Manual of

Security Sensitive
Microbes and Toxins

Manual of

Security Sensitive Microbes and Toxins

Edited by
Dongyou Liu

CRC Press
Taylor & Francis Group
Boca Raton London New York

CRC Press is an imprint of the
Taylor & Francis Group, an **informa** business

The views expressed in this book represent the opinions of individual authors and not necessarily those of their respective institutions.

Neither the editor, the authors and their respective institutions, nor the publisher take the legal responsibility for any personal injury and property damage that may potentially result from the improper use of the information contained in this book.

CRC Press
Taylor & Francis Group
6000 Broken Sound Parkway NW, Suite 300
Boca Raton, FL 33487-2742

First issued in paperback 2019

ISBN-13: 978-1-4665-5396-5 (hbk)
ISBN-13: 978-0-367-37874-5 (pbk)

Visit the Taylor & Francis Web site at
http://www.taylorandfrancis.com

and the CRC Press Web site at
http://www.crcpress.com

Contents

SECTION II Microbes and Toxins Affecting Humans and Animals: Bacteria

SECTION III Microbes and Toxins Affecting Humans and Animals: Fungus and Parasite

SECTION IV   Microbes and Toxins Affecting Humans and Animals: Toxins

SECTION V   Microbes Affecting Animals: Viruses

SECTION VI   Microbes Affecting Animals: Bacteria

SECTION VII   Microbes Affecting Plants

Preface

Security sensitive microbes (viruses, bacteria, fungi, and parasites) and toxins (which are also known as the select agents and toxins in the United States) have the capacity to cause serious illness (illthrift) or death in humans, animals, or plants. Due to their ready availability (commonly occurring in nature), ease of spread/dissemination (through air, water, or in food), and difficulty in identification, these agents have the potential to be used in biological weapons for bioterrorism, creating widespread panic and disruption to a state or a country beyond the actual physical damage they are capable of causing.

This book aims to provide comprehensive coverage on security sensitive microbes and toxins, with an emphasis on the state of the art in the field. Each chapter comprises an authoritative review of one security sensitive microbe or toxin with respect to its classification, biology, epidemiology, pathogenesis, identification, diagnosis, treatment, and prevention followed by a discussion on the limitations of our current knowledge and challenges relating to its improved detection and control.

Each chapter has been written by scientists with expertise in their respective fields of research and provides an all-encompassing, state-of-the-art reference for clinical, veterinary, and industrial scientists involved in the identification/diagnosis and control of security sensitive microbes and toxins. The book is an indispensable roadmap for any upcoming or experienced microbiologist wishing to update their knowledge on security sensitive microbes and toxins and will be useful for prospective undergraduate and graduate students intending to pursue a career in medical, veterinary, and food microbiology.

A comprehensive and inclusive book such as this clearly demands a concerted effort. Toward this end, I am extremely fortunate and honored to have a group of international scientists as chapter contributors, whose generosity to share their in-depth knowledge and technical insights on security sensitive microbes and toxins has made this book a reality. Additionally, the professionalism and dedication of executive editor Barbara Norwitz and senior project coordinator Jill Jurgensen at CRC Press have greatly enhanced its presentation. Finally, without the understanding and support of my family, Liling Ma, Brenda, and Cathy, the compilation of this comprehensive volume would have been impossible.

Editor

Dongyou Liu, PhD, undertook veterinary science education at Hunan Agricultural University, China, and pursued postgraduate training in veterinary parasitology at the University of Melbourne, Victoria, Australia. Over the past two decades, he has worked in several research and clinical laboratories in Australia and the United States, with a focus on molecular characterization and virulence determination of microbial pathogens such as ovine footrot bacterium (*Dichelobacter nodosus*), dermatophyte fungi (*Trichophyton, Microsporum,* and *Epidermophyton*), and listeriae (*Listeria* spp.) as well as security sensitive biological agents. He is the primary author of more than 50 original research and review articles in various international journals and the editor of seven recent biomedical books—*Handbook of* Listeria monocytogenes (2008), *Handbook of Nucleic Acid Purification* (2009), *Molecular Detection of Foodborne Pathogens* (2009), *Molecular Detection of Human Viral Pathogens* (2010), *Molecular Detection of Human Bacterial Pathogens* (2011), *Molecular Detection of Human Fungal Pathogens* (2011), and *Molecular Detection of Human Parasitic Pathogens* (2012)—all of which were published by CRC Press.

Contributors

Mohammad Yazid Abdad
Australian Rickettsial Reference
   Laboratory
Barwon Biomedical Research
Barwon Health
Geelong, Victoria, Australia

and

Environmental and Emerging
   Diseases Unit
Papua New Guinea Institute of Medical
   Research
Goroka, Eastern Highlands Province,
   Papua New Guinea

Katharina Achazi
German Consultant Laboratory for
   Tick-borne Encephalitis, ZBS 1
Robert Koch Institute
and
Institute of Laboratory Medicine,
   Clinical Chemistry and
   Pathobiochemistry
Charité—Universitätsmedizin Berlin
Berlin, Germany

Diallo Adama
Animal Production and Health
   Laboratory
FAO/IAEA Agriculture and
   Biotechnology Laboratory
IAEA Laboratories, International
   Atomic Energy Agency (IAEA)
Vienna, Austria

Claudio L. Afonso
Southeast Poultry Research Laboratory
Agricultural Research Service
United States Department of
   Agriculture
Athens, Georgia

Hiroomi Akashi
Department of Veterinary
   Microbiology
Graduate School of Agricultural and
   Life Sciences
The University of Tokyo
Tokyo, Japan

Sascha Al Dahouk
Federal Institute for Risk Assessment
Division of Hygiene and Microbiology
Department of Biological Safety
Berlin, Germany

Jeremy G. Allen
Department of Agriculture and Food
Animal Health Laboratories
South Perth, Western Australia,
   Australia

Rodrigo P.P. Almeida
Department of Environmental Science,
   Policy and Management
University of California, Berkeley
Berkeley, California

Emilio Aranda
Food Science and Technology
School of Agricultural Engineering
Badajoz, Spain

Agustín Ariño
Faculty of Veterinary Science
Department of Animal Production and
   Food Science
University of Zaragoza
Zaragoza, Spain

Isao Arita
Kumamoto Medical Center
National Hospital Organization
Kumamoto, Japan

Frank W. Austin
College of Veterinary Medicine
Mississippi State University
Mississippi State, Mississippi

Rosa Aznar
Department of Microbiology and
   Ecology
University of Valencia
and
Institute of Agrochemistry and Food
   Technology
Spanish Council for Scientific Research
Valencia, Spain

Vinayagamurthy Balamurugan
Project Directorate on Animal Disease
   Monitoring and Surveillance
Virology Unit
Bangalore, India

Michael J. Banazis
School of Veterinary and Life
   Sciences
Murdoch University
Murdoch, Western Australia, Australia

Norasuthi Bangphoomi
Department of Veterinary
   Microbiology
Graduate School of Agricultural and
   Life Sciences
The University of Tokyo
Tokyo, Japan

Ashley C. Banyard
Wildlife Zoonoses and Vector Borne
   Disease Research Group
Animal Health and Veterinary
   Laboratories Agency
New Haw, Surrey, United Kingdom

Martin J. Barbetti
School of Plant Biology
and
Institute of Agriculture
University of Western Australia
Crawley, Western Australia, Australia

Armanda D.S. Bastos
Department of Zoology and
   Entomology
University of Pretoria
Pretoria, South Africa

María J. Benito
Food Science and Technology
School of Agricultural Engineering
Badajoz, Spain

Mark D. Bennett
School of Veterinary and Life
   Sciences
Murdoch University
Murdoch, Western Australia, Australia

Veerakyathappa Bhanuprakash
FMD Laboratory
Indian Veterinary Research Institute
Bangalore, India

Bishnupriya Bhattacharya
Faculty of Infectious and Tropical Diseases
London School of Hygiene and
   Tropical Medicine
London, United Kingdom

Jon-Paul Bingham
Department of Molecular Biosciences
   and Bioengineering
University of Hawaii
Honolulu, Hawaii

J. Bires
State Veterinary and Food
   Administration
Bratislava, Slovakia

Joseph E. Blaney
Emerging Viral Pathogens Section
National Institute of Allergy and
   Infectious Diseases
National Institutes of Health
Fort Detrick, Maryland

Timothy R. Bowden
Australian Animal Health Laboratory
Division of Animal, Food and Health
   Sciences
Commonwealth Scientific and
   Industrial Research Organisation
Geelong, Victoria, Australia

**Raimunda Sâmia Nogueira
Brilhante**
Specialized Medical Mycology Center
Federal University of Ceará
Fortaleza, Brazil

Anke Brüning-Richardson
Translational Neuro-Oncology Group
Section of Oncology and Clinical
   Research
Leeds Institute of Cancer Studies and
   Pathology
St James' Hospital
Leeds University
Leeds, United Kingdom

Cristina Cano-Gómez
Center for Research in Animal Health,
   National Institute for Agricultural
   and Food Research and Technology
Valdeolmos, Spain

Rossana Cavallo
Virology Unit
Department of Public Health and
   Microbiology
University of Turin
Turin, Italy

Rocky Chau
School of Biotechnology and
   Biomolecular Sciences
University of New South Wales
Sydney, New South Wales, Australia

R. Chaudhry
Department of Microbiology
All India Institute of Medical
   Sciences
New Delhi, India

Rama P. Cherla
Department of Microbial and
   Molecular Pathogenesis
College of Medicine
Texas A&M Health Science Center
Bryan, Texas

Helvécio D. Coletta-Filho
Citriculture Center "Sylvio Moreira"
Agronomic Institute-IAC
Cordeiropolis
São Paulo, Brazil

Rossana de Aguiar Cordeiro
Specialized Medical Mycology Center
Federal University of Ceará
Fortaleza-Ceará, Brazil

Juan J. Córdoba
Food Hygiene and Safety
Faculty of Veterinary Science
University of Extremadura
Cáceres, Spain

María G. Córdoba
School of Agricultural Engineering
Food Science and Technology
Badajoz, Spain

Cristina Costa
Virology Unit
Department of Public Health and
   Microbiology
University of Turin
Turin, Italy

Elena Crotti
Department of Food, Environmental
   and Nutritional Sciences
University of Milan
Milan, Italy

Daniele Daffonchio
Department of Food, Environmental
   and Nutritional Sciences
University of Milan
Milan, Italy

Cristina Domingo
German Consultant Laboratory for
   Tick-borne Encephalitis, ZBS 1
Robert Koch Institute
Berlin, Germany

Oliver Donoso-Mantke
German Consultant Laboratory for
   Tick-borne Encephalitis, ZBS 1
Robert Koch Institute
and
GBD Gesellschaft für
   Biotechnologische Diagnostik mbH
Berlin, Germany

Sylviane Dragacci
French Agency for Food, Environmental
   and Occupational Health and Safety
   (Anses)-Food Safety Laboratory
Maisons-Alfort, France

Patricia Elízaquível
Department of Microbiology and
   Ecology
University of Valencia
Valencia, Spain

Ayse Erbay
Faculty of Medicine
Department of Infectious Diseases and
   Clinical Microbiology
Bozok University
Yozgat, Turkey

Camille Escadafal
German Consultant Laboratory for
   Tick-borne Encephalitis, ZBS 1
Robert Koch Institute
Berlin, Germany

Jennifer S. Evans
Wildlife Zoonoses and Vector Borne
   Disease Research Group
Animal Health and Veterinary
   Laboratories Agency
New Haw, Surrey, United Kingdom

Joseph Fair
METABIOTA
GVF Inc.
Washington, DC

Folorunso O. Fasina
Department of Zoology and
    Entomology
University of Pretoria
Pretoria, South Africa

and

Department of Production Animal
    Studies
University of Pretoria
Onderstepoort, South Africa

M.K. Fatica
Department of Food Science and
    Human Nutrition
University of Florida
Gainesville, Florida

Fabio Fava
Industrial and Environmental
    Biotechnologies and Fluid-
    Dynamics Research Unit
Department of Civil, Chemical,
    Environmental and Materials
    Engineering
University of Bologna
Bologna, Italy

Mark Fegan
Biosciences Research Division
Department of Primary Industries
Bundoora, Victoria, Australia

Yaoyu Feng
School of Resource and Environmental
    Engineering
East China University of Science and
    Technology
Shanghai, Xuhui, People's Republic of
    China

Paloma Fernández-Pacheco
Center for Research in Animal Health,
    National Institute for Agricultural
    and Food Research and Technology
Valdeolmos, Spain

Anthony R. Fooks
Wildlife Zoonoses and Vector Borne
    Disease Research Group
Animal Health and Veterinary
    Laboratories Agency
New Haw, Surrey, United Kingdom

Donald Francis
Global Solutions for Infectious
    Diseases
San Francisco, California

Michelle E. Gahan
National Centre for Forensic Studies
University of Canberra
Canberra, Australia

László Galgóczy
Faculty of Science and Informatics
Department of Microbiology
University of Szeged
Szeged, Hungary

Eric A.E. Garber
Center for Food Safety and Applied
    Nutrition
Food and Drug Administration
College Park, Maryland

Libeau Geneviève
Biological Systems
    Department-CIRAD
Control of Exotic and Emerging
    Animal Diseases (UMR15)
Montpellier, France

Jean-Paul Gonzalez
Health Department
Institut de recherche pour le
developpement
Marseille, France

and

METABIOTA
GVF Inc.
Washington, DC

Marc Grandadam
Arboviruses and Emerging Viral
    Diseases Laboratory
Institut Pasteur du Laos
Vientiane, Lao PDR

Peter Hagedorn
German Consultant Laboratory for
    Tick-borne Encephalitis, ZBS 1
Robert Koch Institute
Berlin, Germany

Mohan Kumar Haleyur Giri Setty
Laboratory of Molecular Virology
Food and Drug Administration
Center for Biologics Evaluation &
    Research
Bethesda, Maryland

Zan A. Halford
Department of Molecular Biosciences
    and Bioengineering
University of Hawaii
Honolulu, Hawaii

Glen L. Hartman
Agricultural Research Service
United States Department of
    Agriculture
and
Department of Crop Sciences
National Soybean Research Center
University of Illinois
Urbana, Illinois

Joshua S. Hawley
Department of Medicine
Tripler Army Medical Center
Honolulu, Hawaii

Alan Christopher Hayward
School of Chemistry and Molecular
    Biosciences
University of Queensland
Brisbane, Queensland, Australia

Jacques-Antoine Hennekinne
French Agency for Food,
    Environmental and Occupational
    Health and Safety (Anses)-Food
    Safety Laboratory
Paris, France

Marta Herrera
Faculty of Veterinary Science
Department of Animal Production and
    Food Science
University of Zaragoza
Zaragoza, Spain

Indira K. Hewlett
Laboratory of Molecular Virology
Food and Drug Administration
Center for Biologics Evaluation &
    Research
Bethesda, Maryland

Michael R. Holbrook
Integrated Research Facility at Fort
    Detrick
National Institute of Allergy and Infectious
    Diseases
National Institutes of Health
Frederick, Maryland

M. Hosamani
FMD Laboratory
Indian Veterinary Research Institute
Bangalore, India

Shifeng Hu
College of Veterinary Medicine
Hunan Agricultural University
Changsha, Hunan, People's Republic of
China

M. Hutber
EpiVet
Winchester, United Kingdom

Antje Hüther
German Consultant Laboratory for
Tick-borne Encephalitis, ZBS 1
Robert Koch Institute
Berlin, Germany

Miguel Ángel Jiménez-Clavero
Center for Research in Animal Health,
National Institute for Agricultural
and Food Research and Technology
Valdeolmos, Spain

Reed F. Johnson
Emerging Viral Pathogens Section
National Institute of Allergy and
Infectious Diseases
National Institutes of Health
Fort Detrick, Maryland

Colleen B. Jonsson
Department of Microbiology and
Immunology
Center for Predictive Medicine for
Biodefense and Emerging Infectious
Diseases
University of Louisville
Louisville, Kentucky

John A. Kalaitzis
School of Biotechnology and
Biomolecular Sciences
University of New South Wales
Sydney, New South Wales, Australia

Lyudmila S. Karan
Laboratory of Epidemiology of
Zoonoses
Central Research Institute of
Epidemiology
Moscow, Russia

Donald P. King
The Pirbright Institute
Pirbright, United Kingdom

Luke C. Kingry
Division of Vector-Borne Diseases
Centers for Disease Control and
Prevention
Fort Collins, Colorado

Laura D. Kramer
Division of Infectious Diseases
Wadsworth Center
New York State Department of Health
and
Department of Biomedical Sciences
School of Public Health
University at Albany
Albany, New York

László Kredics
Faculty of Science and Informatics
Department of Microbiology
University of Szeged
Szeged, Hungary

Goro Kuno
Fort Collins, Colorado

Benjamin Lamp
Department of Pathobiology
Institute of Virology
University of Veterinary Medicine
Vienna, Austria

K.A. Lampel
Center for Food Safety and Applied
Nutrition
Food and Drug Administration
College Park, Maryland

Moo-Seung Lee
Infection and Immunity Research Center
Korea Institute of Bioscience and
Biotechnology
Daejeon, South Korea

Thierry Lefrançois
Control of Exotic and
Emerging Animal Diseases
Unit Research
Biological Systems Department
Agricultural Research Centre for
International Development (CIRAD)
Montpellier, France

Robert K. Likeman
Department of Defense
Canberra, Australian Capital Territory,
Australia

Baochuan Lin
Center for Bio/Molecular Science and
Engineering
United States Naval Research
Laboratory
Washington, DC

Dongyou Liu
Biosecurity Quality Assurance
Program
Royal College of Pathologists of
Australasia
St Leonards, New South Wales,
Australia

João R.S. Lopes
Departmento Entomologia e Acarologia
Escola Superior de Agricultura "Luiz
de Queiroz"
Universidade de São Paulo
Piracicaba, Brazil

Anthony P. Malanoski
Center for Bio/Molecular Science and
Engineering
United States Naval Research
Laboratory
Washington, DC

L. Manso-Silván
CIRAD-UMR15
Control of Exotic and Emerging
Animal Diseases
and
National Institute of Agricultural
Research UMR1309 CMAEE
Montpellier, France

Nicholas J. Mantis
Division of Infectious Diseases
Wadsworth Center
New York State Department of Health
and
Department of Biomedical Sciences
School of Public Health
University at Albany
Albany, New York

Isabel Marcelino
Institute of Experimental and
Technological Biology (IBET)
and
Institute of Chemical and Biological
Technology (ITQB)
New University of Lisbon
Oeiras, Portugal

Dominique Martinez
Agricultural Research Centre for
    International Development (CIRAD)
Guadeloupe, France

B.J. McCluskey
Animal and Plant Health Inspection
    Service
United States Department of
    Agriculture
Fort Collins, Colorado

Patti J. Miller
Southeast Poultry Research Laboratory
Agricultural Research Service
United States Department of
    Agriculture
Athens, Georgia

Brett A. Neilan
School of Biotechnology and
    Biomolecular Sciences
University of New South Wales
Sydney, New South Wales, Australia

Heinrich Neubauer
Institute of Bacterial Infections and
    Zoonoses
Friedrich-Loeffler-Institut
Jena, Germany

Matthias Niedrig
German Consultant Laboratory for
    Tick-borne Encephalitis, ZBS 1
Robert Koch Institute
Berlin, Germany

Gene G. Olinger
Virology Division
United States Army Medical Research
    Institute of Infectious Diseases
Fort Detrick, Maryland

Takashi Onodera
Research Center for Food Safety
School of Agricultural and Life Sciences
The University of Tokyo
Tokyo, Japan

Satya Parida
Pirbright Institute
Surrey, United Kingdom

Pranav Patel
German Consultant Laboratory for
    Tick-borne Encephalitis, ZBS 1
Robert Koch Institute
Berlin, Germany

Priyabrata Pattnaik
Merck Millipore
Biomanufacturing Sciences and
    Training Centre
Merck Pte Ltd
Singapore, Singapore

Leanne A. Pearson
School of Biotechnology and
    Biomolecular Sciences
University of New South Wales
Sydney, New South Wales, Australia

Kirsten S. Pelz-Stelinski
Entomology and Nematology
    Department
Citrus Research and Education
    Center
University of Florida
Lake Alfred, Florida

Jeannine M. Petersen
Division of Vector-Borne Diseases
Centers for Disease Control and
    Prevention
Fort Collins, Colorado

E. Pilipcinec
Rector's Office
University of Veterinary Medicine
Košice, Slovakia

Michel R. Popoff
Unité des Bactéries anaérobies et
    Toxines
Institut Pasteur
Paris, France

Ann M. Powers
Division of Vector-Borne Diseases
Centers for Disease Control and
    Prevention
Fort Collins, Colorado

K. Prathyusha
Department of Microbiology
All India Institute of Medical
    Sciences
New Delhi, India

Jarosław Przetakiewicz
Laboratory of Quarantine Organisms
Department of Plant Pathology
Plant Breeding and Acclimatization
    Institute
National Research Institute
Radzików, Poland

Noura Raddadi
Industrial and Environmental
    Biotechnologies and Fluid-
    Dynamics Research Unit
Department of Civil, Chemical,
    Environmental and Materials
    Engineering
University of Bologna
Bologna, Italy

Pradeep B. J. Reddy
Department of Biomedical and
    Diagnostic Sciences
College of Veterinary Medicine
University of Tennessee
Knoxville, Tennessee

Ian T. Riley
School of Agriculture Food and Wine
University of Adelaide
Urrbrae, South Australia, Australia

Marcos Fábio Gadelha Rocha
Specialized Medical Mycology Center
Federal University of Ceará
and
Faculty of Veterinary
Postgraduate Program in Veterinary
    Science
State University of Ceará
Fortaleza, Brazil

Alicia Rodríguez
Food Hygiene and Safety
Faculty of Veterinary Science
University of Extremadura
Cáceres, Spain

Mar Rodríguez
Food Hygiene and Safety
Faculty of Veterinary Science
University of Extremadura
Cáceres, Spain

Paul E. Roffey
Forensics
Australian Federal Police
Canberra, Australia

Polly Roy
Faculty of Infectious and Tropical
    Diseases
London School of Hygiene and
    Tropical Medicine
London, United Kingdom

Till Rümenapf
Department of Pathobiology
Institute of Virology
University of Veterinary Medicine
Vienna, Austria

Janice M. Rusnak
Goldbeltraven LLC
Fort Detrick, Maryland

George C. Russell
Moredun Research Institute
Midlothian, United Kingdom

Daniel Růžek
Department of Virology,
Veterinary Research Institute
Brno, Czech Republic

and

Institute of Parasitology
Biology Centre of the Academy of
    Sciences of the Czech Republic
České Budějovice, Czech Republic

Masayuki Saijo
Department of Virology 1
National Institute of Infectious
    Diseases
Tokyo, Japan

Akikazu Sakudo
Faculty of Medicine
Laboratory of Biometabolic
    Chemistry
School of Health Sciences
University of the Ryukyus
Okinawa, Japan

M.D. Salman
Animal Population Health Institute
College of Veterinary Medicine and
    Biomedical Sciences
Colorado State University
Fort Collins, Colorado

Gloria Sánchez
Institute of Agrochemistry and Food
    Technology
Spanish Council for Scientific
    Research
Valencia, Spain

Andrea Sanchini
German Consultant Laboratory for
    Tick-borne Encephalitis, ZBS 1
Robert Koch Institute
Berlin, Germany

and

European Public Health Microbiology
Training Programme
European Centre for Disease
    Prevention and Control
Stockholm, Sweden

Frank Sauvage
Laboratoire de Biométrie et Biologie
    Evolutive
Université de Lyon
Villeurbanne, France

K.R. Schneider
Department of Food Science and
    Human Nutrition
University of Florida
Gainesville, Florida

Randal J. Schoepp
Diagnostic Systems Division
United States Army Medical Research
    Institute of Infectious Diseases
Fort Detrick, Maryland

Francesca Sidoti
Virology Unit
Department of Public Health and
    Microbiology
University of Turin
Turin, Italy

José Júlio Costa Sidrim
Specialized Medical Mycology Center
Federal University of Ceará
Fortaleza, Brazil

Raj Kumar Singh
National Research Centre on Equines
Haryana, India

Ina L. Smith
Australian Animal Health Laboratory
Commonwealth Scientific and
    Industrial Research Organisation
Geelong, Victoria, Australia

Leonard A. Smith
United States Army Medical Research
    Institute of Infectious Diseases
Fort Detrick, Maryland

Lisa D. Sprague
Institute of Bacterial Infections and
    Zoonoses
Friedrich-Loeffler-Institut
Jena, Germany

Ambuj Srivastava
Division of Virology
Defence Research and Development
    Establishment
Gwalior, India

Lukasz L. Stelinski
Entomology and Nematology
    Department
Citrus Research and Education Center
University of Florida
Lake Alfred, Florida

Sacha Stelzer-Braid
Virology Research Laboratory
Prince of Wales Hospital
Randwick, New South Wales, Australia

John Stenos
Australian Rickettsial Reference
    Laboratory
Barwon Biomedical Research
Barwon Health
Geelong, Victoria, Australia

Norma P. Tavakoli
Division of Genetics
Wadsworth Center
New York State Department of Health
and
Department of Biomedical Sciences
School of Public Health
University at Albany
Albany, New York

Vernon L. Tesh
Department of Microbial and
    Molecular Pathogenesis
College of Medicine
Texas A&M Health Science Center
Bryan, Texas

F. Thiaucourt
CIRAD-UMR15
Control of Exotic and Emerging
    Animal Diseases
and
National Institute of Agricultural
    Research UMR1309 CMAEE
Montpellier, France

Sergey E. Tkachev
Institute of Chemical Biology and
    Fundamental Medicine
Siberian Branch of the Russian
    Academy of Sciences
Novosibirsk, Russia

Akiko Uema
Department of Infection Control
Disease Prevention Graduate School
    of Agricultural and Life Sciences
The University of Tokyo
Tokyo, Japan

Nathalie Vachiéry
Control of Exotic and Emerging
    Animal Diseases Unit
    Research
Biological Systems Department
Agricultural Research Centre for
    International Development (CIRAD)
Guadeloupe, France

Csaba Vágvölgyi
Faculty of Science and Informatics
Department of Microbiology
University of Szeged
Szeged, Hungary

Gnanavel Venkatesan
Division of Virology
Indian Veterinary Research
    Institute
Uttarakhand, India

Estelle H. Venter
Department of Veterinary Tropical
    Diseases
Faculty of Veterinary Science
University of Pretoria
Pretoria, South Africa

Máté Virágh
Faculty of Science and Informatics
Department of Microbiology
University of Szeged
Szeged, Hungary

B.R. Warren
Land O'Lakes, Inc.,
Arden Hills, Minnesota

Lihua Xiao
Division of Foodborne, Waterborne and
    Environmental Diseases
Centers for Disease Control and Prevention
Atlanta, Georgia

Valeriy V. Yakimenko
Omsk Research Institute of Natural
    Foci Infections
Omsk, Russia

Ruifu Yang
Beijing Institute of Microbiology and
    Epidemiology
Beijing, People's Republic of China

Chong Yin
College of Veterinary Medicine
Hunan Agricultural University
Changsha, Hunan, People's Republic of
    China

Peter Y.C. Yu
Department of Molecular Biosciences
    and Bioengineering
University of Hawaii
Honolulu, Hawaii

Dongsheng Zhou
Beijing Institute of Microbiology and
    Epidemiology
Beijing, People's Republic of China

1 Introductory Remarks

Dongyou Liu

CONTENTS

Security sensitive microbes and toxins (also known as select agents and toxins) are noted for their ability to incapacitate and decimate human, animal, and plant hosts. The veracity and pathogenicity of these biological agents are undoubtedly enhanced by their common occurrence, ease of dissemination, and difficulty in identification. Not surprisingly, throughout history, security sensitive microbes and toxins have been exploited in one form or another as biowarfare and bioterror agents that create fear and panic well beyond the actual physical damages they are capable of causing. In the following sections, a brief overview is presented on the historical aspects of security sensitive microbes and toxins. This is followed by a concise summary of the current status in relation to the regulation of security sensitive microbes and toxins. Finally, the future development concerning these biological agents is discussed.

1.1 SECURITY SENSITIVE MICROBES AND TOXINS: THE PAST

From time immemorial, humans have become acquainted with security sensitive microbes and toxins through numerous outbreaks and epidemics, small and large, that have claimed countless lives and created untold miseries. Being on top of the list of most intelligent creatures on the Earth, humans have gradually learned to exploit these biological agents for both beneficial and destructive purposes. While aboriginal use of curare and amphibian-derived toxins as arrow poisons exemplifies their destructive role, the cosmetic application of botulinum toxin (botox) for wrinkle prevention highlights their beneficial utility. Indeed, the modern term "toxin" was derived from the ancient Greek word for arrow poison (*toxon* = bow, arrow). However, more often than not, security sensitive microbes and toxins have been deployed intentionally or unintentionally as biowarfare and biocrime agents throughout documented history, contributing to the loss of innocent lives and recurrent human sufferings (Table 1.1).

The apparent horrors of biowarfare led to the Hague Conventions of 1899 and 1907 that outlawed the use of "poison or poisoned arms." In addition, the Geneva Protocol formulated in 1925 prohibits the use in war of asphyxiating, poisonous or other gases, and of bacteriological methods of warfare. Nonetheless the 1925 Geneva Protocol had failed to fully prevent the proliferation of biological weapons. The next step toward control of biowarfare agents was the opening of the Biological and Toxin Weapons Convention (BWC) (or the Convention on the Prohibition of the Development, Production and Stockpiling of Bacteriological [Biological] and Toxin Weapons and on Their Destruction) in 1972, which currently involves 170 states parties. The BWC appears to be largely effective in controlling the proliferation of biological weapons at state level; however, it is powerless in dealing with biocrimes committed by disgruntled individuals or fanatic extremists (Table 1.1).

1.2 SECURITY SENSITIVE MICROBES AND TOXINS: THE PRESENT

Although >1200 biological agents are pathogenic to humans and animals to some extent [1], only a small proportion possess the necessary characteristics to be regarded as security sensitive microbes and toxins. Given the enormous potential damages security sensitive microbes and toxins may cause to humans, animals, and plants, guidelines have been established by many countries to regulate their possession, transfer, and use. In the United States, the Select Agent Regulations were promulgated by the U.S. Department of Health and Human Services (HHS) and the U.S. Department of Agriculture (USDA) in 1997 to establish and maintain a list of biological select agents and toxins (BSAT, or select agents for short) that have the potential to pose a severe threat to public health and safety, which has ultimately become known as the select agent list [2,3].

The criteria for the inclusion of human pathogens and toxins in the list consist of (1) the effect on human health from exposure to the agent; (2) the degree of contagiousness of the agent and the methods by which the agent is transferred to humans; and (3) the availability and effectiveness of immunizations to prevent and therapies to treat any illness resulting from exposure to the agent. Criteria for the inclusion of animal pathogens consist of (1) availability and effectiveness of therapies and prophylaxis to treat

1

TABLE 1.1
Historical Incidences of Intentional and Unintentional Use of Security Sensitive Microbes and Toxins in Biowarfare and Biocrimes

Time	Agent	Incidence
184 BC	Serpent toxins	In the naval battle of the Eurymedon, the Carthaginian army under the leadership of Hannibal used earthen pots filled with serpents to attack the Pergamene ship of King Eumenes. The panic and chaos created by live serpents among the enemy sailors ensured the Carthaginian victory.
1340	Plague	Attackers used catapult to hurl dead horses and other animals at the castle of Thun L'Eveque in Hainault (northern France). Unable to endure the stink and the abominable air, the defender negotiated a truce.
1343	Plague	The Mongols used trebuchet to hurl "mountains of dead" over the city wall of Caffa, a Genoese colony in the Crimea. This led to an outbreak of plague in that city.
1422	Infectious agents	Hussite attackers used catapults to throw the decaying cadavers of men killed in battle and dung over the castle walls of Karlstein in Bohemia. However, the defenders held fast, and the siege was abandoned after 5 months.
1520	Smallpox	The Narváez expedition of 1520 unintentionally brought smallpox to the Aztec empire and surrounding areas with catastrophic consequences. Subsequent smallpox epidemics contributed to the demise of the Aztec and the Inca empires.
1763	Smallpox	Near the end of the French and Indian War in 1763, the Native Americans attacked British forts along the western frontier. William Trent, the local militia leader, used smallpox as a biological weapon against the Native Americans. At Fort Pitt on the Pennsylvania frontier, British Gen. Jeffery Amherst gave blankets and handkerchiefs from smallpox patients to Delaware Indians at a peace-making parley.
1785	Plague	Tunisian attackers employed plague-tainted clothing as a weapon in the siege of La Calle.
1796	Smallpox	Smallpox employed by the British forced the Continental Army to retreat from Quebec.
1915	Glanders	During World War I, Anton Dilger, a German operative, infected horses to be shipped to Britain with glanders (*Burkholderia mallei*).
1932–1945	*Yersinia pestis, Vibrio cholerae, Bacillus anthracis, Shigella,* and *Neisseria meningitidis*	During World War II, the Japanese military waged biowarfare involving *Y. pestis, V. cholerae, B. anthracis, Shigella* sp., *N. meningitidis*, and possibly other agents on a mass scale against China, with a notorious division of the Imperial Army called Unit 731 spearheading this operation.
1964–1966	*Salmonella typhi*	Dr. Mitsuru Suzuki allegedly contaminated food items, medications, barium contrast, and a tongue depressor with *S. typhi* in Japan, resulting in >120 cases of infection and 4 deaths.
1979	Anthrax	An outbreak of anthrax disease in the city of Sverdlovsk killed nearly 70 people, which was probably linked to secret weapons work at a nearby Soviet army laboratory.
1984	*Salmonella typhimurium*	In Oregon, USA, followers of Bhagwan Shree Rajneesh infected salad bars in restaurants, produce in grocery stores, doorknobs, and other public domains with *S. typhimurium* bacteria, in an attempt to influence the election. The attack resulted in 751 cases of typhoid fever (enteritis), 43 of which were hospitalized.
1993–1995	Anthrax and sarin gas	The apocalyptic religious group Aum Shinrikyo released *B. anthracis* in a Tokyo building in 1993, resulting in 12 deaths and thousands seeking emergency care. Additionally, the release of the nerve agent sarin in the subway system of Tokyo by this group in 1995 killed 13 people, severely injured 50, and caused temporary vision problems for nearly 1000 others.
2001	Anthrax	Letters laced with anthrax spores were mailed to news media offices and the U.S. Congress, resulted in 22 cases of infection, 5 deaths, and approximately 10,000 individuals being offered postexposure prophylaxis.

Sources: Christopher, G.W. et al., *JAMA*, 278, 412, 1997; Tucker, J.B., *Emerg. Infect. Dis.*, 5, 498, 1999; Wheelis, M., A short history of biological warfare and weapons, in: Chevrier, M.I. et al., eds., *The Implementation of Legally Binding Measures to Strengthen the Biological and Toxin Weapons Convention*, Springer, Dordrecht, the Netherlands, 2004; Martin, J.W. et al., History of biological weapons: From poisoned darts to intentional epidemics, in: Dembek, Z.F., ed., *Medical Aspects of Biological Warfare*, Office of the Surgeon General, Borden Institute, Washington, DC, 2007; History of Biowarfare, http://www.pbs.org/wgbh/nova/bioterror/hist_nf.html/

and prevent any illness; (2) economic impact; inclusion in the Office International des Epizooties (OIE) A and B lists; and (3) presence in the Australia Group List. Criteria for the inclusion of plant pathogens consist of (1) the effect of exposure to the agent on plant health and on the production and marketability of plant products; (2) the ability to detect the agent and diagnose the infection during its early stages;

(3) whether the agent is nonnative or exotic; and (4) the economic importance of the host plant [2].

The select agents are divided into three categories: (1) HHS select agents and toxins (affecting humans); (2) USDA select agents and toxins (affecting agriculture); and (3) overlap select agents and toxins (affecting both) (Table 1.2). Those agents and toxins that are in both HHS and USDA lists are

TABLE 1.2
List of HHS and USDA Select Agents and Toxins

HHS Select Agents and Toxins

Abrin
Botulinum neurotoxins[a]
Botulinum neurotoxin producing species of *Clostridium*[a]
Conotoxins (short, paralytic alpha conotoxins containing the amino acid sequence $X_1CCX_2PACGX_3X_4X_5X_6CX_7$)
Coxiella burnetii
Crimean-Congo hemorrhagic fever virus
Diacetoxyscirpenol
Eastern equine encephalitis virus[b]
Ebola virus[a]
Francisella tularensis[a]
Lassa fever virus
Lujo virus
Marburg virus[a]
Monkeypox virus[b]
Reconstructed replication competent forms of the 1918 pandemic influenza virus containing any portion of the coding regions of all eight gene segments (reconstructed 1918 Influenza virus)
Ricin
Rickettsia prowazekii
SARS-associated coronavirus (SARS-CoV)
Saxitoxin
South American hemorrhagic fever viruses:
　Chapare
　Guanarito
　Junin
　Machupo
　Sabia
Staphylococcal enterotoxins A, B, C, D, E subtypes
　T-2 toxin
　Tetrodotoxin
Tick-borne encephalitis complex (flavi) viruses:
　Far Eastern subtype
　Siberian subtype
Kyasanur Forest disease virus
Omsk hemorrhagic fever virus
Variola major virus (Smallpox virus)[a]
Variola minor virus (Alastrim)[a]
Yersinia pestis[a]

Overlap Select Agents and Toxins

Bacillus anthracis[a]
B. anthracis Pasteur strain
Brucella abortus
Brucella melitensis
Brucella suis
Burkholderia mallei[a]
Burkholderia pseudomallei[a]
Hendra virus
Nipah virus
Rift Valley fever virus
Venezuelan equine encephalitis virus[b]

USDA Select Agents and Toxins

African horse sickness virus
African swine fever virus
Avian influenza virus[b]
Classical swine fever virus
Foot-and-mouth disease virus[a]
Goat pox virus
Lumpy skin disease virus
Mycoplasma capricolum[b]
Mycoplasma mycoides[b]
Newcastle disease virus[b,c]
Peste des petits ruminants virus
Rinderpest virus[a]
Sheep pox virus
Swine vesicular disease virus

USDA Plant Protection and Quarantine (PPQ) Select Agents and Toxins

Peronosclerospora philippinensis (Peronosclerospora sacchari)
Phoma glycinicola (formerly *Pyrenochaeta glycines*)
Ralstonia solanacearum
Rathayibacter toxicus
Sclerophthora rayssiae
Synchytrium endobioticum
Xanthomonas oryzae

Source: CDC. 2012. List of select agents and toxins, http://www.selectagents.gov/select%20agents%20and%20toxins%20list.html (accessed on June 30, 2013).

[a] Denotes Tier 1 agent.

[b] Select agents that meet any of the following criteria are excluded from the requirements of this part: any low pathogenic strains of avian influenza virus, South American genotype of eastern equine encephalitis virus, West African clade of Monkeypox viruses, any strain of Newcastle disease virus which does not meet the criteria for virulent Newcastle disease virus, all subspecies *M. capricolum* except subspecies *capripneumoniae* (contagious caprine pleuropneumonia), all subspecies *M. mycoides* except subspecies *mycoides* small colony (Mmm SC) (contagious bovine pleuropneumonia), any subtypes of Venezuelan equine encephalitis virus except for subtypes IAB or IC, and vesicular stomatitis virus (exotic): Indiana subtypes VSV-IN2, VSV-IN3, provided that the individual or entity can verify that the agent is within the exclusion category. For further details on other exclusions from the requirements of the Select Agent Regulations, refer http://www.selectagents.gov/Select%20Agents%20and%20Toxins%20Exclusions.html.

[c] A virulent Newcastle disease virus (avian paramyxovirus serotype 1) has an intracerebral pathogenicity index in day-old chicks (*Gallus gallus*) of 0.7 or greater or has an amino acid sequence at the fusion (F) protein cleavage site that is consistent with virulent strains of Newcastle disease virus. A failure to detect a cleavage site that is consistent with virulent strains does not confirm the absence of a virulent virus.

called "overlap agents and toxins" and are jointly regulated by both HHS and USDA. Currently, the Division of Select Agents and Toxins at the Centers for Disease Control and Prevention (CDC) administers HHS select agent regulations; the Animal and Plant Health Inspection Service (APHIS) administers USDA select agent regulations [2].

The list of select agents and toxins is not static and is subject to changes after biannual review undertaken by CDC. In the most recent edition (December 4, 2012), Chapare virus from South America, Lujo virus from Africa, and SARS (severe acute respiratory syndrome)-coronavirus (SARS-CoV) have been added to the list of HHS select agents and toxins, while 23 items including *Coccidioides* and the Shiga toxins have been dropped. The main reason for removing *Coccidioides* from the list is the availability of a variety of treatments for its infections, while Shiga toxins are taken out in view of the difficulty to aerosolize them for weapon use [4]. Nonetheless, these topics remain relevant and important to science community at large and are thus covered in this book for completeness. Similarly, although some agents (e.g., lyssaviruses, menangle virus and *Salmonella* Typhi) are not listed as the select agents and toxins, they are included in this book due to the potential threat they pose to human and animal populations outside of the United States.

The select agents and toxins that present the greatest risk of deliberate misuse with the most significant potential for mass casualties or devastating effects on the economy, critical infrastructure, or public confidence are designated as "Tier 1" agents. The Tier 1 agents consist of Ebola virus, *Francisella tularensis* (tularemia), Marburg virus, variola major and minor viruses (smallpox), *Yersinia pestis* (plague), *Clostridium botulinum* and botulinum toxin (botulism), *Bacillus anthracis* (anthrax), *Burkholderia mallei* (glanders), and *Burkholderia pseudomallei* (melioidosis) (Table 1.2).

Other countries also have similar, but less extensive, lists of security sensitive microbes and toxins that are regulated in relation to their possession, use, and transfer.

1.3　SECURITY SENSITIVE MICROBES AND TOXINS: THE FUTURE

Because security-sensitive microbes and toxins have the potential to pose a severe threat to public health and safety, to animal health or animal products, and to plant health and plant products, it is critical that necessary research is conducted with the goal of developing reliable detection technologies and effective medical countermeasures and vaccines. As these endeavors are dependent on the availability of related biological materials, promulgation and imposition of regulations and guidelines for the select agents should be based on the key premise of ensuring protection of the public without encumbering legitimate scientific and medical research as well as academic freedom [5,6]. Similarly, it is crucial to keep a balanced approach between ensuring secrecy to protect national security and accommodating the right of local communities

to have access to safety information regarding select agents and laboratory-acquired infections [7,8]. Furthermore, there is tangible evidence that microbial collections are destroyed in the United States with the implementation of the regulations relating to the Select Agents and Toxins List. Given the rapid evolution of microbial strains, the destruction of archival collections is a potentially irretrievable loss of biological diversity. Therefore, government agencies should develop plans to ensure that microbial collections are preserved when considering future additions to microbial threat lists [9].

Future research on security-sensitive microbes and toxins should focus on (i) development of rapid, sensitive, and specific techniques (including whole genome sequencing) for identifying and tracking biological agents; (ii) optimization of decontamination technologies to restore facilities without causing unnecessary environmental concerns; and (iii) design of effective vaccines against bioweapon threats (including passive immunization, which provide immediate protection, whereas a protective response generated by a vaccine is not immediate) [10–18]. Moreover, improved education on security sensitive microbes and toxins, their epidemiology, disease mechanisms, control and prevention measures will help ease the public's fear and panic of these agents in the event of a bioterror incident occurring.

Along with the continuing advances in molecular biology and proteomics, the costs for in vitro synthesis of microbial genes/genomes and toxins are much reduced. This offers an increasingly affordable supply of noninfectious and stable materials for research and development without having to directly acquire and handle security sensitive microbes and toxins. One example of the benefits deriving from these technical advances is the generation of in vitro RNA transcripts of high purity and quantity from recombinant plasmids containing target gene of RNA virus for quality assurance testing of nucleic acid–based diagnostic assays. Another is the production of toxins from expression plasmids containing relevant genes for vaccine development and other applications. Future updates on the regulations concerning the select agents should therefore consider the impact of these technical advances that may bypass the existing guidelines on their acquisition, use, and transfer.

REFERENCES

1. Cleaveland SC, Laurenson MK, Taylor LH. Diseases of humans and their domestic mammals; pathogen characteristics, host range and the risk of emergence. *Philos Trans R Soc Lond B Biol Sci* 2001; 356:991–999.
2. Centers for Disease Control and Prevention (CDC), List of select agents and toxins, 2012. http://www.selectagents.gov/select%20agents%20and%20toxins%20list.html (accessed on June 30, 2013).
3. Centers for Disease Control and Prevention (CDC), List of excluded agents and toxins, 2012. http://www.selectagents.gov/Select%20Agents%20and%20Toxins%20Exclusions.html. (accessed on June 30, 2013).

4. Centers for Disease Control and Prevention (CDC), Department of Health and Human Services (HHS). Possession, use, and transfer of select agents and toxins; biennial review. Final rule. *Fed Regist* 2012; 77:61083–61115.

5. Keel BA. Protecting America's secrets while maintaining academic freedom. *Acad Med* 2004; 79:333–342.

6. Pastel RH et al. Clinical laboratories, the select agent program, and biological surety (biosurety). *Clin Lab Med* 2006; 26:299–312.

7. Kahn LH. Biodefense research: Can secrecy and safety coexist? *Biosecur Bioterror* 2004; 2:81–85.

8. Matthews S. Select-agent status could slow development of anti-SARS therapies. *Nat Med* 2012; 18:1722.

9. Casadevall A, Imperiale MJ. Destruction of microbial collections in response to select agent and toxin list regulations. *Biosecur Bioterror* 2010; 8:151–154.

10. Kadlec RP, Zelicoff AP, Vrtis AM. Biological weapons control: Prospects and implications for the future. *JAMA* 1997; 278:351–356.

11. Ashford DA. Planning against biological terrorism: Lessons from outbreak investigations. *Emerg Infect Dis* 2003; 9:515–519.

12. Buehler JW et al. Syndromic surveillance and bioterrorism-related epidemics. *Emerg Infect Dis* 2003; 9:1197–1204.

13. Dworkin MS, Xinfang M, Golash RG. Fear of bioterrorism and implications for public health preparedness. *Emerg Infect Dis* 2003; 9:503–505.

14. Hoffman RE. Preparing for a bioterrorist attack: Legal and administrative strategies. *Emerg Infect Dis* 2003; 9:241–245.

15. Slezak T et al. Comparative genomics tools applied to bioterrorism defence. *Brief Bioinform* 2003; 4:133–149.

16. Steinbrook R. Biomedical research and biosecurity. *N Engl J Med* 2005; 353:2212–2214.

17. Roberts M. Role of regulation in minimizing terrorist threats against the food supply: Information, incentives, and penalties. *Minn J L Sci. Technol* 2006; 8:199–223.

18. Pohanka M, Kuca K. Biological warfare agents. *EXS* 2010; 100:559–578.

19. Christopher GW et al. Biological warfare: A historical perspective. *JAMA* 1997; 278:412–417.

20. Tucker JB. Historical trends related to bioterrorism: An empirical analysis. *Emerg Infect Dis* 1999; 5:498–504.

21. Wheelis M. A short history of biological warfare and weapons. In: Chevrier MI et al., eds., *The Implementation of Legally Binding Measures to Strengthen the Biological and Toxin Weapons Convention*. Dordrecht, the Netherlands: Springer, 2004.

22. Martin JW, Christopher GW, Eitzen EM Jr. History of biological weapons: From poisoned darts to intentional epidemics. In: Dembek ZF, ed., *Medical Aspects of Biological Warfare*. Washington, DC: Office of the Surgeon General, Borden Institute, 2007.

23. Lewis SK, History of biowarfare, 2009. http://www.pbs.org/wgbh/nova/bioterror/hist_nf.html/ (accessed on June 30, 2013).

Section I

Microbes and Toxins Affecting Humans and Animals: Viruses

2 Arenaviruses

Frank Sauvage, Joseph Fair, and Jean-Paul Gonzalez

CONTENTS

2.1 INTRODUCTION

Classified in the Arenaviridae family arenaviruses naturally and chronically infect rodents, which serve as a reservoir host and carry the virus in an asymptomatic manner. Rodent species appear to be specifically and chronically infected by a virus species and represent a model of virus–host coevolution (Table 2.1).[1] One exception is the Tacaribe virus (TACV), an arenavirus that has been found naturally and specifically infecting chiropterans; however, a recent work raised doubts about the classically identified bat reservoir.[2,3] Recently a virus has been isolated from reptiles (Boinae subfamily), and identified as an arenavirus; if it is confirmed as a new arenavirus species, it will reveal additional insight into the arenavirus family, in relation to its potential to adapt from one host to another, to coevolve, and to pose zoonotic risk in other host species.[4]

Several arenaviruses are known to infect humans accidentally and cause a range of zoonotic diseases, from mild to severe illness and, eventually, death. Although the arenavirus prototype species, the *Lymphocytic Choriomeningitis Virus* (LCMV) of mice, is responsible for a neurological syndrome in humans, viral hemorrhagic fever (VHF) represents the main arenavirus-associated severe syndrome in humans. Bleeding tendencies are often recorded but not always life threatening, with a mortality rate of 30% during epidemics. Nonhuman primates can be infected experimentally, but there is no evidence that these viruses are pathogenic for domestic animals (e.g., livestock, cats, and dogs), although exotic pets (hamster, mice, etc.) represent a potential source of infection, in particular for the LCMV.[5,6]

The Arenaviridae family consists to date of a unique *Arenavirus* genus including 24 recognized virus species (Table 2.1) and two pending species not yet registered.[7] Among the 24 arenavirus species, 7 are highly pathogenic for humans and responsible for causing hemorrhagic fever. These include Machupo virus (MACV) and Chapare virus

(CHAV) responsible for the Bolivian hemorrhagic fever (BHF); Junin virus (JUNV) causing the Argentine hemorrhagic fever (AHF); Guanarito virus (GTOV), which is the etiologic agent of the Venezuelan hemorrhagic fever (VeHF); Sabia virus (SABV) that causes the Brazilian hemorrhagic fever (BrHF); Lassa virus (LASV) associated with the Lassa fever (LF) in West Africa; and, more recently, the Lujo virus (LUJV) responsible for an hemorrhagic fever syndrome in Southern Africa.[8] Also, four other arenaviruses including Flexal virus (FLEV), Pichinde virus (PICV), TACV, and White Water Arroyo virus (WWAV) have the potential to infect humans: FLEV has resulted in two nonfatal severe infections among laboratory workers but needs more study to evaluate its pathogenicity; several seroconversions have been reported with PICV without clinical manifestations; TACV caused a single case involving central nervous system signs of a mild febrile syndrome; and WWAV has been incriminated in three fatal cases in California presenting a severe respiratory distress syndrome and hemorrhagic fever signs.[1,9–11]

Thus since the discovery of the first arenavirus, the LCMV, in the 1930s, new species constantly emerged, with some being highly pathogenic to humans and others posing potential health risk to the public (Figure 2.1). As rodents are the main virus reservoir, the zoonotic niche appears spatially limited by the availability of rodent species and also restricted to a domain of transmission fitness between virus and rodent species.[12,13a,13b] Human infection accidentally occurs when the risk of encountering (i.e., proximity) with infected rodent species increases.

In the present chapter, as the LCMV is not considered a potential biological threat, its pathology (neurological syndrome) and ecology (ubiquity) will not be discussed in detail. However, given its similarity to the other arenaviruses that are primarily responsible for VHFs in terms of having an eco-epidemiology spatially reduced to foci of zoonotic transmission within the reservoir's distribution area, a phenomenon

TABLE 2.1
Arenavirus

Virus Species	Human Pathogenicity Cardinal Signs[a]	Mortality% (Mean%)/ Seroprevalence	Geographic Distribution	Primary Natural Host Reservoir	Acronym
Old World Arenaviruses (Afrotropic)					
Ippy virus	No evidence	—	Central African Republic	*Arvicanthis niloticus*	IPPYV
Lassa virus	HF[a], sore throat, nausea	1–50 (17)/0%–55%[b]	West Africa (Sierra Leone, Nigeria, and Guinea)	*Mastomys* sp.	LASV
Lujo virus	HF, diarrhea, vomiting, sore throat	4/5 (–)/___[c]	Zambia	Unknown	LUDV
Mobala virus	No evidence	—	Central African Republic	*Praomys* sp.	MOBV
Mopeia virus	No evidence	—	Mozambique, Zimbabwe	*Mastomys natalensis*	MOPV
Old World Arenaviruses (Palearctic, Nearctic, Neotropic)					
Lymphocytic choriomeningitis virus	Meningitis or encephalitis, biphasic fever, nausea, teratogenic	<1 (2–5)/1–4	Europe, America (USA, Argentina)	*Mus musculus*	LCMV
New World Arenaviruses of North and Central America (Nearctic)					
Bear Canyon virus	No evidence	—	USA	*Peromyscus californicus*	BCNV
Tamiami virus	No evidence	—	Florida, USA	*Sigmodon hispidus*	TAMV
Whitewater Arroyo virus	HF, respiratory distress, liver failure	3/4	New Mexico, USA	*Neotoma albigula*	WWAV
New World Arenaviruses of South America (Neotropic)					
Allpahuayo virus	No evidence	—	Peru	*Oecomys bicolor*	ALLV
Amapari virus	No evidence	—	Brazil	*Oryzomys capito*	AMAV
Chapare virus	HF, headache, body aches	1/?	Bolivia	Unknown	CHPV
Cupixi virus	No evidence	—	Brazil	*Oryzomys capito*	CPXV
Flexal virus	Fever	0/2	Brazil	*Oryzomys capito*	FLEV
Guanarito virus	HF	1/2	Venezuela	*Zygodontomys brevicauda*	GTOV
Junín virus	HF	20–30	Argentina	*Calomys musculinus*	JUNV
Latino virus	No evidence	—	Bolivia	*Calomys callosus*	LATV
Machupo virus	HF	5–30	Bolivia	*Calomys callosus*	MACV
Oliveros virus	No evidence	—	Argentina	*Bolomys obscurus*	OLVV
Paraná virus	No evidence	—	Paraguay	*Oryzomys buccinatus*	PARV
Pichinde virus	Subclinical infection	0%	Colombia	*Oryzomys albigularis*	PICV
Pirital virus	No evidence	—	Venezuela	*Sigmodon alstoni*	PIRV
Sabiá virus	HF	1/2	Brazil	Unknown	SABV
Tacaribe virus	Fever, mild neurological signs	0/1	Trinidad	*Artibeus* sp.	TCRV

Note: ? indicates no data is available.
[a] Major clinical signs including HF = hemorrhagic fever clinical (petechial, fever, etc.) and biological (thrombocytopenia, etc.) signs.
[b] Mortality (mean mortality)/seroprevalence in endemic area.
[c] Number of deceased patients/number reported.

quoted as "nidality",[13b] we will evoke here the LCMV as a classical model of the arenavirus general biology.

2.2   CLASSIFICATION AND MORPHOLOGY

Arenaviruses are single stranded RNA viruses with a genome of two ambisense segments named "L" for large (N7.2 kb) and "S" for small (3.5 kb). The L genomic segment encodes a viral RNA (vRNA)-dependent RNA polymerase and also a zinc-binding protein. The S genomic segment encodes the nucleocapsid and also the envelope glycoproteins in

nonoverlapping open reading frames (ORFs) of opposite polarities. An intergenic noncoding region separates the S and L segments with potential for forming hairpin nucleotide secondary configurations. Nucleocapsid antigens are shared by most of the arenavirus and contribute to their common antigenicity. Virions are pleomorphic, 50–300 nm in diameter, with a dense lipid envelope, and a surface layer covered with characteristic club-shaped projections approximately 9 nm in length. Within a virus particle, RNA segments (S and L) are organized in closed circles of nucleocapsid of 450–1300 nm. Each segment codes for two proteins with ORFs in opposite

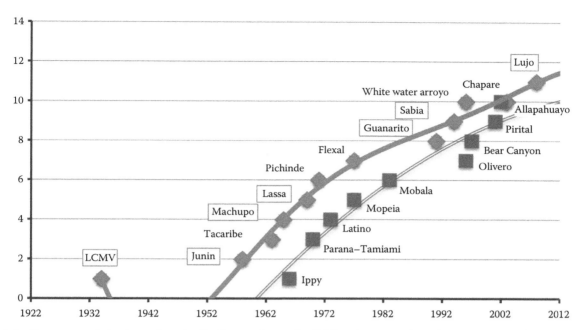

FIGURE 2.1 Emergence of the Arenavirus family. Square dots and solid line, human pathogenic virus species; framed name, highly human pathogenic (P4); Lozenge dot and double line, nonhuman pathogenic virus species.

sense. One protein's sequence is directly coded in the actual sequence of the single strand of RNA of the viral genome; the second is coded in the opposite sense and is coded by the sequence complementary of the vRNA (i.e., viral ambisense strategy). To distinguish between both sequences, one is called the vRNA, the sequence of the RNA actually found in the virions, and the other one is the viral-complementary RNA (vcRNA). The L genomic segment (~7.2 kb) encodes the vRNA-dependent RNA polymerase (L protein) and a zinc-binding protein (Z protein). The S genomic segment (~3.5 kb) encodes the nucleocapsid protein (N protein) and the envelope glycoproteins (GPC protein) in nonoverlapping ORFs of opposite polarities. The genes, of both S and L segments, are separated by an intergenic noncoding region with the potential of forming one or more hairpin configurations. The 5' and 3' untranslated ending sequences of each RNA segment possess a relatively conserved reverse complementary sequence spanning 19 nucleotides at each extremity.

Also a variable number of ribosomal ARN of 20–25 nm, originally captured from the host cell, are present in the virion and give the "sandy" (lat.: *arena*) aspect of it, when visualized by electron microcopy.

Regarding the arenavirus specificities and their shape at the molecular level, they do not differ from one another except for the intergenic hairpin(s) that can have an original tertiary configuration. However, genotypes and proteins (i.e., amino acids) have a unique arrangement with a maximum of 98% of homology among the entire arenavirus family.

Therefore, arenaviruses have often been viewed as relatively stable genetically with amino acid sequence identities of 90%–95% among different strains of LCMV and of 44%–63% for homologous proteins of different arenaviruses.[14] However, considerable variation in biological properties among LCMV

strains has become apparent, with dramatic phenotypic differences among closely related LCMV isolates.[15,16] These data provide strong evidence of viral quasispecies involvement in arenavirus adaptability and pathogenesis.[17]

Arenaviridae family contains a single *Arenavirus* genus originally including two serogroups, later confirmed by phylogenetic analyses, which are (1) The Old World complex (i.e., serogroup) with LCMV (responsible for viral encephalitis and meningitis), LASV, and LUJV, the latter two being responsible for VHF in Africa, and three others nonhuman pathogenic, namely, Ippy virus (IPPV), Mobala virus, (MOBV), and Mopeia virus (MOPV); (2) The New World complex (alias Tacaribe complex serogroup) that contains endemic arenavirus of the Western Hemisphere organized among three phylogenetic clades, which are Clade A with Pirital virus (PIRV), PICV, FLEXV, Parana virus (PARV), and Allpahuayo virus (ALLV); Clade B with five of the most pathogenic arenavirus for humans, JUNV, MACV, GTOV, CHAPV, and SABV, and three less- or nonpathogenic arenavirus for humans, Amapari virus (AMAV), TACV, and Cupixi virus (CUPV); Clade C that contains two nonhuman pathogenic arenavirus, namely, Oliveros virus (OLV) and Latino virus (LATV). Finally, one can identify a fourth cluster of genetic recombinant between parental species of Clades A and B including WWAV, Bear Canyon virus (BCNV), and Tamiami virus (TAMV) (Figure 2.2).[17,18]

2.3 BIOLOGY AND EPIDEMIOLOGY

Almost all arenavirus reservoir hosts belong to rodent species with the exception of TACV isolated one time only and solely from fruit-eating bats (Table 2.1)[2] and the, yet to be confirmed, arenavirus potentially infecting snakes.[2,4]

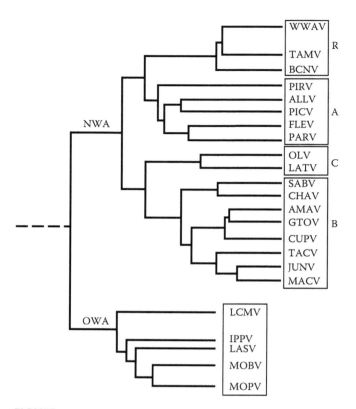

FIGURE 2.2 Arenavirus phylogeny. Cladogram using nucleoprotein amino acid sequences. NWA, New World Arenavirus; OWA, Old World Arenavirus; R, recombinant; A, B, and C, clades A, B, and C. (From Salvato, M. et al., Report of the International Committee on Taxonomy of Viruses, The Arenavirus Study Group, http://www.Ictvonline.Org/Subcommittee.Asp?Committee=3, 2012; Charrel, R.N. et al., *Virology*, 283, 161, 2001.)

The comparison of rodent hosts and arenavirus phylogenies suggests a long association and even a coevolution process, which started by the split of the two rodent groups—Murids vs. Cricetids—35 million years before present (mybp), and a virus–host cospeciation with an ancestral arenavirus, a process that conversely accompanied the present biodiversity of both hosts and viruses. The spread and speciation of the rodent hosts across the continents 10–8.6 mybp ago completed the species radiation. Thus, the biodiversity of arenaviruses is the result of a long-term, shared evolutionary relationship (cospeciation) between the Arenaviridae family and the Muridae family (Table 2.1).[19–21] Hence, the epidemiology of arenaviruses essentially depends on the rodent host's biology. The LCMV prototype arenavirus found in mice and the African viruses, both Old World arenaviruses, are mainly associated with the Murinae subfamily. The New World arenaviruses are found linked to Cricetidae subfamily. Although, besides the specific association between an arenavirus species and a rodent species, the geographic range of an arenavirus—ecologic niche—appears much more restricted than the one of its rodent reservoir host. Beyond the unique worldwide distribution of LCMV, all other arenaviruses have a limited spread within their host biotope territory, often limited by natural barriers for the rodents (river, elevation,

climate, and access to food). This appears to be one of the major characteristics of the epidemiological and dispersion patterns of the arenaviruses (i.e., associated VHFs). Other factors may explain this "natural nidality" as, for example, an independent monophyletic lineage of the reservoir rodent, *Calomys callosus*, in the case of the MACV.[13b]

Most of the arenaviruses infecting their natural rodent hosts induce an asymptomatic infection. Such types of infections are generally suspected to be associated to an insufficient or inappropriate host immune response.[7] Also, rodent asymptomatic infection leads to a chronic viremia and/or viruria, resulting in the shedding of the virus in the environment through urine or droppings. Virus transmission within rodent populations can occur through vertical (mother to progeny) or horizontal routes (directly through bites or indirectly by contacts with urine or feces). Such a persistent infection may be the result of a crucial adaptation for the long-term persistence of arenaviruses in nature. Exceptionally, such chronic infection could have a deleterious effect on their reservoir's fitness, which reduces rodent host fertility. JUNV- and MACV-infected neonatal female rodents are subfertile, while an increased horizontal transmission compensates and favors virus persistence among the rodent populations.[22]

Arenavirus transmission from natural host to humans occurs through contacts with infected rodent biological fluids (i.e., blood, saliva, or urine) when people (through rodent bites, trapping, or eating rodents) are directly exposed to the infected rodent or, indirectly, when exposed to rodent urine contaminated food or environment or by inhalation of infected rodent excreta. Also, human-to-human transmission can occur and arenaviruses can be transmitted through aerosolized particles and sperm fluid. Moreover, transmission to humans may occur by accidental inoculation with infected body fluids and through tissue transplantation.[8,17,23–27]

CHAV was isolated from a fatal human case of hemorrhagic fever during a unique reported outbreak constituted by a small cluster of hemorrhagic fever cases. Outbreak occurred in January 2003 in the village of Samuzabeti near the Chapare River region close to Cochabamba in Bolivia, the original setting of MACV emergence.[28] There is no information concerning an eventual natural rodent host.

Venezuela HF emerged in 1989, with several cases that occurred in the central plains of Venezuela. A new virus was isolated and designated GTOV after the region where the first outbreak occurred.[29] The main affected population was settlers moving into cleared forest areas to practice small agriculture. GTOV has been isolated from several rodent species but, as for the majority of the arenaviruses, only one species appears to be the principal host, that is, the reservoir host of GTOV, a Cricetid, *Zygodontomys brevicauda*.

Although AHF was identified in the early 1940s in Argentina and described in the early 1950s in the rural area of Buenos Aires Province,[30] JUNV was isolated and characterized in 1958. At the time of discovery, JUNV was confined to circulate in an area of around 15,000 km². At the beginning of the year 2000, the JUNV distribution had expanded to 150,000 km². The natural hosts of JUNV are rodents, particularly *Mus musculus*, *Calomys* spp., and *Akodon azarae*. Direct rodent-to-human transmission only happens when contact is made with excrement of an infected rodent. This commonly occurs via ingestion of contaminated food or water, inhalation of rodent urine–infected particles or via direct contact of broken skin with rodent excrement. Currently, AHF remains a major and severe enzootic disease of public importance in Argentina with an endemic risk of crossing the natural barrier of the Rio Paraná and spill over to the closest neighboring countries of the endemic area of Santa Fe to Uruguay.

LF occurs in rural West Africa and appears hyperendemic in Sierra Leone with an antibody prevalence of 8%–52%, also in Guinea (4%–55%) and Nigeria (21%). Natural transmission of LASV occurs from its ubiquitous, prolific, and common multimmate rodent reservoir, *Mastomys natalensis*. LASV is transmitted through rodent-infected dejections, droppings, and urine, which can be easily transmitted by a direct contact to humans or indirectly by contaminated food, based on the multimmate rodent and its primarily domestic habits. Transmission from a person to another has been early established during the first outbreaks of LF and is a risk for health-care workers. LASV can be contracted by an airborne route or by direct contact with infected human blood, urine, or semen and with the late (up to 3 months after) clinical recovery.

Transmission through breast milk has also been observed. Also, LASV is a prominent threat outside the area of endemicity with several imported cases in Europe, Japan, the United States, and the Middle East.[31–36]

About 80% of patients with the LASV experience a mild or asymptomatic infection. Also, LASV has a relatively low mortality (~1%–5%), LF accounts for up to one third of deaths of hospitalized cases. Among the endemic countries, it is estimated that LF is responsible for about 5000 deaths a year. Women in the third trimester of pregnancy and babies have the greatest risk. After an incubation period of up to 3 weeks, an acute illness develops while LASV infects almost every tissue in the human body from the mucosa, intestine, lungs, and urinary system and then progresses to the vascular system. Nonspecific symptoms include fever, facial swelling, and muscle fatigue, as well as conjunctivitis and mucosal bleeding. Other symptoms arise from the affected organs, including gastrointestinal tract bleeding, nausea and bloody vomiting, dysphagia, stomachache, melena, accompanied with cough, dyspnea worsening to cardiovascular system dysfunctions (pericarditis, tachycardia) and hepatitis, and ultimately hearing deficit, meningoencephalitis, seizures, and death. Due to multiorgan involvement and nonspecific clinical signs, LF infections are difficult to distinguish from other VHFs such as Ebola or Marburg, or other common illnesses such as malaria or influenza.[37]

LUJV was associated with an outbreak of five cases of VHF in 2008, four had a fatal issue and the fifth case was treated with Ribavirin® early after the onset of clinical disease and survived. After LASV, LUJV is the second arenavirus

found to be human pathogenic in the African continent. It has only been reported from a few patients from Zambia and from a subsequent nosocomial outbreak in South Africa, the natural reservoir has not yet been identified.[8]

MACV was isolated in 1963 in the Beni Province of Bolivia after several outbreaks of BHF, referred in that time to an hemorrhagic clinical syndrome named "Black Typhus" which occurred since 1959. The incidence increases from April to July (late rainy and early dry season), but the dominant feature of the epidemiological pattern is the small outbreaks in different villages and ranches with several years of dormancy thereafter. *C. callosus*, the natural host of MACV, invades houses during floods of the rainy season, which results in human contaminations. BHF cases occur at the same rate in men, women, and children in small towns, while in remote rural area, adult male patients predominate.[24]

SABV was isolated from one fatal case of BrHF, initially thought to be a case of yellow fever. The transmission naturally occurred in a woman working temporarily in the village of Sabia, outside of Sao Paulo, Brazil, in 1990. It was considered as the first case of the emerging and newly named BrHF.[38] It was then followed by two other nonlethal accidental infections.[39] Lately, CHAV-infected patients were also clinically considered as BrHF cases. SABV and CHAV viruses do not have an identified reservoir; however, like the other arenaviruses, they naturally appear to have a limited geographical area. Another arenavirus reported in 1975 from *Oryzomys* spp. of Brazil, FLEV, has been reported to infect laboratory workers several times.[40,41]

WWAV appears to be endemic of Southwestern United States and hosted by several species of the *Neotoma* genus (woodrats). Infected rodents have been reported from New Mexico, where occurred the initial recognition of the WWAV infecting *Neotoma albigula* and, afterward, identified from Oklahoma, Utah, and Texas in other *Neotoma* spp.[9,42]

2.4 INFECTION FEATURES AND PATHOGENESIS IN HUMANS

LASV, JUNV, MACV, GTOV, and SABV are known to cause a severe HF, in western Africa, Argentina, Bolivia, Venezuela, and Brazil, respectively.[40] More recently, CHAV and LUJV joined this group of highly pathogenic arenaviruses responsible for VHFs. Infection route to humans is known to principally occur through skin lesions or epithelia (lungs, cornea) and aerosol of dried rodent excreta. Arenavirus HF pathogenesis comes from direct viral infection of endothelial cells, resulting in a vascular dysfunction and shock. High levels of circulating endogenous interferon-alpha and pro-inflammatory cytokines are present that appear more detrimental rather than beneficial for the patient.[43]

All known VHFs caused by arenaviruses from South America (SAHFs) present the same clinical course. On average, Argentine, Bolivian, Venezuelan, and Brazilian arenavirus HF exhibit a mortality rate of 22% ± 7%.[44–46] Generally, SAHFs range in severity from mild febrile infections to severe hemorrhagic syndrome. Vascular leak, shock, and multiorgan dysfunction are prominent features. The first symptoms shared by all SAHFs are fever, headache, myalgia, conjunctival suffusion, bleeding, and abdominal pain. There is a pronounced thrombocytopenia and leukopenia, and bone marrow cells can be destroyed. Some complement components are consumed and a progressive alteration in vascular permeability occurs, leading to the hemorrhagic feature of these diseases.[47] Necrosis may appear in several organs like the liver or kidneys, and inflammation of the central nervous system and myocardium was reported.[48] Also, neurological symptoms such as tremor, alterations in consciousness, and seizures are often observed. In the severely ill patients shock develops 7–9 days after onset. Hemorrhages and shock herald a pessimistic prognosis.

Nevertheless, SABV was responsible for single natural infections and the number of CHAV infection cases and clinical records are not totally assessed, one can retain some clinical aspects: in one instance, for one fatal case of CHAV, a 22-year-old male patient, clinical course included fever, headache, arthralgia, myalgia, and vomiting followed by multiple hemorrhagic signs and death 2 weeks after onset.[28,49,52] Regarding symptoms severity and clinical presentation, as it was for SABV, the patient was initially suspected of yellow fever infection.[28]

GTOV causes a severe disease, the VeHF characterized by fever, malaise, sore throat, headache, arthralgia, vomiting, followed by abdominal pain, diarrhea, convulsions, and a variety of hemorrhagic manifestations. Patients also had leukopenia and thrombocytopenia. The overall fatality rate among 165 cases was 33.3%, despite intensive care.[50] The disease affects mostly agricultural male workers, between 14 and 54 years of age. Since its first recognition in 1989 up to 1997, 220 cases have been reported, with a fatality rate of 33%. VeHF has a cyclic performance, with epidemic periods of high incidence between November and January, every 4–5 years during the period of high agricultural activity involving mostly male agricultural workers.

JUNV infection, AHF, classically leads to major alterations within the vascular, neurological, and immune systems with a mortality rate of 20%–30%. Major symptoms are conjunctivitis, purpura, petechia, and occasional sepsis.

For the BHF, after an incubation period of 1–2 weeks, patients infected with MACV develop a slow onset of an influenza-like syndrome with fever, malaise, and fatigue, followed by the onset of headache, dizziness, myalgias, and severe lower back pains. Prostration, abdominal pain, anorexia, tremors, and hemodynamic instability may be followed, in some patients, by an hemorrhagic phase, including petechia on the upper body and bleeding from nasal mucosa, gums, and the gastrointestinal, genitourinary, and bronchopulmonary tracts.[51] A few patients develop neurological symptoms. Death can occur between a few hours and a few days after onset. For patients who recovered, the acute disease lasted 2–3 weeks.

The first case of BrHF, a young healthy woman staying in the village of Sabia experienced at first; high fever, headache, myalgia, nausea, vomiting, weakness, and pronounced sore throat were symptoms exhibited. Additional symptoms such as conjunctivitis, diarrhea, epigastria, and bleeding gums were observed. Symptoms lasted approximately for

15 days. Gastro intestinal hemorrhage was marked, though generalized hemorrhagic fever and severe liver damage led to an initial diagnosis of yellow fever. The patient did not survive. The technician responsible for SABV isolation and identification also contracted the disease during the diagnostic process, and fortunately survived. Four years later, while working under biohazard level 3 conditions, a researcher at Yale New Haven Medical School was accidentally exposed to the virus and successfully treated with Ribavirin. All of the three aforementioned patients had leukopenia, severe thrombocytopenia, and proteinuria.[29,52]

LCMV infection results in an acute central nervous system syndrome and newborn congenital malformations.[53,54]

PICV has resulted in numerous seroconversions without any notable clinical significance.[55]

FLEV has resulted in two symptomatic laboratory infections, and should then be considered as potentially dangerous.[40,41]

TACV has resulted in a single case of febrile disease with mild central nervous system symptoms.[40]

In terms of immunopathology, ultimately, arenaviruses constitute a relevant model system to study virus–host relationships. The interaction with the cellular receptor and subsequent entry into the host cell differs between Old World and New World arenaviruses that use α-dystroglycan and human transferrin receptor 1, respectively, as main receptors for entry into human cells.[56–58] Works from Reignier et al. also questioned the ability of the WWAV to be pathogenic for humans as it appears as a Clade A/Clade B recombinant not using the transferrin receptor 1.58. Emonet et al. presented in their review the LCMV infection of the mouse as the Rosetta stone of the virus–host interaction.[17] Given the absence of proof-reading and post replicative repair mechanisms, RNA viruses in general produce error copies during replication. Hence, even if the consensus genomic sequence of an RNA virus remains invariant (i.e., a collection of mutant genomes) during replication, the population may still be highly dynamic, with mutants (i.e., quasispecies) arising at all times.[59a,59b] The virus is "mutating toward itself" as claimed by Emonet et al.[17] Hence, a large available pool of genomes increases the ability of a population to face environmental changes. The virus in vivo in its reservoir, the domestic mouse, is noncytolytic, and can cause either acute or persistent infections. When immunocompetent adult mice are injected with LCMV, they generate a robust immune response that results in virus clearance, which is mainly mediated by major histocompatibility complex.[60,61] By contrast, mice infected neonatally or in uterus with LCMV become persistently infected for life. In the course of such persistent infections, distinct viral variants can be isolated from the brain and lymphoid tissue, and these variants have different biological properties that correlate with the tissue that they have been isolated from.[62–66] Brain isolates are similar to the parental Armstrong (Arm) strain used to infect the mice, and the infection is cleared within 2 weeks. In contrast, isolates derived from the lymphoid tissue cause chronic infections.[67] The emergence of cell-specific viral variants is explained by the different conditions encountered in the various cell types resulting in the selection of specific viral variants that have a growth advantage in the given cell types.[68]

2.5 IDENTIFICATION AND DIAGNOSIS

Initial stages of the infection are often indistinguishable from other common viral diseases, including the prevalent yellow fever or rising dengue fever and dengue hemorrhagic fever. Therefore, virology testing is undeniably necessary in any cases of suspected arenavirus infection. A rapid diagnosis is also necessary because of the rapid evolution of any infection by the New World arenaviruses, and as death may sometimes occur even before antibodies are detected.

Thus preliminary tests can be targeted at a specific virus according to the natural nidality (i.e., geographically limited enzootic area) of the arenaviruses. The diagnosis of arenavirus infection in nature should be directed by the identification of a suspected area for the contamination overlapping with known endemic areas of the pathogenic arenaviruses, and identification of other risk factors such as potential contacts of the patient with rodents and/or activities in an environment infested with rodents (e.g., agricultural occupation, rodent trapping). Another route of contamination that may drive the diagnosis is a recent contact with a patient with hemorrhagic fever.

In any case of a suspected acute human infection by an arenavirus, all laboratory investigations must be carried out on high security conditions (BSL4). Also, secondarily, containment level can be downgraded using appropriate virus inactivation and process in order to carry out further investigations with noninfectious material (http://www.bt.cdc.gov/lrn/).

For *virus isolation* attempt from a patient, samples have to be taken at the early stage of infection (i.e., before the pick of fever) when viremia occurs. After that stage, only free RNA will be recovered by PCR analysis. Also antigen detection can be carried out on tissue samples after 2–3 weeks after onset, depending on the virus species.

The virus sample collection from wild potential hosts has to be done on blood and organs/tissues that will be used for virus detection (PCR) and isolation (cell culture), identification (genetics), and characterization (proteins, nucleotide) of the etiologic agent. Also, molecular epidemiology (sampling) can be carried out on the preserved RNA sample (PCR/Phylogeny).

One can distinguish basically (1) direct methods for diagnosis that detect the virions, their genome, or some of their antigens, or (2) indirect methods that highlight a past contact with the pathogen through antibody detection. The main advantage of direct detection is that they demonstrate the actual infection and presence of the virus in the patient at the time of sample collection, while the interest of tests based on antibody detection is that the work will be done with inactivated viral antigens that do not require specific high protection, taking into account that field samples need to be handled with specific safety rules and good laboratory practices for potentially highly infectious agents.[69]

Biochemical diagnostic testing of blood and urine samples for the main hemorrhagic syndrome involves several tests; these include coagulation components such as white blood count, platelet count, prothrombin time and partial thromboplastin time, bleeding time, liver function tests, fibrin split products, fibrinogen, urea acid and blood urea nitrogen/creatinine, electrolytes, glucose, pH, and bicarbonate levels. However, bleeding patients need a special care preventing any unnecessary extemporaneous venipuncture and the risk of bleeding.

Specific diagnostic testing can be carried out by a limited number of specialized laboratories dedicated for highly pathogenic agents (P3 to P4 level of protection including sample virus inactivation before use) including antigen-capture ELISA, RT-PCR (most useful clinically), IgM by antibody-capture ELISA. Viral isolation and or vRNA extraction from acute samples requires a high security P4 laboratory. Convalescence for IgG serology in survivors (retrospective studies) has to be carefully managed under biosafety level 3 conditions with potential virus inactivation. Also, in order to avoid any contact with potential infectious material, several approaches can be developed to protect the operator, as for example, recombinant LASV proteins and monoclonal antibodies for diagnostic applications have been developed that can be safely used on inactivated samples; or, unbiased pyrosequencing of LUJV RNA extracts from serum and tissues of outbreak victims that enables identification and detailed phylogenetic characterization within 72 h of sample receipt.[27,70]

Differential diagnosis from other diseases with hemorrhagic signs, including dengue hemorrhagic fever, viral hepatitis, Gram-negative sepsis, toxic shock syndrome, meningococcemia and other bacterial sepsis, leptospirosis, malaria, hemorrhagic smallpox, among others, has to be considered regarding the clinical stage of VHF, epidemiological data (virus reservoir host potential contact) and environmental factors (endemic area).

2.6 TREATMENT AND PREVENTION

Patient isolation: Strict VHF-specific barrier precautions must be initiated on any suspicion of VHF disease.[51,71] Airborne precautions and negative-pressure isolation room are recommended. Close medical surveillance is emphasized for all those with close or high-risk contact or blood exposure within 21 days of a patient's onset of symptoms. Also convalescent patients should abstain from sexual activity for 3 months as reproductive cells can retain live viral particles for several weeks after patient recovery.

There is no specific treatment except for Ribavirin an analogic of purine nucleotide, nor vaccine, except for the live attenuated Candid #1 directed against AHF. Besides, no antiviral preparation has been approved yet or commercially available, and no other vaccines are marketed. Therefore, provision of supportive care blood pressure monitoring and special attention to fluid and electrolytic balance are critical to ensure patient's recovery.

Ribavirin is an antiviral drug, a prodrug that is administered in an inactive form, and subsequently converted into an active drug through normal metabolic processes by resembling a purine RNA nucleotide (nucleoside analog) that interferes with RNA viral replication. Also, Ribavirin has been proved to be an efficient treatment against arenavirus if administered at the early stage of LF and, in some cases, could be effective against other arenavirus including BHF, SABV (BrHF), or LUJV. Also, it has been shown to be effective in the advanced stage of LASV infection, reducing the virus load.[24,27,52,72,73] Also, as mentioned by Charrel et al. there are several antiviral molecules under development and the most promising are directed to interfere with the membrane fusion of the arenavirus cell entry process.[74–76]

Although hyperimmune sera treatment has been effectively used in several instances, clinical experiences are limited and only circumstantial reports are available. Hyper immune sera treatments have been used successfully for AHF patients and plasma banks were developed in Argentina.[77] Also, neutralizing antibodies contained in human immune plasma appears to be effective on patients with BHF, reducing the viremia. However, LASV infection produces a limited neutralizing antibody reaction, and hyperimmune sera treatment is not applicable.[78]

The "Candid vaccine" (Candid #1) was created in 1985 by Argentine virologist Julio Barrera Oro and manufactured by the Salk Institute in the United States, and then became available in Argentina in 1990. The 95.5% protective efficacy for prevention of any illness associated with JUNV infection was clearly demonstrated. No serious adverse events were attributed to vaccination. Candid #1 was, and still is, the first safe and highly efficacious vaccine for the prevention of illness caused by an arenavirus.[79] Candid #1 is now locally manufactured in Argentina; it was finally registered in 2010. Other promising vaccines are at the experimental stage on animals, including DNA vector vaccine and live vaccine. Live-attenuated Lassa vaccine candidates have been investigated and are under development, including a reassortant of Lassa and Mopeia viruses, and a recombinant between yellow fever vaccine (YF17D) and the Lassa glycoprotein.[80] Moreover, another original and challenging approach has been recently developed using a cell-mediated vaccine strategy to develop a cross-protective vaccine against several species of arenaviruses.[81]

Altogether, specific treatments (i.e., vaccine and antiviral drug) against arenavirus HF remain scarce. The Candid #1 vaccine reduced the AHF incidence by more than 1500 to less than 100 cases a year. However, in Africa, the LF remains highly incident with an estimate of annual number of cases across the endemic countries of Sierra Leone, the Republic of Guinea, Nigeria, and Liberia, greater than 450,000 infections with a minimum mortality rate of 10%. Arenavirus diseases have to be considered as truly neglected diseases and much need to be done regarding the Arenaviridae family model that has served important knowledge acquisition and discovery in the field of virology and immunology sciences.

Arenaviruses are inactivated by most of the detergents and disinfectants including the classical 1% sodium hypochlorite and 2% glutaraldehyde. They are also altered by ultraviolet light gamma irradiation and can be inactivated by temperature greater than 56°C and by a pH less than 5.5 or greater than 8.5.

For decontamination of environmental surfaces and equipment, 1:10 to 1:100 dilution of sodium hypochlorite or other EPA-registered disinfectant should be used. Linens should be handled as per CDC guidelines.[51]

2.7 BIOSECURITY

Biosecurity covers firstly intentional exposure to a population with a harmful agent of biological origin, but it also considers the natural and accidental exposure to such agents. Natural exposure refers to the epidemiology of the agent responsible for HF with a special interest in its epidemiological patterns of high infectivity and propensity for rapid diffusion to a naive population. Accidental exposures are essentially due to an exposition of a person attending a patient or during laboratory handling of infectious material (i.e., diagnostic or research).

With consideration to intentional exposure, VHFs are considered to be a significant threat if used as biological weapons due to several of their characteristics including highly infectious low doses; a potential of person-to-person spread; a dissemination ability through aerosols with high rates of morbidity and mortality; a tendency to present significant hemorrhagic syndrome with a bleeding diathesis from petechiae to profuse bleeding leading to death; clinical and epidemiological patterns often causing fear and panic in the general public with respect to the dramatic clinical presentation of the hemorrhagic syndrome; and, ultimately, that they can be readily produced in large quantities. Moreover, effective vaccines are not available or limited and, in 1999, the US Centers for Disease Control and Prevention classified most of the VHF agents as "category A" bioweapon agents.[10] All VHF etiological agents belong to the RNA viruses including essentially four viral families: Bunyaviridae (Rift Valley fever virus, Crimean Congo hemorrhagic fever virus, Hantaan virus), Filoviridae (Ebola virus and Marburg virus), Flaviviridae (yellow fever virus, Omsk hemorrhagic fever, and Kyasanur Forest disease virus), and Arenaviridae families. Half of the 14 viruses clearly identified as potential bioweapon agents belong to the Arenaviridae family, with one third of them belonging to the Arenaviridae family, which includes LASV, MACV, GTOV, and SABV, while the more recently isolated CHAV, WWAV, and LUJV are under consideration. All of them potentially present characteristics of bioweapon agents.[71] Moreover, animal studies, using nonhuman primates, have shown that clinical infection can be caused by aerosolized preparations of arenavirus including Junín virus and LASV. Both can be released as low virus concentration aerosols and, consequently, besides their other characteristics (i.e., infectiosity and lethality), they have the suitable potential to serve as biological weapons.[82–84]

Also, it is of importance to understand that known and basic laboratory safety procedures, hospital- and clinical-based good practices applied for infectious diseases, are highly efficient and can ultimately halt epidemics by patient isolation and standard infection control (i.e., barrier nursing procedures). Furthermore, human infection by aerosol can cause mass casualties if deliberately disseminated and requires significant public health preparedness.[71,85]

LSAV can also be an important threat outside its endemic areas while viremic patients within the subclinical incubation period are traveling and ultimately seeking for care. Subsequently, several imported cases have been recorded throughout the world.[31–36] Moreover, other arenaviruses can also be a threat for the populations outside but geographically close to the endemo-enzootic area as demonstrated by the emergence of a nosocomial outbreak of a novel arenavirus (LUJV) imported by a patient from Zambia and hospitalized in South Africa.[8]

2.8 CONCLUSIONS AND FUTURE PERSPECTIVES

Rodents, as well as chiropterans, are the most widespread mammals with a unique biodiversity, a worldwide distribution. Also, both present high density of population and a migratory behavior. Altogether, rodents are the main arenavirus reservoir and, if they had coevolved for a long time, it is understandable that new virus species will be isolated time to time as much as we study and understand the ecology and pathology of their natural reservoir. Also, the known genome plasticity of the arenavirus, their ability to jump from one rodent species to another, and their potentiality for ARN recombination represent potential mechanisms that can lead to a high risk of new pathogen species emergence. Indeed arenaviruses emergence occurs mostly when a newly recognized arenavirus infects humans (human pathogenic) or when the strategy of virus discovery is applied targeting one or another micromammal species (nonhuman pathogenic). Arenavirus species continue and will continue to emerge; also any rare, unique, and inaugural case that occurs, for example, by the past for SABV, LUJV, or TACV, will need to draw all our attention in order to understand the fundamentals of arenavirus emergence.

In addition, if the acquisition of knowledge on the mechanisms of arenavirus emergence remains a fundamental quest, there is a need to favor research on active molecules against the associated HF. Respectively, antiviral drugs or vaccine need to be produced at a low cost, taken orally, and able to withstand tropical climates where most of these infections occur. For this reason, as proposed by Lee et al., drug discovery based on screening large collections of synthetic small molecules could be an immediate action to take and be an answer to finding a better remedy for neglected diseases.[86–88]

REFERENCES

1. Gonzalez, J.P., Pourrut, X., and Leroy, E. Ebolavirus and other filoviruses. Wildlife and emerging zoonotic diseases: The biology, circumstances and consequences of cross-species transmission. *Current Topics in Microbiology and Immunology* 315, 363–387 (2007).
2. Downs, W.G. et al. Tacaribe virus, a new agent isolated from *Artibeus* bats and mosquitoes in Trinidad, West Indies. *American Journal of Tropical Medicine and Hygiene* 12, 640–646 (1963).
3. Cogswell-Hawkinson, A. et al. Tacaribe virus causes fatal infection of an ostensible reservoir host, the Jamaican fruit bat. *Journal of Virology* 86, 5791–5799 (2012).
4. Stenglein, M.D. et al. Identification, characterization, and in vitro culture of highly divergent arenaviruses from Boa Constrictors and annulated tree boas: Candidate etiological agents for snake inclusion body disease. *Mbio* 3, 12 (2012).
5. Maetz, H.M., Sellers, C.A., Bailey, W.C., and Hardy, G.E. Lymphocytic choriomeningitis from pet hamster exposure—Local public-health experience. *American Journal of Public Health* 66, 1082–1085 (1976).
6. Ceianu, C. et al. Lymphocytic choriomeningitis in a pet store worker in Romania. *Clinical and Vaccine Immunology* 15, 1749 (2008).
7. Salvato, M. et al. Report of the International Committee on Taxonomy of Viruses. The Arenavirus Study Group (2012). http://www.Ictvonline.Org/Subcommittee.Asp?Committee = 3 (accessed on June 30, 2013).
8. Paweska, J.T. et al. Nosocomial outbreak of novel arenavirus infection, Southern Africa. *Emerging Infectious Diseases* 15, 1598–1602 (2009).
9. Fulhorst, C.F. et al. Isolation and characterization of Whitewater Arroyo virus, a novel North American arenavirus. *Virology* 224, 114–120 (1996).
10. Byrd, R.G. et al. Fatal illnesses associated with a New World arenavirus-California, 1999–2000. MMWR 49, 709–711 (2000). http://www.cdc.gov/mmwr/preview/mmwrhtml/mm4931a1.htm.
11. Enserink, M. Emerging diseases—New arenavirus blamed for recent deaths in California. *Science* 289, 842–843 (2000).
12. Pavlovsky, E.N. *Natural Nidality of Transmissible Diseases with Special Reference to the Landscape Epidemiology of Zooanthroponoses*. University of Illinois Press, Champaign, IL, 261pp. (1966).
13a. Salazar-Bravo, J., Ruedas, L.A., and Yates, T.L. Mammalian reservoirs of arenaviruses. In: Oldstone, M.B.A. (ed.), *Arenaviruses I: The Epidemiology, Molecular and Cell Biology of Arenaviruses*, Vol. 262, pp. 25–63. Springer, Berlin, Germany (2002).
13b. Salazar-Bravo, J. et al. Natural nidality in Bolivian hemorrhagic fever and the systematics of the reservoir species. *Infection Genetic and Evolution* 3, 191–199 (2002).
14. Southern, P.J. and Bishop, D.H.L. Sequence comparison among arenaviruses. *Current Topics in Microbiology and Immunology* 133, 19–39 (1987).
15. Sevilla, N., Domingo, E., and De La Torre, J.C. Contribution of LCMV towards deciphering biology of quasispecies in vivo. *Current Topics in Microbiology and Immunology* 263, 197–220 (2002).
16. Sevilla, N. and De La Torre, J.C. Arenavirus diversity and evolution: Quasispecies in vivo. *Current Topics in Microbiology and Immunology* 299, 315–335 (2006).
17. Emonet, S.F., De La Torre, J.C., Domingo, E., and Sevilla, N. Arenavirus genetic diversity and its biological implications. *Infection Genetics and Evolution* 9, 417–429 (2009).
18. Charrel, R.N., De Lamballerie, X., and Fulhorst, C.F. The Whitewater Arroyo virus: Natural evidence for genetic recombination among Tacaribe serocomplex viruses (Family Arenaviridae). *Virology* 283, 161–166 (2001).
19. Johnson, K.M., Mackenzi, R.B., Webb, P.A., and Kuns, M.L. Chronic infection of rodents by Machupo virus. A model of coevolution between rodents and arenaviruses. *Science* 150, 1618–1619 (1965).

20. Gonzalez, J.P. and McCormick, J.B. Proposal of a model of coevolution between arenavirus and rodents (in French). *Mammalia*. 50, 425–438 (1987).

21. Bowen, M.D., Peters, C.J., and Nichol, S.T. Phylogenetic analysis of the Arenaviridae: Patterns of virus evolution and evidence for cospeciation between arenaviruses and their rodent hosts. *Molecular Phylogenetics and Evolution* 8, 301–316 (1997).

22. Webb, P.A., Justines, G., and Johnson, K.M. Infection of wild and laboratory-animals with Machupo and Latino viruses. *Bulletin of the World Health Organization* 52, 493–499 (1975).

23. Peters, C.J. et al. Hemorrhagic fever in Cochabamba, Bolivia, 1971. *American Journal of Epidemiology* 99, 425–433 (1974).

24. Kilgore, P.E. et al. Treatment of Bolivian hemorrhagic fever with intravenous Ribavirin. *Clinical Infectious Diseases* 24, 718–722 (1997).

25. Fischer, S.A. et al. Transmission of lymphocytic choriomeningitis virus by organ transplantation. *New England Journal of Medicine* 354, 2235–2249 (2006).

26. Palacios, G. et al. A new arenavirus in a cluster of fatal transplant-associated diseases. *New England Journal of Medicine* 358, 991–998 (2008).

27. Briese, T. et al. Genetic detection and characterization of Lujo virus, a new hemorrhagic fever-associated arenavirus from Southern Africa. *PLoS Pathogens* 5, e1000455 (2009).

28. Delgado, S. et al. Chapare virus, a newly discovered arenavirus isolated from a fatal hemorrhagic fever case in Bolivia. *PLoS Pathogens* 4, e1000047 (2008).

29. Salas, R. et al. Venezuelan hemorrhagic fever. *Lancet* 338, 1033–1036 (1991).

30. Agnese G. "Una enfermedad alarma a la modesta poblacion de o'higgins" Análisis Del Discurso De La Prensa Escrita Sobre La Epidemia De Fiebre Hemorrágica Argentina De 1958. *Revista De Historia & Humanida Des Médicas* 3, 1958 (2007).

31. Van Der Heide, R.M. A patient with Lassa fever from the Upper Volta, diagnosed in the Netherlands. *Nederlands Tijdschrift Voor Geneeskunde* 126, 566–569 (1982).

32. Cummins, D. Lassa fever. *British Journal of Hospital Medicine* 43, 186–188 (1990).

33. Gunther, S. et al. Imported Lassa fever in Germany: Molecular characterization of a new Lassa virus strain. *Emerging Infectious Diseases* 6, 466–476 (2000).

34. Hirabayashi, Y. et al. An imported case of Lassa fever with late appearance of polyserositis. *Journal of Infectious Diseases* 158, 872–875 (1988).

35. Holmes, G.P. et al. Lassa fever in the United-States—Investigation of a case and new guidelines for management. *New England Journal of Medicine* 323, 1120–1123 (1990).

36. Schlaeffer, F., Barlavie, Y., Sikuler, E., Alkan, M., and Keynan, A. Evidence against high contagiousness of Lassa fever. *Transactions of the Royal Society of Tropical Medicine and Hygiene* 82, 311 (1988).

37. Yun, N.E. and Walker, D.H. Pathogenesis of Lassa fever. *Viruses* 9, 2031–2048 (2012).

38. Coimbra, T.L.M. et al. New arenavirus isolated in Brazil. *Lancet* 343, 391–392 (1994).

39. Gandsman, E.J., Aaslestad, H.G., Ouimet, T.C., and Rupp, W.D. Sabia virus incident at Yale University. *American Industrial Hygiene Association Journal* 58, 51–53 (1997).

40. Peters, C.J., Buchmeier, M., Rollin, P.E., and Ksiazek, T.G. Arenaviruses. In: Fields, B.N. et al. (eds.), *Fields Virology*, 3rd edn. Lippincott-Raven Publishers, Philadelphia, PA, pp. 1521–1551 (1996).

41. Peters, C.J., Jahrling, P.B., and Khan, A.S. Patients infected with high-hazard viruses: Scientific basis for infection control. *Archives of Virology* 11, 141–168 (1996).

42. Fulhorst, C.F. et al. Geographic distribution and genetic diversity of Whitewater Arroyo virus in the Southwestern United States. *Emerging Infectious Diseases* 7, 403–407 (2001).

43. Fan, L., Briese, T., and Lipkin, W.I. Z proteins of New World arenaviruses bind rig-I and interfere with type I interferon induction. *Journal of Virology* 84, 1785–1791 (2010).

44. Stinebau, B.J. et al. Bolivian hemorrhagic fever. A report of four cases. *American Journal of Medicine* 40, 217–230 (1966).

45. Sabattini, M.S. and Maiztegui, J.I. Argentine hemorrhagic fever. *Medicina* 30, S111–S128 (1970).

46. Maiztegui, J.I. et al. Ultrastructural and immunohistochemical studies in 5 cases of Argentine hemorrhagic fever. *Journal of Infectious Diseases* 132, 467–475 (1975).

47. Rimoldi, M.T. and Deedebracco, M.M. In vitro inactivation of complement by a serum factor present in Junin virus infected guinea pigs. *Immunology* 39, 159–164 (1980).

48. Walker, D.H. and Murphy, F.A. Pathology and pathogenesis of arenavirus infections. *Current Topics in Microbiology and Immunology* 133, 89–113 (1987).

49. Cajimat, M.N.R. et al. Genetic diversity among Bolivian arenaviruses. *Virus Research* 140, 24–31 (2009).

50. De Manzione, N. et al. Venezuelan hemorrhagic fever: Clinical and epidemiological studies of 165 cases. *Clinical Infectious Diseases* 26, 308–313 (1998).

51. Anon. Management of patients with suspected viral hemorrhagic fever. *MMWR* 37(S-3), 1–16 (1988). http://www.cdc.gov/mmwr/preview/mmwrhtml/00037085.htm

52. Barry, M. et al. Brief report—Treatment of a laboratory-acquired Sabia virus-infection. *New England Journal of Medicine* 333, 294–296 (1995).

53. Barton, L.L. et al. Congenital lymphocytic choriomeningitis virus-infection in twins. *Pediatric Infectious Disease Journal* 12, 942–946 (1993).

54. Barton, L.L. and Hyndman, N.J. Lymphocytic choriomeningitis virus: Reemerging central nervous system pathogen. *Pediatrics* 105, E35 (2000).

55. Buchmeier, M.J., Adam, E., and Rawls, W.E. Serological evidence of infection by Pichinde virus among laboratory workers. *Infection Immunology* 9, 821–823 (1974).

56. Reignier, T. et al. Receptor use by pathogenic arenaviruses. *Virology* 353, 111–120 (2006).

57. Radoshitzky, S.R. et al. Transferrin receptor 1 is a cellular receptor for New World haemorrhagic fever arenaviruses. *Nature* 446, 92–96 (2007).

58. Flanagan, M.L. et al. New World clade B arenaviruses can use transferrin receptor 1 (Tfr1)-dependent and -independent entry pathways, and glycoproteins from human pathogenic strains are associated with the use of Tfr1. *Journal of Virology* 82, 938–948 (2008).

59a. Domingo, E., Sabo, D., Taniguchi, T., and Weissmann, C. Nucleotide-sequence heterogeneity of an RNA phage population. *Cell* 13, 735–744 (1978).

59b. Reignier, T. et al. Receptor use by the Whitewater Arroyo virus glycoprotein. *Virology* 371, 439–446 (2008).

60. Byrne, J.A. and Oldstone, M.B.A. Biology of cloned cytotoxic lymphocytes-T specific for lymphocytic choriomeningitis virus—Clearance of virus in vivo. *Journal of Virology* 51, 682–686 (1984).

61. Fungleung, W.P., Kundig, T.M., Zinkernagel, R.M., and Mak, T.W. Immune-response against lymphocytic choriomeningitis virus-infection in mice without Cd8-expression. *Journal of Experimental Medicine* 174, 1245–1249 (1991).

62. Ahmed, R., Salmi, A., Butler, L.D., Chiller, J.M., and Oldstone, M.B.A. Selection of genetic-variants of lymphocytic choriomeningitis virus in spleens of persistently infected mice—Role in suppression of cytotoxic lymphocyte-T response and viral persistence. *Journal of Experimental Medicine* 160, 521–540 (1984).

63. Parekh, B.S. and Buchmeier, M.J. Proteins of lymphocytic choriomeningitis virus—Antigenic topography of the viral glycoproteins. *Virology* 153, 268–278 (1986).

64. Borrow, P. and Oldstone, M.B.A. Characterization of lymphocytic choriomeningitis virus-binding protein(S)—A candidate cellular receptor for the virus. *Journal of Virology* 66, 7270–7281 (1992).

65. Tishon, A., Borrow, P., Evans, C., and Oldstone, M.B.A. Virus-induced immunosuppression; age at infection relates to a selective or generalized defect. *Virology* 195, 397–405 (1993).

66. Evans, C.F., Borrow, P., Delatorre, J.C., and Oldstone, M.B.A. Virus-induced immunosuppression—Kinetic-analysis of the selection of a mutation associated with viral persistence. *Journal of Virology* 68, 7367–7373 (1994).

67. Ahmed, R. and Oldstone, M.B.A. Organ-specific selection of viral variants during chronic infection. *Journal of Experimental Medicine* 167, 1719–1724 (1988).

68. Sevilla, N. et al. Immunosuppression and resultant viral persistence by specific viral targeting of dendritic cells. *Journal of Experimental Medicine* 192, 1249–1260 (2000).

69. U.S. Department of Health and Human Services, Centers for Disease Control and Prevention, & National Institutes of Health. *Biosafety in Microbiological and Biomedical Laboratories (BMBL)*, 4th edn. J. Y. Richmond and R. W. McKinney (eds.). Washington, DC: U.S. Government Printing Office (1999). Available at: www.cdc.gov/od/ohs/biosfty/bmbl4/bmbl4toc.htm

70. Fair, J. Development and characterization of recombinant Lassa virus proteins and monoclonal antibodies for diagnostic applications. PhD dissertation, Tulane University, New Orleans, LA, 173pp. (2007).

71. Borio, L. et al. Hemorrhagic fever viruses as biological weapons—Medical and public health management. *JAMA* 287, 2391–2405 (2002).

72. Mccormick, J.B. et al. Lassa fever—Effective therapy with Ribavirin. *New England Journal of Medicine* 314, 20–26 (1986).

73. Enria, D.A. et al. Tolerance and antiviral effect of Ribavirin in patients with Argentine hemorrhagic fever. *Antiviral Research* 7, 353–359 (1987).

74. Charrel, R.N. and De Lamballerie, X. Zoonotic aspects of arenavirus infections. *Veterinary Microbiology* 140, 213–220 (2010).

75. Larson, R.A. et al. Identification of a broad-spectrum arenavirus entry inhibitor. *Journal of Virology* 82, 10768–10775 (2008).

76. York, J., Dai, D., Amberg, S.M., and Nunberg, J.H. pH-induced activation of arenavirus membrane fusion is antagonized by small-molecule inhibitors. *Journal of Virology* 82, 10932–10939 (2008).

77. Maiztegui, J.I., Fernandez, N.J., and Dedamilano, A.J. Efficacy of immune plasma in treatment of Argentine hemorrhagic-fever and association between treatment and a late neurological syndrome. *Lancet* 2(8154), 1216–1217 (1979).

78. Aguilar, P.V. et al. Reemergence of Bolivian hemorrhagic fever, 2007–2008. *Emerging Infectious Diseases* 15, 1526–1528 (2009).

79. Maiztegui, J.I. et al. Protective efficacy of a live attenuated vaccine against Argentine hemorrhagic fever. *Journal of Infectious Diseases* 177, 277–283 (1998).

80. Lukashevich, I.S. et al. A live attenuated vaccine for Lassa fever made by reassortment of Lassa and Mopeia viruses. *Journal of Virology* 79, 13934–13942 (2005).

81. Kotturi, M.F. et al. A multivalent and cross-protective vaccine strategy against arenaviruses associated with human disease. *PLoS Pathogens* 5(12), e1000695 (2009).

82. Mckee, K.T. et al. Infection of *Cebus* monkeys with Junin virus. *Medicina-Buenos Aires* 4, 144–152 (1985).

83. Kenyon, R.H. et al. Aerosol infection of rhesus macaques with Junin virus. *Intervirology* 33, 23–31 (1992).

84. Stephenson, E.H., Larson, E.W., and Dominik, J.W. Effect of environmental-factors on aerosol-induced Lassa virus-infection. *Journal of Medical Virology* 14, 295–303 (1984).

85. Bray, M. Defense against filoviruses used as biological weapons. *Antiviral Research* 57, 53–60 (2003).

86. Lee, A.M. et al. Unique small molecule entry inhibitors of hemorrhagic fever arenaviruses. *Journal of Biological Chemistry* 283, 18734–18742 (2008).

87. Botten, J. et al. A multivalent vaccination strategy for the prevention of Old World arenavirus infection in humans. *Journal of Virology* 84, 9947 (2010).

88. Radoshitzky, S.R. et al. Drug discovery technologies and strategies for Machupo virus and other New World arenaviruses. *Expert Opinion on Drug Discovery* 7, 613–632 (2012).

3 Avian Influenza Virus (Highly Pathogenic)

Dongyou Liu and Sacha Stelzer-Braid

CONTENTS

3.1 INTRODUCTION

The genus *Influenzavirus A* contains a number of segmented, single-stranded, negative sense RNA viruses that are further separated into 17 hemagglutinin (H or HA) subtypes and 9 neuraminidase (N or NA) subtypes. As members of the genus *Influenzavirus A* are adapted in wild aquatic birds (natural reservoir host), they are commonly known as avian influenza viruses (or "avian influenza," "bird flu"). Among these viruses, several subtypes (e.g., H1N1, H2N2, H3N2, H5N1) are highly pathogenic, with the tendency to induce significant mortality in domestic fowls as well as serious diseases in humans. Hence, they are frequently referred to as highly pathogenic avian influenza viruses (HPAI).

In particular, highly pathogenic avian influenza A (H5N1) virus (HPAI H5N1, sometimes shortened to H5N1) is highly contagious and deadly to domestic poultry. Being capable of direct transmission from poultry to humans, and airborne transmission between mammals, this virus has been responsible for causing sporadic, severe illness in humans and many animal species. After first appearing in Asia in 1997, HPAI H5N1 has created global concern as a potential pandemic threat. Between 2003 and 2011, this virus was responsible for a total of 566 confirmed human cases and 332 deaths. On the other hand, avian influenza subtype H7N9 virus is generally considered as of low pathogenicity affecting birds only and has not been implicated in human outbreaks. However, between February and November 2013, 138 patients have been confirmed in eastern China provinces with avian influenza virus subtype H7N9 infection, resulting in 45 casualties.

This chapter focuses on HPAI H5N1 as well as the newly emerged H7N9, in relation to its classification, morphology, biology, epidemiology, clinical features, pathogenesis, identification, diagnosis, treatment, and prevention. The other important influenza A virus H1N1, which caused "Spanish flu" in 1918 and "swine flu" in 2009, is discussed in Chapter 11.

3.2 CLASSIFICATION, MORPHOLOGY, AND GENOME ORGANIZATION

3.2.1 CLASSIFICATION

The family Orthomyxoviridae (orthos means "straight," myxa means "mucus" in Greek) covers a number of enveloped, segmented, negative-strand RNA viruses that are separated taxonomically into five genera: *Influenzavirus A*, *Influenzavirus B*, *Influenzavirus C*, *Isavirus*, and *Thogotovirus* (Figure 3.1) [1]. A sixth genus has been proposed recently to accommodate three previously unclassified viruses that were originally isolated from ticks and masked weaver bird in the 1950s–1960s [2].

The genus *Influenzavirus A* consists of a single species influenza A virus, which has been subdivided previously into 17 H subtypes (or serotypes) (H1–H17) and 9 N subtypes (N1–N9) based on the antigenicity of viral surface glycoproteins HA or H and NA or N. It is notable that influenza A virus H antigen type 17 (H17) was only recently identified from fruit bats in 2012 [3]. Each influenza A virus has one type of HA and one type of NA antigen; and specific strain/isolate is named by a standard nomenclature specifying virus type,

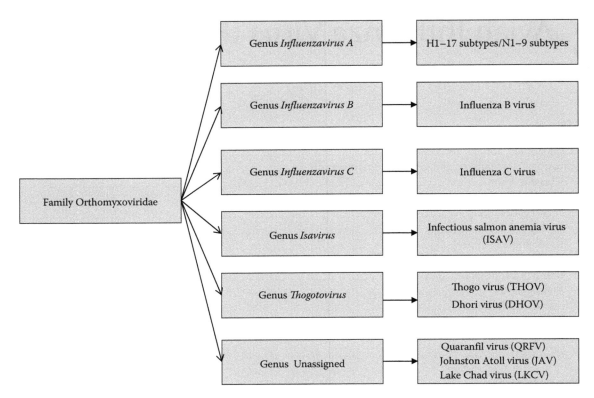

FIGURE 3.1 Classification of the family Orthomyxoviridae.

geographical location where it was first isolated, sequential number of isolation, year of isolation, and HA and NA subtype. For example, A/Brisbane/59/2007 (H1N1) denotes that the virus is of Influenza virus A, which was first isolated in Brisbane with a strain number of 59 in the year of 2007, and belongs to subtype H1N1. The genomes of influenza A viruses comprise 8 RNA segments (1–8) encoding at least 11 proteins (HA, NA, PB1, PB2, PB1-F2, PA, NP, M1, M2, NS1, and NS2).

Commonly present in wild aquatic birds, in which they rarely induce overt clinical symptoms, influenza A viruses can be highly pathogenic to domestic fowls and mammals including humans. The influenza A virus serotypes that have caused serious flu pandemics in humans include H1N1 ("Spanish flu" in 1918 and "swine flu" in 2009), H2N2 ("Asian flu" in 1957), H3N2 ("Hong Kong flu" in 1968), H5N1 (pandemic flu or HPAI), H1N2 (endemic in humans and pigs), and more recently H7N9 (bird flu in eastern China in 2013) [4].

Analyses of the HA gene sequences revealed that HPAI H5N1 viruses can be separated into 10 clades, with the main clades consisting of clade 0 (progenitor viruses from Hong Kong and China, 1996–2002), clade 2.1 (avian and human isolates from Indonesia, 2003–2007), clade 2.2 (2005 progenitors from Qinghai Lake and Mongolia and avian and human 2005–2007 isolates from Europe, the Middle East, and Africa), clade 2.3 (avian and human isolates from China, Hong Kong, Vietnam, Thailand, Laos, and Malaysia, 2003–2006), clade 2.4 (avian isolates from Yunnan and Guangxi, China, 2002–2005), and clade 2.5 (avian isolates from Korea, Japan, and China, 2003/2004, and from Shantou, China, 2006) [5].

Furthermore, based on the NA gene sequences, HPAI H5N1 isolates can be divided into two lineages: 1997 human and poultry isolates and 2005 Vietnamese waterfowl isolates, the latter of which is further divided into two sublineages—GS/GD/1/96-like (containing a full-length NA stalk) or GS/GD/1/96-derived (with a 20-amino acid deletion at amino acids 49–68 in the NA stalk). In addition, phylogenetic comparisons of the PB2, PB1, PA, and NP and M genes of Chinese HPAI H5N1 viruses allow their division into two lineages; and the NS genes of Chinese HPAI H5N1 duck viruses separate them into two alleles, A and B. Combined examination of the HA, N, PB2, PB1, PA, NP, M, and NS gene sequences facilitates the differentiation of HPAI H5N viruses into 13 nodes: I [Gs/GD], II [X-series], III [clades 1, 2, 8, 9], IV [clade 1], V [Vietnam, Thailand, Malaysia (VTM)+ precursor], VI [VTM], VII [Indonesia+precursor], VIII [clade 2.1 (Indonesia)], IX [clade 2.2 (Qinghai lineage)], X [clade 2.3], XI [clades 2.3.1, 2.3.2], XII [clades 2.3.3, 2.3.4], and XIII [clade 2.3.4 (Fujian-like)] [6].

The genus *Influenzavirus B* consists of a single species influenza B virus, with only one influenza B serotype. It is mostly a human pathogen, with seal being the only other animal known to be susceptible. Given its reduced rate of antigenic change, and its limited host range (which inhibits cross-species antigenic shift), influenza B virus is less common than influenza A virus and has not been responsible for any pandemics in humans. Similar to that of influenza A viruses, the genome of influenza B virus harbors 8 RNA segments, which encode at least 11 proteins.

The genus *Influenzavirus C* comprises a single species influenza C virus that is infective to humans and pigs. Being less common than the other types, influenza C virus usually causes mild diseases in children. The genome of influenza C virus has seven RNA segments and encodes nine proteins.

The genus *Isavirus* contains a single species infectious salmon anemia virus (ISAV), which is responsible for causing ISAV in Atlantic salmon, Pacific salmon, brown trout (*Salmo trutta*), and rainbow trout (*Oncorhynchus mykiss*). Several distinct strains of the virus have been identified, with a European strain and a North American strain being the most common. The infected fish develop pale gills and tend to swim close to the water surface, gulping for air. Some fish (notably Pacific salmon, brown trout, and rainbow trout) may show no external signs of illness and thus act as carriers of the virus.

The genus *Thogotovirus* consists of two species: Thogotovirus (THOV) and Dhori virus (DHOV). THOV has been isolated from ticks in Africa and southern Europe and is known to infect humans. Phylogenetic analysis of four THOV RNA segments uncovered the existence of a Euro-Asian lineage and an African lineage. DHOV has been isolated from ticks in India, eastern Russia, Egypt, and southern Portugal. The virus is infective to humans, causing febrile illness and encephalitis. DHOV possesses seven RNA segments.

The sixth yet to be named genus of the family Orthomyxovirus currently comprises Quaranfil virus (QRFV) (which was isolated from ticks *Argas* [*Persicargus*] *arboreus* collected near Cairo, Egypt, in 1953), Johnston Atoll virus (JAV) (which was isolated from ticks [*Ornithodoros capensis*] in Johnston Atoll, in the central Pacific in 1964), and Lake Chad virus (LKCV) (which was isolated from a masked weaver bird, *Ploceus vitellinus*, at Lake Chad, Nigeria, in 1969) [2]. QRFV is known to cross-react serologically with two other unclassified viruses, JAV and LKCV. While QRFV and JAV PB1 and HA share 80% and 70% amino acid identity to each other, respectively, LKCV PB1 shares 83% amino acid identity with the corresponding region of QRFV PB1. QRFV, JAV, and LKCV are lethal to newborn mice via intracerebral (i.c.) inoculation [2].

3.2.2 MORPHOLOGY AND GENOME ORGANIZATION

The virions of influenza A viruses are pleomorphic, mostly spherical or ovoid (50–120 nm in diameter) and filamentous (20 nm in diameter and 200–300 nm in length), with a lipoprotein envelope. The envelope supports about 500 rod-shaped spikelike surface projections (made up of glycoproteins HA and NA), each projecting 10–14 nm from the surface, with the HA protein interposed irregularly by clusters of the NA, or sialidase, protein in a ratio of 4–5 to 1 (HA to NA). Underneath the envelope are matrix (M) proteins (forming capsid) and nucleoproteins (NPs) that enclose the viral genome consisting of eight single-stranded, negative sense RNA segments (ranging from about 900 to 2,500 nt per segment, with the total genome length of about 13,500 nt). The segments all have highly conserved terminal repeats of 9–13 nt at their 5′- and 3′-ends. The longest RNA segment is closely associated with the NP (of 50–130 nm in length and 9–15 nm in diameter) to form helical symmetry.

The eight single-stranded, negative sense RNA segments of the influenza A virus genome encode 11 proteins. Namely, the first segment encodes polymerase basic 2 (PB2); the second segment encodes polymerase basic 1 (PB1) and PB1-F2 by using alternative reading frames; the third segment encodes polymerase acidic (PA); the fourth segment encodes HA; the fifth segment encodes NP; the sixth segment encodes NA; the seventh segment encodes M proteins M1 and M2 by using alternative reading frames; and the eighth segment encodes nonstructural proteins NS1 and nuclear export protein (NEP) (formally referred to as NS2) due to alternatively splicing events.

Among the proteins encoded by influenza A virus genome, the HA protein is a lectin that mediates the viral binding to the host sialic acid receptors on respiratory epithelia, the fusion of the viral and endosomal membranes, and the subsequent entry to target cells. The HA protein is synthesized as a single polypeptide chain (HA0), which is posttranslationally cleaved by cellular proteases. HA cleavage exposes the hydrophobic N-terminus of HA2, which is critical for the fusion of the viral and endosomal membranes to occur. Interestingly, the HA0 precursor protein of low-pathogenic avian influenza viruses (LPAI) contains a single arginine at the cleavage site and another basic amino acid at position -3 or -4 from the cleavage site, which displays limited cleavage by extracellular host proteases and restricts the viral replication at sites in the host where such enzymes are found, that is, the respiratory and intestinal tracts. On the other hand, the HA0 precursor protein of HPAI possesses multiple basic amino acids (arginine and lysine), which are cleavable by intracellular ubiquitous proteases, enabling the viruses to replicate throughout the bird, causing damages in vital organs and tissues, that lead to severe disease and death. The NA is an enzyme (sialidase) that cleaves sialic acid residues on host cells, resulting in the release of newly produced viral particles from infected cells. NP binds to and encapsidates viral RNA in the infected cell nucleus. M constitutes a proton channel in the virus surface that is targeted by the adamantane class of drugs.

What makes influenza A viruses a constant threat to domestic fowls, humans, and other mammals is their uncanny ability to evolve through genetic alteration and recombination (reassortment) that enhance their capacity to undergo interspecies transmission and/or evade host immune responses. Especially, when two subtypes of influenza A viruses co-occur in a single host, new pandemic viruses may emerge as a result of gene reassortments among RNA segments from different subtypes. In theory, reassortments of influenza A viral RNA segments can lead to at least 256 (2^8) different genotypes, providing a potentially unlimited supply of new pandemic viruses [7,8].

Indeed, HPAI H5N1 subtype viruses were shown to harbor an HA segment from A/goose/Guandong/1/96 (H5N1; GS/GD/1/96-like virus), with the remaining seven segments originating from A/teal/Hong Kong/W312/97 (H6N1; W312-like virus). As the HA gene of GS/GD/1/96-lineage has a multibasic sequence (MBS) (RRKKR) at the cleavage site that is targeted by intracellular endoproteases, including PC6 and furin, the virus containing this gene has an increased ability to infect various cell types and cause systemic infections.

The amino acid changes identified in H5N1 subtype viruses may have also underscored its increased pathogenicity in humans [9,10]. Notable variations in H5N1 subtype viruses were found in the NA protein (H274Y for increased oseltamivir resistance), the NS1 protein (P42S for increased IFN antagonism, D92E for reduced sensitivity to IFN and TNFα, deletion from 85 to 94 for impaired inhibition of IFN production), and the PB2 protein (T271A for increased polymerase activity in mammalian cells, E627K for increased replication in mammalian respiratory tract, D701N for increased ability to replicate in mice) [11–13]. In addition to viral evolution in nature, the introduction of mutations into the H5N1 genome by site-directed mutagenesis and passage in ferrets resulted in a novel, highly contagious strain of H5N1 that has the ability to be transmitted via aerosols among mammals [14].

Similarly, the newly emerged H7N9 viruses from China may be a triple reassortant that has been possibly formed by A/brambling/Beijing/16/2012-like viruses (H9N2) incorporating the gene encoding HA from A/duck/Zhejiang/12/2011 (H7N3, subtype ZJ12) and the gene encoding NA protein from A/wild bird/Korea/A14/2011 (H7N9, subtype KO14) [15–19]. Additionally, a number of amino acid changes have been detected in the H7N9 viruses causing epidemics in China in 2013. These include (1) Q226L substitution in the receptor binding region of the HA protein, (2) 69–73 deletion and R294K substitution (for increased resistance to oseltamivir) in the NA protein, (3) L89V and E627K substitutions (for enhanced polymerase activity and increased virulence in mice) in the PB2 protein, (4) I368V substitution (for increased transmissibility in ferrets) in the PB2 protein, (5) N30D and T215A substitutions (for increased virulence in mice) in the M1 protein, (6) S31N substitution (for increased resistance to amantadine) in the M2 protein, and (7) P42S substitution (for increased virulence in mice) [15–19]. Together, these changes have been responsible for transforming a H7N9 virus of low pathogenicity into a highly virulent one, facilitating its jump from avian to human hosts [20].

3.3 BIOLOGY AND EPIDEMIOLOGY

3.3.1 BIOLOGY

The biology of influenza A virus is a complicated process that relies on the involvement of a number of virus- and host-derived molecules. Showing a predilection for respiratory system, influenza A virus employs its HA glycoprotein to bind to sialic acid sugars on the surfaces of epithelial cells in the lung and throat. Assisted by endocytosis, the virus gains entry into the cell and subsequently moves inside the acidic endosome, where with the help of its HA protein, the viral envelope fuses with the vacuole's membrane and the viral contents (viral RNA molecules, accessory proteins, and RNA-dependent RNA polymerase) are released into the cytoplasm. Forming a complex, viral RNA and proteins are transported into the cell nucleus, where the RNA-dependent RNA polymerase transcribes complementary positive-sense RNA (cRNA). The cRNA binds to newly-synthesized viral proteins to form new viral genome particles, which together with viral core proteins leave the nucleus and enter the membrane protrusion formed by HA and NA molecules. The mature virus buds off from the cell and acquires HA and NA as well as host phospholipid membrane coat. The viruses utilizes its HA to adhere to the cell and its NA to cleave sialic acid residues from the host cell, permitting the mature viruses to detach from the cell. The host cell dies once new influenza virus is released.

Influenza A viruses can remain infectious for 1 week at human body temperature, over 30 days at 0°C, and indefinitely at very low temperatures. Off the host, the viruses can survive in mucus for several hours, in feces on cages for up to 2 weeks, in distilled water for 100 days at room temperature, 200 days at 17°C, and indefinitely when frozen. Influenza viruses are susceptible to disinfectants and detergents (e.g., bleach, 70% ethanol, aldehydes, oxidizing agents, and quaternary ammonium compounds) and are inactivated by heat of 56°C for 60 min or longer, as well as by low pH < 2.

3.3.2 EPIDEMIOLOGY

Influenza A viruses are widespread in nature and commonly occur in wild aquatic birds (of >105 species belonging to 26 families), which, as their natural reservoirs, do not present any noticeable clinical signs [21]. However, when transmitted to domestic fowls such as chicken and ducks, some influenza viruses (so-called HPAI) may cause significant morbidity and mortality, while others (so-called LPAI) only induce a mild respiratory disease, along with a decreased egg production or depression [22].

Transmission of influenza A viruses from aquatic birds to other animals, including domestic poultry and mammals, is often via bird droppings that contaminate the water and food. The virus can be also spread by animal to animal contact, bites and scratches, and the movement of infected live birds, poultry products, or contaminated feed, equipment, and materials. In addition, transmission between infected mammals including humans is possible through aerosols created by coughs or sneezes or through contact with saliva, nasal secretions, feces, and blood [23,24].

Because pigs possess cell surface receptors for both human and avian influenza viruses in their trachea, they provide a milieu conducive to viral replication and genetic reassortment for influenza strains that usually infect three different species, pigs, birds, and humans. Thus, the pigs may act as an "intermediate host" where influenza viruses (e.g., H1N1, H1N2, H2N3, H3N1, and H3N2) might exchange genes, producing new and dangerous strains.

HPAI H5N1 virus, with its first outbreak in 1987 and its first lethal human infection in Hong Kong in 1997, represents the first direct transmission of avian influenza A viruses to humans, without prior reassortment in pigs. There is evidence that the 1918 H1N1 influenza strain may have been the result of an avian virus adapting to humans; however, further sequence analyses are required for confirmation (see Chapter 11). Close contact with sick or dead poultry through slaughter, food preparation, and defeathering has accounted for a majority of human cases of H5N1 virus infection [25,26]. Since November 2003, >600 human HPAI H5N1 cases have been reported to World Health Organization (WHO) from 15 countries in Asia, Africa, the Pacific, Europe, and the Near East, of which approximately 60% patients have died [27].

H5N1 is a flu virus strain that is capable of infecting more species than any previously known strain and is deadlier than any previously known strain [28]. Besides poultry and humans, HPAI H5N1 virus has been also detected in pigs in China, Indonesia, and Vietnam; domestic cats in Germany, Austria, Thailand, and Iraq; dogs in Thailand; wild stone marten (a weasellike mammal) in Germany; wild civet cat in Vietnam; and tigers and leopards at zoos in Thailand [29–35].

The recently emerged human pathogenic H7N9 virus is another example of avian influenza A viruses transmitted directly from poultry to humans without apparent involvement of swine. In comparison with H5N1, which is highly pathogenic to both chicken and humans, H7N9 virus appears to possess molecular markers suggestive of "low pathogenicity" in chickens and is able to cause mild or no clinical disease in poultry but severe illness in humans [18,36–40].

3.4 CLINICAL FEATURES AND PATHOGENESIS

3.4.1 CLINICAL FEATURES

In humans, infection with HPAI H5N1 virus causes typical flu-like symptoms, with gastrointestinal and respiratory involvements, ranging from (1) cough (dry or productive); (2) diarrhea, vomiting, and abdominal pain; (3) shortness of breath; (4) fever (>38°C); (5) headache; (6) malaise; (7) muscle aches; (8) runny nose; to (9) sore throat. Complications include (1) acute respiratory distress; (2) organ failure; (3) pneumonia; and (4) sepsis. In most cases, fatality (exceeding 60%) results largely from progressive respiratory failure, particularly among children and adults younger than 40 years old [41,42].

Human infections with avian influenza H7N9 subtype show characteristics of influenza, with clinical presentations ranging from fever, cough, shortness of breath, fatigue, muscle aches, hemoptysis, gastrointestinal symptoms, fever, to rapidly progressive pneumonia that does not respond to antibiotics. Encephalopathy and conjunctivitis are uncommon. Laboratory findings consist of normal white cell count, leukocytopenia, lymphocytopenia, thrombocytopenia, and mildly elevated liver enzymes. Most cases are severe, and a few cases deteriorate rapidly within 1–2 days of hospitalization to acute respiratory failure, leading to refractory hypoxemia, disseminated intravascular coagulation with disease progression, impaired liver or renal function, and death. Sputum specimens are more likely to test positive for the H7N9 virus than samples from throat swabs [15,43]. Men and women are equally represented in the youngest age category 20–34 years, but men are two- to threefold more frequent than women in older age groups [44].

In domestic birds, the clinical signs of H5N1 infection include (1) sudden death in several birds; (2) ruffled feathers;

(3) unusual head or neck posture; (4) inability to walk or stand; (5) reluctance to move, eat, or drink; (6) droopy appearance; (7) respiratory distress; (8) diarrhea; (9) swollen head, wattle, or comb; and (10) a drop in egg production. On the other hand, H7N9 infection in poultry is largely asymptomatic.

3.4.2 PATHOGENESIS

The primary target of influenza infection is airway epithelial cells. The infection begins when the HA protein on the viral surface binds to sialic acid residues expressed on the host cell surface. While avian influenza strains (e.g., H5N1) preferentially bind to α2,3 sialic linkages (SAα2,3Gal) (which are found abundantly on the epithelial cells of duck intestine), human influenza strains (e.g., H1N1) bind to α2,6 linkages (which are commonly present on airway epithelial cells) [45,46].

The HPAI H5N1 viruses of human origin have avian-type receptor SAα 2,3Gal specificity [47]. Given that avian-type receptors exist on the epithelial cells of the lower and, to a lesser extent, upper human respiratory tract as well as on the ciliated cells of *in vitro* differentiated human epithelial cells from tracheal/bronchial tissues, it is no surprise that H5N1 viruses can be directly transmitted from birds to humans and cause serious lower respiratory tract damage in humans. In addition, H5N1 virus is able to infect also nasopharyngeal and oropharyngeal epithelia that apparently do not express SAa2-3Gal receptor [48–50]. Similarly, the recently reported H7N9 subtype of human origin contains Gln226Leu and Gly186Val substitutions in the HA protein (which are associated with increased affinity for α-2,6-linked sialic acid receptors) and Asp701Asn mutation in the PB2 protein (which is associated with mammalian adaptation) [15].

Following viral endocytosis and through pattern recognition receptors, influenza virus activates the endosomal Toll-like receptors (TLRs, specifically TLR3 and TLR7) and cytoplasmic retinoic acid-induced gene 1-like receptor (RIG-1), melanoma differentiation-associated gene 5, and NOD2, triggering type I interferon (IFN) secretion (e.g., IFN-α and IFN-β). IFN-β stimulates the transcription of antiviral genes, including RNA-dependent protein kinase (PKR), 2′-5′-oligoadenylate synthetases, Mx protein, and GTPases in neighboring cells through binding of the IFNAR1/IFNAR2 receptor and activation of the IFN-stimulated gene factor 3 transcription factor. Whereas TLR3 stimulation strongly induces NF-κB-dependent inflammatory responses, RIG-I induces antiviral responses, NOD2 activates the inflammasome complex, and TLR7 regulates the induced B-cell response. Apart from the direct damages caused by the virus replication in respiratory and nonrespiratory tissues, an intense inflammatory reaction, possibly enhanced by virus-induced cytokine dysregulation, may add to the severity of pathological changes [51–58].

Among the viral proteins of avian influenza A viruses, the NS-1 protein blocks key pathways of the innate immune response and suppresses the production of type I IFN by both airway epithelial cells and dendritic cells, thus delaying the induction of immune responses that hampers the viral establishment. NS1 also binds and inhibits the antiviral function of protein kinase R (PKR) by downregulating the translation of the viral mRNA [59–63].

3.5 IDENTIFICATION AND DIAGNOSIS

In view of the nonspecific, flu-like symptoms elicited by avian influenza viruses, the diagnosis of human cases on the clinical presentations is difficult. Laboratory tests should be performed on people presenting flu-like signs with a higher risk for developing the avian flu: (1) farmers and others who work with poultry; (2) travelers visiting affected countries; (3) individuals who have touched an infected bird; (4) individuals who eat raw or undercooked poultry meat, eggs, or blood from infected birds; (5) healthcare workers and household contacts of patients with avian influenza; and (6) individuals with an acute respiratory infection and clinical, radiological, or histopathological evidence of pulmonary parenchymal disease (e.g., pneumonia or acute respiratory distress syndrome [ARDS]) and a history of close contact with a laboratory-confirmed case of avian influenza virus infection 2 weeks before illness.

Traditionally, laboratory identification of avian influenza viruses relies on virus isolation in specific pathogen-free (SPF) eggs or in cell cultures. Although virus isolation in fowl's eggs is highly sensitive and remains the gold standard for avian influenza virus detection, it is costly and time-consuming (taking 3–7 days).

The application of enzyme-linked immunosorbent assay (ELISA), hemagglutination inhibition (HI), neuraminidase inhibition (NI), and neutralization (VN) tests has enabled differentiation of avian influenza viruses into HA and NA subtypes [64–66]. Microneutralization (MN) assay remains a valuable test for the detection of avian influenza subtype-specific antibodies in humans. However, serological tests are generally unsuitable for identifying newly emerging strains [67].

In recent years, molecular techniques have been increasingly utilized for rapid, sensitive, and specific detection and characterization of avian influenza virus RNA and proteins [68–72]. In particular, reverse transcriptase PCR (RT-PCR) has shown potential for accurate identification, subtyping, and quantitation of avian influenza viral RNA in both cultured and clinical specimens [73–106]. Other molecular procedures used for avian influenza detection include nucleic acid sequencing-based amplification (NASBA), reverse transcription LAMP (RT-LAMP), microarray, and pyrosequencing [107–116].

The common gene targets for avian influenza virus subtyping and specific detection comprise the conserved regions located in the genes encoding for the M proteins (M1&2) or the NP and the variable regions located in the genes encoding for the HA and NA proteins [117].

Considering that the first 12 nucleotides of the 3′ terminus (Uni12) and the first 13 nucleotides of 5′ terminus (Uni13) are conserved within avian influenza viruses, RT-PCR using Uni12 and Uni13 primers offers a powerful tool for detecting these viruses in human clinical specimens such as nasal swabs [118,119].

Degenerate primers are useful for the detection of unknown or newly emerging subtypes of avian influenza viruses. The use of five sets of degenerated primers covering the HA0 cleavage sites has enabled the detection of 16 HA subtypes of avian influenza viruses.

3.6 TREATMENT AND PREVENTION

Currently, two classes of antiviral agents are approved for the treatment and prevention of influenza: the M2 inhibitors (amantadine and rimantadine) and the neuraminidase inhibitors (oseltamivir, laninamivir, peramivir and zanamivir). The M2 inhibitors (amantadine and rimantadine) inhibit the M2 proton channel that allows for uncoating of the virus in the endosome, thereby hampering the replication of susceptible influenza A viruses and slowing down the progression to pneumonia. The neuraminidase inhibitors (oseltamivir [GS4104; Tamiflu®], zanamivir [GG167; Relenza®], laninamivir [CS08958; Inavir] and peramivir [BCX-1812 and previously RWJ-270201; Rapiacta and Peramiflu]) inhibit the virus neuraminidase and thereby prevent destruction of sialic acid-bearing receptors that are recognized by influenza virus hemagglutinins, thus stopping the virus from chemically cutting ties with its host cell.

H5N1 influenza viruses are generally susceptible to oseltamivir and zanamivir. Early treatment (within 48 h of symptoms appearing) with the antiviral medication oseltamivir (Tamiflu) or zanamivir (Relenza) may help reduce the disease severity. Oseltamivir may also be used as prophylaxis for persons in close contact with patients diagnosed with avian flu.

H5N1 strains causing human avian flu appear to be resistant to amantadine and rimantadine, which should not be used in case of an H5N1 outbreak. Oseltamivir (Tamiflu)-resistant H5N1 strains (harboring H274Y and N294S mutations in the NA protein) have been detected since 2005, although they remain sensitive to Relenza [120–122].

Most H7N9 viruses appear to be susceptible to the neuraminidase inhibitors (e.g., oseltamivir and zanamivir), but resistant to the adamantanes (amantadine and rimantadine). Therefore, amantadine and rimantadine are not recommended for treatment of H7N9 virus infection.

The Centers for Disease Control and Prevention (CDC) guidance recommends treatment for all hospitalized H7N9 cases, and for confirmed and probable outpatient H7N9 cases. In addition, outpatient cases under investigation who have had recent close contact with a confirmed H7N9 case should receive antiviral treatment, whereas outpatients meeting only the travel exposure criteria for a case under investigation are not recommended to receive antiviral treatment.

For preventive purposes, people working with birds that might be infected should use protective clothing and special breathing masks; avoiding undercooked or uncooked meat reduces the risk of exposure to avian flu and other foodborne diseases; travelers should avoid visits to live-bird markets in areas with an avian flu outbreak; and patients with severe unexplained acute respiratory disease should be tested [123,124].

As a flu vaccine (containing antigens from three or four influenza virus strains-one influenza type A subtype H1N1 virus strain, one influenza type A subtype H3N2 virus strain, and either one or two influenza type B virus strains) is available, its use will help reduce potential infection with avian influenza viruses in humans [125].

3.7 CONCLUSION

Members of the genus *Influenzavirus A* are segmented, single-stranded, negative sense RNA viruses that are currently distinguished into 17 H subtypes and 9 N subtypes. Due to their common occurrence in wild aquatic birds (in which no clinical symptoms are observed), influenza A viruses are often referred to as avian influenza viruses (or "avian influenza," "bird flu"). While some avian influenza virus subtypes (e.g., H1N1, H2N2, H3N2, H5N1) are highly pathogenic and cause significant mortality in domestic fowls and serious diseases in humans, other subtypes are of low virulence. Given their tendency to undergo genetic alterations and reassortments, novel avian influenza virus subtypes that show increased capacity to cross interspecies barriers and induce severe diseases will emerge from time to time and render the control and prevention strategies that have been put in place for the existing virus subtypes ineffective and/or redundant.

The appearance of avian influenza A virus subtype H5N1 in Asia in 1997 represents an excellent example of this organism's ability to incorporate genetic changes to outsmart host's immune responses, resulting significant mortalities in domestic poultry and sporadic and severe diseases in humans. Between 2003 and mid-2013, H5N1 has been responsible for causing a total of 630 confirmed human cases and 375 deaths. Another example is the recently emerged avian influenza virus subtype H7N9 that has been responsible for 138 confirmed human cases in eastern China provinces, including 45 casualties between February–November 2013 [126]. Although avian influenza virus subtype H7N9 is generally considered as of low pathogenicity affecting birds only, and has not been implicated in human outbreaks until 2013, the emergence of this novel H7N9 virus reinforces this organism's capacity to evolve and become a potential threat to human well-being. What is even more remarkable is that this novel H7N9 virus causes silent, asymptomatic infections in domestic poultry and provides no forewarning for human epidemics before the infection is established and much damage has been done [127,128].

To keep an upper hand over the pandemic threats posed by avian influenza viruses such as the highly pathogenic H5N1 and H7N9 subtypes, it is important that rapid, sensitive, and specific molecular techniques are adopted and applied in reference and clinical laboratories for identification and surveillance purposes. Further research to unravel the mechanisms of avian influenza viruses to evade host immune systems will lead to the development of new immunotherapeutics that provide enhanced protection against this organism. In addition, the continued collaboration and knowledge sharing among international and regional frameworks is critical to ensure any future avian influenza outbreaks that will be dealt with in a timely fashion and appropriate manner.

REFERENCES

1. Wright F et al. Orthomyxoviruses. In: Fields B N, Knipe DM, Howley PM (eds.), *Fields Virology*, pp. 1691–1740. Wolters Kluwer/Lippincott Williams & Wilkins, Philadelphia, PA, 2007.
2. Presti RM et al. Quaranfil, Johnston Atoll, and Lake Chad viruses are novel members of the family Orthomyxoviridae. *J Virol* 83, 11599–11606, 2009.
3. Tong S et al. A distinct lineage of influenza a virus from bats. *Proc Natl Acad Sci USA* 109, 4269–4274, 2012.
4. Gambotto A et al. Human infection with highly pathogenic H5N1 influenza virus. *Lancet* 371, 1464–1475, 2008.
5. Chen H et al. Establishment of multiple sublineages of H5N1 influenza virus in Asia: Implications for pandemic control. *Proc Natl Acad Sci USA* 103, 2845–2850, 2006.
6. Vijaykrishna D et al. Evolutionary dynamics and emergence of panzootic H5N1 influenza viruses. *PLoS Pathog* 4, e1000161, 2008.
7. Lei F and Shi W. Prospective of genomics in revealing transmission, reassortment and evolution of wildlife-borne avian influenza A (H5N1) viruses. *Curr Genomics* 12, 466–474, 2011.
8. Mak PW, Jayawardena S, and Poon LL. The evolving threat of influenza viruses of animal origin and the challenges in developing appropriate diagnostics. *Clin Chem* 58, 1527–1533, 2012.
9. Yamada S et al. Haemagglutinin mutations responsible for the binding of H5N1 influenza A viruses to human-type receptors. *Nature* 444, 378–382, 2006.
10. Neumann G et al. H5N1 influenza viruses: Outbreaks and biological properties. *Cell Res* 20, 51–61, 2010.
11. Hatta M et al. Molecular basis for high virulence of Hong Kong H5N1 influenza A viruses. *Science* 293, 1840–1842, 2001.
12. Guan Y et al. H5N1 influenza: A protean pandemic threat. *Proc Natl Acad Sci USA* 101, 8156–8161, 2004.
13. Yin J, Liu S, and Zhu Y. An overview of the highly pathogenic H5N1 influenza virus. *Virol Sin* 28, 3–15, 2013.
14. Herfst S et al. Airborne transmission of influenza A/H5N1 virus between ferrets. *Science* 336(6088), 1534–1541, 2012.
15. Chen Y et al. Human infections with the emerging avian influenza AH7N9 virus from wet market poultry: Clinical analysis and characterisation of viral genome. *Lancet* 381, 1916–1925, 2013.
16. Gao R et al. Human infection with a novel avian-origin influenza A (H7N9) virus. *N Engl J Med* 368, 1888–1897, 2013.
17. Kageyama T et al. Genetic analysis of novel avian A(H7N9) influenza viruses isolated from patients in China, February to April 2013. *Euro Surveill* 18, 20453, 2013.
18. Li Q et al. Preliminary report: Epidemiology of the avian influenza A (H7N9) outbreak in China. *N Engl J Med* 2013, Apr 24 [Epub ahead of print].
19. Liu D et al. Origin and diversity of novel avian influenza A H7N9 viruses causing human infection: Phylogenetic, structural, and coalescent analyses. *Lancet* 381, 1926–1932, 2013.
20. Uyeki TM and Cox NJ. Global concerns regarding novel influenza A (H7N9) virus infections. *N Engl J Med* 368, 1862–1864, 2013.
21. Olsen B et al. Global patterns of influenza A virus in wild birds. *Science* 312, 384–388, 2006.
22. Beato MS and Capua I. Transboundary spread of highly pathogenic avian influenza through poultry commodities and wild birds: A review. *Rev Sci Tech* 30, 51–61, 2011.

23. Yang Y et al. Detecting human-to-human transmission of avian influenza A (H5N1). *Emerg Infect Dis* 13, 1348–1353, 2007.

24. Wang H et al. Probable limited person-to-person transmission of highly pathogenic avian influenza A (H5N1) virus in China. *Lancet* 371, 1427–1434, 2008.

25. Subbarao K et al. Characterization of an avian influenza A (H5N1) virus isolated from a child with a fatal respiratory illness. *Science* 279, 393–396, 1998.

26. Alexander DJ. An overview of the epidemiology of avian influenza. *Vaccine* 25, 5637–5644, 2007.

27. Amendola A et al. Is avian *influenza virus A* (H5N1) a real threat to human health? *J Prev Med Hyg* 52, 107–110, 2011.

28. Watanabe Y et al. The changing nature of avian influenza A virus (H5N1). *Trends Microbiol* 20, 11–20, 2012.

29. Puthavathana P et al. Molecular characterization of the complete genome of human influenza H5N1 virus isolates from Thailand. *J Gen Virol* 86, 423–433, 2005.

30. Songserm T et al. Fatal avian influenza A H5N1 in a dog. *Emerg Infect Dis* 12, 1744–1747, 2006.

31. Thiry E et al. Highly pathogenic avian influenza H5N1 virus in cats and other carnivores. *Vet Microbiol* 122, 25–31, 2007.

32. Mushtaq MH et al. Complete genome analysis of a highly pathogenic H5N1 influenza A virus isolated from a tiger in China. *Arch Virol* 153, 1569–1574, 2008.

33. Desvaux S et al. Highly pathogenic avian influenza virus (H5N1) outbreak in captive wild birds and cats, Cambodia. *Emerg Infect Dis* 15, 475–478, 2009.

34. Qi X et al. Molecular characterization of highly pathogenic H5N1 avian influenza A viruses isolated from raccoon dogs in China. *PLoS One* 4, e4682, 2009.

35. Nidom CA et al. Influenza A (H5N1) viruses from pigs, Indonesia. *Emerg Infect Dis* 16, 1515–1523, 2010.

36. Kahn RE and Richt JA. The novel H7N9 influenza A virus: Its present impact and indeterminate future. *Vector Borne Zoonotic Dis* 13, 347–348, 2013.

37. Parry J. H7N9 virus is more transmissible and harder to detect than H5N1, say experts. *BMJ* 346, f2568, 2013.

38. Ranst MV and Lemey P. Genesis of avian-origin H7N9 influenza A viruses. *Lancet* 381, 1883–1885, 2013.

39. Tang RB and Chen HL. An overview of the recent outbreaks of the avian-origin influenza A (H7N9) virus in the human. *J Chin Med Assoc* 76, 245–248, 2013.

40. Zhang W et al. Epidemiological characteristics of cases for influenza A (H7N9) virus infections in China. *Clin Infect Dis* 57, 619–620, 2013.

41. de Jong MD et al. Fatal outcome of human influenza A (H5N1) is associated with high viral load and hypercytokinemia. *Nat Med* 12, 1203–1207, 2006.

42. Abdel-Ghafar AN et al. Update on avian influenza A (H5N1) virus infection in humans. *N Engl J Med* 358, 261, 2008.

43. Lu SH et al. Analysis of the clinical characteristics and treatment of two patients with avian influenza virus (H7N9). *Biosci Trends* 7, 109–112, 2013.

44. Xiong C et al. Evolutionary characteristics of A/Hangzhou/1/2013 and source of avian influenza virus H7N9 subtype in China. *Clin Infect Dis* 57, 622–624, 2013.

45. Fukuyama S and Kawaoka Y. The pathogenesis of influenza virus infections: The contributions of virus and host factors. *Curr Opin Immunol* 23, 481–486, 2011.

46. Kido H et al. Role of host cellular proteases in the pathogenesis of influenza and influenza-induced multiple organ failure. *Biochim Biophys Acta* 1824, 186–194, 2012.

47. Matrosovich MN et al. Avian influenza A viruses differ from human viruses by recognition of sialyloligosaccharides and gangliosides and by a higher conservation of the HA receptor-binding site. *Virology* 233, 224–234, 1997.

48. Gu J et al. H5N1 infection of the respiratory tract and beyond: A molecular pathology study. *Lancet* 370, 1137–1145, 2007.

49. Korteweg C et al. Pathology, molecular biology and pathogenesis of avian influenza A (H5N1) infection in humans. *Am J Pathol* 172, 1155–1170, 2008.

50. Nicholls JM, Peiris JS, and Guan Y. Sialic acid and receptor expression on the respiratory tract in normal subjects and H5N1 and non-avian influenza patients. *Hong Kong Med J* 15, 16, 2009.

51. Chan MC et al. Proinflammatory cytokine responses induced by influenza A (H5N1) viruses in primary human alveolar and bronchial epithelial cells. *Respir Res* 6, 135, 2005.

52. Le Goffic R et al. Cutting edge: Influenza A virus activates TLR3-dependent inflammatory and RIG-I-dependent antiviral responses in human lung epithelial cells. *J Immunol* 178, 3368–3372, 2007.

53. Kuiken T and Jeffery K. Taubenberger. Pathology of human influenza revisited. *Vaccine* 26S, D59–D66, 2008.

54. Guarner J and Falcon-Escobedo R. Comparison of the pathology caused by H1N1, H5N1 and H3N2 influenza viruses. *Arch Med Res* 40, 655–661, 2009.

55. Hui KP et al. Induction of proinflammatory cytokines in primary human macrophages by influenza A virus (H5N1) is selectively regulated by IFN regulatory factor 3 and p38 MAPK. *J Immunol* 182, 1088–1098, 2009.

56. Lee SM et al. Systems-level comparison of host-responses elicited by avian H5N1 and seasonal H1N1 influenza viruses in primary human macrophages. *PLoS One* 4, E8072, 2009.

57. Mok KP et al. Viral genetic determinants of H5N1 influenza viruses that contribute to cytokine dysregulation. *J Infect Dis* 200, 1104–1112, 2009.

58. Ramos I and Fernandez-Sesma A. Innate immunity to H5N1 influenza viruses in humans. *Viruses* 4, 3363–3388, 2012.

59. Fernandez-Sesma A et al. Influenza virus evades innate and adaptive immunity via the NS1 protein. *J Virol* 80, 6295–6304, 2006.

60. Hale BG et al. The multifunctional NS1 protein of influenza A viruses. *J Gen Virol* 89, 2359–2376, 2008.

61. Haye K et al. The NS1 protein of a human influenza virus inhibits type I interferon production and the induction of antiviral responses in primary human dendritic and respiratory epithelial cells. *J Virol* 83, 6849–6862, 2009.

62. Jia D et al. Influenza virus non-structural protein 1 (NS1) disrupts interferon signaling. *PLoS One* 5, E13927, 2010.

63. Kuo RL et al. Influenza A virus strains that circulate in humans differ in the ability of their NS1 proteins to block the activation of IRF3 and interferon-β transcription. *Virology* 408, 146–158, 2010.

64. Tsuda Y et al. Development of an immunochromatographic kit for rapid diagnosis of H5 avian influenza virus infection. *Microbiol Immunol* 51, 903–907, 2007.

65. Miyagawa E et al. Development of a novel rapid immunochromatographic test specific for the H5 influenza virus. *J Virol Methods* 173, 213–219, 2011.

66. Wada A et al. Development of a highly sensitive immunochromatographic detection kit for H5 influenza virus hemagglutinin using silver amplification. *J Virol Methods* 178, 82–86, 2011.

67. Alexander DJ. Avian influenza—Diagnosis. *Zoonoses Public Health* 55, 16–23, 2008.

68. Michael K et al. Diagnosis and strain differentiation of avian influenza viruses by restriction fragment mass analysis. *J Virol Methods* 158, 63–69, 2009.

69. Sakai-Tagawa Y et al. Sensitivity of influenza rapid diagnostic tests to H5N1 and 2009 pandemic H1N1 viruses. *J Clin Microbiol* 48, 2872–2877, 2010.

70. Wang R and Taubenberger JK. Methods for molecular surveillance of influenza. *Expert Rev Anti Infect Ther* 8, 517–527, 2010.

71. Chandler DP et al. Rapid, simple influenza RNA extraction from nasopharyngeal samples. *J Virol Methods* 183, 8–13, 2012.

72. Sakurai A and Shibasaki F. Updated values for molecular diagnosis for highly pathogenic avian influenza virus. *Viruses* 4, 1235–1257, 2012.

73. Starick E, Römer-Oberdörfer A, and Werner O. Type- and subtype-specific RT-PCR assays for avian influenza A viruses (AIV). *J Vet Med B Infect Dis Vet Public Health* 47, 295–301, 2000.

74. Poddar SK. Influenza virus types and subtypes detection by single step single tube multiplex reverse transcription-polymerase chain reaction (RT-PCR) and agarose gel electrophoresis. *J Virol Methods* 99, 63–70, 2002.

75. Spackman E et al. Development of a real-time reverse transcriptase PCR assay for type A influenza virus and the avian H5 and H7 hemagglutinin subtypes. *J Clin Microbiol* 40, 3256–3260, 2002.

76. Dybkaer K et al. Application and evaluation of RT-PCR-ELISA for the nucleoprotein and RT-PCR for detection of low-pathogenic H5 and H7 subtypes of avian influenza virus. *J Vet Diagn Invest* 16, 51–56, 2004.

77. Phipps LP, Essen SC, and Brown IH. Genetic subtyping of influenza A viruses using RT-PCR with a single set of primers based on conserved sequences within the HA2 coding region. *J Virol Methods* 122, 119–122, 2004.

78. Cattoli G and Capua I. Molecular diagnosis of avian influenza during an outbreak. *Dev Biol (Basel)* 124, 99–105, 2006.

79. Di Trani L et al. A sensitive one-step real-time PCR for detection of avian influenza viruses using a MGB probe and an internal positive control. *BMC Infect Dis* 6, 87, 2006.

80. Ng LF et al. Specific detection of H5N1 avian influenza A virus in field specimens by a one-step RT-PCR assay. *BMC Infect Dis* 2, 6, 2006.

81. Payungporn S et al. Single step multiplex real-time RT-PCR for H5N1 influenza A virus detection. *J Virol Methods* 131, 143–147, 2006.

82. Pourmand N et al. Rapid and highly informative diagnostic assay for H5N1 influenza viruses. *PLoS One* 20, 1, e95, 2006.

83. Xie Z et al. A multiplex RT-PCR for detection of type A influenza virus and differentiation of avian H5, H7, and H9 hemagglutinin subtypes. *Mol Cell Probes* 20, 245–249, 2006.

84. Agüero M et al. A real-time TaqMan RT-PCR method for neuraminidase type 1 (N1) gene detection of H5N1 Eurasian strains of avian influenza virus. *Avian Dis* 51(Suppl 1), 378–381, 2007.

85. Ellis JS et al. Design and validation of an H5 TaqMan real-time one-step reverse transcription-PCR and confirmatory assays for diagnosis and verification of influenza A virus H5 infections in humans. *J Clin Microbiol* 45, 1535–1543, 2007.

86. Hoffmann B et al. Rapid and highly sensitive pathotyping of avian influenza A H5N1 virus by using real-time reverse transcription-PCR. *J Clin Microbiol* 45, 600–603, 2007.

87. Rossi J, Cramer S, and Laue T. Sensitive and specific detection of influenza virus A subtype H5 with real-time PCR. *Avian Dis* 51(Suppl 1), 387–389, 2007.

88. Slomka MJ et al. Validated H5 Eurasian real time reverse transcriptase-polymerase chain reaction and its application in H5N1 outbreaks in 2005–2006. *Avian Dis* 51(Suppl 1), 373–377, 2007.

89. Thontiravong A et al. The single-step multiplex reverse transcription-polymerase chain reaction assay for detecting H5 and H7 avian influenza A viruses. *Tohoku J Exp Med* 211, 75–79, 2007.

90. Alvarez AC et al. A broad spectrum, one-step reverse-transcription PCR amplification of the neuraminidase gene from multiple subtypes of influenza A virus. *Virol J* 5, 77, 2008.

91. Chang HK et al. Development of multiplex RT-PCR assays for rapid detection and subtyping of influenza type A viruses from clinical specimens. *J Microbiol Biotechnol* 18, 1164–1169, 2008.

92. Chantratita W et al. Qualitative detection of avian influenza A (H5N1) viruses: A comparative evaluation of four real-time nucleic acid amplification methods. *Mol Cell Probes* 22, 287–293, 2008.

93. Li PQ et al. Development of a multiplex real-time polymerase chain reaction for the detection of influenza virus type A including H5 and H9 subtypes. *Diagn Microbiol Infect Dis* 61, 192–197, 2008.

94. Lu YY et al. Rapid detection of H5 avian influenza virus by TaqMan-MGB real-time RT-PCR. *Lett Appl Microbiol* 46, 20–25, 2008.

95. Monne I et al. Development and validation of a one step real time PCR assay for the simultaneous detection of H5, H7 and H9 subtype avian influenza viruses. *J Clin Microbiol* 46, 1769–1773, 2008.

96. Suwannakarn K et al. Typing (A/B) and subtyping (H1/H3/H5) of influenza A viruses by multiplex real-time RT-PCR assays. *J Virol Methods* 152, 25–31, 2008.

97. Wu C et al. A multiplex real-time RT-PCR for detection and identification of influenza virus types A and B and subtypes H5 and N1. *J Virol Methods* 148, 81–88, 2008.

98. Chaharaein B et al. Detection of H5, H7 and H9 subtypes of avian influenza viruses by multiplex reverse transcription-polymerase chain reaction. *Microbiol Res* 164, 174–179, 2009.

99. He J et al. Rapid multiplex reverse transcription-PCR typing of influenza A and B virus, and subtyping of influenza A virus into H1, 2, 3, 5, 7, 9, N1 (human), N1 (animal), N2, and N7, including typing of novel swine origin influenza A (H1N1) virus, during the 2009 outbreak in Milwaukee, Wisconsin. *J Clin Microbiol* 47, 2772–2778, 2009.

100. Qiu BF et al. A reverse transcription-PCR for subtyping of the neuraminidase of avian influenza viruses. *J Virol Methods* 155, 193–198, 2009.

101. Thanh TT et al. A real-time RT-PCR for detection of clade 1 and 2 H5N1 influenza A virus using locked nucleic acid (LNA) TaqMan probes. *Virol J* 7, 46, 2010.

102. Sakurai A et al. Rapid typing of influenza viruses using super high-speed quantitative real-time PCR. *J Virol Methods* 178, 75–81, 2011.

103. Tsukamoto K et al. SYBR green-based real-time reverse transcription-PCR for typing and subtyping of all hemagglutinin and neuraminidase genes of avian influenza viruses and comparison to standard serological subtyping tests. *J Clin Microbiol* 50, 37–45, 2012.

104. Corman VM et al. Specific detection by real-time reverse-transcription PCR assays of a novel avian influenza A(H7N9) strain associated with human spillover infections in China. *Euro Surveill* 18, 20461, 2013.

105. Eurosurveillance Editorial Team. Joint ECDC CNRL and WHO/Europe briefing note on diagnostic preparedness in Europe for detection of avian influenza A(H7N9) viruses. *Euro Surveill* 18, 20466, 2013.

106. Wong CK et al. Molecular detection of human H7N9 influenza A virus causing outbreaks in China. *Clin Chem* 59, 1062–1067, 2013.

107. Collins RA et al. A NASBA method to detect high- and low-pathogenicity H5 avian influenza viruses. *Avian Dis* 47(Suppl 3), 1069–1674, 2003.

108. Kessler N et al. Use of the DNA flow-through chip, a three-dimensional biochip, for typing and subtyping of influenza viruses. *J Clin Microbiol* 42, 2173–2185, 2004.

109. Lau LT et al. Nucleic acid sequence-based amplification methods to detect avian influenza virus. *Biochem Biophys Res Commun* 313, 336–342, 2004.

110. Lodes M et al. Use of semiconductor-based oligonucleotide microarrays for influenza A virus subtype identification and sequencing. *J Clin Microbiol* 44, 1209–1218, 2006.

111. Dawson ED et al. Identification of A/H5N1 influenza viruses using a single gene diagnostic microarray. *Anal Chem* 79, 378–384, 2007.

112. Huang Y et al. Multiplex assay for simultaneously typing and subtyping influenza viruses by use of an electronic microarray. *J Clin Microbiol* 47, 390–396, 2009.

113. Moore C et al. Development and validation of a commercial real-time NASBA assay for the rapid confirmation of influenza A H5N1 virus in clinical samples. *J Virol Methods* 170, 173–176, 2010.

114. Postel A et al. Evaluation of two commercial loop-mediated isothermal amplification assays for detection of avian influenza H5 and H7 hemagglutinin genes. *J Vet Diagn Invest* 22, 61–66, 2010.

115. Dinh DT et al. An updated loop-mediated isothermal amplification method for rapid diagnosis of H5N1 avian influenza viruses. *Trop Med Health* 39, 3–7, 2011.

116. Ryabinin VA et al. Universal oligonucleotide microarray for sub-typing of Influenza A virus. *PLoS One* 6, e17529, 2011.

117. Fouchier RA et al. Detection of influenza A viruses from different species by PCR amplification of conserved sequences in the matrix gene. *J Clin Microbiol* 38, 4096–4101, 2000.

118. Hoffmann E et al. Universal primer set for the full-length amplification of all influenza A viruses. *Arch Virol* 146, 2275–2289, 2001.

119. Gall A et al. Universal primer set for amplification and sequencing of HA0 cleavage sites of all influenza A viruses. *J Clin Microbiol* 46, 2561–2567, 2008.

120. de Jong MD et al. Oseltamivir resistance during treatment of influenza A (H5N1) infection. *N Engl J Med* 353, 2667–2672, 2005.

121. Le QM et al. Avian flu: Isolation of drug-resistant H5N1 virus. *Nature* 437, 1108, 2005.

122. Järhult JD. Oseltamivir (Tamiflu®) in the environment, resistance development in influenza A viruses of dabbling ducks and the risk of transmission of an oseltamivir-resistant virus to humans—A review. *Infect Ecol Epidemiol* 2012;2. doi: 0.3402/iee.v2i0.18385.(Epub 2012 June 21).

123. Du L et al. Potential strategies and biosafety protocols used for dual-use research on highly pathogenic influenza viruses. *Rev Med Virol* 22, 412–419, 2012.

124. WHO. Interim WHO surveillance recommendations for human infection with avian influenza A(H7N9) virus. May 10, 2013. www.who.int/influenza/human_animal_interface/influenza_h7n9/InterimSurveillanceRecH7N9_10May13.pdf/ (accessed November 12, 2013).

125. Banner D and Kelvin AA. The current state of H5N1 vaccines and the use of the ferret model for influenza therapeutic and prophylactic development. *J Infect Dev Ctries* 6, 465–469, 2012.

126. WHO website. http://www.who.int/influenza/human_animal_interface/EN_GIP_20130604CumulativeNumberH5N1cases.pdf (accessed 30 June, 2013).

127. Liu Q et al. Genomic signature and protein sequence analysis of a novel influenza A (H7N9) virus that causes an outbreak in humans in China. *Microbes Infect* 15, 432–439, 2013.

128. Nicoll A and Danielsson N. A novel reassortant avian influenza A (H7N9) virus in China—What are the implications for Europe. *Euro Surveill* 18, 20452, 2013.

4 Cercopithecine herpesvirus 1 (B Virus)

Dongyou Liu

CONTENTS

4.1 INTRODUCTION

Cercopithecine herpesvirus 1 (CeHV-1), also known as *Herpesvirus simiae*, monkey B virus, herpesvirus B, herpes B virus, or simply B virus, was first identified from a human patient in 1932. The patient, a young physician named William Brebner, became infected after being bitten by a monkey during research on poliomyelitis-causing virus, with symptoms ranging from localized erythema, lymphangitis, lymphadenitis, to transverse myelitis. Following his death, an ultrafilterable agent showing similarity to herpes simplex virus (HSV) in cell culture was isolated from neurologic tissues of the patient and was initially named "W virus," and subsequently "B virus" [1]. By 2011, about 50 human cases of B virus infections had been documented, with a high proportion of these patients succumbing to the disease in a relatively short period after acquiring the virus.

4.2 CLASSIFICATION AND GENOME ORGANIZATION

B virus is a double-stranded DNA virus belonging to the genus *Simplexvirus*, subfamily Alphaherpesvirus, family Herpesviridae (Figure 4.1) The family name originated from the Greek word *herpein* ("to creep"), referring to the latent, recurring infection typical of this group of viruses, although Herpesviridae can also cause lytic (symptomatic) infections [2]. The only other known species within the genus *Simplexvirus* is human herpesvirus 1 (or herpes simplex 1, HSV 1), which mainly produces cold sores in humans.

Herpesviruses possess relatively large, linear, dsDNA genomes of 100–200 kb, encoding 100–200 genes. The genome is encased within an icosahedral protein called the capsid, which is in turn covered in a lipid bilayer membrane called the envelope. Similar to that of HSV-1 and other human herpesviruses, the genome of B virus is about 157 kb in length with 74.5% G+C and contains unique long (U_L) and unique short (U_S) segments flanked by inverted long (R_L) and short (R_S) repeat sequences that are covalently joined in four possible isomeric configurations [3–5]. Owing to tandem duplication of both oriL and oriS regions, six origins of DNA replication exist in the B virus genome. Of the 74 genes identified in the B virus genome, 73 encode proteins sharing sequence homology to those in herpes simplex viruses. The glycoproteins of B virus, including gB, gC, gD, gE, and gG, demonstrate about 50% homology with that of HSV, with a slightly higher predilection toward HSV-2 over HSV-1 [6]. Thus, B virus and HSV types 1 and 2 may have evolved from a common ancestor. However, B virus lacks a homolog of the HSV gamma(1)34.5 gene, which encodes a neurovirulence factor. This suggests that B virus may utilize distinct mechanisms to sustain efficient replication in neuronal cells in comparison with HSV.

In addition to genetic similarity, B virus also displays strong serological cross-reactivity with HSV. Indeed, B virus may be copresent with HSV in some patients; therefore, accurate diagnosis of B virus infection in both human and the natural B virus host requires a specific assay to distinguish B virus from the closely related HSV [7].

4.3 BIOLOGY AND EPIDEMIOLOGY

Infection begins when a B virus particle binds to specific types of cell membrane receptors via viral envelope glycoproteins. After internalization, the virion releases viral DNA that then migrates to the cell nucleus, where the viral DNA is transcribed to RNA. During symptomatic primary infection, lytic viral genes are transcribed, leading to a self-limited period of clinical illness. In some host cells, a small number of viral gene products termed latency-associated transcript

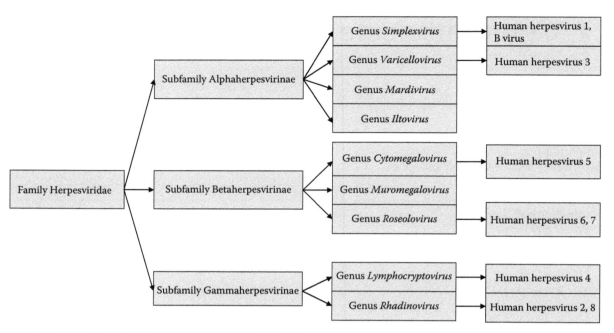

FIGURE 4.1 Human pathogenic species in the family Herpesviridae.

(LAT) are generated, which allow the virus to persist in the cell (and thus the host) indefinitely, without inducing any clinical symptoms [8].

B virus can survive in monkey central nervous system (CNS) tissue and saliva and in monkey kidney cell cultures and can remain viable at 37°C for 7 days, at 4°C for weeks and at −70°C for a long period. However, the virus is susceptible to 1% sodium hypochlorite, 70% ethanol, 2% glutaraldehyde, and formaldehyde. It is also vulnerable to treatment at 50°C–60°C for 30 min or longer, lipid solvents, acidic pH, and detergents.

Macaques (genus *Macaca*) are the most widely distributed nonhuman primates (NHPs) that have gray, brown, or black fur, and tend to be heavily built and medium to large in stature. Macaques are native to Asia and northern Africa, but thousands are housed in research facilities, zoos, wildlife, or amusement parks and are kept as pets in private homes throughout the world. Free-ranging feral populations of macaque species have also been established in Texas and Florida. Of at least 19 species of macaques, rhesus (*Macaca mulatta*), Japanese, cynomolgus (*Macaca fascicularis*), pigtailed, and stump-tailed macaques are most commonly used in biomedical research.

B virus is an infectious agent that is highly prevalent (80%–90%) and relatively benign in its natural macaque monkeys, including rhesus macaques, pig-tailed macaques, and cynomolgus monkeys, but not other NHPs [9]. In captive animals, B virus is often acquired as an oral infection by infants and juveniles or as a genital infection in sexually mature animals [10–12]. The biology of B virus resembles that of HSV in humans and *Cercopithecine herpesvirus 16* (herpesvirus papio 2; HVP2) in baboons (*Papio* spp.) [13–15]. Similar to HSV infection in humans, B virus infections in macaques are lifelong, with intermittent reactivation

and shedding of the virus in saliva, conjunctival fluid, or urogenital secretions [7,16–20]. Virus shedding is more frequent during the mating season (roughly March to June) and when an animal is ill, under stress, or immunosuppressed, although no signs of viral shedding are apparent. As B virus travels within hosts along the peripheral nerves, this neurotropic virus is not found in the blood [21–23].

B virus targets the CNS of its natural host, macaque monkeys, and establishes latent infections subsequently without severely damaging the host. An initial acute phase is generally present when the virus replicates in peripheral tissue of the host [24], which induces a series of specific immune responses as markers of infection. However, when B virus infects other hosts (e.g., humans), severe pathogenesis is observed.

NHPs harboring infectious agents with zoonotic potential come into contact with humans in a variety of contexts (e.g., global travel, tourism, and medical research) [25–29]. B-virus infection in humans usually results from macaque bites or scratches or exposure to the tissues or secretions of macaques (e.g., direct contact, splashes, or needle-stick injuries). In contrast to macaque monkeys, humans infected with B virus develop into a rapidly ascending encephalomyelitis, with an estimated 80% of untreated patients dying of complications associated with the infection [30]. The extreme severity of B virus infections has made it the only biosafety level 4 (BSL-4) herpesvirus and a potential bioterrorism weapon.

Persons at greatest risk for B virus infection include veterinarians, laboratory workers, and others who have close contact with Old World macaques or monkey cell cultures. Therefore, it is important that macaque handlers are provided with comprehensive personal protective equipment (PPE) including appropriate goggles (with antifog lenses) to protect the eyes against splash hazards in combination with a mask (face shields) before entering areas containing macaques, conducting captures, and transporting caged macaques. The fully awake macaques should not be handled, and animals with oral lesions suggestive of active B virus infection or under treatments that suppress immune functions leading to enhanced virus shedding should be quarantined to reduce the risk of virus transmission to workers and other macaques. In addition, macaque handlers need to be aware of the early symptoms of B virus infection and the need to report injuries and/or symptoms suggestive of B virus infection immediately. They also need to know the fact that some medication or underlying medical conditions may heighten the risk for B virus infection. Cages and other equipment for macaques should be free of sharp edges and corners that may cause scratches or wounds to workers. Cages should be designed and arranged in animal housing areas so that the risk of workers being accidentally grabbed or scratched is minimized. All bite or scratch wounds resulting in bleeding and abraded skin from macaques or from cages that might be contaminated with macaque secretions should be immediately and thoroughly scrubbed and cleansed with soap and water. Following an eye exposure, immediate flushing of the eye for at least 15 min is recommended. Further medical attention may then be sought [31–34].

4.4 CLINICAL FEATURES AND PATHOGENESIS

Humans become infected with B virus through macaque bites or scratches, injuries from needles used near a macaque's mucous membranes or CNS, or contact with infectious products from the macaques, or bodily fluids from an infected person [35]. Incubation periods may be as short as 2 days, but more commonly 2–5 weeks. Typically appearing 5 days to 1 month following exposure, symptoms comprise vesicular skin lesions at or near the exposure site, aching, chills and other flu-like symptoms, persistent fever, nausea, lethargy, chest pain and difficulty in breathing, and neurological symptoms (e.g., itching or tingling at or near the exposure site, numbness, dizziness, double vision, difficulty in swallowing, and confusion). Without prompt intervention and treatment, severe meningoencephalitis, coma, and respiratory failure may lead to permanent neurological dysfunction or death (with mortality approaching 80%).

In the macaque host, the virus exhibits pathogenesis similar to that of HSV in humans. These may include oral or genital lesions from where B virus may be discharged, although virus may be also shed in the absence of lesions. After initial infection, B virus can remain latent in the dorsal root of spinal nerves serving the region of exposure or cranial ganglia.

Both B virus and HSV are neurotropic that tend to establish latency in the sensory nerve ganglia of their natural foreign hosts [21]. Reactivation of these viruses from the latent state is induced by stress and possibly other diseases (e.g., shingles, pityriasis rosea) [36]. Following activation, viral genes transit from LAT to multiple lytic genes, leading to enhanced replication and virus production. Infectious viruses then begin shedding from mucosal tissue. Often, lytic activation leads to cell death. Clinically, lytic activation is often accompanied by emergence of nonspecific symptoms (e.g., low-grade fever, headache, sore throat, malaise, and rash) and specific signs (e.g., swollen or tender lymph nodes), as well as immunological findings (e.g., reduced levels of natural killer cells) [37].

4.5 IDENTIFICATION AND DIAGNOSIS

While B virus (CeHV-1) infection in primates is almost always benign, its infection in humans is deadly without prompt antiviral therapy in the early stages of infection. Therefore, development and application of laboratory means to differentiate meningoencephalitis caused by B virus and other pathogens are essential for its early diagnosis and prevention [38–40].

Virus isolation in cell lines has been the traditional technique for identification of B virus infection. Given the highly infectious nature of B virus, a level 3 or higher biosafety containment facility is necessary for safe handling of specimens containing this virus. In addition, virus isolation requires lengthy incubation (several days) and its sensitivity may be inadequate. For this reason, serologic assays have been employed as an alternative method for identifying and diagnosing B virus infection [27,41–43]. A Triton X-100 extract of HVP-2-infected Vero cells was shown to be useful as antigen for B virus antibody detection. B virus could also be detected with B-virus specific monoclonal antibodies in a competitive radioimmunoassay [43] or by using the Enzygnost® Anti-HSV/IgG Test Kit (DADE Chiron, Marburg, Germany) [20].

The recent advances in nucleic acid amplification and detection technologies (especially polymerase chain reaction [PCR]) have facilitated the improved laboratory diagnosis of B virus infection [44–55]. This not only enhances the sensitivity and specificity of B virus testing, it also reduces the risk during the handling of virus-contaminated specimens and shortens the testing time.

Scinicariello et al. [45,46] developed a PCR targeting a 128 bp product from B virus. Subsequent restriction enzyme *Sac*II digestion of the amplicon resulted in the formation of the 72 and 56 bp fragments. The assay has proven valuable for sensitive and specific detection of B virus from both human and monkey specimens. Slomka et al. [47] employed PCR to amplify a specific 188 bp fragment from B virus only, but not Epstein–Barr virus, cytomegalovirus, varicella–zoster virus, HSV types 1 and 2, with a superior sensitivity over virus culture. Hirano et al. [50,51] showed that B virus could be differentiated from other closely related primate alphaherpesviruses (e.g., simian agent 8 of green monkeys and HVP2 of baboons or the human herpes simplex viruses types 1 and 2) through PCR amplification of a DNA segment within the glycoprotein G gene of B virus. DNA polymerase (DPOL) gene was also targeted by PCR for discrimination of B virus from HSV-1 and HSV-2 and other herpesviruses [55]. By using primers targeting the US5 gene encoding glycoprotein J, it was not only possible to distinguish B virus from HSV-1 and HSV-2, but also enabled assessment of the frequency and the titer of shed viral DNA in the clinical specimens [56].

A PCR–microplate hybridization assay with the primer pair HB2A and HB2B targeting the C region was also reported for the identification of B virus. This assay offered a valuable tool for detecting unknown or new B virus genotypes in both natural and human hosts and for quantifying the B virus genome [57]. Furthermore, a TaqMan based real-time PCR assay was reported for rapid detection and quantitation of B virus (CeHV 1) in clinical samples. The assay exploited the nonconserved region of the gG gene to identify B virus from closely related alphaherpesviruses, with a detection limit of 50 copies of B virus DNA [56].

4.6 TREATMENT AND PREVENTION

B virus infection in humans is often fatal unless treatment is undertaken. In addition to the rapid and accurate identification and diagnosis, immediate application of first aids and implementation of treatment measures are the keys to the effective control of human B virus infection [32–34,58–66].

For a human individual with potential exposure to B virus, the first aids comprise (1) immediate cleansing of the exposed area by thoroughly washing and scrubbing the area or wound with soap, concentrated solution of detergent, povidone-iodine, or chlorhexidine and water (these solutions can destroy the viral lipid envelope and inactivate the virus on the skin but are too harsh to use on the eye or mucous membranes); (2) irrigation of the eye or mucous membranes as well as the washed area with sterile saline or running water for 15 min; and (3) avoidance of collecting specimen from the wound area prior to washing the site for PCR testing (as this could force the virus more deeply into the wound and thus reduce the effectiveness of the cleansing protocol), but a serum specimen is obtained from the patient to provide a baseline antibody level [32–34,64–66].

Treatment is recommended when human individuals are exposed to bites, scratches, or excretion of monkeys of the macaque family, which serve as the natural reservoir for B virus. In particular, patients with deep puncture wounds and wounds to the head, neck, or torso (which provide potentially rapid access to the CNS) require timely antiviral treatment. Other scenarios that require treatment include (1) skin exposure (with loss of skin integrity) or mucosal exposure to a high-risk source (e.g., a macaque that is known to be shedding virus or has lesions compatible with B virus infection); (2) inadequately cleansed skin exposure (with loss of skin integrity) or mucosal exposure; (3) needlestick involving tissue or fluid from the nervous system, lesions suspicious for B virus, eyelids, or mucosa; and (4) postcleansing culture is positive for B virus. However, treatment is generally not recommended in the following cases: (1) skin exposure where the skin remains intact and (2) exposure associated with nonmacaque species of NHPs [32–34,64–66].

For prophylaxis of B virus exposure, the recommended antiviral therapy is valacyclovir (1 g per os 3 times daily for 14 days), which is the 6-valine ester of acyclovir, or acyclovir (800 mg per os 5 times daily for 14 days). Famciclovir, an esterified form of penciclovir, may be also considered for postexposure

prophylaxis (500 mg per os 3 times daily for 14 days). For treatment of B virus infection, acyclovir (12.5–15 mg/kg i.v. every 8 h), ganciclovir (5 mg/kg i.v. every 12 h) without CNS symptoms, or ganciclovir (5 mg/kg i.v. every 12 h) with CNS symptoms may also be used. Follow-up appointments should be made at 1, 2, and 4 weeks after the exposure and at any time when there is a change in the clinical status of the exposed primate worker. Serologic testing should be performed 3–6 weeks after the initial exposure [32–34,64–66].

The prevention of human B virus infection should center on the adherence to appropriate laboratory and animal facility protocols, including (1) use of humane restraint methods to reduce the potential for bites and scratches when working with B virus–susceptible monkeys; (2) use of proper PPE (e.g., a lab coat, gloves, and a face shield); (3) immediate cleansing of bites, scratches, or exposure to the tissues or secretions of macaques; and (4) prompt testing of B virus samples from both the exposed human and the implicated macaque [32–34,64–66].

4.7 CONCLUSION

B virus (*Cercopithecine herpesvirus* 1) is a member of the herpesviruses that is enzootic in rhesus (*M. mulatta*), cynomolgus (*M. fascicularis*), and other Asiatic monkeys of the genus *Macaca*. While primary B virus infection in macaques may result in gingivostomatitis with characteristic buccal mucosal lesions, more often it remains latent without producing such symptoms. The virus may reactivate spontaneously or be reactivated in times of stress, resulting in shedding of virus in saliva and/or genital secretions. The virus is transmitted to humans through exposure to monkey saliva (bites or scratches) and tissues, with vesicular skin lesions at or near the site of inoculation, localized neurologic symptoms, and encephalitis [64,67]. Although virus isolation and serological tests are useful for B virus detection, the development of PCR assays (single round, nested, or real time) has made rapid, sensitive, and specific discrimination of this deadly pathogen a reality.

REFERENCES

1. Palmer, A.E., B virus, *Herpesvirus simiae*: Historical perspective. *J. Med. Primatol.*, 16, 99, 1987.
2. Eberle, R. and Hilliard, J., The simian herpesviruses. *Infect. Agents Dis.*, 4, 55, 1995.
3. Perelygina, L. et al., Complete sequence and comparative analysis of the genome of herpes b virus (*Cercopithecine herpesvirus* 1) from a rhesus monkey. *J. Virol.*, 77, 6167, 2003.
4. Ohsawa, K. et al., Sequence and genetic arrangement of the UL region of the monkey B virus (*Cercopithecine herpesvirus* 1) genome and comparison with the UL region of other primate herpesviruses. *Arch. Virol.*, 148, 989, 2003.
5. Ohsawa, K. et al., Sequence and genetic arrangement of the US region of the monkey B virus (*Cercopithecine herpesvirus* 1) genome and primate herpesviruses. *J. Virol.*, 76, 1516, 2003.
6. Harrington, L., Wall, L.V.M., and Kelly, D.C., Molecular cloning and physical mapping of the genome of simian herpes B virus and comparison of genome organization with that of herpes simplex virus type 1. *J. Gen. Virol.*, 73, 1217, 1992.

7. Smith, A.L., Black, D.H., and Eberle, R., Molecular evidence for distinct genotypes of monkey B virus (*Herpesvirus simiae*) which are related to the macaque host species. *J. Virol.*, 72, 9224, 1998.

8. Vizoso, A.D., Recovery of *Herpes simiae* (B virus) from both primary and latent infections in rhesus monkeys. *Br. J. Exp. Pathol.*, 56, 485, 1975.

9. Thompson, S.A. et al., Retrospective analysis of an outbreak of B virus infection in a colony of DeBrazza's monkeys (*Cercopithecus neglectus*). *Comp. Med.*, 50, 649, 2000.

10. Zwartouw, H.T. and Boulter, E.A. Excretion of B virus in monkeys and evidence of genital infection. *Lab. Anim.*, 18, 65, 1984.

11. Zwartouw, H.T. et al., Transmission of B virus infection between monkeys especially in relation to breeding colonies. *Lab. Anim.*, 18, 125, 1984.

12. Weigler, B.J., Scinicariello, F., and Hilliard, J.K., Risk of venereal B virus (*Cercopithecine herpesvirus* 1) transmission in rhesus monkeys using molecular epidemiology. *J. Infect. Dis.*, 171, 1139, 1995.

13. Ritchey, J.W. et al., Comparative pathology of infections with baboon and African green monkey alpha-herpesviruses in mice. *J. Comp. Pathol.*, 127, 150, 2002.

14. Orcutt, R.P. et al., Multiple testing for the detection of B virus antibody in specially handled rhesus monkeys after capture from virgin trapping grounds. *Lab. Anim. Sci.*, 26, 70, 1976.

15. Weigler, B.J., Biology of B virus in macaque and human hosts: A review. *Clin. Infect. Dis.*, 14, 555, 1992.

16. Weigler, B.J. et al., Epidemiology of *Cercopithecine herpesvirus* 1 (B virus) infection and shedding in a large breeding cohort of rhesus macaques. *J. Infect. Dis.*, 167, 257, 1993.

17. Lees, D.N. et al., *Herpesvirus simiae* (B virus) antibody response and virus shedding in experimental primary infection of cynomolgus macaques. *Lab. Anim. Sci.*, 41, 360, 1991.

18. Weir, E.C. et al., Infrequent shedding and transmission of *Herpesvirus simiae* from seropositive macaques. *Lab. Anim. Sci.*, 43, 541, 1993.

19. Carlson, C.S. et al., Fatal disseminated *Cercopithecine herpesvirus* 1 (herpes B infection in cynomolgus monkeys (*Macaca fascicularis*). *Vet. Pathol.*, 34, 405, 1997.

20. Coulibaly, C. et al., A natural asymptomatic herpes B virus infection in a colony of laboratory brown capuchin monkeys (*Cebus apella*). *Lab. Anim.*, 38, 432, 2004.

21. Gosztonyi, G., Falke, D., and Ludwig, H., Axonal and transsynaptic (transneuronal) spread of *Herpesvirus simiae* (B virus) in experimentally infected mice. *Histol. Histopathol.*, 7, 63, 1992.

22. Rogers, K.M. et al., Neuropathogenesis of herpesvirus papio 2 in mice parallels infection with *Cercopithecine herpesvirus* 1 (B virus) in humans. *J. Gen. Virol.*, 7, 267, 2006.

23. Oya, C. et al., Prevalence of herpes B virus genome in the trigeminal ganglia of seropositive cynomolgus macaques. *Lab. Anim.*, 42, 99, 2008.

24. Hilliard, J.K. et al., *Herpesvirus simiae* (B virus): Replication of the virus and identification of viral polypeptides in infected cells. *Arch. Virol.*, 93, 185, 1987.

25. Engel, G.A. et al., Human exposure to herpes virus B-seropositive macaques, Bali, Indonesia. *Emerg. Infect. Dis.*, 8, 789, 2002.

26. Huff, J.L. and Barry, P.A., B-virus (*Cercopithecine herpesvirus* 1) infection in humans and macaques: Potential for zoonotic disease. *Emerg. Infect. Dis.*, 9, 246, 2003.

27. Schillaci, M.A. et al., Prevalence of enzootic simian viruses among urban performance monkeys in Indonesia. *Trop. Med. Int. Health*, 10, 1305, 2005.

28. Jones-Engel, L. et al., Temple monkeys and health implications of commensalism, Kathmandu, Nepal. *Emerg. Infect. Dis.*, 12, 900, 2006.

29. Freifeld, A.G. et al., A controlled seroprevalence survey of primate handlers for evidence of asymptomatic herpes B virus infection. *J. Infect. Dis.*, 171, 1031, 1995.

30. Holmes, G.P. et al., B-virus (*Herpesvirus simiae*) infection in humans: Epidemiologic investigations of a cluster. *Ann. Intern. Med.*, 112, 833, 1990.

31. Boulter, E.A. et al., Successful treatment of experimental B virus (*Herpesvirus simiae*) infection with acyclovir. *Br. Med. J.*, 280, 681, 1980.

32. The B Virus working group. Guidelines from prevention of *Herpesvirus simiae* (B virus) infection in monkey handlers. *J. Med. Primatol.*, 17, 77, 1988.

33. Holmes, G.P. et al., Guidelines for the prevention and treatment of B-virus infections in exposed persons. The B virus Working group. *Clin. Infect. Dis.*, 20, 421, 1995.

34. Cohen, J.I. et al., Recommendations for prevention of and therapy for exposure to B virus (*Cercopithecine herpesvirus* 1). *Clin. Infect. Dis.*, 35, 1191, 2002.

35. Artenstein, A.W. et al., Human infection with B virus following a needlestick injury. *Rev. Infect. Dis.*, 13, 288, 1991.

36. Chellman, G.J. et al., Activation of B virus (*Herpesvirus simiae*) in chronically immunosuppressed cynomolgus monkeys. *Lab. Anim. Sci.*, 42, 146, 1992.

37. Ritchey, J.W., Payton, M.E., and Eberle, R., Clinicopathological characterization of monkey B virus (*Cercopithecine herpesvirus* 1) infection in mice. *J. Comp. Pathol.*, 132, 202, 2005.

38. Davenport, D.S. et al., Diagnosis and management of human B virus (*Herpesvirus simiae*) infections in Michigan. *Clin. Infect. Dis.*, 19, 33, 1994.

39. Ward, J.A. and Hilliard, J.K., B virus-specific pathogen-free (SPF) breeding colonies of macaques: Issues, surveillance, and results in 1992. *Lab. Anim. Sci.*, 44, 222, 1994.

40. Hilliard, J.K. and Ward, J.A., B-virus specific-pathogen-free breeding colonies of macaques (*Macaca mulatta*): Retrospective study of seven years of testing. *Lab. Anim. Sci.*, 49, 144, 1999.

41. Ohsawa, K. et al., Detection of a unique genotype of monkey B virus (*Cercopithecine herpesvirus* 1) indigenous to native Japanese macaque (*Macaca fuscata*). *Comp. Med.*, 52, 555, 2002.

42. Kessler, M.J. and Hilliard, J.K., Seroprevalence of B virus (*Herpesvirus simiae*) in a naturally formed group of rhesus macaques. *J. Med. Primatol.*, 19, 155, 1990.

43. Norcott, J.P. and Brown, D.W., Competitive radioimmunoassay to detect antibodies to herpes B virus and SA8 virus. *J. Clin. Microbiol.*, 31, 931, 1993.

44. Black, D.H. and Eberle, R., Detection and differentiation of primate alpha-herpesviruses by PCR. *J. Vet. Diagn. Invest.*, 9, 225, 1997.

45. Scinicariello, F., Eberle, R., and Hilliard, J.K., Rapid detection of B virus (*Herpesvirus simiae*) DNA by polymerase chain reaction. *J. Infect. Dis.*, 168, 747, 1993.

46. Scinicariello, F., English, W.J., and Hilliard, J.K., Identification by PCR of meningitis caused by herpes B virus. *Lancet*, 341, 1660, 1993.

47. Slomka, M.J. et al., Polymerase chain reaction for detection of *Herpesvirus simiae* (B virus) in clinical specimens. *Arch. Virol.*, 131, 89, 1993.

48. Weigler, B.J. et al., A cross sectional survey for B virus antibody in a colony of group housed rhesus macaques. *Lab. Anim. Sci.*, 40, 257, 1990.

49. Van Devanter, D.R. et al., Detection and analysis of diverse herpesviral species by consensus primer PCR. *J. Clin. Microbiol.*, 34, 1666, 1996.

50. Hirano, M. et al., Rapid discrimination of monkey B virus from human herpes simplex viruses by PCR in the presence of betaine. *J. Clin. Microbiol.*, 38, 1255, 2000.

51. Hirano, M. et al., One-step PCR to distinguish B virus from related primate alphaherpesviruses. *Clin. Diagn. Lab. Immunol.*, 9, 716, 2002.

52. Johnson, G. et al., Comprehensive PCR-based assay for detection and species identification of human herpesviruses. *J. Clin. Microbiol.*, 38, 3274, 2000.

53. Johnson, G. et al., Detection and species-level identification of primate herpesviruses with a comprehensive PCR test for human herpesviruses. *J. Clin. Microbiol.*, 41, 1256, 2003.

54. Huff, J.L. et al., Differential detection of B virus and rhesus cytomegalovirus in rhesus macaques. *J. Gen. Virol.*, 84, 83, 2003.

55. Miranda, M.B., Handermann, M., and Darai, G., DNA polymerase gene locus of *Cercopithecine herpesvirus* 1 is a suitable target for specific and rapid identification of viral infection by PCR technology. *Virus Genes*, 30, 307, 2005.

56. Perelygina, L. et al., Quantitative real-time PCR for detection of monkey B virus (*Cercopithecine herpesvirus* 1) in clinical samples. *J. Virol. Methods*, 109, 245, 2003.

57. Oya, C. et al., Specific detection and identification of herpes B virus by a PCR-microplate hybridization assay. *J. Clin. Microbiol.*, 42, 1869, 2004.

58. Ostrowski, S.R. et al., B-virus from pet macaque monkeys: An emerging threat in the United States? *Emerg. Infect. Dis.*, 4, 117, 1998.

59. Remé, T. et al. Recommendation for post-exposure prophylaxis after potential exposure to herpes b virus in Germany. *J. Occup. Med. Toxicol.*, 4, 29, 2009.

60. Elmore, D. and Eberle, R., Monkey B virus (*Cercopithecine herpesvirus* 1). *Comp. Med.*, 58, 11, 2008.

61. Karsten Tischer, B. and Osterrieder, N., Herpesviruses-A zoonotic threat? *Vet. Microbiol.*, 140, 266, 2010.

62. Estep, R.D., Messaoudi, I., and Wong, S.W., Simian herpes viruses and their risk to humans. *Vaccine* 28, S2, B78, 2010.

63. Tyler, S. et al. Structure and sequence of the saimiriine herpesvirus 1 genome. *Virology*, 410, 181, 2011.

64. Hilliard, J., Monkey B virus. In: Arvin, A. et al. (eds.), *Human Herpesviruses: Biology, Therapy, and Immunoprophylaxis*. Cambridge, MA: Cambridge University Press, Chapter 57, 2007.

65. Keller, C.E. B virus (*Cercopithecine herpesvirus* 1) therapy and prevention recommendations. *US Army Med. Dept. J.*, Jan-Mar, 46–50, 2009.

66. Newton, F., United States Armed Forces. Monkey bite exposure treatment protocol. *J. Spec. Oper. Med.*, 10, 48, 2010.

67. Tregle, R.W. Jr. et al. *Cercopithecine herpesvirus* 1 risk in a child bitten by a bonnet macaque monkey. *J. Emerg. Med.*, 41, e89, 2011.

5 Crimean-Congo Hemorrhagic Fever Virus

Ayse Erbay

CONTENTS

5.1 INTRODUCTION

Crimean-Congo hemorrhagic fever (CCHF) virus is the most extensive tick-borne virus causing a fatal viral infection in humans. CCHF has been known for a long time in Central Asia; a case reported of a hemorrhagic syndrome from where Tajikistan is in present day in the twelfth century may have been the first known case of CCHF [1,2]. In 1944, an acute febrile illness, accompanied with severe bleeding in over 200 Soviet military personnel who were bitten by ticks, in the Steppe region of western Crimea was first described as Crimean hemorrhagic fever, but the virus could not be isolated at that time. Later, it was shown that a viral strain isolated from a patient suffering from Crimean hemorrhagic fever was antigenically identical to Congo virus that was isolated from a febrile patient in Democratic Republic of Congo in 1956 and the combined name Crimean-Congo hemorrhagic fever virus (CCHFV) has been used since the late 1970s [1,3].

5.2 CLASSIFICATION AND MORPHOLOGY

CCHFV is a member of the genus *Nairovirus*, within the Bunyaviridae family. The *Nairovirus* genus includes 34 viruses and is divided into 7 distinct serogroups, which are all believed to be transmitted by either ixodid or argasid ticks. The most important serogroups are the CCHF group (which includes CCHFV and Hazara virus) and the Nairobi sheep disease group (which includes Nairobi sheep disease and Dugbe viruses). Only three viruses are known to cause disease; CCHFV, Dugbe virus, and Nairobi sheep disease virus. Dugbe virus causes a mild febrile illness and thrombocytopenia in humans while Nairobi sheep disease virus is primarily a pathogen of sheep and goats [4,5].

CCHFV is a spherical, enveloped, negative-sense, single-stranded RNA virus with a genome of 17,100–22,800 nucleotides. Virions of CCHFV are spherical, approximately 100 nm in diameter, and have a host cell–derived lipid bilayered envelope approximately 5–7 nm thick, through which protrude glycoprotein spikes 8–10 nm in length. The genome contains three segments: small (S), medium (M), and large (L), all having terminal inverted repeat sequences essential for replication and packaging. The large segment (L segment, of 11,000–14,000 nucleotides) encodes an RNA-dependent RNA polymerase (RdRp), the medium segment (M segment, of 4,400–6,300 nucleotides) is a precursor of glycoproteins G_N and G_C, and the smallest segment (S segment, of 1,700–2,100 nucleotides) has a nucleocapsid protein (NP) (Figure 5.1) [6–9].

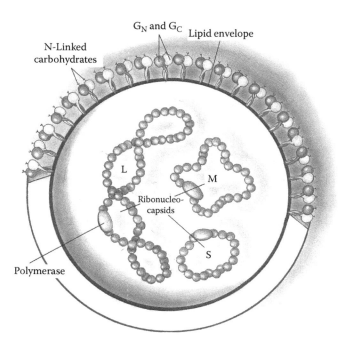

FIGURE 5.1   Schematic cross section of a CCHFV virion.

Their respective plus-strand complements, each with a characteristically short 5′ noncoding region (NCR), encode the nucleocapsid (N) protein, the precursor of two surface glycoproteins (G_N and G_C), and a polyprotein that includes an RdRp component [8,10,11].

The CCHFV glycoproteins show several unusual structural features and undergo several processing events. CCHFV glycoproteins contain, on average, 78–80 cysteine residues, suggesting the presence of an exceptionally large number of disulfide bonds and a complex secondary structure. G_N precursor protein (Pre-G_N) contains a highly variable domain at its amino terminus, which is composed of a high proportion of serine, threonine, and proline residues, and it is predicted to be heavily O glycosylated, thus resembling a mucin-like domain present in other viral glycoproteins [12].

The viral envelope glycoproteins G_N and G_C interact with specific receptors on the host cell. After attachment, the virus enters the cell by receptor-dependent endocytosis. G_C plays an important role in virus entry. Replication takes place in the cytoplasm followed by association of nucleocapsid protein with viral RNA to yield ribonucleoparticles [6]. The N protein interacts with newly synthesized viral RNAs, forming the ribonucleocapsids, which, in turn, interact with the cytoplasmic part of G_N and G_C. These processes trigger the budding of virions into the Golgi compartment [13].

The CCHFV genome M segment encodes an unusually large polyprotein (1684 amino acids in length), which undergoes complex proteolytic processing. The CCHFV mature 37 kDa G_N and 75 kDa G_C proteins have been shown to be processed from the 140 kDa PreGn and 85 kDa PreGc precursors by SKI-1 and SKI-1-like proteases, respectively [14,15].

CCHFV glycoprotein processing is rather unique among RNA viruses in that, in addition to cotranslational cleavage by

signalase, it involves posttranslational cleavage by at least two additional classes of proteases, namely furin/PCs and SKI-1 [16].

In the endoplasmic reticulum (ER), cotranslational cleavage of the precursor polyprotein by signalase and N-glycosylation is expected to occur. As a result of signalase cleavage of the full-length polyprotein, PreG$_N$, which contains mucin, GP38, and G_N and PreG$_C$ are generated. Mature G_C is processed from PreG$_C$ by cleavage at an RKPL1040 motif in the ER. Mature G_C is predominantly localized in the ER and requires G_N to allow its transport to the Golgi analogous. The C terminus of PreG$_N$ is predicted to span the membrane four times based on the presence of hydrophobic amino acid stretches. In the ER/*cis* Golgi, mature G_N is processed from PreG$_N$ at the motif RRLL519 by SKI-1 [15].

G_N predominantly located with Golgi marker. G_C was transported to the Golgi apparatus only in the presence of G_N. Both proteins remained endo-ß-*N*-acetylglucosaminidase H sensitive, showing that the CCHFV glycoproteins are targeted to the *cis* Golgi apparatus [12,16]. Golgi-targeting information partly exists within the G_N ectodomain because a soluble version of G_N lacking its transmembrane and cytoplasmic domains also localizes to the Golgi apparatus. Coexpression of soluble versions of G_N and G_C also results in localization of soluble G_C to the Golgi apparatus, indicating that the ectodomains of these proteins are adequate for the interactions needed for Golgi targeting [16].

Once the mature G_N is generated, it may directly or indirectly interact with G_C and transfer to the cellular compartment where virus assembly occurs. After SKI-1-mediated cleavage, the N-terminal part of PreGn composed of the mucin domain and GP38 is released. Substantial O-glycans are accompanied with the mucin portion of the glycoprotein during its transport through the secretory pathway. In the *trans*-Golgi network, the RSKR247 motif is recognized by furin or furin-like PCs, resulting in the cleavage of the mucin domain and GP38. The processed G_N and G_C, in conjunction with NP, L, and the RNA genome, can initiate virus assembly and can be transported to the plasma membrane for releasing into the medium. GP38, the mucin protein, and the uncleaved forms GP85 and GP160 are released to the medium in significant quantities (Figure 5.2). Function of these three proteins is unknown, but they are secreted in significant amounts [16].

Although the genetic heterogeneity of tick-borne RNA viruses is generally low, CCHFV shows a high level of genetic heterogeneity up to 22% nucleotide variations and 27% amino acid variations based on the S fragment, which is reflected into six genetically distinct clades [17]. Reassortment of L and S segments appears to be largely restricted within phylogenetic groups while M recombination can occur between groups, potentially resulting in a new virus subtype [1]. There is extensive genetic diversity among the strains of CCHFV within the S and M segments. Thirteen full-length CCHF genomes were completely sequenced and reported in Genbank [17–20]. Genetic sequencing and phylogenetic analysis of the virus genomes has identified six distinct lineages of S and L segments and seven M segment lineages [17,18,21]. CCHF virus strains cluster in six groups according

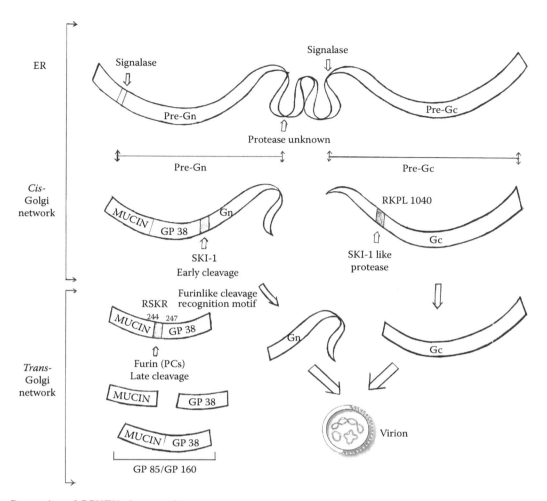

FIGURE 5.2 Processing of CCHFV glycoproteins.

to S segment correlating with the region of origin: European strains from Bulgaria, Albania, Kosovo–Yugoslavia, South West Russia, and Turkey, the nonpathogenic AP92 strain from Greece, West Africa, Democratic Republic of Congo, South/West Africa, and Asia/Middle East strains (Figure 5.3) [21,22]. Recent phylogenetic analyses based on L-RNA segment sequences showed that the L tree topology was similar to the S tree topology [17,18]. On the other hand, the phylogenetic topology based on M-RNA segment sequences of CCHF viruses is different from that based on S-RNA segments [17,18,23–28].

The genetic reassortment may occur in ticks coinfected with different types of CCHF viruses since the virus persists for long periods in ticks. The reason why M segment reassortment is more frequently observed is not clear, but it is probably due to the strong interrelation between the N protein encoded in the S-RNA segment and RNA polymerase encoded in the L-RNA segment that may be required to produce viable virus, when confronted with reassortment opportunities with other CCHF viruses. In addition, the virus is thought to be highly adapted to a particular species of host ticks in endemic region, and the S- and L-RNA segments may have evolved together in a particular tick. In contrast, the M-RNA segment sequence may not be restricted to

a particular tick species; thus, the reassortment event is frequently observed in the M-RNA segment [1].

5.3 BIOLOGY AND EPIDEMIOLOGY

The natural cycle of CCHF virus includes transovarial and transstadial transmission among ticks and a tick–vertebrate host cycle involving wild and domestic animals. CCHF virus circulates in nature in an enzootic tick–vertebrate–tick cycle [5,29]. All members of the genus *Nairovirus* seem to be transmitted mainly by hard ticks (family Ixodidae). Even though CCHFV has been detected in or isolated from more than 30 species of ticks throughout the world, the principal vector is *Hyalomma* ticks, especially *Hyalomma marginatum*, followed by *Rhipicephalus* and *Dermacentor* spp. [5,30,31]. The biological role of ticks is important not only as virus vectors but also as reservoirs of the virus in nature [2,32]. Ticks are able to become infected when co-feeding with virus-infected ticks on the same vertebrate host, even if the vertebrate does not develop a detectable viremia [30]. The geographic distribution of CCHF is closely related with the distribution of *Hyalomma* ticks [2,32]. Virus isolation and illness have been documented in an expanding geographic area that currently includes more than 30 countries

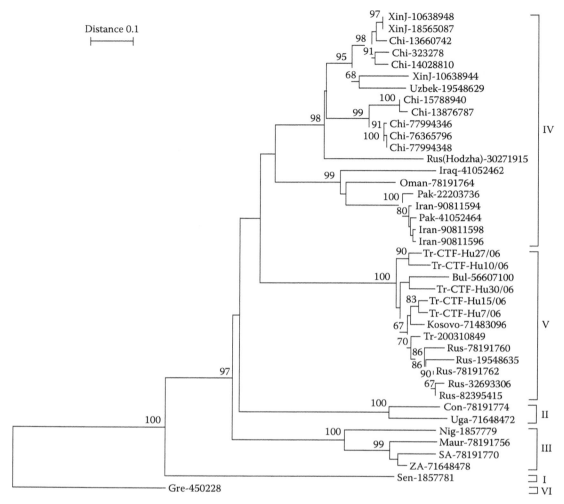

FIGURE 5.3 Phylogenetic tree for the CCHF strains according to S segment. I, West Africa; II, Democratic Republic of Congo; III, South/West Africa; IV, Asia/Middle East; V, Europe/Turkey; VI, Greece. (From Midilli, K. et al., *BMC Infect. Dis.*, 7, 54, 2007.)

in Africa, Central and Southwestern Asia, the Middle East, and Southeastern Europe. Nowadays, CCHFV is known to be widely distributed throughout large areas of sub-Saharan Africa, the Balkans, Northern Greece, European Russia, Pakistan, Xinjiang Province of Northwest China, India, the Arabian Peninsula, Turkey, Iraq, and Iran [32–47]. CCHFV infection was first defined in Turkey in 2003 from persons who became sick during 2002 [45]. Since 2002, when the largest epidemic occurred in Turkey, there had been 6545 confirmed cases and 330 deaths till the end of 2011 with a fatality rate of 4.9% [48].

CCHFV, similar to other zoonotic agents, appears to produce little or no disease in its natural hosts, but causes severe disease in humans [29]. Viremia or antibody production due to CCHFV has been documented through in numerous domestic and wild vertebrates, including cattle, goats, sheep, horses, pigs, hares, ostriches, camels, donkeys, hedgehogs, mice, giraffes, buffalos, rhinoceroses, rodents, and domestic dogs in endemic areas [29,32,37,44,49,50]. CCHF IgG antibody was present in 67% of goats and 86% of sheep in endemic areas in Turkey [51]. Large herbivores are the most common hosts of adult *Hyalomma* ticks with the largest

prevalence of CCHFV antibodies. Small herbivores serve as amplifying hosts; they are infested during the nymphal stages of ticks, which transmit the virus transstadially [52]. Except ostriches, reptiles and birds are refractory to CCHFV infection [45]. However, many migrating bird species are known to be infested by immature ticks such as *H. marginatum marginatum* in Eurasia, and *H. marginatum rufipes, H. truncum*, and some other *Hyalomma* spp. in Africa [52]. Migrating birds can carry infected ticks and thus the birds may play an important role in virus dissemination [44,53].

The virus may be spread into other geographical regions via infected livestock [54]. Importing livestock from endemic to nonendemic areas can cause the transfer of infected ticks [37,55,56]. Even though a variety of animals were demonstrated to be infected, CCHFV can only cause disease in humans and newborn mice [44].

Infection in humans is caused mainly through direct contact with blood or tissues of viremic hosts, or through tick bite or crushing infected ticks with unprotected hands [57]. Cattle, sheep, and goats do not become ill after infection but are viremic for about 1 week. During this period of time, the virus may be transmitted to humans that have close contact to these animals [58]. In endemic areas, high-risk groups are persons in occupational contact with livestock and other animals, including farmers, livestock owners, abattoir workers, and veterinarians [32,36,40,44,57,59–61]. Recreational activities such as hiking and camping in endemic areas are also risk factors for tick bite. As CCHF virus is destroyed by tissue acidification and would not survive after cooking, meat consumption is safe [57]. An epidemiological model found that the ratio of subclinical to clinical CCHF cases is approximately 5:1; 80% of infections are asymptomatic [62].

The overall tick-bite frequency was 62% among persons at high risk and has been reported among 40%–60% of CCHFV patients in Turkey [63]. In a seroprevalence study among 40 veterinarians at the beginning of the CCHF outbreak in Turkey, 2.5% were found to be seropositive [64]. In another seroprevalence study, it was reported that the seroprevalence of CCHFV is higher in persons living in rural areas than in urban areas of the CCHFV epicenter in Turkey (12.8% versus 2.0%). It was also determined that the occupations of animal husbandry and farming were significantly associated with CCHFV seropositivity [60]. The CCHFV IgG positivity was found to be 3.4% in Yunnan Province, China, in relation with history of tick bite, living in a rural area, and being aged >30 years [65]. Within the CCHF endemic areas, there are hyperendemic areas where one out of every five residents and one out of every two residents with a history of tick bite have antibodies against CCHF virus [66].

In temperate zones, CCHF cases occur between spring and early autumn when tick activity is high [40,41,67]. In tropical and subtropical areas, CCHF demonstrates various patterns of seasonality depending on local temperature and humidity [42,68,69]. Most CCHF cases occur sporadically, but climate and environmental changes may influence CCHF epidemiology through facilitating survival of *Hyalomma* ticks and may lead to outbreaks [57]. Mild winters preceded the onset of CCHF outbreaks in Kosovo and Turkey [40,57]. Initial cessation of agricultural activities and hare hunting, followed by reinitiation of agricultural activities, has been reported to be associated with CCHF outbreaks in the former Soviet Union, Bulgaria, Kosovo, and Turkey [57].

Nosocomial transmission is an important route of acquirement of CCHFV. Healthcare workers (HCWs) caring for patients with CCHF are a major risk group [22,42,49,57]. A case of nosocomial transmission of CCHF from patient to patient who had shared the same room as well as the same toilet was also described [70]. The risk of person-to-person transmission is highest during the later stages of the disease. CCHF has not been reported in persons who had an exposure to an infected patient during the incubation period [42]. Direct transmission is thought to occur through contact of viremic blood or other fluids with broken skin. Interventions for gastrointestinal bleeding, surgical operations on unsuspected cases, needlestick injuries, and unprotected handling of infected materials are high-risk activities [22,42,49,57]. Healthcare workers who have had contact with tissue or blood from patients with suspected or confirmed CCHF should be followed up with daily temperature and symptoms monitoring

for at least 14 days after the exposure. Case fatality rates among nosocomial cases tend to be higher than those recorded among community-acquired, and this may be related to viral inoculum [22]. In a hospital outbreak, all five healthcare workers (a nurse, two washermen, and two sweepers) who were exposed to the blood, vomits, sweat, saliva, urine, and stool of the CCHF patients died [42]. The transmission of CCHFV and resultant death among HCWs has been reported since 1976 [71–73]. It was reported that 8.7% of HCWs who had been exposed to infected blood and 33% of HCWs who had had a needlestick injury developed the disease [74].

Airborne transmission of the CCHFV was suspected but could not be documented [57]. Horizontal transmission of the CCHF infection from mother to child was also reported [75]. Viral genomes were detected in saliva and urine [76], and it was assumed that visually nonbloody fluids such as saliva, respiratory secretions, and urine could be a source of CCHV infection [75]. Vertical transmission is possible in CCHF. Vertical transmission of CCHF virus could be either intra-uterine or perinatal. CCHF virus infection among pregnant women may result in abortion or neonatal complications and death. The result is closely related to the term of the pregnancy, as well as the severity of the illness of the mother. In a report consisting of three pregnant women with CCHF, two delivered their babies at term, and the babies died because of massive bleeding within the first 5 days after delivery. CCHF virus was confirmed by PCR and enzyme-linked immunosorbent assay (ELISA) in one baby [77]. In another report, four out of six patients with CCHF had an abortion or stillbirth [78].

CCHF virus RNA was not found in breast milk, but due to the possibility of virus transmission via asymptomatic mastitis, minor areolar lesions, and close maternal–infant contact, it was suggested to stop breastfeeding during the course of CCHF [79].

5.4 CLINICAL FEATURES AND PATHOGENESIS

The typical course of CCHF was described through four distinct phases: incubation, prehemorrhagic, hemorrhagic, and convalescence periods [57]. However, the duration and symptoms of these phases could show variations. The incubation period is usually 1–7 days after a tick bite, but it may last longer, depending on the route of exposure and inoculated viral dose [32,80]. In South Africa, the time to onset of disease after exposure to tick bite was 3.2 days, contact to blood or tissue of livestock was 5 days, and to blood and fomites of human cases was 5.6 days [32]. The incubation period is shorter in fatal cases probably due to higher viral inoculum [36].

After the incubation period, the prehemorrhagic period begins by a sudden onset of fever, chills, headache, lack of appetite, dizziness, photophobia, and myalgia. Additional symptoms such as nausea, vomiting, and diarrhea can be seen [68,80,81]. Fever is often very high up to 40°C and lasts approximately 4–5 days [80]. The duration of the prehemorrhagic period is 1–5 days with an average of 3 days [44,57]. The disease is limited in the prehemorrhagic period in some of the patients and can be misdiagnosed due to nonspecific symptoms [80,82]. Unfortunately, some of the patients can be treated for respiratory tract infection and gastroenteritis during the prehemorrhagic period [82].

The hemorrhagic period develops usually 3–6 days after the onset of the disease [22]. Clinically, CCHF is associated with the most severe bleeding and ecchymoses among the hemorrhagic fevers. Ecchymoses are often large and pressure-linked [83]. The most frequently observed hemorrhagic manifestations are epistaxis, petechiae, ecchymosis, melena, gingival bleeding, hematemesis, and hematoma [81]. Hemoptysis, hematuria, vaginal bleeding, and cerebral hemorrhage have also been reported [32,81,84]. Bleeding to any space in the body is possible [80]. Ocular findings such as subconjunctival hemorrhage and retinal hemorrhage are present in most of the patients without visual complaints [85]. Hepatomegaly is present in 20%–40% [80,81] and splenomegaly in 10%–23% of the patients [45,61,81]. Hepatorenal insufficiency was reported from South Africa [32]. Jaundice may be seen [61,81]. Cardiovascular changes, including bradycardia and low blood pressure, can be seen [68]. Impaired cardiac functions such as lower left ventricular ejection fraction and higher systolic pulmonary artery pressure and pericardial effusion are also reported [86].

The convalescence period among the survivors begins about 10–20 days after the onset of illness [80]. The duration of hospitalization was approximately 8 days (range 2–19) in nonfatal cases [81]. In the convalescence period, labile pulse, tachycardia, temporary complete loss of hair, polyneuritis, difficulty in breathing, xerostomia, poor vision, loss of hearing, and loss of memory were present in the early reports in the Soviet literature [2], although none of these findings were mentioned in the recent reports.

The clinical course of CCHF among children seems to be milder than in adults. Fever, malaise, nausea, vomiting, diarrhea, tonsillopharyngitis, headache, and myalgia are the most common initial symptoms, where diarrhea is more frequent compared with the adult age group [87].

The evaluation of the blood count and biochemical test results provides important clues in the early diagnosis of CCHF. The diagnostic microbiologic methods to detect the virus might not be available in all healthcare settings. Therefore, blood count and biochemical tests have significant values for the suspicion of the diagnosis. Thrombocytopenia is characteristic of CCHF, while most patients also have leucopenia and elevated levels of aspartate aminotransferase (AST), alanine aminotransferase (ALT), lactate dehydrogenase (LDH), and creatine phosphokinase (CPK) [32,45,61,81]. Liver enzyme elevations with AST being higher than ALT are a diagnostic hallmark [54]. Coagulation tests such as prothrombin time (PT) and activated partial thromboplastin time (aPTT) are prolonged, and international normalized ratio (INR) is increased [32,45,61,81]. Low fibrinogen levels might be detected, and fibrin degradation products could be increased [32,80]. Laboratory tests, including complete blood count and biochemical tests, usually returned to normal levels within 5–9 days among the surviving patients [80].

Various fatality rates were reported in the literature, 2.8%–80% in CCHF [42,44,59,68,81,88]. The fatality rate differences may be due to phylogenetic variation of the virus, transmission route, and different treatment facilities. Deaths have been reported to occur on days 3–19 of the disease [81]. Any of the following clinical pathologic values during the first 5 days of illness were found to be >90% predictive of fatal outcome in a series of South African CCHF patients: leukocyte counts $>10 \times 10^9$/L, platelet counts $<20 \times 10^9$/L, AST >200 U/L, ALT >150 U/L, aPTT >60 s, and fibrinogen <110 mg/L [32]. Other case series have confirmed that levels of AST and ALT are significantly higher and platelet counts are significantly lower among severe cases [45,61,81,84]. Independent risk factors for fatality in CCHF were defined as existence of somnolence and melena, prolonged aPTT ($\geq$60 s), and decreased platelet count ($\leq$20,000/mm^3) [81].

The level of viremia has been shown to have prognostic significance in CCHF. High viral load tended to indicate fatal outcome, and viral load is a useful predictor of clinical progress [89–91]. It was reported that the patients with $\geq 1 \times 10^9$ RNA copies/mL had most of the previously reported severity criteria and patients who had RNA titers $\geq 10^9$ RNA copies/mL within 7 days after the onset of symptoms were most likely fatal outcome [90]. In patients with fatal outcome, detectable antibody response was weaker [91].

CCHF pathogenesis could not be studied in detail because of the following reasons: the virus is classified as a Biosafety Level (BSL) 4 pathogen, lack of a suitable animal model to study the disease, CCHF outbreaks are usually sporadic, and very few of the endemic countries have the required safety facilities for working with infectious material. Most of what is known today is established from human case studies, *in vitro* experiments, and inference from other viral hemorrhagic fevers [92].

A common pathogenic feature of viral hemorrhagic fevers is the ability of the etiologic agent to disable the host immune response by attacking and manipulating the cells that initiate the antiviral response [93]. All viral hemorrhagic fever viruses are capable of replicating to high titer in macrophages at their point of entry into the body, resulting in viremia and infection of similar cells in lymphoid organs and other tissues throughout the body. This capacity for rapid dissemination suggests that the viral hemorrhagic fever viruses are able to block human type I interferon (IFN) responses [94]. Viral hemorrhagic fever viruses infect endothelial cells lining blood vessels and destroy these vessels [83]. The role of the endothelium in viral hemorrhagic fevers has not been clearly defined, but it can be targeted in two ways; either by direct viral infection or indirectly by activation of immunological and inflammatory pathways [95]. Impairment of endothelial cell function can cause a wide range of vascular effects that lead to changes in vascular permeability or hemorrhage. Endothelial damage contributes to hemostatic failure by stimulating platelet aggregation and degranulation, with subsequent activation of the intrinsic coagulation cascade [93].

In CCHF patients, progressive lymphopenia develops, and postmortem tissue examination revealed lymphoid depletion, suggesting that extensive loss of lymphocytes through programmed cell death also occurs in this disease [94]. In the examination of bone marrow, reactive hemophagocytosis was detected, which suggested that hemophagocytosis can play a role in the cytopenia observed during CCHF infection [33].

Examination of the tissues collected from CCHF patients at autopsy has shown the presence of viral antigen and RNA within endothelial cells. Immunohistochemistry and *in situ* hybridization analyses revealed that the mononuclear phagocytes, endothelial cells, and hepatocytes are main targets of infection. Association of parenchymal necrosis in the liver with viral infection suggests that cell damage may be mediated by a direct viral cytopathic effect [96].

CCHFV causes extensive infection of hepatocytes, with an increase in circulating liver enzymes, swelling, and necrosis [94]. The ability of CCHFV to replicate in liver cells was demonstrated *in vivo* in mice experiments, and the liver had the highest levels of CCHFV load [97,98]. In an *in vitro* study, CCHFV-infected human hepatocyte Huh7 cell line, replicated to high titers, activated inflammatory mediator and modulated both intrinsic and extrinsic pathways of apoptosis and induced a cytopathic effect [99].

The activation of endothelial cells increases vascular permeability, initiates inflammatory responses, and recruits leukocytes by the upregulation of the leukocyte adhesion molecules, intercellular adhesion molecule 1 (ICAM1), vascular cell adhesion molecule 1 (VCAM1), and E-selectin [100]. The direct activation of endothelial cells in CCHF infection was demonstrated by upregulation of mRNA levels for E-selectin, VCAM1, and ICAM1, and it was determined that the induction of ICAM1 cell surface expression occurred in a virus dose–dependent manner [101].

Elevated interleukin-10 (IL-10), IL-6, gamma interferon (IFN-γ), TNF-α, soluble intercellular adhesion molecule-1 (sICAM-1), soluble vascular adhesion molecule-1 (sVCAM-1), and Vascular endothelial growth factor A (VEGF-A) levels were detected in sera of CCHFV-infected patients and this was correlated to severity of the disease [91,102–104]. It was assumed that CCHF could be a result of a delayed and downregulated immune response caused by IL-10, which leads to an increased replication and spread of CCHFV throughout the body. This consequently activates increased production of IFN-γ and TNF-α, cytokines mediating vascular dysfunction, disseminated intravascular coagulation, organ failure, and shock [91].

In an experimental study, consistently higher levels of TNF-α, IL-6, and IL-10 were measured in supernatants from infected monocyte-derived dendritic cells (moDCs) compared to uninfected cells. However, there was no difference between infected and uninfected cells with regard to IL-8, IL-19, and IL-1β release. Also, it was shown that supernatants from infected moDCs activate endothelial cells by upregulating ICAM-1 expression. The ICAM-1 upregulation is thought to be a virally induced soluble mediator from moDCs that activates endothelial cells [92]. In another experimental study, it was found that CCHFV was able to trigger moDCs upregulation of CD-86 and CD-83, but had a moderate effect on CD-40 and no effect on either CD-80 or HLA-DR, and was also shown to activate the secretion of IL-6 and IL-8, but not TNF-α. On the contrary, in macrophages, CCHFV infection developed a high IL-6 and TNF-α response and a moderate chemokine response [105].

The effect of CCHFV infection on tight junctions (TJs) was investigated to clarify the virus effect, as TJs play an important role in vascular homeostasis and can cause leakage upon deregulation. Infection of CCHFV did not cause any cytopathic effect in Madin–Darby canine kidney 1 (MDCK-1) cells, and no effect on TJs could be either visualized or measured by transepithelial electrical resistance, demonstrating that there is no direct viral effect on TJs in the MDCK-1 epithelial cells and it was assumed that the bleeding observed in patients was possibly due to immune-mediated mechanisms [6,13].

It was shown that replicating CCHFV delays the interferon production in infected cells and that biologically active interferon is induced 48 h after infection. In addition, a significant reduction in CCHFV viral titers by one log step or more when cells have been treated with IFN-α 24 h prior to infection was observed [106]. The host cells need to have an established IFN response in order to inhibit CCHFV infection. The virus delays the early immune responses for the time required for the replication machinery to operate, and once at that point, the IFN induced has little or no effect on virus replication. Pretreatment of cells with IFN-α 24 h before the infection resulted in a reduction of viral titers by approximately two logs. The titers also reduced markedly when cells were treated with IFN-α 2 h before or 1 h after the infection, but not as significantly as after 24 h pretreatment. However, when IFN-α was given to CCHFV-infected cells 6 h postinfection, no effect was observed on the viral titers. Once the virus is replicating, virus replication is more or less insensitive to the antiviral effects induced by the interferon [107].

Several domestic and laboratory animals have been tested as potential animal models since the first isolation of the virus [29]. Experimentally infected animals, such as calves, horses, hamsters, and rabbits, generate little or no viremia and high levels of neutralizing antibodies but do not exhibit clinical signs of disease [108]. Contrary to many hemorrhagic fever viruses, CCHFV has not been found to cause disease in commonly used species of nonhuman primates. Adult immunocompetent mice are not susceptible to CCHFV infection and show no signs of disease. Anyhow, an infant mouse infection model has been described with high virus titers in blood and liver [109]. While there is evidence of systemic viral dissemination and infection of macrophages in mice, which is compatible with the pattern of viral hemorrhagic fever, virus could not be isolated from the spleen where large numbers of mononuclear phagocytes exist [110]. The results of *in vitro* studies have shown that the IFN response plays an essential role in controlling CCHFV replication, and a recent study established that IFN receptor knockout (KO) mice are highly susceptible to CCHFV infection [97]. Just a while ago, an animal model of CCHFV infection was assessed in mice lacking the STAT-1 signaling molecule, a central component of the IFN signaling pathways. In the absence of the IFN response, CCHFV infection of STAT129 mice causes promptly disease and death. After CCHFV administration, mice showed fever, leukopenia, thrombocytopenia, and highly elevated liver enzymes. CCHFV-infected STAT129 mice encounter an early peak of viral replication in the blood on the first day of infection, and in the liver and the spleen on the second day, which are the main places of replication. STAT129 mice respond to CCHFV infection with high levels of IFN-γ, IL-1β, IL-6, IL-10, CCL-2, and TNF on days 2 and 3 of infection [98].

5.5 IDENTIFICATION AND DIAGNOSIS

Early diagnosis is crucial both for the treatment of the patient and to prevent further transmission of the disease, as CCHFV has the potential for nosocomial spread [44,111,112]. Patients' history, particularly history of traveling to endemic areas and history of tick bite or exposure to blood or tissues of livestock or human patients, along with clinical symptoms, blood count, and biochemical test results, is the first indicator of CCHF [44]. In the absence of bleeding or organ manifestations, the diagnosis of CCHF is clinically difficult, and the various etiologic agents can hardly be distinguished by clinical tests [20,34,82]. The differential diagnosis should include rickettsiosis, leptospirosis, and borreliosis. Additionally, other infections (e.g., Alkhurma and Rift Valley fever, Omsk hemorrhagic fever, Kyasanur Forest disease, hantavirus hemorrhagic fever, Lassa, Ebola, Marburg, yellow fever, dengue, malaria, meningococcal infections, sepsis, toxic shock syndrome, typhoid fever, viral hepatitis A, B, E, and brucellosis) and noninfectious reasons (e.g., vitamin B12 deficiency, febrile neutropenia, HELLP syndrome, and drug side effects) should also be considered in the differential diagnosis [44,54,80,112,113].

The definitive diagnosis of CCHF mainly depends on laboratory testing. Direct and indirect approaches for the diagnosis of CCHFV include virus culture, antigen-specific enzyme-linked immunoassay, antibody-specific enzyme-linked immunoassay, and reverse transcription-PCR (RT-PCR). Virus detection in the acute stage of disease is essential, and RT-PCR provides the best sensitivity [44,112,114–116]. In general, viremia is demonstrated in the first 9 days from the onset of the disease, while antibodies are detected in 7 days from the onset of the disease [57].

Special attention is needed for transportation and handling of the CCHF-suspected specimens. Transportation of the specimens must be done in clearly labeled (Infectious Risk) triple-package containers. Body secretions and excretions, blood and tissue specimens contain virus and are highly infectious. Procedures involving CCHFV should be carried out in a BSL-4 laboratory due to high risks for laboratory-acquired infections and the lack of a specific safe vaccine [112]. In laboratories without such a facility, viral inactivation of the blood samples is needed as these samples pose a serious health risk. In order to protect laboratory workers, serum or blood has to be treated with heat, γ-irradiation,

β-propriolactone, or Triton X-100 to inactivate HFV [117]. Serum specimens can be inactivated by heat treatment at 60°C for 60 min. Treatment in acetone (85%–100%), glutaraldehyde (1% or greater), or 10% buffered formalin for 15 min is suitable before handling [112].

5.5.1 CONVENTIONAL TECHNIQUES

5.5.1.1 Virus Isolation

Isolation and culture of the CCHFV should only be performed in a maximum biocontainment laboratory BSL-4 [44,92]. The most frequently used isolation procedure is intracranial inoculation of the sample (e.g., blood from an acute-phase patient or ground tick pools) into newborn suckling mice [44,112]. Isolation in cell cultures has considerable advantages over mouse inoculation, especially since it provides a more rapid result, but generally it is considered to be less sensitive and can only allow detection of the relatively high viremia [44,118].

A large variety of cells could be used for *in vitro* viral cultivation: primary chicken embryo, human embryo, primary green monkey kidney cells, or continuous cell lines as CF-1 (*Cercopithecus aethiops*), Vero or VeroE6 (African green monkey kidney, *C. aethiops*), SW13 (human small cell carcinoma of adrenal cortex), LLC-MK2 (rhesus monkey, *Macaca mulatta*), or CER (derived from hamster) [112]. Usually CCHFV strains produce little or no cytopathic effect, and viral replication is detected in 2–6 days by indirect immunofluorescence assay (IFA) using a specific CCHF mouse hyperimmune ascitic fluid (MHIAF) or monoclonal antibodies against the nucleocapsid NP [44,112,118]. The plaque assay in CER cells was found to be of similar sensitivity to the fluorescence focus assay, but both were found to be 10- to 100-fold less sensitive than the mouse inoculation. From the specimens of 26 CCHF patients in South Africa, which were collected before day 8 of disease, CCHFV was isolated from 20 patients by mouse inoculation and from 11 both by cell cultures and mouse inoculation. Isolation of CCHFV in mice was detected in a mean of 7.7 days, whereas the mean time for positive cell cultures was 3.3 days [118]. However, isolation of some CCHF viral strains from field-collected ticks can only be obtained by suckling mice inoculation [112]. It has been observed that CCHFV can be isolated from the blood of acutely ill patients for 8 days and occasionally for up to 12 days after the onset of disease [118]. Infectivity of the blood in newborn mice remains for 10 days when stored at 4°C. Blood and plasma of the patients and autopsy materials such as lungs, liver, spleen, bone marrow, kidney, and brain can be used for CCHFV isolation. Postmortem material should be secured within 11 h after death [58].

5.5.1.2 Antigen Detection

The detection of CCHFV antigen can be used for the diagnosis of acute infections [112]. The viral antigen can be detected by immunocapture ELISA or reverse passive hemagglutination (RPHA). The RPHA and ELISA have similar sensitivities for detection of cumulative CCHF antigen in infected mouse brain and cell culture supernatants and extracts. However, ELISA is superior to RPHA in the detection of viral antigen in viremic sera from mice and humans [119]. For immunocapture of CCHFV antigen, plates are coated with CCHFV MHIAF or purified monoclonal antibodies to CCHFV recombinant nucleoprotein (rNP) [119,120]. Antigen-capture ELISA was shown to be useful for the diagnosis during the acute phase of disease. It was reported that none of the nested RT-PCR-positive and antibody-positive serum samples reacted positively by antigen-capture ELISA, suggesting that this method is useful for testing serum samples collected during the acute phase of illness before antibody responses are detected [120].

5.5.1.3 Antibody Detection

Earlier serologic tests for the detection of CCHFV antibodies, such as complement fixation, immunodiffusion, and hemagglutination inhibition, suffered from a lack of sensitivity and reproducibility [44]. Nowadays, ELISA and immunofluorescence tests are used for the detection of IgM and IgG antibodies on days 7–9 of illness [57,121]. Specific IgM declines to undetectable levels by 4 months after infection, but IgG remains demonstrable for at least 5 years. An antibody response is rarely detectable in fatal cases. In a study, in 7 out of 11 fatal cases no serum IgM or IgG was detected although the virus was identified by PCR [90]. Recent or current infection is confirmed by demonstrating seroconversion, or a fourfold or greater increase in antibody titer in paired serum samples, or IgM antibody in a single sample [44]. The indirect IFA is useful for a rapid serodiagnosis of the disease; however, ELISA is found to be more specific and sensitive than IFAs and neutralization tests [115].

Serological methods have been developed for the diagnosis of CCHFV using either inactivated virus or extracts from an infected suckling mouse brain [115]. Nevertheless, all of the serological methods require at one stage that live virus should be manipulated, which obligates the use of BSL-4 laboratory [122]. Because of this, recombinant antigens were produced, replacing native Ag in ELISA and IFA tests. Since the nucleoprotein N is recognized as the predominant antigen inducing a high immune response, the rNP of CCHFV has been produced via recombinant baculoviruses or expressed constitutively in established cell lines and tested using ELISA and indirect IFA [122–125].

ELISA is the most regularly used technique for CCHFV antibody detection, and it is more sensitive than IFA [116]. A common source of CCHFV antigen had been the sucrose–acetone–treated suckling mouse brain suspension or a crude suckling mouse brain suspension inactivated by beta propiolactone or heating [112]. ELISA tests using recombinant proteins as antigen have also been developed [54]. For IgM detection, inactivated native CCHFV antigens grown in Vero E6 cells are generally used by immunocapture on plates coated with specific anti-μ serum [33,112]. Then the serum sample, CCHFV antigen, specific CCHFV antibody (MHIAF), antispecies conjugate, and a chromogenic substrate are added successively [112]. For the detection of

CCHFV IgG, usually a sandwich ELISA with capture of the CCHFV antigen in plates coated with a CCHFV MHIAF is used [112]. An IgG ELISA system developed with the recombinant CCHFV NP is a valuable tool for diagnosis and epidemiological investigations of CCHFV infections [122,126,127]. The derived recombinant assays by ELISA G and M (sandwich) or IFA have high sensitivity and specificity for detecting CCHFV antibodies [122,127]. When tested with laboratory animal sera representing all seven serogroups of nairoviruses using recombinant assays, the only reactive sera were those raised to CCHFV, and a weak reaction to Hazara virus was observed [123].

5.5.2 Molecular Techniques

As the specific IgM antibodies against the CCHFV are first detectable about 7 days after the onset of illness, a rapid and accurate diagnosis of CCHF can be made only by an adequate molecular method [121]. Molecular diagnostic tools such as RT-PCR allow rapid detection of CCHFV RNA [54]. The high specificity of RT-PCR makes a probable diagnosis of CCHF without the need to culture the virus feasible [128]. In addition, due to the high sensitivity of RT-PCR, positive results can often be obtained from samples that are culture-negative [128]. Retrospective studies by the stored serum samples can also be done with RT-PCR [45,128]. Usually serum or plasma is used for PCR; however, viral genomes were also detected in the saliva and urine [76].

Techniques usually combine the reverse transcription step with specific amplification, minimizing the risks of contaminations. For diagnostic purposes, assays are typically based on consensus nucleotide sequences, primarily on the S segment, which is best characterized among the three genomic segments [54,112].

RT-PCR also allows for molecular epidemiology to be elucidated. Amplified viral complementary DNA (cDNA) can be sequenced and subjected to phylogenetic analysis, and phylogenetically distinct viral variants can be identified [44,55]. The analysis of sequences on the S segment allows determining the origin and possible source of infections [1,55,129].

Several RT-PCR protocols have been reported; however, most of the published assays are time-consuming as they include a separate cDNA synthesis step prior to PCR, agarose gel analysis of PCR products, and, in some instances, a second round of nested amplification or Southern hybridization. Moreover, post-PCR processing or nested PCR steps increase the risk of false-positive results due to carryover contamination [54,114,128]. The high mortality rate and high risk for nosocomial infection, along with widespread geographic distribution of CCHFV with sporadic outbreaks, prompt the need for a rapid and specific presumptive diagnostic assay against numerous strains [20]. Therefore, one-step real-time RT-PCR assays have been developed for diagnosing CCHF [20,130–132]. Because these assays are more rapid than the conventional RT-PCR, they provide a sensitivity that is comparable to that of the nested PCR with a low risk of contamination, allowing the quantification of viral load. When using the real-time assay, the results of the presence of CCHFV could be obtained within approximately 2 h; however, when using the nested RT-PCR, the results could be available in approximately 4–5 h, excluding gel electrophoresis [131].

5.6 TREATMENT AND PREVENTION

5.6.1 TREATMENT

Treatment options for CCHF are limited. Supportive therapy is the most essential part of case management, and most CCHF patients receive only supportive therapy. Currently, there is no specific antiviral therapy for CCHF approved for use in humans. Although ribavirin has been employed for over 25 years for the prophylaxis and therapy of CCHF, its efficacy remains controversial [44,133].

5.6.1.1 Supportive Treatment

Supportive treatment includes careful attention to fluid balance and correction of electrolyte abnormalities, oxygenation and administration of platelets, fresh frozen plasma, and erythrocyte preparations and appropriate treatment of secondary infections [44,61].

CCHF patients must be monitored closely for effective support. Blood count and hemostasis should be assessed daily or two or three times a day in patients with bleeding, according to platelet count. Blood urea nitrogen, creatinine, ALT, AST, and bilirubin should be examined daily or every other day, depending on the presence of organ failure. In case of disseminated intravascular coagulation, arterial blood gases, fibrinogen, D-dimer, and fibrin degradation products can be tested [134].

Intramuscular injections and invasive interventions should be avoided. Nonsteroidal antiinflammatory drugs are contraindicated [57]. Oral or intravenous paracetamol can be administered for pain and fever. Proton pump inhibitors are recommended for the prophylaxis of gastrointestinal system bleeding. In women, oral progesterone can be used for menstrual bleeding prophylaxis or to control vaginal bleeding [134].

Data are not available to support the use of corticosteroid treatment in CCHF; there is only one report that high-dose corticosteroid treatment in five pediatric CCHF cases led to increased platelet counts and reductions in fever [135].

Even though in a clinical study rapid improvement of hematological parameters was detected in 8 patients treated with intravenous immunoglobulin (IVIG) compared with 14 control patients, there are not enough data to recommend its use in patients with CCHF [134].

Plasma exchange transfusion is a method used for virus inactivation or elimination in patients with VHF, particularly in the case of DF and Lassa fever [136,137]. There is one report of a CCHF case that was treated successfully with a combination of plasma exchange and ribavirin therapy [138].

5.6.1.2 Ribavirin

Ribavirin is a purine nucleoside analogue with broad-spectrum antiviral activity against RNA viruses [139]. Ribavirin has been shown to inhibit *in vitro* viral replication in Vero cells [140] and reduce the mean time to death in a suckling mouse model of CCHF [109]. Also, ribavirin inhibited CCHFV replication according to plaque-reduction assays, and no significant differences in drug sensitivity of different viral isolates were observed [141]. The protective activity of ribavirin has also been demonstrated in CCHFV-infected STAT-1 knockout mice in a recent *in vivo* experiment. In this experiment, 60% of the animals survived in the group in which ribavirin was given 1 h postinfection, whereas none of the animals survived in the group in which ribavirin was given 24 h postinfection, and it was hypothesized that the initial phase following CCHFV infection is crucial in controlling virus replication [98].

The World Health Organization recommends ribavirin as a potential therapeutic drug for CCHF and has added the drug to the essential drug list, mainly based on its *in vitro* effect [142,143].

In the literature, some case reports, observational studies, and one randomized trial exist about efficacy of ribavirin in CCHF, but the results of these studies are conflicting.

The first reported clinical use of ribavirin in CCHF was during a nosocomial outbreak in a South African hospital in 1985. Six of nine healthcare workers with penetrating injuries from CCHFV-contaminated needles were treated with ribavirin. One of the patients on ribavirin had a mild clinical course while five others who received the drug developed neither clinical CCHF nor antibodies to the virus. Two of the three needle contacts not treated with ribavirin had a severe clinical course [74].

Mardani et al. compared the fatality rate among confirmed cases of CCHF who received oral ribavirin treatment and those who did not. The fatality rate in 69 treated patients was 11.6%. Historical cohort consisted of 12 patients and the fatality rate was 58.3% [144]. In another study from Iran in 2006 about the efficacy of oral ribavirin, it was found that 15.7% of treated patients and 63.2% of untreated patients died [145]. Ergonul et al. evaluated the efficacy of ribavirin therapy in 35 confirmed CCHF patients. Eight patients were given ribavirin and all survived. In this study, one patient in the untreated group had a fatal outcome [59]. Ozkurt et al. described the efficacy of ribavirin therapy for CCHF among 60 confirmed CCHF patients in Eastern Turkey. The ean recovery time was shorter in the cases treated with ribavirin than those of control. But the need for blood and blood product, mean hospitalization duration, fatality rates (9.0% in treated and 10.5% in the untreated group), and hospital expenditure values were not significantly different between the two groups of patients who received ribavirin or not [61]. Recently, 218 Turkish CCHF patients were evaluated retrospectively for clinical outcome based on oral ribavirin treatment. The case-fatality rate was 7.1% in the treated group and 11.9% in the untreated group. The average interval between disease onset and ribavirin administration was 4.4 days among fatal and 5.8 days in nonfatal cases in the treated group [146]. Sheikh et al. treated 83 confirmed CCHF patients with ribavirin, and the fatality rate was 9.6%. A remarkable feature of this study was the death of all five healthcare workers who were exposed to the blood and secretions of patients with CCHF, although they had been given the recommended postexposure prophylaxis with ribavirin [42]. The effect of oral ribavirin on viral load and disease progression was evaluated in a retrospective case–control study, which included 10 patients who received ribavirin for 10 days and 40 who received only supportive therapy. No statistically significant differences were found in the decrease in viral load, reduction in liver enzyme concentrations, increase in platelet count, or the case fatality rate [147]. Ribavirin appears to be more effective when given early in the course of illness; it was reported that the fatality rate was 5% when the ribavirin was used within 4 days after the onset of symptoms whereas 10% after fifth day and 27% in untreated patients [148]. In a case–control study in 2006 from Iran, 84% of the patients who were treated with oral ribavirin within the first 72 h of illness onset recovered from the disease, whereas the survival rate of the patients whose treatment began after 72 h was 74.8%; it was concluded that oral ribavirin is an effective treatment for patients with CCHF, especially when it is used within 72 h of the onset of disease and as soon as it is possible [149]. In another Iranian study, it was found that ribavirin therapy for patients who survived had begun on average approximately 24 h earlier than the initiation of ribavirin therapy in the cases of death. It was suggested that the interval between the onset of disease or hemorrhage and the initiation of ribavirin administration was the most important variable correlated with survival [150]. A recent study showed that the early beginning of therapy (days 1–4 after the onset of the disease) reduced the number of severe cases and manifestations of hemorrhagic syndrome [151]. The only randomized clinical trial on efficiency of ribavirin in treatment of CCHF was from Turkey. In this study, 138 Turkish CCHF patients were randomized, such that group A (n = 64) received oral ribavirin and supportive therapy while group B (n = 72) received only supportive therapy. There were no statistically significant differences between the two groups in the incubation period, clinical presentation, laboratory findings, time of hospitalization, requirement for platelet infusions, and time needed for normalization of platelet counts or mortality (6.3% among ribavirin-treated patients versus 5.6% among untreated patients) [152].

5.6.1.3 Specific Immunoglobulin

Specific antibodies against CCHFV are present 5–9 days after the onset of illness, and patients who die do not usually develop a measurable antibody response [67]. This finding encourages the therapeutic use of antibodies derived from recovered patients or animals [2]. Bulgarian investigators reported the prompt recovery of seven severely ill patients treated with two specific immunoglobulin preparations CCHF-bulin (for intramuscular use) and CCHF-venin (for intravenous use) obtained from the plasma of CCHF survivors boosted with one dose of CCHF vaccine [153]. In a recent study from Turkey, prompt administration of CCHFV hyperimmunoglobulin was suggested as an alternative treatment

approach, especially for high-risk individuals with a viral load of 10^8 copies/mL or more [154]. Further studies with larger sample size and more detailed design are necessary for the therapeutic use of CCHFV hyperimmunoglobulin.

5.6.2 Prevention

Prevention and control of CCHF should be applied both at the community level and in the hospitals.

At the community level, measures should be taken to prevent human contact with livestock and minimize the tick burden in these vertebrate hosts. Personal protection against tick bites includes avoiding of areas where ticks are extensively present, use of light-colored fully covered clothing, regular examination of the body surfaces for ticks, and use of tick repellents. While handling livestock or domesticated animals, repellents should be used on skin and clothing, and protective clothing to prevent skin contact with the infected materials. Unpasteurized milk and uncooked meat should not be eaten [155,156].

As mentioned previously, healthcare workers is an important risk group at the hospital settings. Infected patients should be under barrier nursing techniques and isolation. Strict universal precautions are necessary; healthcare workers should wear protective clothing such as disposable gowns, gloves, masks, and goggles or face shields. In procedures that may cause aerosols, an N95 mask should be worn. All items should be removed on leaving the patient's room and be safely disposed of or disinfected. Safety-engineered devices should be available [156,157].

5.6.2.1 Postexposure Management

In CCHF, use of oral ribavirin as postexposure prophylaxis was well described as effective and beneficial drug [158]. Ribavirin prophylaxis is generally well tolerated, potentially useful, and should therefore be recommended for high risk of exposures such as percutaneous injuries or splash of contaminated blood or body fluid to the face or mucosal surfaces [157].

5.6.2.2 Vaccines

At this point, there is no commercially available vaccine. The only currently available vaccine is that produced in Bulgaria, which is made from a suckling mouse brain (inactivated by chloroform, heated at 58°C, and adsorbed on aluminum hydroxide). In Bulgaria, it has been used since 1974 to immunize mainly healthcare workers, livestock breeders (for professionalists that grow sheep, goats, and cattle), mowers, military personnel, forest workers, geologists, and people that go on camping trips. The initial immunization dose includes two applications of 1 mL in an interval of 30–45 days. First re-immunization (booster dose) is performed with one application of 1 mL 1 year after the initial dose, and a second re-immunization should be carried out after 5 years. It was reported that a fourfold reduction in the number of reported CCHF cases in Bulgaria was observed over a 21-year period following initiation of vaccination [159,160].

In a recent study, the effectiveness of Bulgarian CCHF vaccine was investigated. Individuals immunized with the Bulgarian CCHFV vaccine develop specific T-cell activities, and this activity was higher in individuals vaccinated several times than in those who were vaccinated only once. Although the individuals demonstrated high levels of CCHFV-specific antibodies, the neutralization activity was low and it was concluded that the vaccine lacks clinical efficacy [161].

5.7 CONCLUSION

CCHF is a potentially fatal infection. Although it has been almost 70 years since recognition of the virus, the genetic sequencing and phylogenetic analysis of the virus genomes have not been elucidated for all CCHFV strains yet.

So far, the pathogenesis of the CCHF has not been completely determined, and advanced molecular studies are still needed. The molecular mechanisms of viral uptake or entry into target cells require to be explored. Also, it should be clarified whether CCHFV utilizes different cell surface receptors and attachment factors on different cell types. Further, the role of systemic inflammation in the pathogenesis of CCHF should be defined.

Two small animal models, STAT-1 knockout mice and knockout mice lacking the type 1 IFN receptor, that have recently been reported could potentially be used in future studies for antiviral drug development and evaluation and for vaccine design. Understanding mechanisms of viral replication of CCHFV will lead to the invention of new drugs and vaccines. Effective and efficient disease control can be constructed once the virus host–cell interactions are identified.

The definitive diagnosis of CCHF is mainly based on laboratory tests. The accepted approach for CCHF diagnosis combines the detection of the viral RNA genome and the detection of specific IgM antibodies in serum or blood. Early and accurate diagnosis of CCHF is crucial for the treatment and prevention of further transmission. In many endemic areas, the conditions of the laboratories are limited for the diagnosis of the CCHFV infections. Molecular-based diagnostic assays are usually the first choice in the diagnosis of CCHF. Currently, one-step real-time RT-PCR for the detection of CCHFV is accepted as a rapid, specific, and sensitive assay. There is a need for the development of simple and rapid diagnostic tests that do not require high-level laboratories, especially for the endemic areas.

Treatment options for CCHF are limited. There is an urgent need to establish effective treatment for CCHF. Current treatment consists of supportive therapy and ribavirin, but the efficacy of the ribavirin for the prophylaxis and therapy is still on debate. Promising results were mainly associated with early treatment. Currently, ribavirin is used in most endemic countries, and ethical issues about using a placebo-control group in studies have been raised. Given the high fatality rates associated with CCHF, a well-designed multicenter trial taking into account severity criteria is urgently needed in order to provide evidence-based data about ribavirin efficacy.

REFERENCES

1. Chamberlain, J. et al., Co-evolutionary patterns of variation in small and large RNA segments of Crimean-Congo hemorrhagic fever virus, *J. Gen. Virol.*, 86, 3337, 2005.
2. Hoogstraal, H., The epidemiology of tick-borne Crimean–Congo hemorrhagic fever in Asia, Europe, and Africa, *J. Med. Entomol.*, 15, 307, 1979.
3. Casals, J., Antigenic similarities between the virus causing Crimean hemorrhagic fever and Congo virus, *Proc. Soc. Exp. Biol. Med.*, 131, 233, 1969.
4. Burt, F.J. et al., Investigation of tick-borne viruses as pathogens of humans in South Africa and evidence of Dugbe virus infection in a patient with prolonged thrombocytopenia, *Epidemiol. Infect.*, 116, 353, 1996.
5. Flick, R. et al., Reverse genetics for Crimean-Congo hemorrhagic fever virus, *J. Virol.*, 77, 5997, 2003.
6. Connolly-Andersen, A.M., Magnusson, K.E., and Mirazimi, A., Basolateral entry and release of Crimean-Congo hemorrhagic fever virus in polarized MDCK-1 cells, *J. Virol.*, 81, 2158, 2007.
7. Aitichou, M. et al., Identification of Dobrava, Hantaan, Seoul, and Puumala viruses by one-step real-time RT-PCR, *J. Virol. Methods*, 124, 21, 2005.
8. Honig, J.E., Osborne, J.C., and Nichol, S.T., Crimean-Congo hemorrhagic fever virus genome L RNA segment and encoded protein, *Virology*, 321, 29, 2004.
9. Kinsella, E. et al., Sequence determination of the Crimean-Congo hemorrhagic fever virus L segment, *Virology*, 321, 23, 2004.
10. Clerx, J.P., Casals, J., and Bishop, D.H., Structural characteristics of nairoviruses (genus *Nairovirus*, Bunyaviridae), *J. Gen. Virol.*, 55, 165, 1981.
11. Sanchez, A.J., Vincent, M.J., and Nichol, S.T., Characterization of the glycoproteins of Crimean-Congo hemorrhagic fever virus, *J. Virol.*, 76, 7263, 2002.
12. Bertolotti-Ciarlet, A. et al., Cellular localization and antigenic characterization of Crimean-Congo hemorrhagic fever virus glycoproteins, *J. Virol.*, 79, 6152, 2005.
13. Weber, F. and Mirazimi, A., Interferon and cytokine responses to Crimean Congo hemorrhagic fever virus; an emerging and neglected viral zoonosis, *Cytokine Growth Factor Rev.*, 19, 395, 2008.
14. Altamura, L.A. et al., Identification of a novel C-terminal cleavage of Crimean-Congo hemorrhagic fever virus PreGN that leads to generation of an NSM protein, *J. Virol.*, 81, 6632, 2007.
15. Vincent, M.J. et al., Crimean-Congo hemorrhagic fever virus glycoprotein proteolytic processing by subtilase SKI-1, *J. Virol.*, 77, 8640, 2003.
16. Sanchez, A.J. et al., Crimean-Congo hemorrhagic fever virus glycoprotein precursor is cleaved by furin-like and SKI-1 proteases to generate a novel 38-kilodalton glycoprotein, *J. Virol.*, 80, 514, 2006.
17. Deyde, V.M. et al., Crimean-Congo hemorrhagic fever virus genomics and global diversity, *J. Virol.*, 80, 8834, 2006.
18. Hewson, R. et al., Evidence of segment reassortment in Crimean-Congo haemorrhagic fever virus, *J. Gen. Virol.*, 85, 3059, 2004.
19. Yashina, L. et al., Genetic variability of Crimean-Congo haemorrhagic fever virus in Russia and Central Asia, *J. Gen. Virol.*, 84, 1199, 2003.
20. Garrison, A.R. et al., Development of a TaqMan®–minor groove binding protein assay for the detection and quantification of Crimean-Congo hemorrhagic fever virus, *Am. J. Trop. Med. Hyg.*, 77, 514, 2007.

21. Midilli, K. et al., Imported Crimean-Congo hemorrhagic fever cases in Istanbul, *BMC Infect. Dis.*, 7, 54, 2007.
22. Vorou, R., Pierroutsakos, I.N., and Maltezou, H.C., Crimean-Congo hemorrhagic fever, *Curr. Opin. Infect. Dis.*, 20, 495, 2007.
23. Morikawa, S. et al., Genetic diversity of the M RNA segment among Crimean-Congo hemorrhagic fever virus isolates in China, *Virology*, 296, 159, 2002.
24. Ahmed, A.A. et al., Presence of broadly reactive and group-specific neutralizing epitopes on newly described isolates of Crimean–Congo hemorrhagic fever virus, *J. Gen. Virol.*, 86, 3327, 2005.
25. Yashina, L. et al., Genetic analysis of Crimean-Congo hemorrhagic fever virus in Russia, *J. Clin. Microbiol.*, 41, 860, 2003.
26. Papa, A. et al., Genetic characterization of the M RNA segment of Crimean Congo hemorrhagic fever virus strains China, *Emerg. Infect. Dis.*, 8, 50, 2002.
27. Meissner, J.D. et al., The complete genomic sequence of strain ROS/HUVLV-100, a representative Russian Crimean Congo hemorrhagic fever virus strain, *Virus Genes*, 33, 87, 2006.
28. Papa, A. et al., Genetic characterization of the MRNA segment of a Balkan Crimean–Congo hemorrhagic fever virus strain, *J. Med. Virol.*, 75, 466, 2005.
29. Nalca, A. and Whitehouse, C.A., Crimean-Congo hemorrhagic fever virus infection among animals, in *Crimean Congo Hemorrhagic Fever: A Global Perspective*, Ergonul, O. and Whitehouse, C.A. (eds.), Springer, Dordrecht, the Netherlands, 2007, p. 155.
30. Turell, M.J., Role of ticks in the transmission of Crimean Congo hemorrhagic fever virus, in *Crimean Congo Hemorrhagic Fever: A Global Perspective*, Ergonul O. and Whitehouse C.A. (eds.), Springer, Dordrecht, the Netherlands, 2007, p. 143.
31. Logan, T.M. et al., Experimental transmission of Crimean-Congo hemorrhagic fever virus by *Hyalomma truncatum* Koch, *Am. J. Trop. Med. Hyg.*, 40, 207, 1989.
32. Swanepoel, R. et al., Epidemiologic and clinical features of Crimean–Congo hemorrhagic fever in southern Africa, *Am. J. Trop. Med. Hyg.*, 36, 120, 1987.
33. Karti, S.S. et al., Crimean-Congo hemorrhagic fever in Turkey, *Emerg. Infect. Dis.*, 10, 1379, 2004.
34. Drosten, C. et al., Crimean-Congo hemorrhagic fever in Kosovo, *J. Clin. Microbiol.*, 40, 1122, 2002.
35. Nabeth, P. et al., Human Crimean-Congo hemorrhagic fever Senegal, *Emerg. Infect. Dis.*, 10, 1881, 2004.
36. Yesilyurt, M. et al., The early prediction of fatality in Crimean Congo hemorrhagic fever patients, *Saudi Med. J.*, 32, 742, 2011.
37. El-Azazy, O.M. and Scrimgeour, E.M., Crimean-Congo haemorrhagic fever virus infection in the western province of Saudi Arabia, *Trans. R. Soc. Trop. Med. Hyg.*, 91, 275, 1997.
38. Williams, R.J. et al., Crimean-Congo haemorrhagic fever: A seroepidemiological and tick survey in the Sultanate of Oman, *Trop. Med. Int. Health*, 5, 99, 2000.
39. Burney, M.I. et al., Nosocomial outbreak of viral hemorrhagic fever caused by Crimean Hemorrhagic fever-Congo virus in Pakistan, January 1976, *Am. J. Trop. Med. Hyg.*, 29, 941, 1980.
40. Papa, A. et al., Crimean-Congo hemorrhagic fever in Albania, 2001, *Eur. J. Clin. Microbiol. Infect. Dis.*, 21, 603, 2002.
41. Papa, A. et al., Crimean-Congo hemorrhagic fever in Bulgaria, *Emerg. Infect. Dis.*, 10, 1465, 2004.
42. Sheikh, A.S. et al., Bi-annual surge of Crimean-Congo haemorrhagic fever (CCHF): A five-year experience, *Int. J. Infect. Dis.*, 9, 37, 2005.
43. Sun, S. et al., Epidemiology and phylogenetic analysis of Crimean-Congo hemorrhagic fever viruses in Xinjiang, China, *J. Clin. Microbiol.*, 47, 2536, 2009.
44. Whitehouse, C.A., Crimean-Congo hemorrhagic fever, *Antiviral Res.*, 64, 145, 2004.
45. Bakir, M. et al., Crimean-Congo haemorrhagic fever outbreak in Middle Anatolia: A multicentre study of clinical features and outcome measures, *J. Med. Microbiol.*, 54, 385, 2005.
46. Mishra, A.C. et al., Crimean-Congo haemorrhagic fever in India, *Lancet*, 378, 372, 2011.
47. Majeed, B. et al., Morbidity and mortality of Crimean-Congo hemorrhagic fever in Iraq: Cases reported to the National Surveillance System, 1990–2010, *Trans. R. Soc. Trop. Med. Hyg.*, 106, 480, 2012.
48. The reports of the Communicable Diseases Department of the Ministry of Health of Turkey [in Turkish]. Ankara, Turkey, 2012. http://www.kirim-kongo.saglik.gov.tr
49. Van Eeden, P.J. et al., A nosocomial outbreak of Crimean-Congo haemorrhagic fever at Tygerberg Hospital. Part I. Clinical features, *S. Afr. Med. J.*, 68, 711, 1985.
50. Nabeth, P. et al., Crimean-Congo hemorrhagic fever, Mauritania, *Emerg. Infect. Dis.*, 10, 2143, 2004.
51. Albayrak, H., Ozan, E., and Kurt, M., Serosurvey and molecular detection of Crimean-Congo hemorrhagic fever virus (CCHFV) in northern Turkey, *Trop. Anim. Health. Prod.*, 44, 1667, 2012.
52. Morikawa, S., Saijo, M., and Kurane, I., Recent progress in molecular biology of Crimean-Congo hemorrhagic fever, *Comp. Immunol. Microbiol. Infect. Dis.*, 30, 375, 2007.
53. Jameson, L.J. et al., Importation of *Hyalomma marginatum*, vector of Crimean-Congo haemorrhagic fever virus, into the United Kingdom by migratory birds, *Ticks Tick Borne Dis.*, 3, 95, 2012.
54. Drosten, C. et al., Molecular diagnostics of viral hemorrhagic fevers, *Antiviral Res.*, 57, 61, 2003.
55. Rodriguez, L.L. et al., Molecular investigation of a multi-source outbreak of Crimean-Congo hemorrhagic fever in the United Arab Emirates, *Am. J. Trop. Med. Hyg.*, 57, 512, 1997.
56. Khan, A.S. et al., An outbreak of Crimean-Congo hemorrhagic fever in the United Arab Emirates, 1994–1995, *Am. J. Trop. Med. Hyg.*, 57, 519, 1997.
57. Ergonul, O., Crimean-Congo haemorrhagic fever, *Lancet Infect. Dis.*, 6, 203, 2006.
58. Charrel, R.N. et al., Tick-borne virus diseases of human interest in Europe, *Clin. Microbiol. Infect.*, 10, 1040, 2004.
59. Ergonul, O. et al., Characteristics of patients with Crimean-Congo hemorrhagic fever in a recent outbreak in Turkey and impact of oral ribavirin therapy, *Clin. Infect. Dis.*, 39, 284, 2004.
60. Gunes, T. et al., Crimean-Congo hemorrhagic fever virus in high-risk population, Turkey, *Emerg. Infect. Dis.*, 15, 461, 2009.
61. Ozkurt, Z. et al., Crimean-Congo hemorrhagic fever in Eastern Turkey: Clinical features, risk factors and efficacy of ribavirin therapy, *J. Infect.*, 52, 207, 2006.
62. Goldfarb, L.G. et al., An epidemiological model of Crimean hemorrhagic fever, *Am. J. Trop. Med. Hyg.*, 29, 260, 1980.
63. Vatansever Z. et al., Crimean-Congo hemorrhagic fever in Turkey, in *Crimean Congo Hemorrhagic Fever: A Global Perspective*, Ergonul, O. and Whitehouse, C.A. (eds.), Springer, Dordrecht, the Netherlands, 2007, pp. 59–74.

64. Ergonul, O. et al., Zoonotic infections among veterinarians in Turkey: Crimean-Congo hemorrhagic fever and beyond, *Int. J. Infect. Dis.*, 10, 465, 2006.

65. Xia, H. et al., Epidemiological survey of Crimean-Congo hemorrhagic fever virus in Yunnan, China, 2008, *Int. J. Infect. Dis.*, 15, e459, 2011.

66. Estrada-Peña, A. et al., Modeling the spatial distribution of Crimean Congo hemorrhagic fever outbreaks in Turkey, *Vector Borne Zoonotic Dis.*, 7, 667, 2007.

67. Papa, A. et al., Genetic detection and isolation of Crimean-Congo hemorrhagic fever virus, Kosovo, Yugoslavia, *Emerg. Infect. Dis.*, 8, 852, 2002.

68. Schwarz, T.F., Nsanze, H., and Ameen, A.M., Clinical features of Crimean-Congo haemorrhagic fever in the United Arab Emirates, *Infection*, 25, 364, 1997.

69. Athar, M.N. et al., Short report: Crimean-Congo hemorrhagic fever outbreak in Rawalpindi, Pakistan, February 2002, *Am. J. Trop. Med. Hyg.*, 69, 284, 2003.

70. Gürbüz, Y. et al., A case of nosocomial transmission of Crimean-Congo hemorrhagic fever from patient to patient. *Int. J. Infect. Dis.*, 13, e105, 2009.

71. Shepherd, A.J. et al., A nosocomial outbreak of Crimean-Congo haemorrhagic fever at Tygerberg Hospital. Part V. Virological and serological observations, *S. Afr. Med. J.*, 68, 733, 1985.

72. Aradaib, I.E. et al., Nosocomial outbreak of Crimean-Congo hemorrhagic fever, Sudan, *Emerg. Infect. Dis.*, 16, 837, 2010.

73. Naderi, H.R. et al., Nosocomial outbreak of Crimean-Congo haemorrhagic fever, *Epidemiol. Infect.*, 139, 862, 2011.

74. van de Wal, B.W. et al., A nosocomial outbreak of Crimean-Congo haemorrhagic fever at Tygerberg Hospital. Part IV. Preventive and prophylactic measures, *S. Afr. Med. J.*, 68, 729, 1985.

75. Saijo, M. et al., Possible horizontal transmission of Crimean-Congo hemorrhagic fever virus from a mother to her child, *Jpn. J. Infect. Dis.*, 57, 55, 2004.

76. Bodur, H. et al., Detection of Crimean-Congo hemorrhagic fever virus genome in saliva and urine, *Int. J. Infect. Dis.*, 14, e247, 2010.

77. Ergonul, O. et al., Pregnancy and Crimean-Congo haemorrhagic fever, *Clin. Microbiol. Infect.*, 16, 647, 2010.

78. Mood, B.S., Mardani, M., and Metenat, M., Clinical manifestations, laboratory findings and clinical outcome in 6 pregnant women with Crimean-Congo hemorrhagic fever, *Iran. J. Clin. Infect. Dis.*, 2, 193, 2007.

79. Erbay, A. et al., Breastfeeding in Crimean-Congo haemorrhagic fever, *Scand. J. Infect. Dis.*, 40, 186, 2008.

80. Ergonul, O., Clinical and pathologic features of Crimean-Congo hemorrhagic fever, in *Crimean Congo Hemorrhagic Fever: A Global Perspective*, Ergonul, O. and Whitehouse, C.A. (eds.), Springer, Dordrecht, the Netherlands, 2007, p. 207.

81. Cevik, M.A. et al., Clinical and laboratory features of Crimean-Congo haemorrhagic fever: Predictors of fatality, *Int. J. Infect. Dis.*, 12, 374, 2008.

82. Tasdelen Fisgin, N. et al., Initial high rate of misdiagnosis in Crimean Congo haemorrhagic fever patients in an endemic region of Turkey, *Epidemiol. Infect.*, 7, 1, 2009.

83. Franchini, G., Ambinder, R.F., and Barry, M., Viral disease in hematology, *Hematology, Am. Soc. Hematol. Educ. Program*, 1, 409, 2000.

84. Ergonul, O. et al., Analysis of risk-factors among patients with Crimean-Congo haemorrhagic fever virus infection: Severity criteria revisited, *Clin. Microbiol. Infect.*, 12, 551, 2006.

85. Engin, A. et al., Ocular findings in patients with Crimean-Congo hemorrhagic fever, *Am. J. Ophthalmol.*, 147, 634, 2009.

86. Engin, A. et al., Crimean-Congo hemorrhagic fever: Does it involve the heart? *Int. J. Infect. Dis.*, 13, 369, 2009.

87. Tezer, H. et al., Crimean-Congo hemorrhagic fever in children, *J. Clin. Virol.*, 48, 184, 2010.

88. Knust, B. et al., Crimean-Congo hemorrhagic fever, Kazakhstan, 2009–2010, *Emerg. Infect. Dis.*, 18, 643, 2012.

89. Duh, D. et al., Viral load as predictor of Crimean-Congo hemorrhagic fever outcome, *Emerg. Infect. Dis.*, 13, 1769, 2007.

90. Cevik, M.A. et al., Viral load as a predictor of outcome in Crimean-Congo hemorrhagic fever, *Clin. Infect. Dis.*, 45, e96, 2007.

91. Saksida, A. et al., Interacting roles of immune mechanisms and viral load in the pathogenesis of Crimean-Congo hemorrhagic fever, *Clin. Vac. Immunol.*, 17, 1086, 2010.

92. Connolly-Andersen, A.M. et al., Crimean Congo hemorrhagic fever virus infects human monocyte-derived dendritic cells, *Virology*, 390, 157, 2009.

93. Geisbert, T.W. and Jahrling, P.B., Exotic emerging viral diseases: Progress and challenges, *Nat. Med.*, 10, S110, 2004.

94. Bray, M., Comparative pathogenesis of Crimean-Congo hemorrhagic fever and ebola hemorrhagic fever, in *Crimean Congo Hemorrhagic Fever: A Global Perspective*, Ergonul, O. and Whitehouse, C.A. (eds.), Springer, Dordrecht, the Netherlands, 2007, p. 221.

95. Schnittler, H.J. and Feldmann, H., Viral hemorrhagic fever—A vascular disease? *Thromb. Haemost.*, 89, 967, 2003.

96. Burt, F.J. et al., Immunohistochemical and in situ localization of Crimean-Congo hemorrhagic fever (CCHF) virus in human tissues and implications for CCHF pathogenesis, *Arch. Pathol. Lab. Med.*, 121, 839, 1997.

97. Bereczky, S. et al., Crimean-Congo hemorrhagic fever virus infection is lethal for adult type I interferon receptor-knockout mice, *J. Gen. Virol.*, 91, 1473, 2010.

98. Bente, D.A. et al., Pathogenesis and immune response of Crimean-Congo hemorrhagic fever virus in a STAT-1 knockout mouse model, *J. Virol.*, 84, 11089, 2010.

99. Rodrigues R. et al., Crimean-Congo hemorrhagic fever virus-infected hepatocytes induce ER-stress and apoptosis crosstalk, *PLoS One*, 7, e29712, 2012.

100. Vestweber, D., Adhesion and signaling molecules controlling the transmigration of leukocytes through endothelium, *Immunol. Rev.*, 218, 178, 2007.

101. Connolly-Andersen, A.M. et al., Crimean-Congo hemorrhagic fever virus activates endothelial cells, *J. Virol.*, 85, 7766, 2011.

102. Ergonul, O. et al., Evaluation of serum levels of interleukin (IL)-6, IL-10, and tumor necrosis factor-alpha in patients with Crimean-Congo hemorrhagic fever, *J. Infect. Dis.*, 193, 941, 2006.

103. Papa, A. et al., Cytokine levels in Crimean-Congo hemorrhagic fever, *J. Clin. Virol.*, 36, 272, 2006.

104. Ozturk, B. et al., Evaluation of the association of serum levels of hyaluronic acid, sICAM-1, sVCAM-1, and VEGF-A with mortality and prognosis in patients with Crimean-Congo hemorrhagic fever, *J. Clin. Virol.*, 47, 115, 2010.

105. Peyrefitte, C.N. et al., Differential activation profiles of Crimean-Congo hemorrhagic fever virus- and Dugbe virus-infected antigen-presenting cells, *J. Gen. Virol.*, 91, 189, 2010.

106. Andersson, I. et al., Type I interferon inhibits Crimean-Congo hemorrhagic fever virus in human target cells, *J. Med. Virol.*, 78, 216, 2006.

107. Andersson, I. et al., Crimean-Congo hemorrhagic fever virus delays activation of the innate immune response, *J. Med. Virol.*, 80, 1397, 2008.
108. Shepherd, A.J., Leman, P.A., and Swanepoel, R., Viremia and antibody response of small African and laboratory animals to Crimean-Congo hemorrhagic fever virus infection, *Am. J. Trop. Med. Hyg.*, 40, 541, 1989.
109. Tignor, G.H. and Hanham, C.A., Ribavirin efficacy in an in vivo model of Crimean-Congo hemorrhagic fever virus (CCHF) infection, *Antiviral Res.*, 22, 309, 1993.
110. Gowen, B.B. and Holbrook, M.R., Animal models of highly pathogenic RNA viral infections: Hemorrhagic fever viruses, *Antiviral Res.*, 78, 79, 2008.
111. Jamil, B. et al., Crimean-Congo hemorrhagic fever: Experience at a tertiary care hospital in Karachi, Pakistan, *Trans. R. Soc. Trop. Med. Hyg.*, 99, 577, 2005.
112. Zeller, H., Laboratory diagnosis of Crimean-Congo hemorrhagic fever, in *Crimean Congo Hemorrhagic Fever: A Global Perspective*, Ergonul, O. and Whitehouse, C.A. (eds.), Springer, Dordrecht, the Netherlands, 2007, p. 233.
113. Ergonul, O., Crimean-Congo hemorrhagic fever virus: New outbreaks, new discoveries, *Curr. Opin. Virol.*, 2, 215, 2012.
114. Drosten, C. et al., Rapid detection and quantification of RNA of Ebola and Marburg viruses, Lassa virus, Crimean-Congo hemorrhagic fever virus, Rift Valley fever virus, dengue virus, and yellow fever virus by real-time reverse transcription-PCR, *J. Clin. Microbiol.*, 40, 2323, 2002.
115. Burt, F.J. et al., Serodiagnosis of Crimean-Congo haemorrhagic fever, *Epidemiol. Infect.*, 113, 551, 1994.
116. Burt, F.J., Swanepoel, R., and Braack L.E., Enzyme-linked immunosorbent assays for the detection of antibody to Crimean-Congo haemorrhagic fever virus in the sera of livestock and wild vertebrates, *Epidemiol. Infect.*, 111, 547, 1993.
117. Loutfy, M.R. et al., Effects of viral hemorrhagic fever inactivation methods on the performance of rapid diagnostic tests for *Plasmodium falciparum*, *J. Infect. Dis.*, 178, 1852, 1998.
118. Shepherd, A.J. et al., Comparison of methods for isolation and titration of Crimean–Congo hemorrhagic fever virus, *J. Clin. Microbiol.*, 24, 654, 1986.
119. Shepherd, A.J., Swanepoel, R., and Gill, D.E., Evaluation of enzyme-linked immunosorbent assay and reversed passive hemagglutination for detection of Crimean-Congo hemorrhagic fever virus antigen, *J. Clin. Microbiol.*, 26, 347, 1988.
120. Saijo, M. et al., Antigen-capture enzyme-linked immunosorbent assay for the diagnosis of Crimean-Congo hemorrhagic fever using a novel monoclonal antibody, *J. Med. Virol.*, 77, 83, 2005.
121. Shepherd, A.J., Swanepoel, R., and Leman, P.A., Antibody response in Crimean-Congo hemorrhagic fever, *Rev. Infect. Dis.*, 11, S801, 1989.
122. Saijo, M. et al., Recombinant nucleoprotein-based enzyme-linked immunosorbent assay for detection of immunoglobulin G antibodies to Crimean-Congo hemorrhagic fever virus, *J. Clin. Microbiol.*, 40, 1587, 2002.
123. Marriott, A.C. et al., Detection of human antibodies to Crimean-Congo haemorrhagic fever virus using expressed viral nucleocapsid protein, *J. Gen. Virol.*, 75, 2157, 1994.
124. Saijo, M. et al., Immunofluorescence technique using HeLa cells expressing recombinant nucleoprotein for detection of immunoglobulin G antibodies to Crimean-Congo hemorrhagic fever virus, *J. Clin. Microbiol.*, 40, 372, 2002.
125. Tang, Q. et al., A patient with Crimean-Congo hemorrhagic fever serologically diagnosed by recombinant nucleoprotein-based antibody detection systems, *Clin. Diagn. Lab. Immunol.*, 10, 489, 2003.
126. Saijo, M. et al., Recombinant nucleoprotein based serological diagnosis of Crimean-Congo hemorrhagic fever virus infections, *J. Med. Virol.*, 75, 295, 2005.
127. Garcia, S. et al., Evaluation of a Crimean-Congo hemorrhagic fever virus recombinant antigen expressed by Semliki forest suicide virus for IgM and IgG antibody detection in human and animal sera collected in Iran, *J. Clin. Microbiol.*, 35, 154, 2006.
128. Burt, F.J. et al., The use of a reverse transcription-polymerase chain reaction for the detection of viral nucleic acid in the diagnosis of Crimean-Congo haemorrhagic fever, *J. Virol. Methods*, 70, 129, 1998.
129. Schwarz, T.F. et al., Polymerase chain reaction for diagnosis and identification of distinct variants of Crimean–Congo hemorrhagic fever virus in the United Arab Emirates, *Am. J. Trop. Med. Hyg.*, 55, 190, 1996.
130. Yapar, M. et al., Rapid and quantitative detection of Crimean-Congo hemorrhagic fever virus by one-step real-time reverse-transcriptase PCR, *Jpn. J. Infect. Dis.*, 58, 358, 2005.
131. Duh, D. et al., Novel one-step real-time RT-PCR assay for rapid and specific diagnosis of Crimean-Congo hemorrhagic fever encountered in the Balkans, *J. Virol. Methods*, 133, 175, 2006.
132. Wölfel, R. et al., Low-density macroarray for rapid detection and identification of Crimean-Congo hemorrhagic fever virus, *J. Clin. Microbiol.*, 47, 1025, 2009.
133. Keshtkar-Jahromi, M. et al., Crimean-Congo hemorrhagic fever: Current and future prospects of vaccines and therapies, *Antiviral Res.*, 90, 85, 2011.
134. Leblebicioglu, H. et al., Case management and supportive treatment for patients with Crimean-Congo hemorrhagic fever, *Vector Borne Zoonotic Dis.*, 12, 805, 2012.
135. Dilber, E. et al., High-dose methylprednisolone in children with Crimean-Congo haemorrhagic fever, *Trop. Doct.*, 40, 27, 2010.
136. Cummins, D., Bennett, D., and Machin, S.J., Exchange transfusion of a patient with fulminant Lassa fever, *Postgrad. Med. J.*, 67, 193, 1991.
137. Xie, Y.W. et al., Clearance of dengue virus in the plasma-derived therapeutic proteins, *Transfusion*, 48, 1342, 2008.
138. Kurnaz, F. et al., A case of Crimean-Congo haemorrhagic fever successfully treated with therapeutic plasma exchange and ribavirin, *Trop. Doct.*, 41, 181, 2011.
139. Graci, J.D. and Cameron, C.E., Mechanisms of action of ribavirin against distinct viruses, *Rev. Med. Virol.*, 16, 37, 2006.
140. Watts, D.M. et al., Inhibition of Crimean-Congo hemorrhagic fever viral infectivity yields in vitro by ribavirin, *Am. J. Trop. Med. Hyg.*, 41, 581, 1989.
141. Paragas, J. et al., A simple assay for determining antiviral activity against Crimean-Congo hemorrhagic fever virus, *Antiviral Res.*, 62, 21, 2004.
142. WHO. Crimean-Congo haemorrhagic fever. http://www.who.int/mediacentre/factsheets/fs208/en/ (accessed November 19, 2013).
143. WHO. Application for inclusion of ribavirin in the WHO Model List of Essential Medicines. http://archives.who.int/eml/expcom/expcom15/applications/newmed/ribaravin/ribavirin.pdf (accessed November 19, 2013).
144. Mardani, M. et al., The efficacy of oral ribavirin in the treatment of Crimean–Congo hemorrhagic fever in Iran, *Clin. Infect. Dis.*, 36, 1613, 2003.

145. Alavi-Naini, R. et al., Crimean-Congo hemorrhagic fever in Southeast of Iran, *J. Infect.*, 52, 378, 2006.

146. Elaldi, N. et al., Efficacy of oral ribavirin treatment in Crimean-Congo haemorrhagic fever: A quasi-experimental study from Turkey, *J. Infect.*, 58, 238, 2009.

147. Bodur, H. et al., Effect of oral ribavirin treatment on the viral load and disease progression in Crimean-Congo hemorrhagic fever. *Int. J. Infect. Dis.*, 15, e44, 2011.

148. Tasdelen Fisgin, N. et al., The role of ribavirin in the therapy of Crimean-Congo hemorrhagic fever: Early use is promising, *Eur. J. Clin. Microbiol. Infect. Dis.*, 28, 929, 2009.

149. Metanat, M. et al., Clinical outcomes in Crimean-Congo hemorrhagic fever: A five-years experience in the treatment of patients in oral ribavirin, *Int. J. Virol.*, 2, 21, 2006.

150. Izadi, S. and Salehi, M., Evaluation of the efficacy of ribavirin therapy on survival of Crimean-Congo hemorrhagic fever patients: A case–control study, *Jpn. J. Infect. Dis.*, 62, 11, 2009.

151. Cherenov, I.V. et al., Efficacy of antiviral agents in the treatment of Crimean hemorrhagic fever, *Klin. Med. (Mosk.)*, 90, 59, 2012.

152. Koksal, I. et al., The efficacy of ribavirin in the treatment of Crimean-Congo hemorrhagic fever in Eastern Black Sea region in Turkey, *J. Clin. Virol.*, 47, 65, 2010.

153. Vasilenko, S.M. et al., Specific intravenous immunoglobulin for Crimean–Congo haemorrhagic fever, *Lancet*, 335, 791, 1990.

154. Kubar, A. et al., Prompt administration of Crimean-Congo hemorrhagic fever (CCHF) virus hyperimmunoglobulin in patients diagnosed with CCHF and viral load monitorization by reverse transcriptase-PCR. *Jpn. J. Infect. Dis.*, 64, 439, 2011.

155. Appannanavar, S.B. and Mishra, B., An update on Crimean Congo hemorrhagic fever, *J. Glob. Infect. Dis.*, 3, 285, 2011.

156. Zavitsanou, A., Babatsikou, F., and, Koutis, C., Crimean Congo hemorrhagic fever: An emerging tick-borne disease, *Health Sci. J.*, 3, 10, 2009.

157. Tarantola, A., Ergonul, O., and Tattevin, P., Estimates and prevention of Crimean Congo hemorrhagic fever risks for health care workers. In *Crimean Congo Hemorrhagic Fever: A Global Perspective*, Ergonul, O. and Whitehouse, C.A. (eds.), Springer, Dordrecht, the Netherlands, 2007, p. 281.

158. Ergonul, O., Treatment of Crimean–Congo hemorrhagic fever, *Antiviral Res.*, 78, 125, 2008.

159. Kalvatchev, N. and Christova, I., Current state of Crimean-Congo hemorrhagic fever in Bulgaria, *Biotechnol. Biotechnol. Equip.*, 26, 3079, 2012.

160. Christova, I. et al., Vaccine against Congo-Crimean haemorrhagic fever virus-Bulgarian input in fighting the disease, *Probl. Infect. Parasit. Dis.*, 37, 7, 2010.

161. Mousavi-Jazi, M. et al., Healthy individuals' immune response to the Bulgarian Crimean-Congo hemorrhagic fever virus vaccine, *Vaccine*, 28, 30, 6225, 2012.

6 Eastern Equine Encephalitis Virus

Laura D. Kramer and Norma P. Tavakoli

CONTENTS

6.1 INTRODUCTION

Eastern equine encephalitis virus (EEEV; Togaviridae: *Alphavirus*) is a highly pathogenic mosquito-borne virus that is amplified in an enzootic cycle between mosquitoes and avian hosts. It produces severe fatal encephalitis in horses and humans and neurologic symptoms in certain other mammals, such as deer, swine, cervids, and camelids, all of which experience high mortality following infection. EEEV is considered the most virulent of the encephalitic alphaviruses for horses and humans with a case fatality of 90%–95% in equines [1,2] and 30%–70% in humans, with children and the elderly experiencing the highest mortality [3,4]. Neurologic sequelae are commonly experienced by survivors. The ecologic and evolutionary factors that cause an enzootic EEEV cycle to become an epizootic are unknown and are critical for predictive modeling and public health control. The virus presents an important public health problem and has the potential for use as a biological weapon.

6.2 CLASSIFICATION AND MORPHOLOGY

EEEV is a member of the *Alphavirus* genus, family Togaviridae. According to the most recent report of the International Committee on Taxonomy of Viruses (ICTV), the Togaviridae family is comprised of 2 genera, *Alphavirus* (29 species) and *Rubivirus* (1 species) [5]. The alphaviruses include at least seven antigenic complexes, two of which are Western equine encephalitis virus (WEEV) and EEEV, the latter of which will be reviewed in this chapter; the other five are Middelburg, Ndumu, Semliki Forest, Venezuelan equine encephalitis virus (VEEV), and Barmah Forest viruses [6,7]. While EEEV is the sole species in the EEE antigenic complex, North and South American antigenic varieties can be distinguished serologically [8] and ecologically [9]. EEEV strains have been grouped into four subtypes, I–IV. Lineage I is found mainly in the Eastern United States, Canada, and the Caribbean islands [10], and the isolates from North America form a highly conserved lineage. Lineages II–IV have been isolated mainly in Central and South America, that is, lineage II strains are found in Brazil, Guatemala, and Peru; lineage III strains have been isolated in Argentina, Brazil, Colombia, Ecuador, Guiana, Panama, Peru, Trinidad, and Venezuela; and lineage IV have been found in Brazil [11]. Nucleotide sequencing and phylogenetic analyses revealed additional genetic diversity within the South American variety; three major South/Central American lineages were identified including one represented by a single isolate from eastern Brazil and two lineages with more widespread distributions in Central and South America [12]. In addition, there are old reports of EEEV isolations made in the Philippines, Thailand, the Asiatic region of Russia, and the former Czechoslovakia [13]. Antigenic differentiation should be further examined to determine whether viruses in these distinct lineages with separate ecologies represent different species [14].

The alphavirus virion is spherical, roughly 60–65 nm in diameter, and composed of three structural proteins. The nucleocapsid contains single-stranded positive-sense ribonucleic acid (RNA), approximately 11,700 nucleotides in length. Four nonstructural proteins, nsp1–4, make up the 5′ two-thirds of the genome and are involved in viral replication. A subgenomic RNA is produced that is homologous to the 3′ one-third of the genome. This RNA is translated directly into a structural polyprotein that is proteolytically cleaved into the capsid, E2, and E1 envelope glycoproteins [12].

6.3 BIOLOGY AND EPIDEMIOLOGY

In North America, EEEV is maintained in an enzootic life cycle involving wild birds and ornithophilic mosquito vectors [15], predominantly *Culiseta melanura* (Figure 6.1). Bridge vectors including *Coquillettidia perturbans* and *Aedes* spp., as well as *Cs. melanura*, carry the virus from viremic birds in the enzootic cycle to humans and equines. In the Caribbean, the most important vector has been reported to be *Culex taeniopus* [9]. In South America, less is known about the transmission cycle of EEEV, where rodents appear to play a more important role than birds in the enzootic cycle [16], and the critical vectors are in the *Culex* (*Melanoconion*) subgenus [17]. Human cases have not been observed in South America, although equine cases have been noted.

The intensity of yearly transmission of EEEV is highly variable (Figure 6.2), which has been hypothesized to result from the introduction of new viral genotypes, climatic and ecological changes that affect vector populations, viral replication rates, and/or changes in the population of avian reservoir hosts that cause increased amplification of the virus and subsequent spillover of infection to incidental hosts such as horses and humans [18–20]. Although many hypotheses have been put forth, previous studies have been limited in scope and are therefore unable to parse out the role of each factor in EEEV transmission.

The distribution of human and nonhuman EEE cases in the New World is depicted in Figure 6.3. In the United States, virus activity in mosquitoes, equines, and birds historically has been restricted to the eastern, southeastern, and some southern states as well as in the upper midwestern states of Ohio, Michigan, and Wisconsin. The virus also has been isolated from Arkansas, Minnesota, South Dakota, and Texas. The cumulative number of EEE cases reported in the United States since 1964 is listed in Table 6.1. The North American EEEV variant also occurs in Eastern Canada.

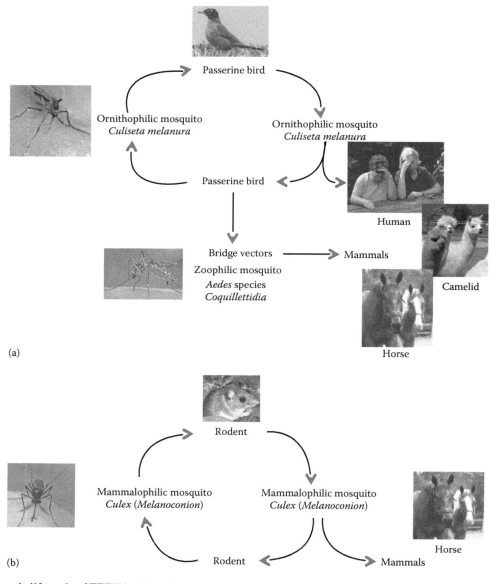

FIGURE 6.1 Enzootic life cycle of EEEV in (a) North and (b) South America.

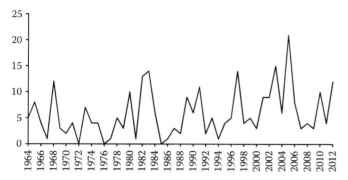

FIGURE 6.2 Eastern equine encephalitis neuroinvasive disease cases in the United States from 1964 to 2012. (Adapted from Centers for Disease Control and Prevention (CDC), Eastern Equine Encephalitis, Epidemiology and geographic distribution, http://www.cdc.gov/EasternEquineEncephalitis/tech/epi.html, accessed November 8, 2012.)

The South American variant occurs in parts of Central and South America, particularly along the Gulf Coast. Most isolates in the Caribbean belong to the North American EEEV group, but the South American variant can also be found in some locations. The virus most likely remains active year round in the tropics, including Florida, but is seasonal in more temperate regions. Yearly enzootic activity may be the result of annual reintroduction of virus from birds migrating from the north/south, or alternatively, perpetuation in situ from overwintering in mosquitoes or resident birds, or possibly reptiles [21,22]. Attempts to isolate virus from *Culiseta melanura* larvae and early season mosquitoes have not been successful [23,24]. However, recrudescence of virus in a relevant avian host, the gray catbird, was not observed after experimental infection: all birds became viremic initially, but no viral RNA was detected in the blood or bodily tissues 19 weeks following infection [25]. On the other hand, garter snakes infected in the laboratory were able to maintain viremia after 30 days forced hibernation at 7°C, but upon emergence, viremia decreased rapidly to undetectable levels [21].

It is not clear how environmental factors such as temperature and rainfall impact virus activity. Molecular epidemiological data provide evidence that some EEEV epizootics follow viral introduction from remote locations [18,26,27]. Introduction may be followed by establishment in the enzootic focus through amplification in the transmission cycle and then spillover of infection to secondary hosts including humans and horses. Before 2003 in New York State, a single dominant genotype circulated in enzootic foci for 1–2 years before becoming undetectable [18]. Since then, EEEV remained strongly active through 2011, with very minimal activity detected in 2012. In other states, such as Massachusetts and Vermont, in 2012, EEEV was very highly active with seven and two human cases, respectively.

Previous studies of mosquito vector competence and components of vectorial capacity (host choice, infection prevalence, longevity) in the Northeast and Mid-Atlantic states have implicated *Cs. melanura* as the dominant enzootic vector with other mosquito species such as *Culiseta morsitans*,

Coquillettidia perturbans, and members of the genera *Aedes*, *Ochlerotatus*, and *Culex* serving as additional potential enzootic or bridge vectors, that is, mosquitoes that feed on birds and mammals [28–30]. Recent evidence suggests *Cs. melanura* also may act as a bridge vector [31]. Overall, EEEV has been isolated from more than 25 species of mosquitoes. Field studies in New York State have implicated passeriform birds, wood thrushes, and American robins in particular as the primary avian hosts that fed on by *Cs. melanura*, the enzootic vector [32]. Ten species of birds (including eight passeriforms) have been tested for EEEV reservoir competence, that is, intensity and duration of infection and ability to infect feeding mosquitoes [33]. It was found that viremias were longer lasting and higher in starlings than in robins and other birds, and starlings frequently died while other birds generally survived.

EEEV primarily causes neurologic disease in horses and humans although other mammals, such as deer, dogs, and others, have occasionally become symptomatic. Swine and camelids (llamas and alpacas) commonly demonstrate neurologic symptoms following infection. Mammals are generally considered to be incidental (dead end) hosts, but some horses develop a transient viremia sufficient to infect a mosquito. Reptiles have been demonstrated to potentially serve as competent hosts, with circulation of levels of virus that appear infectious for mosquitoes over 7 days [21] and no evidence of pathology. Two species of wild-caught vipers, the cottonhead and the cottonmouth, were found to be positive for EEEV RNA by quantitative reverse transcription-polymerase chain reaction (qRT-PCR) [22].

6.4 CLINICAL FEATURES AND PATHOGENESIS

EEE virus is considered an emerging and reemerging human and veterinary pathogen and has the potential for use as a biological weapon [34]. EEE is of significant public health concern, due to the high mortality rates observed in infected humans, equines, and game birds. It is therefore a nationally notifiable disease in the United States.

With an overall case fatality rate of approximately 30%–70% [3,4,11,13,35,36], EEE is the most severe of the arboviral encephalitides. Both systemic and encephalitic forms of the disease are observed. The systemic disease occurs abruptly and is characterized by malaise, arthralgia, and myalgia [3,4,11,13,35,37]. Chills and muscular shaking may last for a few days, and maximum temperature may reach 104°F. The illness does not affect the central nervous system (CNS) and can last 1–2 weeks after which the patient recovers with no sequelae [36]. The encephalitic form, however, is more severe. Onset in infants is abrupt, but in older children and adults the incubation period can exceed 1 week. Symptoms of encephalitis include fever up to 106.4°F, irritability, restlessness, drowsiness, anorexia, vomiting, diarrhea, headache, convulsions, and coma [4,35,37–40]. Some patients also suffer from muscle twitching, tremors, and continuous neck rigidity. Once patients develop neurologic manifestations, their condition deteriorates rapidly [4].

Patients infected with EEEV commonly exhibit cerebrospinal fluid (CSF) abnormalities. An increase in pressure, as

FIGURE 6.3   Geographic distribution of eastern equine encephalitis in the Americas.

TABLE 6.1

Cumulative US Cases of EEE from 1964 to 2012

State	Case Count
Alabama	3
Delaware	3
Florida	71
Georgia	27
Indiana	3
Louisiana	17
Maryland	4
Massachusetts	43
Michigan	13
Mississippi	6
Missouri	2
North Carolina	15
New Hampshire	12
New Jersey	20
New York	5
Rhode Island	6
South Carolina	13
Texas	2
Vermont	2
Virginia	5
Wisconsin	2
Total	274

Source: Adapted from US Department of the Interior/US Geological Survey Disease Maps, Human eastern equine encephalitis, http://diseasemaps.usgs.gov/eee_us_human.html, accessed November 6, 2012.

well as pleocytosis, is observed in the CSF [35,36]. Glucose levels remain normal, but protein levels are generally elevated [3,4]. The white cell count ranges from 200 to 2000 cells, with 60%–90% neutrophils [35,36]. High CSF white cell count or severe hyponatremia is a predictor of a poor outcome [13]. Children younger than 15 years of age and individuals older than 55 are most at risk of serious disease. The age of the patient as well as other host factors determine the course of the disease.

During epidemics, approximately 70% of infected humans die 2–10 days after onset of illness [35]. Of those who survive long term, 35%–80% show progressive mental and physical sequelae [13,36,37,41]. The financial burden of the disease is significant and has been estimated to be in excess of $1M per patient mainly due to the requirement for costly permanent institutional care [37,42].

Short-term death is due to encephalitis, in some cases with evidence of myocardial insufficiency and pulmonary involvement. The observed encephalomyelitis is characterized by intense vascular engorgement that is perivascular and parenchymous, in the cortex, midbrain, and brainstem. Experimental infection of cynomolgus macaques with EEEV resulted in elevated serum levels of blood urea nitrogen, sodium, and alkaline phosphatase and prominent leukocytosis. The leukocytes were primarily granulocytes [43]. In addition, the onset and severity of neurological signs were similar to what is seen in human EEEV cases and also in cynomolgus macaques experimentally infected with WEEV [43].

Horses infected with EEEV may or may not be symptomatic. In cases where symptoms are apparent, they include fever, depression, ataxia, paralysis, anorexia, or stupor [37]. Horses that survive infection frequently exhibit neurological sequelae [9]. With all equine encephalitis viruses (EEE, VEE, WEE), disease in horses can be severe and result in encephalitis and a high mortality rate of up to 90% [44,45]. Pheasants are particularly susceptible to infection with EEEV, although in domestic fowl and wild birds, the infection may or may not cause morbidity and mortality [46,47]. Younger cases among birds and humans have a greater cerebral tissue sensitivity to the virus, are more viremic, and are therefore more susceptible to CNS disease.

In horses, the course of EEE can be swift, with death occurring 2–3 days after onset of clinical signs despite intensive care. Surviving horses may have long-lasting impairments and neurologic problems. Clinical signs of EEE include moderate to high fever, depression, lack of appetite, cranial nerve deficits (facial paralysis, tongue weakness, difficulty swallowing), behavioral changes (aggression, self-mutilation, or drowsiness), gait abnormalities, or severe CNS signs, such as head pressing, circling, blindness, and seizures. According to the US Department of Agriculture (USDA) Animal and Plant Health Inspection Service, 177 equine EEE cases were confirmed in the United States in 2012, but as with the disease in humans, the number of yearly cases has ranged widely, from 60 in 2011 to 712 in 2003 [48].

6.5 IDENTIFICATION AND DIAGNOSIS

Laboratory testing is critical in the diagnosis of EEEV infections, because generally arboviral infections display no specific clinical symptoms or physical findings that would permit unambiguous identification. In a potential biological warfare scenario, whenever epidemic febrile illness progressing to neurological disease is encountered, alphaviruses should be included in the differential diagnosis. In addition, sick and dying equines near an epidemic of febrile illness would suggest alphavirus exposure. Other potential agents that could cause similar symptoms include *Brucella* species, *Yersinia pestis*, *Salmonella typhi*, *Coxiella burnetii*, and *Clostridium botulinum*. In addition, other viral etiologies, which present with similar clinical features, a noninclusive list which is included in Table 6.2, should also be considered [49]. It is important to perform continuous surveillance of bioterrorism-related conditions such as EEE so that information is available for geographic, demographic, and seasonal distribution of these endemic but rare diseases. Such baseline information will assist in rapidly detecting changes in incidence and identifying potential bioterrorism events [50].

In North America, there is little geographic overlap between EEEV and WEEV, but both viruses exist sympatrically with Saint Louis encephalitis virus (SLEV),

TABLE 6.2
Some Important Viral Causes[a] of Endemic Encephalomyelitis

	Virus Family	Genus	Species
DNA viruses	Herpesviridae	*Simplex virus*	Herpes simplex virus 1 and 2
		Varicellovirus	Varicella zoster virus
		Lymphocryptovirus	Epstein–Barr virus
		Cytomegalovirus	Human cytomegalovirus
		Roseolovirus	Human herpesvirus 6
	Adenoviridae	*Mastadenovirus*	Adenovirus
RNA viruses	Togaviridae	*Alphavirus*	EEEV
			WEEV
			VEEV
	Flaviviridae	*Flavivirus*	SLEV
			Murray Valley encephalitis virus
			WNV
			Japanese encephalitis virus
			Dengue virus
			Tick-borne encephalitis virus
			Powassan virus
	Bunyaviridae	*Orthobunyavirus*	La Crosse encephalitis virus
		Phlebovirus	Rift Valley fever virus
			Toscana virus
	Paramyxoviridae	*Rubulavirus*	Mumps virus
		Morbillivirus	Measles virus
		Henipavirus	Hendra virus
			Nipah virus
	Arenaviridae	*Arenavirus*	Lymphocytic choriomeningitis virus
			Machupo virus
			Junin virus
	Picornaviridae	*Enterovirus*	Poliovirus
			Coxsackievirus
			Echovirus
	Reoviridae	*Coltivirus*	Colorado tick fever virus
	Rhabdoviridae	*Lyssavirus*	Australian bat lyssavirus
			Rabies virus
	Retroviridae	*Lentivirus*	Human immunodeficiency virus

Source: Modified from Chaudhuri, A. and Kennedy, P.G.E., *Postgrad. Med. J.*, 78, 575, 2002.

[a] Not all-inclusive.

California serogroup viruses, and Powassan virus as well. It is therefore important to consider multiple viruses in a differential diagnosis of a suspected arboviral infection (Table 6.2). In addition, any diagnostic tests for EEEV should distinguish this virus from Highlands J virus (HJV) even though HJV is not known to cause disease in humans. This is because HJV may be mistaken for EEEV as their geographic distributions overlap.

6.5.1 Containment Requirements

EEE virus is a select agent requiring registration with the Centers for Disease Control and Prevention (CDC) and/or USDA for possession, use, storage, and/or transfer [51,52]. In addition, there are strict containment requirements when working with the virus. Diagnostic or research procedures involving the propagation and manipulation of EEE virus are required to be performed in biosafety level 3 (BSL-3) containment. It is recommended that laboratorians use respiratory protection, gloves, shoe covers, and solid front gowns. If the exhaust air does not circulate to other areas of the building and is vented away from public areas and supply air intake vents, the air in the room where the work is performed is not required to be exhausted through high-efficiency particulate air (HEPA) filtration.

Work that may result in the aerosolization of EEE virus must be conducted under BSL-3 containment within a biosafety cabinet (BSC), class III cabinet, or in a bubble chamber that is HEPA filtered if the air is exhausted into the room. Personal protective equipment (PPE) should consist of respiratory protection such as N95/N100 mask or powered air-purifying respirator (PAPR), gloves, a solid front gown, scrubs or Tyvek suits, and shoe covers.

Research or diagnostic work with EEE virus involving animals must be performed in animal BSL-3 (ABSL-3) containment. HEPA filtration of exhaust air is required. Animal manipulation must be conducted in a BSC or class III cabinet. Depending on the animal host, HEPA-filtered containment cages may be required. PPE must be N95/N100 masks or PAPR, gloves, a solid front gown, scrubs or Tyvek suits, and shoe covers. Entities working with arthropods and EEE virus are required to use arthropod containment level 3 (ACL-3) containment. Appropriate filter/barriers are required to prevent the escape of arthropods.

6.5.2 Conventional Techniques for Diagnosis

The diagnosis of arboviral infections is generally based on serological assays because virus isolation from clinical specimens is labor-intensive, challenging, and time sensitive due to the process of viral clearance within days of the onset of symptoms. The most reliable sampling for virus isolation in cell culture or virus detection by RT-PCR assay is the mosquito vector. Most arboviruses can be amplified on Vero cells, which are also easy to cell culture. Therefore, virus isolation is generally performed on Vero cells.

Virus-positive samples are identified further by indirect immunofluorescence assay and/or RT-PCR [53].

The immunoglobulin M (IgM) antibody capture enzyme-linked immunosorbent assay (MAC-ELISA) is a method for rapid screening of serum or CSF samples in the diagnosis of acute arboviral infections [54]. The interpretation of MAC-ELISA results is made in the context of the specimen-collection and illness-onset dates, and the corresponding immunoglobulin G (IgG) results from a convalescent specimen. In most cases, IgM is detectable 8 days after the onset of symptoms. Positive specimens are confirmed by the plaque reduction neutralization test (PRNT). Confirmatory diagnostic testing by PRNT using wild-type (wt) EEEV poses challenges in public health laboratories because live virus is required. The US Department of Health and Human Services (HHS) has classified North American wt EEEV as a select agent, and it can be used only in registered BSL-3 containment facilities by security risk assessment (SRA)-approved individuals [55]. This prevents non-select-agent-registered diagnostic laboratories and those with only BSL-2 facilities from confirming EEEV by PRNT. To overcome this obstacle, chimeric viruses have been used for PRNTs for arboviral disease diagnostics [56].

Microsphere immunoassays (MIAs) for the detection of antibodies to EEEV utilize the Luminex (Austin, TX) laboratory multianalyte profiling (LabMAP) technology. The method is based on color coding of microsphere beads, to generate 100 different bead sets. Each bead set can be coated with a specific probe that will recognize and detect a particular target. Laser technology is then used to detect the identity of each bead, and hence the probe linked to the bead, and also the fluorescence of the reporter dye captured during the assay. It is theoretically feasible to perform a multiplex MIA in one tube that tests for EEEV and other arboviruses found in the region, as long as the viral protein targets are specific and conjugated to microspheres with different fluorescent signals. The interpretation of MIA results is made in the context of the specimen-collection and illness-onset dates and the MIA and/or MAC-ELISA results from convalescent specimens. If the MIA result is positive, the specimen is submitted for confirmatory testing by PRNT. MIAs have been successfully developed for a number of viruses, including West Nile virus (WNV) [57]. Advantages of the Luminex-based immunoassay over the ELISA include multiplexed capabilities, small sample volume, less reagent consumption, ease of use, speed, and sensitivity.

The PRNT is considered the gold standard procedure for the identification of arboviral antibody, and protocols are well established [58,59]. The test can help to distinguish false-positive results arising from ELISAs and other serologic assays. Ideally, the PRNT is performed on paired specimens consisting of acute (0–45 days after onset) and convalescent (3–7 weeks after the acute) serum samples. A fourfold rise in titer from former to latter is indicative of a current infection. The PRNT is also useful for epidemiological studies that address antibody seroprevalence in a population. Since the test is costly and time-consuming, requires training to perform and interpret, and must be performed in a BSL-3 facility for certain viruses, it usually is not employed as a screening tool. Neutralization titers are reported as the highest dilution of test serum that inhibits formation of at least 90% of the plaques, as compared with the virus control. A fourfold difference in PRNT titers between related viruses, as well as a fourfold rise in titer between paired acute and convalescent sera, is required for confident determination of etiology of disease. If paired acute and convalescent sera are not included in the PRNT, it is impossible to determine whether neutralizing antibody detected by the assay is due to an ongoing or past infection.

6.5.3   Molecular Techniques for Diagnosis

Surveillance to detect virologic activity in natural hosts, as well as monitoring of disease in humans and horses, is essential for early detection of outbreaks and rapid implementation of vector-control measures. Molecular detection by RT-PCR is an ideal method for detection of arboviruses in hosts, since it is rapid, sensitive, specific, reproducible, and amenable to automation. The method is being used increasingly as an adjunct to serology for the diagnosis of arboviruses. RT-PCR has been used for the detection of arboviruses at the group level [60–63] and for specific detection of individual arboviruses [64–67] as well as for a sequential detection at the two levels [68].

There are limitations to the use of molecular methods for the detection of arboviruses. Viral nucleic acids are generally only present in the host's blood 2–6 days after infection, that is, during the viremic stage. The probability of obtaining a positive result outside this time period is poor. Therefore, an alternative detection method, in particular serology, is advisable. In addition, variability in the genome of RNA viruses can result in strains of viruses whose nucleic acids are not amplified by specific primers, as a result of mutations in the primer binding sites in these strains. Where possible, primers should be designed in highly conserved regions of the genome. Furthermore, given the susceptibility of RNA to degradation, care must be taken to ensure that specimens are correctly handled and transported.

Various methods have been used for the molecular detection of EEEV. Standard RT-PCR assays specifically for the detection of EEEV [64–66,69] as well as group-specific RT-PCR assays for the detection of members of the *Alphavirus* genus followed by sequence analysis for the identification of members of the group have been reported [60,63]. However, these assays may not be as sensitive as real-time RT-PCR assays [69].

Armstrong and colleagues reported the use of a colorimetric dot assay to detect EEE viral RNA in mosquitoes, following polymerase chain reaction (PCR) amplification [70]. Regions of the gene encoding the capsid protein are targeted for PCR. Although the assay is sensitive and specific, it is laborious and time-consuming, due to the additional steps entailed by performing the dot assay. Lee and colleagues modified the EEEV PCR assay; incorporated it into

a conventional multiplex RT-PCR assay for the detection of EEE, La Crosse, and SLE viruses; and used it to monitor the activities of the three viruses in mosquito populations in the southeastern United States [71]. PCR and sequencing of three regions of the EEEV genome (C-terminal region of the nsp4 gene, a portion of the E2 gene, and the 3′ untranslated genome region) have also been carried out in phylogenetic analyses of representatives of the North and South American antigenic varieties [11] as well as isolates from New York (NY) [18] and Connecticut (CT) [27]. The North American isolates were found to be highly conserved whereas the existence of three major lineages was discerned in South America.

Linssen and colleagues reported the development of a combination RT-PCR-nested-PCR assay for the detection of equine encephalitis viruses (EEEV, WEEV, and VEEV), with a sensitivity of 30 RNA molecules (copies) for EEEV and 20 RNA copies for WEEV [66]. In a different multiplex assay, RT-PCR-ELISA was used to identify human pathogenic alphaviruses that included EEEV, WEEV, VEEV, and Mayaro virus [72]. That assay combined an RT-PCR, targeting the nsp1 gene, with an ELISA, in which a sequence-specific biotin-labeled probe was targeted against sequences specific to each virus. The assay was reported to be specific, sensitive, and rapid, with a turnaround time of 6–7 h.

A duplex real-time PCR assay for the detection of EEEV as well as SLEV has been reported for the detection of EEEV RNA in patient specimens, but can also be used for vector surveillance [73]. The EEEV assay, which can also be performed as a singleplex assay, targets the E1 gene that encodes a surface glycoprotein. The E1 gene is mostly conserved among alphaviruses [7] although there is sufficient dissimilarity that allowed primers and probes to be selected, which are specific to EEEV. The assay detected 12 different strains representing all four lineages of EEEV and was reported as being sensitive, specific, rapid, and amenable to high throughput [73].

6.6   TREATMENT AND PREVENTION

There is no specific treatment or postexposure prophylaxis for EEE. Treatment is primarily supportive and may involve hospitalization, respiratory support, and IV fluids. In addition, patients are prescribed with analgesics to relieve pain, corticosteroids to prevent swelling of the brain, anticonvulsants to prevent seizures, sedatives for restlessness and irritability, and antibiotics to treat secondary infections. Patient care centers on treating symptoms and complications.

The prevention of disease caused by encephalitic arboviruses centers on mosquito control and vaccination of horses. During outbreaks of EEE, humans and horses must be protected from mosquito bites. The likelihood of mosquito bites can be reduced by avoiding outdoor activities during dawn and dusk when mosquitoes are most active. Insect repellant sprays and wipes that include *N,N*-diethyl-meta-toluamide (DEET), picaridin, and oil of eucalyptus should be applied according to the label directions. When outdoors, wearing shoes, socks, long-sleeve shirts, and long pants will help prevent mosquito bites. Tight fitting screens should be fitted on doors and windows to prevent mosquitoes from entering homes. Horses should be covered with fly sheets and fans should be used in buildings. Any stagnant water that supports mosquito breeding including water in tires, wheelbarrows, bird baths, buckets, and other containers should be emptied. Any ditches that collect dirty stagnant water should be filled in. In barns and outbuildings, gutters should be cleared of debris. Lights should be turned off at night near horse barns to prevent mosquitoes from gathering near barns. Reducing the number of mosquitoes near homes and barns and taking personal precautions to prevent mosquito bites are key in preventing mosquito-borne illnesses.

There is no EEE vaccine for the general public although vaccines have been developed for horses. USDA-approved veterinary vaccines for EEE, WEE, and WNV are used widely in the United States for equines and are highly effective when administered correctly. Various pharmaceutical companies produce commercial multivalent vaccines that protect against EEEV as well as other agents such as WEEV, WNV, VEEV, tetanus, equine influenza virus, and equine herpes virus. For example, the *3-way* combination vaccine, or 3 in 1, protects against EEE, WEE, and tetanus. The *4-way* combination vaccine, or 4 in 1, protects against EEE, WEE, tetanus, and influenza. Depending on the presence of mosquitoes, horses should be vaccinated once or twice per year. The latter in areas where mosquitoes are active year round. The combination vaccines are thought to protect for 6–8 months, so vaccination in April to protect the horse until October is usually recommended.

Although there are no licensed EEE, WEE, or VEE vaccines marketed for the general public, for over 25 years, an experimental vaccine has been available at the US Army Medical Research Institute of Infectious diseases (USAMRIID) for researchers who are working with the virus in a laboratory. The EEE vaccine for humans is an inactivated PE-6, which is used under investigational new drug (IND) status. The PE-6 strain of EEE virus was passed in primary chick-embryo cell cultures and then was formalin-treated and lyophilized to produce an inactivated vaccine for EEE [74,75]. This vaccine is administered as a 0.5 mL dose subcutaneously on days 0 and 28, with 0.1 mL intradermal booster doses given as needed to maintain neutralizing antibody titers. Mild reactions to the vaccine were observed, and immunogenicity was demonstrated in initial clinical trials [75]. The vaccine was given to 896 at-risk laboratory workers between 1976 and 1991. No significant clinical reactions have been observed. A long-term follow-up study of 573 recipients indicated a 58% response rate after the primary series and a 25% chance of failing to maintain adequate titers for 1 year. Response rates and persistence of titers increased with the administration of additional booster doses [76]. Human sera of vaccinated individuals has been shown to display high neutralizing titers against the North American isolates examined, but only negligible neutralizing titers were obtained against South American isolates [77]. These data suggest that immunized individuals would mount an effective antibody response against infection with North American strains of EEE virus, but that further investigation is warranted to assess the protective capability of the vaccine against infection with South American strains.

In addition to vaccinating humans and equids, the formalin-inactivated vaccine has been used to vaccinate other domestic and wild animals affected by the virus such as pheasants, whooping cranes, pigs, and emus [78–83]. However, immunity is weak and short-lived, and residual live virus can cause severe disease, as exemplified in a vaccinated horse in California [84]. Safety is a concern with inactivated-virus vaccines because they have the potential to cause disease if the virus used in the manufacture of the vaccine is not completely inactivated. In addition, alphavirus vaccines that rely on point mutations that accumulate over cell passages may not have a consistent attenuated phenotype possibly due to high rates of reversion. Efforts are being made to develop recombinant vaccines, which are safer to use and which induce rapid and effective humoral and cellular immunity [85]. In addition, the potential use of recombinant, chimeric EEEV and, for example, WEEV or Sindbis virus as the basis for vaccines is under investigation [86,87].

Rapid reporting of suspected EEE and other arboviral cases will help in initiating mosquito control activities that could protect other horses and humans. In general, state and local government agencies would make the decision to use insecticides or other chemical and biological methods of containment for mosquito control. The most common products are synthetic pyrethroid insecticides, which are sprayed at dusk when mosquitoes are most active. The United States Environmental Protection Agency (USEPA) reviews and approves insecticides and their labeling to ensure minimal risk to humans and the environment.

6.7 CONCLUSIONS AND FUTURE PERSPECTIVES

Bioterrorism incidents have increased in the United States since the 1980s with two significant periods in 1998 and 2001. In 1998, a number of anthrax threats occurred in states including California and Indiana, and in 2001, following the September 11 attacks, a number of letters containing anthrax spores were mailed to two senators and a number of news media offices [88]. In the latter events, 5 people were killed and 17 people were infected. Years after the attack, several infected individuals continued to report health problems. It is estimated that cleanup costs exceeded 1 billion dollars. Such threats and incidents also create considerable panic and social disruption. In the past decade, there has been a tremendous drive, especially by the federal government, to improve infrastructure, public health response and preparedness, and funding of basic research to medically defend against bioterrorist threats. Among the agents identified as a bioterrorism threat is EEEV, which, although rare, is endemic in the eastern United States and southeast Canada and every year causes significant morbidity and mortality in equines as well as severe disease in humans. There are a number of reasons why EEEV is a candidate for bioterrorism;

the virus can be produced fairly easily and inexpensively in large amounts, it is stable and highly infectious when aerosolized, it has a high mortality rate, it is amenable to genetic manipulation, and there is no treatment for it.

A vigilant surveillance program is required to detect activity of EEEV as well as other arboviruses in order to establish background levels if a potential EEEV bioterrorism event were to occur. Data on host populations, that is, changes in abundance and/or composition of vector and vertebrate host species, and the presence of virus, as established by surveillance testing, are used to determine geographical areas of high risk, to monitor the efficacy of control measures, and to better understand the transmission cycles. Any divergence from what would be considered a "natural event" for the virus would quickly alert public health authorities to investigate an intentional event. An intentional event has to be differentiated from a spontaneous outbreak of a new or reemerging disease, a spontaneous outbreak following an accidental release, or an outbreak of an endemic disease. In addition to recognizing the event, populations at risk have to be identified, preventive measures such as vaccinations must be targeted, the spread of disease must be tracked and limited, and postexposure prophylaxis must be provided.

In addition to aerosolization, an intentional event could involve the release of infected mosquitoes. It would therefore be important to protect humans and other animals from infectious bites. In the United States, state and local government agencies oversee mosquito control programs, which can include both chemical and biological methods of containment. However, issues regarding exposure and risks to humans and the environment must be considered.

The availability of licensed vaccines for EEEV and other viral encephalitides not only would be valuable in cases of intentional biothreat events but would also be beneficial for at-risk individuals who routinely work with these viruses. There are currently inactivated vaccines being used for VEE, EEE, and WEE available under the FDA's investigational drug status. However, research is ongoing to develop safe and effective alternatives including chimeric vaccines [64,86,87]. There is an urgent need to develop rapid diagnostic methods for the detection of arboviruses in human and other animal hosts for testing of vector mosquito populations. Due to the small sample volume in some cases, multiplexed testing would be advantageous. Multiplexing would reduce hands-on time, numbers of analytical instruments required, and reagent costs. The types of technology that allow multiplexing include multiplex real-time RT-PCR, microarrays, and LabMAP. Currently, the feasible level of multiplexing for real-time RT-PCR assays is four to six agents. Higher multiplexing is possible with LabMAP technology targeting antibody/antigens or genomic nucleic acid. However, further refinement is needed to improve the sensitivity of the assays and to reduce interference and background fluorescence. Microarrays remain expensive and have not yet been adopted by most diagnostic laboratories that perform routine screening. As the methods continue to be improved, more and more assays are being developed that use these technologies.

Future bioterrorism threats are difficult to predict. However, it is clear that continued and preferably increased rigorous surveillance activity is needed, and research into developing diagnostic methods, vaccines, and treatment methods will be crucial in protecting public health and minimizing the potential risks of bioterrorism events.

ACKNOWLEDGMENTS

The authors thank Dr. Betsy Kauffman and Ms. Mary Franke for their valuable assistance in preparing this manuscript.

REFERENCES

1. Waldridge, B.M. et al., Serologic responses to eastern and western equine encephalomyelitis vaccination in previously vaccinated horses, *Vet. Ther.*, 4, 242, 2003.
2. Weaver, S.C. et al., Molecular epidemiological studies of veterinary arboviral encephalitides, *Vet. J.*, 157, 123, 1999.
3. Calisher, C.H., Medically important arboviruses of the United States and Canada, *Clin. Microbiol. Rev.*, 7, 89, 1994.
4. Deresiewicz, R.L. et al., Clinical and neuroradiographic manifestations of eastern equine encephalitis, *N. Engl. J. Med.*, 336, 1867, 1997.
5. King, A.M.Q., Lefkowitz, E., Adams, M.J., and Carstens, E.B., eds., *Virus Taxonomy: Ninth Report of the International Committee on Taxonomy of Viruses*, 1st edn. Elsevier, Waltham, MA, 2011.
6. Calisher, C.H. et al., Reevaluation of the western equine encephalitis antigenic complex of alphaviruses (family Togaviridae) as determined by neutralization tests, *Am. J. Trop. Med. Hyg.*, 38, 447, 1988.
7. Powers, A.M. et al., Evolutionary relationships and systematics of the alphaviruses, *J. Virol.*, 75, 10118, 2001.
8. Casals, J., Antigenic variants of Eastern equine encephalitis virus, *J. Exp. Med.*, 119, 547, 1964.
9. Scott, T.W. and Weaver, S.C., Eastern equine encephalomyelitis virus: Epidemiology and evolution of mosquito transmission, *Adv. Virus Res.*, 37, 277, 1989.
10. Weaver, S.C. et al., Evolution of alphaviruses in the eastern equine encephalomyelitis complex, *J. Virol.*, 68, 158, 1994.
11. Brault, A.C. et al., Genetic and antigenic diversity among eastern equine encephalitis viruses from North, Central, and South America, *Am. J. Trop. Med. Hyg.*, 61, 579, 1999.
12. Brault, A.C. et al., Positively charged amino acid substitutions in the E2 envelope glycoprotein are associated with the emergence of Venezuelan equine encephalitis virus, *J. Virol.*, 76, 1718, 2002.
13. von Sprockhoff, H. and Ising, E., On the presence of viruses of the American equine encephalomyelitis in Central Europe. Review, *Arch. Gesamte Virusforsch.*, 34, 371, 1971.
14. Griffin, D.E., Alphaviruses. In *Field's Virology*, Knipe, D.M. et al. (eds.). Lippincott, Williams & Wilkins, Philadelphia, PA, 2007, p. 1023.
15. Freier, J.E., Eastern equine encephalomyelitis, *Lancet*, 342, 1281, 1993.
16. Arrigo, N.C. et al., Evolutionary patterns of eastern equine encephalitis virus in North versus South America suggest ecological differences and taxonomic revision, *J. Virol.*, 84, 1014, 2010.

17. Weaver, S.C. et al., Alphaviruses: Population genetics and determinants of emergence, *Antiviral Res.*, 94, 242, 2012.

18. Young, D.S. et al., Molecular epidemiology of eastern equine encephalitis virus, New York, *Emerg. Infect. Dis.*, 14, 454, 2008.

19. Day, J.F. and Stark, L.M., Transmission patterns of St. Louis encephalitis and eastern equine encephalitis viruses in Florida: 1978–1993, *J. Med. Entomol.*, 33, 132, 1996.

20. Crans, W.J. et al., Eastern equine encephalomyelitis virus in relation to the avian community of a coastal cedar swamp, *J. Med. Entomol.*, 31, 711, 1994.

21. White, G. et al., Competency of reptiles and amphibians for eastern equine encephalitis virus, *Am. J. Trop. Med. Hyg.*, 85, 421, 2011.

22. Bingham, A.M. et al., Detection of eastern equine encephalomyelitis virus RNA in North American snakes, *Am J. Trop. Med. Hyg.*, 87(6):1140–1144, 2012. Published online October 1, 2012, doi:10.4269/ajtmh.2012.12-0257.

23. Morris, C.D. and Srihongse, S., An evaluation of the hypothesis of transovarial transmission of eastern equine encephalomyelitis virus by *Culiseta melanura*, *Am. J. Trop. Med. Hyg.*, 27, 1246, 1978.

24. Watts, D.M. et al., Ecological evidence against vertical transmission of eastern equine encephalitis virus by mosquitoes (Diptera: Culicidae) on the Delmarva Peninsula, USA, *J. Med. Entomol.*, 24, 91, 1987.

25. Owen, J.C. et al., Test of recrudescence hypothesis for overwintering of eastern equine encephalomyelitis virus in gray catbirds, *J. Med. Entomol.*, 48, 896, 2011.

26. White, G.S. et al., Phylogenetic analysis of eastern equine encephalitis virus isolates from Florida, *Am. J. Trop. Med. Hyg.*, 84, 709, 2011.

27. Armstrong, P.M. et al., Tracking eastern equine encephalitis virus perpetuation in the northeastern United States by phylogenetic analysis, *Am. J. Trop. Med. Hyg.*, 79, 291, 2008.

28. Turell, M.J. et al., Experimental transmission of eastern equine encephalitis virus by strains of *Aedes albopictus* and *A. taeniorhynchus* (Diptera: Culicidae), *J. Med. Entomol.*, 31, 287, 1994.

29. Vaidyanathan, R. et al., Vector competence of mosquitoes (Diptera: Culicidae) from Massachusetts for a sympatric isolate of eastern equine encephalomyelitis virus, *J. Med. Entomol.*, 34, 346, 1997.

30. Sardelis, M.R. et al., Experimental transmission of eastern equine encephalitis virus by *Ochlerotatus j. japonicus* (Diptera: Culicidae), *J. Med. Entomol.*, 39, 480, 2002.

31. Molaei, G. et al., Molecular identification of blood-meal sources in *Culiseta melanura* and *Culiseta morsitans* from an endemic focus of Eastern Equine Encephalitis virus in New York, *Am. J. Trop. Med. Hyg.*, 75, 1140, 2006.

32. Howard, J.J. et al., Antibody response of wild birds to natural infection with alphaviruses, *J. Med. Entomol.*, 41, 1090, 2004.

33. Komar, N. et al., Eastern equine encephalitis virus in birds: Relative competence of European starlings (*Sturnus vulgaris*), *Am. J. Trop. Med. Hyg.*, 60, 387, 1999.

34. Hawley, R.J. and Eitzen, E.M. Jr., Biological weapons—A primer for microbiologists, *Annu. Rev. Microbiol.*, 55, 235, 2001.

35. Farber, S. et al., Encephalitis in infants and children caused by the virus of the eastern variety of equine encephalitis, *J. Am. Med. Assoc.*, 114, 1725, 1940.

36. Przelomski, M.M. et al., Eastern equine encephalitis in Massachusetts: A report of 16 cases, 1970–1984, *Neurology*, 38, 736, 1988.

37. Morris, C.D., Eastern equine encephalomyelitis. In *The Arboviruses: Epidemiology and Ecology*, Monath, T.P. (ed.). CRC Press, Inc., Boca Raton, FL, 1988, p. 1.

38. Feemster, R.F., Equine encephalitis in Massachusetts, *N. Engl. J. Med.*, 257, 701, 1957.

39. Hart, K.L. et al., An outbreak of Eastern equine encephalomyelitis in Jamaica, West Indies. I. Description of human cases, *Am. J. Trop. Med. Hyg.*, 13, 331, 1964.

40. Johnston, R.E. and Peters, C.J., Alphaviruses. In *Fields Virology*, Fields, B.N. et al. (eds.). Lippincott Williams & Wilkins, Philadelphia, PA, 1996, p. 843.

41. Clarke, D.H., Two nonfatal human infections with the virus of eastern encephalitis, *Am. J. Trop. Med. Hyg.*, 10, 67, 1961.

42. Villari, P. et al., The economic burden imposed by a residual case of eastern encephalitis, *Am. J. Trop. Med. Hyg.*, 52, 8, 1995.

43. Reed, D.S. et al., Aerosol exposure to western equine encephalitis virus causes fever and encephalitis in cynomolgus macaques, *J. Infect. Dis.*, 192, 1173, 2005.

44. Acha, P.N. and Szyfres, B., Venezuelan equine encephalitis. In *Zoonoses and Communicable Diseases Common to Man and Animals*. Pan American Health Organization, Washington, DC, 2003, p. 333.

45. Summers, B.A., Cummings, J.F., and de Lahunta, A., *Veterinary Neuropathology*, 1st edn. Mosby-Year Book Inc., St. Louis, MO, 1995.

46. Kissling, R.E. et al., Studies on the North American arthropod-borne encephalitides. III. Eastern equine encephalitis in wild birds, *Am. J. Hyg.*, 62, 233, 1955.

47. Luginbuhl, R.E. et al., Investigation of eastern equine encephalomyelitis. II. Outbreaks in Connecticut pheasants, *Am. J. Hyg.*, 67, 4, 1958.

48. USDA Animal and Plant Health Inspection Service, Eastern and western equine encephalitis surveillance data, http://www.aphis.usda.gov/vs/nahss/equine/ee/index.htm (accessed November 5, 2013).

49. Chaudhuri, A. and Kennedy, P.G.E., Diagnosis and treatment of viral encephalitis, *Postgrad. Med. J.*, 78, 575, 2002.

50. Chang, M.H. et al., Endemic, notifiable bioterrorism-related diseases, United States, 1992–1999, *Emerg. Infect. Dis.*, 9, 556, 2003.

51. Pastel, R.H. et al., Clinical laboratories, the select agent program, and biological surety (biosurety), *Clin. Lab. Med.*, 26, 299, vii, 2006.

52. Gonder, J.C., Select agent regulations, *ILAR. J.*, 46, 4, 2005.

53. Kauffman, E.B. et al., Virus detection protocols for West Nile virus in vertebrate and mosquito specimens, *J. Clin. Microbiol.*, 41, 3661, 2003.

54. Martin, D.A. et al., Standardization of immunoglobulin M capture enzyme-linked immunosorbent assays for routine diagnosis of arboviral infections, *J. Clin. Microbiol.*, 38, 1823, 2000.

55. U.S. Departments of Health and Human Services (HHS), Select agents and toxins list, http://www.selectagents.gov/Select%20Agents%20and%20Toxins%20List.html (accessed November 5, 2013).

56. Johnson, B.W. et al., Use of sindbis/eastern equine encephalitis chimeric viruses in plaque reduction neutralization tests for arboviral disease diagnostics, *Clin. Vaccine Immunol.*, 18, 1486, 2011.

57. Wong, S.J. et al., Immunoassay targeting nonstructural protein 5 to differentiate West Nile virus infection from dengue and St. Louis encephalitis virus infections and from flavivirus vaccination, *J. Clin. Microbiol.*, 41, 4217, 2003.

58. Beaty, B.J. et al., Arboviruses. In *Diagnostic Procedures for Viral, Rickettsial and Chlamydial Infections*, Schmidt, N.J. and Emmons, R.W. (eds.). American Public Health Association, Washington, DC, 1989, p. 797.

59. Calisher, C.H. et al., Complex-specific immunoglobulin M antibody patterns in humans infected with alphaviruses, *J. Clin. Microbiol.*, 23, 155, 1986.

60. Sanchez-Seco, M.P. et al., A generic nested-RT-PCR followed by sequencing for detection and identification of members of the *Alphavirus* genus, *J. Virol. Methods*, 95, 153, 2001.

61. Bronzoni, R.V. et al., Multiplex nested PCR for Brazilian *Alphavirus* diagnosis, *Trans. R. Soc. Trop. Med. Hyg.*, 98, 456, 2004.

62. Scaramozzino, N. et al., Comparison of flavivirus universal primer pairs and development of a rapid, highly sensitive hemi-nested reverse transcription-PCR assay for detection of flaviviruses targeted to a conserved region of the NS5 gene sequences, *J. Clin. Microbiol.*, 39, 1922, 2001.

63. Pfeffer, M. et al., Genus-specific detection of alphaviruses by a semi-nested reverse transcription-polymerase chain reaction, *Am. J. Trop. Med. Hyg.*, 57, 709, 1997.

64. O'Guinn, M.L. et al., Field detection of eastern equine encephalitis virus in the Amazon Basin region of Peru using reverse transcription-polymerase chain reaction adapted for field identification of arthropod-borne pathogens, *Am. J. Trop. Med. Hyg.*, 70, 164, 2004.

65. Lee, J.H. et al., Identification of mosquito avian-derived blood meals by polymerase chain reaction-heteroduplex analysis, *Am. J. Trop. Med. Hyg.*, 66, 599, 2002.

66. Linssen, B. et al., Development of reverse transcription-PCR assays specific for detection of equine encephalitis viruses, *J. Clin. Microbiol.*, 38, 1527, 2000.

67. Lanciotti, R.S. and Kerst, A.J., Nucleic acid sequence-based amplification assays for rapid detection of West Nile and St. Louis encephalitis viruses, *J. Clin. Microbiol.*, 39, 4506, 2001.

68. de Morais Bronzoni, R.V. et al., Duplex reverse transcription-PCR followed by nested PCR assays for detection and identification of Brazilian alphaviruses and flaviviruses, *J. Clin. Microbiol.*, 43, 696, 2005.

69. Lambert, A.J. et al., Detection of North American eastern and western equine encephalitis viruses by nucleic acid amplification assays, *J. Clin. Microbiol.*, 41, 379, 2003.

70. Armstrong, P. et al., Sensitive and specific colorimetric dot assay to detect eastern equine encephalitis viral RNA in mosquitoes (Diptera: Culicidae) after polymerase chain reaction amplification, *J. Med. Entomol.*, 32, 42, 1995.

71. Lee, J.H. et al., Simultaneous detection of three mosquito-borne encephalitis viruses (eastern equine, La Crosse, and St. Louis) with a single-tube multiplex reverse transcriptase polymerase chain reaction assay, *J. Am. Mosq. Control Assoc.*, 18, 26, 2002.

72. Wang, E. et al., Reverse transcription-PCR-enzyme-linked immunosorbent assay for rapid detection and differentiation of alphavirus infections, *J. Clin. Microbiol.*, 44, 4000, 2006.

73. Hull, R. et al., A duplex real-time reverse transcriptase polymerase chain reaction assay for the detection of St. Louis encephalitis and eastern equine encephalitis viruses, *Diagn. Microbiol. Infect. Dis.*, 62, 272, 2008.

74. Maire, L.F., III et al., An inactivated eastern equine encephalomyelitis vaccine propagated in chick-embryo cell culture. I. Production and testing, *Am. J. Trop. Med. Hyg.*, 19, 119, 1970.

75. Bartelloni, P.J. et al., An inactivated eastern equine encephalomyelitis vaccine propagated in chick-embryo cell culture. II. Clinical and serologic responses in man, *Am. J. Trop. Med. Hyg.*, 19, 123, 1970.

76. McKinney, R.W., Inactivated and live VEE vaccines: A review. In *Proceedings of the Workshop Symposium on Venezuelan Encephalitis Virus, Washington, D.C., 14–17 September 1971.* Pan American Health Organization, Washington, DC, 1972, p. 1369.

77. Strizki, J.M. and Repik, P.M., Differential reactivity of immune sera from human vaccinees with field strains of eastern equine encephalitis virus, *Am. J. Trop. Med. Hyg.*, 53, 564, 1995.

78. Tengelsen, L.A. et al., Response to and efficacy of vaccination against eastern equine encephalomyelitis virus in emus, *J. Am. Vet. Med. Assoc.*, 218, 1469, 2001.

79. Jochim, M.M. et al., Venezuelan equine encephalomyelitis: Antibody response in vaccinated horses and resistance to infection with virulent virus, *J. Am. Vet. Med. Assoc.*, 162, 280, 1973.

80. Snoeyenbos, G.H. et al., Immunization of pheasants for Eastern encephalitis, *Avian Dis.*, 22, 386, 1978.

81. Clark, G.G. et al., Antibody response of sandhill and whooping cranes to an eastern equine encephalitis virus vaccine, *J. Wildl. Dis.*, 23, 539, 1987.

82. Elvinger, F. et al., Prevalence of exposure to eastern equine encephalomyelitis virus in domestic and feral swine in Georgia, *J. Vet. Diagn. Invest.*, 8, 481, 1996.

83. Olsen, G.H. et al., Efficacy of eastern equine encephalitis immunization in whooping cranes, *J. Wildl. Dis.*, 33, 312, 1997.

84. Franklin, R.P. et al., Eastern equine encephalomyelitis virus infection in a horse from California, *Emerg. Infect. Dis.*, 8, 283, 2002.

85. Wu, J.Q. et al., Complete protection of mice against a lethal dose challenge of western equine encephalitis virus after immunization with an adenovirus-vectored vaccine, *Vaccine*, 25, 4368, 2007.

86. Schoepp, R.J. et al., Recombinant chimeric western and eastern equine encephalitis viruses as potential vaccine candidates, *Virology*, 302, 299, 2002.

87. Wang, E. et al., Chimeric alphavirus vaccine candidates for chikungunya, *Vaccine*, 26, 5030, 2008.

88. Calisher, C.H. et al., Identification of arboviruses and certain rodent-borne viruses: Reevaluation of the paradigm, *Emerg. Infect. Dis.*, 7, 756, 2001.

89. Centers for Disease Control and Prevention (CDC), Eastern Equine Encephalitis, Epidemiology and geographic distribution, http://www.cdc.gov/EasternEquineEncephalitis/tech/epi.html (accessed November 8, 2012).

90. U.S. Department of the Interior/U.S. Geological Survey Disease Maps, Human eastern equine encephalitis, http://diseasemaps.usgs.gov/eee_us_human.html (accessed November 6, 2012).

7 Filoviruses

Randal J. Schoepp and Gene G. Olinger

CONTENTS

7.1   INTRODUCTION

Filoviruses, specifically the marburgviruses and ebolaviruses, cause hemorrhagic fevers with significant mortality in humans in sub-Saharan Africa. Outside of Africa, the viruses are less of a medical concern, but the potential for utilizing these viruses in biowarfare or bioterrorism has heightened the interests in understanding the viruses, the disease they cause, and its prevention. In developed countries, the dramatic disease and high lethality of filoviruses with the serious social consequences of the ensuing anxiety and panic of an outbreak makes these viruses a major public health priority. Given the very rapid spread of the disease and the lack of a vaccine or effective therapy, the only medical countermeasures remain the prevention and rapid control of outbreaks.

7.2   CLASSIFICATION AND MORPHOLOGY

7.2.1   TAXONOMY

The family Filoviridae is comprised of enveloped, non-segmented, negative-stranded RNA viruses and includes two genera, *Marburgvirus* and *Ebolavirus*.[1] A third genus,

Cuevavirus, is proposed for Lloviu virus (LLOV) identified in bats from Spain.[2] Filoviruses are members of the order Mononegavirales, which includes other families whose members have similar genomic characteristics and organization. The genus *Ebolavirus* has five members based on genetic divergence: Ebola virus (EBOV), Sudan virus (SUDV), Taï Forest virus (TAFV), Reston virus (RESTV), and Bundibugyo virus (BDBV). Each virus is represented by multiple variants that emerged during outbreaks. Unlike ebolaviruses, which are quite divergent, there are only two marburgviruses, Marburg virus (MARV) and Ravn virus (RAVV).[3,4] The taxonomy of the family Filoviridae continues to evolve since the first characterizations of MARV (1947) and EBOV (1976). The naming conventions for the viruses and taxa are a struggle between the laboratory virologists and the International Committee on Taxonomy of Viruses (ICTV), resulting in confusing and frequently improper taxonomic usage. Today, there is an attempt to come to a compromise between the logic of the ICTV and the needs of the laboratory virologist community.[2,5] A comparison of the old and new filovirus taxonomy can be found in Table 7.1. In this chapter, we will use the new taxonomy that is accepted by the ICTV.

TABLE 7.1

Comparison of Filoviridae Taxonomy Used among Laboratory Virologists and the ICTV

Nomenclature Laboratory Virologists	Outdated Taxonomy (Eighth ICTV Report)	Approved New Taxonomy (Ninth ICTV Report)
Order Mononegavirales	Order Mononegavirales	Order Mononegavirales
Family Filoviridae	Family Filoviridae	Family Filoviridae
Genus *Marburgvirus*	Genus *Marburgvirus*	Genus *Marburgvirus*
	Species *Lake Victoria marburgvirus*	Species *Marburg marburgvirus*
Virus: Marburg virus (MARV)	Virus: Lake Victoria marburgvirus (MARV)	Virus 1: Marburg virus (MARV)
		Virus 2: Ravn virus (RAVV)
Genus *Ebolavirus*	Genus *Ebolavirus*	Genus *Ebolavirus*
	Species *Zaire ebolavirus*	Species *Zaire ebolavirus*
Virus: Ebola virus (Zaire) (EBOV)	Virus: Zaire ebolavirus (ZEBOV)	Virus: Ebola virus (EBOV)
	Species *Sudan ebolavirus*	Species *Sudan ebolavirus*
Virus: Ebola virus (Sudan) (EBOV)	Virus: Sudan ebolavirus (SEBOV)	Virus: Sudan virus (SUDV)
	Species *Cote d'Ivoire ebolavirus* [sic]	Species *Taï Forest ebolavirus*
Virus: Ebola virus (Ivory Coast) (EBOV)	Virus: Cote d'Ivoire ebolavirus (CIEBOV) [sic]	Virus: Taï Forest virus (TAFV)
	Species *Reston ebolavirus*	Species *Reston ebolavirus*
Virus: Ebola virus (Reston) (EBOV)	Virus: Reston ebolavirus (REBOV)	Virus: Reston virus (RESTV)
	[Species *Bundibugyo ebolavirus*][a]	Species *Bundibugyo ebolavirus*
Virus: Ebola virus (Bundibugyo) (EBOV)	[Virus: Bundibugyo virus (BEBOV)]	Virus: Bundibugyo virus (BDBV)
		Genus *Cuevavirus* (provisional)
		Species *Lloviu cuevavirus* (provisional)
		Virus: Lloviu virus (LLOV)

Sources: Feldmann, H. et al., Family Filoviridae, in Fauquet, C.M., Mayo, M.A., Maniloff, J., Desselberger, U., and Ball, L.A. (eds.), *Virus Taxonomy: VIIIth Report of the International Committee on Taxonomy of Viruses*, Fauquet, Elsevier/Academic Press, London, U.K., 2005, pp. 645–653; Kuhn, J.H. et al., *Arch. Virol.*, 155, 2083, 2010; Kuhn, J.H. et al., Virus taxonomy—Ninth report of the International Committee on Taxonomy of Viruses, in A.M.Q. King, M.J. Adams, E.B. Carstens, and E.J. Lefkowitz (eds.), *Family Filoviridae*, Elsevier/Academic Press, London, U.K., 2011, pp. 665–671; Adams, M.J. and Carstens, E.B., *Arch. Virol.*, 157, 1411, 2012.

[a] Not addressed in the Eighth ICTV Report.

7.2.2 VIRUS STRUCTURE

The Filoviridae family derives its name from the characteristic long flexible, filamentous particles (Latin filum, *thread*) or shorter pleomorphic virions forming U- or six-shaped structures that measure 80 nm in diameter with lengths varying up to 1400 nm.[1,6–9] Filovirus virions consist of a single linear RNA molecule surrounded by a helical ribonucleoprotein (RNP) complex composed of the nucleoprotein (NP), structural proteins (VP30 and VP35), and RNA-dependent RNA polymerase (L protein). Two proteins, VP40 and possibly VP24, are matrix proteins. The virion is surrounded by a host cell-derived lipid envelope studded with distinctive knob-shaped peplomers that form the glycoprotein (GP) spikes (Figure 7.1a).

The filovirus genome is single-stranded, negative-sense RNA, approximately 19 kb nucleotides long with seven genes sequentially arranged along its length (Figure 7.1b).[1,10] The gene order is 3′-NP-VP35-VP40-GP/sGP-VP30-VP24-L-5′ with each gene defined by a highly conserved transcriptional start site at their 3′ end and a stop codon at their 5′ end. The genes are separated by intergenic regions varying in length and nucleotide composition. Gene overlap is characteristic of the filoviruses; in the marburgvirus genome, VP30–VP24 overlap, while in the ebolavirus genome, there are several overlapping genes (Figure 7.1b). The functional significance of the overlaps is unclear. Unique to ebolaviruses, the GP gene consists of two overlapping open reading frames (ORFs) that give rise to two gene products through transcriptional editing. The first ORF through unedited transcription gives rise to soluble, secreted GP (sGP) that is released from the infected cell. The membrane-bound GP results from an insertion of an adenosine residue during transcription causing a reading frame shift. Most genes possess long, noncoding sequences at their 3′ and/or 5′ ends, which contain the signals for replication and encapsidation.[11–14] The filovirus genomes are complementary at the very extreme ends and may form stem-loop structures.[15–18] These sequences contain the encapsidation signals, the replication origin, and the transcription promoter.[10,19]

The seven structural proteins of filoviruses can be divided into the RNP complex, composed of the NP, viral proteins VP35 and VP30, and the RNA-dependent RNA polymerase or L protein and those that are associated with the lipid envelope, the GP, and two matrix proteins, VP40 and VP24. Three RNP complex proteins, NP, VP35, and L, are essential for the transcription and replication of the marburgvirus RNA, with ebolavirus requiring the additional presence of VP30.[10,12,19] The NP is the major phosphoprotein surrounding

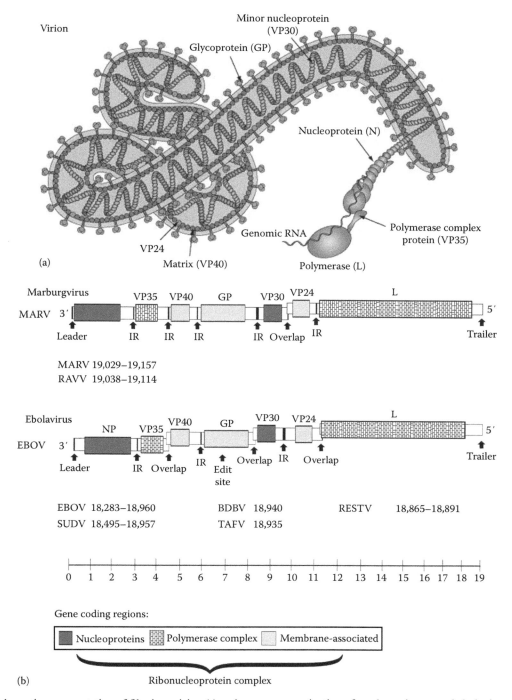

FIGURE 7.1 Schematic representation of filovirus virion (a) and genome organization of marburgviruses and ebolaviruses (b). The virion is represented as a full-length particle with internal and external proteins depicted. Representative virus genome structures (EBOV and MARV) illustrate gene coding regions and important elements. Different genome lengths (in nucleotides) are depicted for each marburgvirus and ebolavirus. EBOV, Ebola virus; SUDV, Sudan virus; BDBV, Bundibugyo virus; TAFV, Taï Forest virus; RESTV, Reston virus; IR, intergenic region. (a-Reproduced with permission from ViralZone, Swiss Institute of Bioinformatics. All rights reserved.)

the viral RNA to form the nucleocapsid. VP35 serves as a bridge that connects NP and L and serves as a polymerase cofactor.[19,20] The largest protein of the RNP complex is the RNA-dependent RNA polymerase or L protein. VP30 functions as activator and regulator of ebolavirus transcription and is involved in nucleocapsid assembly.[10] The role of VP30 in marburgvirus transcription and replication is not

completely understood; it is a minor phosphoprotein in the RNP that interacts with the NP.[21] Of the envelope-associated proteins, GP is the structural protein composing the spike inserted into the lipid envelopes and mediates binding to cellular receptors and fusion with cellular membranes for viral entry. Marburgvirus GP is transcribed from a single ORF and has only a single transmembrane form in contrast to ebolavirus, which has a transmembrane form and a secreted form (sGP) that results from RNA editing. VP40 is the most abundant protein in the virion and is a matrix protein that plays a role in viral assembly and budding.[22–24] VP24 is a minor matrix protein and plays an unclear role in assembly and budding.[25,26]

7.3 EPIDEMIOLOGY AND ECOLOGY

To date, no state-sponsored or terrorist organization has deployed a filovirus as a biological weapon or bioweapon. The consideration of these viruses as bioweapons is based on the lethality observed in sporadic human outbreaks during the past three decades. The epidemiology of human filovirus infections that give rise to natural outbreaks is unknown. Regardless of the viral source of the index case, the outbreak is maintained by person-to-person transmission through contact with infected blood or secretions during the care for the sick or preparation of the dead.

7.3.1 FILOVIRUS EPIDEMIOLOGY

Marburg hemorrhagic fever occurs in the arid woodland regions of Eastern and Southern Africa, described in the countries of Kenya, Uganda, Zimbabwe, Democratic Republic of the Congo (DRC), and Angola (Figure 7.2a). Ebola hemorrhagic fever occurs in the humid rain forests or nearby savannas of Central and Western Africa, described in the countries of Sudan, DRC, Uganda, Republic of Congo (RC), Gabon, and Côte d'Ivoire (Ivory Coast)[27] (Figure 7.2b). Incidence of hemorrhagic fever in these areas is compiled from outbreaks of the disease and limited seroprevalence studies, which may not accurately describe the geographical

extent of the true public health concern.[28–30] Ecological niche modeling describes a broader potential distribution for marburgviruses and ebolaviruses that generally spans the humid African tropics.[27,31] For marburgviruses, the most recent model correlates well with the countries in which outbreaks have occurred and expands the geographic potential to Burundi, Ethiopia, Malawi, Mozambique, Rwanda, Tanzania, Zambia, and a small region in northern Cameroon (Figure 7.2a).[31] The predicted regions for ebolaviruses correspond more closely to areas of known outbreaks and extend it to a few small, disjoint areas in West Africa.[27]

7.3.1.1 Marburgviruses

Marburg hemorrhagic fever was the first recognized filovirus outbreak, occurring in 1967 almost simultaneously in two places—Marburg and Frankfurt, Germany, and Belgrade, Yugoslavia (presently Serbia)—among laboratory staff working with blood and tissues from African green monkeys (*Cercopithecus aethiops*) imported from Uganda.[32–34] Eventually, 31 humans were infected resulting in seven deaths. The virus name Marburg comes from the city of Marburg where the virus was first isolated.

Since the initial discovery, Marburg hemorrhagic fever remains relatively rare, occurring sporadically in sub-Saharan Africa, causing isolated cases in travelers to endemic regions or in miners entering underground mines[35–41] (Figure 7.2a; Table 7.2). The first large, natural outbreak of Marburg hemorrhagic fever occurred in 1998 in the DRC (formerly known as Zaire) and was associated with gold miners in the area of the villages of Durba and Watsa.[42] Most of the cases were implicated with unauthorized mining activities at the Goroumbwa mine. Cases of Marburg hemorrhagic fever continued through September 2000 when the mine flooded

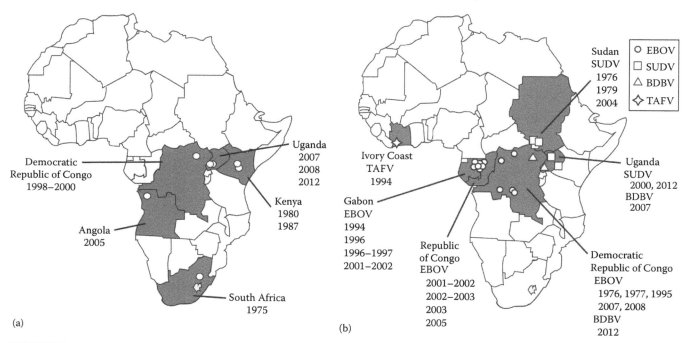

(a)   (b)

FIGURE 7.2 Filovirus outbreaks in Africa. Outbreaks of hemorrhagic fever caused by marburgviruses (a) and ebolaviruses (b) organized by the country and year of occurrence and coded by virus-specific points of reference. Outbreaks as noted in Table 7.2.

TABLE 7.2

Filovirus Outbreaks Listed with Corresponding Date, Location, Number of Clinical Cases, Fatality Rate, and References

Virus	Year	Country	Number of Human Cases	Case Fatality (%)	References
Marburg virus (MARV or RAVV)	1967	Germany—Yugoslavia	31	23	Kissling et al. [138]
	1975	South Africa	3	33	Gear et al. [35]
	1980	Kenya	2	50	Smith et al. [37]
	1987	Kenya	1	100	Johnson et al. [38]
	1998–2000	DRC	154	83	Bausch et al. [28]
	2004	Angola	252	92	Ligon [139]
	2007	Uganda	4	25	Towner et al. [46]
	2007	Uganda	1	0	Towner et al. [46]
	2008	Uganda	1	100	Hartman et al. [47]
	2012	Uganda	20	45	WHO [41]
Ebola virus (EBOV)	1976	DRC	318	88	Johnson et al. [140]
	1977	DRC	1	100	Heymann et al. [51]
	1994	Gabon	49	65	Georges et al. [141]
	1995	DRC	315	81	Khan et al. [142]
	1996	Gabon	37	57	Georges et al. [141]
	1996–1997	Gabon	60	75	Georges et al. [141]
	2001–2002	Gabon/RC	123	79	Leroy et al. [143]
	2002–2003	RC	143	90	Formenty et al. [144]
	2003	RC	35	83	WHO [145]
	2005	RC	10	83	Nkoghe et al. [146]
	2007	DRC	264	71	Grard et al. [147]
	2008	DRC	32	47	Grard et al. [147]
Sudan virus (SUDV)	1976	Sudan	284	53	WHO [49]
	1979	Sudan	34	65	Baron et al. [52]
	2000	Uganda	425	53	Okware et al. [148]; Lamunu et al. [149]
	2004	Sudan	17	41	WHO [150]
	2012	Uganda	1	100	Shoemaker et al. [151]
	2012	Uganda	24	71	WHO [54]
Taï Forest virus (TAFV)	1994	Ivory Coast	1	0	Formenty et al. [152]
Bundibugyo virus (BDBV)	2007	Uganda	102	42	Towner et al. [56]
	2012	DRC	62	55	ProMED [57]
Reston virus (RESTV)	1989	United States	0	0	Jahrling et al. [58]
	1990	United States	0	0	Groseth et al. [11]
	1992	Italy	0	0	WHO [62]
	1996	United States	0	0	Rollin et al. [153]

Abbreviations: DRC, Democratic Republic of the Congo; RC, Republic of Congo.

and halted all operations and access to the mines. A total of 154 cases were identified with 128 deaths, a case fatality rate of 83%. Interestingly, secondary transmission was rare. Most infections were the result of multiple introductions of MARV into the population as documented by isolation of multiple genetic lineages of virus circulating during the outbreak.

The largest recorded outbreak of Marburg hemorrhagic fever probably began in the Uige Province of northern Angola in October 2004 but remained undiagnosed until March 2005 and continued until the last confirmed case in July 2005.[43] This Marburg outbreak was uncharacteristic of previous outbreaks, occurring for the first time in West Africa,

in an urban setting, and tending to infect young children, yet the virus was genetically similar to other East African viruses.[44] This massive outbreak rivaled those of EBOV. The severity of the disease and the proximity to previous Ebola outbreaks left many to assume the cause was EBOV.[45] By the time the epidemic was declared over in November 2005, 374 cases were reported, resulting in 329 deaths, a case fatality rate of 88%. Since the outbreak proceeded unrecognized for some time, determining the exact origin of the MARV infection and/or the reservoir host was not possible, but the close genetic similarity to earlier Marburg strains suggested that the reservoir was similar. In the subsequent years, there were small sporadic outbreaks occurring in western Uganda among miners in the Kitaka Cave near Ibanda.[46–48] The latest reported outbreak occurred in October 2012 in four districts in Uganda[41] (Table 7.2).

7.3.1.2 Ebolaviruses

Ebola hemorrhagic fever was first recognized in 1976 in two almost simultaneous outbreaks in the DRC (formerly known as Zaire) and Sudan caused by two distinct ebolavirus species, *Zaire ebolavirus* (caused by EBOV) and *Sudan ebolavirus* (caused by SUDV), respectively[49,50] (Table 7.2; Figure 7.2b). The DRC outbreak in Yambuku resulted in 318 human cases and 280 deaths (88% case fatality); the Sudan outbreak was centered in Nzara and Maridi causing 284 human cases with 151 deaths (53% case fatality). In 1977, there was an unconfirmed, fatal case of EBOV in the DRC.[51] Following another 1979 SUDV outbreak, again in the Nzara and Maridi region of Sudan that caused 34 cases and 22 deaths (65%), there was a 15-year lull in which no ebolavirus outbreaks were observed.[52] EBOV reemerged in 1994 with outbreaks in Gabon in 1994 and 1996, in DRC in 1995 and 2007, in the border region of Gabon and the RC in 2001–2002, and in the RC in 2002–2003. Outbreaks of SUDV occurred in Uganda in 2000 and in Sudan in 2004. In 2012, there was an SUDV outbreak in Kibaale region, Uganda, that caused 24 cases with 17 deaths and was declared over in October.[53] In November, another ebolavirus outbreak was declared in the Luwero district of Uganda; no ebolavirus species was identified as yet.[54]

In 1994, a new ebolavirus species, *Taï Forest ebolavirus*, emerged and caused a single nonfatal human case contracted during the autopsy of a wild chimpanzee in the Taï Forest, Côte-d'Ivoire (Ivory Coast).[55] This was the first case of any ebolavirus in West Africa and Taï Forest ebolavirus (TAFV) has not occurred anywhere since.

In 2007, BDBV was isolated from a human blood sample in an outbreak in the Bundibugyo district in Western Uganda. The virus, representing a new ebolavirus species, *Bundibugyo ebolavirus*, caused 149 suspected cases with 37 deaths (25% case fatality).[56] In August 2012, a BDBV outbreak in Orientale Province, DRC, caused 62 cases of hemorrhagic disease resulting in 34 deaths before officially ending in November 2012.[57]

RESTV, representing a new ebolavirus species, *Reston ebolavirus*, was first isolated in an animal quarantine facility in Reston, Virginia, United States from dead and dying cynomolgus macaques (*Macaca fascicularis*).[58] The hemorrhagic disease was originally thought to be simian hemorrhagic fever virus (SHFV), an arterivirus, but was later attributed to RESTV. The source of the virus was traced back to the Philippine animal exporter where serologic evidence of past infection was demonstrated.[59] The virus was pathogenic for nonhuman primates, however, it appeared to infect humans and cause subclinical disease. None of the animal handlers at the quarantine facility or the export facility reported any significant disease, but four employees at the quarantine facility and 21 employees at the Philippine export facility had antibodies to RESTV, indicating a previous infection.[60]

Other outbreaks of RESTV in nonhuman primates occurred in the United States in 1990 and 1996, in Italy in 1992, and in the Philippines in 1992 and 1996, all of which were attributed to nonhuman primates imported from the Philippines.[59,61,62] In 2008, RESTV host tropism was expanded when the virus was detected in domestic swine from the Philippines that were coinfected with porcine reproductive and respiratory syndrome virus (PRRSV), an arterivirus.[63] Evidence was originally based on polymerase chain reaction (PCR) results but was eventually confirmed by virus isolation.[64] Serological evidence of pig-to-human transmission was suggested by demonstration of immunoglobulin G (IgG) antibodies to RESTV among Philippine pig farm and slaughterhouse workers, yet there was no report of human disease.[65]

7.3.1.3 Cuevaviruses

A third Filoviridae genus, *Cuevavirus* (provisional), was proposed with the discovery of a novel filovirus in Schreiber's bats (*Miniopterus schreibersii*) dying in Cueva del Lloviu, Spain.[66] The virus, provisionally named LLOV, was detected by PCR analysis in bats dying of viral pneumonia. The virus genotypic and phenotypic characteristics differ significantly from marburgviruses and ebolaviruses sufficiently to propose a separate genus.[2]

7.3.2 FILOVIRUS ECOLOGY

Little is known about the enzootic cycle maintaining filoviruses in nature. Humans are infected when they come in contact with the as yet unknown reservoir host or a closely associated intermediary host. Clearly, nonhuman primates can be a source of transmission to humans as demonstrated in the initial Marburg outbreak in 1967 or ebolavirus infections associated with bush meat[32,67]; however, the high mortality in nonhuman primates would imply that they are not the reservoir.[27] Since the filoviruses were first described, attempts to identify the reservoir host have centered on mammals, but arthropods, arachnids, amphibians, and birds have also been investigated.[36,68–72] Small mammals, particularly bats, are considered to be one of the most logical reservoir candidates because they occur in large numbers, have constant recruitment to their population, and can presumably be infected and amplify the virus.[68,70] The first indirect evidence of bats as filovirus reservoirs came from the 1998 marburgvirus outbreak in miners in the Durba and Watsa region of DRC (discussed earlier). During the course of the 2 year outbreak, infections resulted from repeated exposures acquired in a gold mine infested with bats.[42] Similarly, infections of tourists visiting bat-inhabited caves have resulted in the export of marburgvirus disease to countries outside of Africa.[39,40] Possible evidence for direct bat-to-human transmission occurred during the 2007 EBOV outbreak in DRC in which infection was traced back to an individual that purchased bat bush meat during a large seasonal migration.[73] Even though indirect evidence implicates bats as possible reservoirs of filoviruses and experimental evidence indicates that they can be infected, produce high-titered viremias, and have subclinical infections, direct evidence of their role was lacking.[74]

Recent studies have demonstrated the first direct evidence of marburgvirus infection in wild bats.[46,68,69] Egyptian fruit bats (*Rousettus aegyptiacus*) collected in Gabon contained marburgvirus-specific RNA and IgG antibodies, suggesting that this common bat species may play a role as reservoir and extended the range of marburgvirus to Gabon.[69] Similarly, virus-specific RNA and IgG antibodies in the same fruit bat species, *R. aegyptiacus*, and two insectivorous bat species, *Rhinolophus eloquens* and *Miniopterus inflatus*, collected from the Goroumbwa mine site of the Durba, DRC, outbreak were demonstrated.[68] The most convincing evidence comes from the isolation of genetically diverse marburgviruses from the tissues of Egyptian fruit bats, *R. aegyptiacus*, collected from the Kitaka Cave, Uganda.[46] This cave was the site of a Marburg hemorrhagic fever outbreak in miners in 2007, and the viruses isolated from the fruit bats closely matched those isolated from the miners. Isolation of virus over 9 months in a bat colony numbering over 100,000 demonstrated that bats can represent a major viral reservoir and a source for human infection. While evidence strongly suggests that bats play a role as reservoirs for filoviruses, additional field and laboratory studies must be done, since it is possible that bats could be an intermediary host, and the reservoir remains undiscovered.

7.4 CLINICAL FEATURES AND PATHOGENESIS

Infection with filovirus leads to a severe disease with high morbidity and mortality. The clinical features of the disease during outbreaks and the apparent ease of transmission from human to human instill fear and anxiety in the public and likely contribute heavily to the consideration of these viruses as weapons.

7.4.1 TRANSMISSION

Filovirus outbreaks are explosive and sporadic making them a source of alarm and anxiety. During outbreaks, transmission is known to occur by contact in three ways: (1) transmission from sick individuals between family members, close contacts, and caregivers; (2) transmission during preparation of the dead and funeral proceedings; and (3) transmission during patient care due to breaches in barrier nursing and contaminated medical equipment.[75] Once infected, most cases are symptomatic and cause severe clinical manifestations. Serological surveys indicate that asymptomatic cases are possible; however, there is limited data available.[76–78] In symptomatic human infections, the onset of clinical signs and symptoms of disease can vary but is generally 3–9 days (range 3–21 days). During the incubation period, the concentration of virus in blood and tissue is relatively low to nonexistent. As clinical signs of disease present, virus concentrations in the blood increase significantly and virus can be found in multiple tissues. Following disease onset, person-to-person transmission is highly probable and is associated with contact of blood or other body fluids (feces, vomit, urine, saliva, and respiratory secretions). Transmission via sexual activity may be possible since virus is detected in semen up to 7 weeks after clinical recovery. Filovirus outbreaks can be amplified through nosocomial or hospital-acquired infections. Early diagnosis and patient isolation is critical to prevent transmission. The cycle can be easily broken by patient isolation, strict barrier nursing methods, and safe burial procedures.[79] Aerosol transmission of filoviruses is possible; however, there is no direct evidence that this occurs in natural infections.[80] Experimentally, aerosol infections of guinea pigs and nonhuman primates are documented.[81,82] The possibility of aerosol infections of filoviruses and their use as biowarfare or bioterrorism weapons resulted in their classification as biological safety level-4 (BSL-4) agents.[80]

Patients become increasingly more infectious as their illness progresses. For health-care personnel and family members, contact during this period as a result of patient care in the home or in a hospital setting and preparation of the body for burial is commonly associated with transmission. Reuse of needles and syringes contaminated during patient care often can maintain or amplify outbreaks. Transmission through contaminated needles and needlestick injuries is associated with more severe disease, rapid deterioration, and possibly higher fatality. Fomite and aerosol route of transmission is reported in animal models of disease. Experimentally, transmission is demonstrated through fomites, oral and conjunctiva routes in animal models of disease.[83–85] Aerosol transmission is suggested in at least one case of human disease and is of concern from a biological terrorism or warfare perspective. All age groups are susceptible to infection, but most cases have occurred in adults.

7.4.2 CLINICAL PRESENTATION

Filovirus infection initially presents abruptly with nonspecific flulike symptoms such as fever (as high as 39°C), myalgia (generalized muscle and joint pain), and malaise (discomfort) that includes chills, headache, and sore throat.[35,86–89] Vomiting, abdominal pain, watery diarrhea, anorexia (poor appetite), dyspnea (shortness of breath), and dysphagia (difficulty in swallowing) are often reported in patients.[90] Early symptoms are similar to other common diseases and have limited diagnostic value. Furthermore, in endemic areas of Africa, outbreaks may occur with preexisting and concomitant infections. Illness is usually recognized by the existence of a high fever, followed by progressive and rapid debilitation to the more severe symptoms associated with hemorrhagic fever. In severe cases, patients may exhibit severe bleeding and coagulation abnormalities, including gastrointestinal bleeding, rash, and a range of hematological irregularities, such as lymphopenia and neutrophilia. Cytokines are released during infection, which may exaggerate inflammatory responses and cause some of the symptoms associated with the disease. Diarrhea, vomiting, and the inability to hydrate can lead to patients being described as being "ghost-like" with drawn features, deep-set eyes, expressionless faces, and extreme lethargy. A maculopapular rash may be visible on the axilla (armpit), groin, forehead, and trunk (chest, back, stomach). As the disease progresses, the gastrointestinal tract, respiratory tract, vascular, and neurologic systems are affected. Many, but not all, patients develop severe hemorrhagic manifestations 5–7 days from onset. A higher proportion of fatal cases are associated with some form of bleeding, often from multiple sites. Damage to the liver, combined with massive viremia, leads to disseminated intravascular coagulopathy. The virus eventually infects microvascular endothelial cells and compromises vascular integrity, resulting in bleeding characteristic in hemorrhagic fever patients. Fresh blood found in vomitus and feces or bleeding from the nose, gums, and vagina is of clinical significance. Controlling bleeding following medical procedures such as blood collection is paramount at this stage of disease. During the severe phase of illness, patients continue to sustain high fevers and may exhibit spontaneous bleeding at venipuncture sites. Involvement of the central nervous system can result in confusion, irritability, and aggression. Orchitis (inflammation of the testicles) is reported occasionally in the late phase of disease.

In terminal stages, death occurs most often between 2 and 16 days after symptom onset, usually preceded by hypotensive shock and diffuse bleeding. Clinical signs of disease and symptoms increase in severity and may include sustained high fever, development of prostration, tachypnea (rapid breathing), anuria (inability to urinate), and jaundice possibly from liver destruction. Laboratory blood chemistry may reveal elevated levels of transaminase, amylase, creatinine, and blood urea nitrogen indicating kidney damage. Massive bleeding from the gastrointestinal tract occurs; lymphopenia, thrombocytopenia, and coagulopathies such as disseminated intravascular coagulation are described.[35,91,92] Central nervous system involvement may occur resulting in confusion, irritability, delirium, and convulsions. Pulmonary edema and pleural effusions result in respiratory failure. Pericardial and retroperitoneal bleeding is observed in some patients. Death occurs during the second week with a median interval of 8 days (range 2–16 days). In most cases, the cause of death is associated with cardiovascular failure and hypovolemic shock. The fatality rate ranges from 20% to 90%, but the availability of adequate supportive health care can improve the chance of survival.

7.4.3 Pathogenesis

Filovirus infections culminate in multiple organ dysfunction and hemorrhagic shock that often results in death. At the molecular level, filovirus infection occurs when the virus binds to receptors, ligands, or coreceptors on macrophages, monocytes, dendritic cells, hepatocytes, and other tissues. With only a few exceptions (i.e., lymphocytes), all cells are susceptible to infection. Viral entry leads to membrane fusion and the release of RNA; the RNP complex disassembles and viral replication occurs relatively unimpeded. Dendritic cells are key sentinel antigen-presenting cells that activate both humoral and T-cell-mediated immune responses and are

often the first cells to be infected by the virus.[93,94] Infected dendritic cells fail to mature and thus are incapable of producing appropriate cytokines necessary for T-cell activation and signaling. Infected macrophages begin to secrete inappropriate cytokines that may result in induced anergy and apoptosis of T-lymphocytes in the tissues that are responsible for the acquired immune response. Costimulatory molecules such as CD40, CD80, and CD86 are impaired. A substantial disruption in cytokine release occurs leading to inflammation and may contribute to the T-cell anergy (immune unresponsiveness) and exhaustion observed in lethal cases. An accelerated release of proinflammatory cytokines, such as tumor necrosis factor (TNF)α, disrupts critical cellular processes resulting in tissue and vascular endothelium damage. As virus continues to replicate unabated, CD8+ T-lymphocytes undergo induced apoptosis that further impairs the host functional immunity leading to viral resistance. The adaptive immune response is defective due to interferon (IFN) antagonism in dendritic cells and monocytes as well as inappropriate response to proinflammatory cytokines and costimulatory and coinhibitory molecules. Innate immunity, which initially compensated for the failing adaptive immunity, eventually fails as the viral load increases. Viral proteins such as VP35 and VP24 disrupt the immune response thus limiting the ability of the host to respond to the infection. VP35 interferes with the production of Type 1 IFNs (IFNα and IFNβ) and VP24 prevents cells from responding to IFNs (IFNα, IFNβ, and IFNγ).[95–97] For ebolavirus, secreted forms of the viral GPs (sGP) are produced and are hypothesized to sequester antibodies that would otherwise protect against the virus. While the ebolavirus and marburgvirus proteins are thought to exhibit similar responses in hosts, the proteins vary considerably between viruses and the proteins may have distinct, unique properties, and yet to be identified roles in pathogenesis.[98] Circulating, infected monocytes express large amounts of tissue factor and initiate disseminated intravascular coagulation with its corresponding tissue damage. This continues as the viral load increases during the course of the disease and the patient is soon overwhelmed and can no longer contain the infection. In nonfatal cases, the host immune response, both innate and adaptive, leads to the development of both antibody and cellular immune responses that are virus specific.[99]

By the end of the first week, symptomatic patients that will survive the infection begin to improve, and signs of coagulopathy generally are limited to conjunctival hemorrhages, easy bruising, and bleeding from venipuncture sites; viremia also begins to diminish.[100] Recovery from the disease is prolonged and can be marked by inflammation or infection of various organs. Secondary infections are observed in animal models of disease that are likely due to impaired immune responses during infection. Viral RNA can be detected for several weeks after the illness has subsided. Typically, ebolavirus and marburgvirus infections will resolve within 14–21 days of onset. For patients that survive infection, there are social, economical, and clinical sequelae that can be severe and life changing. Survivors are typically ostracized by their community and are more likely to have lost family members and close acquaintances during the outbreak. Clinical outcomes observed in survivors are associated with long-term clinical sequelae including hearing loss, motor impairment, and cognitive defects. In Africa, the long-term medical consequences from the infection and the social stigmatism may be associated with economical implications since survivors may not be able to work to sustain themselves and their family.

7.5 IDENTIFICATION AND DIAGNOSIS

Marburgviruses and ebolaviruses cause serious disease with high morbidity and mortality. Rapid, sensitive, specific, and reliable diagnostics are essential in suspected viral hemorrhagic fever cases. This is especially true of filovirus infections whether they are naturally occurring in sub-Saharan Africa or as an act of bioterrorism in a developed country. While the need to direct early treatment decisions and manage outbreak responses is the same, the diagnostic options are different. Laboratory diagnoses of viral hemorrhagic fevers are generally done in national and international reference laboratories. These laboratories are usually equipped with current diagnostic assays and instrumentation to detect and identify the pathogens, in contrast to field diagnostic or civil support units where diagnostic options are more limited.

7.5.1 IDENTIFICATION

Detection and identification of Marburg or Ebola hemorrhagic fever always begins with a clinical assessment of the signs and symptoms. The virus is endemic in sub-Saharan Africa where assessment is complicated by a variety of similar early-stage presentations of other pathogens, most commonly malaria and typhoid fever.[101] The disease manifests with fever, chills, headache, abdominal pain, chest pain, sore throat, and nausea. Maculopapular rash is more commonly seen in filovirus infections but can also occur with dengue hemorrhagic fever and Lassa fever.[102] Clinical presentation and residence in or travel to an endemic area is the best indication of marburgvirus or ebolavirus infection. In a bioterrorism scenario, clinical assessment early in the infection would still be confounded by other diseases with general malaise and influenza-like presentation. Of particular concern would be the medical personnel's lack of experience with or likelihood to suspect viral hemorrhagic fever.

7.5.2 DIAGNOSIS

Virus isolation is the diagnostic gold standard, but is not suited to the field due to the need for BSL-4 containment; therefore, it is confined to specially equipped confirmatory laboratories. During the acute phase of the disease, virus isolation from blood, serum, or other clinical samples can be attempted by infecting Vero (*C. aethiops*, African green monkey kidney) cells, most commonly the E6 clone. Other continuous cell lines such as MA-104 (*Macaca mulatta*, rhesus monkey kidney) and SW13 (human adrenal carcinoma)

cells or permissive primary cell cultures can also be used for virus isolation efforts. Since field isolates do not always demonstrate cytopathic effects, additional cell passages or inoculation into guinea pigs may be required.

Whether in the field or confirmatory laboratory, diagnosis can be accomplished by detection of some viral component such as proteins or nucleic acids or detection of host antibodies to the infection. During the acute phase when virus is present, antigen capture enzyme-linked immunosorbent assay (ELISA) and reverse transcription polymerase chain reaction (RT-PCR) are most commonly used for detecting viral antigens or RNA, respectively.[103] During the convalescent phase, marburgvirus- or ebolavirus-specific antibodies are commonly detected by capture IgM ELISA and direct IgG and IgM ELISA; IgM antibodies are produced later in the acute phase and as antibody class switching occurs are replaced by rising IgG antibody titers. Generally, an orthogonal approach is used to provide the greatest confidence in a diagnostic result. A combination of immunological assays detecting virus antigens and/or virus-specific antibodies and nucleic acid assays detecting virus genomic material is utilized. Nucleic acid detection takes advantage of the exquisite sensitivity and specificity of PCR while the less-sensitive but more broadly reactive antibody-based immunological assays ensure that genetic variants are detected. The utility of the orthogonal approach is best illustrated by the identification of the newest *ebolavirus* species, *Bundibugyo*.[56] During the outbreak, clinical presentation of the infected individuals suggested a viral hemorrhagic fever. Samples initially tested with highly sensitive real-time RT-PCR assay to EBOV and SUDV were negative. Ebolavirus infection was ultimately detected using a broadly reactive ebolavirus antigen capture and a cross-reactive EBOV IgM capture ELISAs. Due to the genetic variation of the new *Bundibugyo ebolavirus* species, the more specific RT-PCR assays failed to detect the variant. Only by using an orthogonal approach was the virus detected and eventually identified as a new *ebolavirus species*, BDBV.

Antibody detection assays that rely on accurately reproduced agent-specific antigen are being revolutionized by recombinant DNA technology to clone and express recombinant antigens of consistently high quality in large quantities. Compared to traditional methods of reagent preparation involving growth and inactivation of virus, these recombinants can be produced less expensively, in a shorter amount of time, and with greater safety. Recombinant marburgvirus NP expressed in *Escherichia coli* was used as antigen in a direct IgG ELISA to detect IgG antibodies in human sera from marburgvirus-infected patients.[104] The IgG ELISA using the recombinant NP showed high sensitivity and specificity in detecting marburgvirus-specific antibodies.

Molecular diagnostics detect virus-specific genomic material that is contained in a wide variety of complex sample matrices, like blood, serum, or sputum. Detection is most commonly accomplished by real-time RT-PCR. The extraction of the relevant nucleic acids from other inhibitory components that may negatively impact downstream assays is vital to diagnostic success.[105,106] Sample processing for filoviruses requires two considerations, inactivation of viral infectivity and stabilization of nucleic acids for detection. Inactivation is critical given the infectious nature of the virus and the severity of the disease. Several methods for inactivation of filoviruses in samples are available, with selection of the method contingent on the downstream diagnostic applications. Gamma-irradiation and TRIzol/chaotrope immersion are the most commonly used techniques. Gamma-irradiation inactivation uses Cobalt 60 radionuclide gamma radiation to irreversibly damage the viral genome.[107,108] While applicable for immune-based downstream detection of filovirus proteins or antibodies, this technique requires expensive and bulky machinery that prohibits use in field applications. In contrast, TRIzol/chaotrope inactivation is relatively easy to use and is amenable to use in a field laboratory setting. Treatment of filovirus samples with TRIzol or other chaotropic solutions effectively eliminates infectivity.[109] However, chaotropic degradation of proteins and other macromolecules is not compatible with immune-based detection, thus limits use of this inactivation method to samples for nucleic acid-based diagnostics. Detergents/surfactant solubilization can also be used to inactivate the filoviruses[110,111]; however, this method is not as efficacious as the preceding techniques.

Real-time RT-PCR amplification of marburgvirus or ebolavirus sequences requires isolation and stabilization of the target nucleic acids. In this context, primary consideration is not for the organism but the surrounding sample matrix. Isolating amplifiable filovirus nucleic acid is relatively easy compared to removal of the PCR inhibitors inherent in blood, tissue, or other matrices. Chaotropic silica adsorption or organic solvents can be used to both inactivate viral infectivity while isolating nucleic acids from inhibitors for downstream detection. Column-based silica adsorption or TRIzol-based extraction of filovirus genomic RNA are commonly used for both field- and lab-based samples.[112–114] Automated processing of filovirus samples can be used to reduce potential risk for infection of laboratory personnel and sample cross-contamination.[115]

7.6 TREATMENT AND PREVENTION

Currently, there are no US Food and Drug Administration (FDA)- or European Union (EU)-approved treatments for filovirus infections. Control of outbreaks is primarily achieved through diagnosis, case management, and preventative measures such as universal precautions and barrier nursing procedures. When properly identified, transmission can be limited by patient isolation and case contact medical surveillance. While virus isolates and host genetics are important to the case fatality and transmission rates, the ability to quickly identify and manage cases and case contacts is paramount to containing infections. In the recent SUDV outbreak in Uganda, guided by the coordination and collaboration of the government, the outbreak response consisting of both local and international members implemented public health safety measures that reduced transmission.[54,116] Multidisciplinary teams consisting of traditional health-care

workers, epidemiologists, and infectious disease scientists supplemented with nontraditional team members such as anthropologists, sociologists, public officials, and the press were utilized to provide effective outbreak plans and communication during and after the outbreak. The World Health Organization outbreak communication and planning guide is a useful resource that can guide these responses.[117]

7.6.1 TREATMENT

Supportive care of patients is the objective of medical interventions. Supportive care can vary depending on the location and resources available and may include oral fluid rehydration, medications, nutritional supplementation, and psychosocial support. Oral and intravenous (IV) medications include drugs that alleviate symptoms of disease, such as nausea, vomiting, dyspepsia, anxiety, agitation, confusion, and pain.[118] Prevention and treatment of dehydration via IV fluids and nasogastric delivery of nutritional and vitamin supplementation remains a high priority but is difficult to achieve in resource-limited settings.[43,118] The benefits of fluid replacement were evaluated in the rhesus macaque (*M. mulatta*) model of the disease, but did not provide significant benefit to survival. However, the study did demonstrate that fluid replacement could benefit by limiting severe renal dysfunction and may yield further benefit in larger studies or in conjunction with developing therapeutic interventions.[119] Further prospective and retrospective studies are needed to fully evaluate the benefit of rehydration. While the clinical value is unknown, supportive care and fluid management remains the standard of care during filovirus outbreaks. The remote locations and limited resources available in some of the regions where these infections occur make this level of care a challenge.

7.6.2 DISEASE PREVENTION

During the past two decades, the development of medical countermeasures for filoviruses has received considerable attention and vaccines and therapeutic options are now on the horizon. Prophylactic and postexposure (similar to rabies vaccination following exposure) vaccination has shown value in animal models of disease and was used in a human infection following laboratory exposure.[120] Vaccination is the primary strategy to protect against infection. There are a variety of vaccines developed to protect against filovirus infection, including subunit vaccines, viruslike particles, vectored systems, DNA vaccines, and live-attenuated virus systems that express the ebolavirus and/or marburgvirus proteins.[121] Vaccination with the viral GP is sufficient to provide protection in animal models of disease. Unfortunately, limited cross-protection is achieved, thus requiring the vaccine to express multiple virus proteins to achieve protection. Currently, an adenovirus and DNA vaccine are in clinical safety assessments in humans and several vaccines are in preclinical development. Given sufficient resources and programmatic commitment, a vaccine for human use is possible. Deployment of the vaccine to aid in preventing human and other incidental host infections (great apes and other nonhuman primates) remains a significant hurdle given the cost to manufacture the vaccines and the limited commercial utility of the vaccines.

Therapeutic interventions in filovirus infections may have advantages in development and deployment over vaccines to treat infected or exposed individuals. Small molecules and antibodies directed against the virus or critical host proteins or pathways associated with pathogenesis were successfully employed in treating or preventing disease in animal models of filovirus disease. The two most advanced treatments are the use of silencing RNA technologies or the use of hyperimmune sera or other antibody sources as an immunotherapeutic.[122–124] Several investigational new drug (IND) applications are submitted to the US FDA for the silencing RNA technologies targeting ebolavirus or marburgvirus. These drugs have shown clinical benefit in macaque models of disease for animals exposed to the virus (1–48 h following infection) and are being prepared for early human clinical safety assessments.[125]

Passive immunity provided by the transfer of sera or monoclonal antibodies is considered a potential intervention for filovirus infections. In the 1995 outbreak, patients were treated with whole blood from convalescent patients, and while the study was not well controlled, a reduction in morbidity was observed in the limited number of cases (8 patients, 12.5% vs. 80%).[126] Similarly, hyperimmune goat, equine, and nonhuman primate sera were shown to protect animals when administered after infection.[124] More recently, monoclonal antibody mixtures were used to successfully treat infected animals.[127,128] An equine formulation of the equine hyperimmune serum was developed in the former Soviet Union, tested in a limited number of human volunteers for safety, and utilized during laboratory accidents. Given the history of immune therapy for other infectious diseases, there remains a significant interest in this therapeutic approach.

There is a rich pipeline of candidate drugs discovered over the past decade. These candidate drugs have demonstrated the ability to alter disease outcome in cell culture or in animal models of disease. Numerous small molecules and other compounds have demonstrated antiviral activity by directly inhibiting the virus or virus–host interactions. Other compounds that are identified prevent or limit viral pathogenesis during infection. Many of the unapproved drugs and new indications for existing drugs will require full preclinical and clinical development before use in humans; however, some of the drugs approved for other clinical indications could be considered in the event of an outbreak where ethical use of the drugs could be justified. More research will be needed to translate these findings into approved treatments.

Furthermore, the ability to utilize the drugs following the onset of clinical signs of disease is being studied in animals. The window of opportunity for therapeutic value following infection remains an area of concern because the filovirus disease progresses so rapidly. The ability to provide a beneficial therapeutic value following onset of clinical diseases remains a significant barrier for the current treatments. Moreover, the quality of life of patients following infection and treatment may require additional development efforts or the combination of multiple therapeutic approaches (combination therapy). Similarly, if approved treatment modalities for human virus infections are modeled, the ability to combine treatments to increase effectiveness will be an area of interest during the next decade.

Given the sporadic nature of the infections and the relatively small numbers of patients affected, any countermeasure will be licensed using the FDA animal rule.[129] To date, few antiviral candidates have demonstrated a benefit following early infection (first hours to 48 h post infection). In studies with animals, clinical onset and diagnostic methods used to detect infection are unable to detect infection until 72–96 h post infection. Thus, most of the countermeasures developed provide only a postexposure protection. Similarly, a significant degree of morbidity is observed in both humans and animals that survive symptomatic infection. Combined, the ability to provide a treatment that will increase survival and reduce long-term pathologies following infection is desired. Given the long path to licensure and financial commitment required, Emergency Use Authorization (EUA) may be necessary to consider during outbreaks. Lastly, since zoonotic transmission is central to the sporadic human outbreaks, development of veterinary vaccines and therapies could generate a barrier to disease in humans and protect wild animals impacted by filoviruses.[130]

7.7 CONCLUSIONS AND FUTURE PERSPECTIVES

Filoviruses naturally occur in sub-Saharan Africa causing sporadic outbreaks of hemorrhagic fever with high mortality. The diseases, by nature of their virulence, cause significant fear and anxiety; however, in the normal disease landscape for the average African, they infect a relatively small number of people and are of less concern than more commonly occurring pathogens such as malaria or acquired immune deficiency syndrome (AIDS). Outside of Africa, the viruses are less of a medical concern but have captured the attention of the public, dramatized in popular books such as *The Hot Zone*[131] and Hollywood movies like *Outbreak* as highly contagious and virulent diseases.

The US Centers for Disease Control and Prevention (CDC) classifies filoviruses as Category A biological pathogens due to their high virulence, ability to infect by aerosol in the laboratory, and the fear and anxiety they can cause in natural outbreaks and possible bioterrorism scenarios. Category A agents are given the highest priority for preparedness in the United States and other developed countries. This prioritization is less based on the likelihood of their use as biowarfare or bioterrorism weapons but more on the probability of their impact on public health and the disruption to social, economic, and political aspects of society.[132]

The filovirus Category A classification is partially determined by their consideration as biological warfare agents and bioterrorism agents. After the Biological Weapons Convention in 1972, significantly limiting the development of biological weapons, the Soviet Union continued a secret effort to develop MARV and EBOV as an offensive weapon.[133–135] At the end of 1983, attempts to develop MARV as a bioweapon were initiated in the Soviet Union.[136] The program focus was on a military use of MARV as a denial agent, an agent that would contaminate a specific geographical area for a short period of time. Person-to-person transmission was considered to be less effective for MARV than EBOV, but other characteristics prioritized its development. The weaponization process was to serve as a model for other viruses, including Ebola. Two phases of development were used to achieve a Marburg bioweapon. The first phase used guinea pigs as living "bioreactors" where spleen and liver preparations were blended, lyophilized, and stored for use as a suspended aerosol. The second phase utilized cell culture virus generation to achieve large-scale preparation of the agent without the laborious and dangerous processing required for the tissue-derived material. While the Marburg weapon was valued, the lack of an effective countermeasure to protect the Soviet Union personnel during attacks, and costs associated with production likely impacted the concepts of operation intended for the weapon. Efforts to develop an EBOV bioweapon began about 2 years after the program started for MARV. The operational issues encountered in the Marburg program led the Ebola program to divide its focus between production methods and the development of an effective countermeasure to protect personnel.

Filoviruses are a poor choice as a biological warfare or terrorism agents compared to other more easily obtainable and deliverable pathogens such as *Bacillus anthracis* or *Yersinia pestis*. The primary reason a terrorist organization would want to use filoviruses is their immediately induced fear and anxiety and the future psychological impact on the population. However, the use of these viruses remains dangerous, expensive, and technically demanding for mass production. However, reports of terrorists attempting to attain virus during outbreaks have been reported and small-scale attacks remain a potential threat. Filoviruses used as biowarfare or bioterrorism threat agents would be most effectively delivered as an aerosol.[134] Scenarios involving contamination of food, water, or objects would have a lower probability of success due to the sensitivity of the viruses to environmental factors. Experimentally filoviruses are highly infectious by the aerosol route.[80,84,137] Outside the laboratory, there is no evidence that aerosol transmission occurs as a significant route of infection in a natural setting. In natural outbreaks, transmission of virus occurs through direct contact with infected body fluids. For a successful aerosol delivery of filoviruses, there are many technological challenges to overcome: production of sufficient amounts of highly infectious material, equipment to produce an infecting aerosol, and favorable conditions of wind, temperature, and humidity.

A biological attack of limited success, which results in only a few infections would still achieve the desired result of social and economic disruptions due to the fear and panic that ebolaviruses and marburgviruses induce in the population. If such an attack went unnoticed, infected individuals would begin to develop symptoms a week or more after the initial infection.

The early presentation of the viral hemorrhagic fever would be difficult to differentiate from other commonly occurring diseases such as influenza. Since the medical personnel would not be expecting a viral hemorrhagic fever, concerns would not elevate until the more serious symptoms arise such as hemorrhage. In a developed country such as the United States, once viral hemorrhagic fever was suspected, confirmatory tests at the equivalent of the CDC or US Army Medical Research Institute of Infectious Diseases (USAMRIID) would detect and identify a filovirus infection. Without an effective antiviral therapy for filoviruses, treatment would consist mainly of supportive care. From that point, the outbreak would be effectively managed through standard patient isolation and barrier nursing procedures. However, the psychological damage would far outweigh the cost of medical management of the disease with the "worried well" potentially overwhelming an already reeling medical system.

ACKNOWLEDGMENTS

This work was funded in part by the US Department of Defense, Division of GEIS Operations at the Armed Forces Health Surveillance Center, Research Plan C0602_12_RD, and the Defense Threat Reduction Agency, Research Plan CBCALL12-THRFDA1-2-0135.

DISCLAIMER

Opinions, interpretations, conclusions, and recommendations are those of the authors and are not necessarily endorsed by the U.S. Army.

REFERENCES

1. Feldmann, H. et al. Family Filoviridae. In Fauquet, C.M., Mayo, M.A., Maniloff, J., Desselberger, U., and Ball, L.A. (eds.) *Virus Taxonomy: VIIIth Report of the International Committee on Taxonomy of Viruses*, pp. 645–653 (Elsevier/Academic Press, London, U.K., 2005).
2. Kuhn, J.H. et al. Proposal for a revised taxonomy of the family Filoviridae: Classification, names of taxa and viruses, and virus abbreviations. *Arch Virol* 155, 2083–2103 (2010).
3. Kuhn, J.H. et al. Virus taxonomy—Ninth report of the International Committee on Taxonomy of Viruses. In King, A.M.Q., Adams, M.J., Carstens, E.B., and Lefkowitz, E.J. (eds.) *Family Filoviridae*, pp. 665–671 (Elsevier/Academic Press, London, U.K., 2011).
4. Adams, M.J. and Carstens, E.B. Ratification vote on taxonomic proposals to the International Committee on Taxonomy of Viruses (2012). *Arch Virol* 157, 1411–1422 (2012).
5. Kuhn, J.H. et al. Virus nomenclature below the species level: A standardized nomenclature for natural variants of viruses assigned to the family Filoviridae. *Arch Virol* 158, 301–311 (2013).
6. Peters, D., Muller, G., and Slenczka, W. Morphology, development and classification of the Marburg virus. In Martini, G. and Siegert, R. (eds.) *Marburg Virus Disease*, pp. 68–83 (Springer-Verlag, New York, 1971).
7. Geisbert, T.W. and Jahrling, P.B. Differentiation of filoviruses by electron microscopy. *Virus Res* 39, 129–150 (1995).

8. Murphy, F.A., van der Groen, G., Whitflied, S.G., and Lange, J.V. Ebola and Marburg virus morphology and taxonomy. In Pattyn, S. (ed.) *Ebola Virus Haemorrhagic Fever*, pp. 61–84 (Elsevier/North Holland, Amsterdam, the Netherlands, 1978).

9. Sanchez, A., Geisbert, T.W., and Feldmann, H. Filoviridae: Marburg and Ebola viruses. In D.M. Knipe and Peter M. Howley (eds.) *Field's Virology*, Vol. 5, pp. 1409–1448 (Lippincott Williams & Wilkins, Philadelphia, PA, 2007).

10. Mühlberger, E. Filovirus replication and transcription. *Future Virol* 2, 205–215 (2007).

11. Groseth, A., Stroher, U., Theriault, S., and Feldmann, H. Molecular characterization of an isolate from the 1989/90 epizootic of Ebola virus Reston among macaques imported into the United States. *Virus Res* 87, 155–163 (2002).

12. Muhlberger, E., Lotfering, B., Klenk, H.D., and Becker, S. Three of the four nucleocapsid proteins of Marburg virus, NP, VP35, and L, are sufficient to mediate replication and transcription of Marburg virus-specific monocistronic minigenomes. *J Virol* 72, 8756–8764 (1998).

13. Neumann, G., Feldmann, H., Watanabe, S., Lukashevich, I., and Kawaoka, Y. Reverse genetics demonstrates that proteolytic processing of the Ebola virus glycoprotein is not essential for replication in cell culture. *J Virol* 76, 406–410 (2002).

14. Volchkov, V.E. et al. Recovery of infectious Ebola virus from complementary DNA: RNA editing of the GP gene and viral cytotoxicity. *Science* 291, 1965–1969 (2001).

15. Feldmann, H. and Klenk, H.D. Filoviruses (Chapter 72). In Baron, S. (ed.) *Medical Microbiology*, Vol. 4 (The University of Texas Medical Branch, Galveston, TX, 1996).

16. Volchkov, V.E. et al. Characterization of the L gene and 5′ trailer region of Ebola virus. *J Gen Virol* 80, 355–362 (1999).

17. Sanchez, A. and Rollin, P.E. Complete genome sequence of an Ebola virus (Sudan species) responsible for a 2000 outbreak of human disease in Uganda. *Virus Res* 113, 16–25 (2005).

18. Crary, S.M., Towner, J.S., Honig, J.E., Shoemaker, T.R., and Nichol, S.T. Analysis of the role of predicted RNA secondary structures in Ebola virus replication. *Virology* 306, 210–218 (2003).

19. Muhlberger, E., Weik, M., Volchkov, V.E., Klenk, H.D., and Becker, S. Comparison of the transcription and replication strategies of Marburg virus and Ebola virus by using artificial replication systems. *J Virol* 73, 2333–2342 (1999).

20. Becker, S., Rinne, C., Hofsass, U., Klenk, H.D., and Muhlberger, E. Interactions of Marburg virus nucleocapsid proteins. *Virology* 249, 406–417 (1998).

21. Enterlein, S. et al. Rescue of recombinant Marburg virus from cDNA is dependent on nucleocapsid protein VP30. *J Virol* 80, 1038–1043 (2006).

22. Feldmann, H. et al. Marburg virus, a filovirus: Messenger RNAs, gene order, and regulatory elements of the replication cycle. *Virus Res* 24, 1–19 (1992).

23. Sanchez, A., Kiley, M.P., Holloway, B.P., and Auperin, D.D. Sequence analysis of the Ebola virus genome: Organization, genetic elements, and comparison with the genome of Marburg virus. *Virus Res* 29, 215–240 (1993).

24. Bavari, S. et al. Lipid raft microdomains: A gateway for compartmentalized trafficking of Ebola and Marburg viruses. *J Exp Med* 195, 593–602 (2002).

25. Bamberg, S., Kolesnikova, L., Moller, P., Klenk, H.D., and Becker, S. VP24 of Marburg virus influences formation of infectious particles. *J Virol* 79, 13421–13433 (2005).

26. Han, Z. et al. Biochemical and functional characterization of the Ebola virus VP24 protein: Implications for a role in virus assembly and budding. *J Virol* 77, 1793–1800 (2003).

27. Peterson, A.T., Bauer, J.T., and Mills, J.N. Ecologic and geographic distribution of filovirus disease. *Emerg Infect Dis* 10, 40–47 (2004).

28. Bausch, D.G. et al. Risk factors for Marburg hemorrhagic fever, Democratic Republic of the Congo. *Emerg Infect Dis* 9, 1531–1537 (2003).

29. Monath, T.P. Ecology of Marburg and Ebola viruses: Speculations and directions for future research. *J Infect Dis* 179(Suppl 1), S127–S138 (1999).

30. Leroy, E.M., Gonzalez, J.P., and Baize, S. Ebola and Marburg haemorrhagic fever viruses: Major scientific advances, but a relatively minor public health threat for Africa. *Clin Microbiol Infect* 17, 964–976 (2011).

31. Peterson, A.T., Lash, R.R., Carroll, D.S., and Johnson, K.M. Geographic potential for outbreaks of Marburg hemorrhagic fever. *Am J Trop Med Hyg* 75, 9–15 (2006).

32. Siegert, R., Shu, H.L., and Slenczka, W. Detection of the "Marburg virus" in patients. *Ger Med Mon* 13, 521–524 (1968).

33. Martini, G.A., Knauff, H.G., Schmidt, H.A., Mayer, G., and Baltzer, G. A hitherto unknown infectious disease contracted from monkeys. "Marburg-virus" disease. *Ger Med Mon* 13, 457–470 (1968).

34. Stille, W., Bohle, E., Helm, E., van, R.W., and Siede, W. An infectious disease transmitted by *Cercopithecus aethiops* ("Green monkey disease"). *Ger Med Mon* 13, 470–478 (1968).

35. Gear, J.S. et al. Outbreak of Marburg virus disease in Johannesburg. *Br Med J* 4, 489–493 (1975).

36. Conrad, J.L. et al. Epidemiologic investigation of Marburg virus disease, Southern Africa, 1975. *Am J Trop Med Hyg* 27, 1210–1215 (1978).

37. Smith, D.H. et al. Marburg-virus disease in Kenya. *Lancet* 1, 816–820 (1982).

38. Johnson, E.D. et al. Characterization of a new Marburg virus isolated from a 1987 fatal case in Kenya. *Arch Virol Suppl* 11, 101–114 (1996).

39. World Health Organization. Case of Marburg haemorrhagic fever imported into the Netherlands from Uganda. In *International Travel and Health*, July 10, 2008 ed. (2008). http://www.who.int/ith/updates/2008_07_10/en/

40. World Health Organization. Case of Marburg haemorrhagic fever imported into the United States. In *International Travel and Health*, February 5, 2009 ed. (2009). http://www.who.int/ith/updates/2009_02_05_MHF/en/

41. World Health Organization. Marburg haemorrhagic fever in Uganda—Update. In *Global Alert and Response (GAR)* (2012). http://www.who.int/csr/don/2012_11_23_update/en/

42. Bausch, D.G. et al. Marburg hemorrhagic fever associated with multiple genetic lineages of virus. *N Engl J Med* 355, 909–919 (2006).

43. Jeffs, B. et al. The Medecins Sans Frontieres intervention in the Marburg hemorrhagic fever epidemic, Uige, Angola, 2005. I. Lessons learned in the hospital. *J Infect Dis* 196(Suppl 2), S154–S161 (2007).

44. Towner, J.S. et al. Marburgvirus genomics and association with a large hemorrhagic fever outbreak in Angola. *J Virol* 80, 6497–6516 (2006).

45. Geisbert, T.W. et al. Marburg virus Angola infection of rhesus macaques: Pathogenesis and treatment with recombinant nematode anticoagulant protein c2. *J Infect Dis* 196(Suppl 2), S372–S381 (2007).

46. Towner, J.S. et al. Isolation of genetically diverse Marburg viruses from Egyptian fruit bats. *PLoS Pathog* 5, e1000536 (2009).

47. Hartman, A.L., Towner, J.S., and Nichol, S.T. Ebola and Marburg hemorrhagic fever. *Clin Lab Med* 30, 161–177 (2010).

48. Adjemian, J. et al. Outbreak of Marburg hemorrhagic fever among miners in Kamwenge and Ibanda Districts, Uganda, 2007. *J Infect Dis* 204(Suppl 3), S796–S799 (2011).

49. World Health Organization. Ebola haemorrhagic fever in Sudan, 1976: Report of a WHO/International Study Team. *Bull World Health Organ* 56, 247–270 (1978).

50. World Health Organization. Ebola haemorrhagic fever in Zaire, 1976. Report of an International Commission. *Bull World Health Organ* 56, 271–293 (1978).

51. Heymann, D.L. et al. Ebola hemorrhagic fever: Tandala, Zaire, 1977–1978. *J Infect Dis* 142, 372–376 (1980).

52. Baron, R.C., McCormick, J.B., and Zubeir, O.A. Ebola virus disease in southern Sudan: Hospital dissemination and intrafamilial spread. *Bull World Health Organ* 61, 997–1003 (1983).

53. World Health Organization. End of Ebola outbreak in Uganda. In *Global Alert and Response (GAR)*, October 4, 2012 ed. (2012). http://www.who.int/csr/don/2012_10_04/en/index.html

54. World Health Organization. Ebola in Uganda—Update. In *Global Alert and Response (GAR)*, November 30, 2012 ed. (2012). http://www.who.int/csr/don/2012_11_30_ebola/en/

55. Le Guenno, B. et al. Isolation and partial characterisation of a new strain of Ebola virus. *Lancet* 345, 1271–1274 (1995).

56. Towner, J.S. et al. Newly discovered ebola virus associated with hemorrhagic fever outbreak in Uganda. *PLoS Pathog* 4, e1000212 (2008).

57. ProMED-mail. Ebola virus disease—Democratic Republic Congo (24): (Oriental Province). November 28, 2012 ed. (2012). http://www.promedmail.org/direct.php?id=20121128.1426607

58. Jahrling, P.B. et al. Preliminary report: Isolation of Ebola virus from monkeys imported to USA. *Lancet* 335, 502–505 (1990).

59. Hayes, C.G. et al. Outbreak of fatal illness among captive macaques in the Philippines caused by an Ebola-related filovirus. *Am J Trop Med Hyg* 46, 664–671 (1992).

60. Kurosaki, Y. and Yasuda, J. Ebola viruses. In Lui, D. (ed.) *Molecular Detection of Human Viral Pathogens*, pp. 559–572 (CRC Press, Boca Raton, FL, 2011).

61. Miranda, M.E. et al. Epidemiology of Ebola (subtype Reston) virus in the Philippines, 1996. *J Infect Dis* 179(Suppl 1), S115–S119 (1999).

62. World Health Organization. Viral haemorrhagic fever in imported monkeys. *Wkly Epidemiol Rec* 67, 142–143 (1992).

63. Barrette, R.W. et al. Discovery of swine as a host for the *Reston ebolavirus*. *Science* 325, 204–206 (2009).

64. Miranda, M.E. and Miranda, N.L. *Reston ebolavirus* in humans and animals in the Philippines: A review. *J Infect Dis* 204(Suppl 3), S757–S760 (2011).

65. World Health Organization. *WHO Experts Consultation on Ebola Reston Pathogenicity in Humans* (World Health Organization, Geneva, Switzerland, 2009).

66. Negredo, A. et al. Discovery of an ebolavirus-like filovirus in Europe. *PLoS Pathog* 7, e1002304 (2011).

67. Formenty, P. et al. Ebola virus outbreak among wild chimpanzees living in a rain forest of Cote d'Ivoire. *J Infect Dis* 179(Suppl 1), S120–S126 (1999).

68. Swanepoel, R. et al. Studies of reservoir hosts for Marburg virus. *Emerg Infect Dis* 13, 1847–1851 (2007).

69. Towner, J.S. et al. Marburg virus infection detected in a common African bat. *PLoS ONE* 2, e764 (2007).

70. Peterson, A.T., Carroll, D.S., Mills, J.N., and Johnson, K.M. Potential mammalian filovirus reservoirs. *Emerg Infect Dis* 10, 2073–2081 (2004).

71. Leirs, H. et al. Search for the Ebola virus reservoir in Kikwit, Democratic Republic of the Congo: Reflections on a vertebrate collection. *J Infect Dis* 179(Suppl 1), S155–S163 (1999).

72. Leroy, E.M. et al. Fruit bats as reservoirs of Ebola virus. *Nature* 438, 575–576 (2005).

73. Leroy, E.M. et al. Human Ebola outbreak resulting from direct exposure to fruit bats in Luebo, Democratic Republic of Congo, 2007. *Vector Borne Zoonotic Dis* 9, 723–728 (2009).

74. Swanepoel, R. et al. Experimental inoculation of plants and animals with Ebola virus. *Emerg Infect Dis* 2, 321–325 (1996).

75. Macneil, A. and Rollin, P.E. Ebola and Marburg hemorrhagic fevers: Neglected tropical diseases? *PLoS Negl Trop Dis* 6, e1546 (2012).

76. Tomori, O. et al. Serologic survey among hospital and health center workers during the Ebola hemorrhagic fever outbreak in Kikwit, Democratic Republic of the Congo, 1995. *J Infect Dis* 179(Suppl 1), S98–S101 (1999).

77. Leroy, E.M., Baize, S., Debre, P., Lansoud-Soukate, J., and Mavoungou, E. Early immune responses accompanying human asymptomatic Ebola infections. *Clin Exp Immunol* 124, 453–460 (2001).

78. Leroy, E.M. et al. Human asymptomatic Ebola infection and strong inflammatory response. *Lancet* 355, 2210–2215 (2000).

79. Fisher-Hoch, S.P. Lessons from nosocomial viral haemorrhagic fever outbreaks. *Br Med Bull* 73–74, 123–137 (2005).

80. Leffel, E.K. and Reed, D.S. Marburg and Ebola viruses as aerosol threats. *Biosecur Bioterror* 2, 186–191 (2004).

81. Bazhutin, N.B. et al. The effect of the methods for producing an experimental Marburg virus infection on the characteristics of the course of the disease in green monkeys. *Vopr Virusol* 37, 153–156 (1992).

82. Lub, M.I. et al. Certain pathogenetic characteristics of a disease in monkeys in infected with the Marburg virus by an airborne route. *Vopr Virusol* 40, 158–161 (1995).

83. Jaax, N.K. et al. Lethal experimental infection of rhesus monkeys with Ebola-Zaire (Mayinga) virus by the oral and conjunctival route of exposure. *Arch Pathol Lab Med* 120, 140–155 (1996).

84. Jaax, N. et al. Transmission of Ebola virus (Zaire strain) to uninfected control monkeys in a biocontainment laboratory. *Lancet* 346, 1669–1671 (1995).

85. Weingartl, H.M. et al. Transmission of Ebola virus from pigs to non-human primates. *Sci Rep* 2, 811 (2012).

86. Borio, L. et al. Hemorrhagic fever viruses as biological weapons: Medical and public health management. *JAMA* 287, 2391–2405 (2002).

87. Gear, J.H. Clinical aspects of African viral hemorrhagic fevers. *Rev Infect Dis* 11(Suppl 4), S777–S782 (1989).

88. Egbring, R., Slenczka, W., and Baltzer, G. Clinical manifestations and mechanism of the haemorrhagic diathesis in Marburg virus disease. In Martini, G.A. and Siegert, R. (eds.) *Marburg Virus Disease*, pp. 41–49 (Springer-Verlag, New York, 1971).

89. Martini, G.A. Marburg virus disease. Clinical syndrome. In Martini, G.A. and Siegert, R. (eds.) *Marburg Virus Disease*, pp. 1–9 (Springer-Verlag, New York, 1971).

90. Sureau, P.H. Firsthand clinical observations of hemorrhagic manifestations in Ebola hemorrhagic fever in Zaire. *Rev Infect Dis* 11(Suppl 4), S790–S793 (1989).

91. Rollin, P.E., Bausch, D.G., and Sanchez, A. Blood chemistry measurements and D-Dimer levels associated with fatal and nonfatal outcomes in humans infected with Sudan Ebola virus. *J Infect Dis* 196(Suppl 2), S364–S371 (2007).

92. Hutchinson, K.L. and Rollin, P.E. Cytokine and chemokine expression in humans infected with Sudan Ebola virus. *J Infect Dis* 196(Suppl 2), S357–S363 (2007).

93. Geisbert, T.W. et al. Pathogenesis of Ebola hemorrhagic fever in cynomolgus macaques: Evidence that dendritic cells are early and sustained targets of infection. *Am J Pathol* 163, 2347–2370 (2003).

94. Mohamadzadeh, M., Chen, L., and Schmaljohn, A.L. How Ebola and Marburg viruses battle the immune system. *Nat Rev Immunol* 7, 556–567 (2007).

95. Cardenas, W.B. et al. Ebola virus VP35 protein binds double-stranded RNA and inhibits alpha/beta interferon production induced by RIG-I signaling. *J Virol* 80, 5168–5178 (2006).

96. Bosio, C.M. et al. Ebola and Marburg viruses replicate in monocyte-derived dendritic cells without inducing the production of cytokines and full maturation. *J Infect Dis* 188, 1630–1638 (2003).

97. Reid, S.P. et al. Ebola virus VP24 binds karyopherin alpha1 and blocks STAT1 nuclear accumulation. *J Virol* 80, 5156–5167 (2006).

98. Sullivan, N., Yang, Z.Y., and Nabel, G.J. Ebola virus pathogenesis: Implications for vaccines and therapies. *J Virol* 77, 9733–9737 (2003).

99. Baize, S. et al. Defective humoral responses and extensive intravascular apoptosis are associated with fatal outcome in Ebola virus-infected patients. *Nat Med* 5, 423–426 (1999).

100. Mahanty, S. and Bray, M. Pathogenesis of filoviral haemorrhagic fevers. *Lancet Infect Dis* 4, 487–498 (2004).

101. Gear, J.H. Hemorrhagic fevers, with special reference to recent outbreaks in southern Africa. *Rev Infect Dis* 1, 571–591 (1979).

102. Beer, B., Kurth, R., and Bukreyev, A. Characteristics of Filoviridae: Marburg and Ebola viruses. *Naturwissenschaften* 86, 8–17 (1999).

103. Grolla, A., Lucht, A., Dick, D., Strong, J.E., and Feldmann, H. Laboratory diagnosis of Ebola and Marburg hemorrhagic fever. *Bull Soc Pathol Exot* 98, 205–209 (2005).

104. Saijo, M. et al. Enzyme-linked immunosorbent assays for detection of antibodies to Ebola and Marburg viruses using recombinant nucleoproteins. *J Clin Microbiol* 39, 1–7 (2001).

105. Rantakokko-Jalava, K. and Jalava, J. Optimal DNA isolation method for detection of bacteria in clinical specimens by broad-range PCR. *J Clin Microbiol* 40, 4211–4217 (2002).

106. Wilson, I.G. Inhibition and facilitation of nucleic acid amplification. *Appl Environ Microbiol* 63, 3741–3751 (1997).

107. Elliott, L.H., McCormick, J.B., and Johnson, K.M. Inactivation of Lassa, Marburg, and Ebola viruses by gamma irradiation. *J Clin Microbiol* 16, 704–708 (1982).

108. Hall, E.J. and Giaccia, A.J. *Radiobiology for the Radiologist*, pp. 1–656 (Lippincott Williams & Wilkins, Philadelphia, PA, 2006).

109. Blow, J.A., Mores, C.N., Dyer, J., and Dohm, D.J. Viral nucleic acid stabilization by RNA extraction reagent. *J Virol Methods* 150, 41–44 (2008).

110. Chepurnov, A.A. et al. Inactivation of Ebola virus with a surfactant nanoemulsion. *Acta Trop* 87, 315–320 (2003).

111. Kallstrom, G. et al. Analysis of Ebola virus and VLP release using an immunocapture assay. *J Virol Methods* 127, 1–9 (2005).

112. Drosten, C. et al. Rapid detection and quantification of RNA of Ebola and Marburg viruses, Lassa virus, Crimean-Congo hemorrhagic fever virus, Rift Valley fever virus, dengue virus, and yellow fever virus by real-time reverse transcription-PCR. *J Clin Microbiol* 40, 2323–2330 (2002).

113. Morvan, J.M. et al. Identification of Ebola virus sequences present as RNA or DNA in organs of terrestrial small mammals of the Central African Republic. *Microbes Infect* 1, 1193–1201 (1999).

114. Leroy, E.M. et al. Diagnosis of Ebola haemorrhagic fever by RT-PCR in an epidemic setting. *J Med Virol* 60, 463–467 (2000).

115. Towner, J.S., Sealy, T.K., Ksiazek, T.G., and Nichol, S.T. High-throughput molecular detection of hemorrhagic fever virus threats with applications for outbreak settings. *J Infect Dis* 196(Suppl 2), S205–S212 (2007).

116. World Health Organization. Ebola in Uganda. In *Global Alert and Response (GAR)*, July 29, 2012 ed. (2012). http://www.who.int/csr/don/2012_07_29/en/index.html

117. World Health Organization. *World Health Organization Outbreak Communication Planning Guide*, p. 30 (World Health Organization, Geneva, Switzerland, 2008).

118. Roddy, P. et al. Filovirus hemorrhagic fever outbreak case management: A review of current and future treatment options. *J Infect Dis* 204(Suppl 3), S791–S795 (2011).

119. Kortepeter, M.G. et al. Real-time monitoring of cardiovascular function in rhesus macaques infected with *Zaire ebolavirus*. *J Infect Dis* 204(Suppl 3), S1000–S1010 (2011).

120. Marzi, A., Feldmann, H., Geisbert, T.W., and Falzarano, D. Vesicular stomatitis virus-based vaccines for prophylaxis and treatment of filovirus infections. *J Bioterror Biodef* S1, 2157–2526 (2011).

121. Falzarano, D., Geisbert, T.W., and Feldmann, H. Progress in filovirus vaccine development: Evaluating the potential for clinical use. *Expert Rev Vaccines* 10, 63–77 (2011).

122. Aman, M.J. et al. Development of a broad-spectrum antiviral with activity against Ebola virus. *Antiviral Res* 83, 245–251 (2009).

123. Bausch, D.G., Sprecher, A.G., Jeffs, B., and Boumandouki, P. Treatment of Marburg and Ebola hemorrhagic fevers: A strategy for testing new drugs and vaccines under outbreak conditions. *Antiviral Res* 78, 150–161 (2008).

124. Kudoyarova-Zubavichene, N.M., Sergeyev, N.N., Chepurnov, A.A., and Netesov, S.V. Preparation and use of hyperimmune serum for prophylaxis and therapy of Ebola virus infections. *J Infect Dis* 179(Suppl 1), S218–S223 (1999).

125. Warren, T.K. et al. Advanced antisense therapies for postexposure protection against lethal filovirus infections. *Nat Med* 16, 991–994 (2010).

126. Mupapa, K. et al. Treatment of Ebola hemorrhagic fever with blood transfusions from convalescent patients. International Scientific and Technical Committee. *J Infect Dis* 179(Suppl 1), S18–S23 (1999).

127. Olinger, G.G. Jr. et al. Delayed treatment of Ebola virus infection with plant-derived monoclonal antibodies provides protection in rhesus macaques. *Proc Natl Acad Sci U S A* 109, 18030–18035 (2012).

128. Qiu, X. et al. Successful treatment of ebola virus-infected cynomolgus macaques with monoclonal antibodies. *Sci Transl Med* 4, 138ra81 (2012).

129. Gronvall, G.K. et al. The FDA animal efficacy rule and biodefense. *Nat Biotechnol* 25, 1084–1087 (2007).

130. Wong, G. and Kobinger, G. A strategy to simultaneously eradicate the natural reservoirs of rabies and Ebola virus. *Expert Rev Vaccines* 11, 163–166 (2012).

131. Preston, R. *The Hot Zone*, pp. 1–300 (Random House, New York, 1994).

132. Rotz, L.D., Khan, A.S., Lillibridge, S.R., Ostroff, S.M., and Hughes, J.M. Public health assessment of potential biological terrorism agents. *Emerg Infect Dis* 8, 225–230 (2002).

133. Davis, C.J. Nuclear blindness: An overview of the biological weapons programs of the former Soviet Union and Iraq. *Emerg Infect Dis* 5, 509–512 (1999).

134. Bray, M. Defense against filoviruses used as biological weapons. *Antiviral Res* 57, 53–60 (2003).

135. Alibek, K. and Handelman, S. *Biohazard* (Random House, New York, 1999).

136. Leitenberg, M. and Zilinskas, R.A. *The Soviet Biological Weapons Program. A History*, pp. 216–221 (Harvard University Press, Cambridge, MA, 2012).

137. Johnson, E., Jaax, N., White, J., and Jahrling, P. Lethal experimental infections of rhesus monkeys by aerosolized Ebola virus. *Int J Exp Pathol* 76, 227–236 (1995).

138. Kissling, R.E., Robinson, R.Q., Murphy, F.A., and Whitfield, S.G. Agent of disease contracted from green monkeys. *Science* 160, 888–890 (1968).

139. Ligon, B.L. Outbreak of Marburg hemorrhagic fever in Angola: A review of the history of the disease and its biological aspects. *Semin Pediatr Infect Dis* 16, 219–224 (2005).

140. Johnson, K.M., Lange, J.V., Webb, P.A., and Murphy, F.A. Isolation and partial characterisation of a new virus causing acute haemorrhagic fever in Zaire. *Lancet* 1, 569–571 (1977).

141. Georges, A.J. et al. Ebola hemorrhagic fever outbreaks in Gabon, 1994–1997: Epidemiologic and health control issues. *J Infect Dis* 179(Suppl 1), S65–S75 (1999).

142. Khan, A.S. et al. The reemergence of Ebola hemorrhagic fever, Democratic Republic of the Congo, 1995. Commission de Lutte contre les Epidemies a Kikwit. *J Infect Dis* 179 (Suppl 1), S76–S86 (1999).

143. Leroy, E.M., Souquiere, S., Rouquet, P., and Drevet, D. Re-emergence of ebola haemorrhagic fever in Gabon. *Lancet* 359, 712 (2002).

144. Formenty, P. et al. Outbreak of Ebola hemorrhagic fever in the Republic of the Congo, 2003: A new strategy?. *Med Trop (Mars)* 63, 291–295 (2003).

145. World Health Organization. Ebola haemorrhagic fever in the Republic of the Congo—Update 6. In *Global Alert and Response (GAR)*, January 6, 2004 ed. (2004). http://www. who.int/csr/don/2004_01_06/en/

146. Nkoghe, D., Kone, M.L., Yada, A., and Leroy, E. A limited outbreak of Ebola haemorrhagic fever in Etoumbi, Republic of Congo, 2005. *Trans R Soc Trop Med Hyg* 105, 466–472 (2011).

147. Grard, G. et al. Emergence of divergent Zaire ebola virus strains in Democratic Republic of the Congo in 2007 and 2008. *J Infect Dis* 204(Suppl 3), S776–S784 (2011).

148. Okware, S.I. et al. An outbreak of Ebola in Uganda. *Trop Med Int Health* 7, 1068–1075 (2002).

149. Lamunu, M. et al. Containing a haemorrhagic fever epidemic: The Ebola experience in Uganda (October 2000-January 2001). *Int J Infect Dis* 8, 27–37 (2004).

150. World Health Organization. Ebola haemorrhagic fever— Fact sheet revised in May 2004. *Wkly Epidemiol Rec* 79, 433–440 (2004).

151. Shoemaker, T. et al. Reemerging Sudan Ebola virus disease in Uganda, 2011. *Emerg Infect Dis* 18, 1480–1483 (2012).

152. Formenty, P. et al. Human infection due to Ebola virus, subtype Cote d'Ivoire: Clinical and biologic presentation. *J Infect Dis* 179(Suppl 1), S48–S53 (1999).

153. Rollin, P.E. et al. Ebola (subtype Reston) virus among quarantined nonhuman primates recently imported from the Philippines to the United States. *J Infect Dis* 179(Suppl 1), S108–S114 (1999).

8 Hantaviruses

Colleen B. Jonsson

CONTENTS

8.1 INTRODUCTION

Hantaviruses cause two very different illnesses in humans when they spillover into humans from their rodent reservoirs, hemorrhagic fever with renal syndrome (HFRS) and hantavirus pulmonary syndrome (HPS).[1–3] Western medicine first recognized HFRS in humans in the early 1950s during the Korean War when United Nations troops fell ill. During that period, the illness was referred to as Korean hemorrhagic fever.[4,5] In 1978, the etiological agent for this disease, Hantaan virus (HTNV), and its reservoir, the striped field mouse (*Apodemus agrarius*), were reported.[6] These pioneering discoveries were followed by the recognition of additional HFRS-related viruses in many parts of Asia and Europe.[4,7–9] In 1993, an outbreak of severe, unexplained acute respiratory distress in the Four Corners region of the United States led to the discovery of the New World hantaviruses causing HPS. The first virus recognized to cause HPS, Sin Nombre virus (SNV), was shown to be harbored by *Peromyscus maniculatus*, a common rodent species within North America.[10] In a relatively short time following the outbreak in the United States, additional New World hantaviruses were discovered in outbreaks of HPS throughout the Americas.

Hantaviruses are commonly referred to as the Old World and New World hantaviruses, which in general reflect the geographic distribution of their rodent hosts and the type of illness that manifests upon transmission to humans. The Old World hantaviruses have been detected in the Murinae and Arvicolinae subfamilies, while all of the New World viruses are harbored by the Sigmodontinae subfamily (Table 8.1). In general, viruses are named by their geographic location. Transmission of hantaviruses to humans is thought to occur primarily through the inhalation of rodent excreta. However, not all species within the *Hantavirus* genus cause disease in humans. Interestingly, hantaviruses do not cause disease in their rodent hosts and suggest that hantaviruses have evolved a mechanism to disarm the rodent immune system.[11,12]

As potential biological terror threats, hantaviruses have been on and off the Centers for Disease Control and Prevention (CDC) category A list, reflecting ambiguity as to the threat posed by these viruses. When hantaviruses are viewed individually, rather than as a genus, it becomes obvious why certain hantaviruses pose a greater bioterror threat than others. A partial listing of the reasons why certain South American hantaviruses (e.g., Andes virus [ANDV]) are of particular concern includes (1) high lethality (30%–50% case fatality rate), (2) potential for person-to-person transmission,[13] (3) rapid disease course that can occur within 48 h, (4) lack of vaccine or antiviral drug, (5) lack of population base with preexisting immunity, (6) susceptibility of all ages to lethal infection, and (7) growth to relatively high titers in cell culture supernatants and in rodent organs.

8.2 CLASSIFICATION AND MORPHOLOGY

The genus *Hantavirus*, family Bunyaviridae, was created in 1983 to distinguish it from other genetically related genera within the Bunyaviridae, a large family of over 300 viruses that infect animals, plants, humans, and insects.[14–16] Similar to other members in the family, hantaviruses have three, negative-sense, single-stranded RNA genomes (M, L, and S) that encode for two glycoproteins (GPs) (Gn, Gc), a polymerase (L protein or RNA-dependent RNA polymerase, [RdRp]), and a nucleocapsid (N) protein, respectively. With the exception of the *Hantavirus* genus, all the viruses within the Bunyaviridae are arboviruses (arthropod borne). These viruses are maintained by sylvatic transmission between various arthropods, mosquitoes, ticks, sand flies, or thrips and susceptible vertebrate or plant hosts. In contrast, hantaviruses are harbored by rodents, shrews, moles, and one bat species. While many groups have surveyed insects, sylvatic transmission cycles have never been detected.

Hantavirus virions are generally spherical in nature, and electron microscopy suggested an average diameter

TABLE 8.1

Representative Old and New World Viruses of the Genus *Hantavirus* Harbored by Rodents

Genus	Notable Members	Geographic Distribution	Rodent Host	Disease
		Murinae—subfamily		
Old World	Hantaan	China, Korea, Russia	*A. agrarius*	HFRS
	Dobrava–Belgrade	Germany, Slovakia, Russia, Hungary, Slovenia, Croatia, Estonia, Turkey	*A. flavicollis*	HFRS
			A. agrarius	
			Apodemus ponticus	
	Seoul	Worldwide	*Rattus*	HFRS
	Amur	Far East Russia	*Apodemus peninsulae*	HFRS
		Arvicolinae—subfamily		
	Puumala	Europe, Asia and Americas	*Clethrionomys glareolus*	HFRS/NE
	Prospect Hill	Maryland	*Microtus pennsylvanicus*	Unknown
	Tula	Russia/Europe	*Microtus arvalis*	Unknown
		Sigmodontinae—subfamily		
New World	Sin Nombre	North America	*Peromyscus maniculatus*	HPS
	Black Creek Canal	SE USA	*Sigmodon hispidus*	HPS
	Bayou	SE USA	*Oryzomys palustris*	HPS
	Andes	Argentina, Chile, Uruguay	*Oligoryzomys longicaudatus*	HPS
	Laguna Negra	Paraguay, Bolivia	*Calomys laucha*	HPS
		N. Argentina	*Calomys callosus*	HPS
	Jabora	Paraguay, Brazil	*Akodon montensis*	Unknown
	Araraquara	Brazil	*Bolomys lasiurus*	HPS
	Caño Delgadito	Venezuela	*Sigmodon alstoni*	Unknown

of approximately 80–120 nm.[17–22] Recent ultrastructural studies of Tula and Hantaan (HTN) viruses suggest the diameter of the virion is larger, ranging from 120 to 160 nm (Figure 8.1a).[23,24] The grid-like pattern of the virion surface appears distinct from other genera in the Bunyaviridae.[18,19,21,23–25] A 3D icosahedral reconstruction of Rift Valley fever virus virions[26–28] shows a more ordered virion surface with quasi-sixfold symmetry than the hantaviruses.[23,24] The pattern of the outer surface reflects the GP projections from the lipid bilayer, which extend ~10 nm from the lipid bilayer (Figure 8.1b).[23,24] The images suggest that each GP projection is a dimer within ~14 nm domains. Biochemical studies confirm that the GP projections are composed of heterodimers of the two GPs, Gn (formerly G1) and Gc (formerly G2),[29] which are embedded within a host-derived lipid bilayer presumably derived from the Golgi apparatus.

The interior makeup of a virion particle consists of three tightly packed ribonucleocapsids (RNPs).[30] The first molecular analyses of HTNV genome revealed three negative-sense, single-stranded RNAs with a common 3′ terminal sequence of the three genome segments.[31] It is widely held for all of the viruses in the Bunyaviridae that each genomic RNA can form a circular molecule by base pairing between these inverted complementary terminal sequences.[32] The total size of the

RNA genome is 11,845 nt for HTNV. Each viral RNA segment forms a complex with the N protein to form the three ribonucleoprotein (RNP) structures.[33,34] These RNP complexes are believed to be the source of the virion's internal filamentous appearance.[35] Unlike other Bunyaviridae, hantaviruses do not have a nonstructural (NSs) protein.[30] Hantaviruses also lack a matrix protein, and therefore, the N protein may provide this function to facilitate physical interaction with the GP projections on the inner leaf of the lipid membrane and the RNPs.

Hantaviruses share similar physical properties to those of other viruses in the family Bunyaviridae. The density of the HTNV virion can range from 1.16 to 1.18 g/cm³ in sucrose and 1.20–1.21 g/cm³ in CsCl. Treatment of the HTNV with nonionic detergents releases the three RNPs that sediment to densities of 1.18 and 1.25 g/cm³ in sucrose and CsCl, respectively, using rate-zonal centrifugation methods.[36] Studies show that the HTNV can remain viable for 30 min in buffers from pH 6.6 to pH 8.8, but in the presence of 10% fetal bovine serum (FBS), the virus can remain viable in a wider range of pH 5.8–9.0.[37] HTNV is infectious at temperatures ranging from 4°C to 42°C in the presence or absence of serum and can remain infectious for 1–3 days when at least 10% serum is present in dried samples. Sucrose-purified HTN virions retain infectivity for at least 4 h with heat treatments of 65°C or 2 h at 80°C (Yong-Kyu Chu, unpublished data).

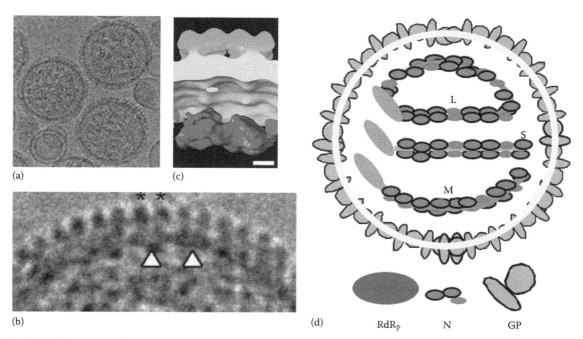

FIGURE 8.1 HTNV images and illustration. (a) Vitrified HTNV virions. Particles are typically round with diameters in the 120–160 nm range; however, elongated particles are sometimes seen. Gn and Gc (GP) spikes extend ~10–12 nm from the membrane. In some places, the spikes seem to form a regular repeating pattern (see b). The image was taken with accelerating voltage of 300 kV and a defocus of ~6 μm. (b) GP spikes. In many images, the spike densities repeat at regular intervals suggesting local order. In this merged image, the spikes appear dimeric. Dimers are separated by ~14 nm. (c) A 3D isosurface rendering of the twofold averaged reconstruction. The map was contoured at 1.5 and tilted to show the cytoplasmic densities. The internal ceiling of density represents the boundary between the cytoplasmic components of the GP spikes and the lower-density interior of the virus, presumably the RNP. Bar, 5 nm. (d) Illustration of virion based on images showing potential assignments of density to the nucleocapsid (N), polymerase (RdRp), and glycoproteins (GP). (a through c: Courtesy of Anthony J. Battisti and Paul R. Chipman, Purdue University, West Lafayette, IN.)

8.3 BIOLOGY AND EPIDEMIOLOGY

Despite their differences in pathogenesis, the Old World and New World hantaviruses share high homology in the organization of their nucleic sequences and exhibit similar aspects of their life cycle.[38] The L and N proteins lack transmembrane domains and are translated on free ribosomes and remain within the cytoplasm. The GP Gn/Gc precursor is translated on membrane-bound ribosomes and co-translated into the rough endoplasmic reticulum (ER). The precursor is proteolytically processed into two transmembrane polypeptides, Gn and Gc, during import into the ER.[39,40] In the HTNV GP polypeptide, a conserved amino acid motif, WAASA, located at the end of Gn, is presumed to be the proteolytic cleavage site.[41] The Gn and Gc proteins are glycosylated in the ER and subsequently transported to the Golgi complex.[29,37,39,42,43]

Soon after the initial round of transcription, the virus switches to replication of the viral genomic RNA (vRNA). The newly synthesized vRNAs are encapsidated by N proteins to form the RNPs.[16] It is unclear if the L protein or RdRp is part of the RNP complex. The RNPs are thought to target the Golgi in order to be enveloped by membranes containing Gn and Gc proteins in the final virion. These particles bud into the Golgi to produce the virion and exit the Golgi through the formation of an additional membrane that surrounds the hantavirus particles. The mechanisms by which these events occur have not been demonstrated experimentally. To date, there is little known regarding how the N protein, RNPs, or L protein targets or associates with the Golgi. One finding is that the HTNV N protein traffics to the ER-Golgi Intermediate Compartment (ERGIC) via microtubules.[44,45] At the Golgi, the N protein could target cellular matrix proteins or transmembrane proteins that reside in the Golgi. Both have been shown to cycle between the Golgi and the ERGIC and could facilitate the coordinated trafficking of N and Gn/Gc proteins to the same compartments. Possibly, the localization of the N protein at the ER, ERGIC, and the Golgi could suggest a backflow mechanism, where uncoupled N protein may recycle back from the Golgi to the ERGIC or the ER to increase the probability of its interaction with G proteins to form complexes required for envelopment. Whether the N protein contains viral RNA prior to arrival at the ERGIC and Golgi is not known. If not, this may suggest that this region is also the site for viral RNA replication.

Hantaviruses enter host epithelial cells via interaction of the larger viral Gn with the host's cell surface receptor(s): β1 and β3 integrins.[46,47] HTNV entry is mediated by clathrin-coated pits, followed by movement to early endosomes, and subsequent delivery to late endosomes or lysosomes.[48] Within the endolysosomal compartments, the virus is uncoated to liberate the three RNP complexes into the cytoplasm that contain the negative-sense, single-stranded RNA segments complexed with N protein and, presumably,

L protein. The L protein initiates primary transcription to give rise to the three viral mRNAs. Of these, the mRNA for the N protein is the most abundant, followed by the M and L transcripts. The abundance of the N mRNA and protein makes it a good target for diagnostics.

Immediately following the initial discovery of the HTNV,[6,49] epidemiological studies of HFRS progressed in both human and rodent populations. Initially, studies showed that farmers, soldiers, and inhabitants of endemic regions were most likely to fall victim to HFRS. Further, it was initially believed that HFRS only occurred in rural areas of Eurasia, specifically China, Korea, Eastern Russia, and Northern Europe.[4] Prudent surveillance demonstrated that HFRS could also occur in urbanized cities and in many parts of the world.[50] These studies show that the distribution of HFRS cases east of the Ural mountains, China, and Korea is caused primarily by HTNV and Seoul virus (SEOV). The severe form of HFRS caused by HTNV occurs primarily in Korea, China, Mongolia, Russia, Hong Kong, and Myanmar, while the moderate form of HFRS caused by SEOV occurs in Japan, Korea, China, and Southeast Asia. In Europe, several variants of Dobrava–Belgrade virus (DOBV) harbored by *A. agrarius* and *Apodemus flavicollis* have been reported to cause severe cases of HFRS in Serbia, Croatia, Slovenia, Bosnia–Herzegovina, Hungary, Greece, Lithuania, Czech Republic, Estonia, and Albania.[51–58] Nephropathia epidemica is a milder form of HFRS caused by Puumala virus (PUUV), which is found throughout Europe in the bank vole, *Myodes glareolus* (previously known as *Clethrionomys glareolus*).[59] This virus affects urban and rural communities of Sweden and Finland. The number of cases in rural areas usually peaks in November and January, while most urban cases occur during August.

Worldwide, up to 150,000–200,000 cases of HFRS can occur annually although this estimate is probably much lower in recent years.[60–62] A large number of HFRS cases in China may be attributed to the HTNV, where recently about 40,000–60,000 cases have been reported each year.[61,62] The total estimated number of cases in China from 1950 to 1997 was greater than 1.2 million and resulted in 44,304 deaths.[61] Approximately 300–900 annual HFRS cases are estimated to occur each year in Korea and in Eastern Russia.[63,64] In all of Europe, approximately 5000 cases occur each year.[65,66] HTNV-related HFRS has been demonstrated to have a mortality rate of 10%–15%. Most cases of HFRS occur in adult men ranging from 20 to 50 years of age.[4,67] Although over 100 cases of HFRS have been reported in large metropolitan areas of Seoul and other larger cities of Korea where patients had direct contact with the domestic rat, HFRS caused by SEOV is rare outside of China and Korea.[68] The seasonality of SEOV outbreaks is somewhat different from the HTNV, with most outbreaks occurring in October–December in Korea and January–May in China.[69] The higher prevalence of HFRS in men than in women directly correlates with the rural breakdown of duties of men working outside in the fields, while most women work in the home.

There have been greater than 3000 individual and small clusters of cases of HPS since 1993 in the Americas with nearly 20% of those cases in the United States.[70,71] Brazil represents greater than one-third of all case (>1100)[72] and Argentina (>710).[73] Additional cases have been reported in Bolivia, Canada, Chile, Ecuador, Paraguay, Panama and Uruguay, and Venezuela.

Surveillance of rodent populations near HPS cases has revealed numerous hantaviral lineages or genotypes within Central and South American rodents, only some of which are listed in Table 8.1.[2,74] New World rodent species harbor various lineages of hantaviruses in Argentina,[73,75] Bolivia,[76,77] Brazil,[78–80] Chile,[81] Costa Rica,[82] Mexico,[83] Panama,[84] Paraguay,[76,85,86] Peru,[87] and Venezuela.[88,89] In general, the Sigmodontinae rodents and their associated hantaviruses exhibit a pattern that suggests cospeciation. According to the International Committee on the Taxonomy of Viruses (2011), only Andes (AND), Laguna Negra, Rio Mamoré (*Oligoryzomys microtis*, Peru), and Caño Delgadito viruses are recognized as distinct hantaviral species in South America. Further efforts to resolve the taxonomy of the complex systematics of many sigmodontinae rodents in Central and South America will be helpful in resolving the taxonomy of hantaviruses.

Rodent ecology largely drives the risk of exposure and the seasonality associated with human cases whether HFRS or HPS. For example, the striped field mouse that harbors HTNV, *A. agrarius*, is naturally found in rural areas[50] and very common in agricultural fields.[90] Rodents are known to invade houses in search of food and shelter. Rural cases seem to occur biannually, with two seasonal peaks of cases occurring during the late spring and in the fall seasons. There are several risk factors that could possibly play a role in the increased incidences of HFRS during the fall months, which may be attributed to the harvesting of crops and possibly the relocation of rodents to indoors in preparation for the winter.[91] Military exercises played a key role in the discovery of the pathogens and present a high degree of risk. Since the Korean War, there have been at least 34 cases in US soldiers from 1987 to 2005 that have been treated for HFRS.[92] Furthermore, the population size of rodents may help to escalate human incidences. For example, the *A. agrarius* is high in Asia during the late spring and early fall seasons. Similarly, increased incidence usually correlates with high number of bank vole populations. For example, in 1993, an outbreak PUU virus-associated HFRS was recorded in southern Belgium, which directly correlated with increased populations of bank voles.[93] Other outbreaks of HFRS have been observed in northeastern France and Germany that also correlate with the increase in the population of bank voles.[94]

Epidemiological surveillances should take into account other Old World viruses that might be endemic in the geographic region. For example, the common domestic rat, *Rattus norvegicus* and *Rattus rattus*, can be found in most regions throughout the world. These rodents harbor SEOV that is the primary agent responsible for HFRS cases that occur in urbanized areas.[7,95]

8.4 CLINICAL FEATURES AND PATHOGENESIS

More than 20 unique hantaviruses have been identified that cause illness in humans ranging from proteinuria to pulmonary edema and frank hemorrhage illnesses when transmitted from their rodent reservoirs to humans. HFRS and HPS target the renal and pulmonary systems, respectively, each causing vascular leakage in endothelial tissues (the major target of the viral infections).[3,70] Studies suggest that some percentage of hantaviral infections may be asymptomatic.

HFRS is characterized by fever, vascular leakage resulting in hemorrhagic manifestations, and renal failure with case fatality ranging from <0.1% to 12%. The clinical course of disease depends on the viral species. The four Old World hantaviruses associated with HFRS include PUUV, DOBV, SEOV, and HTNV. Death occurs in less than 0.1% in patients infected with the PUUV, whereas fatalities as high as 15% have been observed for HFRS patients infected with the HTNV. SEOV is associated with a mortality rate of less than 1%. The progression of HFRS following exposure shows an incubation period of approximately 7–21 days before the development of illness. However, this may vary from 4 to 42 days. Clinical progression and manifestations of disease caused by HFRS are outlined by five overlapping stages: febrile, hypotensive, oliguric, diuretic, and convalescent (Table 8.2).[68,96,97] Approximately 11%–40% of persons with febrile illness develop hypotension and approximately 40%–60% develop oliguria.

Common clinical symptoms begin with influenza-like symptoms, headache, backache, fever, and chills. Once the infection has entered the febrile phase, lasting approximately 3–6 days, slight hemorrhage manifestations are evident in the conjunctiva. This stage is characterized by headache, fever, vertigo, nausea, and myalgia. The clinical symptoms of the febrile stage are eventually augmented to the hypotensive

stage characterized by thirst, restlessness, nausea, and vomiting, each lasting hours or days. Approximately one-third of all patients suffering during the hypotensive stage of HFRS develop shock and mental confusion.[98] Symptoms of vascular leakage, abdominal pain, and tachycardia are observed within this stage. Oliguria is urine output of less than 400 mL/day and lasts 3–7 days. The oliguric phase lasts from 1 to 16 days for HFRS, in contrast to 4–24 h for the New World HPS. Conjuctival, cerebral, and gastrointestinal hemorrhage occurs in about one-third of all patients.[98] The oliguric stage accounts for approximately one-half of all hantavirus-related deaths. In this stage, patients are at risk for hypertension, pulmonary edema, and complications of renal insufficiency. Dialysis is required in approximately 40% of HTNV and 20% of SEOV patients. Death is usually due to complications from renal insufficiency, shock, or hemorrhage. Although a patient may completely recover from HFRS after several weeks to months of convalescence, renal or pulmonary dysfunction may persist for the life of the patient.

HPS is characterized by four phases with fever, vascular leakage within the lungs leading to a pulmonary syndrome often accompanied by cardiovascular failure (Table 8.2).[99–102] Case fatality ranges from <18% to 56% for SNV infections. Some viruses that cause HPS have a much lower fatality (e.g., Laguna Negra virus[103]). As with HFRS, during the incubation period, the patient is asymptomatic and has no detectable antibodies. In contrast to HFRS, the onset of illness is rapid and the patient develops a febrile prodrome with fever, myalgia, and progressively worsening thyrombocytopenia. This can be accompanied by headache, back pain, abdominal pain, and diarrhea. Both IgM and IgG antibodies can typically be detected at or shortly after the onset of the prodrome. The cardiopulmonary phase typically begins with cough,

TABLE 8.2
Clinical Manifestations of Infections Caused by HFRS and HPS

Phase of Illness	HFRS	HPS
Febrile	Duration: 3–6 days Headaches, fever, dizziness, myalgia, nausea, anorexia	Onset, 1 day; duration, 3–6 days Headaches, fever, dizziness, myalgia, nausea, gastrointestinal
Hypotensive (HFRS)	Duration: 24 h to 5 days	Onset, 4–8 days; duration, 1–16 days (combined with oliguric)
Cardiopulmonary (HPS)	Hypotension, shock (1/3 of patients) lumbar, backache, abdominal pain, visceral hemorrhage, tachycardia, and mental confusion	Pulmonary edema, hypoxemia, cough, pleural effusion, gastrointestinal symptoms, tachypnea, tachycardia, myocardial depression, and cardiogenic shock
Oliguric	Duration: 1–16 days Hiccups, hemorrhage, vomiting, rare central nervous system (CNS) bleeding	
Polyuric (HFRS)/diuresis (HPS)	Duration: 9–14 days	Onset, 9–14 days; duration, 7–14 days Clearance of pulmonary edema, resolution of fever and shock
Convalescent	Duration: 3–12 weeks Patients regaining weight, myalgia, polyuria, and hyposthenuria	Onset: 15–21 days Duration: 4–8 weeks

shortness of breath, and the development of bilateral pulmonary infiltrates. Patients with mild HPS may only require supplemental oxygen. Patients with severe disease progress over hours to respiratory failure with or without shock. Deaths, almost invariably due to shock, may occur within hours of the onset of the cardiopulmonary phase, and in the absence of extracorporeal membrane oxygenation, almost all deaths occur within 24–48 h of the onset of this phase. After several days, surviving patients enter a diuretic phase. Improvement is usually rapid, and most ventilator-dependent patients can be extubated within a day or two of the onset of the diuretic phase. The convalescent phase, which is characterized by weakness, fatigue, and abnormal diffusion capacity, may persist for months or years.

8.5 IDENTIFICATION AND DIAGNOSIS

Clinical cases of HFRS are confirmed by a variety of laboratory analyses such as the verification of the presence of specific hantavirus antibodies, viral antigens, or viral RNA. These same techniques can be applied to rodent surveillance. Four basic technology platforms exist for the detection of hantaviruses in human, animal, and environment samples. These include conventional techniques based on immunofluorescence detection, immunoassays (enzyme-linked immunosorbent assay [ELISA], lateral-flow immunoassay), plaque/plaque reduction neutralization test (PRNT), and modern techniques such as the reverse transcription polymerase chain reaction (RT-PCR).[104,105] A great deal of research effort has focused on the development of serological tests or RT-PCR to facilitate rapid diagnosis and strain identification due to the quick progression of disease and requirement for early therapeutic intervention. In general, diagnostic reagents and tests have been developed by scientists who study these viruses and not pharmaceutical companies. However, in recent years, commercially available immunoassay kits have become available.[106] Progen (Heidelberg, Germany) markets an immunofluorescence assay (IFA) and ELISA for PUUV and HTNV, and MRL Diagnostics (Cypress, CA, USA) markets an enzyme immunoassay (EIA). However, most laboratories still use in-house diagnostic methods such as IFA, Western blot (WB), or ELISA.

Specific antibodies to hantaviral proteins are consistently present in acute HFRS and HPS patients as well as in convalescents and survivors. At the onset of symptoms, virtually all acute cases contain IgM and IgG antibodies to the N protein. Due to this fact, immunoassays have become the pillar in laboratory diagnosis. Immunoassays have primarily been developed using purified N protein. N protein can be expressed and purified from a number of recombinant expression systems including bacteria,[107,108] baculovirus,[107,109] insect,[110] yeast[111,112] plants,[113,114] and mammalian cells.[115] The high antigenicity of the N protein maps to its amino terminus, and there have been efforts to use this region to create strain-specific diagnostics.[116–118] For IFA and ELISA, one can also use virus-infected cells as will be discussed in more detail in the following sections.

ELISA tests are being used extensively for routine diagnosis of HFRS and HPS. A rapid IgM capture ELISA for hantaviruses has been developed and proven to be sensitive for detection of HFRS and HPS patients by the US Army Medical Research Institute of Infectious Diseases (USAMRIID) and the CDC.[119] Although this test is highly sensitive, false-positive reactions in malaria-positive individuals have been detected. Other disadvantages of ELISA are the requirement for specialized equipment, personnel, and incubation conditions that most often are not present in field or satellite laboratories. For both ELISA and WB formats, the N protein is commonly used in these serologic formats since it has been proven to be the most antigenic and cross-reactive of the four hantaviral proteins.

A newer method, the strip immunoblot assay (SIA), that can rival the ELISA in sensitivity and specificity has been developed.[120,121] This format has great advantages over ELISA, as it requires a minimal amount of effort, equipment, and expertise. In addition, it can easily be performed in field conditions. From one single run, the strip assay can analyze the reaction to multiple homologous antigens, which is ideal when dealing with min quantities of sample. Therefore, it can be designed to be highly specific and confers the formidable advantage of distinguishing distinct hantavirus serotypes by analysis of band intensities and titer. Although the SIA assay has been proven to be very sensitive, a major disadvantage is the antigens required for the test must be highly purified for human serosurveys. Specifically, all possible traces of *Escherichia coli* proteins must be removed to decrease the possibility of false positives due to nonspecific antibody reaction to these contaminants. This does not seem to be a problem in rodent serosurveys.

Lateral-flow assays have been considered the ideal field-usable diagnostic platform since they are simple to perform, rapid, stable at field conditions, and portable. These assays use specialized membrane-based methods where the reactants move by capillary action along a narrow rectangular strip. The sample is applied at one end and traverses the strip, coming into contact first with detecting antigen and subsequently with capture antibodies that have been dried onto the membranes. If the sample contains antibodies to hantaviral antigens, it will form two visible lines as it accumulates at the position of the detecting antigen and capture antibodies. If the sample does not contain antibody, it will form a visible line only at the position of capture antibody. This assay promises to be easy to perform in field settings, economic and highly sensitive for detection of viruses in patients with HFRS. Currently, the system has been shown to be useful for detection of IgG and IgM for HTNV, PUUV, and DOBV with a sensitivity and specificity of 96%–100%.[122] Cross-reactivity occurred between HTNV and DOBV.

Plaque reduction neutralization test (PRNT) is the most definitive method to facilitate the identification and differentiation of hantaviruses.[123,124] It is a specific test that can detect and measure neutralizing antibodies. Cross-PRNT has permitted serotypic classification of hantavirus infection in rodents and humans.[123,125,126] Although the assay is highly specific and is capable of distinguishing hantaviruses with serum from experimentally infected animals, it was shown to be less specific when human acute sera from HFRS and HPS patients were used.[125] A major disadvantage of PRNT is that its use is confined to very specialized research laboratories. PRNT necessitates the need for a biosafety level 3 containment laboratory and the specialized training of staff in the use of such a facility. In addition, the amount of time and effort to perform these tests requires the expertise of highly trained personnel and is not amenable for diagnostics. Furthermore, PRNT necessitates the existence of a virus that can be propagated in cell culture, and hantaviruses are known to have a slow growth rate, which present low titers when grown in these conditions. In order to visualize the plaques, isolates must be generally passed several times in cell culture and require careful optimization of plaque assay parameters. In summary, the PRNT is not a practical model for diagnostic purposes due to the length of time required to perform a test but is a powerful confirmatory test.

Immunohistochemistry (IHC) is a very simple confirmatory test that permits visual identification of viral antigens in tissues utilizing a specific antibody, but these tests can only be performed in postmortem individuals. Although this tool has been proven to be very useful, IHC cannot identify the specific strain of hantavirus present, due to the cross-reactivity between closely related hantavirus N protein antigens.[127] Virus-specific diagnosis can be confirmed by IHC, but the test requires the use of monoclonal antibodies to hantaviral antigens which are not generally available.

The RT-PCR is used for primary amplification of hantaviral RNA from cell culture and tissue samples.[128,129] The levels of viral RNA present in human and rodent tissue samples usually require nested RT-PCR techniques. Through the course of disease, patients under the early acute phase have suitable detectable levels of viral RNA at the onset of pulmonary edema but afterward clear virus rapidly from circulation.[130,131] Therefore, this method has been more effective in patients with acute HPS and not HFRS. Recent advances in quantitative real-time PCR (qPCR) have proved promising for a number of pathogens for measuring the pathogen load in the patient upon entry into the hospital and over the course of illness (e.g., see Refs. [132–134]). qPCR differs from RT-PCR in that reaction curves are quantified *in real time* during cycling rather than the amount of target accumulated after a fixed number of cycles. The detection of fluorescence signal generated by the qPCR reaction is proportional to the known amount of targets that will be added to the reaction mixture. qPCR diagnostics require advanced training to identify the linear range, accuracy and precision, limits of detection, efficiency, and the specificity of the method.

The PCR–EIA is a colorimetric hybridization assay that is simple and a highly sensitive and specific method for detecting and differentiating hantaviruses,[135] which combines the specificity of PCR with the sensitivity of enzymatic detection. Within this assay, digoxigenin-labeled PCR products are amplified with degenerate primers that correspond to a highly conserved sequence of viral S segment. The amplicons are mixed with biotinylated type-specific probes that permit the binding of the probe-DNA hybrids to streptavidin plates. Subsequently, hybrids are detected colorimetrically as in ELISA. This assay has proven to be highly sensitive, as little as 1.5 PFU of virus produced a positive signal. The assay is highly specific and can correctly identify virus types with no cross-reactivity. Although this test is highly promising and could replace PRNT for strain identification of hantaviruses, specific capture probes must be available. This is a major disadvantage, because as the number of discovered hantaviruses increases, additional capture probes will need to be designed and added to the panel. Overall, the procedure of PCR–EIA is much simpler and faster than PRNT and shows equal potential for typing hantaviruses in cell cultures, rodent samples, and clinical samples.

Given the cocirculation of some hantaviruses and their global presence, a diagnostic assay that contains the major antigens of all critical hantaviruses would be ideal. The requirement for a field-friendly diagnostic test is based on the location of many of the HFRS/HPS outbreaks, which have occurred in rural areas where adequate laboratory facilities are often lacking. Point-of-care diagnostics would increase the likelihood that prompt care, and management, which is vital for the survival of the patients, would be administered in a timely fashion. The storage and stability of the diagnostic reagent is also of concern given the unpredictable, episodic nature of these viruses. The availability of rapid diagnostics and reference laboratories that could facilitate rapid dissemination of kits to outbreak areas would be of great benefit for these lethal and unpredictable viruses.

8.6   TREATMENT AND PREVENTION

Current therapeutic efforts for HFRS and HPS are generally limited to supportive care as there are no FDA-approved antivirals, therapeutics, or vaccines. The antiviral ribavirin (1-β-dribofuranosyl-1,2,4-triazole-3 carboxamide) has been shown to be effective in reducing renal insufficiency if given early.[92] In this report by Rusnak et al., early treatment with intravenous ribavirin reduces the occurrence of oliguria and the severity of renal insufficiency.[92] Earlier studies performed in China on HFRS patients suggest that intravenous ribavirin when administered by the end of the first week of illness reduces the severity of clinical manifestations and death.[136,137]

In addition to small molecule therapeutics, human neutralizing antibodies administered during the acute phase might prove effective for the treatment and/or prophylaxis of hantaviral infections as shown for other viral diseases such as rabies, hepatitis A and B, and varicella virus. Studies in hamsters and rats show that passive transfer of neutralizing mAbs or polyclonal sera to HTNV or ANDV can passively protect animals from challenge with hantaviruses.[138–141] Furthermore, Gc-specific neutralizing mAbs, administered up to 4 days after challenge with virus, have been shown to cure hamsters and mice from infection.[141–143]

Vaccine efforts have focused on the M and S segment products that elicit a protective neutralization response.[144-146] During early efforts, a vaccinia virus-vectored vaccine containing the M and S genes of HTNV was developed.[147] This was tested in humans and found to elicit neutralizing antibodies; however, preexisting immunity to vaccinia (smallpox vaccination) mitigated the efficacy of the vaccine.[148] In another approach, HTNV GPs were used to pseudotype vesicular stomatitis virus (VSV), and this vaccine elicited neutralizing antibodies in mice and protected against HTNV infection.[149] More recently, plasmid DNA approaches have been developed using the full-length SEOV and HTNV M genes.[150,151] Both the SEOV and HTNV M gene-based DNA vaccines administered by gene gun elicited high-titer neutralizing antibodies in macaques.[151] The first DNA vaccine to elicit high-titer neutralizing antibodies against HPS hantaviruses contained the full-length ANDV M gene.[141] A plasmid containing both the HTNV and ANDV full-length M genes elicits antibodies that neutralized both HFRS- and HPS-associated hantavirus, albeit with lower titers than the single-gene constructs.[152] More recently, a PUUV M gene-based DNA vaccine was produced, and this construct, combined with the HTNV M gene-based DNA vaccine, forms an HFRS DNA vaccine that the USAMRIID is currently in phase II clinical trials. Finally, the N protein delivered as a recombinant protein (e.g., made in baculovirus, yeast, or *E. coli* systems or on hepatitis B virus core particles) provides protection against challenge in small animal models against the challenge with the Old World and New World hantaviruses.[153-158] The mechanism of protection is unknown but presumably N promotes a protective cellular immune response.

8.7 CONCLUSION

Hantaviruses represent an important and growing source of global disease emergence to society.[1,2] Hantaviruses also represent a threat to military troops that operate in areas endemic for hantaviral rodent reservoirs as demonstrated in the Korean and Balkan Wars. Retrospective analyses of other wars such as the American Civil War, First World War, and Second World War suggest HFRS and HFRS-like illness has long been a military problem.[159] Further, the high rates of morbidity and mortality qualify hantaviruses as category A agents by NIAID and category C agents by CDC and underscore the need for their continued surveillance in the environment. The New World viruses have been listed as potential biological weapons because of their lethality to humans and high infectivity by the aerosol route. Combined with the severity and mortality associated with HFRS and HPS, a critical gap remains in the availability of cost-effective, rapid diagnostics and treatments.

REFERENCES

1. Schmaljohn, C. and Hjelle, B. Hantaviruses: A global disease problem. *Emerg Infect Dis* 3, 95–104 (1997).
2. Jonsson, C.B., Figueiredo, L.T., and Vapalahti, O.A global perspective on hantavirus ecology, epidemiology, and disease. *Clin Microbiol Rev* 23, 412–441 (2010).
3. Peters, C.J., Simpson, G.L., and Levy, H. Spectrum of hantavirus infection: Hemorrhagic fever with renal syndrome and hantavirus pulmonary syndrome. *Annu Rev Med* 50, 531–545 (1999).
4. Lee, H.W. Korean hemorrhagic fever. *Prog Med Virol* 28, 96–113 (1982).
5. Lee, M. Coagulopathy in patients with hemorrhagic fever with renal syndrome. *J Korean Med Sci* 2, 201–211 (1987).
6. Lee, H.W., Lee, P.W., and Johnson, K.M. Isolation of the etiologic agent of Korean hemorrhagic fever. *J Infect Dis* 137, 298–308 (1978).
7. Lee, H.W., Baek, L.J., and Johnson, K.M. Isolation of Hantaan virus, the etiologic agent of Korean hemorrhagic fever, from wild urban rats. *J Infect Dis* 146, 638–644 (1982).
8. Childs, J.E. et al. Epizootiology of Hantavirus infections in Baltimore: Isolation of a virus from Norway rats, and characteristics of infected rat populations. *Am J Epidemiol* 126, 55–68 (1987).
9. LeDuc, J.W. et al. Global survey of antibody to Hantaan-related viruses among peridomestic rodents. *Bull World Health Organ* 64, 139–144 (1986).
10. Nichol, S.T. et al. Genetic identification of a hantavirus associated with an outbreak of acute respiratory illness. *Science* 262, 914–917 (1993).
11. Easterbrook, J.D., Zink, M.C., and Klein, S.L. Regulatory T cells enhance persistence of the zoonotic pathogen Seoul virus in its reservoir host. *Proc Natl Acad Sci USA* 104, 15502–15507 (2007).
12. Schountz, T. et al. Regulatory T cell-like responses in deer mice persistently infected with Sin Nombre virus. *Proc Natl Acad Sci USA* 104, 15496–15501 (2007).
13. Wells, R.M. et al. An unusual hantavirus outbreak in southern Argentina: Person-to-person transmission? Hantavirus Pulmonary Syndrome Study Group for Patagonia. *Emerg Infect Dis* 3, 171–174 (1997).
14. Fenner, F. The classification and nomenclature of viruses. Summary of results of meetings of the International Committee on Taxonomy of Viruses in Madrid, September 1975. *Intervirology* 6, 1–12 (1975).
15. Bishop, D.H. et al. Bunyaviridae. *Intervirology* 14, 125–143 (1980).
16. Schmaljohn, C.S. and Hooper, J.W. Bunyaviridae: The viruses and their replication. In Fields, B.N., Knipe, D.M., and Howley, P.M. (eds.) *Virology*, Vol. 2, pp. 1581–1602 (Lippincott-Raven, Philadelphia, PA, 2001).
17. Lee, H.W. et al. Observations on natural and laboratory infection of rodents with the etiologic agent of Korean hemorrhagic fever. *Am J Trop Med Hyg* 30, 477–482 (1981).
18. Hung, T. et al. Morphology and morphogenesis of viruses of hemorrhagic fever with renal syndrome (HFRS). I. Some peculiar aspects of the morphogenesis of various strains of HFRS virus. *Intervirology* 23, 97–108 (1985).
19. Martin, M.L. et al. Distinction between Bunyaviridae genera by surface structure and comparison with Hantaan virus using negative stain electron microscopy. *Arch Virol* 86, 17–28 (1985).
20. McCormick, J.B. et al. Morphological identification of the agent of Korean haemorrhagic fever (Hantaan virus) as a member of the Bunyaviridae. *Lancet* 1, 765–768 (1982).
21. White, J.D. et al. Hantaan virus, aetiological agent of Korean haemorrhagic fever, has Bunyaviridae-like morphology. *Lancet* 1, 768–771 (1982).
22. Schmaljohn, C.S. and Nichol, S.T. Bunyaviridae. In Knipe, D. (ed.) *Virology*, Vol. 2, pp. 1741–1789 (Lippincott-Raven, Philadelphia, PA, 2006).

23. Battisti, A.J. et al. Structural studies of Hantaan virus. *J Virol* 85, 835–841 (2011).

24. Huiskonen, J.T. et al. Electron cryotomography of Tula hantavirus suggests a unique assembly paradigm for enveloped viruses. *J Virol* 84, 4889–4897 (2010).

25. Overby, A.K. et al. Insights into bunyavirus architecture from electron cryotomography of Uukuniemi virus. *Proc Natl Acad Sci USA* 105, 2375–2379 (2008).

26. Sherman, M.B. et al. Single-particle cryo-electron microscopy of Rift Valley fever virus. *Virology* 387, 11–15 (2009).

27. Huiskonen, J.T. et al. Electron cryo-microscopy and single-particle averaging of Rift Valley fever virus: Evidence for GN-GC glycoprotein heterodimers. *J Virol* 83, 3762–3769 (2009).

28. Freiberg, A.N. et al. Three-dimensional organization of Rift Valley fever virus revealed by cryoelectron tomography. *J Virol* 82, 10341–10348 (2008).

29. Antic, D., Wright, K.E., and Kang, C.Y. Maturation of Hantaan virus glycoproteins G1 and G2. *Virology* 189, 324–328 (1992).

30. Schmaljohn, C.S. et al. Coding strategy of the S genome segment of Hantaan virus. *Virology* 155, 633–643 (1986).

31. Schmaljohn, C.S. and Dalrymple, J.M. Analysis of Hantaan virus RNA: Evidence for a new genus of bunyaviridae. *Virology* 131, 482–491 (1983).

32. Hewlett, M.J., Pettersson, R.F., and Baltimore, D. Circular forms of Uukuniemi virion RNA: An electron microscopic study. *J Virol* 21, 1085–1093 (1977).

33. Obijeski, J.F. et al. Segmented genome and nucleocapsid of La Crosse virus. *J Virol* 20, 664–675 (1976).

34. Dahlberg, J.E., Obijeski, J.F., and Korb, J. Electron microscopy of the segmented RNA genome of La Crosse virus A absence of circular molecules. *J Virol* 22, 203–209 (1977).

35. Donets, M.A. et al. Physicochemical characteristics, morphology and morphogenesis of virions of the causative agent of Crimean hemorrhagic fever. *Intervirology* 8, 294–308 (1977).

36. Schmaljohn, C.S. et al. Characterization of Hantaan virions, the prototype virus of hemorrhagic fever with renal syndrome. *J Infect Dis* 148, 1005–1012 (1983).

37. Schmaljohn, C. Molecular biology of hantaviruses. In Elliot, R. (ed.) *The Bunyaviridae*, pp. 63–90 (Plenum Press, New York, 1996).

38. Jonsson, C.B. and Schmaljohn, C.S. Replication of hantaviruses. *Curr Top Microbiol Immunol* 256, 15–32 (2001).

39. Ruusala, A. et al. Coexpression of the membrane glycoproteins G1 and G2 of Hantaan virus is required for targeting to the Golgi complex. *Virology* 186, 53–64 (1992).

40. Spiropoulou, C.F. Hantavirus maturation. *Curr Top Microbiol Immunol* 256, 33–46 (2001).

41. Lober, C. et al. The Hantaan virus glycoprotein precursor is cleaved at the conserved pentapeptide WAASA. *Virology* 289, 224–229 (2001).

42. Vapalahti, O. et al. Human B-cell epitopes of Puumala virus nucleocapsid protein, the major antigen in early serological response. *J Med Virol* 46, 293–303 (1995).

43. Ravkov, E.V. et al. Role of actin microfilaments in Black Creek Canal virus morphogenesis. *J Virol* 72, 2865–2870 (1998).

44. Ramanathan, H.N. et al. Dynein-dependent transport of the hantaan virus nucleocapsid protein to the endoplasmic reticulum-Golgi intermediate compartment. *J Virol* 81, 8634–8647 (2007).

45. Ramanathan, H.N. and Jonsson, C.B. New and Old World hantaviruses differentially utilize host cytoskeletal components during their life cycles. *Virology* 374, 138–150 (2008).

46. Gavrilovskaya, I.N. et al. beta3 Integrins mediate the cellular entry of hantaviruses that cause respiratory failure. *Proc Natl Acad Sci USA* 95, 7074–7079 (1998).

47. Gavrilovskaya, I.N. et al. Cellular entry of hantaviruses which cause hemorrhagic fever with renal syndrome is mediated by beta3 integrins. *J Virol* 73, 3951–3959 (1999).

48. Jin, M. et al. Hantaan virus enters cells by clathrin-dependent receptor-mediated endocytosis. *Virology* 294, 60–69 (2002).

49. Lee, H.W. and Johnson, K.M. Korean hemorrhagic fever: Demonstration of causative antigen and antibodies. *Korean J Intern Med* 19, 371 (1976).

50. Lee, H.W. and van der Groen, G. Hemorrhagic fever with renal syndrome. *Prog Med Virol* 36, 62 (1989).

51. Avsic-Zupanc, T. et al. Genetic and antigenic properties of Dobrava virus: A unique member of the *Hantavirus* genus, family Bunyaviridae. *J Gen Virol* 76 (Pt 11), 2801–2808 (1995).

52. Avsic-Zupanc, T. et al. Genetic analysis of wild-type Dobrava hantavirus in Slovenia: Co-existence of two distinct genetic lineages within the same natural focus. *J Gen Virol* 81, 1747–1755 (2000).

53. Golovljova, I. et al. Puumala and Dobrava hantaviruses causing hemorrhagic fever with renal syndrome in Estonia. *Eur J Clin Microbiol Infect Dis* 19, 968–969 (2000).

54. Jakab, F. et al. Detection of Dobrava hantaviruses in *Apodemus agrarius* mice in the Transdanubian region of Hungary. *Virus Res* 128, 149–152 (2007).

55. Klempa, B. et al. Central European Dobrava hantavirus isolate from a striped field mouse (*Apodemus agrarius*). *J Clin Microbiol* 43, 2756–2763 (2005).

56. Klempa, B. et al. Hemorrhagic fever with renal syndrome caused by 2 lineages of Dobrava hantavirus, Russia. *Emerg Infect Dis* 14, 617–625 (2008).

57. Klingstrom, J., Hardestam, J., and Lundkvist, A. Dobrava, but not Saaremaa, hantavirus is lethal and induces nitric oxide production in suckling mice. *Microbes Infect* 8, 728–737 (2006).

58. Papa, A., Bojovic, B., and Antoniadis, A. Hantaviruses in Serbia and Montenegro. *Emerg Infect Dis* 12, 1015–1018 (2006).

59. Clement, J. et al. Hantavirus infections in Europe. *Lancet Infect Dis* 3, 752–753; discussion 753–754 (2003).

60. Lee, H. Epidemiology and pathogenesis of haemorrhagic fever with renal syndrome. In Elliot, R. (ed.) *The Bunyaviridae*, pp. 253–267 (Plenum Press, New York, 1996).

61. Song, G. Epidemiological progresses of hemorrhagic fever with renal syndrome in China. *Chin Med J (Engl)* 112, 472–477 (1999).

62. Bi, Z., Formenty, P.B.H., and Roth, C.E. Hantavirus infection: A review and global update. *J Infect Dev Countries* 2, 3–23 (2008).

63. Cho, H.W., Howard, C.R., and Lee, H.W. Review of an inactivated vaccine against hantaviruses. *Intervirology* 45, 328–333 (2002).

64. Yashina, L., Mishin, V., Zdanovskaya, N., Schmaljohn, C., and Ivanov, L. A newly discovered variant of a hantavirus in *Apodemus peninsulae*, far Eastern Russia. *Emerg Infect Dis* 7, 912–913 (2001).

65. Mailles, A. et al. Larger than usual increase in cases of hantavirus infections in Belgium, France and Germany, June 2005. *Euro Surveill* 10, E050721.4 (2005).

66. Mailles, A. et al. Increase of Hantavirus infections in France, 2003. *Med Mal Infect* 35, 68–72 (2005).

67. Chen, H.X. and Qiu, F.X. Epidemiologic surveillance on the hemorrhagic fever with renal syndrome in China. *Chin Med J (Engl)* 106, 857–863 (1993).

68. Lee, H.W. Hemorrhagic fever with renal syndrome in Korea. *Rev Infect Dis* 11, S864–S867 (1989).

69. Chen, H.X. et al. Epidemiological studies on hemorrhagic fever with renal syndrome in China. *J Infect Dis* 154, 394–398 (1986).

70. Macneil, A., Nichol, S.T., and Spiropoulou, C.F. Hantavirus pulmonary syndrome. *Virus Res* 162, 138–147 (2011).

71. MacNeil, A., Ksiazek, T.G., and Rollin, P.E. Hantavirus pulmonary syndrome, United States, 1993–2009. *Emerg Infect Dis* 17, 1195–1201 (2011).

72. Raboni, S.M. et al. Phylogenetic characterization of hantaviruses from wild rodents and HPS cases in the state of Parana (Southern Brazil). *J Gen Virol* 90, 2166–2171 (2009).

73. Martinez, V.P. et al. Hantavirus pulmonary syndrome in Argentina, 1995–2008. *Emerg Infect Dis* 16, 1853–1860 (2010).

74. Palma, R.E. et al. Ecology of rodent-associated hantaviruses in the southern cone of South America: Argentina, Chile, Paraguay and Uruguay. *J Wildl Dis* 48, 267–281 (2012).

75. Bohlman, M.C. et al. Analysis of hantavirus genetic diversity in Argentina: S segment-derived phylogeny. *J Virol* 76, 3765–3773 (2002).

76. Johnson, A.M. et al. Laguna Negra virus associated with HPS in western Paraguay and Bolivia. *Virology* 238, 115–127 (1997).

77. Carroll, D.S. et al. Hantavirus pulmonary syndrome in Central Bolivia: Relationships between reservoir hosts, habitats, and viral genotypes. *Am J Trop Med Hyg* 72, 42–46 (2005).

78. Figueiredo, L.T. et al. Hantaviruses in Sao Paulo State, Brazil. *Emerg Infect Dis* 9, 891–892 (2003).

79. Oliveira, R.C. et al. Genetic characterization of a Juquitiba-like viral lineage in *Oligoryzomys nigripes* in Rio de Janeiro, Brazil. *Acta Trop* 112, 212–218 (2009).

80. Figueiredo, L.T. et al. Hantavirus pulmonary syndrome, central plateau, southeastern, and southern Brazil. *Emerg Infect Dis* 15, 561–567 (2009).

81. Medina, R.A. et al. Ecology, genetic diversity, and phylogeographic structure of Andes virus in humans and rodents in Chile. *J Virol* 83, 2446–2459 (2009).

82. Hjelle, B. et al. Prevalence and geographic genetic variation of hantaviruses of New World harvest mice (*Reithrodontomys*): Identification of a divergent genotype from a Costa Rican *Reithrodontomys mexicanus*. *Virology* 207, 452–459 (1995).

83. Chu, Y.K. et al. Genetic characterization and phylogeny of a hantavirus from Western Mexico. *Virus Res* 131, 180–188 (2008).

84. Vincent, M.J. et al. Hantavirus pulmonary syndrome in Panama: Identification of novel hantaviruses and their likely reservoirs. *Virology* 277, 14–19 (2000).

85. Chu, Y.K. et al. Phylogenetic and geographical relationships of hantavirus strains in eastern and western Paraguay. *Am J Trop Med Hyg* 75, 1127–1134 (2006).

86. Chu, Y.K. et al. The complex ecology of hantavirus in Paraguay. *Am J Trop Med Hyg* 69, 263–268 (2003).

87. Powers, A.M. et al. Isolation and genetic characterization of a hantavirus (Bunyaviridae: *Hantavirus*) from a rodent, *Oligoryzomys microtis* (Muridae), collected in northeastern Peru. *Am J Trop Med Hyg* 61, 92–98 (1999).

88. Fulhorst, C.F. et al. Maporal virus, a hantavirus associated with the fulvous pygmy rice rat (*Oligoryzomys fulvescens*) in western Venezuela. *Virus Res* 104, 139–144 (2004).

89. Fulhorst, C.F. et al. Isolation, characterization and geographic distribution of Cano Delgadito virus, a newly discovered South American hantavirus (family Bunyaviridae). *Virus Res* 51, 159–171 (1997).

90. Chernukha, Y.G., Evdokimova, O.A., and Cheechovich, A.V. Results of karyologic and immunobiological studies of the striped field mouse (*Apodemus agrarius*) from different areas of its range. *Zool J* 65, 471–475 (1986).

91. LeDuc, J.W. Epidemiology of Hantaan and related viruses. *Lab Anim Sci* 37, 413–418 (1987).

92. Rusnak, J.M. et al. Experience with intravenous ribavirin in the treatment of hemorrhagic fever with renal syndrome in Korea. *Antiviral Res* 81, 68–76 (2009).

93. Clement, J., Colson, P., and McKenna, P. Hantavirus pulmonary syndrome in New England and Europe. *N Engl J Med* 331, 545–546; author reply 547–548 (1994).

94. Pilaski, J. et al. Genetic identification of a new Puumala virus strain causing severe hemorrhagic fever with renal syndrome in Germany. *J Infect Dis* 170, 1456–1462 (1994).

95. Sugiyama, K. et al. Four serotypes of haemorrhagic fever with renal syndrome viruses identified by polyclonal and monoclonal antibodies. *J Gen Virol* 68(Pt 4), 979–987 (1987).

96. Sheedy, J.A. et al. The clinical course of epidemic hemorrhagic fever. *Am J Med* 16, 619–628 (1954).

97. Powell, G.M. Hemorrhagic fever: A study of 300 cases. *Medicine (Baltimore)* 33, 97–153 (1954).

98. Lee, H.W. Clinical manifestations of HFRS. In Calisher, C., and Schmaljohn, C. (eds.) *Manual of Hemorrhagic Fever with Renal Syndrome*, pp. 19–38 (WHO Collaborating Center for Virus Reference and Research [Hantaviruses], ASAN Institute for Life Sciences, Seoul, Korea, 1989).

99. Moolenaar, R.L., Breiman, R.F., and Peters, C.J. Hantavirus pulmonary syndrome. *Semin Respir Infect* 12, 31–39 (1997).

100. Khan, A.S., Kitsutani, P.T., and Corneli, A.L. Hantavirus pulmonary syndrome in the Americas: The early years. *Semin Respir Crit Care Med* 21, 313–322 (2000).

101. Jonsson, C.B., Hooper, J., and Mertz, G. Treatment of hantavirus pulmonary syndrome. *Antiviral Res* 78, 162–169 (2008).

102. Riquelme, R. et al. Hantavirus pulmonary syndrome, southern Chile. *Emerg Infect Dis* 9, 1438–1443 (2003).

103. Williams, R.J. et al. An outbreak of hantavirus pulmonary syndrome in western Paraguay. *Am J Trop Med Hyg* 57, 274–282 (1997).

104. Lee, H.W., Calisher, C.H., and Schmaljohn, C. *Manual of Hemorrhagic Fever with Renal Syndrome and Hantavirus Pulmonary Syndrome* (WHO Collaborating Center for Virus Reference and Research [Hantaviruses], Asan Institute for Life Sciences, Seoul, Korea, 1998).

105. Vaheri, A., Vapalahti, O., and Plyusnin, A. How to diagnose hantavirus infections and detect them in rodents and insectivores. *Rev Med Virol* 18, 277–288 (2008).

106. Koraka, P. et al. Evaluation of two commercially available immunoassays for the detection of hantavirus antibodies in serum samples. *J Clin Virol* 17, 189–196 (2000).

107. Kallio-Kokko, H. et al. Antigenic properties and diagnostic potential of recombinant dobrava virus nucleocapsid protein. *J Med Virol* 61, 266–274 (2000).

108. Jonsson, C.B. et al. Purification and characterization of the Sin Nombre virus nucleocapsid protein expressed in *Escherichia coli*. *Protein Expr Purif* 23, 134–141 (2001).

109. Schmaljohn, C.S. et al. Baculovirus expression of the small genome segment of Hantaan virus and potential use of the expressed nucleocapsid protein as a diagnostic antigen. *J Gen Virol* 69(Pt 4), 777–786 (1988).

110. Vapalahti, O. et al. Antigenic properties and diagnostic potential of Puumala virus nucleocapsid protein expressed in insect cells. *J Clin Microbiol* 34, 119–125 (1996).

111. Razanskiene, A. et al. High yields of stable and highly pure nucleocapsid proteins of different hantaviruses can be generated in the yeast *Saccharomyces cerevisiae. J Biotechnol* 111, 319–333 (2004).

112. Schmidt, J. et al. Nucleocapsid protein of cell culture-adapted Seoul virus strain 80–39: Analysis of its encoding sequence, expression in yeast and immuno-reactivity. *Virus Genes* 30, 37–48 (2005).

113. Khattak, S., Darai, G., Sule, S., and Rosen-Wolff, A. Characterization of expression of Puumala virus nucleocapsid protein in transgenic plants. *Intervirology* 45, 334–339 (2002).

114. Kehm, R. et al. Expression of immunogenic Puumala virus nucleocapsid protein in transgenic tobacco and potato plants. *Virus Genes* 22, 73–83 (2001).

115. Billecocq, A. et al. Expression of the nucleoprotein of the Puumala virus from the recombinant Semliki Forest virus replicon: Characterization and use as a potential diagnostic tool. *Clin Diagn Lab Immunol* 10, 658–663 (2003).

116. Elgh, F. et al. A major antigenic domain for the human humoral response to Puumala virus nucleocapsid protein is located at the amino-terminus. *J Virol Methods* 59, 161–172 (1996).

117. Kang, J.I. et al. A dominant antigenic region of the hantaan virus nucleocapsid protein is located within a aminoterminal short stretch of hydrophilic residues. *Virus Genes* 23, 183–186 (2001).

118. Lindkvist, M. et al. Cross-reactive and serospecific epitopes of nucleocapsid proteins of three hantaviruses: Prospects for new diagnostic tools. *Virus Res* 137, 97–105 (2008).

119. Feldmann, H. et al. Utilization of autopsy RNA for the synthesis of the nucleocapsid antigen of a newly recognized virus associated with hantavirus pulmonary syndrome. *Virus Res* 30, 351–367 (1993).

120. Hjelle, B. et al. Rapid and specific detection of Sin Nombre virus antibodies in patients with hantavirus pulmonary syndrome by a strip immunoblot assay suitable for field diagnosis. *J Clin Microbiol* 35, 600–608 (1997).

121. Yee, J. et al. Rapid and simple method for screening wild rodents for antibodies to Sin Nombre hantavirus. *J Wildl Dis* 39, 271–277 (2003).

122. Hujakka, H. et al. Diagnostic rapid tests for acute hantavirus infections: Specific tests for Hantaan, Dobrava and Puumala viruses versus a hantavirus combination test. *J Virol Methods* 108, 117–122 (2003).

123. Chu, Y.K. et al. Serological relationships among viruses in the *Hantavirus* genus, family Bunyaviridae. *Virology* 198, 196–204 (1994).

124. Schmaljohn, C.S. et al. Antigenic and genetic properties of viruses linked to hemorrhagic fever with renal syndrome. *Science* 227, 1041–1044 (1985).

125. Chu, Y.K. et al. Cross-neutralization of hantaviruses with immune sera from experimentally infected animals and from hemorrhagic fever with renal syndrome and hantavirus pulmonary syndrome patients. *J Infect Dis* 172, 1581–1584 (1995).

126. Lee, P.W. et al. Serotypic classification of hantaviruses by indirect immunofluorescent antibody and plaque reduction neutralization tests. *J Clin Microbiol* 22, 940–944 (1985).

127. Zaki, S.R. et al. Retrospective diagnosis of hantavirus pulmonary syndrome, 1978–1993: Implications for emerging infectious diseases. *Arch Pathol Lab Med* 120, 134–139 (1996).

128. Giebel, L.B. et al. Rapid detection of genomic variations in different strains of hantaviruses by polymerase chain reaction techniques and nucleotide sequence analysis. *Virus Res* 16, 127–136 (1990).

129. Hjelle, B. Virus detection and identification with genetic tests. In Lee, H.W., Calisher, C.H., and Schmaljohn, C.S. (eds.) *Manual of Hemorrhagic Fever with Renal Syndrome and Hantavirus Pulmonary Syndrome*, pp. 132–137 (WHO Collaborating Center for Virus Reference and Research [Hantaviruses], Asan Institute for Life Sciences, Seoul, Korea, 1998).

130. Hjelle, B. et al. Detection of Muerto Canyon virus RNA in peripheral blood mononuclear cells from patients with hantavirus pulmonary syndrome. *J Infect Dis* 170, 1013–1017 (1994).

131. Terajima, M. et al. High levels of viremia in patients with the Hantavirus pulmonary syndrome. *J Infect Dis* 180, 2030–2034 (1999).

132. Bustin, S.A. and Mueller, R. Real-time reverse transcription PCR (qRT-PCR) and its potential use in clinical diagnosis. *Clin Sci (Lond)* 109, 365–379 (2005).

133. Ng, E.K. and Lo, Y.M. Molecular diagnosis of severe acute respiratory syndrome. *Methods Mol Biol* 336, 163–175 (2006).

134. Deback, C. et al. Use of the Roche LightCycler 480 system in a routine laboratory setting for molecular diagnosis of opportunistic viral infections: Evaluation on whole blood specimens and proficiency panels. *J Virol Methods* 159, 291–294 (2009).

135. Dekonenko, A., Ibrahim, M.S., and Schmaljohn, C.S. A colorimetric PCR-enzyme immunoassay to identify hantaviruses. *Clin Diagn Virol* 8, 113–121 (1997).

136. Huggins, J.W. Prospects for treatment of viral hemorrhagic fevers with ribavirin, a broad-spectrum antiviral drug. *Rev Infect Dis* 11(Suppl 4), S750–S761 (1989).

137. Huggins, J.W. et al. Prospective, double-blind, concurrent, placebo-controlled clinical trial of intravenous ribavirin therapy of hemorrhagic fever with renal syndrome. *J Infect Dis* 164, 1119–1127 (1991).

138. Arikawa, J. et al. Protective role of antigenic sites on the envelope protein of Hantaan virus defined by monoclonal antibodies. *Arch Virol* 126, 271–281 (1992).

139. Schmaljohn, C.S. et al. Antigenic subunits of Hantaan virus expressed by baculovirus and vaccinia virus recombinants. *J Virol* 64, 3162–3170 (1990).

140. Zhang, X.K., Takashima, I., and Hashimoto, N. Characteristics of passive immunity against hantavirus infection in rats. *Arch Virol* 105, 235–246 (1989).

141. Custer, D.M. et al. Active and passive vaccination against hantavirus pulmonary syndrome with Andes virus M genome segment-based DNA vaccine. *J Virol* 77, 9894–9905 (2003).

142. Liang, M., Chu, Y.K., and Schmaljohn, C. Bacterial expression of neutralizing mouse monoclonal antibody Fab fragments to Hantaan virus. *Virology* 217, 262–271 (1996).

143. Xu, Z. et al. The in vitro and in vivo protective activity of monoclonal antibodies directed against Hantaan virus: Potential application for immunotherapy and passive immunization. *Biochem Biophys Res Commun* 298, 552–558 (2002).

144. Hooper, J.W. and Li, D. Vaccines against hantaviruses. *Curr Top Microbiol Immunol* 256, 171–191 (2001).

145. Schmaljohn, C. Vaccines for hantaviruses. *Vaccine* 27(Suppl 4), D61–D64 (2009).

146. Schmaljohn, C.S. Vaccines for hantaviruses: Progress and issues. *Expert Rev Vaccines* 11, 511–513 (2012).

147. Chu, Y.K., Jennings, G.B., and Schmaljohn, C.S. A vaccinia virus-vectored Hantaan virus vaccine protects hamsters from challenge with Hantaan and Seoul viruses but not Puumala virus. *J Virol* 69, 6417–6423 (1995).

148. McClain, D.J. et al. Clinical evaluation of a vaccinia-vectored Hantaan virus vaccine. *J Med Virol* 60, 77–85 (2000).

149. Lee, B.H. et al. A pseudotype vesicular stomatitis virus containing Hantaan virus envelope glycoproteins G1 and G2 as an alternative to hantavirus vaccine in mice. *Vaccine* 24, 2928–2934 (2006).

150. Hooper, J.W. et al. DNA vaccination with hantavirus M segment elicits neutralizing antibodies and protects against Seoul virus infection. *Virology* 255, 269–278 (1999).

151. Hooper, J.W. et al. A lethal disease model for hantavirus pulmonary syndrome. *Virology* 289, 6–14 (2001).

152. Hooper, J.W. et al. Hantaan/Andes virus DNA vaccine elicits a broadly cross-reactive neutralizing antibody response in nonhuman primates. *Virology* 347, 208–216 (2006).

153. Dargeviciute, A. et al. Yeast-expressed Puumala hantavirus nucleocapsid protein induces protection in a bank vole model. *Vaccine* 20, 3523–3531 (2002).

154. de Carvalho Nicacio, C. et al. Cross-protection against challenge with Puumala virus after immunization with nucleocapsid proteins from different hantaviruses. *J Virol* 76, 6669–6677 (2002).

155. Klingstrom, J. et al. Vaccination of C57/BL6 mice with Dobrava hantavirus nucleocapsid protein in Freund's adjuvant induced partial protection against challenge. *Vaccine* 22, 4029–4034 (2004).

156. Maes, P. et al. Truncated recombinant Puumala virus nucleocapsid proteins protect mice against challenge in vivo. *Viral Immunol* 21, 49–60 (2008).

157. Maes, P. et al. Hantaviruses: Immunology, treatment, and prevention. *Viral Immunol* 17, 481–497 (2004).

158. Geldmacher, A. et al. A hantavirus nucleocapsid protein segment exposed on hepatitis B virus core particles is highly immunogenic in mice when applied without adjuvants or in the presence of pre-existing anti-core antibodies. *Vaccine* 23, 3973–3983 (2005).

159. Lee, H.W. Epidemiology and epizoology. In Lee, H.W., Calisher, C.H., and Schmaljohn, C. (eds.) *Manual of Hemorrhagic Fever with Renal Syndrome and Hantavirus Pulmonary Syndrome*, pp. 39–73 (WHO Collaborating Center for Virus Reference and Research [Hantaviruses] Asan Institute for Life Sciences, Seoul, Korea, 1998).

9 Henipaviruses

Ina L. Smith

CONTENTS

9.1   INTRODUCTION

Hendra virus (HeV) and Nipah virus (NiV) are of public health and economic concern as they have a broad host range, causing respiratory and/or neurological disease in humans and domesticated animals such as horses, pigs, dogs, and cats, and infection is often associated with high mortality. In their reservoir host, the flying fox, HeV and NiV show no clinical symptoms of infection[1] suggesting that they have coexisted with flying foxes for a long time.

9.2   CLASSIFICATION AND MORPHOLOGY

HeV and NiV are the prototype viruses of the genus *Henipavirus* in the subfamily Paramyxovirinae and family Paramyxoviridae.[2] HeV and NiV are zoonotic viruses transmitted to humans from pteropid bats (genus *Pteropus*) (commonly referred to as flying foxes or fruit bats) via horses or pigs, respectively.[3,4] In some cases, transmission of NiV has been documented from bats to human via contaminated food.[5,6]

HeV and NiV are classified as Biosafety Level 4 (BSL 4) agents due to their broad host range causing fatal disease in animals and humans, high virulence, and unique genetic makeup. There are no vaccines or therapeutics currently available against henipaviruses. NiV is designated a category C priority agent in the National Institute of Allergy and Infectious Diseases Biodefense Research Agenda.[7]

The virions are enveloped, pleomorphic (38–600 nm), and covered in 10–18 nm surface projections.[8] HeV has a double-fringed envelope when viewed by negative contrast electron microscopy[8] and can be differentiated from NiV that has a single fringe.[9,10] The nucleocapsid (N) protein, phosphoprotein (P), and polymerase together with the viral ribonucleic acid (RNA) make up the nucleocapsid core. The glycoprotein (G), fusion (F) protein, and the matrix (M) protein form the outer surface of the virion.[2]

The genomes of the henipaviruses are some of the largest of the paramyxoviruses. The NiV strains, NiV Bangladesh and NiV Malaysia, are 18,252 and 18,246 nucleotides in length, respectively, and HeV has a genome of 18,234 nucleotides in length. The non-segmented single-stranded negative sense RNA genomes have a gene order of N, P, M, F, G, and large (L)-5′ proteins[11] and conform to the *rule of six*, with the length being divisible by six. A highly conserved tri-nucleotide 3′-CTT-5′ intercistronic sequence borders each gene.[12] Gene transcription initiation and termination sequences are conserved among viruses in the Paramyxoviridae family.[13] Henipaviruses have unique complementary genome terminal sequences and large 3′-untranslated regions (UTR) in comparison to other

TABLE 9.1
Gene Features of HeV and NiV

Gene	Protein	Virus	Length of Gene Element or Product			
			Protein (aa)	ORF (nt)	5′-UTR (nt)	3′-UTR (nt)
N	Nucleocapsid	Hendra	532	1,599	57	568
		Nipah Bangladesh	532	1,599	57	586
		Nipah Malaysia	532	1,599	57	586
P	Phosphoprotein	Hendra	707	2,124	105	469
		Nipah Bangladesh	709	2,130	105	469
		Nipah Malaysia	709			
V	V	Hendra	457	—	—	—
		Nipah Bangladesh	459			
		Nipah Malaysia	456			
W	W	Hendra	448	—	—	—
		Nipah Bangladesh	450			
		Nipah Malaysia	450			
C	C	Hendra	166	—	—	—
		Nipah Bangladesh	166			
		Nipah Malaysia	166			
M	Matrix	Hendra	352	1,059	100	200
		Nipah Bangladesh	352	1,059	100	200
		Nipah Malaysia	352	1,059	100	200
F	Fusion	Hendra	546	1,641	272	418
		Nipah Bangladesh	546	1,641	290	412
		Nipah Malaysia	546	1,641	284	412
G	Glycoprotein	Hendra	604	1,815	233	516
		Nipah Bangladesh	602	1,809	233	504
		Nipah Malaysia	602	1,809	233	504
L	RNA polymerase	Hendra	2,244	6,735	153	67
		Nipah Bangladesh	2,244	6,735	153	67
		Nipah Malaysia	2,244	6,735	153	67
	Genome length	Hendra	18,234			
		Nipah Bangladesh	18,252			
		Nipah Malaysia	18,246			

Abbreviations: ORF, open reading frame; UTR, untranslated region; nt, nucleotide; aa, amino acid.[61]

viruses within the Paramyxovirinae subfamily.[11] All gene lengths and other features are summarized in Table 9.1.

The N protein is the most abundant structural protein. The N protein associates with the P protein and the RNA polymerase (L) protein to protect the nucleic acid in the virion.[11,13] The central domain (aa 171–383) is the most conserved region in the N protein and is thought to be involved in N–N, N–P, and N–L interactions. The carboxy terminal is thought to interact with the M protein during virus assembly. Sequencing of the carboxy terminal of the N gene from multiple isolates of HeV has revealed considerable nucleotide variation in this region.[14] This is consistent with other findings for paramyxoviruses that have shown that this region contains antigenic sites and is hypervariable.[13] The N, P, and L proteins form the ribonucleoprotein (RNP) complex with the genomic RNA that functions in the transcription and replication of the virus.[2]

The P protein is part of the RNP and thus has a crucial role in virus replication. The P gene codes for 707 amino acids in

HeV and 709 amino acids in NiV and is the least conserved gene. In addition to coding for the P protein, the P gene encodes three other nonstructural proteins C, V, and W, which interact with host cellular factors.[15] The first open reading frame (ORF) on the P gene produces the full-length P protein of 707 amino acids. An alternate ORF transcribes the C protein. The V and W proteins share the same N-terminal. A highly conserved AG-rich sequence expresses the V and W proteins by RNA editing.[16,17] At the transcriptional editing site, the insertion of one extra G nucleotide produces the cysteine-rich V protein C-terminus.[12] The insertion of two extra G nucleotides produces the C-terminus for the W protein.[17] Another ORF between the C and V proteins of HeV, but not NiV, putatively encodes for a putative small binding (SB) protein.[16]

The P, V, W, and C proteins are interferon antagonists that work to inhibit the host antiviral innate immune response and thus have a role in pathogenicity (reviewed by [18,19]).

The M protein is a basic protein of 352 amino acids. In early stages of infection, the NiV M protein is trafficked through the nucleus and is then localized in the cytoplasm.[20] The nuclear localization sequences and nuclear export sequences identified in the NiV M protein are also present in HeV. The M protein plays a role in viral assembly and budding with the YPLGVG sequence being required for budding in NiV.[20,21]

The F protein, a type I membrane protein, is involved in virus entry into the cell.[22] The F gene transcribes the precursor F_0 protein, which is cleaved at amino acid K109 to the active F_1 and F_2 by the host cellular protease cathepsin L after endocytosis of the virion.[22-24] Both the F and G proteins are required for fusion.[25]

The G gene transcribes the attachment glycoprotein, a type II membrane glycoprotein, which lacks both hemagglutination and neuraminidase functions.[26,27] The viral glycoprotein binds to the host cellular receptor ephrin-B2 ligand or, to a lesser extent, ephrin-B3 to facilitate entry into the host cell.[26,28-30] These cell receptors are widely expressed in various tissues and highly conserved among mammalian species, enabling the henipaviruses to exhibit broad species and cellular tropisms.[31]

The largest protein at 2244 amino acids is the L protein, which functions as the RNA-dependent RNA polymerase.[11] This protein is involved in the enzymatic process of the viruses such as mRNA transcription and replication of the genome.[2]

9.3 BIOLOGY AND EPIDEMIOLOGY

9.3.1 HENDRA VIRUS

HeV was first identified in an outbreak in 1994 in Queensland, Australia, that resulted in the deaths of 20 horses and one human and the infection of another. Due to biosafety issues and public health concern, all surviving horses were destroyed.[4,32] Although originally called equine morbillivirus (EMV), genetic analysis revealed the virus to be distinct from other paramyxoviruses, and it was subsequently renamed HeV, after the Brisbane suburb of Hendra in which the outbreak occurred.[11,33]

Following the outbreaks in 1994, serological studies of the surrounding wildlife and domestic animals found all four species of Australian flying foxes had antibody to HeV.[34] The seroprevalence of anti-HeV antibodies among flying foxes was found to be 47%.[35] HeV was isolated from *Pteropus poliocephalus* and *Pteropus alecto* bat fetal tissue and the reproductive tract.[36] Experimentally, no clinical disease was observed in flying foxes when infected with HeV; however, seroconversion was detected.[1,37]

The exact mode of transmission of HeV from flying foxes to horses has not been fully elucidated, and further investigation is required. The most likely route of transmission from bats to horses is through the ingestion of grass or partially eaten fruit contaminated with bat urine, saliva, or other fluids.[38] HeV has been isolated in cell culture directly from the urine of flying foxes.[14] The Hendra virion is sensitive to changes in temperature, pH, and desiccation and rapidly inactivated following excretion from horses.[37,38]

HeV has a wide host range that distinguishes this virus from other Paramyxoviridae viruses that generally have narrow host specificity.[27] Laboratory experiments have revealed that HeV has low transmissibility and close contact is required for transmission to occur. The virus has been detected in the urine of experimentally infected horses, cats, and guinea pigs.[39] Experimentally, transmission between horses or from bats to horses has not been demonstrated, although transmission from cats to a horse was observed.[39] Laboratory experiments have shown that cats, guinea pigs, monkeys, pigs, mice, and hamsters are susceptible to HeV infection.[40-45] Experimentally, pigs have been found to be susceptible to HeV infection and therefore could play a role in transmission to humans if they were naturally infected.[44]

HeV infects cells from a wide range of hosts, and thus, a wide range of cell lines are susceptible to infection, including cells derived from mammals (RK13 [rabbit kidney], MDBK [bovine kidney], LLC-MK2 [monkey kidney], BHK [baby hamster kidney] Hep-2, and HeLa cells), reptiles, and amphibians, as well as embryonated chicken eggs.[32,42,46,47]

Transmission of HeV from horses to humans is thought to occur via droplets, cuts, or abrasions. In all cases of human HeV infection, there has been close contact with infected horses, which have acted as intermediate hosts.[34,36] Human infections with HeV have been acquired through performing or assisting in horse necropsies[48,49] via close contact with horses during husbandry procedures[4,32] or while performing veterinary procedures, such as nasal lavage of an infected horse.[50]

There have been 40 known outbreaks of HeV (Table 9.2) resulting in the infection of 80 horses, 7 humans, and 1 dog. An unprecedented number of recognized spillovers occurred in 2011, with 18 recorded outbreaks resulting in infection of 24 horses. In 2012, there have been 8 outbreaks resulting in the deaths of 11 horses.

9.3.2 NIPAH VIRUS

The first recognized outbreak of NiV infection occurred in Malaysia and then Singapore between September 1998 and June 1999.[51,52] There were 283 human cases of NiV infection and 109 deaths (case fatality of 39%) recognized in Malaysia.[53] NiV was isolated from the cerebrospinal fluid (CSF) of a patient from Sungai Nipah Village, Selangor State in Malaysia.[51]

In Malaysia, the outbreak of a deadly viral infection affecting pig farmers was initially thought to be caused by the arbovirus Japanese encephalitis virus, and control measures for this virus were implemented. Agricultural intensification contributed to the outbreak of NiV whereby fruiting orchards in close proximity to piggeries attracted flying foxes to the farm.[54] Infection of pigs was thought to have occurred via the ingestion of partially eaten fruit contaminated with bat saliva or ingestion via urine or feces.[55] Repeated introduction of NiV was suggested to have primed for a persistent infection to establish itself on the farm.[54] The high rate of

TABLE 9.2

Outbreaks of HeV in Humans and Horses

Date	Location	Horse Cases	Human Cases
August 1994	Mackay, Queensland (QLD)	2 [49]	1 (1 death)
September 1994	Hendra, QLD	20 [4,32]	2 (1 death)
January 1999	Cairns, QLD	1 [96]	
October 2004	Gordonvale, near Cairns, QLD	1 [4,160]	1
December 2004	Townsville, QLD	1 [160]	
June 2006	Peachester, QLD	1 [48]	
November 2006	Near Murwillumbah, New South Wales (NSW)	1 [161]	
June 2007	Peachester, QLD	1 [162]	
July 2007	Clifton Beach, QLD	1 [163]	
July 2008	Thornlands, Redlands, QLD	5	2 (1 death) [50,98]
July 2008	Proserpine, QLD	4 [163]	
August 2009	Cawarral, QLD	4	1 (1 death) [164]
September 2009	Bowen, QLD	2 [165]	
May 2010	Tewantin, QLD	1 [166]	
July 2011	Mt. Alford, QLD	3 (infection of one dog) [167]	
June 2011	Logan Reserve, QLD	1 [168]	
June 2011	Kerry, QLD	1 [169]	
June 2011	McLeans Ridges, NSW	2 [169]	
July 2011	Utungun, NSW	1 [170]	
July 2011	Park Ridge, QLD	1 [171]	
July 2011	Kuranda, QLD	1 [172]	
July 2011	Hervey Bay, QLD	1 [173]	
July 2011	Corndale, NSW	1 [174]	
July 2011	Boondall, QLD	1 [173]	
July 2011	Chinchilla, QLD	1 [175]	
July 2011	Mullumbimby, NSW	1 [176]	
August 2011	Newrybar, NSW	1 [177]	
August 2011	Pimlico, NSW	2 [178]	
August 2011	Mullumbimby, NSW	1 [178]	
August 2011	Currumbin Valley, QLD	1 [179]	
August 2011	Tintenbar, NSW	1 [180]	
October 2011	Beachmere, QLD	3 [181]	
January 2012	Near Townsville, QLD	1 [182]	
May 2012	Near Rockhampton, QLD	2 [183]	
May 2012	Ingham, QLD	1 [184]	
June 2012	Mackay, QLD	1 [185]	
July 2012	Rockhampton, QLD	3 [186]	
July 2012	Cairns, QLD	1 [187]	
September 2012	Near Port Douglas, QLD	1 [188]	
November 2012	Ingham, QLD	1	
Total		80	7 (4 deaths)

transmission combined with the high density of pigs on the farm and the movement of pigs between farms contributed to the outbreak spreading.[35] Pigs acted as amplifying hosts for the transmission of NiV to humans. The importation of live pigs from Malaysia to Singapore led to infection of abattoir workers with 11 cases and one death identified[56] (Table 9.3).

The reservoir hosts of NiV were identified as *Pteropus vampyrus* and *Pteropus hypomenalus* based on serological evidence for infection in these species,[57] and NiV was subsequently isolated from the urine of *P. hypomenalus* bats on Tioman Island, Malaysia.[58] NiV RNA has also been detected in the liver of a *Pteropus giganteus* bat in West Bengal, India.[59] Rahman and coworkers[60] have suggested the NiV can undergo recrudescence in infected bats, thus suggesting a mechanism for maintenance of viral infections in flying fox populations.

There have been no further spillovers of NiV identified in Malaysia; however, there have been multiple spillover events occurring in Bangladesh and India from 2001 (Table 9.3). In these countries, numerous spillover events involving direct transmission from bats to humans have been identified.[61–68] There have been two recognized outbreaks resulting in 96 cases and 50 deaths in India and 14 outbreaks resulting in 206 cases and 160 deaths in Bangladesh. There have been higher mortalities in outbreaks of NiV in Bangladesh with an overall mortality from NiV in Bangladesh of 77.7% compared to 39% in Malaysia and 52% in India (Table 9.3). Consumption of raw date palm sap contaminated with bat excreta has been identified as a major source of human infection.[68] Infrared cameras have been used to study the *Pteropus* bats visiting the date palms where collection of sap was occurring.[5,69] Infection of humans is also thought to occur via infected animals that have become infected through the consumption of food *presumably* contaminated with bat saliva or urine.[70]

Person-to-person transmission of NiV in Bangladesh was associated with contact with infectious respiratory secretions,[71] with patients with respiratory distress being more likely to transmit the virus.[70] Social factors, such as close contact due to family members nursing infected patients, have been implicated in person-to-person transmission in Bangladesh.[68,71–73] In one outbreak, a religious leader infected with NiV transmitted the disease to 22 others, representing 61% of cases in the outbreak.[74] Further details on the transmission of NiV have been reviewed by numerous authors.[70,75–77]

A comparison of the NiV Malaysia and Bangladesh strains in ferrets found there were significantly higher levels of RNA present in oral secretions from ferrets infected with the Bangladesh strain, which reflects the propensity of this strain to be transmitted from human to human during outbreaks.[78]

Serological studies have shown that dogs,[79] horses, and cats[3,80,81] were naturally infected with NiV during the outbreak in Malaysia, and so these animals could act as intermediary hosts for transmission to humans. In Bangladesh, goats have also been implicated in the transmission of NiV to humans.[70]

TABLE 9.3
NiV Outbreaks in Humans

Date	Location	Cases	Deaths	Case Fatality (%)
September 1998–May 1999	Malaysia (Perak, Selangor, and Negeri Sembilan)[53]	283	109	39
March 1999	Singapore[64]	11	1	9
February 2001	India (Siliguri)[64,189]	66	45	68
April–May 2001	Bangladesh (Meherpur)[64]	13	9	69
January 2003	Bangladesh (Naogaon)[64]	12	8	67
January–April 2004	Bangladesh (Goalanda, Rajbari)[67,190,191]	29	22	76
April 2004	Bangladesh (Faridpur)[64]	36	27	75
January–March 2005	Bangladesh (Tangail)[64]	12	11	92
January–February 2007	Bangladesh (Thakurgaon)[64,73]	7	3	43
March–April 2007	Bangladesh (Kushtia)[192]	8	5	63
April 2007	India (Nadia)[68,193]	30	5	17
February 2008	Bangladesh (Manikganj and Rajbari)[66,191,194]	11	9	82
April 2008	Bangladesh (Satkhira and Jessore)[64]	2	1	50
January 2009	Bangladesh (Gaibandha, Rangpur, and Nilphamari)[191]	3	0	0
January 2009	Bangladesh (Rajbari)[191]	1	1	100
December 2009–April 2010	Bangladesh (Faridpur)[191]	16	14	87.5
January–February 2011	Bangladesh (Lalmonirhat)[191]	44	40	91
February 2012	Bangladesh (Joypurhat, Rajshahi, Natore, Rajbari, and Gopalganj)[191,195]	12	10	83
Total		596	320	54

The reservoir of NiV in Bangladesh and India has been identified as *P. giganteus* based on serological evidence for infection in this species,[82,83] while in Cambodia, NiV has also been isolated from *Pteropus lylei*.[84] There is also serological evidence of NiV in *P. hypomelanus*, *P. vampyrus*, *P. lylei*, and *Hipposideros larvatus*[85] and RNA evidence in *P. lylei* and *H. larvatus*[86] in Thailand. There is serological evidence of henipavirus infection in Papua New Guinea in fruit bats *Pteropus conspicillatus*, *P. alecto*, and *Dobsonia magna* and in *P. vampyrus* from Indonesia.[87]

9.3.3 OTHER HENIPAVIRUSES

Recently, a new henipavirus named Cedar virus was isolated from *Pteropus* bat urine in Australia. Initial characterization of Cedar virus has demonstrated that it displays no pathogenicity in laboratory animals.[88] Henipavirus antibodies and PCR-amplified products have been identified in *Eidolon helvum* fruit bats from Ghana[89,90] and from bat bush meat from the Republic of Congo.[91] Serological evidence has been found in Madagascan fruit bats *Pteropus rufus*, *Eidolon dupreanum*, and *Rousettus madagascariensis*[92] and in *Rousettus leschenaulti* fruit bats and *Myotis* spp. microbats in China.[93] Thus, the henipaviruses appear to be widespread in bats from Australia to Asia to Africa.

9.4 CLINICAL FEATURES AND PATHOGENESIS

HeV and NiV are vasotropic and neurotropic viruses; infection can present with a range of symptoms that include respiratory signs and acute or relapsing encephalitis.

The cellular receptors ephrin-B2 and ephrin-B3[26,28,29,94,95] are used by NiV as well to enable the virus to infect a wide range of cells. Employing a hamster model, Guillaume and coworkers[40] demonstrated that the pathogenesis of HeV was similar to NiV.

9.4.1 HENDRA VIRUS

HeV infects endothelial cells, resulting in vasculitis affecting multiple organs, including the kidneys, brain, lungs, and heart, and the formation of multinucleated endothelial syncytia. Vasculitis in turn produces thromboses and micro-infarctions. In addition, HeV directly infects parenchymal cells, including those in the central nervous system.[40]

Infection in horses may be asymptomatic in some cases. Horses with clinical disease have presented with respiratory and neurological signs.[8,32,96,97] Clinical signs include high fevers, increased heart rate and respiratory rate, facial swelling, depression, anorexia, respiratory distress, and frothy nasal discharge and neurological signs such as head pressing and ataxia.[32] Severe pulmonary edema, congestion, and dilated pulmonary lymphatics are commonly seen, with the appearance of characteristic endothelial syncytial cells in capillaries and arterioles observed histologically. Neurological involvement in addition to respiratory disease is less commonly seen; however, in the 2008 HeV outbreak, only neurological signs were observed that included ataxia, head tilt, and facial nerve paralysis.[49,98] Recent studies have shown that HeV is excreted in nasopharyngeal secretions of experimentally infected horses at least 2 days before clinical signs of infection.[99]

There have been seven known human cases of HeV infection with 57% mortality. The incubation period in humans and horses is up to 16 days.[8,50] The first recognized case of human HeV infection occurred in a stable hand. His symptoms included *myalgia, headaches, lethargy, and vertigo* with an absence of respiratory symptoms.[4] The first fatal case of HeV infection in a human occurred in a horse trainer in the same outbreak. He had an acute respiratory infection without apparent neurological involvement; however, recent revision of archival tissues has shown that this case also had acute encephalitis and systemic infection.[100] In this case, nausea, vomiting, fevers, hypoxemia, and renal failure progressed to multiple-organ failure and death. At autopsy, hemorrhage and edema of the lungs and, histologically, necrotizing alveolitis with syncytia and viral inclusions were observed. Syncytia within the blood vessels indicated vascular tropism of the virus. Other findings included a pulmonary embolism, myocarditis, and inflammation with necrosis of the kidney.[4]

The second fatality and third case of HeV occurred in a horse breeder who had assisted his wife, a veterinarian, with the necropsies of two horses. He initially displayed mild encephalitis with headache, drowsiness, and neck stiffness. He succumbed to a relapsing HeV infection 13 months later with an encephalitis-like illness that included seizures and fevers.[101,102] Retrospectively, both horses were found to be infected with HeV by reverse transcription polymerase chain reaction (RT-PCR) and immunofluorescence assay (IFA). A relapsing encephalitis syndrome has also been observed with NiV infections[103] including a relapse recognized 11 years after the initial NiV infection.[104,105]

The fourth case of HeV occurred in October 2004 and involved a veterinarian who performed a necropsy on a terminally ill horse using minimal personal protective equipment (PPE). A week later, she became ill with a sore throat, dry cough, and a fever that lasted 4 days, at which point her symptoms resolved. It was reported that she remained clinically well 2 years post-infection despite a rise in HeV antibody titer a year post-infection.[48]

The next two HeV infections in humans occurred in the Queensland suburb of Thornlands, in the shire of Redlands (near Brisbane) in July 2008. Infection occurred in a veterinarian and veterinary nurse wearing no PPE during the nasal lavage of an infected horse. In this outbreak, horses displayed neurological signs including ataxia, depression, disorientation, and facial nerve paralysis,[98] and HeV was not initially considered as a potential cause for the outbreak. Both human cases initially presented with an influenza-like illness, then showed improvement before the development of neurological symptoms 1–4 days later.[50] The veterinarian exhibited mild confusion, ataxia, and ptosis on day 5 of illness, and HeV RNA was detected in his CSF by RT-PCR. Neurological signs along with high fever progressed to seizures and the requirement for mechanical ventilation. He died on day 40 of his illness and no autopsy was performed. The veterinary nurse had an influenza-like illness with fever, and HeV RNA was detected in serum and nasopharyngeal

aspirate (NPA) on day 3 of her illness. Encephalitic symptoms developed 4 days following abatement of her fever (on day 12 of the illness), and she displayed a worsening of neurological symptoms for the following 12 days before she stabilized.[50]

More recently, a 55-year-old male veterinarian succumbed to HeV infection following exposure at Cawarral, in central Queensland in August 2009. Infection resulted following exposure to nasal fluids during examination of an infected horse; no PPE was worn. His symptoms included fever, headache, mild confusion, and seizures. Hyperintense lesions were visualized throughout the brain by MRI and renal failure accompanied his death 14 days after presentation. At autopsy, generalized vasculitis and endothelialitis were observed in numerous organs including the coronary arteries, lungs, mesentery, and kidneys. Foci of interstitial inflammation surrounding small involved renal arteries were also noted. Additionally, multinucleated syncytial cells were noted in the lymph nodes, lungs, and renal glomeruli (G. Playford and K. Urankar, Queensland Health).

9.4.2 Nipah Virus

The symptoms of NiV infections range from mild to severe respiratory and neurological disease, often resulting in death. Human patients presented with symptoms that included headache, sore throat, cough, vomiting, difficulty breathing, atypical pneumonia, convulsions, and fever sometimes with severe encephalitis.[56,106–108] Brain stem involvement was associated with a poor prognosis.[107]

In Malaysia, the incubation period for NiV infection was usually less than 2 weeks[107] while patients in Bangladesh had a median incubation period of 9 days, with a range of between 6 and 11 days.[106] In Bangladesh, a greater incidence of respiratory symptoms was observed and there was a higher fatality rate compared to the Malaysia outbreak, most likely due to differences in the level of health care available to infected patients.[106]

In Malaysia, patients that survived the initial clinical disease but developed ongoing neurological sequelae associated with NiV infection[107] commonly experienced fatigue and functional impairment.[109] In approximately 10% of cases, patients that had an initial asymptomatic or nonencephalitic infection suffered relapsing encephalitis or late-onset encephalitis.[103,108,110] The longest time to relapse following initial exposure has reported to occur in a 35-year-old woman after 11 years.[104] It is thought that relapsing encephalitis may be due to recrudescence of virus that has replicated at low levels within the brain after initial infection.[103,105]

In cases of fatal NiV infection, disseminated microinfarctions associated with vasculitis were common pathological findings, especially in the nervous system. Parenchymal lesions and vasculitis were observed in the major organs, with the lung being most severely affected. Multinucleated syncytia were found in lymphoid tissue (spleen and lymph node), lungs, and kidneys.[107]

In the Malaysia outbreak, symptoms in pigs included fever, agitation, difficulty breathing, drooling, and a nonproductive barking cough. However, most pigs had a mild illness or were asymptomatically infected.[111] Under experimental conditions, asymptomatic infections were observed in some inoculated and in-contact pigs.[112] In pigs with pneumonia, multinucleated cells with syncytia were observed.

In the hamster model, the clinical outcome of infection with HeV or NiV was found to be dose related with high doses resulting in respiratory disease, while low doses of virus produced neurological disease.[113]

9.5 IDENTIFICATION AND DIAGNOSIS

The diagnosis of HeV and NiV infections can be made by serology, virus isolation, or molecular detection of viral nucleic acids. Procedures for diagnosis of HeV and NiV in animals have been detailed in the *Manual of Diagnostic Tests and Vaccines for Terrestrial Animals*.[114]

Careful consideration should be given to the collection and handling of samples so that the risk of human exposure is minimized.

9.5.1 MOLECULAR ASSAYS

Henipavirus RNA can be extracted from serum, NPA, CSF, urine, or tissues with confirmation of a positive result by reextraction and retesting of the sample. Failure to detect viral RNA does not rule out an infection. Henipavirus RNA can be detected following extraction using conventional RT-PCR[8,12,71,86,115,116] or by the real-time detection system such as TaqMan-based assays. Real-time RT-PCR is a rapid and highly sensitive method for the detection of HeV and NiV and is therefore the assay of choice. The TaqMan-based assay developed by Smith and coworkers[117] has been found to be a reliable assay for the detection of HeV. There have been numerous real-time assays developed for the detection of NiV.[115,118–121] Henipavirus assays that detect both Hendra and Nipah viruses have also been developed[121] and provide added tools for the diagnosis of Henipaviruses. Sequencing can be performed to determine variability and relationships between isolates of HeV and NiV.[14,86,116] Genotyping of HeV to differentiate strains has focused on the hypervariable carboxy terminal of the N protein (HeV genome positions 1500–2400).[14] Genotyping of NiV has been performed utilizing partial-sequence data from a 357 nucleotide region coding for the carboxy terminus of N (NiV genome positions 1197–1553)[86] and also a 729 nucleotide region in the amino terminal region of the N gene ORF (N ORF nt 123–852, NiV genome positions 236–964).[116]

9.5.2 ISOLATION

The isolation of HeV and NiV requires the highest level of containment, BSL 4. The isolation of virus can be attempted from serum, urine, NPA, kidney, spleen, lung, and brain in African green monkey kidney cells (Vero E6 [ATCC C1008] or Vero [ATCC CCL81]). HeV has been isolated from humans, from kidney,[4] and from an NPA.[50] In horses,

isolation has been successful from nasal swabs, throat swabs, blood, urine, lung, liver, spleen, kidney, and lymph nodes.[122] NiV has been isolated from humans from CSF, urine, saliva, nasal and pharyngeal secretions.[77,123] Following inoculation onto Vero cells, henipaviruses produce large syncytia containing multiple nuclei in the monolayer, and the presence of virus can be confirmed by IFA, TaqMan RT-PCR, and/or electron microscopy.

9.5.3 SEROLOGY

A recent henipavirus infection is confirmed by the testing of acute and convalescent serum samples from exposed patients or horses in parallel and demonstrating a fourfold rise in antibody titer. Antihenipavirus antibodies can be detected by IFA, enzyme-linked immunosorbent assays (ELISAs), neutralization assays, or the recently developed microsphere immunoassays.[50,124]

Both ELISA and IFA assays use inactivated antigen and therefore can be performed at lower containment levels. The IFA is more suited to testing of small numbers of samples and is less laborious than the ELISA. The ELISA is well suited to screening large numbers of samples; however, some nonspecific reactions have been observed giving false positives.[122] Further testing employing neutralization assays is required for confirmation of positive ELISAs. ELISAs using antigen expressed in *Escherichia coli*, baculovirus, and yeast have been developed.[122,125,126] An immune plaque assay that is more sensitive than the ELISA and that uses methanol fixed HeV-infected monolayers and 5-bromo-4-chloro-3-indolyl phosphate and *p*-nitro blue tetrazolium substrate has also been used to detect antibodies.[127]

More recently, a microsphere immunoassay (Luminex) has been successfully adapted from Bossart et al.[124] using biotinylated anti-human IgM and IgG for the detection of human antibodies to HeV.[50] A surrogate neutralization assay has also been developed whereby the biotinylated henipavirus receptor ephrin-B2 is used in a blocking assay to measure the presence of neutralizing antibodies. In this assay, henipavirus antibodies present in the sample attach to a soluble form of the viral glycoprotein (sG) coated to the beads leading to a reduction or total blocking of the signal.[124] The requirement for physical containment 4 (PC4) is eliminated as recombinant antigen is used in these assays.

Neutralization assays are performed under PC4 conditions for serological confirmation of infection and can also be used for differentiating infections with related viruses such as HeV and NiV. Other neutralization protocols have been published.[114,122]

In order to alleviate the need for performing neutralization assay in high containment laboratories, researchers have developed surrogate neutralization assays utilizing vesicular stomatitis virus (VSV) or lentivirus pseudotyped particles where the F and G proteins of NiV and/or HeV express either green fluorescent protein (GFP) or luciferase.[128–130] More recently, Kaku and coworkers[131] replaced the GFP with secreted alkaline phosphatase allowing the assay to be read using an ELISA plate reader. These assays offer the ability to perform a neutralization assay with comparable sensitivity at much lower containment levels and are thus safer to perform.[128,131]

9.5.4 IMMUNOHISTOCHEMISTRY

Immunohistochemistry has been a useful technique for the detection of viral antigen in formalin-fixed tissues and has been used in the retrospective diagnosis of HeV.[97] Anti-HeV and NiV antibodies are used with the biotin–streptavidin peroxidase-linked staining system or the anti-rabbit/mouse dextran polymer conjugated with alkaline phosphatase to determine the areas of replication of virus in tissues.[114,122]

9.5.5 ELECTRON MICROSCOPY

Negative contrast electron microscopy has provided a valuable technique for the discovery of both HeV and NiV.[8,132] Immunoelectron microscopy has also been employed to show the location of the virus within tissues and cells.[10,133,134]

9.6 TREATMENT AND PREVENTION

Currently, no vaccines or antiviral treatments are licensed for henipavirus infections in humans or animals. However, recently, there have been some very promising developments in the area of treatment and prevention of henipavirus infection.[30,135]

9.6.1 THERAPEUTICS

The most promising candidate for treatment of henipavirus infection and as a postexposure prophylaxis is a recombinant human neutralizing monoclonal antibody (mAb) that was developed following selection by panning a human antibody phage display library using the sG of HeV. This mAb, m102, maps to the ephrin receptor binding site and displays strong neutralization activity against HeV and NiV.[136–138] In animal trials, administration of m102 protected ferrets and African green monkeys against lethal disease following HeV and NiV challenge.[43,139] This mAb has been administered to humans following high risk exposures to HeV.[140,141]

Ribavirin is a ribonucleoside that was used to treat humans in the Malaysian NiV outbreak, where it was found to reduce morbidity and mortality.[142] However, HeV patients treated with ribavirin did not display any benefits from this drug.[50] Experimentally, ribavirin delayed the onset of clinical disease and death in hamsters infected with NiV and delayed death by 1–2 days in African green monkeys exposed to HeV[143] but did not affect the overall mortality associated with infection in either case. In light of these animal studies and the experience of the treatment of HeV-infected patients, the efficacy of ribavirin is questionable.

Chloroquine, an antimalarial drug, was found to inhibit infection with HeV and NiV in vitro.[144] However, when used either in combination with ribavirin or alone, chloroquine did not display a therapeutic effect against HeV and NiV infection in hamsters or ferrets.[145–147]

Poly(I)–poly(C_{12}U), a strong inducer of interferon, has shown promise for the treatment of henipavirus infections, with the prevention of mortality in the majority of animals tested.[146] Another possible candidate for development as an antiviral therapeutic is RNA interference (RNAi) that has been found to inhibit the replication of henipaviruses in vitro.[148]

Heptad peptides that correspond to the C-terminal heptad repeat of the F protein and block the formation of the six-helix bundle structures have been shown to block HeV and NiV infection in vitro.[149] With the addition of cholesterol to the peptides, they were able to cross the blood-brain barrier to act on NiV in vivo in golden hamsters[150] and are showing potential as preventive and therapeutic agents. However, further testing in vivo is required before they can be utilized.

Combination therapies may prove to be a successful strategy in the management of henipavirus infections.

9.6.2 PREVENTION

Understanding the cycle of transmission of henipaviruses to humans and their domesticated animals and the ecology of the virus in the host has enabled better education of populations at risk from exposure, so that control measures can be implemented to significantly reduce the likelihood of infection. In the case of HeV, preventive measures include the wearing of PPE such as gloves, safety glasses, and face shields when attending to potentially infected horses. In the case of NiV in Bangladesh and India, using traditional covers on date palm pots to prevent bats from accessing the date palm sap during the collection process will reduce the risk of bat to human transmission.[151–153] Improved barrier nursing such as wearing of face masks and gloves will reduce the risk of human-to-human transmission.

Vaccination of the intermediary host, such as horses in the case of HeV or pigs in the case of NiV, would interrupt the cycle of transmission from domesticated animals to humans. A recombinant sG of HeV has been developed as a potential vaccine against HeV and NiV. Vaccination with the sG of HeV and NiV elicited high levels of antibodies that were protective against challenge[118,154] with NiV in cats and in ferrets prevented a productive infection clinical disease following exposure to HeV.[155] Recently, the recombinant HeV sG subunit vaccine was found to protect African green monkeys against challenge with NiV[156] and therefore could be developed as a human vaccine to protect against HeV and NiV. The sG is currently in trials as a vaccine against HeV in horses and could also be used in pigs to protect against NiV.

Virus-like particles composed of the M, F, and G proteins of NiV have been found to be immunogenic in mice and offer a potential vaccine source.[157] Replication-defective VSV particles expressing the F and G proteins of NiV have also been found to induce neutralizing antibodies in mice.[158]

A recombinant canarypox expressing the F and G protein of NiV has been shown to protect pigs against challenge and prevent shedding of virus in immunized pigs.[159] The canary-pox-based HeV G protein vaccine is also under development. Therefore, the vaccination of pigs would be possible should NiV reemerge in pigs.

9.7 CONCLUSION AND FUTURE PERSPECTIVES

Henipavirus infections in humans, horses, and pigs display a wide spectrum of disease ranging from asymptomatic to respiratory and encephalitic manifestations. Therefore, awareness will remain the key and be the greatest challenge to detecting future outbreaks and minimizing harm. As henipaviruses occur naturally in flying fox colonies, spillovers into animals and humans will be likely in the future. More research on the ecology of the virus in flying foxes will provide a better understanding that will allow for improved management of spillovers and hence decrease the risk of outbreaks.

There is currently a wide range of diagnostic assays to detect HeV and NiV infections. The implementation of recombinant antigens may help reduce the numbers of false positives in ELISA. Microsphere immunoassays show promise in the serological diagnosis of HeV infections in humans and will provide improved serological diagnostic reagents in the future. Surrogate neutralization assays will allow for the confirmation of infection without the need for high containment, thus reducing risk to laboratory workers and also allowing neutralization assays to be performed at lower biosafety levels making the assays more widely available.

With the release of a vaccine against HeV to be used in horses, the risk of infection from horses should be eliminated. However, there is an ongoing need for the further development of therapeutics for the treatment of infections resulting from henipaviruses as NiV continues to reemerge. In addition, new henipaviruses may emerge in the future to cause zoonotic disease outbreaks that will require further develop of therapeutics.

ACKNOWLEDGMENTS

Thanks to Dr. Bronwyn Clayton and Reuben Klein for critical reading of the manuscript.

REFERENCES

1. Halpin, K. et al. Pteropid bats are confirmed as the reservoir hosts of henipaviruses: A comprehensive experimental study of virus transmission. *Am J Trop Med Hyg* 85, 946–951 (2011).
2. Eaton, B.T., Mackenzie, J.S., and Wang, L.-F. Henipaviruses. In Knipe, D.M. and Howley, P.M. (eds.), *Fields Virology*, Lippincott Williams & Wilkins, Philadelphia, PA, pp. 1587–1600 (2007).
3. Chua, K.B. et al. Nipah virus: A recently emergent deadly paramyxovirus. *Science* 288, 1432–1435 (2000).
4. Selvey, L.A. et al. Infection of humans and horses by a newly described morbillivirus. *Med J Aust* 162, 642–645 (1995).
5. Rahman, M.M.A. et al. Date palm sap linked to Nipah virus outbreak in Bangladesh, 2008. *Vector Borne Zoonotic Dis* 12, 65–72 (2012).
6. Luby, S.P. et al. Foodborne transmission of Nipah virus, Bangladesh. *Emerg Infect Dis* 12, 14–16 (2006).
7. National Institutes of Health. NIAID Biodefense Research Agenda for Category B and C priority pathogens. U.S. Department of Health and Human Services, Bethesda, MD, NIH Publication No. 03-5315, pp. 1–66 (2003).
8. Murray, K. et al. A morbillivirus that caused fatal disease in horses and humans. *Science* 268, 94–97 (1995).
9. Hyatt, A.D. and Selleck, P.W. Ultrastructure of equine morbillivirus. *Virus Res* 43, 1–15 (1996).
10. Hyatt, A.D. et al. Ultrastructure of Hendra virus and Nipah virus within cultured cells and host animals. *Microbes Infect* 3, 297–306 (2001).
11. Wang, L.F. et al. The exceptionally large genome of Hendra virus: Support for creation of a new genus within the family Paramyxoviridae. *J Virol* 74, 9972–9979 (2000).
12. Gould, A.R. Comparison of the deduced matrix and fusion protein sequences of equine morbillivirus with cognate genes of the Paramyxoviridae. *Virus Res* 43, 17–31 (1996).
13. Yu, M. et al. Sequence analysis of the Hendra virus nucleoprotein gene: Comparison with other members of the subfamily Paramyxovirinae. *J Gen Virol* 79, 1775–1780 (1998).
14. Smith, I. et al. Identifying Hendra virus diversity in pteropid bats. *PLoS One* 6, e25275 (2011).
15. Sleeman, K. et al. The C, V and W proteins of Nipah virus inhibit minigenome replication. *J Gen Virol* 89, 1300–1308 (2008).
16. Wang, L.F. et al. A novel P/V/C gene in a new member of the Paramyxoviridae family, which causes lethal infection in humans, horses, and other animals. *J Virol* 72, 1482–1490 (1998).
17. Harcourt, B.H. et al. Molecular characterization of Nipah virus, a newly emergent paramyxovirus. *Virology* 271, 334–349 (2000).
18. Shaw, M.L. Henipaviruses employ a multifaceted approach to evade the antiviral interferon response. *Viruses* 1, 1190–1203 (2009).
19. Basler, C.F. Nipah and Hendra virus interactions with the innate immune system. *Curr Top Microbiol Immunol* 359, 123–152 (2012). doi:10.1007/82_2012_209.
20. Wang, Y.E. et al. Ubiquitin-regulated nuclear-cytoplasmic trafficking of the Nipah virus matrix protein is important for viral budding. *PLoS Pathog* 6, e1001186 (2010).
21. Patch, J.R. et al. The YPLGVG sequence of the Nipah virus matrix protein is required for budding. *Virol J* 5, 137 (2008).
22. Meulendyke, K.A. et al. Endocytosis plays a critical role in proteolytic processing of the Hendra virus fusion protein. *J Virol* 79, 12643–12649 (2005).
23. Pager, C.T. and Dutch, R.E. Cathepsin L is involved in proteolytic processing of the Hendra virus fusion protein. *J Virol* 79, 12714–12720 (2005).
24. Michalski, W.P. et al. The cleavage activation and sites of glycosylation in the fusion protein of Hendra virus. *Virus Res* 69, 83–93 (2000).
25. Bossart, K.N. et al. Functional expression and membrane fusion tropism of the envelope glycoproteins of Hendra virus. *Virology* 290, 121–135 (2001).
26. Bonaparte, M.I. et al. Ephrin-B2 ligand is a functional receptor for Hendra virus and Nipah virus. *Proc Natl Acad Sci USA* 102, 10652–10657 (2005).
27. Yu, M. et al. The attachment protein of Hendra virus has high structural similarity but limited primary sequence homology compared with viruses in the genus *Paramyxovirus*. *Virology* 251, 227–233 (1998).

28. Bossart, K.N. et al. Functional studies of host-specific ephrin-B ligands as henipavirus receptors. *Virology* 372, 357–371 (2008).

29. Negrete, O.A. et al. Single amino acid changes in the Nipah and Hendra virus attachment glycoproteins distinguish ephrin B2 from ephrin B3 usage. *J Virol* 81, 10804–10814 (2007).

30. Steffen, D.L. et al. Henipavirus mediated membrane fusion, virus entry and targeted therapeutics. *Viruses* 4, 280–308 (2012).

31. Aljofan, M. et al. Characteristics of Nipah virus and Hendra virus replication in different cell lines and their suitability for antiviral screening. *Virus Res* 142, 92–99 (2009).

32. Murray, K. et al. A novel morbillivirus pneumonia of horses and its transmission to humans. *Emerg Infect Dis* 1, 31–33 (1995).

33. Murray, K. et al. Flying foxes, horses, and humans: A zoonosis caused by a new member of the paramyxoviridae. *Emerg Infect* 1, 43–58 (1998).

34. Young, P.L. et al. Serologic evidence for the presence in *Pteropus* bats of a paramyxovirus related to equine morbillivirus. *Emerg Infect Dis* 2, 239–240 (1996).

35. Field, H. et al. The natural history of Hendra and Nipah viruses. *Microbes Infect* 3, 307–314 (2001).

36. Halpin, K. et al. Isolation of Hendra virus from pteropid bats: A natural reservoir of Hendra virus. *J Gen Virol* 81, 1927–1932 (2000).

37. Williamson, M.M. et al. Experimental Hendra virus infection in pregnant guinea-pigs and fruit Bats (*Pteropus poliocephalus*). *J Comp Pathol* 122, 201–207 (2000).

38. Fogarty, R. et al. Henipavirus susceptibility to environmental variables. *Virus Res* 132, 140–144 (2008).

39. Williamson, M.M. et al. Transmission studies of Hendra virus (equine morbillivirus) in fruit bats, horses and cats. *Aust Vet J* 76, 813–818 (1998).

40. Guillaume, V. et al. Acute Hendra virus infection: Analysis of the pathogenesis and passive antibody protection in the hamster model. *Virology* 387, 459–465 (2009).

41. Hooper, P.T., Westbury, H.A., and Russell, G.M. The lesions of experimental equine morbillivirus disease in cats and guinea pigs. *Vet Pathol* 34, 323–329 (1997).

42. Westbury, H.A. et al. Equine morbillivirus pneumonia: Susceptibility of laboratory animals to the virus. *Aust Vet J* 72, 278–279 (1995).

43. Bossart, K.N. et al. A neutralizing human monoclonal antibody protects African green monkeys from Hendra virus challenge. *Sci Transl Med* 3, 105ra103 (2011).

44. Li, M., Embury-Hyatt, C., and Weingartl, H.M. Experimental inoculation study indicates swine as a potential host for Hendra virus. *Vet Res* 41, 33 (2010).

45. Dups, J. et al. A new model for hendra virus encephalitis in the mouse. *PLoS One* 7, e40308 (2012).

46. Aljofan, M. et al. Development and validation of a chemiluminescent immunodetection assay amenable to high throughput screening of antiviral drugs for Nipah and Hendra virus. *J Virol Methods* 149, 12–19 (2009).

47. Crameri, G. et al. Establishment, immortalisation and characterisation of pteropid bat cell lines. *PLoS One* 4, e8266 (2009).

48. Hanna, J.N. et al. Hendra virus infection in a veterinarian. *Med J Aust* 185, 562–564 (2006).

49. Rogers, R.J. et al. Investigation of a second focus of equine morbillivirus infection in coastal Queensland. *Aust Vet J* 74, 243–244 (1996).

50. Playford, E.G. et al. Human Hendra virus encephalitis associated with equine outbreak, Australia, 2008. *Emerg Infect Dis* 16, 219–223 (2010).

51. Chua, K.B. et al. Fatal encephalitis due to Nipah virus among pig-farmers in Malaysia. *Lancet* 354, 1257–1259 (1999).

52. CDC Update. Outbreak of Hendra-like virus—Malaysia and Singapore, 1998–1999. *Morb Mortal Wkly Rep* 48, 1998–1999 (1999).

53. Chua, K.B. Nipah virus outbreak in Malaysia. *J Clin Virol* 26, 265–275 (2003).

54. Pulliam, J.R.C. et al. Agricultural intensification, priming for persistence and the emergence of Nipah virus: A lethal bat-borne zoonosis. *J R Soc Interface* 9, 89–101 (2012).

55. Wong, K.T. et al. Nipah virus infection, an emerging paramyxoviral zoonosis. *Springer Semin Immunopathol* 24, 215–228 (2002).

56. Paton, N.I. et al. Outbreak of Nipah-virus infection among abattoir workers in Singapore. *Lancet* 354, 1253–1256 (1999).

57. Yob, J.M. et al. Nipah virus infection in bats (order Chiroptera) in peninsular Malaysia. *Emerg Infect Dis* 7, 439–441 (2001).

58. Chua, K.B. A novel approach for collecting samples from fruit bats for isolation of infectious agents. *Microbes Infect* 5, 487–490 (2003).

59. Yadav, P.D. et al. Short report: Detection of Nipah virus RNA in fruit bat (*Pteropus giganteus*) from India. *Am J Trop Med Hyg* 87, 576–578 (2012).

60. Rahman, S.A. et al. Characterization of Nipah virus from naturally infected *Pteropus vampyrus* bats, Malaysia. *Emerg Infect Dis* 16, 1990–1993 (2010).

61. Harcourt, B.H. et al. Genetic characterization of Nipah virus, Bangladesh, 2004. *Emerg Infect Dis* 11, 1594–1597 (2005).

62. Luby, S.P. et al. Recurrent zoonotic transmission of Nipah virus into humans, Bangladesh, 2001–2007. *Emerg Infect Dis* 15, 1229–1235 (2009).

63. Harit, A.K. et al. Nipah/Hendra virus outbreak in Siliguri, West Bengal, India in 2001. *Indian J Med Res* 123, 553–560 (2006).

64. World Health Organization. Regional office for South East Asia Nipah virus infection. WHO, Geneva, Switzerland, pp. 1–9 (2009). http://www.searo.who.int/LinkFiles/CDS_Nipah_Virus.pdf (accessed June 4, 2012).

65. ICDDRB. Nipah outbreak in Faridpur District, Bangladesh, 2010. *ICDDR, B Health Sci Bull* 8, 6–11 (2010).

66. ICDDRB. Outbreaks of Nipah virus in Rajbari and Manikgonj, February 2008. *ICDDR, B Health Sci Bull* 6, 12–13 (2008).

67. ICDDRB. Nipah encephalitis outbreak over wide area of western Bangladesh, 2004. *ICDDR, B Health Sci Bull* 2, 7–11 (2004).

68. Arankalle, V. et al. Genomic characterization of Nipah virus, West Bengal, India. *Emerg Infect Dis* 17, 907–909 (2011).

69. Khan, M.S.U. et al. Use of infrared camera to understand bats' access to date palm sap: Implications for preventing Nipah virus transmission. *Ecohealth* 7, 517–525 (2010).

70. Luby, S.P., Gurley, E.S., and Hossain, M.J. Transmission of human infection with Nipah virus. *Clin Infect Dis* 49, 1743–1748 (2009).

71. Gurley, E.S. et al. Risk of nosocomial transmission of Nipah virus in a Bangladesh hospital. *Infect Control Hosp Epidemiol* 28, 740–742 (2007).

72. Blum, L.S. et al. In-depth assessment of an outbreak of Nipah encephalitis with person-to-person transmission in Bangladesh: Implications for prevention and control strategies. *Am J Trop Med Hyg* 80, 96–102 (2009).

73. Homaira, N. et al. Nipah virus outbreak with person-to-person transmission in a district of Bangladesh, 2007. *Epidemiol Infect* 138, 1630–1636 (2010).

74. Gurley, E.S. et al. Person-to-person transmission of Nipah virus in a Bangladeshi community. *Emerg Infect Dis* 13, 1031–1037 (2007).

75. Luby, S.P. and Gurley, E.S. Epidemiology of henipavirus disease in humans. *Curr Top Microbiol Immunol* 359, 25–40 (2012). doi:10.1007/82_2012_207.

76. Clayton, B.A., Wang, L.F., and Marsh, G.A. Henipaviruses: An updated review focusing on the Pteropid reservoir and features of transmission. *Zoonoses Public Health* 60, 69–83 (2013). doi:10.1111/j.1863-2378.2012.01501.x.

77. Chua, K.B. Epidemiology, surveillance and control of Nipah virus infections in Malaysia. *Malays J Pathol* 32, 69–73 (2010).

78. Clayton, B. et al. Transmission routes for nipah virus from Malaysia and Bangladesh. *Emerg Infect* 18, 1983–1993 (2012).

79. Mills, J.N. et al. Nipah virus infection in dogs, Malaysia, 1999. *Emerg Infect Dis* 15, 950–952 (2009).

80. Hooper, P.T. and Williamson, M.M. Hendra and Nipah virus infections. *Emerg Infect Dis* 16, 597–603 (2000).

81. Hooper, P., Zaki, S., Daniels, P., and Middleton, D. Comparative pathology of the diseases caused by Hendra and Nipah viruses. *Microbes Infect* 3, 315–322 (2001).

82. Epstein, J.H. et al. Henipavirus infection in fruit bats (*Pteropus giganteus*), India. *Emerg Infect Dis* 14, 1309–1311 (2008).

83. Hsu, V.P. et al. Nipah virus encephalitis reemergence, Bangladesh. *Emerg Infect Dis* 10, 2082–2087 (2004).

84. Reynes, J.M. et al. Nipah virus in Lyle's flying foxes, Cambodia. *Emerg Infect Dis* 11, 1042–1047 (2005).

85. Wacharapluesadee, S. et al. Bat Nipah virus, Thailand. *Emerg Infect Dis* 11, 1949–1951 (2005).

86. Wacharapluesadee, S. and Hemachudha, T. Duplex nested RT-PCR for detection of Nipah virus RNA from urine specimens of bats. *J Virol Methods* 141, 97–101 (2007).

87. Sendow, I. et al. Screening for Nipah virus infection in West Kalimantan province, Indonesia. *Zoonoses Public Health* 57, 499–503 (2010).

88. Marsh, G.A. et al. Cedar virus: A novel henipavirus isolated from Australian bats. *PLoS Pathogens* 8, e1002836 (2012).

89. Hayman, D.T. et al. Evidence of henipavirus infection in West African fruit bats. *PLoS One* 3, e2739 (2008).

90. Drexler, J.F. et al. Henipavirus RNA in African bats. *PLoS One* 4, e6367 (2009).

91. Weiss, S. et al. Henipavirus-related sequences in fruit bat bushmeat, Republic of Congo. *Emerg Infect Dis* 18, 1536–1537 (2012).

92. Iehle, C. et al. Henipavirus and Tioman virus antibodies in pteropodid bats, Madagascar. *Emerg Infect Dis* 13, 159–161 (2007).

93. Li, Y. et al. Antibodies to Nipah or Nipah-like viruses in bats, China. *Emerg Infect Dis* 14, 1974–1976 (2008).

94. Negrete, O.A. et al. Ephrin B2 is the entry receptor for Nipah virus, an emergent deadly paramyxovirus. *Nature* 436, 401–405 (2005).

95. Negrete, O.A. et al. Two key residues in ephrin B3 are critical for its use as an alternative receptor for Nipah virus. *PLoS Pathog* 2, e7 (2006).

96. Field, H.E. et al. A fatal case of Hendra virus infection in a horse in north Queensland: Clinical and epidemiological features. *Aust Vet J* 78, 279–280 (2000).

97. Hooper, P.T. et al. The retrospective diagnosis of a second outbreak of equine morbillivirus infection. *Aust Vet J* 74, 244–245 (1996).

98. Field, H.E. et al. Hendra virus outbreak with novel clinical features, Australia. *Emerg Infect Dis* 16, 338–340 (2010).

99. Marsh, G.A. et al. Experimental infection of horses with Hendra virus/Australia/horse/2008/Redlands. *Emerg Infect Dis* 17, 2232–2238 (2011).

100. Wong, K.T. et al. Human Hendra virus infection causes acute and relapsing encephalitis. *Neuropathol Appl Neurobiol* 35, 296–305 (2009).

101. O'Sullivan, J.D. et al. Fatal encephalitis due to novel paramyxovirus transmitted from horses. *Lancet* 345, 93–95 (1997).

102. Allworth, A., O'Sullivan, J., and Selvey, L.A. Equine morbillivirus in Queensland. *Commun Dis Intell* 19, 575 (1995).

103. Tan, C.T. et al. Relapsed and late-onset Nipah encephalitis. *Ann Neurol* 51, 703–708 (2002).

104. Abdullah, S. et al. Late-onset Nipah virus encephalitis 11 years after the initial outbreak: A case report. *Neurol Asia* 17, 71–74 (2012).

105. Wong, K.T. and Tan, C.T. Clinical and pathological manifestations of human henipavirus infection. *Curr Top Microbiol Immunol* 359, 95–104 (2012). doi:10.1007/82_2012_205.

106. Hossain, M.J. et al. Clinical presentation of Nipah virus infection in Bangladesh. *Clin Infect Dis* 46, 977–984 (2008).

107. Goh, K.J. et al. Clinical features of Nipah virus encephalitis among pig farmers in Malaysia. *New Engl J Med* 342, 1229–1235 (2000).

108. Chong, H.T. et al. Nipah encephalitis outbreak in Malaysia, clinical features in patients from Seremban. *Can J Neurol Sci* 29, 83–87 (2002).

109. Sejvar, J.J. et al. Long-term neurological and functional outcome in Nipah virus infection. *Ann Neurol* 62, 235–242 (2007).

110. Tan, C.T. and Chua, K.B. Nipah virus encephalitis. *Curr Infect Dis Rep* 10, 315–320 (2008).

111. Mohd, N., Gan, C.H., and Ong, B.L. Nipah virus infection of pigs in peninsular Malaysia. *Rev Sci Tech* 19, 160–165 (2000).

112. Middleton, D.J. et al. Experimental Nipah virus infection in pigs and cats. *J Comp Pathol* 126, 124–136 (2002).

113. Rockx, B. et al. Clinical outcome of henipavirus infection in hamsters is determined by the route and dose of infection. *J Virol* 85, 7658–7671 (2011).

114. OIE. Hendra and Nipah virus diseases. *Manual of Diagnostic Tests and Vaccines for Terrestrial Animals*, Chapter 2.9.6, OIE, Paris, France, (2013). http://www.oie.int/fileadmin/Home/eng/Health_standards/tahm/2.09.06_HENDRA_&_NIPAH_FINAL.pdf (accessed November 8, 2013).

115. Chang, L.Y. et al. Quantitative estimation of Nipah virus replication kinetics in vitro. *Virol J* 3, 47 (2006).

116. Lo, M.K. et al. Characterization of Nipah virus from outbreaks in Bangladesh, 2008–2010. *Emerg Infect Dis* 18, 248–255 (2012).

117. Smith, I.L. et al. Development of a fluorogenic RT-PCR assay (TaqMan) for the detection of Hendra virus. *J Virol Methods* 98, 33–40 (2001).

118. Mungall, B.A. et al. Feline model of acute Nipah virus infection and protection with a soluble glycoprotein-based subunit vaccine. *J Virol* 80, 12293–12302 (2006).

119. Guillaume, V. et al. Nipah virus: Vaccination and passive protection studies in a hamster model. *J Virol* 78, 834–840 (2004).

120. Chen, J.M. et al. A stable and differentiable RNA positive control for reverse transcription-polymerase chain reaction. *Biotechnol Lett* 28, 1787–1792 (2006).

121. Feldman, K.S. et al. Design and evaluation of consensus PCR assays for henipaviruses. *J Virol Methods* 161, 52–57 (2009).

122. Daniels, P., Ksiazek, T.G., and Eaton, B.T. Laboratory diagnosis of Nipah and Hendra virus infections. *Microbes Infect* 3, 289–295 (2001).

123. Chua, K.B. et al. The presence of Nipah virus in respiratory secretions and urine of patients during an outbreak of Nipah virus encephalitis in Malaysia. *J Infect* 42, 40–43 (2001).

124. Bossart, K.N. et al. Neutralization assays for differential henipavirus serology using Bio-Plex protein array systems. *J Virol Methods* 142, 29–40 (2007).

125. Chen, J.-M.M. et al. Expression of truncated phosphoproteins of Nipah virus and Hendra virus in *Escherichia coli* for the differentiation of henipavirus infections. *Biotechnol Lett* 29, 871–875 (2007).

126. Juozapaitis, M. et al. Generation of henipavirus nucleocapsid proteins in yeast *Saccharomyces cerevisiae*. *Virus Res* 124, 95–102 (2007).

127. Crameri, G. et al. A rapid immune plaque assay for the detection of Hendra and Nipah viruses and anti-virus antibodies. *J Virol Methods* 99, 41–51 (2002).

128. Tamin, A. et al. Development of a neutralization assay for Nipah virus using pseudotype particles. *J Virol Methods* 160, 1–6 (2009).

129. Kaku, Y. et al. A neutralization test for specific detection of Nipah virus antibodies using pseudotyped vesicular stomatitis virus expressing green fluorescent protein. *J Virol Methods* 160, 7–13 (2009).

130. Khetawat, D. and Broder, C.C. A functional henipavirus envelope glycoprotein pseudotyped lentivirus assay system. *Virol J* 7, 312 (2010).

131. Kaku, Y. et al. Second generation of pseudotype-based serum neutralization assay for Nipah virus antibodies: Sensitive and high-throughput analysis utilizing secreted alkaline phosphatase. *J Virol Methods* 179, 226–232 (2012).

132. Chua, K.B. Introduction: Nipah virus-discovery and origin. *Curr Top Microbiol Immunol* 359, 1–9 (2012). doi:10.1007/82_2012_218.

133. Chow, V.T. et al. Diagnosis of Nipah virus encephalitis by electron microscopy of cerebrospinal fluid. *J Clin Virol* 19, 143–147 (2000).

134. Chua, K.B. et al. Role of electron microscopy in Nipah virus outbreak investigation and control. *Med J Malaysia* 62, 139–142 (2007).

135. Broder, C.C. Henipavirus outbreaks to antivirals: The current status of potential therapeutics. *Curr Opin Virol* 2, 176–187 (2012).

136. Zhu, Z. et al. Exceptionally potent cross-reactive neutralization of Nipah and Hendra viruses by a human monoclonal antibody. *J Infect Dis* 197, 846–853 (2008).

137. Zhu, Z. et al. Development of human monoclonal antibodies against diseases caused by emerging and biodefense-related viruses. *Expert Rev Anti Infect Ther* 4, 57–66 (2006).

138. Zhu, Z. et al. Potent neutralization of Hendra and Nipah viruses by human monoclonal antibodies. *J Virol* 80, 891–899 (2006).

139. Bossart, K.N. et al. A neutralizing human monoclonal antibody protects against lethal disease in a new ferret model of acute Nipah virus infection. *PLoS Pathog* 5, e1000642 (2009).

140. Miles, J. and Williams, B. Horse dies from Hendra virus infection, Cairns property now quarantined. *Courier Mail*, July 27 (2012).

141. Promed. Hendra virus, equine-Australia (05):(QL) human exposure Archive Number: 20100527.1761 (2010).

142. Chong, H.T. et al. Treatment of acute Nipah encephalitis with ribavirin. *Ann Neurol* 49, 810–813 (2001).

143. Rockx, B. et al. A novel model of lethal Hendra virus infection in African green monkeys and the effectiveness of ribavirin treatment. *J Virol* 84, 9831–9839 (2010).

144. Porotto, M. et al. Simulating henipavirus multicycle replication in a screening assay leads to identification of a promising candidate for therapy. *J Virol* 83, 5148–5155 (2009).

145. Pallister, J. et al. Chloroquine administration does not prevent Nipah virus infection and disease in ferrets. *J Virol* 83, 11979–11982 (2009).

146. Georges-Courbot, M.C. et al. Poly(I)-poly(C12U) but not ribavirin prevents death in a hamster model of Nipah virus infection. *Antimicrob Agents Chemother* 50, 1768–1772 (2006).

147. Freiberg, A.N. et al. Combined chloroquine and ribavirin treatment does not prevent death in a hamster model of Nipah and Hendra virus infection. *J Gen Virol* 91, 765–772 (2010).

148. Mungall, B.A. et al. Inhibition of henipavirus infection by RNA interference. *Antiviral Res* 80, 324–331 (2008).

149. Bossart, K.N. et al. Inhibition of henipavirus fusion and infection by heptad-derived peptides of the Nipah virus fusion glycoprotein. *Virol J* 2, 57 (2005).

150. Porotto, M. et al. Inhibition of Nipah virus infection in vivo: Targeting an early stage of paramyxovirus fusion activation during viral entry. *PLoS Pathog* 6, e1001168 (2010).

151. Nahar, N. et al. Date palm sap collection: Exploring opportunities to prevent Nipah transmission. *Ecohealth* 7, 196–203 (2010).

152. Nahar, N. et al. Piloting the use of indigenous methods to prevent Nipah virus infection by interrupting bats' access to date palm sap in Bangladesh. *Health Promot Int* 28, 378–386 (2013). doi:10.1093/heapro/das020.

153. Khan, S.U. et al. A randomized controlled trial of interventions to impede date palm sap contamination by bats to prevent Nipah virus transmission in Bangladesh. *PLoS One* 7, e42689 (2012).

154. McEachern, J.A. et al. A recombinant subunit vaccine formulation protects against lethal Nipah virus challenge in cats. *Vaccine* 26, 3842–3852 (2008).

155. Pallister, J. et al. A recombinant Hendra virus G glycoprotein-based subunit vaccine protects ferrets from lethal Hendra virus challenge. *Vaccine* 29, 5623–5630 (2011).

156. Bossart, K.N. et al. A Hendra virus G glycoprotein subunit vaccine protects African green monkeys from Nipah virus challenge. *Sci Transl Med* 4, 146ra107 (2012).

157. Walpita, P. et al. Vaccine potential of Nipah virus-like particles. *PLoS One* 6, e18437 (2011).

158. Chattopadhyay, A. and Rose, J.K. Complementing defective viruses that express separate paramyxovirus glycoproteins provide a new vaccine vector approach. *J Virol* 85, 2004–2011 (2011).

159. Weingartl, H.M. et al. Recombinant Nipah virus vaccines protect pigs against challenge. *J Virol* 80, 7929–7938 (2006).

160. Murray, G. Miscellaneous: Hendra virus findings in Queensland, Australia. World Organisation for Animal Health (OIE), Paris, France (2006). http://web.oie.int/eng/info/hebdo/AIS_12.HTM#Sec8 (accessed June 4, 2012).

161. Arthur, R. Hendra virus. *Anim Health Surveil* 11, 8 (2006).

162. Promed. Hendra virus, human, equine—Australia (QLD) (03): corr. Archive Number: 20070903.2897 (2007).

163. Promed. Hendra virus, human, equine—Australia (02): (QLD, NSW). Archive Number: 20080717.2168 (2008).

164. Promed. Hendra virus, human, equine—Australia (04): (QL) fatal. Archive Number: 20090903.3098 (2009).

165. Promed. Hendra virus, human, equine—Australia (05): (QL). Archive Number: 20090910.3189 (2009).

166. Promed. Hendra virus, human, equine—Australia (05): (QL). Archive Number: 20090910.3189 (2010).

167. Promed. Hendra virus, equine—Australia (21): (QL) canine. Archive Number: 20110802.2324 (2011).

168. Biosecurity Queensland. Biosecurity Queensland adds Logan result to confirmed Hendra cases. (2011). http://www.facebook.com/notes/biosecurity-queensland/biosecurity-queensland-adds-logan-result-to-confirmed-hendra-cases/194626020590612 (accessed June 4, 2012).

169. Promed. Hendra virus, equine—Australia (06): (QL,NS) human exposure. Archive number 20110705.2036 (2011).

170. Promed. Hendra virus, equine—Australia (08): (QL,NS). Archive Number: 20110710.2084 (2011).

171. Promed. Hendra virus, equine—Australia (07): (QL,NS). Archive Number: 20110706.2045 (2011).

172. Promed. Hendra virus, equine—Australia (10): (QL,NS) human exposure. Archive Number: 20110713.2110 (2011).

173. Promed. Hendra virus, equine—Australia (14): (QL, NSW). Archive Number: 20110716.2158 (2011).

174. Promed. Hendra virus, equine—Australia (15): (QL, NS). Archive Number: 20110717.2167 (2011).

175. Promed. Hendra virus, equine—Australia (17): (QLD, NSW). Archive Number: 20110725.2243 (2011).

176. Promed. Hendra virus, equine—Australia (20): (QL, NS) canine. Archive Number: 20110729.2274 (2011).

177. Primary Industries, NSW. New Hendra virus case at Ballina (2011). http://www.dpi.nsw.gov.au/__data/assets/pdf_file/0013/404113/Biosecurity-Bulletin-for-veterinarians-17-August-2011.pdf (accessed June 4, 2012).

178. Promed. Hendra virus, equine—Australia (23): (NSW). Archive Number: 20110818.2512 (2011).

179. Promed. Hendra virus, equine—Australia (24): (QL). Archive Number: 20110823.2570 (2011).

180. Promed. Hendra virus, equine—Australia (26): (NSW). Archive Number: 20110830.2666 (2011).

181. Promed. Hendra virus, equine—Australia (29): (QL). Archive Number: 20111014.3075 (2011).

182. Promed. Hendra virus, equine—Australia: (QL). Archive Number: 20120106.1001359 (2012).

183. Promed. Hendra virus, equine—Australia (05) (QL): canine exposure. Archive Number: 20120606.1157585 (2012).

184. Promed. Hendra virus, equine—Australia (03): (QL) human exposure. Archive Number: 20120531.1151213 (2012).

185. Promed. Hendra virus, equine—Australia (07): (QL). Archive Number: 20120629.1184444 (2012).

186. Promed. Hendra virus, equine—Australia (08): (QL). Archive Number: 20120720.1208397 (2012).

187. Biosecurity Queensland. New Hendra virus case in Cairns area. (2012). https://www.facebook.com/notes/biosecurity-queensland/new-hendra-virus-case-in-cairns-area/434789053240973 (accessed July 28, 2012).

188. Promed. Hendra virus, equine—Australia (10): (QL). Archive Number: 20120907.1284588 (2012).

189. Chadha, M.S. et al. Nipah virus-associated encephalitis outbreak, Siliguri, India. *Emerg Infect Dis* 12, 235–240 (2006).

190. Montgomery, J.M. et al. Risk factors for Nipah virus encephalitis in Bangladesh. *Emerg Infect Dis* 14, 1526–1532 (2008).

191. World Health Organization Regional Office for South East Asia. Nipah virus outbreaks in the WHO South-East Asia Region. (2012). http://www.searo.who.int/entity/emerging_diseases/links/nipah_virus_outbreaks_sear/en/index.html (accessed November 8, 2013).

192. Homaira, N. et al. Cluster of Nipah virus infection, Kushtia District, Bangladesh, 2007. *PLoS One* 5, e13570 (2010).

193. World Health Organization. Regional Office for South East Asia Nipah outbreak in India and Bangladesh. *Commun Dis Newsl* 4(2) (2007). http://209.61.208.233/en/Section10/Section372_13452.htm (accessed November 8, 2013).

194. Promed. Nipah virus, fatal—Bangladesh (03): (Dhaka). Archive Number: 20080311.0979 (2008).

195. Promed. Nipah encephalitis, human—Bangladesh (03): (JI). Archive Number: 20120212.1040138 (2012).

10 Human Immunodeficiency Virus

Mohan Kumar Haleyur Giri Setty and Indira K. Hewlett

CONTENTS

10.1 INTRODUCTION

Acquired immunodeficiency syndrome (AIDS) is a disease of the human immune system caused by the human immunodeficiency virus (HIV),[1] which destroys CD4+ T lymphocytes of the immune system that prevent infections. HIV is transmitted in humans through specific body fluids—blood, semen, genital fluids, and breast milk. Having unprotected sex and sharing drug needles with an HIV-infected individual are the most common ways by which the virus is transmitted. Some body fluids, including saliva, sweat, or tears, have not been shown to transmit HIV.[2,3] AIDS has had a major impact on society, both as an illness and as a source of discrimination. The disease has also had significant economic impact.

AIDS is one of the most serious global health problems of unprecedented dimensions and is one of the greatest modern pandemics. At the end of 2010, an estimated 34 million people were living with HIV globally, including 3.4 million children under 15 years of age. There were 2.7 million new HIV infections in 2010, including 390,000 among children less than 15 years.[4]

Since the recognition of AIDS in 1981 and the discovery of HIV as the causative agent in 1983, 60 million people have become infected with HIV, 25 million of whom have since died.[5–8] More people than ever are living with HIV today, largely due to greater access to treatment. The number of people becoming infected with HIV is continuing to decline in some countries more rapidly than others.

Thirty years after AIDS was first reported, HIV continues to spread. Although the annual number of people newly infected with HIV has declined 20% from the global epidemic peak in 1998, an estimated 2.7 million people acquired the virus in 2010 alone. Existing prevention efforts are often insufficient or inadequately tailored to local epidemics. This requires stronger in-country surveillance systems, especially among key populations at higher risk of HIV infection, greater commitment to implementing evidence-informed programmers, and the development of new prevention approaches and improved tools to strengthen national responses and accelerate progress toward achieving the Millennium Development Goals.

10.2 CLASSIFICATION AND MORPHOLOGY

A key feature of HIV infection is extensive viral diversity. At the population level, two types of HIV (HIV-1 and HIV-2) have been identified, as well as multiple subtypes that may differ by ~35% in their nucleotide sequences.[9] The current HIV classification based on phylogenetics was developed in the late 1990s in response to characterization of many new isolates.[10] HIV-2 is morphologically and biologically similar to HIV-1 but differs in the envelope glycoprotein and other antigenic epitopes.[11] It also contains the gene *Vpx* that serves a critical nuclear import function in lieu of the HIV-1 *Vpu* gene.[12,13] Independent zoonotic transmission events from nonhuman primates to humans have generated several HIV lineages: HIV-1 groups M, N, O, and P and HIV-2 groups A–H. The size of the epidemic caused by each group varies considerably. HIV-1 group M is responsible for the global HIV pandemic (approximately 33 million infected individuals), group O is associated with only a few thousand infections in West–Central Africa, group N has been found in a limited number of individuals in Cameroon, and group P was recently identified in two individuals originating from Cameroon.[14–18] It is conceivable that further HIV lineages in humans will be discovered in the future, as all HIV lineages may not yet have been identified and new cross-species transmissions may take place in the future.

10.2.1 HIV-1 Groups and Subtypes

During the course of the HIV epidemics in human, sequences of the different HIV-1 groups have diversified in the population, which has prompted further classifications. Within the HIV-1 group M, nine subtypes are recognized, designated by the letters A–D, F–H, J, and K. Within a subtype, variation at the amino acid level is in the order of 8%–17% but can be as high as 30%, whereas variation between subtypes is usually between 17% and 35% but can be up to 42%, depending on the subtypes and genome regions examined.[9] Intrasubtype diversity continues to increase over time[19] and evolution rates may differ according to subtype.[20] Group O sequence diversity is high, which has led to a classification of sequences into clades I–V, which are as genetically distant from each other as are group M subtypes, consistent with the tMRCA

of group O being similar to group M. However, there is less subtype-like signal in group O compared to group M because group O has not spread much beyond its origin in West–Central Africa.[21–23] All group N viruses found in humans are closely related, as are the two group P sequences.[14,16]

10.2.2 HIV-1 Recombinants: CRFs, URFs, SGRs

New HIV strains arise due to recombination between two strains in an individual when two or more heterologous strains infect either simultaneously or sequentially. Viral genome sequence analyses have shown recombination between strains as a common occurrence. Recombination has been demonstrated between strains from different groups in HIV-1 (groups M and O) as well as between and within group M subtypes.[24,25] Intrasubtype recombination was found to be very extensive within group M subtype C, although such data are lacking for other subtypes.

Recombinants between different HIV-1 group M subtypes are designated as either circulating recombinant forms (CRFs) if fully sequenced and found in three or more epidemiologically unlinked individuals or as unique recombinant forms (URFs) if not meeting these criteria.[26] Up to 55 CRFs have been described so far.[26,27] Some CRFs have recombined further with other subtypes or CRFs giving rise to so-called second-generation recombinants (SGRs). With increased global availability of sequencing techniques, it has become apparent that recombinant forms are widespread and increasing as a proportion of the total number of HIV-1 infections. Recombinants are currently estimated to be responsible for at least 20% of HIV-1 infections worldwide.[28] HIV viral diversity has its impact on disease progression and transmission, diagnosis and measurement of viral load (VL), response to highly active antiretroviral therapy (HAART) and drug resistance, and the immune response and vaccine development.

HIV-2 also exhibits significant genetic variability, and HIV-2 is separated into eight groups, of which A and B are the most prominent, and it is worth noting that HIV-2 CRFs have been reported only once.[29]

10.2.3 HIV Genome Structure and Organization

HIV-1 is a complex retrovirus encoding 15 distinct proteins. The four Gag proteins, MA (matrix), CA (capsid), NC (nucleocapsid) and p6, and the two Env proteins, SU (surface or gp120) and TM (transmembrane or gp41), are structural components that make up the core of the virion and outer membrane envelope. The three Pol proteins, PR (protease), RT (reverse transcriptase), and IN (integrase), provide essential enzymatic functions and are also encapsulated within the particle. Six additional HIV-1 proteins known as the regulatory proteins Tat and Rev and the accessory proteins Nef, Vif, Vpr, and Vpu have important functions during the early stages of the viral life cycle and are essential for efficient viral replication.[30]

HIV is a spherical, membrane-enveloped, pleomorphic virion, 1000–1500 Å in diameter, having two copies of its

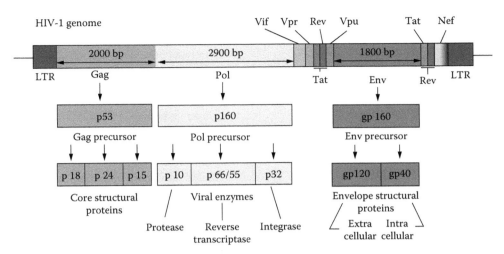

FIGURE 10.1 HIV genome organization from http://www.yale.edu/bio243/HIV/genome.html.

single-stranded and positive-sense RNA genome. The HIV genome is encapsulated in a spherical protein shell composed of around 2500 copies of the virally encoded Gag polyprotein. The Gag molecules are radially arranged, adherent to the inner viral membrane, and closely associated as a hexagonal, paracrystalline lattice. Gag comprises three major structural domains called MA, CA, and NC. For immature virions to become infectious, they must undergo a maturation process that is initiated by proteolytic processing of Gag by the viral protease. The new Gag-derived proteins undergo dramatic rearrangements to form the mature virus. The mature MA protein forms a "matrix" layer and remains attached to the viral envelope, NC condenses with the genome, and approximately 1500 copies of CA assemble into a new cone-shaped protein shell, called the mature capsid, which surrounds the genomic ribonucleoprotein complex. The HIV capsid forms a fullerene shell, in which the CA subunits form about 250 CA hexamers arrayed on a variably curved hexagonal lattice, which is closed by incorporation of exactly 12 pentamers, 7 pentamers at the wide end and 5 at the narrow end of the cone.

HIV belongs to the family Retroviridae and genus *Lentivirus*. The structure of HIV follows the typical pattern of the retrovirus family, comprising a single-stranded, positive-sense ribonucleic acid genome of about 9.7 kb. There are two strands of HIV-positive RNA and each strand has a copy of the virus's nine genes. The RNA is surrounded by a cone-shaped capsid, which consists of approximately 2000 copies of the p24 viral protein. Surrounding the capsid is the viral envelope. The viral envelope is composed of a lipid bilayer membrane, formed from the cellular membrane of the host cell during budding of the newly formed virus particle. Host-cell proteins, such as the major histocompatibility complex antigens and actin, remain embedded within the viral envelope, along with the viral envelope protein. Each envelope subunit consists of two noncovalently linked membrane proteins: glycoprotein 120 (gp120), the outer envelope protein, and gp41, the transmembrane protein that anchors the glycoprotein complex to the surface of the virion. The envelope

protein is the most variable component of HIV, although gp120 itself is structurally divided into highly variable (V) and more constant (C) regions. The variability of V regions may be a product of envelope functionality as has been especially well described in V3, where amino acid changes alter coreceptor usage.[31]

HIV-1 and HIV-2 are similar in most aspects yet differ in many aspects of infection and transmission having distinct genomes. Both viruses have a typical retrovirus structure, which includes long terminal repeats (LTRs) at the 5′ and 3′ ends and the Gag–Pol and Env genes. The general genome organization is similar to lentiviruses; however, there are several additional genes called accessory genes. Five (Vif, Vpr, Tat, Rev, and nef) are common to both HIV-1 and HIV-2. The Vpu exists only in HIV-1, whereas Vpx occurs only in HIV-2 (Figure 10.1).[32,33]

10.3 BIOLOGY AND EPIDEMIOLOGY

The first sign of the HIV global epidemic came in June 1981, when the Centers for Disease Control and Prevention reported five young male homosexuals in Los Angeles had developed *Pneumocystis carinii* pneumonia, a rare disease.[34] The global epidemic was not immediately recognized until the mid-1980s, it rapidly spread in Central Africa and many reports from Europe and other parts of the world came.[35] Soon the causative agent was identified as HIV.[6,36] In 1986, a novel, but related, retrovirus was isolated from an immunocompromised individual who lived in Western Africa and it was named HIV-2.[37] Extensive research was undertaken by many international research organizations, and an understanding of its mode of transmission, pathogenesis, and its biology was evaluated. There were many drugs developed to treat AIDS, but most have failed to completely eliminate the virus and drug resistance has remained a major constraint to eradicate AIDS. Yet the amazing ability to mutate and escape immune response and its capacity to hide are yet to be understood. The recent focus is on understanding latency and elimination of reservoir.

10.3.1   HIV Life Cycle

HIV is an obligate virus that enters the cell by binding to receptors, especially CD4, and other chemokines like CCR5 and CXCR4. The interaction requires the recognition of two host-cell surface-receptor proteins by the viral gp120 envelope protein. The presence or absence of these cellular proteins restricts the range of host-cell types that are susceptible to infection by a strain of HIV.[38,39]

Binding ability and tropism of the virus is dependent on the protein structure of gp120. Particular patterns of sequence of the V3 and V4 variable regions and other regions of gp120 relate to CD4 binding and differential coreceptor affinity. In general, viral strains that bind to CCR5 (R5 strains) infect macrophages and T cells and are characterized by less aggressive growth *in vitro*. Strains that recognize CXCR4 (X4 strains), by contrast, infect only T cells and T cell lines. The growth of X4 strains *in vitro* is characterized by high viral titer and the presence of syncytial cells, which are formed by the fusion of multiple infected cells and can be observed by light microscopy.[40–44]

Preintegration complex: Following infection of a new host cell, in the cytoplasm, the RNA genome is first reverse transcribed into single-stranded DNA that is then further transcribed to double-stranded DNA. These two polymerase steps are performed by viral reverse transcriptase, which is copackaged in the viral particle. Self-priming of the single-stranded RNA and DNA and removal of the transcribed RNA strand occur by a complex series of steps dependent upon interactions between the viral LTR and host-cell enzymes. The double-stranded DNA genome forms a complex with host-cell and viral proteins (including matrix, integrase, and Vpr) that is actively transported to the nucleus.[45,46] During the early steps of the HIV-1 replication cycle, the virus counteracts specific host proteins that have evolved to limit retroviral replication.

Integration and transcription: The double-stranded HIV genome transcribed is then either randomly integrated into the host-cell genome by means of DNA splicing, performed by the viral integrase, or forms stable DNA circles.[47] The integrated form of HIV is known as the provirus and takes the form shown in Figure 10.1, with identical LTR copies flanking the coding regions. Proviral DNA is replicated as part of the normal cell genome and may persist in this form for long periods and through many rounds of mitotic cell division. The 5′ end of the LTR now functions as a promoter, regulating the production of RNA transcripts dependent on the presence of host-cell transcription factors (such as promoter-specific transcription factor, SP1, and nuclear factor kappa beta [NFκB]) and the viral protein Tat.[48] The transcribed HIV RNA molecules may be either spliced in preparation for translation of viral proteins or exported from the nucleus in an unspliced form for packaging into newly produced virions. Nuclear export of spliced RNA is assisted by the viral protein Rev.

HIV assembly and release: Immature viral polypeptides are processed into their functional forms by the enzyme protease and assembled with full-length HIV RNA transcripts into nascent viral particles. The structural immature viral polypeptide Gag encodes the majority of such proteins. Gag itself uses cellular proteins to make its way to the plasma membrane for assembly into the new progeny. It is now known that the p6 protein located at the C-terminus of the Gag polypeptide interacts with tumor suppressor gene 101, an interaction that is critical for the release of the newly assembled particles.[49] Immature viral polypeptides together with the full-length HIV RNA transcripts are initially assembled into immature particles. During budding from the plasma membrane, viral proteins within these particles are processed into their functional forms by the enzyme protease and rearranged into mature particles. The Vpu protein facilitates virion release from the cell membrane in the late stage of the replication cycle.[50] Vpu interacts with a host-cell factor named tetherin, an interferon-alpha (IFN-α)-induced human protein.[51] Tetherin is an endogenous membrane-associated protein that blocks release of viral particles. Without Vpu, HIV-1 particles become tethered to the cell membrane and cannot be released.

10.3.2   Molecular Epidemiology

Advances in DNA sequencing as well as the availability of new bioinformatics tools to analyze data have allowed an assessment of the impact of epidemiological factors on the evolutionary dynamics of HIV at global (worldwide), regional (single epidemics), local (transmission chains), and individual (intrahost) levels by examining genetic variation across its genome and over geographic space and time.[52] HIV-2 has been identified mostly in Western Africa. In turn, HIV-1 is distributed worldwide, group M being the one that accounts for most infections, while groups O, N, and P appear to be concentrated in West–Central Africa. Epidemiological and subtyping studies primarily focus on the capsid proteins encoded within the Env reading frame.

Group M causes most HIV-1 infections, owing to its high numbers of subtypes and CRFs. These subtypes form phylogenetic clusters with amino acid differences of 25%–30% in Env, 20% in Gag, and 10% in Pol. Some subtypes are linked geographically. Variation within group M is greatest in the Congo River basin, which is probably the site of initial zoonotic transmissions and regional diversification. Initial subtype distribution indicated dominance of subtype B in the western world and of subtype A in sub-Saharan Africa. Subtype A was eventually identified in the former Soviet Union. In the past 15 years, however, the rapid emergence of new subtypes and intermixing of strains has altered the geographical distribution of subtypes and intersubtype recombinants. In addition, some preexisting subtypes, such as A and F, have continued to evolve into sub-subtypes—for instance, A1–A4 and F1–F2 that form distinct lineages within a given subtype but have lesser degrees of genetic divergence. The greatest diversity and prevalence of HIV-1 continue to be mostly in sub-Saharan Africa. The epidemics in West and Central Africa seem to have stabilized in prevalence, but these regions, along with the Congo River basin, continue to be hot spots for HIV diversity. Most, if not all, subtypes, sub-subtypes, and CRFs have been reported in the Democratic Republic of Congo (DRC) and Cameroon.

HIV-1 strains in Angola are highly diverse, and classification into sub-subtypes A5 and A6 might be required. From Cameroon, moving on westward to Nigeria, HIV-1 diversity decreases, with the dominance of CRF02_AG (subtypes A and G). This trend continues with western migration to Côte d'Ivoire, Ghana, Senegal, and Mali, where CRF02_AG predominates, with reports of isolated cases of CRF06_cpx (complex subtype). Finally, CRF06_cpx and the SGRs CRF02_AG/CRF06_cpx dominate the epidemic in Burkina Faso, with subtypes A and G rarely being reported. Although HIV-1 genetic diversity is high in West and Central Africa, HIV-1 prevalence remains surprisingly lower than in most other regions of sub-Saharan Africa. The highest prevalence shifted in the late 1990s from East Africa (Uganda, Kenya, and Tanzania) to the Southern African region. On an average, close to 20% of the human population in South Africa, Lesotho, Botswana, and Zimbabwe are thought to be infected with HIV-1. This shift provides strong evidence for the founder-effect theory (a single introduction followed by a rapid spread) since the Southern African epidemic is due almost entirely to the spread of HIV-1 subtype C. Independent, rapid spread of subtype C in East Asia has also contributed to this subtype being responsible for more than 51% of all HIV-1 infections worldwide. Although initially absent, HIV-1 subtype C now circulates at low levels as a pure (nonrecombinant) subtype or a recombinant form in Kenya and Uganda.

The epidemic in Asia is due to multiple founder events, subtype B being introduced in the mid-1980s and seen mainly in China, India, and Thailand. The subtype B has been in Thailand, referred to as Thai B that is distinct from the subtype B found in the Americas. The spread and divergence of subtype B coincides with introduction of CRF01_AE to Thailand. CRF01_AE has now gained dominance in Thailand; likewise, subtype C is currently dominant in most East Asian countries. Subtype C was reportedly introduced into India from South Africa and later into China. Subtypes B and C have recombined to form CRF07_B′C and CRF08_B′C in China. The northern triangle of Burma seems to have the greatest HIV-1 diversity in Asia, with seeding of CRF01_AE and subtype B′ from Thailand, possibly subtypes A, B, and C from India, and CRF07 and CRF08_B′C from Yunnan province in China.

In Eastern Europe, the major HIV-1 infection is due to the spread of sub-subtype A1, which has been linked to intravenous drug use, and of subtype B (and to a lesser extent, CRF03_AB), mainly through sexual transmission. In Western Europe, as in North America and Australia, subtype B predominates. The prevalence of non-B strains has, however, increased, owing to the influx of immigrants from Africa and Asia. Portugal has the highest prevalence of HIV-2 and other non-B subtypes in Europe. In South America, prevalence and diversity of HIV-1 are highest in Brazil and

Argentina, with substantial circulation of subtypes B, C, and F, and BC and BF recombinants. Two reports showed close relations between the subtype C viruses in South America and those in Kenya, Ethiopia, and Burundi. The subtype C epidemic is estimated to have originated in 1958 and was first reported in Ethiopia in 1990 from where it spread to Israel; the virus also spread from Eastern Africa to Brazil and then to Argentina and Uruguay. The subtype C epidemic is now the fastest emerging epidemic in South America: in the Rio do Sul region of Southern Brazil, subtype C or a BC recombinant accounts for nearly 70%. The HIV-1 epidemic in the Caribbean might be the clearest reflection of different founder events, given notable immigration, travel, and trade.

Longstanding links between the DRC and Haiti and between Angola and Cuba might explain the presence and age of certain HIV-1 strains in the Caribbean. Gilbert and coworkers[121] applied a relaxed molecular clock to sequences derived from a 1982 Haitian DNA sample and calculated that subtype B was introduced into Haiti between 1962 and 1970, probably from the DRC, and from Haiti into the United States in 1972. The epidemic in Cuba is diversified, with CRF18_cpx and CRF19_cpx and subtype C strains circulating, but is marked mainly by the CRF20, CRF23, and CRF24, for which subtypes B and G are parental. HIV-1 was probably introduced into Cuba via troops who fought in the Angolan War of Independence. Relative to Cuba and Haiti, HIV-1 has recently arrived through true founder event in other Caribbean islands, such as Trinidad and Tobago and the Dominican Republic.[53]

10.4 PATHOGENESIS AND CLINICAL FEATURES

In the early phase of HIV infection, many people do not develop clinical symptoms. However, some people develop a flu-like illness within several days to weeks after exposure to the virus accompanied by fever, headache, tiredness, and enlarged lymph glands in the neck. Signs and symptoms of late-stage HIV infection may include blurred vision, diarrhea, dry cough, fever, night sweats, permanent tiredness, shortness of breath, swollen glands, and white spots on the tongue or mouth. AIDS is the more advanced stage of HIV infection characterized by a decline in CD4 T cell numbers to a very low level resulting in an impaired immune system and the ability to fight infection. Individuals with suffering from AIDS are highly susceptible to a variety of infections and develop certain types of cancers. Tuberculosis is the most common opportunistic infection associated with HIV and a leading cause of death among people living with AIDS in the developing world. Other common opportunistic coinfections are salmonellosis, cytomegalovirus, candidiasis, cryptococcal meningitis, toxoplasmosis, cryptosporidiosis, human T-lymphotropic virus (HTLV), *Trypanosoma cruzi*, hepatitis B virus (HBV), hepatitis C virus (HCV), and human simplex virus (HSV). Some cancers, such as Kaposi's sarcoma, invasive cervical cancer, lung cancer, rectal carcinomas, hepatocellular carcinomas, head and neck cancers, and cancers of the immune system known as lymphomas, are common in AIDS patients. Other AIDS-related complications include neurological, renal, wasting syndrome, low bone mineral density, and lipodystrophy.

HIV infection is generally characterized by an acute phase of intense viral replication and dissemination to lymphoid tissues; a chronic, often asymptomatic phase of sustained immune activation and viral replication; and an advanced phase of marked depletion of CD4+ T cells that leads to acquired immune deficiency syndrome.[54] Following virus infection, viral RNA in plasma remains undetectable by current nucleic acid assays for up to 7 days, a phase described as the eclipse phase. Antibodies generally appear around 21 days of infection.[55]

After HIV infection, there is local replication in the mucosa; the target cells are CD4+ T cells and Langerhans cells. The virus is then transported to draining lymph nodes where further replication occurs. This is the end of the eclipse phase with the first detection of HIV RNA in blood and the spread of the virus to other lymphoid tissues, particularly gut-associated lymphoid tissue (GALT), resulting peak viremia and in the depletion of approximately 80% of CD4+ T cells in GALT. While viral replication occurs in GALT and other lymphoid tissues, plasma HIV RNA increases exponentially. Following the peak in viremia, VL declines to a more stable level known as the viral set point. This decline is associated with antiviral responses of the innate and adaptive immune system. The early events of intense HIV replication and dissemination lead to the establishment of stable viral reservoirs in lymphoid tissues and, at the cellular level, in the form of latently infected resting CD4+ T cells. These stable reservoirs are major impediments to eradication of the virus.

Signs and symptoms: The acute retroviral syndrome. Acute HIV infection (AHI) is often asymptomatic, but sometimes, it presents with serious manifestations requiring hospital admission. There is a wide spectrum between complete absence of symptoms during the time of seroconversion and severe disease. Symptoms typically start 2–4 weeks after infection and in exceptional cases, as early as 5 days or as late as several months after infection. The median duration of symptoms is difficult to quantify and ranges between 12 and 28 days. Moderate and subjective symptoms, such as fatigue, may persist for months, although almost all patients eventually enter an asymptomatic phase lasting years. None of the features of the acute retroviral syndrome are specific for HIV infection. Coinfection with other pathogens might enhance the proliferation of HIV and are presented with severe symptoms. Progression to AIDS among HIV-infected individuals is highly heterogeneous due to host and viral factors ranging from 3 years in rapid progressors (RPs) to 10 years in long-term nonprogressors (LTNPs). Usually, LTNPs show undetectable or controlled (2000 copies/mL) HIV replication; however, a reduced number of LTNPs show uncontrolled VL (2000 copies/mL) with asymptomatic HIV infection over almost 10 years after seroconversion, and they are called elite controllers.

Elite controllers are a subgroup of HIV-infected individuals characterized by the ability to spontaneously maintain virologic control. Elite control of HIV infection is established soon after seroconversion and occurs in less than 1% of HIV-infected individuals. Elite controllers are demographically heterogeneous with diverse racial backgrounds and modes of HIV transmission, though genetic studies demonstrate an overrepresentation of protective HLA alleles. Elite controllers typically have elevated CD4 cell counts, stable CD4 trajectories, and more favorable clinical outcomes compared with viremic patients. A proportion of elite controllers, however, may experience HIV disease progression with loss of virologic control, CD4 cell declines, and rarely AIDS-defining events. Long-term elite controllers who maintain both virus undetectability and normal CD4 counts for at least 10 years appear to represent a very promising model. In the majority of those cases, a peculiar host's genetic profile induces strong antiviral cell-mediated immunity, which imprints the HIV reservoir by preserving the long-lived CD4 T cells.[56]

Furthermore, a really limited group of HIV-infected individuals show a particular discordant profile with high VL (10,000 copies/mL) in the absence of quantitative immune defects (viremic nonprogressors [VNPs]). This fact is paradoxical, as HIV-infected CD4 T lymphocytes have a shortened life span due to direct cytopathic effects of HIV or lysis by immune cells. Moreover, the number of dying cells in infected individuals greatly exceeds the number of HIV-infected cells due to detrimental effects of immune activation, HIV proteins, or abortive infection on the bystander uninfected CD4 T cell population. Among viral determinants, the envelope glycoprotein (gp120/gp41, Env), which defines HIV tropism for CCR5 or CXCR4, can influence CD4 T cell decline *in vitro* and *in vivo*. Furthermore, Env is a major determinant of viral pathogenicity, which is related to the fusogenic activity of gp41 and affects both infected and bystander CD4 T cells.[57]

Cellular and humoral immune responses are generated against HIV during the acute and early phases of infection, yet they fail to restrict HIV replication in the majority of infected individuals. There is strong evidence for selection of viral mutants that escape these cellular and humoral responses. The immune systems of patients with HIV infection are characterized by an immunodeficiency occurring in the setting of immune activation. The CD4 T cell pool declines while the CD8 T cell pool expands. Homeostatic cytokines such as interleukin 7 (IL-7) and proinflammatory cytokines such as IFN-α are elaborated and may play a significant role in some of the pathologic aspects of HIV infection. An interaction between IL-7 and IFN-α signaling may be responsible for the death of CD4 T cells. Immune cell dysfunction, manifested by poor immune responses against HIV and other pathogens, is observed in the majority of untreated HIV-infected individuals. Chronic HIV viremia leads to increased cell turnover and changes in cellular phenotype and function that are consistent with increased immune activation, differentiation, and exhaustion.

Implementation of HAART has deeply changed the landscape of HIV-associated malignancies. Some AIDS-defining tumors, namely, primitive lymphoma of the central nervous system (CNS), have drastically declined, whereas a steady increase has been observed for non-AIDS-defining tumors, maybe due to longer survival of HIV-infected people. Easier immune restoration, subsequent to availability of a number of drugs targeting HIV at different points, has decreased opportunistic infections that hampered treatment of HIV-associated cancers. Consistently, procedures that have been so far precluded to HIV+ subjects, such as transplant of hematopoietic stem cells (HSCs), either autologous or allogenic, and liver transplant, are expected to be performed more and more extensively in this population. In fact, if assimilation of HIV patients with cancer and the general population is a remarkable achieved goal, uniqueness of HIV infection in terms of immune status still makes HIV-associated cancer a unique chapter in the setting of oncology.[58]

10.5 IDENTIFICATION AND DIAGNOSIS

HIV testing plays an important role in HIV prevention in that knowledge of HIV status has both individual and public health benefits. Early and accurate detection of HIV infection is important to public health because this stage is one of high infectiousness and appears to account for a disproportionate amount of HIV transmission and reduction of virus in plasma and genital secretions by use of and adherence to antiretroviral therapy (ART). The individual benefits of HIV testing are primarily associated with individuals accessing care and treatment. Individuals entering care and treatment have a substantial reduction in adverse health outcomes and increased life expectancy. The first diagnostic assay in the United States to screen blood donations was licensed in 1985. There has been considerable progress since then in terms of number of new assays coupled with improvements in the sensitivity and specificity of these assays.

10.5.1 INITIAL DIAGNOSIS OF HIV

There are a number of tests available to determine whether a person is infected with HIV, the virus that causes AIDS. These include HIV antibody tests (measured in blood, saliva, or urine), p24 antigen tests, and polymerase chain reaction (PCR) tests. HIV rapid tests generally come in the form of lateral flow strips or cassettes, which are convenient, self-contained tools for HIV serologic testing. They are generally easy to use, can usually be performed on fingerstick blood, contain built-in quality controls, and can be administered by technicians and nontechnicians alike, including community health workers.

Immunoassays: First-generation immunoassays detect IgG antibodies to HIV using whole viral lysate as the antigen in a standard indirect immunoassay format. These assays detect HIV infection in the same time frame as the Western blot (WB), approximately 45–60 days following infection.[55,59] Second-generation immunoassays also detect IgG in an indirect format but were designed to increase specificity by incorporating recombinant proteins or peptides as the antigens for detection. Second-generation immunoassays detect HIV

infection approximately 5–7 days sooner than first-generation assays. The third-generation immunoassays could detect both IgG and IgM using peptides and recombinant proteins in an antigen sandwich format and improve detection of recent HIV infection. These assays generally detect infection within about 20–25 days after infection.[59,60] Most third-generation immunoassays marketed in the United States and around the world detect HIV-2 in addition to HIV-1.[60] The latest laboratory immunoassays to come to the market, fourth-generation immunoassays, detect p24 antigen in addition to detecting anti-HIV, IgM, and IgG. Fourth-generation or combination antigen/antibody assays have been approved and used in many countries since the late 1990s.[61–65] These assays detect p24 antigen at the level of 11–18 pg/mL,[66] which is equivalent to approximately 30,000–50,000 copies/mL of HIV RNA.[67] Recently, two such assays (Abbott ARCHITECT HIV Ag/Ab Combo and Bio-Rad GS HIV Combo Ag/Ab EIA) have been approved for use in the United States. Several studies have been conducted with these assays, and the data indicate they have similar performance characteristics as those marketed elsewhere[68–70] and detect p24 approximately 5–7 days after the appearance of nucleic acid.[71] Importantly, the data published to date indicate the assays available in the United States are capable of detecting AHI in greater than 80% of individuals that are nucleic acid amplification test (NAAT) positive but nonreactive or indeterminate in antibody-only assays.[68–70]

Nanoparticle labels conjugated with biomolecules have been used in a variety of different assay applications. Nanoparticles offer adjustable and expandable reactive surface area compared to the more traditional solid-phase forms utilized in bioaffinity assays due to the high surface/volume ratio. Signal enhancement by conjugating nanoparticles with fluorescent, luminescent, and other measurable properties has enhanced detection limit by severalfold. Multiplexing capabilities, ability to increase sensitivity and specificity without using enzymes, have increased the use of nanoparticles in immunoassays for early detection.

CD4 counting: Enumeration of CD4 lymphocytes is an essential diagnostic tool for initiating therapy and monitoring its efficacy. CD4 testing typically relies on complex flow cytometry equipment that requires infrastructure and technical skills that are commonly unavailable at rural and remote clinics.[60] Currently, there are a good number of technology choices for CD4 testing in resource-limited settings. Most of these are laboratory-based platforms using proven flow cytometry methodologies. In reference laboratory settings with well-trained technicians, these technologies function well and can be cost-effective. Many, but not all, of these CD4 testing platforms, including BD FACSCalibur and FACSCount, have been the subject of independent evaluations and have performed well, within the recognized limitations, both physiological and technical, of CD4 performance. There are many new point-of-care (POC) devices coming to market that are affordable, disposable, and easy to use and read the results.[72]

Western blot: WB tests are regarded as the most reliable and sensitive form of confirmatory testing. In a WB assay, a sample containing antigen is separated using gel electrophoresis with SDS-PAGE. Although the WB assay is a highly sensitive platform, the technique again requires complex and expensive equipment. The use of this equipment also requires considerable technical training. These conditions represent significant challenges to resource-poor settings such as those seen in the developing world. Furthermore, WB tests can often produce inconclusive results.

10.5.2 Viral Loading Testing Technologies

The first molecular assay for quantifying HIV viral RNA was approved by the United States in 1999. Since then, a number of assays have been developed and will be considered here in some detail.

Quantitative NAAT detects HIV RNA approximately 10 days after infection and has typically been used to determine viral burden and monitor response to therapy.[55,73,74] In 2006, a qualitative NAAT (Gen-Probe APTIMA) gained approval as a supplemental assay for diagnosing HIV infection. This assay allows detection of HIV infection prior to the appearance of HIV-specific antibody.[60,71,75] However, the current NAATs (quantitative and qualitative) that are on the market have several limitations, including the need to draw blood, extraction of nucleic acid, cost, processing time, and the technical skill required to perform the tests. To address some of the limitations associated with NAAT, there is considerable effort being spent to simplify NAAT and potentially make it feasible for POC testing.[76] Many of these technologies rely on isothermal amplification techniques and include loop-mediated isothermal amplification,[77] helicase-dependent amplification,[78] and a simplified amplification-based assay that incorporates a visual dipstick detection device.[79] There is also considerable work being done to decrease the assay time for real-time PCR assays and package them so they can be used in point-of-contact settings.[80–82] Furthermore, although NAAT is ideal for detecting AHI, it has been demonstrated that there is a risk of false-negative NAAT results in individuals with established HIV infection, and this can be up to 3.7% in them.[60,75]

Diagnostic challenges: HIV diversity. The enormous diversity found in HIV as discussed earlier underscores its extraordinary ability to evolve continuously. The high level of genetic heterogeneity of HIV-1 and the emergence of recombinant strains of the virus complicate VL assay development. In an ideal world, VL assays would detect and quantify all known HIV-1 subtypes as well as intersubtype recombinants and emerging variations thereon. But currently, that is not the case, although the assays are able to recognize most HIV-1 subtypes. Therefore, it is important to consider the prevalence of HIV-1 and HIV-2 groups and subtypes in a particular geographical region when choosing a VL assay.

10.5.3 Early Infant Diagnostics for HIV

Early infant diagnostics and treatments for HIV still lag far behind those for adults. The usual serological tests for HIV cannot be used in the first 18 months of a child's life because maternal antibodies remain in fetal blood. These antibodies unequivocally generate false positives. Accordingly, the use of more appropriate diagnostic tools is vital for infants, particularly because up to half of HIV-infected infants under the age of 18 months will die. Viral DNA detection by PCR is the most prevalent technique to diagnose HIV in infants under this age in developing countries. More recently, the use of dried blood spots (DBSs) in lieu of plasma samples has been shown to be a successful collecting method that simplifies diagnostics for infants.[83,84] Blood samples from the infant are taken, absorbed onto a filter paper, and then set out to dry. The filter paper can be directly transported by local mail services or regular ground transport to the testing facility for HIV RNA or DNA PCR analysis. DBS samples remain stable up to 10 weeks with desiccants and do not require refrigeration. The main limitation to these quantitative assays is the lower specificity due to the fact that low levels of RNA in serum may not always be detected. This could be improved by changing the threshold of detection. On the whole, these results are very promising for pediatric diagnosis since HIV RNA assays can be more readily accessible than PCR machines in some areas.

10.6 TREATMENT AND PREVENTION

The implementation of HAART has been a milestone in the treatment of HIV-infected individuals, improving both their life expectancy and quality of life, transforming an otherwise lethal disease into a chronic illness. Nevertheless, HIV infection is a major public health issue as a high number of cases are still diagnosed every year, especially in developing countries where access to HAART is still limited due to financial constraints.

Vaccines have eradicated and prevented other epidemics with much less cost and allowing them to be used in large populations. Effective HIV vaccine remains elusive and many clinical trials have failed. Therefore, other form of prevention and development of effective drug combinations have gained more thrust recently. The ART has evolved from monotherapy with azidothymidine (AZT) to a cocktail or combination of antiretroviral agents reducing the plasma VL to below the limits of detection of the most sensitive clinical assays. However, development of drug resistance due to the HIV to evolve and diverge has posed a major challenge to drug and vaccine development. Understanding of the virus replication cycle has been instrumental in developing specific targeted drugs at every stage of its replication cycle. Whereas the HIV-1 life cycle presents many potential opportunities for therapeutic intervention, only a few have been exploited. Major steps in viral replication include virus entry, nuclear import, reverse transcription, genomic integration, and viral maturation, and each of these steps can be targeted by different classes of drugs. Anti-HIV medications (also called antiretrovirals) are grouped into six drug classes according to how they fight HIV. The six classes are nonnucleoside reverse transcriptase inhibitors (NNRTIs), nucleoside reverse transcriptase inhibitors

(NRTIs), protease inhibitors (PIs), fusion inhibitors, CCR5 antagonists, and integrase inhibitors.

10.6.1 Treatment of HIV Infection

10.6.1.1 Nonnucleoside Reverse Transcriptase Inhibitors

NNRTIs inhibit HIV-1 RT by binding to reverse transcriptase and inducing the changes in the spatial conformation of the substrate-binding site and reduce polymerase activity.[85] Unlike NRTIs, these non-/uncompetitive inhibitors do not inhibit the RT of other lentiviruses such as HIV-2 and simian human immunodeficiency viruses (SHIV). NNRTIs are an older class of antiretroviral drug—the first was approved in 1996. Currently, there are four approved NNRTIs: etravirine (ETV), delavirdine, efavirenz, and nevirapine, and several in development, including rilpivirine in phase. Because these three drugs work in a very similar way, once HIV develops resistance to one of them, then the others are often ineffective as well. Newer NNRTIs are designed to work differently so as to avoid this problem of cross-resistance. NNRTIs are an integral part of initial treatment regimen along with one or two NRTIs and a PI.

Nevirapine is safe in pregnancy and has been extensively used to prevent vertical transmission. However, nevirapine-based regimen has been reported to cause fatal cutaneous hypersensitivity and hepatotoxicity. Efavirenz is now preferred over nevirapine in initial regimen as it is available as once a day formulation or along with other drug combinations. Efavirenz has short-term CNS side effects and is also teratogenic. Moreover, both the drugs are a substrate for cytochrome enzyme (CYP 3A4) that results into frequent interactions with drugs metabolized through the same pathway. Like NRTIs, NNRTIs also have low genetic barrier to drug resistance. Even a single mutation may result into cross-resistance. Interestingly, the majority of NNRTI-resistant mutations selected under NNRTI treatment are commonly found as wild-type sequence in HIV-1 group O and HIV-2.

ETV, a second-generation NNRTI, has been approved by United States Food and Drug Administration (USFDA). An evaluation of its efficacy, safety, and tolerability in treatment-experienced patients has been promising. It has significant activity against first-generation NNRTI-resistant HIV-1 virus with some activity against HIV-2.[86] The drug is specially developed with a high genetic barrier to resistance with unique genotypic resistance profile.[87] It is specially recommended for patients with documented NNRTI resistance.[86] The long half-life, high barrier to resistance, and good antiviral efficacy of ETV make it a major force in this class of drugs. However, it is also a substrate and inhibitor of several CYP 3A4 enzymes and therefore contraindicated for use with anticonvulsants, antimycobacterials, and other NNRTIs. It also requires dosage adjustment if used with drugs, metabolized by CYP enzyme system.

There are many drugs that are in the clinical trials like rilpivirine, which is being tested for convenient dosage schedule, high genetic barrier to drug resistance, effectiveness against HIV strains resistant to conventional NNRTIs.

Its lack of antagonism with other drugs as well as its absence of adverse reactions represent important characteristics of rilpivirine.[88] RDEA806 is a new NNRTI that has shown activity against HIV-1-resistant mutant in phase II trials. Unlike other NNRTIs, it does not inhibit or induce cytochrome enzyme system.

10.6.1.2 Nucleoside Reverse Transcriptase Inhibitors

NRTIs were the first class of drugs administered as prodrugs, which require host-cell entry and phosphorylation to become active drugs. Lack of a 3'-hydroxyl group at the sugar (2'-deoxyribosyl) moiety of the NRTIs prevents the formation of a 3'-5'-phosphodiester bond between the NRTIs and incoming 5'-nucleoside triphosphates, resulting in termination of the growing viral DNA chain. Chain termination can occur during RNA-dependent DNA or DNA-dependent DNA synthesis, inhibiting production of either the (+) or (−) strands of the HIV-1 proviral DNA.[89]

These drugs have favorable pharmacokinetic profile, especially long intracellular half-life, high oral bioavailability and administration without regard to food, availability as fixed dose combinations with convenient once or twice daily dosage schedule, and low risk for drug–drug interactions.[90] However, NRTIs have low genetic barrier for drug resistance, and continued treatment is reported to accumulate mutations that causes resistance and cross-resistance to agents within the class. Moreover, the current drugs in this class are associated with bone marrow suppression and high mitochondrial toxicity.[91] The potential to cause serious and irreversible toxicity increases with long-term use. Currently, there are eight NRTIs: abacavir, didanosine, emtricitabine, lamivudine, stavudine, zalcitabine, zidovudine, and tenofovir disoproxil fumarate (TDF), a nucleotide RT inhibitor. The new drugs that are being investigated are Apricitabine, Elvucitabine, and Amdoxovir.

10.6.1.3 Protease Inhibitors

The HIV-1 protease is the enzyme responsible for the cleavage of the viral Gag and Gag–Pol polyprotein precursors during virion maturation. PIs have high genetic barrier for drug resistance having great plasticity with polymorphism observed in 49 of the 99 codons and more than 20 substitutions known to be associated with resistance.

All PIs share relatively similar chemical structures and cross-resistance is commonly observed. For most PIs, primary resistance mutations cluster near the active site of the enzyme, at positions located at the substrate/inhibitor-binding site (e.g., D30N, G48V, I50V, V82A, or I84V, among others). These amino acid changes usually have a deleterious effect on the replicative fitness. In addition to mutations in the protease gene, changes located within eight major protease cleavage sites (i.e., Gag and Pol genes) have been associated with resistance to PIs. Cleavage site mutants are better substrates for the mutated protease and thus partially compensate for the resistance-associated loss of viral fitness.[89] The limitations of PIs include insulin resistance, dyslipidemia, hypertriglyceridemia, high risk of coronary artery disease, and clinically significant interactions with antifungals, antimycobacterials, hormonal contraceptives, HMG-coenzyme reductase inhibitors, antihistaminics, anticonvulsants, psychotropics, ergot alkaloids, and sedatives.[92]

10.6.1.4 Integrase Inhibitors

Integrase inhibitors inhibit strand transfer of viral DNA to host-cell DNA by targeting the strand transfer reaction and are thus referred to as either INIs or, more specifically, integrase strand transfer inhibitors (InSTIs). Raltegravir, MK-0518, was approved in 2007, and other integrase inhibitors, including Elvitegravir (GS-9137) are progressing through clinical development. Raltegravir has potential for use at all stages of HIV treatment in treatment-experienced and naïve patients due to significant antiviral activity, short-term safety, and tolerability. Dolutegravir (DTG), currently in phase III clinical trials, is effective at reducing HIV VL and raising CD4 counts in integrase-naive patients. There is no evidence of emergent resistance to DTG in virologic failure. DTG plus NRTIs could be an option for first-line HIV treatment.[93]

Entry inhibitors: A range of hematopoietic cells, including monocyte macrophages, B lymphocytes, eosinophils, and dendritic cells, as well as columnar epithelial cells, have been found to be infected by HIV-1, but the CD4-positive helper T lymphocyte has been identified as the primary target for HIV-1 infection.[94] HIV-1 enters CD4-positive T cells through direct interaction of the viral envelope gp120 with the D1 region of the CD4 receptor on the cell surface of target cells. The interaction of gp120 with CD4 causes a conformational change in the viral envelope gp120, resulting in exposure of the gp41 transmembrane envelope protein that subsequently inserts into and fuses with the target cell membrane. HIV-1 envelope proteins interact with coreceptor molecules on the surface of CD4-positive cells, typically either the α-chemokine receptor CXCR4 or the β-chemokine receptor CCR5 or both, to trigger the fusion of the viral and cellular membranes. To stop the entry of virus, new fusion inhibitors and chemokine receptor antagonists are there in the market.

10.6.1.5 Chemokine Receptor Inhibitors

The HIV coreceptors belong to the family of chemokine receptors—seven-transmembrane G-protein-coupled receptors with prominent roles in immune cell trafficking. These receptors have three extracellular and intracellular loops, extracellular N-termini, and intracellular C-termini. While several different chemokine receptors are capable of mediating HIV entry *in vitro*, current evidence suggests that the CCR5 and CXCR4 chemokine receptors are the most frequently utilized *in vivo*.[95–98] Viruses capable of utilizing CCR5 alone, CXCR4 alone, or both coreceptors are labeled R5-tropic, X4-tropic, and RX54 or dual-tropic viruses, respectively. CCR5 is the primary coreceptor for the majority of HIV-1 isolates and is expressed on CD4+ T cell subsets, macrophages, and dendritic cells, while CXCR4 is less commonly used but is expressed on a wide variety of cells both within and outside the immune system.[99] An interaction

between HIV-1 and CCR5 can be prevented by small-molecule antagonists, monoclonal antibodies, and modified natural CCR5 ligands:

1. *CCR5 receptor inhibitors*: Aplaviroc was the first CCR5 inhibitor and has been followed by maraviroc and vicriviroc. Further evaluation of aplaviroc and vicriviroc has been discontinued due to adverse reactions and poor antiviral activity, respectively. Maraviroc was used for combination therapy in 2007 that blocks the HIV-1 gp120 protein from associating with the CCR5 coreceptor. The drug is well-tolerated with minimal side effects like abdominal pain, asthenia, and postural hypotension. Maraviroc is a substrate for the CYP 3A4, and therefore, dose adjustment is necessary when coadministered with other drugs that use the same pathway. A tropism testing is also necessary prior to use of maraviroc as patients with CXCR4 or mixed HIV-1 do not respond well. In addition, resistance patterns have already been observed with CCR5 inhibitors, either through selection of minority variants of CXCR4 or a dual/mixed tropic virus or by development of mutations.[100] This is potentially dangerous as CXCR4 tropic virus is associated with greater and faster decline in CD4 counts.

2. *CXCR4 chemokine receptor inhibitors*: A variety of CXCR4 antagonists have also been developed, and while they exhibit potent anti-HIV activity *in vitro* (against X4 but not R5 virus strains), administration of these drugs *in vivo* results in mobilization of HSCs from the bone marrow to the peripheral blood, highlighting the important role of CXCR4 in HSC homing.[101] Although this side effect limits their use in HIV-infected individuals, the CXCR4 antagonist *plerixafor* is currently used to mobilize HSCs for subsequent autologous transfer in patients with non-Hodgkin lymphoma and multiple myeloma.[102] AMD-070 is an investigational compound with potent *in vitro* antiviral activity against wild-type X4 virus but with no action against R5-tropic virus.

10.6.1.6 Fusion Inhibitors

The first generation of fusion inhibitors was enfuvirtide, a peptidic mimic of the HR2 domain of the HXB2 isolate, and acts by competitively binding to HR1 domain of gp41, there by blocking formation of the six-helix bundle blocking formation of the six-helix bundle. Enfuvirtide is a synthetic peptide and is structurally similar to a section of gp41. It blocks the conformational changes in gp41 and hence blocks virus entry into the host cell. An advantage of enfuvirtide over CCR5 inhibitors is that it targets both R5- and X4-tropic viruses.[103] Parenteral administration of the drug, local site reactions, cost, and inconvenience associated with its use place the drug in a reserve category for patients when all treatment fails. Resistance to enfuvirtide develops within a few weeks due to mutation of HR1 region of gp41 as well as HR2, which indicates low genetic barrier.[104] The other promising drugs that are at different stages of development are sifuvirtide and T1249.

10.6.1.7 Other Viral and Host Targets

Considerable research progress has been made to understand and target other viral proteins. They include proteins like maturation inhibitors, capsid assembly, nucleocapsid protein, and HIV-1 RNase H and HIV-1 regulatory and accessory proteins like Tat, Rev, nef, Vpu, Vpr, and nif. The host factors also play a major role, which is restricting the viral replication. Knowledge of the mechanisms by which restriction factors interfere with HIV-1 replication and how their effects are avoided by HIV-1 in human cells could allow for novel forms of therapeutic intervention. The host targets include tetherin, APOBEC3G, lens epithelium-derived growth factor (LEDGF/p75), tumor susceptibility gene 101 (TSG101), DEAD-box RNA helicase DDX3, the NFκB transcription factor, extracellular signal-regulated kinases (ERK), mitogen-activated protein kinase (MAPK), and prostaglandin synthetase 2 (PTGS2) that control the viral replication.

10.6.1.8 Problem of Latent Reservoirs

HIV-1 is known to establish latent reservoirs where the virus is maintained for long periods of time in an essentially quiescent state. Low-rate viral replication also comes from anatomical sites, such as the brain, where drug penetration is limited and only suboptimal drug concentration can be achieved.[105] The long-lived HIV-1 reservoirs like cells from monocyte–macrophage lineage and resting memory CD4+ T cells that are established early during primary infection constitute a major obstacle to virus eradication. Eradication of latently infected cells by targeting with antiretroviral agents and other methods is evaluated to get functional cure. A new strategy so-called shock and kill has been recently proposed to eradicate the virus from infected patients. The main objective is to facilitate the reactivation of viruses from the latent reservoirs, naturally (via host immune system or viral cytopathic effects), which are then destroyed by HAART.[106,107] Many factors have been involved in reactivation including physiological stimuli, chemical compounds (phorbol esters), histone deacetylase inhibitors (HDACIs), P-TEFb activators, and activation with antibodies (anti-CD3).[108]

10.6.1.9 Immunotherapy

Strengthening immune response by immune modulators is another strategy for treatment of HIV-1 infection. Several immune modulators are in the preclinical stage of development. ALFERON, human leukocyte-derived interferon alfa-n3 developed by Hemispherx Biopharma, is currently in phase III clinical trials.[109] Recombinant human IL-7 called CYT107, developed by Cytheris, is in phase II clinical trials with raltegravir and maraviroc, with the hope of improving T cell counts in patients classified as immunological nonresponders on antiviral therapy.[110] IL-7 has a crucial stimulatory effect on T lymphocyte development and on homeostatic expansion of peripheral T cells. Tarix Pharmaceuticals is

developing TXA127, angiotensin 1–7, to stimulate bone marrow production of progenitor cells.[111] TXA127 is currently in phase I studies. Several immune modulators designed to signal immune cells to respond to infection have antiviral activity. They act through different mechanisms such as downregulation of cell surface antigen expression, thus reducing virus replication through an entry blocking mechanism. Many immune modulators will either inhibit or induce cellular proliferation of specific cell types.

10.6.1.10 Gene Therapy

Anti-HIV/AIDS gene therapy may be one of the most promising strategies, although challenging, to eradicate HIV-1 infection. In fact, genetic modification of HSCs with one or multiple therapeutic genes is expected to generate blood cell progenies resistant to viral infection and thereby eventually eliminate infected unprotected cells. Ultimately, protected cells will reestablish a functional immune system able to control HIV-1 replication. Sixty gene therapy clinical trials against AIDS employing different viral vectors and transgenes are currently ongoing worldwide.[112]

Novel gene therapy strategies may have the potential to reverse the infection by eradicating HIV-1. For example, expression of LTR-specific recombinase (Tre-recombinase) has been shown to result in chromosomal excision of proviral DNA and, as a consequence, in the eradication of HIV-1 from infected cell cultures. However, the delivery of Tre-recombinase currently depends on the genetic manipulation of target cells, a process that complicates such therapeutic approaches and, thus, might be undesirable in a clinical setting. It has been demonstrated that *Escherichia coli* expressed Tre-recombinases, tagged either with the protein transduction domain (PTD) from the HIV-1 Tat transactivator or the translocation motif (TLM) of the HBV PreS2 protein; were able to translocate efficiently into cells; and showed significant recombination activity on HIV-1 LTR sequences. Furthermore, the TLM-tagged enzyme was able to excise the full-length proviral DNA from chromosomal integration sites of HIV-1-infected HeLa and CEM-SS cells. This study showing Tre-recombinase activity on integrated HIV-1 can provide the basis for the nongenetic transient application of engineered recombinases, which may be a valuable component of future HIV eradication strategies.[113]

Genetic knockout of CCR5: The success story of bone marrow transplant to HIV patients having cancer eliminating HIV completely showed great promise and created a hope for HIV cure. This prompted several researchers to modify CCR5 gene. Several studies have successfully decreased expression of CCR5 in primary human HSCs, CD4 T cells, and macrophages *in vitro* using a variety of unique genetic approaches. Dominant negative forms of CCR5, anti-CCR5 intracellular antibodies (intrabodies), and a number of RNA-based approaches including RNA interference (RNAi), short-hairpin RNAs (shRNA), and ribozymes have been all been shown to decrease CCR5 expression.[114–116] In one of the recent studies simultaneously targeting CCR5, the viral genes *Tat* and *Rev* and the viral transactivating region (TAR) resulted in the so-called triple-R phase I trial where four patients with AIDS-related lymphoma who were scheduled to receive autologous HSC transplants had a fraction of their infused stem cells modified by the triple-R construct.[117] The study was successfully completed without toxicity related to the genetic modification of HSCs suggesting that modification of these long-term progenitor cells is a safe and viable option for future genetic therapies aimed at eliminating HIV infection. However, achieving lasting stable expression remains a problem. Despite their promise, genetic approaches targeting CCR5 are currently incapable of achieving the complete genetic knockout of CCR5 that exists in *ccr5Δ32* homozygous individuals and was accomplished in the Berlin patient. Whether the incomplete ablation of CCR5 provided by these genetic approaches will be able to provide long-term or indeed any control of established HIV infection remains to be seen. Considerable research is in progress to introduce somatic mutations into the *ccr5* gene with the ultimate goal of phenocopying the natural *ccr5Δ32* mutation by rendering a fraction of cells *ccr5* negative.

10.6.2 Prevention of HIV Infection

A number of HIV prevention methods are available, including male and female condoms, voluntary medical male circumcision, prevention of mother-to-child HIV transmission (PMTCT), and harm-reduction strategies such as provision of sterile injecting equipment and opiate substitution therapy for people who inject drugs.

The field of HIV prevention, until recently, experienced years of disappointment, as the search for potential vaccines and non-antiretroviral microbicides has yielded little result. Now, however, a promising new approach has emerged: the use of antiretroviral drugs for HIV prevention, both for those uninfected and for those already living with HIV. These recommendations have been developed specifically to address the daily use of antiretrovirals in HIV-uninfected people to block the acquisition of HIV infection. This prevention approach is known as preexposure prophylaxis (PrEP). All these have contributed to a leveling of the rate of new infections in some countries. Elsewhere, however, the momentum of the epidemic remains strong. In 2010 alone, an estimated 2.7 million people became newly infected with HIV. Additional safe and effective approaches to HIV prevention are urgently needed.

10.6.2.1 Vaccines

Vaccines are the most safe, effective, and inexpensive way of preventing any epidemic and allow their use in large population. There has been remarkable progress in understanding HIV biology and development of antiretroviral drugs and other preventive mechanisms including HIV vaccines. Despite all the progress, effective HIV prevention is still a challenge as there are no good drugs or vaccines. Vaccine challenges include the genetic diversity and mutability of HIV-1, which create a plethora of constantly changing antigens, the structural features of the viral envelope glycoprotein that disguise conserved receptor-binding sites from the immune system, and the presence of carbohydrate moieties that shield potential epitopes from antibodies.

However, there has been significant scientific progress in recent years. In 2009, a large-scale clinical trial known as RV144 demonstrated that a HIV-1 vaccine could modestly reduce the incidence of HIV-1 infection. Further, the identification of broadly neutralizing monoclonal antibodies (such as VRC01, a human monoclonal antibody capable of neutralizing over 90% of natural HIV-1 isolates, as well as PG and PGT antibodies that recognize conserved glycopeptide epitopes) has revealed new opportunities for vaccine design.[118]

Recent progress: In 2010, scientists discovered two potent human antibodies that can stop more than 90% of known global HIV strains from infecting human cells in the laboratory. The antibodies, known as VRC01 and VRC02, are naturally occurring and were found using a novel molecular approach that honed in on the specific cells that make antibodies against HIV. Both VRC01 and VRC02 were found to neutralize more HIV strains with greater overall strength than previously known antibodies to the virus. The atomic-level structure of VRC01 when attached to HIV also was determined, helping define precisely where and how the antibody attaches to the virus. With this knowledge, scientists have begun to design components of a candidate vaccine that could teach the human immune system to make antibodies similar to VRC01 and that might prevent infection by the vast majority of HIV strains worldwide.

Other approaches to generating broadly neutralizing antibodies: The IAVI Neutralizing Antibody Center has developed antibodies directed to the V1 and V2 region of the HIV-1 gp120 named PG9 and PG16, capable of achieving neutralization up to 70%–80% of circulating HIV-1 isolates. Additionally, scientists at the Center for HIV/AIDS Vaccine Immunology (CHAVI) have identified a clonal lineage of four V1V2-directed broadly neutralizing antibodies (CH01–CH04) from a Tanzanian elite controller.

Exciting new technologies and the recent identification of broadly neutralizing HIV-1 antibodies enable us to approach HIV-1 vaccine development in ways that were previously unanticipated. It is clear, for example, that broadly neutralizing antibodies are not as rare as previously thought and that the immune system can effectively target highly conserved regions on diverse strains of HIV-1.

CHAVI scientists also characterized the critical role of the T cell immune response in early virus control. Through analysis of host immune responses to HIV infection, they showed that the first CD8+ T cells, despite limited breadth and very rapid virus escape, suppressed HIV as the amount of HIV in the blood was declining from a peak level. This implies that vaccine-induced HIV-specific T cells could contribute to the control of acute viremia (amount of HIV in the blood) if they are present before or early in HIV infection.

Another important study found that a new HIV vaccination strategy using a "mosaic" design could expand the breadth and depth of immune responses in rhesus monkeys. The mosaic vaccine was designed through computational methods that created small sets of highly variable artificial viral proteins. When combined, these proteins theoretically could provide nearly optimal coverage of the diverse forms of HIV circulating in the world. Vaccine Research Center (VRC) researchers also have developed a new "scaffold" strategy, which would prime the immune system to recognize certain protein structures on the viral surface and produce antibodies that bind to those structures and neutralize HIV. Several recently published HIV vaccine trials have shown that different HIV vaccine prime/boost combinations can greatly affect the immune response generated, but mechanistic insights into their modes of action are lacking.[118]

10.6.2.2 Preexposure Prophylaxis

Preexposure chemoprophylaxis (now commonly referred to as PrEP) of HIV infection has gained increased momentum, concomitantly with the successful use of combination drug regimens for the treatment of AIDS. A pivotal component in the current drug combination regimens for the treatment of AIDS as well as the ongoing PrEP trials is TDF (Viread1) and its combination with emtricitabine (FTC). The combination of TDF with FTC has been marketed as Truvada1. TDF and TDF/FTC have proven effective, if orally administered daily or intermittently, in the prevention of rectal SHIV infection in macaques.

Topical tenofovir gel has proven effective in the prevention of HIV infection in women in South Africa. Oral TDF/FTC has proven effective in the prevention of HIV infection in men having sex with men, and recent data indicate that oral TDF/FTC is also effective in preventing HIV infection in serodiscordant couples in Botswana, Kenya, and Uganda. Other PrEP studies are still ongoing. Available data point to the efficacy and safety of TDF with or without FTC in the prophylaxis of HIV infection (AIDS).

People at highest risk for getting HIV include injection drug users, infants born to mothers with HIV who did not receive HIV therapy during pregnancy, and people who have unprotected sex. HIV spreads through sexual contact including oral, vaginal, and anal sex; through blood transfusions and needle sharing; and from mother to fetus through their shared blood circulation, or a nursing mother can pass it to her baby in her breast milk. Appropriate prevention strategies at every possible transmission stage will reduce HIV incidence.[119,120]

10.7 FUTURE PERSPECTIVES

The *XIX International AIDS Conference 2012* held in Washington, DC, called for AIDS-free generation adopting the theme "Turning the Tide Together," which emphasized "treatment for prevention," more broadly the "strategic use of antiretrovirals (ARVs)." Future strategies for HIV prevention and control are likely to involve enhanced use of PrEP, new combination ARV therapies, targeted treatment of individuals identified as recently infected, eradication of the latent reservoirs in chronically infected individuals, and development of new preventive and therapeutic vaccines. These strategies will require a global collaborative effort to minimize infection of high-risk populations and effectively treat already infected individuals in order to achieve the future goal of an AIDS-free generation.

DISCLAIMER

The writing and conclusions in this chapter have not been formally disseminated by the Food and Drug Administration and should not be construed to represent any agency determination or policy.

REFERENCES

1. Sepkowitz, K.A. AIDS—The first 20 years. *N Engl J Med* 344, 1764–1772 (2001).
2. Rom, W.N. and Markowitz, S.B. *Environmental and Occupational Medicine*, 4th edn. Philadelphia, PA: Wolters Kluwer/Lippincott Williams & Wilkins, p. 745 (2007).
3. HIV and Its Transmission. Centers for Disease Control and Prevention (2006). http://www.hivlawandpolicy.org/resources/hiv-and-its-transmission-centers-disease-control-and-prevention
4. GLOBAL HIV/AIDS RESPONSE. Epidemic update and health sector progress towards Universal Access—Progress Report (2010). www.unaids.org/documents/20101123_globalreport_em.pdf
5. Gallo, R.C. et al. Isolation of human T-cell leukemia virus in acquired immune deficiency syndrome (AIDS). *Science* 220, 865–867 (1983).
6. Barre-Sinoussi, F. et al. Isolation of a T-lymphotropic retrovirus from a patient at risk for acquired immune deficiency syndrome (AIDS). *Science* 220, 868–871 (1983).
7. Gottlieb, M.S. et al. *Pneumocystis carinii* pneumonia and mucosal candidiasis in previously healthy homosexual men: Evidence of a new acquired cellular immunodeficiency. *N Engl J Med* 305, 1425–1431 (1981).
8. UNAIDS. Report on the Global AIDS Epidemic (2010). http://www.unaids.org/documents/20101123_globalreport_em.pdf
9. Korber, B. et al. Evolutionary and immunological implications of contemporary HIV-1 variation. *Br Med Bull* 58, 19–42 (2001).
10. Gao, F. et al. Evidence of two distinct subsubtypes within the HIV-1 subtype A radiation. *AIDS Res Hum Retroviruses* 17, 675–688 (2001).
11. Clavel, F. et al. Molecular cloning and polymorphism of the human immune deficiency virus type 2. *Nature* 324, 691–695 (1986).
12. Ueno, F. et al. Vpx and Vpr proteins of HIV-2 up-regulate the viral infectivity by a distinct mechanism in lymphocytic cells. *Microbes Infect* 5, 387–395 (2003).
13. Fletcher, T.M. et al. Nuclear import and cell cycle arrest functions of the HIV-1 Vpr protein are encoded by two separate genes in HIV-2/SIV(SM). *EMBO J* 15, 6155–6165 (1996).
14. Vallari, A. et al. Confirmation of putative HIV-1 group P in Cameroon. *J Virol* 85, 1403–1407 (2011).
15. Plantier, J.C. et al. A new human immunodeficiency virus derived from gorillas. *Nat Med* 15, 871–872 (2009).
16. Roques, P. et al. Phylogenetic characteristics of three new HIV-1 N strains and implications for the origin of group N. *AIDS* 18, 1371–1381 (2004).
17. Ayouba, A. et al. HIV-1 group O infection in Cameroon, 1986 to 1998. *Emerg Infect Dis* 7, 466–467 (2001).
18. Simon, F. et al. Identification of a new human immunodeficiency virus type 1 distinct from group M and group O. *Nat Med* 4, 1032–1037 (1998).

19. Rambaut, A. et al. Human immunodeficiency virus. Phylogeny and the origin of HIV-1. *Nature* 410, 1047–1048 (2001).

20. Abecasis, A.B. et al. Quantifying differences in the tempo of human immunodeficiency virus type 1 subtype evolution. *J Virol* 83, 12917–12924 (2009).

21. Worobey, M. et al. Direct evidence of extensive diversity of HIV-1 in Kinshasa by 1960. *Nature* 455, 661–664 (2008).

22. Yamaguchi, J. et al. HIV infections in northwestern Cameroon: Identification of HIV type 1 group O and dual HIV type 1 group M and group O infections. *AIDS Res Hum Retroviruses* 20, 944–957 (2004).

23. Yamaguchi, J. et al. Evaluation of HIV type 1 group O isolates: Identification of five phylogenetic clusters. *AIDS Res Hum Retroviruses* 18, 269–282 (2002).

24. Rousseau, C.M. et al. Extensive intrasubtype recombination in South African human immunodeficiency virus type 1 subtype C infections. *J Virol* 81, 4492–4500 (2007).

25. Peeters, M. et al. Characterization of a highly replicative intergroup M/O human immunodeficiency virus type 1 recombinant isolated from a Cameroonian patient. *J Virol* 73, 7368–7375 (1999).

26. Robertson, D.L. et al. HIV-1 nomenclature proposal. *Science* 288, 55–56 (2000).

27. Database, H. http://www.hiv.lanl.gov/content/sequence/HIV/CRFs/CRFs.html (2012).

28. Hemelaar, J. et al. Global trends in molecular epidemiology of HIV-1 during 2000–2007. *AIDS* 25, 679–689 (2011).

29. Cheng, Z. et al. A genome-wide comparison of recent chimpanzee and human segmental duplications. *Nature* 437, 88–93 (2005).

30. Leonard, J.T. and Roy, K. The HIV entry inhibitors revisited. *Curr Med Chem* 13, 911–934 (2006).

31. Chiu, I.M. et al. Nucleotide sequence evidence for relationship of AIDS retrovirus to lentiviruses. *Nature* 317, 366–368 (1985).

32. Vogt, P.K. *Historical Introduction to the General Properties of Retroviruses*. New York: Cold Spring Harbor Laboratory Press (1997).

33. Wain-Hobson, S., Alizon, M., and Montagnier, L. Relationship of AIDS to other retroviruses. *Nature* 313, 743 (1985).

34. Centers for Disease Control and Prevention. Pneumocystis pneumonia–Los Angeles. *Morb Mortal Wkly Rep* 30, 250–252 (1981).

35. Quinn, T.C. et al. AIDS in Africa: An epidemiologic paradigm. *Science* 234, 955–963 (1986).

36. Gallo, R.C. and Montagnier, L. AIDS in 1988. *Sci Am* 259, 41–48 (1988).

37. Gallo, R. et al. HIV/HTLV gene nomenclature. *Nature* 333, 504 (1988).

38. Borsetti, A. et al. CD4-independent infection of two CD4(−)/CCR5(−)/CXCR4(+) pre-T-cell lines by human and simian immunodeficiency viruses. *J Virol* 74, 6689–6694 (2000).

39. McDougal, J.S. et al. Binding of the human retrovirus HTLV-III/LAV/ARV/HIV to the CD4 (T4) molecule: Conformation dependence, epitope mapping, antibody inhibition, and potential for idiotypic mimicry. *J Immunol* 137, 2937–2944 (1986).

40. Cho, M.W. et al. Identification of determinants on a dualtropic human immunodeficiency virus type 1 envelope glycoprotein that confer usage of CXCR4. *J Virol* 72, 2509–2515 (1998).

41. De Jong, J.J. et al. Minimal requirements for the human immunodeficiency virus type 1 V3 domain to support the syncytium-inducing phenotype: Analysis by single amino acid substitution. *J Virol* 66, 6777–6780 (1992).

42. Groenink, M. et al. Relation of phenotype evolution of HIV-1 to envelope V2 configuration. *Science* 260, 1513–1516 (1993).

43. Hoffman, N.G. et al. Variability in the human immunodeficiency virus type 1 gp120 Env protein linked to phenotype-associated changes in the V3 loop. *J Virol* 76, 3852–3864 (2002).

44. Rizzuto, C.D. et al. A conserved HIV gp120 glycoprotein structure involved in chemokine receptor binding. *Science* 280, 1949–1953 (1998).

45. Fouchier, R.A. and Malim, M.H. Nuclear import of human immunodeficiency virus type-1 preintegration complexes. *Adv Virus Res* 52, 275–299 (1999).

46. Gallay, P. et al. HIV-1 infection of nondividing cells through the recognition of integrase by the importin/karyopherin pathway. *Proc Natl Acad Sci USA* 94, 9825–9830 (1997).

47. Bushman, F.D. et al. Retroviral DNA integration directed by HIV integration protein in vitro. *Science* 249, 1555–1558 (1990).

48. Nekhai, S. and Jeang, K.T. Transcriptional and post-transcriptional regulation of HIV-1 gene expression: Role of cellular factors for Tat and Rev. *Future Microbiol* 1, 417–426 (2006).

49. Bieniasz, P.D. Late budding domains and host proteins in enveloped virus release. *Virology* 344, 55–63 (2006).

50. Strebel, K. et al. A novel gene of Hiv-1, Vpu, and its 16-kilodalton product. *Science* 241, 1221–1223 (1988).

51. Neil, S.J., Zang, T., and Bieniasz, P.D. Tetherin inhibits retrovirus release and is antagonized by HIV-1 Vpu. *Nature* 451, 425–430 (2008).

52. Castro-Nallar, E. et al. Genetic diversity and molecular epidemiology of HIV transmission. *Future Virol* 7, 239–252 (2012).

53. Tebit, D.M. and Arts, E.J. Tracking a century of global expansion and evolution of HIV to drive understanding and to combat disease. *Lancet Infect Dis* 11, 45–56 (2011).

54. Moir, S. et al. Pathogenic mechanisms of HIV disease. *Annu Rev Pathol* 6, 223–248 (2011).

55. Busch, M.P. and Satten, G.A. Time course of viremia and antibody seroconversion following human immunodeficiency virus exposure. *Am J Med* 102, 117–124; discussion 125–126 (1997).

56. Okulicz, J.F. and Lambotte, O. Epidemiology and clinical characteristics of elite controllers. *Curr Opin HIV AIDS* 6, 163–168 (2011).

57. Curriu, M. et al. Viremic HIV infected individuals with high CD4 T cells and functional envelope proteins show anti-gp41 antibodies with unique specificity and function. *PLoS One* 7, e30330 (2012).

58. Catalfamo, M. et al. The role of cytokines in the pathogenesis and treatment of HIV infection. *Cytokine Growth Factor Rev* 23, 207–214 (2012).

59. Fiebig, E.W. et al. Dynamics of HIV viremia and antibody seroconversion in plasma donors: Implications for diagnosis and staging of primary HIV infection. *AIDS* 17, 1871–1879 (2003).

60. Owen, S.M. et al. Alternative algorithms for human immunodeficiency virus infection diagnosis using tests that are licensed in the United States. *J Clin Microbiol* 46, 1588–1595 (2008).

61. Ly, T.D. et al. Evaluation of the sensitivity and specificity of six HIV combined p24 antigen and antibody assays. *J Virol Methods* 122, 185–194 (2004).

62. Ly, T.D. et al. Seven human immunodeficiency virus (HIV) antigen-antibody combination assays: Evaluation of HIV seroconversion sensitivity and subtype detection. *J Clin Microbiol* 39, 3122–3128 (2001).

63. Murphy, G. and Aitken, C. HIV testing—The perspective from across the pond. *J Clin Virol* 52, S71–S76 (2011).

64. Weber, B. Screening of HIV infection: Role of molecular and immunological assays. *Expert Rev Mol Diagn* 6, 399–411 (2006).

65. Weber, B. et al. Multicenter evaluation of a new automated fourth-generation human immunodeficiency virus screening assay with a sensitive antigen detection module and high specificity. *J Clin Microbiol* 40, 1938–1946 (2002).

66. Ly, T.D. et al. Could the new HIV combined p24 antigen and antibody assays replace p24 antigen specific assays? *J Virol Methods* 143, 86–94 (2007).

67. Layne, S.P. et al. Factors underlying spontaneous inactivation and susceptibility to neutralization of human immunodeficiency virus. *Virology* 189, 695–714 (1992).

68. Bentsen, C. et al. Performance evaluation of the Bio-Rad Laboratories GS HIV Combo Ag/Ab EIA, a 4th generation HIV assay for the simultaneous detection of HIV p24 antigen and antibodies to HIV-1 (groups M and O) and HIV-2 in human serum or plasma. *J Clin Virol* 52, S57–S61 (2011).

69. Chavez, P. et al. Evaluation of the performance of the Abbott ARCHITECT HIV Ag/Ab Combo Assay. *J Clin Virol* 52, S51–S55 (2011).

70. Pandori, M.W. et al. Assessment of the ability of a fourth-generation immunoassay for human immunodeficiency virus (HIV) antibody and p24 antigen to detect both acute and recent HIV infections in a high-risk setting. *J Clin Microbiol* 47, 2639–2642 (2009).

71. UNITAID HIV/AIDS Diagnostic Technology Landscape 2nd edn., (2012).

72. Cohen, M.S. et al. The detection of acute HIV infection. *J Infect Dis* 202, S270–S277 (2010).

73. Clinical and Laboratory Standards Institute (CLSI). Criteria for laboratory testing and diagnosis of human immunodeficiency virus infection approved guideline. M53A, 31, 13 (2011).

74. Stekler, J.D. et al. HIV testing in a high-incidence population: Is antibody testing alone good enough? *Clin Infect Dis* 49, 444–453 (2009).

75. Schito, M.L. et al. Challenges for rapid molecular HIV diagnostics. *J Infect Dis* 201, S1–S6 (2010).

76. Curtis, K.A. et al. Rapid detection of HIV-1 by reverse-transcription, loop-mediated isothermal amplification (RT-LAMP). *J Virol Methods* 151, 264–270 (2008).

77. Tang, W. et al. Nucleic acid assay system for tier II laboratories and moderately complex clinics to detect HIV in low-resource settings. *J Infect Dis* 201, S46–S51 (2010).

78. Lee, H.H. et al. Simple amplification-based assay: A nucleic acid-based point-of-care platform for HIV-1 testing. *J Infect Dis* 201, S65–S72 (2010).

79. Pau, C.P. et al. Development of a simple, rapid and inexpensive method for the qualitative detection of HIV-1 RNA. *Proceedings of the 2010 HIV Diagnostics Conference*, Orlando, FL (2010).

80. Lilian, R.R. et al. Early diagnosis of human immunodeficiency virus-1 infection in infants with the NucliSens EasyQ assay on dried blood spots. *J Clin Virol* 48, 40–43 (2010).

81. Lofgren, S.M. et al. Evaluation of a dried blood spot HIV-1 RNA program for early infant diagnosis and viral load monitoring at rural and remote healthcare facilities. *AIDS* 23, 2459–2466 (2009).

82. Mazzola, L.T. et al. Innovative point-of-care HIV viral load detection in RLS. *Proceedings of the 2010 HIV Diagnostics Conference*, Orlando, FL (2010).

83. Early infant diagnosis of HIV through dried blood spot testing. Kenya's prevention of mother to child transmission project, Pathfinder International (2007). http://www.pathfinder.org/publications-tools/Early-Infant-Diagnosis-of-HIV-through-Dried-Blood-Spot-Tesing-Pathfinder-InternationalKenyas-Prevention-of-Mother-to-Child-Transmission-Project.html

84. Sherman, G.G. et al. Oral fluid tests for screening of human immunodeficiency virus-exposed infants. *Pediatr Infect Dis J* 29, 169–172 (2010).

85. Kohlstaedt, L.A. et al. Crystal structure at 3.5 A resolution of HIV-1 reverse transcriptase complexed with an inhibitor. *Science* 256, 1783–1790 (1992).

86. Vingerhoets, J. et al. TMC125 displays a high genetic barrier to the development of resistance: Evidence from in vitro selection experiments. *J Virol* 79, 12773–12782 (2005).

87. Poveda, E. et al. Prevalence of etravirine (TMC-125) resistance mutations in HIV-infected patients with prior experience of non-nucleoside reverse transcriptase inhibitors. *J Antimicrob Chemother* 60, 1409–1410 (2007).

88. de Bethune, M.P. et al. A new potent NNRTI, with an increased barrier to resistance and good pharmacokinetic profile. *12th Conference on Retroviruses and Opportunistic Infections* Boston, MA (2005).

89. Arts, E.J. and Hazuda, D.J. HIV-1 Antiretroviral drug therapy. *Cold Spring Harbor Perspect Med* 2, a007161 (2012).

90. Back, D.J. et al. The pharmacology of antiretroviral nucleoside and nucleotide reverse transcriptase inhibitors: Implications for once-daily dosing. *J AIDS* 39, S1–S23, quiz S24–S25 (2005).

91. Carr, A. et al. A syndrome of peripheral lipodystrophy, hyperlipidaemia and insulin resistance in patients receiving HIV protease inhibitors. *AIDS* 12, F51–F58 (1998).

92. Shah, S.M.J. et al. Identification of drug interactions involving ART in New York City HIV specialty clinics. *Abstracts of the Fourteenth Conference on Retroviruses and Opportunistic Infections*, February 25–28, 2007, Los Angeles, CA, p. 277. Abstract 573 (2007).

93. Shamroe, C.L. et al. Update on raltegravir and the development of new integrase strand transfer inhibitors. *Southern Med J* 105, 370–378 (2012).

94. Hladik, F. and McElrath, M.J. Setting the stage: Host invasion by HIV. *Nat Rev Immunol* 8, 447–457 (2008).

95. Edinger, A.L. et al. Chemokine and orphan receptors in HIV-2 and SIV tropism and pathogenesis. *Virology* 260, 211–221 (1999).

96. Gilliam, B.L. et al. Clinical use of CCR5 inhibitors in HIV and beyond. *J Transl Med* 9, S9 (2011).

97. Jiang, C. et al. Primary infection by a human immunodeficiency virus with atypical coreceptor tropism. *J Virol* 85, 10669–10681 (2011).

98. Zhang, Y.J. et al. Use of coreceptors other than CCR5 by non-syncytium-inducing adult and pediatric isolates of human immunodeficiency virus type 1 is rare in vitro. *J Virol* 72, 9337–9344 (1998).

99. Bleul, C.C. et al. The HIV coreceptors CXCR4 and CCR5 are differentially expressed and regulated on human T lymphocytes. *Proc Natl Acad Sci USA* 94, 1925–1930 (1997).

100. Daar, E.S. Emerging resistance profiles of newly approved antiretroviral drugs. *Topics HIV Med* 16, 110–116 (2008).

101. Liles, W.C. et al. Augmented mobilization and collection of CD34+ hematopoietic cells from normal human volunteers stimulated with granulocyte-colony-stimulating factor by single-dose administration of AMD3100, a CXCR4 antagonist. *Transfusion* 45, 295–300 (2005).

102. Brave, M. et al. FDA review summary: Mozobil in combination with granulocyte colony-stimulating factor to mobilize hematopoietic stem cells to the peripheral blood for collection and subsequent autologous transplantation. *Oncology* 78, 282–288 (2010).

103. Berkhout, B. et al. Is there a future for antiviral fusion inhibitors? *Curr Opin Virol* 2, 50–59 (2012).

104. Molto, J. et al. Increased antiretroviral potency by the addition of enfuvirtide to a four-drug regimen in antiretroviral-naive, HIV-infected patients. *Antiviral Ther* 11, 47–51 (2006).

105. Tuomela, M. et al. Validation overview of bio-analytical methods. *Gene Ther* 12, S131–S138 (2005).

106. Geeraert, L. et al. Hide-and-seek: The challenge of viral persistence in HIV-1 infection. *Annu Rev Med* 59, 487–501 (2008).

107. Savarino, A. et al. "Shock and kill" effects of class I-selective histone deacetylase inhibitors in combination with the glutathione synthesis inhibitor buthionine sulfoximine in cell line models for HIV-1 quiescence. *Retrovirology* 6, 52 (2009).

108. Lafeuillade, A. Eliminating the HIV reservoir. *Curr HIV/AIDS Rep* 9, 121–131 (2012).

109. Ptak, R.G. et al. Cataloguing the HIV type 1 human protein interaction network. *AIDS Res Hum Retroviruses* 24, 1497–1502 (2008).

110. Tavel, J.A. et al. Effects of intermittent IL-2 alone or with peri-cycle antiretroviral therapy in early HIV infection: The STALWART study. *PLoS One* 5, e9334 (2010).

111. Anon. CYT107 enters Phase II clinical trial in HIV-infected patients. *Immunotherapy* 2, 755 (2010).

112. Bovolenta, C. et al. Therapeutic genes for anti-HIV/AIDS gene therapy. *Curr Pharm Biotechnol* (2012) (epub).

113. Mariyanna, L. et al. Excision of HIV-1 proviral DNA by recombinant cell permeable tre-recombinase. *PLoS One* 7, e31576 (2012).

114. Anderson, J. et al. Safety and efficacy of a lentiviral vector containing three anti-HIV genes—CCR5 ribozyme, tat-rev siRNA, and TAR decoy—In SCID-hu mouse-derived T cells. *Mol Ther* 15, 1182–1188 (2007).

115. Didigu, C.A. and Doms, R.W. Novel approaches to inhibit HIV entry. *Viruses* 4, 309–324 (2012).

116. Kim, S.S. et al. RNAi-mediated CCR5 silencing by LFA-1-targeted nanoparticles prevents HIV infection in BLT mice. *Mol Ther* 18, 370–376 (2010).

117. DiGiusto, D.L. et al. RNA-based gene therapy for HIV with lentiviral vector-modified CD34(+) cells in patients undergoing transplantation for AIDS-related lymphoma. *Sci Transl Med* 2, 36ra43 (2010).

118. The Jordan Report. Accelerated development of vaccines. NIAID, NIH, USA (2012).

119. Baeten, J. and Celum, C. Systemic and topical drugs for the prevention of HIV infection: Antiretroviral pre-exposure prophylaxis. *Annu Rev Med* 64, 219–232 (2013).

120. Fernandez-Montero, J.V. et al. Antiretroviral drugs for pre-exposure prophylaxis of HIV infection. *AIDS Rev* 14, 54–61 (2012).

121. Gilbert, M.T.P. et al. The emergence of HIV/AIDS in the Americas and beyond. *Proc Natl Acad Sci USA* 104, 18566–18570 (2007).

11 Influenza Virus (Reconstructed 1918 Influenza Virus)

Cristina Costa, Francesca Sidoti, and Rossana Cavallo

CONTENTS

11.1 INTRODUCTION

Influenza is one of the most ancient and deadliest infectious diseases known to man. Currently, seasonal human influenza epidemics account for over 250,000 deaths worldwide and more than three million cases with severe illness requiring hospitalization each year.[1–4]

In the past century, four influenza pandemics occurred in 1918, 1957, 1968, and 2009.[5] With the possible exception of the 1918 pandemic, three of these events have been shown to be associated to reassorted strains. In the presence of a simultaneous infection of a host with two or more strains derived from different animal species, reassortment may occur with the generation of progeny viruses that contain genes derived from two or more parent strains. This event results in the production of significant changes in the virus antigenic profile and poses relevant problems in terms of epidemiological consequences due to the lack of specific immunity of the host against the novel strain.[6,7] Global human pandemics in 1957 and 1968 originated from reassortment among avian and human type A influenza viruses that produced novel H2N2 and H3N2 strains, respectively.[8,9] The 2009 pandemic caused by the type A H1N1 influenza virus derived from the reassortment between a Eurasian swine virus and a triple reassortant North American swine virus of avian, human, and swine origin.[10] Overall, these pandemics resulted in tens of million deaths worldwide.

The influenza pandemic of 1918 was exceptional and is estimated to have caused the death of up to 50 million people worldwide, including 675,000 in the United States.[11] A particular feature of the 1918 pandemic influenza virus was the unusually high rate of death among young otherwise healthy adults (aged 15–34 years), thus consequently having determined a lower average life expectancy in the United States by more than 10 years.[12] No other previous or subsequent influenza A virus pandemics or epidemics have resulted in a similarly high death rate in this age group.[13]

Genomic RNA of the 1918 influenza virus was extracted from archived formalin-fixed lung autopsy tissues and from frozen unfixed lung tissue of an Alaskan infected subject who was buried in permafrost in November of 1918.[14,15] The reconstruction of the 1918 virus by the synthesis of the complete coding sequences of all eight viral RNA subunits, the analysis of these sequences, and the generation of the virus have provided information about the nature and origin of the viral agent, potentially disclosing new insights in terms of virulence and pathogenicity, and the development of new therapies and vaccines to protect against another such pandemic.[11,16] Plasmid-based reverse genetics has allowed for the generation of influenza viruses bearing specific gene segments in order to study its property. This technique has been used[17,18] for the generation of recombinant viruses containing the 1918 hemagglutinin (HA or H) with or without the 1918 neuraminidase (NA or N) recovered from the contemporary human H1N1 or H1N2 influenza viruses; these resulting strains were demonstrated to cause mortality in mice only at high infection doses. With the availability of the complete 1918 influenza virus coding sequence, reverse genetic has been used by Tumpey et al.[11] to generate an influenza virus bearing all eight subunits of the pandemic virus in order to define the features associated to its exceptional virulence. These studies provided evidence that, in contrast to contemporary human influenza

A H1N1 viruses, the 1918 pandemic pathogen was able to replicate in the absence of trypsin, caused death in mice (no other human influenza viruses that have been tested display a similar pathogenicity for mice 3–4 days after infection) and embryonated chicken eggs, and displayed a high-growth phenotype in human lung epithelial cells, with infectivity titers of the 1918 virus significantly higher, regardless of the presence or absence of trypsin. For this reason, the reconstructed 1918 influenza virus is included in the list of select agents and toxins by the Department of Health and Human Services of the United States.

It is notable that anti-influenza drugs, oseltamivir and amantadine, have been shown to be effective against viruses bearing the 1918 HA and the 1918 M gene, respectively,[19] and that vaccines containing the 1918 HA and NA were protective for mice.[20]

11.2 CLASSIFICATION AND MORPHOLOGY

11.2.1 CLASSIFICATION

Influenza viruses are classified as members of the family Orthomyxoviridae, which encompasses five genera: *Influenzavirus A*, *Influenzavirus B*, *Influenzavirus C*, *Thogotovirus*, and *Isavirus*. *Thogotovirus* includes *Thogoto virus* and *Dhori virus*, whereas *Isavirus* includes infectious salmon anemia virus (ISAV). Three types of human influenza viruses have been recognized (types A, B, and C), on the basis of their type-specific nucleoprotein (NP) and matrix protein antigens. Moreover, type A influenza viruses are further classified into subtypes based on the antigenic properties of the HA and NA glycoproteins expressed on the surface of the virus. The present nomenclature system of influenza viruses includes the type of virus, the species from which the virus was isolated (except if human), the location of isolate, the number of the isolate, and the year of isolation; in the case of influenza A viruses, the nomenclature includes also the HA and NA subtypes. For example, the 15th isolate of an H1N1 subtype of influenza A virus isolated from pigs in Iowa in 1930 is designated as influenza A/swine/Iowa/15/30 (H1N1). Currently, 16 HA subtypes (H1–H16) and 9 NA subtypes (N1–N9) are recognized in the nomenclature system for influenza A viruses recommended by the World Health Organization.[21–23] All these subtypes have been found circulating in wild and domestic birds. Thereby, avian hosts are the major reservoirs for all subtypes; so far, only three types of HA (H1, H2, and H3) and two types of NA (N1 and N2) have been widely prevalent in humans (H1N1, H2N2, and H3N2). Only two of these viruses (H1N1 and H3N2) are currently circulating as seasonal influenza. H2N2 has not circulated in humans since 1968.

11.2.2 MORPHOLOGY

Influenza A viruses are pleomorphic particles with a diameter of approximately 100 nm; filamentous particles with elongated viral structures (approximately 300 nm) have also been

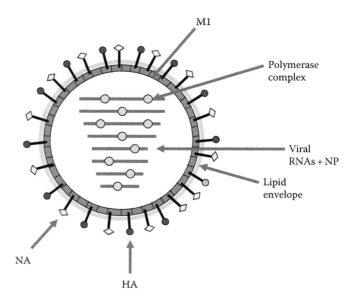

FIGURE 11.1 Schematic figure of human influenza virus particle.

observed, particularly in fresh clinical isolates.[24] Particles of influenza virus are enclosed by a lipid envelope, derived from the plasma membrane of the host cell, to which the HA, NA, and the M2 proteins are attached and from which they project. Therefore, the morphology of these particles is characterized by distinctive spikes (HA and NA) with lengths of about 10–14 nm, observable by electronic microscope. Within the lipid envelope lies the matrix (M1) protein, just beneath the envelope. The core of the virus particle is comprised of the ribonucleoprotein (RNP) complex, consisting of the viral RNA segments, the polymerase proteins (polymerase basic protein 1 [PB1], polymerase basic protein 2 [PB2], and polymerase acidic protein [PA]), and the NP. Two virus-encoded nonstructural proteins (NS1 and NS2) are found in infected cells.[25] Finally, a nuclear export protein (NEP/NS2) is also associated with the virus.[26] The overall composition of virus particles is about 1% RNA, 5%–8% carbohydrate, 20% lipid, and 70% protein.[27–29] Virion structure is reproduced in Figure 11.1.

Influenza B viruses are mostly indistinguishable from the A viruses by electron microscopy. They present four proteins inserted in their lipid envelope (the HA, NA, nucleotide binding [NB], and BM2[30–32]), the matrix protein, the RNP complex, the nonstructural proteins NS1, and the NEP/NS2 are also associated with the virus.

Influenza C viruses possess a single surface glycoprotein (referred to as HEF because it contains the viral hemagglutinating, esterase, and fusion (HEF) activities) instead of the two HA and NA found on the type A and B influenza viruses. The virus also contains a core of three polymerase proteins (PB2, PB1, and P3), the NP (which is associated with RNA segments), the nonstructural proteins NS1 and the NEP/NS2, the M1 protein, and the glycosylated CM2, which is structurally analogous to the M2 protein of influenza A viruses and the NB of influenza B viruses.[33–35]

The genome of influenza viruses types A and B consists of 8 single-stranded negative-sense RNA segments that encode

10 or 11 viral proteins, depending on the strain. Influenza virus type C has only seven genome segments because it possesses a single surface glycoprotein with HEF activities. The three largest RNAs of influenza A virus code for the polymerase proteins (PB2, PB1, and PA), the fourth RNA codes for the HA, and the fifth and sixth RNAs code for the NP and NA, respectively. The seventh RNA codes for the M1 and M2 proteins, the latter via a spliced messenger RNA (mRNA). The eighth RNA codes for the NS1 protein and, via a spliced mRNA, for the NEP/NS2 protein.[25,36] Recently, an 11th protein, the PB1-F2, has been found; in this case, an alternate open reading frame gives rise to the PB1-F2 polypeptide.[37]

The influenza B virus genome is similar to that of influenza A virus. Interestingly, the sixth NA segment contains an additional open reading frame resulting in the NB protein.[38] Therefore, the NA gene codes for the NB protein and the NA. The seventh RNA encodes the M1 protein and the BM2 protein.[39]

The genome of influenza C viruses has only seven RNA segments: the three largest RNAs code for the polymerase proteins (PB2, PB1, and P3).[40] The fourth RNA codes for the HEF protein,[41] whereas the fifth segment codes for the NP. Finally, the sixth M segment encodes the CM1 and CM2 proteins,[35,42] and the seventh RNA codes for the NS1 protein and, via a spliced mRNA, for the NEP/NS2 protein.[34,43,44]

11.3 BIOLOGY AND EPIDEMIOLOGY

11.3.1 BIOLOGY

The HA glycoprotein of influenza virus plays an essential role in initiation of infection and is the main target for immunity by neutralizing antibodies. It is a trimer built of two structurally distinct regions: a triple-stranded coiled-coil region and a globular region that contains the receptor binding site. The HA monomer is synthesized as a single polypeptide chain that undergoes posttranslational cleavage at three places. An N-terminal signal sequence is removed, and depending on the host cell and virus strain, the molecule is cleaved to give two polypeptide chains: HA1 and HA2. Cleavage of HA is essential for the fusing capacity and for the infectivity of the virus. Cellular proteases are involved in this reaction of activation, and depending on the presence of an appropriate enzyme in a certain cell, virus particles with cleaved or with uncleaved HA may be formed.[45,46] Proteases of different specificities are able to cleave HA; however, activation is observed only after cleavage with trypsin or trypsin-like enzymes.[46,47] The HA glycoprotein mediates attachment and entry of the virus by binding to the sialic acid receptors on the cell surface. The binding affinity of the HA to the host sialic acid allows for the host specificity of influenza virus.[48,49] Avian influenza subtypes prefer to bind to sialic acid linked to galactose by α-2,3 linkages, frequent in avian respiratory and intestinal epithelium.[50] On the contrary, human virus subtypes bind to α-2,6 linkages frequent in human respiratory epithelium.[50,51] Swine has both α-2,3 and α-2,6 linkages in its respiratory epithelium, allowing

for easy coinfection with both human and avian subtypes.[52] Humans have been found to contain both α-2,3 and α-2,6 linkages in their lower respiratory tract and conjunctivae, which allows for human infections by avian subtypes.[51,53,54] Subsequently, influenza virus requires a low pH to initiate fusion and is therefore internalized by endocytic compartments. Clathrin-mediated endocytosis has traditionally been the model for influenza virus entry[55]; however, a nonclathrin-mediated internalization mechanism has also been described for influenza virus.[56] After binding to the cell surface and endocytosis, the low pH of the endosome activates fusion of the viral membrane with that of the endosome. This fusion activity is induced by a structural change in the HA of influenza virus: conformational change exposes the fusion peptide of the HA2 subunit, enabling it to interact with the membrane of the endosome. Then, the structural change of several HA glycoproteins opens up a pore that releases the contents of the virion into the cytoplasm of the cell. Effective uncoating also depends on the presence of the M2 protein, which has ion channel activity.[57] In particular, M2 protein allows the influx of the H+ ions from the endosome into the virus particle causing the release of RNP complex free of the M1 protein into the cytoplasm.[58] Subsequently, viral RNP complex is imported by an energy-dependent process into the nucleus where viral RNA synthesis occurs. Replication and transcription of the influenza virus genome are catalyzed by the same viral polymerase complex, although distinct functions of each subunit are employed at different steps. The influenza virus RNA genome exists as RNP complex with viral RNA polymerases (PB1, PB2, and PA) and NP. PB1 binds to the vRNA (negative-sense viral RNA genome) and cRNA (full-sized complementary copy of vRNA) promoters and functions as a polymerase catalytic subunit for the sequential addition of nucleotides to the nascent RNA chains. PB2 is responsible for the recognition and binding of the cap structure of host pre-mRNAs. PA is involved in virus genome replication and transcription,[59,60] but recent reports showed that PA is involved also in the assembly of a functional polymerase.[61] Finally, NP is required for vRNA synthesis possibly by assembling functional RNP complex.[62,63]

Transcription of the viral genome is initiated using the oligonucleotide containing the cap-1 structure derived from cellular pre-mRNAs. The capped oligonucleotide is generated through recognition of the cap structure by PB2 and endonucleolytic cleavage by PB1.[64] The elongation of the mRNA chain proceeds until the viral polymerase reaches a polyadenylation signal.[65] Therefore, the viral polymerase generates a poly(A) tail at the end of the viral mRNA. As concerns the genome replication, this reaction generates full-length positive-sense cRNA from vRNA, and progeny vRNA is in turn copied from the cRNA. Newly formed RNP complex is assembled in the nucleus, and the nuclear export of this newly synthesized viral RNP complex into the cytoplasm is mediated by the viral NEP/NS2 and the M1 protein.

Influenza virus assembles and buds from the apical plasma membrane of infected cells. Viral envelope proteins (HA, NA, and M2) are seen to accumulate at the same polar surface where virus budding occurs.[66] Following synthesis on membrane-bound ribosomes, the three integral membrane proteins—HA, NA, and M2—enter the endoplasmic reticulum (ER), where they are folded and glycosylated (except for M2) and where HA is assembled into a trimer and NA and M2 into tetramers. Then, they are transported to the Golgi apparatus where cleavage of HA into HA1 and HA2 subunits may occur.[67] From here, HA, NA, and M2 are all directed to the virus assembly site on the apical plasma membrane. Little is known about how the remaining viral components reach the assembly site. Only the association of the M1 protein with the RNP–NEP/NS2 complex is well documented[68]; M1 protein is therefore proposed to play a vital role in assembly by recruiting the viral components to the site of assembly at the plasma membrane. As concerns the mechanism of packaging of the RNA segments, this is not well known. In fact, two different models have been proposed: the first named *random incorporation model* suggests the random packaging of vRNA segments into budding particle,[69] whereas the second named *selective incorporation* model indicates the selective packaging of each independent RNA segment.[70] Initiation of bud process requires outward curvature of the plasma membrane. The virus bud is then extruded, and finally, the budding process is completed when the membranes fuse at the base of the bud and the enveloped virus particle is released. The essential role of NA in particle release has been demonstrated.[71] The enzymatic activity of the NA protein is required to remove the sialic acid and thereby to release the virus from the infected cell and also is required to remove sialic acid from the carbohydrates present on the viral glycoproteins themselves, so the individual virus particles do not aggregate. The absence of NA enzymatic activity was seen to cause viral particles amassing at the cell surface resulting in a loss of infectivity.

11.3.2 Epidemiology

The epidemiology of human influenza reflects the peculiar characteristics of the virus genome (segmented single-stranded RNA), as well as the diversity and host range of the viruses. The outstanding feature of human influenza virus consists in its capability of evading host immunity and causing recurrent annual epidemics of disease and, at infrequent intervals, major worldwide pandemics due to the introduction of antigenically novel viruses into an immunologically naïve human population. Influenza viruses present two different mechanisms that allow them to reinfect humans and cause disease: antigenic drift and antigenic shift. It is now known that small changes in antigenicity (antigenic drift) are the result of a gradual accumulation of point mutations, while the complete change in antigenic properties (antigenic shift) involves the replacement of the gene coding for one HA with that for another. Antigenic shift is then derived from reassortment of gene segments between viruses. This may or may not be accompanied by the replacement of the NA gene. Moreover, antigenic shift occurs only in influenza type A virus, whereas type B and C have not been shown to undergo

antigenic shift, probably because they lack the extensive animal reservoir of type A virus. However, like type A virus, they can undergo less drastic antigenic drift due to point mutation in the relevant genes. Therefore, influenza type A and B viruses are responsible for recurrent annual epidemics. They cocirculate and either may predominate in a particular influenza season. An increased incidence of influenza B frequently follows a peak of influenza A activity. Moreover, in recent years, human influenza B virus has tended to be prominent every 2–3 years. Although influenza B virus has been responsible for severe epidemics, the impact of influenza A virus is greater in terms of annual epidemics as well as the infrequent more devastating pandemics. Whereas influenza type B virus infects predominantly humans, type A virus is especially an avian virus that periodically transmits to other species, including mammals. Moreover, influenza A virus comprises a large variety of antigenically distinct subtypes, with different combinations of 16 HA and 9 NA subtypes, that replicate in the intestine of aquatic birds and constitute a large reservoir of potential pandemic viruses.[22] Historical evidence suggests that pandemics have occurred at 10–40-year intervals since the sixteenth century, originating mainly in Asia. In the twentieth century, there were three overwhelming pandemics, in 1918, 1957, and 1968, caused by H1N1 (Spanish flu) that claimed an estimated 50 million lives and H2N2 (Asian flu) and H3N2 (Hong Kong flu) that each resulted in 1–2 million deaths, respectively.[72] In 1957 and 1968, excess mortality was noted in infants, the elderly, and persons with chronic diseases, similar to what occurred during interpandemic periods. On the contrary, in 1918, there was one distinct peak of excess death in young adults aged between 20 and 40 years old. Acute pulmonary edema and hemorrhagic pneumonia contributed to rapidly lethal outcome in young adults.

Influenza viruses are maintained in human populations by direct person-to-person spread during acute infections. In the Northern Hemisphere, influenza activity is generally seasonal: it increases during the cooler months and peaks from January to April but may flare up as early as December or as late as May. In the Southern Hemisphere, outbreaks occur between May and September, whereas seasonality in tropical and subtropical climates is believed to coincide with the onset of the rainy season. The incubation period is about 3 days for influenza type A virus and 4 days for influenza type B virus. The most effective means of spread among humans are aerosols. Human influenza invades the epithelial cells of the upper respiratory tract. Viral replication leads to the secretion of proinflammatory cytokines and the necrosis of ciliated epithelial cells.[73] Small respiratory droplets are generated when humans exhale or talk, but these are generally less than 1 μm.[74] The transmission of influenza is primarily from person to person by large droplets (>5 μm) that are generated when infected persons cough. These droplets settle on the mucosal surfaces of the upper respiratory tracts of susceptible persons. Finally, contact transmission may play a role. In fact, infected people will often touch mucous membranes before direct interpersonal contact (handshaking) or indirect contact, such as touching common surfaces. Influenza virus has been detected in over 50% of the fomites tested in homes and day-care centers during influenza season.[75] As concerns the morbidity and mortality, influenza virus is estimated to cause about 50 million illnesses annually in the United States. Moreover, seasonal influenza causes more than 200,000 hospitalizations and 41,000 deaths in the United States every year, thus representing the seventh leading cause of death.[76] Direct costs, including hospitalizations, medical fees, drugs, and testing, were estimated in 1986 to be about $1 billion annually, while indirect costs such as loss of productivity reach $2–$4 billion annually.

11.4 CLINICAL FEATURES AND PATHOGENESIS

Influenza is an acute respiratory disease characterized by the sudden onset of high fever, myalgia, sore throat, coryza, prostration, malaise, nonproductive cough, and inflammation of the upper respiratory tree and trachea. Acute symptoms and fever often persist for 7–10 days. Weakness and fatigue may linger for weeks. However, only about 50% of infected persons present with these classic symptoms. Additional symptoms may include rhinorrhea, headache, nausea, and diarrhea. Although most influenza is associated with a mild acute self-limited illness, more severe manifestations can occur. Influenza virus infections can present as a typical community-acquired pneumonia with fever, cough, hypoxemia, and leukopenia. Influenza virus is the etiological agent of 5%–10% of community-acquired pneumonias.[77] The incidence is slightly higher in pediatric (12%) and immunosuppressed individuals (11%).[78] More severe diseases are generally seen in young children, people older than 65 years, and persons with underlying pathological conditions.[79]

Immunocompetent adults are infectious from the day before symptoms begin until about 5 days after, and their respiratory symptoms include primarily sneezing, nasal obstruction, rhinorrhea, sore throat, hoarseness, and dry cough.[80] Conjunctival inflammation and excessive tearing may also occur. These symptoms probably result from damage produced by viral replication in the upper and lower respiratory tract.[81] The upper respiratory tract pathology shows an inflammatory response and desquamation of ciliated and basal cells in which virus replication has occurred. Although damage is predominantly confined to the upper respiratory tract, tracheitis, bronchitis, and impaired lower respiratory function may also occur.[82] Moreover, constitutional effects like headache, myalgia, shivering, listlessness, nausea, vomiting, anorexia, and high fever have been attributed either to the toxic effects of products (viral or host) from cells destroyed by viral replication or to complement activation by antigen–antibody complexes of viral components.[81]

Influenza is a common respiratory infection of young children for whom its effects are generally mild. However, febrile convulsions, croup, otitis media, and myositis are common, and bronchiolitis and pneumonia can occur, sometimes with fatal consequences.[83] Croup (laryngotracheobronchitis) can occur predominantly in children younger than

1 year, whereas gastrointestinal manifestations such as nausea, vomiting, diarrhea, and abdominal pain are much more frequent in children than adults, especially in children less than 3 years old.

There are a number of individuals who are at increased risk for complicated influenza infection, in particular,

1. People with pulmonary diseases such as cystic fibrosis, asthma, cor pulmonale, and bronchopulmonary dysplasia or chronic cardiac disorders
2. People with chronic medical conditions such as diabetes mellitus, renal insufficiency, hemoglobinopathy, immunodeficiency, and immunosuppression
3. Women in the second or third trimester of pregnancy
4. People older than 65 years or younger than 2 years

Three distinct syndromes of severe pneumonia can follow influenza infection in these people categories: primary viral pneumonia, combined viral–bacterial pneumonia, and secondary bacterial pneumonia. Primary viral pneumonia occurs predominantly in high-risk patients (the elderly or patients with cardiopulmonary disease) but has been occasionally described in otherwise healthy adults. Presentation is often abrupt and dramatic, progressing within 6–24 h to a severe pneumonia with rapid respiration rate, tachycardia, cyanosis, high fever, hypotension, respiratory failure, and shock. The illness may rapidly progress to hypoxemia and death in 1–4 days. Mortality is in the order of 10%–20%. Combined viral–bacterial pneumonia is at least three times more common than viral pneumonia, from which it is clinically indistinguishable. The bacteria most often involved are *Streptococcus pneumoniae*, *Staphylococcus aureus*, and *Haemophilus influenzae*. The case-fatality rate for combined viral–bacterial pneumonia is 10%–12%, but the coinfection with influenza and *S. aureus* can have a fatality rate of up to 42%.[84] Clinically, secondary bacterial pneumonia may be easier to differentiate from combined viral–bacterial pneumonia, as patients typically develop shaking chills, pleuritic chest pain, and an increase in cough productive of bloody or purulent sputum. Mortality is about 7%.

Extrapulmonary complications are not frequent, and they include the following:

1. *Myositis.* Acute myositis may present with generalized pain and extreme tenderness of the affected muscles, most commonly in the legs. Markedly elevated serum levels of muscle enzymes, myoglobinemia, and myoglobinuria are seen, and acute renal failure has been reported.[85]
2. Central nervous system involvement, including encephalitis, transverse myelitis, aseptic meningitis, and Guillain–Barrè syndrome.[86] Psychiatric complications such as irritability, drowsiness, confusion, depression, psychosis, delirium, and coma have been recognized.
3. *Reye syndrome.* Reye syndrome is a rapidly progressive noninflammatory encephalopathy and fatty

infiltration of the viscera (especially the liver) that results in severe hepatic dysfunction. Children seem preferentially affected. An epidemiological association with aspirin use has been described.[87] The case-fatality rate varied between 22% and 42%.[88]

4. *Cardiac involvement.* Myocarditis and pericarditis were reported in the 1918 influenza pandemic. However, during the Asian epidemic in 1957, signs of focal or diffuse myocarditis were found in a third of autopsies.

Symptoms of human influenza virus infection can vary widely, ranging from a minor upper respiratory illness to the classic febrile respiratory disease of abrupt onset accompanied by systemic symptoms such as headache, myalgias, extreme weakness, and malaise. As the symptoms of influenza are not readily distinguishable from those caused by other respiratory pathogens including respiratory syncytial virus, parainfluenza virus, adenovirus, rhinovirus, and coronavirus, it is impossible to differentiate clinically one virus infection from another. Therefore, influenza cannot be diagnosed on clinical grounds alone, although during a well-defined outbreak or epidemic, influenza is responsible for a high proportion of acute respiratory illnesses.[89] There is also no consistent clinical basis on which to differentiate between type A, B, and C influenza virus infections, although the symptoms of type C virus are almost always milder than those caused by type A or B.[90] For these reasons, the laboratory identification of influenza virus is fundamental. Since the clinical presentation of numerous illnesses may resemble influenza, diagnosis can be confirmed only by laboratory tests.

11.5 IDENTIFICATION AND DIAGNOSIS

Over the past two decades, several approaches have been used for diagnosing influenza virus infection. Virus isolation and serology have been the principal reference of the clinical laboratory for diagnosing respiratory virus infections. Initially, virus isolation was performed using a modest number of cell lines and, together with embryonated chicken eggs, provided the means for isolating influenza viruses. A variety of serological tests including the hemagglutination inhibition (HAI) test, complement fixation, and enzyme immunoassay (EIA) were used for testing paired acute- and convalescent-phase sera for diagnosing infections. Moreover, HAI was the reference subtyping method able to subtype the influenza virus as being H1 or H3 using specific antisera. More rapid subtyping techniques using monoclonal antibodies that differentiate between influenza A/H1N1, A/H3N2, and B virus subtypes have been developed in rapid culture assays and directly on clinical specimens.[91] In the early 1990s, tube cultures were replaced by shell vial culture that, using specific monoclonal antibodies, could detect specific viral antigens in 1–2 days instead of 7–10 days for tube culture. Direct fluorescent antibody (DFA) staining of cells derived from nasopharyngeal swabs or nasopharyngeal aspirates became the mainstay for many laboratories and provided a rapid test result in about

3–4 h. Also EIAs were introduced in the 1980s and 1990s, but these tests lacked sensitivity and are no longer used in the clinical laboratory to diagnose influenza virus infection. A wide variety of rapid viral diagnostic tests are now available that greatly facilitate the diagnosis of influenza. Direct testing of sputum and nasal washes for influenza antigen permits a rapid diagnosis in a variety of settings. Commercially available rapid antigen testing kits vary by their complexity, storage conditions, and reporting metrics, but assay features (sensitivity and specificity) are largely similar. Generally, these tests are very specific (95%–100%), although sensitivity is modest, especially in adults (50%–70%).[92,93] Higher sensitivity is reported in children compared with adults.[94] These rapid viral diagnostic tests are regularly summarized by the Centers for Disease Control and Prevention (CDC) at www.cdc.gov/flu/professionals/labdiagnosis.htm.

Influenza virus genome may be detected in clinical samples by nucleic acid testing molecular techniques such as reverse transcription-polymerase chain reaction (RT-PCR).[95–99] Depending on primer selection, these assays may be type- or subtype-specific. The sequencing of amplified HA and NA genes is an important subtyping method, as it allows the rapid identification of novel or highly divergent strains, the analysis of strain variation, and the determination of the origin of outbreaks.[100] DNA microarrays have been also used to detect type- and subtype-specific amplification sequences.[101] Although some studies reported a similar sensitivity of nucleic acid testing in comparison to cell culture, others have reported 5%–15% more influenza virus detections using RT-PCR.[95,97,98,102] Moreover, specimen quality, timing, and transportation conditions may be less critical for nucleic acid testing than for culture or antigen detection, as viable virus and intact infected cells do not need to be preserved. Therefore, when specimen quality is limited, yields from RT-PCR are significantly higher than from cell culture.[95,97,103] Multiplex RT-PCR assays that can simultaneously detect influenza and other viral respiratory pathogens directly in clinical specimens have been developed.[99,104] Some of these multiplex RT-PCR assays allow simultaneous testing of multiple viral, bacterial, mycobacterial, and fungal agents. Nested RT-PCR assays have been developed and in some cases provide an increased sensitivity over that of non-nested RT-PCR[105]; however, most clinical laboratories will not use nested RT-PCR because the amplification workload is doubled and the risk for PCR contamination is dramatically increased. Real-time RT-PCR assays for influenza virus infection offer results more quickly than end-point assays, and their sensitivity is similar or better than cell culture.[99] Recently, other nucleic acid testing assays have been proposed for the detection of influenza virus, such as nucleic acid sequence-based amplification (NASBA) and loop-mediated isothermal amplification (LAMP).

The key to community management of influenza is the collection of good-quality respiratory tract specimens for laboratory testing. Specimens should be collected early in the clinical illness (within the first 96 h, during maximal viral shedding), transported to the laboratory at 4°C for virus isolation or room temperature for other assays, and processed as rapidly as possible. Preferred respiratory samples for influenza testing include nasopharyngeal or nasal swab, nasal wash or aspirate, and bronchoalveolar lavages, depending on which type of test is used. Single-use swabs containing viral transport media are commercially available. The most practical samples to collect from adults are combined nose and throat swabs, whereas nasopharyngeal aspirates are the specimen of choice from children younger than 3 years. In general, the recovery of virus from nasopharyngeal aspirates, nasal washes, and bronchoalveolar lavages is superior to that from nasopharyngeal and throat swabs and expectorated sputum, as the latter generally contain less columnar and more squamous epithelial cells.[106] The likelihood of successful isolation will also depend on the interval between the onset of symptoms and the procurement of the specimen, the temperature and duration of specimen storage. Specimens from infected patients are most likely to yield virus when they are obtained within 4 days of symptom onset and stored at 4°C for less than 48 h or at −70°C for prolonged storage. The composition and other characteristics of the collecting medium, including pH 7.0, the presence of broad-spectrum antibiotics, and the absence of serum (which contains non-specific HA inhibitors), also influence the success of virus isolation.[107] Specimen requirements for antigen and nucleic acid testing are similar to those for virus isolation. Serum specimens for serology should be collected during the acute phase within 7–10 days of symptom onset and the convalescent phase 14–21 days after symptom onset and tested in parallel.

For virus isolation, influenza viruses are first grown in embryonated eggs but can be grown in a variety of cell culture systems.[108] Currently, Madin–Darby canine kidney (MDCK) cells and African green monkey kidney (Vero) cells are generally used for the isolation of influenza viruses from human clinical specimens.[109] Influenza viruses also replicate in a number of primary cell cultures, including monkey, calf, hamster, and chicken kidney cells, as well as in chicken embryo fibroblasts and primary human epithelial cells. With the exception of primary kidney cells, most other cell culture systems require the addition of trypsin to cleave the HA protein of human viruses, a prerequisite for efficient replication. Cytopathic effect (CPE) consistent with influenza virus can be visualized by light microscopy but is variable depending on cell types.[110] Many, but not all, strains of human influenza A viruses can be isolated in the allantoic cavity of embryonated eggs, although some human influenza A and B viruses must first be isolated in the chicken embryo amniotic cavity and subsequently adapted to growth in the allantoic cavity. In all cases, the isolated viruses must be characterized serologically to confirm the diagnosis, since several potential respiratory virus pathogens, especially the paramyxoviruses, can also cause hemadsorption or hemagglutination or both. In the application of these techniques, the collection of appropriate clinical samples, the measures to maintain the infectious titer of virus, and the quality of cells and reagents used are crucial to obtain the sensitivity and specificity required.

Neither viral culture nor shell vial culture (described in the following) can detect inactivated virus. Nevertheless, viral culture is considered the most accurate method for identifying specific viral strains and subtypes, it recovers novel or highly divergent strains missed by other tests, it provides an isolate for subsequent characterization and consideration as potential vaccine strains, and finally, it allows the simultaneous recovery of other respiratory viruses if an appropriate range of cell lines is used. Moreover, viral culture is usually more sensitive than the rapid culture and antigen detection assays. However, the disadvantage of virus isolation is the time needed to obtain a positive result, usually 7–10 days.[111] Furthermore, viral culture is costly (about $100 per test) and requires special laboratory equipment and procedures, as well as skilled technical expertise for correct performance.

Shell vial culture is a centrifuge-enhanced tissue culture assay that has revolutionized viral culturing in terms of rapidity. The technique involves inoculation of the clinical specimens onto preformed cell monolayers on a cover slip in an appropriate tube, followed by centrifugation and overnight incubation. This system is based on the principle that centrifugation enhances viral infectivity to the susceptible cells. Viral antigens are produced in the cells within a few hours, so that specific monoclonal antibodies directly conjugated to a fluorescent dye (DFA) or staining with antibodies to the virus and second conjugated antibodies directed at the firsts (indirect fluorescent antibody) can be used to reveal the infection. Commercially available cell mixture, such as R-Mix cells (a mixture of A549 and Mink lung cells), has been used by clinical laboratories.[112] The advantage of shell vial culture is that influenza virus can be identified in 1–2 days instead of 7–10 days for conventional cell culture technique.

Serologic diagnosis is based on the fact that recovery from influenza virus infection is accompanied by the development of demonstrable antibodies to the virus. The antibodies may be detected as early as 4–7 days after symptom onset and reach their peak after 14–21 days. Since a significant proportion of the population, especially adults, may already possess strain-specific antibodies as a result of previous exposure to related strains, it is essential to obtain two serum specimens, one in the acute phase and another in convalescence, for a comparative titration of antibody levels.[107] A fourfold rise in antibody titer against a specific type or strain of influenza virus can be considered diagnostic. A variety of influenza-specific assays are available, including HAI test, complement fixation, immunofluorescence and EIA, neutralization (plaque reduction assay), single radial hemolysis, and hemadsorption inhibition. The traditional "gold standard" techniques for detecting influenza-specific antibodies are the neutralization and the HAI assays, as they can differentiate subtype-specific and strain-specific serological responses. Complement fixation is more commonly used as it is easier to perform than the HAI assay and neutralization, but it does not distinguish between subtypes. Serology can be used when specimens for virus isolation or antigen detection are negative, inadequate, or unavailable. However, routine serological testing does not provide information on the antigenic composition of circulating strains and may help with clinical decision making. Moreover, it is only available at limited number of public health or research laboratories and is not generally recommended, except for research and epidemiological investigations. Serological testing results for human influenza on a single serum specimen are not interpretable and are not recommended.

Commercial rapid diagnostic tests are available that can detect influenza viruses within 15 min or less. Rapid diagnostic tests can help in the diagnosis and management of patients who present with signs and symptoms compatible with influenza. These rapid tests differ in the types of influenza viruses they can detect and whether they can distinguish between influenza types. Typically, these tests produce a visual result on an immunochromatographic strip using influenza A or B NP-specific monoclonal antibodies within about 15 min. Specimen quality is a major determinant of their performance. The specificity and, in particular, the sensitivity of rapid tests are lower than for viral culture and vary by test. In general, sensitivities are approximately 50%–70% when compared with viral culture or RT-PCR, and specificities are approximately 90%–95%. Because of the lower sensitivity of the rapid tests, physicians should consider confirming negative tests with viral culture, molecular techniques, or other means because of possibility of false-negative rapid test results, especially during periods of peak community influenza activity. In contrast, false-positive rapid test results are less likely but can occur during periods of low influenza activity. Different rapid tests for detecting influenza virus A and/or B are commercially available, as reported in Table 11.1. Several rapid influenza tests have been approved by the US Food and Drug Administration (FDA), and the cost per test varies from $12 to $24.[113–117]

Molecular tests for influenza virus RNA detection are gaining widespread use due to the versatility while maintaining high sensitivity and specificity. Molecular assays, in fact, have a sensitivity and specificity approaching 100%, and sometimes the sensitivity may exceed virus isolation.[102,118,119] Moreover, molecular techniques are less affected by specimen quality and transport and provide an objective interpretation of results. Different RT-PCR assays for influenza virus detection have been reported, and several gene targets have been used for amplification including the matrix, HA, and NS protein genes.[120,121] All of these targets have both conserved and unique sequences, permitting their use in either consensus or subtype-specific (H1 and H3) virus detection assays. Different targets are required for the detection of influenza B virus. Although the turnaround time for nucleic acid testing is intermediate between virus isolation and direct antigen detection, newer techniques can reduce this to 4–5 h or less.[99] Real-time RT-PCR technology, with specific detection of the product by fluorescent probes that combines nucleic acid amplification with amplicon detection, provides results more quickly than conventional RT-PCR and, in some cases, has shown improved sensitivity and specificity.[99] Moreover, it provides a uniform platform for quantifying both single and multiple pathogens in a single sample.[122,123] Although reagent and instrument costs are higher for real-time RT-PCR technology than conventional virus isolation, real-time RT-PCR requires less hands-on time per specimen than virus isolation, which is labor-intensive. Cost-effective implementation of molecular testing in routine diagnostics requires further attention. Automation of the extraction process and the use of real-time RT-PCR reduce the hands-on time in the laboratory. Additional cost benefits may result from the more rapid diagnosis in reduced time of hospitalization, decreased nosocomial spread, and decreased use of antibiotics.[124,125] Moreover, real-time RT-PCR assays offer advantages over conventional PCR by providing lower risk of false-positive results due to amplicon contamination, identification of the etiologic agent in a clinically relevant time

TABLE 11.1

Commercial Rapid Viral Tests for Influenza Virus A and/or B

Rapid Diagnostic Tests	Influenza Types Detected	Time for Results (min)
3M Rapid Detection Flu A + B Test	A + B	15
Directigen Flu A, Directigen Flu A + B, and Directigen EZ Flu A + B (Becton Dickinson)	A; A + B; A + B	<15
BinaxNOW Influenza A & B (Inverness)	A + B	<15
Influ A + B Dipstick (Biolife Italiana)	A + B	15
Influ-A + B-Respi-Dipstick	A + B	<15
OSOM® Influenza A&B (Genzyme)	A + B	<15
QuickVue Influenza Test, QuickVue Influenza A + B (Quidel)	A + B	<15
SAS FluAlert (SA Scientific)	A + B	<15
TRU FLU, ImmunoCard STAT! Flu A and B Plus Test (Meridian Bioscience)	A + B; A + B	15; 15–20
Xpect Flu A&B (Remel)	A + B	<15
ZstatFlu Test (ZymeTx., Inc.)	A + B	20–25
FLU OIA (BioStar Inc.)	A + B	15–20
ESPLINE Influenza A&B	A + B	15

period, and quantification of viral load. Therefore, real-time RT-PCR assays are useful tools for further investigations on the epidemiology and disease of influenza viruses and will provide information to better understand the relationship between illness and the quantity of virus being shed.

Influenza multiplex assays have been reported that can differentiate influenza A from B and influenza A H3 from H1, reducing both time and overall costs of diagnosis.[96,118,126] Multiplex RT-PCR for clinical diagnosis has a significant advantage, as it permits simultaneous amplification of several influenza viruses in a single reaction mixture, facilitating cost-effective diagnosis (Table 11.1).[127,128] The disadvantage of RT-PCR methods, compared to direct PCR of DNA targets, is that the RT step is often performed separately from the PCR, increasing both assay time and the risk of contamination.

11.6 TREATMENT AND PREVENTION

Antiviral agents active against influenza viruses belong to two categories of drugs that can be used for treatment as well as chemoprophylaxis purposes: inhibitors of protein M2, amantadine and rimantadine, and NA inhibitors, oseltamivir and zanamivir. Amantadine and rimantadine block the activity of membrane protein M2 that facilitates the entry of hydrogen ions into the endosomes, thus obstaculating the capsid removal within the infected cell. Rimantadine is rarely used because of the potential development of relevant side effects regarding the central nervous system. Amantadine, although more used than rimantadine, also displays important limitations, including activity only against influenza A virus, side effects particularly in elderly, and rapid development of resistant strains (sometimes even within 2–3 days). These resistant strains can be easily transmitted to other subjects. Oseltamivir and zanamivir are chemically related drugs that are active against both influenza A and B viruses. They are approved by the FDA for the treatment and chemoprophylaxis of influenza A and B; for details, see Table 11.2. Clinical data evidence that early antiviral therapy is able to shorten the duration of fever and illness and reduce the risk of influenza-related complications, including otitis media, pneumonia, respiratory insufficiency, and death, as well as the length of hospital stay. Therapy is recommended as early as possible for patients that are hospitalized; present severe, complicated, or progressive disease; or are at high risk of complications (children <2 years, adults >65 years, subjects with chronic diseases, immunosuppressed patients, subjects resident in nursing institutes or chronic care facilities). Two main limitations characterize NA inhibitors: they are effective only when administered during the acute phase, while efficacy rapidly diminishes after 48 h from the symptom onset; resistant strains have been isolated *in vitro* and *in vivo*. Phenotype of resistant strains seems to be associated to mutations regarding both the active site of NA (that alters the sensibility of the enzyme to inhibitors) and the HA (that alters the affinity of HA for cellular receptor, thus allowing the virus to exit from the cell without the intervention of NA). All the antiviral medications are able to prevent the illness in 50%–90% of the cases, although they did not alter the possibility of causing subclinical infections that elicit a specific immune response.

During the flu season, influenza viruses are circulating in the population. The administration of annual seasonal flu vaccine is the best method to reduce the risk of illness and spread to other subjects. Two main types of influenza vaccines are available: the "flu shot" (an inactivated vaccine that is approved for use in people older than 6 months) and the nasal spray flu vaccine (a vaccine made with live, attenuated viruses that is approved for use in people aged 2–49 years [not pregnant women]). Given the potential risks associated to the use of a live vaccine, the flu shot is far more used. The subunit vaccine is constituted only by the viral proteins, extremely purified; this product contains only the peptides necessary to stimulate an antibody response. In this vaccine, HA and NA of different strains of influenza A and B are usually present in order to confer the widest protection. Another similar type of vaccine is constituted by fragmented different strains of the influenza viruses and is commonly named "split vaccine." This vaccine may be conjugated with an adjuvant in order to amplify the development of the immune response. More recently, a virosomal vaccine is constituted by liposomal particles that incorporate in their phospholipidic bilayer the viral proteins, thus mimicking the natural antigenic structure of the protein that is present in the virus in the absence of an infecting agent. For vaccine preparation, it should be considered the circulating strains in the specific flu season based on the data supplied by the surveillance systems of the World Health Organization. This vaccine is effective for the prevention of the viruses belonging to the same strains of those used for vaccine preparation, while it is not protective against other strains. During the 2011–2012 flu season, the intradermal

TABLE 11.2
Antiviral Agents Recommended for the Treatment and Chemoprophylaxis of Influenza

Antiviral Agent	Treatment	Chemoprophylaxis	Side Effects
Oseltamivir (Tamiflu®)	Influenza A and B (≥1 year)	Influenza A and B (≥1 year)	Nausea, vomiting, and neuropsychiatric events
Zanamivir (Relenza®)	Influenza A and B (≥7 years)	Influenza A and B (≥5 years)	Oropharyngeal or facial edema, nausea, diarrhea, sinusitis, bronchitis, cough, headache, and nasopharyngeal infections

Source: Modified from Centers for Disease Control and Prevention (CDC), www.cdc.gov/flu.

vaccine Fluzone Intradermal® was first made available. It is a shot that is injected into the skin instead of the muscle; it requires less antigen to be as effective as the regular influenza shot. This vaccine is approved by the FDA for use in adults 18–64 years of age and represents another option in this category of patients and seems to provide an immune response similar to the regular flu shot. Common reactions include redness, swelling, toughness, pain, and itching; moreover, also headache, muscle ache, and asthenia have been reported, with spontaneous disappearing within 3–7 days.

11.7 CONCLUSIONS AND FUTURE PERSPECTIVES

Since the first isolation of human influenza viruses in 1933, they have been studied extensively. Much progress has been made in elucidating the components of the virus and in understanding the clinical consequences of an influenza virus infection. Today, the intensity of influenza virus research has not diminished and yields approximately 40,000 entries in a PubMed search. Many approaches that served us well in the past have now been superseded by newer molecular techniques that have allowed us to obtain an excellent understanding of the influenza virus on a molecular level and to learn how it has changed over the years. The reconstruction of the 1918 virus by the synthesis of all the eight subunits with generation of infectious virions has posed a relevant ethical problem.[129] As appropriately appointed by Sharp,[129] should the sequence of the 1918 virus have been published, given its potential use by terrorists? Taking into consideration all the possible issues, the publication of such data is correct in terms of national security and public health. The availability of this information could allow for the development of new therapeutic approaches or procedures to manage future pandemics. For examples, recent studies have demonstrated that the HA protein of the 1918 virus and 2009 H1N1 pandemic virus shares cross-reactive antigenic determinants and that the 2010–2011 seasonal trivalent vaccine induces neutralizing antibodies that cross-react with the reconstructed 1918 influenza virus in ferrets. The immunized ferrets infected with the 1918 virus evidenced a significant reduction in illness and virus shedding in comparison to nonimmunized control animals.[130]

What are the challenges for the future? With the threat of another pandemic influenza virus emerging, a detailed molecular understanding of virus–host interactions is sorely needed in order to know how best to disable the virus. In a pandemic outbreak, the availability of new diagnostic tests will be imperative.

REFERENCES

1. Lun, A.T. et al., FluShuffle and FluResort: New algorithms to identify reassorted strains of the influenza virus by mass spectrometry, *BMC Bioinform.*, 13, 208, 2012.
2. Wilschut, J.C., McElhaney, J.E., and Palache, A.M. (eds.), *Influenza Rapid Reference*, 2nd ed. Amsterdam, the Netherlands: Elsevier, 2006.
3. Van-Tam, J. and Sellwood, C., *Introduction to Pandemic Influenza*. Wallingford, U.K.: CAB International, 2010.
4. Nelson, M.I. and Holmes, E.C., The evolution of epidemic influenza, *Nat. Rev. Genet.*, 8, 196, 2007.
5. Kilbourne, E.D., Influenza pandemics of the 20th century, *Emerg. Infect. Dis.*, 12, 9, 2006.
6. Nguyen-Van-Tam, J.S. and Hampson, A.W., The epidemiology and clinical impact of pandemic influenza, *Vaccine*, 21, 1762, 2008.
7. Nelson, M.L. et al., Multiple reassortment events in the evolutionary history of H1N1 influenza A virus since 1918, *PLoS Pathog.*, 4, e1000012, 2008.
8. Schäfer, J.R. et al., Origin of the pandemic 1957 H2 influenza A virus and the persistence of its possible progenitors in the avian reservoir, *Virology*, 194, 781, 1993.
9. Fang, R. et al., Complete structure of A7duck/Ukraine/63 influenza hemagglutinin gene: Animal virus as progenitor of human H3 Hong Kong 1968 influenza hemagglutinin, *Cell*, 25, 315, 1981.
10. Smith, G.J. et al., Origins and evolutionary genomics of the 2009 swine-origin H1N1 influenza A epidemic, *Nature*, 459, 1122, 2009.
11. Tumpey, T.M. et al., Characterization of the reconstructed 1918 Spanish Influenza pandemic virus, *Science*, 310, 77, 2005.
12. Glezen, W.P., Emerging infections: Pandemic influenza, *Epidemiol. Rev.*, 18, 64, 1996.
13. Kilbourne, E.D. Epidemiology of Influenza. In: *The Influenza Viruses and Influenza*, Kilbourne, E.D. (ed.). New York: Academic Press, p. 483, 1975.
14. Taubenberger, J.K. et al., Initial genetic characterization of the 1918 "Spanish" influenzavirus, *Science*, 275, 1793, 1997.
15. Reid, A.H. et al., Origin and evolution of the 1918 "Spanish" influenza virus hemagglutinin gene, *Proc. Natl. Acad. Sci. USA*, 96, 1651, 1999.
16. Taubenberger, J. et al., Characterization of the 1918 influenza virus polymerase genes, *Nature*, 437, 889, 2005.
17. Kobasa, D. et al., Enhanced virulence of influenza A viruses with the haemagglutinin of the 1918 pandemic virus, *Nature*, 431, 703, 2004.
18. Tumpey, T.M. et al., Pathogenicity of influenza viruses with genes from the 1918 pandemic virus: Functional roles of alveolar macrophages and neutrophils in limiting virus replication and mortality in mice, *J. Virol.*, 79, 14933, 2005.
19. Tumpey, T.M. et al., Existing antivirals are effective against influenza viruses with genes from the 1918 pandemic virus, *Proc. Natl. Acad. Sci. USA*, 99, 13849, 2002.
20. Tumpey, T.M. et al., Pathogenicity and immunogenicity of influenza viruses with genes from the 1918 pandemic virus, *Proc. Natl. Acad. Sci. USA*, 101, 3166, 2004.
21. World Health Organization Expert Committee, A revision of the system of nomenclature for influenza viruses: A WHO Memorandum, *Bull. WHO*, 58, 585, 1980.
22. Webster, R.G. et al., Evolution and ecology of influenza A viruses, *Microbiol. Rev.*, 56, 152, 1992.
23. Fouchier, R.A.M. et al., Characterization of a novel influenza A virus hemagglutinin subtype (H16) obtained from black-headed gulls, *J. Virol.*, 79, 2814, 2005.
24. Chu, C.M., Dawson, I.M., and Elford, W.J., Filamentous forms associated with newly isolated influenza virus, *Lancet*, 1, 602, 1949.
25. Lamb, R.A. and Choppin, P.W., Identification of a second protein (M2) encoded by RNA segment 7 of influenza virus, *Virology*, 112, 729, 1981.

26. Richardson, J.C. and Akkina, R.K., NS2 protein of influenza virus is found in purified virus and phosphorylated in infected cells, *Arch. Virol.*, 116, 69, 1991.

27. Ada, G.L. and Perry, B.T., The nucleic acid content of influenza virus, *Aust. J. Exp. Biol. Med. Sci.*, 32, 453, 1954.

28. Frommhagen, L.H., Knight, C.A., and Freeman, N.K., The ribonucleic acid, lipid, and polysaccharide constituents of influenza virus preparations, *Virology*, 8, 176, 1959.

29. Compans, R.W., Meier-Ewert, H., and Palese, P., Assembly of lipid-containing viruses, *J. Supramol. Struct.*, 2, 496, 1974.

30. Betakova, T., Nermut, M.V., and Hay, A.J., The NB protein is an integral component of the membrane of influenza B virus, *J. Gen. Virol.*, 77, 2689, 1996.

31. Brassard, D.L., Leser, G.P., and Lamb, R.A., Influenza B virus NB glycoprotein is a component of the virion, *Virology*, 220, 350, 1996.

32. Odagiri, T., Hong, J., and Ohara, Y., The BM2 protein of influenza B virus is synthesized in the late phase of infection and incorporated into virions as a subviral component, *J. Gen. Virol.*, 80, 2573, 1999.

33. Nakada, S. et al., Influenza C virus hemagglutinin: Comparison with influenza A and B virus hemagglutinins, *J. Virol.*, 50, 118, 1984.

34. Nakada, S. et al., Influenza C virus RNA 7 codes for a nonstructural protein, *J. Virol.*, 56, 221, 1985.

35. Pekosz, A. and Lamb, R.A., The CM2 protein of influenza C virus is an oligomeric integral membrane glycoprotein structurally analogous to influenza A virus M2 and influenza B virus NB proteins, *Virology*, 237, 439, 1997.

36. Lamb, R.A. and Choppin, P.W., Segment 8 of the influenza virus genome is unique in coding for two polypeptides, *Proc. Natl. Acad. Sci. USA*, 76, 4908, 1979.

37. Chen, W. et al., A novel influenza A virus mitochondrial protein that induces cell death, *Nat. Med.*, 7, 1306, 2001.

38. Racaniello, V.R. and Palese, P., Influenza B virus genome: Assignment of viral polypeptides to RNA segments, *J. Virol.*, 29, 361, 1979.

39. Horvath, C.M., Williams, M.A., and Lamb, R.A., Eukaryotic coupled translation of tandem cistrons: Identification of the influenza B virus BM2 polypeptide, *EMBO J.*, 9, 2639, 1990.

40. Yamashita, M., Krystal, M., and Palese, P., Comparison of the three large polymerase proteins of influenza A, B, and C viruses, *Virology*, 171, 458, 1989.

41. Herrler, G. et al., The glycoprotein of influenza C virus is the haemagglutinin, esterase, and fusion factor, *J. Gen. Virol.*, 69, 839, 1988.

42. Yamashita, M., Krystal, M., and Palese, P., Evidence that the matrix protein of influenza C virus is coded for by a spliced mRNA, *J. Virol.*, 62, 3348, 1988.

43. Hongo, S. et al., Cloning and sequencing of influenza C/Yamagata/1/88 virus NS gene, *Arch. Virol.*, 126, 343, 1992.

44. Alamgir, A.S. et al., Phylogenetic analysis of influenza C virus nonstructural (NS) protein genes and identification of the NS2 protein, *J. Gen. Virol.*, 81, 1933, 2000.

45. Klenk, H.-D. et al., Activation of influenza A viruses by trypsin treatment, *Virology*, 68, 426, 1975.

46. Lazarowitz, S.G. and Choppin, P.W., Enhancement of the infectivity of influenza A and B viruses by proteolytic cleavage of the hemagglutinin polypeptide, *Virology*, 68, 440, 1975.

47. Klenk, H.-D., Rott, R., and Orlich, M., Further studies on the activation of influenza virus by proteolytic cleavage of the hemagglutinin, *J. Gen. Virol.*, 36, 151, 1977.

48. Ito, T. et al., Molecular basis for the generation in pigs of influenza A viruses with pandemic potential, *J. Virol.*, 72, 7367, 1998.

49. Gambaryan, A.S. et al., Differences between influenza virus receptors on target cells of duck and chicken and receptor specificity of the 1997 H5N1 chicken and human influenza viruses from Hong Kong, *Avian Dis.*, 47, 1154, 2003.

50. Couceiro, J.N., Paulson, J.C., and Baum, L.G., Influenza virus strains selectively recognize sialyloligosaccharides on human respiratory epithelium: The role of the host cell in selection of hemagglutinin receptor specificity, *Virus Res.*, 29, 155, 1993.

51. Matrosovich, M.N. et al., Human and avian influenza (AI) viruses target different cell types in cultures of human airway epithelium, *Proc. Natl. Acad. Sci. USA*, 101, 4620, 2004.

52. Matrosovich, M.N. et al., The surface glycoproteins of H5 influenza viruses isolated from humans, chickens, and wild aquatic birds have distinguishable properties, *J. Virol.*, 73, 1146, 1999.

53. Shinya, K. et al., Avian flu: Influenza virus receptors in the human airway, *Nature*, 440, 43, 2006.

54. van Riel, D. et al., H5N1 virus attachment to lower respiratory tract, *Science*, 312, 399, 2006.

55. Matlin, K.S. et al., Infectious entry pathway of influenza virus in a canine kidney cell line, *J. Cell. Biol.*, 91, 601, 1981.

56. Sieczkarski, S.B. and Whittaker, G.R., Influenza virus can enter and infect cells in the absence of clathrin-mediated endocytosis, *J. Virol.*, 76, 10455, 2002.

57. Pinto, L.H., Holsinger, L.J., and Lamb, R.A., Influenza virus M2 protein has ion channel activity, *Cell*, 69, 517, 1992.

58. Zhirnov, O.P. and Grigoriev, V.B., Disassembly of influenza C viruses, distinct from that of influenza A and B viruses requires neutral-alkaline pH, *Virology*, 200, 284, 1994.

59. Fodor, E. et al., A single amino acid mutation in the PA subunit of the influenza virus RNA polymerase inhibits endonucleolytic cleavage of capped RNAs, *J. Virol.*, 76, 8989, 2002.

60. Hara, K. et al., Amino acid residues in the N-terminal region of the PA subunit of influenza A virus RNA polymerase play a critical role in protein stability, endonuclease activity, cap binding, and virion RNA promoter binding, *J. Virol.*, 80, 7789, 2006.

61. Kawaguchi, A., Naito, T., and Nagata, K., Involvement of influenza virus PA subunit in assembly of functional RNA polymerase complexes, *J. Virol.*, 79, 732, 2005.

62. Honda, A. et al., RNA polymerase of influenza virus: Role of NP in RNA chain elongation, *J. Biochem. (Tokyo)*, 104, 1021, 1988.

63. Shapiro, G.I. and Krug, R.M., Influenza virus RNA replication in vitro: Synthesis of viral template RNAs and virion RNAs in the absence of an added primer, *J. Virol.*, 62, 2285, 1988.

64. Fechter, P. et al., Two aromatic residues in the PB2 subunit of influenza A RNA polymerase are crucial for cap binding, *J. Biol. Chem.*, 278, 20381, 2003.

65. Poon, L.L. et al., Direct evidence that the poly(A) tail of influenza A virus mRNA is synthesized by reiterative copying of a U track in the virion RNA template, *J. Virol.*, 73, 3473, 1999.

66. Boulan, E.R. and Pendergast, M., Polarized distribution of viral envelope proteins in the plasma membrane of infected epithelial cells, *Cell*, 20, 45, 1980.

67. Stieneke-Grober, A. et al., Influenza virus hemagglutinin with multibasic cleavage site is activated by furin, a subtilisin-like endoprotease, *EMBO J.*, 11, 2407, 1992.

68. Palese, P. and Shaw, M.L., Orthomyxoviridae: The viruses and their replication. In: *Fields Virology*, Knipe, D.M. and Howley, P.M. (eds.). Philadelphia, PA: Lippincott Williams & Wilkins, Chapter 47, 2007.

69. Bancroft, C.T. and Parslow, T.G., Evidence for segment-nonspecific packaging of the influenza A virus genome, *J. Virol.*, 76, 7133, 2002.

70. Mindich, L., Packaging, replication and recombination of the segmented genome of bacteriophage Phi6 and its relatives, *Virus Res.*, 101, 83, 2004.

71. Luo, C., Nobusawa, E., and Nakajima, K., An analysis of the role of neuraminidase in the receptor-binding activity of influenza B virus: The inhibitory effect of Zanamivir on haemadsorption, *J. Gen. Virol.*, 80, 2969, 1999.

72. Pyle, G.F., *The Diffusion of Influenza: Patterns and Paradigms.* Totowa, NJ: Rowman & Littlefield, 1986.

73. Adachi, M. et al., Expression of cytokines on human bronchial epithelial cells induced by influenza virus A, *Int. Arch. Allergy Immunol.*, 113, 307, 1997.

74. Papineni, R.S. and Rosenthal, F.S., The size distribution of droplets in the exhaled breath of healthy human subjects, *J. Aerosol Med.*, 10, 105, 1997.

75. Boone, S.A. and Gerba, C.P., The occurrence of influenza A virus on household and day care center fomites, *J. Infect.*, 51, 103, 2005.

76. Dushoff, J. et al., Mortality due to influenza in the United States—An annualized regression approach using multiple-cause mortality data, *Am. J. Epidemiol.*, 163, 181, 2006.

77. Lauderdale, T.L. et al., Etiology of community acquired pneumonia among adult patients requiring hospitalization in Taiwan, *Respir. Med.*, 99, 1079, 2005.

78. Numazaki, K. et al., Etiological agents of lower respiratory tract infections in Japanese children, *In Vivo*, 18, 67, 2004.

79. de Roux, A. et al., Viral community-acquired pneumonia in nonimmunocompromised adults, *Chest*, 125, 1343, 2004.

80. Douglas, R.G., Influenza in man. In: *The Influenza Viruses and Influenza*, Kilbourne, E.D. (ed.). London, U.K.: Academic Press, Inc., p. 395, 1975.

81. Fenner, F. et al., *The Biology of Animal Viruses*, 2nd ed. London, U.K.: Academic Press, Inc., 1974.

82. Walsh, J.J. et al., Bronchotracheal response in human influenza, *Arch. Int. Med.*, 108, 376, 1961.

83. Paisley, J.W. et al., Type A_2 influenza viral infections in children, *Am. J. Dis. Child.*, 132, 34, 1978.

84. Robertson, L., Caley, J.P., and Moore, J., Importance of *Staphylococcus aureus* in pneumonia in the 1957 epidemic of influenza A, *Lancet*, 2, 233, 1958.

85. Dell, K.M. and Schulman, S.L., Rhabdomyolysis and acute renal failure in a child with influenza A infection, *Pediatr. Nephrol.*, 11, 363, 1997.

86. Fujimoto, S. et al., PCR on cerebrospinal fluid to show influenza-associated acute encephalopathy or encephalitis, *Lancet*, 352, 873, 1998.

87. Waldman, R.J. et al., Aspirin as a risk factor in Reye's syndrome, *JAMA*, 247, 3089, 1982.

88. Hurwitz, E.S. et al., National surveillance for Reye syndrome: A five-year review, *Pediatrics*, 70, 895, 1982.

89. Nicholson, K.G. et al., Acute upper respiratory tract viral illness and influenza immunization in homes for the elderly, *Epidemiol. Infect.*, 105, 609, 1990.

90. Noble, G.R., Epidemiological and clinical aspects of influenza. In: *Basic and Applied Influenza Research*, Beare, A.S. (ed.). Boca Raton, FL: CRC Press, p. 11, 1982.

91. Tkácová, M. et al., Evaluation of monoclonal antibodies for subtyping of currently circulating human type A influenza viruses, *J. Clin. Microbiol.*, 35, 1196, 1997.

92. Hurt, A.C. et al., Performance of six influenza rapid tests in detecting human influenza in clinical specimens, *J. Clin. Virol.*, 39, 132, 2007.

93. Smit, M. et al., Comparison of the NOW Influenza A & B, NOW Flu A, NOW Flu B, and Directigen Flu A + B assays, and immunofluorescence with viral culture for the detection of influenza A and B viruses, *Diagn. Microbiol. Infect. Dis.*, 57, 67, 2007.

94. Steininger, C. et al., Effectiveness of reverse transcription-PCR, virus isolation, and enzyme-linked immunosorbent assay for diagnosis of influenza A virus infection in different age groups, *J. Clin. Microbiol.*, 40, 2051, 2002.

95. Ellis, J.S., Fleming, D.M., and Zambon, M.C., Multiplex reverse transcription-PCR for surveillance of influenza A and B viruses in England and Wales in 1995 and 1996, *J. Clin. Microbiol.*, 35, 2076, 1997.

96. Stockton, J. et al., Multiplex PCR for typing and subtyping influenza and respiratory syncytial viruses, *J. Clin. Microbiol.*, 36, 2990, 1998.

97. Schweiger, B. et al., Application of a fluorogenic PCR assay for typing and subtyping of influenza viruses in respiratory samples, *J. Clin. Microbiol.*, 38, 1552, 2000.

98. Herrmann, B., Larsson, C., and Zweygberg, B.W., Simultaneous detection and typing of influenza viruses A and B by a nested reverse transcription-PCR: Comparison to virus isolation and antigen detection by immunofluorescence and optical immunoassay (FLU OIA), *J. Clin. Microbiol.*, 39, 134, 2001.

99. Van Elden, L.J.R. et al., Simultaneous detection of influenza viruses A and B using real-time quantitative PCR, *J. Clin. Microbiol.*, 39, 196, 2001.

100. Young, L.C. et al., Summer outbreak of respiratory disease in an Australian prison due to an influenza A/Fujian/411/2002(H3N2)-like virus, *Epidemiol. Infect.*, 133, 107, 2005.

101. Li, J., Chen, S., and Evans, D.H., Typing and subtyping using DNA microarrays and multiplex reverse transcriptase PCR, *J. Clin. Microbiol.*, 39, 696, 2001.

102. Zitterkopf, N.L. et al., Relevance of influenza A virus detection by PCR, shell vial assay, and tube cell culture to rapid reporting procedures, *J. Clin. Microbiol.*, 44, 3366, 2006.

103. Carman, W.F. et al., Rapid virological surveillance of community influenza infection in general practice, *BMJ*, 321, 736, 2000.

104. Playford, E.G. and Dwyer, D.E., Laboratory diagnosis of influenza virus infection, *Pathology*, 34, 115, 2002.

105. Zambon, M. et al., Diagnosis of influenza in the community: Relationship of clinical diagnosis to confirmed virological, serologic, or molecular detection of influenza, *Arch. Intern. Med.*, 161, 2116, 2001.

106. Schmid, M.L. et al., Prospective comparative study of culture specimens and methods in diagnosing influenza in adults, *BMJ*, 316, 275, 1998.

107. Kendal, A.P. and Harmon, M.W., Influenza virus. In: *Laboratory Diagnosis of Infectious Disease—Principles and Practice. Viral, Rickettsial, and Chlamydial Diseases*, Lennette, E.H., Halonen, P., and Murphy, F.M. (eds.). Berlin, Germany: Springer-Verlag, p. 612, 1988.

108. Burnet, F.M., Influenza virus on the developing egg. I. Changes associated with the development of an egg-passage strain of virus, *Br. J. Exp. Pathol.*, 17, 282, 1936.

109. Clinical and Laboratory Standards Institute., Viral culture: Approved guideline, CLSI document M41-A. Wayne, PA: Clinical and Laboratory Standards Institute, 2006.

110. Atmar, R.L., Influenza viruses. In: *Manual of Clinical Microbiology*, Murray, P.R. et al. (eds.). Washington, DC: ASM Press, p. 1340, 2007.

111. Leland, D.S. and Ginocchio, C.C., Role of cell culture for virus detection in the age of technology, *Clin. Microbiol. Rev.*, 20, 49, 2007.

112. Weinberg, A. et al., Evaluation of R-mix shell vials for the diagnosis of viral respiratory tract infections, *J. Clin. Virol.*, 30, 100, 2004.

113. BD Directigen Flu A [package insert]. Sparks, MD.: Becton, Dickinson and Co., August 2000.

114. Newton, D.W., Treanor, J.J., and Menegus, M.A., Clinical and laboratory diagnosis of influenza virus infections, *Am. J. Manag. Care*, 6, S265, 2000.

115. BD Directigen Flu A+B [package insert]. Sparks, MD: Becton, Dickinson and Co., March 2001.

116. QuickVue [package insert]. San Diego, CA: Quidel Corp., March 2002.

117. Centers for Disease Control and Prevention (CDC). Laboratory diagnostic procedures for influenza. Retrieved December 11, 2002, from www.cdc.gov/ncidod/diseases/flu/flu_dx_table.htm.

118. Ellis, J.S. and Zambon, M.C., Molecular diagnosis of influenza, *Rev. Med. Virol.*, 12, 375, 2002.

119. Harnden, A. et al., Near patient testing for influenza in children in primary care: Comparison with laboratory test, *BMJ*, 326, 480, 2003.

120. Cherian, T. et al., Use of PCR-enzyme immunoassay for identification of influenza A virus matrix RNA in clinical samples negative for cultivable virus, *J. Clin. Microbiol.*, 32, 623, 1994.

121. Atmar, R.L. et al., Comparison of reverse transcription-PCR with tissue culture and other rapid diagnostic assays for detection of type A influenza virus, *J. Clin. Microbiol.*, 34, 2604, 1996.

122. Dagher, H. et al., Rhinovirus detection: Comparison of real-time and conventional PCR, *J. Virol. Methods*, 117, 113, 2004.

123. Poon, L.L. et al., Detection of SARS coronavirus in patients with severe acute respiratory syndrome by conventional and real-time quantitative reverse transcription-PCR assays, *Clin. Chem.*, 50, 67, 2004.

124. Adcock, P.M. et al., Effect of rapid viral diagnosis on the management of children hospitalized with lower respiratory tract infection, *Pediatr. Infect. Dis. J.*, 16, 842, 1997.

125. Woo, P.C. et al., Cost-effectiveness of rapid diagnosis of viral respiratory tract infections in pediatric patients, *J. Clin. Microbiol.*, 35, 1579, 1997.

126. Hibbitts, S. and Fox, J.D., The application of molecular techniques to diagnosis of viral respiratory tract infections, *Rev. Med. Microbiol.*, 13, 177, 2002.

127. Liolios, L. et al., Comparison of a multiplex reverse transcription-PCR-enzyme hybridization assay with conventional viral culture and immunofluorescence techniques for the detection of seven viral respiratory pathogens, *J. Clin. Microbiol.*, 39, 2779, 2001.

128. Coiras, M.T. et al., Simultaneous detection of influenza A, B, and C viruses, respiratory syncytial virus, and adenoviruses in clinical samples by multiplex reverse transcription nested-PCR assay, *J. Med. Virol.*, 69, 132, 2003.

129. Sharp, P.A., 1918 Flu and responsible science, *Science*, 310, 17, 2005.

130. Pearce, M.B. et al., Seasonal trivalent inactivated influenza vaccine protects against 1918 Spanish influenza virus infection in ferrets, *J. Virol.*, 86, 7118, 2012.

12 Japanese Encephalitis Virus

Dongyou Liu

CONTENTS

12.1 INTRODUCTION

First isolated in 1935 from a 19 year old who died of encephalitis in Tokyo, Japan, Japanese encephalitis virus (JEV) is an arthropod-borne virus (or arbovirus) that involves mosquitoes, pigs, and water birds during its life cycle. Besides causing reproductive disorder in pigs (amplifying host), JEV is also responsible for inducing encephalitis in humans (dead-end host, due to the low level and transient viremia, which is insufficient for the virus to spread to the next host through mosquito bites). Despite the availability of vaccines, JEV remains a serious public health concern in eastern and southern Asia as a result of increased population density, deforestation, and the expanding irrigation of agricultural areas. It is estimated that JEV is responsible for >50,000 cases annually, including at least 10,000 deaths and 15,000 cases with neuropsychiatric sequelae [1].

12.2 CLASSIFICATION, MORPHOLOGY, AND GENOME STRUCTURE

12.2.1 CLASSIFICATION

JEV is an enveloped, positive sense, single-stranded RNA virus belonging to the JEV serogroup, mosquito-borne virus group, genus *Flavivirus*, family Flaviviridae (*flavus* means yellow in Latin, which comes from the type virus of the family, the yellow fever virus) [2].

Of the three genera within the family Flaviviridae, *Flavivirus* is arthropod-borne while *Hepacivirus* and *Pestivirus* are nonarthropod-borne. The genus *Flavivirus* encompasses 73 known viruses, including 34 mosquito-borne, 17 tick-borne, and 22 zoonotic agents with no known arthropod vector. About 40 of the 73 recognized viruses in this genus are associated with human diseases, among which dengue, yellow fever, Japanese encephalitis, tick-borne encephalitis, and West Nile encephalitis viruses are responsible for extensive morbidity and mortality in human populations worldwide [1,2].

Upon phylogenetic analysis of the *NS5* gene, the genus *Flavivirus* is separated into 12 serogroups (or antigenic complexes): seven mosquito-borne serogroups (i.e., Aroa, dengue, Japanese encephalitis, Kokobera, Ntaya, Spondweni, and yellow fever virus groups), two tick-borne serogroups (mammalian tick-borne and seabird tick-borne groups), and three serogroups with no known arthropod vector (Entebbe, Modoc, and Rio Bravo) (Figure 12.1). Within the JEV serogroup, at least 10 virus strains/isolates are recognized: Cacipacore, Ilheus, Koutango, Japanese encephalitis, Murray Valley encephalitis (including a distinct subtype Alfuy virus), Rocio, St. Louis encephalitis, Usutu, West Nile (including a distinct subtype Kunjin virus), and Yaounde (Figure 12.1). In addition, examination of structural genes (i.e., prM and E) facilitates the identification of five genotypes (I–IV) within JEV [3].

The genus *Hepacivirus* contains a single member, hepatitis C virus and its relatives, and the genus *Pestivirus* comprises four serogroups, that is, bovine viral diarrhea virus 1 (BVDV-1), bovine viral diarrhea virus 2 (BVDV-2), border disease virus (BDV), and classical swine fever virus (CSFV), as well as a tentative species (pestivirus of giraffe). Furthermore, several pestiviruses have been identified and suggested for recognition as novel subgroups/species [4].

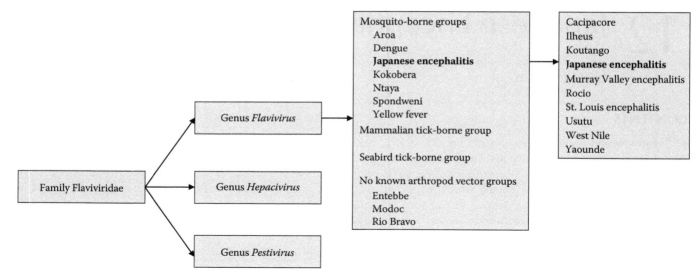

FIGURE 12.1 Classification of JEV.

12.2.2 MORPHOLOGY AND GENOME STRUCTURE

The virion of JEV appears spherical and measures approximately 50 nm in diameter. Its external surface is covered (enveloped) by a host-derived lipid bilayer that is formed by 180 copies of the envelope (E) and membrane (M) proteins arranged in a herringbone pattern. Underneath the envelope is a nucleocapsid (C) core enclosing a single-stranded, positive sense RNA genome of approximately 11 kb. Flanked by the 5′ and 3′ untranslated region (UTR), the genome consists of a single open reading frame (ORF), which is translated as a single polyprotein of three structural (capsid [C], envelope [E], and membrane [M] proteins) and seven nonstructural proteins (NS1, NS2A, NS2B, NS3, NS4A, NS4B, and NS5) organized in the order of 5′UTR-C-prM/M-E-NS1-NS2A-NS2B-NS3-NS4A-NS4B-NS5-3′UTR. Subsequent cleavages by viral and cellular proteases yield 10 individual proteins [2].

Being a well-studied virus, over 140 full-length genomic sequences of JEV have become available in GenBank [5–11]. As exemplified by JEV_CNS769_Laos_2009, a JEV genotype I isolated from a patient with encephalitis in Laos, the complete genome is 10,965 nucleotides (nt) in length, including the 5′ and 3′ UTRs of 96 and 570 nt, respectively. The ORF (10,299 nt) encodes a polyprotein processed into three structural proteins [capsid (C) (127 amino acids [aa]), premembrane/membrane (prM/M) (167 aa), and envelope (E) (500 aa)] and seven nonstructural proteins [NS1 (352 aa), NS2A (227 aa), NS2B (131 aa), NS3 (619 aa), NS4A (149 aa), NS4B (255 aa), and NS5 (905 aa)] [12].

The premembrane/membrane protein (prM/M) is a 20–25 kDa precursor to M protein. During virus maturation and upon exposure to low pH in the late endosomes, prM is cleaved by the cellular protease, furin, at a highly conserved cleavage site, resulting in the release of the singly glycosylated pr protein [13].

The envelope protein (E) is a 50 kDa protein belonging to the type I membrane protein group. It is a monomer containing 3 structural domains that are held together by 12 conserved cysteine residues forming 6 disulfide bonds. While domain I (DI) features an eight-stranded β barrel, domain II (DII) contains the highly conserved fusion peptide and domain III (DIII) is an immunoglobulin-like structure that is involved in receptor binding and fusion with the host membrane. Being the main focus of host antibody response, the E protein is extensively targeted in the serological assays such as hemagglutination inhibition, neutralization, and IgM capture enzyme-linked immunosorbent assays (ELISA) to diagnose flavivirus infections. Further, neutralizing antibodies raised against the E protein following vaccination provide protection from JEV infection.

Nonstructural protein 1 (NS1) is a glycosylated protein of about 48 kDa, with a highly conserved structure that is stabilized by six disulfide bridges. Multiple forms of NS1 exist and are found intracellularly (cell membrane bound), on the cell surface, or as a soluble secreted hexameric lipoparticle. The intracellular form of NS1 is colocalized with the double-stranded RNA and other components of the viral replication complex and is involved in RNA replication. The secreted form of NS1 is capable of attenuating complement activation and is implicated in immune evasion strategies. Secreted and cell surface–associated NS1 are highly immunogenic with the ability to elicit a strong, protective, but nonneutralizing, antibody response. These forms of NS1 offer useful targets for serological diagnosis [14].

12.3 BIOLOGY AND EPIDEMIOLOGY

JEV enters the host through a mosquito bite and reaches the central nervous system (CNS) via leukocytes (probably T lymphocytes). After binding to the endothelial surface of the CNS with its E protein, the virus is internalized via receptor-mediated endocytosis. Subsequently, the viral nucleocapsid releases the viral RNA in the cytoplasm, and the genomic RNA is translated into viral proteins in one ORF. Individual proteins are then cleaved by cellular proteases and a viral serine protease. In addition to neuronal cells, JEV also infects astrocytes (which constitute a component of the blood–brain barrier [BBB]), which facilitate transmission of JEV from peripheral tissues to the cerebrospinal fluid (CSF) [15].

Although the first Japanese encephalitis (JE) case was reported in Japan in 1871, it was not until 1935 when JEV was successfully isolated. Phylogenetic investigations suggested that JEV had evolved in the mid-1500s from its ancestral flavivirus genotype IV in the Indonesia–Malaysia region and diverged into genotypes V, III, II, and I. From there, it has spread northward and westward to other parts of Asia (including China, Japan, Korea, Taiwan, India, and Southeast Asia) and recently to northern Australia, correlating with the population explosion, globalization, and changes in global climatic condition as a result of industrialization and deforestation [16–22].

Of the five JEV genotypes identified to date, genotypes I and III occur primarily in temperate, epidemic areas, whereas genotypes II, IV, and V are found in tropical, endemic areas [23,24]. Interestingly, all five genotypes are present in Indonesia, including genotype IV (the oldest), which is confined to the Indonesia–Malaysia region; genotype III is the only type found in India; genotype V has been isolated on a few occasions only in Malaysia, China, and South Korea [25–28]. On the other hand, genotypes I–III are distributed throughout Asia and Australasia. It is notable that in Indonesia, JEV isolates collected between 1974 and 1980 belonged to genotype II, those collected between 1980 and 1981 belonged to genotype IV, and those collected in 1987 belonged to genotype III [29].

JE is a disease occurring predominantly in rural and suburban areas where rice culture and pig farming coexist. Water birds act as a natural reservoir; pigs and ardeid birds such as pond heron (*Ardeola grayii*) and cattle egret (*Bubulcus ibis*) function as amplifying hosts; *Culex* mosquitoes (especially *Culex tritaeniorhynchus* and, to a lesser extent, *Culex vishnui*, *Culex Gelidus*, and *Culex annulirostris*) are the principal vectors for JEV [30,31].

Culex tritaeniorhynchus mosquitoes feed on large domestic animals, birds, and occasionally humans outdoors, especially after sunset. *Culex tritaeniorhynchus* larvae are present in flooded rice fields, marshes, and stagnant water. Vertical transmission of JEV has been observed in *Culex tritaeniorhynchus*, *Culex pipiens*, *Culex annulus*, *Culex quinquefasciatus*, *Aedes albopictus*, *Aedes togoi*, and *Armigeres subalbatus*.

Because of their close proximity to human dwellings and their rapid population turnover, pigs are the main contributors in the transmission of JEV to humans. Although JEV infection generally results in a transient febrile illness in older (nonpregnant) pigs, it can cause reproductive defects (e.g., abortions and stillborn fetuses) in pregnant pigs and encephalitis and death in young pigs. The level of viremia in pigs is sufficient to allow passage of the virus onto feeding mosquitoes [32–34].

However, while JEV infections in horses, donkeys, cattle, and water buffaloes can be symptomatic, these animals do not produce high virus titers for the virus to spread to other hosts via mosquito feedings. These animals are therefore considered dead-end hosts for JEV. In horses, the clinical symptoms of JEV infection range from fever, anorexia, lethargy, neck rigidity, falling or staggering, hyperexcitability, disordered behavior, blindness, sweating, collapse, to death [35].

Similar to situation in horses, JEV infection in humans does not generate a sufficient level of viremia. Thus, humans also represent a dead-end host for JEV. A number of host (e.g., age, general health, and genetic makeup) and viral factors (e.g., strain virulence, route of entry, and amount of virus inoculated) may influence the outcome of JEV infection relating to the absence/presence of clinical symptoms. JE is largely a disease of children in the endemic areas, but it may affect both the adults and children in the newly invaded areas due to the absence of protective antibodies [36].

In the subtropics and tropics, JEV transmission can take place year-round, particularly during the rainy season. In temperate areas, JEV transmission is seasonal, with the human disease usually peaking in the summer and fall. Countries with major JEV epidemics in the past include China, Korea, Japan, Taiwan, and Thailand, but use of vaccines has reduced the disease prevalence in these regions [37–46]. However, periodic JEV epidemics continue to occur in Vietnam, Cambodia, Myanmar, India, Nepal, and Malaysia [47–49].

12.4 CLINICAL FEATURES AND PATHOGENESIS

12.4.1 CLINICAL FEATURES

With an incubation period of 5–15 days, Japanese encephalitis is asymptomatic in the vast majority of cases, with only a small number (<1%) developing into encephalitis and meningitis [50]. The onset of JE in humans is marked with the sudden and abrupt appearance of fever (38°C–41°C), chills, headache, and malaise in mild cases followed by nausea, vomiting, abdominal pain, diarrhea, photophobia, rigor, neck stiffness, stupor, disorientation, cachexia, hemiparesis, convulsions, poliomyelitis-like flaccid paralysis, and coma (alpha, theta, and delta coma) in more severe cases [51]. Some patients may show parkinsonian syndrome (encompassing dull, flat, mask-like face with wide, unblinking eyes, tremor, generalized hypertonia, cogwheel rigidity, and choreoathetoid movements) and dystonia (limb, axial, orofacial). Symptoms of brainstem infection consist of changed respiratory pattern, flexor and extensor posturing, and abnormalities in the papillary and oculocephalic reflexes. Up to 30% of JE cases admitted to hospital die and 50% of the survivors suffer from permanent neurological or psychiatric sequelae (e.g., seizures, upper and lower motor neuron weakness, cerebellar and extrapyramidal signs, flexion deformities of the arms, hyperextension of the legs, cognitive deficits, language impairment, learning difficulties, deafness, and emotional liability) [52,53]. Pathological changes in fatal cases include vascular congestion, microglial proliferation, formation of gliomesenchymal nodules, focal or confluent areas of cystic necrosis, cerebral edema, and transcompartmental shift in thalamus, corpus striatum, brainstem, spinal cord, and other parts of the brain [54,55].

Ankur Nandan et al. [56] documented a JE case involving a 13-year-old adolescent boy, who presented with fever and acute onset paraparesis with urinary retention initially, progressed to quadriparesis and then followed by headache and altered sensorium. Brain magnetic resonance imaging (MRI) showed bilateral basal ganglia that were grossly swollen with vasogenic edema tracking along the internal capsule and midbrain. Adjacent ventrolateral thalamus and internal capsule also displayed mild abnormal intensities. Spinal screening revealed abnormal cord intensities in the entire cord with gross edema in cervical and conus regions. The patient had elevated IgM titers against JE virus in the CSF. After treatment with intravenous methyl prednisolone, he regained near normal power at 3 months follow-up, but still had hesitancy, dysarthria, and slowness of movement.

12.4.2 Pathogenesis

The ability of JEV to cross the BBB is important for the pathogenesis and clinical outcome of the neurotropic infection [57]. JEV is known to utilize macrophages as a "Trojan horse" mechanism to infiltrate into the CNS. Because JEV-infected monocytes do not undergo apoptosis, they provide a means of transporting the virus across BBB. The newly synthesized virus particles inside the infected monocytes induce deformation of tight junctions and dissociation of endothelial cells facilitating the viral entry and leading to the infection of nearby CNS cells [58–61].

At the molecular level, JEV upregulates matrix metalloproteinase 9 (MMP-9) on endothelial cells, resulting in the degradation of components of the basal lamina and disruption of BBB. Inside neuronal cells, JEV elicits strong proinflammatory and antiviral responses, leading to apoptosis. Neuronal cell death resulting from JEV infection can be due to (1) direct killing, where viral multiplication within neuronal cells causes apoptosis, and (2) indirect killing, where massive inflammatory response upregulates reactive oxygen species and cytokines such as tumor necrosis factor α (TNFα), contributing to neuronal death. Neuronal cell damage or death induces the subsequent astroglial and microglial activation and release of further proinflammatory mediators (e.g., interleukin 6 [IL-6], TNFα monocyte chemotactic protein 1 [MCP1], and RANTES [regulated upon activation, normal T cell expressed and secreted]) [62–64]. This promotes massive leukocyte migration and infiltration into the brain. Thus, the intense cascade of proinflammatory mechanism taking over the homeostatic mechanism underlines the pathogenesis of encephalitis [65–74].

Furthermore, JEV appears to have the ability to modulate histone deacetylases (HDACs), leading to the suppressed expression of proinflammatory cytokines and chemokines and increased expression of anti-inflammatory molecules in monocyte/macrophages [75]. Other immune evasion strategies of flaviviruses including JEV are the inhibition of interferon, complement, natural killer (NK) cell, B cell, and T cell responses in the host cells [76].

12.5   IDENTIFICATION AND DIAGNOSIS

Patient showing evidence of a neurologic infection (e.g., encephalitis, meningitis, or acute flaccid paralysis) who has recently returned from endemic regions should be suspected of JE. Given the nonspecific signs induced by JEV, diagnosis of JE by clinical symptoms alone is difficult. In particular, JE needs to be differentiated from other encephalitis-causing microbes such as herpesviruses, bacterial meningitis, malaria, and other arboviruses (e.g., Murray Valley encephalitis, West Nile/Kunjin and dengue viruses) [77].

Clinical laboratory investigations often reveal moderately elevated white blood cell count, mild anemia, hyponatremia, thrombocytopenia, and elevated hepatic enzymes in the blood and lymphocytic pleocytosis with moderately elevated protein in the CSF. MRI of the brain helps uncover abnormal changes in the thalamus, basal ganglia, midbrain, pons, and medulla. However, laboratory confirmation of human JEV infection relies on the demonstration of JE virus–specific IgM in the CSF and serum [78–81]. Although isolation of the virus in cell culture or the detection of nucleic acid by reverse transcription polymerase chain reaction (RT-PCR) provides definitive evidence of JEV infection, these methods lack the sensitivity needed for routine diagnosis considering that humans often have low or undetectable levels of viremia by the time distinctive clinical symptoms are observable.

JEV-specific IgM antibodies are often detected by using JE virus immunoglobulin M capture enzyme-linked immunosorbent assay (MAC ELISA). In general, such antibodies are present in patient CSF and serum 4–7 days after onset of symptoms. In particular, detection of JEV-specific IgM antibodies in the CSF indicates recent CNS infection. This helps distinguish clinical disease attributed to JEV from previous vaccination. Nevertheless, cross-reaction with other arboviruses may occur occasionally, and interpretation of the MAC ELISA results should correlate with clinical and epidemiologic information [82–87].

Plaque reduction neutralization test (PRNT) measures virus-specific neutralizing antibodies against viral envelope protein and is useful for confirming recent JEV infection or discriminating between cross-reacting antibodies and other flaviviral infections. Usually, a fourfold or greater rise in virus-specific neutralizing antibodies between acute- and convalescent-phase serum specimens collected 2–3 weeks apart is indicative of JEV infection. Again, interpretation of PRNT results should consider vaccination history, date of onset of symptoms, and information about other arboviruses known to circulate in the geographic area [88].

Other serological tests useful for confirmation of JEV infection include virus neutralization test (VNT, which shows the absence or decrease in visible cytopathic effect in the cell monolayer due to patient serum), hemagglutination and hemagglutination inhibition assays (HI, which detect the envelope glycoprotein of arthropod-borne viruses and the antibodies developed to this virus protein, respectively), immunofluorescence assay (IFA, which can differentiate the

IgM and IgG responses to flaviviral infection), and Western blot (which uses lysates of flavivirus-infected cell monolayers to differentiate flaviviral infections) [89–92].

Although it is not a sensitive procedure to be useful for routine diagnosis of JEV infection, in vitro isolation of the virus from serum, CSF, or a 10% brain tissue homogenate provides a means to further characterize the pathogen. The following cell lines may be employed for this purpose: C6/36 (*A. albopictus*) cells, baby hamster kidney (BHK), African green monkey kidney (Vero), or porcine stable equine kidney (PSEK) cells. In addition, suckling mice (24–48 h old) may be utilized for the isolation of virus.

The development of nucleic acid amplification techniques (especially RT-PCR) has offered additional tools for the diagnosis and genotyping of JEV [93–115]. Sequence analysis of the resulting PCR products permits identification as well as phylogenetic investigation. In addition, application of real-time PCR enables simultaneous amplification and detection of target organism in a sensitive and specific manner with a decreased risk of contamination. It also allows easy quantification, standardization, and high-throughput capacity. Although nucleic acid amplification techniques are not recommended for ruling out JE infection in humans, they are valuable for the detection of the virus in the amplifying hosts and in mosquitoes, contributing to improved surveillance activities and control measures [116,117].

12.6 TREATMENT AND PREVENTION

12.6.1 TREATMENT

As no specific treatment for Japanese encephalitis is available, care of JE patients is centered on supportive measures, with assistance given for feeding, breathing, or seizure control as required. Raised intracranial pressure may be managed with mannitol. Since there is no transmission from person to person, JE patients do not need to be isolated.

Currently, treatment options for JEV are largely in developmental stages. Several natural occurring compounds such as arctigenin (a phenylpropanoid dibenzylbutyrolactone lignin) and rosmarinic acid (a phenolic compound found in Labiatae herbs) have been shown to markedly decrease JEV-induced neuronal apoptosis, enhance microglial activation and caspase activity, and induce proinflammatory mediators in the brains of experimental mice. In addition, curcumin appears to decrease cellular reactive oxygen species level, restore cellular membrane integrity, decrease proapoptotic signaling molecules, and modulate cellular levels of stress-related proteins, providing in vitro neuroprotection against JEV infection. Similarly, minocycline (a member of the broad-spectrum antibiotic tetracycline group) also helps reduce neuronal apoptosis, microglial activation, active caspase activity, proinflammatory mediators, and viral titer as well as prevent BBB damage in mice challenged with JEV infection [118]. Further, *N*-methylisatin-β-thiosemicarbazone derivative (which inhibits virus replication) and *N*-nonyldeoxynojirimycin (which blocks the trimming step of

N-linked glycosylation) also demonstrate the potential to eliminate the production of endoplasmic reticulum–budding viruses. Finally, baicalein, a purportedly antiviral bioflavonoid has been shown to exhibit direct extracellular virocidal activity on JEV [119].

12.6.2 PREVENTION

As Japanese encephalitis is transmitted via mosquito bites, an important aspect of JE prevention is vector control [120,121]. The use of larvicides and insecticides is useful for controlling mosquitoes in paddy fields, which are breeding grounds for mosquitoes and also attract birds. Growing larvivorous fish and using neem cakes represent alternative, ecologically friendly ways to control mosquitoes in paddy fields. Vaccinating pigs also helps break the mosquito–pig–human transmission cycle of JEV.

To prevent human JEV infection, precaution should be taken to reduce exposure by using insect repellent, wearing permethrin-impregnated clothing, sleeping under permethrin-impregnated bed nets, and staying in accommodations with screens. Another potential source of JEV exposure is working with the virus in a laboratory setting, where the virus can be transmitted through needlesticks and through mucosal or inhalational accidental exposures.

Since infection with JEV confers lifelong immunity, vaccination represents a most effective way to prevent JE [122]. Travelers to the rural areas in endemic regions such as Southeast Asia and India during JEV transmission season are recommended to have JE vaccine [123–127]. Vaccination is also recommended for all laboratory workers with a potential for exposure to infectious JEV [128].

JEV vaccines are mostly based on the genotype III virus, although they seem to elicit protective levels of neutralizing antibodies against heterologous strains of genotypes I–IV [129]. A formalin-inactivated mouse brain–derived vaccine was first produced in Japan in the 1930s, which has contributed to the control of JE in Japan, Korea, Taiwan, and Singapore. A live-attenuated SA14-14-2 JE vaccine, licensed in 1988 in China, is available in Nepal, Sri Lanka, South Korea, and India. A derivative of SA 14-14-2 strain was developed in September 2012 by an Indian firm Biological E Limited [130,131].

Currently, two JE vaccines are in common use: (1) an inactivated mouse brain–derived JE vaccine (JE-MB) and (2) an inactivated Vero cell culture–derived Japanese encephalitis vaccine (JE-VC) [132–134].

JE-MB (also known as JE-VAX) is an inactivated vaccine against JE that is prepared by inoculating mice intracerebrally with the JEV Nakayama-NIH strain. This vaccine contains porcine gelatin as stabilizer and thimerosal as preservative. It is injected via subcutaneous route at days 0, 7, and 30. Licensed since 1992, this vaccine is used to prevent JE in persons aged >1 year traveling to JE-endemic countries. The adverse effects of JE-MB include allergic hypersensitivity reactions (e.g., general urticaria and angioedema of the extremities, face, and oropharynx), bronchospasm, respiratory distress, and hypotension, which may be relieved with antihistamines or corticosteroids [135].

JE-VC (also known as IXIARO, which is marketed by Intercell AG, Vienna, Austria) is an inactivated, Vero cell-derived vaccine based on SA14-14-2 strain against JE. This vaccine contains aluminum hydroxide as adjuvant and is injected via intramuscular route at days 0 and 28. Introduced in March 2009, this vaccine is highly immunogenic, showing significantly higher geometric mean antibody titers and is used in persons aged >17 years, especially travelers to rural areas of Asia. The main side effects of IXIARO are headache and myalgia [122,131,136].

12.7 CONCLUSION

JEV is a mosquito-borne virus that is responsible for causing JE in many parts of Asia, with significant morbidity and mortality. With rapid globalization and climatic shift, JEV has begun to emerge in other areas and is poised to become a global pathogen and a potential cause of worldwide pandemics [137]. Given the tendency of JEV to induce transient and low-level viremia in humans, serological assays such as the IgM capture ELISA for detection of the virus in serum and CSF remain the preferred method for diagnosis of JEV infections. Nonetheless, molecular assays provide an indispensible tool for surveillance activities in pigs and mosquitoes and facilitate timely implementation of public health responses and control measures against this disease [138]. In endemic regions, prevention should be focused on breaking up the JEV infection cycle, involving mosquitoes, pigs, and humans, by controlling mosquito larvae, vaccinating pigs, and reducing human exposure to mosquitoes. The use of vaccines in travelers to endemic regions and workers in laboratories dealing with JEV provides another useful means to mitigate this disease.

REFERENCES

1. Campbell GL et al. Estimated global incidence of Japanese encephalitis: A systematic review. *Bull World Health Organ* 2011;89:766–774.
2. Gubler DJ, Kuno G, and Markoff L. Flaviviruses, in *Fields Virology*, 5th ed., Knipe DM and Howley PM (eds.), Lippincott Williams & Wilkins, Philadelphia, PA, p. 1153, 2007.
3. Mackenzie JS, Barrett AD, and Deubel V. The Japanese encephalitis serological group of flaviviruses: A brief introduction to the group. *Curr Top Microbiol Immunol* 2002;267:1–10.
4. Kuno G et al. Phylogeny of the genus *Flavivirus*. *J Virol* 1998;72:73–83.
5. Sumiyoshi H et al. Complete nucleotide sequence of the Japanese encephalitis virus genome RNA. *Virology* 1987;161:497–510.
6. Chambers TJ et al. Flavivirus genome organization, expression, and replication. *Annu Rev Microbiol* 1990;44:649–688.
7. Yun SI et al. Molecular characterization of the full-length genome of the Japanese encephalitis viral strain K87P39. *Virus Res* 2003;96:129–140.
8. Unni SK et al. Japanese encephalitis virus: From genome to infectome. *Microbes Infect* 2011;13:312–321.
9. Carney J et al. Recombination and positive selection identified in complete genome sequences of Japanese encephalitis virus. *Arch Virol* 2012;157:75–83.
10. Singha H et al. Complete genome sequence analysis of Japanese encephalitis virus isolated from a horse in India. *Arch Virol* 2013;158:113–122.
11. Xu LJ et al. Genomic analysis of a newly isolated of Japanese encephalitis virus strain, CQ11-66, from a pediatric patient in China. *Virol J* 2013;10:101.
12. Aubry F et al. Complete genome of a genotype I Japanese encephalitis virus isolated from a patient with encephalitis in Vientiane, Lao PDR. *Genome Announc* 2013;1(1).
13. Li L et al. The flavivirus precursor membrane-envelope protein complex: Structure and maturation. *Science* 2008;319:1830–1834.
14. Roehrig JT. Antigenic structure of flavivirus proteins. *Adv Virus Res* 2003;59:141–175.
15. Mackenzie JS, Gubler DJ, and Petersen LR. Emerging flaviviruses: The spread and resurgence of Japanese encephalitis, West Nile and dengue viruses. *Nat Med* 2004;10:S98–109.
16. Rosen L. The natural history of Japanese encephalitis virus. *Annu Rev Microbiol* 1986;40:395–414.
17. Hanna JN et al. Japanese encephalitis in north Queensland, Australia, 1998. *Med J Aust* 1999;170:533–536.
18. Williams DT et al. Molecular characterization of the first Australian isolate of Japanese encephalitis virus, the FU strain. *J Gen Virol* 2000;81:2471–2480.
19. Solomon T et al. Origin and evolution of Japanese encephalitis in southeast Asia. *J Virol* 2003;77:3091–3098.
20. Erlanger TE et al. Past, present, and future of Japanese encephalitis. *Emerg Infect Dis* 2009;15:1–7.
21. van den Hurk AF, Ritchie SA, and Mackenzie JS. Ecology and geographical expansion of Japanese encephalitis virus. *Annu Rev Entomol* 2009;54:17–35.
22. Chen SP. Molecular phylogenetic and evolutionary analysis of Japanese encephalitis virus in China. *Epidemiol Infect* 2012;140:1637–1643.
23. Morita K. Molecular epidemiology of Japanese encephalitis in East Asia. *Vaccine* 2009;27:7131–7132.
24. Pan XL et al. Emergence of genotype I of Japanese encephalitis virus as the dominant genotype in Asia. *J Virol* 2011;85:9847–9853.
25. Lindenbach BD and Rice CM. Molecular biology of flaviviruses. *Adv Virus Res* 2003;59:23–61.
26. Li M-H et al. Genotype V Japanese encephalitis virus is emerging. *PLoS Negl Trop Dis* 2011;5:e1231.
27. Mohammed MA et al. Molecular phylogenetic and evolutionary analyses of Muar strain of Japanese encephalitis virus reveal it is the missing fifth genotype. *Infect Genet Evol* 2011;11:855–862.
28. Takhampunya R et al. Emergence of Japanese encephalitis virus genotype V in the Republic of Korea. *Virol J* 2011;8:449.
29. Schuh AJ et al. Genetic diversity of Japanese encephalitis virus isolates obtained from the Indonesian Archipelago between 1974 and 1987. *Vector Borne Zoonotic Dis* 2013;13(7):479–488.
30. Endy TP and Nisalak A. Japanese encephalitis virus: Ecology and epidemiology. *Curr Top Microbiol Immunol* 2002;267:11–48.
31. Kim HC et al. Japanese encephalitis virus in culicine mosquitoes (Diptera: Culicidae) collected at Daeseongdong, a village in the demilitarized zone of The Republic of Korea. *J Med Entomol* 2011;48:1250–1256.

32. Nidaira M et al. Detection of Japanese encephalitis virus genome in Ryukyu wild boars (Sus scrofa riukiuanus) in Okinawa, Japan. *Jpn J Infect Dis* 2008;61:164.

33. Kurane I et al. The effect of precipitation on the transmission of Japanese encephalitis (JE) virus in nature: A complex effect on antibody-positive rate to JE virus in sentinel pigs. *Int J Environ Res Public Health* 2013;10(5):1831–1844.

34. Zheng H et al. Molecular characterization of Japanese encephalitis virus strains prevalent in Chinese swine herds. *J Vet Sci* 2013;14:27–36.

35. Lian WC et al. Diagnosis and genetic analysis of Japanese encephalitis virus infected in horse. *J Vet Med B* 2002;49:361.

36. Libraty DH et al. Clinical and immunological risk factors for severe disease in Japanese encephalitis. *Trans R Soc Trop Med Hyg* 2002;96:173–178.

37. Wang LH et al. Japanese encephalitis outbreak, Yuncheng, China, 2006. *Emerg Infect Dis* 2007;13:1123–1125.

38. Wang JL et al. Japanese encephalitis viruses from bats in Yunnan, China. *Emerg Infect Dis* 2009;15:939–942.

39. Yun S-M et al. Molecular epidemiology of Japanese encephalitis virus circulating in South Korea, 1983–2005. *Virol J* 2010;7:127.

40. Nitatpattana N et al. Change in Japanese encephalitis virus distribution, Thailand. *Emerg Infect Dis* 2008;14:1762.

41. Wang L et al. Identification and isolation of genotype-I Japanese encephalitis virus from encephalitis patients. *Virol J* 2010;7:345.

42. Cao QS et al. Isolation and molecular characterization of genotype 1 Japanese encephalitis virus, SX09S-01, from pigs in China. *Virol J* 2011;8:472.

43. Sarkar A et al. Molecular evidence for the occurrence of Japanese encephalitis virus genotype I and III infection associated with acute encephalitis in patients of West Bengal, India, 2010. *Virol J* 2012;9:271.

44. Feng Y et al. High incidence of Japanese encephalitis, Southern China. *Emerg Infect Dis* 2013;19:672–673.

45. Hu Q et al. Recurrence of Japanese encephalitis epidemic in Wuhan, China, 2009–2010. *PLoS One* 2013;8(1):e52687.

46. Kuwata R et al. Surveillance of Japanese encephalitis virus infection in mosquitoes in Vietnam from 2006 to 2008. *Am J Trop Med Hyg* 2013;88:681–688.

47. Parida MM et al. Development and evaluation of reverse transcription-loop-mediated isothermal amplification assay for rapid and real-time detection of Japanese encephalitis virus. *J Clin Microbiol* 2006;44:4172–4178.

48. Misra UK and Kalita J. Overview: Japanese encephalitis. *Prog Neurobiol* 2010;91:108–120.

49. Kumari R et al. First indigenous transmission of Japanese Encephalitis in urban areas of National Capital Territory of Delhi, India. *Trop Med Int Health* 2013;18:743–749.

50. Kuwayama M et al. Japanese encephalitis virus in meningitis patients, Japan. *Emerg Infect Dis* 2005;11:471–473.

51. Murgod UA et al. Persistent movement disorders following Japanese encephalitis. *Neurology* 2001;57:2313–2315.

52. Ding D et al. Long-term disability from acute childhood Japanese encephalitis in Shanghai, China. *Am J Trop Med Hyg* 2007;77:528–533.

53. Joshi R et al. Clinical presentation, etiology, and survival in adult acute encephalitis syndrome in rural Central India. *Clin Neurol Neurosurg* 2013;115:1753–1761.

54. Ghosh D and Basu A. Japanese encephalitis—A pathological and clinical perspective. *PLoS Negl Trop Dis* 2009;3(9):e437.

55. Basumatary LJ et al. Clinical and radiological spectrum of Japanese encephalitis. *J Neurol Sci* 2013;325:15–21.

56. Ankur Nandan V et al. Acute transverse myelitis (ascending myelitis) as the initial manifestation of Japanese encephalitis: A rare presentation. *Case Rep Infect Dis* 2013;2013:487659.

57. Sips GJ, Wilschut J, and Smit JM. Neuroinvasive flavivirus infections. *Rev Med Virol* 2012;22:69–87.

58. Liu TH et al. The blood-brain barrier in the cerebrum is the initial site for the Japanese encephalitis virus entering the central nervous system. *J Neurovirol* 2008;14:514–521.

59. Thongtan T et al. Highly permissive infection of microglial cells by Japanese encephalitis virus: A possible role as a viral reservoir. *Microbes Infect* 2010;12:37–45.

60. Thongtan T et al. Characterization of putative Japanese encephalitis virus receptor molecules on microglial cells. *J Med Virol* 2012;84:615–623.

61. Thongtan T, Thepparit C, and Smith DR. The involvement of microglial cells in Japanese encephalitis infections. *Clin Dev Immunol* 2012;2012:890586.

62. Winter PM et al. Proinflammatory cytokines and chemokines in humans with Japanese encephalitis. *J Infect Dis* 2004;190:1618–1626.

63. Swarup V et al. Japanese encephalitis virus infection decrease endogenous IL-10 production: Correlation with microglial activation and neuronal death. *Neurosci Lett* 2007;420:144–149.

64. Bardina SV and Lim JK. The role of chemokines in the pathogenesis of neurotropic flaviviruses. *Immunol Res* 2012;54:121–132.

65. Mishra MK, Kumawat KL, and Basu A. Japanese encephalitis virus differentially modulates the induction of multiple pro-inflammatory mediators in human astrocytoma and astroglioma cell-lines. *Cell Biol Int* 2008;32:1506–1513.

66. Pierson TC and Diamond MS. Molecular mechanisms of antibody-mediated neutralisation of flavivirus infection. *Expert Rev Mol Med* 2008;10:e12.

67. Gupta N et al. Chemokine profiling of Japanese encephalitis virus-infected mouse neuroblastoma cells by microarray and real-time RT-PCR: Implication in neuropathogenesis. *Virus Res* 2010;147:107–112.

68. Kaushik DK et al. NLRP3 inflammasome: Key mediator of neuroinflammation in murine Japanese encephalitis. *PLoS One* 2012;7:e32270.

69. Ariff IM et al. Japanese encephalitis virus infection alters both neuronal and astrocytic differentiation of neural stem/progenitor cells. *J Neuroimmune Pharmacol* 2013;8(3):664–676.

70. Kalia M et al. Japanese encephalitis virus infects neuronal cells through a clathrin-independent endocytic mechanism. *J Virol* 2013;87:148–162.

71. Katoh H et al. Japanese encephalitis virus core protein inhibits stress granule formation through an interaction with Caprin-1 and facilitates viral propagation. *J Virol* 2013;87(1):489–502.

72. Larena M, Regner M, and Lobigs M. Cytolytic effector pathways and IFN-γ help protect against Japanese encephalitis. *Eur J Immunol* 2013;43:1789–1798.

73. Yiang GT et al. The NS3 protease and helicase domains of Japanese encephalitis virus trigger cell death via caspase-dependent and-independent pathways. *Mol Med Rep* 2013;7:826–830.

74. Zhang LK et al. Identification of host proteins involved in Japanese encephalitis virus infection by quantitative proteomics analysis. *J Proteome Res* 2013b;12(6):2666–2678.

75. Adhya D et al. Histone deacetylase inhibition by Japanese encephalitis virus in monocyte/macrophages: A novel viral immune evasion strategy. *Immunobiology* 2013;218:1235–1247.

76. Ye J et al. Immune evasion strategies of flaviviruses. *Vaccine* 2013;31:461–471.

77. Solomon T et al. A cohort study to assess the new WHO Japanese encephalitis surveillance standards. *Bull WHO* 2008;86:178.

78. Johnson AJ et al. Detection of anti-arboviral immunoglobulin G by using a monoclonal antibody-based capture enzyme-linked immunosorbent assay. *J Clin Microbiol* 2000;38(5):1827–1831.

79. World Health Organization. *Manual for the Laboratory Diagnosis of Japanese Encephalitis Virus Infection*, 2007. http://www.who.int/immunization_monitoring/Manual_lab_diagnosis_JE.pdf

80. Chiou SS. High antibody prevalence in an unconventional ecosystem is related to circulation of a low-virulent strain of Japanese encephalitis virus. *Vaccine* 2007;25(8):1437–1443.

81. Hobson-Peters J. Approaches for the development of rapid serological assays for surveillance and diagnosis of infections caused by zoonotic flaviviruses of the Japanese encephalitis virus serocomplex. *J Biomed Biotechnol* 2012;2012:379738.

82. Chanama S et al. Detection of Japanese encephalitis (JE) virus-specific IgM in cerebrospinal fluid and serum samples from JE patients. *Jpn J Infect Dis* 2005;58:294–296.

83. Taylor C et al. Development of immunoglobulin M capture enzyme-linked immunosorbent assay to differentiate human flavivirus infections occurring in Australia. *Clin Diagn Lab Immunol* 2005;12:371.

84. Ravi V et al. Development and evaluation of a rapid IgM capture ELISA (JEV-Chex) for the diagnosis of Japanese encephalitis. *J Clin Virol* 2006;35:429–434.

85. Robinson JS et al. Evaluation of three commercially available Japanese encephalitis virus IgM enzyme-linked immunosorbent assays. *Am J Trop Med Hyg* 2010;83(5):1146–1155.

86. Mansfield KL et al. Flavivirus-induced antibody cross-reactivity. *J Gen Virol* 2011;92:2821–2829.

87. Mei L et al. Development and application of an antigen capture ELISA assay for diagnosis of Japanese encephalitis virus in swine, human and mosquito. *Virol J* 2012;9:4.

88. Maeda A and Maeda J. Review of diagnostic plaque reduction neutralization tests for flavivirus infection. *Vet J* 2013;195:33–40.

89. Koraka P et al. Reactivity of serum samples from patients with a flavivirus infection measured by immunofluorescence assay and ELISA. *Microbes Infect* 2002;4:1209–1215.

90. Kuno G. Serodiagnosis of flaviviral infections and vaccinations in humans. *Adv Virus Res* 2003;61:3–65.

91. Oceguera LF III et al. Flavivirus serology by western blot analysis. *Am J Trop Med Hyg* 2007;77:159–163.

92. Shrivastva A et al. Comparison of a dipstick enzyme-linked immunosorbent assay with commercial assays for detection of Japanese encephalitis virus-specific IgM antibodies. *J Postgrad Med* 2008;54:181–185.

93. Fulop L et al. Rapid identification of flaviviruses based on conserved NS5 gene sequences. *J Virol Methods* 1993;44:179.

94. Tanaka M. Rapid identification of flavivirus using the polymerase chain reaction. *J Virol Methods* 1993;41:311.

95. Chang GJ et al. An integrated target sequence and signal amplification assay, reverse transcriptase-PCR-enzyme-linked immunosorbent assay, to detect and characterize flaviviruses. *J Clin Microbiol* 1994;32:477.

96. Pierre V, Drouet MT, and Deubel V. Identification of mosquito-borne flavivirus sequences using universal primers and reverse transcription/polymerase chain reaction. *Res Virol* 1994;145:93.

97. Puri B. A rapid method for detection and identification of flaviviruses by polymerase chain reaction and nucleic acid hybridization. *Arch Virol* 1994;134:29.

98. Meiyu F et al. Detection of flaviviruses by reverse transcriptase-polymerase chain reaction with the universal primer set. *Microbiol Immunol* 1997;41:209.

99. Kuno G. Universal diagnostic RT-PCR protocol for arboviruses. *J Virol Methods* 1998;72:27.

100. Paranjpe S and Banerjee K. Detection of Japanese encephalitis virus by reverse transcription/polymerase chain reaction. *Acta Virol* 1998;42:5.

101. Scaramozzino N et al. Comparison of flavivirus universal primer pairs and development of a rapid, highly sensitive heminested reverse transcription-PCR assay for detection of flaviviruses targeted to a conserved region of the NS5 gene sequences. *J Clin Microbiol* 2001;39:1922.

102. Huang JL et al. Sensitive and specific detection of strains of Japanese encephalitis virus using a one-step TaqMan RT-PCR technique. *J Med Virol* 2004;74:589.

103. Pyke AT et al. Detection of Australasian Flavivirus encephalitic viruses using rapid fluorogenic TaqMan RT-PCR assays. *J Virol Methods* 2004;117:161.

104. Yang DK et al. TaqMan reverse transcription polymerase chain reaction for the detection of Japanese encephalitis virus. *J Vet Sci* 2004;5:45.

105. Shirato K et al. Detection of West Nile virus and Japanese encephalitis virus using real-time PCR with a probe common to both viruses. *J Virol Methods* 2005;126:119.

106. Parida M. Japanese encephalitis outbreak, India, 2005. *Emerg Infect Dis* 2006;12:1427.

107. Toriniwa H and Komiya T. Rapid detection and quantification of Japanese encephalitis virus by real-time reverse transcription loop-mediated isothermal amplification. *Microbiol Immunol* 2006;50:379.

108. Chao DY et al. Development of multiplex real-time reverse transcriptase PCR assays for detecting eight medically important flaviviruses in mosquitoes. *J Clin Microbiol* 2007;45:584.

109. Santhosh SR et al. Development and evaluation of SYBR Green I-based one-step real-time RT-PCR assay for detection and quantitation of Japanese encephalitis virus. *J Virol Methods* 2007;143:73.

110. Maher-Sturgess SL et al. Universal primers that amplify RNA from all three flavivirus subgroups. *Virol J* 2008;5:16.

111. Swami R et al. Usefulness of RT-PCR for the diagnosis of Japanese encephalitis in clinical samples. *Scand J Infect Dis* 2008;40:815–820.

112. Liu Y et al. In situ reverse-transcription loop-mediated isothermal amplification (in situ RT-LAMP) for detection of Japanese encephalitis viral RNA in host cells. *J Clin Virol* 2009;46:49.

113. Saxena V, Mishra VK, and Dhole TN. Evaluation of reverse transcriptase PCR as a diagnostic tool to confirm Japanese encephalitis virus infection. *Trans R Soc Trop Med Hyg* 2009;103:403.

114. Patel P et al. Development of one-step quantitative reverse transcription PCR for the rapid detection of flaviviruses. *Virol J* 2013;10:58.

115. Seo HJ et al. Molecular detection and genotyping of Japanese encephalitis virus in mosquitoes during a 2010 outbreak in the Republic of Korea. *PLoS One* 2013;8(2):e55165.

116. Chen YY et al. Detection and differentiation of genotype I and III Japanese encephalitis virus in mosquitoes by multiplex reverse transcriptase-polymerase chain reaction. *Transbound Emerg Dis* 2012; November 16 [Epub ahead of print].

117. Lindahl JF et al. Circulation of Japanese encephalitis virus in pigs and mosquito vectors within Can Tho city, Vietnam. *PLoS Negl Trop Dis* 2013;7:e2153.

118. Mishra MK and Basu A. Minocycline neuroprotects, reduces microglial activation, inhibits caspase 3 induction, and viral replication following Japanese encephalitis. *J Neurochem* 2008;105:1582–1595.

119. Johari J et al. Antiviral activity of baicalein and quercetin against the Japanese encephalitis virus. *Int J Mol Sci* 2012;13:16785–16795.

120. Ritchie SA. Operational trials of remote mosquito trap systems for Japanese encephalitis virus surveillance in the Torres Strait, Australia. *Vector Borne Zoonotic Dis* 2007;7:497.

121. van den Hurk AF et al. Evolution of mosquito-based arbovirus surveillance systems in Australia. *J Biomed Biotechnol* 2012;2012:325659.

122. Larena M et al. JE-ADVAX vaccine protection against Japanese encephalitis virus mediated by memory B cells in the absence of CD8(+) T cells and pre-exposure neutralizing antibody. *J Virol* 2013;87:4395–4402.

123. Cutfield NJ et al. Japanese encephalitis acquired during travel in China. *Intern Med J* 2005;35:497–498.

124. Buhl M and Lindquist L. Japanese encephalitis in travelers: Review of cases and seasonal risk. *J Travel Med* 2009;16:217–219.

125. Hatz C et al. Japanese encephalitis: Defining risk incidence for travelers to endemic countries and vaccine prescribing from the UK and Switzerland. *J Travel Med* 2009;16:200–203.

126. Hills S et al. Japanese encephalitis among travelers from non-endemic countries, 1973–2008. *Am J Trop Med Hyg* 2010;82:930–936.

127. Ratnam I et al. Low risk of Japanese encephalitis in short-term Australian travelers to Asia. *J Travel Med* 2013;20(3):206–208.

128. Beasley DW, Lewthwaite P, and Solomon T. Current use and development of vaccines for Japanese encephalitis. *Expert Opin Biol Ther* 2008;8:95–106.

129. Erra EO et al. Cross-protective capacity of Japanese encephalitis (JE) vaccines against circulating heterologous JE virus genotypes. *Clin Infect Dis* 2013;56:267–270.

130. Wiwanitkit V. Development of a vaccine to prevent Japanese encephalitis: A brief review. *Int J Gen Med* 2009;2:195–200.

131. McCallum AD and Jones ME. Allergy to IXIARO and BIKEN Japanese encephalitis vaccines. *J Travel Med* 2013;20:60–62.

132. Halstead SB and Thomas SJ. New Japanese encephalitis vaccines: Alternatives to production in mouse brain. *Expert Rev Vac* 2011;10:355–364.

133. Appaiahgari MB and Vrati S. Clinical development of IMOJEV®—A recombinant Japanese encephalitis chimeric vaccine (JE-CV). *Expert Opin Biol Ther* 2012;12:1251–1263.

134. McArthur MA and Holbrook MR. Japanese encephalitis vaccines. *J Bioterror Biodef* 2011;S1:2.

135. Heinz FX and Stiasny K. Flaviviruses and flavivirus vaccines. *Vaccine* 2012;30:4301–4306.

136. Jelinek T. IXIARO® updated: Overview of clinical trials and developments with the inactivated vaccine against Japanese encephalitis. *Expert Rev Vac* 2013; May 29 [Epub ahead of print].

137. Weaver SC and Reisen WK. Present and future arboviral threats. *Antiviral Res* 2010;85:328–345.

138. Wong SC et al. A decade of Japanese encephalitis surveillance in Sarawak, Malaysia: 1997–2006. *Trop Med Int Health* 2008;13:52.

13 Kyasanur Forest Disease Virus

Pradeep B.J. Reddy, Ambuj Srivastava, and Priyabrata Pattnaik

CONTENTS

13.1 INTRODUCTION

Kyasanur forest disease (KFD), also called as monkey fever, is an infectious hemorrhagic disease caused by a highly pathogenic virus, namely, Kyasanur forest disease virus (KFDV).[1] KFDV falls under the genus *Flavivirus* within the family Flaviviridae. There are about 70 identified viruses in the Flaviviridae family[2] and most of them are arthropod-borne, which are transmitted to vertebrates by either infected mosquitoes or ticks.[3] Dengue virus, yellow fever virus, and Japanese encephalitis virus are the few important flaviviruses transmitted by mosquitoes, whereas KFDV, Omsk hemorrhagic fever virus (OHFV), and Alkhurma virus (ALKV) are the few viruses transmitted by ticks in this genus.[4] Generally these tick-borne encephalitis viruses (TBEV) are defined geographically with KFD being restricted to the Indian subcontinent. KFD is a febrile illness and it was initially identified in the Shimoga district of the Karnataka state in 1957 (Figure 13.1).[1,5] Initially, the symptoms were similar to that of typhoid fever when the outbreak occurred; however, the negative results in Widal test ruled out the possibility.[1] With numerous fatal cases in the black-faced langur and the red-faced bonnet monkeys occurring parallelly, it was suspected to be the dawn of yellow fever virus infection in India.[1] Later, the causative virus was isolated from the postmortem blood and visceral organs of dead langurs and bonnet monkeys, and the same virus was also isolated from the *Haemaphysalis* ticks.[1] The disease was named KFD and the causative agent KFDV after the name of the forest where it was identified for the first time. KFD epidemics have occurred every year in this region with an average of about 400–500 cases annually, reflecting the prevalence of active foci of infection in the forest region. The highest recorded incidence was in 1983, with 1555 cases and 150 deaths.[6] KFD has so far been localized largely to the Karnataka state, although it was found occasionally in some other parts of India. KFDV commonly infects monkeys, that is, black-faced langur (*Presbytis entellus*) and red-faced bonnet monkeys (*Macaca radiata*), and KFDV gets transmitted by the bite of infective ticks (*Haemaphysalis spinigera*), especially at its nymphal stage, and the ticks remain infectious throughout their life.[7] KFDV also circulates in small animals such as rodents, shrews, and birds.[7–9] KFDV-transmitting adult ticks (*H. spinigera*) commonly feed on the large animals, and neutralizing antibodies have been found in large animals like cattle, buffaloes, goats, and wild bears and also in a number of avian species, but hardly have they suffered from KFD (Table 13.1).[10] Direct transmission of the virus from rodents to humans has been documented; however, person-to-person transmission has not been known yet. During 1994, a variant of KFD was isolated from a butcher living in the Mecca province of Saudi Arabia, which was named as Alkhurma virus (ALKV).[11] Due to the strong genetic similarity between KFDV and ALKV, the latter was considered to be a subtype of KFDV introduced into Saudi Arabia.[12]

13.2 CLASSIFICATION AND MORPHOLOGY

KFDV is a member of the mammalian tick-borne virus group (previously referred to as the tick-borne encephalitis serogroup) of the genus *Flavivirus* and family Flaviviridae.[3] Although isolated cases of KFD have been reported from several parts of India, epidemics of KFD mostly remain confined to few districts of Karnataka, a southern Indian state. The first recorded epidemic of KFD in humans occurred between January and May 1956, when four villages of Belthangady taluk of Kanara district were affected.[1] The following year, KFD spread to more than 20 villages, and by 2003, it was prevalent in at least 70 villages in four districts adjacent to Shimoga in the Karnataka state of India. The exact causes of the sudden emergence of KFDV in India during the late

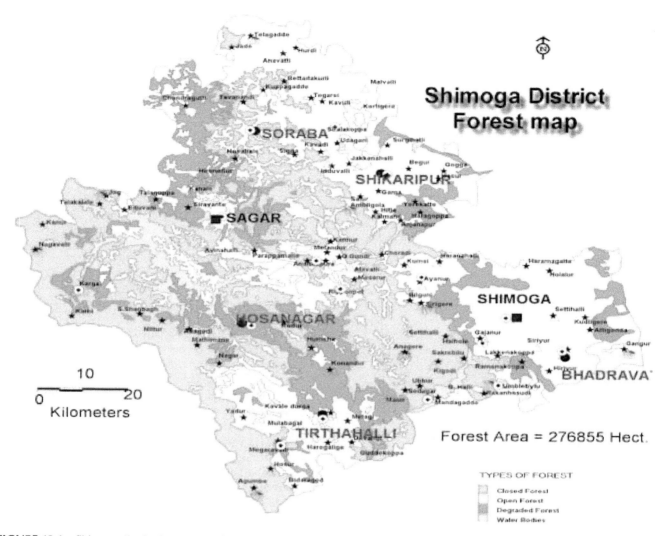

FIGURE 13.1   Shimoga district forest map. (Courtesy of Forest Department, Government of India.)

1950s and the subsequent localization to Karnataka state alone are not clear. Despite its high pathogenicity and potential epidemiological importance, there have been relatively few detailed antigenic and molecular studies on KFDV. The only molecular information available is the nucleotide sequences of genes encoding the structural proteins.[13] KFDV is a spherical, enveloped virus measuring 45 nm in diameter. It has a single-stranded, positive-sense RNA genome with 10,774 nucleotides encased in an icosahedral nucleocapsid. The genome of KFDV encodes for a polyprotein that is cleaved by viral and cellular proteases into three structural (C, prM/M, and E) and seven nonstructural (NS) proteins (NS1, NS2A, NS2B, NS3, NS4, NS4B, and NS5).[14,15]

13.3   BIOLOGY AND EPIDEMIOLOGY

KFD in humans is characterized by an incubation period of approximately 3–8 days, followed by chills, frontal headache, body ache, and high fever for 5–12 days with a case fatality rate of 2%–10%.[6] Subsequent to fever, hemorrhagic manifestations, including intermittent epistaxis, hematemesis,

melena, and blood in the stools, are observed.[6] Following KFDV infection, viral titers remain high at least 10 days following the onset of symptoms.[16] In another report,[17] it was demonstrated that viremia in patients lasted for 12–13 days and, unlike other flaviviruses, remains high during the first 3–6 days with titers as high as 3.1×10^6.[16] Development of central nervous system abnormalities is rarely observed after an afebrile period of 1–2 weeks. It has been demonstrated that IgE plays a critical in role in the immunopathology of KFD and possibly of other hemorrhagic fevers.[18] Although disseminated intravascular coagulation is suspected, the exact cause of hemorrhage observed in KFD is not well known. OHFV and KFDV are two important viruses within the genus *Flavivirus* that are genetically closely related to TBEV but produce hemorrhagic fever instead of encephalitis.[6] A variant of KFDV known as ALKV that is genetically similar to KFDV has been recently identified in Saudi Arabia. Because of the highly pathogenic nature, KFDV is accepted as level 4 virus based on international biosafety rules. Upon inoculation through intracerebral or intraperitoneal route, KFDV is lethal to infant mice. KFDV induces cytopathic effect in

TABLE 13.1

Hosts Known to Be Susceptible to KFDV/Carry KFDV-Specific Neutralizing Antibodies

Species: Scientific Name (Common Name)	Experimental Transmission	Experimental Infection	Neutralization Antibody Positive	HI Antibody Positive	Virus Isolation
Rattus blanfordi (white-tailed rat)	√		√		
Rattus rattus wroughtoni (field rat)			√		√
Funambulus tristriatus tristriatus (striped squirrel)	√		√		
Vandeleuria oleracea		√			√
Suncus murinus (shrew)	√		√		
Petaurista petaurista philippensis (giant flying squirrel)		√			
Cynopterus sphinx (frugivorous bat)		√	√		
Golunda ellioti			√		
Mus booduga (field mouse)			√		
Mus platythrix		√	√		
Lepus nigricollis (black-naped hare)	√				
R. rattus rufescens			√		
F. tristriatus numarius			√		
Funambulus pennanti (northern palm squirrel)			√		
Tatera indica (Indian gerbil)			√		
Miniopterus schreibersii (insectivorous bat)			√		
Rousettus leschenaultia (frugivorous bat)			√		
Eonycteris spelaea (frugivorous bat)			√		
Hipposideros lankadiva (insectivorous bat)			√		
Rhinolophus rouxi (insectivorous bat)			√		
Hipposideros speoris (insectivorous bat)			√		
Tephrodornis virgatus				√	
Megalaima zeylanica				√	
Chalcophaps indica				√	
Treron pompadora				√	
Rhoppocichla atriceps				√	

chick embryo and hamster kidney cells and produces plaques in monkey kidney cell cultures. However, it replicates without cytopathic effect in a continuous tick cell line from *H. spinigera*.[15]

KFDV was first isolated from sick monkeys captured in the Kyasanur forest of the Shimoga district of the Karnataka state of India. Shimoga is located at 571 m altitude, between 13°27′ and 14°39′ north longitude, and between 74°38′ and 76°4′ latitude (Figure 13.1). The Shimoga district of the Karnataka state is full of closed forests and few open forests. Black-faced langur and red-faced bonnet monkeys that are commonly found in the forest localities get infected by KFDV through infected tick (*H. spinigera*) bites. Apart from *H. spinigera*, KFDV has also been isolated from more than 16 different types of ticks (Table 13.2). Upon biting, highly anthropophilic unfed ticks are capable of transovarial transmission of KFDV to humans. Nymphs are active during the months of October to December, and hence, frequent transmission of KFDV has been documented during this period. Fed female adult ticks lay a large number of eggs that hatch to larvae under the leaves in the forests. The ticks survive in the vegetation/grasses/undergrowth (in the forests) and feed

on animals that move across the forest floor. The animals are exposed to the ticks when they roam the ground that has leaves, grasses, undergrowth, etc., that maintain the correct conditions for the survival of the ticks (i.e., in terms of temperature and humidity). The ticks feed on small mammals and monkeys, so also larvae and later mature to nymphs, and the cycle is repeated (Figure 13.2). It has been demonstrated in an experimental model that larvae of *Rhipicephalus haemaphysaloides* that feed on KFDV-infected viremic rodents carry the infection to the next generation. In this study, the fed larvae, nymphs, unfed adults, fed adult males and females after oviposition were infected, while the unfed larvae were free from infection. Nymphs and adults transmitted the infection by biting on rodents and rabbits, respectively, and this rodent–tick cycle continues for more than one life cycle. The viral titers measures up to 3.5×10^6 of mouse LD50/30 µL in an infected tick, and viral titers persist in the ticks up to 245 days post infection.[19] The presence of KFDV-specific neutralizing antibodies in fed blood has no effect on survival of KFDV in these ticks.

Shimoga, the central focus of the KFD, is located in great part due to the deep encroachment of human colonization on

TABLE 13.2
Vectors Known to Transmit KFDV

Insect Species	Virus Isolation	Experimental Transmission
H. spinigera	√	√
Haemaphysalis turturis	√	√
Haemaphysalis papuana kinneari	√	√
Haemaphysalis minuta	√	√
Haemaphysalis cuspidata	√	
Haemaphysalis kyasanurensis	√	√
Haemaphysalis bispinosa	√	
Haemaphysalis wellingtoni	√	√
Haemaphysalis aculeata	√	
Rh. haemaphysaloides		√
Hyalomma marginatum isaac		√
Ornithodoros crosi		√
Argas persicus		√
Ixodes petauristae	√	√
Dermacentor auratus		√
Ixodes ceylonensis		√

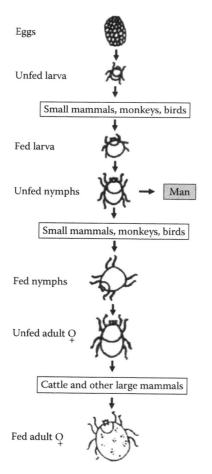

FIGURE 13.2 Stages of development of *H. spinigera* (tick) responsible for the transmission of KFDV to humans.

a primitive sylvan territory. Originally, most of the vegetation consists of tall rain forest interspersed with deciduous and semideciduous forest on the slopes and on the ridges of rolling hills with mixed bamboo and shrub jungle at the edges. Optimal humidity and fertile bottoms between the terrain's undulations are suitable for the cultivation of paddy fields. High humidity generated from the cultivated fields is adequate for the maintenance of the ticks throughout the year, and availability of nearby forest sustains a large population of wild monkeys. Such ecological specificity is seldom observed in any other parts of India. The combination of these climatic conditions could be the strong reason for geographical localization of KFDV only to the Karnataka state of India. Local villagers staying in and around the forest area frequently visit the forest for collection of firewood and get infected through the bite of a tick. Generally females mostly visit forest for firewood collection, whereas children do not. Males are in between. So females have the highest opportunity to become infected, whereas children have the lowest.[20] The presence of adequate animal reservoir in this area, a large population of carriers having consistent susceptibility, the ability to circulate KFDV and the capability to infect an adequate number of vectors in Shimoga area are thought to be important for the sustenance of KFDV. Seasonal fluctuations of phase density of ticks in these areas are so timed that transmission of KFDV through consecutive seasons is ascertained, independent of individual variations in vector density.[21]

Black-faced langur and red-faced bonnet monkey commonly residing in the Kyasanur forest succumb to the natural infection with KFDV. During October 1964 and September 1973, a total of 1046 monkeys (860 black-faced langurs and 186 red-faced bonnet monkeys) succumbed to death, and KFDV was successfully isolated from 118 black-faced langurs and 13 red-faced bonnet monkeys. High mortality of monkeys was generally observed during December to May, which coincides with the seasonal activity of immature stages of *Haemaphysalis* ticks, the vectors of KFDV. Epizootiology of KFD in wild monkeys of Shimoga, specifically the death of monkeys in dry seasons (February and March), correlates well with human cases of KFD (Figure 13.3).[22] Several studies have confirmed the susceptibility of *M. radiata* (bonnet macaques) to KFD and demonstrated KFDV-specific gastrointestinal and lymphoid lesions.[23] The epidemic period correlated well with the period of greatest human activity in forest, that is, during June–September when the monsoon was over, November to December for harvesting paddy, and December to May gathering firewood and other forest products.[24]

KFD epizootic is initially characterized by the spread of the disease to the areas very close to the original focus of infection. Since the first record, the epidemics of KFD have occurred repeatedly in the Shimoga district of Karnataka and its adjoining areas. During the epizootic period of 1964–1965, monkey deaths occurred only within the previously known infected area. During 1965–1966, the epizootic extended towards a contiguous forest southeast of Sagar town

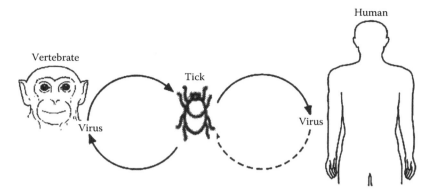

FIGURE 13.3 Circulation of KFDV among different species.

involving an area covering approximately 30 km². The epizootics that appeared within the original infected area between October 1966 and September 1969 showed a tendency to move northwest of Sorab town. During the epizootic season of 1969–1970 and 1970–1971, monkey deaths occurred in Yakshi and Gudavi forests located further northwest of Sorab town. During 1971–1972, epizootics continued to occur in the old focus of Sagar and Sorab taluk and occurred in new focus at Gadgeri and a nearby valley. By the end of year 1973, epizootics and epidemics were recognized in several new foci, distant from the original focus. The new foci were Aramanekoppa in Hosanagara taluk of the Shimoga district and Kodani in Honnavar taluk of North Kanara district. It was also observed in certain localities of original focus (Barur, Jambai, Maisavi, and Padavagodu) that the KFDV activity persisted over several years.[25] During 1975, epizootics of KFD spread to Mandagadde area in Thirthahalli taluk, Shimoga district, approximately 50 km southwest of the periphery of Aramanekoppa focus. Later during 1982, the disease appeared at Patrame area in Belthangady taluk, South Kanara district, approximately 80 km south of the periphery of Mandagadde focus.[25] Other than the Karnataka state, the primary site of introduction of KFDV, antibodies against KFDV have also been detected in humans in several parts of Kutch and Saurashtra of the Gujarat state (semiarid area, around 1200 km away from main focus of KFD) and isolated localities like Ramtek (near Nagpur), Kingaon, and Parbatpur of the West Bengal state of India.[26] The human population of the Andaman and Nicobar islands of India was reported to have the highest prevalence of hemagglutination-inhibition antibodies against KFDV. Neutralizing antibodies to KFDV were detected in Middle Andaman.[27] Considering the previous findings, Andaman and Nicobar islands of India might be a major silent focus of KFDV activity, but there are no corroborative data from case reports of KFD.

13.3.1 GEOGRAPHICAL DISTRIBUTION

KFDV has never been reported from any country other than India. However, recently, Wang et al.[28] determined that Nanjianyin virus, isolated from the serum of a febrile patient

(a 38-year-old woman) from the Hengduan Mountain region of Yunnan Province, China, in 1989, is highly homologous to that of KFDV, hence a variant of KFDV. Results of a 1987–1990 seroepidemiologic investigation in Yunnan Province had shown that residents of the Hengduan Mountain region had been infected with the Nanjianyin virus. A previous serosurvey demonstrated that KFD exposure had occurred (Lushui and Eryuan counties).[29,30] This report indicates that humans and animals in the Hengduan Mountain region of the Yunnan Province of China have been infected with KFDV since the 1980s. This report also corroborates the finding of KFD antibodies in human and bird sera in the Chinese provinces of Guangdong, Guangxi, Guizhou, Hubei, Henan, Xinjiang, and Qinghai in 1983.[31] Migratory birds are known to play a critical role in the epidemiology and spread of arboviruses[32,33] since they frequently pass through Yunnan Province during their migration from south India and the Indian Ocean islands to Mongolia and Siberia. Hence, it is postulated that KFDV was likely carried to the region by these migratory birds and their parasitic ticks. A previous report also indicates that Hengduan Mountain in Yunnan Province provides a suitable habitat for *H. spinigera*, the vector for KFDV.[34,35]

A variant genotype of KFDV, characterized serologically and genetically as ALKV, has been identified in Jeddah, Saudi Arabia.[12] ALKV has been recommended for inclusion as a subtype of KFDV.[12] ALKV was first isolated in the 1990s from the blood of a butcher. Since then, a total of 24 cases were reported in the last 10 years and most of these cases originated from Mecca or Jeddah, Saudi Arabia.[36] Thirty-seven cases have been reported from February 2001 to February 2003 from Mecca, Saudi Arabia. Until 1999, not more than three cases of Alkhurma hemorrhagic fever (AHF) per year had been reported. Recent increase of AHF cases in Saudi Arabia may be due to better reporting and early identification of the disease. The transmission of most of the reported AHV cases was via wound, camel raw milk, tick/mosquito bite, or direct contact with the animals. Although the sources of AHF infection are very diverse, most of the cases were localized in either Mecca or Jeddah. Any widespread epidemic of AHF has not occurred in Saudi Arabia perhaps due to the scarce distribution of the virus. In contrast, India has experienced several KFD epidemics, although they have been very much confined to Karnataka state.

Nucleotide sequence analysis of KFDV isolates from India differed from Alkhurma isolates obtained from Saudi Arabia by approximately 8%–9%.[37] However, the Nanjianyin isolate[28] from China differed only by a single nucleotide in reference to the 1957 KFDV isolate raising concern regarding the authenticity of the isolate. The striking similarity of the Chinese variant isolated after 32 years and 3000 km apart from their Indian counterpart raises doubts on the possibility of contamination with the Indian isolate used for analysis.[38]

Recent reports of the Italian tourists being infected with ALKV in Southeastern Egypt proved the evolutionary origin of these tick-borne flaviviruses.[39] ALKV was initially isolated from camel ticks[12] and it has close resemblance to KFDV that was first reported in India. When KFDV was subsequently isolated in China,[28] it was inferred that migratory birds were responsible for carrying the infected ticks to China from India. Later, it was found that large numbers of animals are transported annually from Africa and other countries to Saudi Arabia both for transport and also food. These animals were potent source of ticks that were infected with ALKV and thus causing infections in humans. This theory is in content with the phylogenetic evidence that tick-borne encephalitic flavivirus serocomplex originated in Africa and gradually evolved and dispersed across the northern hemisphere of the Old World.[40,41] This evidence also supports the African evolutionary origin for these viruses.[42]

KFDV existed as a part of the ecosystem in South Kanara. Outbreaks of KFDV were always preceded by deforestation. One known report of KFD outbreak in Belthangady occurred after the clearance of forests to make way for the cashew plantations. Following the deforestation, the black-faced and rhesus monkeys were forced to roam on the ground and hence very much prone to the vector. Cattle rearing has played an essential role in the transmission of KFD to human settlements, the reason being the less susceptibility of cattle to KFDV, and at the same time, cattle act as a reservoir for maintaining and propagating the KFDV vector. Mature ticks lay eggs under the leaves, and nymphs emerge after the rainy season that coincides with the KFDV epidemic with peak incidence in January and February. Another striking feature of the epidemic is their localization; the disease spreads only in few villages every season, which could be due to the elevated immune responses in those unaffected villages.

The factors responsible for the sudden emergence of KFDV and ALKV in India and Saudi Arabia, respectively, are not well understood. Increased human populations in the Shimoga districts of Karnataka during the 1950s could be one of the reasons for the emergence of KFD during 1957. Rapid population growths lead to increased deforestation and extension of grazing fields into previously forest areas. Increase in cattle population that harbors the ticks infected with KFDV could be another reason for the increased spread in the KFDV.[21]

13.3.2 ETIOLOGICAL MYTH

Before the KFDV epidemic in the Karnataka state of India during 1957, a large tree (*Pterocarpus marsupium*), locally called as Benga tree, the abode of a local deity, was cut by a timber contractor. People believed that the red sap seeped out of the cut tree as blood. Within a short span of time, the contractor succumbed to death and so did few of the workers who cut the tree. This episode was followed with an epidemic, and local people believed that the epidemic resulted because of the wrath of the deity in the cut tree. By 1984, local villagers in Shimoga district learnt from the health workers that ticks from the forest carried KFDV, which is the cause of epidemic.[43]

13.4 CLINICAL FEATURES AND PATHOGENESIS

KFDV infection leads to a febrile and occasionally to a fatal illness among people living in the vicinity of Shimoga district of Karnataka concurrently with the KFDV isolation and identification. The incubation period varies between 3 and 8 days with the appearance of sudden manifestations of KFD. The clinical symptoms of KFDV infection follow a biphasic course with flu-like symptoms, and over half of the infected people may progress into a second phase of illness. Relative bradycardia is frequently seen along with inflammation of the conjunctiva. A small proportion of patients develop coma or bronchopneumonia prior to death. The acute phase of illness lasts for ~2 weeks. Mortality rates in KFDV infections range between 2% and 10% in humans and up to 85% in monkeys.[44] Human mortality rates are significantly lower than in Alkhurma virus infection where case fatality has been reported to be 25%.[45] Convalescence in KFD survivors is generally prolonged, up to 4 weeks. Viremia in KFD patients lasts for 12–13 days of illness, and unlike most other arboviruses, KFDV is easily isolated from human sera. The level of virus in blood circulation is considerably high (3.1×10^6), especially during 3–6 days after the onset of illness.[17] Leukopenia is a constant feature in KFD patients, and in most cases, neutrophil count drops below 2000 cells/µL. Lymphopenia was usually observed within the first week of illness and significant eosinopenia during the first or early in the second week. In several patients, lymphocytosis was also observed between the third and the fifth week.[18] Prominent thrombocytopenia and neutropenia in KFD are probably mediated through antibodies to platelets and leukocytes.[46] However, no experimental evidence is available in support of this postulation.

Pathologic manifestations of KFD in human patients include parenchymal degeneration of the liver and kidney, hemorrhagic pneumonitis, and a moderate to marked prominence of the reticuloendothelial elements in the liver and spleen, with marked erythrophagocytosis.[47] Pathologic and hematologic investigations emphasized similarities between KFD and Omsk hemorrhagic fever. Later, there was a shift in clinical emphasis from hemorrhagic to neurologic complications.[48,49] During KFDV infection, interferonemia is concomitant with the viremic phase. IgE has also been implicated as a cofactor in the immunopathology of KFD and possibly of other hemorrhagic fevers.[18] Neutralization studies indicated that the endogenous (circulating) interferon (IFN) was antigenically similar to the acid-stable form of IFN-α. Sathe et al.[50] monitored endogenous IFN levels in acute[51] and convalescent phase[29] sera collected from KFD patients and reported that the levels of circulating IFN in the acute samples (GM 216.3 ± 8.7) collected during 4–7 post-onset day (POD) were significantly higher (P less than 0.001) than the convalescent samples (GM 13.19 ± 1.6) collected between 30 and 90 POD. Clinical or postmortem biopsies of various organs suggested that KFD could pass through four stages each lasting for about a week, that is, a prodromal stage with fever, hypotension, and hepatomegaly, a stage of complication characterized by hemorrhage, neurological manifestation, or bronchopneumonia, and a stage of recovery followed by a stage of fever in some cases. Adhikary Prabha and coworkers[52] clinically studied 100 cases of KFD and conducted biopsies of various organs. They interpreted that hypotension in KFD could be of myocardial origin, whereas encephalopathy could be due to a metabolic cause probably of hepatic origin and lung signs due to intra-alveolar hemorrhage and secondary infection.[52]

Clinical manifestations caused by the Alkhurma virus (ALKV), a variant of KFDV, include fever, headache, retro-orbital pain, joint pain, generalized muscle pain, anorexia, and vomiting associated with leukopenia, thrombocytopenia, and elevated levels of liver enzymes, such as alanine transferase and aspartate transferase.[36] Like KFDV, ALKV also produces encephalitic features such as confusion, disorientation, drowsiness, coma, neck stiffness, hemiparesis, paraparesis, or convulsion and hemorrhagic manifestations such as ecchymosis, purpura, petechiae, gastrointestinal bleeding, epistaxis, bleeding from puncture sites, or menorrhagia.[45] Like other hemorrhagic fever viruses, both KFDV and ALKV are zoonotic.[53] Unlike other hemorrhagic fever viruses classified under the Arenaviridae (Lassa and Junin viruses), Bunyaviridae (Hantaviruses), and Filoviridae (Marburg and Ebola viruses), KFDV and ALKV classified under the Flaviviridae are vector-borne with less intensity of hemorrhagic manifestations than the others. Although ALKV and KFDV are somewhat similar in clinical presentations, high mortality (25%) reported for ALKV infection is surprising. However, this may be due to limited numbers of actually reported cases. As more cases get reported, the high mortality indexes may drop for the ALKV infections.

In general, the antibodies raised against viral infections are maintained for years at high levels due to repeated reexposure to the virus in the form of subclinical infections or due to exposure to closely related viruses. Even in the absence of reexposure, antibodies may persist for many years, as has been shown in several arboviruses.[54,55] Like other arboviruses, Haemagglutination inhibition (HI) and complement fixing (CF) antibodies begin to rise in the first week after the onset of KFD. Neutralizing antibodies appear during the second week, and during the third to fourth week, the titers

reach the peak. Antibody-mediated immunity to KFDV persists for a decade even in the absence of reexposure.[56]

13.4.1 Mechanism of Immune Evasion by KFDV

During evolution, the host develops various mechanisms to encounter the pathogens, and at the same time, pathogens develop various mechanisms to evade the host immune system[57] and increase their ability to transmit to a fresh host. Following an infection, the host's innate immune system is the first to respond and eventually triggers the adaptive immune responses.[58] Once the pathogen is cleared from the host, pathogen-specific memory response is established in the host.[59] Different viruses employ various mechanisms to evade the host immune response.[60] Dissection of these immune evasion mechanisms could lead to the identification of new targets and consequently better therapies.

Type 1 IFNs (IFN-α and IFN-β) are generally produced by various innate immune cells following the viral infections[61] and induce an antiviral state in uninfected cells in their vicinity.[62] IFNs induce the expression of numerous genes called interferon-stimulated genes (ISGs) that play an essential role in the maintenance of the antiviral state in the naïve cells. Inhibition of the type I IFNs is one of the well-known immune evasion mechanisms employed especially by the members of the flavivirus group.[63] Different viruses like dengue, Japanese encephalitis (JE), West Nile Virus (WNV), and TBEV exploit their NS proteins to inhibit the IFN signaling.[63–66] Cook et. al.,[67] with the help of reverse genetics system, demonstrated that KFDV also targets the type I IFN system with their NS5 protein. Specifically the RNA-dependent RNA polymerase (RdRp) region of the NS5 played a critical role in the inhibition of the type I IFN signaling. Definitely, this report has shed light on the mechanism of immune evasion by KFDV and hopefully aid in the designing of new therapies for KFD in the future.

13.5 IDENTIFICATION AND DIAGNOSIS

Diagnosis in most of the infections depends on demonstration of the infecting virus or its products in acute serum samples. KFDV was reported to circulate in the blood until 10 days after the onset of the disease. The clinical manifestations of KFDV infection do not usually suggest a specific etiological agent, so it is critical that compatible epidemiologic factors be aligned with clinical features in order to make the diagnosis of KFD, which can then only be confirmed by conventional diagnostic tests like serological tests, virus isolation, and polymerase chain reaction (PCR). Reliable and rapid diagnosis is critical to confirm the suspected KFD cases. The conventional diagnostic methods[68] were discontinued in India during 1970 because the virus was known to cause aerosol infections in field and laboratory workers. Serologic tests showing an antibody rise is most practical for diagnosis. Fevers that are accompanied by arthralgia are mostly confused with dengue, Chikungunya, and West Nile virus diseases. Yellow fever must be differentiated from viral hepatitis, falciparum malaria, or drug-induced hepatic injury. Laboratory tests include hemagglutination-inhibition, immunofluorescence, and neutralization tests. Neutralization test is the most useful test for the detection of KFDV-specific antibody. Especially during the early phase of the disease, KFDV can be isolated from blood, organs of sick persons employing animal cell culture, or experimental animals (new born mice).

The common method of preparation of KFDV antigen is from mouse brain preparations. To prepare mouse brain antigen (MB Ag), KFDV is inoculated by intracranial route in 24–48-hour-old suckling mice. On fourth to fifth postinoculation day (PID), the sick animals are sacrificed by anesthesia with ether and the brain materials are scooped out. A 20% suspension is prepared in sterile phosphate-buffered saline (PBS). The suspension is then clarified by centrifugation for removal of cell debris. To one part of this preparation, 2% bovine serum albumin (BSA) is added. This is aliquoted and stored at −70°C for future use as seed virus. The other part (without BSA) is generally exposed to sonic disintegration at 40 W for 2 min. After clarification by low-speed centrifugation, this is used as MB Ag for immunization as well as in diagnostic tests.

Interpretation of serologic data obtained by hemagglutination-inhibition, complement-fixation, and fluorescent antibody tests is difficult in most tropical areas where several flaviviruses are endemic. In primary infections, the virus neutralization test provides virus-specific confirmation. If a patient has had previous flavivirus infections, cross-reactions make even neutralization test results difficult or impossible to interpret. Demonstration of specific IgM antibodies in the cerebrospinal fluid by immunoassay can be an excellent way to diagnose flavivirus encephalitis. A recent report described about the diagnosis of KFDV by detecting IgM antibodies by IgM antibody capture ELISA (MAC-ELISA).[69] The specificity of this method was further evaluated by using known positive samples of other flaviviruses like JE, WNV, and dengue. Additionally, the method was used to evaluate the suspected KFD samples collected from the Karnataka state of India, and the results were in accord with the PCR-based methods.

Reverse transcriptase ploymerase chain reaction (RT-PCR) methods of pathogen detection are generally highly sensitive and, at the same time, can involve some artifacts or contamination. Till recent times, the rapid methods for the detection of KFDV were not available. However, recent reports proposed the novel and rapid methods of KFD diagnosis by using nested RT-PCR (nRT-PCR) and TaqMan-based real-time RT-PCR.[69] These real-time PCR methods were specific to the nonstructural 5 (NS5) gene that is highly conserved among flaviviruses.[69] This group typically focused on the areas of mismatch among other flaviviruses in order to design the KFDV-specific primers. The nRT-PCR reactions yielded two PCR products, 756 and 355 base pairs in size, and further, the specificity was confirmed by DNA sequencing. The sensitivity of this method was impressive with a detection of 2.67 pfu/mL. One-step and two-step TaqMan RT-PCRs were also performed by the same group to compare the sensitivity

of detection. They could successfully detect 38 RNA copies of RNA by a two-step RT-PCR, which was tenfold more sensitive than one-step RT-PCR, which could detect 380 copies of RNA. Additionally, these techniques were successfully validated by using clinical samples collected during 2006–2007 from the Karnataka state of India. Additionally, MAC-ELISA and RT-PCR methods were sensitive to differentiate between KFDV and ALKV. These novel methods of KFD diagnosis are rapid, sensitive, and specific and are definitely safe and better alternatives for the existing conventional methods.

13.6   TREATMENT AND PREVENTION

As of now, there is no specific treatment available for KFDV infection, but supportive therapy is critical for KFD-infected patients. Immediate hospitalization of the patient and careful maintenance of fluid balance is very much essential. There are no specific treatments available for KFDV; hence, the strategies to control and prevent this highly infectious disease transmitted by tick bites are through the development and application of effective vaccines. Control of ticks on wild reservoir hosts is another attractive strategy to control the spread of KFD.

The very first vaccine tested to control KFD in the area of Shimoga district, Mysore state of India, was a 5%–10% suspension of formalin-inactivated RSSE virus (mouse brain preparation) produced by the Walter Reed Army Institute of Research laboratory, Washington, United States. It was injected subcutaneously, two doses a week apart followed by a third dose 5 weeks after the second shot. About 4000 vaccine recipients displayed no unwanted reaction such as allergic and febrile reactions.[70] The vaccine induced weak HI antibody response but stimulated no CF antibody response. The vaccine failed to evoke booster response in many individuals with previous KFDV infection. The RSSE vaccine thus was found to be ineffective in reducing the attack rate of KFD or in modifying the disease course.[71,72]

A concerted effort to produce KFDV vaccine was made late in 1965 by growing KFDV in brains of infant Swiss albino mice and subsequent inactivation by formalin. The vaccine produced retained potency up to 6 months storage in a refrigerator. It induced neutralizing antibodies in mice.[73] An experimental KFDV vaccine was also produced by growing the virus in chick embryo, but the product was poorly immunogenic and failed to evoke neutralizing antibody response in mice.[74] Later in the year 1966, formalinized experimental vaccine was prepared by growing KFDV in chick embryo fibroblast cultures. The vaccine was found to be immunogenic, potent, stable, and safe.[75,76] Field trials with this tissue culture vaccine yielded satisfactory results, raising neutralizing antibodies in 50% of the vaccinees and inducing protective response in ~23% vaccinees.[51] Efforts were also made to develop a live attenuated KFDV vaccine by attenuating the P9605 strain through a serial tissue culture passage. Langurs (*P. entellus*) vaccinated with this attenuated KFDV generated neutralizing antibodies and resisted the challenge infection.[77] In another experiment, two dosages of formalin-killed KFDV administered to langurs by subcutaneous route induced neutralizing antibodies. The antibodies, though transient, could be detected up to 15 months after first vaccination. The vaccine did not protect the langurs from challenge infection. However, it did prevent death.[78] A surveillance study indicated that the formalinized KFDV vaccine has some but not absolute protective effect on the vaccinees in Sagar–Sorab taluks of Shimoga district.[79] An attenuated Langat virus vaccine was also tested and showed protection against KFDV.[80] Kayser et al.[81] studied the human antibody response to immunization with 17D yellow fever virus and inactivated TBEV vaccine. They reported that vaccinees producing HI titers ≥20 against TBE showed cross-reaction with KFD.

The formalin-inactivated KFDV vaccine produced in chick embryo fibroblasts has been licensed and is currently in use in the endemic areas in the Karnataka state of India. The places for vaccination are selected on the basis of prevalence of KFDV activity in the previous years, including the villages from which mortality in monkeys was reported and those adjacent to the KFDV-affected areas. The efficacy of the vaccine was satisfactory, exerting a highly significant protective effect.[82] Coverage of vaccine is fairly good. Almost all the individuals including children are being routinely vaccinated by local government authorities. However, the occurrence of KFD cases, despite vaccination, has suggested some changes in virus antigenic determinants in due course. The KFDV strain currently used for vaccine preparation was isolated late in the 1950s. Thereafter, KFDV strains have not been well characterized at the molecular level; hence, possible antigenic drifts and diversity since its first introduction in India during the late1950s remain poorly understood. Increasing trend of KFD cases in the Karnataka state warrants development of a new vaccine preparation that includes currently circulating KFDV. Improper storage of vaccine and lack of maintenance of cold chain result in inactivation of the vaccine and could be another reason for the emergence of KFD despite routine vaccination.[20]

Ticks are the most significant vectors of animal diseases after mosquitoes in the number of human diseases they transmit.[83] Ticks transmit a variety of pathogens like protozoans, rickettsiae, spirochetes, and viruses. Tick control represents another effective method of disease prevention. Despite routine vaccination programs, the Indian state of Karnataka has an increasing number of KFD cases every year that imply insufficient efficacy of vaccine protocol and that call for improved tick control methods as well. Presently, the effective tick control method is the application of acaricides. However, this approach is associated with a number of disadvantages such as chemical pollution of the food chain and the environment as well as the development of resistance against acaricides by ticks.[84] Effective reduction in the transmission of tick-borne diseases by vaccination against vectors was shown to be effective.[83,85–87] This strategy of developing a vaccine against the KFD vector *H. spinigera* could be an effective method of tick control and in turn KFD spread. An interesting approach suggested recently[88] by administering oral vaccines using baits could be an option to immunize the monkeys, the amplifier host of KFDV. The programs focusing on the development of tick vaccines should also be encouraged in countries endemic for the tick-borne diseases. A combination of the KFD vaccines and also the tick vaccines could be more effective in containing the disease rather than using one of the previously mentioned vaccines.

13.7 CONCLUSIONS AND FUTURE PERSPECTIVES

KFD is a zoonotic disease caused by KFDV, which constitutes a member of the tick-borne encephalitis virus serocomplex within the genus *Flavivirus*, family Flaviviridae. While the disease is localized largely to the Karnataka state of India, viruses related to KFDV have also been reported recently in other parts of the world such as China and Saudi Arabia. The main clinical manifestations of KFD include fever, hemorrhage, and encephalitis, with a fatality rate of 2%–10%.

KFDV, despite regional significance, carries much importance related to origin, evolution, dispersal, and antigenic diversity of flaviviruses. It is one of the few flaviviruses that lead to hemorrhagic manifestations. Ecology and epidemiology of KFDV are very unique with distinct clinical symptoms and pathological manifestations. ALKV and KFDV share a high sequence identity despite their differences in geographic distribution.[38,89] Both the viruses possess a positive-sense RNA genome with 11 kb in length, coding for a polyprotein that is posttranscriptionally cleaved into three structural (C, M, and E) and seven NS proteins (NS1, NS2a, NS2b, NS3, NS4a, NS4b, and NS5). With the striking similarities between the two viruses, it has been proposed that ALKV is a descendant of KFDV, and studies on envelope and NS3 segments proved the recent divergence of ALKV and KFDV during 1828 and 1942.[38] Understanding the complete history of evolution of these two closely related viruses will shed light on the circumstances that lead to their emergence. Recent reports on full-length sequencing analysis of ALKV and KFDV proved a much prior divergence of ALKV from KFDV, approximately 700 years ago, raising a question about the existence of closely related but undiscovered virus variants in the regions between Saudi Arabia and India.[89] This speculation of existence of related viruses is further supported by the presence of competent ticks in these regions.[90] Hence, further phylogenetic studies using numerous isolates of KFDV and ALKV could unravel the mystery of the emergence of this deadly pathogen.

Earlier studies indicated that the antigenic structure of KFDV could be very different from that of TBEV. However, considering the positionally conserved cysteines and a similar structural feature of domain III of the E protein to those of other known flaviviruses, it could be inferred that KFDV may have similar receptor–ligand interaction.[91] A common mechanism of virus–cell fusion in virus entry, which is shared by KFDV and other flaviviruses, has also been suggested by amino acid sequence comparison and structural modeling of the E proteins. Although the available vaccine has once successfully controlled the KFD, the increasing trend of the disease in the last 5 years irrespective of routine vaccination is alarming. Such a trend may be indicative of the possible mutations in exposed antigenic domains of the virus, thereby allowing the virus to evade the immune response generated by the formalin-inactivated vaccine.

Grard et al.[37] reported complete characterization of an old strain P9605 of KFDV (GenBank accession number AY323490) and proposed significant taxonomic improvements. However, lack of sequence data of recent isolates of KFDV restricts our understanding on the relationship of sequence diversity and antigenic diversity of this virus. We speculate that KFDV may show an altered degree of virulence due to many changing factors such as changes of social behavior of humans like urbanization, rapid transport, and migration of people or of vectors, large-scale changes in ecology due to deforestation or building of dams or canals, and changing agricultural practices.

Given the requirement of developing safer and more effective vaccines in general, it is important to make an effort to develop an alternative vaccine version for KFD. A recombinant subunit vaccine or recombinant chimeric live vaccine could be the better options. The development of rapid and easy-to-use diagnostic system is also needed for the field surveillance of KFD. Recent developments in the field of diagnosis[69] and also the elucidation of immune evasion mechanism[67] prove that the measures to encounter a lethal pathogen like KFD are in progress. Homology modeling of KFDV envelope (E) protein exhibited a structure similar to those of other flaviviruses, suggesting a common mechanism of virus–cell fusion. The possible mechanism of receptor–ligand interaction involved in infection by KFDV may resemble that of other flavivirses.[91] A present understanding is that KFDV may be persisting silently in several regions of India and that antigenic and structural differences from other tick-borne viruses may be related to the unique host specificity and pathogenicity of KFDV. From January 1999 through January 2005, an increasing number of KFD cases have been detected in the Karnataka state of the Indian subcontinent despite routine vaccination, suggesting insufficient efficacy of the current vaccine protocol. The exact cause of increase of KFD cases needs further investigation. The changing ecology of the prime focus of the KFD also warrants attention, as it may lead to the establishment of the disease in newer localities never reported earlier.

REFERENCES

1. Work, T.H. and Trapido, H. Summary of preliminary report of investigations of the Virus Research Centre on an epidemic disease affecting forest villagers and wild monkeys of Shimoga District, Mysore. *Indian J Med Sci* 11, 341–342 (1957).
2. Francki, R.I.B. et al. Fifth report of the International Committee on Taxonomy of Viruses. *Arch Virol* (Suppl. 2) 2, 1–450 (1991).
3. Thiel, H.J. et al., Family Flaviviridae. In: Fauquet C.M. et al., eds. *Virus Taxonomy: Classification and Nomenclature, Eighth Report of the International Committee on the Taxonomy of Viruses*. Amsterdam, the Netherlands: Elsevier Academic Press, p. 981 (2005).
4. Calisher, C.H. et al. Antigenic relationships between flaviviruses as determined by cross-neutralization tests with polyclonal antisera. *J Gen Virol* 70 (Pt 1), 37–43 (1989).
5. Work, T.H., Roderiguez, F.R., and Bhatt, P.N. Virological epidemiology of the 1958 epidemic of Kyasanur Forest disease. *Am J Public Health Nations Health* 49, 869–874 (1959).
6. Banerjee, K. Kyasanur Forest disease. In: Monath, T.P., ed. *The Arboviruses: Epidemiology and Ecology*. Boca Raton, FL: CRC Press, pp. 93–116 (1988).
7. Trapido, H., Rajagopalan, P.K., Work, T.H., and Varma, M.G. Kyasanur Forest disease. VIII. Isolation of Kyasanur Forest disease virus from naturally infected ticks of the genus *Haemaphysalis*. *Indian J Med Res* 47, 133–138 (1959).
8. Goverdhan, M.K. and Anderson, C.R. The reaction of *Funambulus tristriatus tristriatus Rattus blanfordi* and *Suncus murinus* to Kyasanur Forest disease virus. *Indian J Med Res* 74, 141–146 (1981).
9. Goverdhan, M.K. and Anderson, C.R. Reaction of *Rattus rattus wroughtoni* to Kyasanur Forest disease virus. *Indian J Med Res* 67, 5–10 (1978).
10. Anderson, C.R. and Singh, K.R. The reaction of cattle to Kyasanur Forest disease virus. *Indian J Med Res* 59, 195–198 (1971).
11. Zaki, A.M. Isolation of a flavivirus related to the tick-borne encephalitis complex from human cases in Saudi Arabia. *Trans R Soc Trop Med Hyg* 91, 179–181 (1997).
12. Charrel, R.N. et al. Complete coding sequence of the Alkhurma virus, a tick-borne flavivirus causing severe hemorrhagic fever in humans in Saudi Arabia. *Biochem Biophys Res Commun* 287, 455–461 (2001).
13. Venugopal, K., Gritsun, T., Lashkevich, V.A., and Gould, E.A. Analysis of the structural protein gene sequence shows Kyasanur Forest disease virus as a distinct member in the tick-borne encephalitis virus serocomplex. *J Gen Virol* 75 (Pt 1), 227–232 (1994).
14. Harris, E., Holden, K.L., Edgil, D., Polacek, C., and Clyde, K. Molecular biology of flaviviruses. *Novartis Found Symp* 277, 23–39; discussion 40, 71–73, 251–253 (2006).
15. Lindenbach, B.D., Heinz-Jurgen, T., and Rice, C.M. In: Howley, P.M. and Knipe, D.M., eds. *Fields' Virology*, 5th ed. Philadelphia, PA: Lippincott Williams & Wilkins, Chapter 33 (Flaviviruses), pp. 1043–1127, (2007).
16. Bhatt, P.N. et al. Kyasanur Forest diseases. IV. Isolation of Kyasanur Forest disease virus from infected humans and monkeys of Shimoga district, Mysore state. *Indian J Med Sci* 20, 316–320 (1966).
17. Upadhyaya, S., Narasimha Murthy, D.P., and Yashodhara Murthy, B.K. Viraemia studies on the Kyasanur Forest disease human cases of 1966. *Indian J Med Res* 63, 950–953 (1975).
18. Pavri, K. Clinical, clinicopathologic, and hematologic features of Kyasanur Forest disease. *Rev Infect Dis* 11 (Suppl 4), S854–S859 (1989).
19. Bhat, H.R., Naik, S.V., Ilkal, M.A., and Banerjee, K. Transmission of Kyasanur Forest disease virus by *Rhipicephalus haemaphysaloides* ticks. *Acta Virol* 22, 241–244 (1978).
20. Pattnaik, P. Kyasanur Forest disease: An epidemiological view in India. *Rev Med Virol* 16, 151–165 (2006).

21. Boshell, J. Kyasanur Forest disease: Ecologic considerations. *Am J Trop Med Hyg* 18, 67–80 (1969).

22. Goverdhan, M.K. et al. Epizootiology of Kyasanur Forest disease in wild monkeys of Shimoga district, Mysore State (1957–1964). *Indian J Med Res* 62, 497–510 (1974).

23. Kenyon, R.H., Rippy, M.K., McKee, K.T., Zack, P.M., and Peters, C.J. Infection of *Macaca radiata* with viruses of the tick-borne encephalitis group. *Microb Pathog* 13, 399–409 (1992).

24. Upadhyaya, S., Murthy, D.P., and Anderson, C.R. Kyasanur Forest disease in the human population of Shimoga district, Mysore State, 1959–1966. *Indian J Med Res* 63, 1556–1563 (1975).

25. Sreenivasan, M.A., Bhat, H.R., and Rajagopalan, P.K. The epizootics of Kyasanur Forest disease in wild monkeys during 1964 to 1973. *Trans R Soc Trop Med Hyg* 80, 810–814 (1986).

26. Sarkar, J.K. and Chatterjee, S.N. Survey of antibodies against arthropod-borne viruses in the human sera collected from Calcutta and other areas of West Bengal. *Indian J Med Res* 50, 833–841 (1962).

27. Padbidri, V.S. et al. A serological survey of arboviral diseases among the human population of the Andaman and Nicobar Islands, India. *Southeast Asian J Trop Med Public Health* 33, 794–800 (2002).

28. Wang, J. et al. Isolation of Kyasanur forest disease virus from febrile patient, Yunnan, China. *Emerg Infect Dis* 15, 326–328 (2009).

29. Hou, Z.L. et al. Study of the serologic epidemiology of tick-borne viruses in Yunnan, China [in Chinese]. *Chin J Vector Biol Control* 3, 173–176 (1992).

30. Yang, Q.R., Liu, X.Z., Zhang, J.Y., Zi, D.Y., and Zhang, H.L. A study of arbovirus antibodies in birds of the Niao-Diao mountain area of Eryuan County in Yunnan Province [in Chinese]. *Chin J Endemiol* 9, 150–153 (1988).

31. Chen, B.Q., Liu, Q.Z., and Zhou, G.F. Investigation of arbovirus antibodies in serum from residents of certain areas of China [in Chinese]. *Chin J Endemiol* 4, 263–266 (1983).

32. Ghosh, S.N., Rajagopalan, P.K., Singh, G.K., and Bhat, H.R. Serological evidence of arbovirus activity in birds of KFD epizootic—Epidemic area, Shimoga District, Karnataka, India. *Indian J Med Res* 63, 1327–1334 (1975).

33. Venugopal, K., Buckley, A., Reid, H.W., and Gould, E.A. Nucleotide sequence of the envelope glycoprotein of Negishi virus shows very close homology to Louping ill virus. *Virology* 190, 515–521 (1992).

34. Gong, Z.D. and Hai, B.Q. Investigation of small animals in the Gaoli Mountain region [in Chinese]. *J Vet Med* 24, 28–32 (1989).

35. Gong Z.D., Zi, D.Y., and Feng, X.G. Composition and distribution of ticks in the Hengduan Mountain region of western Yunnan, China [in Chinese]. *Chin J Pest Control* 2, 13–15 (2001).

36. Charrel, R.N. et al. Low diversity of Alkhurma hemorrhagic fever virus, Saudi Arabia, 1994–1999. *Emerg Infect Dis* 11, 683–688 (2005).

37. Grard, G. et al. Genetic characterization of tick-borne flaviviruses: New insights into evolution, pathogenetic determinants and taxonomy. *Virology* 361, 80–92 (2007).

38. Mehla, R. et al. Recent ancestry of Kyasanur Forest disease virus. *Emerg Infect Dis* 15, 1431–1437 (2009).

39. Carletti, F. et al. Alkhurma hemorrhagic fever in travelers returning from Egypt, 2010. *Emerg Infect Dis* 16, 1979–1982 (2010).

40. Gould, E.A., de Lamballerie, X., Zanotto, P.M., and Holmes, E.C. Evolution, epidemiology, and dispersal of flaviviruses revealed by molecular phylogenies. *Adv Virus Res* 57, 71–103 (2001).

41. Gould, E.A., de Lamballerie, X., Zanotto, P.M., and Holmes, E.C. Origins, evolution, and vector/host coadaptations within the genus *Flavivirus*. *Adv Virus Res* 59, 277–314 (2003).

42. Charrel, R.N. and Gould, E.A. Alkhurma hemorrhagic fever in travelers returning from Egypt, 2010. *Emerg Infect Dis* 17, 1573–1574; author reply 1574 (2011).

43. Nichter, M. Kyasanur Forest disease: An ethnography of a disease of development. *Medical Anthropology Quarterly* 1, 406–423 (1987).

44. Dobler, G. Zoonotic tick-borne flaviviruses. *Vet Microbiol* 140, 221–228 (2010).

45. Madani, T.A. Alkhurma virus infection, a new viral hemorrhagic fever in Saudi Arabia. *J Infect* 51, 91–97 (2005).

46. Chatterjea, J.B., Swarup, S., Pain, S.K., and Laxmana Rao, R. Haematological and biochemical studies in Kyasanur Forest disease. *Indian J Med Res* 51, 419–435 (1963).

47. Iyer, C.G., Laxmana Rao, R., Work, T.H., and Narasimha Murthy, D.P. Kyasanur Forest Disease VI. Pathological findings in three fatal human cases of Kyasanur Forest Disease. *Indian J Med Sci* 13, 1011–1022 (1959).

48. Webb, H.E. and Rao, R.L. Kyasanur forest disease: A general clinical study in which some cases with neurological complications were observed. *Trans R Soc Trop Med Hyg* 55, 284–298 (1961).

49. Wadia, R.S. Neurological involvement in Kyasanur Forest disease. *Neurol India* 23, 115–120 (1975).

50. Sathe, P.S., Dandawate, C.N., Sharadamma, K., and Ghosh, S.N. Circulating interferon-alpha in patients with Kyasanur forest disease. *Indian J Med Res* 93, 199–201 (1991).

51. Banerjee, K., Dandawate, C.N., Bhatt, P.N., and Rao, T.R. Serological response in humans to a formalized Kyasanur Forest disease vaccine. *Indian J Med Res* 57, 969–974 (1969).

52. Adhikari Prabha, M.R., Prabhu, M.G., Raghuveer, C.V., Bai, M., and Mala, M.A. Clinical study of 100 cases of Kyasanur Forest disease with clinicopathological correlation. *Indian J Med Sci* 47, 124–130 (1993).

53. LeDuc, J.W. Epidemiology of hemorrhagic fever viruses. *Rev Infect Dis* 11 (Suppl 4), S730–S735 (1989).

54. Sabin, A.B. and Blumberg, R.W. Human infection with Rift Valley fever virus and immunity twelve years after single attack. *Proc Soc Exp Biol Med* 64, 385–389 (1947).

55. Price, W.H. Studies on the immunological overlap among certain arthropod-borne viruses. II. The role of serological relationships in experimental vaccination procedures. *Proc Natl Acad Sci USA* 43, 115–121 (1957).

56. Achar, T.R., Patil, A.P., and Jayadevaiah, M.S. Persistence of humoral immunity in Kyasanur forest disease. *Indian J Med Res* 73, 1–3 (1981).

57. Woolhouse, M.E., Webster, J.P., Domingo, E., Charlesworth, B., and Levin, B.R. Biological and biomedical implications of the co-evolution of pathogens and their hosts. *Nat Genet* 32, 569–577 (2002).

58. Iwasaki, A. and Medzhitov, R. Toll-like receptor control of the adaptive immune responses. *Nat Immunol* 5, 987–995 (2004).

59. Kaech, S.M., Wherry, E.J., and Ahmed, R. Effector and memory T-cell differentiation: Implications for vaccine development. *Nat Rev Immunol* 2, 251–262 (2002).

60. Alcami, A. and Koszinowski, U.H. Viral mechanisms of immune evasion. *Trends Microbiol* 8, 410–418 (2000).

61. Takeuchi, O. and Akira, S. Innate immunity to virus infection. *Immunol Rev* 227, 75–86 (2009).

62. Pestka, S. The interferons: 50 years after their discovery, there is much more to learn. *J Biol Chem* 282, 20047–20051 (2007).

63. Robertson, S.J., Mitzel, D.N., Taylor, R.T., Best, S.M., and Bloom, M.E. Tick-borne flaviviruses: Dissecting host immune responses and virus countermeasures. *Immunol Res* 43, 172–186 (2009).

64. Muñoz-Jordan, J.L., Sánchez-Burgos, G.G., Laurent-Rolle, M., and García-Sastre, A. Inhibition of interferon signaling by dengue virus. *Proc Natl Acad Sci USA* 100, 14333–14338 (2003).

65. Laurent-Rolle, M. et al. The NS5 protein of the virulent West Nile virus NY99 strain is a potent antagonist of type I interferon-mediated JAK-STAT signaling. *J Virol* 84, 3503–3515 (2010).

66. Liu, W.J. et al. A single amino acid substitution in the West Nile virus nonstructural protein NS2A disables its ability to inhibit alpha/beta interferon induction and attenuates virus virulence in mice. *J Virol* 80, 2396–2404 (2006).

67. Cook, B.W., Cutts, T.A., Court, D.A., and Theriault, S. The generation of a reverse genetics system for Kyasanur Forest disease virus and the ability to antagonize the induction of the antiviral state in vitro. *Virus Res* 163, 431–438 (2012).

68. Pavri, K.M. and Anderson, C.R. Serological response of man to Kyasanur Forest disease. *Indian J Med Res* 58, 1587–1607 (1970).

69. Mourya, D.T. et al. Diagnosis of Kyasanur Forest disease by nested RT-PCR, real-time RT-PCR and IgM capture ELISA. *J Virol Methods* 186, 49–54 (2012).

70. Aniker, S.P. et al. The administration of formalin-inactivated RSSE virus vaccine in the Kyasanur Forest disease area of Shimoga District, Mysore State. *Indian J Med Res* 50, 147–152 (1962).

71. Pavri, K.M., Gokhale, T., and Shah, K.V. Serological response to Russian spring-summer encephalitis virus vaccine as measured with Kyasanur Forest disease virus. *Indian J Med Res* 50, 153–161 (1962).

72. Shah, K.V. et al. Evaluation of the field experience with formalin-inactivated mouse brain vaccine of Russian spring-summer encephalitis virus against Kyasanur Forest disease. *Indian J Med Res* 50, 162–174 (1962).

73. Mansharamani, H.J., Dandawate, C.N., Krishnamurthy, B.G., Nanavathi, A.N., and Jhala, H.I. Experimental vaccine against Kyasanur forest disease (KFD) virus from mouse brain source. *Indian J Pathol Bacteriol* 12, 159–177 (1965).

74. Dandawate, C.N., Mansharamani, H.J., and Jhala, H.I. Experimental vaccine against Kyasanur Forest Disease (KFD) virus from embryonated eggs. I. Adaptation of the virus to developing chick embryo and preparation of formalized vaccines. *Indian J Pathol Bacteriol* 8, 241–260 (1965).

75. Mansharamani, H.J., Dandawate, C.N., and Krishnamurthy, B.G. Experimental vaccine against Kyasanur Forest disease (KFD) virus from tissue culture source. I. Some data on the preparation and antigenicity tests of vaccines. *Indian J Pathol Bacteriol* 10, 9–24 (1967).

76. Mansharamani, H.J. and Dandawate, C.N. Experimental vaccine against Kyasanur Forest disease (KFD) virus from tissue culture source. II. Safety testing of the vaccine in cortisone sensitized Swiss albino mice. *Indian J Pathol Bacteriol* 10, 25–32 (1967).

77. Bhatt, P.N. and Anderson, C.R. Attenuation of a strain of Kyasanur Forest disease virus for mice. *Indian J Med Res* 59, 199–205 (1971).

78. Bhatt, P.N. and Dandawate, C.N. Studies on the antibody response of a formalin inactivated Kyasanur Forest Disease Virus vaccine in langurs *Presbytis entellus*. *Indian J Med Res* 62, 820–826 (1974).

79. Upadhyaya, S., Dandawate, C.N., and Banerjee, K. Surveillance of formalized KFD virus vaccine administration in Sagar-Sorab talukas of Shimoga district. *Indian J Med Res* 69, 714–719 (1979).

80. Thind, I.S. Attenuated Langat E5 virus as a live virus vaccine against Kyasanur Forest disease virus. *Indian J Med Res* 73, 141–149 (1981).

81. Kayser, M. et al. Human antibody response to immunization with 17D yellow fever and inactivated TBE vaccine. *J Med Virol* 17, 35–45 (1985).

82. Dandawate, C.N., Desai, G.B., Achar, T.R., and Banerjee, K. Field evaluation of formalin inactivated Kyasanur forest disease virus tissue culture vaccine in three districts of Karnataka state. *Indian J Med Res* 99, 152–158 (1994).

83. Mulenga, A. et al. Molecular characterization of a *Haemaphysalis longicornis* tick salivary gland-associated 29-kilodalton protein and its effect as a vaccine against tick infestation in rabbits. *Infect Immun* 67, 1652–1658 (1999).

84. Shapiro, S.Z., Voigt, W.P., and Fujisaki, K. Tick antigens recognized by serum from a guinea pig resistant to infestation with the tick *Rhipicephalus appendiculatus*. *J Parasitol* 72, 454–463 (1986).

85. Elvin, C.M. and Kemp, D.H. Generic approaches to obtaining efficacious antigens from vector arthropods. *Int J Parasitol* 24, 67–79 (1994).

86. Willadsen, P. Immunity to ticks. *Adv Parasitol* 18, 293–311 (1980).

87. Labuda, M. et al. An antivector vaccine protects against a lethal vector-borne pathogen. *PLoS Pathog* 2, e27 (2006).

88. Ghosh, S., Azhahianambi, P., and Yadav, M.P. Upcoming and future strategies of tick control: A review. *J Vector Borne Dis* 44, 79–89 (2007).

89. Dodd, K.A. et al. Ancient ancestry of KFDV and AHFV revealed by complete genome analyses of viruses isolated from ticks and mammalian hosts. *PLoS Negl Trop Dis* 5, e1352 (2011).

90. Kolonin, G.V. *World Distribution of Ixodid Ticks (Genus Haemaphysalis)*. Moscow, Russia: Nauka, 70 pp. (1978).

91. Pattnaik, P. et al. Cloning, expression, purification and homology modeling of envelope protein of Kyasanur forest disease virus (KFDV). In *16th Annual Convention of Indian Virological Society and Symposium on Management of Vector-Borne Viruses*. International Crop Research Institute for Semi-Arid Tropics (ICRISAT), Hyderabad, India, p. 49, February 7–10, 2006.

14 Lyssaviruses

Jennifer S. Evans, Anthony R. Fooks, and Ashley C. Banyard

CONTENTS

14.1 INTRODUCTION

The lyssaviruses constitute an important group of viruses that are of significance to both human and animal health. All viruses within this genus are highly neurotropic and are capable of causing a fatal encephalitis, known as rabies. The prototypic lyssavirus is rabies virus (RABV), which is primarily transmitted through the bite of an infected dog and causes a higher burden of disease in both humans and animals than any other lyssavirus. Indeed, the term "lyssa" is of Greek origin and is believed to refer to a "raging dog."[1] Of the remaining lyssaviruses, all have been detected in bats with only two exceptions to this, Mokola virus (MOKV) and Ikoma lyssavirus (IKOV). Despite safe and effective prophylactic and postexposure tools being available, RABV remains endemic across much of Africa and the Indian subcontinent. Members of the *Lyssavirus* genus are able to infect and cause disease in a very wide host range, with theoretically all mammalian species being susceptible to infection. Susceptibility to infection with these viruses does vary with several restrictions being hypothesized to contribute to incubation period, progression of disease, and overall outcome of infection

including virus dose, infecting strain, immunocompetence of the host, and site of wound. Despite the availability of tools to control rabies infection, the presence of virus in sylvatic populations has made elimination in certain mammalian species difficult and eradication problematic. From a human perspective, variation in incubation period following infection can mean that appropriate prophylactic tools are not sought and fatalities occur as where clinical disease develops, infection with lyssaviruses is invariably fatal.

14.2 CLASSIFICATION AND MORPHOLOGY

The lyssaviruses all possess a single strand of negative-sense, nonsegmented RNA as their genomic material, and as such, these viruses are classified within the family Rhabdoviridae within the order Mononegavirales. The *Lyssavirus* genus contains 14 distinct viral species with one further virus awaiting classification (Table 14.1). Within almost all of the currently defined lyssavirus species, variants have been described where viruses have been isolated from different mammalian species. This observation suggests coevolution of certain lineages of virus within

TABLE 14.1

Currently Defined Lyssavirus Species, Their Phylogroups, Geographical Location, and the Species from Which Each Has Been Isolated

Phylogroup	Lyssavirus Species	Species Isolated	Location
I	RABV	All mammals susceptible	Global
I	ARAV	Bat	Eurasian
I	KHUV	Bat	Eurasian
I	BBLV	Bat	European
I	EBLV-2	Human, bat	European
I	ABLV	Human, bat	European
I	IRKV	Human, bat	Eurasian
I	EBLV-1	Human, bat, cat, sheep, stone marten	European
I	DUVV	Human, bat	Africa
II	LBV	Bat, dog, cat, water mongoose	Africa
II	SHIBV	Bat	Africa
II	MOKV	Human, dog, cat, shrews, rodent species	Africa
III	WCBV	Bat	Eurasia
III (tentative)	IKOV	African civet	Africa

particular host reservoirs.[2] The species can be further divided into at least two and possibly a third phylogroup based on antigenicity and cross protection afforded by current rabies biologics[3–5] (Figure 14.1). It has been suggested, due to genetic and antigenic divergence, that West Caucasian bat virus (WCBV) may belong to a tentatively proposed phylogroup III as it is highly divergent from the phylogroup II viruses.[6] By this reasoning, and due to its genetic divergence from the other lyssaviruses, IKOV may become the second member of phylogroup III or may constitute a fourth phylogroup. Further studies are required to characterize this novel virus and Lleida bat lyssavirus.[7,8]

All lyssaviruses are enveloped, deriving their envelope from the plasma membrane of the infected host cell upon budding, and exhibit the classic bullet-like morphology common to all rhabdoviruses. They generally measure ~75 nm by 180 nm.[9] All lyssavirus genomes consist of between 11,000 and 12,000 nucleotides and contain open reading frames that encode five proteins, namely, the nucleoprotein (N), the phosphoprotein (P), the matrix protein (M), the glycoprotein (G), and the large polymerase protein (L).[10] The N protein encapsidates the RNA genome and in combination with the L and P proteins forms the ribonucleoprotein (RNP) complex, which is the minimal essential replication unit.[11] The M protein forms a protective coating surrounding the RNP and interacts with the cytoplasmic tail of the viral G protein at the cell membrane. The G protein is the only viral protein exposed on the virion surface and as such plays a crucial role in attachment and cell entry[12] as well as being the main target of neutralizing antibodies.[10]

14.3 BIOLOGY AND EPIDEMIOLOGY

14.3.1 TRANSMISSION

The most common route of infection with a lyssavirus is via the mechanistic action of a bite from an infected animal that contains live virus in its saliva. Historically, this mechanism was first described over 200 years ago by experimental work in 1804 by Zinke. Alongside this early observation, it was later determined that these viruses are unable to cross an intact dermal barrier. However, if the skin is broken, the risk of infection is greatly increased. Lyssaviruses can also infect hosts via mucous membranes including the nasal lining, oral cavity, conjunctivae, external genital organs, and anus. There have also been reports of infection via inhalation of virus via the aerosol route[13,14] although conclusive evidence to validate this route of infection is lacking.

For human exposure, human-to-human transmission has only been reported very rarely with transmission of bodily fluids being implicated as the source of virus.[15] Alongside this, there have been occurrences of infection through transplantation of infected tissue. A case in 2004 resulted in three rabies fatalities when an infected donor supplied a liver section, a kidney, and a section of artery to three separate recipients. Within 30 days of transplantation, all three recipients had returned to the hospital with rapidly progressing neurological disease, which was diagnosed postmortem as rabies infection.[16] A further case in Germany resulted in the deaths of three individuals following organ donation.[17] Corneal transplantation has also facilitated infection in the recipient following surgery.[18]

14.3.2 RECEPTORS

The lyssavirus G protein is the main viral protein that enables virion attachment and entry into host cells. Currently, while there are postulated to be as yet undefined receptors utilized by lyssaviruses for cell entry, several receptor molecules have been proposed to be involved in virus binding and entry.[19] Once the virus has entered the host, it may either enter the nervous system directly or initially replicate in the muscle tissue surrounding the locus of infection before entering the nervous system. The exact sites of virus replication prior to entering the central nervous system (CNS) are yet to be defined; however, the suggestion of low-level replication in the nonneuronal tissue following infection may provide some explanation for the long incubation periods that have been reported.[19] It has been demonstrated *in vitro* that the virus is able to enter motor nerves via neuromuscular junctions (NMJs). As well as this, viral antigen has been shown to be present in proprioceptors, sensory spindles, and stretch receptors *in vitro*.[20]

The virus is able to enter neural tissues via one/a combination of three proposed receptors that are present in all mammals: the neural cell adhesion molecule (NCAM) that is present in NMJs at postsynaptic membranes,[21] the nicotinic

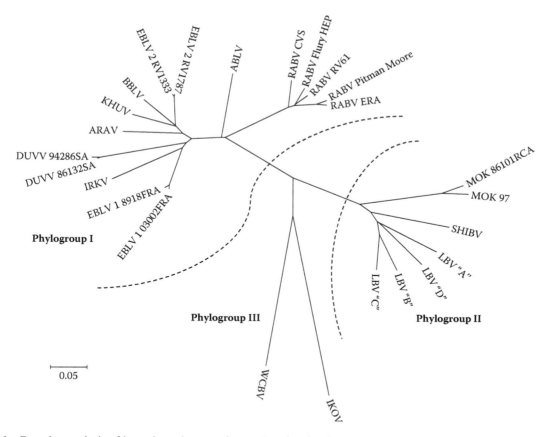

FIGURE 14.1 Bayesian analysis of lyssavirus glycoprotein ectodomains, implemented in vector NTI. The minimum evolution tree was chosen using MEGA 4. The scale bar represents substitutions per site. Accession numbers for each of the lyssavirus species analyzed are as follows: RABV CVS EU352767, RABV Flury-HEP GU565704, RABV RV61 JQ685981, RABV Pitman Moore AJ871962, RABV ERA J02293, ABLV AF801020, EBLV-2 RV1787 EU352769, EBLV-2 RV1333 EF57977, BBLV JF311903, KHUV EF614261, ARAV EF614259, DUVV 94286SA EU293120, DUVV 86132SA EU293119, IRKV EF614260, EBLV-1 8918FRA EU293112, EBLV-1 03002FRA EU293109, MOKV 86101RCA EU293118, MOKV 97 NC_006429, SHIBV GU170201, LBV "A" EF547432, LBV "D" GU170202, LBV "B" EF547431, LBV "C" EF547425, WCBV EF614258, and IKOV JN800509.

acetylcholine receptor (nAchR) that is involved in inter-neuronal signaling within both the peripheral and central nervous system (P/CNS),[22] and p75NTR that is a neurotrophin receptor responsible for apoptosis, axonal elongation, and synaptic transmission.[23] Each of these three proposed receptors is not only expressed in the CNS but can be expressed on cells in the dermis; NCAM and p75NTR are expressed from dermal fibroblasts associated with hair follicles,[24,25] while nAchR has been detected in the squamous epithelium.[26] This suggests that they may be available around sites of infection as a route to the CNS for the virus; however, the extent to which rabies virus and the lyssaviruses utilize each of these proposed receptors is undefined. Upon attachment of the virus to a permissive cell, it is thought that entry occurs by receptor-mediated endocytosis or by fusion of the virion membrane with the cell membrane. This results in the release of the RNP into the cell cytoplasm, which initiates transcription and replication to produce new virus particles.

Once within the nervous system of the host, virus particles are able to travel along axons in a retrograde fashion at a rate of 50–100 mm/day[27] until they reach the CNS.

The mechanism of axonal transport may in part be due to interactions between the P protein and actin and microtubule networks including the microtubule motor molecule, dynein LC8.[28,29] The virus is able to spread rapidly within the CNS due to the ubiquitous nature of the proposed receptors while movement in the periphery appears limited, which restricts the spread of virus in these areas.[30]

14.3.3 INCUBATION PERIOD

The incubation period of RABV is widely accepted to be between 20 and 90 days; however, there have been reports of extreme cases of incubation periods ranging between 14 and 19 years although in these cases, the possibility of a secondary infection cannot be discounted.[31] Regardless, it is clear that the incubation period following infection can vary greatly depending on a range of factors. Due to the neurotropic nature of the virus, a bite/site of infection that is highly innervated (e.g., face and hands) can lead to a much shorter incubation period or faster progression of disease than infection at a less densely innervated area.

Alongside this, the presence/absence of permissive receptors in the tissue surrounding the infection site can affect viral entry and spread, hence increasing the incubation period. The dose of infective virus also plays a role in determining incubation period.[30] As well as this, the immune status and age of an individual can influence incubation as incubation periods in pediatric cases are shorter than those in adults, presumably due to their smaller stature.[32]

Incubation periods following exposure to nonrabies lyssaviruses are less well defined as possible exposure times are often difficult to determine. In particular, it is recognized that often bat bites may go totally unnoticed leading to cryptic cases of rabies.[33] A case of European bat lyssavirus 1 (EBLV-1) infection in a human recorded an incubation period of 45 days.[34] Further to this, a case of EBLV-2 infection in a bat handler reported a 19-week incubation period; however, the patient had a documented history of bat bites that may have confounded determination of the incubation period.[35] Duvenhage virus (DUVV) infection has been reported to have an incubation period of around 4 weeks,[36,37] whereas Australian bat lyssavirus (ABLV) may have a much longer incubation period as a case in 1998 reported onset of clinical disease 20 months after exposure to a bat.[38,39]

14.3.4 GENOMIC VARIATION WITHIN LYSSAVIRUSES

Two other elements of lyssavirus virology that remain poorly understood are both the role of defective interfering (DI) particles and the role of viral quasispecies. Virus quasispecies are variant RNA species present in all RNA virus populations and are generated as a result of the error-prone nature of the RNA-dependent RNA polymerase.[40] Virus species that contribute to a virus population may include those with single mutations, those with multiple mutations, and those that have had complete deletions of stretches of the genome: DI particles.[41,42] The latter are believed to only be generated in isolates following extensive *in vitro* passage but can contribute significantly to pathogenesis.[43] The former will, no doubt, affect viral fitness, but as yet determining the dynamics of how this occurs remains to be studied. In particular, the role of virus mutation leading to cross species transmission (CST) events remains poorly understood so requires further research.[44]

The role of single/multiple mutations within dominant virus populations has been investigated in several studies. Growth of a human RABV isolate in cell culture was studied and significant intrinsic heterogeneity within the glycoprotein sequence was observed.[45] When comparing a dog-derived street RABV isolate with Challenge Virus Standard (CVS) strain, a 10% divergence in amino acid composition was observed; however, in the ectodomain, there was only a 6% divergence, which may suggest that some structural features may restrain divergence in this region. Additional evidence comes from sequencing of viral clones; only one-third of clones isolated from the brain of a rabid dog exhibited the consensus sequence, and the remaining two-thirds showed between one and three amino acid substitutions. It is suggested that the nature of these heterogeneous populations enables the virus to rapidly adapt to changes in the environment. Deep sequencing techniques are increasingly being used to analyze the full range of sequences present in a viral population to further our understanding of variation among viral quasispecies.

Importantly, and most commonly, transmission of RABV by hematophagous bats has shown that virus transmission from bats to terrestrial species occurs. For non-RABV lyssaviruses, several "spillover" events have occurred whereby the virus has been transmitted from a bat species into nonhuman terrestrial species to cause disease.[46–49] However, CST events that have resulted in continued transmission in the recipient species are rare[50,51] but it is these events that are of most significance to human and animal health as they represent mechanisms of the establishment of an endemic cycle within a new population.[44] Theoretically, all events where viruses cross a species barrier may result in a true CST event occurring should the new ecosystem favor virus replication. Ultimately, evolutionary constraints on viral host range have indicated that host species barriers may overwhelm the intrinsic mutability of RNA viruses in determining the fate of emerging host–virus interactions.[44] Most recently, host ecology has been suggested to play a key role in shaping the rate of evolution in viruses that are found in multiple hosts.[52] However, understanding the genetic changes or microscopic environmental factors that enable maintenance and spread of virus following a spillover event that then leads to a CST is of most interest as such data may enable prediction of CST capabilities.[53]

14.3.5 EPIDEMIOLOGY

RABV is distributed globally with only a few, principally island nations being free of disease. Concerted vaccination campaigns with domestic animals across both the United States and Western Europe have now eliminated the burden of rabies from domestic terrestrial carnivore populations. In the United States, sylvatic rabies remains a constant threat to human populations, while extensive wildlife vaccination campaigns have eliminated sylvatic rabies from Western Europe. However, globally, the problem of lyssaviruses circulating in the chiropteran reservoir remains and so a complete eradication of RABV is thought to be unrealistic.[54] Within the United States, bat species maintain RABV, while, in contrast across the Old World, classical RABV appears to be totally absent from bat populations, while several nonrabies lyssaviruses persist. These populations of bats are very difficult to access and, across the EU, are protected by law, making eradication of lyssaviruses an unfeasible target.

Within Europe, there are four lyssaviruses: the EBLVs (EBLV-1 and EBLV-2) that circulate in chiropteran hosts, the latter being endemic within UK bat populations[55]; Bokeloh bat lyssavirus (BBLV) that has been detected in Natterer's bats in Germany and France[56,108]; and WCBV that was isolated from a bat in the Caucasus mountains[57] and is genetically, highly divergent from the currently defined lyssaviruses.[58,59] Genomic material of a potential fifth European lyssavirus has also recently been reported and named Lleida bat lyssavirus although this requires further clarification.[60] Alongside

European viruses, three Eurasian viruses have been characterized, all isolated from bats: Aravan virus (ARAV) in Kyrgystan,[57] Khujand virus (KHUV) isolated in Tajikistan,[61] and Irkut virus (IRKV) isolated in Eastern Siberia.

Alongside these viruses, there are also a number of African lyssaviruses including Lagos bat virus (LBV), which is the most regularly isolated of the African lyssaviruses and circulates in a range of bats across sub-Saharan Africa[6]; MOKV, which is prevalent across Africa and has been associated with two recorded human cases[62]; DUVV, which was initially isolated from a fatal human bat bite case in Kenya in 1971[63] and subsequently from a number of fruit bats in South Africa; and Shimoni bat virus (SHIBV), which was first isolated in 2009[8] from a Commerson's leaf-nosed bat in Kenya.[6] Most recently, IKOV was isolated from an African civet in Tanzania in 2009[8] and, although discovered in a terrestrial carnivore, is believed to be of bat origin.

Within Australasia, only ABLV has been reported and has been isolated from five different bat species since its initial isolation in 1996.[64]

14.4 CLINICAL FEATURES AND PATHOGENESIS

The principal mechanisms of lyssavirus pathogenesis within the P/CNS remain poorly defined and human rabies patients show little morphological damage with almost no neuronal cell death in the brain. There are, however, a number of suggestions of mechanisms behind the development of disease. These include the ability of the virus to spread within peripheral neuronal cells to reach the CNS, which depends on receptor load and promiscuity. Once in the CNS, the virus may alter neurotransmission and normal neural function. It has been demonstrated that there is a reduction in the expression of housekeeping genes in RABV-infected neurons. This results in a reduction of protein synthesis and subsequently affects neuronal activity. It has also been suggested that the presence of the G protein on the surface of infected cells may disrupt ion channels, supported by the observation of electrophysiological disturbances in RABV-infected murine brains.[65]

14.4.1 HUMAN INFECTION

In human infection, where disease progression has been characterized, the development of clinical disease following lyssavirus infection is indistinguishable from that caused by RABV. Indeed, in areas where RABV is endemic, a patient presenting with idiopathic neurological symptoms should not have rabies ruled out and an absence of bite history should not rule out the possibility of rabies in a differential diagnosis. Though infection through contact with saliva from a rabid animal is the most common route of infection, other routes must be considered; infection may occur through exposed scratches or an open wound as well as exposure with mucous membranes. A selection of documented human cases of infection are displayed in Table 14.2.

TABLE 14.2

Documented Human Cases of Lyssavirus Infection Including the Results of Diagnostic Investigation and Clinical Outcome

Source of Infection	Lyssavirus Species	Specific Antibodies	Virus/Antigens Detected	Clinical Outcome	Reference
Bat, United States	RABV	Detected at clinical presentation	No virus isolated	Survived	[84]
Bat, United States	RABV	Detected at clinical presentation	Skin and saliva RT-PCR and DFA positive	Deceased, 34 days after symptoms	[100]
Bat, United States	RABV	Detected on day 12	Saliva RT-PCR positive, corneal impressions DFA positive	Deceased, 29 days after symptoms	[100]
Dog, Thailand	RABV	None detected	Hair follicle, nucleic acid, sequence-based amplification positive	Deceased, 8 days after symptoms	[101]
Dog, Germany	RABV	Detected from day 11	Saliva and corneal swab RT-PCR positive	Deceased, 31 days after symptoms	[102]
Bat, Canada	RABV	IgG and IgM detected	Skin and saliva RT-PCR positive, FAT positive	Deceased, 65 days after symptoms	[103]
Bat, Kenya	DUVV	None detected	Saliva and nuchal skin biopsy RT-PCR positive	Deceased, 45 days post contact with bat	[104]
Dog, Equatorial Guinea	RABV	Detected in serum on day 18 and in cerebral spinal fluid (CSF) on day 20	Saliva RT-PCR positive, skin biopsy MIT positive	Deceased—attributed to renal failure and malnutrition	[105]
Dog, Belfast	RABV	Detected in serum and CSF from day 8	Skin, CSF, and sputum PT-PCR positive, RTCIT positive	Deceased, 35 days after symptoms	[106]

The 2–20-day prodromal phase of a RABV or lyssavirus infection is easy to confuse with almost any other disease as the patient often presents with weakness, malaise, fever, headaches, urinary problems, and paresthesia at the site of infection that leads to ascending paralysis. For this reason, other diseases such as Guillain–Barré syndrome are included in a differential diagnosis. From the point of clinical onset, the likelihood of recovery is effectively zero. Eighty percent of individuals then go on to develop the classic "furious" form of rabies that can be characterized by hydrophobia, respiratory convulsions, and madness, while the remainder may develop a "dumb" or "paralytic" form of disease. The reason for these different clinical courses is not yet defined but may be related to the source of infection, that is, a low-level dermal bite from a bat versus a deep intramuscular bite from a dog. Death most often results from respiratory failure as a result of the spasms brought on by hydrophobia.

14.4.2 Animal Infection

It appears that there are no species-specific pathognomonic symptoms, but similar to human infection, there is a prodromal phase where the animal appears lethargic and exhibits some weakness. It has been observed that domesticated animals may show less interest in their owners, whereas wild animals may actively seek contact with humans. Animal infection also follows the dumb or furious forms of disease but individuals may present with features of both. Clinical progression may be linked to the site of viral replication within the CNS, which is linked to the site of infection and the dose of virus transmitted.[66] In animals suffering with furious rabies, there is often a short prodromal phase where the animal may appear to have heightened alertness, and this is followed by a neurological phase with signs similar to those in human infections. Rabid animals have been observed to snap at any objects within reach, and it is presumed that animals also experience paresthesia at the site of infection as excessive grooming and scratching has been observed, which can lead to self-mutilation and, in extreme cases, to self-consumption. Death results from organ failure and respiratory arrest. Paralytic rabies in both domesticated and wild dogs starts with a clinical presentation of lethargy and a lack of coordination. This leads to weakness in the hind limbs and subsequent motor paralysis that spreads eventually to the mandible, which results in cessation of the swallowing reflex. This is the final stage of disease before death.

14.5 DIAGNOSIS

14.5.1 Histopathology

Historically, histopathological methods were the first reliable option for accurate rabies diagnosis. In 1903, Negri described the presence of inclusion bodies in the cytoplasm of nerve cells in the hippocampus of rabies-infected brains.[67] These "Negri" bodies are composed of a reticulogranular matrix that contains tubular structures 4–5 µm in length.

They are most commonly found in the neurons and ganglioneurons of the cerebellum, Ammon's horn on the hippocampus, the medulla, and the motor section of the cerebral cortex. Despite this apparent specificity, the absence of Negri bodies does not exclude rabies infection.

There are now more sensitive and specific histopathological techniques available including a digoxigenin-labeled riboprobe that is able to detect lyssavirus RNA in infected mouse brain samples. This technique can be used to detect messenger RNA (mRNA), which indicates the presence of replicating virus and hence a current lyssavirus infection. This also has some utility in the potential differentiation of infection with the EBLV from a classic RABV infection.[68]

14.5.2 Antigen Detection

The direct fluorescent antibody test (FAT), developed nearly 40 years ago,[69] remains the World Health Organization (WHO) and the World Animal Health Organisation (OIE) gold standard test for lyssavirus antigen detection. It utilizes fluorescently labeled antibodies directed against the viral N protein. This antibody can be applied to clinical brain smears and *intravitam* skin biopsies or to detect virus in tissue culture. With these types of sample, it is able to give 95%–99% accurate diagnosis of rabies infection within hours, but the sensitivity of the test is affected by the region of the brain submitted and the quality of tissue. In addition to the standard FAT, the direct rapid immunohistochemical test (dRIT) also utilizes antibodies that target the virus nucleoprotein. In contrast to the FAT, however, the dRIT requires low-cost histological stains and can be read on a light microscope, making this method a financially viable alternative to the FAT in resource-limited settings.[70]

Another diagnostic test, the rapid rabies enzyme immunodiagnosis (RREID) test is able to detect RABV in brain tissue, even tissue that is partially decomposed, without the need for a microscope. However, the sensitivity of this test, compared to the FAT, is low.[71] Most recently, pen-side tests have been developed using lateral flow devices coated with RABV antigen-trapping antibodies.[72] These tests and other similar formats have now been tested in multiple settings and are appropriate for field testing of samples.[73,74]

14.5.3 Virus Isolation

The OIE guidance for a rabies diagnosis recommends an FAT on postmortem material and subsequent virus isolation to confirm the diagnosis. Historically, *in vitro* tissue culture isolation of virus was problematic as lyssaviruses rarely produce cytopathic effect observable under light microscope, though some strains produce large fusions of cells termed syncytia. Application of the anti-N antibody used in the FAT to tissue culture cells has enabled the development of the rabies tissue culture inoculation test (RTCIT) to detect antigen generated by replicating virus in tissue culture cells. The RTCIT can be used to detect virus antigen and ultimately the presence of live virus 2–4 days post infection.

The *in vivo* alternative to the RTCIT is the mouse inoculation test (MIT), which is the most widely used *in vivo* virus isolation technique. It involves homogenization of brain material and subsequent inoculation of homogenate intracerebrally into mice. The mice are then observed over a 28-day period for signs of clinical disease, and an FAT is performed following a 28-day incubation or on material following the development of clinical disease. With classic RABV infection, disease in some mice can be observed as early as 6 days post infection. However, there is substantial variation in onset of clinical disease with this test.[75] Also the test is time consuming and ethically challenging so the WHO has recommended that the MIT be replaced with an alternative *in vitro* isolation test, for example, the RTCIT, where *in vitro* virus isolation techniques have been validated within diagnostic laboratories.[76]

14.5.4 MOLECULAR TECHNIQUES FOR DETECTION OF VIRAL NUCLEIC ACID

The development of polymerase chain reaction (PCR) revolutionized diagnosis of pathogens and PCR is now widely used as a diagnostic technique for both postmortem and *intravitam* samples. However, due to the potential for contamination and the requirement of highly skilled staff and expensive equipment, currently, PCR techniques are not recommended for routine postmortem diagnosis by the WHO although these techniques are utilized in numerous laboratories that have these capabilities.[77] There are many advantages to the use of reverse transcriptase PCR (RT-PCR) as this can detect viral RNA in a range of biological samples, including skin biopsies and saliva.[78] Additionally, real-time PCR is able to differentiate between viral species in real time, which makes diagnosis rapid and accurate.[79] Sequencing of PCR products is also a useful tool for virus characterization as it is able to differentiate between all lyssavirus species and phylogenetically characterize isolates.[80]

14.5.5 SEROLOGICAL ASSAYS

The use of serological assays for the detection of a current rabies infection is redundant as lyssavirus infection is invariably fatal and postinfective antibodies are unlikely to be detected in easy to obtain samples, for example, blood prior to death. They are, however, very effective for monitoring vaccination status in susceptible populations.

The OIE recommends the rapid fluorescent focus inhibition test (RFFIT) and the fluorescent antibody virus neutralization (FAVN) test for monitoring of antibodies to allow both international travel in companion animals and evaluation of human vaccination. Both of these tests measure the ability of antibodies in suspect sera to neutralize live virus. Once the virus is applied to the sera, fluorescent antibody staining is used to detect any virus that has not been neutralized by the sera. In order to quantify the virus-neutralizing antibody, standardized reference sera from the WHO and OIE

are used as a comparison. These tests are widely considered as the gold standard in serological assays, but these have some logistical restrictions as the use of live virus requires containment facilities that may not be available in resource-limited settings. This has led to efforts to develop serological assays that do not require the use of live virus.

Novel enzyme-linked immunosorbent assay (ELISA) techniques are available that are capable of the detection of rabies-specific antibodies in serum from both humans and animals.[81] However, the current ELISAs have a major drawback in that they are unable to conclusively show the presence of neutralizing antibodies within a sample.

The development of lyssavirus pseudotype virus particles as an alternative neutralization assay to the FAVN has been reported. It is possible to express the glycoproteins of all currently known lyssaviruses from a lentiviral vector. The lentiviral gag–pol complex along with a suitable reporter gene, commonly luciferase, green fluorescent protein (GFP), or the gene encoding β-galactosidase, is cotransfected with a plasmid containing the lyssavirus glycoprotein to produce the pseudotype particles. The virus can be detected by assaying for the reporter gene activity.[82] These viruses are replication deficient, which means they are safe to use outside of containment, which makes them a promising approach for classic RABV and nonrabies lyssavirus detection, especially in areas where high security facilities are limited, for example, Africa.[83]

14.6 TREATMENT AND PREVENTION

14.6.1 TREATMENT

The lyssaviruses represent the only group of viruses for which, following the development of clinical disease, the outcome of infection is invariably fatal; however, a notable dichotomy between bat-derived infections is becoming more apparent. This feature of these viruses has made RABV one of the most feared infections in human history. Interestingly, the development of a medical intervention that ultimately involves the induction of a therapeutic coma has resulted in a small number of cases where individuals have survived infection, albeit with permanent neurological sequelae. This treatment, termed the Milwaukee protocol (Table 14.3) where success has been seen, appears to work by reducing the activity of the patient's nervous system, which is hypothesized to enable the immune system to catch up with the progress of the virus through the nervous system and to neutralize the spread of virus.[84] While initially heralded as a breakthrough in the battle against RABV, the procedure has so far had a low survival rate.[85]

14.6.2 PREVENTION

14.6.2.1 Human Intervention

One of the most important methods of intervention is vaccination. Historically, nerve tissue-based vaccines (NTVs) were used that were generated from desiccated brain and spinal cord material from infected animals.

TABLE 14.3

Basic Cocktail of Drugs Used in the Milwaukee Protocol and the Intended Function of Each Drug

Drug	Function
Ketamine	Inhibition of viral genome transcription inside host cells
Midazolam	Sedative to suppress brain activity
Phenobarbital	Sedative to suppress brain activity
Amantadine	Antiviral agent that blocks neuronal N-methyl-D-aspartate (NMDA) receptors preventing transport of the virus
Ribavirin	Antiviral agent thought to inhibit viral transcription and translation

Source: Willoughby, R.E., *Sci. Am.*, 296, 88, 2007.

The issue with these vaccines was neuroallergenic responses in vaccines though later, these live vaccines were inactivated with β-propiolactone to reduce reactogenicity.[86] Some reports suggest that these vaccines are still in use in some rabies-endemic countries[87] though the WHO recommended termination of the use of these vaccines in 1973. The advent of tissue culture techniques for the generation of vaccines led to the development of the human diploid cell vaccine (HDCV),[88] which involved adapting classic RABV strains to grow in human cell culture then subsequent inactivation. This vaccine was able to produce a highly protective immune response in most vaccinees but had high production costs, which inhibited its use in many of the developing countries that needed it most. Purified chick embryo cell vaccines (PCECVs) and purified Vero rabies vaccines (PVRVs) have replaced HDCV in areas where production costs preclude its use. The PVRV is now licensed across most of the world and is commonly used in Europe and Latin America.

These available vaccines can be used as preexposure immunization for individuals at high risk of infection including veterinarians, laboratory workers, and those traveling to rabies-endemic countries. The OIE and WHO specify that scientists working with RABV must have a minimum antibody titer of 0.5 IU/mL post immunization in order to be safe while working with the viruses. If the titer falls lower, a booster injection is required. Additionally, for other at-risk groups including bat workers, antibody tests and boosters are recommended every 3–5 years.

One potential drawback with the current RABV vaccines, however, is that they are unable to induce neutralizing antibody responses that can protect against infection with all of the lyssaviruses. Indeed, the current RABV vaccines for both human and animal use are based on inactivated preparations of phylogroup I rabies strains. Studies have shown that with these vaccines, little to no cross protection is afforded against phylogroup II or III viruses, meaning that for a number of lyssaviruses, there is no vaccine protection.[3,59,89] A number of studies have looked into viable cross protective vaccine strategies, for example, it has been shown using recombinant vaccinia viruses expressing the MOKV and WCBV G proteins that protection can be obtained for both viruses, both singly and in combination.[90] This shows that there may be potential to generate a pan-lyssavirus vaccine in the future.[59,91] In addition to pan-lyssavirus vaccines, there is also ongoing development of live attenuated vaccines for use in human postexposure prophylaxis (PEP) in order to avoid the need for multiple vaccinations before a protective titer is reached. An attenuated triple G variant of rabies has been generated, which could potentially be used for human PEP. The risk of the live vaccine reverting to virulence is negligible due to the inherent safety of the triple apathogenic G, which has been shown to be dominant over the pathogenic G in coinfection studies.[92] In the meantime, while these vaccines are under development, intradermal regimens are available, which utilize the existing rabies vaccines but use 60% less vaccine, and are safer and more practicable than intramuscular methods.[93]

Another feature of the existing rabies vaccines is that they can be used to protect against disease after exposure to the virus but before the development of clinical disease. The regimen used once an individual has experienced a potential exposure to virus is termed PEP. This requires that the intervention be administered as soon after the exposure as possible, to reduce the likelihood of the virus having established a productive infection. Initially, in the case of a bite or scratch, the WHO recommends washing the wound thoroughly with soap and water; in some cases, it is thought that this simple action can increase survival by 50%.[94] Once the wound has been thoroughly washed, it is advisable to soak the site in 70% ethanol or iodine. It is important that the wound is not sutured as the blood flow stimulated to the area may increase the likelihood of carrying virus away from the site of infection and toward nerve endings. The next step is to administer PEP. Rabies immunoglobulins are applied to the wound and any nearby mucous membranes. The immunoglobulins are obtained either from vaccinated human rabies immunoglobulin (HRIG) or from vaccinated equine rabies immunoglobulin (ERIG). Once the RIG has been applied, a shot of rabies vaccine is administered at a site distant from the wound so that the RIG does not neutralize the virus in the vaccine. There are two main regimens for administration of RIG: the Essen regimen, which is most popular in Europe and North America and consists of five intramuscular doses of vaccine, and the Zagreb 2-1-1 regimen, which is more common in some parts of Europe as the number of doses of vaccine is reduced compared to the Essen.[76] Despite PEP being effective in reducing incidence of disease, it is estimated that it is only used in 5% of cases due to its lack of availability, which is related to its almost prohibitive cost.[95] There are, however, efforts to develop more affordable antibody cocktails to replace RIG, including mouse-derived monoclonal antibody cocktails[96] and plant-derived antibodies (plantibodies).[95]

14.6.2.2 Animal Vaccines

The vaccines used in animals, mainly for dogs as these are the species most associated with human infection, have been derived in a similar way to human counterparts. The first vaccines were derived from neural tissue, but the reactogenicity of these vaccines drove the development of embryonic chicken egg-derived vaccines. These egg-derived vaccines were effective in adult dogs but could cause clinical rabies in young dogs, cats, and even cattle.[97] As well as vaccines based on fixed virus strains, there are a number of genetically engineered vaccines designed to control rabies in wildlife. A successful oral vaccine composed of vaccinia virus expressing the RABV glycoprotein helped reduce concerns over the safety of live attenuated vaccines.[98] In Canada, the United States, and across much of western Europe, the application of oral rabies vaccination campaigns has successfully eliminated terrestrial rabies, which has, in turn, reduced the number of human cases in these areas.[99] However, within the United States, RABV continues to circulate in certain wildlife populations, and the development of suitable bait formulations to try and eliminate these pockets of infection continues.

14.7 CONCLUSIONS AND FUTURE PERSPECTIVES

Rabies remains a significant threat to both human and animal populations across the globe. Despite the availability of vaccines and, in the event of exposure, PEP tools, the virus causes numerous human and untold animal fatalities annually. From a bioterror perspective, the mere mention of rabies is able to strike terror into the human population because of the historical association of the virus with human deaths[109]. Even in the event of eradication of rabies from terrestrial carnivore species, virus can still be propagated from bat species and introduced into areas from which it has been eradicated. The application of reverse genetics techniques to rabies and related viruses opens up the potential for the development of a virus that could spread readily by an aerosol route, which could have devastating effects. Furthermore, the identification and characterization of lyssaviruses that are not neutralized by antibodies generated by standard rabies vaccination highlights another potential area whereby these viruses could be used against humanity. Work is currently being done to generate pan-lyssavirus vaccines to protect against such possibilities and to generally protect those at occupational risk from these pathogens. Indeed, until suitable antiviral molecules are developed to combat these viruses, infection will remain invariably fatal in those exposed.

REFERENCES

1. Neville, J. Rabies in the ancient world. In *Historical Perspective of Rabies in Europe and the Mediterranean Basin* (eds. King, A.A., Fooks, A.R., Aubert, M., and Wanderler, A.I.), pp. 1–12. OIE Publications, Paris, France, 2004.
2. Blancou, J. and Wandeler, A. Rabies virus and its vectors in Europe. *Scientific and Technical Review of the Office International des Epizooties* 8, 927–929 (1989).
3. Badrane, H., Bahloul, C., Perrin, P., and Tordo, N. Evidence of two Lyssavirus phylogroups with distinct pathogenicity and immunogenicity. *Journal of Virology* 75, 3268–3276 (2001).
4. Hanlon, C.A. et al. Efficacy of rabies biologics against new lyssaviruses from Eurasia. *Virus Research* 111, 44–54 (2005).
5. Kuzmin, I.V., Wu, X., Tordo, N., and Rupprecht, C.E. Complete genomes of Aravan, Khujand, Irkut and West Caucasian bat viruses, with special attention to the polymerase gene and non-coding regions. *Virus Research* 136, 81–90 (2008).
6. Kuzmin, I.V. et al. Shimoni bat virus, a new representative of the *Lyssavirus* genus. *Virus Research* 149, 197–210 (2010).
7. Marston, D.A. et al. Complete genomic sequence of Ikoma Lyssavirus. *Journal of Virology* 86, 10242–10243 (2012).
8. Marston, D.A. et al. Ikoma lyssavirus, highly divergent novel lyssavirus in an African civet. *Emerging Infectious Diseases* 18, 664–667 (2012).
9. Warrell, M.J. and Warrell, D.A. Rabies and other lyssavirus diseases. *Lancet* 363, 959–969 (2004).
10. Dietzschold, B., Li, J., Faber, M., and Schnell, M. Concepts in the pathogenesis of rabies. *Future Virology* 3, 481–490 (2008).
11. Schnell, M.J., McGettigan, J.P., Wirblich, C., and Papaneri, A. The cell biology of rabies virus: Using stealth to reach the brain. *Nature Reviews Microbiology* 8, 51–61 (2010).
12. Rupprecht, C.E., Hanlon, C.A., and Hemachudha, T. Rabies re-examined. *The Lancet Infectious Diseases* 2, 327–343 (2002).
13. Constantine, D.G. Rabies transmission by nonbite route. *Public Health Reports* 77, 287–289 (1962).
14. Johnson, N., Phillpotts, R., and Fooks, A.R. Airborne transmission of lyssaviruses. *Journal of Medical Microbiology* 55, 785–790 (2006).
15. Fekadu, M. et al. Possible human-to-human transmission of rabies in Ethiopia. *Ethiopian Medical Journal* 34, 123–127 (1996).
16. Burton, E.C. et al. Rabies encephalomyelitis: Clinical, neuroradiological, and pathological findings in 4 transplant recipients. *Archives of Neurology* 62, 873–882 (2005).
17. Johnson, N., Brookes, S.M., Fooks, A.R., and Ross, R.S. Review of human rabies cases in the UK and in Germany. *Veterinary Record* 157, 715 (2005).
18. Javadi, M.A., Fayaz, A., Mirdehghan, S.A., and Ainollahi, B. Transmission of rabies by corneal graft. *Cornea* 15, 431–433 (1996).
19. Lafon, M. Rabies virus receptors. *Journal of Neurovirology* 11, 82–87 (2005).
20. Lewis, P., Fu, Y., and Lentz, T.L. Rabies virus entry at the neuromuscular junction in nerve-muscle cocultures. *Muscle and Nerve* 23, 720–730 (2000).
21. Thoulouze, M.I. et al. The neural cell adhesion molecule is a receptor for rabies virus. *Journal of Virology* 72, 7181–7190 (1998).
22. Gastka, M., Horvath, J., and Lentz, T.L. Rabies virus binding to the nicotinic acetylcholine receptor alpha subunit demonstrated by virus overlay protein binding assay. *Journal of General Virology* 77(Pt 10), 2437–2440 (1996).

23. Dechant, G. and Barde, Y.A. The neurotrophin receptor p75(NTR): Novel functions and implications for diseases of the nervous system. *Nature Neuroscience* 5, 1131–1136 (2002).

24. Muller-Rover, S., Peters, E.J., Botchkarev, V.A., Panteleyev, A., and Paus, R. Distinct patterns of NCAM expression are associated with defined stages of murine hair follicle morphogenesis and regression. *Journal of Histochemistry and Cytochemistry* 46, 1401–1410 (1998).

25. Botchkareva, N.V., Botchkarev, V.A., Chen, L.H., Lindner, G., and Paus, R. A role for p75 neurotrophin receptor in the control of hair follicle morphogenesis. *Developmental Biology* 216, 135–153 (1999).

26. Arredondo, J. et al. Central role of alpha7 nicotinic receptor in differentiation of the stratified squamous epithelium. *Journal of Cell Biology* 159, 325–336 (2002).

27. Tsiang, H., Ceccaldi, P.E., and Lycke, E. Rabies virus infection and transport in human sensory dorsal root ganglia neurons. *Journal of General Virology* 72(Pt 5), 1191–1194 (1991).

28. Jacob, Y., Badrane, H., Ceccaldi, P.E., and Tordo, N. Cytoplasmic dynein LC8 interacts with lyssavirus phosphoprotein. *Journal of Virology* 74, 10217–10222 (2000).

29. Raux, H., Flamand, A., and Blondel, D. Interaction of the rabies virus P protein with the LC8 dynein light chain. *Journal of Virology* 74, 10212–10216 (2000).

30. Ugolini, G. Use of rabies virus as a transneuronal tracer of neuronal connections: Implications for the understanding of rabies pathogenesis. *Development in Biologicals (Basel)* 131, 493–506 (2008).

31. Fishbein, D.B. Rabies in humans. In *The Natural History of Rabies* (ed. Baer, G.M.), pp. 519–549. CRC Press, Boca Raton, FL, 1991.

32. Warrell, M.J. Emerging aspects of rabies infection: With a special emphasis on children. *Current Opinion in Infectious Diseases* 21, 251–257 (2008).

33. Messenger, S.L., Smith, J.S., Orciari, L.A., Yager, P.A., and Rupprecht, C.E. Emerging pattern of rabies deaths and increased viral infectivity. *Emerging Infectious Diseases* 9, 151–154 (2003).

34. Botvinkin, A., Selnikova, O.P., Anotonova, L.A., Moiseeva, A.B., and Nesterenko, E.Y. New human rabies case caused from a bat bite in the Ukraine. *Rabies Bulletin Europe* 3, 5–7 (2006).

35. Fooks, A.R. et al. Case report: Isolation of a European bat lyssavirus type 2a from a fatal human case of rabies encephalitis. *Journal of Medical Virology* 71, 281–289 (2003).

36. Meredith, C.D., Prossouw, A.P., and Koch, H.P. An unusual case of human rabies thought to be of chiropteran origin. *South African Medical Journal* 45, 767–769 (1971).

37. Paweska, J.T. et al. Fatal human infection with rabies-related Duvenhage virus, South Africa. *Emerging Infectious Diseases* 12, 1965–1967 (2006).

38. Hanna, J.N. et al. Australian bat lyssavirus infection: A second human case, with a long incubation period. *Medical Journal of Australia* 172, 597–599 (2000).

39. Johnson, N., Fooks, A., and McColl, K. Human rabies case with long incubation, Australia. *Emerging Infectious Diseases* 14, 1950–1951 (2008).

40. Vignuzzi, M., Stone, J.K., Arnold, J.J., Cameron, C.E., and Andino, R. Quasispecies diversity determines pathogenesis through cooperative interactions in a viral population. *Nature* 439, 344–348 (2006).

41. Wunner, W.H. and Clark, H.F. Regeneration of DI particles of virulent and attenuated rabies virus: Genome characterization and lack of correlation with virulence phenotype. *Journal of General Virology* 51, 69–81 (1980).

42. Finke, S. and Conzelmann, K.K. Virus promoters determine interference by defective RNAs: Selective amplification of mini-RNA vectors and rescue from cDNA by a 3′ copy-back ambisense rabies virus. *Journal of Virology* 73, 3818–3825 (1999).

43. Perrault, J. Origin and replication of defective interfering particles. *Current Topics in Microbiology and Immunology* 93, 151–207 (1980).

44. Streicker, D.G. et al. Host phylogeny constrains cross-species emergence and establishment of rabies virus in bats. *Science* 329, 676–679 (2010).

45. Benmansour, A. et al. Rapid sequence evolution of street rabies glycoprotein is related to the highly heterogeneous nature of the viral population. *Virology* 187, 33–45 (1992).

46. Stougaard, E. and Ammendrup, E. Rabies in individual countries: Denmark. *Rabies Bulletin Europe* 22, 6 (1998).

47. Muller, T. et al. Spill-over of European bat lyssavirus type 1 into a stone marten (*Martes foina*) in Germany. *Journal of Veterinary Medicine Series B: Infectious Diseases and Veterinary Public Health* 51, 49–54 (2004).

48. Ronsholt, L. A new case of European bat lyssavirus (EBL) infection in Danish sheep. *Rabies Bulletin Europe* 26, 15 (2002).

49. Tjornehoj, K., Ronsholt, L., and Fooks, A.R. Antibodies to EBLV-1 in a domestic cat in Denmark. *Veterinary Record* 155, 571–572 (2004).

50. Leslie, M.J. et al. Bat-associated rabies virus in Skunks. *Emerging Infectious Diseases* 12, 1274–1277 (2006).

51. Daoust, P.Y., Wandeler, A.I., and Casey, G.A. Cluster of rabies cases of probable bat origin among red foxes in Prince Edward Island, Canada. *Journal of Wildlife Diseases* 32, 403–436 (1996).

52. Streicker, D.G. et al. Ecological and anthropogenic drivers of rabies exposure in vampire bats: Implications for transmission and control. *Proceedings of the Royal Society of London. Series B: Biological Sciences* 279, 3384–3392 (2012).

53. Wood, J.L. et al. A framework for the study of zoonotic disease emergence and its drivers: Spillover of bat pathogens as a case study. *Philosophical Transactions of the Royal Society of London B: Biological Science* 367, 2281–2292 (2012).

54. Rupprecht, C.E. et al. Can rabies be eradicated? *Developmental Biology (Basel)* 131, 95–121 (2008).

55. Fooks, A.R., Brookes, S.M., Johnson, N., McElhinney, L.M., and Hutson, A.M. European bat lyssaviruses: An emerging zoonosis. *Epidemiology and Infection* 131, 1029–1039 (2003).

56. Freuling, C.M. et al. Novel lyssavirus in Natterer's bat, Germany. *Emerging Infectious Diseases* 17, 1519–1522 (2011).

57. Botvinkin, A.D. et al. Novel lyssaviruses isolated from bats in Russia. *Emerging Infectious Diseases* 9, 1623–1625 (2003).

58. Horton, D.L. et al. Quantifying antigenic relationships among the lyssaviruses. *Journal of Virology* 84, 11841–11848 (2010).

59. Fooks, A. The challenge of new and emerging lyssaviruses. *Expert Review of Vaccines* 3, 333–336 (2004).

60. Aréchiga, N., Vázquez-Morón, S., Berciano, J., Nicolás, O., Aznar, C., Juste, J., Rodriguez, C., Aguilar, A., and Echevarria, J. Novel lyssavirus from a *Miniopterus schreibersii* bat in Spain. In *23rd Rabies in the Americas Meeting*, São Paulo, Brazil, October 14–19, 2012.

61. Kuzmin, I.V., Botvinkin, A.D., and Khabilov, T.K. The lyssavirus was isolated from a whiskered bat in northern Tajikistan. *Plecotus et al* 4, 75–81 (2001).

62. Familusi, J.B. and Moore, D.L. Isolation of a rabies related virus from the cerebrospinal fluid of a child with 'aseptic meningitis'. *The African Journal of Medical Sciences* 3, 93–96 (1972).

63. Tignor, G.H. et al. Duvenhage virus: Morphological, biochemical, histopathological and antigenic relationships to the rabies serogroup. *The Journal of General Virology* 37, 595–611 (1977).

64. Thornber, P., Rooney, J., Longbottom, H., and Crerar, S. Human health aspects of a possible lyssavirus in a black flying fox. *Communicable Diseases Intelligence* 20, 325 (1996).

65. Dietzschold, B., Schnell, M., and Koprowski, H. Pathogenesis of rabies. *Current Topics in Microbiology and Immunology* 292, 45–56 (2005).

66. Healy, D.M., Brookes, S.M., Banyard, A.C., Núñez, A., Cosby, S.L., and Fooks, A.R. Pathobiology of rabies virus and the European bat lyssaviruses in experimentally infected mice. *Virus Research* 172(1–2), 46–53 (2013).

67. Schaaf, J. Technic and reliability of microscopic diagnosis of rabies. I. Demonstration of Negri bodies in smears and sections. *Journal of Veterinary Medicine. Series B.* 15, 241–248 (1968).

68. Finnegan, C.J., Brookes, S.M., Johnson, L., and Fooks, A.R. Detection and strain differentiation of European bat lyssaviruses using in situ hybridisation. *Journal of Virological Methods* 121, 223–229 (2004).

69. Dean, D.J. and Abelseth, M.K. Laboratory techniques in rabies: The fluorescent antibody test. *Monograph Series. World Health Organ* 23, 73–84 (1973).

70. Durr, S. et al. Rabies diagnosis for developing countries. *PLoS Neglected Tropical Diseases* 2, e206 (2008).

71. Perrin, P., Rollin, P.E., and Sureau, P. A rapid rabies enzyme immuno-diagnosis (RREID): A useful and simple technique for the routine diagnosis of rabies. *Journal of Biological Standardization* 14, 217–222 (1986).

72. Kang, B. et al. Evaluation of a rapid immunodiagnostic test kit for rabies virus. *Journal of Virological Methods* 145, 30–36 (2007).

73. Markotter, W. et al. Evaluation of a rapid immunodiagnostic test kit for detection of African lyssaviruses from brain material. *Onderstepoort Journal of Veterinary Research* 76, 257–262 (2009).

74. Servat, A. et al. Evaluation of a rapid immunochromatographic diagnostic test for the detection of rabies from brain material of European mammals. *Biologicals: Journal of the International Association of Biological Standardization* 40, 61–66 (2012).

75. Johnson, N. et al. Isolation of a European bat lyssavirus type 2 from a Daubenton's bat in the United Kingdom. *Veterinary Record* 152, 383–387 (2003).

76. WHO. WHO Expert Consultation on rabies. World Health Organization Technical Report Series 931, pp. 1–88, Back Cover (2005).

77. Fooks, A.R. OIE Support for World Rabies Day. *OIE Bulletin* 3, 32 (2012).

78. Heaton, P.R., McElhinney, L.M., and Lowings, J.P. Detection and identification of rabies and rabies-related viruses using rapid-cycle PCR. *Journal of Virological Methods* 81, 63–69 (1999).

79. Wakeley, P.R. et al. Development of a real-time, TaqMan reverse transcription-PCR assay for detection and differentiation of lyssavirus genotypes 1, 5, and 6. *Journal of Clinical Microbiology* 43, 2786–2792 (2005).

80. Hoffmann, B. et al. Improved safety for the molecular diagnosis of classical rabies viruses using a TaqMan(C) real-time RT-PCR 'double check' strategy. *Journal of Clinical Microbiology* 48, 3970 (2010).

81. Servat, A. et al. A quantitative indirect ELISA to monitor the effectiveness of rabies vaccination in domestic and wild carnivores. *Journal of Immunological Methods* 318, 1–10 (2007).

82. Wright, E. et al. Investigating antibody neutralization of lyssaviruses using lentiviral pseudotypes: A cross-species comparison. *Journal of General Virology* 89, 2204–2213 (2008).

83. Wright, E. et al. A robust lentiviral pseudotype neutralisation assay for in-field serosurveillance of rabies and lyssaviruses in Africa. *Vaccine* 27, 7178–7186 (2009).

84. Willoughby, R.E. Jr. et al. Survival after treatment of rabies with induction of coma. *New England Journal of Medicine* 352, 2508–2514 (2005).

85. Willoughby, R.E. Jr. "Early death" and the contraindication of vaccine during treatment of rabies. *Vaccine* 27, 7173–7177 (2009).

86. Bugyaki, L., Costy, F., De Bruycker, M., and Marchal, A. Rabies in Belgium. *Archives of Belgium Medical Society* 37, 465–479 (1979).

87. Parviz, S., Chotani, R., McCormick, J., Fisher-Hoch, S., and Luby, S. Rabies deaths in Pakistan: Results of ineffective post-exposure treatment. *International Journal of Infectious Diseases* 8, 346–352 (2004).

88. Wiktor, T.J., Fernandes, M.V., and Koprowski, H. Cultivation of rabies virus in human diploid cell strain WI-38. *Journal of Immunology* 93, 353–366 (1964).

89. Hanlon, C.A., Niezgoda, M., Morrill, P.A., and Rupprecht, C.E. The incurable wound revisited: Progress in human rabies prevention? *Vaccine* 19, 2273–2279 (2001).

90. Weyer, J., Kuzmin, I.V., Rupprecht, C.E., and Nel, L.H. Cross-protective and cross-reactive immune responses to recombinant vaccinia viruses expressing full-length lyssavirus glycoprotein genes. *Epidemiology and Infection* 136, 670–678 (2008).

91. Evans, J.S., Horton, D.L., Easton, A.J., Fooks, A., and Banyard, A.C. Rabies virus vaccines: Is there a need for a pan-lyssavirus vaccine? *Vaccine* 30, 7447–7454 (2012).

92. Faber, M. et al. Effective preexposure and postexposure prophylaxis of rabies with a highly attenuated recombinant rabies virus. *Proceedings of the National Academy of Sciences of the United States of America* 106, 11300–11305 (2009).

93. Warrell, M.J. et al. A simplified 4-site economical intradermal post-exposure rabies vaccine regimen: A randomised controlled comparison with standard methods. *PLoS Neglected Tropical Diseases* 2, e224 (2008).

94. Rupprecht, C.E. and Gibbons, R.V. Clinical practice. Prophylaxis against rabies. *New England Journal of Medicine* 351, 2626–2635 (2004).

95. Both, L. et al. Passive immunity in the prevention of rabies. *The Lancet Infectious Diseases* 12, 397–407 (2012).

96. Muller, T. et al. Development of a mouse monoclonal antibody cocktail for post-exposure rabies prophylaxis in humans. *PLoS Neglected Tropical Diseases* 3, e542 (2009).

97. Bunn, T.O. Canine and feline vaccines, past and present. In *The Natural History of Rabies* (ed. Baer, G.M.). CRC Press, Boca Raton, FL, 1991.

98. Pastoret, P.P. et al. Experience with antirabies vaccination of foxes using the oral route coordinated among several European countries and perspectives on the use of recombinant vaccinia-rabies virus. *Parassitologia* 30, 149–154 (1988).

99. Krebs, J.W., Mandel, E.J., Swerdlow, D.L., and Rupprecht, C.E. Rabies surveillance in the United States during 2004. *Journal of the American Veterinary Medical Association* 227, 1912–1925 (2005).

100. US Centres for Disease Control and Prevention. Human rabies—Indiana and California, 2006. *Morbidity and Mortality Weekly Report* 56, 361–365 (2007).

101. Hemachudha, T., Sunsaneewitayakul, B., Desudchit, T., Suankratay, C., Sittipunt, C., Wacharapluesadee, S., Khawplod, P., Wilde, H., and Jackson, A.C. Failure of therapeutic coma and ketamine for therapy of human rabies. *Journal of Neurovirology* 12, 407–409 (2006).

102. Schmiedel, S., Panning, M., Lohse, A., Kreymann, K.G., Gerloff, C., Burchard, G., and Drosten, C. Case report on fatal human rabies infection in Hamburg, Germany, March 2007. *Euro Surveillance* 12, E070531 5 (2007).

103. McDermid, R.C., Saxinger, L., Lee, B., Johnstone, J., Gibney, R.T., Johnson, M., and Bagshaw, S.M. Human rabies encephalitis following bat exposure: Failure of therapeutic coma. *CMAJ* 178, 557–561 (2008).

104. van Thiel, P.P., de Bie, R.M., Eftimov, F., Tepaske, R., Zaaijer, H.L., van Doornum, G.J., Schutten, M., Osterhaus, A.D., Majoie, C.B., Aronica, E., Fehlner-Gardiner, C., Wandeler, A.I., and Kager, P.A. Fatal human rabies due to Duvenhage virus from a bat in Kenya: Failure of treatment with coma-induction, ketamine, and antiviral drugs. *PLoS Neglected Tropical Diseases* 3, e428 (2009).

105. Rubin, J., David, D., Willoughby, Jr. R.E., Rupprecht, C.E., Garcia, C., Guarda, D.C., Zohar, Z., and Stamler, A. Applying the Milwaukee protocol to treat canine rabies in Equatorial Guinea. *Scandinavian Journal of Infectious Diseases* 41, 372–375 (2009).

106. Hunter, M., Johnson, N., Hedderwick, S., McCaughey, C., Lowry, K., McConville, J., Herron, B. et al. Immunovirological correlates in human rabies treated with therapeutic coma. *Journal of Medical Virology* 82, 1255–1265 (2010).

107. Willoughby, R.E., A cure for rabies? *Scientific American*, 296, 88–95 (2007).

108. Picard-Meyer, E., Servat, A., Robardet, E., Moinet, M., Borel, C., and Cliquet, F. Isolation of Bokeloh bat lyssavirus in Myotis nattereri in France. *Archives of Virology* 158(11):2333–2340 (2013).

109. Fooks, A.R., Johnson, N., and Rupprecht, and C.E. Rabies. In: Vaccines for Biodefense and Emerging and Neglected Diseases, A.D.T. Barrett and L.R. Stanberry (eds.), Chapter 33, pp. 609–630 (2009).

15 Menangle Virus

Timothy R. Bowden

CONTENTS

15.1 INTRODUCTION

Menangle virus, a recent addition to the family *Paramyxoviridae*, subfamily *Paramyxovirinae*, was isolated in 1997 from stillborn piglets at a large commercial piggery in New South Wales, Australia, during the investigation of an outbreak of severe reproductive disease, which persisted from April to September of that year.[1] The index piggery, which housed 2,600 sows in four separate breeding units, was located approximately 60 km southwest of Sydney on a property that was adjacent to the Nepean River.[2,3] The disease was characterized by a reduction in both the farrowing rate and the number of live piglet births per litter, occasional abortions, and an increase in the proportion of mummified and stillborn piglets, some of which had deformities.[1,2,4] Although Menangle virus was only ever isolated from affected stillborn piglets, subsequent seroepidemiological investigations resulted in the implementation of a successful eradication program,[3] the identification of a likely natural host (fruit bats, also known as flying foxes, in the genus *Pteropus*),[1] and the unexpected realization that the virus had infected two piggery workers, causing severe influenza-like illness and a rash.[5]

15.2 CLASSIFICATION AND MORPHOLOGY

Paramyxoviruses are a diverse group of enveloped viruses that infect vertebrates, primarily mammals and birds, but also reptiles and fish. The family *Paramyxoviridae* is divided into two subfamilies, *Paramyxovirinae* and *Pneumovirinae*, the member species of which can be distinguished based on ultrastructure, genome organization, sequence relatedness of the encoded proteins, antigenic cross-reactivity, and

biological properties of the attachment proteins (presence or absence of hemagglutinating and neuraminidase activities).[6,7] *Pneumovirinae* contains two genera, *Pneumovirus* and *Metapneumovirus*, whereas *Paramyxovirinae* is divided into five genera, *Rubulavirus*, *Avulavirus*, *Respirovirus*, *Henipavirus*, and *Morbillivirus*, the type species of which are mumps, Newcastle disease, Sendai, Hendra, and measles viruses, respectively.

Following its original isolation and propagation in cell culture, Menangle virus was shown to be a member of the subfamily *Paramyxovirinae* by virtue of its ultrastructure as determined by electron microscopy (Figure 15.1).[1] Virions in this subfamily are typically 150–350 nm in diameter and are usually spherical in shape, although pleomorphic and filamentous forms are often observed.[6,7] Within each virion is the ribonucleoprotein core, a single strand of nonsegmented, negative-sense RNA, which is tightly bound along its entire length by nucleocapsid (N) proteins such that the genome is rendered insensitive to attack by nucleases.[8–10] The ribonucleoprotein complex, or nucleocapsid, is 19 ± 4 nm in diameter and has a helical symmetry with a pitch of 5.8 ± 0.4 nm,[1] thereby resulting in its characteristic "herringbone" appearance. Also associated with the nucleocapsid is the viral RNA-dependent RNA polymerase, which consists of two protein subunits, namely, L and P. Surrounding the core is a lipid envelope containing two surface glycoproteins, which are visualized by electron microscopy as an external fringe of projections 17 ± 4 nm long.[1] One glycoprotein—HN for rubulaviruses, avulaviruses, and respiroviruses, H for morbilliviruses, and G for henipaviruses—mediates attachment of the virion to the cellular receptor; the other glycoprotein, F, is responsible for mediating fusion between the virion

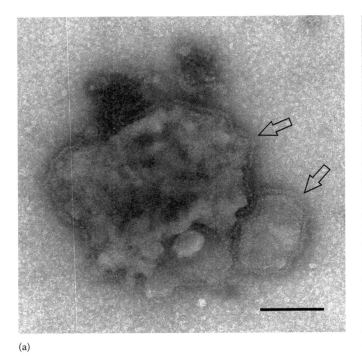

(a)

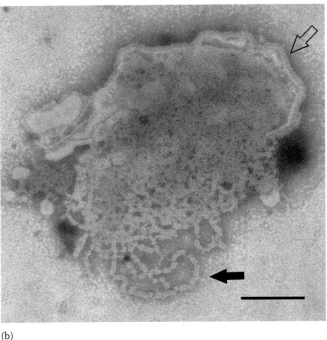

(b)

FIGURE 15.1 Transmission electron micrographs of Menangle virus negatively stained with 2% phosphotungstic acid (pH 6–8): (a) depicts an intact virion with single fringe of surface projections (open arrows) extending from the viral envelope; bar, 100 nm and (b) shows a disrupted virion, viral envelope with surface projections (open arrow), and ribonucleoprotein complex or nucleocapsid (solid arrow); bar, 100 nm. (Adapted from Philbey, A.W. et al., *Emerg. Infect. Dis.*, 4, 269, 1998. Electron micrographs kindly provided by the AAHL Biosecurity Microscopy Facility, CSIRO Animal, Food and Health Sciences, Australia.)

envelope and the cell plasma membrane, a process that delivers the viral genome into the cell cytoplasm where viral transcription and replication occur.[7] Additionally, located on the inner surface of the viral envelope is a nonglycosylated matrix (M) protein, which is thought to play a central role in coordinating virion assembly and budding.[7]

Characterization and analysis of the complete genome sequence (15,516 nucleotides), and the encoded proteins, subsequently established that Menangle virus should be classified as a new member of the genus *Rubulavirus*.[11,12]

15.3 BIOLOGY AND EPIDEMIOLOGY

As for most other members of the genus *Rubulavirus*, there are six transcriptional units, in the order 3′-N-V/P-M-F-HN-L-5′, which are separated by intergenic regions of variable length and sequence (Figure 15.2). Although believed to transcribe each gene sequentially from the 3′ end of the

genome, the viral polymerase does not always reinitiate RNA synthesis at every gene junction following release of the newly synthesized mRNA. Instead, the polymerase detaches from the genome template, at which point it must reenter the genome at its 3′ terminus for further transcription to take place. Consequently, an mRNA gradient is produced such that genes located closest to the 3′ terminus of the genome are expressed at greater levels than their downstream neighbors.[7,13,14]

In contrast to the N, M, F, HN, and L genes, all of which encode single proteins using individual open reading frames, the V/P gene encodes at least two different proteins from two or more overlapping reading frames using a mechanism known as RNA editing, during which the polymerase adds one or more nontemplated G residues, cotranscriptionally, to a percentage of the newly synthesized mRNAs.[7] As such, two or more different transcripts are synthesized, which code for proteins that possess common N-termini

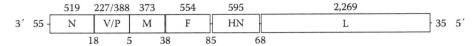

FIGURE 15.2 Genome organization of Menangle virus. The full-length genome (15,516 nucleotides) comprises six genes, each of which is drawn to scale. Numbers above each gene indicate the length, in amino acids, of the encoded proteins. Numbers below the gene boundaries indicate the length, in nucleotides, of the intergenic regions. Numbers immediately to the left and right of the genome indicate the length, in nucleotides, of the extragenic leader and trailer sequences, respectively. Note that unedited V/P gene transcripts encode the V protein, whereas insertion of two nontemplated G residues at the editing site (not shown) results in synthesis of the P gene mRNA. Note also that the negative-sense genome is depicted rather than the complementary antigenome, which is in the same sense as the viral mRNA molecules. GenBank accession number: AF326114.

but different C-termini, resulting from a translational shift downstream of the editing site. For Menangle virus, as for other rubulaviruses, faithful gene transcripts encode the V protein, whereas the predominant editing event (the insertion of two nontemplated G residues at the editing site) yields the P mRNA, which codes for the P protein.[11] Significantly, many of the *Paramyxovirinae* accessory gene products, which are generated using a variety of distinct mechanisms, have important roles such as regulating viral RNA synthesis, mediating virulence *in vivo*, and/or counteracting the induction of an interferon-mediated antiviral state by host cells following infection.[7]

Several novel findings were also apparent, the most significant of which would appear to be the limited sequence homology of the deduced Menangle virus HN protein when compared to attachment proteins of other *Paramyxovirinae*.[11] Although the predicted structure of Menangle virus HN was most similar to other rubulavirus HN proteins, it lacked the majority of the amino acids that are considered critical determinants of both sialic acid binding and hydrolysis. This unexpected finding is unique among all known rubulavirus, avulavirus, and respirovirus HN proteins, with the exception of that from the closely related Tioman virus,[15] another rubulavirus that was later isolated from fruit bats in Malaysia during the search for the natural reservoir of Nipah virus.[16] This lack of conservation in functional amino acids suggests that the Menangle and Tioman virus attachment proteins, in contrast to those from all other known members of the rubulavirus, avulavirus, and respirovirus genera, are unlikely to use sialic acid as a cellular receptor, a prediction which in part would explain the observation, made at the time of its initial isolation, that Menangle virus is nonhemadsorbing and nonhemagglutinating using erythrocytes from various species.[1] The uniqueness of the Menangle and Tioman virus HN proteins is further highlighted by the apparent marked divergence in their evolutionary development in comparison to the N, P, M, F, and L proteins, the significance of which is not yet understood.[11,12,15]

The isolation of Menangle virus in cell culture provided a means of detecting Menangle virus-specific neutralizing antibodies in sera, thus enabling seropositive animals to be identified. Sera from pigs that had been collected during and prior to 1996, as well as samples collected as late as February 1997, before the onset of clinical disease, were all seronegative for antibodies against the new virus. However, 58 of 59 samples collected from 3 of the 4 breeding units in late May, immediately preceding the onset of disease in these units, had high neutralizing antibody titers (256–4,096) against Menangle virus.[3] Subsequent epidemiological investigations using archival sera collected during separate national surveys for unrelated infectious disease agents, trace forward and trace back piggeries, and pigs with and without reproductive disease elsewhere in Australia, revealed that infection with Menangle virus was confined to the index piggery and two associated piggeries, also within New South Wales, which were located several hundred kilometers away at Young and Trunkey Creek.[3] Neither of these piggeries was a breeding

establishment. Instead, each was used for rearing batches of 12- to 14-week-old pigs, originating from the index piggery, until slaughter at 24 weeks of age.

Although the reproductive parameters had returned to normal by mid-September 1997, two separate cross-sectional studies, the first conducted during August and September 1997, the other in March 1998, subsequently revealed that the virus was being maintained in the affected piggeries by infection of successive batches of growing pigs as they lost their protective maternal antibody at about 12–14 weeks of age.[3] Whereas colostral antibodies had declined to almost undetectable levels by this age, 95% of pigs subsequently developed neutralizing antibody titers $\geq$128 by the time of slaughter, at 24–26 weeks of age, or by the time they entered the breeding herd as replacement gilts at 28 weeks of age.[3] Significantly, this immunity appeared to be protective because no further reproductive disease was seen after September 1997 and sows, which had produced affected litters during the outbreak, farrowed normal litters subsequently.[2,3]

Menangle virus was only ever isolated from the brain, heart, and lungs of affected stillborn piglets.[1,2] Thus, it was not determined how the virus was spread between pigs. However, serological monitoring of various groups of sentinel pigs provided circumstantial evidence regarding likely modes of transmission. Sentinel pigs placed into the grower and grower–finisher areas of one of the units during September and October 1997, respectively, when infection was known to be active, took up to 1 month to seroconvert, despite the fact that the sentinel pigs in the grower–finisher area were allowed free access to walkways between pens to maximize their exposure to other pigs.[3] This suggested that transmission between pigs was relatively slow and required close contact, and thus probably resulted from viral excretion in feces or urine rather than in respiratory aerosols.[3] Furthermore, 180 gilts and sows, which had been introduced into the breeding herd in November 1997, were seronegative to Menangle virus when tested in February 1999. This finding demonstrated that Menangle virus infection was not active in the breeding herd after the period of reproductive disease and that persistent shedding did not occur.[3] Finally, a large group of 8-week-old pigs that was moved into an uncleaned building, which until 3 days earlier was occupied by pigs known to be infected with Menangle virus, was seronegative at slaughter age, suggesting that survival of the virus in the environment was not prolonged and that maternally derived antibodies were protective in weaned pigs less than 8 weeks of age.[3] The realization that active Menangle virus infection was confined to the grower–finisher age group enabled a strategy for virus eradication to be developed. Implementation of this program, together with depopulation of the piggery at Young, resulted in successful elimination of the virus from the remaining two piggeries by February 1999.[3]

Determining the probable source of the outbreak was facilitated by the fact that fruit bats had previously been identified as the likely natural host of two newly described zoonotic viruses, Hendra virus,[17] a paramyxovirus that had caused the death of two humans in Queensland within the

3 years preceding the discovery of Menangle virus,[18–23] and Australian bat lyssavirus, a rhabdovirus that, although first identified in a sick fruit bat,[24–26] had subsequently caused the death of a bat carer.[27,28] A large breeding colony of 20,000–30,000 fruit bats, which consisted primarily of grey-headed fruit bats (*Pteropus poliocephalus*), but also little red fruit bats (*Pteropus scapulatus*), roosted annually, from October to April, within 150–200 m of the nearest breeding unit, and had done so for at least 30 years.[1,3] Serum samples were initially collected from this colony, as well as from fruit bats elsewhere in New South Wales and Queensland. Forty-two of 125 samples were positive in a virus neutralization test with titers against Menangle virus ranging from 16 to 256.[1] Considering that antibodies were not detected in a range of wild and domestic species found within the vicinity of the piggery, including rodents, birds, cattle, sheep, cats, and a dog, fruit bats were implicated as the likely natural host of Menangle virus.[1] More comprehensive serological surveys have since demonstrated that antibodies to Menangle virus are widespread in at least three of the four pteropid species endemic to Australia.[3,29] However, despite finding paramyxovirus-like particles in fruit bat feces collected from the colony nearby the index piggery, attempts to isolate Menangle virus from these samples were unsuccessful.[3,29] Nevertheless, following the discovery of Menangle virus, two other paramyxoviruses were identified in Malaysia, namely, Nipah and Tioman viruses, both of which were isolated from fruit bat urine.[16,30] Using similar sampling procedures, and newly established primary cell lines of bat origin, Menangle virus has now also been isolated from fruit bat (*Pteropus alecto*) urine samples that were collected in southeast Queensland.[31] It would therefore seem probable that transmission of Menangle virus from fruit bats to pigs occurred indirectly, perhaps through ingestion of fruit bat excreta, and this may have been associated with the movement of pigs between buildings using uncovered walkways, a management practice that had only been implemented 2 years before the outbreak.[3]

Prior to the isolation of Menangle virus from pigs, only two paramyxoviruses were considered to be zoonotic, namely, Newcastle disease virus, which in humans can cause conjunctivitis,[32,33] and Hendra virus, which was known to have infected three people, two of whom subsequently died.[18–23] Considering that the relationship between Hendra and Menangle viruses had not been clearly established, but that both were considered to have "jumped" from fruit bats to infect other species (horses and humans in the case of Hendra virus), a serological survey was conducted to assess the zoonotic potential of the new virus. A total of 251 people, all of whom were considered to have been exposed to potentially infected pigs, were tested. This included all staff at the index ($n = 33$) and grower ($n = 5$) piggeries, abattoir workers ($n = 142$), researchers and animal handlers ($n = 41$), veterinarians and pathology laboratory workers ($n = 24$), and others ($n = 6$).[5] Two piggery staff, one from the index piggery and the other from one of the two associated grower piggeries, were shown to be seropositive to Menangle virus with neutralizing antibody titers of 128 and 512, respectively.[5] Further

investigations revealed that both workers had contracted very similar illnesses in early June 1997, during the period when reproductive disease in pigs was present. The first worker had regular prolonged contact with farrowing pigs, which resulted in splashing of amniotic fluid and blood to his face, and he also frequently received minor wounds to his hands and forearms.[5] Although the second worker had no exposure to birthing pigs, he performed necropsies on pigs without wearing protective eyewear or gloves and reported that exposure to pig secretions, including feces and urine, was common.[5] Neither worker had any contact with bats. Extensive testing of both men in September 1997 failed to identify an alternative cause for the illnesses, which were therefore considered to have resulted from infection with Menangle virus. Although the mode of transmission from pigs to humans was not established, the available evidence suggested that Menangle virus did not readily transmit to people and that infection probably required parenteral or permucosal exposure to infectious materials.[5]

Experimental infection of weaned pigs, by intranasal inoculation, has since demonstrated that Menangle virus is shed in nasal and oral secretions, feces, and urine, typically for periods of less than 1 week, following an incubation period of only a few days.[34] Although the stability of Menangle virus in these samples was not evaluated, these findings suggested that transmission is possible not only by direct contact between pigs, by exposure to infectious nasal and oral secretions, but also by indirect means, following contamination of the environment, food, and water with virus. Furthermore, the presence of infectious virus in nasal and oral secretions indicates that spread by aerosol droplet might also play a role in transmission. Nevertheless, the low concentrations of virus shed in each sample type would appear to correlate with the slow rate of spread observed in the index piggery, which led to the hypothesis that Menangle virus was more likely to be spread by fecal or urinary excretion than by respiratory aerosols.[2,3]

15.4 CLINICAL FEATURES AND PATHOGENESIS

15.4.1 Pigs

In naive pigs, infection with Menangle virus causes a marked decline in reproductive performance, characterized by a reduction in both the farrowing rate and the number of live piglet births per litter, occasional abortions, and an increase in the proportion of mummified and stillborn piglets, some of which have deformities; there are otherwise no clinical signs attributable to infection with the virus in postnatal, growing, or adult pigs of any age.[1–4]

The first affected litter, detected during the week commencing April 21, 1997, consisted of one stillborn piglet and seven mummified fetuses.[2] The proportion of affected litters (those containing fewer than six live born piglets) in this unit increased significantly in the following weeks and peaked at 64.3% 5 weeks later, at which time only 17 of 45 mated sows farrowed.[2] During the period of disease in this unit, which lasted for 15 weeks, the farrowing rate decreased from a pre-outbreak average of 80.2%–63.2% and the number of piglets born alive per litter decreased from an average of 9.6 ± 2.9 to 8.3 ± 3.9.[2] The disease occurred sequentially in the remaining three breeding units, commencing 7, 8, and 11 weeks after the disease was initially recognized in the first affected unit, and continued for periods of 12, 12, and 11 weeks, respectively. The reproductive performance in all four units had returned to normal by mid-September 1997, 7 months after the virus was considered to have first entered the breeding herd.[2]

The composition of affected litters varied markedly throughout the Menangle virus outbreak. Some litters contained only one or two mummified or stillborn fetuses, suggesting that infection had occurred early during gestation, with subsequent death and resorption of the embryos.[2] In contrast, others contained a mixture of mummified and stillborn piglets in various stages of development, as well as apparently normal fetuses. Like porcine parvovirus,[35] this suggested that transplacental infection of fetuses was somewhat variable, and that subsequent spread of the virus from infected to neighboring fetuses, in utero, was possible. The size of mummified fetuses, and the stage of development at which pathology was induced, also indicated that infection prior to day 70 of gestation was required for fetuses to be adversely affected,[2] which is consistent with other infectious causes of reproductive failure in swine.

Pathologic changes in affected stillborn piglets were characterized by severe degeneration, or even absence, of the brain and spinal cord (particularly the cerebral hemispheres, cerebellum, and brain stem), arthrogryposis, brachygnathia, and kyphosis,[1,2,4] although the pattern of central nervous system lesions observed within affected litters appeared to change over time.[4] Occasionally, marked pulmonary hypoplasia was observed and, in 41% of stillborn piglets examined, an excessive amount of straw-colored fluid, sometimes fibrinous, was present in body cavities.[1,4] Histologically, pathology was most marked in the central nervous system and was characterized by extensive degeneration and necrosis of grey and white matter, in association with infiltrates of macrophages and lymphocytes.[1,4] In some lesions within the brain and spinal cord, intranuclear and intracytoplasmic inclusion bodies were present in neurons and, occasionally, nonsuppurative myocarditis and leptomeningitis were also observed.[1,4]

Following experimental challenge of weaned pigs intranasally, a viremia of short duration (lasting only a matter of days) and low titer was detected.[34] Menangle virus was shed at low to moderate concentrations in nasal and oral secretions, feces, and urine, typically for periods of less than 1 week. Infectious virus and viral RNA were detected in a wide range of solid tissues, including tonsil, mandibular lymph node, jejunum, ileum, and mesenteric lymph node, indicating that secondary lymphoid organs and gastrointestinal tissues were major sites of viral replication and dissemination. This was subsequently confirmed by abundant immunolabeling of viral nucleocapsid protein in lymphoid tissues and throughout the intestinal epithelium, particularly within jejunum, which would explain the presence of virus in feces.[34]

15.4.2 Humans

Only two people are suspected to have been infected with Menangle virus, both of whom experienced a severe influenza-like illness with a rash. In each case, the diagnosis was made retrospectively after demonstrating that the men had high neutralizing antibody titers to the virus, with no other likely cause identified. The first patient experienced sudden onset of malaise and chills, followed by drenching sweats and fever, and was confined to bed with severe headaches and myalgia for 10 days.[5] By the fourth day, he had developed a spotty red rash and, on examination, exhibited upper abdominal tenderness, lymphadenopathy, and a rubelliform rash. He was absent from work for 14 days, tired easily on his return, and reported weight loss of 10 kg (~22 lb) due to this illness. The second patient experienced fever, chills, rigors, drenching sweats, marked malaise, back pain, severe frontal headache of 4–5 days duration, and photophobia.[5] A spotty, red, nonpruritic rash developed on his torso 4 days after the illness started and persisted for 7 days. He had essentially recovered after 10 days, having lost 3 kg (~6.6 lb).[5] Neither patient had cough, vomiting, or diarrhea.

15.4.3 Fruit Bats

Infection of pteropid bats with Menangle virus has not been associated with disease,[29] and this is consistent with the perceived role of bats as reservoir hosts of not only this and several other paramyxoviruses that have emerged in the Australasian region since 1994, namely, Hendra, Nipah, Tioman, and Cedar viruses, but also many other viruses from a diverse spectrum of viral families.[31,36–39]

15.5 LABORATORY DIAGNOSIS

Suspicion or confirmation of Menangle virus infection must be reported to relevant government authorities. In the United States, Menangle virus is listed as a US Department of Agriculture select agent on the National Select Agent Registry. Although the recommended sample types for confirming disease in humans will need to be guided by future characterization of the infection in people, should this opportunity arise, multiple laboratory tests have been developed and used, both during and since the outbreak, to increase understanding of the epidemiology and pathogenesis of Menangle virus, principally in pigs.

15.5.1 Conventional Techniques

Virus isolation: Menangle virus will grow in a wide range of cell types, including cells of porcine, human, and pteropid bat origin, and possesses neither hemadsorbing nor hemagglutinating activities using erythrocytes from several species.[1,31,34] The cytopathic effect is similar to that caused by other paramyxoviruses and is characterized by vacuolation of cells, formation of syncytia, and, eventually, lysis of the cell monolayer. During the outbreak, however, the virus was only isolated from lung ($n = 9$), brain ($n = 9$), and heart ($n = 5$) of 10 affected, stillborn piglets, after 3–5 blind passages in baby hamster kidney (BHK_{21}) cells.[1,2] Considering that 170 tissues (including lung, brain, heart, kidney, and spleen from 57 affected piglets) were tested, the virus was isolated from fewer than 1 in 7 samples. It has since been determined that virus is more likely to be isolated from piglets that exhibit gross or histological pathology of the central nervous system.[4]

Serology: Detection of Menangle virus-specific antibodies is undertaken using the virus neutralization test.[2,34] This test requires the use of live virus and cell culture and routinely takes 5–7 days to complete. Neutralizing antibody titers >16 are considered positive in humans and pigs,[2,3] whereas titers ≥8 are considered positive in fruit bats.[3,29] The development of a prototype enzyme-linked immunosorbent assay (ELISA), based on recombinant Menangle virus nucleocapsid proteins, offers the future prospect of a rapid and convenient serological test with no requirement for infectious reagents.[40]

Immunohistochemistry: In formalin-fixed pig tissues, the demonstration of viral antigen in lesions can assist with diagnosis, either at the time of an outbreak investigation or retrospectively.[41] Following experimental infection of weaned pigs with Menangle virus, by intranasal inoculation, viral antigen was readily detected in lymphoid tissues and intestinal mucosa, and in limited amounts in other tissues, using a rabbit hyperimmune serum raised against purified recombinant viral nucleocapsid protein.[34]

Electron microscopy: Negative contrast[1] or ultrathin section[42] electron microscopical analysis of virus propagated in cell culture reveals the presence of pleomorphic enveloped virions and nucleocapsids, the ultrastructures of which are consistent with viruses in the family *Paramyxoviridae*, subfamily *Paramyxovirinae*.

15.5.2 MOLECULAR TECHNIQUES

15.5.2.1 Real-Time Reverse Transcriptase-Polymerase Chain Reaction

During the Menangle virus outbreak, molecular tests such as the reverse transcriptase-polymerase chain reaction (RT-PCR) assay were not available to assist with disease investigation and control activities. Several attempts to amplify segments of the Menangle virus genome using primers based on conserved regions of other viruses in the subfamily *Paramyxovirinae* were unsuccessful and, it was not until later, when a PCR-based cDNA subtraction strategy was used to isolate, clone, and sequence viral cDNAs from Menangle virus-infected cells, that the viral genome sequence was determined.[11,12] This led to the development of a multiplexed real-time RT-PCR TaqMan assay that, to date, has only been evaluated using clinical specimens (blood and solid tissues) collected from experimentally infected pigs.[34] The virus-specific primers and probe were designed to amplify and detect a short cDNA target of 66 nucleotides in length (corresponding to nucleotides 478–543 of the viral antigenome, within the N gene of Menangle virus). Their sequences are

as follows: MP80F (5′-CGGATTTGAGCCTGGTACGT-3′); MP81R (5′-ACCTCTCCATTTGTCATCGGA-3′); and MT01-FAM (5′-FAM-TTCTCGCATTTGCCCTTAGCCGG-TAMRA-3′). Using random hexamer primers to initiate cDNA synthesis enables viral genome, antigenome, and N mRNA (the most abundant viral transcript) to be amplified and detected, thereby maximizing test sensitivity. Simultaneous detection and quantitation of an endogenous control, 18S rRNA, offers a means of excluding the presence of RT or PCR inhibitors in every sample tested, thereby providing a convenient and simple method of excluding false-negative results. The 18S rRNA primers and probe, the sequences of which are proprietary, are supplied in the TaqMan ribosomal RNA control reagents kit (Applied Biosystems). Notably, however, despite the N genes of the pig and bat isolates of Menangle virus sharing 95% nucleotide identity, two mutations in the 3′ terminal region of the reverse primer would prevent the newly identified bat variant from being detected by this assay. Development of additional virus-specific real-time RT-PCR assays will therefore be required to ensure that the currently known pig[11,12] and bat[31] isolates of Menangle virus are detectable.

In experimentally infected pigs, viral RNA was detected in blood (for up to 21 days following challenge) and in solid tissues, including jejunum, ileum, nasal mucosa, lung, duodenum, colon, cecum, rectum, and spleen, all of which could conceivably be used to confirm recent infection with Menangle virus.[34] However, in a disease investigation scenario, brain, lung, and heart from affected stillborn piglets would be the samples of choice, since these are the tissues from which infectious virus was isolated during the outbreak in 1997.[1,2] Prior to the development of neutralizing antibodies in naive pigs, it is also likely that infectious virus would be present in nasal and oral secretions, as well as in feces and urine,[34] although the kinetics of viral RNA shedding in these samples has not been evaluated.

In a diagnostic sense, several significant advantages of real-time PCR technologies are readily apparent, compared to conventional PCR: there is no requirement for any post-PCR processing, such as agarose gel electrophoresis or DNA sequencing, to confirm the identity of amplicons, and this greatly enhances the efficiency of sample testing; the sensitivity of the system is at least as good as conventional nested PCR tests; detection is performed in a closed-tube system, which greatly reduces the risk of carryover contamination, either by previously amplified products or by samples containing high endogenous concentrations of target sequences; the methodologies are readily automated to increase sample throughput, if required; and finally, because all TaqMan assays conform to universal cycling conditions, multiple assays targeting different pathogens can be run simultaneously in the same 96- or 384-well plate. Other improvements include hot-start modifications to increase fidelity[43] and the inclusion of uracil N-glycosylase[44] for the prevention of carryover contamination. Nevertheless, separation of work space into designated areas exclusively for reagent preparation, sample processing (nucleic acid extraction), PCR amplification, and post-PCR sample analysis, should this be required, is always recommended.[45]

15.6 TREATMENT AND PREVENTION

There are no vaccines or therapeutics currently available, or under development, for use in pigs, humans, or fruit bats. Fortunately, it would appear that spillover of Menangle virus from the reservoir host to pigs is a rare event, with only one such occurrence documented to date.[1–5] In addition, it would seem that transmission to humans likely requires intense occupational exposure to recently infected pigs, including contamination of cuts and abrasions or mucosal surfaces, such as conjunctivae, with infectious materials.[5] Despite the lack of knowledge regarding the ecology of the virus, basic controls that restrict direct and indirect exposure of fruit bats to pigs should prevent future disease outbreaks in intensively managed piggeries that have effective biosecurity measures in place. An understanding of how and when the virus is shed from the natural host will help to define the risk factors associated with spillover to pigs, including extensively farmed and feral species. Increased awareness and recognition of Menangle virus as a potential cause of serious reproductive disease in swine, and influenza-like illness, with a rash, in humans, should also ensure its prompt consideration in the differential diagnosis of similar disease episodes in the future.

15.7 CONCLUSIONS AND FUTURE PERSPECTIVES

Menangle virus is one of five novel paramyxoviruses of fruit bat origin to be identified in the Australasian region within the last two decades. However, while it has caused disease in people, it is currently not in the same category as Hendra and Nipah viruses (see Chapter 9), which have caused serious illness and multiple fatalities in humans. Isolated in 1997 from stillborn piglets at a large commercial piggery in New South Wales, Australia, Menangle virus is a newly described cause of serious reproductive disease in pigs and may also cause severe influenza-like illness, with a rash, in humans. Although successfully eradicated from the affected piggeries within 2 years of entering the breeding herd of the index farm, many of the details relating to its pathogenesis and epidemiology, not only in pigs, but particularly in fruit bats and humans, remain unknown. Menangle virus probably remains endemic in fruit bats, the likely natural host of the virus, and therefore a continuing risk of reinfection exists.

Also intriguing is the existence, in Malaysia, of Tioman virus, which cross-reacts with Menangle virus-specific antisera[16,46] and is closely related genetically.[11,12,15,16] Considering that antibodies to Tioman virus, or a Tioman-like virus, have since been detected in fruit bats in Madagascar,[47] novel rubulaviruses have been identified in, and isolated from, fruit bats in China[48] and Ghana,[49,50] and an estimated 66 new paramyxoviruses have been discovered in a worldwide surveillance study targeting 119 bat and rodent species,[38] it would appear that related viruses with an unknown propensity to cause disease in domestic animal species or humans are far more abundant and widespread than previously thought.

Although Tioman virus has not been associated with disease in pigs or humans, a recent study has established that weaned pigs are susceptible to experimental infection.[51] Whether the virus is pathogenic in pregnant pigs, however, is not known. Furthermore, low neutralizing antibody titers to Tioman virus, or a closely related virus, have been detected in sera from 3 of 169 villagers on the island from which it was originally isolated, suggesting possible prior infection by this or a similar virus.[52] In the absence of any identifiable link to pigs, it was postulated that human infection might have occurred through consumption of fruit that had been partially eaten by bats, a potential risk factor common to 32 of the surveyed villagers, including 2 of the 3 who had detectable neutralizing antibodies to the virus.[52] It would not seem unreasonable to presume, therefore, that direct transmission of Menangle virus from fruit bats to people, such as wildlife carers, might be possible without prior amplification in pigs.

Many fundamental questions remain to be answered, such as how Menangle virus is maintained in fruit bats and over what geographic range it should be considered endemic, as well as which cellular receptor is used for viral attachment and how it relates to pathogenesis in each susceptible host species. Obtaining answers to complex questions such as these will be challenging and will require a multidisciplinary approach. However, until such time as Menangle virus infection in humans is better characterized, the true zoonotic potential of this recently emerged paramyxovirus will remain unclear.

The author would like to acknowledge that certain parts of this chapter, including the figures, were previously published in *Molecular Detection of Human Viral Pathogens*, 2010, edited by Dongyou Liu, CRC Press, Boca Raton, FL. With permission. ISBN: 9781439812365.

REFERENCES

1. Philbey, A.W. et al., An apparently new virus (family *Paramyxoviridae*) infectious for pigs, humans, and fruit bats, *Emerg. Infect. Dis.*, 4, 269, 1998.
2. Love, R.J. et al., Reproductive disease and congenital malformations caused by Menangle virus in pigs, *Aust. Vet. J.*, 79, 192, 2001.
3. Kirkland, P.D. et al., Epidemiology and control of Menangle virus in pigs, *Aust. Vet. J.*, 79, 199, 2001.
4. Philbey, A.W. et al., Skeletal and neurological malformations in pigs congenitally infected with Menangle virus, *Aust. Vet. J.*, 85, 134, 2007.
5. Chant, K. et al., Probable human infection with a newly described virus in the family *Paramyxoviridae*, *Emerg. Infect. Dis.*, 4, 273, 1998.
6. Wang, L.F. et al., Family *Paramyxoviridae*, in *Virus Taxonomy: Classification and Nomenclature of Viruses. Ninth Report of the International Committee on Taxonomy of Viruses*, eds. A.M.Q. King et al. (Elsevier/Academic Press, San Diego, CA, 2012), p. 672.
7. Lamb, R.A. and Parks, G.D., *Paramyxoviridae*: The viruses and their replication, in *Fields Virology*, eds. D.M. Knipe et al. (Lippincott Williams & Wilkins, Philadelphia, PA, 2007), p. 1449.

8. Kingsbury, D.W. and Darlington, R.W., Isolation and properties of Newcastle disease virus nucleocapsid, *J. Virol.*, 2, 248, 1968.

9. Compans, R.W. and Choppin, P.W., The nucleic acid of the parainfluenza virus SV5, *Virology*, 35, 289, 1968.

10. Heggeness, M.H., Scheid, A., and Choppin, P.W., Conformation of the helical nucleocapsids of paramyxoviruses and vesicular stomatitis virus: Reversible coiling and uncoiling induced by changes in salt concentration, *Proc. Natl. Acad. Sci. USA*, 77, 2631, 1980.

11. Bowden, T.R. et al., Molecular characterization of Menangle virus, a novel paramyxovirus which infects pigs, fruit bats, and humans, *Virology*, 283, 358, 2001.

12. Bowden, T.R. and Boyle, D.B., Completion of the full-length genome sequence of Menangle virus: Characterisation of the polymerase gene and genomic 5′ trailer region, *Arch. Virol.*, 150, 2125, 2005.

13. Sedlmeier, R. and Neubert, W.J., The replicative complex of paramyxoviruses: Structure and function, *Adv. Virus Res.*, 50, 101, 1998.

14. Curran, J. and Kolakofsky, D., Replication of paramyxoviruses, *Adv. Virus Res.*, 54, 403, 1999.

15. Chua, K.B. et al. Full length genome sequence of Tioman virus, a novel paramyxovirus in the genus *Rubulavirus* isolated from fruit bats in Malaysia, *Arch. Virol.*, 147, 1323, 2002.

16. Chua, K.B. et al., Tioman virus, a novel paramyxovirus isolated from fruit bats in Malaysia, *Virology*, 283, 215, 2001.

17. Young, P.L. et al., Serologic evidence for the presence in *Pteropus* bats of a paramyxovirus related to equine morbillivirus, *Emerg. Infect. Dis.*, 2, 239, 1996.

18. Murray, K. et al., A novel morbillivirus pneumonia of horses and its transmission to humans, *Emerg. Infect. Dis.*, 1, 31, 1995.

19. Murray, K. et al., A morbillivirus that caused fatal disease in horses and humans, *Science*, 268, 94, 1995.

20. Selvey, L.A. et al., Infection of humans and horses by a newly described morbillivirus, *Med. J. Aust.*, 162, 642, 1995.

21. Rogers, R.J. et al., Investigation of a second focus of equine morbillivirus infection in coastal Queensland, *Aust. Vet. J.*, 74, 243, 1996.

22. Hooper, P.T. et al., The retrospective diagnosis of a second outbreak of equine morbillivirus infection, *Aust. Vet. J.*, 74, 244, 1996.

23. O'Sullivan, J.D. et al., Fatal encephalitis due to novel paramyxovirus transmitted from horses, *Lancet*, 349, 93, 1997.

24. Crerar, S. et al., Human health aspects of a possible lyssavirus in a black flying fox, *Commun. Dis. Intell.*, 20, 325, 1996.

25. Fraser, G.C. et al., Encephalitis caused by a lyssavirus in fruit bats in Australia, *Emerg. Infect. Dis.*, 2, 327, 1996.

26. Hooper, P.T. et al., A new lyssavirus—The first endemic rabies-related virus recognized in Australia, *Bull. Inst. Pasteur.*, 95, 209, 1997.

27. Allworth, A., Murray, K., and Morgan, J., A human case of encephalitis due to a lyssavirus recently identified in fruit bats, *Commun. Dis. Intell.*, 20, 504, 1996.

28. Samaratunga, H., Searle, J.W., and Hudson, N., Non-rabies Lyssavirus human encephalitis from fruit bats: Australian bat Lyssavirus (pteropid Lyssavirus) infection, *Neuropathol. Appl. Neurobiol.*, 24, 331, 1998.

29. Philbey, A.W. et al., Infection with Menangle virus in flying foxes (*Pteropus* spp.) in Australia, *Aust. Vet. J.*, 86, 449, 2008.

30. Chua, K.B. et al., Isolation of Nipah virus from Malaysian Island flying-foxes, *Microbes Infect.*, 4, 145, 2002.

31. Barr, J.A. et al., Evidence of bat origin for Menangle virus, a zoonotic paramyxovirus first isolated from diseased pigs, *J. Gen. Virol.*, 93, 2590, 2012.

32. Trott, D.G. and Pilsworth, R., Outbreaks of conjunctivitis due to the Newcastle disease virus among workers in chicken-broiler factories, *Br. Med. J.*, 5477, 1514, 1965.

33. Mustaffa-Babjee, A., Ibrahim, A.L., and Khim, T.S., A case of human infection with Newcastle disease virus, *Southeast Asian J. Trop. Med. Public Health*, 7, 622, 1976.

34. Bowden, T.R. et al., Menangle virus, a pteropid bat paramyxovirus infectious for pigs and humans, exhibits tropism for secondary lymphoid organs and intestinal epithelium in weaned pigs, *J. Gen. Virol.*, 93, 1007, 2012.

35. Mengeling, W.L., Porcine parvovirus, in *Diseases of Swine*, eds. B.E. Straw et al. (Blackwell Publishing, Ames, IA, 2006), p. 373.

36. Calisher, C.H. et al., Bats: Important reservoir hosts of emerging viruses, *Clin. Microbiol. Rev.*, 19, 531, 2006.

37. Marsh, G.A. et al., Cedar virus: A novel Henipavirus isolated from Australian bats, *PLoS Pathog.*, 8, e1002836, 2012.

38. Drexler, J.F. et al., Bats host major mammalian paramyxoviruses, *Nat. Commun.*, 3, 796, 2012.

39. Wong, S. et al., Bats as a continuing source of emerging infections in humans, *Rev. Med. Virol.*, 17, 67, 2007.

40. Juozapaitis, M. et al., Generation of Menangle virus nucleocapsid-like particles in yeast *Saccharomyces cerevisiae*, *J. Biotechnol.*, 130, 441, 2007.

41. Hooper, P.T. et al., Immunohistochemistry in the identification of a number of new diseases in Australia, *Vet. Microbiol.*, 68, 89, 1999.

42. Yaiw, K.C. et al., Viral morphogenesis and morphological changes in human neuronal cells following Tioman and Menangle virus infection, *Arch. Virol.*, 153, 865, 2008.

43. Birch, D.E. et al., Simplified hot start PCR, *Nature*, 381, 445, 1996.

44. Pang, J., Modlin, J., and Yolken, R., Use of modified nucleotides and uracil-DNA glycosylase (UNG) for the control of contamination in the PCR-based amplification of RNA, *Mol. Cell Probes*, 6, 251, 1992.

45. Storch, G.A., Diagnostic virology, in *Fields Virology*, eds. D.M. Knipe et al. (Lippincott Williams & Wilkins, Philadelphia, PA, 2007), p. 565.

46. Petraityte, R. et al., Generation of Tioman virus nucleocapsid-like particles in yeast *Saccharomyces cerevisiae*, *Virus Res.*, 145, 92, 2009.

47. Iehle, C. et al., Henipavirus and Tioman virus antibodies in pteropodid bats, Madagascar, *Emerg. Infect. Dis.*, 13, 159, 2007.

48. Lau, S.K. et al., Identification and complete genome analysis of three novel paramyxoviruses, Tuhoko virus 1, 2 and 3, in fruit bats from China, *Virology*, 404, 106, 2010.

49. Baker, K.S. et al., Co-circulation of diverse paramyxoviruses in an urban African fruit bat population, *J. Gen. Virol.*, 93, 850, 2012.

50. Baker, K.S. et al., Novel potentially-zoonotic paramyxoviruses from the African straw-colored fruit bat, *Eidolon helvum*, *J. Virol.*, 87, 1348, 2013.

51. Yaiw, K.C. et al., Tioman virus, a paramyxovirus of bat origin, causes mild disease in pigs and has a predilection for lymphoid tissues, *J. Virol.*, 82, 565, 2008.

52. Yaiw, K.C. et al., Serological evidence of possible human infection with Tioman virus, a newly described paramyxovirus of bat origin, *J. Infect. Dis.*, 196, 884, 2007.

16 Monkeypox Virus

Joseph E. Blaney and Reed F. Johnson

CONTENTS

16.1 INTRODUCTION

Monkeypox virus (MPXV) is a zoonotic pathogen endemic to Central and West Africa that causes disease resembling human discrete ordinary-type smallpox but with a lower case fatality rate. The virus is maintained in several species of rodents in Africa and routinely enters the human population causing sporadic disease outbreaks. Fortunately, human-to-human transmission of MPXV is relatively low when compared to variola virus, the causative agent of smallpox, so disease outbreaks are limited. The threat posed to public health outside of Africa was illustrated by the recent introduction of MPXV to the United States in 2003 through importation of rodents for exotic pets. Thirty-seven individuals were confirmed to have been infected and several were hospitalized. While this event demonstrated that MPXV could spread to nonendemic regions by normal commerce or travel, there are also significant concerns that MPXV could be used for bioterrorism. While not as deadly or easily transmitted as variola virus, monkeypox would be considerably easier for a malefactor to obtain. A deliberate introduction of a virulent MPXV strain would likely cause significant morbidity and considerable social and economic interruption. Infection with MPXV results in disease characterized by the pathognomonic rash, fever, malaise, and lymphadenopathy. The case fatality rate has been reported to be between 1% and 10%. Currently, there is no approved countermeasure for treatment of MPXV disease. However, immunization with smallpox vaccine (vaccinia virus) does confer protective immunity to MPXV. Here, we describe the salient features of MPXV biology, epidemiology, clinical manifestations, diagnosis, and treatment with a focus on recent investigations. Remaining questions that confront the medical and research community are identified and proposed for future study.

16.2 CLASSIFICATION, MORPHOLOGY, AND MOLECULAR VIROLOGY

MPXV is classified in the *Orthopoxvirus* genus of the subfamily Chordopoxvirinae along with vaccinia virus, variola virus (smallpox), and cowpox virus.[1] Orthopoxviruses range from 200 to 400 nm in length and are described as either oval- or brick-shaped depending upon the viral species. The orthopoxvirus virion is composed of three major substructures: (1) an outer lipid bilayer membrane, (2) a biconcave core, and (3) lateral bodies located in the concavities of the core. Poxviruses produce two types of infectious particles termed extracellular enveloped virus (EEV) and intracellular mature virions (IMV). Therefore, the composition and complexity of the outer lipid membrane varies depending upon the infectious particle produced. The EEV is the viral particle that is thought to initiate the infectious process in a new host and the IMV is thought to play a role in dissemination of the virus within the host.[2] The lipid membrane for both particles is derived from the Golgi of infected cells. The core of the virus contains the nucleoprotein complex that forms an insoluble structural network.[3] The core wall is composed of two layers approximately 18–19 nm thick. The inner layer is continuous with channels and its density is consistent with a lipid membrane. The outer layer is described as a palisade structure made of T-shaped spikes anchored in the lower membrane.[3] The dense layer under the core wall has a fiber-like morphology and is considered the nucleoprotein complex.

The MPXV genome is composed of linear double-stranded DNA of 197 kb that is covalently linked at each end. Similar to other orthopoxviruses, the central portion of the genome is composed of genes necessary for replication with the left and right side of the genome composed of immunomodulatory and host range genes. The termini of the DNA strands are inverted tandem repeats, and for MPXV, the left side of the genome is composed of four left terminal open reading frames (ORFs) duplicated from the right side of the genome. The inverted tandem repeats have been characterized for vaccinia virus and are A+T-rich regions that allow for incomplete base-pairing to presumably maintain a double-stranded configuration.[4]

MPXV is closely related to variola virus, cowpox virus, and vaccinia virus although the extent of the relationship depends upon the methodology used and the region investigated. The central regions of the genomes of variola and MPXV are 96.3% identical, while the ends vary greatly.[5] Further analysis suggests that variola is more closely related to vaccinia than either cowpox GRI strain or MPXV Zaire strain. Based on overall nucleotide sequence homology, Gubser demonstrated that MPXV is more distantly related to variola than cowpox.[6] Xing and colleagues analyzed the 49 immunomodulatory and host range determining genes common between MPXV, variola, and cowpox by two methodologies.[7] Using the most parsimonious trees method, MPXV and cowpox were more closely related to each other than either was to variola. By the neighbor joining tree method of the same 49 genes, MPXV was more closely related to variola than was cowpox. Given the organization of poxvirus genomes, the wide variation observed within the left- and right-hand sides of the genomes is expected.[7] The varying severity of disease and host range of these viruses indicates that determining the function of the immunomodulatory and host range genes provides an opportunity to determine the mechanism of pathogenesis and is an active area of investigation. A detailed comparison of the genetic differences between variola, vaccinia, and MPXV is provided in Weaver and Isaacs, 2008.[8]

MPXV can be divided into two clades, West African and Central African. West African strains of MPXV are considered to be less virulent than Central African MPXV strains, which is discussed in detail in Section 16.4. Central African strains are associated with a 10% case fatality rate, while no fatalities have been reported for West African MPXV.[9] There is a 0.56% nucleotide difference between the genomes.[8] There are 173 functional genes in MPXV Zaire (a Central African strain) and 170 for MPXV Sierra Leone (a West African strain). The shared genes are 99.4% identical and no significant differences can be found between genes that regulate transcription and replication. Of the 56 pox virulence genes, 53 are in both strains and there is high amino acid similarity between them.

Poxviruses have a typical viral infection cycle: entry, transcription, replication, assembly, and egress.[1] Due to restrictions associated with variola and MPXV research, vaccinia has served as the prototype orthopoxvirus for modeling the steps of the virus life cycle, and unless specified, the step of the life cycle described has been investigated for vaccinia and it is assumed that MPXV uses similar mechanisms. As with any virus, specific targeting of any of the viral proteins involved in the basic life cycle steps may provide an opportunity to develop inhibitors with high effective concentrations and low toxicity. Such targets are summarized in Pritchard and Kern[10] and Table 16.1.

TABLE 16.1
Select Genes Involved in the Viral Life Cycle That May Serve as Therapeutic Targets

Genes	Function	Therapeutic	References
A50R	DNA ligase	Novantrone	[98,115]
B1R	Ser/Thr kinase packaged in the virion and need for early events	siRNA	[116]
G7L	Morphogenesis	siRNA	[116]
C7L	Interferon antagonism	Not developed	[117]
A28	Membrane fusion	Not developed	[118]
006	NF-kappaB signaling suppressor	Not developed	[119]
F13L	Virus egress	ST-246	[31]

Note: Additional targets are listed in Pritchard and Kern [10].

As stated previously, orthopoxviruses exist in two infectious particle types, EEV and IMV. Either particle can initiate the infectious cycle. The mechanism used for entry by either particle is incompletely understood but is thought to be different due to the variation in the expression of glycoproteins and membranes associated with each particle. Viral proteins are thought to bind to glycosaminoglycans or components of the extracellular matrix to initiate this process.[2] IMV can enter the cell by direct fusion with the plasma membrane or macropinocytosis. Virus proteins D8, A27, and H3 are thought to be necessary for IMV entry. D8 binds chondroitin sulfates on the cell surface, while A27 and H3 bind heparin sulfate, which is also readily available on the surface of most cell types. Vaccinia relies on the entry–fusion complex, which likely forms a pore for core entry into the cytoplasm. Twelve proteins form the complex (A16, A21, A28, F9, G3, G9, A2, I2, J5, L1, L5, and O3); of these, A16, A21, A28, G3, G9, H2, J5, L5, and O3 are considered integral because the absence of one destabilizes the entire complex.[11,12] Physical and functional interactions of these proteins have not been definitively determined. However, it has been demonstrated that A28 interacts with H2, A16 interacts with G9, G3 interacts with L5, and it is possible that G3 interacts with G9.[11] EEV entry is thought to be mediated by A34 and B5.[13] A34 recruits B5 into the EEV membrane and B5 regulates EEV membrane rupture upon exposure to glycosaminoglycans,[14] which then presumably allows for IMV membrane fusion to the host plasma membrane and release of the core into the cytoplasm.

Transcription is functionally separated into early, intermediate, and late genes. Once the core is released into the cytoplasm, the early transcription events begin to occur and can be detected within minutes of infection. The core contains all machinery necessary for capping, including the DNA-dependent RNA polymerase, RAP94, vaccinia early transcription factor (VETF), poly(A) polymerase, nucleotide phosphohydrolase 1, and a topoisomerase. The early genes include proteins necessary for interaction with the host, viral DNA synthesis, and proteins needed for uncoating and intermediate gene expression. Termination of early gene expression coincides with uncoating. Intermediate genes are expressed between the cessation of replication, but prior to late gene expression. They mainly function in interacting with newly produced viral genome and are DNA binding, packaging, and nonenzymatic core-associated proteins. Intermediate genes are regulated by a specific promoter and include 21 genes (for vaccinia, unknown for MPXV).[15] Late genes are defined as those that occur after intermediate transcription and persist to the end of the life cycle.[3] Yang and colleagues identified 38 genes from vaccinia as late genes. These genes encode redox disulfide enzymes, proteins involved in morphogenesis, crescent formation, mature virion membrane proteins, and entry–fusion complex proteins.[15]

After 1–2 h postinfection replication begins, approximately 10,000 gene copies are produced per cell and about half are packaged.[3] Initiation and the mechanism of genome replication are not well understood and evidence is available that replication occurs by either a strand displacement mechanism or through lagging strand synthesis. About half of the genome is transcribed prior to replication and takes place between early transcription and intermediate transcription. DNA replication occurs exclusively in the cytoplasm, and the virus encodes its own DNA-dependent DNA polymerase complex. Orthopoxviruses also encode proteins that increase the pool of available deoxyribonucleotides such as thymidine kinase, which can be targeted for therapeutic intervention.

As with any virus, effective countermeasures could be targeted to the replication machinery as the mechanism usually requires less host involvement and thus increases specificity of the inhibitor and may minimize toxicity to the host. For example, cidofovir was identified as an inhibitor of poxvirus replication by Neyts et al. in 1993.[16] The mechanism of action was thought to be similar to its mechanism against other virus and essentially act as a chain terminator[17,18] for

the polymerase complex. Further study has determined that cidofovir allosterically repositions the 3′ nucleophile of the terminal and short + strand synthesis products that leads to aberrant chain extension.[19,20] Cidofovir can also inhibit the 3′–5′ exonuclease activity at the +1 position relative to the terminus, which would also inhibit chain elongation. Resistance can occur and a nucleotide comparison between wild-type and resistant strains indicated 55 single nucleotide polymorphisms, 44 nucleic acid substitutions, and 10 intergenic and 45 intragenic single nucleotide polymorphism (SNPs) that resulted in 17 synonymous and 26 nonsynonymous substitutions, and one tandem repeat contraction was identified.

Virions assemble in virus factories within the cytoplasm. The first structures are electron-dense viroplasms composed of viral core proteins. Single lipid bilayer crescent-shaped membranes (Figure 16.1) that are stabilized by VACV D13 develop at the edge of the viroplasms and engulf the cores to form immature virions. The viral genome is encapsidated prior to complete encirclement by the crescent membrane and this structure is referred to as the immature virus (IV). IVs mature into the characteristic oval/brick shape and are termed intracellular mature virions (IMVs)[21] (Figure 16.1). The EEV is formed by further wrapping of the IMV with host-derived membranes to form intracellular enveloped virions (IEVs). IEVs travel to the plasma membrane by a kinesin–microtubule transport system and fuse with the plasma membrane to form cell-associated enveloped virions (CEVs). CEV initiates actin polymerization that moves the CEV toward an adjacent cell. A CEV that detaches from the tip of an actin tail or from the membrane is termed an EEV.[22]

As expected, orthopoxviruses recruit host pathways to promote viral survival. Reeves et al. demonstrated that vaccinia CEVs use Abl and Src family tyrosine kinases for actin motility and that the release of CEV from the cell requires Abl but not Src family tyrosine kinases.[23] A follow-up study with variola and MPXV indicated that variola and MPXV release CEV

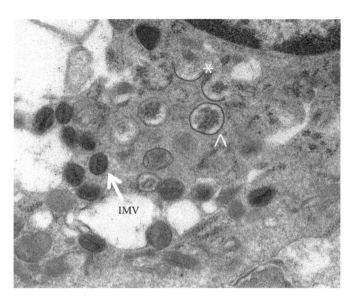

FIGURE 16.1 Electron micrograph of a spleen sample from a cynomolgus macaque infected with MPXV Zaire V79. Various stages of the virion life cycle are present. An IMV is indicated. * Indicates crescents and ∧ indicates immature virion.

in an Abl and Src family tyrosine kinase–dependent fashion.[24] Fusion of the IEV with the plasma membrane results in the formation of CEV. CEVs are able to activate Src and Abl family kinases to induce phosphorylation of vaccinia A36 that is localized in the plasma membrane beneath the viral particle. A36 phosphorylation leads to the recruitment of a signaling complex consisting of the host proteins Grb2, Nck, IP, and N-WASP that activate Arp2/3 to stimulate actin tail polymerization and virus motility leading to cell-to-cell spread of the virus.[2,23] In the absence of cell lysis, the release of infectious virus is limited by its size and is an active process mediated by VACV F11 protein, which binds RhoA (RhoA is a regulator of the actin cytoskeleton and can change cell morphology) and inhibits RhoA signaling.

Therapeutics targeting virus assembly and release have been rejuvenated since the 2003 US MPXV outbreak and will be further discussed in the succeeding text. One of the most promising is ST-246 from SIGA Technologies. ST-246 was initially identified in a large-scale screen of low-molecular-weight compounds to identify vaccinia virus replication inhibitors and was found to be effective against the orthopoxviruses.[25–28] The mechanism of action of ST-246 is not precisely understood. However, it is apparent that ST-246 blocks the function of the vaccinia F13L gene product and its orthologues. F13L encodes a phospholipase that is essential for maturation of viral particles from IMV to EEV. F13L colocalizes to the Golgi, plasma, and endosomal membranes and shuttles between these compartments by clathrin-mediated trafficking.[29,30] The F13L gene product contains a conserved histidine–lysine–aspartate phospholipase (HKD), and mutagenesis of this motif has abrogated phospholipase activity and subcellular localization.[30–33] The resulting effect of F13L's inhibition is that EEV production is impaired, which inhibits viral spread thus reducing disease[27] and allowing the host immune system to respond and develop protection.

16.3 BIOLOGY AND EPIDEMIOLOGY

MPXV was initially identified in 1958 at the State Serum Institute in Denmark in cynomolgus macaques with vesicular lesions and since then, several outbreaks in nonhuman primate colonies across the globe have been reported, but there are no reports of human infections from these episodes. The first human case was identified in 1970 in the Democratic Republic of the Congo (DRC) (formerly Zaire).[34] It is likely that smallpox overshadowed MPXV infections until it was controlled by vaccination. The emergence of MPXV in humans and investigations into the disease were intertwined with the smallpox eradication program in the 1970s. As part of the program to confirm smallpox eradication, rare cases of pox-like disease in Central and West Africa were investigated in areas where smallpox cases had recently been interrupted, and 58 MPXV cases were reported between 1970 and 1979.[35] Nearly 80% of cases were reported in the DRC, while cases also occurred in Cameroon, Ivory Coast, Liberia, Nigeria, and Sierra Leone. Cases were predominantly observed in children and associated with animal contact but

limited person-to-person transmission was also observed. No deaths were reported in West Africa, but a case fatality rate of approximately 20% was reported in the DRC.[36,37] Approximately 90% of cases in the DRC were attributed to animal contact, while 10% were believed to be secondary human-to-human transmission.

After the declaration of smallpox eradication in 1980, the WHO instituted a special surveillance program to examine MPXV epidemiology and ecology from 1981 to 1986. There was legitimate concern that MPXV cases and transmission might increase once smallpox vaccination waned resulting in MPXV becoming a more profound public health threat. Between 1981 and 1986, 338 cases were reported in the regions of the DRC that were covered by health institution–based surveillance representing 89% of total reported MPXV cases in the DRC.[38] The increase in prevalence was attributed to the enhanced surveillance. More than 90% of cases were found in small villages associated with tropical rain forests. Contact with animals was the suspected infection source in 72% of cases, while human-to-human transmission was suspected in the remaining 28% of cases. The average annual incidence rate in the DRC for this period was 0.63 cases per 10,000 and the case fatality rate was 10%. Children were considerably more susceptible to disease potentially because of discontinuation of smallpox vaccination. Despite the MPXV disease reported in this heightened surveillance period, the secondary attack rate (9%) of unvaccinated household members was considerably less than observed for smallpox transmission. Analysis of these data and mathematical modeling predicted that MPXV would be unlikely to spread with the ferocity of smallpox even in unvaccinated human population.[39] Luckily, our experience over the last 25 years has indicated that MPXV continues to be a rare zoonosis with limited human-to-human transmission. Interestingly, unlike the situation in Central Africa, MPXV disease all but disappeared from West Africa, which will be discussed later.

Not surprisingly, investigation into MPXV epidemiology waned after the intense studies of the early 1980s, and few cases were reported but intermittent cases in the DRC were reported as well as suspected cases in the Republic of Congo and Sudan in 2003 and 2005, respectively.[37,40] However, the most recent investigation into MPXV surveillance in the DRC by Rimoin and colleagues has yielded some dramatic results on MPXV incidence.[41,42] Active disease surveillance in nine health zones of the Sankuru District of the DRC was performed from November 2005 to November 2007. A total of 760 laboratory-confirmed cases (MPXV genome detection by polymerase chain reaction [PCR] analysis of serum or scab) were identified. The average annual cumulative incidence among health zones was 5.53 per 10,000 with a range of 2.18–14.42. As had been found in the 1980s, forested areas had considerably more disease burden. Other risk factors included male gender (ages 5–14) and no history of smallpox vaccination. The comparison of incidence data of similar areas between 2005 and 2007 and the early 1980s indicated that incidence may have increased up to 20-fold. Increased surveillance would appear to be called for despite the still relatively low disease burden compared to other infectious diseases that Africa confronts. The results of this recent surveillance also suggest that as vaccination coverage wanes in the population, we can expect even more MPXV infections to affect Central Africa with the concomitant risk of accidental export to other areas of the globe.

The risk that MPXV might spread to other regions was illustrated by the unexpected outbreak of MPXV disease in the Midwestern United States in 2003.[43] The source of infection was tracked to Ghana in West Africa by way of imported rodents destined for the exotic pet market in the United States. Transmission to humans was actually mediated by newly acquired prairie dogs purchased as pets that had become infected through contact with the imported rodents.[44] Of 72 potential cases, 37 were laboratory-confirmed MPXV infections.[37,45] Although 18 patients were hospitalized,[46] the disease manifestations of this outbreak were relatively mild and no fatalities were reported, which is consistent with previous findings that West African MPXV strains may be of low virulence. The clinical manifestations of the US outbreak will be described later. It is highly likely that the severity of disease and potential for secondary transmission would have been far greater if a Central African strain would have been introduced into the United States. Luckily, it does not appear that MPXV has become endemic in the United States as no cases have been reported since 2003. It is important to remember that numerous rodents and other wildlife in the United States are likely susceptible to MPXV infection and could potentially serve as reservoirs.

Considerable research has aimed at defining MPXV ecology in Africa and specifically the animal species responsible for maintenance of MPXV and transmission to humans since the initial reports that human infection was associated with animal contact. It is now understood that both humans and monkeys are incidental hosts for MPXV in contrast to smallpox, which was strictly a human pathogen with no animal reservoir. As mentioned previously, 72% of the cases identified in the DRC between 1980 and 1986 were likely attributed to animal contact.[38] Nonhuman primates (*Cercopithecus, Colobus, Cercocebus*) were the suspected animal source of infection in 55% of these cases.[38] Arboreal (*Funisciurus, Heliosciurus*) and terrestrial (primarily *Cricetomys*) rodents were associated with 31% of cases. Fourteen percent of cases were linked to contact with duikers (*Cephalophus*) or gazelles (*Gazella*). Patient histories indicated that 90% of the cases were associated with contact with a healthy animal, while the remaining cases were linked with ill or dead wildlife. Natural history studies of MPXV infection have primarily relied on serosurveys as isolation of virus from wildlife has rarely been reported. Two reviews have recently summarized the diverse array of species in Africa that show some level of seropositivity for MPXV.[47,48] Based on the diversity of species affected and high seroprevalence, arboreal and semiterrestrial squirrels are believed to be the top candidate for the MPXV reservoir in Africa, but there is no unequivocal evidence.

16.4 CLINICAL FEATURES AND PATHOGENESIS

MPXV infection presents in a similar manner as ordinary discrete smallpox although the presence of lymphadenopathy is a distinguishing characteristic. However, while smallpox spread was primarily attributed to upper respiratory invasion by virus from oropharyngeal droplets of an individual with a rash, MPXV exposure is less clearly understood. Exposure to MPXV-infected wildlife is responsible for the vast majority of cases, but it is unclear if infection is primarily mediated by abrasions, oropharyngeal or nasopharyngeal contact, or even consumption. Like smallpox, MPXV is secreted in the upper respiratory tract in humans and human-to-human transmission can occur by respiratory exposure in addition to physical contact with lesions or lesion exudates. The clinical characteristics of MPXV infection are summarized in Table 16.2.

Jezek and Fenner carefully described MPXV clinical disease observed in the DRC in the early 1980s, which is summarized in the succeeding text.[38] The incubation period (time from exposure to fever) was reported to be 7–19 days with 72% of cases having incubation periods between 10 and 13 days. The prodromal phase (time between onset of fever and appearance of rash) lasted for 1–3 days in 84% of cases. In addition to high fever, early pre-rash symptoms included severe headache, backache, malaise, prostration, and lymphadenopathy. Lymphadenopathy was commonly observed and associated with firm nodes, tenderness, and pain. Enlargement of submaxillary and cervical lymph nodes (neck area) were commonly observed early in disease, but diffuse lymphadenopathy developed in more than half of cases.

The first signs of lesional rash typically occurred on the face followed rapidly by appearance on the body. Lesions evolved as in classical smallpox from macules to papules to vesicles to pustules prior to umbilication, drying, and desquamation. Lesion number ranged from a few to several thousand in severe MPXV cases that resembled semiconfluent smallpox. Lesion size was typically 0.5 cm but ranged up to 1.0 cm. Representative lesions are shown in Figure 16.2. Vaccination was associated with less severe lesional disease.

TABLE 16.2
Characteristics of Monkeypox Virus Disease in Humans

Transmission	Contact with infected wildlife; limited human-to-human
Incubation period	6–16 days
Rash	Resembles smallpox rash; lesions evolve over approximately 10 days from macropapules to vesicles to pustules to crusts
Other clinical signs	Fever, lymphadenopathy, headache, myalgia, asthenia, upper/lower respiratory
Duration	14–28 days
Complications	Secondary bacterial infections, bronchopneumonia
Countermeasures	No approved antiviral; vaccinia immunization is preventative
Case fatality rate	Up to 10%
Long-term sequelae	Scarring, blindness (rare)

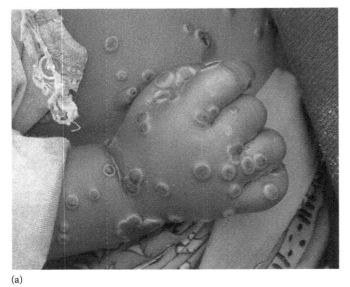

(a)

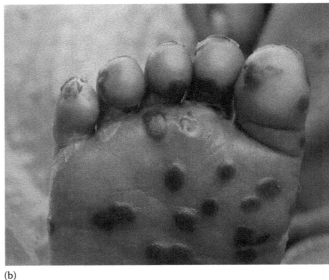

(b)

FIGURE 16.2 Lesions resembling a smallpox rash present on the (a) hand and (b) foot of children infected with MPXV in the DRC. (From Reynolds, M.G. and Damon, I.K., *Trends Microbiol.*, 20, 80, 2012.)

Lesions were commonly observed in the oropharynx and associated with painful sore throat and coughing. Ocular involvement was also quite common with conjunctivitis and eyelid edema. Of the 338 cases, 11% were characterized as mild illness with less than 25 lesions and no incapacitation, 19% had moderate illness with 25–99 lesions and incapacitation with nursing care required, and 70% had severe illness necessitating intensive medical attention with more than 100 lesions. The disease course typically lasted 2–4 weeks. The case fatality rate was nearly 10% with fatalities highest among unvaccinated patients and children of age 4 years or younger. Complications associated with death were bronchopneumonia (primarily), encephalitis, and septicemia.

As mentioned earlier, the disease in the US outbreak in 2003 was considerably milder than that observed in the Central African cases.[49,50] The average incubation period was approximately 12 days. The most common clinical signs of 34 patients were rash (97% of cases), fever (85%), chills (71%), lymphadenopathy (71%), headache (65%), and myalgia (56%). Common abnormal laboratory findings included elevated transaminase levels, low blood urea nitrogen levels, hypoalbuminemia, leukocytosis, and thrombocytopenia. Many patients developed only a single or few lesions, reminiscent of cowpox virus exposure, at the suspected inoculation site mediated by handling or bite/scratch of a prairie dog. Approximately 20% of patients had lesion counts of 100 or greater. Despite the generalized mild disease, two children did present with severe disease including encephalitis, a rare complication of MPXV infection, in one child and serious cervical and tonsillar lymphadenopathy with numerous lesions on the body and oropharynx. Interestingly, complex exposure (invasive bite or scratch) was associated with more severe disease and requirement for hospitalization.[50]

After the US outbreak of 2003, considerable efforts have focused on increasing our understanding of MPXV

pathogenesis and specifically the difference in disease manifestation between West and Central (Congo Basin) African strains. Cases of MPXV infection in West African countries such as Côte d'Ivoire, Liberia, Nigeria, and Sierra Leone are far less common than in the DRC and are characterized by mild disease, very limited human-to-human transmission, and a lack of fatalities.[51] Restriction fragment length polymorphism (RFLP) and sequence analysis of viral isolates indicated that the geographically isolated strains found in West and Central Africa group into two separate clades suggesting that there may be genetic basis underlying the difference in virulence.[43,51,52] The nucleotide homology of virus strains within the central and western clades is approximately 99%, while strains across the clades have approximately 95% homology. Two investigations compared the amino acid differences and gene truncations or deletions between the two MPXV clades and related these modifications to known gene functions among MPXV and the related orthopoxviruses including cowpox, vaccinia, and smallpox viruses.[51,52] Several genes related to either poxvirus replication or host response modification were identified as candidates for virulence determinants including D10L, D14L, B10R, B14R, and B19R. The potential contribution of D14L to virulence was particularly interesting because West African strains completely lacked this immunomodulatory gene while Central African strains possessed a truncated but functional gene. The D14L gene encodes for the MPXV orthologue of smallpox inhibitor of complement enzymes (SPICEs), termed monkeypox inhibitor of complement enzymes (MOPICE) or complement control protein (CCP) for MPXV strains. The protein product of this poxvirus orthologue disrupts the classical and alternative complement pathways preventing complement-mediated antiviral activity in the host. Thus, West African strains may lack this ability to escape complement-mediated protection in the host potentially explaining their decreased virulence.

Recently, studies have attempted to determine if the virulence differences observed in humans for the two MPXV clades are reflected in animal models of disease so that the genetic determinants of virulence could be assayed in vivo. Chen and colleagues compared a West African MPXV strain, COP-58, with a Central African strain, Zaire V79, in an aerosol model of infection in cynomolgus macaques.[52] As expected, the Central African strain was 100% lethal in three animals, but the West African strain was not lethal although it did induce clinical signs including lesional disease. Saijo reported that intranasal inoculation of cynomolgus macaques with a West and Central African strain resulted in virulence differences. The West African strain induced a 10-fold reduction in viremia and lower mortality (33% vs. 75% for Central African strain).[53] Thus, in the model of disease that best reflects human MPXV infection, there is evidence that western strains are less virulent. Additional studies have been conducted in rodents. Ground squirrels (13-lined, *Ictidomys tridecemlineatus*) were administered subcutaneous injections of the US 2003 strain and MPXV Zaire V79.[54] Both virus strains caused lethal disease, but infection with the Central African strain was associated with an earlier time of death, higher viral loads, and more frequent hemorrhagic manifestations than squirrels infected with the western strain. Hutson and colleagues inoculated black-tailed prairie dogs (*Cynomys ludovicianus*), the species responsible for transmission in the US outbreak, by the intranasal or intradermal route with the US 2003 strain or a Central African strain from the Republic of Congo.[55–57] The prairie dog model demonstrates several clinical signs of human MPXV including a 6–12-day incubation period, rash, oral lesions, and occasional lymphadenopathy. The Central African strain had a 22-fold reduction in virulence as determined by lethal dose—50% by the intranasal route.

The study of the MPXV clades in various animal models has culminated in two recent studies employing recombinant viruses to try to discern the genetic basis for the observed virulence differences. Estep and colleagues generated a recombinant MPXV Zaire V79 with a deletion in the gene encoding the D14L ORF termed MOPICE or CCP.[58] As mentioned previously, this gene is absent in less virulent West African strains but present in virulent Central African strains. Rhesus macaques were infected by the intrabronchial route with 2×10^5 PFU of wild-type or deletion virus. The deletion of the gene did not appear to significantly attenuate disease when compared to wild-type virus although the viruses were administered at low, sublethal doses so comparison was difficult. In fact, all animals infected with wild-type virus survived, while one of the four animals infected with the deletion virus succumbed to disease. Viremia was surprisingly slightly higher in the MOPICE deletion virus. Therefore, while these results suggest that MOPICE may play a role in virulence, its activity does not appear to explain virulence differences between the two clades. In a separate study, Hudson et al. introduced MOPICE into a West African recombinant virus and deleted the gene from a Central African recombinant virus.[59] Analysis of these viruses in the intranasal prairie dog model demonstrated that the introduced expression of MOPICE did not greatly affect the virulence of the West African strain. However, the deletion of MOPICE from the Central African strain reduced mortality from 100% in wild-type control animals to 0% in animals infected with the deletion virus. Taken together, these results indicate that MOPICE may play a role in the virulence differences between the MPXV clades but that the differing virulence is almost certainly a result of multiple virulence genes. With a genome of nearly 200,000 nucleotides, this is not a surprising outcome.

16.5 IDENTIFICATION AND DIAGNOSIS

A combination of clinical, epidemiological, and laboratory analyses are usually required to identify MPXV infections. Pre-eruptive infection cannot be distinguished from various other acute febrile infections, but the development of pox-like rash is usually pathognomonic. However, using clinical signs alone, MPXV infection can be difficult to distinguish from related orthopoxvirus infections or other infectious disease where rash and fever are present. Lymphadenopathy is likely the sole distinguishing characteristic between MPXV and smallpox infection of humans. Enlargement of submaxillary and cervical lymph nodes (neck area) is commonly observed early in disease, but diffuse lymphadenopathy develops in most patients characterized by firm lymph nodes, tenderness, and pain.[38] Lymphadenopathy can develop at the onset of fever or rash but typically is observed on the second or third day after fever. Jezek noted that in the DRC, the differential diagnosis of MPXV included chicken pox, measles, tanapox, syphilis, scabies, and drug allergy eruptions.[38] Chicken pox was typically the most difficult to differentiate.[60] The causative agent of chicken pox is actually an alpha-herpesvirus, varicella zoster virus, which produces fever and rash, but the rash is characterized by a heterogeneous nature with multiple lesion forms present at a given time. In contrast, orthopoxvirus lesions are usually homogeneous in form throughout the stages of infection. Chicken pox lesions are typically denser on the trunk, while MPXV lesions are more dispersed with more characteristic lesion involvement of the face, palms, and soles. Most importantly, for MPXV differential diagnosis, lymphadenopathy is not usually associated with chicken pox.

Over the years, several laboratory methods have been developed to identify orthopoxvirus infections. If available, electron microscopy is a reliable method to diagnose during active infection although distinguishing the various orthopoxviruses is not possible. Electron micrographs of lesional biopsies can be used to identify brick-shaped mature virions with dumbbell cores and round immature virion forms such as in Figure 16.1. In the smallpox eradication era, it became important to provide a definitive diagnosis of MPXV ruling out smallpox infection in suspected human cases. Thus, much research focused on laboratory methods to distinguish the poxviruses. An early but effective method was evaluation of pock formation on the chorioallantoic membrane (CAM) of chick embryos. Each orthopoxvirus induces a characteristic pock on the CAM.[38] MPXV induces small and opaque hemorrhagic pocks, while variola virus pocks are small and opaque but white. Furthermore, MPXV pock formation is less temperature sensitive than variola virus with a ceiling temperature of 39°C versus 37.5°C–38.5°C for variola virus. Additionally, MPXV was found to produce characteristic hemorrhagic lesions upon intradermal inoculation of rabbits. Another laboratory method used to identify cases in the early 1980s in the DRC was the analysis of restriction fragment length polymorphism (RFLPs).[61,62] Digested viral DNA genomes isolated from purified virions or homogenates of cell culture infections exhibit characteristic electrophoretic patterns that can be used to differentiate the various orthopoxviruses and even West or Central African strains of MPXV. Serological techniques such as enzyme-linked immunosorbent assay (ELISA), hemagglutinin inhibition assays, and radioimmunoassays were also used for diagnosis, but were not MPXV-specific.

The Midwestern US outbreak in 2003 brought MPXV diagnosis into the twenty-first century. In addition to light and electron microscopy, immunohistochemical analysis, and virus isolation by cell culture, investigators used modern genetic methods of detection such as PCR assays to identify MPXV as the causative agent.[43] Histological analysis of lesion biopsies at the pustular stage demonstrated degenerating and necrotic keratinocytes and a moderate inflammatory infiltrate with multinucleated cells and viral inclusion bodies. Orthopoxvirus antigen was detected by immunohistochemical analysis with an antivaccinia polyclonal antibody preparation (ViroStat) in primarily degenerating keratinocytes and the follicular epithelium. Transmission and negative-stain electron microscopy were utilized to observe virions with the characteristics of the orthopoxviruses including immature and mature forms with dumbbell-shaped cores in lesion biopsies. Tissue samples from human lesions or a prairie dog lymph node were inoculated onto various cell lines permissive for orthopoxvirus growth, and cytopathic effects were observed including plaques of elongated and rounded cells and formation of syncytia.

For confirmation that an orthopoxvirus was the causative agent of the outbreak and specific determination of the viral species, molecular genetic techniques were employed.[43] Isolated DNA from tissue biopsies, swabs, and infected cell cultures was analyzed first by a PCR assay specific for the DNA polymerase gene, E9L, of all Old World orthopoxviruses except variola virus. As anticipated, the clinical samples tested positive for the presence of orthopoxvirus DNA. PCR amplification of the hemagglutinin and A-type inclusion genes followed by RFLP analysis demonstrated a banding pattern consistent with known West African strains of MPXV. Finally, sequence analysis of the hemagglutinin gene confirmed the cause of the outbreak as a West African strain of MPXV.

For the more recent 2005–2007 epidemiological analysis of MPXV incidence in the DRC described earlier, PCR-based assays were utilized to confirm MPXV as the causative agent in suspected cases.[63] DNA was isolated from scabs or vesicle fluid, and the presence of orthopoxvirus DNA was confirmed by a pan-orthopoxvirus PCR assay against the strongly conserved 14 kDa fusion protein gene or other than *orthopoxvirus* genes.[64,65] The confirmation of MPXV infection was performed by real-time PCR assays specific for MPXV targeting the F3L or N3R genes using TaqMan® MGB probes. Kulesh et al. had previously validated the assay to demonstrate MPXV specificity by evaluating the assay against numerous strains of MPXV, variola virus, vaccinia virus, cowpox, camelpox, and several herpesviruses including varicella zoster virus.[66] The assay specifically detected all West and Central African MPXV strains but none of the other viruses. Research and development of improved molecular genetic assays for orthopoxvirus identification continues aiming for enhanced speed, specificity, or breadth of detection.[67–69] For example, a multiplex real-time PCR assay has just been reported that can differentiate MPXV, variola virus, vaccinia virus, or cowpox virus in a single reaction using four virus-specific primer sets and TaqMan probes.[67]

In addition to the strides in molecular diagnostics, studies of the US outbreak also incorporated new immunological techniques to characterize infection. Serological diagnosis of orthopoxvirus infection is hampered by the cross-reactivity of the response, but the advantage over PCR-based methods is that serological techniques are not dependent on catching the infection as it occurs with virus still present. Karem et al. developed an IgM sandwich ELISA using vaccinia antigen that is not specific for MPXV, but was able to detect acute, active orthopoxvirus infection.[70] The presence of orthopoxvirus-specific IgM indicates the likely presence of an acute-phase humoral response indicating recent infection, whereas IgG ELISA results are difficult to interpret in vaccinated individuals that would likely already have virus-specific IgG responses. Separate investigations have attempted to generate MPXV-specific ELISAs using MPXV peptides.[71,72]

16.6 TREATMENT AND PREVENTION

There is no specific countermeasure that has been developed for MPXV disease. However, immunization with the smallpox vaccine is known to confer protective immunity. Furthermore, two antivirals that have activity against MPXV have recently been developed in response to the smallpox bioterrorism threat. Next, we describe the historical and current status of both vaccination and antiviral development.

Several lines of evidence exist to support the ability of smallpox vaccination to confer protection from MPXV. The vaccine utilized to eradicate smallpox consists of a live replication-competent orthopoxvirus, vaccinia virus, which induces strong cellular and humoral immunity that induces cross-reactivity and protection. Jezek and colleagues reported that history of smallpox vaccination was associated with less severe lesional disease in the DRC in their post-eradication studies in the 1980s.[35,38,73] Greater than 50% of vaccinated patients but less than 25% of unvaccinated patients had fewer than 100 lesions. Furthermore, a confluent rash was seen in 10% of unvaccinated patients but only in 2% of vaccinated individuals. When overall disease was scored, severe illness was reported in 40% and 79% of vaccinated and unvaccinated patients, respectively. Of the 33 deaths among 338 patients, all were in unvaccinated children. An additional critical benefit of smallpox vaccination was illustrated by the finding that secondary attack rate among vaccinated family members was reduced eightfold in comparison to unvaccinated household contacts. Fine et al. reported that smallpox immunization provided approximately 85% protection against MPXV infection by comparison of attack rates among vaccinated versus unvaccinated individuals.[39]

More recent epidemiological studies by Rimoin et al. also demonstrated smallpox vaccination as a critical protective factor in the DRC.[42] In analysis of individuals born before 1980, vaccinated individuals had a greater than fivefold lower risk of MPXV disease when compared to unvaccinated individuals. Vaccination was determined to confer approximately 80% protection from disease at time points of 25 years or greater postimmunization. This remarkable durability of the immune response induced by smallpox vaccination has been characterized by ex vivo studies where more than 90% of individuals vaccinated 25–75 years ago had detectable neutralizing antibodies and/or cellular immunity to vaccinia virus.[74] In the 2003 outbreak in the United States, Centers for Disease Control (CDC) investigators did not determine a protective effect of vaccination in terms of disease severity or hospitalization.[49] A subsequent study by Hammarlund et al. indicated that additional cases of inapparent infection in vaccinated individuals may have occurred indicating long-term cross protection, but it currently remains unclear what level of protection vaccination conferred in the US outbreak.[72,75] Nevertheless, the culmination of the studies described earlier produces strong evidence that smallpox vaccination induces protection from severe MPXV disease. In addition to the clinical and epidemiological investigations that have reported that vaccinia immunization protects against MPXV, it should be noted that vaccination studies using nonhuman primate models of MPXV infection have confirmed this cross-protective response.[76–79]

While evidence exists that MPXV incidence is growing in Central Africa, there are no plans currently to reinstate vaccinia immunization to protect the at-risk population.[41,42] Several reasons underlie the lack of support for vaccination against MPXV at this time. First, the impact of MPXV is relatively low when compared to other diseases and public health resources must target areas where impact can be greatest. Second, vaccinia immunization is fairly reactogenic and contraindicated in individuals with certain health conditions including immunosuppression. Therefore, many individuals in the affected regions of Africa may be at risk for adverse reactions to immunization. Third, human-to-human transmission of MPXV is still relatively low and unless it were to significantly increase in scope and impact, vaccination to prevent major outbreaks is not necessary. Despite these conditions, vaccination of healthcare workers or other individuals with a high risk of MPXV exposure should still be considered by public health officials in the most affected regions of Africa.

The US CDC has issued recommendations on vaccination for protection against MPXV.[80] Guidelines for preexposure vaccination include healthcare workers, investigators of outbreaks, individuals who come in contact with infected animals, and laboratory workers who handle MPXV samples. The CDC also recommends postexposure vaccination for up to 4 days after MPXV exposure, and it should be considered for up to 14 days postexposure in individuals with no contraindications.

While the use of the live vaccinia virus vaccine resulted in one of the great public health accomplishments with the eradication of smallpox, there are serious adverse events that can be caused by vaccination.[81,82] As such, the modern list of contraindications for vaccination is considerable. The CDC lists the following contraindications for potential vaccinees and their household contacts because of the risk of exposure: (1) eczema or atopic dermatitis and other acute, chronic, or exfoliative skin conditions; (2) diseases or conditions that cause immunodeficiency or immunosuppression including HIV/AIDS, malignancy, autoimmune conditions, and genetic immunodeficiency; (3) treatments associated with immunodeficiency or immunosuppression including radiation, corticosteroids, or organ transplant medications; and (4) pregnancy.[83] Additional contraindications for individual vaccinees include presence of moderate or severe acute illness, age of less than 1 year, breastfeeding, or heart disease.[83] While adherence to the list of contraindications may decrease the risk of serious adverse events in target populations, it also significantly decreases the pool of potential vaccinees in the population. For example, the US military reported in 2003 on their experience with a vaccinia immunization program in response to bioterrorism fears of smallpox.[84] Contraindications due to personal or household contacts resulted in vaccination deferments in 11%–34% of eligible personnel. In deployed personnel where household contacts were not an issue, deferments ranged from 4.9% to 7.8%.

Concern for the considerable potential of adverse outcomes associated with the current smallpox vaccine and the large number of individuals with contraindications has resulted in substantial investigation into new vaccine candidates that might offer an improved balance between reactogenicity and immunogenicity. While this effort has largely stemmed from bioterrorism concerns associated with smallpox, this new era of orthopoxvirus vaccine development is also applicable to efforts to prevent MPXV disease. The new vaccine strategies have recently been reviewed.[85] The smallpox vaccine that was most widely used for immunization in the eradication era and up until recently was dryvax. Wyeth (now part of Pfizer) manufactured the vaccine, which consisted of freeze-dried calf lymph obtained from lesions on cow skin formed by inoculation with the vaccinia virus (New York City Board of Health Strain). In an effort to make a smallpox vaccine using modern cell culture technology and potentially lessen vaccine reactogenicity, Acambis (now part of Sanofi Pasteur) isolated individual virus clones with low reactogenicity from dryvax and propagated a new vaccine on

Vero cells. This vaccine termed ACAM2000 was licensed by the US Food and Drug Administration (FDA) in 2007 and has replaced dryvax.[86] ACAM2000 has been shown to induce humoral and cellular immunity levels comparable to dryvax, but unfortunately, vaccine reactogenicity has not substantially improved.[85]

While multiple, modern strategies including DNA vaccines and subunit vaccines are under development, the new smallpox vaccine candidates that are farthest along the development pipeline include two further attenuated vaccinia virus vaccines, MVA or LC16m8.[87,88] MVA (modified vaccinia Ankara) was generated in Germany in the mid-twentieth century by extensive serial passage of CVA (chorioallantois vaccinia Ankara) in chicken embryo fibroblast cells. The serial passage introduced several large deletions in the genome of MVA as well as numerous deletions, insertions, or mutations in the genome resulting in highly restricted replication on mammalian cell culture and strong attenuation in animal models.[87] MVA was used to immunize greater than 120,000 humans in the 1960s and had an excellent safety profile and provided protection from intradermal challenge with live vaccinia virus. MVA was never tested during the smallpox epidemic, but it has shown efficacy in animal models of MPXV infection.[77,78,89] Clinical trials of MVA candidates are ongoing.[90,91] A second vaccine candidate that has progressed into recent clinical trials is LC16m8, which was generated by serial passage of the Lister strain in rabbit kidney cells at 30°C.[88,92] LC16m8 was shown to have reduced neurotoxicity in animal models and had an excellent safety profile when administered to 90,000 individuals in Japan in 1974–1975. Current studies of MVA and LC16m8 focus on dosing regimen and showing equivalent immunogenicity between the attenuated strains and dryvax or ACAM2000.

Antiviral development against the orthopoxviruses has also seen great advances as bioterrorism fears have increased. While many antiviral approaches have been pursued against MPXV and variola virus,[93–99] two strategies have progressed the farthest, ST-246 and CMX001.[100,101] ST-246 (tecovirimat) was identified in a high-throughput screening assay to identify small molecule inhibitors of vaccinia replication that could potentially be developed as broad inhibitors of the orthopoxviruses.[100] In vitro activity was shown against MPXV, vaccinia, camelpox, cowpox, ectromelia (mousepox), and variola viruses.[25] Efficacy in animal models has also been rigorously pursued. Oral administration of 100 mg/kg ST-246 in mice has conferred protection from vaccinia, cowpox, and ectromelia viruses at up to 72 h postchallenge. Sbrana et al. demonstrated that ST-246 had potent antiviral activity against MPXV in the ground squirrel model.[102] Oral administration of drug at 100 mg/kg conferred complete protection from lethal disease if started on days 0, 1, 2, or 3. Furthermore, clinical disease was absent in the treated animals. Finally, ST-246 has been shown to confer protection in nonhuman primate models of MPXV and variola virus infection.[28,103] Drug treatment up to day 3 post i.v. challenge with MPXV was associated with no

lethal outcomes and reduced lesion development and viremia. ST-246 has progressed into phase I clinical trials and proven to be safe and well tolerated. Recently, the drug was administered to two patients with vaccinia virus complications and may have contributed to the positive outcomes of both cases although this cannot be fully discerned because cidofovir or CMX001 and vaccinia immune globulin was also administered.[100,104] An additional antiviral, CMX001, has been developed. CMX001 is a safer, orally bioavailable lipid conjugate of cidofovir, an FDA-approved drug used for cytomegalovirus retinitis. CMX001 has activity against multiple orthopoxviruses including MPXV and variola virus and has shown efficacy in rodent or nonhuman primate models of MPXV, vaccinia, cowpox, ectromelia, and variola viruses.[101] The drug has progressed to clinical trials and has proven to be safe and well tolerated.

16.7 CONCLUSIONS AND FUTURE PERSPECTIVES

One of the most interesting and outstanding questions with regard to MPXV is whether the virus will ever develop greater potential for human-to-human transmission resulting in establishing humans as a maintenance host rather than an incidental host. Commentaries describing conditions that might increase widespread human disease have been reported.[48,105] It is believed that viral evolution resulting in enhanced transmissibility and/or changes in MPXV ecology would be required to increase MPXV prevalence whether in a vaccinated or unvaccinated population. As mentioned earlier, the secondary attack rate (approximately 10%) of MPXV is considerably less than observed for smallpox transmission (approximately 60%). Viral evolution resulting in MPXV strains with enhanced transmissibility might occur if virus titers were increased in magnitude or duration in the oropharyngeal secretions. If ecological changes, such as MPXV becoming established in mice or rats with greater human contact, resulted in increased human infections, this too could serve to increase the potential frequency of viral evolution and increased prevalence of human disease. Further research on virulence factors in MPXV and specifically transmission potential is required. However, the work is primarily performed in animal models and thus extrapolation of results to human disease is difficult.

Since the US outbreak of MPXV in 2003 and heightened concerns of orthopoxvirus bioterrorism since 2001, several laboratories have investigated animal models of MPXV infection in primarily nonhuman primates and also mice.[106–110] The studies have been conducted not only to increase our basic understanding but to develop models that satisfy the FDA animal rule. The FDA animal rule was established to provide a mechanism for licensure of medical countermeasures (MCMs) for pathogens that could not be assayed in normal clinical trials in humans. To fulfill the requirements of the "animal rule," an MCM must demonstrate efficacy in at least two animal models in which the route and dose of virus administration, time of onset of disease, and time course/progression of infection optimally reflect human disease. Considerable research has focused on respiratory exposure since it is the primary means of human transmission, and aerosol, intratracheal, and intrabronchial inoculation results in disease resembling human MPXV infection.[106,108–110] Positron emission tomography/computed tomography (PET/CT) imaging was used to quantitate lymphadenopathy in nonhuman primates infected by the intravenous and intrabronchial route.[106] Studies continue to characterize viral pathogenesis and evaluate countermeasures in these established models.

An additional area of active MPXV research is the characterization of viral and host gene and protein expression and function by modern molecular techniques.[111–114] In addition to increasing understanding of molecular viral pathogenesis, the studies are intended to identify viral or host targets for countermeasures. Rubins and colleagues used microarrays targeting all predicted ORFs in the MPXV genome as well as >18,000 human genes to characterize the viral and host gene expression patterns in infected human monocytes, fibroblasts, and HeLa cells.[113] The temporal progression of both characterized and undefined viral genes was reported including the first reported expression of some immunomodulatory genes. Interestingly, MPXV infection was found to selectively inhibit the expression of genes involved in activation of the innate immune response.[114] Alkhalil et al. reported changes in transcriptional regulation of MPXV-infected cells including genes associated with histones, actin dynamics, cell cycle regulators, and ion channels. A recent study characterized the global proteome response in bronchoalveolar lavage fluid of macaques infected with MPXV or vaccinia virus by the respiratory route.[111] In addition to modulation of host defense proteins, MPXV infection was preferentially associated with a decrease of structural and metabolic proteins in lavage fluids when compared to vaccinia infection suggesting that degradation of structural integrity of the lungs is an important factor in pathogenesis. Finally, a recent study using global systems kinomics demonstrated that Central African MPXV infection selectively downregulates host responses as compared with West African MPXV, including growth factor- and apoptosis-related responses.[112] In addition, differentially phosphorylated kinases were identified and validated as potential targets for inhibition including Akt S473 and p53 S15.

The natural history of MPXV and the emergence of cases in Africa and the United States has been an interesting scientific story with implications for smallpox eradication, bioterrorism, and our awareness of emerging infectious diseases. Current studies should emphasize our need to obtain a better understanding of MPXV epidemiology in Africa since there is recent evidence of increased disease prevalence. In addition, focus on viral pathogenesis in animal models and molecular virology should continue so that additional countermeasures can be developed against MPXV and the other orthopoxviruses.

ACKNOWLEDGMENTS

This work was supported by the NIAID Division of Intramural Research. We would like to thank Dr. Reynolds and Dr. Damon from the US CDC for granting permission to use the artworks in Figure 16.2.

REFERENCES

1. Moss, B. Poxviridae: The viruses and their replication. In *Fields Virology*, 5th ed., Vol. 2 (eds. Knipe, D.M., Howley, P.M., and Griffin, D.E.), pp. 2905–2945 (Lippincott Williams & Wilkins, Philadelphia, PA, 2007).
2. McFadden, G. Poxvirus tropism. *Nat. Rev. Microbiol.* 3, 201–213 (2005).
3. Fields, B.N., Knipe, D.M., and Howley, P.M. *Fields Virology* (Wolters Kluwer Health/Lippincott Williams & Wilkins, Philadelphia, PA, 2007).
4. Garon, C.F., Barbosa, E., and Moss, B. Visualization of an inverted terminal repetition in vaccinia virus DNA. *Proc. Natl. Acad. Sci. USA* 75, 4863–4867 (1978).
5. Shchelkunov, S.N. et al. Human monkeypox and smallpox viruses: Genomic comparison. *FEBS Lett.* 509, 66–70 (2001).
6. Gubser, C. et al. Poxvirus genomes: A phylogenetic analysis. *J. Gen. Virol.* 85, 105–117 (2004).
7. Xing, K. et al. Genome-based phylogeny of poxvirus. *Intervirology* 49, 207–214 (2006).
8. Weaver, J.R. and Isaacs, S.N. Monkeypox virus and insights into its immunomodulatory proteins. *Immunol. Rev.* 225, 96–113 (2008).
9. Jezek, Z. et al. Human monkeypox: Disease pattern, incidence and attack rates in a rural area of northern Zaire. *Trop. Geograph. Med.* 40, 73–83 (1988).
10. Prichard, M.N. and Kern, E.R. Orthopoxvirus targets for the development of antiviral therapies. *Curr. Drug Targets Infect. Disord.* 5, 17–28 (2005).
11. Wolfe, C.L. and Moss, B. Interaction between the G3 and L5 proteins of the vaccinia virus entry-fusion complex. *Virology* 412, 278–283 (2011).
12. Senkevich, T.G., Ward, B.M., and Moss, B. Vaccinia virus entry into cells is dependent on a virion surface protein encoded by the A28L gene. *J. Virol.* 78, 2357–2366 (2004).
13. McIntosh, A.A. and Smith, G.L. Vaccinia virus glycoprotein A34R is required for infectivity of extracellular enveloped virus. *J. Virol.* 70, 272–281 (1996).
14. Roberts, K.L. et al. Acidic residues in the membrane-proximal stalk region of vaccinia virus protein B5 are required for glycosaminoglycan-mediated disruption of the extracellular enveloped virus outer membrane. *J. Gen. Virol.* 90, 1582–1591 (2009).
15. Yang, Z. et al. Expression profiling of the intermediate and late stages of poxvirus replication. *J. Virol.* 85, 9899–9908 (2011).
16. Neyts, J. and De Clercq, E. Efficacy of (S)-1-(3-hydroxy-2-phosphonylmethoxypropyl)cytosine for the treatment of lethal vaccinia virus infections in severe combined immune deficiency (SCID) mice. *J. Med. Virol.* 41, 242–246 (1993).
17. Neyts, J., Snoeck, R., Balzarini, J., and De Clercq, E. Particular characteristics of the anti-human cytomegalovirus activity of (S)-1-(3-hydroxy-2-phosphonylmethoxypropyl) cytosine (HPMPC) in vitro. *Antiviral Res.* 16, 41–52 (1991).
18. De Clercq, E. Cidofovir in the treatment of poxvirus infections. *Antiviral Res.* 55, 1–13 (2002).

19. Magee, W.C., Hostetler, K.Y., and Evans, D.H. Mechanism of inhibition of vaccinia virus DNA polymerase by cidofovir diphosphate. *Antimicrob. Agents Chemother.* 49, 3153–3162 (2005).

20. Magee, W.C. et al. Cidofovir and (S)-9-[3-hydroxy-(2-phosphonomethoxy) propyl]adenine are highly effective inhibitors of vaccinia virus DNA polymerase when incorporated into the template strand. *Antimicrob. Agents Chemother.* 52, 586–597 (2008).

21. Meng, X. et al. Vaccinia virus A6 is essential for virion membrane biogenesis and localization of virion membrane proteins to sites of virion assembly. *J. Virol.* 86, 5603–5613 (2012).

22. Roberts, K.L. and Smith, G.L. Vaccinia virus morphogenesis and dissemination. *Trends Microbiol.* 16, 472–479 (2008).

23. Reeves, P.M. et al. Disabling poxvirus pathogenesis by inhibition of Abl-family tyrosine kinases. *Nat. Med.* 11, 731–739 (2005).

24. Reeves, P.M. et al. Variola and monkeypox viruses utilize conserved mechanisms of virion motility and release that depend on abl and SRC family tyrosine kinases. *J. Virol.* 85, 21–31 (2011).

25. Yang, G. et al. An orally bioavailable antipoxvirus compound (ST-246) inhibits extracellular virus formation and protects mice from lethal orthopoxvirus challenge. *J. Virol.* 79, 13139–13149 (2005).

26. Grosenbach, D.W. et al. Efficacy of ST-246 versus lethal poxvirus challenge in immunodeficient mice. *Proc. Natl. Acad. Sci. USA* 107, 838–843 (2010).

27. Berhanu, A. et al. ST-246 inhibits in vivo poxvirus dissemination, virus shedding, and systemic disease manifestation. *Antimicrob. Agents Chemother.* 53, 4999–5009 (2009).

28. Huggins, J. et al. Nonhuman primates are protected from smallpox virus or monkeypox virus challenges by the antiviral drug ST-246. *Antimicrob. Agents Chemother.* 53, 2620–2625 (2009).

29. Husain, M., Weisberg, A., and Moss, B. Topology of epitope-tagged F13L protein, a major membrane component of extracellular vaccinia virions. *Virology* 308, 233–242 (2003).

30. Husain, M. and Moss, B. Similarities in the induction of post-Golgi vesicles by the vaccinia virus F13L protein and phospholipase D. *J. Virol.* 76, 7777–7789 (2002).

31. Grosenbach, D.W. and Hruby, D.E. Analysis of a vaccinia virus mutant expressing a nonpalmitylated form of p37, a mediator of virion envelopment. *J. Virol.* 72, 5108–5120 (1998).

32. Roper, R.L. and Moss, B. Envelope formation is blocked by mutation of a sequence related to the HKD phospholipid metabolism motif in the vaccinia virus F13L protein. *J. Virol.* 73, 1108–1117 (1999).

33. Sridhar, P. and Condit, R.C. Selection for temperature-sensitive mutations in specific vaccinia virus genes: Isolation and characterization of a virus mutant which encodes a phosphonoacetic acid-resistant, temperature-sensitive DNA polymerase. *Virology* 128, 444–457 (1983).

34. Ladnyj, I.D., Ziegler, P., and Kima, E. A human infection caused by monkeypox virus in Basankusu Territory, Democratic Republic of the Congo. *Bull. WHO* 46, 593–597 (1972).

35. Arita, I., Jezek, Z., Khodakevich, L., and Ruti, K. Human monkeypox: A newly emerged orthopoxvirus zoonosis in the tropical rain forests of Africa. *Am. J. Trop. Med. Hyg.* 34, 781–789 (1985).

36. Breman, J.G. et al. Human monkeypox, 1970–79. *Bull. WHO* 58, 165–182 (1980).

37. Damon, I.K. Status of human monkeypox: Clinical disease, epidemiology and research. *Vaccine* 29(Suppl 4), D54–D59 (2011).

38. Jezek, Z. and Fenner, F. Human monkeypox. *Monogr. Virol.* 17, 1–134 (1988).

39. Fine, P.E., Jezek, Z., Grab, B., and Dixon, H. The transmission potential of monkeypox virus in human populations. *Int. J. Epidemiol.* 17, 643–650 (1988).

40. Reynolds, M.G. and Damon, I.K. Outbreaks of human monkeypox after cessation of smallpox vaccination. *Trends Microbiol.* 20, 80–87 (2012).

41. Rimoin, A.W. and Graham, B.S. Whither monkeypox vaccination. *Vaccine* 29(Suppl 4), D60–D64 (2011).

42. Rimoin, A.W. et al. Major increase in human monkeypox incidence 30 years after smallpox vaccination campaigns cease in the Democratic Republic of Congo. *Proc. Natl. Acad. Sci. USA* 107, 16262–16267 (2010).

43. Reed, K.D. et al. The detection of monkeypox in humans in the Western Hemisphere. *New Engl. J. Med.* 350, 342–350 (2004).

44. Hutson, C.L. et al. Monkeypox zoonotic associations: Insights from laboratory evaluation of animals associated with the multi-state US outbreak. *Am. J. Trop. Med. Hyg.* 76, 757–768 (2007).

45. Nalca, A. et al. Reemergence of monkeypox: Prevalence, diagnostics, and countermeasures. *Clin. Infect. Dis.* 41, 1765–1771 (2005).

46. Update: Multistate outbreak of monkeypox—Illinois, Indiana, Kansas, Missouri, Ohio, and Wisconsin, 2003. *MMWR* 52, 589–590 (2003).

47. Essbauer, S., Pfeffer, M., and Meyer, H. Zoonotic poxviruses. *Vet. Microbiol.* 140, 229–236 (2010).

48. Parker, S. et al. Human monkeypox: An emerging zoonotic disease. *Future Microbiol.* 2, 17–34 (2007).

49. Huhn, G.D. et al. Clinical characteristics of human monkeypox, and risk factors for severe disease. *Clin. Infect. Dis.* 41, 1742–1751 (2005).

50. Reynolds, M.G. et al. Clinical manifestations of human monkeypox influenced by route of infection. *J. Infect. Dis.* 194, 773–780 (2006).

51. Likos, A.M. et al. A tale of two clades: Monkeypox viruses. *J. Gen. Virol.* 86, 2661–2672 (2005).

52. Chen, N. et al. Virulence differences between monkeypox virus isolates from West Africa and the Congo basin. *Virology* 340, 46–63 (2005).

53. Saijo, M. et al. Virulence and pathophysiology of the Congo Basin and West African strains of monkeypox virus in nonhuman primates. *J. Gen. Virol.* 90, 2266–2271 (2009).

54. Sbrana, E. et al. Comparative pathology of North American and central African strains of monkeypox virus in a ground squirrel model of the disease. *Am. J. Trop. Med. Hyg.* 76, 155–164 (2007).

55. Hutson, C.L. et al. Monkeypox disease transmission in an experimental setting: Prairie dog animal model. *PLoS One* 6, e28295 (2011).

56. Hutson, C.L. et al. Dosage comparison of Congo Basin and West African strains of monkeypox virus using a prairie dog animal model of systemic orthopoxvirus disease. *Virology* 402, 72–82 (2010).

57. Hutson, C.L. et al. A prairie dog animal model of systemic orthopoxvirus disease using West African and Congo Basin strains of monkeypox virus. *J. Gen. Virol.* 90, 323–333 (2009).

58. Estep, R.D. et al. Deletion of the monkeypox virus inhibitor of complement enzymes locus impacts the adaptive immune response to monkeypox virus in a nonhuman primate model of infection. *J. Virol.* 85, 9527–9542 (2011).

59. Hudson, P.N. et al. Elucidating the role of the complement control protein in monkeypox pathogenicity. *PLoS One* 7, e35086 (2012).

60. Jezek, Z. et al. Human monkeypox: Confusion with chickenpox. *Acta Trop.* 45, 297–307 (1988).

61. Esposito, J.J., Obijeski, J.F., and Nakano, J.H. Orthopoxvirus DNA: Strain differentiation by electrophoresis of restriction endonuclease fragmented virion DNA. *Virology* 89, 53–66 (1978).

62. Esposito, J.J. and Knight, J.C. Orthopoxvirus DNA: A comparison of restriction profiles and maps. *Virology* 143, 230–251 (1985).

63. Rimoin, A.W. et al. Endemic human monkeypox, Democratic Republic of Congo, 2001–2004. *Emerg. Infect. Dis.* 13, 934–937 (2007).

64. Kulesh, D.A. et al. Smallpox and pan-orthopox virus detection by real-time 3′-minor groove binder TaqMan assays on the roche LightCycler and the Cepheid smart Cycler platforms. *J. Clin. Microbiol.* 42, 601–609 (2004).

65. Olson, V.A. et al. Real-time PCR system for detection of orthopoxviruses and simultaneous identification of smallpox virus. *J. Clin. Microbiol.* 42, 1940–1946 (2004).

66. Kulesh, D.A. et al. Monkeypox virus detection in rodents using real-time 3′-minor groove binder TaqMan assays on the Roche LightCycler. *Lab. Invest.* 84, 1200–1208 (2004).

67. Shchelkunov, S.N. et al. Species-specific identification of variola, monkeypox, cowpox, and vaccinia viruses by multiplex real-time PCR assay. *J. Virol. Methods* 175, 163–169 (2011).

68. Aitichou, M. et al. Dual-probe real-time PCR assay for detection of variola or other orthopoxviruses with dried reagents. *J. Virol. Methods* 153, 190–195 (2008).

69. Putkuri, N. et al. Detection of human orthopoxvirus infections and differentiation of smallpox virus with real-time PCR. *J. Med. Virol.* 81, 146–152 (2009).

70. Karem, K.L. et al. characterization of acute-phase humoral immunity to monkeypox: Use of immunoglobulin M enzyme-linked immunosorbent assay for detection of monkeypox infection during the 2003 North American outbreak. *Clin. Diagn. Lab. Immunol.* 12, 867–872 (2005).

71. Dubois, M.E., Hammarlund, E., and Slifka, M.K. Optimization of peptide-based ELISA for serological diagnostics: A retrospective study of human monkeypox infection. *Vector Borne Zoonotic Dis.* 12, 400–409 (2012).

72. Hammarlund, E. et al. Multiple diagnostic techniques identify previously vaccinated individuals with protective immunity against monkeypox. *Nat. Med.* 11, 1005–1011 (2005).

73. Jezek, Z. et al. Human monkeypox: A study of 2,510 contacts of 214 patients. *J. Infect. Dis.* 154, 551–555 (1986).

74. Hammarlund, E. et al. Duration of antiviral immunity after smallpox vaccination. *Nat. Med.* 9, 1131–1137 (2003).

75. Karem, K.L. et al. Monkeypox outbreak diagnostics and implications for vaccine protective effect. *Nat. Med.* 12, 495–496; author reply 496–497 (2006).

76. McConnell, S. et al. Protection of rhesus monkeys against monkeypox by vaccinia virus immunization. *Am. J. Vet. Res.* 25, 192–195 (1964).

77. Earl, P.L. et al. Immunogenicity of a highly attenuated MVA smallpox vaccine and protection against monkeypox. *Nature* 428, 182–185 (2004).

78. Earl, P.L. et al. Rapid protection in a monkeypox model by a single injection of a replication-deficient vaccinia virus. *Proc. Natl. Acad. Sci. USA* 105, 10889–10894 (2008).

79. Saijo, M. et al. LC16m8, a highly attenuated vaccinia virus vaccine lacking expression of the membrane protein B5R, protects monkeys from monkeypox. *J. Virol.* 80, 5179–5188 (2006).

80. CDC. Fact sheet: Smallpox vaccine and monkeypox. http://www.cdc.gov/ncidod/monkeypox/smallpoxvaccine_mpox.htm (2003) (accessed on October, 2012).

81. Lane, J.M. and Goldstein, J. Adverse events occurring after smallpox vaccination. *Semin. Pediatr. Infect. Dis.* 14, 189–195 (2003).

82. Lane, J.M. and Goldstein, J. Evaluation of 21st-century risks of smallpox vaccination and policy options. *Ann. Intern. Med.* 138, 488–493 (2003).

83. CDC. Fact sheet: Smallpox (vaccinia) vaccine contraindications. http://www.emergency.cdc.gov/agent/smallpox/vaccination/contraindications-clinic.asp (2003) (accessed on October, 2012).

84. Grabenstein, J.D. and Winkenwerder, W., Jr. US military smallpox vaccination program experience. *JAMA* 289, 3278–3282 (2003).

85. Golden, J.W. and Hooper, J.W. The strategic use of novel smallpox vaccines in the post-eradication world. *Expert Rev. Vaccines* 10, 1021–1035 (2011).

86. Nalca, A. and Zumbrun, E.E. ACAM2000: The new smallpox vaccine for United States Strategic National Stockpile. *Drug Des. Devel. Ther.* 4, 71–79 (2010).

87. McCurdy, L.H. et al. Modified vaccinia Ankara: Potential as an alternative smallpox vaccine. *Clin. Infect. Dis.* 38, 1749–1753 (2004).

88. Kenner, J. et al. LC16m8: An attenuated smallpox vaccine. *Vaccine* 24, 7009–7022 (2006).

89. Stittelaar, K.J. et al. Modified vaccinia virus Ankara protects macaques against respiratory challenge with monkeypox virus. *J. Virol.* 79, 7845–7851 (2005).

90. Wilck, M.B. et al. Safety and immunogenicity of modified vaccinia Ankara (ACAM3000): Effect of dose and route of administration. *J. Infect. Dis.* 201, 1361–1370 (2010).

91. von Krempelhuber, A. et al. A randomized, double-blind, dose-finding phase II study to evaluate immunogenicity and safety of the third generation smallpox vaccine candidate IMVAMUNE. *Vaccine* 28, 1209–1216 (2010).

92. Kennedy, J.S. et al. Safety and immunogenicity of LC16m8, an attenuated smallpox vaccine in vaccinia-naive adults. *J. Infect. Dis.* 204, 1395–1402 (2011).

93. Smee, D.F., Bray, M., and Huggins, J.W. Antiviral activity and mode of action studies of ribavirin and mycophenolic acid against orthopoxviruses in vitro. *Antivir. Chem. Chemother.* 12, 327–335 (2001).

94. Alkhalil, A. et al. Inhibition of monkeypox virus replication by RNA interference. *Virol. J.* 6, 188 (2009).

95. De Clercq, E. and Neyts, J. Therapeutic potential of nucleoside/nucleotide analogues against poxvirus infections. *Rev. Med. Virol.* 14, 289–300 (2004).

96. Jin, Y.H. et al. Practical synthesis of D- and l-2-cyclopentenone and their utility for the synthesis of carbocyclic antiviral nucleosides against orthopox viruses (smallpox, monkeypox, and cowpox virus). *J. Organ. Chem.* 68, 9012–9018 (2003).

97. Altmann, S.E. et al. Antiviral activity of the EB peptide against zoonotic poxviruses. *Virol. J.* 9, 6 (2012).

98. Altmann, S.E. et al. Inhibition of cowpox virus and monkeypox virus infection by mitoxantrone. *Antiviral Res.* 93, 305–308 (2012).

99. Johnston, S.C. et al. in vitro inhibition of monkeypox virus production and spread by Interferon-beta. *Virol. J.* 9, 5 (2012).

100. Jordan, R. et al. Development of ST-246(R) for treatment of poxvirus infections. *Viruses* 2, 2409–2435 (2010).

101. Lanier, R. et al. Development of CMX001 for the treatment of poxvirus infections. *Viruses* 2, 2740–2762 (2010).

102. Sbrana, E. et al. Efficacy of the antipoxvirus compound ST-246 for treatment of severe orthopoxvirus infection. *Am. J. Trop. Med. Hyg.* 76, 768–773 (2007).

103. Jordan, R. et al. ST-246 antiviral efficacy in a nonhuman primate monkeypox model: Determination of the minimal effective dose and human dose justification. *Antimicrob. Agents Chemother.* 53, 1817–1822 (2009).

104. Lederman, E.R. et al. Progressive vaccinia: Case description and laboratory-guided therapy with vaccinia immune globulin, ST-246, and CMX001. *J. Infect. Dis.* 206, 1372–1385 (2012).

105. Reynolds, M.G., Carroll, D.S., and Karem, K.L. Factors affecting the likelihood of monkeypox's emergence and spread in the post-smallpox era. *Curr. Opin. Virol.* 2, 335–343 (2012).

106. Dyall, J. et al. Evaluation of monkeypox disease progression by molecular imaging. *J. Infect. Dis.* 204, 1902–1911 (2011).

107. Earl, P.L., Americo, J.L., and Moss, B. Lethal monkeypox virus infection of CAST/EiJ mice is associated with a deficient gamma interferon response. *J. Virol.* 86, 9105–9112 (2012).

108. Goff, A.J. et al. A novel respiratory model of infection with monkeypox virus in cynomolgus macaques. *J. Virol.* 85, 4898–4909 (2011).

109. Johnson, R.F. et al. Comparative analysis of monkeypox virus infection of cynomolgus macaques by the intravenous or intrabronchial inoculation route. *J. Virol.* 85, 2112–2125 (2011).

110. Nalca, A. et al. Experimental infection of cynomolgus macaques (*Macaca fascicularis*) with aerosolized monkeypox virus. *PloS One* 5, e12880 (2010).

111. Brown, J.N. et al. Characterization of macaque pulmonary fluid proteome during monkeypox infection: Dynamics of host response. *Mol. Cell. Proteomics* 9, 2760–2771 (2010).

112. Kindrachuk, J. et al. Systems kinomics demonstrates Congo Basin monkeypox virus infection selectively modulates host cell signaling responses as compared to West African monkeypox virus. *Mol. Cell. Proteomics* 11, M111 015701 (2012).

113. Rubins, K.H. et al. Comparative analysis of viral gene expression programs during poxvirus infection: A transcriptional map of the vaccinia and monkeypox genomes. *PLoS One* 3, e2628 (2008).

114. Rubins, K.H. et al. Stunned silence: Gene expression programs in human cells infected with monkeypox or vaccinia virus. *PLoS One* 6, e15615 (2011).

115. Lin, Y.C. et al. Vaccinia virus DNA ligase recruits cellular topoisomerase II to sites of viral replication and assembly. *J Virol* 82, 5922–5932 (2008).

116. Vigne, C.S. et al. Inhibition of vaccinia virus replication by two small interfering RNAs targeting B1R and G7L genes and their synergistic combination. *Antimicrobial Agents and Chemotherapy* 53(6), 2579–2588 (2009). doi:10.1128/AAC.01626-08.

117. Meng, X., Embry, A., Rose, L., Yan, B., Xu, C., and Xiang, Y. Vaccinia virus A6 is essential for virion membrane biogenesis and localization of virion membrane proteins to sites of virion assembly. *Journal of Virology* 86(10), 5603–5613 (2012). doi:10.1128/JVI.00330-12.

118. Laliberte, J.P., Wesberg, A.S., and Moss, B. The membrance fusion step of vaccinia virus entry is cooperatively mediated by multiple viral proteins and host cell components. *Plos Pathogens* 12 (2011).

119. Mohamed, M.R., Rahman, M.M., Rice, A., Moyer, R.W., Werden, S.J., and McFadden, G. Cowpox virus expresses a novel ankyrin repeat NF-kappaB inhibitor that controls inflammatory cell influx intoe virus-infected tissues and is critical for virus patheogenesis. *Journal of Virology* 83(18), 9223–9236 (2009).

17 Omsk Hemorrhagic Fever Virus

*Daniel Růžek, Michael R. Holbrook, Valeriy V. Yakimenko,
Lyudmila S. Karan, and Sergey E. Tkachev*

CONTENTS

17.1 INTRODUCTION

Omsk hemorrhagic fever (OHF) was first described as a new disease in 1945–1946, when physicians in the northern-lake steppe and forest-steppe areas of the Omsk oblast (an administrative unit similar to county or province) of Russia recorded sporadic cases of an acute febrile disease with abundant hemorrhagic signs (i.e., hemorrhaging from the nose, mouth, uterus, and skin and hemorrhagic rash) and leukopenia.[1,2] In 1947, an expedition including Russian scientists M.P. Chumakov, A.P. Belyayeva, A.V. Gagarina, and their coworkers arrived in the Omsk oblast to investigate this new disease and identify its causative agent and mode of transmission. During the expedition, a new virus, Omsk hemorrhagic fever virus (OHFV), was isolated from a human patient and later from a pool of *Dermacentor reticulatus* ticks. The *D. reticulatus* tick was subsequently identified as the principal arthropod vector of OHFV.[1,2] Further investigation also identified a predominant pattern of direct contact with muskrats (*Ondatra zibethicus*) among new cases. An increased incidence of OHF was found among muskrat hunters and their family members, who participated in muskrat skinning and preparing skins.[3] In order to capture muskrats, hunters destroy muskrat lodges with their bare hands and seize the animals. The killing of muskrats is carried out by stretching the rodent, which causes large vessels to rupture. The capture and processing of muskrat carcasses represent a high-risk activity for potential infection with OHFV by either aerosol or contact exposure.

Muskrats are an alien animal species in Russia that were introduced to Siberia from Canada for industrial fur production.[4] In 1935, the first muskrats were imported into the Novosibirsk oblast. Muskrats are prolific breeders, and once introduced into Russia, they propagated quickly and rapidly increased their range. The breeding of muskrats, however, did not reach its economic potential in Siberia because of fatal epizootics within the muskrat population. It is probable that OHFV was endemic in Siberia prior to the release of the nonnative muskrat. The introduction of the highly susceptible muskrat appears to have greatly amplified infection rates in other animals, including human beings.[5,6] In the years 1946–1970, 76 different epizootics of OHF were recorded within the muskrat population in Tyumen, Kurgan, Omsk, and Novosibirsk oblasts; the epizootics were followed by human cases of OHF in Omsk and Novosibirsk oblasts. More recently, epizootic activity of OHFV was recorded beginning in the late 1980s in Tyumen (1987), Omsk (1988, 1999–2007), Novosibirsk (1989–present), and Kurgan (1992) oblasts. Between 1946 and 1958, a total of 972 human cases of OHF were officially recorded. In 1960–1970, the incidence of OHF decreased significantly. Between 1990 and the present time (2012), cases of OHF were reported only in the Novosibirsk oblast with the highest number of cases in 1990 (29 cases) and 1991 (38 cases).[6,7]

17.2 CLASSIFICATION AND MORPHOLOGY

OHFV is a member of the tick-borne encephalitis (TBE) serocomplex of flaviviruses (family Flaviviridae, genus *Flavivirus*). OHFV is closely related phylogenetically to TBE virus (TBEV). Using classical serological methods, such as neutralization tests (NTs) or complement fixation (CF) tests, for the initial identification of OHFV, it was very difficult to differentiate between OHFV and TBEV, which had previously been identified in the region. This serological cross-reactivity suggested a close antigenic relationship between these viruses. OHFV morphology, structural features, and mode of replication are the same as those of TBEV.[8] However, the genetic factors that determine the virus association with a hemorrhagic manifestation rather than encephalitis are unknown, but are under intensive investigation. OHFV is also closely related to other tick-borne flaviviruses, which include Kyasanur Forest disease virus, louping ill virus, Powassan virus, and Langat virus.

Virions of flaviviruses, including OHFV, are spherical particles, approximately 50–60 nm in diameter with an electron-dense core forming a nucleocapsid enclosed in capsid (C) protein and surrounded by a host cell-derived lipid bilayer.[9] Membrane (M) and envelope (E) proteins are integrated in the bilayer. It has been shown that E protein shares N-glycosylation sites, cysteine residues, the fusion peptide, and a hexapeptide with other tick-borne flaviviruses.[10] A sequence of three amino acids (AQN; amino acids 232–234), which was previously shown to be specific for the TBEV, is altered to MVG or MMG in OHFV. It is predicted that it has a higher hydrophobicity than AQN sequence, and this may have significant implications for the phenotypic characteristics of OHFV. However, several other unique amino acid substitutions were identified in positions that may have significance for OHF pathogenesis.[10] M protein, produced during the maturation of nascent virus particles within the secretory pathway, is a small proteolytic fragment of the precursor prM protein. Glycoprotein E, the major protein component of the virus surface, mediates receptor binding and fusion activity after uptake by a receptor-mediated endocytosis.[11] E protein is the main target of neutralizing antibodies and induces protective immunity in infected organisms. For its functional importance, it is believed that E protein is also an important determinant of virulence.[12]

OHFV has a (+)ssRNA genome of 10,787 bases in length with an open reading frame (ORF) of 10,242 nucleotides encoding 3,414 amino acids. Like all flaviviruses, the ORF is flanked by 5′ and 3′ untranslated regions (UTRs). The 5′ UTR of OHFV contains a 5′-methylG-cap. The structure of the 5′ UTR is considerably different from other TBE complex viruses through an approximately 30-nucleotide stretch, while the remainder of the 5′ UTR is highly homologous. As this difference does not necessarily define tissue tropism, it may determine replication efficiency in a cell or tissue once infection has occurred.[13] The 3′ UTR lacks 3′-poly(A) tail-like other flaviviruses and is slightly shorter in comparison to other TBEV, but is otherwise similar to the 3′ UTR observed in Far Eastern and Siberian strains of TBEV.[13] The viral ORF encodes a large polyprotein that is co- and posttranslationally cleaved by cellular and viral proteases into three structural proteins (C, prM, and E) and seven nonstructural proteins (NS1, NS2A, NS2B, NS3, NS4A, NS4B, and NS5). Viral nonstructural proteins have several functions during virus replication in the host cells. For example, they form the RNA-dependent RNA–polymerase replication complex, provide a serine protease needed to cleave the polyprotein, and may also play a role in regulating the host innate immune response following infection.[14]

In the 1960s, strains of OHFV isolated from various sources (e.g., ixodid ticks, muskrats, and blood and organs of sick people) were differentiated into two groups based on biological and serological properties.[15,16] Based on the comparison of the nucleotide sequence of viral E protein genes, it was demonstrated that OHFV is most closely related to the western subtype of TBEV, rather than to far eastern and Siberian subtypes. Phylogenetic analysis of E and NS5 genes of various OHFV strains revealed two genotypes.[17] The first genotype is formed predominantly by strains from Novosibirsk and the most strains from Omsk oblasts. The second genotype includes strains from the Kurgan oblast and two strains from the western part of Omsk oblast. Within the first genotype, the level of homology of E gene is 96.7% while the second genotype has a homology of 98.1%–100%.[18] The homology between genotypes is 87.2%–89.0%. Homology with the most closely related TBEV is 81.7%–83.4% (genotype I), and 81.9%–82.1% (genotype II), respectively.[18] In the case of NS5 gene, the homology between strains within genotype I is 96.4%–99.8%, and 88.6%–89.5% between the two genotypes. The homology between OHFV and the most closely related TBEV in NS5 is 83.2%–83.9%.[18] When the complete genomes of OHFV strains Kubrin (genotype I) and Bogoluvovska (genotype II) were compared, only six nucleotide differences were identified between these two strains throughout the entire viral genome, and they encoded four amino acid changes including three in the E protein.[17]

17.3 BIOLOGY AND EPIDEMIOLOGY

Human beings can be infected with OHFV by transmissive (i.e., via feeding of infected tick) or nontransmissive (e.g., direct contact with carcasses of infected animals, respiratory or alimentary) routes. There is no evidence of direct transmission of OHFV or any other tick-borne flavivirus, between people. No cases of transmission were reported within the same hospital, and none was noted among members of the same family. People who took care of the sick persons or who came in close contact with them did not contract the disease.[1]

At the time of the discovery of OHF, researchers focused on the investigation of transmissive infections as most of the initial cases had transmissive features (i.e., via vector). The only exceptions were cases in a village Gornostayevka in 1945 (Sargat district, Omsk oblast), where potential nontransmissive OHF cases were reported. In the active OHFV foci, a high abundance of *D. reticulatus* (*pictus*) ticks was observed, suggesting a connection between these ticks and the new disease. This connection was subsequently confirmed. The narrow-skulled vole (*Microtus gregalis*) was predominant in the area and was considered the most likely intermediate rodent host of OHFV with *D. reticulatus* ticks serving as the vector as well as the main reservoir of the virus.[1] Low levels of transovarial transmission of OHFV in *D. reticulatus* suggested that this mode does not play a crucial role in virus maintenance in the natural focus.[18] Adult *D. reticulatus* ticks feed on wild ungulates and humans, whereas immature forms feed mainly on water voles (*M. gregalis*) in forest-steppe habitats. Vole populations are cyclic, and expansion of the virus-infected tick population coincides with increases in vole populations.[19,20] In the steppe regions of southern and western Siberia, the virus is transmitted mainly by the

Dermacentor marginatus tick. Gamasid mites and the taiga tick (*Ixodes persulcatus*) are believed to be involved in the sylvatic cycle of OHFV.[21] Although OHFV was also isolated from several species of mosquitoes (*Ochlerotatus excrucians*, *Coquilletidia richiardii*, and *Oligoryzomys flavescens*), their role in the circulation of OHFV in nature is unclear.[3]

As early as 1946–1948, cases of OHF were reported in laboratory workers following contact with infected muskrats. In 1954, OHFV was isolated from the brain of a dead muskrat for the first time. It was also demonstrated experimentally that muskrats are highly sensitive to OHFV infection.[3]

During the years 1946–1970, 76 OHF epizootics in muskrats were reported. OHF natural foci are situated in Omsk, Tyumen, Novosibirsk, and Kurgan oblasts of Russia. Natural foci are typical in areas with forests, steppe with multiple marshes, lakes, and reed thickets (Figure 17.1). Development and establishment of natural foci for OHFV are favored by a combination of landscape, climatic, and biotic factors. The development of an epizootic is usually slow, not occurring simultaneously in the whole lake area. Initially, there is a local focus of OHFV; the development of the full epizootic takes one to three years. From 1971 to 1989, there was no documented outbreak of OHF in muskrats (Figure 17.1). The apparent disappearance of OHF might have been the result of changes in agricultural activities, leading to decreased numbers of *D. reticulatus* and narrow-skulled voles as the main host of the preimaginal stages of these ticks. However, it was also suggested

that selection of resistant muskrat populations may have been responsible for the disappearance of OHF.[22] Currently, there is virtually no OHFV detected in ixodid ticks in Siberia. As a consequence, the numbers of transmissive human OHF cases is very low (less than 7.4% of all OHF cases).[23]

The key role of muskrats in the dissemination of OHFV in the endemic regions is well accepted. The introduction of muskrats to Siberia provided a susceptible virus shedding host for OHFV and stimulated the emergence of this disease. Urine and other excreta from infected muskrats contain high titers of OHFV. In water, the virus remains stable for more than 2 weeks in summer and for 3.5 months in winter providing ample opportunity for the infection of aquatic animal species.[3] Animals living in or near water may become infected through contact with, or consumption of, water contaminated by muskrat corpses, urine or feces. Seroepidemiological studies suggest that many animal species come in contact with OHFV, including rodents, insectivores, birds, ungulates, and domestic animals. Some wild hosts develop latent chronic infections, others develop acute disease (e.g., root vole [*Microtus oeconomus*], narrow skulled vole [*M. gregalis*], red-cheeked suslik [*Citellus erythrogenys*], hedgehog, etc.), and in some cases, fatal infections. Multiple hosts are, therefore, involved in the OHFV natural foci, in particular water voles (*Arvicola amphibius*) and narrow-skulled voles (*M. gregalis*). Some small mammal hosts have been shown to disseminate the

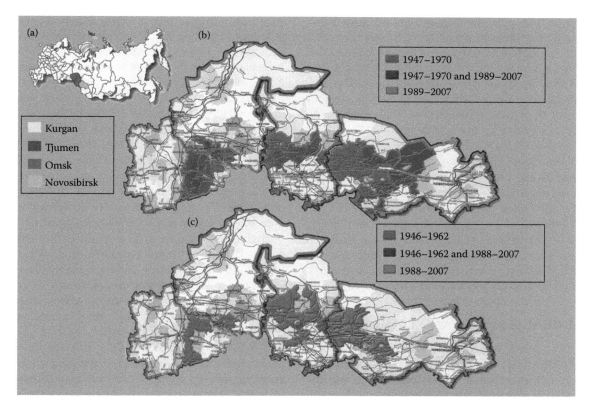

FIGURE 17.1 (a) OHF endemic administrative regions of Russia, that is, Kurgan, Tyumen, Omsk, and Novosibirsk regions. (b) Geographical distribution of OHF epizootics in the muskrat population. (c) Geographical distribution of the areas of OHF human morbidity since 1946. (From Růžek, D. et al., Omsk hemorrhagic fever virus, in: Singh, S.K. and Růžek, D., eds., *Viral Hemorrhagic Fevers*, CRC Press, Boca Raton, FL, 2013, 553–561.)

virus in their excreta.[3] The virus transmission can also occur following direct contact between infected animals and possibly via an aerosol.[3] Nontransmissive infection of humans with OHFV occurs mainly after close contact with infected muskrats; that is, the patients are generally rural residents, agricultural workers, or people involved in hunting and skinning these animals.

Seasonal morbidity of OHF in humans has two peaks. The first peak in May and June correlates with the activity of *D. marginatus* (in the southern and western areas of Siberia) or *D. reticulatus* (in the northern forest-steppe regions of Siberia). The second peak of OHF morbidity occurs between August and September. This correlates with the muskrat hunting season where most of the patients are infected following contact with the muskrats harboring OHFV.

Between the years 1946 and 1958, 972 OHF cases were officially recorded with a case fatality rate of 1%–2%. However, it is assumed that the number of cases was much higher, because mild or subclinical cases were not reported. The patients were local residents, who come in contact with infected ticks.[23] The established zone of OHFV distribution at the time encompassed the entire forest-steppe region of the Western Siberian lowland.[24] In December 1961, an outbreak of OHF was reported in the Zdvinskiy Rayon of the Novosibirsk oblast among hunters and family members who had direct contact with muskrats. However, during the next 10 years, the incidence of OHF decreased markedly, and only few cases were reported. In November–December 1962 and in February 1963 in the Veselovskiy region of the Novosibirsk oblast, acute febrile diseases, similar to OHF, were recorded in nine persons. All the affected persons had hunted muskrats and had noted a large number of dead and sick animals. The decrease in OHF incidence during 1960–1970 probably correlated with the decrease in OHFV natural foci, decrease in the numbers of *D. reticulatus*, and narrow-skulled vole as the main host of the preimaginal stages of these ticks.[18] From 1990 until the present time (2012), OHF cases were reported only in the Novosibirsk region. The last large outbreaks of OHF were in 1990 (29 cases) and 1991 (38 cases). Most of the recent OHF cases represent muskrat hunter and poachers. Only about 10% of cases were associated with tick bites. In 1998, seven OHF cases were reported, one was fatal and three were severe. However, the exact numbers of OHF cases are not available, since this disease is frequently misdiagnosed or mild cases are not reported.[25] Humans are susceptible to the infection at any age, but 40- to 50-year-old patients predominate.

17.4 CLINICAL FEATURES AND PATHOGENESIS

OHF typically presents as a nonspecific febrile illness where the clinical picture is a combination of various symptoms. In the majority of cases, apparent nasal hemorrhage in the first days of illness, hemoptysis, petechial eruption on the skin, hemorrhagic enanthema on the mucous cavity of the mouth, and other hemorrhagic manifestations can be seen.

The incubation period of OHF is 3–7 days, usually without prodromal signs, but sometimes malaise, aches, and pains

are recorded. There is sudden onset of headache and hemorrhagic signs (e.g., bleeding from the nose, mouth, and uterus). Cough, muscle pain, gastrointestinal symptoms, and dehydration are also frequent symptoms. Fever is high (39°C–40°C), lasting 8–15 days, and frequently accompanied by a chill. Hypotension and bradycardia; hyperemia of the face, neck, and breast; acute scleral injection; bright colorization and light edema of the tunica mucosa in the mouth and throat; unusual dryness of mucous membranes, especially on the tongue; putrid odor from the mouth; and most predominantly an enlarged liver are seen. The face can also become slightly puffy, and labial fissures and crusts appear.[6,26–29] By the third and fourth days of clinical disease, the signs described above typically progress further; for example, face hyperemia and sclera injection are pronounced more intensely, the face becomes puffy, and pharynx hyperemia is intensified (looks like "flaming"). Mucous membranes become increasingly dry, and labial fissure and crusts appear. Also, permanent gingivitis without pronounced stomatorrhagia, tonsils, and soft palate hyperemia and uvula edema (without inflammatory changes) are observed. In rare cases, surface necroses in the pharynx (which usually disappears rapidly) may be observed. Poignant skin hyperesthesia and muscle pain may be recorded. The skin is very sensitive to the touch. Patients typically exhibit arterial hypotension and bradycardia. In severe cases of OHF, patients may develop hemorrhagic manifestations that are an important marker of the disease. Hemorrhagic signs may include hemorrhaging on mucous membranes, uterine bleeding, bleeding at venipuncture sites, gastrointestinal bleeding, and, in very severe cases, pulmonary bleeding. Petechial rash was documented in up to 22% of cases during the most recent outbreaks. Typhoid maculopapular rash on the skin of abdomen, and upper and lower extremities is seen in some cases of OHF, but is very rare.[6,26–28]

After 1–2 weeks of symptomatic disease, some patients recover without any complications, but in 30%–50% of cases, a second phase occurs. The second phase is characterized by fever and signs of (meningo)encephalitis. The duration of the relapse typically ranges from 5 to 14 days. Common complaints include permanent headache, meningism, nausea, chill, reddening of the face and sclerae, nasal and gingival bleeding, hematuria, and uterine bleeding. Sometimes, petechial rash may appear, with bruises at the site of pressure or injections. Laboratory analysis of blood has shown leukopenia, a neutrophil shift up to stabnuclear during the acute period of the disease, thrombocytopenia, and plasmacytosis, and some patients have pathology in internal organs (pneumonia and nephrosis).[6,26–28]

Recovery from OHF is usually without sequelae after quite a long period of asthenia. While permanent sequelae are rare, they may include weakness, hearing loss, hair loss, and behavioral, psychological, or psychiatrical difficulties associated with the loss of neurological function (e.g., poor memory, reduced ability to concentrate, and reduced ability to work).[30] Pareses are not seen in patients with OHF.[6,26–28]

There are no pathomorphological changes in internal organs typical exclusively for OHF. In brain, focal diffuse degenerative necrotic and encephalitic inflammatory changes, proliferation of microglial cells, and perivascular cell infiltrates and edema can be seen. The blood–brain barrier is usually severely affected.[31] In the internal organs, there is stagnant plethora, dilatation of capillaries, diapedetic hemorrhages, toxic edema, and degeneration of parenchymatous elements (liver, heart, and kidneys).[2]

The prognosis of OHF is usually favorable. The case fatality rate of OHF is quite low, ranging from 0.4% to 2.5%. In fatal cases, the patients die either in the period of disease with a rapid increase in hemorrhagic signs (i.e., gastric and intestinal bleeding) or in the second phase of the disease as a result of sepsis (i.e., suppurative bilateral parotiditis and empyema). The patients are highly sensitive to secondary infections.[6,28]

Apart from the classical form of OHF, asymptomatic or influenza-like courses without hemorrhagic signs of OHF have also been reported. In the literature, subfebrile cases with equivocal clinical signs, but with a characteristic blood picture have been described.[23,28,29] This atypical clinical course of OHF has been seen more frequently in patients infected after contact with an ill muskrat and has been the predominant course of disease during more recent outbreaks. Typical cases with clear hemorrhagic signs were seen in less than 20% of the patients.[6,23,28,29]

OHFV possesses a wide range of pathogenicity. The experimental evaluation of OHF pathogenesis has been investigated primarily using laboratory mice and nonhuman primates.[32,33] Laboratory mice are sensitive to infection with OHFV and develop fatal neuroinfection. Signs of disease in OHFV-infected mice include spasms, paresis, and paralysis. The animals are weakened, lose mobility and appetite, lie in a corner, and become hyperpneic. Mice usually die within a few hours, up to 1 day, after the onset of the disease. Regardless of the method of virus challenge (i.e., intracerebral, subcutaneous, intraperitoneal, intranasal, and intracutaneous challenge), viral antigen has been found to accumulate in the cerebellum and brain hemispheres. However, similar virus titers were also found in lungs, kidney, blood, and feces. A lower titer was observed in the spleen and the least in the liver.[34] In another study, OHFV infection of mice did not cause paralysis or significant infection of the cerebrum but showed marked involvement of the cerebellum.[32] In mice, OHFV can also be detected in the urine. The tissue of the cerebellar cortex contains foci of altered and destroyed neurons. Viral antigen accumulation has been found in the cytoplasm of the perikarya, on the cell surface, in axons, and in the intercellular space.[34] Many damaged Purkinje cells have been observed in the cerebellar cortex. Early and prominent induction of IL-1α, TNF-α, IL-12p10, MCP-1, MIP-1α, and MIP-1β in the spleen of infected mice has also been observed.[35]

Experimentally infected macaques (*Macaca radiata*) do not develop any signs of clinical disease, and no virus can be isolated from tissues or blood. However, the animals seroconvert and elevated levels of serum aminotransferase can be detected.[33]

17.5 IDENTIFICATION AND DIAGNOSIS

Diagnosis of OHF is usually based on clinical and epidemiological observations. Laboratory diagnosis is based on serological methods, which represent the gold standard in OHF diagnostics. Antibodies to OHFV can be detected by ELISA, and seroconversion with paired sera is investigated with hemagglutination inhibition (HI), CF, and NTs.

OHFV is a biosafety level (BSL)-3 or BSL-4 agent in several countries, and therefore, all work with the live virus or with samples potentially containing live virus must be done in appropriate facilities.[36] Tissues for viral isolation, electron microscopy, and assays for the direct detection of antigen and genomic sequences should be collected aseptically. Samples for virus isolation should be kept frozen at −70°C or on dry ice continuously, avoiding freeze–thaw cycles, which can inactivate the virus. Tissues collected for electron microscopy should be minced and placed directly in glutaraldehyde (2.5%). Because autolytic changes occur rapidly, the tissues should be fixed as quickly as possible.[37]

For virus isolation, tissue homogenates, fluids, and serum collected in the acute phase of the disease should be inoculated into cultures of several cell lines and intracerebrally into 2- to 4-day-old suckling mice.[3,38] Serum samples should be inoculated undiluted and in 10^{-1} and 10^{-2} dilutions. Intracerebrally inoculated mice should be observed twice daily for up to 2 weeks for signs of illness (neurological signs, apathy, mice out of the nest, etc.) and death. Moribund or dead mice should be dissected immediately or frozen at −70°C until their brains are removed and processed further.[37]

After inoculation, cell cultures should be observed daily for the presence of cytopathic effect. However, OHFV often replicates without any cytopathic effect, or the cytopathic effect is only weak.[39] Therefore, the replication of OHFV in each cell line must also be evaluated by immunofluorescence assay or by RT-PCR.

Antibodies against OHFV can be detected by enzyme-linked immunosorbent assay (ELISA). Serum samples should be collected aseptically and stored either refrigerated or frozen. Freeze–thaw cycles may reduce antibody titers and should be therefore avoided. This is best done by comparing antibody titers in serum samples drawn during the first week of illness and 2–3 weeks later.[37]

HI antibody titers rise rapidly within the first week of illness and are long-lived. The use of goose (*Anser cinereus*) erythrocytes is preferred.[40] Inactivated sucrose–acetone or acetone and ether extracts of infected mouse brains provide a high-titer source of viral antigen. The phenomenon of hemagglutination, however, is markedly dependent on pH, does not occur with most flaviviruses above pH 7, and is usually maximal at a slightly acidic pH. It has been reported that for OHFV, the pH for hemagglutinating activity is between 5.95 and 6.28 (optimum pH 6.08–6.15) at 4°C; however, a preliminary HI test should be performed to identify a specific pH optimum for a specific lot of antigen and to determine the antigen titer. Arbovirus catalog indicates the HI pH range of 6.6–7.0 with an optimum of 6.8.[41] The optimal pH is dependent on the method of antigen preparation, the presence or absence of bovine albumin in the diluent, and the temperature of incubation. A fourfold change between acute- and convalescent-phase samples is indicative of a recent infection. A titer of >1:80 indicates presumptive recent infection, and a titer >2560 represents obvious recent secondary antibody response to flavivirus infection. The problem of this assay lies in the fact that HI antibodies tend to be broadly cross-reactive to common epitopes among the flaviviruses, especially with viruses within the TBE serocomplex.[37]

The CF test is moderately specific. Because this test is relatively insensitive, it should be used in combination with some other procedure. A fourfold change is indicative of a recent infection. It is important to note that some individuals fail to produce detectable CF antibodies. Elderly individuals, in particular, have a delayed or undetectable response.[37]

The NT is considered to be the most specific serological test for the identification of arbovirus infections. Neutralizing antibodies usually become detectable within the first week after the onset of the disease and can persist for years, or over a lifetime. However, it is important to note that antibodies to other tick-borne flaviviruses also have the ability to cross-neutralize OHFV.[38] In studies evaluating the cross-neutralization of OHFV by heterologous serum from related tick-borne flaviviruses, it was found that TBEV antiserum was the most highly cross-reactive while those from more distantly related viruses, including Langat, Kyasanur Forest disease, Negishi, and louping ill viruses, had lower neutralization titers. In plaque reduction assays, the serum samples are diluted in 1:5 and with an addition of an equal volume of the virus, giving an initial dilution 1:10. OHFV forms plaques in the culture of PS cells under carboxymethylcellulose overlay, or in HeLa, and hamster kidney cells, mouse embryo cells (MEC-1), and some other cell lines under agar overlay.[3,42] Fourfold changes in titer are interpreted as an evidence of a recent infection. Titer >1:80 indicates a presumptive recent infection.[37]

Detection procedures based on reverse transcription-polymerase chain reaction (RT-PCR): It is believed that PCR-based methods are not suitable for laboratory diagnosis of OHF. The viremic phase of OHFV infection appears to occur during the prodromal phase, or before the onset of clinical signs of disease. Subsequently, the virus is typically cleared by the time a patient reports to hospital, making molecular detection of the virus likely to be unsuccessful. Molecular methods are suitable for the screening of ticks and animals for the presence of OHFV or for postmortem investigations.

1. Standard RT-PCR protocol: Two primer pairs OHF-E1F/OHF-E2R, and OHF-E3F and OHF-E4R were derived from the highly conserved regions of OHFV envelope gene (GenBank accession No. AY438626). As these two primer sets cover the whole E gene, they can be also used for the preparation of templates for sequence analyses and subsequent phylogeny in molecular-epidemiological surveys.

Primer	Sequence (5′–3′)	Nucleotide Positions[a]	Expected Product (bp)
OHF-E1F	ACCAGGATTGTCATC-GTGTCAGCA	922–945	769
OHF-E2R	GTTCAGCATTGTTC-CAACCCACCAT	1690–1666	
OHF-E3F	CACGGCATGGCAG-GTTCACAGAGAT	1602–1626	696
OHF-E4R	GTTCCATTCTTTCAGT-GTCCACAGCACAT	2497–2469	

[a] Based on the genome sequences of OHFV strain Kubrin (GenBank accession No. AY438626).

The expected PCR product is 769 bp with primers OHF-E1F and OHF-E2R, and 696 bp with primers OHF-E3F and OHF-E4R. The RT-PCR can be used for viral RNA detection in brains of mice and supernatants cell cultures inoculated with clinical samples. In addition, despite the fact that this method has not been evaluated clinically, it may be considered for a direct detection of viral RNA in blood samples at the first days of the disease or in postmortem tissues.[7]

2. Real time RT-PCR protocol: Primers OHF-d1F and OHF-d2R from the conserved envelope gene sequence (GenBank accession No. X66694) can be used in combination with a TaqMan probe for specific detection of OHFV RNA via a real-time RT-PCR method.[7]

Primers/Probe	Sequence (5′-3′)	Nucleotide Positions[a]	Expected Product (bp)
OHF-s1F	GGCACARACCGTT-GTTCTTGAGCT	582–605	139
OHF-d2R	GCGTTCWGCATT-GTTCCAWCCCAC-CAT	720–694	
DNA probe	JOE-AGGTGTTCT-GCTGTCTTGTC-GAGCACCT-BQH1	626–603	

[a] Based on the nucleotide sequence of OHFV envelope gene (GenBank accession No. X66694).

The RT-PCR and real-time RT-PCR assays exhibit high specificity. More than 50 strains of other flaviviruses, for example, TBEV, Langat virus, Powassan virus, West Nile virus, Japanese encephalitis virus, louping ill virus, and Greek goat encephalitis virus, were tested using these assays, and no cross-reactivity was observed (Karan, unpublished data).

17.6 TREATMENT AND PREVENTION

No specific therapy against OHFV is available. The disease is generally self-limiting, but patients must be kept on strict bed rest. The main focus of the treatment is to control symptoms. Abundant intake of liquids and a nutritious diet is recommended. Administration of potassium chloride, glucose, and vitamins K and C is believed to be beneficial for the patients. A long convalescent period is necessary to reduce any permanent complications. Any complications that might occur during OHF, like pneumonia, cardiac manifestations, bacterial infections, etc., should be treated accordingly. Transfusions are indicated in cases with severe blood loss. Hemostatic drugs that strengthen vascular walls can be used.[43]

Several compounds exhibit anti-OHFV activity during experiments in cell culture or in laboratory animals. Recombinant human interferon α-2b completely inhibits OHFV reproduction in cell culture.[44] Ribavirin and interferon inducers larifan and rifastin exhibit moderate virus growth inhibition in cell culture; larifan has the highest antiviral activity when tested in laboratory animals infected with OHFV. However, no data on the effectiveness of the treatment in human OHF patients with these compounds are available.[44]

A vaccine against OHFV was developed as early as in 1948. This vaccine was highly effective, but its production was discontinued owing to the adverse reactions to the mouse-brain components of the vaccine and because the disease incidence has dropped.[45] Vaccines against TBEV can be used to prevent infection with OHFV, since there is high antigenic similarity between these two viruses.[15] These vaccines were used as a preventive measure against OHFV infection during the 1991 outbreak. Recently, the European TBEV vaccines were shown to be highly protective against OHFV challenge.[46,47]

17.7 CONCLUSIONS

OHF is a reemerging disease in the endemic region in Russia. After decades of decreased OHFV activity, a new epizootic and partially also epidemic activity is seen in the natural foci. However, this disease is relatively unstudied. The dynamics of the enzootic environment and the transmission cycle of the OHFV need to be further detailed. We know very little about the molecular biology of OHFV and interaction of OHFV with host cells, but also pathogenesis of OHF in reservoir animals, muskrats, and humans. New data on viral replication offer substantial potential for the development of new drugs. Understanding mechanisms of OHFV infection will assist in the research of other hemorrhagic fevers, interaction of hemorrhagic viruses with vascular endothelium, and will help in the development of new therapeutic approaches in the area of viral hemorrhagic fevers.

ACKNOWLEDGMENTS

This chapter represents an updated version of previously published chapters (Růžek et al., 2011, 2013).[7,48] The authors acknowledge financial support by the Czech Science Foundation project No. P302/10/P438 and No. P502/11/2116, and grant Z60220518 from the Ministry of Education, Youth, and Sports of the Czech Republic, AdmireVet project No. CZ.1.05./2.1.00/01.006. The founders had no role in study design, data collection and analysis, decision to publish, or preparation of the manuscript. MRH was supported by a contract to Battelle Memorial Institute with NIAID No. HHSN272200200016I. SET was partially supported by Integration interdisciplinary project grant No. 135 from the Siberian Branch of the Russian Academy of Sciences.

REFERENCES

1. Chumakov, M.P., Results of a study made of Omsk hemorrhagic fever (OL) by an expedition of the Institute of Neurology. *Vestn. Acad. Med. Nauk SSSR*, 2, 19, 1948.
2. Mazbich, I.B. and Netsky, G.I., Three years of study of Omsk hemorrhagic fever (1946–1948). *Proc. Omsk Inst. Epidemiol. Microbiol. Gigien.*, 1, 51, 1952.
3. Kharitonova, N.N. and Leonov, Yu.A., *Omsk Hemorrhagic Fever. Ecology of the Agent and Epizootiology.* Amerind, New Delhi, India, 230pp., 1985.
4. Neronov, V.M. et al., Alien species of mammals and their impact on natural ecosystems in the biosphere reserves of Russia. *Integr. Zool.*, 3, 83, 2008.
5. Korsh, P.V., Epizootic characteristics of Omsk hemorrhagic fever natural foci at the south of Western Siberia. In: *The Problems of Infectious Pathology.* Russian Academy of Sciences, Omsk, Russia, p. 70, 1971.
6. Růžek, D. et al., Omsk haemorrhagic fever. *Lancet*, 376, 2104, 2010.
7. Růžek, D. et al., Omsk hemorrhagic fever virus. In: Liu, D. (ed.), *Molecular Detection of Human Viral Pathogens.* CRC Press, Boca Raton, FL, p. 231, 2011.
8. Shestopalova, N.M. et al., Electron microscope studies into the morphology and localization of Omsk hemorrhagic fever virus in infected tissue culture cells. *Vopr. Virusol.*, 4, 425, 1965.
9. Heinz, F.X. and Stiasny, K., Flaviviruses and flavivirus vaccines. *Vaccine*, 30, 4301, 2012.
10. Gritsun, T.S., Lashkevich, V.A., and Gould, E.A., Nucleotide and deduced amino acid sequence of the envelope glycoprotein of Omsk haemorrhagic fever virus; comparison with other flaviviruses. *J. Gen. Virol.*, 74, 287, 1993.
11. Stiasny, K. et al., Molecular mechanisms of flavivirus membrane fusion. *Amino Acids*, 41, 1159, 2011.
12. Lindenbach, B.D. and Rice, C.M., Molecular biology of flaviviruses. *Adv. Virus Res.*, 59, 23, 2003.
13. Lin, D. et al., Analysis of the complete genome of the tick-borne flavivirus Omsk hemorrhagic fever virus. *Virology*, 313, 81, 2003.
14. Best, S.M. et al., Inhibition of interferon-stimulated JAK-STAT signaling by a tick-borne flavivirus and identification of NS5 as an interferon antagonist. *J. Virol.*, 79, 12828, 2005.
15. Clarke, D.H., Further studies on antigenic relationships among the viruses of the group B tick-borne complex. *Bull. World Health Organ.*, 31, 45, 1964.
16. Kornilova, E.A., Gagarina, A.V., and Chumakov, M.P., Comparison of the strains of Omsk hemorrhagic fever virus isolated from different objects of a natural focus. *Vopr. Virusol.*, 15, 232, 1970.

17. Li, L. et al., Molecular determinants of antigenicity of two subtypes of the tick-borne flavivirus Omsk haemorrhagic fever virus. *J. Gen. Virol.*, 85, 1619, 2004.

18. Yakimenko, V.V., Omsk hemorrhagic fever virus: Epidemiological and clinical aspects. In: *Tick-Borne Infections in Siberia Region*. Siberian Branch of the Russian Academy of Sciences, Novosibirsk, Russia, p. 279, 2011.

19. Hoogstraal, H., Argasid and Nuttalliellid ticks as parasites and vectors. *Adv. Parasitol.*, 24, 135, 1985.

20. Estrada-Peña, A. and Jongejan, F., Ticks feeding on humans: A review of records on human-biting Ixodoidea with special reference to pathogen transmission. *Exp. Appl. Acarol.*, 23, 685, 1999.

21. Kondrashova, Z.N., Study of Omsk hemorrhagic fever virus preservation in *Ixodes persulcatus* ticks during mass dosated infection. *Med. Parasitol. Parasit. Dis.*, 39, 274, 1970.

22. Gritsun, T.S., Nuttall, P.A., and Gould E.A., Tick-borne flaviviruses. *Adv. Virus Res.*, 61, 317, 2003.

23. Busygin, G.G., Omks hemorrhagic fever—Current status of the problem. *Vopr. Virusol.*, 45, 4, 2000.

24. Fedorova, T.N. and Sizemova, G.A., Omsk hemorrhagic fever incidence in a man and muskrats in winter. *Zh. Mikrobiol. Epidemiol. Immunol.*, 11, 134, 1964.

25. Netesov, S.V. and Conrad, J.L., Emerging infectious diseases in Russia, 1990–1999. *Emerg. Infect. Dis.*, 7, 1, 1991.

26. Akhrem-Akhremovich, R.M., Spring-autumn fever in Omsk region. *Proc. Omsk Inst. Epidemiol. Microbiol. Gigien.*, 13, 3, 1948.

27. Akhrem-Akhremovich, R.M., Problems of hemorrhagic fevers. *Proc. Omsk Inst. Epidemiol. Microbiol. Gigien.*, 25, 107, 1959.

28. Lebedev, E.P., Sizemova, G.A., and Busygin, F.F., Clinical and epidemiological characteristics of Omsk hemorrhagic fever. *Zh. Mikrobiol.*, 11, 132, 1975.

29. Belov, G.F. et al., Clinico-epidemiological characterization of Omsk hemorrhagic fever at the period of 1988–1992. *Zh. Mikrobiol. Epidemiol. Immunobiol.*, 4, 88, 1995.

30. Jelínková-Skalová, E. et al., Laboratory infection with virus of Omsk hemorrhagic fever with neurological and psychiatric symptomatology. *Čs. Epidemiol. Mikrobiol. Immunol.*, 23, 290, 1974.

31. Novitskiy, V.S., Pathologic anatomy of spring-summer fever in Omsk region. *Proc. Omsk Inst. Epidemiol. Microbiol. Gigien.*, 13, 97, 1948.

32. Holbrook, M.R. et al., An animal model for the tickborne flavivirus—Omsk hemorrhagic fever virus. *J. Infect. Dis.*, 191, 100, 2005.

33. Kenyon, R.H. et al., Infection of *Macaca radiata* with viruses of the tickborne encephalitis group. *Microbiol. Pathogen.*, 13, 399, 1992.

34. Shestopalova, N.M. et al., Electron microscopic study of the central nervous system in mice infected by Omsk hemorrhagic fever (OHF) virus. *J. Ultrastruct. Res.*, 40, 458, 1972.

35. Tigabu, B., Juelich, T., and Holbrook, M.R., 2010. Comparative analysis of immune responses to Russian spring-summer encephalitis and Omsk hemorrhagic fever viruses in mouse models. *Virology*, 408, 57, 2010.

36. Von Lubitz, D.K.J.E., *Bioterrorism. Field Guide to Disease Identification and Initial Patient Management.* CRC Press, Washington, DC, 175pp., 2004.

37. Tsai, T.F., Arboviruses. In: Murray, P.R., Baron, E.J., Pfaller, M.A., Tenover, F.C., and Yolken, R.H. (eds.), *Manual of Clinical Microbiology*, 7th ed. ASM Press, Washington, DC, p. 1107, 1999.

38. Calisher, C.H. et al., Antigenic relationships between flaviviruses as determined by cross-neutralization tests with polyclonal antisera. *J. Gen. Virol.*, 70, 37, 1989.

39. Libíková, H., Viruses of the tick-borne encephalitis group in HeLa cells, *Acta Virol.*, 3, 41, 1959.

40. Porterfield, J.S., Use of goose cells in haemagglutination tests with arthropod-borne viruses, *Nature*, 180, 1201, 1957.

41. Karabatsos, N. (ed.), *International Catalogue of Arboviruses Including Certain Other Virus of Vertebrates*, 3rd ed. American Society of Tropical Medicine and Hygiene, San Antonio, TX, 1985.

42. Porterfield, J.S., Plaque production by arboviruses. *Ann. Microbiol.*, 11, 221, 1963.

43. Gaidamovich, S.Ya., Tick-borne flavivirus infections. In: Porterfield, J.S. (ed.), *Exotic Viral Infections*. Chapman and Hall, London, U.K., p. 203, 1995.

44. Loginova, S.Ia. et al., Effectiveness of virazol, realdiron and interferon inductors in experimental Omsk hemorrhagic fever. *Vopr. Virusol.*, 47, 27, 2002.

45. Stephenson, J.R., Flavivirus vaccines. *Vaccine*, 6, 471, 1988.

46. Orlinger, K.K. et al., A tick-borne encephalitis virus vaccine based on the European prototype strain induces broadly reactive cross-neutralizing antibodies in humans. *J. Infect. Dis.*, 203, 1556, 2011.

47. Ngulube, C.N., Yoshii, K., and Kariwa, H. Evaluation of antigenic cross-reactivity of tick-borne encephalitis virus and Omsk hemorrhagic fever virus. Abstract No. P1-7. *The Fourth International Young Researcher Seminar for Zoonosis Control.* Graduate School of Veterinary Medicine, Hokkaido University, Sapporo, Japan, September 19–20, 2012.

48. Růžek, D. et al., Omsk hemorrhagic fever virus. In: Singh, S.K. and Růžek, D. (eds.), *Viral Hemorrhagic Fevers.* CRC Press, Boca Raton, FL, 2013, pp. 553–561.

18 Rift Valley Fever Virus

Marc Grandadam

CONTENTS

18.1 INTRODUCTION

Rift Valley fever virus (RVFV) (family Bunyaviridae, genus *Phlebovirus*) is an arthropod-transmitted virus that causes high morbidity and mortality in ruminants and humans. Among arthropod-borne viruses, RVFV is probably one of those infecting the wider spectrum of vertebrates through the most diversified range of vectors. RVFV primarily infects ruminants in either wildlife or livestock in eastern and sub-Saharan Africa. Since its discovery in 1931, the virus continuously spread in Africa, but a profound epidemiological switch occurred in 2000 when the virus spread out of the African continent to the Arabian Peninsula and Indian Ocean islands. Outbreaks are in most cases explosive especially in domestic ruminants, provoking high abortion and mortality rates. Humans are also susceptible to RVFV infection, which may lead to severe hemorrhagic forms.

Recent progression of the virus out of its historical endemic area raises the question of a possible introduction of the virus into new geographic areas including northern and western territories.

Whereas arthropods play the role of viral reservoir and initiate active transmission to vertebrates, viral spreading in humans and animals mainly occurs through alternative routes. Ingestion, transcutaneous passage, and exposure to aerosol are predominant alternative mechanisms of viral dispersion that, in addition to the dramatic epidemic potential of RVFV, pose an important problem of professional exposure. Inactivated and live attenuated vaccines, available for preventive immunization of livestock and humans, are at the center of disease control plans. Development of predictive models is a major area of research, but their validation requires entomological and epidemiological field investigations based on sensitive and specific virological methods.

The high incidence rate of RVFV in animals and humans and the variety of infection routes represent an important health and economical threat. Risk evaluations of RVFV spreading integrate both accidental and purposeful means. International sanitary organizations set up and revised guidelines to regulate livestock trading in order to limit the viral extension. Research efforts are ongoing in different fields such as outbreak modeling, vaccine, and antiviral agent development to improve capabilities to survey and prevent RVFV extension.

18.2 CLASSIFICATION AND MORPHOLOGY

18.2.1 VIROLOGICAL ASPECTS

The shape of RVFV particle observed by negative staining appears roughly spherical with an average diameter of 95 nm.[1,2] The enveloped virions contain three independent nucleocapsids giving a filamentous appearance to the interior of the particles. The viral genome consists of three single-stranded negative-sense RNA segments denoted L (large),

M (medium), and S (small). The mean size of the L, M, and S segments is 6400, 3890, and 1690 nucleotides, respectively (Figure 18.1). The L and M segments encode a unique open reading frame encoding, respectively, the RNA-dependent RNA polymerase and the precursor to the envelop glycoproteins GN and GC.[3] Posttranslational cleavages of the M derivated polyprotein also release a nonstructural protein (NSm) of unidentified function. Both L and M segments are of negative polarity. Transcription of S segment relies on an ambisense strategy, allowing the synthesis of the nonstructural protein (NSs) in sense orientation, whereas the nucleoprotein N is expressed in the antisense orientation. An intergenic region separates N and NSs open reading frames and plays a crucial role in transcription termination.[4,5] The genome

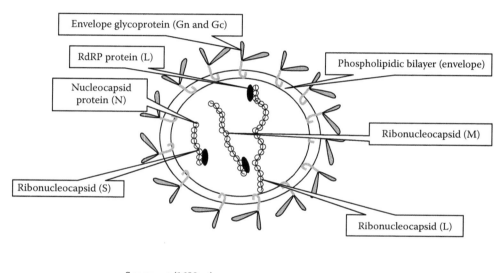

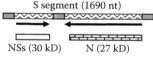

S segment (1690 nt)

NSs (30 kD)   N (27 kD)

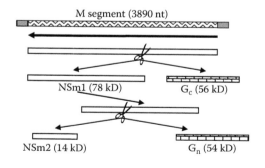

M segment (3890 nt)

NSm1 (78 kD)   G_c (56 kD)

NSm2 (14 kD)   G_n (54 kD)

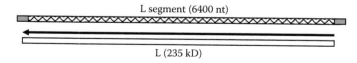

L segment (6400 nt)

L (235 kD)

FIGURE 18.1 Structure of viral particles and genome organization of RVFV. RdRP, RNA-dependent RNA polymerase. ▨▨▨▨, viral RNA segments; ▥▥▥▥, structural proteins; ▭▭, noncoding regions; ▭▭, nonstructural proteins. Bold arrows indicate open reading frames.

transcription, replication, encapsidation, and packaging steps are driven by *cis*-acting elements contained within the non-coding regions flanking the viral genes. Highly conserved complementary nucleotide stretches are located at each segments ends. The partial double-stranded RNA structure provides a functional promoter for the viral polymerase. All steps of the viral cycle occur in the cytoplasm. However, it has been demonstrated that the phosphoprotein NSs accumulates and forms filamentous structure in the nucleus of infected cells.[3–6] Despite the dispensable characteristic of nonstructural proteins NSs and NSm in the replication cycle, both proteins may have virulence properties.[4] Interaction of GC and GN glycoproteins with the Golgi membranes determines the site of viral particles budding. A direct interaction between glycoprotein cytoplasmic domain and the N protein within ribonucleocapsids complexes seems to play an important role in the genome packaging.[2]

18.2.2 Position of RVFV in Viral Taxonomy

The general features presented previously characterize the Bunyaviridae family that covers nearly 300 viruses affecting a wide diversity of hosts including animals, humans, and plants. Within this family, viruses are classified in five genera, that is, *Orthobunyavirus*, *Phlebovirus*, *Nairovirus*, *Hantavirus*, and *Tospovirus*.[1,7] RVFV groups in the *Phlebovirus* genus. The primary vectors of that genus are phlebotomine sand flies with two main exceptions, the RVFV, mainly transmitted by mosquitoes, and the tick-borne Uukuniemi-related viruses.

Members of the genus *Phlebovirus* are unrelated at the antigenic level to other members of the Bunyaviridae family but cross-react serologically with each other at different degrees. Attempt to discriminate phleboviruses on their antigenic properties revealed nine complexes of viruses.[8] Intriguingly, RVFV was first found to be more closely related to Belterra and Icoaraci, two South American phleboviruses, than to Old World phleboviruses. In the past sero-prevalence studies, false IgG positivity in enzyme-linked immunosorbent assays (ELISAs) might result from possible cross-reactions with antigenic-related phleboviruses. Recent data suggested a revision of the classification where RVFV would be the only member of Rift Valley fever complex.[8] The independent position of RVFV is subsequently confirmed as no consistent serological cross-reaction between RVFV and African or Mediterranean phleboviruses has been found yet.[8,9] Use of plaque reduction neutralization test further verified this specificity.[9,10] Antigenic composition seemed to be highly conserved among geographically distant strains; thus, all RVFV identified are gathered in a unique serotype.

Phylogenetic analysis based on partial sequences of the three genome segments of Old World and New World phleboviruses permitted identification of five main lineages, which correlated with antigenic classification.[10,11] Nucleotide sequence comparison clearly showed that RVFV represents an independent species within the *Phlebovirus* genus. Comparison of NSs genes indicated that RVFV strains from different geographic origins and hosts diverged of less than 10%.[12] Recent comparison of RVFV isolates allowed determining the mean evolutionary rate of RVFV.[13] The numbers of nucleotide substitution per site per year for the first, second, and third codon positions were 6.5×10^{-5}, 2.7×10^{-5}, and 6.3×10^{-4}, respectively. The mean molecular evolution rates calculated for each segment were estimated to be 2.35×10^{-4}, 2.42×10^{-4}, and 2.78×10^{-4}, respectively for S, M, and L segments.[13] These data are consistent with a high degree of constraint at the protein coding level. The overall high conservation of RVFV that is maintained among geographically and timely distant isolates could be explained either by a low tolerance for mutation or by recent divergence of current strains from a common ancestor. Comparison of RVFV substitution rates to those of mammalian-adapted viruses and detailed Bayesian analyses sustains the recent ancestor hypothesis. Depending on the number of strains and the genome segments, the most recent ancestor of RVFV was dated around 1880 or 1892.[13,14] History of RVF disease supports the argument of a contemporary origin of RVFV as clinical presentations compatible with RVF recorded in animals were observed in the early 1900s.[5] The report documented an epizootic of hepatitis in sheep, which occurred in Western Kenya coincidently to the importation of European sheep and cattle.[5] Changes in agricultural practices may have offered to RVFV ancestor the opportunity to emerge and colonized this new ecological niche.

18.2.3 Biosafety Level Classification

RVFV is a cause of hemorrhages both in animals and humans, and all detected cases are of mandatory declaration at the national level in a number of countries. The high potential of viral spreading and the lack of commercially available vaccine and specific antiviral treatment raise question about the possible misappropriate usage of the virus for purposeful dissemination of the virus. All these reasons justify applying specific measures to control as much as possible the viral spreading and to reinforce global health security. Thus, the virus has been classified as a Category A pathogen by the Centers for Diseases Control and registered in the list of biological agents for export control (Australia group) that also includes Ebola virus, Marburg virus, Crimean–Congo virus, or Kyasanur Forest virus (http://www.australiagroup.net/en/biological_agents.html). In respect to the International Sanitary Regulation, all confirmed cases should be reported without delay to international health organizations such as the World Health Organization and the World Organization for Animal Health (OIE).

Direct diagnosis and experimentation on RVFV require biosafety level 3 or 4 facilities and highly qualified personnel. Specific measures regarding biosafety and biosecurity are needed and could be submitted to specific audit by governmental agencies.

18.3   BIOLOGY AND EPIDEMIOLOGY

18.3.1   Vectors and Viral Maintenance

The first description of an RVFV outbreak is attributed to Montgomery and Stordy in 1913 who reported an undiagnosed epidemic of hepatitis among cases in sheep, cattle, and human near Lake Naivasha in Kenya. The virus was isolated from human and sheep during a second outbreak that occurred 20 years later in the same region.[15] Since 1977, RVFV emerged in different African countries but also spread out of the continent (Figure 18.2).[16,17] The diversity of the macroenvironment supporting these emergences suggested a slight adaptation of the natural transmission cycle to local conditions including adaptation to local vector species. Transmission of the virus by arthropod vectors was demonstrated by Smithburn in 1948.[18,19]

Following this first description, a number of arthropod species, including *Aedes*, *Anopheles*, *Culex*, *Eretmapodites*,

and *Mansonia*, have been recognized as potential vectors for RVFV.[20] Competence of different African phlebotomine species has been experimentally demonstrated raising the question of the role of sand flies in natural transmission and maintenance in natural conditions.[9,21,22] However, field collection and further experimental competence assays are needed to verify the involvement of sand flies in RVFV epidemiological cycle.[9,23,24] RVFV could also be isolated from ticks. These exoparasites may serve as possible vector for the virus and may favor its spreading as they can be transported over long distances by different vertebrate hosts.[25] The wide diversity of arthropods from which the virus has been successfully isolated highlights the complexity of the enzootic/endemic cycle of RVFV. Seroprevalence studies and experimental vector transmission of RVFV to wild vertebrates suggested that some rodents species including *Rattus rattus* may play the role of viral reservoirs participating to the viral persistence in environment.[26,27,28]

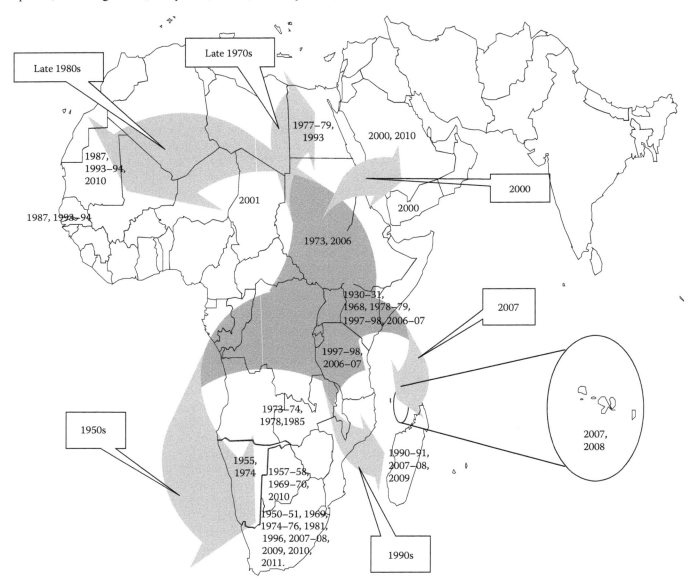

FIGURE 18.2   History of Rift Valley fever outbreaks and routes of dispersal. Outbreak years are indicated within the country borders. Labels referred to the period or years of probable emergence. Gray arrows represent the ways of spreading of RVFV.

18.3.2   Epidemic Transmission

RVFV outbreaks coincidentally occurred with heavy rainfalls sometimes simultaneously over geographic areas distant by hundreds of kilometers.[28,29] Human activities such as construction and flooding of dams were also associated with huge epidemics in Egypt and in Senegal.[3] In Eastern and Southern Africa, epizootics and epidemics have been clearly linked to large-scale meteorological events such as El Niño Southern Oscillation.[30,31] Thus, satellite surveillance has an increasing interest for the development of predictive models and mapping risk zones.[32,33] RVFV epizootics/epidemics are the consequence of two distinct but overlapping transmission cycles. An epizootic cycle takes place near shallow stream less depression called "dambos" in Kenya or broad "vleis" in Zimbabwe.[23,28] Recovery of RVFV from *Aedes* males and females collected as larvae in flooded ground pools also called "dambos" suggested a possible transovarial transmission of the virus.[23,34] This mechanism is the most likely mean for maintaining the virus during dry seasons when environmental conditions are unsuitable for active transmission between competent vectors and susceptible hosts. Dambos may be kept dry for years thus may not support active vector breeding except during the short periods of time after heavy rainfalls. Desiccation-resistant mosquito eggs hatch and may lead to very high vector density compatible with an efficient transmission of RVFV.[24] The presence of naïve livestock near recently flooded dambos is in most cases the starting point of epizootics. Studies on host feeding pattern of mosquito species demonstrated that *Aedes* and *Culex* species captured near dambos preferentially took their blood meal on cattle.[24] Thus, trophic habits of potential vectors and high density of animal populations may explain the low rate of transmission to humans. Studies in different Western and Central African countries raised the question of alternative mechanisms for viral maintenance. During rainy seasons, a permanent horizontal transmission takes place between mosquitoes and ruminants, whereas in dry seasons, maintenance occurs through vertical transmission in *Aedes* species.[35] In forestry environment, the transmission could be uninterrupted as suggested by serological studies. Finally, RVFV transmission may also occur in regions where ruminant density is very low.[36]

Migration of herds favors the spreading of RVFV from enzootic to enzootic and epidemic foci in urban or rural periurban areas.[37] In mountainous regions of Madagascar, where environmental conditions are a priori not suitable for viral maintenance, recurrent circulation is nonetheless observed. Replacement of local herds by importing domestic ruminants from other parts of the great island is the most probable mechanism underlying the reemergence of the virus.[35]

Human exposure results from either bites of infected mosquitoes, inhalation of infectious aerosols, ingestion of uncooked milk or meat, or transcutaneous infection during slaughtering or manipulation of carcasses.[38–42] More recently, mother to child transmission cases have been reported.[43,44]

18.3.3   Professional Exposure to RVFV

RVFV virus represents an important risk for different categories of professionals that may enter into contact with a contaminated source of the virus. In endemic countries, shepherds could potentially be exposed to the virus by three different routes. Mosquito-borne infection can occur near the water points where vectors of the virus are proliferating. The breeders could get the virus by handling contaminated tissues during dropping or manipulating abortion products and slaughtering. At last, consumption of insufficiently cooked meat or crude milk has been described as an efficient source for human infections. Slaughterhouse personnel and butchers might get infected by injuries with contaminated knifes, direct contact of infected tissues with broken skin, and aerosol exposure.[45]

Veterinarians have multiple risks of infection during animal care and of course during outbreak investigations during clinical or postmortem diagnosis.[46]

Whereas no human-to-human transmission has ever been documented, health workers should be aware and trained for the management of suspected cases. High levels of viremia have been recorded in humans raising the question of a possible parenteral transmission during patients sampling procedures.[47,48] Free access specific guidelines are regularly updated (http://www.nicd.ac.za/; http://www.who.int/mediacentre/factsheets/fs207/en/). Arboviral infections of laboratory workers have been reported since the 1960s. A report published in 1967 documented the increasing number of infected lab workers directly linked to the intensification of research on arboviruses since 1950.[49,50] Analysis of the sources of contamination revealed that animal experimental infection, aerosol exposure, and agent or biological samples handling were the main causes of infection. RVFV has been at the origin of a number of accidents. Even though mortality is low, the number of laboratory exposures illustrates the potential risk of handling this pathogen or infected biological material and justifies strict experimental procedures and highly trained personnel.

18.3.4   Significance of RVFV for Intentional Dissemination

Studies were conducted for all yet identified viral hemorrhagic fevers to determine their potential to militarization.[51,52] Large-scale dissemination by aerosol spreading has been retained as a major parameter in evaluation risk. Transmission of RVFV to humans by aerosolization has been proven during outbreaks and by multiple laboratory accidents. High mortality and morbidity rates could be recorded after RVFV infection, and even if it existed, vaccines are only available in limited quantity and only recommended for professional prevention. In culture, RVFV grows rapidly and produces high titers. Limiting access to viral strain and a strict control of transfer of biological material produced by laboratories or collected on from infected hosts are key measures to lower the risk of intentional dissemination.

18.4 CLINICAL FEATURES AND PATHOGENESIS

18.4.1 INFECTION IN DOMESTIC ANIMALS

RVFV dramatically hits domestic species playing a central economical role and affecting the survival of human rural populations in African countries. Sheep, cattle, and goats are highly susceptible to RVFV infection with abortion and mortality rates ranging from 10% to 100% and from 10% to 70%, respectively (Table 18.1).[3,5,15,28,32] Susceptibility decreases with age as suggested by the lower mortality rates in calves and adults compared to neonates. Typically, acute phase is short ranging from 12 to 24 h in young individuals and from 24 to 72 h in adults. Symptom onset is marked by an acute fever and a rapid progression to death in newborns, whereas adults are affected by lethargy, acute hepatitis, and hemorrhages.[53] Serological studies in camels revealed high prevalence ranging from 3% to 45% in endemic regions such as Kenya and Egypt.[53] RVFV may cause extensive abortion outbreaks in camels and a mortality rate of 20% among calves (Table 18.1).

Abnormal abortions were observed among domestic pigs during an RVFV outbreak in South Africa, but no evidence of the virus was found in necropsies.[54] Experimental infections demonstrated a dose-dependent resistance of pigs to RVFV.[55] Altogether, the data are too scarce to draw conclusion on the participation of *Suidae* in RVFV transmission cycle.

18.4.2 INFECTION IN WILDLIFE

Evidence of exposition of wild animals to RVFV has been established by seroprevalence studies in different species.[53] Some mammal species appeared to be highly exposed to RVFV, whereas birds seem to be refractory to either natural or experimental infection. However, pathogenicity among the suspected mammal species is not fully documented yet.[28] Serological investigations of primates captured in the wild revealed rare positivity suggesting that the exposition should be very rare. Experimental infection of different species showed that a viremia may persist for several days after inoculation.[56] However, only minor clinical signs, in most cases limited to fever, were recorded in infected primates. Altogether, even if the participation of primates is not fully excluded, they probably play a marginal role in natural transmission and maintenance of the virus in the environment.

Experimental infection of pregnant African buffaloes led to abortion within 16 days after inoculation. A positive viremia with a mean titer of 10^4 TCID$_{50}$/mL could be detected in 80% of infected animals. Rodents and chiropterans have been suspected to participate in the maintenance of RVFV during interepizootic periods.[28] Experimental infections demonstrated the susceptibility of some bats to RVFV with prolonged shedding of virus in urine, but no clinical manifestation has been reported. Rodents were early suspected to participate in RVFV transmission cycle. Daubney and Hudson reported a high mortality among unstriped grass mice (*Arvicanthis abyssinicus*) and rats (*R. rattus*) during the first RVFV outbreak in Kenya.[57] Serological investigations supported a possible link between RVFV and rodents, but the effective role of rodents in the maintenance of RVFV remains unclear.[28] These data support the circulation of RVFV in wild fauna, but further investigations are required to fully establish the role of wildlife in RVFV maintenance and reemergence, as well as spread of RVFV to new areas.[28]

TABLE 18.1
Clinical Features of RVFV in Susceptible Hosts

Susceptible Hosts	Transmission	Specific Epidemiological Features: Symptoms
Human	C; V; I; A; M–C	Self-limiting dengue-like syndrome, with malaise, arthralgia, myalgia; 1%–2% severe forms: hemorrhagic fever, acute hepatitis/hepatic necrosis, loss of visual and hearing acuities. Max. mortality: 50%
Cattle	V	Calves: febrile syndrome; rapid progression to death. Max. mortality: 70%
		Adults: fever, lethargy, hematochezia, epistaxis, lactation decrease. Max mortality: 10%
Sheep	V	Lambs: febrile syndrome, rapid progression to death. Mortality: >90%
		Adults: fever, lethargy, hematemesis, nasal discharge. Mortality: up to 30%; abortion >90%
Goats	V	Fever, inappetence, lethargy, abortion. Max mortality: 48%
Horses	V	Asymptomatic infections
Camels	V	Calves: febrile syndrome. Mortality: 20%
		Adults: fever, lethargy, hematochezia, epistaxis. Max mortality: 10%

Note: V, vector transmission; C, transmission by contact; I, transmission by ingestion; A, aerosol transmission; M–C, mother to child.

18.4.3 HUMAN INFECTION

According to studies implemented during recent outbreaks in Africa and the Arabian Peninsula, only a small proportion of infected subject experienced clinical symptoms. It must be pointed out that no correlation could be established between the lack of symptoms and the infection route. Asymptomatic infection may represent up to 80% of human cases. Clinical phase of RVFV virus infection in human is preceded by a short incubation period ranging from 3 to 7 days. No correlation could be established so far between the length of the incubation phase, the viral dose, and the mean of virus transmission (vector borne, mechanical, aerosol, etc.,).

RVFV infection in humans leads to four recognized clinical forms. In most cases, an uncomplicated influenza-like syndrome is depicted.[40] Febrile syndrome associated with hemorrhagic symptoms with liver involvement, jaundice, thrombocytopenia, and bleeding tendencies represents about 20%–30% of RVFV infections. This clinical presentation is rapidly self-limiting and recovery occurs in general within 10 days without complications and sequelae. At the biochemical level, RVFV infection is associated with a profound leukopenia, elevated liver enzymes, and thrombocytopenia.[58] Up to 8% of symptomatic patients progressed to severe acute hepatitis or encephalitis with confusion and coma.[47,58,59] Neurological sequelae are not infrequent. Ocular involvement characterized by retinal hemorrhages and macular edema may lead to blurred vision and sometime permanent loss of visual acuity.[47] Case fatality rate in human is generally low, ranging from 1% to 3% but may be as high as 50% in patients with severe forms (Table 18.1).

18.5 IDENTIFICATION AND DIAGNOSIS

Early detection of suspected cases and rapid diagnosis are crucial for an efficient containment of RVFV outbreaks. Clinical diagnosis is difficult; some bacterial or viral infections such as brucellosis, vibriosis, trichromaniasis, Nairobi sheep disease, bluetongue, Wesselsbron and Middelburg virus infections, or chemical poisoning are also responsible for high abortion rates in livestock.[42] In this context, biological tools play a central role for surveillance and differential diagnosis.[5] Despite the importance of RVF in human and animal health, commercial diagnosis kits are poorly developed. Some reference laboratories are involved in the organization of field sample collection and subsequent testing. However, the laboratory coverage of endemic regions is heterogeneous; thus, diagnosis may be delayed sometimes for several weeks. Ideally, the biological diagnosis may rely on a combination of serological and molecular tools. Indeed, the kinetics of the virological markers of RVFV infection highlight the benefit of direct and indirect approaches (Figure 18.3). In some cases, genome detection with appropriate primers helps to discriminate naturally infected to vaccinated seropositive individuals. RVFV induces specific histopathological lesions in tissues of

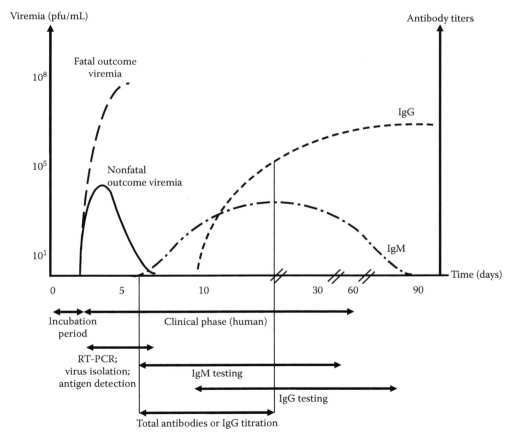

FIGURE 18.3 Kinetic of virological markers of RVFV infection.

infected animals. Anatomicopathological investigations coupled with antigen detection by immunohistochemistry allow specific postmortem diagnosis. Molecular methods are also relevant for necropsy investigation. Efforts for standardization of RVFV diagnosis are necessary. Reference methods are depicted in a reference manual edited by OIE available online at www.oie.int/eng/Normes/mmanual/A_00031.htm.

18.5.1 INDIRECT DIAGNOSIS

Serological investigations are commonly used for RVFV. A number of serological tests, including virus neutralization, ELISA, and hemagglutination inhibition assays (HIA), have been widely applied. Others, such as indirect immunofluorescence, complement fixation, radioimmunoassay, and immunodiffusion, are nowadays used less frequently. Cross-reactions with other phleboviruses may interfere with serological tests excepted for virus neutralization.

Historically, serological investigations of suspected cases were based on HIA. Preparation of viral antigens compatible with hemagglutination tests requires specific methods such as suckling mouse inoculation and sucrose–acetone extraction.[60] Preparation of erythrocytes is also a determinant step for HIA effectiveness. Incubation temperature and pH variations may have a dramatic impact on the test. As for ELISA, comparison of paired sera allows infection dating. A correlation has been established between antibody titers and the nature of the immune stimulation. Indeed, vaccinated animals have significant lower antibody titers compared to naturally infected individuals.[61] Whereas HIA is of major interest for RVFV diagnosis, biosafety and practicability limitations restricted its field application.

ELISAs have been developed for the serological diagnosis of RVFV infection in animals or humans. Monoclonal IgM capture (MAC-ELISA) and sandwich ELISA allow sensitive and specific detection of anti-RVFV IgM and IgG, respectively. Crude inactivated antigens prepared from cells culture or recombinant N protein are used as source of antigen. The major limitation of these techniques is the specie specificity of the test due to the natures of the monoclonal antibodies used for IgM capture and of the immunoconjugates used for IgG detection.

A competitive-based ELISA has been developed as a universal serological method.[62] Polyclonal anti-RVFV sheep antibodies are passively coated on ELISA microplates. Suspected sera mixed with a known amount of RVFV antigen (crude inactivated or recombinant N protein) are incubated onto the plates.[63,64] An anti-RVFV-positive mouse serum is then added on the plates and detected by an antimouse horseradish peroxidase conjugate. Direct incubation of mouse sera serves as control. Optical density (OD) of the mixed sera is compared to the OD of the reference serum alone. Reduction of OD in the mix revealed the presence of anti-RVF antibodies in the sample. This competitive assay detects total anti-RVFV antibodies and thus does not discriminate recent infections (IgM or IgM plus IgG) from past infections (IgG alone) when analyzing a unique sample. Rising of at least

four times of the antibodies titers between acute and convalescent sera indicates a recent infection (Figure 18.3). Like almost serological tests, discrimination between naturally infected and vaccinates individuals is not possible.

18.5.2 DIRECT DIAGNOSIS

RVFV can be isolated or detected by reverse transcriptase-polymerase chain reaction (RT-PCR) from the blood collected during the febrile phase of illness. Viral titers in tissues are often high, thus allowing antigen detection on histological preparations or impression smears or on tissue extract.

Histopathological examination of liver of affected animal reveals characteristic lesions.[42,65] The presence of viral antigens can easily be confirmed by immunostaining methods using hyperimmune mouse ascitic fluids or monoclonal antibodies. Agar gel diffusion is also a valuable method for antigen detection in tissues extracts.

Experimental infection or isolation of RVFV can be performed in various animal models. Hamsters, adult or suckling mice, and embryonic or 2-day-old chicken eggs are highly susceptible to the virus. Several cell lines including baby hamster kidney cells, monkey kidney (Vero) cells, chicken embryo reticulum, and primary cultures from cattle or sheep had significantly improved viral isolation and are more adapted for the treatment of large sample series than laboratory animals. Virus may be isolated from a wide panel of samples including blood or serum collected during the acute phase of the disease or from tissues (liver, spleen, brain) of died animals or aborted fetuses (Table 18.1). RVFV induces a characteristic cytopathic effect generally in <5 days followed by complete destruction of the cell monolayer within 12–24 h. Virus identification in infected cells or tissues remains on antigen detection by means of indirect immunofluorescence assay using specific hyperimmune ascitic fluids or monoclonal antibodies.

A number of RT-PCR or RT-PCR-derived methods have been proposed for the diagnosis of RVFV in human and livestock. Detection of viral genome during the acute phase is easy due to the high level of the viremia that occurs either in human and animal hosts (Figure 18.3). Moreover, the viremia titration may have a predictive value of the outcome of RVFV infection.[47,48,66] Thus, molecular methods and particularly "real-time" RT-PCR are of major interest. Despite their high sensitivity, requiring no postamplification analysis, and their adaptation to high throughput, these tests have limitation due to the short duration of RVFV viremia and their cost-effectiveness. Durable deployment of these techniques is still challenging.

Recently, loop-mediated isothermal amplification (LAMP) has been developed for the rapid detection of a wild range of pathogens including arboviruses.[67] LAMP amplifies specific nucleic acids sequences using a set of six primers by strand displacement activity of a DNA polymerase. The amplification yield leads to the formation of a DNA precipitate easily detectable even with the naked eye. The specificity of the technique relies on the combination of multiple primers; thus, no postamplification analysis is required. Moreover, the isothermal process (60°C–65°C) avoids the requirement of complex equipments making LAMP of particular interest for field diagnosis.

18.5.3 IMPORTANCE OF MOLECULAR EPIDEMIOLOGY

Retrospective analysis of viral genomes evidenced different variability features depending on the epidemiological situation. In most epizootics/epidemics, a single viral genotype was found with minor nucleotide differences over the whole genomes (less than 0.35%) regardless whatever was the host from which the virus could be isolated.[13,68] By contrast, RVFV strains isolated during periods of low-level enzootic activity or very located human or animal outbreaks were remarkably highly divergent. In some cases, viral strains falling in different RVFV genotypes could be evidenced.

Moreover, reassortant strains support the probable cocirculation of different genotypes over restricted geographic areas and within short periods of time.[13,69]

Recent progresses and the increasing number of sequenced strains allow better documenting the genome plasticity of RVFV. The mean diversities recorded for the S, M, and L segments were 4%, 5%, and 4% at the nucleotide level and 1%, 2%, and 1% at the amino acid level.[6,13,14,69] Sequence comparison identifies up to 15 lineages (Table 18.2) and helps to draw some hypothesis on viral dispersion and origin of some outbreaks.[6] Hypothesis could be proposed about the origin of RVFV. It is assumed that the virus emerged from an African reservoir contemporarily to the introduction of large numbers of naïve and highly sensitive sheep and cattle. This may explain the low level of genetic diversity observed between the different RVFV lineages. Permanent sequence database update helps identifying the geographical origins of

TABLE 18.2
Evolution of Rift Valley Fever Lineage Definition and Geographic Distribution

LD	Sall, 1999 (*n* = 18; NSs)	LD	Bird, 2007 (*n* = 33; Full Genome)	LD	Grobbelaar, 2011 (*n* = 203; Partial M)
Cluster I		A	Zimbabwe: 1974; Egypt: 1977, 1978, 1979, 1993; Madagascar: 1979	A	Zimbabwe: 1974; Egypt: 1977–1978–1979, 1993; Madagascar: 1979
1a	Uganda: 1955; Uganda, 1944 (SNS); Madagascar, 1991; Mauritania, 1987 (HD47502); Senegal: 1993 (ArD104769); CAR: 1969	B	Kenya: 1983, 1998; CAR: 1973; Saudi Arabia: 2000	B	Kenya, 1972
1b	Burkina Faso: 1984; Mauritania: 1987 (HD47311, 47408, 48255); Senegal: 1993 (AnD106417); Guinea: 1984	C	Guinea: 1981, 1984; CAR: 1969, 1973, 1974; Zimbabwe: 1978	C	Zimbabwe: 1976, 1978, 1979, 1998; Kenya: 1977, 1983, 1997, 1998, 2007; Saudi Arabia: 2000; Somalia: 1998; SA: 1999, 2008, 2009; Madagascar: 1991, 2008; Mauritania: 2003
Cluster II		D	Kenya: 1965; Zimbabwe, 1970, SA: 1975; Burkina Faso: 1983; Mauritania: 1987	D	CAR: 1973
	Egypt: 1977, 1993; Senegal: 1984; Madagascar: 1979	E	Uganda: 1944; Zimbabwe: 1974 (strain 2373/74)	E	Zimbabwe: 1975, 1978; CAR: 1973, 1974; Zambia: 1985
		F	Zimbabwe: 1974 (strain 2250/74)	F	SA: 1981
		G	SA: 1951	G	CAR: 1969, 1985; Guinea: 1981, 1984; Senegal: 1993; Zimbabwe: 1978
				H	SA: 2009, 2010
				I	SA: 1955, 1956
				J	Zimbabwe: 1974 (strain 2269/74)
				K	Uganda: 1944; Kenya: 1951, 1962, 1963; S.A.: 2010; Zimbabwe: 1974 (strain 2373/74); SNS; vaccine strain 95EG
				L	Egypt: 1995; Kenya: 1951, 1962, 1963; SA: 1971, 1974, 1975; Zimbabwe: 1969, 1970
				M	Uganda: 1955; SA: 1955
				N	Mauritania: 1987, 1988; Senegal: 1975, 1983, 1993; Burkina Faso 1983
				O	SA, 1951

Note: LD, lineage designation; SA, South Africa; CAR, Central African Republic; SNS, Smithburn neurotropic strain.

the strains and provides crucial information for maintaining the performances of molecular detection tools.[6,12,13,68,70]

18.6 TREATMENT AND PREVENTION

18.6.1 TREATMENT

Treatment of RVFV infections in humans or animals is symptomatic. At the moment, there is no specific antiviral available.[4] *In vitro* and *in vivo* activities of ribavirin and poly-riboinosinic acid against RVFV have been demonstrated.[71] However, the limited penetration of ribavirin through the blood–brain barrier and the adverse effects limited the utility of this molecule in the treatment of hemorrhagic and neurological forms. At present, ribavirin is no longer recommended for the treatment of RVFV infection. However, as clinical presentation does not discriminate RVFV and Crimean–Congo hemorrhagic fever virus, ribavirin might be administered in the early phase until the virological confirmation in geographical areas where the two viruses are cocirculating. New broad-spectrum antiviral compounds with proven *in vitro* activity on RVFV and on other closely related bunyaviruses are in development, but their *in vivo* effectiveness still remains uncertain.[4] In hamsters, a preventive effect was obtained against Punta Toro virus, another member of the *Phlebovirus* genus, after intranasal administration of pyrazine derivates.[71] Whether such effect remains to be demonstrated on RVFV, this result opens new hopes to prevent aerosol contamination especially for professional.

Studies in different animal models demonstrated the high sensitivity of RVFV to interferon-alpha (IFN-α) or to IFN inducers, but the rapid degradation of the molecules limits their therapeutic potential.[49,72] To circumvent the short half-life and the cost of IFN-α derivates, a recombinant adenovirus expressing consensus IFN-α was used to induce constitutive expression of cIFN-α in transduced cells. Intranasal delivery allows obtaining a rapid and high serum level concentration of cIFN-α in a murine model and demonstrates a preventive effect against Punta Toro challenge.[73] Such strategy offers multiple perspectives for prevention plans both for targeted groups such as health workers and for large-scale risk containment after intentional release of pathogens.

18.6.2 SURVEILLANCE

Human cases are generally preceded by RVFV outbreaks in animals. Thus, an active surveillance of herd's health is essential in providing early warning for both veterinary and human health organization. World organization for animal health (OIE) and the Food and Agriculture Organization (FAO) defined rules for the management of RVFV outbreaks. Quarantine of suspected herds has been proposed to limit the virus spreading during transhumances from infected to uninfected areas, but this measure is almost never implemented. Live attenuated and inactivated vaccines have been developed and licensed for veterinary use under the OIE's recommendations. Early detection and confirmation of suspected

cases are crucial to limit the geographic range of outbreaks. Increasing diagnostic capacities and capabilities in endemic regions is a main challenge. It appears impossible to eradicate RVFV from countries in which it has become established. Thus, definition of high-risk areas is a major concern for countries bordering the endemic regions but also for far distant zones where RVFV could be introduced through livestock trading, vector spreading, or bioterrorism acts.[74]

18.6.3 VECTOR CONTROL

The impact for insect vector control programs as components of an RVF campaign is very limited. Control of adult vectors by mass insecticide spraying is in most cases impractical and expensive and may have environmental consequences. Treatment of well-defined mosquito breeding and resting sites may be effective in some circumstances. Thus, preventing animal exposure to mosquito bites by moving livestock away from infested areas or stabling herds in specific facilities to protect them from mosquito bites is largely nonapplicable in endemic countries. The best vector control strategy is through larvicidal treatment of potential mosquito breeding sites. Because RVFV outbreaks occur after flooding, conventional antilarval preventive treatments are of low interest. Tentative treatments of potential breeding sites before flooding are still experimental, but promising results were obtained with toxins derived from *Bacillus thuringiensis* and *Bacillus sphaericus* and methoprene, a chemical larval growth inhibitor.[42]

18.6.4 VACCINES

Since the identification of RVFV in 1931, various vaccine strategies have been developed in order to limit the viral activity in endemic countries and to contain this rapidly expanding anthropozoonosis considered as a worldwide threat. In spite of the unique serotype of RVFV, vaccine development faced some hitches.

The first live attenuated vaccine has been derivated from the mosquito strain Entebbe isolated in 1944 in Uganda.[28] The so-called Smithburn strain has been neuroadapted by serial passages on suckling mouse brain. The modified Smithburn strain was submitted to further passages on embryonated eggs to obtain a vaccine preparation. Single injection of Smithburn vaccine confers high level of protective immunity among vaccinated animals in different species.[75] However, adverse effects have been observed during livestock immunization campaigns. Up to 28% of abortion or teratogenic effects in pregnant animals have been reported demonstrating the partial attenuation of the vaccine strain.[4,65,76] The question of a possible reversion of the vaccine strain to a highly virulent phenotype excluded this vaccine for a use in nonendemic countries.[76]

A naturally attenuated RVFV strain has been proposed as an alternative to the Smithburn strain. Clone 13, isolated from a mild human case, displays a large mutation in the gene encoding the nonstructural protein NSs.[77] NSs protein interplays with the type 1 IFN pathway, suggesting that RVFV virulence relies at least in part on NSs anti-IFN activity.[71,78] Genetic reassortment has been demonstrated within Bunyaviridae including RVFV for which such mechanism may occur naturally.[79] Thus, reversion to a virulent phenotype is raised for all attenuated RVFV by both mutation and reassortment. However, the stable large deletion within the NSs gene of Clone 13 and the overall low frequency of reassortment renders the risk of reversion quite unlikely. Immunization trials performed on rodents demonstrated that a single injection of Clone 13 vaccine elicited a high rate of immunization and effective protective immune response against wild-type RVFV strain.[4,77] Protection studies of natural hosts have been undertaken with this promising vaccine candidate.

Chemical attenuation of the reference RVFV strain ZH548 has been obtained after serial passages of the virus in the human diploid fibroblast cells MRC5 in the presence of 5-fluorouracil. The MP12 attenuated clone elicited a protective immune response in livestock.[80,81] Moreover, neutralizing antibodies produced by ewes could be transmitted to neonates by colostrum. However, teratogenic effects of the vaccine have been evidenced in ewes.[82] Clinical Phase II trials in humans demonstrated that a single injection of MP12 induced a seroconversion associated with significant neutralization titers in 95% of volunteers.[4] No evidence of reversions of the vaccine strain could be evidenced in vaccinated individuals.

In order to reinforce the safety of attenuated vaccine strains, a reassortant that combines the attenuation markers of Clone 13 and MP12 has been generated.[4] Preliminary trials of single-dose injection of the reassortant R566 in cattle and pregnant ewes in Senegal didn't show any adverse effect (i.e., signs of illness, abortion). The dose response study demonstrated that all vaccinated animals with 10^5 pfu developed neutralizing antibodies.[4] Clone 13 and R566 are now considered as potential candidates by OIE for field studies in livestock.

Despite the low risk of reversion of attenuated strains, inactivated vaccines are of interest for preventive vaccination of livestock in nonendemic areas or to protect personnel with high occupational risk of exposure to RVFV such as laboratory workers or vets.

Whereas a live attenuated RVFV has been tested for human immunization, only a formalin-inactivated vaccine is presently available but only dispensed to exposed lab workers or professionals potentially exposed to infected animals.[83]

Potential reversion and restrictive rules for the use of live attenuated vaccines and lack of long-lasting protective immunity conferred by inactivated vaccines highlight the need for continuous research to overcome these limitations. Different fields of vaccine development have been open by recent progresses in molecular biology.[4] RVFV Gn and Gc were genetically introduced in the genome of viral vectors such as lumpy skin disease virus and alphavirus replicons. Both recombinant viral vectors were shown to elicit neutralizing antibodies in mouse models and total protection of challenged mice.[4] Expression of RVFV glycoproteins in recombinant baculovirus gave promising results in terms of immunization. Moreover, this expression system is of particular interest for large-scale production. However, combination of the recombinant proteins with complete Freund's adjuvant is required to obtain a 100% protection rate of challenged mice. Baculovirus has also been used as a platform for virus-like particles (VLPs). Coexpression of RVFV nucleocapsid and envelope glycoproteins was successfully achieved. VLPs could be derived from human embryonic kidney cells transfected with M and L segment of RVFV genome.

More recently, clinical trials in animals were performed to evaluate the efficacy of new generations of recombinant vaccine candidates based on the structural glycoproteins. A significant neutralizing antibody response could be induced within three weeks after vaccination and after administration of a single dose of the different candidates that conferred protection to challenged lambs.[84] These encouraging results open new perspectives for large-scale vaccination of herds in low-income countries.

18.7 CONCLUSIONS AND PERSPECTIVES

RVFV is considered as a major animal and human health concern for decades. Imported livestock from northern hemisphere in the African continent revealed that European species were significantly more susceptible to RVFV infection compared to local species. The spreading of the virus out of the sub-Saharan Africa regions since 1977 demonstrated that various biotopes gathered conditions compatible with an efficient transmission cycle. High mortality rates among livestock, the various routes of contamination, the severity of the disease in humans, and the dramatic economical impact of the outbreaks led to consider RVFV as a potential agent for bioterrorism. The emergence of RVFV in agriculture systems of industrialized countries may lead to economical crisis. Thus, surveillance networks with reference laboratories are implemented outside endemic countries. Competent mosquito species have still been identified in Europe and the United States. Efforts for the development of new vaccines more adapted to the different epidemiological contexts are in progress. New vaccine strategies are built up with the perspectives to be able to discriminate naturally infected and vaccinated animals and to be adapted to wide vaccination programs of human populations.

Molecular epidemiology helps to better understand the means of virus spreading and allows establishing the genetic evolutionary rate of RVFV. These data are of major interest to draw containment measures and to take the viral variability into account for the development of new vaccines and molecular diagnostic tools. Early diagnosis is essential for the recognition of RVFV outbreaks. Multidisciplinary surveillance systems have been implemented in endemic countries. Diagnostic procedures are crucial for the efficiency of these networks. Highly sensitive and specific molecular tools have been developed and applied for the detection of the RVFV genome in samples collected within the different susceptible hosts of the virus including mosquitoes. RVFV viremia titration by real-time RT-PCR may have a predictive value for the patients' outcome and by the way an interest for the management of antiviral treatments. Efforts are now needed to fully adapt these technologies to field conditions and to standardize the diagnosis algorithm to reinforce international surveillance of RVFV.

RVF is an international mandatory animal disease. Thus, all cases should be reported to the FAO, the world organization for animal health ("Organisation Internationale des epizooties" [OIE]), and the World Health Organization. Standardization efforts for diagnosis are held by OIE through a network of reference laboratories. Revised documentation on outbreak management and diagnosis is available on the following websites: www.oie.int/eng/normes/mmanual/A_summry.htm and www.who.int/en and www.fao.org.

History of RVFV outbreaks highlights the predominant role of animal movements of viral extension or reintroduction. Thus, control measures of herds trading for the management of RVFV are a main economical concern in developing countries but also in possible emergence areas. Training of veterinarians to diagnose RVF disease in endemic zones but also in countries commercially connected is required for early detection and alert. The role of wildlife in the maintenance and spreading of RVFV is not fully elucidated. Thus, special attention should also be focused on new companion animals especially because of illegal trafficking of protected animal species.

REFERENCES

1. Schmaljohn, C.S. and Nichol, S.T. Bunyaviridae. In: Knipe, D.M. and Howley, P., eds. *Fields Virology*, 5th ed. Philadelphia, PA: Lippincott, Williams & Wilkins, pp. 1741–1789, 2007.
2. Freiberg, A.N. et al. Three-dimensional organization of Rift Valley fever virus revealed by cryoelectron tomography. *J. Virol.* 2008;82:10341–10348.
3. Flick, R. and Bouloy, M. Rift Valley fever virus. *Curr. Mol. Med.* 2005;5:827–834.
4. Bouloy, M. and Flick, R. Reverse genetics technology for Rift Valley fever virus: Current and future applications for the development of therapeutics and vaccines. *Antiviral Res.* 2009;84:101–118.
5. Pepin, M. et al. Rift Valley fever virus (Bunyaviridae: *Phlebovirus*): An update on pathogenesis, molecular epidemiology, vectors, diagnostics and prevention. *Vet. Res.* 2010;41:61.
6. Ikegami, T. Molecular biology and genetic diversity of Rift Valley fever virus. *Antiviral Res.* 2012;95:293–310.
7. Nichol, S.T. et al. Family Bunyaviridae. In: Fauquet, C.M., Mayo, M.A., Maniloff, J., Desselberger, U., and Ball, L.A., eds. *Virus Taxonomy: Eighth Report of the International Committee on Taxonomy of Virus*. San Diego, CA: Elsevier Academic Press, pp. 695–716, 2005.
8. Xu, F. et al. Antigenic and genetic relationships among Rift Valley fever virus and other selected members of the genus *Phlebovirus* (Bunyaviridae). *Am. J. Trop. Med. Hyg.* 2007;76:1194–1200.
9. Zeller, H.G. et al. Enzootic activity of Rift Valley fever virus in Senegal. *Am. J. Trop. Med. Hyg.* 1997;56:265–272.
10. Gonzalez, J.P. et al. Serological evidence in sheep suggesting phlebovirus circulation in a Rift Valley fever enzootic area in Burkina Faso. *Trans. R. Soc. Trop. Med. Hyg.* 1992;86:680–682.
11. Liu, D.Y. et al. Phylogenetic relationships among members of the genus *Phlebovirus* (Bunyaviridae) based on partial M segment sequence analyses. *J. Gen. Virol.* 2003;84:465–473.
12. Sall, A.A. et al. Variability of the NS(S) protein among Rift Valley fever virus isolates. *J. Gen. Virol.* 1997;78:2853–2858.
13. Bird, B.H. et al. Complete genome analysis of 33 ecologically and biologically diverse Rift Valley fever virus strains reveals widespread virus movement and low genetic diversity due to recent common ancestry. *J. Virol.* 2007;81:2805–2816.
14. Grobbelaar, A.A. et al. Molecular epidemiology of Rift Valley fever virus. *Emerg. Infect. Dis.* 2011;17:2270–2276.
15. Daubney, R., Hudson, J.R., and Garnham, P.C. Enzootic hepatitis or Rift Valley fever. An undescribed virus disease of sheep, cattle and man from East Africa. *J. Pathol. Bacteriol.* 1931;34:545.
16. Abdo-Salem, S. et al. Descriptive and spatial epidemiology of Rift valley fever outbreak in Yemen 2000–2001. *Ann. N.Y. Acad. Sci.* 2006;1081:240–242.
17. Sissoko, D. et al. Rift Valley fever, Mayotte, 2007–2008. *Emerg. Infect. Dis.* 2009;15:568–570.
18. Smithburn, K.C., Haddow, A.J., and Gillett, J.D. Rift Valley fever; isolation of the virus from wild mosquitoes. *Br. J. Exp. Pathol.* 1948;29:107–121.
19. Fontenille, D. et al. New vectors of Rift Valley fever in West Africa. *Emerg. Infect. Dis.* 1998;4:289–293.
20. Turell, M.J. and Perkins, P.V. Transmission of Rift Valley fever virus by the sand fly, *Phlebotomus duboscqi* (Diptera: Psychodidae). *Am. J. Trop. Med. Hyg.* 1990;42:185–188.
21. Dohm, D.J. et al. Laboratory transmission of Rift Valley fever virus by *Phlebotomus duboscqi, Phlebotomus papatasi, Phlebotomus sergenti*, and *Sergentomyia schwetzi* (Diptera: Psychodidae). *J. Med. Entomol.* 2000;37:435–438.
22. Linthicum, K.J. et al. Rift Valley fever virus (family Bunyaviridae, genus *Phlebovirus*). Isolations from Diptera collected during an inter-epizootic period in Kenya. *J. Hyg. (Lond.)* 1985;95:197–209.
23. Meegan, J.M. and Bailey, C.L. Rift Valley fever. In: Monath, T.P., ed. *The Arboviruses: Epidemiology and Ecology*, vol. 4. Boca Raton, FL: CRC Press, pp. 51–76, 1989.
24. Linthicum, K.J. et al. Transstadial and horizontal transmission of Rift Valley fever virus in *Hyalomma truncatum. Am. J. Trop. Med. Hyg.* 1989;41:491–496.
25. McIntosh, B.M., Dickinson, D.B., and dos Santos, I. Rift Valley fever. 3. Viraemia in cattle and sheep. 4. The susceptibility of mice and hamsters in relation to transmission of virus by mosquitoes. *J. S. Afr. Vet. Assoc.* 1973;44:167–169.
26. Pretorius, A. et al. Rift Valley fever virus: A seroepidemiologic study of small terrestrial vertebrates in South Africa. *Am. J. Trop. Med. Hyg.* 1997;57:693–698.
27. Olive, M.M., Goodman, S.M., and Reynes, J.M. The role of wild mammals in the maintenance of Rift Valley fever virus. *J. Wildl. Dis.* 2012;48:241–266.
28. Swanepoel, R. Observations on Rift Valley fever in Zimbabwe. *Contrib. Epidemiol. Biostat.* 1981;3:1549–1555.
29. Davies, F.G., Linthicum, K.J., and James, A.D. Rainfall and epizootic Rift Valley fever. *Bull. World Health Organ.* 1985;63:941–943.
30. Anyamba, A., Linthicum, K.J., and Tucker, C.J. Climate-disease connections: Rift Valley Fever in Kenya. *Cadernos de Saude Publica* 2001;17(Suppl):133–140.
31. Anyamba, A. et al. Prediction of a Rift Valley fever outbreak. *Proc. Natl. Acad. Sci. USA* 2009;106:955–959.
32. Clements, A.C. et al. A Rift Valley fever atlas for Africa. *Prev. Vet. Med.* 2007;82:72–82.

33. Tourre, Y.M. et al. Mapping of zones potentially occupied by *Aedes vexans* and *Culex poicilipes* mosquitoes, the main vectors of Rift Valley fever in Senegal. *Geospat. Health* 2008;3:69–79.

34. Logan, T.M. et al. Egg hatching of *Aedes* mosquitoes during successive floodings in a Rift Valley fever endemic area in Kenya. *J. Am. Mosq. Control Assoc.* 1991;7:109–112.

35. Chevalier, V. et al. An unexpected recurrent transmission of Rift Valley fever virus in cattle in a temperate and mountainous area of Madagascar. *PLoS Negl. Trop. Dis.* 2011;5:e1423.

36. Pourrut, X. et al. Rift Valley fever virus seroprevalence in human rural populations of Gabon. *PLoS Negl. Trop. Dis.* 2010;4:e763.

37. Gad, A.M. et al. A possible route for the introduction of Rift Valley fever virus into Egypt during 1977. *J. Trop. Med. Hyg.* 1986;89:233–236.

38. Meegan, J.M. Rift Valley fever in Egypt: An overview of the epizootics in 1977 and 1978. *Contrib. Epidemiol. Biostat.* 1981;3:100–113.

39. Meegan, J.M., Watten, R.H., and Laughlin, L.W. Clinical experience with Rift Valley fever in humans during the 1977 Egyptian epizootic. *Contrib. Epidemiol. Biostat.* 1981;3:114–123.

40. van Velden, D.J. et al. Rift Valley fever affecting humans in South Africa: A clinicopathological study. *S. Afr. Med. J.* 1977;51:867–871.

41. Hoogstraal, H. et al. The Rift Valley fever epizootic in Egypt 1977–78. 2. Ecological and entomological studies. *Trans. R. Soc. Trop. Med. Hyg.* 1979;73:624–629.

42. Gerdes, H. Rift Valley fever. *Rev. Sci. Tech. Off. Int. Epiz.* 2004;23:613–623.

43. Arishi, H.M., Aqeel, A.Y., and Al Hazmi, M.M. Vertical transmission of fatal Rift Valley fever in a newborn. *Ann. Trop. Paediatr.* 2006;26:251–253.

44. Adam, I. and Karsany, M.S. Case report: Rift Valley Fever with vertical transmission in a pregnant Sudanese woman. *J. Med. Virol.* 2008;80:929.

45. Abu-Elyazeed, R. et al. Prevalence of anti-Rift-Valley-fever IgM antibody in abattoir workers in the Nile delta during the 1993 outbreak in Egypt. *Bull. World Health Organ.* 1996;74:155–158.

46. Archer, B.N. et al. Outbreak of Rift Valley fever affecting veterinarians and farmers in South Africa, 2008. *S. Afr. Med. J.* 2011;101:263–266.

47. Njenga, M.K. et al. Using a field quantitative real-time PCR test to rapidly identify highly viremic Rift Valley fever cases. *J. Clin. Microbiol.* 2009;47:1166–1171.

48. Garcia, S. et al. Quantitative real-time PCR detection of Rift Valley fever virus and its application to evaluation of antiviral compounds. *J. Clin. Microbiol.* 2001;39:4456–4461.

49. Hanson, R.P. et al. Arbovirus infections of laboratory workers. Extent of problem emphasizes the need for more effective measures to reduce hazards. *Science* 1967;158:1283–1286.

50. The Subcommittee on Arbovirus Laboratory Safety of the American Committee on Arthropods-Borne Virus. Laboratory safety for arboviruses and certain other viruses of vertebrates. *Am. J. Trop. Med. Hyg.* 1980;29:1359–1381.

51. Borio, L. et al. Hemorrhagic fever viruses as biological weapons: Medical and public health management. *JAMA* 2002;287:2391–2405.

52. Bossi, P. et al. Bichat guidelines for the clinical management of haemorrhagic fever viruses and bioterrorism-related haemorrhagic fever viruses. *Euro Surveill.* 2004;9:E11–E12.

53. Bird, B.H. et al. Rift Valley fever virus. *J. Am. Vet. Med. Assoc.* 2009;234:883–893.

54. Gear, J. et al. Rift valley fever in South Africa. 2. The occurrence of human cases in the Orange Free State, the North-Western Cape Province, the Western and Southern Transvaal. B. Field and laboratory investigation. *S. Afr. Med. J.* 1951;25:908–912.

55. Scott, G.R. Pigs and Rift Valley fever. *Nature* 1963;200:919–920.

56. Davies, F.G., Clausen, B., and Lund, L.J. The pathogenicity of Rift Valley fever virus for the baboon. *Trans. R. Soc. Trop. Med. Hyg.* 1972;66:363–365.

57. Daubney, J.R. Hudson Rift Valley fever. *Lancet* 1932;219:611–612.

58. Madani, T.A. et al. Rift Valley fever epidemic in Saudi Arabia: Epidemiological, clinical, and laboratory characteristics. *Clin. Infect. Dis.* 2003;37:1084–1092.

59. Alrajhi, A.A., Al-Semari, A., and Al-Watban, J. Rift Valley fever encephalitis. *Emerg. Infect. Dis.* 2004;10:554–555.

60. Clarke, D.H. and Casals, J. Techniques for hemagglutination and hemagglutination-inhibition with arthropod-borne viruses. *Am. J. Trop. Med. Hyg.* 1958;7:561–573.

61. Office International des Epizooties (World Organization for Animal Health). Chapter 2.1.14—Rift Valley fever. In: *Manual of Diagnostic Tests and Vaccines for Terrestrial Animals.* 2013. Available online at www.oie.int/en/international-standard-setting/terrestrial-manual/access-online/.

62. Paweska, J.T. et al. An inhibition enzyme-linked immunosorbent assay for the detection of antibody to Rift Valley fever virus in humans, domestic and wild ruminants. *J. Virol. Methods* 2005;127:10–18.

63. Paweska, J.T., Jansen van Vuren, P., and Swanepoel, R. Validation of an indirect ELISA based on a recombinant nucleocapsid protein of Rift Valley fever virus for the detection of IgG antibody in humans. *J. Virol. Methods* 2007;146:119–124.

64. Jansen van Vuren, P. and Paweska, J.T. Laboratory safe detection of nucleocapsid protein of Rift Valley fever virus in human and animal specimens by a sandwich ELISA. *J. Virol. Methods* 2009;157:15–24.

65. Kamal, S.A. Pathological studies on postvaccinal reactions of Rift Valley fever in goats. *Virol. J.* 2009;6:94.

66. Sall, A.A. et al. Single tube and nested reverse transcriptase-polymerase chain reaction for the detection of Rift Valley fever virus in human and animal sera. *J. Virol. Methods* 2001;91:85–92.

67. Peyrefitte, C.N. et al. Real-time reverse-transcription loop-mediated isothermal amplification for rapid detection of Rift Valley fever virus. *J. Clin. Microbiol.* 2008;46:3653–3659.

68. Shoemaker, T. et al. Genetic analysis of viruses associated with emergence of Rift Valley fever in Saudi Arabia and Yemen, 2000–01. *Emerg. Infect. Dis.* 2002;8:1415–1420.

69. Bird, B.H. et al. Multiple virus lineages sharing recent common ancestry were associated with a Large Rift Valley fever outbreak among livestock in Kenya during 2006–2007. *J. Virol.* 2008;82:11152–11166.

70. Cêtre-Sossah, C. et al. Genome analysis of Rift Valley fever virus, Mayotte. *Emerg. Infect. Dis.* 2012;18:969–971.

71. Gowen, B.B. et al. Efficacy of favipiravir (T-705) and T-1106 pyrazine derivatives in phlebovirus disease models. *Antiviral Res.* 2010;86:121–127.

72. Billecocq, A. et al. NSs protein of Rift Valley fever virus blocks interferon production by inhibiting host gene transcription. *J. Virol.* 2004;78:9798–9806.

73. Gowen, B.B. et al. Extended protection against phlebovirus infection conferred by recombinant adenovirus expressing consensus interferon (DEF201). *Antimicrob. Agents Chemother.* 2012;56:4168–4174.

74. Pfeffer, M. and Dobler, G. Emergence of zoonotic arboviruses by animal trade and migration. *Parasit. Vectors* 2010;3:35.

75. Botros, B. et al. Immunological response of Egyptian fat-tail sheep to inactivated and live attenuated Smithburn Rift Valley fever vaccines. *J. Egypt. Vet. Med. Assoc.* 1995;55:895–907.

76. Botros, B. et al. Adverse response of non-indigenous cattle of European breeds to live attenuated Smithburn Rift Valley fever vaccine. *J. Med. Virol.* 2006;78:787–791.

77. Muller, R. et al. Characterization of clone 13, a naturally attenuated avirulent isolate of Rift Valley fever virus, which is altered in the small segment. *Am. J. Trop. Med. Hyg.* 1995;53:405–411.

78. Bouloy, M. et al. Genetic evidence for an interferon-antagonistic function of rift valley fever virus nonstructural protein NSs. *J. Virol.* 2001;75:1371–1377.

79. Bowen, M.D. et al. A reassortant bunyavirus isolated from acute hemorrhagic fever cases in Kenya and Somalia. *Virology* 2001;291:185–190.

80. Morrill, J.C. et al. Further evaluation of a mutagen-attenuated Rift Valley fever vaccine in sheep. *Vaccine* 1991;9:35–41.

81. Morrill, J.C., Mebus, C.A., and Peters, C.J. Safety and efficacy of a mutagen-attenuated Rift Valley fever virus vaccine in cattle. *Am. J. Vet. Res.* 1997;58:1104–1109.

82. Hunter, P., Erasmus, B.J., and Vorster, J.H. Teratogenicity of a mutagenised Rift Valley fever virus (MVP 12) in sheep. *Onderstepoort J. Vet. Res.* 2002;69:95–98.

83. Pittman, P.R. et al. Immunogenicity of an inactivated Rift Valley fever vaccine in humans: A 12-year experience. *Vaccine* 1999;18:181–189.

84. Kortekaas, J. et al. Efficacy of three candidate Rift Valley fever vaccines in sheep. *Vaccine* 2012;30: 3423–3429.

19 SARS Coronavirus

Baochuan Lin and Anthony P. Malanoski

CONTENTS

19.1 INTRODUCTION

Severe acute respiratory syndrome coronavirus (SARS CoV) emerged in 2003 as a global health threat that causes a life-threatening infectious respiratory disease. The first SARS outbreak occurred in November 2002 in Guangdong Province, China, and the epidemic quickly spread to more than 30 countries. There were more than 8000 cases and approximately 800 deaths during the outbreak in 2002–2003.[1] Fortunately, the outbreak triggered a successful global response, and the spread of the disease was stopped at the end of June 2003 and did not reach the scale of an influenza pandemic. Before the appearance of SARS CoV, human coronaviruses, such as 229E and OC43, were relatively obscure and usually associated with a mild upper respiratory tract infection presenting symptoms of the common cold. In contrast, animal coronaviruses are responsible for a variety of severe diseases in domesticated animals.[2–8] The severity and sudden onset of symptoms of atypical pneumonia with dry cough and persistent high fever, which is drastically different from symptoms of the previously known human pathogenic coronaviruses, suggested that SARS CoV emerged through interspecies transmission.[1]

After the outbreak, the origin of the SARS CoV was traced back to exotic animals, such as the Himalayan palm civets (*Paguma larvata*) and raccoon dogs (*Nyctereutes procyonoides*), sold in Chinese wet markets.[9] Subsequent studies found the presence of SARS CoV in various animals including Chinese ferret-badger (*Melogale moschata*) and several bat species. Furthermore, these studies also suggested that although *P. larvata* may have been the source of animal-to-human interspecies transmission that precipitated the SARS outbreak, these animals only serve as intermediate hosts while bats, Chinese horseshoe bats (*Rhinolophus sinicus*) in particular, are the likely animal reservoir of SARS CoV.[9–11] The zoonotic origin of SARS CoV from an identified animal reservoir that has ongoing circulation of SARS CoV via several different intermediate animal host places this organism in the list of pathogens that pose a continuous threat to the public health and economic welfare of many nations. It is easily possible for the current strain to be reintroduced into the human population from its animal reservoir or even for a new variant, mutation or related variant in the animal reservoir, to be introduced initiating a new outbreak and possible epidemic. Thus, SARS outbreak is another representative case of a relatively benign organism in its host reservoir becoming a serious health threat for the human population as contact and interaction with the animal reservoir or the environment that the animal lives in has increased. Nature remains the best source for potentially apocalyptic biological agents.[12,13] In this chapter, we will discuss the current understanding of SARS CoV related to its classification, biology, epidemiology, pathogenesis, identification/diagnosis, treatment, and prevention, as well as future perspectives in improving detection and control of SARS CoV.

19.2 CLASSIFICATION AND MORPHOLOGY

SARS CoV belongs to the Coronaviridae family, which is a family of large enveloped positive-stranded RNA viruses with the largest known RNA viral genome (~27–33 kb) that are capped and polyadenylated.[1,6–8,14] Coronaviruses are divided into three serological groups based on their natural hosts, nucleotide sequences, and serological relationship.[2,6] Group I and II viruses mainly infect mammals, while group III viruses are found in birds.[4,6] So far, five human coronaviruses, 229E, OC43, NL63, SARS, and HKU1, have been identified. The first two human coronaviruses HCoV-229E (group I) and HCoV-OC43 (group II) identified in the mid-1960s cause common cold symptoms.[6,15,16] Following the outbreak and identification of SARS CoV (group IIb), two other strains HCoV-NL63 (group I) and HCoV-HKU1 (group II) have since been identified.[2,4,5,15–26] Phylogenetic analysis of SARS CoV showed a marked degree of divergence from all other known coronaviruses.[20,27] Further detection and phylogenetic analysis found one other strain, SARS-like CoV or SARS CoV-like virus (SL-CoV) in wild animals, that has a

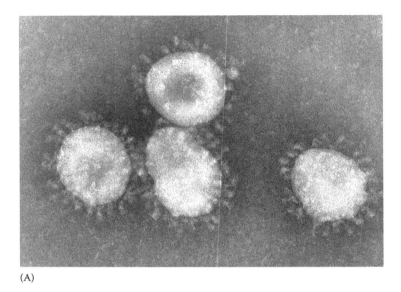

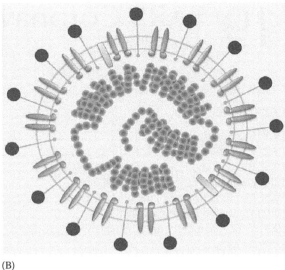

(A)   (B)

FIGURE 19.1   SARS CoV. (A) Negative-contrast electron microscopy image. The large petal-shaped spikes consisting of S glycoprotein seen on the envelopes of the viruses are a distinct feature of the coronavirus. (Courtesy of Dr. Charles D. Humphrey, Centers for Disease Control and Prevention, Atlanta, GA.) (B) Structure model of the SARS CoV virion. The S protein was represented with a dark circle attached to a gray line, the light gray rectangles embedded within membrane represent E proteins, paired M proteins were represented with ellipsoid shapes, protruding from the membrane and dark gray circles represent genomic RNA. (Modified from Holmes, K. V., *N. Engl. J. Med.*, 348, 1948–1951, 2003; Bárcena M. et al., *Proc. Natl. Acad. Sci. USA*, 106, 582–587, 2009.)

close genetic and antigenic relationship to SARS CoV. This further analysis shows that the two strains, SARS CoV and SL-CoV, form a distinct cluster (group IIb) that is distantly related to other group II coronavirus.[28–31]

SARS CoV has a distinct virion morphology similar to all the other known coronaviruses in the Coronaviridae family. The viral particles, having an average diameter of 80–140 nm, are enveloped with extended spike membrane proteins producing a crown-like structure (Figure 19.1). The three main structural proteins are spike (S), membrane (M), and envelope (E) proteins (Figure 19.1). Inside the virion, a coronavirus possesses a helically symmetric ribonucleocapsids core formed by the association of nucleocapsid (N) proteins with genomic RNA. The ribonucleocapsids core is enclosed by a lipoprotein envelope composed of M glycoprotein and has a diameter of 65 nm. When released from disrupted viral particles, the N proteins appear as a threadlike structure with a diameter of 14–16 nm and hollow core of 3–4 nm. SARS CoV is classified as a group IIb coronavirus, a distinct subgroup, because it lacks the hemagglutinin–esterase (HE) glycoprotein that is characteristic of group II coronaviruses.[2,6,32–34]

19.3   BIOLOGY AND EPIDEMIOLOGY

The near absence of antibodies against SARS CoV in the general population before 2003 suggested the zoonotic origin of SARS CoV.[1,34–36] This finding prompted the investigation into animal origins of SARS CoV. In addition to likely animal-to-human infection sources such as the Himalayan palm civets, raccoon dogs, and Chinese ferret-badger, SARS CoVs were found in several bat species. Further surveillance of coronavirus in wildlife indicated there was no widespread

infection of SARS CoV in wild or farmed civets, which suggested that civet cats, raccoon dogs, and ferret-badgers from wet markets were only vectors for SARS CoV outbreaks in 2002 and 2003. On the other hand, studies of several species of bats have found very diverse assemblages of coronaviruses. A group of SL-CoVs shares 88%–92% nucleotide identity with SARS CoV also identified from bats suggesting that they may be the natural reservoir of SARS CoV and other Coronaviridae family members.[9,10,28,29,31,37–40]

The genome sequence of SARS CoV was completed in 2003 by two independent groups (accession no. NC_004718, AY278741).[20,27] The genome size of SARS CoV is 29.75 kb and has similar organization to other known coronaviruses with a conserved structure order of 5′ cap, leader sequence (72 nucleotide long), untranslated region (5′ UTR), and replicase, followed by S, E, M, and N proteins and 3′UTR (Figure 19.2A).[2,5,6,20,27] The replicase gene consists of two overlapping open reading frames ORF1a and ORF1b, approximately two-thirds of coronavirus genome size (19.2 kb for SARS CoV) from the 5′-end, and encodes proteins necessary for viral RNA synthesis. ORF1a encodes one or two papain-like protease (PLpro), a picornavirus 3C-like protease (3CLpro), which processes two large precursor proteins (pp1a and pp1ab) into the mature replicase proteins, and putative ADP-ribose-1′-phosphatase. ORF1b encodes RNA-dependent RNA polymerase (RdRp), helicase, putative 3′ to 5′ exonuclease, poly(U)-specific endoribonuclease, and putative S-adenosylmethionine-dependent ribose 2′-O-methyltransferase.[8] SARS CoV does not encode HE glycoprotein that is unique to other group II coronaviruses as reflected in its classification as group IIb rather than just group II. Interspersed among ORF1 and structure protein

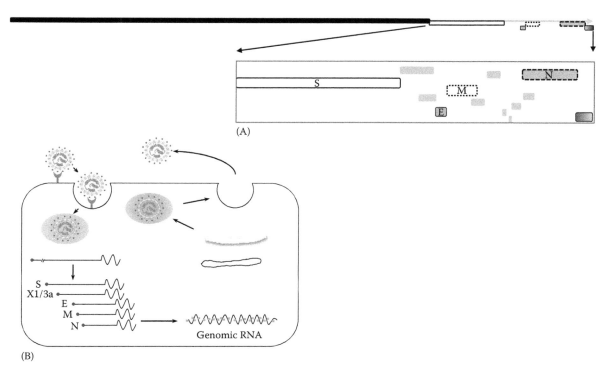

FIGURE 19.2 (A) Genome structure of SARS CoV. ■, ORF1; ▬, accessory proteins; the rest of the proteins are marked as S (spike), E (envelop), M (membrane), and N (nucleocapsid). The genome structures are constructed based on GenBank sequence access number NC_004718. (B) Life cycle of SARS CoV. *Note*: DMV, double membrane vesicles; ERGIC, ER–Golgi intermediate compartment. (Modified from de Haan, C.A. and Rottier, P.J., *Cell. Microbiol.*, 8, 1211–1218, 2006. With permission.)

genes are nonstructural genes (accessory proteins), which vary in number and position for different coronaviruses. SARS CoV has eight potential nonstructural proteins, two located between S and E proteins (3a and 3b), five located between M and N proteins (ORF6, 7a, 7b, 8a, and 8b), and one within N genes (9b).[5,8,20,27] Unlike other coronaviruses, the functions of the accessory proteins of SARS CoV have been investigated due to the SARS epidemic. These studies implicated the involvement of accessory proteins in SARS pathogenesis and evasion of host immune responses (detailed in the following).

Like other coronavirus, the S glycoprotein of SARS CoV mediate binding with its functional receptor, angiotensin-converting enzyme 2 (ACE2), of host cells that led to the endocytosis and entry of viral particles into the cytoplasm.[41] Once inside the cells, viral genomic RNA was used as a template for translation of the replicase genes that produces pp1a and pp1ab. The pp1a and pp1ab are then cleaved by viral proteinases, PLpro, and 3CLpro, into 16 subunits, and collectively constitute the functional replication–transcription complexes that mediate both genome replication and transcription of subgenomic mRNAs downstream of OFR1b. The replication–transcription complexes mediate viral RNA replication through the synthesis of a negative-strand RNA that in turn is the template for the synthesis of progeny virus genomes. The subgenomic RNAs are translated into structural proteins and accessory proteins through a cap-dependent ribosomal scanning mechanism. The membrane-bound structure proteins, M, S, and E, are inserted into the endoplasmic reticulum

(ER) membrane and transited to ER–Golgi-associated complexes. The N proteins encapsulate progeny genomic RNA and assemble to form helical N proteins, which then combine with membrane-bound components, forming viral particles by budding into the ER–Golgi-associated complexes. Finally, progeny virions are released through an exocytosis-like mechanism by fusion of smooth-walled virion-containing vesicles with the plasma membrane (Figure 19.2B).[2,6,41]

The SARS epidemic in 2003 mainly occurred in Southeast Asia with the highest incidence in China (~2700) followed by Hong Kong (~1000), Taiwan (~220), and Singapore (~160). In addition, through international air travel, visitors that were infected by the first index case resulted in the widespread of SARS cases to 28 other countries.[33,42] Retrospective analysis indicated that health-care facilities were the major transmission setting as health workers accounted for 21% of all SARS cases and a high percentage of patients in the same ward as SARS patients and hospital visitors were infected. The second most common transmission setting was the household of a previously infected SARS patient as surveys in Hong Kong and Singapore indicated 6%–8% of SARS cases were in this category.[43,44] Health-care workers and people in close contact with SARS patients have the highest risk of contracting SARS infection.

The possible transmission modes of SARS CoV are droplets, direct contact, and fomite (indirect contact), which is similar to other human coronaviruses.[6,33,45] The incubation period of human SARS CoV ranges from 2 to 10 days with median range of 4–6 days, and the symptoms can last from 3 to 18 days with a mean of 7 days.[5,38] The peak virus excretion of SARS CoV from the respiratory tract occurs on ca. day 10 of illness and declines thereafter to a low level at ca. day 23. During the peak viral load, the patients are highly infectious. In addition to respiratory secretion, SARS CoV can also be detected in urine and stools, which seem to start later with a peak between days 12 and 14 and a slower decline thereafter.[33,46] It was further found that fecal samples remained infectious for up to 4 days from patients with diarrhea (with higher pH). This is consistent with the animal origin of SARS CoV since animal coronaviruses are primarily spread through the fecal–oral route.[38] Fecal–oral transmission of SARS CoV has been proposed as a possible means of infection; however, there was insufficient evidence from the 2003 outbreak to conclusively support the fecal–oral route of transmission for SARS CoV infection; but, it also could not be ruled out as a route of transmission.[47] So the significance of shedding SARS CoV in urine and stool samples is not clear at this time.

19.4 CLINICAL FEATURES AND PATHOGENESIS

In contrast to other known human coronaviruses that only induce mild to moderate common cold symptoms, SARS CoV infections cause serious respiratory symptoms with a high fatality rate, including upper respiratory infection, bronchiolitis, and pneumonia in the elderly and in patients with underlying diseases, such as diabetes mellitus, heart disease, and hepatitis B infection.[42,48] The SARS CoV infection also leads to viral pneumonia with rapid respiratory deterioration, fever, chills, myalgia, malaise, and intestinal complications in adults while causing milder symptoms in children. During the first 10 days of illness, pneumocyte proliferation and desquamation, hyaline-membrane formation, mixed inflammatory infiltrate, intra-alveolar edema, and increased numbers of interstitial and alveolar macrophages, with focal hemophagocytosis in interstitial macrophage, are common. Diffuse alveolar damage, squamous metaplasia, and multinucleated giant cells macrophage are observed after longer duration of illness.[1,33] Lower respiratory tract symptoms are common and typically included a nonproductive cough with later onset of dyspnea. Diarrhea is the most common extrapulmonary manifestation, concurrent with viral replication in small and large intestines.[8,38,49] Leucopenia, lymphopenia, and thrombocytopenia are also common features in SARS CoV infection.[33,42,47,50] Acute respiratory distress, multiorgan failure, thromboembolic complication, secondary infections, and septic shock are the cause of death of patients suffering SARS CoV infection later in the course of the illness.[33] The overall case-fatality rate is approximately 10% and increases to >50% in people older than 60 years of age.[1,33,51]

Clinical manifestation of SARS can be defined in two phases, which are correlated with major mechanisms that contribute to the pathogenesis of SARS. The initial phase of clinical symptoms, such as high fever and myalgia, is correlated with the increase of viral replication and direct lytic effects of host cells. During phase 2, patients experience recurrence of fever, onset of diarrhea, and oxygen

desaturation, which is correlated with falls in viral load and rise of host immune response to the infection. Most patients recovered after phase 2, while 20% progressed to acute respiratory distress syndrome (ARDS) characterized by pulmonary edema, accumulation of inflammatory cells, and severe hypoxia.[52,53]

The entry route for SARS CoV into host cells is by receptor-mediated endocytosis via the interaction of S glycoprotein and ACE2 receptor. Once infection is established, the pathogenesis of SARS CoV can be separated into direct viral damage and indirect consequences of host immune response to infection.[49] Direct viral damage results from viral replication within host cells, which become overwhelmed by virus multiplication, leading to a variety of effects, that is, necrosis, lysis, or apoptosis of host cells to release virus particles. In consequence, patients infected by SARS CoV show respiratory epithelial cell necrosis and suffer lymphopenia and liver impairment due to the depletion of T lymphocytes and hepatocytes by apoptosis.[7] Viral replication relies on host machinery; thus, viral-specific synthesis downregulates genes involved in host cell translation and upregulates genes related to viral protein synthesis. In addition to shutting off host cell translation machinery, viral replication also downregulates genes related to the maintenance of cytoskeletal structure; the disruption of these microtubule networks is conducive to persistent viral infection and subsequent release. Furthermore, viral replication also upregulates genes involved in stress response, proapoptosis, proinflammatory response, and procoagulation. The balance between up- and downregulated genes plays a role in the severity of illness.[54]

While the contribution of direct viral damage is straightforward to enumerate, the hyperimmune response is dependent on host factors, that is, the ability to trigger and regulate innate and adaptive inflammatory responses, resulting in great variation in the severity of symptoms caused by SARS CoV infections. The modulation of host immune response is a common survival strategy of highly pathogenic viruses, and SARS CoV is no exception. Various aspects of the immune response to SARS CoV have been studied. Studies of cytokine and chemokine profiles of SARS patients suggested deregulation of proinflammatory cytokines, such as evaluated interferon (IFN)γ-inducible protein-10 (IP-10/CXCL10), IL-6, and lacking of IL-10 production, probably contributed to the severe lung injury in SARS CoV-infected patients. Elevated levels of other inflammatory mediators, including interleukin 8 (IL-8), CCL2/monocyte chemoattractant protein (MCP)-1, and CXCL9/monokine induced by IFNγ (MIG), were also reported. Other cytokines and chemokines were also studied but with controversial results (Table 19.1). The discrepancies of the results were probably caused by differences in assay sensitivity and the time frames (relative to onset of the infection) of the sample collection.[55–70]

Further studies using cell culture or animal models to confirm these findings and clarify some of the controversial results, however, resulted in more discrepancies. The results of animal and *in vitro* studies are even more complex and need to take into account the response of different species,

TABLE 19.1
Cytokines/Chemokines Profiles of SARS Patients

Sample	Cytokines/Chemokines Profiles	N	References
Serum	IL-2↑, IL-10↑, IL-12↑	35	[55]
Plasma	IFNγ↑, IL-1β↑, IL-6↑, IL-8↑, IL-12↑, MCP-1↑, IP-10↑ TNFα, IL-2, IL-4, IL-10	20	[56]
Plasma	IL-8↑, TNFα↑, IL-1β, IL-2, IL-4, IL-10, IL-12p70, IFNγ	24	[57]
Plasma	IL-8↑, IP-10↑, MIG↑, MCP-1, CCL5	255	[58]
Serum	IL-2↑, IL-6↑, IP-10↑, MIG↑, IL-5↑, IFNγ, IL-4, IL-8, IL-10, MCP-1, TNFα	14	[59]
Serum	IL-6↑, IL-8↑, IP-10↑, MCP-1↑, MIP-1α↑, TNFα, IL-1β, IL-2, IL-4, IL-10, IL-12, IFNγ, GM-CSF, CCL5	23	[60]
T$_{helper}$ cells	IL-1↑, IL-6↑, IL-8↑, IL-12↑, MCP-1↑, IP-10↑, TNFα, IL-2, IL-4, IL-10	20	[61]
Serum	IFNγ↑, IL-6↑, IL-8↑, IL-18↑, TGFβ↑, IP-10↑, MCP-1↑, MIG↑, TNFα, IL-2, IL-4, IL-10, IL-13, TNFRI	88	[62]
Serum	IL-6↑, IL-10↑, IL-4↓, IL-8↓, IFNγ↓, TGFβ↓, IL-1, TNFα	228	[63]
Serum	IL-1β↑, IL-2↑, IL-4↑, IL-6↑, IL-8↑, IL-10↑, TNFα↑	46	[64]
Plasma	IL-1β↑, IL-6↑, TNFα↑, IL-8, IL-10, IL-12p70	8	[65]
Serum	IFNγ↑, IL-4↑, IL-10↑, IL-12p70↑, CXCL8↑, IP-10↑, CCL5↑	9	[66]
Serum	IL-6↑, IL-8↑, IL-13↑, IL-16↑, IL-18↑, TNFα↑, TGFβ1↑	61	[67,68]
Serum	IL-6↑, IL-8↑, TNFα↑, IL-1β	14	[69]
Blood/plasma	CCL2, CXCL14↓, CCL19↑, IFNα↑, IFNγ↑, IP-10↑, MCP-1↑	50	[70]

Note: The underlined cytokines/chemokines listed were studied but no significant elevation or decrease was observed. CXCL, chemokine (C-X-C motif); CCL, chemokine (C-C motif); CCL5/RANTES, regulated on activation, normal T expressed and secreted; TNF, tumor necrosis factor; GM-CSF, granulocyte–macrophage colony-stimulating factor; IFN, interferon; IL, interleukin; IP-10 (CXCL-10), IFN-inducible protein-10; MCP, monocyte chemoattractant protein; MIG (CXCL9), monokine induced by IFNγ; TNFRI, tumor necrosis factor receptor I; ↑, indicated elevated level; ↓, indicated decreased level. N column indicated number of patients included in the study.

cell types' origins, and how the samples were prepared (Table 19.2).[71–96] The complex picture of how SARS CoV modulates host immune response remains confusing and requires further clarification; however, the basic framework is that SARS CoV evades the type I IFN production and type 1 IFN-mediated immunity by lowering production of antiviral cytokines, that is, IFNs, and elevating the proinflammatory and inflammatory cytokines, that is, IL-6, MIP, and CCL2, which helps them evade antiviral response evoked by the host and cause significant pathological damages.

Substantial research efforts in SARS CoV have gained insight to the putative roles of most of SARS proteins in the aforementioned pathogenic mechanisms. The S and N proteins are the main antigenic determinants; in addition, the accessory protein 3a of SARS CoV is also highly immunogenic and can induce neutralizing antibodies.[38] Besides being one of the main antigenic determinants, S proteins also play a critical role in SARS pathogenesis by downregulating ACE2 expression, triggering ACE2 signaling, and activating fibrosis-associated chemokine (C-C motif) ligand 2 (CCL2) expression through the Ras–ERK–AP-1 pathway, which in turn causes severe lung injury and acute lung failure.[97–100]

Although not immunogenic, several proteins of SARS CoV, including E, M, and several accessory proteins, can induce cell apoptosis and modulate host immune response. Alterations in cell cycle and apoptosis pathways of host cells have been used by viruses for their own replication advantage as well as modulating host immune response as a survival strategy of highly pathogenic viruses. SARS CoV also follows this behavior, which directly leads to the pathology of disease. E protein was implicated in downregulating the stress response and reducing virus-induced inflammation in virus-infected cells, which in turn minimizes the innate immune responses.[101] M protein was suggested to induce cell apoptosis through Akt prosurvival pathway and mitochondrial cytochrome c release.[102] Nsp1 was indicated in inhibiting host protein synthesis by blocking translation of preexisting transcripts, increasing RNA degradation, and inducing the secretion of chemokines, such as CCL5, CXCL10, and CCL3 through the nuclear factor kappa B (NF-κB) signaling pathway.[103,104] Nsp3, PLpro, was also indicated in counteracting IFN-induced process by blocking the IFN induction through IFN regulatory factor 3 (IRF3) and NF-κB.[105–112] NSP5, 3CLpro, was implicated in inducing host cell growth arrest and apoptosis through activation of caspase-3 and caspase-9 activities, as well as mitochondrial-mediated apoptosis.[113,114] NSP-10 probably enhances the cytopathic effect of SARS CoV-infected cells by impairing the oxidoreductase system in the mitochondria

TABLE 19.2

Cytokines/Chemokines Profiles of SARS-Infected Cells/Animals

Model	Cytokines/Chemokines Profiles	Reference
C57BL/6Mice Lung	CCL1↑, CCL2↑, CCL3↑, CCL5↑, CXCL-1↑, MIG↑, IP-10↑, XCL-1↑, CCL7↓, CCL4, IFNγ, IL-12 p70, IL-4, IL-10, TNFα	[71]
DCs	IL-6↑, TNFα↑, MIP1α↑, CCL5↑, IP-10↑, MCP-1↑, IFNα, IFNβ, IFNγ, IL12p40	[72]
Macrophages	CCL5↑, CCL7↑, CCL8↑, CCL20↑, CXCL2↑, MIG↑, IP-10↑, MIP1α, IFNβ	[73]
PBMCs	CXCL1↑, IL-1↑, IL-8↑, IL-17↑, IL-18↑, TNFα↑, CCL5↓, IP-10↓, CXCL11↓	[74]
A549 cells macrophages DCs		[75]
PBMC	No induction of cytokines	[92]
ACE2+ cells	IL-1β↑, IL-6↑, MCP-1↑, TGFβ↑, TNFα↑	[76]
Caco-2 cells	IL-6↑, IL-8↑, IP-10↑, IFNα, IFNβ, IFNγ, CCL5	[77]
HEK293 cells	IFNα↑, IL6↑, IL-8↑, IP-10↑, IFNβ, IFNγ, CCL5	[77]
A549 cells	IL-8↑, MCP-1↑, CXCL1↑, MCP-1↑, TNFβ↑,	[78]
THP-1 cells	MCP-1↑, IL-8↑, CCL3↑, IP-10↑, CCL4↑, CCL5↑	[78]
Macaques lung	IL-1β↑, IL-6↑, IL-8↑, MCP-1↑, IP-10↑, IFNα↑, IFNβ↑, IFNγ↑, CXCL1↑, 3↑, 9↑, 10↑, 11↑, 13↑; CCL3↑, 4↑, 7↑, 11↑, 13↑, 19↑, 20↑	[79]
RAW264.7 cells	IL-6↑, TNFα↑, IL-8	[80]
DCs	IL-6↑, TNFα↑, IL-1β, IL-8	[81]
BALB/c mice young	MCP-1↑, MIP-1α↑, IP-10↑, MIG↑, KC↑, VEGF↑, IFNγ↑, IL-1 α↑, IL-2↑, IL-4↑, IL-10↑, IL-12p40/p70↑, IL-13↑, BFGF, GM-CSF, IL-1α, IL-1β, IL-4, IL-5, IL-6, IL-17, KC, TNF-α	[82]
BALB/c mice adult	MCP-1↑, MIP-1α↑, IP-10↑, IL-1α↑, IL-1β↑, TNFα↑, IL-6↑, IL-12p40/p70↑, IL-13↑, BFGF, GM-CSF, IFNγ, IL-2, IL-4, IL-5, IL-10, IL-17, KC, MIG, VEGF	[82]
PBMC	Low induction of IFNα, β, and γ in comparison to other viral infection	[83]
BABL/c mice	IL-1α↑, IL-6↑, MIP-1α↑, MCP1↑, CCL5↑, IL-2, IL-3, IL-4, IL-5, IL-9, IL-10, TNFα, GM-CSF	[84]
NCI-H1650 6 h	IL-6↑, IL-8↑, IL-18↑, IP-10↑, MIG↑, CCL20↑, TNFα↑, IFNγ, IL-10, IL-12β	[85]
NCI-H1650 24 h	IL-6↑, IL-8↑, IL-10↑, IL-12β↑, IL-18↑, IFNγ↑, IP-10↑, MIG↑, TNFα↑, CCL20↑,	[85]
LOVO 6 h	IL-8↑, IL-18↑, IP-10↑, MIG↑, TNFα↑, IFNγ↑, IL-6, IL-10, IL-12β, CCL20	[85]
LOVO 24 h	IL-6↑, IL-8↑, IL-10↑, IL-12β↑, IFNγ↑, IP-10↑, MIG↑, CCL20↑, TNFα↑, IL-18↓	[85]
NP460 6 h	IP-10↑, MIG↑, IL-8↑, CCL20↑, IL-10, IL-12β, IL-18, IL-6, IFNγ, TNFα	[85]
NP460 24 h	IP-10↑, MIG↑, IL-8↓, CCL20↓, IL-6, IL-10, IL-12β, IL-18, IFNγ, TNFα	[85]
NP69 6 h	IL-8↑, IL-18↑, IP-10↑, MIG↑, CCL20↑, IL-6, IL-10, IL-12β, IFNγ, TNFα	[85]
NP69 24 h	IL-18↑, IP-10↑, MIG↑, IL-6, IL-8, IL-10, IL-12β, CCL20, IFNγ, TNFα	[85]
AC22 mice lung	IL-1α↑, IL-1β↑, IL-6↑, KC↑, IL12p40↑, CCL5↑, MCP-1↑, IL-2, IL-3, IL-4, IL-5, IL-9, IL-10, IL-13, IL-17, IFNγ, TNFα	[86]
AC70 mice lung	IL12p40↑, KC↑, CCL5↑, MCP-1↑, IL-1α, IL-1β, IL-6, IL-2, IL-3, IL-4, IL-5, IL-9, IL-10, IL-13, IL-17, IFNγ, TNFα	[86]
AC22 and AC70 mice brain	IL-1α↑, IL-1β↑, IL-6↑, IL-8↑, IL-9↑, IL-10↑, IL12p40↑, MIP1α↑, MIP1β↑, MCP-1↑, eotaxin↑, G-CSF↑, CCL5↑	[86]
Young macaques Lung	IL-6↑, CCL2↑, CCL3↑, CXCL11↑, CCL11↑, CCL8↑, IP-10↑, CXCL9, IFNβ1, TNFRSF21, TNFSF13B, IL-1RN	[87]
Aged adult macaques lung	IL-6↑, IL-8↑, CCL2↑, CCL3↑, CCL8↑, CXCL11↑, IP-10↑, TNFSF13B↑, CXCL1/CXCL3↑, CXCL6↓, IL29, CCL4L1, CCL19, CXCL9,	[87]
2B4 cells	IL-1α↑, IL-6↑, IL-8↑, IP-10↑, CCL5↑, IFN-β1↑, IFN-λ1↑, IFN-λ2↑, CXCL1↑, PDGF BB↑, TRAIL↑	[88]
Ferret	CCL2↑, CCL4↑, CCL14↑, CCL19↑, CCL25↑	[89]
Macaques lung	IL-2↑, IL12/23 (p40)↑, IL-13↑, MCP-1↑, MIP-1α↑, CCL8↑, IL29↑, IP-10↑, CXCL11↑	[90]
African green monkey	CXCL11↑, CXCL13↑, IL-1β↑, IFNγ↑, IL-12↑, CCL3↑, IL-2↑, IL-8, CCL19, CXCL12, IL-18, IL-15, IL-6, CCL21, TGF-β, CCL2, CCL5	[91]
PBMC	IFNα↑, IFNγ↑	[93]
Huh7 cells	IL-8↑, TGF β2↑, CXCL1↑, 2↑, 3↑, 5↑, 6↑; IP-10	[94]

TABLE 19.2 (continued)
Cytokines/Chemokines Profiles of SARS-Infected Cells/Animals

Model	Cytokines/Chemokines Profiles	Reference
Caco2, CL-14	CCL3↑, 20↑; CXCL1↑, 2↑, 3↑, 11↑; IL-1↑, IL-8↑, IP-10↑, PDGFB↑, TGFβ↑, IL-18↓, MIF↓	[95]
Caco2	IL-6↑, IL-8↑, IFNα↑, IFNγ↑, IFNβ↑, CCL5↑	[96]

Note: Caco2 and CL-14, human colon carcinoma cell lines; DCs, dendritic cells; Huh7, human hepatoma cell line; NCI-H1650, human lung cell line; LOVO, human colon cell line; NP460 and NP69, human nasopharyngeal cell line; PBMCs, peripheral blood mononuclear cells; A549 cells, lung epithelial cells; Caco-2, colon carcinoma cell line; HEK293 cells, human embryonic kidney cell line; THP-1, human monocytic cell line; RAW264.7 cells, murine macrophage cell line; 2B4 cells, a cloned bronchial epithelial cell line derived from Calu-3 cells; BFGF, basic fibroblast growth factor; G-CSF, granulocyte colony-stimulating factor; GM-CSF, granulocyte–macrophage colony-stimulating factor; IFN, interferon; IL, interleukin; IP-10 (CXCL-10), IFN-inducible protein-10; KC, keratinocyte chemoattractant; MCP, monocyte chemoattractant protein; MIG (CXCL9), monokine induced by IFNγ; CCL5/RANTES, regulated on activation, normal T expressed and secreted; PDGF BB, platelet-derived growth factor (PDGF) BB; TNF, tumor necrosis factor; TRAIL (TNFSF10); VEGF, vascular endothelial growth factor; CXCL, chemokine (C-X-C motif); CCL, chemokine (C-C motif); ↑, indicated moderate elevated level; ↓, indicated decreased level.

and is responsible for the cytopathogenicity induced by SARS CoV infection.[115] Accessory proteins 3a and 7a were shown to upregulate NF-κB, which in turn activates proinflammatory genes, and c-Jun N-terminal kinase (JNK), as well as chemokines, such as IL-8 and RANTES.[116] Accessory proteins 3a, 3b, and 7a were indicated to induce cell cycle arrest and apoptosis *in vitro*.[7,117,118] Evidence also suggested that accessory proteins 3b and ORF6 prevent IFN production and signaling, which is essential for the antiviral innate immune response. The interference with the innate immune response reduces the host's ability to clear the infection and enables SARS CoV to evade the host immune response.[119,120]

19.5 IDENTIFICATION AND DIAGNOSIS

The severity and the widespread of SARS infection from Asia to other continents through international travel in a short period of time has given rise to intense research efforts in developing identification and diagnosis tools for SARS CoVs. Almost all methods (traditional or molecular) that have been used for identification and diagnosis of other viruses were also applied to SARS CoV. In the following is a summary of what is available today. Besides the innate sensitivity and specificity of the particular diagnostic tool, the variable viral load in clinical specimens particularly at the early stage of infection needs to be taken into account when evaluating the clinical sensitivity of diagnostic assays. The low viral load in upper respiratory tract at the early stage of infection affects the detection sensitivity, whereas lower respiratory tract specimens, such as sputum and bronchoalveolar lavage fluid, provide better specimens for SARS CoV diagnosis. The collection of multiple specimens of different types can increase the overall clinical sensitivity when using any diagnostic technique and in particular the molecular-based methods.[33,42]

Despite the severity of SARS infection, differential diagnosis is still difficult due to the fact that SARS infection exhibits clinical features similar to other respiratory infections.[45]

Virus isolation using tissue culture or organ culture was the first method developed to identify the etiology agents of SARS infection. In contrast to other human coronaviruses, SARS CoV can be cultured easily in a variety of cell lines, including fetal rhesus monkey kidney (FRhK) and African green monkey (Vero E6) cells, with distinct cytopathic effects. However, propagation of SARS CoV is not recommended due to the high risk associated with infections and is not suitable for diagnosis in outbreak settings.[12,19,20,27,34,121] Electronic microscope examination of viral particles, immunohistochemical analysis, and *in situ* hybridization have been used for pathological examination but are not suitable for routine diagnosis although they are useful for studying the distribution of viruses in clinical specimens.[122–126]

Quite a few antibody-based assays, such as enzyme-linked immunosorbent assay (ELISA), immunofluorescence assay, and indirect fluorescent antibody (IFA), were developed for detecting SARS infection in clinical specimens. The targets for these antibody-based assays are S and N proteins since they are the antigenic determinants of SARS CoV.[14,127–148] Based on a similar principle, protein microarrays that contain peptides derived from SARS CoVs were also developed as a diagnostic tool to detect the presence of antibodies to SARS CoV.[149,150] Although sensitive, the antibody-based assays do not provide early diagnostic information since seroconversion is often delayed until 2–3 weeks after infection, which is not useful for diagnosis of early stage of infection.[45]

Molecular assays, such as RT-PCR, real-time RT-PCR, nucleic acid sequence-based amplification tests, rolling circle amplification, and loop-mediated isothermal amplification (LAMP), are more versatile for SARS CoV diagnosis. Several RT-PCR (one-step or two-step), nested PCR, and real-time RT-PCR using SYBR green or probes (TaqMan, molecular beacon, or hybridization probes) were developed for SARS CoV detection.[45,151–168] A few LAMP assays were also developed.[162,169,170] The main target for primers' design is polymerase 1b region of the 5'-replicase gene.[151–153,156,158,160,163,164,169] Other targets include N gene, 3′-noncoding region, NSP1, and NSP14, and later on, S, M, and E were also used for developing PCR assays.[154,155,157,159] The detection sensitivity of these assays is around 10–100 genome copies per reaction. There are commercially available detection assays, RealArt HPA CoV RT-PCR assay (now SARS CoV RT-PCR kit, Qiagen), EraGen SARS CoV NP and POL Multicode-RT kit (EraGen Biosciences, discontinued), and LightCycler SARS CoV kit (Roche). The performances of these commercial kits were not vigorously tested under real conditions due to the successful control of the disease.[45] The performances of some of these nucleic acid-based assays were reviewed by Mahony and Richardson.[45] Microarray-based approaches were also developed for SARS CoV identification. Whole-genome resequencing microarrays were developed to track the genetic diversity of SARS CoV, which will facilitate contact tracing and identifying the infectious source.[171,172] A SARS-specific detection macroarray in conjunction with multiplex RT-PCR was developed to detect multiple genomic sequences of SARS CoV.[173] Other microarrays designed for multiple human respiratory pathogens detection, which included coronaviruses, were also used for SARS CoV identification.[173–179]

19.6 TREATMENT AND PREVENTION

During SARS outbreak, broad-spectrum antibiotics and antiviral drugs (mainly ribavirin) and the combination of ribavirin and steroid were the treatment options; however, the effectiveness of these treatment options remains questionable since multicenter, randomized, controlled interventional trials were not possible.[180,181] Research efforts on developing new treatment options against SARS infection with *in vitro* methods and animal models have derived three types of potential therapies based on the principles of (1) blocking viral entry, (2) inhibition of viral replication, and (3) modulating immune response.[35,181–183]

Viral entry of SARS CoV into host cells is through the interaction of S glycoprotein with ACE2 receptor, which induces conformational changes in S glycoprotein followed by cathepsin L proteolysis and causes fusion between the viral and cellular membranes. Thus, peptides and small compounds that bind to ACE2, soluble form of the ACE2, ACE-2 inhibitor, antibodies against S protein or ACE2, and cathepsin L inhibitors are all potential antiviral agents for the treatment and prevention of SARS. Indeed, several treatment options based on all aspects of blocking viral entry have been developed (for details, see Tong, T.R., 2009, part 1).[181,182]

The life cycle of SARS CoV has been extensively studied (see Section 19.3 for a detailed description of viral replication). The main treatment targets during viral replication are two proteases, PLpro and 3CLpro, which process the formation of replication–transcription complex, RdRp, as well as E protein, which mediates viral assembly and form cation-selective ion channels. Small molecules and peptide-like protease inhibitors against 3CLpro and PLpro, nucleoside analog inhibitors against RdRp, and ion channels blockers that inhibit E protein

were also generated for treating SARS infection. In addition, mimicking cells' self-defense mechanisms besides immune systems against viral infections, such as RNA interference (RNAi) and intracellular molecules that retard viral processes, was also investigated as potential treatment options for SARS infection. Several siRNAs targeting replicase, RdRp, 5′-leader sequence, and 3′UTR, as well as short oligonucleotides disrupting base pairing or subgenomic mRNA synthesis, causing frame shift, were designed and tested against SARS replication (for details, see Tong, part 2).[181,183]

SARS CoV uses various strategies to evade host immune responses (see Section 19.4); thus, immunomodulating agents such as IFN inducers, 2′-5′-oligoadenylate synthetase (intracellular antiviral effector), and CpG single-strand oligonucleotide were also potential candidates for treating SARS infection. However, the caveat of the aforementioned treatment options is that whether any of these treatments benefited patients during the SARS outbreak is not possible to determine right now.

For prevention purposes, several different types of SARS vaccines, including inactivate and live attenuated viruses, subunit or expression proteins, viral-vectored vaccines, and DNA vaccines, have been developed. The inactivated viruses have been tested in monkey, mice, and 30 human volunteers and showed good induction of neutralizing antibodies against SARS CoVs. The attenuated virus has been evaluated in hamster model with promising results. Subunit vaccines, such as S, N, and M proteins, have been expressed and used in mice to induce the production of neutralizing antibodies with promising results; however, no *in vivo* protection experiments were performed with these vaccines with the exception of S protein. DNA vaccines expressing S and N proteins were also tested in mice and showed protective immunity. Several viral-vectored vaccines expressing S, N, and M proteins were described with successful results in animal models (for details, see Enjuanes et al.).[181,184,185]

19.7 CONCLUSION AND FUTURE PERSPECTIVE

The large genome size of SARS CoVs enables the sophisticated mechanisms it employs to evade the host defense system and successfully propagate itself. The very quick response at the start of the outbreak represents a wonderful success for human health management, but it also means that less data are now available in estimating the risks this disease presents. A primary gap is the uncertainty surrounding the transmission mechanism from human to human. It appeared from the case studies that relatively close contact or extended contact was most often required to enable transmission, but several different routes are still possible. Oxford et al. argued that SARS did not spread rapidly in comparison to influenza; SARS outbreaks in other countries besides Southeast Asia were due to contact with patients in this area; the high density of living and population may have allowed virus spread in this area, but careful barrier control can reduce transmissibility.[186] The fact that the disease transmission did not appear to be as rapid and simple as in say influenza may have contributed to the successful containment of the outbreak. Without better understanding of the transmission modes, it will not be possible to determine if SARS CoV might ever change to produce a more deadly transmission mode. Another potential risk is that although several different treatment and prevention methods (such as vaccine development) have been developed since the outbreak, the effectiveness of these treatment options in human remains questionable.

Control of this agent is as difficult as any other that exists in a natural host reservoir. SL-CoVs have ongoing circulation in bats and several different intermediate animal hosts present a challenge in deriving an effective control measurement. Although coronaviruses normally have a very narrow host range, SARS CoVs can efficiently infect domestic cats, ferrets, and several laboratory animal models.[2,5] Even though transmission of the virus from the domestic cats and ferrets to human was not shown, SL-CoVs can replicate in a number of animals, and species variations in host cell factors impose selection pressure on SL-CoVs that can potentially allow successful subsequent intraspecies transmission.[187–189] A further concern is that unlike most human coronaviruses, SARS CoV can be easily cultured in a variety of cell lines and replicates to high titers in these cells. Mutations can arise during adaptation from one cell line to another that can contribute to the virus evolution to potential human pathogenic strains.[12]

To derive better guidelines on the best preventative procedures in the case of a new outbreak, the doctrine of transmission modes, intra- and interspecies, will need to be understood. In addition, considering the ease of cross-species transmission by SL-CoVs, the wet markets that trade live game animals provide a venue for the animal SL-CoVs to amplify and transmit to new hosts, including human. Better rules and regulations of wet market practices that currently contain large numbers and varieties of wild game mammals in overcrowded cages can also provide some preventative measure in limiting disease transmission.

REFERENCES

1. Peiris, J.S. et al. The severe acute respiratory syndrome. *N Engl J Med* 349, 2431–2441 (2003).
2. Lai, M.M.C. and Holmes, K.V. Coronaviridae: The viruses and their replication. In *Fields Virology* (eds. Knipe, D.M. and Howley, P.M.), pp. 1163–1185 (Lippincott Williams & Wilkins, Philadelphia, PA, 2001).
3. Pyrc, K., Berkhout, B., and van der Hoek, L. The novel human coronaviruses NL63 and HKU1. *J Virol* 81, 3051–3057 (2007).
4. van der Hoek, L., Pyrc, K., and Berkhout, B. Human coronavirus NL63, a new respiratory virus. *FEMS Microbiol Rev* 30, 760–773 (2006).
5. Holmes, K.V. Coronaviruses. In *Fields Virology* (eds. Knipe, D.M. and Howley, P.M.), pp. 1187–1203 (Lippincott Williams & Wilkins, Philadelphia, PA, 2001).
6. Masters, P.S. The molecular biology of coronaviruses. *Adv Virus Res* 66, 193–292 (2006).
7. Enjuanes, L. et al. Biochemical aspects of coronavirus replication and virus-host interaction. *Annu Rev Microbiol* 60, 211–230 (2006).
8. Weiss, S.R. and Navas-Martin, S. Coronavirus pathogenesis and the emerging pathogen severe acute respiratory syndrome coronavirus. *Microbiol Mol Biol Rev* 69, 635–664 (2005).
9. Guan, Y. et al. Isolation and characterization of viruses related to the SARS coronavirus from animals in southern China. *Science* 302, 276–278 (2003).
10. Li, W. et al. Bats are natural reservoirs of SARS-like coronaviruses. *Science* 310, 676–679 (2005).
11. Poon, L.L. et al. The aetiology, origins, and diagnosis of severe acute respiratory syndrome. *Lancet Infect Dis* 4, 663–671 (2004).
12. Sims, A.C. et al. SARS-CoV replication and pathogenesis in an in vitro model of the human conducting airway epithelium. *Virus Res* 133, 33–44 (2008).
13. Gottschalk, R. and Preiser, W. Bioterrorism: Is it a real threat? *Med Microbiol Immunol* 194, 109–114 (2005).
14. Woo, P.C. et al. Longitudinal profile of immunoglobulin G (IgG), IgM, and IgA antibodies against the severe acute respiratory syndrome (SARS) coronavirus nucleocapsid protein in patients with pneumonia due to the SARS coronavirus. *Clin Diagn Lab Immunol* 11, 665–668 (2004).
15. Tyrrell, D.A. and Bynoe, M.L. Cultivation of a novel type of common-cold virus in organ cultures. *Br Med J* 1, 1467–1470 (1965).
16. Hamre, D. and Procknow, J.J. A new virus isolated from the human respiratory tract. *Proc Soc Exp Biol Med* 121, 190–193 (1966).
17. Pyrc, K., Berkhout, B., and van der Hoek, L. Identification of new human coronaviruses. *Expert Rev Anti Infect Ther* 5, 245–253 (2007).
18. Woo, P.C. et al. Characterization and complete genome sequence of a novel coronavirus, coronavirus HKU1, from patients with pneumonia. *J Virol* 79, 884–895 (2005).
19. Drosten, C. et al. Identification of a novel coronavirus in patients with severe acute respiratory syndrome. *N Engl J Med* 348, 1967–1976 (2003).
20. Rota, P.A. et al. Characterization of a novel coronavirus associated with severe acute respiratory syndrome. *Science* 300, 1394–1399 (2003).
21. Pyrc, K. et al. Genome structure and transcriptional regulation of human coronavirus NL63. *Virol J* 1, 7 (2004).
22. van der Hoek, L. et al. Identification of a new human coronavirus. *Nat Med* 10, 368–373 (2004).
23. Almeida, J.D. and Tyrrell, D.A. The morphology of three previously uncharacterized human respiratory viruses that grow in organ culture. *J Gen Virol* 1, 175–178 (1967).
24. Bradburne, A.F., Bynoe, M.L., and Tyrrell, D.A. Effects of a "new" human respiratory virus in volunteers. *Br Med J* 3, 767–769 (1967).
25. McIntosh, K., Becker, W.B., and Chanock, R.M. Growth in suckling-mouse brain of "IBV-like" viruses from patients with upper respiratory tract disease. *Proc Natl Acad Sci USA* 58, 2268–2273 (1967).
26. McIntosh, K. et al. Recovery in tracheal organ cultures of novel viruses from patients with respiratory disease. *Proc Natl Acad Sci USA* 57, 933–940 (1967).
27. Marra, M.A. et al. The genome sequence of the SARS-associated coronavirus. *Science* 300, 1399–1404 (2003).
28. Lau, S.K. et al. Severe acute respiratory syndrome coronavirus-like virus in Chinese horseshoe bats. *Proc Natl Acad Sci USA* 102, 14040–14045 (2005).
29. Ren, W. et al. Full-length genome sequences of two SARS-like coronaviruses in horseshoe bats and genetic variation analysis. *J Gen Virol* 87, 3355–3359 (2006).

30. Snijder, E.J. et al. Unique and conserved features of genome and proteome of SARS-coronavirus, an early split-off from the coronavirus group 2 lineage. *J Mol Biol* 331, 991–1004 (2003).

31. Yip, C.W. et al. Phylogenetic perspectives on the epidemiology and origins of SARS and SARS-like coronaviruses. *Infect Genet Evol* 9, 1185–1196 (2009).

32. Koning, R.I. et al. Cryo electron tomography of vitrified fibroblasts: Microtubule plus ends in situ. *J Struct Biol* 161, 459–468 (2008).

33. Parashar, U.D. and Anderson, L.J. Severe acute respiratory syndrome: Review and lessons of the 2003 outbreak. *Int J Epidemiol* 33, 628–634 (2004).

34. Ksiazek, T.G. et al. A novel coronavirus associated with severe acute respiratory syndrome. *N Engl J Med* 348, 1953–1966 (2003).

35. Anderson, L.J. and Tong, S. Update on SARS research and other possibly zoonotic coronaviruses. *Int J Antimicrob Agents* 36(Suppl 1), S21–S25.

36. Chan, K.H. et al. Detection of SARS coronavirus in patients with suspected SARS. *Emerg Infect Dis* 10, 294–299 (2004).

37. Poon, L.L. et al. Identification of a novel coronavirus in bats. *J Virol* 79, 2001–2009 (2005).

38. Cheng, V.C. et al. Severe acute respiratory syndrome coronavirus as an agent of emerging and reemerging infection. *Clin Microbiol Rev* 20, 660–694 (2007).

39. Dong, B.Q. et al. Detection of a novel and highly divergent coronavirus from Asian leopard cats and Chinese ferret badgers in Southern China. *J Virol* 81, 6920–6926 (2007).

40. Shi, Z. and Hu, Z. A review of studies on animal reservoirs of the SARS coronavirus. *Virus Res* 133, 74–87 (2008).

41. de Haan, C.A. and Rottier, P.J. Hosting the severe acute respiratory syndrome coronavirus: Specific cell factors required for infection. *Cell Microbiol* 8, 1211–1218 (2006).

42. Christian, M.D. et al. Severe acute respiratory syndrome. *Clin Infect Dis* 38, 1420–1427 (2004).

43. Lau, J.T. et al. Probable secondary infections in households of SARS patients in Hong Kong. *Emerg Infect Dis* 10, 235–243 (2004).

44. Goh, D.L. et al. Secondary household transmission of SARS, Singapore. *Emerg Infect Dis* 10, 232–234 (2004).

45. Mahony, J.B. and Richardson, S. Molecular diagnosis of severe acute respiratory syndrome: The state of the art. *J Mol Diagn* 7, 551–559 (2005).

46. Anderson, R.M. et al. Epidemiology, transmission dynamics and control of SARS: The 2002–2003 epidemic. *Philos Trans R Soc Lond B Biol Sci* 359, 1091–1105 (2004).

47. Samaranayake, L.P. and Peiris, M. Severe acute respiratory syndrome and dentistry: A retrospective view. *J Am Dent Assoc* 135, 1292–1302 (2004).

48. Knudsen, T.B. et al. Severe acute respiratory syndrome— A new coronavirus from the Chinese dragon's lair. *Scand J Immunol* 58, 277–284 (2003).

49. Hui, D.S. and Chan, P.K. Clinical features, pathogenesis and immunobiology of severe acute respiratory syndrome. *Curr Opin Pulm Med* 14, 241–247 (2008).

50. Groneberg, D.A. et al. Severe acute respiratory syndrome: Global initiatives for disease diagnosis. *QJM* 96, 845–852 (2003).

51. Chan-Yeung, M. et al. Severe acute respiratory syndrome. *Int J Tuberc Lung Dis* 7, 1117–1130 (2003).

52. Peiris, J.S. et al. Clinical progression and viral load in a community outbreak of coronavirus-associated SARS pneumonia: A prospective study. *Lancet* 361, 1767–1772 (2003).

53. Wang, H., Rao, S., and Jiang, C. Molecular pathogenesis of severe acute respiratory syndrome. *Microbes Infect* 9, 119–126 (2007).

54. Leong, W.F. et al. Microarray and real-time RT-PCR analyses of differential human gene expression patterns induced by severe acute respiratory syndrome (SARS) coronavirus infection of Vero cells. *Microbes Infect* 7, 248–259 (2005).

55. Li, Z. et al. The relationship between serum interleukins and T-lymphocyte subsets in patients with severe acute respiratory syndrome. *Chin Med J* 116, 981–984 (2003).

56. Lam, C.W., Chan, M.H., and Wong, C.K. Severe acute respiratory syndrome: Clinical and laboratory manifestations. *Clin Biochem Rev* 25, 121–132 (2004).

57. Xie, J. et al. Dynamic changes of plasma cytokine levels in patients with severe acute respiratory syndrome. *Zhonghua Nei Ke Za Zhi* 42, 643–645 (2003).

58. Tang, N.L. et al. Early enhanced expression of interferon-inducible protein-10 (CXCL-10) and other chemokines predicts adverse outcome in severe acute respiratory syndrome. *Clin Chem* 51, 2333–2340 (2005).

59. Chien, J.Y. et al. Temporal changes in cytokine/chemokine profiles and pulmonary involvement in severe acute respiratory syndrome. *Respirology* 11, 715–722 (2006).

60. Jiang, Y. et al. Characterization of cytokine/chemokine profiles of severe acute respiratory syndrome. *Am J Respir Crit Care Med* 171, 850–857 (2005).

61. Wong, C.K. et al. Plasma inflammatory cytokines and chemokines in severe acute respiratory syndrome. *Clin Exp Immunol* 136, 95–103 (2004).

62. Huang, K.J. et al. An interferon-gamma-related cytokine storm in SARS patients. *J Med Virol* 75, 185–194 (2005).

63. Zhang, Y. et al. Analysis of serum cytokines in patients with severe acute respiratory syndrome. *Infect Immun* 72, 4410–4415 (2004).

64. Duan, Z.P. et al. Clinical characteristics and mechanism of liver injury in patients with severe acute respiratory syndrome. *Zhonghua Gan Zang Bing Za Zhi* 11, 493–496 (2003).

65. Ng, P.C. et al. Inflammatory cytokine profile in children with severe acute respiratory syndrome. *Pediatrics* 113, e7–e14 (2004).

66. Ward, S.E. et al. Dynamic changes in clinical features and cytokine/chemokine responses in SARS patients treated with interferon alfacon-1 plus corticosteroids. *Antivir Ther* 10, 263–275 (2005).

67. Wang, C. and Pang, B.S. Dynamic changes and the meanings of blood cytokines in severe acute respiratory syndrome. *Zhonghua Jie He He Hu Xi Za Zhi* 26, 586–589 (2003).

68. Beijing Group of National Research Project for SARS. Dynamic changes in blood cytokine levels as clinical indicators in severe acute respiratory syndrome. *Chin Med J (Engl)* 116, 1283–1287 (2003).

69. Sheng, W.H. et al. Clinical manifestations and inflammatory cytokine responses in patients with severe acute respiratory syndrome. *J Formos Med Assoc* 104, 715–723 (2005).

70. Cameron, M.J. et al. Interferon-mediated immunopathological events are associated with atypical innate and adaptive immune responses in patients with severe acute respiratory syndrome. *J Virol* 81, 8692–8706 (2007).

71. Glass, W.G. et al. Mechanisms of host defense following severe acute respiratory syndrome-coronavirus (SARS-CoV) pulmonary infection of mice. *J Immunol* 173, 4030–4039 (2004).

72. Law, H.K. et al. Chemokine up-regulation in SARS-coronavirus-infected, monocyte-derived human dendritic cells. *Blood* 106, 2366–2374 (2005).

73. Cheung, C.Y. et al. Cytokine responses in severe acute respiratory syndrome coronavirus-infected macrophages in vitro: Possible relevance to pathogenesis. *J Virol* 79, 7819–7826 (2005).

74. Yu, S.Y. et al. Gene expression profiles in peripheral blood mononuclear cells of SARS patients. *World J Gastroenterol* 11, 5037–5043 (2005).

75. Ziegler, T. et al. Severe acute respiratory syndrome coronavirus fails to activate cytokine-mediated innate immune responses in cultured human monocyte-derived dendritic cells. *J Virol* 79, 13800–13805 (2005).

76. He, L. et al. Expression of elevated levels of pro-inflammatory cytokines in SARS-CoV-infected ACE2+ cells in SARS patients: Relation to the acute lung injury and pathogenesis of SARS. *J Pathol* 210, 288–297 (2006).

77. Spiegel, M. and Weber, F. Inhibition of cytokine gene expression and induction of chemokine genes in non-lymphatic cells infected with SARS coronavirus. *Virol J* 3, 17 (2006).

78. Yen, Y.T. et al. Modeling the early events of severe acute respiratory syndrome coronavirus infection in vitro. *J Virol* 80, 2684–2693 (2006).

79. de Lang, A. et al. Functional genomics highlights differential induction of antiviral pathways in the lungs of SARS-CoV-infected macaques. *PLoS Pathog* 3, e112 (2007).

80. Wang, W. et al. Up-regulation of IL-6 and TNF-alpha induced by SARS-coronavirus spike protein in murine macrophages via NF-kappaB pathway. *Virus Res* 128, 1–8 (2007).

81. Ma, C.L., Yao, K., and Zhou, F. Effect on mRNA and secretion levels of proinflammatory cytokines in DC infected by SARS-CoV N gene recombinant adenovirus. *Zhonghua Shi Yan He Lin Chuang Bing Du Xue Za Zhi* 22, 431–433 (2008).

82. Nagata, N. et al. Mouse-passaged severe acute respiratory syndrome-associated coronavirus leads to lethal pulmonary edema and diffuse alveolar damage in adult but not young mice. *Am J Pathol* 172, 1625–1637 (2008).

83. Scagnolari, C. et al. Severe acute respiratory syndrome coronavirus elicits a weak interferon response compared to traditional interferon-inducing viruses. *Intervirology* 51, 217–223 (2008).

84. Day, C.W. et al. A new mouse-adapted strain of SARS-CoV as a lethal model for evaluating antiviral agents in vitro and in vivo. *Virology* 395, 210–222 (2009).

85. To, K.F. and Chan, P.K. Identification of human cell line model of persistent SARS coronavirus infection and studies of the response to cytokines and chemokines. *Hong Kong Med J* 15(Suppl 6), 39–43 (2009).

86. Yoshikawa, N. et al. Differential virological and immunological outcome of severe acute respiratory syndrome coronavirus infection in susceptible and resistant transgenic mice expressing human angiotensin-converting enzyme 2. *J Virol* 83, 5451–5465 (2009).

87. Smits, S.L. et al. Exacerbated innate host response to SARS-CoV in aged non-human primates. *PLoS Pathog* 6, e1000756 (2010).

88. Yoshikawa, T. et al. Dynamic innate immune responses of human bronchial epithelial cells to severe acute respiratory syndrome-associated coronavirus infection. *PLoS One* 5, e8729 (2010).

89. Danesh, A. et al. Early gene expression events in ferrets in response to SARS coronavirus infection versus direct interferon-alpha2b stimulation. *Virology* 409, 102–112 (2011).

90. Rockx, B. et al. Comparative pathogenesis of three human and zoonotic SARS-CoV strains in cynomolgus macaques. *PLoS One* 6, e18558 (2011).

91. Clay, C. et al. Primary SARS-CoV infection limits replication but not lung inflammation upon homologous rechallenge. *J Virol* 86, 4234–4244 (2012).

92. Reghunathan, R. et al. Expression profile of immune response genes in patients with severe acute respiratory syndrome. *BMC Immunol* 6, 2 (2005).

93. Castilletti, C. et al. Coordinate induction of IFN-alpha and -gamma by SARS-CoV also in the absence of virus replication. *Virology* 341, 163–169 (2005).

94. Tang, B.S. et al. Comparative host gene transcription by microarray analysis early after infection of the Huh7 cell line by severe acute respiratory syndrome coronavirus and human coronavirus 229E. *J Virol* 79, 6180–6193 (2005).

95. Cinatl, J. Jr. et al. Infection of cultured intestinal epithelial cells with severe acute respiratory syndrome coronavirus. *Cell Mol Life Sci* 61, 2100–2112 (2004).

96. Okabayashi, T. et al. Cytokine regulation in SARS coronavirus infection compared to other respiratory virus infections. *J Med Virol* 78, 417–424 (2006).

97. Kuba, K. et al. Lessons from SARS: Control of acute lung failure by the SARS receptor ACE2. *J Mol Med (Berl)* 84, 814–820 (2006).

98. Kuba, K. et al. Trilogy of ACE2: A peptidase in the renin-angiotensin system, a SARS receptor, and a partner for amino acid transporters. *Pharmacol Ther* 128, 119–128 (2010).

99. Haga, S. et al. Modulation of TNF-alpha-converting enzyme by the spike protein of SARS-CoV and ACE2 induces TNF-alpha production and facilitates viral entry. *Proc Natl Acad Sci USA* 105, 7809–7814 (2008).

100. Chen, I.Y. et al. Upregulation of the chemokine (C-C motif) ligand 2 via a severe acute respiratory syndrome coronavirus spike-ACE2 signaling pathway. *J Virol* 84, 7703–7712 (2010).

101. DeDiego, M.L. et al. Severe acute respiratory syndrome coronavirus envelope protein regulates cell stress response and apoptosis. *PLoS Pathog* 7, e1002315 (2011).

102. Chan, C.M. et al. The SARS-coronavirus membrane protein induces apoptosis through modulating the Akt survival pathway. *Arch Biochem Biophys* 459, 197–207 (2007).

103. Kamitani, W. et al. Severe acute respiratory syndrome coronavirus nsp1 protein suppresses host gene expression by promoting host mRNA degradation. *Proc Natl Acad Sci USA* 103, 12885–12890 (2006).

104. Law, A.H. et al. Role for nonstructural protein 1 of severe acute respiratory syndrome coronavirus in chemokine dysregulation. *J Virol* 81, 416–422 (2007).

105. Frieman, M. et al. Severe acute respiratory syndrome coronavirus papain-like protease ubiquitin-like domain and catalytic domain regulate antagonism of IRF3 and NF-kappaB signaling. *J Virol* 83, 6689–6705 (2009).

106. Ratia, K. et al. A noncovalent class of papain-like protease/deubiquitinase inhibitors blocks SARS virus replication. *Proc Natl Acad Sci USA* 105, 16119–16124 (2008).

107. Lindner, H.A. et al. Selectivity in ISG15 and ubiquitin recognition by the SARS coronavirus papain-like protease. *Arch Biochem Biophys* 466, 8–14 (2007).

108. Barretto, N. et al. Deubiquitinating activity of the SARS-CoV papain-like protease. *Adv Exp Med Biol* 581, 37–41 (2006).

109. Ratia, K. et al. Severe acute respiratory syndrome coronavirus papain-like protease: Structure of a viral deubiquitinating enzyme. *Proc Natl Acad Sci USA* 103, 5717–5722 (2006).

110. Barretto, N. et al. The papain-like protease of severe acute respiratory syndrome coronavirus has deubiquitinating activity. *J Virol* 79, 15189–15198 (2005).

111. Lindner, H.A. et al. The papain-like protease from the severe acute respiratory syndrome coronavirus is a deubiquitinating enzyme. *J Virol* 79, 15199–15208 (2005).

112. Sulea, T., Lindner, H.A., Purisima, E.O., and Menard, R. Deubiquitination, a new function of the severe acute respiratory syndrome coronavirus papain-like protease? *J Virol* 79, 4550–4551 (2005).

113. Lin, C.W. et al. Severe acute respiratory syndrome coronavirus 3C-like protease-induced apoptosis. *FEMS Immunol Med Microbiol* 46, 375–380 (2006).

114. Lai, C.C. et al. Proteomic analysis of up-regulated proteins in human promonocyte cells expressing severe acute respiratory syndrome coronavirus 3C-like protease. *Proteomics* 7, 1446–1460 (2007).

115. Li, Q. et al. The interaction of the SARS coronavirus non-structural protein 10 with the cellular oxido-reductase system causes an extensive cytopathic effect. *J Clin Virol* 34, 133–139 (2005).

116. Kanzawa, N. et al. Augmentation of chemokine production by severe acute respiratory syndrome coronavirus 3a/X1 and 7a/X4 proteins through NF-kappaB activation. *FEBS Lett* 580, 6807–6812 (2006).

117. Weiss, S.R. and Leibowitz, J.L. Coronavirus pathogenesis. *Adv Virus Res* 81, 85–164 (2011).

118. Narayanan, K., Huang, C., and Makino, S. SARS coronavirus accessory proteins. *Virus Res* 133, 113–121 (2008).

119. Frieman, M. et al. Severe acute respiratory syndrome coronavirus ORF6 antagonizes STAT1 function by sequestering nuclear import factors on the rough endoplasmic reticulum/Golgi membrane. *J Virol* 81, 9812–9824 (2007).

120. Kopecky-Bromberg, S.A. et al. Severe acute respiratory syndrome coronavirus open reading frame (ORF) 3b, ORF 6, and nucleocapsid proteins function as interferon antagonists. *J Virol* 81, 548–557 (2007).

121. Bermingham, A. et al. Laboratory diagnosis of SARS. *Philos Trans R Soc Lond B Biol Sci* 359, 1083–1089 (2004).

122. Nicholls, J.M. et al. Time course and cellular localization of SARS-CoV nucleoprotein and RNA in lungs from fatal cases of SARS. *PLoS Med* 3, e27 (2006).

123. Shieh, W.J. et al. Immunohistochemical, in situ hybridization, and ultrastructural localization of SARS-associated coronavirus in lung of a fatal case of severe acute respiratory syndrome in Taiwan. *Hum Pathol* 36, 303–309 (2005).

124. Ding, Y. et al. Organ distribution of severe acute respiratory syndrome (SARS) associated coronavirus (SARS-CoV) in SARS patients: Implications for pathogenesis and virus transmission pathways. *J Pathol* 203, 622–630 (2004).

125. Chong, P.Y. et al. Analysis of deaths during the severe acute respiratory syndrome (SARS) epidemic in Singapore: Challenges in determining a SARS diagnosis. *Arch Pathol Lab Med* 128, 195–204 (2004).

126. Nakajima, N. et al. SARS coronavirus-infected cells in lung detected by new in situ hybridization technique. *Jpn J Infect Dis* 56, 139–141 (2003).

127. Gimenez, L.G. et al. Development of an enzyme-linked immunosorbent assay-based test with a cocktail of nucleocapsid and spike proteins for detection of severe acute respiratory syndrome-associated coronavirus-specific antibody. *Clin Vaccine Immunol* 16, 241–245 (2009).

128. Lee, H.K. et al. Detection of antibodies against SARS-Coronavirus using recombinant truncated nucleocapsid proteins by ELISA. *J Microbiol Biotechnol* 18, 1717–1721 (2008).

129. Yu, F. et al. Recombinant truncated nucleocapsid protein as antigen in a novel immunoglobulin M capture enzyme-linked immunosorbent assay for diagnosis of severe acute respiratory syndrome coronavirus infection. *Clin Vaccine Immunol* 14, 146–149 (2007).

130. Guo, Z.M. et al. Comparison of effectiveness of whole viral, N and N199 proteins by ELISA for the rapid diagnosis of severe acute respiratory syndrome coronavirus. *Chin Med J (Engl)* 120, 2195–2199 (2007).

131. Li, Y.H. et al. Detection of the nucleocapsid protein of severe acute respiratory syndrome coronavirus in serum: Comparison with results of other viral markers. *J Virol Methods* 130, 45–50 (2005).

132. Woo, P.C. et al. Differential sensitivities of severe acute respiratory syndrome (SARS) coronavirus spike polypeptide enzyme-linked immunosorbent assay (ELISA) and SARS coronavirus nucleocapsid protein ELISA for serodiagnosis of SARS coronavirus pneumonia. *J Clin Microbiol* 43, 3054–3058 (2005).

133. He, Q. et al. Characterization of monoclonal antibody against SARS coronavirus nucleocapsid antigen and development of an antigen capture ELISA. *J Virol Methods* 127, 46–53 (2005).

134. Zhao, J. et al. Development and evaluation of an enzyme-linked immunosorbent assay for detection of antibodies against the spike protein of SARS-coronavirus. *J Clin Virol* 33, 12–18 (2005).

135. Saijo, M. et al. Recombinant nucleocapsid protein-based IgG enzyme-linked immunosorbent assay for the serological diagnosis of SARS. *J Virol Methods* 125, 181–186 (2005).

136. Di, B. et al. Monoclonal antibody-based antigen capture enzyme-linked immunosorbent assay reveals high sensitivity of the nucleocapsid protein in acute-phase sera of severe acute respiratory syndrome patients. *Clin Diagn Lab Immunol* 12, 135–140 (2005).

137. Chan, P.K. et al. Evaluation of a recombinant nucleocapsid protein-based assay for anti-SARS-CoV IgG detection. *J Med Virol* 75, 181–184 (2005).

138. Lai, S.C. et al. Characterization of neutralizing monoclonal antibodies recognizing a 15-residues epitope on the spike protein HR2 region of severe acute respiratory syndrome coronavirus (SARS-CoV). *J Biomed Sci* 12, 711–727 (2005).

139. Shao, P.L. et al. Development of immunoglobulin G enzyme-linked immunosorbent assay for the serodiagnosis of severe acute respiratory syndrome. *J Biomed Sci* 12, 59–64 (2005).

140. Carattoli, A. et al. Recombinant protein-based ELISA and immuno-cytochemical assay for the diagnosis of SARS. *J Med Virol* 76, 137–142 (2005).

141. Zhao, J.C. et al. Prokaryotic expression, refolding, and purification of fragment 450–650 of the spike protein of SARS-coronavirus. *Protein Expr Purif* 39, 169–174 (2005).

142. Mo, H.Y. et al. Evaluation by indirect immunofluorescent assay and enzyme linked immunosorbent assay of the dynamic changes of serum antibody responses against severe acute respiratory syndrome coronavirus. *Chin Med J (Engl)* 118, 446–450 (2005).

143. Woo, P.C. et al. False-positive results in a recombinant severe acute respiratory syndrome-associated coronavirus (SARS-CoV) nucleocapsid enzyme-linked immunosorbent assay due to HCoV-OC43 and HCoV-229E rectified by Western blotting with recombinant SARS-CoV spike polypeptide. *J Clin Microbiol* 42, 5885–5888 (2004).

144. Wang, Y.D. et al. Detection of antibodies against SARS-CoV in serum from SARS-infected donors with ELISA and Western blot. *Clin Immunol* 113, 145–150 (2004).

145. Lau, S.K. et al. Detection of severe acute respiratory syndrome (SARS) coronavirus nucleocapsid protein in sars patients by enzyme-linked immunosorbent assay. *J Clin Microbiol* 42, 2884–2889 (2004).

146. Woo, P.C. et al. Detection of specific antibodies to severe acute respiratory syndrome (SARS) coronavirus nucleocapsid protein for serodiagnosis of SARS coronavirus pneumonia. *J Clin Microbiol* 42, 2306–2309 (2004).

147. Guan, M. et al. Evaluation and validation of an enzyme-linked immunosorbent assay and an immunochromatographic test for serological diagnosis of severe acute respiratory syndrome. *Clin Diagn Lab Immunol* 11, 699–703 (2004).

148. Hsueh, P.R. et al. SARS antibody test for serosurveillance. *Emerg Infect Dis* 10, 1558–1562 (2004).

149. Zhu, H. et al. Severe acute respiratory syndrome diagnostics using a coronavirus protein microarray. *Proc Natl Acad Sci USA* 103, 4011–4016 (2006).

150. Lu, D.D. et al. Screening of specific antigens for SARS clinical diagnosis using a protein microarray. *Analyst* 130, 474–482 (2005).

151. Chantratita, W. et al. Development and comparison of the real-time amplification based methods—NASBA-Beacon, RT-PCR taqman and RT-PCR hybridization probe assays—For the qualitative detection of sars coronavirus. *Southeast Asian J Trop Med Public Health* 35, 623–629 (2004).

152. Drosten, C. et al. Evaluation of advanced reverse transcription-PCR assays and an alternative PCR target region for detection of severe acute respiratory syndrome-associated coronavirus. *J Clin Microbiol* 42, 2043–2047 (2004).

153. Escutenaire, S. et al. SYBR Green real-time reverse transcription-polymerase chain reaction assay for the generic detection of coronaviruses. *Arch Virol* 152, 41–58 (2007).

154. Hadjinicolaou, A.V. et al. Development of a molecular-beacon-based multi-allelic real-time RT-PCR assay for the detection of human coronavirus causing severe acute respiratory syndrome (SARS-CoV): A general methodology for detecting rapidly mutating viruses. *Arch Virol* 156, 671–680 (2011).

155. Houng, H.S. et al. Development and evaluation of an efficient 3'-noncoding region based SARS coronavirus (SARS-CoV) RT-PCR assay for detection of SARS-CoV infections. *J Virol Methods* 120, 33–40 (2004).

156. Hourfar, M.K. et al. Comparison of two real-time quantitative assays for detection of severe acute respiratory syndrome coronavirus. *J Clin Microbiol* 42, 2094–2100 (2004).

157. Huang, J.L. et al. Rapid and sensitive detection of multiple genes from the SARS-coronavirus using quantitative RT-PCR with dual systems. *J Med Virol* 77, 151–158 (2005).

158. Hui, R.K. et al. Reverse transcriptase PCR diagnostic assay for the coronavirus associated with severe acute respiratory syndrome. *J Clin Microbiol* 42, 1994–1999 (2004).

159. Inoue, M. et al. Performance of single-step gel-based reverse transcription-PCR (RT-PCR) assays equivalent to that of real-time RT-PCR assays for detection of the severe acute respiratory syndrome-associated coronavirus. *J Clin Microbiol* 43, 4262–4265 (2005).

160. Keightley, M.C. et al. Real-time NASBA detection of SARS-associated coronavirus and comparison with real-time reverse transcription-PCR. *J Med Virol* 77, 602–608 (2005).

161. Peiris, J.S. and Poon, L.L. Detection of SARS coronavirus in humans and animals by conventional and quantitative (real time) reverse transcription polymerase chain reactions. *Methods Mol Biol* 454, 61–72 (2008).

162. Poon, L.L. et al. Evaluation of real-time reverse transcriptase PCR and real-time loop-mediated amplification assays for severe acute respiratory syndrome coronavirus detection. *J Clin Microbiol* 43, 3457–3459 (2005).

163. Poon, L.L. et al. A one step quantitative RT-PCR for detection of SARS coronavirus with an internal control for PCR inhibitors. *J Clin Virol* 30, 214–217 (2004).

164. Poon, L.L. et al. Early diagnosis of SARS coronavirus infection by real time RT-PCR. *J Clin Virol* 28, 233–238 (2003).

165. Yam, W.C. et al. Clinical evaluation of real-time PCR assays for rapid diagnosis of SARS coronavirus during outbreak and post-epidemic periods. *J Clin Virol* 33, 19–24 (2005).

166. Yam, W.C. et al. Evaluation of reverse transcription-PCR assays for rapid diagnosis of severe acute respiratory syndrome associated with a novel coronavirus. *J Clin Microbiol* 41, 4521–4524 (2003).

167. Mahony, J.B. et al. Performance and cost evaluation of one commercial and six in-house conventional and real-time reverse transcription-PCR assays for detection of severe acute respiratory syndrome coronavirus. *J Clin Microbiol* 42, 1471–1476 (2004).

168. Wang, B. et al. Rapid and sensitive detection of severe acute respiratory syndrome coronavirus by rolling circle amplification. *J Clin Microbiol* 43, 2339–2344 (2005).

169. Hong, T.C. et al. Development and evaluation of a novel loop-mediated isothermal amplification method for rapid detection of severe acute respiratory syndrome coronavirus. *J Clin Microbiol* 42, 1956–1961 (2004).

170. Poon, L.L. et al. Rapid detection of the severe acute respiratory syndrome (SARS) coronavirus by a loop-mediated isothermal amplification assay. *Clin Chem* 50, 1050–1052 (2004).

171. Sulaiman, I.M. et al. Evaluation of affymetrix severe acute respiratory syndrome resequencing GeneChips in characterization of the genomes of two strains of coronavirus infecting humans. *Appl Environ Microbiol* 72, 207–211 (2006).

172. Wong, C.W. et al. Tracking the evolution of the SARS coronavirus using high-throughput, high-density resequencing arrays. *Genome Res* 14, 398–405 (2004).

173. Juang, J.L. et al. Coupling multiplex RT-PCR to a gene chip assay for sensitive and semiquantitative detection of severe acute respiratory syndrome-coronavirus. *Lab Invest* 84, 1085–1091 (2004).

174. Kistler, A. et al. Pan-viral screening of respiratory tract infections in adults with and without asthma reveals unexpected human coronavirus and human rhinovirus diversity. *J Infect Dis* 196, 817–825 (2007).

175. Long, W.H. et al. A universal microarray for detection of SARS coronavirus. *J Virol Methods* 121, 57–63 (2004).

176. Wang, D. et al. Microarray-based detection and genotyping of viral pathogens. *Proc Natl Acad Sci USA* 99, 15687–15692 (2002).

177. Wang, D. et al. Viral discovery and sequence recovery using DNA microarrays. *PLoS Biol* 1, E2 (2003).

178. Lin, B. et al. Using a resequencing microarray as a multiple respiratory pathogen detection assay. *J Clin Microbiol* 45, 443–452 (2007).

179. Wang, Z. et al. Resequencing microarray probe design for typing genetically diverse viruses: Human rhinoviruses and enteroviruses. *BMC Genome* 9, 577 (2008).

180. Stockman, L.J., Bellamy, R., and Garner, P. SARS: Systematic review of treatment effects. *PLoS Med* 3, e343 (2006).

181. Groneberg, D.A. et al. Treatment and vaccines for severe acute respiratory syndrome. *Lancet Infect Dis* 5, 147–155 (2005).

182. Tong, T.R. Therapies for coronaviruses. Part I of II—Viral entry inhibitors. *Expert Opin Ther Pat* 19, 357–367 (2009).

183. Tong, T.R. Therapies for coronaviruses. Part 2: Inhibitors of intracellular life cycle. *Expert Opin Ther Pat* 19, 415–431 (2009).

184. Enjuanes, L. et al. Vaccines to prevent severe acute respiratory syndrome coronavirus-induced disease. *Virus Res* 133, 45–62 (2008).

185. Gillim-Ross, L. and Subbarao, K. Emerging respiratory viruses: Challenges and vaccine strategies. *Clin Microbiol Rev* 19, 614–636 (2006).

186. Oxford, J.S., Bossuyt, S., and Lambkin, R. A new infectious disease challenge: Urbani severe acute respiratory syndrome (SARS) associated coronavirus. *Immunology* 109, 326–328 (2003).

187. Martina, B.E. et al. Virology: SARS virus infection of cats and ferrets. *Nature* 425, 915 (2003).

188. Li, W. et al. Animal origins of the severe acute respiratory syndrome coronavirus: Insight from ACE2-S-protein interactions. *J Virol* 80, 4211–4219 (2006).

189. Groneberg, D.A., Hilgenfeld, R., and Zabel, P. Molecular mechanisms of severe acute respiratory syndrome (SARS). *Respir Res* 6, 8 (2005).

190. Holmes K.V., SARS-associated coronavirus. *N Engl J Med* 15;348(20), 1948–1951 (2003).

191. Bárcena M. et al., Cryo-electron tomography of mouse hepatitis virus: Insights into the structure of the coronavirion. *Proc Natl Acad Sci USA* 106(2), 582–587 (2009).

20 Tick-Borne Encephalitis Viruses

Oliver Donoso-Mantke, Camille Escadafal, Andrea Sanchini,
Cristina Domingo, Peter Hagedorn, Pranav Patel, Katharina Achazi,
Antje Hüther, and Matthias Niedrig

CONTENTS

20.1 INTRODUCTION

Tick-borne encephalitis (TBE) is a life-threatening neurological infection affecting humans.[1] Tick-borne encephalitis virus (TBEV) is the etiological agent of TBE. It was discovered in 1937 in Russia during an expedition led by Lev Zilber searching for the etiological agent of acute encephalitis associated with tick bites.[1] Subsequently, three subtypes of TBEV have been identified using serological techniques: the European virus TBEV-Eu (previously Central European encephalitis [CEE] virus), the Far Eastern virus TBEV-Fe (previously Russian spring–summer encephalitis [RSSE] virus), and the Siberian virus TBEV-Sib (previously west Siberian virus).[2] Because of their pathogenic potential in humans, these TBEV subtypes were included in the HHS list of select agents and toxins by Centers for Disease Control and Prevention (CDC) of the United States.[2a] In the ruling dated October 5, 2012, the European virus TBEV-Eu (CEE virus) was removed from the HHS list of select agents and toxins due to its relatively low virulence in humans in comparison with other TBEV subtypes.[2b]

Over the last decades, TBE has become a growing public health concern in Europe and Asia and is the most important viral tick-borne disease in Europe.[2] The increase of TBE incidence can be explained by the recent changing of climate that affects the spread of the tick vector.[3] A vaccine is available for TBE, and vaccination is recommended for persons who live/work in or visit TBE endemic areas with an increased risk of tick bites.[3]

20.2 CLASSIFICATION AND MORPHOLOGY

TBEV is a member of the genus *Flavivirus* of the family Flaviviridae, which comprises several other important arboviral human pathogens as West Nile virus, dengue virus, or yellow fever virus.[4,5] Most individuals exposed to these viruses show asymptomatic or subclinical infection, while a minority develops hemorrhagic fever, permanent neurological damages (meningitis, encephalitis), or death.[4]

TBEV is included in the mammalian tick-borne flavivirus serogroup, distinct from the seabird tick-borne flavivirus serogroup.[2,6] The mammalian tick-borne flavivirus serogroup includes other viruses genetically and antigenically related to TBEV, such as louping ill virus, Powassan virus, Omsk hemorrhagic fever virus, Alkhurma hemorrhagic fever virus, Kyasanur Forest disease virus, Langat virus, Karshi virus, Royal Farm virus, and Gadgets Gully virus, all known as the TBEV serocomplex.[7] The large homology between all these viruses has practical implications for laboratory differential diagnostics due to cross-reactivity.[8]

There are three subtypes of TBEV distinguished by serological analyses: the European virus TBEV-Eu (previously CEE virus), the Far Eastern virus TBEV-Fe (previously RSSE virus), and the Siberian virus TBEV-Sib (previously west Siberian virus).[2] The analysis of amino-acid sequences of the major envelope (E) protein (see in the following) showed that the variation is 5%–6% between the subtypes.[9,10] Furthermore, European TBEV isolates have shown fairly homogeneous sequences, without a clear geographical clustering, and do not undergo significant antigenic variation.[11] On the other hand, the diversity of Siberian and Far Eastern TBEV subtypes is much higher.[9] Recently, at least two groups in the Siberian genotype were identified (European and Asian groups, separated by the Ural Mountains).[12] However, immunization of persons with a vaccine based on the TBEV-Eu subtype provides cross-protection against the other two subtypes.[13,14]

TBEV, such as the other flaviviruses, is a small, icosahedral, lipid-enveloped RNA virus with a diameter of approximately 50 nm.[10,15] The genome comprises a single-stranded, positive-sense RNA of about 11,000 nucleotides in length, coding for one unique open reading frame that is flanked by 5′- and 3′-untranslated regions, containing a 5′-cap and lacking a 3′-polyadenylate tail.[7,16] The 5′-cap is important for mRNA stability and translation, while the untranslated regions form conserved secondary stem-loops that probably serve as *cis*-acting elements for genome amplification, translation, and packaging.[7,17]

Translation yields a polyprotein of 3414 amino acids that is glycosylated by cellular glycosyltransferases and then cleaved by viral and cellular proteases into three structural proteins, the capsid (C) protein, the membrane (M) protein (which is derived from its larger precursor prM) and the E glycoprotein, along with seven nonstructural proteins (NS1, NS2A, NS2B, NS3, NS4A, NS4B, and NS5).[17,18] The viral RNA genome *per se* is infectious and would act as an mRNA producing progeny virus if introduced into susceptible cells.[19]

The C protein, along with the viral RNA, forms the spherical 30 nm nucleocapsid of the virus, which is covered by a host cell–derived lipid bilayer with two surface proteins, prM and E, which possess double membrane anchors.[10] The E protein is involved in receptor binding, membrane fusion, and virus assembly and is thus responsible for both the replication and the maturation process of the virus.[17] The E protein contains the most important antigenic and virulence determinant of TBEV.[15] Many studies have focused on this protein, which is formed of three distinct domains.[20] In the mature virus, the E protein appears as flat dimers extending in a direction parallel to the viral membrane without forming particular spikes so that the fusion peptide in the tip of the distal domain is hidden under the proximal part of the dimer partner.[20]

After binding to a poorly characterized cell surface receptor on cells, internalization of the virus occurs through receptor-mediated clathrin-dependent endocytosis.[20] The low pH within the endosomal vesicle changes the conformation of the E protein and rearranges its dimers to trimeric forms with spikes and subsequently exposes the fusion peptide at the tip toward the endosome membrane. These changes result in fusion of the viral envelope and the membrane of the endosomal vesicle and the release of the viral nucleocapsid into the cytoplasm.[10,21] Furthermore, the second transmembrane region of the E protein is known to be important for virion formation, and E protein forms together with prM-enveloped virus-like particles.[21]

The seven nonstructural proteins regulate viral translation, transcription, and replication and also attenuate the host antiviral immune response.[22] NS1 is a glycoprotein acting as cofactor for the viral replicase; it is secreted from infected cells and has complement inhibitory activity.[23] The NS3 has protease, NTPase, and helicase activities and requires the protein NS2B as cofactor.[24] The proteins NS2A, NS4A, and NS4B are involved in the function as a replicative complex, and an overexpression of NS4A induces membrane rearrangements observed in flavivirus-infected cells.[25] The NS5 is an RNA-dependent RNA polymerase and a methyltransferase.[26]

Apparently, the ability to use multiple receptors can be responsible for the very wide host range of flaviviruses, which replicate in arthropods and in a broad range of vertebrates. In vertebrate cells, flavivirus RNA traffics to the rough endoplasmic reticulum, where it is translated, and leads to formation of immature virions that contain the proteins C, prM, and E. These immature particles are transported through the cellular secretory pathway, and shortly before release, prM—which acts as a chaperon for the E protein—is cleaved by cellular protease furin in the acidic compartment of the trans-Golgi network to yield mature and fully infectious virions.[27] However, the TBEV maturation process in tick cells is completely different than in the cells of vertebrate hosts. In cell lines derived from ticks infected with TBEV, nucleocapsids occur in cytoplasm, and the envelope is acquired by budding on cytoplasmic membranes or into cell vacuoles.[20,28] Studies focusing on the adaptation of TBEV to ticks and mammals described the presence of respective adaptive mutations within the second domain of the E protein.[29]

20.3 BIOLOGY AND EPIDEMIOLOGY

According to the concept of Pavlovskij, in nature, TBEV is propagated in a cycle involving ticks and wild vertebrate animals under certain botanical, zoological, climatical, and geo-ecological conditions.[30] The development of a TBE natural focus depends on the coincidence of all these factors.[14,31–33] The main vector and reservoir of the European TBEV subtype is *Ixodes ricinus*, a hard tick predominant across Europe.[34] *I. ricinus* ticks are mainly found in areas with high relative humidity like the dense undergrowth of forests.[14] TBEV from Far Eastern and Siberian subtypes are transmitted predominantly by *Ixodes persulcatus* ticks, which preferentially live in taiga-type forests.[35] Although the virus had also been isolated from several other tick species,[1,36,37] only the two mentioned ixodid tick species appear to play an important role in virus maintenance.[38]

The life cycle of ticks consists of three developmental stages: the larva, nymph, and adult. A tick can become chronically infected for the duration of its life by feeding on infected animal reservoir hosts (Figure 20.1). Therefore, the virus is transmitted from one developmental stage of the tick to the next (transstadial transmission). In the period that precedes molting, the virus multiplies in the tick and invades almost all the tick's organs.[39] TBEV can also be transmitted transovarially,[39] and despite the fact that the percentage of transovarial transmission of members of the European TBEV subtype in *I. ricinus* is much lower than that of Siberian and Far Eastern strains in *I. persulcatus*, it is sufficient under certain conditions to ensure the continuity of virus population. Furthermore, cofeeding of infected ticks together with naïve ticks on the same host allows TBEV transmission even in the absence of systemic viremia.[40,41]

Horizontal TBEV transmission between ticks and their vertebrate reservoir host is necessary for virus endemism.[42] Natural reservoir hosts are sensitive to the virus, exhibiting viremia for a long period without presenting any symptoms of illness. Reservoir hosts of TBEV are mainly small rodent species but also include insectivore and carnivore species.[14,43]

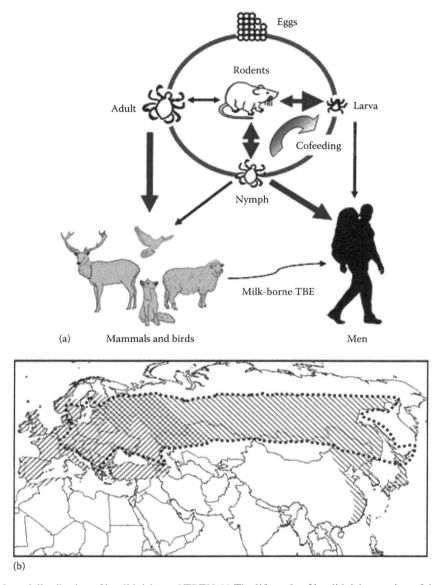

FIGURE 20.1 Life cycle and distribution of ixodid ticks and TBEV. (a) The life cycle of ixodid ticks consists of three developmental stages (larva, nymph, and adult). Each stage needs to take a blood meal on a suitable vertebrate host, usually for a period of a few days, to develop into the next stage. Further, adult female ticks need a blood meal for egg production. Each stage of *I. ricinus* takes approximately 1 year to develop into the next stage. Thus, it takes 3 years on average for the completion of the shortest life cycle. TBEV infects ticks chronically for the duration of their life while feeding on infected animal reservoir hosts. It can be transmitted transstadially (from one stage to a subsequent one), transovarially, or primarily from nymphs to naïve larvae while cofeeding on the same rodent host. TBEV can be transmitted to man or other animal hosts by all three tick stages with different prevalence. (b) Different shadings show the western distribution of *I. ricinus* and eastern distribution of *I. persulcatus*. The distribution of both vectors overlaps in the checkered area. The dotted line shows the border of the TBE endemic area. Note that TBE is distributed in an endemic pattern of the so-called natural foci. The distribution of ixodid ticks in China is uncertain and, therefore, not shown in detail.

Humans are accidental hosts of TBEV, that is, they can develop a disease with viremia but they do not contribute to the circulation of the virus in nature.

Most frequently, TBEV infection in humans occurs following the bite of an infected tick, which is unnoticed in about a third of cases.[44] The tick usually attaches itself to humans that are walking in the dense vegetation of forests. The virus is transmitted from the saliva within the first minutes of feeding, so early removal of ticks is not a guarantee for preventing disease. On humans, ticks prefer to attach themselves to the hair-covered portion of the head, to the arm and knee bends, hand, feet, and ears, as well as the gluteal and genital regions. The incidence of human TBE cases correlates with the period of activity of ticks. The seasonal activity of *I. ricinus* reaches its maximum in April–May and in September–October, while the activity of *I. persulcatus* reaches only one peak lasting from end of April to beginning of June.[14] However, alimentary routes of TBEV transmission by raw milk consumption have also been described in humans several times in the past years.[45–50] Single cases of laboratory-acquired TBE infections from needlestick injuries or associated with aerosol infection were also described.[51–53]

The epidemiology of TBE is closely related to the distribution, ecology, and biology of the ixodid tick vectors. While *I. ricinus* occurs in most parts of Europe, and the distribution extends to the southeast (Turkey, Northern Iran, and Caucasus),[32] *I. persulcatus* is found in a belt extending from Eastern Europe to China and Japan (Figure 20.1).[54] Parallel occurrence of both tick species was reported in North–Eastern Europe and the east of Estonia and Latvia as well as in several European regions of Russia.[55–57] The prevalence of TBEV-infected *I. ricinus* ticks varies from 0.5% to 5%, whereas in *I. persulcatus* in certain regions of Russia, prevalence up to 40% was recorded.[58] Several other studies on TBEV prevalence in ticks, as well as in livestock and domestic animals, have been conducted in the last years.[59–62] However, it should be noted that methods for measuring virus prevalence in ticks or animal reservoirs are not yet standardized and that there is still a need for reliable tools to translate epizootic prevalence data into infection risk for humans.

TBE occurs in many parts of Central Europe and Scandinavia, particularly in Austria, Czech Republic, Estonia, Finland, Germany, Hungary, Latvia, Lithuania, Poland, Russia, Slovak Republic, Slovenia, Sweden, Switzerland, and also Northern Asia (Figure 20.1).[63–65] Further, new TBE foci are emerging and latent ones reemerging in a number of other European countries.[66] In Russia, the highest TBE incidence is reported in Western Siberia and Ural.[31] Sporadic TBE occurrence has been observed in Bulgaria, Croatia, Denmark, France, Greece, Italy, Norway, Romania, Serbia, China, and Japan, while no TBE cases have been reported, for example, in Great Britain, Ireland, Iceland, Belgium, the Netherlands, Luxemburg, Spain, and Portugal. In general, an increase of TBE incidence has been observed in the last decade.[66,67] This is due to a complex interrelation of several factors, such as ecological (e.g., effect of climate change on vector distribution), social (e.g., human leisure activities and sociopolitical changes), and technological factors (e.g., medical awareness and advanced diagnostics).[68,69]

Although TBE is a growing concern in Europe, surveillance and notification schemes are not uniform and not always mandatory within countries. Because of the lack of a Europe-wide standard case definition, varying diagnostic procedures, and wide differences in the quality of national surveillance of TBE cases, surveillance data from different countries are difficult to compare.[63,64]

20.4 CLINICAL FEATURES AND PATHOGENESIS

The main route by which TBEV enters the body is by an *Ixodes* tick bite. Upon inoculation of virus into the skin, initial infection and replication occur in dendritic cells (Langerhans cells) localized in the skin and in neutrophils. Dendritic cells are thought to transport the virus to nearby lymph nodes. The virus further replicates in T and B cells as well as in macrophages of the lymph nodes, thymus, and spleen. After replication in the lymphatic organs, TBEV spreads through efferent lymphatics and the thoracic duct to produce viremia.[70] During the viremic phase, many extraneural tissues are infected, and the release of the virus from these tissues enables the viremia to continue for several days. The virus probably reaches the brain via the blood vessels. High production of the virus in the primarily affected organs is a prerequisite for the virus to cross the blood–brain barrier because the capillary endothelium is not easily infected. Once it has invaded these endothelial cells from the lumen, the virus replicates and enters the central nervous system (CNS) by seeding through the capillary endothelium into the brain tissue. However, individual reports correlating tick bites of the upper trunk to the development of localized shoulder girdle paralysis and paresis suggest direct entry of the virus via peripheral nerves as well.[71] Neuroinvasion of TBEV in humans has been well reported, and the primary targets of infection are neurons. A number of human neural cell lines have been demonstrated to undergo apoptosis as well as necrosis upon infection with TBEV *in vitro*.[72] However, the mechanisms by which the virus enters and damages the CNS are not defined.

There are a number of animal models employed to examine the neuropathogenicity of TBEV. The mouse is the most commonly used model, as it is susceptible to TBEV-induced disease unlike other wild and domestic animals.[73] Wild-type strains of TBEV are generally neuropathogenic when experimentally inoculated into mice (intracranially or peripherally), usually resulting in a lethal infection depending on the age of the animal. An apparent feature of TBEV is its ability to cause persistent infections in experimental animals and humans.[1] Animal models have also been used to demonstrate degenerative changes in the CNS following infection with TBEV, that is, by intracranial or subcutaneous inoculation of hamsters[74] and intranasal or intracerebral inoculation of monkeys.[73] The pathogenic process following experimental infection with TBEV is fundamentally similar to that of other encephalitic flaviviruses, including Japanese encephalitis, West Nile virus, and Murray Valley virus. The CNS pathology of TBEV involves two distinct features, that is, neuroinvasiveness and neurovirulence. Direct intracerebral infection usually results in high mortality rates, and 50% lethal doses (LD_{50}) are often below 1 PFU (plaque-forming unit). Therefore, it is generally believed that after the CNS is entered by viruses, the host develops lethal encephalitis. Thus, mortality rates following direct intracerebral infection represent neurovirulence, whereas mortality following peripheral infection represents neuroinvasiveness.[27] However, CNS pathology is the consequence of viral infection and the resulting inflammatory responses in the CNS. Direct viral infection of neurons is considered to be the major cause of neurological disease because viral infections cause apoptosis or degeneration of neurons *in vivo* and *in vitro*. Although critical for controlling viral infection in the CNS, the host immune response has been implicated in contributing to neuropathology.[75] In addition, recent studies have demonstrated that TBEV induces widespread inflammatory changes, characterized by diffuse inflammatory infiltrates in combination with astrogliosis, the formation of microglial nodules, neuronophagia, and varying degrees of neuronal loss.[76]

The most obvious feature of TBE in patients and in experimentally infected monkeys is ataxia, followed by paresis or paralysis of one or more limbs.[77] These and other neurological symptoms of TBE can be explained by the affinity of TBEV for certain regions of the CNS. Postmortem examination of the brain and spinal cord from patients with a lethal course of TBE and from monkeys, which were infected experimentally with TBEV, show similar findings.[78] Cerebral and spinal meninges usually show diffuse infiltration with lymphocytes and sometimes leukocytes. The most extensive area of meningitis is around the cerebellum. The brain is edematous and hyperemic. Microscopic lesions are present in almost all parts of the CNS but particularly in the medulla oblongata, pons, cerebellum, brainstem, basal ganglia, thalamus, and spinal cord. The lesions are localized in the gray matter and consist of lymphocytic perivascular infiltrations, accumulation of glial cells, necrosis of nerve cells, and neuronophagia. In particular, Purkinje cells in the cerebellum and the anterior horn cells in the spinal cord are frequently attacked. Infiltration and rarefaction of cells are also noted in the mesencephalon and diencephalon. Changes in the cerebral cortex are almost invariably restricted to the motor area with degeneration and necrosis of the pyramidal cells and lymphocytic accumulation and glial proliferation near the surface.[78] However, the understanding of the pathogenic mechanism of TBE is incomplete. The difficulties associated with detecting viral RNA in cerebrospinal fluid (CSF) during the encephalitic phase strongly suggest that virus replication may be inhibited or reduced when neutralizing antibodies appear in serum and CSF, although the virus may be located in neurons.[76] Immunohistochemical experiments have demonstrated that many large neurons within affected areas contain viral antigen, although an inverse correlation between the number of infected neurons and magnitude of the infiltrating immune response was observed, suggesting underlying immunopathogenic mechanism.[76] A thorough examination of the specific immune response in postmortem tissue sections indicated that CD8+ granzyme B-releasing cytotoxic T cells might significantly contribute to neuronal damage *in vivo* via the introduction of bystander damage.[78] Human TBEV patients have elevated CCL3, CXCL10, and CXCL13 levels in the serum; CCL2, CCL3, CCL5, and CXCL10-13 have been detected in CSF, possibly implicating the role of a Th1-mediated response and the trafficking of T-lymphocytes into the CNS.[79] Moreover, CCL2 and CCL5 were found to be elevated in the CSF of TBE patients even after the acute symptoms subsided, and unlike other neurotropic flaviviruses, TBEV induces CXCL13.[80] In a recent study, it has been suggested that the loss of CCR5 may predispose individuals to a severe outcome of a TBEV infection.[81] Low levels of neutralizing serum antibodies correlate with severe course of the disease. The theory of antibody-dependent enhancement, which applies for dengue severity studies, has been considered, but there are no laboratory data indicating a similar phenomenon in human TBE.

While courses and symptoms are quite similar in the early stage of disease, human infections with different subtypes of TBEV may result in the development of clinical manifestations of varying severity. Human infections with the Far Eastern subtype result in the most severe forms of CNS disorder, with a tendency for the patient to develop focal meningoencephalitis or polyencephalitis accompanied by loss of consciousness and prolonged feelings of fatigue during recovery. A fatality rate of 5%–35% and usually absence of chronic forms have been reported. In contrast, TBEV infections with the Siberian subtype present a less severe pattern (lethality: 1%–3%), but chronic forms seem to be possible more frequently pointing toward the importance of specific viral factors in the pathogenesis of chronic flaviviral CNS disease.[35] With the European subtype, the one more thoroughly analyzed in this chapter, one-third of the patients develop symptomatic disease, and a biphasic course of illness occurs in 72%–87% of the patients.[1,82] In general, the case fatality rate is approximately 1%–2% following TBEV infection with the European subtype,[7] with residual sequelae in 25%–50% of the patients.[1]

The incubation period ranges from 4 to 28 days but in average is 7–14 days. After exposure to infected milk, there is a shorter incubation period of 3–4 days. There is no correlation between the length of the incubation period and the severity of subsequent illness. In the biphasic disease, the first viremic stage usually presents with muscle pain, fatigue (63%), headache (54%), and fever (99%) lasting 2–10 days, which peaks at a temperature of 37.5°C–39°C. There are no signs or symptoms of meningoencephalitis during this phase. Thrombocytopenia, leukopenia, hyperalbuminorrachia, sometimes the unique finding in CSF, and elevated liver enzymes are common,[83] while leukocytosis is frequent in the second stage. Compared to other forms of viral meningitis, the white blood cell count in CSF is low in TBE. The first phase is followed by an afebrile and relatively asymptomatic period lasting about 7 days (range: 1–33).

The second febrile phase correlates with virus invasion into the CNS, where viral replication is associated with inflammation, lysis, and dysfunction of the cells.[82] This phase is characterized by fever that is usually 1°C–2°C higher than the peak temperature in the first phase. Flushing of the face and neck, conjunctival injection, headache, somnolence, nausea, vomiting, dizziness, and myalgia are common findings. Hyperesthesia, hyperacusia, and increased sensitivity to smells occur.

Different clinical presentations can occur in the second phase. TBE presents as meningitis in about 50% of the patients, as meningoencephalitis in about 40%, and as meningoencephalomyelitis in about 10%.[70] Meningitis is frequently accompanied by high fever, headache, nausea, vomiting, and vertigo. Signs of meningeal irritation usually occur, but may not be pronounced. All patients exhibit CSF pleocytosis, with two-thirds of the patients having 100 leukocytes per ml or less. An initial predominance of polymorphonuclear cells is later changed to almost 100% mononuclear cell dominance. Two-thirds have a moderate increased CSF albumin, peaking at a median day 9. Objective meningeal signs could be absent in about 10% despite CSF pleocytosis. Therefore, patients presenting only with fever as the prominent symptom without encephalitic signs could be suspected as having other infectious syndromes. Abnormalities on magnetic resonance imaging (MRI) are seen in up to 18% with lesions usually confined to the thalamus, cerebellum, brainstem, and nucleus caudatus. Electroencephalogram (EEG) is abnormal in 77%. Both MRI and EEG abnormalities are unspecific and not diagnostic.[84]

Meningoencephalitis presentation is variable. The fever is usually high and sustained in the face of antipyretics. Fever in the last phase of the disease mostly lasts 4–10 days but can even last up to 1 month. Meningeal signs are usually present, and patients are somnolent or unconscious. Severe tremors of extremities and fasciculations of the tongue, profuse sweating, asymmetrical paresis of cranial nerves, and nystagmus are common symptoms. In some patients, delirium and psychosis may develop rapidly. Encephalitis is characterized by a disturbance of consciousness ranging from somnolence to stupor and, in rare cases, coma. Other symptoms include restlessness, hyperkinesia of limb and face muscles, lingual tremor, convulsions, vertigo, and speech disorders. The cranial nerves may be involved; mainly the ocular, facial, and pharyngeal muscles are affected.[85]

Meningoencephalomyelitis is the most severe form of the disease. It is characterized by paresis that usually develops 5–10 days after the remission of fever. Severe pain in the arms, back, and legs occasionally precedes the development of paresis. The upper extremities are affected more frequently than the lower ones, and the proximal segments more often than the distal ones.[85] Rarely, myelitis occurs as the sole manifestation of the disease; mostly brainstem encephalitis is prominent.[86] Involvement of cranial nerve nuclei and motor neurons of the spinal cord produces a flaccid poliomyelitis-like paralysis that, unlike polio, usually affects the neck and upper extremity muscles. Involvement of the medulla oblongata and the central portions of the brainstem is associated with a poor prognosis due to substantial respiratory and circulatory failure. Death usually occurs within 5–10 days of the onset of the neurologic signs and is most commonly secondary to bulbar involvement or diffuse brain edema. Occasionally, TBE can be associated with autonomic dysfunction including reduced heart rate variability and tachycardia. Apart from myelitis, TBE can develop into a myeloradiculitic form typically presenting a few days after defervescence and could be accompanied by severe pain in the back and limbs, weak muscle reflexes, and sensory disturbances. Paralyses could develop that, compared with myelitis, have a more favorable prognosis.

The severity of TBE increases with age. Severe forms of encephalitis were seen in 44%–55% of adults, while severe disease associated with TBE in children <3 years of age is rare. In children and adolescents, meningitis is the predominant form of the disease, which is why the infection usually takes a milder course with a better prognosis than in adults. There is a clear tendency for a more severe course of TBE

above the age of 7 years. A substantial increase in morbidity in elderly people makes them a special target group for immunization. Age, severity of illness in the acute stage, and low-titer neutralizing antibodies at onset are associated with severe forms of the disease, along with low early CSF IgM response. The degree of virus neutralizing capacity can determine the degree of viremia that is experimentally associated with development of disease.[3]

20.5 IDENTIFICATION/DIAGNOSIS

As the clinical features of TBEV infections in patients are nonspecific and similar to other CNS diseases, reliable laboratory methods are required to monitor the clinical outcome of TBE in patients. Furthermore, these methods can also be used for epidemiological studies, as well as for research purposes. In the first (viremic) phase of the disease, TBEV can be isolated from blood, serum, liquor, or organs, and the viral RNA can be detected by polymerase chain reaction (PCR)-based methods. For epidemiological studies, these methods also could be used to identify the TBEV in ticks and host animals.[87] In the later stages, when CNS symptoms are manifest, the serological methods based on the detection of TBE-specific antibodies are used to detect the immune response of the host (human or animal) and confirm the diagnosis of TBEV infection.[88–90]

The TBEV can be isolated by virus cultivation using mammalian cell cultures susceptible to TBEV (e.g., VeroE6, BHK-21, PS, HEK 293T, or A549 cells) in the first phase of infection. For virus isolation, patient's serum is first incubated with the susceptible cells until a cytopathic effect can be observed, typically 5–7 days. Finally, the infection of cells by TBEV can be tested by specific PCR or immunofluorescence assay (IFA).[91] However, virus isolation is not always successful, as the viral titer in human blood is low and viral particles are already cleared from the blood when the first CNS symptoms arise and patients are usually admitted to the hospital.[89] Furthermore, virus isolation is time-consuming and has to be performed under BSL-3 conditions by specially trained persons.

In the last years, PCR-based molecular techniques have become increasingly popular as a standard laboratory method for the diagnosis of flaviviral diseases during the viremic phase.[88,89] The reverse transcriptase-PCR (RT-PCR) is widely described as a sensitive and specific tool to detect virus-specific RNA in a wide range of samples. The major advantages of molecular techniques for diagnosis of TBEV infections are the following: (1) detection of viruses in the early phase of the disease and before the appearance of antibodies; (2) use of a wide range of samples such as a blood, serum and CSF samples (of TBE patients and infected mammals), infected ticks, and tissue samples (e.g., brain or spleen of infected humans and animals); (3) quantification of viral load; and (4) differentiation between the three TBEV subtypes.

Various RT-PCR assays ranging from the conventional RT-PCR format to the quantitative real-time RT-PCR format have been described for the detection of TBEV previously. Saksida et al.[92] described an RT-PCR assay to detect the TBEV in blood and serum samples of patients with febrile illness following a tick bite. Růžek et al.[93] developed an easy multiplex RT-PCR assay, which is able to detect and to differentiate the three TBEV subtypes. As RNA amounts could be rather low, these RT-PCR assays are more suitable for surveillance studies in host animals than for diagnostic purposes.

The first TBEV real-time RT-PCR assay was published in 2003 by Schwaiger and Cassinotti.[94] In a quality control assessment for the molecular diagnosis of TBEV infections performed in 2007, this method showed the best performance when compared with other techniques.[95] This assay detects down to 10 copies of virus genome providing a good sensitivity for molecular diagnostics of TBEV infection. However, this assay cannot differentiate between the three TBEV subtypes.[94] An improvement for diagnostic as well as for surveillance is provided by the most recently published TBEV quantitative real-time RT-PCR assay followed by pyrosequencing.[91] This method combines the advantages of both assays previously mentioned, the quantitative and sensitive detection, and the possibility of differentiating between the subtypes, with a fast and easy handling.

As the majority of patients come to the hospital in the second phase of illness when neurological symptoms are manifest, it is the experts' current opinion that virus isolation and RT-PCR at this time are less important for the diagnosis of TBE. At the beginning of the second phase of illness, the virus itself is only rarely detectable in blood and CSF. Therefore, the diagnosis of TBE is mainly done by serological methods, usually enzyme-linked immunosorbent assay (ELISA) based on purified virions or recombinant virus-like particles, IFA, and neutralization test (NT), which have been developed toward higher specificity and sensitivity in the last decade.[89,96–98]

Serological methods are based on the detection of IgM and IgG antibodies against the TBEV in the sera of patients and animals. One characteristic of these methods is that they can be only used in the second phase of infection when a humoral immune response has started and antibodies are produced in response to the TBEV infection. Within the first 6 days of CNS symptoms, the level of TBE-specific IgM raises. Typically, a decrease of the IgM level is observed after 6 weeks.[99] The TBE-specific IgG antibodies are detectable simultaneously or a few days after the appearance of IgM antibodies. As the IgM and IgG antibodies are usually detectable from the beginning of the second phase over a longer time, the serological methods work well for this phase of infection.

An early diagnosis by detecting only IgM is sometimes questionable, since IgM antibodies can persist for up to 10 months in vaccinees or individuals who acquired the infection naturally. Therefore, confirmation by detection of specific IgG is recommended but may turn out negative at the beginning of the second phase. Therefore, it is necessary to monitor an increase of the titers of IgG antibodies against TBE 1–2 weeks later, but this testing is rarely done. Moreover, a major problem when using ELISA and IFA lays in the high cross-reactivity of the flaviviral antigenic structure. Possible diagnostic difficulties may arise due to cross-reactions of antibodies elicited by other flavivirus infections or vaccinations. This could happen in areas where other flaviviruses cocirculate (e.g., West Nile virus in the southern parts of the TBE endemic areas of Europe), in patients recently returned from areas endemic for other flaviviruses (e.g., dengue virus) or in individuals being vaccinated against TBEV, Japanese encephalitis virus, or yellow fever virus.[89,90]

An external quality assurance study has shown that the assays based on detection of IgM were less sensitive than the assays based on detection of IgG, whereas the IgG assays showed cross-reactivity with other flaviviruses.[100] Nevertheless, ELISA methods based on the detection of IgM/IgG antibodies are still the method of choice for TBE diagnostic. ELISAs are often used due to their simple performance, the ease of automation, and the availability of commercial kits with high sensitivity and specificity.

Recently, the development of a new immune complex (IC) ELISA, detecting antibodies against domain III of the TBEV E protein (EDIII), was published. This assay offers a high sensitivity comparable with the whole tissue culture virus (TCV) ELISA.[98] The IC ELISA showed a higher specificity than the TCV ELISA and might be helpful to reduce false positive results from cross-reactivity to other flaviviruses in the future.

Other techniques like NT and IFA are rarely used in clinical diagnostics.[100–102] The NT is laborious and expensive and limited to laboratories equipped with BSL-3 facilities due to the use of infectious viruses. The IFA is done using commercial tests but needs experienced persons for interpretation. Both methods still have their range of application in research laboratories and are useful to carry out detailed diagnostics in specialized laboratories.

Both serological and molecular detection methods for TBE are useful as a single application or in combination for clinical diagnosis, immunity testing, and epidemiological surveillance studies of virus prevalence in ticks and vertebrate hosts.

General trends in molecular diagnostics indicate that more real-time RT-PCR assays in multiplex format will be available, which will allow the identification of several flaviviral or tick-borne diseases simultaneously. Concerning serological testing, recombinant TBEV particles or antigens are being developed in order to establish new simple and cost-effective methods that could overcome cross-reactivity issues.

20.6 TREATMENT AND PREVENTION

There is currently no specific therapeutic for the clinical treatment of TBE patients, and recommended measures are only symptomatic with strict bed rest, usually in an intensive care unit, and administration of analgesics and antipyretics. The administration of hyperimmunoglobulin for a passive postexposure prophylaxis (PEP) is highly questionable concerning the virtue and controversial due to concerns about antibody-dependent enhancement of infection.[44,103–105] Therefore, hyperimmunoglobulin preparations for PEP against TBE were withdrawn from the European market in the late 1990s and are also not recommended—as well as vaccination—after a tick bite. In contrast, such preparations are sometimes used for early treatment in the Russian Federation.[106] Besides general preventive measures, like wearing appropriate clothing or checking the skin for attached ticks, active immunization offers the most effective protection against TBE.[107–110]

Currently, four vaccines are widely used and of assured quality.[111] Two vaccines based on Far Eastern TBE subtype viruses are mainly licensed for use in Russia: the TBE-Moscow vaccine (Chumakov Institute of Poliomyelitis and Viral Encephalitides, Moscow, Russia; also licensed in Kazakhstan and Ukraine) with strain Sofjin and EnceVir (NPO Microgen, Tomsk, Russia) with strain 205. In Europe, two vaccines are available that are based on European TBEV strains: FSME-IMMUN (Baxter Bioscience, Vienna, Austria; also licensed for use in Canada and Russia) with strain Neudörfl and Encepur (Novartis Vaccines, Marburg, Germany; also available in Russia) with strain K23. These vaccines are all based on cell-cultured, formalin-inactivated, and purified whole TBE viral strains, adjuvanted by aluminum hydroxide, and produced according to the WHO's Good Manufacturing Practice requirements with some differences in the production method, the excipients, and the distributed product (Table 20.1).[111] The large E protein induces the production of neutralizing antibodies important for the protective immunity. Due to the highly conserved structure of this antigen, broad cross-protection by the vaccines could be shown against TBE viruses of all three known subtypes.[9,13,112–116]

The conventional vaccination schedules for primary immunization are similar for both vaccines manufactured in Europe, with three intramuscular doses given on 0, 1–3, and 5–12 months, respectively (see Table 20.1). Besides vaccines for adults, both European vaccine manufacturers offer pediatric vaccine formulations containing half the dose of viral antigen of the adult ones to improve tolerability in children (FSME-IMMUN Junior for children ≤15 years; Encepur Children for children ≤11 years). Both vaccines induce antibody concentrations that are believed to be protective in over 90% (seroconversion) of children and adults after the second dose.[10,111] Due to the high homology of the antigens and demonstrated cross-boostering in clinical studies, it is accepted that both vaccines can be used interchangeably.[117,118] However, when possible, it is recommended to use the same vaccine throughout the primary immunization series. So far, the protective amount of antibodies is not clearly defined and standardized for both vaccines, and little information is available on the persistence of postimmunization antibodies. Pending more information, regular boosters are recommended by the manufacturers every 5 years for age groups ≤49 years of age (except for the first booster after 3 years). In age groups >49 years of age, a 3-year booster interval is recommended due to the significant gradual decline of postimmunization antibodies.[119–121] Despite, in Switzerland, booster

intervals have been extended to 10 years in order to improve acceptability of vaccination.[122]

Concerning safety aspects, moderate local reactions at injection site are commonly associated with TBE vaccines, but severe adverse reactions have not been casually linked. There are no specific contraindications to TBE vaccination except allergy to vaccine components and severe acute infections.[111] As for all other inactivated vaccines, pregnancy is considered to be a relative contraindication, and pregnant women at risk can be vaccinated after an individual risk–benefit assessment. Regardless of the duration of the delay, interrupted immunization schedules should be completed without repeating previous doses. The principle here is "Each vaccination dose counts!" On the other hand, the recommended booster intervals as specified by technical information and official immunization plans should be maintained. Vaccine breakthroughs have been occasionally reported, especially in persons >50 years.[123–125] Although there are no reports of interference between TBE vaccines and other simultaneously administered vaccines, more information is needed on the immune response to the vaccine in individuals who have been previously immunized against yellow fever or Japanese encephalitis.[111]

The current Russian vaccine formulations are licensed for adults and children ≥3 years (see Table 20.1). The TBE-Moscow vaccine is given in two doses over 1–7 months with the revaccination after 1 year and then every 3 years. EnceVir vaccine is also administered in two doses over 5–7 months, with the revaccination after 1 year and every 3 years thereafter. Both Russian vaccines have shown safety and efficacy profiles similar to those seen for the European vaccines. However, the body of evidence is significantly larger for the European than for the Russian vaccines. An increased incidence of high fever and allergic reactions following vaccination with some lots of EnceVir, particularly in children, resulted in withdrawal for pediatric use (children ≤17 years). A pediatric formulation of this vaccine is under development.[111]

Rapid immunization schedules have been introduced, by both European vaccine producers (for FSME-IMMUN and Encepur) and by the Russian manufacturer of EnceVir, for people who require immunity at short notice, such as persons traveling to TBE endemic areas or when the tick season has already started.[111] However, since the experience with TBE vaccines is mainly based on the conventional immunization schedules, these should be always applied wherever possible.

In areas where the disease is highly endemic (i.e., where the average prevaccination incidence of clinical disease is ≥5 cases/100,000 population per year), implying that there is a high individual risk of infection, WHO recommends that vaccination should be offered to all age groups, including children. Inclusion of vaccination against TBE into immunization programs at regional level or national level should be considered, depending on the epidemiological situation. Where the prevaccination incidence of the disease is moderate or low (i.e., the annual average during a 5-year period is < 5/100,000) or is limited to particular geographical locations

TABLE 20.1
Description of Available TBE Vaccines (Overview)[a]

Vaccine	FSME-IMMUN	Encepur	TBE-Moscow	EnceVir
Manufacturer	Baxter Bioscience, Vienna, Austria	Novartis Vaccines, Marburg, Germany	Chumakov Institute of Poliomyelitis and Viral Encephalitides, Moscow, Russia	NPO Microgen, Tomsk, Russia
Vaccine Strain	Neudörfl (European subtype)	K23 (European subtype)	Sofjin (Far Eastern subtype)	205 (Far Eastern subtype)
Production Method	Produced on chick embryonic fibroblast cells, purified after formaldehyde inactivation by continuous-flow zonal ultracentrifugation		Propagated in primary chicken embryo cells, purified and concentrated after formaldehyde inactivation, treated with protamine sulfate, subsequently gel-filtrated	Propagated in primary chicken embryo cells, purified and concentrated after formaldehyde inactivation, treated with protamine sulfate; ultrafiltration by gel-permeated chromatography
Excipients	Aluminum hydroxide, human serum albumin	Aluminum hydroxide, sucrose	Aluminum hydroxide (final formulation), human serum albumin, gelatin, sucrose, bovine serum albumin (<0.5 μg/dose)	Aluminum hydroxide, human serum albumin, sucrose
Distributed Product	Stored as liquid formulation in prefilled syringe		Lyophilized in excipient, with saline containing aluminum hydroxide as gel	Stored as liquid formulation
Shelf Life (2°C–8°C)	2 years		3 years	2 years
Availability in/ Licensed for	Throughout Europe, Russia, and Canada (licensed using purely national routes)	Throughout Europe and Russia	Russia, Kazakhstan, and Ukraine	Russia
Pediatric Formulation	For children ≤15 years old.	For children ≤11 years old.	Formulation is licensed for all persons ≥3 years old.	Formulation is licensed for all persons ≥3 years old.
Conventional Immunization Schedule	0, 1–3 months, 5–12 months Regular boosters are recommended by the manufacturer: ≤49 years old, first booster after 3 years, subsequently 5-year intervals >49 years old, 3-year intervals	0, 1–3 months, 9–12 months Regular boosters are recommended by the manufacturer: ≤49 years old, first booster after 3 years, subsequently 5-year intervals >49 years old, 3-year intervals	0, 1–7 months with the revaccination after 1 year, and then every 3 years	0, 5–7 months with the revaccination after 1 year, and then every 3 years
Immunogenicity (Seroconversion) Conventional	In children: 95%–100% after 2 or 3 vaccine doses In adults: 96.6% after 2 vaccinations	In children: 97%–100% after 2 or 3 vaccine doses In adults: 92%–95% after 2 vaccinations	In children: 89%–96% after 2 vaccine doses In adults: 84%–93% after 2 vaccinations	In children: 84%–97% after 2 vaccine doses In adults: 82%–89% after 2 vaccinations
Accelerated Immunization Schedule	Day 0, day 14, 5–12 months Booster doses as above	Day 0, day 7, day 21 First booster at 12–18 months, subsequent booster doses as above	—	Day 0, 21–35* days, 42–70* days First booster at 5–12 months, subsequent boosters as above

(*continued*)

TABLE 20.1 (continued)

Description of Available TBE Vaccines (Overview)[a]

Vaccine	FSME-IMMUN	Encepur	TBE-Moscow	EnceVir
Safety	Moderately reactogenic (redness, pain at injection site in up to 45%, fever in up to 5%–6% of vaccinees) but no severe adverse events.		No randomized, controlled trials published. The Russian National Regulatory Authority has assessed both Russian vaccines and concluded that they are both safe and well tolerated (with some local and systematic reactogenicity). However, in 2010 and 2011, some lots of EnceVir associated with frequent high fever and allergic reactions in children were withdrawn. EnceVir is not recommended for use in children aged 3–17 years.	

Note: *, double doses.

[a] According to WHO publication and relevant background documents.[111]

or certain outdoor activities, immunization should target individuals in the most severely affected cohorts, including travelers from nonendemic to endemic areas if their visits will include extensive outdoor activities.[111]

In order to characterize the epidemiological situation, which might influence the quality of vaccine recommendations, improved TBE surveillance and reporting is critical. Therefore, standardization of case definitions, laboratory diagnostics, and surveillance/reporting activities is needed.[64,126] Also, in order to improve vaccination coverage, awareness of this tick-transmissible disease needs to be increased by providing relevant information in all endemic areas, for example, in schools, medical offices, and tourist information materials.

20.7 CONCLUSIONS AND FUTURE PERSPECTIVES

According to the increasing number of surveillance studies in the recent years, it became obvious that the areas for the distribution of the TBEV are expanding westward.[127] Since this is a sneaking process depending on regional, environmental, and climate factors, the appearance of new TBE cases in previously nonendemic areas has to be seriously investigated. Therefore, serological studies in wild animals like roe deer and rodents, and domestic animals like sheep or goats, should be initiated to inform us on the presence of TBEV in a certain area.[62,87,128] To evaluate the risk for TBE infection in an area with low endemicity, the screening of suitable surrogate marker is an important task in order to provide clear advice for public health measures. This preparedness also comprises the diagnostic confirmation of suspected TBE cases in patients with clinical symptoms like encephalitis. Such evidence is required to launch further public health measures such as recommending vaccination of risk groups or providing specific information to physicians in the exposed area. Since September 2012, TBE is a notifiable disease and is part of the list of diseases to be covered by epidemiological surveillance within the European community.[129] This will help to get a better picture on the epidemiological situation in the affected countries and will help to follow the further distribution of the infection.

The diagnostic of TBEV infection still struggles with the high cross-reactivity of IgG antibodies within the Flaviviridae family. Since the detection of the TBEV genome in the acute phase often fails, due to undetectable viremia or lack of early sampling, the diagnostic has to rely on serology. Even though IgM detection is less prone to cross-reactivity than IgG antibodies, only a fourfold increase of the titer in consecutive sera samples is a clear indication for a TBEV infection. Hopefully, more specific methods than the currently used ELISA will soon enter the market and contribute to a more accurate and simple diagnostic in the future.

Vaccines of *Flavivirus* origin (yellow fever, Japanese encephalitis) are administered more and more frequently. Nevertheless, there is a lack of studies investigating the role and/or interaction of cross-reactive antibodies during the immune response.[130,131] So far, it is unclear whether a second flavivirus vaccine boosters the previous immune response by increasing the antibody titer of the first vaccine or whether preexisting cross-reactive antibodies hinder the virus propagation of the second flavivirus vaccine. Upcoming new flavivirus-based vaccines (e.g., against dengue and West Nile) will enhance the complexity of such mixed immune response in multivaccinated vaccinees.[132] Therefore, it is crucial to address these questions in clinical studies including all existing flavivirus vaccines and including gradually upcoming new vaccines.

REFERENCES

1. Gritsun, T.S., Lashkevich, V.A., and Gould, E.A., Tick-borne encephalitis, *Antiviral Res.*, 57, 129, 2003.
2. Süss, J., Tick-borne encephalitis 2010: Epidemiology, risk areas, and virus strains in Europe and Asia—An overview, *Ticks Tick Borne Dis.*, 2, 2, 2011.
2a. CDC National Select Agents Registry. Select Agents and Toxins List. http://www.selectagents.gov/resources/List%20 of%20Select%20Agents%20and%20Toxins%2009-19-2011. pdf (accessed on June 30, 2013).
2b. CDC National Select Agents Registry. Possession, use, and transfer of select agents and toxins; biennial review. Final rule. http://www.selectagents.gov/resources/CDC%20Select%20 Agent%20Biennial%20Review%20Final%20Rule%20 10%2005%202012.pdf (accessed on June 30, 2013).

3. Sips, G.J., Wilschut, J., and Smit, J.M., Neuroinvasive flavivirus infections, *Rev. Med. Virol.*, 22, 69, 2012.

4. Turtle, L., Griffiths, M.J., and Solomon, T., Encephalitis caused by flaviviruses, *QJM*, 105, 219, 2012.

5. Weissenböck, H. et al., Zoonotic mosquito-borne flaviviruses: Worldwide presence of agents with proven pathogenicity and potential candidates of future emerging diseases, *Vet. Microbiol.*, 140, 271, 2010.

6. Heinz, F.X., Molecular aspects of TBE virus research, *Vaccine*, 21, S3, 2003.

7. Mansfield, K.L. et al., Tick-borne encephalitis virus—A review of an emerging zoonosis, *J. Gen. Virol.*, 90, 1781, 2009.

8. Lobigs, M. and Diamond, M.S., Feasibility of cross-protective vaccination against flaviviruses of the Japanese encephalitis serocomplex, *Expert Rev. Vaccines*, 11, 177, 2012.

9. Ecker, M. et al., Sequence analysis and genetic classification of tick-borne encephalitis viruses from Europe and Asia, *J. Gen. Virol.*, 80, 179, 1999.

10. Lindquist, L. and Vapalahti, O., Tick-borne encephalitis, *Lancet*, 371, 1861, 2008.

11. Grard, G. et al., Genetic characterization of tick-borne flaviviruses: New insights into evolution, pathogenetic determinants and taxonomy, *Virology*, 36, 80, 2007.

12. Pogodina, V.V. et al., Evolution of tick-borne encephalitis and a problem of evolution of its causative agent, *Vopr. Virusol.*, 52, 16, 2007.

13. Hayasaka, D. et al., Evaluation of European tick-borne encephalitis virus vaccine against recent Siberian and far-eastern subtype strains, *Vaccine*, 19, 4774, 2001.

14. Süss, J., Epidemiology and ecology of TBE relevant to the production of effective vaccines, *Vaccine*, 21, S19, 2003.

15. Lindenbach, B.D. and Rice, C.M., Molecular biology of flaviviruses, *Adv. Virus Res.*, 59, 23, 2003.

16. Wengler, G., Wengler, G., and Gross, H.J., Studies on virus-specific nucleic acids synthesized in vertebrate and mosquito cells infected with flaviviruses, *Virology*, 89, 423, 1978.

17. Chambers, T.J. et al., Flavivirus genome organization, expression, and replication, *Ann. Rev. Microbiol.*, 44, 649, 1990.

18. Heinz, F.X. and Allison, S.L., Flavivirus structure and membrane fusion, *Adv. Virus Res.*, 59, 63, 2003.

19. Mandl, C.W. et al., Infectious cDNA clones of tick-borne encephalitis virus European subtype prototypic strain Neudorfl and high virulence strain Hypr, *J. Gen. Virol.*, 78, 1049, 1997.

20. Kaufmann, B. and Rossmann, M.G., Molecular mechanisms involved in the early steps of flavivirus cell entry, *Microbes Infect.*, 13, 1, 2011.

21. Orlinger, K.K. et al., Construction and mutagenesis of an artificial bicistronic tick-borne encephalitis virus genome reveals an essential function of the second transmembrane region of protein E in flavivirus assembly, *J. Virol.*, 80, 12197, 2006.

22. Diamond, M.S., Mechanisms of evasion of the type I interferon antiviral response by flaviviruses, *J. Interferon Cytokine Res.*, 29, 521, 2009.

23. Chung, K.M. et al., West Nile virus nonstructural protein NS1 inhibits complement activation by binding the regulatory protein factor H, *Proc. Natl. Acad. Sci. USA*, 103, 19111, 2006.

24. Xu, T. et al., Structure of the Dengue virus helicase/nucleoside triphosphatase catalytic domain at a resolution of 2.4 A, *J. Virol.*, 79, 10278, 2005.

25. Miller, S., Sparacio, S., and Bartenschlager, R., Subcellular localization and membrane topology of the Dengue virus type 2 non-structural protein 4B, *J. Biol. Chem.*, 281, 8854, 2006.

26. Yap, T.L. et al., Crystal structure of the dengue virus RNA-dependent RNA polymerase catalytic domain at 1.85-angstrom resolution, *J. Virol.*, 81, 4753, 2007.

27. Mandl, C.W., Steps of tick-borne encephalitis virus replication cycle that affect neuro-pathogenesis, *Virus Res.*, 111, 161, 2005.

28. Šenigl, F., Grubhoffer, L., and Kopecký, J., Differences in maturation of tick-borne encephalitis virus in mammalian and tick cell line, *Intervirology*, 49, 236, 2006.

29. Romanova, L. et al., Microevolution of tick-borne encephalitis virus in course of host alternation, *Virology*, 362, 75, 2007.

30. Pavlovskij, E.N., On natural focality of infectious and parasitic diseases, *Vestn. Akad. Nauk. SSSR*, 10, 98, 1939.

31. Korenberg, E.I. and Kovalevskij, Y.V., Main features of tick-borne encephalitis eco-epidemiology in Russia, *Zent. Bakteriol.*, 289, 525, 1999.

32. Nuttall, P.A. and Labuda, M., Tick-borne encephalitides. In *Zoonoses*, pp. 469–486. Palmer, S.R., Soulsby, L., Simpson, D.I.H. (eds.), Oxford, U.K.: Oxford University Press, 1998.

33. Spielman, A. et al., Issues in public health entomology, *Vector Borne Zoonotic Dis.*, 1, 3, 2001.

34. Rampas, J. and Gallia, F., Isolation of tick-borne encephalitis virus from ticks *Ixodes ricinus*, *Čas. Lék. Čes.*, 88, 1179, 1949.

35. Gritsun, T.S. et al., Characterization of a siberian virus isolated from a patient with progressive chronic tick-borne encephalitis, *J. Virol.*, 77, 25, 2003.

36. Krivanec, K. et al., Isolation of TBE virus from the tick *Ixodes hexagonus*. *Folia Parasitol. (Praha)*, 35, 273, 1988.

37. Grešíková, M. and Kaluzová, M., Biology of tick-borne encephalitis virus, *Acta Virol.*, 41, 115, 1997.

38. Labuda, M. and Randolph, S.E., Survival strategy of tick-borne encephalitis virus: Cellular basis and environmental determinants, *Zentralbl. Bakteriol.*, 289, 513, 1999.

39. Benda, R., The common tick "*Ixodes ricinus*" as a reservoir and vector of tick-borne encephalitis. I. Survival of the virus (strain B3) during the development of ticks under laboratory condition, *J. Hyg. Epidemiol. (Prague)*, 2, 314, 1958.

40. Labuda, M. et al., Efficient transmission of tick-borne encephalitis virus between cofeeding ticks, *J. Med. Entomol.*, 30, 295, 1993.

41. Labuda, M. et al., Tick-borne encephalitis virus transmission between ticks cofeeding on specific immune natural rodent hosts, *Virology*, 235, 138, 1997.

42. Nuttall, P.A. and Labuda, M., Dynamics of infection in tick vectors and at the tick-host interface, *Adv. Virus. Res.*, 60, 233, 2003.

43. Kozuch, O. et al., The role of small rodents and hedgehogs in a natural focus of tick-borne encephalitis, *Bull. World Health Organ.*, 36(Suppl. 1), 61, 1967.

44. Kaiser, R., The clinical and epidemiological profile of tick-borne encephalitis in southern Germany 1994–1998: A prospective study of 656 patients, *Brain*, 122, 2067, 1999.

45. Gresikova, M. et al., Sheep milk-borne epidemic of tick-borne encephalitis in Slovakia, *Intervirology*, 5, 57, 1975.

46. Vereta, L.A. et al., The transmission of the tick-borne encephalitis virus via cow's milk, *Med. Parazitol. (Mosk)*, May–Jun(3), 54, 1991.

47. Kerbo, N. et al., Tick-borne encephalitis outbreak in Estonia linked to raw goat milk, May–June 2005, *Euro Surveill.*, 10, E050623.2, 2005.

48. Holzmann, H. et al., Tick-borne encephalitis from eating goat cheese in a mountain region of Austria, *Emerg. Infect. Dis.*, 15, 1671, 2009.

49. Kriz, B., Benes, C., and Daniel, M., Alimentary transmission of tick-borne encephalitis in the Czech Republic (1997–2008), *Epidemiol. Mikrobiol. Imunol.*, 58, 98, 2009.

50. Balogh, Z. et al., Tick-borne encephalitis outbreak in Hungary due to consumption of raw goat milk, *J. Virol. Methods*, 163, 481, 2010.

51. Molnar, E. and Fornosi, F., Accidental laboratory infection with the Czechoslovakian strain of tick encephalitis, *Orv. Hetil.*, 93, 1032, 1952.

52. Bodemann, H.H. et al., Tick-borne encephalitis (FSME) as laboratory infection, *Med. Welt.*, 28, 1779, 1977.

53. Avšič-Županc, T. et al., Laboratory acquired tick-borne meningoencephalitis: Characterization of virus strains, *Clin. Diagn. Virol.*, 4, 51, 1995.

54. Jaenson, T. et al., Geographical distribution, host association, and vector roles of ticks (Acari: Ixodidae, Argasidae) in Sweden, *J. Med. Entomol.*, 31, 240, 1994.

55. Golovljova, I. et al., Characterization of tick-borne encephalitis virus from Estonia, *J. Med. Virol.*, 74, 580, 2004.

56. Haglund, M. et al., Characterisation of human tick-borne encephalitis virus from Sweden, *J. Med. Virol.*, 71, 610, 2003.

57. Bormane, A. et al., Vectors of tick-borne diseases and epidemiological situation in Latvia in 1993–2002, *Int. J. Med. Microbiol.*, 293, S36, 2004.

58. Charrel, R.N. et al., Tick-borne virus diseases of human interest in Europe, *Clin. Microbiol. Infect.*, 10, 1040, 2004.

59. Gaumann, R. et al., High-throughput procedure for tick surveys of tick-borne encephalitis virus and its application in a national surveillance study in Switzerland, *Appl. Environ. Microbiol.*, 76, 4241, 2010.

60. Jaaskelainen, A.E. et al., Tick-borne encephalitis virus in ticks in Finland, Russian Karelia and Buryatia, *J. Gen. Virol.*, 91, 2706, 2010.

61. Beugnet, F. and Marie, J.L., Emerging arthropod-borne diseases of companion animals in Europe, *Vet. Parasitol.*, 163, 298, 2009.

62. Klaus, C. et al., Goats and sheep as sentinels for tick-borne encephalitis (TBE) virus—Epidemiological studies in areas endemic and non-endemic for TBE virus in Germany, *Ticks Tick Borne Dis.*, 3, 27, 2012.

63. Donoso Mantke, O. et al., A survey on cases of tick-borne encephalitis in European countries, *Euro Surveill.*, 13, pii: 18848, 2008.

64. Donoso Mantke, O. et al., Tick-borne encephalitis in Europe, 2007 to 2009, *Euro Surveill.*, 16, pii: 19976, 2011.

65. Süss, J., Tick-borne encephalitis in Europe and beyond—The epidemiological situation as of 2007, *Euro Surveill.*, 13, pii: 18916, 2008.

66. Bröker, M. and Gniel, D., New foci of tick-borne encephalitis virus in Europe: Consequences for travellers from abroad, *Travel Med. Infect. Dis.*, 1, 181, 2003.

67. Petri, E., Gniel, D., and Zent, O., Tick-borne encephalitis (TBE) trends in epidemiology and current and future management, *Travel Med. Infect. Dis.*, 8, 233, 2010.

68. Randolph, S.E., To what extent has climate change contributed to the recent epidemiology of tick-borne diseases? *Vet. Parasitol.*, 167, 92, 2010.

69. Randolph, S.E. et al., Human activities predominate in determining changing incidence of tick-borne encephalitis in Europe, *Euro Surveill.*, 15, 24, 2010.

70. Haglund, M. and Gunther, G., Tick-borne encephalitis—Pathogenesis, clinical course and long-term follow-up, *Vaccine*, 21, S11, 2003.

71. Malkova, D. and Frankova, V., The lymphatic system in the development of experimental tick-borne encephalitis in mice, *Acta Virol.*, 3, 210, 1959.

72. Růžek, D. et al., Morphological changes in human neural cells following tick-borne encephalitis virus infection, *J. Gen. Virol.*, 9, 1649, 2009.

73. Zlotnik, G.D.P. and Carter, G.B., Experimental infection of monkeys with viruses of the tick-borne encephalitis complex: Degenerative cerebellar lesions following inapparent forms of the disease or recovery from clinical encephalitis, *Br. J. Exp. Pathol.*, 57, 200, 1976.

74. Andzhaparidze, O.G. et al., Morphological characteristics of the infection of animals with tick-borne encephalitis virus persisting for a long time in cell cultures, *Acta Virol.*, 22, 218, 1978.

75. Toporkova, M.G. et al., Serum levels of interleukin 6 in recently hospitalized tick-borne encephalitis patients correlate with age, but not with disease outcome, *Clin. Exp. Immunol.*, 152, 517, 2008.

76. Gelpi, E. et al., Visualization of Central European tick-borne encephalitis infection in fatal human cases, *J. Neuropathol. Exp. Neurol.*, 64, 506, 2005.

77. Duniewicz, M., Clinical picture of Central European tick-borne encephalitis, *MMW Munch. Med. Wochenschr.*, 118, 1609, 1976.

78. Gelpi, E. et al., Inflammatory response in human tick-borne encephalitis: Analysis of post-mortem brain tissue, *J. Neurovirol.*, 12, 322, 2006.

79. Zajkowska, J. et al., Evaluation of CXCL10, CXCL11, CXCL12 and CXCL13 chemokines in serum and cerebrospinal fluid in patients with tick-borne encephalitis (TBE), *Adv. Med. Sci.*, 56, 311, 2011.

80. Bardina, S.V. and Lim, J.K., The role of chemokines in the pathogenesis of neurotropic flaviviruses, *Immunol. Res.*, 54, 121, 2012.

81. Kindberg, E. et al., A deletion in the chemokine receptor 5 (CCR5) gene is associated with tick-borne encephalitis, *J. Infect. Dis.*, 197, 266, 2008.

82. Dumpis, U., Crook, D., and Oksi, J., Tick-borne encephalitis, *Clin. Infect. Dis.*, 28, 882, 1999.

83. Lotric-Furlan, S. and Strle, F., Thrombocytopenia, leukopenia and abnormal liver function tests in the initial phase of tick-borne encephalitis, *Zentralbl. Bakteriol.*, 282, 275, 1995.

84. Marjelund, S. et al., Magnetic resonance imaging findings and outcome in severe tick-borne encephalitis. Report of four cases and review of the literature, *Acta Radiol.*, 45, 88, 2004.

85. Kaiser, R., Tick-borne encephalitis, *Infect. Dis. Clin. North Am.*, 22, 561, 2008.

86. Fauser, S., Stich, O., and Rauer, S., Unusual case of tick-borne encephalitis with isolated myeloradiculitis, *J. Neurol. Neurosurg. Psychiatry*, 78, 909, 2007.

87. Achazi, K. et al., Rodents as sentinels for the prevalence of tick-borne encephalitis virus, *Vector Borne Zoonotic Dis.*, 11, 641, 2011.

88. Donoso Mantke, O., Achazi, K., and Niedrig, M., Serological versus PCR methods for the detection of tick-borne encephalitis virus infections in humans, *Future Virol.*, 2, 565, 2007.

89. Holzmann, H., Diagnosis of tick-borne encephalitis, *Vaccine*, 21, S36, 2003.

90. Niedrig, M. et al., Comparison of six different commercial IgG-ELISA kits for the detection of TBEV-antibodies, *J. Clin. Virol.*, 20, 179, 2001.

91. Achazi, K. et al., Detection and differentiation of tick-borne encephalitis virus subtypes by a reverse transcription quantitative real-time PCR and pyrosequencing, *J. Virol. Methods*, 171, 34, 2011.

92. Saksida, A. et al., The importance of tick-borne encephalitis virus RNA detection for early differential diagnosis of tick-borne encephalitis, *J. Clin. Virol.*, 33, 331, 2005.

93. Růžek, D. et al., Rapid subtyping of tick-borne encephalitis virus isolates using multiplex RT-PCR, *J. Virol. Methods*, 144, 133, 2007.

94. Schwaiger, M. and Cassinotti, P., Development of a quantitative real-time RT-PCR assay with internal control for the laboratory detection of tick-borne encephalitis virus (TBEV) RNA, *J. Clin. Virol.*, 27, 136, 2003.

95. Donoso Mantke, O. et al., Quality control assessment for the PCR diagnosis of tick-borne encephalitis virus infections, *J. Clin. Virol.*, 38, 73, 2007.

96. Sonnenberg, K. et al., State-of-the-art serological techniques for detection of antibodies against tick-borne encephalitis virus, *IJMM*, 293(Suppl. 37), 148, 2004.

97. Gunther, G. and Haglund, M., Tick-borne encephalopathies: Epidemiology, diagnosis, treatment and prevention, *CNS Drugs*, 19, 1009, 2005.

98. Ludolfs, D., Reinholz, M., and Schmitz, H., Highly specific detection of antibodies to tick-borne encephalitis (TBE) virus in humans using a domain III antigen and a sensitive immune complex (IC) ELISA, *J. Clin. Virol.*, 45, 125, 2009.

99. Gunther, G. et al., Tick-borne encephalitis in Sweden in relation to aseptic meningo-encephalitis of other etiology: A prospective study of clinical course and outcome, *J. Neurol.*, 244, 230, 1997.

100. Niedrig, M. et al., Quality control assessment for the serological diagnosis of tick-borne encephalitis virus infections, *J. Clin. Virol.*, 38, 260, 2007.

101. Vene, S. et al., A rapid fluorescent focus inhibition test for detection of neutralizing antibodies to tick-borne encephalitis virus, *J. Virol. Methods*, 73, 71, 1998.

102. Danielova, V., Holubova, J., and Daniel, M., Tick-borne encephalitis virus prevalence in *Ixodes ricinus* ticks collected in high risk habitats of the south-Bohemian region of the Czech Republic, *Exp. Appl. Acarol.*, 26, 145, 2002.

103. Waldvogel, K. et al., Severe tick-borne encephalitis following passive immunization, *Eur. J. Pediatr.*, 155, 775, 1996.

104. Jones, N. et al., Tick-borne encephalitis in a 17-day-old newborn resulting in severe neurologic impairment, *Pediatr. Infect. Dis. J.*, 26, 185, 2007.

105. Bröker, M. and Kollaritsch, H., After a tick bite in a tick-borne encephalitis virus endemic area: Current positions about post-exposure treatment, *Vaccine*, 26, 863, 2008.

106. Onischenko, G.G., Fedorov, Y.M., and Pakskina, N.D., Organization of supervision of tick-borne encephalitis and ways of its prevention in the Russian Federation, *Vopr. Virusol.*, 52, 8, 2007.

107. Kunz, C., TBE vaccination and the Austrian experience, *Vaccine*, 21, S50, 2003.

108. Heinz, F.X. and Kunz, C., Tick-borne encephalitis and the impact of vaccination, *Arch. Virol. Suppl.*, 18, 201, 2004.

109. Heinz, F.X. et al., Field effectiveness of vaccination against tick-borne encephalitis, *Vaccine*, 25, 7559, 2007.

110. Romanenko, V.V., Esiunina, M.S., and Kiliachina, A.S., Experience in implementing the mass immunization program against tick-borne encephalitis in the Sverdlovsk Region, *Vopr. Virusol.*, 52, 22, 2007.

111. WHO Publication, Vaccines against tick-borne encephalitis: WHO position paper, *Wkly. Epidemiol. Rec.*, 86, 241, 2011.

112. Klockmann, U. et al., Protection against European isolates of tick-borne encephalitis virus after vaccination with a new tick-borne encephalitis vaccine, *Vaccine*, 9, 210, 1991.

113. Holzmann, H. et al., Molecular epidemiology of tick-borne encephalitis virus: Cross-protection between European and Far Eastern subtypes, *Vaccine*, 10, 345, 1992.

114. Leonova, G.N. and Maistrovskaya, O.S., Viremia in patients with tick-borne encephalitis and in patients with sucking ixodidae ticks, *Vopr. Virusol.*, 5, 224, 1996.

115. Orlinger, K.K. et al., A tick-borne encephalitis virus vaccine based on the European prototype strain induces broadly reactive cross-neutralizing antibodies in humans, *J. Infect. Dis.*, 203, 1556, 2011.

116. Fritz, R. et al., Quantitative comparison of the cross-protection induced by tick-borne encephalitis virus vaccines based on European and Far Eastern virus subtypes, *Vaccine*, 30, 1165, 2012.

117. Rendi-Wagner, P. et al., Immunogenicity and safety of a booster vaccination against tick-borne encephalitis more than 3 years following the last immunisation, *Vaccine*, 23, 427, 2004.

118. Bröker, M. and Schöndorf, I., Are tick-borne encephalitis vaccines interchangeable?, *Expert Rev. Vaccines*, 5, 461, 2006.

119. Girgsdies, O.E. and Rosenkranz, G., Tick-borne encephalitis: Development of a paediatric vaccine. A controlled, randomized, double-blind and multicentre study, *Vaccine*, 14, 1421, 1996.

120. Hainz, U. et al., Insufficient protection for healthy elderly adults by tetanus and TBE vaccines, *Vaccine*, 23, 3232, 2005.

121. Stiasny, K. et al., Age affects quantity but not quality of antibody responses after vaccination with inactivated flavivirus vaccine against tick-borne encephalitis, *PLoS ONE*, 7, e34145, 2012.

122. Bundesamt für Gesundheit, Recommendations on TBE vaccination in Switzerland from the Federal Office of Public Health, Section Vaccinations, *Bull. BAG*, 13, 225, 2006.

123. Bender, A. et al., Two severe cases of tick-borne encephalitis despite complete active vaccination—The significance of neutralizing antibodies, *J. Neurol.*, 251, 353, 2004.

124. Kleiter, I. et al., Delayed humoral immunity in a patient with severe tick-borne encephalitis after complete active vaccination, *Infection*, 35, 26, 2007.

125. Andersson, C.R. et al., Vaccine failures after active immunization against tick-borne encephalitis, *Vaccine*, 28, 2827, 2010.

126. Stefanoff, P. et al., Reliable surveillance of tick-borne encephalitis in European countries is necessary to improve the quality of vaccine recommendations, *Vaccine*, 29, 1283, 2011.

127. Linden, A. et al., Tick-borne encephalitis virus antibodies in wild cervids in Belgium, *Vet. Rec.*, 170, 108, 2012.

128. Kiffner, C. et al., Tick-borne encephalitis virus antibody prevalence in roe deer (*Capreolus capreolus*) sera, *Exp. Appl. Acarol.*, 51, 405, 2010.

129. 2012/492/EU: Commission Decision of 3 September 2012 amending Decision 2000/96/EC as regards tick-borne encephalitis and the category of vector-borne communicable diseases (notified under document C(2012) 3241).

130. Litzba, N. et al., Evaluation of serological diagnostic test systems assessing the immune response to Japanese encephalitis vaccination, *PLoS Negl. Trop. Dis.*, 4, e883, 2010.

131. Niedrig, M. et al., Evaluation of an indirect immunofluorescence assay for detection of immunoglobulin M (IgM) and IgG antibodies against yellow fever virus, *Clin. Vaccine Immunol.*, 15, 177, 2008.

132. Kohler, S. et al., The early cellular signatures of protective immunity induced by live viral vaccination, *Eur. J. Immunol.*, 42, 2363, 2012.

21 Variola Viruses

Isao Arita, Donald Francis, and Masayuki Saijo

CONTENTS

Recently, on a bus in the western area of Japan, a young mother with a baby boarded the bus and was seated next to one of the authors (Isao Arita).

"Your baby is pretty! How old is your baby?" Arita asked.

"Thank you. He is now 6 months old." She replied smiling.

"Perhaps your baby just finished his routine vaccinations?" Arita asked.

"Yes, diphtheria, whooping cough, and tetanus." Then, after a pause, she added, "also polio."

To see what her response would be, Arita asked, "No smallpox vaccination?"

"What do you mean smallpox? What is that?" She obviously did not know what smallpox was. She was extremely embarrassed as if she had missed an important issue for her baby. Arita then explained the situation and she understood. Japan had removed the recommendation for routine smallpox vaccination in 1976—probably a decade before the mother was born. Her reaction to the question is likely similar to any mother anywhere in the world these days.

21.1 INTRODUCTION

The last naturally transmitted smallpox occurred in Somalia in October 1977. The last laboratory-associated smallpox occurred in the United Kingdom in August 1978. Considering the large number of laboratories with live virus at that time, the latter incident was rare but it highlighted the risk of laboratory infections. Thus, 1979 was the first year when no human smallpox was reported worldwide.

The total interruption of smallpox transmission was the result of a large global effort under the auspices of the World Health Organization (WHO). All nations worked together despite differences of government structure, race, religions, cultures, and relative wealth. In May 1980, after a 2-year certification period that assured the absence of the disease, the WHO assembly in Geneva declared smallpox eradicated in its Resolution WHA 33.3 and urged, with Resolution WHA 33.4, the immediate discontinuation of routine smallpox vaccination and the use of the international certificates of vaccination.

Since that time, more than 30 years have elapsed without a single report of smallpox in the world. And the world has clearly changed. With the cessation of routine smallpox vaccination beginning in 1980, the proportion of smallpox-susceptible people has been rapidly increasing worldwide. By 2080, virtually no one younger than 100 years of age will have been vaccinated against smallpox. Moreover, in many nations, especially those with a high proportion of youth (like many African nations), the proportion susceptible is even much greater.

Now, consider what would happen if smallpox were introduced into such a population today?

The projected outcome of a single person-to-person exposure is straightforward and easy to estimate. What is less straightforward is estimating the risk of a purposeful or accidental introduction from a clandestine production lab. Recent assessments conclude that the threat of bioterrorism or unintentional release of smallpox virus has not been decreasing. Rather, considering the 9/11 incident, many feel that the risk has been increasing since smallpox eradication 30 years ago. With declining global economies, highlighted by Stiglitz in his 2002 book *Globalization and Its Discontent*, media reports documenting increasing dissatisfaction of younger populations in certain nations, and occasional military conflicts in countries like Afghanistan and elsewhere, have led us to be highly concerned about the world's insecurity [1]. We cannot deny that bioterrorism is a threat. Nor can we deny the horrors that would occur if variola virus were used as a bioterrorist weapon.

This chapter will start with a review of the virology of variola virus and the past epidemics it caused. This will be followed by a description of the program of smallpox eradication—the first successful human effort in history of public health to actually eradicate a human disease. Next, there will be a discussion on how to deal with the threat of a purposeful reintroduction of smallpox using the experience and lessons learned from the previous smallpox control and eradication efforts. Last, it will stress the indispensable role of the WHO for all phases of planning and implementation of an effective response. A smallpox bioterrorist introduction will not be a single-country issue. It will be a global issue. Thus, WHO will have to continue to be front and center on this issue.

21.2 CLASSIFICATION AND MORPHOLOGY OF SMALLPOX VIRUS

Smallpox virus (variola virus) causes a systematic infection in man. It belongs to the genus *Orthopox* that includes vaccinia, cowpox, monkeypox, camelpox, and others. It is a DNA virus having the dimensions of $300 \times 250 \times 200$ nm. It is brick shaped and easy to identify being morphologically different from other viruses causing exanthema in man such as measles and varicella. Infection by the most common smallpox virus (known as variola major) carried a 30%–50% case fatality, whereas a group of smallpox viruses known as variola minor or alastrim carried a fatality rate closer to 1%. In addition, there was also variola intermediate carrying a case fatality between minor and major. Although the genome of the variola major is known and published [2], the genetic understanding of the reasons for the different clinical expression is not well understood. Clinically, smallpox is generally very distinct and easy to diagnose. But there are severe forms of smallpox, termed flat smallpox or hemorrhagic smallpox, that are usually rapidly fatal and difficult to diagnose. These might be misdiagnosed as Stevens–Johnson syndrome or erythema multiforme [3]. The clinical resemblance to disease caused by monkeypox is discussed in the succeeding text.

21.3 HISTORY OF SMALLPOX

Humans were the only reservoir of smallpox virus. Thus, it has had to survive on this planet through infecting one human after another for centuries. There is some evidence that Ramses V probably died of smallpox in 1175 BC in Egypt.

Smallpox virus infection expanded as human populations expanded. Around 5000–6000 years ago, humans started to live together in small villages of perhaps 500 or more inhabitants. Before that time, humans survived by hunting and gathering where it was not practical to live together in a large numbers. But later, when the first agriculture techniques were developed in Africa, villages emerged. This "urbanization" facilitated smallpox virus transmission with expansion of the virus's sole natural reservoir. As humans moved, they carried smallpox virus down the Nile to the North African coast, then to Asia minor and the Eurasian continent. Thus, smallpox can be traced back probably 4000–5000 years [4].

From that period up to now, there have been several studies on the history of smallpox (Donald Hopkins' 1983 book, University of Chicago Press, *Princes and Peasants: Smallpox in History*, Ian and Jennifer Glynn's 2004 book, Profile Books, *The Life and Death of Smallpox*). Epidemics occurred in France and Italy in 570 AD when it was first named as variola. In the eighth century, there is evidence that smallpox moved from East Asia via Korea to Japan. Finally, it moved from Europe to the Americas in the sixteenth century. During that time, Spanish adventurers used smallpox-contaminated materials, such as blankets, as bioweapons that caused smallpox epidemics among the native Americans. These epidemics, plus other naturally occurring imported diseases, clearly contributed to the fall of Aztec and Incan empires. At that time, Spaniards were immune to smallpox, almost all of them having been infected with endemic smallpox at home. In contrast, the indigenous people in the New World had never experienced the disease before and were totally susceptible to its wrath. From that time forward, the disease affected essentially all people throughout the world until it was eliminated late in the twentieth century.

Before the development of smallpox vaccination by Edward Jenner in the late eighteenth century, almost everyone was infected by smallpox at some time in their life. People had to accept the hazard of smallpox infection—namely, 30%–40% of family contacts would develop smallpox, and if the virus was variola major, 40%–50% of those would die [5]. For smallpox, prevention by vaccination was the only effective tool. Its effect on the world was striking. Indeed, when nations in Europe introduced smallpox vaccination, their life expectancy was prolonged 10–20 years [6]. Only now are therapeutic antiviral agents showing some promise as discussed in the succeeding text.

Between the early nineteenth century and the mid-twentieth century, public health measures had gradually been developed to effectively diagnose and prevent smallpox. These included vaccine standardization and its quality control (virus titer, purity, and stability), laboratory diagnosis of smallpox virus, and measures such as obligatory reporting of cases, isolation facilities, and quarantine measures including international certificate of smallpox vaccination for international travelers. These measures were effective and smallpox transmission became increasingly circumscribed to those nations where vaccination programs and other preventive measures were not effectively implemented. In 1967, when WHO's intensified smallpox eradication program was initiated, the disease was endemic only in less developed nations, mainly in Africa and South Asia. At that time, there were 41 countries reporting smallpox including those endemic countries as well as those infected through importations. A decade later, no country had smallpox.

Since then, the global population returned to an era without natural exposure. What then does the future hold? This is the issue discussed in this chapter.

21.4 PATHOLOGY, CLINICAL FEATURES, AND EPIDEMIOLOGY OF SMALLPOX

The acutely infected human releases smallpox virus from upper respiratory droplets, and after the rash has matured, additional virus is released from the patient's skin lesions. The infection spreads most commonly to close contacts within distance of 2 m.

After infection, an incubation period of 7–17 days (most frequently 12–14 days) passes before clinical signs and symptoms develop. Then the prodromal symptoms of headache, fever, prostration, and backache appear. Finally, skin eruptions begin. The first is erythema, followed by papules, vesicle, and, finally, pustules—each stage lasting for about 2 days. At any given time, the rash, anywhere on the body, is typically at the same development stage. The rash distribution is unique for smallpox being centripetal, predominating on the face and extremities—being more dense on the extensor side of arms and legs (Figure 21.1).

After a week, the pustular eruption moves to the crust stage. This lasts for another week and is followed by the desquamation stage for the final week. Thus, within 3 weeks, all stages of the rash have passed leaving the whitish spots and discoloration. Within a few additional weeks, the whitish spots darken. With time, the lesions become scarred, known as pockmarks, leaving 80% of smallpox survivors with clear evidence that the area has been smallpox infected—very useful for surveillance.

With the exception of fatality rate, variola major and variola minor (or intermediate) follow similar courses. In fact, it is of no value for smallpox control to differentiate major from minor. As an example, the world's last case of naturally transmitted smallpox was Ali Maomalin in Somalia in 1977. He was recorded later as a variola minor or attenuated variola major, but during the operation, WHO and all the program staff paid little attention to this: "It was a case of smallpox. Period!"

21.5 IDENTIFICATION, DIAGNOSIS, AND SURVEILLANCE

As seen in Figure 21.1, the clinical manifestations of smallpox are distinct from any other exanthematous disease including varicella, measles, or allergic exanthem. As a result, the disease was well known by the public in any endemic area. Indeed, in all endemic areas, local names were common: boshonto in Bangladesh, furka in Somalia, and heaven's flower (English translation) in China. Only varicella (chicken pox) was sometimes misdiagnosed as smallpox even though the varicella exanthema has a mixture of development stages (size, papule, vesicle, pustule, and scab) at any given time. Furthermore, the distribution of the chicken pox rash is generally concentrated on the central body, while for smallpox, the rash is peripheral being more concentrated on the face and extensor sides of arms and legs. Moreover, the mortality of varicella is minuscule compared with smallpox. The unique challenge of hemorrhagic smallpox was mentioned in the previous section.

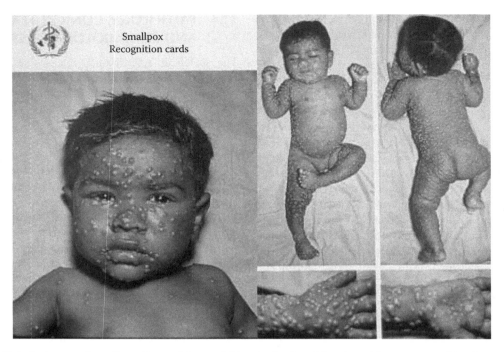

FIGURE 21.1   Clinical pictures of smallpox. Recognition card (showing back and front) used by teams searching for cases.

Monkeypox is a rare pox virus infection of humans occurring in tropical rain forest areas of Central and West Africa where the natural reservoir of monkeypox virus, rodents or squirrels, exists. From these animals, occasionally monkeys or humans become infected. Human monkeypox resembles smallpox in terms of clinical appearance of rash, but, uniquely, there is also distinct swelling caused by lymphadenitis, especially of the neck and inguinal lymph nodes. In the United States in 2003, there was an outbreak of human monkeypox. In this instance, imported giant rats from Africa infected prairie dogs that were sold as pets. These prairie dogs infected humans. Laboratory testing, using recently developed polymerase chain reaction (PCR) method, was helpful to rapidly identify the causative agent as monkeypox virus [7] (also see Chapter 16).

Techniques for the laboratory diagnosis of smallpox have evolved over time. Up to the period of worldwide eradication, the most common method used for laboratory diagnosis involved observing the morphology of pocks grown on chorioallantoic membranes (CAMs) of chick embryos. Clinical specimens were usually taken from skin lesions. Electron microscopic examination was helpful in emergency situations. But this CAM test needed time: for preparation of eggs, virus inoculation, growth, and neutralization test with smallpox antibody. It required several days. Moreover, it was not usually available in countries with limited health resources—most notably in Africa, Asia, and South America.

During the intensified WHO smallpox eradication program, control measures, as well as smallpox reporting, were exclusively based on the clinical observations and epidemiological findings. Laboratory confirmation was only initiated when the program was reaching the final phases in 1975. The only exception took place in nonendemic countries that had

highly developed health structures. In those areas, a report of suspected smallpox was a national emergency requiring definitive laboratory confirmation.

In recent years, PCR technology has become common to confirm viral infections. The method is in common use in industrialized countries as well as in a few nations in Africa and South Asia (personal communication, Inger K. Damon). As discussed in the last section, if suspect smallpox is found, specimens should be sent to WHO collaborating centers (Atlanta, United States, and/or VECTOR, Novosibirsk, Russia), with simultaneous notification to the local health authority as well as WHO. It is essential that the utmost precautions be followed to handle specimens in case of a suspected smallpox case. During all steps, from collection of specimens from patients to shipping to collaborating centers, extreme cautions will be necessary to prevent any transmission to others.

21.6   VACCINIA VIRUS AND VACCINATION

The control or prevention of smallpox was not possible before smallpox vaccine was introduced by Edward Jenner in 1798. Around that time, virology, bacteriology, and epidemiology of infectious diseases, including smallpox, were in their infancy. When and how cowpox virus was replaced with vaccinia virus is not clear. The first Jenner vaccine was produced using cowpox virus. The change of vaccine virus strain as such took place sometime during long history of smallpox vaccine production—between Jenner and the early twentieth century.

The principal difficulty in using smallpox vaccine has been its adverse side effects. Vaccine complications include vaccinal encephalitis, eczema vaccinatum, vaccinia necrosum, generalized vaccinia, and accidental vaccinia inoculations. In a 1968 study in the United States, it was estimated that nine deaths (from encephalitis, eczema vaccinatum, and vaccinia necrosum) would occur per one million vaccinations [8]. Importantly, the rate of serious complications varies according to both the strain of vaccine virus used and age of vaccinees. Svetlana Marennikova at the Research Institute of Viral Preparation, a WHO collaborating center, then in the USSR, studied the human pathogenicity of different vaccine production strains. She reported that the strain-specific pathogenicity varied significantly in terms of certain laboratory markers. With this information, WHO, during the smallpox eradication program, advised national vaccine producers regarding what strains (Lister strain or New York Board of Health) were appropriate for local vaccine production.

Since that time, there has been considerable progress in developing safer vaccine strains. Up to 1980, when routine vaccination ceased, more than 30 vaccinia production strains had been isolated in more than 40 production laboratories. At that time, vaccine production used skin inoculation of live calves where a certain level of bacterial contamination was allowed. Subsequently, modern tissue culture systems were developed. Moreover, the development of attenuated vaccinia strains was also a main priority for research. These included the CVI-78 and the River's strains in the early twentieth century. Other efforts focused on improving vaccine shelf life and finding better inoculation methods (discussed in the next section).

21.7 TREATMENT AND PREVENTION

21.7.1 TREATMENT

Recently, with help from the WHO Variola Research Committee, studies of therapeutic agents have progressed. These studies include some promising agents such as ST246 and cidofovir (CCMX001). Such studies have included evaluations using reliable animal models including the nonhuman primate/monkeypox virus model, which is the best available model presently. Final recommendations will have to include the best available practical method of administration, side effects, storage conditions, and price so that its use would be feasible in countries having limited resources [2].

21.7.2 PREVENTION

The eradication of smallpox was possible with the use of the key preventive tool—vaccination. Quarantine, with isolation and restriction of movement of patients, was also important, but without the vaccine, it would not have been possible to reach the ambitious target of complete eradication. In the following section, we will describe how the effective immunization efforts succeeded in eradication of smallpox. This description has both historical interest and current importance—especially for a possible return of smallpox in the future.

21.8 WHO INTENSIFIED SMALLPOX ERADICATION PROGRAM 1967–1980

The program had its preparatory phase from 1958 to 1966, but due to lack of financial support as well as less-than-optimal strategy, the program was failing. Not until 1967 when the intensified global program was initiated did the program receive a regular WHO budget and international financial cooperation. This consumed about $300 million over the next 13 years. The program operation was simple and practical with the small number of medical officers at the smallpox eradication unit, WHO headquarters, and four WHO regional offices and at national level assignments in endemic states (Figure 21.2).

The operation was simple consisting of a combination of national and WHO staff members focused on smallpox-infected districts. Free of bureaucracy, innovative approaches were applied locally. The results were dramatic. There was a rapid reduction of the number of smallpox cases. In 1967, there were 131,892 reported cases that declined to 3,234 in 1977 and, finally, endemic smallpox was eliminated in 1977 (except for the 2 laboratory-associated cases in 1978). Experience showed that there was extremely weak case reporting during the early phase of the program with estimates that only about 1% of all cases being reported in many endemic nations. In the end, the WHO certification committee verified that the zero cases having been reported in December 1979 was, indeed, correct. Why was this program so successful?

21.8.1 DETECTING AND REPORTING ALL CASES

The clinical manifestations of smallpox are clinically distinct. Surveillance of smallpox was stimulated by the distribution of a few million recognition cards (Figure 21.1) to local health services in nations undertaking eradication. Also, rewards for reporting smallpox were announced at national and later global levels. This encouraged the lay public to report hidden smallpox foci. During the last phase of the program, Arita recalls that the reports were coming from major international airports such as New York or Paris as well as from very remote villages in Guinea, West Africa, and Sumatra, South Asia.

21.8.2 ENSURING VACCINE QUALITY AND ADMINISTRATION

WHO requested all the smallpox vaccine producers, more than 30 throughout the world, to submit their vaccines for testing by WHO collaborating centers in Bilthoven, the Netherlands, and Toronto, Canada. Surprisingly, in 1967, only one-third of the producers' samples met with the WHO standard. Most failed potency and/or heat stability tests. Over the ensuing 3 years, the situation improved with full involvement of the WHO collaborating centers including the provision of a special production manual. Site training was set up and quality testing continued as long as it was required. Only the

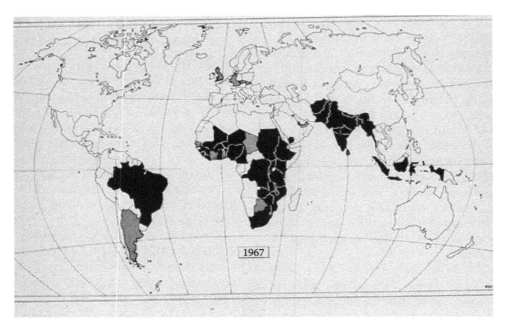

FIGURE 21.2 Map of smallpox-endemic nations in 1967.

freeze-dried vaccine was used. A major delivery advantage was provided by the use of bifurcated needles. These needles assured a good take with 15 punctures using a markedly reduced vaccine volume—1/5 of the previous method. The needles were patented by Wyeth Laboratory, United States, which donated its technology when they were produced for WHO's eradication program.

21.8.3 SURVEILLANCE AND CONTAINMENT STRATEGY

The US bilateral collaborating program for measles control and smallpox eradication targeted 19 countries in West and Central Africa between 1966 and 1970. That program greatly contributed to designing the strategy, termed "search and containment" that led to the ultimate interruption of smallpox transmission. Initially, Rafe Henderson (United States) and M. Yakpe (Benin) investigated a smallpox outbreak in a village in Dahomey and found that smallpox spread slowly, needing very close contact and resulting in spread in a very narrow geographical locale [3]. Then William Foege was working in Northern Nigeria when supplies of vaccine were very limited. As a result of these limits, he stopped mass vaccination and, instead, concentrated the vaccination only in affected villages. This strategy led to a surprisingly rapid elimination of smallpox [10]. The US bilateral program for smallpox eradication shifted to this strategy instead of the previous mass vaccination campaign approach. The result was zero case by 1970 in West and Central Africa.

21.8.4 AUTUMN CAMPAIGN IN INDIA

The search and containment strategy was, in fact, employed by the global WHO smallpox eradication program in all smallpox-endemic WHO regions. Notably, in September 1973, the eradication program in India developed a special

campaign, following WHO advice, called the autumn campaign. In this strategy, all health centers conducted a special smallpox campaign for 1 week every month. The health center teams visited every village to search for cases of smallpox. If cases were found, 100% of the people living in the surrounding 50 households were vaccinated. The results were dramatic: 88,114 cases in 1973, 188,003 cases in 1974, and 1,428 cases in 1975. After May 1975, no smallpox was discovered in India. All Southern Asia nations, including Pakistan, Indonesia, Bangladesh, and others, succeeded in stopping smallpox transmission. The last case of smallpox occurred in Bangladesh in October 1975. Surveillance and containment was a principal strategy but the exact methods of delivery differed among the nations.

21.8.5 LAST BATTLE IN THE HORN OF AFRICA

In the autumn of 1976, Ethiopia was the only nation reporting smallpox in the world. The assumption was that Ethiopia would be the world's last smallpox reporting nation. But WHO's wishful thinking was soon shattered. At that time, smallpox transmission was occurring in the northern areas of the Ogaden desert where there had been political conflict between Ethiopia and Somalia. Then, in September 1976, WHO found a smallpox case in Mogadishu, the capital of Somalia. Retrospectively, there were hidden smallpox foci in the Ogaden desert shared by Ethiopia, Kenya, and Somalia. But the major foci were mainly in Somalia. The lesson here was clear. Political conflict/war is the major detrimental factor threatening any eradication effort. Eradication requires an all-out effort and such conflicts severely hamper the ability to deliver such an effort. Now, let us move to the period September 1976–October 1977 when the world's last smallpox case was discovered in Somalia.

21.8.6 FINAL SOMALIA CAMPAIGN

The situation in the Horn of Africa was complex. Nomads from Ethiopia, Somalia, and Kenya were traveling in the Ogaden desert. Experience showed that smallpox was maintained over 6 months among these traveling nomads. Once discovered, WHO emergency actions were considerable. They consisted of (1) mobilization of an emergency fund of $470,000 constructed both from voluntary contributions and from the UN disaster relief operation (UNDRO); (2) swift transfer of WHO staff from India to Somalia to work in Ogaden; (3) formation of a special task force (national and WHO); and (4) air delivery of 16 land rovers from Copenhagen, Denmark, to Mogadishu. There were a few exciting episodes. One involved WHO staff in Ogaden who, despite the availability of car and helicopter, chose to walk because of fear of missing cases if they use such transport. As a result, three WHO staff members were captured by Somali guerrillas. But, typical of smallpox workers, they persuaded the guerrillas to be vaccinated. Fortunately, as the guerrillas realized the importance of smallpox eradication program, they released their captives. In total for Somalia in 1977, 3229 cases occurred. The last case there, in fact the last case of smallpox in the world, occurred on October 26, at Merca, a small seaside Somali town.

21.8.7 WAR AND SMALLPOX ERADICATION

Smallpox eradication efforts developed during the peak of the cold war period. Fortunately, there were only two major military conflicts during that time—namely, the Indo/Pakistan war in 1970 and the Somali/Ethiopian war in 1977. The major threat to smallpox eradication occurred in Bangladesh (then East Pakistan). That country, with the application of new surveillance and containment strategy, was about to reach the zero case point, when multiple importations came via smallpox-infected refugees from India. Massive smallpox epidemics followed. It wasn't until 1975 that the last case was found. It was the last case in Asia.

21.8.8 GLOBAL CERTIFICATION OF SMALLPOX ERADICATION

How does one prove that smallpox is really gone? WHO established the WHO Commission for Certification of Smallpox Eradication that decided to certify eradication only after 2 years had passed without a reported case. The methods included house-to-house visits to search cases, laboratory confirmation of suspected cases, pockmark surveys of young children, independent assessment of national surveillance records, and announcement of cash rewards for discovering any new smallpox case. The commission visited the assigned geographical regions and concluded in 1979 that the global smallpox eradication was, indeed, achieved. Various additional recommendations were also made. In the end, one recommendation, made in May 1980 by the World Health Assembly (WHA), called for the cessation of routine smallpox vaccination. It called for all member states to implement the recommendation.

21.8.9 MYSTERY (OR CONUNDRUM) OF VARIOLA VIRUS ISOLATES

The commission's report indicated that "monkeypox virus does not reconstitute a threat to the performance of smallpox eradication." Indeed, it recognized that person-to-person transmission of monkeypox virus is difficult [11]. More confusing were reports that variola virus had in the past caused some mysterious episodes of disease. During intensive passage work in a few laboratories, including a WHO collaborating laboratory, there were reports that variola-like pock lesions on CAM appeared either during monkeypox passage work or in healthy monkey kidney cell cultures. These were termed "white pox virus" since the specimen records lacked any smallpox source.

The white pox virus isolates are believed to have come from more than 10 laboratories. Fortunately, follow-up studies indicated that the mutation of monkeypox virus to variola was not genetically possible [12]. In the end, it was strongly suspected that white pox virus were the result of cross contamination as these laboratories were the very labs where variola work was frequently conducted.

21.8.10 SEARCH FOR VARIOLA VIRUS STOCKS

During the final phase of eradication, there was a special WHO program to identify the laboratories that possessed variola virus stocks. Using available records, laboratories that published variola virus studies since 1960 were either visited or referred to by the independent teams—either national or WHO. Unnecessary virus stocks were either destroyed or transferred to one of the two WHO collaborating centers in the United States and Russia. For variolation materials, some proved to have no viability and others were destroyed.

21.8.11 SUMMING UP OF PREVENTION OF VARIOLA

Eradication is the best outcome to deal with any infectious disease. Yet, its success requires a combination of multiple factors. Looking back, smallpox eradication was very complex. It had to deal with not only public health matters but also social, economic, and political matters. The lead group in WHO consisted of a small number of workers who worked in a system that was remarkably flexible and free from the usual bureaucracy. This allowed for the rapid and practical decisions that proved essential for success. In 1997, the Dahlem Workshop was held in Berlin to discuss the eradication of infectious diseases with a focus on future programs. In this occasion, Arita indicated that the eradication program was "a concentrated global effort which can function best if such a special program is developed in WHO or elsewhere in the UN." Following this proposal, William Foege advised that future eradication work would be best carried out "by an ad hoc organization" not a grandiose organization lacking flexibility [13]. Perhaps, through good administration or good fortune (or both), the smallpox eradication team accomplished the task within the WHO system. In this case, WHO was successful and, indeed, essential for the project's eventual success.

For our current issue, preparedness to control smallpox bioterrorism, we may require a similar approach.

21.9 PREPAREDNESS FOR SMALLPOX BIOTERRORISM

Recently, the world has been changing. Molecular biology has made great progress over the last three decades. World peace is becoming frazzled due to the world's economic decline. Could these and other issues drive some to use smallpox virus as a weapon? Let us start this discussion of whether the variola virus stocks in two WHO collaborating laboratories should be destroyed and later deal with the various issues of smallpox as a bioterrorist threat.

21.9.1 SHOULD WE DESTROY THE VARIOLA VIRUS STOCKS IN THE TWO LABORATORIES?

Concern has been raised about accidental release from the officially maintained stocks. Thus, the 2012 WHO assembly recommended that the stocks be destroyed on the date, which will be decided in 2015. Currently, there remains some concern regarding this decision. Some wonder if destruction is really necessary considering the current possibility of de novo synthesis of variola virus [14]. Others consider the issue of dual use of research as exemplified by the synthesis of ectromelia virus [15]. Moreover, there still remain important research questions such as the following: Why is smallpox so lethal to humans? What is the role of cytokine storm? What determines the host tropism for humans? [16]. These studies would be very important to address from the vantage of both treatment and prevention should we find ourselves confronted with reintroduction. Finally, the ultimate question is as follows: Are these two variola virus stocks the only ones that really exist on the planet? A compromised solution would be to place the virus stocks into UN facilities under the strict security with a title—"Symbol of victory over a vicious enemy, variola virus." Any newly found variola virus stocks would then be placed into the facilities [17].

Now let's move on to a discussion of preparedness for smallpox bioterrorism.

21.9.2 BRIEF HISTORY OF BIOTERRORISM

In the WHO's 2007 annual report, experts highlighted the serious risk of smallpox bioterrorism "in the absence of global capacity to contain an outbreak rapidly, smallpox might reestablish endemicity, undoing one of public health greatest achievements."

Throughout history, extremist political or religious cults have inspired humans to do despicable acts of evil. The incident committed by the Aum Shinrikyo religious cult in the middle of 1990s resulted in 11 deaths and more than 6000 injuries in the Kanto area of Japan [18]. Six years later, in 2001, another incident followed—the 9/11 terrorist attacks in the United States where the death toll reached nearly 3000. Subsequently, another biological terrorist act took place in the United States—the posting of multiple anthrax-containing envelopes in the mail. This attack, originally aimed at political groups, caused several deaths. Here, the terrorist is thought to have been a government researcher [19].

These events have led some to consider the possibility of a deliberate reintroduction of smallpox. Indeed, a few months before the 9/11 incident, the US government conducted an exercise concerning a simulated smallpox bioterrorist attack termed Dark Winter. It modeled simultaneous US smallpox epidemics involving three million cases during four generations of disease as well as exportation to other nations. The conclusion of this exercise was as follows: "Our lack of preparation is a real emergency" [20].

The problem is clear. If a smallpox epidemic were to get out of control and spread widely, it could be a huge threat to the world. In this article, we will review the threat of a smallpox bioterrorist attack and how we should prepare to prevent a possible pandemic. It is highly likely that if a smallpox bioterrorist event hits a single nation, it will lead to widespread smallpox importations. Some of these will likely be to nations having very limited health services—especially in areas in sub-Saharan Africa and South Asia. As physicians experienced in both the past smallpox program and the current polio eradication efforts, we are of the opinion that the world's antismallpox terrorist strategy must be one of a global nature.

21.9.3 MANUALS FOR BIOTERRORISM

Excellent manuals have been produced detailing how to cope with a smallpox bioterrorist event. One, a book entitled *Bioterrorism, Guidelines for Medical and Public Health Management*, was edited by DA Henderson et al., with foreword by Anthony S. Fauci, director, US NIAID. A review of this book was published in the *Journal of American Medical Association* in 2002—the year after the 9/11 incident occurred. In great technical detail, the book gives excellent advice and guidelines on how to implement and manage control strategies. These include guidance for preexposure preventive vaccination, postexposure therapy, infection control, vaccine administration, and complications—detailed on page 99 of the book.

Not wanting to duplicate what has been presented in the aforementioned reference, this section will specifically focus on the global aspects of the response strategy. Here, we will stress: how best to ascertain if smallpox has been reintroduced by terrorist, what are the emergency responses that need to be implemented, and, at the same time, what massive efforts will be required to prevent pandemic spread. Special attention will be given to the strategy required to promote effective surveillance and response in developing countries where health services will need to be strengthened as emergency operations are launched. Such strengthening will require a large international effort of the UN, the WHO, and other appropriate organizations.

21.9.4 ESTABLISHMENT OF A SPECIAL WHO ADVISORY COMMITTEE ON VARIOLA VIRUS RESEARCH

In the late 1990s, WHO set up the WHO Advisory Committee on Variola Virus Research. The committee, with responsibility for all serious pox viruses including monkeypox, made considerable progress. While they are working to develop practical tools to cope with smallpox bioterrorism, we focus here on the arena of vaccines [21].

Any smallpox vaccine to be used in the future should have both an excellent protective efficacy against variola and, at the same time, carry very low adverse health risks. Renewed vaccination using vaccines of the past would cause a great number of severe adverse effects for people living in both poor and rich areas of the world. In this respect, WHO's variola research committee has been working to further the development of both a safer vaccine and therapeutic antiviral agents. In the vaccine arena, the LC16 m8 (LC) vaccine, an attenuated smallpox vaccine, studied since the mid-1970s, is a success story. Its safety has been well documented and it can be administered using the bifurcated needle. Being produced during the final phases of smallpox eradication, there was no opportunity to test it in the face of actual smallpox outbreaks. In place of actual efficacy testing, extensive laboratory testing has been undertaken, the results of which strongly support the efficacy of the LC vaccine to prevent smallpox. The vaccine was licensed in Japan in the mid-1970s.

In addition, the IMVAMUNE vaccine is also an attenuated vaccine. It has been reported to be safe and immunogenic for individuals having medical conditions that put them at risk to receive other smallpox vaccines. This vaccine is not freeze-dried and, as we understand, is to be given in two subcutaneous doses.

Both the LC and IMVAMUNE vaccines were under discussion during 12th and 13th WHO Advisory Committee on Variola Virus Research, in 2010 and 2011 [22,23].

Although additional information will be collected in due course, at this time, the authors are of the opinion that LC should be the vaccine of choice for an emergency today. The considerable clinical data available suggest that using the LC vaccine for large vaccination programs over wide areas will carry very low complication rates. Moreover, its simple, bifurcated needle administration method, its stability and simple storage requirements, and its presumed low cost make it an attractive alternative to any other available product.

21.9.5 BIOTERRORIST THREAT

A worrisome revelation came forth in 1992. Ken Alibek, the deputy chief of the USSR agency for biological weapon production, defected to the US and reported that the former USSR developed a laboratory called Biopreparat where extensive efforts were made, beginning in 1970, to weaponize dangerous pathogens. Smallpox virus (variola major) was one of the priority pathogens, and they grew large stockpiles from which to produce virus aerosols for biological warfare attacks [24]. Although the then Russian president Boris Yeltsin declared in 1992 that his nation had discontinued the activities at Biopreparat, there is no guarantee that such an event may not or will not be occurring elsewhere.

21.9.6 KNOWN SOURCES OF VARIOLA VIRUS

As of 2012, there are only two laboratories known to possess smallpox virus. One is at the Centers for Disease Control and Prevention (CDC), Atlanta, United States. This laboratory worked as a WHO smallpox surveillance laboratory during the eradication program. The second laboratory is located at Koltsovo, Siberia, in the Russian Federation. This laboratory was once situated in Moscow, then known as the Research Institute of Viral Preparation. During the smallpox eradication program, it also served as a WHO collaborating center. Sometime in the 1980–1990 timeframe, the work was transferred to Koltsovo, Siberia.

The WHO committee on variola virus research indicated that since the genome of variola virus is now available, it could be possible to synthesize the virus in the laboratory [25].

21.9.7 SMALLPOX BIOTERRORISM: INTENTIONAL RELEASE

Although somewhat challenging, there is theoretical possibility that terrorists could obtain smallpox virus and, with considerable technical assistance, produce enough aerosol virus to conduct a deliberate release to targeted populations.

In order to make definitive response plans to counteract a deliberate release of smallpox virus, we need to predict what would actually happen epidemiologically if such a release were to occur. That is, given the susceptibility of today's populations and the known ability of smallpox virus to spread once introduced, what can we predict would happen?

In 1980, as mentioned earlier, with WHO's declaration of smallpox eradication, all nations discontinued smallpox vaccination. Thus, with today's global population of 7 billion, 3.55 billion people 30 years of age and below have never been vaccinated against smallpox and are, therefore, susceptible. Looking ahead, as the immune cohort continues to age, the proportion susceptible to smallpox will continue to increase (Figure 21.3). Although the exact proportion of unvaccinated people in different countries will vary according to both life span and birth cohort, we can predict that by 2080, essentially the entire global population will be susceptible to variola virus.

A bioterrorist-associated smallpox outbreak could either be an accidental infection of those possessing limited laboratory skills who try to grow the virus or a deliberate release of the virus into the population by those with greater laboratory abilities. A critical point with the latter would be whether the deliberate introduction would target a single location

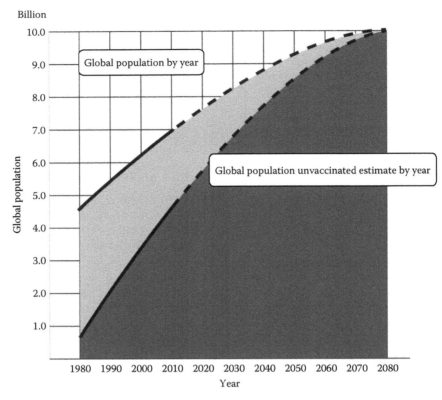

FIGURE 21.3 Proportion of unvaccinated population worldwide, 1980–2080. The number of vaccinated population worldwide will reach virtually zero in 2080 since smallpox vaccination discontinued in 1980 onward: those who were born in or before 1980 will live only in small number for the age of 100 years (From WHO 2011).

causing a single outbreak or target multiple locations causing multiple outbreaks. Regardless of where the initial outbreak takes place, in view of ever-increasing speed of global traffic, exportations would likely occur not only to adjacent countries but also to the countries on faraway continents and different WHO regions.

The ultimate result of such an introduction would clearly depend on our ability to respond quickly—both the early recognition of the outbreak and the rapid delivery of preventative measures. Unfortunately, if past history is an indicator of the future, the initial cases may well be misdiagnosed and, by the time the correct diagnosis is made, the transmission will likely be extensive—as was seen with hospital infections during the eradication program. In such a case, it would be hard to prevent the exportation of cases via today's busy and fast intercontinental traffic.

To get a "best-case" estimate of the effectiveness of smallpox control, we can look back to when smallpox was endemic and the world's physicians were fully capable of diagnosing it and when public health practitioners were fully capable of controlling it. For this purpose, we examined outbreaks detected from 1970 to 1978 during the time of WHO's smallpox eradication program. During that period, there were 25 smallpox exportations from 18 endemic countries (Table 21.1).

Such exportation occurred despite ongoing intensive surveillance and containment efforts at a time when physicians were familiar with the disease and capable of making the diagnosis. In addition, containment efforts included isolation

of smallpox patients as well as requirements for international travelers to have up-to-date vaccination certificates. Moreover, at that time, there were extensive efforts for routine vaccination. Despite these facts, in the last exportation of smallpox into Europe in 1972, 175 cases with 35 deaths occurred in Yugoslavia. That outbreak was controlled after a massive effort of case detection, quarantine, and mass vaccination of 20 million people [26].

TABLE 21.1

Frequency of Importation during the Final Phase of Smallpox Eradication[a]

WHO Regions	Number of Endemic Countries	Number of Imported Outbreak Episodes in Nonendemic Countries
WPRO	0	2
SEARO	4	3
EURO	0	7
EMRO	3	8
AMRO	1	1
AFRO	10	4
Total	18	25

Source: The global eradication of smallpox. WHO Report 1980.
[a] Importations of smallpox infection 1970–1978.

Perhaps, the most detailed study of exported smallpox was undertaken in Bihar (population 56 million, 1971), India. There, even in the face of active eradication activities in 1974, a year before last case in May 1975, the exportation of smallpox was extensive through India's well-developed train network (Figure 21.4).

These exportations occurred despite the strong interest of all parties, including national governments, in smallpox eradication. Contrast this with today, where a few nations, especially those in sub-Saharan Africa, contain large numbers of susceptible people yet, as discussed in the next section, have very limited abilities to cope with a smallpox introduction.

21.9.8   SURVEILLANCE AND CONTAINMENT STRATEGY

It should be noted that during smallpox eradication, the surveillance and containment strategy was effective even in densely populated areas. However, at that time, the populations had high levels of preexisting vaccine and/or natural epidemic-induced immunity. But, now and in the future, with the high proportion of unvaccinated people, this approach may not be effective enough to interrupt transmission. It is likely that we will need to use widespread mass vaccination. This certainly needs some additional study.

21.9.9   VACCINE AVAILABILITY

As early as 2007, WHO, in its annual report entitled "Global Public Health Security in the 21st Century," indicated that "in the absence of global capacity to contain an outbreak rapidly, smallpox might be reestablished endemicity...." Absolutely essential to effectively cope with a potential return of smallpox will be sufficient supplies of good smallpox vaccine. To this end, the latest information indicates that WHO has established a stock of 30.5 million doses of vaccine. Additionally, through a "virtual stockpile," France, Germany, New Zealand, United Kingdom, and United States have pledged 31 million vaccine doses to WHO [21]. Even with a moderate size outbreak of smallpox, this quantity will clearly not be sufficient for mass vaccination campaigns.

The epidemiological modeling of smallpox transmission among our current and/or projected susceptible population indicates the potential disaster before us should smallpox be introduced in the absence of sufficient vaccine supplies. Fortunately, it is not difficult to produce and stockpile sufficient supplies of smallpox vaccine, and the LC16m8 vaccine has an almost infinite shelf life at −20°C [22,23]. In 1980, the WHA endorsed the recommendations of the global commission of smallpox eradication on the need for reserve stocks of vaccine. They are listed as follows, with the author's updates of the situation as of 2011 in parentheses:

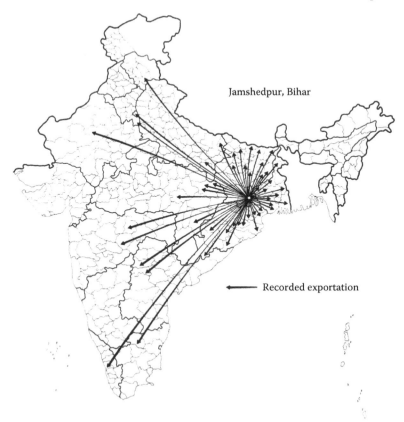

FIGURE 21.4   Smallpox exportation from Bihar to all India in 1971. Nearly 300 outbreaks of smallpox in other states, most of which had passed through the Tatanagar railway station. The worst of these exportations was to Bilaspur, Madhya Pradesh, where a single case from Tatanagar railway station led to 464 additional cases and 76 deaths affecting 72 villages and 18 PHCs within the district and at least 3 other districts and 2 other states.

(i) Sufficient freeze-dried smallpox vaccine to vaccinate 200 million people. The recommendation: Resolution WHA 33.4. (*As of 2011, an urgent need to consult with member states was recognized.*)

(ii) Periodic testing of vaccine quality in stock. (*As of 2011, this was being done, but results were not available outside of WHO.*)

(iii) To maintain the seed lots in designated WHO collaborating centers. (*In 2011, we believe this was being done.*)

(iv) National health authorities should be asked to inform WHO of the amount of vaccine maintained and their production capacity to follow up the emergence of smallpox epidemics. (*This was partially done.*)

At the time of writing of this manuscript, further action is expected through 2012.

21.9.10 CLINICAL AND LABORATORY DIAGNOSIS

After three decades without smallpox, medical and public health professionals will be challenged to make a clinical diagnosis of smallpox. The ability to accurately diagnose orthopox-related diseases, with special emphasis on variola virus, will be essential if outbreaks are to be detected early. Here again, we can use our experience from the smallpox eradication program. In this instance, a picture is worth a thousand words. During the program, a few million post-card-size photos of smallpox patients were distributed worldwide (Figure 21.1). That approach could be extremely valuable today. Clearly, early detection will lead to early implementation of appropriate containment measures. In addition, the recent development of rapid laboratory tests, such as PCR, should speed accurate laboratory diagnosis. Its effective use should be developed with the new WHO collaborating center network.

21.10 LOOKING AHEAD: PREPARATIONS REQUIRED

So how do we prepare ourselves to effectively respond to an outbreak of smallpox? The Dark Winter exercise indicated what a challenge it will be. To what extent it is feasible to maintain isolation facilities and the like in the absence of smallpox is not clear. This issue, incorporated into the general requirements for emergency medical and public health practice, will require considerable discussion elsewhere. Note that during eradication program, varicella cases misdiagnosed as smallpox were isolated with smallpox cases and, as a result, were infected with smallpox.

Perhaps, at a minimum, routine smallpox vaccination should be continued or newly initiated for a very limited group such as first-responder personnel. Such vaccine could use the currently available attenuated vaccines, LC16m8 or IMVAMUNE.

Most importantly, as mentioned earlier, we must immediately construct large stocks of safe and effective smallpox vaccine. As a first step, WHO's regional offices should request national health administrations to supply them with information regarding (1) the amount and types of smallpox vaccine stocks currently available, (2) any plans/capabilities to produce vaccine in the case of an emergency, and (3) any capacity to donate vaccine to WHO for the international stockpile. The quality of the vaccine for donations must meet WHO standards as tested by WHO collaborating laboratories, which will have to be established shortly.

There is also a problem with supporting smallpox vaccine production. It is not feasible to support the production costs through standard market mechanisms. An alternative approach to consider would involve transferring the technical production know-how and intellectual property to a number of low-cost manufacturing facilities throughout the world. The technology transfer could be provided free of charge with the proviso that once produced, some of the product would be donated to UN or WHO stores. Another alternative, suggested by Arita recently [17], would be for the UN or WHO to consider setting up a UN vaccine production center. In this respect, the LC16m8 vaccine production process could be rapidly transferred.

21.11 CONCLUSIONS AND FUTURE PERSPECTIVES

In this chapter, we discuss variola virus (smallpox and alastrim virus) including its biology, clinical manifestation, epidemiology, and the human efforts that were put forth to eradicate it under the leadership of the WHO in 1980. A brief historical view of the disease is also added to highlight the unprecedented human effort put forth to eliminate this vicious natural enemy from the planet. However, even with eradication, the story has not ended. We cannot ignore the possibility that through intentional or unintentional release from known or unknown virus stocks, variola could return. If it did return, the ever-increasing susceptibility of global population will allow smallpox to rapidly spread. It is important to note that a number of practical manuals/guidelines have been published over the last three decades. While these are excellent tools necessary to cope with the possibility, this manuscript stresses the global efforts that will have to be launched to counteract such an epidemic.[8,9]

We conclude that there is an urgent need to make preparations to respond appropriately to a possible smallpox bioterrorist event. In view of the epidemiology of smallpox, an effective response will have to involve all nations regardless of race, religion, civil structure, and economic status. Critical to our ability to detect and respond to an outbreak of smallpox will be the rapid access to both advanced diagnostic laboratories and safe and efficacious smallpox vaccine. For the world to produce enough smallpox vaccine to cope with a potential smallpox terrorist attack, we cannot rely on standard commercial or market forces. Instead, we will have to employ a UN-/WHO-coordinated system with direct financial support. If we employ such a system effectively and produce large quantities (200 million doses) of smallpox vaccine and if we

TABLE 21.2

Worldwide Requirements of Smallpox Vaccine Stocks

Smallpox Vaccine Stocks

We need a 200-million-dose stockpile as recommended in 1980.

We assume that in two WHO regions, Americas and Europe, most nations will likely have their own supply capability. The rest of the four WHO regions, India, Pakistan, Indonesia, and China, and certain other nations may have the ability to obtain the vaccine through direct purchase or national production. Thus, the areas truly needing a WHO international stockpile are illustrated in the succeeding text.

	Proportion Needing Vaccine from Stockpile	Required Doses (Millions)
African region	Almost all	800
East Mediterranean	1/2	300
Southeast Asia	1/3	600
Western Pacific	1/3	600
Total		2300

It is estimated that a total of 2300 million doses are needed in the four WHO regions. But actual needs may be even greater due to wastage of vaccine during operation.

Let's consider a hypothetical situation: Bioterrorists attack three large cities having populations ranging from 10 to 30 million. This would mean that a total of 30 to 90 million people would immediately need containment vaccination. We can assume that at least three exportations to nearby or distant areas would occur within a few weeks. Each of these areas would require an additional 10 million doses. Thus, 60–120 million doses would be needed for the initial containment from the 200-million-dose stockpile. The remaining 140 to 80 million doses can be reserved and, as needed, be provided for emergency vaccination campaigns by selecting priority area populations.

At the same time, WHO will discuss with nations having self-sufficient vaccine production capacity as to how their vaccine stores could be partly used as their emergency vaccine production is activated to fill the supply gap in the future. In addition, while doing this supply action, some selected nations will start to newly produce the vaccine needed according to the epidemiological situation. Here, WHO may have to take a major leadership role. In this model, it is estimated that the 200-million-dose stockpile may have sufficient flexibility to cope with regional and global situation caused by the smallpox bioterrorism. In addition, it will be essential to have a sufficient supply of bifurcated needles to deliver the vaccine.

Source: WHO report to 57th WHA, Smallpox, Global Smallpox Vaccine Reserve, April 7, 2005.

remain alert so we can effectively diagnose smallpox quickly, we are confident that we can minimize the damage of a reintroduction of smallpox (Table 21.2). It would clearly be wise for nations to continue (or initiate) a practice to vaccinate a very limited number of selected staff. These people would undertake, as needed, the initial investigation of suspected smallpox cases. To avoid unnecessary vaccine-associated risks, they should be vaccinated using the attenuated vaccine.

Smallpox eradication is the first time in history of biological evolution that the human species successfully eliminated a disease. We cannot imagine that a terrorist would want to join forces with such a vicious enemy, the variola virus, and once again reintroduce it to the people of this planet. They must consider that such a plan would be a tragedy to all humans including the terrorist's friends, relatives, and families. We hope that such an "unimaginable evil" will be outside the bounds for even those who want to cause people great harm.

ACKNOWLEDGMENTS

We are grateful to all of our colleagues who worked for smallpox eradication. Specifically, we acknowledge DA Henderson and Haruo Watanabe for their valuable technical information and to Hanako Ushizima for her skilled assistance. The authors are solely responsible for the opinions expressed in the paper. We are grateful to the editor of *Japan Journal of Infectious Diseases* for giving permission to use some portions of Arita's publication, Ref. [9] "A personal recollection of smallpox eradication…." We are also thankful to WHO for the use of the WHO figures and table.

REFERENCES

1. Stiglitz, J. 2002. *The Globalization and Its Discontents.* Penguin Books, London, U.K.
2. WHO Advisory Committee on Variola Virus Research (ACVVR). 2010. Report of the 12th Meeting, Geneva, Switzerland, November 17–18, 2010.
3. Fenner, F., Henderson, D.A., Arita I., Jezek, Z., and Ladny, I.D. 1988. *Smallpox and Its Eradication.* WHO, Geneva, Switzerland, pp. 59, 62, and 204.
4. Arita, I. 1979. *Smallpox Eradication: Target Zero.* Mainichi Press, Kumamoto, Japan (in Japanese).
5. Rao, A.R. 1972. *Smallpox, Transmission of Infection.* Kothari Book Depot, Bombay, India, pp. 93–34.
6. Fenner, F., Henderson, D.A., Arita, I., Jezek, Z., and Ladny, I.D. 1988. Smallpox and its eradication. WHO, Geneva, Switzerland, p. 262.
7. Guarner, J., Johnson, B.J., Paddock, C.D. et al. March 2004. Monkeypox transmission and pathogenesis in prairie dogs. *Emerg. Infect. Dis.,* 10: 426–431. http://wwwnc.cdc.gov/eid/article/10/3/03-0878.htm.
8. WHO Advisory Committee Report. 12th Meeting Report, November 17–18, 2010 and 13th Meeting Report (draft), October 31–November 1, 2011.

9. WHO Advisory Committee on Variola Virus Research Report of the Twelfth Meeting, Geneva, Switzerland, 17–18 November 2010 WHO reference number: WHO/HSE/GAR/BDP/2010.5.

10. Foege, W.H., Millar, J.D., and Lane, J.M. 1971. Selective epidemiologic control in smallpox eradication. *Am. J. Epidemiol.*, 94(4): 311–315.

11. Jezek, Z. and Fenner, F. 1988. *Human Monkeypox*. Monographs in Virology No. 17, pp. 1–140. S. Karger Publisher, New York.

12. Esposito, J.J., Nakano, J.H., and Obijeski, J.F., 1985. Can variola like virus be derived from monkeypox virus? An investigation based on DNA mapping. *Bull. WHO*, 63: 695–703.

13. Dowdle, W.R. and Hopkins, D.R. (eds.) 1997. *The Eradication of Infectious Diseases* (Report of Dahlem Workshop on the Eradication of Infectious Diseases, Berlin, Germany) Arita, I., Chapter 15 (Are There Better Mechanism for Formulating, Implementing, and Evaluating Progress?); Foege, W., Chapter 16 (Thoughts on Organization for Disease Eradication), John Wiley & Son, New York.

14. Wimmer, E. 2006. The test tube synthesis of a chemical called polio virus. *EMBO Rep.*, 17(Special Issue): S3–S9.

15. Interview by Selgelid, M.J. and Lorna, W. 2010. The mousepox experience. *EMBO Rep.*, 11: 18–24.

16. McFadden, G. January 2010. Killing a killer: What next for smallpox? *PLOS Pathog.*, 6: e1000727.

17. Arita, I. 2011. Smallpox: Should we destroy the last stockpile? *Exp. Rev. Anti Infect. Ther.*, 9(10): 837–839.

18. The Supreme Court Decision on Aum Cult Members. November 18, 2011. Asahi Press, Japan.

19. Bhattacharjee, Y. February 19, 2010. FBI closes anthrax case. Says Bruce Ivins was sole culprit behind letter attacks. *Science*. http://news.sciencemag.org/policy/2010/02/fbi-closes-anthrax-case-says-bruce-ivins-was-sole-culprit-behind-letter-attacks.

20. http://www.terrorisminfo.mipt.org/dark-winter.asp. 2009.

21. WHO Advisory Committee Reports. 2011. 12th Meeting Report, November 17–18, 2010, and 13th Meeting Report (Draft), October 31–November 1.

22. Saijio, M. et al. 2011. National Institute of Infectious Diseases, Japan: The view on smallpox vaccine LC 16 m, September 22, 2011, Submitted to the Meeting of 13th WHO Advisory Committee on Variola Virus Research, October 31–November 1, 2011, Geneva, Switzerland.

23. Kennedy, J.S., Gurwith, M., Dekker, C.L. et al. 2011. Safety and immunology of LC 16 m 8, an attenuated smallpox vaccine in vaccinia naïve adults. *J. Infect. Dis.*, 204(9): 1395–1402.

24. Alibek, K. 1999. *Biohazard*. Arrow Books, London, U.K.

25. Report of the 10th Meeting, WHO Advisory Committee on Variola Virus Research, 11. November 19–20, 2008. Synthesis of Variola Virus, Geneva, Switzerland.

26. Lane, J.M., Brandling-Bennett, D., Francis, D.P. et al. 1972. *Smallpox in Yugoslavia*. Memorandum to the Director, CDC (EPI-SEP-72-91-2). Centers for Disease Control, Atlanta, GA.

FURTHER READINGS

Arita, I. 2010. Forward by Heymann, D. *Smallpox Eradication Saga: An Insider's View*, 197pp. Oriental Black Swan, Hyderabad, India.

Fenner, F., Henderson, D.A., Arita, I., Jezek, Z., and Ladny, I. 1988. *Smallpox and Its Eradication*, 1500pp. WHO Publication, Geneva, Switzerland.

Henderson, D.A. 2009. Foreword by Preston, R. *Smallpox: The Death of a Disease*, p. 334. Prometheus, New York.

The second book is an encyclopedia to find any information relevant to the disease. The others are a type of personal history by those who closely worked for the program.

22 Venezuelan Equine Encephalitis Virus

Ann M. Powers

CONTENTS

22.1 INTRODUCTION

The alphaviruses are a group of antigenically related arthropod-borne viruses (arboviruses) that were first isolated in the 1930s. Based upon the results of hemagglutination–inhibition (HI) tests, the alphaviruses were originally designated as group A viruses, thus distinguishing them from other arboviruses such as flaviviruses and bunyaviruses.[1,2] Additional serological and subsequent molecular testing further separated these viruses and led to the serological antigenic complexes still recognized within the *Alphavirus* genus.[3,4] Alphaviruses can be divided serologically into 10 antigenic complexes. Three of these serocomplexes of bioterrorism importance are represented by eastern equine encephalitis (EEE), western equine encephalitis (WEE), and Venezuelan equine encephalitis (VEE) viruses. In both humans and equines, these viruses cause the most severe manifestation associated with alphaviruses infection (encephalitis) and are of significant concern during either a natural or intentional release outbreak.

This chapter will focus on the viruses of the VEE complex that, in Central and South America, have been the cause of focal outbreaks that occur periodically with rare large regional epizootics involving thousands of equine cases and deaths. These epizootic/epidemic viruses that are the likely candidates as biological terrorism agents are theorized to emerge periodically from mutations occurring in the continuously circulating enzootic viruses in northern South America. In the United States, the classical epizootic varieties of the virus are not present; however, if a subtype that was not previously present in the United States were to be introduced during a bioterrorist attack, it is possible that the virus could establish a natural cycle involving local mosquitoes and small vertebrates or that human to mosquito to human transmission could occur.

The weaponization of VEE viruses (VEEVs) by various countries around the world has made these priority agents. There are a number of characteristics of VEEVs that set them apart from other viral BT agents:

- Unlike smallpox and the viral hemorrhagic fevers, VEEVs are readily available from natural sources.
- VEEV is fully infectious via aerosol and requires enhanced BSL-3 containment (HEPA-filtered exhaust air) for laboratory investigations.
- Unlike smallpox, vaccines are not readily available for VEEVs, and there are currently no therapeutics for these viruses.
- The properties of these viruses to be transmitted and maintained independently of humans in the mosquito and small animal reservoir make them unique. Once released, it is possible that they will develop stable enzootic transmission cycles, resulting in prolonged epidemics and epizootics.
- Infectious and defective/helper expression systems of VEEV have been developed, and it is feasible that these constructs could be modified confusing the diagnostic picture.

Because there has been no VEEV epidemic activity documented for years, large clusters of cases, especially in urban settings, geographic areas not normally associated with the natural distribution of these viruses, or times of the year not normally associated with the natural transmission, should arouse suspicion of a possible bioterrorism event.

Thus, an understanding of the epidemiology, clinical presentation, detection, and control options is important from a public health perspective.

22.2 CLASSIFICATION AND MORPHOLOGY

VEEV is a member of the family *Togaviridae*, genus *Alphavirus*.[5] Within the VEE complex, there are eight viral species with multiple distinct varieties of VEEV (Figure 22.1 and Table 22.1).[6] The VEEV strains isolated during major outbreaks are referred to as epizootic or epidemic and typically belong to subtypes IAB and IC. The remaining subtypes (ID–IE) are considered enzootic strains.[7,8] The only other genus within *Togaviridae*, *Rubivirus* genus, contains a single species, *Rubella virus*.[5]

Like all alphaviruses, VEEV structurally is a 60–70 nm icosahedral virion with T = 4 symmetry.[9,10] The nucleocapsid core, which consists of 240 copies of the capsid protein combined with one molecule of genomic RNA, is encased in a lipid envelop acquired by budding from the host cell plasma membrane. This membrane contains heterodimers of two virally encoded membrane glycoproteins, E1 and E2, that are arranged as trimers on the virion surface. Alphavirus structural studies indicate that the E2 protein forms spikes on the surface of the virion, while the E1 protein lies adjacent to the host cell-derived lipid envelope.[9]

The genome consists of single-stranded, positive-sense RNA of approximately 11.5 kb (Figure 22.2).[11] Four nonstructural proteins (designated nsP1-4) are encoded in the 5′ two-thirds of the genome. These proteins participate in genome replication and viral protein processing in the host cell cytoplasm with each protein having a specific function. The nsP1 protein is required for initiation of synthesis of minus strand RNA and also functions as a methyltransferase to cap the genomic and subgenomic RNAs during transcription. The second gene, nsP2, encodes a protein that has RNA helicase activity in its N-terminus, while the C-terminal domain functions as a proteinase for the alphavirus nonstructural polyprotein. The nsP3 gene encodes a protein consisting of two domains: a widely conserved N-terminal domain and a hypervariable carboxyl terminus. The C-terminal region has been shown to tolerate numerous mutations, including large deletions or insertions, and still produce viable viral particles in vertebrate cells. While the functions of these distinct elements are not yet fully understood, significant research has recently focused on the role of nsP3 in replication.[12] The final nonstructural protein, nsP4, contains the characteristic GDD motif of an RNA-dependent RNA polymerase (RdRP).[11,13]

The 3′ one-third of the genome encodes the structural proteins. These are generated by a subgenomic message that is translated to produce the three major structural proteins (the capsid and the E1 and E2 envelope glycoproteins) and two peptides (6K and E3). Functionally, the E2 protein has been

FIGURE 22.1 Distribution of enzootic VEE antigenic complex viruses. Note the focal characteristic of these strains due to ecological and susceptibility parameters of the vectors and reservoir hosts. Epizootic activity (attributed to varieties IAB and IC) is dispersed throughout northern South America (including Colombia, Venezuela, Peru, and Ecuador), Mexico, and the southern United States.

TABLE 22.1
VEE Antigenic Complex Viruses

Species	Geographic Location
Cabassou virus	French Guiana
Everglades virus	Southern Florida
Mosso das Pedras virus	Brazil
Mucambo virus	South America
Pixuna virus	Brazil
Rio Negro virus	Northern Argentina
Tonate virus	South/North America
VEEV (epizootic IAB and IC varieties)[a]	Americas
VEEV (enzootic ID and IE varieties)	South/Central America

[a] Complex viruses of bioterrorism concern.

found to be an important determinant of antigenicity and cell receptor binding in both the vertebrate host and the insect vector, while the other major structural protein, E1, has been found to contain domains associated with membrane fusion. Neither 6K nor E3 has been identified with the final intact VEEV virion.[9,11]

The 5′ end of the genome has a 7-methylguanosine cap and the 3′ end possesses a polyadenylated tail. Just upstream of the poly(A) tract is a noncoding region of varying lengths (depending upon the virus). This region contains specific repeat elements that are distinctly associated with each of the different viruses. There are secondary structures associated with this noncoding region that may be involved in replication, host specificity, or virulence patterns.[14]

22.3 BIOLOGY AND EPIDEMIOLOGY

Geographically, VEEV is found primarily in Central and South America (Figure 22.1). Distribution of a particular VEEV complex virus is tied to its vertebrate reservoir and invertebrate vector availability. Initially, VEEV strains were only isolated during equine epizootics and human epidemics as there was no awareness of the maintenance cycle of the virus during interepidemic periods. Then, beginning in the late 1950s, other viruses in the complex (i.e., Everglades, Mucambo) were detected in sylvatic habitats in Central America, South America, Mexico, Florida, and Colorado.[15–21] However, there was an absence of equine disease associated with these strains. While there typically was no equine disease associated with these

enzootic strains, humans were shown to become infected with some of these strains when they contacted enzootic foci of transmission.

Each individual VEE complex virus tends to be transmitted enzootically by only a single or small number of invertebrate species; these very specific host–virus interactions are related to geography and ecological dynamics associated with the mosquito vectors. The majority of the viruses within the VEE complex circulate continuously in enzootic habitats between small vertebrate rodent hosts and mosquitoes of the subgenus *Culex* (*Melanoconion*).[22,23] For example, VEEV subtype IE viruses are transmitted only by *Culex* (*Mel.*) *taeniopus*, Everglades virus is vectored exclusively by *Culex* (*Mel.*) *cedecei*, while VEEV subtype ID viruses are transmitted by several melanoconions including *Culex* (*Mel.*) *aikenii s.l.* (*ocossa, panocossa*), *Culex* (*Mel.*) *vomerifer, Culex* (*Mel.*) *pedroi*, and *Culex* (*Mel.*) *adamesi*.[24–27] Many adult females, especially sylvatic vectors in the Spissipes section of the *Melanoconion* subgenus, feed primarily on small sylvatic mammals as would be expected for these zoonotic cycles. However, some species, including proven VEEV vectors, exhibit more opportunistic feeding behavior and readily bite humans. This characteristic suggests how humans can occasionally become infected when they enter the vector habitat. Because small rodents that do not travel great distances serve as the vertebrate reservoirs, transmission can be extremely focal.[28] Finally, it is believed that there are no or minimal adverse effects due to viral infection in either the reservoir hosts or the zoonotic vectors.

In contrast to enzootic maintenance, epidemic or epizootic outbreaks may utilize numerous species of mammalophilic mosquitoes in the genera *Aedes* and *Psorophora*, but transmission by these species rarely continues once the outbreak subsides.[29–31] In laboratory studies, many of these species have been found to be poorly susceptible or almost completely refractory to infection with VEEV. However, ecological and behavioral traits such as longevity, host preference, survival, and population size are probably more important than susceptibility for VEEV. Because equines develop extremely high-titered viremias, some species that appear to be only moderately susceptible to infection are able to become infected after biting equines and have been incriminated as important vectors during outbreaks.

Curiously, the most closely related virus to VEEV, EEE virus (EEEV), utilizes extremely different invertebrate vectors in its transmission cycles. EEEV is maintained in North America in a transmission cycle including *Culiseta*

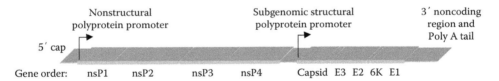

FIGURE 22.2 Organization of the VEEV genome showing genes, promoter elements, and noncoding regions. Proteins involved in viral replication that are translated from viral genomic RNA directly are encoded by the nonstructural genes. Proteins involved in viral attachment, entry, and encapsidation of the viral RNA are coded for by the structural genes.

melanura mosquitoes and avians.[32] However, this species of mosquito rarely feeds on mammals so epidemic or epizootic transmission to humans or equines typically involves multiple other species of mosquitoes in several genera as is found with VEEV.[33]

VEEV was first recognized as a disease of horses, mules, and donkeys in Colombia and Venezuela during the 1930s, and the virus was first isolated in Venezuela in 1938 from the brain of a horse during an epizootic.[34,35] VEE outbreaks continued sporadically for most of the twentieth century primarily in northern South America but occasionally in Central America; Mexico and the US focal outbreaks occurred occasionally; but, infrequently, large regional epizootics were documented with thousands of equine cases and deaths. One epizootic that began in Peru and Ecuador in 1969 reached Texas in 1971.[7] During the course of this extensive outbreak, it was estimated that over 200,000 horses died prior to the eventual control of the epizootic by a massive equine vaccination program using the live-attenuated TC-83 vaccine.[36,37] Additionally, there were several thousand human infections. However, after this epizootic ended, there was no confirmed VEE activity leading to speculation that the epizootic strains had become extinct. However, the 1990s saw a resurgence and changing of VEEV activity causing concern among public health officials.

In 1992, several dozen human and equine cases were documented in western Venezuela. A few months later, additional cases were reported on the western shore of Lake Maracaibo. Viruses isolated from humans and horses of this small outbreak represented a genetically novel subtype IC strain distinct from those isolated during all previous epizootics. The degree of genetic identity in comparison to enzootic, subtype ID strains from the same region of western Venezuela suggested that this outbreak resulted from the recent evolution of the IC strain from continuously circulating, enzootic, subtype ID progenitors in western Venezuela.[22,38] This indicated that the epizootic virus, while perhaps not maintained in nature, could reemerge in epizootic fashion at any time with little or no warning.

Two more small outbreaks in 1993 and 1996 in southern Mexico generated additional cause for concern. Pacific coastal areas of Chiapas State in southern Mexico reported a small equine outbreak in 1993, while in 1996, another equine focus was identified in coastal Oaxaca State. The virus responsible for these outbreaks was found to belong to subtype IE, a subtype that had never before been associated with equine disease.[39] Genetic studies again supported the hypothesis that these epizootic viruses had recently gained the equine virulence phenotype from local, enzootic progenitors.[40] The Mexican subtype IE epizootic strains were found to differ fundamentally from epizootic strains (IAB and IC) isolated during more extensive outbreaks because they did not replicate to high titers in equines as IAB and IC strains had been shown to do. Recent genetic studies have demonstrated the continued activity of IE viruses including infection of both equines and humans[41]; however, the lower titers in these affected horses probably limited the magnitude of the outbreak in Mexico and prevented spread to Central America and the United States.[42]

Another major VEE epidemic occurred in the fall of 1995 in Venezuela and Colombia with estimates of over 75,000 human infections.[43,44] This subtype IC outbreak was found to be genetically virtually identical to samples recovered from a 1962 to 1964 epizootic in the same geographic region. Curiously, a reference strain isolated in 1963 contained the predicted ancestral sequence of the 1995 outbreak suggesting a possible laboratory source for the 1995 epizootic.[45] While there has been no further documented outbreak activity since 1996, viruses continue to be isolated from nature (unpublished data), and these periodic emergence events via a small number of genetic mutations suggest that further epidemics are indeed possible.

The epidemiological patterns of the alphaviruses are as diverse as their geography and ecological characteristics. Generally, outbreaks of human or animal illness due to alphavirus infections coincide with peak mosquito seasons in temperate zones, while viruses such as VEEV that exist in tropical climates occur year-round. The magnitude of each outbreak can vary dramatically depending upon whether the outbreak was localized to urban or rural settings. However, in both ecologies, the attack rates can be significant if a virulent strain is the etiological agent and the seropositivity rates often correlate with the vector infectivity and transmissibility rates as well as host preference of the mosquito. Occupational exposure differences to mosquitoes are also likely to affect incidence rates.

22.4 CLINICAL FEATURES AND PATHOGENESIS

Alphaviruses can be broadly categorized into three distinct groups based upon the type of illness they produce in humans and/or animals. Disease patterns include (1) febrile illness associated with a severe and prolonged arthralgia, (2) encephalitis, or (3) no apparent or unknown clinical illness. VEEV, as its name indicates, can cause severe encephalitis that can lead to death in a small percentage of cases. Infection of man is less severe than with the closely related EEE and WEE viruses, and fatalities are rare with VEEV infections.[7] In general, disease is often more severe in young children and rates of immunity are lower in young children. This indicates that infections produce long-lived antibodies that are presumably protective for life. Adults usually develop only an influenza-like illness, while overt encephalitis is usually confined to children. However, clinical illness is also related to subtype or variety; some of the subtypes and varieties of VEEV cause only mild febrile illness or are not known to cause human illness. Similarly, many of the other viral species within the genus also cause only mild disease.[46]

Many VEEV infections are clinically silent (asymptomatic) but may result in illnesses of variable severity sometimes associated with central nervous system (CNS) involvement. When the CNS is affected, clinical syndromes ranging from febrile headache to aseptic meningitis to encephalitis may occur. These syndromes are usually indistinguishable from similar symptoms caused by other viruses. The presentation may include the following syndromes: (1) confusion, stupor, or coma; (2) aphasia or mutism; (3) convulsions; (4) hemiparesis with asymmetric deep tendon reflexes and positive Babinski sign; (5) ataxia, myoclonus, and involuntary movements; (6) cranial nerve dysfunctions producing facial weakness, nystagmus, and ocular palsies; (7) vomiting; and (8) involvement of meninges producing a stiff neck.

In general, arboviral meningitis is characterized by fever, headache, stiff neck, and pleocytosis, while arboviral encephalitis is characterized by fever, headache, and altered mental status ranging from confusion to coma with or without additional signs of brain dysfunction (e.g., paresis or paralysis, cranial nerve palsies, sensory deficits, abnormal reflexes, generalized convulsions, or abnormal movements).[47–50]

One further clinical consideration with VEEV is the possibility of alternate transmission routes. In the Americas, naturally acquired VEE viral infections result from the bites of infected mosquitoes. But as a potential intentional release agent, VEEV could be dispersed as an infectious aerosol. Consequently, the clinical presentation of individuals receiving inhalational exposure to VEEV could be different from the presentations of mosquito-borne exposure.[23] Unlike naturally acquired infection, inhalational exposure to VEEV could result in direct viral invasion of the olfactory nerve and the pulmonary alveolar epithelium. Laboratory animal studies have demonstrated that aerosol exposure to VEEV can result in attachment to olfactory nerve endings, in direct invasion of the CNS, and a high incidence of CNS disease.[51] This suggests that in contrast to mosquito-borne disease, VEE resulting from an intentional aerosol release would be likely to result in rapid CNS involvement and increased morbidity and mortality. In this setting, seizures and profound neurologic dysfunction could be much more common than other systemic manifestations. Following an intentional aerosol release of VEEV, cases of mild or moderate VEE viral illness commonly seen in mosquito-borne outbreaks might not be as frequent.

22.5 IDENTIFICATION AND DIAGNOSIS

Any presumptive clinical diagnosis of VEEV infection must be confirmed with laboratory testing. Typically, this confirmation consists of (1) demonstration of specific viral antigen or genomic sequences in tissue, blood, or cerebrospinal fluid (CSF); (2) isolation of virus from tissue, blood, CSF, or other body fluid; (3) fourfold or greater change in virus-specific serum antibody titer; (4) virus-specific immunoglobulin M (IgM) antibodies demonstrated in CSF by IgM antibody capture enzyme-linked immunosorbent assay (MAC-ELISA); or (5) virus-specific IgM antibodies demonstrated in serum by MAC-ELISA and confirmed by demonstration of virus-specific serum immunoglobulin G (IgG) antibodies in the same or a later specimen by another serological assay (e.g., IgG-ELISA, neutralization, or HI).

The first step in accurate diagnosis is correct collection and handling of specimens. In putative VEEV cases, a lumbar puncture to obtain CSF and venipuncture to collect serum and whole, anticoagulated blood should be obtained on every person with symptoms suggesting encephalitis or meningitis. Specific testing that could be performed on each of these samples is listed in the following.

Laboratory diagnosis of human VEEV infections has changed greatly over the last few years. In the past, the identification of VEEV-specific antibody relied on four tests: HI, complement fixation, plaque reduction neutralization test, or the indirect fluorescent antibody (IFA) test.[52] With the advent of solid-phase antibody-binding assays, the diagnostic algorithm for identification of VEEV has changed. Rapid serological assays such as MAC-ELISA are now utilized early in infection.[53,54] This has the advantage of being able to provide a confirmed result without the need to wait for a convalescent sample. For example, a positive MAC-ELISA result on an acute CSF sample is considered confirmatory. Furthermore, IgM antibody obtained early in infection is more specific, while the IgG antibody detected later in infection is more cross-reactive.[55]

Virus isolation and identification have also been useful in defining VEEV infection using serum or CSF. While virus isolation still depends upon growth of an unknown virus in cell culture or neonatal mice, virus identification has been greatly facilitated by the availability of virus-specific monoclonal antibodies as well as rapid sequencing methods for use in identification assays.[56–58] Additional numerous molecular techniques including standard reverse-transcription polymerase chain reaction (RT-PCR) and quantitative RT-PCR have dramatically increased the speed of laboratory confirmation of VEEV infections. Unfortunately, rapid point-of-care diagnostics for VEEV are not readily available to most laboratories.

22.5.1 CONVENTIONAL TECHNIQUES

One of the most popular currently used conventional diagnostic methods for VEEV is the ELISA. Depending upon the format of the ELISA, specific antibody isotypes can be detected. The MAC-ELISA provides rapid and early documentation of an antibody response in specimens or cultures from presumptive VEEV cases. Assays that detect virus-reactive IgM are advantageous because they detect antibodies produced 2–10 days after onset of clinical symptoms in a primary infection, possibly obviating the need for convalescent-phase specimens in many cases. Because the MAC-ELISA utilizes early infection material, it is important to note the potentially infectious nature of the serum specimens involved, and this assay should be performed in microbiology laboratories that are biological safety level 2 (BSL-2) and practice BSL-3 safety procedures. An annually certified class II biological safety cabinet is recommended.

If the acute sample is negative, the IgG-ELISA provides a useful alternative to immunofluorescence for presumptive identification of a serological response, particularly in convalescent samples. IgG antibody is less virus-specific than IgM, appears in serum slightly later in the course of infection than IgM, and remains detectable until long after IgM ceases to be present. Using the IgG-ELISA in parallel with the MAC-ELISA, the relative rises and falls in antibody levels in paired serum samples can be noted. This simple and sensitive test is applicable to serum specimens but not generally to CSF samples.

The neutralization of viral infectivity (neutralization test [NT] assay) is the most sensitive and specific method for determining the identity of an isolate and for determining the presence of specific antibodies in a patient's serum. The serum dilution-plaque reduction procedure performed in cell culture is the standard method. In this assay, if neutralizing antibody is present, virus cannot attach to cells, and infectivity is blocked.[59] For VEEV, this method is able to distinguish the various subtypes and varieties that may be cross-reactive in other immunologic tests. Unfortunately, the procedure is expensive and time-consuming and requires specialized laboratories if using live VEEV. However, the development of chimeric alphaviruses such as Sindbis/VEEV that have a reduced pathogenicity compared with wild-type VEEV makes the possibility of NT assays more broadly available to laboratories with only BSL-2 capability.[60]

22.5.2 Molecular Techniques

Infection with VEEV often produces a viremia of sufficient magnitude and duration that the viruses can be detected directly from blood during the acute phase of illness (which is typically 0–5 days after onset). Virus can be also be detected in biopsy or autopsy tissue of the brain as well as in the upper respiratory tract of patients with acute VEEV infection.[61] The most common approach to direct viral detection and diagnosis is molecular detection typically using RT-PCR. There are many variations on the RT-PCR approaches mostly based on the variety of kits and instruments available.[22,62,63] However, all can rapidly, and with a high degree of sensitivity, detect VEEV nucleic acid from human cases. Specimens for any molecular detection assay for VEEV include acute human serum samples or ground tissues. One of the few limitations of this approach is that if the sample is not collected during the viremic phase, detection is unlikely.

The earliest detection techniques utilized traditional RT-PCR that could involve an either one- or two-step process followed by visualization of amplification products on agarose gels. Products could be identified by size of the amplicons, presence of specific banding patterns, or sequencing of amplification material. Sequencing of VEEV amplicons is a common practice as many investigators use the resulting data for epidemiological analysis as well as a diagnostic tool. An example is the detection of active and ongoing transmission of VEEV in coastal Mexico based upon sequencing of clinical samples.[41] Subsequent development of real-time detection instruments led to even more rapid detection of VEEV as not only was the assay time reduced but the need for follow-up sequencing was eliminated. Sensitivity of real-time methods is typically 10–100-fold greater than traditional RT-PCR. There are other molecular detection approaches as well

including methods such as NASBA, LAMP, or bead-based detection assays.[64–66] However, a detailed description of all these approaches is beyond the scope of this report.

22.6 TREATMENT AND PREVENTION

Like most arthropod-borne viruses, there are no vaccines or therapeutics commercially available for humans. Live-attenuated and inactivated vaccines are both considered investigational new drugs (INDs), but it is unlikely that these will come to licensure for human use. Additional vaccine options have been and are being developed for VEEV including further attenuated live strains, chimeric alphaviruses, DNA vaccines, and viruslike replicon particles.[67–70] None of these are currently available for human use. There are currently no treatment options available for VEEV infection either, but new arenas of investigation have suggested there might be therapeutics available in the future. One example is the development of humanized anti-VEEV antibodies[71] that could rapidly be deployed in the event of an outbreak.

Fortunately, in the case of VEEV, there is an excellent equine vaccine that is reported to have been responsible for halting the 1972/1973 epidemic that reached the southern United States.[72] If evidence of infected horses, mosquitoes, or sentinel animals from surveillance programs is detected, a multiple-component intervention is preferred to control the outbreak. The components are (1) mosquito control (including aerial ULV adulticide, larval management, and local source reduction at affected farms), (2) topical application of pesticides to horses (which can be conducted during site surveillance), (3) individual protection against mosquito bites (such as wearing long-sleeved clothing and insect repellent), and (4) vaccination of horses in affected and surrounding areas expanding outward using mosquito densities, habitat structure, and distribution of horses as a guide for prioritizing the applications.

22.7 CONCLUSIONS AND FUTURE PERSPECTIVES

In Central and South America where VEEVs circulate enzootically in continuous transmission cycles, epizootic/epidemic VEE transmission primarily involves horses and a variety of mosquito species. Because VEE is not native to North America and the only US outbreak that afforded opportunity to observe how this virus would behave occurred in a limited area of south Texas during 1971–1972 where a variety of *Psorophora* and *Aedes* mosquito species appeared to be involved, there is little background information available to infer what consequences may result from an intentional release of VEE in most areas of the United States. However, we can assume that a release in an area where horses and horse-feeding mosquito species are present may result in establishment of mosquito to horse transmission cycles combined with the associated risk of transmission to humans.

Since there are no commercially available human VEEV vaccines, rapid detection and control efforts (equine vaccination and mosquito control) are essential. Because VEEV is a zoonotic virus, outbreaks will continue to occur at least sporadically. Ideally, these emergence events will be rapidly detected and characterized because of effective surveillance programs combined with rapid and reliable diagnostic methodologies. In addition to refining existing diagnostic methods, the continued development of novel diagnostic approaches, new vaccines, and possible therapeutics will ensure that public health officials are as prepared as possible for preventing and/or controlling any human VEEV epidemics.

REFERENCES

1. Casals, J. Viruses: The versatile parasites; the arthropod-borne group of animal viruses. *Trans N Y Acad Sci* 19, 219–235 (1957).

2. Casals, J. and Brown, L.V. Hemagglutination with arthropod-borne viruses. *J Exp Med* 99, 429–449 (1954).

3. Calisher, C.H. and Karabatsos, N. Arbovirus serogroups: Definition and geographic distribution. In *The Arboviruses: Epidemiology and Ecology*, Vol. I, Monath, T.P. (ed.), pp. 19–57. CRC Press, Boca Raton, FL (1988).

4. Calisher, C.H. et al. Proposed antigenic classification of registered arboviruses. *Intervirology* 14, 229–232 (1980).

5. Powers, A. et al. Togaviridae. In *Virus Taxonomy: Ninth Report of the International Committee on Taxonomy of Viruses*, King, A.M.Q., Adams, M.J., Carstens, E.B., and Lefkowtiz, E.J. (eds.), pp. 1103–1110. Elsevier Academic Press, Amsterdam, the Netherlands (2012).

6. Powers, A.M. et al. Evolutionary relationships and systematics of the alphaviruses. *J Virol* 75, 10118–10131 (2001).

7. Walton, T.E. and Grayson, M.A. Venezuelan equine encephalomyelitis. In *The Arboviruses: Epidemiology and Ecology*, Vol. IV, Monath, T.P. (ed.), pp. 203–233. CRC Press, Boca Raton, FL (1988).

8. Young, N.A. and Johnson, K.M. Antigenic variants of Venezuelan equine encephalitis virus: Their geographic distribution and epidemiologic significance. *Am J Epidemiol* 89, 286–307 (1969).

9. Jose, J., Snyder, J.E., and Kuhn, R.J. A structural and functional perspective of alphavirus replication and assembly. *Future Microbiol* 4, 837–856 (2009).

10. Mukhopadhyay, S. et al. Mapping the structure and function of the E1 and E2 glycoproteins in alphaviruses. *Structure* 14, 63–73 (2006).

11. Strauss, J.H. and Strauss, E.G. The alphaviruses: Gene expression, replication, and evolution. *Microbiol Rev* 58, 491–562 (1994).

12. Aaskov, J. et al. Lineage replacement accompanying duplication and rapid fixation of an RNA element in the nsP3 gene in a species of alphavirus. *Virology* 410, 353–359 (2011).

13. Powers, A.M. Togaviruses: Alphaviruses. In *Encyclopedia of Virology*, 3rd ed., Vol. 5, Mahy, B.W. and van Regenmortel, M.H.V. (eds.), pp. 96–100. Elsevier, Oxford, U.K. (2009).

14. Pfeffer, M., Kinney, R.M., and Kaaden, O.R. The alphavirus 3'-nontranslated region: Size heterogeneity and arrangement of repeated sequence elements. *Virology* 240, 100–108 (1998).

15. Causey, O.R., Causey, C.E., Maroja, O.M., and Macedo, D.G. The isolation of arthropod-borne viruses, including members of hitherto undescribed serological groups, in the Amazon region of Brazil. *Am J Trop Med Hyg* 10, 227–249 (1961).

16. Johnson, K.M. et al. Recovery of Venezuelan equine encephalomyelitis virus in Panama. A fatal case in man. *Am J Trop Med Hyg* 17, 432–440 (1968).

17. Scherer, W.F. et al. Venezuelan equine encephalitis virus in Veracruz, Mexico, and the use of hamsters as sentinels. *Science* 145, 274–275 (1963).

18. Scherer, W.F., Dickerman, R.W., and Ordonez, J.V. Discovery and geographic distribution of Venezuelan encephalitis virus in Guatemala, Honduras, and British Honduras during 1965–68, and its possible movement to Central America and Mexico. *Am J Trop Med Hyg* 19, 703–711 (1970).

19. Shope, R.E., Causey, O.R., Homobono Paes de Andrade, A., and Theiler, M. The Venezuelan equine encephalomyelitis complex of group A arthropod-borne viruses, including Mucambo and Pixuna from the Amazon region of Brazil. *Am J Trop Med Hyg* 13, 723–727 (1964).

20. Ventura, A.K. and Ehrenkranz, N.J. Detection of Venezuelan equine encephalitis virus in rural communities of Southern Florida by exposure of sentinel hamsters. *Am J Trop Med Hyg* 24, 715–717 (1975).

21. Monath, T.P. et al. Recovery of Tonate virus ("Bijou Bridge" strain), a member of the Venezuelan equine encephalomyelitis virus complex, from Cliff Swallow nest bugs (*Oeciacus vicarius*) and nestling birds in North America. *Am J Trop Med Hyg* 29, 969–983 (1980).

22. Powers, A.M. et al. Repeated emergence of epidemic/epizootic Venezuelan equine encephalitis from a single genotype of enzootic subtype ID virus. *J Virol* 71, 6697–6705 (1997).

23. Weaver, S.C. et al. Genetic determinants of Venezuelan equine encephalitis emergence. *Arch Virol Suppl* 18, 43–64 (2004).

24. Chamberlain, R.W. et al. Arbovirus studies in south Florida, with emphasis on Venezuelan equine encephalomyelitis virus. *Am J Epidemiol* 89, 197–210 (1969).

25. Cupp, E.W., Scherer, W.F., and Ordonez, J.V. Transmission of Venezuelan encephalitis virus by naturally infected *Culex (Melanoconion) opisthopus*. *Am J Trop Med Hyg* 28, 1060–1063 (1979).

26. Ferro, C. et al. Natural enzootic vectors of Venezuelan equine encephalitis virus, Magdalena Valley, Colombia. *Emerg Infect Dis* 9, 49–54 (2003).

27. Galindo, P. and Grayson, M.A. *Culex (Melanoconion) aikenii*: Natural vector in Panama of endemic Venezuelan encephalitis. *Science* 172, 594–595 (1971).

28. Franck, P.T. and Johnson, K.M. An outbreak of Venezuelan encephalitis in man in the Panama Canal Zone. *Am J Trop Med Hyg* 19, 860–865 (1970).

29. Kramer, L.D. and Scherer, W.F. Vector competence of mosquitoes as a marker to distinguish Central American and Mexican epizootic from enzootic strains of Venezuelan encephalitis virus. *Am J Trop Med Hyg* 25, 336–346 (1976).

30. Sudia, W.D. et al. Epidemic Venezuelan equine encephalitis in North America in 1971: Vector studies. *Am J Epidemiol* 101, 17–35 (1975).

31. Turell, M.J., Ludwig, G.V., and Beaman, J.R. Transmission of Venezuelan equine encephalomyelitis virus by *Aedes sollicitans* and *Aedes taeniorhynchus* (Diptera: Culicidae). *J Med Entomol* 29, 62–65 (1992).

32. Hachiya, M. et al. Human eastern equine encephalitis in Massachusetts: Predictive indicators from mosquitoes collected at 10 long-term trap sites, 1979–2004. *Am J Trop Med Hyg* 76, 285–292 (2007).

33. Molaei, G. et al. Molecular identification of blood-meal sources in *Culiseta melanura* and *Culiseta morsitans* from an endemic focus of eastern equine encephalitis virus in New York. *Am J Trop Med Hyg* 75, 1140–1147 (2006).

34. Beck, C.E. and Wyckoff, R.W.G. Venezuelan equine encephalomyelitis. *Science* 88, 530 (1938).

35. Kubes, V. and Rios, F.A. The causative agent of infectious equine encephalomyelitis in Venezuela. *Science* 90, 20–21 (1939).

36. Sudia, W.D. et al. Epidemic Venezuelan equine encephalitis in North America in 1971: Vertebrate field studies. *Am J Epidemiol* 101, 36–50 (1975).

37. Baker, E.F., Jr. et al. Venezuelan equine encephalomyelitis vaccine (strain TC-83): A field study. *Am J Vet Res* 39, 1627–1631 (1978).

38. Rico-Hesse, R. et al. Emergence of a new epidemic/epizootic Venezuelan equine encephalitis virus in South America. *Proc Natl Acad Sci USA* 92, 5278–5281 (1995).

39. Oberste, M.S. et al. Association of Venezuelan equine encephalitis virus subtype IE with two equine epizootics in Mexico. *Am J Trop Med Hyg* 59, 100–107 (1998).

40. Brault, A.C. et al. Positively charged amino acid substitutions in the e2 envelope glycoprotein are associated with the emergence of Venezuelan equine encephalitis virus. *J Virol* 76, 1718–1730 (2002).

41. Adams, A.P. et al. Venezuelan equine encephalitis virus activity in the gulf coast region of Mexico, 2003–2010. *PLoS Negl Trop Dis* 6, e1875 (2012).

42. Gonzalez-Salazar, D. et al. Equine amplification and virulence of subtype IE Venezuelan equine encephalitis viruses isolated during the 1993 and 1996 Mexican epizootics. *Emerg Infect Dis* 9, 161–168 (2003).

43. Rivas, F. et al. Epidemic Venezuelan equine encephalitis in La Guajira, Colombia, 1995. *J Infect Dis* 175, 828–832 (1997).

44. Weaver, S.C. et al. Re-emergence of epidemic Venezuelan equine encephalomyelitis in South America. VEE Study Group. *Lancet* 348, 436–440 (1996).

45. Brault, A.C. et al. Potential sources of the 1995 Venezuelan equine encephalitis subtype IC epidemic. *J Virol* 75, 5823–5832 (2001).

46. Weaver, S.C. et al. Molecular epidemiological studies of veterinary arboviral encephalitides. *Vet J* 157, 123–138 (1999).

47. Watts, D.M. et al. Venezuelan equine encephalitis febrile cases among humans in the Peruvian Amazon River region. *Am J Trop Med Hyg* 58, 35–40 (1998).

48. Calisher, C.H. Medically important arboviruses of the United States and Canada. *Clin Microbiol Rev* 7, 89–116 (1994).

49. Hommel, D. et al. Association of Tonate virus (subtype IIIB of the Venezuelan equine encephalitis complex) with encephalitis in a human. *Clin Infect Dis* 30, 188–190 (2000).

50. Talarmin, A. et al. Tonate virus infection in French Guiana: Clinical aspects and seroepidemiologic study. *Am J Trop Med Hyg* 64, 274–279 (2001).

51. Ryzhikov, A.B. et al. Spread of Venezuelan equine encephalitis virus in mice olfactory tract. *Arch Virol* 140, 2243–2254 (1995).

52. Tsai, T.F. and Monath, T.P. Alphaviruses. In *Clinical Virology*, Richman, D.D., Whitley, R.J., and Hayden, F.G. (eds.). Churchill Livingstone, New York (1997).

53. Martin, D.A. et al. Standardization of immunoglobulin M capture enzyme-linked immunosorbent assays for routine diagnosis of arboviral infections. *J Clin Microbiol* 38, 1823–1826 (2000).

54. Calisher, C.H. et al. Complex-specific immunoglobulin M antibody patterns in humans infected with alphaviruses. *J Clin Microbiol* 23, 155–159 (1986).

55. Johnson, A.J. et al. Detection of anti-arboviral immunoglobulin G by using a monoclonal antibody-based capture enzyme-linked immunosorbent assay. *J Clin Microbiol* 38, 1827–1831 (2000).

56. Rico-Hesse, R., Roehrig, J.T., and Dickerman, R.W. Monoclonal antibodies define antigenic variation in the ID variety of Venezuelan equine encephalitis virus. *Am J Trop Med Hyg* 38, 187–194 (1988).

57. Roehrig, J.T. and Bolin, R.A. Monoclonal antibodies capable of distinguishing epizootic from enzootic varieties of subtype I Venezuelan equine encephalitis viruses in a rapid indirect immunofluorescence assay. *J Clin Microbiol* 35, 1887–1890 (1997).

58. Meissner, J.D. et al. Sequencing of prototype viruses in the Venezuelan equine encephalitis antigenic complex. *Virus Res* 64, 43–59 (1999).

59. Bowen, G.S. and Calisher, C.H. Virological and serological studies of Venezuelan equine encephalomyelitis in humans. *J Clin Microbiol* 4, 22–27 (1976).

60. Ni, H. et al. Recombinant alphaviruses are safe and useful serological diagnostic tools. *Am J Trop Med Hyg* 76, 774–781 (2007).

61. Charles, P.C. et al. Mucosal immunity induced by parenteral immunization with a live attenuated Venezuelan equine encephalitis virus vaccine candidate. *Virology* 228, 153–160 (1997).

62. Linssen, B. et al. Development of reverse transcription-PCR assays specific for detection of equine encephalitis viruses. *J Clin Microbiol* 38, 1527–1535 (2000).

63. Pfeffer, M. et al. Genus-specific detection of alphaviruses by a semi-nested reverse transcription-polymerase chain reaction. *Am J Trop Med Hyg* 57, 709–718 (1997).

64. Lambert, A.J., Martin, D.A., and Lanciotti, R.S. Detection of North American eastern and Western equine encephalitis viruses by nucleic acid amplification assays. *J Clin Microbiol* 41, 379–385 (2003).

65. Parida, M.M. Rapid and real-time detection technologies for emerging viruses of biomedical importance. *J Biosci* 33, 617–628 (2008).

66. Johnson, A.J. et al. Duplex microsphere-based immunoassay for detection of anti-West Nile virus and anti-St. Louis encephalitis virus immunoglobulin m antibodies. *Clin Diagn Lab Immunol* 12, 566–574 (2005).

67. Konopka, J.L. et al. Acute infection with Venezuelan equine encephalitis virus replicon particles catalyzes a systemic antiviral state and protects from lethal virus challenge. *J Virol* 83, 12432–12442 (2009).

68. Fine, D.L. et al. Venezuelan equine encephalitis virus vaccine candidate (V3526) safety, immunogenicity and efficacy in horses. *Vaccine* 25, 1868–1876 (2007).

69. Dupuy, L.C. et al. Directed molecular evolution improves the immunogenicity and protective efficacy of a Venezuelan equine encephalitis virus DNA vaccine. *Vaccine* 27, 4152–4160 (2009).

70. Paessler, S. et al. Recombinant Sindbis/Venezuelan equine encephalitis virus is highly attenuated and immunogenic. *J Virol* 77, 9278–9286 (2003).

71. Hunt, A.R. et al. Treatment of mice with human monoclonal antibody 24h after lethal aerosol challenge with virulent Venezuelan equine encephalitis virus prevents disease but not infection. *Virology* 414, 146–152 (2011).

72. Zehmer, R.B. et al. Venezuelan equine encephalitis epidemic in Texas, 1971. *Health Serv Rep* 89, 278–282 (1974).

23 Yellow Fever Virus

Goro Kuno

CONTENTS

23.1 INTRODUCTION

Yellow fever (YF) is a viral disease transmitted to subhuman primates and humans by the bite of mosquitoes. Until the development of an effective vaccine in the 1930s, for several centuries, YF had been one of the most feared infectious diseases of the world not only for its high mortality but for the enormous social and economic impacts. The virulent trait was even exploited in bioterrorism as early as 1894; and during the Cold War, an aerial dissemination trial of uninfected *Aedes aegypti* mosquitoes was conducted to determine their survival, with an ultimate objective of developing YF virus (YFV)-infected mosquitoes for biological warfare.

Although the oldest accounts of YF-like disease outbreaks were recorded in 1649 in the West Indies, in 1686 in Pernambuco, Brazil, and in 1768 in Senegal, it is generally accepted that Africa is the birthplace of YFV and that both YFV and its main vector, *Ae. aegypti*, subsequently dispersed to the New World presumably through slave trade. The development of an effective vaccine in 1937 clearly contributed to the dramatic decline of YF cases; and the spectacular advancement in virology helped deciphering the viral genes and functions at molecular levels. YF outbreak in major urban centers in North America and in South America has not occurred since 1905 and 1928–1929, respectively.

From these medical advancements, however, arose a false sense of security even among many medical professionals that YF was a disease of the past. In reality, YF still poses a serious threat because of totally inadequate vaccination programs in too many endemic countries, lack of progress in controlling domesticated vectors in urban environments, and increased international travel by jets that brings exotic disease agents, such as YFV, rapidly to susceptible human populations far away.

23.2 CLASSIFICATION AND MORPHOLOGY

Among more than 70 viruses (hereafter called flaviviruses) that belong to the genus *Flavivirus* in the family Flaviviridae, YFV is a member of the mosquito-borne subgroup transmitted between mosquitoes (vectors) and vertebrates.

23.2.1 VIRION

Flaviviral virions are icosahedral particles with a diameter of around 50 nm containing an electron-dense core (around 30 nm in diameter), which is surrounded by an envelope. The envelope contains two viral structural proteins (envelope [E] and membrane [M]) and a small amount of lipids. Cryo-electron microscopy reveals head-to-tail configuration of E protein dimers lying parallel to the lipid bilayer and 90 E protein dimers tightly packed in an unusual herring pattern of icosahedral symmetry. In infected hosts, the virus also releases small (around 14 nm in diameter) noninfectious particles (made of E and M proteins), which are called "slowly sedimenting hemagglutinin" (SHA). Like complete virions, SHAs aggregate red blood cells at low pH. The size of recombinant subviral particles (RSP) used in serodiagnosis, research, and as candidate for a potential vaccine is around 30 nm in diameter.

23.2.2 NUCLEIC ACID, GENOME ORGANIZATION, AND GENE FUNCTIONS

The nucleic acid of YFV is a positive-sense, single-strand RNA. The genome generates a single open-reading frame (ORF) flanked by 5′ and 3′ untranslated regions (UTRs) of 121 nt and 440–660 nt, respectively. The genome organization and functions are shown in the direction of 5′ head to 3′ tail in Table 23.1. For detailed information about molecular biology of replication and gene functions of YFV, readers are recommended to consult with a publication by Lindenbach et al.[1]

23.2.3 VIRULENCE, VISCEROTROPISMS, NEUROTROPISM, HOST RANGE, AND EPIZOOTIC STRAIN

In flaviviruses, multiple genes are believed to be involved in each of these phenotypic expressions. However, it is more likely that some genes (or genomic segments) are more important than the others. For example, more important molecular determinants of host range are suspected to reside in multiple NS genes, although prM and E genes are also involved.[2] As for virulence determinants, one or more mutations in the NS protein genes and/or 3′UTR of YFV are most likely involved.

23.2.4 PHYLOGENETIC RELATIONSHIP AMONG FLAVIVIRUSES CLOSELY RELATED TO YFV

A group of about 10 flaviviruses phylogenetically clustered in one major branch as YFV lineage contain more viruses with distribution in Africa than elsewhere. This has been used as a supporting evidence for the African origin of YFV. Nonetheless, the distributions of Sepik virus (SEPV), the virus very close to YFV, only in New Guinea and of another YFV lineage virus, Edge Hill virus (EHV), only in Australia (Figure 23.1) are puzzling.[3]

23.2.5 GENOTYPES OF YFV

Phylogenetic studies of the YFV strains isolated in Africa and the Americas revealed five genotypes in Africa (African genotype I [Ivory Coast]; African genotype II [Nigeria, Senegal, Gambia]; Central–South African genotype [Angola]; East African genotype [Uganda]; East–Central African genotype [Ethiopia]) and two genotypes in the Americas (Americas genotype I [Brazil, Colombia, Ecuador, Trinidad, Venezuela]; Americas genotype II [primarily Bolivia, Ecuador, and Peru but including sporadic isolates in Brazil]). Phylogenetic studies have suggested that YFV originated in East or Central Africa and was subsequently introduced to West Africa and South America approximately three to four centuries ago and that the two genotypes of the Americas had been derived from a shared origin in Brazil.

23.3 BIOLOGY AND EPIDEMIOLOGY

23.3.1 BIOLOGICAL TRANSMISSION

In this mode of transmission in nature, YFV is maintained in mosquito vectors that transmit the virus to subhuman primates (monkeys) through blood feeding. Uninfected female mosquitoes feeding on infected monkeys (with the virus being

TABLE 23.1

Organization of YFV Genome and Functions of the Genome Segments or Genes

Region or Gene	Length (AAs or nt)	Gene or Protein Function
5′UTR	120 nt	Involved in translation, cyclization of viral RNA, replication.
Capsid (C)	121 AAs	Together with genomic RNA, constitutes nucleocapsids.
prM (M)	164 AAs	M is a small proteolytic fragment of the precursor prM protein that is generated during maturation of nascent viral particles.
		Involved in maintenance of a proper configuration of E glycoprotein and prevents E from undergoing acid-catalyzed rearrangement to fusogenic form.
E	493 AAs	Is a glycoprotein and the major antigenic determinant in immune response, mediates binding to cell receptors and intracellular fusion during the cell entry process.
Nonstructural protein—NS1	352 AAs	Is a glycoprotein likely involved intracellularly in RNA replication. The function of the extracellular form of NS1 remains unknown.
NS2A	224 AAs	Is involved in viral assembly and functions as interferon antagonist.
NS2B	130 AAs	Is involved in autoproteolytic cleavage at NS2B/NS3 junction.
NS3	623 AAs	Is a serine protease to cleave at NS2A/NS2B, NS2B/NS3, NS3/NS4A, and NS4B/NS5 junctions. Contains two functional domains: RNA helicase and RNA triphosphatase.
NS4A	126 AAs	Involved in RNA replication.
2K	23 AAs	Signal peptide.
NS4B	250 AAs	IFN signaling antagonist.
NS5	905 AAs	Contains two functional domains: methyltransferase, RNA-dependent RNA polymerase (RdRp).
3′UTR	Variable length (444–660 nt)	Involved in RNA cyclization, translation, and replication.
ORF	3411 AAs	

circulated in blood or in viremia) acquire the virus through blood feeding. The virus replicates in multiple organs of the vectors; and when viral concentration in the salivary gland rises to a sufficient level, the mosquitoes are ready to transmit the virus back to susceptible monkeys when they feed, thus completing the biological transmission cycle. Humans are normally accidental dead-end hosts in sylvan environments or savannah. Mosquitoes, once infected, remain infected for the rest of their life. Although the efficacy is low, YFV is transmitted vertically to the following generations of mosquitoes via infected eggs. Thus, mosquito vectors are the true reservoir of the virus in nature.

In monkeys, the infectious virions persist only for short periods after infection. The outcome of infection is either elimination of the virus from the vertebrate hosts by host's immune responses or death of the hosts. Either way, the size of susceptible primate host population is reduced, leading to a lower frequency of transmission, unless high reproductive rate of the vertebrate hosts compensates the loss or immigration of susceptible monkeys to the endemic area occurs or vectors move away to find a new susceptible host

population elsewhere. This partly explains somewhat cyclical epizootics in monkey populations. In urban environments, transmission between *Ae. aegypti* mosquitoes and humans occurs when YFV is introduced by infected mosquitoes or humans.

While YFV and dengue virus (DENV) can be transmitted in urban environments by the same vector, there exists a marked contrast between the two viruses. While persistent endemic urban dengue transmission is common in large population centers in the tropics, this does not happen in YF. When YF epidemic occurred in urban centers in parts of Africa, it typically occurred in urban areas socioeconomically and ecologically mixed with or adjacent very close to rural YF-endemic environments. This is partly because, unlike DENV, YFV genotypes have not fully adapted to perpetual transmission strictly in urban environments and still depend on association with sylvan environments. Thus, the mechanism involved in the YF outbreaks in urban areas in the past has been sporadic incursion of sylvan strains that typically died out at the end of outbreak regardless of the size of human population.[4]

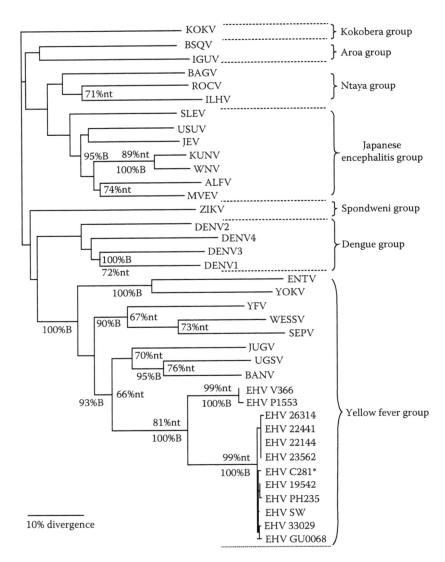

FIGURE 23.1 Phylogenetic relationships of YFV in the mosquito-borne subgroup flaviviruses. (From Macdonald, J. et al., *Evol. Bioinform.*, 6, 91, 2010.)

23.3.1.1 Vectors

The vector species involved vary depending on location. The species involved in four types of transmission cycles are as follows: (1) savannah–sylvan cycle in West Africa (*Ae. africanus, Ae. furcifer, Ae. taylori, Ae. opok, Ae. vittatus*); (2) Central–East African cycle (*Ae. africanus* and *Ae. opok* serving as bridge vectors and *Ae. bromeliae* and *Ae. africanus* serving as peridomestic vectors); (3) New World jungle cycle (*Haemagogus* and *Sabethes* spp.); (4) urban cycle in both Africa and the New World (*Ae. aegypti*). *Ae. albopictus* of Asian origin has spread to tropical Americas and Africa, but little is known about its competence in YF transmission in nature, although it was found competent under laboratory conditions.

23.3.1.2 Vertebrate Hosts

Many species of cercopithecid (i.e., *Cercopithecus* spp.) and colobid (i.e., *Colubus* spp.) monkeys are the primary hosts in Africa, while patas monkeys (*Erythrocebus* spp.)

are widespread in dry savannah regions, savannah woodlands, and gallery forests. Baboons (*Papio* spp.) have also been implicated, while the involvement of galagos in Africa is questionable. In the Americas, although howler monkeys (*Alouatta* spp.) are the most important species, spider monkeys (*Ateles* spp.) and owl monkeys (*Aotus* spp.) are also hosts of YFV, developing often fatal infection. Capuchin monkeys (*Cebus* spp.) and wooly monkeys (*Lagothrix* spp.) develop viremia, but do not develop clinical signs or symptoms. A few species of marmosets are apparently involved in jungle transmission.

23.3.1.3 Animal Model

Although African monkeys are infected, they do not present the typical YF syndrome, while the monkeys from India (i.e., *Macaca mulatta*) and howler monkeys of the Americas demonstrate severe and fatal syndrome. Golden hamsters (*Mesocricetus auratus*) have been recently proposed to be an ideal model because of hemorrhage and clinical characteristics in blood and histological alterations similar to those in the liver of humans.

23.3.2 DIRECT TRANSMISSION

The Walter Reed Commission at the turn of the twentieth century conclusively established that YF is not transmitted by fomites (beds, clothing, pillows, etc., used by patients). But, contact of normal skin (without an abrasion) with YFV may result in infection. Intranasal infection through respiration of airborne YFV has been suspected strongly in multiple reports of infections that occurred in the absence of an invasive accident in mosquito-free laboratories in association with autopsy or handling of infectious clinical specimens. In fact, nasopharyngeal, direct transmission (intranasal and oral) of YFV has been experimentally confirmed in monkeys and bats.[5] One case of transmission to infant by breast feeding in a woman shortly after vaccination was also reported. In a laboratory-confirmed case of perinatal transmission, a woman in late pregnancy was naturally infected. An infant girl born 3 days after the onset of illness of the mother became symptomatic on the 3rd day of life, presenting severe bleeding, oliguria, hypotension, hepatomegaly, jaundice, and tachycardia, and died on the 12th day of life.[6]

23.3.3 EPIDEMIOLOGY

23.3.3.1 Endemicity of YF

As of 2012, YF is endemic in 35 countries (including the newly independent Republic of South Sudan) stretching from Senegal to Kenya, 11 countries in South America, and Trinidad in the Caribbean (Figure 23.2). The other areas of the world infested by *Ae. aegypti* are considered "receptive areas" because the potential of YF outbreak exists. Ecologically, the endemic regions in Africa are rainforest, gallery forest, humid savannah, and dry savannah. In the Americas, YF occurs typically in or around the forest, feral, and/or rural environments. YF outbreak in major urban centers is always a concern, but the last such outbreak in the Americas occurred in 1929. One major medical puzzle has been absence of YF in Asia and the Pacific infested by *Ae. aegypti* despite proven vector competence and lack of racial difference in human susceptibility. In fact, none of nearly a dozen hypotheses proposed has gained a general acceptance. The only exceptional case of YF in the Pacific occurred in Hawaii in 1911. Also, the scarcity of YF outbreak in East Africa had been enigmatic until an outbreak occurred in Kenya in 1993.

23.3.3.2 Pattern of Outbreak Occurrence

The pattern of occurrence of YF outbreak and the determinants behind it have been the major focus of research for many decades. Accumulated records reveal that periods of increased activity lasting as long as several years are followed by variable lengths of relative inactivity, in both Africa and South America. In the savannah of Western Africa, cases appear often in the midst of rainy season (August) and peak

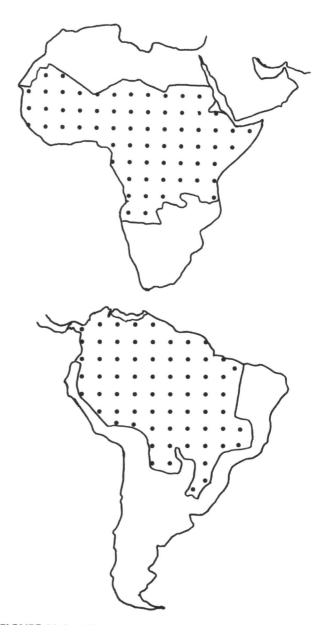

FIGURE 23.2 YF-endemic areas of the world. The dotted areas represent YF-endemic areas in Africa and South America. Trinidad in the Caribbean, where YF is endemic, is not shown.

during the early dry season (October). In Brazil, between 1954 and 2008, the patterns in central–western and northern regions roughly fit the 7- and 14-year cycles, respectively. This corresponded to the depletion of susceptible monkeys and subsequent rise of YF activity associated with rebounding of monkey populations.

23.3.3.3 Morbidity and Mortality

The number of human cases per annum and case fatality rate (CFR) vary a great deal from year to year and are unpredictable. Nonetheless, annual morbidity and mortality worldwide are estimated to be around 200,000 cases (with 90% of which occurring in Africa) and approximately 30,000 cases, respectively. The CFRs in Africa range from 15% to 35%, while in

South America, they range from 25% to over 50%. Although both sexes are susceptible, mortality is higher in males than in females because of gender difference in outdoor activities (and hence in exposure rate). Thus, in Brazil, more than 80% of the cases occurred in males.

23.3.3.4 Risk Factors

Sporadic cases typically occur nonclustered in time and place in unvaccinated individuals who are exposed to the vectors in endemic areas for a variety of reasons (i.e., logging, road or dam construction, mining, food gathering, hunting or fishing, recreation). The risks (number of cases per 100,000) of contracting YF or death due to YF for the travelers from nonrisk countries are estimated to be 50 and 10 in West Africa, while in South America, they are 5 and 1, respectively. Infants and elderly persons are more often associated with manifestation of more severe syndrome. Little is known if genetic factors of human contribute to variation in severity, although CFRs in Africa have been generally lower than in South America.

23.4 CLINICAL FEATURES AND PATHOGENESIS

23.4.1 CLINICAL DIAGNOSIS

The classic triad of symptoms (jaundice, hemorrhage, and intense albuminuria) is manifested only in a small proportion of infections.[7] Thus, 50%–85% of the infected individuals are totally asymptomatic or only very mildly symptomatic.

23.4.1.1 Incubation Period

The length of incubation period in humans varies. It is usually 3–6 days after infection but may be longer in cases of accidental direct transmission. During this period, YFV is believed to replicate first in dendritic cells, leading to expression of CD38 and later in macrophages and antigen-presenting lymphoid cells. Thereafter, YFV disseminates first to Kupffer cells of the liver within 24 h. In the following 1–2 days, infection spreads to multiple organs. In defense, natural killer cells in the liver release IFN-γ; and activated CD4$^+$ T cells release IFN-γ and TFN-α. CD8$^+$ T cells lyse infected cells, releasing virions, which are neutralized by specific antibodies released by B cells.

23.4.1.2 Three Phases of YF

Initial phase: Prodromal symptoms are usually absent, and symptoms emerge suddenly. This phase, which may last up to 48 h, is characterized by flushed face or neck, fever (>39°C), malaise, generalized myalgia, nausea, vomiting, irritability, sleeplessness, and/or dizziness. Mild epistaxis and Faget's sign (slow pulse associated with high fever) may be observed in some patients. The margins and tip of the tongue are bright red with the center furred, and the gums are congested. Laboratory tests often reveal leucopenia and elevated levels of serum transaminases. In very mild cases, the only symptoms (fever and headache) disappear within a day or so. In a small number of abortive infections, symptoms may be severe at the onset, but patients recover rapidly.

Period of remission: In this phase, temperature and blood pressure drop and symptoms subside considerably for a few hours to 1–2 days. In many cases of abortive infections, anicteric patients recover at this phase.

Period of intoxication: This phase is characterized by a return of fever, vomiting, epigastric pain, prostration, dehydration, sudden increase in albumin in the urine, bleeding diathesis (including gastric bleeding due to ecchymosis of the mucosa and coffee-ground hematemesis [or "black vomit"] and melena), albuminuria, and oliguria. Multiorgan system failure involving the liver, the heart, the kidneys, the blood vessels, and the vessels of the brain is common. In the blood, serum transaminase levels are proportional to the severity of disease. Approximately, 15%–30% of symptomatic cases in West Africa (but 35%–65% in South America) develop moderate to severe syndrome characterized by jaundice, which stresses the fact that jaundice is often not a prominent symptom of YF.[7] The period of intoxication is usually 3–4 days but may be as long as 2 weeks in exceptional cases. In severe cases, coma may set in 1–2 days before death. Fatal outcome occurs usually between the 7th and 10th day after onset and is characterized by intensified jaundice with extensive degeneration of hepatocytes, hemorrhages, bradycardia, and/or hypotension. Infrequently, fulminant and fatal cases occur as early as the third day after onset without a period of remission. Shortly before death, many patients become delirious and wildly agitated, most likely as a result of perivascular hemorrhages in the brain. Myocardial failure may occur in some fully recovered patients, although no YF-associated sequelae have been recognized.

23.4.1.3 Blood Analyses

Leucopenia is a fairly consistent finding. Serum transaminase levels (AST and ALT) rise on the second and third day, AST level being higher than ALT level, which is different from viral hepatitis in which ALT levels are far higher than AST levels.[8] Albuminuria is also common. Coagulation defects are manifested in prolonged prothrombin, partial thromboplastin, and clotting times; reduced levels of Factors II, V, VII, VIII, IX, and X and fibrinogen; and thrombocytopenia. Increased fibrin split products are signs of disseminated intravascular coagulation (DIC). Decreased synthesis of vitamin K-dependent coagulation factors by the liver, platelet dysfunction, and DIC together contribute to the bleeding diathesis.

23.4.2 PATHOLOGIES IN HUMANS

Liver: The pathologies of the liver typically include the damages to hepatocytes, such as steatosis, apoptosis, and necrosis. Hepatic failure itself is not necessarily the primary cause of death. Despite the severity of jaundice, the liver is normal or slightly enlarged and appears reddish or yellow. Microscopic examination reveals cloudy swelling and coagulative necrosis of hepatocytes in the midzonal area of the lobule. However, the distribution of the necrotic process is not uniform, and the proportion of the lobule in the necrotic process varies from 5% to 100%, with an average of about 60%.[9] In most cases, a ring of relatively normal cells remains around the central veins and portal areas. Eosinophilic signs of degeneration of hepatocytes (Councilman bodies) are present, and occasionally intranuclear eosinophilic granular inclusions (Torres bodies) are observed. Apoptosis of Kupffer cells and hepatocytes may be seen. Fatty infiltration is a consistent finding, but inflammation is either absent or minimal.[8] Like in leptospirosis, TNF-α and IFN-γ are expressed, but TGF-β expression in YF is more intense, suggesting both Th1 and Th3 patterns of immune response. Also, the immune response to hepatocytic lesions includes expression of CD45RO, CD4, CD8, CD20, S100, CD57, and CD68. CD4+ T cells in the presence of CD8+ T cells, natural killers (CD57), macrophages, and antigen-presenting cells (S100) are commonly observed.[8]

Heart: On gross examination, the heart is flabby with a variable amount of subserous petechial hemorrhage and moderate dilation of the right ventricle. Icterus is often visible. Upon sectioning, the myocardium is cloudy, icteric, and soft. Patchy subendocardial fatty changes extend deeply into the myocardium. The nuclei of the muscle fibers tend to be enlarged and often show necrosis with karyolysis and vacuolization of the cytoplasm.[9]

Kidneys: On gross examination, the kidney is plump and turgid, with a variable amount of icterus visible, with or without subcapsular hemorrhages. Upon sectioning, the kidney is found cloudy, swollen, and gray pink in color. The capsular space is distended, and the cells lining Bowman's capsule reveal desquamation and massive necrosis. Small globules of fat are found in the capsular epithelium. The tubular epithelium shows severe changes.[9]

Brain: On gross examination, no obvious abnormality is evident, except for a small number of hemorrhagic spots. The pathologies include petechiae, perivascular hemorrhage, and edema. Cerebral edema may be an important component of the encephalopathy observed during the terminal phase of infection. Although paralysis, cranial nerve palsy, optic neuritis, and other encephalitic syndromes have been observed, they are the manifestations of encephalopathy, because true YFV infection of the central nervous system has not been definitively established.

23.4.3 Vaccine-Associated Adverse Responses

The rare but severe adverse reactions to YF vaccination have become a hot issue lately. However, actually, they have been reported for many years since the development of YF vaccine. Three kinds of such severe reactions are vaccine-associated neurotropic disease (YEL-AND) characterized by encephalitis syndrome, YF vaccine-associated viscerotropic disease (YEL-AVD) characterized by jaundice and multiorgan failure, and hypersensitivity reactions (such as

anaphylaxis and urticaria) in the vaccine recipients allergic to eggs or gelatin. The incidence of YEL-AND for all age groups is one case per 150,000–250,000 doses, while the incidence of YEL-AVD is one case per 200,000–300,000 doses; but for the elderly group (>60 year), it is one case per 40,000–50,000 doses. YEL-AVD occurs more frequently in younger women with defects in innate immunity for whom CFR is higher.

23.4.4 Pathologies in Animal Models

In nature, only nonhuman primates (monkeys) develop viscerotropic infections (hepatitis). Although subhuman primates develop viremia, most African monkeys do not present comparable YF syndrome; while monkeys from India (such as rhesus monkey) and several species in South America present severe YF syndrome including development of fulminant hepatitis, renal failure, coagulation, and shock. Generally, death occurs more rapidly (3–6 days) in monkeys than in humans. DIC also occurs in rhesus monkeys. As in human infections, CNS signs (such as depression, stupor, coma) occur during the terminal phase and appear to be metabolic in origin. In rodents (i.e., mice, hamsters, and guinea pigs), the virus is principally neurotropic. The only extraneural organ showing viral replication is the adrenal gland. Rodents develop encephalitis only after intracerebral, intraocular, or intranasal inoculation. The time from infection to death is typically 5–9 days, depending on virus strain and passage history. The only nonprimate species that develops pathologies in response to experimental infection is the European hedgehog, but recently golden hamster was found to produce histopathologies, hemorrhage, and thrombocytopenia similar to those in humans and rhesus monkeys. In immunosuppressed hamsters, fatal histopathologies comparable to hepatic changes of severe YF in humans (such as inflammation, hepatocyte necrosis, and stenosis) develop by intraperitoneal inoculation of wild-type YFV strains.

23.5 IDENTIFICATION AND DIAGNOSIS

23.5.1 Considerations in Differential Diagnosis

Most ideally, a case of YF should be definitively confirmed by a combination of clinical diagnosis, laboratory diagnosis, latest epidemiologic condition, and medical and travel histories of the patient. Case definitions typically provided by miscellaneous international health-care organizations including the World Health Organization and governmental agencies are largely based on manifestation of clinical syndrome and epidemiologic information alone. They are useful for an outbreak monitoring once the cause of the outbreak is definitively identified on at least several cases. However, they should not be strongly emphasized for sporadic cases or at the early phase of an outbreak when the number of patients with similar syndrome is small. Such a precaution is based on the facts that (1) all mild cases cannot be clinically diagnosed with confidence; (2) "typical signs and symptoms" are not manifested in all YF patients; and (3) differential diagnosis is difficult because of the presence of multiple agents presenting syndromes indistinguishable from that observed in the early phase of YF.

Because of the difficulty of differential diagnosis of YF, at a minimum, a possibility of the involvement of the following hemorrhagic viruses must be considered initially, in particular when the occurrence of suspected cases is sporadic: filoviruses (Ebola and Marburg viruses); arenaviruses (Chapare, Guanarito, Junin, Lassa fever, Machupo, Sabia viruses); bunyaviruses (Crimean–Congo hemorrhagic fever, Hanta, and Rift Valley fever viruses); and flaviviruses (Alkhurma, dengue, and Kyasanur forest viruses).

23.5.2 Importance of Laboratory Diagnosis

Laboratory diagnosis is indispensable in YF case diagnosis. Selection of a technique depends on the kind and quality of the specimens and at which stage of infection they were obtained. The traditional procedure consists of two components: virus isolation from blood in acute phase of illness and serologic tests. In current practices, molecular diagnosis constitutes the third component. Because no single technique is perfect in all respects under all circumstances, it is imperative that diagnosticians be fully knowledgeable regarding the limitations, advantages, and drawbacks of each technique used and when a particular technique is applicable under a particular circumstance. As reported in a recent international quality assessment study, variation of accuracy and false result in YF diagnosis among laboratories of multiple countries are the persistent problems.[10]

23.5.2.1 Epidemiologic Information and Personal Medical History

For each suspected case, the latest epidemiologic information regarding occurrence of infectious diseases in the areas where the suspected case occurred or exposure to YFV by mosquito bite could have occurred, personal history of related syndrome, vaccination history against any flaviviruses, and recent travel record of the patient are critically important for assessing the significance of the clinical and laboratory diagnostic data obtained.

23.5.2.2 Virus Isolation

The combination of virus isolation and compatible clinical syndrome observed in a patient is generally regarded a strong indication of the etiology. With the advent of molecular techniques, which are generally far superior in sensitivity and generate results faster, fewer people employ virus isolation today on the mistaken belief that molecular techniques replaced cumbersome, less sensitive, and time-consuming virus isolation. However, molecular techniques can never replace virus isolation. It should be noted that, once isolated, virus strains can be propagated and used in all laboratory experiments including viral characterization, histopathological studies in an animal model, production of monoclonal antibodies, and molecular epidemiology. Furthermore, they can be shared with other institutions. On the other hand, the application of the data generated by molecular techniques (such as real-time RT-PCR and microarray) is severely limited because these techniques are basically only detection–identification techniques. The data generated by these techniques, while providing identification of etiologic agents very quickly and being highly valuable, cannot be used in any virologic experimentation or for further characterization of the virus. Thus, virus isolation in conjunction with other techniques should be attempted as much as possible. It should be noted that YFV is not as readily isolated from acute phase (≤ 3 days after onset) blood specimens as in dengue, even though YFV was isolated in one case from blood as late as the seventh day after onset of illness. This virus is more often isolated from vector mosquitoes captured in the area of YF cases or epidemic than from patients. YFV is generally isolated in susceptible cell cultures, such as AP-61 cell line derived from *Aedes pseudoscutellaris* or C6/36 clone of *Ae. albopictus*. Cell lines derived from monkeys, such as Vero, are also used, even though the sensitivities are lower, in particular with wild strains of YFV. The total length of time between sample collection and virus identification takes more than 4–5 days, a serious disadvantage when immediate diagnosis is required.

23.5.2.3 Serologic Techniques

The old methods, such as hemagglutination inhibition test (HI) and complement fixation test (CF), rely on significant (>4-fold) rise in titer of reacting antibodies in blood between the acute phase (usually within 7 days after onset) and convalescent phase (usually 10–14 days after onset) of illness. Although relatively inexpensive, these techniques are rarely used today because of multiple disadvantages: (1) that results are not made available in acute phase when they are needed urgently, because second blood samples in convalescent phase must be obtained, and (2) that etiology is established only as an unidentified member of flavivirus group but not as a specific flavivirus. CF tests are more virus-specific than HI tests, but CF antibodies are generally detectable only for less than a few months after onset of illness. Furthermore, in many patients, CF antibodies are not induced sufficiently, contributing to false-negative results. Also, the procedure is too cumbersome for many laboratorians to use routinely.[11]

Plaque reduction neutralization test (PRNT) is the most virus-specific serologic test in primary flaviviral infections. However, like HI and CF tests, it requires paired serum specimens and cannot be used in rapid diagnosis in acute phase of illness. Furthermore, in secondary flaviviral infections or in patients who had a prior flaviviral vaccination, identification of the etiologic agents is difficult due to induction of high neutralizing antibody (Nt) titers to multiple mosquito-borne flaviviruses.[11]

Today, in most laboratories, enzyme-linked immunosorbent assay (ELISA) is the serologic test of choice. Direct antibody-capture methods are far more common than indirect methods. Among multiple classes of specific antibodies, IgM (rather than IgG or IgA) detection on single specimens is most popular. IgM detection on single specimens alone does not always establish a definitive diagnosis because in some YF vaccines, specific IgM persists in blood for as long as 18 months. However, if a reliable, relative ratio of IgG and IgM titers is used as a criterion, accuracy of etiologic identification in ELISA tests is improved because specific IgM is generally detectable only for a few months after the onset of illness. ELISA tests may be performed using multiwell plastic plates or membrane strips precoated with reactants (enzyme and chromogen) as in immunochromatographic kits. As of 2012, the only commercial diagnostic kit on the market specifically for YF diagnosis is an IgG/IgM immunofluorescence kit (EUROIMMUN, Lübeck, Germany).

23.5.2.4 Electron Microscopy

Transmission electron microscopy is useful to rapidly narrow down the range of possible groups of viruses so that appropriate reagents, primers, and/or procedures can be selected for a definitive identification of the unknown viruses in subsequent tests. Negative staining and immune electron microscopy are the two techniques most often applied for etiologic identification. The provisional identification of many viral agents is based on a combination of morphologic traits of virus with other aforementioned sets of clinical and epidemiologic information.

23.5.2.5 Molecular Diagnostic Techniques

Most rapid (real-time) molecular techniques are designed to detect viral genome (or its fragments) through nucleic acid amplification within 1–2 h. Thus, these techniques are most suitable for rapid diagnosis under emergency conditions. The clinical specimens used for these techniques include blood, cerebrospinal fluid, pleural fluid, liver biopsy, urine, and saliva specimens. Selected YFV-specific primers and probes used in RT-PCR and real-time RT-PCR are shown in Table 23.2.[12–20] When absolutely nothing is known about the etiologic agent involved, it is prudent to apply a variety of techniques designed specifically for unknowns.[12–14] Before application of these methods, cells and mitochondria are removed and virus particles are isolated by centrifugation. Cellular DNA is then removed by DNase. The five types of these methods are (1) sequence-independent single primer amplification (SISPA), (2) random PCR, (3) microarray, (4) panmicrobial array, and (5) next-generation sequencing (NGS).

SISPA requires the direct ligation of an asymmetric oligonucleotide linker or primer onto the target population of blunt-ended DNA molecules. The method is useful when viremia levels are high. A primer that is complementary to the common end sequence is used in subsequent PCR amplification. Amplified DNA bands in gel are excised from the gel and sequenced. Generally, the disadvantage of the technique is that numerous sequencing attempts are necessary before the unknown is identified conclusively because the sequences obtained are short. The detailed procedure is found in the article by Ambrose and Clewley.[13]

Random PCR: In this method, sample preparation and nucleic acid extraction processes are the same as in SISPA. A primer with a known specific 5′-end sequence is used to combine with a degenerate 3′-end sequence. The degenerate 3′-end sequence usually consists of a hexamer attached to a 15-mer of any nucleotide base. A mixture of primers will bind stochastically across the nucleic acid of unknown sequence, and the nucleic acid is subsequently linear-amplified with a suitable polymerase. The PCR products are recovered from gel and sequenced. Unknown viral sequence is obtained from a large number of such products. The sequences of random primers are shown in Table 23.2. Large sizes of fragments amplified (>1000 bp) are advantageous, but it is often hard to distinguish viral nucleic acid from contaminating DNA.

Microarrays: The methods are useful when the agent of unknown etiology is strongly suspected to be related to a family of viruses already characterized. In some commercial products, nearly 15,000 viral oligonucleotides derived from GenBank and other databanks are employed for a hybridization with unknown. If hybridization is successful, the nucleic acid of the unknown is amplified by random PCR, sequenced, and unknown identified.

Panmicrobial array: This variation of microarray designed for a rapid, differential diagnosis of infectious diseases employs nearly 30,000 sixty-mer oligonucleotide probes for vertebrate viruses (including YFV), bacteria, fungi, and parasites. The procedure entails nucleic acid preparation and random-primed PCR amplification.[12]

NGS: In this technique, total RNA is extracted from patients, purified, and treated with DNase. cDNA is generated using random octamer primer with a 17-mer anchor sequence and a reverse transcriptase and incubating in a specific cycle amplification program. Residual RNA is removed by RNase H. The cDNA library thus obtained is amplified with random octamer primer and a primer with a 17-mer anchor sequence with an additional CGCC 5′ extension, followed by further amplification. After removal of unincorporated nucleotides and small fragments, library fragments are directly ligated to adaptors for 454 NGS (Roche Diagnostics, Indianapolis, United States). Emulsion-based PCR and pyrosequencing are carried out, and the sequence reads are analyzed, using a series of bioinformatics tools.[14]

23.6 TREATMENT AND PREVENTION

23.6.1 ANTIVIRAL DRUGS

Currently, there is no effective drug for the treatment of YF. Thus far, nearly all potentially useful antiviral drugs

TABLE 23.2

Characteristics of Selected Primers and Probes for the Detection and/or Identification of YFV

Specificity	Amplification Format	Primer Sequence (5'>3')	Reference
Flavivirus group-specific	RT-PCR	MA(f) catgatgggRaaRagRgaRRAg	
		cFD2(r) gtgtcccagccggcggtgtcatcagc	Kuno [15]
Flavivirus group-specific	RT-PCR	FS778(f) aaRggHagYMcDgcHatHtggt	
		MAMD(f) aacatgatgggRaaRagRgaRaa	
		cFD2(r) as above	Scaramozzino et al. [16]
Flavivirus group-specific	RT-PCR	PF1s (f) tgYRtBtaYaacatgatggg	Crochu et al. [17]
		PF2 bis(r) gtgtcccaIccNgcNgtRtc	Morreau et al. [18]
YFV-specific	Two-step RT-PCR	YF1955F2 agccccctgcaggattccagtgatag	
		YF2200R cgttccacgccctttcatggtctgact	Kuno [15]
YFV-specific	Real-time RT-PCR	YFS aatcgagttgctaggcaataaacac	
		YFSs tccctgagctttacgaccaga	
		YFP FAM-atcgttcgttgagcgattagcag-TX	Drosten et al. [19]
YFV-specific	Real-time RT-PCR	mFU1 tacaacatgatgggaaagcgagagaaaaa	
		cFD2 as above	
		YFP TX-tcagagacctggctgcaatggatggt-BHQ2	Chao et al. [20]
For unknown etiology	Random PCR	Primer A gtttcccagtcacgatgNNNNNNNN	
		Primer B gtttcccagtcacgatc	Ambrose and Clewley [13]
For unknown etiology	NGS	The outline of the NGS technique using a random octamer primer is found in the text. For more details, see the original publication	McMullan et al. [14]

have been discovered based on experiments using monkeys, hamsters, and other rodents. The popular objective of research for therapeutic intervention has been easing systemic inflammatory syndrome induced by cytokine activation. Antibody, interferon-α, poly-CLC, ribavirin, carboxamide drugs, tiazofurin, and iminocyclitol compounds have been investigated.[21] At least, one company has been pursuing development of a drug specifically against YF. Others have focused on intervention of human gene activation. For example, viral entry process alone activates nearly 300 human genes. This approach is sound, because viruses cannot develop resistance against the drugs targeting the hosts.

In other investigations, the major focus of research has been interference in the functions of genes, domains, or genomic segments responsible for translation in the synthesis of key viral proteins, such as helicase domain of NS3, RdRp domain of NS5, proteases encoded in NS2, and three genes as well as 3' terminal UTR. Some drugs have been found promising, but thus far no drug has been approved. Treatment is basically tailored to control symptoms and signs. Rest, fluids, analgesics, and antipyretics may relieve

fever and aching to some extent. Care should be taken to avoid certain medications, such as aspirin and nonsteroidal anti-inflammatory drugs, which may increase the risk of bleeding.

23.6.2 Vaccination

In the absence of any effective therapeutic drug to ameliorate the YF syndrome and because of the enormous difficulty of devising an effective vector control strategy in most YF-at risk and receptive areas, the best strategy for prevention is vaccination. The live, attenuated YF vaccine currently in use derived from the 17D strain developed in 1937 through repeated passages of a wild strain in the laboratory of the Rockefeller Institute for Medical Research. Many millions of people have been safely vaccinated since. It is safe, effective, sustainable, and economic, if managed properly. Although revaccination is strongly recommended after 10 years, actually for many vaccines, protective antibodies have persisted for many more years. However, in too many endemic countries, the implementation of a vaccination program has met many socioeconomic, infrastructural, or administrative problems. In fact, in 32 of 45 endemic countries where YF vaccination is conducted, actually vaccination rate per country has been far less than 50%. Regarding contraindication, the vaccine should not be used for infants less than 9 months, pregnant women, the elderly (>60 years), and people with thymus disorder. More recently, inactivated vaccines prepared with β-propiolactone or by hydrostatic pressure have been evaluated as potential alternative vaccine in the future. The vaccine candidate inactivated by the former method induced Nt antibodies in 100% of the subjects.

23.6.3 VECTOR CONTROL

Because YFV is maintained in the jungle, eradication of this virus is *impossible*. Accordingly, vector control for YF refers to *Ae.* aegypti control in urban environments. In the urban environments, control of this vector has been difficult. Nonetheless, during an urban outbreak, a speedy insecticide (such as malathion) application is imperative for quickly reducing vector population. No solution has been found for three basic issues: high cost of insecticide for many countries, labor intensiveness of fogging operation, and difficulty of facilitating insecticide penetration deep inside houses for domestic vector control. More recently, new strategies based on the release of genetically modified mosquitoes have been developed. However, even if such a strategy was found effective, simple to apply, and economical under experimental conditions, its actual efficacy, practicality, and economic cost in urban environments in the tropics are totally unknown at this moment. In the meanwhile, for personal protection, use of a repellent, DEET, is strongly recommended.

23.7 CONCLUSIONS AND FUTURE PERSPECTIVES

Despite the availability of a very effective vaccine, YF still presents a serious public health concern for a variety of reasons. First, as mentioned earlier, in too many endemic countries, vaccination is difficult due to socioeconomic, administrative, and even political reasons. Regrettably, this problem is expected to persist. Second, while the proposals of new vector control strategies including release of genetically modified mosquitoes are encouraging, they should not be expected to become available or to be a panacea any time soon. Control of domesticated mosquitoes, unlike control of outdoor pests, requires collaboration, dedication, and strong will stressing self-reliance and perseverance for many years by the residents in urban environments. Thus far, all educational and behavioral change or motivational programs initiated have failed. Until we find an effective and simple way of cultivating the aforementioned set of human behaviors and attitude for a long-term battle for vector control in democratic societies, effective control of domesticated *Ae. aegypti* mosquitoes in urban areas will remain difficult. In contrast, thanks to intensified research

by multiple groups, availability of licensed therapeutic drugs will be strongly expected in the near future. Another bright spot is rapid progress in molecular techniques for diagnosis. With further improvement in real-time, on-site diagnosis, automated result analysis, and instant interpretation, coupled with rapid transmission of digitized data via smartphones (i.e., iPhone, Android) and geographic locator program (i.e., Google Map), an outbreak of YF will be detected far more accurately and reported for a proper action to be taken.

REFERENCES

1. Lindenbach, B.D., Thiel, H.-J., and Rice, C.M., Flaviviridae: The viruses and their replication. In: *Fields Virology*, 5th ed., Vol. 1, D.M. Knipe and P.M. Howley (eds.). Lippincott, Williams & Wilkins, Philadelphia, PA, pp. 1101–1152; 2005.
2. Kuno, G., Chang, G.-J.J., and Chien, L.-J., Correlation of phylogenetic relation with host range, length of ORF or genes, organization of conserved sequences in the 3′ non-coding region, and viral classification among the members of the genus *Flavivirus*. In: *Viral Genomes: Diversity, Properties and Parameter*, Z. Feng and M. Long (eds.). Nova Science Publishers, Inc., Hauppauge, New York, pp. 1–33; 2009.
3. Macdonald, J. et al. Molecular phylogeny of Edge Hill virus supports its position in the yellow fever virus group and identifies a new genetic variant. *Evol. Bioinform. Online* 6: 91–96; 2010.
4. Kuno, G. and Chang, G.-J. J., Biological transmission of arboviruses: Reexamination and new insights into components, mechanisms, and unique traits as well as their evolutionary trends. *Clin. Microbiol. Rev.* 18: 608–637; 2005.
5. Kuno, G. Transmission of arboviruses without involvement of arthropod vectors. *Acta Virol.* 45: 139–150; 2001.
6. Bentlin, M.R. et al. Perinatal transmission of yellow fever, Brazil, 2009. *Emerg. Infect. Dis.* 17: 1779–1780; 2009.
7. Kerr, J.A. The clinical aspects and diagnosis of yellow fever. In: *Yellow Fever*, G.K. Strode (ed.). McGraw-Hill Book Co., New York, pp. 385–425; 1951.
8. Monath, T.P. Yellow fever. In: *Vaccines*, 3rd ed., A. Plotkin and W.A. Orenstein (eds.). W.B. Saunders Co., Philadelphia, PA, pp. 815–879; 1999.
9. Bugher, J.C. The pathology of yellow fever. In: *Yellow Fever*, G.K. Strode (ed.). McGraw-Hill Book Co., New York, pp. 137–163; 1951.
10. Domingo, C. et al. First international external quality assessment study on molecular and serological methods for yellow fever diagnosis. *PloS One* 7: e36291; 2012.
11. Kuno, G., Serodiagnosis of flaviviral infections and vaccinations in humans. *Adv. Virus Res.* 61: 3–65; 2003.
12. Palacios, G. et al. Panmicrobial oligonucleotide array for diagnosis of infectious diseases. *Emerg. Infect. Dis.* 13: 73–81; 2007.
13. Ambrose, H.E. and Clewley, J.P., Sequence-independent virus detection and discovery. In: *Molecular Detection of Human Viral Pathogens*, D. Liu (ed.). CRC Press, Boca Raton, FL, pp. 1109–1121; 2011.
14. McMullan, L.K. et al. Using next generation sequencing to identify yellow fever virus in Uganda. *Virology* 422: 1–5; 2012.

15. Kuno, G. Universal diagnostic RT-PCR protocol for arboviruses. *J. Virol. Methods* 72: 27–41; 1998.

16. Scaramozzino, N. et al. Comparison of flavivirus universal primer pairs and development of rapid, highly sensitive heminested reverse transcription-PCR assay for detection of flaviviruses targeted to a conserved region of the NS5 gene sequences. *J. Clin. Microbiol.* 39: 1922–1927; 2001.

17. Crochu, S. et al. Sequences of flavivirus-related RNA viruses persist in DNA form integrated in the genome of *Aedes* spp. mosquitoes. *J. Gen. Virol.* 85: 1971–1980; 2004.

18. Morreau, G. et al. A real-time RT-PCR method for the universal detection and identification of flaviviruses. *Vector Borne Zoonotic. Dis.* 7: 467–477; 2007.

19. Drosten, C. et al. Rapid detection and amplification of RNA of Ebola and Marburg viruses, Lassa virus, Crimean-Congo hemorrhagic fever virus, Rift Valley fever virus, dengue viruses, and yellow fever virus by real-time reverse transcription-PCR. *J. Clin. Microbiol.* 40: 2323–2330; 2002.

20. Chao, D.Y., Davis, B.S., and Chang, G.J., Development of multiplex real-time reverse transcriptase PCR assays for detecting eight medically important flaviviruses in mosquitoes. *J. Clin. Microbiol.* 45: 584–589; 2007.

21. Novartis Foundation. *New Treatment Strategies for Dengue and Other Flaviviral Diseases*. Novartis Foundation Symposium 277. John Wiley & Sons, Hoboken, NJ; 2006.

Section II

Microbes and Toxins Affecting Humans and Animals: Bacteria

24 *Bacillus anthracis*

Noura Raddadi, Elena Crotti, Fabio Fava, and Daniele Daffonchio

CONTENTS

24.1 INTRODUCTION

The genus *Bacillus* encompasses rod-shaped Gram-positive and spore-forming bacteria that are characterized by high diversity in terms of physiology, ecological niche, gene sequences, and regulation. These bacteria have an important impact on human activity varying from probiotic effect[1–3] to severe pathogenicity. Species within the genus have been emerged as new human pathogens that are associated with foodborne diseases and can cause severe and even fatal infections. These include *Bacillus cereus*, *B. weihenstephanensis*, *B. pumilus*, *B. mojavensis*, *B. licheniformis*, *B. subtilis*, and *B. circulans*.[4] However, the oldest, most infectious, and potentially lethal human disease is the anthrax that is caused by the species *Bacillus anthracis*, a member of the *B. cereus* group. The infection with the bacterium often leads to systemic anthrax, defined as the invasive infection associated with bacterial dissemination or toxin-mediated multiorgan dysfunction, and which may be secondary to any of the four well-described forms of the disease.[5]

Before giving details on the clinical features and pathogenesis, diagnosis, and methods of detection of the bacterium and the treatment and prevention of the disease, we will briefly introduce the complex ecology and the identification challenges that characterize bacteria in the *B. cereus* group as well as the anthrax epidemiology worldwide.

24.2 CLASSIFICATION, MORPHOLOGY, AND BIOLOGY OF *Bacillus cereus* GROUP

The *B. cereus* group encompasses five other species in addition to *B. anthracis*. *B. cereus* is the most known foodborne pathogen within the group, which has been associated with food-poisoning illnesses as well as other kinds of clinical infections.[4,6] *B. thuringiensis* is an insect pathogen widely used as biopesticide. It is different from *B. cereus* in that it has a very useful impact on human activities,

being widely used in agriculture for insect pest biocontrol. *B. weihenstephanensis*, a psychrotolerant species capable of growing at temperatures as low as 4°C–6°C is implicated in food spoilage.[7] Two other species, *B. mycoides* and *B. pseudomycoides* that are typically isolated from soil and plant rhizosphere, are also members of the *B. cereus* group.

The six species are known to be strictly related phylogenetically, according to DNA–DNA hybridization studies[8–10] and the sequencing of the ribosomal RNA genes.[11–13] However, a marked variability is always observed when large collections of strains are examined by DNA fingerprinting methods that target the whole genome[14–18] and/or discrete genes.[19–23] Hence, the phylogenetic and taxonomic relationship among these species are far to be clear. The genetic variability observed within/between these species has raised several interesting questions and posed a challenge to microbiologists: (1) How did the species evolve their evolutionary pathway and how are they differentiated during evolution? (2) What is the genetic baseline driving the different ecotypes and the virulence?

Also, the ecology of these bacteria has puzzled microbiologists for a long time and is still debated, despite several analyses showing that a major environmental niche for these bacteria is the invertebrate intestinal tract.[24–26] In a general model, these species commonly live associated with the intestine of invertebrates assuming a symbiotic lifestyle, but occasionally they escape such an ecotype becoming invasive for other animal hosts, which are in specific insects, arthropods and nematodes for *B. thuringiensis*, mammals and humans for *B. anthracis* and *B. cereus*. It has been shown that with respect to other soil inhabiting bacteria, like those of the *B. subtilis* group that have genomes harboring many genetic determinants for the metabolism of plant-derived carbohydrates, bacteria of the *B. cereus* group share a high number of genes for protein metabolism. This suggests that they evolved following selection by nutrient (protein)-rich environments like animal guts or animal tissues and fluids rather than the plant environment.[27,28] Specialization on different "animal" niches has led to different evolution of the species in the *B. cereus* group.

For instance, *B. anthracis* has been proposed to diverge from the other members of the group by specializing as a lethal mammal pathogen. It has been supposed that *B. anthracis* has evolved along two possible models[29]: (1) In the first proposed model, *B. anthracis* is considered a relatively ancient organism with a low growth rate, determined by the previously mentioned ecological constraints and evolving separately from a common ancestor with *B. cereus* and other relatives; (2) in a second model, it has been supposed to be derived relatively recently from *B. cereus*, through the acquisition and rearrangement of plasmids resulting in the actual pXO plasmids responsible for lethal virulence.

The discrimination between the species in the *B. cereus* group is an important point to be addressed, considering the differences in the virulence potential of these microorganisms. Typically, *B. anthracis* can be differentiated from most of *B. cereus* and *B. thuringiensis* strains by specific tests such as penicillin G and gamma phage sensitivity, lack of motility, and of beta hemolysis on blood agar. However, a problem still exists for borderline isolates between the species such as the pathogenic *B. cereus* G9241, for which these tests failed to recognize the pathogenic potential.[30] This strain has been confirmed to be a *B. cereus* harboring a virulence plasmid very similar to the *B. anthracis* pXO1 plasmid and a second plasmid encoding for a capsule synthesis. However, such a second plasmid and the capsule coding genes completely differed from those in the pXO2 plasmid and the typical capsule of *B. anthracis*.[30] Thus, considering such a virulence potential of strains genetically near neighbor of *B. anthracis*, approaches that might rapidly identify these strains are of great importance. From a safety perspective, these strains could represent alternative hosts for *B. anthracis* toxin genes.[30] Indeed, according to DNA fingerprinting, several other *B. cereus* strains resulted strictly related to *B. anthracis*. For example, Keim et al.[29] and Radnedge et al.[30] identified *B. cereus* and *B. thuringiensis* strains closely related to *B. anthracis* by amplified fragment length polymorphism (AFLP). Apart from AFLP, several other methods based on whole genome fingerprinting have been used for typing *B. cereus/B. thuringiensis* and the identification of *B. anthracis* borderline strains. Among others, there are rep-PCR[31] and multilocus sequence typing (MLST),[32] or, more recently, microarray analysis and comparative genomics.[33,34] Alternative approaches have been based on length or sequence polymorphisms in variable-number tandem repeats in multiple loci (multilocus variable-number tandem repeats [VNTR] analysis [MLVA])[35] or on signature single nucleotide polymorphisms (SNPs) in the genome.[36] MLVA has been used as a gold standard for subtyping *B. anthracis* isolated worldwide (see among others[35,37–39]). Together with MLVA, SNP analysis has greatly contributed in typing and tracing *B. anthracis* isolates, especially by the whole genome SNP analysis.[40] Besides those identified in the whole genome,

SNPs in housekeeping genes have been nicely exploited to identify strains related to certain species or genetic types. For example, Prüss et al.[39] showed that certain nucleotides in the 16S rRNA gene and their relative prevalence among the different ribosomal operons in the genome correlate with the psychrotolerance of *B. cereus* strains and hence are signature of the species *B. weihenstephanensis*. Gierczyński et al.[41] developed a simple and cost-effective restriction site insertion-PCR (RSI-PCR)-based assay that precisely detects the *B. anthracis*–specific nonsense mutation in *plcR* gene by restriction digestion with the endonuclease *Ssp*I.

24.3 ANTHRAX EPIDEMIOLOGY

Anthrax is classified into four different syndromes in relation to the primary site of infections: cutaneous, gastrointestinal, inhalational, and injectional.[42] It has been estimated that approximately 20,000–100,000 annual cases of human accidental infections with *B. anthracis* occurred worldwide.[43] More than 95% of these cases are largely due to cutaneous form, and particularly they were related to occupational exposure in meat-packing, leather-tanning, bone meal–processing, and hair-/wool-sorting industries, with the remaining cases mainly related to the agricultural exposure. The employment of vaccines in high-risk populations of animal and humans and the improvement in hygiene practices reduced the occurrence of *B. anthracis* infections in industrialized countries. However, anthrax infections are still a serious problem especially in low-income countries.[44,45]

Several outbreaks have been documented, for instance and among others, the largest infection in United States in a goat hair–processing plant in 1957; the accidental release of aerosolized *B. anthracis* spores from a military laboratory in Sverdlovsk, Russia, in 1979, which caused 68 deaths; the infections and deaths in Zimbabwe between 1979 and 1985, in Tibet in 1989, and in China in 1996 and 1997; and the increased anthrax incidence reported in Spain, Central America, and Africa between 1991 and 1995.[44]

Among the different anthrax syndromes, the one with the higher mortality rate is the inhalational one with 5% of the incidence of reported anthrax cases, whereas approximately 95% of anthrax cases occur through the cutaneous route, mainly in Africa, Asia, and Eastern Europe where animal and human vaccination is inadequate. Generally, cutaneous syndrome resolves spontaneously, but, when infections are not treated adequately, the mortality rate is estimated to be between 5% and 20%. When antibiotic treatments are applied, the mortality rate falls down to less than 1%.[42,45]

In inhalational anthrax, *B. anthracis* spores are inhaled and deposited in the alveoli, where they are then phagocytized by macrophages and carried to local mediastinal lymph nodes where after germination they produce hemorrhagic mediastinitis.[46] When not treated, infections result in meningitis, gastrointestinal involvement, and refractory shock. Due to the earlier diagnosis, better supportive care, and multidrug antibiotic therapy, the mortality rate of inhalational anthrax decreased over time: from 94% for naturally occurring cases before 1976, to 86% in Sverdlovsk in 1979, and to 46% in the outbreak in the United States in 2001.[47–49]

Gastrointestinal anthrax is due to the ingestion of *B. anthracis*–contaminated meat. Its incidence is reported to be rare, but it is often underestimated due to the infection's nonspecific symptoms. The mortality rate accounts for 25%–60%, and severe cases are frequently complicated by shock.[50] Outbreaks due to the ingestion of spore-contaminated meat have been reported in Thailand in the 1980s and in Russia in the late 1990s and recently (2009–2010) in Bangladesh.[45,50,51]

Injectional anthrax, firstly reported in Norway in a heroin user, is due to the subcutaneous injection of drugs contaminated with *B. anthracis* spores and results to be different from the cutaneous syndrome.[52] In comparison to cutaneous syndrome, injectional anthrax has an increased risk of shock and a higher mortality rate in spite of antibiotic therapy (34% for injectional one, less than 1% for cutaneous one). Recently, it has been reported a severe outbreak of injectional anthrax in the United Kingdom with 47 recognized cases and 13 deaths in Scotland and 5 cases and 4 deaths in England.[42,53–56] Three cases (and two deaths) among which one in June 2012 have been reported in Germany.[42,53–56]

Outbreaks in animals are more widespread. The common victims are cattle and sheep with a mortality of 80%; horses and swine can also be affected, together with wild animals like deer, toads, many rodents, and cats. Frogs, rats, and adult dogs are largely resistant. Recent examples of animal infections are the outbreak in Iran that caused the death of 1 million sheep in 1945; the death of wild animals in South Africa and Namibia between 1984 and 1989; the outbreak in Nepal in 1996; and the infections of cattle in China in 1996, in Australia in 1997, and in India between 1991 and 1996.[44]

While on one hand there are the improvements made to minimize the accidental infections with anthrax spores, on the other hand *B. anthracis* remains a serious threat as a bioterrorism and a biowarfare agent. During the twentieth century, several episodes indicated its applicability as bioweapon, and it is supposed that at least 17 countries have biological weapon capabilities.[44,57–59] Anthrax is currently considered one of the most serious bioterrorism threats. Recent events include the attacks perpetrated in the United States via the mail, with 22 cases of inhalational and cutaneous anthrax, and 5 people died from inhalational anthrax, and the aerosolized spores dispersed several times in Tokyo in the mid-1990s by a Japanese terrorist sect.[44]

24.4 CLINICAL FEATURES AND PATHOGENESIS OF *Bacillus anthracis*

B. anthracis has been well studied for over 150 years and continues to be a threat as a bioweapon.[59] This bacterium forms spores (1–2 μm in size) able to withstand several kinds of stresses that are usually lethal to vegetative cells, such as

ultraviolet and ionizing radiation, heat, various chemicals,[60] and oxidative stress.[61] Anthrax is primarily a disease affecting herbivore animals, but it can be transmitted to all mammals including humans. Typically, animals become infected with *B. anthracis* by ingesting spores while grazing on contaminated soils or feed but also through the skin by biting insects.[62] Transmission to humans occurs through different routes of entry of the spores[42]: (1) cutaneously by direct contact with infected animals or by handling infected animal products, (2) respiratory by inhalation of sufficient quantity of spores, (3) oropharyngeal and/or gastrointestinal by the ingestion of undercooked contaminated meat, and (4) recently, a fourth syndrome, referred to as "injectional anthrax," has emerged in injection drug users with the first case described in 2000 in a heroin user in Norway.[52]

All forms of anthrax could be fatal if not treated in due time, but the cutaneous form is often self-limiting.[56] The cutaneous anthrax infection results in primary skin lesion corresponding to a painless, pruritic papule that appears at the following 3–5 days. The lesion develops in a vesicle that undergoes central necrosis and drying, leaving a characteristic black eschar surrounded by edema. After 1–2 weeks, the eschar dries and sloughs off.[42]

In inhalational anthrax, *B. anthracis* spores are inhaled and deposited in the alveoli, where they are then phagocytized by macrophages and carried to local mediastinal lymph nodes. Further, they germinate into vegetative forms, replicate, and produce hemorrhagic mediastinitis.[46] Inhalational anthrax has a biphasic clinical course: flu-like symptoms with fever and nonproductive cough and myalgias that are developed by patients following an incubation period of approximately 4 days. If not treated in time, the disease enters a second fulminant phase, characterized by hypotension and dyspnea and which may progress to death within the following 24 h.[63]

In oropharyngeal anthrax, spores reach the pharyngeal area where they produce ulcerative lesions, whereas in intestinal one, spores settle and cause ulcers anywhere from the jejunum to the cecum. Most patients in an outbreak reported from Thailand developed the oropharyngeal form presented with fever and neck swelling with symptoms developed in a mean incubation time of 42 h. In intestinal anthrax, patients frequently present with nonspecific gastrointestinal symptoms (i.e., nausea, vomiting, abdominal pain, or diarrhea). In more severe cases, fever, ascites, increased abdominal girth, and shock develop. Also, flu-like symptoms followed by nausea, vomiting, and abdominal cramps has been described in the one confirmed case in the Unites States.[50]

The "injectional anthrax," characterized by severe soft tissue infections observed in 1–10 days following drug injection, appeared different from cutaneous disease in terms of mortality rate (34% versus <1%) and absence of the development of the typical papules, vesicles, and eschars, although development of edema at the injection site is common. Injectional anthrax has been considered to be the result of the germination of spores in the inoculation site followed by a capsule-facilitated local spread of the bacteria. It is not known, however, whether disseminated disease results from intravascular injection of spores or spread of local soft tissue infection. Surgical exploration revealed a diffuse capillary bleeding and necrosis of the adipose tissue but without production of the characteristic purulent material observed in case of necrotizing fasciitis.[54,64]

It is believed that, once in the host, the germination of *B. anthracis* spores to vegetative cells occurs inside host macrophages following the interaction of germinant nutrient receptors located on the inner membrane of the spore with amino acids and purine nucleosides.[65] Following germination, vegetative cells multiply in their way to the regional lymph nodes.[66] Along with other mechanisms, two spore-bound superoxide dismutases (SOD15 and SODA1) enzymes, which are capable of detoxifying oxygen radicals,[67] were recently reported to play a crucial role in protecting the spores against the oxidative burst of macrophages in mouse model.[61] Following spore germination and bacterial proliferation, the capacity of the lymph nodes is exceeded; hence, the vegetative bacteria enter the blood stream, which leads to bacteremia and the distinctive pathology of edema, hemorrhage, and tissue necrosis.[44] Dissemination of *B. anthracis* by hematogenous or lymphogenous spread from a cutaneous, mediastinal, or gastrointestinal focus can lead to infection of other sites like the brain, allowing entry into the central nervous system (CNS) and development of anthrax meningitis.[68] Another form of anthrax meningitis in which no primary focus is found has been reported and is known as "primary anthrax meningitis."[69]

The pathogenicity of *B. anthracis* depends on two major factors: a poly-γ-D-glutamic acid capsule and the three-component secreted anthrax toxin, codified by plasmids pXO2 and pXO1, respectively. In addition to protecting the bacteria from the phagocytosis, the capsule allows bacterial proliferation and subsequent toxemia to develop in the blood stream due to its poor immunogenic properties. The anthrax toxin is composed of three proteins known as protective antigen (PA), lethal factor (LF), and edema factor (EF). Once in the body, the PA binds to receptors present on most cells including macrophages, and after proteolytic activation, it combines with LF and EF allowing their delivery inside the cell.[44] Transcription of the genes *pag*, *lef*, and *cya* encoding the three toxin proteins PA, LF, and EF, respectively, is regulated by a gene known as the anthrax toxin activator (*atxA*), whereas the capsule proteins, codified by the *capBCADE* operon, are regulated by AcpA and AcpB.[70,71]

24.5 IDENTIFICATION AND DIAGNOSIS

The diagnosis of a human illness is based on the combination of different clues including the symptoms of the illness, the history of the patient, the possible risk factors, and the detection and identification of the suspected agent and/or its pathogenicity factors. *B. anthracis* identification is challenged by the presence of *B. cereus* closely related strains that own numerous phenotypical and genotypical characteristics in common with *B. anthracis*. Moreover, the ubiquitous presence of these close neighbors in the environment makes more difficult the identification of *B. anthracis*.[72] Conventional, biochemical, and affinity- and nucleic acid–based identification methods are currently available for *B. anthracis* detection.

Conventional diagnosis procedures: A timely and adequate treatment of a patient foresees a fast and reliable identification of the pathogenic microorganism involved in the illness. Normally, *B. anthracis* can be differentiated from other *Bacillus* spp. by classical microbiological tests that are recommended by the World Health Organization (WHO) and by the Centers for Disease Control and Prevention (CDC).

Like other *B. cereus* group members, *B. anthracis* is a Gram-positive spore former, but differently from other nonanthrax group members (*B. cereus* and *B. thuringiensis*), it produces a polypeptide capsule in vivo and in vitro when grown under suitable conditions, it is sensitive to penicillin G and susceptible to lysis by gamma phage, and it lacks motility and β-hemolytic activity on sheep or horse blood agar medium.[73] In addition to blood agar medium, *B. anthracis* can be also isolated on the selective polymyxin-lysozyme-EDTA-thallous acetate (PLET) agar[74] or on anthracis chromogenic agar (ACA[75]). On PLET, the growth requires 1–2 days, with *B. anthracis* colonies that appear small, white, domed, and circular. This method must be then subject to other confirmations. ACA is a chromogenic medium that contains 5-bromo-4-chloro-3-indoxyl-cholinephosphate that is hydrolyzed by phosphatidylcholine-specific phospholipase C (also known as lecithinase C). Upon incubation at 35°C–37°C, *B. cereus* and *B. thuringiensis* colonies are dark teal blue within 24 h, whereas in the case of *B. anthracis*, the colonies need at least 48 h to develop the color due to the slower synthesis rate of this enzyme as a result of the truncated *plcR* gene and PlcR regulator. PLET and ACA are suitable for the isolation of *B. anthracis* by trained and expert personnel, from samples with high numbers of bacteria.[76] On the other side, the blood agar medium must be used when *B. anthracis* is isolated from samples with a small number of background bacilli being moreover easier and more cost-effective than PLET and ACA.

Capsulated *B. anthracis* cells can be observed with McFadyean's capsule-staining test[77] or stained with India ink.[77,78] The virulence of the cultured bacteria might be proved by inoculation in guinea pigs or mice, followed by the observation of the animal death.[73,79]

Classical identification is straightforward for an experimented bacteriologist in a biological safety level 3 (BSL-3) working condition, but it is time-consuming requiring at least 24–48 h to be accomplished. Moreover, there are some other challenges in the detection of *B. anthracis*: (1) Not all the *B. anthracis* isolates fit all the previously described characteristics, and atypical *B. anthracis* isolates lacking pXO2 (and hence capsule) have been described; (2) pathogenic capsule-producing *B. cereus* strains harboring plasmids similar to those of *B. anthracis* have been described associated to human fatal pneumonia and from carcasses of great apes in Africa[80]; (3) blood cultures do provide *B. anthracis* identification, while cultures from skin or lesions cannot be reliable as they yield positive results in only 60%–65% of cases especially after antibiotic treatment.[81]

Biochemical diagnosis procedures: Biochemical identification methods are mainly used as confirmatory methods, performed on isolated colonies already suspected to be *B. anthracis*. These rely on automatic identification systems that are based on the analysis of the metabolic features of the isolates and on their susceptibility to antimicrobial agents.[82] Some researchers suggest the use of intact cell mass spectrometry (MS) for the detection of microbial protein patterns that have to be compared to the profiles deposited in the reference library.[83] Another biochemical technique is based on the employment of gas chromatography in order to determine the cellular fatty acid profile, even if the evidence of a clear discrimination between *B. anthracis* and the other nonanthrax species is not always possible.[84] As previously reported in the case of conventional identification methods, also these assays require well-trained personnel and are difficult to use on environmental samples because a culturing phase is needed in order to obtain the bacterial isolates to be processed.

Affinity-based diagnosis procedures: The aforementioned limitations of culture- and biochemical-based diagnosis of *B. anthracis* make necessary the application of quicker and more specific approaches for the detection of the bacterium. Several antibody-based tests targeting spores, vegetative cells, and anthrax toxin proteins have been developed for the specific detection of *B. anthracis*.[44,72,85–87] Recently, several other approaches targeting the *B. anthracis* capsular antigen detection by latex agglutination[88] or the entire organism mediating phage bioluminescence[89] have been reported.

Several detection techniques are available, such as enzyme-linked immunosorbent assay (ELISA), immunoradiometric assays (IRMAs), flow cytometry, direct fluorescent antibody (DFA) assays, and electrochemiluminescence (ECL).[44] For instance, Morel et al.[90] produced 48 monoclonal antibodies against *B. anthracis* spore surface and developed colorimetric and ECL immunoassays to detect *B. anthracis*. A detection limit between 30 and 100 spores per test has been documented according to the considered *B. anthracis* strain. In complex matrices, such as soil and various suspensions of domestic powders, they were able to detect anthrax both through ELISA and ECL assay, in a sensitive manner: the signal detected for the spores spiked into the different matrices has never reported to decrease less than three-fold when compared to the control.

Recently, a microarray-based method has been developed to detect *B. anthracis* antibodies. Briefly, bacterial "signature" carbohydrates have been immobilized onto glass slides, allowing differentiating *B. anthracis* antibodies in infected patients and infected or vaccinated animals.[91]

Immunomagnetic capture of *B. anthracis* spores has been applied to capture and concentrate anthrax spores from several food matrices, such as milk, juices, processed meat, water, and salad. Immunomagnetic beads from different sources and with different chemical properties have been functionalized with monoclonal and polyclonal antibodies. Interesting results have been reported warranting efficient capture of anthrax spores at very low concentrations.[92]

Nucleic acid–based diagnosis procedures: Other *B. anthracis* diagnosis techniques include the nucleic acid–based molecular approaches. DNA isolation from clinical specimens is a critical step in molecular biology–based diagnosis techniques due to the presence of high level of inhibitors and host DNA in relation to the amount of DNA from the pathogenic bacterium and especially in the first step of the infection. Also, safety is an important consideration for personnel handling specimens, which are suspected to contain infectious *B. anthracis* spores known to be difficult to inactivate. Several extraction kits are commercially available for the rapid isolation of DNA to be used in nucleic acid–based detection assays. In particular, the UltraClean Microbial DNA Isolation Kit (Mo Bio Laboratories, Inc.) was shown to offer the best method for depleting viable *B. anthracis* spores from samples.[93] Dauphin et al.[94] reported that gamma irradiation of DNA preparations allowed inactivation of residual *B. anthracis* spores; their complete removal could be reached by centrifugal filtration with 0.1 μm filter unit.[95]

Actually, there is a wide variety of molecular-based techniques for the detection of *B. anthracis*, including PCR, multiplex PCR, and real-time PCR.[44] All these techniques are based on the specific amplification of the genes *pag*, *lef*, and *cap*. However, all these genetic determinants reside on plasmids, and hence *B. anthracis* isolates that lack one of the plasmids could not be identified easily based on these techniques. For this reason, chromosomal-based *B. anthracis*–specific markers such as BA813, *rpoB*, *gyrA*, *gyrB*, *saspB*, *plcR*, and BA5345 has been identified.[39,96,97] Recently, Antwerpen et al.[98] reported the development of a TaqMan real-time-based assay that targets the locus BA5345 and that detected specifically *B. anthracis* as was shown by testing 328 *Bacillus* strains. Moreover, TaqMan array cards targeting plasmid sequences have been applied for simultaneous detection of 21 respiratory agent pathogens, including *B. anthracis*[99]: when multiplex PCR is performed with 500 fg of DNA, 100% presence of *B. anthracis* is found, whereas with 100 fg level of DNA, *B. anthracis* was detected in 71% of the times.

Another molecular assay that could be applied for the rapid and sensitive/quantitative detection of *B. anthracis* has been reported recently.[100] It is based on the combination of molecular beacons and real-time PCR technologies. Molecular beacons are highly specific nucleic acid probes that, when bound to their target DNA sequences, act as switches by emitting fluorescence as a result of the alteration of their conformation.[100,101] The molecular beacon–based real-time PCR assay was performed by designing five molecular beacons targeting *B. anthracis capA, capB, capC, lef,* and *pag* alleles that were used in five uniplex assays for the detection of these genes in a broad range of samples.[100]

Specific PCR coupled with electrospray ionization MS (PCR/ESI-MS) has been used for the detection of *B. anthracis* and other potential bioterrorism-relevant microorganisms.[102] Briefly, following PCR amplifications, PCR products are analyzed by MS, allowing the calculation of the base compositions of the PCR products that are then compared to a database for identification of the pathogens. In the tested conditions, identification rate of 100% has been found for *B. anthracis.*

Molecular methods as MLST, multiple-locus variable-number tandem repeat analysis (MLVA), and next generation sequencing (NGS) methods have been recently used for the discrimination of *B. anthracis* strains.[34,72] MLST and MLVA allowed a clear-cut discrimination between *B. anthracis* and other isolates of *B. cereus* group, but they revealed to be not sensitive enough to discriminate *B. anthracis* isolates. Instead researchers showed that NGS can be used for the discrimination of isolates from *B. anthracis* group and moreover can be useful for the discovery of novel strain-specific candidate SNPs.[72]

24.6 TREATMENT AND PREVENTION

The treatment of anthrax must be considered in the case of active infection diagnosis, of suspected infection, or of potential exposure even if the patients did not express yet clinical symptoms. Although the treatment recommendations are based on the clinical form of the disease in the case of an active infection, they can be classified mainly as antimicrobial and antitoxin therapies.[42,54,63,103]

The antimicrobial therapy involves the use of antimicrobials with activity against *B. anthracis.* In the case of inhalational anthrax, current treatment recommendations by the CDC involve the use, for a minimum of 60-day period, of ciprofloxacin or doxycycline in combination with at least one or two other agents against which the organism is typically susceptible, such as vancomycin, rifampin, penicillin, clindamycin, ampicillin, chloramphenicol, tetracycline, erythromycin, streptomycin, fluoroquinolones, and cefazolin. Intravenous antibiotic therapy is recommended at the onset of the treatment with ciprofloxacin as primary antimicrobial agent in the case of severe (stage 2) inhalational anthrax. Oral antibiotics administration is possible once the patient is clinically stable. Ciprofloxacin is also recommended as a component of combination therapy in pregnant women and children; but if ciprofloxacin is contraindicated, doxycycline is used independently of the patient' category. Conversely, many later-generation cephalosporins, such as cefuroxime, cefotaxime, ceftazidime, aztreonam, and trimethoprim–sulfamethoxazole to which *B. anthracis* is resistant, are contraindicated. Another important measure in the management of inhalational anthrax

consists in the continued chest drainage or intermittent thoracentesis, which proved to be effective and improved survival.[54]

With regard to cutaneous anthrax, oral ciprofloxacin or doxycycline for 60 days or intravenous multidrug therapy is recommended in case of stage 1 cutaneous anthrax or complicated severe infection, respectively.[54–63]

Actual recommendations for antibiotic therapy of injectional anthrax include a treatment that covers *B. anthracis* and other more common causes of soft tissue infection, when an injectional anthrax with soft tissue infection is suspected. Intravenous ciprofloxacin and clindamycin are used in combination with other antibiotics like penicillin, flucloxacillin, and metronidazole. Moreover, debridement is also necessary when clinical control of tissue infection is needed.[104] If injectional anthrax is suspected but the drug user does not present soft tissue infection symptoms, a treatment as for inhalational disease with possible meningeal involvement including at least one agent with adequate CNS penetration like ampicillin, penicillin, meropenem, or vancomycin is recommended.

In addition to antimicrobial therapy, treatment of active anthrax infection could be performed through passive immunotherapy. Other new approaches including the use of drugs approved for other diseases and which show activities against anthrax toxins or of agents that can interfere at different phases of anthrax infection process are under investigation and will be more detailed in the perspectives section of this chapter.[5]

The passive immunotherapy consists in the administration of toxin-neutralizing antibodies to patients with active anthrax disease, either in combination with antibiotic therapy or as an alternative to it in those not responsive to antimicrobials. Preparation of polyclonal antibodies derived from the pooled plasma of healthy, anthrax-vaccinated donors known as anthrax immune globulin (AIG) has been used in the United States and the United Kingdom for the treatment of different forms of anthrax, including the injectional form.[54] Another antibody-based preparation is raxibacumab, a human monoclonal antibody generated against PA and produced by Human Genome Sciences, which is a component of the US Strategic National Stockpile.[5,54] In addition to AIG and raxibacumab, other preparations with antibodies directed against other pathogenicity factors of *B. anthracis* such as LF and edema are under studies.

The treatment of anthrax could also be performed through the use of agent approved for the treatment of other diseases and which has been shown to be active against anthrax toxins. These include, among others, the antimalarial chloroquine, the antiprotozoal quinacrine, and dantrolene used for the treatment of malignant hyperthermia.[5]

The prevention measures of anthrax include widespread livestock/wild animal vaccination where the disease is endemic, burning or cremation of anthrax-infected animals or humans, vaccination of people with high exposure risk, as well as following guidelines of anthrax postexposure prophylaxis. With reference to these guidelines, both antibiotic treatment and vaccination aimed at accelerating host immune response are recommended. Following potential inhalation exposure to *B. anthracis* spores, ciprofloxacin and doxycycline are recommended as first-line antimicrobial agents for 60 days. Vaccine treatment includes the administration of AVA (anthrax vaccine adsorbed) in the United States or AVP (anthrax vaccine precipitated) in the United Kingdom[54] at 10 days, 2 and 4 weeks after exposure.

Moreover, it is also important to consider the event of a terroristic attack when defining prevention measures. In this frame, stable production and supply of vaccines as well as provision of stockpiles of antimicrobials effective against *B. anthracis* are required. In addition, development of rapid and sensitive detection methods are crucial in order to allow for a focused prevention and therapeutic procedure and consequently to limit early morbidity and mortality.[59]

24.7 CONCLUSIONS AND FUTURE PERSPECTIVES

Although natural anthrax infection is rare in developed countries, the potential for large outbreaks persists, whether related to the dissemination of the bacterium through intended release as a biological weapon or its propagation through drug/heroin distribution networks.[56] In addition to its infrequency, the nonspecific early symptomatology of the gastrointestinal, inhalational, and injectional forms of the disease constitutes a big challenge in the treatment of patients, which may present with advanced disease in the case of an outbreak. This makes imperative the development of effective management strategies at large scale based, among others, on understanding/defining the mechanisms underlying advanced anthrax stage.

To date, the treatment of suspected anthrax infection is based on antibiotic therapy and mainly the administration of ciprofloxacin or doxycycline as primary antimicrobial agents in combination with other antibiotics for at least 60 days. Together with antibiotics, antitoxin therapy has been recommended by the CDC. When the patient presents the active anthrax infection, based on the disease form and stage, antibiotic therapy should be coupled with passive immunotherapy, that is, administration of anti-PA mono- or polyclonal antibodies.

In addition to PA-targeting antibodies, several studies/clinical tests are being carried out in order to find alternatives or at least to find other anthrax pathology elements that could be used as target for the development of new strategies for anthrax treatment. These include the production of antibodies against LF and/or EF as well as proteases that are able to inhibit furin, a key protein involved in the processing of PA.[5,105] Moreover, the risk of the release of antibiotic-resistant *B. anthracis* in the frame of a bioterrorism action makes necessary the search of other antimicrobials. In this context,

bacteriophage-encoded lysins are being studied in depth in order to overcome this challenge.[106,107] Other antimicrobial strategies aiming at supplementing or replacing the existing ones are being developed and include molecules able to inhibit quorum-sensing, spore germination, or PA assemblage or binding.[5,108]

REFERENCES

1. Raddadi, N. et al., The most important *Bacillus* species in biotechnology. In Sansinenea Royano, E. (ed.), *Bacillus thuringiensis Biotechnology*, Chapter 17, Springer, Dordrecht, the Netherlands, 2012.
2. Kim, K.M. et al., Characterization of *Bacillus polyfermenticus* KJS-2 as a probiotic, *J. Microbiol. Biotechnol.*, 19, 1013, 2009.
3. Sánchez, B. et al., Identification of surface proteins involved in the adhesion of a probiotic *Bacillus cereus* strain to mucin, *Microbiology*, 155, 1708, 2009.
4. Raddadi, N. et al., *Bacillus*. In Liu, D. (ed.), *Molecular Detection of Foodborne Pathogens*, 1st ed., Chapter 10, Taylor & Francis Group, New York, 2010.
5. Artenstein, A.W. and Opal, S.M., Novel approaches to the treatment of systemic anthrax. *Clin. Infect. Dis.*, 54, 1148, 2012.
6. Dierick, K. et al., Fatal family outbreak of *Bacillus cereus*-associated food poisoning, *J. Clin. Microbiol.*, 43, 4277, 2005.
7. Baron, F. et al., Isolation and characterization of a psychrotolerant toxin producer, *Bacillus weihenstephanensis*, in liquid egg products, *J. Food Prot.*, 70, 2782, 2007.
8. Nakamura, L.K., DNA relatedness among *Bacillus thuringiensis* serovars, *Int. J. Syst. Bacteriol.*, 44, 125, 1994.
9. Nakamura, L.K. and Jackson, M.A., Clarification of the taxonomy of *Bacillus mycoides*, *Int. J. Syst. Bacteriol.*, 45, 46, 1995.
10. Nakamura, L.K., *Bacillus pseudomycoides* sp. nov., *Int. J. Syst. Bacteriol.*, 48, 1031, 1998.
11. Ash, C. et al., Phylogenetic heterogeneity of the genus *Bacillus* revealed by comparative analysis of small-subunit-ribosomal RNA sequences, *Lett. Appl. Microbiol.*, 13, 202, 1991.
12. Ash, C. et al., Comparative analysis of *Bacillus anthracis*, *Bacillus cereus*, and related species on the basis of reverse transcriptase sequencing of 16S rRNA, *Int. J. Syst. Bacteriol.*, 41, 343, 1991.
13. Ash, C. and Collins, M.D., Comparative analysis of 23S ribosomal RNA gene sequences of *Bacillus anthracis* and emetic *Bacillus cereus* determined by PCR-direct sequencing, *FEMS Microbiol. Lett.*, 94, 75, 1992.
14. Brousseau, R. et al., Arbitrary primer polymerase chain reaction, a powerful method to identify *Bacillus thuringiensis* serovars and strains, *Appl. Environ. Microbiol.*, 59, 114, 1993.
15. Carlson, C.R., Grønstad, A., and Kolstø, A.-B., Physical maps of the genomes of three *Bacillus cereus* strains, *J. Bacteriol.*, 174, 3750, 1992.
16. Carlson, C.R. and Kolstø, A.-B., A complete physical map of a *Bacillus thuringiensis* chromosome, *J. Bacteriol.*, 175, 1053, 1993.
17. Carlson, C.R., Caugant, D.A., and Kolstø, A.-B., Genotypic diversity among *Bacillus cereus* and *Bacillus thuringiensis* strains, *Appl. Environ. Microbiol.*, 60, 1719, 1994.

18. Carlson, C.R., Johansen, T., and Kolstø, A.-B., The chromosome map of *Bacillus thuringiensis* subsp. *canadensis* HD224 is highly similar to that of the *Bacillus cereus* type strain ATCC 14579, *FEMS Microbiol. Lett.*, 141, 163, 1996.

19. Daffonchio, D. et al., Restriction site insertion-PCR (RSI-PCR) for rapid discrimination and typing of closely related microbial strains, *FEMS Microbiol. Lett.*, 180, 77, 1999.

20. Daffonchio, D. et al., A randomly amplified polymorphic DNA marker specific for the *Bacillus cereus* group is diagnostic for *Bacillus anthracis*, *Appl. Environ. Microbiol.*, 65, 1298, 1999.

21. Kim, Y.R., Czajka, J., and Batt, C.A., Development of a fluorogenic probe-based PCR assay for detection of *Bacillus cereus* in nonfat dry milk, *Appl. Environ. Microbiol.*, 66, 1453, 2000.

22. Schraft, H. and Griffith, M.W., Specific oligonucleotide primers for detection of lecithinase-positive *Bacillus* spp. by PCR, *Appl. Environ. Microbiol.*, 61, 98, 1995.

23. Schraft, H. et al., Epidemiological typing of *Bacillus* spp. isolated from food, *Appl. Environ. Microbiol.*, 62, 4229, 1996.

24. Margulis, L. et al., The Arthromitus stage of *Bacillus cereus*: Intestinal symbionts of animals, *Proc. Natl. Acad. Sci. USA*, 95, 1236, 1998.

25. Jensen, G.B. et al., The hidden lifestyles of *Bacillus cereus* and relatives, *Environ. Microbiol.*, 5, 631, 2003.

26. Swiecicka, I. and Mahillon, J., Diversity of commensal *Bacillus cereus* sensu lato isolated from the common sow bug (*Porcellio scaber*, Isopoda), *FEMS Microbiol. Ecol.*, 56, 132, 2006.

27. Ivanova, N. et al., Genome sequence of *Bacillus cereus* and comparative analysis with *Bacillus anthracis*, *Nature*, 423, 87, 2003.

28. Gohar, M. et al., A comparative study of *Bacillus cereus*, *Bacillus thuringiensis* and *Bacillus anthracis* extracellular proteomes, *Proteomics*, 5, 3696, 2005.

29. Keim, P. et al., Molecular evolution and diversity in *Bacillus anthracis* as detected by amplified fragment length polymorphism markers, *J. Bacteriol.*, 179, 818, 1997.

30. Radnedge, L. et al., Genome differences that distinguish *Bacillus anthracis* from *Bacillus cereus* and *Bacillus thuringiensis*, *Appl. Environ. Microbiol.*, 69, 2755, 2003.

31. Cherif, A. et al., Genetic relationship in the 'Bacillus cereus group' by rep-PCR fingerprinting and sequencing of a *Bacillus anthracis*-specific rep-PCR fragment, *J. Appl. Microbiol.*, 94, 1108, 2003.

32. Helgason, E. et al., Multilocus sequence typing scheme for bacteria of the *Bacillus cereus* group, *Appl. Environ. Microbiol.*, 70, 191, 2004.

33. Zwick, M.E. et al., Microarray-based resequencing of multiple *Bacillus anthracis* isolates, *Genome Biol.*, 6, R10, 2005.

34. Keim, P. et al., Multiple-locus variable-number tandem repeat analysis reveals genetic relationships within *Bacillus anthracis*, *J. Bacteriol.*, 182, 2928, 2000.

35. Read, T.D. et al., Comparative genome sequencing for discovery of novel polymorphisms in *Bacillus anthracis*, *Science*, 296, 2028, 2002.

36. Gierczyński, R. et al., Intriguing diversity of *Bacillus anthracis* in eastern Poland—The molecular echoes of the past outbreaks, *FEMS Microbiol. Lett.*, 239, 235, 2004.

37. Merabishvili, M. et al., Diversity of *Bacillus anthracis* strains in Georgia and of vaccine strains from the former Soviet Union, *Appl. Environ. Microbiol.*, 72, 5631, 2006.

38. Ciammaruconi, A. et al., Fieldable genotyping of *Bacillus anthracis* and *Yersinia pestis* based on 25-loci multi locus VNTR analysis, *BMC Microbiol.*, 8, 21, 2008.

39. Prüss, B.M. et al., Correlation of 16S ribosomal DNA signature sequences with temperature-dependent growth rates of mesophilic and psychrotolerant strains of the *Bacillus cereus* group, *J. Bacteriol.*, 181, 2624, 1999.

40. Van Ert, M.N. et al., Global genetic population structure of *Bacillus anthracis*, *PLoS One*, 2, e461, 2007.

41. Gierczyński, R. et al., Specific *Bacillus anthracis* identification by a plcR–targeted restriction site insertion-PCR (RSI-PCR) assay, *FEMS Microbiol. Lett.*, 272, 55, 2007.

42. Sweeney, D.A. et al., Anthrax infection, *Am. J. Respir. Crit. Care Med.*, 184, 1333, 2011.

43. Swartz, M.N., Recognition and management of anthrax—An update, *N. Engl. J. Med.*, 345, 1621, 2001.

44. Edwards, K.A., Clancy, H.A., and Baeumner, AJ., *Bacillus anthracis*: Toxicology, epidemiology and current rapid-detection methods, *Anal. Bioanal. Chem.*, 384, 73, 2006.

45. Chakraborty, A. et al., Anthrax outbreaks in Bangladesh, 2009–2010, *Am. J. Trop. Med. Hyg.*, 86, 703, 2012.

46. Ross, J., The pathogenesis of anthrax following the administration of spores by the respiratory route, *J. Pathol. Bacteriol.*, 73, 485, 1957.

47. Abramova, F.A. et al., Pathology of inhalational anthrax in 42 cases from the Sverdlovsk outbreak of 1979, *Proc. Natl. Acad. Sci. USA*, 90, 2291, 1993.

48. Kyriacou, D.N. et al., Clinical predictors of bioterrorism-related inhalational anthrax, *Lancet*, 364, 449, 2004.

49. Holty, J.E. et al., Systematic review: A century of inhalational anthrax cases from 1900 to 2005, *Ann. Intern. Med.*, 144, 270, 2006.

50. Sirisanthana, T. and Brown, A.E., Anthrax of the gastrointestinal tract, *Emerg. Infect. Dis.*, 8, 649, 2002.

51. Hugh-Jones, M., 1996-97 global anthrax report, *J. Appl. Microbiol.*, 87, 189, 1999.

52. Ringertz, S.H. et al., Injectional anthrax in a heroin skin-popper, *Lancet*, 356, 1574, 2000.

53. Radun, D. et al., Preliminary case report of fatal anthrax in an injecting drug user in North-Rhine-Westphalia, Germany, December 2009, *Euro Surveill.*, 15, 19464, 2010.

54. Hicks, C.W. et al., An overview of anthrax infection including the recently identified form of disease in injection drug users, *Intensive Care Med.*, 38, 1092, 2012.

55. Holzmann, T. et al., Fatal anthrax infection in a heroin user from southern Germany, June 2012, *Euro Surveill.*, 17, 20204, 2012.

56. Price, E.P. et al., Molecular epidemiologic investigation of an anthrax outbreak among heroin users, Europe, *Emerg. Infect. Dis.*, 18, 1307, 2012.

57. Mourez, M., Anthrax toxins, *Rev. Physiol. Biochem. Pharmacol.*, 152, 135, 2004.

58. Cole, L.A., The specter of biological weapons, *Sci. Am.*, 275, 60, 1996.

59. Fowler, R.A. and Shafazand, S., Anthrax bioterrorism: Prevention, diagnosis and management strategies, *J. Bioterror. Biodef.*, 2, 2, 2011.

60. Nicholson, W.L. et al., Resistance of *Bacillus endospores* to extreme terrestrial and extraterrestrial environments, *Microbiol. Mol. Biol. Rev.*, 64, 548, 2000.

61. Cybulski, R.J. et al., Four superoxide dismutases contribute to *Bacillus anthracis* virulence and provide spores with redundant protection from oxidative stress, *Infect. Immun.*, 77, 274, 2009.

62. Hugh-Jones, M. and Blackburn, J., The ecology of *Bacillus anthracis*, *Mol. Aspects Med.*, 30, 356, 2009.

63. Inglesby, T.V. et al., Working Group on Civilian Biodefense. Anthrax as a biological weapon, 2002: Updated recommendations for management, *JAMA*, 287, 2236, 2002.

64. Booth, M.G. et al., Health Protection Scotland Anthrax Clinical Network. Anthrax infection in drug users, *Lancet*, 375, 1345, April 17, 2010.

65. Cote, C.K., Welkos, S.L., and Bozue J., Key aspects of the molecular and cellular basis of inhalational anthrax, *Microbes Infect.*, 13, 1146, 2011.

66. Sanz, P. et al., Detection of *Bacillus anthracis* spore germination in vivo by bioluminescence imaging, *Infect. Immun.*, 76, 1036, 2008.

67. Fang, F.C., Antimicrobial reactive oxygen and nitrogen species: Concepts and controversies, *Nat. Rev. Microbiol.*, 2, 820, 2004.

68. van Sorge, N.M. et al., Anthrax toxins inhibit neutrophil signaling pathways in brain endothelium and contribute to the pathogenesis of meningitis, *PLoS One*, 3, e2964, 2008.

69. Sejvar, J.J., Tenover, F.C., and Stephens, D.S., Management of anthrax meningitis, *Lancet Infect. Dis.*, 5, 287, 2005.

70. Drysdale, M., Bourgogne, A., and Koehler, T.M. Transcriptional analysis of the *Bacillus anthracis* capsule regulators, *J. Bacteriol.*, 187, 5108, 2005.

71. Fouet, A. and Mock, M. Regulatory networks for virulence and persistence of *Bacillus anthracis*, *Curr. Opin. Microbiol.*, 9, 160, 2006.

72. Irenge, L.M. and Gala, J.L., Rapid detection methods for *Bacillus anthracis* in environmental samples: A review, *Appl. Microbiol. Biotechnol.*, 93, 1411, 2012.

73. Turnbull, P.C., Definitive identification of *Bacillus anthracis*—A review, *J. Appl. Microbiol.*, 87, 237, 1999.

74. Knisely, R.F., Selective medium for *Bacillus anthracis*, *J. Bacteriol.*, 92, 784, 1966.

75. Juergensmeyer, M.A. et al., A selective chromogenic agar that distinguishes *Bacillus anthracis* from *Bacillus cereus* and *Bacillus thuringiensis*, *J. Food Prot.*, 69, 2002, 2006.

76. Marston, C.K. et al., Evaluation of two selective media for the isolation of *Bacillus anthracis*, *Lett. Appl. Microbiol.*, 47, 25, 2008.

77. Spencer, R.C., *Bacillus anthracis*, *J. Clin. Pathol.*, 56, 182, 2003.

78. Hoffmaster, A.R. et al., Identification of anthrax toxin genes in a *Bacillus cereus* associated with an illness resembling inhalation anthrax, *Proc. Natl. Acad. Sci. USA*, 101, 8449, 2004.

79. Titball, R.W., Turnbull, P.C., and Hutson, R.A., The monitoring and detection of *Bacillus anthracis* in the environment, *Soc. Appl. Bacteriol. Symp. Ser.*, 20, 9S, 1991.

80. Klee, S.R. et al., Characterization of *Bacillus anthracis*-like bacteria isolated from wild great apes from Cote d'Ivoire and Cameroon, *J. Bacteriol.*, 188, 5333, 2006.

81. Dixon, T. et al., Anthrax, *N. Engl. J. Med.*, 341, 815, 1999.

82. Baillie, L.W., Jones, M.N., Turnbull, P.C., and Manchee, R.J., Evaluation of the Biolog system for the identification of *Bacillus anthracis*, *Lett. Appl. Microbiol.*, 20, 209, 1995.

83. Keys, C.J. et al., Compilation of a MALDI-TOF mass spectral database for the rapid screening and characterisation of bacteria implicated in human infectious diseases, *Infect. Genet. Evol.*, 4, 221, 2004.

84. AOAC International, Initiative yields effective methods for anthrax detection; RAMP and MIDI, Inc., methods approved, *Inside Lab. Manage.*, 10, 3, 2004.

85. Hao, R. et al., Rapid detection of *Bacillus anthracis* using monoclonal antibody functionalized QCM sensor, *Biosens. Bioelectron.*, 24, 1330, 2009.

86. Tang, S. et al., Detection of anthrax toxin by an ultrasensitive immunoassay using europium nanoparticles, *Clin. Vaccine Immunol.*, 16, 408, 2009.

87. Duriez, E. et al., Femtomolar detection of the anthrax edema factor in human and animal plasma, *Anal. Chem.*, 81, 5935, 2009.

88. AuCoin, D.P. et al., Rapid detection of the poly-γ-D-glutamic acid capsular antigen of *Bacillus anthracis* by latex agglutination, *Diagn. Microbiol. Infect. Dis.*, 64, 229, 2009.

89. Schofield, D.A. and Westwater, C. Phage-mediated bioluminescent detection of *Bacillus anthracis*, *J. Appl. Microbiol.*, 107, 1468, 2009.

90. Morel, N. et al., Fast and sensitive detection of *Bacillus anthracis* spores by immunoassay, *Appl. Environ. Microbiol.*, 78, 6491, 2012.

91. Parthasarathy, N. et al., Application of carbohydrate microarray technology for the detection of *Burkholderia pseudomallei*, *Bacillus anthracis* and *Francisella tularensis* antibodies, *Carbohydr. Res.*, 343, 2783, 2008.

92. Shields, M.J. et al., Immunomagnetic capture of *Bacillus anthracis* spores from food, *J. Food Prot.*, 75, 1243, 2012.

93. Dauphin, L.A., Moser, B.D., and Bowen, M.D., Evaluation of five commercial nucleic acid extraction kits for their ability to inactivate *Bacillus anthracis* spores and comparison of DNA yields from spores and spiked environmental samples, *J. Microbiol. Methods*, 76, 30, 2009.

94. Dauphin, L.A. et al., Gamma irradiation can be used to inactivate *Bacillus anthracis* spores without compromising the sensitivity of diagnostic assays, *Appl. Environ. Microbiol.*, 74, 4427, 2008.

95. Dauphin, L.A. and Bowen, M.D., A simple method for the rapid removal of *Bacillus anthracis* spores from DNA preparations, *J. Microbiol. Methods*, 76, 212, 2009.

96. Ramisse, V. et al., The Ba813 chromosomal DNA sequence effectively traces the whole *Bacillus anthracis* community, *J. Appl. Microbiol.*, 87, 224, 1999.

97. Qi, Y. et al., Utilization of the rpoB gene as a specific chromosomal marker for real-time PCR detection of *Bacillus anthracis*, *Appl. Environ. Microbiol.*, 67, 3720, 2001.

98. Antwerpen, M.H. et al., Real-time PCR system targeting a chromosomal marker specific for *Bacillus anthracis*, *Mol. Cell. Probes*, 22, 313, 2008.

99. Rachwal, P.A. et al., The potential of TaqMan Array Cards for detection of multiple biological agents by real-time PCR, *PLoS One*, 7, e35971, 2012.

100. Hadjinicolaou, A.V. and Victoria, L., Use of molecular beacons and multi-allelic real-time PCR for detection of and discrimination between virulent *Bacillus anthracis* and other *Bacillus* isolates, *J. Microbiol. Methods*, 78, 45, 2009.

101. Li, Y., Zhou, X., and Ye, D., Molecular beacons: An optimal multifunctional biological probe, *Biochem. Biophys. Res. Commun.*, 373, 457, 2008.

102. Jacob, D. et al., Rapid and high-throughput detection of highly pathogenic bacteria by Ibis PLEX-ID technology, *PLoS One*, 7, e39928, 2012.

103. Athamna, A. et al., In vitro susceptibility of *Bacillus anthracis* to various antibacterial agents and their time-kill activity, *J. Antimicrob. Chemother.*, 53, 247, 2004.

104. Jallali, N. et al., The surgical management of injectional anthrax, *J. Plast. Reconstr. Aesthet. Surg.*, 64, 276, 2011.

105. Jean, F. et al., Alpha antitrypsin Portland, a bioengineered serpin highly selective for furin: Application as an antipathogenic agent, *Proc. Natl. Acad. Sci. USA*, 95, 7293, 1998.

106. Schuch, R., Nelson, D., and Fischetti, V.A., A bacteriolytic agent that detects and kills *Bacillus anthracis*, *Nature*, 418, 884, 2002.

107. Kikkawa, H.S. et al., Characterization of the catalytic activity of the gamma-phage lysin, PlyG, specific for *Bacillus anthracis*, *FEMS Microbiol. Lett.*, 286, 236, 2008.

108. Jones, M.B. et al., Inhibition of *Bacillus anthracis* growth and virulence-gene expression by inhibitors of quorum-sensing, *J. Infect. Dis.*, 191, 1881, 2005.

25 Brucella (B. abortus, B. melitensis, and B. suis)

Sascha Al Dahouk, Heinrich Neubauer, and Lisa D. Sprague

CONTENTS

25.1 INTRODUCTION

From the numerous agents selectable for an effective attack against society, brucellae are certainly among the most suitable. The resulting disease, brucellosis, is a member of the so-called dirty dozen, which reads like a "who is who" of highly infectious and incapacitating diseases. This illustrious circle further includes plague, tularemia, glanders/melioidosis, Q-fever, anthrax, small pox, viral encephalitis, viral hemorrhagic fevers, ricin, botulinum toxins, and enterotoxin B from *Staphylococcus aureus*.

Brucella spp. have always been in the focus of military decision makers[1] mainly due to the fact that the organism can be easily transmitted via aerosols to humans and animals. Given the very low infective dose, corresponding to 10 bacteria for humans and its moderate to high attack rate, 50%–80% of the persons exposed will develop clinical disease, making the agent an effective weapon. According to a WHO expert committee, the deployment of 50 kg aerosolized *B. melitensis* upstream from a city with 500,000 inhabitants would not only result in 150–500 fatalities and 27,000–100,000 diseased persons but also entail the loss of $26.2 billion per 100,000 exposed inhabitants.[2] The applicability of weaponized *Brucella* spp. in attacking livestock was extensively tested by the United States between 1942 and 1967 but was later abandoned with the ratification of the Biological Weapons Convention.[3] However, following the dissolution of the former USSR and the end of the cold war between NATO and the Warsaw Pact, the clearly defined and controlled threat scenario collapsed and was replaced by the global threat of internationally operating terror groups committed to destruction on an unprecedented scale.[4]

Confronted with this new challenge and in view of possible acts of bioterrorism, army and public health experts tried to evaluate the genuine risks posed by the use of various B agents against the civilian population. Special emphasis was placed on the influence of a large-scale outbreak on public health and medical infrastructure. Based on the criteria of public health impact, the delivery potential to large populations, public perception, and special public health preparedness needs, brucellosis was ranked in category B due to its weaker medical and public impact.[5]

It must be stressed, however, that the conclusions drawn were founded on the local conditions prevailing in the United States and can therefore not be transferred as such to other countries without careful reflection. Therefore, local conditions, that is, endemicity, awareness of the disease by first responders, public health systems in action, and preparedness in the respective country should be reevaluated by native experts. In countries where brucellosis has been successfully eradicated, even small-scale outbreaks will cause major problems since the first responders, for example, family physicians, are usually unfamiliar with the clinical picture. This can result in a significant diagnostic delay (up to 6 months) and above-average mortality.[6]

Despite brucellosis being generally considered a disease with lower-ranking medical and public health impact, it actually lends itself superbly to agroterrorism, a fact frequently ignored by political decision makers. Agroterrorism is the deliberate tampering with and/or contamination of the food supply with the intent of adversely affecting the social, economic, physical, and psychological well-being of society.[7] Agroterrorism is believed to be more attractive to terrorists because it carries less risk to the terrorist, can be carried out more covertly, and does not need the sophisticated skills of weaponization. The bacterium can be readily acquired from endemic areas and production in sufficient quantity is easy to accomplish in low-tech facilities. Moreover, the prolonged incubation time will make tracking of the executer difficult.[8] Vulnerable targets include farm animals (cattle, swine, sheep, horses, poultry, and fish) and processed food and storage facilities.[9]

In veterinary public health, brucellosis has always been a disease subject to scrutiny due to its considerable impact on animal health and international trade. An outbreak can significantly obstruct international trade of animals and the products thereof and lead to major socioeconomic and public health-related problems not only in the country of origin but also on a global scale. Numerous factors in today's intensive mass animal farming, such as monitoring of animal health only at flock level, rapid movement of animals and their products over long distances, and insufficient farm- and food-related security, multiply the impact of an attack. Inefficient disease reporting systems and limited knowledge of the first responding veterinarian can also delay the installation of appropriate countermeasures. Agroterroristic attacks are believed to cause immediate economic disruption of markets due to eradication procedures (mass culling); considerable

loss of jobs; difficulties in sustaining adequate food supplies; increased consumer costs; indirect multiplier effects, that is, compensation for the animal owners; and, finally, international trade embargos.[10] Countermeasures will also cause restrictions of civil liberties[8] for fear of human causalities, which are likely to occur if *B. melitensis*, *B. abortus*, and *B. suis* biovar (bv) 1 and 3 are used. The reputation of the government and its administration, including the public health and veterinary public health sector, can be seriously damaged if the public does not feel sufficiently protected and provided for.[10]

25.2 CLASSIFICATION AND MORPHOLOGY

Brucellae are facultative intracellular coccobacilli belonging to the order of Rhizobiales of the α-2 subgroup of Proteobacteria. The class Alphaproteobacteria comprises organisms that are mammalian or plant pathogens and symbionts.[11] Within the family Brucellaceae, *Ochrobactrum* is the closest phylogenetic neighbor of *Brucella*. Historically, brucellae are differentiated by host tropism, pathogenicity, and phenotypic traits. Presently, the genus consists of the six "classical" species (Table 25.1), that is, *B. melitensis* bvs 1–3 (mainly isolated from sheep and goats); *B. abortus* bvs 1–6 and 9 (from cattle and other Bovidae); *B. suis* bvs 1–3 (from pigs), bv 4 (from reindeer), and

TABLE 25.1
The "Classical" *Brucella* Species and the "New" Atypical Strains Including Their Pathogenicity for Humans

Brucella (*B.*) Species	Biovars	Animal Host	Human Disease
B. abortus	1–6, 9	Cattle, bison, buffalo, elk, yak, camel	Yes
B. melitensis	1–3	Sheep, goat, cow, camel	Yes
	3	Nile catfish, dog	
B. suis	1	Horse	Yes (bvs 1, 3, 4)
	1–3	Pig, wild boar	
	2	European hare	
	4	Caribou, reindeer	
	5	Rodents	
B. canis		Canines	Yes (rarely)
B. ovis		Ram	Not reported
B. neotomae		Rodents	Not reported
B. ceti		Whale, dolphin, porpoise	Yes (rarely)
B. pinnipedialis		Seal	Not reported
B. microti		Common vole, red fox, (soil)	Not reported
B. inopinata		Unknown	Yes
BO2		Unknown	Yes
Baboon isolates		Baboons	Not reported
Rodent isolates		Rodents	Not reported
Fox isolates		Red fox	Not reported
Frog isolates		African bullfrog	Not reported

Source: Modified according to Sprague, L.D. et al., *Pathog. Glob. Health*, 106, 144, 2012.

bv 5 (from small rodents); *B. canis* (from dogs); *B. ovis* (from sheep); and *B. neotomae* (from desert wood rats). However, the majority of notified cases are still caused by *B. melitensis*, *B. abortus*, and *B. suis* in descending order of occurrence and rarely by *B. canis*.

In the last decade, several "new" species (Table 25.1), that is, *B. pinnipedialis* (isolated from seals) and *B. ceti* (from whales and dolphins),[12] *B. microti* (isolated from the common vole and red foxes),[13,14] and *B. inopinata* (isolated from a human breast implant and an unknown animal reservoir),[15] have been described. Moreover, numerous "atypical" strains have been found possibly representing novel species or lineages. These strains have been isolated from nonhuman primates,[16] from wild native rodents in North Queensland, Australia,[17] red foxes in Eastern Austria,[18] and most recently from African bullfrogs[19] (Table 25.1). At present, the pathogenicity of these newly described species and atypical strains for humans is unknown and needs to be investigated. However, human pathogenicity can be assumed since several reports not only described human infections with marine mammal strains[20,21] and the isolation of *B. inopinata* (BO1) and BO2 from human samples[22,23] but also confirmed the pathogenicity of *B. microti* for mammalian hosts.[24]

Brucellae are very small (0.5–0.7 × 0.6–1.5 μm), faintly stained gram-negative coccoid rods, which microscopically look like "fine sand." Primary culturing reveals punctate, nonpigmented, and nonhemolytic colonies. Colonies of smooth (S) *Brucella* strains are raised, convex, circular, translucent, and 0.5–1 mm in diameter. After subcultivation or prolonged culture (>4 days), the colony morphology of the bacteria may tend toward less convex and more opaque colonies with a dull, dry, yellowish-white granular appearance. These changes are caused by the dissociation of brucellae from smooth to rough (R) forms.

25.3 PATHOGENICITY AND EPIDEMIOLOGY

25.3.1 PATHOGENICITY

The pathogenesis of brucellosis and the virulence factors of *Brucella* are unique. Classical virulence factors commonly recognized by innate immunity, for example, capsules, fimbriae, and pili, are missing. Brucellae are able to survive and replicate inside professional and nonprofessional phagocytic host cells without affecting cellular viability.[25] The pathogenic potential differs between the various *Brucella* species; *B. microti*, for instance, replicates more efficiently in human and murine macrophages than the virulent species *B. suis*.[24] *Brucella* uses a surreptitious strategy that enables the bacterium to evade host defense mechanisms, penetrate host cells, alter intracellular trafficking, avoid killing in lysosomes, and modulate the intracellular environment.[26,27] Contrariwise, cellular immunity is absolutely necessary for the host to control and resolve *Brucella* infections.[28]

Unlike the endotoxins of other gram-negative bacteria, the lipid A moiety of the lipopolysaccharide (LPS) of *Brucella* induces a reduced and delayed inflammatory response in infected hosts. *Brucella* LPS inhibits complement and cationic peptide-mediated killing and prevents the synthesis of immune mediators.[29] The O-polysaccharide chain of the LPS moreover mediates the entry of *Brucella* into macrophages.[30] The LPS also interferes with MHC II antigen presentation resulting in the alteration of *Brucella*-specific CD4+ T cell activation.[31] This alteration of T cell function may thus play a key role in the development of chronic and relapsing brucellosis.[32]

Brucella entry into host cells is also dependent on the two-component regulatory system *Bvr*R/*Bvr*S,[33] which controls the expression of various cell-surface proteins (e.g., Omp 3 family), enabling *Brucella* to escape from the lysosomal pathway. After internalization, virulent brucellae inhibit the fusion of their phagosomes with lysosomal compartments and modulate the phagosomal composition in order to evade intracellular degradation. The type IV secretion system encoded by the *Vir*B operon is also a major virulence factor and plays a key role in intracellular trafficking.[34] Effector molecules secreted by the type IV secretion machinery are responsible for maintaining the interactions of the *Brucella*-containing vacuole with the endoplasmic reticulum of the host cell until the replicative niche of *Brucella* is fully developed.[35] Brucellae are able to survive and even replicate within this compartment, also known as the "brucellosome", without restricting basic functions of the host cells. Intracellular bacteria have to resist severe nutrient deprivation and microaerobic conditions.[36,37] A considerable degree of metabolic versatility is required for long-term survival and replication under the harsh conditions in this compartment.[38] The majority of proteins regulated at this stage of infection are involved in the primary metabolism of *Brucella*, confirming a "covert" adaptation of the pathogen to the host cell environment.[39]

25.3.2 EPIDEMIOLOGY

Bovine brucellosis (*B. abortus*) has been successfully eradicated in Canada, Australia, Japan, and Northern Europe by means of vaccination schemes and rigorous test-and-slaughter programs. According to the European Commission Decision 2003/467/EC (last update May 11, 2011), Austria, Belgium, the Czech Republic, Denmark, Estonia, Finland, France, Germany, Ireland, Luxembourg, the Netherlands, Norway, Poland, Slovakia, Slovenia, Sweden, and Switzerland are "officially free from bovine brucellosis". This does not however apply to numerous southern European countries (http://www.oie.int/wahis/public.php). While bovine brucellosis has been successfully contained in the last decades, the control of brucellosis in small ruminants continues to be a serious problem. Ovine/caprine brucellosis (*B. melitensis*) is endemic in countries surrounding the Mediterranean Sea, highly prevalent in the Persian Gulf region, and has also been reported from several African countries (http://www.oie.int/wahis/public.php). Endemic regions are moreover found in South and Central Asia and in parts of Central and

South America, whereby Mexico and northern Argentina are the countries most seriously affected. In contrast, *B. melitensis* is not known to be enzootic in the United States, Canada, Northern Europe, Australia, New Zealand, and Southeast Asia, and only sporadic incursions have been notified in these regions.

Despite ovine/caprine brucellosis having a limited geographic distribution, *B. melitensis* is the main cause of clinically apparent human disease.[40,41] Currently, about half a million human cases worldwide are reported annually but the estimated number of unreported cases due to the unspecific clinical symptoms of the disease is assumed to be ten times higher. In endemic countries, prevalence rates frequently exceed 10 cases per 100,000 population[42] and the incidence of human *B. melitensis* infections correlates directly with the distribution of the disease in sheep and goat populations.[43] Of note is that living in the proximity of an infected flock does not constitute a significant danger, provided that direct contact with contaminated tissues or secretions is avoided.[44]

Porcine brucellosis (*B. suis*) is present in most pig breeding areas and can be encountered in domestic, feral, and wild pigs. The zoonotic potential of the different *B. suis* biovars differs significantly, for example, bv 2, the most common causative agent of porcine brucellosis in Europe, is known to only exceptionally infect humans.[45,46] The human pathogenic biovars, *B. suis* bv 1 and bv 3, occur mainly in Asia, South America, in the southeastern states of the United States, and in Queensland, Australia.[47,48]

In the United States, brucellosis used to be an occupational disease and up to 1960 *B. abortus* was the most frequently reported cause of infection. By the early 1970s, infections with *B. suis* predominated and currently cases of *B. melitensis* infections are on the rise and increasingly found in the general population.[49] Major risk factors accountable for contracting human brucellosis in the United States are Hispanic ethnicity, recent travel to endemic areas in Mexico, and ingestion of nonpasteurized dairy products.[49–52] An association of brucellosis with the immigrant population has also been described for other nonendemic countries, for example, Denmark[53] and Germany.[6] Individuals with a migration background usually acquire the disease in their native country or through the consumption of illegally imported and contaminated food products.[54] In spite of sanitary, socioeconomic, and political measures as well as efficient disease control in many developed countries, brucellosis still remains a major public health concern. Within Europe, for example, human disease is mostly limited to the consumption of contaminated food products and to traveling since even short-distance trips can be accompanied by a significant health risk.[55] In addition, spillover of brucellosis from wildlife to livestock and globalization of animal trade continues to promote the local and worldwide dissemination of animal and human disease.[42,47] Finally, novel species, that is, *B. microti*, have emerged from an environmental reservoir that may pose a new public health risk, leading to the recurrence of brucellosis in nonendemic regions.[56]

25.4 CLINICAL FEATURES AND PATHOGENESIS

Brucellosis is a classic zoonotic disease transmitted from animal to man. Human infection usually occurs after the consumption of unpasteurized dairy products or less frequently after consumption of raw meat. However, some occupational groups with close animal contacts such as veterinarians, farmers, butchers, meat packers, and abattoir workers have an above-average exposure risk. *Brucella* spp. are highly contagious and may also enter the body as aerosol via the mucous membranes of the respiratory tract. Brucellosis is therefore among the most frequent laboratory-acquired bacterial infections worldwide.[57]

Blood transfusion,[58] bone marrow transplantation,[59] breast-feeding,[60,61] sexual intercourse,[62–64] and congenital infection[65] have been described as potential but marginal routes of human-to-human transmission. However, the isolation of patients is not necessary.[66]

Clinical signs and symptoms of brucellosis in patients are independent of the route of transmission[66]; airborne transmission does not increase the number of respiratory infections,[67] whereas those infected by other routes may show coughing and pneumonia.

After infection, bacteremia leads to the spreading of the organism to various organs, mainly to the reticuloendothelial tissues, that is, liver, spleen, skeletal, and hematopoietic system.

While asymptomatic infections may occur in humans, the frequently multisystemic disease presents as fever of unknown origin with a wide range of symptoms.[68]

Brucellosis usually begins as an acute febrile illness with nonspecific flu-like symptoms such as headaches, arthralgia and myalgia, fatigue, malaise, weight loss, chills, and sweating. Undulant fever may develop that can last for several months. The persistence of brucellae in the mononuclear phagocytic system is responsible for the high frequency of chronic diseases, complications, and relapses. Clinical manifestations are protean. Cutaneous, hematologic, gastrointestinal, genitourinary, respiratory, osteoarticular, cardiovascular, as well as neurologic disorders may occur.[69,70] Hepatomegaly and splenomegaly are major clinical findings due to the invasion of the reticuloendothelial system by *Brucella*.[71] The most frequent focal complications are spondylitis, sacroiliitis, and peripheral arthritis.[69,70] These osteoarticular manifestations can be severely debilitating and are difficult to treat.[72] The majority of fatal outcomes are due to neurobrucellosis and endocarditis. Despite of the high frequency of focal complications, the case fatality rate of human brucellosis is low (<1%).

Congenital brucellosis can lead to preterm birth, with signs of septicemia, but asymptomatic or cases with mild symptoms have also been reported. Whether brucellosis can lead to spontaneous abortion in humans is still a matter of debate; however, reports describing an increased risk of spontaneous abortion, premature delivery, and intrauterine fetal death after infection with *Brucella* during pregnancy exist.[73,74] In terms of the rate and severity of complications and the response to treatment, the clinical course of childhood brucellosis appears to be less severe than in adults.

25.5 IDENTIFICATION AND DIAGNOSIS

Artificial outbreaks among humans are difficult to recognize since brucellosis lacks distinctive clinical features and the incubation period varies widely, ranging from a few days to several months. A further major drawback when trying to determine a deliberate release of *Brucella* spp. very early on is the lack of an immediate rise in cases. However, a marked increase of brucellosis cases within a short period of time in a nonendemic country without specific risk factors, such as traveling to an endemic region, consumption of unpasteurized milk or raw met, direct animal contacts, and working in a microbiology laboratory, should raise suspicion of an attack, and target-oriented laboratory diagnosis and appropriate antibiotic treatment have to be initiated.[75]

Although various microbiological, serological, and molecular methods for direct or indirect detection of *Brucella* infections exist, the laboratory diagnosis of human brucellosis is still a challenge. The rapid laboratory diagnosis confirming a clinical suspicion is most important for a successful treatment outcome. Currently, the isolation of the disease-causing pathogen by culture methods is the gold standard in the diagnosis of an active infection.[76] However, primary cultures and subsequent phenotypic characterization of the isolate are time-consuming. In addition, the sensitivity of culture methods varies widely depending on the stage of disease (acute > chronic), previous use of antibiotics, sample material (bone marrow > blood > tissue/body fluids), and the technical approach.

Brucella spp. grow on most standard media, for example, blood agar, chocolate agar, trypticase soy agar (TSA), and serum-dextrose agar (SDA). Bovine or equine serum (2%–5%) is routinely added to the basal medium because some strains need it for growth. The inoculated agar plates should be incubated at 35°C–37°C in 5%–10% CO_2. The primary culturing of the fastidious bacterium from clinical specimens may take several days or even weeks. In acute cases, the isolation rate from blood samples may vary from 80% to 90%, whereas only 30%–70% of the chronic cases can be confirmed bacteriologically.[77,78]

The recovery rate from clinical specimens has been progressively maximized by various technical improvements, for example, the biphasic Castañeda method, automated systems, and yield-optimizing methods such as lysis centrifugation or blood clot culture.[78–80] Automated blood culture systems, such as the BACTEC™ (Becton Dickinson Diagnostic Systems, Sparks, Maryland, United States), the BacT/ALERT™ (bioMerieux Inc., Durham, North Carolina, United States) (which continuously monitor the CO_2 release of potentially growing microorganisms), and the BACTEC™ Myco/F Lytic system integrating lytic activity and automation,[81] have reduced the time to detection significantly. Oxidase- and urease-positive bacteria must raise suspicion of *Brucella*. The suspicious colonies can be rapidly confirmed by the slide agglutination test using undiluted polyvalent *Brucella* antiserum (anti-S[smooth] serum) mixed with a saline suspension of the bacteria. Elaborate microbiological methods testing for CO_2 requirement, H_2S production, urease activity, agglutination with monospecific sera (A and M), selective inhibition of growth on media containing dyes such as thionin or basic fuchsine, and phage typing further identify *Brucella* species and biovars.[76,82] These methods are not suited for routine clinical microbiology laboratories because they are time-consuming, hazardous, and subject to variable interpretation. Results of commercially available biochemical identification systems such as the API 20 NE® (BioMerieux, Nürtingen, Germany) have to be carefully assessed, since *Brucella* spp. are frequently misidentified as *Psychrobacter phenylpyruvicus* (formerly *Moraxella phenylpyruvica*)[83] or *Ochrobactrum anthropi*.[84] A feasible alternative is the semiautomated metabolic biotyping system (Micronaut™) based on a selection of 93 different substrates.[85] In comparison to phenotypic techniques or molecular methods, matrix-assisted laser desorption ionization–time-of-flight mass spectrometry (MALDI–TOF MS) could be a rapid, precise, and cost-effective tool for the identification of brucellae in clinical microbiology laboratories.[86,87] However, the access to comprehensive protein profile databases is severely hampered by the classification of the microorganism as a category B bioterrorism agent.

Automated, continuous monitoring blood culture systems are preferable for culturing clinical samples suspicious for brucellosis in order to avoid spreading within laboratory facilities. Since brucellae are considered class 3 organisms, live *Brucella* cultures as well as presumptive organisms should be handled in a biosafety cabinet. Until *Brucella* has been ruled out, laboratory staff should strictly adhere to standard precautions and to good laboratory practice. The high frequency of accidental exposures to brucellae in clinical laboratories of industrialized countries indicates a lack of alertness when dealing with such potential bioterrorist agents.[57] Although culture and isolation of *Brucella* is time-consuming, labor-intensive, and hazardous and does not seem appropriate for the fast and reliable detection of an outbreak, a thorough analysis of the live pathogen is essential for identifying genetic modifications in virulence and antibiotic resistance indicative of a deliberate release.

Because of the diagnostic delay related to classical microbiological methods and the low isolation rate in chronic courses, most physicians rely on high or rising titers of anti-*Brucella* antibodies. Serologic tests can only indirectly prove brucellosis but agglutination titers ≥1:160 or a fourfold rise of titers in follow-up sera are considered to be indicative for active infections. After successful therapy, diagnostic titers have been detected for months or even years despite negative blood cultures.[88,89] In addition, antibody titers may persist due to an ongoing exposure to *Brucella* in endemic regions. Therefore, the significance of anti-*Brucella* antibody detection has to be evaluated with careful consideration of clinical signs and symptoms, a history of potential exposure, and the background prevalence in healthy individuals.

Serologic tests are not only suitable for the initial diagnosis of human brucellosis but also for treatment follow-up. However, the monitoring of treatment response requires serial testing. An effective antibiotic therapy is usually accompanied by a rapid decline of antibody titers, whereas persisting high and rising IgG titers can be an indication of treatment failure and relapse, respectively.[90,91]

Numerous serological tests have been described so far,[92] but they have several drawbacks. In serum agglutination tests (SATs), false-negative results may occur due to the prozone effect, blocking antibodies, or delayed antibody generation during the incubation period. False-positive results due to cross-reactive anti-Brucella LPS antibodies can be encountered in patients suffering from other gram-negative bacterial infections such as salmonellosis, yersiniosis, cholera, and tularemia.[92] Nevertheless, the SAT is still considered to be the method of reference in the serological diagnosis of human brucellosis. For routine laboratories with a high turnover of clinical samples practicable formats such as slide, plate and card agglutination have replaced the labor-intensive and time-consuming classical tube agglutination test (Wright test). The Rose Bengal test (RBT) is an example of a card agglutination method that is widely used for rapid screening in endemic countries although its performance is poor in patients formerly and/or repeatedly exposed to the agent.[93] Therefore, diagnostic titers have to be confirmed by more specific serologic tests or, alternatively, diluted sera can be tested by the RBT in high-risk populations.[94] In chronic courses and during relapse, SAT results are frequently negative or inconclusive and the classical Coombs test (CT) may help to complement SAT by detecting incomplete, blocking, or nonagglutinating antibodies.[90] Since both CT and SAT are labor-intensive and time-consuming, the Brucellacapt® (Vircell, Santa Fé, Granada, Spain), a single-step immuno-capture assay, is recommended for the detection of total anti-Brucella antibodies. The Brucellacapt titers are a valuable marker of infection activity independent of disease stage.[90,95]

Although SAT is considerably cheaper, the serological diagnosis of human brucellosis in nonendemic countries is usually based on commercial enzyme-linked immunoassay (ELISA) kits due to the significantly shorter turnaround time.[96] Anti-Brucella antibodies are reliably detected by ELISA and results are consistent with SAT and CT.[97,98] A combined analysis of IgM and IgG results increases ELISA sensitivity and allows the staging of the disease.[96,99]

Independent of the disease stage, polymerase chain reaction (PCR) assays are more sensitive than blood cultures and more specific than serologic tests. Various PCR assays have been developed in the last years for the rapid detection of brucellae in clinical and environmental samples.[100,101] Especially real-time PCR techniques allow a fast online diagnosis and are extremely valuable for screening a larger number of samples that may accumulate in a deliberate outbreak. Moreover, the analytical sensitivity of real-time PCR assays is very high and as few as five bacteria per reaction can be detected.[102,103] Nevertheless, the low number of bacteria in clinical specimens, such as body fluids or tissues and inhibitory effects

arising from matrix components, are still a challenge for the direct detection of Brucella DNA in brucellosis patients.[104] Inhibitory effects can be diminished by using commercial DNA preparation kits, for example, the QIAamp™ DNA Mini Kit (Qiagen Inc., Valencia, California, United States) and the UltraClean™ DNA BloodSpin Kit (MO BIO Laboratories Inc., Carlsbad, California, United States).[105–107] In complex matrices, residual PCR inhibition can be easily unmasked by using an internal amplification control.[103]

Genus-specific PCR assays are appropriate for diagnosing human brucellosis.[75] The bcsp31 gene, coding for a 31-kDa immunogenic outer membrane protein conserved among all Brucella spp., is a well-established molecular target for clinical applications.[108] PCR assays targeting the IS711 insertion element that is found in multiple copies within Brucella chromosomes can enhance analytical sensitivity.[109,110] Genus-specific PCR assays help to avoid false-negative results when atypical species and biovars are tested. However, a second gene target is absolutely necessary to confirm the primary molecular diagnosis[103]; 16S rRNA gene sequencing[111] or differential PCR assays, for example, the conventional Bruce-ladder PCR[112–114] or species-specific real-time PCR assays,[103] can be applied for this purpose.

Finally, the sole detection of Brucella-specific gene targets does not provide evidence for an active infection but quantitative real-time PCR might offer the tool to differentiate active from past Brucella infections.[115]

After a deliberate release of brucellae, genotyping of bacterial isolates may help to trace an infection and to identify a potential aggressor. A multiple-locus variable number of tandem repeats (VNTR) analysis assay based on 16 markers (MLVA-16) has been designed for this particular purpose.[116–118]

25.6 TREATMENT AND PREVENTION

Although naturally occurring Brucella strains only rarely reveal antibiotic resistance, the recommended antimicrobial therapies are often followed by primary treatment failures and relapses. The negative therapeutic outcome is a consequence of the intracellular lifestyle of brucellae and the acidic environment within the phagosome.[119,120] Therefore, the selected antibiotics should be able to penetrate cellular membranes and remain effective despite the low pH within the phagosome. Therapeutic regimens are commonly based on various combinations of four antibiotic drugs. These are doxycycline, rifampin, streptomycin (or other aminoglycosides), and trimethoprim–sulfamethoxazole (co-trimoxazole). Prolonged chemotherapy using at least two synergistic antibiotics is required to control acute brucellosis and to prevent complications and relapse. Monotherapies have shown unacceptably high rates of relapse and should therefore not be used.[121] Since successful treatment of human brucellosis needs long-term application of multiple drugs with a multitude of adverse effects, a confirmative laboratory diagnosis is obligatory.[75]

According to the latest therapeutic guidelines, the treatment of choice for acute nonfocal brucellosis includes oral doxycycline 100 mg twice a day over a 6-week course and streptomycin i.m. (15 mg/kg/day) for 2–3 weeks.[121–123] Streptomycin may be replaced by other aminoglycosides, for example, gentamicin (5–6 mg/kg/day i.m. once daily for the first 7 days), which has proven to be equally effective and has shown fewer adverse effects.[123,124] An alternative first-line therapy is the combination of doxycycline (200 mg/day) with rifampin 600–900 mg/day (15 mg/kg/day) in a single oral dose for 6 weeks. The latter regimen, however, leads to higher relapse rates (about 10% of the patients),[124–126] but the compliance of patients can be increased by an exclusively oral administration of antibiotic drugs. In acute nonfocal disease, treatment periods shorter than 6 weeks yield poor therapeutic outcomes, whereas longer treatment periods do not necessarily constitute an advantage.[123] Focal complications, such as *Brucella* endocarditis, neurobrucellosis, and spondylitis, necessitate triple and quadruple combinations of effective antimicrobial agents. Depending on the individual clinical course, the duration of antibiotic therapy has to be prolonged accordingly (>45 days).[127,128] Surgical procedures, for example, prosthetic heart valve replacement, may, however, still be required. Relapses can be treated with the same regimens used for first-line therapy since brucellae usually do not develop antibiotic resistance.[120,128,129] Susceptibility testing is therefore unnecessary as it does not provide additional information with regard to treatment measures. Fluoroquinolones can be used as a secondary alternative therapy but they should always be used in combination with other antibiotics, such as doxycycline or rifampin.[122] However, a clear recommendation of their use is missing and requires further studies.[123]

Because of undesired side effects, not all of the aforementioned antibiotics are suitable during pregnancy and childhood. In neonates and children younger than 8 years of age, tetracyclines are generally contraindicated so that either rifampin or co-trimoxazole, if necessary, a combination of both drugs eventually supplemented by an aminoglycoside, is recommended.[124,127,128] During pregnancy, rifampin and/or co-trimoxazole is the drug of choice.[121,127] However, one has to bear in mind that sulfonamides such as co-trimoxazole may induce kernicterus in the newborn when taken by the mother.[127]

Surveillance and control of brucellosis in livestock has significantly reduced human exposure risks and undoubtedly limited the development of human brucellosis vaccines.[130] Nevertheless, strategies directed at protecting public health against the potential use of *Brucella* in bioterrorism or biowarfare may still legitimate the vaccination of an exposed population. However, one has to bear in mind that the development of novel vaccines based on immunogenic antigen fractions or live attenuated, noninfectious *Brucella* mutants are not capable of protecting the exposed population after deliberate release because a preemptive vaccine administration would be necessary. The live attenuated vaccine strains (*B. abortus* S19 and RB51, *B. melitensis* Rev1), which are widely used to prevent animal brucellosis and spreading of the pathogen within livestock, are not applicable in humans as they elicit clinical disease comparable to natural infections with wild-type strains.[131,132] Heat-killed bacteria or subcellular fractions might be safer for human application[133] but they are frequently associated with reduced efficacy. *Brucella* mutants including deletions in genes required for intracellular survival are much more protective against a subsequent challenge but the level of attenuation needs the appropriate adjustment to provide a protective immune response while maintaining safety.[130] A further influencing factor of vaccine efficacy is the pharmaceutical formulation since adjuvants and particle formation can enhance immune stimulatory effects of a vaccine. Currently, no reliable and safe human vaccines are available and antibiotic therapies are frequently ineffective or prone to primary treatment failures and relapses.

After a deliberate release that most likely will involve the use of more virulent strains, postexposure prophylaxis appears reasonable, albeit controlled studies to assess the value of postexposure prophylaxis to persons at risk have not been conducted for ethical reasons. A combination of oral doxycycline 100 mg twice a day plus rifampin 600 mg daily for 3–6 weeks[66,134] or a 6-week course of doxycycline[122] is generally recommended. Postexposure prophylaxes in natural infected single cases or in natural outbreaks are not to be recommended as they are heterogeneous with respect to the causative agent, the source of infection, the ways of transmission, etc., and the number of cases is usually so low that no meta-analysis can be done. Moreover, there is some evidence that the application of antibiotic drugs in the early course of brucellosis (<10 days of symptomatic disease) might be a risk factor for relapse.[135] Instead, the clinical and serological long-term follow-up is recommended to detect late infections after a prolonged incubation period.

25.7 CONCLUSIONS AND FUTURE PERSPECTIVES

The world has become safer in the last decades due to the intensive efforts of the global community to effectively ban the use of weapons of mass destruction. However, bio- and agroterrorism, especially the attack against the agricultural infrastructure, is considered to be a permanent danger.[10]

Brucella was one of the first agents included in offensive research programs and still belongs to category B pathogens. Brucellae can be easily aerosolized, the infectious dose is low, and the pathogen induces a severe debilitating disease with vague clinical signs and symptoms. Because of the minimal mortality of brucellosis, the protracted incubation period, and the availability of several treatment options, nowadays, its usefulness as a potential biological weapon is mainly of historical significance.[136] However, its global abundance and high impact on veterinary and public health makes *Brucella* still interesting as an agent of bioterrorism. As such, the use of *Brucella* through the food chain will be more feasible in the future than airborne transmission after aerolization, although food has to be contaminated after the pasteurization process. As a consequence, research on airborne transmission of brucellae[137–140] does not seem to be relevant anymore, especially since reliable data on the infectious oral doses are still not available.

Since only a few expert laboratories experienced in dealing with serological diagnosis, cultivation, or even agent identification exist in both developed and less developed countries, there is a worldwide need for improvement of public health, that is, medical awareness, surveillance, and laboratory diagnostic capabilities. Accurate international surveillance of animal brucellosis, both in livestock and wildlife, and online tracing of human disease to disease hotspots in animal and environmental reservoirs are required to follow regional trends and to detect outbreaks with high impact on animal and human health. However, control strategies for brucellosis are not easy to develop, both because of the complex spread dynamics of zoonotic diseases and because of the need for implementing an extensive multispecies surveillance.[141] Only a diligent comparative and preventive medicine in all affected species may help to substantially reduce the zoonotic burden of brucellosis and map background prevalence necessary to detect a deliberate release. However, the adequate allocation of countermeasures remains a challenge and still requires a profound knowledge about the pathogen and its reservoirs as well as the animal–human interface.

REFERENCES

1. Christopher, G.W. et al., History of U.S. military contributions to the study of bacterial zoonoses, *Mil. Med.*, 170, 39, 2005.
2. World Health Organization (WHO), *Health Aspects of Chemical and Biological Weapons*, Report of a WHO group of consultants. Annex 3 Bases of quantitative estimates, pp. 84–100, World Health Organization, Geneva, Switzerland, 1970.
3. Sidell, F.R., Takafuji, E.T., and Franz, D.R., *Medical Aspects of Chemical and Biological Warfare*, TMM Publications, Washington, DC, 1997.
4. Rosand, E., *Global Terrorism: Multilateral Responses to an Extraordinary Threat. Coping with Crisis*, International Peace Academy, New York, 2007.
5. Rotz, L.D. et al., Public health assessment of potential biological terrorism agents, *Emerg. Infect. Dis.*, 8, 225, 2002.
6. Al Dahouk, S. et al., Changing epidemiology of human brucellosis, Germany, 1962–2005, *Emerg. Infect. Dis.*, 13, 1895, 2007.
7. Levin, J. et al., Agroterrorism workshop: Engaging community preparedness, *J. Agromed.*, 10, 7, 2005.
8. Wilson, T.M. et al., Agroterrorism, biological crimes, and biowarfare targeting animal agriculture. The clinical, pathologic, diagnostic, and epidemiologic features of some important animal diseases, *Clin. Lab. Med.*, 21, 549, 2001.
9. Lutz, B.D. and Greenfield, R.A., Agroterrorism, *J. Okla. State Med. Assoc.*, 96, 259, 2003.
10. Crutchley, T.M. et al., Agroterrorism: Where are we in the ongoing war on terrorism? *J. Food Prot.*, 70, 791, 2007.
11. Ficht, T.A., *Brucella* taxonomy and evolution, *Future Microbiol.*, 5, 859, 2010.
12. Foster, G., Osterman, B.S., Godfroid, J., Jacques, I., and Cloeckaert, A., *Brucella ceti* sp. nov. and *Brucella pinnipedialis* sp. nov. for *Brucella* strains with cetaceans and seals as their preferred hosts, *Int. J. Syst. Evol. Microbiol.*, 57, 2688, 2007.
13. Scholz, H.C. et al., *Brucella microti* sp. nov., isolated from the common vole *Microtus arvalis*, *Int. J. Syst. Evol. Microbiol.*, 58, 375, 2008.
14. Scholz, H.C. et al., Isolation of *Brucella microti* from mandibular lymph nodes of red foxes, *Vulpes vulpes*, in lower Austria, *Vector Borne Zoonotic Dis.*, 9, 153, 2009.
15. Scholz, H.C. et al., *Brucella inopinata* sp. nov., isolated from a breast implant infection, *Int. J. Syst. Evol. Microbiol.*, 60, 801, 2010.
16. Schlabritz-Loutsevitch, N.E. et al., A novel *Brucella* isolate in association with two cases of stillbirth in non-human primates—First report, *J. Med. Primatol.*, 38, 70, 2009.
17. Tiller, R.V. et al., Characterization of novel *Brucella* strains originating from wild native rodent species in North Queensland, Australia, *Appl. Environ. Microbiol.*, 76, 5837, 2010.
18. Hofer, E. et al., A potential novel *Brucella* species isolated from mandibular lymph nodes of red foxes in Austria, *Vet. Microbiol.*, 155, 93, 2012.
19. Eisenberg, T. et al., Isolation of potentially novel *Brucella* spp. from frogs, *Appl. Environ. Microbiol.*, 78, 3753, 2012.
20. Sohn, A.H. et al., Human neurobrucellosis with intracerebral granuloma caused by a marine mammal *Brucella* spp., *Emerg. Infect. Dis.*, 9, 485, 2003.
21. McDonald, W.L. et al., Characterization of a *Brucella* sp. strain as a marine-mammal type despite isolation from a patient with spinal osteomyelitis in New Zealand, *J. Clin. Microbiol.*, 44, 4363, 2006.
22. De, B.K. et al., Novel *Brucella* strain (BO1) associated with a prosthetic breast implant infection, *J. Clin. Microbiol.*, 46, 43, 2008.
23. Tiller, R.V. et al., Identification of an unusual *Brucella* strain (BO2) from a lung biopsy in a 52 year-old patient with chronic destructive pneumonia, *BMC Microbiol.*, 10, 23, 2010.
24. Jiménez de Bagüés, M.P. et al., The new species *Brucella microti* replicates in macrophages and causes death in murine models of infection, *J. Infect. Dis.*, 202, 3, 2010.
25. Martirosyan, A., Moreno, E., and Gorvel, J.P., An evolutionary strategy for a stealthy intracellular *Brucella* pathogen, *Immunol. Rev.*, 240, 211, 2011.
26. Moreno, E. and Moriyón, I., *Brucella melitensis*: A nasty bug with hidden credentials for virulence, *Proc. Natl. Acad. Sci. USA*, 99, 1, 2002.
27. Barquero-Calvo, E. et al., *Brucella abortus* uses a stealthy strategy to avoid activation of the innate immune system during the onset of infection, *PLoS One*, 2, 631, 2007.
28. Roop, R.M. 2nd, Bellaire, B.H., Valderas, M.W., and Cardelli, J.A., Adaptation of the brucellae to their intracellular niche, *Mol. Microbiol.*, 52, 621, 2004.
29. Lapaque, N. et al., *Brucella* lipopolysaccharide acts as a virulence factor, *Curr. Opin. Microbiol.*, 8, 60, 2005.
30. Porte, F., Naroeni, A., Ouahrani-Bettache, S., and Liautard, J.P., Role of the *Brucella suis* lipopolysaccharide O antigen in phagosomal genesis and in inhibition of phagosome-lysosome fusion in murine macrophages, *Infect. Immun.*, 71, 1481, 2003.
31. Forestier, C. et al., *Brucella abortus* lipopolysaccharide in murine peritoneal macrophages acts as a down-regulator of T cell activation, *J. Immunol.*, 165, 5202, 2000.
32. Cannella, A.P. et al., Antigen-specific acquired immunity in human brucellosis: Implications for diagnosis, prognosis, and vaccine development, *Front. Cell. Infect. Microbiol.*, 2, 1, 2012.
33. von Bargen, K., Gorvel, J.P., and Salcedo, S.P., Internal affairs: Investigating the *Brucella* intracellular lifestyle, *FEMS Microbiol. Rev.*, 36, 533, 2012.

34. Boschiroli, M.L. et al., The *Brucella suis virB* operon is induced intracellularly in macrophages, *Proc. Natl. Acad. Sci. USA*, 99, 1544, 2002.

35. Celli, J. and Gorvel, J.P., Organelle robbery: *Brucella* interactions with the endoplasmic reticulum, *Curr. Opin. Microbiol.*, 7, 93, 2004.

36. Köhler, S. et al., The analysis of the intramacrophagic virulome of *Brucella suis* deciphers the environment encountered by the pathogen inside the macrophage host cell, *Proc. Natl. Acad. Sci. USA*, 99, 15711, 2002.

37. Köhler, S. et al., What is the nature of the replicative niche of a stealthy bug named *Brucella*? *Trends Microbiol.*, 11, 215, 2003.

38. Al Dahouk, S. et al., Proteomic analysis of *Brucella suis* under oxygen deficiency reveals flexibility in adaptive expression of various pathways, *Proteomics*, 9, 3011, 2009.

39. Al Dahouk, S. et al., Quantitative analysis of the intramacrophagic *Brucella suis* proteome reveals metabolic adaptation to late stage of cellular infection, *Proteomics*, 8, 3862, 2008.

40. Corbel, M.J., Brucellosis: An overview, *Emerg. Infect. Dis.*, 3, 213, 1997.

41. Pappas, G. et al., Brucellosis, *N. Engl. J. Med.*, 352, 2325, 2005.

42. Pappas, G. et al., The new global map of human brucellosis, *Lancet Infect. Dis.*, 6, 91, 2006.

43. De Massis, F. et al., Correlation between animal and human brucellosis in Italy during the period 1997–2002, *Clin. Microbiol. Infect.*, 11, 632, 2005.

44. Piffaretti, J.C., Staedler, P., and Beretta-Piccoli, C.F., Risk of infection by *Brucella melitensis* for people living near infected goats, *J. Infect.*, 15, 177, 1987.

45. Teyssou, R. et al., About a case of human brucellosis due to *Brucella suis* biovar 2, *Méd. Mal. Infect.*, 19, 160, 1989.

46. Paton, N.I. et al., Visceral abscesses due to *Brucella suis* infection in a retired pig farmer, *Clin. Infect. Dis.*, 32, 129, 2001.

47. Godfroid, J., Brucellosis in wildlife, *Rev. Sci. Tech.*, 21, 277, 2002.

48. Godfroid, J. et al., From the discovery of the Malta fever's agent to the discovery of a marine mammal reservoir, brucellosis has continuously been a re-emerging zoonosis, *Vet. Res.*, 36, 313, 2005.

49. Fosgate, G.T. et al., Time-space clustering of human brucellosis, California, 1973–1992, *Emerg. Infect. Dis.*, 8, 672, 2002.

50. Chomel, B.B. et al., Changing trends in the epidemiology of human brucellosis in California from 1973 to 1992: A shift toward foodborne transmission, *J. Infect. Dis.*, 170, 1216, 1994.

51. White, Jr, A.C. and Atmar, R.L., Infections in hispanic immigrants, *Clin. Infect. Dis.*, 34, 1627, 2002.

52. Troy, S.B., Rickman, L.S., and Davis, C.E., Brucellosis in San Diego: Epidemiology and species-related differences in acute clinical presentations, *Medicine (Baltimore)*, 84, 174, 2005.

53. Eriksen, N., Lemming, L., Højlyng, N., and Bruun, B., Brucellosis in immigrants in Denmark, *Scand. J. Infect. Dis.*, 34, 540, 2002.

54. Memish, Z.A. and Balkhy, H.H., Brucellosis and international travel, *J. Travel Med.*, 11, 49, 2004.

55. Field, V. et al., EuroTravNet network, Travel and migration associated infectious diseases morbidity in Europe, 2008, *BMC Infect. Dis.*, 10, 330, 2010.

56. Al Dahouk, S. et al., Intraspecies biodiversity of the genetically homologous species *Brucella microti*, *Appl. Environ. Microbiol.*, 78, 1534, 2012.

57. Yagupsky, P. and Baron, E.J., Laboratory exposures to brucellae and implications for bioterrorism, *Emerg. Infect. Dis.*, 11, 1180, 2005.

58. Doganay, M., Aygen, B., and Eşel, D., Brucellosis due to blood transfusion, *J. Hosp. Infect.*, 49, 151, 2001.

59. Ertem, M. et al., Brucellosis transmitted by bone marrow transplantation, *Bone Marrow Transplant.*, 26, 225, 2000.

60. Palanduz, A. et al., Brucellosis in a mother and her young infant: Probable transmission by breast milk, *Int. J. Infect. Dis.*, 4, 55, 2000.

61. Tikare, N.V., Mantur, B.G., and Bidari, L.H., Brucellar meningitis in an infant—Evidence for human breast milk transmission, *J. Trop. Pediatr.*, 54, 272, 2008.

62. Ruben, B., Band, J.D., Wong, P., and Colville, J., Person-to-person transmission of *Brucella melitensis*, *Lancet*, 337, 14, 1991.

63. Vigeant, P., Mendelson, J., and Miller, M.A., Human to human transmission of *Brucella melitensis*, *Can. J. Infect. Dis.*, 6, 153, 1995.

64. Kato, Y. et al., Brucellosis in a returned traveler and his wife: Probable person-to-person transmission of *Brucella melitensis*, *J. Travel Med.*, 14, 343, 2007.

65. Mesner, O. et al., The many faces of human-to-human transmission of brucellosis: Congenital infection and outbreak of nosocomial disease related to an unrecognized clinical case, *Clin. Infect. Dis.*, 45, 135, 2007.

66. Bossi, P. et al., Task Force on Biological and Chemical Agent Threats, Public Health Directorate, European Commission, Luxembourg, Bichat guidelines for the clinical management of brucellosis and bioterrorism-related brucellosis, *Euro Surveill.*, 9, 1, 2004.

67. Pappas, G. et al., Brucellosis and the respiratory system, *Clin. Infect. Dis.*, 37, 95, 2003.

68. Saltoglu, N. et al., Fever of unknown origin in Turkey: Evaluation of 87 cases during a nine-year-period of study, *J. Infect.*, 48, 81, 2004.

69. Colmenero, J.D. et al., Complications associated with *Brucella melitensis* infection: A study of 530 cases, *Medicine (Baltimore)*, 75, 195, 1996.

70. Gür, A. et al., Complications of brucellosis in different age groups: A study of 283 cases in southeastern Anatolia of Turkey, *Yonsei Med. J.*, 44, 33, 2003.

71. Buzgan, T. et al., Clinical manifestations and complications in 1028 cases of brucellosis: A retrospective evaluation and review of the literature, *Int. J. Infect. Dis.*, 14, 469, 2010.

72. Pappas, G. et al., Treatment of *Brucella* spondylitis: Lessons from an impossible meta-analysis and initial report of efficacy of a fluoroquinolone-containing regimen, *Int. J. Antimicrob. Agents*, 24, 502, 2004.

73. Khan, M.Y., Mah, M.W., and Memish, Z.A., Brucellosis in pregnant women, *Clin. Infect. Dis.*, 32, 1172, 2001.

74. Elshamy, M. and Ahmed, A.I., The effects of maternal brucellosis on pregnancy outcome, *J. Infect. Dev. Ctries*, 2, 230, 2008.

75. Al Dahouk, S. and Nöckler, K., Implications of laboratory diagnosis on brucellosis therapy, *Expert Rev. Anti Infect. Ther.*, 9, 833, 2011.

76. Al Dahouk, S. et al., Laboratory-based diagnosis of brucellosis—A review of the literature. Part I: Techniques for direct detection and identification of *Brucella* spp., *Clin. Lab.*, 49, 487–505, 2003.

77. Franco, M.P., Mulder, M., Gilman, R.H., and Smits, H.L., Human brucellosis, *Lancet Infect. Dis.*, 7, 775, 2007.

78. Espinosa, B.J. et al., Comparison of culture techniques at different stages of brucellosis, *Am. J. Trop. Med. Hyg.*, 80, 625, 2009.

79. Yagupsky, P., Detection of brucellae in blood cultures, *J. Clin. Microbiol.*, 37, 3437, 1999.

80. Mantur, B.G. et al., Diagnostic yield of blood clot culture in the accurate diagnosis of enteric fever and human brucellosis, *Clin. Lab.*, 53, 57, 2007.

81. Mantur, B.G. and Mangalgi, S.S., Evaluation of conventional Castañeda and lysis centrifugation blood culture techniques for diagnosis of human brucellosis, *J. Clin. Microbiol.*, 42, 4327, 2004.

82. Alton, G.G., Jones, L.M., Angus, R.D., and Verger, J.M., *Techniques for the Brucellosis Laboratory*, Institut National de la Recherche Agronomique, Paris, France, 1988.

83. Barham, W.B., Church, P., Brown, J.E., and Paparello, S., Misidentification of *Brucella* species with use of rapid bacterial identification systems, *Clin. Infect. Dis.*, 17, 1068, 1993.

84. Elsaghir, A.A. and James, E.A., Misidentification of *Brucella melitensis* as *Ochrobactrum anthropi* by API 20NE, *J. Med. Microbiol.*, 52, 441, 2003.

85. Al Dahouk, S. et al., Differential phenotyping of *Brucella* species using a newly developed semi-automated metabolic system, *BMC Microbiol.*, 10, 269, 2010.

86. Ferreira, L. et al., Identification of *Brucella* by MALDI-TOF mass spectrometry. Fast and reliable identification from agar plates and blood cultures, *PLoS One*, 5, 14235, 2010.

87. Lista, F. et al., Reliable identification at the species level of *Brucella* isolates with MALDI-TOF-MS, *BMC Microbiol.*, 11, 267, 2011.

88. Ariza, J. et al., Specific antibody profile in human brucellosis, *Clin. Infect. Dis.*, 14, 131, 1992.

89. Roushan, M.R. et al., Follow-up standard agglutination and 2-mercaptoethanol tests in 175 clinically cured cases of human brucellosis, *Int. J. Infect. Dis.*, 14, 250, 2010.

90. Casanova, A. et al., Brucellacapt versus classical tests in the serological diagnosis and management of human brucellosis, *Clin. Vac. Immunol.*, 16, 844, 2009.

91. Bosilkovski, M. et al., The role of Brucellacapt test for follow-up patients with brucellosis, *Comp. Immunol. Microbiol. Infect. Dis.*, 33, 435, 2010.

92. Al Dahouk, S. et al., Laboratory-based diagnosis of brucellosis—A review of the literature. Part II: Serological tests for brucellosis, *Clin. Lab.*, 49, 577, 2003.

93. Ruiz-Mesa, J.D. et al., Rose Bengal test: Diagnostic yield and use for the rapid diagnosis of human brucellosis in emergency departments in endemic areas, *Clin. Microbiol. Infect.*, 11, 221, 2005.

94. Díaz, R. et al., The Rose Bengal Test in human brucellosis: A neglected test for the diagnosis of a neglected disease, *PLoS Negl. Trop. Dis.*, 5, 950, 2011.

95. Mantur, B.G. et al., Comparison of a novel immunocapture assay with standard serological methods in the diagnosis of brucellosis, *Clin. Lab.*, 57, 333, 2011.

96. Fadeel, M.A. et al., Comparison of four commercial IgM and IgG ELISA kits for diagnosing brucellosis, *J. Med. Microbiol.*, 60, 1767, 2011.

97. Araj, G.F. et al., Evaluation of the PANBIO *Brucella* immunoglobulin G (IgG) and IgM enzyme-linked immunosorbent assays for diagnosis of human brucellosis, *Clin. Diagn. Lab. Immunol.*, 12, 1334, 2005.

98. Prince, H.E. et al., Performance characteristics of the Euroimmun enzyme-linked immunosorbent assay kits for *Brucella* IgG and IgM, *Diagn. Microbiol. Infect. Dis.*, 65, 99, 2009.

99. Mantur, B.G. et al., ELISA versus conventional methods of diagnosing endemic brucellosis, *Am. J. Trop. Med. Hyg.*, 83, 314, 2010.

100. Al Dahouk, S., Nöckler, K., and Tomaso, H., *Brucella* (Chapter 23). In Liu, D. (ed.), *Molecular Detection of Foodborne Pathogens*, pp. 317–330, Taylor & Francis, CRC Press, Boca Raton, FL, 2010.

101. Al Dahouk, S., Neubauer, H., and Tomaso, H., *Brucella* (Chapter 54). In Liu, D. (ed.), *Molecular Detection of Human Bacterial Pathogens*, pp. 629–646, Taylor & Francis, CRC Press, Boca Raton, FL, 2011.

102. Navarro, E. et al., Use of real-time quantitative polymerase chain reaction to monitor the evolution of *Brucella melitensis* DNA load during therapy and post-therapy follow-up in patients with brucellosis, *Clin. Infect. Dis.*, 42, 1266, 2006.

103. Al Dahouk, S. et al., Evaluation of genus-specific and species-specific real-time PCR assays for the identification of *Brucella* spp., *Clin. Chem. Lab. Med.*, 45, 1464, 2007.

104. Queipo-Ortuño, M.I. et al., Preparation of bacterial DNA template by boiling and effect of immunoglobulin G as an inhibitor in real-time PCR for serum samples from patients with brucellosis, *Clin. Vac. Immunol.*, 15, 293, 2008.

105. Maas, K.S. et al., Evaluation of brucellosis by PCR and persistence after treatment in patients returning to the hospital for follow-up, *Am. J. Trop. Med. Hyg.*, 76, 698, 2007.

106. Queipo-Ortuño, M.I. et al., Comparison of seven commercial DNA extraction kits for the recovery of *Brucella* DNA from spiked human serum samples using real-time PCR, *Eur. J. Clin. Microbiol. Infect. Dis.*, 27, 109, 2008.

107. Vrioni, G. et al., An eternal microbe: *Brucella* DNA load persists for years after clinical cure, *Clin. Infect. Dis.*, 46, 131, 2008.

108. Baily, G.G., Krahn, J.B., Drasar, B.S., and Stoker, N.G., Detection of *Brucella melitensis* and *Brucella abortus* by DNA amplification, *J. Trop. Med. Hyg.*, 95, 271, 1992.

109. Hinić, V. et al., Novel identification and differentiation of *Brucella melitensis*, *B. abortus*, *B. suis*, *B. ovis*, *B. canis*, and *B. neotomae* suitable for both conventional and real-time PCR systems, *J. Microbiol. Methods*, 75, 375, 2008.

110. Bounaadja, L. et al., Real-time PCR for identification of *Brucella* spp.: A comparative study of *IS711*, *bcsp31* and *per* target genes, *Vet. Microbiol.*, 137, 156, 2009.

111. Gee, J.E. et al., Use of 16S rRNA gene sequencing for rapid confirmatory identification of *Brucella* isolates, *J. Clin. Microbiol.*, 42, 3649, 2004.

112. López-Goñi, I. et al., Evaluation of a multiplex PCR assay (Bruce-ladder) for molecular typing of all *Brucella* species, including the vaccine strains, *J. Clin. Microbiol.*, 46, 3484, 2008.

113. Mayer-Scholl, A. et al., Advancement of a multiplex PCR for the differentiation of all currently described *Brucella* species, *J. Microbiol. Methods*, 80, 112, 2010.

114. López-Goñi, I. et al., New Bruce-ladder multiplex PCR assay for the biovar typing of *Brucella suis* and the discrimination of *Brucella suis* and *Brucella canis*, *Vet. Microbiol.*, 154, 152, 2011.

115. Queipo-Ortuño, M.I. et al., Usefulness of a quantitative real-time PCR assay using serum samples to discriminate between inactive, serologically positive and active human brucellosis, *Clin. Microbiol. Infect.*, 14, 1128, 2008.

116. Le Flèche, P. et al., Evaluation and selection of tandem repeat loci for a *Brucella* MLVA typing assay, *BMC Microbiol.*, 6, 9, 2006.

117. Al Dahouk, S. et al., Evaluation of *Brucella* MLVA typing for human brucellosis, *J. Microbiol. Methods*, 69, 137, 2007.

118. Gwida, M. et al., Cross-border molecular tracing of brucellosis in Europe, *Comp. Immunol. Microbiol. Infect. Dis.*, 35, 181, 2012.

119. Akova, M. et al., *In vitro* activities of antibiotics alone and in combination against *Brucella melitensis* at neutral and acidic pHs, *Antimicrob. Agents Chemother.*, 43, 1298, 1999.

120. Al Dahouk, S. et al., Failure of a short-term antibiotic therapy for human brucellosis using ciprofloxacin. A study on *in vitro* susceptibility of *Brucella* strains, *Chemotherapy*, 51, 352, 2005.

121. Ariza, J. et al., International Society of Chemotherapy; Institute of Continuing Medical Education of Ioannina, Perspectives for the treatment of brucellosis in the 21st century: The Ioannina recommendations, *PLoS Med.*, 4, 1872, 2007.

122. Corbel, M.J., Treatment of brucellosis in humans. Post-exposure prophylaxis (Chapter 5). In Corbel, M.J., Elberg, S.S., and Cosivi, O. (eds.), *Brucellosis in Humans and Animals*, p. 40, World Health Organization, Geneva, Switzerland, 2006.

123. Solís García del Pozo, J. and Solera, J., Systematic review and meta-analysis of randomized clinical trials in the treatment of human brucellosis, *PLoS One*, 7, 32090, 2012.

124. Skalsky, K. et al., Treatment of human brucellosis: Systematic review and meta-analysis of randomised controlled trials, *BMJ*, 336, 701, 2008.

125. Ariza, J. et al., Treatment of human brucellosis with doxycycline plus rifampin or doxycycline plus streptomycin. A randomized, double-blind study, *Ann. Intern. Med.*, 117, 25, 1992.

126. Solera, J. et al., Doxycycline-rifampin versus doxycycline-streptomycin in treatment of human brucellosis due to *Brucella melitensis*. The GECMEI Group. Grupo de Estudio de Castilla-la Mancha de Enfermedades Infecciosas, *Antimicrob. Agents Chemother.*, 39, 2061, 1995.

127. Al-Tawfiq, J.A., Therapeutic options for human brucellosis, *Expert Rev. Anti Infect. Ther.*, 6, 109, 2008.

128. Solera, J., Update on brucellosis: Therapeutic challenges, *Int. J. Antimicrob. Agents*, 36, 18, 2010.

129. Ariza, J. et al., Relevance of *in vitro* antimicrobial susceptibility of *Brucella melitensis* to relapse rate in human brucellosis, *Antimicrob. Agents Chemother.*, 30, 958, 1986.

130. Ficht, T.A. et al., Brucellosis: The case for live, attenuated vaccines, *Vaccine*, 27, 40, 2009.

131. Ollé-Goig, J.E. and Canela-Soler, J., An outbreak of *Brucella melitensis* infection by airborne transmission among laboratory workers, *Am. J. Public Health*, 77, 335, 1987.

132. Wallach, J.C. et al., Occupational infection due to *Brucella abortus* S19 among workers involved in vaccine production in Argentina, *Clin. Microbiol. Infect.*, 14, 805, 2008.

133. Oliveira, S.C., Giambartolomei, G.H., and Cassataro, J., Confronting the barriers to develop novel vaccines against brucellosis, *Expert Rev. Vac.*, 10, 1291, 2011.

134. Robichaud, S., Libman, M., Behr, M., and Rubin, E., Prevention of laboratory-acquired brucellosis, *Clin. Infect. Dis.*, 38, 119, 2004.

135. Solera, J. et al., Multivariate model for predicting relapse in human brucellosis, *J. Infect.*, 36, 85, 1998.

136. Pappas, G., Panagopoulou, P., Christou, L., and Akritidis, N., *Brucella* as a biological weapon, *Cell. Mol. Life Sci.*, 63, 2229, 2006.

137. Mense, M.G., Borschel, R.H., Wilhelmsen, C.L., Pitt, M.L., and Hoover, D.L., Pathologic changes associated with brucellosis experimentally induced by aerosol exposure in rhesus macaques (*Macaca mulatta*), *Am. J. Vet. Res.*, 65, 644, 2004.

138. Smither, S.J. et al., Development and characterization of mouse models of infection with aerosolized *Brucella melitensis* and *Brucella suis*, *Clin. Vac. Immunol.*, 16, 779, 2009.

139. Yingst, S.L. et al., A rhesus macaque (*Macaca mulatta*) model of aerosol-exposure brucellosis (*Brucella suis*): Pathology and diagnostic implications, *J. Med. Microbiol.*, 59, 724, 2010.

140. Teske, S.S. et al., Animal and human dose-response models for *Brucella* species, *Risk Anal.*, 31, 1576, 2011.

141. Roy, S., McElwain, T.F., and Wan, Y., A network control theory approach to modeling and optimal control of zoonoses: Case study of brucellosis transmission in sub-Saharan Africa, *PLoS Negl. Trop. Dis.*, 5, 1259, 2011.

142. Sprague, L.D., Al Dahouk, S., and Neubauer, H., A review on camel brucellosis: A zoonosis sustained by ignorance and indifference, *Pathog. Glob. Health*, 106, 144, 2012.

26 Burkholderia (B. mallei and B. pseudomallei)

Dongyou Liu, Shifeng Hu, and Chong Yin

CONTENTS

26.1 INTRODUCTION

The genus *Burkholderia* consists of >60 Gram-negative β-proteobacterial species that are ubiquitously distributed in various environments such as water, sewage, soil, and plants. Among them, *Burkholderia cepacia* complex (BCC), *Burkholderia gladioli*, *Burkholderia mallei*, and *Burkholderia pseudomallei* are of clinical importance. In particular, *B. mallei* and *B. pseudomallei*, which cause glanders and melioidosis, respectively, and have a history of being on the list of potential biological warfare agents, demonstrate high infectivity, low lethal dosage, easy aerosolization, resistance to commonly used antibiotics, and lack of effective vaccines. Indeed, *B. mallei* causing glanders was intentionally released by German agents to infect animals and humans in the allied and nonaligned countries in the 1910s (World War I). The bacterium was also used by the Japanese biological warfare research unit 731 to deliberately infect horses, civilians, and prisoners of war and test its effectiveness in contaminating water supplies in the 1940s (World War II). In addition, the Russian army might have employed *B. mallei* during their Afghanistan war in the 1980s [1]. The melioidosis-causing *B. pseudomallei*, from which *B. mallei* has evolved, has the potential to be developed as a biological weapon [2,3]. As both *B. mallei* and *B. pseudomallei* are classified as Category B biological agents by the Centers for Disease Control and Prevention (CDC), United States, this chapter will focus on these two species, reviewing their classification, morphology, biology, epidemiology, clinical features, pathogenesis, diagnosis, treatment, and prevention.

26.2 *Burkholderia mallei*

26.2.1 CLASSIFICATION AND MORPHOLOGY

Classification: *B. mallei* (synonyms: *Glanders bacillus, Bacillus mallei, Actinobacillus mallei, Pfeifferella mallei, Malleomyces mallei, Loefferella mallei, Acinetobacter mallei, Pseudomonas mallei*) is classified in the genus *Burkholderia*, family Burkholderiaceae, order Burkholderiales, class Betaproteobacteria, phylum Proteobacteria, kingdom Bacteria.

The genus *Burkholderia* (named after W. H. Burkholder, who originally isolated *B. cepacia* from onions with sour skin rot in 1950) was created in 1992 to cover a number of Gram-negative bacteria that were previously considered as *Pseudomonas*, including *Pseudomonas cepacia* (*B. cepacia*), *Pseudomonas gladioli* (*B. gladioli*), *P. mallei* (*B. mallei*), and *Pseudomonas pseudomallei* (*B. pseudomallei*) [4,5]. Currently, the genus comprises over 60 recognized species, of which BCC, *B. gladioli*, *B. mallei*, and *B. pseudomallei* are clinically important.

Forming a subgroup of the *Burkholderia* genus, BCC encompasses at least 17 distinct genomic species or genomovars on the basis of molecular phylogenetic characteristics: *B. cepacia*, *B. multivorans*, *B. cenocepacia*, *B. stabilis*, *B. vietnamiensis*, *B. dolosa*, *B. ambifaria*, *B. anthina*, *B. pyrrocinia*, *B. latens*, *B. diffusa*, *B. arboris*, *B. seminalis*, *B. metallica*, *B. ubonensis*, *B. contaminans*, and *B. lata*. BCC species are associated with plant rhizospheres (of sugarcane, rice, corn, wheat, soybean, alfalfa, bananas, mushrooms, onions, etc.) and have been also isolated from sewage, surface water sources, medical-grade water, and drinking water. BCC infections in humans commonly occur in patients with cystic fibrosis (CF; due to mutation in the CF transmembrane conductance regulator gene that leads to thick mucus buildup in the lungs and susceptibility to bacterial infection) and chronic granulomatous disease (CGD; resulting from mutations in phagocytic NADPH oxidase gene components that render patients' susceptibility to pathogens).

B. gladioli is separated into four pathovars: *gladioli*, *alliicola*, *agaricicola*, and *cocovenenans*. Whereas pathovars *gladioli*, *alliicola*, and *agaricicola* are primarily phytopathogens, pathovar *cocovenenans* is present in tempe bongkrek (an Indonesian fermented coconut cake) and fermented corn flour (for making breads and noodles in China). *B. gladioli* pathovars *gladioli*, *alliicola*, and *agaricicola* have been shown to infect immune-suppressed patients that have CF, CGD, and acquired immune deficiency syndrome (AIDS).

B. mallei, the causative agent of glanders (from Middle English *glaundres* or Old French *glandres*, both meaning glands; less common names include equinia, malleus, droes, or farcy, when the skin is involved), is closely related to melioidosis-causing *B. pseudomallei* [6]. Recent molecular evidence suggests that *B. mallei* has evolved from *B. pseudomallei* by selective reductions and deletions from the *B. pseudomallei* genome, and thus *B. mallei* represents a clone (subspecies) of *B. pseudomallei* [7,8]. However, for historical considerations, *B. mallei* and *B. pseudomallei* will be dealt as separate identities here.

Morphology: *B. mallei* is a Gram-negative, nonflagellated, nonmotile, bipolar aerobic bacterium (in contrast to *B. pseudomallei* and other genus members that are motile) and measures 1.5–3 μm in length and 0.5–1 μm in diameter with rounded ends.

Glanders in horses was first described by Aristotle in 330 BC, who named it malleus, meaning hammer or mallet. The bacterium was first isolated in 1882 from the liver and spleen of a glanderous horse. Unlike other members of the *Burkholderia* genus, indeed the Burkholderiaceae family, which mostly lives in soil, *B. mallei* is an obligate, intracellular mammalian pathogen, which is transmitted directly from one host to another [6].

To date, four *B. mallei* strains have been sequenced. These include ATCC 23344 (which is the type strain, alternatively referred to as NBL 7, China 7, 3873, EY 2233, RH627; whose genome contains a 3,510,148 bp chromosome I and a 2,325,379 bp chromosome II) [9], NCTC 10229 (with a 3,458,208 bp chromosome I and a 2,284,095 bp chromosome II), NCTC 10247 (with a 3,495,687 bp chromosome I and a 2,352,693 bp chromosome II), and SAVP1 (which is a nonvirulent strain with a 3,497,479 bp chromosome I and a 1,734,922 bp chromosome II) [10]. *B. mallei* chromosome 1 encompasses genes relating to metabolism, capsule formation, and lipopolysaccharide (LPS) biosynthesis; and its chromosome 2 contains genes relating to secretion systems and virulence-associated genes. The presence of a polysaccharide capsule in *B. mallei* indicates its potential as a pathogen.

Compared to the genome of *B. pseudomallei* strain K96243 (a clinical isolate from Thailand), which harbors two chromosomes of 4,074,542 bp and 3,173,005 bp, respectively [11], the genome of *B. mallei* type strain ATCC 23344 is 1.4 Mb smaller, resulting in the reductions and alterations of approximately 1000 genes that are present in *B. pseudomallei* K96243 genome [9–14]. As a consequence, *B. mallei* has lost its ability to survive in a soil environment (including the capacity to protect against bactericidals, antibiotics, and antifungals) and adapted into an obligate intracellular pathogen of mammals [6,8,10,15].

26.2.2 Biology and Epidemiology

Biology: As a causal agent for glanders (or farcy), *B. mallei* is an obligate animal pathogen with limited ability to survive outside its host and has been only isolated from humans, animals (e.g., horses, mules, and donkeys), and animal-associated environments (e.g., troughs and sewages). *B. mallei* is highly infective and pathogenic, and inhalation of only a few organisms (1–10) may cause disease in humans, equids, and other susceptible species.

The bacterium is easily aerosolized and susceptible to drying, heat, and sunlight. In addition, it is sensitive to a number of disinfectants (e.g., benzalkonium chloride, iodine, mercuric chloride, potassium permanganate, 1% sodium hypochlorite, and ethanol) and antibiotics (e.g., streptomycin, amikacin, tetracycline, doxycycline, carbapenems, ceftazidime, amoxicillin/clavulanic acid, piperacillin, chloramphenicol, and sulfathiazole). The organism can survive for 1 month in water at room temperature and for a few months in warm and moist environments (e.g., stable bedding, manure, feed and water troughs, wastewater, and enclosed equine transporters). In nature, the bacterium tends to lose its infectivity within 3 weeks and its viability within 3 months.

Epidemiology: The natural reservoir of *B. mallei* is horse (which is prone to develop chronic and latent disease); other animals may also act as the amplifying host such as donkeys, which tend to develop acute forms of glanders, and mules (a crossbred between horse and donkey), which are susceptible to acute, chronic, and latent disease. Field mice, goats, cats, and guinea pigs are also susceptible to *B. mallei* infection. Humans represent an accidental host for this bacterium.

The organism is usually transmitted by direct contact (involving abraded or lacerated skin in the arms, hands, face, and neck) or by aerosol inhalation (involving the nasal, oral, and conjunctival mucous membranes, as well as lungs). Handling contaminated fomites (e.g., grooming tools, hoof-trimming equipment, harnesses, tack, feeding and husbandry equipment, bedding, and veterinary equipment) may also play a role in transmission. *B. mallei* transmission among equids may involve oronasal mucous membrane exposure, inhalation, mastication of skin exudates, and respiratory secretions of infected animals, as well as sharing feed and water troughs facilitates, grooming and snorting. Carnivores may acquire *B. mallei* infection after consuming contaminated carcasses and meat.

Individuals whose occupation, hobby, and lifestyle involve frequent contact with horses may be prone to acquire *B. mallei* infection. These include veterinarians, veterinary students, farriers, flayers (hide workers), transport workers, soldiers, slaughterhouse personnel, farmers, horse fanciers and caretakers, and stable hands [16]. Occasionally, *B. mallei* infection may be acquired through ingestion of contaminated food and water. In addition, *B. mallei* culture aerosols are highly infectious to laboratory workers. Close contact with glanders-infected individuals may also lead to infection of other family members.

The incubation period of *B. mallei* infection in humans ranges from a day to several weeks, with cutaneous and mucous membrane exposure generally requiring 3–5 days and inhalational exposure taking 7–21 days for symptoms to appear. On the other hand, the incubation period for glanders in equids varies from days to months, typically 2–6 weeks.

B. mallei was once widespread throughout the globe. However, with extensive control and prevention campaigns, the bacterium was eradicated in nearly 20 countries, including Australia (last reported in 1891), Egypt (1928), United Kingdom (1928), Japan (1935), United States (1942, except for eight laboratory-related cases between 1944 and 2000), South Africa (1945), Germany (1955), France (1965), and India (1988). Glanders, caused by *B. mallei*, is notifiable to the 164-member Office International des Epizooties (OIE). To date, glanders in livestock has been reported in a number of countries in northern Africa, the Middle East, South America, and Eastern Europe, such as Bolivia, Belarus, Brazil, Eritrea, Ethiopia, Iran, Latvia, Mongolia, Myanmar, Pakistan, and Turkey; and glanders in humans was described in Cameroon, Curaçao, Sri Lanka, Turkey, and the United States (laboratory-acquired).

26.2.3 CLINICAL FEATURES AND PATHOGENESIS

26.2.3.1 Clinical Features

B. mallei infection in humans: Being an accidental host, humans are occasionally infected with *B. mallei* by direct contact with infected animals (through skin abrasions and nasal and oral mucosal surfaces or by inhalation). Clinically, several forms of infection may be observed, including localized (skin or nasal), pulmonary, septicemic, disseminated, and chronic infections. In general, glanders may be characterized by ulcerating granulomatous lesions of the skin and mucous membranes, accompanied by a range of generalized symptoms (e.g., fever or low-grade fever in the afternoon to evening, chills with or without rigors, headache, malaise, myalgia [of the limbs, joints, neck, and back], dizziness, nausea, vomiting, diarrhea, tachypnea, diaphoresis [with night sweats], altered mental status, fatigue, and weight loss). Other nonspecific signs may include tender lymph nodes, sore throat, chest pain, blurred vision, splenomegaly, abdominal pain, photophobia, and marked lacrimation.

In case of cutaneous exposure, the most characteristic feature of the disease is glanders nodes and small papular to egg-sized abscesses (blisters). The nodular abscesses (farcy buds) then become ulcerated together with thickened and indurated regional cutaneous lymphatic pathways (farcy pipes), producing a glanders-typical yellow-green gelatinous pus (farcy oil) for long periods of time, with accompanying pain and swelling. Lymphangitis or regional lymphadenopathy may emerge in the lymphatic pathways (which are palpable as firm, ropy cords) that drain the infection site. If the bacterium infects through the eyes, nose, and mouth, the affected membranes become infected, swell, and weep a serosanguineous to mucopurulent discharge within 1–5 days, leading to fever, rigors, septicemic or disseminated infections (affecting the lungs, liver, or spleen), septic shock, and death. The disease has a fatality rate of 95% within a week to 10 days if left untreated and a 50% fatality rate in individuals treated with antibiotics. The surviving individuals may develop a protracted infection with alternating remissions and exacerbations.

A relatively recent incidence of laboratory-acquired glanders highlights the danger of *B. mallei* to humans, especially those with underlying diseases that render them vulnerable to microbial infections. In March 2000, a young microbiologist with type 1 diabetes experienced enlargement of the lymph nodes, fever, fatigue, rigors, night sweats, and loss of weight. The patient has worked in a US research facility dealing with *B. mallei* for 2 years, without always wearing gloves (and in the absence of cut or skin abrasion) while conducting his research. After treatment with clarithromycin, his symptoms seemed to disappear and then reappeared after cessation of the medication. Cultures of the researcher's blood and a biopsied portion of a liver abscess led to the isolation and identification of *B. mallei*. With another course of antibiotics (imipenem and doxycycline) for 6 months, the patient made a full recovery [17].

B. mallei infection in animals: Being the natural host for *B. mallei*, horses may develop either acute or chronic form of glanders disease. In acute infection, horses tend to show a high fever, coughing, loss of fat or muscle, erosion of the surface of the nasal septum, hemorrhaging, and the release of an infectious nasal discharge, followed by septicemia and death within days. In chronic infection (lasting longer than

6 months), horses typically develop mucus containing nasal discharge, nodular lesions in the lungs, ulceration of the mucous membranes in the upper respiratory tract, and nodules around the liver or spleen, with death occurring within months, while survivors acting as carriers. The infection also occurs in mules, donkeys, dogs, cats, and goats, usually by ingestion of contaminated food or water.

26.2.3.2 Pathogenesis

B. mallei usually gains entry into the host through skin exposure or inhalation. Covered with a tenacious capsule, the bacterium is protected from phagocytosis. After lysis of the vacuole, the organism moves by means of bacterial protein dependent actin-based motility to the cytosol for efficient replication. By initiating host cell fusion that results in multinucleated giant cells (MNGCs), the organism efficiently spreads to neighboring cells. This helps its evasion from host's immune responses and protects it against host's lysosomal defensins and other pathogen killing agents. With an affinity for the lymphatics, *B. mallei* travels through lymph channels, first to regional lymph nodes, causing irritation (lymphangitis, lymphadenitis), and then via the bloodstream to other parts of the body. It induces the formation of glanderous nodes even deep within the musculature. By producing an endotoxin that affects smooth muscle cells of various organs, the bacterium causes mucous membrane erosions, which slows down the healing processes of local infections [18–21].

26.2.4 Identification and Diagnosis

Diagnosis of glanders on the basis of clinical presentations is often inconclusive and presumptive, as *B. mallei* infection tends to generate nonspecific signs. In case of inhalation exposure in humans, headache, chest pain, fever, rigors, night sweats, fatigue, cough, and nasal discharge may be observed. Application of diagnostic imaging may reveal localized or lobar pneumonia, bronchopneumonia, miliary nodules, lobar infiltrative pneumonia, and consolidation (early) or cavitating (later) pulmonary lesions. In case of cutaneous or percutaneous exposure in humans, lymphadenopathy with or without ulceration and single or multiple slow-healing cutaneous eruptions (particularly along lymphatic pathways) may be observed. Imaging examinations may reveal the presence of characteristic but not diagnostic honeycomb abscesses in the liver or prostatic abscesses.

In equids, symptoms of glanders include fever, cough, weakness, and emaciation, together with white-to-greenish viscous unilateral or bilateral nasal exudates (which form thin yellowish crusts along the external nares); irregularly shaped abscesses on the nasal septum; regional lymphadenopathy; boil-like lesions with thick, ropy lymphatic pathways; swelling of the limbs; and dull hair coat.

Definitive diagnosis of glanders relies on laboratory isolation and identification of *B. mallei* from nodules, abscesses, lesions, sputum, blood, etc. Since *B. mallei* has a tendency for aerosol or droplet production, BSL-3 personnel and primary

containment precautions are needed in order to avoid laboratory-acquired infection.

Typically, samples are inoculated on meat infusion nutrient media (or nutrient agar enhanced with 1%–5% glucose and/or glycerol) that contain supplements (e.g., crystal violet, proflavine, penicillin) to inhibit the growth of Gram-positive organisms [22,23]. After incubation at 37°C for 2 days, *B. mallei* forms smooth and pinpoint colonies of 1 mm in width, which are white (turning yellow with age) and semitranslucent and viscid on Loeffler's serum agar and blood agar. Colonies show a clear honey-like layer by day 3, later darkening to brown or reddish brown when grown on glycerin-potato medium. To differentiate *B. mallei* from *B. pseudomallei* and *Pseudomonas aeruginosa*, it is notable that *B. mallei* does not grow at 42°C, but *B. pseudomallei* and *P. aeruginosa* do; that *B. mallei* does not grow at 21°C, but *P. aeruginosa* does; and that *B. mallei* does not grow in 2% sodium chloride solution, but *B. pseudomallei* and *P. aeruginosa* do. In addition, *B. pseudomallei* forms colonies with white-to-yellow pigmentation and metallic sheen.

B. mallei is a small, nonmotile, nonsporulating, nonencapsulating aerobic Gram-negative bacillus 2–4 μm in length and 0.5–1 μm in width. It is facultatively anaerobic in the presence of nitrate. *B. mallei* from young cultures and fresh exudate or tissue samples typically stains in a bipolar fashion with Wright's stain and methylene blue. Organisms from older cultures may be pleomorphic. *B. mallei* is often extracellular in vivo. The lack of motility helps differentiate *B. mallei* from *B. pseudomallei*.

Other phenotypic techniques such as bacteriophage plaque assays, latex agglutination, indirect immunofluorescence assays, complement fixation test (CFT), enzyme-linked immunosorbent assay (ELISA), and mallein test as well as commercially available biochemical identification systems may be used for *B. mallei* identification [24]. Gas–liquid chromatography of cellular fatty acids has proven valuable for helping identify the organism to the genus level.

More recently, genotypic methods (e.g., pulsed-field gel electrophoresis [PFGE], ribotyping, polymerase chain reaction [PCR], multilocus sequence typing [MLST], variable amplicon typing [VAT], multiple-locus variable-number tandem repeat analysis [MLVA], DNA microarrays, and DNA sequencing of 16S and 23S ribosomal RNA [rRNA] genes) have been employed for specific identification of *B. mallei* [7,25–27]. A single-nucleotide polymorphism at position 75 in the 16S rRNA differentiates *B. mallei* from *B. pseudomallei*, as *B. mallei* has a thymidine and *B. pseudomallei* has a cytidine at this position [28]. Furthermore, *B. mallei* has a thymidine, while *B. pseudomallei* has a cytidine at position 2143 in the 23S rRNA [29]. An MLST targeting the genes *ace* (acetyl coenzyme A reductase), *gltB* (glutamate synthase), *gmhD* (ADP glycerol-mannoheptose epimerase), *lepA* (GTP-binding elongation factor), *lipA* (lipoic acid synthetase), *narK* (nitrite extrusion protein), and *ndh* (NADH dehydrogenase) provides an efficient approach for the differentiation of *B. mallei* from other related bacteria [7].

26.2.5 Treatment and Prevention

Antibiotic treatment for *B. mallei* infection is feasible as the organism is susceptible to a number of antibiotics in vitro, including amikacin, netilmicin, gentamicin, streptomycin, tobramycin, azithromycin, novobiocin, piperacillin, imipenem, ceftazidime, tetracycline, oxytetracycline, minocycline, doxycycline, ciprofloxacin, norfloxacin, ofloxacin, erythromycin, sulfadiazine, and amoxicillin–clavulanate. However, *B. mallei* is resistant to amoxicillin, ampicillin, penicillin G, bacitracin, chloromycetin, carbenicillin, oxacillin, cephalothin, cephalexin, cefotetan, cefuroxime, cefazolin, ceftriaxone, metronidazole, and polymyxin, while its susceptibility to streptomycin and chloramphenicol is variable. Because *B. mallei* is a facultative intracellular pathogen, antibiotics (e.g., aminoglycosides) that are incapable of penetrating host cells are probably not useful in vivo. Sulfadiazine, Aureomycin, streptomycin, imipenem, doxycycline, azithromycin, and doxycycline have been applied successfully for treatment of glanders in human patients. Currently, oral doxycycline and trimethoprim–sulfamethoxazole (TMP-SMX) plus chloramphenicol are recommended for mild disease. Either intravenous (IV) ceftazidime, imipenem, meropenem, or TMP-SMX is suggested for severe disease. Surgical drainage is useful for prostatic abscesses, septic arthritis, and parotid abscesses, but not for hepatosplenic abscesses. Patients should be checked regularly for at least 5 years after recovery.

Prophylaxis with doxycycline or ciprofloxacin may be helpful to prevent infection in individuals with potential exposure, including emergency responders, as previous infection or vaccination does not provide adequate immunity against glanders [30,31].

To prevent human-to-human transmission nosocomially and via close personal contact, health staff should use disposable gloves, face shields, surgical masks, and surgical gowns to protect mucous membranes and skin. Family members should be advised of blood and body fluid precautions for patients recovering at home, and measures are adapted to reduce exposure of mucous membranes, cuts and sores, skin abrasions (genital, oral, nasal), and body fluids.

Decontamination procedures for the patient include the removal and containment of outer clothing and management of all waste in accordance with BSL-3 containment protocols. Appropriate barriers to direct skin contact with the organism are mandatory for laboratory and health personnel.

Environmental decontamination consists of disinfection with 1% sodium hypochlorite, 5% calcium hypochlorite, 70% ethanol, 2% glutaraldehyde, 70 to 150 ppm iodine, 0.05% benzalkonium chloride, 1% potassium permanganate, 3% solution of alkali, or 3% sulfur–carbolic solution; heating at >55°C for 10 min; and ultraviolet (UV) irradiation.

26.3 *Burkholderia pseudomallei*

26.3.1 Classification and Morphology

B. pseudomallei (synonyms: *Bacillus pseudomallei*, *Bacterium whitmori*, *Malleomyces pseudomallei*, *Loefferella pseudomallei*, *Pfeiferella pseudomallei*, *Actinobacillus pseudomallei*, *Flavobacterium pseudomallei*, *P. pseudomallei*) belongs to the genus *Burkholderia*, family Burkholderiaceae, order Burkholderiales, class Betaproteobacteria, phylum Proteobacteria, kingdom Bacteria.

B. pseudomallei is a Gram-negative, bipolar, aerobic, motile rod-shaped bacterium 1.5 μm in length and 0.4–0.8 μm in diameter. It possesses a polar tuft of two to four flagella, which assist its self-propulsion and mobility, and exhibits bipolar staining with a "safety pin" appearance. *B. pseudomallei* tends to demonstrate variable interstrain and medium-dependent colony morphology [32]. On Ashdown's selective medium, *B. pseudomallei* displays two distinct colony phenotypes, due probably to its differential uptake of crystal violet and neutral red or its differential production of ammonia and oxalic acid. After 2–3 days of incubation at 37°C, most strains produce lavender colonies, while some isolates form deep purple colonies. After 5 days at 37°C, the colonies become dull and wrinkled and generate a distinctive sweet earthy smell. The bacterium is capable of infecting humans and animals resulting in melioidosis (also known as pseudoglanders, Whitmore's disease [after Alfred Whitmore, who first described the disease], Nightcliff gardener's disease [Nightcliff is a suburb of Darwin, Australia, where melioidosis is endemic], Paddy-field disease, and Morphia injector's septicemia). It is also infective to plants.

Recently, the complete genome sequences of four *B. pseudomallei* strains have become available: K96243 (chromosome I, 4,074,542 bp; chromosome II, 3,173,005 bp), 1710b (chromosome I, 4,126,292 bp; chromosome II, 3,181,762 bp), 1106a (chromosome I, 3,988,455 bp; chromosome II, 3,100,794 bp), and 668 (chromosome I, 3,912,947 bp; chromosome II, 3,127,456 bp) [9,11,33]. *B. pseudomallei* strain K96243 is a clinical isolate from a diabetic patient in Thailand. Its genome is predicted to encode 5855 proteins, with chromosome 1 encoding a high proportion of core housekeeping functions (DNA replication, transcription, translation, amino acid and nucleotide metabolism, basic carbohydrate metabolism, and cofactor synthesis) and chromosome 2 encoding a high proportion of accessory functions (adaptation to atypical conditions, osmotic protection, and secondary metabolism) and essential functions (transfer RNA [tRNA], amino acid biosynthesis, and energy metabolism) as well as plasmid-like replication genes and mobile genetic elements (e.g., prophages, insertion sequences), indicative of its plasmid (or megaplasmid) origin. Similar to the related species *B. mallei*, the *B. pseudomallei* K96243 genome contains 16 genomic islands and a number of small sequence repeats that may promote antigenic variations, suggesting extensive horizontal gene transfer among *Burkholderia* species. The strain contains three different type III secretion systems (TTSSs), which contribute to pathogenicity, with two being similar to plant pathogenic TTSSs and the third resembling the *Salmonella* pathogenicity island. Other virulence determinants consist of multidrug efflux pumps, secreted toxins and proteases, adhesions, and capsular polysaccharide, the latter of which plays a role in protecting the organism from host defense mechanisms [34,35].

26.3.2 Biology and Epidemiology

B. pseudomallei was first isolated by Whitmore and Krishnaswami in 1911 from postmortem tissue samples of patients in Rangoon, Burma, with a previously unrecognized disease (resembling disseminated or localized, suppurative infection) that appeared similar to glanders, a zoonotic disease of equines caused by *B. mallei* [36]. *B. pseudomallei* was distinguished from *B. mallei* by its motility, luxuriant growth on peptone agar, and wrinkled colony morphology. The disease became known as "Whitmore's disease" or "morphine injector's septicemia." In 1921, a bacterium sharing identity with *B. pseudomallei* was identified from a guinea pig colony in Kuala Lumpur with a septicemic disease, which was named "melioidosis" (melis meaning "a distemper of asses," -oid meaning "similar to," and -osis meaning "a condition" in Greek; together it means a condition similar to glanders). Since then, "melioidosis" has become a preferred term to describe the disease caused by *B. pseudomallei* [37].

The nonsporulating, Gram-negative bacillus is a saprophyte that occurs in surface waters and wet soils in tropical and subtropical regions [38–43]. Being metabolically versatile, *B. pseudomallei* can grow on numerous carbon sources and a variety of microbial media, with Ashdown's selective medium being often used for its isolation from environmental and clinical specimens. In the presence of nitrate or arginine, it can undergo anaerobic growth. By accumulating intracellular stores of poly-β-hydroxybutyric acid, the bacterium is capable of surviving in distilled water for 11 years [44]. In general, *B. pseudomallei* strains are able to ferment sugars without gas formation (glucose and galactose, older cultures may also metabolize maltose and starch). Although the optimal growth temperature is between 24°C and 32°C, *B. pseudomallei* is capable of enduring temperatures up to 42°C. *B. pseudomallei* is sensitive to disinfectants (e.g., benzalkonium chloride, iodine, mercuric chloride, potassium permanganate, 1% sodium hypochlorite, 70% ethanol, 2% glutaraldehyde, and phenolic preparations) and effectively killed by the commercial disinfectants PeraSafe and Virkon as well as by heating to above 74°C for 10 min or by UV irradiation [45].

Many mammalian species are susceptible to melioidosis, including sheep, goats, horses, swine, cattle, dogs, and cats [46,47]. The median lethal dose (LD50) for hamsters and ferrets is <10^2 bacteria, while for rats, pigs, and rhesus monkeys, >10^6 bacteria. As an environmental organism that has no requirement to pass through an animal host in order to replicate, *B. pseudomallei* is transmitted from the contaminated source (e.g., soil and water) to humans through broken skin, by inhalation or by ingestion [48]. The major risk factors include diabetes, alcoholism, chronic renal disease, thalassemia, CF, and steroid abuse, with up to 50% of melioidosis sufferers having diabetes mellitus; individuals (especially males and females and those over 45 years of age) working in muddy soil (e.g., rice paddy) without good hand and foot protection may also acquire the infection [49,50].

Melioidosis caused by *B. pseudomallei* is endemic in Thailand, Burma, Cambodia, Laos, Vietnam, southern China, Hong Kong, Taiwan, Malaysia, Singapore, Brunei, and northern Australia as well as other regions located between 20°N and 20°S latitude [51–57]. In northeast Thailand, 80% of children have been tested positive for antibodies against *B. pseudomallei* by the age of 4; and the number of melioidosis cases increased from 11.5/100,000 inhabitants in 1997 to 21.3/100,000 in 2006 throughout the country [58,59]. Melioidosis was also reported in the South Pacific, Africa, India, the Middle East, Aruba, Brazil, Mexico, Panama, Ecuador, Haiti, Peru, and Guyana [60–63]. Recent travel to endemic areas may account for melioidosis cases that occur in temperate regions. The number and severity of melioidosis cases tend to peak during rainy season; heavy rainfall may play a role in shifting from percutaneous inoculation to inhalation as the primary mode of infection, leading to more severe illness [64].

26.3.3 Clinical Features and Pathogenesis

26.3.3.1 Clinical Features

Although most *B. pseudomallei* infections in humans are asymptomatic or mild, they may assume either acute melioidosis (acute localized suppurative soft tissue infection, acute pulmonary infection, acute fulminant infection [bloodstream septicemia]) or chronic melioidosis (suppurative/pyogenic infection), which are typically characterized by the formation of abscesses and depend on both patient's preexisting risk factors and the route of inoculation [65]. The incubation periods of melioidosis range from 2 days to many years (with the longest period between presumed exposure and clinical presentation being 62 years), and the mortality may be 20%–50% even with treatment. Because melioidosis is associated with a wide spectrum of symptoms, the disease is sometimes referred to as "the great mimicker" [54,66].

Acute melioidosis: Acute melioidosis may be localized, pulmonary, or fulminant (bloodstream-related), with the incubation period being about 9 days (range 1–21 days). However, fulminant infection, particularly associated with near drowning, may present with severe symptoms within hours.

Patients with active melioidosis usually present with fever, cough, pleuritic chest pain (pneumonia), bone or joint pain (osteomyelitis, septic arthritis, or cellulitis), and abscesses (in liver, spleen, skeletal muscle, prostate, and kidney) that tend to show a characteristic hypoechoic, multiseptate, multiloculate structure, similar to "honeycomb" or "Swiss cheese" on computed tomography (CT) [67–70]. The disease demonstrates some regional variations: Acute suppurative parotitis (parotid abscesses) commonly occurs in Thailand, typically presenting with fever, pain, and swelling over the parotid/salivary gland without other evidence of underlying predisposing conditions; encephalomyelitis syndrome (with peripheral motor weakness, cerebellar signs, flaccid paraparesis, and cranial nerve palsies) tends to appear in northern Australia due probably to multiple focal *B. pseudomallei* microabscesses in the brain stem and spinal cord; and prostatic abscesses are found in up to 20% of Australian males with *B. pseudomallei* infection [53]. Patients with the acute septic form of melioidosis are often severely ill with signs of sepsis and possibly septic shock.

Less common presentations of acute melioidosis consist of intravascular infection, lymph node abscesses, pyopericardium, myocarditis, pericardial fluid collections, mediastinal masses, adrenal, thyroid or scrotal abscesses, and uncomplicated infections of the skin, subcutaneous tissues, or the eye. Corneal trauma is vulnerable to *B. pseudomallei* infection, leading to destructive ulceration.

Chronic melioidosis: Chronic melioidosis (suppurative/ pyogenic infection) is defined by a duration of symptoms greater than 2 months that occur in approximately 10% of patients [71]. The clinical presentations of chronic melioidosis are that protean and various organs (e.g., the skin, brain, lungs, liver, spleen, and bones) are affected, including chronic skin infection, skin ulcers, and lung nodules or chronic pneumonia, which closely mimic tuberculosis (sometimes called "Vietnamese tuberculosis") or tuberculous pericarditis [72–74].

In a recent report, Rossi et al. [75] documented two cases of melioidosis and hairy cell leukemia in French travelers returning from Thailand. The first patient was a 48-year-old man showing with fever, asthenia, chills, and pancytopenia after returning from a 1-week visit to flooded regions in Thailand. CT revealed multiple liver, spleen, and lung abscesses. Culture of a CT scan-guided liver abscess puncture specimen was positive for *B. pseudomallei* after 12 days of antimicrobial drug treatment. Treatment with ceftazidime (120 mg/kg/day) and TMP/SMX (10/50 mg/kg/day) and oral doxycycline (200 mg/day) for 3 weeks led to a complete remission of melioidosis.

The second patient was a 64-year-old man with persistent fever, chills, abdominal pain, and cough 16 days after his return from Thailand. Clinical examination showed left lung crackles and chest and abdomen CT images unveiled a focus of lung consolidations, left pleural effusion, pericarditis, and spleen abscesses. Blood cultures were positive for *B. pseudomallei*. The patient recovered after treatment with IV TMP/ SMX (10 mg/50 mg/kg/day), oral doxycycline (200 mg/day), and IV ceftazidime (120 mg/kg/day), followed by a 6-month course of oral TMP/SMX (50 mg/10 mg/kg/day) and doxycycline (4 mg/kg/day) [75].

26.3.3.2 Pathogenesis

B. pseudomallei is a facultative intracellular pathogen that produces a number of molecules to assist its entry to host cell, replication in cytosol, migration inside the cell, and spread from cell to cell, causing cell fusion and the formation of MNGCs [76–81]. The bacterium has the ability to evade the humoral immune surveillance and may remain in a dormant stage in macrophages for months or years [21].

To date, the following molecules have been implicated in *B. pseudomallei* virulence: (1) capsule (which is a 200 kDa homopolymer, contributing to survival in serum by reducing

complement factor C3b deposition), (2) TTSS (the effector proteins of TTSS 3 facilitate the invasion of epithelial cells and escape from endocytic vesicles), (3) quorum-sensing molecules (*B. pseudomallei* encodes three *luxI* homologues that produce at least three quorum-sensing molecules and five *luxR* homologues to sense these signals), (4) LPS O-antigen (which promotes survival in serum by preventing killing by the alternative pathway of complement); (5) flagellin (a surface-associated 43 kd protein required for motility), (6) type II secretion (which is involved in the production of protease, lipase, and phospholipase C), (7) type IV pili (which plays a part in cell attachment), (8) biofilm formation (which may play a minor role in virulence); (9) malleobactin (which is a water-soluble siderophore of the hydroxamate class capable of scavenging iron from both lactoferrin and transferrin in vitro), (10) exopolysaccharide (EPS) (which may be a virulence determinant), (11) endotoxin (which is lipid A portion of *B. pseudomallei* LPS), and (12) actin-based motility (after gaining access to the host cell cytoplasm, *B. pseudomallei* replicates and exploits actin-based motility for cell-to-cell spread and evasion of the humoral immune response) [21,82].

26.3.4 IDENTIFICATION AND DIAGNOSIS

Any severely ill febrile patient with an associated risk factor, who has traveled to and/or stayed at endemic areas, should be suspected of melioidosis. Due to the protean nature of its clinical manifestations, a definitive diagnosis of melioidosis requires the isolation and identification of *B. pseudomallei* from clinical specimens, considering that it never forms part of the normal human flora. A complete set of samples (including blood, sputum, urine, abscess fluid, throat and rectal swabs), instead of specimen from the affected site only, should be screened on patients suspected of melioidosis [83–86].

Being a nonfastidious organism, *B. pseudomallei* grows well on a variety of laboratory media (e.g., blood agar and Ashdown's medium) [23,87–89]. Ashdown's medium is a crystal violet and gentamicin-containing medium that permits selective growth of *B. pseudomallei*, especially in clinical samples obtained from nonsterile sites [22]. *B. pseudomallei* typically becomes positive in 24−48 h (in contrast to *B. mallei*, which takes a minimum of 72 h to grow). However, its growth is notably slower than other bacteria present in clinical specimens. *B. pseudomallei* colonies are large and wrinkled, with a metallic appearance and an earthy odor. It is a Gram-negative rod with a characteristic "safety pin" appearance (bipolar staining). *B. pseudomallei* is innately resistant to a large number of antibiotics including colistin and gentamicin (in contrast to *B. mallei*, which is sensitive to a large number of antibiotics). *B. pseudomallei* is differentiated from the nonpathogenic *Burkholderia thailandensis* using an arabinose test, and *B. thailandensis* is never isolated from clinical specimens.

The use of biochemical identification systems is helpful for *B. pseudomallei* identification, with the API20NE system (bioMérieux) and the automated VITEK 1 system (bioMérieux) accurately identifying *B. pseudomallei* in 99% of cases but the automated VITEK 2 system (bioMérieux) requiring correct software to analyze the data for its identification [24,90–93].

B. pseudomallei is intrinsically resistant to aminoglycosides and polymyxins (gentamicin and colistin) but susceptible to amoxicillin–clavulanate [94]. This antibiotic pattern helps distinguish it from *P. aeruginosa*.

Serologic tests (e.g., indirect hemagglutination test [IHA], latex agglutination assay, immunofluorescence assay, and ELISA) have been employed for the diagnosis of melioidosis [95–98]. A *B. pseudomallei* identification algorithm consisting of screening tests (Gram stain, oxidase test, gentamicin, and polymyxin susceptibility testing) together with monoclonal antibody agglutination testing and gas–liquid chromatography analysis of bacterial fatty acid methyl esters has proven to be highly sensitive [99].

Considering the time and technical expertise required for in vitro isolation and subsequent macroscopic and microscopic identification, MLST, fluorescence in situ hybridization (FISH), and PCR-based methods have been developed and applied for improved detection and typing of *B. pseudomallei* and *B. mallei* [7,100–112]. Recently, Janse et al. [113] reported a single-reaction quadruplex qPCR assay for rapid detection and differentiation of *B. mallei* and *B. pseudomallei* strains. The assay targeted a 150 bp region within the transposase ISBma2, which is shared between *B. mallei* and *B. pseudomallei*: a *B. pseudomallei* signature sequence psu encoding for a putative acetyltransferase (which is part of the TTSS-associated gene cluster) and a *B. mallei* signature sequence mau encoding for a phage integrase family protein, permitting specific detection of less than 1 genome equivalent per reaction [113].

26.3.5 TREATMENT AND PREVENTION

B. pseudomallei is intrinsically resistant to aminoglycosides, cephalosporins, rifamycins, and nonureidopenicillins (including gentamicin and colistin) as well as quinolones and macrolides [114–117], while it is sensitive to kanamycin, chloramphenicol, doxycycline, and co-trimoxazole [94,118]. The antibiotic of choice is ceftazidime. Treatment of melioidosis usually begins with an IV therapy for at least 2 weeks (e.g., ceftazidime 2 g IV every 6 h, up to 8 g/day, or imipenem/cilastatin 1 g IV every 6 h, or meropenem 1 g IV every 8 h, combined with TMP-SMX 320 mg/1600 mg IV or by mouth every 12 h). Meropenem may be preferred to imipenem/cilastatin due to its fewer neurological side effects. In general, it takes an average 10 days for the resolution of fever, but patients with large abscesses or empyema may have fluctuating fevers for > 1 month. The IV therapy is not usually stopped until the patient's temperature has returned to normal for more than 48 h. This is followed by an eradication/maintenance oral therapy (e.g., chloramphenicol 40 mg/kg/day, doxycycline 4 mg/kg/day, and TMP-SMX 10 mg/50 mg/kg/day or TMP-SMX 8 mg/40 mg/kg up to 320/1600 mg by mouth twice daily together with doxycycline)

for a minimum of 3 months to prevent disease relapse [119]. Chemotherapy is also helpful for isolation of *B. pseudomallei* from clinical specimens of patients.

As there is no licensed vaccine available to prevent human melioidosis, avoidance of *B. pseudomallei* in the environment by individuals with known risk factors (e.g., diabetes mellitus, chronic renal failure, and chronic lung disease) is critical. Doxycycline 100 mg (or ciprofloxacin 500 mg), by mouth twice daily, may be recommended to individuals with risk factors and exposure to *B. pseudomallei*. In endemic areas, people (e.g., rice-paddy farmers) are advised to avoid direct contact with soil, mud, and surface water. Laboratory staff should handle *B. pseudomallei* specimens under BSL-3 conditions [50,120,121].

Currently, a number of vaccine candidates (e.g., live attenuated, heterologous, acellular, and subunit vaccines) are shown to provide significant protection in animals but are not far from being ready for human application because of their failure to elicit a sterilizing immunity [122,123]. Further research is clearly warranted in this area, with the goal to generate an effective vaccine against human melioidosis.

26.4 CONCLUSION AND FUTURE PERSPECTIVES

B. mallei and *B. pseudomallei* are highly infective, zoonotic pathogens capable of causing glanders and melioidosis, respectively, with severe consequence in the absence of prompt antibiotic treatment and intervention. Given their high infectiveness, low lethal dosage, ease of dissemination, and resistance to common antibiotics, as well as the lack of effective vaccines, *B. mallei* and *B. pseudomallei* pose serious risks to humans if they are intentionally used as biowarfare and bioterror weapons.

While *B. mallei* is nonflagellated and nonmotile with a relatively low growth rate and an obligate intracellular lifestyle, *B. pseudomallei* is flagellated and motile with a fast growth rate and a lifestyle that alternates between saprophyte and pathogen [124,125]. Despite their apparent morphological, biological, and behavioral discrepancy, *B. mallei* and *B. pseudomallei* are closely related genetically. Indeed, molecular evidence suggests that *B. mallei* has evolved from *B. pseudomallei* by gene reductions that abrogate its ability to survive in soil environment and rely on direct-host transmission to perpetuate its intracellular lifestyle.

Because the clinical presentations due to *B. mallei* and *B. pseudomallei* are generally nonspecific (e.g., fever, myalgia, headache, fatigue, diarrhea, and weight loss as well as lesions in the skin and mucous membranes), a definitive diagnosis of glanders and melioidosis depends on the isolation and identification of the organisms concerned from clinical specimens through in vitro culture and morphological characterization. In recent years, a number of PCR-based procedures have been reported for rapid and reliable detection and identification of *B. mallei* and *B. pseudomallei*. Future adoption of these new generational techniques in clinical laboratories will facilitate prompt implementation of treatment and control measures in the event of glanders and melioidosis outbreaks. In view of the current lack of effective vaccines against glanders and melioidosis, continued research efforts on *B. mallei* and *B. pseudomallei*'s ability to evade host's immune surveillance will contribute to the further advances in this area.

REFERENCES

1. Fong, I.W. and Alibek, K. (2005). *Bioterrorism and Infectious Agents: A New Dilemma for the 21st Century*. Springer, New York, pp. 99–145.
2. Wheelis, M. (1998). First shots fired in biological warfare. *Nature*, 395: 213.
3. Rotz, L.D. et al. (2002). Public health assessment of potential biological terrorism agents. *Emerg. Infect. Dis.*, 8: 225.
4. Burkholder, W.H. (1950). Sour skin, a bacterial rot of onion bulbs. *Phytopathology*, 40: 115.
5. Yabuuchi, E. et al. (1992). Proposal of *Burkholderia* gen. nov. and transfer of seven species of the genus *Pseudomonas* homology group II to the new genus, with the type species *Burkholderia cepacia* (Palleroni and Holmes 1981) comb. nov. *Microbiol. Immunol.*, 36: 1251.
6. Whitlock, G.C., Estes, D.M., and Torres, A.G. (2007). Glanders: Off to the races with *Burkholderia mallei*. *FEMS Microbiol. Lett.*, 277: 115–122.
7. Godoy, D. et al. (2003). Multilocus sequence typing and evolutionary relationships among the causative agents of melioidosis and glanders, *Burkholderia pseudomallei* and *Burkholderia mallei*. *J. Clin. Microbiol.*, 41: 2068.
8. Song, H. et al. (2010). The early stage of bacterial genome-reductive evolution in the host. *PLoS Pathog.*, 6: e1000922.
9. Nierman, W.C. et al. (2004). Structural flexibility in the *Burkholderia mallei* genome. *Proc. Natl. Acad. Sci. USA*, 101: 14246.
10. Losada, L. et al. (2010). Continuing evolution of *Burkholderia mallei* through genome reduction and large—Scale rearrangements. *Genome Biol. Evol.*, 2: 102–116.
11. Holden, M.T. et al. (2004). Genomic plasticity of the causative agent of melioidosis, *Burkholderia pseudomallei*. *Proc. Natl. Acad. Sci. USA*, 101: 14240–14245.
12. Moore, R.A. et al. (2004). Contribution of gene loss to the pathogenic evolution of *Burkholderia pseudomallei* and *Burkholderia mallei*. *Infect. Immun.*, 72: 4172–4187.
13. Ong, C. et al. (2004). Patterns of large-scale genomic variation in virulent and avirulent *Burkholderia* species. *Genome Res.*, 14: 2295–2307.
14. Fushan, A. et al. (2005). Genome-wide identification and mapping of variable sequences in the genomes of *Burkholderia mallei* and *Burkholderia pseudomallei*. *Res. Microbiol.*, 156: 278–288.
15. Kim, H.S. et al. (2005). Bacterial genome adaptation to niches: Divergence of the potential virulence genes in three *Burkholderia* species of different survival strategies. *BMC Genomics*, 6: 174.
16. Srinivasan, A. et al. (2001). Glanders in a military research microbiologist. *N. Engl. J. Med.*, 345: 256.
17. Centers for Disease Control and Prevention (CDC). (2000). Laboratory-acquired human glanders—Maryland, May 2000. *MMWR*, 49: 532–535.
18. DeShazer, D. et al. (2001). Identification of a *Burkholderia mallei* polysaccharide gene cluster by subtractive hybridization and demonstration that the encoded capsule is an essential virulence determinant. *Microb. Pathog.*, 30: 253–269.
19. Schell, M.A. et al. (2007). Type VI secretion is a major virulence determinant in *Burkholderia mallei*. *Mol. Microbiol.*, 64: 1466–1485.
20. Schell, M.A., Lipscomb, L., and DeShazer, D. (2008). Comparative genomics and an insect model rapidly identify novel virulence genes of *Burkholderia mallei*. *J. Bacteriol.*, 190: 2306.
21. Galyov, E.E., Brett, P.J., and DeShazer, D. (2010). Molecular insights into *Burkholderia pseudomallei* and *Burkholderia mallei* pathogenesis. *Ann. Rev. Microbiol.*, 64: 495–517.
22. Ashdown, L.R. (1979). Identification of *Pseudomonas pseudomallei* in the clinical laboratory. *J. Clin. Pathol.*, 32: 500–504.
23. Glass, M.B. et al. (2009). Comparison of four selective media for the isolation of *Burkholderia mallei* and *Burkholderia pseudomallei*. *Am. J. Trop. Med. Hyg.*, 80: 1023.
24. Glass, M.B. and Popovic, T. (2005). Preliminary evaluation of the API 20NE and RapID NF plus systems for rapid identification of *Burkholderia pseudomallei* and *B. mallei*. *J. Clin. Microbiol.*, 43: 479.
25. Ulrich, R.L. et al. (2006). Development of a polymerase chain reaction assay for the specific identification of *Burkholderia mallei* and differentiation from *Burkholderia pseudomallei* and other closely related Burkholderiaceae. *Diagn. Microbiol. Infect. Dis.*, 55: 37.
26. Scholz, H.C. et al. (2006). Detection of the reemerging agent *Burkholderia mallei* in a recent outbreak of glanders in the United Arab Emirates by a newly developed *fliP*-based polymerase chain reaction assay. *Diagn. Microbiol. Infect. Dis.*, 54: 241.
27. Spilker, T. et al. (2009). Expanded multilocus sequence typing for *Burkholderia* species. *J. Clin. Microbiol.*, 47: 2607.
28. Gee, J.E. et al. (2003). Use of 16S rRNA gene sequencing for rapid identification and differentiation of *Burkholderia pseudomallei* and *B. mallei*. *J. Clin. Microbiol.*, 41: 4647.
29. Bauernfeind, A. et al. (1998). Molecular procedure for rapid detection of *Burkholderia mallei* and *Burkholderia pseudomallei*. *J. Clin. Microbiol.*, 36: 2737.
30. Healey, G.D. et al. (2005). Humoral and cell-mediated adaptive immune responses are required for protection against *Burkholderia pseudomallei* challenge and bacterial clearance postinfection. *Infect. Immun.*, 73: 5945–5951.
31. Rowland, C.A. et al. (2010). Protective cellular responses to *Burkholderia mallei* infection. *Microbes Infect.*, 12: 846–853.
32. Chantratita, N. et al. (2007). Biological relevance of colony morphology and phenotypic switching by *Burkholderia pseudomallei*. *J. Bacteriol.*, 189: 807–817.
33. Ou, K. et al. (2005). Integrative genomic, transcriptional, and proteomic diversity in natural isolates of the human pathogen *Burkholderia pseudomallei*. *J. Bacteriol.*, 187: 4276–4285.
34. Chaiyaroj, S.C. et al. (1999). Differences in genomic macrorestriction patterns of arabinose-positive (*Burkholderia thailandensis*) and arabinose-negative *Burkholderia pseudomallei*. *Microbiol. Immunol.*, 43: 625–630.
35. Sim, S.H. et al. (2008). The core and accessory genomes of *Burkholderia pseudomallei*: Implications for human melioidosis. *PLoS Pathog.*, 4: e1000178.
36. Whitmore, A. and Krishnaswami, C.S. (1912). An account of the discovery of a hitherto undescribed infective disease occurring among the population of Rangoon. *Indian Med. Gaz.*, 47: 262–267.
37. White, N.J. (2003). Melioidosis. *Lancet*, 361: 1715–1722.

38. Wuthiekanun, V., Smith, M.D., and White, N.J. (1995). Survival of *Burkholderia pseudomallei* in the absence of nutrients. *Trans. R. Soc. Trop. Med. Hyg.*, 89: 491.

39. Tong, S. et al. (1996). Laboratory investigation of ecological factors influencing the environmental presence of *Burkholderia pseudomallei*. *Microbiol. Immunol.*, 40: 451.

40. Zanetti, F., De Luca, G., and Stampi, S. (2000). Recovery of *Burkholderia pseudomallei* and *B. cepacia* from drinking water. *Int. J. Food. Microbiol.*, 59: 67.

41. Howard, K. and Inglis, T.J.J. (2005). Disinfection of *Burkholderia pseudomallei* in potable water. *Water Res.*, 39: 1085–1092.

42. Palasatien, S. et al. (2008). Soil physicochemical properties related to the presence of *Burkholderia pseudomallei*. *Trans. R. Soc. Trop. Med. Hyg.*, 102: S5.

43. Lee, Y.H. et al. (2010). Identification of tomato plant as a novel host model for *Burkholderia pseudomallei*. *BMC Microbiol.*, 10: 28.

44. Pumpuang, A. et al. (2011). Survival of *Burkholderia pseudomallei* in distilled water for 16 years. *Trans. R. Soc. Trop. Med. Hyg.*, 105: 598–600.

45. Howard, K. and Inglis, T.J.J. (2003). The effect of free chlorine on *Burkholderia pseudomallei* in potable water. *Water Res.*, 37: 4425–4432.

46. Sprague, L.D. and Neubauer, H. (2004). Melioidosis in animals: A review on epizootiology, diagnosis and clinical presentation. *J. Vet. Med. B Infect. Dis. Vet. Public Health*, 51: 305–320.

47. Parkes, H.M. et al. (2009). Primary ocular melioidosis due to a single genotype of *Burkholderia pseudomallei* in two cats from Arnhem Land in the Northern Territory of Australia. *J. Feline Med .Surg.*, 11: 856–863.

48. Kunakorn, M., Jayanetra, P., and Tanphaichitra, D. (1991). Man-to-man transmission of melioidosis. *Lancet*, 337: 1290–1291.

49. Suputtamongkol, Y. et al. (1999). Risk factors for melioidosis and bacteremic melioidosis. *Clin. Infect. Dis.*, 29: 408–413.

50. Cheng, A.C. and Currie, B.J. (2005). Melioidosis: Epidemiology, pathophysiology, and management. *Clin. Microbiol. Rev.*, 18: 383–416.

51. Vuddhakul, V. et al. (1999). Epidemiology of *Burkholderia pseudomallei* in Thailand. *Am. J. Trop. Med. Hyg.*, 60: 458–461.

52. Yang, S. (2000). Melioidosis research in China. *Acta Trop.*, 77: 157–165.

53. Currie, B.J. et al. (2004). Melioidosis epidemiology and risk factors from a prospective whole-population study in northern Australia. *Trop. Med. Int. Health*, 9: 1167.

54. Currie, B.J. (2008). Advances and remaining uncertainties in the epidemiology of *Burkholderia pseudomallei* and melioidosis. *Trans. R. Soc. Trop. Med. Hyg.*, 102: 225.

55. Inglis, T.J., Rolim, D.B., and De Queroz Sousa, A. (2006). Melioidosis in the Americas. *Am. J. Trop. Med. Hyg.*, 75: 947–954.

56. Chen, Y.S. et al. (2010). Distribution of melioidosis cases and viable *Burkholderia pseudomallei* in soil: Evidence for emerging melioidosis in Taiwan. *J. Clin. Microbiol.*, 48: 1432–1434.

57. Baker, A. et al. (2011). Molecular phylogeny of *Burkholderia pseudomallei* from a remote region of Papua New Guinea. *PLoS One*, 6: e18343.

58. Limmathurotsakul, D., Wongratanacheewin, S., and Teerawattanasook, N. (2010). Increasing incidence of human melioidosis in Northeast Thailand. *Am. J. Trop. Med. Hyg.*, 82: 1113–1117.

59. Apisarnthanarak, A., Khawcharoenporn, T., and Mundy, L.M. (2012). Flood-associated melioidosis in a non-endemic region of Thailand. *Int. J. Infect. Dis.*, 16: e409–e410.

60. Raja, N.S., Ahmed, M.Z., and Singh, N.N. (2005). Melioidosis: An emerging infectious disease. *J. Postgrad. Med.*, 51: 140–145.

61. Kite-Powell, A. et al. (2006). Imported melioidosis—South Florida, 2005. *Morb. Mortal. Wkly. Rep.*, 55: 873–876.

62. Corkeron, M.L., Norton, R., and Nelson, P.N. (2010). Spatial analysis of melioidosis distribution in a suburban area. *Epidemiol. Infect.*, 22: 1–7.

63. Inglis, T.J. et al. (2011). The aftermath of the Western Australian melioidosis outbreak. *Am. J. Trop. Med. Hyg.*, 84: 851–857.

64. Athan, E. et al. (2005). Melioidosis in tsunami survivors. *Emerg. Infect. Dis.*, 11: 1638.

65. Barnes, J.L. and Ketheesan, N. (2005). Route of infection in melioidosis. *Emerg. Infect. Dis.*, 11: 638–639.

66. Brilhante, R.S. et al. (2012). Clinical-epidemiological features of 13 cases of melioidosis in Brazil. *J. Clin. Microbiol.*, 50: 3349–3352.

67. Muttarak, M. et al. (2008). Spectrum of imaging findings in melioidosis. *Br. J. Radiol.*, 82: 514–521.

68. Laopaiboon, V. et al. (2009). CT findings of liver and splenic abscesses in melioidosis: Comparison with those in non-melioidosis. *J. Med. Assoc. Thai.*, 92(11): 1476–1484.

69. Chong, V.H. (2011). Changing spectrum of microbiology of liver abscess: Now *Klebsiella*, next *Burkholderia pseudomallei*. *J. Emerg. Med.* 41: 676–677.

70. Lim, K.S. and Chong, V.H. (2010). Radiological manifestations of melioidosis. *Clin. Radiol.*, 65: 66–72.

71. Chlebicki, M.P. and Tan, B.H. (2006). Six cases of suppurative lymphadenitis caused by *Burkholderia pseudomallei* infection. *Trans. R. Soc. Trop. Med. Hyg.*, 100: 798–801.

72. Ngauy, V. et al. (2005). Cutaneous melioidosis in a man who was taken as a prisoner of war by the Japanese during World War II. *J. Clin. Microbiol.*, 43: 970–972.

73. Vidyalakshmi, K. et al. (2008). Tuberculosis mimicked by melioidosis. *Int. J. Tuberc. Lung Dis.*, 12: 1209–1215.

74. Chetchotisakd, P. et al. (2010). Melioidosis pericarditis mimicking tuberculous pericarditis. *Clin. Infect. Dis.*, 51: e46–e49.

75. Rossi, B. et al. (2013). Melioidosis and hairy cell leukemia in 2 travelers returning from Thailand. *Emerg. Infect. Dis.*, 19: 503–505.

76. Haase, A. et al. (1997). Toxin production by *Burkholderia pseudomallei* strains and correlation with severity of melioidosis. *J. Med. Microbiol.*, 46: 557–563.

77. Simpson, A.J.H. et al. (2000). Differential antibiotic-induced endotoxin release in severe melioidosis. *J. Infect. Dis.*, 181: 1014–1019.

78. Wiersinga, W.J. et al. (2006). Melioidosis: Insights into the pathogenicity of *Burkholderia pseudomallei*. *Nat. Rev. Microbiol.*, 4: 272.

79. Shalom, G., Shaw, J.G., and Thomas, M.S. (2007). In vivo expression technology identifies a type VI secretion system locus in *Burkholderia pseudomallei* that is induced upon invasion of macrophages. *Microbiology*, 153: 2689–2699.

80. Kespichayawattana, W. et al. (2000). *Burkholderia pseudomallei* induces cell fusion and actin-associated membrane protrusion: A possible mechanism for cell-to-cell spreading. *Infect. Immun.*, 68: 5377–5384.

81. Cruz-Migoni, A. et al. (2011). A *Burkholderia pseudomallei* toxin inhibits helicase activity of translation factor *eIF4A*. *Science*, 334: 821–824.

82. Reckseidler, S.L. et al. (2001). Detection of bacterial virulence genes by subtractive hybridization: Identification of capsular polysaccharide of *Burkholderia pseudomallei* as a major virulence determinant. *Infect. Immun.*, 69: 34–44.

83. Dance, D.A. et al. (1990). The use of bone marrow culture for the diagnosis of melioidosis. *Trans. R. Soc. Trop. Med. Hyg.*, 84: 585–587.

84. Walsh, A.L. and Wuthiekanun, V. (1996). The laboratory diagnosis of melioidosis. *Br. J. Biomed. Sci.*, 53: 249–253.

85. Wuthiekanun, V. et al. (2001). Value of throat swab in diagnosis of melioidosis. *J. Clin. Microbiol.*, 39: 3801–3802.

86. Limmathurotsakul, D. et al. (2005). Role and significance of quantitative urine cultures in diagnosis of melioidosis. *J. Clin. Microb.*, 43: 2274–2276.

87. Peacock, S.J. et al. (2005). Comparison of ashdown's medium, *Burkholderia cepacia* medium, and *Burkholderia pseudomallei* selective agar for clinical isolation of *Burkholderia pseudomallei*. *J. Clin. Microbiol.*, 43: 5359–5361.

88. Francis, A. et al. (2006). An improved selective and differential medium for the isolation of *Burkholderia pseudomallei* from clinical specimens. *Diagn. Microbiol. Infect. Dis.*, 55: 95–99.

89. Roesnita, B. et al. (2012). Diagnostic use of *Burkholderia pseudomallei* selective media in a low prevalence setting. *Trans. R. Soc. Trop. Med. Hyg.*, 106: 131–133.

90. Inglis, T.J. et al. (1998).Potential misidentification of *Burkholderia pseudomallei* by API 20NE. *Pathology*, 30: 62–64.

91. Lowe, P., Engler, C., and Norton, R. (2002). Comparison of automated and non automated systems for identification of *Burkholderia pseudomallei*. *J. Clin. Microbiol.*, 40: 4625–4627.

92. Amornchai, P. et al. (2007). Accuracy of *Burkholderia pseudomallei* identification using the API 20NE system and a latex agglutination test. *J. Clin. Microbiol.*, 45: 3774–3776.

93. Deepak, R.N., Crawley, B., and Phang, E. (2008). *Burkholderia pseudomallei* identification: A comparison between the API 20NE and VITEK 2 GN systems. *Trans. R. Soc. Trop. Med. Hyg.*, 102: S42.

94. Cheng, A.C. et al. (2005). Short report: Consensus guidelines for dosing of amoxicillin-clavulanate in melioidosis. *Am. J. Trop. Med. Hyg.*, 78: 208–209.

95. Wuthiekanun, V. et al. (2005). Rapid immunofluorescence microscopy for diagnosis of melioidosis. *Clin. Diag. Lab. Immunol.*, 12: 555–556.

96. Limmathurotsakul, D. et al. (2011). Enzyme-linked immuno sorbent assay for the diagnosis of melioidosis: Better than we thought. *Clin. Infect. Dis.*, 52: 1024–1028.

97. Puthucheary, S.D., Anuar, A.S., and Tee, T.S. (2010). *Burkholderia thailandensis* whole cell antigen cross-reacts with *B. pseudomallei* antibodies from patients with melioidosis in an immunofluorescent assay. *Southeast Asian J. Trop. Med. Public Health*, 41: 397–400.

98. Peacock, S.J. et al. (2011). The use of positive serological tests as evidence of exposure to *Burkholderia pseudomallei*. *Am. J. Trop. Med. Hyg.*, 84: 1021–1022.

99. Sheridan, E.A. et al. (2007). Evaluation of the Wayson stain for the rapid diagnosis of melioidosis. *J. Clin. Microbiol.*, 45: 1669–1670.

100. Gal, D. et al. (2005). Short report: Application of a polymerase chain reaction to detect *Burkholderia pseudomallei* in clinical specimens from patients with suspected melioidosis. *Am. J. Trop. Med. Hyg.*, 73: 1162–1164.

101. Liu, Y. et al. (2006). Rapid molecular typing of *Burkholderia pseudomallei*, isolated in an outbreak of melioidosis in Singapore in 2004, based on variable-number tandem repeats. *Trans. R. Soc. Trop. Med. Hyg.*, 100: 687–692.

102. Wattiau, P. et al. (2007). Identification of *Burkholderia pseudomallei* and related bacteria by multiple-locus sequence typing-derived PCR and real-time PCR. *J. Clin. Microbiol.*, 45: 1045–1048.

103. Hodgson, K. et al. (2009). A comparison of routine bench and molecular diagnostic methods in the identification of *Burkholderia pseudomallei*. *J. Clin. Microbiol.*, 47: 1578–1580.

104. Hagen, R.M. et al. (2011). Rapid identification of *Burkholderia pseudomallei* and *Burkholderia mallei* by fluorescence in situ hybridization (FISH) from culture and paraffin-embedded tissue samples. *Int. J. Med. Microbiol.*, 301: 585–590.

105. Bartpho, T. et al. (2012). Genomic islands as a marker to differentiate between clinical and environmental *Burkholderia pseudomallei*. *PLoS One*, 7: e37762.

106. Kaestli, M. et al. (2012). Comparison of Taq Man PCR assays for detection of the melioidosis agent *Burkholderia pseudomallei* in clinical specimens. *J. Clin. Microbiol.*, 50: 2059–2062.

107. Koh, S.F. et al. (September 2012). Development of a multiplex PCR assay for rapid identification of *Burkholderia pseudomallei*, *Burkholderia thailandensis*, *Burkholderia mallei* and *Burkholderia cepacia* complex. *J. Microbiol. Methods*, 90: 305–308.

108. Marques, M.A. et al. (December 2012). Evaluation of six commercial DNA extraction kits for recovery of *Burkholderia pseudomallei* DNA. *J. Microbiol. Methods*, 91: 487–489.

109. Nandagopal, B. et al. (January 2012). Application of polymerase chain reaction to detect *Burkholderia pseudomallei* and *Brucella* species in buffy coat from patients with febrile illness among rural and peri-urban population. *J. Glob. Infect. Dis.*, 4: 31–37.

110. Price, E.P. et al. (2012). Development and validation of *Burkholderia pseudomallei*-specific real-time PCR assays for clinical, environmental or forensic detection applications. *PLoS One*, 7: e37723.

111. Richardson, L.J. et al. 2012. Towards a rapid molecular diagnostic for melioidosis: Comparison of DNA extraction methods from clinical specimens. *J. Microbiol. Methods*, 88: 179–181.

112. Zhang, B. et al. (2012). Development of hydrolysis probe-based real-time PCR for identification of virulent gene targets of *Burkholderia pseudomallei* and *B. mallei*—A retrospective study on archival cases of service members with melioidosis and glanders. *Mil. Med.*, 177: 216–221.

113. Janse, I. et al. (2013). Multiplex qPCR for reliable detection and differentiation of *Burkholderia mallei* and *Burkholderia pseudomallei*. *BMC Infect. Dis.*, 13: 86.

114. Moore, R.A. et al. (1999). Efflux-mediated amino glycoside and macrolide resistance in *Burkholderia pseudomallei*. *Antimicrob. Agents Chemother.*, 43: 465–470.

115. Viktorov, D.V. et al. (2008). High-level resistance to fluoro-quinolones and cephalosporins in *Burkholderia pseudomallei* and closely related species. *Trans. R. Soc. Trop. Med. Hyg.*, 102: S103.

116. Norris, M.H. et al. (2009). Glyphosate resistance as a novel select-agent-compliant, non-antibiotic-selectable marker in chromosomal mutagenesis of the essential genes *asd* and *dapB* of *Burkholderia pseudomallei*. *Appl. Environ. Microbiol.*, 75: 6062–6075.

117. Wuthiekanun, V. et al. (2010). Perasafe, Virkon and bleach are bactericidal for *Burkholderia pseudomallei*, a select agent and the cause of melioidosis. *J. Hosp. Infect.*, 77: 183–184.

118. Chaowagul, W. (2000). Recent advances in the treatment of severe melioidosis. *Acta Trop.*, 74: 133–137.

119. Chetchotisakd, P. et al. (2001). Maintenance therapy of meli-oidosis with ciprofloxacin plus azithromycin compared with cotrimoxazole plus doxycycline. *Am. J. Trop. Med. Hyg.*, 64: 24–27.

120. Peacock, S.J. et al. (2008). Management of accidental labo-ratory exposure to *Burkholderia pseudomallei* and *B. mallei*. *Emerg. Infect. Dis.*, 14: e2.

121. Wuthiekanun, V. and Peacock, S.J. (2006). Management of melioidosis. *Expert Rev. Anti Infect. Ther.*, 4: 445–455.

122. Nelson, M. et al. (2004). Evaluation of lipopolysaccha-ride and capsular polysaccharide as subunit vaccines against experimental melioidosis. *J. Med. Microbiol.*, 53: 1177–1182.

123. Bondi, S.K. and Goldberg, J.B. (2008). Strategies toward vac-cines against *Burkholderia mallei* and *Burkholderia pseudo-mallei*. *Pubmed Central*, 7: 1357–1365.

124. Currie, B.J., Dance, D.A.B., and Cheng, A.C. (2008). The global distribution of *Burkholderia pseudomallei* and melioi-dosis: An update. *Trans. R. Soc. Trop. Med. Hyg.*, 102: S1.

125. Nandi, T. et al. (2010). A genomic survey of positive selection in *Burkholderia pseudomallei* provides insights into the evo-lution of accidental virulence. *PLoS Pathog.*, 6: e1000845.

27 *Chlamydophila (Chlamydia) psittaci*

Dongyou Liu

CONTENTS

27.1 INTRODUCTION

Chlamydophila (Chlamydia) psittaci is an obligate intracellular pathogen with the ability to efficiently colonize mucosal surfaces and thrive within a wide variety of animal hosts. As the pathogenic agent of a primarily avian respiratory disease called ornithosis or psittacosis (psittakos, Greek word for parrot), *C. psittaci* exerts sizeable impact on poultry farming and bird breeding economic returns. Given its ability to transmit from birds to other animals (e.g., cattle, horses, and pigs) as well as humans through respiratory route, and cause significant morbidity and mortality, *C. psittaci* is regarded as a category B bioterrorism agent by Centers for Disease Control and Prevention (CDC) in the United States. Indeed, psittacosis (parrot fever) pandemic across the United States and Europe in the winter of 1929–1930 produced a mortality rate of 20% and up to 80% in pregnant women. The aftermath of this outbreak and how it was handled played a crucial role in the founding of the National Institutes of Health in the United States.

27.2 CLASSIFICATION, MORPHOLOGY, AND GENOME ORGANIZATION

Classification: Chlamydophila (Chlamydia) psittaci is one of the nine gram-negative species classified in the genus *Chlamydia* (from Greek *khlamus* mantle or cloak), family Chlamydiaceae, order Chlamydiales, class Chlamydiae, phylum Chlamydiae, and domain Bacteria. The other species in the genus *Chlamydia* include *Chlamydia trachomatis, Chlamydia muridarum, Chlamydia pneumoniae, Chlamydia pecorum, Chlamydia suis, Chlamydia abortus, Chlamydia felis*, and *Chlamydia caviae*. This classification is based largely on the morphology of inclusions, sensitivity to sulfadiazine, ability to synthesize and accumulate glycogen in chlamydial inclusions, and estimates of the DNA–DNA homology. The detection of the family-specific lipopolysaccharide epitope

αKdo-(2→8)-αKdo-(2→4)-αKdo (previously regarded as the genus-specific epitope) has also played a role in the determination of the family Chlamydiaceae [1].

Upon phylogenetic analyses of the 16S and 23S ribosomal RNA genes, the nine species in the genus *Chlamydia* were separated into two groups (which display <95% sequence identity), forming distinct genera *Chlamydia* and *Chlamydophila* [2]. While the genus *Chlamydia* contains three species (*Chlamydia trachomatis, Chlamydia muridarum*, and *Chlamydia suis*), the genus *Chlamydophila* comprises six species (*Chlamydophila pneumonia, Chlamydophila pecorum, Chlamydophila psittaci, Chlamydophila abortus, Chlamydophila felis*, and *Chlamydophila caviae*) [2,3].

However, the separation of the genus *Chlamydia* into two genera was shown to be inconsistent with the natural history of the organisms and thus not widely accepted by the *Chlamydia* research community. Therefore, the genera *Chlamydia* and *Chlamydophila* were proposed to be reunited into a single genus *Chlamydia* [4,5]. Nonetheless, recent phylogenetic analyses of concatenated protein sequences, the numbers of shared orthologous genes, and the level of synteny between genomes indicated that the separation of the genus *Chlamydia* into *Chlamydia* and *Chlamydophila* remains useful for delineation of evolutionarily distinct branches within the family Chlamydiaceae [6].

Of the nine *Chlamydia* species, three (*C. trachomatis, C. pneumonia*, and *C. psittaci*) are medically important. *Chlamydia trachomatis* is a causative agent of human urogenital infections, trachoma, conjunctivitis, infant pneumonia, and lymphogranuloma venereum (LGV). Being infective only to human epithelial cells (apart from one strain that can infect mice), *C. trachomatis* is a leading cause of sexually transmitted bacterial diseases and ocular infections (trachoma) in humans. According to its host or tissue specificity, *C. trachomatis* is separated into two biovars (biological variants): trachoma and LGV. Currently, the trachoma biovar consists of 14 serovars, which primarily infect epithelial cells of mucous

313

membranes. The LGV biovar consists of four serovars, L1, L2, L2a, and L3, which can invade lymphatic tissue.

Chlamydophila (Chlamydia) pneumonia is linked to atypical pneumonia (or walking pneumonia) and other acute respiratory illnesses with persistent cough and malaise. Additionally, it may be involved in human sinusitis, pharyngitis, bronchitis, and atherosclerosis. Originally called the TWAR strain from the names of the two initial isolates, Taiwan (TW-183) and AR-39 (an acute respiratory isolate), the organism is now considered a separate species, that is, *Chlamydia (Chlamydophila) pneumonia* [7].

Chlamydophila (Chlamydia) psittaci is a causal agent of endemic chlamydiosis in birds, epizootic outbreaks in mammals (cattle, pigs, sheep, and horses), and pneumonia (psittacosis) in humans. Because of its epidemiological link to birds, psittacosis (or parrot fever) has also been referred to as ornithosis (from Greek word for "bird"). Sequencing analysis of the major outer membrane protein (*ompA*) gene differentiates *C. psittaci* into nine genotypes (A to F, E/B, M56, and WC), with genotypes A to F being commonly present in birds, and genotypes WC and M56 occurring in mammals. More specifically, genotypes A (reference strain 84/55) and B (reference strain CP3) are associated with psittacine birds (cockatoos, parrots, parakeets, and lories) and pigeons, respectively. Genotype C (reference strain GR9) resides in ducks and geese, genotype D (reference strain NJ1) exists mainly in turkeys, and genotype E (reference strain MN) is found in pigeons, ratites, ducks, turkeys, and occasionally humans. Genotype F (reference strain VS225) infects psittacine birds and turkeys, and genotype E/B (reference strain WS/RT/E30) occurs mainly in ducks. Genotypes WC (reference strain WC) and M56 (reference strain M56) exist in cattle and muskrats, respectively [8]. However, all *C. psittaci* genotypes have the potential to infect humans. Due to its infectiousness as an aerosol agent and pathogenicity, *C. psittaci* was developed as a bioweapon by the United States during the first half of the twentieth century. This led to its classification as a biothreat agent by CDC.

Chlamydia muridarum was created out of the former mouse pneumonitis biovar of *C. trachomatis*. Of the two *C. muridarum* strains, MoPn and SFPD, isolated from mice and hamsters, MoPn may cause asymptomatic infection or pneumonia in mice, while SFPD is a causative agent of proliferative ileitis.

Chlamydia suis causes conjunctivitis, enteritis, pneumonia, polyarthritis, polyserositis, abortion, orchitis, or asymptomatic infection in swine.

Chlamydophila (Chlamydia) pecorum is associated with pneumonia, polyarthritis, conjunctivitis, abortion, encephalomyelitis, metritis, salpingitis, enteritis, and diarrhea in mammals including cattle, sheep and goats, koala, and swine.

Chlamydophila (Chlamydia) caviae causes conjunctivitis in guinea pigs.

Chlamydophila (Chlamydia) felis is linked to pneumonia and conjunctivitis in cats.

Chlamydophila (Chlamydia) abortus is a cause of abortion and fetal loss in ruminants (sheep, cattle, and goats) such as ovine enzootic abortion (OEA) or enzootic abortion of ewes (EAE).

Morphology and genome organization: Members of the Chlamydiaceae family are obligate intracellular parasites that were once considered to be viruses because they are small enough to pass through 0.45 μm filters [9]. Subsequent demonstration of DNA, RNA, and prokaryotic ribosomes in Chlamydiae put beyond doubt their true bacterial identity. Chlamydiae possess an inner and outer membrane similar to gram-negative bacteria and a lipopolysaccharide, but do not have a peptidoglycan layer [10].

Chlamydiae are gram-negative bacteria of about 0.25–0.5 μm in diameter with a genome of about 1–1.2 Mb. They undergo obligately intracellular growth in eukaryotic host cells during their biphasic developmental life cycle, with the formation of metabolically inactive infectious elementary body (EB), metabolically active noninfectious reticulate body (RB), and intermediate body (IB) [11].

The EB is a small, spherical, occasionally pear-shaped body (of 0.2–0.3 μm in diameter), which is characterized by highly electron-dense nucleoid, comprising tightly coiled genomic DNA covered by histone-like proteins, located at the periphery and separated from an electron-dense cytoplasm. It possesses a rigid outer membrane that enables the bacterium to withstand harsh environmental conditions and to survive outside of the eukaryotic host cells. Being the infectious stage of chlamydial developmental cycle, the EB binds to receptors on susceptible cells (e.g., columnar epithelial cells as well as macrophages) and enters via endocytosis and/or phagocytosis. Within the host cell endosome, the EB replicates by binary fission and reorganizes into the RB.

The RB is of 0.5–2.0 μm in diameter and has a fragile membrane that lacks the extensive disulfide bonds characteristic of the EB. Being metabolically active and containing ribosomes in the cytoplasm, which are required for protein synthesis, the RB replicates exclusively in an intracellular vacuole termed inclusion, and the resulting inclusions may contain 100–500 progeny. As the RB begins to differentiate into the EB at the end of the chlamydial developmental cycle, recondensation of nucleic acid appears in the inclusion's cytoplasm. In the maturing inclusion, chlamydial particles pack tightly within the inclusion membrane. Nutrient deficiencies (e.g., low glucose, iron, or amino acid levels) may delay chlamydial development and lead to the formation of aberrant chlamydial organisms within the inclusion. These aberrant inclusions may contribute to the maintenance of the infection in asymptomatic hosts. Eventually, the cells and inclusions lyse (*C. psittaci*) or the inclusion is extruded by reverse endocytosis (*C. trachomatis* and *C. pneumoniae*).

The IB (about 0.3–1.0 μm in diameter) representing the transition from the EB to the RB may be observed occasionally.

The genomes of Chlamydiae show a conserved gene order (synteny) with few pseudogenes and a notable absence of disruptive mobile elements [12–14]. As exemplified by the 6BC strain, *C. psittaci* genome contains a single circular chromosome of 1.172 Mb (with 967 coding sequences and a GC content of 39.06%) and a plasmid of 7553 bp (with eight coding sequences) [15]. *C. psittaci* carries 36 tRNA genes and

1 rRNA operon. Most of the *C. psittaci*-specific sequences are located in the plasticity zone (PZ), which includes a 9 kb coding sequence (CDS) predicted to encode a cytotoxin and also contains the intact *guaBA-add* operon encoding proteins involved in purine nucleotide interconversion. *C. psittaci* lacks the tryptophan biosynthesis operon. Phylogenetic comparison reveals the close relationship of *C. psittaci* with the three chlamydial species originally considered as the "mammalian" *Chlamydia psittaci* abortion, feline, and Guinea pig strains (i.e., *C. abortus*, *C. felis*, and *C. caviae*; here referred to as "*psittaci* group"). Whole-genome comparisons with the other Chlamydiaceae genomes show that the *C. psittaci* 6BC genome is essentially syntenic to sequences from *C. abortus*, *C. felis*, and *C. pneumoniae* [6,16]. In addition, comparison of 20 *C. psittaci* genomes from diverse strains representing the 9 known serotypes revealed a core genome of 911 genes. Molecular clock analysis suggested that psittacosis is a recently emerged disease originating in South America and that *C. psittaci* strains have a history of frequently switching hosts and undergoing recombination [17].

27.3 BIOLOGY AND EPIDEMIOLOGY

Due to their inability to produce metabolic energy and synthesize ATP, Chlamydiae are dependent on the host cells (from single-celled organisms such as amoebae to other multicellular organisms) for energy including adenosine triphosphate (ATP) and nicotinamide adenine dinucleotide (NAD+).

Similar to other Chlamydiae, *C. psittaci* life cycle includes several developmental stages. With the ability to infect new hosts, but not replicate, the EB travels from an infected bird to the lungs of an uninfected bird or person in small droplets and initiates infection. Once inside the lungs, the EB enters into the endosome through phagocytosis, where it transforms into the RB, which is capable of replicating, but is unable to cause new infection, and begins replication. Two to three days after infection, the RBs convert back to the EBs, which are released back into the lung following the death of the host cell, to infect new cells, either in the same host or in a new host.

Psittacosis, the disease caused by *C. psittaci*, was first noted in Switzerland in 1879 from patients with pneumonia after exposure to tropical pet birds. In the winter of 1929–1930, outbreak of parrot fever (psittacosis) caused high mortality (20%) and led to the isolation of *C. psittaci* in Europe and the United States.

C. psittaci infection is commonly present in a wide range of domesticated and wild bird species (e.g., parrots, parakeets, canaries, turkeys, pigeons, and ducks), which may be acute, protracted, chronic, or subclinical, with intermittent shedding of the organism [18–31]. Clinical signs of psittacosis in birds (also referred to as avian Chlamydiosis) include conjunctivitis or keratoconjunctivitis, dyspnea, rales, coryza, sinusitis, diarrhea and polyuria diarrhea, lethargy, loss of appetite, weight loss, dehydration, ruffled plumage, the voiding of greenish-yellow gelatinous droppings, and reduced egg production [32,33]. It is noticeable that some distinct symptoms may be observed in different bird species, including

reduced anorexia and droopiness in parrots, parakeets, and other cage birds; rhinitis, bronchitis, and inflammation of the air sac in racing pigeons; and serous or pussy nasal and ocular discharges and encrusted eyes and nostrils in ducks. Crowding, handling, shipping, and changes in ambient conditions may activate or aggravate shedding of the organisms through feces and nasal discharges by infected birds and increase the prevalence of the disease.

A recent investigation on *C. psittaci* outbreak in Magellanic penguins (*Spheniscus magellanicus*) showed that affected penguins developed inappetence, lethargy, light green urates, and seizures. Necropsy revealed hepatomegaly and splenomegaly, and histologic lesions included necrotizing hepatitis, splenitis, and vasculitis. Pathologic examination uncovered renal and visceral gout and cardiac insufficiency [34]. In cattle, pigs, sheep, and horses, *C. psittaci* is responsible for ocular, respiratory, intestinal, and arthritic diseases as well as abortion [35,36].

C. psittaci is found in tissues, feces, and feathers of infected birds and is transmitted via the respiratory system to other animals (e.g., cats and cattle) and humans [30,37–43]. Persons (e.g., veterinarians, zoo keepers, pet shop workers, poultry processing workers, and laboratory workers) with regular contact with birds are at great risk for developing psittacosis after inhalation of *C. psittaci* from aerosolized dried avian excreta or respiratory secretions from sick birds [44,45]. Upon entering the respiratory epithelial cells, *C. psittaci* spreads via the blood stream to the reticuloendothelial system. The subsequent development of secondary bacteremia leads to lung infection. The mortality rate of *C. psittaci* infection was approximately 15%–20% prior to the advent of antimicrobial treatment and <1% with appropriate antibiotic therapy.

27.4 CLINICAL FEATURES AND PATHOGENESIS

Clinical features: *Chlamydia psittaci* is an obligate intracellular bacterium causing pulmonary psittacosis in humans and chlamydiosis in the birds, with clinical features varying from no apparent symptoms to pneumonitis, pericarditis, conjunctivitis, air sacculitis, peritonitis, and hepatosplenomegaly [46–49].

In general, human psittacosis results from exposure to infected animals. With an incubation period of 5–14 days, *C. psittaci* infection in humans may be asymptomatic or appear as an influenza-like illness that evolves into a life-threatening pneumonia [50,51]. In the first week of psittacosis, the symptoms may range from prostrating high fevers, severe headache, chill, a nonproductive cough, sore throat and mild pharyngitis, chest and abdominal pain, nausea and vomiting, diarrhea, photophobia, agitation, lethargy, arthralgias, conjunctivitis, epistaxis, leukopenia, splenomegaly, stupor, to coma. Facial rash (or Horder's spots), erythema multiforme, and erythema nodosum may also appear. The second week of psittacosis, the symptoms include continuous high fevers, cough, dyspnoea, pulmonary embolism, and infarction, with x-rays showing patchy infiltrates or a diffuse whiteout of lung fields.

In complicated cases, the disease may induce other conditions, including endocarditis, pericarditis, myocarditis, hepatitis, keratoconjunctivitis, encephalitis, Guillain–Barré syndrome, anemia secondary to hemolysis, disseminated intravascular coagulation, acute glomerulonephritis, tubulointerstitial nephritis, polyarticular arthritis, acute respiratory failure, septic shock, convulsions, seizures, coma, and death [52]. In addition, *C. psittaci* has been linked to the development and maintenance of ocular adnexal marginal zone B-cell lymphoma (OAMZL) in some geographic regions [53] and may also act as a trigger for psoriasis [54].

Petrovay et al. [50] documented two fatal cases of psittacosis that highlighted the danger of *C. psittaci* exposure to poultry-processing-plant workers. The first case involved a 69-year-old woman who had worked in a poultry processing plant for 6 months. Before hospital admission, the patient had symptoms of a bad cough, difficulty in breathing, and general malaise for 5 days, followed by fever (39°C) and diarrhea for 3 days. A chest radiograph revealed extensive pleuropneumonia in the left lung and right lower lobe infiltrates. Biochemical tests showed white blood cell count of 9.97×10^9 cells per liter, comprising 86.2% neutrophils, 8.9% lymphocytes, and 4.4% monocytes. Despite receiving intranasal oxygen and ciprofloxacin (200 mg) intravenously (i.v.), her dyspnoea deteriorated and required mechanical ventilation. The patient was then treated with antibiotic levofloxacin administered via a nasogastric tube (500 mg once daily), together with preventative anticoagulant, corticosteroids, and analgesic drugs. However, she became bradycardic and succumbed to the disease shortly afterward. Examination of postmortem samples showed an immunoglobulin M (IgM) titer of 1:16 and elevated titers of IgG (1:512) and IgA (1:64) to *C. psittaci* in serum using a microimmunofluorescence (MIF) assay and *C. psittaci* DNA in lung tissue using a touchdown nested polymerase chain reaction (PCR) method targeting the *ompA* gene. The autopsy showed extensive bronchopneumonia in the left lung, small disseminated pneumonic foci in the right lower lobe, and a massive pulmonary embolism originating from the thrombosis of pelvic veins, which might be the ultimate cause of death in this patient.

In the second case, a 48-year-old poultry-processing-plant female worker with a 2-week history of fever (38°C) and nonproductive cough was admitted to the hospital. A chest radiograph showed a homogeneous left lower lobe infiltrate and an inhomogeneous infiltrate in the whole of the right lobe. The elevated white blood cell count was 1.193×10^{10} cells/L, comprising 93.2% neutrophils, 5.4% lymphocytes, and 1.0% monocytes. Antibiotic therapy with moxifloxacin (400 mg orally once daily) and co-amoxiclav (1 g i.v. 3 times daily) was prescribed, along with intranasal oxygen, a bronchodilator, and corticosteroids. However, her conditions worsened, with the development of severe respiratory failure and hypoxia (respiratory rate 50 *breaths per min*, oxygen saturation 26%), and she required mechanical ventilation. Her serum had an IgG titer of 1:256 and an IgA titer of 1:32 to *C. psittaci*; however, no specific IgM was detected. A PCR performed on the bronchial fluid sample was positive for *C.*

psittaci DNA. Despite the broad-spectrum antibiotic treatment (moxifloxacin, rifampicin, clindamycin, ceftriaxone, and doxycycline), the patient developed toxic symptoms, multiorgan failure, extreme tachycardia, and hypoxia and died shortly. The autopsy revealed confluent pneumonia, a septic spleen and kidneys, and an alcoholic fatty liver [50].

Pathogenesis: *C. psittaci* gains entry into the human host via inhalation and infects mucosal epithelial cells and macrophages of the respiratory tract. From the lungs, the bacterium migrates to the blood stream and is transported to the liver and spleen, where it replicates. A lymphocytic inflammatory response in the alveoli and interstitial spaces induces edema, infiltration of macrophages, necrosis, and hemorrhage. Mucus plugs that develop in the alveoli may cause cyanosis and anoxia.

Several chlamydial genes and related proteins are implicated in the pathogenesis of psittacosis. These include (1) the polymorphic membrane proteins (pmps), which are important in adhesion of the EB to the host cell, molecular transport, and cell wall–associated functions; (2) genes located in the chlamydial "plasticity zone"; (3) the type III effector proteins, which help translocate effector proteins into the host cytoplasm to mediate colonization and parasitation of susceptible hosts; and (4) other genes (e.g., *incA*, *groEL*, *cpaf*, and *ftsW*) related to virulence [26,55].

Cell-mediated immunity plays a major part in immune protection and disease development of chlamydial infection. In experimental mouse model, *C. psittaci* virulent strain causes an influx of nonactivated macrophages, neutrophils, and significant end organ damage. It appears that *C. psittaci* modulates virulence by alteration of host immunity, which is conferred by small differences in the chromosome [56]. There is evidence that upon infection of blood monocytes/macrophages, *C. psittaci* may deliberately limit its replication and immediately arms itself to infect other cells elsewhere in the host, whilst using the monocytes/macrophages as a quick transport vehicle [57].

In OAMZL, *C. psittaci* may be involved in clonal selection on induced mucosa-associated lymphoid tissue with subsequent lymphoma development. Additionally, heat-shock proteins produced by *C. psittaci* may trigger both humoral and cell-mediated immune responses, generating a cross-reactivity against human proteins and other self-antigens and contributing to the altered local tolerance and chronic antigenic stimulation. The chronic antigenic stimulation associated with *C. psittaci* infection could then promote genetic instability with subsequent chromosomal abnormalities as a consequence of induced proliferation and DNA oxidative damage [58].

27.5 IDENTIFICATION AND DIAGNOSIS

As *C. psittaci* induces many nonspecific symptoms, exposure history plays a paramount role in clinical diagnosis of psittacosis. The most common finding of psittacosis in chest radiography is unilateral, lower-lobe dense infiltrate/consolidation, which may be resolved within an average of

6 weeks (range 3–20 week). The characteristic intracytoplasmic inclusions of Chlamydiae may be visualized under light microscopy in smears and tissue samples using modified Machiavello, modified Giménez, Giemsa, or modified Ziehl–Neelsen stain. In particular, the modified Ziehl–Neelsen stain makes small coccoid elementary bodies appear red/pink against a counterstained blue or green cellular background. Histologic examination may reveal tracheobronchitis and interstitial pneumonitis with air space involvement and predominant mononuclear cell infiltration. Other findings may include macrophages containing cytoplasmic inclusion bodies (Levinthal-Coles-Lillie or LCL bodies) within macrophages in bronchial alveolar lavage (BAL) fluid, focal necrosis of hepatocytes along with Kupffer cell hyperplasia, and hepatic noncaseating granulomata in the liver.

Differential diagnoses include typhus (*Rickettsia typhi* or *Rickettsia prowazekii*), typhoid fever (*Salmonella typhi*), Mycoplasma, Legionella, Q fever (*Coxiella burnetii*), brucellosis, *Chlamydophila pneumonia*, tularemia (*Francisella tularensis*), and infective endocarditis, as well as pneumonia due to other bacterial, fungal, and viral pathogens.

Laboratory confirmation of psittacosis has been largely reliant on in vitro culture and serological tests. In fact, CDC criteria for *C. psittaci* infection consist of the following: (1) confirmed cases give a positive culture result for *C. psittaci* from respiratory secretions, (2) a fourfold increase in antibody titer in two serum samples obtained by complement fixation (CF) or MIF 2 weeks apart or IgM antibodies against *C. psittaci* with a titer of 16 by MIF, and (3) possible cases show the presence of antibodies against *C. psittaci* with titers of 1:32 by CF or MIF.

It should be noted that in vitro culture of *C. psittaci* is hazardous and is only carried out in biosafety laboratories. Because *C. psittaci* is an obligate intracellular bacterium, use of cell lines (e.g., McCoy, buffalo green monkey kidney, HeLa 229, and Hep-2) and embryonated hens' eggs is necessary for its cultivation. Given the safety concern and time-consuming nature of in vitro culture techniques, serological tests have been developed for diagnosis and typing of *C. psittaci* organisms [59–62]. In general, MIF tests are more sensitive and specific than CF tests, and the use of MIF assays helps differentiate between human chlamydial species. However, cross-reactions among Chlamydial species sometimes occur due to antigenic homology [63].

In recent years, molecular methods based on the amplification and detection of *C. psittaci* nucleic acids have been increasingly applied for rapid, sensitive, and specific identification and genotyping of *C. psittaci* bacteria [3,39,64–80]. The adoption of multiplex and real-time formats further permits simultaneous differentiation of *C. psittaci* and related bacteria with instant result availability [81–84]. For example, Mitchell et al. [85] described a real-time PCR assay targeting the *ompA* gene for rapid detection and genotyping of *C. psittaci*. This test gives valuable insight into the distribution of *C. psittaci* genotype among specific hosts and provides epidemiological and epizootiological data in human and mammalian/avian cases. Pantchev et al. [82] utilized species-specific

real-time PCR assays to identify *C. psittaci*, *C. abortus*, *C. felis*, *C. caviae*, and *C. suis*. In addition, Nordentoft et al. [86] developed SYBR green-based real-time assays to detect all members of Chlamydiaceae and to differentiate the most prevalent veterinary species: *C. psittaci*, *C. abortus*, *C. felis*, and *C. caviae*. The assays proved to be more sensitive than traditional microscopic examination of stained tissue smears for identification of Chlamydiaceae.

27.6 TREATMENT AND PREVENTION

Treatment: *C. psittaci* infection in humans is treatable with antibiotics. While doxycycline hyclate and tetracycline hydrochloride are the drugs of choice for psittacosis, other antibiotics such as chloramphenicol palmitate and erythromycin may be used. Ibuprofen or acetaminophen and fluids are also administered. However, in children aged <9 years or pregnant women, tetracycline should not be prescribed, and erythromycin should be given instead. Because relapse can occur, antibiotic treatment of psittacosis must continue for at least 10–14 days after fever abates. It should be noted that chloramphenicol may cause agranulocytosis, and failures with macrolide (such as erythromycin) and quinolone have been observed occasionally. Furthermore, doxycycline is not recommended for children because it may cause tooth discoloration [87].

To treat chlamydiosis in the birds, tetracycline and enrofloxacin are often utilized. In addition, ovotransferrin (ovoTF), a natural antimicrobial protein, has been investigated for prevention of experimental *C. psittaci* infection in specific pathogen free (SPF) turkeys [88,89].

Prevention: Ventilation, cleaning, hand hygiene, and personal protective equipment represent the most effective protective measures to limit and control exposure to *C. psittaci*. High-risk individuals (such as poultry workers, zoo workers, and pet shop owners) are advised to become aware of the symptoms of psittacosis (e.g., ocular or nasal discharge, diarrhea, or low body weight) and to avoid handling newly imported or sick birds [90,91].

At the moment, no commercial vaccine against psittacosis is available. Recent studies indicated that DNA vaccines expressing the major outer membrane protein helped reduce clinical signs, macroscopic lesions, pharyngeal and cloacal excretion, and chlamydial replication in experimental birds with *C. psittaci* challenge [92–95].

27.7 CONCLUSION

Chlamydophila (Chlamydia) psittaci is an obligate intracellular bacterium that is endemic in at least 153 psittacine and 322 nonpsittacine bird species. The organism is shed in bird droppings and respiratory tract excretions and can remain infectious in litter and feces for many months. Transmission to humans and animals is mainly through inhalation of contaminated aerosols originating from feathers (wing flapping), excreta (dried), or environment (sand and cages). Handling and processing of the plumage, carcasses, and tissues from infected birds pose a significant zoonotic risk. Considering the variable and nonspecific clinical presentations of *C. psittaci* infection (e.g., high fever, chills, headache, myalgia, non-productive cough, pharyngitis, dyspnea, rash, splenomegaly, and pulse–temperature dissociation), it is important that correct diagnosis is achieved and prompt antibiotic therapy is implemented. This is crucial to decrease the mortality of psittacosis and its potential spread to the wider community. In addition, promotion of increased awareness among professional health-care workers and the general public, a broader application of new diagnostic methods, and development of effective vaccine will undoubtedly contribute to the reduction of the economic burden of this highly infectious, potentially life-threatening disease.

REFERENCES

1. Bush, R.M. and Everett, K.D.E., Molecular evolution of the *Chlamydiaceae. Int. J. Syst. Evol. Microbiol.*, 51, 203, 2001.
2. Everett, K.D.E., Bush, R.M., and Andersen, A.A., Emended description of the order *Chlamydiales*, proposal of *Parachlamydiaceae* fam. nov. and *Simkaniaceae* fam. nov., each containing one monotypic genus, revised taxonomy of the family *Chlamydiaceae*, including a new genus and five new species, and standards for the identification of organisms. *Int. J. Syst. Bacteriol.*, 49, 415–440, 1999.
3. Everett, K.D. and Andersen, A.A., Identification of nine species of the *Chlamydiaceae* using PCR-RFLP. *Int. J. Syst. Bacteriol.*, 49, 803, 1999.
4. Stephens, R.S. et al., Divergence without difference: Phylogenetics and taxonomy of *Chlamydia* resolved. *FEMS Immunol. Med. Microbiol.*, 55, 115, 2009.
5. Greub, G., International Committee on Systematics of Prokaryotes. Subcommittee on the taxonomy of the *Chlamydiae*: Minutes of the inaugural closed meeting, March 21, 2009, Little Rock, AR. *Int. J. Syst. Evol. Microbiol.*, 60, 2691–2694, 2010.
6. Voigt, A., Schöfl, G., and Saluz, H.P., The *Chlamydia psittaci* genome: A comparative analysis of intracellular pathogens. *PLoS One*, 7, e35097, 2012.
7. Grayston, J.T. et al., A new *Chlamydia psittaci* strain, TWAR, isolated in acute respiratory tract infections. *N. Engl. J. Med.*, 315, 161, 1986.
8. Van Lent, S. et al., Full genome sequences of all nine *Chlamydia psittaci* genotype reference strains. *J. Bacteriol.*, 194, 6930–6931, 2012.
9. Eaton, M.D., Beck, M.D., and Pearson, H.E., A virus from cases of atypical pneumonia: Relation to the viruses of meningopneumonitis and psittacosis. *J. Exp. Med.*, 73, 641–654, 1941.
10. Horn, M. et al., Illuminating the evolutionary history of *Chlamydiae. Science*, 304, 728–730, 2004.
11. Harkinezhad, T. et al., Prevalence of *Chlamydophila psittaci* infections in a human population in contact with domestic and companion birds. *J. Med. Microbiol.*, 58, 1207–1212, 2009.
12. Collingro, A. et al., Unity in variety—The pan-genome of the *Chlamydiae. Mol. Biol. Evol.*, 28, 3253–3270, 2011.
13. Schöfl, G. et al., Complete genome sequences of four mammalian isolates of *Chlamydophila psittaci. J. Bacteriol.*, 193, 4258, 2011.
14. Seth-Smith, H.M. et al., Genome sequence of the zoonotic pathogen *Chlamydophila psittaci. J. Bacteriol.*, 193, 1282–1283, 2011.
15. Grinblat-Huse, V. et al., Genome sequences of the zoonotic pathogens *Chlamydia psittaci* 6BC and Cal10. *J. Bacteriol.*, 193, 4039–4040, 2011.
16. Voigt, A. et al., Full-length de novo sequence of the *Chlamydophila psittaci* type strain, 6BC. *J. Bacteriol.*, 193, 2662–2663, 2011.
17. Read, T.D. et al., Comparative analysis of *Chlamydia psittaci* genomes reveals the recent emergence of a pathogenic lineage with a broad host range. *MBio*, 4, e00604–e00612, 2013.
18. Kaleta, E.F. and Taday, E.M., Avian host range of *Chlamydophila* spp. based on isolation, antigen detection and serology. *Avian Pathol.*, 32, 435, 2003.
19. Raso, T.de.F. et al., An outbreak of chlamydiosis in captive blue-fronted Amazon parrots (*Amazona aestiva*) in Brazil. *J. Zoo. Wildl. Med.*, 35, 94–96, 2004.
20. Laroucau, K., Chlamydial infections in duck farms associated with human cases of psittacosis in France. *Vet. Microbiol.*, 135, 82, 2009.
21. Pannekoek, Y. et al., *Chlamydophila psittaci* infections in the Netherlands. *Drugs Today (Barc).*, 45(Suppl B), 151–157, 2009.
22. Sharples, E. and Baines, S.J., Prevalence of *Chlamydophila psittaci* positive cloacal PCR tests in wild avian casualties in the UK. *Vet. Rec.*, 164, 16–17, 2009.
23. Beeckman, D.S. and Vanrompay, D.C., Biology and intracellular pathogenesis of high or low virulent *Chlamydophila psittaci* strains in chicken macrophages. *Vet. Microbiol.*, 141, 342–353, 2010.
24. Arraiz, N. et al., Evidence of zoonotic *Chlamydophila psittaci* transmission in a population at risk in Zuliastate, Venezuela. *Rev. Salud Publica (Bogota)*, 14, 305–314, 2012.
25. Blomqvist, M. et al., *Chlamydia psittaci* in Swedish wetland birds: A risk to zoonotic infection? *Avian Dis.*, 56, 737–740, 2012.
26. Braukmann, M. et al., Distinct intensity of host-pathogen interactions in *Chlamydia psittaci-* and *Chlamydia abortus-*infected chicken embryos. *Infect. Immun.*, 80, 2976–2988, 2012.
27. Geigenfeind, I., Vanrompay, D., and Haag-Wackernagel, D., Prevalence of *Chlamydia psittaci* in the feral pigeon population of Basel, Switzerland. *J. Med. Microbiol.*, 61, 261–265, 2012.
28. McGuigan, C.C., McIntyre, P.G., and Templeton, K., Psittacosis outbreak in Tayside, Scotland, December 2011 to February 2012. *Euro. Surveill.*, 17, 20186, 2012.
29. Piasecki, T., Chrząstek, K., and Wieliczko, A., Detection and identification of *Chlamydophila psittaci* in asymptomatic parrots in Poland. *BMC Vet. Res.*, 8, 233, 2012.
30. Sachse, K. et al., More than classical *Chlamydia psittaci* in urban pigeons. *Vet. Microbiol.*, 157, 476–480, 2012.
31. Gartrell, B.D. et al., First detection of *Chlamydia psittaci* from a wild native passerine bird in New Zealand. *N. Z. Vet. J.*, 61, 174–176, 2013.
32. Vanrompay, D., Ducatelle, R., and Haesebrouck, F., Pathogenicity for turkeys of *Chlamydia psittaci* strains belonging to the avian serovars A, B and D. *Avian Pathol.*, 23, 247–262, 1994.
33. Pilny, A.A. et al., Evaluation of *Chlamydophila psittaci* infection and other risk factors for atherosclerosis in pet psittacine birds. *J. Am. Vet. Med. Assoc.*, 240, 1474–1480, 2012.

34. Jencek, J.E. et al., An outbreak of *Chlamydophila psittaci* in an outdoor colony of Magellanic penguins (*Spheniscus magellanicus*). *J. Avian Med. Surg.*, 26, 225–231, 2012.

35. Theegarten, D. et al., *Chlamydophila* spp. infection in horses with recurrent airway obstruction: Similarities to human chronic obstructive disease. *Respir. Res.*, 9, 14, 2008.

36. Lenzko, H. et al., High frequency of *Chlamydial* co-infections in clinically healthy sheep flocks. *BMC Vet. Res.*, 7, 29, 2011.

37. Longbottom, D. and Coulter, L.J., Animal chlamydioses and zoonotic implications. *J. Comp. Pathol.*, 128, 217–244, 2003.

38. Heddema, E.R. et al., An outbreak of psittacosis due to *Chlamydophila psittaci* genotype A in a veterinary teaching hospital. *J. Med. Microbiol.*, 55, 1571–1575, 2006.

39. Vanrompay, D. et al., *Chlamydophila psittaci* transmission from pet birds to humans. *Emerg. Infect. Dis.*, 13, 1108–1110, 2007.

40. Reinhold, P., Sachse, K., and Kaltenboeck, B., *Chlamydiaceae* in cattle: Commensals, trigger organisms, or pathogens? *Vet. J.*, 189, 257–267, 2011.

41. Reinhold, P. et al., A bovine model of respiratory *Chlamydia psittaci* infection: Challenge dose titration. *PLoS One*, 7, e30125, 2012.

42. Gacouin, A. et al., Distinctive features between community-acquired pneumonia (CAP) due to *Chlamydophila psittaci* and CAP due to Legionella pneumophila admitted to the intensive care unit (ICU). *Eur. J. Clin. Microbiol. Infect. Dis.*, 31, 2713–2718, 2012.

43. Osman, K.M. et al., Prevalence of *Chlamydophila psittaci* infections in the eyes of cattle, buffaloes, sheep and goats in contact with a human population. *Transbound Emerg. Dis.*, 60, 245–251, 2013.

44. Gaede, W. et al., *Chlamydophila psittaci* infections in humans during an outbreak of psittacosis from poultry in Germany. *Zoonoses Public Health*, 55, 184, 2008.

45. Pospischil, A., From disease to etiology: Historical aspects of *Chlamydia*-related diseases in animals and humans. *Drugs Today (Barc)*, 45(Suppl B), 141–146, 2009.

46. Crosse, B.A., Psittacosis: A clinical review. *J. Infect.*, 21, 251–259, 1990.

47. Eidson, M., Psittacosis/avian chlamydiosis. *J. Am. Vet. Med. Assoc.*, 221, 1710–1712, 2002.

48. Cunha, B.A., The atypical pneumonias: Clinical diagnosis and importance. *Clin. Microbiol. Infect.*, 12(Suppl 3), 12–24, 2006.

49. Harkinezhad, T., Geens, T., and Vanrompay, D., *Chlamydophila psittaci* infections in birds: A review with emphasis on zoonotic consequences. *Vet. Microbiol.*, 135, 68–77, 2009.

50. Petrovay, F. and Balla, E., Two fatal cases of psittacosis caused by *Chlamydophila psittaci*. *J. Med. Microbiol.*, 57, 1296–1298, 2008.

51. Fraeyman, A. et al., A typical pneumonia due to *Chlamydophila psittaci*: 3 Case reports and review of literature. *Acta Clin. Belg.* 65, 192–196, 2010.

52. Fabris, M. et al., *Chlamydophila psittaci* subclinical infection in chronic polyarthritis. *Clin. Exp. Rheumatol.*, 29(6), 977–982, 2011.

53. Ferreri, A.J. et al., Chlamydial infection: The link with ocular adnexal lymphomas. *Nat. Rev. Clin. Oncol.*, 6, 658–669, 2009.

54. Stinco, G. et al., Detection of DNA of *Chlamydophila psittaci* in subjects with psoriasis: A casual or acausal link? *Br. J. Dermatol.* 167, 926–928.

55. Millman, K.L., Tavaré, S., and Dean, D., Recombination in the ompA gene but not the omcB gene of *Chlamydia* contributes to serovar-specific differences in tissue tropism, immune surveillance, and persistence of the organism. *J. Bacteriol.*, 183, 5997–6008, 2001.

56. Miyairi, I. et al., *Chlamydia psittaci* genetic variants differ in virulence by modulation of host immunity. *J. Infect. Dis.*, 204, 654–663, 2011.

57. Beeckman, D.S. and Vanrompay, D.C., Zoonotic *Chlamydophila psittaci* infections from a clinical perspective. *Clin. Microbiol. Infect.*, 15, 11–17, 2009.

58. Collina, F. et al., *Chlamydia psittaci* in ocular adnexa MALT lymphoma: A possible role in lymphoma genesis and a different geographical distribution. *Infect. Agent Cancer*, 7, 8, 2012.

59. Andersen, A.A., Serotyping of *Chlamydia psittaci* isolates using serovar-specific monoclonal antibodies with the microimmunofluorescence test. *J. Clin. Microbiol.*, 29, 707–711, 1991.

60. Wood, M.M. and Timms, P., Comparison of nine antigen detection kits for diagnosis of urogenital infections due to *Chlamydia psittaci* in koalas. *J. Clin. Microbiol.*, 30, 3200, 1992.

61. Vanrompay, D. et al., Evaluation of five immunoassays for detection of *Chlamydia psittaci* in cloacal and conjunctival specimens from turkeys. *J. Clin. Microbiol.*, 32, 1470, 1994.

62. Toyokawa, M. et al., Severe *Chlamydophila psittaci* pneumonia rapidly diagnosed by detection of antigen in sputum with an immunochromatography assay. *J. Infect. Chemother.*, 10, 245–249, 2004.

63. She, R.C. et al., Correlation of *Chlamydia* and *Chlamydophila* spp. IgG and IgM antibodies by microimmunofluorescence with antigen detection methods. *J. Clin. Lab. Anal.*, 25, 305–308, 2011.

64. Kaltenboeck, B., Kousoulas, K.G., and Storz, J., Detection and strain differentiation of *Chlamydia psittaci* mediated by a two-step polymerase chain reaction. *J. Clin. Microbiol.*, 29, 1969, 1991.

65. Tong, C.Y. and Sillis, M., Detection of *Chlamydia pneumoniae* and *Chlamydia psittaci* in sputum samples by PCR. *J. Clin. Pathol.*, 46, 313, 1993.

66. Yoshida, H. et al., Differentiation of *Chlamydia* species by combined use of polymerase chain reaction and restriction endonuclease analysis. *Microbiol. Immunol.*, 42, 411, 1998.

67. Madico, G. et al., Touchdown enzyme time release-PCR for detection and identification of *Chlamydia trachomatis*, *C. pneumoniae*, and *C. psittaci* using the 16S and 16S–23S spacer rRNA genes. *J. Clin. Microbiol.*, 38, 1085, 2000.

68. DeGraves, F.J. et al., Quantitative detection of *Chlamydia psittaci* and *C. pecorum* by high-sensitivity real-time PCR reveals high prevalence of vaginal infection in cattle. *J. Clin. Microbiol.*, 41, 1726, 2003.

69. Sachse, K. et al., Detection of *Chlamydia suis* from clinical specimens: Comparison of PCR, antigen ELISA, and culture. *J. Microbiol. Methods*, 54, 233, 2003.

70. Sachse, K. et al., Recent developments in the laboratory diagnosis of chlamydial infections. *Vet. Microbiol.*, 135, 2, 2009.

71. Geens, T. et al., Sequencing of the *Chlamydophila psittaci* ompA gene reveals a new genotype, E/B, and the need for a rapid discriminatory genotyping method. *J. Clin. Microbiol.*, 43, 2456–2461, 2005.

72. Geens, T. et al., Development of a *Chlamydophila psittaci* species-specific and genotype-specific real-time PCR. *Vet. Res.*, 36, 787, 2005.

73. Chahota, R. et al., Genetic diversity and epizootiology of *Chlamydophila psittaci* prevalent among the captive and feral avian species based on VD2 region of ompA gene. *Microbiol. Immunol.*, 50, 663–678, 2006.

74. Borel, N. et al., Direct identification of *Chlamydiae* from clinical samples using a DNA microarray assay—A validation study. *Mol. Cell. Probes*, 22, 55, 2008.

75. Robertson, T. et al., Characterization of *Chlamydiaceae* species using PCR and high resolution melt curve analysis of the 16 SrRNA gene. *J. Appl. Microbiol.*, 107, 2017–2028, 2009.

76. Van Droogenbroeck, C. et al., Simultaneous zoonotic transmission of *Chlamydophila psittaci* genotypes D, F and E/B to a veterinary scientist. *Vet. Microbiol.*, 135, 78–81, 2009.

77. Van Droogenbroeck, C. et al., Evaluation of bioaerosol sampling techniques for the detection of *Chlamydophila psittaci* in contaminated air. *Vet. Microbiol.*, 135, 31–37, 2009b.

78. Dickx, V. et al., *Chlamydophila psittaci* in homing and feral pigeons and zoonotic transmission. *J. Med. Microbiol.*, 59, 1348–1353, 2010.

79. Pannekoek, Y. et al., Multi locus sequence typing of *Chlamydia* reveals an association between *Chlamydia psittaci* genotypes and host species. *PLoS One*, 5, e14179, 2010.

80. Frutos, M.C. et al., Genotyping of *C. psittaci* in central area of Argentina. *Diagn. Microbiol. Infect. Dis.*, 74, 320–322, 2012.

81. Menard, A. et al., Development of a real-time PCR for the detection of *Chlamydia psittaci*. *J. Med. Microbiol.*, 55, 471, 2006.

82. Pantchev, A. et al., Detection of all *Chlamydophila* and *Chlamydia* spp. of veterinary interest using species-specific real-time PCR assays. *Comp. Immunol. Microbiol. Infect. Dis.*, 33, 473–484, 2010.

83. Okuda, H. et al., Detection of *Chlamydophila psittaci* by using SYBR green real-time PCR. *J. Vet. Med. Sci.*, 73(2), 249–254, 2011.

84. Tramuta, C. et al., Development of a set of multiplex standard polymerase chain reaction assays for the identification of infectious agents from aborted bovine clinical samples. *J. Vet. Diagn. Invest.*, 23, 657–664, 2011.

85. Mitchell, S.L. et al., Genotyping of *Chlamydophila psittaci* by real-time PCR and high-resolution melt analysis. *J. Clin. Microbiol.*, 47, 175–181, 2009.

86. Nordentoft, S., Kabell, S., and Pedersen, K., Real-time detection and identification of *Chlamydophila* species in veterinary specimens by using SYBR green-based PCR assays. *Appl. Environ. Microbiol.*, 77, 6323–6330, 2011.

87. Ferreri, A.J. et al., *Chlamydophila psittaci* eradication with doxycycline as first-line targeted therapy for ocular adnexae lymphoma: Final results of an international phase II trial. *J. Clin. Oncol.*, 30, 2988–2994, 2012.

88. Van Droogenbroeck, C. et al., Evaluation of the prophylactic use of ovotransferrin against chlamydiosis in SPF turkeys. *Vet. Microbiol.*, 132, 372–378, 2008.

89. Van Droogenbroeck, C. and Vanrompay, D., Use of ovotransferrin on a turkey farm to reduce respiratory disease. *Vet. Rec.*, 172, 71, 2013.

90. Smith, K.A. et al., Compendium of measures to control *Chlamydophila psittaci* (formerly *Chlamydia psittaci*) infection among humans (psittacosis) and pet birds. *J. Am. Vet. Med. Assoc.*, 226, 532–539, 2005.

91. Deschuyffeleer, T.P. et al., Risk assessment and management of *Chlamydia psittaci* in poultry processing plants. *Ann. Occup. Hyg.*, 56, 340–349, 2012.

92. Verminnen, K. et al., Protection of turkeys against *Chlamydophila psittaci* challenge by DNA and rMOMP vaccination and evaluation of the immunomodulating effect of 1 alpha, 25-dihydroxyvitamin D(3). *Vaccine*, 23, 4509–4516, 2005.

93. Verminnen, K. and Vanrompay, D., *Chlamydophila psittaci* infections in turkeys: Overview of economic and zoonotic importance and vaccine development. *Drugs Today (Barc).*, 45(Suppl B), 147–150, 2009.

94. Verminnen, K. et al., Vaccination of turkeys against *Chlamydophila psittaci* through optimised DNA formulation and administration. *Vaccine*, 28, 3095–3105, 2010.

95. Harkinezhad, T., Schautteet, K., and Vanrompay, D., Protection of budge rigars (*Melopsittacus undulatus*) against *Chlamydophila psittaci* challenge by DNA vaccination. *Vet. Res.*, 40, 61, 2009.

28 Clostridium (C. botulinum and C. perfringens)

Mar Rodríguez, Emilio Aranda, María G. Córdoba,
María J. Benito, Alicia Rodríguez, and Juan J. Córdoba

CONTENTS

28.1 INTRODUCTION

Species of *Clostridium* genus are widely present in the environment (e.g., soil, sewage, and marine sediments) and in the gastrointestinal tract of humans and domestic and feral animals. Although most *Clostridium* species are saprophytes, 34 species have been considered pathogenic to man and animals. Among the main pathogenic species are *C. botulinum* and *C. perfringens*. *C. botulinum* forms the highly potent botulinum neurotoxin (BoNT) that is responsible for botulism, a severe disease with a high fatality rate. There are three major types of botulism in humans: foodborne botulism, infant/intestinal (adult) botulism, and wound botulism.[1] *C. botulinum* is a group of four physiologically and phylogenetically distinct clostridia that share the common feature of producing neurotoxins. Particularly two members of this group, proteolytic type A, B, and F and nonproteolytic *C. botulinum* type B, E, and F, cause botulism food poisoning.[2] The foodborne botulism is an intoxication caused by the consumption of preformed toxin, while infant/intestinal (adult) botulism and wound botulism are infections involving toxin formation in situ.[1] *C. perfringens* is considered to be one of the common microorganisms that causes human and veterinary diseases.[3] In addition, *C. perfringens* produces around 15 different toxins.

28.2 CLASSIFICATION AND MORPHOLOGY

The *Clostridium* genus comprises a highly heterogeneous group of Gram-positive, endospore-producing, rod-shaped anaerobes, very diverse in both their physiology and genetics.[4] The species of this genus are grouped in 19 different clusters.[5] More than half of the pathogenic species are members of cluster I, including those considered the major pathogenic agents, that is, *C. botulinum* and *C. perfringens*. Although most *Clostridium* species are saprophytes, the genus includes several infamous human and veterinary pathogens that secrete various toxin types responsible of dreaded disease as botulism and food poisoning.

Species of *Clostridium* genus are ubiquitously distributed in the environment (e.g., soil, sewage, and marine sediments) and exist in the gastrointestinal tract of humans and domestic and feral animals.[6,7]

Clostridium species are classified based on the shape of vegetative cells, cell wall structure, endospore formation, biochemical properties, 16S rRNA sequence homology, mol% G + C content of DNA, and polymerase chain reaction (PCR) amplification of spacer regions of 16S and 23S rRNA genes. Genome sequences of various species including *C. botulinum* and *C. perfringens* have been completed, which will help understand the molecular nature of these organisms.[8–10]

TABLE 28.1

Characteristic of the Neurotoxinogenic Clostridia

Species	Group	Toxin Type	Proteolysis	Glucose	Lipase
C. botulinum	I	A, B, F	+	+	+
C. botulinum	II	B, E, F	–	+	+
C. botulinum	III	C, D	+/–	–	–
C. argentinense	IV	G	+	–	–
C. butyricum	V	E	–	–	–
C. baratii	VI	F	–	+	–

Source: Quinn, C.P. and Minton, N.P., Clostridial neurotoxins, in *Clostridia: Biotechnology and Medical Applications*, Bahl, H. and Dürre, P., eds., Wiley-VCH Verlag GmbH, Weinheim, Germany, 2001.

TABLE 28.2

Toxins of *C. perfringens* Active on the Gastrointestinal Tract

Toxin	Molecular Action	Associated Diseases
CPA	Phospholipase C	Gas gangrene
CPB	Pore former	Necrotizing enteritis
CPB2	Cytotoxin	Animal enterotoxemias
ETX	Cytotoxin	Animal enterotoxemias
ITX	ADP ribosylates actin	Animal enterotoxemias
NetB	Pore former	Necrotizing enteritis
CPE	Affect tight junctions	Food poisoning

Source: McClane, B.A. and Rood, J.I., Clostridial toxins involved in human enteric and histotoxic infections, in *Clostridia: Biotechnology and Medical Applications*, Bahl, H. and Dürre, P., eds., Wiley-VCH Verlag GmbH, Weinheim, Germany, 2001.

BoNT is produced by some species of the *Clostridium* genus, particularly *C. botulinum* but also some strains of *C. baratii* and *C. butyricum*. The BoNTs intoxicate the peripheral nervous system where they inhibit calcium-dependent secretion of acetylcholine at the neuromuscular junction.[11] BoNTs can be separated into seven distinct serotypes (A–G) on the basis of the immunological characteristics. Moreover, four immunologically or genetically distinct BoNT/A subtypes, five BoNT/B subtypes, and six BoNT/E subtypes have been identified.[12,13] Some strains of *C. butyricum* and *C. baratii* have been shown to produce type E and F neurotoxins, respectively. *C. botulinum* is subdivided into four distinct groups based on physiological and genetic criteria (Table 28.1); only group I (proteolytic *C. botulinum*) strains and group II (nonproteolytic *C. botulinum*) strains cause botulism in humans, while group III is involved in animal botulism. Group IV strains (named *C. argentinense*) have not generally associated with illness.[14]

Phenotypically, groups I and II *C. botulinum* strains differ from each other significantly. Group I organisms seem to be more terrestrial origin and are present in temperature climates, whereas group II strains, particularly those BoNT/E-producing strains, are frequently found in aquatic environments in the Northern Hemisphere.[14]

The strains of group I *C. botulinum* are proteolytic and capable of utilizing amino acids as an energy source. The minimum growth temperature of group I *C. botulinum* is 13°C–16°C and the optimum one is 37°C–42°C.[15] Under otherwise favorable conditions, growth is typically inhibited by a water activity of 0.94, corresponding to approximately 10% NaCl (w:v) in brine. Growth does not occur in pH values lower than 4.5. Group II *C. botulinum* is a nonproteolytic and saccharolytic microorganism. The minimum growth temperature is 3°C.[16] The inhibitory water activity is 0.97, corresponding to 5% NaCl (w:v) in brine. The spores of group I *C. botulinum* possess a very high heat resistance higher than that of group II *C. botulinum*. As groups I and II strains differ in their epidemiologies, discrimination between the two groups should be done whenever an outbreak strain is isolated.

C. perfringens (formerly called *C. welchii*) produces clostridial toxins activated on the gastrointestinal tract (Table 28.2). Among the 15 toxins produced by *C. perfringens*, alpha (CPA), beta (CPB), epsilon (ETX), and iota (ITX) toxins are the four major toxins present in five different toxinotypes (A–E) of the bacterium.[16,17] In addition, *C. perfringens* may produce other toxins such as *C. perfringens* enterotoxin (CPE), *C. perfringens* beta2 toxin (CPB2), and *C. perfringens* NetB toxin.[18,19] CPE causes *C. perfringens* type A food poisoning as well as antibiotic-associated diarrhea and sporadic diarrheas in humans. This toxin is also responsible for enteric diseases and enterotoxemias in animals.[20,21]

Growth occurs over the temperature range from 12°C to 50°C although it is very slow below about 20°C. At its temperature optimum, 43°C–47°C, growth is extremely rapid with a generation time of only 7.1 min at 41°C.[22] Under otherwise favorable conditions, the minimum water activity for growth ranges between 0.95 and 0.97, and it will not grow in the presence of 6% NaCl (w:v). Growth does not occur in pH values lower than five. The heat resistance of spores at 100°C shows a wide interstrain variation with recorded values from 0.31 min to more than 38 min. This may, in part, be due to factors that cause spore wet heat resistance.[23]

28.3 BIOLOGY AND EPIDEMIOLOGY

Diseases caused by *C. botulinum* are prevalent all over the world. In 1999/2000, more than 2500 cases of foodborne botulism were reported in Europe, with a high incidence in Armenia, Azerbaijan, Belarus, Georgia, Poland, Russia, Turkey, and Uzbekistan.[24] A smaller but significant number of cases are reported annually in France, Germany, Italy, China, and the United States.[24] It should be noted that the true incidence of foodborne botulism is likely to be much higher, with underreporting an issue. Foodborne botulism is not reportable in all countries and the efficiency of investigating potential outbreaks also varies from country to country.[25] Outbreaks of foodborne botulism have involved both home-prepared foods and commercially prepared foods, with proteolytic *C. botulinum* most frequently implicated. Outbreaks of foodborne botulism have been produced more frequent both at home and food commercially prepared contaminated with proteolytic strains of *C. botulinum* or nonproteolytic *C. botulinum*.[1]

Foodborne botulism involving proteolytic *C. botulinum* most frequently involves type A or type B neurotoxin. In March 2006, a very large outbreak involving 209 cases (134 hospitalized, 42 required mechanical ventilation) was associated with consumption of home-canned bamboo shoots in Thailand.[26,27] An inadequate heat treatment was given, permitting survival of spores of proteolytic *C. botulinum* type A, that was followed by growth and neurotoxin formation during ambient storage. Temperature abuse of foods intended to be stored chilled has also been associated with foodborne botulinum. In 2006, a severe outbreak in Canada and the United States was associated with commercial chilled carrot juice. Six people were affected; all required mechanical ventilation, one of whom died, and two were still dependent on mechanical ventilation 1 year after the initial intoxication.[28,29] In August 2008, a severe outbreak in France was associated with temperature-abused commercial chicken enchiladas. It is estimated that just 10 mg of food may have constituted a lethal dose and both people affected required mechanical ventilation.[30]

Foodborne botulism involving nonproteolytic *C. botulinum* most frequently involves type B or type E neurotoxin. The original strain isolated by van Ermengem in 1895 was probably nonproteolytic *C. botulinum* type B, and many type B cases in Europe are still due to strains of nonproteolytic *C. botulinum*.[31,32] Recorded outbreaks of botulism involving nonproteolytic *C. botulinum* have been most frequently associated with meat, fish, and homemade foods prepared. One common cause is time and/or temperature abuse of homemade or commercial refrigerated foods.[1] In 1991, a large outbreak in Egypt associated with commercially produced uneviscerated salted fish ("faseikh") affected more than 91 people (18 fatally).[33] It is likely that temperature abuse of the fish, prior to salting, enabled growth and neurotoxin formation by nonproteolytic *C. botulinum* type E. In September 2009, three members of the same family were affected by botulism in southeast France.[34] All of them had consumed commercial vacuum-packed hot-smoked whitefish. Nonproteolytic *C. botulinum* type E had grown and formed neurotoxin in the product, probably during temperature abuse.[34]

The *C. perfringens*- and CPE-mediated food poisoning is among the most common foodborne illnesses in the industrialized countries.[35] Outbreaks typically involve a large number of victims and are associated with temperature-abused meat or poultry dishes. Optimal conditions for food poisonings arise when food contaminated with CPE-positive *C. perfringens* spores is slowly chilled or held or served at a temperature range of 10°C–54°C, allowing germination and rapid growth of *C. perfringens*.[36] Abdominal cramps and diarrhea appear within 8–12 h of ingestion, followed by recovery within 24 h. Fatalities are rare but possible in elderly

or debilitated humans.[35] Regan et al.[37] described an outbreak of *C. perfringens* food poisoning in two hospitals. They observed a statistically significant association between the consumption of pork roast and disease. *C. perfringens* type A was isolated from samples of precooked vacuum-sealed pork supplied by a local meat producer. Failures were observed in the food production process in the factory. The cuts of meat were too large and equipment to ensure rapid cooling of cooked meat had not been used. Similarly in 1989, there was an outbreak of diarrhea in 58 elderly residents in a hospital and two deaths. The causal agent was identified as *C. perfringens* type A, serotype TW23, and the source of the outbreak was found to be reheated insufficiently minced beef.[38]

Sudden infant death syndrome (SIDS) accounts for unexpected deaths in infants under the age of 1 year.[39] Strains of *C. perfringens* type A have been discovered to be commonly present in the intestines of babies dying with SIDS.[40–42] The enterotoxin produced by *C. perfringens* in the intestine is absorbed into the systemic circulation and leads to rapid death in SIDS.[43]

28.4 CLINICAL FEATURES AND PATHOGENESIS

C. botulinum produces BoNT, the most poisonous toxin known to man and the causative agent of botulism, a severe disease of humans and animals. BoNT is a single polypeptide chain of around 150 kDa. After nicking by bacterial protease (or gastric proteases), it turns into a light chain of 50 kDa and a heavy chain of 100 kDa. Being specific for peripheral nerve endings at the point where a motor neuron stimulates a muscle, BoNT binds to the neuron and prevents the release of acetylcholine across the synaptic cleft. The heavy chain mediates binding to presynaptic receptors via its carboxy terminus and forms a channel through the membrane of the neuron via its amino terminus, facilitating entrance of the light chain. Inside a neuron, the toxin specifically cleaves SNARE complex proteins (soluble NSF-attachment protein receptors; NSF stands for *N*-ethylmaleimide-sensitive fusion protein), synaptobrevin (VAMP), synaptosomal protein (SNAP-25), and syntaxin, which are involved in neurotransmitter release from synaptic vesicles.[16] This abolishes the ability of affected cells to release a neurotransmitter, paralyzing thus the motor system.

In structural and functional terms, BoNT is similar to tetanus toxin, as both are zinc-dependent endopeptidases that cleave a set of proteins in the synaptic vesicles of neurons involved in the excretion of neurotransmitters. However, while BoNT displays a preference for stimulatory motor neurons at a neuromuscular junction in the peripheral nervous system, causing weakness or flaccid paralysis, tetanus toxin (tetanospasmin) acts on inhibitory motor neurons of the central nervous system, causing rigidity and spastic paralysis.

Clinically, botulism often shows one of the four following manifestations: (1) foodborne botulism, which is caused by the ingestion of foods containing BoNT (as little as 30 ng of neurotoxin can be fatal[24]); (2) infant botulism, which occurs in infants under 1 year of age after ingestion of *C. botulinum* spores that then colonize the intestinal tract and produce BoNT; (3) wound botulism, which is caused by infection of *C. botulinum* in a wound, with BoNT being produced and spread to the body via the bloodstream; and (4) intestinal colonization of *C. botulinum* in adults, which resembles infant botulism in its etiology and occurs when competing bacteria in the normal intestinal microbiota have been suppressed (e.g., by antibiotic treatment). Foodborne botulism is most commonly caused by group I and group II *C. botulinum*, while infant and wound botulism are most frequently caused by group I *C. botulinum*.[44]

The paralysis typically starts in ocular and facial nerves, then proceeds to the throat, chest, and extremities, potentially leading to respiratory difficulty and death by asphyxia. The fatal paralytic condition in humans and various animal species is known as "botulism."

C. perfringens type A causes gas gangrene (myonecrosis), diarrhea, and foodborne illness in humans,[17] whereas type B and type D strains are responsible for fatal enterotoxemia in domestic animals and occasionally in humans[45] and type C strains induce severe necrotic enteritis in humans. The type E strains of *C. perfringens* have rarely been isolated in humans.[45] The role of CPE in foodborne illness has been well established. The production of enterotoxins is linked to the sporulation process, which occurs in the small intestine following the consumption of large numbers of cells from temperature-abused foods.[46] These enterotoxins produced a severe pathogenic effect due to their ability to generate pore in eukaryotic cells, which causes changes in the cellular membrane permeability, resulting in death of the cells via either the apoptotic or oncotic pathways due to calcium influx through CPE pores.[47] Enterotoxins produced by *C. perfringens* are constituted by a single polypeptide of 35 kDa (containing 319 amino acids), which binds to protein receptors belonging to the claudin family (proteins involved in tight junction structure and function) to form a large complex of 155 kDa. The continuing presence of these toxins in intoxicated cells leads to the formation of another large complex of 200 kDa together with a tight junction protein named occludin. These enterotoxin-containing complexes are responsible for the mentioned changes in cellular membrane permeability, resulting in death of the cells.

28.5 IDENTIFICATION AND DIAGNOSIS

C. botulinum produces well-defined and very distinctive symptoms of *C. perfringens* and its differentiation relies on the use of laboratory procedures. Various physiological and morphological methods have been traditionally used in taxonomic determination of *Clostridium*. However, traditional identification methods based on phenotypic properties are laborious and lack discriminatory power, and misidentification may occur occasionally. Molecular assays have become widely available for diagnostic microbiology, spurred by technological developments and commercial profit motives, and let results in time to be of clinical value in decision-making for the patient or for infection prevention.[48] Progress in

the molecular biology in the last decade has opened up possibilities of characterizing *Clostridium* at the genomic level.[49–51] While DNA sample preparation methods have improved over the last few decades, current methods are still time-consuming and labor intensive. Strotman et al.[52] described a method that replaces the multiple wash and centrifugation steps required by traditional DNA sample preparation methods with a single step.

28.5.1 *C. BOTULINUM*

For implementing early supportive care in the treatment of BoNTs outbreaks such as early administration of antitoxin, it is necessary to have rapid and accurate methods to detect *C. botulinum* and BoNTs.

28.5.1.1 Conventional Techniques

The standard detection and isolation of *C. botulinum* is based on the presumptive identification of the toxigenic strains through the mouse bioassay injected with culture supernatant of cultured microorganisms or with samples from patients exhibiting symptoms of botulism.[7,14] Although this method is highly sensitive and specifically requires specialized facilities and staff. In addition, it is time-consuming and expensive, raises ethical concerns due to numbers of animals, and can be carried out only by a select number of laboratories. In addition, to determine the serotype, a second, independent neutralization assay is required,[53] indicating that the antigenicity of BoNT/A partially disappeared with formalin treatment and reactivity of the heavy chain to some monoclonal antibodies was influenced by nontoxic components. In the event of a suspected BoNT contamination event, the mouse bioassay, while extremely sensitive, does not meet the need of emergency responders.[48] All these disadvantages are described by Sibley et al.[51]

Culture of *C. botulinum* requires strict anaerobic conditions. All culture media must be deoxygenated by heating in a boiling water bath or by a continuous flow of an anaerobic gas mixture, and the use of anaerobic jars and anaerobic workstations is necessary for successful diagnostics.[14] Reddy et al.[54] studied the effect of media, additives, and incubation conditions on the recovery of *C. botulinum* spores, and Mato Rodriguez and Alatossava[55] studied the effects of copper on germination, growth, and sporulation of *Clostridium* spp. The detection of BoNTs production in vitro is, therefore, a preferred approach for the identification of *C. botulinum*, including its distinction from genetically close relatives such as *C. sporogenes* and *C. novyi*.[56]

To enhance the isolation of *C. botulinum* from clinical samples (e.g., serum and feces), various sample pre-preparation steps may be used including (1) ethanol pretreatment to recover spores and eliminate vegetative bacteria; (2) heating to discard non-spore-forming bacteria (i.e., 80°C for 10 min for group I spores or 60°C for 10–20 min for group II spores); and (3) treatment with lysozyme (5 µg/mL) or other heat-resistant lytic enzymes to facilitate germination of heat-stressed spores.[14,57] In the germination of spores, the effect of the spore density system or proximity on the time of germination could be of great utility.[57]

The cultivation of *C. botulinum* in liquid media could be carried out in chopped-meat–glucose–starch medium; cooked-meat medium; broths containing various combinations of tryptone, peptone, glucose, yeast extract, and trypsin; reinforced clostridial medium; and fastidious anaerobe broth.[54,58–60] Pretreatment with some of the previously mentioned liquid culture media is of great importance to assure viability of spores, as it has been reported by Stringer et al.[61] However, all of these media are nonselective and thus allow the growth of a range of other bacteria.

To identify *C. botulinum*, unselective plating media such as blood agar and egg yolk agar (EYA)[62] are commonly used, since it could enable the typical lipase produced by this microorganism. However, other clostridia species have lipase and may therefore confuse the identification.[63] EYA medium alone does not contain any inhibitory compounds, but it has been reported as supplemented medium with cycloserine, sulfamethoxazole, and trimethoprim for identifying select group I of *C. botulinum*.[64,65]

The selection of the correct incubation temperature is essential to differentiate strains from the different physiologies of group I and II of *Clostridium*. Group I strains grow optimally at 35°C–37°C, whereas temperatures of 25°C–30°C or even lower favor growth of group II strains.[66]

The quantification of *C. botulinum* by the use of plate count procedures in samples containing other bacteria is difficult, since the prevalence of *C. botulinum* in naturally contaminated samples is generally low (10–1000 spores/kg) and proper selective media are not available.[14]

Detection of *C. botulinum* by culture media should be complemented by the detection of the BoNT toxin. To detect the BoNT, several procedures such as enzyme-linked immunosorbent assay (ELISA)[67,68] or mass spectrometric (MS)-based endopeptidase methods[69] have been reported. Recently, surface plasmon resonance (SPR) sensors have been also applied for detecting BoNTs serotypes A, B, and F.[70,71] This method shows a label-free biosensor assay for BoNT/B in food and human serum based on protein chip assay. Lastly, Ching et al.[72] have reported an immunochromatographic test strip for the detection of BoNT serotypes A and B, and Liu et al.[73] have developed an immunobiochemical assay to detect the BoNT. In addition, an ELISA and Endopep-MS analyses using fluorogenic reporters to detect BoNT in drinking water have been reported.[74,75]

28.5.1.2 Molecular Techniques

PCR represents the most popular molecular technique for microbial detection as the sensitivity of PCR can be as low as one copy of the target sequence by using two oligonucleotide primers.[76,77] Fakruddin et al.[49] described improvements and alternatives to the PCR such as loop-mediated isothermal amplification (LAMP), nucleic acid sequence-based amplification (NASBA), self-sustained sequence replication (3SR), and rolling circle amplification (RCA).

PCR detection of *C. botulinum* often targets the BoNT genes, although other types of toxin genes are also of diagnostic value.[14] Use of multiplex PCR assays enables the simultaneous detection of two or more types of BoNT genes.[78–80] However, PCR detection of the neurotoxin genes does not provide details on the physiological group and epidemiology of *C. botulinum* isolates. For differentiation between group I (proteolytic) and group II (nonproteolytic) strains of *C. botulinum*, several molecular typing methods are useful.[81–83] Furthermore, Janda and Whitehouse[84] have developed a multilocus PCR followed by electrospray ionization MS to identify *C. botulinum*, and Fach et al.[85] have described an innovative molecular detection tool based on the GeneDisc cycler for tracking all types A, B, E, F, and other BoNTs.

PCR offers the possibility to detect any of toxic strain regardless type of *C. botulinum* and BoNT produced. This strategy is possible using universal primers that recognize all BoNT genes, but the nucleotide diversity among the BoNT gene may present a potential problem.[86] Nonetheless, BoNTs are generated as part of a progenitor toxin complex and a conserved component among serotypes is the nontoxic nonhemagglutinin (NTNH). East and Collins[87] demonstrated that the gene encoding NTNH reveals a high level of similarity and is present in all strains that produce BoNTs and absent from strains that are nontoxic. Aranda et al.[76] developed a PCR protocol to detect all BoNTs-producing strains using a single set of degenerate primers. This protocol yields a single PCR product of 1.1 kbp in agarose gel providing a more specific detection than hybridization with a BoNT probe.

Several protocols such as ELISA and immuno-PCR based on protein antibody microarray have been recently reported to detect BoNT neurotoxins A, B, C, D, E, and F in complex clinical, food, and environmental samples at higher sensitivity than mouse bioassay.[88,89]

Real-time PCR, biosensors, and DNA microarray offer the possibility of continuous and real-time monitoring of the environment for the presence of infectious agents. Therefore, several real-time PCR protocols[90] along with biosensor technology[91] and DNA microarray[92] have been reported to detect *Clostridium*. More recently, Fenicia et al.[93] proposed a real-time PCR method for detecting and typing BoNT-producing *C. botulinum* types A, B, E, and F in clinical, food, and environmental samples and thus support its use as an international standard method.

To avoid problems of false positives in the detection of naked DNA derived from dead and degrading cells, reverse-transcription PCR (RT-PCR) could be used as an alternative approach to ensure that only live cells are detected. RT-PCR protocols for detecting *C. botulinum* have been described.[94–96] Furthermore, Chung et al.[97] reported a multiplex ligation-dependent probe amplification (MLPA), a PCR-based method involving in electrophoretic separation of amplicons, which has been proposed as an alternative method of genetic marker detection.

To type *C. botulinum* isolates, other PCR procedures such as PCR-RFLP, RAPD, Q-PCR, and Rep-PCR are used.[14] Other nonamplified molecular procedures for the characterization of *C. botulinum* consist of DNA sequencing, pulsed-field gel electrophoresis (PFGE), and ribotyping.[14]

These methods are only used to type *C. botulinum* isolates, but they do not allow detecting this microorganism directly from clinical or food samples.

28.5.2   *C. PERFRINGENS*

The detection of genes encoding *C. perfringens* toxins or the corresponding toxins is the most accepted criterion in establishing a definitive diagnosis of this microorganism.[3,98] Several procedures including conventional and molecular methods have been described.

28.5.2.1   Conventional Techniques

To culture and detect *C. perfringens*, blood agar plates are used. In this medium, this microorganism produces flat, rough, and translucent colonies, with regular or irregular margins, an inner zone of hemolysis, and an outer zone of less complete clearing. Occasionally, strains without inner zone hemolysis are seen.[77] On Nagler agar containing 5%–10% egg yolk, *C. perfringens* generates a characteristic white precipitate as a result of its alpha toxin interacting with the lipids in egg yolk.[16]

Furthermore, several solid media have been reported as able to detect and isolate *C. perfringens* from different environmental samples. These techniques include neomycin blood agar, Shahidi–Ferguson *perfringens* (SFP) agar, sulfite polymyxin sulfadiazine (SPS) agar, oleandomycin polymyxin sulfadiazine *perfringens* (OPSP) agar, tryptone sulfite neomycin (TSN) agar, and tryptose sulfite cycloserine (TSC) agar with or without egg yolk.[99–101] Isolation of typical colonies from media is followed by confirmation of *C. perfringens* colonies with different tests to reveal phenotypic characteristics. Observation of nitrate reduction, gelatin liquefaction, lactose fermentation, or lack of motility in suspected *C. perfringens* isolates is usually enough to differentiate *C. perfringens* from other microorganism.[63,77] Identification of *C. perfringens* may be also carried out by using biochemical test kits such as API.[101] Several standard international methods are available to differentiate *C. perfringens*, including methods published by the Association of Official Analytical Chemists (AOAC), the International Standardization Organization (ISO), and the Nordic Committee on Food Analysis (NCFA).

Given that not all *C. perfringens* produce toxins and consequently not all strains are related to diseases or cause food poisonings, the confirmation of toxins productions is the most accepted criterion in conventional diagnosis.[102] Among the techniques for detection of the major toxins of *C. perfringens* are the mouse neutralization test (MNT),[102] ELISAs,[103–105] counterimmnunoelectrophoresis,[106] and latex agglutination test.[107,108] Lastly, Seyer et al.[109] reported a method to detect *C. perfringens* toxins in complex foods and biological matrixes by immunopurification and ultraperformance liquid chromatography–tandem MS.

28.5.2.2   Molecular Techniques

Many PCR protocols focus on individual *C. perfringens* toxin genes,[45] in particular the *cpe* gene[110,111] and *cpa* gene[112]

have been reported. In addition, multiplex PCR protocols targeting several genes have also been reported.[113–116] Baums et al.[117] developed a reliable species-specific multiplex PCR for the detection of the *cpa, cpb, cpb2, cpe, etx,* and *iap* genes of *C. perfringens* isolates without DNA purification. Furthermore, Joshy et al.[118] reported a multiplex PCR to simultaneously detect the enterotoxin gene of *C. perfringens* and BoNT genes of *C. botulinum.* In addition to PCR, other approaches such as DNA microarray have also been reported for detecting toxin genes of *C. perfringens.*[3]

The real-time PCR allows quantifying the copy numbers of plasmid-borne toxin genes.[119] Gurjar et al.[120] designed a dual-labeled fluorescence hybridization probe (TaqMan®)–based real-time multiplex PCR assay for the detection of toxin genes α (*cpa*), β (*cpb*), ι (*iap*), ε (*etx*), and β 2 (*cpb2*) and enterotoxin (*cpe*) of *C. perfringens* directly from cattle feces. Similarly, three real-time fluorogenic (TaqMan®) multiplex PCRs have been reported for the detection of six toxic genes of *C. perfringens.*[121,122] Recently, Chon et al.[123] developed an accurate real-time PCR for the detection and quantification of *C. perfringens* in foods.

To type *C. perfringens*, several methods can be used. Thus, PCR–denaturing gradient gel electrophoresis (DGGE) of the ribosomal DNA is a variation of PCR used for the study of microbial ecology. DGGE is based on the separation of PCR amplicons of the same size but different sequences. This method has been used for the differentiation of *C. perfringens* from other composition of ileal bacterial microbiota of broiler chicken.[119] Other molecular typing tools for *C. perfringens* include DNA sequencing, PFGE, plasmid profiling, and ribotyping.[124–127] These methods are only used to type *C. perfringens* isolates, but do not allow detecting this microorganism directly from clinical or food samples.

28.6   PREVENTION OF INTOXICATION BY BoNT AND *C. perfringens* TOXINS

C. botulinum, especially those of group II (nonproteolytic), poses a safety hazard in modern food processing, which consists of mild pasteurization treatments, anaerobic packaging, extended shelf lives, and chilled storage.[128] These authors stated that spore heat resistance should be investigated for each food in order to determine the efficiency of industrial heat treatments to prevent this hazard. Similarly, *C. perfringens* poses a health hazard by consumption of the kind of foods previously mentioned. To reduce risk for intoxication by BoNT and *C. perfringens* toxins, effective food safety programs for the prevention of growth of *C. botulinum* and toxins *C. perfringens* strains in foods should be implemented throughout the entire food chain. This safety program should include an effective and running hazard analysis and critical control point (HACCP) in the food industry.

In the HACCP, several methods reported as appropriated to control *C. botulinum* and *C. perfringens* growth and sporulation could be used as preventive actions. Therefore, in the *C. botulinum* case, when a 6-log reduction of spore number is not guaranteed by the corresponding food processing, inhibition of germination of spore should be assured. No germination of spores occurs when foods are preserved at temperature under 3°C or foods reach NaCl content of 5% (w/v), water activity below 0.97, or pH values lower than 5.0.[128] To avoid germination of *C. perfringens* spores, foods should be preserved under 7°C or reach water activity values below 0.95.

Anderson et al.[129] analyzed the food safety objective (FSO) for controlling the most concerns of the aforementioned microorganism (*C. botulinum*) in commercially sterile foods. These authors stated that the risk-based framework of *C. botulinum* should encourage development of innovative technologies that result in microbial safety levels equivalent to those achieved with traditional processing methods. Thus, it has been reported that package of food products under modified atmospheres package (MAP) reduces spore germination and toxin production of the aforementioned microorganism. Therefore, fresh mussels packed under MAP packaging inoculated with *C. botulinum* did not produce toxin, even at an abusive storage temperature and when held beyond their shelf life.[130] Several studies have stated the inactivation of spores of *C. botulinum* by high-pressure processing (HPP) as an alternative method for preservation of foods, using various combinations of high pressure and elevated temperatures.[54,131] These innovative methods may be also used to control *C. perfringens* growth and toxin production in foods.

On the other hand, the development of vaccines against BoNT may be useful to control BoNT intoxication. Investigation about vaccines against BoNT was intensified during the Second World War because of concerns that the toxin might be used as a biological weapon against allied forces, and methods for preparing alum-precipitated type A and B toxoids for use in animals and humans were developed by Nigg and coworkers in the United States and by Rice and coworkers in Canada.[132] Today, recombinant subunit vaccines are in development and a bivalent (AB) Hc vaccine (rBV A/B [*Pichia pastoris*]) has been developed and has performed well in clinical trials by Smith.[132] Recent advances have facilitated the development of the next generation of BoNT vaccines, utilizing noncatalytic full-length BoNT or a subunit vaccine composed of the receptor-binding domain of BoNT as immunogens.[133] It is expected that these and other new vaccines will continue onto the corresponding licensure to be used in the prevention of BoNT intoxication.

28.7 CONCLUSIONS AND FUTURE PERSPECTIVES

Species of *Clostridium* genus are widely present in the environment (e.g., soil, sewage, and marine sediments) and in the gastrointestinal tract of humans and domestic and feral animals. *C. botulinum* and *C. perfringens* have been considered pathogenic to man and animals. These pathogenic species often secrete an array of invasins, exotoxins, and other extracellular enzymes, and various toxins are responsible for such human diseases as botulism, foodborne illness, and diarrheas. Although in some species such as *C. botulinum*, symptoms are well defined and very distinctive, other species

as *C. perfringens* could exhibit similar symptoms and molecular differentiation would be needed. Different physiological and morphological methods have been traditionally used for taxonomic differentiation of *Clostridium*. However, several studies have shown that traditional identification methods, based on phenotypic properties, are laborious and lack discriminatory power and misidentification occurs frequently. Progress in the molecular biology in the last decade has opened up possibilities of characterizing *Clostridium* at the genomic level. The application of molecular procedures has greatly improved laboratory diagnostics of these important bacterial pathogens. We here review the main conventional methods, as well as the molecular techniques, for detecting the main human pathogenic *Clostridium* species (i.e., *C. botulinum* and other BoNT-producing species such as some strains of *C. baratii*, *C. butyricum*, and *C. perfringens*). Sample processing and molecular methods for their rapid detection and identification are presented, such as a PCR for *C. botulinum* encoding BoNT A, B, E, and F and a multiduplex real-time PCR for the detection of the toxin genes encoding for the toxins of *C. perfringens*. There is no doubt that continuing refinement in the sample processing procedures and automation in the target amplification and detection will make molecular identification of *C. botulinum* and *C. perfringens* an indispensable tool in clinical testing laboratories and further strengthen our capacity in the tracking and illnesses prevention resulting from these pathogenic clostridia. In addition, new strategies based in the prevention of intoxication by BoNT and *C. perfringens* toxins should be implemented to reduce risk of both types of intoxication.

ACKNOWLEDGMENTS

Authors thank the Spanish MICINN (Carnisenusa CSD2007-00016 Consolider Ingenio 2010 and Projects AGL2007-64639), the governments of Extremadura (GRU10162), and FEDER.

REFERENCES

1. Peck, M.W., Stringer, S.C., and Carter, A.T. *Clostridium botulinum* in the post-genomic era. *Food Microbiol.*, 28, 183, 2011.
2. Smelt, J.P., Stringer, S.C., and Brul, S. Behaviour of individual spores of non proteolytic *Clostridium botulinum* as an element in quantitative risk assessment. *Food Control*, 29, 358, 2013.
3. Al-Khaldi, S.F. et al. Identification and characterization of *Clostridium perfringens* using single target DNA microarray chip. *Int. J. Food Microbiol.*, 91, 289, 2004.
4. Berezina, O.V. et al. Extracellular glycosyl hydrolase activity of the *Clostridium* strains producing acetone, butanol, and ethanol. *Appl. Biochem. Microbiol.*, 44, 42, 2008.
5. Collins, M.D. et al., The phylogeny of the genus *Clostridium*: Proposal of five new genera and eleven new species combination. *Int. J. Syst. Bacteriol.*, 44, 812, 1994.
6. Hatheway, C.L. and Johnson E.A. *Clostridium*: The spore bearing anaerobes. In: *Microbiology and Microbial Infections*. Balows, A. and Duerden, B.I., (eds.) 9th ed. Topley and Wilson's, Arnold, London, U.K., 1998.

7. Lauro, F.M., Bertoloni, G., and Obraztsova, A. Pressure effects on *Clostridium* strains isolated from a cold deep-sea environment. *Extremophiles*, 8, 169, 2004.

8. Sebaihia, M. et al. Genome sequence of a proteolytic (Group I) *Clostridium botulinum* strain Hall A and comparative analysis of the clostridial genomes. *Genome Res.*, 17, 1082, 2007.

9. Skarin, H. et al. *Clostridium botulinum* group III: A group with dual identity shaped by plasmids, phages and mobile elements. *BMC Genomics*, 12, 185, 2011.

10. Shimizu, T. et al. Complete genome sequence of *Clostridium perfringens*, an anaerobic flesh-eater. *PNAS*, 99, 996, 2002.

11. Quinn, C.P. and Minton, N.P. Clostridial neurotoxins. In: *Clostridia: Biotechnology and Medical Applications*. Bahl, H. and Dürre, P., (eds.) Wiley-VCH Verlag GmbH, Weinheim, Germany, 2001.

12. Hill, K.K. et al. Genetic diversity among botulinum neurotoxin-producing clostridial strains. *J. Bacteriol.*, 189, 818, 2007.

13. Chen, Y. et al. Sequencing the botulinum neurotoxin gene and related genes in *Clostridium botulinum* type E strains reveals orfx3 and a novel type E neurotoxin subtype. *J. Bacteriol.*, 189, 8643, 2007.

14. Lindström, M. and Korkeala, H. Laboratory diagnostics of botulism, *Clin. Microbiol. Rev.*, 19, 298, 2006.

15. Hinderink, K., Lindström, M., and Korkeala, H. Group I *Clostridium botulinum* strains show significant variation in growth at low and high temperature. *J. Food Prot.*, 72, 375, 2008.

16. Heikinheimo, A. et al. *Clostridium*. In: *Molecular Detection of Foodborne Pathogens*. Liu, D., (ed.) Taylor & Francis, Boca Raton, FL, p. 145, 2009.

17. Hatheway, C.L. Toxigenic clostridia. *Clin. Microbiol. Rev.*, 3, 66, 1990.

18. Smedley, J.G. et al. The enteric toxins of *Clostridium perfringens*. *Rev. Physiol. Biochem. Pharmacol.*, 152, 183, 2005.

19. Keyburn, A.L. et al. NetB, a pore-forming toxin from necrotic enteritis strains of *Clostridium perfringens*. *Toxins*, 2, 1913, 2008.

20. Songer, J.G. and Uzal, F.A. Clostridial enteric infections in pigs. *J. Vet. Diagn. Invest.*, 17, 528, 2005.

21. McClane, B.A. and Rood, J.I. Clostridial toxins involved in human enteric and histotoxic infections. In: *Clostridia: Biotechnology and Medical Applications*. Bahl, H. and Dürre, P., (eds.) Wiley-VCH Verlag GmbH, Weinheim, Germany, 2001.

22. Adams, M.R. and Moss, M.O. Bacterial agents of foodborne illness. In: *Food Microbiology*. The Royal Society of Chemistry, Cambridge, U.K., 2005.

23. Wang, G. et al. Effects of wet heat treatment on the germination of individual spores of *Clostridium perfringens*. *J. Appl. Microbiol.* 113, 824, 2012.

24. Peck, M.W. *Clostridium botulinum* and the safety of minimally heated, chilled foods: An emerging issue. *J. Appl. Microbiol.*, 101, 556, 2006.

25. Therre, H. Botulism in the European Union. *Euro Surveill.*, 4, 2, 1999.

26. Ungchusak, K. et al. The need for global planned mobilization of essential medicine: Lessons from a massive Thai outbreak. *Bull. World Health Organ.*, 85, 238, 2007.

27. Wongtanate, M. et al. Signs and symptoms predictive of respiratory failure in patients with foodborne botulism in Thailand. *Am. J. Trop. Med. Hyg.*, 77, 386, 2007.

28. CDC (Centers for Disease Control and Prevention). Botulism associated with commercial carrot juice e Georgia and Florida, September 2006. *MMWR—Morb. Mortal. Wkly. Rep.*, 55, 1098, 2006.

29. Sheth, A.N. et al. International outbreak of severe botulism with prolonged toxemia caused by commercial carrot juice. *Clin. Infect. Dis.*, 47, 1245, 2008.

30. King, L.A. Two severe cases of botulism associated with industrially produced chicken enchiladas, France, August. *Euro Surveill.*, 13, 1, 2008.

31. Lücke, F.K. Psychrotrophic *Clostridium botulinum* strains from raw hams. *System. Appl. Microbiol. J.* 5, 274, 1984.

32. Hauschild, A.H.W. Epidemiology of foodborne botulism. In: *Clostridium botulinum. Ecology and Control in Foods.* Hauschild, A.H.W. and Dodds, K.L., (eds.) Marcel Dekker, New York, p. 69, 1993.

33. Weber, J.T. et al. A massive outbreak of type E botulism associated with traditional salted fish in Cairo. *J. Infect. Dis.* 167, 51, 1993.

34. King, L.A. et al. Botulism and hot-smoked whitefish: A family cluster of type E botulism in France, September 2009. *Euro Surveill.*, 14, 7, 2009.

35. Lindström, M. et al. Novel insights into the epidemiology of *Clostridium perfringens* type A food poisoning. *Food Microbiol.*, 28, 192, 2011.

36. Li, J. and McClane, B.A. Comparative effects of osmotic, sodium nitrite-induced, and pH-induced stress on growth and survival of *Clostridium perfringens* type A isolates carrying chromosomal or plasmid-borne enterotoxin genes. *Appl. Environ. Microbiol.* 72, 7620, 2006.

37. Regan, C.M., Syedt, V.Q., and Tunstall P.J. A hospital outbreak of *Clostridium perfringens* food poisoning- implications for food hygiene review in hospitals. *J. Hosp. Infect.*, 29, 69, 1995.

38. Pollock, A.M. and Whitty, P. Outbreak of *Clostridium perfringens* food poisoning. *J. Hosp. Infect.*, 17, 179, 1991.

39. Mage, D.T., Cohen, M., and Donner, M. Sudden infant death syndrome. *N. Engl. J. Med.*, 24, 2581, 2009.

40. Murrell, T.G. et al. A hypothesis concerning *Clostridium perfringens* type A enterotoxin (CPE) and sudden infant death syndrome (SIDS). *Med. Hypotheses*, 22, 401, 1987.

41. Murrell, W.G. et al. Enterotoxigenic bacteria in the sudden infant death syndrome. *J. Med. Microbiol.* 39, 114, 1993.

42. Lindsay, J.A. et al. *Clostridium perfringens* type A cytotoxic-enterotoxin(s) as triggers for death in the sudden infant death syndrome: Development of a toxico-infection hypothesis. *Curr. Microbiol.* 27, 51, 1993.

43. Siarakas, S., Damas, E., and Murrell, W.G. The effect of enteric bacterial toxins on the catecholamine levels of the rabbit. *Pathology* 29, 278, 1997.

44. Peck, M.W. *Clostridium botulinum*. In: *Understanding Pathogen Behaviour*. Griffiths, M. (ed.) Woodhead Press, Cambridge, U.K., p. 531, 2005.

45. Wu, J. et al. Detection and toxin typing of *Clostridium perfringens* in formalin-fixed, paraffin-embedded tissue samples by PCR. *J. Clin. Microbiol.*, 47, 807, 2009.

46. Labbé, R. and Juneja, V. *Clostridium perfringens* gastroenteritis. In: *Foodborne Infections and Intoxications*. Cliver, D. and Rieimann, H. (eds.) 3rd ed. Elsevier, New York, p. 136, 2006.

47. Smedley, J.G. et al. Identification of a prepore large-complex stage in the mechanism of action of *Clostridium perfringens* enterotoxin. *Infect. Immun.*, 75, 2381, 2007.

48. Baron, E.J. Conventional versus molecular methods for pathogen detection and the role of clinical microbiology in infection control. *J. Clin. Microbiol.*, 49, S43, 2011.

49. Fakruddin, M., Chowdhury, A., and Hossain, Z. Competitiveness of polymerase chain reaction to alternate amplification methods. *Am. J. Biochem. Mol. Biol.*, 3, 71, 2013.

50. Janvilisri, T. et al. Development of a microarray for identification of pathogenic *Clostridium* spp. *Diagn. Microbiol. Infectious Dis.*, 66, 140, 2010.
51. Sibley, C.D., Peirano, G., and Church, D.L. Molecular methods for pathogen and microbial community detection and characterization: Current and potential application in diagnostic microbiology. *Infection Gen. Evol.*, 12, 505, 2012.
52. Strotman, L.N. et al. Facile and rapid DNA extraction and purification from food matrices using IFAST (immiscible filtration assisted by surface tension). *Analyst*, 137, 4023, 2012.
53. Zhao, H., et al. Characterization of the monoclonal antibody response to botulinum neurotoxin type A in the complexed and uncomplexed forms. *Jpn. J. Infect. Dis.*, 65, 138, 2012.
54. Reddy, N.R., Tetzloff, R.C., and Skinner, G.E. Effect of media, additives, and incubation conditions on the recovery of high pressure and heat-injured *Clostridium botulinum* spores. *Food Microbiol.*, 27, 613, 2010.
55. Mato Rodriguez, L. and Alatossava, T. Effects of copper on germination, growth and sporulation of *Clostridium tyrobutyricum*. *Food Microbiol.*, 27, 434, 2010.
56. Collins, M.D. and East, A.K. Phylogeny and taxonomy of the food-borne pathogen *Clostridium botulinum* and its neurotoxins. *J. Appl. Microbiol.*, 84, 5, 1998.
57. Webb, M.D. et al. Does proximity to neighbours affect germination of spores of non-proteolytic *Clostridium botulinum*? *Food Microbiol.*, 32, 104, 2012.
58. Lilly, T. et al. An improved medium for detection of *Clostridium botulinum* type E. *J. Milk Food Technol.*, 34, 492, 1971.
59. Quagliaro, D.A. An improved cooked meat medium for the detection of *Clostridium botulinum*. *J. AOAC*, 60, 563, 1977.
60. Saeed, E.M.A. Studies on isolation and identification of *Clostridium botulinum* investigating field samples specially from equine grass sickness cases. Doctoral thesis, University of Göttingen, Göttingen, Germany, 2005.
61. Stringer, S.C., Webb, M.D., and Peck, M.W. Lag time variability in individual spores of *Clostridium botulinum*. *Food Microbiol.*, 28, 228, 2011.
62. Hauschild, A.H.W. and Hilsheimer, R. Enumeration of *Clostridium botulinum* spores in meats by a pour-plate procedure. *Can. J. Microbiol.*, 23, 829, 1977.
63. Cato, E., George, W.L., and Finegold, S. Genus *Clostridium*. In: *Bergey's Manual of Systematic Bacteriology*. Sneath, P.H.A. et al., (eds.) Williams & Wilkins, Baltimore, MD, 1986.
64. Mills, D.C., Midura, T.F., and Arnon, S.S. Improved selective medium for the isolation of lipase-positive *Clostridium botulinum* from feces of human infants. *J. Clin. Microbiol.*, 2, 947, 1985.
65. Silas, J.C. et al. Selective and differential medium for detecting *Clostridium botulinum*. *Appl. Environ. Microbiol.*, 50, 1110, 1985.
66. Smith, L.D.S. and Sugiyama, H. *Botulism. The Organism, Its Toxins, the Disease*. Charles C. Thomas, Springfield, IL, 1988.
67. Ferreira, J.L. et al. Detection of botulinal neurotoxins A, B, E, and F by amplified enzyme-linked immunosorbent assay: Collaborative study. *J. AOAC Int.*, 86, 314, 2003.
68. Grate, J.W. et al. Advances in assays and analytical approaches for botulinum-toxin detection. *TrAC—Trend Anal. Chem.*, 29, 1137, 2010.
69. Kalb, S.R. et al. The use of Endopep–MS for the detection of botulinum toxins A, B, E, and F in serum and stool samples. *Anal. Biochem.*, 351, 84, 2006.
70. Ladd, J. et al. Detection of botulinum neurotoxins in buffer and honey using a surface plasmon resonance (SPR) sensor. *Sens. Actuat. B*, 130, 129, 2008.
71. Ferracci, G. et al. A label-free biosensor assay for botulinum neurotoxin B in food and human serum. *Anal. Biochem.*, 410, 281, 2011.
72. Ching, K.H. et al. Rapid and selective detection of botulinum neurotoxin serotype-A and -B with a single immunochromatographic test strip. *J. Immunol. Methods*, 380, 23, 2012.
73. Liu, Y.Y.B. et al. A functional dual-coated (FDC) microtiter plate method to replace the botulinum toxin LD50 test. *Anal. Biochem.*, 425, 28, 2012.
74. Ruge, D.R. et al. Detection of six serotypes of botulinum neurotoxin using fluorogenic reporters. *Anal. Biochem.* 411, 200, 2011.
75. Raphael, B.H. et al. Ultrafiltration improves ELISA and Endopep MS analysis of botulinum neurotoxin type A in drinking water. *J. Microbiol. Methods*, 90, 267, 2012.
76. Aranda, E. et al. Detection of *Clostridium botulinum* types A, B, E, and F in foods by PCR and DNA probe. *Lett. Appl. Microbiol.*, 25, 186, 1997.
77. Córdoba, J.J. et al. *Clostridium*. In: *Molecular Detection of Human Bacterial Pathogens*. Liu, D. (ed.) CRC Press, Taylor & Francis Group, Boca Raton, FL, p. 367, 2011.
78. Lindström, M. et al. Multiplex PCR assay for detection and identification of *Clostridium botulinum* types A, B, E, and F in food and fecal material. *Appl. Environ. Microbiol.*, 67, 5694, 2001.
79. Kasai, Y. et al. Quantitative duplex PCR of *Clostridium botulinum* types A and B neurotoxin genes. *J. Food Hyg. Soc. Jpn.*, 48, 19, 2007.
80. Sakuma, T. et al. Rapid and simple detection of *Clostridium botulinum* types A and B by loop-mediated isothermal amplification. *J. Appl. Microbiol.*, 106, 1252, 2009.
81. Hyytiä, E. et al. Characterisation of *Clostridium botulinum* groups I and II by randomly amplified polymorphic DNA analysis and repetitive element sequence-based PCR. *Int. J. Food Microbiol.*, 48, 179, 1999.
82. Paul, C. et al. A unique restriction site in the flaA gene allows rapid differentiation of group I and group II *Clostridium botulinum* strains by PCR-Restriction fragments length polymorphism analysis. *J. Food Prot.*, 70, 2133, 2007.
83. Dahlsten, E. et al. PCR assay for differentiating between Group I (proteolytic) and Group II (nonproteolytic) strains of *Clostridium botulinum*. *Int. J. Food Microbiol.*, 124, 108, 2008.
84. Janda, J.M. and Whitehouse, C.A. Usefulness of multilocus polymerase chain reaction followed by electrospray ionization mass spectrometry to identify a diverse panel of bacterial isolates. *Diagn. Microbiol. Infec. Dis.*, 63, 403, 2009.
85. Fach, P. et al. An innovative molecular detection tool for tracking and tracing *Clostridium botulinum* types A, B, E, F and other botulinum neurotoxin producing *Clostridia* based on the GeneDisc cycler. *Int. J. Food Microbiol.*, 145, S145, 2011.
86. Hauser, D. et al. Botulinal neurotoxin C1 complex genes, clostridial neurotoxin homology, and genetic transfer in *Clostridium botulinum*. *Toxicon*, 33, 515, 1995.
87. East, A.K. and Collins, M.D. Conserved structure of genes encoding components of botulinum neurotoxin complex M and the sequence of the gene coding for the nontoxic component in nonproteolytic *Clostridium botulinum* type F. *Curr. Microbiol.*, 29, 69, 1994.

88. Rajkovic, A. et al. Detection of *Clostridium botulinum* neurotoxins A and B in milk by ELISA and immuno-PCR at higher sensitivity than mouse bio-assay. *Food Anal. Methods*, 5, 319, 2012.

89. Zhang, Y. et al. Simultaneous and sensitive detection of six serotypes of botulinum neurotoxin using enzyme-linked immunosorbent assay-based protein antibody microarrays. *Anal. Biochem.*, 430, 185, 2012.

90. Raphael, B.H. and Andreadis, J.D. Real-time PCR detection of the nontoxic nonhemagglutinin gene as a rapid screening method for bacterial isolates harboring the botulinum neurotoxin (A–G) gene complex. *J. Microbiol. Methods*, 71, 343, 2007.

91. Dover, J.E. et al. Recent advances in peptide probe-based biosensors for detection of infectious agents. *J. Microbiol. Methods*, 78, 10, 2009.

92. Yoo, M. et al. High-throughput identification of clinically important bacterial pathogens using DNA microarray. *Mol. Cell. Probe*, 23, 171, 2009.

93. Fenicia, L. et al. Towards an international standard for detection and typing botulinum neurotoxin-producing *Clostridia* types A, B, E and F in food, feed and environmental samples: A European ring trial study to evaluate a real-time PCR assay. *Int. J. Food Microbiol.*, 145, S152, 2011.

94. Lövenklev, M. et al. Relative neurotoxin gene expression in *Clostridium botulinum* type B, determined using quantitative reverse transcription-PCR. *Appl. Environ. Microbiol.*, 70, 2919, 2004.

95. McGrath, S., Dooley, J.S.G., and Haylock, R.W. Quantification of *Clostridium botulinum* toxin gene expression by competitive reverse transcription-PCR. *Appl. Environ. Microbiol.*, 66, 1423, 2000.

96. Chen Y. et al. Quantitative real-time reverse transcription-PCR analysis reveals stable and prolonged neurotoxin cluster gene activity in a *Clostridium botulinum* type E strain at refrigeration temperature. *Appl. Environ. Microbiol.*, 74, 6132, 2008.

97. Chung, B. et al. Multiplex quantitative foodborne pathogen detection using high resolution CE-SSCP coupled stuffer-free multiplex ligation-dependent probe amplification. *Electrophoresis*, 33, 1477, 2012.

98. Uzal, F.A. Diagnosis of *Clostridium perfringens* intestinal infections in sheep and goats. *Anaerobe*, 10, 135, 2004.

99. Harmon, S.M., Kautter, D.A., and Peeler, J.T. Improved medium for enumeration of *Clostridium perfringens*, *Appl. Microbiol.*, 22, 688, 1971.

100. Hauschild, A.H.W. and Hilsheimer, R. Enumeration of food borne *Clostridium perfringens* in egg yolk free tryptose sulphite cycloserine agar. *J. Appl. Microbiol.*, 27, 521, 1974.

101. Labbé, R. *Clostridium perfringens*. In: *The Microbiological Safety and Quality of Food*. Aspen Publishers, Gaithersburg, MD, 2000.

102. Sterne, M. and Batty, I. Criteria for diagnosing clostridial infection. In: *Pathogenic clostridia*. Sterne, M. and Batty, L., (eds.) Butterworths, London, U.K., p. 79, 1975.

103. Piyankarage, R.H. et al. Sandwich enzyme-linked immunosorbent assay by using monoclonal antibody for detection of *Clostridium perfringens* enterotoxin. *J. Vet. Med. Sci.*, 61, 45, 1999.

104. El Idrissi, A.H. and Ward, G.E. Development of double sandwich ELISA for *Clostridium perfringens* beta and epsilon toxins. *Vet. Microbiol.*, 31, 89, 1992.

105. Nagahama, M. et al. Enzyme-linked immunosorbent assay for rapid detection of toxins from *Clostridium perfringens*. *FEMS Microbiol Lett.*, 84, 41, 1991.

106. Petit, L., Gibert, M., and Popoff, M.R. Detection of enterotoxin of *Clostridium perfringens*. In: *Encyclopedia of Food Microbiology*. Batt, C.A., Patel, P., and Robinson, R.K., (eds.) Academic Press, London, U.K., 1999.

107. Marks, S.L. et al. Evaluation of methods to diagnose *Clostridium perfringens* associated diarrhea in dogs. *J. Am. Vet. Med. Assoc.*, 214, 357, 1999.

108. Berry, P.R. et al. Evaluation of ELISA, RPLA, and Vero cell assays for detecting *Clostridium perfringens* enterotoxin in faecal specimens. *J. Clin. Pathol.*, 41, 458, 1988.

109. Seyer, A. et al. Rapid quantification of clostridial epsilon toxin in complex food and biological matrixes by immunopurification and ultraperformance liquid chromatography-tandem mass spectrometry. *Anal. Chem.*, 84, 5103, 2012.

110. Nakamura, M. et al. PCR identification of the plasmid-borne enterotoxin gene (cpe) in *Clostridium perfringens* strains isolated from food poisoning outbreaks. *Int. J. Food Microbiol.*, 294, 261, 2004.

111. Tang, Y. et al. Detection, cloning, and sequencing of the enterotoxin gene of *Clostridium perfringens* type C isolated from goat. *Turkish J. Vet. Anim. Sci.*, 36, 153, 2012.

112. Shanmugasamy, M. and Rajeswar, J. Alpha toxin specific PCR for detection of toxigenic strains of *Clostridium perfringens* in Poultry. *Vet. World*, 5, 365, 2012.

113. Meer, R. and Songer, G. Multiplex polymerase chain reaction assay for genotyping *Clostridium perfringens*. *Am. J. Vet. Res.*, 58, 702, 1997.

114. Heikinheimo, A. and Korkeala, H. Multiplex PCR assay for toxinotyping *Clostridium perfringens* isolates obtained from Finnish broiler chickens. *Lett. Appl. Microbiol.*, 40, 407, 2005.

115. Uzal, F.A. and Songer J.G. Diagnosis of *Clostridium perfringens* intestinal infections in sheep and goats. *J. Vet. Diag. Invest.*, 20, 253, 2008.

116. Goncuoglu, E.M. et al. Molecular typing of *Clostridium perfringens* isolated from turkey meta by multiplex PCR. *Lett. Appl. Microbiol.*, 47, 31, 2009.

117. Baums, C.G. et al. Diagnostic multiplex PCR for toxin genotyping of *Clostridium perfringens* isolates. *Vet. Microbiol.*, 100, 11, 2004.

118. Joshy, L., Chaurdhry, R., and Chandel, D.S. Multiplex PCR for the detection of *Clostridium botulinum* and *C. perfringens* toxin genes. *Indian J. Med. Res.*, 128, 206, 2008.

119. Feng, Y. et al. Identification of changes in the composition of ileal bacterial microbiota of broiler chickens infected with *Clostridium perfringens*. *Vet. Microbiol.*, 140, 116, 2010.

120. Gurjar, A.A. et al. Real-time multiplex PCR assay for rapid detection and toxintyping of *Clostridium perfringens* toxin producing strains in feces of dairy cattle. *Mol. Cell. Probe.*, 22, 90, 2008.

121. Albini, S. et al. Real-time multiplex PCR assays for reliable detection of *Clostridium perfringens* toxin genes in animal isolates. *Vet. Microbiol.*, 127, 179, 2008.

122. Abildgaard, L. et al. Sequence variation in the a-toxin encoding plc gene of *Clostridium perfringens* strains isolated from diseased and healthy chickens. *Vet. Microbiol.*, 136, 293, 2009.

123. Chon, J. et al. Development of real-time PCR for the detection of *Clostridium perfringens* in meats and vegetables. *J. Microbiol. Biotechnol.*, 22, 530, 2012.

124. Ridell, J. et al. Prevalence of the enterotoxin gene and clonality of *Clostridium perfringens* strains associated with food-poisoning outbreaks. *J. Food Prot.*, 61, 240, 1998.

125. Maslanka, S.E. et al. Molecular subtyping of *Clostridium perfringens* by pulsed-field gel electrophoresis to facilitate food-borne-disease outbreak investigations. *J. Clin. Microbiol.*, 37, 2209, 1999.

126. Eisgruber, H., Wiedmann, M., and Stolle, A. Use of plasmid profiling as a typing method for epidemiologically related *Clostridium perfringens* isolates from food poisoning cases and outbreaks. *Lett. Appl. Microbiol.*, 20, 290, 1995.

127. Schalch, B. et al. Ribotyping for strain characterization of *Clostridium perfringens* isolates from food poisoning cases and outbreaks. *Appl. Environ. Microbiol.*, 63, 3992, 1997.

128. Lindström, M., Kiviniemi, K., and Korkeala, H. Hazard and control of group II (non-proteolytic) *Clostridium botulinum* in modern food processing. *Int. J. Food Microbiol.* 108, 92, 2006.

129. Anderson, N.M. et al. Food safety objective approach for controlling *Clostridium botulinum* growth and toxin production in commercially sterile foods. *J. Food Prot.*, 74, 1956, 2011.

130. Newell, C.R., Ma, L., and Doyle, M. Botulism challenge studies of a modified atmosphere package for fresh mussels: Inoculated pack studies. *J. Food Prot.* 75, 1157, 2012

131. Juliano, P. et al. *C. botulinum* inactivation kinetics implemented in a computational model of a high-pressure sterilization process. *Biotechnol. Prog.*, 25, 163, 2009.

132. Smith, L.A. Botulism and vaccines for its prevention. *Vaccine*, 27, D33, 2009

133. Karalewitz, A.P. and Barbieri, J.P. Vaccines against botulism. *Curr. Opin. Microbiol.*, 15, 317, 2012.

29 *Coxiella burnetii*

Mark D. Bennett and Michael J. Banazis

CONTENTS

29.1 INTRODUCTION

Coxiella burnetii was recognized as a potential bioterrorism agent only two decades after it was first described. In 1950s America, Operation Whitecoat used conscientious objectors in human experiments to investigate infectious dose, vaccine efficacy, and antibiotic therapies for *C. burnetii*.[1]

Operation Whitecoat was a response to the Soviet Union biological warfare program that may have deliberately released *C. burnetii* during World War II.[2]

Although Q fever is rarely fatal, *C. burnetii* has traits that make it suitable for weaponization. Its low infectious dose, its ease of spread by aerosol, and its extreme resilience and persistence in the environment[1] simplify the problems of storage, transportation, and dispersal.[2] Due to its environmental stability, ease of transmission to humans, extremely low infectious dose, high morbidity, ability to incapacitate large numbers of people, and prior history of weaponization, *C. burnetii* was included and retained in the United States Federal Government Department of Health and Human Services (HHS) list of select agents and toxins by the Centers for Disease Control and Prevention (CDC), United States.[2a,2b]

While *C. burnetii* is considered an "incapacitating" rather than a "lethal" bioweapon,[1] infection would likely cause both immediate and long-term negative health outcomes for an exposed population. The World Health Organization's (WHO) modeling predicted 125,000 cases of acute Q fever, 9,000 cases of chronic Q fever, and 150 deaths in an urban population of 500,000 exposed by aerosol to 50 kg of *C. burnetii*.[3] More recent modeling suggests 50% of people exposed to such infectious aerosols could become sick, 13% seriously ill, and 0.6% could die.[4] Following the Dutch outbreaks that started in 2007,[5] data from the field showed that 52% of Q fever patients had severe fatigue 6 months after their primary illness.[6]

Some properties of *C. burnetii* make it potentially attractive to achieve terrorist objectives. *C. burnetii* is found virtually everywhere around the globe[7] and therefore is relatively easily obtained.[2] Also, Q fever is difficult to clinically differentiate from other naturally occurring influenza-like diseases. These traits would hamper attempts to accurately identify and trace a deliberate release and propagate a major aim of bioterrorism: panic.[2] *C. burnetii* is certain to remain a potential bioterrorism threat until diagnostic and treatment capabilities exist that can rapidly meet the challenges posed by the misuse of this unique bacterium.

29.2 CLASSIFICATION AND MORPHOLOGY

C. burnetii is classified in the class Gammaproteobacteria and the order Legionellales.[8–10] While *C. burnetii* is the type species of the genus, other (unofficial) congeners include a bacterium of freshwater crayfish, *Coxiella cheraxi*, and various partially characterized endosymbionts of ticks.[11–23] The molecular detection of *Coxiella*-like DNA sequences in occasional clinical[24–26] and environmental samples[27,28] hints at a potentially wide distribution of as yet unknown or incompletely characterized relatives of *C. burnetii*.

During investigations of a mysterious febrile condition of abattoir workers in Brisbane, Australia, in the mid-1930s, it was initially suspected that the agent causing the query, or "Q," fever was a filterable virus.[29,30] This was later disproved by Burnet and Freeman who determined that the Q fever agent was in fact bacterial and named it *Rickettsia burneti*.[8]

Almost simultaneously, in the United States, Drs. Davis and Cox of the Rocky Mountain Laboratory detected a tick-borne transmissible febrile disease in a guinea pig upon which ticks collected from Nine Mile Creek had fed.[31] Serendipitously, a laboratory accident led to a human infection with this agent, and the unfortunate patient developed

Q fever-like symptoms.[32] Cox proposed the name *Rickettsia diaporica* (to indicate the organism's ability to pass through a filter) for the Nine Mile Creek agent.[33]

It was soon realized that *R. burneti* and the Nine Mile Creek agent were closely related, if not the same organism,[34] and, furthermore, that the classification of the Q fever agent within the genus *Rickettsia* was inappropriate. Thus, the subgenus *Coxiella* in which *R. burneti* had been temporarily placed was elevated to full genus status and *R. burneti* was renamed *C. burnetii*.[35] The organism's name acknowledges the contributions of Drs. Cox and Burnet in pioneering *C. burnetii* research.

The US government lists all *C. burnetii* strains except *C. burnetii* Nine Mile phase II (NMII) plaque purified clone 4 on their National Select Agent Registry under category B.[36] All *C. burnetii* strains, with the exception of NMII clone 4, must be handled in appropriately certified laboratories under biosafety level 3 (BSL-3) conditions.[37] In Australia and New Zealand, no distinction is drawn between *C. burnetii* strains: all must be handled under physical containment level 3 conditions.[38]

C. burnetii is an acidophilic and microaerophilic diderm pleomorphic coccobacillus that inhabits the phagolysosomes of eukaryotic cells.[39,40] Although generally considered Gram negative, the Gram reaction for *C. burnetii* may vary.[40,41] At least two morphological forms of *C. burnetii* may be detected within infected phagolysosomes: the more metabolically active large cell variant (LCV) and the relatively quiescent small cell variant (SCV) (see Figure 29.1a).[40,42] The LCV can be differentiated from the SCV by its size (≥ 1 μm vs. 0.2–0.5 μm), shape (rounded, pleomorphic vs. rod), chromatin (dispersed vs. condensed), and distinct outer membrane, periplasmic space, and inner membrane (see Figure 29.1b).[40,42] In contrast, the more compact SCV has an inner and outer membrane apposed to an electron-dense, peptidoglycan-rich periplasm (see Figure 29.1c).[40,42] The SCV has a higher proportion of peptidoglycan relative to its mass (32% [w/w]) than the LCV (2% [w/w]), and this may contribute to the SCVs' robustness.[43]

Under both *in vitro* and *in vivo* conditions, *C. burnetii* LCVs may be observed to contain an electron-dense endogenous spore-like polar structure.[40,44,45] This is remarkable, because endogenous spore formation is a feature of certain monoderm (Gram positive) bacteria. Traditional stains for bacterial spores did not stain *C. burnetii* spores,[46] so this morphological form is known as the "pseudospore." The pseudospore may give rise to the SCV, but this remains unproven.

A variant of the SCV has been proposed: the small dense cell (SDC), which is distinguishable from the SCV by its increased resistance to immense pressures (e.g., 20,000 psi [138 MPa]).[40] The distinct morphological variants of *C. burnetii* have different protein expression profiles as was recently reviewed.[40]

29.3 BIOLOGY AND EPIDEMIOLOGY

29.3.1 GENOME AND EVOLUTIONARY HISTORY

C. burnetii Nine Mile phase I encodes its genome on a circular 1,995,275-bp chromosome (G+C content, 42.6%) and a 37,393-bp plasmid (QpH1).[47,48] Other *C. burnetii* isolates possess

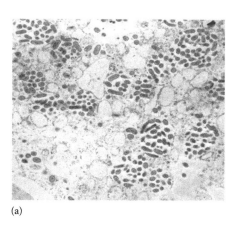 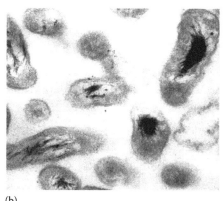 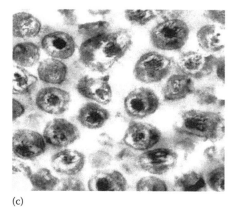

(a) (b) (c)

FIGURE 29.1 Transmission electron micrographs of *C. burnetii*. (a) *C. burnetii* infected murine macrophages, showing LCVs and SCVs within PVs. (b) *C. burnetii* LCVs purified by gradient ultracentrifugation. (c) *C. burnetii* SCVs purified by gradient ultracentrifugation. (From electron micrographs kindly supplied by Dr. Kati Kiss, Texas A&M System Health Science Center, Microbial and Molecular Pathogenesis.)

genomes of similar length, but may bear alternate plasmids, designated QpDG, QpDV, and QpRS.[49,50] The plasmids all share significant regions of homology, but there are also plasmid-specific sequence differences.[51,52] Some strains of *C. burnetii* lack plasmids; however, plasmid-less strains carry plasmid-homologous sequences integrated into their chromosomes.[53,54]

C. burnetii is divided into six genomic groups: I–III are associated with acute Q fever; IV and V, mainly with chronic Q fever; and VI with low-virulence rodent-derived isolates.[55–57]

Like *Yersinia pestis*, which has relatively recently adapted to life within eukaryotic cells, *C. burnetii* carries many copies of insertion sequence (IS) elements. These are thought to help in mediating genome plasticity.[55,58] Furthermore, compared to many intracellular pathogens, the proportion of pseudogenes in *C. burnetii* is low.[47] Together, these observations indicate that *C. burnetii* may have evolved a parasitic lifestyle relatively recently in its evolutionary history.[47]

29.3.2 RESCUE FROM THE HOST CELL

C. burnetii was long considered to be an obligate intracellular bacterium. This is no longer the case. An axenic growth medium capable of supporting the *in vitro* growth of *C. burnetii* has been developed.[59,60] Acidified citrate cysteine medium-2 (ACCM-2) contains a citrate buffer at pH 4.75 with interstitial fluid-like concentrations of Na^+, K^+, and Cl^-; a complex mixture of casamino acids, peptides, vitamins, minerals, and nucleotides; methyl-β-cyclodextrin; and supraphysiological levels of L-cysteine. ACCM-2 agarose is also available for the cultivation of *C. burnetii* on semisolid agar, under microaerophilic conditions (2.5% oxygen).[59,60]

29.3.3 PHASE VARIATION

The lipopolysaccharide (LPS) of *C. burnetii* is a well-known virulence factor that on serial passage in embryonated eggs or tissue culture undergoes an antigenic change associated with reduction in virulence called phase variation.[61–63] This antigenic change has been compared to the smooth–rough transition observed in Gram-negative bacteria.[64] Wild-type, virulent phase I *C. burnetii*, isolated from natural sources or infections (e.g., ticks, ruminant placentas), produces full-length LPS. Serial *in vitro* passage in embryonated eggs or tissue culture produces bacteria with significantly truncated forms of LPS, termed phase II *C. burnetii*.[61,63,65]

Phase variation is apparently a unidirectional and permanent adaptation to life under *in vitro* conditions.[63,66,67] Phase II organisms are approximately tenfold more infectious for cultured cells than phase I organisms[61] but are approximately 3000-fold less infectious for immunocompetent mammalian hosts[68] and far less virulent.[61,69]

Phase I *C. burnetii* LPS can resist lysis by complement, sterically inhibit antibody binding, mask Toll-like receptor (TLR) ligands, and may inhibit CR3 binding on macrophages.[70–73] The chemical differences between phase I and phase II LPS lie in the O-antigen and outer core regions.[74–76] In particular, two uncommon methylated sugars, virenose and dihydrohydroxystreptose, are lost from the O-antigen during phase transition.[65] Although phase transition in the Nine Mile strain of *C. burnetii* has been associated with deletion mutations in chromosomal genes essential for carbohydrate metabolism, LPS and lipooligosaccharide synthesis, the same mutations have not been detected in all phase II *C. burnetii* isolates.[77] Rather, there is a considerable variety in the size and location of chromosomal genetic changes associated with phase variation, and the LPS-synthesis genes are not always deleted.[66] A notable case in point is the Nine Mile *C. burnetii* clone RSA 514, or "crazy" isolate, which has an intermediate-length LPS compared with phase I and phase II *C. burnetii* and also has an intermediate virulence.[61]

29.3.4 LIFE CYCLE: OVERVIEW

C. burnetii SCVs are internalized by cells via microfilament-dependent endocytosis;[78–80] however, different molecular mechanisms are involved, depending on the host cell type and the phase of the invading *C. burnetii* LPS.

Phagocytosis and subsequent acidification of the nascent phagolysosome (parasitophorous vacuole [PV]) trigger SCVs to differentiate into LCVs.[39] During the 1–2-day lag phase of bacterial growth, there is no increase in bacterial number, merely a morphological change from SCVs to LCVs.[81] LCVs predominate during the subsequent 4 days of exponential growth, consistent with their role as the most metabolically active life cycle stage of *C. burnetii*, responsible for intracellular replication.

Intracellular growth is relatively slow: doubling time estimates range from 5 to 45 h.[82–84] As the bacteria pass from the exponential growth phase to the stationary phase at around 6 days postinfection,[81] the proportion of SCVs compared to LCVs increases, with a concomitant decrease in *C. burnetii* metabolic activity.

Cell lysis, or possibly exocytosis, releases the bacteria into the extracellular medium,[85] allowing the next generation of *C. burnetii* organisms to infect cells.

29.3.5 Survival in the Parasitophorous Vacuole

The precise mechanisms that allow *C. burnetii* to thrive within the austere conditions of the PV are not well understood. *C. burnetii* requires low pH for *in vivo* growth and replication.[39] The bacterium's surface passively resists attack by cationic peptides and acid hydrolases,[86] and *C. burnetii* peptidoglycan, imbued with protease-resistant proteins, may provide additional protection from lysosomal digestion.[87] *C. burnetii* prevents the respiratory burst in infected phagocytes,[88–90] by inhibiting NADPH oxidase assembly.[89,91,92] *C. burnetii* chooses to inhabit niches low in iron concentration and encodes Fe–Mn and Cu–Zn superoxide dismutases, glutaredoxin, thioredoxin reductases, peroxiredoxin (bacterioferritin comigratory protein), peroxide-scavenging alkyl hydroperoxide reductases, DNA repair genes, and catalase, all of which help protect it from reactive oxygen and nitrogen species and contribute to intracellular persistence.[93–96]

Just which subcellular elements *C. burnetii* recruits to form the PV has been the subject of much research and confusion.[97] The current paradigm suggests that the PV (or *Coxiella*-containing vacuole [CCV]) can be decorated with markers indicative of the secretory, autophagic, and phagolysosomal compartments,[97,98] and the array of markers depends on both the host cell and the *C. burnetii* phase.[97,99] In nonmicrobiocidal cells (e.g., epithelial cells, trophoblasts), phase I *C. burnetii* resides within phagolysosomes, whereas in microbiocidal cells (e.g., monocytes, macrophages), phase I *C. burnetii* resides within acidic late phagosomes.[97,99,100] Phase II *C. burnetii*, when engulfed by a monocyte/macrophage, are destroyed, whereas within nonmicrobiocidal cells, phase II *C. burnetii* survive in a phagolysosome–autophagosome hybrid compartment.[97,99,101,102] As *C. burnetii* adapts to life *in vitro*, its phase transition manifests as a change in the structure of its LPS and also in the ways in which it manipulates cellular phagocytic and autophagocytic pathways.[103]

The PV membrane is cholesterol-rich, and the chemical inhibition of host cholesterol metabolism and host cell kinase activity retards PV biogenesis and *C. burnetii* replication.[99,104–106] *C. burnetii* lacks enzymes for *de novo* cholesterol biosynthesis; however, it encodes putative eukaryote-like Δ24 and Δ7 sterol reductase homologs. The roles of these enzymes are currently unclear,[105,107] but it seems *C. burnetii* may have acquired these genes through horizontal transfer from amoebae.[107]

29.3.6 Type IVB Secretion System

Legionella pneumophila and *C. burnetii* have distinct but evolutionarily related type IVB secretion systems (T4BSSs) that are essential for their survival within cells.[108–110] Their T4BSSs are sufficiently similar that it is possible to use the machinery of one to test the substrates of the other,[111–112] and this has facilitated research into the functional properties of the suite of effector proteins secreted by the *C. burnetii* T4BSS.[110] By synthesizing and secreting effector proteins through its T4BSS, *C. burnetii* interferes with host cell apoptosis,[113] autophagy, cytoskeletal organization, immunity and development, and maintenance of the mature vacuole, thus constructing for itself a spacious and comfortable PV within which to replicate.[99,110,114–116]

29.3.7 Survival in the Environment

The *C. burnetii* SCV is the environmentally resistant life cycle stage. SCVs are relatively metabolically quiescent and can resist heat, desiccation, extreme pH, disinfectants, chemical products, and UV radiation.[117–121]

29.3.8 Tissue Tropism

When infection occurs via the respiratory route, pulmonary alveolar macrophages are among the first cells to be infected. Following engulfment, *C. burnetii* are transported systemically,[122] with histological evidence of *C. burnetii* infection most often found in the lungs, liver, spleen, and other sites of the "reticuloendothelial system."[122]

29.3.9 Geographic Distribution

C. burnetii is endemic around the globe, except at the poles, and possibly New Zealand.[123,124] The prevalence of Q fever may be increasing, perhaps due to increasing development of previously rural areas.[125] Reported national seroprevalences vary from about 1% to >48% (see Table 29.1). The incidence of Q fever varies between 0.028 and 3.3 per 100,000 population (see Table 29.1), although this is likely an underestimate.[126] The wide range of prevalence and incidence estimates is partially explained by inconsistencies between immunological tests[127] and differences in surveillance effort expended at particular locales.[7]

29.3.10 Risk Factors

The most widely recognized risk factor for contracting Q fever is contact with domestic ruminants, particularly newborn or pregnant animals.[128] Bacterial shedding is demonstrable

TABLE 29.1

Human Seroprevalence of *C. burnetii* in 17 Countries

Country	Notifications (per 100,000)	Seroprevalence (Sample Size)	References
Algeria	—	15.5% (729)	[342]
Australia	1.7–4.9	—	[313]
Canada	—	1.0% (250)	[340]
Croatia	3.3	—	[344]
Egypt	—	20.0% (358)	[338]
France	—	Acute, 2.1%; chronic, 0.9% (179,794)	[290]
Germany	—	7.5% (1,036)	[172]
Greece	—	7.5% (1,007)	[345]
Italy	—	13.6% (280)	[142]
Netherlands[a]	—	2.4% (5,654)	[335]
Northern Ireland	—	12.8% (2,394)	[336]
People's Republic of China	—	6.4% (220)	[346]
Senegal	—	3.7% (241)–24.8% (238)	[343]
Spain	—	23.1% (863)–48.6% (595)	[141,172]
Tunisia	—	26.0% (500)	[341]
Turkey	—	32.3% (601)	[339]
United States	0.028	3.1% (4,437)	[137,140,337]

[a] Recorded in 2006–2007 prior to Q fever outbreaks beginning in 2007.

in feces, urine, products of parturition, and milk.[129–131] While direct contact is the strongest risk factor, merely living or working in rural areas is also significantly associated with *C. burnetii* exposure.[124,128,132–135]

Q fever is an occupational hazard for people working with or around domestic livestock,[136] and veterinarians have a higher average seroprevalence than the general population.[137,138]

The Q fever risk is 5 times greater in individuals aged >15 years compared with those aged <15 years,[139] and the 30–69-year-old age group is most at risk[128] with risk generally increasing proportionally with age.[140–142] Men are over twice as likely to contract *C. burnetii* infection compared to women, and this may be related to sex hormone influences.[143,144] It is widely held that Q fever is a significant disease of pregnant women and that its incidence is underestimated in this group.[145,146] However, the clinical consequences of *C. burnetii* infection in pregnant women remain uncertain.[5,146–148]

29.3.11 Reservoirs

The most epidemiologically significant hosts of *C. burnetii* are domestic ruminants (sheep, cattle, and goats), although a wide variety of vertebrate hosts, from marsupials to birds to rodents and domestic pets, is well known.[149–155]

Direct transmission from ticks is not thought to be a significant source of human infection; however, ticks have been implicated for cycling *C. burnetii* in wild vertebrates[156] and maintenance through vertical transmission.[20,157]

Haemaphysalis humerosa was first implicated in the transmission of *C. burnetii* not long after the bacterium was discovered,[8,158] and since then, various studies have indicated that arthropods are almost universally susceptible.[159–162]

C. burnetii is also capable of infecting and persisting in amoebae,[163,164] like its close relative *L. pneumophila*.

29.3.12 Transmission

Inhalation of bacteria via infectious aerosols or dust contaminated with birthing fluids of domestic ruminants is the commonest mode of *C. burnetii* transmission.[165–169] Modeling suggests that ≥200 cells/m^3 of air are required to establish human infections,[170] possibly equating to as few as 1 viable bacterium inoculated into a lung.[61,171] Aerosolized bacteria can spread up to 5 km.[172] Ruminant products of parturition can have particularly high concentrations of bacteria,[173] and contaminated environmental matrices may remain infectious for years.[174]

Less common routes of infection include ingestion of unpasteurized milk products,[175–178] tick bites,[123,164,179–183] transplacental congenital infection,[184] blood[185,186] and bone marrow transplantation,[187] and sexual transmission.[188–190]

Despite the concern over blood donations following the Q fever outbreaks in the Netherlands starting in 2007, the European Commission Directorate-General for Health and Consumers concluded that the risk of transmission of infection from a blood or organ transplant was less than the risk of acquiring an infection from the environment of regions affected by the epidemic.[172]

29.4 CLINICAL FEATURES AND PATHOGENESIS

29.4.1 ACUTE Q FEVER

Up to 60% of human *C. burnetii* infections are asymptomatic.[143] Acute Q fever manifests in approximately 38% of human *C. burnetii* cases as a mild to moderate, self-limiting influenza-like illness.[143,164] Symptoms are typically nonspecific and may include fever of approximately 2-week duration, severe headache, chills, fatigue, and sweats. The incubation period varies from a few days to several weeks, and the disease severity varies proportionally with the infecting dose in otherwise healthy individuals.[191–193]

Severe acute Q fever requiring patient hospitalization occurs in 2%–5% of cases and is associated with greater exposure to *C. burnetii* organisms. Fatalities are rare (<1%), with myocarditis as the leading proximate cause of death.[29,143,194] The most consistently reported symptoms in cases of acute Q fever in descending order of frequency are fever, fatigue, arthromyalgia, hepatitis, chills, headache, sweats, gastrointestinal symptoms, pneumonia, skin rash, and neck stiffness.[164,195]

Some regional variation in clinical presentation exists: for instance, in Greece, Spain, Japan, Italy, Germany, and the Netherlands, atypical pneumonia is most often reported; in France, acute hepatitis is the most common presentation, while prolonged fever of unknown origin is most common in Andalusia, California, and Australia.[196]

In pregnant women, Q fever can lead to spontaneous abortion, intrauterine fetal death (IUFD), premature delivery, or intrauterine growth retardation.[197] Transplacental *in utero* infection of the fetus has also been reported.[198–200] Q fever is associated with oligoamnios, a recognized cause of neonatal morbidity and mortality, and first-trimester Q fever is especially associated with undesirable outcomes.[197] After infection, breast feeding is, of course, contraindicated.[199] Despite this, data collected during the 2007–2010 Dutch Q fever outbreak found no evidence of adverse pregnancy outcomes in women with asymptomatic *C. burnetii* infection acquired in early pregnancy.[5]

29.4.2 CHRONIC Q FEVER

Up to 9% of patients fail to completely recover from acute Q fever and develop chronic Q fever[201] that is defined as an infection persisting for >6 months with high IgG antiphase I antibodies and most often presenting as infective endocarditis.[196] Chronic Q fever can arise years after the initial presentation of acute disease[193] and can lead to spontaneous abortions.[198,202,203] The immunocompromised or those with prior cardiovascular disease are at increased risk for developing chronic Q fever.[204] Valvular endocarditis, particularly of the aortic and mitral valves,[205] is the most common manifestation of chronic Q fever and is diagnosed in 60%–70% of cases.[206] Approximately 90% of Q fever endocarditis patients have preexisting valvular heart disease.[207] Up to one third of acute Q fever patients with cardiac valve abnormalities may develop endocarditis.[208] The most common presentations of

chronic disease include vascular graft infections, aneurism infections,[209] aortic and/or mitral valve signs, cardiac failure, and arterial embolism.[210] Other symptoms include osteomyelitis,[211] digital clubbing and purpuric rash,[210] chronic hepatitis, septic arthritis, interstitial pneumonia, and chronic fatigue syndrome.[29,212–214]

Rarely, *C. burnetii* may infect the gall bladder[215,216], pericardium[217], uvea[218,219], and subcutis.[220]

29.4.3 Q FEVER FATIGUE SYNDROME

Q fever fatigue syndrome (QFS) is a manifestation of chronic Q fever clinically similar to idiopathic chronic fatigue syndrome. *C. burnetii* DNA can be recovered from QFS sufferers many years after their primary infection, but infectious/viable bacterial cells cannot.[221] Noninfective, nonbiodegradable, antigenic *C. burnetii* material including DNA may induce an aberrant immune response, including the release of cytokines,[222,223] which could mediate the development of QFS. Patient genetic factors may also play a role.[224]

29.4.4 *C. BURNETII* INFECTION IN ANIMALS

In contrast to human Q fever, animal infection with *C. burnetii* is usually devoid of any detectable clinical signs.[123] The term "coxiellosis" has been introduced to refer to animal infection with *C. burnetii*.[225] *C. burnetii* localizes in the uterus and mammary glands of infected ruminants.[214] Abortion may occur with *C. burnetii* infection in ruminants: aborted fetuses usually appear grossly unremarkable; however, there may be exudative placentitis with intercotyledonary fibrosis.[226] Following parturition, metritis may result.[7] *C. burnetii* can also be recovered from milk for up to 42 months.[174] Goats shed *C. burnetii* in feces before and after kidding, and the mean duration of excretion is 20 days.[130] High concentrations of *C. burnetii* organisms can be detected in placentas and amniotic fluid of infected ruminants.[227,228]

29.4.5 PATHOGENESIS

Crucial to the pathogenesis of Q fever is the competence and nature of the host immune response. Immature dendritic cells (DCs) are likely to be among the first to internalize *C. burnetii* during natural infections.[229] When phase I *C. burnetii* infect human DCs, there is little evidence of DC maturation or inflammatory cytokine production.[230,231] In contrast, phase II bacteria trigger marked DC maturation and inflammatory cytokine expression.[229] Full-length *C. burnetii* LPS is required to avoid marked DC stimulation,[230,231] which suggests phase I LPS may act as a shield to hide *C. burnetii* surface molecules from pattern recognition receptors.[229] Muted DC maturation and proinflammatory signaling in response to virulent *C. burnetii* would presumably result in the presentation of bacterial antigens in such a way as to induce tolerance to *C. burnetii* antigen *in vivo*, thereby contributing to the establishment of persistent infection.[229]

Using TLR knockout mouse models, it has been shown that both TLR2 and TLR4 influence inflammatory responses to *C. burnetii*, and both are dispensable for bacterial clearance.[232–234] Given that the infectious dose of *C. burnetii* is so low,[61,171] the innate immune system appears to be incapable of stifling *C. burnetii* infection by itself,[229] so the adaptive immune response is central to determining the outcome of *C. burnetii* infections.

Passive transfer of antibody can protect immunocompetent animals against subsequent challenge with *C. burnetii*,[235–237] but this fails for SCID mice, which proves the importance of the cell-mediated arm of the adaptive immune response in clearing this organism.[236] There are conflicting literature reports on the role of antibody opsonization in the pathogenesis of natural *C. burnetii* infection, but it seems that opsonization of *C. burnetii* does not hinder and may actually potentiate the infection.[238,239]

Cell-mediated immunity is clearly important in controlling *C. burnetii* infection.[229] B cell-deficient mice display more severe pathology than wild-type mice exposed to *C. burnetii*, indicating a potential role for anti-*C. burnetii* antibodies in minimizing tissue damage. However, these antibodies do not appear essential for the resolution of the primary infection.[240,241] T cells are vital in controlling *C. burnetii* in experimental mice, with CD8+ T cells of greater use than CD4+ T cells, though either subset is sufficient to control the infection.[241] The production of IFN-γ is likely to be one key factor that makes T cells indispensable in the effective control of *C. burnetii* infections.[229,231,240]

The immunopathogenesis of chronic Q fever is thought to depend on deficiencies in cell-mediated immunity.[229] Monocytes from chronic Q fever patients are unable to kill *C. burnetii*.[242,243] This manifests as reduced granuloma formation in chronic Q fever patients compared with those who experience acute Q fever only.[143] The suppression of cell-mediated immunity in mice, whether due to pregnancy, corticosteroids, or gamma irradiation, permits the reactivation of persistent *C. burnetii* infections.[244–246] Patients with chronic Q fever endocarditis had significantly higher concentrations of Foxp3+ regulatory T cells[247] and were more likely to have mutations in their IL-10 promoter microsatellites R and G and to have the TNF-α receptor II 196R polymorphism.[224]

The living bacterium itself stimulates macrophages to adopt an atypical M2 activation program that facilitates intracellular survival.[248] M2 macrophages secrete a range of antiinflammatory cytokines including IL-10, and it is well established that chronic Q fever patients overproduce IL-10.[248–250] Indeed, IL-10 appears to be a key cytokine in the immunopathogenesis of chronic Q fever.[251–254]

29.5 IDENTIFICATION AND DIAGNOSIS

29.5.1 CULTURE

Animal inoculation remains the gold standard for determining the viability of *C. burnetii*,[173] and despite the development of plaque assays[68] and other culture-based methods,[255] results

from *in vitro* culture have correlated poorly with titrations in animals.[256,257] Culture from clinical samples is not done routinely for diagnosis due to inherent infection risks[258] and its significantly lower sensitivity than PCR and IFA.[146] Recently, scientists discovered how to culture *C. burnetii* in an axenic medium,[59] but the risk of laboratory infection may prevent the use of this method for routine diagnosis.

29.5.2 MOLECULAR METHODS

PCR can be used to detect *C. burnetii* DNA in clinical samples, environmental samples including dust and air, animal waste products, and milk products.[129,165,259–264]

PCR is clinically useful for diagnosis early in infection, before a consistent immune response has developed[265–268] and in blood culture-negative infective endocarditis.[269] Rapid typing of *C. burnetii* isolates may be used in epidemiological studies to trace the source of an outbreak[270,271] and therefore inform the selection of appropriate control measures. High sensitivity, good interlaboratory reproducibility,[272] and the potential for quantitative results[273] are all advantages of PCR. Depending on the detection chemistry used, the genetic target, and the sample type being tested, PCR may be able to detect as few as three *C. burnetii* genomes per reaction.[274] High copy number targets such as the *IS1111a* repetitive element are often used due to the potential for increased sensitivity.[272] However, the copy number per genome of the *IS* repetitive element can vary substantially between strains, with reported variation ranging from as few as 7 copies to as many as 110.[273]

Several PCR assays may be combined into multiplex microarrays, enabling simultaneous detection of multiple organisms. Microarray assays have been developed that target multiple potential bioterror agents, including *C. burnetii*, and the hope is to reliably accelerate the detection of these agents in field samples.[275–277]

PCR-based assays may be inhibited by various substances in certain samples[278] and poor performance in some clinical samples,[146,147] and the relevance of pathogen DNA detection to the presence of viable bacteria remains debatable.[221,279] For these reasons and others, PCR remains a diagnostic tool only applicable to reference and research laboratories.[280]

29.5.3 IMMUNOLOGICAL METHODS

Indirect immunofluorescence (IFA) and enzyme-linked immunosorbent assay (ELISA) are commercially available for Q fever diagnosis; however, wide interlaboratory variability indicates the need for better standardized tests.[281,282] Despite significant limitations, measuring the host antibody response to *C. burnetii* is currently the mainstay of Q fever diagnosis and patient monitoring.

29.5.3.1 Indirect Immunofluorescence Assay

IFA is considered the reference diagnostic method for Q fever,[283,284] although it is not a true "gold standard."[127] IFA is superior to complement fixation[285,286] and can detect IgG,

IgM, or IgA antiphase I and II *C. burnetii* antibodies.[258] Acute Q fever is diagnosed when two samples, taken ≥8 days apart, show a fourfold increase in antiphase II IgG antibodies to a dilution of ≥1 in 1024. Alternatively, an antiphase II IgG titer ≥1 in 1024 accompanied by an antiphase II IgM antibody titer ≥1 in 256 is also considered diagnostic of acute Q fever. If the accompanying antiphase II IgM titer is <1 in 256, this is considered to indicate previous infection.[287] Chronic Q fever is diagnosed when the patient's antiphase I IgG titer is >1:1600.[288–290] Some suggest an "equivocal" zone should be employed, rather than strictly applied cutoffs, with any "equivocal" results being retested with a different immunological assay.[291]

Despite its status as the reference diagnostic method, the IFA is flawed. It is more laborious than ELISA,[127] relies on subjective interpretation, and is not easily standardized,[283,292,293] and the concordance between laboratories has been poor for diagnosing chronic disease.[294]

29.5.3.2 Enzyme-Linked Immunosorbent Assay

Compared to IFA, ELISA is better suited to automation[127] and therefore may be a more useful epidemiological tool.[286] Comparisons between ELISA and IFA have given inconsistent results,[265,293] but any superiority in sensitivity and/or specificity of IFA should be balanced against the assay's cost, subjectivity, incompatibility with automation, and potential for operator error.[293] ELISA is also superior to the complement fixation test (CFT) in regard to both sensitivity and specificity.[286,295,296]

The ELISA also has disadvantages and remains the second choice for routine serological diagnosis. For instance, commercially available tests only detect IgG,[296] the source of the antigen may affect test results,[295] and considerable technical proficiency is required to perform and interpret the test.[258]

29.5.3.3 Complement Fixation Test

The CFT has limited utility in human samples due to the availability of more sensitive and specific tests, but can be used for surveillance in animals.[297] When used for veterinary surveillance, CFT has been found inferior to commercially available ELISAs.[295,296] Furthermore, complement-fixing antibodies are often not present during early infection but can persist for long periods after primary infection, thus complicating interpretation of results.[297]

29.5.3.4 Immunohistochemistry

Immunohistochemistry (IHC) is relatively safe since live bacteria are not required and tissues are fixed prior to staining and analysis. This test has found use in the confirmation of Q fever endocarditis through biopsy of cardiac valves,[298] aortic grafts,[299] and liver[300] and may be particularly useful where blood cultures are negative.[298] However, IHC is not suitable for large epidemiological studies and has been used more commonly in animal studies.[85]

29.6 TREATMENT AND PREVENTION

29.6.1 Antibiotics

Postexposure prophylactic therapy for Q fever may be effective if initiated within 12 days of infection.[4] Current guidelines for medical treatment of Q fever (acute or chronic) include doxycycline (100 mg twice daily) and hydroxychloroquine (200 mg three times daily) for ≥18 months.[164] For Q fever endocarditis, quinolone/doxycycline therapy for ≥3 years is recommended.[301] Serum concentrations of doxycycline and phase I antibodies should be measured and the dose adjusted to achieve a sufficient effective antibiotic concentration.[302] Following antibiotic therapy for chronic Q fever, successful treatment is best detected serologically, with a decrease in phase I IgG and IgA of ≥2 dilutions in a year.[303]

Doxycycline and chloroquine are contraindicated in pregnant women; therefore, co-trimoxazole for >5 weeks may be used instead. After the birth, doxycycline–hydroxychloroquine therapy can be reinstituted.[197]

For chronic Q fever endocarditis with substantial valvular damage, surgery is indicated. Q fever endocarditis patients with prosthetic valves should ideally have the valve replaced,[304] although treatment with doxycycline and hydroxychloroquine for >2 years may be considered. Due to the risk of relapse, patients should be monitored serologically for ≥5 years.[305]

29.6.2 Prevention

29.6.2.1 Behavioral

In the event of an outbreak of Q fever, adjacent areas should be notified and awareness campaigns instituted among health-care providers[306] and at-risk individuals.[167] For workers in outbreak areas, personal protective equipment (filtering face pieces, masks, gloves, overalls, and hairnets) have proven ineffective in preventing seroconversion and therefore vaccination is recommended.[307]

People from an epidemic area should defer donating blood for 6 weeks, and individuals with acute Q fever should defer donating blood for 2 years following confirmed cure.[306]

29.6.2.2 Vaccination

Early efforts to produce an effective, safe vaccine to protect against Q fever have been reviewed.[308] An attenuated live vaccine was tested in Russia but was abandoned due to safety concerns; a chloroform–methanol residue (CMR) extracted vaccine was trialed in animals but not humans[136]; and Q-Vax[123] has been available in Australia since 1989.[136] Phase I trials of a CMR vaccine have been conducted,[309] but currently, Q-Vax is the only human vaccine available commercially and it is only approved for use in Australia.

Evaluations of Q-Vax have shown it to confer a 97% protection rate in those at risk of occupational exposure to *C. burnetii*.[310] In 2006, Q-Vax had been administered to approximately 49,000 Australians (predominantly abattoir workers) and appears to have significantly reduced Q fever notification rates.[136]

A licensed vaccine is not available in the United States[164] or Europe, but following the Netherlands outbreak, the EDCD recommended that Q-Vax be made available in Europe until a new vaccine is developed.[172] The major limitation restricting the widespread deployability of Q-Vax is the two-step screening and administration process used to minimize the risk of a serious adverse event, such as injection site granulomas and necrotic lesions.[311–313]

29.6.2.3 Reservoir and Environmental Control

The extreme environmental resistance of *C. burnetii*[45] makes it difficult to control. However, through surveillance testing of bulk tank milk,[314] dairy farms with infected herds can be identified and appropriate control measures implemented. Effective control measures include crutch clipping ewes prior to lambing, confinement housing of animals at lambing in winter and spring,[315] ectoparasite control,[123] animal vaccination, hygiene measures,[166] and, in extreme circumstances, culling animals.[166,314]

29.6.2.4 Decontamination

The SCV of *C. burnetii* is extremely resistant to high temperature, high pressure, desiccation, and osmotic shock.[117,119,316] However, *C. burnetii* is highly sensitive to oxidative stress: relatively low concentrations of hydrogen peroxide (10 µmol/L H_2O_2) completely inhibited its growth.[59]

High-risk materials such as aborted fetuses, placentas, and contaminated bedding should be buried with lime or incinerated, while feces from infected herds should be treated with lime or CaCN before using it as a fertilizer.[1] The UK Health Protection Agency advises that decontaminating surfaces with 2% formaldehyde, 1% Lysol, 5% hydrogen peroxide, 70% ethanol, or 5% chloroform are "thought to be effective." For spills or contaminated material, it is suggested that hypochlorite (5000 ppm available chlorine), 5% peroxide, or phenol-based solutions should be used immediately.[317] However, in liquid waste contaminated with organic matter, the disinfection efficacy of hypochlorite and peroxide may be significantly reduced.[318,319] Heat, such as is applied in pasteurization, also effectively inactivates *C. burnetii*.[117]

29.6.2.5 Vaccination and Antibiotic Treatment of Reservoirs

Young animals infected with *C. burnetii* shed bacteria profusely and contribute significantly to environmental contamination.[320,321] Antibiotic therapy reduces the incidence of abortions and the quantity of organisms shed at parturition[123] but may have little effect on the number of animals shedding overall.[322] In contrast, vaccination reduces shedding and abortions[123] but has reduced efficacy under conditions of high exposure.[320,323] It has been suggested that to achieve a significant effect, vaccination programs need to continue for >4 years.[323,324] Recently, a phase I vaccine for animal use has become available (Coxevac, CEVA Santè Animale)[124] and was successfully used in the Netherlands outbreak.[314]

29.7 CONCLUSIONS AND FUTURE PERSPECTIVES

The single most important aspect of preparedness for intentional release of *C. burnetii* is the capacity to promptly identify the agent involved.[325] New technologies, such as protein microarrays[229,326,327] and mass spectrometry,[328,329] are being used to hasten the identification of candidate antigens for use in serological diagnostic assays. DNA microarray technologies and molecular typing methods also promise to facilitate the rapid identification of a bioweapon with no prior knowledge of the organism used.[275,276,330]

The most significant recent breakthrough in *C. burnetii* research has been the development of an axenic growth medium. This discovery, in conjunction with recent technical breakthroughs that now permit molecular dissection of virulence factors and genetic engineering of *C. burnetii*, has reinvigorated the *C. burnetii* research community and promises to provide new insights into virulence factors and pathogenesis mechanisms.[331–334] While this new age of *C. burnetii* research may aid preventative detection and treatment approaches, forthcoming discoveries could also conceivably provide solutions to the technical barriers that make weaponizing *C. burnetii* difficult.[1]

REFERENCES

1. Oysten, P.C.F. and Davies, C., Q fever: The neglected biothreat agent, *J. Med. Microbiol.*, 60, 9, 2011.
2. Madariaga, M.G. et al., Q fever: A biological weapon in your backyard, *Lancet Infect. Dis.*, 3, 709, 2003.
2a. CDC National Select Agents Registry. HHS and USDA Select Agents and Toxins, 7 CFR Part 331, 9 CFR Part 121, and 42 CFR Part 73. http://www.selectagents.gov/resources/List_of_Select_Agents_and_Toxins_2013-09-10.pdf (accessed on November 5, 2013).
3. World Health Organization; Group of Consultants, *Health Aspects of Chemical and Biological Weapons*. World Health Organization, Geneva, Switzerland, 1970.
4. Moodie, C.E. et al., Prophylaxis after exposure to *Coxiella burnetii*, *Emerg. Infect. Dis.*, 14, 1558, 2008.
5. Van der Hoek, W. et al., Epidemic Q fever in humans in the Netherlands. In: Toman, R. et al. (eds.), *Coxiella burnetii: Recent Advances and New Perspectives in Research of the Q Fever Bacterium*, *Adv. Exp. Med. Biol.*, 984, 329, 2012.
6. Limonard, G.J.M. et al., Detailed analysis of health status of Q fever patients 1 year after the first Dutch outbreak: A case-control study, *QJM*, 103, 953, 2010.
7. Arricau-Bouvery, N. and Rodolakis, A., Is Q fever an emerging or re-emerging zoonosis?, *Vet. Res.*, 36, 327, 2005.
8. Derrick, E.H., *Rickettsia burneti*: The cause of 'Q' fever, *Med. J. Aust.*, 1, 14, 1939.
9. Weisburg, W.G. et al., Phylogenetic diversity of the Rickettsiae, *J. Bacteriol.*, 171, 4202, 1989.
10. Waag, D.M. and Thompson, H.A., Pathogenesis of and immunity to *Coxiella burnetii*. In: Linder, L.E., Lebeda, F.J., and Korch, G.W. (eds.), *Biological Weapons Defense: Infectious Diseases and Counter Bioterrorism*, p. 185. Humana Press, Totowa, NJ, 2005.

11. Zhong, J., *Coxiella*-like endosymbionts. In: Toman, R. et al. (eds.), *Coxiella burnetii: Recent Advances and New Perspectives in Research of the Q Fever Bacterium*, Adv. Exp. Med. Biol., 984, 365, 2012.

12. Taylor, M. et al., Endosymbiotic bacteria associated with nematodes, ticks and amoebae, *FEMS Immunol. Med. Microbiol.*, 64, 21, 2012.

13. Andreotti, R. et al., Assessment of bacterial diversity in the cattle tick *Rhipicephalus (Boophilus) microplus* through tag-encoded pyrosequencing, *BMC Microbiol.*, 11, 6, 2011.

14. Noda, H., Munderloh, U.G., and Kurtti, T.J., Endosymbionts of ticks and their relationship to *Wolbachia* spp. and tick-borne pathogens of humans and animals, *Appl. Environ. Microbiol.*, 63, 3926, 1997.

15. Mediannikov, O. et al., Molecular evidence of *Coxiella*-like microorganism harboured by *Haemaphysalis concinna* ticks in the Russian far east, Rickettsiology: Present and Future Directions, *Ann. N. Y. Acad. Sci.*, 990, 226, 2003.

16. Klyachko, O. et al., Localization and visualization of a *Coxiella*-type symbiont within the lone star tick, *Amblyomma americanum*, *Appl. Environ. Microbiol.*, 73, 6584, 2007.

17. Clay, K. et al., Microbial communities and interactions in the lone star tick, *Amblyomma americanum*, *Mol. Ecol.*, 17, 4371, 2008.

18. Machado-Ferreira, E. et al., *Coxiella* symbionts in the cayenne tick *Amblyomma cajennense*, *Microb. Ecol.*, 62, 134, 2011.

19. Reeves, W.K. Molecular evidence for a novel *Coxiella* from *Argas monolakensis* (Acari: Argasidae) from Mono Lake, California, USA, *Exp. Appl. Acarol.*, 44, 57, 2008.

20. Reeves, W.K. et al., Molecular and biological characterization of a novel *Coxiella*-like agent from *Carios capensis*, *Ann. N. Y. Acad. Sci.*, 1063, 343, 2005.

21. Satta, G. et al., Pathogens and symbionts in ticks: A survey on tick species distribution and presence of tick–transmitted microorganisms in Sardinia, Italy, *J. Med. Microbiol.*, 60, 63, 2011.

22. Jasinskas, A., Zhong, J., and Barbour, A.G., Highly prevalent *Coxiella* sp. bacterium in the tick vector *Amblyomma americanum*, *Appl. Environ. Microbiol.*, 73, 334, 2007.

23. Heise, S.R., Elshahed, M.S., and Little, S.E., Bacterial diversity in *Amblyomma americanum* (Acari: Ixodidae) with a focus on members of the genus *Rickettsia*, *J. Med. Entomol.*, 47, 258, 2010.

24. Shivaprasad, H.L. et al., *Coxiella*-like infection in psittacines and a toucan, *Avian Dis.*, 52, 426, 2008.

25. Vapniarsky, N., Barr, B.C., and Murphy, B., Systemic *Coxiella*-like infection with myocarditis and hepatitis in an eclectus parrot (*Eclectus roratus*), *Vet. Pathol.*, 49, 717, 2012.

26. Woc-Colburn, A.M. et al., Fatal coxiellosis in Swainson's Blue Mountain rainbow lorikeets (*Trichoglossus haematodus moluccanus*), *Vet. Pathol.*, 45, 247, 2008.

27. Kim, E. et al., Complex array of endobionts in *Petalomonas sphagnophila*, a large heterotrophic euglenid protist from *Sphagnum*-dominated peatlands, *ISME J.*, 4, 1108, 2010.

28. Michaud, L. et al. Phylogenetic characterization of the heterotrophic bacterial communities inhabiting a marine recirculating aquaculture system, *J. Appl. Microbiol.*, 107, 1935, 2009.

29. Derrick, E.H., "Q" fever, new fever entity: Clinical features, diagnosis, and laboratory investigation, *Med. J. Aust.*, 2, 281, 1937.

30. Burnet, F.M. and Freeman, M., Experimental studies on the virus of "Q" fever, *Med. J. Aust.*, 2, 299, 1937.

31. Davis, G.E. and Cox, H.R., A filter-passing infectious agent isolated from ticks. I. Isolation from *Dermacentor andersoni*, reactions in animals, and filtration experiments, *Public Health Rep.*, 53, 2259, 1938.

32. Dyer, R.E., A filter-passing infectious agent isolated from ticks. IV. Human infection, *Public Health Rep.*, 53, 2277, 1938.

33. Cox, H.R., Studies of a filter-passing infectious agent isolated from ticks. V. Further attempts to cultivate in cell-free media. Suggested classification, *Public Health Rep.*, 54, 1822, 1939.

34. Dyer, R.E., Similarity of Australian "Q" fever and a disease caused by an infectious agent isolated from ticks in Montana, *Public Health Rep.*, 54, 1229, 1939.

35. Philip, C.B., Comments on the name of the Q fever organism, *Public Health Rep.*, 63, 58, 1948.

36. Atlas, R.M., Bioterrorism and biodefence research: Changing the focus of microbiology, *Nat. Rev. Microbiol.*, 1, 70, 2003.

37. Hackstadt, T., Biosafety concerns and *Coxiella burnetii*, *Trends Microbiol.*, 4, 341, 1996.

38. Anon. Australian/New Zealand Standard. Safety in laboratories. Part 3: Microbiological safety and containment (AS/NZS 2243.3:2010), SAI Global Limited, Sydney, Australia, 2010.

39. Hackstadt, T. and Williams, J.C., Biochemical stratagem for obligate parasitism of eukaryotic cells by *Coxiella burnetii*, *Proc. Natl. Acad. Sci. USA*, 78, 3240, 1981.

40. Minnick, M.F. and Raghavan, R., Developmental biology of *Coxiella burnetii*. In: Toman, R. et al. (eds.), *Coxiella burnetii: Recent Advances and New Perspectives in Research of the Q Fever Bacterium*, Adv. Exp. Med. Biol., 984, 231, 2012.

41. Giménez, D.F., Gram staining of *Coxiella burnetii*, *J. Bacteriol.*, 90, 834, 1965.

42. McCaul, T.F. and Williams, J.C., Developmental cycle of *Coxiella burnetii*: Structure and morphogenesis of vegetative and sporogenic differentiations, *J. Bacteriol.*, 147, 1063, 1981.

43. Amano, K. et al., Biochemical and immunological properties of *Coxiella burnetii* cell wall and peptidoglycan-protein complex fractions, *J. Bacteriol.*, 160, 982, 1984.

44. McCaul, T.F. et al., *In vivo* endogenous spore formation by *Coxiella burnetii* in Q fever endocarditis, *J. Clin. Pathol.*, 47, 978, 1994.

45. Heinzen, R.A., Hackstadt, T., and Samuel, J.E., Developmental biology of *Coxiella burnetii*, *Trends Microbiol.*, 7, 149, 1999.

46. McCaul, T.F., The developmental cycle of *Coxiella burnetii* in Q fever. In: Williams, J.C. and Thompson, H.A. (eds.), *The Biology of Coxiella burnetii*. CRC Press, Boca Raton, FL, 1991, p. 149.

47. Seshadri, R. et al., Complete genome sequence of the Q-fever pathogen *Coxiella burnetii*, *Proc. Natl. Acad. Sci. USA*, 100, 5455, 2003.

48. Samuel, J.E. et al., Isolation and characterization of a plasmid from phase I *Coxiella burnetii*, *Infect. Immun.*, 41, 488, 1983.

49. Valková, D. and Kazár, J., A new plasmid (QpDV) common to *Coxiella burnetii* isolates associated with acute and chronic Q fever, *FEMS Microbiol. Lett.*, 125, 275, 1995.

50. Mallavia, L.P., Genetics of rickettsiae, *Eur. J. Epidemiol.*, 7, 213, 1991.

51. Jäger, C. et al., *Coxiella burnetii* plasmid types QpDG and QpH1 are closely related and likely identical, *Vet. Microbiol.*, 89, 161, 2002.

52. Lautenschläger, S. et al., Sequencing and characterization of the cryptic plasmid QpRS from *Coxiella burnetii*, *Plasmid*, 44, 85, 2000.

53. Willems, H. et al., Plasmid-homologous sequences in the chromosome of plasmidless *Coxiella burnetii* Scurry Q217, *J. Bacteriol.*, 179, 3293, 1997.

54. Savinelli, E.A. and Mallavia, L.P., Comparison of *Coxiella burnetii* plasmids to homologous chromosomal sequences present in plasmidless endocarditis-causing isolate, *Ann. N.Y. Acad. Sci.*, 590, 523, 1990.

55. Beare, P.A. et al., Comparative genomics reveal extensive transposon-mediated genomic plasticity and diversity among potential effector proteins within the genus *Coxiella*, *Infect. Immun.*, 77, 642, 2009.

56. Sekeyová, Z., Roux, V., and Raoult, D. et al., Intraspecies diversity of *Coxiella burnetii* as revealed by *com1* and *mucZ* sequence comparison, *FEMS Microbiol. Lett.*, 180, 61, 1999.

57. Stoenner, H.G. and Lackman, D.B., The biologic properties of *Coxiella burnetii* isolated from rodents collected in Utah, *Am. J. Hyg.*, 71, 45, 1960.

58. Parkhill, J. et al., Genome sequence of *Yersinia pestis*, the causative agent of plague, *Nature*, 413, 523, 2001.

59. Omsland, A. and Heinzen, R.A., Life on the outside: The rescue of *Coxiella burnetii* from its host cell, *Annu. Rev. Microbiol.*, 65, 111, 2011.

60. Omsland, A., Axenic growth of *Coxiella burnetii*. In: Toman, R. et al. (eds.), *Coxiella burnetii: Recent Advances and New Perspectives in Research of the Q Fever Bacterium*, *Adv. Exp. Med. Biol.*, 984, 215, 2012.

61. Moos, A. and Hackstadt, T., Comparative virulence of intra- and interstrain lipopolysaccharide variants of *Coxiella burnetii* in the guinea pig model, *Infect. Immun.*, 55, 1144, 1987.

62. Hackstadt, T. et al., Lipopolysaccharide variation in *Coxiella burnetii*: Intrastrain heterogeneity in structure and antigenicity, *Infect. Immun.*, 48, 359, 1985.

63. Narasaki, C.T. and Toman, R., Lipopolysaccharide of *Coxiella burnetii*. In: Toman, R. et al. (eds.), *Coxiella burnetii: Recent Advances and New Perspectives in Research of the Q Fever Bacterium*, *Adv. Exp. Med. Biol.*, 984, 65, 2012.

64. Stoker, M.G.P. and Fiset, P., Phase variation of the Nine Mile and other strains of *Rickettsia burneti*, *Can. J. Microbiol.*, 2, 310, 1956.

65. Ftácek, P., Skultéty, L., and Toman, R., Phase variation of *Coxiella burnetii* strain Priscilla: Influence of this phenomenon on biochemical features of its lipopolysaccharide, *J. Endotoxin Res.*, 6, 369, 2000.

66. Vodkin, M.H. and Williams, J.C., Overlapping deletion in two spontaneous phase variants of *Coxiella burnetii*, *J. Gen. Microbiol.*, 132, 2587, 1986.

67. Burton, P.R. et al. Some ultrastructural effects of persistent infections by the rickettsia *Coxiella burnetii* in mouse L cells and green monkey kidney (Vero) cells, *Infect. Immun.*, 21, 556, 1978.

68. Ormsbee, R. et al., Limits of rickettsial infectivity, *Infect. Immun.*, 19, 239, 1978.

69. Andoh, M. et al., Comparative virulence of phase I and II *Coxiella burnetii* in immunodeficient mice, *Ann. N. Y. Acad. Sci.*, 1063, 167, 2005.

70. Shannon, J.G., Howe, D., and Heinzen, R.A., Virulent *Coxiella burnetii* does not activate human dendritic cells: Role of lipopolysaccharide as a shielding molecule, *Proc. Natl. Acad. Sci. USA*, 102, 8722, 2005.

71. Capo, C. et al., Subversion of monocyte functions by *Coxiella burnetii*: Impairment of the cross-talk between αvβ3 integrin and CR3, *J. Immunol.*, 163, 6078, 1999.

72. Vishwanath, S. and Hackstadt, T., Lipopolysaccharide phase variation determines the complement-mediated serum susceptibility of *Coxiella burnetii*, *Infect. Immun.*, 56, 40, 1988.

73. Hackstadt, T., Steric hindrance of antibody binding to surface proteins of *Coxiella burnetii* by phase I lipopolysaccharide, *Infect. Immun.*, 56, 802, 1988.

74. Schramek, S. and Mayer, H., Different sugar compositions of lipopolysaccharides isolated from phase I and pure phase II cells of *Coxiella burnetii*, *Infect. Immun.*, 38, 53, 1982.

75. Schramek, S., Radziejewska-Lebrecht, J., and Mayer, H., 3-C branched aldoses in lipopolysaccharide of phase I *Coxiella burnetii* and their role as immunodominant factors, *Eur. J. Biochem.*, 148, 455, 1985.

76. Amano, K. et al., Structure and biological relationships of *Coxiella burnetii* lipopolysaccharides, *J. Biol. Chem.*, 262, 4740, 1987.

77. Denison, A.M., Massung, R.F., and Thompson, H.A., Analysis of the O-antigen biosynthesis regions of phase II isolates of *Coxiella burnetii*, *FEMS Microbiol. Lett.*, 267, 102, 2007.

78. Baca, O.G., Klassen, D.A., and Aragon, A.S., Entry of *Coxiella burnetii* into host cells, *Acta Virol.*, 37, 143, 1993.

79. Meconi, S. et al., *Coxiella burnetii* induces reorganization of the actin cytoskeleton in human monocytes, *Infect. Immun.*, 66, 5527, 1998.

80. Capo, C. et al., *Coxiella burnetii* avoids macrophage phagocytosis by interfering with spatial distribution of complement receptor 3, *J. Immunol.*, 170, 4217, 2003.

81. Coleman, S.A. et al., Temporal analysis of *Coxiella burnetii* morphological differentiation, *J. Bacteriol.*, 186, 7344, 2004.

82. Baca, O.G. and Paretsky, D., Q fever and *Coxiella burnetii*: A model for host-parasite interactions, *Microbiol. Rev.*, 47, 127, 1983.

83. Mertens, K. and Samuel, J.E. Bacteriology of *Coxiella*. In: *Rickettsial Diseases*. Information Healthcare, New York, 2007, p. 257.

84. Afseth, G. and Mallavia, L.P., Copy number of the 16S rRNA gene in *Coxiella burnetii*, *Eur. J. Epidemiol.*, 13, 729, 1997.

85. Porter, S.R. et al., Q fever: Current state of knowledge and perspectives of research of a neglected zoonosis, *Int. J. Microbiol.*, Article ID 248418, 2011.

86. Howe, D. et al., Fusogenicity of the *Coxiella burnetii* parasitophorous vacuole, *Ann. N. Y. Acad. Sci.*, 990, 556, 2003.

87. Amano, K. and Williams, J.C., Chemical and immunological characterization of lipopolysaccharides from phase I and phase II *Coxiella burnetii*. *J. Bacteriol.*, 160, 994, 1984.

88. Baca, O.G., Li, Y.P., and Kumar, H., Survival of the Q fever agent *Coxiella burnetii* in the phagolysosome, *Trends Microbiol.*, 2, 476, 1994.

89. Siemsen, D.W. et al., Inhibition of the human neutrophil NADPH oxidase by *Coxiella burnetii*, *Microbes Infect.*, 11, 671, 2009.

90. Akporiaye, E.T. et al., *Coxiella burnetii* fails to stimulate human neutrophil superoxide anion production, *Acta Virol.*, 34, 64, 1990.

91. Hill, J. and Samuel, J.E. *Coxiella burnetii* acid phosphatase inhibits the release of reactive oxygen intermediates in polymorphonuclear leukocytes, *Infect. Immun.*, 79, 414, 2011.

92. Li, Y.P. et al., Protein-tyrosine phosphatase activity of *Coxiella burnetii* that inhibits human neutrophils, *Acta Virol.*, 40, 263, 1996.

93. Mertens, K. and Samuel, J.E. Defense mechanisms against oxidative stress in *Coxiella burnetii*: Adaptation to a unique intracellular niche. In: Toman, R. et al. (eds.), *Coxiella burnetii: Recent Advances and New Perspectives in Research of the Q Fever Bacterium*, *Adv. Exp. Med. Biol.*, 984, 39, 2012.

94. Heinzen, R.A., Frazier, M.E., and Mallavia, L.P., *Coxiella burnetii* superoxide dismutase gene: Cloning sequencing, and expression in *Escherichia coli*, *Infect. Immun.*, 60, 3814, 1992.

95. Hicks, L.D. et al., A DNA-binding peroxiredoxin of *Coxiella burnetii* is involved in countering oxidative stress during exponential-phase growth, *J. Bacteriol.*, 192, 2077, 2010.

96. Park, S.H., Lee, H.-W., and Cao, W., Screening of nitrosative stress resistance genes in *Coxiella burnetii*: Involvement of nucleotide excision repair, *Microb. Pathog.*, 49, 323, 2010.

97. Barry, A.O., Mege, J.L., and Ghigo, E., Hijacked phagosomes and leukocyte activation: An intimate relationship, *J. Leukoc. Biol.*, 89, 373, 2011.

98. Campoy, E.M., Zoppino, F.C.M., and Colombo, M.I., The early secretory pathway contributes to the growth of the *Coxiella*-replicative niche, *Infect. Immun.*, 79, 402, 2011.

99. Ghigo, E., Colombo, M.I., and Heinzen, R.A., The *Coxiella burnetii* parasitophorous vacuole. In: Toman, R. et al. (eds.), *Coxiella burnetii: Recent Advances and New Perspectives in Research of the Q Fever Bacterium, Adv. Exp. Med. Biol.*, 984, 141, 2012.

100. Ghigo, E. et al., *Coxiella burnetii* survival in THP-1 monocytes involves the impairment of phagosome maturation: IFN-γ mediates its restoration and bacterial killing, *J. Immunol.*, 169, 4488, 2002.

101. Howe, D. and Heinzen, R.A., *Coxiella burnetii* inhibits a cholesterol-rich vacuole and influences cellular cholesterol metabolism, *Cell. Microbiol.*, 8, 496, 2006.

102. Berón, W. et al., *Coxiella burnetii* localizes in a Rab7-labeled compartment with autophagic characteristics, *Infect. Immun.*, 70, 5816, 2002.

103. Romano, P.S. et al., The autophagic pathway is actively modulated by phase II *Coxiella burnetii* to efficiently replicate in the host cell, *Cell. Microbiol.*, 9, 891, 2007.

104. Howe, D. and Heinzen, R.A., Replication of *Coxiella burnetii* is inhibited in CHO K-1 cells treated with inhibitors of cholesterol metabolism, *Ann. N. Y. Acad. Sci.*, 1063, 123, 2005.

105. Gilk, S.D., Beare, P.A., and Heinzen, R.A., *Coxiella burnetii* expresses a functional Δ24 sterol reductase, *J. Bacteriol.*, 192, 6154, 2010.

106. Hussain, S.K. et al., Host kinase activity is required for *Coxiella burnetii* parasitophorous vacuole formation, *Front. Microbiol.*, 1, 137, 2010.

107. Gilk, S.D., Role of lipids in *Coxiella burnetii* infection. In: Toman, R. et al. (eds.), *Coxiella burnetii: Recent Advances and New Perspectives in Research of the Q Fever Bacterium, Adv. Exp. Med. Biol.*, 984, 199, 2012.

108. Juhas, M., Crook, D.W., and Hood, D.W. Type IV secretion systems: Tools of bacterial horizontal gene transfer and virulence, *Cell. Microbiol.*, 10, 2377, 2008.

109. Nagai, H. and Kubori, T. Type IVB secretion systems of *Legionella* and other Gram-negative bacteria, *Front. Microbiol.*, 2, 136, 2011.

110. McDonough, J.A., Newton, H.J., and Roy, C.R., *Coxiella burnetii* secretion systems. In: Toman, R. et al. (eds.), *Coxiella burnetii: Recent Advances and New Perspectives in Research of the Q Fever Bacterium, Adv. Exp. Med. Biol.*, 984, 171, 2012.

111. Carey, K.L. et al., The *Coxiella burnetii* Dot/Icm system delivers a unique repertoire of type IV effectors into host cells and is required for intracellular replication, *PLoS Pathog.*, 7, e1002056, 2011.

112. Chen, C. et al., Large-scale identification and translocation of type IV secretion substrates by *Coxiella burnetii*, *Proc. Natl. Acad. Sci. USA*, 107, 21755, 2010.

113. Lührmann, A. et al., Inhibition of pathogen-induced apoptosis by a *Coxiella burnetii* type IV effector protein, *Proc. Natl. Acad. Sci. USA*, 107, 18997, 2010.

114. Voth, D.E. and Heinzen, R.A., Lounging in a lysosome: The intracellular lifestyle of *Coxiella burnetii*, *Cell Microbiol.*, 9, 829, 2007.

115. Mahapatra, S., Ayoubi, P., and Shaw, E.I., *Coxiella burnetii* Nine Mile II proteins modulate gene expression of monocytic host cells during infection, *BMC Microbiol.*, 10, 244, 2010.

116. Hussain, S.K. and Voth, D.E., *Coxiella* subversion of intracellular host signalling. In: Toman, R. et al. (eds.), *Coxiella burnetii: Recent Advances and New Perspectives in Research of the Q Fever Bacterium, Adv. Exp. Med. Biol.*, 984, 131, 2012.

117. Ransom, S.E. and Huebner, R.J., Studies on the resistance of *Coxiella burnetii* to physical and chemical agents, *Am. J. Hyg.*, 53, 110, 1951.

118. Mitscherlich, E. and Marth, E.H., *Microbial Survival in the Environment—Bacteria and Rickettsiae Important in Human and Animal Health*. Springer-Verlag, New York, 1984.

119. McCaul, T.F., Hackstadt, T., and Williams, J.C., Ultrastructural and biological aspects of *Coxiella burnetii* under physical disruptions. In: Burgdorfer, W. and Anacker, R.L. (eds.), *Rickettsiae and Rickettsial Diseases*. Harcourt Brace Jovanovich, New York, 1981.

120. Samuel, J.E., Developmental cycle of *Coxiella burnetii*. In: Brun, Y.V. and Shimkets, L.J. (eds.), *Procaryotic Development*. ASM Press, Washington, DC, 2000, p. 427.

121. Babudieri, B. and Moscovici, C., Research on the behavior of *Coxiella burnetii* in relation to various physical and chemical agents, *Rend. Inst. Sup. Sanit.*, 13, 739, 1950.

122. Russell-Lodrigue, K.E. et al., Clinical and pathologic changes in a guinea pig aerosol challenge model of acute Q fever, *Infect. Immun.*, 74, 6085, 2006.

123. Angelakis, E. and Raoult, D., Q fever, *Vet. Microbiol.*, 140, 297, 2010.

124. Guatteo, R. et al., Prevalence of *Coxiella burnetii* infection in domestic ruminants: A critical review, *Vet. Microbiol.*, 149, 1, 2011.

125. Hellenbrand, W., Breuer, T., and Petersen, L. et al., Changing epidemiology of Q fever in Germany, 1947–1999, *Emerg. Infect. Dis.*, 7, 789, 2001.

126. Van Der Hoek, W. et al., Relation between Q fever notifications and *Coxiella burnetii* infections during the 2009 outbreak in The Netherlands, *Euro Surveill.*, 17, 20058, 2012.

127. Blaauw, G.J. et al., The application of an enzyme-linked immunosorbent assay or an immunofluorescent assay test leads to different estimates of seroprevalence of *Coxiella burnetii* in the population, *Epidemiol. Infect.*, 140, 36, 2012.

128. Raoult, D. et al., Q fever 1985–1998. Clinical and epidemiologic features of 1,383 infections, *Medicine* (Baltimore), 79, 109, 2000.

129. Berri, M., Laroucau, K., and Rodolakis, A. et al., The detection of *Coxiella burnetii* from ovine genital swabs, milk and fecal samples by the use of a single touchdown polymerase chain reaction, *Vet. Microbiol.*, 72, 285, 2000.

130. Arricau-Bouvery, N. et al., Experimental *Coxiella burnetii* infection in pregnant goats: Excretion routes, *Vet. Res.*, 34, 423, 2003.

131. Guatteo, R. et al., Shedding routes of *Coxiella burnetii* in dairy cows: Implications for detection and control, *Vet. Res.*, 37, 827, 2006.

132. Garner, M.G. et al., A review of Q fever in Australia 1991–1994, *Aust. N. Z. J. Public Health*, 21, 722, 1997.

133. Sting, R. et al., The occurrence of *Coxiella burnetii* in sheep and ticks of the genus *Dermacentor* in Baden-Wuerttemberg, *Dtsch. Tierarztl. Wochenschr.*, 111, 390, 2004.

134. Schulz, J. et al., Detection of *Coxiella burnetii* in the air of a sheep barn during shearing, *Dtsch. Tierarztl. Wochenschr.*, 112, 470, 2005.

135. Anderson, A.D. et al., Q fever and the US military, *Emerg. Infect. Dis.*, 11, 1320, 2005.
136. Chiu, C.K. and Durrheim, D.N., A review of the efficacy of human Q fever vaccine registered in Australia, *N. S. W. Public Health Bull.*, 18, 133, 2007.
137. Whitney, E.A.S. et al., Seroepidemiologic and occupational risk survey for *Coxiella burnetii* antibodies among US veterinarians, *Clin. Infect. Dis.*, 48, 550, 2009.
138. de Rooij, M.M.T. et al., Risk factors of *Coxiella burnetii* (Q fever) seropositivity in veterinary medicine students, *PLOS ONE*, 7, e32108, 2012.
139. Dupuis, G. et al., An important outbreak of human Q fever in a Swiss alpine valley, *Int. J. Epidemiol.*, 16, 282, 1987.
140. Anderson, A.D. et al., Seroprevalence of Q fever in the United States, 2003–2004, *Am. J. Trop. Med. Hyg.*, 81, 691, 2009.
141. Pascual-Velasco, F. et al., High seroprevalence of *Coxiella burnetii* infection in Eastern Cantabria (Spain), *Int. J. Epidemiol.*, 27, 142, 1998.
142. Monno, R. et al., Seroprevalence of Q fever, brucellosis and leptospirosis in farmers and agricultural workers in Bari, southern Italy, *Ann. Agric. Environ. Med.*, 16, 205, 2009.
143. Raoult, D., Marrie, T.J., and Mege, J.L., Natural history and pathophysiology of Q fever, *Lancet Infect. Dis.*, 5, 219, 2005.
144. Textoris, J. et al., Sex-related differences in gene expression following *Coxiella burnetii* infection in mice: Potential role of circadian rhythm, *PLOS ONE*, 5, e12190, 2010.
145. Baud, D. et al., Seroprevalence of *Coxiella burnetii* and *Brucella abortus* among pregnant women, *Clin. Microbiol. Infect.*, 15, 499, 2009.
146. Vaidya, V.M. et al., Comparison of PCR, immunofluorescence assay, and pathogen isolation for diagnosis of Q fever in humans with spontaneous abortions, *J. Clin. Microbiol.*, 46, 2038, 2008.
147. Munster, J.M. et al., Placental histopathology after *Coxiella burnetii* infection during pregnancy, *Placenta*, 33, 128, 2012.
148. Nielsen, S.Y. et al., Presence of antibodies against *Coxiella burnetii* and risk of spontaneous abortion: A nested case-control study, *PLOS ONE*, 7, e31909, 2012.
149. Kersh, G.J. et al., *Coxiella burnetii* infection of marine mammals in the Pacific Northwest, 1997–2010, *J. Wildl. Dis.*, 48, 201, 2012.
150. Cooper, A. et al., Determination of *Coxiella burnetii* seroprevalence in macropods in Australia, *Vet. Microbiol.*, 155, 317, 2012.
151. Buhariwalla, F., Cann, B., and Marrie, T.J., A dog-related outbreak of Q fever, *Clin. Infect. Dis.*, 23, 753, 1996.
152. Marrie, T.J. et al., Exposure to parturient cats: A risk factor for acquisition of Q fever in Maritime Canada, *J. Infect. Dis.*, 158, 101, 1988.
153. Psaroulaki, A. et al., Rats as indicators of the presence and dispersal of six zoonotic microbial agents in Cyprus, an island ecosystem: A seroepidemiological study, *Trans. R. Soc. Trop. Med. Hyg.*, 104, 733, 2010.
154. Reusken, C. et al., *Coxiella burnetii* (Q fever) in *Rattus norvegicus* and *Rattus rattus* at livestock farms and urban locations in the Netherlands; could *Rattus* spp. represent reservoirs for (re)introduction? *Prev. Vet. Med.*, 101, 124, 2011.
155. Ioannou, I. et al., Carriage of *Rickettsia* spp., *Coxiella burnetii* and *Anaplasma* spp. by endemic and migratory wild birds and their ectoparasites in Cyprus, *Clin. Microbiol. Infec.*, 15, 158, 2009.
156. Herrin, B. et al., Growth of *Coxiella burnetii* in the *Ixodes scapularis*-derived IDE8 tick cell line, *Vector Borne Zoonotic Dis.*, 11, 917, 2011.
157. Aitken, I.D. et al., Q fever in Europe: Current aspects of aetiology, epidemiology, human infection, diagnosis and therapy, *Infection*, 15, 323, 1987.
158. Smith, D.J.W. and Derrick, E.H., Studies in the epidemiology of Q fever. The isolation of six strains of *Rickettsia burneti* from the tick *Haemaphysalis humerosa*, *Aust. J. Exp. Biol. Med. Sci.*, 18, 99, 1940.
159. Široký, P. et al., Tortoise tick *Hyalomma aegyptium* as long term carrier of Q fever agent *Coxiella burnetii*—Evidence from experimental infection, *Parasitol. Res.*, 107, 1515, 2010.
160. Nelder, M.P. et al., *Coxiella burnetii* in wild-caught filth flies. *Emerg. Infect. Dis.*, 14, 1002, 2008.
161. Pope, J.H., Scott, W., and Dwyer, E. et al., *Coxiella burnetii* in kangaroos and kangaroo ticks in western Queensland, *Aust. J. Exp. Biol. Med. Sci.*, 38, 17, 1960.
162. Parola, P. and Raoult, D., Ticks and tickborne bacterial diseases in humans: An emerging infectious threat, *Clin. Infect. Dis.*, 32, 897, 2001.
163. La Scola, B. and Raoult, D., Survival of *Coxiella burnetii* within free-living amoeba *Acanthamoeba castellanii*, *Clin. Microbiol. Infect.*, 7, 75, 2001.
164. Maurin, M. and Raoult, D., Q fever, *Clin. Microbiol. Rev.*, 12, 518, 1999.
165. Yanase, T. et al., Detection of *Coxiella burnetii* from dust in a barn housing dairy cattle, *Microbiol. Immunol.*, 42, 51, 1998.
166. Dijkstra, F. et al., The 2007–2010 Q fever epidemic in the Netherlands: Characteristics of notified acute Q fever patients and the association with dairy goat farming, *FEMS Immunol. Med. Microbiol.*, 64, 3, 2012.
167. Wallensten, A. et al., Q fever outbreak in Cheltenham, United Kingdom, in 2007 and the use of dispersion modelling to investigate the possibility of airborne spread, *Euro Surveill.*, 15, pii:19521, 2010.
168. DeLay, P.D., Lennette, E.H., and Deome, K.B., Q fever in California. II. Recovery of *Coxiella burnetii* from naturally infected airborne dust, *J. Immunol.*, 65, 211, 1959.
169. Tissot-Dupont, H. et al., Hyperendemic focus of Q fever related to sheep and wind, *Am. J. Epidemiol.*, 150, 67, 1999.
170. Vorobeychikov, E. et al., Evaluation of low concentration aerosol for infecting humans with the Q fever pathogen, *Ann. N. Y. Acad. Sci.*, 1063, 466, 2005.
171. Tigertt, W.D., Benenson, A.S., and Gochenour, W.S. et al., Airborne Q fever, *Bacteriol. Rev.*, 25, 285, 1961.
172. Forland, F. et al., ECDC technical report: Risk assessment on Q fever. European Centre for Disease Prevention and Control, Stockholm, Sweden, 2010.
173. Arricau-Bouvery, N. et al., Effect of vaccination with phase I and phase II *Coxiella burnetii* vaccines in pregnant goats, *Vaccine*, 23, 4392, 2005.
174. PHAC., Material safety data sheet: Infectious agent; *Coxiella burnetii*. C. Public Health Agency of Canada, Ottawa, Canada, 2001.
175. Rodolakis, A., Q fever in dairy animals, *Ann. N. Y. Acad. Sci.*, 1166, 90, 2009.
176. Fishbein, D.B. and Raoult, D., A cluster of *Coxiella burnetii* infections associated with exposure to vaccinated goats and their unpasteurized dairy products, *Am. J. Trop. Med. Hyg.*, 47, 35, 1992.
177. Hatchette, T.F. et al., Goat-associated Q fever: A new disease in Newfoundland, *Emerg. Infect. Dis.*, 7, 413, 2001.
178. Maltezou, H.C. et al., Q fever in children in Greece, *Am. J. Trop. Med. Hyg.*, 70, 540, 2004.
179. Sprong, H. et al., Prevalence of *Coxiella burnetii* in ticks after a large outbreak of Q fever, *Zoonoses Public Health*, 59, 69, 2012.

180. Kazar, J., Q fever. In: Kazar, J. and Toman, R. (eds.), *Rickettsiae and Rickettsial Diseases*. Slovak Academy of Sciences, Bratislava, Slovakia, p. 353, 1996.

181. Lang, G.H., *Coxiellosis (Q Fever) in Animals*. In: Marrie, T.J. (ed.) Q Fever Volume I: The Disease, p. 23. CRC Press, Boca Raton, FL, 1990.

182. Marrie, T.J. et al., Truckin' pneumonia—An outbreak of Q fever in a truck repair plant probably due to aerosols from clothing contaminated by contact with newborn kittens, *Epidemiol. Infect.*, 102, 119, 1989.

183. Bennett, M.D. et al., *Coxiella burnetii* in western barred bandicoots (*Perameles bougainville*) from Bernier and Dorre Islands in Western Australia, *EcoHealth*, 8, 519, 2011.

184. Carcopino, X. et al., Q fever during pregnancy: A cause of poor fetal and maternal outcome, *Ann. N. Y. Acad. Sci.*, 1166, 79, 2009.

185. Pantanowitz, L., Telford, S.R., and Cannon, M.E., Tick-borne diseases in transfusion medicine, *Transfus. Med.*, 12, 85, 2002.

186. Hogema, B.M. et al., *Coxiella burnetii* infection among blood donors during the 2009 Q-fever outbreak in the Netherlands, *Transfusion*, 52, 144, 2012.

187. Kanfer, E. et al., Q fever following bone marrow transplantation, *Bone Marrow Transplant.*, 3, 165, 1988.

188. Kruszewska, D., Lembowicz, K., and Tylewska-Wierzbanowska, S., Possible sexual transmission of Q fever among humans, *Clin. Infect. Dis.*, 22, 1087, 1996.

189. Miceli, M.H. et al., A case of person-to-person transmission of Q fever from an active duty serviceman to his spouse, *Vector Borne Zoonotic Dis.*, 10, 539, 2010.

190. Oliphant, J.W. and Gordon, D.A. et al., Q fever in laundry workers, presumably transmitted from contaminated clothing, *Am. J. Hyg.*, 49, 76, 1949.

191. Sawyer, L.A., Fishbein, D.B., and McDade, J.E., Q fever: Current concepts, *Rev. Infect. Dis.*, 9, 935, 1987.

192. Tiggert, W.D. and Benenson, A.S., Studies on Q fever in man. *Trans. Assoc. Am. Physicians*, 69, 98, 1956.

193. Waag, D.M. *Coxiella burnetii*: Host and bacterial responses to infection, *Vaccine*, 25, 7288, 2007.

194. Fournier, P.E. et al., Myocarditis, a rare but severe manifestation of Q fever: Report of 8 cases and review of the literature, *Clin. Infect. Dis.*, 32, 1440, 2001.

195. Bellazreg, F. et al., Acute Q fever in hospitalised patients in Central Tunisia: Report of 21 cases, *Clin. Micro. Infect.*, 15, 138, 2009.

196. Gikas, A., Kokkini, S., and Tsioutis, C., Q fever: Clinical manifestations and treatment, *Expert Rev. Anti. Infect. Ther.*, 8, 529, 2010.

197. Carcopino, X. et al., Managing Q fever during pregnancy: The benefits of long-term cotrimoxazole therapy, *Clin. Infect. Dis.*, 45, 548, 2007.

198. Raoult, D. and Stein, A., Q fever during pregnancy—A risk for women, fetuses and obstetricians, *N. Engl. J. Med.*, 330, 371, 1994.

199. Raoult, D., Fenollar, F., and Stein, A., Q fever during pregnancy: Diagnosis, treatment, and follow-up, *Arch. Intern. Med.*, 162, 701, 2002.

200. Kaplan, B. et al., An isolated case of Q-fever during pregnancy, *Acta Obstet. Gynecol. Scand.*, 74, 848, 1995.

201. Raoult, D. et al., Chronic Q fever: Diagnosis and follow-up, *Ann. N. Y. Acad. Sci.*, 590, 51, 1990.

202. Stein, A. and Raoult, D., Q fever during pregnancy: A public health problem in southern France, *Clin. Infect. Dis.*, 27, 592, 1998.

203. Langley, J.M. et al., *Coxiella burnetii* seropositivity in parturient women is associated with adverse pregnancy outcomes, *Am. J. Obstet. Gynecol.*, 189, 228, 2003.

204. Heard, S.R., Ronalds, C.J., and Heath, R.B., *Coxiella burnetii* infection in immunocompromised patients, *J. Infect.*, 11, 15, 1985.

205. Tobin, M.J. et al., Q fever endocarditis, *Am. J. Med.*, 72, 396, 1982.

206. Fenollar, F., Fournier, P.E., and Raoult, D., Molecular detection of *Coxiella burnetii* in the sera of patients with Q fever endocarditis or vascular infection, *J. Clin. Microbiol.*, 42, 4919, 2004.

207. Brouqui, P. et al., Chronic Q fever: Ninety-two cases from France, including 27 cases without endocarditis, *Arch. Intern. Med.*, 153, 642, 1993.

208. Fenollar, F. et al., Risks factors and prevention of Q fever endocarditis, *Clin. Infect. Dis.*, 33, 312, 2001.

209. Wegdam-Blans, M.C.A. et al., Vascular complications of Q-fever infections, *Eur. J. Vasc. Endovasc. Surg.*, 42, 384, 2011.

210. Williams, J.C. and Sanchez, V., Q fever and Coxiellosis. In: Beran, G.W. Steele, J.H. (eds.), *Handbook of Zoonoses* section A: Bacterial, rickettsial, chlamydial, and mycotic, p. 429. CRC Press, Boca Raton, FL, 1994.

211. Acquacalda, E. et al., A case of multifocal chronic Q fever osteomyelitis, *Infection*, 39, 167, 2011.

212. Hickie, I. et al., Post-infective and chronic fatigue syndromes precipitated by viral and non-viral pathogens: Prospective cohort study, *BMJ*, 333, 575, 2006.

213. Wildman, M.J. et al., Chronic fatigue following infection by *Coxiella burnetii* (Q fever): Ten-year follow-up of the 1989 UK outbreak cohort. QJM, 95, 527, 2002.

214. Babudieri, B. Q fever. A zoonosis, *Adv. Vet. Sci.*, 5, 82, 1959.

215. Figtree, M. et al., Q fever cholecystitis in an unvaccinated butcher diagnosed by gallbladder polymerase chain reaction, *Vector Borne Zoonotic Dis.*, 10, 421, 2010.

216. Rolain, J.M. et al., Acute acalculous cholecystitis associated with Q fever: Report of seven cases and review of the literature, *Eur. J. Clin. Microbiol. Infect. Dis.*, 22, 222, 2003.

217. Levy, P.Y. et al., Diagnosis of *Coxiella burnetii* pericarditis by using a systematic prescription kit in cases of pericardial effusion: An 8-year experience, *Clin. Microbiol. Infect. Dis.*, 15, 173, 2009.

218. Mantonti, F. et al., Uveitis in the course of Q-fever, *Clin. Microbiol. Infect. Dis.*, 15, 176, 2009.

219. Udaondo, P. et al., Q fever: A new ocular manifestation, *Clin. Ophthalmol.*, 5, 1273, 2011.

220. Soulard, R. et al., Histopathology of a granulomatous lobar panniculitis in acute Q fever: A case report, *J. Cutan. Pathol.*, 37, 870, 2010.

221. Marmion, B.P. et al., Q fever: Persistence of antigenic non-viable cell residues of *Coxiella burnetii* in the host—implications for post-Q-fever infection fatigue syndrome and other chronic sequelae, *QJM*, 102, 673, 2009.

222. Sukocheva, O.A. et al., Long-term persistence after acute Q fever of non-infective *Coxiella burnetii* cell components, including antigens, *QJM*, 103, 847, 2010.

223. Peuttila, I.A. et al., Cytokine dysregulation in the post-Q-fever fatigue syndrome, *QJM*, 91, 549, 1998.

224. Helbig, K. et al., Immune response genes in the post-Q-fever fatigue syndrome, Q fever endocarditis and uncomplicated acute primary Q fever, *QJM*, 98, 565, 2005.

225. Lang, G.H., Q fever, *Vet. Rec.*, 123, 582, 1988.

226. To, H. et al., Prevalence of *Coxiella burnetii* infection in dairy cattle with reproductive disorders, *J. Vet. Med. Sci.*, 60, 859, 1998.

227. Oda, H. and Yoshiie, K., Q fever and *Coxiella burnetii*, *Nihon Saikingaku Zasshi*, 50, 703, 1995.

228. Hirai, K. and To, H., Advances in the understanding of *Coxiella burnetii* infection in Japan, *J. Vet. Med. Sci.*, 60, 781, 1998.

229. Shannon, J.G. and Heinzen, R.A., Adaptive immunity to the obligate intracellular pathogen *Coxiella burnetii*, *Immunol. Res.*, 43, 138, 2009.

230. Shannon, J.G., Howe, D., and Heinzen, R.A., Lack of dendritic cell maturation following infection by *Coxiella burnetii* synthesizing different lipopolysaccharide chemotypes, *Ann. N. Y. Acad. Sci.*, 1063, 154, 2005.

231. Capo, C. and Mege, J.L., Role of innate and adaptive immunity in the control of Q fever. In: Toman, R. et al. (eds.), *Coxiella burnetii: Recent Advances and New Perspectives in Research of the Q Fever Bacterium*, *Adv. Exp. Med. Biol.*, 984, 273, 2012.

232. Meghari, S. et al., TLR2 is necessary to inflammatory response in *Coxiella burnetii* infection, *Ann. N. Y. Acad. Sci.*, 1063, 161, 2005.

233. Zamboni, D.S. et al., Stimulation of toll-like receptor 2 by *Coxiella burnetii* is required for macrophage production of pro-inflammatory cytokines and resistance to infection, *J. Biol. Chem.*, 279, 54405, 2004.

234. Honstettre, A. et al., Lipopolysaccharide from *Coxiella burnetii* is involved in bacterial phagocytosis, filamentous actin reorganization, and inflammatory responses through Toll-like receptor 4, *J. Immunol.*, 172, 3695, 2004.

235. Burnet, F.M. and Freeman, M. Q fever: Factors affecting the appearance of Rickettsiae in mice, *Med. J. Aust.*, 2, 1114, 1938.

236. Zhang, G. et al., Mechanisms of vaccine-induced protective immunity against *Coxiella burnetii* infection in BALB/c mice, *J. Immunol.*, 179, 8372, 2007.

237. Humphres, R.C. and Hinrichs, D.J., Role of antibody in *Coxiella burnetii* infection, *Infect. Immun.*, 31, 641, 1981.

238. Shannon, J.G. et al., Antibody-mediated immunity to the obligate intracellular bacterial pathogen *Coxiella burnetii* is Fc receptor- and complement-independent, *BMC Immunol.*, 10, 26, 2009.

239. Desnues, B. et al., Role of specific antibodies in *Coxiella burnetii* infection of macrophages, *Clin. Microbiol. Infect.*, 15, 161, 2009.

240. Andoh, M. et al., T cells are essential for bacterial clearance, and gamma interferon, tumor necrosis factor alpha, and B cells are crucial for disease development in *Coxiella burnetii* infection in mice, *Infect. Immun.*, 75, 3245, 2007.

241. Read, A.J., Erickson, S., and Harmsen, A.G., Role of CD4+ and CD8+ T cells in clearance of primary pulmonary infection with *Coxiella burnetii*, *Infect. Immun.*, 78, 3019, 2010.

242. Dellacasagrande, J. et al., *Coxiella burnetii* survives in monocytes from patients with Q fever endocarditis: Involvement of tumor necrosis factor, *Infect. Immun.*, 68, 160, 2000.

243. Delaby, A. et al., Defective monocyte dynamics in Q fever granuloma deficiency, *J. Infect. Dis.*, 205, 1086, 2012.

244. Sidwell, R.W., Thorpe, B.D., and Gebhardt, L.P., Studies of latent Q fever infections. 2. Effects of multiple cortisone injections, *Am. J. Hyg.*, 79, 320, 1964.

245. Sidwell, R.W., Thorpe, B.D., and Gebhardt, L.P., Studies of latent Q fever infections. 1. Effects of whole body X-irradiation upon latently infected guinea pigs, white mice and deer mice, *Am. J. Hyg.*, 79, 113, 1964.

246. Sidwell, R.W. and Gebhardt, L.P., Studies of latent Q fever infections. 3. Effects of parturition upon latently infected guinea pigs and white mice, *Am. J. Epidemiol.*, 84, 132, 1966.

247. Layez, C. et al., Foxp3[+]CD4[+]CD25[+] regulatory T cells are increased in patients with *Coxiella burnetii* endocarditis, *FEMS Immunol. Med. Microbiol.*, 64, 137, 2012.

248. Benoit, M. et al., *Coxiella burnetii*, the agent of Q fever, stimulates an atypical M2 activation program in human macrophages, *Eur. J. Immunol.*, 38, 1065, 2008.

249. Capo, C. et al., Upregulation of tumor necrosis factor alpha and interleukin-1 beta in Q fever endocarditis, *Infect. Immun.*, 64, 1638, 1996.

250. Honstettre, A. et al., Dysregulation of cytokines in acute Q fever: Role of interleukin-10 and tumor necrosis factor in chronic evolution of Q fever, *J. Infect. Dis.*, 187, 956, 2003.

251. Ghigo, E. et al., Intracellular life of *Coxiella burnetii* in macrophages: An update, *Ann. N. Y. Acad. Sci.*, 1166, 55, 2009.

252. Meghari, S. et al., Persistent *Coxiella burnetii* infection in mice overexpressing IL-10: An efficient model for chronic Q fever pathogenesis, *PLoS Pathog.*, 4, e23, 2008.

253. Ghigo, E. et al., Link between impaired maturation of phagosomes and defective *Coxiella burnetii* killing in patients with chronic Q fever, *J. Infect. Dis.*, 190, 1767, 2004.

254. Amara, A.B., Bechah, Y., and Mege, J.-L., Immune response and *Coxiella burnetii* invasion. In: Toman, R. et al. (eds.), *Coxiella burnetii: Recent Advances and New Perspectives in Research of the Q Fever Bacterium*, *Adv. Exp. Med. Biol.*, 984, 287, 2012.

255. Schneider, W., Titration of *Coxiella burnetii* in Buffalo green monkey (BGM) cell cultures, *Zentralbl. Bakteriol.*, 271, 77, 1989.

256. McDade, J.E. and Gerone, P.J., Plaque assay for Q fever and scrub typhus rickettsiae, *Appl. Microbiol.*, 19, 963, 1970.

257. Wike, D.A. et al., Studies of the rickettsial plaque assay technique, *Infect. Immun.*, 5, 715, 1972.

258. Fournier, P.E., Marrie, T.J., and Raoult, D. et al., Diagnosis of Q fever, *J. Clin. Microbiol.*, 36, 1823, 1998.

259. Wattiau, P. et al., Q fever in woolsorters, Belgium, *Emerg. Infect. Dis.*, 17, 2368, 2011.

260. Hirai, A. et al., Development of a method for detecting *Coxiella burnetii* in cheese samples, *J. Vet. Med. Sci.*, 74, 175, 2012.

261. Lockhart, M.G. et al., A comparison of methods for extracting DNA from *Coxiella burnetii* as measured by a duplex qPCR assay, *Lett. Appl. Microbiol.*, 52, 514, 2011.

262. Muramatsu, Y. et al., Detection of *Coxiella burnetii* in cow's milk by PCR-enzyme-linked immunosorbent assay combined with a novel sample preparation method, *Appl. Environ. Microbiol.*, 63, 2142, 1997.

263. Issartel, B. et al., Clinically and histologically silent Q fever endocarditis accidentally diagnosed by PCR, *Clin. Microbiol. Infect.*, 8, 113, 2002.

264. Brennan, R.E. and Samuel, J.E., Evaluation of *Coxiella burnetii* antibiotic susceptibilities by real-time PCR assay, *J. Clin. Microbiol.*, 41, 1869, 2003.

265. Boden, K. et al., Diagnosis of acute Q fever with emphasis on enzyme-linked immunosorbent assay and nested polymerase chain reaction regarding the time of serum collection, *Diagn. Microbiol. Infect. Dis.*, 68, 110, 2010.

266. Denman, J. and Woods, M., Acute Q fever in pregnancy: Report and literature review, *Intern. Med. J.*, 39, 479, 2009.

267. Ughetto, E. et al., Three years experience of real-time PCR for the diagnosis of Q fever, *Clin. Microbiol. Infec. Dis.*, 15, 200, 2009.

268. Schneeberger, P.M. et al., Real-time PCR with serum samples is indispensable for early diagnosis of acute Q fever, *Clin. Vaccine Immunol.*, 17, 286, 2010.

269. Cotar, A.I. et al., Q fever endocarditis in Romania: The first cases confirmed by direct sequencing, *Int. J. Mol. Sci.*, 12, 9504, 2011.

270. Sidi-Boumedine, K. et al., Evaluation of randomly amplified polymorphic DNA (RAPD) for discrimination of *Coxiella burnetii* ruminant strains isolated in France, *Clin. Microbiol. Infec. Dis.*, 15, 194, 2009.

271. Huijsmans, C.J.J. et al., Single-nucleotide-polymorphism genotyping of *Coxiella burnetii* during a Q fever outbreak in the Netherlands, *Appl. Environ. Microbiol.*, 77, 2051, 2011.

272. Jones, R.M. et al., Interlaboratory comparison of real-time polymerase chain reaction methods to detect *Coxiella burnetii*, the causative agent of Q fever, *J. Vet. Diagn. Invest.*, 23, 108, 2011.

273. Klee, S.R. et al., Highly sensitive real-time PCR for specific detection and quantification of *Coxiella burnetii*, *BMC Microbiol.*, 6, 2, 2006.

274. Tilburg, J.J.H.C. et al., Interlaboratory evaluation of different extraction and real-time PCR methods for detection of *Coxiella burnetii* DNA in serum, *J. Clin. Microbiol.*, 48, 3293, 2010.

275. Janse, I. et al., Development and comparison of two assay formats for parallel detection of four biothreat pathogens by using suspension microarrays, *PLOS ONE*, 7, e31958, 2012.

276. Leski, T.A. et al., Application of a broad-range resequencing array for detection of pathogens in desert dust samples from Kuwait and Iraq, *Appl. Environ. Microbiol.*, 77, 4285, 2011.

277. Curran, T. et al., Development of a novel DNA microarray to detect bacterial pathogens in patients with chronic obstructive pulmonary disease (COPD), *J. Microbiol. Methods*, 80, 257, 2010.

278. Jiang, J. et al., Development of procedures for direct extraction of *Cryptosporidium* DNA from water concentrates and for relief of PCR inhibitors, *Appl. Environ. Microbiol.*, 71, 1135, 2005.

279. Lockhart, M. et al., Detecting and measuring small numbers of viable *Coxiella burnetii*, *FEMS Immunol. Med. Microbiol.*, 64, 61, 2012.

280. Parker, N.R., Barralet, J.H., and Bell, A.M., Q fever, *Lancet*, 367, 679, 2006.

281. Raoult, D., Reemergence of Q fever after 11 September 2001, *Clin. Infect. Dis.*, 48, 558, 2009.

282. Kowalczewska, M. et al., Protein candidates for Q fever serodiagnosis, *FEMS Immunol. Med. Microbiol.*, 64, 140, 2012.

283. Field, P.R. et al., Comparison of a commercial enzyme-linked immunosorbent assay with immunofluorescence and complement fixation tests for detection of *Coxiella burnetii* (Q fever) immunoglobulin M, *J. Clin. Microbiol.*, 38, 1645, 2000.

284. Office International Des Epizooties (OIE), Q fever. In: *World Organisation for Animal Health, Manual of Standards for Diagnostic Tests and Vaccines*, 4th ed. OIE, Paris, France, 2000, p. 822.

285. Ascher, M.S. et al., A rapid immunofluorescent procedure for serodiagnosis of Q fever in mice, guinea pigs, sheep and humans, *Diagn. Immunol.*, 1, 33, 1983.

286. Péter, O. et al., Comparison of enzyme-linked immunosorbent assay and complement fixation and indirect fluorescent-antibody tests for detection of *Coxiella burnetii* antibody, *J. Clin. Microbiol.*, 25, 1063, 1987.

287. Bacci, S. et al., Epidemiology and clinical features of human infection with *Coxiella burnetii* in Denmark during 2006–07, *Zoonoses Public Health*, 59, 61, 2012.

288. Dupont, H.T., Thirion, X., and Raoult, D., Q fever serology: Cutoff determination for microimmunofluorescence, *Clin. Diagn. Lab. Immunol.*, 1, 189, 1994.

289. Fournier, P.E. et al., Modification of the diagnostic criteria proposed by the Duke endocarditis service to permit improved diagnosis of Q fever endocarditis, *Am. J. Med.*, 100, 629, 1996.

290. Frankel, D. et al., Q fever in France, 1985–2009, *Emerg. Infect. Dis.*, 17, 350, 2011.

291. Setiyono, A. et al., New criteria for immunofluorescence assay for Q fever diagnosis in Japan, *J. Clin. Microbiol.*, 43, 5555, 2005.

292. Péter, O. et al., Enzyme-linked immunosorbent assay for diagnosis of chronic Q fever, *J. Clin. Microbiol.*, 26, 1978, 1988.

293. Meekelenkamp, J.C.E. et al., Comparison of ELISA and indirect immunofluorescent antibody assay detecting *Coxiella burnetii* IgM phase II for the diagnosis of acute Q fever, *Eur. J. Clin. Microbiol. Infect. Dis.*, 31, 1267, 2012.

294. Healy, B. et al., Chronic Q fever: Different serological results in 3 countries—Results of a follow-up study 6 years after a point source outbreak, *Clin. Infect. Dis.*, 52, 1013, 2011.

295. Horigan, M.W. et al., Q fever diagnosis in domestic ruminants: Comparison between complement fixation and commercial enzyme-linked immunosorbent assays, *J. Vet. Diagn. Invest.*, 23, 924, 2011.

296. Kittelberger, R. et al., Comparison of the Q-fever complement fixation test and two commercial enzyme-linked immunosorbent assays for the detection of serum antibodies against *Coxiella burnetii* (Q-fever) in ruminants: Recommendations for use of serological tests on imported animals in New Zealand, *N. Z. Vet. J.*, 57, 262, 2009.

297. Péter, O. et al., Evaluation of the complement fixation and indirect immunofluorescence tests in the early diagnosis of primary Q fever, *Eur. J. Clin. Microbiol.*, 4, 394, 1985.

298. Lepidi, H. et al., Autoimmunohistochemistry: A new method for the histologic diagnosis of infective endocarditis, *J. Infect. Dis.*, 193, 1711, 2006.

299. Lepidi, H. et al., Immunohistochemical detection of *Coxiella burnetii* in an aortic graft, *Clin. Microbiol. Infect.*, 15, 171, 2009.

300. Lepidi, H., Gouriet, F., and Raoult, D., Immunohistochemical detection of *Coxiella burnetii* in chronic Q fever hepatitis, *Clin. Microbiol. Infect.*, 15, 169, 2009.

301. Levy, P.Y. et al., Comparison of different antibiotic regimens for therapy of 32 cases of Q fever endocarditis, *Antimicrob. Agents Chemother.*, 35, 533, 1991.

302. Lecaillet, A. et al., Therapeutic impact of the correlation of doxycycline serum concentrations and the decline of phase I antibodies in Q fever endocarditis, *J. Antimicrob. Chemother.*, 63, 771, 2009.

303. Rolain, J.M. et al., Correlation between ratio of serum doxycycline concentration to MIC and rapid decline of antibody levels during treatment of Q fever endocarditis, *Antimicrob. Agents Chemother.*, 49, 2673, 2005.

304. Krol, V., Kogan, V., and Cunha, B.A., Q fever bioprosthetic aortic valve endocarditis (PVE) successfully treated with doxycycline monotherapy, *Heart Lung*, 37, 157, 2008.

305. Million, M. et al., Long-term outcome of Q fever endocarditis: A 26-year personal survey, *Lancet Infect. Dis.*, 10, 527, 2010.

306. Forland, F. et al., Applicability of evidence-based practice in public health: Risk assessment on Q fever under an ongoing outbreak, *Euro Surveill.*, 17, 20060, 2012.

307. Whelan, J. et al., Q fever among culling workers, the Netherlands, 2009–2010, *Emerg. Infect. Dis.*, 17, 1719, 2011.

308. Hendrix, L.R. and Chen, C., Antigenic analysis for vaccines and diagnostics. In: Toman, R. et al. (eds.), *Coxiella burnetii: Recent Advances and New Perspectives in Research of the Q Fever Bacterium, Adv. Exp. Med. Biol.*, 984, 299, 2012.

309. Waag, D.M. et al., Low-dose priming before vaccination with the phase I chloroform-methanol residue vaccine against Q fever enhances humoral and cellular immune responses to *Coxiella burnetii*, *Clin. Vaccine Immunol.*, 15, 1505, 2008.

310. Gefenaite, G. et al., Effectiveness of the Q fever vaccine: A meta-analysis, *Vaccine*, 29, 395, 2011.

311. Marmion, B.P. et al., Vaccine prophylaxis of abattoir-associated Q fever, *Lancet*, 2, 1411, 1984.

312. Fairweather, P., O'Rourke, T., and Strutton, G. et al., Rare complication of Q fever vaccination, *Australas. J. Dermatol.*, 46, 124, 2005.

313. Gidding, H.F. et al., Australia's national Q fever vaccination program, *Vaccine*, 27, 2037, 2009.

314. Hermans, M.H.A. et al., *Coxiella burnetii* DNA in goat milk after vaccination with Coxevac®, *Vaccine*, 29, 2653, 2011.

315. Lang, G.D., Waltner-Toews, D., and Menzies, P. et al., The seroprevalence of coxiellosis (Q fever) in Ontario sheep flocks, *Can. J. Vet. Res.*, 55, 139, 1991.

316. Cerf, O. and Condron, R., *Coxiella burnetii* and milk pasteurization: An early application of the precautionary principal? *Epidemiol. Infect.*, 134, 946, 2006.

317. Health Protection Agency Centre for Infections, Guidelines for action in the event of a deliberate release: Q fever. Health Protection Agency, U.K., Version 1.7, 2010. http://www.hpa.org.uk/webc/HPAwebFile/HPAwec_C/1194947387885 (accessed on November 5, 2013).

318. Scott, G.H. and Williams, J.C., Susceptibility of *Coxiella burnetii* to chemical disinfectants, *Ann. N. Y. Acad. Sci.*, 590, 291, 1990.

319. Baldry, M.G., The bactericidal, fungicidal and sporicidal properties of hydrogen peroxide and peracetic acid, *J. Appl. Bacteriol.*, 54, 417, 1983.

320. de Cremoux, R. et al., *Coxiella burnetii* vaginal shedding and antibody responses in dairy goat herds in a context of clinical Q fever outbreaks, *FEMS Immunol. Med. Microbiol.*, 64, 120, 2012.

321. Rousset, E. et al., Efficiency of a phase 1 vaccine for the reduction of vaginal *Coxiella burnetii* shedding in a clinically affected goat herd, *Clin. Micro. Infect.*, 15, 188, 2009.

322. Astobiza, I. et al., Kinetics of *Coxiella burnetii* excretion in a commercial dairy sheep flock after treatment with oxytetracycline, *Vet. J.*, 184, 172, 2010.

323. Astobiza, I. et al., Four-year evaluation of the effect of vaccination against *Coxiella burnetii* on reduction of animal infection and environmental contamination in a naturally infected dairy sheep flock, *Appl. Environ. Microbiol.*, 77, 7405, 2011.

324. Courcoul, A. et al., Modelling effectiveness of herd level vaccination against Q fever in dairy cattle, *Vet. Res.*, 42, 68, 2011.

325. WHO Group of Consultants, *Public Health Response to Biological and Chemical Weapons—WHO Guidance*, 2nd ed. World Health Organization, Geneva, Switzerland, 1970.

326. Beare, P.A. et al., Candidate antigens for Q fever serodiagnosis revealed by immunoscreening of a *Coxiella burnetii* protein microarray, *Clin. Vaccine Immunol.*, 15, 1771, 2008.

327. Vigil, A. et al., Genome-wide profiling of humoral immune response to *Coxiella burnetii* infection by protein microarray, *Proteomics*, 10, 2259, 2010.

328. Papadioti, A. et al., A proteomic approach to investigate the differential antigenic profile of two *Coxiella burnetii* strains, *J. Proteomics*, 74, 1150, 2011.

329. Deringer, J.R. et al., Immunoreactive *Coxiella burnetii* Nine Mile proteins separated by 2D electrophoresis and identified by tandem mass spectrometry, *Microbiology*, 157, 526, 2011.

330. Massung, R.F., Cutler, S.J., and Frangoulidis, D., Molecular typing of *Coxiella burnetii* (Q Fever). In: Toman, R. et al. (eds.), *Coxiella burnetii: Recent Advances and New Perspectives in Research of the Q Fever Bacterium*, *Adv. Exp. Med. Biol.*, 984, 381, 2012.

331. Beare, P.A. et al., Advances in genetic manipulation of obligate intracellular bacterial pathogens, *Front. Microbiol.*, 2, 97, 2011.

332. Beare, P.A. et al., Dot/Icm type IVB secretion system requirements for *Coxiella burnetii* growth in human macrophages. *MBio*, 2, e00175-11, 2011.

333. Beare, P.A. et al., Characterization of a *Coxiella burnetii* *ftsZ* mutant generated by *Himar1* transposon mutagenesis, *J. Bacteriol.*, 191, 1369, 2009.

334. Beare, P.A., Genetic manipulation of *Coxiella burnetii*. In: Toman, R. et al. (eds.), *Coxiella burnetii: Recent Advances and New Perspectives in Research of the Q Fever Bacterium*, *Adv. Exp. Med. Biol.*, 984, 249, 2012.

335. Schimmer, B. et al., Low seroprevalence of Q fever in The Netherlands prior to a series of large outbreaks, *Epidemiol. Infect.*, 140, 27, 2012.

336. McCaughey, C. et al., Human seroprevalence to *Coxiella burnetii* (Q fever) in Northern Ireland, *Zoonoses Public Health*, 55, 189, 2008.

337. McQuiston, J.H. et al., National surveillance and the epidemiology of human Q fever in the United States, 1978–2004, *Am. J. Trop. Med. Hyg.*, 75, 36, 2006.

338. Botros, B.A. et al., *Coxiella burnetii* antibody prevalences among human populations in north-east Africa determined by enzyme immunoassay, *J. Trop. Med. Hyg.*, 98, 173, 1995.

339. Kilic, S. et al., Prevalence of *Coxiella burnetii* antibodies in blood donors in Ankara, Central Anatolia, Turkey, *New Microbiol.*, 31, 527, 2008.

340. Campagna, S. et al., Seroprevalence of 10 zoonotic infections in 2 Canadian Cree communities, *Diagn. Microbiol. Infect. Dis.*, 70, 191, 2011.

341. Letaief, A.O. et al., Seroepidemiological survey of rickettsial infections among blood donors in central Tunisia, *Trans. R. Soc. Trop. Med. Hyg.*, 89, 266, 1995.

342. Lacheheb, A. and Raoult, D., Seroprevalence of Q-fever in Algeria, *Clin. Microbiol. Infect.*, 15, 167, 2009.

343. Mediannikov, O. et al., *Coxiella burnetii* in humans and ticks in rural Senegal, *PLoS Negl. Trop. Dis.*, 4, e654, 2010.

344. Morović, M. et al., Q fever in Dalmatia, Croatia, *Clin. Microbiol. Infect.*, 15, 181, 2009.

345. Pape, M. et al., Seroprevalence of *Coxiella burnetii* in a healthy population from Northern Greece, *Clin. Microbiol. Infect.*, 15, 148, 2009.

346. Zhang, L. et al., Rickettsial seroepidemiology among farm workers, Tianjin, People's Republic of China, *Emerg. Infect. Dis.*, 14, 938, 2008.

30 *Escherichia coli* O157:H7

Patricia Elízaquível, Gloria Sánchez, and Rosa Aznar

CONTENTS

30.1 INTRODUCTION

Escherichia coli O157 was first reported in 1982 in outbreaks of severe bloody diarrhea in the United States. Such outbreaks increased dramatically and became widespread in the following years. In Europe, the first recognized outbreak of *E. coli* O157:H7 occurred in the United Kingdom in the summer of 1985, and further outbreaks and sporadic cases have been reported throughout Europe ever since. Nowadays, *E. coli* O157:H7 is an established human pathogen of great significance that accounts for an estimated 73,000 cases per year, 2,000 hospitalizations, and 60 deaths each year.[1] According to the cost list of foodborne illness in the United States, *E. coli* O157:H7 is ranked in sixth position, with an estimated cost of $993 million a year.[2] The pathogenicity of *E. coli* O157:H7 is primarily due to the bacterium's ability to produce Shiga toxin that may cause hemolytic uremic syndrome (HUS), a complication in which the red blood cells are destroyed and the kidneys fail.

In the early outbreaks, the sources were most often found to be contaminated beef meat, frequently minced, since the organism is widespread in the guts of asymptomatic cattle. As a reflection of the seriousness of such outbreaks, extensive legislation, new food handling practices, and food-producer education have been introduced and implemented over the last 20 years. Despite these, large outbreaks still occur mainly due to producer incompetence; for instance, an outbreak among schoolchildren in south Wales in 2005 involved 157 cases, with the death of a 5-year-old schoolboy. At the time, the requirements for food hygiene should have been sufficient to prevent it. Therefore, it seems that even though lessons have been learnt and processes changed accordingly, outbreaks cannot be avoided. Besides, risk factor studies indicate a shift from contaminated meat products toward environmental transmission in some countries, such as via contact with farm animals, gardening, or produce.[3]

In September 2006, there was an outbreak caused by *E. coli* O157:H7 found in uncooked spinach that involved 26 US states, 199 infected people, including 3 deaths and 31 who suffered HUS. The outbreak was traced to organic bagged fresh spinach and wild pigs on a ranch and the proximity of surface waterways to irrigation wells as "potential environmental risk factors." As a result, California's farm industry announced the adoption of a set of "good agricultural practices" to reduce the risk of *E. coli* contamination for leafy green vegetables. In addition, challenges for public health interventions of this potentially serious disease should also consider that there are over 200 serotypes of Shiga toxin-producing *E. coli* (STEC) that are commonly referred to as non-O157 STEC. It is estimated that non-O157 STEC double the prevalence of *E. coli* O157:H7 in foods but accounts for half the clinical cases with HUS being encountered. The potential health risk that these strains represent in food depends on the presence of virulence factors. In fact, a recent outbreak of STEC in northern Germany in June 2011 had serious health implications, causing a total of 3,816 cases (including 54 deaths), 845 (22%) of which involved HUS.[3] The strain causing the outbreak was identified as *E. coli* O104:H4, *stx*2 positive, *eae* negative, *iha* positive, and ESBL positive, but gentamicin and fluoroquinolone susceptible. Whole-genome sequencing revealed a unique combination of virulence traits that makes this German outbreak clone a unique hybrid of different *E. coli* pathovars.[4,5]

Risk evaluation studies will be necessary to make decisions about which pathogenic *E. coli* and on which foodstuff, water, or facilities should be surveyed to improve food safety measures.

30.2　CLASSIFICATION AND MORPHOLOGY

E. coli (obsolete synonym: *Bacillus coli communis* [Escherich 1885]) is one of the six species in the genus *Escherichia* that includes *E. adecarboxylata*, *E. blattae*, *E. coli*, *E. fergusonii*, *E. hermannii*, and *E. vulneris*. Named after Theodor Escherich (the person who first discovered this bacterium), the genus *Escherichia* is classified taxonomically in the family Enterobacteriaceae (consisting of Gram-negative bacteria that are motile via peritrichous flagella, grow well at 37°C, are oxidase negative and catalase positive, and reduce nitrates), order Enterobacteriales (rod-shaped facultatively anaerobic Gram-negative bacteria), class Gammaproteobacteria (facultatively anaerobic Gram-negative bacteria), phylum Proteobacteria (Gram-negative bacteria with an outer membrane composed primarily of lipopolysaccharides), and domain Bacteria.

Of the six *Escherichia* species, *E. coli* is the most frequently isolated and the best understood and characterized living organism. As Gram-negative, rod-shaped, and nonspore-forming bacilli, *E. coli* cells are about 2 μm in length and 0.5 μm in diameter. It is a facultative anaerobe, capable of reducing nitrates to nitrites, and produces acid and gas when growing fermentatively on glucose or other carbohydrates. Most strains are oxidase, citrate, urease, and hydrogen sulfide negative. They are positive for indole production and the methyl red test and can be differentiated from the closely related *Shigella* and *Salmonella* by its ability to ferment lactose.[6]

30.3　BIOLOGY AND EPIDEMIOLOGY

Members of *E. coli* are common inhabitants of the lower intestinal tract of humans and other vertebrates, where they often represent the most abundant facultative anaerobes. In fact, healthy humans typically carry more than a billion *E. coli* cells. They form a beneficial symbiotic relationship with mammal host, providing nutrients and immune regulation against other pathogens.[7] However, some *E. coli* strains have diverged to a more pathogenic nature causing serious diseases both within the intestinal tract and elsewhere within the host. They can also be present in soil and water as the result of fecal contamination, and therefore, their presence has classically been used as an indicator of poor water and food quality.

High genetic heterogeneity of *E. coli* has been reported, with only half of the genome conserved among the different strains.[8] The virulence genes responsible for the clinical capabilities are carried on plasmids, lysogenic bacteriophages, or large chromosomal insertions known as "pathogenicity-associated islands" (PAIs).[9] As a result, different disease-causing genotypes are recognized and referred to

as pathotypes or categories. Typically, each pathotype has a different pathogenesis and comprises a different combination of O (somatic), H (flagellar), and K (capsular) surface antigen profiles. So far, more than 700 antigenic types (serotypes) of *E. coli* are recognized based on these antigens.

Pathogenic *E. coli* strains are mainly responsible for three types of infections in humans: urinary tract infections (UTIs), neonatal meningitis, and intestinal diseases (gastroenteritis) that, in some instances, can evolve to HUS. Currently, there are six pathotypes associated with intestinal diseases that are classified according to their unique interaction with eukaryotic cells as enterotoxigenic *E. coli* (ETEC), enteroinvasive *E. coli* (EIEC), enterohemorrhagic *E. coli* (EHEC), enteropathogenic *E. coli* (EPEC), enteroaggregative *E. coli* (EAEC), and diffusely adhering *E. coli* (DAEC).[10] Of these, only the first four pathotypes have been associated with food- or waterborne illness. Moreover, two extraintestinal pathotypes, the uropathogenic *E. coli* (UPEC) and the meningitis-associated *E. coli* (MAEC), cause UTIs, ranging from asymptomatic bacteriuria to urosepsis and neonatal septicemia/meningitis, respectively.

E. coli accounts for the majority of enteric disease cases and is the major cause of traveler's diarrhea. The most severe human enteric infections are due to EHEC strains. However, most of the reported human EHEC outbreaks and sporadic EHEC infections are associated with a minor number of O:H serotypes. Out of 100 serotypes that have been implicated in cases of human disease, the O157 serotype is the most frequently reported.[11,12] In addition, serotypes O26, O103, O111, and O145 have increasingly been isolated from clinical cases and outbreaks as well as from animals and environmental sources.[13–16] As a consequence, the European Food Safety Authority (EFSA) has requested the monitoring of the "top-five" serogroups (O26, O103, O111, O145, and O157) that are most often encountered in foodborne diseases. Furthermore, the recent *E. coli* O104:H4 outbreak in the European Union (EU) (Germany, Spring 2011), caused by the unusual *E. coli* strain O104:H4, highlighted the need for rapid methods to approach characterization of foodborne pathogens, especially in such emergency situations. Indeed, the *E. coli* O104:H4 strain has been reported as a new pathotype, enteroaggregative hemorrhagic *E. coli* (EAHEC) that possess characteristics of both the EHEC and EAEC groups.[5]

Human infection may be acquired through the consumption of contaminated food or water, by direct transmission from person to person, or from infected animals to humans. EHEC (including serotype O157) are zoonotic pathogens associated with food- and waterborne illness around the world and have been isolated from many different animal species. Of them, the gastrointestinal tract of healthy ruminants seems to be the foremost important reservoir for EHEC. The sources of contamination include mainly cattle and food of bovine origin, with undercooked meat or ground beef being the major sources of human infections.[17] Other important food sources, which have been implicated in foodborne infections, include cider, lettuce, spinach, sprouts, and recreational water.[3,18]

The significance of many EHEC types that can be isolated from animals and foodstuffs for infections in humans is, however, not yet clear. Data on the epidemiology of these organisms are often influenced by the ability to differentiate between the pathogenic groups and serotypes. On the one hand, the enteric disease symptoms frequently derive from coinfection with ETEC and other enteric pathogens that may lead to problems in the accurate diagnosis.[19] On the other hand, in contrast to serotype O157, only a few laboratories screen for the presence of non-O157 serotypes. Therefore, the relevance of non-O157 EHEC is still not well understood and might be underestimated.[13] In the last years, the incidence of EHEC O157 infection has declined by 44% in the United States. Data from 2010 (FoodNet) on foodborne diseases revealed that the incidence per 100,000 population was 0.9 for O157 and 1.0 for non-O157.[20] In the EU, the incidence of *E. coli* O157 per 100,000 population was 0.9 according to the last EFSA report[21] based on data from 2010 where the total number of confirmed EHEC cases in the EU was 4,000, representing a 12.0% increase compared to 2009. The increase was mostly attributed to the German outbreak linked to the *E. coli* O104:H4 strain. Almost half of the reported O serogroups were O157 (41.1%). This represents an 18.8% decrease compared to the reported cases associated with the O157 serogroup in 2009.

Overall, more than one-half of the reported EHEC cases occur in 0–4-year-old children, the highest percentage of hospitalized persons age > 50 years, and the case fatality rate is among children aged <4 years. In the German outbreak, cases of the HUS occurred predominantly in adults, with a preponderance of cases occurring in women. The HUS developed in more than 20% of the identified cases.[3]

A marked seasonality in human EHEC[12] and ETEC[16] cases is observed, with most cases being reported during summer and autumn. This is linked, for instance, to the fact that fecal prevalence of *E. coli* O157:H7 in ruminants is the highest in summer, decreasing to low or undetectable levels during winter.[22]

30.4   CLINICAL FEATURES AND PATHOGENESIS

With a large range of pathologies, pathogenic *E. coli* strains are a major cause of human morbidity and mortality around the world. Of these, the EHEC including *E. coli* O157:H7 are also known as verotoxigenic *E. coli* (VTEC) since they are characterized by the production of toxin(s) that are designated verotoxin(s) (VT), VT1 and VT2, which cause a cytopathic effect on Vero cells. They are also referred to as STEC because their cytotoxins closely resemble the Shiga toxins produced by *Shigella dysenteriae* 1 (Stx1 and Stx2). Shiga toxin genes are ubiquitous among *E. coli* strains due to their transmission as part of lambdoid phages. Human pathogenic EHEC usually harbor additional virulence factors that are important for the development of disease in humans. The O157:H7 strain was probably originated by the acquisition of Shiga toxin-encoding prophages and the large virulence plasmid pO157 common in EHEC strains.[23,24]

In general, *E. coli* O157:H7 infection develops within 3–4 days of eating contaminated food, with a range of 1–10 days. A low infection dose of 10–100 organisms is sufficient to cause disease, which indicates a high acid-tolerance ability of this microorganism compared to other enteric bacterial pathogens.[6] Symptoms vary from a simple watery diarrhea to a hemorrhagic colitis that consists in bloody stools with ulcerations of the bowel. In most events, symptoms disappear in around 5 days, with a low rate of hospitalizations. However, in individuals at risk such as children and the elderly, infection can result in HUS that is characterized by acute renal failure, hemolytic anemia, and thrombocytopenia. It develops in up to 10% of patients infected with *E. coli* O157:H7 and is the leading cause of acute renal failure in young children. The incidence of HUS in children who have EHEC infection is 8%–10%. Of these, 3%–5% will die, and 12%–30% will have severe long-term effects including renal impairment, hypertension, or central nervous system manifestations.

30.5   IDENTIFICATION AND DIAGNOSIS

Pathogenic *E. coli* are identified based on its virulence factors; therefore, the analytical procedure for these pathogens generally requires the isolation and identification of the organisms as *E. coli* before testing for the specific virulence traits. Whereas the identification of *E. coli* is easily performed by traditional culture tests, differentiation between pathotypes requires molecular-based techniques to detect the virulence factors or their genes, since there are limited examples where pathogenic *E. coli* strains differ from the commensals in their metabolic abilities. For instance, commensal *E. coli* strains generally use sorbitol, but *E. coli* O157:H7 does not. Most diarrheagenic strains cannot utilize D-serine as a carbon and nitrogen source, but uropathogenic and commensal fecal strains can use it.[6]

30.5.1   CONVENTIONAL TECHNIQUES

E. coli detection is typically performed according to conventional recommended culture-based methods, that is, FDA's Bacteriological Analytical Manual (http://www.fda.gov/Food/ScienceResearch/LaboratoryMethods/BacteriologicalAnalyticalManualBAM/ucm070080.htm). *E. coli* can be recovered easily from clinical specimens on general or selective lactose-containing media, usually MacConkey or eosin methylene-blue agar, at 37°C under aerobic conditions. *E. coli* isolation from environmental or food samples normally requires previous steps such as filtration or pre-enrichment. A wide variety of selective and differential media are available to facilitate selection of presumptive colonies. Further, identification is based on biochemical tests generally performed using commercially available strips, such as API tests, allowing quick and satisfactory results. Besides lactose fermentation, other specific biochemical traits have been applied to design rapid procedures for the isolation and identification of *E. coli*. For instance, it can be identified within 1 h by oxidase, indole, lactose, and beta-glucuronidase tests following growth on MacConkey agar. Moreover, the presence of the enzyme beta-glucuronidase can be tested using broth or agar medium containing the substrate 4-methylumbelliferyl-β-D-glucuronide (MUG). This compound is colorless, but following hydrolysis by the glucuronidase, the cleaved moiety is fluorescent under long-wave ultraviolet light. The incorporation of MUG into MacConkey agar isolation plates allows direct determination of beta-glucuronidase activity, and identification can be completed in 5 min.

The discrimination of pathogenic *E. coli* from non-pathogenic strains is a key issue, which complicates the diagnosis because they have to be tested for the specific virulence factors. Serotyping has extensively been used to identify pathogenic strains. However, correlation between serotypes (combination of O and H antigens) and specific diarrheagenic *E. coli* strains is not frequent, an exception of the serotype O157:H7. The recognized threat to human health of these pathogens has led to the development of selective–differential media to aid in detection, isolation, and presumptive identification of EHEC strains, particularly serotype O157:H7. Nearly all isolates of this serotype ferment D-sorbitol slowly or not at all. This unique phenotypic feature is exploited by substituting the carbohydrate sorbitol for lactose in MacConkey agar to develop the Sorbitol-MacConkey (SMAC) agar. Sorbitol-negative colonies will appear colorless on SMAC. CHROMagar O157™ is also based on sorbitol fermentation but using a highly specific chromogenic reaction, where *E. coli* O157 shows mauve colonies and *E. coli* non-O157 blue colonies. Cefixime and tellurite can be added for an increased selectivity (i.e., CT-SMAC, T-CHROMagar O157). Rainbow Agar O157 contains chromogenic substrates that are specific for two *E. coli*-associated enzymes: beta-galactosidase (a blue-black chromogenic substrate) and beta-glucuronidase (a red chromogenic substrate). The O157:H7 strains are typically glucuronidase negative and form black or gray colonies. Following, presumptive colonies that tested positive for agglutinating in O157 antiserum or O157 latex reagent should be identified biochemically as *E. coli*, since strains of several species cross-react with O157 antiserum. The confirmation of *E. coli* O157:H7 requires the identification of the H7 flagellar antigen, which requires special expertise. In addition, isolates that are nonmotile or that are negative for the H7 antigen should be tested for the production of the Shiga-like toxins to identify pathogenic strains. The specificity of selective–differential media can be increased, including a previous step by enriching the sample by immunocapture with magnetic beads coated with antibodies against *E. coli* O157.

In addition to *E. coli* O157:H7, atypical sorbitol-positive strains have increasingly been isolated from clinical cases stressing the importance of screening for both sorbitol-positive and sorbitol-negative O157. Improved recovery of non-O157 EHEC serotypes (O26, O103, O111, and O145) has been approached by Possé et al.[15] using selective–differential media that combine the use of a chromogenic compound

to signal beta-galactosidase activity and one or more fermentative carbon sources with a pH indicator and several inhibitory components.

Nevertheless, isolates have to be confirmed as toxigenic strains either by serotyping or testing for the presence of virulence factors using tissue culture methods. Alternatively, PCR-based molecular methods, using serotype-specific PCR and PCR-based virulence typing (Table 30.1), have gained acceptance in the last decade.

30.5.2 IMMUNOASSAYS

Traditionally, antibodies have been used for *E. coli* serotyping by mixing them with a culture of suspected *E. coli*. When antibodies react with the specific antigen, agglutination of antigens can be visualized on a slide. A number of serological tests (enzyme-linked immunosorbent assay [ELISA], radioimmunoassay, immunoblot, and reversed-passive latex agglutination [RPLA]) have extensively been used for the detection of *E. coli* or *E. coli* toxins. Among them, ELISA test is the most popular format where antibodies are fixed on a solid support, for example, wells of microtiter plate. Nowadays, many of these immunological tests are commercially available, which is an advantage for routine laboratories. However, it is essential that an international and independent organization, for example, AOAC, validates or evaluates them prior to become an official method. In addition, some companies have completely automated the entire procedure. For instance, the Vitek Immunodiagnostic Assay System (VIDAS) (bioMérieux) is widely used for the detection of *E. coli* O157. This system consists on a reagent strip in which the enriched broth is added. Thereafter, the strip is introduced in the multiparameter automated immunoanalyzer. The analytical module automatically performs all stages of the analysis, including printing out the full results. The immunoanalyzer contains five sections, each accepting 6 tests; thus, it can handle 30 tests per hour. This format is designed as a "sandwich" assay where a specific antibody is bound to a solid matrix that captures the target pathogen from enrichment cultures. Then, a second antibody conjugated to an enzyme fixes itself to the captured antigen. The detection takes place by measuring the intensity of the

TABLE 30.1
Most Frequent Target Genes for the Detection of *E. coli* O157:H7 by PCR or qPCR

Target Gene	Virulence Factor	References
eae	Intimin	[10,57,58]
*rfb*E	O157 antigen	
*fli*C	H7 antigen	
*stx*1	Shiga toxin 1	
*stx*2	Shiga toxin 2	
wzx	Flippase	
*ihp*1	Inserted hypothetical protein 1	
*uid*A	Glucuronidase	
*hly*A	Hemolysin	

fluorescence emitted by an enzymatic reaction. Currently, there are two different VIDAS assays for the detection of *E. coli*, the VIDAS-ICE for the generic detection of *E. coli* and the VIDAS-ECO for the specific detection of serotype O157. In both cases, to confirm the presence of *E. coli* O157:H7, all presumptive positive results have to be confirmed culturally by streaking onto CT-SMAC and SMAC agar plates.

Besides the VIDAS system, the lateral flow technology is a simple, easy, and rapid method that allows taking the analysis out of the laboratory. Briefly, lateral flow tests are immunochromatographic assays, like the pregnancy tests, adapted to operate along a single axis to suit the test strip format. A number of variations of the technology have been developed into commercial products to detect *E. coli* O157, but they all operate according to the same basic principle. It is also a "sandwich" procedure but, instead of enzyme conjugates, the detection antibody is coupled to colored latex beads or to colloidal gold. In this way, the result is obtained within 10 min after cultural enrichment.

Another innovative development is the immunomagnetic separation system, where antibodies are coupled to magnetic beads, which can easily be recovered. Thereafter, cells can be detected by direct plating, ELISA, PCR, or other procedures. This technique allows not only the detection but also the concentration of the pathogen in the sample. It can be applied instead of the selective enrichment, thus preventing from the use of growth inhibitors that can cause cells stress injury. Several systems based on magnetic beads have been developed for the isolation of *E. coli*, and some of them are currently commercially available.[25] Moreover, fully automated immunomagnetic separation systems have also been developed for *E. coli* O157:H7 concentration,[26,27] reducing hands-on manipulations.

30.5.3 PCR-Based Techniques

Molecular methods are the most widely used techniques for differentiating pathogenic from nonpathogenic *E. coli* isolates. They mostly focus on the detection of EHEC by searching Shiga toxins production or virulence-associated genes. The most common methods currently used to identify *E. coli* in clinical or food samples involve PCR-based techniques. Primers used in PCR have successfully been applied to detect all six intestinal categories (Table 30.1). Moreover, in the recent years, real-time PCR (qPCR) assays have revolutionized nucleic acid detection by the high speed, sensitivity, reproducibility, and minimization of cross contamination. Besides, it also offers the possibility to quantify the targeted pathogen in the sample. The qPCR technology is based on the detection of a fluorescent signal coming either from dsDNA-specific dyes (e.g., SYBR Green I) or from specific probes that include hydrolysis probes (e.g., TaqMan or TaqMan-MGB) and hybridization probes (e.g., molecular beacons or Scorpion primers).

Theoretically, as few as three copies of the targeted gene per PCR can be detected.[28] However, due to the small sample volume (1–10 µL) analyzed, the reality is that for most PCRs,

the limit of detection is 1–1000 cells per reaction that translates to $1000–1 \times 10^6$ cells per mL. Assuming that the pathogen might be present at lower levels, that is, water or food samples, initial cell numbers could be risen up including an enrichment step or by concentrating the initial sample. Nevertheless the concentration step, other than immunoconcentration, may enhance the presence of PCR inhibitors. Therefore, inhibition of PCR or qPCR should be checked by dilution of the sample and using an internal amplification control (IAC).

Presently, many qPCR assays for the detection of pathogenic *E. coli* have been described; most of them are designed for serotype O157:H7 and targeting a single gene. For screening purposes, this represents a limitation since only a specific serotype is investigated. To overcome this issue, multiplex PCR assays allow simultaneous detection of various *E. coli* serotypes/pathotypes. Moreover, the combination of multiplex SYBR® Green Real-time PCR methods and high-resolution melting (HRM) analysis allows the discrimination among strains possessing similar virulence traits. Using this approach, Guion et al.[29] developed a multiplex qPCR for the detection of all six intestinal *E. coli* categories. Kagkli et al.[30] developed a multiplex qPCR capable of rapidly detecting the presence of the toxin genes together with intimin (eae) (in the case of VTEC) or aggregative protein (aggR) (in the case of the O104:H4 strain) responsible for the outbreak in Germany in 2011. The targeted pathogenic *E. coli* include the "top-five" serogroups (O26, O103, O111, O145, and O157) requested for monitoring by the EFSA. On the other hand, generic primers for the detection of *E. coli* have been described[31] that provide a rapid tool for *E. coli* monitoring.

Nevertheless, the major obstacle with PCR assays is how to distinguish between DNA from dead cells and DNA from live cells. Intact DNA can be present although the organisms are dead. This is particularly relevant for qPCR quantification that may overestimate the number of live cells due to the relatively long persistence of DNA after cell death. The most commonly used strategy to avoid this overestimation relies on the detection of mRNA that is a direct indicator of the active bacterial metabolism. Thus, reverse transcription (RT)-PCR and nucleic acid sequence amplification (NASBA) have been applied for the detection of viable *E. coli* cells.[32–34] Both can be used as quantitative procedures, RT-qPCR and NASBA-qPCR, respectively. Other alternative assays consist on the use of DNA-binding molecules, like ethidium monoazide (EMA) or propidium monoazide (PMA), as a sample treatment previous to the DNA extraction. Both dyes intercalate with DNA after a photolysis process using visible light inhibiting the PCR process. This treatment can be combined with the qPCR for the quantitative detection of viable cells, as recently showed for *E. coli* O157:H7 detection.[35]

30.5.4 Next-Generation Assays

These include microarrays, biosensors, DNA chips, ion semiconductor sequencing, and other upcoming new technologies.

Microarrays contain hundreds or thousands of micrometer-diameter spots, which can contain antibodies or probes to capture the target of interest. In the first case, for detection, a second labeled antibody is then used and the latter generally involves gene amplification followed by hybridization analysis. DNA arrays or DNA chips are available in formats ranging from high-density chips with over a hundred thousand gene probes to low-density membranes with several genes only. Microarrays allow the screening of thousands of samples in a single run, combined with high-throughput analysis. Microarrays based on antibodies and different types of probes have successfully been applied to detect *E. coli* O157:H7.[36–38] Moreover PCR or multiplex PCR can be combined with a microarray hybridization step for the detection of different *E. coli* virulence genes. In addition, mRNA of *E. coli* has also been targeted, allowing the detection of the viable cells.[35]

Biosensors are devices that detect biological or chemical recognition complexes in the form of antigen–antibody, enzyme–substrate, or receptor–ligand, placed in proximity to a transducer that generates a signal. They are self-contained integrated devices capable of providing specific quantitative or semiquantitative analytical information. A biosensor should be clearly distinguished from a bioanalytical system, which requires additional processing steps, such as reagent addition.[39] The advantages of biosensors are the possibility of portability, miniaturization and working on-site, and the ability to measure pathogens in complex matrices with minimal sample preparation. Different types of biosensors have been reported for *E. coli* detection, including optical, microgravimetric, and electrochemical devices.[40–42] Moreover, biosensors have also been assayed targeting mRNA allowing the detection of *E. coli* viable cells.[43]

Pyrosequencing technology is a real-time DNA-sequencing method, based upon the release of pyrophosphate when any nucleotide triphosphate is successfully added to the growing polynucleotide chain, that is, sequencing by synthesis. This technology is particularly advantageous for the fast identification of short DNA sequences and the detection of single-nucleotide polymorphisms. This technique is simple to use and accurate and allows a quantitative analysis of DNA sequences. It has successfully been applied to differentiate bacteria commonly associated with diarrhea by targeting a partial amplicon of the *gyr*B gene as well to differentiate O157 serotype from other *E. coli* serotypes.[44]

Nowadays, with the advent of next-generation sequencing, large amounts of sequence data can be generated very rapidly, that is, from DNA sample to sequence data, in a single day. This technology is based on the use of an integrated semiconductor device enabling nonoptical genome sequencing. Sequence data are obtained by directly sensing the ions produced by template-directed DNA polymerase synthesis using all-natural nucleotides on a massively parallel semiconductor-sensing device or ion chip. This technology has been successfully applied to discover within days a unique combination of virulence traits in the strain causing the German outbreak in 2011.[4]

30.6 TREATMENT AND PREVENTION

30.6.1 TREATMENT

Despite the increased understanding of the pathophysiology of EHEC, treatment modalities have not evolved during the last years. The first symptom showed after an infection is acute bloody diarrhea. As a first treatment, patients showing this pathology should be rehydrated intravenously in order to maintain fluids and avoid kidney damage.[45] However, standard rehydration protocols that consist on the replacement of the liquid losses are not appropriate as they are inaccurate due to the presence of vascular leakage and edemas.[46] Instead, isotonic crystalloid and infusion of higher sodium concentrations are recommended.[47] Once a person is infected with an EHEC, the percentage of patients in whom the infection progresses to HUS depends on the infecting EHEC serotype and was reported for *E. coli* O157:H7 to be 15%.[46] The development of the HUS comprises acute renal failure leading to the perturbation of fluid and electrolyte balance, hemolysis, and disruption of the clotting cascade with thrombocytopenia. This syndrome must be managed and addressed urgently using a multitargeted approach.[48] To that aim, the treatment of Stx-HUS includes supportive measures to mainly manage fluid levels and electrolyte balance that are extremely important in preventing the development of HUS.[45,49] HUS is a life-threatening condition that is usually treated in an intensive care unit. Blood transfusions and kidney dialysis are often required. With intensive care, the death rate for HUS is 3%–5%.

Moreover, other symptoms such as anemia resulting from hemolysis, fluid and electrolyte disturbances, hypertension, and other extrarenal complications must be addressed as well. In this sense, peritoneal dialysis (PD) or hemodialysis has been shown to improve outcomes. Anemia may need correction with transfusions of whole blood or packed red cells. If these approaches are not successful, it may be necessary to perform renal replacement therapy required in 30%–60% of cases.[50]

Nevertheless, most people recover without specific treatment within 5–10 days. Antibiotics should not be used to treat this infection, as they involve bacterial lysis that causes the release of VT.[51–53] It is thought that treatment with some antibiotics could lead to kidney complications. Antidiarrheal agents should also be avoided. Other strategies include the use of ligand mimics of the receptor for Stx in the gastrointestinal tract to prevent the spread of toxin to extraintestinal sites that have been proposed or neutralizing Shiga toxin-specific antibodies.[48] However, it is unlikely that targeting a single pathway with a treatment modality will be sufficiently successful, and thus, a multitargeted approach would seem necessary.

30.6.2 PREVENTION

The prevention of *E. coli* O157:H7 infection is largely dependent on the control of the spread of this bacterium, whose way of transmission is changing over time. On one hand, good hygienic practices are the key measures in food of animal origin. The steps include proper control of fecal soiling of meat during slaughter processing and proper cooking of food products. Internal meat temperature over 68.3°C (155°F) eradicates enterohemorrhagic *E. coli* contamination. Additionally, pasteurization of milk and milk products is another important measure. On the other hand, "good agricultural practices" should also be implemented to reduce the risk of *E. coli* contamination for produce, that is, leafy green vegetables and sprouts. Besides, harmonized recommendations for the monitoring of *E. coli* O157:H7 as well as other EHEC serotypes are necessary for a better understanding of the biology of these pathogens leading to the establishment of the microbiological standards for food safety regulations.

30.7 CONCLUSIONS AND FUTURE PERSPECTIVES

Although *E. coli* O157:H7 is the most predominant pathogenic serotype, *E. coli* constitutes a very versatile and diverse enterobacterial species whose genome is constantly subjected to rearrangements that is frequently the origin of strains with new virulence traits, for example, the recent German outbreak caused by *E. coli* O104:H4. Therefore, despite *E. coli* pathotypes are well defined, the existence of atypical strains showing new combinations of pathogenic traits and new antigenic formula or being nontypeable hampers detection and diagnosis.[54] At present, the assessment of the risk related to a given strain is achieved by determining not only the serogroup but also the simultaneous presence of the virulence-associated genes. Eight real-time PCR methods for the detection of *E. coli* toxin genes and their variants (stx(1), stx(2)), the intimin gene (eae), and five serogroup-specific genes have been proposed by the European Reference Laboratory for VTEC (EURL-VTEC) as a technical specification to the European Normalization Committee (CEN TC275/WG6) and further validated in food matrices.[55] Molecular diagnostic techniques and specially PCR-based methods are undoubtedly the best choice for rapid and reliable detection of pathogenic strains, even if they are atypical or nontypeable. Rapid and accurate identification of causative bacteria with a high degree of specificity and sensitivity is essential to outbreak surveillance and investigation. In addition, it would allow the implementation of proper clinical therapies as well as food ban regulations. Nowadays, high-throughput systems are available for accurate identification by targeting several genes such as qPCR-HRM,[30] multiplex qPCR,[55] microarrays,[56] Pyrosequencing™,[48] and the next-generation sequencing approach,[4] providing rapid and accurate genotypic identification that can also be applied in pathogen detection.

Prevention measures should continue focusing on good hygienic practices, while incorporating harmonized recommendations for the monitoring of EHEC, not only O157:H7 serotype. Besides, molecular detection of virulence factors,

either by PCR or qPCR methods or by the metagenomic approach using the next-generation sequencing, will improve sensitivity and diagnostic accuracy in future surveys.

REFERENCES

1. Mead, P.S. and Griffin, P.M. *Escherichia coli* O157:H7. *Lancet* 352, 1207–1212 (1998).
2. Scharff, R.L. Economic burden from health losses due to foodborne illness in the United States. *J. Food Prot.* 75, 123–131 (2012).
3. Frank, C. et al. Epidemic profile of Shiga-toxin-producing *Escherichia coli* O104:H4 outbreak in Germany. *N. Engl. J. Med.* 365, 1771–1780 (2011).
4. Mellmann, A. et al. Prospective genomic characterization of the German enterohemorrhagic *Escherichia coli* O104:H4 outbreak by rapid next generation sequencing technology. *PLoS One* 6, e22751 (2011).
5. Brzuszkiewicz, E. et al. Genome sequence analyses of two isolates from the recent *Escherichia coli* outbreak in Germany reveal the emergence of a new pathotype: Entero-aggregative-haemorrhagic *Escherichia coli* (EAHEC). *Arch. Microbiol.* 193, 883–891 (2011).
6. Welch, R.A. The genus *Escherichia*, in *The Prokaryotes*, eds. Falkow, S., Rosenberg, E., Schleifer, K.-H., Stackebrandt, E., and Dworkin, M., pp. 60–71, Springer, New York (2006).
7. Yan, F. and Polk, D.B. Commensal bacteria in the gut: Learning who our friends are. *Curr. Opin. Gastroenterol.* 20, 565–571 (2004).
8. Rasko, D.A. et al. The pangenome structure of *Escherichia coli*: Comparative genomic analysis of *E. coli* commensal and pathogenic isolates. *J. Bacteriol.* 190, 6881–6893 (2008).
9. Hacker, J. et al. Pathogenicity islands of virulent bacteria: Structure, function and impact on microbial evolution. *Mol. Microbiol.* 23, 1089–1097 (1997).
10. Nataro, J.P. and Kaper, J.B. Diarrheagenic *Escherichia coli*. *Clin. Microbiol. Rev.* 11, 142–201 (1998).
11. Anonymous. Annual listing of foodborne disease outbreaks, United States. CDC Report (2006).
12. European Food Safety Authority. The community summary report on trends and sources of zoonoses, zoonotic agents, antimicrobial resistance and foodborne outbreaks in the European Union in 2006. *EFSA J.* 130, 3–352 (2007).
13. Bettelheim, K.A. The non-O157 shiga-toxigenic (verocytotoxigenic) *Escherichia coli*; under-rated pathogens. *Crit. Rev. Microbiol.* 33, 67–87 (2007).
14. Eklund, M., Scheutz, F., and Siitonen, A. Clinical isolates of non-O157 shiga toxin-producing *Escherichia coli*: Serotypes, virulence characteristics, and molecular profiles of strains of the same serotype. *J. Clin. Microbiol.* 39, 2829–2834 (2001).
15. Possé, B. et al. Quantitative isolation efficiency of O26, O103, O111, O145 and O157 STEC serotypes from artificially contaminated food and cattle faeces samples using a new isolation protocol. *J. Appl. Microbiol.* 105, 227–235 (2008).
16. Bettelheim, K.A. Non-O157 verotoxin-producing *Escherichia coli*: A problem, paradox, and paradigm. *Exp. Biol. Med. (Maywood)* 228, 333–344 (2003).
17. Karmali, M.A., Gannon, V., and Sargeant, J.M. Verocytotoxin-producing *Escherichia coli* (VTEC). *Vet. Microbiol.* 140, 360–370 (2010).
18. Rangel, J.M. et al. Epidemiology of *Escherichia coli* O157:H7 outbreaks, United States, 1982–2002. *Emerg. Infect. Dis.* 11, 603–609 (2005).

19. Qadri, F. et al. Enterotoxigenic *Escherichia coli* in developing countries: Epidemiology, microbiology, clinical features, treatment, and prevention. *Clin. Microbiol. Rev.* 18, 465–483 (2005).

20. MMWR. http://www.cdc.gov/mmwr/preview/mmwrhtml/mm5813a2.htm?s_cid = mm5813 a2e 58, 333–337 (2009).

21. Anonymous. The Community summary report on trends and sources of zoonoses, zoonotic agents and food-borne outbreaks in the European Union in 2008. *EFSA J.* 1496, 1–288 (2010).

22. Edrington, T.S. et al. Seasonal shedding of *Escherichia coli* O157:H7 in ruminants: A new hypothesis. *Foodborne Pathog. Dis.* 3, 413–421 (2006).

23. Reid, S.D. et al. Parallel evolution of virulence in pathogenic *Escherichia coli*. *Nature* 406, 64–67 (2000).

24. Lathem, W.W. et al. Acquisition of stcE, a C1 esterase inhibitor-specific metalloprotease, during the evolution of *Escherichia coli* O157:H7. *J. Infect. Dis.* 187, 1907–1914 (2003).

25. Lund, A., Wasteson, Y., and Olsvik, O. Immunomagnetic separation and DNA hybridization for detection of enterotoxigenic *Escherichia coli* in a piglet model. *J. Clin. Microbiol.* 29, 2259–2262 (1991).

26. Chandler, D.P. et al. Automated immunomagnetic separation and microarray detection of *E. coli* O157:H7 from poultry carcass rinse. *Int. J. Food Microbiol.* 70, 143–154 (2001).

27. Prentice, N. et al. Rapid isolation and detection of *Escherichia coli* O157:H7 in fresh produce. *J. Rapid Methods Autom. Microbiol.* 14, 299–308 (2007).

28. Bustin, S.A. et al. The MIQE guidelines: minimum information for publication of quantitative real-time PCR experiments. *Clin. Chem.* 55, 611–622 (2009).

29. Guion, C.E. et al. Detection of diarrheagenic *Escherichia coli* by use of melting-curve analysis and real-time multiplex PCR. *J. Clin. Microbiol.* 46, 1752–1757 (2008).

30. Kagkli, D.M. et al. Towards a pathogenic *Escherichia coli* detection platform using multiplex SYBR(R) Green real-time PCR methods and high resolution melting analysis. *PLoS One* 7, e39287 (2012).

31. Bernasconi, C. et al. Use of the tna Operon as a new molecular target for *Escherichia coli* detection. *Appl. Environ. Microbiol.* 73, 6321–6325 (2007).

32. Liu, Y. et al. Electronic deoxyribonucleic acid (DNA) microarray detection of viable pathogenic *Escherichia coli*, *Vibrio cholerae*, and *Salmonella typhi*. *Anal. Chim. Acta* 578, 75–81 (2006).

33. McIngvale, S.C., Elhanafi, D., and Drake, M.A. Optimization of reverse transcriptase PCR to detect viable Shiga-toxin-producing *Escherichia coli*. *Appl. Environ. Microbiol.* 68, 799–806 (2002).

34. Cook, N. The use of NASBA for the detection of microbial pathogens in food and environmental samples. *J. Microbiol. Methods* 53, 165–174 (2003).

35. Wang, L., Li, Y., and Mustapha, A. Detection of viable *Escherichia coli* O157:H7 by ethidium monoazide real-time PCR. *J. Appl. Microbiol.* 107, 1719–1728 (2009).

36. Gehring, A.G. et al. Antibody microarray detection of *Escherichia coli* O157:H7: Quantification, assay limitations, and capture efficiency. *Anal. Chem.* 78, 6601–6607 (2006).

37. Kim, H. et al. A molecular beacon DNA microarray system for rapid detection of *E. coli* O157:H7 that eliminates the risk of a false negative signal. *Biosens. Bioelectron.* 22, 1041–1047 (2007).

38. Willenbrock, H. et al. Design of a seven-genome *Escherichia coli* microarray for comparative genomic profiling. *J. Bacteriol.* 188, 7713–7721 (2006).

39. Rodriguez-Mozaz, S., Lopez de Alda, M.J., and Barcelo, D. Biosensors as useful tools for environmental analysis and monitoring. *Anal. Bioanal. Chem.* 386, 1025–1041 (2006).

40. Wang, L. et al. A novel electrochemical biosensor based on dynamic polymerase-extending hybridization for *E. coli* O157:H7 DNA detection. *Talanta* 78, 647–652 (2009).

41. Pyun, J.C. et al. Development of a biosensor for *E. coli* based on a flexural plate wave (FPW) transducer. *Biosens. Bioelectron.* 13, 839–845 (1998).

42. Zhu, P. et al. Detection of water-borne *E. coli* O157 using the integrating waveguide biosensor. *Biosens. Bioelectron.* 21, 678–683 (2005).

43. Baeumner, A.J. et al. RNA biosensor for the rapid detection of viable *Escherichia coli* in drinking water. *Biosens. Bioelectron.* 18, 405–413 (2003).

44. Hou, X.L. et al. Pyrosequencing analysis of the *gyrB* gene to differentiate bacteria responsible for diarrheal diseases. *Eur. J. Clin. Microbiol. Infect. Dis.* 27, 587–596 (2008).

45. Ake, J.A. et al. Relative nephroprotection during *Escherichia coli* O157:H7 infections: Association with intravenous volume expansion. *Pediatrics* 115, e673–e680 (2005).

46. Tarr, P.I., Gordon, C.A., and Chandler, W.L. Shiga-toxin-producing *Escherichia coli* and haemolytic uraemic syndrome. *Lancet* 365, 1073–1086 (2005).

47. Tarr, P.I. and Neill, M.A. *Escherichia coli* O157:H7. *Gastroenterol. Clin. North Am.* 30, 735–751 (2001).

48. Goldwater, P.N. and Bettelheim, K.A. Treatment of enterohemorrhagic *Escherichia coli* (EHEC) infection and hemolytic uremic syndrome (HUS). *BMC Med.* 10, 12 (2012).

49. Hickey, C.A. et al. Early volume expansion during diarrhea and relative nephroprotection during subsequent hemolytic uremic syndrome. *Arch. Pediatr. Adolesc. Med.* 165, 884–889 (2011).

50. Bitzan, M., Schaefer, F., and Reymond, D. Treatment of typical (enteropathic) hemolytic uremic syndrome. *Semin. Thromb. Hemost.* 36, 594–610 (2010).

51. Cimolai, N., Morrison, B.J., and Carter, J.E. Risk factors for the central nervous system manifestations of gastroenteritis-associated hemolytic-uremic syndrome. *Pediatrics* 90, 616–621 (1992).

52. Bell, B.P. et al. Predictors of hemolytic uremic syndrome in children during a large outbreak of *Escherichia coli* O157:H7 infections. *Pediatrics* 100, E12 (1997).

53. Wong, C.S. et al. The risk of the hemolytic–uremic syndrome after antibiotic treatment of *Escherichia coli* O157:H7 infections. *N. Engl. J. Med.* 342, 1930–1936 (2000).

54. Hernandes, R.T. et al. An overview of atypical enteropathogenic *Escherichia coli*. *FEMS Microbiol. Lett.* 297, 137–149 (2009).

55. Kagkli, D.M. et al. Application of the modular approach to an in-house validation study of real-time PCR methods for the detection and serogroup determination of verocytotoxigenic *Escherichia coli*. *Appl. Environ. Microbiol.* 77, 6954–6963 (2011).

56. Salehi, T.Z. et al. Genetic characterization of *Escherichia coli* O157:H7 strains isolated from the one-humped camel (*Camelus dromedarius*) by using microarray DNA technology. *Mol. Biotechnol.* 51, 283–288 (2012).

57. Beutin, L., Jahn, S., and Fach, P. Evaluation of the 'GeneDisc' real-time PCR system for detection of enterohaemorrhagic *Escherichia coli* (EHEC) O26, O103, O111, O145 and O157 strains according to their virulence markers and their O- and H-antigen-associated genes. *J. Appl. Microbiol.* 106, 1122–1132 (2009).

58. Fu, Z., Rogelj, S., and Kieft, T.L. Rapid detection of *Escherichia coli* O157:H7 by immunomagnetic separation and real-time PCR. *Int. J. Food Microbiol.* 99, 47–57 (2005).

31 *Francisella tularensis*

Luke C. Kingry and Jeannine M. Petersen

CONTENTS

31.1 INTRODUCTION

Present throughout the Northern Hemisphere, the gram-negative organism *Francisella tularensis* was first isolated in 1912 from rodents suffering from a "plague"-like disease in Tulare County, California, United States.[1,2] Descriptions of similar illnesses, affecting lemmings in Norway and humans with rabbit contact in Japan, date back to the seventeenth century, suggesting recognition of the disease caused by *F. tularensis* decades prior to its isolation.[3,4] The first diagnosed (bacteriologically confirmed) human case due to infection with *F. tularensis* was reported in 1914 in a meat cutter from Ohio, United States.[5] Edward Francis spent his career studying the bacterium and coined the term "tularemia" to describe the resulting disease, after determining deer-fly fever and the "plague"-like disease of rodents observed in Tulare County were both caused by *F. tularensis*.[6] In 1947, the genus name *Francisella* was chosen to commemorate Edward Francis' numerous contributions to the field.[7,8]

During the first half of the twentieth century, the incidence of tularemia in the Northern Hemisphere was higher than present day. Between 1927 and 1948, over 22,000 cases were reported in the United States.[9,10] In the former Soviet Union, substantial numbers of tularemia cases occurred leading up to and during World War II.[11] Most famously, a tularemia outbreak, which affected both German and Soviet soldiers at the Siege of Stalingrad, resulted in hundreds of thousands of cases between 1942 and 1943.[12] This outbreak has been attributed to an intentional release as well as to wartime conditions, which led to a breakdown in public health infrastructure, unharvested crops, and large numbers of rodents in trenches and dugouts.[12] Regardless of the cause, the enormity of the outbreak established the epidemic potential of *F. tularensis*.

31.2 *F. tularensis* AS A BIOLOGICAL WEAPON

F. tularensis was examined by multiple countries for its possible use as both an offensive and retaliatory biological weapon in the years leading up to and following World War II.[13] In 1937, Japan began studying numerous biological agents, including *F. tularensis*, for potential weaponization, with some of these agents tested on prisoners of war.[13,14] The former Soviet Union developed vaccine and antibiotic-resistant strains of *F. tularensis* as part of their extensive bioweapons research program.[13,15] In 1942, the United States agreed to only use biological weapons in retaliation while continuing to engage in biological weapons research.[13] By 1955, the Pine Bluff Arsenal in Arkansas, United States was producing *F. tularensis* for use as a retaliatory biological warfare agent, with advances made in drying concentrated suspensions of the bacterium in order to improve long-term storage, stability, and ease of dispersal.[16,17] In 1972, the United States opened a biological defense facility at Fort Detrick in Maryland, United States, where botulism and anthrax were the primary focus, but brucellosis, tularemia, and glanders were also studied.[13,18,19] Vaccines for *F. tularensis* were studied extensively at Fort Detrick using human subjects exposed to aerosols of *F. tularensis* in the one million liter spherical aerosol chamber nicknamed the "8-ball."[16]

Decades later, concern regarding intentional use of *F. tularensis* as a bioweapon was renewed. A marked increase in bioterrorism research occurred after the deliberate distribution of anthrax spores, immediately following the September 11, 2001, terrorist attacks in the United States. In 2002, *F. tularensis* was classified as a category A select agent by the US Centers for Disease Control and Prevention (CDC), one of four bacterial agents with highest potential for misuse.[20] Likewise, the World Health Organization (WHO) and European Centre for Disease Prevention and

Control (ECDC) categorized *F. tularensis* as a top priority pathogen of concern with respect to intentional misuse.[21,22] In October 2012, the US Department of Health and Human Services released 42 CFR Part 73 Final Rule reclassifying biological agents or toxins that "present the greatest risk of deliberate misuse with the most significant potential for mass casualties or devastating effects to the economy, critical infrastructure, or public confidence" as Tier 1 agents.[23] Designation of *F. tularensis* as a Tier 1 pathogen is likely due to a number of criteria including, but not limited to, its history of weaponization, its low infectious dose, its ability to cause pneumonic disease upon inhalation, its potential for causing morbidity and mortality in humans, the relative ease in growing the organism, and the predicted economic impact of an outbreak.[19,24,25]

F. tularensis is one of the most infectious bacterial agents known. Experimental studies performed in human volunteers in the 1970s, for the purpose of vaccine efficacy testing, provided evidence that humans were regularly infected by as few as 25 organisms when aerogenically exposed to bacteria.[26] Clinically overt disease in these individuals occurred quickly, within 3–5 days following aerosol exposure.[26] More recently, mathematical modeling of the minimal infectious dose, based on several historical studies, concluded that a single *F. tularensis* cell would result in a 40%–50% infectivity rate ($ID_{50} = 1$ bacteria).[27]

A report from the WHO in 1970 estimated the economic impact of an *F. tularensis* attack and concluded that dispersal over a population of 5 million people could render 250,000 people extremely ill and cause approximately 19,000 fatalities.[28] More recently, predictive modeling was used to estimate the medical and economic impact due to a release of *F. tularensis* on a large suburban population. The study concluded that 82,500 pneumonic or typhoidal tularemia cases would result, with a 7.5% death rate per 100,000 persons exposed to an aerosol of *F. tularensis*.[25] In addition, this attack scenario was projected to cost 4–5.5 billion dollars in medical costs (hospitalized and outpatient) for treatment, prophylaxis, and loss of productivity for patients and hospital staff.[25]

31.3 CLASSIFICATION AND MORPHOLOGY

F. tularensis was originally assigned to the *Pasteurella* genus (*Pasteurella tularensis*).[8] In 1947, it was proposed that the organism be separated from *Pasteurella* and placed within a unique genus, termed *Francisella*, due to the bacterium's comparatively small size and specific nutritional requirements for growth on microbiological media.[8] DNA:DNA hybridization studies performed in the 1960s with other bacterial species confirmed that *F. tularensis* warranted classification within a unique genus and family of the gamma-proteobacteria.[29–31]

F. tularensis is a gram-negative and nonmotile pathogen.[32] Cells are pleomorphic coccobacilli that range from 0.2 to 0.7 μm in size.[33] The surface structure of *F. tularensis* is characterized by a unique lipopolysaccharide (LPS) that is tetra-acylated with hydroxyl fatty acids opposed to the

hexa-acylated form displayed on the surface of *Escherichia coli*.[34,35] The LPS is also monophosphorylated on the glycosyl dimer of the lipid A anchor; this modification is associated with the generally poor immunogenic properties of *F. tularensis* LPS.[35] The O-antigen of *F. tularensis* LPS is present in both free and lipid A–anchored forms and makes up a polysaccharide capsule on the bacterial cell surface.[36]

F. tularensis is comprised of three subspecies, (subsp. *tularensis*, *holarctica*, *mediasiatica*). Other members of the *Francisella* genus include *Francisella philomiragia*, *Francisella noatunensis*, *Francisella novicida*, and *Francisella hispaniensis*,[37,38] all of which either do not or very rarely cause disease in humans. *F. noatunensis* is a significant pathogen of fish that has had a profound economic impact on the wild and farmed fishing industries around the world.[39–44]

31.4 PATHOGENESIS

F. tularensis is an obligate intracellular pathogen *in vivo*, infecting and replicating within host cells. A hallmark of natural and experimental tularemia infections in human and animal models of disease is an initial delay in the innate immune response upon infection, followed by rapid dissemination of the bacteria to secondary sites within the body.[6,45,46] This evasion of the host response has been shown to be an important aspect of *F. tularensis* pathogenesis (reviewed in [47]), as mutants that stimulate a robust innate immune response are impaired in their ability to cause disease. In disseminated disease, the lung, spleen, liver, and kidneys are all target organs, with the bacterium causing substantial tissue damage at these sites.[6,48–50]

F. tularensis enters host cells via a unique, looping phagocytic mechanism; once in the cell, it rapidly escapes the phagosome and replicates in the host cell cytoplasm.[51–53] Early understanding of molecular mechanisms of pathogenesis utilized by *F. tularensis* came with the discovery of a mutant unable to replicate in host macrophages.[54] The mutation was mapped to a gene that shared 20% homology to the nutrient starvation transcriptional regulator from *E. coli*, SspA,[54] which interacts with the alpha-subunit of RNA polymerase to regulate transcription of stationary-phase genes during stringent response to nutrient starvation.[55] The *F. tularensis* gene was named macrophage growth locus A (*mglA*) and is expressed as a bicistronic transcript with macrophage growth locus B (*mglB*), which is also involved in phagosomal escape and intracellular replication.[54]

Virulence of *F. tularensis* is tightly linked to the 30-kilobase *Francisella* pathogenicity island (FPI), which contains 16 open reading frames organized in two operons.[56,57] Eight of these genes share homology to the type VI secretion system encoded by *Pseudomonas aeruginosa* and *Vibrio cholerae*.[58] The most studied gene within the FPI is intracellular growth locus C (*iglC*), which was identified as the most abundant bacterial protein present in *F. tularensis*–infected macrophages.[59] Studies to identify the function of IglC have shown its role in phagosomal escape and the prevention of phagosome–lysosome fusion.[60] Results also indicate that IglC is important for cytosolic growth, macrophage apoptosis, and the disruption of host Toll-like receptor (TLR) signaling; however, it is unclear if these are direct effects of the protein or polar effects of the gene knockout.[60]

The *F. tularensis* genome encodes a relatively large number of genes, annotated to be associated with the bacterial cell surface, compared to other intracellular pathogens.[61] It has therefore been hypothesized that the cell surface of *F. tularensis* is important for its pathogenesis and interaction with its host.[62] As previously discussed, *F. tularensis* LPS is weakly immunogenic and therefore an important factor in the evasion of the initial innate immune response. In addition, the O-antigen/polysaccharide complex is important for allowing the bacteria to evade serum killing by complement proteins.[63] Disruption of capsule biogenesis leads to an avirulent phenotype that is sensitive to serum killing.[63] *F. tularensis* also encodes type IV pilin genes within its genome that share homology and structural arrangement with pilin genes of *Neisseria meningitidis* and *P. aeruginosa*.[64] Bacterial pilin machinery is commonly associated with pathogenesis including adhesion to host cells, biofilm formation, motility, and protein secretion.[58] Indeed, mutations in the *F. tularensis* pilin gene, *pilA*, lead to a severely attenuated phenotype, consistent with a role in pathogenesis.[65]

31.5 EPIDEMIOLOGY AND CLINICAL DISEASE

F. tularensis is a zoonotic pathogen, infecting a large number of hosts within the animal kingdom. The bacterium has been isolated from a number of vertebrates, including mammals, amphibians, and aves, as well as from invertebrate arthropods.[66,67] In both vertebrates and invertebrates, infection with *F. tularensis* can be lethal.[66,68] The organism is most often associated with rodents and mammals (particularly hares/rabbits); however, evidence is lacking to suggest that *F. tularensis* may be adapted to a particular host species.[67]

Humans are incidental hosts and acquire tularemia primarily by handling infected animals or bites from infected animals or arthropods.[69] Infections in humans can also occur via inhalation or ingestion of *F. tularensis* from contaminated environmental sources.[69–71] Most cases of tularemia are sporadic, although outbreaks can occur, and may often follow tularemia epizootics in animals.[24] Tularemia in humans is predominantly a rural disease, with urban cases rare.[24] Hunters and other people who spend a great deal of time outdoors are at a risk of exposure to tularemia. Additional risk factors include landscaping and laboratory work with cultures of *F. tularensis*.[24]

Despite broad geographic distribution of *F. tularensis* across the Northern Hemisphere, overall disease occurrence is suspected to be low; however, worldwide incidence is not known.[24] In the last decade, an average of 0.04 and 3.5 cases per 100,000 individuals per year was reported in the United States and Sweden, respectively.[69,72,73] Within endemic regions, distinct disease foci exist where infection prevalence is much higher. For example, in Örebro, Sweden, from the time period 2002–2011, an average of 17.2 cases was reported per 100,000 inhabitants, and in Arkansas, United States, an average of 5.5 cases was reported per 100,000 inhabitants from 2001 to 2010.[72,73]

Tularemia worldwide is caused by the *F. tularensis* subspecies, *tularensis* (type A) and *holarctica* (type B), which differ from each other with respect to glycerol fermentation, disease outcomes in humans, and virulence in laboratory animals.[74] Type A is the more virulent subspecies and causes infections only in North America, whereas type B infections have been reported throughout the Northern Hemisphere and more recently in Tasmania.[74,75] Type A strains have been further differentiated via molecular typing into three clades, termed A1a, A1b, and A2.[76] Infections caused by A1 strains are more common in the eastern United States, whereas A2 infections occur primarily in the western United States.[76] For culture-confirmed cases in the United States, fatality rates due to infection by an A1b, A1a, or A2 strain differ; A1b strains are associated with the highest mortality (24%), followed by A1a (4%) and A2 (0%) strains.[76] Similarly, in mice, the time to death following infection with A1b strains is significantly shorter as compared to infection with A1a or A2 strains.[77]

In addition to strain type, the severity of human illness is based on the route of inoculation, the infecting dose, as well as the patient's immune status and age. Upon infection, the period for disease onset is typically 3–5 days but can range from 1 to 21 days.[24,69] Disease presentation (e.g., symptoms) correlates with the route of bacterial entry, with all forms of tularemia accompanied by fever.[69,78] Ulceroglandular tularemia is characterized by a punched-out ulcer at the inoculum site and enlarged lymph nodes and is the most common form of disease worldwide. Glandular tularemia occurs when the organism is directly inoculated into the bloodstream and is similar in presentation to ulceroglandular tularemia but lacking an ulcer. Entry of the bacterium through the eye causes oculoglandular tularemia. Symptoms include inflammation of the eye and swollen preauricular lymph nodes. Ingestion or inhalation of the organism results in oropharyngeal tularemia, characterized by enlarged cervical lymph nodes and respiratory tularemia, respectively. Respiratory tularemia is the most severe form of the disease and is the primary concern with respect to an intentional event.[24] Person to person spread of tularemia via the respiratory route is generally not thought to be a concern and has never been documented.

The third subspecies of *F. tularensis*, *mediasiatica*, has only been isolated from regions of Central Asia, and a description of human illness due to this subspecies is lacking in the literature.[74] Experimental studies in rabbits indicate virulence comparable to *F. tularensis* subsp. *holarctica*.[74] The reassignment of *F. novicida* as a subspecies of *F. tularensis* has been proposed based on the high degree of genome similarity observed between *F. tularensis* and *F. novicida* via DNA:DNA hybridization studies.[79,80] Several findings, however, support the original classification of these organisms as two separate species, with *F. novicida* therefore not considered a biothreat agent. *F. novicida* (a) has never been isolated from animals or arthropod vectors; (b) infections in humans are extremely rare, associated with compromised individuals, and do not display the characteristic features of tularemia; and (c) infectious dose in animals is higher as compared to *F. tularensis*.[81] Consistent with these findings, whole genome analyses indicate that *F. tularensis* and *F. novicida* represent distinct population lineages that evolved separately.[82]

31.6 IDENTIFICATION AND DIAGNOSIS

The low incidence of disease, sporadic nature of cases, and ease in mistaking tularemia symptoms for more common illnesses all contribute to the difficulty in physician diagnosis. Isolation of *F. tularensis* in culture is the definitive means to confirm the diagnosis of tularemia.[69] Appropriate specimens for culture include swabs of skin lesions or ulcers, lymph node aspirates or biopsies, pharyngeal washings, sputum, pleural fluid, or gastric aspirates, depending on the form of illness.[69] Blood cultures are also appropriate, in cases where the patient is febrile.[69]

Care must be taken when handling *F. tularensis* cultures in the laboratory, given the low infectious dose of the organism via the respiratory route.[26,83] *F. tularensis* has a long history of causing disease in laboratory workers, particularly in individuals with aerosol exposure to the organism.[49,84] Diagnostic procedures with clinical materials can be performed using biosafety level 2 precautions.[85] All work with suspect cultures of *F. tularensis* should be done in a biological safety cabinet to limit exposure and risk of inhalation. Manipulation of cultures and other procedures that might produce aerosols should be conducted using biosafety level 3 practices.[85]

Although *F. tularensis* grows relatively slowly (1–2 h doubling time *in vitro*), the organism can be readily cultivated on microbiologic agar and in liquid broth. A defined medium supporting the growth of *F. tularensis* contains cysteine (required for growth), other amino acids, vitamins, iron, and ions.[86] CO_2 can increase growth rate but is not required for growth. Optimum growth of the bacterium occurs within the temperature range of 35°C–37°C. Addition of IsoVitalex™ to culture media, including Mueller-Hinton and Brain-Heart Infusion, supports the growth of *F. tularensis*.[69,87] Several commercially available microbiologic agars can be used for cultivation of *F. tularensis*, including chocolate, buffered charcoal yeast extract, and Thayer-Martin. Specialized media for growth and propagation of *F. tularensis* include cysteine heart agar with 9% chocolatized sheep blood (CHAB) and glucose-cysteine agar (GC II) with 1% IsoVitalex and 1% hemoglobin.[69] On most agar formulations, *F. tularensis* colonies are typically pinpoint in size after 24 h of growth. When grown on CHAB, colonies are usually green and shiny; this distinctive colony morphology can be used to aid in identification of the organism.[69] Selective media, containing antimicrobials, are required for isolation of *F. tularensis* from contaminated specimens as other organisms such as *Staphylococcus*, *E. coli*, and *Pseudomonas* can inhibit its growth.[88,89]

The diagnosis of tularemia can also be established serologically by demonstrating a fourfold or significant change in specific antibody titers or response between acute and convalescent sera.[69] Serologic testing is not useful for clinical management as antibodies take 7–10 days to develop and convalescent sera are best drawn at least 4 weeks after illness onset.[90] Additionally, IgM responses may remain high for sustained periods and therefore not always indicative of recent infection.[69] A rapid presumptive diagnosis of tularemia may be made via testing of clinical specimens using antigen (e.g., direct fluorescent antibody, antigen capture) or molecular detection methods (e.g., PCR).[69,91]

31.7 TREATMENT AND PREVENTION

Tularemia can be treated with various classes of antibiotics, including aminoglycosides, tetracyclines, and fluoroquinolones.[69,78] Streptomycin was previously considered the drug of choice for treatment of tularemia based on experience, efficacy, and US Food and Drug Administration (FDA) approval.[24,78] Due to its potential to cause vestibular toxicity, it is no longer readily available and another aminoglycoside, gentamicin, is considered an acceptable alternative. Tetracyclines, primarily doxycycline, have been widely used in the treatment of tularemia. Due to their bacteriostatic mode of action, there is a risk of relapse; therefore, this class of antibiotics is typically administered for longer periods of time.[24] Ciprofloxacin and other fluoroquinolones are not FDA-approved for treatment of tularemia; however, they have shown very good efficacy *in vitro*, in animals, and in humans.[69,78]

Naturally occurring resistance in *F. tularensis* to aminoglycosides, tetracyclines, and fluoroquinolones has never been shown.[69,78] The risk for development of antibiotic resistance in clinical disease is low as tularemia is an end-stage disease and also is not transmitted person to person.[69] Treatment failures in tularemia patients correlate with a delay in the initiation of antibiotics with respect to symptom onset.[92–94] Lymph node suppuration can develop in these cases, which is nonresponsive to all classes of antibiotics and requires surgical drainage. Considerably longer recovery times are observed in these cases and can be greater than 70 days.[94]

β-Lactam antibiotics are not used for treatment of tularemia, as *F. tularensis* strains encode a class A β-lactamase, FTU-1.[95] Resistance, due to FTU-1, appears limited to penicillins; *in vitro* MIC determinations indicate that FTU-1 does not confer resistance to other beta-lactams, including first- and second-generation cephalosporins.[95] Although third-generation cephalosporins have been shown to be active against *F. tularensis in vitro*, clinical experience indicates that ceftriaxone is not effective for treatment of tularemia.[96] Macrolides are also not recommended for treatment of tularemia, as type B strains from Central and Eastern Europe and Asia are naturally resistant to erythromycin and other macrolides.[97]

Management options for laboratory workers exposed to *F. tularensis* include a "fever watch" or antimicrobial prophylaxis.[98] Considerations in choosing the method of exposure management include the nature and potential dose of the exposure, whether the incubation period is already passed, and the level of concern. For fever watch, temperature is monitored and treatment obtained if a fever develops (usually defined as a single oral temperature above 101°F or 38.5°C). For prophylaxis in adults, doxycycline (100 mg orally BID × 14 days) is generally recommended.

Prevention of tularemia involves implementing protective measures to reduce the risk of infection. Insect repellant is recommended for preventing insect bites. To reduce risk in tick-infested areas, individuals can also wear long-sleeved shirts and long pants, tuck pants into socks, check skin and clothing frequently for ticks, and remove ticks promptly if found. For individuals who hunt, trap, or skin animals or handle/dispose of animals found dead, gloves should be used, especially when handling rabbits and other rodents, and all game meat should be cooked thoroughly before eating. When mowing or landscaping, aerosol exposure can be diminished by not mowing over sick or dead animals and using dust masks. All surface water should be treated, to avoid exposure via ingestion of the organism. A vaccine, the attenuated *F. tularensis* live vaccine strain (LVS), was previously offered to protect laboratorians.[99] This vaccine, however, is under review by the US FDA and no longer available as a preventative measure.[24]

31.8 CONCLUSIONS AND FUTURE PERSPECTIVES

Due to the categorization of *F. tularensis* as a biothreat agent, an explosion in both diagnostic and basic research has occurred over the last decade, leading to the development of improved diagnostics, the ability to distinguish between strain types worldwide, and an improved understanding of the molecular mechanisms of pathogenesis utilized by *F. tularensis*. Because tularemia is a rare disease and presents with varying symptoms, diagnosis of the disease remains a challenge, owing largely to a lack of clinical suspicion. Currently, the most commonly performed unbiased diagnostic testing is culture of clinical specimens and subsequent identification of recovered isolates. In suspect biothreat situations or outbreaks of unknown etiology, specialized and reference laboratories worldwide now have the capability to directly detect *F. tularensis* DNA and/or antigen in primary samples, both clinical and environmental. For naturally occurring cases of tularemia, however, the cost and time required to perform *F. tularensis*–specific molecular tests on diagnostic specimens, with no prior suspicion of tularemia, remain prohibitive. Future development of low-cost broadscale approaches that can be applied to differing specimen types (e.g., wound samples, respiratory samples, aspirate samples, and blood samples) will be important for improving diagnoses due to infections caused by rare or unusual pathogens. This cost-efficient testing strategy does not require prior disease suspicion yet can lead to a rapid presumptive

diagnosis, thereby allowing for diagnostic-driven treatment. Ultimately these changes should decrease mortality as well as improve tularemia surveillance worldwide.

REFERENCES

1. McCoy, G.W. and Chapin, C.C. Studies of plague. A plague-like disease and tuberculosis among rodents in California. *Journal of Infectious Disease* VI, 170–180 (1912).
2. McCoy, G.W. Some features of the squirrel plague problem. *California State Journal of Medicine* 9, 105–109 (1911).
3. Ohara, S. Studies on yato-byo (Ohara's disease, tularemia in Japan). I. *The Japanese Journal of Experimental Medicine* 24, 69–79 (1954).
4. Omland, T. et al. A survey of tularemia in wild mammals from Fennoscandia. *Journal of Wildlife Diseases* 13, 393–399 (1977).
5. Wherry, W.B. and Lamb, B.H. Infection of man with *Bacterium tularense*. *Journal of Infectious Disease* 15, 331–340 (1914).
6. Francis, E. Tularemia. *Journal of the American Medical Association* 84, 1243–1250 (1925).
7. Sjöstedt, A. Tularemia: History, epidemiology, pathogen physiology, and clinical manifestations. *Annals of the New York Academy of Sciences* 1105, 1–29 (2007).
8. Olsufjev, N.G. Taxonomy and characteristic of the genus *Francisella* Dorofeev, 1947. *Journal of Hygiene, Epidemiology, Microbiology, and Immunology* 14, 67–74 (1970).
9. Larson, C. Tularemia. In *Tice's Practice of Medicine*, Vol. 3, pp. 663–676. Harper & Row Publishers Inc., Hagerstown, MD, 1970.
10. Sanford, J.P. Landmark perspective: Tularemia. *Journal of the American Medical Association* 250, 3225–3226 (1983).
11. Tärnvik, A., Priebe, H.S., and Grunow, R. Tularaemia in Europe: An epidemiological overview. *Scandinavian Journal of Infectious Diseases* 36, 350–355 (2004).
12. Croddy, E. and Krčálová, S. Tularemia, biological warfare, and the battle for Stalingrad (1942–1943). *Military Medicine* 166, 837–838 (2001).
13. Evans, F. Tularemia. In *Textbook of Military Medicine Aspects of Chemical and Biological Warfare*. Medical Research Institute of Chemical Defense, Washington, DC, 1997.
14. Harris, S. Japanese biological warfare research on humans: A case study of microbiology and ethics. *Annals of the New York Academy of Sciences* 666, 21–52 (1992).
15. Frischknecht, F. The history of biological warfare. Human experimentation, modern nightmares and lone madmen in the twentieth century. *EMBO Reports* 4 Spec. No. S47–S52 (2003).
16. Franz, D.R., Parrot, C.D., and Takafuji, E.T. The U.S. biological warfare and biological defense programs. In *Textbook of Military Medicine*, Part 1 (eds. Sidell, F.R., Takafuji, E.T., and Franz, D.R.), pp. 425–436. TMM Publications, Washington, DC, 1997.
17. Smart, J.K. The U.S. biological warfare and biological defense programs. In *Textbook of Military Medicine*, Part 1 (eds. Sidell, F.R., Takafuji, E.T., and Franz, D.R.), pp. 425–436. TMM Publications, Washington, DC, 1997.
18. Christopher, G.W. et al. Biological warfare. A historical perspective. *Journal of the American Medical Association* 278, 412–417 (1997).
19. Franz, D.R. et al. Clinical recognition and management of patients exposed to biological warfare agents. *Journal of the American Medical Association* 278, 399–411 (1997).
20. Federal Register. Possession, use, and transfer of select agents and toxins. *Federal Register* 240, 76886–76905 (2002).

21. Tegnell, A. et al. The European commission's task force on bioterrorism. *Emerging Infectious Diseases* 9, 1330–1332 (2003).

22. World Health Organization. Public health response to biological and chemical weapons. In *Health Aspects of Chemical and Biological Weapons*, 2nd ed. World Health Organization, Geneva, Switzerland, 2003.

23. Federal Register. Possession, use, and transfer of select agents and toxins. *Federal Register* 77, 61084–61115 (2012).

24. Dennis, D.T. et al. Tularemia as a biological weapon: Medical and public health management. *Journal of the American Medical Association* 285, 2763–2773 (2001).

25. Kaufmann, A.F., Meltzer, M.I., and Schmid, G.P. The economic impact of a bioterrorist attack: Are prevention and post-attack intervention programs justifiable? *Emerging Infectious Diseases* 3, 83–94 (1997).

26. McCrumb, F.R. Aerosol infection of man with *Pasteurella tularensis*. *Bacteriological Reviews* 25, 262–267 (1961).

27. Jones, R.M. et al. The infectious dose of *Francisella tularensis* (Tularemia). *Applied Biosafety* 10, 227–239 (2005).

28. World Health Organization. *Health Aspects of Chemical and Biological Weapons*, pp. 75–76. World Health Organization, Geneva, Switzerland, 1970.

29. Ritter, D.B. and Gerloff, R.K. Deoxyribonucleic acid hybridization among some species of the genus *Pasteurella*. *Journal of Bacteriology* 92, 1838–1839 (1966).

30. Olsufiev, N.G., Emelyanova, O.S., and Dunayeva, T.N. Comparative study of strains of *B. tularense* in the old and new world and their taxonomy. *Journal of Hygiene, Epidemiology, Microbiology, and Immunology* 3, 138–149 (1959).

31. Owen, C.R. et al. Lack of demonstrable enhancement of virulence of *Francisella tularensis* during animal passage. *Zoonoses Research* 1, 75–85 (1961).

32. Sjöstedt, A. Family XVII. *Francisellaceae*, Genus I. *Francisella*. In *Bergey's Manual of Systemic Bacteriology* (ed. Brenner, D.J.). Springer-Verlag, New York, 2005.

33. Evans, M.E. *Francisella tularensis*. *Infection Control* 6, 381–383 (1985).

34. Nichols, P.D. et al. Determination of monounsaturated double-bond position and geometry in the cellular fatty acids of the pathogenic bacterium *Francisella tularensis*. *Journal of Clinical Microbiology* 21, 738–740 (1985).

35. Phillips, N.J. et al. Novel modification of lipid A of *Francisella tularensis*. *Infection and Immunity* 72, 5340–5348 (2004).

36. Apicella, M.A. et al. Identification, characterization and immunogenicity of an O-antigen capsular polysaccharide of *Francisella tularensis*. *PLOS One* 5, e11060 (2010).

37. Ahlinder, J. et al. Increased knowledge of *Francisella* genus diversity highlights the benefits of optimised DNA-based assays. *BMC Microbiology* 12, 220 (2012).

38. Sjödin, A. et al. Genome characterisation of the genus *Francisella* reveals insight into similar evolutionary paths in pathogens of mammals and fish. *BMC Genomics* 13, 268 (2012).

39. Birkbeck, T.H., Bordevik, M., Froystad, M.K., and Baklien, A. Identification of *Francisella* sp. from Atlantic salmon, *Salmo salar* L., in Chile. *Journal of Fish Diseases* 30, 505–507 (2007).

40. Mauel, M.J. et al. Occurrence of piscirickettsiosis-like syndrome in tilapia in the continental United States. *Journal of Veterinary Diagnostic Investigation* 17, 601–605 (2005).

41. Mauel, M.J. et al. A piscirickettsiosis-like syndrome in cultured Nile tilapia in Latin America with *Francisella* spp. as the pathogenic agent. *Journal of Aquatic Animal Health* 19, 27–34 (2007).

42. Ottem, K.F. et al. Characterization of *Francisella* sp., GM2212, the first *Francisella* isolate from marine fish, Atlantic cod (*Gadus morhua*). *Archives of Microbiology* 187, 343–350 (2007).

43. Nylund, A. et al. *Francisella* sp. (Family *Francisellaceae*) causing mortality in Norwegian cod (*Gadus morhua*) farming. *Archives of Microbiology* 185, 383–392 (2006).

44. Olsen, A.B. et al. A novel systemic granulomatous inflammatory disease in farmed Atlantic cod, *Gadus morhua* L., associated with a bacterium belonging to the genus *Francisella*. *Journal of Fish Diseases* 29, 307–311 (2006).

45. Chase, J.C., Celli, J., and Bosio, C.M. Direct and indirect impairment of human dendritic cell function by virulent *Francisella tularensis* Schu S4. *Infection and Immunity* 77, 180–195 (2009).

46. Bosio, C.M., Bielefeldt-Ohmann, H., and Belisle, J.T. Active suppression of the pulmonary immune response by *Francisella tularensis* Schu4. *Journal of Immunology* 178, 4538–4547 (2007).

47. Jones, C.L. et al. Subversion of host recognition and defense systems by *Francisella* spp. *Microbiology and Molecular Biology Reviews* 76, 383–404 (2012).

48. Kingry, L.C. et al. Genetic identification of unique immunological responses in mice infected with virulent and attenuated *Francisella tularensis*. *Microbes and Infection* 13, 261–275 (2011).

49. Overholt, E.L. et al. An analysis of forty-two cases of laboratory-acquired tularemia. Treatment with broad spectrum antibiotics. *The American Journal of Medicine* 30, 785–806 (1961).

50. Wickstrum, J.R. et al. *Francisella tularensis* induces extensive caspase-3 activation and apoptotic cell death in the tissues of infected mice. *Infection and Immunity* 77, 4827–4836 (2009).

51. Clemens, D.L., Lee, B.Y., and Horwitz, M.A. *Francisella tularensis* enters macrophages via a novel process involving pseudopod loops. *Infection and Immunity* 73, 5892–5902 (2005).

52. Golovliov, I. et al. An attenuated strain of the facultative intracellular bacterium *Francisella tularensis* can escape the phagosome of monocytic cells. *Infection and Immunity* 71, 5940–5950 (2003).

53. Santic, M. et al. *Francisella tularensis* travels a novel, twisted road within macrophages. *Trends in Microbiology* 14, 37–44 (2006).

54. Baron, G.S. and Nano, F.E. MglA and MglB are required for the intramacrophage growth of *Francisella novicida*. *Molecular Microbiology* 29, 247–259 (1998).

55. Williams, M.D., Ouyang, T.X., and Flickinger, M.C. Glutathione S-transferase-sspA fusion binds to *E. coli* RNA polymerase and complements delta sspA mutation allowing phage P1 replication. *Biochemical and Biophysical Research Communications* 201, 123–127 (1994).

56. Nano, F.E. et al. A *Francisella tularensis* pathogenicity island required for intramacrophage growth. *Journal of Bacteriology* 186, 6430–6436 (2004).

57. Larsson, P. et al. The complete genome sequence of *Francisella tularensis*, the causative agent of tularemia. *Nature Genetics* 37, 153–159 (2005).

58. Alvarez-Martinez, C.E. and Christie, P.J. Biological diversity of prokaryotic type IV secretion systems. *Microbiology and Molecular Biology Reviews* 73, 775–808 (2009).

59. Golovliov, I. et al. Identification of proteins of *Francisella tularensis* induced during growth in macrophages and cloning of the gene encoding a prominently induced 23-kilodalton protein. *Infection and Immunity* 65, 2183–2189 (1997).

60. Lindgren, H. et al. Factors affecting the escape of *Francisella tularensis* from the phagolysosome. *Journal of Medical Microbiology* 53, 953–958 (2004).

61. Titball, R.W. and Petrosino, J.F. *Francisella tularensis* genomics and proteomics. *Annals of the New York Academy of Sciences* 1105, 98–121 (2007).

62. Forsberg, A. and Guina, T. Type II secretion and type IV pili of *Francisella. Annals of the New York Academy of Sciences* 1105, 187–201 (2007).

63. Sandström, G., Lofgren, S., and Tärnvik, A. A capsule-deficient mutant of *Francisella tularensis* LVS exhibits enhanced sensitivity to killing by serum but diminished sensitivity to killing by polymorphonuclear leukocytes. *Infection and Immunity* 56, 1194–1202 (1988).

64. Gil, H., Benach, J.L., and Thanassi, D.G. Presence of pili on the surface of *Francisella tularensis. Infection and Immunity* 72, 3042–3047 (2004).

65. Forslund, A.L. et al. Direct repeat-mediated deletion of a type IV pilin gene results in major virulence attenuation of *Francisella tularensis. Molecular Microbiology* 59, 1818–1830 (2006).

66. Hopla, C.E. The ecology of tularemia. *Advances in Veterinary Science and Comparative Medicine* 18, 25–53 (1974).

67. Mörner, T. The ecology of tularaemia. *Revue Scientifique et Technique* 11, 1123–1130 (1992).

68. Reese, S.M. et al. Transmission dynamics of *Francisella tularensis* subspecies and clades by nymphal *Dermacentor variabilis* (Acari: Ixodidae). *The American Journal of Tropical Medicine and Hygiene* 83, 645–652 (2010).

69. World Health Organization. *WHO Guidelines on Tularemia.* WHO Press, Geneva, Switzerland, 2007.

70. Feldman, K.A. et al. An outbreak of primary pneumonic tularemia on Martha's Vineyard. *The New England Journal of Medicine* 345, 1601–1606 (2001).

71. Syrjälä, H. et al. Airborne transmission of tularemia in farmers. *Scandinavian Journal of Infectious Diseases* 17, 371–375 (1985).

72. Centers for Disease Control and Prevention (CDC). Tularemia Statistics. Available from: http://www.cdc.gov/tularemia/statistics/ (accessed on November, 2012).

73. Swedish Institute for Communicable Disease Control. Statistik for Harpest. Available from: http://www.smittskyddsinstitutet.se/statistik/harpest/ (accessed on November, 2012).

74. Olsufjev, N.G. and Meshcheryakova, I.S. Infraspecific taxonomy of tularemia agent *Francisella tularensis* McCoy et Chapin. *Journal of Hygiene, Epidemiology, Microbiology, and Immunology* 26, 291–299 (1982).

75. Jackson, J. et al. *Francisella tularensis* subspecies *holarctica*, Tasmania, Australia, 2011. *Emerging Infectious Diseases* 18, 1484–1486 (2012).

76. Kugeler, K.J. et al. Molecular epidemiology of *Francisella tularensis* in the United States. *Clinical Infectious Diseases* 48, 863–870 (2009).

77. Molins, C.R. et al. Virulence differences among *Francisella tularensis* subsp. *tularensis* clades in mice. *PLOS One* 5, e10205 (2010).

78. Tärnvik, A. and Chu, M.C. New approaches to diagnosis and therapy of tularemia. *Annals of the New York Academy of Sciences* 1105, 378–404 (2007).

79. Huber, B. et al. Description of *Francisella hispaniensis* sp. nov., isolated from human blood, reclassification of *Francisella novicida* (Larson et al. 1955) Olsufiev et al. 1959 as *Francisella tularensis* subsp. *novicida* comb. nov. and emended description of the genus *Francisella. International Journal of Systematic and Evolutionary Microbiology* 60, 1887–1896 (2010).

80. Hollis, D.G. et al. *Francisella philomiragia* comb. nov. (formerly *Yersinia philomiragia*) and *Francisella tularensis* biogroup *novicida* (formerly *Francisella novicida*) associated with human disease. *Journal of Clinical Microbiology* 27, 1601–1608 (1989).

81. Johansson, A. et al. Objections to the transfer of *Francisella novicida* to the subspecies rank of *Francisella tularensis. International Journal of Systematic and Evolutionary Microbiology* 60, 1717–1718; author reply 1718–1720 (2010).

82. Larsson, P. et al. Molecular evolutionary consequences of niche restriction in *Francisella tularensis*, a facultative intracellular pathogen. *PLoS Pathogens* 5, e1000472 (2009).

83. Saslaw, S., Eigelsbach, H.T., Prior, J.A., Wilson, H.E., and Carhart, S. Tularemia vaccine study. II. Respiratory challenge. *Archives Internal Medicine* 107, 702–714 (1961).

84. Pike, R.M. Laboratory-associated infections: Summary and analysis of 3921 cases. *Health Laboratory Science* 13, 105–114 (1976).

85. Center for Disease Control. *Biosafety in Microbiological and Biomedical Laboratories.* Department of Health and Human Services, Bethesda, MA, 2009.

86. Chamberlain, R.E. Evaluation of live tularemia vaccine prepared in a chemically defined medium. *Applied Microbiology* 13, 232–235 (1965).

87. McGann, P. et al. A novel brain heart infusion broth supports the study of common *Francisella tularensis* serotypes. *Journal of Microbiological Methods* 80, 164–171 (2010).

88. Petersen, J.M. et al. Direct isolation of *Francisella* spp. from environmental samples. *Letters in Applied Microbiology* 48, 663–667 (2009).

89. Petersen, J.M. et al. Methods for enhanced culture recovery of *Francisella tularensis. Applied and Environmental Microbiology* 70, 3733–3735 (2004).

90. Koskela, P. and Salminen, A. Humoral immunity against *Francisella tularensis* after natural infection. *Journal of Clinical Microbiology* 22, 973–979 (1985).

91. Petersen, J.M. Schriefer, M.E., and Araj, G.F. *Francisella and Brucella.* ASM Press, Washington, DC, 2011.

92. Celebi, G. et al. Tularemia, a reemerging disease in northwest Turkey: Epidemiological investigation and evaluation of treatment responses. *Japanese Journal of Infectious Diseases* 59, 229–234 (2006).

93. Kaya, A. et al. Treatment failure of gentamicin in pediatric patients with oropharyngeal tularemia. *Medical Science Monitor: International Medical Journal of Experimental and Clinical Research* 17, CR376–CR380 (2011).

94. Meric, M. et al. Evaluation of clinical, laboratory, and therapeutic features of 145 tularemia cases: The role of quinolones in oropharyngeal tularemia. *Acta Pathologica, Microbiologica, et Immunologica Scandinavica* 116, 66–73 (2008).

95. Antunes, N.T. et al. The class A beta-lactamase FTU-1 is native to *Francisella tularensis. Antimicrobial Agents and Chemotherapy* 56, 666–671 (2012).

96. Cross, J.T. and Jacobs, R.F. Tularemia: Treatment failures with outpatient use of ceftriaxone. *Clinical Infectious Diseases* 17, 976–980 (1993).

97. Kudelina, R.I. and Olsufiev, N.G. Sensitivity to macrolide antibiotics and lincomycin in *Francisella tularensis holarctica. Journal of Hygiene, Epidemiology, Microbiology, and Immunology* 24, 84–91 (1980).

98. Centers for Disease Control and Prevention (CDC). Managing Potential Laboratory Exposures to *Francisella.* Available from: http://www.cdc.gov/tularemia/laboratoryexposure/ (accessed on November 10, 2013).

99. Sandström, G. The tularaemia vaccine. *Journal of Chemical Technology and Biotechnology* 59, 315–320 (1994).

32 *Rickettsia*

Mohammad Yazid Abdad and John Stenos

CONTENTS

32.1 INTRODUCTION

Zoonotic rickettsial infections occur naturally in both urban and rural communities worldwide. Transmitted via vectors such as ticks, fleas, mites, and lice, rickettsial infections are often viewed as mild or are misdiagnosed as common ailments such as the common cold and flu. Through lack of understanding of the actual cause of disease, complications leading to organ failure and possibly even death.

The genus *Rickettsia* was first described by Howard Ricketts in 1906 and 1907, where he investigated an outbreak of Rocky Mountain spotted fever (RMSF).[1] His work was instrumental in paving the way for modern rickettsiology. Most significant were the roles that ectoparasites were found to play as vectors of rickettsial infection and reservoirs of rickettsial organisms in an endemic area.[1] Today, we know that the rickettsial life cycle involves both vertebrate and invertebrate hosts. Not only do hematophagous arthropods play an important role as vectors, they are also a primary reservoir and amplifying host. In some instances, small mammals such as rats and opossums also act as hosts.[2]

Rickettsiae are obligate intracellular Gram-negative parasitic bacteria. Their ability to grow within the cytoplasm and nucleus of the eukaryotic host cell differentiates them from other obligate intracellular bacteria of the genera *Coxiella*, *Ehrlichia*, and *Chlamydia*. Rickettsial entry into host cells is by induced phagocytosis.[3]

32.2 TAXONOMY

The original classification of species within the genus *Rickettsia* into the spotted fever group (SFG) and typhus group (TG) depended on a variety of characteristics including intracellular localization, optimal growth temperature, and the cross-reaction of sera from an infected patient with somatic antigens of three strains of *Proteus* (Weil–Felix test).[4] The advent of modern molecular and serological techniques both reinforced and disputed original classification of members of the genus, and as such, the last couple of decades have seen the restructuring of rickettsial species. Genomic analyses have now proposed the creation of two new groups within the genus *Rickettsia*: ancestral group (AG) consisting of *Rickettsia bellii* and *Rickettsia canadensis*[5] and transitional group (TRG) consisting of *Rickettsia akari* and *Rickettsia felis*.[6] Currently, there are at least 17 known pathogenic SFG rickettsiae (Table 32.1). TG consists of *Rickettsia prowazekii* and *Rickettsia typhi*. Another notable rickettsia that has been reclassified is *Rickettsia tsutsugamushi*, which has been placed in its own genus, *Orientia* (*Orientia tsutsugamushi*), and is in the same α-1 subgroup as the genus *Rickettsia*.[7] More recently, *Orientia chuto* has been proposed as the second member of the *Orientia* genus.[8] *Coxiella burnetii* (originally *Rickettsia burnetii*) has also been reclassified due to the findings that its 16S rRNA sequence is more similar to members of the γ-subgroup and it is no longer viewed as a true rickettsia.[9]

32.3 EPIDEMIOLOGY AND TRANSMISSION

Rickettsial organisms are found worldwide with some species endemic to most continents (e.g., *R. felis* and *R. typhi*) while others are restricted to specific regions due to climatic, geographic and host constraints (e.g., *Rickettsia japonica* in Japan and *Rickettsia rickettsii* and *Rickettsia parkeri* in North and Central America).[2,10,11]

TABLE 32.1
Known Pathogenic SFG Rickettsiae

Rickettsia Species	Known or Potential Vectors	Year of First Identification in Vectors	Disease (Year of First Clinical Description)	References
R. rickettsii	*Dermacentor andersoni*	1906	RMSF (1899)	[59]
	Dermacentor variabilis			
	Rhipicephalus sanguineus			
	Amblyomma cajennense			
	Amblyomma aureolatum			
Rickettsia conorii subsp. *conorii*	*Rh. sanguineus*	1932	Mediterranean spotted fever (1910)	[60]
Rickettsia sibirica subsp. *sibirica*	*Dermacentor nuttalli*	Unknown	Siberian tick typhus (1934)	[61]
	Dermacentor marginatus			
	Dermacentor silvarum			
	Haemaphysalis concinna			
R. parkeri	*Amblyomma maculatum*	1939	Unnamed (2004)	[62]
	Amblyomma americanum			
	Amblyomma triste			
R. conorii subsp. *indica*	*Rh. sanguineus*	1950	Indian tick typhus	[63]
Rickettsia slovaca	*D. marginatus*	1968	Tick-borne lymphadenopathy (1997)	[64]
	Dermacentor reticulatus			
			Dermacentor-borne necrosis and lymphadenopathy (1997)	
R. conorii subsp. *israelensis*	*Rh. sanguineus*	1974	Israeli spotted fever (1940)	[65,66]
Rickettsia australis	*Ixodes holocyclus*	1974	Queensland tick typhus (1946)	[67]
	Ixodes tasmani			
R. sibirica subsp. *mongolotimonae*	*Dermacentor sinicus*	1974	North Asian tick typhus (1960)	[68]
	Hyalomma asiaticum	1991	Lymphangitis-associated rickettsiosis (1996)	[69]
	Hyalomma truncatum			
Rickettsia heilongjiangensis	*D. silvarum*	1982	Far Eastern spotted fever (1992)	[35]
Rickettsia africae	*Amblyomma hebraeum*	1990	African tick bite fever (1934)	[70,71]
	Amblyomma variegatum			
R. conorii subsp. *caspia*	*Rh. sanguineus*	1992	Astrakhan fever (1970s)	[72,73]
	Rhipicephalus pumilio			
Rickettsia massiliae	*Rh. sanguineus*	1992	Unnamed (2005)	[74,75]
	Rhipicephalus turanicus			
	Rhipicephalus muhsamae			
	Rhipicephalus lunulatus			
	Rhipicephalus sulcatus			
Rickettsia honei	*Bothriocroton (Aponomma) hydrosauri*	1993	Flinders Island spotted fever (1991)	[76,77]
	A. cajennense			
	Ixodes granulatus			
R. japonica	*Ixodes ovatus*	1996	Oriental or Japanese spotted fever (1984)	[78]
	Dermacentor taiwanensis			
	Haemaphysalis longicornis			
	Haemaphysalis flava			
Rickettsia aeschlimannii	*Hyalomma marginatum marginatum*	1997	Unnamed (2002)	[79]
	Hyalomma marginatum rufipes			
	Rhipicephalus appendiculatus			
R. honei strain *marmionii*	*Haemaphysalis novaeguineae*	2005	Australian spotted fever (2005)	[80]

All pathogenic rickettsiae require arthropods for transmission; however, how the rickettsia is transmitted from the arthropod to humans can differ. The vector for most rickettsiae (especially SFG rickettsia) is ticks, and the species of tick varies between each species of rickettsiae (Table 32.1).

Other vectors have been implicated with rickettsial transmission, fleas for *R. felis* (cat flea typhus) and *R. typhi* (murine/endemic typhus), mites for *R. akari* (rickettsial pox) and *O. tsutsugamushi* (scrub typhus), and lice for *R. prowazekii* (epidemic typhus).[2] Most rickettsiae are transmitted by a single type of host, *R. felis*, which has been found to be unique in that regard due to its large ectoparasite host range that includes ticks, fleas, mites, and lice.[10]

Tick- and mite-borne rickettsiae rely on the transfer of rickettsiae during feeding. Rickettsiae are transferred to the host via the tick or mite saliva at the bite site. Very little is known with regard to rickettsial transmission efficacy via tick bite and attachment duration; however, it has been demonstrated to be a major factor with ehrlichiosis and Lyme disease.[12,13]

Cases of transmission of rickettsiae via the conjunctiva have occurred with exposure of infected tick hemolymph on fingers after crushing an infected tick. Flea- and louse-borne rickettsiae are transmitted via bite site or cuts on the skin. Inoculation via inhalation is possible and has happened in the laboratory.[14] Although target cells at the site of inoculation are currently unknown, rickettsiae possibly infect dermal cells such as vascular endothelium (main cell infected), fibroblasts, lymphatic endothelium, and macrophages.[15]

32.4 RICKETTSIAE–HOST INTERACTION

The advent of modern methods in cell culture and imagery has provided a better understanding of the interactions between rickettsiae and their hosts. Severity of rickettsial disease can be attributed to patient age as the main consistent factor in almost all cases. Initial hypotheses for rickettsioses having a higher fatality rate in males (esp. RMSF) have been disproved by recent studies.[16] Other factors affecting rickettsiae–host interaction would be underlying patient diseases and enhanced oxidative stress increasing the severity of illness. Mediterranean spotted fever shows greater virulence in patients with diabetes mellitus.[14]

32.4.1 ROUTES OF SPREAD

Rickettsial infection starts at the site of inoculation with nucleated cells such as vascular and lymphatic endothelium, dermal dendritic cells, fibroblasts, and macrophages likely to be targeted.[14] Upon initiation of infection, spread through the host is facilitated via the lymphatic and vascular systems. Antigenic spread by the lymphatic system can be observed by the presence of lymphadenopathy in the same region as the inoculation site. The vascular system has been implicated with the spread of rickettsial organisms to susceptible cells through the action of rickettsial adhesions on host cell receptors.[14,17]

Intracellular movement of rickettsiae is facilitated by actin polymerization, shown to be present in all SFG rickettsiae except *Rickettsia peacockii*. Actin polymerization is activated on one pole of the bacterium, and the continuous conversion of globular actin to filamentous actin propels the rickettsia through the cytosol, eventually hitting the host cell membrane.[18,19] Upon collision with the cell membrane, there are two possible outcomes: (1) bouncing off in the inner surface of the cell membrane or (2) deformation of the membrane and formation of a filopodium. The formation of a filopodium

may result in two further possibilities depending on what is on the other side of the cell membrane; the rickettsia either enters the adjacent cell or exits as a luminal filopodium into the bloodstream.[20] The protein RickA is responsible for the actin-based mobility observed in SFG rickettsiae. The gene *rickA* is however absent in TG rickettsiae (*R. prowazekii* and *R. typhi*), which is consistent with the lack of actin polymerization observed and the use of a burst release model instead. Erratic actin-based mobility has been observed in *R. typhi*, suggesting that members of the genus use multiple redundant mechanisms for actin polymerization, which has yet to be clarified.[21]

32.4.2 PATHOLOGY

Increased vascular permeability as a result of rickettsial infection has been associated with bacterial load and tumor necrosis factor (TNF) α causing disrupted endothelial cell junctions.[22,23] Perivascular T lymphocytes and macrophages are heavily involved in the clearance of rickettsial infections. As most rickettsial infection is localized and not widespread, the result of vasculitis is not usually severe except in the most serious of cases. Previous lymphohistiocytic infiltrate has been observed in the organs affected by rickettsial infection including the brain, lung, heart, kidney, skin, gastrointestinal tract, pancreas, gall bladder, skeletal muscle, and tests.[24]

Destruction of endothelial cells as a result of multifocal rickettsial infection usually results in the impairment of normal anticoagulation functions. The occurrence of fibrin–platelet thrombi in both living and dead patients is rare.[25] Hemostatic plugs are only observed in the foci of severe rickettsial-inflicted vascular injury. In spite of the general hypothesis that thrombosis is the cause of organ infarction and tissue damage among patients, pathological and experimental evidence proves otherwise. Although infarction of organs seldom occurs, it has been observed in the white matter of the brain. Tissue damage to kidneys is usually the result of poor perfusion. Gangrene and cutaneous necrosis have also been observed in some cases of severe rickettsioses resulting in extensive damage to the vascular microcirculation of the affected area.[26]

32.5 WEAPONIZATION HISTORY

Rickettsial organisms are unstable in storage and transmission from human to human has not been reported. Stable storage is only achieved by freezing at under −20°C for short periods of time and −140°C for long periods. Thus, this makes it a poor choice as a bioweapon. However, in the past, various nations have experimented with various rickettsiae (both *Rickettsia* and *Orientia*) to determine their potential for warfare. Canada, United States, Germany, Soviet Union, and Japan have made attempts in weaponizing rickettsiae in the early twentieth century. Of note was the Soviet Biopreparat program in 1928 to weaponize *R. prowazekii*. The most recent report of a bioweapon program involving rickettsiae was North Korea's production of weaponized *R. prowazekii* in the 1980s.

Only *R. prowazekii* and *R. typhi* can cause infection by inhalation, in addition to a bite from their vector. However, between the two, only *R. prowazekii* is known to cause severe disease in humans and its aerosolization has been demonstrated to have caused disease.[27] *R. prowazekii* is also the only rickettsia in recorded history to have caused disease among human populations at epidemic levels with more than 30 million cases during and after World War I and a mortality rate of 10%.[27] Even at the cusp of crossing over into the twenty-first century, epidemic typhus still occur when there is a breakdown in the living and social conditions of a population as witnessed in Burundi with about 30,000 cases reported (10%–15% mortality rate).[28]

To date, there has, so far, only been one attempt to potentially transmit tick typhus using ticks as the vector. In 1976, envelopes containing live ticks or letter bombs were mailed to the director and other members of the Federal Bureau of Investigation (United States). The WHO estimates that the hypothetical aerosolization of 50 kg of *R. prowazekii* would result in 104,000 casualties of which 19,000 dead and 85,000 incapacitated.[29,30]

Due to the physical and logistical limitations with regard to transmitting rickettsiae to humans and ease of treatment, the weaponization of *Rickettsia* and *Orientia* is believed to be impractical, where other more efficient bio-agents can be utilized with more devastating effect.[31] However, *R. prowazekii* and *R. rickettsii* continue to be on the Center of Disease Control and Prevention category B list for pathogens of bioterrorism importance.

32.6 CLINICAL PRESENTATION AND TREATMENT

Diagnosis of rickettsial infections can be hard without appropriate diagnostic assays. Prior to the advent of modern diagnostic methods, identifying rickettsial infections and prompt treatment relied heavily on the physician's knowledge of endemic rickettsial diseases in their locality. Unfortunately, to some extent, this is still the case in suburban and rural areas in various locations worldwide today.

A classic indicator of rickettsial disease is the presence of an attached arthropod or inflamed arthropod bite on the patient together with classical rickettsial symptoms. The two classic symptoms that usually differentiate rickettsioses from other ailments are the presence of an eschar at the arthropod bite site and a rash. Other major symptoms as a result of rickettsiosiae such as myalgia, fever, headaches, and local lymphadenopathy could also be attributed to a myriad of other diseases.[14] There have been reports of "spotless" spotted fever and "escharless" rickettsioses, major symptoms are similar for most rickettsial diseases, and variations that do occur can vary greatly from classic symptoms.[32] Thus, there is a need for prompt and accurate laboratory tests to assist physicians with the diagnosis of rickettsial disease.

Variability of endemic areas and host between species is also reflected by the severity of disease they cause in humans. Due to the high variability observed in rickettsiosis between species, it is of no surprises that disease names are equally varied (Table 32.1). Rickettsiosis in some cases is a self-limiting disease that can be of minor clinical significance or all together asymptomatic and not require treatment. However, cases of severe rickettsioses worldwide are not uncommon.[14] RMSF had a high mortality rate (63%) before the introduction of antibiotics.[33] The mortality rate of RMSF in the later part of the twentieth century has dropped to about 2%–6%.[34]

Treatment with antibiotics such as tetracycline or chloramphenicol is essential to avoid morbidity and mortality.[35,36] These drugs however do cause significant side effects in children such as staining of teeth by tetracyclines and bone marrow toxicity by chloramphenicol. Symptoms relating to rickettsial infection usually resolve within 48–72 h of treatment regardless of severity. Treatment regimens last between 5 and 14 days depending on the severity of the disease.[36,37]

32.7 DIAGNOSTICS

Due to the complexity of diagnosis by clinical symptoms common with various other ailments, the only way to confirm rickettsial infection is through the use of robust and rapid diagnostic tools. Blood samples can be screened for rickettsiae by serology or polymerase chain reaction (PCR). Culture or PCR can also be performed on eschar biopsy. Though uncommon, screening the biting arthropod directly by either PCR or culture for evidence of rickettsiae is another alternative.[38]

One of the first serological tests used for rickettsial diagnostics was the Weil–Felix test.[39] However, its poor specificity and sensitivity are well documented, even so it is still in use today in rudimentary clinical laboratories.[4,33,40,41] The current gold standard for diagnosis of rickettsioses is immunofluorescence assay (IFA).[42] The use of IFA allows for a fast and accurate diagnosis of rickettsioses, but differentiation between species of rickettsiae as the cause of infection is often difficult due to cross-reacting antibodies. This is further complicated by the operator subjectivity when determining antibody titer. To reduce false-positives, the Australian Rickettsial Reference Laboratory (Geelong, Australia) recommends that the cutoff value of antibody titer of 1:128.[43] Current cutoff values used by the Unité des Rickettsies (Marseille, France) are 1:64 for IgG and 1:32 for IgM.[44]

PCR has revolutionized the diagnosis of diseases and in no small part the detection of rickettsial infection in humans and animals, allowing for diagnosis before seroconversion and growth in culture.[45] Samples tested are usually fresh and can range from biopsy of eschars to buffy coat; additionally, paraffin-preserved tissue can also be tested for rickettsiae.[46] A recent publication demonstrated diagnosis of rickettsiosis through detection of the etiologic agent from swabs of patients' eschars.[47] Various assays have been developed allowing for rapid detection of *Rickettsia* sp. with most targeting well-described genes such as citrate synthase (*glt*A),[48] OmpA (*omp*A),[49] OmpB (*omp*B),[50] 16S rRNA,[51] and geneD (*sca*4).[52] The implementation of real-time PCR (qPCR) in rickettsial detection further increased the sensitivity of PCR assays and allows for the semi quantitation of rickettsiae in a sample.[53]

32.8 GENETIC MANIPULATION

The advancements in molecular biology in the last decade have allowed the genetic manipulation of various organisms to further benefit humanity through recombinant DNA technology. However, the same technology can be used for more nefarious means. The bioengineering of organisms to make them resistant to known treatments with increased virulence is a real danger. Very little hurdles exist in using various methods to introducing DNA or inducing mutations in rickettsiae. Other than the clonal nature of rickettsial growth, requiring a host cell and the absence of a known rickettsial bacteriophage, the recent discovery of rickettsial plasmids in various species of rickettsiae has further facilitated studies into molecular manipulation of the rickettsial genome.[6,54] The first successful genetic transformation performed on *R. prowazekii* demonstrated the ability to induce allelic transfer between two strains, inducing rifampin resistance in the receiving strain.[55] More studies on genetic transfer in *Rickettsia* with the use of plasmids as a vector have been performed since, including the introduction of plasmids consisting of fluorescent reporter genes, giving the means to study rickettsia–host cell interaction by visualizing the expressed fluorescence.[56] Recent development of shuttle vectors has further revolutionized the field, allowing the transformation of rickettsiae with greater ease, facilitating the genetic manipulation of rickettsiae.[57]

32.9 CONCLUSIONS

Should rickettsiae pose a major public health problem especially in endemic areas or the threat of release with bioweapons, the prospects of vaccine creation are excellent.[58] Coupled with the ease of treatment with tetracyclines, any outbreaks, if responded to promptly, would be short lived. Many factors pertaining to rickettsial pathophysiology, transmission, and disease treatment make it a poor candidate as a weapon for bioterrorism. However, in underdeveloped parts of the world with a poor public health system, the result of a man-made or natural outbreak can still be devastating to the social and economic structure. Continued vigilance by public health authorities in susceptible communities such as the outbreak in Burundi in 1997 would help negate the risk of any significant natural or man-made rickettsial outbreaks.

REFERENCES

1. Ricketts, H.T., *Contributions to Medical Sciences by Howard Taylor Ricketts, 1870–1910*, University of Chicago Press, Chicago, IL, 1911.
2. Telford, 3rd, S.R. and Parola, P. Chapter 3: Arthropods and rickettsiae. In *Rickettsial Diseases*, Raoult, D. and Parola, P. (Eds.). Informa Healthcare, New York, 2007.
3. Walker, T.S., Rickettsial interactions with human endothelial cells in vitro: Adherence and entry. *Infect. Immun.*, 44, 205–210, 1984.
4. Hechemy, K.E. et al., Discrepancies in Weil–Felix and microimmunofluorescence test results for Rocky Mountain spotted fever. *J. Clin. Microbiol.*, 9, 292–293, 1979.
5. Stothard, D.R., Clark, J.B., and Fuerst, P.A., Ancestral divergence of *Rickettsia bellii* from the spotted fever and typhus groups of *Rickettsia* and antiquity of the genus *Rickettsia*. *Int. J. Syst. Bacteriol.*, 44, 798–804, 1994.
6. Gillespie, J.J. et al., Plasmids and rickettsial evolution: Insight from *Rickettsia felis*. *PLoS One*, 2, e266, 2007.
7. Tamura, A. et al., Classification of *Rickettsia tsutsugamushi* in a new genus, *Orientia* gen. nov., as *Orientia tsutsugamushi* comb. nov. *Int. J. Syst. Evol. Microbiol.*, 45, 589–591, 1995.
8. Izzard, L. et al., Isolation of a novel *Orientia* species (*O. chuto* sp. nov.) from a patient infected in Dubai. *J. Clin. Microbiol.*, 48, 4404–4409, 2010.
9. McDade, J.E. Historical aspects of Q fever. In *Q Fever, Volume I: The Disease*, Marrie, T.J. (Ed.). CRC Press, Boca Raton, FL, 1990.
10. Abdad, M.Y., Stenos, J., and Graves, S., *Rickettsia felis*, an emerging flea-transmitted human pathogen. *Emerg. Health Threats J.*, 4, 7168, 2011.
11. Parola, P., Paddock, C.D., and Raoult, D., Tick-borne rickettsioses around the world: Emerging diseases challenging old concepts. *Clin. Microbiol. Rev.*, 18, 719–756, 2005.
12. Katavolos, P. and Armstrong, P.M., Duration of tick attachment required for transmission of granulocytic ehrlichiosis. *J. Infect. Dis.*, 177, 1422–1425, 1998.
13. Sood, S.K. et al., Duration of tick attachment as a predictor of the risk of Lyme disease in an area in which Lyme disease is endemic. *J. Infect. Dis.*, 175, 996–999, 1997.
14. Walker, D.H. et al., Chapter 2: Pathogenesis, immunity, pathology, and pathophysiology in rickettsial diseases. In *Rickettsial Diseases*, Raoult, D. and Parola, P. (Eds.). Informa Healthcare, New York, 2007.
15. Walker, D.H., Valbuena, G.A., and Olano, J.P., Pathogenic mechanisms of diseases caused by *Rickettsia*. *Ann. N.Y. Acad. Sci.*, 990, 1–11, 2003.
16. Walker, D.H. and Fishbein, D.B., Epidemiology of rickettsial diseases. *Eur. J. Epidemiol.*, 7, 237–245, 1991.
17. Martinez, J.J. et al., Ku70, a component of DNA-dependent protein kinase, is a mammalian receptor for *Rickettsia conorii*. *Cell*, 123, 1013–1023, 2005.
18. Heinzen, R.A. et al., Directional actin polymerization associated with spotted fever group *Rickettsia* infection of Vero cells. *Infect. Immun.*, 61, 1926–1935, 1993.
19. Teysseire, N., Chiche-Portiche, C., and Raoult, D., Intracellular movements of *Rickettsia conorii* and *R. typhi* based on actin polymerization. *Res. Microbiol.*, 143, 821–829, 1992.
20. Heinzen, R.A. et al., Dynamics of actin-based movement by *Rickettsia rickettsii* in vero cells. *Infect. Immun.*, 67, 4201–4207, 1999.
21. Jeng, R.L. et al., A *Rickettsia* WASP-like protein activates the Arp2/3 complex and mediates actin-based motility. *Cell. Microbiol.*, 6, 761–769, 2004.
22. Valbuena, G. and Walker, D.H., Changes in the adherens junctions of human endothelial cells infected with spotted fever group rickettsiae. *Virchows Arch.*, 446, 379–382, 2008.
23. Woods, M.E. and Olano, J.P., Host defenses to *Rickettsia rickettsii* infection contribute to increased microvascular permeability in human cerebral endothelial cells. *J. Clin. Immunol.*, 28, 174–185, 2008.
24. Walker, D.H. et al., The pathology of fatal Mediterranean spotted fever. *Am. J. Clin. Pathol.*, 87, 669–672, 1987.
25. Davidson, M.G. et al., Vascular permeability and coagulation during *Rickettsia rickettsii* infection in dogs. *Am. J. Vet. Res.*, 51, 165–170, 1990.

26. Walker, D.H. and Mattern, W.D., Acute renal failure in Rocky Mountain spotted fever. *Arch. Intern. Med.*, 139, 443–448, 1979.

27. Raoult, D., Woodward, T., and Dumler, J.S., The history of epidemic typhus. *Infect. Dis. Clin. North Am.*, 18, 127–140, 2004.

28. World Health Organisation, A large outbreak of epidemic louse-borne typhus in Burundi. *Wkly. Epidemiol. Rec.*, 72, 152–153, 1997.

29. Kelly, D.J. et al., The past and present threat of rickettsial diseases to military medicine and international public health. *Clin. Infect. Dis.*, 34, S145–S169, 2002.

30. Nettleman, M.D., Biological warfare and infection control. *Infect. Control Hosp. Epidemiol.*, 12, 368–372, 1991.

31. Azad, A.F., Pathogenic rickettsiae as bioterrorism agents. *Clin. Infect. Dis.*, 45, S52–S55, 2007.

32. Raoult, D. et al., Spotless rickettsiosis caused by *Rickettsia slovaca* and associated with *Dermacentor* ticks. *Clin. Infect. Dis.*, 34, 1331–1336, 2002.

33. Ormsbee, R.A., Review: Studies in *Pyroplasmosis hominis* ('spotted fever' or 'tick fever' of the Rocky Mountains) by Louis B. Wilson and William M. Chowning, published in *The Journal of Infectious Diseases* 1, 31–57, 1904. *Rev. Infect. Dis.*, 1, 559–562, 1979.

34. Paddock, C.D. et al., Assessing the magnitude of fatal Rocky Mountain spotted fever in the United States: Comparison of two national data sources. *Am. J. Trop. Med. Hyg.*, 67, 349–354, 2002.

35. Fournier, P.-E. et al., Gene sequence-based criteria for identification of new rickettsia isolates and description of *Rickettsia heilongjiangensis* sp. nov. *J. Clin. Microbiol.*, 41, 5456–5465, 2003.

36. Gikas, A. et al., Comparison of the effectiveness of five different antibiotic regimens on infection with *Rickettsia typhi*: Therapeutic data from 87 cases. *Am. J. Trop. Med. Hyg.*, 70, 576–579, 2004.

37. Chapman, A.S. et al., Diagnosis and management of tickborne rickettsial diseases: Rocky Mountain spotted fever, ehrlichioses, and anaplasmosis—United States: A practical guide for physicians and other health-care and public health professionals. *MMWR Morb. Mortal. Wkly. Rep.*, 55, 1–27, 2006.

38. Fenollar, F., Fournier, P.E., and Raoult, D., Chapter 23: Diagnostic strategy of rickettsiosis and ehrlichiosis. In *Rickettsial Diseases*, Raoult, D. and Parola, P. (Eds.). Informa Healthcare, New York, 2007.

39. Weil, E. and Felix, A., Zur serologischen Diagnose des Fleckfiebers. *Wien Klin Wochenschr*, 29, 33–35, 1916.

40. Brown, G.W. et al., Diagnostic criteria for scrub typhus: Probability values for immunofluorescent antibody and Proteus OXK agglutinin titers. *Am. J. Trop. Med. Hyg.*, 32, 1101–1107, 1983.

41. Raoult, D. et al., Mediterranean Boutonneuse fever. Apropos of 154 recent cases. *Ann. Dermatol. Vener.*, 110, 909–914, 1984.

42. La Scola, B. and Raoult, D., Laboratory diagnosis of rickettsioses: Current approaches to diagnosis of old and new rickettsial diseases. *J. Clin. Microbiol.*, 35, 2715–2727, 1997.

43. Unsworth, N.B. et al., Three rickettsioses, Darnley Island, Australia. *Emerg. Infect. Dis.*, 13, 1105–1107, 2007.

44. Fournier, P.-E. et al., Kinetics of antibody responses in *Rickettsia africae* and *Rickettsia conorii* infections. *Clin. Diagn. Lab. Immun.*, 9, 324–328, 2002.

45. Anderson, B.E. and Tzianabos, T., Comparative sequence analysis of a genus-common rickettsial antigen gene. *J. Bacteriol.*, 171, 5199–5201, 1989.

46. Stein, A. and Raoult, D., A simple method for amplification of DNA from paraffin-embedded tissues. *Nucleic Acids Res.*, 20, 5237–5238, 1992.

47. Mouffok, N. et al., Diagnosis of rickettsioses from eschar swab samples, Algeria. *Emerg. Infect. Dis.*, 17, 1968–1969, 2011.

48. Roux, V. et al., Citrate synthase gene comparison, a new tool for phylogenetic analysis, and its application for the rickettsiae. *Int. J. Syst. Bacteriol.*, 47, 252–261, 1997.

49. Fournier, P.-E., Roux, V., and Raoult, D., Phylogenetic analysis of spotted fever group rickettsiae by study of the outer surface protein rOmpA. *Int. J. Syst. Bacteriol.*, 48, 839–849, 1998.

50. Roux, V. and Raoult, D., Phylogenetic analysis of members of the genus *Rickettsia* using the gene encoding the outer-membrane protein rOmpB (ompB). *Int. J. Syst. Evol. Microbiol.*, 50, 1449–1455, 2000.

51. Roux, V. and Raoult, D., Phylogenetic analysis of the genus *Rickettsia* by 16S rDNA sequencing. *Res. Microbiol.*, 146, 385–396, 1995.

52. Sekeyova, Z., Roux, V., and Raoult, D., Phylogeny of *Rickettsia* spp. inferred by comparing sequences of 'gene D', which encodes an intracytoplasmic protein. *Int. J. Syst. Evol. Microbiol.*, 51, 1353–1360, 2001.

53. Paris, D.H. et al., Real-time multiplex PCR assay for detection and differentiation of rickettsiae and orientiae. *Trans. R. Soc. Trop. Med. Hyg.*, 102, 186–193, 2008.

54. Gillespie, J.J. et al., A *Rickettsia* genome overrun by mobile genetic elements provides insight into the acquisition of genes characteristic of an obligate intracellular lifestyle. *J. Bacteriol.*, 194, 376–394, 2012.

55. Rachek, L.I. et al., Transformation of *Rickettsia prowazekii* to rifampin resistance. *J. Bacteriol.*, 180, 2118–2124, 1998.

56. Baldridge, G.D. et al., Analysis of fluorescent protein expression in transformants of *Rickettsia monacensis*, an obligate intracellular tick symbiont. *Appl. Environ. Microbiol.*, 71, 2095–2105, 2005.

57. Burkhardt, N.Y. et al., Development of shuttle vectors for transformation of diverse *Rickettsia* species. *PLoS One*, 6, e29511, 2011.

58. Walker, D., The realities of biodefense vaccines against *Rickettsia*. *Vaccine*, 27, D52–D55, 2009.

59. Ricketts, H.T., A microorganism which apparently has a specific relationship to Rocky Mountain spotted fever. *J. Am. Med. Assoc.*, 52, 379, 1909.

60. Letaïef, A., Epidemiology of rickettsioses in North Africa. *Ann. N.Y. Acad. Sci.*, 1078, 34–41, 2006.

61. Rolain, J.M., Shpynov, S., and Raoult, D., Spotted-fever-group rickettsioses in north Asia. *Lancet*, 362, 1939, 2003.

62. Paddock, C.D. et al., *Rickettsia parkeri*: A newly recognized cause of spotted fever rickettsiosis in the United States. *Clin. Infect. Dis.*, 38, 805–811, 2004.

63. Rovery, C. and Raoult, D., Chapter 10: *Rickettsia conorii* infections (Mediterranean Spotted Fever, Israeli Spotted Fever, Indian Tick Typhus, Astrakhan Fever). In *Rickettsial Diseases*, Raoult, D. and Parola, P. (Eds.). Informal Healthcare, New York, 2007.

64. Sekeyová, Z. et al., *Rickettsia slovaca* sp. nov., a member of the spotted fever group rickettsiae. *Int. J. Syst. Bacteriol.*, 48, 1455–1462, 1998.

65. Goldwasser, R.A. et al., The isolation of strains of rickettsiae of the spotted fever group in Israel and their differentiation from other members of the group by immunofluorescence methods. *Scand. J. Infect. Dis.*, 6, 53–62, 1974.

66. Valero, A., Rocky Mountain spotted fever in Palestine. *Harefuah*, 36, 99–101, 1949.

67. Campbell, R.W. and Domrow, R., Rickettsioses in Australia: Isolation of *Rickettsia tsutsugamushi* and *R. australis* from naturally infected arthropods. *Trans. R. Soc. Trop. Med. Hyg.*, 68, 397–402, 1974.

68. Walker, D.H., Rickettsioses of the spotted fever group around the world. *J. Dermatol.*, 16, 169–177, 1984.

69. Yu, X. et al., Genotypic and antigenic identification of two new strains of spotted fever group rickettsiae isolated from China. *J. Clin. Microbiol.*, 31, 83–88, 1993.

70. Jensenius, M. et al., African tick bite fever. *Lancet Infect. Dis.*, 3, 557–564, 2003.

71. Kelly, P.J. et al., *Rickettsia africae* sp. nov., the etiological agent of African tick bite fever. *Int. J. Syst. Bacteriol.*, 46, 611–614, 1996.

72. Tarasevich, I.V. et al., Studies of a "new" rickettsiosis "Astrakhan" spotted fever. *Eur. J. Epidemiol.*, 7, 294–298, 1991.

73. Eremeeva, M.E. et al., Proteinic and genomic identification of spotted fever group rickettsiae isolated in the former USSR. *J. Clin. Microbiol.*, 31, 2625–2633, 1993.

74. Beati, L. and Raoult, D., *Rickettsia massiliae* sp. nov., a new spotted fever group *Rickettsia. Int. J. Syst. Bacteriol.*, 43, 839–840, 1993.

75. Vitale, G. et al., *Rickettsia massiliae* human isolation. *Emerg. Infect. Dis.*, 12, 174–175, 2006.

76. Robertson, R.G. and Wisseman, C.L.J., Tick-borne rickettsiae of the spotted fever group in West Pakistan. II. Serological classification of isolates from West Pakistan and Thailand: Evidence for two new species. *Am. J. Epidemiol.*, 97, 55–64, 1973.

77. Graves, S.R. et al., Spotted Fever Group rickettsial infection in south-eastern Australia: Isolation of rickettsiae. *Comp. Immunol. Microbiol.*, 16, 223–233, 1993.

78. Uchida, T. et al., Spotted fever group rickettsiosis in Japan. *Jpn. J. Med. Sci. Biol.*, 38, 151–153, 1985.

79. Beati, L. et al., *Rickettsia aeschlimannii* sp. nov., a new spotted fever group rickettsia associated with *Hyalomma marginatum* ticks. *Int. J. Syst. Bacteriol.*, 47, 548–554, 1997.

80. Unsworth, N.B. et al., Flinders Island spotted fever rickettsioses caused by "marmionii" strain of *Rickettsia honei*, Eastern Australia. *Emerg. Infect. Dis.*, 13, 566–573, 2007.

33 *Salmonella* Typhi

K. Prathyusha and R. Chaudhry

CONTENTS

33.1 INTRODUCTION

Salmonella enterica subspecies *enterica* serotype Typhi causes typhoid fever leading to approximately 17 million cases annually [1]. Typhoid fever is an important cause of morbidity and mortality in crowded and unsanitary settings. Most cases are confined to the developing world. Cases in the West are usually associated with history of travel to endemic regions [2]. Typhoid is contracted by ingestion of food or water contaminated by fecal or urinary carriers excreting *Salmonella* Typhi. Though amenable to treatment with antimicrobials, emergence of drug resistance has resulted in a major setback in the management of typhoid.

Salmonella species belongs to category B in the critical biological agents list. Though nontyphoidal *Salmonella* have been used as agents of terror, *S.* Typhi has seldom found a similar application. Laboratory infections by *S.* Typhi have been reported with few fatalities too [3]. Some cases have also occurred in individuals not working in a microbiology laboratory, like relatives and close acquaintances of microbiologists. Most lab-acquired cases were associated with mouth pipetting and handling proficiency test strains [4].

Though efforts have been undertaken to prevent *S.* Typhi through vaccination and adequate hygiene measures like hazard analysis control critical point (HACCP), the multidrug-resistant strains resistant to even ciprofloxacin and ceftriaxone are always a cause for concern. These strains, when in wrong hands, can easily be introduced into food or water sources leading to widespread morbidity and mortality.

Hence, understanding the epidemiology and pathogenesis of *S.* Typhi and developing an efficient vaccine is all the more important in the present world.

33.2 CLASSIFICATION AND MORPHOLOGY

33.2.1 CLASSIFICATION

Genus *Salmonella* belongs to the family Enterobacteriaceae comprising a large number of medically important pathogens [5]. Several proposals were suggested by various scientists for the nomenclature of *Salmonella*. Kauffmann proposed a one-serotype–one-species concept on the basis of the serologic identification of O (somatic) and H (flagellar) antigens [6]. Schemes of nomenclature were also suggested based on biochemical characteristics that divide the serotypes into subgenera and clinical role of a strain and on genomic relatedness. The nomenclature system currently used at the CDC for the genus *Salmonella* is based on recommendations from the WHO Collaborating Centre [7,8]. According to the CDC system, the genus *Salmonella* contains two species: *S. enterica* (the type species) and *Salmonella bongori*. *S. enterica* is divided, on the basis of biochemical characteristics and genomic relatedness, into six subspecies, namely, *S. enterica* subsp. *enterica* (I), *S. enterica* subsp. *salamae* (II), *S. enterica* subsp. *arizonae* (IIIa), *S. enterica* subsp. *diarizonae* (IIIb), *S. enterica* subsp. *houtenae* (IV), and *S. enterica* subsp. *indica* (VI) [9,10]. Using the Kauffmann–White scheme, members of the genus are classified into more than

2500 serotypes based on differences between the O (cell wall) and H (flagella) antigens [9,10]. CDC uses names for serotypes in subspecies I and antigenic formulas for unnamed serotypes in subspecies II, IV, and VI and in *S. bongori* [11]. The serotypes were named according to the disease and/or the animal from which the organism was isolated, such as *S.* Typhi and *Salmonella* Typhimurium, or by the geographic area where the strain was first isolated, for example, *Salmonella panama* [12]. For named serotypes, to emphasize that they are not separate species, the serotype name is not italicized and the first letter is capitalized [13].

At the first citation of a serotype, the genus name is given followed by the word "serotype" or the abbreviation "ser." and then the serotype name (e.g., *Salmonella* serotype or ser. Typhimurium). Serotype names designated by antigenic formulas include the following: (1) subspecies designation (subspecies I through VI), (2) O (somatic) antigens followed by a colon, (3) H (flagellar) antigens (phase 1) followed by a colon, and (4) H antigens (phase 2, if present). The antigenic formula for *S.* Typhi is *S.* I 9,12[Vi]:d [13].

33.2.2 MORPHOLOGY

The organism is a Gram-negative bacillus of 2–4 μm × 0.6 μm size, nonacid fast, noncapsulated, nonsporulating, and motile with peritrichous flagella. The strains possess type 1 fimbriae associated with mannose-sensitive adhesive properties [5,14].

33.3 BIOLOGY AND EPIDEMIOLOGY

33.3.1 BIOLOGY

S. Typhi expresses three antigens on the cell surface:

1. O antigens that are heat-stable polysaccharides and are part of the cell wall lipopolysaccharide (LPS)
2. H antigens that are heat-labile proteins of the flagella
3. Vi antigen—surface polysaccharide

The O polysaccharides have a core structure common to all Enterobacteriaceae, and side chains of sugars attached to the core determine O specificity. Rough mutant strains are present, which lack the O-specific side chains and show a progressive loss of sugar constituents from core. The O antigen specificities may also change with lysogenic conversion by phages [5].

The flagellar filaments have highly antigenic components called flagellins that are encoded by chromosomally located genes, namely, *fliC* (phase I) and *fljB* (phase II).

Vi is a capsular polysaccharide of α-(1–4)-linked *N*-acetyl-D-galactosaminouronic acid. It inhibits the binding of C3b and antibodies to LPS. The rate of synthesis is controlled by the *viaA* and *viaB* genes [5].

Strains also express other antigens like fimbrial antigens associated with type 1 fimbriae, K or capsular antigens, and M antigen, which is a loose extracellular polysaccharide called colanic acid [5,15].

The genome of *S.* Typhi is approximately five million base pairs long and codes for around 4000 genes [16]. The complete genome sequences of two *S.* Typhi isolates, CT18 and Ty2, are available in public databases, and various studies have concluded that the genomes of *S. enterica* serotype Typhi CT18, *S. enterica* serotype Typhimurium LT2, and *Escherichia coli* are essentially collinear [17–19]. At the same time, the analysis of MLST patterns, the examination of the DNA sequence, and the rate of change of single-nucleotide polymorphisms suggest that *S.* Typhi may be as young as 30,000 years old indicating that its divergence occurred significantly later [20,21].

S. Typhi has several large insertions in its genome thought to be important for survival in the host, and these are called Salmonella Pathogenicity Islands (SPI). The genome also has multiple smaller gene insertions that may also play a part in pathogenicity [22].

The major pathogenicity islands, SPI-1 and SPI-2, encode type three secretion system (T3SS). The function of T3SS is to translocate proteins, especially those involved in virulence, from the bacterial cytoplasm into the host cell. They are often referred to as "molecular syringes" [23]. The base structure of the T3SS complex spans the cell membrane and the cell wall of *Salmonella*. A needle structure protrudes from the base that interacts with the host cells. An inner rod, within, forms the conduit between the bacterial cytoplasm and the host cell membrane. On the cytoplasmic side of the TTSS structure, there is a set of export machinery that contains an ATPase complex that facilitates the transport of effector molecules through the inner rod to a translocase structure in the host cell membrane [24].

The SPI-1 secretes two effectors, SopE and SopE2, which act as guanine nucleotide exchange factors (GEFs) for small GTPases [25]. Two other effectors translocated by SPI-1 are SipA and SipC that affect actin dynamics during *Salmonella* invasion. SipA binds and stabilizes actin and SipC, which forms part of the TTSS delivery pore, nucleates, and bundles actin while anchored in the host cell membrane [26].

Invasive *Salmonella* infections also require expression of SPI-2. The pathogenicity island encodes a second T3SS, effector proteins, molecular chaperones, and a two-component regulatory system and is required for intracellular survival and replication at systemic sites of infection [25].

Other PIs like SPI-3, SPI-4, and SPI-5 have also been identified. Studies suggest that these PIs also have a role in virulence. When genes associated with the SPI-3, SPI-4, and SPI-5 together with SPI-1 and SPI-2 are inactivated, *S.* Typhi loses the ability to express several virulence-associated traits [27].

The pathogenicity island, SPI-7, is responsible for the production of the Vi polysaccharide capsule. The island encodes the Vi locus, a phage encoding the SopE effector protein of PSI-1, a type IV pilus, and a putative type IV secretion system. The type IV pilus is involved in aiding attachment to eukaryotic cells, and the SopE prophage harbors a gene encoding an effector protein secreted through the TTSS. SPI-7 has a viaB operon that encodes the gene responsible for the synthesis and transport of the virulence antigen Vi [28].

S. Typhi also harbors plasmids that favor antimicrobial resistance like the pHCM1 plasmid that encodes resistance to chloramphenicol (*cat*I), ampicillin (TEM-1, *bla*), trimethoprim (*dhfr* 1b), sulfonamides (*sul* II), and streptomycin (*str*AB) [29].

33.3.2 EPIDEMIOLOGY

There are no animal reservoirs of *S.* Typhi; humans are the only source. Since the organism is shed in stool, the infection is transmissible via the fecal–oral route. Foodborne or waterborne transmission occurs as a result of fecal contamination by patients with active disease or by asymptomatic chronic carriers [30]. Mary Mallon or Typhoid Mary, as she was better known, was the first person in the United States identified as an asymptomatic carrier of *S.* Typhi. She was presumed to have infected some 51 people, three of whom died, over the course of her career as a cook. Waterborne transmission involves the ingestion of fewer microorganisms and, hence, has a longer incubation period and lower attack rate as compared to foodborne transmission. Flies may also transmit the infection through their role as mechanical vectors. In addition, fomites are important in disease transmission because *S.* Typhi may survive for several weeks within water, milk, or dust. Although direct person-to-person transmission is uncommon, *S.* Typhi can be transmitted sexually, mainly by anal and oral sex [30]. Health-care and laboratory workers can acquire the disease from infected patients as a result of poor hand hygiene or handling laboratory specimens [4].

Enteric fever is endemic in areas where overcrowding, poor sanitation, and lack of clean running water are the norm. The highest incidence of the infection is seen in areas of poor sanitation especially during the hot and dry season during which the availability of potable water is limited. Rise in the number of cases is also seen during heavy rains when flooding of sewer lines contaminate water supply [31].

Reported risk factors also include food and drinks purchased from street vendors, raw fruits, and vegetables grown in fields fertilized with sewage, ill contacts in the household, lack of hand washing, and lack of hygienic toilets. Gastric acidity is one of the primary defense mechanisms against *S.* Typhi [30]. Hence, conditions associated with chronic reduced gastric acidity like the frequent use of antacids or proton pump inhibitors, gastric achlorhydria, history of gastric surgery, and evidence of prior *Helicobacter pylori* infection also predispose to enteric fever [32,33]. Patients suffering from malaria, sickle cell disease, schistosomiasis, and cellular deficiencies like HIV also show increased susceptibility to typhoid fever [30,34,35].

Until the early twentieth century, enteric fever had a worldwide distribution. Historical surveillance indicates that the infection was endemic in Western Europe and North America [36]. But these countries have witnessed a massive decline in the incidence of typhoid in the past half century due to changes in sanitation and hygiene like introduction of safe municipal water and strict food hygiene practices including pasteurization [37]. The annual incidence of typhoid globally is approximately 17 million cases and 600,000 deaths [1]. In 2000, there were 22 million new cases of typhoid fever and 210,000 typhoid fever-related deaths [38]. In 2010, there were an estimated 13.5 million typhoid fever episodes [39].

Most infections in the present day occur in developing countries in the Asian, African, and Latin American subcontinent where sanitary conditions remain poor and water supplies are not adequately treated. The greatest burden of disease is in Asia, where 13 million cases are assumed with 400,000 deaths annually. The incidence is highest among infants, children, and adolescents [38]. An issue of major concern in the developing world is that the public health figures may be underestimating the actual number of enteric fever cases. The diagnosis is usually clinical on the basis of which the patient is given an empirical therapy. Most cases are not confirmed by blood culture, and surveillance techniques face a lot of logistic challenges [40].

The incidence of typhoid fever varies substantially in Asia. Very high typhoid fever incidence has been found in India and Pakistan [41]. In comparison, typhoid fever frequency was moderate in Vietnam and China and intermediate in Indonesia [42].

In India, the disease is seen to be prevalent mostly in urban areas. In the endemic pockets, the annual incidence approaches around one percent of the population with most of the cases being reported in children less than 15 years of age [43]. A large-scale community study performed in an Indian urban slum showed incidence as high as 2 per 1000 population per year for children under 5 years and 5.1 per 1000 populations per year for children under 10 years [44]. Another study in Northern India showed that the majority of cases occurred in children aged 5–12 years and 24.8% of cases were in children up to 5 years of age. The peak incidence is seen from July to September, coinciding with the rainy season [45].

The incidence of enteric fever is poorly characterized in subSaharan Africa [46]. But frequent outbreaks do occur and many patients present with intestinal perforation [47].

There is a decline in the disease burden of typhoid in Latin America. The decrease in the region is attributed to improved economic conditions and implementation of sanitation and hygiene like safe potable water. These safety measures were undertaken to control the cholera menace, but this is an ideal model that shows that the burden of typhoid can be curbed with the correct interventions [38].

In developed countries, enteric fever is predominantly associated with travel to an endemic region. The risk for travellers appears to vary by the geographic region visited. Several reports indicate that the Indian subcontinent has the higher risk for acquiring typhoid fever [48]. Among travellers from the USA, a visit to six countries, namely India, Mexico, Philippines, Pakistan, El Salvador, and Haiti account for 80% of cases of typhoid [2]. The overall risk from travel to the Indian subcontinent was higher than from travel to any other geographic region.

Outbreaks are uncommon in the developed countries. They are mostly associated with asymptomatic carriers who handle food and due to breaches in the safety of food and water supply [37]. Outbreaks have been traced back to potato salads, egg salads, and shrimp salads, all of which were also handled by carriers. A recent outbreak occurred in Florida that was traced back to imported mamey oranges in a fruit shake [49]. In 1964, 507 cases of typhoid fever in Scotland were linked to canned Argentinean corned beef that was cooled in contaminated river water [50]. A 1997 French outbreak and a 1998 Danish outbreak of multiresistant *S.* Typhi were associated with ingestion of contaminated pork [51].

Between 1970 and 1989, several strains of multidrug-resistant *S.* Typhi emerged, which were showing plasmid-mediated resistance to chloramphenicol, ampicillin, and trimethoprim. The strains were isolated from all over the world, especially from the Indian subcontinent [50]. Multidrug-resistant outbreaks have been reported in 1995 from Bangalore, India, with 76% resistance to ampicillin, 64% to chloramphenicol, and 75% to tetracycline [51]. Development of resistance to common first-line antimicrobials leads to an increased use of fluoroquinolones for the treatment of multidrug-resistant typhoid in the 1990s. This was soon followed by chromosomal and plasmid-mediated fluoroquinolone resistance.

Strains that are multidrug resistant and show resistance to nalidixic acid with reduced susceptibility to fluoroquinolones constitute 80% of the world's typhoid burden [52]. Although multidrug-resistant strains remain common in many areas of Asia, in some areas, antimicrobial-susceptible strains have reemerged.

33.4 PATHOGENESIS AND CLINICAL FEATURES

33.4.1 PATHOGENESIS

In studies involving the administration of laboratory *Salmonella* strains to healthy human volunteers, the infectious dose of *S.* Typhi varies between 1000 and 1 million organisms [53]. The typhoid bacilli that survive the gastric acid barrier reach the small intestine. *Salmonella* strains adhere to the apical surface of intestinal epithelial cells and disrupt the normal brush border to induce the formation of membrane ruffles that reach out and enclose the adherent bacteria in large vesicles. This morphologically distinct process is termed *bacteria-mediated endocytosis* [54].

It has been postulated that M cells, specialized epithelial cells overlying Peyer's patches, are the principal portal of entry in enteric fever. Once internalized, the typhoid bacilli are phagocytosed by antigen presenting cells. *Salmonella* can survive inside the phagocytes within a vacuolar compartment that fuses with lysosomes. Resistance to a variety of vacuolar bactericidal activities is essential to pathogenesis including resistance to antimicrobial peptides, nitric oxide, and oxidative killing [55].

The infected phagocytes are organized in discrete foci that become pathological lesions, surrounded by normal tissue and are called typhoid nodules.

The intestinal pathological stages include (1) hyperplasia in the first week within typhoidal nodules in the Peyer's patches and the lymphoid follicles of the cecum, (2) necrosis of the mucosa during the second week, and (3) sloughing of the necrotic mucosa to form ulcerations that may bleed or perforate. Perforation and bleeding typically occur on the antimesenteric border of the intestine near the ileocecal valve [56].

Typhoid bacilli translocate from the intestinal lymphoid follicles to the draining mesenteric lymph nodes into reticuloendothelial cells. At a critical point, depending on factors such as the number of bacilli, their virulence, and the host immune response, *S.* Typhi escapes from their intracellular sequestered havens and disseminates into the blood stream. They seed secondary sites of infection like the liver, spleen, bone marrow, gallbladder, and Peyer's patches in the terminal ileum. Kupffer cells in the liver try to neutralize the bacilli with free radicals and enzymes. The bacilli that survive this onslaught invade hepatocytes and cause apoptosis [55].

The systemic manifestations of typhoid fever are attributed to endotoxins, immune complexes, and disseminated intravascular coagulation (DIC). The current hypothesis is that interactions between *S.* Typhi and macrophages lead to the production of such cytokines as tumor necrosis factor-α, interleukin-1, and interferon-α and interferon-β, leading in turn to fever and other constitutional symptoms [56].

33.4.2 CLINICAL FEATURES

The symptoms of typhoid fever appear after an incubation period of 5–21 days. The cytokines released into the bloodstream cause fever, which is initially low grade but becomes high and sustained by the second week. The fever is said to have a characteristic "step ladder" pattern. Nonspecific symptoms like headache, anorexia, sore throat, myalgia, and mental confusion have also been observed. Besides fever, relative bradycardia, tender abdomen, hepatomegaly, and splenomegaly are also common. During the second week of illness, blanching maculopapular erythematous lesions called rose spots appear on the chest and abdomen in 5%–30% of the cases. Rose spots are clearly visible on fair-skinned individuals only and hence are not a significant feature in typhoid fever in the endemic regions. In a small number of patients, constipation or diarrhea is seen. If prompt and accurate antibiotic therapy is not instituted, the illness progresses into the third week. Ileocecal hyperplasia and ulceration leading to gastrointestinal bleed and perforation occurs. Due to death of hepatocytes, elevated liver enzymes and bilirubin may also be observed. Though recovery occurs by the fourth week, weakness and weight loss can persist for months [57,58]. Rarely complications like septicemia, endocarditis, pneumonia, pyogenic meningitis, and septic meningitis are also seen [59]. Unusual presentations of *S.* Typhi infection like breast abscess, splenic abscess, and liver abscess have also been reported [60,61].

33.5 IDENTIFICATION AND DIAGNOSIS

Enteric fever should be differentiated from other febrile illness like malaria, tuberculosis, leptospirosis, amoebic liver abscess, and visceral leishmaniasis since these are also common in regions endemic for typhoid. Several algorithms have been developed to diagnose typhoid in endemic areas. For patients in countries where typhoid is not endemic, a travel history is crucial. In endemic settings, physicians rely on their clinical judgement and prescribe broad-spectrum antibiotics that take care of most of the etiological agents of febrile illness. Patients also self-medicate with over-the-counter antimicrobials since strict pharmaceutical laws are seldom followed in resource-limited countries [62,63]. This common usage of antimicrobials alters the classical presentation of typhoid when the patients seek treatment in health centers after failure to respond to self-medication. Moreover, self-treatment and broad-spectrum antibiotics without proper patient compliance have also led to the menace of drug resistance.

The standard diagnostic method is isolation of the organism from blood or bone marrow. Snips of skin from rose spots and samples from other sterile sites may also be used for isolation of *S.* Typhi. Culture is considered as the gold standard test [64]. The major advantage is that isolation of the organism can be followed up with antimicrobial susceptibility tests and typing (phage typing and genotyping). Typing methods are useful to identify the source during outbreaks.

Blood culture is positive in 50%–90% of patients, during the first week of illness before starting antibiotics. After the first week, stool and urine cultures may be positive in 50%–75% and 5%–10% of patients, respectively. The most sensitive test is a bone marrow culture [90%–95%], which may remain positive even after antibiotic administration [65,67]. Other diagnostic tests include a duodenal string test during the third week of illness (70% sensitivity) and rose spot culture (up to 60%) [65,66].

The variable sensitivity of blood culture is attributed to various factors such as volume of blood sample collected, time of sample collection, type of culture media used for isolation, and pretreatment with antimicrobials. The volume of blood taken is critical and relates directly to the number of bacteria in the blood. The ideal period of time for blood culture would be during the first week of illness when typhoid bacilli are circulating in blood. Several studies have indicated that bile containing oxgall media is ideal for typhoid bacilli isolation. Oxgall inhibits antibacterial activity of fresh blood. Other media include those containing tryptone soya broth or brain–heart infusion broth, with additional sodium polyethanol sulfonate [68]. Automated blood culture systems like BACTEC (Becton Dickinson, United Kingdom) and BacTAlert (bioMérieux, France) increase the speed of recovery of the organisms. Cultures from buffy coat of blood and streptokinase-treated blood clot have also increased the sensitivity of culture.

Viable typhoid bacilli concentrate more in the bone marrow, the sensitivity of bone marrow culture is highest even after pretreatment with antimicrobials. Since bone marrow collection is invasive and painful, the method is generally not preferred [69].

Cultures of feces and urine in the third and fourth week of illness, respectively, are also employed for isolation of typhoid bacilli. But culture positivity may indicate either acute infection or chronic carriage.

The major limitations of culture are the time taken for isolation, identification, and susceptibility testing of the organism and lack of adequate microbiology laboratory facilities in endemic countries.

Serological diagnosis of enteric fever is another modality to identify *S.* Typhi. Widal test measures agglutinating antibodies against somatic O antigen and flagellar H antigen of *S.* Typhi in the serum of individuals with suspected enteric fever. The test is simple and cheap to perform and is available in the slide and tube format. Ideally paired acute and convalescent sera are required for accurate diagnosis; a fourfold rise or fall in titer indicates a positive Widal [70]. In endemic regions, when paired samples are difficult to obtain, a cutoff titer is determined based on baseline level of anti *S.* Typhi antibodies in the healthy population [71]. Limitations of Widal are due to the lack of standardization of test procedure and result interpretation and presence of cross reacting antibodies to surface antigens on other *Enterobacteriaceae*.

ELISA-based studies use LPS and flagella as the antigens to detect circulating antibodies to *S.* Typhi [72]. Though reported to be more sensitive than Widal, it lacks in specificity. ELISA for the detection of antibodies against Vi antigen is helpful to identify carriers, especially in outbreak situations.

Clinicians have always laid emphasis on point-of-care diagnostic tests for enteric fever. "Point of care tests" or rapid tests are those that make the results available within an hour after initiation of the assay. They should also be simple to perform, yet sensitive and specific. Typhidot is an immunodot that detects specific IgM and IgG against a 50 kD *S.* Typhi outer membrane protein [73]. TUBEX is a latex agglutination test that detects *S.* Typhi-specific anti-LPS antibodies. The analyte is mixed with reagent latex particles and visual results are interpreted. Since the reaction involves colored indicator particles, visual interpretation is relatively easy [74]. When compared with blood culture-positive typhoid cases, sensitivities for TUBEX vary between 56% and 100% and between 67% and 98% for Typhidot and specificities from 58% to 100% for TUBEX and 73% to 100% for Typhidot [75–77].

Molecular diagnostic tests, both polymerase chain reaction (PCR) and real-time PCR, are available for the detection of *S.* Typhi, the common targets being Hd flagella gene (*fliC-d*) [78], the Ha flagella gene (*fliC-a*), the Vi capsular gene *viaB* [79], the tyvelose epimerase gene (*tyv*, previously *rfbE*), the paratose synthase gene (*prt*, previously rfbS), *groEL* [80], and the 16sRNA gene [81]. Though variable sensitivities have been reported in blood cultures, specificities are always 100%. The author has developed an in-house PCR for the rapid detection of *S.* Typhi in clinical samples. The assay specifically amplifies the VI–VIII region of the Hd flagellin gene. The assay has a sensitivity and specificity of

93.58% and 87.19%, respectively, with blood culture as the gold standard. The positive and negative predictive values of the assay are 43.45% and 99.23%, respectively [82].

33.6 TREATMENT AND PREVENTION

33.6.1 TREATMENT

Enteric fever should be promptly treated with appropriate antibiotics and diet. Patients should be under constant monitoring to detect life-threatening complications. The antibiotic of choice for typhoid fever is ciprofloxacin, 500 mg twice daily for 7–10 days. This regimen has a high cure rate (>97%) and has a very low relapse and carrier rate (<1%) [83]. Other fluoroquinolones like norfloxacin, ofloxacin, and levofloxacin can also be used. Alternatively, intravenous ceftriaxone, 2 g daily for 14 days, is also effective [84]. In pregnant women and children, the drug of choice is ceftriaxone since quinolones are associated with cartilage-related damage in young children [85]. Azithromycin (1000 mg/day for 1 day, then 500 mg/day for 6 days) may also be effective [86]. Exclusion of carrier state requires three consecutive negative stool cultures, collected one month apart at least 48 hours after cessation of antibiotic therapy. For treatment of carriers, a 4-week course of ciprofloxacin, 750 mg orally twice a day, proves to be effective [87].

Historically, typhoid fever was treated with chloramphenicol, ampicillin, or trimethoprim–sulfamethoxazole (co-trimoxazole). Resistance to these three antibiotics is termed multidrug resistance in *S.* Typhi [88]. Plasmid-mediated chloramphenicol resistance emerged in the early 1970s spreading across to as far as Mexico, India, Vietnam, Chile, and Bangladesh [89]. These strains harbored a plasmid of the H1 incompatibility group and were also resistant to streptomycin, sulfonamides, and tetracycline [90]. Chloramphenicol resistance was due to acetylation and presence of efflux pumps. The presence of beta lactamases like TEM-1 leads to ampicillin resistance. The emergence of these MDR strains resulted in the widespread use of fluoroquinolones. Very soon, strains were isolated that seemed to show susceptibility to ciprofloxacin by MIC values (though on the upper limit) but showed clinical resistance and resistance to nalidixic acid disc. These were referred to as NARSTs or nalidixic acid-resistant strains. These strains showed decreased susceptibility to ciprofloxacin as shown by their MIC values that were creeping up toward resistance [91]. These strains were mostly seen in endemic settings. A study conducted in our tertiary institute from 2000 to 2003 showed a gradual increase in the NARSTs, over the years. These strains had a higher range of MIC of ciprofloxacin ranging from 0.023 to 1.0 µg/mL [92]. Besides these, quinolone-resistant strains are also isolated with point mutation in the quinolone resistance-determining region (QRDR) of DNA gyrase. Recently, sporadic cases of high-level ceftriaxone resistance with *S.* Typhi strains expressing CTX-M-15 and SHV-12 extended-spectrum beta lactamases have been reported [93,94]. A ray of hope seems to be the reemergence of chloramphenicol-susceptible strains in areas where its use was prohibited. The concept of antibiotic recycling of older antimicrobials may prove to be useful in the future [95].

33.6.2 PREVENTION

In spite of the drug resistance, enteric fever is still treatable with appropriate antibiotics. But in endemic areas, an efficient typhoid vaccine is a more cost-effective option. Current licensed vaccines are a capsular polysaccharide vaccine and a live-attenuated vaccine.

The earliest typhoid vaccines were whole-cell vaccines that were heat killed, phenol preserved, or acetone inactivated. The latter was better since it included the Vi capsular component. These vaccines were injectable and offered 51% to 88% protection to children and young adults, lasting for up to 12 years. But they also caused local inflammation, pain, fever, and malaise in 9%–34% of the recipients [96].

The licensed polysaccharide vaccine is formulated from the Vi capsular component and is available in the injectable form for intramuscular delivery. The vaccination is recommended for use in adults and in children at least 2 years of age. Booster doses are required every 2–3 years. Immunoconversion is seen in 85% of the adults and provides protection for up to 3 years in 70% of adults [97]. The limitations of the Vi polysaccharide vaccine are the short duration of protection and hence the need for multiple booster doses. The vaccine is also not immunogenic for children under 2 years of age. The most important limitation is that there are several strains of *S.* Typhi that are Vi negative, and the vaccine cannot protect patients from infection by these strains. There is also the possibility that continuous use of this vaccine, especially in endemic settings, would lead to the emergence and persistence of Vi-negative strains [98].

Ty21a is a live vaccine made from a highly attenuated Vi-negative *S.* Typhi strain. The vaccine strain is a *gal* E mutant having a deficiency of galactose epimerase enzyme. The strain is unable to synthesize a complete lipopolysaccharide in the absence of galactose in the medium [99].

Ty21a is formulated as enteric-coated capsules, and three doses every other day in endemic settings and 3–4 doses as a travel vaccine are recommended [100]. Ty21a vaccination results in robust cell-mediated immune responses, immunologic memory, and prolonged protection of 60%–70% over 7 years [101]. The vaccine also confers cross protection against *S.* Paratyphi B but not *S.* Paratyphi A [102]. Revaccination is recommended at 5–7-year intervals. A major limitation of the vaccine is its thermal instability with dependence on cold chain to maintain viability [96].

The highest priority in endemic settings is to make available a vaccine that can even be administered to children less than 2 years old. Approaches are being used to increase the immunogenicity of the Vi vaccine by protein conjugation. A Vi conjugate vaccine linked to a recombinant *Pseudomonas aeruginosa* exoprotein A has been formulated and is called Vi-rEPA.

TABLE 33.1

Dosage and Schedule for Typhoid Fever Vaccine

Vaccine	Age (Years)	Dose and Mode of Administration	No. of Doses	Boosting Interval
Live-attenuated Ty21a vaccine				
Primary	≥6	1 capsule (oral)	4	NA
Booster	≥6	1 capsule (oral)	4	Every 5 years
Vi capsular polysaccharide vaccine				
Primary	≥2	0.50 mL (intramuscular)	1	NA
Booster	≥2	0.50 mL (intramuscular)	1	Every 2 years

Source: Adapted from Centers for Disease Control and Prevention, Typhoid and paratyphoid fever, in: US Department of Health and Human Services, ed., *CDC Health Information for International Travel 2010*, Centers for Diseases Control and Prevention, Atlanta, GA, 2009.

In 2–4-year-olds, Vi-rEPA showed more than 90% efficacy over 27 months and 82% efficacy at 46 months. The vaccine is very immunogenic in infants and young children [103]. Vi conjugate vaccines are also being developed using tetanus toxoid, diphtheria toxoid, or CRM197 as the carrier protein [96]. The most advanced, a Vi–tetanus toxoid conjugate produced by Bharat Biotech (India), has completed phase 2 evaluations. A phase 3 study in infants, children, and adults comparing the immunogenicity and safety of Vi–tetanus toxoid with Bharat's unconjugated Vi polysaccharide is ongoing (CTRI/2011/08/001957) (Table 33.1).

CDC has redefined vaccination recommendations for travel and now enlists pretravel vaccination for 195 global destinations [104]. But vaccination alone cannot offer complete protection. Sanitary and dietary precautions like drinking boiled water and washing fruits and vegetables are equally important for prevention of enteric fever. Hygiene measures all along the food chain through HACCP approach are of utmost importance. So are measures like hand washing, sanitary disposal of feces, control of flies, good cooking practices, and pasteurization of milk and dairy products.

33.7 CONCLUSIONS AND FUTURE PERSPECTIVES

Typhoid fever has a global epidemiology and outbreaks are related to breach in safe water supply and food handling. Patients present with fever, malaise, and abdominal pain. Complications like gastric ulcer and perforation, meningitis, and septicemia may occur. Clinical presentation may vary when patients try to self-treat themselves. Though blood culture is the gold standard of diagnostics, rapid tests are the need of the hour. Ciprofloxacin and ceftriaxone are the current choices to treat typhoid. But management has suffered a setback thanks to the emerging and rampant drug-resistant strains. The currently licensed vaccines are not adequately immunogenic and do not protect infants and young children. The new conjugate vaccines are currently in various stages of development.

The drug-resistant strains of *S.* Typhi have the potential to be used as weapons of terror. The procurement of strains is easy; they can even be isolated from the source. The strains can be easily maintained in culture. The emergence of strains of *S.* Typhi that are multidrug resistant with resistance to fluoroquinolones and third-generation cephalosporins has made typhoid fever less amenable to treatment. There are no effective vaccines available to provide protection during outbreaks. When spread through food or water, they can lead to widespread morbidity and mortality.

Such a critical situation could be ideally managed using accurate rapid diagnostic tests and effective oral antibiotics. Strict and thorough measures of sanitation and hygiene should be enforced. Newer drugs should be evaluated. Combination chemotherapy and antibiotic recycling may help tackle the drug resistance. Newer and more immunogenic vaccines and their widespread use in endemic settings are also of considerable importance. Strategies to avert such a critical event capable of mass discomfort should be thought about seriously by world bodies like WHO and CDC.

REFERENCES

1. Edelman R, Levine MM: Summary of an international workshop on typhoid fever. *Rev. Infect. Dis.* 1986; 8:329–349.
2. Connor BA, Schwartz E: Typhoid and paratyphoid fever in travellers. *Lancet Infect. Dis.* 2005; 5:623–628.
3. Pike RM: Laboratory-associated infections: Incidence, fatalities, causes and prevention. *Annu. Rev. Microbiol.* 1979; 33:41–66.
4. Sewell DL: Laboratory associated infections and biosafety. *Clin. Microbiol. Rev.* 1995; 8:389–405.
5. Mahy BWJ, Meulen V, Borriello SP et al.: *Topley and Wilson's Microbiology and Microbial Infections*, 10th ed. Wiley.
6. Kauffmann, F: *The Bacteriology of Enterobacteriaceae.* Copenhagen, Denmark: Munksgaard, 1966.
7. Popoff MY, Le Minor L: *Antigenic Formulas of the Salmonella Serovars*, 8th revision. Paris, France: World Health Organization Collaborating Centre for Reference and Research on *Salmonella*, Pasteur Institute, 2001.
8. Popoff MY, Bockemuhl J, Gheesling LL: Supplement 2002 (no. 46) to the Kauffmann–White scheme. *Res. Microbiol.* 2004; 155:568–570.
9. Reeves MW, Evins GM, Heiba AA, Plikaytis BD, Farmer JJ III: Clonal nature of *Salmonella* Typhi and its genetic relatedness to other salmonellae as shown by multilocus enzyme electrophoresis and proposal of *Salmonella bongori* comb. nov. *J. Clin. Microbiol.* 1989; 27:313–320.
10. Popoff MY, Le Minor L: *Antigenic Formulas of the Salmonella Serovars*, 7th revision. Paris, France: World Health Organization Collaborating Centre for Reference and Research on *Salmonella*, Pasteur Institute, 1997.
11. Le Minor L, Popoff MY: Request for an opinion. Designation of *Salmonella enterica* sp. nov., nom. rev., as the type and only species of the genus *Salmonella*. *Int. J. Syst. Bacteriol.* 1987; 37:465–468.

12. Su L-H, Chiu C-H: *Salmonella*: Clinical importance and evolution of Nomenclature. *Chang Gung Med. J.* 2007; 30:3.

13. Brenner FW, Villar RG, Angulo FJ, Tauxe R, Swaminathan B: *Salmonella* nomenclature. *J. Clin. Microbiol.* 2000; 38(7):2465.

14. Murray PR, Baron EJ, Landry ML, Jorgensen JH, Pfaller MA, Eds.: *Manual of Clinical Microbiology*, 9th ed. Washington, DC: ASM Press.

15. Collee JG, Marimon BP, Fraser AG, Simmons A: *Mackie and McCartney Practical Medical Microbiology*, 14th ed. Churchill.

16. Baker S, Dougan G: The genome of *Salmonella enterica* serovar Typhi. *Clin. Infect. Dis.* 2007; 45:S29–S33.

17. Parkhill J, Dougan G, James KD et al.: Complete genome sequence of a multiple drug resistant *Salmonella enterica* serovar Typhi CT18. *Nature* 2001; 413:848–852.

18. McClelland M, Sanderson KE, Spieth J et al.: The complete genome sequence of *Salmonella enterica* serovar Typhimurium LT2. *Nature* 2001; 413:852–856.

19. Blattner FR, Plunkett G III, Bloch CA et al.: The complete genome sequence of *Escherichia coli* K-12. *Science* 1997; 277:1453–1474.

20. Selander RK, Beltran P, Smith NH et al.: Evolutionary genetic relationships of clones of *Salmonella* serovars that cause human typhoid and other enteric fevers. *Infect. Immun.* 1990; 58:2262–2275.

21. Kidgell C, Reichard U, Wain J et al.: *Salmonella* Typhi, the causative agent of typhoid fever is approximately 50,000 years old. *Infect. Genet. Evol.* 2002; 2:39–45.

22. Hacker J, Blum-Oehler G, Muhldorfer I, Tschape H. Pathogenicity islands of virulent bacteria: Structure, function and impact on microbial evolution. *Mol. Microbiol.* 1997; 23:1089–1097.

23. Hueck CJ: Type III protein secretion systems in bacterial pathogens of animals and plants. *Microbiol. Mol. Biol. Rev.* 1998; 62:379–433.

24. Gala'n JE, Wolf-Watz H: Protein delivery into eukaryotic cells by type III secretion machines. *Nature* 2006; 444:567–573.

25. Thomson N, Baker S, Pickard D et al.: The role of prophage-like elements in the diversity of *Salmonella enterica* serovars. *J. Mol. Biol.* 2004; 339:279–300.

26. Fu Y, Galán JE: A *Salmonella* protein antagonizes Rac-1 and Cdc42 to mediate host-cell recovery after bacterial invasion. *Nature* 1999; 401:293–297.

27. Zhang X-L, Jeza VT, Pan Q: *Salmonella* Typhi: From a human pathogen to a vaccine vector. *Cell. Mol. Immunol.* 2008; 5:91–97.

28. Seth-Smith HMB: SPI-7: *Salmonella*'s Vi-encoding pathogenicity island. *J. Infect. Dev. Countries* 2008; 2(4):267–271.

29. Taylor DE, Chumpitaz JC, Goldstein F: Variability of IncHI1 plasmids from *Salmonella* Typhi with special reference to Peruvian plasmids encoding resistance to trimethoprim and other antibiotics. *Antimicrob. Agents Chemother.* 1985; 28:452–455.

30. Mandell GL, Bennett JE, Dolin R, Eds.: *Principles and Practices of Infectious Diseases*, 7th ed. Churchill Livingstone.

31. Crum NF: Current trends in typhoid fever. *Curr. Gastroenterol. Rep.* 2003; 5:279–286.

32. Blaser MJ, Rafuse EM, Wells JG et al.: An outbreak of salmonellosis involving multiple vehicles. *Am. J. Epidemiol.* 1981; 114:663–670.

33. Waddell WR, Kunz LJ: Association of *Salmonella enteritis* with operation of stomach. *N. Engl. J. Med.* 1956; 255:555–559.

34. Barrett-Conner E: Bacterial infection and sickle cell anemia: An analysis of 250 infections in 166 patients and a review of the literature. *Medicine* 1971; 50:97–112.

35. Neves J, Raso P, Marinko PP: Prolonged septicemic salmonellosis intercurrent with *Schistosomiasis mansoni* infection. *J. Trop. Med. Hyg.* 1971; 74:9.

36. Osler W: *The Principles and Practice of Medicine: Designed for the Use of Practitioners and Students of Medicine*, 8th ed. New York: D. Appleton, 1912:1–46.

37. Crump JA, Griffin PM, Angulo FJ: Bacterial contamination of animal feed and its relationship to human foodborne illness. *Clin. Infect. Dis.* 2002; 35:859–865.

38. Crump JA, Luby SP, Mintz ED: The global burden of typhoid fever. *Bull. World Health Organ.* 2004; 82:346–353.

39. Buckle GC, Walker CL, Black RE: Typhoid fever and paratyphoid fever: Systematic review to estimate global morbidity and mortality for 2010. *J. Glob. Health.* 2012; 2:10401.

40. Archibald LK, Reller LB: Clinical microbiology in developing countries. *Emerg. Infect. Dis.* 2001; 7:302–305.

41. Siddiqui FJ, Rabbani F, Hasan R, Nizami SQ, Bhutta ZA: Typhoid fever in children: Some epidemiological considerations from Karachi, Pakistan. *Int. J. Infect. Dis.* 2006; 10:215–222.

42. Ochiai RL, Acosta CJ, Danovaro-Holliday MC, Baiqing D, Bhattacharya SK, Domi Typhoid Study Group: A study of typhoid fever in five Asian countries: Disease burden and implications for controls. *Bull. World Health Organ.* 2008; 86:260–268.

43. Bhan MK, Bahl R, Bhatnagar S: Typhoid and paratyphoid fever. *Lancet* 2005; 366(9487):749–762.

44. Sur D, von Seidlein L, Manna B et al.: The malaria and typhoid fever burden in the slums of Kolkata, India: Data from a prospective community based study. *Trans. R. Soc. Trop. Med. Hyg.* 2006; 100:725–733.

45. Walia M, Gaind R, Paul P et al.: Age related clinical and microbiological characteristics of enteric fever in India. *Trans. R. Soc. Trop. Med. Hyg.* 2006; 100:942–948.

46. Mweu E, English M: Typhoid fever in children in Africa. *Trop. Med. Int. Health* 2008; 13:1–9.

47. Muyembe-Tamfum JJ, Veyi J, Kaswa M, Lunguya O, Verhaegen J, Boelaert M: An outbreak of peritonitis caused by multidrug-resistant *Salmonella* Typhi in Kinshasa, Democratic Republic of Congo. *Travel. Med. Infect. Dis.* 2009; 7:40–43.

48. Mermin JH, Townes JM, Gerber M et al.: Typhoid fever in the United States, 1985–1994. *Arch. Intern. Med.* 1998; 158:633–638.

49. Katz DJ, Cruz MA, Trepka MJ et al.: An outbreak of typhoid fever in Florida associated with an imported frozen fruit. *J. Infect. Dis.* 2002; 186:234–239.

50. Mirza SH, Beeching NJ, Hart CA: Multi-drug resistant typhoid: A global problem. *J. Med. Microbiol.* 1996; 44:317–319.

51. Rathish KC, Chandrashekar MR, Nagesha CN: An outbreak of multidrug resistant typhoid fever in Bangalore. *Ind. J. Pediatr.* 1995; 62:445–448.

52. Chau TT, Campbell JI, Galindo CM et al.: Antimicrobial drug resistance of *Salmonella enterica* serovar Typhi in Asia and molecular mechanism of reduced susceptibility to the fluoroquinolones. *Antimicrob. Agents Chemother.* 2007; 51:4315–4323.

53. Hornick RB, Greisman SE, Woodward TE, DuPont HL, Dawkins AT, Snyder MJ. Typhoid fever: Pathogenesis and immunologic control. *N. Engl. J. Med.* 1970; 283:686–691, 739–746.

54. Francis CL, Starnbach MN, Falkow S: Morphological and cytoskeletal changes in epithelial cells occur immediately upon interaction with *Salmonella* Typhimurium grown under low-oxygen conditions. *Mol. Microbiol.* 1992; 6:3077–3087.

55. House D, Bishop A, Parry C et al.: Typhoid fever: Pathogenesis and disease. *Curr. Opin. Infect. Dis.* 2001; 14:573–578.

56. Everest P, Wain J, Roberts M et al.: The molecular mechanisms of severe typhoid fever. *Trends Microbiol.* 2001; 9:316–320.

57. Stuart BM, Pullen RL: Typhoid: Clinical analysis of three hundred and sixty cases. *Arch. Intern. Med.* 1946; 78:629–661.

58. Huckstep RL: *Typhoid Fever and Other Salmonella Infections.* Edinburgh, U.K.: E.&S. Livingstone, 1962.

59. Khan M, Coovadia Y, Connolly C et al.: Risk factors predicting complications in blood culture-proven typhoid fever in adults. *Scand. J. Infect. Dis.* 2000; 32:201–205.

60. Mahajan RK, Duggal S, Chande DS, Duggal N, Hans C, Chaudhry R: *Salmonella enterica* serotype Typhi from a case of breast abscess. *J. Commun. Dis.* 2007; 39:201–204.

61. Chaudhry R, Mahajan RK, Diwan A, Khan S, Singhal R, Chandel DS, Hans C: Unusual presentation of enteric fever: Three cases of splenic and liver abscesses due to *Salmonella* Typhi and *Salmonella* Paratyphi A. *Trop. Gastroenterol.* 2003; 24:198–199.

62. Brooks WA, Hossain A, Goswami D et al.: Bacteremic typhoid fever in an urban slum, Bangladesh. *Emerg. Infect. Dis.* 2005; 11:326–329.

63. Siddiqui FJ, Rabbani F, Hasan R, Nizami SQ, Bhutta ZA: Typhoid fever in children: Some epidemiological considerations in Karachi, Pakistan. *Int. J. Infect. Dis.* 2006; 10:215–222.

64. World Health Organization. Background document: The diagnosis, treatment and prevention of typhoid fever. WHO/V&B/03.07. Geneva, Switzerland: WHO, 2003.

65. Gilman RH, Terminel M, Levine MM, Hernandez-Mendoza P, Hornick RB: Relative efficacy of blood, urine, rectal swab, bone marrow, and rose-spot cultures for recovery of *Salmonella* Typhi in typhoid fever. *Lancet* 1975; 1:1211–1213.

66. Vallenas C, Hernandez H, Kay B, Black R, Gotuzzo E: Efficacy of bone marrow, blood, stool and duodenal contents cultures for bacteriologic confirmation of typhoid fever in children. *Pediatr. Infect. Dis.* 1985; 4:496–498.

67. Guerra-Caceres JG, Gotuzzo-Herencia E, Crosby-Dagninio E, Miro-Quesada M, Carrillo-Parodi C: Diagnostic value of bone marrow culture in typhoid fever. *Trans. R. Soc. Trop. Med. Hyg.* 1979; 73:680–683.

68. Escamilla J, Santiago LT, Sangalang RP, Ranoa CP, Cross JH. Comparative study of three blood culture systems for isolation of enteric fever *Salmonella. Southeast Asian J. Trop. Med. Public Health* 1984; 15:161–166.

69. Wain J, Diep TS, Bay PV et al.: Specimens and culture media for the laboratory diagnosis of typhoid fever. *J. Infect. Dev. Countries* 2008; 2:469–474.

70. House D, Chinh NT, Diep TS et al.: Use of paired serum samples for serodiagnosis of typhoid fever. *J. Clin. Microbiol.* 2005; 43:4889–4990.

71. Levine MM, Grados O, Gilman RH, Woodward WE, Solis-Plaza R, Waldman W: Diagnostic value of the Widal test in areas endemic for typhoid fever. *Am. J. Trop. Med. Hyg.* 1978; 27:795–800.

72. Nardiello S, Pizzella S, Russo M, Galanti B: Serodiagnosis of typhoid fever by enzyme-linked immunosorbent assay determination of anti-*Salmonella* Typhi lipopolysaccharide antibodies. *J. Clin. Microbiol.* 1984; 20:718–721.

73. Choo KE, Oppenheimer SJ, Ismail AB, Ong KH: Rapid serodiagnosis of typhoid fever by dot enzyme immunoassay in an endemic area. *Clin. Infect. Dis.* 1994; 19:172–176.

74. Lim PL, Tam FCH, Cheong YM, Jegathesan M: One-step 2-minute test detect typhoid specific antibodies based on particle separation in tubes. *J. Clin. Microbiol.* 1998; 36:2271–2278.

75. Olsen SJ, Pruckler J, Bibb W et al.: Evaluation of rapid diagnostic tests for typhoid fever. *J. Clin. Microbiol.* 2004; 42:1885–1889.

76. Kawano RL, Leano SA, Agdamag DMA: Comparison of serological test kits for diagnosis of typhoid fever in the Philippines. *J. Clin. Microbiol.* 2007; 45:246–247.

77. Dutta S, Sur D, Manna B, Sen B, Deb AK, Deen JL, Wain J, Von Seidlein L, Ochiai L, Clemens JD, Kumar Bhattacharya S. Evaluation of new-generation serologic tests for the diagnosis of typhoid fever: data from a community-based surveillance in Calcutta, India. *Diagn. Microbiol. Infect. Dis.* 2006; 56:359–365.

78. Song JH, Cho H, Park MY, Na DS, Moon HB, Pai CH. Detection of *Salmonella* Typhi in the blood of patients with typhoid fever by polymerase chain reaction. *J. Clin. Microbiol.* 1993; 31:1439–1443.

79. Nizami SQ, Bhutta ZA, Siddiqui AA, Lubbard L: Enhanced detection rate of typhoid fever in children in a periurban slum in Karachi, Pakistan using polymerase chain reaction technology. *Scand. J. Clin. Lab. Invest.* 2006; 66:429–436.

80. Ali A, Haque A, Haque A et al.: Multiplex PCR for differential diagnosis of emerging typhoidal pathogens directly from blood samples. *Epidemiol. Infect.* 2009; 137:102–107.

81. Zhu Q, Lim CK, Chan YN: Detection of *Salmonella* Typhi by polymerase chain reaction. *J. Appl. Bacteriol.* 1996; 80:244–251.

82. Chaudhry R, Chandel DS, Verma N, Singh N, Singh P, Dey AB: Rapid diagnosis of typhoid fever by an in house flagellin PCR. *J. Med. Microbiol.* 2010; 59:1391–1393.

83. Akalin HE: Quinolones in the treatment of typhoid fever. *Drugs* 1999; 58:52–54.

84. Wallace MR, Yousif AA, Mahroos GA et al.: Ciprofloxacin versus ceftriaxone in the treatment of multidrug resistant typhoid fever. *Eur. J. Clin. Microbiol. Infect. Dis.* 1993; 12:907–910.

85. Doherty CP, Saha SK, Cutting WAM: Typhoid fever, ciprofloxacin and growth in young children. *Ann. Trop. Paediatr.* 2000; 20:297–303.

86. Butler T, Sridhar CB, Daga MK et al.: Treatment of typhoid fever with azithromycin versus chloramphenicol in a randomized multicentre trial in India. *J. Antimicrob. Chemother.* 1999; 44:243–250.

87. Ferreccio C, Morris JG, Valdivieso C et al.: Efficacy of ciprofloxacin in the treatment of chronic typhoid carriers. *J. Infect. Dis.* 1988; 157:1235–1239.

88. Wain J, Kidgell C: The emergence of multidrug resistance to the antimicrobial agents for the treatment of typhoid fever. *Trans. R. Soc. Trop. Med. Hyg.* 2004; 98:423–430.

89. Brown JD, Duong Hong M, Rhoades ER: Chloramphenicol resistant *Salmonella* Typhi in Saigon. *JAMA* 1975; 231:162–166.

90. Rowe B, Ward LR, Threlfall EJ: Multidrug-resistant *Salmonella* Typhi: A worldwide epidemic. *Clin. Infect. Dis.* 1997; 24(Suppl. 1):S106–S109.

91. Crump JA, Barrett TJ, Nelson JT, Angulo FJ: Reevaluating fluoroquinolone breakpoints for *Salmonella enterica* serotype Typhi and for non-Typhi salmonellae. *Clin. Infect. Dis.* 2003; 37:75–81.

92. Renuka K, Kapil A, Kabra SK et al.: Reduced susceptibility to ciprofloxacin and *gyra* gene mutation in North Indian strains of *Salmonella enterica* serotype Typhi and serotype Paratyphi A. *Microb. Drug Resist.* 2004; 10:146–153.

93. Pfeifer Y, Matten J, Rabsch W: *Salmonella enterica* serovar Typhi with CTX-M β-lactamase, Germany. *Emerg. Infect. Dis.* 2009; 15:1533–1534.

94. Al Naiemi N, Zwart B, Rijnsburger MC et al.: Extended-spectrum-beta-lactamase production in a *Salmonella enterica* serotype Typhi strain from the Philippines. *J. Clin. Microbiol.* 2008; 46:2794–2795.

95. Harish BN, Menezes GA: Preserving efficacy of chloramphenicol against typhoid fever in a tertiary care hospital, India. *Reg. Health Forum*, WHO South-East Asia Region 2011; 15:92–96.

96. Marathe SA, Lahiri A, Negi VD, Chakravortty D: Typhoid fever and vaccine development: A partially answered question. *Indian J. Med. Res.* 2012; 135:161–169.

97. Tacket CO, Ferreccio C, Robbins JB et al.: Safety and immunogenicity of two *Salmonella* Typhi Vi capsular polysaccharide vaccines. *J. Infect. Dis.* 1986; 154:342–345.

98. Saha MR, Ramamurthy T, Dutta P, Mitra U: Emergence of *Salmonella* Typhi Vi antigen-negative strains in an epidemic of multidrug-resistant typhoid fever cases in Calcutta, India. *Natl. Med. J. India* 2000; 13:164.

99. Levine MM, Ferreccio C, Black RE, Tacket CO, Germanier R: Progress in vaccines against typhoid fever. *Rev. Infect. Dis.* 1989; 11(Suppl. 3):S552–S567.

100. Levine MM, Ferreccio C, Cryz S, Ortiz E: Comparison of enteric-coated capsules and liquid formulation of Ty21a typhoid vaccine in randomised controlled field trial. *Lancet* 1990; 336:891–894.

101. Levine MM, Ferreccio C, Abrego P et al.: Duration of efficacy of Ty21a, attenuated *Salmonella* Typhi live oral vaccine. *Vaccine* 1999; 17(Suppl. 2): S22–S27.

102. Wahid R, Simon R, Zafar SJ et al.: Live oral typhoid vaccine Ty21a induces cross reactive humoral immune responses against *S.* Paratyphi A and *S.* Paratyphi B in humans. *Clin. Vaccine Immunol.* 2012; 19:825–834.

103. Lin FY, Ho VA, Khiem HB et al.: The efficacy of a *Salmonella* Typhi Vi conjugate vaccine in two-to-five-year-old children. *N. Engl. J. Med.* 2001; 344:1263–1269.

104. Centers for Disease Control and Prevention. Typhoid and Paratyphoid Fever. In: US Department of Health and Human Services, Ed., *CDC Health Information for International Travel 2010.* Atlanta, GA: Centers for Diseases Control and Prevention, 2009.

34 *Shigella*

K.R. Schneider, M.K. Fatica, K.A. Lampel, and B.R. Warren

CONTENTS

34.1 INTRODUCTION

Shigella spp. are the causative agent of shigellosis or "bacillary dysentery" and have been implicated in many worldwide foodborne outbreaks. In the United States, it is estimated that these pathogens cause over 400,000 cases of illness, 5,400 hospitalizations, and 38 deaths each year [1]. Humans have been the only identified hosts of this pathogen, although higher primates in close proximity to humans have also been infected. Transmission occurs via the fecal–oral route, with a common route in foods prepared by an infected food handler who practices poor personal hygiene. Shigellosis outbreaks usually involve large numbers of individuals due to its low infectious dose and ease of spreading through populations. In some cases, outbreaks can linger due to the carriers of *Shigella* [2]. *Shigella* carriers asymptomatically harbor the pathogen. In an area of Bangladesh with endemic shigellosis, a study of young children under the age of 5 years was performed where 249 ill children without diarrheal symptoms and 699 healthy children were screened for *Shigella*; 6.4% of the ill children and 2.1% of the healthy group were positive carriers of *Shigella* [3].

Shigella is acid resistant and salt tolerant and can survive at infective levels in many types of foods including fruits and vegetables, low-pH foods, prepared foods, and foods held in modified atmosphere or vacuum packaging or held at refrigeration temperatures. Identification and diagnostic methods for *Shigella* include conventional culture methods, immunological methods, and molecular-based methods. Fluoroquinolone-based antibiotics are the current antibiotic treatment of choice; however, fluoroquinolone-resistant *Shigella* spp. have been isolated in several countries. The increase in antibiotic resistance in *Shigella* spp. has become a significant public health concern worldwide. This chapter reviews *Shigella* classification, morphology, epidemiology, clinical features, and pathogenesis characteristics of *Shigella*, along with methods of diagnosis and the treatment and prevention of shigellosis.

34.2 CLASSIFICATION AND MORPHOLOGY

Over 100 years ago, Kiyoshi Shiga discovered the etiological agent that caused bacillary dysentery, which was different than the pathogen causing amoebic dysentery [4]. In total, three additional distinct species were isolated in 1950 and the genus *Shigella* was established. *Shigellae* are members of the family Enterobacteriaceae, which are nearly genetically identical to *Escherichia coli* and are closely related to *Salmonella* and *Citrobacter* spp. [5]. *Shigella* spp. are Gram-negative, facultatively anaerobic, nonspore-forming, nonmotile rods that typically do not ferment lactose. Other important biochemical characteristics include that they are lysine decarboxylase, acetate, and mucate negative and do not produce gas from glucose (some *Shigella flexneri* 6 serotypes may produce gas) [6,7].

There are four serogroups of *Shigella*: *Shigella dysenteriae* (serogroup A) serotypes 1–15, *S. flexneri* (serogroup B) serotypes 1–8, *Shigella boydii* (serogroup C) serotypes 1–19, and *Shigella sonnei* (serogroup D) serotype 1. These serotypes are

TABLE 34.1

Percentage of *Shigella* Isolates in the United States Reported by the National *Shigella* Surveillance System from 1995 to 2009

Serotype	*S. sonnei* (%)	*S. flexneri* (%)	*S. boydii* (%)	*S. dysenteriae* (%)	Ungrouped (%)	Total Isolates
1995	76.6	15.6	1.2	0.5	6.1	19,330
1996	72.9	19.2	2.0	0.7	5.2	14,071
1997	71.5	20.9	2.1	0.6	4.9	12,314
1998	75.2	17.7	1.7	0.7	4.8	12,485
1999	73.0	20.1	1.6	0.5	4.8	10,084
2000	79.6	13.6	1.4	0.4	5.0	12,732
2001	77.3	15.7	1.2	0.5	5.3	10,598
2002	83.5	12.2	0.8	0.3	3.2	12,992
2003	80.2	14.4	1.1	0.4	3.9	11,552
2004	68.9	17.2	1.8	0.4	11.7	9,343
2005	74.4	13.6	1.2	0.5	10.3	10,484
2006	72.3	14.3	1.1	0.5	11.9	10,336
2007	77.5	9.3	0.6	0.2	12.5	10,996
2008	76.1	7.5	0.6	0.2	15.6	14,805
2009	80.4	10.6	0.7	0.4	8.0	10,173

Source: Centers for Disease Control and Prevention (CDC), National *Shigella* Surveillance System, National *Shigella* Annual Summaries, 2012, http://www.cdc.gov/nationalsurveillance/shigella_surveillance.html (accessed July 21, 2012).

mainly differentiated based on O (lipopolysaccharide (LPS)) antigen and biochemical characteristics, as well as varying epidemiology [8]. *S. dysenteriae* is primarily associated with epidemics and serotype 1 associated with the highest fatality rate (5%–15%) [8,9]. *S. flexneri* predominates in areas of endemic infection, while *S. sonnei* has been implicated in source outbreaks in developed countries [10]. In particular, *S. sonnei* accounts for over two-thirds of shigellosis cases in the United States. *S. boydii* is predominantly isolated within the Indian subcontinent, but source outbreaks have occurred in Central and South America and is rarely isolated in North America [9]. Table 34.1 shows the percentage of the four *Shigella* serogroups and total number of isolates reported by the National *Shigella* Surveillance System in the United States each year from 1995 to 2009 [11]. The average number of *Shigella* isolates associated with each serogroup remains relatively unchanged throughout the 14-year span as do the total number of isolates (Table 34.1). The disparity between the number of estimated cases of shigellosis and the number of isolates depicted in Table 34.1 may be due to patients not seeking medical care, lack of illness reporting, or lack of specimen collection and testing at medical facilities, among other factors.

34.3 BIOLOGY AND EPIDEMIOLOGY

Shigella has been classically characterized as a waterborne pathogen [12], and outbreaks have occurred from contaminated community water sources that were insufficiently chlorinated [13–15]. Outbreaks of *Shigella* contributed to contaminated foods are those subjected to processing or preparation by hand, are exposed to a limited heat treatment,

or are served/delivered raw to the consumer [16]. Examples of food products from which *Shigella* has been isolated include potato salad, ground beef, bean dip, raw oysters, fish, and raw vegetables.

34.3.1 SHIGELLA SURVIVAL AND RESISTANCE

Shigella spp. are generally heat sensitive, acid resistant, salt tolerant, and can withstand low levels of organic acids [17–20]. Acid resistance under various temperature conditions was demonstrated using *S. flexneri* strain 5348. The culture was prepared in brain-heart infusion (BHI) broth at pH 5 and incubated at 19°C, 28°C, and 37°C. However, bacterial counts declined over time when the culture was held at 12°C or lower in the BHI broth at pH 5. Population numbers also decreased at all temperatures tested in BHI broths adjusted to pH 4, 3, and 2 [18]. This trend was also observed in food products with growth studies of *Shigella* in low-pH foods held under refrigeration, such as fruit juices [21,22]. *S. sonnei* and *S. flexneri* survived in apple juice (pH 3.3–3.4) and tomato juice (pH 3.9–4.1) when held at 7°C for 14 days. No reduction was observed in the tomato juice, while a 1.2–3.1 $\log_{10}$ reduction was observed in apple juice over the 14 days [21].

S. flexneri survival was also assessed when inoculated into prepared salads that were acidic. *S. flexneri* did survive in carrot salad (pH 2.7–2.9), potato salad (pH 3.3–4.4), coleslaw (pH 4.1–4.2), and crab salad (pH 4.4–4.5) held at 4°C. After 11 days, a decrease from 4.3×10^6 to 4.2×10^2 CFU/g was observed in the carrot salad and a decrease from 1.32×10^6 to 8.5×10^2 CFU/g in the potato salad. Sampling of the coleslaw and the crab salads ceased due to product spoilage on days 13 and 20, respectively. However, the number of recovered *S. flexneri* was 2.16×10^4 and 2.4×10^5 CFU/g in the respective salads. These studies indicate that *Shigella* can be found and recovered in foods held at refrigeration temperatures, even in the presence of background microflora and low pH [22].

Salt tolerance of *Shigella* has also been reported. In one study, *S. flexneri* strain 5348 grew in BHI broth, pH 6, supplemented with ≤6% NaCl when held at 19°C and 37°C or ≤7% NaCl when held at 28°C. Growth of *S. flexneri* was also observed in BHI broth, pH 5, containing ≤2%, ≤4%, ≤4%, and ≤0.5% NaCl when held at 37°C, 28°C, 19°C, and 12°C, respectively [19]. However, *S. flexneri* populations gradually declined in BHI broth, pH 4, regardless of incubation temperatures and percentage of NaCl. Results from this study suggest that *S. flexneri* is salt tolerant and may survive in saline foods including pickled vegetables, caviar, pickled herring, dry cured ham, and certain cheeses for extended periods of time [19].

S. flexneri survival was also studied in BHI broth supplemented with organic acids commonly found in fruits and vegetables (citric, malic, and tartaric acid) or fermentation acids commonly used as preservatives (acetic and lactic acid) at 0.04 M and adjusted to pH 4 [20]. Acetic and lactic acids had a greater effect on survival than citric, malic, and tartaric acids. When incubated at 37°C, *S. flexneri* survived for 1–2 days in the presence of each organic acid tested. However, at 4°C, *S. flexneri* survived in the presence of all the organic acids tested for longer than 55 days [20]. This study suggests that organic acids may aid in the inactivation of *Shigella* at specific temperatures; however, foods with low levels of acids stored at low temperatures may support the survival of the bacterium for extended periods of time [20].

This long-term survival of *Shigella* has been exhibited on produce commodities. Fresh-cut papaya, jicama, and watermelon were artificially inoculated with *S. sonnei*, *S. flexneri*, or *S. dysenteriae*, and growth was observed within 6 h at room temperature [23]. Rafi and Lunsford inoculated raw cabbage, onion, and green pepper with *S. flexneri*, and although counts decreased slightly at 4°C, survival was observed on onion and green pepper until day 12, when spoilage of the produce halted further analysis [22]. *S. sonnei* survival was observed on chopped and whole parsley for 14 days at 4°C, with growth observed on chopped parsley held at 21°C [16]. *Shigella* is also able to survive on produce packaged under vacuum or modified atmosphere. The survival of *S. sonnei* in shredded cabbage packaged under vacuum or in a modified atmosphere of nitrogen and carbon dioxide was assessed when stored at room temperature or under refrigeration. At room temperature, *S. sonnei* populations remained stable for 3 days, at which time population declines were observed. At refrigeration conditions, *S. sonnei* populations were stable for 7 days. Differences in stability are attributed to decline in cabbage pH at room temperature storage with little pH change under refrigeration storage [24]. These studies demonstrate *Shigella* survival on refrigerated produce for periods of time that may exceed expected shelf life.

Temperature is also an important factor in the survival of *Shigellae*. Freezing (−20°C) and refrigeration (4°C) temperatures support survival, but not growth, of *Shigella*. Elevated temperatures are less permissive for *Shigella* survival, with traditional pasteurization and cooking temperatures being sufficient for inactivation. Sublethal heat exposure can sensitize *Shigella* to selective components of microbiological media. Tollison and Johnson demonstrated that *S. flexneri* sublethally heat-stressed by exposure to 50°C for 30 min in phosphate buffer became sensitive to 0.85% bile salts and 0.50% sodium deoxycholate. Since these compounds are ingredients in several enrichment and isolation media used for detection of *Shigella*, the thermal history of the food sample to be analyzed should be known [25].

The growth and survival of *S. flexneri* was studied in boiled rice, lentil soup, milk, cooked beef, cooked fish, mashed potato, mashed brinjal, and raw cucumber. All food samples, except raw cucumber, were autoclaved prior to inoculation. Growth of *S. flexneri* was observed in all food products within 6–18 h after inoculation at 25°C and 37°C. Survival of initial inoculum was observed through 72 h in all foods, with a slight decrease in population levels in the rice and milk [26]. Initial inoculum levels were maintained throughout the 72 h holding period for all foods, except rice and milk. These results demonstrate the ability of *Shigella* to grow and survive in a variety of prepared foods that may be contaminated during cooking or preparation by an infected food handler.

In addition to foods, *Shigella* is also capable of long-term survival in water and on inanimate objects (also known as fomites). *Shigella* can survive in water with little decline in population. Distilled water was artificially inoculated with *S. flexneri* at 2.8×10^8 CFU/mL and held the samples at 4°C. After 26 days, 9.2×10^7 CFU/mL of *S. flexneri* was recovered. This high survival rate of *S. flexneri* in water supports the historical association of shigellosis outbreaks and unchlorinated water sources [22]. Inanimate objects can also serve as vectors for transmission of *Shigella*. *S. sonnei* survival was observed on cotton, glass, wood, paper, and metal at various temperatures (−20°C to 45°C). When held at −20°C, most of the strains survived for more than 14 days on each surface; however, surfaces held at 45°C did not support survival of most strains, supporting the increased survival of *Shigella* at low temperatures [27].

34.3.2 SHIGELLA EPIDEMIOLOGY

The epidemiology of the four subgroups of *Shigella* differs, with *S. dysenteriae* being primarily associated with epidemics [8]. *S. flexneri* predominates in areas of endemic infection, while *S. sonnei* has been implicated in source outbreaks in developed countries [10]. *S. boydii* has been associated with source outbreaks in Central and South America but is most commonly restricted to the Indian subcontinent. *S. boydii* is rarely isolated in North America; however, slight increases in the numbers of isolates had been observed in both 2003 and 2004 [9,28].

The demographic variability of the four *Shigella* subgroups also differs based on the FoodNet data on the US shigellosis cases from 1996 to 1999 [29]. The overall incidence of shigellosis was highest among the following groups: children aged 1–4 years, male patients, blacks, Hispanics, and Native Americans [29]. There were also marked demographic differences between infection with *S. sonnei* and *S. flexneri* with respect to age, sex, and race. While the incidence of both *S. sonnei* and *S. flexneri* were higher among those aged 1–4 years, there was a second peak of *S. flexneri* infection among those aged 30–39 years [29]. The incidence of *S. sonnei* among men and women was similar; however, the incidence of *S. flexneri* among men was almost twice that of women. In addition, the incidence of *S. sonnei* among blacks and whites was higher than that of *S. flexneri*, while the incidence of *S. flexneri* was higher among Native Americans [29].

The estimated number of illnesses caused by *Shigella* per year in the United States is over 400,000, with the number of reported laboratory-confirmed cases being over 14,500 [1]. According to the Centers for Disease Control and Prevention (CDC) Emerging Infections Program, Foodborne Diseases Active Surveillance Network (FoodNet), *Shigella* was the third most reported foodborne bacterial pathogen in 2010, with the highest incidence in children age 4 and younger [30,31]. Of 19,089 laboratory-diagnosed cases in 2010, *Shigella* accounted for 1780 cases (9.3% of total cases) behind only *Salmonella* (8256 cases) and *Campylobacter* (6365 cases) [30]. The same trend was observed in 2006; with *Shigella* being the third most reported foodborne pathogen behind *Salmonella* and *Campylobacter* [31,32]. From 1996–1998 to 2009, the incidence of *Shigella* has decreased 55%, with a 27% decrease in the number of *Shigella* cases in 2009 from those of 2006–2008 [31]. The incidence of each serogroup in the United States has remained relatively constant, with serogroup D (*S. sonnei*) accounting for the majority of all isolates, followed by serogroup C (*S. flexneri*), serogroup B (*S. boydii*), and serogroup A (*S. dysenteriae*) [11]. Despite overall decreasing incidence of *Shigella* cases, it remains a foodborne bacterial pathogen of concern worldwide.

Characteristic of foodborne shigellosis, several outbreaks have been associated with foods consumed raw and processed or prepared by hand (Table 34.2). In 2000, a multistate outbreak of shigellosis was traced to a commercially prepared five-layer bean dip [33]. In this outbreak, there were 406 reported cases of illness, resulting in 14 hospitalizations and zero deaths across 10 states. After extensive epidemiological investigation, numerous problems were identified in the manufacturing process and investigators determined that the source of the outbreak was most likely an infected food handler [33]. Another large outbreak of shigellosis occurred in 2010 where over 300 cases of illness occurred. The source of the outbreak was found to be a food product served at a national sandwich chain store in the Chicago area of Illinois [34]. The investigation identified 140 laboratory-confirmed cases of *S. sonnei*. Those who consumed tomatoes and the nine-grain bread at the location had a higher risk of

TABLE 34.2
Selected Foodborne Outbreaks Associated with *Shigella*

Year	Serogroup	Food Product(s) Implicated	References
1983	*S. sonnei*	Tossed salad	[35]
1986	*S. sonnei*	Shredded lettuce	[36]
1986	*S. sonnei*	Raw oysters	[37]
1987	*S. sonnei*	Watermelon	[38]
1988	*S. sonnei*	Uncooked tofu salad	[39,40]
1989	*S. flexneri* 4a	German potato salad	[41]
1992	*S. flexneri* 2	Tossed salad	[42]
1994	*S. sonnei*	Iceberg lettuce	[43–45]
1995–1996	*S. sonnei*	Fresh pasteurized milk cheese	[46]
1996	*S. flexneri*	Salad vegetables	[47]
1998	*S. sonnei*	Uncooked, chopped curly parsley	[48]
1998	*S. flexneri*	Restaurant-associated, source unknown	[49]
1999	*S. boydii* 18	Bean salad (parsley or cilantro)	[50]
2000	*S. sonnei*	Five-layer bean dip	[33]
2001	*S. sonnei*	Raw oysters	[51]
2001	*S. flexneri*	Tomatoes	[52]
2002	*Shigella* spp.	Greek-style pasta salad	[53]
2004	*S. sonnei*	Raw carrots	[54]
2004	*S. flexneri*	Macaroni salad, coleslaw, potato salad	[55]
2004	*S. flexneri*	Tamale	[55]
2005	*S. sonnei*	Chicken dishes	[55]
2005	*S. flexneri*, serotype 2a	Beef, other	[55]
2006	*S. sonnei*	Beef, picadillo	[55]
2006	*S. sonnei*	Beans, unspecified	[55]
2006	*S. sonnei*	Salad, unspecified	[55]
2006	*S. sonnei*	Salads, lettuce based	[55]
2007	*S. sonnei*	Raw baby corn	[56]
2008	*S. sonnei*	Salad	[57]
2009	*S. sonnei*	Sugar peas	[58]
2010	*S. sonnei*	Infected food handler at restaurant	[59]
2010	*S. sonnei*	Restaurant-associated	[34]
2011	*S. sonnei*	Fresh basil	[60]

illness. Employee illness at the location was also discovered, but it remains unclear if a food product was contaminated upon arrival and caused the employee illness or if an infected food handler was the source of the outbreak [34].

The 2000 and 2010 shigellosis outbreaks, among other studies, demonstrate the importance of proper food handling and the role of food handlers in the transmission of *Shigella*. For instance, a study investigating the presence of enteropathogens among food handlers in Irbid, Jordan, showed that *Shigella* was isolated from the stools of four out of 283 examined food handlers [61]. Mensah et al. evaluated 511 food items from the streets of Accra, Ghana, from which *S. sonnei* was isolated from one sample of macaroni. It was noted that the macaroni was served using bare hands instead of clean utensils, which may have led to the *S. sonnei* contamination [62]. Wood et al. examined foods from Mexican homes, commercial sources in Guadalajara, Mexico, and from restaurants in Houston, TX, for contamination with bacterial enteropathogens [63]. While no *Shigella* was isolated from foods sampled from 12 Houston restaurants or from food commercially prepared in Guadalajara, Mexico, four isolates were obtained from meals prepared in Mexican homes.

As discussed in the previous section, *Shigella* can also persist on produce commodities. The US Food and Drug Administration (FDA) has investigated the presence of foodborne pathogens, including *Shigella*, on imported and domestic produce [64,65]. An FDA survey of imported broccoli, cantaloupe, celery, parsley, scallions, loose-leaf lettuce, and tomatoes found *Shigella* contamination in nine of 671 total samples: 3 of 151 cantaloupe samples, 2 of 84 celery samples, 1 of 116 lettuce samples, 1 of 84 parsley samples, and 2 of 180 scallion samples [65]. Another FDA survey of domestically grown fresh cantaloupe, celery, scallions, parsley and tomatoes found *Shigella* contamination in 5 of 665 total samples: 1 of 164 cantaloupe samples, 3 of 93 scallion samples, and 1 of 90 parsley samples [64].

34.4 CLINICAL FEATURES AND PATHOGENESIS

There are over 400,000 estimated cases of shigellosis per year in the United States, but approximately only 20,000 cases are actually reported [1,11]. Most cases are due to poor personal hygiene; however, shigellosis can be associated with contaminated food and water and international travel [9]. Each year, approximately 20% of Americans become infected with shigellosis due to international travel to developing countries [66].

34.4.1 CLINICAL FEATURES OF *SHIGELLA*

Shigella is an exclusively human disease transmitted indirectly through contaminated food, water, or fomites or directly through the fecal–oral route. *Shigella* is highly contagious and is capable of causing disease with only low cell counts; the infective dose ranges from 10 cells for *S. dysenteriae* to 500 cells for *S. sonnei* [66]. Children under the age of 5 years, the elderly, and immunocompromised individuals may be at greater risk for *Shigella* infection due to increased susceptibility. *Shigella* is an important cause of morbidity of children and is challenging to control, particularly in daycare settings due to common person-to-person transmission where toddlers are especially susceptible and commonly practice poor personal hygiene [67]. Shigellosis manifests in a spectrum of symptoms with two basic clinical presentation, watery diarrhea and severe dysentery [67]. All subgroups of *Shigella* are capable of causing dysentery, but *S. sonnei* generally results in milder symptoms than *S. dysenteriae* and *S. flexneri* [69,70].

The symptoms of infection include bloody diarrhea, abdominal pain, fever, and malaise. The acute inflammation and mucosal destruction from the bacterial invasion into the colonic and rectal mucosa result in bloody and mucoid stools in the infected patients [71]. Although the mechanism is unknown, seizures have also been reported in 5.4% of shigellosis cases involving children [72]. Chronic sequelae from infections of *S. dysenteriae* serotype 1 and other *Shigella* that encode the genetic information for the Shiga toxin (STX), can include hemolytic uremic syndrome (HUS) and toxic megacolon, while *S. flexneri* infections are associated with later development of reactive arthritis, especially in persons with the genetic marker HLA-B27 [71,73]. Reactive arthritis is characterized by joint pain and eye irritation [73].

34.4.2 PATHOGENESIS OF *SHIGELLA*

The invasion of colonic (large intestine) epithelial cells by *Shigella* involves four steps: entry into epithelial cells, intracellular multiplication, intra- and intercellular spreading, and killing of the host cell [74]. The genetic machinery responsible for invasion is encoded on the 220 kDa virulence plasmid. The virulence plasmid contains invasion plasmid antigen (*ipa*) genes, which encode four highly immunogenic polypeptides: IpaA, IpaB, IpaC, and IpaD. Bacterial strains containing Tn5 insertions affecting the expression of the *ipa* genes revealed that expression of *ipaB*, *ipaC*, and *ipaD* is strongly associated with cellular entry, while an *ipaA* mutant is invasive but at reduced ability as compared to wild type [74].

Along with the *ipa* genes, the type III secretory apparatus carries out invasion of *Shigella* into the host. *Shigella* can enter the host intestinal epithelial cells by secreting virulence factors directly into the epithelial host cells using type III secretion [75]. The secretion system is composed of nearly 50 proteins, along with the chaperones IpgA, IpgC, IpgE, and Spa15, and several transcription activators encoded on the large virulence plasmid harbored by all four pathogenic strains. The complex mechanism that involves the assembly of the type III secretory needle and the subsequent translocation of the effector protein molecules is controlled by a cascade of regulatory activators and partially activated when the pathogen is in contact with the host cell [75].

The *virF* gene, located on the virulence plasmid, and the *virR* gene, located on the chromosome, also mediate invasion. The *virF* gene encodes a 30 kDa protein that positively regulates the expression of the *ipa* genes and *icsA* (also known as *virG*), another virulence plasmid gene that encodes intra- and intercellular spread. Environmental factors that affect the expression of virF are not known. The *virR* gene is a repressor of the plasmid invasion genes in a temperature-dependent manner [74]. When *Shigellae* are grown at 30°C, they do not express the Ipa polypeptides and are therefore noninvasive at this temperature. However, *Shigellae* grown at 37°C are fully invasive and all virulence plasmid polypeptides are encoded [74].

Once ingested, *Shigellae* transit through the gastrointestinal tract to the colon where they translocate the epithelial barrier via M cells that overlay the solitary lymphoid nodules [76]. Upon reaching the basolateral side of the M cells, *Shigella* infects macrophages and induces cell apoptosis [76]. Infected macrophages release interleukin-1β, which elicits a strong inflammatory response [77]. Once released from the macrophage, *Shigella* enters neighboring epithelial cells (enterocytes), through a process known as macropinocytosis. This involves the host cell extending its membrane into formations known as pseudopodia, which engulf large volumes and close around them forming a vacuole. Macropinocytosis requires actin polymerization and the presence of myosin, an actin-binding protein [78]. Common stimuli that induce macropinocytosis include cytokines and bacterial antigens. In response to invasion, epithelial cells produce proinflammatory cytokines that contribute to inflammation of the colon [75].

Once engulfed by epithelial cells, *Shigellae* immediately disrupt the vacuole, releasing them to the host cell cytoplasm where they multiply rapidly. Sansonetti et al. observed the generation time of *S. flexneri* in HeLa cells to be approximately 40 min [79]. Efficient intracellular growth requires the acquisition of host cell nutrients. Since little free iron exists within mammalian host cells, *Shigella* spp. must also express high-affinity iron acquisition systems. In order to obtain iron, *Shigella* synthesizes and transports the siderophores aerobactin and enterobactin and utilizes a receptor/transport system in which iron is obtained from heme [80]. Siderophores are low molecular weight, iron-binding compounds that remove iron from host proteins. Enterobactin is produced by some but not all *Shigella* spp., while aerobactin is synthesized by *S. flexneri*, *S. boydii*, and some *S. sonnei* [81–84]. Headley et al. demonstrated that aerobactin systems, although active in extracellular environments, are not expressed intracellularly. This suggests that siderophore-independent iron acquisition systems can provide essential iron during intracellular multiplication [85].

The capacity for *Shigella* to spread intracellularly and infect adjacent epithelial cells is critical in the infection process [74]. Intra- and intercellular spreading is controlled by the *icsA* (*virG*) gene located on the virulence plasmid. The *icsA* gene encodes the protein IcsA, which enables actin-based motility and intercellular spread [86,87]. IcsA is a surface-exposed outer membrane protein consisting of three distinctive domains: a 52-amino-acid N-terminal signal sequence, a 706-amino-acid α-domain, and a 344-amino-acid C-terminal β-core [88–90]. The α-domain is the exposed portion, while the β-core is embedded in the outer membrane [90].

IcsA is distributed at only one pole of the outer membrane surface. This asymmetrical distribution allows the polar formation of actin tails and thus directional movement of *Shigella* within host cell cytoplasm. The polar localization of IcsA is primarily affected by two events: (1) the rate of diffusion of outer membrane IcsA [91–94] and (2) the specific cleavage of IcsA by the protease IcsP (SopA) [95–97]. The rate of diffusion of outer membrane IcsA is directly affected by the O-side chains of the membrane LPS. Sandlin et al. demonstrated this relationship with a *S. flexneri* LPS mutant, which did not produce O antigen. Without O-side chains, the

LPS mutant polymerized actin in a nonpolar fashion [92]. In addition to O-side chains, the correct composition of the IcsA C-terminus α domain is also required for polar localization and polar movement of *S. flexneri*. A *S. flexneri* mutant, in that a segment of the IcsA C-terminus α domain had been deleted, was unable to polymerize actin in a polar fashion or move unidirectionally [98].

The *icsP* gene encodes the outer membrane protease IcsP (also called SopA) that cleaves laterally diffused IcsA, thus promoting polar localization [96]. An *E. coli* K-12 strain, engineered to express the *icsA* gene, was shown to diffuse IcsA along its outer membrane [99]. When the same *E. coli* K-12 strain was engineered to express the *icsP* gene with the *icsA* gene, the number of bacteria that polymerized actin at one pole increased [99]. The reader is directed to the cited references for information on molecular mechanisms of actin assembly in *Shigella* [76,88,100,101].

Intercellular spreading is dependent upon an actin-based motility mechanism [99]. *Shigella* cells first form a membrane-bound protrusion into an adjacent cell. This protrusion must distend two membranes: one from the donor cell and another from the recipient cell [103]. As the protrusion pushes further into the recipient cell, it is taken up by the recipient cell resulting in the bacteria enclosed in a double-membrane vacuole [99]. Intercellular spread is completed when *Shigella* rapidly escapes from the double-membrane vacuole, releasing it into the cytosol of the secondary cell. Monack and Theriot observed intercellular spread of an *E. coli* K-12 strain expressing the *icsA* and *icsP* genes in HeLa cells [99]. After invading adjacent cells, the *E. coli* K-12 strain was observed both enclosed in a double-membrane vacuole and free in the cytosol of host cells.

The early killing of host cells by *Shigella* requires invasion and is mediated by the genes found on the virulence plasmid. Sansonetti demonstrated the inability of noninvasive *S. flexneri* to kill host cells, whereas the invasive phenotype killed efficiently and rapidly [74]. Early killing of host cells involves metabolic events that rapidly drop the intracellular concentration of ATP, increase pyruvate, and arrest lactate production [104]. STX, a potent toxin produced by *S. dysenteriae* serotype 1, does not play a role in the early killing of host cells. Fontaine et al. constructed a Tox⁻ mutant strain of *S. dysenteriae serotype 1* and found that the mutant killed as efficiently as the wild-type strain [105].

34.4.3 SHIGELLA TOXINS

Shigella dysenteriae serotype 1 is the predominant *Shigella* sp. that produces the potent STX. Although STX is not necessary to sustain an infection, though toxin expression increases the severity of disease. Three biologic activities associated with STX are cytotoxicity, enterotoxicity, and neurotoxicity, while the one known biochemical effect is the inhibition of protein synthesis [106]. STX is considered the prototype to a family of toxins known as Shiga-like toxins (SLTs), which are similar in structure and function and share the same receptor sites. Perhaps the most widely known human pathogen that produces SLTs is *E. coli* O157:H7, which has two toxin variants (SLT I and SLT II).

STX is composed of two polypeptides: an A subunit (32,225 kDa) and five B subunits (7691 kDa each) [106]. The B subunits mediate binding to cell surface receptors, which have been identified as glycolipids containing terminal galactose-α(1–4)galactose disaccharides, such as galabiosylceramide and globotriaosylceramide (Gb3) [107,108]. The A subunit, once inside the host cell cytoplasm, acts enzymatically to cleave the N-glycosidic bond of adenine at nucleotide position 4324 in the 28S rRNA of the 60S ribosomal unit [106].

Recently, two additional enterotoxins, *Shigella* enterotoxin 1 (ShET 1) and *Shigella* enterotoxin 2 (ShET 2), have been characterized and are believed to play a role in the clinical manifestation of shigellosis [109]. ShET 1, which is chromosomally encoded, was only prevalent in isolates of *S. flexneri* 2a. ShET 2, however, is encoded on the virulence plasmid and was detected in all *Shigella* isolates tested, except several isolates that had been cured of the virulence plasmid.

34.5 DIAGNOSIS AND IDENTIFICATION

A patient presenting with severe diarrhea, fever, and possible bloody and/or mucoid stool is likely to be diagnosed with infectious bacterial colitis [110]. There are numerous pathogens that result in gastroenteritis and diarrhea symptoms, so the disease setting and symptoms should be taken into account by a clinician to help determine the specific bacterial etiology [111]. When a bacterial enteropathogen is suspected as the cause of illness, a stool sample should be obtained. The stool sample should be tested for *Salmonella*, *Campylobacter*, and *Shigella*, the three most predominately reported laboratory strains causing gastroenteritis in the United States per year [31]. The samples should also be analyzed with a toxin assay upon the occurrence of bloody stools. The four major bacterial pathogens that cause bloody stool in the United States are *Shigella*, *Campylobacter*, *Salmonella*, and Shiga toxin-producing *E. coli* (STEC) [112]. Stools cultures are often not collected in the case of watery diarrhea due to the low concentration of pathogens [111].

Initial isolation of the bacterial agent is often performed on selective and differential media. The patient sample is streaked onto selective and/or differential media and incubated in an aerobic environment to inhibit growth of anaerobic gut microbiota. There are a variety of media that can be used for the isolation of *Shigella* including MacConkey, Hektoen enteric (HE) agar, xylose lysine deoxycholate (XLD) agar, and *Salmonella–Shigella* (SS) agar [113]. If necessary, an enrichment media may be used to increase population levels before isolation on agar plates. Upon isolation of putative *Shigella* colonies, the cultures should be used to inoculate Kligler's iron agar or triple sugar iron agar slants for presumptive identification. The production of an alkaline slant, acid butt, and no gas bubble formation in either agar is considered a presumptive positive *Shigella* isolate [113].

Further, traditional microbiological assays for the identification of *Shigella* include agglutination assays. The O antigen is a component of the LPSs of Gram-negative bacteria, which is also used to differentiate bacteria into serogroups. As previously stated, *Shigella* is divided into four serogroups, *S. dysenteriae* (serogroup A), *S. flexneri* (serogroup B), *S. boydii* (serogroup C), and *S. sonnei* (serogroup D). Agglutination assays utilize polyclonal antisera specific to group A, B, C, or D, where cell clumping indicates a positive identification of the culture subgroup [114]. Agglutination assays are often necessary to identify *Shigella* strains because of its close relatedness to *E. coli*.

Further methods of analysis for the identification of *Shigella* from human samples include toxin assays and polymerase chain reaction (PCR). Enzyme-linked immunosorbent assays (ELISA) have been developed to target and identify the presence of the Shigella toxin using prepared antitoxin serum [115]. The bacterial agent should not be positively identified as *Shigella* until an agglutination assay is performed since the ELISA assay may not distinguish between the *Shigella* toxin and the SLTs produced by some STEC strains. The utilization of PCR is also a rapid method of *Shigella* detection. PCR assays for *Shigella* spp. have targeted the invasion-associated locus (*ial*), the *virA* gene, or the *ipaH* gene for amplification [26,116–122]. The same PCR primers are used to detect each of the serogroups of *Shigella* and enteroinvasive *E. coli*. The *ial* (now referred to as the *mxi/spa* genes) and *virA* genes are located on the large virulence plasmid (sometimes referred to as the invasion plasmid); however, sequencing of the *S. flexneri* genome revealed the *ipaH* gene to be encoded multiple times on both the chromosome and the large virulence plasmid [123]. Thus, the major advantage of using *ipaH* genes as targets for PCR is twofold; first, there are multiple genes, nine in total, which from the onset acts as an amplification system, and secondly, there are five copies of the *ipaH* gene in the chromosome that can be amplified in the event the virulence plasmid is lost in the pathogen either during prolonged storage or during cultivation [116]. Various forms of PCR, including multiplex PCR, nested PCR, immuno-capture PCR, Rep-PCR, and ELISA-PCR, have been described for rapid identification of *Shigella* [124].

This discussion is limited to the general identification and diagnosis of *Shigella* from clinical samples; however, more detailed methods of detection of *Shigella* in clinical and food samples appear in the literature. While this was not discussed, in the case of a foodborne illness, it is also important to isolate and detect the pathogen from the food source. Although *Shigella* is readily isolated from clinical samples, food samples are more problematic. Isolation of *Shigella* from food samples can be inhibited by indigenous microflora, especially the coliform bacteria and *Proteus* spp. [7]. *Shigella* identification through ELISA assays can be disrupted by the presence of proteases or antiglobulin antibodies [115]. PCR methods can also be unsatisfactory in distinguishing between *Shigella* and *E. coli* or discriminating between the species of *Shigella* [123].

34.6 TREATMENT AND PREVENTION

34.6.1 TREATMENT OF SHIGELLOSIS

Once a diagnosis of shigellosis is determined, attention to fluid and electrolyte replacement is important, as in all cases of diarrhea. Along with rehydration, feeding is encouraged during and after shigellosis. There is no current data on the effect in adults, but one study supported the feeding of children suffering from acute diarrhea [125]. Shigellosis can be self-limiting with the proper hydration and feeding during the onset and throughout the illness. The severity of disease, age of the patient, and likelihood of further transmission of shigellosis should be considered in deciding any further treatment options [113]. The use of antibiotic treatments serves to alleviate dysenteric syndromes including fever and abdominal cramps and decreases the average duration of pathogen excretion, therefore reducing the risk of transmission and/or lethal complications [126]. Effective and prompt antibiotic treatments can reduce the average illness from 5–7 days to 3 days [113]. Unfortunately, resistance to multiple antibiotics among Shigella strains has progressively increased [127].

In the past, ampicillin and trimethoprim–sulfamethoxazole (TMP-SMX) were the antibiotic treatments preferred for the treatment of shigellosis. The antibiotics could be safely administered to ill children, were inexpensive, and attained concentrations in the serum and gut lumen that were capable of inhibiting Shigella proliferation [128]. Unfortunately, antibiotic resistance of Shigella has become an increasing problem in both undeveloped and developed countries. Among Shigella isolates from five Canadian provinces (1997–2000), high rates of resistance to ampicillin (65%) and TMP-SMX (70%) were observed [129]. Similarly, Shigella isolates in the United States (1999–2002) were resistant to ampicillin (78%) and TMP-SMX (46%) [130].

In the United States, Shigella isolates remain susceptible to ciprofloxacin and ceftriaxone [130]. The fluoroquinolone-based antibiotics are not widely approved for use in children and are more costly than the previously used antibiotics, and indiscriminate use of the antibiotics could result in rapid development of antibiotic resistance in Shigella [128]. The WHO has recommended ciprofloxacin as the first-line antibiotic for treatment of shigellosis, although fluoroquinolone-resistant Shigella strains have been documented in several countries [131]. The rise in fluoroquinolone-resistant Shigella is relatively low in the American and European regions, but resistance to ciprofloxacin in the Asia and African regions has progressively increased, reaching 29.1% in 2007–2009 [132]. In 2010, up to 40% of tested Shigella isolates were found to be resistant to three or more classes of antibiotics [133]. This reinforces the importance of responsible use of antibiotic treatments and monitoring antibiotic resistances in Shigella spp. There is need for a Shigella vaccine. There are both polysaccharide conjugate and live attenuated vaccines currently under advanced development; however, progress is limited in part due to the lack of a sufficient animal model for vaccine analysis [134].

34.6.2 PREVENTION OF SHIGELLOSIS

Shigella is an exclusively human disease transmitted directly through the fecal–oral route or indirectly through contaminated food, water, or fomites. The exclusive spread of the pathogen highlights the importance of personal hygiene. Perhaps the most effective way shigellosis can be directly prevented is by diligent adherence to proper hand washing. This is especially important in day-care settings or where large groups are in close quarters, such as cruise ships or institutional living facilities. These dangers have been documented in the previous shigellosis outbreaks occurring in these settings [41,135]. In the case of contacting Shigella through an indirect source, Shigella is capable of survival on produce commodities, in packaged produce, and in fruit juices along with prepared foods at both room and refrigeration temperatures [16–27]. Preventative measures against shigellosis include thoroughly washing fruits and vegetables, properly storing and refrigerating raw and prepared foods, purging foods past their expiration date, thoroughly cooking and reheating prepared foods, and frequently washing hands with soap and water, especially when preparing or consuming foods. Furthermore, someone ill with shigellosis or any diarrheal disease should not be preparing food. If a water source is suspected of containing Shigella, the water should be boiled before use or consumption. Chlorination and treatment of water also reduces Shigella contamination [13–15]. The survival of Shigella on fomites again highlights the importance of frequent hand washing throughout the day and minimizing hand-to-face contact.

34.7 CONCLUSION

Shigella spp. continues to be a significant agent of foodborne illness in the United States. Shigella has been classically characterized as a waterborne pathogen, with outbreaks occurring from contaminated community water sources that were insufficiently chlorinated. Along with the connections to water and foods, there is some information indicating that the prevalence of Shigella on food or among food handlers remains a concern in regard to food safety. Several examples of food products from which Shigella has been isolated include potato salad, ground beef, bean dip, raw oysters, fish, and raw vegetables. The low infectious dose of Shigella also allows common person-to-person transmission, especially in community settings. Community-wide infection is a concern in areas where food is distributed to a large group or in facilities such as day-care centers where fecal contact is facilitated through diaper changes. Despite the multiple routes of Shigella contamination, the best prevention against illness is to remain diligent with personal hygiene and to understand the sources and risks of illness. The increasing antibiotic resistance of Shigella is of grave concern, but a profound reduction of foodborne shigellosis may depend on the education of employees and consumers as to proper personal hygiene.

ACKNOWLEDGMENTS

Portions of the material presented in this chapter have been previously published by Taylor & Francis (www.informaworld.com) in Warren, B. R., Parish, M. E., and K. R. Schneider. 2006. *Shigella* as a foodborne pathogen and current methods for detection. *Crit. Rev. Food. Sci. Nutr.* 46:551–567, and by CRC Press (www.crcnetbase.com) in Schneider, K. R., Strawn, L. K., Lampel, K. K., and B. R. Warren. 2011. *Shigella*. In: *Molecular Detection of Human Bacterial Pathogens*, ed. Liu, D., 1049–1068. CRC Press.

REFERENCES

1. Scallan, E., Hoekstra, R.M., Angulo, F.J. et al. 2011. Foodborne illness acquired in the United States—Major pathogens. *Emerg. Infect. Dis.* 17:7–15.
2. Levine, M.M., DuPont, H.L., Khodabandelou, M. et al. 1973. Long-term *Shigella*-carrier state. *N. Engl. J. Med.* 288:1169–1171.
3. Hossain, M.A., Hasan, K.Z., and Albert, M.J. 1994. *Shigella* carriers among non-diarrhoeal children in an endemic area of shigellosis in Bangladesh. *Trop. Geogr. Med.* 46:40–42.
4. Shiga, K. 1898. Ueber den Dysenterie-*Bacillus* (*Bacillus dysenteriae*). *Zentralbl. Bakteriol. Orig.* 24:913–918.
5. Lampel, K.A. 2001. *Shigella*. In: *Compendium of Methods for the Microbiological Examination of Foods*, 4th ed., Downes, F.P. and Ito, K. (eds.), pp. 381–385. Washington, DC: American Public Health Association.
6. Echeverria, P., Sethabutr, O., and Pitarangsi, C. 1991. Microbiology and diagnosis of infections with *Shigella* and enteroinvasive *Escherichia coli*. *Rev. Infect. Dis.* 13:S220–S225.
7. International Commission on Microbiological Specifications for Foods (ICMSF). 1996. *Shigella*. In: *Microorganisms in Foods*, 5th ed., Roberts, T.A., Baird-Parker, A.C., and Tompkins, R.B. (eds.), pp. 280–298. London, U.K.: Chapman & Hall, Blackie Academic & Professional.
8. Ingersoll, M., Groisman, E.A., and Zychlinsky, A. 2002. Pathogenicity islands of *Shigella*. *Curr. Top. Microbiol. Immunol.* 264:49–65.
9. Centers for Disease Control and Prevention (CDC). 2009. Disease information: Shigellosis. http://www.cdc.gov/nczved/divisions/dfbmd/diseases/shigellosis/technical.html (accessed July 21, 2012).
10. Hale, T. 1991. Genetic basis of virulence in *Shigella* species. *Microbiol. Rev.* 55:206–224.
11. Centers for Disease Control and Prevention (CDC). 2012. National *Shigella* Surveillance System. National *Shigella* annual summaries. http://www.cdc.gov/nationalsurveillance/shigella_surveillance.html (accessed July 21, 2012).
12. Smith, J.L. 1987. *Shigella* as a foodborne pathogen. *J. Food Prot.* 50:788–810.
13. Blostein, J. 1991. Shigellosis from swimming in a park pond in Michigan. *Public Health Rep.* 106:317–322.
14. Centers for Disease Control and Prevention (CDC). 2001. Shigellosis outbreak associated with an unchlorinated fill-and-drain wading pool—Iowa. *Morb. Mortal. Wkly. Rep.* 50:797–800.
15. Fleming, C.A., Caron, D., Gunn, J. et al. 2000. An outbreak of *Shigella sonnei* associated with a recreational spray fountain. *Am. J. Public Health.* 90:1641–1642.
16. Wu, F.M., Doyle, M.P., Beuchat, L.R. et al. 2000. Fate of *Shigella sonnei* on parsley and methods of disinfection. *J. Food Prot.* 63:568–572.
17. Evans, D.A., Hankinson, D.J., and Litsky, W. 1970. Heat resistance of certain pathogenic bacteria in milk using a commercial plate heat exchanger. *J. Dairy Sci.* 53:1659–1665.
18. Zaika, L. 2001. The effect of temperature and low pH on survival of *Shigella flexneri* in broth. *J. Food Prot.* 64:1162–1165.
19. Zaika, L. 2002. The effect of NaCl on survival of *Shigella flexneri* in broth as affected by temperature and pH. *J. Food Prot.* 65:774–779.
20. Zaika, L. 2002. Effect of organic acids and temperature on survival of *Shigella flexneri* in broth at pH 4. *J. Food Prot.* 65, 1417–1421.
21. Bagamboula, C., Uyttendaele, M., and Debevere, J. 2002. Acid tolerance of *Shigella sonnei* and *Shigella flexneri*. *J. Appl. Microbiol.* 93:479–486.
22. Rafi, F. and Lunsford, P. 1997. Survival and detection of *Shigella flexneri* in vegetables and commercially prepared salads. *J. AOAC Int.* 80:1191–1197.
23. Escartin, E.F., Ayala, A.C., and Lozano, J.S. 1989. Survival and growth of *Salmonella* and *Shigella* on sliced fresh fruit. *J. Food Prot.* 52:471–472.
24. Satchell, F.B., Stephenson, P., Andrews, W.H. et al. 1990. The survival of *Shigella sonnei* in shredded cabbage. *J. Food Prot.* 53:558–562.
25. Tollison, S.B. and Johnson, M.G. 1985. Sensitivity to bile salts of *Shigella flexneri* sublethally heat stressed in buffer or broth. *Appl. Environ. Microbiol.* 50:337–341.
26. Islam, D., Tzipori, S., Islam, M. et al. 1993. Rapid detection of *Shigella dysenteriae* and *Shigella flexneri* in faeces by an immunomagnetic assay with monoclonal antibodies. *Eur. J. Clin. Microbiol. Infect. Dis.* 12:25–32.
27. Nakamura, M. 1962. The survival of *Shigella sonnei* on cotton, glass, wool, paper and metal at various temperatures. *J. Hyg.* 60:35–39.
28. Centers for Disease Control and Prevention (CDC). 2004. National *Shigella* Surveillance Data. *Shigella* annual summary, 2004. http://www.cdc.gov/ncidod/dbmd/phlisdata/shigella.htm (accessed July 22, 2012).
29. Shiferaw, B., Shallow, S., Marcus, R. et al. 2004. Trends in population-based active surveillance for shigellosis and demographic variability in FoodNet sites, 1996–1999. *Clin. Infect. Dis.* 38:S175–S180.
30. Centers for Disease Control and Prevention (CDC). 2011. Vital signs: Incidence and trends of infection with pathogens transmitted commonly through food—Foodborne disease active surveillance network, 10 U.S. sites, 1996–2010. *Morb. Mortal. Wkly. Rep.* 60:749–755.
31. Centers for Disease Control and Prevention (CDC). 2010. Preliminary FoodNet data on the incidence of infection with pathogens transmitted commonly through food—10 States, United States, 2009. *Morb. Mortal. Wkly. Rep.* 59:418–422.
32. Centers for Disease Control and Prevention (CDC). 2006. Preliminary FoodNet data on the incidence of infection with pathogens transmitted commonly through food—10 States, United States, 2005. *Morb. Mortal. Wkly. Rep.* 55:392–395.
33. Kimura, A.C., Johnson, K., Palumbo, M.S. et al. 2004. Multistate shigellosis outbreak and commercially prepared foods, United States. *Emerg. Infect. Dis.* 10:1147–1149.
34. DuPage County Health Department (DCHD). 2010. Subway *Shigella sonnei* outbreak: Investigation Report IL2010-060. http://www.shigellablog.com/uploads/image/DUPAGE%20 Outbreak%20Report.pdf (accessed on July 28, 2012).

35. Martin, D.L., Gustafson, T.L., Pelosi, J.W. et al. 1986. Contaminated produce—A common source for two outbreaks of *Shigella* gastroenteritis. *Am. J. Epidemiol.* 124:299–305.

36. Davis, H., Taylor, J.P., Perdue, J.N. et al. 1988. A shigellosis outbreak traced to commercially distributed shredded lettuce. *Am. J. Epidemiol.* 128:1312–1321.

37. Reeve, G., Martin, D.L., Pappas, J. et al. 1989. An outbreak of shigellosis associated with the consumption of raw oysters. *N. Engl. J. Med.* 321:224–227.

38. Fredlund, H., Back, E., Sjoberg, L. et al. 1987. Water-melon as a vehicle of transmission of shigellosis. *Scand. J. Infect. Dis.* 19:219–221.

39. Lee, L.A., Ostroff, S.M., McGee, H.B. et al. 1991. An outbreak of shigellosis at an outdoor music festival. *Am. J. Epidemiol.* 133:608–615.

40. Yagupsky, P., Loeffelholz, M., Bell, K. et al. 1991. Use of multiple markers for investigation of an epidemic of *Shigella sonnei* infections in Monroe County, New York. *J. Clin. Microbiol.* 29:2850–2855.

41. Lew, J.F., Swerdlow, D.L., Dance, M.E. et al. 1991. An outbreak of shigellosis aboard a cruise ship caused by a multiple-antibiotic-resistant strain of *Shigella flexneri*. *Am. J. Epidemiol.* 134:413–420.

42. Dunn, R.A., Hall, W.N., Altamirano, J.V. et al. 1995. Outbreak of *Shigella flexneri* linked to salad prepared at a central commissary in Michigan. *Public Health Rep.* 110:580–586.

43. Frost, J.A., McEvoy, M.B., Bentley, C.A. et al. 1995. An outbreak of *Shigella sonnei* infection associated with consumption of iceberg lettuce. *Emerg. Infect. Dis.* 1:26–29.

44. Kapperud, G., Rorvik, L.M., Hasseltvedt, V. et al. 1995. Outbreak of *Shigella sonnei* infection traced to imported iceberg lettuce. *J. Clin. Microbiol.* 33:609–614.

45. Long, S.M., Adak, G.K., O'Brien, S.J. et al. 2002. General outbreaks of infectious intestinal disease linked with salad vegetables and fruit, England and Wales, 1992–2000. *Commun. Dis. Public Health.* 5:101–105.

46. Garcia-Fulgueiras, A. 2001. A large outbreak of *Shigella sonnei* gastroenteritis associated with consumption of fresh pasteurized milk cheese. *Eur. J. Epidemiol.* 17:533–538.

47. Public Health Laboratory Service (PHLS). 1997. Outbreaks of foodborne illness in humans, England and Wales: Quarterly Report. *CDR Wkly.* 7:207–216.

48. Centers for Disease Control and Prevention (CDC). 1999. Outbreaks of *Shigella sonnei* infection associated with eating fresh parsley—United States and Canada, July–August 1998. *Morb. Mortal. Wkly. Rep.* 48:285–289.

49. Trevejo, R.T., Abbott, S.L., Wolfe, M.I. et al. 1999. An untypeable *Shigella flexneri* strain associated with an outbreak in California. *J. Clin. Microbiol.* 37:2352–2353.

50. Chan, Y.C. and Blaschek, H.P. 2005. Comparative analysis of *Shigella boydii* 18 foodborne outbreak isolate and related enteric bacteria: Role of rpoS and adiA in acid stress response. *J. Food Prot.* 68:521–527.

51. Terajima, J., Tamura, K., Hirose, K. et al. 2004. A multi-prefectural outbreak of *Shigella sonnei* infections associated with eating raw oysters in Japan. *Microbiol. Immunol.* 48:49–52.

52. Reller, M., Nelson, J.M., Molbak, K. et al. 2006. A large, multiple-restaurant outbreak of infection with *Shigella flexneri* serotype 2a traced to tomatoes. *Clin. Infect. Dis.* 42:163–169.

53. Public Health Agency of Canada (PHAC). 2009. Canadian integrated surveillance report: *Salmonella, Campylobacter*, verotoxigenic *E. coli* and *Shigella*, from 2000 to 2004. *CCDR.* 35-S3:35–40.

54. Gaynor, K., Park, S.Y., Kanenaka, R. et al. 2009. International foodborne outbreak of *Shigella sonnei* infection in airline passengers. *Epidemiol. Infect.* 137:335–341.

55. Center for Science in the Public Interest (CSPI). 2009. Outbreak alert database. http://www.cspinet.org/foodsafety/outbreak/pathogen.php (accessed July 28, 2012).

56. Lewis, H.C., Ethelberg, S., Olsen, K.E. et al. 2009. Outbreaks of *Shigella sonnei* infections in Denmark and Australia linked to consumption of imported raw baby corn. *Epidemiol. Infect.* 137:326–334.

57. Hung-Wei, K., Kasper, S., Jelovcan, S. et al. 2009. A foodborne outbreak of *Shigella sonnei* gastroenteritis, Austria, 2008. *Middle Eur. J. Med.* 121:157–163.

58. Heier, B.T., Nygard, K., Kapperud, G. et al. 2009. *Shigella sonnei* infections in Norway associated with sugar peas, May–June 2009. *Euro Surveill.* 14:1–2.

59. Gutierrez Garitano, I., Naranjo, M., Forier, A. et al. 2011. Shigellosis outbreak linked to canteen-food consumption in a public institution: A matched case–control study. *Epidemiol. Infect.* 139:1956–1964.

60. Guzman-Herrador, B., Vold, L., Comelli, H. et al. 2011. Outbreak of *Shigella sonnei* infection in Norway linked to consumption of fresh basil, October 2011. *Euro Surveill.* 16:1–2.

61. al-Lahham, A.B., Abu-Saud, M., and Shehabi, A.A. 1990. Prevalence of *Salmonella, Shigella* and intestinal parasites in food handlers in Irbid, Jordan. *J. Diarrhoeal Dis. Res.* 8:160–162.

62. Mensah, P., Yeboah-Manu, D., Owusu-Darko, K. et al. 2002. Street foods in Accra, Ghana: How safe are they? *Bull. World Health Organ.* 80:546–554.

63. Wood, L.V., Ferguson, L.E., Hogan, P. et al. 1983. Incidence of bacterial enteropathogens in foods from Mexico. *Appl. Environ. Microbiol.* 46:328–332.

64. Food and Drug Administration (FDA). 2003. FDA survey of domestic fresh produce, FY 2000/2001 field assignment. http://www.fda.gov/Food/FoodSafety/Product-SpecificInformation/FruitsVegetablesJuices/GuidanceComplianceRegulatoryInformation/ucm118306.htm (accessed July 28, 2012).

65. Food and Drug Administration (FDA). 2001. FDA survey of imported fresh produce, FY 1999 field assignment. http://www.fda.gov/Food/FoodSafety/Product-SpecificInformation/FruitsVegetablesJuices/GuidanceComplianceRegulatoryInformation/ucm118891.htm (accessed July 28, 2012).

66. Kothary, M.H. and Babu, U.S. 2001. Infective dose of foodborne pathogens in volunteers: A review. *J. Food Saf.* 21:49–68.

67. Acheson, D.W. and Keusch, G.T. 2002. *Shigella* and enteroinvasive *Escherichia coli*. In: *Infections of the Gastrointestinal Tract*, Blaser, M., Smith, P., and Raudin, J. (eds.), pp. 763–784. New York: Raven Press.

68. Keusch, G.T. and Bennish, M.L. 1989. Shigellosis: Recent progress, persisting problems and research issues. *J. Pediatr. Infect Dis.* 8:713–719.

69. Keusch, G.T., Formal, S.B., and Bennish, M.L. 1990. Shigellosis. In: *Tropical and Geographical Medicine*, Warren, K.S. and Mahmoud, A.A.F. (eds.), pp. 762–776. New York: McGraw-Hill.

70. Stoll, B.J., Glass, R.I., Huq, M.I. et al. 1982. Epidemiologic and clinical features of patients infected with *Shigella* who attended a diarrheal disease hospital in Bangladesh. *J. Infect. Dis.* 146:177–183.

71. Phalipon, A. and Sansonetti, P.J. 2007. *Shigella*'s way of manipulating the host intestinal innate and adaptive immune system: A tool box for survival? *Immunol. Cell Biol.* 85:119–129.

72. Galanakis, E., Tzoufi, M., Charisi, M. et al. 2002. Rate of seizures in children with shigellosis. *Acta Paediatr.* 91:101–102.

73. Colmegna, I., Cuchacovish, R., and Espinoza, L.R. 2004. HLA-B27-Associated reactive arthritis: Pathogenetic and clinical considerations. *Clin. Microbiol. Rev.* 17:348–369.

74. Sansonetti, P.J. 1991. Genetic and molecular basis of epithelial cell invasion by *Shigella* species. *Rev. Infect. Dis.* 13:S285–S292.

75. Tamano, K., Aizawa, S., Katayama, E. et al. 2002. Supramolecular structure of the *Shigella* type III secretion machinery: The needle part is changeable in length and essential for delivery of effectors. *EMBO J.* 19:3876–3887.

76. Suzuki, T., and Sasakawa, C. 2001. Molecular basis of the intracellular spreading of *Shigella*. *Infect. Immun.* 69:5959–5966.

77. Zychlinsky, A., Fitting, C., Cavaillon, J. et al. 1994. Interleukin 1 is released by murine macrophages during apoptosis induced by *Shigella flexneri*. *J. Clin. Invest.* 94:1328–1332.

78. Stendahl, O.I., Hartwig, J.H., Brotschi, E.A. et al. 1980. Distribution of actin-binding protein and myosin in macrophages during spreading and phagocytosis. *J. Cell. Biol.* 84: 215–224.

79. Sansonetti, P.J., Ryter, A., Clerc, P. et al. 1986. Multiplication of *Shigella flexneri* within HeLa cells: Lysis of the phagocytic vacuole and plasmid-mediated contact hemolysis. *Infect. Immun.* 54:461–469.

80. Vokes, S.A., Reeves, S.A., Torres, A.G. et al. 1999. The aerobactin iron transport system genes in *Shigella flexneri* are present within a pathogenicity island. *Mol. Microbiol.* 33:63–73.

81. Lawlor, K.M. and Payne, S.M. 1984. Aerobactin genes in *Shigella* spp. *J. Bacteriol.* 160:266–272.

82. Payne, S.M. 1988. Iron and virulence in the family Enterobacteriaceae. *CRC Crit. Rev. Microbiol.* 16:81–111.

83. Payne, S.M., Niesel, D.W., Peixotto, S.S. et al. 1983. Expression of hydroxamate and phenolate siderophores by *Shigella flexneri*. *J. Bacteriol.* 155:949–955.

84. Perry, R.D. and San Clemente, C.L. 1979. Siderophore synthesis in *Klebsiella pneumoniae* and *Shigella sonnei* during iron deficiency. *J. Bacteriol.* 140:1129–1132.

85. Headley, V., Hong, M., Galko, M. et al. 1997. Expression of aerobactin genes by *Shigella flexneri* during extracellular and intracellular growth. *Infect. Immun.* 65:818–821.

86. Bernardini, M.L., Mounier, J., d'Hauteville, H. et al. 1989. Identification of icsA, a plasmid locus of *Shigella flexneri* that governs bacterial intra- and intercellular spread though interaction with F-actin. *Proc. Natl. Acad. Sci. USA* 86:3867–3871.

87. Makino, S., Sasakawa, C., Kamata, K. et al. 1986. A genetic determinate required for continuous reinfection of adjacent cells on a large plasmid of *Shigella flexneri* 2a. *Cell.* 46:551–555.

88. Goldberg, M.B., Barzu, O., Parsot, C. et al. 1993. Unipolar localization and ATPase activity of IcsA, a *Shigella flexneri* protein involved in intracellular movement. *J. Bacteriol.* 175:2189–2196.

89. Lett, M.C., Yang, X.F., Coster, T.S. et al. 1989. *virG*, a plasmid-coded virulence gene of *Shigella flexneri*: Identification of the *virG* protein and determination of the complete coding sequence. *J. Bacteriol.* 171:353–359.

90. Suzuki, T., Lett, M.C., and Sasakawa, C. 1995. Extracellular transport of VirG protein in *Shigella*. *J. Biol. Chem.* 270:30874–30880.

91. Sandlin, R.C. and Maurelli, A.T. 1999. Establishment of unipolar localization of IcsA in *Shigella flexneri* 2a is not dependent on virulence plasmid determinants. *Infect. Immun.* 67:350–356.

92. Sandlin, R.C., Goldberg, M.B., and Maurelli, A.T. 1996. Effect of O side-chain length and composition on the virulence of *Shigella flexneri* 2a. *Mol. Microbiol.* 22:63–73.

93. Sandlin, R.C., Lampel, K.A., Keasler, S.P. et al. 1995. Avirulence of rough mutants of *Shigella flexneri*: Requirement of O antigen for correct unipolar localization of IcsA in the bacterial outer membrane. *Infect. Immun.* 63:229–237.

94. Robbins, J.R., Monack, D., McCallum, S.J. et al. 2001. The making of a gradient: IcsA (VirG) polarity in *Shigella flexneri*. *Mol. Microbiol.* 41:861–872.

95. d'Hauteville, H., Dufourcq Lagelouse, R., Nato, F. et al. 1996. Lack of cleavage of IcsA in *Shigella flexneri* causes aberrant movement and allows demonstration of a cross-reactive eukaryotic protein. *Infect. Immun.* 64:511–517.

96. Egile, C., d'Hauteville, H., Parsot, C. et al. 1997. SopA, the outer membrane protease responsible for polar localization of IcsA in *Shigella flexneri*. *Mol. Microbiol.* 23:1063–1073.

97. Steinhauer, J., Agha, R., Pham, T. et al. 1999. The unipolar *Shigella* surface protein IcsA is targeted directly to the bacterial old pole: IcsP cleavage of IcsA occurs over the entire bacterial surface. *Mol. Microbiol.* 32:367–377.

98. Suzuki T., Saga, S., and Sasakawa, C. 1996. Functional analysis of *Shigella* VirG domains essential for interaction with vinculin and actin-based motility. *J. Biol. Chem.* 271:21878–21885.

99. Monack, D.M. and Theriot, J.A. 2001. Actin-based motility is sufficient for bacterial membrane protrusion formation and host cell uptake. *Cell. Microbiol.* 3:633–647.

100. Loisel, T.P., Boujemaa, R., Pantaloni, D. et al. 1999. Reconstitution of actin-based motility of *Listeria* and *Shigella* using pure proteins. *Nature.* 401:613–616.

101. Miki, H., Miura, K., and Takenawa, T. 1996. N-WASP, a novel actin-depolymerizing protein, regulates the cortical cytoskeletal rearrangement in a PIP2-dependent manner downstream of tyrosine kinases. *EMBO J.* 15:5326–5335.

102. Shibata, T., Takeshima, F., Chen, F. et al. 2002. Cdc42 facilitates invasion but not the actin-based motility of *Shigella*. *Curr. Biol.* 12:341–345.

103. Parsot, C. and Sansonetti, P.J. 1996. Invasion and the pathogenesis of *Shigella* infections. *Curr. Top. Microbiol. Immunol.* 209:25–42.

104. Sansonetti, P.J. and Mounier, J. 1987. Metabolic events mediating early killing of host cells infected by *Shigella flexneri*. *Microbiol. Pathog.* 3:53–61.

105. Fontaine, A., Arondale, J., and Sansonetti, P.J. 1988. Role of shiga toxin in the pathogenesis of bacillary dysentery studied using a tox- mutant of *Shigella dysenteriae* 1. *Infect. Immun.* 56:3099–3109.

106. Donohue-Rolfe, A., Acheson, D.W., and Keusch, G.T. 1991. Shiga toxin: Purification, structure, and function. *Rev. Infect. Dis.* 13:S293–S297.

107. Brown, J.E., Echeverria, P., and Lindberg, A.A. 1991. Digalactosyl-containing glycolipids as cell surface receptors for shiga toxin of *Shigella dysenteriae* 1 and related cytotoxins of *Escherichia coli*. *Rev. Infect. Dis.* 13:S298–S303.

108. Keusch, G.T., Jacewicz, M., Mobassaleh, M. et al. 1991. Shiga toxin: Intestinal cell receptors and pathophysiology of enterotoxic effects. *Rev. Infect. Dis.* 13:S304–S310.

109. Yavzori, M., Cohen, D., and Orr, N. 2002. Prevalence of the genes for *Shigella* enterotoxins 1 and 2 among clinical isolates of *Shigella* in Israel. *Epidemiol. Infect.* 128:533–535.

110. Pfeiffer, M.L., DuPont, H.L., and Ochoa, T.J. 2012. The patient presenting with acute dysentery—A systematic review. *J. Infect.* 64:374–386.

111. DuPont, H.L. 2009. Bacterial diarrhea. *N. Engl. J. Med.* 361:1560–1569.

112. Talan, D., Morgan, G.J., Newdow, M. et al. 2001. Etiology of bloody diarrhea among patients presenting to United States emergency departments: Prevalence of *Escherichia coli* O157:H7 and other enteropathogens. *Clin. Infect. Dis.* 32:573–580.

113. Hale, T.L. and Keusch, G.T. 1996. *Shigella*. In: *Medical Microbiology*, 4th ed., Baron, S. (ed.). Galveston, TX: University of Texas Medical Branch.

114. Lefebvre, J., Gosselin, F., Ismail, J. et al. 1995. Evaluation of commercial antisera for *Shigella* serogrouping. *J. Clin. Microbiol.* 33:1997–2001.

115. Donohue-Rolfe, A., Kelley, M.A., Bennish, M. et al. 1986. Enzyme-linked immunosorbent assay for *Shigella* toxin. *J. Clin. Microbiol.* 24:65–68.

116. Lampel, K.A. and Orlandi, P.A. 2002. Polymerase chain reaction detection of invasive *Shigella* and *Salmonella enterica* in food. *Methods Mol. Biol.* 179:235–244.

117. Lindqvist, R. 1999. Detection of *Shigella* spp. in food with a nested PCR method—Sensitivity and performance compared with a conventional culture method. *J. Appl. Microbiol.* 86:971–978.

118. Sethabutr, O., Venkatesan, M., Murphy, G.S. et al. 1993. Detection of Shigellae and enteroinvasive *Escherichia coli* by amplification of the invasion plasmid antigen H DNA sequence in patients with dysentery. *J. Infect. Dis.* 167:458–461.

119. Sethabutr, O., Venkatesan, M., Yam, S. et al. 2000. Detection of PCR products of the *ipaH* gene from *Shigella* and enteroinvasive *Escherichia coli* by enzyme linked immunosorbent assay. *Diagn. Microbiol. Infect. Dis.* 37:11–16.

120. Theron, J., Morar, D., Du Preez, M. et al. 2001. A sensitive seminested PCR method for the detection of *Shigella* in spiked environmental water samples. *Water Res.* 35:869–874.

121. Vantarakis, A., Komninou, G., Venieri, D. et al. 2000. Development of a multiplex PCR detection of *Salmonella* spp. and *Shigella* spp. in mussels. *Lett. Appl. Microbiol.* 31:105–109.

122. Villalobo, E. and Torres, A. 1998. PCR for the detection of *Shigella* spp. in mayonnaise. *Appl. Environ. Microbiol.* 64:1242–1245.

123. Jin, Q., Yuan, Z., Xu, J. et al. 2002. Genome sequence of *Shigella flexneri* 2a: Insights into pathogenicity through comparison with genomes of *Escherichia coli* K12 and O157. *Nucleic Acids Res.* 30:4432–4441.

124. Levin, R.E. 2009. Molecular methods for detecting and discriminating *Shigella* associated with foods and human clinical infections—A review. *Food Biotechnol.* 23:214–228.

125. Brown, K.H., Gastanaduy, A.S., Saavedra, J.M. et al. 1988. Effect of continued oral feeding on clinical and nutritional outcomes of acute diarrhea in children. *J. Pediatr.* 112:191–200.

126. Salam, M.A. and Bennish, M.L. 1991. Antimicrobial therapy for shigellosis. *Rev. Infect. Dis.* 13:S332–S341.

127. Woodward, D.L. and Rodgers, F.G. 2000. Surveillance of antimicrobial resistance in *Salmonella*, *Shigella* and *Vibrio cholerae* in Latin America and the Caribbean: A collaborative project. *Can. J. Infect. Dis.* 11:181–186.

128. Bennish, M.L. and Salam, M.A. 1992. Rethinking options for the treatment of shigellosis. *J. Antimicrob. Chemother.* 30:243–247.

129. Martin, L.J., Flint, J., Ravel, A. et al. 2006. Antimicrobial resistance among *Salmonella* and *Shigella* isolates in five Canadian provinces (1997 to 2000). *Can. J. Infect. Dis. Med. Microbiol.* 17:243–250.

130. Sivapalasingham, S., Nelson, J.M., Joyce, K. et al. 2006. High prevalence of antimicrobial resistance among *Shigella* isolates in the United States tested by the National Antimicrobial Resistance Monitoring System from 1999 to 2002. *Antimicrob. Agents Chemother.* 50:49–54.

131. Niyogi, S.K. 2007. Increasing antimicrobial resistance—An emerging problem in the treatment of shigellosis. *Clin. Microbiol. Infect. Dec.* 13:1141–1143.

132. Gu, B., Cao, Y., Pan, S. et al. 2012. Comparison of the prevalence and changing resistance to nalidixic acid and ciprofloxacin of *Shigella* between Europe–America and Asia–Africa from 1998 to 2009. *Int. J. Antimicrob. Agents* 40:9–17.

133. Centers for Disease Control and Prevention (CDC). 2010a. National antimicrobial resistance monitoring system (NARMS): Enteric bacteria. Human isolates final report—2010. http://www.cdc.gov/narms/pdf/2010-annual-report-narms.pdf (accessed September 15, 2012).

134. Kweon, M.N. 2008. Shigellosis: The current status of vaccine development. *Curr. Opin. Infect. Dis.* 21:313–318.

135. Centers for Disease Control and Prevention (CDC). 1992. Shigellosis in child day care centers—Lexington–Fayette County, Kentucky. *Morb. Mortal. Wkly. Rep.* 41:440–442.

35 *Vibrio cholerae*

Dongyou Liu

CONTENTS

35.1 INTRODUCTION

Vibrio cholerae, the causal agent of cholera, was first isolated by Filippo Pacini in 1854. Transmitted via ingestion of contaminated seafood or water, the bacterium (called "vibrion" by its discoverer due to its motility) causes acute diarrheal infection with an incubation period of 2 h to 5 days. Spreading from its original reservoir in the Ganges delta in India during the nineteenth century, *V. cholerae* has killed millions of people across all continents in subsequent pandemics. In spite of continuous control efforts, cholera is still endemic in many countries in Africa, Asia, Middle East, and South and Central America, with an estimated 3–5 million cases annually, including >100,000 deaths.

35.2 CLASSIFICATION AND MORPHOLOGY

The genus *Vibrio* is classified taxonomically in the family Vibrionaceae, order Vibrionales, class Gammaproteobacteria, domain Bacteria. Currently, the family Vibrionaceae consists of three genera: *Vibrio* (formerly genus I), *Photobacterium* (formerly genus II), and *Salinivibrio* (formerly genus III). In turn, the genus *Vibrio* contains >70 recognized species, of which *V. cholerae*, *V. parahaemolyticus*, and *V. vulnificus* are important human foodborne pathogens. Other human pathogenic *Vibrio* species in the genus include *V. alginolyticus*, *V. fluvialis*, *V. furnissii*, *V. metchnikovii*, and *V. hollisae* [1,2].

Based on the differences in the sugar composition of the surface somatic "O" antigen, *V. cholerae* is separated into 206 "O" serogroups. Of these, two serogroups, O1 and O139, are associated with epidemic cholera, with *V. cholerae* O1 causing the majority of outbreaks worldwide and O139 being confined to Southeast Asia. The O1 serogroup can be further distinguished into two biotypes: classical and El Tor (the latter is a more common epidemic biotype). In addition, the variations in antigenic composition allow differentiation of the *V. cholerae* O1 serogroup into three serotypes: Ogawa, Inaba, and Hikojima. While the Ogawa serotype expresses the A and B antigens and a small amount of C antigen, Inaba serotype expresses only the A and C antigens, and rare and unstable Hikojima serotype expresses all three antigens. Moreover, depending on whether it produces toxins or not, *V. cholerae* O1 can be subdivided into toxigenic or nontoxigenic strains. *V. cholerae* O139 shares some common characteristics with the O1 El Tor biotype but differs in its polysaccharide surface antigen [3,4].

Non-O1 and non-O139 *V. cholerae* strains (which do not agglutinate with "O" antiserum) are also capable of causing mild diarrhea but not epidemics [5]. Non-O1 strains of *V. cholerae* are occasionally isolated from extraintestinal infections, wounds, the ear, sputum, urine, and cerebrospinal fluid [6].

Morphologically, *V. cholerae* is a Gram-negative, facultatively anaerobic, nonspore forming curved rod of approximately 1.4–2.6 μm in length and 0.5 μm in width. The bacterium is motile by means of a single, sheathed, polar flagellum. It is capable of respiratory (oxygen-utilizing) and fermentative metabolism, produces oxidase and catalase, reduces nitrate, and ferments glucose without generating gas. While its growth is stimulated by addition of 1% sodium chloride, *V. cholerae* differs from other *Vibrio* spp. by its ability to grow in nutrient broth without added NaCl. Most *Vibrio* spp. can be isolated using standard growth media such as nutrient agar/broth or media supplemented with required concentrations of NaCl (0.5%–3%) and/or a mixture of salts (preferably sea salts) [2].

Pulse-field gel electrophoresis (PFGE) analysis showed that *V. cholerae* genome consists of two unique and circular replicons [7]. Sequencing examination of *V. cholerae* El Tor N16961 revealed that the two circular chromosomes of 2,961,146 bp and 1,072,314 bp contain 3,885 open reading frames (ORF) together. While the genes for essential cell functions (e.g., DNA replication, transcription, translation, and cell-wall biosynthesis) and pathogenicity (e.g., toxins, surface antigens, and adhesins) are mostly located on large chromosomes, a gene capture system (the integron island) and host "addiction" genes (that are typically found on plasmids) as well other many other hypothetic genes are found on small chromosome [8–11]. According to a recent study on *V. cholerae* O1 strain G4222, a clinical isolate from South Africa, chromosome I (of 3,139,654 bp in length) has an average G+C content of 47.72% and 2,809 coding sequences, while chromosome II (of 1,061,058 bp in length) has a G+C content of 46.88% and 1,051 coding sequences. A ~150 kb integrative conjugative element (ICE), belonging to the SXT family, is located on chromosome I and carries the genes implicated in multiple-drug resistance [12].

35.3 BIOLOGY AND EPIDEMIOLOGY

Members of the genus *Vibrio* normally exist in aquatic environment (freshwater, estuarine and marine) as free-living organisms or in association with aquatic animals and plants, although a number of vibrios are pathogenic to humans or aquatic (primarily marine) animals such as fish and eels [13].

As aquatic sources (e.g., brackish water and estuaries, especially those associated with algal blooms) are its main reservoirs, transmission of *V. cholerae* to humans is largely through the ingestion of raw or poorly cooked seafood, drinking of contaminated potable fresh water, or exposure of warm coastal waters to cut and bruises on the skin leading to septicemia. Thus, two conditions are sufficient for a cholera outbreak to occur: (1) presence of cholera in the population and (2) breaches in the water, sanitation, and hygiene infrastructure, which facilitate large-scale exposure of population to food or water contaminated with *V. cholerae* organisms [14–17].

Typical areas at risk are peri-urban slums and over-crowded camps for displaced people or refugees, where clean water and sanitation are below minimum requirements [18]. Immunocompromised individuals, particularly those with liver diseases, are also susceptible to the infection. Individuals who produce less stomach acid (e.g., young children and older people) and who take drugs that reduce stomach acid (e.g., proton pump inhibitors such as omeprazole and histamine-2 blockers such as ranitidine) have an increased tendency to contact the infection [13,19].

V. cholerae was known to cause cholera disease in India (including present day Bangladesh and West Bengal) in the early 1800s. *V. cholerae* serogroup O1 biotype classical was responsible for six pandemic outbreaks between 1817 and 1923 in Asia, Africa, Europe, and America [20–22]. The seventh pandemic due to the *V. cholerae* El Tor biotype began in Sulawesi (Celebes), Indonesia, in 1961, spread to Africa in 1971 and reached South America in 1991 through ship ballast waters and oceanic current and zooplanktons [23,24]. In 1992, *V. cholerae* O139 biotype was first identified as the cause of another outbreak (the eighth cholera pandemic) in Madras, India [25–27]. Despite numerous control attempts, the incidences of human cholera continued to remain high [28–30]. The postflooding cholera outbreak in Bangladesh in 2004 involved >17,000 cases, with the isolation of *V. cholerae* O1 Ogawa and O1 Inaba. For the year 2011 alone, a total of 589,854 cases (including 7816 deaths) were notified to WHO from 58 countries. It was especially notable that Haiti experienced a cholera outbreak following a devastating earthquake in early 2011, claiming thousands of lives and affecting populations in nearby Dominican Republic, Venezuela, and the United States [31].

35.4 CLINICAL FEATURES AND PATHOGENESIS

35.4.1 Clinical Features

V. cholerae O1 and O139 are primarily causative agents for cholera disease in humans. While a majority (75%) of people infected with *V. cholerae* does not develop any overt clinical symptoms, they may shed the bacteria in feces for 7–14 days after infection, and represent a source of potential infection to other people. Of the remaining 25% of infected people, 18% may show mild illness, 5% have moderate illness, and 2% develop acute watery diarrhea (with the stool appearing gray and containing flecks of mucus—the so-called "rice water stool," resulting in the loss of >1 L of water and salts per hour), vomiting, intense thirst, muscle cramps, loss of body weight, loss of normal skin turgor, dry mucous membranes, sunken eyes, lethargy, anuria, weak pulse, kidney failure, shock, coma, and death. People with suppressed immune functions (e.g., malnourished children or people living with HIV) are at a greater risk of death if infected. In surviving cholera patients, symptoms usually subside in 3–6 days and bacteria are cleared in 2 weeks.

On the other hand, *V. cholerae* non-O1, non-O139 strains are noncholeragenic, and only cause a milder form of gastroenteritis than *V. cholerae* O1 and O139 strains. They are often associated with sporadic cases and small outbreaks rather than with epidemics and pandemics. It is of interest to note that while *V. cholerae* O1 rarely causes bacteremia or invasive extraintestinal disease, non-O1/non-O139 *V. cholerae* can invade the bloodstream, leading to bacteremia and septicemia.

35.4.2 Pathogenesis

Cholera disease results from fluid imbalance that contributes to dehydration, hypokalemia (loss of potassium), metabolic acidosis (bicarbonate loss), and renal failure. The outcome of the disease largely reflects the capacity of the individual host to initiate various defensive responses (both immunological and molecular) to the invading *V. cholerae* bacteria [32–35].

V. cholerae expresses a number of microbial factors that are implicated in its virulence and disease processes. These include cholera toxin (a T-cell-dependent antigen inducing intestinal secretion of electrolytes and water), toxin coregulated pilus (TCP), lipopolysaccharide (a T-cell-independent antigen and a serogrouping/serotyping determinant), O-specific polysaccharide (a T-cell-independent antigen and a determinant of serogroup and serotype), etc. [36–38].

The cholera toxin (ctx) is a protein exotoxin encoded by the ctxA and ctxB genes that form part of a filamentous bacteriophage called CTX phage. The toxin is composed of one A subunit associated with five B subunits. The B subunit pentamer binds to the ganglioside GM1 on eukaryotic cells, whereas the A subunit is translocated intracellularly and acts enzymatically to activate cytoplasmic adenylate cyclase of the intestinal epithelial cells by adenosine diphosphate (ADP)-ribosylation of the stimulatory G protein. The presence of high concentrations of intracellular cyclic adenosine monophosphate (cAMP) activates the cystic fibrosis transmembrane conductance regulator, causing a dramatic efflux of ions and water from infected enterocytes and thus watery diarrhea [39].

The TCP is a T-cell-dependent group of antigens that is encoded by the tcpA gene forming part of the *V. cholerae* pathogenicity island [40,41]. In addition to be involved in intestinal colonization, it acts as a receptor for the CTX phage. In response to the environment at the intestinal wall, *V. cholerae* produces the TcpP/TcpH proteins, which along with the ToxR/ToxS proteins, activate the expression of the ToxT regulatory protein. Subsequently, ToxT activates expression of toxin-producing virulence genes, facilitating bacterial colonization of the intestine and development of diarrhea in the infected person.

35.5 IDENTIFICATION AND DIAGNOSIS

Isolation and identification of *V. cholerae* O1 and O139 from a patient's stool provide a definitive diagnosis of cholera. Culture isolation of *Vibrio* organisms normally involves the enrichment of the sample and plating on *Vibrio*-specific agar plates (e.g., thiosulfate-citrate-bile salts-sucrose [TCBS] agar). Through the inclusion of alkaline pH (8.6), ox bile, and NaCl (1%), TCBS suppresses the growth of nontarget bacteria, such as members of the family Enterobacteriaceae, the genus *Pseudomonas*, and Gram-positive bacteria. On TCBS agar, *V. cholerae* colonies appear yellow, whereas *V. vulnificus* and *V. parahaemolyticus* colonies are green. Subsequent examination using a series of biochemical and serologic tests enables their correct identification [42–44]. For example, use of oxidase test, string test, glucose and sucrose fermentation, salt broths, hemolysis test, and antimicrobial susceptibility test facilitates differentiation of vibrios from other bacteria, and also pandemic from non-pandemic *Vibrio* species/serogroups. In addition, *Vibrio* isolates obtained from TCBS can be differentiated by a commercially available API 20E Biolog system. Application of serological tests such as the coagglutination test (COAT), the Institute Pasteur (IP)

cholera dipstick, the sensitive membrane antigen rapid test (SMART), the IP dipstick and Medicos also enables rapid discrimination of *Vibrio* species [45–49].

In recent years, nucleic acid amplification techniques such as polymerase chain reaction (PCR), loop-mediated isothermal amplification (LAMP), and nucleic acid sequence-based amplification (NASBA) are utilized to improve the sensitivity, specificity, and speed of laboratory detection and epidemiological tracking of *V. cholerae* [50–64]. The common gene targets for *V. cholerae* identification include the *ctx* (cholera toxin), *hlyA* (hemolysin), *ompU*, *toxR*, *tcpAI*, and *lolB* genes [65–67]. The development of multiplex PCR allows simultaneous detection of *V. cholerae* O1, O139, non-O1 and non-O139 strains, differentiation between toxigenic and non-toxigenic strains, as well as discrimination among *V. cholerae*, *V. parahaemolyticus*, and *V. vulnificus* [68–71]. Furthermore, use of real time PCR platforms enables rapid and sensitive detection of *V. cholerae* bacteria from clinical, food, and environmental specimens [56,58,59,68].

35.6 TREATMENT AND PREVENTION

35.6.1 TREATMENT

As cholera is a severe acute dehydrating diarrheal disease that contributes to dehydration and loss of electrolytes, the first line of treatment is rapid fluid and electrolyte replacement, optimally in the form of oral rehydration solution containing salts, sugar, and water (e.g., WHO/UNICEF ORS standard sachet). Use of rice-based oral rehydration solution also helps decrease volume of stools and is indicated for patients of 6 months and older. Intravenous administration of isotonic fluids is necessary for those with severe dehydration or unable to tolerate oral therapy. Antibiotic use (e.g., tetracycline) is essential for disease treatment in severe cholera, as it will shorten the duration of *V. cholerae* excretion and reduce the volume of rehydration fluids needed. However, widespread application of antibiotic therapy should be avoided due to the possible emergence of antibiotic resistant *V. cholerae* [72].

35.6.2 PREVENTION

Effective prevention measures for mitigating cholera outbreaks, controlling cholera in endemic areas, and reducing deaths encompass personal hygiene, water treatment, and emergency responses.

For personal hygiene, use of soap and hand washing promotion can achieve a significant decrease in the incidence of diarrhea in endemic areas. Water treatment includes boiling the water for drinking, washing, and cooking purposes; and treatment of sewages and drainage systems. Proper disposal of infected materials (e.g., waste products, clothing, and beddings of cholera victims) either by boiling or by using chlorine bleach is also crucial. Adequate emergency responses are dependent on the provision of manpower, equipment, drugs, and consumables, along with improved surveillance systems, communication, and transport [73].

Immunization with cholera vaccines also forms part of the recommended control measures in areas where cholera is endemic or at risk of outbreaks [74,75]. Currently, two oral cholera vaccines are WHO-prequalified and licensed in over 60 countries [76,77]. One contains killed whole cells of *V. cholerae* classical and El Tor O1, supplemented with recombinant CtxB subunit (WC-rBS; Dukoral, Crucell, Sweden); the other is bivalent, containing killed *V. cholerae* classical and El Tor O1 as well as O139, without CtxB supplementation (WC; Shanchol, Shantha Biotechnics, India). Dukoral has been shown to provide short-term (4–6 months) protection of 85%–90% against *V. cholerae* O1 among all age groups, while Shanchol provides longer-term protection against *V. cholerae* O1 and O139 in children <5 years of age.

35.7 CONCLUSION

Cholera is an acute diarrheal infection resulting from ingestion of food or water contaminated with choleragenic *V. cholerae* O1 and O139 [78]. Epidemiologically, the disease is characterized by its tendency to appear in explosive outbreaks and its predisposition to causing pandemics that may spread to other countries and continents. Despite concerted control and prevention efforts, the disease continues to cause a major public health problem in developing countries, especially those in Africa, parts of Asia, the Middle East, and South and Central America. Disruptions of public sanitation services by civil unrest, natural disasters (e.g., earthquake, tsunami, volcanic eruptions, landslides, and floods) are important contributing factors for the outbreaks of cholera disease in endemic areas. Application of rapid diagnostic tests such as nucleic acid-based techniques is critical for prompt identification and confirmation of cholera epidemics. Because cholera is a severe acute dehydrating diarrheal disease, rapid fluid and electrolyte replacement represents an effective first-line treatment, and use of antibiotics such as tetracycline in severe cholera helps shorten the duration of illness and lessen the amount of fluid replacement. Prevention of cholera outbreaks relies on personal hygiene, water treatment, and emergency responses as well as vaccination.

REFERENCES

1. Daniels NA, Shafaie A. A review of pathogenic *Vibrio* infections for clinicians, *Infect. Med.* 17, 665, 2000.
2. Farmer JJ, Janda M. Vibrionaceae. In: *Bergey's Manual of Systematic Bacteriology*, 2nd ed., Vol. 2, The Proteobacteria, p. 491, Garrity, G.M. et al. (eds.), Springer-Verlag, New York, 2005.
3. Berche P et al. The novel epidemic strain O139 is closely related to the pandemic strain O1 of *Vibrio cholerae*, *J. Infect. Dis.* 170, 701, 1994.
4. Li BS et al. Phenotypic and genotypic characterization *Vibrio cholerae* O139 of clinical and aquatic isolates in China. *Curr. Microbiol.* 62, 950–955, 2011.
5. Sack DA et al. Cholera. *Lancet* 363, 223–233, 2004.
6. Cariri FAMO et al. Characterization of potentially virulent non-O1/non-O139 *Vibrio cholerae* strains isolated from human patients. *Clin. Microbiol. Infect.* 16, 62–67, 2010.
7. Trucksis M et al. The *Vibrio cholerae* genome contains two unique circular chromosomes. *Proc. Natl. Acad. Sci. USA* 95, 14464–14469, 1998.
8. Heidelberg JF et al. DNA sequence of both chromosomes of the cholera pathogen *Vibrio cholerae*. *Nature* 406, 477–483, 2000.
9. Faruque SM et al. Genomic analysis of the Mozambique strain of *Vibrio cholerae* O1 reveals the origin of el Tor strains carrying classical CTX prophage. *Proc. Natl. Acad. Sci. USA* 104, 5151–5156, 2007.
10. Chun J et al. Comparative genomics reveals mechanism for short-term and long-term clonal transitions in pandemic *Vibrio cholerae*. *Proc. Natl. Acad. Sci. USA* 106, 15442–15447, 2009.
11. Grim CJ et al. Genome sequence of hybrid *Vibrio cholerae* O1 MJ-1236, B-33, and CIRS101 and comparative genomics with *V. cholerae*. *J. Bacteriol.* 192, 3524–3533, 2010.
12. le Roux WJ et al. Genome sequence of *Vibrio cholerae* G4222, a South African clinical isolate. *Genome Announc.* 1, e0004013, 2013.
13. Kaper JB, Morris JG, Levine MM. Cholera. *Clin. Microbiol. Rev.* 8, 48–86, 1995.
14. Faruque SM, Albert MJ, Mekalanos JJ. Epidemiology, genetics, and ecology of choleragenic *Vibrio cholerae*. *Microbiol. Mol. Biol. Rev.* 62, 1301–1314, 1998.
15. Faruque SM, Nair GB. Molecular ecology of toxigenic *Vibrio cholerae*. *Microbiol. Immunol.* 46, 59–66, 2002.
16. Lipp EK, Huq A, Colwell RR. Effects of global climate on infectious disease: The cholera model. *Clin. Microbiol. Rev.* 15, 757–770, 2002.
17. Ranjbar R et al. A cholera outbreak associated with drinking contaminated well water. *Arch. Iran. Med.* 14, 339–340, 2011.
18. Talavera A, Pérez EM. Is cholera disease associated with poverty? *J. Infect. Dev. Countries* 3, 408–411, 2009.
19. Sigman M, Luchette FA. Cholera: Something old, something new. *Surg. Infect. (Larchmt)* 13, 216–222, 2012.
20. Lan R, Reeves PR. Pandemic spread of cholera: Genetic diversity and relationships within the seventh pandemic clone of *Vibrio cholerae* determined by amplified fragment length polymorphism. *J. Clin. Microbiol.* 40, 172–181, 2002.
21. Cho YJ et al. Genomic evolution of *Vibrio cholerae*. *Curr. Opin. Microbiol.* 13, 646–651, 2010.
22. Safa A, Nair GB, Kong RY. Evolution of new variants of *Vibrio cholerae* O1. *Trends Microbiol.* 18, 46–54, 2010.
23. Kirigia JM et al. Economic burden of cholera in the WHO African region. *BMC Int. Health Hum. Rights* 9, 1–14, 2009.
24. Adagbada AO et al. Cholera epidemiology in Nigeria: An overview. *Pan. Afr. Med. J.* 12, 59, 2012.
25. Ramamurthy T et al. Emergence of novel strain of *Vibrio cholerae* with epidemic potential in southern and eastern India. *Lancet* 341, 703, 1993.
26. Morris JG Jr. *Vibrio cholerae* O139 Bengal: Emergence of a new epidemic strain of cholera. *Infect. Agents. Dis.* 4, 41, 1995.
27. Mishra A et al. Amplified fragment length polymorphism of clinical and environmental *Vibrio cholerae* from a freshwater environment in a cholera-endemic area, India. *BMC Infect. Dis.* 11, 249, 2011.
28. Emch M et al. Seasonality of cholera from 1974 to 2005: A review of global patterns. *Int. J. Health Geogr.* 7, 31, 2008.

29. Harris JB et al. Susceptibility to *Vibrio cholerae* infection in a cohort of household contacts of patients with cholera in Bangladesh. *PLoS Negl. Trop. Dis.* 2, e221, 2008.

30. Chowdhury F et al. Impact of rapid urbanization on the rates of infection by *Vibrio cholerae* O1 and enterotoxigenic *Escherichia coli* in Dhaka, Bangladesh. *PLoS Negl. Trop. Dis.* 5, e999, 2011.

31. Barzilay EJ et al. Cholera surveillance during the Haiti epidemic—The first 2 years. *N. Engl. J. Med.* 368, 599–609, 2013.

32. Zhang D et al. The *Vibrio* pathogenicity island-encoded Mop protein modulates the pathogenesis and reactogenicity of epidemic *Vibrio cholerae*. *Infect. Immun.* 71, 510–515, 2003.

33. Weil AA et al. Memory T-cell responses to *Vibrio cholerae* O1 infection. *Infect. Immun.* 77, 5090–5096, 2009.

34. Kuchta A et al. *Vibrio cholerae* O1 infection induces pro-inflammatory CD4+ T-cell responses in blood and intestinal mucosa of infected humans. *Clin. Vaccine Immunol.* 18, 1371–1377, 2011.

35. Leung DT et al. Immune responses to cholera in children. *Exp. Rev. Anti Infect. Ther.* 10, 435–444, 2012.

36. DiRita VJ et al. Regulatory cascade controls virulence in *Vibrio cholerae*. *Proc. Natl. Acad. Sci. USA* 88, 5403–5407, 1991.

37. Flach CF et al. Broad up-regulation of innate defense factors during acute cholera. *Infect. Immun.* 75, 2343–2350, 2007.

38. Bhuiyan TR et al. Cholera caused by *Vibrio cholerae* O1 induces T-cell responses in the circulation. *Infect. Immun.* 77, 1888–1893, 2009.

39. Qadri F et al. Acute dehydrating disease caused by *Vibrio cholerae* serogroups O1 and O139 induce increases in innate cells and inflammatory mediators at the mucosal surface of the gut. *Gut* 53, 62–69, 2004.

40. Karaolis DK et al. A *Vibrio cholerae* pathogenicity island associated with epidemic and pandemic strains. *Proc. Natl. Acad. Sci. USA* 95, 3134–3139, 1998.

41. Tay CY, Reeves PR, Lan R. Importation of the major pilin *TcpA* gene and frequent recombination drive the divergence of the *Vibrio* pathogenicity island in *Vibrio cholerae*. *FEMS Microbiol. Lett.* 289, 210–218, 2008.

42. Bhuiyan NA et al. Use of dipsticks for rapid diagnosis of cholera caused by *Vibrio cholerae* O1 and O139 from rectal swabs. *J. Clin. Microbiol.* 41, 3939–3941, 2003.

43. Kalluri P et al. Evaluation of three rapid diagnostic tests for cholera: Does the skill level of the technician matter? *Trop. Med. Int. Health* 11, 49–55, 2006.

44. Roozbehani AD et al. A rapid and reliable species-specific identification of clinical and environmental isolates of *Vibrio cholerae* using a three-test procedure and recA polymerase chain reaction. *Indian J. Med. Microbiol.* 30, 39–43, 2012.

45. Nato F et al. One-step immunochromatographic dipstick tests for rapid detection of *Vibrio cholerae* O1 and O139 in stool samples. *Clin. Diagn. Lab. Immunol.* 10, 476–478, 2003.

46. Wang X et al. Field evaluation of a rapid immunochromatographic dipstick test for the diagnosis of cholera in a high-risk population. *BMC Infect. Dis.* 6, 17, 2006.

47. Harris JR et al. Field evaluation of Crystal VC Rapid Dipstick test for cholera during a cholera outbreak in Guinea-Bissau. *Trop. Med. Int. Health* 14, 1117–1121, 2009.

48. Pengsuk C et al. Differentiation among the *Vibrio cholerae* serotypes O1, O139, O141 and non-O1, non-O139, non-O141 using specific monoclonal antibodies with dot blotting. *J. Microbiol. Methods* 87, 224–233, 2011.

49. Dick MH et al. Review of two decades of cholera diagnostics—How far have we really come? *PLoS Negl. Trop. Dis.* 6, e1845, 2012.

50. Koch WH et al. Rapid polymerase chain reaction method for detection of *Vibrio cholerae* in foods. *Appl. Environ. Microbiol.* 59, 556–560, 1993.

51. Olsvik O et al. Use of automated sequencing of PCR-generated amplicons to identify three types of cholera toxin subunit B in *Vibrio cholerae* O1 strains. *J. Clin. Microbiol.* 31, 22–25, 1993.

52. Karunasagar I et al. Rapid detection of *Vibrio cholerae* contamination of seafood by polymerase chain reaction. *Mol. Mar. Biol. Biotechnol.* 4, 365–368, 1995.

53. Chaicumpa W et al. Rapid diagnosis of cholera caused by *Vibrio cholerae* O139. *J. Clin. Microbiol.* 36, 3595–3600, 1998.

54. Rivera IN et al. Genotypes associated with virulence in environmental isolates of *Vibrio cholerae*. *Appl. Environ. Microbiol.* 67, 2421, 2001.

55. Rivera IN et al. Method of DNA extraction and application of multiplex polymerase chain reaction to detect toxigenic *Vibrio cholerae* O1 and O139 from aquatic ecosystems. *Environ. Microbiol.* 5, 599, 2003.

56. Blackstone GM et al. Use of a real time PCR assay for detection of the ctxA gene of *Vibrio cholerae* in an environmental survey of Mobile Bay. *J. Microbiol. Methods* 68, 254, 2006.

57. González-Escalona N et al. Quantitative reverse transcription polymerase chain reaction analysis of *Vibrio cholerae* cells entering the viable but non-culturable state and starvation in response to cold shock. *Environ. Microbiol.* 8, 658, 2006.

58. Fedio W et al. Rapid detection of the *Vibrio cholerae* ctx gene in food enrichments using real-time polymerase chain reaction. *J. AOAC Int.* 90, 1278, 2007.

59. Fykse EM et al. Detection of *Vibrio cholerae* by real-time nucleic acid sequence-based amplification. *Appl. Environ. Microbiol.* 73, 1457–1456, 2007.

60. Yamazaki W et al. Sensitive and rapid detection of cholera toxin-producing *Vibrio cholerae* using a loop-mediated isothermal amplification. *BMC Microbiol.* 8, 94, 2008.

61. Srisuk C et al. Rapid and sensitive detection of *Vibrio cholerae* by loop-mediated isothermal amplification targeted to the gene of outer membrane protein ompW. *Lett. Appl. Microbiol.* 50, 36, 2010.

62. Shuan JTC et al. Comparative PCR-based fingerprinting of *Vibrio cholerae* isolated in Malaysia. *J. Gen. Appl. Microbiol.* 57, 19–26, 2011.

63. Naha A et al. Development and evaluation of a PCR assay for tracking the emergence and dissemination of Haitian variant ctx Bin *Vibrio cholerae* O1 strains isolated from Kolkata, India. *J. Clin. Microbiol.* 50, 1733–1736, 2012.

64. Spagnoletti M, Ceccarelli D, Colombo MM. Rapid detection by multiplex PCR of Genomic Islands, prophages and Integrative Conjugative Elements in *V. cholerae* 7th pandemic variants. *J. Microbiol. Methods* 88, 98–102, 2012.

65. Lalitha P et al. Analysis of *lolB* gene sequence and its use in the development of a PCR assay for the detection of *Vibrio cholerae*. *J. Microbiol. Methods* 75, 142–144, 2008.

66. Gardès J, Croce O, Christen R. In silico analyses of primers used to detect the pathogenicity genes of *Vibrio cholerae*. *Microbes Environ.* 27, 250–256, 2012.

67. Cho MS et al. A novel marker for the species-specific detection and quantitation of *Vibrio cholerae* by targeting an outer membrane lipoprotein *lolB* gene. *J. Microbiol. Biotechnol.* 23, 555–559, 2013.

68. Lyon WJ, TaqMan PCR for detection of *Vibrio cholerae* O1, O139, non-O1, and non-O139 in pure cultures, raw oysters, and synthetic seawater. *Appl. Environ. Microbiol.* 67, 4685, 2001.

69. Chua AL et al. Development of a dry reagent-based triplex PCR for the detection of toxigenic and non-toxigenic *Vibrio cholerae*. *J. Med. Microbiol.* 60, 481–485, 2011.

70. Izumiya H et al. Multiplex PCR assay for identification of three major pathogenic *Vibrio* spp., *Vibrio cholerae*, *Vibrio parahaemolyticus*, and *Vibrio vulnificus*. *Mol. Cell. Probes* 25, 174–176, 2011.

71. Mehrabadi JF et al. Detection of toxigenic *Vibrio cholerae* with new multiplex PCR. *J. Infect. Public Health* 5, 263–267, 2012.

72. Rashed SM et al. Genetic characteristics of drug-resistant *Vibrio cholerae* O1 causing endemic cholera in Dhaka, 2006–2011. *J. Med. Microbiol.* 61, 1736–1745, 2012.

73. Zuckerman JN, Rombo L, Fisch A. The true burden and risk of cholera: Implications for prevention and control. *Lancet Infect. Dis.* 7, 521–530, 2007.

74. Kanungo S et al. Immune responses following one and two doses of the reformulated, bivalent, killed, whole-cell, oral cholera vaccine among adults and children in Kolkata, India: A randomized, placebo-controlled trial. *Vaccine* 27, 6887–6893, 2009.

75. Verma R, Khanna P, Chawla S. Cholera vaccine: New preventive tool for endemic countries. *Hum. Vaccin. Immunother.* 8, 682–684, 2012.

76. López-Gigosos RM et al. Vaccination strategies to combat an infectious globe: Oral cholera vaccines. *J. Glob. Infect. Dis.* 3, 56–62, 2011.

77. Saha A et al. Safety and immunogenicity study of a killed bivalent (O1 and O139) whole-cell oral cholera vaccine Shanchol, in Bangladeshi adults and children as young as 1 year of age. *Vaccine* 29, 8285–8292, 2011.

78. Kaysner CA, DePaola A Jr. In: *Bacteriological Analytical Manual*, 8th edn., Chapter 9, Hammack T, et al. (eds.), Food and Drug Administration, Washington, DC, 2004. http://www.fda.gov/Food/ScienceResearch/LaboratoryMethods/BacteriologicalAnalyticalManualBAM/ucm070830.htm.

36 *Yersinia pestis*

Dongsheng Zhou and Ruifu Yang

CONTENTS

36.1 INTRODUCTION

Plague, caused by *Yersinia pestis*, is one of the most devastating bacterial infections in human history. Plagues have claimed millions of lives during the three recorded pandemics.[1] Plague circulates naturally among rodent reservoirs and flea vectors in various enzootic foci.[2] Humans typically acquire plague infections via the bite of an infected flea or via direct contact with infected rodents or plague patients. Enzootic plague foci exist throughout the world, especially in Eurasia, Africa, and America. Although the occurrence of large-scale human plague epidemics is minimally probable at present, human cases of plague are reported annually. Moreover, *Y. pestis* can be used as a biowarfare or bioterrorism agent.[3] It was employed during warfare in the fourteenth century. The notorious Japanese Unit 731 conducted plague warfare from 1937 to 1945 in China.[4] This nefarious pathogen was classified as a category A bioterrorism agent by the Centers for Disease Control and Prevention (CDC), United States.[5]

36.2 CLASSIFICATION AND MORPHOLOGY

Y. pestis is a Gram-negative bacterium that was first isolated and correctly described by Dr. Alexandre Émile Jean Yersin from the Pasteur Institute, Vietnam, in 1894 during a plague outbreak in Hong Kong, although Dr. Shibasaburo Kitasato from Japan also contributed to the first isolation of this pathogenic bacterium.[6] The previously reported synonyms for *Y. pestis* include *Bacterium pestis* Lehmann and Neumann 1896, *Bacillus pestis* (Lehmann and Neumann 1896) Migula 1900, *Pasteurella pestis* (Lehmann and Neumann 1896) Bergey et al. 1923, and *Pestisella pestis* (Lehmann and Neumann 1896) Dorofeev 1947.[7,8] According to the current prokaryote classification, the genus *Yersinia* belongs to the family Enterobacteriaceae, order Enterobacteriales, class Gammaproteobacteria, division/phylum Proteobacteria, and domain/empire Bacteria (http://www.bacterio.cict.fr/classificationsz.html).

The genus *Yersinia* has 18 species and 3 subspecies (http://www.bacterio.cict.fr/xz/yersinia.html), including three human pathogenic species, namely, *Y. pestis*, *Y. enterocolitica*, and *Y. pseudotuberculosis*. The high DNA similarity (83%) of *Y. pestis* to *Y. pseudotuberculosis* led to its reclassification as a subspecies of the latter.[9] However, the two bacteria cause very distinct diseases; thus, the name *Y. pseudotuberculosis* subsp. *pestis* was rejected to avoid confusion.[10]

36.2.1 BIOTYPING

Y. pestis is a homogenous species with only one serotype and one phage type.[1] Only a few antibiotic-resistant strains were found.[11,12] Biotyping systems based on biochemical features are widely used in plague research. The nature of plague and its causative agent hinders exchanging bacterial strains between plague researchers. Other typing systems, such as ecotyping and subspecies classification system, are only studied and used in certain countries.[2,13,14] Although these traditional methods have insufficient discriminatory power, show poor reproducibility, and are affected by physiologic factors, and their specific reagents lack of availability, they have greatly contributed to plague prevention and control and to our understanding of *Y. pestis*.

Early investigators utilized its capacity to ferment glycerol to classify *Y. pestis* strains into glycerol-positive and glycerol-negative strains.[2] The negative strains were called the oceanic type because they were usually isolated from rats in seaports; the positive ones were called continental because they were isolated from "wild" rodents, susliks (ground squirrels), gerbils, and natural plague foci.[2]

Y. pestis can be classified into four biovars: *Antiqua* (glycerol positive, arabinose positive, and nitrate positive), *Medievalis* (glycerol positive, arabinose positive, and nitrate negative), *Orientalis* (glycerol negative, arabinose positive, and nitrate positive), and *Microtus* (glycerol positive, arabinose negative, and nitrate negative).[15] The first three biovars were linked to the first, second, and third pandemic of human plague, respectively,[16] whereas the fourth is avirulent to humans, naturally causing only *Microtus* plague and its epidemic. Some natural plague foci have been found in Russia, where no human cases of plague have been reported. It is possible that additional new biovars may be found in the future. However, biovar characteristics are unstable and one strain can undergo spontaneous phenotypic variation that may cause it to be classified into another biovar.[2] For example, the Nicholisk 51 strain, an isolate from Manchuria, was classified into biovar *Orientalis* via the IS*100* genotype and the presence of specific a phage remains, but it is glycerol positive and should be classified into biovar *Antiqua*.[17] The authors considered Nicholisk 51 as an ancestor of biovar *Orientalis* or a variant of this biovar that had undergone phenotypic reversion to the glycerol-positive phenotype.[17]

36.2.2 Ecotyping

Researchers have developed an ecotyping system that exploits several biochemical features such as glycerin, rhamnose, maltose, melibiose, and arabinose fermentation; nitrate reduction; amino acid utilization; mutation rate from Pgm+ to Pgm−; and water-soluble protein patterns on SDS-PAGE to group Chinese isolates of *Y. pestis* into 18 ecotypes.[13,14,18,19] Each of the ecotypes is located in a particular geographic region. Most of the plague foci with different primary reservoirs have unique ecotypes.[19] For the other plague foci, more than one ecotype occurs in a single focus with a single primary reservoir, and each of the ecotypes corresponds to a unique set of natural landscapes and primary vector(s). The ecotyping system has been used as a framework for ecological and epidemiological analysis of plague in China. It provides a preliminary explanation of the relationship of the ecotypes of *Y. pestis*, natural environment, reservoirs, and vectors.

TABLE 36.1

Relationship between *Y. pestis* Biovar and Subspecies

Biovar	Subspecies	Other Nomenclatures
Antiqua	*pestis*	Main subspecies
	caucasica	Nonmain subspecies
Medievalis	*altaica*	Pestoides
	hissarica	
	ulegeica	
	talassica	
Orientalis	*pestis*	
Microtus	*xilingolensis*	Once classified as biovar *Medievalis*
	qinghaiensis	

36.2.3 Subspecies Classification

The subspecies classification system was proposed by Russian scientists, and the following description about this classification scheme was derived from Anisimov's review.[2] *Y. pestis* isolated from the Former Soviet Union (FSU) and Mongolia were classified into six "subspecies," including *Y. pestis* subsp. *pestis* (sometimes referred to as the "main" subspecies), *Y. pestis* subsp. *altaica*, *Y. pestis* subsp. *caucasica*, *Y. pestis* subsp. *hissarica*, *Y. pestis* subsp. *ulegeica*, and *Y. pestis* subsp. *talassica* based on the numerical analysis of 60 phenotypic features. The last five subspecies are sometimes referred to as the "nonmain" subspecies and have been referred to as the "pestoides" group of *Y. pestis* isolates (Table 36.1). All pestoides strains ferment rhamnose and are less virulent to guinea pigs, but they are highly virulent to mice. These strains cause occasional human and animal plague cases, but they have rarely been associated with epizootics of plague.[2,20] Recent studies indicate that *Y. pestis* strains from *Microtus* in Inner Mongolia and Qinghai Province of China should be classified into *Y. pestis* subsp. *xilingolensis* and *qinghaiensis* in accordance with the previous nomenclature.[21,22]

36.2.4 Genotyping

Several genotyping methods have been described for *Y. pestis*, including ribotyping,[23] different region (DFR) analysis,[24] multilocus variable number tandem repeat analysis (MLVA),[21] repetitive DNA,[25] clustered regularly interspaced short palindromic repeats (CRISPRs),[22] and single-nucleotide polymorphisms (SNPs).[26,27] They have all been successfully used for genotyping *Y. pestis* in spite of their profound differences in discriminatory power. For a specific investigation of a plague outbreak, they are all applicable. However, for effective source tracing, MLVA and SNP analysis demonstrate a higher discriminatory power. MLVA was successfully employed to trace the source of a primary pneumonic outbreak in Qinghai Province of China in 2009.[28] SNP analysis and whole genome sequencing have also been used to confirm that the ancient Black Death in Europe in the fourteenth century was caused by *Y. pestis*.[29,30]

36.3 BIOLOGY AND EPIDEMIOLOGY

Y. pestis is a Gram-negative, rod-shaped, nonmotile, non-spore-forming, bipolar-staining coccobacillus (0.5–0.8 μm in diameter and 1–3 μm in length) (giving it a safety pin appearance).[1] *Y. pestis* grows optimally at 28°C. Biochemically, *Y. pestis* is unable to ferment lactose, sucrose, rhamnose, and melibiose; it does not produce hemolysin, urease, and indole, but it is catalase positive. *Y. pestis* can survive for hours after drying on different environmental surfaces and for at least 5 days with the addition of nutrient media onto the surfaces.[31] Thus, public areas could be contaminated with *Y. pestis* even after the infected individual or source has left.

Enzootic plague foci exist throughout the world, especially in Eurasia, Africa, and America. Plague is primarily an enzootic disease transmitted among rodents by the bite of infected fleas. *Y. pestis* is a multihost and multivector pathogen that involves more than 200 wild rodent species as reservoirs and over 80 flea species as vectors.[2] The maintenance of plague in nature is almost absolutely dependent on the cyclic transmission between fleas and rodents in various enzootic foci,[32] although *Y. pestis* has a limited ability to live in environments such as soil.[33,34] In this case, an enzootic plague focus can be considered as a well-balanced terrestrial ecosystem composed of spatial contacts and food chains among *Y. pestis*, its animal hosts, and the environment.

The natural environment, including biotic and abiotic factors, in enzootic plague foci is a crucial habitat for *Y. pestis* and its rodent reservoirs and flea vectors.[35] Biotic factors include animals, plants, microorganisms, soil, and vegetation, whereas abiotic factors consist of longitude and latitude, topography, geomorphology, geology, and climate (temperature, humidity, sunlight, and environmental chemical factors). Compared with biotic factors, abiotic ones are relatively stable. The natural environment determines geographic zoning, regional distribution, habitats, and population abundance of both rodent reservoirs and flea vectors. The natural environment in enzootic plague foci determines the geographic distribution, habitats, and population abundance of both rodent reservoirs and flea vectors.

The rodent reservoirs of *Y. pestis* are usually classified as major, minor, and accidental reservoirs.[19] Major reservoirs are typically the predominant rodent species in the plague focus. They are the primary carriers of *Y. pestis* and are essential to the long-term survival of *Y. pestis* in nature. Major reservoirs are generally highly susceptible to *Y. pestis*. However, the infected individuals that are somewhat tolerant to the pathogen challenge will not die within a short period. The mammalian bacteremia continues for a certain period with high numbers of *Y. pestis* in the blood, which facilitates pathogen transmission by fleabites. Moreover, some of the rodents may be highly tolerant to *Y. pestis* in nature, even with asymptomatic infections. The dynamics of population abundance in major reservoirs coincides with the epidemic situation of animal plague in the enzootic focus.[36] Major reservoirs determine the type and geographic distribution of a plague focus, as well as

the epidemiological pattern of the plague in the focus. The enzootic properties of plague disappear without major reservoirs. Minor plague reservoirs have a very high mortality and usually die shortly after infection without bacteremia or with temporary bacteremia. Minor reservoirs are insufficient for the long-term survival of *Y. pestis* in nature, although plague epidemics occasionally occur violently in their populations. Accidental reservoirs are highly resistant to the pathogen, but they usually catch *Y. pestis* because of frequent contact with major and minor reservoirs. Human contact with any of the *Y. pestis*–infected reservoirs is likely to cause the incidence of human plague.

Although *Y. pestis* is highly pathogenic to rodent reservoirs, it has a limited effect on the entire rodent population. Even during the period of animal plague epidemics, the abundance of rodents remains high. In specific plague foci or parts of foci, the unique natural environment ultimately molds a distinct food chain-based relationship among *Y. pestis*, reservoirs, and vectors. The resulting specific complex interaction among the environment, hosts, and *Y. pestis*, termed as "host–niche interaction," determines the presence and types of *Y. pestis*.[37] The survival of *Y. pestis* in nature primarily depends on rodents and fleas, whereas fleas parasitize rodents and act as vectors for bacterial transmission.[37,38] Natural environments in various plague foci remodel distinct sets of rodents and fleas. The accumulation of functional genetic variations promotes the parallel diversification of *Y. pestis* in different plague foci, as reflected by the expansion of various plague foci in nature.[37,38] This adaptive evolution is likely determined by the complex interaction among the environment, hosts, and pathogen.

As an enzootic pathogen, *Y. pestis* has the potential to infect humans. Humans typically acquire plague infections via the bite of infected fleas or through direct contact with infected rodents, yet this deadly disease can be transmitted from person to person by the respiratory route.[32] In most cases, humans are a biological dead end for *Y. pestis*. Thus, the long-term maintenance of *Y. pestis* in nature is generally independent of humans.

Three major global plague pandemics have been recorded in history, namely, the Justinian plague, the Black Death, and the modern plague.[1] The third plague pandemic was believed to have originated from Yunnan, China, in 1855 and then spread around the world through modern transportation.[37] A recent study revealed that the Black Death was caused by *Y. pestis*[29] and might have spread along the ancient Silk Road.[26]

36.4 CLINICAL FEATURES, PATHOGENESIS, AND TRANSMISSION

36.4.1 MAJOR CLINICAL FEATURES

Clinically, plague has different forms. The three common forms, namely, bubonic plague, septicemic plague, and pneumonic plague, are mentioned in textbooks and the literature. Other kinds of plague have been reported, including skin plague, plague encephalitis, tonsillar plague, eye plague, and intestinal plague. In epidemic areas, asymptomatic plague has also been detected.[39]

Bubonic plague: This is the most common form clinically. After a bite from an infected flea, the patient usually develops swollen lymph nodes, which become buboes, the classic sign of bubonic plague. The affected lymph node depends on the bite site. Inguinal nodes are the most frequently involved because the legs are the most common site of fleabites. The incubation period is 2–6 days. Flu-like symptoms, such as a sudden onset of fever, weakness, headache, and chills, are the primary clinical manifestations at the beginning of the disease. When buboes develop, the local subcutaneous tissues swell with the enlarged node(s), which range from 1 cm × 1 cm for small ones to 5 cm × 7 cm for larger ones. The enlarged nodes coalesce and adhere to the surrounding tissues with unclear edges. The nodes are firm and the patient feels severe pain when touched, which constrains posture.

Pneumonic plague: It is the most contagious form of plague because *Y. pestis* spreads from person to person by respiratory droplets. Pneumonic plague is classified as primary or secondary. Primary pneumonic plague is caused by direct inhalation of *Y. pestis*, and secondary pneumonic plague is caused by blood spread from bubonic or septicemic plague. Patients develop other kinds of plague before they develop secondary pneumonic plague. Primary pneumonic plague usually presents with severe pulmonary signs in addition to flu-like symptoms, with a very short incubation period (2–4 days).[28] The patients generally display high fever (39°C–41°C), chills, bloody and suppurative coughing, weakness, chest pain, dyspnea, hemoptysis, lethargy, hypotension, and shock. Due to the severe respiratory distress, the patients develop hypoxia, which manifests as lip or even body cyanosis. Thus, the disease plague epidemics were called Black Death in medieval times.

Septicemic plague: It is the most severe form of plague clinically. A large quantity of *Y. pestis* is present in the patient's blood and could become new sources of pathogens for fleabites. If bubonic plague and other kinds of plague are not treated properly, the disease worsens to septicemic plague. Patients have systemic symptoms, including high fever, severe headaches, hypotension, seizures in children, hepatosplenomegaly, severe arrhythmia or extremely weak pulse, delirium, and shock, with subcutaneous or mucus membrane bleeding, bloody diarrhea, and vomiting. The patient usually dies within 1–3 days after the onset of these symptoms if specific treatment is not administered on time. The patient may even die before any symptoms appear.

36.4.2 Pathogenesis

The transmission of *Y. pestis* relies primarily on the bite of flea vectors, although infections can occur through direct contact or inhalation. The development of heavy bacteremia

in rodents is crucial to reliably infect fleas, which can then transmit the disease by biting new animal hosts.[40] The infected fleas regurgitate about 25,000–100,000 *Y. pestis* into the host's skin when they bite people.[41]

Y. pestis synthesizes biofilms to attach onto the surface of proventricular spines, and the heavy bacterial proliferation in the biofilms promotes the blockage of the gut of fleas.[42] Blockage of fleas inhibits feeding and makes them feel hungry and repeatedly attempt to feed, during which the ingested blood is regurgitated back into the bite sites, causing *Y. pestis* to infect the new hosts.

Yersinia biofilms contain bacterial colonies that are embedded in the self-synthesized extracellular matrix, and the matrix is primarily composed of exopolysaccharides, homopolymers of *N*-acetyl-D-glucosamine.[43] The *hmsHFRS* operon is responsible for the synthesis and translocation of biofilm exopolysaccharides across the cell envelope.[43,44] The signaling molecule 3′,5′-cyclic diguanylic acid (c-di-GMP) is a central positive allosteric activator of the enzymes that catalyzes the production of biofilm exopolysaccharides.[45] c-di-GMP is produced from guanosine triphosphate (GTP) by diguanylate cyclases and is degraded by phosphodiesterases. HmsT[46,47] and YPO0449[48,49] are the only two diguanylate cyclases in *Y. pestis* that synthesize c-di-GMP. The predominant effect of HmsT is in vitro biofilm formation, whereas the role of YPO0449 in biofilm production is much greater in fleas than in vitro.[48] HmsP exhibits c-di-GMP-specific phosphodiesterase activity and is involved in c-di-GMP degradation; therefore, it negatively affects biofilm formation.[46,50]

In contrast to its progenitor, *Y. pestis*, *Y. pseudotuberculosis* is transmitted via the food-borne route. *Y. pseudotuberculosis* harbors all of the known structural genes required for biofilm formation, but it typically cannot synthesize adhesive biofilms and create blockages in fleas.[51] The action of multiple antibiofilm factors such as NghA[52] and RcsAB[53] produces a tight biofilm-negative phenotype in typical *Y. pseudotuberculosis*. By contrast, *Y. pestis* has lost the function of NghA[52] and RcsA[53] during its evolution. Moreover, *Y. pestis* has acquired an additional factor Ymt, which promotes bacterial survival in fleas.[54] The aforementioned evolutionary events enable *Y. pestis* to survive in fleas and to synthesize adhesive biofilms in flea proventriculi to cause blockage, which results in efficient arthropod-borne transmission.[55]

After a bite from a blocked flea, the organisms migrate through cutaneous lymphatic vessels to regional lymph nodes. Once in the lymph nodes, most of the organisms are phagocytosed and killed by polymorphonuclear leukocytes that are attracted to the infection site in large numbers.[56] However, a few bacilli are taken up by macrophages that are unable to kill them. After the *Y. pestis* bacilli grow inside phagocytes, they develop the ability to resist subsequent phagocytosis.[57] Based on in vitro experiments and in vivo experiments in rodent peritoneal cavities, the infected macrophages provide a protected environment for the pathogens to proliferate intracellularly and synthesize their capsule and other virulence determinants, thereby enabling the bacteria to resist phagocytes[55,56] and to annihilate the host immune response.[58] After initial subcutaneous and intradermal colonization, the bacteria migrate into regional lymph nodes, resulting in inflammation and cellulitis, and occasionally large carbuncles developing around the bubo (bubonic plague).[59] Without timely and effective treatment, the bacteria rapidly escape from containment in the lymph nodes and systemically spread through the blood to various organs, including the spleen, liver, lungs, and brain, which initiates an immunologic cascade that leads to disseminated intravascular coagulation, which in turn results in bleeding, as well as skin and tissue necrosis.[56] High-density bacteremia is characteristic of moribund patients with plague.

Secondary pneumonic plague could result from hematogenous spread from the buboes to the lungs, which presents in patients as severe bronchopneumonia, cavitation, or consolidation with production of bloody or purulent sputum.[56] Primary pneumonic plague could be directly caused by the inhalation of infectious droplets or aerosols, with symptoms such as acute pneumonia, intra-alveolar hemorrhage and edema, profound lobular exudation, fibrin deposition, and bacillary aggregation.[60] The pneumonic form of the disease is the most feared because of the rapidity of its development (1–3 days) and its high mortality rate among infected individuals (approaching 100%) without timely effective treatment. Coughing results in the production of airborne droplets that contain bacteria, which can be inhaled by susceptible individuals and lead to the rapid airborne transmission of diseases among close contacts.

The pathogenicity of *Y. pestis* involves a diverse array of virulence determinants that are coordinately expressed during different stages of infection, including colonization and invasion, early intracellular growth, avoidance of host defense, and extracellular proliferation (Table 36.2). *Y. pestis* must survive during infection against different host-responding milieus that make bacterial living conditions far from optimal through appropriate adaptive/protective responses that are primarily reflected by changes in the expression of specific sets of genes. Thus, the regulatory networks that govern a complex cascade of cellular pathways facilitate the *Y. pestis* pathogenic mechanisms that operate in a concerted manner.[61]

36.4.3 Transmission

Rodents and humans acquire *Y. pestis* through infected fleabites, contact with infected tissues, or the inhalation of respiratory droplets or aerosols.[1] In addition to acquiring *Y. pestis* from wild animals, domestic animals such as cats, dogs, and guinea pigs have also been reported to be infectious sources of plague.[28]

Bubonic plague is transmitted by bites from blocked fleas. However, blockage development takes a relatively long time (about 2 weeks), which might be insufficient for explaining the rapid spread that typifies plague epidemics.[32,62] Infected fleas are immediately infectious and efficiently transmit the microorganisms for at least 4 days post infection; the mode of "early-phase transmission" by unblocked fleas has been proposed accordingly.[62,63] During the testing of an early-phase

transmission model, defects in *Y. pestis* biofilm formation did not prevent flea-borne transmission, whereas biofilm overproduction inhibited efficient early-phase transmission.[64] Unlike traditional blockage-dependent plague transmission models, early-phase transmission occurs when a flea takes its first blood meal after initial infection by feeding on a bacteremic host, which may explain the rapid spread of disease from fleas to mammalian hosts during epizootic outbreak.[62,63]

Primary or secondary pneumonic plague could be transmitted from person to person by respiratory droplets. Primary pneumonic plague outbreaks rarely happen in modern times,[28,65–69] but they have been recorded in history. The outbreaks in Oakland in 1919 and in Los Angeles in 1924,[70–72] and in Manchuria from 1910 to 1911[73,74] are the most cited instances. According to estimates from data from historical outbreaks, the basic reproduction number is 2.8–3.5.[73,75] R_0 measures the transmission potential of a pathogen; it is the average number of secondary cases that spread from the introduction of a single primary case into an otherwise fully susceptible population. However, the transmissibility of pneumonic plague also depends on the environment where the patients live. If the patients stay in poorly ventilated rooms and other people get close contact with the patients without any preventive measures, more people will be infected.

36.5 IDENTIFICATION AND DIAGNOSIS

The CDC of the United States provides case classifications based on clinical symptoms and laboratory results.[76] A suspected case is defined as a clinically compatible case that lacks presumptive or confirmatory laboratory diagnosis results. A presumptive case is defined as a clinically compatible case with available presumptive laboratory diagnosis results. A confirmed case is a clinically compatible case with the available confirmatory laboratory diagnosis results. A suspected case can be diagnosed quickly based on the symptoms and the epidemiological features, including the environmental exposure history. When a case of plague is suspected, clinical specimens can immediately be collected. Diagnostic samples include blood and appropriate site-specific samples, such as aspirates from suspected buboes, pharyngeal swabs, sputum, and endotracheal washings from patients suspected of plague pharyngitis or pneumonia, and cerebrospinal fluid from those with suspected meningitis.[76] Presumptive laboratory diagnosis may also be performed by observing a fourfold or greater increase in the serum antibody titers against the *Y. pestis* F1 antigen without a history of plague vaccinations, as determined by enzyme-linked immunosorbent assays (ELISA) and older, less-sensitive passive hemagglutination assays. Confirmatory laboratory diagnoses are traditionally based on the isolation of pure *Y. pestis* from clinical specimens together with staining and microscopy observation and phage lysis assay with *Y. pestis*-specific bacteriophages. Tiny, 1–3 mm "beaten-copper" colonies appear on blood agar after 48 h of cultivation. Gram's staining confirms the presence of Gram-negative rods and, in some cases, the identification of the double-curved shapes.

TABLE 36.2
Major Virulence Determinants of *Y. pestis*

Function	Gene IDs	Description	References
Colonization and dissemination			
Pla	YPPCP1.07	Plasminogen activator promoting bacterial in vivo dissemination	[113,114]
Ail	YPO2905	Invasin adhesin and invasin	[115]
YadBC	YPO1387-1388	Invasin adhesin and invasin	[116]
YapC	YPO2796	Autotransporter adhesin	[117]
YapE	YPO3984		[118]
Intracellular growth			
RipA	YPO1926	Putative acetyl coenzyme A transferase	[119]
Ugd	YPO2174	LPS modification	[120]
MgtCB	YPO1660-1661	Magnesium uptake	[120]
Yfe	YPO2439-2442	ABC-type iron transporter	[121]
FeoBA	YPO0132-0133	Ferrous iron transporter	[121]
Annihilation of hose immune response			
T3SS		Annihilation of innate immune cells	[122]
F1 capsule	YPMT1.81c-1.84	Resistance to phagocytosis	[123]
pH6 antigen	YPO1301-1305	Resistance to phagocytosis	[124]
Ail	YPO2905	Serum resistance	[115]
Tc proteins		Toxicity to mammalian cells	[125]
O-antigen genes		Lack of O-antigen is essential for Pla function	[126]
Iron uptake			
Yfe	YPO2439-2442	ABC-type iron transporters	[127]
HPI	YPO1906-1916	Siderophore yersiniabactin-based iron acquisition system	

However, other *Yersinia* species may exhibit very similar appearances. In recent years, numerous rapid tests for detecting *Y. pestis* based on different variants of polymerase chain reaction have been developed.[77–80] The continued development and implementation of DNA-based methods with increased sensitivity and defined specificity are particularly useful for detecting residual *Y. pestis* DNA in situations where *Y. pestis* cannot be cultured from clinical specimens. In addition, immunochromatographic strips and biosensors have also been developed for plague diagnosis.[81–84] Their applications will greatly help in the rapid diagnosis of plague.

36.6 TREATMENT AND PREVENTION

36.6.1 ANTIMICROBIAL THERAPY

If a case of plague is suspected or diagnosed, specific antimicrobial therapy should be started immediately. The suspected or diagnosed plague patients and their direct contacts should be arranged individually in separate wards or isolation rooms. Patients suspected of pneumonic plague

should be managed with respiratory droplet precautions. All patients should be isolated for the first 48 h after treatment initiation. If pneumonic plague is present, then respiratory isolation procedures should be enforced strictly and rigidly, including the use of gowns, gloves, and eye protection. Patients with pneumonia must be isolated until they have completed at least 4 days of antibiotic therapy. Timely antibiotic treatment is effective against plague. When administered during the early phase of the disease, antibiotic treatment effectively reduces overall human mortality ranging from 5% to 14%. However, when left untreated, the mortality rate is between 50% and 90%.[85] Streptomycin, chloramphenicol, and tetracycline, alone or in combination, are the reference drugs for treating plague, and streptomycin is the most preferred.[86,87] Antibiotics should be given intramuscularly at a dose of 30 mg/kg/day in two divided doses. Antibiotics should be administered intramuscularly for 10 days or until 3 days after the patient's temperature returns to normal. Chloramphenicol should be used in cases of plague meningitis.[76] Streptomycin or gentamicin can be used to treat plague in children. Gentamicin is preferred among pregnant women because of its safety.[76] Tetracycline is contraindicated in pregnant women and in children less than 7 years of age because of possible staining of developing teeth. Alternatively, antibiotics may be applied.[88]

The successful treatment of septicemic and pneumonic plague with antibiotics is less likely because the disease rapidly develops and treatment must commence during the early stages of the infection. Fulminant plague is especially difficult to treat because of the possibility of bacteriolysis, which subsequently releases large amounts of endotoxin. Early treatment with sufficient quantities of effective antibiotics is critical for saving lives from primary pneumonic plague.[28] Supportive care is an important part of management, which may include immediate fluid resuscitation, vasopressors, hemodynamic monitoring, and respiratory care, including ventilator support. The following symptomatic treatments may be needed: (1) patients with dysphoria or severe pain can be given sedatives and painkillers, (2) patients with heart failure can be given cardiac stimulants, and (3) patients with toxic shock can be given timely antishock treatment.

36.6.2 Immunization and Prevention

The recent emergence of multiple antibiotic-resistant *Y. pestis* strains[89,90] and their possible use as bioweapons and bioterrorism agents means that the long-term potential for the use of antibiotics to treat plague is less certain and that a vaccine effective against plague is urgently needed. The individuals at high risk of plague should be immunized, including military troops and other field personnel working in plague-endemic areas where exposure to rats and fleas cannot be controlled. Laboratory personnel working with *Y. pestis*, people who reside in enzootic or epidemic plague areas, and those whose vocations bring them into regular contact with wild animals, particularly rodents and rabbits, should also be vaccinated.

Both killed and live whole-cell plague vaccines have saved thousands of human lives in the twentieth century and have continually been used in a few countries where the threat of plague is imminent.[91,92] A formaldehyde-killed whole bacilli vaccine is the only US-licensed vaccine for plague and requires a series of injections. Unfortunately, the vaccine does not protect against primary pulmonary plague, and its manufacture was discontinued in 1999. The EV (initials from the plague patient's name) vaccine was initially developed in the early 1900s and has been used since then in some parts of the world. However, the vaccine strain is not avirulent, and its safety in humans has been questioned. An ideal vaccine candidate should have an important role in pathogenicity and/or survival of pathogens during infection. The current interest is in developing plague vaccines that consist of purified protein subunits, with improved protection and reduced side effects.[91–93] The subunit vaccines, which consist of a capsular subunit F1 and/or the major virulence V antigen, provide excellent long-lasting protection in a number of animal models, including nonhuman primates.[94] The efficacy of this formulation for human use remains to be determined. The identification of novel protective antigens and their combined use with F1 and V antigens[92,95,96] as multicomponent subunit vaccines have been widely considered as one of the leading strategies for plague vaccine development in the future.[97] Thus, a safe, effective, and licensed vaccine for plague prevention is still unavailable.

Other measures should be taken to prevent plague during outbreaks. For bubonic plague, patients should be isolated for timely medical treatment. The possible site where the patient acquired the infection should be monitored for potential infected individuals. If more people are afflicted, the infectious area should be decontaminated by controlling fleas and mice using pesticides. For primary and secondary pneumonic plague, aside from patient isolation, all close contacts should be isolated for medical observation.[28] Sometimes, simple preventive measures could effectively prevent the spread of plague from person to person, including the wearing of masks and avoiding close patient contact (more than 2 m away).[28]

36.7 CONCLUSIONS AND FUTURE PERSPECTIVES

Y. pestis has caused tragedies in human history and has reshaped our civilization. It is one of the most dangerous bacterial pathogens because it has the capacity to spread from person to person; thus, it has been listed as a category A bioterrorism agent. *Y. pestis* is a typical zoonotic pathogen, with rodents as reservoirs and fleas as vectors. Its life cycle involves the interaction between among, reservoirs, vectors, and the environment. Therefore, it could be used as a model for studying pathogen evolution.[98–102]

Although *Y. pestis* was identified more than 100 years ago, many of its facets remain unclear, including its pathogenesis, host immune responses, its evolution, and its natural life cycles. The application of newly developed technologies, such as omics-based methods,[103] to clarify these basic issues, will help find targets for developing novel diagnostic procedures, vaccines, and drugs. Considering *Y. pestis* as a dangerous bioterrorism agent, techniques for the rapid, highly specific, and sensitive detection of this pathogen on-site that are easy to use are urgently needed, although a number of immunologic and nucleic acid-based assays have been developed to date.[77,82–84,104–111] Vaccine development is still a continuous and difficult task for preventing *Y. pestis* infections. *Y. pestis* could also be employed as a model for developing strategies for handling bioterrorism events because it is a typical pathogen with zoonotic and epidemic features, and it has historically been associated with bioterrorism or biowarfare. After the "911" anthrax spore letter bioterrorism in the United States in 2001, a new field, microbial forensics, was proposed for the source tracing of *Bacillus anthracis*.[112] A key step for the success of microbial forensics is the development of an international database for source-tracing analysis. *Y. pestis* represents one of the best model organisms for developing such a database.[112]

ACKNOWLEDGMENTS

We thank members of our laboratory for helpful comments and suggestions. *Y. pestis* works in our laboratory have been supported by grants from the National Basic Research Program of China (2009CB522600), the National Natural Science Foundation of China (Nos. 30930001, 30430620, 30371284, 30471554, 31000015, and 31071111), and the National Science Foundation of China for Distinguished Young Scholars (No. 30525025).

REFERENCES

1. Perry, R.D. and Fetherston, J.D. *Yersinia pestis*—Etiologic agent of plague. *Clin. Microbiol. Rev.* 10, 35–66 (1997).
2. Anisimov, A.P., Lindler, L.E., and Pier, G.B. Intraspecific diversity of *Yersinia pestis. Clin. Microbiol. Rev.* 17, 434–464 (2004).
3. Inglesby, T.V. et al. Plague as a biological weapon: Medical and public health management. Working Group on Civilian Biodefense. *JAMA* 283, 2281–2290 (2000).
4. Drea, E. et al. *Researching Japanese War Crimes Records: An Introductory Essay.* Nazi War Crimes and Japanese Imperial Government Records Interagency Working Group, ISBN: 1-880875-28-4 (2007).
5. Louie, A. et al. Use of an in vitro pharmacodynamic model to derive a moxifloxacin regimen that optimizes kill of *Yersinia pestis* and prevents emergence of resistance. *Antimicrob. Agents Chemother.* 55, 822–830 (2011).
6. Bibel, D.J. and Chen, T.H. Diagnosis of plaque: An analysis of the Yersin-Kitasato controversy. *Bacteriol. Rev.* 40, 633–651 (1976).
7. Skerman, V.B.D., McGowan, V., and Sneath, P.H.A. Approved lists of bacterial names. *Int. J. Syst. Bacteriol.* 30, 225–420 (1980).
8. Van Loghem, J.J. The classification of the plague-bacillus. *Antonie Van Leeuwenhoek* 10, 15 (1944).
9. Bercovier, H. et al. Intra- and interspecies relatedness of *Yersinia pestis* by DNA hybridization and its relationship to *Yersinia pseudotuberculosis. Curr. Microbiol.* 4, 225–229 (1980).
10. Judicial Commission of ICSB: Rejection of the name *Yersinia pseudotuberculosis* subsp. *pestis* (van Loghem) Bercovier et al. 1981 and conservation of the name *Yersinia pestis* (Lehmann and Neumann) van Loghem 1944 for the plague bacillus. *Int. J. Syst. Bacteriol.* 35, 540 (1985).
11. Anisimov, A.P. and Dyatlov, I.A. A novel mechanism of antibiotic resistance in plague. *J. Med. Microbiol.* 46, 887–889 (1997).
12. Dennis, D.T. and Hughes, J.M. Multidrug resistance in plague. *N. Engl. J. Med.* 337, 702–704 (1997).
13. Ji, S. et al. A study on typing of *Yersinia pestis* in China and its ecologioepidmiologica significance (in Chinese). *Chin. J. Epidemiol.* 11, 60–66 (1990).
14. Ji, S. et al. The discovery and research of plague natural foci in China (in Chinese). *Chin. J. Epidemiol.* 11(Suppl), 1–41 (1990).
15. Zhou, D. et al. Genetics of metabolic variations between *Yersinia pestis* biovars and the proposal of a new biovar, microtus. *J. Bacteriol.* 186, 5147–5152 (2004).
16. Devignat, R. Varieties of *Pasteurella pestis*; new hypothesis. *Bull. World Health Organ.* 4, 247–263 (1951).
17. Motin, V.L. et al. Genetic variability of *Yersinia pestis* isolates as predicted by PCR-based IS100 genotyping and analysis of structural genes encoding glycerol-3-phosphate dehydrogenase (*glpD*). *J. Bacteriol.* 184, 1019–1027 (2002).
18. Hai, R. et al. Molecular biological characteristics and genetic significance of *Yersinia pestis* in China. *Zhonghua Liu Xing Bing Xue Za Zhi* 25, 509–513 (2004).
19. Zhou, D. et al. Comparative and evolutionary genomics of *Yersinia pestis. Microbes Infect.* 6, 1226–1234 (2004).
20. Golubov, A. et al. Structural organization of the pFra virulence-associated plasmid of rhamnose-positive *Yersinia pestis. Infect. Immun.* 72, 5613–5621 (2004).
21. Li, Y. et al. Genotyping and phylogenetic analysis of *Yersinia pestis* by MLVA: Insights into the worldwide expansion of Central Asia plague foci. *PLoS One* 4, e6000 (2009).
22. Cui, Y. et al. Insight into microevolution of *Yersinia pestis* by clustered regularly interspaced short palindromic repeats. *PLoS One* 3, e2652 (2008).
23. Guiyoule, A. et al. Plague pandemics investigated by ribotyping of *Yersinia pestis* strains. *J. Clin. Microbiol.* 32, 634–641 (1994).
24. Williams, G.J. and Stickler, D.J. Effect of triclosan on the formation of crystalline biofilms by mixed communities of urinary tract pathogens on urinary catheters. *J. Med. Microbiol.* 57, 1135–1140 (2008).
25. Kingston, J.J. et al. Genotyping of Indian *Yersinia pestis* strains by MLVA and repetitive DNA sequence based PCRs. *Antonie Van Leeuwenhoek* 96, 303–312 (2009).
26. Morelli, G. et al. *Yersinia pestis* genome sequencing identifies patterns of global phylogenetic diversity. *Nat. Genet.* 42, 1140–1143 (2010).
27. Achtman, M. et al. Microevolution and history of the plague bacillus, *Yersinia pestis. Proc. Natl. Acad. Sci. U.S.A.* 101, 17837–17842 (2004).
28. Wang, H. et al. A dog-associated primary pneumonic plague in Qinghai Province, China. *Clin. Infect. Dis.* 52, 185–190 (2011).
29. Bos, K.I. et al. A draft genome of *Yersinia pestis* from victims of the Black Death. *Nature* 478, 506–510 (2011).
30. Drancourt, M. et al. *Yersinia pestis* Orientalis in remains of ancient plague patients. *Emerg. Infect. Dis.* 13, 332–333 (2007).
31. Rose, L.J. et al. Survival of *Yersinia pestis* on environmental surfaces. *Appl. Environ. Microbiol.* 69, 2166–2171 (2003).
32. Eisen, R.J. and Gage, K.L. Adaptive strategies of *Yersinia pestis* to persist during inter-epizootic and epizootic periods. *Vet. Res.* 40, 1 (2008).
33. Eisen, R.J. et al. Persistence of *Yersinia pestis* in soil under natural conditions. *Emerg. Infect. Dis.* 14, 941–943 (2008).
34. Ayyadurai, S. et al. Long-term persistence of virulent *Yersinia pestis* in soil. *Microbiology* 154, 2865–2871 (2008).
35. Liu, P. *The Atlas of Plague and Its Environment in the People's Republic of China* (in Chinese and English). Science Press, Beijing, China (2000).
36. Davis, S. et al. The abundance threshold for plague as a critical percolation phenomenon. *Nature* 454, 634–637 (2008).
37. Zhou, D. et al. DNA microarray analysis of genome dynamics in *Yersinia pestis*: Insights into bacterial genome microevolution and niche adaptation. *J. Bacteriol.* 186, 5138–5146 (2004).
38. Li, Y. et al. Different region analysis for genotyping *Yersinia pestis* isolates from China. *PLoS One* 3, e2166 (2008).
39. Li, M. et al. Asymptomatic *Yersinia pestis* infection, China. *Emerg. Infect. Dis.* 11, 1494–1496 (2005).
40. Lorange, E.A. et al. Poor vector competence of fleas and the evolution of hypervirulence in *Yersinia pestis. J. Infect. Dis.* 191, 1907–1912 (2005).
41. Reed, W.P. et al. Bubonic plague in the Southwestern United States. A review of recent experience. *Medicine (Baltimore)* 49, 465–486 (1970).
42. Darby, C. Uniquely insidious: *Yersinia pestis* biofilms. *Trends Microbiol.* 16, 158–164 (2008).

43. Hinnebusch, B.J. and Erickson, D.L. *Yersinia pestis* biofilm in the flea vector and its role in the transmission of plague. *Curr. Top. Microbiol. Immunol.* 322, 229–248 (2008).

44. Bobrov, A.G. et al. Insights into *Yersinia pestis* biofilm development: Topology and co-interaction of Hms inner membrane proteins involved in exopolysaccharide production. *Environ. Microbiol.* 10, 1419–1432 (2008).

45. Cotter, P.A. and Stibitz, S. c-Di-GMP-mediated regulation of virulence and biofilm formation. *Curr. Opin. Microbiol.* 10, 17–23 (2007).

46. Kirillina, O. et al. HmsP, a putative phosphodiesterase, and HmsT, a putative diguanylate cyclase, control Hms-dependent biofilm formation in *Yersinia pestis. Mol. Microbiol.* 54, 75–88 (2004).

47. Simm, R. et al. Phenotypic convergence mediated by GGDEF-domain-containing proteins. *J. Bacteriol.* 187, 6816–6823 (2005).

48. Sun, Y.C. et al. Differential control of *Yersinia pestis* biofilm formation in vitro and in the flea vector by two c-di-GMP diguanylate cyclases. *PLoS One* 6, e19267 (2011).

49. Bobrov, A.G. et al. Systematic analysis of cyclic di-GMP signalling enzymes and their role in biofilm formation and virulence in Yersinia pestis. *Mol. Microbiol.* 79, 533–551 (2011).

50. Bobrov, A.G., Kirillina, O., and Perry, R.D. The phosphodiesterase activity of the HmsP EAL domain is required for negative regulation of biofilm formation in *Yersinia pestis. FEMS Microbiol. Lett.* 247, 123–130 (2005).

51. Erickson, D.L. et al. Serotype differences and lack of biofilm formation characterize *Yersinia pseudotuberculosis* infection of the *Xenopsylla cheopis* flea vector of *Yersinia pestis. J. Bacteriol.* 188, 1113–1119 (2006).

52. Erickson, D.L. et al. Loss of a biofilm-inhibiting glycosyl hydrolase during the emergence of *Yersinia pestis. J. Bacteriol.* 190, 8163–8170 (2008).

53. Sun, Y.C., Hinnebusch, B.J., and Darby, C. Experimental evidence for negative selection in the evolution of a *Yersinia pestis* pseudogene. *Proc. Natl. Acad. Sci. U.S.A.* 105, 8097–8101 (2008).

54. Hinnebusch, B.J. et al. Role of *Yersinia* murine toxin in survival of *Yersinia pestis* in the midgut of the flea vector. *Science* 296, 733–735 (2002).

55. Zhou, D. and Yang, R. Formation and regulation of *Yersinia* biofilms. *Protein Cell* 2, 173–179 (2011).

56. Zhou, D., Han, Y., and Yang, R. Molecular and physiological insights into plague transmission, virulence and etiology. *Microbes Infect.* 8, 273–284 (2006).

57. Cavanaugh, D.C. and Randall, R. The role of multiplication of *Pasteurella pestis* in mononuclear phagocytes in the pathogenesis of flea-borne plague. *J. Immunol.* 83, 348–363 (1959).

58. Lukaszewski, R.A. et al. Pathogenesis of *Yersinia pestis* infection in BALB/c mice: Effects on host macrophages and neutrophils. *Infect. Immun.* 73, 7142–7150 (2005).

59. Sebbane, F. et al. Kinetics of disease progression and host response in a rat model of bubonic plague. *Am. J. Pathol.* 166, 1427–1439 (2005).

60. Lathem, W.W. et al. Progression of primary pneumonic plague: A mouse model of infection, pathology, and bacterial transcriptional activity. *Proc. Natl. Acad. Sci. U.S.A.* 102, 17786–17791 (2005).

61. Zhou, D. and Yang, R. Molecular Darwinian evolution of virulence in *Yersinia pestis. Infect. Immun.* 77, 2242–2250 (2009).

62. Eisen, R.J. et al. Early-phase transmission of *Yersinia pestis* by unblocked fleas as a mechanism explaining rapidly spreading plague epizootics. *Proc. Natl. Acad. Sci. U.S.A.* 103, 15380–15385 (2006).

63. Eisen, R.J. et al. Early-phase transmission of *Yersinia pestis* by cat fleas (*Ctenocephalides felis*) and their potential role as vectors in a plague-endemic region of Uganda. *Am. J. Trop. Med. Hyg.* 78, 949–956 (2008).

64. Vetter, S.M. et al. Biofilm formation is not required for early-phase transmission of *Yersinia pestis. Microbiology* 156, 2216–2225 (2010).

65. Wong, D. et al. Primary pneumonic plague contracted from a mountain lion carcass. *Clin. Infect. Dis.* 49, e33–e38 (2009).

66. Joshi, K. et al. Epidemiological features of pneumonic plague outbreak in Himachal Pradesh, India. *Trans. R. Soc. Trop. Med. Hyg.* 103, 455–460 (2009).

67. Bubonic and pneumonic plague—Uganda, 2006. *MMWR Morb. Mortal. Wkly. Rep.* 58, 778–781 (2009).

68. Gupta, M.L. and Sharma, A. Pneumonic plague, northern India, 2002. *Emerg. Infect. Dis.* 13, 664–666 (2007).

69. Begier, E.M. et al. Pneumonic plague cluster, Uganda, 2004. *Emerg. Infect. Dis.* 12, 460–467 (2006).

70. Kellogg, W.H. An epidemic of pneumonic plague. *Am. J. Public Health (N.Y.)* 10, 599–605 (1920).

71. Bogen, E. The pneumonic plague in Los Angeles. *Cal. West Med.* 23, 175–176 (1925).

72. Meyer, K.F. Pneumonic plague. *Bacteriol. Rev.* 25, 249–261 (1961).

73. Nishiura, H. Epidemiology of a primary pneumonic plague in Kantoshu, Manchuria, from 1910 to 1911: Statistical analysis of individual records collected by the Japanese Empire. *Int. J. Epidemiol.* 35, 1059–1065 (2006).

74. Wu, L.-T., Chun, J.W.H., and Pollitzer, R. *Plague: A Manual for Medical and Public Health Workers.* National Quarantine Service, Shanghai, China (1936).

75. Nishiura, H. et al. Transmission potential of primary pneumonic plague: Time inhomogeneous evaluation based on historical documents of the transmission network. *J. Epidemiol. Community Health* 60, 640–645 (2006).

76. Koirala, J. Plague: Disease, management, and recognition of act of terrorism. *Infect. Dis. Clin. North Am.* 20, 273–287, viii (2006).

77. Qu, S. et al. Ambient stable quantitative PCR reagents for the detection of *Yersinia pestis. PLoS Negl. Trop. Dis.* 4, e629 (2010).

78. Griffin, K.A. et al. Detection of *Yersinia pestis* DNA in prairie dog-associated fleas by polymerase chain reaction assay of purified DNA. *J. Wildl. Dis.* 46, 636–643 (2010).

79. Franklin, H.A., Stapp, P., and Cohen, A. Polymerase chain reaction (PCR) identification of rodent blood meals confirms host sharing by flea vectors of plague. *J. Vector Ecol.* 35, 363–371 (2010).

80. Bakanidze, L. et al. Polymerase chain reaction assays for the presumptive identification of *Yersinia pestis* strains in Georgia. *Adv. Exp. Med. Biol.* 529, 333–336 (2003).

81. Hong, W. et al. Development of an up-converting phosphor technology-based 10-channel lateral flow assay for profiling antibodies against *Yersinia pestis. J. Microbiol. Methods* 83, 133–140 (2010).

82. Wei, H. et al. Sensitive detection of antibody against antigen F1 of *Yersinia pestis* by an antigen sandwich method using a portable fiber optic biosensor. *Sens. Actuators B* 127, 525–530 (2007).

83. Wei, H. et al. Direct detection of *Yersinia pestis* from the infected animal specimens by a fiber optic biosensor. *Sens. Actuators B* 123, 204–210 (2007).

84. Yan, Z.Q. et al. Rapid quantitative detection of *Yersinia pestis* by lateral-flow immunoassay and up-converting phosphor technology-based biosensor. *Sens. Actuators B* 119, 656–663 (2006).

85. Roussos, D. Plague. *Prim. Care Update Ob./Gyns.* 9, 125–128 (2002).

86. Meyer, K.F. Modern therapy of plague. *J. Am. Med. Assoc.* 144, 982–985 (1950).

87. Zaini, K.R. Role of streptomycin in plague. *Antiseptic* 49, 560–562 (1952).

88. Anisimov, A.P. and Amoako, K.K. Treatment of plague: Promising alternatives to antibiotics. *J. Med. Microbiol.* 55, 1461–1475 (2006).

89. Galimand, M. et al. Multidrug resistance in *Yersinia pestis* mediated by a transferable plasmid. *N. Engl. J. Med.* 337, 677–680 (1997).

90. Guiyoule, A. et al. Transferable plasmid-mediated resistance to streptomycin in a clinical isolate of *Yersinia pestis*. *Emerg. Infect. Dis.* 7, 43–48 (2001).

91. Williamson, E.D. Plague vaccine research and development. *J. Appl. Microbiol.* 91, 606–608 (2001).

92. Titball, R.W. and Williamson, E.D. Vaccination against bubonic and pneumonic plague. *Vaccine* 19, 4175–4184 (2001).

93. Titball, R.W. and Williamson, E.D. *Yersinia pestis* (plague) vaccines. *Expert Opin. Biol. Ther.* 4, 965–973 (2004).

94. Wang, Z. et al. Long-term observation of subunit vaccine F1-rV270 against *Yersinia pestis* in mice. *Clin. Vaccine Immunol.* 17, 199–201 (2010).

95. Williamson, E.D. et al. A new improved sub-unit vaccine for plague: The basis of protection. *FEMS Immunol. Med. Microbiol.* 12, 223–230 (1995).

96. Grosfeld, H. et al. Vaccination with plasmid DNA expressing the *Yersinia pestis* capsular protein F1 protects mice against plague. *Adv. Exp. Med. Biol.* 529, 423–424 (2003).

97. Matson, J.S. et al. Immunization of mice with YscF provides protection from *Yersinia pestis* infections. *BMC Microbiol.* 5, 38 (2005).

98. Ben Ari, T. et al. Plague and climate: Scales matter. *PLoS Pathog.* 7, e1002160 (2011).

99. Davis, S. et al. Predictive thresholds for plague in Kazakhstan. *Science* 304, 736–738 (2004).

100. Easterday, W.R. et al. An additional step in the transmission of *Yersinia pestis*? *ISME J.* 6, 231–236 (2012).

101. Samia, N.I. et al. Dynamics of the plague–wildlife–human system in Central Asia are controlled by two epidemiological thresholds. *Proc. Natl. Acad. Sci. U.S.A.* 108, 14527–14532 (2011).

102. Stenseth, N.C. et al. Plague dynamics are driven by climate variation. *Proc. Natl. Acad. Sci. U.S.A.* 103, 13110–13115 (2006).

103. Teiten, M.H. et al. OMICS, a multidisciplinary friendship. *Cell Death Dis.* 3, e267 (2012).

104. Anderson, G.P., Breslin, K.A., and Ligler, F.S. Assay development for a portable fiberoptic biosensor. *Asaio J.* 42, 942–946 (1996).

105. Cao, L.K. et al. Detection of *Yersinia pestis* fraction 1 antigen with a fiber optic biosensor. *J. Clin. Microbiol.* 33, 336–341 (1995).

106. Ghindilis, A.L. et al. CombiMatrix oligonucleotide arrays: Genotyping and gene expression assays employing electrochemical detection. *Biosens. Bioelectron.* 22, 1853–1860 (2007).

107. Meyer, M.H. et al. Magnetic biosensor for the detection of *Yersinia pestis*. *J. Microbiol. Methods* 68, 218–224 (2007).

108. Weller, S.A. et al. Evaluation of two multiplex real-time PCR screening capabilities for the detection of *Bacillus anthracis*, *Francisella tularensis*, and *Yersinia pestis* in blood samples generated from murine infection models. *J. Med. Microbiol.* 61, 1546–1555 (2012).

109. Rachwal, P.A. et al. The potential of TaqMan array cards for detection of multiple biological agents by real-time PCR. *PLoS One* 7, e35971 (2012).

110. Malou, N. et al. Immuno-PCR—A new tool for paleomicrobiology: The plague paradigm. *PLoS One* 7, e31744 (2012).

111. Matero, P. et al. Rapid field detection assays for *Bacillus anthracis*, *Brucella* spp., *Francisella tularensis* and *Yersinia pestis*. *Clin. Microbiol. Infect.* 17, 34–43 (2011).

112. Yang, R.F. and Keim, P. Microbial forensics: A powerful tool for pursuing bioterrorism perpetrators and the need for an international database. *J. Bioterr. Biodef.* S3:007 (2012). doi:10.4172/2157-2526.S3-007.

113. Lathem, W.W. et al. A plasminogen-activating protease specifically controls the development of primary pneumonic plague. *Science* 315, 509–513 (2007).

114. Sebbane, F. et al. Role of the *Yersinia pestis* plasminogen activator in the incidence of distinct septicemic and bubonic forms of flea-borne plague. *Proc. Natl. Acad. Sci. U.S.A.* 103, 5526–5530 (2006).

115. Kolodziejek, A.M. et al. Phenotypic characterization of OmpX, an Ail homologue of *Yersinia pestis* KIM. *Microbiology* 153, 2941–2951 (2007).

116. Forman, S. et al. yadBC of *Yersinia pestis*, a new virulence determinant for bubonic plague. *Infect. Immun.* 76, 578–587 (2008).

117. Felek, S., Lawrenz, M.B., and Krukonis, E.S. The *Yersinia pestis* autotransporter YapC mediates host cell binding, auto-aggregation and biofilm formation. *Microbiology* 154, 1802–1812 (2008).

118. Lawrenz, M.B., Lenz, J.D., and Miller, V.L. A novel autotransporter adhesin is required for efficient colonization during bubonic plague. *Infect. Immun.* 77, 317–326 (2008).

119. Pujol, C. et al. Replication of *Yersinia pestis* in interferon gamma-activated macrophages requires *ripA*, a gene encoded in the pigmentation locus. *Proc. Natl. Acad. Sci. U.S.A.* 102, 12909–12914 (2005).

120. Grabenstein, J.P. et al. Characterization of phagosome trafficking and identification of PhoP-regulated genes important for survival of *Yersinia pestis* in macrophages. *Infect. Immun.* 74, 3727–3741 (2006).

121. Perry, R.D., Mier, I., Jr., and Fetherston, J.D. Roles of the Yfe and Feo transporters of *Yersinia pestis* in iron uptake and intracellular growth. *Biometals* 20, 699–703 (2007).

122. Ramamurthi, K.S. and Schneewind, O. Type III protein secretion in *Yersinia* species. *Annu. Rev. Cell Dev. Biol.* 18, 107–133 (2002).

123. Du, Y., Rosqvist, R., and Forsberg, A. Role of fraction 1 antigen of *Yersinia pestis* in inhibition of phagocytosis. *Infect. Immun.* 70, 1453–1460 (2002).

124. Huang, X.Z. and Lindler, L.E. The pH 6 antigen is an antiphagocytic factor produced by *Yersinia pestis* independent of *Yersinia* outer proteins and capsule antigen. *Infect. Immun.* 72, 7212–7219 (2004).

125. Hares, M.C. et al. The *Yersinia pseudotuberculosis* and *Yersinia pestis* toxin complex is active against cultured mammalian cells. *Microbiology* 154, 3503–3517 (2008).

126. Kukkonen, M. et al. Lack of O-antigen is essential for plasminogen activation by *Yersinia pestis* and *Salmonella enterica*. *Mol. Microbiol.* 51, 215–225 (2004).

127. Bearden, S.W. and Perry, R.D. The Yfe system of *Yersinia pestis* transports iron and manganese and is required for full virulence of plague. *Mol. Microbiol.* 32, 403–414 (1999).

Section III

Microbes and Toxins Affecting Humans and Animals: Fungus and Parasite

37 Coccidioides (C. posadasii and C. immitis)

Rossana de Aguiar Cordeiro, Marcos Fábio Gadelha Rocha,
Raimunda Sâmia Nogueira Brilhante, and José Júlio Costa Sidrim

CONTENTS

37.1 INTRODUCTION

Coccidioides spp. are dimorphic fungi that cause coccidioidomycosis, a deep-seated infection that occurs in humans and several mammal species. The disease is caused by two nearly identical species, *Coccidioides immitis* and *Coccidioides posadasii*, which share many biological traits regarding immunogenicity, pathogenicity, and virulence. Both species are soil-inhabiting fungi that occur in arid and semiarid areas in the Americas. *Coccidioides* spp. are considered the most virulent fungal pathogens of humans and animals,[1] being classified as biosafety level 3 microorganisms by regulatory agencies around the world. *Coccidioides* are considered important laboratory hazard and potential bioweapons.[2]

37.2 CLASSIFICATION, BIOLOGY, AND MORPHOLOGY

Coccidioides spp. are mitosporic fungi (phylum Ascomycota), classified into the order Onygenales, which encompasses a broad range of fungal species, including pathogens of immunocompetent animals, such as *Histoplasma capsulatum* and *Paracoccidioides brasiliensis*, as well as plant pathogens and saprobes.[3] *C. immitis* and *C. posadasii* are heterothallic species that have a mating-type sexual reproduction system, *MAT1-1* and *MAT1-2*, even though the sexual phase of both species is unknown in nature.[4]

Coccidioides spp. are telluric organisms that form hyaline hyphae during the filamentous phase of the biological cycle. During their maturation, the hyphae undergo enterothallic conidiogenesis, originating from arthroconidia interspaced with cells lacking cytoplasmatic material, called disjunctor cells. The arthroconidia are easily detached from the vegetative mycelium and contain at their extremities remnants of the disjunctor cell walls. This facilitates their dispersion through air currents.[5] After being inhaled by a susceptible host, the arthroconidia undergo morphological changes, evolving into spherules that release a great number of endospores after reaching cell maturity.[5,6] When these endospores reach the soil under adequate conditions, they grow into the filamentous form, thus guaranteeing the continuity of the microorganism's biological cycle.

Dimorphic conversion in vitro depends on some nutritional requirements and suitable temperature, but in general, the full conversion of cells is considered a difficult task. Parasitic spherules may be attained at 40°C in 20% CO_2 atmosphere[7] in a synthetic medium supplemented with the dispersant N-Tamol.[8]

37.3 EPIDEMIOLOGY

Coccidioidomycosis occurs exclusively in the Western Hemisphere, at latitudes between 40°N and 40°S, from California to the south of Argentina.[9,10] However, autochthone cases have been described only in the United States, Mexico, Honduras, Guatemala, Colombia, Venezuela, Bolivia, Paraguay, Argentina, and Brazil.[10–13] *C. immitis* seems to be restricted to California, but it may also occur in Arizona and Mexico. *C. posadasii* is found in a broader area, being reported throughout the Americas.[14] The exact incidence of coccidioidomycosis in the Americas is unknown because approximately 60% of the infected individuals are asymptomatic. In addition, South American countries do not have a compulsory notification of coccidioidomycosis cases. Nevertheless, it is estimated that around 150,000 new cases of coccidioidomycosis occur annually in the United States.[15]

Since the respiratory route is the main route source of coccidioidomycosis,[15] occupational activities with dust exposition—such as archeological digging[16] and military training in open natural fields[17]—present a high risk of infection. The increase in the incidence of coccidioidomycosis in recent years in the United States is likely related to the progressive growth of immunocompromised people living in high-risk areas, as well as by the rising in the number of immigrants from nonendemic areas.[10,18]

The disease occurs more frequently in males, which is likely due to occupational exposure to contaminated dust. However, laboratory studies have shown that human sexual hormones may stimulate spherule maturation rates and endospore release,[19] which may explain the sex- and pregnancy-related tendency of dissemination.[10,20]

Epidemiological data reveal that coccidioidomycosis occurs in all age groups, but infants and older people are at higher risk for the severe forms of the disease, including disseminated illness and chronic pulmonary infection.[10,21,22] The risk factors for the disseminated disease also include genetic/racial heritage (African Americans and Filipinos are 10–175 times more likely to develop severe forms of the disease), immunosuppressive conditions such as HIV infection, cancer (particularly Hodgkin's lymphoma), and organ transplantation.[23–25] In addition, pregnancy (mainly in the third trimester) and postpartum are also considered risk conditions for disseminated coccidioidomycosis.[10,20]

Epidemiological studies involving cutaneous tests with spherulin and/or coccidioidin have shown a great variability in the incidence of coccidioidomycosis in endemic areas. In the United States, reactivity indexes of 64% have already been detected,[26] but these values can vary depending on the population studied. The estimated coccidioidomycosis incidence varies, respectively, in Mexico (10%–93%), Honduras (2.5%–50%), Guatemala (0.5%–42%), Paraguay (0.9%–50%), Argentina (7.3%–40%), and Brazil (0%–12%).[10,13,27–30]

37.4 CLINICAL FEATURES AND PATHOGENESIS

Coccidioidomycosis can be classified as asymptomatic, acute, or chronic pneumonia of disseminated or primary cutaneous type.[15,22] It is estimated that up to 60% of individuals exposed to infective conidia are asymptomatic, with the presence of the fungus only being detected through serological inquiry or cutaneous tests. Nearly all of these individuals develop permanent immunity and are therefore protected against second infections.[22] In approximately 40% of individuals, pneumonia can occur within 1–3 weeks after infection.[22] However, during epidemic outbreaks caused by earthquakes or participation in large construction projects or excavation of archeological sites, the rate of symptomatic individuals can reach up to 90%.[31,32]

The most common clinical syndrome is pneumonia,[33] characterized by cough, dyspnea and thoracic pain, fever, fatigue, anorexia, and arthralgia.[34] The acute pulmonary primary form can regress spontaneously, after a few months, even without specific therapy.[25] Among patients with pulmonary disease, approximately 5% develop erythema nodosum or erythema multiform.[6,35] Other cutaneous manifestations, such as acute exanthema, Sweet's syndrome, and interstitial granulomatous dermatitis, can also be observed.[36]

Approximately 5% of patients with primary pneumonia do not get well spontaneously, and the disease can evolve to chronic pulmonary infection, characterized by nodular lesions or fibro-cavitary pulmonary disease.[25] Patients with chronic pulmonary coccidioidomycosis present nocturnal sudoresis, muscular fatigue, weight loss, chronic cough, and hemoptysis.[22,25] Due to clinical, radiographic, and histopathological resemblances, progressive pulmonary coccidioidomycosis can be mistaken for pulmonary tuberculosis.[37]

Disseminated coccidioidomycosis can affect 1%–5% of infected individuals,[25] and it can occur even when there are no clinical or radiological signs of pulmonary disease.[6] Disseminated coccidioidomycosis most frequently affects immunocompromised individuals.[15] Fungal dissemination occurs through hematogenic or lymphatic viae and can reach several sites, such as lymph nodes, skin, bones, central nervous system, joints, and genital–urinary apparatus.[15,25] Coccidioidal meningitis is the most severe form of the disseminated disease and has a high mortality index when it is not diagnosed and properly treated.[25] The most common complications associated with coccidioidal meningitis are hydrocephalus, vasculitis, and parenchymal abscesses.[38]

Although not common, pericardiac involvement in coccidioidomycosis can result from contiguous pleural–pulmonary lesions or from lymphohematogenous dissemination. In these patients, clinical complaints are not very different from those of patients with pulmonary involvement, which are thoracic pain, cough, and dyspnea. Coccidioidal pericarditis usually presents an unfavorable prognosis, with high rates of morbidity and mortality.[39,40]

Primary cutaneous disease is rare and results from traumatic inoculation of the microorganism from an environmental source or through laboratory manipulation.[6,15,34] Clinical signs include papules, nodules, and verrucose plaques that can evolve to the formation of ulcers and abscesses.[25] Due to the clinical diversity of the lesions, cutaneous coccidioidomycosis can be mistaken for several other diseases, making a laboratory confirmation necessary.[25]

37.5 LABORATORY DIAGNOSIS

Coccidioidomycosis does not have pathognomonic clinical–radiological findings, and it can be mistaken for other infectious diseases.[12,41] Thus, a laboratory confirmation is essential for correct diagnosis.

Direct examination of the clinical specimen must be performed through optical microscopy with slide preparations and smears. Depending on the clinical presentation, several biological specimens may be sent to mycology laboratories for analysis of *Coccidioides*, such as sputum, bronchoalveolar wash, cerebrospinal fluid (CSF), aspirates of osteoarticular lesions, pleural fluid, and skin and pulmonary tissue samples for biopsy.[12,39,42–44]

Under direct examination, using wet mounts on glass slides, clinical specimens are initially clarified by KOH 10%–40% or calcofluor white fluorescent stain (CFW) to facilitate spherule visualization. Smears and imprints can also be stained by Grocott's methenamine silver (GMS). Microscopic examination shows spherules of up to 100 μm diameter, with birefringent walls, containing endospores of 2–5 μm.[42,43] This image is considered pathognomonic for the parasitic phase of *Coccidioides* (Figure 37.1a). Immature spherules can be mistaken for other agents of endemic mycosis, such as *Blastomyces dermatitidis*[42,44] and *P. brasiliensis*, respectively. Some authors describe the morphological similarity between *Coccidioides* spherules and those produced

in vivo by *Rhinosporidium seeberi*, *Emmonsia* spp.,[45] and *Prototheca* spp.[46] Free endospores can be mistaken for conidia of *H. capsulatum*, *Cryptococcus* spp., or *Candida* spp.[42,43] Although cheap and feasible when performed by experienced mycologists, direct examination has low sensitivity. In patients with acute pulmonary disease, positive rates ranging from 30% to 40% have been described.[47]

Rarely, arthroconidiated filamentous structures can be formed in CSF, during CNS infections[42,43] and in pleural fluid.[46] The presentation of atypical forms in respiratory specimens can be four times more frequent in patients with cavitary lesions who have type 2 diabetes when compared with patients who are not affected by this comorbidity.[48] This finding is of great importance, since these atypical parasitic structures are highly infectious and can cause infection in poorly trained lab technicians.

Despite the advances provided by molecular and immunological techniques in recent years, isolation and growth of

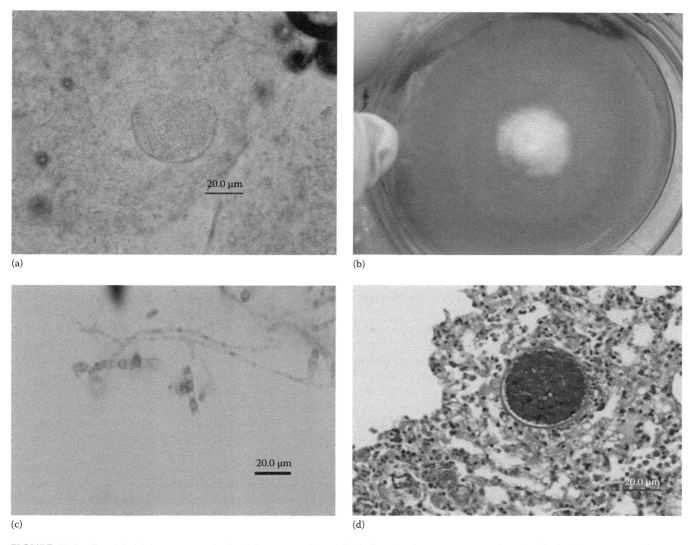

(a)

(b)

(c)

(d)

FIGURE 37.1 Coccidioidal structures obtained from experimentally infected mice. (a) Mature spherule filled with endospores in lung tissue (KOH 10%); (b) macroscopic aspect of *C. posadasii* culture in Sabouraud dextrose agar; (c) hyaline hyphae with arthroconidia and disjunctor cells; and (d) histopathology of lungs revealing parasitic coccidioidal spherule.

Coccidioides is still the gold standard for laboratory diagnosis of coccidioidomycosis.[44] Primary cultures obtained from clinical specimens develop quickly, within approximately 5 days with incubation at 25°C–30°C. The microorganism grows on most of the media routinely used in mycology laboratories, such as 2% Sabouraud dextrose agar, BHI agar, and potato dextrose agar. However, fungal recovery from clinical specimens can sometimes take 3 weeks or longer, especially when the patient is using antimicrobial drugs.[39,44] Usually, young colonies, with up to 5 days, are apiculated, glabrous, white colored, and do not exhibit arthroconidia. As the colonies mature, they become apiculated, velvety or cottony, and white, cream, or gray colored (Figure 37.1b), and under microscopy, they exhibit septate hyaline hyphae having arthroconidia interspaced by disjunctor cells lacking cytoplasmatic material, which are 5–7.5 μm long and 2.5–5 μm wide[49] (Figure 37.1c). In mature colonies, arthroconidia may be easily detached from the hyphae, becoming airborne and spreading rapidly by air flow. Because of this ease of dissemination in the environment, *Coccidioides* filamentous cultures must be manipulated inside a biological safety cabinet Class II B2 in a biological safety level 3 laboratory (BSL3). Petri dishes are not recommended for routine diagnosis. *Coccidioides* can also be isolated on bacteriologic media such as blood agar, chocolate agar, agar-buffered charcoal–yeast extract, Bordet–Gengou agar, and Regan–Lowe agar.[42,43]

In histopathological exams, *Coccidioides* can be identified through several staining techniques, such as hematoxylin–eosin (HE), which, although not allowing easy visualization of fungal structures, continues to be the most frequent technique used in laboratories. Other options include periodic acid-Schiff (PAS), which allows better spherule (Figure 37.1d), endospore, and hyphal individualization, and GMS, which is considered the most sensitive technique.[42,43] Other options, such as Giemsa, Papanicolaou, mucicarmine, and Gram, present low efficiency and should not be employed as primary stains for laboratory diagnosis of coccidioidomycosis.[42,43] Through histopathological examination, it is possible to observe spherules in different stages of maturation.

Cellular immune response during coccidioidomycosis can be demonstrated through positive results for the cutaneous test with the antigens coccidioidin or spherulin, which are produced from filamentous or parasitic forms of the fungus, respectively.[50–52] The use of these tests to establish diagnosis and prognosis of coccidioidomycosis is very limited, since reactivity to fungal antigens may be of little value for disease detection, because positive results can be associated with current or past infections or even cross-reactions. However, the conversion of a negative result can be useful for identifying recent fungal infections. Patients with positive intradermal reaction for coccidioidin and/or spherulin have a better prognosis, while cases where cutaneous tests are negative during the infectious process generally evolve to the disseminated disease.[6,50]

In coccidioidomycosis, the produced antibodies do not have a protective activity, but serve as tools for diagnosis, prognosis, and follow-up of patients.[53] The search for specific antibodies can be performed in various clinical specimens, such as serum, CSF, pleural, peritoneal, and synovial fluids,[53] depending on the standardization of the technique used. Tests can also be used in immunocompromised patients, though in some cases, it is necessary to combine more than one diagnostic technique.[54]

For immunodiagnosis of coccidioidomycosis, there are several commercially available diagnostic kits: immunodiffusion IDTP and IDCF, complement fixation (IMMY Immuno-Mycologics, Inc., USA), latex agglutination (IMMY Immuno-Mycologics, Inc., USA; Miravista Diagnostics, USA), ELISA for the qualitative detection of IgM and IgG antibodies (Meridian Bioscience, Inc., USA), and quantitative antigen enzyme immune assay (Miravista Diagnostics, USA). An early detection of IgM can be performed within the first to third weeks, with a positivity of 50% and 90% of the cases, respectively.[55] Approximately 90% of the individuals show reactivity to tests aimed at detecting IgG, around the fourth week, after the appearance of symptoms.[56] Regardless of the method chosen, immunological diagnosis of the disease has the advantage that it can be performed in level 2 biosafety laboratories.

A laboratory diagnosis of coccidioidomycosis based on molecular techniques is very promising, since it allows fungal detection from contaminated clinical samples and from mixed colonies.[57–59] In fact, it is the only strategy so far that allows distinguishing between the species *C. immitis* and *C. posadasii*. Besides this, the use of molecular techniques for diagnosis reduces the inherent risk of handling *Coccidioides* cultures in their filamentous phase. However, few studies have investigated the potential of molecular techniques for the diagnosis of coccidioidomycosis. Although the first techniques used for diagnosing the disease were based on hybridization reactions, currently, PCR-based protocols are preferred because they are easily executed. The first studies involving molecular detection of *Coccidioides* based on PCR were performed by Pan and Cole,[60] who established a protocol for the amplification of the 520 bp sequence from the CSA gene, which codifies a thermostable antigen of 19 kDa, directly from fungal cultures. Subsequently, several PCR-based tests were developed to confirm *Coccidioides* from cultures and clinical specimens. In this context, the gene clusters from ribosomal DNA and partial sequences from internal transcribed spacers (ITSs) of rDNA have been used to detect *Coccidioides* in cultures,[61] as well as in blood samples from patients and experimentally infected animals.[59] Through real-time PCR, partial sequences from the ITS regions can also be used as targets for *Coccidioides* detection in several clinical specimens, such as respiratory samples or fresh biopsy material fixed in formaldehyde or paraffinized.[62] The detection of *Coccidioides* in filamentous cultures or in skin and lung biopsy material can also be performed through amplification of the partial sequence of the codifying gene for the molecule "antigen 2" or "PRA" following protocols based on nested PCR.[63] Lindley et al.[64] developed a method based on PCR-enzyme immunoassay (PCR-EIA) for the detection of *Coccidioides* and other dimorphic fungal species. Even though the method allows the amplification and detection of small quantities of DNA, cross-reactivity was observed against *B. dermatitidis*.

Although the distinction of the species *C. immitis* and *C. posadasii* was originally performed through an analysis of microsatellite loci *GAC2* and *621*[65] and single nucleotide polymorphisms,[14] these methods are of limited value for routine diagnosis since they are expensive and require laborious techniques.[66] Currently, the identification of the etiological agent can be performed through only one PCR[66] or amplification of the ITS region and posterior sequencing or RFLP of the complete sequence.[67]

37.6 TREATMENT AND PREVENTION

The treatment of coccidioidomycosis basically depends on three factors: the severity of the pulmonary infection, the presence of disseminated disease, and the patient's individual risk factors.[15,68] The majority of cases of pulmonary coccidioidomycosis are self-limited, and according to some authors, there is no need for therapeutic intervention. In mild forms of the disease, patients do not need antifungal treatment, but they should be clinically and radiographically monitored for up to 2 years to document infection resolution.[23] In contrast, experts recommend the use of systemic antifungal therapy for all symptomatic patients, with fluconazole or itraconazole for up to 6 months.[34]

More severe, chronic, or disseminated clinical cases require a specific therapeutic approach. The drug of choice depends on the clinical manifestations and host immunological status.[24] Currently, amphotericin B and its lipid formulations are reserved for severe cases of the disease. Azole derivatives, especially fluconazole and itraconazole, are used as sequential or combined treatment. In cases of meningitis, fluconazole is preferred because of its good penetration in the CSF. However, due to its fungistatic activity, it demands long-lasting therapy, possibly throughout life, especially in immunocompromised patients.[68]

Recently obtained data show that voriconazole and posaconazole are efficient for the treatment of chronic pulmonary coccidioidomycosis and coccidioidal meningitis.[69–74] The clinical efficacy of caspofungin was reported in two cases of disseminated coccidioidomycosis;[75,76] a combined therapy with caspofungin and fluconazole was reported as an alternative treatment for diffuse coccidioidal pneumonia.[77]

Azole chemoprophylaxis must be considered for organ transplant recipients and HIV patients who live in endemic areas, due to high risk of coccidioidomycosis in these populations.[78,79]

In the past years, new therapeutic strategies have been proposed for the treatment of coccidioidomycosis, including the combined use of antifungal drugs[75,77,80] and the use of IFN-γ for the treatment of disseminated disease refractory to conventional antifungal agents.[81]

37.7 SPECIAL REMARKS REGARDING BIOSAFETY

37.7.1 WHY ARE *COCCIDIOIDES* SPP. CONSIDERED SELECT AGENTS?

In the United States, *Coccidioides* spp. are enrolled on the biological select agent list, which comprises biological pathogens that have the potential to pose a severe threat to humans and animals. In that country, the possession, use, and transfer of *Coccidioides* spp. cultures and derived products are strictly regulated by the Department of Health and Human Services (HHS)/Centers for Disease Control and Prevention (CDC) and the US Department of Agriculture (USDA)/Animal and Plant Health Inspection Service (APHIS).[82] Although *C. posadasii* is recognized as a BL3 microorganism in endemic areas of Latin America, unfortunately, there are no official guidelines and laws regulating the access of cultures and products derived from this fungus.

According to the US federal authorities, the inclusion of *Coccidioides* spp. on the select agent list was based on the following criteria:

(i) *The effect on human (lab workers and the community in general) and animal health after exposure of infectious conidia.*

Coccidioides spp. are true primary pathogens, being able to cause disease to immunocompetent individuals. Besides humans, an extended list of susceptible animals has been continually updated throughout the years and includes both domestic and wild animals, such as dogs, cats, horses, sea lions, coyotes, llamas, armadillos, bats, and even snakes. A symptomatic infection has rarely been detected in cattle and sheep, although lung lesions may be found in these animals. Coccidioidomycosis does not seem to occur in birds.

The outcome of the encounter of a susceptible host with *Coccidioides* conidia is determined by the genetic background and the general health of the host and the virulence of the fungal strain. Studies have shown that polymorphisms on HLA genes and the presence of blood group antigen B are associated with severe, disseminated disease. Ethnic groups with such genetic markers, such as Filipinos and African Americans, have an increased susceptibility to coccidioidomycosis and are more likely to develop the disseminated disease, in comparison to Caucasians, for example.[83]

Besides genetically determined susceptibility, the risk for disseminated coccidioidomycosis is also related to age, with those younger than 5 years or older than 50 years being significantly more vulnerable to severe disease. Apparently, males are more susceptible than females, except during pregnancy, when women are at high risk of severe coccidioidomycosis, due either to the immunosuppressive state that occurs during gestation or by the presence of progesterone and 17β-estradiol, which may stimulate the growth of the spherule/endospore phase. Patients with immunosuppressive diseases or conditions, such as HIV infection, malignant neoplasias, recipients of immunosuppressive drug therapy, and organ transplant recipients, also show a high risk of disseminated coccidioidomycosis with high levels of mortality.[10,20,23–25]

(ii) *The infectivity and pathogenicity of the fungi, including the means of transmission of their propagules.*

Experimental studies have shown that *Coccidioides* spp. are pathogens requiring a low infectious dose, roughly estimated to be around 1–10 arthroconidia,[84] with truly virulence factors.[85] Nevertheless, the infectious dose, defined as the number of cells sufficient to establish an infection, is clearly dependent on the route of infection and the genetic susceptibility of the host. In general, intraperitoneal, intravenous, and intranasal infections seem to require a higher number of arthroconidia than intracranial infections.[86]

However, it is important to note that as airborne pathogens, *Coccidioides* spp. represent a real threat, so it is necessary to establish rigorous biosafety measures to control their spread.[1,2]

(iii) *The availability of efficiency therapy for treating the disease and the existence of proper vaccines to prevent any illness resulting from infection.*

In general, the management of patients with primary coccidioidomycosis without risk factors for dissemination relies on continuous reevaluation of symptoms and radiographic findings to guarantee clinical resolution.[68] However, patients with disseminated disease frequently need antifungal therapy, surgical debridement, or both.[68] However, patients with disseminated disease frequently require antifungal therapy. According to several authors, the management of coccidioidal meningitis—the most threatening form of disseminated disease—is challenging even for experienced clinicians, as the disease can be refractory to antifungal therapies. For severe forms of coccidioidomycosis, therapy can range from many months to years, and for some patients, a lifelong antifungal treatment is necessary.

Despite the progress in the study of the immunology of coccidioidomycosis in the past 30 years,[6,87,88] a commercial vaccine is not available yet.

37.7.2 COULD *COCCIDIOIDES* SPP. BE CONSIDERED BIOLOGICAL WEAPONS?

Each microorganism presents some weapon potential, which can range from high to null.[89] However, the inclusion of *Coccidioides* in the select agent list has been criticized by the American Society for Microbiology (ASM).[90] According to ASM experts, *Coccidioides* do not fit the criteria for biological weapons, as they are pathogens widely found in nature. The questions raised by the ASM experts also included the following:

(i) *Degree of pathogenicity.*

Although coccidioidomycosis is considered a common cause of community-acquired pneumonia in the southwestern United States (around 150,000 new cases per year are expected to occur in that area), only 40% of the infected individuals typically develop a clinical disease with moderate respiratory and/or systemic symptoms. It is estimated that only 1%–5% of the infected people will develop disseminated disease, which is strongly related to the immunological status of the host and/or a high inocula containing infectious arthroconidia.[6]

The susceptibility to coccidioidomycosis varies among animal species and strains. In a seminal study performed by Kirkland and Fierer,[91] it was shown that DBA mice are 1000 times more resistant to intraperitoneal coccidioidomycosis than BALB/c mice. Dissemination of *C. immitis* to the lungs also occurred frequently in BALB/c but not in DBA mice, which clearly shows the importance of the host's genetic background to prevent coccidioidomycosis.

Apart from the issues regarding susceptibility of each animal species, it is believed that fungal virulence is a strain-specific trait. In an experiment conducted by Cox and Magee,[6] groups of 10 Balb/c mice were infected by intranasal route with strains of *C. posadasii* and then monitored for survival over 45 days. According to the authors, strains could be classified as displaying high, intermediate, and low virulence. Mice infected with 29 arthroconidia of a highly virulent *C. posadasii* strain began to die on day 9, with a median of 50% survival on day 10.5. When animals were infected with 27 arthroconidia of an intermediately virulent *C. posadasii* strain, death was first recorded on day 13, with a median of 50% survival on day 14. On the other hand, infection with 35 arthroconidia of a weakly virulent *C. posadasii* strain resulted in 100% survival for over 45 days after the challenge.

(ii) *Communicability.*

Communicability is a measure of transmissibility and contagiousness.[89] There is no evidence that coccidioidomycosis can be directly transmitted person to person. Only one case of coccidioidomycosis in a transplant recipient acquired from an infected donor was reported to date.[92] Presumed transmission from animals was described by Kohn et al.[93] in a veterinarian with disseminated disease, probably by inhalation of endospores aerosolized during the autopsy of an infected horse. Recently, Gaidici and Saubolle[94] reported a case of cutaneous coccidioidomycosis in a patient that had been bitten by a cat with disseminated disease. However, these are unusual and isolated cases.

(iii) *Ease of dissemination.*

Although the infectious conidia are easily airborne, continuous wind activity—as during dust storms—is necessary to their dissemination. Infectious arthroconidia cannot persist in still air for prolonged periods of time. Therefore, increasing rates of coccidioidomycosis in the United States are frequently associated with extended drought periods, dust storms, earthquakes, or soil disturbance caused by construction projects.

(iv) *Route of exposure.*

Unlike other BSL3 pathogens, *Coccidioides* spp. have a preferential route of transmission, which is via the respiratory tract. Respiratory route of exposure depends on environmental factors, such as continuous wind activity and dry air, and, therefore, tends to be less efficient than other routes, for example, percutaneous or oral ingestion. Coccidioidomycosis resulting from direct inoculation is a very rare condition and is associated with lab accidents.

(v) *Environmental stability.*

Although *Coccidioides* have a very strong cell wall and can resist environmental droughts and high temperatures, their survival depends on many ecological conditions. Even in endemic areas, the isolation of *Coccidioides* from soil is rarely successful. Apparently, the distribution of *Coccidioides* in these environments is irregular and sporadic.[61] A study by Elconin et al.[95] who analyzed more than 5000 soil samples from endemic areas for the disease during 8 years showed that the isolation rates for *Coccidioides* ranged from 0% to 43%. Greene et al.[61] reported an isolation rate lower than 1% after analyzing 720 soil samples from hyperendemic areas. Theoretically, the natural sites for *Coccidioides* can be of two types: growth sites, where physical, chemical, and biological conditions are adequate for the microorganism's complete development, and accumulation sites, which are areas where infective arthroconidia can be found after being carried away from growth sites by wind, water, animals, or anthropogenic disturbances.[96]

Several biotic and abiotic factors, still not completely understood, influence the discontinuous distribution of *Coccidioides* in soil. Apparently, the most important factors are temperature, amount and timing of rainfall, and conditions that are directly related to soil physicochemical characteristics, such as low moisture, texture, alkaline pH, salinity, and levels of exposure to sun and/or ultraviolet light. In addition, *Coccidioides* are poor competitors for nutrients in comparison with other telluric microorganisms.[96] *Coccidioides* are not common saprophytes of farm soils and are not considered contaminants of soils in urbanized areas, even in endemic regions.

(vi) *Ease of production.*

In contrast to many biological threats, *Coccidioides* can be recovered from environmental sources—like soil from rodent burrows—without the need for specialized devices.

(vii) *Difficulty of genetic manipulation.*

Like many filamentous fungi, the genetic transformation of *Coccidioides* is a difficult task. At present, the genetic manipulation of *Coccidioides* depends on protoplast formation[97] or infection with *Agrobacterium tumefaciens* bacteria.[98] High numbers of transformant cells are not expected to happen.

(viii) *Long-term effects.*

Only a small number of infected people (1%–5%) are expected to develop disseminated coccidioidomycosis, with long-lasting monitoring necessary. The majority of symptomatic patients present a manageable disease in which antifungal therapy is fairly successful. Patients without risk factors for disseminated disease can present spontaneous remission even without antifungal therapy.

(ix) *Acute morbidity and mortality.*

As previously stated, only 40% of infected individuals present symptomatic disease. The acute pulmonary primary form can regress spontaneously, after a few months, even without specific therapy.[25] In immunocompetent patients, mortality is a rare outcome of acute pneumonia, and suitable antifungal treatment can reduce successfully the mortality rate. Higher inocula are necessary to increase the morbidity and mortality rates among immunocompetent individuals.

(x) *Availability of therapy.*

Although coccidioidal meningitis is considered an unmanageable disease, treatment of other clinical forms of coccidioidomycosis can be performed with amphotericin B, triazoles, and echinocandins. To date, at least five different antifungal drugs have been employed successfully in the treatment of coccidioidomycosis: amphotericin B, itraconazole, fluconazole, voriconazole, and caspofungin.

(xi) *Immunity.*

Nearly 60% of infected individuals develop permanent immunity and are therefore protected against second infections with *Coccidioides*.[15] Apparently, infection with *C. immitis* leads to cross-protection against *C. posadasii* and vice versa.

(xii) *Vulnerable populations.*

Although some ethnic groups, infants, and older people, as well as immunosuppressed patients, are at greater risk to develop severe disseminated disease, they would be a small percent of the total population exposed to a terrorist attack. Immediate antifungal therapy should reduce the fungal burden.

(xiii) *Impact on health-care systems.*

ASM experts believe that the use of *Coccidioides* cultures by terrorists would cause an impact on health services similar to outbreaks previously experienced in endemic regions. Although the cost of antifungal medication can be as high as $20,000 per year,[68] only few patients would require therapy. Furthermore, these costs can be lowered by using cheaper alternative drugs, such as generic fluconazole.

37.7.3 BIOSAFETY RECOMMENDATIONS FOR SAFE WORK PRACTICES WITH *COCCIDIOIDES* CULTURES

Coccidioides cultures pose a hazard to microbiologists, lab technicians, and researchers. Filamentous cultures and respiratory specimens from patients with type 2 diabetes mellitus and chronic and cavitary pulmonary coccidioidomycosis—as they are four times more likely than nondiabetic patients to develop parasitic mycelial forms during the disease[48]—must be manipulated inside a biological safety cabinet (minimum Class II, type B2), without inside air recirculation, in a BSL3. This requires the establishment of proper policies, practices, and procedures for work:

(i) *General principles and responsibilities*

A BSL3 is able to manipulate and carry out the storage of Level 3 pathogens, defined as agents that can be transmitted by the respiratory route and can cause serious infections to humans and animals. Personnel working with Level 3 pathogens must be instructed about the specific BSL3 practices and should understand that they have an active role in keeping the containment conditions required for safe work. The main investigator or laboratory director must schedule continuous training sessions for all laboratory staff. The main investigator or laboratory director is also responsible for controlling access to the laboratory; he has the final responsibility for assessing each circumstance and determining who may enter or work in the laboratory, as well as for defining which activities will be performed by each person. Pregnant women, immunocompromised individuals, or people under 18 years old should not be allowed to enter in a BLS3. Formal written consent documents notifying the risks associated with BLS3 are required for each lab worker/researcher.

(ii) *Personal protective equipment*

For the manipulation of *Coccidioides* cultures, lab workers must use special protective clothing, masks (N95 or N99 respirator is required), gloves, and eye protection. Face shields should be used when appropriate. Lab workers may not be allowed to leave BSL3 wearing any of these items.

(iii) *Material waste and decontamination*

All kinds of waste from BSL3 (including fungal cultures, clinical specimens, animal bodies, and animal excreta) must be decontaminated, preferably by autoclaving, before disposal. A BLS3 must be provided with an autoclave for decontamination within the laboratory.

Nondisposable materials and surfaces of the room—including floor, ceiling, and walls—should be decontaminated with highly efficient disinfectants, such as bleach at a 1:10 dilution or undiluted commercial products in water.[99] Each worker should decontaminate his or her own work bench immediately after each use and immediately after contamination with viable material.

(iv) *Medical surveillance*

Any lab worker having occupational exposure to *Coccidioides* must be evaluated periodically by serological and/or skin tests. Accidental exposure must be reported immediately to the main investigator or laboratory director. Workers should be treated with either itraconazole or fluconazole orally (400 mg daily) for 6 weeks, as a prophylaxis measure.[99]

37.8 CONCLUSIONS

C. immitis and *C. posadasii* are pathogenic species, able to cause severe disease, especially among the immunocompromised. They possess several intrinsic virulence factors, such as dimorphism and the ability to survive within the phagocyte or to escape phagocytosis. Infectious conidia can persist on inanimate surfaces for a long time and can resist environmental stress. They are considered potential biological weapons, although some experts disagree with this classification. The manipulation of *Coccidioides* cultures and potentially contaminated clinical specimens must be performed only by well-trained personnel inside BSL3 containment.

REFERENCES

1. Dixon, D.M., *Coccidioides immitis* as a select agent of bioterrorism. *J Appl Microbiol.* 2001; 91(4):602–605.
2. Derensinski, S., *Coccidioides immitis* as a potential bioweapon. *Semin Respir Infect.* 2003; 18(3):216–219.
3. Sharpton, T.J. et al., Comparative genomic analyses of the human fungal pathogens *Coccidioides* and their relatives. *Genome Res.* 2009; 19(10):1722–1731.
4. Mandel, M.A. et al., Genomic and population analyses of the mating type loci in *Coccidioides* species reveal evidence for sexual reproduction and gene acquisition. *Eukaryot Cell.* 2007; 6(7):1189–1199.
5. Huppert, M. et al., Morphogenesis throughout saprobic and parasitic cycles of *Coccidioides immitis. Mycopathologia.* 1982; 78(2):107–122.
6. Cox, R. and Magee, D.M., Coccidioidomycosis: Host response and vaccine development. *Clin Microbiol Rev.* 2004; 17(4): 804–839.
7. Sun, S.H. et al., Rapid in vitro conversion and identification of *Coccidioides immitis. J Clin Microbiol.* 1976; 3(2):186–190.

8. Petkus, A.F. et al., Pure spherules of *Coccidioides immitis* in continuous culture. *J Clin Microbiol.* 1985; 22(2):165–167.

9. Bonifaz, A. et al., Endemic systemic mycoses: Coccidioidomycosis, histoplasmosis, paracoccidioidomycosis and blastomycosis. *J Dtsch Dermatol Ges.* 2011; 9(9):705–714.

10. Laniado-Laborín, R. Coccidioidomycosis and other endemic mycoses in Mexico. *Rev Iberoam Micol.* 2007; 24(4): 249–258.

11. Canteros, C.E. et al., Coccidioidomycosis in Argentina. *Rev Argent Microbiol.* 2010; 42(4):261–268.

12. de Aguiar Cordeiro, R. et al., Twelve years of coccidioidomycosis in Ceará State, Northeast Brazil: Epidemiologic and diagnostic aspects. *Diagn Microbiol Infect Dis.* 2010; 66(1):65–72.

13. Negroni, R., Historical evolution of some clinical and epidemiological knowledge of coccidioidomycosis in the Americas. *Rev Argent Microbiol.* 2008; 40(4):246–256.

14. Fisher, M.C. et al., Biogeographic range expansion into South America by *Coccidioides immitis* mirrors New World patterns of human migration. *Proc Natl Acad Sci USA.* 2001; 98(8):4558–4562.

15. Galgiani, J.N. et al., Coccidioidomycosis. *Clin Infect Dis.* 2005; 41(9):1217–1223.

16. Petersen, L.R. et al., Coccidioidomycosis among workers at an archeological site, northeastern Utah. *Emerg Infect Dis.* 2004; 10(4):637–642.

17. Crum, N.F. et al., Seroincidence of coccidioidomycosis during military desert training exercises. *J Clin Microbiol.* 2004; 42(10):4552–4555.

18. Centers for Disease Control and Prevention (CDC), Increase in coccidioidomycosis—California, 2000–2007. *MMWR.* 2009; 58(5):105–109.

19. Drutz, D.J. et al., Human sex hormones stimulate the growth and maturation of *Coccidioides immitis. Infect Immun.* 1981; 32(2):897–907.

20. Bercovitch, R.S. et al., Coccidioidomycosis during pregnancy: A review and recommendations for management. *Clin Infect Dis.* 2011; 53(4):363–368.

21. Blair, J.E. et al., Coccidioidomycosis in elderly persons. *Clin Infect Dis.* 2008; 47(12):1513–1518.

22. Chiller, T.M. et al., Coccidioidomycosis. *Infect Dis Clin North Am.* 2003; 17(1):41–57.

23. Ruddy, B.E. et al., Coccidioidomycosis in African Americans. *Mayo Clin Proc.* 2011; 86(1):63–69.

24. Parish, J.M. and Blair, J.E., Coccidioidomycosis. *Mayo Clin Proc.* 2008; 83(3):343–348.

25. Crum, N.F. et al., Coccidioidomycosis: A descriptive survey of a reemerging disease. Clinical characteristics and current controversies. *Medicine (Baltimore).* 2004; 83(3):149–175.

26. Levine, H.B. et al., Dermal sensitivity to different doses of spherulin and coccidioidin. *Chest.* 1974; 65(5):530–533.

27. Mondragón-González, R. et al., Detection of *Coccidioides immitis* infection in Coahuila, Mexico. *Rev Argent Microbiol.* 2005; 37(3):135–138.

28. Padua y Gabriel, A. et al., Prevalence of skin reactivity to coccidioidin and associated risks factors in subjects living in a northern city of Mexico. *Arch Med Res.*1999; 30(5):388–392.

29. Diógenes, M.J.N. et al., Inquérito epidemiológico com esferulina em Jaguaribara-CE, Brasil, 1993. *An Bras Dermatol.* 1995; 70(6):525–529.

30. Lacaz, C.S. et al., Inquérito imunoalérgico com esferulina em um hospital geral de São Paulo. *Rev Assoc Med Bras.* 1978; 24(12):403–404.

31. Cairns, L. et al., Outbreak of coccidioidomycosis in Washington state residents returning from Mexico. *Clin Infect Dis.* 2000; 30(1):61–64.

32. Schneider, E. et al., A coccidioidomycosis outbreak following the Northridge, Calif, earthquake. *JAMA.* 1997; 277(11):904–908.

33. Valdivia, L. et al., Coccidioidomycosis as a common cause of community-acquired pneumonia. *Emerg Infect Dis.* 2006; 12(6):958–962.

34. DiCaudo, D.J., Coccidioidomycosis: A review and update. *J Am Acad Dermatol.* 2006; 55(6):929–942.

35. Ampel, N.M., The complex immunology of human coccidioidomycosis. *Ann N Y Acad Sci.* 2007; 1111:245–258.

36. DiCaudo, D.J. et al., Sweet syndrome (acute febrile neutrophilic dermatosis) associated with pulmonary coccidioidomycosis. *Arch Dermatol.* 2005; 141(7):881–884.

37. Castañeda-Godoy, R. and Laniado-Laborín, R., Coexistencia de tuberculosis y coccidioidomicosis. Presentacion de dos casos clínicos. *Rev Inst Nal Enf Resp Mex.* 2002; 15(2):98–101.

38. Blair, J.E., Coccidioidal meningitis: Update on epidemiology, clinical features, diagnosis, and management. *Curr Infect Dis Rep.* 2009; 11(4):289–295.

39. Brilhante, R.S. et al., Coccidioidal pericarditis: A rapid presumptive diagnosis by an in-house antigen confirmed by mycological and molecular methods. *J Med Microbiol.* 2008; 57(Pt 10):1288–1292.

40. Arsura, E.L. et al., Coccidioidal pericarditis: A case presentation and review of the literature. *Int J Infect Dis.* 2005; 9(2):104–109.

41. Cadena, J. et al., Coccidioidomycosis and tuberculosis coinfection at a tuberculosis hospital: Clinical features and literature review. *Medicine (Baltimore).* 2009; 88(1):66–76.

42. Saubolle, M.A., Laboratory aspects in the diagnosis of coccidioidomycosis. *Ann N Y Acad Sci.* 2007; 1111:301–314.

43. Saubolle, M.A. et al., Epidemiologic, clinical, and diagnostic aspects of coccidioidomycosis. *J Clin Microbiol.* 2007; 45(1):26–30.

44. Sutton, D.A., Diagnosis of coccidioidomycosis by culture: Safety considerations, traditional methods, and susceptibility testing. *Ann N Y Acad Sci.* 2007; 1111:315–325.

45. Pfaller, M.A. and Diekema, D.J., Unusual fungal and pseudofungal infections of humans. *J Clin Microbiol.* 2005; 43(4):1495–504.

46. Kaufman, L. et al., Misleading manifestations of *Coccidioides immitis* in vivo. *J Clin Microbiol.* 1998; 36(12):3721–3723.

47. DiTomasso, J.P. et al., Bronchoscopic diagnosis of pulmonary coccidioidomycosis. Comparison of cytology, culture, and transbronchial biopsy. *Diagn Microbiol Infect Dis.* 1994; 18(2):83–87.

48. Muñoz-Hernández, B. et al., Mycelial forms of *Coccidioides* spp. in the parasitic phase associated to pulmonary coccidioidomycosis with type 2 diabetes mellitus. *Eur J Clin Microbiol Infect Dis.* 2008; 27(9):813–820.

49. Cordeiro, R.A. et al., Phenotypic characterization and ecological features of *Coccidioides* spp. from Northeast Brazil. *Med Mycol.* 2006; 44(7):631–639.

50. Ampel, N.M., Measurement of cellular immunity in human coccidioidomycosis. *Mycopathologia.* 2003; 156(4):247–262.

51. Ampel, N.M. et al., An archived lot of coccidioidin induces specific coccidioidal delayed-type hypersensitivity and correlates with in vitro assays of coccidioidal cellular immune response. *Mycopathologia.* 2006; 161(2):67–72.

52. Woodruff, W.W. 3rd et al., Reactivity to spherule-derived coccidioidin in the southeastern United States. *Infect Immun.* 1984; 43(3):860–869.

53. Pappagianis, D., Serologic studies in coccidioidomycosis. *Semin Respir Infect.* 2001; 16(4):242–250.

54. Blair, J.E. et al., Serologic testing for symptomatic coccidioidomycosis in immunocompetent and immunosuppressed hosts. *Mycopathologia.* 2006; 162(5):317–324.

55. Blair, J.E. and Currier, J.T., Significance of isolated positive IgM serologic results by enzyme immunoassay for coccidioidomycosis. *Mycopathologia.* 2008; 166(2):77–82.

56. Pappagianis, D. and Zimmer, B.L., Serology of coccidioidomycosis. *Clin Microbiol Rev.* 1990; 3(3):247–268.

57. de Aguiar Cordeiro, R. et al., Rapid diagnosis of coccidioidomycosis by nested PCR assay of sputum. *Clin Microbiol Infect.* 2007; 13(4):449–451.

58. Bialek, R. et al., Coccidioidomycosis and blastomycosis: Advances in molecular diagnosis. *FEMS Immunol Med Microbiol.* 2005; 45(3):355–360.

59. Johnson, S.M. et al., Amplification of coccidioidal DNA in clinical specimens by PCR. *J Clin Microbiol.* 2004; 42(5):1982–1985.

60. Pan, S. and Cole, G.T., Molecular and biochemical characterization of a *Coccidioides immitis*–specific antigen. *Infect Immun.* 1995; 63(10):3994–4002.

61. Greene, D.R. et al., Soil isolation and molecular identification of *Coccidioides immitis. Mycologia.* 2000; 92(3):406–410.

62. Binnicker, M.J. et al., Detection of *Coccidioides* species in clinical specimens by real-time PCR. *J Clin Microbiol.* 2007; 45(1):173–178.

63. Bialek, R. et al., PCR assays for identification of *Coccidioides posadasii* based on the nucleotide sequence of the antigen 2/proline-rich antigen. *J Clin Microbiol.* 2004; 42(2):778–783.

64. Lindley, M.D. et al., Rapid identification of dimorphic and yeast-like fungal pathogens using specific DNA probes. *J Clin Microbiol.* 2001; 39(10):3505–3511.

65. Fisher, M.C. et al., Molecular and phenotype description of *Coccidioides posadasii* sp. nov., previously recognized as the non-Californian population of *Coccidioides immitis. Mycologia.* 2002; 94(1):73–84.

66. Umeyama, T. et al., Novel approach to designing primers for identification and distinction of the human pathogenic fungi *Coccidioides immitis* and *Coccidioides posadasii* by PCR amplification. *J Clin Microbiol.* 2006; 44(5):1859–1862.

67. Tintelnot, K., Taxonomic and diagnostic markers for identification of *Coccidioides immitis* and *Coccidioides posadasii. Med Mycol.* 2007; 45(5):385–393.

68. Galgiani, J.N. et al., Practice guideline for the treatment of coccidioidomycosis. Infectious Diseases Society of America. *Clin Infect Dis.* 2000; 30(4):658–661.

69. Catanzaro, A. et al., Safety, tolerance, and efficacy of posaconazole therapy in patients with nonmeningeal disseminated or chronic pulmonary coccidioidomycosis. *Clin Infect Dis.* 2007; 45(5):562–568.

70. Stevens, D.A. et al., Posaconazole therapy for chronic refractory coccidioidomycosis. *Chest.* 2007; 132(3):952–958.

71. Anstead, G.M. et al., Refractory coccidioidomycosis treated with posaconazole. *Clin Infect Dis.* 2005; 40(12):1770–1776.

72. Prabhu, R.M. et al., Successful treatment of disseminated nonmeningeal coccidioidomycosis with voriconazole. *Clin Infect Dis.* 2004; 39(7):e74–e77.

73. Proia, L.A. and Tenorio, A.R., Successful use of voriconazole for treatment of *Coccidioides* meningitis. *Antimicrob Agents Chemother.* 2004; 48(6):2341.

74. Cortez, K.J. et al., Successful treatment of coccidioidal meningitis with voriconazole. *Clin Infect Dis.* 2003; 36(12):1619–1622.

75. Antony, S., Use of the echinocandins (caspofungin) in the treatment of disseminated coccidioidomycosis in a renal transplant recipient. *Clin Infect Dis.* 2004; 39(6):879–880.

76. Hsue, G. et al., Treatment of meningeal coccidioidomycosis with caspofungin. *J Antimicrob Chemother.* 2004; 54(1):292–294.

77. Park, D.W. et al., Combination therapy of disseminated coccidioidomycosis with caspofungin and fluconazole. *BMC Infect Dis.* 2006; 15(6):26.

78. Blair, J.E. et al., Early results of targeted prophylaxis for coccidioidomycosis patients undergoing orthotopic liver transplantation within an endemic area. *Transpl Infect Dis.* 2003; 5(1):3–8.

79. Blair, J.E. et al., The prevention of recrudescent coccidioidomycosis after solid organ transplantation. *Transplantation.* 2007; 83(9):1182–1187.

80. Antony, S.J. et al., Successful use of combination antifungal therapy in the treatment of coccidioides meningitis. *J Natl Med Assoc.* 2006; 98(6):940–942.

81. Kuberski, T.T. et al., Successful treatment of a critically ill patient with disseminated coccidioidomycosis, using adjunctive interferon-gamma. *Clin Infect Dis.* 2004; 38(6):910–912.

82. Department of Health and Human Services, Centers for Disease Control and Prevention: OMB approval data collection notice. *Federal Register.* 2002; 67:51063.

83. Louie, L. et al., Influence of host genetics on the severity of coccidioidomycosis. *Emerg Infect Dis.* 1999; 5(5):672–680.

84. Kruse, R.H. et al., Infection of control monkeys with *Coccidioides immitis* by caging with inoculated monkeys. In *Proceedings of the 2nd Coccidioidomycosis Symposium.* Tucson, University of Arizona Press, Phoenix, AR, 1967, pp. 387–396.

85. Hung, C.Y. et al., Virulence mechanisms of *Coccidioides. Ann N Y Acad Sci.* 2007; 1111:225–235.

86. Clemons, K.V. et al., Experimental animal models of coccidioidomycosis. *Ann N Y Acad Sci.* 2007; 1111:208–224.

87. Hung, C.Y. et al., Vaccine immunity to coccidioidomycosis occurs by early activation of three signal pathways of T helper cell response (Th1, Th2, and Th17). *Infect Immun.* 2011; 79(11):4511–4522.

88. Hurtgen, B.J., Construction and evaluation of a novel recombinant t cell epitope-based vaccine against coccidioidomycosis. *Infect Immun.* 2012; 80(11):3960–3974.

89. Casedevall, A. and Pirofski, L., The weapon potential of a microbe. *Trends Microbiol.* 2004; 12(6):259–263.

90. American Society of Microbiology. Comments on the changes to the HHS list of select agents and toxins. http://www.selectagents.gov/resources/American%20Society%20for%20Microbiology.pdf (2010) (accessed on October 17, 2012).

91. Kirkland, T.N. and Fierer, J., Inbred mouse strains differ in resistance to lethal *Coccidioides immitis* infection. *Infect Immun.* 1983; 40(3):912–916.

92. Brugière, O. et al., Coccidioidomycosis in a lung transplant recipient acquired from the donor graft in France. *Transplantation.* 2009; 88(11):1319–1320.

93. Kohn, G.J. et al., Acquisition of coccidioidomycosis at necropsy by inhalation of coccidioidal endospores. *Diagn Microbiol Infect Dis.* 1992; 15(6):527–530.

94. Gaidici, A. and Saubolle, M.A., Transmission of coccidioi-domycosis to a human via a cat bite. *J Clin Microbiol.* 2009; 47(2):505–506.

95. Elconin, A.F. et al., Significance of soil salinity on the ecology of *Coccidioides immitis*. *J Bacteriol.* 1964; 87: 500–503.

96. Fisher, F.S. et al., *Coccidioides* niches and habitat parameters in the southwestern United States. *Ann N Y Acad Sci.* 2007; 1111:47–72.

97. Hung, C.J. et al., Gene disruption in *Coccidioides* using hygromycin or phleomycin resistance markers. *Methods Mol Biol.* 2012; 845:131–147.

98. Aboudeh, R.O., Genetic transformation of *Coccidioides immitis* facilitated by *Agrobacterium tumefaciens*. *J Infect Dis.* 2000; 181(6):2106–2110.

99. Stevens, D.A., Expert opinion: What to do when there is *Coccidioides* exposure in a laboratory. *Clin Infect Dis.* 2009; 49(6):919–923.

38 Cryptosporidium

Lihua Xiao and Yaoyu Feng

CONTENTS

38.1 INTRODUCTION

Cryptosporidium spp. inhabit the brush borders of the gastrointestinal and respiratory epithelium of various vertebrates, causing enterocolitis, diarrhea, and cholangiopathy in humans.[1] Immunocompetent children and adults with cryptosporidiosis usually have a short-term illness accompanied by watery diarrhea, nausea, vomiting, and weight loss. In immunocompromised persons, however, the infection can be protracted and life-threatening.[2] In animals, cryptosporidiosis is an important cause of diarrhea in neonatal animals.[3]

Cryptosporidiosis is a notifiable disease in most industrialized nations. There were 2769–3787 annual cases of reported human cryptosporidiosis in the United States during 1999–2002, 3505–8269 during 2003–2005, 6479–11,657 during 2006–2008, and 7656–8951 during 2009–2010.[4–7] The Centers for Disease Control and Prevention estimates that there are about 748,123 cases of cryptosporidiosis in the United States each year, with a 98.6% underdiagnosis of cases.[8]

Cryptosporidium spp. are well recognized water- and foodborne pathogens, having caused many outbreaks of human illness in the United States and other countries.[9] Because of the ubiquitous nature of *Cryptosporidium* spp., *Cryptosporidium* oocysts are found in surface and drinking water source at high frequency.[10] Its persistence in the environment, resistance to chlorine and many other disinfectants and drinking water treatment procedures, low infective dose, lack of effective therapeutic treatments, and demonstrated ability to cause massive outbreaks make *Cryptosporidium* a biodefense Category B priority pathogen in the United States (http://www.niaid.nih.gov/topics/biodefenserelated/biodefense/pages/cata.aspx).

38.2 CLASSIFICATION AND MORPHOLOGY

Cryptosporidium spp. belong to the family Cryptosporidiidae, which is a member of the phylum Sporozoa (syn. Apicomplexa). The exact placement of Cryptosporidiidae in Sporozoa is uncertain. It was long considered a member of the class Coccidia, in the order of Eimeriida.[3] Recent phylogenetic studies, however, indicate that *Cryptosporidium* spp. are more related to gregarines.[11] Putative extracellular gregarine-like reproductive stages were described.[12] Thus, *Cryptosporidium* spp. are no longer considered classic coccidian parasites.

The taxonomy of *Cryptosporidium* has gone through revisions as the result of extensive molecular genetic studies and biologic characterizations of parasites from various animals.[1,13] The validity of several early-described species

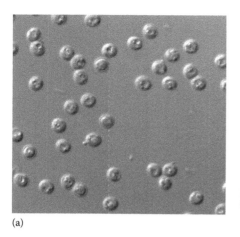
(a)

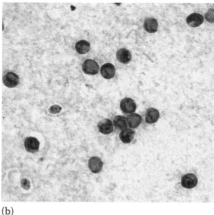
(b)

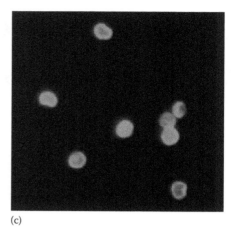
(c)

FIGURE 38.1 Oocysts of *C. hominis* under DIC (a), modified Ziehl–Neelsen acid-fast (b), and immunofluorescence (c) microscopy.

has been established, such as *C. parvum* in ruminants and humans; *C. muris* in rodents; *C. wrairi* in guinea pigs; *C. felis* in cats; *C. cuniculus* in rabbits; *C. meleagridis*, *C. baileyi*, and *C. galli* in birds; and *C. varanii* and *C. serpentis* in reptiles. Many new species have been named, such as *C. hominis* and *C. viatorum* in humans; *C. andersoni*, *C. bovis*, and *C. ryanae* in weaned calves and adult cattle; *C. xiaoi* in sheep; *C. ubiquitum* in ruminants, rodents, and primates; *C. canis* in dogs; *C. suis* and *C. scrofarum* in pigs; *C. tyzzeri* in rodents; *C. fayeri* and *C. macropodum* in marsupials; *C. molnari* and *C. scophthalmi* in fish; and *C. fragile* in amphibians. Thus, there are currently 26 established *Cryptosporidium* species in vertebrates. There are also many host-adapted *Cryptosporidium* genotypes that do not yet have species names, such as *Cryptosporidium* horse, hamster, ferret, skunk, squirrel, bear, deer, fox, mongoose, hedgehog, wildebeest, duck, woodcock, seal, snake, tortoise, mouse II, goose I and II, muskrat I and II, opossum I and II, chipmunk I–III, rat I–IV, deer mouse I–IV, avian I–V, and several piscine genotypes.[1,13] These species biologically, morphologically, and phylogenetically belong to two large groups—intestinal species and gastric species—although piscine species/genotypes are clearly different from those in other vertebrates.[14]

Currently, nearly 20 *Cryptosporidium* species and genotypes have been reported in humans, including *C. hominis*, *C. parvum*, *C. meleagridis*, *C. felis*, *C. canis*, *C. ubiquitum*, *C. cuniculus*, *C. viatorum*, *C. muris*, *C. suis*, *C. andersoni*, *C. tyzzeri*, *C. fayeri*, *C. bovis*, *C. scrofarum*, and *Cryptosporidium* horse, skunk genotypes, and chipmunk genotype I. Humans are most frequently infected with *C. hominis* and *C. parvum*. The former almost exclusively infects humans, thus is considered an anthroponotic parasite, whereas the latter mostly infects humans and ruminants, thus is considered a zoonotic pathogen. Other species, such as *C. meleagridis*, *C. felis*, *C. canis*, *C. ubiquitum*, *C. cuniculus*, and *C. viatorum*, are less common. The remaining species and genotypes have been only found in a few human cases.[15–18] These *Cryptosporidium* spp. infect both immunocompetent and immunocompromised persons. The distribution of these species in humans is different among geographic areas and

socioeconomic conditions, probably as a reflection of differences in infection sources and transmission routes.[15] Among all *Cryptosporidium* spp., *C. parvum* is the most likely biodefense agent because of its high infection rates and shedding intensity in preweaned dairy calves and common availability.

The infectious stage of *Cryptosporidium* is the oocyst (Figure 38.1). This is also the stage shed in the environment and targeted by most diagnostic assays. Oocysts of intestinal species are generally spherical and 4–6 μm in size. In contrast, oocysts of gastric species are more elongated and 6–9 μm in size. Each oocyst contains four banana-shaped sporozoites. *Cryptosporidium* oocysts are acid-fast, stained red in various modified acid-fast stains (Figure 38.1).

38.3 BIOLOGY AND EPIDEMIOLOGY

38.3.1 BIOLOGY AND LIFE CYCLES

Cryptosporidium spp. are intracellular parasites that primarily infect epithelial cells of the stomach, intestine, and biliary ducts. In birds and in severely immunosuppressed persons, the respiratory tract is sometimes involved. The infection site varies according to species, but almost the entire development of *Cryptosporidium* spp. occurs between the two lipoprotein layers of the membrane of the epithelial cells,[3] with the exception in *C. molnari*, *C. scophthalmi*, and other piscine genotypes, for which oogonial and sporogonial stages are located deep within the epithelial cells.[19]

Cryptosporidium infections in humans or other susceptible hosts start with the ingestion of viable oocysts (Figure 38.2). Upon contact with gastric and duodenal fluid, four sporozoites are liberated from each excysted oocyst, invade the epithelial cells, and develop to trophozoites surrounded by a parasitophorous vacuole. Within the epithelial cells, trophozoites undergo 2–3 generations of asexual amplification called merogony, leading to the formation of different types of meronts containing 4–8 merozoites. The latter differentiate into sexually distinct stages called macrogamonts and microgamonts (containing microgametes) in the process of gametogony. New oocysts are formed in the

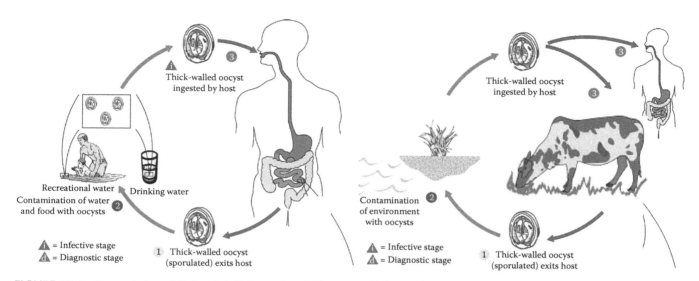

FIGURE 38.2 Transmission of *C. hominis* (left panel) and *C. parvum* (right panel). (Image courtesy of DPDx [http://dpd.cdc.gov/dpdx] from the Centers for Disease Control and Prevention.)

epithelial cells from the fusion of macrogamonts and microgametes, sporulate in situ in the process of sporogony, and contain four sporozoites. It is believed that about 20% of oocysts are thin-walled and may excyst within the digestive tract of the host, leading to the infection of new cells (autoinfection). The remaining 80% of oocysts are excreted into the environment; are resistant to low temperature, high salinity, and most disinfectants; and can initiate infection in a new host upon ingestion without further development. The time from ingestion of infective oocysts to the completion of the endogenous development and excretion of new oocysts varies with species, hosts, and infection doses; it is usually between 4 and 10 days for intestinal species.[3,20]

The only extracellular stages in the *Cryptosporidium* life cycle are oocysts, which are the environmental stage of the parasite, and released sporozoites, merozoites, and microgametes, which are briefly in the lumen of the digestive tract.[3] However, recently, a gregarine-like extracellular stage was supposedly found in *C. andersoni* and *C. parvum*, which might go through multiplication via syzygy, a sexual reproduction process involving the end-to-end fusion of two or more parasites.[12]

38.3.2 SUSCEPTIBLE POPULATIONS

In developing countries, human *Cryptosporidium* infection occurs mostly in children younger than 5 years, with peak occurrence of infections and diarrhea in children less than 2 years of age.[20,21] In developed countries, pediatric cryptosporidiosis occurs in children later than in developing countries, probably due to delayed exposures to contaminated environments as a result of better hygiene.[4,6,7,18,22] Cryptosporidiosis is also common in elderly people attending nursing homes where person-to-person transmission probably plays a major role in the spread of *Cryptosporidium* infections.[23] In the general population, a substantial number of adults are susceptible to *Cryptosporidium* infection, as sporadic infections occur in all age groups in the United

States and United Kingdom, and traveling to developing countries and consumption of contaminated food or water can frequently lead to infection.[9,24–26]

Cryptosporidiosis is common in immunocompromised persons, including AIDS patients, persons with primary immunodeficiency, and cancer and transplant patients undergoing immunosuppressive therapy.[2,27–30] Hemodialysis patients with chronic renal failure and renal transplant patients commonly develop cryptosporidiosis.[30–34] In HIV+ persons, the occurrence of cryptosporidiosis increases as the CD4+ lymphocyte cell counts fall, especially below 200 cells/μL.[28]

38.3.3 TRANSMISSION ROUTES

Humans can acquire cryptosporidiosis through several transmission routes, such as direct contact with infected persons or animals and consumption of contaminated water (drinking or recreational) or food.[2] The transmission routes differ between the two dominant species in humans, *C. parvum* and *C. hominis* (Figure 38.2). The relative role of each transmission route in the epidemiology of cryptosporidiosis and the infection sources are frequently unclear for a particular country or area.

38.3.4 ANTHROPONOTIC VERSUS ZOONOTIC TRANSMISSION

Studies in the United States and Europe have shown that cryptosporidiosis is more common in homosexual men than persons in other HIV-transmission categories,[35] indicating that direct person-to-person or anthroponotic transmission of cryptosporidiosis is common. Contact with persons with diarrhea has been identified as a major risk factor for sporadic cryptosporidiosis in industrialized countries.[25,36–39] This is exemplified by the high prevalence of cryptosporidiosis in childcare facilities, nursing home, and mothers with young children in these countries.

Only a few case–control studies assessed the role of zoo-notic transmission in the acquisition of cryptosporidiosis in humans. In industrialized countries, contact with farm animals (especially cattle) is a major risk factor in sporadic cases of human cryptosporidiosis.[9,25,36,37,40,41] Contact with pigs, dogs, or cats was a risk factor for cryptosporidiosis in children in Guinea-Bissau and Indonesia in one study.[42,43] A weak association was observed between the occurrence of cryptosporidiosis in HIV+ persons and contact with dogs in another study.[44] In other studies, no increased risk in the acquisition of cryptospo-ridiosis was associated with contact with companion animals.[45]

The distribution of *C. parvum* and *C. hominis* in humans is probably a good indicator of the transmission routes. Thus far, studies conducted in developing countries have shown a pre-dominance of *C. hominis* in children or HIV+ adults. This is also true for most areas in the United States, Canada, Australia, and Japan. In Europe and New Zealand, however, several stud-ies have shown an almost equal prevalence of *C. parvum* and *C. hominis* in both immunocompetent and immunocompro-mised persons.[15] In contrast, children in the Middle East are mostly infected with *C. parvum*.[46] The differences in the dis-tribution of *Cryptosporidium* genotypes in humans are consid-ered an indication of differences in infection sources[15,41,47,48]; the occurrence of *C. hominis* in humans is most likely due to anthroponotic transmission, whereas *C. parvum* in a popula-tion can be the result of both anthroponotic and zoonotic trans-mission. Thus, in most developing countries, it is possible that anthroponotic transmission of *Cryptosporidium* plays a major role in human cryptosporidiosis, whereas in Europe, New Zealand, and rural areas of the United States, both anthro-ponotic and zoonotic transmissions are important.

Recent subtyping studies based on sequence analyses of the 60 kDa glycoprotein (gp60) gene have shown that many *C. parvum* infections in humans are not results of zoonotic transmission.[15] Among several *C. parvum* subtype families identified, IIa and IIc are the two most common families. The former has been identified in both humans and rumi-nants, thus can be a zoonotic pathogen, whereas the latter has only been seen in humans,[14,15,49] thus is an anthroponotic pathogen. In developing countries, most *C. parvum* infections in children and HIV+ persons are caused by the subtype fam-ily IIc, with IIa largely absent, indicating that anthroponotic transmission of *C. parvum* is common in these areas.[14,15] In contrast, both IIa and IIc subtype families are seen in humans in developed countries. Even in the United Kingdom where zoonotic transmission is known to play a significant role in the transmission of human cryptosporidiosis, anthroponotic transmission of *C. parvum* is also common.[50] Another *C. par-vum* subtype family commonly found in sheep and goats, IId, is the dominant *C. parvum* subtype family in humans in Mideast countries.[46] Results of multilocus subtyping support the conclusions of gp60 subtyping studies.[51,52]

38.3.5 Waterborne Transmission

Epidemiologic studies have frequently identified water as a major route of *Cryptosporidium* transmission in disease-endemic areas. In most tropical countries, *Cryptosporidium* infections in children usually peak dur-ing the rainy season; thus, waterborne transmission probably plays a role in the transmission of cryptosporidiosis in these areas.[53–55] Seasonal variations in the incidence of human *Cryptosporidium* infection in industrialized nations have also been partially attributed to waterborne transmission.[9,24,25,56] In the United States, there is a late summer peak in sporadic cases of cryptosporidiosis.[7,9,25] It is generally accepted that the late summer peak of cryptosporidiosis cases is due to rec-reational activities such as swimming and water sports.[57] In a Canadian study, swimming in a lake or river was identified as a risk factor.[38]

The role of drinking water in sporadic *Cryptosporidium* infection is not clear. In Mexican children living near the US border, cryptosporidiosis is associated with consumption of municipal water instead of bottled water.[58] In England, there is an association between the numbers of glasses of tap water drunk at home each day and the occurrence of sporadic cryptosporidiosis.[36] In the United States, drinking untreated surface water was identified as a risk factor for the acquisi-tion of *Cryptosporidium* in a small case–control study,[59] and residents living in cities with surface-derived drinking water generally have higher blood antibody levels against *Cryptosporidium* than those living in cities with groundwa-ter as drinking water, indicating that drinking water plays a role in the transmission of human cryptosporidiosis.[60] Nevertheless, case–control studies conducted in the United States and Canada have failed to show a direct linkage of *Cryptosporidium* infection to drinking water.[61–63]

Numerous waterborne outbreaks of cryptosporidi-osis have occurred in the United States, Canada, United Kingdom, France, Australia, Japan, and other industrialized nations.[64–67] These include outbreaks associated with both drinking water and recreational water (swimming pools and water parks). After the massive cryptosporidiosis outbreak in Milwaukee in 1993, the Environmental Protection Agency (EPA) revised its drinking water regulations and the water industry adopted these more stringent regulations. Currently, the number of drinking water-associated outbreaks is in decline in the United States and United Kingdom, and most outbreaks in the United States are associated with recre-ational water.[9,57] Even though five *Cryptosporidium* species are commonly found in humans, *C. parvum* and *C. hominis* are responsible for most cryptosporidiosis outbreaks, with *C. hominis* responsible for more outbreaks than *C. parvum*.[15] This is even the case for the United Kingdom, where *C. parvum* and *C. hominis* are both common in the general pop-ulation. Recently, there was one drinking water-associated cryptosporidiosis outbreak caused by *C. cuniculus*.[68]

38.3.6 Foodborne Transmission

The role of food in the transmission of cryptosporidiosis is much less clear. *Cryptosporidium* oocysts have been isolated from several foodstuffs and these have mainly been associated with fruits, vegetables, and shellfish.[69–71]

Direct contamination of food by fecal materials from animals or food handlers has been implicated in several foodborne outbreaks of cryptosporidiosis in industrialized nations. In most instances, human infections were acquired through the consumption of contaminated fresh produce and unpasteurized apple cider or milk.[72–77]

Very few case–control studies have examined the role of contaminated food as a risk factor in the acquisition of *Cryptosporidium* infection in endemic areas. A pediatric study in Brazil failed to show any association between *Cryptosporidium* infection and diet or type of food hygiene.[45] Case–control studies conducted in the United States, United Kingdom, and Australia have actually shown lower prevalence of *Cryptosporidium* infection in immunocompetent persons with frequent consumption of raw vegetables.[25,36,39,78] Nevertheless, it is estimated that about 8% of *Cryptosporidium* infections in the United States are foodborne.[8]

38.4 CLINICAL FEATURES AND PATHOGENESIS

In developing countries, frequent symptoms of cryptosporidiosis in children include diarrhea, abdominal cramps, nausea, vomiting, headache, fatigue, and low-grade fever. The diarrhea can be voluminous and watery, but usually resolves within 1–2 weeks without treatment. Not all infected children have diarrhea or other gastrointestinal symptoms, and the occurrence of diarrhea in children with cryptosporidiosis can be as low as 30% in community-based studies.[79] Even subclinical cryptosporidiosis exerts a significant adverse effect on child growth, as infected children with no clinical symptoms experience growth faltering, both in weight and in height.[2] Children can have multiple episodes of cryptosporidiosis, implying that the anti-*Cryptosporidium* immunity in children is short-lived or incomplete.[80] Cryptosporidiosis has been associated with increased mortality in developing countries in hospitalized children.[53]

Unlike in developing countries, immunocompetent persons with sporadic cryptosporidiosis in industrialized nations usually have diarrhea.[2,26,78] The median number of stools per day during the worst period of the infection is 7–9.5.[78] Other common symptoms include abdominal pain, nausea, vomiting, and low-grade fever.[26,78] The duration of illness has a mean or median of 9–21 days, with a median of five lost work or study days and hospitalization of 7%–22% of patients.[24,26,78] Patients infected with *C. hominis* are more likely to have joint pain, eye pains, recurrent headache, dizziness, and fatigue than those infected with *C. parvum*.[81] There are significant differences among different *Cryptosporidium* species and *C. hominis* subtype families in the clinical manifestations of pediatric cryptosporidiosis.[80]

Cryptosporidiosis in immunocompromised persons, including AIDS patients, is frequently associated with chronic, life-threatening diarrhea.[28] Before the era of highly active antiretroviral therapy (HAART), Manabe et al. described four clinical forms of diarrhea caused by cryptosporidiosis in AIDS patients in the United States: chronic diarrhea (36% of patients), cholera-like disease (33%), transient diarrhea (15%),

and relapsing illness (15%).[82] Sclerosing cholangitis and other biliary involvements are also common in AIDS patients with cryptosporidiosis.[83] Symptoms of cryptosporidiosis in AIDS patients vary in severity, duration, and responses to treatment. Much of this variation can be explained by the degree of immunosuppression. In addition, variations in the infection site (gastric infection, proximal small intestine infection, ileo-colonic infection, versus pan-enteric infection) have been seen in AIDS patients with cryptosporidiosis, which may contribute to the differences in disease severity and survival.[84,85] Likewise, different *Cryptosporidium* species and *C. hominis* subtype families are associated with different clinical manifestations in HIV+ persons in developing countries.[86] Cryptosporidiosis in AIDS patients is associated with increased mortality and shortened survival.[82]

Cryptosporidium spp. commonly infect the colonic and ileal mucosa. However, developing stages may be found in the entire gastrointestinal tract in immunocompromised persons. Histologically, mononuclear cell infiltration in the lamina propria, mucosal cell apoptosis, mucosal inflammation with villus blunting, and cryptitis are usually seen, leading to the loss of barrier function and malabsorption. Illness severity correlates with the extent of infection and host immunity. Infection of the crypts usually leads to more severe disease, as does infection of the proximal small bowel with villus flattening.[2,83,84,87] The voluminous diarrhea is accompanied with chloride secretion and impaired glucose absorption. It is probably mediated by substance P, a gastrointestinal neuropeptide.[88] CD4+ T lymphocytes and Th-1 immune responses play a key role in acquired immunity against cryptosporidiosis, but CD8+ T lymphocytes contribute to the clearance of the parasite from the intestine.[87,89] These protective immune responses appear to be mediated through TLR4/NF-kappaB-dependent nitric oxide production.[90,91]

38.5 TREATMENT AND PREVENTION

Numerous pharmaceutical compounds have been screened for anti-*Cryptosporidium* activities in vitro or in laboratory animals.[92–94] Some of those showing promises have been used in the experimental treatment of cryptosporidiosis in humans, but few have been proven effective in controlled clinical trials.[28,95] Oral or intravenous rehydration and antimotility drugs are widely used in treating the severe diarrhea associated with cryptosporidiosis. Nitazoxanide (NTZ) is the only FDA-approved drug for the treatment of cryptosporidiosis in immunocompetent persons. Clinical trials have demonstrated that NTZ can shorten clinical disease and reduce parasite loads.[93,96] This drug, however, is not effective in the treatment of *Cryptosporidium* infections in immunodeficient patients.[95,96] For this population, paromomycin and spiramycin have been used in the treatment of some patients, but their efficacy remains unproven.[93–95]

In industrialized nations, the most effective treatment and prophylaxis for cryptosporidiosis in AIDS patients is the use of HAART.[93,94,97] It is probably also an effective prevention for cryptosporidiosis in HIV+ persons in developing countries.[98]

It is believed that the eradication and prevention of the infection are related to the replenishment of CD4+ cells in treated persons and the antiparasitic activities of the protease inhibitors (such as indinavir, nelfinavir, and ritonavir) used in HAART.[94,97] A relapse of cryptosporidiosis is common in AIDS patients who have stopped taking HAART.[99] In developing countries, protease inhibitors are generally not included in the HAART regimens. Thus, the effect of HARRT on the occurrence of cryptosporidiosis there remains unclear.[100] Limited reports have shown that cryptosporidiosis is still common in HIV-positive patients receiving HAART in developing countries, although at lower frequencies than those generally reported in untreated HIV patients.[98,101,102]

As for any pathogens that are transmitted by the fecal-oral route, good hygiene is key in preventing the acquisition of *Cryptosporidium* infection.[103] Immunosuppressed persons especially should take necessary precautions to prevent the occurrence of cryptosporidiosis.[104] This includes washing hands before eating and after going to the bathroom, changing diapers, and contacting pets or soil (including gardening); avoiding drinking water from lakes and rivers, swallowing water in recreational activities, and unpasteurized milk, milk products, and juices; and following safe-sex practices (avoiding oral–anal contact). During cryptosporidiosis outbreaks or when a community advisory to boil water is issued, individuals should boil water for one minute to kill the parasite or use a tap water filter capable of removing particles less than one micron in diameter. Immunosuppressed persons also should avoid eating raw shellfish and should not eat uncooked vegetables and unpeeled fruits when traveling to cryptosporidiosis-endemic areas (such as developing countries).[104]

38.6 IDENTIFICATION AND DIAGNOSIS

At present, almost all active *Cryptosporidium* infections are diagnosed by the analysis of stool specimens. Stool specimens are usually collected fresh or in fixative solutions such as 10% buffered formalin and polyvinyl alcohol (PVA).[105] However, stool specimens fixed in formalin and mercury-based preservatives (such as LV-PVA) cannot be used for polymerase chain reaction (PCR), which requires the use of fresh or frozen stool specimens or stools preserved in TotalFix, zinc PVA, or 2.5% potassium dichromate. It is recommended that whenever possible, multiple specimens (three specimens passed at intervals of 2–3 days) from each patient should be examined if *Cryptosporidium* infection is suspected and the examination of the initial stool specimen is negative. This is because carriers with low oocyst shedding are common, and examination of individual specimens can lead to the detection of only 53% of infections.[106] In clinical laboratories, *Cryptosporidium* spp. in stool specimens are commonly detected by microscopic examinations of oocysts or immunologic detection of antigens.[107] PCR-based typing methods, however, are increasingly used in investigations of outbreaks, surveillance, and diagnosis of cryptosporidiosis.[15]

Examination of intestinal or biliary biopsy materials is sometimes used in the diagnosis of cryptosporidiosis in AIDS patients.[84] However, the sensitivity of the diagnosis depends on the location of tissues examined; the duodenum is usually infected with *Cryptosporidium* only in severe infection, and the terminal ileum has significantly higher detection rates than the duodenum.[106] Thus, upper endoscopic biopsies are much less sensitive than lower endoscopic biopsies in diagnosing cryptosporidiosis. However, lower endoscopy is generally considered too invasive and risky for many AIDS patients. In such cases, colonoscopy should be used in biopsy.

38.6.1 MICROSCOPY

Stool specimens can be examined directly for *Cryptosporidium* oocysts by microscopy of direct wet mount or stained fecal materials if the number of oocysts in specimens is high. *Cryptosporidium* oocysts in humans are generally 4–6 μm in size. Oocysts present are often concentrated using either traditional ethyl acetate or Weber-modified ethyl acetate concentration methods.[107] Concentrated stool specimens can be examined by microscopy in several ways. *Cryptosporidium* oocysts can be detected by bright-field microscopy in direct wet mounts. This allows the observation of oocyst morphology and a more accurate measurement of oocysts, which is frequently needed in biologic studies. Differential interference contrast (DIC) can be used in microscopy, which produces better images and visualization of internal structures of oocysts (Figure 38.1). Most *Cryptosporidium* species look similar under microscopes and have similar morphometric measurements.[13]

More often, *Cryptosporidium* oocysts in concentrated stool specimens are detected by microscopy after staining of the fecal smears. Many special stains have been used in the detection of *Cryptosporidium* oocysts, but modified acid-fast stains are the most commonly used,[107] especially in developing countries, because of their low cost, ease of use, no need for special microscopes, and simultaneous detection of several other pathogens such as *Cystoisospora* and *Cyclospora*. Two stains widely used are the modified Ziehl–Neelsen acid-fast stain and the modified Kinyoun's acid-fast stain.[107] Oocysts are stained bright red to purple against a blue or green background (Figure 38.1).

Direct immunofluorescence assays (DFAs) have been used increasingly in *Cryptosporidium* oocyst detection by microscopy, especially in industrialized nations. Compared to acid-fast staining, DFA has higher sensitivity and specificity.[108] Many commercial DFA kits are marketed for the diagnosis of *Cryptosporidium*, most of which include reagents allowing simultaneous detection of *Giardia* cysts. Oocysts appear apple green against a dark background in immunofluorescence microscopy (Figure 38.1). Because of the high sensitivity and specificity, DFA has been used by some as the gold standard or reference test.[108,109] It has been shown that most antibodies in commercial DFA kits react with oocysts of almost all *Cryptosporidium* species, making species diagnosis impossible.[110,111]

The sensitivity of most microscopic methods is low. The detection limit for the combination of ethyl acetate concentration and DFA was shown to be 10,000 oocysts per gram of liquid stool and 50,000 oocysts per gram of formed stool.[112,113] A similar sensitivity was achieved with fecal specimens from dogs.[114] The sensitivity of modified acid-fast staining was tenfold lower than DFA,[113] probably because acid-fast stains do not consistently stain all oocysts.[115] The sensitivity of the DFA can be significantly improved by the incorporation of an oocyst isolation step using an expensive immunomagnetic separation technique.[116]

38.6.2 FECAL ANTIGEN DETECTION

Cryptosporidium infection can also be diagnosed by the detection of *Cryptosporidium* antigens in stool specimens by immunoassays.[107] Antigen capture-based enzyme immunoassays (EIAs) have been used in the diagnosis of cryptosporidiosis since 1990. In recent years, EIAs have gained popularity because they do not require experienced microscopists and can be used to screen a large number of samples.[117] Several commercial EIA kits are commonly used in clinical laboratories. High specificity (99%–100%) has been generally reported for these EIA kits.[108,109] Sensitivities, however, have been reported to range from 70%[108] to 94%–100%.[109,118,119] Occasional false positivity of EIA kits is known to occur in the detection of *Cryptosporidium*,[120,121] and manufacturer's recalls of EIA kits have occurred because of high nonspecificity.[122] If a patient is in the carrier state or during self-cure, the number of parasites may drop below the sensitivity levels of these kits.[108] Most EIA kits have been evaluated only with human stool specimens presumably from patients infected with *C. hominis* or *C. parvum*. Their usefulness in the detection of *Cryptosporidium* spp. in animals may be compromised by the specificity of the antibodies.

In recent years, several lateral flow immunochromatographic assays have been marketed for rapid detection of *Cryptosporidium* in stool specimens. In the few evaluation studies conducted, these assays have been shown to have high specificities (>90%) and sensitivity (98%–100%).[108,123–127] However, sensitivities of 68%–75% were shown in some studies for some assays.[108,128] These rapid assays have also been affected by quality control problems and were subjected to several manufacturer's recalls because of false-positive results.[129,130] High false-positive rates of several rapid assays have been recently reported in clinical diagnosis of cryptosporidiosis in the Unites States. It has also been shown that some rapid assay kits have low sensitivity in detecting some *Cryptosporidium* species other than *C. hominis* and *C. parvum*.

38.6.3 GENOTYPING AND SUBTYPING

Molecular techniques, especially PCR and PCR-related methods, have been developed and used in the detection and differentiation of *Cryptosporidium* spp. for many years. Several genus-specific PCR-RFLP-based genotyping tools have been developed for the detection and differentiation of *Cryptosporidium* at the species level.[131–135] Most of these techniques are based on the small-subunit (SSU) rRNA gene. Other genotyping techniques are designed mostly for the differentiation of *C. parvum* and *C. hominis*, thus cannot detect and differentiate other *Cryptosporidium* spp. or genotypes.[15] Their usefulness in the analysis of human stool specimens is compromised by their inability to detect *C. canis*, *C. felis*, *C. suis*, *C. muris*, and other species/genotypes divergent from *C. parvum* and *C. hominis*.[136]

Several subtyping tools have also been developed to characterize the diversity within *C. parvum* or *C. hominis*.[15] One of the most commonly used techniques is DNA sequence analysis of the gp60 gene (also known as gp15/45/60, gp40/15).[22,49,137,138] Most of the genetic heterogeneity in this gene is present in the number of a trinucleotide repeat (TCA, TCG, or TCT), although extensive sequence differences are also present between families of subtypes. Multilocus mini- and microsatellite subtyping tools for *C. parvum* and *C. hominis* have also been developed.[52,139–142] The usefulness of subtyping tools has been demonstrated by the analysis of samples from foodborne and waterborne outbreaks of cryptosporidiosis.[39,76,143–152]

38.6.4 CASE REPORTING

Cryptosporidiosis is a notifiable disease in most industrialized countries. Thus, the detection of the pathogen in stools or tissues from humans should be reported to the local health department in addition to the physicians. The reporting of a significant number of cases above background levels in industrialized nations indicates the likely occurrence of outbreaks of cryptosporidiosis or the false positivity of diagnostic kits.[121,122,129,130,153] In situations like this, it is crucial to have the test results verified by a confirmatory test such as DFA or PCR, and report them to the state or local public health department. Residual specimens from these cases should be submitted to public health laboratories for genotyping and subtyping, especially when an outbreak of cryptosporidiosis is suspected.

38.6.5 DETECTION OF *CRYPTOSPORIDIUM* OOCYSTS IN FOOD AND WATER

The detection of *Cryptosporidium* oocysts in water samples is now mostly done using the standard EPA Method 1623 and its equivalent in other countries.[154,155] It involves the concentration of oocysts in 10–100 L of water, immunomagnetic isolation of oocysts from water concentrates, immunofluorescence and DAPI staining of isolated oocysts, and examination of oocysts under epifluorescence and DIC microscopy. Method 1623 has four certified sample concentration systems to capture oocysts in water, including Envirochek filters (Pall Corporation, Ann Arbor, MI), Envirochek HV filters (Pall Corporation), Filta-Max filters (IDEXX, Westbrook, MA), or continuous flow centrifugation (Haemonetics, Braintree, MA). More recently, in response to the need for simultaneous detection of viral pathogens of biodefense importance, hollow-fiber ultrafiltration has been used as an alternative oocyst concentration technique in the conventional EPA Method 1623, using either the tangential flow or the dead-end filtration format.[156–158]

Because most *Cryptosporidium* oocysts in surface water samples are not from human-pathogenic species, genotyping *Cryptosporidium* oocysts has been used in Method 1623 downstream analysis. This allows the assessment of the human infective potential of oocysts in source or drinking water and contamination sources in watersheds.[159,160] This can be done using either residual water concentrates or microscopy-positive slides from Method 1623.[161,162] In resource-limited countries, calcium carbonate flocculation can be used to replace the filtration in oocyst concentration, and PCR analysis of DNA extraction from water concentrates can be used as an alternative to immunomagnetic separation and fluorescence microscopy to reduce the high cost associated with Method 1623.[163]

No standard methods are available for the detection of *Cryptosporidium* oocysts in food.[10] Current methods in research laboratories for the detection of oocysts in fresh produces, meat products, and other food generally include steps for oocyst elution, concentration, separation, and detection, some of which are adapted from Method 1623 for water samples.[164–166] The recovery rates for these methods are reportedly high, but have not been verified through vigorous interlaboratory trials.

38.7 CONCLUSIONS AND FUTURE PERSPECTIVES

Since the massive waterborne outbreak of cryptosporidiosis in Milwaukee in 1993, in which an estimated 403,000 people were infected,[167] *Cryptosporidium* has been regulated by drinking water standards of many industrialized nations.[168,169] Tremendous progress has been made to better understand its basic biology, transmission routes/infection sources, and environmental ecology and to improve drinking water treatment processes. This has led to the reduction of drinking water-associated outbreaks of cryptosporidiosis in the United States and some other industrialized nations. Nevertheless, the number of reported cryptosporidiosis cases has increased significantly in the United States in recent years,[4,6,7] where recreational water is now a major transmission route.[57] This has been accompanied with the emergence of a new *C. hominis* subtype.[170] The recent Global Enteric Multicenter Study in seven developing countries in Asia and Africa has also identified cryptosporidiosis as the No. 2 cause for diarrhea in children less than 2 years in age.

More research on *Cryptosporidium* biology is needed in response to these new public health challenges. Currently, our understanding of the basic biology of *Cryptosporidium* spp., such as invasion mechanisms, virulence determinants, and protective immune responses, lags far behind other related apicomplexan parasites such as *Plasmodium* and *Toxoplasma*. This is partially due to the lack of effective research tools, such as efficient laboratory animal and cell culture models and genetic manipulation techniques. The completion of the whole genome sequencing of *C. hominis* and *C. parvum* in 2004 improved our understanding of the biochemistry and metabolism of *Cryptosporidium* spp. and provided some optimism in the development of new intervention tools. However, 8 years later, we still have no fully effective drugs and promising vaccines against cryptosporidiosis and have sequenced the partial genome of only two other *Cryptosporidium* isolates.

Increased research funding is clearly needed to combat cryptosporidiosis, which is causing an estimated 750,000 cases per year in the United States. This includes an estimated 2725 annual hospitalizations costing $45.7 million in direct costs.[171] The reduced drinking water-associated transmission of cryptosporidiosis and the availability of HAART as a therapeutic and preventive measure against cryptosporidiosis in AIDS patients have created some complacency in funding agencies and the water and pharmaceutical industry. Thus, research on *Cryptosporidium* peaked in the United States in 2000 and in the world in 2007, as indicated by the number of publications per year in a recent search of the Web of Knowledge database. Likewise, a recent search of the National Institutes of Health (NIH) database has identified only three new and completing R01 grants and two other smaller grants on *Cryptosporidium* during 2009–2011.[172] The USEPA and water industry used to be a driving force for *Cryptosporidium* research but now offer few research grants on *Cryptosporidium*. Despite the availability of many potential targets in *Cryptosporidium* metabolism identified through whole genome sequencing,[173,174] few pharmaceutical companies are actively pursuing the development of effective drugs or vaccines against cryptosporidiosis. Stimulation of *Cryptosporidium* research activities will require the concerted efforts of governmental agencies, private industry, nonprofit organizations, and research scientists.

DISCLAIMER

The findings and conclusions in this report are those of the authors and do not necessarily represent the views of the Centers for Disease Control and Prevention.

REFERENCES

1. Fayer, R. Taxonomy and species delimitation in *Cryptosporidium*. *Exp Parasitol* 124, 90–97 (2010).
2. Chalmers, R.M. and Davies, A.P. Minireview: Clinical cryptosporidiosis. *Exp Parasitol* 124, 138–146 (2010).
3. Fayer, R. Introduction. In *Cryptosporidium and Cryptosporidiosis*, 2nd ed., Fayer, R. and Xiao, L. (eds.), pp. 1–42 (Taylor & Francis, Boca Raton, FL, 2008).
4. Yoder, J.S. and Beach, M.J. Cryptosporidiosis surveillance—United States, 2003–2005. *MMWR Surveill Summ* 56, 1–10 (2007).
5. Hlavsa, M.C., Watson, J.C., and Beach, M.J. Cryptosporidiosis surveillance—United States 1999–2002. *MMWR Surveill Summ* 54, 1–8 (2005).

6. Yoder, J.S., Harral, C., and Beach, M.J. Cryptosporidiosis surveillance—United States, 2006–2008. *MMWR Surveill Summ* 59, 1–14 (2010).

7. Yoder, J.S. et al. Cryptosporidiosis surveillance—United States, 2009–2010. *MMWR Surveill Summ* 61, 1–12 (2012).

8. Scallan, E. et al. Foodborne illness acquired in the United States—Major pathogens. *Emerg Infect Dis* 17, 7–15 (2011).

9. Yoder, J.S. and Beach, M.J. *Cryptosporidium* surveillance and risk factors in the United States. *Exp Parasitol* 124, 31–39 (2010).

10. Smith, H.V. and Nichols, R.A. *Cryptosporidium*: Detection in water and food. *Exp Parasitol* 124, 61–79 (2010).

11. Leander, B.S. Marine gregarines: Evolutionary prelude to the apicomplexan radiation? *Trends Parasitol* 24, 60–67 (2008).

12. Rosales, M.J. et al. Extracellular like-gregarine stages of *Cryptosporidium parvum*. *Acta Trop* 95, 74–78 (2005).

13. Xiao, L. et al. *Cryptosporidium* taxonomy: Recent advances and implications for public health. *Clin Microbiol Rev* 17, 72–97 (2004).

14. Xiao, L. and Feng, Y. Zoonotic cryptosporidiosis. *FEMS Immunol Med Microbiol* 52, 309–323 (2008).

15. Xiao, L. Molecular epidemiology of cryptosporidiosis: An update. *Exp Parasitol* 124, 80–89 (2010).

16. Elwin, K. et al. *Cryptosporidium viatorum* n. sp. (Apicomplexa: Cryptosporidiidae) among travellers returning to Great Britain from the Indian subcontinent, 2007–2011. *Int J Parasitol* 42, 675–682 (2012).

17. Elwin, K. et al. The epidemiology of sporadic human infections with unusual cryptosporidia detected during routine typing in England and Wales, 2000–2008. *Epidemiol Infect* 140, 673–683 (2012).

18. Nichols, G.L. et al. Cryptosporidiosis: A report on the surveillance and epidemiology of *Cryptosporidium* infection in England and Wales. Drinking Water Directorate Contract Number DWI 70/2/201. Drinking Water Inspectorate, London, U.K., 142 (2006).

19. Alvarez-Pellitero, P. et al. *Cryptosporidium* scophthalmi n. sp. (Apicomplexa: Cryptosporidiidae) from cultured turbot *Scophthalmus maximus*. Light and electron microscope description and histopathological study. *Dis Aquat Organ* 62, 133–145 (2004).

20. Leitch, G.J. and He, Q. Cryptosporidiosis—An overview. *J Biomed Res* 25, 1–16 (2012).

21. Mor, S.M. and Tzipori, S. Cryptosporidiosis in children in sub-Saharan Africa: A lingering challenge. *Clin Infect Dis* 47, 915–921 (2008).

22. Sulaiman, I.M. et al. Unique endemicity of cryptosporidiosis in children in Kuwait. *J Clin Microbiol* 43, 2805–2809 (2005).

23. Mor, S.M. et al. Cryptosporidiosis in the elderly population of the United States. *Clin Infect Dis* 48, 698–705 (2009).

24. Dietz, V. et al. Active, multisite, laboratory-based surveillance for *Cryptosporidium parvum*. *Am J Trop Med Hyg* 62, 368–372 (2000).

25. Roy, S.L. et al. Risk factors for sporadic cryptosporidiosis among immunocompetent persons in the United States from 1999 to 2001. *J Clin Microbiol* 42, 2944–2951 (2004).

26. Goh, S. et al. Sporadic cryptosporidiosis, North Cumbria, England, 1996–2000. *Emerg Infect Dis* 10, 1007–1015 (2004).

27. McLauchlin, J. et al. Polymerase chain reaction-based diagnosis of infection with *Cryptosporidium* in children with primary immunodeficiencies. *Pediatr Infect Dis J* 22, 329–335 (2003).

28. Hunter, P.R. and Nichols, G. Epidemiology and clinical features of *Cryptosporidium* infection in immunocompromised patients. *Clin Microbiol Rev* 15, 145–154 (2002).

29. Wolska-Kusnierz, B. et al. *Cryptosporidium* infection in patients with primary immunodeficiencies. *J Pediatr Gastroenterol Nutr* 45, 458–464 (2007).

30. Krause, I. et al. Cryptosporidiosis in children following solid organ transplantation. *Pediatr Infect Dis J* 31, 1135–1138 (2012).

31. Tran, M.Q. et al. *Cryptosporidium* infection in renal transplant patients. *Clin Nephrol* 63, 305–309 (2005).

32. Seyrafian, S. et al. Prevalence rate of *Cryptosporidium* infection in hemodialysis patients in Iran. *Hemodial Int* 10, 375–379 (2006).

33. Bonatti, H. et al. *Cryptosporidium* enteritis in solid organ transplant recipients: Multicenter retrospective evaluation of 10 cases reveals an association with elevated tacrolimus concentrations. *Transpl Infect Dis* 14, 635–648 (2012).

34. Bandin, F. et al. Cryptosporidiosis in paediatric renal transplantation. *Pediatr Nephrol* 24, 2245–2255 (2009).

35. Hellard, M. et al. Risk factors leading to *Cryptosporidium* infection in men who have sex with men. *Sex Transm Infect* 79, 412–414 (2003).

36. Hunter, P.R. et al. Sporadic cryptosporidiosis case-control study with genotyping. *Emerg Infect Dis* 10, 1241–1249 (2004).

37. Pollock, K.G. et al. Spatial and temporal epidemiology of sporadic human cryptosporidiosis in Scotland. *Zoonoses Public Health* 57, 487–492 (2010).

38. Pintar, K.D. et al. A modified case-control study of cryptosporidiosis (using non-*Cryptosporidium*-infected enteric cases as controls) in a community setting. *Epidemiol Infect* 137, 1789–1799 (2009).

39. Valderrama, A.L. et al. Multiple risk factors associated with a large statewide increase in cryptosporidiosis. *Epidemiol Infect* 137, 1781–1788 (2009).

40. Lake, I.R. et al. Case-control study of environmental and social factors influencing cryptosporidiosis. *Eur J Epidemiol* 22, 805–811 (2007).

41. Snel, S.J., Baker, M.G., and Venugopal, K. The epidemiology of cryptosporidiosis in New Zealand, 1997–2006. *N Z Med J* 122, 47–61 (2009).

42. Molbak, K., Aaby, P., Hojlyng, N., and da Silva, A.P. Risk factors for *Cryptosporidium* diarrhea in early childhood: A case–control study from Guinea-Bissau, West Africa. *Am J Epidemiol* 139, 734–740 (1994).

43. Katsumata, T. et al. Cryptosporidiosis in Indonesia: A hospital-based study and a community-based survey. *Am J Trop Med Hyg* 59, 628–632 (1998).

44. Glaser, C.A. et al. Association between *Cryptosporidium* infection and animal exposure in HIV-infected individuals. *J AIDS Hum Retrovirol* 17, 79–82 (1998).

45. Pereira, M.D. et al. Intra-familial and extra-familial risk factors associated with *Cryptosporidium parvum* infection among children hospitalized for diarrhea in Goiania, Goias, Brazil. *Am J Trop Med Hyg* 66, 787–793 (2002).

46. Nazemalhosseini-Mojarad, E., Feng, Y., and Xiao, L. The importance of subtype analysis of *Cryptosporidium* spp. in epidemiological investigations of human cryptosporidiosis in Iran and other Mideast countries. *Gastroenterol Hepatol Bed Bench* 5, 67–70 (2012).

47. Learmonth, J.J. et al. Genetic characterization and transmission cycles of *Cryptosporidium* species isolated from humans in New Zealand. *Appl Environ Microbiol* 70, 3973–3978 (2004).

48. Chalmers, R.M. et al. Long-term *Cryptosporidium* typing reveals the aetiology and species-specific epidemiology of human cryptosporidiosis in England and Wales, 2000 to 2003. *Euro Surveill* 14(pii), 19086 (2009).

49. Widmer, G. Meta-analysis of a polymorphic surface glycoprotein of the parasitic protozoa *Cryptosporidium parvum* and *Cryptosporidium hominis*. *Epidemiol Infect* 137, 1800–1808 (2009).

50. Hunter, P.R. et al. Subtypes of *Cryptosporidium parvum* in humans and disease risk. *Emerg Infect Dis* 13, 82–88 (2007).

51. Mallon, M.E. et al. Multilocus genotyping of *Cryptosporidium parvum* Type 2: Population genetics and sub-structuring. *Infect Genet Evol* 3, 207–218 (2003).

52. Grinberg, A. et al. Host-shaped segregation of the *Cryptosporidium parvum* multilocus genotype repertoire. *Epidemiol Infect* 136, 273–278 (2008).

53. Tumwine, J.K. et al. *Cryptosporidium parvum* in children with diarrhea in Mulago Hospital, Kampala, Uganda. *Am J Trop Med Hyg* 68, 710–715 (2003).

54. Peng, M.M. et al. Molecular epidemiology of cryptosporidiosis in children in Malawi. *J Eukaryot Microbiol* 50(Suppl), 557–559 (2003).

55. Bhattacharya, M.K. et al. *Cryptosporidium* infection in children in urban Bangladesh. *J Trop Pediatr* 43, 282–286 (1997).

56. McLauchlin, J. et al. Molecular epidemiological analysis of *Cryptosporidium* spp. in the United Kingdom: Results of genotyping *Cryptosporidium* spp. in 1,705 fecal samples from humans and 105 fecal samples from livestock animals. *J Clin Microbiol* 38, 3984–3990 (2000).

57. Hlavsa, M.C. et al. Surveillance for waterborne disease outbreaks and other health events associated with recreational water—United States, 2007–2008. *MMWR Surveill Summ* 60, 1–32 (2011).

58. Leach, C.T. et al. Prevalence of *Cryptosporidium parvum* infection in children along the Texas-Mexico border and associated risk factors. *Am J Trop Med Hyg* 62, 656–661 (2000).

59. Gallaher, M.M. et al. Cryptosporidiosis and surface water. *Am J Public Health* 79, 39–42 (1989).

60. Frost, F.J. et al. Serological responses to *Cryptosporidium* antigens among users of surface- vs. ground-water sources. *Epidemiol Infect* 131, 1131–1138 (2003).

61. Khalakdina, A. et al. Is drinking water a risk factor for endemic cryptosporidiosis? A case–control study in the immunocompetent general population of the San Francisco Bay Area. *BMC Public Health* 3, 11 (2003).

62. Sorvillo, F. et al. Municipal drinking water and cryptosporidiosis among persons with AIDS in Los Angeles County. *Epidemiol Infect* 113, 313–320 (1994).

63. Pintar, K.D. et al. Considering the risk of Infection by *Cryptosporidium* via consumption of municipally treated drinking water from a surface water source in a Southwestern Ontario community. *Risk Anal* 32, 1122–1138 (2012).

64. Fayer, R. *Cryptosporidium*: A water-borne zoonotic parasite. *Vet Parasitol* 126, 37–56 (2004).

65. Baldursson, S. and Karanis, P. Waterborne transmission of protozoan parasites: Review of worldwide outbreaks—An update 2004–2010. *Water Res* 45, 6603–6614 (2011).

66. Karanis, P., Kourenti, C., and Smith, H. Waterborne transmission of protozoan parasites: A worldwide review of outbreaks and lessons learnt. *J Water Health* 5, 1–38 (2007).

67. Semenza, J.C. and Nichols, G. Cryptosporidiosis surveillance and water-borne outbreaks in Europe. *Euro Surveill* 12, E13–E14 (2007).

68. Chalmers, R.M. et al. *Cryptosporidium* sp. rabbit genotype, a newly identified human pathogen. *Emerg Infect Dis* 15, 829–830 (2009).

69. Robertson, L.J. and Gjerde, B. Occurrence of parasites on fruits and vegetables in Norway. *J Food Prot* 64, 1793–1798 (2001).

70. Fayer, R., Dubey, J.P., and Lindsay, D.S. Zoonotic protozoa: From land to sea. *Trends Parasitol* 20, 531–536 (2004).

71. Budu-Amoako, E. et al. Foodborne illness associated with *Cryptosporidium* and *Giardia* from livestock. *J Food Prot* 74, 1944–1955 (2011).

72. Millard, P.S. et al. An outbreak of cryptosporidiosis from fresh-pressed apple cider [published erratum appears in *JAMA* 273, 776, 1995]. *JAMA* 272, 1592–1596 (1994).

73. Quiroz, E.S. et al. An outbreak of cryptosporidiosis linked to a foodhandler. *J Infect Dis* 181, 695–700 (2000).

74. Ponka, A. et al. A foodborne outbreak due to *Cryptosporidium parvum* in Helsinki, November 2008. *Euro Surveill* 14(pii), 19269 (2009).

75. Ethelberg, S. et al. A foodborne outbreak of *Cryptosporidium hominis* infection. *Epidemiol Infect* 137, 348–356 (2009).

76. Blackburn, B.G. et al. Cryptosporidiosis associated with ozonated apple cider. *Emerg Infect Dis* 12, 684–686 (2006).

77. Yoshida, H. et al. An outbreak of cryptosporidiosis suspected to be related to contaminated Food, October 2006, Sakai City, Japan. *Jpn J Infect Dis* 60, 405–407 (2007).

78. Robertson, B. et al. Case-control studies of sporadic cryptosporidiosis in Melbourne and Adelaide, Australia. *Epidemiol Infect* 128, 419–431 (2002).

79. Bern, C. et al. Epidemiologic differences between cyclosporiasis and cryptosporidiosis in Peruvian children. *Emerg Infect Dis* 8, 581–585 (2002).

80. Cama, V.A. et al. *Cryptosporidium* species and subtypes and clinical manifestations in children, Peru. *Emerg Infect Dis* 14, 1567–1574 (2008).

81. Hunter, P.R. et al. Health sequelae of human cryptosporidiosis in immunocompetent patients. *Clin Infect Dis* 39, 504–510 (2004).

82. Manabe, Y.C. et al. Cryptosporidiosis in patients with AIDS—Correlates of disease and survival. *Clin Infect Dis* 27, 536–542 (1998).

83. De Angelis, C. et al. An update on AIDS-related cholangiopathy. *Minerva Gastroenterol Dietol* 55, 79–82 (2009).

84. Clayton, F., Heller, T., and Kotler, D.P. Variation in the enteric distribution of cryptosporidia in acquired immunodeficiency syndrome. *Am J Clin Pathol* 102, 420–425 (1994).

85. Lumadue, J.A. et al. A clinicopathologic analysis of AIDS-related cryptosporidiosis. *AIDS* 12, 2459–2466 (1998).

86. Cama, V.A. et al. Differences in clinical manifestations among *Cryptosporidium* species and subtypes in HIV-infected persons. *J Infect Dis* 196, 684–691 (2007).

87. Pantenburg, B. et al. Intestinal immune response to human *Cryptosporidium* sp. infection. *Infect Immun* 76, 23–29 (2008).

88. Hernandez, J. et al. Substance P is responsible for physiological alterations such as increased chloride ion secretion and glucose malabsorption in cryptosporidiosis. *Infect Immun* 75, 1137–1143 (2007).

89. Pantenburg, B. et al. Human CD8(+) T cells clear *Cryptosporidium parvum* from infected intestinal epithelial cells. *Am J Trop Med Hyg* 82, 600–607 (2010).

90. Costa, L.B. et al. *Cryptosporidium*—Malnutrition interactions: Mucosal disruption, cytokines and TLR signaling in a weaned murine model. *J Parasitol* 97, 1113–1120 (2011).

91. Zhou, R. et al. miR-27b targets KSRP to coordinate TLR4-mediated epithelial defense against *Cryptosporidium parvum* infection. *PLoS Pathog* 8, e1002702 (2012).

92. Gargala, G. Drug treatment and novel drug target against *Cryptosporidium*. *Parasite* 15, 275–281 (2008).

93. Rossignol, J.F. *Cryptosporidium* and *Giardia*: Treatment options and prospects for new drugs. *Exp Parasitol* 124, 45–53 (2010).

94. Pantenburg, B., Cabada, M.M., and White, A.C., Jr. Treatment of cryptosporidiosis. *Expert Rev Anti Infect Ther* 7, 385–391 (2009).

95. Abubakar, I. et al. Treatment of cryptosporidiosis in immunocompromised individuals: Systematic review and meta-analysis. *Br J Clin Pharmacol* 63, 387–393 (2007).

96. Bailey, J.M. and Erramouspe, J. Nitazoxanide treatment for giardiasis and cryptosporidiosis in children. *Ann Pharmacother* 38, 634–640 (2004).

97. Zardi, E.M., Picardi, A., and Afeltra, A. Treatment of cryptosporidiosis in immunocompromised hosts. *Chemotherapy* 51, 193–196 (2005).

98. Werneck-Silva, A.L. and Prado, I.B. Gastroduodenal opportunistic infections and dyspepsia in HIV-infected patients in the era of highly active antiretroviral therapy. *J Gastroenterol Hepatol* 24, 135–139 (2009).

99. Maggi, P. et al. Effect of antiretroviral therapy on cryptosporidiosis and microsporidiosis in patients infected with human immunodeficiency virus type 1. *Eur J Clin Microbiol Infect Dis* 19, 213–217 (2000).

100. Ajjampur, S.S., Sankaran, P., and Kang, G. *Cryptosporidium* species in HIV-infected individuals in India: An overview. *Natl Med J India* 21, 178–184 (2008).

101. Certad, G. et al. Cryptosporidiosis in HIV-infected Venezuelan adults is strongly associated with acute or chronic diarrhea. *Am J Trop Med Hyg* 73, 54–57 (2005).

102. Tuli, L. et al. Correlation between CD4 counts of HIV patients and enteric protozoan in different seasons—An experience of a tertiary care hospital in Varanasi (India). *BMC Gastroenterol* 8, 36 (2008).

103. Juranek, D.D. Cryptosporidiosis: Sources of infection and guidelines for prevention. *Clin Infect Dis* 21, S57–S61 (1995).

104. Kaplan, J.E. et al. Guidelines for prevention and treatment of opportunistic infections in HIV-infected adults and adolescents: Recommendations from CDC, the National Institutes of Health, and the HIV Medicine Association of the Infectious Diseases Society of America. *MMWR Recomm Rep* 58, 1–207; quiz CE1-4 (2009).

105. Garcia, L.S. et al. Techniques for the recovery and identification of *Cryptosporidium* oocysts from stool specimens. *J Clin Microbiol* 18, 185–190 (1983).

106. Greenberg, P.D., Koch, J., and Cello, J.P. Diagnosis of *Cryptosporidium parvum* in patients with severe diarrhea and AIDS. *Digest Dis Sci* 41, 2286–2290 (1996).

107. Smith, H.V. Diagnostics. In *Cryptosporidium and Cryptosporidiosis*, 2nd ed., Fayer, R. and Xiao, L. (eds.), pp. 173–207 (CRC Press, Boca Raton, FL, 2008).

108. Johnston, S.P. et al. Evaluation of three commercial assays for detection of *Giardia* and *Cryptosporidium* organisms in fecal specimens. *J Clin Microbiol* 41, 623–626 (2003).

109. Garcia, L.S. and Shimizu, R.Y. Evaluation of nine immunoassay kits (enzyme immunoassay and direct fluorescence) for detection of *Giardia lamblia* and *Cryptosporidium parvum* in human fecal specimens. *J Clin Microbiol* 35, 1526–1529 (1997).

110. Graczyk, T.K., Cranfield, M.R., and Fayer, R. Evaluation of commercial enzyme immunoassay (EIA) and immunofluorescent antibody (FA) test kits for detection of *Cryptosporidium* oocysts of species other than *Cryptosporidium parvum*. *Am J Trop Med Hyg* 54, 274–279 (1996).

111. Yu, J.R. et al. A common oocyst surface antigen of *Cryptosporidium* recognized by monoclonal antibodies. *Parasitol Res* 88, 412–420 (2002).

112. Webster, K.A. et al. Detection of *Cryptosporidium parvum* oocysts in faeces: Comparison of conventional coproscopical methods and the polymerase chain reaction. *Vet Parasitol* 61, 5–13 (1996).

113. Weber, R. et al. Threshold of detection of *Cryptosporidium* oocysts in human stool specimens: Evidence for low sensitivity of current diagnostic methods. *J Clin Microbiol* 29, 1323–1327 (1991).

114. Rimhanen-Finne, R. et al. Evaluation of immunofluorescence microscopy and enzyme-linked immunosorbent assay in detection of *Cryptosporidium* and *Giardia* infections in asymptomatic dogs. *Vet Parasitol* 145, 345–348 (2007).

115. Garcia, L.S., Brewer, T.C., and Bruckner, D.A. Fluorescence detection of *Cryptosporidium* oocysts in human fecal specimens by using monoclonal antibodies. *J Clin Microbiol* 25, 119–121 (1987).

116. Robinson, G., Watkins, J., and Chalmers, R.M. Evaluation of a modified semi-automated immunomagnetic separation technique for the detection of *Cryptosporidium* oocysts in human faeces. *J Microbiol Methods* 75, 139–141 (2008).

117. Church, D. et al. Screening for *Giardia/Cryptosporidium* infections using an enzyme immunoassay in a centralized regional microbiology laboratory. *Arch Pathol Lab Med* 129, 754–759 (2005).

118. Bialek, R. et al. Comparison of fluorescence, antigen and PCR assays to detect *Cryptosporidium parvum* in fecal specimens. *Diagn Microbiol Infect Dis* 43, 283–288 (2002).

119. Srijan, A. et al. Re-evaluation of commercially available enzyme-linked immunosorbent assay for the detection of *Giardia lamblia* and *Cryptosporidium* spp from stool specimens. *Southeast Asian J Trop Med Public Health* 36(Suppl 4), 26–29 (2005).

120. Chapman, P.A., Rush, B.A., and McLauchlin, J. An enzyme immunoassay for detecting *Cryptosporidium* in faecal and environmental samples. *J Med Microbiol* 32, 233–237 (1990).

121. Doing, K.M. et al. False-positive results obtained with the Alexon ProSpecT *Cryptosporidium* enzyme immunoassay. *J Clin Microbiol* 37, 1582–1583 (1999).

122. Anonymous. False-positive laboratory tests for *Cryptosporidium* involving an enzyme-linked immunosorbent assay—United States, November 1997–March 1998. *MMWR* 48, 4–8 (1999).

123. Katanik, M.T. et al. Evaluation of ColorPAC *Giardia/Cryptosporidium* rapid assay and ProSpecT *Giardia/Cryptosporidium* microplate assay for detection of *Giardia* and *Cryptosporidium* in fecal specimens. *J Clin Microbiol* 39, 4523–4525 (2001).

124. Garcia, L.S. and Shimizu, R.Y. Detection of *Giardia lamblia* and *Cryptosporidium parvum* antigens in human fecal specimens using the ColorPAC combination rapid solid-phase qualitative immunochromatographic assay. *J Clin Microbiol* 38, 1267–1268 (2000).

125. Garcia, L.S. et al. Commercial assay for detection of *Giardia lamblia* and *Cryptosporidium parvum* antigens in human fecal specimens by rapid solid-phase qualitative immunochromatography. *J Clin Microbiol* 41, 209–212 (2003).

126. Abdel Hameed, D.M., Elwakil, H.S., and Ahmed, M.A. A single-step immunochromatographic lateral-flow assay for detection of *Giardia lamblia* and *Cryptosporidium parvum* antigens in human fecal samples. *J Egypt Soc Parasitol* 38, 797–804 (2008).

127. Regnath, T., Klemm, T., and Ignatius, R. Rapid and accurate detection of *Giardia lamblia* and *Cryptosporidium* spp. antigens in human fecal specimens by new commercially available qualitative immunochromatographic assays. *Eur J Clin Microbiol Infect Dis* 25, 807–809 (2006).

128. Weitzel, T. et al. Evaluation of seven commercial antigen detection tests for *Giardia* and *Cryptosporidium* in stool samples. *Clin Microbiol Infect* 12, 656–659 (2006).

129. Anonymous. Manufacturer's recall of rapid assay kits based on false positive *Cryptosporidium* antigen tests—Wisconsin, 2001–2002. *MMWR* 51, 189 (2002).

130. Anonymous. Manufacturer's recall of rapid cartridge assay kits on the basis of false-positive *Cryptosporidium* antigen tests—Colorado, 2004. *MMWR* 53, 198 (2004).

131. Xiao, L. et al. Phylogenetic analysis of *Cryptosporidium* parasites based on the small-subunit rRNA gene locus. *Appl Environ Microbiol* 65, 1578–1583 (1999).

132. Sturbaum, G.D. et al. Species-specific, nested PCR-restriction fragment length polymorphism detection of single *Cryptosporidium parvum* oocysts. *Appl Environ Microbiol* 67, 2665–2668 (2001).

133. Amar, C.F., Dear, P.H., and McLauchlin, J. Detection and identification by real time PCR/RFLP analyses of *Cryptosporidium* species from human faeces. *Lett Appl Microbiol* 38, 217–222 (2004).

134. Nichols, R.A., Campbell, B.M., and Smith, H.V. Identification of *Cryptosporidium* spp. oocysts in United Kingdom noncarbonated natural mineral waters and drinking waters by using a modified nested PCR-restriction fragment length polymorphism assay. *Appl Environ Microbiol* 69, 4183–4189 (2003).

135. Coupe, S. et al. Detection of *Cryptosporidium* and identification to the species level by nested PCR and restriction fragment length polymorphism. *J Clin Microbiol* 43, 1017–1023 (2005).

136. Jiang, J. and Xiao, L. An evaluation of molecular diagnostic tools for the detection and differentiation of human-pathogenic *Cryptosporidium* spp. *J Eukaryot Microbiol* 50(Suppl), 542–547 (2003).

137. Alves, M. et al. Subgenotype analysis of *Cryptosporidium* isolates from humans, cattle, and zoo ruminants in Portugal. *J Clin Microbiol* 41, 2744–2747 (2003).

138. Feng, Y. et al. Subtypes of *Cryptosporidium* spp. in mice and other small mammals. *Exp Parasitol* 127, 238–242 (2011).

139. Mallon, M. et al. Population structures and the role of genetic exchange in the zoonotic pathogen *Cryptosporidium parvum*. *J Mol Evol* 56, 407–417 (2003).

140. Gatei, W. et al. Development of a multilocus sequence typing tool for *Cryptosporidium hominis*. *J Eukaryot Microbiol* 53, S43–S48 (2006).

141. Tanriverdi, S. et al. Inferences about the global population structure of *Cryptosporidium parvum* and *Cryptosporidium hominis*. *Appl Environ Microbiol* 74, 7227–7234 (2008).

142. Tanriverdi, S. et al. Emergence of distinct genotypes of *Cryptosporidium parvum* in structured host populations. *Appl Environ Microbiol* 72, 2507–2513 (2006).

143. Glaberman, S. et al. Three drinking-water-associated cryptosporidiosis outbreaks, Northern Ireland. *Emerg Infect Dis* 8, 631–633 (2002).

144. Leoni, F. et al. Molecular epidemiological analysis of *Cryptosporidium* isolates from humans and animals by using a heteroduplex mobility assay and nucleic acid sequencing based on a small double-stranded RNA element. *J Clin Microbiol* 41, 981–992 (2003).

145. Chalmers, R.M. et al. Direct comparison of selected methods for genetic categorisation of *Cryptosporidium parvum* and *Cryptosporidium hominis* species. *Int J Parasitol* 35, 397–410 (2005).

146. Xiao, L. and Ryan, U.M. Molecular epidemiology. In *Cryptosporidium and Cryptosporidiosis*, Fayer, R. and Xiao, L. (eds.), pp. 119–171 (CRC Press and IWA Publishing, Boca Raton, FL, 2008).

147. Leoni, F. et al. Multilocus analysis of *Cryptosporidium hominis* and *Cryptosporidium parvum* from sporadic and outbreak-related human cases and *C. parvum* from sporadic cases in livestock in the UK. *J Clin Microbiol* 45, 3286–3294 (2007).

148. Waldron, L.S. et al. Molecular epidemiology and spatial distribution of a waterborne cryptosporidiosis outbreak in Australia. *Appl Environ Microbiol* 77, 7766–7771 (2011).

149. Mayne, D.J. et al. A community outbreak of cryptosporidiosis in Sydney associated with a public swimming facility: A case–control study. *Interdiscip Perspect Infect Dis* 2011, 341065 (2011).

150. Ng, J.S. et al. Molecular characterisation of *Cryptosporidium* outbreaks in Western and South Australia. *Exp Parasitol* 125, 325–328 (2010).

151. Chalmers, R.M. et al. Detection of *Cryptosporidium* species and sources of contamination with *Cryptosporidium hominis* during a waterborne outbreak in north west Wales. *J Water Health* 8, 311–325 (2010).

152. Feng, Y. et al. Extended outbreak of cryptosporidiosis in a pediatric hospital, China. *Emerg Infect Dis* 18, 312–314 (2012).

153. Robinson, T.J. et al. Evaluation of the positive predictive value of rapid assays used by clinical laboratories in Minnesota for the diagnosis of cryptosporidiosis. *Clin Infect Dis* 50, e53–e55 (2010).

154. USEPA. Method 1623: *Cryptosporidium* and *Giardia* in water by filtration/IMS/FA (EPA 821-R-01-025. Office of Water, U.S. Environmental Protection Agency, Washington, DC, 2001).

155. Weintraub, J.M. Improving *Cryptosporidium* testing methods: A public health perspective. *J Water Health* 4(Suppl 1), 23–26 (2006).

156. Hill, V.R. et al. Comparison of hollow-fiber ultrafiltration to the USEPA VIRADEL technique and USEPA method 1623. *J Environ Qual* 38, 822–825 (2009).

157. Rhodes, E.R. et al. A modified EPA Method 1623 that uses tangential flow hollow-fiber ultrafiltration and heat dissociation steps to detect waterborne *Cryptosporidium* and *Giardia* spp. *J Vis Exp* 65(pii), 4177 (2012).

158. Smith, C.M. and Hill, V.R. Dead-end hollow-fiber ultrafiltration for recovery of diverse microbes from water. *Appl Environ Microbiol* 75, 5284–5289 (2009).

159. Kothavade, R.J. Potential molecular tools for assessing the public health risk associated with waterborne *Cryptosporidium* oocysts. *J Med Microbiol* 61, 1039–1051 (2012).

160. Ruecker, N.J. et al. Molecular and phylogenetic approaches for assessing sources of *Cryptosporidium* contamination in water. *Water Res* 46, 5135–5150 (2012).

161. Ruecker, N.J. et al. Tracking host sources of *Cryptosporidium* spp. in raw water for improved health risk assessment. *Appl Environ Microbiol* 73, 3945–3957 (2007).

162. Yang, W. et al. *Cryptosporidium* source tracking in the Potomac River watershed. *Appl Environ Microbiol* 74, 6495–6504 (2008).

163. Feng, Y. et al. Occurrence, source, and human infection potential of *Cryptosporidium* and *Giardia* spp. in source and tap water in Shanghai, China. *Appl Environ Microbiol* 77, 3609–3616 (2011).

164. Shields, J.M., Lee, M.M., and Murphy, H.R. Use of a common laboratory glassware detergent improves recovery of *Cryptosporidium parvum* and *Cyclospora cayetanensis* from lettuce, herbs and raspberries. *Int J Food Microbiol* 153, 123–128 (2012).

165. Robertson, L.J. and Huang, Q. Analysis of cured meat products for *Cryptosporidium* oocysts following possible contamination during an extensive waterborne outbreak of cryptosporidiosis. *J Food Prot* 75, 982–988 (2012).

166. Rzezutka, A. et al. *Cryptosporidium* oocysts on fresh produce from areas of high livestock production in Poland. *Int J Food Microbiol* 139, 96–101 (2010).

167. MacKenzie, W.R. et al. Massive outbreak of waterborne *Cryptosporidium* infection in Milwaukee, Wisconsin: Recurrence of illness and risk of secondary transmission. *Clin Infect Dis* 21, 57–62 (1995).

168. Gostin, L.O. et al. Water quality laws and waterborne diseases: *Cryptosporidium* and other emerging pathogens. *Am J Public Health* 90, 847–853 (2000).

169. USEPA. National primary drinking water regulations: Long Term 2 Enhanced Surface Water Treatment Rule. *Fed Regist* 71, 654–786 (2006).

170. Xiao, L. et al. Subtype analysis of *Cryptosporidium* specimens from sporadic cases in Colorado, Idaho, New Mexico, and Iowa in 2007: Widespread occurrence of one *Cryptosporidium hominis* subtype and case history of an infection with the *Cryptosporidium* horse genotype. *J Clin Microbiol* 47, 3017–3020 (2009).

171. Collier, S.A. et al. Direct healthcare costs of selected diseases primarily or partially transmitted by water. *Epidemiol Infect* 140, 2003–2013 (2012).

172. Sinai, A.P. et al. The state of research for AIDS-associated opportunistic infections and the importance of sustaining smaller research communities. *Eukaryot Cell* 11, 90–97 (2012).

173. Yu, Y., Zhang, H., and Zhu, G. Plant-type trehalose synthetic pathway in *Cryptosporidium* and some other apicomplexans. *PLoS One* 5, e12593 (2010).

174. Fritzler, J.M. and Zhu, G. Novel anti-*Cryptosporidium* activity of known drugs identified by high-throughput screening against parasite fatty acyl-CoA binding protein (ACBP). *J Antimicrob Chemother* 67, 609–617 (2012).

Section IV

Microbes and Toxins Affecting
Humans and Animals: Toxins

39 Abrin

Dongyou Liu and Eric A.E. Garber

CONTENTS

39.1 INTRODUCTION

Abrin is a toxic protein and is found in the seeds of a plant called *Abrus precatorius* (*Abrus* means beautiful). Native to Asia, Africa, Australia, and the Western Pacific, *A. precatorius* thrives in tropical and subtropical climates throughout the world. Given its ability to grow over and kill other native species, it has the tendency to become weedy and invasive.

The seeds of *A. precatorius* have been recognized since times immemorial for their toxicity. Ground *Abrus* seeds, dried paste, or extract from the plant has been used as arrow poison, in fish poisoning, and in folk medicine and extensively reviewed by Ross [1]. For example, *Abrus* leaves have been made into tea to treat fevers, coughs, colds, and diarrhea; extract from the seeds with antimicrobial, antioxidant, anti-inflammatory, and analgesic potential has been employed to heal minor skin irritations (e.g., scratches, sores, wounds), chronic eye diseases (e.g., trachoma), kidney inflammation, pain relief, leukoderma, tetanus, rabies, worms, and diabetes; the products of *A. precatorius* capable of causing reversible alterations in the estrous cycle pattern have been used orally to quicken labor and induce abortion and as an oral contraceptive; and oil prepared from the seeds has been claimed to be an aphrodisiac [1–3].

With a glossy appearance (being red with a black spot, resembling ladybug), the seeds of *A. precatorius* have been used as beads in rosaries (precatorius or precare means to pray), necklaces and bracelets, as well as percussion instruments. Despite their frequent use and apparent toxicity, intoxication by the seeds of *A. precatorius* in humans has been rare, unless the seeds are chewed and swallowed. One likely reason for this is that with a hard shell, the seeds of *A. precatorius* are not readily digested and usually passed through the intestinal tract intact without the toxin being released. This, along with poor uptake from the gut and being heat-labile explain the low number of intoxications reported despite abrin's being 5- to 10-fold more toxic than ricin [4–6]. Though there are no published records on the weaponization of abrin, its toxicity, availability, and ease of production make it a potential health risk and justify its classification as a select agent [7].

39.2 CLASSIFICATION, MORPHOLOGY, PURIFICATION, AND BIOLOGY

39.2.1 CLASSIFICATION

Commonly referred to as rosary pea, jequirity pea, bead vine, black-eyed Susan, coral-bead plant, crab's eye, Indian licorice (due to its use as a licorice substitute), John Crow bead, jumbie bead, lucky bean, or precatory bean, *A. precatorius* (synonyms: *Abrus cyaneus* and *Glycine abrus*) is classified taxonomically in the genus *Abrus*, family Fabaceae (pea), order Fabales, class Magnoliopsida (dicotyledons), division Magnoliophyta (flowering plants), superdivision Spermatophyta (seed plants), subkingdom Tracheobionta (vascular plants), kingdom Plantae (plants).

The genus *Abrus* currently consists of 20 recognized species: *Abrus aureus* (Madagascar), *Abrus baladensis* (Somalia), *Abrus bottae* (Saudi Arabia, Yemen), *Abrus canescens* (Africa), *Abrus cantoniensis* Hance (China), *Abrus diversifoliatus* (Madagascar), *Abrus diversifoliolatus*, *Abrus fruticulosus* (India), *Abrus gawenensis* (Somalia), *Abrus laevigatus* (Southern Africa), *Abrus longibracteatus* (Laos, Vietnam), *Abrus lusorius*, *Abrus madagascariensis* (Madagascar), *Abrus parvifolius* (Madagascar), *A. precatorius* (Africa, Australia, Southeast Asia), *Abrus pulchellus* (Africa), *Abrus sambiranensis* (Madagascar), *Abrus schimperi* (Africa), *Abrus somalensis* (Somalia), and *Abrus wittei* (Zaire). In turn, the species *A. precatorius* can be further separated into two subspecies: *A. precatorius* subsp. *africanus* and *A. precatorius* subsp. *precatorius*. Whereas *A. precatorius* subsp. *africanus* produces smaller pods (2–3.5 cm long) with a rough surface texture (tuberculate), *A. precatorius* subsp. *precatorius* produces larger pods (3–5 cm long) with a relatively smooth surface texture. Detailed surveys for the presence of class II ribosome inhibiting protein (RIP II) toxins have not been conducted for all members of the *Abrus* genus, but a toxin analogous to abrin was observed in *A. pulchellus* var. *tenuiflorus* [8].

Abrus seeds may be dispersed by birds, as well as spread along waterways during floods and in dumped garden waste. *A. precatorius* plants may also establish in forest after fires, thus accounting for its ubiquitous presence.

39.2.2 MORPHOLOGY

A. precatorius is a perennial, deciduous climbing plant with slender twining stems that twines around trees, shrubs, and hedges and can reach up to 10 m or more in height. It shows a preference for shady areas and moist soil. While younger stems are slender, twining, woody, greenish in color, smooth, and covered in tiny hairs (pubescent), older stems are covered in a smooth-textured or wrinkled brown bark. The plant generates alternately compound leaves (pinnate) of 5–12 cm in length, with 5–15 pairs of green, oblong leaflets (0.5–2.5 cm long and 0.2–0.8 cm wide) on each pinna, which are mostly hairless (glabrous) and have rounded tips (obtuse apices). There is a conspicuous lack of a terminal leaflet on the compound leaves. The flowers are pea-shaped (about 1 cm long), whitish, pale pink, lavender, mauve, or purplish in color and densely clustered in racemes (peduncles) that emerge from the leaf axils (axillary racemes). Individual flowers have five small green sepals, that is, a large upper petal (standard), two side petals (wings), and two lower petals that are fused together at the base into a short tube (calyx tube, or a keel). The flowers are followed by oblong, flat, truncate, gray-brown, hairy or silky-textured pods (3–5 cm long and 1.2–1.5 cm wide) with a sharp point (beak). When mature, the brown pods split open and curl back to reveal several (usually 3–5) oval-shaped (ellipsoid), hard, shiny, red, black-capped, clinging seeds (5–7 mm long and 4–5 mm wide). The seeds are usually scarlet and black, with a large black spot surrounding the hilum, which is the point of attachment

FIGURE 39.1   Seeds of *A. precatorius*.

(Figure 39.1). The seed coat (testa) is smooth and glossy, becomes hard when the seed matures, and generally remains on the plant for several months. Some *A. precatorius* varieties may produce seeds of black, yellow, white, and green [9–11].

Being very uniform in size (each weighing 1/10th of a gram), hard, and durable, *A. precatorius* seeds have been used as a measure (ratti, where 8 rattis equal to 1 masha; 12 mashas equal to 1 tola or 11.6 g) to weigh gold and jewels (1 carat = 2 seeds) in India and Southeast Asia.

Although all parts of the plant are toxic to humans and livestock (e.g., cattle and horses), the highest concentrations of cytotoxic compounds are found in the seeds. Apart from small amounts of glycyrrhizin and uric acid, *Abrus* seeds contain two major toxins: abrin and *Abrus* agglutinin. Abrin is a 63 kDa heterodimeric glycoprotein, whereas agglutinin is a heterotetrameric glycoprotein with a molecular weight of 134 kDa. The agglutinin consists of two A chains (molecular weight 30 kDa) and two B chains (molecular weight 31 kDa), displaying 67% and 80% homology, respectively, to the comparable subunits in abrin [12]. As observed with ricin, most antibodies derived against the toxin also recognize the agglutinin. The agglutinin is considerably less toxic than abrin and is not classified as a select agent [12–14].

39.2.3 PURIFICATION

Two different methods are routinely used for the purification of abrin. Both methods separate the abrin isomers into fractions or groups. One method, described by Hegde et al. [5], entails affinity chromatography using a lactamyl-Sepharose® column followed by Sephadex® G100 and Diethylaminoethyl–Sephacel® (DEAE-Sephacel®) chromatography. The abrin isomer fractions are designated as fractions I, II, and III based on elution from the DEAE-Sephacel column. The second procedure described by Wei et al. [15] generates primarily two products designated abrin A and abrin C, and a third less abundant isolate-labeled abrin B was noted. Fractionation of these isomers is based on DEAE-Sephadex A-50 followed by carboxymethyl (CM) cellulose and DEAE cellulose chromatography with abrin A and C displaying significant differences in Sepharose 4B affinity. The complexity of the isomeric pattern is consistent with DNA studies that have successfully cloned three isomers of the A chain [16,17]. Though routinely stored as concentrated solutions at –80°C, powdered abrin is also stable. The yield of abrin using various purification and extraction methods varies from 0.05% to 0.5% [5,15,18–20].

The application of abrin as a pharmaceutical and immunoconjugate has been examined and shows promise in the treatment of several important human diseases such as cancer [21–29].

39.2.4 Biology

The proteinaceous nature of abrin was first noted in the 1880s where it played a critical role in the development of immunology and the concepts of specificity and cross-reactivity [30,31]. It was not until early 1970s that the mechanisms of abrin intoxication began to emerge. The initial experiments showed that abrin targeted protein synthesis rather than RNA and DNA synthesis. Because the structure of the polyribosomes remained preserved, abrin was suspected of preventing elongation of the already initiated polypeptide chain.

Abrin is a member of the binary, A–B toxin family consisting of an enzymatically active A-moiety and a receptor-binding B-moiety. The A–B toxin motif is common and associated with bacteria such as diphtheria, *Pseudomonas*, cholera, Shiga, and anthrax toxins. More specifically, as a member of RIP II, abrin, like ricin, consists of two subunits (termed A and B).

The abrin B chain was characterized as a lectin in the 1970s after demonstration of its capacity to bind to Sepharose and β-D-galactopyranoside moieties and subsequent elution with galactose or lactose [32]. While the underlying gene for abrin B chain encodes three galactose-binding peptides (termed α, β, and γ), only two of these structures (1α and 2γ) are able to bind galactose [33]. It is believed that the B chain facilitates toxin uptake by binding to cell-surface receptors (e.g., galactosyl residues and mannose receptors on reticuloendothelial cells) leading to the toxin's uptake by endocytosis [34]. Consistent with the critical role ascribed to the B chain are results showing that isolated A chains are not toxic [35] and that CHO-resistant cells display a reduced affinity by the holotoxin while purified ribosomes from the resistant cells are readily deadenylated by abrin [36]. Consistent with this model are thermodynamic studies showing an increased stabilization of the toxin upon lactose binding [37].

Temperature and kinetic studies of abrin and ricin binding to erythrocytes and HeLa cells and inhibition of rabbit reticulocyte lysate protein synthesis support a model in which the two toxins function by similar mechanisms [35,38,39] with differences in temperature sensitivity in cellular binding and enzymatic activity consistent with the greater toxicity of abrin [28]. Detailed x-ray structure [40,41] and thermodynamic studies of abrin [39] have tried to identify possible sources of the differences observed between the two proteins. The observations that the A chain of abrin displays a greater activity than the A chain of ricin in *in vitro* assays

while the binding of abrin is less sensitive to the presence of lactose [35] are consistent with the differences between the two toxins reflecting differences in both the A and B chains.

Once inside, the toxin is transported to endosomes, and a small fraction is translocated to the cytosol. The A chain, reductively cleaved from the B chain, with N-glycosidase activity (rRNA N-glycohydrolase, EC 3.2.2.22), deadenylates the highly conserved adenine-4324 at the end of the 28S rRNA hairpin loop of the 60S subunit of ribosomes [39,42]. The deadenylation inhibits EF-1 binding and protein synthesis, leading to cell death by both necrosis and apoptosis [43]. The A chain of abrin and ricin is similar to the RIP I proteins common in many plants. A third type of RIP (RIP III) was recently identified in which an additional peptide must be removed for enzymatic activity [44]. Though abrin and ricin only cleaves the adenine at position 4324 of 28S rRNA, other RIPs (saporin-R2 and the RIP from lamjapin) have been shown to deadenylate at multiple sites [43,45].

Abrin inhibition of protein synthesis initiates within 90 min of administration though death is usually not until 15–30 h post exposure. Thus, protection necessitates almost immediate treatment or prior immunization [46]. Indeed, one A chain molecule has the capacity to inactivate 2000 ribosomes per minute [39]. At this speed, more ribosomes are inactivated than the cell can make new ones, abrogating the capacity of 60S ribosomes to support protein synthesis with cell death being the inevitable outcome [12,33]. Using rodent models, prior immunization with abrin successfully protected the animals from intoxication, though prior immunization with ricin had no effect [30,46,47].

X-ray crystallographic analysis of abrin has been reported for several crystalline forms with the A chain refined to a resolution of 2.1 Å [40,48,49].

39.3 CLINICAL FEATURES AND PATHOGENESIS

39.3.1 Clinical Features

Abrin is extremely toxic and 3 μg can be lethal. However, the toxicity of abrin depends on the route of administration (inhalation, ingestion, absorption, or injection), dose, and length of exposure to the substance [50].

39.3.1.1 Inhalation

Inhalation of abrin by humans may lead to respiratory distress, fever, cough, nausea, and tightness in the chest within 8 h of exposure. This is followed by heavy sweating and pulmonary edema. Excess fluid in the lungs (which is diagnosable by x-ray or by listening to the chest with a stethoscope) tends to make breathing more difficult, with the skin turning blue due to lack of oxygen. Low blood pressure and respiratory failure may eventually result in death. Critical to the inhalation toxicity of abrin is the ability of the particles to reach the inner lung; this requires a particle size less than 5 μm [51,52]. The toxicity of particles capable of reaching the inner lung is comparable to that of injection. Griffiths et al. [46] studied the inhalation toxicity of rats using abrin particles of a diameter of 0.9 μm (geometric standard deviation, σ_g 1.6).

Rats exposed to abrin by inhalation may remain clinically well for 18–24 h, before showing general malaise, lethargy, anorexia, piloerection, and respiratory difficulties. In severe cases, blood-stained fluid exudes from the nostrils. Upon postmortem examination, inhaled abrin causes pulmonary edema, alveolitis, diffuse airway inflammation, necrosis, and diffuse alveolar flooding.

39.3.1.2 Ingestion

Most research on the oral toxicity of abrin has focused on rodents with LD_{50} values approximately sevenfold more toxic than ricin [30] or 1 mg/kg body weight [6,20]. A detailed understanding of human toxicity has been limited to anecdotal stories complicated by inconclusive information regarding the digestion of the hard peas in the gastrointestinal tract. Death has been ascribed to the chewing and swallowing of 2–4 seeds [53,54] while a recent attempted suicide that involved the ingestion of 10 seeds failed [4]. Ingestion of abrin may lead to nausea, severe vomiting, bloody diarrhea, abdominal pain, and colic in 6 h to 3 days. This is followed by gastrointestinal bleeding, hematemesis, dehydration, low blood pressure, tachycardia, headaches, dilated pupils, irrationality, hallucinations, drowsiness, weakness, tetany, tremors, seizures, fever, and dysrhythmias [54]. Pulmonary edema and hypertension have also been reported [55]. The liver, spleen, and kidneys may stop functioning, with death occurring up to 4 days after the onset of symptoms. If death does not occur in 3–5 days, the victim usually recovers. Postmortem findings may include hemorrhage, erythema, and edema in the gastrointestinal tract.

39.3.1.3 Absorption

Symptoms of skin and eye exposure to abrin powder or mist include redness, severe irritation, pain, inflammation of the conjunctiva (conjunctivitis), localized necrosis, permanent damage to the cornea, and occasionally blindness. As with ricin, skin exposure is the least efficient route of administration [56].

39.3.1.4 Injection

Parenteral injection of abrin is more toxic than oral administration. LD_{50} values for abrin fractions I, II, and III are 22, 2.4, and 10 μg/kg-bw, 0.55, 0.06, and 0.25 refer to μg per mouse, methods state weight 25 ± 1 g [5]. Human intravenous (i.v.) injection has an estimated lethal dose of 0.1–1 μg/kg body weight [50]. Enzymatic assays of abrin-catalyzed deadenylation of nucleic acid substrates and cell culture studies are consistent with the uptake of abrin from the gut being the primary cause of the 1000-fold difference between oral and intraperitoneal (i.p.)/i.v. toxicity [57]. The i.v. administration of abrin to mice and dogs causes weakness, shivering, anorexia, weight loss, edema, ascites, and rectal bleeding. Subcutaneous injection of abrin into mice, guinea pigs, and rats causes liver damage [58,59].

39.3.2 Pathogenesis

Abrus agglutinin has an affinity for the red blood cell and other parenchyma cells (e.g., liver and kidney cells). In experimental rodents, abrin is mainly detected in the liver, blood, lungs, spleen, kidneys, and heart following i.v. injection; and the toxin is found predominantly in the liver, kidneys, and blood after i.p. injection. Due in part to its molecular weight (63 kDa) and glycosylation, abrin is poorly absorbed from the intestine. However, even a small amount of undigested toxin is sufficient to cause severe complications and death in human. Abrin is eliminated almost exclusively via renal route as low molecular weight degradation products.

Abrin increases capillary permeability causing fluid and protein leakage and tissue edema, resulting in endothelial cell damage or vascular leak syndrome. The abrin A chain also causes vascular disturbances and organ and tissue lesions. Primary toxicity of abrin to the epithelial membrane lies in the direct injury to the type I and II pneumocytes it induces [60].

Abrin triggers apoptosis in mammalian hosts via caspase-3 activation and follows the intrinsic mitochondrial pathway involving potential damage to the mitochondrial membrane and reactive oxygen species production [61–66]. The apoptosis process in leukemic cells takes place 1 h after abrin exposure (before it penetrates into the cells), with PSer translocation from the inner to the outer monolayer of plasma membrane, accompanied by caspase activation on the first to second hour and DNA fragmentation on the fourth to sixth hour. It is likely that the B chain may trigger the apoptosis, while the A chain expedites the progress [67].

Lectins are polyclonal activators of lymphocytes, capable of inducing/enhancing the production of a battery of cytokines [68–70]. Both native and heat-denatured *Abrus* agglutinin are shown to activate splenocytes, natural killer (NK) cells, innate effector cells (e.g., macrophages); increase the expression of activation markers (CD25, CD71) in B and T cells; and induce the production of cytokines such as interleukin-2 (IL-2), interferon-γ (IFN-γ), and tumor necrosis factor-α (TNF-α), suggestive of a Th1 type of immune response. In addition, agglutinin increases production of nitric oxide and hydrogen peroxide, enhances phagocytic and bactericidal activity of macrophages, and augments the humoral and cell-mediated immune response of the host. Similarly, abrin stimulates specific humoral responses and increases the total leukocyte count, lymphocytosis, the weights of the spleen and thymus, circulating antibody titer, antibody forming cells, bone marrow cellularity, and α-esterase-positive bone marrow cell in experimental mice [71].

The ability to elicit the Th1 type of cytokine (IL-2 and IFN-γ) response and augment the humoral and cell-mediated immune response of the host has been exploited for antitumor therapies. Abrin derived peptide (ABR) treatment of tumor-bearing mice showed enhanced antibody-dependent cellular cytotoxicity (ADCC), which links humoral and cell-mediated immunity, indicating its potential use as an immunoadjuvant. Tryptic-digested agglutinin and abrin-derived peptides (of 5–15 kDa in molecular weight) induce mitochondria-dependent cell death [22,72].

39.4 DETECTION, IDENTIFICATION, AND DIAGNOSIS

Considering the historic roles played by ricin and abrin in the development of immunology, it is not surprising that antibody-based methods for detecting these toxins are common with limits of detection often <1 ng/mL. Indeed, one of the first methods developed to detect abrin sought to exploit the classic experiment of Paul Ehrlich and employed a bioassay whereby immunized and nonimmunized mice were exposed to suspect samples [47].

A variety of immunological and chromatographic techniques have been used to detect abrin in food and other matrices ranging from enzyme-linked immunosorbent assays (ELISAs) [73–75], lateral flow devices (LFDs) [76], electrochemiluminescence [73], surface plasmon resonance (SPR) [77,78], and multiplex bead-based assays [79,80] to a novel silver-enhanced LFD with a detection limit of 0.1 ng/mL [81]. Commercial antibody-based ELISAs and LFDs are also available [82] with a new ELISA format displaying an interpolated limit of detection (LOD) of 0.10 ± 0.02 ng/mL.

Associated with the various antibody-based assays was the use of novel methods to generate the capture and detector antibodies. In developing the SPR assay, Zhou et al. [77] used phage display to select for human monoclonal antibodies with optimal sensitivity. The procedure resulted in antibodies that displayed LOD values of 35 and 75 ng/mL, considerably less than a human lethal dose. Goldman et al. [78] used llama-derived single-domain antibodies.

Molecular methods have also been reported using aptamer and nucleic acid technology to detect abrin and abrin extracts. Tang et al. [83] used a novel molecular light switching reagent $[Ru(phen)_2(dppz)]^{2+}$ to generate aptamers displaying an LOD of 1 nM abrin (equivalent to approximately 60 ng/mL). Felder et al. [84] developed a duplex real-time polymerase chain reaction (real time-PCR) assay for simultaneous detection of abrin and ricin DNA based on 5′-nuclease technology and the OmniMix HS bead PCR reagent mixture. The assay was capable of detecting 1.2 genomes per reaction for abrin DNA and 3 genomes per reaction for ricin DNA and was applied successfully to the detection of abrin and ricin contaminations in a food matrix [84].

An alternative nontoxic biomarker used for the detection of *Abrus* contamination is abrine, Nα-methyl-L-tryptophan. Abrine is present in *A. precatorius* at levels comparable to that of abrin. Liquid chromatography-tandem mass spectrometry (LC/MS/MS) was successfully applied to the detection of abrine following Strata-X solid phase extraction (SPE) sample cleanup with recoveries ranging from 88% to 111% from milk, cola, juice, tea, and water with an LOD of 0.025 μg/mL and a linear dynamic range of 0.05–10 μg/mL [85]. An alternative method developed for the analysis of urine samples after solid phase extraction and addition of $^{13}C_1{}^2H_3$-L-abrine as an internal standard displayed an LOD of 0.09 ng/mL following high performance liquid chromatography (HPLC) fractionation and electrospray ionization in a triple-quadrupole mass spectrometer [86]. Inasmuch as abrine does

not only occur solely in *A. precatorius* but may be associated with other members of the genus, its presence should be interpreted as a warning of a possible health risk.

Several enzymatic assays have been developed for the detection of RIPs. Of these, a bead-based electrochemiluminescence assay developed by Keener et al. [87] was successfully adapted to enable distinguishing between different RIPs (i.e., abrin, ricin, and saporin) by comparing the activity with alternate substrates [88]. This assay was subsequently extended to a plate-based instrument and for the analysis of food [89].

39.5 TREATMENT, PREVENTION, AND DECONTAMINATION

Currently, no treatment is available for abrin poisoning and its management centers on symptomatic and supportive approaches to minimize the effects of poisoning. This includes decontamination to rid the body of the toxin and prevent further exposure as quickly as possible. The toxin is readily deactivated by treatment with bleach (sodium hypochlorite), pH, or heat. Since the antigenicity of abrin with the commonly used commercially available antibody-based assays is less sensitive than the cytotoxicity of the toxin to these decontamination measures, these antibody-based assays can be used as an indicator of effective decontamination [90].

In cases of gastrointestinal exposure, treatment has been proposed involving vigorous gastric decontamination with activated charcoal along with use of cathartics such as magnesium citrate and replacement of gastrointestinal fluid losses [91]. In cases of aerosol or inhalant exposure, intensive respiratory therapy, fluid and electrolyte replacement, anti-inflammatory agents, and analgesics have been recommended. In cases of ocular or dermal exposure, washing eyes and skin with saline is recommended. Other types of supportive medical care include treatments for seizure and low blood pressure.

Given the absence of an antidote for abrin, the most effective way to prevent abrin poisoning is to avoid abrin exposure. Considering the high antigenicity of abrin and the historic studies immunizing mice against abrin, it is not surprising that efforts are underway to develop abrin vaccines. Recent advances and efforts to develop an abrin vaccine make the possibility of such in the next few years likely [92–94]. Efforts have focused on mutating two key amino acid residues, Glu164 and Arg167, of abrin A chain to generate a less toxic immunogen. BALB/c mice given three doses of the recombinant expressed in *Escherichia coli* cells displayed a complete protective immune response against i.p. administration of $10 \times LD_{50}$ of native abrin. Additionally, the sera from immunized mice offered full passive protection in naive mice.

39.6 CONCLUSION

Abrin is a highly toxic protein from the seeds of *A. precatorius*, which has the capacity to induce endothelial cell damage, leading to an increase in capillary permeability with consequent fluid and protein leakage and tissue edema (vascular leak syndrome), and ultimately cell death.

As a member of the large family of ribosome-inactivating proteins, containing two disulfide-linked heterodimeric chains (A and B), the abrin A chain is an N-glycosidase that can irreversibly inactivate ribosomes by cleaving a specific adenine at a highly conserved site in the 28S ribosomal RNA in the cytoplasm of eukaryotic cells. This prevents the binding of elongation factor 2 and blocks protein synthesis. The abrin B chain is a lectin that binds to galactosyl residues of cell membranes and assists the toxin heterodimer enter target cells.

Human exposure to abrin can be via inhalation, ingestion, absorption, and injection. Given its high toxicity, ready availability, and ease of production, abrin poses a serious health risk. In view of the current lack of an antidote for abrin, further research toward the development of an effective vaccine is anticipated.

REFERENCES

1. Ross IA. *Abrus precatorius* L. In: *Medicinal Plants of the World, Vol. 1: Chemical Constituents, Traditional and Modern Uses*, 2nd ed. Springer, Humana Press, Totowa, NJ, 2003; pp. 15–31.
2. Saganuwan SA and Onyeyili PA. Biochemical effects of aqueous leaf extract of *Abrus precatorius* (jequirity bean) in Swiss albino mice. *Herba Palonica* 2010;56(3). http://www.herbapolonica.pl/magazines-files/5524294-art.6-3-2010.pdf. Accessed March 23, 2013.
3. Khan FZ, Saeed MA, and Ahmad E. Oxytoxic activity and toxic effects of globulins of *Abrus* seeds (scarlet variety) in rabbits. *J Islamic Acad Sci*. 1993;6:108–113.
4. Jang DH, Hoffman RS, and Nelson LS. Attempted suicide, by mail order: *Abrus precatorius*. *J Med Toxicol*. 2010;6:427–430.
5. Hedge R, Maiti TK, and Podder SK. Purification and characterization of three toxins and two agglutinins from *Abrus precatorius* seed by using lactamyl-Sepharose affinity chromatography. *Anal Biochem*. 1991;194:101–109.
6. Garber EAE. Toxicity and detection of ricin and abrin in beverages. *J Food Prot*. 2008;71:1875–1883.
7. National Select Agent Registry Updated: Monday, January 7, 2013. http://www.selectagents.gov/select%20agents%20and%20toxins%20list.html. Accessed March 23, 2013.
8. Ramos MV et al. Isolation and partial characterisation of highly toxic lectins from *Abrus pulchellus* seeds. *Toxicon*. 1998;36:477–484.
9. Royal Botanic Gardens, Kew and Missouri Botanical Garden. The Plant List (2010). Version 1. http://www.theplantlist.org/browse/A/Leguminosae/Abrus. Accessed November 7, 2013.
10. Anonymous. *Abrus precatorius* L. rosarypea. Plants profile. http://plants.usda.gov/java/profile?symbol=ABPR3. Natural Resources Conservation Service, United States Department of Agriculture (USDA), Washington, DC, 2006. Accessed November 7, 2013.
11. Anonymous. *Abrus precatorius* subsp. *precatorius*. New South Wales Flora Online. PlantNET—The Plant Information Network System of Botanic Gardens Trust. http://plantnet.rbgsyd.nsw.gov.au (Accessed November 7, 2013). Royal Botanic Gardens and Domain Trust, Sydney, New South Wales, Australia, 2006.
12. Liu CL et al. Primary structure and function analysis of the *Abrus precatorius* agglutinin A chain by site-directed mutagenesis. Pro(199) of amphiphilic alpha-helix H impairs protein synthesis inhibitory activity. *J Biol Chem*. 2000;275:1897–1901.
13. Bagaria A et al. Structure-function analysis and insights into the reduced toxicity of *Abrus precatorius* agglutinin I in relation to abrin. *J Biol Chem*. 2006;281(45):34465–34474.
14. Cheng J et al. A biophysical elucidation for less toxicity of agglutinin than abrin-a from the seeds of *Abrus precatorius* in consequence of crystal structure. *J Biomed Sci*. 2010;17:34.
15. Wei CH et al. Purification and characterization of two major toxic proteins from seeds of *Abrus precatorius*. *J Biol Chem*. 1974;249:3061–3067.
16. Evensen G, Mathiesen A, and Sundan A. Direct molecular cloning and expression of two distinct abrin A-chains. *J Biol Chem*. 1991;266:6848–6852.
17. Hung CH et al. Cloning and expression of three abrin A-chains and their mutants derived by site-specific mutagenesis in *Escherichia coli*. *Eur J Biochem*. 1994;219:83–87.
18. Olsnes S. Toxic and nontoxic lectins from *Abrus precatorius*. *Methods Enzymol*. 1978;50:323–330.
19. Griffiths GD, Hartman FC, and Upshall DG. Examination of the toxicity of several protein toxins of plant origin using bovine pulmonary endothelial cells. *Toxicology*. 1994;90:11–27.
20. Garber EAE. Effects of pasteurization on detection and toxicity of the beans from *Abrus precatorius*. In: Al-Taher F, Jackson L, and DeVries J (eds.), *Intentional and Unintentional Contaminants in Food and Feed*. ACS Symposium Series, Vol. 1020. ACS Publications, Washington, DC, 2009; pp. 143–151.
21. Arora R et al. Phytopharmacological evaluation of ethanolic extract of the seeds of *Abrus precatorius* linn. *J Pharmacol Toxicol*. 2011;6:580–588.
22. Bhutia SK et al. Inhibitory effect of *Abrus* abrin-derived peptide fraction against Dalton's lymphoma ascites model. *Phytomedicine*. 2009;16:377–385.
23. Okoko II et al. Antiovulatory and anti-implantation potential of the methanolic extract of seeds of *Abrus precatorius* in the rat. *Endocr Pract*. 2010;16:554–560.
24. Ramnath V, Kuttan G, and Kuttan R. Antitumour effect of abrin on transplanted tumours in mice. *Indian J Physiol Pharmacol*. 2002;46:69–77.
25. Ramnath V et al. Regulation of caspase-3 and Bcl-2 expression in Dalton's lymphoma ascites cells by abrin. *Evid Based Complement Altern Med*. 2009;6:233–238.
26. Smagur A et al. Chimeric protein ABRaA-VEGF121 is cytotoxic towards VEGFR-2-expressing PAE cells and inhibits B16-F10 melanoma growth. *Acta Biochim Pol*. 2009;56:115–124.
27. Verma D et al. Pharmacognostical evaluation and phytochemical standardization of *Abrus precatorius* L. seeds. *Nat Product Sci*. 2011;17:51–57.
28. Sivam G et al. Immunotoxins to a human melanoma-associated antigen: Comparison of gelonin with ricin and other A chain conjugates. *Cancer Res*. 1987;47:3169–3173.
29. Wawrzynczak EJ et al. Molecular and biological properties of an abrin A chain immunotoxin designed for therapy of human small cell lung cancer. *Br J Cancer*. 1992;66:361–366.
30. Ehrlich P. Experimentelle untersuchungen uber immunitat I Ueber Ricin. *Deutsche Medizinische Wochenschrift DMW*. 1891;17:976–979.
31. Silverstein AM. *Paul Ehrlich's Receptor Immunology: The Magnificent Obsession*. Academic Press, Orlando, FL, 2002.

32. Sandvig K, Olsnes S, and Pihl A. Interactions between *Abrus* lectins and Sephadex particles possessing immobilized desialylated fetuin. Model studies of the interaction of lectins with cell surface receptors. *Eur J Biochem.* 1978;88:307–313.

33. Olsnes S. The history of ricin, abrin and related toxins. *Toxicon.* 2004;44:361–370.

34. Wu AM et al. Carbohydrate specificity of a toxic lectin, abrin A, from the seeds of *Abrus precatorius* (jequirity bean). *Life Sci.* 2001;69:2027–2038.

35. Olsnes S, Refsnes K, and Pihl A. Mechanism of action of the toxic lectins abrin and ricin. *Nature.* 1974;249:627–631.

36. Ko J-L and Lin J-Y. Establishment and characterization of an abrin-resistant cell line. *Cell Biol Toxicol.* 1997;13:75–81.

37. Krupakar J et al. Calorimetric studies on the stability of the ribosome-inactivating protein abrin II: Effects of pH and ligand binding. *Biochem J.* 1999;338:273–279.

38. Sandvig K, Olsnes S, and Pihl A. Kinetics of binding of the toxic lectins abrin and ricin to surface receptors of human cells. *J Biol Chem.* 1976;251:3977–3984.

39. Olsnes S et al. Rates of different steps involved in the inhibition of protein synthesis by the toxic lectins abrin and ricin. *J Biol Chem.* 1976;251:3985–3992.

40. Tahirov TH et al. Crystal structure of abrin-a at 2.14 A. *J Mol Biol.* 1995;250:354–367.

41. Weston SA et al. X-ray structure of recombinant ricin A-chain at 1.8 A resolution. *J Mol Biol.* 1994;244:410–422.

42. Endo Y et al. The mechanism of action of ricin and related toxic lectins on eukaryotic ribosomes. The site and the characteristics of the modification in 28S ribosomal RNA caused by the toxins. *J Biol Chem.* 1987;262:5908–5912.

43. Liu YH, Peck K, and Lin JY. Involvement of prohibitin upregulation in abrin-triggered apoptosis. *Evid Based Complement Altern Med.* 2012;2012:605154.

44. Wu JH et al. Recognition intensities of submolecular structures, mammalian glyco-structural units, ligand cluster and polyvalency in abrin-a–carbohydrate interactions. *Biochimie.* 2010;92:147–156.

45. Barbieri L et al. Some ribosome-inactivating proteins depurinate ribosomal RNA at multiple sites. *Biochem J.* 1992;286:1–4.

46. Griffiths GD et al. Protection against inhalation toxicity of ricin and abrin by immunisation. *Hum Exp Toxicol.* 1995;14:155–164.

47. Clarke EGC and Humphreys DJ. The detection of abrin. *J Forensic Sci Soc.* 1971;11:109–112.

48. McPherson A Jr and Rich A. Studies on crystalline abrin: X-ray diffraction data, molecular weight, carbohydrate content and subunit structure. *FEBS Lett.* 1973;35:257–261.

49. Wei CH. Two phytotoxic anti-tumor proteins: Ricin and abrin. Isolation, crystallization, and preliminary x-ray study. *J Biol Chem.* 1973;248:3745–3747.

50. Dickers KJ, Bradberry SM, Rice P, Griffiths GD, and Vale JA. Abrin poisoning. *Toxicol Rev.* 2003;22:137–142.

51. Collins EJ et al. Primary amino acid sequence of alpha-trichosanthin and molecular models for abrin A-chain and alpha-trichosanthin. *J Biol Chem.* 1990;265:8665–8669.

52. Eitzen Jr EM. Chapter 20: Use of biological weapons. In: Sidell FR, Takafuji ET, and Franz DR (eds.), *Medical Aspects of Chemical and Biological Warfare.* Borden Institute, Walter Reed Army Medical Center, Washington, DC, 1997; pp. 437–450.

53. Gunsolus JM. Toxicity of jequirity beans. *J Am Med Assoc.* 1955;157:779.

54. Sahni V et al. Acute demyelinating encephalitis after jequirity pea ingestion (*Abrus precatorius*). *Clin Toxicol (Phila).* 2007;45:77–79.

55. Fernando C. Poisoning due to *Abrus precatorius* (jequirity bean). *Anaesthesia.* 2001;56:1178–1180.

56. Franz DR and Jaax KK. Chapter 32: Ricin toxin. In: Sidell FR, Takafuji ET, and Franz DR (eds.), *Medical Aspects of Chemical and Biological Warfare.* Borden Institute, Walter Reed Army Medical Center, Washington, DC, 1997; pp. 631–642.

57. Hughes JN, Lindsay CD, and Griffiths GD. Morphology of ricin and abrin exposed endothelial cells is consistent with apoptotic cell death. *Hum Exp Toxicol.* 1996;15:443–451.

58. Fodstad O, Olsnes S, and Pihl A. Toxicity, distribution and elimination of the cancerostatic lectins abrin and ricin after parenteral injection into mice. *Br J Cancer.* 1976;34:418–425.

59. Fodstad O et al. Toxicity of abrin and ricin in mice and dogs. *J Toxicol Environ Health.* 1979;5:1073–1084.

60. Mensah AY, Bonsu AS, and Fleischer TC. Investigation of the bronchodilator activity of *Abrus precatorius*. *Int J Pharmaceut Sci Rev Res.* 2011;6:9–13.

61. Narayanan S, Surolia A, and Karande AA. Ribosome-inactivating protein and apoptosis: Abrin causes cell death via mitochondrial pathway in Jurkat cells. *Biochem J.* 2004;377:233–240.

62. Ohba H et al. Plant-derived abrin-a induces apoptosis in cultured leukemic cell lines by different mechanisms. *Toxicol Appl Pharmacol.* 2004;195:182–193.

63. Qu X and Qing L. Abrin induces HeLa cell apoptosis by cytochrome *c* release and caspase activation. *J Biochem Mol Biol.* 2004;37:445–453.

64. Bhutia SK et al. *Abrus* abrin derived peptides induce apoptosis by targeting mitochondria in HeLa cells. *Cell Biol Int.* 2009;33:720–727.

65. Bora N, Gadadhar S, and Karande AA. Signaling different pathways of cell death: Abrin induced programmed necrosis in U266B1 cells. *Int J Biochem Cell Biol.* 2010;42:1993–2003.

66. Liu YH, Peck K, and Lin JY. Involvement of prohibitin upregulation in abrin-triggered apoptosis. *Evid Based Complement Altern Med.* 2012;2012:605154.

67. Bhaskar AS et al. Abrin induced oxidative stress mediated DNA damage in human leukemic cells and its reversal by *N*-acetylcysteine. *Toxicol In Vitro.* 2008;22:1902–1908.

68. Moriwaki S et al. Biological activities of the lectin, abrin-a, against human lymphocytes and cultured leukemic cell lines. *J Hematother Stem Cell Res.* 2000;9:47–53.

69. Shih SF et al. Abrin triggers cell death by inactivating a thiol-specific antioxidant protein. *J Biol Chem.* 2001;276:21870–21877.

70. Ramnath V, Kuttan G, and Kuttan R. Effect of abrin on cell-mediated immune responses in mice. *Immunopharmacol Immunotoxicol.* 2006;28:259–268.

71. Bhaskar AS, Gupta N, and Rao PV. Transcriptomic profile of host response in mouse brain after exposure to plant toxin abrin. *Toxicology.* 2012;299:33–43.

72. Bhutia SK, Mallick SK, and Maiti TK. In vitro immunostimulatory properties of *Abrus* lectins derived peptides in tumor bearing mice. *Phytomedicine.* 2009;16:776–782.

73. Garber EAE, Walker JL, and O'Brien TW. Detection of abrin in food using enzyme-linked immunosorbent assay and electrochemiluminescence technologies. *J Food Prot.* 2008;71:1868–1874.

74. Li XB et al. Preparation and identification of monoclonal antibody against abrin-a. *J Agric Food Chem.* 2011;59:9796–9799.

75. Zhou Y et al. Development of a monoclonal antibody-based sandwich-type enzyme-linked immunosorbent assay (ELISA) for detection of abrin in food samples. *Food Chem.* 2012;135:2661–2665.

76. Gao S et al. Colloidal gold-based immunochromatographic test strip for rapid detection of abrin in food samples. *J Food Prot.* 2012;75:112–117.

77. Zhou H et al. Selection and characterization of human monoclonal antibodies against abrin by phage display. *Bioorg Med Chem Lett.* 2007;17:5690–5692.

78. Goldman ER et al. Llama-derived single domain antibodies specific for *Abrus* agglutinin. *Toxins (Basel).* 2011;3:1405–1419.

79. Pauly D et al. Simultaneous quantification of five bacterial and plant toxins from complex matrices using a multiplexed fluorescent magnetic suspension assay. *Analyst.* 2009;134:2028–2039.

80. Garber EAE, Venkateswaran KV, and O'Brien TW. Simultaneous multiplex detection and confirmation of the proteinaceous toxins abrin, ricin, botulinum toxins, and *Staphylococcus* enterotoxins A, B, and C in food. *J Agric Food Chem.* 2010;58:6600–6607.

81. Yang W et al. A colloidal gold probe-based silver enhancement immunochromatographic assay for the rapid detection of abrin-a. *Biosens Bioelectron.* 2011;26:3710–3713.

82. Tetracore, Inc. Current BioThreat Alert® kits of lateral flow immunoassay test strips. http://www.tetracore.com/elisa-kits/ and http://www.tetracore.com/bio-warfare/index.html. Accessed March 23, 2013.

83. Tang J et al. In vitro selection of DNA aptamer against abrin toxin and aptamer-based abrin direct detection. *Biosens Bioelectron.* 2007;22:2456–2463.

84. Felder E et al. Simultaneous detection of ricin and abrin DNA by real-time PCR (qPCR). *Toxins.* 2012;4:633–642.

85. Owens J and Koester C. Quantitation of abrine, an indole alkaloid marker of the toxic glycoproteins abrin, by liquid chromatography/tandem mass spectrometry when spiked into various beverages. *J Agric Food Chem.* 2008;56:11139–11143.

86. Johnson RC et al. Quantification of L-abrine in human and rat urine: A biomarker for the toxin abrin. *J Anal Toxicol.* 2009;33:77–84.

87. Keener WK et al. An activity-dependent assay for ricin and related RNA *N*-glycosidases based on electrochemiluminescence. *Anal Biochem.* 2006;357:200–207.

88. Keener WK et al. Identification of the RNA N-glycosidase activity of ricin in castor bean extracts by an electrochemiluminescence-based assay. *Anal Biochem.* 2008;378:87–89.

89. Cho CY, Keener WK, and Garber EAE. Application of deadenylase electrochemiluminescence assay for ricin to foods in a plate format. *J Food Prot.* 2009;72:903–906.

90. Tolleson WH et al. Chemical inactivation of protein toxins on food contact surfaces. *J Agric Food Chem.* 2012;60:6627–6640.

91. Sahoo R et al. Acute demyelinating encephalitis due to *Abrus precatorius* poisoning—Complete recovery after steroid therapy. *Clin Toxicol (Phila).* 2008;46:1071–1073.

92. Surendranath K and Karande AA. A neutralizing antibody to the A chain of abrin inhibits abrin toxicity both in vitro and in vivo. *Clin Vaccin Immunol.* 2008;15:737–743.

93. Wang LC et al. Abrin-a A chain expressed as soluble form in *Escherichia coli* from a PCR-synthesized gene is catalytically and functionally active. *Biochimie.* 2004;86:327–333.

94. Han YH et al. A recombinant mutant abrin A chain expressed in *Escherichia coli* can be used as an effective vaccine candidate. *Hum Vaccin.* 2011;7:838–844.

40 Botulinum Neurotoxins from *Clostridium botulinum*

Janice M. Rusnak and Leonard A. Smith

CONTENTS

40.1 INTRODUCTION

Botulinum neurotoxins (BoNTs) are produced by gram-positive, anaerobic, spore-forming bacilli known as *Clostridium botulinum* (and uncommonly by *Clostridium baratii* and *Clostridium butyricum*) and may cause a severe neuroparalytic illness known as botulism.[1–3] BoNTs cause paralysis by preventing the transmission of nerve impulses to muscles by inhibiting acetylcholine release at the presynaptic nerve terminals of voluntary motor and autonomic cholinergic neuromuscular junctions (NMJs).[4,5] There are seven antigenically distinct BoNTs, which have been designated toxin serotypes A through G (BoNT A–G).[1] Most human cases of botulism are due to BoNTs A, B, and E (rarely F), but all seven serotypes may potentially cause disease in humans based on animal studies.[6]

Botulism in humans may be acquired after ingestion of food containing preformed toxin (food botulism) or from colonization of the intestinal tract (infant or adult intestinal botulism [AIB]) or a wound (wound botulism [WB]) by toxin-producing clostridia.[7–9] Nonnatural acquisition of botulism may occur from aerosol exposure to the toxin (i.e., bioterrorism event, laboratory exposure) or by injection of the toxin (mainly iatrogenic botulism). Intoxication generally manifests as a rapidly descending, flaccid paralysis, regardless of the route of exposure. Death is most commonly from respiratory failure, with mortality reported as high as 60% without medical intervention.[10,11] Supportive intensive care (including mechanical ventilation, when necessary) and early treatment with antitoxin have resulted in the reduction of mortality to less than 10%.[11–13] Due to its lethality and capability for aerosol dissemination, BoNT is designated a major biological threat agent.[6]

40.2 CLASSIFICATION, MICROPHYSIOLOGY, AND MORPHOLOGY

Clostridia are gram-positive, anaerobic, spore-forming bacilli (generally range from 0.5 to 2.0 μm in width and 1.6 to 20 μm in length) that possess a peritrichous flagellum and may form subterminal ellipsoidal spores.[1,3,14] There are four groups of *C. botulinum* with differences in phylogenetic and physiological properties (designated Groups I–IV) that produce the seven BoNT serotypes, designated A–G (Table 40.1). Most human cases of botulism are caused by *C. botulinum* in Groups I and II. Group I *C. botulinum* are proteolytic (based on the ability to digest complex proteins such as casein, meat, and coagulated egg or serum) and produce BoNTs A, B, and E that are fully activated upon release, whereas toxins produced by the nonproteolytic Group II organisms (BoNTs B, E, and F) require exogenous proteolysis for activation.[1,28,29] Group I organisms are also more adaptive to growth in humans as they grow optimally near human body temperature (37°C); Group II organisms grow optimally at lower temperatures (between 25°C and 30°C and as low as 3°C) and are less adaptive to colonization and growth in humans.[3,17,18,20,28–30] Group I strains are also more heat resistant and salt tolerant than Group II strains.[28] Group III *C. botulinum* strains produce toxin serotypes C and D, which cause botulism mainly in nonhuman mammals and birds (only a few sporadic cases reported in humans).[17,31–33] Group IV *C. botulinum* strains (renamed *Clostridium argentinense*) produce toxin serotype G, but naturally occurring type G botulism has yet to be confirmed in humans or animals.[25–27] Two other groups of clostridia, *C. butyricum* and *C. baratii*, may produce BoNTs (serotypes E and F, respectively), but are relatively uncommon causes of human botulism. While most *C. botulinum* strains produce only one toxin serotype, dual-producing toxin strains (Af, Bf, Ab, and Ba) and strains carrying a silent gene for a second serotype have been reported.[9,34,35]

40.3 BIOLOGY AND EPIDEMIOLOGY

Botulism may occur after ingestion of food contaminated with preformed toxin (foodborne botulism [FB]), by clostridia colonizing the intestinal tract and producing toxin (resulting mainly in infant botulism [IB] and uncommonly AIB), and by clostridia contaminating a wound and producing toxin (WB).[8] Botulism may also be acquired by nonnatural routes, such as aerosol exposure or injection of toxin.

The spores of *C. botulinum* may be found worldwide in soil and aquatic sediments.[1,36] The incidence of botulism and the various forms of botulism may vary throughout the world and may be higher in some areas of the world due to ingestion of certain indigenous cultural foods or may be factitiously lower in some areas due to the underreporting or lack of resources to confirm a diagnosis.[9,37–40] In the United States, IB (initially recognized in 1976) has been the most commonly reported form of botulism since 1979, with a mean of 94 cases (range 82–112) reported annually (from 2001 to 2010), compared to 18 cases (range 8–33) of FB and 26 cases (range 17–45) of WB.[5,41] Nearly all cases of WB (a rare form of botulism until 1994) were associated with injecting drug usage (IDU).[41,42] AIB remains a rare form of botulism.[8]

TABLE 40.1

Groups and Characteristics of Clostridia Capable of Producing BoNT

Group	I	II	III	IVa	*C. butyricum*	*C. baratii*
Toxins	A, B, F	B, E, F	C, D	G	E	F
Botulism in humans	Yes	Yes	Very rareb	Noa	Rare	Rare
Proteolyticc	Yes	No	No/slight	Yes	No	No
Liquification of gelatin	Yes	Yes	Yes	Yes	No	No
Lipase	Yes	Yes	Yes	No	No	No
Lecithinase	No	No	No	No	No	Yes
Fermentation						
Glucose	Yes	Yes	Yes	No	Yes	Yes
Sucrose	No	Yes	No	No	Yes	Yes
Lactose	No	No	No	No	Yes	Yes
Temperature growth (°C)						
Minimal growth	10–12	3	15	12	10	—
Optimal growth	35–40 (37)	18–25(30)d	37–45 (40)	25–45 (37)	30–45	30–37
Spore heat resistance (°C)	121	82.2–85	104	104	—	—
Inhibition NaCl (%)	10	5	3	>3	—	—
pH (inhibition)	4.6	5.0	5.1e	—	4.6	—
Similar nontoxigenic organism	*C. sporogenes*	No species name assigned	*C. novyi*	*C. subterminale*	Nontoxigenic *C. butyricum*	Nontoxigenic *C. baratii*
References	[3,14–16]	[3,14–19]	[3,14,15]	[3,14,15]	[1,18,20–22]	[1,14,23,24]

Source: Adaptation of Table 1 in With kind permission from Springer Science+Business Media: *Curr. Top. Microbiol. Immmunol.*, Botulism: The present status of the diseases, 195, 1995, 55–75, Hatheway, C.

a Also known as *C. argentinense* (in soil samples in Argentina and Switzerland).[25] No confirmed naturally occurring cases in humans or animals (found on autopsy of adults and infants with sudden death but causal relationship unclear).[26,27]

b Causes botulism mostly in animals (i.e., cattle, horse, sheep, mink) and birds.[14,17]

c Proteolysis based on the ability to digest complex proteins such as casein, meat particles in a cooked meat medium, coagulated egg white, and coagulated serum and not on ability to liquefy gelatin.[28]

d Optimal growth from 25 to 30.[20]

e Limiting pH of 5.1 for marine strains and 5.6 for terrestrial strains.[3]

Most cases of human botulism are due to BoNTs A, B, and E (and rarely F) produced by *C. botulinum*. BoNTs A and B are responsible for most FB cases in the mainland United States (BoNT A more predominant on the west coast and BoNT B produced by proteolytic Group I clostridia on the east coast), China, and many countries in Europe (nonproteolytic BoNT B is the predominant serotype).[13,15,16,39] Serotype E-producing *C. botulinum* strains are found mainly in aquatic environments in the Northern Arctic and subarctic regions and are responsible for most cases of botulism in Alaska, Canada, Scandinavia, and Japan (associated with ingestion of freshwater and marine products).[1,3,37,43,44] *C. botulinum* type E botulism outbreaks not associated with seafood ingestion have been reported in the noncoastal regions of northwest China, and *C. butyricum* type E outbreaks from the ingestion of legume products have been reported in eastern China and India.[13,45-47] Also, international travel and commerce may result in imported botulism with serotypes not common to the country (i.e., type E botulism in France due to imported marine products that were inadequately refrigerated).[48,49]

IB (>98% cases), WB (>99% cases), and AIB are caused mainly by toxin serotypes A and B produced by Group I *C. botulinum* and rarely by type E produced by Group II *C. botulinum* strains (only one reported case each of wound and IB) that are poorly adapted for growth and toxin production at the higher body temperature of humans.[9,50,51] Infant and AIB may also be caused uncommonly by Group I serotype F-producing *C. botulinum* or by *C. butyricum* or *C. baratii* (serotype E and F botulism, respectively).[9,23,24,45,52-57] *C. butyricum* has been found in soil, the rumen of calves, and the intestines of both healthy and ill neonates; the presumed source of *C. baratii* is soil.[36,45,58] Only rare cases of types C and D botulism have been reported in humans (mainly occurs in nonhuman mammals and birds), and serotype G botulism has been confirmed only in laboratory animals.[6,31,32]

40.3.1 FOODBORNE BOTULISM

FB in the United States (1990–2000) has been most commonly associated with improperly canned or home-prepared foods and less frequently from contaminated commercially produced or restaurant-prepared foods.[7] Implicated sources of botulism have generally involved foods with low salt, sugar, and acid (pH > 4.1) and higher water (a_w > 0.955) content, which were stored under anaerobic conditions.[8,15,40,59] Such conditions promote the germination of clostridia spores. In 2006, the failure to refrigerate a commercial carrot juice product (a nonacidic juice with low salt and sugar content) resulted in a botulism outbreak in the United States and Canada that was associated with extremely high serum toxin levels (10 times higher than any previous case) and prolonged toxemia (as long as 25 days).[40,60] This resulted in a change in the Food and Drug Administration (FDA) Guidelines for Industry to require nonacidic juices to have an additional safety factor other than refrigeration

(i.e., acidification) to prevent *C. botulinum* growth.[40,60,61] Ready-to-eat meals that are only minimally heated and then chilled (and often vacuum- or modified atmosphere-packed) may not be sterile from *C. botulinum*. The safety of these foods depends on a combination of a mild heat treatment, adequate refrigeration, a restricted shelf life, and occasionally intrinsic preservatives.[30,40,61,62] Most botulism cases associated with these products have been attributed to inadequate refrigeration and often involved Group II clostridia that may grow at chilled temperatures as low as 3°C (i.e., type E botulism from inadequate refrigeration of vacuum-packed fish).[19,30,40,48,49,62] In contrast, botulism due to improperly home-canned vegetables is more likely to occur with the more heat-resistant Group I clostridia.[29] Implicated sources of FB in recent years have included inadequately prepared or refrigerated salsa, home-canned green beans, baked potato wrapped in foil, inadequately refrigerated sausage in China, home-prepared native Alaskan foods (often inadequately fermented, smoked, salted, and/or refrigerated marine and freshwater products), home brew made in a Utah prison, and bamboo shoots in Thailand.[7,63-65]

40.3.2 INFANT AND ADULT INTESTINAL BOTULISM

Since the initial report of IB in the United States in 1976, nearly 3000 cases of IB have been reported in 26 countries and 5 continents.[9] The absence of IB in Africa and only two cases reported in China are attributed to underreporting or underdiagnosis, as *C. botulinum* spores are present worldwide. IB is acquired from inhalation or ingestion of *C. botulinum* spores. Infants less than 12 months of age are at risk for IB, due to the absence of competitive bowel flora that allows for the colonization of *C. botulinum*. Risk factors for IB include exposure to soil and dust that may carry spores (i.e., sweeper dust, increased soil exposure due to construction, or father working in construction, nursery, or farming).[66,67] While honey ingestion was associated with 40% of IB cases in the 1970s, honey is now implicated in less than 5% of cases in California due to public health education.[9,68] Powdered infant formula was linked to a case of IB serotype B in the United Kingdom, but *C. botulinum* spores were not detected in infant formula in the United States.[69,70] *C. botulinum* was found in small quantities on herbs (i.e., chamomile, linden flower) in Argentina and raised caution for feeding these herbs to infants.[71,72]

AIB occurs rarely, as the normal endogenous microflora of the adult intestinal tract is poorly conducive even for Group I *C. botulinum* that may grow optimally near the body temperature of humans. *C. baratii* and *C. butyricum* appear to be somewhat more adaptive to growth in the adult intestine, particularly in the presence of underlying gastrointestinal pathology that may provide a localized site or altered gut flora to aid in growth (i.e., Meckel's diverticulum, recent gastrointestinal procedure, recent antibiotic usage).[52,53,73] *C. baratii* was isolated from 8 of 10 type F AIB cases in the United States from 1981 to 2002.[52]

40.3.3 Wound Botulism

WB is caused by the contamination of a wound with *C. botulinum*, with subsequent absorption of the locally produced toxin into the bloodstream. Historically, WB was a rare form of botulism associated with traumatic wounds or abdominal surgery.[42,74] The increase in WB associated with IDU in the United States is mainly from subcutaneous injection (skin popping) of black tar heroin imported from Mexico (contamination likely during cutting of heroin).[42,75,76] In Europe, WB is associated with injection of powdered heroin imported from Asia.[58,76–80] Intranasal cocaine use (with or without associated sinusitis) has been implicated in botulism.[75,81]

40.3.4 Iatrogenic Botulism

Botulism or botulism-like symptoms have been reported uncommonly after receiving injections with FDA-approved toxin serotype A or B BoNT products for treating spasmodic disorders.[4,82–87] Botulism symptoms have been reported after the initial one to four injections, but not until after 2.5–5 years of injections in other cases. Botulism-like serious adverse events have not yet been confirmed with the use of Botox Cosmetic (Allergan, Inc., Irvine, CA), which contains a lower toxin dose.[87,88] An unlicensed, highly concentrated BoNT preparation (not intended for human use) administered for cosmetic purposes resulted in severe botulism (four cases), with serum toxin levels 21–43 times the estimated human lethal injection dose.[89]

40.3.5 Inhalational-Acquired Botulism

BoNT is considered a biological threat agent due to the ability to aerosolize and weaponize the toxin and to the public health impact and panic that would result from dissemination of this highly lethal toxin. Based on an estimated LD_{50} of 0.7–0.9 μg by aerosol exposure, 1 g of aerosolized toxin has the potential to kill over a million humans.[90] BoNT has been weaponized (but never released) by state-sponsored programs in the past. During the Gulf War, Iraq was reported to have produced over 19,000 L of concentrated BoNT.[90] A Japanese cult attempted to aerosolize BoNT (although unsuccessfully) on three occasions from 1990 to 1995. The only reported cases of inhalational-acquired botulism occurred in three laboratory workers (3 days after performing a necropsy) who inhaled residual toxin from the fur of animals (from an aerosol challenge with lyophilized toxin); all three individuals survived.[91]

40.4 PATHOGENESIS AND CLINICAL FEATURES

40.4.1 Structure and Mechanism of Action

BoNTs target the NMJ where they exert their potent and long-lasting effects. The NMJ is the synapse or junction of the axon terminal of a motor neuron with the motor end plate, the highly excitable region of muscle fiber plasma membrane responsible for the initiation of action potentials across the muscle's surface, ultimately causing the muscle to contract. In vertebrates, the signal passes through the NMJ via the neurotransmitter acetylcholine. To understand how the toxin blocks neurotransmitter release from the NMJ, it is important to understand the structure of the neurotoxins.

BoNTs are synthesized as single-chain protoxins that are posttranslationally processed by endogenous proteases to form di-chains consisting of a C-terminal 100 kDa heavy chain and an N-terminal 50 kDa light chain (LC). The heavy chain is composed of a C-terminal receptor-binding domain (fragment C or H_c) and an N-terminal translocation domain (H_n). The LC is a zinc-dependent endoprotease that specifically inactivates neuronal proteins required for neuroexocytosis and release of neurotransmitter from motor end plates. The extent of protease nicking varies from completely nicked (serotype A) to completely unnicked (serotype E), and the di-chains remain covalently attached by a single disulfide bond.[92] Toxin serotypes that are not activated by the proteases from the *C. botulinum* bacteria (e.g., serotype E) can be activated by the proteases in the gastrointestinal tract of humans and animals. While in the single-chain form, toxins have significantly lower toxicity as neurotoxins compared to their nicked counterparts.

The nucleotide and deduced amino acid sequence comparisons for all seven structural genes encoding BoNT serotypes A–G with consensus sequences between the genes have been reported.[93–97] The regions of sequence homology among the serotypes along with the x-ray crystallographic structures of their proteins[98–102] support the hypothesis that all serotypes employ a similar mode of action in causing blockade of neurotransmitter release from nerve cells.[103] Nerve intoxication results from the interplay of three key events: (1) the binding of toxin to cell surface receptors on motor neurons, (2) the trafficking of the endoprotease to the cytoplasmic compartment, and (3) the intracellular inactivation of the neuroexocytosis apparatus by the endoprotease (Figure 40.1). The three functional domains of BoNT (LC, H_n, and H_c) are endowed with intrinsic activities responsible for each of the consecutive phases leading to neuromuscular paralysis.[103–107]

First, the carboxy half of the heavy chain (fragment C or H_c) is required for receptor-specific binding primarily but not exclusively on cholinergic nerve cells.[108,109] Binding of BoNTs to nerve cell membranes is believed to occur in two steps: (1) an initial binding to a cell surface ganglioside and (2) followed by binding to a coreceptor protein component. With the exception of serotype D that interacts with a phospholipid,[110] BoNTs are known to interact with gangliosides.[111] The crystal structure of BoNT type A in complex with the ganglioside cell surface coreceptor ganglioside trisialoganglioside 1b (GT1b) leads to the supposition that GT1b mediates the initial contact between toxin and the neuronal membrane, resulting in an increase in the local toxin concentration at the membrane surface, permitting the toxin to diffuse in the plane of the membrane and bind its protein receptor.[112]

The protein component of the receptor for BoNT type A was previously reported as secretory vesicle 2 (SV2) protein.[113,114] During exocytosis, SV2 proteins become exposed to the surface, allowing botulinum toxin type A to bind. More recently, fibroblast growth factor receptor-3 (FGFR3) was shown to be

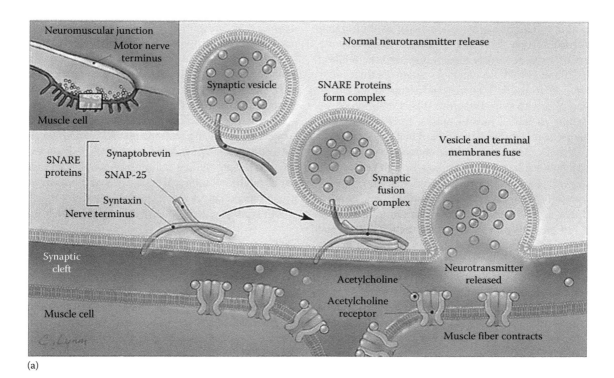

(a)

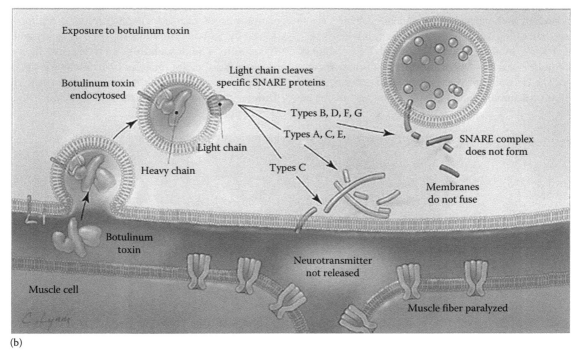

(b)

FIGURE 40.1 Mechanism of botulinum action. (a) Release of acetylcholine at the NMJ is mediated by the assembly of a synaptic fusion complex that allows the membrane of the synaptic vesicle containing acetylcholine to fuse with the neuronal cell membrane. The synaptic fusion complex is a set of SNARE proteins, which include synaptobrevin, SNAP-25, and syntaxin. After membrane fusion, acetylcholine is released into the synaptic cleft and then bound by receptors on the muscle cell. (b) Botulinum toxin binds to the neuronal cell membrane at the nerve terminus and enters the neuron by endocytosis. The LC of botulinum toxin cleaves specific sites on the SNARE proteins, preventing complete assembly of the synaptic fusion complex and thereby blocking acetylcholine release. Botulinum toxins types B, D, F, and G cleave synaptobrevin; types A, C, and E cleave SNAP-25; and type C cleaves syntaxin. Without acetylcholine release, the muscle is unable to contract. SNARE indicates soluble NSF-attachment protein receptor; NSF, *N*-ethylmaleimide-sensitive fusion protein; and SNAP-25, synaptosomal-associated protein of 25 kDa. (Reproduced from Arnon, S.S. et al., *JAMA*, 285, 1059, 2001. With permission of the American Medical Association.)

the high-affinity receptor for BoNT/A in neuronal cells.[115] Synaptotagmins I and II have been identified as the protein receptors for BoNT type B and G.[116,117] Synaptotagmins are associated with synaptic vesicle membranes, and the binding of serotypes B and G to these receptors promotes their internalization into neurons.[116,118] The crystal structure of BoNT type B in complex with synaptotagmin II was reported.[116,119]

After binding to cell surface receptors, BoNT is internalized into the nerve cell through the inward budding of plasma membrane vesicles (formation of the endosome), a process known as receptor-mediated endocyctosis.[120–122] The internalization process is energy dependent and a required step for the toxin's activity.[108] Upon acidification of the endosome, it is believed that a pH-dependent change in the translocation domain (H_n) of the heavy chain facilitates the movement and transfer of the zinc-dependent endoprotease (LC) from the inside of the endosomal vesicle across the endosomal membrane into the cytosol of the nerve cell, a process called translocation. The precise mechanism of this translocation process is not known, but it has been speculated that the heavy chain can form a pore through which the LC can pass.[122] Recent data indicate that acidification does not trigger substantial structural changes to the botulinum toxin protein as previously thought, but instead may eliminate repulsive electrostatic interactions between the translocation domain and the membrane, leading to the protein's translocation.[123]

The final event of intoxication involves the catalytic hydrolysis of the toxin's molecular targets, the SNARE (soluble N-ethylmaleimide-sensitive factor-attachment protein receptor) proteins. The LC domains selectively cleave and inactivate three essential proteins involved in the docking and fusion of acetylcholine-containing synaptic vesicles to the plasma membrane, thus preventing neurotransmitter release.[124–134] The LCs of BoNT serotypes A, C_1, and E cleave SNAP-25 (synaptosomal-associated protein of 25 kDa); serotypes B, D, F, and G cleave VAMP/synaptobrevin (synaptic vesicle-associated membrane protein); and serotype C_1 cleaves syntaxin. Inactivation of SNAP-25, VAMP, or syntaxin by the toxin's endoprotease prevents acetylcholine-containing synaptic vesicles from fusing with the plasma membrane and, in doing so, inhibiting the release of acetylcholine from the presynaptic NMJ. The clinical manifestations of botulism are dominated by the neurological signs and symptoms resulting from a toxin-induced blockade of the voluntary motor and autonomic cholinergic junctions and are described in the following.

40.4.2 CLINICAL FEATURES

The classic triad of a symmetric descending flaccid paralysis with prominent bulbar palsies in an afebrile patient with a clear sensorium should suggest a diagnosis of botulism (Figure 40.2). Gastrointestinal symptoms are an early occurrence in FB (generally absent with other exposure routes) and are soon followed by an acute, symmetrical descending flaccid paralysis. Cranial nerves are affected initially, with patients often presenting with ocular symptoms of blurred vision, diplopia, and/or ptosis, followed by bulbar symptoms of dysarthria, dysphasia, and/or dysphonia. The descending paralysis may then progress to involve the skeletal muscles controlling the neck and upper extremities and lastly the lower extremities. Respiratory failure may occur due to the paralysis of respiratory

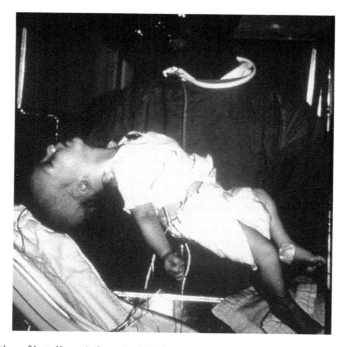

FIGURE 40.2 Clinical presentation of botulism. A 6-week-old infant with botulism, showing a marked loss of muscle tone, especially in the region of the head and neck. (Reproduced from Centers for Disease Control and Prevention Public Health Image [PHI] Library website, PHI 1935, Available at http://phil.CDC.gov.)

muscles (diaphragm and accessory breathing muscles) or from upper airway obstruction (due to the paralysis of posterior pharyngeal muscles, epiglottis, and tongue).[68,135] Symptoms of autonomic dysfunction include dry mouth, constipation, nonreactive dilated pupils, urinary retention, and orthostatic hypotension. Sensory symptoms are absent in botulism. Disease severity and onset of symptoms may vary with the exposure dose and toxin serotype. Botulism due to BoNT A is generally more severe than BoNT B and E and also has a longer recovery time compared to serotypes B, E, and F.[52,74,136-138] Mortality is generally less than 10% with intensive supportive care (including mechanical ventilation) and antitoxin therapy.[11-13,139]

The recovery of normal muscle function after botulism may take months (muscular weakness and autonomic systems often persist >24 months), with clinical recovery likely to correlate with the formation of new presynaptic end plates and NMJs.[5,136,140,141] Chronic symptoms (fatigue, limitation in physical activity, and emotional problems) persisting years later were reported by 68% of individuals with a history of FB in the Republic of Georgia[142]:

1. *FB.* The pentad of dysphagia, diplopia, dry mouth, fixed and dilated pupils, and nausea or vomiting was not sensitive in detecting most cases of FB in the Republic of Georgia, with only 2% of cases meeting all five criteria and 68% meeting three criteria.[143] Initial symptoms are often manifested within 12–36 h after the ingestion of toxin (range 2 h to 10 days).[137,144] Intubation was more likely to be required for botulism due to BoNT A (67%–85% of cases) than BoNT B (24%–59%) and BoNT E (33%–39%).[7,137,138]

2. *WB.* The onset of symptoms of WB is generally a mean of 5–7 days (range 2–18 days) in wounds not associated with IDU (i.e., traumatic wounds) and ranged from 2 to 14 days in IDU-associated WB.[42,74,79] Initial presenting symptoms of blurred vision, diplopia, and/or ptosis in injecting drug users should prompt the consideration of a botulism diagnosis.[42] A grossly infected wound (most commonly an abscess) was present in most cases, usually located at an injecting site on an extremity or buttock.[42,75] A close exam of all injecting sites is required as (1) the infected site(s) may occur at unusual locations (i.e., the base of the tongue or vulva) and (2) the injecting site may not be grossly infected in 15%–50% of cases (only minimal swelling or erythema at the site).[42,75,79,145] In spite of the wound infection, fever is generally not present in IDU-associated WB. Delays in treatment (up to 64 days) have occurred due to misdiagnosis, often attributing symptoms to a drug overdose or intoxication (particularly when the opiate test is positive), to pneumonia (fever with respiratory failure), or to other medical conditions because of atypical presentations (asymmetric weakness, dysphasia

as initial symptom).[42,79,145-148] Mortality has been reported as high as 29% (4 of 15 cases) in WB associated with trauma, but generally lower with IDU-associated WB.[41,78,79,149]

3. *IB.* Early symptoms of IB include constipation or change in stool pattern, poor feeding, irritability, lethargy, and signs of cranial nerve dysfunction (i.e., weak cry, poor suck or gag reflex, ocular findings of sluggish and incomplete pupillary response, ophthalmoplegia, ptosis).[67] Cranial nerve abnormalities may be mild and overlooked if repetitive testing (for 1–3 min) to induce muscle fatigue is not performed, as cranial nerve abnormalities may initially be observed only after fatigability of muscles.[68] Mortality from IB is less than 1% with the availability of supportive intensive care (even before antitoxin for infants was available).[150] *C. botulinum* intestinal colonization has been found at autopsy in a small percentage (3.5%) of infants with unexplained sudden death.[27,68]

4. *Iatrogenic and inhalational botulism (IHB).* As the clinical presentation of IHB is similar to FB (except for the absence of initial gastrointestinal symptoms), IHB may be considered in localized outbreaks that cannot be attributed to a common food source, particularly if an unusual serotype (i.e., C, D, G, or F) is involved. The time of onset of illness and death after aerosol exposure is dose dependent.[151,152] Death in nonhuman primates occurred 2–4 days after aerosol challenge (5–10 times the median lethal dose), generally within 12–18 h after symptom onset.[153] The onset of symptoms in iatrogenic botulism from therapeutic BoNT products was within hours to 3 weeks after the last injection.[83,85]

40.5 DIAGNOSIS

A clinical syndrome consistent with botulism may be the initial basis for the decision to treat with antitoxin, as antitoxin therapy is most effective when given early in illness and laboratory confirmation may not be available for days. The epidemiologic history (i.e., ongoing foodborne outbreak, IDU) may help support a botulism diagnosis. Differential diagnosis includes other causes of paralysis (i.e., Guillain–Barré syndrome [GBS], myasthenia gravis, tick paralysis, Eaton–Lambert syndrome, or a central nervous system [CNS] etiology).[8] GBS and tick paralysis generally present as an ascending symmetrical paralysis with paresthesias and without cranial nerve involvement (except for the Miller Fisher variant of GBS that may present with a triad of ophthalmoplegia, sensory ataxia, and hyperreflexia).[42] Eaton–Lambert syndrome does not generally involve cranial nerves. Paralysis from CNS lesions is generally asymmetric, and brain imaging scans will often identify the lesion (i.e., stroke, tumor).

Electromyography (EMG) studies demonstrating facilitation (incremental response to repetitive stimulation at 20–50 Hz) may be helpful in distinguishing botulism from GBS and myasthenia gravis, but not Eaton–Lambert syndrome.[5,80] However, facilitation may not be demonstrated in approximately one third of botulism cases.[138] As EMG findings may not appear until a few days after onset of peripheral muscle weakness, testing should be repeated the following day if normal.[74,80] A diagnosis of botulism may also be suggested by a small evoked muscle action potential response to a single supramaximal nerve stimulus (present in approximately 85% of botulism cases).[5,74,138] Sensory nerve function and motor nerve conduction velocity are normal in botulism.[80] Edrophonium chloride testing (Tensilon test) resulting in a dramatic improvement in strength may not necessarily distinguish myasthenia gravis from botulism, as 26% of botulism cases may exhibit a mild, nondramatic response.[138,154] Cerebrospinal fluid (CSF) analysis is normal in botulism (slightly elevated CSF protein has been reported in a few cases).[138] Routine chemistries and renal, hepatic, and hematological studies are also normal unless complications of botulism are present (i.e., secondary infections, severe hypoxia).[138]

A diagnosis of botulism may be laboratory-confirmed (1) by the detection of toxin from the serum, stool, gastric contents, or wound site or (2) by the culture of *C. botulinum* from stool, tissue, or wound. Detection of toxin in the implicated food is highly supportive for a diagnosis of botulism in an individual with symptoms consistent with botulism.[155] However, a positive culture of food in the absence of detectable toxin must be interpreted in context with the epidemiological findings, as *C. botulinum* is widely distributed in nature. Reporting of suspected cases of botulism and diagnostic laboratory testing should follow the requirements/policies of the public health department.

BoNT confirmation requires in vivo testing, using a mouse neutralization bioassay. The mouse bioassay is highly sensitive (0.01 ng/mL detection limit) and involves the use of serotype-specific antitoxin that will protect mice against injection with the respective toxin serotype.[18,155] Polymerase chain reaction (PCR) has correlated with results of the mouse neutralization toxin assay in recent cases.[155–159] PCR has the advantage of providing earlier results (usually within 24 h) than mouse bioassays that may take up to 4 days (range 6–96 h) and cultures that may take 7–10 days (range 5–21 days).[90,155,156] PCR should only be used in conjunction with the mouse bioassay for a diagnosis of botulism, as PCR detects fragments of *C. botulinum* (not active BoNT) and has not been validated for toxin confirmation on food and clinical materials.[7,8,155,156,160] Serology is not helpful in diagnosing botulism as the small amount of toxin needed to cause symptoms does not result in an antibody response.[6,161] While more rapid detection methods for BoNT are available (i.e., enzyme-linked immunosorbent assay [ELISA], lateral flow immunoassays, endopeptidase activity assays), the mouse bioassay remains the standard procedure for toxin detection (Table 40.2).[18,155,162–188]

Toxin assays in FB of serum and stool specimens are most commonly positive if obtained within 3 days after toxin ingestion (44% cases), but stool cultures may remain positive in 40% of cases even 7–9 days after ingestion (Table 40.2).[137] As toxin assays and stool cultures may be negative in approximately one third of FB cases, negative test results do not exclude a diagnosis of botulism.[137] The confirmation of infant and AIB is usually made by toxin assay or culture of stool specimens.[67] Toxin was identified in the serum in only 13% (9/67) of stool culture-positive IB cases.[189] In WB, serum toxin assay was sensitive for diagnosis in one US cohort (positive in 95% IDU-related and 83% non-IDU-related cases) but reported less sensitive in later reports (38%–68% IDU-related cases).[42,77,159] Wound specimens were positive by culture in 65% of cases (61% IDU-associated cases and 79% non-IDU cases) and by toxin assay in one third of cases.[42] In botulism acquired by inhalation or injection of toxin, toxin assay of the serum should be obtained (and perhaps gastric contents and stool with IHB); *C. botulinum* is not likely to be present.[90] Although not validated, a nasal swab for PCR/ELISA may be considered if within 24 h of exposure for support of inhalational exposure.[6]

40.6 TREATMENT AND PREVENTION

40.6.1 TREATMENT

Treatment of botulism requires supportive care (may require mechanical ventilation) and botulinum antitoxin. As respiratory failure may occur suddenly, the vital capacity and maximal inspiratory force should be closely monitored. Intubation is recommended for respiratory failure before the onset of hypercarbia or hypoxia (generally recommended when the vital capacity falls below 12 mL/kg or below 30% of the predicted vital capacity) or for signs of upper airway compromise.[80] The absence of gastric motility may require parenteral nutritional support. WB requires antibiotic therapy and debridement with irrigation of the infected site to eliminate the source of continued toxin production (preferably after antitoxin administration as wound manipulation may cause increased toxin release into the bloodstream).[42] *C. botulinum* is generally sensitive to penicillin and Flagyl (sensitivity testing should be performed). Aminoglycosides, clindamycin, and magnesium containing medications should be avoided as they may potentiate neuromuscular blockade.[190]

Mortality from adult botulism has been reduced to less than 10% with the use of botulinum antitoxin and supportive care (including mechanical ventilation).[11–13] Botulinum antitoxin is most effective in preventing paralysis when given early in the illness as the antitoxin only neutralizes circulating toxin and has no effect on toxin already bound to nerve terminals (does not reverse paralysis).[12,63] Early administration of antitoxin (particularly within 24 h of illness onset but even as late as 4 days) has been associated with a decrease in the duration of illness, number of days of mechanical ventilation, requirement for mechanical ventilation, and length of hospital stay.[12,63,150,191,192] An effect from the antitoxin may

TABLE 40.2
Diagnostic Assays for Botulinum Toxin

A. Laboratory Tests Used in Confirmation of Botulism Diagnosis[8,16]

Source	Test[a–c]	Amount	Form of Botulism	Comment
Serum	Toxin assay	10 mL minimum	All forms of botulism	≥30 mL preferred. Obtain before antitoxin given. Use earliest serum sample.
Stool	Toxin assay, culture	25 g minimum	FB, IB, AIB, IHB,[d] unknown	Sterile water enema may be needed.
Wound	Toxin assay, culture	Exudates, tissue	WB	Transport in anaerobic sterile container.
Gastric	Toxin assay, culture	20 cm³	FB	Vomitus or secretions. Within 72 h onset of symptoms.
Food	Toxin assay, culture	—	FB	Keep suspect food in original container.
Nasal	Toxin assay[d]	Nasal swab[d]	IHB	Up to 24 h after toxin inhalation.

B. Newer Diagnostic Assays for Botulinum Toxin (Compared to Mouse Bioassay)

Assay	Sensitivity (ng/mL)	Comments
Mouse bioassay	0.01	Standard and only accepted method for toxin confirmation by the Association of Analytical Communities.[18,155] May inject food extracts, serum, feces, vomitus, gastric secretions, wound tissue/pus, and culture supernatants (may dilute sample in phosphate buffer). Laborious, expensive, requires animals. Results often <48 h (range 6–96 h). Note: 1 mouse LD_{50} is equivalent to ~10–20 pg BoNT.
ELISA	0.2 (BoNT A/B); 0.5 (BoNT E); 2 (BoNT F)[165,167] (buffer, sera)	Experience mostly in food, *C. botulinum* cultures, purified toxin (less experience in serum/fecal samples).[165–167] False-positives (<1% cross-reactivity with other serotypes).[165,167] Positive test requires mouse bioassay confirmation.[155,162] Results <8 h.
Amplified ELISA	0.1 (BoNT A/B/E); 1.0 BoNT F (pure toxin)	BoNTs ABEF; for presumptive diagnosis in foods/environmental samples. High sensitivity in cultures containing ≥100 MLD/mL toxin (100 MLD/mL sample contains ~2 ng for BoNT A); positive test requires mouse bioassay confirmation (false positives mainly due to cross-reactivity between toxin strains—ranged from 1.5% to 28.6% for the 4 BoNTs).[168–170]
Digoxigenin (DIG)-ELISA	0.06–0.176 (casein buffer)	BoNTs ABEF; sensitive and specific ELISA (<0.82% cross-reactivity with other toxin serotypes). Uses toxin serotype-specific polyclonal antibodies for toxin capture and DIG-labeled toxin serotype-specific secondary antibodies (less background problems as DIG is only found in digitalis plants). Less sensitive (2–5 ng/mL) in food samples (~30 various liquid/semisolid/solid foods tested).[171]
ECLA ELISA	0.005–0.01	BoNTs ABE; most sensitive ELISA; complex and costly test (sophisticated amplification system that utilizes snake venom coagulation factor); contains biotinylated and chicken antibodies that preclude use in foods with chicken meat, milk, eggs, or biotin.[172–174]
PCR	—	PCR screens for botulinum gene (not toxin); positive test requires mouse bioassay confirmation.[8,156,160,162] Pentaplexed real-time PCR detects gene variations of 20 BoNT-ABEF sero- or subtypes.[175] Two-step "universal PCR" involves (1) detection of nontoxin-nonhemagglutinin gene (in all *C. botulinum*) and (2) quantitative PCR with primers/probes for BoNT A–G serotypes (tested on 22/26 BoNT subtypes, food, stool); portable, high throughput.[176] Cannot determine toxin production. Results <24 h.
Lateral flow detection kits	10–100	Immunochromatographic assay. Simple test may be adapted for field settings (i.e., red line appears if positive). Low sensitivity and false positives support limitation to presumptive screening (i.e., food samples) with mouse bioassay to confirm positive results; variability between kits.[177–179] Results <30 min.
Electrochemiluminescence immunoassay	0.05–0.1	BoNT ABEF; commercially available kit; detects BoNTs via antibodies attached to magnetic beads; tested in various food matrices, human sera, assay buffer.[180] Results <2.5 h.
Endopeptidase immunosorbent assay	0.005 (BoNT B)	Endopeptidase activity assay. Detects proteolytic activity of toxin cleavage of SNARE complex proteins by immunological detection of cleaved peptide or fluorescence release associated with cleavage.[181] Mainly evaluated for food. Results 4–6 h.
Forster resonance energy transfer (FRET)	0.045 (BoNTA/E) 0.45 (BoNT B/F) 4.5 (BoNT D/E)	Uses optimization of fluorogenic reporters that detect proteolytic activity using FRET. Suitable for high-throughput screening, quantitative; tested in serum, carrot juice, and whole blood.[182] Results 4–20 h.

(continued)

TABLE 40.2 (continued)

Diagnostic Assays for Botulinum Toxin

B. Newer Diagnostic Assays for Botulinum Toxin (Compared to Mouse Bioassay)

Assay	Sensitivity (ng/mL)	Comments
Endopep-MS	Serum: 20, 1, 0.2, and 1 MLD_{50}/mL (BoNTs A/B/E/F). Stool: 200, 10, 1, and 10 MLD_{50}/mL (BoNT A/B/E/F). Buffer: 0.02–0.16 MLD_{50}/mL	Detects endopeptidase activity (ABEF) followed by mass spectrometry to confirm peptide products.[183] Determines serotype, quantitative. Protease inhibitors with assay in clinical samples. Concentration of stool samples (extraction of toxin using Ab-coated magnetic beads) to avoid protease interference associated with decreased sensitivity due to endopeptide inhibition by antibodies. Requires specialized equipment and personnel; expensive. Results 4–8 h.
Protein chip	0.01 MLD_{50}/mL	Sensor chip allows for direct label-free detection of molecular interactions from SNARE complex (VAMP2) cleavage, using surface plasmon resonance (SPR); measures vesicle capture by anti-VAMP antibodies coupled to microchips (capture inhibited when epitopes are clipped by BoNT B); tested in apple/carrot juice, milk, and serum.[184] Protein chip technology with SPR also sensitive for detecting BoNT A/B in cultured neuron cells (neuronal cells increase sensitivity close to mouse bioassay).[185,186] Results 4–5 h.

Notes: AIB, adult intestinal botulism; ECLA, enzyme-linked coagulation assay; ELISA, enzyme-linked immunosorbent assay; FB, foodborne botulism; IB, infant botulism; IHB, inhalational botulism; MLD_{50}, mouse 50% lethal dose; WB, wound botulism.

[a] Mouse lethality bioassay is the standard procedure for toxin detection.[18,155]

[b] Anaerobic cultures of stool, wound, and food samples generally in enrichment broth media (i.e., nonselective liquid media such as cooked meat media or tryptone–peptone–glucose–yeast extract [TGBY] broth) or on agar using nonselective plating media (i.e., blood agar, egg yolk agar [EYA]). EYA screens (causes iridescent film surrounding colonies) for lipase-positive *C. botulinum* (Groups I–III).[1,163] Selective media supplemented with cycloserine, sulfamethoxazole, and trimethoprim used for stool cultures in IB (limited use in FB as antibiotics suppress Group II *C. botulinum* growth).[164]

[c] ELISA and PCR may be used to screen for toxin (positive results must be confirmed by mouse bioassay).[8,160,162,170]

[d] IHB may have inadvertent swallowing of toxin; use of PCR/ELISA for toxin detection in nares not validated.

not be seen until 12 h after administration (as long as 24 h), as neurological signs may not appear until hours after binding of toxin to nerve terminals. Generally, only one dose of antitoxin is needed. However, a second antitoxin dose may be required should symptoms continue to progress or relapse, as may occur with exposure to high toxin doses, incomplete wound debridement in WB, or AIB (persistent intestinal colonization).[42,60,193] Toxin assay 24 h after antitoxin administration is recommended in such cases.

In 2010, equine botulinum antitoxins at the CDC were replaced with a despeciated equine heptavalent (A-G) antitoxin (Cangene Corporation, Winnepeg, CN; FDA-approved in 2013).[194] Despeciated equine products are made by cleaving the species-specific F_c fragment from the horse immunoglobulin G product.[11,195] The resulting product ($F(ab')_2$ and Fab' fragments) contains less than 2% horse protein, which may potentially reduce the risk of serum sickness and hypersensitivity reactions to horse protein (1% risk with a single vial of whole equine antitoxin and 9% with two to four vials).[8,11,194–196] Rebound intoxication in an AIB serotype F case 10–12 days after receiving HBAT was attributed to the more rapid clearance of HBAT from the circulation in conjunction with continued intraintestinal toxin production (estimated 12–24 h HBAT serum half-life is shorter than the 5- to 7-day half-life of whole equine immune globulin).[193,197,198]

For infants, two human botulinum antitoxin products are available at the California Department of Health Services: (1) a bivalent (BoNT AB) FDA-approved product known as Baby-BIG IV (botulism immune globulin intravenous) and (2) a monovalent (BoNT E) IND product. The use of human antitoxin in infants decreases the risk of hypersensitivity reactions and avoids the risk of possible lifelong hypersensitivity to equine antigens. A single infusion of Baby-BIG is estimated to neutralize circulatory toxin for at least 6 months due to its prolonged half-life of approximately 28 days.[55,150] Antibodies in Baby-BIG may interfere with the immune response to live viral vaccines if given shortly before or within 5 months of Baby-BIG.[55] For nonserotype A and B IB (or if Baby-BIG is unavailable), HBAT may be administered to infants.[199] Also, a retrospective review of 31 cases of IB in Argentina treated with equine botulinum antitoxin noted a statistically significant reduction in hospital stay and tube feedings by 24 days and mechanical ventilation by 11 days, a 47% reduction in sepsis, and no serious hypersensitivity reactions (compared to 18 untreated controls).[200]

40.6.2 PREVENTION

Proper food handling and preparation may prevent FB.[40,59,201] BoNT may be inactivated by heating foods to a core temperature $\geq 85°C$ for at least 5 min, boiling, or sterilization.[7,16,17,62,202] The CDC suggests boiling or cooking home-canned foods for 10 min before ingesting as an added safety measure to kill toxin.[203] Warming foods in a microwave without stirring may not uniformly heat the food and result in cold spots with non-inactivated toxin.[59,204] Spores may be inactivated (i.e., in home-canning foods) by heating to $121°C$ in a pressure cooker (under pressure of 15–20 lb/in.^{2}) for at least 20 min.[7,16,201] Food industry safety guidelines employ multiple measures for inhibiting *C. botulinum* growth (i.e., combination of storage temperature conditions [i.e., refrigeration], limited storage times, acidification [generally pH $\leq$ 4.5], food preservatives, redox potential, competing organisms, and/or water activity).[7,15,16,62] Commercial or home-canned products with bulging lids should not be opened, and spoiled or expired food should not be eaten.[40,59] Infants (<12 months age) should not be fed with honey.[20,55,203]

Standard precautions are recommended for hospitalized patient with botulism (no person-to-person transfer reported).[8,203] If exposed to aerosolized BoNT, clothing and skin (toxin does not penetrate intact skin) should be washed with soap and water.[90] Surfaces and objects may be decontaminated using 0.1% hypochlorite bleach solution.[203] Aerosolized toxin decays naturally within hours to days (estimated decay rate from <1% to 4% per minute); sunlight may inactivate BoNTs within 1–3 h.[90,205] Free active chlorine (3 mg/L) in water supplies may inactivate >99.7% BoNT within 20 min.[205]

There is currently no FDA-approved vaccine for botulism. The Investigational New Drug (IND) formalin-inactivated pentavalent (A–E) botulinum toxoid (manufactured over 30 years ago) is no longer available to at-risk laboratory workers due to declining immunogenicity and increased local reactinogenicity.[11,206–210] A quadrivalent (ABEF) botulinum toxoid in Japan may be available.[211] A recombinant bivalent (AB) vaccine (based on heavy-chain fragments of BoNT) for aerosol protection against BoNT has completed phase 2 trials (phase 3 trials are planned in the near future).[208,209,212,213]

40.7 CONCLUSIONS AND FUTURE PERSPECTIVES

Since the original clinical recognition of botulism (then known as "sausage poisoning") nearly 200 years ago, the science and medical knowledge of BoNT has evolved to identify nonfoodborne forms of botulism and the use of BoNT for medicinal and cosmetic purposes.[1] Improvements in medical care (mechanical ventilation, botulinum antitoxin) and public health interventions have contributed to decreases in mortality and prevention of botulism. However, the threat of BoNT use as a biological weapon has necessitated real-time diagnostic tests that are less cumbersome than the mouse bioassay, a vaccine to replace the aging PBT, the expansion of antitoxin therapy for all seven BoNT serotypes, and a goal to develop equine-free treatment products.

DISCLAIMER

Opinions, interpretations, conclusions, and recommendations are those of the author and are not necessarily endorsed by the US Army. This article was written by an officer or employee of the US Government as part of his or her official duties and therefore is not subject to US copyrights. Research on human subjects was conducted in compliance with the US Department of Defense, federal, and state statutes and regulations relating to the protection of human subjects and adheres to the principles identified in the Belmont Report (1979). All data and human subjects' research were gathered and conducted for this publication under institutional review board-approved protocols.

REFERENCES

1. Hatheway, C.L., Toxigenic clostridia, *Clin. Microbiol. Rev.*, 3, 66, 1990.
2. Hatheway, C.L., *Clostridium botulinum* and other clostridia that produce botulinum neurotoxin. In *Clostridium botulinum. Ecology and Control in Foods*, eds. A.H.W. Hauschild and K.L. Dodds, pp. 3–20, New York: Marcel Dekker, 1992.
3. Smith, L.D.S. and Sugiyama, H., *Botulism: The Organism, Its Toxin, the Disease*, 2nd ed., pp. 23–37, Springfield, IL: C.C. Thomas, 1988.
4. Girlanda, P.G. et al., Botulinum toxin therapy: Distant effects on neuromuscular transmission and autonomic nervous system, *J. Neurol. Neurosurg. Psychiatry*, 55, 844, 1992.
5. Shapiro, R.L., Hatheway, C., and Swerdlow, D.L., Botulism in the United States: A clinical and epidemiologic review, *Ann. Intern. Med.*, 129, 221, 1998.
6. Middlebrook, J.L. and Franz, D.R., Botulism toxins. In *Medical Aspects of Chemical and Biological Warfare*, eds. F.R. Sidell, E.T. Takafugi, and D.R. Franz, pp. 243–254, Washington, DC: TMM Publications and Bethesda, MD: Office of the Surgeon General and Borden Institute.
7. Sobel, J. et al., Foodborne botulism in the United States, 1990–2000, *Emerg. Infect. Dis.*, 10, 1606, 2004.
8. Sobel, J., Botulism, *Clin. Infect. Dis.*, 41, 1167, 2005.
9. Koepke, R.J. et al., Global occurrence of infant botulism, 1976–2006, *Pediatrics*, 122, e72, 2008.
10. Burke, G.S., Notes on *Bacillus botulinus*, *J. Bacteriol.*, 4, 555, 1919.
11. Rusnak, J.M., Harper, I.M., and Abbassi, I., Laboratory exposures to botulinum toxins: Review and updates of therapeutics for the occupational health provider. In *Anthologies in Biosafety XII. Managing Challenges for Safe Operations of BSL-3/ABSL-3 Facilities*, eds. J.Y. Richmond and K.B. Meyers, pp. 123–138, Mundelein, IL: American Biological Safety Association, 2011.
12. Tacket, C.O. et al., Equine antitoxin use and other factors that predict outcome in type A foodborne botulism, *Am. J. Med.*, 76, 794, 1984.
13. Ying, S. and Shuyan, C., Botulism in China, *Rev. Infect. Dis.*, 8, 984, 1986.
14. Lund, B.M. and Peck, M.W., *Clostridium botulinum*. In *The Microbiological Safety and Quality of Food*, Vol. II, eds. B.M. Lund, T.C. Baird-Parker, and G.W. Gould, pp. 1057–1109, Gaithersburg, MD: Aspen Publishers, Inc., 2000.

15. International Commission on Microbiological Specifications for Foods, *Clostridium botulinum*. In *Micro-Organisms in Foods 5: Characteristics of Microbial Pathogens*, pp. 65–111, London, U.K.: Blackie Academic & Professional, 1996.

16. Centers for Disease Control and Prevention, Botulism in the United States, *Handbook for Epidemiologists, Clinicians and Laboratory Workers, 1998*. Atlanta, GA: U.S. Department of Health and Human Services, Public Health Service, 1899–1998.

17. Lindstrom, M. et al., *Clostridium botulinum* in cattle and dairy products, *Clin. Rev. Food Sci. Nutr.*, 50, 281, 2010.

18. Sharma, S.K. and Whiting, R.C., Methods for detection of *Clostridium botulinum* toxin in foods, *J. Food Prot.*, 68, 1256, 2005.

19. Graham, A.F. and Lund, B.M., The combined effect of sub-optimal temperature and sub-optimal pH on growth and toxin formation from spores of *Clostridium botulinum*, *J. Appl. Bacteriol.*, 63, 387, 1997.

20. World Health Organization, International programme on chemical safety poisons information monograph 858 bacteria, 2000. http://www.who.int/csr/delibepidemics/clostridiumbotulism.pdf, 1–32 (accessed April 26, 2012).

21. Anniballi, F. et al., Influence of pH and temperature on the growth of and toxin production by neurotoxigenic strains of *Clostridium butyricum* type E, *J. Food. Prot.*, 65, 1267, 2002.

22. McCroskey, L.M. et al., Characterization of an organism that produces type E botulinal toxin but which resembles *Clostridium butyricum* from the feces of an infant with type E botulism, *J. Clin. Microbiol.*, 23, 201, 1986.

23. Hall, J.D. et al., Isolation of an organism resembling *Clostridium baratii* which produces type F botulinal toxin from an infant with botulism, *J. Clin. Microbiol.*, 21, 654, 1985.

24. McCroskey, L.M. et al., Type F botulism due to neurotoxigenic *Clostridium baratii* from an unknown source in an adult, *J. Clin. Microbiol.*, 29, 2618, 1991.

25. Suen, J.C. et al., *Clostridium argentinense* sp. nov.: A genetically homogeneous group composed of all strains of *Clostridium botulinum* toxin type G and some nontoxigenic strains previously identified as *Clostridium subterminale* or *Clostridium hastiforme*, *Int. J. Syst. Biol.*, 38, 375, 1988.

26. Sonnebend, W.F. et al., Isolation of *Clostridium botulinum* type G from Swiss soil specimens using sequential steps in an acidification scheme, *Appl. Environ. Microbiol.*, 53, 1880, 1987.

27. Sonnabend, O. et al., Isolation of *Clostridium botulinum* type G and identification of type G botulinal toxin in humans: Report of five sudden unexpected deaths, *J. Infect. Dis.*, 143, 22, 1981.

28. Lynt, R.K. et al., Difference and similarities among proteolytic and nonproteolytic strains of *Clostridium botulinum* types A, B, E, and F: A review, *J. Food Prot.*, 45, 466, 1982.

29. Hatheway, C.L., Botulism: The present status of the disease, *Curr. Top. Microbiol. Immunol.*, 195, 55, 1995.

30. Lindstrom, M., Kiviniemi, K., and Korkeala, H., Hazard and control of group II (non-proteolytic) *Clostridium botulinum* in modern food processing, *Int. J. Food Microbiol.*, 108, 92, 2006.

31. Oguma, K. et al., Infant botulism due to *Clostridium botulinum* type C toxin, *Lancet*, 336, 1449, 1990.

32. Demarchi, J. et al., Existence du botulisme humain de type D, *Bull. Acad. Nat. Med.*, 142, 580, 1958.

33. Prevot, A.R. et al., Existence en France du botulisme humain de type C, *Bull. Acad. Nat. Med.*, 139, 355, 1955.

34. Barash, J.R. and Arnon, S.S., Dual toxin-producing strain of *C. botulinum* type Bf isolated from a California patient with infant botulism, *J. Clin. Microbiol.*, 42, 1713, 2004.

35. Gimenez, D.F., *Clostridium botulinum* subtype Ba, *Zentbl. Bakteriol. Parasitenkd. Infektkrankh. Hyg. Abt. 1 Orig. Reighe A*, 257, 68, 1984.

36. Haushchild, A.H.W., *Clostridium botulinum*. In *Foodborne Bacterial Pathogens*, ed. M.P. Doyle, pp. 112–189, New York: Marcel Dekker, Inc., 1989.

37. Horowitz, B.Z., Type E botulism, *Clin. Toxicol.*, 48, 880, 2010.

38. Reller, M.E. et al., Wound botulism acquired in the Amazonian rain forest of Ecuador, *Am. J. Trop. Med. Hyg.*, 74, 628, 2006.

39. Therre, H., Botulism in the European Union, *Euro. Surveill.*, 4, 2, 1999.

40. Peck, M.W., *Clostridium botulinum* and the safety of minimally heated, chilled foods: An emerging issue? *J. Appl. Microbiol.*, 101, 556, 2006.

41. Centers for Diseases Control, National enteric disease surveillance: Botulism annual summary, 2000 to 2010. http://www.cdc.gov/nationalsurveillance/PDFs/BotulismCSTE (accessed January 22, 2012).

42. Werner, S.B. et al., Wound botulism in California, 1951–1998: Recent epidemic in heroin injectors, *Clin. Infect. Dis.*, 31, 1018, 2000.

43. Dolman, C.E., Human botulism in Canada (1919–1973), *Can. Med. Assoc. J.*, 110, 191, 1974.

44. Austin, J.W. and Leclair, D., Botulism in the north: A disease without borders, *Clin. Infect. Dis.*, 52, 593, 2011.

45. Meng, X.T. et al., Characterization of a neurotoxigenic *Clostridia butyricum* strain isolated from the food implicated in an outbreak of food-borne type E botulism, *J. Clin. Microbiol.*, 35, 2160, 1997.

46. Fu, S. and Wang, C., An overview of type E botulism in China, *Biomed. Environ. Sci.*, 21, 353, 2008.

47. Chaudhry, R.B. et al., Outbreak of suspected *Clostridium butyricum* botulism in India, *Emerg. Infect. Dis.*, 4, 506, 1998.

48. King, L.A. et al., Botulism and hot-smoked whitefish: A family cluster of type E botulism in France, September 2009, *Euro. Surveill.*, 14(pii), 19394, 2009.

49. Boyer, A.C. et al., Two cases of foodborne botulism type E and review of epidemiology in France, *Eur. J. Clin. Microbiol., Infect. Dis.*, 20, 192, 2001.

50. Artin, I.P. et al., First case of type E wound botulism diagnosed using real-time PCR, *J. Clin. Microbiol.*, 45, 3589, 2007.

51. Luquez, C. et al., First report worldwide of an infant botulism case due to *Clostridium botulinum* type E, *J. Clin. Microbiol.*, 48, 326, 2010.

52. Gupta, A. et al., Adult botulism type F in the United States, 1981–2002, *Neurology*, 65, 1694, 2005.

53. Fenicia, L. et al., Intestinal toxemia botulism in two young people, caused by *Clostridium butyricum* type E, *Clin. Infect. Dis.*, 29, 1381, 1999.

54. Fenicia, L. et al., Intestinal toxemia botulism in Italy. 1984–2005, *Eur. J. Clin. Microbiol. Infect. Dis.*, 26, 385, 2007.

55. Fenicia, L. and Anniballi, F., Infant botulism, *Ann. 1st Super Sanita*, 45, 134, 2009.

56. Aureli, P.L. et al., Two cases of type E infant botulism caused by neurotoxigenic *Clostridium butyricum* in Italy, *J. Infect. Dis.*, 154, 207, 1986.

57. Barash, J.R., Tang, T.W.H., and Arnon, S.S., First case of infant botulism caused by *Clostridium baratii* type F in California, *J. Clin. Microbiol.*, 43, 42880, 2005.

58. Schechter, R. and Arnon, S., Commentary: Where Marco Polo meets Meckel: Type E botulism from *Clostridium butyricum*, *Clin. Infect. Dis.*, 29, 1388, 1999.

59. Date, K. et al., Three outbreaks of foodborne botulism caused by unsafe home canning of vegetables—Ohio and Washington, 2008 and 2009, *J. Food Prot.*, 74, 2090, 2011.

60. Sheth, A.N. et al., International outbreak of severe botulism with prolonged toxemia caused by commercial carrot juice, *Clin. Infect. Dis.*, 47, 1245, 2008.

61. FDA, Guidance for Industry: Refrigerated carrot juice and other refrigerated low-acid juices. Health and Human Services, 2007. http://www.fda.gov.Food/GuidanceComplianceRegulatoryInformation/GuidanceDocuments/Juice/ucm072481.htm (accessed April 26, 2012).

62. FDA, Safety practices for food processes. Chapter III. Potential hazards in cold-smoked fish. *Clostridia botulinum* type E, 2001. http://fda.gov/Food/ScienceResearchAreas/SafePracticesforFoodProcesses/ucm092182.htm (accessed April 17, 2012).

63. Kongsaengdao, S. et al., An outbreak of botulism in Thailand: Clinical manifestations and management of severe respiratory failure, *Clin. Infect. Dis.*, 43, 1247, 2006.

64. Zhang, S. et al., Multilocus outbreak of foodborne botulism linked to contaminated sausage in Hebei Province, China, *Clin. Infect. Dis.*, 51, 323, 2010.

65. McFarland, S., Utah inmates get sick from tainted brew, *The Salt Lake Tribune*, October 6, 2011.

66. Nevas, M. et al., Infant botulism acquired from household dust presenting as sudden infant death syndrome, *J. Clin. Microbiol.*, 43, 511, 2005.

67. Thompson, J.A. et al., Infant botulism in the age of botulism immune globulin, *Neurology*, 64, 2029, 2005.

68. Arnon, S.S. and Chin, J., The clinical spectrum of infant botulism, *Rev. Infect. Dis.*, 4, 614, 1979.

69. Brett, M.M. et al., A case of infant botulism with a possible link to infant formula milk powder: Evidence for the presence of more than one strain of *Clostridium botulinum* in clinical specimens and food, *J. Med. Microbiol.*, 54, 769, 2005.

70. Barash, J.R., Hsia, J.K., and Arnon, S.S., Presence of soil-dwelling *Clostridia* in commercial powdered infant formulas, *J. Pediatr.*, 156, 402, 2010.

71. Bianco, M.I. et al., Linden flower (*Tilia* spp.) as potential vehicle of *Clostridium botulinum* spores in the transmission of infant botulism, *Revista Argentina de Microbiologia*, 41, 232, 2009.

72. Bianco, M.I. et al., Presence of *Clostridium botulinum* spores in *Matricaria chamomilla* (chamomile) and its relationship with infant botulism, *Int. J. Food Microbiol.*, 121, 357, 2008.

73. Sobel, J. et al., Clinical recovery and circulating botulinum toxin type F in adult patient, *Emerg. Infect. Dis.*, 15, 969, 2009.

74. Merson, M.H. et al., Current trends in botulism in the United States, *JAMA*, 229, 1305, 1974.

75. MacDonald, K.L. et al., Botulism and botulism-like illness in chronic drug abusers, *Ann. Int. Med.*, 102, 616, 1985.

76. Passaro, D.J. et al., Wound botulism associated with black tar heroin among injecting drug users, *JAMA*, 279, 859, 1998.

77. Akbulut, D. et al., Wound botulism in injectors of drugs: Upsurge in cases in England during 2004, *Euro. Surveill.*, 10, 172, 2005.

78. Schroeter, M. et al., Outbreak of wound botulism in injecting drug users, *Epidemiol. Infect.*, 137, 1602, 2009.

79. Brett, M.M., Hallas, G., and Mpamugo, O., Wound botulism in the UK and Ireland, *J. Med. Microbiol.*, 53, 555, 2004.

80. Davis, L.E. and Kingk, M.K., Wound botulism from heroin skin popping, *Curr. Neurol. Neurosci. Rep.*, 8, 462, 2008.

81. Roblot, F. et al., Botulism in patients who inhale cocaine: The first cases in France, *Clin. Infect. Dis.*, 43, e51, 2006.

82. Bakheit, A.M.O., Ward, C.D., and MClellan, D.L., Generalized botulism-like syndrome after intramuscular injections of botulinum toxin type A: A report of two cases. Letter to editor, *J. Neurol. Neurosurg. Psychiatry*, 62, 198, 1997.

83. Bhatia, K.P. et al., Generalised muscular weakness after botulinum toxin injections for dystonia: A report of three cases, *J. Neurol. Neurosurg. Psychiatry*, 67, 90, 1999.

84. Coban, A.Z. et al., Iatrogenic botulism after botulinum toxin type A injections, *Clin. Neuropharmacol.*, 33, 158, 2010.

85. Comella, C.L. et al., Dysphagia after botulinum toxin injections for spasmodic torticollis: Clinical and radiologic findings, *Neurology*, 42, 1307, 1992.

86. FDA, Early communication about ongoing safety review of Botox and Botox Cosmetic (*Botulinum* toxin type A) and Myobloc (*Botulinum* toxin type B), 2008. http://www.fda.gov/Drugs/DrugSafety/PostmarketDrugSafetyInformationforpatientsandProviders/DrugSafetyInformationforHealthcareProfessional/ucm070366.htm (accessed July 6, 2010).

87. FDA, FDA Alert [08/2009] Information for healthcare professionals: On a botulinum toxin A (marketed as Botox/Botox Cosmetic), Abo botulinum toxin A (marketed as Dysport) and Rimabotulinum toxin B (marketed as Myobloc), 2009. http://www.fda.gov/Drugs/DrugSafety/PostmarketDrugSafetyInformationforpatientsandProviders/DrugSafetyInformationforHealthcareProfessional/ucml174949.htm (accessed July 6, 2010).

88. Omprakash, H.M. and Rajendran, S.C., Botulinum toxin deaths: What is the fact? *J. Cutan. Aesthet. Surg.*, 1, 95, 2008.

89. Chertow, D.S. et al., Botulism in 4 adults following cosmetic injections with an unlicensed, highly concentrated *Botulinum* preparation, *JAMA*, 296, 76, 2006.

90. Arnon, S.S.R. et al., Botulinum toxin as a biological weapon. Medical and Public Health Management, *JAMA*, 285, 1059, 2001.

91. Holzer, E., *Botulism* caused by inhalation, *Med. Klin.*, 41, 1735, 1962.

92. DasGupta, B.R., Structures of botulinum neurotoxin, its functional domains, and perspectives on the crystalline type A toxin. In *Therapy with Botulinum Toxin*, eds. J. Jankovic and M. Hallen, pp. 15–39, New York: Marcel Dekker, 1994.

93. Binz, T. et al., The complete sequence of botulinum neurotoxin type A and comparison with other clostridial neurotoxins, *J. Biol. Chem.*, 265, 9153, 1990.

94. Thompson, D.E. et al., Nucleotide sequence of the gene coding for *Clostridium baratii* type F neurotoxin: Comparison with other clostridial neurotoxins. *FEMS Microbiol. Lett.*, 108, 175, 1993.

95. Willems, A. et al., Sequence of the gene coding for the neurotoxin of *Clostridium botulinum* type A associated with infant botulism: Comparison with other clostridial neurotoxins, *Res. Microbiol.*, 144, 547, 1993.

96. Hutson, R.A. et al., Nucleotide sequence of the gene coding for non-proteolytic *Clostridium botulinum* type B neurotoxin: Comparison with other clostridial neurotoxins, *Curr. Microbiol.*, 28, 101, 1994.

97. Elmore, M.J. et al., Nucleotide sequence of the gene coding for proteolytic (group I) *Clostridium botulinum* type F neurotoxin: Genealogical comparison with other clostridial neurotoxins, *Syst. Appl. Microbiol.*, 18, 23, 1995.

98. Lacy, D.B. et al., Crystal structure of botulinum neurotoxin type A and implications for toxicity, *Nat. Struct. Biol.*, 5, 898, 1998.

99. Swaminathan, S. and Eswaramoorthy, S., Structural analysis of the catalytic and binding sites of *Clostridium botulinum* neurotoxin, B, *Nat. Struct. Biol.*, 7, 693, 2000.

100. Hasegawa, K. et al., A novel subunit structure of *Clostridium botulinum* serotype D toxin complex with three extended arms, *J. Biol. Chem.*, 24, 24777, 2007.

101. Kumaran, D. et al., Domain organization in *Clostridium botulinum* neurotoxin type E is unique: Its implication in faster translocation, *J. Mol. Biol.*, 386, 233, 2009.

102. Agarwal, R. et al., Mode of VAMP substrate recognition and inhibition of *Clostridium botulinum* neurotoxin type F, *Nat. Struct. Mol. Biol.*, 16, 789, 2009.

103. Montecucco, C. and Schiavo, G., Mechanism of action of tetanus and *Botulinum* neurotoxins, *Mol. Microbiol.*, 13, 1, 1994.

104. Montecucco, C., Papini, E., and Schiavo, G., Bacterial protein toxins penetrate cells via a four-step mechanism, *FEBS Lett.*, 346, 92, 1994.

105. Montecucco, C. and Schiavo, G., Structure and function of tetanus and botulinum neurotoxins, *Q. Rev. Biophys.*, 28, 423, 1995.

106. Halpern, J.L. and Neale, E.A., Neurospecific binding, internalization, and retrograde axonal transport. In *Current Topics in Microbiology and Immunology, Clostridial Neurotoxins: The Molecular Pathogenesis of Tetanus and Botulism*, ed. C. Montecucco, pp. 221–241, Berlin, Germany: Springer, 1995.

107. Simpson, L.L., Identification of the major steps in botulinum toxin action, *Annu. Rev. Pharmacol. Toxicol.*, 44, 167, 2004.

108. Black, J.D. and Dolly, J.O., Interaction of ^{125}I-labeled botulinum neurotoxins with nerve terminals. I. Ultrastructural autoradiographic localization and quantitation of distinct membrane acceptors for types A and B on motor nerves, *J. Cell. Biol.*, 103, 521, 1986.

109. Black, J.D. and Dolly, J.O., Selective location of acceptors for botulinum neurotoxin A in the central and peripheral nervous systems, *Neuroscience*, 23, 767, 1987.

110. Tsukamoto, K. et al., Binding of *Clostridium botulinum* type C and D neurotoxins to ganglioside and phospholipid. Novel insights into the receptor for clostridial neurotoxins, *J. Biol. Chem.*, 280, 35164, 2005.

111. Eidels, L., Proia, R.L., and Hart, D.A., Membrane receptors for bacterial toxins, *Microbiol. Rev.*, 47, 596, 1983.

112. Stenmark, P. et al., Crystal structure of botulinum neurotoxin type A in complex with the cell surface co-receptor GT1b-insight into the toxin–neuron interaction, *PLoS Pathog.*, 4, e1000129, 2008.

113. Mahrhold, S. et al., The synaptic vesicle protein 2C mediates the uptake of botulinum neurotoxin A into phrenic nerves, *FEBS Lett.*, 580, 2011, 2006.

114. Dong, M. et al., SV2 is the protein receptor for botulinum neurotoxin A, *Science*, 312, 592, 2006.

115. Jacky, B.P.S. et al., Identification of fibroblast growth factor receptor 3 (FGFR3) as a protein receptor for botulinum neurotoxin serotype A (BoNT/A), *PLoS One*, 9, e1003369, 2013.

116. Dong, M. et al., Mechanism of botulinum neurotoxin B and G entry into hippocampal neurons, *J. Cell. Biol.*, 179, 1511, 2007.

117. Jin, R. et al., Botulinum neurotoxin B recognizes its protein receptor with high affinity and specificity, *Nature*, 444, 1092, 2006.

118. Dong, M. et al., Synaptotagmins I and II mediate entry of botulinum neurotoxin B into cells, *J. Cell. Biol.*, 162, 1293, 2003.

119. Chai, Q. et al., Structural basis of cell surface receptor recognition by botulinum neurotoxin B, *Nature*, 444, 1096, 2006.

120. Black, J.D. and Dolly, J.O., Interaction of ^{125}I-labelled botulinum neurotoxins with nerve terminals. II. Autoradiographic evidence for its uptake into motor nerves by acceptor-mediated endocytosis, *J. Cell. Biol.*, 103, 535, 1986.

121. Shone, C.C., Hambleton, P., and Melling, J., A 50-kDa fragment from the NH_2-terminus of the heavy subunit of *Clostridium botulinum* type A neurotoxin forms channels in lipid vesicles, *Eur. J. Biochem.*, 167, 175, 1987.

122. Pellizzari, R. et al., Tetanus and botulinum neurotoxins: Mechanism of action and therapeutic uses, *Philos. Trans. R. Soc. Lond. B. Biol. Sci.*, 354, 259, 1999.

123. Galloux, M. et al., Membrane interaction of botulinum neurotoxin A translocation (T) domain. The belt region is a regulatory loop for membrane interaction, *J. Biol. Chem.*, 283, 27668, 2008.

124. Schiavo, G. et al., Tetanus and botulinum-B neurotoxins block neurotransmitter release by proteolytic cleavage of synaptobrevin, *Nature*, 359, 832, 1992.

125. Oguma, K., Fujinaga, Y., and Inoue, K., Structure and function of *Clostridium botulinum* toxins, *Microbiol. Immunol.*, 39, 161, 1995.

126. Schiavo, G. et al., Identification of the nerve terminal targets of botulinum neurotoxin serotypes A, D, and E, *J. Biol. Chem.*, 268, 23784, 1993.

127. Shone, C.C. et al., Proteolytic cleavage of synthetic fragments of vesicle-associated membrane protein, isoform-2 by botulinum type B neurotoxin, *Eur. J. Biochem.*, 217, 965, 1993.

128. Schiavo, G. et al., Tetanus toxin is a zinc protein and its inhibition of neurotransmitter release and protease activity depend on zinc, *EMBO J.*, 11, 3577, 1992.

129. Blasi, J. et al., Botulinum neurotoxin A selectively cleaves the synaptic protein SNAP-25, *Nature*, 365, 160, 1993.

130. Schiavo, G. et al., Botulinum neurotoxin serotypes A and E cleave SNAP-25 at distinct COOH-terminal peptide bonds, *FEBS Lett.*, 335, 99, 1993.

131. Binz, T. et al., Proteolysis of SNAP-25 by types E and A botulinal neurotoxins, *J. Biol. Chem.*, 269, 1617, 1994.

132. Foran, P. et al., Botulinum neurotoxin C1 cleaves both syntaxin and SNAP-25 in intact and permeabilized chromaffin cells: Correlation with its blockade of catecholamine release, *Biochemistry*, 35, 2630, 1996.

133. Niemann, H., Blasi, J., and Jahn, R., Clostridial neurotoxins: New tools for dissecting exocytosis, *Trends Cell. Biol.*, 4, 179, 1994.

134. Blasi, J. et al., Botulinum neurotoxin C1 blocks neurotransmitter release by means of cleaving HPC1/syntaxin, *EMBO J.*, 12, 4821, 1993.

135. Arnon, S.S. et al., Infant botulism, epidemiological, clinical, and laboratory aspects, *JAMA*, 237, 1946, 1977.

136. Keller, J.E., Recovery from botulinum neurotoxin poisoning in vivo, *Neuroscience*, 139, 629, 2006.

137. Woodruff, B.A. et al., Clinical and laboratory comparison of botulism from toxin serotypes A, B, and E in the United States, *J. Infect. Dis.*, 166, 1281, 1992.

138. Hughes, J.M. et al., Clinical features of types A and B foodborne botulism, *Ann. Intern. Med.*, 95, 442, 1981.

139. Iida, H., Specific antitoxin therapy in type E botulism, *Jpn. J. Med. Sci. Biol.*, 16, 311, 1963.

140. Mann, J.M. et al., Patient recovery from type A botulism: Morbidity assessment following a large outbreak, *Am. J. Public Health*, 71, 266, 1981.

141. Duchen, L.W., An electron microscopic study of the changes induced by botulism toxin in the motor end-plates of slow and fast skeletal muscle fibres of the mouse, *J. Neurol. Sci.*, 14, 47, 1971.

142. Gottlieb, S.L. et al., Long-term outcomes of 217 botulism cases in the Republic of Georgia, *Clin. Infect. Dis.*, 45, 174, 2007.

143. Varma, J.K. et al., Signs and symptoms predictive of death in patients with foodborne botulism—Republic of Georgia, 1980–2002, *Clin. Infect. Dis.*, 39, 357, 2004.

144. Aminzadeh, Z. et al., A survey on 80 cases of botulism and its clinical presentations as a public health concern, *Iran. J. Clin. Infect. Dis.*, 2, 77, 2007.

145. Mitchell, P.A. and Pons, P.T., Wound botulism associated with black tar heroin and lower extremity cellulitis, *J. Emerg. Med.*, 20, 371, 2001.

146. Royl, G. et al., Diagnostic pitfall: Wound botulism in an intoxicated intravenous drug abuser presenting with respiratory failure, *Intensive Care Med.*, 33, 1301, 2007.

147. Filozov, A. et al., Asymmetric type F botulism with cranial nerve demyelination, *Emerg. Infect. Dis.*, 18, 102, 2012.

148. Preuss, S.F. et al., A rare differential diagnosis in dysphagia: Wound botulism, *Laryngoscope*, 116, 831, 2006.

149. Weber, J.T. et al., Wound botulism in a patient with a tooth abscess: Case reports and review, *Clin. Infect. Dis.*, 16, 635, 1993. Rev 47 WB cases 1951-90;incub8da(rg4-18).

150. Arnon, S.S. et al., Human botulism immune globulin for the treatment of infant botulism, *N. Eng. J. Med.*, 354, 462, 2006.

151. Sanford, D.C. et al., Inhalational botulism in rhesus macaques exposed to botulinum neurotoxin complex serotype A1 and B1, *Clin. Vaccine Immunol.*, 17, 1293, 2010.

152. Taysse, L. et al., Induction of acute lung injury after intranasal administration of toxin botulinum A complex, *Toxicol. Pathol.*, 33, 336, 2005.

153. Franz, D.R. et al., Efficacy of prophylactic and therapeutic administration of antitoxin for Inhalation botulism. In *Botulinum and Tetanus Neurotoxins, Neurotransmission and Biomedical Aspects*, ed. B.R. DasGupta, pp. 473–476, New York: Plenum Press, 1993.

154. Edell, T.A. et al., Wound botulism associated with a positive Tensilon test, *West. J. Med.*, 139, 218, 1983.

155. Lindstrom, M. and Korkeala, H., Laboratory diagnostics of botulism, *Clin. Microbiol. Rev.*, 19, 298, 2006.

156. Lindstrom, M. et al., Multiplex PCR Assay for detection and identification of *Clostridium botulinum* types A, B, E, and F in food and fecal material, *Appl. Environ. Microbiol.*, 67, 5694, 2001.

157. Korkeala, H. et al., Type E botulism associated with vacuum-packaged hot-smoked whitefish, *Int. J. Food Microbiol.*, 43, 1, 1998.

158. Akbulut, D., Grant, K.A., and McLauchlin, J., Development and application of real-time PCR assays to detect fragments of the *Clostridium botulinum* types A, B, and E neurotoxin genes for investigation of human foodborne and infant botulism, *Foodborne Pathog. Dis.*, 1, 247, 2004.

159. Wheeler, C.G. et al., Sensitivity of mouse bioassay in clinical wound botulism, *Clin. Infect. Dis.*, 48, 1669, 2009.

160. Raphael, B.H. and Andreadis, J.D., Real-time PCR detection of the nontoxic nonhemagglutinin gene as a rapid screening method for bacterial isolates harboring the botulinum neurotoxin (A–G) gene complex, *J. Microbiol. Methods*, 71, 343, 2007.

161. Griffin, P.M. et al., Endogenous antibody production to botulinum toxin in an adult with intestinal colonization botulism and underlying Crohn's disease, *J. Infect. Dis.*, 175, 633, 1997.

162. Maslanka, S.E., Laboratory confirmation of human cases of botulism, 2008. htpp://iccvam.neihs.nih.gov/methods/biologics/botdocs/biolowkshp/Presentations/Session1/SMaslanka.pdf (accessed July 13, 2012).

163. McClung, L.S. and Toabe, R., The egg yolk plate reaction for the presumptive diagnosis of *Clostridium sporogenes* and certain species of the gangrene and botulinum groups, *J. Bacteriol.*, 53, 139, 1947.

164. Glasby, C. and Hatheway, C.L., Isolation and enumeration of *Clostridium botulinum* by direct inoculation of infant fecal specimens on egg yolk agar and *Clostridium botulinum* isolation media, *J. Clin. Microbiol.*, 21, 264, 1985.

165. Dezfulian, M. et al., Enzyme-linked immunosorbent assay for detection of *Clostridium botulinum* type A and type B toxins in stool samples of infants with botulism, *J. Clin. Microbiol.*, 20, 279, 1984.

166. Poli, M.A., Rivera, V.R., and Neal, D., Development of sensitive colorimetric capture ELISAs for *Clostridium botulinum* neurotoxin serotypes E and F, *Toxicon*, 40, 797, 2002.

167. Szilagyi, M. et al., Development of sensitive colorimetric capture ELISAs for *Clostridium botulinum* neurotoxin serotypes A and B, *Toxicon*, 38, 381, 2000.

168. Ferreira, J.L., Comparison of the amplified ELISA and mouse bioassay for determination of botulinum toxins A, B, E, and F, *J. AOAC Int.*, 84, 85, 2001.

169. Ferreira, J.L. et al., Detection of botulinal neurotoxins A, B, E, and F by amplified enzyme-linked immunosorbent assay: Collaborative study, *J. AOAC Int.*, 86, 314, 2003.

170. Ferreira, J.L. et al., Comparison of the mouse bioassay and enzyme-linked immunosorbent assay procedures for the detection of type A botulinal toxin in food, *J. Food. Prot.*, 67, 203, 2004.

171. Sharma, S.K. et al., Detection of type A, B, E, and F *Clostridium botulinum* neurotoxins in foods by using an amplified enzyme-linked immunosorbent assay with digoxigenin-labeled antibodies, *Appl. Environ. Microbiol.*, 72, 1231, 2006.

172. Doellgast, G.J. et al., Sensitive enzyme-linked immunosorbent assay for the detection of *Clostridium botulinum* neurotoxins A, B, and E using signal amplification via enzyme-linked coagulation assay, *J. Clin. Microbiol.*, 31, 2402, 1993.

173. Doellgast, G.J. et al., Enzyme-linked immunosorbent assay and enzyme-linked coagulation assay for detection of *Clostridium botulinum* neurotoxins A, B, and E and solution-phase complexes with dual-label antibodies, *J. Clin. Microbiol.*, 32, 105, 1994.

174. Sharma, S.K. and Whiting, K.C., Methods for detection of *Clostridium botulinum* toxin in foods, *J. Food Prot.*, 68, 1256, 2005.

175. Kirchner, S. et al., Pentaplexed quantitative real-time PCR assay for the simultaneous detection and quantification of botulinum neurotoxin-producing *Clostridia* in food and clinical samples, *Appl. Environ. Microbiol.*, 76, 4387–4395, 2010.

176. Hill, B.J. et al., Universal and specific quantitative detection of botulinum neurotoxin genes, *BMC Microbiol.*, 10, 267, 2010.

177. Sharma, S.K. et al., Evaluation of lateral-flow *Clostridium botulinum* neurotoxin detection kits for food analysis, *Appl. Environ. Microbiol.*, 71, 2935, 2005.

178. Ahn-Yoon, S., De Cory, T.R., and Durst, R.A., Ganglioside-liposome immunoassay for the detection of botulinum toxin, *Anal. Bioanal. Chem.*, 378, 68, 2004.

179. Gessler, F. et al., Evaluation of lateral flow assays for the detection of botulinum neurotoxin type A and their application in laboratory diagnosis of botulism, *Diagn. Microbiol. Infect. Dis.*, 57, 342, 2007.

180. Rivera, V.R. et al., Rapid detection of *Clostridium botulinum* toxins A, B, E, and F in clinical samples, selected foods, matrices, and buffer using paramagnetic bead-based electrochemiluminescent detection, *Anal. Biochem.*, 353, 248, 2006.

181. Wictome, M. et al., Development of an in vitro bioassay for *Clostridium botulinum* type B neurotoxin in foods that is more sensitive than the mouse bioassay, *Appl. Environ. Microbiol.*, 65, 3787, 1999.

182. Ruge, D.R. et al., Detection of six serotypes of botulinum neurotoxin using fluorogenic reporters, *Anal. Biochem.*, 441, 200, 2011.

183. Kalb, S.R. et al., The use of Endopep-MS for the detection of botulinum toxins A, B, E, and F in serum and stool samples, *Anal. Biochem.*, 251, 84, 2006.

184. Ferracci, G. et al., A label-free biosensor assay for botulinum neurotoxin B in food and human serum, *Anal. Biochem.*, 410, 281, 2011.

185. Marconi, S. et al., A protein chip membrane-capture assay for botulinum neurotoxin activity, *Toxicol. Appl. Pharmacol.*, 233, 439, 2008.

186. Pellett, S. et al., Sensitive and quantitative detection of botulinum neurotoxin in neurons derived from mouse embryonic stem cells, *Biochem. Biophys. Res. Commun.*, 404, 388, 2011.

187. Cai, S. and Singh, B.R., Botulism diagnostics: From clinical symptoms to in vitro assays, *Crit. Rev. Microbiol.*, 33, 109, 2007.

188. Shone, C. et al., The 5th international conference in basic and therapeutic aspects of botulinum and tetanus neurotoxins. Workshop review: Assays and detection, *Neurotox. Res.*, 9, 205, 2006.

189. Hatheway, C.L. and McCroskey, L.M., Examination of feces and serum for diagnosis of infant botulism in 336 patients, *J. Clin. Microbiol.*, 25, 2334, 1987.

190. Santos, J.I., Swensen, P., and Glasgow, L.A., Potentiation of *Clostridium botulinum* toxin aminoglycoside antibiotics: Clinical and laboratory observations, *Pediatrics*, 68, 50, 1981.

191. Offerman, S.R. et al., Wound botulism in injection drug users: Time to antitoxin correlates with intensive care unit length of stay, *West J. Emerg. Med.*, 10, 251, 2009.

192. Sandrock, C.E. and Murin, S., Clinical predictors of respiratory failure and long-term outcome in black tar heroin-associated wound botulism, *Chest*, 120, 562, 2001.

193. Fagan, R.P. et al., Initial recovery and rebound of type F intestinal colonization botulism after administration of investigational heptavalent botulinum antitoxin, *Clin. Infect. Dis.*, e125, 1–5, 2011.

194. CDC, Investigational heptavalent botulinum antitoxin (HBAT) to replace licensed botulinum antitoxin AB and investigational botulinum antitoxin E, *MMWR*, 59, 299, 2010.

195. Hibbs, R.G., Experience with the use of an investigational F(ab')₂ heptavalent botulism immune globulin of equine origin during an outbreak of type E botulism in Egypt, *Clin. Infect. Dis.*, 23, 340, 1996.

196. Black, R.E. and Gunn, R.A., Hypersensitivity reactions associated with botulinal antitoxin, *Am. J. Med.*, 69, 567, 1980.

197. Hathaway, C.H. and Snyder, J.D., Antitoxin levels in botulism patients treated with trivalent equine botulism antitoxin to toxin types A, B, and E, J, *Infect. Dis.*, 150, 407, 1984.

198. FDA, Package insert-Botulism Antitoxin Heptavalent (A,B,C,D,E,F,G)-(Equine). 2013. http://www.fda.gov/downloads/BiologicsBloodVaccines/BloodBloodProducts/ApprovedProducts/LicensedProductsBLAs/FractionatedPlasmaProducts/UCM345147.pdf (accessed November 5, 2013).

199. Al-Sayyed, B. et al., A 3-day-old boy with acute flaccid paralysis, *Pediatr. Ann.*, 38, 479, 2009.

200. Vanella de Cuetos, E.E. et al., Equine botulinum antitoxin for the treatment of infant botulism, *Clin. Vaccine Immunol.*, 18, 1845–1849, 2011.

201. USDA, USDA Complete Guide to Home Canning, 2009 revision, 2009. http://nchfp.uga.edu/publications/usda/INTRO%20section%20Home%20Can.pdf (accessed April 23, 2012).

202. Siegel, L.S., Destruction of botulinum toxins in food and water. In *Clostridium botulinum. Ecology and Control in Foods*, eds. A.H.W. Hauschild and K.L. Dodds, pp. 323–332, New York: Marcel Dekker, 1993.

203. CDC, Botulism: Control measures overview for clinicians, June 14, 2006. http://www.bt.cdc.gov/agent/botulism/clinicians/control.asp (accessed April 16, 2012).

204. Smith, K.E. et al., Outbreaks of salmonellosis in Minnesota (1998 through 2006) associated with frozen, microwaveable, breaded, stuffed chicken products, *J. Food Prot.*, 71, 2153, 2008.

205. Burrows, W.D. and Renner, S.E., Biological warfare agents as threats to potable water, *Environ. Health Perspect.*, 107, 975, 1999.

206. Rusnak, J.M. et al., Experience in the medical management of potential laboratory exposures to agents of bioterrorism on the basis of risk assessment at the United States Army Medical Research Institute of Infectious Diseases (USAMRIID), *J. Occup. Environ. Med.*, 46, 801, 2004.

207. Rusnak, J.M. et al., Risk of occupationally acquired illnesses from biological threat agents in unvaccinated laboratory workers, *Biosec. Bioterr.*, 2, 281, 2004.

208. Rusnak, J.M. and Smith, L.A., Botulism neurotoxin vaccines. Past history and recent developments, *Hum. Vaccin.*, 5, 1, 2009.

209. Smith, L.A. and Rusnak, J.M., Botulinum neurotoxin vaccines: Past, present, and future, *Crit. Rev. Immunol.*, 27, 303, 2007.

210. CDC, Notice of CDC's discontinuation of investigational pentavalent (ABCDE) botulinum toxoid vaccine for workers at risk for occupational exposure to botulinum toxins, *MMWR*, 60, 1454, 2011.

211. Torii, Y. et al., Production and immunogenic efficacy of botulinum tetravalent (A, B, E, F) toxoid, *Vaccine*, 20, 2556, 2002.

212. Shearer, J.D. et al., Botulinum neurotoxin neutralizing activity of immune globulin (IG) purified from clinical volunteers vaccinated with recombinant botulinum vaccine (rBV A/B), *Vaccine*, 28, 7313, 2010.

213. Shearer, J.D., Manetz, T.S., and House, R.V., Preclinical safety assessment of recombinant botulinum vaccine A/B (rBV A/B), *Vaccine*, 30, 1917, 2012.

41 Conotoxins

Jon-Paul Bingham, Robert K. Likeman, Joshua S. Hawley,
Peter Y.C. Yu, and Zan A. Halford

CONTENTS

41.1 INTRODUCTION

Cone snails are a group of carnivorous marine gastropods, having perfected venom needed for rapid prey immobilization. This neurotoxic cocktail is skillfully delivered using a hollow, lone, disposable radula harpoon, which is also designed to impale and tether its prey (Figure 41.1a). Armed with this combination, certain cone snail species have been responsible for human fatalities [1–3]. To understand their toxic nature, science has taken these highly prized snails to discover a wealth of biological materials that separate them from most other venomous organisms.

The genus of ~600 species, carrying >100,000 peptides (termed "conotoxins" or "conopeptides"), provides clear indication to immense chemical evolution and diversity. What variation these snails have achieved potentially parallels the diversity of bioactive constituents observed in the microbial world. This evaluation has been recently reiterated examining milked venoms of the piscivorous/fish-eating members of *Conus* [4–7]. Coinciding with their characterization [8–10],

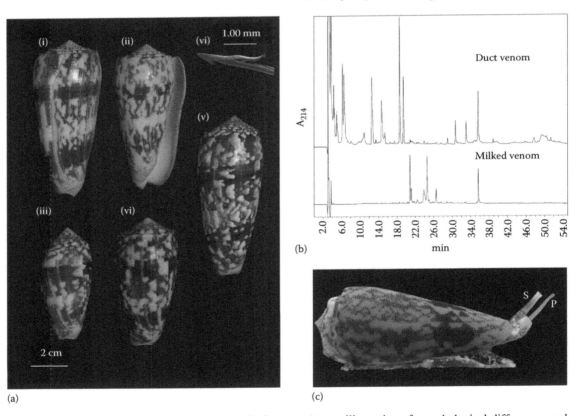

FIGURE 41.1 (a) Representatives of piscivorous cone snail *Conus striatus*—illustration of morphological differences and representative geographic distribution: (i) and (ii) Oahu, Hawaii; (iii) Pago Pago, American Samoa; (iv) Great Barrier Reef, Australia; and (v) Cebu, Philippines. (vi) The radula harpoon from *C. striatus*—each individual snail species has a different radula structure. (b) The reverse-phase high-performance liquid chromatography (RP-HPLC)/ultraviolet detection profile of the duct and milked venom extracts from *C. striatus*, Oahu, Hawaii—Note the dramatic differences in venom composition; some peaks (conotoxins) are common in both extracts. (c) Live *C. striatus* showing siphon (S) and proboscis (P)—the latter holds the radula for venom delivery.

synthesis [11–13], and bioengineering [14–16], these peptides have operated as pharmacological probes to dissect ion channel and receptor functions since the 1970s [17–19] and have now transitioned into analgesic pharmaceuticals [20,21].

On a regulatory basis, conotoxins/conopeptides represent a grey area in their formal classification listing. Exemption status as "select agents" has been provided by the US Department of Agriculture (USDA), Animal and Plant Health Inspection Service (APHIS), and Centers for Disease Control and Prevention (CDC) (see Section 41.8), but they remain as security-sensitive materials with specific biosafety concerns. However, combining the many thousand conotoxins/conopeptides into one general regulatory classification brings about a complex situation, as the various described peptide families demonstrate diversity in their pharmacology, offering differential phyla selectivity in higher organisms (Section 41.2). Synthetic techniques are used in the production of *Conus* peptides to compensate for their limited natural availability [22–24]. To a lesser extent, recombinant protein expression technology may be used [25,26]; this is limited due to the abundance of posttranslational modifications (PTMs). Some of these modifications are required for biological function (Section 41.6). Such synthetic or biotechnological approaches potentially present their own unique regulatory concerns.

Only a few laboratories in the world have access to these native venom extracts; thus far, from those efforts, <2% of materials have been described. Yet, of the described compounds, some have demonstrated unparalleled scientific and medical worth [27,28]. As correctly gauged by the scientific community and federal regulatory agencies, placing restrictive access and practices on these "agents" may actually hinder potential drug development for major human afflictions such as pain, as seen with the analgesic properties of ω-conotoxin MVIIA/Prialt® [29–31]. Thus, establishing a basic understanding of the biology of these venomous organisms (Section 41.2), the chemical and pharmacological diversity of their toxins (Section 41.5), and the practices employed in safely procuring materials and handling (native and synthetic materials) within the confidences of a research laboratory environment becomes essential. To achieve this, we have provided standard operating procedures (SOPs) (Section 41.9), detailed first aid protocols (Box 41.1), and notes to physicians (Box 41.2) to ensure a centralized document to assist forming individual requirements and protocols while dealing with conotoxins and conopeptides as security-sensitive agents.

BOX 41.1   FIRSTHAND ACCOUNT

FIRSTHAND ACCOUNT OF TREATMENT OF A CONE SNAIL STING

In 1974, I was a government medical officer at Kavieng, in New Ireland, which at that time was still a part of the Commonwealth of Australia. Spearfishing on the reefs at night was a popular pastime with the local inhabitants, using a homemade speargun, powered by strong elastic bands, and a waterproof flashlight. They also used to keep an eye open for specimens of *Conus gloriamaris*, which at that time were selling for several hundred dollars each. Although papers had been published many years earlier both in Australia [64] and Papua New Guinea [147] and elsewhere [1] reporting the potential risk to humans from cone snails, both expatriates and indigenous people alike were generally unaware of the danger. Although an amateur shell collector myself, I was equally ignorant of the potential hazards.

The victim in this case was a 35-year-old male, and he was brought to the hospital unconscious at first light. The history was obtained in part from family members and later from the patient himself after recovery. While spearfishing on a reef in the dark, he had seen and picked up a cone snail, which he had placed in the pocket of his shorts. Soon after, he felt a sharp prick in his thigh, but thought nothing of it. While cycling home from his fishing expedition, he began to feel weak and went straight to bed. After a while, his wife with exceptional vigilance observed that his chest wall did not move when he breathed, and on attempting to rouse him, she found that he was unresponsive and paralyzed below the neck. They urgently brought him to the hospital and with commendable diligence also brought the cone snail, which was still very much alive. Examination revealed a flaccid paralysis to the level of T1, with diaphragmatic respiration. No lesion was identified on the thigh where the victim had felt the sting. Rapid consultation of a textbook and identification of the shell as *C. geographus* made the diagnosis all too clear.

There was nothing to be done but to prepare for intubation and mechanical ventilation, which in a small hospital on a Pacific island was a somewhat daunting prospect for a young doctor of only three years experience. Fortunately, the paralysis progressed no further and soon began to resolve. Consciousness gradually returned, and after 24 h, the patient could move his legs. The next day, he was out of bed and two days later, left the hospital, apparently none the worse for his experience.

The cone shell spent a short time in the freezer and was then added to my collection. I developed a healthy respect for cone snails, and, although I never saw another case of envenomation in my subsequent 38 years of medical practice, I have always treated them with great respect and encouraged others to do the same [148]. Prevention as always is better than cure, particularly if there is not one.

Col. Robert Likeman MA BM BCh (Oxon), FRCOG, FRANZCOG, FACTM, FACRRM

BOX 41.2   NOTES TO PHYSICIANS

- All cases, or suspected cases, should be treated as potentially life threatening and moved as quickly as possible to a medical facility, which can provide mechanical respiration for 24 h.
- The only serious consequence of a sting is paralysis of respiratory muscles; however, this paralysis is not permanent and in most cases, resolves in less than 24 h.
- Treatment should be directed at maintaining respiration until the paralysis resolves by itself. Mechanical ventilation is the treatment of choice.

Basic treatment: Establish a patent airway with suction where necessary. Watch for signs of respiratory insufficiency and assist ventilation as necessary. Administer oxygen by non-rebreather mask at 10–15 L/min. Monitor and treat, where necessary, for pulmonary edema. Monitor and treat, where necessary, for shock. Anticipate seizures. DO NOT use emetics.
    NO recurrent or persistent coagulopathy has been reported in humans.
    Direct cardiac toxicity by a cone snail envenomation has not been demonstrated. If cardiac arrest occurs, it is likely secondary to respiratory arrest and secondary cardiac ischemia. Management should focus on airway protection and ventilation, with cardiac interventions (such as chest compressions, antiarrhythmic medications, and cardioconversion) given only as indicated once the respiratory status is stabilized.

Advanced treatment: Consider orotracheal or nasotracheal intubation for airway control in an unconscious patient or where respiratory arrest has occurred. Positive-pressure ventilation using a bag-valve mask or mechanical ventilator might be of use. Monitor and treat, where necessary, for arrhythmias.
    Start an IV and provide intravenous fluid resuscitation as needed, but avoid volume overload. Drug therapy should be considered for pulmonary edema. Hypotension with signs of hypovolemia requires the cautious administration of fluids. Fluid overload might create complications.
    Treat seizures with midazolam.

Laboratory safety: Proparacaine hydrochloride should be used to assist eye irrigation. Peptides from cone snails are also poorly absorbed dermally or via oral route. Treat needle stabs as formerly mentioned.

41.2   BIOLOGY OF *CONUS*

The genus *Conus* owes much of its 60 million years of evolutionary success as marine predators to its distinct venom apparatus. The main structure, a highly convoluted, longitudinal secretory duct, can range from <2 to 80 cm in length [32]. Crude venom extracts can yield a few micrograms to tens of grams of venom, depending on the species. Histological examination of the venom duct's epithelial cells provides clear evidence of their secretory nature [1,33,34]. Different sections of the secretory venom duct possess varying degrees of toxicity, yet these observations remain unexplained [35,36].

The radula harpoon plays a significant role in the process of prey capture and venom delivery, and this tooth completes the terminal interaction between the predator and prey [1,37]. The hollow, chitinous harpoon is a single projectile that is morphologically particular to each *Conus* species [38–40] (Figure 41.1b). Structural features can be clearly correlated to specific feeding types observed in *Conus* [41,42].

Cone snails are either (a) piscivorous/fish-eaters, (b) molluscivorous/mollusk-eaters, (c) vermivorous/worm-eaters, or (d) omnivorous. Although representing a small number of species, the piscivorous cone snails' venoms are considered

life threatening to humans, while molluscivorous feeders are considered dangerous, via their aggressive nature—unconfirmed fatalities have been attributed to some members (see succeeding text). The vermivorous class represents the largest group within the genus; most are considered timid and nonthreatening, unlike their previous counterparts. Lastly, omnivorous species show no preferential feeding behavior. These four feeding classes have a few "general" morphological distinctions that include aperture size, weight, and general pigmentation patterns.

Vermivore shells are generally heavy in comparison to other classes of feeders and possess a narrow parallel aperture. Molluscivores typically tend to have multiple white-tented marks on a brown-background medium-weighted shell with a graduated aperture, while piscivore shells are ornate, are typically lightweight in comparison, and have the widest apertures to accommodate vertebrate prey (Figure 41.1a). Despite these delineations, snail classification is an onerous undertaking due to the morphological differences between the species.

Cone snails are nocturnal but are mainly active around sunset and sunrise. Vermivores impale polychaetes/worms with the radula affixed to the extended, freely mobile proboscis, as prey is drawn into the rostrum [43]. In a second strategy, the foot locates and orientates the prey for consumption.

The foot moves the worm to the rostrum and using a "spaghetti-sucking" approach, with or without active envenomation engulfs the prey [44].

Molluscivores are aggressive hunters, feeding on mitres, trochus, and cowries, including other cone snails [40,45–47]. Stimulated snails use their proboscis to probe for the prey's shell and orifice. Once the prey is found, it is impaled with multiple radulae. At times, a "puff" of surplus venom is observed, and this action may be repeated during subduing strikes [45]. Once paralyzed, the cone snail's rostrum enters the prey's shell cavity so partial digestion and extrapolation can occur. Indigestible material and used radulae are regurgitated after digestion [40,48].

Piscivores also use two feeding approaches [4,40,43]. The first is via "tag and reel" harpooning, in which the extended proboscis and internalized radula impale, inject, and draw the paralyzed fish into the fully expanded rostrum [49,50] (see Figure 41.1b). After digestion, a mass of fish scales, bones, and used radula(e) are regurgitated [51,52]. *Conus geographus* and *Conus tulipa* have exhibited an alternative "passive" strategy. The unsuspecting fish is "netted" by the extended rostrum, and envenomation occurs after engulfment [40,51,53]. However, variations of this behavior have been observed [54].

The milked venom, resulting from piscivore envenomation, possesses a pharmacological cocktail to block pre- and postsynaptic ion channels at neuromuscular junction. Injected venom containing various α- and ω-conotoxins (i.e., by *C. geographus*) [54,55] provides dual assurance in accomplishing rapid prey immobilization. Termination of action potential propagation is achieved by simultaneously inhibiting the presynaptic Ca^{2+} influx through voltage-gated calcium channels (VGCCs) (ω-conotoxins) and postsynaptic muscle-type acetylcholine receptors (AChR) (α-conotoxins). Other synergistic strategies exist, in which *Conus* peptides combine targeting sodium (Na_V) and potassium (K_V) voltage-gated channels to induce an excitotoxic state in prey [56,57]. These combined mechanisms intensify the venom's effectiveness in rapid "lightning-strike cabal" immobilization; Na_V channels are opened simultaneously (via μ-conotoxins) alongside blockage of K_V channels (via κ-conotoxins) [58]. Further discussion on conopeptide pharmacology is presented in Section 41.5.

41.3 VENOM OF *CONUS*

Venom appearance: Dissected venom gland extracts can be opaque, milky white to sulfur yellow in color. The milked venom is clear unless hydrophobic peptides, such as the δ-conotoxins, are present. Milked venom consists to a lesser extent of proteins and low molecular mass organic compounds with peptides being the dominant constituent; the majority of the milked venom volume is equivalent to seawater. Freeze-dried or lyophilized materials, both native (desalted) and synthetic, can be fluffy/velutinous, electrostatic, and hydroscopic in nature.

Venom solubility: Synthetic and extracted conotoxins/conopeptides are soluble in water, producing a slightly translucent solution that may foam if agitated. Native venoms contain small insoluble particles or granules [59–61], these being heavily pronounced in dissected whole duct venom extracts along with other cellular debris. Whole venom extracts require centrifugation and secondary extraction. This process results in a translucent peptide-containing supernatant. To achieve maximal solubility, small amounts of immiscible organic solvent, such as acetonitrile (5% v/v), are added to aqueous solvents; sonication may assist in dissolving peptidic materials.

Venom stability: When stored correctly under laboratory conditions, conotoxins/conopeptides are highly stable identities. The analysis of dissected *C. geographus* duct venom obtained from the late Dr. Robert Endean (University of Queensland, Australia), which being collected, extracted, and freeze-dried in 1962, demonstrated to contain many recognizable peptide molecular masses that correlated to known conotoxins/conopeptides [62]. Due to disulfide bonding, a high level of intrinsic structural stability is present within the conotoxins. Yet, as peptides, they are not resistant to enzymatic digestions, microbial breakdown, disulfide reduction, and/or chemical oxidation. Any chemical modification to the native peptide(s) will typically lead to a decrease or removal of biological activity. Although heating leads to their degradation, prolonged exposure to heat, such as autoclaving alone, is regarded as an ineffective measure to completely remove biological functionality. An effective combined laboratory approach for peptide neutralization is discussed in Section 41.11.

41.4 EPIDEMIOLOGY OF *CONUS*

Cone snail-related deaths have been recorded dating back to 1705, first from a molluscivore, *Conus textile* [63]. Since, there have been around 30 recorded fatalities, principally, by one of the most studied species, *C. geographus* [34,40,64]. In humans, the fatality rate from this species alone is approximately 70% [3].

Shell collectors have the highest risk of cone snail envenomation, whether being a tourist or hobby malacologist. Increased exposure and handling increases the likelihood of envenomation (see Box 41.1). This includes researchers that handle cone snails in the same category. Collecting live cone snails is dangerous! Specimen handling, containment, and separation in the field requires due diligence, behavioral knowledge, and positive identification of species. Removal of cone snails from their natural environment induces a defensive response. When stressed they can eject 5–30 times more venom than during a predatory response; a contribution to their lethal nature. Radula harpoons are fired at a velocity of 200 m/s [64,65] and can penetrate 5 mm neoprene wet suit material [66]. The proboscis has been observed to extend 2–3 shell lengths from the snail, probing for the source of antagonism; careful handling is essential.

Moreover, even cone snails have become "taboo"; certain cultures of the South Pacific islands are long accustomed and wary of the dangers from cone snail envenomation.

TABLE 41.1
Symptoms of Cone Snail Envenomation and First Aid

Cone Snail Envenomation

Symptoms	First Aid
• Mild to sharp burning sensation at the site of the sting. • Sensation of tingling, burning, pricking, or numbness (paresthesia) especially about lips and mouth. • Pruritus—desire to scratch at site of penetration. • Edema at site of penetration; actual puncture wound may not be evident; possible localized discoloration; edema may show effects within the entire limb. • Fatigue and malaise. • Faintness or altered mentation. • Nausea, prolonged stomach cramps. • Facial muscle paralysis. • Ptyalism—drooling/hypersalivation. • Slurred speech and potentially aphonia (total loss of speech). • Blurred vision or diplopia (double vision). • Ptosis—abnormal drooping of the upper eyelid(s). • Progressive muscle paralysis and numbness. • Absence of reflexes in lower legs/limbs and arms. • Dyspnea—difficulty in breathing. • Unconsciousness. • Respiratory arrest (40 min to 5 h after sting). • Cardiac impairment, leading to cardiac arrest. • *DO NOT* cut or excise the bitten or stung area. • *DO NOT* attempt to suck out the venom. • *DO NOT* unnecessarily move the victim. • *DO NOT* submerge limb in hot water or pour hot water, vinegar, denatured alcohol or ethanol, etc., on sting area. • *DO NOT* apply an arterial tourniquet. • *DO NOT* elevate sting site. • *DO NOT* operate vehicle if envenomated.	As there is no antivenom/antivenin for cone snail venom, death is typically due to respiratory failure then cardiac arrest. Initial first responder measures: • General DRABC (*D*anger, *R*esponse, *A*irway, *B*reathing, *C*irculation). • Activate emergency medical services. • Seek assisted medical evacuation. • Go directly to an emergency room—escorted. • Keep victim calm. • Prolonged artificial respiration, even mechanical ventilation, may be required. Be prepared. • Cardiopulmonary resuscitation (CPR) may be required. • If possible and with great caution, retain specimen(s) for identification process, which may aid in prognosis. *Bandage Application*: • Apply a broad pressure bandage directly over the sting area about as tight as an elastic wrap to a sprained ankle. • Bind bandage distally, away from the heart—this avoids venous congestion and discomfort. • Ensure that arterial circulation is not cut off; ensure fingers or toes stay pink and warm. • Immobilize limb using splint; use a sling if the stung area is on the arm or hand. • In cases that involve swelling of the affected area, the compression bandage may need to be more proximally positioned to wrap ahead of the swollen area. • In route, hold the stung site below the rest of the body. • The bandage pressure should be released within 8 h or as soon as medical care is reached.

If an islander has been stung, they resort to "on-the-spot" bloodletting, inflicting numerous deep lacerations on the afflicted limb [67]. Although not recommended, such actions would likely decrease circulatory distribution. Table 41.1 lists reported symptoms observed after a cone snail envenomation, representing details from numerous species [68–71]. Also compiled are first aid measures that should be immediately implemented. These expectations too have equal relevance in the laboratory setting, particularly when involving injectable materials. To present how to handle cone snail envenomation, a list of "interventions to avoid" is also presented; these have no effect in lessening the effect or venom distribution and may cause more harm. An advance note to physicians is also provided (Box 41.2), as a reference if such a case is encountered.

41.5   CONOTOXIN AND CONOPEPTIDE CLASSIFICATION

Bioactive constituents within *Conus* venom demonstrate inter- and intraspecies diversity [72]. An individual duct venom extract can contain ~200 different pharmacological components [18,73,74], while only ~20% of these elements appear in milked venom. The milked venom may also demonstrate differential peptide toxin expression leading to intraspecimen variation [57,75]. This unique ability to vary venom constituents potentially makes production of an effective antivenom/antivenin impossible.

There are ~5500 individual sequences derived from *Conus*, with ~80% originating from genomic origin [76–78]. A majority of these peptides have been organized into seven main superfamilies (A-, M-, O-, P-, S- T-, and I-superfamilies; Table 41.2). These groupings are based on a combination of their pharmacological selectivity and their cysteine frameworks, together with pre-propeptide genetic sequence homology. The naming convention of *Conus* first assigns a Greek letter to indicate the pharmacological targeting; the main families of interest are the α-, ω-, μ-, δ-, and κ-"conotoxins" (see succeeding text), while other bioactive peptides such as conantokins, contulakins, contryphans, and others *Conus* peptides are assigned as "conopeptides." Distinguishing between conotoxin and conopeptide sequences, one or two letters are used to indicate the source species, which the first (capitalized) and second letter (typically a sequential nonvowel character)

TABLE 41.2

Representative Conotoxins and Conopeptides from *Conus*

Superfamily	Disulfide Framework	Family		Target (Receptor/Channel)	Example	Amino Acid Sequence	Pharmacological Affinity IC$_{50}$ [nM] (Organism)	Conus Species	Feeding Type	References
A	CC-C-C	α		nAChR	α-ImI	GCCSDPRCAWRC*	250–500 (frog)	*C. imperialis*	V	[131]
		ρ		nAChR	ρ-TIA	FNWRCCLIPACRRNHKKFC*	150 (rat)	*C. tulipa*	P	[132]
	CC-C-C-C-C	αA		nAChR	αA-PIVA	GCCGSYONAACHOCSCKDROSYCGQ*	350 (fish)	*C. purpurascens*	P	[133]
		κA		K$_v$	κA-SIVA	ZKSLVPSVITTCCGYDOGIMCOOCRCTNSC*	0.050–5 (fish) [LD$_{50}$]	*C. striatus*	P	[134]
	CC-C-C-CC	μ		Na$_v$	μ-GIIIA	RDCCTOOKKCCKDRQCKOQRCCA*	100 (rat skeletal muscle) [K$_D$]	*C. geographus*	P	[84,85]
		ψ		nAChR	ψ-PIIIE	HOCCLYGKCRRYOGCSSASCCQR*	127 (eel skeletal muscle)	*C. purpurascens*	P	[8]
		κM		K$_v$	κM-RIIIK	LOSCCSLNLRLCOVOACKRNOCCT*	20 (fish)	*C. radiatus*	P	[103]
O	C-C-CC-C-C	μO		Na$_v$	μO-MrVIA	ACRKKWEYCIVPIIGFIYCCPGLICGPFVCV	500 (chromaffin cell)	*C. marmoreus*	M	[135]
		δ		Na$_v$	δ-TxVIA	WCKQSGEMCNLLDQNCCDGYCIVLVCT	0.0025 (rat CNS) [K$_D$]	*C. textile*	M	[47,136]
		ω		Ca$_v$	ω-MVIIA	CKGKGAKCSRLMYDCCTGSCRSGKC*	0.03 (rat CNS)	*C. magus*	P	[137]
		κ		K$_v$	κ-PVIIA	CRIONQHCFQHLDDCCSRKCNRFNKCV	360 ± 70 (frog) [K$_D$]	*C. purpurascens*	P	[9,138]
		γ		Voltage-gated pacemaker	γ-PnVIIA	DCTSWFGCTVNSγCCSNSCDQTYCγLYAFOS	0.0632 nM/100 mg body weight (mussel) [ED$_{50}$]	*C. pennaceus*	M	[139]
P	C-C-C-C-C-C	Spastics		Undetermined	Tx9a	GCNNSCQγHSDCγSHCICTFRGCGAVN*	>0.250 nM/g body weight (mouse brain) [LD$_{50}$]	*C. textile*	M	[97,140]
S	C-C-C-C-C-C-C-C	σ		5-HT$_3$	σ-GVIIIA	GCTRTCGGOKCTGTCTCTNSSKCGCRYN-VHPSGwGCGCACS*	53 ± 3 (HEK 293)	*C. geographus*	P	[18,98]
T	CC-CC	τ		Ca^{2+}	τ-TxIX	γCCγDGWCCT†AAO	0.20 ± 0.029 (fish) [ED$_{min}$]	*C. textile*	P	[18,99]
		χ		Norepinephrine transporter	χ-MrIA	NGVCCGYKLCHOC*	500 (rat); 1700 (human)	*C. marmoreus*	M	[141]
I	C-C-CC-CC-C-C	Excitatory		Undetermined	Sr11a	CRTEGMSCγγNQQCCWRSCCRGEC EAPCRFGP*	640 (frog)	*C. spurius*	V	[142,143]
No assignment	C-C	Conopressins		Vasopressin	Lys-conopressin-G	CFIRNCPKG*	0.010 (mouse) [ED$_{min}$]	*C. geographus*	P	[98]
Linear	No cysteines	Contryphans		L-type Ca^{2+} channel	Am975	GCDOWDPWC*	<100 (rat PNS)	*C. amadis*	M	[144]
		Conantokins		Glutamate (NMDA)	Pr1	GEDγYAγGIRγYQLIHGKI	200 (mouse NR2B)	*C. parius*	P	[145]
		Conorfamide		RF amide	CNF-Sr2	GPMγDPLγIIRI*	active range of ±0.25–1.00 (mouse)	*C. spurius*	V	[146]
	O-linked	Contulakins		Neurotensin	Contulakin-G	ZSEEGGSNA7KKPYIL*	960 (human); 250 (mouse)	*C. geographus*	P	[79]

*, amidated C-terminus; O, 4-*trans*-hydroxyproline; Z, pyroglutamate; S̲, O-glycosylated serine; γ, gamma-carboxyglutamic acid; w, L-6-bromotryptohan; T†, O-glycosylated threonine; W̲, D-tryptophan; T̲, threonine

+ β-D-Gal(1→3)α-D-GalNAc; P, piscivorous; M, molluscivorous; V, vermivorous; ED, effective dose.

derived from the species name, that is, Tx = C. *textile*. The use of a single letter assignment is typically reserved for piscivorous species, that is, M = *Conus* **magus**.

Conus nomenclature is partly influenced by disulfide bridges; these are essential in forming and maintaining the necessary 3D structures that convey biological activity. Assigning of a Roman numeral indicates the disulfide framework category, as illustrated in Table 41.2. Lastly, a sequential uppercase letter is used to signify the order of discovery. Usually the initially described peptide, within the individual species, often lacks this designation. A discussion outlining the major described superfamilies is provided in the following:

41.5.1 A-SUPERFAMILY

The A-superfamily is separated into four toxin families: α-, αA-, κA-, and ρ-conotoxins; these consist of ~200 individual sequences (excluding the precursor sequences). The α- and αA-conotoxins are nicotinic AChR (nAChR) antagonists, ρ-conotoxins have structurally similar cysteine arrangement to the α-conotoxins but are $_{1B}$-adrenoceptor antagonists (Table 41.2), while κA-conotoxins target potassium (K+) ion channels and are typically glycosylated and larger in size [79,80]. The classical α-conotoxins contain ~40 individually isolated venom peptides and have been discovered throughout the genus. They range from 12 to 22 amino acids (αα) and contain four cysteine moieties that form two disulfide bonds. Postsynaptic inhibition at the neuromuscular junction by α-conotoxins results in paralysis and death from respiratory failure [3,68,70]. Other α-conotoxins have the ability to distinguish between different AChR subtypes, as observed with α-conotoxin ImI (Table 41.2) that targets neuronal isoforms.

41.5.2 M-SUPERFAMILY

The M-superfamily is composed of three families, ψ-, μ-, and κM-conotoxins [81–83]. About 300 peptide sequences (excluding the precursor sequences) from *Conus* fall into this class, a large majority being described from genetic sources. These peptides structurally contain 14–28 αα residues, six cysteines—forming three disulfide bonds in a CC-C-C-CC framework. Some members contain high concentrations of 4-*trans*-hydroxyproline, a PTM αα (Section 41.6.4). ψ-Conotoxins target nAChRs, noncompetitively, while the μ-conotoxins block Na$_V$ channels in excitable cells of muscle, heart, skeletal, and nerve tissue [81]. Na$_V$ channels modulate rapid electrical signaling in both neuronal and muscle cell types [84]. μ-Conotoxins GIIIA, GIIIB, and GIIIC are examples of conotoxins from *C. geographus* that act upon Na$_V$ channels in muscle [85,86] (Table 41.2). These peptides work in unison with other milked venom constituents to induce rapid paralysis in fish [54]. At a molecular level, they selectively bind to the ion channel pore, blocking the electrochemical flux of Na+ ions into the cell. This feature makes them extremely useful for electrophysiological research.

41.5.3 O-SUPERFAMILY

The O-superfamily is the largest group with ~500 peptide sequences (excluding precursor sequences), in five distinct families: μO-, δ-, ω-, κ-, and γ-conotoxins [87–89]. Structural similarities exist between pharmacologically unrelated conotoxins (Table 41.2). This superfamily is primarily dictated by a highly conserved disulfide framework of C-C-CC-C-C or commonly referred to as a "6/4-Cys loop" pattern, which is observed in the mature peptide sequence [90]. The interrelation of this superfamily is compounded by peptide precursor sequence homology [91]. Pharmacologically, μO- and δ-conotoxin families target Na$_V$ channels but will not compete with μ-conotoxins for the pore-binding site [92,93], while ω-, κ-, and γ-conotoxins target VGCCs, K+ channels, and pacemaker channels, respectively [94]. The δ-conotoxins typically consists of about 30 ααs, retaining three disulfide bridges and demonstrating a characteristic hydrophobic nature. The ω-conotoxins follow these same biochemical properties, but they lack the hydrophobicity. Known as the "shaker peptides," these induce persistent tremors when injected into mice and have drawn much pharmacological interest. ω-Conotoxin MVIIA, an N-type selective VGCC isolated from *C. magus*, was the first conotoxin to be clinically approved for the treatment of chronic intractable pain, mostly targeting patients who have high tolerances to opioids [95]. Other ω-conotoxins are under investigation as therapeutic leads.

41.5.4 P-SUPERFAMILY

The P-superfamily, which consists of nine peptide sequences, has been granted the family name "spastics" because their effects on their molecular target manifest a spastic paralysis in the test organism [96]. Presently, these peptides have only been isolated from molluscivorous and vermivorous cone snails with the majority being derived from genetic sequences. An example of a spastic conotoxin is Tx9a, a PTM peptide found in the duct venom of molluscivorous *C. textile* [97] (Table 41.2). Pharmacological targeting of this unique peptide family remains to be established.

41.5.5 S-SUPERFAMILY

The S-superfamily targets muscle-type nAChR or competitively antagonizes the serotonin 5HT$_3$ receptor and consists of seven peptides (excluding the precursor sequences). Isolated from all feeding types, these sequences are unique, both in size (<40 αα) and disulfide arrangement; these having the highest content observed in any conotoxin/conopeptide with the formation of five individual disulfide bonds [98]. Those isolated from the duct venom of piscivores demonstrate extensive PTM, including bromotryptophan, as seen in σ-conotoxin GVIIIA from *C. geographus* (Table 41.2).

41.5.6 T-SUPERFAMILY

The T-superfamily includes both τ- and χ-conotoxin families and consists of ~110 peptide sequences (excluding the precursor sequences) with most (<95%) being derived from genetic sources representing all feeding groups within the genus. The majority of these peptides have four cysteines in a -CC-CC- framework producing two disulfide bonds [99]. τ-Conotoxins function as presynaptic Ca^{2+} channel blockers [100], while χ-conotoxins inhibit norepinephrine transporters (NETs). χ-Conotoxins are receiving much interest due to their pharmacological transition into clinical drug trials to combat pain [101].

41.5.7 I-Superfamily

A large number of I-superfamily members have been disclosed, ~88 individual sequences (excluding precursor sequences), most representing genetic-derived sequences from vermivorous cone snails. Most of these sequences remain pharmacologically undefined. These peptides possess eight cysteine moieties with a pattern of -C-C-CC-CC-C-C- and from those few isolated native peptides demonstrate to contain multiple PTM ααs. Examples include κ-conotoxin BtX (BeTx), isolated from *Conus betulinus*. Pharmacologically, BtX demonstrated to target the calcium-activated potassium channel [102] without affecting other voltage-gated channels [103]. The pharmacological selectivity and differentiation within the I-superfamily toward K_V channels is illustrated with conotoxin ViTx (*Conus virgo*), which demonstrated specific selectivity inhibiting the K^+ ion channel subtypes $K_V1.1$ and $K_V1.3$, but not $K_V1.2$ [104]. However, the role of PTMs in this class is basically unknown.

41.5.8 Single Disulfide Bonding and Linear Conopeptides

A number of additional classes of conopeptides are gaining momentum together. Most are not classified within the conotoxin superfamily structure, presently. These contain either a single disulfide bridge or, as with the linear conopeptides, possess no cysteine moieties at all. Conopressins and contryphans contain one disulfide bond, which are vasopressin receptor antagonists and target unknown receptor types, respectively [105,106]. Being absent of disulfide bonds, the conantokins are structurally stabilized by the chelating ability γ-carboxy glutamate. These peptides selectively target the *N*-methyl-D-aspartate subtype of glutamate receptors and ligand-gated Ca^{2+} channels involved in seizures of intractable epilepsy [107]. The contulakins target the neurotensin (NT) receptors agonistically [108]. It is a given that the pharmacological repertoire of ion channel and receptor-targeting capabilities within *Conus* is not fully complete. Proposed extensions to these superfamilies is now being undertaken to accommodate the discovery of new conotoxins and conopeptides.

41.6 POSTTRANSLATION MODIFICATION IN CONOTOXINS AND CONOPEPTIDES

Native PTMs in conotoxins and conopeptides play an important biochemical and pharmacological role, as seen with their incorporation and dominance throughout the peptide families [109]. Many PTMs have multiple implications in peptide folding and structural stabilization [110], providing resistance to degradation [111] and importantly increasing pharmacological potency [112]. These will be briefly discussed in the succeeding text but do not represent all PTMs observed in *Conus*.

41.6.1 Disulfide Bond Formation

Biological activity is reflective on a native 3D structure. In conotoxins and some conopeptides, this is conveyed by a specific, directed formation of disulfide bonds between individual cysteine moieties. Each class of native peptides has a specific connectivity pattern. The modification of the resulting structure by changing their connectivity can lead, in most, to biological inactivity [113]. This PTM represents the most commonly observed in *Conus*.

41.6.2 C-Terminal Amidation

C-terminal amidation is common to most conotoxins. Its presence can be genetically predicted with a level of certainty by establishing the rear flanking sequence region of the predicted mature peptide toxin [114]. C-terminal amidation has a role in the disulfide-coupled folding of conotoxins, which can structurally influence the biological activity of the peptide [115] but also provides a level of enzymatic resistance to carboxypeptidase degradation. Its removal from native-like peptides has demonstrated minor decreases in potency [116]. This represents the second most common PTM seen in *Conus*.

41.6.3 N-Terminal Cyclization

A few conotoxins contain an N-terminal glutamic acid; these residues can undergo N-terminal cyclization via glutaminyl cyclase to form pyroglutamic acid [79]. This results in changing the N-terminal charge state that can affect target affinity as well as provide resistance to endopeptidase degradation. This alteration represents one of the least observed αα modifications in *Conus*.

41.6.4 Hydroxylation of Proline

Hydroxylation of proline to 4-*trans*-hydroxyproline is a common PTM undertaken by proline hydroxylase. Observed in various conotoxins and conopeptide classes, it has been implicated in assisting in 3D folding [117]. For the μ-conotoxins, incorporating multiple moieties is important for interacting with the Na_V channel. This represents the third most commonly observed PTM in *Conus*.

41.6.5 Carboxylation of Glutamic Acid

About 10% of conotoxins/conopeptides contain the vitamin K-dependent carboxylation of glutamic acid, which results in γ-carboxyglutamic acid. Structurally, it promotes the formation of helices and Ca^{2+} ion binding [110]. γ-Carboxylation is predominant in the conantokins (Table 41.2). Its removal or exclusion often causes inactivity [110]. It is also seen as a mechanism to increase the chemical/charge state diversity within conotoxins and conopeptides. The presence of γ-carboxyglutamic acid represents the fourth most common PTM seen in *Conus*.

41.6.6 Isomerization of Amino Acids

L-Amino acids are the biologically active and prevalent chemical precursors of proteins, but some ααs undergo epimerization. Although uncommon, a number of D-amino acids have been observed in *Conus*, most being restricted to D-tryptophan, phenylalanine, valine, and leucine [118]. Epimerase enzymes cause their occurrence, and the biological value of such modifications in these peptides remains elusive.

41.6.7 Bromination of Tryptophan

The halogenation of ααs in *Conus* is presently restricted to the bromination of L-tryptophan [119]. Produced by bromoperoxidase, forming the analogue 6-L-bromotryptophan, this unique PTM is present in the I- and S-superfamilies. There is no specific functionality assigned to peptides that contain 6-L-bromotryptophan. This represents one of the more unusually PTMs observed in *Conus*.

41.6.8 O-Glycosylation

Glycosyltransferase enzymes posttranslationally modify serine and threonine residues by O-glycosylation. These modifications can have different sugar interlinkages within different conformations. Contalukin-G contains one of the most common glycan linkages, β-D-Gal(1→3)α-D-GalNAc, which is found attached to threonine. Glycosylation can increase the efficacy of certain conotoxins. Without the glycosylation, the bioactivity is lower or abolished; this is evident with the κ-conotoxins and their actions upon K$_V$ channels [79]. Glycosylation is expected to have a higher occurrence in conotoxins than presently discussed in the current literature. Glycosylation events can be found predominantly in the milked venoms of *Conus* (Bingham unpublished results).

41.6.9 Sulfation of Tyrosine

Tyrosyl sulfotransferase results in the sulfation of the hydroxyl group of tyrosine producing sulfotyrosine. Identified originally in the α-conotoxin EpI, its non-PTM form, EpI[Tyr]^{15}, demonstrated to have similar potency as a competitive nicotinic antagonist on bovine adrenal chromaffin cells [120]. Thus, the biological consequences of this PTM remain undetermined. Although an uncommon PTM found in *Conus*, it reveals a different chemical feature from its hydroxylated counterpart.

41.7 BIOENGINEERED CONOTOXINS

Conotoxins/conopeptides have been subjected to extensive chemical manipulation and bioengineering. These endeavors outweigh the combined efforts undertaken with other peptide toxins derived from scorpion, spider, ant, and bee venoms. The resulting bioactive, nonnative conotoxin/conopeptide hybrids and peptidomimetic analogues have advanced our present understanding of structural biology [121] and drug design [122]. The development of the "conopolytides" hybrids represents a significant step forward in conotoxin/conopeptide bioengineering [123]. This approach incorporates replacements of the peptide backbone and side functionalities to streamline molecular size and make "redundant" or minimize "nonessential" interactions that may deter pharmacological selectivity or directed targeting [123,124].

Other mechanisms in conotoxin/conopeptide bioengineering revolve around the all-essential disulfide bridges, replacing them with organic nonreducing structures. These include cystathionine thioether and dicarba linkages as seen applied to α-conotoxin ImI [125,126]. Here, mimicking structural constraints has pharmaceutical benefit providing potentially great stability *in vivo*. This is also observed with the more recent approach in undertaking peptide backbone cyclization of the N- to C-termini by incorporating a space linker with native chemical ligation for the desired cyclic backbone structure. The bioengineering approaches as applied to Vc1.1 and α-conotoxins AuIB [127], RgIA [128], and MII [129] increase oral bioavailability and systemic stability. This offers advancement in seeing conotoxins/conopeptide leads, increasing their therapeutic potential.

Synthetic bioengineered conotoxins/conopeptides potentially represent a perplexing situation in "select agent" regulation. Here, more resilient procedures may be required in neutralization and decontamination (see Section 41.9). Granted, the complications regarding the bioengineering of orally active "full" or "partial" peptidomimetics represent unique circumstances that need to be evaluated to ensure personal and workplace safety. These measures may include (1) handling procedures, (2) decontamination and neutralization procedures, (3) animal disposal procedures, (4) security measures, and (5) possible reclassification and regulatory delineation between bioengineered and native materials. Present recommended SOPs for conotoxins/conopeptides are detailed in the succeeding text.

41.8 SELECT AGENT CLASSIFICATION OF CONOTOXINS AND CONOPEPTIDES

Preamble. The words conotoxin and conopeptide are typically used interchangeably; however, the term "conotoxin" is usually reserved for their pharmacological-hierarchical classification (see Section 41.5). These peptides vary in molecular mass, solubility, and phylogenetic (dependent) bioactivity and toxicity. General safety precautions in handling and storing these materials must be undertaken in the laboratory, as with any potentially toxic venom constituent(s). Here, we describe the

US National Select Agent Registry (NSAR)'s* regulatory classification regarding conotoxins/conopeptides and provide recommended SOPs for their laboratory use.

41.8.1 SELECT AGENTS EXCLUSION: EFFECT 4-29-2003

For research purposes, the US Departments of Health and Human Services (HHS) and Agriculture (USDA) excludes specifically:

> The class of sodium channel antagonist μ-conotoxins, including GIIIA; the class of calcium channel antagonist ω-conotoxins, including GVIA, GVIB, MVIIA, MVIIC, and their analogs or synthetic derivatives; the class of NMDA-antagonist conantokins, including con-G, con-R, con-T and their analogs or synthetic derivatives; and the putative neurotensin agonist, contulakin-G and its synthetic derivatives … as select agents.[†]

41.8.2 PERMISSIBLE AMOUNTS

Also excluded by the NSAR are permissible amounts <100 mg of any conotoxin/conopeptide under the control of a principal investigator[‡] (PI). These present regulatory exclusions that meet normal research demands are founded on current experimental evidence and scientific literature, which indicates that these conotoxins/conopeptides do not possess sufficient or acute toxicity to pose a significant threat to public health or safety.

41.8.3 "NONFUNCTIONAL" CONOTOXINS AND CONOPEPTIDES

"Nonfunctional" conotoxins/conopeptides are also excluded from these regulations[§]—this includes synthetic peptides in either their on- or off-resin state, which are not folded/oxidized and remain in an inactive/"nonfunctional" state.

41.9 STANDARD OPERATING PROCEDURES FOR LABORATORY USE OF CONOTOXINS AND CONOPEPTIDES

41.9.1 PURPOSE

This SOP describes the techniques and procedures for safe and proper handling, storage and preparation for experimental use and disposal of conotoxins/conopeptides as "select agents."

Note: Not all conotoxins/conopeptides demonstrate toxicity, but the user must regard all materials, independent of source or species, as having the ability to be toxic.

41.9.2 MINIMUM PERSONAL PROTECTIVE EQUIPMENT

At minimum, personal protective equipment (PPE) for handling of conopeptides should include the following:

1. *Eye protection*: Safety glasses or chemical-splash goggles must be worn at all times when handling conotoxins/conopeptides. Adequate safety glasses must meet the requirements of the "Practice of Occupational and Educational Eye and Face Protection" (ANSI Z.87.1 2003).
2. *Face shield*: An optional face shield may be worn in addition to safety glasses or chemical-splash goggles if the potential for splashing exists.
3. *Gloves*: (a) Appropriate gloves shall be worn when handling solutions that contain conotoxins/conopeptides. (b) Due to strength and durability, nitrile gloves are recommended when handling conotoxins/conopeptides. (c) Gloves that protect against the generation of static charges are also preferred, especially when handling dry or powdered forms of conotoxins/conopeptides. This is important when weighing out materials as static charge may cause conotoxins/conopeptides to become motile and airborne.
4. *Respirator*: During the handling of bioengineered, aerosol-derived conopeptides, a respirator should be utilized (e.g., half-mask respirator equipped with high-efficiency particulate air (HEPA) cartridges[¶]). A respirator should be used during events of spill cleanup or decontamination. Potential users should have obtained prior medical evaluation, training, and fit testing before the usage of a respirator.
5. *Protective clothing*: A traditional cotton-polyester lab coat may not provide adequate protection against chemicals. Wear impervious protective wear such as a polyvinylchloride (PVC) or polyethylene (PE) apron, lab coat, or at minimum, a smock with elastic cuffs and high neck collar. This may be required in possible events of decontamination.

41.9.3 HAZARDOUS WARNING SIGNS

Any room associated with the usage of conotoxins/conopeptides should be posted with a sign "Caution! Toxins in Use" (user name/date). Visitors to the laboratory should be notified to take precaution by the researchers present.

41.9.4 DELIVERY OF CONOTOXINS AND CONOPEPTIDES

1. When a shipment of conotoxins/conopeptides is received, the specification sheet is dated and goes in the select agents notebook. Log in the agent's name and quantity shipped on the select agent inventory sheet. Notify the sender within 24 h of receipt of the toxin.

---

* Maintained jointly by the APHIS and the CDC Select Agent Programs.
[†] http://www.selectagents.gov/Select%20Agents%20and%20Toxins%20Exclusions.html.
[‡] http://www.selectagents.gov/Permissible%20Toxin%20Amounts.html.
[§] http://www.selectagents.gov/Guidance_for_the_Inactivation_of_Select_Agents_and_Toxins.html.

[¶] High-efficiency particulate air.

2. Inspect external package for damage during shipment. If damaged, immediately contain and isolate; notify principal scientist/PI, environmental, health, and safety (EH&S), and sender; and undertake decontamination and disposal of the material upon instruction.

3. All primary vials containing select agents should be handled in a glove box or Class III biosafety cabinet—when practical.

4. For storage, the primary vial is placed in a secondary sealed container, upright, within a lockable freezer. The primary vial remains factory sealed until "working solution" stocks are required.

41.9.5 Handling Procedures for Conotoxins and Conopeptides

1. A "buddy system" should be practiced during any high-risk operations. Each must be familiar with the applicable procedures, maintain visual contact with the other, and be ready to assist in the event of an accident.

2. If the conopeptide is received as a powder, place a pad on the inside of a glove box or Class III biosafety cabinet in order to minimize the spread of contamination during the solvation process. Ensure that the ducted biosafety cabinet is working properly with an inward airflow prior to initiating work. All work should be performed within the operationally effective zone of the biosafety cabinet.

3. To avoid static charge buildup on glass or plasticware, use static discharger, that is, Zerostat® 3 antistatic gun; use caution to avoid flammable or combustible solvents/materials during its use.

4. Avoid unnecessary handling and/or transfer of materials by making a stock solution using manufacturers vial and weight specifications; adjust volume addition to achieve desired stock concentration. Carefully suspend the select agent by slow titration, preventing possible foaming or aerosolization. Buffer such as 100 mM NaCl, pH 7.0 (±100 mM glycine), is recommended; the addition of 1 mg/mL BSA can be used to minimize nonspecific binding, if this is suspected.

5. Aliquot stock solution into separate labeled vials (agent name, concentration, buffer type, date) for storage and later concentration readjustments to obtain desired experimental concentration ranges—this is referred to as "working solution." Dispose of used pipette tips into a beaker containing a 10% v/v bleach or alternatively 1% v/v solution of glutaraldehyde.

6. Decontaminate/neutralize the exterior surfaces of all materials leaving the glove box and biosafety cabinet, including the closed aliquot vial(s), with a 10% v/v bleach solution. Decontaminate all work surfaces with a 10% v/v bleach solution. Until the glove box or ducted biosafety cabinet is decontaminated, the equipment should be posted to indicate that select agents are in use, and access to the equipment and apparatus is restricted to authorized personnel.

7. Select agents should be transported in a leak-proof closed secondary container, kept upright, and placed into a lockable freezer best kept at −80°C for longer-term storage, while −20°C is good for short term (1–6 weeks).

8. If primary stock material is in a powder form and requires weighting, transfer material from primary stock vial into preweighted capped vial operating within the inside of a glove box or Class III biosafety cabinet. Wipe the outside of the closed vial with 10% v/v bleach solution and air-dry. Reweigh closed vial and calculate the required buffer addition to achieve working stock volume and then continue as from (3) earlier, with modification. This becomes a "working solution."

41.9.6 "Working Solutions" Storage

While not in use, keep select agents within a lockable storage freezer. Make sure to secure any vial with parafilm in order to prevent any possible spill or leakage. Ensure permanent labeling—with date, amount, concentration, etc. An inventory control system should be in place—detailing inventory, stocks and physical state (i.e., in-solution vs. lyophilized form), and dates received/used/accessed.

41.9.7 "Working Solutions" Usage

Remove required "working solutions" adjusting inventory lists when required. Thaw the vial at 4°C in lockable refrigerator and once in liquid form, mix and centrifuge before opening. Open the vial with caution as the material may be pressurized. It is advisable to plan the use of "working solutions" with care, avoiding prolonged and repeated vial openings, minimizing vial handling and movement. Also of consideration is stability of materials with repeated usage. Avoid repeated freeze-thawing if possible. However, some conotoxins/conopeptides have demonstrated extreme stability if handled and stored correctly. Yet, more conservative considerations are required when using native crude venoms of unknown composition as degradation does occur, which is seen as a decrease in biological activity or chromatographic component abundance.

41.10 ANIMAL HANDLING

Note—practice standard: Sharps must not be used with conotoxins/conopeptides unless specifically approved by EH&S and principal scientist/PI. If it is absolutely necessary to use sharps with conotoxins/conopeptides, sharps with engineering controls must be used (see succeeding text).

41.10.1 USE OF SYRINGES AND NEEDLES

Always wear gloves prior to usage of needles or syringes. Locate a sharps container and bring it to the area of usage. Do not attempt to recap the syringe after usage; instead, dispose of it immediately, sharp end first into the container. Remain stationary; do not walk with any needles. Keep sharp ends away from you at all times.

41.10.2 PREINJECTION

Any animal used as a candidate for conotoxin/conopeptide analysis should be properly obtained and humanely taken care of, complying with approved and current governing animal protocols. It is important to ensure careful experimental planning and organization to decrease stressors that may affect the health of the organism and thus observations and data. Follow guidelines provided in regard to protective wear and safety. Extra precautions should always be taken during the handling of needles and syringes. Engineered controls include a Luer lock with a simple tubular needle guard that can assist leak avoidance and in reproducible needle depth penetration.

41.10.3 INJECTION

If hypodermic needles are used, the injection site should be targeted to an area of high vascular activity. This will ensure maximum efficiency of the toxin and prevent unnecessary pain directed toward the organism. It is recommended to obtain help from a second labmate during this procedure. While one person handles the organism, the other records observations and data. The use of a basic platform to immobilize animals is recommended; with the use of fish, ensure that dose is administered in a quick fashion, returning the animal to its observation habitat.

41.10.4 POSTINJECTION

The use of conotoxins/conopeptides may cause paralysis, which might induce the false appearance of death. Obtain knowledge about each organism used in regard to the most pain-free and efficient way to euthanize, and make sure that each organism is deceased prior to processing/disposal. All small animals used are considered pathological waste and must be incinerated.

41.11 WASTE DISPOSAL AND DECONTAMINATION/NEUTRALIZATION OF CONOTOXINS AND CONOPEPTIDES

All biological liquid waste must be decontaminated/neutralized using a 10% v/v sodium hypochlorite (Clorox®) or solution of sodium hypochlorite and sodium hydroxide (10% w/v) for a 30 min contact time and may be drain disposed with plenty of water. Alternatively, use reactive disinfectants such as glutaraldehyde and formaldehyde (1% v/v solution) for 30 min contact time; however, the resulting material requires biohazards waste disposal.

The majority of native conotoxins/conopeptides are susceptible to (1) reducing agents followed by thiol alkylation (the reducing agents are dithiothreitol [DTT]; β-mercaptoethanol [BME]; and tris(2-carboxyethyl)phosphine [TCEP], 50–100 mM, 65°C–100°C, 15 min; the alkylation agents are 50–100 mM maleimide or substituted maleimides in isopropanol, 65°C, 15 min [130]) and (2) hydrolytic activity (acid or base; 10 M, 100°C, 60 min) leading to their degradation and "nonfunctionality." This approach is usually reserved for destruction of concentrated materials or stock solutions. Dry waste must be placed in double red biohazard bags autoclaved at 121°C for 60 min at 18 psi. Contaminated and potentially contaminated protective clothing and equipment should be decontaminated using methods known to be effective against the conotoxins/conopeptides before removal from the laboratory for disposal, cleaning, or repair. If decontamination is not possible/practical, materials (e.g., used gloves) should be disposed of as hazardous waste.

Note: "Nonfunctional" peptides after disposal/neutralization must represent a zero risk of acquiring toxicity—thus, simple disulfide reduction alone is not an advised process for conotoxins/conopeptide neutralization but combining thiol reduction with thiol alkylation is satisfactory.

41.12 EMERGENCY PROCEDURES

Follow these steps when an emergency involving toxins occurs. Circumstances may include spills, fire and evacuation, personal exposure or injury, and power and ventilation failure.

41.12.1 SPILL

If a spill occurs outside of the fume hood or ducted biosafety cabinet, notify the lab personnel and evacuate immediately. Close the doors as you exit to allow the aerosols to settle. Notify EH&S immediately! Do not reenter the facility. Personnel decontaminating the spill must wear the minimum PPE of safety goggles, disposable lab coat, double gloves, and shoe coverings. Respiratory protection may be necessary based on a risk assessment of the select agent.

Cover the spill area with absorbent material and cover with 10% bleach solution for a 30 min contact time. Avoid raising dust when cleaning up. Ventilate the area and wash the spill site after material pickup is complete. Wash contaminated clothing before reuse. Refer to an MSDS for proper cleanup measures.

Clean up the material and place in double red biohazard bags. Ventilate the area and wash the spill site again after decontamination is complete.

Note: Notify the principal scientist/PI and EH&S of the spill immediately.

41.12.2 FIRE AND EVACUATION

Refer to standard procedures and responsibilities to the emergency, following an established procedure for laboratory/building evacuation—account for all personnel.

41.12.3 PERSONAL INJURY/EXPOSURE, FIRST AID, AND MEDICAL EMERGENCY

Undertake all standard first aid procedures paying particular attention to respiratory support.

A note to physicians: There are no known antidotes/antivenoms for conotoxins/conopeptides. Therapy is supportive. Supplemental oxygen, airway management, and mechanical ventilation may be required (see Box 41.2).

41.12.4 POWER/VENTILATION FAILURE

Undertake the following procedures: (1) Stop the work; (2) secure and cover the toxin; (3) whenever possible, store the toxin(s) in a secure storage location or chemical fume hood; (4) lower the hood's sash completely; (5) post an alert or warning on the sash; and (6) alert all personnel.

41.13 CONTROL, SECURITY, AND TRAINING

41.13.1 EQUIPMENT CONTROL

Engineering controls such as fume hoods and biological safety cabinets must be used as primary containment in order to limit personnel exposure to conotoxins/conopeptides. Engineering controls must be inspected in order to ensure efficient removal of hazard and must have a visual indication of airflow and alarms to indicate that airflow has fallen below acceptable standards. Notification of laboratory personnel must be made prior to any maintenance that will impact the capture velocity of ventilation systems. Local exhaust ventilation systems including laboratory-type chemical fume hoods and biosafety cabinets must be certified upon installation, after maintenance, and annually thereafter.

41.13.2 ADMINISTRATIVE CONTROL

Written SOPs should be developed for all laboratories possessing, using, transferring, or receiving conopeptides in order to ensure consistently safe work practices are being used. SOPs must be reviewed annually or as necessary to reflect changes to procedures. Standard Laboratory Practices for Guidelines for Work with Toxins of Biological Origin are addressed in the BMBL Appendix I (http://www.cdc.gov/biosafety/publications/bmbl5/BMBL.pdf).

41.13.3 SECURITY

Laboratory areas using conotoxins/conopeptides, including those housing cone snails, must be locked at all times. All entries (including entries by visitors, maintenance workers, and others needing one-time or occasional entry) should be recorded with signature in a logbook. Only workers required to perform a job should be allowed in laboratory/holding areas, and workers should be allowed only in areas and at hours required to perform their particular job. Access to areas containing conotoxins/conopeptides should be restricted to those whose work assignments require access. All unknown persons must be politely confronted and questioned to their presence, then escorted to a safe hazard-free area and await consultation. Access for students, visiting scientists, etc., should be limited to hours when regular employees are present. Access for routine cleaning, maintenance, and repairs should be limited to hours when regular employees are present. The laboratory must be cleared of hazardous materials by EH&S prior to maintenance work or cleaning.

41.13.4 TRAINING

All research personnel possessing, using, transferring, or receiving conotoxins/conopeptides must have appropriate training as to (1) the symptoms of conopeptide exposure, (2) postexposure management protocol, and (3) spill cleanup and conotoxin/conopeptide decontamination/neutralization and disposal. Research personnel also require training in the proper use of (1) engineering controls, (2) administrative and work practice controls, (3) PPE, and (4) security requirements for conotoxin/conopeptide possession and use. Research personnel must be trained on the (1) chemical hygiene plan and the (2) conotoxin-/conopeptide-specific SOPs. Training is required before initiation of research involving conotoxins/conopeptides and annually thereafter. PIs are responsible for ensuring that all laboratory workers and visitors understand security requirements and are trained appropriately and equipped to follow established procedures.

41.14 CONCLUSIONS AND FUTURE PERSPECTIVES

The study of venomous organisms or the utilization of their potentially toxic extracts in research comes with inherent risks. Yet, these risks can be minimized with a basic understanding of the organisms' biology, an awareness of their toxicity, and a healthy respect for their pharmacology complexity. Conotoxins/conopeptides are well recognized for advancing the current developments in analgesic medicines and have continued to increase our understanding of receptor–ligand interactions. Cone snails will undoubtedly continue to be a source of novel phyla-selective agents, demonstrating unique isoform pharmacological targeting abilities, but as conotoxins/conopeptides develop and now transition, more so in the synthetic world, examination and reconsideration of these risks becomes an evolving issue. Various regulatory bodies have recognized the scientific communities' need for access to these "agents." In doing so, the agencies have made classification exemptions, providing specific measures and guidelines in their research use. Such access will continue the present momentum in conotoxin and conopeptide drug development.

ACKNOWLEDGMENTS

The authors would like to thank Mr. Jeffery Milisen for photographic assistance, and we wish to acknowledge the continued financial support from the American Heart Association

(Scientist Development Award 0530204N to J.-P.B.), USDA T-STAR (#2009-34135-20067) and HATCH (HAW00595-R) (J.-P.B) that have helped expand our own horizons in conopeptide research.

DISCLAIMER

The views expressed in this manuscript are those of the author(s) and do not reflect the official policy or position of the Department of the Army, Department of Defense, or the US government.

REFERENCES

1. Hermitte, L.C., Venomous marine molluscs of the genus, *Conus*, *Trans. R. Soc. Trop. Med. Hyg.*, 39, 485, 1946.
2. Kohn, A.J., Cone shell stings: Recent cases of human injury due to venomous marine snails of the genus, *Conus*, *Hawaii Med. J.*, 17, 528, 1958.
3. Yoshiba, S., An estimation of the most dangerous species of cone shell *Conus geographus* venom dose in humans, *Jpn. J. Hyg.*, 39, 565, 1984.
4. Hopkins, C. et al., A new family of *Conus* peptides targeted to the nicotinic acetylcholine receptor, *J. Biol. Chem.*, 270, 22361, 1995.
5. Shon, K.J. et al., Purification, characterization, synthesis, and cloning of the lockjaw peptide from *Conus purpurascens* venom, *Biochemistry*, 34, 4913, 1995.
6. Jakubowski, J.A. et al., Intraspecific variation of venom injected by fish-hunting *Conus* snails, *J. Exp. Biol.*, 208, 2873, 2005.
7. Rivera-Ortiz, J.A., Cano, H., and Marí, F., Intraspecies variability and conopeptide profiling of the injected venom of *Conus ermineus*, *Peptides*, 32, 306, 2011.
8. Shon, K.J. et al., A noncompetitive peptide inhibitor of the nicotinic acetylcholine receptor from *Conus purpurascens* venom, *Biochemistry*, 36, 9581, 1997.
9. Shon, K.J. et al., κ-Conotoxin PVIIA is a peptide inhibiting the *Shaker* K+ channel, *J. Biol. Chem.*, 273, 33, 1998.
10. Mitchell, S.S. et al., Three-dimensional solution structure of conotoxin psi-PIIIE, an acetylcholine gated ion channel antagonist, *Biochemistry*, 37, 1215, 1998.
11. Schroeder, C.I. et al., Neuronally micro-conotoxins from *Conus striatus* utilize an alpha-helical motif to target mammalian sodium channels, *J. Biol. Chem.*, 283, 21621, 2008.
12. Martinez, J.S. et al., Alpha-conotoxin EI, a new nicotinic acetylcholine receptor antagonist with novel selectivity, *Biochemistry*, 34, 14519, 1995.
13. Teichert, R.W. et al., Alpha-conotoxin OIVA defines a new alpha-conotoxin subfamily of nicotinic acetylcholine receptor inhibitors, *Toxicon*, 44, 207, 2004.
14. Khoo, K.K. et al., Lactam-stabilized helical analogues of the analgesic μ-conotoxin KIIIA, *J. Med. Chem.*, 54, 7558, 2011.
15. Carstens, B.B. et al., Engineering of conotoxins for the treatment of pain, *Curr. Pharm. Des.*, 17, 4242, 2011.
16. Armishaw, C.J. et al., Improving the stability of α-conotoxin AuIB through N-to-C cyclization: The effect of linker length on stability and activity at nicotinic acetylcholine receptors, *Antioxid. Redox. Signal.*, 14, 65, 2011.
17. Olivera, B.M. et al., Diversity of *Conus* neuropeptides, *Science*, 249, 257, 1990.
18. Terlau, H. and Olivera, B.M., *Conus* venoms: A rich source of novel ion channel-targeted peptides, *Physiol. Rev.*, 84, 41, 2004.
19. Olivera, B.M. et al., Neuronal calcium channel antagonists. Discrimination between calcium channel subtypes using omega-conotoxin from *Conus magus* venom, *Biochemistry*, 26, 2086, 1987.
20. Bingham, J.P., Mitsunaga, E., and Bergeron, Z.L., Drugs from slugs—Past, present and future perspectives of omega-conotoxin research, *Chem. Biol. Interact.*, 183, 1, 2010.
21. Vetter, I. and Lewis, R.J., Therapeutic potential of cone snail venom peptides(conopeptides), *Curr. Top. Med. Chem.*, 12, 1546, 2012.
22. Nishiuchi, Y. et al., Synthesis of gamma-carboxyglutamic acid-containing peptides by the Boc strategy, *Int. J. Pept. Protein Res.*, 42, 533, 1993.
23. Simmonds, R.G., Tupper, D.E., and Harris, J.R., Synthesis of disulfide-bridged fragments of omega-conotoxins GVIA and MVIIA. Use of Npys as a protecting/activating group for cysteine in Fmoc syntheses, *Int. J. Pept. Protein Res.*, 43, 363, 1994.
24. Buczek, P., Buczek, O., and Bulaj, G., Total chemical synthesis and oxidative folding of delta-conotoxin PVIA containing an N-terminal propeptide, *Biopolymers*, 80, 50, 2005.
25. Hernandez-Cuebas, L.M. and White, M.M., Expression of a biologically-active conotoxin PrIIIE in *Escherichia coli*, *Protein Expr. Purif.*, 82, 6, 2012.
26. Bruce, C. et al., Recombinant conotoxin, TxVIA, produced in yeast has insecticidal activity, *Toxicon*, 58, 93, 2011.
27. Olivera, B.M., E.E. Just Lecture, 1996. *Conus* venom peptides, receptor and ion channel targets, and drug design: 50 million years of neuropharmacology, *Mol. Biol. Cell.*, 8, 2101, 1997.
28. Norton, R.S. and Olivera, B.M., Conotoxins down under, *Toxicon*, 48, 780, 2006.
29. Halai, R. and Craik, D.J., Conotoxins: Natural product drug leads, *Nat. Prod. Rep.*, 26, 526, 2009.
30. Han, T.S. et al., *Conus* venoms—A rich source of peptide-based therapeutics, *Curr. Pharm. Des.*, 14, 2462, 2008.
31. Teichert, R.W. and Olivera, B.M., Natural products and ion channel pharmacology, *Future Med. Chem.*, 2, 731, 2010.
32. Hu, H. et al., Elucidation of the molecular envenomation strategy of the cone snail *Conus geographus* through transcriptome sequencing of its venom duct, *BMC Genom.*, 13, 284, 2012.
33. Hinegardner, R.T., The venom apparatus of the cone shell, *Hawaii Med. J.*, 17, 533, 1958.
34. Endean, R. and Duchemin, C., The venom apparatus of *Conus magus*, *Toxicon*, 4, 275, 1967.
35. Whyte, J.M. and Endean, R., Pharmacological investigation of the venoms of the marine snails *Conus textile* and *Conus geographus*, *Toxicon*, 1, 25, 1962.
36. Tayo, L.L. et al., Proteomic analysis provides insights on venom processing in *Conus textile*, *J. Proteome Res.*, 9, 2292, 2010.
37. Taylor, J.D., Kantor, Y.I., and Sysoev, A.V., Foregut anatomy, feeding mechanisms, relationships and classification of the Conoidea (= Toxoglossa) Gastropoda, *Bull. Nat. Hist. Mus. Lond. (Zool.)*, 59, 125, 1993.
38. Endean, R. and Rudkin, C., Further studies of the venoms of *Conidae*, *Toxicon*, 2, 225, 1965.
39. James, M.J., Comparative morphology of radula teeth in *Conus*: Observations with scanning electron microscopy, *J. Moll. Stud.*, 46, 116, 1980.
40. Johnson, C.R. and Stablum, W., Observations on the feeding behaviour of *Conus geographus* (Gastopoda: Toxoglossa), *Pacific Sci.*, 25, 109, 1971.
41. Lim, C.F., Identification of the feeding types in the genus *Conus* Linnaeus, *Veliger*, 12, 160, 1968.

42. Nybakken, J.W., Correlation of radular tooth structure and food habits of three vermivorous species of *Conus*, *Veliger*, 12, 316, 1970.

43. Yoshiba, S., Predatory behavior of *Conus* (*Gastridium*) *tulipa* with comparison to other species, *Jpn. J. Malacol.*, 61, 179, 2002

44. Marsh, H., Preliminary studies of the venoms of some vermivorous *Conidae*, *Toxicon*, 8, 271, 1970.

45. Schoenberg, O., Life and death in the home aquarium, *Hawaiian Shell News*, 29, 4, 1981.

46. Kohn, A.J. and Waters, V., Escape responses to three herbivorous gastropods to the predatory gastropod *Conus textile*, *Anim. Behav.*, 14, 340, 1966.

47. Fainzilber, M. et al., A new neurotoxin receptor site on the sodium channels is identified by a conotoxin that affects sodium channels inactivation in molluscs and acts as an antagonist in rat brain, *J. Biol. Chem.*, 269, 2574, 1994.

48. Endean, R., Venomous cones, *Aust. Nat. Hist.*, 14, 400, 1964.

49. McIntosh, M. et al., Isolation and structure of a peptide toxin from the marine snail *Conus magus*, *Arch. Biochem. Biophys.*, 218, 329, 1982.

50. Olivera, B.M. et al., Conotoxins: Targeted peptide ligands from snail venoms, *Marine Toxins*, American Chemical Society, Washington, DC, 256, 1990.

51. Olivera, B.M. et al., Peptide neurotoxins from fish hunting cone snails, *Science*, 230, 1338, 1985.

52. Terlau, H. et al., Strategy for rapid immobilization of prey by a fish-hunting marine snail, *Nature*, 381, 148, 1996.

53. Le Gall, F. et al., The strategy used by some piscivorous cone snails to capture their prey: The effects of their venoms on vertebrates and on isolated neuromuscular preparations, *Toxicon*, 37, 985, 1999.

54. Bingham, J.P. et al., Analysis of a cone snail's killer cocktail—The milked venom of *Conus geographus*, *Toxicon*, 60, 1166, 2012.

55. Janes, R.W. Nicotinic acetylcholine receptors: α-conotoxins as templates for rational drug design, *Biochem. Soc. Trans.*, 31(3), 634–636, 2003.

56. Bulaj, G. et al., Novel conotoxins from *Conus striatus* and *Conus kinoshitai* selectively block TTX-resistant sodium channels, *Biochemistry*, 44, 7259, 2005.

57. Chun J.B. et al., Cone snail milked venom dynamics—A quantitative study of *Conus purpurascens*, *Toxicon*, 60, 83, 2012.

58. Teichert, R.W. et al., Discovery and characterization of the short κA-conotoxins: A novel subfamily of excitatory conotoxins, *Toxicon*, 49, 318, 2007.

59. Marshall, J. et al., Anatomical correlates of venom production in *Conus californicus*, *Biol. Bull.*, 203, 27, 2002.

60. Maguire, D. and Kwan, J., Cone shell venoms—Synthesis and packaging, *Toxins and Targets* (Watters, D., Lavin, M., and Pearn, J., eds.), Harwood Academic Publishers, Newark, NJ, p. 11, 1992.

61. Marshall, J. et al., Anatomical correlates of venom production in *Conus californicus*, *Biol. Bull.*, 203, 27, 2002.

62. Bingham, J.P. et al., *Conus* venom peptides (conopeptides): Inter-species, intra-species and within individual variation revealed by ionspray mass spectrometry, *Biochemical Aspects of Marine Pharmacology*, Alaken Inc., Fort Collins, CO, 1996.

63. Rumphius, G.E., *The Ambonese Curiosity Cabinet* (translated by Beekman, E.M.), Yale University Press, New Haven, CT, p. 149, 1999.

64. Flecker, H., Cone shell poisoning, with report of a fatal case, *Med. J. Aust.*, 1, 464, 1936.

65. Schulz, J.R., Norton, A.G., and Gilly, W.F., The projectile tooth of a fish-hunting cone snail: *Conus catus* injects venom into fish prey using a high-speed ballistic mechanism, *Biol. Bull.*, 207, 77, 2004.

66. Milton East, personal communication, 1998.

67. Hinde, B., Notes and exhibits, *Proc. Linn. Soc., N.S.W.*, IX, 944, 1885.

68. Fegan, D. and Andresen, D., *Conus geographus* envenomation, *Lancet*, 349, 1672, 1997.

69. Cruz, L.J. and White, J., Clinical toxicology of *Conus* snail stings, *Clinical Toxicology of Animal Venoms* (Meier, J. and White, J., eds.), CRC Press, Boca Raton, FL, p. 117, 1995.

70. Rice, R.D. and Halstead, B.W., Report of fatal cone shell sting by *Conus geographus* Linnaeus, *Toxicon*, 5, 223, 1968.

71. Johnstone, K.Y., Handle with care—The dangerous cone shells, *Collecting Seashells*, Grosst and Dunlap Publishers, New York, Chapter 17, 1970.

72. Bingham, J.P., Novel toxins from the genus *Conus*—From taxonomy to toxins, PhD thesis, University of Queensland, Brisbane, Queensland, Australia, 1998.

73. Olivera, B.M. et al., Diversity of *Conus* neuropeptides, *Science*, 249, 257, 1990.

74. Olivera, B.M. and Cruz, L.J., Conotoxins, in retrospect, *Toxicon*, 39, 7, 2001.

75. Dutertre, S. et al., Dramatic intraspecimen variations within the injected venom of *Conus consors*: An unsuspected contribution to venom diversity, *Toxicon*, 55, 1453, 2010.

76. Conticello, S.G. et al., Mechanisms for evolving hypervariability: The case of conopeptides, *Mol. Biol. Evol.*, 18, 120, 2001.

77. Woodward, S.R. et al., Constant and hypervariable regions in conotoxin propeptides, *EMBO J.*, 9, 1015, 1990.

78. Kaas, Q. et al., ConoServer, a database for conopeptide sequences and structures, *Bioinformatics*, 24, 445, 2008.

79. Craig, G. et al., Contulakin-G, an O-glycosylated invertebrate neurotensin, *J. Biol. Chem.*, 274, 13752, 1999.

80. Kelley, W.P. et al., Two toxins from *Conus striatus* that individually induce tetanic paralysis, *Biochemistry*, 45, 14212, 2006.

81. Van Wagoner, R.M. et al., Characterization and three-dimensional structure determination of psi-conotoxin PIIIF, a novel noncompetitive antagonist of nicotinic acetylcholine receptors, *Biochemistry*, 42, 6353, 2003.

82. French, R.J. et al., The tetrodotoxin receptor of voltage-gated sodium channels—Perspectives from interactions with microconotoxins. *Mar. Drugs*, 13, 2153, 2010.

83. Jacob, R.B. and McDougal, O.M., The M-superfamily of conotoxins: A review, *Cell. Mol. Life. Sci.*, 67, 17, 2010.

84. Cruz, L.J. et al., *Conus geographus* toxins that discriminate between neuronal and muscle sodium channels, *J. Biol. Chem.*, 260, 9280, 1985.

85. Wilson, M.J. et al., μ-Conotoxins that differentially block sodium channels Na$_V$1.1 through 1.8 identify those responsible for action potentials in sciatic nerve, *Proc. Natl. Acad. Sci. USA*, 108, 10302, 2011.

86. Li, R.A. and Tomaselli, G.F., Using the deadly μ-conotoxins as probes of voltage-gated sodium channels, *Toxicon*, 44, 117, 2004.

87. Jimenez, E.C. and Olivera, B.M., Divergent M- and O-superfamily peptides from venom of fish-hunting *Conus parius*, *Peptides*, 31, 1678, 2010.

88. McIntosh, J.M. et al., A new family of conotoxins that blocks voltage-gated sodium channels, *J. Biol. Chem.*, 270, 16796, 1995.

89. Holford, M. et al., Pruning nature: Biodiversity-derived discovery of novel sodium channel blocking conotoxins from *Conus bullatus*, *Toxicon*, 53, 90, 2009.

90. Olivera, B.M. et al., Speciation of cone snails and interspecific hyperdivergence of their venom peptides. Potential evolutionary significance of introns, *Ann. N Y Acad. Sci.*, 870, 223, 1999.

91. Espiritu, D.J. et al., Venomous cone snails: Molecular phylogeny and the generation of toxin diversity, *Toxicon*, 39, 1899, 2001.

92. Bulaj, G. et al., Delta-conotoxin structure/function through a cladistic analysis, *Biochemistry*, 40, 13201, 2001.

93. Bulaj, G. et al., Novel conotoxins from *Conus striatus* and *Conus kinoshitai* selectively block TTX-resistant sodium channels, *Biochemistry*, 44, 7259, 2005.

94. Han, T.S. and Teichert, R.W., *Conus* venoms—A rich source of peptide-based therapeutics, *Pharm. Des.*, 14, 2462, 2008.

95. Pexton, T. et al., Targeting voltage-gated calcium channels for the treatment of neuropathic pain: A review of drug development, *Exp. Opin. Investig. Drugs*, 20, 1277, 2011.

96. McIntosh, J.M. and Jones, R.M., Cone venom—From accidental stings to deliberate injections, *Toxicon*, 39, 1447, 2001.

97. Miles, L.A. et al., Structure of a novel P-superfamily spasmodic conotoxin reveals an inhibitory cystine knot motif, *J. Biol. Chem.*, 277, 43033, 2002.

98. England, L.J. et al., Inactivation of a serotonin-gated ion channel by a polypeptide toxin from marine snails, *Science*, 281, 575, 1998.

99. Walker, C.S. et al., The T-superfamily of conotoxins, *J. Biol. Chem.*, 274, 30664, 1999.

100. Rigby, A.C. et al., A conotoxin from *Conus textile* with unusual posttranslational modifications reduces presynaptic Ca^{2+} influx, *Proc. Natl. Acad. Sci. USA*, 96, 5758, 1999.

101. Paczkowski, F.A. et al., chi-Conotoxin and tricyclic antidepressant interactions at the norepinephrine transporter define a new transporter model, *J. Biol. Chem.*, 282, 17837, 2007.

102. Fan, C.X. et al., A novel conotoxin from *Conus betulinus*, kappa-BtX, unique in cysteine pattern and in function as a specific BK channel modulator, *J. Biol. Chem.*, 278, 12624, 2003.

103. Ferber, M. et al., A novel *Conus* peptide ligand for K^+ channels, *J. Biol. Chem.*, 278, 2177, 2003.

104. Kauferstein, S. et al., A novel conotoxin inhibiting vertebrate voltage-sensitive potassium channels, *Toxicon*, 42, 43, 2003.

105. Cruz, L.J. et al., Invertebrate vasopressin/oxytocin homologs. Characterization of peptides from *Conus geographus* and *Conus striatus* venoms, *J. Biol. Chem.*, 262, 15821, 1987.

106. Jacobsen, R. et al., The contryphans, a D-tryptophan-containing family of *Conus* peptides: Interconversion between conformers, *J. Pept. Res.*, 51, 173, 1998.

107. Ragnarsson, L. et al., Spermine modulation of the glutamate (NMDA) receptor is differentially responsive to conantokins in normal Alzheimer's disease human cerebral cortex, *J. Neurochem.*, 81, 765, 2002.

108. Lewis, R.J., Conotoxins as selective inhibitors of neuronal ion channels, receptors and transporters, *IUBMB Life*, 56, 89, 2004.

109. Craig, A.G., Bandyopadhyay, P., and Olivera, B.M., Post-translationally modified neuropeptides from *Conus* venoms, *Eur. J. Biochem.*, 264, 271, 1999.

110. Layer, R.T., Wagstaff, J.D., and White, H.S., Conantokins: Peptide antagonists of NMDA receptors, *Curr. Med. Chem.*, 11, 3073, 2004.

111. Shon, K.J. et al., mu-Conotoxin PIIIA, a new peptide for discriminating among tetrodotoxin-sensitive Na^+ channel subtypes, *J. Neurosci.*, 18, 4473, 1998.

112. Azam, L. et al., α-Conotoxin BuIA[T5A;P6O]: A novel ligand that discriminates between α6ß4 and α6ß2 nicotinic acetylcholine receptors and blocks nicotine-stimulated norepinephrine release, *FASEB J.*, 24, 5113, 2010.

113. Dutton, J.L. et al., A new level of conotoxin diversity, a nonnative disulfide bond connectivity in alpha-conotoxin AuIB reduces structural definition but increases biological activity, *J. Biol. Chem.*, 277, 48849, 2002.

114. Yuan, D.D. et al., From the identification of gene organization of alpha conotoxins to the cloning of novel toxins, *Toxicon*, 49, 1135, 2007.

115. Kang, T.S. et al., Effect of C-terminal amidation on folding and disulfide-pairing of alpha-conotoxin ImI, *Angew. Chem. Int.*, 44, 6333, 2005.

116. Price-Carter, M., Gray, W.R., and Goldenberg, D.P., Folding of omega-conotoxins 2. Influence of precursor sequences and protein disulfide isomerase, *Biochemistry*, 35, 15547, 1996.

117. Lopez-Vera, E. et al., Role of hydroxyprolines in the in vitro oxidative folding and biological activity of conotoxins, *Biochemistry*, 47, 1741, 2008.

118. Buczek, O. et al., Characterization of D-amino-acid-containing excitatory conotoxins and redefinition of the I-conotoxin superfamily, *FEBS J.*, 272, 4178, 2005.

119. Jimenez, E.C., Watkins, M., and Olivera, B.M., Multiple 6-bromotryptophan residues in a sleep-inducing peptide, *Biochemistry*, 43, 12343, 2004.

120. Loughnan, M. et al., Alpha-conotoxin EpI, a novel sulfated peptide from *Conus episcopatus* that selectively targets neuronal nicotinic acetylcholine receptors, *J. Biol. Chem.*, 273, 15667, 1998.

121. Clark, R.J. et al., Cyclization of conotoxins to improve their biopharmaceutical properties, *Toxicon*, 59, 446, 2012.

122. Yamamoto, T. and Takahara, A., Recent updates of N-type calcium channel blockers with therapeutic potential for neuropathic pain and stroke, *Curr. Top. Med. Chem.*, 9, 377, 2009.

123. Green, B.R. et al., Conotoxins containing nonnatural backbone spacers: Cladistic-based design, Chemical synthesis, and improved analgesic activity, *Chem. Biol.*, 14, 399, 2007.

124. Armishaw, C.J. et al., A synthetic combinatorial strategy for developing alpha-conotoxin analogs as potent alpha7 nicotinic acetylcholine receptor antagonists, *J. Biol. Chem.*, 15, 1809, 2010.

125. Dekan, Z. 4th et al., α-Conotoxin ImI incorporating stable cystathionine bridges maintains full potency and identical three-dimensional structure, *J. Am. Chem. Soc.*, 133, 15866, 2011.

126. MacRaild, C.A. et al., Structure and activity of (2,8)-dicarba-(3,12)-cystino alpha-ImI, an alpha-conotoxin containing a nonreducible cystine analogue, *J. Med. Chem.*, 52, 755, 2009.

127. Lovelace, E.S. et al., Stabilization of α-conotoxin AuIB: Influences of disulfide connectivity and backbone cyclization, *Antioxid. Redox. Signal.*, 14, 87, 2011.

128. Halai, R. et al., Effects of cyclization on stability, structure, and activity of α-conotoxin RgIA at the α9α10 nicotinic acetylcholine receptor and GABA(B) receptor, *J. Med. Chem.*, 54, 6984, 2011.

129. Clark, R.J. et al., The synthesis, structural characterization, and receptor specificity of the alpha-conotoxin Vc1.1, *J. Biol. Chem.*, 281, 23254, 2006.

130. Bingham, J.P. et al., Optimizing the connectivity in disulfide-rich peptides: Alpha-conotoxin SII as a case study, *Anal. Biochem.*, 338, 48, 2005.

131. McIntosh, J.M. et al., A nicotinic acetylcholine receptor ligand of unique specificity, alpha-conotoxin ImI, *J. Biol. Chem.*, 269, 16733, 1994.

132. Sharpe, I.A. et al., Allosteric alpha 1-adrenoceptor antagonism by the conopeptide rho-TIA, *J. Biol. Chem.*, 278, 34451, 2003.

133. Chi, S.W. et al., Solution conformation of alpha A-conotoxin EIVA, a potent neuromuscular nicotinic acetylcholine receptor antagonist from *Conus ermineus*, *J. Biol. Chem.*, 278, 42208, 2003.

134. Craig, A.G. et al., An O-glycosylated neuroexcitatory *Conus* peptide, *Biochemistry*, 37, 16019, 1998.

135. Jimenez, E.C., Sasakawa, N., and Kumakura, K., Effects of sodium channel-targeted conotoxins on catecholamine release in adrenal chromaffin cells, *Phil. J. Sci.*, 137, 127, 2008.

136. Fainzilber, M. et al., A new conotoxin affecting sodium current inactivation interacts with the δ-conotoxin receptor site, *J. Biol. Chem.*, 270, 1123, 1995.

137. McGivern, J.G., Ziconotide: A review of its pharmacology and use in the treatment of pain, *Neuropsychiatr. Dis. Treat.*, 3, 69, 2007.

138. Naranjo, D., Inhibition of single *Shaker* K channels by kappa-conotoxin-PVIIA, *Biophys. J.*, 82, 3003, 2002.

139. Fainzilber, M. et al., γ-Conotoxin-PnVIIA, a γ-carboxyglutamate-containing peptide agonist of neuronal pacemaker cation currents, *Biochemistry*, 36, 1470, 1998.

140. Lirazan, M.B. et al., The spasmodic peptide defines a new conotoxin superfamily, *Biochemistry*, 37, 1583, 2000.

141. Sharpe, I.A. et al., Inhibition of the norepinephrine transporter by the venom peptide chi-MrIA. Site of action, Na$^+$ dependence, and structure–activity relationship, *J. Biol. Chem.*, 278, 40317, 2003.

142. Aguilar, M.B. et al., I-Conotoxins in vermivorous species of the West Atlantic: Peptide sr11a from *Conus*, *Peptides*, 28, 18, 2007.

143. Aguilar, M.B. et al., Peptide sr11a from *Conus spurius* is a novel peptide blocker for Kv1 potassium channels, *Peptides*, 31, 1287, 2010.

144. Sabareesh, V. et al., Characterization of contryphans from *Conus loroisii* and *Conus amadis* that target calcium channels, *Peptides*, 27, 2647, 2006.

145. Teichert, R.W. et al., Novel conantokins from *Conus parius* venom are specific antagonists of *N*-methyl-D-aspartate receptors, *J. Biol. Chem.*, 282, 36905, 2007.

146. Aguilar, M.B. et al., Conorfamide-Sr2, a gamma-carboxyglutamate-containing FMRF amide-related peptide from the venom of *Conus spurius* with activity in mice and mollusks, *Peptides*, 29, 186, 2008.

147. Petrauskas, L.E., A case of cone shell poisoning by "Bite" in Manus Island, *PNG Med. J.*, 1, 67, 1955.

148. Likeman, R.K., Turtle meat and cone shell poisoning, *PNG Med. J.*, 18, 125, 1975.

42 *Clostridium perfringens* Epsilon Toxin

Michel R. Popoff

CONTENTS

42.1 INTRODUCTION

Based on the toxins produced, *Clostridium perfringens* is responsible for diverse pathologies in man and animals, resulting from a gastrointestinal or wound contamination and including food poisoning, enteritis, necrotic enteritis, enterotoxemia, gangrene, and puerperal septicemia. Among the various toxins produced by *C. perfringens*, epsilon toxin (ETX) is one of the most potent toxins known. Its lethal activity ranges just below the botulinum neurotoxins. Indeed, the lethal dose by intraperitoneal injection in mice is 1.2 ng/kg for botulinum neurotoxin A and 70 ng/kg for ETX (Table 42.1).[1,2] For this reason, ETX is considered as a potential biological weapon classified as biological agent of the category B, although very few ETX-mediated natural diseases have been reported in humans.[3] ETX belongs to the family of Aerolysin pore-forming toxins; however, its precise mode of action accounting for its high potency is not yet fully understood.[4] Recent reviews have been focused on ETX.[5,6]

42.2 CHARACTERISTICS OF ETX

42.2.1 CHARACTERISTICS OF THE AGENT

C. perfringens is a Gram-positive, rod-shaped, anaerobic, and sporulating bacterium, which produces the largest number of toxins compared to other bacteria. According to the main lethal toxins (alpha, beta, epsilon, and iota toxins),

C. perfringens is divided into five toxinotypes (A to E) (Table 42.2). ETX is synthesized by toxinotypes B and D. However, the high diversity of toxin combinations, which can be produced by *C. perfringens* strains, makes more complex the classification into five toxinotypes.[7]

C. perfringens is widespread in the environment (soil, dust, sewage, sediments, cadavers, and also the intestinal content of healthy human and animals) and is pathogenic to humans and animals (Table 42.2). *C. perfringens* strains producing only alpha toxin as main toxin are the most common in the environment. The ETX-producing strains are mainly isolated from diseased animals.

C. perfringens grows easily in the complex media containing peptones, preferentially from meat. The multiplication rate is very rapid, and growth temperature ranges from 15°C to 50°C, with an optimum temperature of 40°C–45°C. The generation time (7 min at 41°C in optimum conditions) is one of the shortest reported for any bacterium. Gas is abundantly produced during growth. The colonies on sheep blood agar are characteristics. They are round, smooth, 2–5 mm in diameter, and surrounded with two areas of hemolysis. A large area of partial hemolysis is due to the alpha toxin, and a small area of total hemolysis just around the colonies is caused by the theta toxin. *C. perfringens* is an obligate anaerobic bacterium, but it is aerotolerant. Cultures can start at redox values of E_h of +350 mV which reach after growth until −400 mV. *C. perfringens* is glucidolytic and proteolytic. *C. perfringens*

TABLE 42.1
Toxicity Comparison of Selected Toxins according to References 128,129

Toxin	Origin	Mouse Lethal Dose (µg/kg)
Botulinum toxin A	*Clostridium*	0.0003
Tetanus toxin	*Clostridium*	0.001
C. perfringens epsilon toxin	*Clostridium*	0.1
C. sordellii lethal toxin	*Clostridium*	0.1
Diphtheria toxin	Bacterium (*Corynebacterium*)	0.16 (guinea pig)
C. perfringens beta toxin	*Clostridium*	0.4
C. difficile toxin A	*Clostridium*	0.5
Taipotoxin	Snake	2
Ricin	Plant	3
Conotoxin	Mollusc	5
C. perfringens alpha toxin	Clostridium	5
Tetradotoxin	Fish	8
Saxitoxin	Dinoflagellae	9
Alpha-latrotoxin	Spider	10
Beta-bungarotoxin	Snake	14
C. perfringens theta toxin (perfringolysin)	*Clostridium*	16
Cobratoxin	Snake	75
C. perfringens enterotoxin	*Clostridium*	80
Curare	Plant	500
DFP	Toxic gas	1,000
Sodium cyanide	Chemical	10,000

rarely sporulates on regular culture media. Sporulation is achieved on special sporulation media and is variable according to the strains. Toxins are produced during the exponential growth, except the enterotoxin which is only synthesized during the sporulation.

42.2.2 ETX GENETICS

C. perfringens contains a single circular chromosome, which shows some degree of diversity between strains, with strains from gastrointestinal origin harboring a large number of mobile elements probably acquired by horizontal transfer in the digestive ecosystem.[8] The ETX gene is located on large plasmids in *C. perfringens* like the other main toxin genes (beta toxin and iota toxin genes), which are used for *C. perfringens* typing. This accounts for the great genetic diversity of *C. perfringens* strains, since plasmids can be acquired, rearranged, or lost. Thus, a *C. perfringens* strain can change from one toxinotype to another by acquisition or loss of a toxigenic plasmid. The ETX gene is harbored by diverse plasmids.[9] At least five different plasmids (48–110 kb) in *C. perfringens* type D[10] and a 65 kb plasmid in *C. perfringens* type B have been described.[11] The same strain can contain several toxin plasmids. Indeed, most of *C. perfringens* type B strains carry a 65 kb ETX plasmid, a 90 kb beta toxin plasmid, and a third plasmid with the lambda protease gene, or some *C. perfringens* type D strains contain a ETX plasmid and a smaller one with beta2 toxin gene.[10,11] On the other hand, a single plasmid can harbor several toxin genes. For example, some plasmid from *C. perfringens* type D can encode three toxins, ETX, *C. perfringens* enterotoxin, and beta2 toxin.[10] Additional genes located on these large plasmids encoding for specific metabolism pathways or potential virulence factors such as collagen adhesin or sortase might be responsible for the adaptation of distinct *C. perfringens* toxinotypes to specific ecological niches, for example, *C. perfringens* type D in the digestive tract of ruminants and more specifically of ovine. Moreover, plasmids in *C. perfringens* type B and D contain insertion sequences, which can mobilize toxin genes between different plasmids or between plasmid and chromosome or vice versa like for *C. perfringens* enterotoxin gene.[10–13] Plasmids carrying ETX gene in *C. perfringens* type B and D have probably evolved from a common ancestor plasmid by insertion of mobile genetic elements.[14] In addition, plasmids with an ETX gene from *C. perfringens* type D contain the *tcp* conjugative locus and are conjugative and can be transferred in other *C. perfringens* type strains such as *C. perfringens* type A strain.[9] The horizontal transfer of toxin plasmids contributes to the genetic complexity and plasticity of *C. perfringens* toxinotypes.

TABLE 42.2
Clostridium perfringens Toxinotypes and Associated Diseases

Toxins									
Alpha	Beta	Epsilon	Iota	Enterotoxin (CPE)	Beta2	TpeL	NetB	Typing	Associated Diseases
+	−	−	−	−	+/−	−	−	A	Humans: gangrene
+	−	−	−	+	+/−	−	−	A	Humans: food poisoning, Animals: enteritis
+	−	−	−	−	+/−	−	+	A	Poultry: necrotic enteritis
+	+	+	−	+/−	+/−	+	−	B	Animals: diarrhea, enteritis
+	+	−	−	+/−	+/−	+/−	−	C	Humans and animals: necrotic enteritis
+	−	+	−	+/−	+/−	+/−	−	D	Animals: enterotoxaemia
+	−	−	+	+/−	+/−	−	−	E	Animals: enterotoxaemia

42.2.3 ETX Production

ETX is synthesized during the exponential growth phase of *C. perfringens* as a single protein containing a signal peptide (32 N-terminal amino acids). ETX synthesis has been found to be under the control of an Agr-like quorum sensing system.[15,16] The secreted protein (32,981 Da) is poorly active and it is called prototoxin.[17] The prototoxin is activated by proteases such as trypsin, α-chymotrypsin, and λ-protease, which are produced by *C. perfringens*. Activation by λ-protease is comparable to that obtained with trypsin plus α-chymotrypsin. The λ-protease removes 11 N-terminal and 29 C-terminal residues whereas trypsin plus α-chymotrypsin cleaves 13 N-terminal residues and the same C-terminal amino acids. This results in a reduction of size (28.6 kDa) and an important decrease in p*I* from 8.02 to 5.36, accompanied probably by a conformational change. The charged C-terminal residues could prevent the interaction of the protein with its substrate or receptor.[2]

42.2.4 ETX Structure

At the amino acid sequence level, ETX shows some homology with the *Bacillus sphaericus* mosquitocidal toxins Mtx2 and Mtx3, with 26% and 23% sequence identity, respectively.[17] Mtx2 and Mtx3 are toxins specific of mosquito larvae, which are activated by proteolytic cleavage and which probably act by pore formation.[18] In addition, a hypothetical protein encoded by a gene located in the vicinity of the C2 toxin genes on a large plasmid in *C. botulinum* type D shows a sequence similarity with that of ETX.[19]

ETX retains an elongated form and contains three domains, which are mainly composed of β-sheets[20] (Figure 42.1). Despite poor sequence identity (14%), the ETX overall structure is significantly related to that of the pore-forming toxin aerolysin produced by *Aeromonas* species,[21,22] and to the model of alpha toxin from *Clostridium septicum*, an agent of gangrene.[23] However, ETX is a much more potent toxin with a 100 times more lethal activity in mice, than aerolysin and *C. septicum* alpha toxin.[2,21,24] The main difference between both toxins is that the aerolysin domain I, which is involved in initial toxin interaction with cells, is missing in ETX. Domain 1 of ETX consists in a large α-helix followed by a loop and three short α-helices and is similar to domain 2 of aerolysin, which interacts with the glucosyl phosphatidylinositol (GPI) anchors of proteins. This domain of ETX could have a similar function of binding to the receptor. A cluster of aromatic residues (Tyr49, Tyr43, Tyr42, Tyr209, and Phe212) in ETX domain 1 has been hypothesized to be involved in receptor binding by analogy with other binding domains.[25] Using site-specific mutagenesis, the aromatic amino acids Tyr29, Tyr30, Tyr36, and Tyr129 have been identified to play a critical role in ETX binding to cell membrane.[26] Domain 2 is a β-sandwich structurally related to domain 3 of aerolysin. This domain contains a two-stranded sheet with an amphipathic sequence predicted to be the channel-forming domain (see the following). In contrast to the cholesterol-dependent cytolysins, only one amphipathic β-hairpin from each monomer is involved in the pore structure of ETX and other heptameric β-pore-forming toxins (β-PFTs) like aerolysin. Domain 3 is also a β-sandwich analogous to domain 4 of aerolysin and contains the cleavage site for toxin activation. Domain 3, after removing the C-terminus, is likely involved in monomer–monomer interaction required for oligomerization.[4,20] In addition, ETX is also structurally related to *C. perfringens* enterotoxin, another pore-forming toxin produced by some *C. perfringens* strains, despite a low sequence identity. In contrast to ETX and aerolysin, the receptor binding domain of *C. perfringens* enterotoxin is located in the C-terminal part.

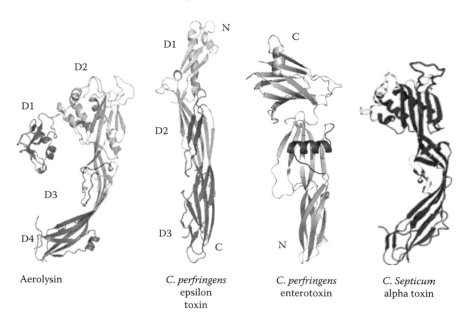

FIGURE 42.1 Structures of aerolysin, *C. perfringens* ETX, *C. perfringens* enterotoxin, and structure model of *C. septicum* alpha toxin. N—N-terminal part: C—C-terminal part.

These toxins probably share a common mechanism of insertion into membranes since the domains involved in pore formation, the ETX and aerolysin C-terminal domains and the corresponding N-terminal domain of *C. perfringens* enterotoxin, share a high level of structure identity.[27]

The pore-forming domain has been identified in domain 2. The segment His151-Ala181 contains alternate hydrophobic–hydrophilic residues, which are characteristic of membrane-spanning β-hairpins, and forms two amphipathic β-strands on ETX structure. Site-directed mutagenesis confirmed that this segment is involved in ETX channel activity in lipid bilayers.[28] Interestingly, the ETX pore-forming domain shows higher sequence similarity to those of the binding components (Ib, C2-II, CDTb, CSTb) of clostridial binary toxins (iota toxin, C2 toxin, *Clostridium difficile* transferase (CDT), *Clostridium spiroforme* toxin (CST), respectively), and to a lesser extent to *B. anthracis* protective antigen (PA, the binding component of anthrax toxins), than with that of aerolysin (Figure 42.2). However, the ETX segment Lys162 to Glu169, which is exposed to the transmembrane side of the channel and forms the loop linking the two β-strands forming the transmembrane β-hairpin, is unrelated at the amino acid sequence level to those of other β-PFTs. The ETX loop is flanked by two charged residues, Lys162 and Glu169, and

contains a proline in the central part, similarly to the sequence of the corresponding aerolysin loop. Binding components share a similar structure organization with that of β-PFTs and notably contain an amphipathic flexible loop that forms a β-hairpin, playing a central role in pore formation.[29,30] This suggests that binding components and β-PFTs have evolved from a common ancestor. However, β-PFTs have acquired a specific function consisting in the translocation of the corresponding enzymatic components of binary toxins through the membrane of endosomes at acidic pH. In contrast, β-PFTs such as ETX and aerolysin can form pores in plasma membrane at neutral pH, which are responsible for cytotoxicity.

Essential amino acids for the lethal activity have been identified by biochemistry and mutagenesis. A previous work with chemical modifications shows that His residues are required for the active site, and Trp and Tyr residues are necessary for the binding to target cells.[31] The molecule contains a unique Trp and two His. Amino acid substitutions showed that His106 is important for the biological activity, whereas His149 and Trp190 probably are involved to maintain the structure of ETX, but they are not essential for the activity.[32]

42.3 MECHANISM OF ACTION AND EPIDEMIOLOGY

42.3.1 MOLECULAR AND CELLULAR MECHANISM OF ACTION OF ETX

Specific activity of ETX is observed in cultured cells. Only very few cell lines including renal cell lines from various species such as Madin–Darby canine kidney (MDCK), mpkCCD$_{cl4}$, and to a lesser extent the human leiomyoblastoma (G-402) cells are sensitive to ETX.[33,34] Surprisingly, kidney cell lines from ETX-susceptible animal species, like from lamb and cattle, are ETX resistant, suggesting that the ETX receptor in primary cells is lost in cultured cell lines (Ref. [33] and unpublished). ETX exhibits a single binding site of high affinity to susceptible cells or synaptosomes with a dissociation constant (K_d) about 3–4 nM.[35,36]

A marked swelling is observed in the first phase of intoxication, followed by mitochondria disappearance, blebbing, and membrane disruption. The cytotoxicity can be monitored by using an indicator of lysosomal integrity (neutral red) or mitochondrial integrity (3-(4,5-dimethylthiazol-2-yl)-2, 5-diphenyltetrazolium bromide MTT).[33,34,37–40]

ETX binds to MDCK cell surface, preferentially to the apical site, and recognizes a specific membrane receptor, which is not present in insensitive cells. Binding of the toxin to its receptor leads to the formation of large membrane complexes which are very stable when the incubation is performed at 37°C. In contrast, the complexes formed at 4°C are dissociated by SDS and heating. This suggests a maturation process like a prepore and then a functional pore formation.[39] Indeed, a prepore stage has been identified, which is followed by the insertion of the prepore into the membrane to form a functional pore.[39] Endocytosis and internalization of the toxin into the cell was not observed, and the toxin remains

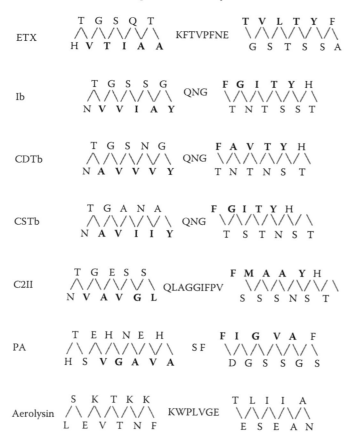

FIGURE 42.2 Sequence alignments of the pore-forming domains of ETX, binding components of the clostridial binary toxins (*C. perfringens* Iota toxin [Ib], *C. difficile* transferase [CDTb], *C. spiroforme* toxin [CSTb], *C. botulinum* C2 toxin [C2-II]), and aerolysin.

associated to the cell membrane throughout the intoxication process.[41] However, a recent report shows that ETX induces cell vacuolation, which is dependent on vacuolar ATPase and could reflect toxin internalization.[42] The ETX large membrane complex in MDCK cells and synaptosomes correspond to the heptamerization of toxin molecules within the membrane and pore formation.[39,43,44] ETX prototoxin is able to bind to sensitive cells but does not oligomerize, in contrast to activated ETX. Thus, the 23 C-terminal residues of the prototoxin control the toxin activity by preventing the heptamerization. These amino acids are removed in the active toxin molecule.[43]

ETX binding to susceptible cells or synaptosomes and subsequent complex formation are prevented by protease treatment but not or weakly by phospholipase C, glycosidases, or neuraminidase, indicating the protein nature of the ETX receptor.[36,39,45,46] In contrast, *C. perfringens* neuraminidase I (NanI) has been reported to increase (about 25%) ETX cytotoxicity on MDCK cells possibly by unmasking receptors on cell surface.[47] The ETX receptor could be related to a 34 or 46 kDa protein or O-glycoprotein in MDCK cells[39,46,48] and to a 26 kDa sialoglycoprotein in the rat brain.[36] Hepatitis A virus cellular receptor 1 (HAVCR1) has been identified to facilitate ETX cytotoxicity in MDCK cells and the human kidney cell line ACHN. ETX binds to HAVCR1 in vitro.[49] However, it is not yet clear whether HAVCR1 is a functional ETX receptor. Moreover, although ETX does not directly interact with a lipid, the lipid environment of the ETX receptor is critical for the binding of ETX to cell surface, since detergent treatment or lipase prevents ETX binding to the cell surface.[36,44,46,50] It is noteworthy that ETX can interact with artificial lipid bilayers and form functional channels, without the requirement of a specific receptor in contrast to cell membrane, albeit less efficiently, compared to MDCK cells. Lipid bilayers have smooth surfaces without any surface structure including the surface-exposed carbohydrates and proteins of biological membranes, which means that the toxins can interact with the hydrocarbon core of the lipid bilayer and can insert without the help of receptors, whereas in general receptors are required to promote such an interaction in cell membrane.[51]

In synaptosomes and MDCK cells, the ETX receptor has been localized in lipid raft microdomains, which are enriched in certain lipids such as cholesterol and sphingolipids as well as in certain proteins like GPI-anchored proteins, suggesting that such a protein could be an ETX receptor.[44,52] However, in contrast to aerolysin and *C. septicum* alpha toxin, ETX does not interact with a GPI-anchored protein as a receptor, since phosphatidylinositol-specific phospholipase C did not impair binding or ETX complex formation.[52] Localization of ETX receptor in lipid microdomains is further supported by the fact that ETX prototoxin and active form bind preferentially to detergent-resistant membrane fractions (DRM) and only activated ETX forms heptamers in DRM.[44] In addition, membrane cholesterol removal with MβCD impairs ETX binding and pore formation.[44,52,53] The composition of lipid rafts in sphingomyelin and gangliosides as well as

membrane fluidity influences ETX binding to sensitive cells, heptamerization, and cytotoxicity.[54,55] Thus, inhibitors of sphingolipid or glycosphingolipid synthesis increase cell susceptibility to ETX, whereas inhibitor of sphingomyelin synthesis or addition of GM1 dramatically decreases ETX binding and subsequent heptamerization.[54] Moreover, phosphatidylcholine (PC) molecules which increase membrane fluidity, facilitate ETX binding and assembly.[55] ETX bound to its receptor shows a confined mobility on cell membrane, probably permitting interaction between ETX monomers and subsequent oligomerization.[56,57] Local lipid composition and membrane fluidity likely control ETX bound to its receptor in cell membrane. In addition, lipids such as diacylglycerol and phosphatidyl ethanolamine, which induce a negative membrane curvature, increase ETX pore formation in liposome, whereas lipids having an opposite effect like lyso-PC impair ETX activity.[55] This is consistent with the model of an ETX prepore formation and subsequent insertion into the membrane to form a functional channel. The structure of an ETX pore has been defined as a cone shape,[58] and thus its insertion in lipid bilayer might be favored by a specific lipid membrane organization. Therefore, although ETX does not directly bind to a lipid receptor, the lipid composition and physical properties of membrane influence ETX access to the receptor, ETX monomer assembly, and insertion of the ETX pore in the membrane. Moreover, other cellular proteins, like caveolin-1 and caveolin-2, seem to interact with ETX and to be involved in toxin oligomerization by a yet unknown mechanism.[59]

The cytotoxicity is associated with a rapid loss of intracellular K^+, and an increase in Cl^- and Na^+, whereas the increase in Ca^{2+} occurs later. In addition, the loss of viability also correlates with the entry of propidium iodide, indicating that ETX forms large pores in the cell membrane. Pore formation is evident in artificial lipid bilayer. ETX induces water-filled channels permeable to hydrophilic solutes up to a molecular mass of 1 kDa, which represent general diffusion pores slightly selective for anions.[51] In polarized MDCK cells, ETX induces a rapid and dramatic increase in permeability. Pore formation in the cell membrane is likely responsible for the permeability change of cell monolayers. Actin cytoskeleton and organization of tight and adherens junctions are not altered, and the paracellular permeability to macromolecules is not significantly increased upon ETX treatment.[52,60] ETX causes a rapid cell death by necrosis characterized by a marked reduction in nucleus size without DNA fragmentation. Toxin-dependent cell signaling leading to cell necrosis is not yet fully understood and includes ATP depletion, AMP-activated protein kinase stimulation, mitochondrial membrane permeabilization, and mitochondrial-nuclear translocation of apoptosis-inducing factor, which is a potent caspase-independent cell death factor (Figure 42.3).[52] The early and rapid loss of intracellular K^+ induced by ETX and also by *C. septicum* alpha toxin seems to be the early event leading to cell necrosis.[61] It is intriguing that ETX, which has a pore-forming activity related to that of aerolysin and *C. septicum* alpha toxin, is much more active.

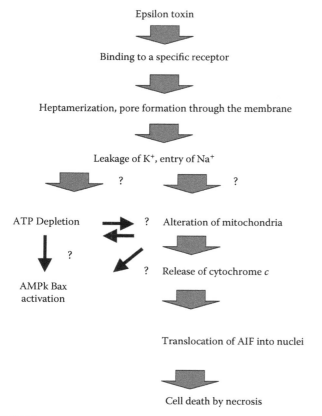

FIGURE 42.3 Main steps of ETX mode of action (? indicates probable but not yet definitively confirmed signaling pathways).

Does ETX induce a specific intracellular signal responsible for a rapid cell death? MβCD, which prevents ETX pore formation in lipid rafts, does not inhibit the sudden decrease in cellular ATP and cell necrosis.[52] A subset of ETX channels unaffected by MβCD might be sufficient to trigger an intracellular signal leading to cell necrosis, excluding the requirement of a large diffusion pore to induce the intracellular toxic program. Therefore, ETX is a very potent toxin which alters the permeability of cell monolayers such as epithelium and endothelium, causing edema and cell death; however, its precise mode of action remains unclear.

42.3.2 EPIDEMIOLOGY

In the natural conditions, ETX-producing *C. perfringens* are essentially associated with diseases in animals. Toxinotype B is the causative agent of lamb dysentery, which is only found in some countries like the United Kingdom, whereas toxinotype D is responsible for a fatal, economically important disease of sheep worldwide, called enterotoxemia. ETX contributes with beta toxin to the pathogenesis of toxinotype B, and it is the causative virulence factor of all symptoms and lesions due to toxinotype D.

The dramatic diseases induced by *C. perfringens* ETX in certain animal species raises the question if humans might also be a target of this toxin? Primary human renal tubular epithelial cells (HRTEC) and a human kidney cell line, G-402, are sensitive to ETX albeit to a lower extent than the highly sensitive

dog kidney cell line MDCK,[34,62] suggesting that human might be susceptible to ETX. However, *C. perfringens* type D disease is extremely rare in humans, even in farmers or other persons in contact with diseased animals or their environment. Two reports mention a *C. perfringens* type D infection in men. One concerned a person with acute intestinal obstruction and subsequent development of *C. perfringens* type D and production of ETX in the intestinal content. A portion of the ileum was gangrenous and bloodstained fluid was present in the peritoneal cavity.[63] A second case hospitalized for treatment of ankylosing spondylitis developed abundant diarrhea and abdominal pain. *C. perfringens* type D was isolated from stool and antibodies against ETX were evidenced in the serum.[64]

42.4   CLINICAL FEATURES AND PATHOGENESIS

42.4.1 ENTEROTOXEMIA

The typical disease caused by ETX is the enterotoxemia which is observed in animals, mainly in lambs.

Enterotoxemia is characterized by a high production of toxin in the intestine, which then passes through the intestinal barrier and disseminates by the blood circulation (toxemia) to several organs, causing a toxic shock and death. The natural habitat of *C. perfringens* type D, like for the other toxinotypes, is the environment: soil, dust, sediment, cadavers, litter, and also the digestive tract of healthy animals. *C. perfringens* is not a usual inhabitant of the digestive tract. However, it can be found in low number ($<10^3$ bacteria/g) in the intestinal content of animals without associated pathology.[65,66]

High production of ETX in the intestine and subsequent disease is conditioned by an overgrowth of ETX-producing *C. perfringens* (more than 10^6 bacteria/g, usually 10^8–10^9 bacteria/g) in the intestinal content, essentially in the small intestine. Rapid multiplication of *C. perfringens* can occur in the digestive tract of very young animals, in which the resident intestinal microflora, which is inhibitory of *C. perfringens* colonization, is not yet developed or not yet functional. This is the case of lamb dysentery due to *C. perfringens* type B, which only occurs in the first days of life. Overeating of a highly concentrated ration or a rapid change to a rich diet such as ration rich in cereal, young cereal crops, or abundant and luxuriant pasture is a common cause of enterotoxemia in older lambs and sheep. Such alimentary conditions induce a perturbation of the microbial balance in the gut and a massive passage into the small intestine of undigested fermentescible carbohydrates, like starch, which are normally metabolized in the rumen and which are an excellent substrate of *C. perfringens* growth. In addition, any cause of intestinal stasis contributes to accumulation of *C. perfringens* and ETX in the intestinal loops.

C. perfringens type D enterotoxemia is very common in lambs, less frequent in sheep and goats, and occasional in other animal species. The most rapidly growing lambs are the most susceptible. This raises the question, which are the host intestinal conditions permitting the selective *C. perfringens* type D overgrowth in the digestive tract of susceptible animals compared to more resistant animal species?

C. perfringens type D enterotoxemia, also called pulpy kidney disease in lambs, is rapidly fatal. The peracute clinical form is characterized by a sudden death without premonitory signs. In the acute form, which is very brief (few minutes to several hours, no more than 12 h), the neurological symptoms of excitatory type are predominant and include violent convulsions, opisthotonos, struggling, nystagmus, bruxism, ataxia, and then lateral recumbency, violent movements of paddling, ptyalism, hyperthermia, and coma. Sheep usually develop a more chronic form, also called focal symmetrical encephalomalacia. Diarrhea might be observed in addition to neurological signs in animals surviving for few days. In contrast to sheep, fibrotic and hemorrhagic enterocolitis in the absence of cerebral lesions is more common in goats.[67–69]

In the peracute form, only a few lesions can be observed such as microscopic brain lesions, edema, and petechia in various organs, including pericardial effusions, subendothelial ecchymoses, and occasionally pulmonary and pleural effusions. Macroscopic brain lesions are more evident in the animals with longer duration of the disease and consist in symmetrical foci of hemorrhagic or gelatinous softenings in the corpus striatum, thalamus, and midbrain cerebellar peduncles. Lambs dying from enterotoxemia show characteristic modifications of the kidneys. Just after death, the kidneys are swollen or only congestive, but they autolyse more rapidly than normal, with the cortical parenchyma totally liquefied. In addition, hyperglycemia and glucosuria are frequently found.[71–75] In mouse kidneys, specific binding sites for ETX have been identified in distal and collecting tubule cells.[48]

42.4.2 Intestinal Absorption of ETX and Dissemination through Blood Circulation

Sheep can support ETX accumulation (10^2–10^3 mouse lethal dose (MLD)/mL) in the intestine without associated symptoms for a few hours. Then, high ETX concentration induces an increase in the intestinal mucosa permeability, mediating its passage into the blood circulation.[66,75–78] In experimental mice and rat intestinal loops, ETX at concentrations of 10^3 MLD/mL and higher causes fluid accumulation into the intestinal lumen, decrease in transepithelial electrical resistance (TER), and increase in the passage of macromolecules through the intestinal barrier.[79–81] The absence of histological and ultrastructural changes in the intestinal epithelium suggests an increased passage through the paracellular pathway. The only observed lesions consist in paravascular edema and apoptotic cells in the lamina propia.[80] Subsequent ETX absorption in the general circulation occurs from small and large intestine but not from the stomach in mice.[82] However, the precise mechanism of ETX-dependent increased permeability of the intestinal barrier remains to be defined.

ETX also influences the gastrointestinal motility. Contradictory results have been obtained in experimental animal models. Indeed, ETX was found to cause contraction of isolated rat ileum, as a consequence of an indirect ETX action via the nervous system.[83] However, ETX administered orally or intravenously in mice reduces the gastrointestinal transit by

a yet undefined mechanism.[84] Inhibition of the gastrointestinal motility is a risk factor, which facilitates bacterial overgrowth and toxin accumulation in the intestinal lumen.

Edema and petechias, which have been observed in various tissues from naturally or experimentally intoxicated animals, indicate that ETX targets endothelial cells and alters the vascular barrier integrity. Indeed, it has been evidenced that ETX efficiently increases the vascular permeability of microvessels of rat mesentery,[85] or that of skin vessels after intradermal ETX injection.[86] These effects seem to result from a direct ETX interaction with endothelial cells and not from an indirect signaling cascade such as an inflammatory response possibly induced by ETX.[85] Fluorescent ETX injected intravenously in mice binds to the luminal surface of endothelium of most blood vessels.[86,87] Observation of necrotic cells and gaps in endothelium indicates that ETX modifies the integrity of the endothelial barrier by destruction of cells rather than by disassembly of intercellular junctions.[85] However, endothelial cell lines from various animal species responsive to *C. perfringens* type D enterotoxemia, which have been tested, are not sensitive to ETX.[88] Possibly cultured cell lines have lost the specific ETX receptor of primary endothelial cells. Moreover, ETX has been reported to increase blood pressure subsequently to vessel contraction in skin by a yet undefined mechanism, possibly including an increased membrane permeability to Na^+ and release of noradrenaline from adrenergic nerve terminals.[89,90] Targeting of endothelial cells and increased endothelial barrier permeability seem to be among the major ETX effects in susceptible animal species. The mechanism and importance of hemodynamic alteration remain to be defined.

42.4.3 ETX and Kidney Disorders

Rapid postmortem autolysis of kidneys is characteristic of lamb enterotoxemia (pulpy kidney disease) and is less evident in sheep and other animal species. At the time of death of lambs intoxicated with ETX, only a variable degree of congestion is observable in kidneys. At 2 h and in a more pronounced manner at 4 h postmortem, kidneys show interstitial hemorrhages between tubules and degeneration of the epithelium of proximal tubules.[70] Similar findings are observed in mice, which show congestion and hemorrhages in the medulla as well as severe degeneration of the distal tubule epithelium.[87] ETX specifically binds to the basolateral side of epithelial cells of distal tubules, in agreement with the degenerative effects in this epithelium, and also to the luminal surface of proximal tubules but in a nonspecific manner, indicating a filtration of the toxin by the glomerules.[87,91] Interestingly, nephrectomy shortens the time to kill the mice injected intravenously with ETX, suggesting that kidneys have a protective role by trapping the toxin from the blood circulation and eliminating it from the organism.[91] In addition, only a few cultured cell lines are sensitive to ETX *in vitro*, and these cell lines are from kidneys such as kidney cells from dog (MDCK), the murine renal cortical collecting duct principal cell line (mpkCCD$_{c14}$), and to a lower extent the renal cell line from human kidney G-402,[33,34,38,52] indicating that kidney is one of the main target organs for ETX. In addition, kidney alteration is also involved in the glucosuria, which is observed in lamb enterotoxemia. However, glucosuria mostly results from a hyperglycemic response which is probably mediated by mobilization of hepatic glycogen subsequently to ETX-dependent vascular endothelial damage.[92]

42.4.4 ETX AND BRAIN DISORDERS

The brain is the second organ, after the kidneys, where ETX massively accumulates (Figure 42.1). In contrast to kidneys, ETX binds to the brain in an exclusively specific manner and with a high affinity (nM range).[36,93] Therefore, this indicates that ETX passes the blood–brain barrier and recognizes specific cells or sites in the brain. Indeed, ETX has been shown to alter the blood–brain barrier integrity, permitting not only its own passage but also that of macromolecules such as horseradish peroxidase or serum albumin.[93–98] A marked extravasation of serum albumin into brain has been evidenced in lambs or rats intoxicated with ETX,[95,99] and extravasation of horseradish peroxidase as measured in the mouse brain after intravenous injection of less than one lethal dose is extremely rapid (about 20 min).[96,98] Such a rapid decrease in blood–brain barrier permeability facilitates a rapid ETX accumulation in the brain. However, the ETX mechanism of blood–brain barrier perturbation is not yet fully understood. ETX in its prototoxin form binds to brain endothelial cells and induces a decreased expression of the endothelial barrier antigen (EBA), which is a specific marker of central nervous system barrier vessels. The ETX-dependent reduction of EBA production in brain endothelial cells by a yet undefined pathway is accompanied by a rapid but mild increase in blood–brain barrier permeability.[97] Impairment of EBA expression might be an early event of ETX effect on the blood–brain barrier. Then, endothelial cells show macroscopic alterations including swelling, abundance of clear vacuoles, and loss of intracellular organelles, as well as protrusions or blebbing of the luminal surface. Then, their cytoplasm is very thin and the nuclei pyknotic leading to a very attenuated capillary endothelium.[100–102]

In mice, ETX causes bilaterally symmetrical lesions in several brain areas, mainly including cerebral cortex, corpus striatum, vestibular area, corpus callosum, lateral ventricles, cerebellum, whereas in lambs or sheep, more restricted areas are concerned such as basal ganglia, thalamus, subcortical white matter, substantia nigra, hippocampus, and cerebellar peduncles.[100,101] The most early and prominent lesions consist in perivascular edema, and they have been described in various animal species including mice,[94,98,103] rats,[99,104,105] sheep,[86,106–108] and calves.[109] A widening of perivascular space is the main early change which is observed as rapidly as 1 h after intraperitoneal injection of a sublethal dose in mice. Then the perivascular edema progresses, even leading to a stenosis of the capillary lumen.[103]

Perivascular edema is mainly distributed in white matter and is accompanied by swelling of perivascular astrocytic cells,[97,100] predominantly in the cerebellum.[70,100] Swelling is also observed in axon terminals and dendrites, and the myelin sheath is distended by edema.[102] Neuronal damages occur in some neurons, and consist in swelling, vacuolation, and necrosis mainly in neurons from certain brainstem nuclei, or in cell shrinkage with hyperchromatosis and nuclear pyknosis most commonly in the cerebral cortex, hippocampus, and thalamus.[99] A consequence of brain vasogenic edema is an overexpression of aquaporin-4 (the most abundant water channel in the central nervous system involved in water homeostasis), mainly in astrocytic cells. Upregulation of aquaporin-4 represents a host response in tempting to resolve the ETX-induced edema.[105]

In the subacute and chronic forms of the disease, the brain lesions ultimately change in foci of necrosis and hemorrhages. Two pathways might account for the generation of necrotic lesions. Impairment of the blood–brain vessels leads to vasogenic edema and reduced perfusion of the tissues and therefore to tissue hypoxia and necrosis. Alternatively, ETX, which diffuses in the brain parenchyma, can directly damage neurons and other cell types. This does not preclude that combination of both processes could be involved. However, the cells directly targeted by ETX remain to be determined.

Using fluorescent ETX injected intravenously, it has been confirmed that both activated and precursor toxin forms bind to the luminal surface of brain vascular endothelial cells.[53,110] In addition, active fluorescent ETX, but not the prototoxin, passes the blood–brain barrier and accumulates in brain tissue, preferentially in cerebellum, cerebellar peduncles, cerebral white matter, hippocampus, thalamus, corpus striatum, olfactory bulb, and colliculi.[53,110] These brain areas of ETX diffusion correspond to regions of the brain that have been already identified as the main sites of ETX-induced histological changes, supporting a direct toxin action on neuronal cells and/or other brain cells. Fluorescent ETX binds in vitro to myelin structure of the central and peripheral nervous system from mice, sheep, cattle, and even humans.[45,53] However, myelin does not seem to be the primary target of ETX since intravenously injected toxin in mice does not show a correlation of ETX staining pattern with that of myelin containing structures.[45] Moreover, it was found that fluorescent ETX binds to only a subset of astrocytes and microglia cells and it is cytotoxic for these cells.[110] However, the pathological significance of ETX on astrocytes and microglia cells is not known. A more detailed analysis of ETX binding to the mouse cerebellum has identified granule cells and oligodendrocytes, but not Purkinje cells and astrocytes, as ETX target cells.[53] ETX binding to myelin probably accounts for ETX staining of oligodendrocytes, which are involved in myelin synthesis in contrast to astrocytes, which participate in blood–brain barrier function, regulation of local pH and electrolytes, and probably in the recapture of neurotransmitter. ETX stains brain white matter, which is enriched in myelinated axons and oligodendrocyte cell bodies. It is noteworthy that ETX binds to the cell body of granule cells or other target cells but not to cell axons or nerve terminals,[45,53] suggesting a specific ETX interaction with a cell body membrane receptor.

42.4.5 ETX STIMULATES GLUTAMATE RELEASE

A direct and rapid ETX effect in brain concerns the stimulation of glutamate release. First, it was found that ETX injected at low doses in rats induces rapidly (4 h) neuronal damages characterized by cell shrinking, vacuolation, and nucleus pyknosis, mainly in the hippocampus and cortex. These effects were neither accompanied of perivascular edema nor of blood flow reduction in the hippocampus and were specifically inhibited by inhibitors of glutamate receptors, indicating that ETX interferes directly with the glutamatergic neurons.[111] Then, it was shown that ETX increases the glutamate release from the mouse hippocampus and not from other brain areas.[112] Thereby, ETX seems to target specifically glutamatergic neurons, stimulating the release of glutamate and then inducing cell alteration (shrinkage, pyknosis).

Glutamate is the most abundant excitatory neurotransmitter in the central nervous system. Its excessive release is probably the main cause of the neurological symptoms of excitation, which are observed in ETX-dependent enterotoxemia. The precise mechanism used by ETX to stimulate the glutamate release is not yet fully understood. Since ETX incubation with synaptosomal fraction containing nerve terminals did not elicit glutamate release, ETX probably interacts directly with the cell bodies of neurons or other cell types in the nervous tissue to induce its effects on neurotransmitter release.[45] Indeed, it was confirmed that in the mouse cerebellum, in addition to oligodendrocytes, ETX only binds to somata of granule cell, which are glutamatergic neurons, and not to nerve terminals, neuronal extensions, or other neuronal cell types like GABAergic Purkinje cells.[53] Alteration of blood circulation in the brain such as ischemia, is among the factors which stimulate the glutamate release. However, no cerebral blood flow modification was observed during the time period of ETX-dependent stimulation of glutamate release,[112] indicating again a direct activity of ETX on neuronal or glial cells. This is further supported by the fact that ETX induces glutamate release from primary and cultured cerebellar granule cells, although it is not ruled out that oligodendrocytes, which are also targeted by ETX, are also not involved. Moreover, as measured by patch clamp, ETX triggers membrane depolarization, leading to a decrease in membrane electrical resistance and rise in intracellular Ca^{2+}. Does ETX mediate glutamate efflux by pore formation through the membrane, plasma membrane disruption, or stimulation of the neuroexocytosis machinery? The observation that the absence of extracellular Ca^{2+} or methyl β-cyclodextrine (MβCD), which sequesters membrane cholesterol and impairs ETX pore activity, prevents ETX-dependent intracellular Ca^{2+}

rise and glutamate release, argues for a ETX activity on the release machinery of glutamate.[53] Although the most direct and prominent ETX effect on the brain is the stimulation of glutamate release, the toxin also induces the release of other neurotransmitters such as dopamine.[113]

42.5 IDENTIFICATION AND DIAGNOSIS

The diagnosis of the natural diseases due to ETX-producing *C. perfringens* is based on the identification of toxigenic *C. perfringens* in the infection site and/or on detection of ETX.

The biological confirmation of animal enterotoxemia is achieved by the isolation and typing of *C. perfringens* from the intestinal content. Since *C. perfringens* can be found in the digestive tract of healthy animals, but at a low number, the enumeration of this bacterium is a decisive factor of the diagnosis. Indeed, during an enterotoxemia the level of *C. perfringens* in the intestinal content is at least 10^6 colonies per g and most commonly ranges between 10^8 to 10^{10} colonies/g. Several culture media have been proposed for the isolation of *C. perfringens* from biological samples. One of the most used culture medium consists in sheep blood agar supplemented with D-cycloserine (400 μg/mL). After overnight incubation in anaerobic conditions, the *C. perfringens* colonies are easily recognizable by their double hemolysis area. It is noteworthy that in the terminal step of the disease, *C. perfringens* disseminates by the blood circulation in all the body, and this bacterium can be isolated from organ samples such as the liver or the spleen. However, the organ sampling has to be performed rapidly after the death, because in cadavers of more than 6–12 h the *C. perfringens* resident in the digestive tract can invade the organism, leading to a misinterpretation of enterotoxemia.

Suspected colonies on isolation medium can be easily identified as *C. perfringens* by evidence of phenotypic characters such as large Gram-positive rod, absence of mobility, production of gas, fermentation of lactose, and sulfite reduction.

The most reliable method of *C. perfringens* toxinotyping is based on genetic detection of toxin genes. Single and multiplex PCR assays have been developed for the identification of all the *C. perfringens* toxin genes. Simultaneous detection of alpha toxin and ETX genes allows the identification of the *C. perfringens* species since alpha toxin gene is conserved in all *C. perfringens* strains except very rare strains, and ETX gene characterizes the toxinotypes which potentially produce this toxin. Further genetic characterization of *C. perfringens* strains includes pulsed-field gel electrophoresis (PFGE), multilocus sequence typing (MLST), and ribotyping.

The presence of ETX gene in a *C. perfringens* strain does not necessarily indicate that the toxin is produced. However, up to now, no cryptic ETX gene has been reported. Therefore, detection of ETX in culture supernatant is required to evidence an ETX-producing strain. The standard method is based on the mouse bioassay. For that, 0.5 mL of supernatant from exponential or stationary culture is intraperitoneally injected into mice. The ETX mediated mouse lethality is identified by seroprotection with specific neutralizing antibodies against ETX. ELISA tests have been developed for ETX detection.[114–117] A monoclonal antibody against ETX is commercially available, but ELISA kit is not yet distributed. Recently, a sensitive method based on mass spectrometry has been developed.[118] The in vitro ETX detection methods can be used with environmental samples regarding a suspicion of bioterrorism threat.

A direct way to establish the diagnosis consists in the identification of ETX in biological samples from intoxicated humans or animals. In cases of sheep enterotoxemia, ETX has been detected by ELISA mainly in ileal content (92% of the samples) and to a lower extent in other biological samples such as duodenal content (64%), colon content (57%), pericardial fluid and aqueous humor (7%), but not in abdominal fluid or urine.[115] In one human natural case, ETX was recovered in the intestinal content as monitored by the mouse bioassay.[63]

A sensitive method based on mass spectrometry has been developed allowing an ETX detection as low as about 5 ng/mL in biological samples like serum and milk.[118]

42.6   TREATMENT AND PREVENTION

No specific treatment of ETX-induced disease is available and no protocol of symptomatic therapy in humans has been developed. In animals, the disease is almost always fatal.

Vaccines against toxins which disseminate through the general circulation and which target organ or tissues at distance of the gastrointestinal tract are among the most efficient. Indeed, vaccines against enterotoxemia due to *C. perfringens* ETX are extensively used in veterinary medicine.[119] Classically, toxin-based vaccines derive from chemically detoxified toxins. New approaches consist in genetically detoxified toxins such as toxin mutants or toxin subunits, which are nonbiologically active but retain the toxin immunogenicity. Recombinant ETX vaccines are under investigation.[32,119–122] For example, cysteine substitutions at Ile51–Ala114 and Val56–Phe118 yield a non-cytotoxic ETX mutant, which could be a vaccine candidate.[123]

To prevent the toxic effects of ETX, polyclonal and monoclonal antibodies have been developed. For example, monoclonal antibodies targeting an epitope close to the pore-forming domain have been found to be efficiently neutralizing.[124] Chemical inhibitors of ETX have been investigated by screening a large compound library. Three compounds, *N*-cycloalkylbenzamide, a furo(2,3-*b*)quinoline, and a 6*H*-anthra(1,9-*cd*)isoxazol, inhibit ETX channel activity and cell death but not ETX binding to cell or ETX oligomerization. These inhibitors possibly block the ETX pore or interfere with an unidentified host factor involved in ETX-dependent cytotoxicity.[125] Interestingly, these inhibitors are specific of ETX and do not prevent aerolysin cytotoxicity, arguing again for a differential mode of action between both pore-forming toxins although they are structurally related and form similar functional pores.

42.7   CONCLUSION AND FUTURE PERSPECTIVES

ETX belongs to the heptameric β-PFTs family including aerolysin and *C. septicum* alpha toxin, which are characterized by the formation of a pore consisting in a β-barrel resulting from the arrangement of 14 amphipatic β-strands.[4] Although these toxins share a similar mechanism of pore formation, ETX is much more potent than aerolysin and *C. septicum* alpha toxin. A main difference is that aerolysin and *C. septicum* alpha toxin recognize GPI-anchored proteins as receptors, whereas the ETX receptor, although localized in lipid rafts, is distinct from GPI-anchored proteins and is distributed in a limited number of cell types. The specific ETX receptor possibly accounts for the high potency of ETX, which also might be dependent on a specific intracellular signaling induced by the toxin. Another particularity of ETX, compared to the other β-PFTs, is its ability to cross the blood–brain barrier, likely mediated by the interaction with its specific receptor. ETX can be considered as a neurotoxin, since it targets specific neurons, which are glutamatergic neurons. In contrast to the other bacterial neurotoxins which inhibit the release of neurotransmitter, ETX has an opposite effect by stimulating the release of glutamate and also acts on other nonneuronal cells. This opens the door to design ETX molecules as a delivery system to address compounds into the central nervous system. Thereby, ETX has been used to facilitate the transport of the drug, bleomycin, through the blood–brain barrier for the treatment of experimental malignant brain tumor in mice.[126] Whether ETX is a powerful toxin, which requires a medical vigilance for the prevention of animals, this toxin also represents a unique tool to vehicle drugs in the central nervous system and/or to target glutamatergic neurons.

REFERENCES

1. Gill, D.M. Bacterial toxins: Lethal amounts, in Laskin, A.I. and Lechevalier, H.A. (eds.) *Toxins and Enzymes*, Vol. 8, pp. 127–135 (CRC Press, Cleveland, OH, 1987).
2. Minami, J., Katayama, S., Matsushita, O., Matsushita, C., and Okabe, A. Lambda-toxin of *Clostridium perfringens* activates the precursor of epsilon-toxin by releasing its N- and C-terminal peptides. *Microbiol. Immunol.* 41, 527–535 (1997).
3. Mantis, N.J. Vaccines against the category B toxins: Staphylococcal enterotoxin B, epsilon toxin and ricin. *Adv. Drug Deliv. Rev.* 57, 1424–1439 (2005).
4. Knapp, O., Stiles, B.G., and Popoff, M.R. The aerolysin-like toxin family of cytolytic, pore-forming toxins. *Open Toxinol. J.* 3, 53–68 (2010).
5. Bokori-Brown, M. et al. Molecular basis of toxicity of *Clostridium perfringens* epsilon toxin. *FEBS J.* 278, 4589–4601 (2011).
6. Popoff, M.R. Epsilon toxin: A fascinating pore-forming toxin. *FEBS J.* 278, 4602–4615 (2011).
7. Petit, L., Gibert, M., and Popoff, M.R. *Clostridium perfringens*: Toxinotype and genotype. *Trends Microbiol.* 7, 104–110 (1999).
8. Myers, G.S. et al. Skewed genomic variability in strains of the toxigenic bacterial pathogen, *Clostridium perfringens*. *Genome Res.* 16, 1031–1040 (2006).

9. Hughes, M.L. et al. Epsilon-toxin plasmids of *Clostridium perfringens* type D are conjugative. *J. Bacteriol.* 189, 7531–7538 (2007).

10. Sayeed, S., Li, J., and McClane, B.A. Virulence plasmid diversity in *Clostridium perfringens* type D isolates. *Infect. Immun.* 75, 2391–2398 (2007).

11. Sayeed, S., Li, J., and McClane, B.A. Characterization of virulence plasmid diversity among *Clostridium perfringens* type B isolates. *Infect. Immun.* 78, 495–504 (2010).

12. Deguchi, A. et al. Genetic characterization of type A enterotoxigenic *Clostridium perfringens* strains. *PLoS One* 4, e5598 (2009).

13. Kobayashi, S. et al. Spread of a large plasmid carrying the *cpe* gene and the tcp locus amongst *Clostridium perfringens* isolates from nosocomial outbreaks and sporadic cases of gastroenteritis in a geriatric hospital. *Epidemiol. Infect.* 137, 108–113 (2009).

14. Miyamoto, K., Li, J., Sayeed, S., Akimoto, S., and McClane, B.A. Sequencing and diversity analyses reveal extensive similarities between some epsilon-toxin-encoding plasmids and the pCPF5603 *Clostridium perfringens* enterotoxin plasmid. *J. Bacteriol.* 190, 7178–7188 (2008).

15. Chen, J., Rood, J.I., and McClane, B.A. Epsilon-toxin production by *Clostridium perfringens* type D strain CN3718 is dependent upon the agr operon but not the VirS/VirR two-component regulatory system. *MBio* 2 (2011).

16. Chen, J. and McClane, B.A. The role of the Agr-like quorum sensing system in regulating toxin production by *Clostridium perfringens* type B strains CN1793 and CN1795. *Infect. Immun.* 80, 3008–3017 (2012).

17. Hunter, S.E., Clarke, I.N., Kelly, D.C., and Titball, R.W. Cloning and nucleotide sequencing of the *Clostridium perfringens* epsilon-toxin gene and its expression in *Escherichia coli. Infect. Immun.* 60, 102–110 (1992).

18. Phannachet, K., Raksat, P., Limvuttegrijeerat, T., and Promdonkoy, B. Production and characterization of N- and C-terminally truncated Mtx2: A mosquitocidal toxin from *Bacillus sphaericus. Curr. Microbiol.* 61, 549–553 (2010).

19. Sakaguchi, Y. et al. Molecular analysis of an extrachromosomal element containing the C2 toxin gene discovered in *Clostridium botulinum* type C. *J. Bacteriol.* 191, 3282–3291 (2009).

20. Cole, A. Structural studies on epsilon toxin from *Clostridium perfringens*, in Duchesnes, C., Mainil, J., Popoff, M.R., and Titball, R. (eds.) *Protein Toxins of the Genus Clostridium and Vaccination*, p. 95 (Presses de la Faculté de Médecine Vétérinaire, Liège, Belgium, 2003).

21. Gurcel, L., Iacovache, I., and van der Goot, F.G. Aerolysin and related *Aeromonas* toxins, in Alouf, J.E. and Popoff, M.R. (eds.) *The Source Book of Bacterial Protein Toxins*, 3rd edition, pp. 606–620 (Elsevier, Academic Press, Amsterdam, the Netherlands, 2006).

22. Parker, M.W. et al. Structure of the *Aeromonas* toxin proaerolysin in its water-soluble and membrane-channel states. *Nature* 367, 292–295 (1994).

23. Melton, J.A., Parker, M.W., Rossjohn, J., Buckley, J.T., and Tweten, R.K. The identification and structure of the membrane-spanning domain of the *Clostridium septicum* alpha toxin. *J. Biol. Chem.* 279, 14315–14322 (2004).

24. Tweten, R.K. *Clostridium perfringens* beta toxin and *Clostridium septicum* alpha toxin: Their mechanisms and possible role in pathogenesis. *Vet. Microbiol.* 82, 1–9 (2001).

25. Cole, A.R. et al. *Clostridium perfringens* ε-toxin shows structural similarity to the pore-forming toxin aerolysin. *Nat. Struct. Mol. Biol.* 11, 797–798 (2004).

26. Ivie, S.E. and McClain, M.S. Identification of amino acids important for binding of *Clostridium perfringens* epsilon toxin to host cells and to HAVCR1. *Biochemistry*, 51, 7588–7595 (2012).

27. Briggs, D.C. et al. Structure of the food-poisoning *Clostridium perfringens* enterotoxin reveals similarity to the aerolysin-like pore-forming toxins. *J. Mol. Biol.* 413, 138–149 (2011).

28. Knapp, O., Maier, E., Benz, R., Geny, B., and Popoff, M.R. Identification of the channel-forming domain of *Clostridium perfringens* epsilon-toxin (ETX). *Biochim. Biophys. Acta* 1788, 2584–2593 (2009).

29. Geny, B. and Popoff, M.R. Bacterial protein toxins and lipids: Pore formation or toxin entry into cells. *Biol. Cell* 98, 667–678 (2006).

30. Schleberger, C., Hochmann, H., Barth, H., Aktories, K., and Schulz, G.E. Structure and action of the binary C2 toxin from *Clostridium botulinum. J. Mol. Biol.* 364, 705–715 (2006).

31. Sakurai, J. Toxins of *Clostridium perfringens. Rev Med. Microbiol.* 6, 175–185 (1995).

32. Oyston, P.C.F., Payne, D.W., Havard, H.L., Williamson, E.D., and Titball, R.W. Production of a non-toxic site-directed mutant of *Clostridium perfringens* e-toxin which induces protective immunity in mice. *Microbiology* 144, 333–341 (1998).

33. Payne, D.W., Williamson, E.D., Havard, H., Modi, N., and Brown, J. Evaluation of a new cytotoxicity assay for *Clostridium perfringens* type D epsilon toxin. *FEMS Microbiol. Lett.* 116, 161–168 (1994).

34. Shortt, S.J., Titball, R.W., and Lindsay, C.D. An assessment of the in vitro toxicology of *Clostridium perfringens* type D epsilon-toxin in human and animal cells. *Hum. Exp. Toxicol.* 19, 108–116 (2000).

35. de la Rosa, C., Hogue, D.E., and Thonney, M.L. Vaccination schedules to raise antibody concentrations against epsilon-toxin of *Clostridium perfringens* in ewes and their triplet lambs. *J. Anim. Sci.* 75, 2328–2334 (1997).

36. Nagahama, M. and Sakurai, J. High-affinity binding of *Clostridium perfringens* epsilon-toxin to rat brain. *Infect. Immun.* 60, 1237–1240 (1992).

37. Heine, K., Pust, S., Enzenmuller, S., and Barth, H. ADP-ribosylation of actin by the *Clostridium botulinum* C2 toxin in mammalian cells results in delayed caspase-dependent apoptotic cell death. *Infect. Immun.* 76, 4600–4608 (2008).

38. Lindsay, C.D., Hambrook, J.L., and Upshall, D.G. Examination of toxicity of *Clostridium perfringens* ε-toxin in the MDCK cell line. *Toxic. In Vitro* 9, 213–218 (1995).

39. Petit, L. et al. *Clostridium perfringens* epsilon-toxin acts on MDCK cells by forming a large membrane complex. *J. Bacteriol.* 179, 6480–6487 (1997).

40. Borrmann, E., Günther, H., and Köhler, H. Effect of *Clostridium perfringens* epsilon toxin on MDCK cells. *FEMS Immunol. Med. Microbiol.* 31, 85–92 (2001).

41. Robertson, S.L., Li, J., Uzal, F.A., and McClane, B.A. Evidence for a prepore stage in the action of *Clostridium perfringens* epsilon toxin. *PLoS One* 6, e22053 (2011).

42. Nagahama, M. et al. Cellular vacuolation induced by *Clostridium perfringens* epsilon-toxin. *FEBS J.* 278, 3395–3407 (2011).

43. Miyata, S. et al. Cleavage of C-terminal peptide is essential for heptamerization of *Clostridium perfringens* ε-toxin in the synaptosomal membrane. *J. Biol. Chem.* 276, 13778–13783 (2001).

44. Miyata, S. et al. *Clostridium perfringens* ε-toxin forms a heptameric pore within the detergent-insoluble microdomains of Madin–Darby canine kidney cells and rat synaptosomes. *J. Biol. Chem.* 277, 39463–39468 (2002).

45. Dorca-Arevalo, J. et al. Binding of epsilon-toxin from *Clostridium perfringens* in the nervous system. *Vet. Microbiol.* 131, 14–25 (2008).

46. Payne, D., Williamson, E.D., and Titball, R.W. The *Clostridium perfringens* epsilon-toxin. *Rev. Med. Microbiol.* 8, S28–S30 (1997).

47. Li, J., Sayeed, S., Robertson, S., Chen, J., and McClane, B.A. Sialidases affect the host cell adherence and epsilon toxin-induced cytotoxicity of *Clostridium perfringens* type D strain CN3718. *PLoS Pathog.* 7, e1002429 (2011).

48. Dorca-Arevalo, J., Martin-Satue, M., and Blasi, J. Characterization of the high affinity binding of epsilon toxin from *Clostridium perfringens* to the renal system. *Vet. Microbiol.* 157, 179–189 (2012).

49. Ivie, S.E., Fennessey, C.M., Sheng, J., Rubin, D.H., and McClain, M.S. Gene-trap mutagenesis identifies mammalian genes contributing to intoxication by *Clostridium perfringens* epsilon-toxin. *PLoS One* 6, e17787 (2011).

50. Dorca-Arevalo, J., Martin-Satue, M., and Blasi, J. Characterization of the high affinity binding of epsilon toxin from *Clostridium perfringens* to the renal system. *Vet. Microbiol.* 157, 179–189 (2012).

51. Petit, L., Maier, E., Gibert, M., Popoff, M.R., and Benz, R. *Clostridium perfringens* epsilon-toxin induces a rapid change in cell membrane permeability to ions and forms channels in artificial lipid bilayers. *J. Biol. Chem.* 276, 15736–15740 (2001).

52. Chassin, C. et al. Pore-forming epsilon toxin causes membrane permeabilization and rapid ATP depletion-mediated cell death in renal collecting duct cells. *Am. J. Physiol. Renal Physiol.* 293, F927–F937 (2007).

53. Lonchamp, E. et al. *Clostridium perfringens* epsilon toxin targets granule cells in the mouse cerebellum and stimulates glutamate release. *PLoS One* 5, e13046 (2010).

54. Shimamoto, S. et al. Changes in ganglioside content affect the binding of *Clostridium perfringens* epsilon-toxin to detergent-resistant membranes of Madin–Darby canine kidney cells. *Microbiol. Immunol.* 49, 245–253 (2005).

55. Nagahama, M., Hara, H., Fernandez-Miyakawa, M., Itohayashi, Y., and Sakurai, J. Oligomerization of *Clostridium perfringens* epsilon-toxin is dependent upon membrane fluidity in liposomes. *Biochemistry* 45, 296–302 (2006).

56. Masson, J.B. et al. Inferring maps of forces inside cell membrane microdomains. *Phys. Rev. Lett.* 102, 048103 (2009).

57. Turkcan, S. et al. Observing the confinement potential of bacterial pore-forming toxin receptors inside rafts with non-blinking eu(3+)-doped oxide nanoparticles. *Biophys. J.* 102, 2299–2308 (2012).

58. Nestorovich, E.M., Karginov, V.A., and Bezrukov, S.M. Polymer partitioning and ion selectivity suggest asymmetrical shape for the membrane pore formed by epsilon toxin. *Biophys. J.* 99, 782–789 (2011).

59. Fennessey, C.M., Sheng, J., Rubin, D.H., and McClain, M.S. Oligomerization of *Clostridium perfringens* epsilon toxin is dependent upon Caveolins 1 and 2. *PLoS One* 7, e46866 (2012).

60. Petit, L. et al. *Clostridium perfringens* epsilon toxin rapidly decreases membrane barrier permeability of polarized MDCK cells. *Cell. Microbiol.* 5, 155–164 (2003).

61. Knapp, O. et al. *Clostridium septicum* alpha-toxin forms pores and induces rapid cell necrosis. *Toxicon* 55, 61–72 (2010).

62. Fernandez Miyakawa, M.E., Zabal, O., and Silberstein, C. *Clostridium perfringens* epsilon toxin is cytotoxic for human renal tubular epithelial cells. *Hum. Exp. Toxicol.* 30, 275–282 (2011).

63. Gleeson-White, M.H. and Bullen, J.J. *Clostridium welchii* epsilon toxin in the intestinal contents of man. *Lancet* 268, 384–385 (1955).

64. Kohn, J. and Warrack, G.H. Recovery of *Clostridium welchii* type D from man. *Lancet* 268, 385 (1955).

65. Bullen, J.J. Enterotoxaemia of sheep: *Clostridium welchii* type D in the alimentary tract of normal animals. *J. Pathol. Bacteriol.* 64, 201–206 (1952).

66. Bullen, J.J. and Battey, I. Enterotoxaemia of sheep. *Vet. Rec.* 69, 1268–1273 (1957).

67. Finnie, J.W. Neurological disorders produced by *Clostridium perfringens* type D epsilon toxin. *Anaerobe* 10, 145–150 (2004).

68. Uzal, F.A. Diagnosis of *Clostridium perfringens* intestinal infections in sheep and goats. *Anaerobe* 10, 135–143 (2004).

69. Uzal, F.A. and Kelly, W.R. Experimental *Clostridium perfringens* type D enterotoxemia in goats. *Vet. Pathol.* 35, 132–140 (1998).

70. Gardner, D.E. Pathology of *Clostridium welchii* type D enterotoxaemia. II. Structural and ultrastructural alterations in the tissues of lambs and mice. *J. Comp. Pathol.* 83, 509–524 (1973).

71. Gardner, D.E. Pathology of *Clostridium welchii* type D enterotoxaemia. I. Biochemical and haematological alterations in lambs. *J. Comp. Pathol.* 83, 499–507 (1973).

72. Filho, E.J. et al. Clinicopathologic features of experimental *Clostridium perfringens* type D enterotoxemia in cattle. *Vet. Pathol.* 46, 1213–1220 (2009).

73. Uzal, F.A. and Kelly, W.R. Enterotoxaemia in goats. *Vet. Res. Commun.* 20, 481–492 (1996).

74. Uzal, F.A. and Songer, J.G. Diagnosis of *Clostridium perfringens* intestinal infections in sheep and goats. *J. Vet. Diagn. Invest.* 20, 253–265 (2008).

75. Bullen, J.J. and Batty, I. Experimental enterotoxaemia of sheep: The effect on the permeability of the intestine and the stimulation of antitoxin production in immune animals. *J. Pathol. Bacteriol.* 73, 511–518 (1957).

76. Bullen, J.J. and Scarisbrick, R. Enterotoxaemia of sheep: Experimental reproduction of the disease. *J. Pathol. Bacteriol.* 73, 495–509 (1957).

77. Fernandez Miyakawa, M.E., Ibarra, C.A., and Uzal, F.A. In vitro effects of *Clostridium perfringens* type D epsilon toxin on water and ion transport in ovine and caprine intestine. *Anaerobe* 9, 145–149 (2003).

78. Fernandez Miyakawa, M.E. and Uzal, F.A. The early effects of *Clostridium perfringens* type D epsilon toxin in ligated intestinal loops of goats and sheep. *Vet. Res. Commun.* 27, 231–241 (2003).

79. Fernandez-Miyakawa, M.E., Jost, B.H., Billington, S.J., and Uzal, F.A. Lethal effects of *Clostridium perfringens* epsilon toxin are potentiated by alpha and perfringolysin-O toxins in a mouse model. *Vet. Microbiol.* 127, 379–385 (2008).

80. Goldstein, J. et al. *Clostridium perfringens* epsilon toxin increases the small intestinal permeability in mice and rats. *PLoS One* 4, e7065 (2009).

81. Batty, I. and Bullen, J.J. The effect of *Clostridium welchii* type D culture filtrates on the permeability of the mouse intestine. *J. Pathol. Bacteriol.* 71, 311–323 (1956).

82. Losada-Eaton, D.M., Uzal, F.A., and Fernandez Miyakawa, M.E. *Clostridium perfringens* epsilon toxin is absorbed from different intestinal segments of mice. *Toxicon* 51, 1207–1213 (2008).

83. Sakurai, J., Nagahama, M., and Takahashi, T. Contraction induced by *Clostridium perfringens* epsilon toxin in the isolated rat ileum. *FEMS Microbiol. Lett.* 49, 269–272 (1989).

84. Losada-Eaton, D.M. and Fernandez-Miyakawa, M.E. *Clostridium perfringens* epsilon toxin inhibits the gastrointestinal transit in mice. *Res. Vet. Sci.* 89, 404–408 (2010).

85. Adamson, R.H. et al. *Clostridium perfringens* epsilon-toxin increases permeability of single perfused microvessels of rat mesentery. *Infect. Immun.* 73, 4879–4887 (2005).

86. Buxton, D. Further studies on the mode of action of *Clostridium welchii* type-D epsilon toxin. *J. Med. Microbiol.* 11, 293–298 (1978).

87. Soler-Jover, A. et al. Effect of epsilon toxin-GFP on MDCK cells and renal tubules in vivo. *J. Histochem. Cytochem.* 52, 931–942 (2004).

88. Uzal, F.A., Rolfe, B.E., Smith, N.J., Thomas, A.C., and Kelly, W.R. Resistance of ovine, caprine and bovine endothelial cells to *Clostridium perfringens* type D epsilon toxin in vitro. *Vet. Res. Commun.* 23, 275–284 (1999).

89. Nagahama, M., Iida, H., and Sakurai, J. Effect of *Clostridium perfringens* epsilon toxin on rat isolated aorta. *Microbiol. Immunol.* 37, 447–450 (1993).

90. Sakurai, J., Nagahama, M., and Fujii, Y. Effect of *Clostridium perfringens* epsilon toxin on the cardiovascular system of rats. *Infect. Immun.* 42, 1183–1186 (1983).

91. Tamai, E. et al. Accumulation of *Clostridium perfringens* epsilon-toxin in the mouse kidney and its possible biological significance. *Infect. Immun.* 71, 5371–5375 (2003).

92. Gardner, D.E. Pathology of *Clostridium welchii* type D enterotoxaemia. 3. Basis of the hyperglycaemic response. *J. Comp. Pathol.* 83, 525–529 (1973).

93. Nagahama, M. and Sakurai, J. Distribution of labeled *Clostridium perfringens* epsilon toxin in mice. *Toxicon* 29, 211–217 (1991).

94. Finnie, J.W. and Hajduk, P. An immunohistochemical study of plasma albumin extravasation in the brain of mice after the administration of *Clostridium perfringens* type D epsilon toxin. *Aust. Vet. J.* 69, 261–262 (1992).

95. Griner, L.A. and Carlson, W.D. Enterotoxemia of sheep. II. Distribution of ^{131}I-radioiodinated serum albumin in brain of *Clostridium perfringens* type D intoxicated lambs. *Am. J. Vet. Res.* 22, 443–446 (1961).

96. Worthington, R.W. and Mulders, M.S. Effect of *Clostridium perfringens* epsilon toxin on the blood brain barrier of mice. *Onderstepoort. J. Vet. Res.* 42, 25–27 (1975).

97. Zhu, C. et al. *Clostridium perfringens* prototoxin-induced alteration of endothelial barrier antigen (EBA) immunoreactivity at the blood–brain barrier (BBB). *Exp. Neurol.* 169, 72–82 (2001).

98. Morgan, K.T., Kelly, B.G., and Buxton, D. Vascular leakage produced in the brains of mice by *Clostridium welchii* type D toxin. *J. Comp. Pathol.* 85, 461–466 (1975).

99. Finnie, J.W., Blumbergs, P.C., and Manavis, J. Neuronal damage produced in rat brains by *Clostridium perfringens* type D epsilon toxin. *J. Comp. Pathol.* 120, 415–420 (1999).

100. Finnie, J.W. Ultrastructural changes in the brain of mice given *Clostridium perfringens* type D epsilon toxin. *J. Comp. Pathol.* 94, 445–452 (1984).

101. Finnie, J.W. Pathogenesis of brain damage produced in sheep by *Clostridium perfringens* type D epsilon toxin: A review. *Aust. Vet. J.* 81, 219–221 (2003).

102. Morgan, K.T. and Kelly, B.G. Ultrastructural study of brain lesions produced in mice by the administration of *Clostridium welchii* type D toxin. *J. Comp. Pathol.* 84, 181–191 (1974).

103. Finnie, J.W. Histopathological changes in the brain of mice given *Clostridium perfringens* type D epsilon toxin. *J. Comp. Pathol.* 94, 363–370 (1984).

104. Ghabriel, M.N. et al. Toxin-induced vasogenic cerebral oedema in a rat model. *Acta Neurochir. Suppl.* 76, 231–236 (2000).

105. Finnie, J.W., Manavis, J., and Blumbergs, P.C. Aquaporin-4 in acute cerebral edema produced by *Clostridium perfringens* type D epsilon toxin. *Vet. Pathol.* 45, 307–309 (2008).

106. Baldwin, E.M., Jr. and Griner, L.A. Clostridia in diarrheal diseases of animals. *Ann. N.Y. Acad. Sci.* 66, 168–175 (1956).

107. Buxton, D. The use of an immunoperoxidase technique to investigate by light and electron microscopy the sites of binding of *Clostridium welchii* type-D epsilon toxin in mice. *J. Med. Microbiol.* 11, 289–292 (1978).

108. Uzal, F.A., Kelly, W.R., Morris, W.E., Bermudez, J., and Baison, M. The pathology of peracute experimental *Clostridium perfringens* type D enterotoxemia in sheep. *J. Vet. Diagn. Invest.* 16, 403–411 (2004).

109. Uzal, F.A., Kelly, W.R., Morris, W.E., and Assis, R.A. Effects of intravenous injection of *Clostridium perfringens* type D epsilon toxin in calves. *J. Comp. Pathol.* 126, 71–75 (2002).

110. Soler-Jover, A. et al. Distribution of *Clostridium perfringens* epsilon toxin in the brains of acutely intoxicated mice and its effect upon glial cells. *Toxicon* 50, 530–540 (2007).

111. Miyamoto, O. et al. Neurotoxicity of *Clostridium perfringens* epsilon-toxin for the rat hippocampus via glutamanergic system. *Infect. Immun.* 66, 2501–2508 (1998).

112. Miyamoto, O. et al. *Clostridium perfringens* epsilon toxin causes excessive release of glutamate in the mouse hippocampus. *FEMS Microbiol. Lett.* 189, 109–113 (2000).

113. Nagahama, M. and Sakurai, J. Effect of drugs acting on the central nervous system on the lethality in mice of *Clostridium perfringens* epsilon toxin. *Toxicon* 31, 427–435 (1993).

114. Moller, K. and Ahrens, P. Comparison of toxicity neutralization-, ELISA-, and PCR tests for typing of *Clostridium perfringens* and detection of the enterotoxin gene by PCR. *Anaerobe* 2, 103–110 (1996).

115. Layana, J.E., Fernandez Miyakawa, M.E., and Uzal, F.A. Evaluation of different fluids for detection of *Clostridium perfringens* type D epsilon toxin in sheep with experimental enterotoxemia. *Anaerobe* 12, 204–206 (2006).

116. Uzal, F.A., Kelly, W.R., Thomas, R., Hornitzky, M., and Galea, F. Comparison of four techniques for the detection of *Clostridium perfringens* type D epsilon toxin in intestinal contents and other body fluids of sheep and goats. *J. Vet. Diagn. Invest.* 15, 94–99 (2003).

117. el Idrissi, A.H. and Ward, G.E. Development of double sandwich ELISA for *Clostridium perfringens* beta and epsilon toxins. *Vet. Microbiol.* 31, 89–99 (1992).

118. Seyer, A. et al. Rapid quantification of clostridial epsilon toxin in complex food and biological matrixes by immunopurification and ultraperformance liquid chromatography-tandem mass spectrometry. *Anal. Chem.* 84, 5103–5109 (2012).

119. Titball, R.W. *Clostridium perfringens* vaccines. *Vaccine* 27(Suppl 4), D44–D47 (2009).

120. Lobato, F.C. et al. Potency against enterotoxemia of a recombinant *Clostridium perfringens* type D epsilon toxoid in ruminants. *Vaccine* 28, 6125–6127 (2010).

121. Mathur, D.D., Deshmukh, S., Kaushik, H., and Garg, L.C. Functional and structural characterization of soluble recombinant epsilon toxin of *Clostridium perfringens* D, causative agent of enterotoxaemia. *Appl. Microbiol. Biotechnol.* 88, 877–884 (2010).

122. Souza, A.M. et al. Molecular cloning and expression of epsilon toxin from *Clostridium perfringens* type D and tests of animal immunization. *Genet. Mol. Res.* 9, 266–276 (2010).

123. Pelish, T.M. and McClain, M.S. Dominant-negative inhibitors of the *Clostridium perfringens* epsilon-toxin. *J. Biol. Chem.* 284, 29446–29453 (2009).

124. McClain, M.S. and Cover, T.L. Functional analysis of neutralizing antibodies against *Clostridium perfringens* epsilon-toxin. *Infect. Immun.* 75, 1785–1793 (2007).

125. Lewis, M., Weaver, C.D., and McClain, M.S. Identification of small molecule inhibitors of *Clostridium perfringens* epsilon-toxin cytotoxicity using a cell-based high-throughput Screen. *Toxins (Basel)* 2, 1825–1847 (2011).

126. Hirschberg, H. et al. Targeted delivery of bleomycin to the brain using photo-chemical internalization of *Clostridium perfringens* epsilon prototoxin. *J. Neurooncol.* 95, 317–329 (2009).

127. Gill, D.M. Bacterial toxins: A table of lethal amounts. *Microbiol. Rev.* 46, 86–94 (1982).

128. Middlebrook, J.L. Cellular mechanism of action of botulism neurotoxin. *J. Toxicol.* 5, 177–180 (1986).

43 Ricin Toxin

Nicholas J. Mantis

CONTENTS

43.1 INTRODUCTION

Ricin toxin is a 65 kDa glycoprotein found in the seeds of the castor bean (or castor oil) plant, *Ricinus communis.* Ricin, in purified or semipurified forms, is extremely toxic to humans (and other mammals) by injection, inhalation, or ingestion [1]. Because castor beans are ubiquitous in tropical and subtropical environments, and the toxin is relatively easy to purify in large amounts from crude bean extracts [1–4], ricin is considered a security concern and a potential biothreat agent by public health and military officials in the United States and abroad [5–7]. In fact, in the United States, the possession, use, and transfer of ricin toxin are regulated by the Office of Public Health Preparedness and Response at the Centers for Disease Control and Prevention (CDC). At the international level, the Organization for the Prohibition of Chemical Weapons (OPCW) provides an oversight regarding the possession and purification of ricin. For more information regarding international regulatory oversight, see a recent excellent review by Worbs et al. [8].

In keeping with the theme of the book, this chapter will focus on aspects of ricin that are most pertinent to its classification as a security sensitive toxin. The reader is referred to a series of recent reviews for more detailed information about aspects of ricin, including incidence of human and animal exposures [8,9], the biology of the toxin at the cellular and subcellular levels [10–12], animal models of ricin intoxication [13], immunity to ricin and vaccine development [14,15], and the design and development of postexposure therapeutics [16].

43.2 CLASSIFICATION AND MORPHOLOGY

The castor bean plant, *R. communis,* is a member of the spurge (Euphorbiaceae) family. The plant originated in Africa but is now found in tropical and subtropical regions around the world [17–19]. While castor bean plants are often propagated as ornamentals in private and public garden spaces, they are primarily cultivated for industrial applications. The oils extracted from the castor beans, notably ricinoleic acid, an unsaturated omega-9 fatty acid, are used in the production of industrial lubricants, cosmetics, and even biofuels [20,21]. It is estimated that more than a million tons of castor beans are produced annually, with cultivation occurring mostly in India and Brazil.

Very recent molecular analysis of the genome of *R. communis* has provided novel insights into the evolution and diversity of the castor bean plant [17,18]. A genome-wide assessment of single nucleotide polymorphisms (SNPs) from *R. communis* plants from 38 different countries revealed relatively low levels of genetic variation [18]. In fact, the authors noted that the molecular variance occurred predominantly within populations (74%), as opposed to among populations or continents. The limited genetic diversity within the castor bean populations was confirmed by a chloroplast genome sequence analysis [19]. The chloroplast genome sequence diversity analysis indicated that there are two major clades of castor beans, each with two distinct subclades, and that modern *Ricinus* populations are a hybrid between two ancestral populations [19].

43.2.1 STRUCTURE AND CLASSIFICATION OF RICIN TOXIN AND ITS SUBUNITS

In the castor bean, ricin is initially synthesized as a 576 amino acid preprotein [22]. Through a circuitous pathway, proricin is eventually processed into its mature, glycosylated (and toxic) form where it accumulates in protein storage vesicles [23,24]. In its mature form, ricin is a 65 kDa glycoprotein consisting of two subunits, RTA and RTB, that are joined by a single disulfide bond (Table 43.1 and Figure 43.1) [25–28].

43.2.2 RTA

RTA is an RNA N-glycosidase that selectively inactivates eukaryotic ribosomes through the cleavage of a universally conserved ribosomal RNA element known as the sarcin–ricin loop (SRL) [29–31]. RTA is 267 amino acids in length and has two N-linked $GlcNAc_2Man_4$ modifications [32]. RTA has two cysteine residues. By SDS-PAGE, RTA generally appears as a doublet of 30,000 and 32,000 molecular weights [2,33]. The higher molecular weight form (A1) has been shown to bind the α-D-mannosyl- and α-D-glucosyl-specific lectin concanavalin A (ConA), while the lower molecular weight species (A2) does not, suggesting the two derivatives of RTA are different in their degree of glycosylation [34].

Structurally, RTA is highly α-helical in nature, consisting of a total of seven α helices encompassing more than a third of the total amino acid residues. RTA can be divided into three folding

domains (FDs) [26,28]. FD I (residues 1–117) is dominated by a five-stranded β-sheet that terminates in a solvent-exposed loop-helix-loop [35]. Domain II (residues 118–210) is dominated by five α-helices that run through the center of RTA. FD III (residues 211–267) forms a protruding element that slides into the cleft between RTB's two domains [26].

43.2.3 RTB

RTB is a 262 amino acid lectin that promotes the attachment, entry, and intracellular trafficking of RTA into mammalian cells. RTB has four intramolecular disulfide bonds, in addition to the single intermolecular disulfide bond that joins it to RTA [36,37]. RTB has two N-linked modifications that have been postulated to interact with mannose-binding protein(s) during ricin toxin intracellular transport and/or influence intracellular stability of RTB [32,38–40]. The removal of RTB's N-linked side chains renders ricin highly attenuated *in vitro* and *in vivo* [39–42].

Structurally, RTB consists of two globular domains with identical folding topologies and has been compared to an elongated "dumbbell" [26]. Each of the two domains (1 and 2) is comprised of three homologous subdomains (α, β, γ) that probably arose by gene duplication from a primordial carbohydrate recognition domain (CRD) [26,43,44]. Only the external subdomains, 1α and 2γ, retain functional carbohydrate recognition activity [44,45]. Subdomain 1α (residues 17–59) is Gal-specific and is considered a "low-affinity" CRD, whereas subdomain 2γ (residues 228–262) binds both Gal and GalNAc and is considered a "high affinity" CRD [46–48]. Although RTB's overall affinity for monosaccharides is quite low (kDa in the range 10^{-3} to 10^{-4}), its affinity for complex sugars on the surface of cells is 3–4 magnitudes greater [49]. The selective ablation of domains 1α and 2γ by genetic or biochemical methods has revealed that both domains must be inactivated to abolish RTB's ability to attach to cells [45,50].

RTB is interesting in a broader sense because the conserved CRD folds of subdomains 1α and 2γ are the prototypes of a superfamily known as the ricin-type (R-type) lectins, which are found in plants, animals, and bacteria [43]. The CRDs of 1α and 2γ each form a shallow pocket created by a sharp bend in the polypeptide backbone associated with the three consecutive residues, Asp, Val, and Arg, plus a more distal fourth variable aromatic residue that provides the binding platform for the sugar [26,44]. Toxin-associated lectins produced by *Campylobacter jejuni*, *Haemophilus ducreyi*, and *Clostridium botulinum* all share RTB's CRD motifs [51–55].

43.2.4 *R. COMMUNIS* AGGLUTININ AND RICIN-LIKE GENES

Castor seeds produce a second lectin that is referred to as *R. communis* agglutinin I or simply RCA-I [56–60]. RCA-I is a ~120,000 MW tetrameric glycoprotein consisting of two identical RTA/RTB-like heterodimers that associate noncovalently with each other. Like ricin, RCA-I is

---

TABLE 43.1

Characteristics of Ricin Toxin's A and B Subunits

Subunit	Residues	Function
RTA	267	RNA N-glycosidases
RTB	262	Gal/GalNAc lectin

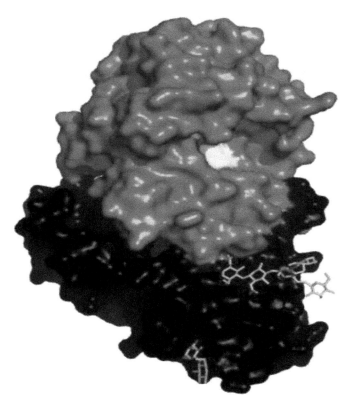

FIGURE 43.1 PyMol surface representation of ricin toxin. RTA (gray) and RTB (black) are joined via a single disulfide bond. RTA's active site is in white. RTB's mannose side chains and one lactose binding site is shown as stick images. (Image from Protein Database [PDB] entry 2AAI.)

initially synthesized as a preprotein [60]. The two preproteins, ricin (576 AA) and RCA-I (564 AA), are 89.6% identical and 95.8% similar. Despite their overwhelming similarities, ricin and RCA-I are functionally different. RCA-I is a strong agglutinin, but a weak or modest ribosome-inactivating protein (RIP). Ricin, on the other hand, is a moderate agglutinin but a potent RIP. Interestingly, a recently completed *R. communis* draft genome sequence revealed 28 members of the ricin/RCA-I gene family, although only seven contain putative full-length RTA- and RTB-like coding regions [17]. These additional ricin/RCA-I genes likely account for the hybrid/variant toxins (e.g., ricin B and E) identified several decades ago [61].

43.2.5   Ricin-Like Toxins in Other Plant Species

Ricin is classified as a member of the so-called RIP family of toxins. Ricin-like RIPs, defined by their RNA *N*-glycosidase activity (EC 3.2.2.22), are found throughout the plant kingdom and are thought to play a role in plant defense against viral and microbial invaders [62–64]. The so-called Type I RIPs are monomeric proteins (~30 kDa) with varying degrees of RNA *N*-glycosidase activity. Type I RIPs are weak cytotoxins due to their inability to penetrate cell membranes. The so-called Type II RIPs, of which ricin is a member, are heterodimers and are generally more potent toxins due to the fact that the binding subunits promote the uptake of the RNA *N*-glycosidase subunit into cells. Despite the fact that Type II RIPs are found in different plant families, they are remarkably conserved at the functional level. For example, the A and B subunits of abrin from the legume *Abrus precatorius* (jequiriti bean) are immunologically distinct from ricin but are similar enough that the subunits from ricin are functionally interchangeable [65,66].

43.3   MECHANISMS OF TOXICITY AND CELL BIOLOGY OF RICIN

43.3.1   Attachment, Entry, and Intracellular Trafficking of Ricin in Mammalian Cells

An overview of ricin's cytotoxic pathway is depicted in Figure 43.2. The attachment of ricin to host cells is mediated by RTB, which binds to glycoproteins and glycolipids displaying terminal galactose and *N*-acetylgalactosamine (Gal/GalNAc) moieties [48,67–70]. As Gal/GAlNAc residues are ubiquitous on mammalian cells, ricin is capable of attaching to virtually all known cell types. It is estimated that there are $>1 \times 10^7$ available RTB binding sites on HeLa cells alone [67]. Binding occurs optimally at pH 7 and is inhibited (and reversed) by molar excess lactose.

Following attachment, ricin is internalized into endosomes by clathrin-dependent and clathrin-independent mechanisms [71,72]. A small fraction (5%) of the toxin is trafficked, by a process known as retrograde transport, to the trans-Golgi network (TGN) and eventually the endoplasmic reticulum (ER) [10,73–75]. In the ER, RTA and RTB separate

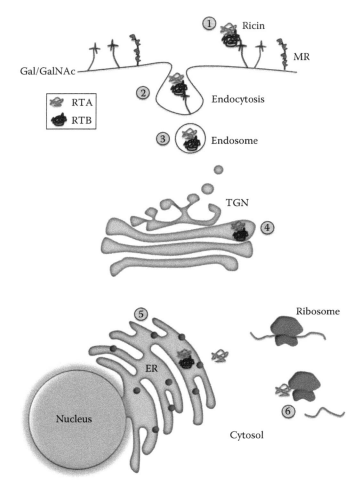

FIGURE 43.2 Retrograde trafficking of ricin. (1) Attachment; (2) endocytosis; (3) transport to TGN; (4) transfer to ER; (5) reduction of disulfide bond linking RTA and RTB, followed by retrotranslocation of RTA into the cytoplasm; (6) RTA "kills" ribosomes and stops protein synthesis.

via a process involving protein disulfide isomerase (PDI) and ER degradation-enhancing α-mannosidase I-like protein 1 (EDEM1) [38,76,77]. RTA's hydrophobic C-terminus becomes exposed following separation from RTB and is proposed to associate with the luminal face of the ER membrane [76,78–81]. RTA is then retrotranslocated across the ER membrane into the cytoplasm where refolding is facilitated by cytoplasmic chaperones (e.g., Hsc70) and possibly ribosomes themselves [10,82–84]. It was recently suggested from studies done in yeast that RTA's N-glycosylation sites are important in egress from the ER [85].

43.3.2 INTERACTIONS OF RTA WITH EUKARYOTIC RIBOSOMES

Once in the cytoplasm, RTA arrests protein synthesis by virtue of its ability to inactivate ribosomes. RTA does not interact favorably with naked rRNA, which was the first clue that the toxin likely interfaces with one or more ribosomal proteins [86]. It is now known that the affinity of RTA for ribosomes is dictated in part by ribosomal stalk proteins P1 and P2 [12,87]. Once associated (albeit transiently) with the ribosome, RTA cleaves the N-glycosidic bond of a conserved adenosine residue (A^{4324} in rat)

within the SRL of 28S rRNA [31,88]. The hydrolysis of the SRL results in an immediate arrest in ribosome progression and a cessation in translation [30], due to the fact that the SRL is involved in the elongation factors, Tu (EF-Tu) and G (EF-G), activation and/or anchoring to the ribosome [89,90].

RTA's active site consists of a large, polar, and solvent-exposed cleft situated on one face of the molecule (Figure 43.1). X-ray crystal structural analysis, coupled with site-directed mutagenesis, identified five residues, Tyr80, Tyr123, Arg180, Glu177, and Trp211, that are integral to RTA's enzymatic activity [28,47,91,92]. RTA's depurination reaction is well understood. RTA residues Tyr80 and Tyr123 play a role in substrate binding through their ability to interact with the first adenosine (GAGA) in the SRL motif [91,93–95]. Glu177 and Arg180 are implicated in transition state stabilization [96,97]. Trp211 lays parallel to Arg180 and may serve a structural role within the active site [98].

43.3.3 RIBOTOXIC STRESS RESPONSE

rRNA depurination by RTA activates the so-called ribotoxic stress response (RSR) [99,100]. Through a mechanism that has yet to be fully elucidated, 28S rRNA damage stimulates cellular stress-activated protein kinases (SAPKs), including p38 mitogen-activated protein kinase (p38 MAPK) and c-Jun N-terminal kinase (JNK) pathways. The activation of these and possibly other SAPKs leads to an increased production of proinflammatory cytokines and apoptosis-mediated cell death [101–103]. The MAP3K, ZAK, has been identified as being responsible for activating the p38 MAPK and JNK pathways in response to ricin [100,101,104].

43.3.4 ALTERNATIVE UPTAKE PATHWAYS FOR RICIN

There is conflicting evidence as to whether ricin exploits the mannose receptor (MR) as a second pathway by which to enter cells and gain access to the cytoplasm. The MR (CD206) is a 175 kDa transmembrane endocytic receptor that recognizes complex oligosaccharides terminating in mannose, fucose, or N-acetylglucosamine [105]. There are numerous in vitro studies that demonstrate convincingly that the MR can recognize ricin holotoxin, as well as RTA, by virtue of the toxin's mannose side chains [42,106–108]. Moreover, MR-mediated uptake of ricin (or RTA) has been shown to result in toxin-induced cell death, indicating that RTA is successfully delivered into the cytoplasm by this route of internalization. If the MR represents a relevant uptake pathway in vivo, then mice lacking MR would be expected to be more resistant to ricin intoxication than wild-type mice. However, a recent study of MR knockout mice revealed that the mice lacking the MR were in fact more sensitive to ricin-induced death than their wild-type counterparts [109], which is consistent with a role for the MR in clearance and degradation of ricin toxin and not its uptake.

43.4 CLINICAL FEATURES AND PATHOGENESIS

43.4.1 INJECTION AND INHALATION

The toxicity of ricin depends on the route of exposure with injection and inhalation being the most severe, as documented in numerous excellent reviews [5,9,110–112]. It should be noted that there are limited examples of human exposure to ricin by injection and no documented cases of aerosolized exposure [8]. Thus, the reported LD_{50} values by these routes (i.e., injection and inhalation) are based on studies performed in rodents, rabbits, or nonhuman primates.

The effects of aerosolized ricin on the respiratory tract are extremely severe, based on challenge studies performed in rodents [113–116], rabbits [117], and rhesus monkeys [118]. In rats, for example, ricin inhalation leads to apoptosis of alveolar macrophages within 6 h of exposure, followed by interalveolar edema, mixed inflammatory cell infiltrates, alveolar flooding, and tissue necrosis 12 h and 15 h later [113]. Similar gross pathological and histological changes occur in rabbits [117] and nonhuman primates. In all cases, death was preceded by widespread necrosis in the airways and alveoli, peribronchovascular edema, mixed inflammatory cell infiltrates, and massive pulmonary alveolar flooding.

43.4.2 INGESTION AND GASTROINTESTINAL TOXICITY

The lethality of ricin by intragastric exposure remains controversial as the actual LD_{50} can vary 100–1000 fold depending on the animal model and whether the toxin is given orally or by gavage [112,119,120]. In general, toxicity of ricin by gavage or ingestion is considered to be 10–100 times less than by injection or inhalation. In humans, there are numerous examples in which the consumption of whole castor beans has led to gastrointestinal and some systemic complications, although death is rare [8].

Nonetheless, the sensitivity of the intestinal epithelium to ricin has been recognized for more than 100 years [121]. Rats challenged with ricin by gavage, for example, develop dose-dependent lesions in the proximal small intestine, including widespread villus atrophy, crypt elongation, sloughing of the epithelium, and occasional infiltration of inflammatory cells, including eosinophils and neutrophils [122–125]. Similar histopathologic changes occur in mice following intragastric ricin challenge [120]. It is postulated that following intestinal exposure, ricin ultimately disseminates from the mucosa into the circulation [126]. Indeed, ricin has been shown to cross polarize epithelial cell monolayers in vitro, although how much toxin crosses the intestinal barrier in vivo remains unknown [127,128].

The toxin directly affects epithelial cells, based on the fact that the application of ricin to the apical surfaces of polarized intestinal epithelial cell monolayers in vitro results in an arrest of protein synthesis within 3–4 h [127]. In addition, ricin activates cellular SAPK pathways and SAPKs in intestinal epithelial cells and induces them to secrete an array of proinflammatory cytokines [101,129–131]. Thus, ricin-induced epithelial destruction may be the consequence of direct cytotoxicity in combination with a local, acute inflammatory response.

43.5 IDENTIFICATION AND DIAGNOSIS

There are an ever-increasing number of reports describing methods to identify ricin toxin in biological and nonbiological samples, a number of which were recently summarized in an excellent review [8]. The most advanced detection methods involve antibody-based detection methods using conventional and single-chain antibodies [132], conventional PCR and immuno-PCR for DNA encoding ricin [133,134], as well as highly sophisticated mass spectroscopy-based methods [135] and others [136,137].

43.5.1 DETECTION OF rRNA CLEAVAGE AND ADENINE RELEASE

One of the earliest methods for measuring the activity of ricin involved extraction of rRNA from ribosomes following incubation with ricin [30,31]. This method is sensitive but is labor intensive and easily affected by RNAse contamination. A primer extension-based method has improved sensitivity of this assay [99] as did a quantitative RT-PCR assay to examine the kinetics of ribosome depurination [138]. Adenine, the by-product of ricin's RNA *N*-glycosidase activity, can be detected directly in reaction supernatants by HPLC or fluorometric methods [139,140] or a modified luciferase-based assay [141].

43.5.2 IN VITRO TRANSLATION ASSAYS

Traditionally, ricin's ribosome-inactivating properties have been measured using an *in vitro* translation (IVT) that relies on the incorporation of radioactive amino acids into newly synthesized proteins [30,94]. As a substitute, we developed a highly reproducible, nonradioactive, luminescence-based IVT assay amenable to high-throughput screening [142–145].

43.5.3 MAMMALIAN CELL-BASED ASSAYS

The activity of ricin has also been studied in cell culture by simply measuring the incorporation of radiolabeled amino acids into nascent polypeptides. We developed and optimized a simpler luciferase-based cytotoxicity assay that proved to be compatible with automated high-throughput screening procedures [144,145].

43.6 TREATMENT AND PREVENTION

There are currently no FDA-approved preventative or therapeutic countermeasures against ricin toxin [146]. Avenues toward the development of countermeasures for use by military personnel, first responders, and research scientists are focused on small molecule inhibitors, vaccines, and immunotherapeutics.

43.6.1 SMALL MOLECULE INHIBITORS OF RICIN TOXIN'S ACTIVE SITE

Virtual screening of chemical libraries has led to the identification of at least three broad classes of small molecules that bind in or near the toxin's active site and thereby interfere with RNA *N*-glycosidase activity [16,145,147]. The first class consists of purine-like compounds that bind within the adenine specificity pocket of RTA. This class consists of substrate analogs such as formycin monophosphate (FMP) and guanosine. While not inhibitors *per se*, FMP and guanosine derivatives have proven invaluable in elucidating the intermolecular bonds involved in RTA's catalytic activity. The second class of inhibitors consists of pterin-like derivatives. These are heterobicyclic compounds that have also been shown by x-ray crystallography to be capable of occupying RTA's active site [95,148]. Both the purine- (class 1) and pterin-type (class 2) inhibitors bind RTA in its "open" form [95,149,150]. The third class of RTA inhibitors consists of single-ring pyrimidine derivatives. Within this class, there is a subset of compounds that bind RTA in its open form (e.g., PBA) and a subset that bind RTA in its closed form (e.g., DDP) [142,149]. Finally, several additional small molecules possibly representing a fourth class of inhibitors have been

described in the past several years [142,144,150–152]. While many of these compounds are predicted to bind RTA's specificity pocket, the x-ray crystal structures of these compounds in complex with RTA have not been solved.

43.6.2 SMALL MOLECULE INHIBITORS OF RICIN INTRACELLULAR TRAFFICKING

The advent of cell-based screening strategies has yielded a new class of ricin inhibitors; small molecules interfere with ricin intracellular trafficking or activation of proapoptotic pathways. A recent high-throughput cell-based screen of >16,500 compounds using a conventional protein synthesis readout from the ChemBridge library uncovered two compounds, including the compound called Retro-2, that blocked retrograde trafficking of ricin at the early endosome–TGN interface [153]. Mice treated with Retro-2 were partially protected against a low-dose ricin challenge, demonstrating for the first time the potential of small molecules to interfere with toxin-induced lethality.

Wahome and colleagues recently identified several small molecules that prolong the viability of ricin- and Stx-treated cells by inhibiting activation and/or catalytic activity of p38 MAPK [145,154]. This screen also uncovered compounds that interfere with the activity of p38 MAPK and caspases 3/7. These results suggest that compounds that reduce and/or inhibit the activation toxin-induced SAPK and apoptosis pathways may be useful as complements of other molecules that target these toxins' active sites.

43.6.3 VACCINES FOR RICIN

Despite ricin's history as an agent of biological warfare and bioterrorism, there is currently no available ricin toxin vaccine [5,155]. As early as the 1940s, the US military focused on the development of a simple formalin-treated holotoxin toxoid vaccine. Although ricin toxoid is highly efficacious in rodents and nonhuman primates, its use in humans was abandoned because of manufacturing problems and safety concerns [146]. For those reasons, current efforts are aimed at the development of a recombinant subunit vaccine. While RTB is an obvious candidate, RTB immunization confers only partial protection against ricin challenge [156,157]. On the other hand, immunization of mice with RTA, or nontoxic derivatives of RTA, is sufficient to protect mice against a $10LD_{50}$ ricin challenge [15,158].

There are currently two RTA-based vaccines under development: RiVax and RVEc. RiVax is a recombinant derivative of RTA with two point mutations: one at position Tyr80 that abolishes the toxin's RNA *N*-glycosidase activity, while the other at Val76 eliminates RTA's ability to elicit vascular leak syndrome (VLS) [159,160]. RVEc is more complicated and was engineered with the primary objective of increasing recombinant RTA's solubility and reducing its propensity to self-aggregate in solution [81,161,162]. RVEc is a recombinant derivative of RTA that lacks a small hydrophobic loop in the *N*-terminus (residues 34–43) and that is truncated at residue 198. RV*Ec* (often referred to as RTA 1–33/44–198) is only 188 residues in length, as compared to RiVax's 267. When described in terms of the three arbitrary FDs, RiVax represents all three domains of RTA, while RV*Ec* essentially consists of FDs 1 and 2 [163]. In mice, RiVax immunization by the intramuscular (i.m.), subcutaneous (s.c.), or intradermal (i.d.) routes elicits toxin-specific serum IgG Abs that are sufficient to confer protection against a lethal dose of ricin [15,119,159,160,164,165]. Phase I clinical trials have demonstrated that RiVax is safe and immunogenic in healthy human volunteers [158,166]. Similarly, RV*Ec* is effective at eliciting toxin-neutralizing antibodies in mice and rabbits [117,167–169] and is now in Phase I clinical trials.

43.6.4 RICIN IMMUNOTHERAPEUTICS

There are dozens of reports demonstrating that antisera and polyclonal antibody preparations derived from different animal species (e.g., mouse, rabbit) and tested on a diversity of cell types (e.g., human, nonhuman primate, and mouse) and animal models (e.g., mice, rats, rabbits) are generally sufficient to neutralize ricin *in vitro* and, in most instances, confer passive immunity *in vivo* [65,116,119,127,156,157,165,170–181]. Abs against either subunit has been shown to be protective, although RTA-specific Abs are generally considered to be more effective than RTB-specific Abs [65,156,157,181,182].

For these reasons, monoclonal antibodies (mAbs) are being pursued as possibly immunotherapies for ricin intoxication [14,183,184]. For example, O'Hara and colleagues recently described the production and characterization of a chimeric derivative of a murine mAb known as GD12 [183]. The chimeric mAb consisted of the murine heavy (V_H), and light chain (V_L) variable regions were fused to a human IgG$_1$ framework. The chimeric mAb was expressed in a *Nicotiana*-based system, which resulted in the rapid production of extremely high amounts of mAbs. It was demonstrated that cGD12 was capable of "rescuing" mice from a 10 LD$_{50}$ challenge with ricin, when the mAb was administered within 6 h following toxin challenge.

43.7 CONCLUSIONS AND FUTURE DIRECTIONS

Ricin toxin remains a security threat and a significant public health concern, and despite enormous advances over the past four decades in the understanding of the molecular and cellular basis of ricin pathogenesis, there are still no FDA-approved vaccines or therapeutics available to treat exposed or at risk individuals. Moreover, no single detection platform has gained widespread acceptance in the public health community, thereby complicating detection and diagnostic efforts.

Looking forward, it is my opinion that the key to effective ricin countermeasure lies in a better understanding of toxin–antibody interactions and in identifying the key determinants involved in toxin neutralization, *in vitro* and *in vivo*. It is becoming increasingly apparent that antibody-mediated toxin neutralization is much more complicated than simply interfering with ricin's ability to attach to host cell receptors. Recent work from our lab and other labs suggests that most potent toxin-neutralizing antibodies are internalized with ricin into host cells and exert their effects on the toxin within one or more intracellular compartments. Identifying exactly how intracellular toxin neutralization occurs will not only shed additional light on the complexity of toxin–host cell interactions but will most certainly reveal more sophisticated avenues to pursue in terms of preventing ricin-induced cell death and toxicity.

REFERENCES

1. Olsnes S. 1978. Ricin and ricinus agglutinin, toxic lectins from castor bean. *Methods Enzymol* 50:330–335.
2. Fulton RJ, Blakey DC, Knowles PP, Uhr JW, Thorpe PE, and Vitetta ES. 1986. Purification of ricin A1, A2, and B chains and characterization of their toxicity. *J Biol Chem* 261:5314–5319.
3. Lappi DA, Kapmeyer W, Beglau JM, and Kaplan NO. 1978. The disulfide bond connecting the chains of ricin. *Proc Natl Acad Sci USA* 75:1096–1100.
4. Thomas TS and Li Steven SL. 1980. Purification and physicochemical properties of ricins and agglutinins from *Ricinus communis*. *Eur J Biochem* 105:453–459.
5. Franz D and Jaax N. 1997. Ricin toxin. In Zajtchuk R and Bellamy RF (eds.), *Textbook of Military Medicine*. pp. 631–642, Borden Institute Books, Washington, DC.
6. Franz DR. 2004. Posting date. Defense against toxin weapons. Virtual Naval Hospital Project, pp. 631–642 [Online].
7. Rotz LD, Khan AS, Lillibridge SR, Ostroff SM, and Hughes JM. 2002. Public health assessment of potential biological terrorism agents. *Emerg Infect Dis* 8:225–230.
8. Worbs S, Kohler K, Pauly D, Avondet MA, Schaer M, Dorner MB, and Dorner BG. 2011. *Ricinus communis* intoxications in human and veterinary medicine—A summary of real cases. *Toxins (Basel)* 3:1332–1372.
9. Griffiths GD. 2011. Understanding ricin from a defensive viewpoint. *Toxins (Basel)* 3:1373–1392.
10. Spooner RA and Lord JM. 2012. How ricin and shiga toxin reach the cytosol of target cells: Retrotranslocation from the endoplasmic reticulum. *Curr Top Microbiol Immunol* 357:19–40.
11. Tesh VL. 2012. The induction of apoptosis by shiga toxins and ricin. *Curr Top Microbiol Immunol* 357:137–178.
12. Tumer NE and Li XP. 2012. Interaction of ricin and shiga toxins with ribosomes. *Curr Top Microbiol Immunol* 357:1–18.
13. Roy CJ, Song K, Sivasubramani SK, Gardner DJ, and Pincus SH. 2012. Animal models of ricin toxicosis. *Curr Top Microbiol Immunol* 357:243-257.
14. O'Hara JM, Yermakova A, and Mantis NJ. 2012. Immunity to ricin: Fundamental insights into toxin–antibody interactions. *Curr Top Microbiol Immunol* 357:209–241.
15. Smallshaw JE and Vitetta ES. 2012. Ricin vaccine development. *Curr Top Microbiol Immunol* 357:259–272.
16. Wahome PG, Robertus JD, and Mantis NJ. 2012. Small-molecule inhibitors of ricin and shiga toxins. *Curr Top Microbiol Immunol* 357:179–207.
17. Chan AP, Crabtree J, Zhao Q, Lorenzi H, Orvis J, Puiu D, Melake-Berhan A et al. 2010. Draft genome sequence of the oilseed species *Ricinus communis*. *Nat Biotechnol* 28:951–956.

18. Foster JT, Allan GJ, Chan AP, Rabinowicz PD, Ravel J, Jackson PJ, and Keim P. 2010. Single nucleotide polymorphisms for assessing genetic diversity in castor bean (*Ricinus communis*). *BMC Plant Biol* 10:13.

19. Rivarola M, Foster JT, Chan AP, Williams AL, Rice DW, Liu X, Melake-Berhan A et al. 2011. Castor bean organelle genome sequencing and worldwide genetic diversity analysis. *PLoS One* 6:e21743.

20. 2007. Final report on the safety assessment of *Ricinus communis* (castor) seed oil, hydrogenated castor oil, glyceryl ricinoleate, glyceryl ricinoleate se, ricinoleic acid, potassium ricinoleate, sodium ricinoleate, zinc ricinoleate, cetyl ricinoleate, ethyl ricinoleate, glycol ricinoleate, isopropyl ricinoleate, methyl ricinoleate, and octyldodecyl ricinoleate. *Int J Toxicol* 26(Suppl 3):31–77.

21. da Silva Nde L, Maciel MR, Batistella CB, and Maciel Filho R. 2006. Optimization of biodiesel production from castor oil. *Appl Biochem Biotechnol* 129–132:405–414.

22. Lamb FI, Roberts LM, and Lord JM. 1985. Nucleotide sequence of cloned cDNA coding for preproricin. *Eur J Biochem* 148:265–270.

23. Lord JM. 1985. Precursors of ricin and *Ricinus communis* agglutinin. Glycosylation and processing during synthesis and intracellular transport. *Eur J Biochem* 146:411–416.

24. Jolliffe NA, Craddock CP, and Frigerio L. 2005. Pathways for protein transport to seed storage vacuoles. *Biochem Soc Trans* 33:1016–1018.

25. Katzin BJ, Collins EJ, and Robertus JD. 1991. Structure of ricin A-chain at 2.5 Å. *Proteins* 10:251–259.

26. Montfort W, Villafranca JE, Monzingo AF, Ernst SR, Katzin B, Rutenber E, Xuong NH, Hamlin R, and Robertus JD. 1987. The three-dimensional structure of ricin at 2.8 Å. *J Biol Chem* 262:5398–5403.

27. Robertus JD, Piatak M, Ferris R, and Houston LL. 1987. Crystallization of ricin A chain obtained from a cloned gene expressed in *Escherichia coli*. *J Biol Chem* 262:19–20.

28. Rutenber E, Katzin BJ, Ernst S, Collins EJ, Mlsna D, Ready MP, and Robertus JD. 1991. Crystallographic refinement of ricin to 2.5 Å. *Proteins* 10:240–250.

29. Olsnes S, Fernandez-Puentes C, Carrasco L, and Vazquez D. 1975. Ribosome inactivation by the toxic lectins abrin and ricin. Kinetics of the enzymic activity of the toxin A-chains. *Eur J Biochem* 60:281–288.

30. Endo Y, Mitsui K, Motizuki M, and Tsurugi K. 1987. The mechanism of action of ricin and related toxins on eukaryotic ribosomes. *J Biol Chem* 262:5908–5912.

31. Endo Y and Tsurugi K. 1987. RNA *N*-glycosidase activity of ricin A-chain. Mechanism of action of the toxic lectin ricin on eukaryotic ribosomes. *J Biol Chem* 262:8128–8130.

32. Kimura Y, Hase S, Kobayashi Y, Kyogoku Y, Ikenaka T, and Funatsu G. 1988. Structures of sugar chains of ricin D. *J Biochem (Tokyo)* 103:944–949.

33. Ramakrishnan S, Eagle MR, and Houston LL. 1982. Radioimmunoassay of ricin A- and B-chains applied to samples of ricin A-chain prepared by chromatofocusing and by DEAE Bio-Gel A chromatography. *Biochim Biophys Acta* 719:341–348.

34. Foxwell BM, Donovan TA, Thorpe PE, and Wilson G. 1985. The removal of carbohydrates from ricin with endoglycosidases H, F and D and alpha-mannosidase. *Biochim Biophys Acta* 840:193–203.

35. Lebeda FJ and Olson MA. 1999. Prediction of a conserved, neutralizing epitope in ribosome-inactivating proteins. *Int J Biol Macromol* 24:19–26.

36. Lewis MS and Youle RJ. 1986. Ricin subunit association. Thermodynamics and the role of the disulfide bond in toxicity. *J Biol Chem* 261:11571–11577.

37. Villafranca JE and Robertus JD. 1981. Ricin B chain is a product of gene duplication. *J Biol Chem* 256:554–556.

38. Slominska-Wojewodzka M, Gregers TF, Walchli S, and Sandvig K. 2006. EDEM is involved in retrotranslocation of ricin from the endoplasmic reticulum to the cytosol. *Mol Biol Cell* 17:1664–1675.

39. Wales R, Richardson PT, Roberts LM, Woodland HR, and Lord JM. 1991. Mutational analysis of the galactose binding ability of recombinant ricin B chain. *J Biol Chem* 266:19172–19179.

40. Zhan J, de Sousa M, Chaddock JA, Roberts LM, and Lord JM. 1997. Restoration of lectin activity to a non-glycosylated ricin B chain mutant by the introduction of a novel N-glycosylation site. *FEBS Lett* 407:271–274.

41. Simeral LS, Kapmeyer W, MacConnell WP, and Kaplan NO. 1980. On the role of the covalent carbohydrate in the action of ricin. *J Biol Chem* 255:11098–11101.

42. Simmons BM, Stahl PD, and Russell JH. 1986. Mannose receptor-mediated uptake of ricin toxin and ricin A chain by macrophages. Multiple intracellular pathways for a chain translocation. *J Biol Chem* 261:7912–7920.

43. Cummings R and Etzler M. 2009. R-type lectins. In Varki A, Cummings R, Esko J, Freeze H, Stanley P, Bertozzi C, Hart G, and Etzler M (eds.), *Essentials of Glycobiology*. Cold Spring Harbor Laboratory Press, Cold Spring Harbor, NY.

44. Rutenber E, Ready M, and Robertus JD. 1987. Structure and evolution of ricin B chain. *Nature* 326:624–626.

45. Swimmer C, Lehar SM, McCafferty J, Chiswell DJ, Blattler WA, and Guild BC. 1992. Phage display of ricin B chain and its single binding domains: System for screening galactose-binding mutants. *Proc Natl Acad Sci USA* 89:3756–3760.

46. Newton DL, Wales R, Richardson PT, Walbridge S, Saxena SK, Ackerman EJ, Roberts LM, Lord JM, and Youle RJ. 1992. Cell surface and intracellular functions for ricin galactose binding. *J Biol Chem* 267:11917–11922.

47. Rutenber E and Robertus JD. 1991. Structure of ricin B-chain at 2.5 Å resolution. *Proteins* 10:260–269.

48. Zentz C, Frenoy JP, and Bourrillon R. 1978. Binding of galactose and lactose to ricin. Equilibrium studies. *Biochim Biophys Acta* 536:18–26.

49. Baenziger JU and Fiete D. 1979. Structural determinants of *Ricinus communis* agglutinin and toxin specificity for oligosaccharides. *J Biol Chem* 254:9795–9799.

50. Sphyris N, Lord JM, Wales R, and Roberts LM. 1995. Mutational analysis of the *Ricinus* lectin B-chains. Galactose-binding ability of the 2 gamma subdomain of *Ricinus communis* agglutinin B-chain. *J Biol Chem* 270:20292–20297.

51. Cao L, Volgina A, Korostoff J, and DiRienzo JM. 2006. Role of intrachain disulfides in the activities of the CdtA and CdtC subunits of the cytolethal distending toxin of *Actinobacillus actinomycetemcomitans*. *Infect Immun* 74:4990–5002.

52. Inoue K, Sobhany M, Transue TR, Oguma K, Pedersen LC, and Negishi M. 2003. Structural analysis by X-ray crystallography and calorimetry of a haemagglutinin component (HA1) of the progenitor toxin from *Clostridium botulinum*. *Microbiology* 149:3361–3370.

53. Lara-Tejero M and Galan JE. 2001. CdtA, CdtB, and CdtC form a tripartite complex that is required for cytolethal distending toxin activity. *Infect Immun* 69:4358–4365.

54. Nesic D, Hsu Y, and Stebbins CE. 2004. Assembly and function of a bacterial genotoxin. *Nature* 429:429–433.

55. Nesic D and Stebbins CE. 2005. Mechanisms of assembly and cellular interactions for the bacterial genotoxin CDT. *PLoS Pathog* 1:e28.

56. Cawley DB, Hedblom ML, and Houston LL. 1978. Homology between ricin and *Ricinus communis* agglutinin: Amino terminal sequence analysis and protein synthesis inhibition studies. *Arch Biochem Biophys* 190:744–755.

57. Ishiguro M, Takahashi T, Funatsu G, Hayashi K, and Funatsu M. 1964. Biochemical studies on ricin. I. Purification of ricin. *J Biochem* 55:587–592.

58. Lin TT and Li SL. 1980. Purification and physicochemical properties of ricins and agglutinins from *Ricinus communis*. *Eur J Biochem* 105:453–459.

59. Nicolson GL, Blaustein J, and Etzler ME. 1974. Characterization of two plant lectins from *Ricinus communis* and their quantitative interaction with a murine lymphoma. *Biochemistry* 13:196–204.

60. Roberts LM, Lamb FI, Pappin DJ, and Lord JM. 1985. The primary sequence of *Ricinus communis* agglutinin. Comparison with ricin. *J Biol Chem* 260:15682–15686.

61. Araki T and Funatsu G. 1987. The complete amino acid sequence of the B-chain of ricin E isolated from small-grain castor bean seeds. Ricin E is a gene recombination product of ricin D and *Ricinus communis* agglutinin. *Biochim Biophys Acta* 911:191–200.

62. Nielsen K and Boston RS. 2001. Ribosome-inactivating proteins: A plant perspective. *Ann Rev Plant Physiol Plant Mol Biol* 52:785–816.

63. Olsnes S and Kozlov JV. 2001. Ricin. *Toxicon* 39:1723–1728.

64. Stirpe F. 2004. Ribosome-inactivating proteins. *Toxicon* 44:371–383.

65. Olsnes S, Pappenheimer AM, Jr., and Meren R. 1974. Lectins from *Abrus precatorius* and *Ricinus communis*. II. Hybrid toxins and their interaction with chain-specific antibodies. *J Immunol* 113:842–847.

66. Pappenheimer AM, Jr., Olsnes S, and Harper AA. 1974. Lectins from *Abrus precatorius* and *Ricinus communis*. I. Immunochemical relationships between toxins and agglutinins. *J Immunol* 113:835–841.

67. Sandvig K, Olsnes S, and Pihl A. 1976. Kinetics of binding of the toxic lectins abrin and ricin to surface receptors of human cells. *J Biol Chem* 251:3977–3984.

68. Sandvig K and Olsnes S. 1979. Effect of temperature on the uptake, excretion and degradation of abrin and ricin by HeLa cells. *Exp Cell Res* 121:15–25.

69. Sandvig K and Olsnes S. 1982. Entry of the toxic proteins abrin, modeccin, ricin, and diphtheria toxin into cells. II. Effect of pH, metabolic inhibitors, and ionophores and evidence for toxin penetration from endocytotic vesicles. *J Biol Chem* 257:7504–7513.

70. Sandvig K and Olsnes S. 1982. Entry of the toxic proteins abrin, modeccin, ricin, and diphtheria toxin into cells. I. Requirement for calcium. *J Biol Chem* 257:7495–7503.

71. Moya M, Dautry-Varsat A, Goud B, Louvard D, and Boquet P. 1985. Inhibition of coated pit formation in Hep2 cells blocks the cytotoxicity of diphtheria toxin but not that of ricin toxin. *J Cell Biol* 101:548–559.

72. Sandvig K, Olsnes S, Petersen OW, and van Deurs B. 1987. Acidification of the cytosol inhibits endocytosis from coated pits. *J Cell Biol* 105:679–689.

73. Rapak A, Falnes PO, and Olsnes S. 1997. Retrograde transport of mutant ricin to the endoplasmic reticulum with subsequent translocation to cytosol. *Proc Natl Acad Sci USA* 94:3783–3788.

74. van Deurs B, Tonnessen TI, Petersen OW, Sandvig K, and Olsnes S. 1986. Routing of internalized ricin and ricin conjugates to the Golgi complex. *J Cell Biol* 102:37–47.

75. van Deurs B, Sandvig K, Petersen OW, Olsnes S, Simons K, and Griffiths G. 1988. Estimation of the amount of internalized ricin that reaches the trans-Golgi network. *J Cell Biol* 106:253–267.

76. Sokolowska I, Walchli S, Wegrzyn G, Sandvig K, and Slominska-Wojewodzka M. 2011. A single point mutation in ricin A-chain increases toxin degradation and inhibits EDEM1-dependent ER retrotranslocation. *Biochem J* 436:371–385.

77. Spooner RA, Watson PD, Marsden CJ, Smith DC, Moore KA, Cook JP, Lord JM, and Roberts LM. 2004. Protein disulphide-isomerase reduces ricin to its A and B chains in the endoplasmic reticulum. *Biochem J* 383:285–293.

78. Chaddock JA, Monzingo AF, Robertus JD, Lord JM, and Roberts LM. 1996. Major structural differences between pokeweed antiviral protein and ricin A-chain do not account for their differing ribosome specificity. *Eur J Biochem* 235:159–166.

79. Day PJ, Pinheiro TJ, Roberts LM, and Lord JM. 2002. Binding of ricin A-chain to negatively charged phospholipid vesicles leads to protein structural changes and destabilizes the lipid bilayer. *Biochemistry* 41:2836–2843.

80. Mayerhofer PU, Cook JP, Wahlman J, Pinheiro TT, Moore KA, Lord JM, Johnson AE, and Roberts LM. 2009. Ricin A chain insertion into endoplasmic reticulum membranes is triggered by a temperature increase to 37°C. *J Biol Chem* 284:10232–10242.

81. Olson MA, Carra JH, Roxas-Duncan V, Wannemacher RW, Smith LA, and Millard CB. 2004. Finding a new vaccine in the ricin protein fold. *Protein Eng Des Sel* 17:391–397.

82. Argent RH, Parrott AM, Day PJ, Roberts LM, Stockley PG, Lord JM, and Radford SE. 2000. Ribosome-mediated folding of partially unfolded ricin A-chain. *J Biol Chem* 275:9263–9269.

83. Li S, Spooner RA, Allen SC, Guise CP, Ladds G, Schnoder T, Schmitt MJ, Lord JM, and Roberts LM. 2010. Folding-competent and folding-defective forms of ricin A chain have different fates after retrotranslocation from the endoplasmic reticulum. *Mol Biol Cell* 21:2543–2554.

84. Spooner RA, Hart PJ, Cook JP, Pietroni P, Rogon C, Hohfeld J, Roberts LM, and Lord JM. 2008. Cytosolic chaperones influence the fate of a toxin dislocated from the endoplasmic reticulum. *Proc Natl Acad Sci USA* 105:17408–17413.

85. Yan Q, Li XP, and Tumer NE. 2012. N-Glycosylation does not affect the catalytic activity of ricin a chain but stimulates cytotoxicity by promoting its transport out of the endoplasmic reticulum. *Traffic* 13:1508–1521.

86. Endo Y and Tsurugi K. 1988. The RNA N-glycosidase activity of ricin A-chain. The characteristics of the enzymatic activity of ricin A-chain with ribosomes and with rRNA. *J Biol Chem* 263:8735–8739.

87. May KL, Li XP, Martinez-Azorin F, Ballesta JP, Grela P, Tchorzewski M, and Tumer NE. 2012. The P1/P2 proteins of the human ribosomal stalk are required for ribosome binding and depurination by ricin in human cells. *FEBS J* 279(20):3925–3936.

88. Gluck A, Endo Y, and Wool IG. 1994. The ribosomal RNA identity elements for ricin and for alpha-sarcin: Mutations in the putative CG pair that closes a GAGA tetraloop. *Nucleic Acids Res* 22:321–324.

89. Moazed D, Robertson JM, and Noller HF. 1988. Interaction of elongation factors EF-G and EF-Tu with a conserved loop in 23S RNA. *Nature* 334:362–364.

90. Shi X, Khade PK, Sanbonmatsu KY, and Joseph S. 2012. Functional role of the sarcin–ricin loop of the 23S rRNA in the elongation cycle of protein synthesis. *J Mol Biol* 419:125–138.

91. Monzingo AF and Robertus JD. 1992. X-ray analysis of substrate analogs in the ricin A-chain active site. *J Mol Biol* 227:1136–1145.

92. Weston SA, Tucker AD, Thatcher DR, Derbyshire DJ, and Pauptit RA. 1994. X-ray structure of recombinant ricin A-chain at 1.8 Å resolution. *J Mol Biol* 244:410–422.

93. Olson MA and Cuff L. 1997. Molecular docking of superantigens with class II major histocompatibility complex proteins. *J Mol Recognit* 10:277–289.

94. Ready MP, Kim Y, and Robertus JD. 1991. Site-directed mutagenesis of ricin A-chain and implications for the mechanism of action. *Proteins* 10:270–278.

95. Yan X, Hollis T, Svinth M, Day P, Monzingo AF, Milne GW, and Robertus JD. 1997. Structure-based identification of a ricin inhibitor. *J Mol Biol* 266:1043–1049.

96. Kim Y, Mlsna D, Monzingo AF, Ready MP, Frankel A, and Robertus JD. 1992. Structure of a ricin mutant showing rescue of activity by a noncatalytic residue. *Biochemistry* 31:3294–3296.

97. Kim Y and Robertus JD. 1992. Analysis of several key active site residues of ricin A chain by mutagenesis and X-ray crystallography. *Protein Eng* 5:775–779.

98. Bradley JL and McGuire PM. 1990. Site-directed mutagenesis of ricin A chain Trp 211 to Phe. *Int J Pept Protein Res* 35:365–366.

99. Iordanov MS, Pribnow D, Magun JL, Dinh TH, Pearson JA, Chen SL, and Magun BE. 1997. Ribotoxic stress response: Activation of the stress-activated protein kinase JNK1 by inhibitors of the peptidyl transferase reaction and by sequence-specific RNA damage to the alpha-sarcin/ricin loop in the 28S rRNA. *Mol Cell Biol* 17:3373–3381.

100. Jandhyala DM, Thorpe CM, and Magun B. 2012. Ricin and shiga toxins: Effects on host cell signal transduction. *Curr Top Microbiol Immunol* 357:41–65.

101. Jandhyala DM, Ahluwalia A, Obrig T, and Thorpe CM. 2008. ZAK: A MAP3 kinase that transduces shiga toxin- and ricin-induced proinflammatory cytokine expression. *Cell Microbiol* 10:1468–1477.

102. Higuchi S, Tamura T, and Oda T. 2003. Cross-talk between the pathways leading to the induction of apoptosis and the secretion of tumor necrosis factor-alpha in ricin-treated RAW 264.7 cells. *J Biochem (Tokyo)* 134:927–933.

103. Smith WE, Kane AV, Campbell ST, Acheson DW, Cochran BH, and Thorpe CM. 2003. Shiga toxin 1 triggers a ribotoxic stress response leading to p38 and JNK activation and induction of apoptosis in intestinal epithelial cells. *Infect Immun* 71:1497–1504.

104. Sauter KA, Magun EA, Iordanov MS, and Magun BE. 2010. ZAK is required for doxorubicin, a novel ribotoxic stressor, to induce SAPK activation and apoptosis in HaCaT cells. *Cancer Biol Ther* 10:258–266.

105. Taylor PR, Gordon S, and Martinez-Pomares L. 2005. The mannose receptor: Linking homeostasis and immunity through sugar recognition. *Trends Immunol* 26:104–110.

106. Magnusson S and Berg T. 1993. Endocytosis of ricin by rat liver cells in vivo and in vitro is mainly mediated by mannose receptors on sinusoidal endothelial cells. *Biochem J* 291(Pt 3):749–755.

107. Magnusson S, Berg T, Turpin E, and Frenoy JP. 1991. Interactions of ricin with sinusoidal endothelial rat liver cells. Different involvement of two distinct carbohydrate-specific mechanisms in surface binding and internalization. *Biochem J* 277(Pt 3):855–861.

108. Magnusson S, Kjeken R, and Berg T. 1993. Characterization of two distinct pathways of endocytosis of ricin by rat liver endothelial cells. *Exp Cell Res* 205:118–125.

109. Gage E, Hernandez MO, O'Hara JM, McCarthy EA, and Mantis NJ. 2011. Role of the mannose receptor (CD206) in immunity to ricin. *Toxins (Basel)* 3(9):1131–1145.

110. Audi J, Belson M, Patel M, Schier J, and Osterloh J. 2005. Ricin poisoning: A comprehensive review. *JAMA* 294:2342–2351.

111. Bradberry SM, Dickers KJ, Rice P, Griffiths GD, and Vale JA. 2003. Ricin poisoning. *Toxicol Rev* 22:65–70.

112. Pincus SH, Smallshaw JE, Song K, Berry J, and Vitetta ES. 2011. Passive and active vaccination strategies to prevent ricin poisoning. *Toxins (Basel)* 3:1163–1184.

113. Brown RF and White DE. 1997. Ultrastructure of rat lung following inhalation of ricin aerosol. *Int J Exp Path* 78:267–276.

114. DaSilva L, Cote D, Roy C, Martinez M, Duniho S, Pitt ML, Downey T, and Dertzbaugh M. 2003. Pulmonary gene expression profiling of inhaled ricin. *Toxicon* 41:813–822.

115. Doebler JA, Wiltshire ND, Mayer TW, Estep JE, Moeller RB, Traub RK, Broomfield CA, Calamaio CA, Thompson WL, and Pitt ML. 1995. The distribution of [^{125}I]ricin in mice following aerosol inhalation exposure. *Toxicology* 98:137–149.

116. Griffiths GD, Phillips GJ, and Bailey SC. 1999. Comparison of the quality of protection elicited by toxoid and peptide liposomal vaccine formulations against ricin as assessed by markers of inflammation. *Vaccine* 17:2562–2568.

117. McLain DE, Lewis BS, Chapman JL, Wannemacher RW, Lindsey CY, and Smith LA. 2011. Protective effect of two recombinant ricin subunit vaccines in the New Zealand white rabbit subjected to a lethal aerosolized ricin challenge: Survival, immunological response and histopathological findings. *Toxicol Sci* 126(1):72–83.

118. Wilhelmsen CL and Pitt ML. 1996. Lesions of acute inhaled lethal ricin intoxication in rhesus monkeys. *Vet Pathol* 33:296–302.

119. Smallshaw JE, Richardson JA, and Vitetta ES. 2007. RiVax, a recombinant ricin subunit vaccine, protects mice against ricin delivered by gavage or aerosol. *Vaccine* 25:7459–7469.

120. Yoder JM, Aslam RU, and Mantis NJ. 2007. Evidence for widespread epithelial damage and coincident production of monocyte chemotactic protein 1 in a murine model of intestinal ricin intoxication. *Infect Immun* 75:1745–1750.

121. Flexner S. 1897. The histological changes produced by ricin and abrin intoxications. *J Exp Med* 2:197–220.

122. Ishiguro M, Harada H, Ichiki O, Sekine I, Nishimori I, and Kikutani M. 1984. Effects of ricin, a protein toxin, on glucose absorption by rat small intestine. (Biochemical studies on oral toxicity of ricin. II.) *Chem Pharm Bull* 32:3141–3147.

123. Ishiguro M, Nakashima H, Tanabe S, and Sakakibara R. 1992. Interaction of toxic lectin ricin with epithelial cells of rat small intestine in vitro. *Chem Pharm Bull* 40:441–445.

124. Ishiguro M, Tanabe S, Matori Y, and Sakakibara R. 1992. Biochemical studies on oral toxicity of ricin. IV. A fate of orally administered ricin in rats. *J Pharmacobiodyn* 15:147–156.

125. Sekine I, Kawase Y, Nishimori I, Mitarai M, Harada H, Ishiguro M, and Kikutani M. 1986. Pathological study on mucosal changes in small intestine of rat by oral administration of ricin. I. Microscopical observation. *Acta Pathol Jpn* 36:1205–1212.

126. Ishiguro M, Matori Y, Tanabe S, Kawase Y, Sekine I, and Sakakibara R. 1992. Biochemical studies on oral toxicity of ricin. V. The role of lectin activity in the intestinal absorption of ricin. *Chem Pharm Bull* 40:1216–1220.

127. Mantis NJ, McGuinness CR, Sonuyi O, Edwards G, and Farrant SA. 2006. Immunoglobulin A antibodies against ricin A and B subunits protect epithelial cells from ricin intoxication. *Infect Immun* 74:3455–3462.

128. van Deurs B, Hansen SH, Petersen OW, Melby EL, and Sandvig K. 1990. Endocytosis, intracellular transport and transcytosis of the toxic protein ricin by a polarized epithelium. *Eur J Cell Biol* 51:96–109.

129. Thorpe CM, Hurley BP, Lincicome LL, Jacewicz MS, Keusch GT, and Acheson DW. 1999. Shiga toxins stimulate secretion of interleukin-8 from intestinal epithelial cells. *Infect Immun* 67:5985–5993.

130. Thorpe CM, Smith WE, Hurley BP, and Acheson DW. 2001. Shiga toxins induce, superinduce, and stabilize a variety of C-X-C chemokine mRNAs in intestinal epithelial cells, resulting in increased chemokine expression. *Infect Immun* 69:6140–6147.

131. Yamasaki C, Nishikawa K, Zeng XT, Katayama Y, Natori Y, Komatsu N, and Oda T. 2004. Induction of cytokines by toxins that have an identical RNA *N*-glycosidase activity: Shiga toxin, ricin, and modeccin. *Biochim Biophys Acta* 1671:44–50.

132. Shia WW and Bailey RC. 2012. Single domain antibodies for the detection of ricin using silicon photonic microring resonator arrays. *Anal Chem* 85(2):805–810.

133. Felder E, Mossbrugger I, Lange M, and Wolfel R. 2012. Simultaneous detection of ricin and abrin DNA by real-time PCR (qPCR). *Toxins (Basel)* 4:633–642.

134. He X, McMahon S, Henderson TD, 2nd, Griffey SM, and Cheng LW. 2010. Ricin toxicokinetics and its sensitive detection in mouse sera or feces using immuno-PCR. *PLoS One* 5:e12858.

135. Schieltz DM, McGrath SC, McWilliams LG, Rees J, Bowen MD, Kools JJ, Dauphin LA et al. 2011. Analysis of active ricin and castor bean proteins in a ricin preparation, castor bean extract, and surface swabs from a public health investigation. *Forensic Sci Int* 209(1–3):70–79.

136. Peruski AH, Johnson LH, 3rd, and Peruski LF, Jr. 2002. Rapid and sensitive detection of biological warfare agents using time-resolved fluorescence assays. *J Immunol Methods* 263:35–41.

137. Webb-Robertson BJ, Kreuzer H, Hart G, Ehleringer J, West J, Gill G, and Duckworth D. 2012. Bayesian integration of isotope ratio for geographic sourcing of castor beans. *J Biomed Biotechnol* 2012:450967.

138. Pierce M, Kahn JN, Chiou J, and Tumer NE. 2011. Development of a quantitative RT-PCR assay to examine the kinetics of ribosome depurination by ribosome inactivating proteins using *Saccharomyces cerevisiae* as a model. *RNA* 17:201–210.

139. Fujimori H, Sasaki T, Hibi K, Senda M, and Yoshioka M. 1990. Direct injection of blood samples into a high-performance liquid chromatographic adenine analyser to measure adenine, adenosine, and the adenine nucleotides with fluorescence detection. *J Chromatogr* 515:363–373.

140. Zamboni M, Brigotti M, Rambelli F, Montanaro L, and Sperti S. 1989. High-pressure-liquid-chromatographic and fluorimetric methods for the determination of adenine released from ribosomes by ricin and gelonin. *Biochem J* 259:639–643.

141. Sturm MB and Schramm VL. 2009. Detecting ricin: Sensitive luminescent assay for ricin A-chain ribosome depurination kinetics. *Anal Chem* 81:2847–2853.

142. Bai Y, Monzingo AF, and Robertus JD. 2009. The x-ray structure of ricin A chain with a novel inhibitor. *Arch Biochem Biophys* 483:23–28.

143. Neal LM, O'Hara J, Brey RN, 3rd, and Mantis NJ. 2010. A monoclonal immunoglobulin G antibody directed against an immunodominant linear epitope on the ricin A chain confers systemic and mucosal immunity to ricin. *Infect Immun* 78:552–561.

144. Wahome PG, Bai Y, Neal LM, Robertus JD, and Mantis NJ. 2010. Identification of small-molecule inhibitors of ricin and shiga toxin using a cell-based high-throughput screen. *Toxicon* 56:313–323.

145. Wahome PG and Mantis NJ. 2013. High-throughput, cell-based screens to identify small-molecule inhibitors of ricin toxin and related category b ribosome inactivating proteins (RIPs). *Curr Protoc Toxicol* Chapter 2:Unit 2.23.

146. Reisler RB and Smith LA. 2012. The need for continued development of ricin countermeasures. *Adv Prev Med* 2012:149737.

147. Barbier J, Bouclier C, Johannes L, and Gillet D. 2012. Inhibitors of the cellular trafficking of ricin. *Toxins (Basel)* 4:15–27.

148. Robertus JD, Yan X, Ernst S, Monzingo A, Worley S, Day P, Hollis T, and Svinth M. 1996. Structural analysis of ricin and implications for inhibitor design. *Toxicon* 34:1325–1334.

149. Miller DJ, Ravikumar K, Shen H, Suh JK, Kerwin SM, and Robertus JD. 2002. Structure-based design and characterization of novel platforms for ricin and shiga toxin inhibition. *J Med Chem* 45:90–98.

150. Pruet JM, Jasheway KR, Manzano LA, Bai Y, Anslyn EV, and Robertus JD. 2011. 7-Substituted pterins provide a new direction for ricin A chain inhibitors. *Eur J Med Chem* 46:3608–3615.

151. Bai Y, Watt B, Wahome PG, Mantis NJ, and Robertus JD. 2010. Identification of new classes of ricin toxin inhibitors by virtual screening. *Toxicon* 56:526–534.

152. Pang YP, Park JG, Wang S, Vummenthala A, Mishra RK, McLaughlin JE, Di R, Kahn JN, Tumer NE, Janosi L, Davis J, and Millard CB. 2011. Small-molecule inhibitor leads of ribosome-inactivating proteins developed using the doorstop approach. *PLoS One* 6:e17883.

153. Stechmann B, Bai SK, Gobbo E, Lopez R, Merer G, Pinchard S, Panigai L et al. 2010. Inhibition of retrograde transport protects mice from lethal ricin challenge. *Cell* 141:231–242.

154. Wahome PG, Ahlawat S, and Mantis NJ. 2012. Identification of small molecules that suppress ricin-induced stress-activated signaling pathways. *PLoS One* 7:e49075.

155. Radosavljevic V and Belojevic G. 2009. A new model of bioterrorism risk assessment. *Biosecur Bioterror* 7:443–451.

156. Maddaloni M, Cooke C, Wilkinson R, Stout AV, Eng L, and Pincus SH. 2004. Immunological characteristics associated with the protective efficacy of antibodies to ricin. *J Immunol* 172:6221–6228.

157. Yermakova A and Mantis NJ. 2011. Protective immunity to ricin toxin conferred by antibodies against the toxin's binding subunit (RTB). *Vaccine* 29:7925–7935.

158. Vitetta ES, Smallshaw JE, and Schindler J. 2012. A small phase IB clinical trial of an alhydrogel-adsorbed recombinant ricin vaccine (RiVax). *Clin Vaccine Immunol* 19(10):1697–1699.

159. Smallshaw JE, Firan A, Fulmer JR, Ruback SL, Ghetie V, and Vitetta ES. 2002. A novel recombinant vaccine which protects mice against ricin intoxication. *Vaccine* 20:3422–3427.

160. Smallshaw JE, Ghetie V, Rizo J, Fulmer JR, Trahan LL, Ghetie MA, and Vitetta ES. 2003. Genetic engineering of an immunotoxin to eliminate pulmonary vascular leak in mice. *Nat Biotechnol* 21:387–391.

161. Carra JH, Wannemacher RW, Tammariello RF, Lindsey CY, Dinterman RE, Schokman RD, and Smith LA. 2007. Improved formulation of a recombinant ricin A-chain vaccine increases its stability and effective antigenicity. *Vaccine* 25:4149–4158.

162. McHugh CA, Tammariello RF, Millard CB, and Carra JH. 2004. Improved stability of a protein vaccine through elimination of a partially unfolded state. *Protein Sci* 13:2736–2743.

163. O'Hara JM, Neal LM, McCarthy EA, Kasten-Jolly JA, Brey RN, 3rd, and Mantis NJ. 2010. Folding domains within the ricin toxin A subunit as targets of protective antibodies. *Vaccine* 28:7035–7046.

164. Marconescu PS, Smallshaw JE, Pop LM, Ruback SL, and Vitetta ES. 2010. Intradermal administration of RiVax protects mice from mucosal and systemic ricin intoxication. *Vaccine* 28:5315–5322.

165. Smallshaw JE, Richardson JA, Pincus S, Schindler J, and Vitetta ES. 2005. Preclinical toxicity and efficacy testing of RiVax, a recombinant protein vaccine against ricin. *Vaccine* 23:4775–4784.

166. Vitetta ES, Smallshaw JE, Coleman E, Jafri H, Foster C, Munford R, and Schindler J. 2006. A pilot clinical trial of a recombinant ricin vaccine in normal humans. *Proc Natl Acad Sci USA* 103:2268–2273.

167. McLain DE, Horn TL, Detrisac CJ, Lindsey CY, and Smith LA. 2011. Progress in biological threat agent vaccine development: A repeat-dose toxicity study of a recombinant ricin toxin a-chain (rRTA) 1-33/44-198 vaccine (RVEc) in male and female New Zealand White rabbits. *Int J Toxicol* 30(2):143–152.

168. Meagher MM, Seravalli JG, Swanson ST, Ladd RG, Khasa YP, Inan M, Harner JC et al. 2011. Process development and cGMP manufacturing of a recombinant ricin vaccine: An effective and stable recombinant ricin a-chain vaccine-RVEc. *Biotechnol Prog* 27(4):1036–1047.

169. Porter A, Phillips G, Smith L, Erwin-Cohen R, Tammariello R, Hale M, and Dasilva L. 2011. Evaluation of a ricin vaccine candidate (RVEc) for human toxicity using an in vitro vascular leak assay. *Toxicon* 58:68–75.

170. Chanh TC, Romanowski MJ, and Hewetson JF. 1993. Monoclonal antibody prophylaxis against the in vivo toxicity of ricin in mice. *Immunol Invest* 22:63–72.

171. Dai J, Zhao L, Yang H, Guo H, Fan K, Wang H, Qian W, Zhang D, Li B, and Guo Y. 2011. Identification of a novel functional domain of ricin responsible for its potent toxicity. *J Biol Chem* 286:12166–12171.

172. Dertzbaugh MT, Rossi CA, Paddle BM, Hale M, Poretski M, and Alderton MR. 2005. Monoclonal antibodies to ricin: In vitro inhibition of toxicity and utility as diagnostic reagents. *Hybridoma (Larchmt)* 24:236–243.

173. Foxwell BM, Detre SI, Donovan TA, and Thorpe PE. 1985. The use of anti-ricin antibodies to protect mice intoxicated with ricin. *Toxicology* 34:79–88.

174. Hazen EL. 1927. General and local immunity to ricin. *J Immunol* 13:171–218.

175. Hewetson JF, Rivera VR, Creasia DA, Lemley PV, Rippy MK, and Poli MA. 1993. Protection of mice from inhaled ricin by vaccination with ricin or by passive treatment with heterologous antibody. *Vaccine* 11:743–746.

176. Griffiths GD, Lindsay CD, Allenby AC, Bailey SC, Scawin JW, Rice P, and Upshall DG. 1995. Protection against inhalation toxicity of ricin and abrin by immunisation. *Hum Exp Toxicol* 14:155–164.

177. Godal A, Fodstad O, and Pihl A. 1983. Antibody formation against the cytotoxic proteins abrin and ricin in humans and mice. *Int J Cancer* 32:515–521.

178. Houston LL. 1982. Protection of mice from ricin poisoning by treatment with antibodies directed against ricin. *J Toxicol Clin Toxicol* 19:385–389.

179. Lemley PV and Wright DC. 1992. Mice are actively immunized after passive monoclonal antibody prophylaxis and ricin toxin challenge. *Immunology* 76:511–513.

180. Olsnes S and Saltvedt E. 1975. Conformation-dependent antigenic determinants in the toxic lectin ricin. *J Immunol* 114:1743–1748.

181. Prigent J, Panigai L, Lamourette P, Sauvaire D, Devilliers K, Plaisance M, Volland H, Creminon C, and Simon S. 2011. Neutralising antibodies against ricin toxin. *PLoS One* 6:e20166.

182. Yermakova A, Vance DJ, and Mantis NJ. 2012. Sub-domains of ricin's B subunit as targets of toxin neutralizing and non-neutralizing monoclonal antibodies. *PLoS One* 7:e44317.

183. O'Hara JM, Whaley K, Pauly M, Zeitlin L, and Mantis NJ. 2012. Plant-based expression of a partially humanized neutralizing monoclonal IgG directed against an immunodominant epitope on the ricin toxin A subunit. *Vaccine* 30:1239–1243.

184. Roche JK, Stone MK, Gross LK, Lindner M, Seaner R, Pincus SH, and Obrig TG. 2008. Post-exposure targeting of specific epitopes on ricin toxin abrogates toxin-induced hypoglycemia, hepatic injury, and lethality in a mouse model. *Lab Invest* 88:1178–1191.

44 Saxitoxin and Related Paralytic Shellfish Poisons

Leanne A. Pearson and Brett A. Neilan

CONTENTS

44.1 INTRODUCTION

Saxitoxin and its analogues are neurotoxic alkaloids produced by certain strains of marine dinoflagellates and freshwater cyanobacteria. Collectively known as paralytic shellfish toxins (PSTs), these heterocyclic guanidine compounds are the causative agents of the condition known as paralytic shellfish poisoning (PSP).

The medical record for PSP dates back over 300 years,[1,2] and outbreaks of epidemic proportions have been recorded since the late 1800s.[3] Human poisonings typically occur via the ingestion of contaminated seafood such as filter-feeding bivalve mollusks and herbivorous finfish that feed on toxic algae.[4] Ingestion of water containing PST-producing cyanobacteria may also lead to saxitoxin poisoning, although this rarely occurs in humans. Intoxication is dose-dependent and symptoms range from a mild tingling of the lips to full cardiopulmonary arrest.[5,6] The LD_{50} parenterally for saxitoxin in mice is around 10 μg/kg[6a] making it one of the most toxic substances ever described. It has therefore been listed among the Schedule 1 substances in the Chemical Weapons Convention and is classified as a select agent by the US Department of Health and Human Services. Similar restrictions are also in place in the United Kingdom under the Anti-terrorism, Crime and Security Act of 2001, which limit the production, use, and transport of the toxin.

It is estimated that around 2000 human cases of PSP occur annually, with a 15% mortality rate.[7] The economic impact of harmful algal blooms is enormous, with an estimated loss of at least $82 million dollars per year in the United States alone.[8] The environmental impacts of PSTs are also devastating with numerous wildlife and livestock fatalities occurring annually.[9–12]

While there is currently no antidote for PSP, the recent genetic characterization of the saxitoxin biosynthesis pathway in cyanobacteria and dinoflagellates will allow for more efficient monitoring of harmful algal blooms, thereby reducing future impacts on human health, industry, and the environment.[13–17] On the other hand, knowledge of the genetic basis for saxitoxin production could be exploited for sinister purposes. While saxitoxin has been recognized as a potential biochemical weapon for decades, its mass production is precluded by the fact that it is difficult to synthesize and purify or alternatively extract from slow-growing producer organisms. However, modern biotechnological advances now permit the cloning and expression of entire toxin gene clusters in fast-growing heterologous hosts such as *Escherichia coli*.[18] Thus, the potential misuse of this toxin as a warfare agent must be reexamined.

44.2 CHEMICAL PROPERTIES AND PRODUCING ORGANISMS

44.2.1 CHEMICAL PROPERTIES

The chemical structure of two saxitoxin analogues, ethyl hemiketal saxitoxin and *p*-bromobenzenesulfonate saxitoxin, was inferred by x-ray diffraction in the mid-1970s.[19,20] Since then, over 57 PST analogues with varying toxicities have been described.[21] PSTs are low-molecular-weight compounds that share a common tricyclic perhydropurine backbone supporting a range of different functional side groups

TABLE 44.1
PSTs

Toxin	R1	R2	R3	R4[a]	R5
Nonsulfated					
Saxitoxin	H	H	H	$OCONH_2$	OH
Neosaxitoxin	OH	H	H	$OCONH_2$	OH
Monosulfated					
Gonyautoxin 1	OH	H	OSO_3^-	$OCONH_2$	OH
Gonyautoxin 2	H	H	OSO_3^-	$OCONH_2$	OH
Gonyautoxin 3	H	OSO_3^-	H	$OCONH_2$	OH
Gonyautoxin 4	OH	OSO_3^-	H	$OCONH_2$	OH
Gonyautoxin 5 (B1)	H	H	H	$OCONHSO_3^-$	OH
Gonyautoxin 6 (B2)	OH	H	H	$OCONHSO_3^-$	OH
Disulfated					
C1	H	H	OSO_3^-	$OCONHSO_3^-$	OH
C2	H	OSO_3^-	H	$OCONHSO_3^-$	OH
C3	OH	H	OSO_3^-	$OCONHSO_3^-$	OH
C4	OH	OSO_3^-	H	$OCONHSO_3^-$	OH
Decarbamoylated (dc)					
dc-Saxitoxin	H	H	H	OH	OH
dc-Neosaxitoxin	OH	H	H	OH	OH
dc-Gonyautoxin 1	OH	H	OSO_3^-	OH	OH
dc-Gonyautoxin 2	H	H	OSO_3^-	OH	OH
dc-Gonyautoxin 3	H	OSO_3^-	H	OH	OH
dc-Gonyautoxin 4	OH	OSO3-	H	OH	OH
Deoxy-decarbamoylated (do)					
do-Saxitoxin	H	H	H	H	OH
do-Gonyautoxin 1	OH	H	OSO_3^-	H	OH
do-Gonyautoxin 1	H	H	OSO_3^-	H	OH
L. wollei toxins (LWTXs)					
LWTX 1	H	H	OSO_3^-	$OCOCH_3$	H
LWTX 2	H	H	OSO_3^-	$OCOCH_3$	OH
LWTX 3	H	OSO_3^-	H	$OCOCH_3$	OH
LWTX 4	H	H	H	H	H
LWTX 5	H	H	H	$OCOCH_3$	OH
LWTX 6	H	H	H	$OCOCH_3$	H
Mono-hydroxy-benzoate analogues					
GC1	H	H	OSO_3^-	OCOPhOH	OH
GC2	H	OSO_3^-	H	OCOPhOH	OH
GC3	H	H	H	OCOPhOH	OH
Di-hydroxyl benzoate (DHB) analogues[b]					
GC1a	H	H	OSO_3^-	DHB	OH
GC2a	H	OSO_3^-	H	DHB	OH
GC3a	H	H	H	DHB	OH
GC4a	OH	H	OSO_3^-	DHB	OH
GC5a	OH	OSO_3^-	H	DHB	OH
GC6a	OH	H	H	DHB	OH

TABLE 44.1 (continued)
PSTs

Toxin	R1	R2	R3	R4[a]	R5
Sulfated benzoate (SB) analogues[b]					
GC1b	H	H	OSO_3^-	SB	OH
GC2b	H	OSO_3^-	H	SB	OH
GC3b	H	H	H	SB	OH
GC3b	H	H	H	SB	OH
GC5b	OH	OSO_3^-	H	SB	OH
GC6b	OH	H	H	SB	OH

[a] R4 group putatively assigned on the basis of major ions obtained via MS.[57]
[b] Adapted from Wiese, M. et al.[21]

R4 group structure:

(Table 44.1). Members of the PST family may be hydrophilic or hydrophobic and are usually classified according to their substituent side chains, which include carbamate, sulfate, hydroxyl, hydroxybenzoate, methyl, and acetate moieties.[21,22] In the case of saxitoxin, neosaxitoxin, and gonyautoxin, the functional side group is carbamate (Figure 44.1). The guanidinium groups on saxitoxin and related PSTs are essential for toxicity as are the hydroxyls on carbon 12.[22,23]

Saxitoxin is a highly polar, water-soluble molecule and is very stable under physiological conditions but degrades slowly under alkaline conditions.[24,25]

44.2.2 Producing Organisms

Interestingly, PSTs are produced by both prokaryotic cyanobacteria and eukaryotic dinoflagellates. Toxic bloom-forming species of cyanobacteria and dinoflagellates are often loosely termed "harmful algae." The cyanobacteria, or "blue-green algae," comprise a diverse group of oxygenic photosynthetic Gram-negative bacteria that possess the ability to synthesize chlorophyll a and several phycobilin proteins. Dinoflagellates, on the other hand, are unicellular protists that possess one to three flagella. About half of all dinoflagellates, including those producing PSTs, are photosynthetic and contain chlorophylls a and $c2$, as well as the carotenoid beta-carotene and a group of xanthophylls that appears to be unique to dinoflagellates.[26]

Most saxitoxin-producing cyanobacteria are filamentous nitrogen-fixing freshwater species that belong to the order Nostocales (e.g., *Anabaena* species, *Aphanizomenon flos-aquae*, *Cylindrospermopsis raciborskii*, and *Scytonema* species).[27–35] However, the toxin has also been isolated from the distantly related Oscillatoriales species, *Lyngbya wollei*.[36,37] Saxitoxin production in dinoflagellates is usually associated with the order Gonyaulacales (e.g., *Alexandrium* and *Pyrodinium* species); however, at least one species of the order Gymnodiniales (*Gymnodinium catenatum*) is also known to produce PSTs.[38,39] These organisms can occur in both temperate and tropical waters. Interestingly, PST production within cyanobacterial and dinoflagellate species is nonuniform, with toxic and nontoxic strains of the same species existing in nature.

44.3 MECHANISM OF TOXICITY AND EPIDEMIOLOGY

44.3.1 Mechanism of Toxicity

Saxitoxin and its analogues elicit their toxicity by binding to and inhibiting the flow of ions through voltage-dependent sodium channels.[40] This biological activity disrupts action potentials in nerve and muscle cells in the peripheral nervous system, leading to paralysis and

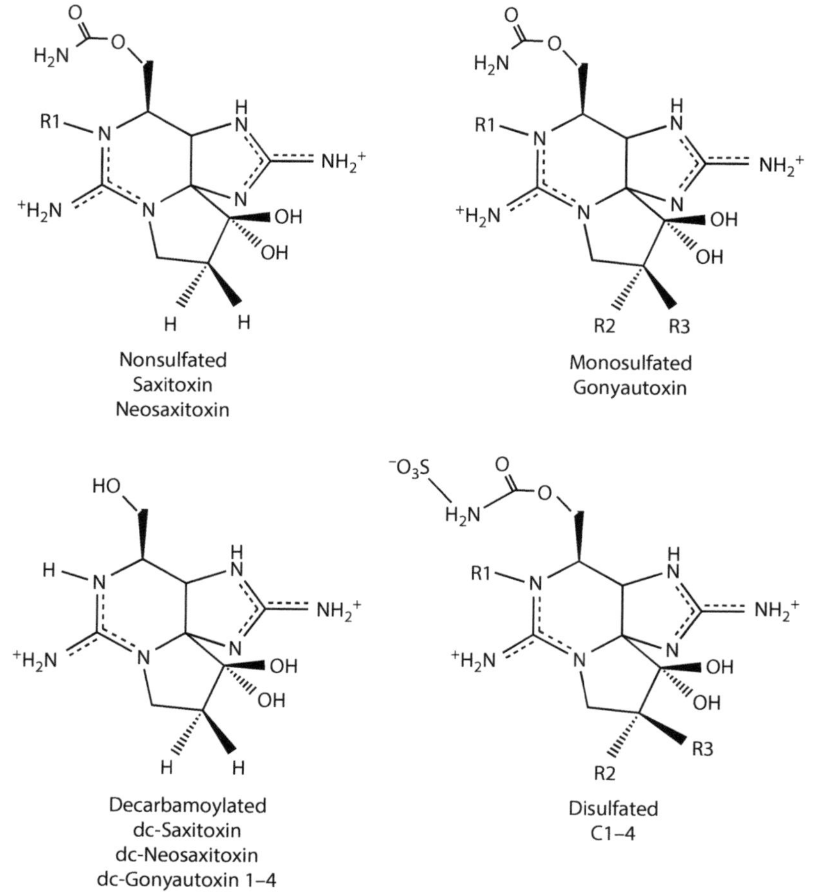

FIGURE 44.1 Structure of saxitoxin and analogous PSTs. Refer to Table 44.1 for assigned R groups.

sensory disturbances. Saxitoxin has also been demonstrated to bind a suite of other proteins, although with reduced potency. These include calcium and potassium channels as well as neuronal nitric oxide synthase and saxitoxin-metabolizing enzymes.[22]

The lethal saxitoxin oral dose in humans is ~1–4 mg, and since, levels up to 100 mg PST/g shellfish have been reported consumption of only a few contaminated shellfish has proven fatal in these rare cases. Furthermore, epidemiological investigations of PSP in Canada have indicated 200–600 μg of poison will produce symptoms in susceptible persons.[40a] The toxicity of saxitoxin derivatives varies greatly. The carbamate toxins (saxitoxin, neosaxitoxin, gonyautoxin-1 to gonyautoxin-4) are 10–200 times more potent than the corresponding N-sulfocarbamoyl derivatives (gonyautoxin-5/gonyautoxin-6 and C-toxins). N-sulfocarbamoyl analogues are labile, however, and may easily be converted to their corresponding carbamates.[41,42] Structure–activity relationships of PSTs are reviewed in detail by Llewellyn.[22]

44.3.2 EPIDEMIOLOGY

Almost 2000 cases of human PST-related poisonings are reported each year around the globe. Of these, around 15% are fatal.[7] Since the late 1800s, outbreaks of PSP have occurred in the United States, Europe, Japan, Southeast Asia, India, and Australia.[7] For example, in California, between 1927 and 1936, over 100 cases of PSP occurred as a result of the consumption of contaminated mussels. Six of these were fatal. In Canada, between 1880 and 1970, 187 cases of PSP were recorded. In the Philippines, between 1988 and 1998, 877 cases of PSP were reported, 44 of which were fatal.[3]

Over the past three decades, the occurrence of harmful algal incidents has increased in many parts of the world, both in frequency and in geographic distribution. It is not known whether this phenomenon is an artifact of increased monitoring programs or if it is an indication that PST-producing organisms are invading new habitats as a result of climate change and/or human activity.[43]

44.4 CLINICAL FEATURES

Symptoms of PSP generally manifest within 30 min to 2 h after consumption of contaminated seafood or water. Early symptoms include tingling and numbness of the lips, which spread to the face and neck in moderate cases. In severe cases, these symptoms spread to the extremities, with concomitant loss of motor control, drowsiness, and incoherence as well as brain stem dysfunction. Gastrointestinal symptoms, such as nausea, vomiting, and diarrhea, may also occur during the early stages of PSP. In severe cases, complete flaccid paralysis and cardiopulmonary arrest may result. PSP has also been associated with hypertension[44] and altered dopamine levels in the brains of human victims and laboratory animals.[44,45]

44.5 DIAGNOSIS

The diagnosis of PSP is based on the presentation of the aforementioned neurological symptoms (Section 44.4) together with a history of ingestion of shellfish or untreated water, often, but not always associated with a recent harmful algal bloom. Identification of the toxin in blood or urine is of little immediate benefit to the patient as there is no specific treatment for saxitoxin poisoning. However, testing potentially contaminated shellfish or water is imperative for preventing repeated poisonings from the same source. This is usually carried out in specialized laboratories rather than hospital clinics, as specialized equipment and expert interpretation of results are required (see Section 44.6).

44.6 TREATMENT AND PREVENTION

44.6.1 TREATMENT

There is currently no antidote or detoxification pathway for PSP, although several experimental trials using the drug, 4-aminopyridine, which selectively blocks Kv1 voltage-activated K^+ channels, have proven effective in reversing saxitoxin-induced lethality in guinea pigs when administered prior to, or immediately following, cardiopulmonary arrest. No harmful side effects were observed in these studies.[46–51]

Early treatment of symptoms is critical for a successful patient outcome. Mechanical ventilation in combination with DL-amphetamine (Benzedrine) is effective in most mild to moderate cases, but may not be sufficient to save the lives of severely poisoned individuals. Other treatments include fluid therapy and periodic monitoring of blood pH, as well as removal of unabsorbed toxin by gastric lavage incorporating activated carbon or diluted bicarbonate solution.[6] If rapid medical care is administered and the patient survives beyond 12 h, a complete recovery is likely.

The primary route of saxitoxin clearance is the kidney, which can remove toxins from the blood in under 24 h, even in patients who have experienced respiratory paralysis.[44,52,53] While some muscle weakness may persist for several weeks following PSP, permanent or chronic effects of saxitoxin poisoning have never been observed.

44.6.2 PREVENTION

PSP can be prevented through strict monitoring of seafood and drinking water in areas where harmful algal blooms are known to occur. Shellfish control authorities may use cell counts of toxin-producing algae in coastal waters to classify safe harvest areas and seasons. Routine detection of PSTs is also paramount, particularly during known bloom seasons. Chemical methods that are based on high-pressure liquid chromatography (HPLC) with either fluorescence (HPLC-FLD) or tandem mass spectroscopic (LC-MS[2]) detection are the methods of choice with regard to sensitivity and specificity and have been used extensively in PST monitoring programs.[54]

The US Food and Drug Administration (FDA) has established an action level of 0.8 ppm (80 µg/100 g) of saxitoxin equivalents in shellfish meat. Until recently, the primary diagnostic method for the detection of PSTs in seafood was the mouse bioassay; however, in 2004 and in response to the summary of actions from the 2003 Interstate Shellfish Sanitation Conference, the FDA concurred that the Jellett Rapid PSP test (JRPT) may also be used for screening shellfish tissues for saxitoxin and related compounds.[55] The JRPT is based on easy-to-use lateral flow immunochromatographic (LFI) test strips that can detect a range of natural PSTs with varying sensitivity. The JRPT is advantageous compared to the mouse bioassay for several reasons; it is rapid, sensitive, and cost-effective and it eliminates the need for laboratory animals. However, there have been concerns related to the test's cross-reactivity, leading to false-positive results.[55]

More recently, DNA-based methods are being employed to detect and monitor toxic algae. For example, Al-Tebrineh et al.[14] developed a quantitative PCR test for saxitoxin-producing cyanobacteria in freshwater bloom samples based on the saxitoxin biosynthesis gene target, *sxtA*. Saxitoxin concentrations correlated positively with *sxtA* gene copy numbers, indicating that the latter can be used as a quantitative measure of potential toxigenicity in cyanobacterial blooms.[14] A similar molecular test for potentially toxic dinoflagellates in marine water samples was later developed by Murray et al.[16] While these protocols have not been validated by government agencies, they are currently being employed by researchers in Australia to monitor harmful algal blooms that threaten the health of humans and the environment.[56]

44.7 CONCLUSIONS AND FUTURE PERSPECTIVES

Saxitoxin and related PSTs are among the most deadly toxins known to man. Produced by both marine and freshwater harmful algae, they constitute a significant hazard to human health and the environment. Monitoring programs have been put in place to reduce the risk of PSP, mainly in the developed world, yet human and animal poisonings from contaminated seafood and drinking water are still a frequent occurrence. Confounding this problem is the lack of a specific treatment for PSP.

A recent breakthrough in the field of harmful algal research was the discovery of the gene clusters for saxitoxin biosynthesis in cyanobacteria[13] and dinoflagellates.[17] These discoveries have paved the way for molecular-based detection methods for potentially harmful algae in fresh and marine waters as well as seafood.[14-16] These PCR-based protocols are advantageous to traditional methods used to detect PSTs insofar as they are more rapid, economically competitive, and incredibly sensitive. Furthermore, they can be adapted to target specific toxigenic strains or a wide range of toxigenic genera. Their ability to detect as few as 10 gene copies per reaction also means that they can be used as an early warning system to identify potentially hazardous water bodies prior to bloom proliferation. Amplified DNA may also be sequenced to provide taxonomic information about the offending organism(s), which may aid in their eradication. While the presence of toxin genes generally correlates well with actual strain toxicity,[14] these methods are best used to assess the potential toxicity of a bloom. Furthermore, they provide no indication of the toxin analogues present in a given sample. As different PSP analogues have incredibly diverse toxicities, molecular methods should therefore be complemented with the more traditional chemoanalytical techniques.

While the aforementioned advances in biotechnology promise to revolutionize food and water testing protocols, the potential misuse of this information should not be overlooked. The saxitoxin (*sxt*) gene cluster in cyanobacteria is around 35 kb in length. Clusters of this size and greater have been cloned and expressed in heterologous hosts such as *Streptomyces* and *E. coli*, which are fast-growing and easy to manipulate.[18] Mass production of saxitoxin via the heterologous expression of the *sxt* gene cluster could, therefore, potentially be achieved by persons with basic molecular biology skills.

While the threat of bioterrorism is ever present, natural exposure to PSTs is presently the greatest challenge. The apparent spread of harmful algal blooms from North America, Europe, and Japan into the previously uncontaminated waters of the Southern Hemisphere, including South America, Australia, Southeast Asia, and India, is a subject of recent debate. Whether this phenomenon is due to the dispersal of toxic algae via the ballast water of ships, the transplantation of shellfish stocks or other human activities is currently under investigation, as is the prospect that the increased distribution, incidence, and severity of these blooms reflect a changing global climate.

REFERENCES

1. Chevalier, A. and Duchesne, E.A. Mémoire sur les empoisonnements par les huïtres, les moules, les crabes, et par certain poissons de mer et de riviere. *Annales d'Hygiene Publique (Paris)* 46, 108–147 (1851).
2. Chevalier, A. and Duchesne, E.A. Mémoire sur les empoisonnements par les huïtres, les moules, les crabes, et par certain poissons de mer et de riviere. *Annales d'Hygiene Publique (Paris)* 45, 387–437 (1851).
3. James, K.J., Carey, B., O'Halloran, J., van Pelt, F.N., and Skrabakova, Z. Shellfish toxicity: Human health implications of marine algal toxins. *Epidemiology and Infection* 138, 927–940 (2010).
4. Deeds, J.R., Landsberg, J.H., Etheridge, S.M., Pitcher, G.C., and Longan, S.W. Non-traditional vectors for paralytic shellfish poisoning. *Marine Drugs* 6, 308–348 (2008).
5. Bower, D.J., Hart, R.J., Matthews, P.A., and Howden, M.E. Nonprotein neurotoxins. *Clinical Toxicology* 18, 813–863 (1981).
6. Kao, C.Y. Paralytic shellfish poisoning, in *Algal Toxins in Seafood and Drinking Water*, pp. 75–86. Academic Press, London, U.K. (1993).
6a. Halstead, B.W. and Schantz, E.J. Paralytic shellfish poisoning. *WHO Offset Publ.* 1984, 1–59.
7. Hallegraeff, G. A review of harmful algal blooms and their apparent global increase. *Phycologia* 32, 79–99 (1993).
8. Hoagland, P. and Scatasta, S. The economic effects of harmful algal blooms, in *Ecology of Harmful Algae*, Graneli, E. and Turner, J (eds.). Springer-Verlag, Dordrecht, the Netherlands (2006).
9. Coulson, J.C., Potts, G.R., Deans, I.R., and Fraser, S.M. Mortality of shags and other sea birds caused by paralytic shellfish poison. *Nature* 220, 23–24 (1968).
10. Hokama, Y. et al. Causative toxin(s) in the death of two Atlantic dolphins. *Journal of Clinical Laboratory Analysis* 4, 474–478 (1990).
11. Anderson, D.M. Red tides. *Scientific American* 271, 62–68 (1994).
12. Negri, A.P., Jones, G.J., and Hindmarsh, M. Sheep mortality associated with paralytic shellfish poisons from the cyanobacterium *Anabaena circinalis*. *Toxicon: Official Journal of the International Society on Toxinology* 33, 1321–1329 (1995).
13. Kellmann, R. et al. Biosynthetic intermediate analysis and functional homology reveal a saxitoxin gene cluster in cyanobacteria. *Applied Environmental Microbiology* 74, 4044–4053 (2008).
14. Al-Tebrineh, J., Mihali, T.K., Pomati, F., and Neilan, B.A. Detection of saxitoxin-producing cyanobacteria and *Anabaena circinalis* in environmental water blooms by quantitative PCR. *Applied and Environmental Microbiology* 76, 7836–7842 (2010).
15. Al-Tebrineh, J., Pearson, L.A., Yasar, S.A., and Neilan, B.A. A multiplex qPCR targeting hepato- and neurotoxigenic cyanobacteria of global significance. *Harmful Algae* 15, 19–25 (2011).
16. Murray, S.A. et al. sxtA-based quantitative molecular assay to identify saxitoxin-producing harmful algal blooms in marine waters. *Applied and Environmental Microbiology* 77, 7050–7057 (2011).
17. Stuken, A. et al. Discovery of nuclear-encoded genes for the neurotoxin saxitoxin in dinoflagellates. *PLoS One* 6, e20096 (2011).
18. Gao, X., Wang, P., and Tang, Y. Engineered polyketide biosynthesis and biocatalysis in *Escherichia coli*. *Applied Microbiology and Biotechnology* 88, 1233–1242 (2010).
19. Bordner, J., Thiessen, W.E., Bates, H.A., and Rapoport, H. The structure of a crystalline derivative of saxitoxin. The structure of saxitoxin. *Journal of the American Chemical Society* 97, 6008–6012 (1975).
20. Schantz, E.J. et al. Letter: The structure of saxitoxin. *Journal of the American Chemical Society* 97, 1238 (1975).
21. Wiese, M., D'Agostino, P.M., Mihali, T.K., Moffitt, M.C., and Neilan, B.A. Neurotoxic alkaloids: Saxitoxin and its analogs. *Marine Drugs* 8, 2185–2211 (2010).
22. Llewellyn, L.E. Saxitoxin, a toxic marine natural product that targets a multitude of receptors. *Natural Product Reports* 23, 200–222 (2006).

23. Strichartz, G. Structural determinants of the affinity of saxitoxin for neuronal sodium channels. Electrophysiological studies on frog peripheral nerve. *The Journal of General Physiology* 84, 281–305 (1984).

24. Rogers, R.C. and Rapoport, H. The pKa's of saxitoxin. *Journal of the American Chemical Society* 102, 7335 (1980).

25. Stafford, R.G. and Hines, H.B. Urinary elimination of saxitoxin after intra-venous injection. *Toxicon* 33, 1501–1510 (1995).

26. Hackett, J.D., Anderson, D.M., Erdner, D.L., and Bhattacharya, D. Dinoflagellates: A remarkable evolutionary experiment. *American Journal of Botany* 91, 1523–1534 (2004).

27. Velzeboer, R.M.A. et al. Geographical patterns of occurrence and composition of saxitoxins in the cyanobacterial genus *Anabaena* (Nostocales, Cyanophyta) in Australia. *Phycologia* 39, 395–407 (2000).

28. Humpage, A.R. et al. Paralytic shellfish poisons from Australian cyanobacterial blooms. *Australian Journal of Marine and Freshwater Research* 45, 761–771 (1994).

29. Ikawa, M., Wegener, K., Foxall, T.L., and Sasner, J.J., Jr. Comparison of the toxins of the blue–green alga *Aphanizomenon flos-aquae* with the gonyaulax toxins. *Toxicon* 20, 747–752 (1982).

30. Dias, E., Pereira, P., and Franca, S. Production of paralytic shellfish toxins by *Aphanizomenon* sp. LMECYA 31 (cyanobacteria). *Journal of Phycology* 38, 705–712 (2002).

31. Sasner, J.J., Jr., Ikawa, M., and Foxall, T.L. Studies on *Aphanizomenon* and *Microcystis* toxins. In ACS Symposium Series No. 262, Seafood Toxins, 391-406. Ragelis, E.; American Chemical Society: Washington, DC (1984).

32. Pereira, P. et al. Paralytic shellfish toxins in the freshwater cyanobacterium *Aphanizomenon flos-aquae*, isolated from Montargil reservoir, Portugal. *Toxicon* 38, 1689–1702 (2000).

33. Lagos, N. et al. The first evidence of paralytic shellfish toxins in the freshwater cyanobacterium *Cylindrospermopsis raciborskii*, isolated from Brazil. *Toxicon* 37, 1359–1373 (1999).

34. Molica, R. et al. Toxins in the freshwater cyanobacterium *Cylindrospermopsis raciborskii* (Cyanophyceae) isolated from Tabocas reservoir in Caruaru, Brazil, including demonstration of a new saxitoxin analogue. *Phycologia* 41, 606–611 (2002).

35. Smith, F.M., Wood, S.A., van Ginkel, R., Broady, P.A., and Gaw, S. First report of saxitoxin production by a species of the freshwater benthic cyanobacterium, *Scytonema* Agardh. *Toxicon* 57, 566–573 (2011).

36. Carmichael, W.W., Evans, W.R., Yin, Q.Q., Bell, P., and Moczydlowski, E. Evidence for paralytic shellfish poisons in the freshwater cyanobacterium *Lyngbya wollei* (Farlow ex Gomont) comb. nov. *Applied and Environmental Microbiology* 63, 3104–3110 (1997).

37. Onodera, H., Satake, M., Oshima, Y., Yasumoto, T., and Carmichael Wayne, W. New saxitoxin analogues from the freshwater filamentous cyanobacterium *Lyngbya wollei*. *Natural Toxins* 5, 146–151 (1997).

38. Oshima, Y., Hasegawa, M., Yasumoto, T., Hallegraeff, G., and Blackburn, S. Dinoflagellate *Gymnodinium catenatum* as the source of paralytic shellfish toxins in Tasmanian shellfish. *Toxicon* 25, 1105–1111 (1987).

39. Negri, A. et al. Three novel hydroxybenzoate saxitoxin analogues isolated from the dinoflagellate *Gymnodinium catenatum*. *Chemical Research in Toxicology* 16, 1029–1033 (2003).

40. Narahashi, T., Haas, H.G., and Therrien, E.F. Saxitoxin and tetrodotoxin: Comparison of nerve blocking mechanism. *Science* 157, 1441–1442 (1967).

40a. The US Food and Drug Administration National Shellfish Sanitation Program Guide (2007). http://www.fda.gov/

41. Cembella, A.D., Shumway, S.E., and Larocque, R. Sequestering and putative biotransformation of paralytic shellfish toxins by the sea scallop *Placopecten magellanicus*—Seasonal and spatial scales in natural populations. *Journal of Experimental Marine Biology and Ecology* 180, 1–22 (1994).

42. Cembella, A.D., Shumway, S.E., and Lewis, N.I. Anatomical distribution and spatio-temporal variation in paralytic shellfish toxin composition in two bivalve species from the Gulf of Maine. *Journal of Shellfish Research* 12, 389–403 (1993).

43. Van Dolah, F.M. Marine algal toxins: Origins, health effects, and their increased occurrence. *Environmental Health Perspectives* 108, 133–141 (2000).

44. Gessner, B.D. et al. Hypertension and identification of toxin in human urine and serum following a cluster of mussel-associated paralytic shellfish poisoning outbreaks. *Toxicon: Official Journal of the International Society on Toxinology* 35, 711–722 (1997).

45. Cervantes Cianca, R.C., Faro, L.R., Duran, B.R., and Alfonso, P.M. Alterations of 3,4-dihydroxyphenylethylamine and its metabolite 3,4-dihydroxyphenylacetic produced in rat brain tissues after systemic administration of saxitoxin. *Neurochemistry International* 59, 643–647 (2011).

46. Benton, B.J., Keller, S.A., Spriggs, D.L., Capacio, B.R., and Chang, F.C. Recovery from the lethal effects of saxitoxin: A therapeutic window for 4-aminopyridine (4-AP). *Toxicon* 36, 571–588 (1998).

47. Benton, B.J., Spriggs, D.L., Capacio, B.R., and Chang, F.-C.T. 4-Aminopyridine antagonizes the lethal effects of saxitoxin (STX) and tetrodotoxin (TTX). In *International Society of Toxicology, 5th Pan American Symposium on Animal, Plant and Microbial Toxins*, p. 217. Frederick, MD (1995).

48. Chang, F.C., Spriggs, D.L., Benton, B.J., Keller, S.A., and Capacio, B.R. 4-Aminopyridine reverses saxitoxin (STX)- and tetrodotoxin (TTX)-induced cardiorespiratory depression in chronically instrumented guinea pigs. *Fundamental and Applied Toxicology: Official Journal of the Society of Toxicology* 38, 75–88 (1997).

49. Chen, H.M., Lin, C.H., and Wang, T.M. Effects of 4-aminopyridine on saxitoxin intoxication. *Toxicology and Applied Pharmacology* 141, 44–48 (1996).

50. Chang, F.C., Bauer, R.M., Benton, B.J., Keller, S.A., and Capacio, B.R. 4-Aminopyridine antagonizes saxitoxin-and tetrodotoxin-induced cardiorespiratory depression. *Toxicon: Official Journal of the International Society on Toxinology* 34, 671–690 (1996).

51. Santamaria, D., Rios, C., Kravzov, J., and Altagracia, M. 4-Aminopyridine antagonizes lethality induced by saxitoxin in mice. *Proceedings of the Western Pharmacology Society* 36, 193–195 (1993).

52. Hines, H.B., Naseem, S.M., and Wannemacher, R.W., Jr. [³H]-Saxitoxinol metabolism and elimination in the rat. *Toxicon* 31, 905–908 (1993).

53. Andrinolo, D., Michea, L.F., and Lagos, N. Toxic effects, pharmacokinetics and clearance of saxitoxin, a component of paralytic shellfish poison (PSP), in cats. *Toxicon: Official Journal of the International Society on Toxinology* 37, 447–464 (1999).

54. van de Riet, J. et al. Liquid chromatography post-column oxidation (PCOX) method for the determination of paralytic shellfish toxins in mussels, clams, oysters, and scallops: Collaborative study. *Journal of AOAC International* 94, 1154–1176 (2011).

55. Oshiro, M. et al. Paralytic shellfish poisoning surveillance in California using the Jellett Rapid PSP test. *Harmful Algae* 5, 69–73 (2006).

56. Al-Tebrineh, J. et al. Community composition, toxigenicity, and environmental conditions during a cyanobacterial bloom occurring along 1,100 kilometers of the Murray River. *Applied and Environmental Microbiology* 78, 263–272 (2012).

57. Vale, P. Complex profiles of hydrophobic paralytic shellfish poisoning compounds in *Gymnodinium catenatum* identified by liquid chromatography with fluorescence detection and mass spectrometry. *Journal of Chromatography A* 1195, 85–93 (2008).

45 Shiga Toxin and Shiga-Like Ribosome-Inactivating Proteins

Moo-Seung Lee, Rama P. Cherla, and Vernon L. Tesh

CONTENTS

45.1 INTRODUCTION

The use of biological agents and toxins as weapons of mass destruction is particularly nefarious given that detection of an attack may not occur until significant numbers of patients are diagnosed. In the case of a highly contagious agent, considerable numbers of secondary cases may occur during the lag time between dissemination of the agent and identification of an attack.

The US Centers for Disease Control and Prevention (CDC) has established three categories of biological agents and toxins that may potentially be used in an attack on civilian populations. Category A agents are most dangerous. They are easily disseminated, frequently via aerosols, and may be readily spread person-to-person. Intentional release of Category A agents would result in high mortality rates leading to significant disruption of normal society. Category B agents are more difficult to disseminate and are frequently incapacitating agents, with moderate morbidity and low mortality. Category C agents are emerging agents requiring additional genetic engineering to improve delivery and/or lethality.[1]

"Food safety threats," which include Shiga toxin-producing bacteria *Shigella dysenteriae* serotype 1 and multiple serotypes of *Escherichia coli*, and ricin, the toxin expressed by the castor bean plant that shares an identical mechanism of action as the Shiga toxins, are currently listed as Category B agents. The USA PATRIOT Act (Uniting and Strengthening America by Providing Appropriate Tools Required to Intercept and Obstruct Terrorism Act of 2001)[2] and the Bioterrorism Act (Public Health Security and Bioterrorism Preparedness and Response Act of 2002)[3] define the regulatory authority of the Department of Health and Human Services (HHS)/CDC and the Department of Agriculture/Animal and Plant Health Inspection Service (APHIS) to restrict access and define safety and security measures for the possession, use, and transport of biological agents and toxins that may constitute a significant bioterrorist threat. The CDC and APHIS share responsibility for 80 agents that potentially threaten humans as well as animals used in food production.

Until recently, Shiga toxin and Shiga-like ribosome-inactivating proteins were on the HHS/CDC Select Agents and Toxins List, requiring registration, review of procedures, and inspection of laboratories that possess, use, and transport the toxins.[4] However, in a ruling dated October 5, 2012, Shiga toxin and Shiga-like ribosome-inactivating proteins were removed from the HHS/CDC Select Agents and Toxins List.[5] This decision was based primarily on considerations dealing with the delivery of the toxins as bioweapons, including the following: (1) introduction of Shiga toxins by the aerosol route has not been reported; (2) Shiga toxins are extremely difficult to synthesize in quantities that are toxic to humans; (3) expression of toxin in bacteria is self-limiting due to inhibitory effects on bacterial cells of overexpressed toxin; (4) limitations to purification and concentration of Shiga toxins make them impractical and ill-suited to methods of dispersal that would require large quantities of toxin for delivery by food, water, or air; (5) difficulty in producing or administering large quantities of toxin via the aerosol route, and their poor environmental stability; (6) lack of significant toxicity seen with oral exposure to purified Shiga toxins (which is the route by which an individual becomes intoxicated by Shiga toxin-expressing bacteria); and (7) the observation that the worst effects seen with intoxication may require the expression of other pathogenic factors by Shiga toxin-producing strains of *E. coli*, which are not regulated.[5]

In spite of the recent ruling, the toxigenic/pathogenic potential of Shiga toxins and Shiga-like ribosome-inactivating proteins to humans cannot be ignored. This chapter will review the genetics, structure, and function of the Shiga toxins, a family of conserved bacterial protein exotoxins that inactivate eukaryotic ribosomes. The natural reservoirs of, the diseases caused by, and treatment options for Shiga toxin-producing *S. dysenteriae* serotype 1 and *E. coli* will be briefly reviewed.

45.2 TOXIN CLASSIFICATION AND STRUCTURE

45.2.1 CLASSIFICATION

45.2.1.1 Shiga Toxins

The bacteriologist Kiyoshi Shiga described an epidemic of bacillary dysentery in Japan in 1896. The organism isolated from these patients was called Shiga's bacillus. In 1903, a lethal toxic activity present in extracts prepared from heat-killed Shiga's bacillus was described (the early history of the characterization of Shiga's bacillus is reviewed in O'Brien et al.[6]). The capacity of *S. dysenteriae* serotype 1, a causative agent of bacillary dysentery, to express a neurotoxic protein was reported in the mid-1950s.[7,8] Based on the capacity of the purified toxin to cause flaccid paralysis in mice and rabbits, the toxin was initially referred to as *Shigella* neurotoxin. These early studies showed that the toxin acted on the neural microvasculature rather than directly damaging neural tissues. When it was subsequently shown that the toxin was capable of causing fluid accumulation in rabbit ligated ileal loops and vascular damage in the intestinal tract of nonhuman primates,[9,10] the term Shiga toxin became widely used.

The genetic locus encoding Shiga toxin was mapped to the *trp-pyrF* region of the *S. dysenteriae* serotype 1 chromosome.[11] Subsequently, the toxin structural genes were cloned and sequenced.[12,13] Two genes encoding Shiga toxin A- and B-subunits are organized in an operon, with the *stxA* gene proximal to the promoter element. The *stx* promoter possesses a Fur operator site conferring transcriptional repression in the presence of iron. The *stxA* and *stxB* genes are separated by 12 untranslated nucleotides. Probes homologous to *E. coli* 16S rRNA were used to define ribosome-binding sites within the *stx* genes. The *stxA* and *stxB* genes encode signal peptides of 22 and 20 amino acids, respectively. The sequence for the mature A-subunit encodes 293 amino acids with a predicted molecular weight of 32,225 Da. The *stxB* gene encodes a

protein comprised of 69 amino acids with a calculated molecular weight of 7691 Da.[6]

45.2.1.2 Shiga-Like Toxins

In 1977, Konowalchuk et al.[14] described a cytotoxic activity present in some culture filtrates prepared from *E. coli* strains isolated from patients with diarrhea. Unlike *E. coli* heat-labile enterotoxins that caused morphological changes in Vero, Y-1, and CHO cells, the novel toxic activity killed Vero cells, but did not kill Y-1 or CHO cells. Based on this observation, the toxin was referred to as Vero cytotoxin or verotoxin. Genes associated with *E. coli* verotoxin activity could be conferred to nontoxigenic *E. coli* recipients by lysogenic conversion.[15,16] In the early 1980s, *E. coli* O157:H7 (then considered a rare serotype of *E. coli*) was identified as the etiologic agent of foodborne outbreaks of bloody diarrhea (hemorrhagic colitis) and acute renal failure (hemolytic uremic syndrome) in the United States and Canada.[17–20] *E. coli* O157:H7 strain 933 was isolated from ground beef implicated as a vehicle in an outbreak of hemorrhagic colitis. The strain was shown to possess toxin activities similar to Shiga toxin; that is, the strain produced exotoxins that were capable of killing Vero and HeLa cells, were paralytic and lethal for mice, and caused fluid accumulation in rabbit ileal loops.[21] Based on these observations, toxins expressed by strain 933 were characterized as Shiga-like toxins. O'Brien et al.[22] subsequently developed a protocol to induce Shiga-like toxin-converting bacteriophages from *E. coli* O157:H7 and other toxin-producing *E. coli* serotypes. Strain 933 was shown to be lysogenized with two Shiga-like toxin-converting bacteriophages designated φ933J and φ933W. The cytotoxicity of the Shiga-like toxin expressed by φ933J was neutralized by antibodies directed against Shiga toxin, suggesting that the toxin encoded by φ933J and Shiga toxin expressed by *S. dysenteriae* serotype 1 were antigenically similar. In contrast, the cytotoxic activity of the φ933W-encoded Shiga-like toxin was not neutralized by antibody directed against Shiga toxin. Antibodies directed against the φ933W-encoded Shiga-like toxin failed to block cytotoxicity mediated by Shiga toxin or the φ933J-encoded toxin. Thus, *E. coli* O157:H7 strain 933 was capable of producing two antigenically distinct but functionally related Shiga-like toxins. The toxin that was neutralized by anti-Shiga toxin antibodies was designated Shiga-like toxin 1, while Shiga-like toxin 2 was not neutralized by anti-Shiga toxin antibodies.[23]

Nucleotide sequencing of the structural genes for the toxins clarified these findings. The deduced amino acid sequence of Shiga toxin and Shiga-like toxin 1 was reported to differ by a single amino acid in the A-subunit.[13,24] In contrast, Shiga-like toxin 2 showed only ~56% homology at the amino acid sequence level compared to Shiga/Shiga-like toxin 1.[25] Based on the genetic and functional similarities among the toxins, it was recommended that the nomenclature of the toxins be changed so that Shiga-like toxin 1 would be referred to as Shiga toxin 1 (Stx1) and Shiga-like toxin 2 would be referred to as Shiga toxin 2 (Stx2). Alternative designations for the toxins are verotoxin 1 (VT1) and verotoxin 2 (VT2).[26]

While the Shiga toxin and Stx1 A and B genes encode proteins containing 293 and 69 amino acids, respectively, *stx2A* and *stx2B* genes encode slightly larger proteins comprised of 296 and 70 amino acids. Like Shiga toxin, Stx1 expression is iron repressed, while the *stx2* promoter lacks the Fur operator sequence.[6] The *stx* genes in *S. dysenteriae* serotype 1 localize to a prophage defective for excision/lysis because of a loss of phage sequences following multiple insertion sequence (IS) element insertions. The Shiga toxin-encoding defective prophage contains regions with homology to φ80, P22, and lambdoid bacteriophages, suggesting that the prophage has undergone extensive gene rearrangement events.[27]

A number of genetic variants of Stx1 and Stx2 may be expressed by Shiga toxin-producing *E. coli* (STEC). As was the case with *E. coli* O157:H7 strain 933, which expressed both Stx1 and Stx2, a single *E. coli* isolate may express multiple toxins and/or toxin variants. Stx1 genetic variants include Stx1c and Stx1d.[28,29] Relatively more Stx2 variants have been defined: Stx2c, Stx2c2, Stx2d, mucus-activatable Stx2d (Stx2d_{act}), Stx2e, Stx2f, and Stx2g. Most of the toxin genetic variants have been isolated from patients with uncomplicated diarrhea. STEC producing Stx2e are primarily associated with edema disease in pigs and are only rarely isolated from humans.[30] Stx2f-producing *E. coli* were initially isolated from pigeons but are isolated with increasing frequency from children with severe diarrhea.[31,32] Stx2g-producing strains have been isolated from the feces of asymptomatic cattle and wild deer and from contaminated water.[33–35]

Interestingly, a recent genetic analysis of 24 Stx2g-expressing STEC strains isolated from human, zoonotic, and environmental sources revealed that they all coexpressed the gene for heat-stable enterotoxin STIa of enterotoxigenic *E. coli* (ETEC), suggesting the emergence of a STEC/ETEC "blended" pathotype.[36] Clinical and epidemiological studies show that three toxin types, Stx2, Stx2c, and Stx2d_{act}, are more frequently isolated from patients with hemorrhagic colitis and/or systemic complications such as the acute renal failure syndrome termed hemolytic uremic syndrome (HUS).[37–40] Stx2c was initially characterized from an *E. coli* O157:H⁻ strain isolated from an HUS patient. This strain expressed both *stx2* and *stx2c* genes.[41] Stx2 and Stx2c show 99% and 96% identity in genes encoding the A- and B-subunits, respectively. One of the amino acid changes in Stx2c maps to a toxin receptor binding site in the B-subunit, which may alter the affinity of toxin binding to its receptor on mammalian cells.[42] Bacteriophages encoding Stx2c integrate within the *E. coli* chromosome at a different locus compared to Stx1- and Stx2-encoding phages.[43] Stx2d_{act} was characterized from *E. coli* O91:H21 isolated from a patient with HUS.[44] The B-subunits of Stx2c and Stx2d_{act} are identical, but Stx2d_{act} possesses unique amino acids in the A-subunit. The cytotoxicity of Stx2d_{act} for Vero cells and the lethality of Stx2d_{act} for mice were shown to be markedly increased when the toxin was preincubated with elastase isolated from human or murine intestinal mucus.[45,46] It was subsequently shown that elastase acts at a trypsin-sensitive site to cleave the Stx2d_{act} A-subunit into a ~27.5 kDa A_1-fragment and a

~4.5 kDa A_2-fragment and cleaves two amino acids from the carboxy-terminus of the Stx2d_{act} A_2-fragment.[47] Precisely how the unique processing of Stx2d_{act} enhances cytotoxicity and murine lethality remains to be fully characterized. STEC that express Stx2d_{act} as the sole toxin have been associated with severe clinical outcome.[39] Sheep and cattle appear to be important natural reservoirs for carriage of Stx2d_{act}-expressing *E. coli*.[48]

45.2.2 STRUCTURE

All members of the Shiga toxin family share structural properties. The toxins possess an AB_5 molecular structure; that is, a single A-subunit noncovalently associates with five identical B-subunits. The pentameric B-subunits form a ring with a central pore with a diameter of approximately 11 Å.[49] The carboxy-terminus of the A-subunit fits within the central pore of B-subunits.[50,51] Other important toxins produced by enteric pathogens share the AB_5 structure, including cholera toxin expressed by *Vibrio cholerae* and heat-labile enterotoxin expressed by ETEC.[52,53] Comparative analyses of the structures of Shiga toxin/Stx1 and Stx2 revealed that the enzymatic site responsible for *N*-glycosidase activity (see Section 45.3) is occluded in the Stx1 holotoxin molecule, while the active site in the Stx2 holotoxin is accessible and shows electronic density for a molecule of formic acid from the crystallization solution bound at the active site.[51] Furthermore, the C-terminus of the Stx2 A-subunit forms a short α-helix after passing through the central pore of pentameric B-subunits. This structure is absent at the C-terminus of the Shiga toxin/Stx1 A-subunit. How differences in active site accessibility and A-subunit C-termini between Shiga toxin/Stx1 and Stx2 affect pathogenesis remain to be fully characterized.

45.3 TOXIN BIOLOGY AND EPIDEMIOLOGY OF DISEASES CAUSED BY SHIGA TOXIN-PRODUCING BACTERIA

45.3.1 TOXIN BIOLOGY

The A-subunits of Shiga toxins are highly specific *N*-glycosidases acting to remove a single adenine base, located at position 4324 in the rat, from the ribose-phosphate backbone of the 28S rRNA component of 60S eukaryotic ribosomes.[54,55] The depurination reaction is directed to an unpaired adenine located within a loop structure created by non-Watson–Crick base pairing. This loop structure is referred to as the α-sarcin/ricin loop (SRL) since the ribosome-inactivating proteins α-sarcin expressed by the fungus *Aspergillus giganteus* and ricin expressed by the castor bean plant (*Ricinus communis*) also act at this rRNA site. Although Shiga toxins and ricin share little overall structural homology, their mechanisms of action are identical, while α-sarcin is an endoribonuclease that cleaves the phosphodiester bond adjacent to the adenine at position 4324.[56,57] Shiga toxin enzymatic activity results in a loss of ribosomal interaction with, and subsequent GTPase function of, elongation factors, leading to protein synthesis inhibition.[58] While the toxins are effective protein synthesis inhibitors, recent studies show that Shiga toxins are multifunctional molecules capable of triggering host cell stress responses.[59] The role of host cell responses in pathogenesis is discussed later (Section 45.4). Epithelial cell lines, such as Vero or HeLa cells, are employed in the laboratory to assess Shiga toxin-mediated protein synthesis inhibition, and 50% cytotoxicity (CD_{50}) values are routinely measured in fg/mL to ng/mL amounts.

The pentameric B-subunits are responsible for toxin binding to target cells. The toxin receptor identified for most Shiga toxins causing disease in humans is the neutral glycolipid globotriaosylceramide (Gb3), although Stx2e, the causative agent of edema disease in swine, binds the glycolipid globotetraosylceramide (Gb4).[60–62] Structure–function studies suggest that there are 15 potential Gb3 binding sites *per* B-subunit pentamer, explaining the high affinity ($K_D \sim 10^{-9}$ M) binding of Shiga toxins with Gb3.[63,64] Thus, the interaction of Shiga toxins with cells expressing Gb3 leads to membrane glycolipid receptor cross-linking and activation of toxin internalization mechanisms that may be clathrin-dependent or clathrin-independent.[65,66] Although Shiga toxins interact with the $Gal(\alpha1\rightarrow4)Gal(\beta1\rightarrow4)Glu$ trisaccharide moiety of Gb3, the lipid component of the receptor plays an important role in toxin binding and internalization. There are multiple isoforms of Gb3 that differ in terms of carbon chain length and degree of saturation and hydroxylation of the ceramide component of the toxin receptor (reviewed in Müthing et al.[67]). The expression of different patterns of Gb3 isoforms in the cell membrane may significantly affect toxin binding, internalization, and intracellular routing. Although the expression of Gb3 is a major factor in the determination of target cell sensitivity, Shiga toxins appear capable of binding additional membrane receptors on the surface of platelets and polymorphonuclear neutrophils (PMNs).[68,69] Although these alternative toxin receptors remain to be definitively characterized, initial studies suggest that the interaction of the toxins with receptors on PMNs does not result in toxin internalization and that PMNs may serve to disseminate the toxins in the bloodstream to target organs containing microvascular endothelial cells rich in membrane Gb3[70] (see Section 45.4).

In order to be an effective inhibitor of protein synthesis, an activated or processed form of the A-subunit must gain access to ribosomes within the target cell cytoplasm. Shiga toxins use host cell intracellular transport apparatus to reach the endoplasmic reticulum (ER), an intracellular compartment rich in membrane-associated ribosomes and the active site of protein synthesis.[71] Using energy-depleted HeLa cells and cytosol-free model membranes, Römer et al.[66] showed that Shiga toxin B-subunit binding was sufficient to induce negative membrane curvature and the generation of toxin-containing invaginations. However, toxin binding also activates membrane-associated signaling molecules that may be necessary for further toxin uptake and intracellular routing (reviewed by Sandvig et al.[72]). Target cell sensitivity is critically dependent upon subsequent intracellular routing of the

toxins. Primary human monocytes express membrane Gb3 but are relatively resistant to killing by the toxins. Shiga toxins are routed to phagolysosomes in human monocytes where the toxin molecules are efficiently degraded.[73–75] In cells that are sensitive to Shiga toxin-mediated protein synthesis inhibition, the partitioning of Gb3 into lipid-enriched microdomains (lipid rafts) appears to be necessary for initiation of retrograde transport leading to the delivery of toxins to the ER. In toxin-sensitive cells, the late endosome pathway is bypassed, so that the toxins utilize a retrograde early/recycling endosome-to-*trans*-Golgi network (TGN) transport process normally used to recycle host cell lipids and proteins.[76,77] The toxins are then transported through Golgi cisternae to reach the lumen of the ER. The reader is referred to several recent reviews of Shiga toxin retrograde transport for additional details.[72,78,79]

How activated or processed Shiga toxins cross the ER membrane to reach ribosomes is an area of active study. Fluorescent microscopy studies utilizing dual-labeled Stx1 A- and B-subunits showed that the AB_5 holotoxin is delivered into the lumen of the ER.[80] During retrograde transport, the A-subunit is cleaved by furin or calpain to form a 27.5 kDa A_1-fragment (amino acids 1–251) and a 4.5 kDa A_2-fragment (amino acids 252–293).[81] The A-subunit fragments remain associated through a disulfide bond that spans the cleavage site. In the ER, the disulfide bond is reduced, freeing the A_1-fragment from $A_2 + B_5$. The enzymatic activity of Shiga toxins resides with the processed A_1-fragment. The next steps in the translocation of the A_1-fragment across the ER membrane remain to be fully characterized. It is known that the reduction of the disulfide bond linking the ricin A-chain with the ricin AB holotoxin results in a conformational change in the A-chain exposing a hydrophobic region previously embedded within the holotoxin molecule. Exposure of this hydrophobic region may be essential for ricin A-chain association with the ER membrane (reviewed by Spooner and Lord).[82] Whether a similar conformational change occurs in the Shiga toxin A_1-fragment following cleavage of the disulfide bond requires additional study. However, a hydrophobic region in the A_1-fragment is necessary for retrotranslocation in yeast ER.[83] Furthermore, synthetic peptides based on the sequence of this A_1-fragment hydrophobic region will readily associate with lipid membranes.[84] Thus, in order to associate with the ER membrane, the Shiga toxin A-subunit appears to require processing, possibly by the enzyme protein disulfide isomerase (PDI).

The ER-associated degradation (ERAD) process is a host cell mechanism by which improperly folded proteins bind chaperones in the ER lumen. The aberrant proteins are then directed to retrotranslocate (or "dislocate") through the ER membrane using membrane-localized ubiquitin ligase complexes that polyubiquitinate misfolded substrates.[85,86] Once in the cytosol, aberrant host proteins interact with the AAA-ATPase p97/Cdc48p complex that routes misfolded proteins to the proteasome for degradation.[87,88] Shiga toxins appear to utilize at least some elements of the ERAD process to access the cytosol. Exposure of a hydrophobic structure within the

A$_1$-fragment may lead to association with ER intralumenal chaperones ERdj3/HEDJ, GRP94, and BiP.[83,89,90] In essence, the A$_1$-fragment may then be sensed by the host cell as an unfolded protein. Although toxin complexed with chaperones has been shown to associate with the Sec61 translocon, attempts to directly co-immunoprecipitate components of the Sec61 translocon with Stx1 A$_1$-fragments were unsuccessful.[80,89,90] Thus, the precise nature of the retrotranslocon (or "dislocon") used to transport Shiga toxin A$_1$-fragments remains to be fully described. However, it has been shown that ricin A-chain uses the yeast Hrd1p E3 ubiquitin ligase complex as a retrotranslocation portal.[91] Ricin A-chain retrotranslocation appears to be independent of Hrd1p complex ubiquitination activity and may be dependent on direct interaction of ricin A-chain with a proteasomal cap protein. In contrast to ricin A-chain, Shiga toxin A$_1$-fragments appear to utilize yeast Hrd1p complex for retrotranslocation in a manner that requires ubiquitination.[92] It is unclear how Shiga toxin A$_1$-fragments escape proteolytic degradation in the cytosol, although inhibition of proteasome activity does increase cytotoxicity, suggesting that some fraction of cytosolic A$_1$-fragments are degraded.[80,92] The A$_1$-fragments contain few (Shiga toxin/Stx1) or no (Stx2) lysine residues,[13,25] suggesting that noncanonical ubiquitination and subsequent deubiquitination may be necessary for toxin disengagement from the ERAD pathway. The association of toxin fragments with cytosolic chaperone proteins may mediate refolding of A$_1$-fragments and subsequent association with ribosomal stalk phosphoproteins (P-proteins). The interaction of A$_1$-fragments with ribosome-associated and cytoplasmic stalk P-proteins facilitates the depurination reaction (reviewed by Tumer and Li[93]). Recently, Stx1 and Stx2 A$_1$-fragments have been shown to differentially interact with ribosomal P-proteins. Using yeast mutants lacking specific ribosomal stalk P-proteins, Chiou et al.[94] showed that Stx2 A$_1$-fragments appear less dependent on interaction with P-proteins for enzymatic activity. Finally, the "highjacking" of the ERAD process for toxin retrotranslocation appears to be conserved by another glycolipid-binding AB$_5$ toxin that undergoes retrograde transport. The A$_1$-fragment of cholera toxin is retrotranslocated across the ER membrane and refolds in the cytosol to form the active ADP-ribosyltransferase that may trigger massive intestinal chloride secretion and secretory diarrhea.[95]

45.3.2 Epidemiology of Diseases Caused by Shiga Toxin-Producing Bacteria

Shigella species are the etiologic agents of bacillary dysentery or shigellosis. The annual incidence of shigellosis has been estimated to be as high as 165 million cases with 1.1 million deaths.[96] Risk factors for death include young age, poor nutritional status, and the development of central nervous system (CNS) complications such as seizures, paralysis, or coma.[97,98] *Shigella* species are human-adapted pathogens that infect via the fecal–oral route. Children between the ages of 1 and 5 years living in conditions of poor sanitary

hygiene are particularly impacted by shigellosis. High attack rates are evident in conditions of crowding, such as child-care centers, nursing homes, and care facilities for the mentally ill. Shigellosis may also be a sexually transmitted disease. Asymptomatic carriers may play an important role in the spread of *Shigella* as ~17% of children continued to excrete *Shigella* in the stool one month after onset of diarrhea. Secondary cases due to person-to-person contact have been as high as 40%. Of the four *Shigella* species, only *S. dysenteriae* serotype 1 expresses Shiga toxin. As a consequence, *S. dysenteriae* serotype 1 causes epidemic outbreaks of shigellosis that are characterized by prolonged and severe illness. HUS and CNS complications have been reported in up to 24% and 45%, respectively, of hospitalized shigellosis cases caused by *S. dysenteriae* serotype 1 (reviewed in Ram et al.[99]). The frequency of *S. dysenteriae* serotype 1 isolation is inversely correlated with *per capita* gross domestic product (GDP), and infection with *S. dysenteriae* serotype 1 remains a major public health concern in sub-Saharan Africa and on the Indian subcontinent.[99]

Unlike shigellosis, which is a disease of the developing world, hemorrhagic colitis and HUS associated with STEC infections occur primarily in developed countries with modern food processing and distribution facilities.[100] Multiple transmission routes for STEC have been established including contact with domesticated ruminants such as cattle, sheep, or goats, ingestion of undercooked contaminated meats derived from ruminant animals, ingestion of vegetables contaminated with manure or irrigated with contaminated water, ingestion of contaminated well water, and person-to-person spread.[101] The capacity of Shiga toxin-encoding bacteriophages to lysogenize *E. coli* serves as a mechanism for lateral spread of the toxin genes, and it is estimated that nearly 500 distinct *E. coli* serotypes may express the toxin genes.[102] Over 150 STEC serotypes have been documented to cause outbreaks or sporadic cases of disease.[103] *E. coli* O157:H7 is the most common cause of hemorrhagic colitis and HUS in the United States, United Kingdom, Canada, and Japan.[104,105] In 1999, Mead et al.[106] estimated that *E. coli* O157:H7 caused approximately 73,000 cases *per* year and non-O157 STEC accounted for approximately 37,000 cases *per* year of diarrheal illness in the United States. More recently, Scallan et al.[107] estimated the annual incidence of STEC infection in the United States as 63,153 cases caused by *E. coli* O157 and 112,752 cases associated with non-O157 STEC, highlighting the emergence of non-O157 STEC as a major public health concern. In a recent 10-year study in Connecticut, the non-O157 serotypes were more commonly isolated (58% of total isolates) compared to *E. coli* O157:H7, although patients with non-O157 STEC infection tended to have less severe disease.[108] The most common non-O157 STEC serotypes isolated in the Connecticut study were O111, O103, O26, and O45. However, the overall estimated incidence of STEC isolation in the state of Connecticut declined from 4.16 isolates *per* 100,000 population in 2000 to 2.93 isolates *per* 100,000 in 2009, suggesting that efforts to educate producers and consumers on the dangers of STEC have been effective. International travel within 7 days of onset of symptoms is a significant risk factor in patients with non-O157 STEC infection.[108,109]

45.4 CLINICAL FEATURES AND PATHOGENESIS

The incubation period for shigellosis varies from 12 h to 5 days. Patients with shigellosis may initially present with fever, malaise, anorexia, myalgia, vomiting, headache, and watery diarrhea. Watery diarrhea may rapidly progress to bloody, mucoid stools that may contain sheaths of denuded epithelium. Patients may experience abdominal cramping and tenesmus (painful straining to pass stools). Fever usually begins within three days of infection. Toxic megacolon and perforation of the colon may occur in the most severe cases.[110] The infectious dose of *S. dysenteriae* serotype 1 may be as low as 10–100 organisms.[111] The organisms are acid resistant, so that after ingestion of small numbers of organisms, the bacteria survive passage through the stomach.[112] All *Shigella* species are invasive microorganisms, and the invasive phenotype is critical for full virulence. The bacteria are internalized by Peyer's patches primarily in the distal colon and rectum. Following translocation across M-cells, the organisms are internalized by macrophages and dendritic cells within lymphoid follicles. The bacteria escape the killing mechanisms of these phagocytic cells and replicate within the cytoplasm to trigger pyroptotic cell death; that is, macrophage and dendritic cell death elicits an inflammatory response. Bacteria released from phagocytes then invade colonic epithelial cells through basolateral membranes. The bacteria rapidly disrupt intracellular vesicles and replicate within the cytoplasm of epithelial cells. By inducing the unipolar polymerization of host cell actin, the bacteria are capable of intracellular motion and intercellular spread. In this manner, *Shigella* may infect adjacent uninfected epithelial cells. Lesions are more common and severe in the distal colon and are progressively less severe in the transverse and ascending colon. The destruction of the colonic epithelium augments the inflammatory response characterized by the robust infiltration of PMNs into the gut, which may contribute to tissue destruction.[113,114]

Infection with STEC may result in different clinical manifestations ranging from uncomplicated watery diarrhea to bloody mucoid diarrhea (hemorrhagic colitis) and life-threatening systemic complications of which acute renal insufficiency or HUS is most common. The infectious dose of *E. coli* O157:H7 may be as low as <100 organisms, although an analysis of data from nine separate outbreaks caused by *E. coli* O157:H7 showed heterogeneity in infectivity; that is, the percentage of individuals infected did not directly correlate with estimated dose of bacteria ingested.[115] These findings suggest that other factors such as the food matrix contaminated with STEC and the immune status of the host may contribute to infectivity. After an incubation period of 3–4 days, patients initially develop watery diarrhea that may be accompanied by abdominal cramping and rarely fever. In culture-confirmed cases, about 90% of patients will develop bloody diarrhea.[116] HUS will occur in approximately 15% of

patients infected with *E. coli* O157:H7 within 5–13 days after the onset of diarrhea. HUS is defined by a triad of clinical findings: microangiopathic hemolytic anemia, thrombocytopenia, and acute renal failure. Clinical criteria used to define HUS include prodromal diarrheal disease, platelet counts of <150,000/mm^3, hemoglobin <10 g/dL with schistocytes (damaged erythrocytes) noted on blood smears, and elevated serum creatinine.[117] Oliguria or anuria develops in 30% to 50% of cases. Extrarenal complications include hepatomegaly, pancreatitis, and CNS abnormalities, including lethargy, seizures, and coma. The development of CNS complications is an indicator of a poor prognosis. Approximately 60% of children with postdiarrheal HUS will recover normal renal function. Approximately 12% of patients will develop end-stage renal disease or die.[118]

Whether an individual will progress from diarrheal disease to HUS is a complex process involving host factors (immune status, age) and virulence factors expressed by the STEC strain. As noted earlier, infections with STEC strains expressing Stx2 and/or Stx2 variants are more likely to result in the development of extraintestinal complications.[119,120] An excellent example of the importance of the expression of multiple virulence factors in determining disease severity is evident in the recent outbreak of hemorrhagic colitis and HUS in Europe, primarily focused in Germany. The outbreak consisted of over 4300 illnesses with a high percentage of cases (approximately 852) progressing to HUS and neurological complications with 50 cases resulting in death.[121,122] In the United States, six cases with one death were reported, associated with travel to Germany.[122] The vehicle of transmission in the outbreak was contaminated fenugreek seeds and the causative agent was *E. coli* O104:H4 that expressed Stx2 but also expressed genes responsible for aggregative adherence to intestinal mucosa characteristic of enteroaggregative *E. coli* (EAEC). All isolates from this outbreak were categorized as sequence type 678 suggesting a clonal source of infection. Isolates from the outbreak had also acquired a plasmid encoding multiple drug resistance. The enhanced capability of the Stx2-expressing *E. coli* O104:H4 ST678 strain to adhere to intestinal epithelial cells may have facilitated toxin uptake into the bloodstream, explaining the increased incidence of HUS and severe neurological complications seen in the outbreak.[123]

Shiga toxins are critical virulence determinants for the development of bloody diarrhea and the systemic complications that may follow. Studies using toxigenic and atoxigenic *S. dysenteriae* isogenic strains in macaque monkeys showed that both strains produced fulminant dysentery. The difference between the strains was the presence of blood in the stools and histopathological evidence of colonic hemorrhage seen in animals given the Shiga toxin-producing strain.[10] Thus, the production of Shiga toxin correlates with the development of colonic vascular damage. Even though the majority of human colonic epithelial cells fail to express membrane Gb3,[124] there is evidence that Shiga toxins are transported across the gut epithelium in a complex process that may involve all of the following: disruption of epithelial tight junctions; PMN extravasation into the lamina propria and subsequent migration through the gut epithelial barrier; and signaling through the GTPase Cdc42 and nonmuscle myosin II to form actin-mediated macropinocytotic membrane blebs distinct from attaching/effacing lesions that serve as sites of toxin uptake.[125–128] Once in the submucosa, Shiga toxins target colonic microvascular endothelial cells, which express the toxin-binding glycolipid Gb3, for damage. The destruction of colonic microvessels may create portals of entry for intestinal contents into the bloodstream, as patients with hemorrhagic colitis may have elevated titers of anti-O157 LPS antibodies in their circulation.[129] However, free Shiga toxins have not been detected in the bloodstream. Flow cytometry studies utilizing fluoresceinated Shiga toxins mixed with human blood showed that the toxins are associated with PMNs and monocytes. Furthermore, labeled toxins could be transferred from PMNs to Gb3-expressing endothelial cells. These findings suggest that PMNs may serve as carrier cells for the transport of Shiga toxins in the bloodstream, delivering the toxins to Gb3-expressing microvascular endothelial cells in the kidneys and brain.[69] Human PMNs do not express Gb3,[130] yet Scatchard analysis suggests ~200,000 toxin-binding sites *per* PMN with a lower affinity for toxin than the Gb3 receptor.[69] A criticism of the PMN carrier model for the hematogenous spread of Shiga toxins is the short half-life of neutrophils in circulation, estimated to be between 6 and 7 h, while extraintestinal complications of hemorrhagic colitis may manifest up to 5–7 days after the onset of diarrhea. However, recent studies involving the coincubation of toxin-laden PMNs with toxin-free PMNs *in vitro* revealed that Shiga toxins may be transferred between neutrophils, and in this manner, Shiga toxins may remain in circulation in a cell-bound form for longer periods of time.[131] In HUS, glomerular endothelial cells appear swollen and detached from the glomerular basement membrane leading to microthrombi deposition.[117] The transfer of toxins from carrier cells to Gb3-expressing glomerular endothelial cells would result in the characteristic thrombotic microangiopathy of HUS.

It has become increasingly apparent that Shiga toxins are multifunctional proteins. In addition to the capacity to enzymatically remove a purine base from the rRNA backbone, the toxins also activate cell stress responses. How these host cell signaling cascades contribute to pathogenesis is incompletely understood. In 1997, Iordanov et al.[132] showed that ribosome-inactivating proteins that act on the peptidyltransferase center (domains V and VI including the SRL) of the 28S rRNA component of eukaryotic ribosomes activate c-Jun NH$_2$-terminal kinase 1 (JNK1). This phenomenon was referred to as the ribotoxic stress response. Subsequently, it was shown that Shiga toxins activate all three major mitogen-activated protein kinase (MAPK) cascades, JNK, p38 MAPK, and extracellular signal-regulated kinases (ERK) to different degrees in different cell types.[133–135] Precisely, how Shiga toxins activate the ribotoxic stress response remains to be fully elucidated. However, the association of Shiga toxin A$_1$-fragments with the ribosomal stalk may induce a sufficient conformational change to activate the serine–threonine

kinase double-stranded RNA-activated protein kinase (PKR).[136] PKR may then activate downstream MAP3Ks and MAP2Ks. Shiga toxin-mediated signaling through the MAPK cascades may lead to the upregulated expression of cytokines and chemokines.[137–140] The cytokines tumor necrosis factor-α (TNF-α) and interleukin-1 have been shown to upregulate the expression of Gb3 on microvascular endothelial cells *in vitro*, thereby increasing toxin binding and sensitizing the target cells to cytotoxicity.[141,142] Furthermore, the administration of recombinant murine TNF-α to mice after the administration of purified Shiga toxins resulted in increased glomerular damage in the animals.[143] The administration of bacterial lipopolysaccharides (LPS) to baboons, a treatment known to activate the expression of TNF-α, increased renal Gb3 expression and decreased the lethal dose of Shiga toxins in the baboon model of HUS.[144] The localized production of chemokines may be essential for the infiltration of PMNs and macrophages into the lamina propria and kidneys.[137,139,145] In addition to activating the ribotoxic stress response, the capacity of processed Shiga toxin A$_1$-fragments to be sensed as unfolded proteins activates the ER stress response in some cell types. Shiga toxins appear capable of activating three ER-membrane-localized sensors of unfolded proteins: PKR-like ER-localized kinases (PERK), inositol-requiring protein 1 (IRE1) and activating transcription factor (ATF6). These activated sensors of unfolded proteins, in turn, upregulate the expression of the transcription factor C/EBP homologous protein (CHOP) and downregulate the expression of the antiapoptotic factor B-cell lymphoma 2 (Bcl-2).[146] While regulated signaling through the ER stress response may be important for cell survival during periods of aberrant protein synthesis, prolonged signaling through the ER stress response triggers apoptosis. Many cell types treated with purified Shiga toxins *in vitro* die by apoptosis, and prolonged activation of ER stress may facilitate apoptotic cell death.[147]

45.5 IDENTIFICATION AND DIAGNOSIS

It is important to screen stool specimens as early as possible after the onset of symptoms since only one-third of cases of diarrhea-associated HUS have positive stool cultures if performed 6 days after the onset of diarrhea.[148] The detection of fecal Shiga toxins may be difficult or impossible as soon as one week after, symptoms manifest as *stx* genes may be lost from STEC strains over time, rendering ineffective assays to detect the toxin genes or phage sequences.[149,150] Finally, the rapid laboratory identification of STEC is critical as clinical data suggest that prompt initiation of intravenous volume expansion improves disease outcome.[151]

The presence of blood or fecal leukocytes in the stool should not be used as criteria for testing since patients may present without bloody diarrhea. *E. coli* O157:H7 can readily be distinguished from commensal *E. coli* in the gut by the inability to ferment sorbitol within a 24 h period. Sorbitol-MacConkey (SMAC) agar with or without antibiotics may be used to detect sorbitol-negative strains in stool samples.

However, the emergence of sorbitol-fermenting *E. coli* O157:H⁻ STEC has complicated the dependence on SMAC agar as the sole differential media. Sorbitol nonfermenting colonies may be tested for reactivity with the O157 antigen using commercially available latex agglutination assays or anti-O157 antiserum. Sorbitol negative colonies that are also O157 negative should be screened by slide agglutination assays using antisera to non-O157 O-antigens. These tests are usually performed in conjunction with automated identification systems or standard biochemical tests (triple sugar iron agar slant, indole reaction, etc.) to confirm *E. coli* isolation. Most non-O157 STEC ferment both sorbitol and lactose. Stool samples and sorbitol-negative or sorbitol-positive colonies may be cultured in nonselective broth media in order to enrich for the presence of STEC. Broth cultures can be tested for Shiga toxin expression using commercially available enzyme immunoassays. Finally, it should be noted that many commonly used enteric media (Hektoen agar, xylose–lysine–deoxycholate [XLD] agar, *Salmonella–Shigella* agar) inhibit the growth of many STEC.[152] Although Vero and HeLa cells are sensitive to the cytotoxic action of Shiga toxins, they are not routinely employed in clinical laboratories because cytotoxicity may require 48–72 h to manifest, and the need for maintenance of sterile tissue culture cells.

Following the presumptive identification of STEC, isolates should be submitted to local or state public health laboratories for additional testing. Pulsed-field gel electrophoresis, detection of the H7 antigen, and identification of the presence of Shiga toxin variants are tests usually performed by public health or reference laboratories. The development of primers to specifically amplify *stx* genes has the potential to improve rapid identification of STEC, although to date, DNA-based methods have primarily been used in research laboratories. PCR amplification of the *eae* gene, encoding the adhesin molecule intimin that is important for intimate adherence and attaching/effacing lesion formation, may be used in conjunction with *stx*-specific primers to identify the virulent *stx⁺/eae⁺* subset of STEC. Recently, Quiñones et al.[153] designed a 30-mer oligonucleotide DNA microarray to detect 11 O-antigen gene clusters encoding LPS of STEC serotypes and 11 virulence genes including the *stx1* and *stx2* genes. The use of microarrays coupled to colorimetric readout systems may allow the rapid genotyping of STEC isolates. Readers are referred to a comprehensive review of diagnostic tests used to identify STEC.[152]

45.6 TREATMENT AND PREVENTION

Most cases of shigellosis resolve within 2–5 days after the onset of diarrhea. However, bacillary dysentery caused by Shiga toxin-producing *S. dysenteriae* serotype 1 is more severe with mortality rates in outbreaks as high as 20%. Antibiotics recommended for treatment of shigellosis include azithromycin, ceftriaxone, and ciprofloxacin.[154] The use of antibiotics may reduce shigellosis severity and duration. The use of some antibiotics to treat hemorrhagic colitis caused by STEC may be contraindicated since they

may induce the SOS response, leading to the induction of phage lysis and increased production of Shiga toxins.[155,156] Additional studies will be necessary to define antibiotic regimens that effectively kill STEC without increasing toxin synthesis.[157,158] Currently, treatment for hemorrhagic colitis and HUS is largely supportive requiring attention to management of fluids and electrolytes. Fluids are administered at a rate replacing insensible loss (400 mL/m³/day) with 5% dextrose containing 0.45% NaCl. Urinary catheters are placed to monitor urine output and diuresis is compensated based on output. In cases of oliguria, diuretics may be administered but should be discontinued if a clinical effect is absent. Patients should be carefully monitored for hyponatremia and hyperkalemia. One-third to one-half of children will require peritoneal dialysis or hemodialysis. Readers are referred to a detailed review of the clinical management of patients with postdiarrheal HUS.[117]

There currently are no efficacious vaccines to prevent hemorrhagic colitis and HUS. However, recent advances in our understanding of the host cell interactions with Shiga toxins have provided potential interventional therapeutic approaches to treat disease caused by the toxins. The finding that toxin B-subunits bound to Gb3 on the surface of target cells led to the development of toxin receptor analogues as potential therapeutic agents. In 1991, Armstrong et al.[159] physically coupled the Galα(1→4)Galβ(1→4)βGlc trisaccharide of Gb3 to inert silicon dioxide particles (Chromosorb) commonly used in liquid chromatography. The concept was that, when fed to patients with bloody diarrhea, the receptor analogue would compete with membrane-bound Gb3 for the toxin and thereby reduce toxin absorption and subsequent pathology. This material, called Synsorb-Pk, showed therapeutic potential in *in vitro* protection experiments as the treatment of Vero cells or human renal adenocarcinoma cells with Synsorb-Pk protected the cells from cytotoxicity caused by Stx1 or Stx2.[159,160] Furthermore, orally administered Synsorb-Pk was well tolerated in phase I clinical trials. Unfortunately, Synsorb-Pk did not prove to be efficacious when tested in children with hemorrhagic colitis.[161,162] The findings that Shiga toxins (1) bind to three Gb3 molecules *per* B-subunit, (2) may be actively transported across the gut epithelium, and (3) may rapidly associate with non-Gb3 receptors on the surface of PMNs led to the refinement of toxin receptor analogues containing clustered Gb3 trisaccharides for oral and intravenous administration. Paton et al.[163] genetically engineered a nonpathogenic *E. coli* strain to express LPS molecules terminating in Galα(1→4)Galβ(1→4)Glc. Feeding mice the recombinant toxin receptor analogue-expressing "probiotic" *E. coli* protected them from challenge with a highly virulent STEC strain. Kitov et al.[164] synthesized a multivalent toxin receptor analogue containing multiple Gb3 trisaccharides. This compound, called STARFISH, bound to two separate B-subunits within a single B-pentamer and effectively protected Vero cells from cytotoxicity *in vitro*. Depending on the orientation of toxin-binding carbohydrates, STARFISH derivatives displayed variable protection of mice against lethality caused by Stx1, Stx2, or Stx2d$_{act}$.[165]

More recently, carbosilane dendrimer biochemistry has been used to create an array of multivalent toxin-receptor analogues. These branched compounds protected mice from the neuropathological and lethal effects of Shiga toxins and appeared to facilitate the uptake and degradation of toxins by macrophages.[166] Although different numbers of Gb3 trisaccharides can be attached to silicon molecules in dendrimeric receptor analogue constructs, more is not necessarily better. For example, dendrimers with 12 or 36 trisaccharides clustered about silicon molecules bound Stx2 with higher affinity compared with a dendrimer expressing 6 trisaccharides. However, the 12 and 36 trisaccharide-containing molecules were not as effective as the 6 trisaccharide-containing dendrimer in protecting mice from intravenously administered Stx2.[167] Structure–function analyses suggest that the presentation of two clusters of Gb3 trisaccharides separated by a hydrophobic linker may be necessary for optimal toxin clearance in vivo.[168]

The characterization of intracellular toxin routing has identified a number of steps that may be inhibited in order to block toxin trafficking to the ER. A peptide library was screened for the ability of cell-permeable peptides to bind to Shiga toxin B-subunits. A tetravalent peptide was identified that when bound to Shiga toxin does not alter internalization but reroutes the toxin molecule into the phagolysosomal pathway. This peptide protects mice from a lethal dose of E. coli O157:H7 and protects baboons from a lethal intravenous challenge with purified Stx2.[169,170] Saenz et al.[171] screened over 14,000 molecules of unknown function for protection of Vero cells from intoxication with Stx1. They identified two compounds that block toxin retrograde transport at different stages in the process. One compound blocked formation of early endosomes and was protective against other toxins, such as diphtheria toxin, that enter the cytosol from an endosomal compartment. The second compound blocked toxin access to the Golgi apparatus and was effective against toxins that are routed to the ER. Some surprising results have come from studies involving toxin transport inhibitors. Exo2, a small molecule inhibitor of anterograde intracellular transport, was shown to have disorganizing effects on the TGN and Golgi apparatus. However, Exo2 did not alter TGN to ER transport of cholera toxin in a human intestinal epithelial cell line, suggesting that an alternative retrograde transport process not requiring functional Golgi apparatus may be active in intracellular routing of cholera toxin.[172] In contrast, Exo2 treatment of HeLa cells altered TGN and Golgi structure and protected the cells from Shiga toxin-induced cytotoxicity. Protection was not due to the aberrant anterograde transport of furin, since Exo2 also protected HeLa cells from cytotoxicity caused by trypsin preactivated toxin. Exo2 appeared to block Shiga toxin intracellular transport at the level of endosome-to-TGN transport.[173] Thus, despite sharing AB_5 structural organization, the requirement for glycolipid binding for toxin uptake, and retrograde transport to the ER, cholera toxin and Shiga toxin appear to utilize different intracellular transport pathways.

45.7 CONCLUSIONS AND FUTURE DIRECTIONS

Shiga toxins are a family of genetically and functionally conserved protein exotoxins expressed by enteric pathogens. The toxins are multifunctional proteins capable of mediating protein synthesis inhibition and activating the ribotoxic stress response and the ER stress response. Together, these phenomena cause cell death in vitro at fg/mL to ng/mL toxin amounts and in vivo in µg/kg amounts. The toxins are probably not useful as biological weapons when administered orally, since the AB_5 holotoxin configuration necessary for cell binding and intracellular transport may be altered in the acidic environment of the stomach. In contrast to ingestion of purified toxins, S. dysenteriae serotype 1 and STEC are extremely effective acid-tolerant delivery agents for the toxins, with oral infectious doses estimated to be in the range of 10–100 organisms. The horizontal transfer of Shiga toxin genes by bacteriophages, and the emergence of STEC with "blended" virulence profiles, for example, STEC/EAEC E. coli O104:H4 ST678, will continue to constitute public health challenges. The capacity to genetically engineer Shiga toxin-producing recombinant bacteria is a significant security concern. Shiga toxins reproduce the systemic complications of natural infection when administered to nonhuman primates via the intravenous route.[174,175] Thus, Shiga toxins may be used as an individualized weapon, as was the case with the 1978 attack on Bulgarian journalist Georgi Markov using ricin.[176] Extreme caution must be used when sharp objects are employed in experiments involving Shiga toxins. A recent literature search using the terms Shiga toxins, verotoxins, and inhalational exposure did not reveal any publications. However, Shiga toxins have been detected in the lungs of mice administered with the toxins and in patients with HUS.[177,178] The toxins bind to Gb3-expressing vascular endothelial cells and pulmonary epithelial cells. Human lung carcinoma cell lines were exquisitely sensitive to the cytotoxic action of Shiga toxins, with CD_{50}s reported between 0.25 and 0.025 pg/mL.[178] In contrast to Shiga toxins, the effects of inhalational delivery of ricin have been examined. The inhalational 50% lethal dose of ricin in mice is 3–5 µg/kg body weight[179]; 60% lethal doses of 21–42 µg/kg were reported in monkeys inhaling ricin coupled to 1–2 µm particles.[180] Until definitively studied in animal models, Shiga toxins should be considered potential inhalational hazards.

REFERENCES

1. Centers for Disease Control and Prevention: Bioterrorism agents/diseases. www.bt.cdc.gov/agent/agentlist-category. asp.
2. United States Government Printing Office, Federal Digital System. http://www.gpo.gov/fdsys/pkg/PLAW-107publ56/pdf/PLAW-107publ56.pdf.
3. United States Government Printing Office, Federal Digital System. http://www.gpo.gov/fdsys/pkg/PLAW-107publ188/pdf/PLAW-107publ188.pdf.

4. CDC National Select Agents Registry. http://www.selecta-gents.gov/resources/List%20of%20Select%20Agents%20and%20Toxins%2009-19-2011.pdf.

5. CDC National Select Agents Registry. http://www.select-agents.gov/resources/CDC%20Select%20Agent%20Biennial%20Review%20Final%20Rule%2010%2005%202012.pdf.

6. O'Brien, A.D. et al. Shiga toxin: Biochemistry, genetics, mode of action and role in pathogenesis. In *Pathogenesis of Shigellosis*, ed. Sansonetti, P.J., pp. 65–94, Springer-Verlag, Berlin, Germany (1992).

7. Bridgwater, F.A.J. et al. The neurotoxin of *Shigella shigae*: Morphological and functional lesions produced in the central nervous system of rabbits. *Br. J. Exp. Pathol.* 36, 447–453 (1955).

8. Howard, J.G. Observations on the intoxication produced in mice and rabbits by the neurotoxin of *Shigella shigae*. *Br. J. Exp. Pathol.* 36, 439–446 (1955).

9. Keusch, G.T. et al. The pathogenesis of *Shigella* diarrhea. I. Enterotoxin production by *Shigella dysenteriae* 1. *J. Clin. Invest.* 51, 1212–1218 (1972).

10. Fontaine, A., Arondel, J., and Sansonetti, P.J. Role of Shiga toxin in the pathogenesis of bacillary dysentery using a tox⁻ mutant of *Shigella dysenteriae* 1. *Infect. Immun.* 56, 3099–3109 (1988).

11. Sekizaki, T., Harayama, S., and Timmis, K.N. Localization of *stx*, a determinant essential for high level production of Shiga toxin by *Shigella dysenteriae* serotype 1, near *pyrF* and generation of *stx* transposon mutants. *Infect. Immun.* 55, 2208–2214 (1987).

12. Kozlov, Y.V. et al. The primary structure of operons coding for *Shigella dysenteriae* toxin and temperate phage H30 Shiga-like toxin. *Gene* 67, 213–221 (1988).

13. Strockbine, N.A. et al. Cloning and sequencing of the genes for Shiga toxin from *Shigella dysenteriae* type 1. *J. Bacteriol.* 170, 1116–1122 (1988).

14. Konowalchuk, J., Speirs, J.I., and Stavric, S. Vero response to a cytotoxin of *Escherichia coli*. *Infect. Immun.* 18, 775–779 (1977).

15. Scotland, S.M. et al. Vero cytotoxin production in a strain of *Escherichia coli* is determined by genes carried on bacteriophage. *Lancet* 322, 216 (1983).

16. Smith, H.W., Green, P., and Parsell, Z. Vero cell toxins in *Escherichia coli* and related bacteria: Transfer by phage and conjugation and toxic action in laboratory animals, chickens and pigs. *J. Gen. Microbiol.* 129, 3121–3137 (1983).

17. Riley, L.W. et al. Hemorrhagic colitis associated with a rare *Escherichia coli* serotype. *N. Engl. J. Med.* 308, 681–685 (1983).

18. Wells, J.G. et al. Laboratory investigation of hemorrhagic colitis outbreaks associated with a rare *Escherichia coli* serotype. *J. Clin. Microbiol.* 18, 512–520 (1983).

19. Karmali, M.A. et al. Sporadic cases of haemolytic–uraemic syndrome associated with faecal cytotoxin and cytotoxin-producing *Escherichia coli* in stools. *Lancet* 321, 619–620 (1983).

20. Karmali, M.A. et al. The association between idiopathic hemolytic uremic syndrome and infection by verotoxin-producing *Escherichia coli*. *J. Infect. Dis.* 151, 775–782 (1985).

21. O'Brien, A.D. et al. Purification of *Shigella dysenteriae* 1 (Shiga)-like toxin from *Escherichia coli* O157:H7 strain associated with haemorrhagic colitis. *Lancet* 322, 573 (1983).

22. O'Brien, A.D. et al. Shiga-like toxin-converting phages from *Escherichia coli* strains that cause hemorrhagic colitis or infantile diarrhea. *Science* 226, 694–696 (1984).

23. Strockbine, N.A. et al. Two toxin-converting phages from *Escherichia coli* O157:H7 strain 933 encode antigenically distinct toxins with similar biologic activities. *Infect. Immun.* 53, 135–140 (1986).

24. Takao, T. et al. Identity of molecular structure of Shiga-like toxin I (VT1) from *Escherichia coli* O157:H7 with that of Shiga toxin. *Microb. Pathog.* 5, 357–369 (1988).

25. Jackson, M.P. et al. Nucleotide sequence analysis and comparison of the structural genes for Shiga-like toxin I and Shiga-like toxin II encoded by bacteriophages from *Escherichia coli* 933. *FEMS Microbiol. Lett.* 44, 109–114 (1987).

26. Calderwood, S.B. et al. Proposed new nomenclature for SLT (VT) family. *ASM News* 62, 118–119 (1996).

27. Greco, K.M., McDonough, M.A., and Butterton, J.R. Variation in the Shiga toxin region of 20th-century epidemic and endemic *Shigella dysenteriae* 1 strains. *J. Infect. Dis.* 190, 330–334 (2004).

28. Zhang, W. et al. Identification, characterization, and distribution of a Shiga toxin 1 gene variant (stx_{1c}) in *Escherichia coli* strains isolated from humans. *J. Clin. Microbiol.* 40, 1441–1446 (2002).

29. Bürk, C. et al. Identification and characterization of a new variant of Shiga toxin 1 in *Escherichia coli* ONT:H19 of bovine origin. *J. Clin. Microbiol.* 41, 2106–2112 (2003).

30. Beutin, L. et al. Evaluation of major types of Shiga toxin 2e-producing *Escherichia coli* bacteria present in food, pigs, and the environment as potential pathogens for humans. *Appl. Environ. Microbiol.* 74, 4806–4816 (2008).

31. Schmidt, H. et al. A new Shiga toxin 2 variant (Stx2f) from *Escherichia coli* isolated from pigeons. *Appl. Environ. Microbiol.* 66, 1205–1208 (2000).

32. Prager, R. et al. *Escherichia coli* encoding Shiga toxin 2f as an emerging human pathogen. *Int. J. Med. Microbiol.* 299, 343–353 (2009).

33. Asakura, H. et al. Phylogenetic diversity and similarity of active sites of Shiga toxin (Stx) in Shiga toxin-producing *Escherichia coli* (STEC) isolates from humans and animals. *Epidemiol. Infect.* 127, 27–36 (2001).

34. Leung, P.H.M. et al. A newly discovered verotoxin variant, VT2g, produced by bovine verocytotoxigenic *Escherichia coli*. *Appl. Environ. Microbiol.* 69, 7549–7553 (2003).

35. Garcia-Aljaro, C. et al. Newly identified bacteriophages carrying the *stx2g* Shiga toxin gene isolated from *Escherichia coli* strains in polluted waters. *FEMS Microbiol. Lett.* 258, 127–135 (2006).

36. Prager, R. et al. Comparative analysis of virulence genes, genetic diversity, and phylogeny of Shiga toxin 2g and heat stable enterotoxin STIa encoding *Escherichia coli* isolates from human, animal and environmental sources. *Int. J. Med. Microbiol.* 301, 181–191 (2011).

37. Friedrich, A.W. et al. *Escherichia coli* harboring Shiga toxin 2 gene variants: Frequency and association with clinical symptoms. *J. Infect. Dis.* 185, 74–84 (2002).

38. Jelacic, J.K. et al. Shiga toxin-producing *Escherichia coli* in Montana: Bacterial genotypes and clinical profiles. *J. Infect. Dis.* 188, 719–729 (2003).

39. Bielaszewska, M. et al. Shiga toxin activatable by intestinal mucus in *Escherichia coli* isolated from humans: Predictor for a severe clinical outcome. *Clin. Infect. Dis.* 43, 1160–1167 (2006).

40. Persson, S. et al. Subtyping method for *Escherichia coli* Shiga toxin (verocytotoxin) 2 variants and correlations to clinical manifestations. *J. Clin. Microbiol.* 45, 2020–2024 (2007).

41. Schmitt, C.K., McKee, M.L., and O'Brien, A.D. Two copies of Shiga-like toxin II-related genes common in enterohemorrhagic *Escherichia coli* strains are responsible for the antigenic heterogeneity of the O157:H⁻ strain E32511. *Infect. Immun.* 59, 1065–1073 (1991).

42. Ling, H. et al. A mutant Shiga-like toxin IIe bound to its receptor Gb$_3$: Structure of a group II Shiga-like toxin with altered binding specificity. *Structure* 8, 253–264 (2000).

43. Strauch, E., Schaudinn, C., and Beutin, L. First-time isolation and characterization of a bacteriophage encoding the Shiga toxin 2c variant, which is globally spread in strains of *Escherichia coli* O157. *Infect. Immun.* 72, 7030–7039 (2004).

44. Ito, H. et al. Cloning and nucleotide sequencing of Vero toxin 2 variant genes from *Escherichia coli* O91:H21 isolated from a patient with the hemolytic uremic syndrome. *Microb. Pathog.* 8, 47–60 (1990).

45. Melton-Celsa, A.R., Darnell, S.C., and O'Brien, A.D. Activation of Shiga-like toxins by mouse and human intestinal mucus correlates with virulence of enterohemorrhagic *Escherichia coli* O91:H21 isolates in orally infected, streptomycin-treated mice. *Infect. Immun.* 64, 1569–1576 (1996).

46. Kokai-Kun, J.F., Melton-Celsa, A.R., and O'Brien, A.D. Elastase in intestinal mucus enhances the cytotoxicity of Shiga toxin 2d. *J. Biol. Chem.* 275, 3713–3721 (2000).

47. Melton-Celsa, A.R., Kokai-Kun, J.F., and O'Brien, A.D. Activation of Shiga toxin type 2d (Stx2d) by elastase involves cleavage of the C-terminal two amino acids of the A$_2$ peptide in the context of the appropriate B pentamer. *Mol. Microbiol.* 43, 207–215 (2002).

48. Tasara, T. et al. Activatable Shiga toxin 2d (Stx2d) in STEC strains isolated from cattle and sheep at slaughter. *Vet. Microbiol.* 131, 199–204 (2008).

49. Stein, P.E. et al. Crystal structure of the cell-binding B oligomer of verotoxin 1 from *E. coli*. *Nature* 355, 748–750 (1992).

50. Fraser, M.E. et al. Crystal structure of the holotoxin from *Shigella dysenteriae* at 2.5 Å resolution. *Nat. Struct. Biol.* 1, 59–64 (1994)

51. Fraser, M.E. et al. Structure of Shiga toxin type 2 (Stx2) from *Escherichia coli* O157:H7. *J. Biol. Chem.* 279, 27511–27517 (2004).

52. Sixma, T.K. et al. Comparison of the B-pentamers of heat-labile enterotoxin and verotoxin-1: Two structures with remarkable similarity and dissimilarity. *Biochemistry* 32, 191–198 (1996).

53. Beddoe, T. et al. Structure, biological function and applications of the AB$_5$ toxins. *Trends Biochem. Sci.* 35, 411–418 (2010).

54. Endo, Y. et al. Site of action of a Vero toxin (VT2) from *Escherichia coli* O157:H7 and of Shiga toxin on eukaryotic ribosomes. RNA *N*-glycosidase activity of the toxins. *Eur. J. Biochem.* 171, 45–50 (1988).

55. Saxena, S.K., O'Brien, A.D., and Ackerman, E.J. Shiga toxin, Shiga-like toxin II variant and ricin are all single-site RNA *N*-glycosidases of 28S RNA when microinjected into *Xenopus* oocytes. *J. Biol. Chem.* 264, 595–601 (1989).

56. Endo, Y. and Tsurugi, K. RNA *N*-glycosidase activity of ricin A chain: Mechanism of action of the toxic lectin ricin on eukaryotic ribosomes. *J. Biol. Chem.* 262, 8128–8130 (1987).

57. Endo, Y., Huber, P.W., and Wool, I.G. The ribonuclease activity of the cytotoxin alpha-sarcin. The characteristics of the enzymatic activity of alpha-sarcin with ribosomes and ribonucleic acids as substrates. *J. Biol. Chem.* 258, 2662–2667 (1983).

58. Vorhees, R.M. et al. The mechanism for activation of GTP hydrolysis on the ribosome. *Science* 330, 835–838 (2010).

59. Tesh, V.L. Activation of cell stress responses by Shiga toxins. *Cell. Microbiol.* 14, 1–9 (2012).

60. Jacewicz, M. et al. Pathogenesis of *Shigella* diarrhea. XI. Isolation of a *Shigella* toxin-binding glycolipid from rabbit jejunum and HeLa cells and its identification as globotriaosylceramide. *J. Exp. Med.* 163, 1391–1404 (1986).

61. Lindberg, A.A. et al. Identification of the receptor glycolipid for Shiga toxin produced by *Shigella dysenteriae* type 1. In *Protein–Carbohydrate Interactions in Biological Systems*, eds. Lark, D.L., Normark, S., Uhlin, B.E., and Wolf-Watz, H., pp. 439–446, Academic Press, London, U.K. (1986).

62. DeGrandis, S. et al. Globotetraosylceramide is recognized by the pig edema disease toxin. *J. Biol. Chem.* 264, 12520–12525 (1989).

63. Ling, H. et al. Structure of the Shiga-like toxin I B-pentamer complexed with an analogue of its receptor Gb3. *Biochemistry* 37, 1777–1788 (1998).

64. Bast, D.J. et al. The identification of three biologically relevant globotriaosyl ceramide receptor binding sites on the verotoxin 1 B subunit. *Mol. Microbiol.* 32, 953–960 (1999).

65. Sandvig, K. et al. Endocytosis from coated pits of Shiga toxin: A glycolipid-binding protein from *Shigella dysenteriae* 1. *J. Cell Biol.* 108, 1331–1343 (1989).

66. Römer, W. et al. Shiga toxin induces tubular membrane invaginations for its uptake into cells. *Nature* 450, 670–675 (2007).

67. Müthing, J. et al. Shiga toxins, glycosphingolipid diversity, and endothelial cell injury. *Thromb. Haemost.* 101, 252–264 (2009).

68. Cooling, L.L. et al. Shiga toxin binds human platelets via globotriaosylceramide (Pk antigen) and a novel platelet glycosphingolipid. *Infect. Immun.* 66, 4355–4366 (1998).

69. te Loo, D.M.W.M. et al. Binding and transfer of verocytotoxin by polymorphonuclear leukocytes in hemolytic uremic syndrome. *Blood* 95, 3396–3402 (2000).

70. Brigotti, M. The interactions of human neutrophils with Shiga toxins and related plant toxins: Danger or safety? *Toxins* 4, 157–190 (2012).

71. Sandvig, K. et al. Retrograde transport of endocytosed Shiga toxin to the endoplasmic reticulum. *Nature* 358, 510–512 (1992).

72. Sandvig, K. et al. Endocytosis and retrograde transport of Shiga toxin. *Toxicon* 56, 1181–1185 (2010).

73. Falguières, T. et al. Targeting of Shiga toxin B-subunit to retrograde transport route in association with detergent-resistant membranes. *Mol. Biol. Cell* 12, 2453–2468 (2001).

74. Lee, M.-S. et al. Shiga toxins induce autophagy leading to differential signalling pathways in toxin-sensitive and toxin-resistant human cells. *Cell. Microbiol.* 13, 1479–1496 (2011).

75. Haicheur, N. et al. The B-subunit of Shiga toxin fused to a tumor antigen elicits CTL and targets dendritic cells to allow MHC class I restricted presentation of peptides derived from exogenous antigens. *J. Immunol.* 165, 3301–3308 (2000).

76. Johannes, L. and Popoff, V. Tracing the retrograde route in protein trafficking. *Cell* 135, 1175–1187 (2008).

77. Jovic, M. et al. The early endosome: A busy sorting station for proteins at the crossroads. *Histol. Histopathol.* 25, 99–112 (2010).

78. Johannes, L. and Römer, W. Shiga toxins—From cell biology to biomedical applications. *Nat. Rev. Microbiol.* 8, 105–116 (2010).

79. Lee, M.-S., Cherla, R.P., and Tesh, V.L. Shiga toxins: Intracellular trafficking to the ER leading to activation of host cell stress responses. *Toxins* 2, 1515–1535 (2010).

80. Tam, P.J. and Lingwood, C.A. Membrane–cytosolic translocation of verotoxin A$_1$ subunit in target cells. *Microbiology* 153, 2700–2710 (2007).

81. Garred, Ø., van Deurs, B., and Sandvig, K. Furin-induced cleavage and activation of Shiga toxin. *J. Biol. Chem.* 270, 10817–10821 (1995).

82. Spooner, R.A. and Lord, J.M. How ricin and Shiga toxin reach the cytosol of target cells: Retrotranslocation from the endoplasmic reticulum. In *Ricin and Shiga Toxins: Pathogenesis, Immunity, Vaccines and Therapeutics*, ed. Mantis, N.J., pp. 19–40, Springer-Verlag, Berlin, Germany (2012).

83. LaPointe, P., Wei, X., and Gariepy, J. A role for the protease-sensitive loop region of Shiga-like toxin 1 in the retrotranslocation of its A1 domain from the endoplasmic reticulum lumen. *J. Biol. Chem.* 280, 23310–23318 (2005).

84. Menikh, A. et al. Orientation in lipid bilayers of a synthetic peptide representing the C-terminus of the A1 domain of Shiga toxin: A polarized ATR-FTIR study. *Biochemistry* 36, 15865–15872 (1997).

85. Carvalho, P., Goder, V., and Rapoport, T. Distinct ubiquitin-ligase complexes define convergent pathways for the degradation of ER proteins. *Cell* 126, 361–373 (2006).

86. Denic, V., Quan, E., and Wiseman, J. A luminal surveillance complex that selects misfolded glycoproteins for ER-associated degradation. *Cell* 126, 349–359 (2006).

87. Bays, N.W. et al. Hrd1/Der3p is a membrane-anchored ubiquitin ligase required for ER-associated degradation. *Nat. Cell Biol.* 3, 24–29 (2001).

88. Ye, Y., Meyer, H.H., and Rapoport, T.A. The AAA ATPase Cdc48/p97 and its partners transport proteins from the ER into the cytosol. *Nature* 414, 652–656 (2001).

89. Yu, M. and Haslam, D.B. Shiga toxin is transported from the endoplasmic reticulum following interaction with the luminal chaperone HEDJ/ERdj3. *Infect. Immun.* 73, 2524–2532 (2005).

90. Falguières, T. and Johannes, L. Shiga toxin B-subunit binds to the chaperone BiP and the nucleolar protein B23. *Biol. Cell* 98, 125–134 (2006).

91. Li, S. et al. Folding-competent and folding-defective forms of ricin A-chain have different fates after retrotranslocation from the endoplasmic reticulum. *Mol. Biol. Cell* 21, 2543–2554 (2010).

92. Li, S. et al. Cytosolic entry of Shiga-like toxin A chain from the yeast endoplasmic reticulum requires catalytically active Hrd1p. *PLoS One* 7, e41119 (2012).

93. Tumer, N.E. and Li, X.-P. Interaction of ricin and Shiga toxins with ribosomes. In *Ricin and Shiga Toxins: Pathogenesis, Immunity, Vaccines and Therapeutics*, ed. Mantis, N.J., pp. 1–18, Springer-Verlag, Berlin, Germany (2012).

94. Chiou, J.C. et al. Shiga toxin 1 is more dependent on the P proteins of the ribosomal stalk for depurination activity than Shiga toxin 2. *Int. J. Biochem. Cell Biol.* 43, 1792–1801 (2011).

95. Cho, J.A. et al. Insights on the trafficking and retrotranslocation of glycosphingolipid-binding bacterial toxins. *Front. Cell. Infect. Microbiol.* 2, e51 (2012).

96. Kotloff, K.L. et al. Global burden of *Shigella* infections: Implications for vaccine development and implementation of control strategies. *Bull. World Health Organ.* 77, 651–666 (1999).

97. Bennish, M.L. and Wojtyniak, B.J. Mortality due to shigellosis: Community and hospital data. *Rev. Infect. Dis.* 13(Suppl. 4), S245–S251 (1991).

98. Islam, S.S. and Shahid, N.S. Morbidity and mortality in a diarrhoeal diseases hospital in Bangladesh. *Trans. Royal Soc. Tropic. Med. Hyg.* 80, 748–752 (1986).

99. Ram, P.K. et al. Analysis of data gaps pertaining to *Shigella* infections in low and medium human development index countries, 1984–2005. *Epidemiol. Infect.* 136, 577–603 (2007).

100. Tesh, V.L. Foodborne enterohemorrhagic *Escherichia coli* infections. In *Preharvest and Postharvest Food Safety: Contemporary Issues and Future Directions*, eds. Beier, R.C., Pillai, S.D., Phillips, T.D., and Ziprin, R.L., pp. 27–42, Blackwell Publishing, Ames, IA (2004).

101. Ferens, W.A. and Hovde, C.J. *Escherichia coli* O157:H7: Animal reservoir and sources of human infection. *Foodborne Pathog. Dis.* 8, 465–487 (2011).

102. Gyles, C.L. Shiga toxin-producing *Escherichia coli*: An overview. *J. Anim. Sci.* 85(Suppl. E), E45–E62 (2007).

103. Johnson, K.E., Thorpe, C.M., and Sears, C.L. The emerging clinical importance of non-O157 Shiga toxin-producing *Escherichia coli*. *Clin. Infect. Dis.* 43, 1587–1595 (2006).

104. Pennington, H. *Escherichia coli* O157:H7. *Lancet* 376, 1428–1435 (2010).

105. Kawano, K. et al. *stx* genotype and molecular epidemiological analyses of Shiga toxin-producing *Escherichia coli* O157:H7/H⁻ in human and cattle isolates. *Eur. J. Clin. Microbiol. Infect. Dis.* 31, 119–127 (2012).

106. Mead, P.S. et al. Food-related illness and death in the United States. *Emerg. Infect. Dis.* 5, 607–625 (1999).

107. Scallan, E. et al. Foodborne illness acquired in the United States—Major pathogens. *Emerg. Infect. Dis.* 17, 7–15 (2011).

108. Hadler, J.L. et al. Ten-year trends and risk factors for non-O157 Shiga toxin-producing *Escherichia coli* found through Shiga toxin testing, Connecticut, 2000–2009. *Clin. Infect. Dis.* 53, 269–276 (2011).

109. Hedican, E. et al. Characteristics of O157 versus non-O157 Shiga toxin-producing *Escherichia coli* infections in Minnesota, 2000–2006. *Clin. Infect. Dis.* 49, 358–364 (2009).

110. Keusch, G.T. Shigellosis. In *Infectious Diseases*, 3rd ed., eds. Gorbach, S.L., Bartlett, J.G., and Blacklow, N.R., pp. 603–607, Lippincott, Williams & Wilkins, Philadelphia, PA (2004).

111. DuPont, H.L. et al. Inoculum size in shigellosis and implications for expected mode of transmission. *J. Infect. Dis.* 159, 1126–1128 (1989).

112. Goh, K. et al. Arginine-dependent acid-resistance pathway in *Shigella boydii*. *Arch. Microbiol.* 193, 179–185 (2011).

113. Phalipon, A. and Sansonetti, P.J. *Shigella*'s ways of manipulating the host intestinal innate and adaptive immune system: A tool box for survival? *Immunol. Cell Biol.* 85, 119–129 (2007).

114. Ashida, H. et al. *Shigella* are versatile mucosal pathogens that circumvent the host innate immune system. *Curr. Opin. Immunol.* 23, 448–455 (2011).

115. Teunis, P.F.M., Ogden, I.D., and Strachan, N.J.C. Hierarchal dose response of *E. coli* O157:H7 from human outbreaks incorporating heterogeneity in exposure. *Epidemiol. Infect.* 136, 761–770 (2008).

116. Karch, H., Tarr, P.I., and Bielaszewska, M. Enterohemorrhagic *Escherichia coli* in human medicine. *Int. J. Med. Microbiol.* 295, 405–418 (2005).

117. Proulx, F. and Tesh, V.L. Renal diseases in the Pediatric Intensive Care Unit: Thrombotic microangiopathy, hemolytic uremic syndrome, and thrombotic thrombocytopenic purpura. In *Pediatric Critical Care Medicine: Basic Science and Clinical Evidence*, eds. Wheeler, D.S., Wong, H.R., and Shanley, T.P., pp. 1189–1203, Springer-Verlag, London, U.K. (2007).

118. Garg, A. et al. Long-term renal prognosis of diarrhea-associated hemolytic uremic syndrome: A systematic review, meta-analysis, and meta-regression. *JAMA* 290, 1360–1370 (2003).

119. Ostroff, S.M. et al. Toxin genotypes and plasmid profiles as determinants of systemic sequelae in *Escherichia coli* O157:H7 infections. *J. Infect. Dis.* 160, 994–998 (1989).

120. Ethelberg, S. et al. Virulence factors for hemolytic uremic syndrome, Denmark. *Emerg. Infect. Dis.* 10, 842–847 (2004).

121. Frank, C. et al. Epidemic profile of Shiga toxin-producing *Escherichia coli* O104:H4 outbreak in Germany. *N. Engl. J. Med.* 365, 1771–1780 (2011).

122. Rump, L.V. et al. Genetic characterization of *Escherichia coli* O104 isolates from different sources in the United States. *Appl. Environ. Microbiol.* 78, 1615–1618 (2012).

123. Bielaszewska, M. et al. Characterisation of the *Escherichia coli* strain associated with an outbreak of haemolytic uraemic syndrome in Germany, 2011: A microbiological study. *Lancet Infect. Dis.* 11, 671–676 (2011).

124. Schüller, S. et al. Shiga toxin binding in normal and inflamed human intestinal mucosa. *Microbes Infect.* 9, 35–39 (2007).

125. Acheson, D.W.K. et al. Translocation of Shiga toxin across polarized intestinal cells in tissue culture. *Infect. Immun.* 64, 3294–3300 (1996).

126. Hurley, B.P., Thorpe, C.M., and Acheson, D.W.K. Shiga toxin translocation across intestinal epithelial cells is enhanced by neutrophil transmigration. *Infect. Immun.* 69, 6148–6155 (2001).

127. Malyukova, I. et al. Macropinocytosis in Shiga toxin 1 uptake by human intestinal epithelial cells and transcellular transcytosis. *Am. J. Physiol. Gastrointest. Liver Physiol.* 296, G78–G92 (2009).

128. Lukyanenko, V. et al. Enterohemorrhagic *Escherichia coli* infection stimulates Shiga toxin 1 macropinocytosis and transcytosis across intestinal epithelial cells. *Am. J. Physiol. Cell. Physiol.* 301, C1140–C1149 (2011).

129. Bitzan, M. et al. High incidence of serum antibodies to *Escherichia coli* O157:H7 lipopolysaccharide in children with hemolytic–uremic syndrome. *J. Pediatr.* 119, 380–385 (1991).

130. Macher, B.A. and Klock, J.C. Isolation and chemical characterization of neutral glycosphingolipids of human neutrophils. *J. Biol. Chem.* 255, 2092–2096 (1980).

131. Brigotti, M. et al. Interactions between Shiga toxins and human polymorphonuclear leukocytes. *J. Leukoc. Biol.* 84, 1019–1027 (2008).

132. Iordanov, M.S. et al. Ribotoxic stress response: Activation of the stress-activated protein kinase JNK1 by inhibitors of the peptidyl transferase reaction and by sequence-specific RNA damage to the alpha-sarcin/ricin loop in the 28S rRNA. *Mol. Cell Biol.* 17, 3373–3381 (1997).

133. Smith, W.E. et al. Shiga toxin 1 triggers a ribotoxic stress response leading to p38 and JNK activation and induction of apoptosis in intestinal epithelial cells. *Infect. Immun.* 71, 1497–1504 (2003).

134. Cameron, P. et al. Verotoxin activates mitogen-activated protein kinase in human peripheral blood monocytes: Role in apoptosis and proinflammatory cytokine release. *Br. J. Pharmacol.* 140, 1320–1330 (2003).

135. Cherla, R.P. et al. Shiga toxin 1-induced cytokine production is mediated by MAP kinase pathways and translation initiation factor eIF4E in the macrophage-like THP-1 cell line. *J. Leukoc. Biol.* 79, 397–407 (2006).

136. Gray, J.S. et al. Double-stranded RNA-activated protein kinase mediates induction of interleukin-8 expression by deoxynivalenol, Shiga toxin 1, and ricin in monocytes. *Toxicol. Sci.* 105, 322–330 (2008).

137. Thorpe, C.M. et al. Shiga toxins induce, superinduce, and stabilize a variety of C-X-C chemokine mRNAs in intestinal epithelial cells, resulting in increased chemokine expression. *Infect. Immun.* 69, 6140–6147 (2001).

138. Harrison, L.M., van Haaften, W.C.E., and Tesh, V.L. Regulation of proinflammatory cytokine expression by Shiga toxin 1 and/or lipopolysaccharides in the human monocytic cell line THP-1. *Infect. Immun.* 72, 2618–2627 (2004).

139. Harrison, L.M. et al. Chemokine expression in the monocytic cell line THP-1 in response to purified Shiga toxin 1 and/or lipopolysaccharides. *Infect. Immun.* 73, 403–412 (2005).

140. Leyva-Illades, D. et al. Regulation of cytokine and chemokine expression by the ribotoxic stress response elicited by Shiga toxin type 1 in human macrophage-like THP-1 cells. *Infect. Immun.* 80, 2109–2120 (2012).

141. van de Kar, N.C.A.J. et al. Tumor necrosis factor and interleukin-1 induce expression of the verocytotoxin receptor globotriaosylceramide on human endothelial cells: Implications for the pathogenesis of the hemolytic uremic syndrome. *Blood* 80, 2755–2764 (1992).

142. Ramegowda, B., Samuel, J.E., and Tesh, V.L. Interaction of Shiga toxins with human brain microvascular endothelial cells: Cytokines as sensitizing agents. *J. Infect. Dis.* 180, 1205–1213 (1999).

143. Lentz, E.K. et al. Role of TNF-α in disease using a mouse model of Shiga toxin-mediated renal damage. *Infect. Immun.* 78, 3689–3699 (2010).

144. Clayton, F. et al. Lipopolysaccharide up-regulates renal Shiga toxin receptors in a primate model of hemolytic uremic syndrome. *Am. J. Nephrol.* 25, 536–540 (2005).

145. Keepers, T.R., Gross, L.K., and Obrig, T.G. Monocyte chemoattractant protein 1, macrophage inflammatory protein 1 alpha, and RANTES recruit macrophages to the kidney in a mouse model of hemolytic–uremic syndrome. *Infect. Immun.* 75, 1229–1236 (2007).

146. Lee, S.-Y. et al. Shiga toxin 1 induces the ER stress response in human monocytes. *Cell. Microbiol.* 10, 770–780 (2008).

147. Tesh, V.L. The induction of apoptosis by Shiga toxins and ricin. In *Ricin and Shiga Toxins: Pathogenesis, Immunity, Vaccines and Therapeutics*, ed. Mantis, N.J., pp. 137–178, Springer-Verlag, Berlin, Germany (2012).

148. Tarr, P.I. et al. *Escherichia coli* O157:H7 and the hemolytic uremic syndrome: Importance of early cultures in establishing etiology. *J. Infect. Dis.* 162, 553–556 (1990).

149. Karch, H. et al. Frequent loss of Shiga-like toxin genes in clinical isolates of *Escherichia coli* upon subcultivation. *Infect. Immun.* 60, 3464–3467 (1992).

150. Feng, P. et al. Isogenic strain of *Escherichia coli* O157:H7 that has lost both Shiga toxin 1 and 2 genes. *Clin. Diagn. Lab. Immunol.* 8, 711–717 (2001).

151. Ake, J.A. et al. Relative nephroprotection during *Escherichia coli* O157:H7 infections: Association with intravenous volume expansion. *Pediatrics* 115, 673–680 (2005).

152. Gould, L.H. et al. Recommendations for diagnosis of Shiga toxin-producing *Escherichia coli* infections by clinical laboratories. *Morbid. Mortal. Wkly. Rep.* 58(RR-12), 1–14 (2009).

153. Quiñones, B. et al. O-antigen and virulence profiling of Shiga toxin-producing *Escherichia coli* by a rapid and cost-effective DNA microarray-colorimetric method. *Front. Cell. Infect. Microbiol.* 2, e61 (2012).

154. DuPont, H.L. Bacterial diarrhea. *N. Engl. J. Med.* 361, 1560–1569 (2009).

155. Walterspiel, J.N. et al. Effect of subinhibitory concentrations of antibiotics on extracellular Shiga-like toxin I. *Infection* 20, 25–29 (1992).

156. Zhang, X. et al. Quinolone antibiotics induce Shiga toxin-encoding bacteriophages, toxin production, and death in mice. *J. Infect. Dis.* 181, 664–670 (2000).

157. Panos, G.Z., Betsi, G.I., and Falagas, M.E. Systematic review: Are antibiotics detrimental or beneficial for the treatment of patients with *Escherichia coli* O157:H7 infection? *Aliment. Pharmacol. Ther.* 24, 731–742 (2006).

158. Rahal, E.A. et al. Decrease in Shiga toxin expression using a minimal inhibitory concentration of rifampicin followed by bactericidal gentamicin treatment enhances survival of *Escherichia coli* O157:H7-infected BALB/c mice. *Ann. Clin. Microbiol. Antimicrob.* 10, 34 (2011).

159. Armstrong, G.D., Fodor, E., and Vanmaele, R. Investigation of Shiga-like toxin binding to chemically synthesized oligosaccharide sequences. *J. Infect. Dis.* 164, 1160–1167 (1991).

160. Takeda, T. et al. In vitro assessment of a chemically synthesized Shiga toxin receptor analogue attached to chromosorb P (Synsorb Pk) as a specific absorbing agent of Shiga toxin 1 and 2. *Microbiol. Immunol.* 43, 331–337 (1999).

161. Armstrong, G.D., McLaine, P.N., and Rowe, P.C. Clinical trials of Synsorb-Pk in preventing hemolytic–uremic syndrome. In *Escherichia coli O157:H7 and Other Shiga Toxin-Producing E. coli Strains*, eds. Kaper, J.B. and O'Brien, A.D., pp. 374–384, ASM Press, Washington, DC (1998).

162. Trachtman, H. et al. Effect of an oral Shiga toxin-binding agent on diarrhea-associated haemolytic uraemic syndrome in children: A randomized controlled trial. *JAMA* 290, 1337–1344 (2003).

163. Paton, A.W., Morona, R., and Paton, J.C. A new biological agent for treatment of Shiga toxigenic *Escherichia coli* infections and dysentery in humans. *Nat. Med.* 6, 265–270 (2000).

164. Kitov, P.I. et al. Shiga-like toxins are neutralized by tailored multivalent carbohydrate ligands. *Nature* 403, 669–672 (2000).

165. Mulvey, G.L. et al. Assessment in mice of the therapeutic potential of tailored, multivalent Shiga toxin carbohydrate ligands. *J. Infect. Dis.* 187, 640–649 (2003).

166. Nishikawa, K. et al. A therapeutic agent with oriented carbohydrates for treatment of infections by Shiga toxin-producing *Escherichia coli* O157:H7. *Proc. Natl. Acad. Sci. USA* 99, 7669–7674 (2002).

167. Nishikawa, K. et al. Identification of the optimal structure for a Shiga toxin neutralizer with oriented carbohydrates to function in the circulation. *J. Infect. Dis.* 191, 2097–2105 (2005).

168. Nishikawa, K. Recent progress of Shiga toxin neutralizer for treatment of infections by Shiga toxin-producing *Escherichia coli. Arch. Immunol. Ther. Exp.* 59, 239–247 (2011).

169. Nishikawa, K. et al. A multivalent peptide library approach identifies a novel Shiga toxin inhibitor that induces aberrant cellular transport of the toxin. *FASEB J.* 20, 2597–2599 (2006).

170. Stearns-Kurosawa, D.J. et al. Rescue from lethal Shiga toxin 2-induced renal failure with a cell-permeable peptide. *Pediatr. Nephrol.* 26, 2031–2039 (2011).

171. Saenz, J.B., Doggett, T.A., and Haslam, D.B. Identification and characterization of small molecules that inhibit intracellular toxin transport. *Infect. Immun.* 75, 4552–4561 (2007).

172. Feng, Y. et al. Retrograde transport of cholera toxin from the plasma membrane to the endoplasmic reticulum requires the *trans*-Golgi network but not the Golgi apparatus in Exo2-treated cells. *EMBO Rep.* 5, 596–601 (2004).

173. Spooner, R.A. et al. The secretion inhibitor Exo2 perturbs trafficking of Shiga toxin between endosomes and the *trans*-Golgi network. *Biochem. J.* 414, 471–484 (2008).

174. Taylor, F.B., Jr. et al. Characterization of the baboon responses to purified Shiga-like toxin: Descriptive study of a new primate model of toxic response to Stx-1. *Am. J. Pathol.* 154, 1285–1299 (1999).

175. Stearns-Kurosawa, D.J. et al. Distinct physiologic and inflammatory responses elicited in baboons after challenge with Shiga toxin type 1 or 2 from enterohemorrhagic *Escherichia coli. Infect. Immun.* 78, 2497–2504 (2010).

176. Audi, J. et al. Ricin poisoning: A comprehensive review. *JAMA* 294, 2342–2351 (2005).

177. Rutjes, N.W. et al. Differential tissue targeting and pathogenesis of verotoxin 1 and verotoxin 2 in the mouse animal model. *Kidney Int.* 62, 832–845 (2002).

178. Uchida, H. et al. Shiga toxins induce apoptosis in pulmonary epithelium-derived cells. *J. Infect. Dis.* 180, 1902–1911 (1999).

179. Roy, C.J. et al. Impact of inhalation exposure modality and particle size on the respiratory deposition of ricin in BALB/c mice. *Inhal. Toxicol.* 15, 619–638 (2003).

180. Wilhelmsen, C.L. and Pitt, M.L. Lesions of acute inhaled lethal ricin intoxication in rhesus monkeys. *Vet. Pathol.* 33, 296–302 (1996).

46 Staphylococcal Enterotoxins

Jacques-Antoine Hennekinne and Sylviane Dragacci

CONTENTS

46.1 INTRODUCTION

The correlation between staphylococci-containing food and symptomatology was not recognized until the twentieth century even if some cases have been described earlier. It was Baerthlein, when reporting on a huge outbreak involving 2000 soldiers of the German army during World War I, who established in 1922 the possible involvement of bacteria contaminating food:

I am going to report the case of an extended demonstration of poisoned sausages (approximately 2,000 cases) held in the spring 1918 during the military campaign of Verdun (France), which would probably have catastrophic military consequences. Early in June 1918, sudden and massive demonstrations that have the appearance of an acute and in some cases severe gastroenteritis, similar to cholera, affected the troops around Verdun; entire companies were disabled except just a few people, and within two days about 2,000 men had been affected. The symptoms were so severe that some troops had to be transferred to field hospitals. The suspicion of food poisoning has been mentioned because, according to reports of the sick, the disease occurred 2 or 3 hours (some of the symptoms appeared after 6 to 8 hours) after eating a dish of sausages. Only troops who did not eat the meal were spared, such as soldiers who had returned to headquarters to receive orders, soldiers who for other reasons had not eaten sausages, and soldiers who were on leave and/or following a different diet. However, it was surprising that among the troops that were not present at the front, such as butchers, who ate the same sausage two days earlier, we did not observe any cases of disease.

46.1.1 CLASSIFICATION OF ONE STAPHYLOCOCCAL TOXIN (ENTEROTOXIN B) AS A BIOLOGICAL WEAPON

Later on, from a military point of view, such powerful bacterial toxins were considered as potentially more effective than classical chemical weapons because of their great potency to cover a larger area with a smaller quantity of agent. As an example, the US Army had developed and stockpiled two kinds of toxin weapons: *botulinum* toxin, a lethal agent, and staphylococcal enterotoxin B (SEB) qualified as an incapacitating agent. Staphylococcal enterotoxins (SEs) exhibit neurotoxic and superantigenic properties. Among SEs, the SEB is nowadays classified as a potential bioweapon as it shares all the characteristics of an ideal biological agent. Indeed, coagulase-positive staphylococci (CPS) including *Staphylococcus aureus* are widespread bacteria commonly found in environment and bear by some animals (mammals and birds). The toxin SEB is very easy to prepare from bacterial cultures needing no specific equipment. SEB is a heat-stable protein that could be easily transformed into an aerosol and air-dispersed. The toxin can produce large and potent multiorgan and system symptoms, and if in high amounts, SEB could provoke up to shock and death. The symptoms developed in humans after an aerosol attack differ from those induced after ingestion of contaminated food. Current symptoms consist of high fever, chills, headache, muscle ache, dry cough, and inflammation of the eyelids. After an aerosol attack, SEB may additionally cause troubles with breathing, chest pains, and the production of fluid in the lungs. Typically, SEB is considered as an incapacitating agent that causes debilitation up to 2 weeks.

This is why SEB was studied, not so much for its mass destruction capability but rather for its ability to incapacitate soldiers so they would be incapable of fighting or defending their posts.[1,2]

In 1943, the United States began research into the offensive use of biological agents. This work was started at Camp Detrick (now Fort Detrick), in response to a perceived German biological warfare (BW) threat as opposed to a Japanese one.

The US bioweapons program studied the SEB toxin intensively and determined that the amount of the toxin (originally named as agent PG and later UC) required to induce incapacitation was considerably less than that of synthesized chemicals. When the toxin and chemicals were compared by expense, time, and complexity of production, SEB was far more cost-effective. A dose of 0.4 ng/kg body weights was estimated to incapacitate 50% of the human population exposed by an aerosol attack, while 200 ng/kg body weights would be lethal for 50% of those exposed.[3,4]

The first military use of SEB probably took place during the D-Day in North Africa where the US Army chemical warfare service has supplied a vial of SEB to Office of Strategic Services agents to incapacitate "Nazi" agent in order to prevent effective handling of intelligence.

After Fidel Castro seized power in Cuba in 1959, Eisenhower administration drew up contingency plans to invade the island and topple the dictator as US officials tolerated with difficulty that a communist administration took place at 350 km of the Florida coast.

Following to his election, President Kennedy carried out on Eisenhower plans and ordered a large force to be assembled for a possible invasion, and the Kennedy administration considered a wide range of military options against Cuba during the crisis. The Pentagon set more than a million people in motion. It estimated that in a conventional assault, the number of Americans wounded or killed in the first 10 days of battle could easily reach 18,500 or roughly 10% of the 180,000 man invasion force.

One of the most secret options was referred to Fort Detrick as the Marshall Plan. This plan called for a huge assault by American troops that would begin with a non-lethal attack against Cuba's soldiers and civilians. Over the years, the work in Fort Detrick included not only close collaborations with Fort Detrick's biological experts but also agent selection, casualty forecasts, and study of the weather patterns over Cuba. There were even drills to see how fast the required germs could be produced. Initially, Fort Detrick could provide mainly lethal agents, such as anthrax. But in the early 1960s, as production increased, Detrick was able to offer an increasingly wide selection of incapacitants meant to leave most of the target population alive. Military biologists and planners argued for using such pathogens and biological toxins if the president ordered an invasion. Incapacitants could immobilize Cuban defenses, significantly reducing projected American combat casualties. Prior to invade Bay of Pigs in the "operation Zapata," the Marshall Plan should douse the island from jet aircraft with a cocktail of SEB and other organisms including Venezuelan equine fever and Q fever.

The toxin of the cocktail was SEB. The germ warriors made it into a weapon by cultivating up trillions of the bacteria and then concentrating the poison. Whoever breathed the vapor would fall ill 3–12 h later. The symptoms included chills, headache, muscle pain, coughing, sudden fever up to 41°C (close to what produces coma, seizures, and death), and, less frequently, nausea, vomiting, and diarrhea. The fever lasted days and the cough weeks.

The virus in the mix caused Venezuelan equine encephalitis. Its incubation period varied from 1 to 5 days, followed by the sudden onset of the nausea and diarrhea often associated with serious infection, as well as spiking fever up to 40.5°C. The acute phase lasted from one to three days, followed by weeks of weakness and lethargy.

The final element was the bug that caused Q fever. Its incubation period was 10–20 days, after which it generally produced up to 2 weeks of debilitating symptoms, including headaches, chills, hallucinations, facial pain, and fevers of up to 40°C. Chronic Q fever, the US Army found, was rare, but if the disease progressed to that stage, it was frequently fatal.

Finally, the Marshall Plan was rejected by President Kennedy, and the operation Zapata was launched on April 17, 1961.

By 1966, the United States and its allies had produced stockpiles of various BW agents, including SEB, and research to establish parameters for SEB's use as an aerosolized bioweapon continued at several facilities in the United States and Great Britain.

On November 25, 1969, the National Security Decision Memorandum 35 contained the formal decision to renounce an offensive bioweapon capability. However, this statement did not clearly mention toxins. In order to clarify the US policy on toxins, the document "review of toxins policy" was issued on December 31, 1969. Then, an interagency debate was initiated, and this one finally led to a three-option proposal on January 21, 1970:

- Option I: Keep entirely open the option to produce and employ toxin warfare agents.
- Option II: Do not produce toxins now, but keep open the possibility of stockpiling them if a method is devised to manufacture them by chemical synthesis, without the need for production in bacteria.
- Option III: Give up toxin weapons entirely, whether produced by biological fermentation or by chemical synthesis, and permit work only on defensive measures against them, such as vaccines or more effective gas masks.

Finally, on February 14, 1970, President Nixon decided to follow option III and stopped by executive order the offensive BW program. Between May 1971 and May 1972, all stockpiles of agents were destroyed.[5,6]

Today, bioterrorism agents are separated into three categories for preparedness purposes depending upon their ease of dissemination and the ability to cause excessive morbidity and mortality.[7] Category A includes agents such as variola major (smallpox) and *Yersinia pestis* (plague) that have been used as a weapon of mass destruction (WMD). Category B agents are easy to disseminate and produce moderate morbidity and low mortality. Category B agents do not meet criteria for use as a WMD, but dispersal of a Category B agent could result in regional disruptions and hysteria. Finally, category C agents include emerging pathogens that could potentially be engineered for future mass dissemination.

From all accounts, SEB meets the criteria for a Category B agent in that it is stable, easy to disseminate, and induces severe emesis and toxic shock. An aerosol of SEB in a crowded area could lead to an incapacitating disease in hundreds of individuals at once. Although mortality would be low, the illness would create a serious public health impact by disrupting normal work days and cause havoc by increasing individual use of emergency rooms.[4] Many bioterrorism agents such as SEB are found in nature, are easy to isolate and produce in mass quantities, and are usually stable in adverse environmental conditions.[3] Because the agent is a common inhabitant in the environment, monitoring the agent becomes more difficult. The fact that there are accidental cases of food poisoning and occasional cases of toxic shock syndrome annually also complicates identifying bioterrorism incidents using SEB. In the final analysis, although SEB may not be the most favored bioterrorism agent, there is always a possibility that it will be used in an attack, and, therefore, mechanisms should be in place for decontamination and treatment.

More than 1200 biological agents could be used to cause illness or death; however, only few of them hold the necessary characteristics to make them the right candidates for BW or terrorism agents. These characteristics are listed in the following text. First of all, the best biological agent should be relatively easy to acquire, process, and use. Then, only small amounts of this one must be needed to incapacitate and/or kill hundreds of thousands of people in a dedicated area. Moreover, ideal BW agents must be easy to hide, difficult to detect, and with no related medical treatment. And finally, this perfect agent should be invisible, odorless, tasteless, and able to be spread silently.

Various dissemination ways could be used: air, water or food, and skin.

By inhalation route: BW must be dispersed as fine particles in a size range necessary for inhalation (from 0.5 to 5 μm) to be an effective biological weapon. Moreover, a person should breathe a sufficient quantity of particles into the lungs to be infected. The use of an explosive device (artillery, missiles, and detonated bombs) to deliver and spread biological agents is not as effective as the delivery by aerosol due to the possible destruction of the agent by the blast, typically leaving less than 5% of the agent capable of causing disease.

By oral route: Contamination of city's water supplies requires an unrealistically large amount of an agent. Moreover, this also needs an introduction of the bioweapon agent into the tap water *after* a facility treatment.

By injection route: This method might be used for assassination, but is not appropriate for mass casualties.

46.1.2 FOODBORNE OUTBREAKS: SOME EXAMPLES

The very first description of foodborne disease involving *S. aureus* was reported by Vaughan and Sternberg in 1884. This food poisoning event happened in Michigan (United States) and was related to the consumption of a cheese contaminated by staphylococci. The authors stated: "It seems not improbable that the poisonous principle is a ptomaine developed in the cheese as a result of the vital activity of the above mentioned Micrococcus or some other microorganisms which had preceded it, and had perhaps been killed by its own poisonous products." Ten years later, in 1894, Denys observed that members of the same family who had consumed meat coming from a cow, which died of vitullary fever, were linked to the presence of pyogenic staphylococci. In 1907, Owen recovered staphylococci from dried beef involved in an outbreak where ill people showed typical symptoms of staphylococcal food poisoning (SFP). The proof of the involvement of staphylococci in food poisoning was first brought by Barber, in 1914. He evidenced that staphylococci were the cause of poisoning by consuming by himself unrefrigerated milk from a cow suffering from mastitis, an inflammation due to staphylococci.

In 1930, Dack found that a sponge cake was responsible for the intoxication of 11 individuals; he highlighted that the disease was probably linked to a toxin called "enterotoxin" produced by yellow hemolytic staphylococcus. Broth culture filtrates of this strain were administrated intravenously to a rabbit and orally to three human volunteers. The rabbit died, after first developing water diarrhea, and the three volunteers developed nausea, chilliness, and vomiting after 3 h.

One of the first well-documented SFP outbreaks (SFPOs) was described by Denison in 1936. This outbreak occurred among high school students after they had eaten tainted cream puffs. He depicted the typical symptoms of 122 cases as follows:

Within 2-4 hours after eating there was first noticed a feeling of nausea. Severe abdominal cramps developed and were quickly followed by vomiting which was severe and continued at 5-20 minute intervals for 1-8 hours […] A diarrhea of 1-7 liquid stools usually began with the vomiting and continued for several hours after its onset […] During the acute stage the temperature was normal or subnormal, the pulse noticeably increased, there were cold sweats, prostration was severe and the patients were very definitely in a state of shock. Headache was mild and of a short duration. Muscular cramping […] was present in the majority. Dehydration was marked in some. While the acute symptoms usually lasted only 1-8 hours, complete recovery […] was delayed for 1-2 days.

The symptoms of SFP have been extensively studied, especially by the US Army. In a report of the US Army regarding naturally occurring outbreaks involving 400 out of 600 soldiers in 1944, DeLay indicated that about 25% of cases were classified as severe or shock cases.[8]

Numerous SFPOs have been described since the end of World War II. For example, Brink and Van Meter from the Institute for Cooperative Research of the University of Pennsylvania wrote a long report on an outbreak related to SE food poisoning that happened in 1960:

> On a Saturday afternoon in the middle of summer, an epidemic of staphylococcal enterotoxin food poisoning occurred at a picnic held two miles from Gabriel, a small Midwestern town (the name of the town and other names in this report are fictitious, in accordance with commitments to Task Surprise respondents.). About 1,700 persons attended the picnic, which is an annual affair sponsored by the Johnson Co., of Croydon, some 60 miles away. Early in the morning, approximately seven hours before the picnic began, an unventilated, unrefrigerated truck containing a large supply of ham sandwiches was parked at the picnic grounds. The truck was exposed to the heat of direct sunlight, while the average ambient temperature for the day was close to 100 degrees Fahrenheit. In this environment, the staphylococcal organisms which elaborate the toxin multiplied rapidly. During the epidemic that followed, approximately 1,100 persons became ill.

As it is one of the main features of outbreaks related to *S. aureus* to involve at one time a large number of people living or eating together (restaurants, troops, schools or corporate catering, family events like wedding banquet), the involved *S. aureus* enterotoxins were named as the "banquet toxin." Indeed, food is often contaminated by toxin-producing *Staphylococcus* when handled or stored in improper hygiene conditions (lack of cleanliness, lack of refrigeration of cooked meals). But, one of the other stories leading to such contaminated food could also be a wrongdoing food process like the one that occurred in Japan, in 2000. The event was called the "the Snow Brand incident" for which 13,420 cases were notified, the highest number ever recorded in Japan. It represents one of the largest and recent *Staphylococcus* foodborne poisoning in the world. The incriminated food was processed milk (low-fat milk and beverage milk), and the process default was an improper control of temperature due to an electric power cut.

When the amount of ingested enterotoxins is high, the staphylococcal poisoning may lead to severe symptoms, making people unable to pursue their (professional) activities. This was illustrated by a violent outbreak happening during air travel and reported by Effersoe and Kjerulf in 1975 (Table 46.1). Three hundred and sixty four passengers and crew members were poisoned by meals containing contaminated ham served just 1 h before landing at Copenhagen (Denmark). Nearly half of the passengers needed to be hospitalized, and two out of the badly touched were discharged from the hospital only on the 10th day but healthy. Numeration of *S. aureus* in meal remnants confirmed later on the involvement of enterotoxins. The teaching, which was retrieved from this toxic episode concerning people grouped in a close area with no ground contact (like in ships or aircraft), was to recommend to embark meals coming from at least two food-delivering centers and to separate people, and especially the crew staff, in two population receiving meals of separate origins.

TABLE 46.1

SFPOs from the Literature

Year	Location	Incriminated Food	Number of Cases
1968	School children, Texas	Chicken salad	1,300
1971	UK Army	Sausages rolls, ham sandwiches	100
1975	Flight from Japan to Denmark	Ham	364
1976	Flight from Rio to NYC	Chocolate eclairs	80
1980	Canada	Cheese curd	62
1982	North Carolina and Pennsylvania	Ham and cheese sandwich; stuffed chicken	121
1983	Caribbean cruise ship	Dessert cream pastry	215
1984	Scotland	Sheep's milk cheese	27
1985	France, United Kingdom, Italy, Luxembourg	Dried lasagna	50
1985	School children, Kentucky	2% chocolate milk	>1,000
1986	Country Club, New Mexico	Turkey, poultry, gravy	67
1989	Various US states	Canned mushrooms	102
1990	Thailand	Eclairs	485
1992	Elementary school, Texas	Chicken salad	1,364
1997	Retirement party, Florida	Precooked ham	18
1998	Minas Gerais, Brazil	Chicken, roasted beef, rice, and beans	4,000
2000	Osaka, Japan	Low-fat milk	13,420
2006	Ile de France area, France	Coconut pearls (Chinese dessert)	17
2007	Scouts' camp, Belgium	Hamburger	15
2007	Elementary school, Austria	Milk, cacao milk, vanilla milk	166
2009	Nagoya University festival, Japan	Crepes	75
2009	Various districts, France	Raw milk cheese	23

The main point highlighted by these reports is that any food that provides a convenient medium for CPS growth may be involved in such SFPOs. The foods most frequently involved differ widely from one country to another, probably due to differing food habits (Table 46.1).[9] For instance, in the United Kingdom or the United States, meat or meat-based products are the food vehicles mostly involved in SFP,[10] although poultry, salads, and cream-filled bakery items are other good examples of foods that have been incriminated.[11] In France, various food types have been associated with SFPOs, but as the consumption of unpasteurized milk cheeses is much more common than in Anglo-Saxon countries, milk-based products are more frequently involved than in other countries.[12]

Although SFPO should or might be reported, depending on national regulations over countries, to the sanitary authorities, they are in fact often underreported, especially when few people were involved and recovered quickly.[12] However, many researchers still consider that SFP is one of the most common foodborne diseases worldwide.[13]

46.2 STAPHYLOCOCCI

46.2.1 PHENOTYPIC AND GENOTYPIC CHARACTERIZATIONS

Staphylococcus is a spherical, nonsporulating, nonmotile bacterium (coccus) that, when observed under the microscope, occurs in pairs, short chains, or grape-like clusters. These facultative aero-anaerobic bacteria are Gram- and catalase-positive. Staphylococci are ubiquitous in the environment and can be found in the air, dust, sewage, water, environmental surfaces, humans, and animals. To date, 52 species and subspecies of staphylococci have been described according to their potential to produce coagulase. Their classification thus distinguishes between coagulase-producing strains, designated as CPS, and non-coagulase-producing strains, called coagulase-negative staphylococci (CNS). However, only CPS strains have been clearly implicated in food poisoning incidents. Among the seven described species belonging to the CPS group, *S. aureus* subsp. *aureus* is the main causative agent described in SFPOs. During processing and storage, temperatures outside the range of 7°C–48°C prevent the growth of *S. aureus*. However, *S. aureus* subsp. *aureus* strains are usually very tolerant to NaCl and grow well in NaCl concentrations of up to 10%; growth is possible, although retarded, even in concentrations of up to 20%.

Various methods have been developed to isolate and/or enumerate staphylococci. Some of them use conventional microbiology-based methods, whereas others are dedicated to a molecular characterization.

Among the first type of methods, some are standardized in order to ensure sanitary rules and mandatory declaration of food poisoning outbreaks; it appears absolutely necessary to use robust and reliable methods for the detection of CPS and of SE. Standardization brings consensual solutions to technical problems in a client–provider perspective. In its international definition, a standard is "a document, established by consensus and approved by a recognized body that provides, for common and repeated use, rules, guidelines or characteristics for activities or their results, aimed at the achievement of the optimum degree of order in a given context." Usually, standards are heavy to implement for operators; standard methods may be supplemented by the so-called alternative methods with the same response characteristics (this however requires a validation procedure against the reference method described in the standard) but that implementation will be easier, faster, and usually cheaper. One of the most used standards methods for CPS enumeration refers to *S. aureus* EN ISO 6888 parts -1, -2, and -3. These are classical techniques of colony-forming unit (cfu) numeration on selective media after direct inoculation of decimal dilutions of food extracts and incubation at 37°C.

The selective medium used in the standard is a modified Giolitti and Cantoni broth whose formula is similar to that of Baird Parker (BP) broth. After enrichment, *S. aureus* is detected by streaking and incubation on agar selective medium. Such standards are based on two types of agar selective media: BP and rabbit plasma fibrinogen-BP (RPF-BP). The standard ISO 6888-3 describes a method of detection and numeration by the technique of the most probable number (MPN) after an enrichment step. Methods using BP medium require a confirmation test of several colonies with characteristic and noncharacteristic aspects, with a coagulase test in vitro. Methods using RPF-BP do not require confirmation test. Whatever the medium used, coagulase-positive colonies are counted, and the result is expressed in cfu of CPS (most likely *S. aureus*) by g or by mL of foodstuff (Figure 46.1).

Molecular-based methods are used for characterization. These methods include the detection of specific genes of CPS such as genes encoding for SEs in order to evaluate the enterotoxigenic potential of CPS isolated from samples. Moreover, other methods are used for the large-scale analysis of the diversity of *S. aureus* isolates, namely, multilocus sequence typing (MLST), spa typing, and multiple locus variable number of tandem repeat (VNTR) analysis (MLVA). In addition, pulsed field gel electrophoresis (PFGE) is still widely considered as the gold standard for typing *S. aureus* isolates both at food and clinical levels.[14]

MLST allows the description of major clonal complexes underlying the *S. aureus* population structure and relationship. However, this method has a moderate discriminatory power. The spa typing is a widely used method in which variations in a highly variable tandem repeat are characterized by sequencing. This method is a very powerful tool offering relevant results when used in first-line assay. MLVA has been recently developed using a variable number of VNTR loci. This method provides rapid and efficient results for the high-resolution genotyping of *S. aureus* isolates.

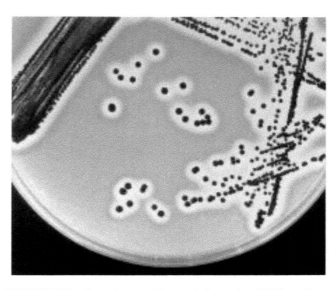

FIGURE 46.1 Coagulase-positive staphylococci on RPFA medium.

46.2.2 Growth and Enterotoxigenic Conditions

Five conditions are required to induce SFPOs: (1) a source containing enterotoxin-producing staphylococci such as raw materials and healthy or infected carrier, (2) transfer of staphylococci from source to food using, for example, dirty tools in food preparation due to poor hygiene practices, (3) food composition with favorable physicochemical characteristics for *S. aureus* growth and toxinogenesis, (4) favorable temperature and sufficient time for bacterial growth and toxin production, and (5) ingestion of food containing sufficient amounts of toxin to rouse symptoms.

Most SFPOs arise due to poor hygiene practices during processing,[15] cooking, or distributing the food product.[16] Staphylococci are commonly found in a wide variety of mammals and birds, and the transfer of *S. aureus* to food has two main sources: human carriage during food processing and dairy animals in case of mastitis.[17]

46.3 STAPHYLOCOCCAL ENTEROTOXINS

46.3.1 Classification and Structure

Since the first characterization of SEA and SEB in 1959 to 1960, 22 different SEs have been described (Figure 46.2; Table 46.2). They are named SEA to SElV2, after the chronological order of their discovery, except for SEF that was later renamed TSST1: enterotoxins A (SEA), B (SEB), C1 (SEC1), C2 (SEC2), C3 (SEC3), D (SED), E (SEE), G (SEG), H (SEH), I (SEI), J (SElJ), K (SElK), L (SElL), M (SElM), N (SElN), O (SElO), P (SElP), Q (SElQ), R (SER), S (SES), T (SET), U (SElU), and U2 and V.[18]

These toxins (enterotoxin and enterotoxin-like) are globular single-chain proteins with molecular weights ranging from 22 to 29 kDa. Moreover, their crystal structures, established for SEA, SEB, SEC, SED, SEH, SElI, and SElK, show significant homology in their secondary and tertiary conformations. However, SEs, SEls, and

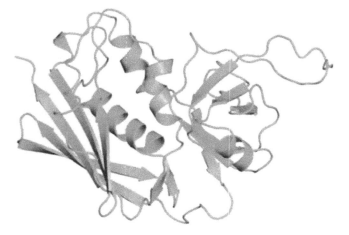

FIGURE 46.2 3D structure of SEB. (This structure was published by Papageorgiou, A.C. et al., *J. Mol. Biol.*, 277, 61, 1998.) This assembly is monomeric and contains one copy of SEB. (From http://www.ebi.ac.uk/pdbe-srv/view/entry/3seb/summary.)

TABLE 46.2
Characteristics of SEs

General Characteristics			Mode of Activity	
Toxin Type	Molecular Weight (Da)	Genetic Basis of SE	Superantigenic Action[a]	Emetic Action[b]
SEA	27,100	Prophage	+	+
SEB	28,336	Chromosome, plasmid, pathogenicity island	+	+
SEC$_{1-2-3}$	≈27,500	Plasmid	+	+
SED	26,360	Plasmid	+	+
SEE	26,425	Prophage	+	+
SEG	27,043	*egc*, chromosome	+	+
SEH	25,210	Transposon	+	+
SEI	24,928	*egc*, chromosome	+	(+)
SElJ	28,565	Plasmids	+	nk
SEK	25,539	Pathogenicity island	+	nk
SElL	25,219	Pathogenicity island	+	–[c]
SElM	24,842	*egc*, chromosome	+	nk
SElN	26,067	*egc*, chromosome	+	nk
SElO	26,777	*egc*, chromosome	+	nk
SElP	26,608	Prophage	+	nk[d]
SElQ	25,076	Pathogenicity island	+	–
SER	27,049	Plasmids	+	+
SES	26,217	Plasmid	+	+
SET	22,614	Plasmid	+	(+)
SElU	27,192	*egc*, chromosome	+	nk
SElU$_2$	26,672	*egc*, chromosome	+	nk
SElV	24,997	*egc*, chromosome	+	nk

[a] +, Positive reaction.

[b] +, Positive reaction; (+), weak reaction; (–), negative reaction; nk, not known.

[c] For SElL, emetic activity was not demonstrated in *Macaca nemestrina* monkey.

[d] For SElP, emetic activity was demonstrated in *Suncus murinus* but not in primate model.

TSST-1 can be divided into four phylogenic groups based on their primary amino acid sequences.

SEs are resistant to environmental conditions (freezing, drying, heat treatment, low pH) that on the opposite destroy the enterotoxin-producing strain. Toxins are also resistant to proteolytic enzymes, and their toxic effects are still operating in the digestive tract after ingestion. Thermal resistance of toxins depends on the relative purity of the SE preparation. Basically, heat treatments commonly used in food processing are not effective for complete destruction of SE found at levels conducting to food poisoning outbreaks (0.5–10 µg/100 mL or 100 g). However, it should be borne in mind that thermal inactivation often resulted in the loss of the serological reactivity of the SE. Biological activity can be lost before the serological activity. On the other hand, some outbreaks occur after eating foods that have been heated *after* SE was produced. Thermal stability of SE depends on the nature of the food, pH level, NaCl content, and on the type of toxin. SEA, for instance, is relatively more stable to heat at pH 6.0 or higher than at pH 4.5–5.5. SED is more stable at pH 4.5–5.5 than pH 6.0 or higher. If SE is not completely inactivated by heat, reactivation may occur under certain conditions like cooking, storage, or incubation.

These proteins have been named according to their emetic activity after oral administration in a primate model. Jarraud et al. renamed some of these toxins as SE-like toxins (SEl), either because no emetic property was detected or because they were not tested in primate models.[19] SEs belong to the broad family of pyrogenic toxin superantigens (SAgs). SAgs, unlike conventional antigens, do not need to be processed by antigen-presenting cells (APCs) before being presented to T cells. They can directly stimulate T cells by cross-linking major histocompatibility complex (MHC) class II molecules on APC with the variable portion of the T cell antigen receptor β chain (TCR Vβ) or the T cell antigen receptor α chain for SE (TCR Vα), thereby inducing polyclonal cell proliferation. SAg binding sites lie outside the peptide-binding groove and therefore do not depend on T cell antigenic specificity but rather on the Vβ and/or Vα region of the TCR. This leads to the activation of a large number of T cells followed by proliferation and massive release of chemokines and proinflammatory cytokines that may possibly result in lethal toxic shock syndrome.[13] The SAgs can interact with epithelial cells leading to their transepithelial transport, cell activation, and induction of inflammatory state. First, most SAgs have dose-dependent capacity to cross the intestinal wall and produce a local and systemic action on the immune system. This transport is favored by the production of proinflammatory cytokine-like elements. Finally, superantigenic stimulation of intestinal epithelial cells induces an inflammatory response.

Although the superantigenic activity of SEs has been well characterized, the mechanisms involved in the emetic activity are less documented. Despite considerable efforts to identify specific amino acids and domains within SEs that may be important for emesis, published data are still limited and controversial. For example, SEIL and SEIQ are nonemetic, whereas SEI displays weak emetic activity. These toxins lack the disulfide loop characteristically found at the top of the N-terminal domain of other SEs. Nonetheless, the loop itself does not appear to be an absolute requirement for emesis, although it may stabilize a crucial conformation important for this activity. Correlation between emetic and T cell stimulatory activities of SEA and SEB where amino acids had been substituted has been studied. In most cases, genetic mutations resulting in a loss of SAg activity also resulted in the loss of emetic activity. However, as there was not a perfect correlation between immunological and emetic activities in all the mutants, this study suggested that both activities could be dissociated.

In contrast to other bacterial enterotoxins, specific cells and receptors in the digestive system have not been clearly linked to oral intoxication by an SE. In 1965, Sugiyama and Hayama suggested that SEs stimulate the vagus nerve in the abdominal viscera, which conducts the signal to the emetic center.[20] Supporting this idea, receptors on vagal afferent neurons are essential for SEA-triggered emesis. In addition, SEs are able to go through the gut lining and activate local and systemic immune responses. The diarrhea sometimes associated with SE intoxication could be due to the inhibition of water and electrolyte reabsorption in the small intestine. In an attempt to link the two separate activities of SEs, that is, superantigenicity and emesis, it has been postulated that enterotoxin activity could facilitate transcitosis that enables the toxin to enter the bloodstream and circulate through the body; this action allowing the interaction with antigen presenting and T cells leads to SAg activity.[13] In this way, circulation of SEs following ingestion of SEs as well as their spread from an *S. aureus* infection site could have more profound effects upon the host than if the toxin remains localized.

Enterotoxin gene locations are numerous. In 2010, Argudin et al. summed up these locations: se genes can be carried by plasmids (*seb, sed, sej, ser, ses, set*), phage (temperate for *sea*, defective for *see*), or by genomic islands (*seb, sec, seg, seh, sei, sek, sel, sem, sen, seo, sep,* and *seq*).[21] Gene encoding for *sec* can be located on a plasmid or a pathogenicity island depending on the origin of the isolate. Jarraud et al. highlighted the existence of an operon, enterotoxin gene cluster (*egc*), encoding for several SEs: SEG, SEI, SEM, SEN, and SEO. The *egc* also contains two pseudogenes (φent1 and φent2). This locus probably plays the role of a nursery for *se* genes, as phenomena of duplication and recombination from a common ancestral gene could explain new forms of toxins. This was demonstrated by the identification of genes encoding for *seu, seu2,* and *sev* within *egc*. The location of *se* genes on mobile genetic elements can result in horizontal gene transfer between strains of *S. aureus*. For example, the *seb* gene is located on the chromosome in some clinical isolates, whereas it has a plasmidic location in other strains of *S. aureus*.

The main regulatory system controlling the expression of virulence factors in *S. aureus* is the accessory gene regulator (*agr*) system. This arrangement works in combination with the staphylococcal accessory regulator (*sar*) system. Most but not all of the expression of SEs are controlled by the *agr* system. For example, expression of *seb, sec,* and *sed* genes is *agr*-dependent, whereas expression of *sea* and *sej* is *agr*-independent. SEB is a negative global regulator of exoprotein gene expression acting through the *agr* system. The expression of *agr* system is closely linked to quorum sensing. Four different patterns of expression using quantitative reverse-transcription PCR have been described. The first pattern for *sea, see, sej, sek, sep,* and *seq* indicated that the abundance of mRNAs was independent of the bacterial growth phases. In the second pattern, the transcript levels for *seg*, *sei*, *sem*, *sen*, *seo*, and *seu* slightly decreased during bacterial growth. The third pattern indicated a huge and rapid induction of *seb*, *sec*, and *seh* at the end of the exponential growth phase, whereas the last highlighted a modest postexponential increase of *sed*, *ser*, and *sel* expression.

To conclude this section, the currently known SEs form a group of serologically distinct, extracellular proteins that share important properties, namely, (1) the ability to cause emesis in primate model, (2) superantigenicity through a noncomplete unspecific activation of T lymphocytes (as each SEs binds to a subset of Vβ chains) followed by cytokine release and systemic shock, (3) resistance to heat and to digestion by pepsin, and (4) structural similarities.

46.3.2 SYMPTOMATOLOGY

SFP is one of the most common foodborne diseases in the world. SFP is caused by the ingestion of SEs, which are produced by enterotoxigenic strains of CPS, mainly *S. aureus*, and very occasionally by other staphylococci species such as *Staphylococcus intermedius*.[22]

The incubation period and severity of symptoms observed are strictly dependent on the amount of enterotoxins ingested and on the susceptibility of each person. Initial symptoms, nausea followed by incoercible characteristic vomiting (in spurts), appear within 30 min to 8 h (3 h on average) after ingesting the contaminated food. This symptomatology is due to the neurotoxic action of the SEB that stimulates the vagus nerve in the abdominal viscera, which transmits the signal to the emetic center. Other commonly described symptoms are abdominal pain, diarrhea, dizziness, shivering, and general weakness, sometimes associated with a moderate fever. In the most severe cases, headaches, prostration, and low blood pressure have been reported. In the majority of cases, recovery occurs within 24–48 h without specific treatment, while diarrhea and general weakness can last 24 h or longer. Death is rare[23] (0.02‰); it occurs in the most susceptible people due to dehydration such as infants and the elderly people and in people affected by an underlying illness.[24]

Regarding the toxin dose, most of the studies referred to SEA. In 1991, Notermans et al.[25] demonstrated the feasibility of a reference material containing about 0.5 μg of staphylococcal enterotoxin A (SEA), as it had been suggested that such a dose can cause symptoms such as vomiting. In 1995, Mossel et al. cited an emetic dose 50 value of about 0.2 μg SE per kg of human body weight.[26] They concluded that an adult would need to ingest about 10–20 μg of SE per capita to suffer symptoms. Other authors considered that less than 1 μg of SE may cause food poisoning symptoms in susceptible individuals: In 1988, Evenson et al. estimated that the amount of SEA needed to cause vomiting and diarrhea was 0.144 μg, which indeed corresponds to the amount recovered from a half-pint (~0.28 L) carton of a 2% chocolate milk.[27] In SFP in Japan, the total intake of SEA in low-fat milk per capita was estimated mostly at approximately 20–100 ng.[15,28] In 2009, in an SFPO involving "coconut pearls" (a Chinese

dessert based on tapioca), Hennekinne et al. estimated the total intake of SEA per capita at around 100 ng.[29] Finally, in 2010, Ostyn et al. investigated an outbreak due to SEE present in raw milk and estimated that the total intake of SEE per capita was 90 ng, a dose in accordance with those previously mentioned.[30]

46.4 DETECTION METHODS

While *S. aureus* is classically enumerated using microbiological techniques with dedicated media such as BP or rabbit plasma fibrinogen agar (RPFA) media, four types of methods are usually performed to detect bacterial toxins in food: bioassays, molecular biology, immunological tools, and mass spectrometry (MS)-based methods.

Bioassays are based on the capacity of an extract of the suspected food to induce symptoms such as vomiting, gastrointestinal symptoms in animals, and/or superantigenic action in cell cultures. Historically, SEs have been detected based on their emetic activity in monkey-feeding and kitten-intraperitoneal tests[31] and, more recently, using animal models such as house musk shrews (*Suncus murinus*).[32] Symptoms of SFP appear if the dose ingested by the animals is above 200 ng, a considerably higher amount than those involved in human food poisoning. Thus, this technique is not appropriate for characterizing SFPOs. More recently, a bioassay to detect the superantigenic activity of SEA has been developed.[33] This method uses SEA's superantigenic activity to induce in cytotoxic T lymphocytes a cytotoxic response against SEA-bound Raji cells. This test can only detect SEA at picomolar concentrations and is thus of little interest for laboratories involved in official controls and SFP testings. In conclusion, in addition to the fact that the use of laboratory animals for testing is now restricted for ethical reasons, bioassays are not sensitive enough to ensure food safety for consumers. Thus, alternative methods for detecting SEs have been developed.

Molecular biology methods often involve the polymerase chain reaction (PCR). These methods usually detect genes encoding enterotoxins in strains of *S. aureus* isolated from contaminated foods. However, these methods display two major limitations: first, staphylococcal strains should be isolated from food, and second, the results inform about the presence or absence of genes encoding SEs, but do not provide any information on the expression of these genes in food, that is, the actual production of toxins in food. This method therefore cannot be the single method to detect SEs in food. However, the PCR approach is a specific, highly sensitive, and rapid method that can characterize the *S. aureus* strains involved in SFPOs, thereby providing highly valuable information.

To improve SFP characterization, very recent efforts have been directed to determine which genes are involved in the biosynthesis of SEs. Following the huge SFP event that occurred in Japan in July 2000 (more than 13,000 people were intoxicated by powdered or liquid milk), authors developed a PCR-based methodology whereby *sea*, *seg*, *seh*, and *sei* genes could be detected in the incriminated powdered skim milk, although cultivable *S. aureus* were not recovered from the sample.[28] Recently, to evaluate the toxic potential of strains isolated from SFPOs, various authors have designed primers to perform PCR and reverse transcription PCR (RT-PCR) for *se* genes.[34,35] These approaches demonstrate possible transcription of mRNA from those genes, but do not indicate whether those strains were able to produce detectable or poisonous levels of toxins in food. For example, in 2009, Derzelle et al. developed an RT-PCR-based procedure to determine the temporal expression of enterotoxin genes, including many of the newly discovered ones, in optimal culture growth conditions.[36] PCR assays that can screen for 18 *se* genes have been developed, and the distribution of these genes was examined on a panel of enterotoxigenic CPS, including reference strains and isolates that have been collected from foods and SFPOs in France since the 1980s. A total of 28 strains displaying multiple enterotoxin genotypes were selected for further mRNA expression kinetics studies. More recently, in 2010, Duquenne et al. developed an efficient method to extract bacterial RNA accessible for RT-quantitative PCR (RT-qPCR) from cheese and adapted a simple, sensitive and reproducible, method for quantifying relative transcript levels to evaluate *S. aureus* enterotoxin gene expression during cheese manufacture.[37]

The third and most commonly used method for detecting SEs in food is based on the use of antienterotoxin polyclonal or monoclonal antibodies. Commercially available kits have been developed according to two different principles: (1) enzyme immunoassay (EIA) comprising enzyme-linked immunosorbent assay (ELISA) and enzyme-linked fluorescent assay (ELFA) and (2) reverse passive latex agglutination (RPLA). It is widely recognized that the use of immunological methods to detect contaminants in food matrices is a difficult task, mainly due to the lack of specificity and sensitivity of the assay. Many drawbacks impair the development and use of these techniques for detecting SEs. First, highly purified toxins are needed to raise specific antibodies to develop an EIA; purified toxins are difficult and expensive to obtain. Moreover, and until very recently, only antibodies against SEA to SEE, SEG, SEH, and SElQ have been available. The ELISA test will not detect the other SEs, which partly explains some discrepancies that have arisen in the analysis of food extracts from SFPOs. Another drawback is the low specificity of some marketed kits, where false-positives may occur depending on food components as it is well known that some proteins, such as protein A, can interfere with binding to the Fc fragment (and, to a lesser extent, Fab fragments) in immunoglobulin G from several animal species, such as mouse or rabbit, but not rat or goat.[38,39] Other interferences are associated with endogenous enzymes, such as alkaline phosphatase or lactoperoxidase. Whatever the detection method used and due to the low amount of SEs present in food, it is crucial to concentrate the extract before performing detection assays. For this purpose, various methodologies have been tested.[40–42] Among them, only extraction followed by dialysis concentration has been approved by

the European Union to extract SEs from food.[43] However, up to now, after enumerating CPS strains, conclusive diagnosis of SFPs has been mainly based on demonstrating the presence of SEs in food using commercial EIA kits designed to detect SEA to SEE or using a confirmatory in-house ELISA method to differentiate and quantify these types of SEs.[18,44] Other authors developed methods to detect SEs in food and clinical or environmental samples based on biochips using specific antibodies.[45,46]

Due to drawbacks and the lack of available antibodies against the newly described SEs, other strategies based on physicochemical techniques have been developed. Among these, MS has recently emerged as an indispensable and suitable technique to analyze protein and peptide mixtures.[47] It is among the most sensitive techniques currently available because it provides specific, rapid, and reliable analytical results. The development of two soft ionization methods, such as electrospray ionization (ESI) and matrix-assisted laser desorption/ionization (MALDI), and the use of appropriate mass analyzers have revolutionized the analysis of biomolecules. Given the wide range of methodologies available, a single MS technique cannot be used for all proteins. The MS method requires the development of a series of techniques, individually suited for each particular case. In the case of food analysis, the situation is complex because the matrix can contain many proteins, lipids, and other molecular species that can interfere with the detection of the targeted toxin and may distort quantification. Sample preparation remains the critical step of the analysis. Several authors have tried to improve this step, by, for example, optimizing digestion parameters[48] or by adding a purification step.[49] The strategy of incorporating an isotopically labeled internal standard into the samples has also been developed. In the case of SE detection, some authors have developed MS tools to detect these toxins in culture supernatants and in spiked samples, such as water or apple juice. For example, in 2002, Bernardo et al. developed a MALDI-TOF method to detect *S. aureus* virulence factors such as enterotoxins and demonstrated that this technique was suitable for detecting SEs other than SEA to SEE in culture supernatants.[50] In contrast, in 2006, Callahan et al. detected and quantified SEB using liquid chromatography coupled to ESI/MS detection in apple juice used as a model food matrix.[51] In this study, enterotoxin types SEA and SEB were detected in spiked cheese. In 2007, Brun et al. developed an MS approach able to perform absolute quantification of SEA and TSST1 in spiked water or urine samples.[52] To improve characterization and absolute quantification of SEs, this latter methodology was successfully used to carry out absolute quantification of SEA in a naturally contaminated cheese sample[53] and in food vehicle of a food poisoning outbreak.[29]

To conclude, an overall approach combining classical microbiology to enumerate CPS strains coupled with immunological techniques, molecular biology, and MS-based methods offers an interesting alternative for assigning outbreaks to SEs. Thus, the development of standards to perform absolute quantification will continue. While the quantitative

MS method overpasses specific technical limitations of existing ELISA methods for detecting and quantifying SEs, its throughput and cost per analysis compare unfavorably with ELISA. For this reason, when the MS-based method becomes available for all SEs involved in SFPOs, it will not be employed for routine analysis, but only in special cases to confirm outbreaks due to SEs.

46.5 STAPHYLOCOCCAL ENTEROTOXIN FOOD POISONING PREVENTION AND MONITORING

46.5.1 Food Poisoning Prevention

The most common way for food to be contaminated with CPS is through contact with food workers who carry the bacteria or through contaminated raw material such as milk. Moreover, due to its properties, CPS are salt tolerant and can grow in a large variety of high salt content food matrices such as salty ham. When CPS are present in the food and if satisfactory conditions are encountered for growth, CPS can multiply and produce toxins that can cause illness. As SEs are heat resistant, they cannot be destroyed by cooking. Foods at highest risk of contamination with CPS and subsequent toxin production are those that are made by hand and require no cooking. Some kinds of food having caused SFP are sliced meat, puddings, some pastries, and sandwiches.

The prevention of SFPO is based on hygiene measures in order to avoid or limit the contamination of food by *S. aureus*. These must include monitoring of the health of animals (e.g., mastitis), good manufacturing practices (GMPs), and cleaning and disinfection of equipments and environmental surfaces. Moreover, for people involved in food processing, the hand care as well as the wear of hair protection covering the whole of the scalp are essential good hygiene practices (GHPs) that must be followed. Food handlers presenting skin lesions must be excluded from handling nonpackaged foods while the lesions are not correctly covered (gloves). Likewise, a mask must be worn when suffering from any throat-type symptom. Regarding preventive measures carried out by at-risk food handlers, it is not necessary to screen *S. aureus* carrier before recruitment, nor to exclude a healthy carrier (nose or throat carriage), provided that these GHPs are applied. Since these measures are not enough to achieve zero contamination, it is necessary to destroy staphylococci through suitable treatment, heat or other, before they have multiplied or to prevent their growth by food storage under +6°C. Complying with the cold chain is vital as far as staphylococci are concerned. All food technological processes applied in a hazardous temperature zone must be short or rely on parameters other than temperature to stop bacterial growth. Since staphylococci are heat sensitive, whereas their enterotoxins are heat resistant, making safe a highly contaminated product by *S. aureus* is not guaranteed by heat treatment. This will destroy the bacteria but not their enterotoxins if these are present. It must be considered that, once preformed in food, enterotoxins cannot be eliminated effectively.

46.5.2 Monitoring

The reporting of foodborne outbreaks has been mandatory for the European Union Member States (EU MS) since 2005. Moreover, since 2007, new harmonized specifications on the reporting of these outbreaks at community level have come into force.[54] However, the foodborne outbreak investigation and reporting systems at national level are not harmonized within the European Union. Therefore, differences in the number of reported outbreaks, the types of outbreaks, and causative agents do not necessarily reflect different levels of food safety between EU MS. The high number of reported outbreaks may reflect the increasing efficiency of the EU-MS systems in investigating and identifying the outbreaks.

For example, in Europe, the European Food Safety Authority (EFSA) is responsible for examining the data on zoonoses, antimicrobial resistance, and foodborne outbreaks submitted by member states in accordance with Directive 2003/99/EC and for preparing the Community Summary Report from the results.[55] Data were produced in collaboration with the European Centre for Disease Control (ECDC), which provides the information on zoonosis cases in humans.

46.6 CONCLUSIONS

SEs, produced by coagulase-positive staphylococcus bacteria, are one of the major active principles of food poisoning outbreaks worldwide. Among these toxin types, the SEB has been extensively studied as this one can be easily produced in large amounts. Although being not primarily lethal, it still proved to be very stable, also causing severe and debilitating symptoms. After September 11, 2001, the CDC redefined bioterrorism agents and separated them into three categories. As previously mentioned, SEB met all the criteria for a category B agent in that it is stable, easy to disseminate, and induces severe emesis and toxic shock. Even if the use of SEB in bioterrorism acts has not been clearly demonstrated, SEB could be used not for its mass destruction capabilities but rather for its ability to incapacitate military or civilian population at large scale.

REFERENCES

1. Croddy, E.C. et al., *Chemical and Biological Warfare* (Google Books), pp. 30–31. Springer, New York, 2002.
2. Hursh, S. et al., Staphylococcal enterotoxin B battlefield challenge modeling with medical and non-medical countermeasures. Science Applications International Corporation, Joppa, MD, Technical Report MB DRP-95-2, 1995.
3. Ahanotu, E. et al., Staphylococcal enterotoxin B as a biological weapon: Recognition, management, and surveillance of staphylococcal enterotoxin. *Appl. Biosafety*, 11, 120, 2006.
4. Ulrich, R.G., Staphylococcal enterotoxin B and related pyrogenic toxins. In *Textbook of Military Medicine. Part I. Warfare, Weaponry and the Casualty*, vol. 3, pp. 621–631. U.S. Government Printing Office, Washington, DC, 1997.
5. Greenfield, R.A. et al., Microbiological, biological, and chemical weapons of warfare and terrorism. *Am. J. Med. Sci.*, 323, 326, 2002.

6. Franz, D.R. et al., The U.S. biological warfare and biological defense programs. In *Textbook of Military Medicine. Part I. Warfare, Weaponry and the Casualty*, vol. 3, pp. 621–631. U.S. Government Printing Office, Washington, DC, 1997.

7. Rotz, L.D. et al., Public health assessment of potential biological terrorism agents. *Emerg. Infect. Dis.*, 8, 225, 2002.

8. DeLay, P.D., Staphylococcal enterotoxin in bread pudding. *Bull. U.S. Army Med. Dep.*, 72, 72, 1944.

9. Le Loir, Y., Baron, F., and Gautier, M., *Staphylococcus aureus* and food poisoning. *Genet. Mol. Res.*, 2, 63, 2003.

10. Genigeorgis, C.A., Present state of knowledge on staphylococcal intoxication. *Int. J. Food Microbiol.*, 9, 327, 1989.

11. Minor, T.E. and Marth, E.H., *Staphylococcus aureus* and staphylococcal food intoxications. A review. IV. Staphylococci in meat, bakery products and other foods. *J. Milk Food. Technol.*, 35, 228, 1972.

12. De Buyser, M.L. et al., Implication of milk and milk products in food-borne diseases in France and indifferent industrialized countries. *Int. J. Food Microbiol.*, 67, 1, 2001.

13. Balaban, N. and Rasooly, A., Staphylococcal enterotoxins. *Int. J. Food Microbiol.*, 61, 1, 2000.

14. Sobral, D. et al., High throughput multiple locus variable number of tandem repeat analysis (MLVA) of *Staphylococcus aureus* from human, animal and food sources. *PLoS One*, 7, 5, e33967, 2012.

15. Asao, T. et al., An extensive outbreak of staphylococcal food poisoning due to low-fat milk in Japan: Estimation of enterotoxin A in the incriminated milk and powdered skim milk. *Epidemiol. Infect.*, 130, 33, 2003.

16. Anonymous, Outbreak of staphylococcal food poisoning associated with precooked ham, Florida, 1997. *CDC MMWR*, 46(50), 1189, 1997.

17. Kerouanton, A. et al., Characterization of *Staphylococcus aureus* strains associated with food poisoning outbreaks in France. *Int. J. Food Microbiol.*, 115, 369, 2007.

18. Hennekinne, J.A. et al., *Staphylococcus aureus* and its food poisoning toxins: Characterization and outbreak investigation. *FEMS Microbiol. Rev.*, 36, 815, 2012.

19. Jarraud, S. et al., egc, a highly prevalent operon of enterotoxin gene, forms a putative nursery of superantigens in *Staphylococcus aureus. J. Immunol.*, 166, 669, 2001.

20. Sugiyama, H. and Hayama, T., Abdominal viscera as site of emetic action for staphylococcal enterotoxin in monkey. *J. Infect. Dis.*, 115, 330, 1965.

21. Argudin, M.A. et al., Food poisoning and *Staphylococcus aureus* enterotoxins. *Toxins*, 2, 1751, 2010.

22. Khambaty, F.M. et al., Application of pulse field gel electrophoresis to the epidemiological characterisation of *Staphylococcus intermedius* implicated in a food-related outbreak. *Epidemiol. Infect.*, 113, 75, 1994.

23. Mead, P.S. et al., Food-related illness and death in the United States. *Emerg. Infect. Dis.*, 5, 607, 1999.

24. Do Carmo, L.S. et al., A case study of a massive staphylococcal food poisoning incident. *Foodborne Pathog. Dis.*, 1, 241, 2004.

25. Notermans, S. et al., Feasibility of a reference material for staphylococcal enterotoxin A. *Int. J. Food Microbiol.*, 14, 325, 1991.

26. Mossel, D.A.A. et al., *Essentials of the Microbiology of Foods. A Textbook for Advanced Studies*, pp. 146–150. Wiley & Sons, Chichester, England, 1995.

27. Evenson, M.L. et al., Estimation of human dose of staphylococcal enterotoxin A from a large outbreak of staphylococcal food poisoning involving chocolate milk. *Int. J. Food Microbiol.*, 7, 311, 1988.

28. Ikeda, T. et al., Mass outbreak of food poisoning disease caused by small amounts of staphylococcal enterotoxins A and H. *Appl. Environ. Microbiol.*, 71, 2793, 2005.

29. Hennekinne, J.A. et al., Innovative contribution of mass spectrometry to characterise staphylococcal enterotoxins involved in food outbreaks. *Appl. Environ. Microbiol.*, 75, 882, 2009.

30. Ostyn, A. et al., First evidence of a food-poisoning due to staphylococcal enterotoxin type E in France. *Eurosurveillance*, 15, 19528, 2010.

31. Surgalla, M. et al., Some observations on the assay of staphylococcal enterotoxin by the monkey feeding test. *J. Lab. Clin. Med.*, 41, 782, 1953.

32. Ono, H.K. et al., Identification and characterization of two novel staphylococcal enterotoxins types S and T. *Infect. Immun.*, 76, 4999, 2008.

33. Hawryluk, T. and Hirshfield, I., A super antigen bioassay to detect staphylococcal enterotoxin A. *J. Food Prot.*, 65, 1183, 2002.

34. Lee, Y.D. et al., Expression of enterotoxin genes in *Staphylococcus aureus* isolates based on mRNA analysis. *J. Microbiol. Biotechnol.*, 17, 461, 2007.

35. Akineden, O. et al., Enterotoxigenic properties of *Staphylococcus aureus* isolated from goats' milk cheese. *Int. J. Food Microbiol.*, 124, 211, 2008.

36. Derzelle, S. et al., Differential temporal expression of the staphylococcal enterotoxins genes during cell growth. *Food Microbiol.*, 26, 896, 2009.

37. Duquenne, M. et al., Tool for Quantification of staphylococcal enterotoxin gene expression in cheese. *Appl. Environ. Microbiol.*, 76, 1367, 2010.

38. Wieneke, A.A., Comparison of four kits for the detection of staphylococcal enterotoxin in foods from outbreaks of food poisoning. *Int. J. Food Microbiol.*, 14, 305, 1997.

39. Park, C.E. et al., Nonspecific reactions of a commercial enzyme-linked immunosorbent assay kit (TECRA) for detection of staphylococcal enterotoxins in foods. *Appl. Environ. Microbiol.*, 58, 2509, 1992.

40. Macaluso, L. et al., Determination of influential factors during sample preparation for staphylococcal enterotoxin detection in dairy products. *Analusis*, 26, 300, 1998.

41. Meyrand, A. et al., Evaluation of an alternative extraction procedure for enterotoxin determination in dairy products. *Lett. Appl. Microbiol.*, 28, 411, 1999.

42. Lapeyre, C. et al., Enzyme immunoassay of staphylococcal enterotoxins in dairy products with cleanup and concentration by immunoaffinity column. *J. AOAC Int.*, 84, 1587, 2001.

43. Anonymous. Commission Regulation. No. 1441/2007 of 5 December 2007. *Off. J. Eur. Union*, L322, 12, 2007.

44. Bennett, R.W. Staphylococcal enterotoxin and its rapid identification in foods by enzyme-linked immunosorbent assay-based methodology. *J. Food Prot.*, 68, 1264, 2005.

45. Rubina, A.Y. et al., Simultaneous detection of seven staphylococcal enterotoxins: Development of hydrogel biochips for analytical and practical application. *Anal. Chem.*, 82, 8881, 2010.

46. Techer, C. et al., Detection and quantification of staphylococcal enterotoxin A in foods with specific and sensitive polyclonal antibodies. *Food Control*, 32:255–261, 2013. http://dx.doi.org/10.1016/j.foodcont.2012.11.021.

47. Mamone, G. et al., Analysis of food proteins and peptides by mass spectrometry based techniques. *J. Chromatogr. A*, 1216, 7130, 2009.

48. Norrgran, J. et al., Optimization of digestion parameters for protein quantification. *Anal. Biochem.*, 393, 48, 2009.

49. Oeljeklaus, S. et al., New dimensions in the study of protein complexes using quantitative mass spectrometry. *FEBS Lett.*, 583, 1674, 2009.

50. Bernardo, K. et al., Identification and discrimination of *Staphylococcus aureus* strains using matrix-assisted laser desorption/ionization-time of flight mass spectrometry. *Proteomics*, 2, 747, 2002.

51. Callahan, J.H. et al., Detection, confirmation, and quantification of staphylococcal enterotoxin B in food matrixes using liquid chromatography-mass spectrometry. *Anal. Chem.*, 78, 1789, 2006.

52. Brun, V. et al., Isotope-labeled protein standards: Toward absolute quantitative proteomics. *Mol. Cell Proteomics*, 6, 2139, 2007.

53. Dupuis, A. et al., Protein Standard Absolute Quantification (PSAQ) for improved investigation of staphylococcal food poisoning outbreaks. *Proteomics*, 8, 4633, 2008.

54. Anonymous. Report of the Task Force on Zoonoses Data Collection on harmonising the reporting of food-borne outbreaks through the Community reporting system in accordance with Directive 2003/99/EC. *EFSA J.*, 123, 1, 2007.

55. Anonymous. Directive 2003/99/EC of the European Parliament and of the Council of 17 November 2003 on the monitoring of zoonoses and zoonotic agents, amending Council Decision 90/424/EEC and repealing Council Directive 92/117/EEC. *Off. J. Eur. Union*, L325, 31, 2003.

56. Papageorgiou, A.C. et al., Crystal structure of microbial superantigen staphylococcal enterotoxin B at 1.5 Å resolution: Implications for superantigen recognition by MHC class II molecules and T-cell receptors. *J. Mol. Biol.*, 277, 61, 1998.

47 T-2, HT-2, and Diacetoxyscirpenol Toxins from *Fusarium*

Alicia Rodríguez, Mar Rodríguez, Marta Herrera,
Agustín Ariño, and Juan J. Córdoba

CONTENTS

47.1 INTRODUCTION

Fusarium species are important plant pathogens that cause significant quality and economic losses in small grain cereals by reducing crop yield and concomitant contamination with mycotoxins, especially trichothecenes. Thus, *Fusarium* genus includes devastating plant pathogenic fungi that cause diseases in plants as diverse as maize, wheat, banana, tomato, and mango around the world. *Fusarium* species are a widespread cosmopolitan group of fungi that commonly colonize aerial and subterranean plant parts, either as primary or secondary invaders. *Fusarium* can be soil-, air-, or water-borne or carried in or on plant residue or seeds, and can be recovered from any part of a plant, for example, from roots, shoots, flowers, and/or cones, as well as seeds.[1]

Many of molds included in the *Fusarium* genus can produce a wide range of biologically active secondary metabolites (e.g., mycotoxins) with an extraordinary chemical diversity. Mycotoxin-producing strains of *Fusarium* are a worldwide problem. When environmental factors such as temperature and humidity are favorable, these fungi produce mycotoxins, which can end up as contaminants in animal feed and food.[2–4] Mycotoxins are compounds produced by naturally occurring fungi, they are not essential for fungal growth, and they have toxic and in some cases carcinogenic properties. *Fusarium* species produce three of the most important types of mycotoxins: trichothecenes, fumonisins, and zearalenone.[5,6] Of them, trichothecenes are the chemically most diverse of *Fusarium* mycotoxins. The trichothecenes are all nonvolatile, low-molecular-weight sesquiterpene epoxides, and can be further classified according to the presence or absence of functional characteristic groups. The C12, 13-epoxide ring, which is necessary for protein synthesis inhibition, is considered to be essential. Most trichothecenes also have a C9–C10 double bond, which is important for their toxicity.[7] This group of toxins includes the quite deadly diacetoxyscirpenol (DAS) and T-2 toxins nivalenol (NIV) and deoxynivalenol (DON).[6] All trichotecenes produced by *Fusarium* spp. are of the simple or nonmacrocyclic type. T-2, HT-2, and DAS are classified as Type A trichothecenes and NIV and DON as Type B. The T-2, HT-2, and DAS are more toxic than NIV and DON; however, these last trichothecenes (Type B) are widely more extended geographically than those of Type A. Due to its widespread occurrence and the potential for economic losses, DON is much more widely regulated than T-2, HT-2, or DAS, even though human or animal intoxications with either DON or NIV are much less likely to be fatal than those with T-2, HT-2 or DAS. The main trichothecenes-producing species of *Fusarium*, *Fusarium langsethiae* and *Fusarium sporotrichioides* have been reported as T-2 and HT-2 toxin producers,[8] but they may not be the only species responsible, because other species such as *Fusarium acuminatum*, *Fusarium oxysporum*, *Fusarium solani*, *Fusarium nivale*, *Fusarium sambucinum*, and *Fusarium poae* were also identified as possible producers of T-2 and HT-2 toxins (T-2 toxin was also found in molds belonging to other genera such as *Trichoderma* sp. and *Myrothecium* sp.).[9] *Fusarium sibiricum*, *Fusarium culmorum*, and *Fusarium graminearum* have also been reported as producers of T-2 and HT-2 toxins.[10]

47.2 CLASSIFICATION AND MORPHOLOGY OF GENUS *FUSARIUM*

The genus *Fusarium* belongs to the *Ascomycete* phylum, Ascomycetes class, Hypocreales order.[11] Although Gräfenhan et al.[12] have gone some way to resolving the generic boundaries of taxa associated with *Fusarium* and have removed a number of species into genera such as *Dialonectria*, *Fusicolla*, and *Microcera*, other teleomorph genera are associated with species of *Fusarium* such as *Gibberella*, *Haematonectria*, and *Albonectria*.[13] A strictly monophyletic definition for *Fusarium* would, most logically, be based on those species that either have a *Gibberella* sexual stage or would have one if they reproduced sexually.

The genus *Fusarium* is well-known for taxonomic difficulties and has been composed of anywhere between 9 and >1000 species, depending on criteria or kind of characterization used. All taxonomic systems developed so far are based on seminal work by Wollenweber and Reinking[14] with several modifications.[15,16] This publication organized the genus in 16 sections including 65 species, 55 varieties, and 22 forms. The taxonomic history of *Fusarium* species has been reviewed in great detail elsewhere.[15–17] Definition and identification of species remains unclear, with problems at both ends of the scale since there are too many things grouped into a single species, for example, *F. oxysporum*[18] and *F. solani*, and the separations have gone too far, for example, the 13 phylogenetic species within *F. graminearum*[19] or the separation of *Fusarium*

brevicatenulatum from *Fusarium pseudoanthophilum*.[20] Traditional classification of *Fusarium* has been based exclusively on morphological characters, but this may not be definitive, while molecular markers need to be checked against other characters since where to draw the line between species when only molecular markers are used is not always clear.[1] This highlights the need to use a combination of morphological, biological, and phylogenetic markers in a polyphasic approach, based on relative weighting of the available data and markers.

Specifically, the identification at specie level of *Fusarium* could be based on important cultural and morphological characters such as pigmentation produced by the fungus in culture media, kind and growth of mycelium, characters of the sporodochia, nature of conidiogenous cells, micro and macroconidia, and presence, absence, and type of chlamydospores.

In addition to a wide range of morphological and cultural diversity, pathogenic variability has been used to characterize the fungus at species, subspecies, and intra-subspecies level. *Fusarium* sp. can be subclassified into *forma specialis* which are not taxonomic species; they are groups of isolates recognized for their ability to cause disease in a specific set of host plant species.[21,22] Thus, over 100 *forma specialis* in *F. oxysporum* have been described from identical number of hosts, or *F. solani* has been classified into different *forma specialis* (e.g., phaseoli, pisi, batatas, or phaseoli), depending on the degree of virulence as well as genetic diversity in these isolates of different origins.[23] Isolates within one *forma specialis* differ in their virulence and they are characterized by assigning pathogenic races. These races have been mostly by use of differential reaction to selected host genotypes. In several cases, these are inadequate or completely lacking and, therefore, precise information on these aspects needs to be generated for elucidation of the extent of pathogenic diversity present in the pathogens.

Progress in molecular biology in the last two decades has opened up possibilities of differentiation between species, and identification of new strain/isolates collected from infected samples at genomic level. Molecular methods have also been used to distinguish between closely related species with few morphological differences and to distinguish strains (or even specific isolates) within a species.[24] Molecular markers monitor the variations in DNA sequences within and between the species and provide accurate identification, for example, the random amplified polymorphic DNA (RAPD) technique has been used successfully to study *forma specialis*, and races of *Fusarium* or the restriction fragment length polymorphisms (RFLP) have been used to identify properly at strain level. The ribosomal DNA (rDNA) based classification is also the method of choice especially when classifying the related species. The internal transcribed spacer (ITS) region has been used successfully as taxonomic discriminator between *Fusarium verticillioides* and *Fusarium proliferatum*[25] and for molecular characterization of pathogenic *Fusarium* species[26] among others.

47.3 BIOLOGY AND EPIDEMIOLOGY

Fusarium species cause an array of diseases that affect agriculture and horticulture in all parts of the world. Several of the diseases, such as head blight of wheat[27] and Panama disease of bananas,[28] have had nearly devastating economic and sociological impacts on the farmers and communities that rely on these crops for their livelihoods.

The plant diseases caused by *Fusarium* species are not restricted to any particular region or cropping scenario. Thus, toxigenic species of *Fusarium* occur worldwide in habitats as diverse as deserts, tidal salt flats, and alpine mountain regions.[29]

Mycotoxin-producing *Fusarium* species are probably the most prevalent fungi of the northern temperature regions and are commonly found on cereals grown in the temperature regions of America, Europe, and Asia.[30] Identifying strains that produce trichothecenes has resulted in many claims in the literature that often fail to hold up under more careful examination. Type B trichothecenes are produced by species that are well known from temperate regions (e.g., *F. culmorum* and *F. graminearum*), while Type A trichothecene producers such as *F. sambucinum*, *F. poae*, and *F. sporotrichioides* are more common in colder climates of Canada, Scandinavia, and Russia.[5,31] Among the Type A of trichothecenes producers, those aspects related to producers of the T-2 and DAS mycotoxins deserve more attention.

47.3.1 T-2 AND HT-2

The T-2 and HT-2 toxins, some of the most poisonous, have been reported to be mainly produced by *F. sporotrichioides*, *F. langsethiae*, *F. poae*, *Fusarium equiseti*, and *F. acuminatum*.[30] These species of *Fusarium* are widespread fungi on a variety of plants in soil throughout the cold-temperature regions.[32] *F. sporotrichioides* of the Sporotrichiella section is the most important producer of T-2 together with *F. langsethiae*, and it is related with cereals as a result of water damage to grains occurring when the cereals remain on the field for extended periods or when the grain is wet during storage.[32] *F. sporotrichioides* grows at −2°C to 35°C and only at high water activities.[30]

47.3.2 DIACETOXYSCIRPENOL

Several *Fusarium* species have been reported as DAS producers, for example, *F. poae*, *Fusarium semitectum*, *F. verticillioides*, *F. sporotrichioides*, *F. acuminatum*, *F. culmorum*, *Fusarium crookwellense*, *Fusarium venenotum*, *F. sambucinum*, *F. equiseti*, *F. graminearum*, *Fusarium avenaceum*, and *F. langsethiae*.[32,33]

DAS is one of the trichothecenes mycotoxins naturally occurring in agricultural products.[34] DAS is abundant in various cereal crops such as corn, barley, mixed feed samples, and other grains from various regions in the world. Coexistence of DAS and T-2 in animal feeds and human foods represents a health threat to humans and animals in some part of the world.[35]

47.4 CHEMISTRY OF T-2 AND DIACETOXYSCIRPENOL TOXINS

All trichothecenes share the tetracyclic, sesquiterpenoid 12,13-epoxytrichothec-9-ene ring system but differ in the type of functional group attached to the carbon backbone.[36] Based on their structure, the group of trichothecenes can be divided into four groups (A–D) according to their different chemical functionalities, being types A and B the most important. The stable epoxide group between C12 and C13 seems to account for many of the typical toxic effects of trichothecenes. Type A trichothecenes include T-2 toxin, HT-2 toxin, and DAS. This group shares some common structural features, such as a double bond between C9 and C10 and an epoxy group at C12 and C13. The structures of T-2 and HT-2 toxins differ only in one functional group: T-2 toxin is acetylated at C-4 whereas HT-2 toxin is not acetylated. T-2 toxin is rapidly metabolized to a large number of products, HT-2 toxin being a major metabolite.[37]

Both T-2 and HT-2 toxins and DAS are nonvolatile, low-molecular-weight compounds, which are stable at neutral and acidic pH. The fewer free hydroxyl groups and the lacking keto group at C8 of type A trichothecenes make them less polar compared with the related type B trichothecenes such as deoxynivalenol, 3-acetyl-deoxynivalenol, 15-acetyldeoxynivalenol, and nivalenol. Therefore, different methods of analysis are usually used to determine type (A or B) of trichothecenes.[37] The solubility of both T-2 and HT-2 toxins is good in most organic solvents including methanol, ethanol, acetone, chloroform, ethyl acetate, diethyl ether, and acetonitrile, but poor in water.

47.5 TOXIC EFFECT, OCCURRENCE, AND DIETARY EXPOSURE OF T-2, HT-2, AND DIACETOXYSCIRPENOL

47.5.1 TOXIC EFFECTS

Trichothecenes mycotoxins are noted for their marked stability under different conditions. Although they are less toxic than other toxins, trichothecene mycotoxins are proven lethal agents in warfare. The typical symptom for acute, high dosage trichothecene ingestion is vomiting and food refusal.[38] Anorexia, emesis, oral and gastro-intestinal lesions, ill-thrift, renal lesions, among others, are also common effects of trichothecene-induced toxicoses to humans and/or animals.[39] Clinical effects produced by trichothecenes can be grouped into four categories: (1) feed refusal, (2) dermal necrosis, (3) gastroenteric effects, and (4) coagulopathy.[32]

At the cellular level, effects on DNA and membrane integrity have been considered as secondary effects of the inhibited protein synthesis.[40] Cytotoxic effects were observed at slightly higher doses of trichothecenes.[41] The myelotoxicity was considered higher for T-2 and HT-2 toxins and lowest for DON and NIV.[42]

Trichothecenes cause the greatest problems in animal health.[43] In acute tests with trichothecenes, type A members such as DAS and T-2 have been found to be more toxic than Type B members such as DON and NIV.[44] Trichothecenes are toxic to many animal species, but the sensitivity varies considerably between species and also between the different trichothecenes.[32]

47.5.1.1 T-2 and HT-2 Toxins

The T-2 cytotoxin has the highest toxicity of all the trichothecenes,[45] although it is less frequently detected compared to the other trichothecenes.[46] The toxicity of the T-2 toxin and related mycotoxins can be affected by a variety of factors such as administration route, time of exposure, the number of exposure, dose, animal's age, sex, and overall health, and presence of other mycotoxins. The 12,13 epoxide ring of T-2 toxin is responsible for its toxic activity. After exposure by the oral, dermal, or inhalation route, T-2 toxin can cause severe effects on various animal organs and tissues. So far, toxic effects have been evidenced in the cells of fungi, protozoa, insects, molds, plants, and different cell cultures.[46] T-2 toxin inhibits DNA, RNA, and protein synthesis in eukaryotic cells and affects the cell cycle.[36,47,48] The chemical structure of T-2 toxin molecule (position of chemical groups on the trichothecene ring) again has an essential role that determines the mode and target of action, because it specifies the interaction with protein molecules. Thus, T-2 toxin inhibits polypeptide chain initiation, while other trichothecenes affect elongation and termination.[49]

The T-2 toxin also induces apoptosis both in vivo and in vitro, depending on the activation of JNK and p38 MAP kinases, but the precise mechanism has not yet been elucidated. In addition, it has an immunomodulatory activity whose mode of action is time- and dose-dependent. Furthermore, it produces toxic effects on almost all cellular processes in the digestive system, and it is a well-known neurotoxin that damages the blood–brain barrier.[49]

Thus, T-2 contaminated products can cause different effects on humans/animals, at the same time they may even result in death.[32] General signs of T-2 include nausea, emesis, dizziness, chills, abdominal pain, diarrhea, dermal necrosis, abortion, irreversible damage to the bone marrow, and reduction in white blood cells, and it is toxic for the hematological and lymphatic systems, producing immunosuppression.[46]

47.5.1.2 DAS Toxin

This toxin like T-2 toxin produces inhibition of the initial step of protein synthesis due to its chemical structure. This toxin is rapidly transformed into four products, including 15-monoacetoxyscirpenol (MAS), scirpenol, and two new compounds identified as 15-acetoxy-3α,4β-dihydroxytrichothecec-9,12-diene (deepoxy MAS) and 3α,4β,15-trihydroxytrichotec-9,12-diene (deepoxy scirpenol). This mycotoxin shows a wide-range biological activity, including toxicity to fungi, plants, animals, and various mammalian tissue cultures.[50] Toxic effects in humans and animals seemed similar such as nausea, vomiting, diarrhea, hypotension, neurological symptoms, chills, and fever.[32]

47.5.2 Maximum Levels and Tolerable Daily Intakes

Worldwide, 13 countries have reported legal maximum levels (MLs) or recommendations for T-2 and HT-2 toxins in food and/or feed products. In the European Union (EU) there is a Commission Recommendation (2013/165/EU) of 27 March 2013 on the presence of T-2 and HT-2 toxin in cereals and cereals products. This recommendation establishes indicative levels for the sum of T-2 and HT-2 (µg/kg) above which further investigation is required, certainly in case of repetitive findings (Table 47.1).

The Scientific Committee for Food[51] established a temporary tolerable daily intake (t-TDI) of 0.06 µg/kg body weight (bw) for the sum of T-2 and HT-2 toxins, which was confirmed in 2002. As new relevant evidence has become available, the Scientific Panel on Contaminants in the Food Chain of the European Food Safety Authority[37] established a full TDI of 100 ng/kg bw for the sum of T-2 and HT-2 toxins.

47.5.3 Occurrence in Food and Feed and Dietary Exposure

T-2 toxin and HT-2 toxin are mycotoxins that may contaminate the harvest grain and feed, and food products thereof, and—after ingestion—affect animal and human health.[36] In a recent survey carried out in 22 European countries during

TABLE 47.1
Indicative Levels (µg/kg) for the Sum of T-2 and HT-2 in Cereals and Cereal Products

Cereals and Cereal Products Intended for Human Consumption	Guidance Value in µg/kg
1. Unprocessed cereals	
1.1. barley (including malting barley) and maize	200
1.2. oats (with husk)	1000
1.3. wheat, rye and other cereals	100
2. Cereal grains for direct human consumption	
2.1. oats	200
2.2. maize	100
2.3. other cereals	50
3. Cereal products for human consumption	
3.1. oat bran and flaked oats	200
3.2. cereal bran except oat bran, oat milling products other than oat bran and flaked oats, and maize milling products	100
3.3. other cereal milling products	50
3.4. breakfast cereals including formed cereal flakes	75
3.5. bread (including small bakery wares), pastries, biscuits, cereal snacks, pasta	25
3.6. cereal-based foods for infants and young children	15
4. Cereal products for feed and compound feed	
4.1. oat milling products (husks)	2000
4.2. other cereal products	500
4.3. compound feed, with the exception of feed for cats	250

2005–2010, a total 20,519 samples of food, feed, and unprocessed grains were collected and analyzed. Overall, 65% of the results were below the limit of detection (LOD) or limit of quantification (LOQ). In the quantified results, HT-2 toxin concentration represents about two thirds of the sum of T-2 and HT-2 toxin concentration. The highest mean concentrations for the total of T-2 and HT-2 toxins were observed in grains and grain milling products, notably in oats and oat products. Levels in unprocessed grains were higher than in grain products for human consumption, suggesting that processing applied to grains results in lower T-2 toxin and HT-2 toxin concentrations. During the milling process T-2 and HT-2 toxins are not destroyed but unevenly redistributed between fractions. Because T-2 and HT-2 toxins are mostly attached to the outer hull of the grain, cleaning, sorting, sieving, and de-hulling of grains lead to marked increases in T-2 and HT-2 toxins in cereal by-products, for example, bran. T-2 and HT-2 toxins are relatively stable compounds during baking and cooking. A recent overview of the occurrence of T-2 toxin and HT-2 toxin in cereals in Europe and derived food products, factors influencing the occurrence, co-occurrence with other trichothecenes, and toxicological effects of T-2 and HT-2 in human has been published by Van der Fels-Klerx and Stratakou.[48]

The Panel on Contaminants in the food chain (CONTAM Panel) estimated total chronic dietary exposures to the sum of T-2 and HT-2 toxins using mean concentrations of the sum of T-2 and HT-2 toxins in foods, and consumption data for different age groups. Grains and grain-based foods, in particular, bread, fine bakery wares, grain milling products, and breakfast cereals, made the largest contribution to the sum of T-2 and HT-2 toxin exposure for humans. Considering a combined TDI of 100 ng/kg bw for the sum of T-2 and HT-2 toxins, the estimates of chronic human dietary exposure to the sum of T-2 and HT-2 toxins were below the TDI for populations of all age groups, and thus not a health concern.

47.6 IDENTIFICATION AND DETECTION OF PRODUCING SPECIES OF *FUSARIUM* AND THE CORRESPONDING TOXINS T-2 AND DIACETOXYSCIRPENOL

47.6.1 DETECTION OF T-2- AND DIACETOXYSCIRPENOL-PRODUCING SPECIES OF *FUSARIUM*

Early and accurate identification/detection of *Fusarium* species producing T-2, HT-2, and DAS toxins occurring in the plant at every step of cereal or other food products processing may allow corrective actions to avoid toxins strains and consequently reduce risk of T-2 and HT-2 toxins accumulation entering the food chain.

Conventional methods to assess the presence of mycotoxigenic mold using mycological keys are relatively simple, but sometimes are laborious, time consuming, and require experienced operators. The identification of *Fusarium* species is particularly difficult and complex because the genus is diverse, presents high intraspecific variability, and conflicting taxonomy, which may depend on external conditions.[17,52]

That is the reason why it is difficult to establish systems of taxonomy for *Fusarium* species and therefore the importance of molecular methods for identification.

Other methodologies based on immunological detection of *Fusarium* species have been described.[53–55] But some of immunoassays developed to detect these species involve cross-reactivity with species of *Aspergillus* and *Penicillium*.[55]

As identification of *Fusarium* species is critical to predict the potential mycotoxigenic risk of the isolates, there is a need for accurate and complementary tools which permit a rapid, sensitive, and reliable specific diagnosis of *Fusarium* species.[56]

Some studies have developed a methodology to detect toxigenic *Fusarium* species based on the volatile compound released[57] and the detection of trichodiene, the volatile intermediate of trichothecenes.[58] These assays were able to differentiate between infected and noninfected wheat grains in the postharvest chain.[57] This methodology could be useful to predict the potential of trichothecene mycotoxin formation, but to date it has been developed only for detecting the presence of *F. graminearum* in wheat cultivars.[58]

Therefore an early and accurate identification of the type A trichothecene-producing *Fusarium* species is critical to prevent mycotoxins entering the food chain. In this way, molecular-markers-based technology for detection and identification of T-2, HT-2, and DAS mycotoxins-producing mold species could be used.

In recent years, different marker systems such as RFLP, RAPD, sequence tagged sites (STSs), amplified fragment length polymorphisms (AFLPs), simple sequence repeats (SSRs), microsatellites, single nucleotide polymorphisms (SNPs) or rolling circle amplification (RCA) have been developed and applied to different fungal species (Figure 47.1).

For species-specific identification, molecular methods such as PCR and real-time PCR (qPCR) are a powerful tool of screening (Figure 47.1). However, what is necessary is a good correlation between DNA and mycotoxin and the absence of false negatives, that is, samples with high mycotoxin and low DNA content.[59] One of the most important factors in the development of PCR and qPCR methods is the reliability of the primer set designed and the targeted DNA sequence of the organism of interest.[60,61] Table 47.2 shows a summary of the main molecular methods and target genes described for the detection of type A trichothecenes.

The use of multicopy target sequences to develop specific primers enhances the sensitivity, in comparison with PCR assays based on single-copy sequences, of the assay since it reduces the amount of DNA templates necessary for PCR amplification and simplifies detection protocols.[62] PCR and qPCR protocols based on spacers of rDNA, intergenic spacer (IGS of rDNA unit), and ITS (of rDNA unit) have been reported for *Fusarium* species; various PCR-based methods have been published for detecting single species or species groups of *Fusarium*.[62–67]

Other PCR assays for the identification of type A trichothecene-producing species of *Fusarium* have been developed based on single-copy genes such as elongation factor 1-α (TEF-1-α), calmodulin, or β-tubulin.[13] The TEF-1-α appears to be consistently in *Fusarium*, and it shows a high level

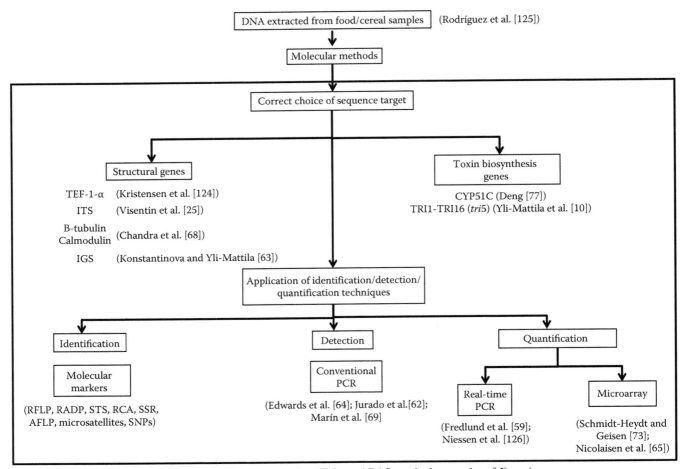

FIGURE 47.1   Flow chart for identification/detection of T-2-, HT-2-, and DAS-producing species of *Fusarium*.

of sequence polymorphism among closely related species, even in comparison to the intron-rich portions of protein-coding genes such as calmodulin, β-tubulin, and histone H3.[68] For these reasons, TEF-1-α has become the marker of choice as a single-locus identification tool in type A trichothecene-producing *Fusarium* species.[69–71]

An alternative is the detection of genes directly involved in mycotoxin biosynthesis such as the TRI1–TRI16 locus that determines type A versus type B trichothecene production.[10] In this sense, the *tri5* gene, one of the key genes of the trichothecene pathway,[72] could be a good target for early detection of type A trichothecene-producing molds (Figure 47.1), because it is active instantly with the growth of mold.[73] Primers based in this target have been used for detecting type A trichothecene-producing *Fusarium* species by PCR and qPCR.[59,73–76]

Recently, a novel market based on the nucleotide sequences of the gene for 14-α-demethylase (CYP51C) has been used. This gene has been identified only in *Fusarium* spp.[77] and has been used for separation of this species according to their synthesized mycotoxin.[78,79]

Microarray-based technologies represent an advance in nucleic acid testing. The goal of this technology is the ability to analyze in a single experiment a high number of target organisms (Figure 47.1). The advantage of microarray-based detection is that it can combine powerful nucleic acid amplification strategies with a massive screening capability, resulting in a high level of sensitivity, specificity, and throughput capacity technology.[80] Various oligonucleotide-microarrays were developed for the identification of *Fusarium* species.[65,71,73] The array could thus distinguish type-B trichothecene producers, type-A trichothecene producers, type-A and -B producers and nontrichothecene producers (*F. avenaceum* and *Fusarium tricinctum*).

47.6.2   Detection of the *Fusarium* Toxins T-2 and Diacetoxyscirpenol

Analytical procedures for type A trichothecenes may be fully quantitative, semiquantitative, or qualitative. Usually these methods determine T-2 and HT-2 toxins at the same time, individually or together as a sum. Several analytical methods applying GC or LC have been validated in-house for applicability in food, beverages, and feedstuffs within the last few years. There is little direct information available on the analysis of DAS toxin alone. Existing data indicate that DAS behave chemically in a similar way to T-2 and HT-2 toxin. Furthermore, T-2 toxin, HT-2 toxin, and DAS share the tetracyclic, sesquiterpenoid 12,13-epoxytrichothec-9-ene ring system as they are all type A trichothecenes. Therefore, the

TABLE 47.2

Molecular Method for Detection of an Identification Type A Trichothecene-Producing *Fusarium*

Method	Target Gene	References
Conventional PCR assays	β-Tubulin (β-Tub)	Edwards et al. [64]
	Calmodulin (CaM)	
	ITS region (ITS)	
	IGS	Yli-Mattila et al. [10] and Jurado et al. [62]
		Marín et al. [69] and Kristensen et al. [70]
	Translocation elongation factor 1-α (TEF)	Yli-Mattila et al. [10]
	Tri 5	Lincy et al. [74] and Niessen et al. [120]
	14-α-Demethylase (CYP51C)	Stakheev et al. [78] and Fernández-Ortuño et al. [79]
Real-time PCR assays	ITS	Frendlud et al. [59], Edwards et al. [64], Yli-Mattila et al. [121], and Nielsen et al. [122]
	Tri 5	Frendlud et al. [59]
Multiplex PCR assays	ITS	Suanthie et al. [66]
	Tri 5	Demeke et al. [75]
Microarray-based assays	ITS	Nicolaisen et al. [65]
	TEF	Kristensen et al. [71]
	Tri 5	Schmidt-Heydt and Geisen [73]
Other methods		
DGGE, ARMS-PCR	β-Tub	Mach et al. [123] Yli-Mattila et al. [67]
UP-PCR	TEF	Kristensen et al. [124]
SNP assays	Tri 5	Wilson et al. [76]
RAPD-PCR	IGS	Konstantinova and Yli-Mattila [63]
RFLP-PCR		

analytical considerations for T-2 and HT-2 toxins can be considered to include DAS, and the results of laboratory studies with T-2 and HT-2 toxins, which are more frequently reported, can be used to approximate the results with DAS toxin.

Accurate quantification of T-2 and HT-2 toxins is mostly carried out by liquid chromatography coupled with (multistage) mass spectrometry often within a multianalyte approach.[81] For rapid screening several immunochemical methods have become available but they may suffer from undesired cross-reactivity with other trichothecenes.

47.6.2.1 Sampling and Storage

Due to the possible inhomogeneous distribution of trichothecenes in batches (of grains, for instance), sampling may contribute to the variability in analytical results. Therefore, prior to the analysis of type A trichothecenes, a representative sample must be provided, as this influences the reliability of the analytical data generated. After taking a sample, it is stored under appropriate conditions (dry, preferably frozen) until being analyzed.

47.6.2.2 Analyte Extraction and Purification

For extraction of type A trichothecene toxins from food and feed, including grains and grain-based products, mostly organic solvent/water mixtures are used. The toxins have been extracted with mixtures of methanol:water, acetonitrile:water, acetone:acetic acid:water, ethyl acetate:formic acid, PBS-buffer,

as well as by microextraction with hollow fiber pieces, MSPD (matrix solid-phase dispersion), ASE (accelerated solvent extraction), and quick, easy, cheap, effective, rugged, and safe (QuEChERS) method.[37,82] The resulting extract is usually further processed to remove impurities/interfering materials and often concentrated to make determination of toxins at low concentrations possible. Rubert et al.[83] have reported a multimycotoxin method in which the analytes are extracted by MSPD followed by liquid chromatography coupled with tandem mass spectrometry (LC-MS/MS) using a hybrid triple quadrupole-linear ion trap mass spectrometer (QTRAP (R)) with extra confirmation by information-dependent acquisition (IDA).

For T-2 and HT-2 toxins, clean-up procedures may involve use of various types of solid phase extraction (SPE) columns,[84] multifunctional columns,[85] and immunoaffinity (IA) columns.[86] For SPE columns various packages are now commercially available, and may contain silica, charcoal, Florisil®, C8, C18, NH$_2$ (amino), and aluminum oxide. There is a trend toward easy-to-use purification techniques with fewer requirements for extensive clean-up procedures.

47.6.2.3 Determination of T-2 and HT-2 Toxins

Physicochemical-based techniques generally include the use of high-performance liquid chromatography (HPLC) or gas chromatography (GC) coupled to various detectors, for example, ultraviolet (UV), diode array (DAD), fluorescence (FLD), electron capture (ECD), or mass spectrometry (MS).[87] The increasing need for reliable and accurate testing methods for the sensitive determination of T-2 and HT-2 toxins has led to a shift from methods using UV, DAD, FLD, and ECD detection to highly sophisticated methods based on liquid chromatography (LC) coupled with multiple-stage mass spectrometry (MS[n]). However, the instrumentation for LC-MS/MS is expensive and the operators require a high level of skill, which often prevents its use for screening purposes. Recently, mostly immunochemically based, rapid methods have been developed for the screening for these type A trichothecenes. Some of the methods in use for the analysis of cereals and cereal products have been recently discussed in an extensive review by Meneely et al.[84]

Chromatographic methods have been developed for the accurate quantification and identification of type A trichothecenes in various matrices including food and feed. GC has largely been the method of choice, in combination with flame ionization detection (FID), ECD, and MS detection. Typical LODs for GC-methods for the analysis of T-2 and HT-2 toxins are close to 2 µg/kg. However, the popularity and use of GC methods for the analysis of trichothecene mycotoxins has declined in recent years.[88] While these GC-based methods provide sensitive and accurate results, the polar compounds require derivatization prior to GC separation. A variety of chemicals have been used for derivatization of type A trichothecenes, and the choice depends on the method of detection employed: trimethylsilylation, fluoroacetylation, derivatizing agents such as anhydrides of trifluoroacetic, heptafluorobutyric acid, 1-anthroylnitrile, and pentafluoropropionic acid. However, GC methods showed higher variations in an intercomparison study on trichothecene determination as compared to LC methods.[89] The major reason for the observed discrepancies was adsorption of derivatized type A trichothecenes to active sites of the GC injector and the upper part of the capillary column. This effect was more pronounced in the absence of the matrix, which led to lower signals for pure calibrants compared to the analyte response in the presence of the matrix.

Within the last decade, HPLC has become the most frequently used method for the determination of type A trichothecenes.[88] In contrast to type B trichothecenes type A trichothecenes lack an exploitable chromophore, so that UV detection is not the method of choice for determination of type A trichothecenes. For this reason, FLD and MS detection methods seem more appropriated. The use of HPLC coupled with FLD has been applied to the determination of T-2 and HT-2 toxins and involves precolumn derivatization of the compounds using different fluorescent labeling reagents.[90–92] Typical LODs for the LC methods for the analysis of T-2 and HT-2 toxins are in the range of 3–5 µg/kg for LC-FLD and from 0.4 to 12 for LC-MS.

LC-MS/MS (also referred to as triple quadrupole MS) methods for the simultaneous determination of multiple groups of mycotoxins, including T-2 toxin, HT-2 toxin, and DAS, in a variety of different foods and feeds have become

very popular during the last few years.[93,94] Recently also high-resolution LC-MS was employed for the quantification of T-2 and HT-2 toxins in beer.[95] Usually, electrospray ionization (ESI) is performed, but some methods (often those with a limited number of measured analytes besides T-2 and HT-2 toxins) rely on atmospheric pressure chemical ionization or even photoionization.[37] While some methods use both positive and negative polarities (either with two consecutive LC runs or fast polarity switching), T-2 and HT-2 toxins are almost exclusively measured in positive ion mode.

An inherent problem with all MS methods is signal suppression or even signal enhancement owing to matrix effects.[96] Possible strategies to cope with these effects include selective clean-up, sample extract dilution, usage of matrix matched standards, standard addition to each sample at multiple levels or internal standards. For the latter, a stable isotope dilution assay for the quantification of T-2 and HT-2 toxins has been described.[97]

Finally, it is worth mentioning that trichothecenes that are bound to polar substances such as sugars, amino acids, or sulfates, may not be detected by most routine methods. Therefore, reliable procedures for their detection and quantification are needed. Veprikova et al.[98] have recently developed a new analytical procedure for monitoring T-2/HT-2 conjugates (glucosides) in cereal samples. Sample preparation is achieved by employing dedicated immunoaffinity cross-reactive cartridges, and the extracts obtained were examined by ultra-high performance liquid chromatography hyphenated to high-resolution tandem mass spectrometry that enabled the confirmation of the presence of conjugated T-2 and HT-2.

47.6.2.4 Immunochemical Methods and Biosensors

Immunochemical methods to determine T-2 and HT-2 toxins are usually employed as screening methods. They include enzyme-linked immunosorbent assays (ELISAs), lateral flow devices, dipstick tests, fluorescence polarization immunoassay, immunofiltration assays, and more recently biosensor assays.[99]

The results generated by ELISAs generally provide low LODs in the range of 20 µg/kg, and can be generated fast and at relatively low cost. One of the limiting factors of immunochemical methods is the fact that structurally related toxins (i.e., other trichothecenes) may lead to cross-reactivities, resulting in overestimation of the toxin content, and also the existence of matrix effects should be carefully considered. The immunochemical methods in use for T-2 and HT-2 toxins are mainly based on competitive ELISA techniques in microtiter plates, though lateral flow devices are becoming more popular since they are extremely rapid to perform and may be used in field conditions.

Biosensors are used in many industrial sectors, including the food sector. They are composed of a biological (often antibody–antigen) recognition element connected to a transducer or sensing device. Both optical and electrochemical sensor-based methods have recently been developed for the determination of type A trichothecenes in cereals and maize-based baby food. Optical biosensors based on the principle of surface plasmon resonance (SPR) have shown excellent in-house performance characteristics and offer possibilities for high throughput analyses, but currently they are costly and can only be used in the laboratory. Meneely et al.[99] described an SPR assay based on an inhibition format employing a monoclonal antibody raised against HT-2 with cross reactivity to T-2. In-house validation has shown LODs of 31, 47, and 36 µg/kg for HT-2 in wheat, breakfast cereal, and maize-based baby food, respectively. A enzyme-linked-immunomagnetic-electrochemical array (ELIME-array) for T-2 and HT-2 toxin detection in food samples based on the use of magnetic beads and screen-printed electrodes has been reported.[100] Advantages of electrochemical measurements over that of spectrophotometric ones include the possibility of increased speed, miniaturization, and multiplexing.

In summary, many different immunochemical platforms exist for the screening of type A trichothecene mycotoxins in food from the low cost ELISAs to the expensive instruments such as optical biosensors based on the phenomenon of SPR. A recent review evaluated and compared a number of these platforms to assess their accuracy and precision when applied to naturally contaminated samples containing HT-2/T-2 mycotoxins.[101]

47.6.2.5 Analytical Quality Assurance: Performance Criteria

In Annex II of the Regulation (EC) No. 401/2006 of February 23, 2006, laying down the methods of sampling and analysis for the official control of the levels of mycotoxins in foodstuffs, criteria for methods of analysis are laid down. Performance criteria for methods of analysis of T-2 and HT-2 toxins used in the official control as mentioned in this regulation are presented in Table 47.3.

47.7 METHODS TO PREVENT GROWTH OF PRODUCING MOLDS

Given that several *Fusarium* species are able to produce T-2, HT-2, and DAS toxins in foods, prevention of growth of these producing species occurring in the plant at every step of their growth, at harvest, or in different stages of cereal or other food products processing is essential to avoid risk

TABLE 47.3
Performance Criteria for T-2 and HT-2 Toxins

Toxin	Level (µg/kg)	RSRr (%)	RSD$_R$ (%)	Recovery (%)
T-2	50–250	≤40	≤60	60–130
T-2	>250	≤30	≤50	60–130
HT-2	100–200	≤40	≤60	60–130
HT-2	>200	≤30	≤50	60–130

Note: RSRr, relative standard deviation under repeatability conditions; RSDR, relative standard deviation under reproducibility conditions.

of mycotoxin accumulation. To prevent growth of T-2, HT-2, and DAS toxins producing molds several strategies pre- and postharvest (during storage and commercialization) of foods could be used.

As a preharvest strategy it has been reported that an *early harvesting* of crops reduces fungal infection in the field before harvest and consequent contamination of harvested produce.[102,103] The importance of assessing mycotoxin risk has been demonstrated on a site-by-site basis to make appropriate decisions on the timing of harvest to minimize mycotoxin levels.[102,104] Other preharvest strategies include several good agricultural practices to reduce crop stress such as *improved irrigation, low plant density, balanced fertilization, use of strains resistant to fungal colonization*, and *genetically modified crops* that inhibit fungal colonization.[105,106]

Fungicide treatments have also been reported as preharvest strategy to inhibit growth of mycotoxin-producing molds, but in most of the cases these treatments did not significantly reduce trichothecenes contamination in products.[107] However, a phytoalexin (stress-induced compound) from *Citrus sinensis* 6,7-dimethoxycoumarin, isolated from *Penicillium digitatum*-infected Valencia fruit, has been reported to confer resistance against the mycotoxigenic fungi *F. verticillioides*, causing a reduction in production of fumonisin B$_1$.[108] This phytoalexin could be used to reduce trichothecenes in the field before harvest.

Use of biological agents by introducing atoxigenic fungal strains is one strategy that has recently gained prominence in literature and may be a good alternative to control growth of trichothecenes-producing fungi pre- and postharvest. Thus, inhibition of growth of producing strains and mycotoxins production was observed when nontoxigenic *F. verticillioides*, as protective culture, was used.[109] In maize residues, inhibition of growth and sporulation of producing strains of *F. verticillioides* and *F. proliferatum* has also been reported.[110] In addition, inhibition of growth of mycotoxin-producing molds by endophytic bacteria such as *Bacillus subtilis* has been widely described.[103,111] Thus Cavaglieri et al.[112] demonstrated the ability of *B. subtilis* to reduce rhizoplane and endorhizosphere of *F. verticillioides* colonization at all inoculum and maize root. Furthermore, use of lactic acid bacteria as protective culture in postharvest products has shown high efficiency to control fungal growth.[113] Thus, the former authors observed inhibition of *F. culmorum* in bread when *Lactobacillus amylovorus* was used as protective culture, due to the production by this microorganism of antifungal compounds such as 3-phenylpropanoic acid, (E)-2-methylcinnamic acid, cytidine, deoxycytidine, sodium decanoate, cyclo(L-Met-L-Pro), cyclo(L-Pro-L-Pro), cyclo(L-Tyr-L-Pro), and cyclo(L-His-L-Pro).

Postharvest strategies include *rapid drying* of agricultural products to low moisture level. Therefore, it creates less favorable conditions for fungal growth and proliferation, mold infestation and helps keep longer.[114] Thus, drying harvested maize to 15.5% moisture content or lower within 24–48 h would reduce the risk of fungal growth and consequent mycotoxins production.[102] In addition, *storage of raw materials at low relative* humidity may be essential to avoid growth of toxigenic molds. Ensuring that cereal grains going into store have "safe" moisture content, efforts should be made to prevent moisture migration into grains through leaking roofs and condensation resulting from inadequate ventilation.[102] In addition, *temperature of storage* has been reported to exert a great impact on gene expression involved in mycotoxins production by *Fusarium* spp.[115–117] These authors demonstrated maximum expression of mycotoxins by *Fusarium* at 20°C–25°C. Consequently, storage of raw materials should be conducted at temperature below 15°C to reduce risk of trichothecenes accumulation.

Postharvest *treatments of cereals with NaCl* at concentration higher than 3.5% NaCl seem to be effective in reducing fungal growth and mycotoxins accumulation. Thus, maize treatment with NaCl at 3.5% concentration seems to inhibit DAS production by *F. tricinctum*.[118] In addition, *treatments with several essential oils from spices* such as clove oil (eugenol), cinnamon (cinnamic aldehyde), oregano (thymol) and carvacol, and mace oils (myristin) have shown inhibitory effect on *Fusarium* spp., regulating production of mycotoxins by these fungi.[119]

47.8 CONCLUSIONS

T-2, HT-2, and DAS are a series of mycotoxins produced by *Fusarium* species, a ubiquitous contaminant of stored cereal grains and derived products. Health risks to humans are still unclear, but by this time we know that consumption of food products contaminated with *Fusarium* spp. has been correlated with an increased risk of human diseases. Furthermore, it causes great economic losses in small grain cereals by reducing crop yield and contamination with mycotoxins. Early and accurate identification/detection of *Fusarium* species producing T-2, HT-2, and DAS toxins occurring in the plant at every step of their growth, at harvest or in different stages of cereal or other food products processing, may allow corrective actions to avoid toxic strains and consequently reduce risk of T-2, HT-2, and DAS toxins accumulation entering the food chain. Several methods to accurate and sensitive detection of T-2-, HT-2-, and DAS-producing strains in foods such as those based on real-time PCR are available. In addition, multiple sensitive methods based on GC, HPLC, LC-MS/MS, and ELISA to accurate detection and quantification of trichothecenes produced by *Fusarium* spp. in food commodities have been reported. Furthermore, biological control and other pre- and postharvest treatments should be used to prevent growth of producing strains of *Fusarium* to reduce risk of mycotoxins accumulation throughout the entire food chain. Detection and quantification of any toxigenic strains, regardless of the genus or the species in which they are included, by accurate techniques such as qPCR and detection of toxins in the contaminated foods and use of corrective and preventive actions in the food chain are the most attractive methods that could be implemented as an integrated strategy to reduce risk of accumulation of T-2, HT-2, and DAS toxins in foods.

ACKNOWLEDGMENTS

Authors thank the Spanish MICINN (Carnisenusa CSD2007-00016 and Projects AGL2007-64639, AGL2011-26808), the Governments of Aragón (Grupo de Investigación Consolidado A01) and Extremadura (GRU10162), FEDER, and the European Social Fund of the EC.

REFERENCES

1. Summerell, B.A., Salleh, B., and Leslie, J.F. A utilitarian approach to *Fusarium* identification. *Plant Dis.*, 87, 117, 2003.
2. World Health Organization (WHO). Selected mycotoxins: Ochratoxin, trichothecenes, ergot. Environmental Health Criteria 105, 1990. Available at http://inchem.org/documents/ehc/ehc105.htm, 2007.
3. Smith, J.E. et al. Mycotoxins in human nutrition and health. European Commission, Brussels. Directorate-General XII Science. Research and Development, Agro-Industrial Research Division. EUR 16048 EN, 1994.
4. Bottalico, A. *Fusarium* diseases of cereals: Species complex and related mycotoxin profiles in Europe. *J. Plant. Pathol.*, 80, 85, 1998.
5. Marasas, W.F.O., Nelson, P.E., and Toussoun, T.A. *Toxigenic Fusarium Species: Identity and Mycotoxicology*. The Pennsylvania State University Press, University Park, PA, 1984.
6. Desjardins, A.E. *Fusarium Mycotoxins: Chemistry, Genetics, and Biology*. APS Press, Eagan, MN, pp. 1–260, 2006.
7. Rocha, O., Ansari, K., and Doohan, F.M. Effects of trichothecene mycotoxins on eukaryotic cells: A review. *Food Addit. Contam.*, 22, 369, 2005.
8. Kokkonen, M., Medina, A., and Magan, N. Comparative study of water and temperature relations of growth and T-2/HT-2 toxin production by strains of *Fusarium sporotrichioides* and *Fusarium langsethiae*. *World Mycotoxin J.*, 5, 365–372, 2012.
9. Schuhmacher-Wolz, U., Heine, K., and Schneider, K. Report on toxicity data on trichothecene mycotoxins HT-2 and T-2 toxins. A scientific report submitted to EFSA (CT/EFSA/CONTAM/2010/03), 2010.
10. Yli-Mattila, T. et al. *Fusarium sibiricum* sp. nov. a novel type A trichothecene-producing *Fusarium* from northern Asia closely related to *F. sporotrichioides* and *F. langsethiae*. *Int. J. Food Microbiol.*, 147, 58, 2011.
11. Kirk, P.M. et al. *Dictionary of the Fungi*, 9th ed. CAB International, Wallingford, U.K., 2001.
12. Gräfenhan, T. et al. An overview of the taxonomy, phylogeny and typification of some nectriaceous fungi classified in *Cosmospora, Acremonium, Fusarium, Stilbella* and *Volutella*. *Stud. Mycol.*, 68, 79, 2011.
13. Moretti, A. and Susca, A. *Fusarium*. In: *Molecular Detection of Foodborne Pathogens*. Liu, D. (ed.), CRC Press, Taylor & Francis Group, Boca Raton, FL, 2009.
14. Wollenweber, H.W. and Reinking, O.A. *Die Fusarien, ihre Beschreibung, Schadwirkung und Bekämpfung*. Paul Parey, Berlin, Germany, 1935.
15. Nelson, P.E. Taxonomy and biology of *Fusarium moniliforme*. *Mycopathologia*, 117, 29, 1992.
16. Nelson, P.E., Tousson, T.A., and Marasas, W.F.O. *Fusarium Species: An Illustrated Manual for Identification*. The Pennsylvania State University Press, University Park, PA, 1983.

17. Leslie, J.F. and Summerell, B.A. *The Fusarium Laboratory Manual*. Blackwell Publishing, Ames, IA, 2006.

18. Laurence, M.H. et al. *Fusarium burgesii* sp. nov. representing a novel lineage in the genus *Fusarium*. *Fungal Divers.*, 49, 101, 2011.

19. O'Donnell, K. et al. Gene genealogies reveal global phylogeographic structure and reproductive isolation among lineages of *Fusarium graminearum*, the fungus causing wheat scab. *PNAS*, 97, 7905, 2000.

20. Amata, R.L. et al. An emended description of *Fusarium brevicatenulatum* and *F. pseudoanthophilum* based on isolates recovered from millet in Kenya. *Fungal Divers.*, 43, 11, 2010.

21. Suga, H. et al. Phylogenetic analysis of the phytopathogenic fungus *Fusarium solani* based on the rDNA-ITS region. *Mycol. Res.*, 104, 1175, 2000.

22. Summerell, B.A. and Leslie, J.F. Fifty years of *Fusarium*: How could nine species have ever been enough? *Fungal Divers.*, 50, 135, 2011.

23. Bahar, M. and Shahab, H. Analysis of Iranian isolates of *Fusarium solani* using morphological, pathogenicity and microsatellite DNA marker characterization. *Afr. J. Biotechnol.*, 11, 474, 2012.

24. Lal, N. and Datta, J. Progress and perspectives in characterization of genetic diversity in plant pathogenic *Fusarium*. *Plant Arch.*, 12, 557, 2012.

25. Visentin, I. et al. The ITS region as a taxonomic discriminator between *Fusarium verticillioides* and *Fusarium proliferatum*. *Mycol. Res.*, 113, 1137, 2009.

26. Chehri, K. et al. Molecular characterization of pathogenic *Fusarium* species in curcubit plants from Kermanshah province, Iran. *Saudi J. Biol. Sci.*, 18, 341, 2011.

27. Windels, C.E. Economic and social impacts of *Fusarium* head blight: Changing farms and rural communities in the Northern Great Plains. *Phytopathology*, 90, 17, 2000.

28. Ploetz, R.C. et al. Importance of *Fusarium* wilt in different banana growing regions. In: *Fusarium Wilt of Banana*. Ploetz, R.C. (ed.), American Phytopathological Society, St. Paul, MN, pp. 9–26, 1990.

29. Wannemacher, R.W. and Wiener, S.L. Trichothecene mycotoxins. In: *Medical Aspects of Chemical and Biological Warfare*. Sidell, F.R., Takafuji, E.T., and Franz, D.R. (eds.), Office of the Surgeon General at TMM Publications, Washington, DC, pp. 655–676, 1997.

30. Creppy, E.E. Update of survey, regulation and toxic effects of mycotoxins in Europe. *Toxicol. Lett.*, 127, 19, 2002.

31. Desjardins, A.E. and Proctor, R.H. Biochemistry and genetics of *Fusarium* toxins. In: *Fusarium*. APS, St. Paul, MN, pp. 50–69, 2001.

32. Yazar, S. and Omurtag, G.Z. Fumonisins, trichothecenes and zearalenone in cereals. *Int. J. Mol. Sci.*, 9, 2062, 2008.

33. Omurtag, G.Z. et al. Occurrence of diacetoxyscirpenol (anguidine) in processed cereals and pulses in Turkey by HPLC. *Food Control*, 18, 970, 2007.

34. Mirocha, C.J. et al. Natural occurrence of *Fusarium* toxins in feedstuff. *Appl. Environ. Microbiol.*, 32, 553, 1976.

35. Wang, J.S., Busby, W.F., and Wogan, G.N. Percutaneous absorption and tissue distribution of [³H]diacetoxyscirpenol (anguidine) in rats and mice. *Toxicol. Appl. Pharmacol.*, 103, 430, 1996.

36. Van der Fels-Klerx, H.J. Occurrence data of trichothecene mycotoxins T-2 toxin and HT-2 toxin in food and feed. A scientific report submitted to EFSA (NP/EFSA/UNIT CONTAM/2010/01), 2010.

37. EFSA (European Food Safety Authority). Scientific Opinion on the risks for animal and public health related to the presence of T-2 and HT-2 toxin in food and feed. *EFSA J.*, 9, 2481, 2011.

38. Pestka, J.J. and Smolinski, A.T. Deoxynivalenol: Toxicology and potential effects on humans. *J. Environ. Sci. Health Pt. B*, 8, 39, 2005.

39. Stenglein, S.A. *Fusarium poae*: A pathogen that need more attention. *J. Plant. Pathol.*, 91, 25, 2009.

40. Eriksen, G.S. and Pettersson, H. Toxicological evaluation of trichothecenes in animal feed. *Anim. Feed Sci. Technol.*, 114, 205, 2004.

41. Hsia, C.C. et al. Natural occurrence and clastogenic effects of nivalenol, deoxynivalenol, 3-acetyl-deoxynivalenol, 15-acetyl-deoxynivalenol, and zearalenone in corn from a high-risk area of esophageal cancer. *Cancer Detect. Prev.*, 13, 79, 1988.

42. Larsen, J.C. et al. Workshop on trichothecenes with a focus on DON: Summary report. *Toxicol. Lett.*, 153, 1, 2004.

43. Widestrand, J. and Pettersson, H. Effect of time, temperature and solvent on the stability of T-2 toxin, HT-2 toxin, deoxynivalenol and nivalenol calibrants. *Food Addit. Contam.*, 18, 987, 2001.

44. D'Mello, J.F.P., Placinta, C.M., and MacDonald, A.M.C. *Fusarium* mycotoxins: A review of global implications for animal health, welfare and productivity. *Anim. Feed Sci. Technol.*, 80, 183, 1999.

45. Sudakin, D.L. Trichothecenes in the environment: Relevance to human health. *Toxicol. Lett.*, 14, 97, 2003.

46. Li, Y. et al. T-2 Toxin, a trichothecene mycotoxin: Review of toxicity, metabolism and analytical methods. *J. Agric. Food Chem.*, 59, 3441, 2011.

47. Doi, K., Shinozuka, J., and Sehata, S. T-2 toxin and apoptosis. *J. Toxicol. Pathol.*, 19, 15, 2006.

48. Van de Fels-Klerx, H.J. and Stratakou, I. T-2 toxin and HT-2 toxin in grain and grain based commodities in Europe: Occurrence, factors affecting occurrence, co-occurrence and toxicological effects. *World Mycotoxin J.*, 3, 349, 2010.

49. Sokolovic, M., Garaj-Vrhovac, V., and Simpraga, B. T-2 toxin: Incidence and toxicity in poultry. *Arch. Ind. Hyg. Toxicol.*, 59, 43, 2008.

50. Bauer, J. et al. Kinetic profiles of diacetoxyscirpenol and two of its metabolites in blood serum of pigs. *Appl. Environ. Microbiol.*, 49, 842, 1985.

51. SCF (Scientific Committee for Food). Opinion of the Scientific Committee for Food on *Fusarium* toxins Part 5: T-2 toxin and HT-2 toxin, adopted on May 30, 2001. Available from http://ec.europa.eu/food/fs/sc/scf/out88_en.pdf, 2001.

52. Seifert, K.A. and Lévesque, C.A. Phylogeny and molecular diagnosis of mycotoxigenic fungi. *Eur. J. Plant. Pathol.*, 11, 449, 2004.

53. Banks, J.N. et al. Specific monoclonal antibodies to *Fusarium* species and *Microdochium nivale*. *Food Agric. Immunol.*, 8, 249, 1996.

54. Gan, Z. et al. The characterization of chicken antibodies raised against *Fusarium* spp. by enzyme-linked immunosorbent assay and immunoblotting. *Int. J. Food Microbiol.*, 38, 191, 1997.

55. Iyer, M.S. and Cousin, M.A. Immunological detection of *Fusarium* species in cornmeal. *J. Food Prot.*, 66, 451, 2003.

56. Sampietro, D.A. et al. A molecular based strategy for rapid diagnosis of toxigenic *Fusarium* species associated to cereal grains from Argentina. *Fungal Biol.*, 114, 74, 2010.

57. Eifler, J. et al. Differential detection of potentially hazardous *Fusarium* species in wheat grains by an electronic nose. *PLoS One*, 6, e21026, 2011.

58. Girotti, J.R. et al. Early detection of toxigenic *Fusarium graminearum* in wheat. *World Mycotoxin J.*, 5, 143, 2012.

59. Fredlund, E. et al. Real-time PCR detection of *Fusarium* species in Swedish oats and correlation to T-2 and HT-2 toxin content. *World Mycotoxin J.*, 3, 77, 2010.

60. Atoui, A. et al. Quantification of *Fusarium graminearum* and *Fusarium culmorum* by real-time PCR system and zearalenone assessment in maize. *Int. J. Food Microbiol.*, 154, 59, 2012.

61. Niessen, L. PCR-based diagnosis and quantification of mycotoxin producing fungi. *Int. J. Food Microbiol.*, 119, 38, 2007.

62. Jurado, M. et al. PCR detection assays for the trichothecene-producing species *Fusarium graminearum, Fusarium culmorum, Fusarium poae, Fusarium equiseti* and *Fusarium sporotrichioides*. *Syst. Appl. Microbiol.*, 28, 562, 2005.

63. Konstantinova, P. and Yli-Mattila, T. IGS-RFLP analysis and development of molecular markers for identification of *Fusarium poae, Fusarium langsethiae, Fusarium sporotrichioides* and *Fusarium kyushuense*. *Int. J. Food Microbiol.*, 95, 321, 2004.

64. Edwards, S.G. et al. Molecular studies to identify the *Fusarium* species responsible for HT-2 and T-2 mycotoxins in UK oats. *Int. J. Food Microbiol.*, 156, 168, 2012.

65. Nicolaisen, M. et al. An oligonucleotide microarray for the identification and differentiation of trichothecene producing and non-producing *Fusarium* species occurring on cereal grain. *J. Microbiol. Methods*, 62, 57, 2005.

66. Suanthie, Y., Cousin, M.A., and Woloshuk, C.P. Multiplex real-time PCR for detection and quantification of mycotoxigenic *Aspergillus, Penicillium* and *Fusarium*. *J. Stored Prod. Res.*, 45, 139, 2009.

67. Yli-Mattila, T. et al. Phylogenetic relationship of *Fusarium langsethiae* to *Fusarium poae* and *Fusarium sporotrichioides* as inferred by IGS, ITS, β-tubulin sequences and UP-PCR hybridization analysis. *Int. J. Food Microbiol.*, 95, 267, 2004.

68. Chandra, N.S. et al. Prospects of molecular markers in *Fusarium* species diversity. *Appl. Microbiol. Biotechnol.*, 90, 1625, 2011.

69. Marín, P. et al. Phylogenetic analyses and toxigenic profiles of *Fusarium equiseti* and *Fusarium acuminatum* isolated from cereals from Southern Europe. *Food Microbiol.*, 31, 229, 2012.

70. Kristensen, R. et al. Phylogeny and toxigenic potential is correlated in *Fusarium* species as revealed by partial translation elongation factor 1 alpha gene sequences. *Mycol. Res.*, 109, 173, 2005.

71. Kristensen, R. et al. DNA microarray to detect and identify trichothecene- and moniliformin-producing *Fusarium* species. *J. Appl. Microbiol.*, 102, 1060, 2007.

72. Hohn, T.M. and Desjardins, A.E. Isolation and gene disruption of the *tox5* gene encoding a trichodiene synthase in *Gibberella pulicaris*. *Mol. Plant–Microbe Interact.*, 5, 249, 1992.

73. Schmidt-Heydt, M. and Geisen, R. A microarray for monitoring the production of mycotoxins in food. *Int. J. Food Microbiol.*, 117, 131, 2007.

74. Lincy, S.V. et al. Detection of toxigenic fungi and quantification of type A trichothecene levels in some food and feed materials from India. *Food Control*, 19, 962, 2008.

75. Demeke, T. et al. Species-specific PCR-based assays for the detection of *Fusarium* species and a comparison with the whole seed agar plate method and trichothecene analysis. *Int. J. Food Microbiol.*, 103, 271, 2005.

76. Wilson, A. et al. Development of PCR assays for the detection and differentiation of *Fusarium sporotrichioides* and *Fusarium langsethiae*. *FEMS Microbiol. Lett.*, 233, 69, 2004.

77. Deng, J. Structural, functional and evolutionary analyses of the rice blast fungal genome, PhD thesis. North Carolina State University, Raleigh, NC, 2006.

78. Stakheev, A.A., Ryazantsev, D.Y., and Zavriev, S.K. Novel DNA markers for taxonomic characterization and identification of *Fusarium* species Russian. *J. Bioorg. Chem.*, 37, 593, 2011.

79. Fernández-Ortuño, D. et al. The *CYP51C* gene, a reliable marker to resolve interspecific phylogenetic relationships within the *Fusarium* species complex and a novel target for species-specific PCR. *Int. J. Food Microbiol.*, 144, 301, 2010.

80. Severgnini, M. et al. Advances in DNA microarray technology for the detection of foodborne pathogens. *Food Bioprocess Technol.*, 4, 936, 2011.

81. Shephard, G.S. et al. Developments in mycotoxin analysis: An update for 2010–2011. *World Mycotoxin J.*, 5, 3, 2012.

82. Shephard, G.S. et al. Developments in mycotoxin analysis: An update for 2009–2010. *World Mycotoxin J.*, 4, 3, 2011.

83. Rubert, J., Soler, C., and Manes, J. Application of an HPLC-MS/MS method for mycotoxin analysis in commercial baby foods. *Food Chem.*, 133, 176, 2012.

84. Meneely, J.P. et al. A rapid optical immunoassay for the screening of T-2 and HT-2 toxin in cereals and maize-based baby food. *Talanta*, 81, 630, 2010.

85. Cano-Sancho, G. et al. Biomonitoring of *Fusarium* spp. mycotoxins: Perspectives for an Individual Exposure Assessment Tool. *Food Sci. Technol. Int.*, 16, 266, 2010.

86. Majerus, P., Hain, J., and Scheer, M. T-2 and HT-2 toxin analysis in cereals and cereal products following IAC cleanup and determination via GC-ECD after derivatization. *Mycotoxin Res.*, 24, 24, 2008.

87. Meneely, J.P. et al. Current methods of analysis for the determination of trichothecene mycotoxins in food. *Trends Anal. Chem.*, 30, 192, 2011.

88. Shephard, G.S. et al. Developments in mycotoxin analysis: An update for 2008–2009. *World Mycotoxin J.*, 3, 3, 2010.

89. Pettersson, H. and Langseth, W. Intercomparison of trichothecene analysis and feasibility to produce certified calibrants and reference material. Final report I, Method Studies. BCR Information, Project Report EUR 20285/1 EN 1-82, 2002.

90. Visconti, A. et al. Analysis of T-2 and HT-2 toxins in cereal grains by immunoaffinity clean-up and liquid chromatography with fluorescence detection. *J. Chromatogr. A*, 1075, 151, 2005.

91. Lippolis, V. et al. Improvement of detection sensitivity of T-2 and HT-2 toxins using different fluorescent labelling reagents by high-performance liquid chromatography. *Talanta*, 74, 1476, 2008.

92. Trebstein, A. et al. Determination of T-2 and HT-2 toxins in cereals including oats after immunoaffinity cleanup by liquid chromatography and fluorescence detection. *J. Agric. Food Chem.*, 56, 4968, 2008.

93. Lattanzio, V.M.T. et al. Simultaneous determination of aflatoxins, ochratoxin A and *Fusarium* toxins in maize by liquid chromatography/tandem mass spectrometry after multitoxin immunoaffinity cleanup. *Rapid Commun. Mass Spectrom.*, 21, 3253, 2007.

94. Spanjer, M.C., Rensen, P.M., and Scholten, J.M. LC-MS/MS multimethod for mycotoxins after single extraction, with validation data for peanut, pistachio, wheat, maize, cornflakes, raisins and figs. *Food Addit. Contam.*, 25, 472, 2008.

95. Zachariasova, M. et al. Analysis of multiple mycotoxins in beer employing (ultra)-high-resolution mass spectrometry. *Rapid Commun. Mass Spectrom.*, 24, 3357, 2010.

96. Rubert, J. et al. Study of mycotoxin calibration approaches on the example of trichothecenes analysis from flour. *Food Chem. Toxicol.*, 50, 2034, 2012.

97. Asam, S. and Rychlik, M. Synthesis of four carbon-13-labeled type a trichothecene mycotoxins and their application as internal standards in stable isotope dilution assays. *J. Agric. Food. Chem.*, 54, 6535, 2006.

98. Veprikova, Z. et al. Occurrence of mono- and di-glycosylated conjugates of T-2 and HT-2 toxins in naturally contaminated cereals. *World Mycotoxin J.*, 5, 231, 2012.

99. Meneely, J.P. et al. Simultaneous screening for T-2/HT-2 and deoxynivalenol in cereals using a surface plasmon resonance immunoassay. *World Mycotoxin J.*, 5, 117, 2012.

100. Romanazzo, D. et al. ELIME (enzyme linked immuno magnetic electrochemical) method for mycotoxin detection. *J. Visual. Exp.*, 32, 1588, 2009.

101. Meneely, J.P. et al. A comparative study of qualitative immunochemical screening assays for the combined measurement of T-2/HT-2 in cereals and cereal-based products. *World Mycotoxin J.*, 4, 385, 2011.

102. Bankole, S.A. and Adebanjo, A. Mycotoxins in food in West Africa: Current situation and possibilities of controlling it. *Afr. J. Biotechnol.*, 2, 254, 2003.

103. Wagacha, J.M. and Muthomi, J.W. Mycotoxin problem in Africa: Current status, implications to food safety and health and possible management strategies. *Int. J. Food Microbiol.*, 124, 1, 2008.

104. Rachaputi, N.R., Wright, G.C., and Kroschi, S. Management practices to minimise pre-harvest aflatoxin contamination in Australian groundnuts. *Aust. J. Exp. Agric.*, 42, 595, 2002.

105. Wild, C.P. and Gong, Y.Y. Mycotoxins and human disease: A largely ignored global health issue. *Carcinogenesis*, 31, 71, 2010.

106. Waśkiewicz, A., Beszterda, M., and Goliński, P. Occurrence of fumonisins in food—An interdisciplinary approach to the problem. *Food Control*, 26, 491, 2012.

107. Blandino, M. et al. Influence of agricultural practices on *Fusarium* infection, fumonisin and deoxynivalenol contamination of maize kernels. *World Mycotoxin J.*, 2, 409, 2009.

108. Mohanlall, V. and Odhav, B. Biocontrol of aflatoxins B1, B2, G1, G2, and fumonisin B1 with 6,7-dimethoxycoumarin, a phytoalexin from *Citrus sinensis. J. Food Prot.*, 69, 9, 2224, 2006.

109. Desjardins, A.E. et al. Distribution of fumonisins in maize ears infected with strains of *Fusarium moniliforme* that differ in fumonisin production. *Plant Dis.*, 82, 953, 1998.

110. Luongo, L. et al. Potential of fungal antagonists for biocontrol of *Fusarium* spp. in wheat and maize through competition in crop debris. *Biocontrol Sci. Technol.*, 15, 229, 2005.

111. Bacon, C.W. et al. Biological control of *Fusarium moniliforme* in maize. *Environ. Health Perspect.*, 109, 325, 2001.

112. Cavaglieri, L. et al. Biocontrol of *Bacillus subtilis* against *Fusarium veticillioides* in vitro and at the maize root level. *Res. Microbiol.*, 156, 748, 2005.

113. Ryan, L.A.M. et al. *Lactobacillus amylovorus* DSM 19280 as a novel food-grade antifungal agent for bakery products. *Int. J. Food Microbiol.*, 146, 276, 2011.

114. Lanyasunya, T.P. et al. The risk of mycotoxins contamination of dairy feed and milk on small holder dairy farms in Kenya. *Pakistan J. Nutr.*, 4, 162, 2005.

115. Jurado, M. et al. Relationship between solute and matric potential stress, temperature, growth and *FUM1* gene expression in two *Fusarium verticillioides* strains from Spain. *Appl. Environ. Microbiol.*, 74, 2032, 2008.

116. Marín, P. et al. Differential effect of environmental conditions on the growth and regulation of the fumonisin biosynthetic gene *FUM1* in the maize pathogens and fumonisin producers *Fusarium verticillioides* and *Fusarium proliferatum. FEMS Microbiol. Ecol.*, 73, 303, 2010.

117. Lazzaro, I. et al. Effects of temperature and water activity on *FUM2* and *FUM21* gene expression and fumonisin B production in *Fusarium verticillioides. Eur. J. Plant. Pathol.*, 134, 685, 2012.

118. Youssef, M.S. Natural occurrence of mycotoxins and mycotoxigenic fungi on Libyan corn with special reference to mycotoxin control. *Res. J. Toxins*, 1, 8, 2009.

119. Juglal, S., Govinden, R., and Odhav, B. Spice oils for the control of co-occurring mycotoxin-producing fungi. *J. Food Prot.*, 65(4), 683, 2002.

120. Niessen, L., Schmidt, H., and Vogel, R.F. The use of *tri5* gene sequences for PCR detection and taxonomy of trichothecene-producing species in the *Fusarium* section *Sporotrichiella. Int. J. Food Microbiol.*, 95, 305, 2004.

121. Yli-Mattila, T. et al. Real-time PCR detection and quantification of *Fusarium poae*, *F. graminearum*, *F. sporotrichioides* and *F. langsethiae* in cereal grains in Finland and Russia. *Arch. Phytopathol. Plant Protect.*, 41, 243, 2008.

122. Nielsen, L.K. et al. *Fusarium* head blight of cereals in Denmark: Species complex and related mycotoxins. *Phytopathology*, 101, 960, 2011.

123. Mach, R.L. et al. Specific detection of *Fusarium langsethiae* and related species by DGGE and ARMS-PCR of a β-tubulin (*tub1*) gene fragment. *Int. J. Food Microbiol.*, 95, 333, 2004.

124. Kristensen, R. et al. Simultaneous detection and identification of trichothecene- and moniliformin-producing *Fusarium* species based on multiplex SNP analysis. *J. Appl. Microbiol.*, 102, 1071, 2007.

125. Rodríguez, A. et al. A comparative study of DNA extraction methods to be used in real-time PCR based quantification of ochratoxin A-producing molds in food products. *Food Control*, 25, 666, 2012.

126. Nicolaisen, M. et al. Real-time PCR for quantification of eleven individual *Fusarium* species in cereals. *J. Microbiol. Methods*, 76, 234, 2009.

48 Tetrodotoxin

Rocky Chau, John A. Kalaitzis, and Brett A. Neilan

CONTENTS

48.1 INTRODUCTION

Tetrodotoxin (TTX; Figure 48.1) is one of the most highly toxic neurotoxins known to humans, and its existence has been known since antiquity. The existence of engravings of puffer fish on a fifth dynasty (2500 BC) Egyptian tomb and references to the toxicity of puffer fish eggs in first and second century BC Chinese texts suggest that both the ancient Egyptians and Chinese knew of the toxic effects of puffer fish. Chinese texts suggest that both the ancient Egyptians and Chinese knew of the toxic effects of puffer fish. Toxic puffer fish are also mentioned in a AD 1600 Chinese *materia medica*. One of the earliest descriptions, by Europeans, of the toxicity of puffer toxin can be found in the journals of James Cook.[1]

In nature, TTX can be found in many phylogenetically diverse organisms[2] where it serves a variety of ecological roles.[3] It is commonly known to be associated with puffer fish (family Tetraodontidae) where it serves as a passive chemical defense system to deter predators. In some puffer species, including *Takifugu niphobles*, TTX has been proposed to function as a sex pheromone. It has been shown that sexually active male *T. niphobles* are attracted by TTX contained in ovulated eggs in female *T. niphobles*, and this is supported by the finding that TTX does not attract female *T. niphobles*.[4] TTX has been reported from the blue-ringed octopus (genus *Hapalochlaena*), where it serves as a predation aid. *Hapalochlaena maculosa* have been observed to inject TTX into its prey to immobilize them and assist capture. Furthermore, many animals, for example, blue-ringed octopus and newts, have been found to invest TTX into their eggs during spawning.[5,6]

Apart from its many physiological and ecological roles in nature, TTX plays a significant role in human health. In minute quantities, TTX causes oral paresthesia, sometimes sought after numbing of the mouth. Therefore, the flesh of TTX-containing animals, especially puffer fish, is considered a delicacy in many countries. The consumption of poorly prepared puffer fish flesh has led to many documented cases of fatal puffer poisoning throughout the world, especially in Asia. The epidemiology of TTX poisoning will be discussed further in Section 48.3.

Due to the high toxicity of TTX in humans, 0.34 mg/kg, it is a controlled substance in many countries. It is listed in the Select Agent Registry of the US Department of Health and Human Services (HHS), and thus, agencies or researchers possessing greater than 100 mg of TTX must be registered with the HHS.

Although the possibility of using TTX as a weapon has long been known, few documented cases of its weaponization can been found. In World War II, a Japanese biological weapons research unit, Unit 731, was believed to have investigated TTX for use as a weapon.[7] As recently as 2012, a man was convicted of acquiring, possessing, and attempting to use TTX as a weapon. He was found to have ordered over 150 mg of TTX, six times the human lethal dose, from a scientific chemicals supplier. He was later sentenced to 7-year imprisonment.[8]

Despite its high toxicity, TTX has many beneficial uses. Due to its ability to selectively block voltage-gated sodium ion channels, it is used extensively in the study of signal conductance in neuronal cells. TTX also has many therapeutic applications. Low doses of TTX have been found to reduce craving and anxiety associated with heroin abstinence in addicts, while having no effect on heart rate or blood pressure.[9] Additionally, TTX has been trialled in the management of moderate to severe cancer pain. An open-label trial was conducted where subcutaneous TTX was administered to 45 patients who failed to respond to opioids and other analgesics.[10] Of these, 47% of patients reported a reduction in pain of over 30%, while 36% of patients reported pain reduction of less than 30%, which was still considered as a meaningful effect upon consultation with a physician.

This chapter will provide an overview of TTX, including the classification and morphology of its producers, the epidemiology of TTX poisoning, and its diagnosis and treatment.

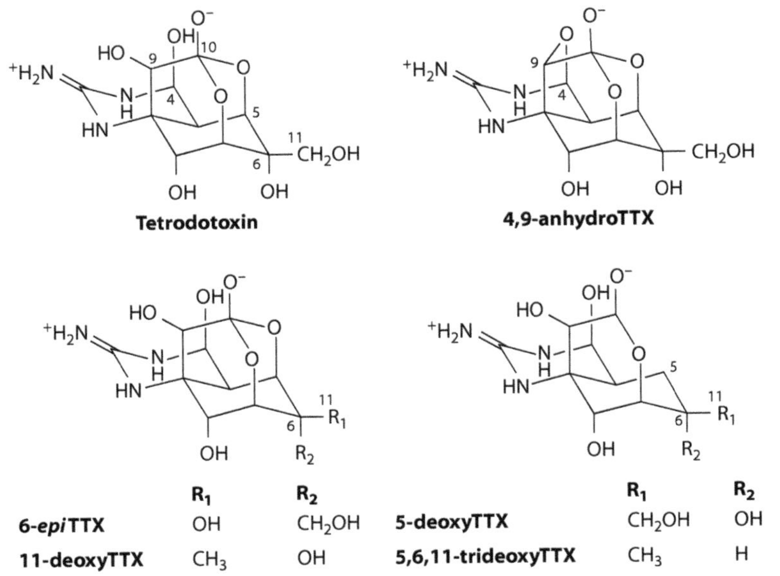

FIGURE 48.1 Structure of TTX. TTX has an unusual oxygenated adamantane backbone. It has many natural analogues possessing varying degrees of affinity to NaV channels. The C9 and C10 groups are important in interacting with the sodium channel.

48.2 CLASSIFICATION AND MORPHOLOGY

TTX is a guanidinium neurotoxin containing a unique oxygenated adamantane carbon backbone that is unusual in natural molecules. The most well-known producers of TTX are Tetraodontidae puffer fish, which are common in waters of Southeast Asia. Many phylogenetically distant organisms, from eight phyla of life, have been reported to contain TTX (Table 48.1). The production of a compound by such a wide distribution of animals is unprecedented. It is unlikely that convergent evolution led to the ability to produce TTX in such a high number of divergent organisms. Hence, it is hypothesized that microbes are the true producers of TTX.[2] It is believed that these microbes are associated with their host organisms via symbiosis or bioaccumulated through the food chain; however, this is still a topic of much controversy. The first bacterial producer of TTX was isolated from the intestines of the reef crab, *Atergatis floridus*, by Noguchi et al. in 1983.[11] Since then, TTX production by bacteria from 19 genera has been documented.[2] Not all instances of TTX production can be attributed to a microbial producer. The newt, *Taricha granulosa*, has been found to contain TTX in its skin, ovaries, and muscles. In order to identify any bacteria present in these tissues, and thus a potential TTX producer, PCR was used to detect bacterial gene sequences; however, this yielded a negative result. The results showed that bacterial DNA was only present in the gastrointestinal tract, which had only low concentrations of TTX.[12] It was concluded that newts were unlikely to contain TTX-producing bacteria. However, there have been many documented cases of TTX producers being isolated from the intestines of marine animals,[13] for example, TTX-producing *Vibrio* species have been isolated from intestines of puffer fish, *Takifugu vermicularis* and *T. niphobles*, and starfish, *Astropecten polyacanthus*.

Over half of the reported TTX-producing bacteria are Gammaproteobacteria, with *Vibrio* being the dominant genus.[2] Due to the diversity of TTX-producing microbes, there is no common characteristic shared between TTX-producing bacteria, apart from the production of TTX. Hence, morphological identification of TTX-producing bacteria relies solely on the detection of TTX in cell cultures. Therefore, accurate methods of detecting TTX are required and these detection methods will be discussed in Section 48.5.

48.3 BIOLOGY AND EPIDEMIOLOGY

The popularity of puffer fish (fugu) as a culinary delicacy in Japan has contributed greatly to Japan having the highest reported incidence rate of TTX intoxication in humans (Figure 48.2). Since 1965, Japan has had over 1300 reported cases of TTX poisonings[14,15] with the highest mortality rates occurring prior to 1980. With an increasing awareness of the dangers of consuming puffer fish, combined with the introduction of regulations on puffer fish preparation, this number has decreased in recent years. Currently, only licensed chefs specially trained in the proper techniques of preparing puffer fish are permitted to serve fugu in restaurants. Chefs in Japan must adhere to regulations regarding the edible parts of a puffer fish and are strictly prohibited to serve puffer liver, which is typically the most toxic part of the fish.[14,15] Additionally, there are now strict regulations in place governing the export of Japanese puffer fish internationally, and this has limited the reported incidences of puffer fish poisoning outside Asia.

In an attempt to reduce puffer-associated TTX poisonings in Japan, aquaculture of nontoxic puffer fish species has been attempted.[16] In the period 2001–2004, approximately 5000 specimens of *Takifugu rubripes* were raised in netcages and land aquaria on TTX-free diets. None of the specimens

TABLE 48.1
Phylogenetic Overview of TTX-Producing Host Organisms

Phylum	Family	Species	References
Chordata (frogs, puffer fish, newts)	Dendrobatidae	*Colostethus inguinalis*	[58]
	Brachycephalidae	*Brachycephalus pernix*	[59]
		Brachycephalus ephippium	[60]
	Bufonidae	*Atelopus oxyrhynchus*	[61]
	Salamandridae	*Notophthalmus viridescens*	[62]
		Taricha torosa	[63]
	Tetraodontidae	*Takifugu xanthopterus*	[64]
		Takifugu niphobles	[65]
		Fugu rubripes[a]	[66,67]
		Fugu vermicularis[a]	[68,69]
		Fugu pardalis[a]	[70]
		Fugu poecilonotus[a]	[70,71]
	Gobiidae	*Gobius criniger*	[72]
Mollusca (gastropods)	Octopodidae	*Hapalochlaena maculosa*[a]	[73,74]
	Nassariidae	*Niotha clathrata*	[75]
		Nassarius semiplicatus[a]	[76]
	Muricidae	*Rapana rapiformis*	[77]
		Rapana venosa venosa	[77]
	Ranellidae	*Charonia sauliae*	[78]
	Buccinidae	*Babylonia japonica*	[79]
	Naticidae	*Polinices didyma*	[80]
	Bursidae	*Tutufa lissostoma*	[81]
	Pleurobranchidae	*Pleurobranchaea maculata*	[50]
Nemertea (nematodes)	Cephalothricidae	*Cephalothrix rufifrons*[a]	[82]
	Lineidae	*Lineus longissimus*[a]	[82]
Echinodermata (starfish)	Astropectinidae	*Astropecten latespinosus*	[83]
		Astropecten polyacanthus[a]	[84]
Chaetognatha (arrow worms)	Sagittidae	*Flaccisagitta enflata*	[85]
		Parasagitta elegans	[85]
	Pterosagittidae	*Zonosagitta nagae*	[85]
	Eukrohniidae	*Eukrohnia hamata*	[85]
Arthropoda (crabs)	Cleroidea	*Carcinoscorpius rotundicauda*	[86,87]
	Carpilioidea	*Lophozozymus pictor*	[88]
	Xanthoidea	*Atergatis floridus*	[11]
Platyhelminthes (flatworms)	Planoceridae	*Planocera multitentaculata*	[89]
Dinoflagellate	Goniodomataceae	*Alexandrium tamarense*	[90]

[a] Bacterial production of TTX has been documented in these species.

grown in aquaculture showed TTX toxicity via mouse bioassay. Although the culturing of nontoxic puffer fish represents a safe and reliable method to reduce the risk of consuming puffer fish, this method has yet to be adopted by the food industry in Japan.

Occurrences of TTX poisoning due to puffer fish consumption is much less in other parts of Asia. There have been 277 reported cases of TTX poisonings for Bangladesh between 2001 and 2008[17–19], and thus, it is the third highest incidence of TTX poisonings worldwide. Although the total number of incidents seems high, the rate of incidents is much lower, compared to Japan. The majority of these cases arise from large-scale outbreaks of puffer fish poisonings in the slum districts, where puffer fish is sold cheaply

and is therefore an attractive food source for poorer families, who have not been educated on the dangers of consuming puffer fish. Many other countries in Asia also have reported occurrences of TTX poisonings from puffer fish, including cases in Thailand,[20] Taiwan,[21] South Korea,[22,23] and Malaysia.[24]

Due to the unavailability of puffer fish and the lack of cultural precedence in puffer cuisine, cases of puffer fish poisoning occurring outside of Asia are uncommon. In Australia, 15 cases of TTX intoxication via puffer fish have been reported,[25,26] and only two cases of TTX intoxication, due to puffer fish consumption, have been documented in Europe.[27] In Italy in 1977, 10 people were hospitalized following the consumption of imported puffer species from

Country	Patients	Deaths
Japan	1344	237
China	309	16
Thailand	288	5
Bangladesh	277	30
Korea	81	0
Taiwan	58	9
Malaysia	34	0
Australia	27	9
Brazil	27	2
USA	17	3
Israel	13	0
Italy	10	3
Mexico	4	0
Spain	1	0
Total	2490	314

FIGURE 48.2 Epidemiology of TTX. A graphical representation of the worldwide distribution of TTX poisoning incidences. Countries are shaded according to the number of reported cases of TTX poisonings.

China, which were labelled incorrectly as anglerfish. The only instance of TTX intoxication from locally caught fish was reported in 2010.[28]

Toxic species of puffer fish are not usually found in the Mediterranean Sea, and thus the limited prevalence of TTX intoxications in the eastern Mediterranean. However, human-induced migrations of *Lagocephalus sceleratus*, a TTX-containing puffer fish species, led to the hospitalization of 13 patients in Israel,[29] who consumed these toxic puffer fish.

There have been 17 cases of TTX poisoning reported in the United States, including seven cases in Hawaii.[30,31] Interestingly, saxitoxin has been implicated as the main toxin present in puffer fish originating from the Indian River Lagoon, Florida, United States.[32] Therefore, the number of puffer fish poisonings attributed to TTX may be actually less than reported. Reports of TTX intoxication in Central and South America are isolated, with cases occurring in Brazil[33] and Mexico.[34]

Apart from puffer fish, many other animals have also been responsible for TTX poisoning. In Thailand, eggs from the horseshoe crab, *Carcinoscorpius rotundicauda*, are used in many Thai foods. Between 1994 and 2006, 280 cases of TTX poisoning due to consumption of *C. rotundicauda* eggs have been reported, of which 5 were fatal.[35] In China and Taiwan, the consumption of toxic marine gastropods has led to 309 and 8 cases of TTX intoxication, respectively.[36,37] Outside of Japan, China has the next highest number of TTX intoxications.

The blue-ringed octopus, *H. maculosa*, has been responsible for 12 cases of TTX intoxication in Australia.[38,39] These cases are unique, as they represent the only documented cases of a nondietary route of TTX intoxication. TTX stored in the salivary glands of *H. maculosa* is released during its bite. *H. maculosa* typically reside in shallow waters of rock pools and possess bright blue marks on its skin. Hence, many envenomation cases involve people who attempt to handle the octopus, which outwardly looks harmless.

48.4 CLINICAL FEATURES AND PATHOGENESIS

As discussed previously, the most common cause of TTX intoxication is via ingestion of TTX-containing organisms, such as puffer fish. However, other routes of intoxication have been reported, including envenomation from the bite of the blue-ringed octopus, *H. maculosa*. Generally, symptoms of TTX intoxication are similar in both cases. However, the onset of symptoms is typically more rapid in envenomation cases. In such cases, onset of intoxication symptoms occurs within 5–10 min of contact with the octopus.

The severity of symptoms varies between patients depending on individual physiology and the amount of toxin that the patient has been exposed to. A study of 53 patients, who were hospitalized due to ingestion of TTX-containing puffer fish flesh, showed that only 3% of patients who ate less than 50 g of puffer fish died. The mortality rate increased to 27% for those who ate between 51 and 100 g of fish. A mortality rate of 50% was recorded for patients who ate greater than 101 g of fish.[17]

Due to the rarity of envenomation cases, little is known or documented regarding the amount of TTX contained in the venom excreted from a *H. maculosa* bite. It has been proposed that the total venom extracted from a 25 g blue-ringed octopus, of which TTX is a major component, could kill 750 kg of rabbits.[40] The precise amount of TTX present in intoxicated patients is not usually recorded, but it has been suggested that 10 μg/kg (intravenous) may be fatal to humans. It has also been suggested that the oral lethal dose of TTX in humans is 344 μg/kg.

The onset of symptoms usually occurs within 10–45 min of intoxication but may also develop as late as 3 h postintoxication.[41] Early symptoms include perioral paresthesia, spreading to the lower and upper limbs, malaise, weakness of upper and lower limbs, incoordination, vertigo, dizziness, blurred vision, sweating, precordial pain, headache, nausea, vomiting, hypersalivation, and diarrhea. Pupil constriction is evident in early-phase TTX poisoning, but the pupils become dilated in late-stage TTX poisoning due to loss of pupillary and corneal muscle control.[41,42]

After the onset of initial symptoms, paresthesia develops into muscle paralysis, immobilizing the patient. Petechial hemorrhages and hematemesis have also been reported, although instances of these are uncommon. During late-stage intoxication, an extreme drop in blood pressure may also be detected. Death is usually due to respiratory failure caused by respiratory muscle paralysis.

TTX is unable to cross the blood–brain barrier;[43] hence, the patient remains conscious and mental faculties are unimpaired, until shortly before death. Depending on the severity of TTX poisoning, symptoms can persist for up to 24 h. Excepting complications arising from ventilatory and cardiovascular support, the prognosis for a patient surviving this period is positive and would be expected to make a full recovery.[42]

Binding of TTX onto voltage-gated sodium ion channels prevents normal neural conduction, leading to the paralytic symptoms associated with TTX intoxication. TTX has a mode of action similar to that of saxitoxin, and both toxins are guanidinium-containing compounds. It has been shown that the guanidinium moiety is responsible for the binding of the toxins to the outer vestibule of voltage-gated sodium ion channels. The binding of TTX onto voltage-gated sodium ion channels has been shown to be reversible; hence, patients fully recover from nonfatal levels of TTX intoxication.

It has been noted that TTX does not affect the resting membrane potential or the resting membrane resistance of neuronal cells. This led to the hypothesis that TTX selectively inhibits the action of sodium channels, with no effect on potassium channels.[44] TTX binds onto the sodium channel by forming a salt bridge between its guanidine moiety and three carboxyl groups on site 1 of the α-subunit of voltage-gated sodium channels.[45] The C9 and C10 hydroxyl groups (Figure 48.1) of TTX are also important in facilitating binding to the sodium channel and form hydrogen bonds

between TTX and the sodium channel site. The importance of these hydroxyl groups is exemplified by anhydrotetrodotoxin, which does not have a C9 hydroxyl and has approximately half the toxicity of TTX.[45]

Mutations in amino acid residues present on site 1 of voltage-gated sodium channels have been shown to have an effect on TTX-binding affinity. Cardiac sodium channels, which have a Cys residue substituted instead of Tyr/Phe, which is present in TTX-sensitive rat skeletal muscle sodium channels, are 25–100 times less sensitive to TTX. Dorsal root ganglion sodium channels have been shown to be even more resistant to TTX and saxitoxin, due to a Tyr/Phe to Ser substitution.[44]

TTX-bearing organisms have been found to have TTX-resistant sodium channels, which protect them from the effects of TTX. In addition, some species have been found to produce TTX-binding compounds that are able to neutralize the effects of TTX.[46]

48.5 IDENTIFICATION AND DIAGNOSIS

Due to the quick onset of symptoms, which can be fatal if untreated, a rapid diagnosis of TTX intoxication must be performed. Diagnosis of TTX poisoning is usually based on clinical symptoms presented by the patient in addition to an interrogation of the patient's recent diet. Such a diagnosis is possible because nearly all cases of TTX poisoning in humans are from food intoxication. The only observed nondietary route to TTX intoxication is envenomation by *H. maculosa*. However, recent contact with any animal in Table 48.2 should be highly regarded in the diagnosis. It is also important to consider whether food was cooked prior to consumption. In nonalkaline conditions, TTX is highly heat stable. Typical heating during cooking is insufficient to inactivate, or reduce the effect of, TTX in contaminated food. This is in contrast to other foodborne neurotoxins such as botulinum toxins, which are heat labile. Other diseases that share similar symptoms to TTX intoxication include Guillain–Barre syndrome, botulism, paralytic shellfish poisoning, tick paralysis, myasthenia gravis, cerebrovascular accident, and atropine.

Initial diagnosis can be confirmed by analytical methods. Until recently, the most common method for TTX detection was the mouse bioassay. The mouse bioassay has been widely used to detect the presence of paralytic shellfish toxins and is suitable for the detection of TTX. However, it cannot unambiguously conclude TTX as the main toxin in foodborne intoxications and is unable to differentiate between TTX and other neurotoxins, such as saxitoxin, which has been detected in TTX-containing animals.[47] Gas chromatography–mass spectrometry (GC–MS) has also been used; however, it has been shown that the alkaline degradation product of TTX has a similar GC–MS profile to peptone, a common bacteriological media constituent. Kawatsu et al.[48] developed an immunoaffinity chromatography method for the detection of TTX in urine samples. Their methodology was highly sensitive

TABLE 48.2

Host Organisms and Their Associated TTX-Producing Bacteria

TTX-Producing Host Organism	TTX-Producing Bacteria	References
Atergatis floridus (reef crab)	*Vibrio* spp.	[11]
Fugu vermicularis (common puffer)	*Vibrio* spp.	[68,69]
Astropecten polyacanthus (comb sea star)	*Vibrio* spp.	[91]
Deep-sea sediment	*Vibrio* spp.	[92]
	Bacillus spp.	
	Alteromonas spp.	
	Aeromonas spp.	
	Micrococcus spp.	
	Acinetobacter spp.	
Chaetognath species (arrow worms)	*Vibrio* spp.	[85]
Freshwater sediment	*Bacillus* spp.	[93]
	Micrococcus spp.	
	Caulobacter spp.	
	Flavobacterium spp.	
Fugu poecilonotus (common puffer)	*Pseudomonas* spp.	[71]
Fugu rubripes (common puffer)	*Nocardiopsis dassonvillei*	[66,67]
	Actinomycete spp.	
	Bacillus spp.	
Hapalochlaena maculosa (blue-ringed octopus)	*Pseudomonas* spp.	[73]
	Bacillus spp.	
	Alteromonas spp.	
	Vibrio spp.	
Jania spp. (red alga)	*Pseudomonas* spp.	[70]
Marine puffer fish	*Microbacterium arabinogalactanolyticum*	[65]
Marine sediment	*Streptomyces* spp.	[94]
Meoma ventricosa (sea urchin)	*Pseudoalteromonas* spp.	[95]
Nassarius semiplicatus (sea snail)	*Marinomonas* spp.	[76]
	Tenacibaculum spp.	
	Vibrio spp.	
Niotha clathrata (marine gastropod)	*Pseudomonas* spp.	[96]
	Aeromonas spp.	
	Plesiomonas spp.	
	Vibrio spp.	
Unidentified puffer fish species	*Vibrio* spp.	[65]
	Serratia marcescens	[65,97]
Seven species of nemertean worms	*Vibrio* spp.	[82]
Takifugu niphobles (common puffer)	*Shewanella putrefaciens*	[98]

and was able to detect TTX at concentrations of 6 ng/mL. TTX was also detected from urine samples collected up to 1-week postintoxication.

One of the most conclusive detection methods for TTX is liquid chromatography–mass spectrometry (LC–MS). Mass spectrometric detection of TTX is currently the most sensitive method for TTX detection. Limits of detection as low as 0.074 ng/mL have been reported.[49] Typically, the *m/z* 320 parent ion of TTX is fragmented to produce daughter ions, *m/z* 162.1 and 60.1 that is confirmatory fingerprint of TTX. Detection of TTX analogues can also be performed via tandem mass spectrometry. Detailed methods for the detection of TTX and analogues can be found in scientific literature.[50,51]

48.6 TREATMENT AND PREVENTION

An antidote for TTX has yet to be developed. Treatment of TTX poisoning is supportive and relies on symptom management.[15] Minor cases of TTX intoxication, resulting in limited paresthesia and weakness, require only monitoring and should be continued until all symptoms recede. More serious cases may require gastrointestinal decontamination followed by life support. If vomiting has not yet occurred, the use of emetics is recommended to remove unabsorbed TTX. Gastric lavage combined with 2% sodium bicarbonate followed by activated charcoal has been reported as effective, if given early enough.[25,41] Ventilatory support should be provided if respiratory paralysis is likely. Light sedation and reassurance may be recommended, as patients usually remain fully conscious throughout their ordeal. As the binding of TTX on voltage-gated sodium channels is reversible, patients make a complete recovery once symptoms have resolved.

Although supportive care is the established treatment for TTX poisoning, numerous other treatments have been proposed. It has been suggested that anticholinesterase drugs are able to competitively reverse the sodium channel blocking action of TTX, by increasing the production of acetylcholine.[41] Kai Chew et al.[52] showed that anticholinesterase drugs were able to rapidly restore muscle power in 16 cases of TTX poisoning. However, this has not been documented in all cases where anticholinesterase drugs were administered.[53,54] The development of a haptenic TTX vaccine has been trialled in mice with success.[55] Immunized mice were intragastrically challenged with increasing doses of TTX from 600 to 2400 μg/kg, with survival rates ranging from 100% to 20%, respectively. All unimmunized mice died at a dose of 600 μg/kg TTX. Furthermore, monoclonal anti-TTX antibodies have been used successfully in the treatment of TTX intoxication in mice. Independent studies performed by Matsumura[56] and Rivera et al.[57] showed that anti-TTX antibodies, administered intravenously, could neutralize lethal doses of TTX.

Unlike other foodborne toxins, such as botulinum toxin, TTX is highly heat stable. Hence, typical preventative measures for foodborne toxins such as heating of food prior to consumption are ineffective. Preventative measures are limited and rely solely on educating the local populace on the dangers of consuming puffer fish. As mentioned in Section 48.3, the culturing of nontoxic puffer fish for the seafood industry is a good preventative measure to reduce TTX poisonings from puffer fish. The lack of perioral paresthesia traditionally associated with consuming minute amounts of TTX in wild puffer fish has made cultured puffer fish an unattractive option to the public.

In areas where TTX-envenoming animals reside, education on the dangers of TTX-bearing organisms has minimized the number of TTX envenomation cases. In Australia, such education campaigns have maintained the low occurrence of *H. maculosa* envenomation. Only 12 cases of *H. maculosa* envenomation, of which 2 were fatal, have been reported since 1950.

48.7 CONCLUSIONS AND FUTURE PERSPECTIVES

The toxicity of TTX and its existence has been long known. The biological mechanisms underpinning the production of TTX remain a mystery. Indeed, it is unknown whether the true producers of TTX are microbes or their host organisms, and the reasons for TTX's production in so many diverse organisms also remain unknown. Although TTX has caused many deaths throughout human civilization, an effective antitoxin has yet to be developed for use in humans. Recent successful trials with anti-TTX antibodies and TTX vaccines in animal models are a positive step toward finding a TTX antitoxin.

Without an antitoxin to TTX, vigilance must be maintained when interacting with known and potential sources of TTX. The most common route to TTX intoxication in humans is via food, and thus, it is vital that the general public continue to be educated about the dangers of eating TTX-containing foods like puffer fish. Such education campaigns have been highly successful in countries such as Japan, which has seen a dramatic decrease in mortality rates from TTX poisonings, from over 50% mortality in the 1960s to 5% in 2007.

Although the greatest threat from TTX is via natural sources, the threat of bioterrorism using TTX is well founded. In areas where TTX-containing organisms are common, TTX is able to be readily purified from these organisms using commonly found chemicals. A criminal case in the United States involving TTX as a weapon highlights the dangers of this toxin, even in areas where tight regulations restrict the distribution of this toxin.

The dangers of TTX are well evidenced. However, its use as a drug in the treatment of pain and drug dependencies shows that the benefits of TTX can be harnessed, while minimizing its dangers.[9,10]

REFERENCES

1. Fuhrman, F.A. Tetrodotoxin, tarichatoxin, and chiriquitoxin: Historical perspectives. *Ann. N.Y. Acad. Sci.* 479, 1–14 (1986).
2. Chau, R., Kalaitzis, J.A., and Neilan, B.A. On the origins and biosynthesis of tetrodotoxin. *Aquat. Toxicol.* 104, 61–72 (2011).
3. Williams, B.L. Behavioral and chemical ecology of marine organisms with respect to tetrodotoxin. *Mar. Drugs* 8, 381–398 (2010).
4. Matsumura, K. Tetrodotoxin as a pheromone. *Nature* 378, 563–564 (1995).
5. Williams, B., Hanifin, C., Brodie, E., and Caldwell, R. Ontogeny of tetrodotoxin levels in blue-ringed octopuses: Maternal investment and apparent independent production in offspring of *Hapalochlaena lunulata*. *J. Chem. Ecol.* 37, 10–17 (2011).
6. Hanifin, C.T. and Brodie, E.D. Tetrodotoxin levels in eggs of the rough-skin newt, *Taricha granulosa*, are correlated with female toxicity. *J. Chem. Ecol.* 29, 1729–1739 (2003).
7. Tu, A. Unit 731, in *Weapons of Mass Destruction: An Encyclopedia of Worldwide Policy, Technology, and History*, vol. 2 (eds. Croddy, E. and Wirtz, J.J.) (ABC-CLIO, Santa Barbara, CA, 2005).
8. Karner, M.T. *Lake in the Hills Man Sentenced to Federal Prison for Acquiring and Possessing Deadly Neurotoxin* (US Department of Justice, Chicago, IL, 2012).
9. Shi, J. et al. Tetrodotoxin reduces cue-induced drug craving and anxiety in abstinent heroin addicts. *Pharmacol. Biochem. Behav.* 92, 603–607 (2009).
10. Hagen, N. et al. A multicentre open-label safety and efficacy study of tetrodotoxin for cancer pain. *Curr. Oncol.* 18, e109 (2011).
11. Noguchi, T. et al. Occurrence of tetrodotoxin as the major toxin in xanthid crab *Atergatis floridus*. *Bull. Jpn. Soc. Sci. Fish.* 49, 1887–1892 (1983).
12. Lehman, E.M., Brodie, E.D., Jr., and Brodie Iii, E.D. No evidence for an endosymbiotic bacterial origin of tetrodotoxin in the newt *Taricha granulosa*. *Toxicon* 44, 243–249 (2004).

13. Yu, C.-F. et al. Two novel species of tetrodotoxin-producing bacteria isolated from toxic marine puffer fishes. *Toxicon* 44, 641–647 (2004).

14. Noguch, T. and Arakawa, O. Tetrodotoxin—Distribution and accumulation in aquatic organisms, and cases of human intoxication. *Mar. Drugs* 6, 220–242 (2008).

15. Noguchi, T. and Ebesu, J.S.M. Puffer poisoning: Epidemiology and treatment. *Toxin Rev.* 20, 1–10 (2001).

16. Noguchi, T., Arakawa, O., and Takatani, T. Toxicity of pufferfish *Takifugu rubripes* cultured in netcages at sea or aquaria on land. *Comp. Biochem. Physiol. Part D Genomics Proteomics* 1, 153–157 (2006).

17. Chowdhury, F. et al. Tetrodotoxin poisoning: A clinical analysis, role of neostigmine and short-term outcome of 53 cases. *Singapore Med. J.* 48, 830–833 (2007).

18. Haque, M.A. et al. Neurological manifestations of puffer fish poisoning and it's outcome: Study of 83 cases. *J. Teachers Assoc.* 21, 121–125 (2009).

19. Islam, Q. et al. Puffer fish poisoning in Bangladesh: Clinical and toxicological results from large outbreaks in 2008. *Trans. R. Soc. Trop. Med. Hyg.* 105, 74–80 (2011).

20. Laobhripatr, S. et al. Food poisoning due to consumption of the freshwater puffer *Tetraodon fangi* in Thailand. *Toxicon* 28, 1372–1375 (1990).

21. Yang, C.C., Liao, S.C., and Deng, J.F. Tetrodotoxin poisoning in Taiwan: An analysis of poison center data. *Vet. Hum. Toxicol.* 38, 282 (1996).

22. Ahn, S.K. et al. Clinical analysis of puffer fish poisoning. *J. Korean Soc. Emerg. Med.* 10, 447–455 (1999).

23. Hyun, S.H. et al. Clinical analysis of puffer fish poisoning cases. *J. Korean Soc. Clin. Toxicol.* 9, 95–100 (2011).

24. Chua, H. and Chew, L. Puffer fish poisoning: A family affair. *Med. J. Malaysia* 64, 181 (2009).

25. Ellis, R. and Jelinek, G.A. Never eat an ugly fish: Three cases of tetrodotoxin poisoning from Western Australia. *Emerg. Med.* 9, 136–142 (1997).

26. Kiernan, M.C. et al. Acute tetrodotoxin-induced neurotoxicity after ingestion of puffer fish. *Ann. Neurol.* 57, 339–348 (2005).

27. Pocchiari, F. Trade of misbranded frozen fish: Medical and public health implications. *Ann. Ist. Super. Sanita* 13, 767 (1977).

28. Fernández-Ortega, J.F. et al. Seafood intoxication by tetrodotoxin: First Case in Europe. *J. Emerg. Med.* 39, 612–617 (2010).

29. Bentur, Y. et al. *Lessepsian* migration and tetrodotoxin poisoning due to *Lagocephalus sceleratus* in the eastern Mediterranean. *Toxicon* 52, 964–968 (2008).

30. Tanner, P. et al. Tetrodotoxin poisoning associated with eating puffer fish transported from Japan–California, 1996. *J. Am. Med. Assoc.* 275, 1631 (1996).

31. Ahmed, F.E. *Seafood Safety* (The National Academies Press, Washington, DC, 1991).

32. Landsberg, J.H. et al. Saxitoxin puffer fish poisoning in the United States, with the first report of *Pyrodinium bahamense* as the putative toxin source. *Environ. Health Perspect.* 114, 1502 (2006).

33. Silva, C.C.P. et al. Clinical and epidemiological study of 27 poisonings caused by ingesting puffer fish (Tetrodontidae) in the states of Santa Catarina and Bahia, Brazil. *Rev. Inst. Med. Trop. Sao Paulo* 52, 51–56 (2010).

34. Sierra-Beltran, A. et al. An overview of the marine food poisoning in Mexico. *Toxicon* 36, 1493–1502 (1998).

35. Kanchanapongkul, J. Tetrodotoxin poisoning following ingestion of the toxic eggs of the horseshoe crab *Carcinoscorpius rotundicauda*, a case series from 1994 through 2006. *Southeast Asian J. Trop. Med. Public Health* 39, 303–306 (2008).

36. Hwang, P.A. et al. Identification of tetrodotoxin in a marine gastropod (*Nassarius glans*) responsible for human morbidity and mortality in Taiwan. *J. Food Prot.* 68, 1696–1701 (2005).

37. Shui, L.M. et al. Tetrodotoxin-associated snail poisoning in Zhoushan: A 25-year retrospective analysis. *J. Food Prot.* 66, 110–114 (2003).

38. Cavazzoni, E. et al. Blue-ringed octopus (*Hapalochlaena* sp.) envenomation of a 4-year-old boy: A case report. *Clin. Toxicol.* 46, 760–761 (2008).

39. Yotsu-Yamashita, M., Mebs, D., and Flachsenberger, W. Distribution of tetrodotoxin in the body of the blue-ringed octopus (*Hapalochlaena maculosa*). *Toxicon* 49, 410–412 (2007).

40. Flachsenberger, W.A. Respiratory failure and lethal hypotension due to blue-ringed octopus and tetrodotoxin envenomation observed and counteracted in animal models. *Clin. Toxicol.* 24, 485–502 (1986).

41. Chew, S. et al. Puffer fish (tetrodotoxin) poisoning: Clinical report and role of anti-cholinesterase drugs in therapy. *Singapore Med. J.* 24, 168–171 (1983).

42. Sims, J.K. and Ostman, D.C. Pufferfish poisoning: Emergency diagnosis and management of mild human tetrodotoxication. *Ann. Emerg. Med.* 15, 1094–1098 (1986).

43. Zimmer, T. Effects of tetrodotoxin on the mammalian cardiovascular system. *Mar. Drugs* 8, 741–762 (2010).

44. Narahashi, T. Pharmacology of tetrodotoxin. *Toxin Rev.* 20, 67–84 (2001).

45. Lipkind, G.M. and Fozzard, H.A. A structural model of the tetrodotoxin and saxitoxin binding site of the Na+ channel. *Biophys. J.* 66, 1–13 (1994).

46. Shiomi, K. et al. Occurrence of tetrodotoxin-binding high molecular weight substances in the body fluid of shore crab (*Hemigrapsus sanguineus*). *Toxicon* 30, 1529–1537 (1992).

47. Nakamura, M., Oshima, Y., and Yasumoto, T. Occurrence of saxitoxin in puffer fish. *Toxicon* 22, 381–385 (1984).

48. Kawatsu, K., Shibata, T., and Hamano, Y. Application of immunoaffinity chromatography for detection of tetrodotoxin from urine samples of poisoned patients. *Toxicon* 37, 325–333 (1999).

49. Nzoughet, J.K. et al. Comparison of sample preparation methods, validation of an UPLC-MS/MS procedure for the quantification of tetrodotoxin present in marine gastropods and analysis of pufferfish. *Food Chem* 136, 1584–1589 (2013).

50. McNabb, P. et al. Detection of tetrodotoxin from the grey side-gilled sea slug—*Pleurobranchaea maculata*, and associated dog neurotoxicosis on beaches adjacent to the Hauraki Gulf, Auckland, New Zealand. *Toxicon* 56, 466–473 (2010).

51. Shoji, Y. et al. Electrospray ionization mass spectrometry of tetrodotoxin and its analogs: Liquid chromatography/mass spectrometry, tandem mass spectrometry, and liquid chromatography/tandem mass spectrometry. *Anal. Biochem.* 290, 10–17 (2001).

52. Kai Chew, S. et al. Anticholinesterase drugs in the treatment of tetrodotoxin poisoning. *Lancet* 324, 108 (1984).

53. Tambyah, P. et al. Central-nervous-system effects of tetrodotoxin poisoning. *Lancet* 343, 538–539 (1994).

54. Tibballs, J. Severe tetrodotoxic fish poisoning. *Anaesth. Intensive Care* 16, 215–217 (1988).

55. Xu, Q.-H. et al. Immunologic protection of anti-tetrodotoxin vaccines against lethal activities of oral tetrodotoxin challenge in mice. *Int. Immunopharmacol.* 5, 1213–1224 (2005).

56. Matsumura, K. In vivo neutralization of tetrodotoxin by a monoclonal antibody. *Toxicon* 33, 1239–1241 (1995).

57. Rivera, V.R., Poli, M.A., and Bignami, G.S. Prophylaxis and treatment with a monoclonal antibody of tetrodotoxin poisoning in mice. *Toxicon* 33, 1231–1237 (1995).

58. Daly, J. et al. First occurrence of tetrodotoxin in a dendrobatid frog (*Colostethus inguinalis*), with further reports for the bufonid genus *Atelopus*. *Toxicon* 32, 279 (1994).

59. Pires, O. et al. Further report of the occurrence of tetrodotoxin and new analogues in the Anuran family Brachycephalidae. *Toxicon* 45, 73–79 (2005).

60. Pires, O. et al. Occurrence of tetrodotoxin and its analogues in the Brazilian frog *Brachycephalus ephippium* (Anura: Brachycephalidae). *Toxicon* 40, 761–766 (2002).

61. Mebs, D. and Schmidt, K. Occurrence of tetrodotoxin in the frog *Atelopus oxyrhynchus*. *Toxicon* 27, 819 (1989).

62. Yotsu-Yamashita, M. and Mebs, D. Occurrence of 11-oxotetrodotoxin in the red-spotted newt, *Notophthalmus viridescens*, and further studies on the levels of tetrodotoxin and its analogues in the newt's efts. *Toxicon* 41, 893–897 (2003).

63. Brown, M.S. and Mosher, H.S. Tarichatoxin: Isolation and purification. *Science* 140, 295–296 (1963).

64. Nagashima, Y. et al. Occurrence of tetrodotoxin-related substances in the nontoxic puffer *Takifugu xanthopterus*. *Toxicon* 39, 415–418 (2001).

65. Yu, C. et al. Two novel species of tetrodotoxin-producing bacteria isolated from toxic marine puffer fishes. *Toxicon* 44, 641–647 (2004).

66. Wu, Z. et al. A new tetrodotoxin-producing actinomycete, *Nocardiopsis dassonvillei*, isolated from the ovaries of puffer fish *Fugu rubripes*. *Toxicon* 45, 851–859 (2005).

67. Wu, Z. et al. Toxicity and distribution of tetrodotoxin-producing bacteria in puffer fish *Fugu rubripes* collected from the Bohai Sea of China. *Toxicon* 46, 471–476 (2005).

68. Lee, M. et al. A tetrodotoxin-producing *Vibrio* strain, LM-1, from the puffer fish *Fugu vermicularis radiatus*. *Appl. Environ. Microbiol.* 66, 1698 (2000).

69. Noguchi, T. et al. *Vibrio alginolyticus*, a tetrodotoxin-producing bacterium, in the intestines of the fish *Fugu vermicularis vermicularis*. *Mar. Biol.* 94, 625–630 (1987).

70. Yasumoto, T. et al. Bacterial production of tetrodotoxin and anhydrotetrodotoxin. *Agric. Biol. Chem.* 50, 793–795 (1986).

71. Yotsu, M. et al. Production of tetrodotoxin and its derivatives by *Pseudomonas* sp. isolated from the skin of a pufferfish. *Toxicon* 25, 225 (1987).

72. Noguchi, T. and Hashimoto, Y. Isolation of tetrodotoxin from a goby *Gobius criniger*. *Toxicon* 11, 305–307 (1973).

73. Hwang, D. et al. Tetrodotoxin-producing bacteria from the blue-ringed octopus *Octopus maculosus*. *Mar. Biol.* 100, 327–332 (1989).

74. Sheumack, D., Howden, M., and Spence, I. Occurrence of a tetrodotoxin-like compound in the eggs of the venomous blue-ringed octopus (*Hapalochlaena maculosa*). *Toxicon* 22, 811 (1984).

75. Jeon, J. et al. Occurrence of tetrodotoxin in a gastropod mollusk, "araregai" *Niotha clathrata*. *Bull. Jpn. Soc. Sci. Fish.* 50, 2099–2102 (1984).

76. Wang, X. et al. Toxin-screening and identification of bacteria isolated from highly toxic marine gastropod *Nassarius semiplicatus*. *Toxicon* 52, 55–61 (2008).

77. Hwang, D., Lu, S., and Jeng, S. Occurrence of tetrodotoxin in the gastropods *Rapana rapiformis* and *R. venosa venosa*. *Mar. Biol.* 111, 65–69 (1991).

78. Narita, H. et al. Occurrence of tetrodotoxin in a trumpet shell, "Boshubora" *Charonia sauliae*. *Bull. Jpn. Soc. Sci. Fish.* 47, 935–941 (1981).

79. Noguchi, T. et al. Occurrence of tetrodotoxin in the Japanese ivory shell *Babylonia japonica*. *Bull. Jpn. Soc. Sci. Fish.* 47, 901–913 (1981).

80. Shih, Y. et al. Occurrence of tetrodotoxin in the causative gastropod *Polinices didyma* and another gastropod *Natica lineata* collected from western Taiwan. *J. Food Drug Anal.* 11, 159–163 (2003).

81. Noguchi, T., Maruyama, J., Narita, H., and Hashimoto, K. Occurrence of tetrodotoxin in the gastropod mollusk *Tutufa lissostoma* (frog shell). *Toxicon* 22, 219 (1984).

82. Carroll, S., McEvoy, E., and Gibson, R. The production of tetrodotoxin-like substances by nemertean worms in conjunction with bacteria. *J. Exp. Mar. Biol. Ecol.* 288, 51–63 (2003).

83. Maruyama, J. et al. Occurrence of tetrodotoxin in the starfish *Astropecten latespinosus*. *Cell. Mol. Life Sci.* 40, 1395–1396 (1984).

84. Miyazawa, K. et al. Occurrence of tetrodotoxin in the starfishes *Astropecten polyacanthus* and *A. scoparius* in the Seto Inland Sea. *Mar. Biol.* 90, 61–64 (1985).

85. Thuesen, E. and Kogure, K. Bacterial production of tetrodotoxin in four species of Chaetognatha. *Biol. Bull.* 176, 191 (1989).

86. Dao, H. et al. Frequent occurrence of the tetrodotoxin-bearing horseshoe crab *Carcinoscorpius rotundicauda* in Vietnam. *Fish. Sci.* 75, 435–438 (2009).

87. Kungsuwan, A. et al. Tetrodotoxin in the horseshoe crab *Carcinoscorpius rotundicauda* inhabiting Thailand. *Nippon Suisan Gakk.* 53, 261–266 (1987).

88. Tsai, Y. et al. Occurrence of tetrodotoxin and paralytic shellfish poison in the Taiwanese crab *Lophozozymus pictor*. *Toxicon* 33, 1669–1673 (1995).

89. Miyazawa, K. et al. Occurrence of tetrodotoxin in the flatworm *Planocera multitentaculata*. *Toxicon* 24, 645–650 (1986).

90. Kodama, M. et al. Occurrence of tetrodotoxin in *Alexandrium tamarense*, a causative dinoflagellate of paralytic shellfish poisoning. *Toxicon* 34, 1101–1105 (1996).

91. Narita, H. et al. *Vibrio alginolyticus*, a TTX-producing bacterium isolated from the starfish *Astropecten polyacanthus*. *Nippon Suisan Gakk.* 53, 617–621 (1987).

92. Do, H., Kogure, K., and Simidu, U. Identification of deep-sea-sediment bacteria which produce tetrodotoxin. *Appl. Environ. Microbiol.* 56, 1162 (1990).

93. Do, H. et al. Presence of tetrodotoxin and tetrodotoxin-producing bacteria in freshwater sediments. *Appl. Environ. Microbiol.* 59, 3934 (1993).

94. Do, H. et al. Tetrodotoxin production of actinomycetes isolated from marine sediment. *J. Appl. Microbiol.* 70, 464–468 (1991).

95. Ritchie, K., Nagelkerken, I., James, S., and Smith, G. Environmental microbiology: A tetrodotoxin-producing marine pathogen. *Nature* 404, 354 (2000).

96. Cheng, C.A. et al. Microflora and tetrodotoxin-producing bacteria in a gastropod, *Niotha clathrata*. *Food Chem. Toxicol.* 33, 929–934 (1995).

97. Yan, Q., Yu, P., and Li, H. Detection of tetrodotoxin and bacterial production by *Serratia marcescens*. *World J. Microbiol. Biotechnol.* 21, 1255–1258 (2005).

98. Matsui, T. et al. Production of tetrodotoxin by the intestinal bacteria of a puffer fish *Takifugu niphobles*. *Nippon Suisan Gakk.* 55, 2199–2203 (1989).

Section V

Microbes Affecting Animals: Viruses

49 African Horse Sickness Virus

Dongyou Liu and Frank W. Austin

CONTENTS

49.1 INTRODUCTION

African horse sickness virus (AHSV) is the causative agent of African horse sickness (AHS), which is a highly infectious and deadly disease of horses, ponies, and European donkeys, as well as mules, camels, goats, and buffalo. The virus is also capable of infecting carnivores such as dogs that ingest contaminated meat. However, African donkeys and zebra are largely refractory to the devastating consequences of AHS and rarely show clinical signs.

Clinically, AHS is characterized by pyrexia, inappetence, edema of subcutaneous and intermuscular tissues and of the lungs, transudation into the body cavities, and hemorrhages (particularly of the serosal surfaces), leading to impaired respiratory and vascular functions and 90% mortality in infected horses.

Although a disease resembling AHS was first recorded in 1327 in Yemen, with the natural zebra population acting as reservoir host, the disease was reported in imported horses in Central and East Africa in 1569 and in southern Africa around 1710s. The first major outbreak of AHS in 1719 claimed the lives of about 1700 animals. Since then, greater than 10 additional outbreaks of AHS were documented on the African continent. Along with the decline in the wild zebra population due to hunting throughout the last century and the introduction of AHS vaccines, the incidence of AHS in southern Africa has been steadily decreasing. Currently, the endemic zones of AHS are restricted to sub-Saharan Africa, Yemen, and the Arabian Peninsula, but the disease has periodically emerged in other parts of the world, including India, Pakistan, Spain (1987 to 1990), and Portugal (1989) [1].

While not directly contagious, AHS is known to be transmitted by insect vectors, such as *Culicoides* midges (particularly *Culicoides* [*Avaritia*] *imicola* and *Culicoides* [*Avaritia*] *bolitinos*), mosquitoes, black flies, sand flies, or ticks. Given the potential of *AHSV* to cause significant mortality and debilitating morbidity in naïve equid populations and its capacity for sudden and rapid expansion, AHS is regarded as a notifiable disease by the World Organisation for Animal Health (OIE) and is included on the list of select agents of the United States Department of Agriculture (USDA).

49.2 CLASSIFICATION AND MORPHOLOGY

49.2.1 CLASSIFICATION

AHSV is a double-stranded RNA virus that belongs to the genus *Orbivirus*, subfamily Sedoreovirinae, family Reoviridae. As one of the two subfamilies (the other being Spinareovirinae) in the family Reoviridae, Sedoreovirinae is divided into six genera: *Cardoreovirus, Mimoreovirus, Orbivirus, Phytoreovirus, Rotavirus,* and *Seadornavirus.* In turn, the genus *Orbivirus* is separated into at least 17 species: *AHSV, Bluetongue virus (BTV), Changuinola virus, Corriparta virus, Epizootic hemorrhagic disease virus (EHDV), Equine encephalosis virus (EEV), Eubenangee virus, Great Island virus, Orungo virus, Palyam virus, Peruvian horse sickness virus (PHSV), St Croix River virus, Umatilla virus, Wallal virus, Warrego virus, Wongorr virus,* and *Yunnan orbivirus* [2]. Of these, *AHSV, BTV, EHDV, EEV, Palyam virus,* and *PHSV* are highly pathogenic and economically important. With the exception of *PHSV* that is likely spread by mosquitoes, most orbiviruses are transmitted by adult females of hematophagous *Culicoides* midges.

At present, nine serotypes of *AHSV* (*AHSV*-1 to *AHSV*-9) are recognized, with the last serotype being identified in 1960. Some cross-relatedness appears to exist between *AHSV* serotypes, notably between serotypes 1 and 2, 3 and 7, 5 and 8, and 6 and 9. Whereas serotypes 1 to 8 are found only in restricted areas of sub-Saharan Africa, serotype 9 has been associated with epidemics outside Africa, with the exception of Spanish–Portuguese outbreaks between 1987 and 1990, which were caused by *AHSV*-4 [3].

42.2.2 Morphology

AHSV virion is an unenveloped particle of about 70 nm in diameter. It is composed of an outer two-layered icosahedral capsid and a core that encloses the genome consisting of 10 double-stranded RNA segments [4,5].

The outer capsid is composed of two viral proteins (VP2 and VP5), with VP5 trimers forming globular domains and VP2 trimers in a three entwined leg formation (triskelions) on the virion surface. VP5 is positioned between the peripentonal VP2 molecules and around VP7 on the threefold axis of symmetry.

The core (inner capsid) comprises two major viral proteins, VP3 and VP7, which are conserved among the nine *AHSV* serotypes and make up the group-specific epitopes, and three minor viral proteins, VP1 (polymerase), VP4 (capping enzyme), and VP6 (helicase). VP3 (a flat, triangular molecule organized in 60 asymmetric dimers) forms the innermost protein layer, with the viral polymerase, capping enzyme, and helicase attached to the inner surface of the VP3 layer, forming a flower-like transcriptase complex. VP7 (780 monomers arranged as 260 trimers) forms a second layer on top of VP3.

The 19.52 kb genome of *AHSV* is composed of 10 double-stranded (ds)RNA segments of different sizes, with segments 1–3 (of 3965, 3203, 2792 bp) being designated L1–L3 (large), segments 4–6 (of 1978, 1748, 1566 bp) designated M4–M6 (medium), and segments 7–10 (of 1179, 1166, 1169, and 758 bp) designated S7–S10 (small). While the segments (L3, S7) coding for the core proteins (VP3, VP7) as well as segments (M5, S8) coding for NS1 and NS2 proteins are highly conserved between the different *AHSV* serotypes, the segment (S10) encoding NS3 and NS3a and the segments (L2 and M6) encoding the outer capsid proteins (VP2 and VP5) are more variable between the virus serotypes [6–8].

42.2.3 Viral Proteins

The 10 segments of *AHSV* genome encode 7 structural proteins (VP1–VP7) and 4 nonstructural proteins (NS1, NS2, NS3, and NS3a), with NS3 and NS3a being synthesized from two inphase overlapping reading frames of segment 10 (S10) [4,9,10].

VP1: Encoded by segment 1 (L1), VP1 is an RNA-dependent RNA polymerase, whose activity is highly dependent upon temperature, with an optimum between 27°C and 42°C and little or no activity below 12°C–15°C. This property permits *AHSV* undertaking viral RNA synthesis and replication within the insect vectors (ectothermic or ambient temperature) and mammalian hosts (about 37°C). As one of the core structural proteins, VP1 is very highly conserved and is unlikely to play a significant role in the specific adaptation of the virus to vectorial transmission [11].

VP2: Encoded by segment 2 (L2), VP2 (124 kDa) is susceptible to proteolytic cleavages (by equine serum proteases and the saliva of the insect host in vivo), which increase the infectivity of the virus in *Culicoides* insect. This may be attributable to a change in VP2-mediated receptor binding or increased efficiency of uncoating and exposure of VP7 during the early stages of infection and cell entry [12]. Similarly, cleavage of VP2 in the *AHSV* outer capsid by trypsin or chymotrypsin generates infectious subviral particles (ISVPs) that demonstrate unchanged infectivity for mammalian cells but greatly enhanced infectivity for *Culicoides sonorensis* cells. Components (VP2 and VP5) of the outer capsid layer are released from the *AHSV* core below pH 6, an event that likely occurs in vivo within endosomes during the processes of infection and cell entry. This results in the activation of the transcriptase activity of the virus core, which is released into the host cell cytoplasm. Therefore, besides its involvement in the initial cell attachment, the activated VP2 also contributes to virus entry in vivo. As a host specificity determinant, VP2 shows remarkable antigenic variability due to selective pressure from the host immune system. This has been exploited for the purpose of serotype determination. Additionally, as a key immunogen and an important target for virus neutralization, VP2 is capable of inducing a protective immune response. VP2 contains 15 antigenic sites, with a major antigenic domain containing 12 of the 15 sites located in the region between residues 223 and 400 and a second domain containing the three remaining sites between residues 568 and 681. Based on in vitro experiments, 3 of the 15 sites are capable of inducing neutralizing antibodies, and two neutralizing epitopes have been defined, "a" and "b," between residues 321 and 339 and 377 and 400, respectively. A combination of peptides representing both sites induces a more effective neutralizing response [5,13,14].

VP3: Encoded by segment 3 (L3), VP3 is a flat, triangular molecule of about 103 kDa that forms the innermost protein layer (with 60 asymmetric dimers) of *AHSV* core (inner capsid).

VP4: Encoded by segment 4 (M4), VP4 is a minor core protein of 642 amino acids (75 kDa) that functions as a capping enzyme.

VP5: Encoded by segment 6 (M6), VP5 (57 kDa) is a major structural protein forming the outer two-layered capsid with VP2. Containing coiled-coil motifs typical of membrane fusion protein, VP5 plays a part in membrane penetration during cell entry as well as introduction of *AHSV*-specific neutralizing antibodies. The most immunodominant region of VP5 is located in the N-terminal 330 residues, defining two antigenic regions (residues 151–200 and residues

83–120), which include 8 antigenic sites, with neutralizing epitopes situated at positions 85–92 (PDPLSPGE) and 179–185 (EEDLRTR) [15,16].

VP6: Encoded by segment 9 (S9), VP6 is a minor structural protein with putative helicase activity. In vitro experiments showed that VP6 binds double- and single-stranded RNA and DNA in an NaCl concentration-sensitive reaction. The binding activity appears to vary with the pH of the binding buffer (ranging from pH 6.0 to 8.0).

VP7: Encoded by segment 7 (S7), VP7 is a 38 kDa molecule (with 780 monomers organized as 260 trimers) forming a second (surface) layer on top of VP3 in the *AHSV* core (inner capsid) [17,18]. Being highly conserved among different serotypes, VP7 may assist cell attachment and virus entry in insect systems, through involvement of a distinct cell surface receptor from that used by VP2 [19].

NS1: Encoded by segment 5 (M5), NS1 forms tubular structures within the cell cytoplasm, which are characteristically produced during virus replication in infected cells. These tubules have an average diameter of 23 ± 2 nm (which is less than half the width of the corresponding *BTV* tubules) and are more fragile at high salt concentrations or pH. NS1 protein is involved in cell lysis, which is the main mode of virus exit from infected mammalian cells and spreading to other cells.

NS2: Encoded by segment 8 (S8), NS2 is a major component of the granular viral inclusion bodies that form within the cytoplasm of infected cells. Sedimentation analysis and nonreducing SDS-PAGE revealed that NS2 is a 7S multimer with both inter- and intramolecular disulfide bonds, probably consisting of six or more NS2 molecules. The 7S NS2 multimer binds ssRNA nonspecifically.

NS3: Encoded by segment 10 (S10), NS3 is a proline-rich protein with strict sequence conservation and two hydrophobic domains (HD1 and HD2), which are linked to NS3 cytotoxicity. Variations in NS3 are found in HD1 and the adjacent variable region between HD1 and HD2. NS3 is associated with the membrane of infected cells and relatively exposed to the host immune system; the latter of which is responsible for its high variability may be partially in attempt to avoid immune recognition. This variability corresponds to genetic differences in the insects that the virus infects. NS3 has the characteristics of lytic viral proteins that modify membrane permeability and contribute to viral pathogenesis. The modification of four amino acids (aa 165–168) from hydrophobic to charged amino residues in a predicted transmembrane region of NS3 alters its ability to interact (anchor) with cellular membrane components and largely abolishes its cytotoxic toxicity [20]. NS3 protein mediates virus budding, which is vital for virus exit from infected insect cells. This is reflected by the fact that NS3 is expressed at higher levels in insect cells than mammalian cells [21–23].

NS3a: Also encoded by segment 10 (S10), but via another inphase overlapping reading frame (compared to NS3), NS3a is a truncated protein localized in areas of plasma membrane disruption and associated with events of viral release from cells [24].

49.3 BIOLOGY AND EPIDEMIOLOGY

49.3.1 BIOLOGY

The establishment of *AHSV* infection and subsequent pathogenic changes in susceptible hosts may likely involve the following series of events: (1) proteolytic cleavage of VP2 in the serum or in midge saliva, (2) interaction with the host cell receptor, (3) entry through an endocytic pathway, (4) activation of VP5 at low pH leading to exposure of the fusion peptide and subsequent insertion into the endosomal membrane, (5) release of the double-shelled core into the cytoplasm, (6) multiplication in the regional lymph nodes, (7) spreading to pulmonary microvascular endothelial cells, (8) breaking up of the endothelial cell barrier (increased vascular permeability resulting in edema, effusion, and hemorrhage), (9) dissemination via the bloodstream (to lungs, spleen, other lymphoid tissues, choroid plexus, pharynx), and (10) replication in these organs producing a secondary viremia. The duration from initial infection to secondary viremia may vary between 2 and 21 days, but it usually takes <9 days.

Being acid sensitive, *AHSV* is inactivated at pH of <6.0 or ≥12.0, but remains relatively stable at pH 7.0–8.5. The virus is vulnerable to treatment with formalin, β-propiolactone, acetylethyleneimine derivatives, or radiation; however, it is resistant to lipid solvents and heat. In the presence of stabilizers such as serum, sodium oxalate, carbolic acid, and glycerine, its infectivity is remarkably stable at 4°C. The virus shows minimal loss of titer after lyophilization or storage at –70°C in Parker Davis medium, but it is labile between –20°C and –30°C [25].

49.3.2 EPIDEMIOLOGY

The natural reservoir host for *AHSV* is zebras (*Equus burchelli*), which display a detectable viremia for up to 40 days postinfection without visible clinical symptoms, while virus could be recovered from tissues and cells for up to 48 days. On the other hand, horses are highly susceptible to *AHSV* infection, with a short period of viremia (4–8 days) and high mortality rates (frequently exceeding 90%). Donkeys typically display a viremia for up to 4 weeks and a mortality rate of around 10% [26]. Other susceptible mammals include mules, asses, elephants, camels, and dogs (after eating infected horsemeat) [27].

The biological vector of *AHSV* is the *Culicoides* (midges), especially *Culicoides imicola*, which measures 1–3 mm, preferring warm, humid conditions, and which is also a transmitting vector for Akabane, bovine ephemeral fever, bluetongue, epizootic hemorrhagic disease, equine encephalosis, Oropouche, and the Palyams. *Culicoides variipennis* (*sonorensis*) (the American *BTV* vector), *Culicoides obsoletus* and *Culicoides pulicaris* (the northern European *BTV* vectors), and *Culicoides bolitinos* (present at higher altitudes in Africa, where *C. imicola* is found) may also have the capacity to transmit *AHSV*. In addition, *AHSV* may also be transmitted by mosquitoes (e.g., *Aedes aegypti*, *Anopheles stephensi*, and *Culex pipiens*) and ticks (e.g., camel tick

Hyalomma dromedarii and dog tick *Rhipicephalus sanguineus*) [28–32]. *AHSV* survives the winter (overwintering) in the adult vector population, via transovarial transmission (vertical transmission of *AHSV* from adult females to their eggs) or retention in adult *Culicoides* that survive the winter drop in temperatures. The occurrence of AHS is often preceded by seasons of heavy rain that alternate with hot and dry climatic conditions.

AHSV serotypes 1–9 are endemic to Eastern and Southern Africa, while serotype 9 predominates the northern parts of sub-Saharan Africa. During 1950–1960, *AHSV* serotype 9 (*AHSV*-9) emerged in the Middle East (Iran, Iraq, Syria, Lebanon, and Jordan), then Cyprus, Afghanistan, Pakistan, India, and Turkey as well as Libya, Tunisia, Algeria, Morocco, and Spain [33,34]. *AHSV* serotype 4 (*AHSV*-4) was reported in Spain (1988–1990), Portugal (1989), and Morocco (1989–1991) following the importation of a number of subclinically infected zebra from Namibia [35–40]. However, AHS has not been reported in the Americas, eastern Asia, or Australasia.

49.4 CLINICAL FEATURES AND PATHOGENESIS

49.4.1 CLINICAL FEATURES

AHSV infections induce four forms of clinical diseases of various severities in horses: acute (pulmonary), subacute (cardiac), mixed, or febrile (horse sickness fever) forms [41].

Pulmonary form. This acute form of AHS is characterized by a short incubation period (3–5 days), high fever (39–41°C, which usually corresponds with the onset of viremia), depression, respiratory distress, dyspnea, spasmodic coughing, dilated nostrils, and sweating. Standing with its legs apart and head extended, the affected animal experiences breathing difficulty (due to lung congestion), coughs frothy fluid from nostril and mouth, and displays signs of pulmonary edema. As the most serious form of AHS, its mortality rate exceeds 95%, with severely affected horses succumbing to the disease (anoxia) within 24 h of infection. The pulmonary form of AHS is also the form most usually seen in dogs. The virus is mainly localized in the cardiovascular and lymphatic system (which is also the case with the cardiac and mixed forms). At necropsy, the lungs are distended and heavy, with edema being visible in the intralobular space, and frothy fluid is present in the trachea, bronchi, and bronchioles. Also observed are edematous thoracic lymph nodes, congested gastric fundus and abdominal viscera, and petechiae in the pericardium.

Cardiac form. This subacute form of AHS has an incubation period longer than that of the pulmonary form, with clinical signs of disease appearing at days 7–12 after infection. The disease is usually characterized by high fever; progressive dyspnea; edema of the head, neck, chest, and supraorbital fossae; petechial hemorrhages in the eyes; ecchymotic hemorrhages on the tongue; and colic (abdominal pain), leading to notable swelling in the supraorbital fossae, eyelids, facial

tissues, neck, thorax, brisket, and shoulders. The mortality rate is between 50% and 70% and surviving animals recover in 7 days. Postmortem examination reveals prominent petechiae and ecchymoses on the epicardium and endocardium; flaccid or slightly edematous lungs; yellow, gelatinous infiltrations of the subcutaneous and intramuscular tissues (especially along the jugular veins and ligamentum nuchae); and other lesions (e.g., hydropericardium, myocarditis, hemorrhagic gastritis, and petechiae on the ventral surface of the tongue and peritoneum).

Mixed form. The mixed form is often the most common form of the disease, with affected horses showing signs of both the pulmonary and cardiac forms of AHS. The mortality rate may exceed 70%, and death often occurs within 3–6 days.

Mild form. The mild (febrile) form (or horse sickness fever) of AHS is a subclinical disease that is seen in zebras and African donkeys as well as mules and partially immune horses. Infected animals may display mild to moderate fever, congested mucous membrane, and some edema of the supraorbital fossae despite having high virus titers in blood. The survival rate is 100%. The virus is concentrated in the spleen and almost absent from the lungs and heart. Interestingly, the virus isolated from the spleen appears to be less virulent than that from the lungs (as in the case of pulmonary and cardiac forms of the disease).

49.4.2 PATHOGENESIS

In a single infection cycle, *AHSV* has to pass through the skin of the equid host twice. Besides forming a mechanical barrier, the skin also represents an immunologically active organ that is capable of initiating an immune response against an invading pathogen. To mitigate the damaging aspects of host response in the skin, arthropod insects such as *Culicoides* secrete saliva with components that inhibit the phagocytic activity of host macrophages at the biting site, suppress the host's inflammatory reactions, prevent blood coagulation, and promote viral replication (the so-called saliva activated transmission) [42].

Given its predilection for the vascular endothelial cells throughout the body, *AHSV* increases vascular permeability, enhances effusions into body cavities and tissues, and causes widespread hemorrhages [43]. The appearance of fever in an infected host may disrupt the barriers to viral replication and dissemination and invite additional insect feeding due to the elevated temperature. The induction of severe clinical signs in the host by *AHSV* will incapacitate the host to mount effective defense against vector attack [44].

49.5 IDENTIFICATION AND DIAGNOSIS

Presumptive diagnosis of AHS is often made on the basis of characteristic clinical signs, postmortem lesions, and presence of competent vectors. Due to its nonspecific clinical signs and macroscopic lesions that can be confused with those caused by *EEV*, purpura hemorrhagica, equine viral arthritis, and babesiosis (*Babesia equi* and *Babesia caballi*), a definitive diagnosis of AHS has traditionally depended on the isolation and identification of the virus from clinical specimens. These include whole blood collected in anticoagulant (preferably EDTA) during the febrile stage of infection and the tissues of the spleen, lung, lymph nodes, and salivary glands collected from newly dead animals. Blood may be preserved in Oxalate/Glycerol (OCG) solution (50% glycerol, 0.5% potassium oxalate, 0.5% phenol), and spleen samples may be preserved in 10% buffered glycerol. Usually, whole blood is washed and lysed and tissues ground in a mortar with sterile sand before inoculation in embryonated hens eggs, 2–4-day-old suckling mice, or in vitro cell culture (e.g., baby hamster kidney [BHK], African green monkey [Vero], monkey kidney [MS], and *Culicoides* [KC] cells) for virus isolation [45].

Further confirmation of AHS is achieved by using immunological techniques such as antigen capture ELISA, complement fixation, agar gel immunodiffusion, immunofluorescence, and virus neutralization for identification and serotyping [46–48]. In recent years, molecular techniques such as PCR have been shown to be valuable for direct detection of viral RNA in whole blood and other tissues within a result turnover time of 24 h [49–58].

49.6 TREATMENT AND PREVENTION

No specific treatment is currently available for animals suffering from AHS, and rest and good husbandry represent the best approach for their recovery. As *AHSV* is only spread via the bites of infected vectors, control and prevention of the disease should center on (1) restricting the movement of infected animals, (2) slaughter of viremic animals, (3) husbandry modification (e.g., installing sand fly netting/mesh on windows and doors), (4) vector control (e.g., use of insecticides such as synthetic pyrethroids, ivermectin, tetrachlorvinphos, 5% temephos granulated with gypsum), and (5) vaccination (including use of polyvalent, monovalent, and monovalent inactivated vaccines) [59–75].

49.7 CONCLUSION AND FUTURE PERSPECTIVES

AHS is an economically important, noncontagious but infectious disease caused by a double-stranded RNA virus—*AHSV*. Transmitted by *Culicoides* midges, the disease severely affects horses, ponies, and European donkeys as well as mules, but has little effect on African donkeys and zebra. Although *AHSV* is largely confined to sub-Saharan Africa, Yemen, and the Arabian Peninsula, it has the capacity for sudden and rapid excursions and causes significant mortality in naïve horse populations in other parts of the world. It is therefore critical that rapid, sensitive, and specific laboratory methods are available for prompt identification and typing of *AHSV* in case of unexpected AHS outbreaks.

In addition, the development of effective vaccines against this deadly disease will also help prevent the serious consequence of a sudden epidemic due to *AHSV*.

REFERENCES

1. Mellor PS and Mertens PPC. African horse sickness viruses. *Encycl. Virol.* 2008;37–43.
2. Taxonomy-UniProt. http://www.uniprot.org/taxonomy/10892. (accessed on June 30, 2013).
3. Calisher CH and Mertens PP. Taxonomy of African horse sickness viruses. *Arch. Virol. Suppl.* 1998;14:3–11.
4. Roy P, Mertens PP, and Casal I. African horse sickness virus structure. *Comp. Immunol. Microbiol. Infect. Dis.* 1994;17:243–273.
5. Manole V et al. Structural insight into African horse sickness virus infection. *J. Virol.* 2012;86:7858–7866.
6. Bremer CW, Huismans H, and Van Dijk AA. Characterization and cloning of the African horse sickness virus genome. *J. Gen. Virol.* 1990;71:793–799.
7. Williams CF et al. The complete sequence of four major structural proteins of African horse sickness virus serotype 6: Evolutionary relationships within and between the orbiviruses. *Virus Res.* 1998;53:53–73.
8. Matsuo E, Celma CC, and Roy P. A reverse genetics system of African horse sickness virus reveals existence of primary replication. *FEBS Lett.* 2010;584:3386–3391.
9. Grubman M and Lewis S. Identification and characterisation of the structural and non-structural proteins of African horse sickness virus and determination of the genome coding assignments. *Virology* 1992;186:444–451.
10. Huismans H et al. A comparison of different orbivirus proteins that could affect virulence and pathogenesis. *Vet. Ital.* 2004;40:417–425.
11. Vreede FT and Huismans H. Sequence analysis of the RNA polymerase gene of African horse sickness virus. *Arch. Virol.* 1998;143:413–419.
12. Marchi PR et al. Proteolytic cleavage of VP2, an outer capsid protein of African horse sickness virus, by species-specific serum proteases enhances infectivity in *Culicoides. J. Gen. Virol.* 1995;76:2607–2611.
13. Burrage TG et al. Neutralizing epitopes of African horse sickness virus serotype 4 are located on VP2. *Virology* 1993;196:799–803.
14. Martinez-Torrecuadrada JL and Casal JI. Identification of a linear neutralization domain in the protein VP2 of African horse sickness virus. *Virology* 1995;210:391–399.
15. Martinez-Torrecuadrada JL et al. Antigenic profile of African horse sickness virus serotype 4 VP5 and identification of a neutralizing epitope shared with bluetongue virus and epizootic hemorrhagic disease virus. *Virology* 1999;257:449–459.
16. Stassen L, Huismans H, and Theron J. Membrane permeabilization of the African horse sickness virus VP5 protein is mediated by two N-terminal amphipathic α-helices. *Arch. Virol.* 2011;156:711–715.
17. Basak AK et al. Crystal structure of the top domain of African horse sickness virus VP7: Comparisons with bluetongue virus VP7. *J. Virol.* 1996;70:3797–3806.
18. Stassen L, Huismans H, and Theron J. Silencing of African horse sickness virus VP7 protein expression in cultured cells by RNA interference. *Virus Genes* 2007;35:777–783.
19. Rutkowska DA et al. The use of soluble African horse sickness viral protein 7 as an antigen delivery and presentation system. *Virus Res.* 2011;156:35–34.

20. Van Niekerk M et al. Membrane association of African horse sickness virus nonstructural protein NS3 determines its cytotoxicity. *Virology* 2001;279:499–508.

21. Stoltz MA et al. Subcellular localization of the nonstructural protein NS3 of African horse sickness virus. *Onderstepoort J. Vet. Res.* 1996;63:57–61.

22. Quan M et al. Molecular epidemiology of the African horse sickness virus S10 gene. *J. Gen. Virol.* 2008;89:1159–1168.

23. Meiring TL, Huismans H, and van Staden V. Genome segment reassortment identifies non-structural protein NS3 as a key protein in African horse sickness virus release and alteration of membrane permeability. *Arch. Virol.* 2009;154:263–271.

24. Martin LA et al. Phylogenetic analysis of African horse sickness virus segment 10: Sequence variation, virulence characteristics and cell exit. *Arch. Virol. Suppl.* 1998;14:281–293.

25. Wellby MP et al. Effect of temperature on survival and rate of virogenesis of African horse sickness virus in *Culicoides variipennis sonorensis* (Diptera: Ceratopogonidae) and its significance in relation to the epidemiology of the disease. *Bull. Entomol. Res.* 1996;86:715–720.

26. Hamblin C et al. Donkeys as reservoirs of African horse sickness virus. *Arch. Virol. Suppl.* 1998;14:37–47.

27. Alexander KA et al. African horse sickness and African carnivores. *Vet. Microbiol.* 1995;47:133–140.

28. Venter GJ, Graham S, and Hamblin C. African horse sickness epidemiology: Vector competence of South African *Culicoides* species for virus serotypes 3, 5 and 8. *Med. Vet. Entomol.* 2000;14:245–250.

29. Venter GJ, Koekemoer JJ, and Paweska JT. Investigations on outbreaks of African horse sickness in the surveillance zone in South Africa. *Rev. Sci. Tech.* 2006;25:1097–1109.

30. Venter GJ and Paweska JT. Virus recovery rates for wild-type and live-attenuated vaccine strains of African horse sickness virus serotype 7 in orally infected South African *Culicoides* species. *Med. Vet. Entomol.* 2007;21:377–383.

31. Venter GJ et al. The oral susceptibility of South African field populations of *Culicoides* to African horse sickness virus. *Med. Vet. Entomol.* 2009;23:367–378.

32. Meiswinkel R and Paweska JT. Evidence for a new field *Culicoides* vector of African horse sickness in South Africa. *Prev. Vet. Med.* 2003;60:243–253.

33. Zientara S et al. Molecular epidemiology of African horse sickness virus based on analyses and comparisons of genome segments 7 and 10. *Arch. Virol. Suppl.* 1998;14:221–234.

34. Aklilu N et al. African horse sickness outbreaks caused by multiple virus types in Ethiopia. *Transbound. Emerg. Dis.* 2012. DOI: 10.1111/tbed.12024.

35. Lubroth J. African horse sickness and the epizootic in Spain 1987. *Equine Pract.* 1988;10:26–33.

36. Rodriguez M, Hooghuis H, and Castano M. African horse sickness in Spain. *Vet. Microbiol.* 1992;33:129–142.

37. Mellor PS et al. Isolations of African horse sickness virus from vector insects made during the 1988 epizootic in Spain. *Epidemiol. Infect.* 1990;105:447–454.

38. Scacchia M et al. African horse sickness: A description of outbreaks in Namibia. *Vet. Ital.* 2009;45:255–264, 265–274.

39. Maclachlan NJ and Guthrie AJ. Re-emergence of blue tongue, African horse sickness, and other orbivirus diseases. *Vet. Res.* 2010;41:35.

40. Thompson GM, Jess S, and Murchie AK. A review of African horse sickness and its implications for Ireland. *Ir. Vet. J.* 2012;65:9.

41. Weyer CT et al. African horse sickness in naturally infected, immunised horses. *Equine Vet. J.* 2013;45:117–119.

42. Burrage TG and Laegreid WW. African horse sickness: Pathogenesis and immunity. *Comp. Immunol. Microbiol. Infect. Dis.* 1994;17:275–285.

43. Laegreid WW et al. Electron microscopic evidence for endothelial infection by African horse sickness virus. *Vet. Pathol.* 1992;29:554–556.

44. Wilson A et al. Adaptive strategies of African horse sickness virus to facilitate vector transmission. *Vet. Res.* 2009;40:16.

45. O'Hara RS et al. Development of a mouse model system and identification of individual genome segments of African horse sickness virus, serotypes 3 and 8 involved in determination of virulence. *Arch. Virol.* 1998;S14:259–279.

46. Clift SJ et al. Standardization and validation of an immunoperoxidase assay for the detection of African horse sickness virus in formalin-fixed, paraffin-embedded tissues. *J. Vet. Diagn. Invest.* 2009;21:655–667.

47. Clift SJ and Penrith ML. Tissue and cell tropism of African horse sickness virus demonstrated by immunoperoxidase labeling in natural and experimental infection in horses in South Africa. *Vet. Pathol.* 2010;47:690–697.

48. Bitew M et al. Serological survey of African horse sickness in selected districts of Jimma zone, Southwestern Ethiopia. *Trop. Anim. Health Prod.* 2011;43:1543–1547.

49. Sailleau C et al. African horse sickness in Senegal: Serotype identification and nucleotide sequence determination of segment S10 by RT-PCR. *Vet. Rec.* 2000;146:107–108.

50. Agüero M et al. Real-time fluorogenic reverse transcription polymerase chain reaction assay for detection of African horse sickness virus. *J. Vet. Diagn. Invest.* 2008;20:325–328.

51. Rodriguez-Sanchez B et al. Novel gel-based and real-time PCR assays for the improved detection of African horse sickness virus. *J. Virol. Methods* 2008;151:87–94.

52. Aradaib IE. PCR detection of African horse sickness virus serogroup based on genome segment three sequence analysis. *J. Virol. Methods* 2009;159:1–5.

53. Fernández-Pinero J et al. Rapid and sensitive detection of African horse sickness virus by real-time PCR. *Res. Vet. Sci.* 2009;86:353–358.

54. Quan M et al. Development and optimisation of a duplex real-time reverse transcription quantitative PCR assay targeting the VP7 and NS2 genes of African horse sickness virus. *J. Virol. Methods* 2010;167:45–52.

55. Maan NS et al. Serotype specific primers and gel-based RT-PCR assays for 'typing' African horse sickness virus: Identification of strains from Africa. *PLoS One* 2011;6:e25686.

56. Monaco F et al. A new duplex real-time RT-PCR assay for sensitive and specific detection of African horse sickness virus. *Mol. Cell Probes* 2011;25:87–93.

57. Scheffer EG et al. Use of real-time quantitative reverse transcription polymerase chain reaction for the detection of African horse sickness virus replication in *Culicoides imicola*. *Onderstepoort J. Vet. Res.* 2011;78:E1–E4.

58. Guthrie AJ et al. Diagnostic accuracy of a duplex real-time reverse transcription quantitative PCR assay for detection of African horse sickness virus. *J. Virol. Methods* 2013;189:30–35.

59. Laegreid WW et al. Characterization of virulence variants of African horse sickness virus. *Virology* 1993;195:836–839.

60. Roy P et al. Recombinant baculovirus-synthesized African horse sickness virus (AHSV) outer-capsid protein VP2 provides protection against virulent AHSV challenge. *J. Gen. Virol.* 1996;77:2053–2057.

61. Stone-Marschat MA et al. Immunization with VP2 is sufficient for protection against lethal challenge with African horse sickness virus type 4. *Virology* 1996;220:219–222.

62. Scanlen M et al. The protective efficacy of a recombinant VP2-based African horse sickness subunit vaccine candidate is determined by adjuvant. *Vaccine* 2002;20:1079–1088.

63. Sánchez-Vizcaíno JM. Control and eradication of African horse sickness with vaccine. *Dev. Biol. (Basel)* 2004;119:255–258.

64. Diouf ND et al. Outbreaks of African horse sickness in Senegal, and methods of control of the 2007 epidemic. *Vet. Rec.* 2013;172:152.

65. MacLachlan NJ et al. Experiences with new generation vaccines against equine viral arteritis, West Nile disease and African horse sickness. *Vaccine* 2007;25:5577–5582.

66. Koekemoer JJ. Serotype-specific detection of African horse sickness virus by real-time PCR and the influence of genetic variations. *J. Virol. Methods* 2008;154:104–110.

67. von Teichman BF and Smit TK. Evaluation of the pathogenicity of African horse sickness (AHS) isolates in vaccinated animals. *Vaccine* 2008;26:5014–5021.

68. von Teichman BF, Dungu B, and Smit TK. In vivo cross-protection to African horse sickness serotypes 5 and 9 after vaccination with serotypes 8 and 6. *Vaccine* 2010;28:6505–6517.

69. Castillo-Olivares J et al. A modified vaccinia ankara virus (MVA) vaccine expressing African horse sickness virus (AHSV) VP2 protects against AHSV challenge in an IFNAR−/− mouse model. *PLoS One* 2011;6:e16503.

70. de Vos CJ, Hoek CA, and Nodelijk G. Risk of introducing African horse sickness virus into the Netherlands by international equine movements. *Prev. Vet. Med.* 2012;106:108–122.

71. El Garch H et al. An African horse sickness virus serotype 4 recombinant canary pox virus vaccine elicits specific cell-mediated immune responses in horses. *Vet. Immunol. Immunopathol.* 2012;149:76–85.

72. Oura CA et al. African horse sickness in The Gambia: Circulation of alive-attenuated vaccine-derived strain. *Epidemiol Infect.* 2012;140:462–465.

73. Pretorius A et al. Virus-specific CD8+ T-cells detected in PBMC from horses vaccinated against African horse sickness virus. *Vet. Immunol. Immunopathol.* 2012;146:81–86.

74. Scheffer EG et al. Comparison of two trapping methods for *Culicoides* biting midges and determination of African horse sickness virus prevalence in midge populations at Onderstepoort, South Africa. *Vet. Parasitol.* 2012;185:265–273.

75. Boom R and Oldruitenborgh-Oosterbaan MM. Can Europe learn lessons from African horse sickness in Senegal? *Vet. Rec.* 2013;172:150–151.

50 African Swine Fever Virus

Armanda D.S. Bastos, Folorunso O. Fasina, and Donald P. King

CONTENTS

50.1 INTRODUCTION

African swine fever virus (ASFV) causes a highly lethal, viral hemorrhagic disease in domestic pigs, which represents a global threat due to the current lack of a vaccine and unavailability of treatment options for a disease that has morbidity and mortality rates of 100%, when highly pathogenic ASFV strains are involved. The virtually worldwide dependence on pork as a source of affordable protein, high virus stability and longevity in inadequately treated meat products, and rapid dissemination of meat products through well-established transport and trading routes, together with pathognomic clinical signs in infected animals can collectively contribute to ill-afforded delays in agent identification and mobilization of essential control measures to limit spread. Countries to which the disease is exotic are particularly prone to unavoidable delays unless veterinary officials and diagnostic laboratory staff are, respectively, primed and prepared to recognize and identify both the disease and the disease-causing agent.

African swine fever (ASF) is a former "List A" disease of the Office International des Epizooties (OIE) and remains a notifiable disease due to its potential for rapid spread and socioeconomic, public health, and trade impact. Whereas recognition of the disease in domestic pigs is relatively recent, being described for the first time by Montgomery in 1921 [1], the association between the virus and its asymptomatic vertebrate (*Phacochoerus* warthogs) and invertebrate (*Ornithodoros* soft tick) hosts in sub-Saharan Africa is an ancient one [2]. The disease, initially termed "East African swine fever," but subsequently renamed "African swine fever" when its broader distribution was realized [3,4], was of local interest and limited concern, until 1957, when it made its first incursion into Europe, with devastating effects [5]. A second incursion, again into Portugal, occurred in 1960 and was followed by the dissemination of the virus throughout the Iberian peninsula,

subsequent exportation to South America and the Caribbean [6], and an apparently reverse introduction in the early 1980s, from these exotic southern hemisphere localities to the West African region [7], from which the virus that was introduced to Europe originated [8]. Shortly thereafter, sporadic outbreaks occurred in Belgium and neighboring Holland. However, for some affected southern European countries, it took more than three decades to eradicate the virus. Portugal, the initial site of the infection, only achieved disease-free status in 1993 but suffered a reemergence in 1999 [9], and Sardinia has yet to achieve this status [10]. After a period of relative quiescence, the virus has made a number of successive incursions, first into the West African region in 1996 where it remains problematic; then, in 1998 and 2007, two sporadic outbreaks occurred for the first time in the Indian Ocean Islands of Madagascar [11] and Mauritius [12], respectively, and further afield, the disease was introduced into Georgia in 2007 [13]. The latter virus subsequently expanded its geographical range to Armenia, Iran [14], the Ukraine, and southern Russia and persists throughout the Caucasus region [9,15]. When one considers that the loss of 300,000 of the Russian Federation's 19 million pigs to ASF came at a cost of around $240 million in 2011, the catastrophic implications of the introduction of the virus to a naïve pig population as large as that of China's, which is estimated to exceed 1 billion, are clear [15].

50.2 CLASSIFICATION AND MORPHOLOGY

ASFV is a double-stranded DNA virus that is the sole member of the *Asfivirus* genus, which in turn is the only representative of the family Asfarviridae, which derives its name from *African swine fever and related viruses* [16]. The unique features of this virus, which was initially classified within the Iridoviridae due to cytoplasmic genome location and structural similarities and later within the Poxviridae on the basis of genome structure and replication, prompted the creation of a new family centered on the virus name. The enveloped icosahedral virions contain approximately 1892 capsomeres and are large, ranging from 175 to 215 nm in diameter. Extracellular virions are composed of five concentric layers, with the fifth lipid envelope layer being acquired when the virion buds from the plasma membrane. The double-stranded DNA genome is also large and variable in size, ranging from 170 to 190 kb in length. The variation in size is mainly due to the length heterogeneity in the regions that flank a central conserved region (CCR) of approximately 125 kb. The left variable region (LVR) ranges from 38 to 47 kb and the right variable region (RVR) ranges from 13 to 16 kb. Variation in genome length between isolates is not only due to length variations in flanking regions but also arises from minor length variations in the CCR [17]. Although negligible in terms of contribution to overall genome length variations, they have been shown to be highly informative targets for molecular epidemiology studies [18,19]. Likewise, the antigenically stable virus protein (VP) 72 (also termed VP73) [20] that makes up more than 30% of the total protein mass of the virion, but represents just one of the more than 150 open reading frames encoded by the genome [21], has been shown to be a valuable molecular diagnostic and epidemiological marker.

50.3 BIOLOGY AND EPIDEMIOLOGY

There are a number of features that make ASFV unusual. First and perhaps foremost is that although antibodies to the virus occur following natural infection, these antibodies fail to neutralize the virus [22]. This, together with the high mortality rate and "bleeding out" symptoms, has resulted in ASFV being likened to Ebola virus [15]. In addition, although antigenically and genetically variable strains are known to occur, it is not possible to assign viruses to a distinct serotype [6], as is the case with many other virus species. Further, ASFV is currently the only known DNA arthropod-borne virus (arbovirus), and it occurs in three epidemiologically distinct cycles. In the ancient sylvatic cycle, it is transmitted between African warthog and soft tick host species. However, the virus also appears capable of cycling between ticks and domestic pigs with the exclusion of wild suids and appears capable of persisting in a cycle that involves domestic pigs alone, with no apparent role for either the soft tick vector or natural sylvatic suid hosts [6]. The ability to survive in three distinct cycles suggests a remarkable adaptability as the latter two cycles are comparatively "young," initially becoming established in Africa on introduction of domestic pigs to the continent. Another atypical feature of this DNA virus is that a number of structural proteins are derived from proteolytic cleavage [17], which, although a common enough strategy for retroviruses and RNA viruses, is unusual for large DNA viruses and suggestive of an expansive gene expression range.

In its endemic sub-Saharan African setting, the infected *Ornithodoros* tick is the primary disease vectoring agent. Outside of this environment, the feeding of infected pork products to pigs has repeatedly been shown to be the main factor precipitating infection in domestic pigs. Once infected, domestic pigs and European wild boars, unlike their African wild suid counterparts, display clinical signs of infections and shed large amounts of virus [2], acting as a source of infection for other pigs in close contact.

In the absence of a serology-based framework for grouping field strains, classification of field strains is reliant on molecular methods when assigning viruses to genotypes. This was initially achieved by whole genome restriction fragment length polymorphism (RFLP) analysis, but has subsequently been replaced by PCR-based methods. Currently, genotyping is achieved by characterization of a C-terminal 478 bp fragment of the *p72* gene [8]. Application of the *p72* genotyping method has revealed the presence of 22 discrete genotypes [8,23–25] and confirmed the identity of the genotypes involved in recent incursions outside of Africa [8,11–13]. Although valuable for genotype assignment, the *p72* gene is too conserved to permit inferences regarding the source and spread of outbreaks. Molecular epidemiology studies are therefore reliant on a three-gene approach that involves characterization of *p54* and CVR genome regions, in addition to *p72*, in order to refine field strain relatedness [19,26].

50.4 CLINICAL FEATURES AND PATHOGENESIS

The process of ASF diagnosis usually begins with field investigation by personnel trained to recognize the typical signs of disease (farmers and/or veterinarians) and is supported by postmortem examination of tissues from affected animals and testing of appropriate samples for the presence of ASFV or ASFV-specific antibodies. Definitive diagnosis hinges on laboratory diagnosis as the range of clinical disease manifestations together with postmortem findings such as splenic enlargement and hemorrhages in lymph nodes and kidney resemble those of classical swine fever (CSF)/hog cholera, as well as a number of infectious (viral and bacterial) and noninfectious (chemical poisoning) diseases.

50.4.1 CLINICAL FEATURES

Following a 2–15 day incubation period, clinical signs of the disease will become apparent. The clinical picture varies from the extremes of peracute disease where animals die suddenly with few signs to chronic cases that are associated with a variety of nonspecific signs such as weight loss, fever, necrosis, and ulceration of the skin, respiratory signs, and arthritis. In the field, the anticipated range of syndromes generally varies with the form of infection, which in turn is dependent on virulence and host factors. In addition to peracute, acute, subacute, and chronic states, apparently healthy carrier forms may also be encountered particularly if subacute and chronically infected animals are not removed from herds through stamping out. The latter two classes of infection should be borne in mind in areas in which the disease persists or has become endemic. Briefly, the four forms of infection can be summarized as follows:

1. *Peracute infection*: Individual pigs within the domestic herds are often found dead without displaying any apparent clinical signs. This type of infection usually occurs in a naïve pig population and/or in a situation where a herd is infected by a highly virulent ASFV.
2. *Acute infection*: Clinical signs will be observed as early as 48 h postinfection. Pigs may present with high fever ($\geq 40.5°C$), thrombocytopenia, leucopenia, reddened skin (especially of the ears, tail, abdomen, ventral thoracic region, and distal extremities in white pigs), anorexia, listlessness, increased pulse and respiratory rates, vomiting, diarrhea, eye discharges, cyanosis, and incoordination, followed by death within 1–2 days. Abortion will be recorded in pregnant sows and death may supervene within 2 or up to 3 weeks in certain instances. Though mortality may approach 100%, survivor pigs often remain carriers for life [6]. This type of infection is caused by a highly virulent ASFV or exposure of a naïve herd.
3. *Subacute infection*: Irregular fever that generally lasts up to a month, but can extend to 45 days, followed by recovery. Anorexia and loss of body condition as well as coughing and dyspnea may be observed. Abortion is an almost-regular feature in pregnant sows and mortality ranges from 30% to 70% depending on associated stressful conditions. This form of infection may be precipitated by a moderate to highly virulent virus but is also observed in herds with previous exposure.
4. *The chronic form* of the disease usually manifests in previously exposed herds and those herds affected by strains of low to mild virulence. While mortality may not occur or in the worst-case scenario be low ($\leq 30\%$), mild to severe weight loss, unthriftiness and severe growth retardation, loss of hair, irregular febrile conditions, undulating respiratory peaks, lameness associated with arthritis, secondary infections, and low-grade perpetual sickness may be found in the herd of pigs. Pigs chronically infected with ASF do not develop conjunctivitis as observed in the case of CSF; this clinical feature may assist in differentiating the two infections.

These different scenarios pose challenges for ASF diagnosis, particularly in endemic regions where a broader range of clinical manifestations can be anticipated due to a broader range in virus virulence and host factors arising from persistent exposure to virus. However, when a virulent strain is introduced to naïve domestic pig herds, ASFV will typically cause high mortality (approaching 100%), as was the case when 90,000 pigs succumbed in the first 6 months following introduction of the virus to Georgia in 2007. Clinically, fever, reddening of the skin (particularly the ears, tail, distal extremities, ventral aspects of chest, and abdomen) that is readily seen in light-skinned pigs, vomiting, diarrhea (sometimes bloody) and eye discharges, anorexia, listlessness, and cyanosis will typically be observed in infected pigs.

Pig-to-pig transmission occurs through the oronasal route. Other potential routes of infection identified include the skin but also intramuscularly, subcutaneously, intraperitoneally, and intravenously as well as from the vector (*Ornithodoros* species) bites. Though the pathogenesis of ASF has not been comprehensively researched in the warthog and bushpig populations, except for the documented lack of clinical signs and low level of viral replication with or without minimal pathological signs [27], its effect on domestic pig populations has been well described.

50.4.2 Pathogenesis

Following exposure of domestic pigs to ASFV, viral penetration into tissues usually occurs through the tonsils or dorsal pharyngeal mucosa, after which the infection spreads to the mandibular/retropharyngeal lymph nodes and later goes to the hematogenous route with subsequent viremia [6]. While the rate of penetration and spread is dependent on the route of infection, ASFV has also been detected in other locations, for example, the gastric lymph nodes [28,29]. After viral infection and replication of the virus in the lymphoid organs and the subsequent hematogenous spread, ASFV titer in the blood may reach 10^8 HAD$_{50}$/mL in acute cases, and the virus is primarily associated with erythrocytes (especially in infections caused by hemadsorbing viruses) and also lymphocytes and neutrophils [30]. A recent study indicates that macropinocytosis may be the principal means of internalization of the virus into the cells [31]. Primary viremia may be observed as early as 8 h postinfection in newborn piglets, with secondary spread to other body organs occurring within a day and recovery of the virus in all tissues occurring 3 days postinfection [32].

Primarily, ASF will infect and replicate *in vivo* in the macrophages and monocytes of the mononuclear phagocytic system. In addition, although there are indications that the virus also infects the megakaryocytes, endothelial cells, mesangial cells of the glomerulus, epithelial cells of the kidney tubes, hepatocytes, thymus reticuloendothelial cells, fibroblast and smooth muscle cells of venules and arterioles, neutrophils, and lymphocytes, no record of viral replication in these cells has been documented [33]. While there were intense debates regarding the infection status of the aforementioned cells, the reasons for the infection of such a wide range of host cells remain unclear, and it is possible that as the disease progresses in a host, these infections supervene at the advanced stage of disease condition. Another hypotheses regarding infection and replication include suggestions that hemorrhage associated with viral replication occurs in the endothelial cells of the interstitial capillaries [34] and initial hemorrhage of the kidney and lymph nodes occurs prior to viral replication in the previously stated cells. It was also recorded that lysosomal proliferations and phagocytic cell debris, capillary fenestrations, necrosis, and loss of endothelial membrane are always associated with hemorrhages.

In many cases where disseminated intravascular coagulopathy (DIC) is observed, these are associated with the release of prostaglandin E_2 by the infected macrophages. It may also be associated with the release of cytokines from infected macrophages including the IL-1 and TNF-α [35]. Late stage of the acute viral infection may present with thrombocytopenia, and the reasons may be due to destruction of thrombocytes during coagulation processes or immune-mediated processes or by the virus effect on megakaryocytes [35].

It is widely accepted that impaired hemostasis is associated with the massive destructions of macrophage in view of the released cytokines, complement factors, and arachidonic acid metabolites [6]. Apoptosis of the thymocytes, decreased blood lymphocytes, and quantitative increase in thymus macrophages has also been demonstrated [36].

The pathogenesis of ASFV infection in the chronic form is not well described. It was suggested that autoimmunity may play a role in these forms and that lesions are associated with immune complexes deposited in tissues (kidneys, lungs, and skin), which subsequently binds to complement [30].

50.5 IDENTIFICATION AND DIAGNOSIS

Accurate and timely diagnosis plays a central role in the effective monitoring and control of ASF. This hinges on differential diagnosis to rule out a number of infectious diseases that may clinically be confused with ASF, notably CSF, but also acute porcine reproductive and respiratory syndrome (PRRS), erysipelas, salmonellosis, Aujeszky's disease (in young animals), pasteurellosis, porcine dermatitis and nephropathy syndrome (PDNS), as well as generalized septicemic or hemorrhagic conditions.

A variety of laboratory tools can be used for the detection of ASFV and ASFV-specific antibodies. A number of these tests are recommended by the World Organisation for Animal Health and are included in the Manual of Diagnostic Tests and Vaccines for Terrestrial Animals. ASFV-specific antibodies can be detected for many years in pigs that have recovered from ASF, even those exposed to avirulent or low-virulent field strains where serological diagnosis can be particularly important. A number of serological assays have been developed to detect these specific antibody responses, including an indirect enzyme-linked immunosorbent assay (ELISA) [37,38], which is the test prescribed by the OIE for international trade. Immunoblotting can be used to support serological diagnosis by confirming the specificity of antibody responses to individual viral proteins [39]. A more thorough understanding of the antigenic determinants of ASFV has led to the development of improved serological tests for ASFV [40]. These new tests exploit advantages of recombinant proteins and provide the prospect for diagnostic kits to be manufactured without the requirement for high-containment laboratories.

Rapid, sensitive, and specific approaches for the detection of ASFV are provided by molecular tests and immunoassays [41]. However, it is likely that the ability to culture ASFV will remain an important component of research efforts to develop vaccines for ASFV. ASFV replicates mainly in porcine cells belonging to the monocyte/macrophage lineage [42]. The established hemadsorption virus isolation (VI) method [43] can be a sensitive approach to isolate ASFV from blood and infected tissue (spleen, lymph nodes, tonsil, and kidney), but it takes several days to obtain a result and is reliant upon the regular sourcing of fresh (and uninfected) pig tissues for the preparation of primary bone marrow cells or pig leukocyte cultures. Furthermore, the emergence of virulent non-hemadsorbing, noncytopathic strains of ASFV [11] raises the potential of this test to generate false-negative results. As an alternative to primary cell culture systems, continuous porcine alveolar macrophage cell lines that can support replication of ASFV have been established using transfection [44]. Although not currently adopted for routine diagnostics, this approach, or alternative strategies to generate myelomonocytic cell lines, may have future application for ASF virus isolation (VI) (ASFV-VI) isolation and culture.

Presumptive diagnosis of ASF can be achieved by visualizing viral antigen in infected tissues or blood smears from affected pigs using the fluorescent antibody test (FAT) with ASFV-specific antibody reagents. Immunocytochemistry as well as in situ hybridization methods can also be used in studies to investigate disease pathogenesis, but these techniques are not ideally suited for routine diagnostics [45]. In order to detect viral antigen, a number of antigen-capture ELISAs (Ag-ELISAs) have been developed. These assays employ specific antibody reagents: polyclonal antisera raised in rabbits and guinea pigs, a combination of rabbit antisera and monoclonal antibodies, or monoclonal antibodies alone. Ag-ELISA tests can detect viral antigen in sera and spleen samples collected from pigs infected with representative field strains of phylogenetically distinct groupings of ASFV [46,47] although these tests can have low sensitivity for diagnosis in the event of subacute and chronic forms of ASF. These and other ASFV-specific antibody reagents are currently being evaluated for incorporation into lateral-flow devices that offer the potential for simple and rapid detection of ASFV and ASFV-specific antibodies in the field.

Molecular assays are now widely used as frontline tests for laboratory diagnosis of ASFV. These tests can be highly sensitive and specific and are particularly appropriate for poor-quality or degraded samples that are unsuitable for testing by VI or antigen detection methods. In order to achieve high diagnostic sensitivity, these assays target ASFV DNA sequences that are conserved across field strains of the virus. Many of the current assays amplify sequences of the *p72* gene, exploiting the large number of sequences that are available in GenBank for this region of the genome. The earliest molecular formats developed for diagnosis used agarose-gel electrophoresis to visualize products generated by polymerase chain reaction (PCR) [48]. One of these methods is still recommended by the OIE for ASFV diagnosis [48] and can be multiplexed with additional primers to provide simultaneous differential diagnosis to CSFV [49]. Alternative multiplex PCR assays [50] for ASFV and CSFV have been expanded to include other porcine viruses (porcine circovirus type 2, PRRS virus, and porcine parvovirus). Furthermore, nested PCR approaches [51] can be adopted for agarose-gel-based PCR assays, allowing more sensitive detection of ASFV to be achieved in ticks (*Ornithodoros erraticus*). However, these relatively labor-intensive procedures have a high risk of generating false-positives due to carry-over of PCR amplicons and are therefore not ideal for routine testing of large numbers of samples. More recently real-time PCR assays have become widely used for routine diagnosis. This approach is a highly sensitive technique enabling simultaneous amplification and quantification of specific nucleic acid sequences. In addition to enhanced sensitivity, the benefits of real-time PCR assays over agarose-gel electrophoresis detection methods include their large dynamic range, a reduced risk of cross contamination, an ability to be scaled up for high-throughput applications, and the potential for accurate target quantification. In addition to the 5′ nuclease (TaqMan®) system that utilizes dual-labelled probes [52–55], ASFV real-time PCR assays have been developed that exploit alternative primer and probe chemistries including molecular beacons [56], minor groove binder probes [57], linear-after-the-exponential PCR [58], and a universal library of locked nucleic acid probes [54].

Although laboratory-based molecular tests to detect ASFV are rapid and can take only a few hours to generate an objective result, the time taken to transport suspect material to a central laboratory can be lengthy. In disease outbreaks where a rapid decision is critical, this delay can preclude laboratory confirmation. Furthermore, in many developing countries in sub-Saharan Africa where ASFV is endemic, access to more sophisticated laboratory equipment is difficult, and as a consequence, real-time surveillance and monitoring of ASF outbreaks is extremely limited. Away from dedicated laboratory facilities, a number of established and emerging technologies can be considered for deployment close to the suspect case of disease. Assays in this format that aim to provide data to support local decision making are frequently referred to as "point-of-care" or "pen-side" tests. A number of different portable PCR platforms have been or are currently under development. These fully integrated platforms can perform all of the assay steps (nucleic acid extraction, PCR setup, amplification and interpretation of the assay results) without user intervention. Proof-of-concept studies to evaluate these PCR platforms for ASF diagnosis have been recently undertaken [59]. For nucleic acid amplification, alternative strategies that utilize "isothermal" approaches show considerable promise for use for ASF diagnostics. Unlike PCR, these assay systems do not require precision (and often expensive) instrumentation to accurately control the different temperature incubation steps in the reaction. Loop-mediated isothermal amplification (LAMP), originally described by Notomi et al. [60], is an approach that allows rapid amplification of target DNA sequences in a highly specific manner under isothermal conditions. LAMP assays that target the topoisomerase II, putative DNA primase, and VP73 genes of ASFV have been

developed (James et al., 2010). Parallel testing of samples by LAMP, real-time PCR, Ag-ELISA, and VI was undertaken using EDTA-blood and tissue samples collected from experimentally infected pigs. These sample types represent the most frequently submitted diagnostic material for suspect cases of swine fever diagnosis. Using these samples, the performance of the LAMP assay was similar to real-time PCR and was superior to Ag-ELISA. LAMP reactions can be performed at a single temperature using a simple water bath or potentially in a disposable thermal device or heat pack. Furthermore, ASFV LAMP amplicons can be dual-labelled with biotin and fluorescein and visualized using novel lateral-flow devices [61]. This assay and detection format represents the first step towards developing a practical, simple-to-use, and inexpensive molecular assay format for ASF diagnosis in the field, which is especially relevant to sub-Saharan Africa where the disease is endemic.

Molecular characterization of field and reference strains plays a central role in the ability of diagnostic laboratories to monitor the distribution and spread of ASFV, as well as providing important information to understand the genetics of interactions of the virus with the host. Prior to the use of nucleotide sequencing methods, analysis of patterns generated by restriction endonuclease (RE) digestion of viral DNA could be used to differentiate ASFV strains. In the past 20 years, nucleotide sequencing has become widely adopted for ASFV strain characterization such that sequences for >1000 genomic fragments of ASFV are now available in GenBank. In view of the relatively large size of ASFV, it is a more challenging exercise to routinely generate complete genome sequences for field strains of the virus. The first complete ASFV genome sequence was generated for a cell culture reference strain of ASFV (BA71V) [21]. This sequence was assembled from plasmid clones generated from RE fragmentation of the ASFV genome using conventional dideoxy terminator (Sanger) sequencing technology. The throughput and robustness of sequencing methods has been improved by the use of fluorescent dyes and capillary separation technologies, such that the pace of ASFV sequencing has increased and has included complete genome analysis of different ASF strains with different phenotypes [62,63]. More recently, next-generation sequencing (NGS) approaches have been utilized for the molecular characterization of ASFV. Clonal emulsion PCR and pyrosequencing (using the 454 platform) have been employed to sequence the genome of a highly virulent isolate that has caused the recent ASF outbreaks in the Caucasus. These new NGS technologies generate much larger quantities of data than was previously possible and have the potential to revolutionize sequencing approaches used for ASFV. However, frame shifts due to homopolymer errors generated during pyrosequencing can pose challenges for accurate genome assembly of ASFV. Therefore, it is likely that robust sequencing protocols will utilize more than one NGS technology to reliably construct complete ASFV genomes. Furthermore, improved bioinformatics tools are also required to align, annotate, and display these data.

50.6 TREATMENT AND PREVENTION

To date, there is no documented treatment protocol for ASF infection since no prophylactic medication exists. Similarly, no vaccine is available for the prevention of ASF as all effort to develop an effective vaccine to date has failed. Quarantine, slaughter policy, and eradication have been adopted by many countries as a means of control. In case of outbreaks, the degree of adoption of the protocol for quarantine, slaughter policy, and eradication differs among countries. While most of the developed economies and countries with principal interest in agriculture will conduct bloodless destruction/stamping out of infected and all in-contact pigs (e.g., through asphyxiation using carbon dioxide gas or anesthetics) and also do preemptive slaughter of pigs in the radius of 5–20 km from infected locations, most of the emerging economies and poorer countries will carry out modified stamping out with the slaughter of infected herds and in-contact pigs only. It is possible that the perpetuation of ASF in African countries is closely related to this lack of complete stamping out policy and the slaughter of sick pigs in the abattoir [64]. However, a recent study on the cost-benefits of biosecurity reveals that adoption of what is generally perceived to be costly or unnecessary measures can have substantial economic benefits for smaller-scale African farmers [65]. Biosecurity is the assurance of biocontainment (restricting a pathogen present in a farm/location to that farm/location only) and bioexclusion (preventing the introduction of new pathogen into a farm/an area). It carries several components, and only the strict enforcement of biosecurity rules, its compliance, and regular evaluation will ensure its effectiveness.

Disinfection is also important in the prevention of ASF, and the use of many chemical substances including the Virkon®, hypochlorite, inactivation by cresol, 2% sodium hydroxide, 1% formalin, 4% anhydrous sodium carbonate plus 0.1% detergent, 10% crystalline sodium carbonate plus 0.1% detergent, detergents (ionic and anionic), 1% phosphorylated iodophors, and chloroform has been effective in containing ASFV. It is advisable to decongest surfaces by scrapping and washing before the application of these disinfectants. This will enhance direct contact with solid surfaces as most of the organic materials present on surfaces reduce the effectiveness of these disinfectants. Heat inactivation at 70°C has also been effective in destruction of the virus.

Management protocols in use per farm have to be reviewed regularly to ensure that it does not constitute a threat to farm health status. It is necessary to avoid the use of swill or cook all swill before feeding to pigs, to carry out effective animal identification systems, and to have in place a good testing protocol (serological/virological surveys) to identify sick animals at the earliest possible time. Intensive farming, prevention of unnecessary animal contact especially if they originate from another farm, removal of slaughter slab from pig communities, rodent control, and animal food control will assist in reducing the risk of ASF.

Although the meat from ASF-infected pigs does not carry a public health threat because of heat inactivation, its ban is related to quarantine and to prevent inadvertent spread of the virus to free areas. In the event of an outbreak, government control ensuring strict importation and testing protocols, monitoring at the live animal markets, pig auctions, and farms, routine national surveillance, prevention of importation of pork and pork product if possible, strict border movement monitoring and quarantine, and development of control zones/compartmentalization, is essential for limiting the spread and impact of the disease.

For countries historically free of the disease, with well-developed pig industries, drastic measures in pursuit of elimination are often advocated. This approach does not translate well to resource-poor countries in which the disease is endemic. As long as there are wildlife reservoirs of infection in sub-Saharan Africa, eradication of ASF is unlikely, and the emphasis must realistically be placed on mitigation, through, for example, improved biosecurity and implementation of locality-specific policies aimed at reducing risk.

50.7 CONCLUSIONS AND FUTURE PROSPECTS

The recent incursions and prolonged field presence/persistence of virus genotypes outside of their traditional ranges, both within and outside of the African continent (Misinzo et al., 2012), necessitate some comment on insights gained retrospectively into factors that facilitated virus persistence in Europe. In particular, establishment of the virus in pig-associated, vector-competent ticks, as was the case with naturally occurring *O. erraticus* in Portugal, means that virus reemergence remains a threat albeit one that can be contained through appropriate legislation [9]. The ubiquitous distribution of diverse long-lived, soft tick species capable of vectoring the disease [66] and the difficulties in eradication once the virus becomes established in these vector-competent hosts make these important aspects to consider once the disease has been eradicated from domestic pigs. Management of this aspect has been facilitated by the development of locality-specific methods that are suited to screening the virus genotype introduced to Europe and the resident *O. erraticus* soft tick species [51,67] and the *Ornithodoros porcinus* sylvatic cycle ticks and their associated genotypes in Africa [68]. The tick-directed methods have been instrumental in identifying residual infected tick populations in areas in exotic localities and in highlighting virus disappearance from previously infected tick populations in endemic areas [69]. Both have economic impacts as appropriate measures can be put in place to prevent outbreaks and to lift unnecessary restrictive control measures, respectively. These and many of the diagnostic advances reviewed in this chapter make rapid diagnosis and effective management of the disease a distinct possibility and crucial first line of defense. However, until an effective vaccine is developed, ASF will remain a significant global threat.

REFERENCES

1. Montgomery RE. 1921. On a form of swine fever occurring in British East Africa (Kenya Colony). *J Comp Pathol.* 34:159–191.
2. Jori F and Bastos ADS. 2009. Role of wild suids in the epidemiology of African swine fever. *EcoHealth.* 6:296–310.
3. Walker J. 1929. Memorandum on research on East African swine fever immunization in Kenya. Paper No. 32. In: *Proceedings of the 1929 Pan African Veterinary Conference,* Kenya, pp. 262–272.
4. Steyn DG. 1932. East African virus disease in pigs. In: *Proceedings of the 18th Report of the Director of Veterinary Services and Animal Industry,* vol. 1. Pretoria, Union of South Africa, pp. 99–109.
5. Wilkinson PJ. 1984. The persistence of African swine fever in Africa and the Mediterranean. *Prev Vet Med.* 2:71–82.
6. Penrith ML, Thomson GR, and Bastos ADS. 2004. African swine fever. In: Coetzer JAW and Tustin RC, eds. *Infectious Diseases of Livestock,* 2nd ed. Oxford, U.K.: Oxford University Press, pp. 1087–1119.
7. Wesley RD and Tuthill AE. 1984. Genome relatedness among African swine fever virus field isolates by restriction endonuclease analysis. *Prev Vet Med.* 2:53–62.
8. Bastos ADS, Penrith M-L, Crucière C, Edrich JL, Hutchings G, Couacy-Hymann E, and Thomson GR. 2003. Genotyping field strains of African swine fever virus by partial *p72* gene characterisation. *Arch Virol.* 148(4):693–706.
9. Boinas FS, Wilson AJ, Hutchings GH, Martins C, and Dixon LJ. 2011. The persistence of African swine fever virus in field-infected *Ornithodoros erraticus* during the ASF endemic period in Portugal. *PLoS One.* 6(5):e20383. doi:10.1371/journal.pone.0020383.
10. Giammarioli M, Gallardo C, Oggiano A, Iscaro C, Nieto R, Pellegrini C, Dei Giudici S, Arias M, and De Mia GM. 2011. Genetic characterisation of African swine fever viruses from recent and historical outbreaks in Sardinia (1978–2009). *Virus Genes.* 42(3):377–387.
11. Gonzague M, Roger F, Bastos A, Burger C, Randriamparany T, Smondack S, and Cruciere C. 2001. Isolation of a non-haemadsorbing, non-cytopathic strain of African swine fever virus in Madagascar. *Epidemiol Infect.* 126(3):453–459.
12. Lubisi BA, Dwarka RM, Meenowa D, and Jaumally R. 2009. An investigation into the first outbreak of African swine fever in the Republic of Mauritius. *Transbound Emerg Dis.* 56(5):178–188.
13. Rowlands RJ, Michaud V, Heath L, Hutchings G, Oura C, Vosloo W, Dwarka R, et al. 2008. African swine fever virus isolate, Georgia, 2007. *Emerg Infect Dis.* 14(12):1870–1874.
14. Rahimi P, Sohrabi A, Ashrafihelan J, Edalat R, Alamdari M, Masoudi M, Mostofi S, and Azadmanesh K. 2010. Emergence of African swine fever virus, northwestern Iran. *Emerg Infect Dis.* 16(12):1946–1948.
15. Callaway, E. 2012. Pig fever sweeps across Russia. *Nature.* 488:565–566.
16. Dixon LK, Escribano JM, Martins C, Rock DL, Salas ML, and Wilkinson PJ. 2005. Asfarviridae. In: Fauquet CM, Mayo MA, Maniloff J, Desselberger U, and Ball LA, eds. *Virus Taxonomy, VIIIth Report of the International Committee on Taxonomy of Viruses.* London, U.K.: Elsevier Academic Press, pp. 135–143.
17. Tulman ER, Delhon GA, Ku BK, and Rock DL. 2009. African swine fever. *Curr Top Microbiol Immunol.* 328:43–87.
18. Irusta PM, Borca MV, Kutish GF, Lu Z, Caler E, Carrillo C, and Rock DL. 1996. Amino acid tandem repeats within a late viral gene define the central variable region of African swine fever virus. *Virology.* 220(1): 20–27.
19. Owolodun O, Bastos ADS, Antiabong JF, Ogedengbe ME, Ekong PS, and Yakubu B. 2010. Molecular characterisation of African swine fever viruses from Nigeria (2003–2006) recovers multiple virus variants and reaffirms CVR epidemiological utility. *Virus Genes.* 41:361–368.
20. Cistué C and Tabarés E. 1992. Expression in vivo and in vitro of the major structural protein (VP73) of African swine fever virus. *Arch Virol.* 123(1–2): 111–124.
21. Yáñez RJ, Rodríguez JM, Nogal ML, Yuste L, Enríquez C, Rodriguez JF, and Viñuela E. 1995. Analysis of the complete nucleotide sequence of African swine fever virus. *Virology.* 208(1):249–278.
22. Martins CL and Leitão AC. 1994. Porcine immune responses to African swine fever virus (ASFV) infection. *Vet Immunol Immunopathol.* 43(1–3):99–106.
23. Lubisi BA, Bastos ADS, Dwarka RM, and Vosloo W. 2005. Molecular epidemiology of African swine fever in East Africa. *Arch Virol.* 150: 2439–2452.
24. Boshoff CI, Bastos ADS, Gerber LJ, and Vosloo W. 2007. Genetic characterisation of African swine fever viruses from outbreaks in southern Africa (1973–1999). *Vet Microbiol.* 121:45–55.
25. Misinzo G, Kasanga CJ, Mpelumbe-Ngeleja C, Masambu J, Kitambi A, and Van Doorsselaere J. 2012. African swine fever virus, Tanzania, 2010–2012. *Emerg Infect Dis.* 18(12):2081–2083.
26. Gallardo C, Mwaengo DM, MacHaria JM, Arias M, Taracha EA, Soler A, Okoth E, et al. 2009. Enhanced discrimination of African swine fever virus isolates through nucleotide sequencing of the p. 54, p. 72, and pB602L (CVR) genes. *Virus Genes* 38: 85–95.
27. Anderson EC, Hutchings GH, Mukarati N, and Wilkinson PJ. 1998. African swine fever virus infection of the bushpig (*Potamochoerus porcus*) and its significance in the epidemiology of the disease. *Vet Microbiol.* 62(1):1–15.
28. Greig A. 1972. Pathogenesis of African swine fever in pigs naturally exposed to the disease. *J Comp Pathol.* 82(1):73–79.
29. Wilkinson PJ. 1989. African swine fever virus. In: Pensaert MC, ed. *Virus Infections of Porcines.* Amsterdam, the Netherlands: Elsevier, pp. 17–32.
30. Plowright W, Thomson GR, and Neser JA. 1994. African swine fever. In: Coetzer JAW, Thomson GR, and Tustin RC, eds. *Infectious Diseases of Livestock,* 1st ed. Oxford, U.K.: Oxford University Press, pp. 568–599.
31. Sánchez EG, Quintas A, Pérez-Núñez D, Nogal M, Barroso S, Carrascosa AL, and Revilla Y. 2012. African swine fever virus uses macropinocytosis to enter host cells. *PLoS Pathog.* 8(6):e1002754. doi:10.1371/journal.ppat.1002754.
32. Colgrove GS, Haelterman EO, and Coggins L. 1969. Pathogenesis of African swine fever in young pigs. *Am J Vet Res.* 30(8):1343–1359.
33. Oura CA, Powell PP, and Parkhouse RM. 1998. Detection of African swine fever virus in infected pig tissues by immunocytochemistry and in situ hybridisation. *J Virol Methods.* 72(2):205–217.
34. Sierra MA, Quezada M, Fernandez A, Carrasco L, Gomez-Villamandos JC, Martin de las Mulas J, and Sanchez-Vizcaino JM. 1989. Experimental African swine fever: Evidence of the virus in interstitial tissues of the kidney. *Vet Pathol.* 26(2):173–176.

35. Gómez-Villamandos JC, Carrasco L, Bautista MJ, Sierra MA, Quezada M, Hervas J, Chacón Mde L, Ruiz-Villamor E, et al. 2003. African swine fever and classical swine fever: A review of the pathogenesis. *Dtsch Tierarztl Wochenschr.* 110(4):165–169.

36. Salguero FJ, Sánchez-Cordón PJ, Sierra MA, Jover A, Núñez A, Gómez-Villamandos JC. 2004. Apoptosis of thymocytes in experimental African swine fever virus infection. *Histol Histopathol.* 19(1):77–84.

37. Wardley RC, Abu Elzein EM, Crowther JR, and Wilkinson PJ. 1979. A solid-phase enzyme linked immunosorbent assay for the detection of African swine fever virus antigen and antibody. *J Hyg (Lond).* 83(2):363–369.

38. Pastor MJ and Escribano JM. 1990. Evaluation of sensitivity of different antigen and DNA-hybridization methods in African swine fever virus detection. *J Virol Methods.* 28(1):67–77.

39. Alcaraz C, De Diego M, Pastor MJ, and Escribano JM. 1990. Comparison of a radio-immunoprecipitation assay to immunoblotting and ELISA for detection of antibody to African swine fever virus. *J Vet Diagn Invest.* 2(3):191–196.

40. Gallardo C, Soler A, Nieto R, Carrascosa AL, De Mia GM, Bishop RP, Martins C, et al. 2013. Comparative evaluation of novel African swine fever virus (ASF) antibody detection techniques derived from specific ASF viral genotypes with the OIE internationally prescribed serological tests. *Vet Microbiol.* 162: 32–43.

41. Crowther JR, Wardley RC, and Wilkinson PJ. 1979. Solidphase radioimmunoassay techniques for the detection of African swine fever antigen and antibody. *J Hyg (Lond).* 83(2):353–361.

42. Carrascosa AL, Bustos MJ, and de Leon P. 2011. Methods for growing and titrating African swine fever virus: Field and laboratory samples. *Curr Protoc Cell Biol.* Chapter 26:Unit 26.14.

43. Malmquist WA, and Hay D. 1960. Hemadsorption and cytopathic effect produced by African swine fever virus in swine bone marrow and buffy coat cultures. *Am J Vet Res,* 21: 104–108.

44. Weingartl HM, Sabara M, Pasick J, van Moorlehem E, and Babiuk L. 2002. Continuous porcine cell lines developed from alveolar macrophages: Partial characterization and virus susceptibility. *J Virol Methods.* 104(2):203–216.

45. Ballester M, Galindo-Cardiel I, Gallardo C, Argilaguet JM, Segalés J, Rodríguez JM, and Rodríguez F. 2010. Intranuclear detection of African swine fever virus DNA in several cell types from formalin-fixed and paraffin-embedded tissues using a new in situ hybridisation protocol. *J Virol Methods.* 168(1–2):38–43.

46. Vidal MI, Stiene M, Henkel J, Bilitewski U, Costa JV, and Oliva AG. 1997. A solid-phase enzyme linked immunosorbent assay using monoclonal antibodies, for the detection of African swine fever virus antigens and antibodies. *J Virol Methods* 66(2):211–8.

47. Hutchings GH, and Ferris NP. 2006. Indirect sandwich ELISA for antigen detection of African swine fever virus: comparison of polyclonal and monoclonal antibodies. *J Virol Methods.* 131:213–7. doi: 10.1016/j.jviromet.2005.08.009.

48. Agüero M, Fernández J, Romero L, Sánchez Mascaraque C, Arias M, and Sánchez-Vizcaíno JM. 2003. Highly sensitive PCR assay for routine diagnosis of African swine fever virus in clinical samples. *J Clin Microbiol.* 41(9):4431–4434.

49. Agüero M, Fernández J, Romero LJ, Zamora MJ, Sánchez C, Belák S, Arias M, and Sánchez-Vizcaíno JM. 2004. A highly sensitive and specific gel-based multiplex RT-PCR assay for the simultaneous and differential diagnosis of African swine fever and Classical swine fever in clinical samples. *Vet Res.* 35(5):551–63.

50. Giammarioli M, Pellegrini C, Casciari C, and De Mia GM. 2008. Development of a novel hot-start multiplex PCR for simultaneous detection of classical swine fever virus, African swine fever virus, porcine circovirus type 2, porcine reproductive and respiratory syndrome virus and porcine parvovirus. *Vet Res Commun.* 32(3):255–262.

51. Basto AP, Portugal RS, Nix RJ, Cartaxeiro C, Boinas F, Dixon LK, Leitão A, and Martins C. 2006. Development of a nested PCR and its internal control for the detection of African swine fever virus (ASFV) in *Ornithodoros erraticus. Arch Virol.* 151(4):819–826.

52. King DP, Reid SM, Hutchings GH, Grierson SS, Wilkinson PJ, Dixon LK, Bastos AD, and Drew TW. 2003. Development of a TaqMan PCR assay with internal amplification control for the detection of African swine fever virus. *J Virol Methods.* 107(1):53–61.

53. Zsak L, Borca MV, Risatti GR, Zsak A, French RA, Lu Z, Kutish GF, et al. 2005. Preclinical diagnosis of African swine fever in contact-exposed swine by a real-time PCR assay. *J Clin Microbiol.* 43(1):112–119.

54. Fernandez-Pinero J, Gallardo C, Elizalde M, Robles A, Gomez C, Bishop R, Heath L, et al. 2013. Molecular diagnosis of African swine fever by a new real-time PCR using universal probe library. *Transbound Emerg Dis.* 60: 48–58. doi:10.1111/j.1865-1682.2012.01317.x

55. Tignon M, Gallardo C, Iscaro C, Hutet E, Van der Stede Y, Kolbasov D, De Mia GM, et al. 2011. Development and interlaboratory validation study of an improved new real-time PCR assay with internal control for detection and laboratory diagnosis of African swine fever virus. *J Virol Methods.* 178(1–2):161–170.

56. McKillen J, Hjertner B, Millar A, McNeilly F, Belák S, Adair B, and Allan G. 2007. Molecular beacon real-time PCR detection of swine viruses. *J Virol Methods.* 140(1–2):155–165.

57. McKillen J, McMenamy M, Hjertner B, McNeilly F, Uttenthal A, Gallardo C, Adair B, and Allan G. 2010. Sensitive detection of African swine fever virus using real-time PCR with a 5′ conjugated minor groove binder probe. *J Virol Methods.* 168(1–2):141–146.

58. Ronish B, Hakhverdyan M, Ståhl K, Gallardo C, Fernandez-Pinero J, Belák S, Leblanc N, and Wangh L. 2011. Design and verification of a highly reliable linear-after-the-exponential PCR (LATE-PCR) assay for the detection of African swine fever virus. *J Virol Methods.* 172(1–2):8–15.

59. Madi M, Mioulet V, King DP, Andreou MP, Millington S, Das P, Wakeley PR, North S. 2012. A hand-held platform to detect trans-boundary livestock disease viruses using loop-mediated isothermal amplification (LAMP). Book of Abstract of the 6th Annual Meeting of EPIZONE, "Viruses on the move". Brighton, UK, June 2012, p. 28.

60. Notomi T, Okayama H, Masubuchi H, Yonekawa T, Watanabe K, Amino N, and Hase T. 2000. Loop-mediated isothermal amplification of DNA. *Nucl. Acids Res.* 28 (12): e63. doi: 10.1093/nar/28.12.e63.

61. James HE, Ebert K, McGonigle R, Reid SM, Boonham N, Tomlinson JA, Hutchings GH, et al. 2010. Detection of African swine fever virus by loop-mediated isothermal amplification. *J Virol Methods.* 164(1–2):68–74.

62. Chapman DA, Tcherepanov V, Upton C, and Dixon LK. 2008. Comparison of the genome sequences of non-pathogenic and pathogenic African swine fever virus isolates. *J Gen Virol.* 89:397–408.

63. de Villiers EP, Gallardo C, Arias M, da Silva M, Upton C, Martin R, and Bishop RP. 2010. Phylogenomic analysis of 11 complete African swine fever virus genome sequences. *Virology.* 400(1):128–136.

64. Fasina FO, Agbaje M, Ajani FL, Talabi OA, Lazarus DD, Gallardo C, Thompson PN, and Bastos AD. 2012a. Risk factors for farm-level African swine fever infection in major pig-producing areas in Nigeria, 1997–2011. *Prev Vet Med.* 107(1–2):65–75.

65. Fasina FO, Lazarus DD, Spencer BT, Makinde AA, and Bastos ADS. 2012b. Cost implications of African swine fever in smallholder farrow-to-finish units: Economic benefits of disease prevention through biosecurity. *Transbound Emerg Dis.* 53:244–255.

66. FAO. 2009. Preparation of African swine fever contingency plans. In: Penrith M-L, Guberti V, Depner K, and Lubroth J, eds. *FAO Animal Production and Health Manual No. 8.* Rome, Italy: FAO.

67. Steiger Y, Ackermann M, Mettraux C, and Kihm U. 1992. Rapid and biologically safe diagnosis of African swine fever virus infection by using polymerase chain reaction. *J Clin Microbiol.* 30(1):1–8.

68. Bastos ADS, Arnot LF, Jacquier MD, and Maree S. 2009. A host species-informative internal control for molecular assessment of African swine fever virus infection rates in the African sylvatic cycle *Ornithodoros* vector. *Med Vet Entomol.* 23(4):399–409.

69. Arnot LF, du Toit JT, and Bastos ADS. 2009. Molecular monitoring of African swine fever virus using surveys targeted at adult *Ornithodoros* ticks: A re-evaluation of Mkuze Game Reserve, South Africa. *Onderstepoort J Vet Res.* 76:385–392.

51 Akabane Virus

Norasuthi Bangphoomi, Akiko Uema, and Hiroomi Akashi

CONTENTS

51.1 INTRODUCTION

Akabane virus (AKAV) causes reproductive and congenital problems in cattle, sheep, and goats, resulting in severe economic losses. The identification of arthropod-borne diseases, such as Akabane disease, was first reported in a case of arthrogryposis and hydranencephaly abnormalities in calves in Australia in 1951–1955.[1] However, viral isolation was not attempted. In 1959, AKAV was first isolated from mosquitoes in Akabane village of the Gunma Prefecture, Japan. This survey collected mosquitoes from animal farms and isolated AKAV JaGAr39 strains from *Aedes* arbovirus.[2,3] *Culicoides brevitarsis* is known to be a vector of this virus in Australia[4] and *Culicoides oxystoma* in Japan.[5] The impact of AKAV was first reported from the outbreak in 1974 in Australia, 1972–1975 in Japan, and 1969–1970 in Israel.[6–8] Previously, an outbreak in Japan caused more than 42,000 congenital abnormalities, abortions, stillbirths, and premature births in cattle.[7] Furthermore, AKAV has been detected in many tropical and subtropical regions[9–13] throughout various countries. Vaccination has reduced the prevalence of Akabane disease; however, cases still occur in Japan and Korea.[14,15] Recently, antigenic and pathogenic variants of AKAV have been isolated.[16–19] These variants may be the cause of disease continuation in areas in which vaccines are administered. Therefore, to elucidate the etiological agent, characterization of both the epidemiology and pathogenesis is required. Moreover, molecular studies will facilitate novel strategies for control of Akabane disease in the future.

51.2 CLASSIFICATION AND MORPHOLOGY

AKAV belongs to the genus *Orthobunyavirus* of the Bunyaviridae family. Based on serology, AKAV is a member of the Simbu serogroup, which consists of 25 viruses distributed globally that cause disease in human and ruminants and are transmitted by biting midges and mosquitoes.[20] Other viruses included in the serogroup are the Aino virus, which affects Japan and Australia[21,22] and Schmallenberg virus, which affects European countries.[23–25] These viruses cause abortion and congenital abnormalities in cattle, sheep, and goats, while Oropouche virus causes febrile disease in humans in South America.[26]

AKAV is an enveloped virus and has a tripartite negative-sense, single-stranded RNA genome. The viral particle is roughly spherical, sometimes pleomorphic, and has a diameter of 70–130 nm (typically 90–100 nm) (Figure 51.1). The virion surface consists of an envelope with a spikelike projection.[27] The envelope is a host cell-derived lipid layer while spikes are comprised of two viral glycoproteins (Gn and Gc). In the *Orthobunyavirus* genus, the virion has three genome segments (L, M, and S segments).[28] Each genome is encapsulated with nucleocapsid proteins (N) and an RNA-dependent RNA polymerase (RdRp), referred to as the ribonucleoprotein (RNP) complex, in a helical configuration. The large (L) segment encodes the large protein (L), which acts as the RdRp. The M segment encodes a precursor polyprotein containing NH_2-Gn-NSm-Gc-COOH. Afterward, the polyprotein is cleaved into two glycoproteins, Gn and Gc, and a nonstructural protein (NSm). Gc is responsible for viral entry,[29] while the function of NSm remains unknown. However, it may

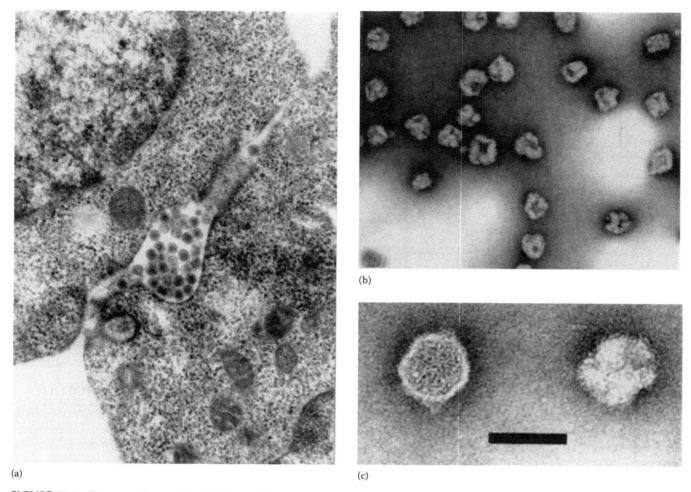

(a)

(b)

(c)

FIGURE 51.1   Electron micrographs of Akabane virus. (a) Virus-infected HmLu-1 cells. Spherical virus particles are seen in the intercellular space. (b, c) Negatively stained virions. The bar represents 100 nm.

be responsible for apoptosis signaling.[30] The S segment encodes N and a nonstructural protein (NSs) in an overlapping open reading frame (ORF). N plays a role in RNP formation, and NSs is responsible for interferon antagonism and regulation of host protein synthesis.[31,32] Furthermore, it has been suggested that the untranslated region (UTR) of each segment plays an essential role in RNA regulation and genome synthesis.[33]

Orthobunyaviruses enter cells by attaching to a cell-surface receptor and penetrating through the endocytic pathway.[34] However, the complete details of viral entry remain unclear. After viral uptake into the cytosol, virus particles fuse with the endosome to gain acidity and release viral genome. Subsequently, primary transcription and translation of viral proteins begins. The new virion assembles and buds into the Golgi compartment, followed by exocytosis from cells.[28,35,36]

51.3 BIOLOGY AND EPIDEMIOLOGY

51.3.1 BIOLOGICAL PROPERTIES

The physicochemical properties of AKAV have been reported previously.[37] The virus is completely inactivated at 56°C within 5 min in an acidic environment at pH 3.0.

This virus is sensitive to lipid solvents (ether and chloroform), ionic detergents (sodium deoxycholate), and protease (trypsin), but is resistant to DNA-suppression agents (5-iodo-2'-deoxyuridine) and DNA-precipitation agents (protamine sulfate). Exposure to gamma radiation of 0.25 megarads decreases viral infectivity.[38]

AKAV agglutinates pigeon, duck, and goose erythrocytes at any temperature and also lyses pigeon erythrocytes significantly at 37°C. The hemagglutination (HA) and hemolytic activities of AKAV are dependent on the pH and NaCl molarity of the diluent.[39,40] The HA titer is markedly increased using a diluent with 0.4 M NaCl and 0.2 M phosphate at pH 6.0–6.2. Analysis of AKAV antigenic properties using monoclonal antibodies showed that Gc possessed HA activity.[16]

51.3.2 GEOGRAPHIC DISTRIBUTION

AKAV has been reported in many tropical and subtropical countries, including East Asia, Southeast Asia, Australia, Africa, and Middle East, by viral isolation and serological surveys. However, pathogenic AKAV in a natural host has been isolated only in Japan, Taiwan, Korea, Australia, Kenya, and Israel.[3,8–11,41–43]

Currently, AKAV is divided into four genetic clusters based on the S RNA segment sequences of field AKAV isolates from Japan, Australia, Taiwan, Israel, and Kenya. The studies showed that two clusters comprised the Japanese, Taiwanese, and Israeli isolates while the Australian and Kenyan isolates were each placed in an individual cluster. Phylogenetic analysis revealed that the Japanese field isolates could be divided into two major clusters (genotypes I and II) and that the second cluster (genotype II) was subdivided into two branches (IIa and IIb) (Figure 51.2). These data suggest that AKAVs have evolved in multiple lineages and may have arrived in Japan from other countries on more than one occasion. Thus, it is suggested that geographic divergence is the main factor driving AKAV variation.[17,44]

51.3.3 HOSTS

AKAV-associated congenital abnormalities have been reported in cattle, sheep, and goats.[7,45–47] In Taiwan, AKAV has been isolated in pigs that showed convulsion and diarrhea.[48] Moreover, horse, donkey, camel, and wild-life animals have been found to be seropositive for AKAV; however, pathogenic lesions have not been reported.[12,49]

Mice are the most commonly used experimental animal for infection with both AKAV and other orthobunyaviruses. Pregnant hamsters and embryonated chicken eggs have also been used as models of AKAV infection. AKAV readily infects mice and causes encephalitis by the intracerebral (i.c.) route. Suckling mice are highly susceptible to AKAV infection not only via the i.c. route but also via the intraperitoneal (i.p.) or subcutaneous (s.c.) route.[50,51] Transplacental

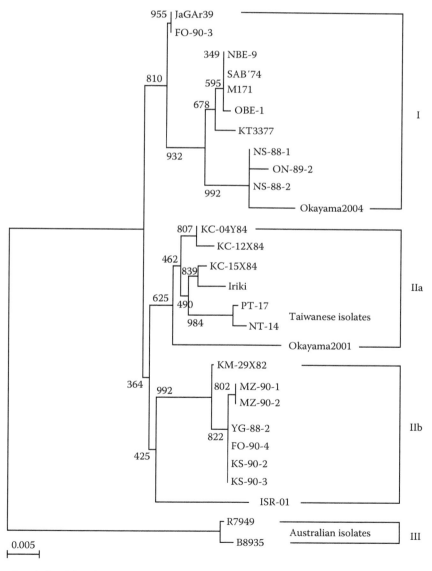

FIGURE 51.2 Phylogenetic relationship of the AKAVs on the basis of their N ORF nucleotide sequences. Phylogenetic analyses of the N ORFs (between nucleotide positions −34 and −735, total 702 nucleotides) show that four groups are extant. Numbers indicate bootstrap percentages out of 1000 replicates. The scale bar represents 0.005% substitution per base.

infections of hamster fetuses were produced by the inoculation of AKAV via either the i.p. or the s.c. route.[52] Moreover, infection of embryonated chicken eggs via the yolk sac with the orthobunyaviruses, including AKAV, resulted in death or congenital deformities, such as arthrogryposis, scoliosis, mandible defects, and retarded development, similar to those observed in infected fetal ruminants. This suggests that the teratogenic potential of AKAV, as well as other orthobunyaviruses, can be assessed using this model system.[51]

51.3.4 TRANSMISSION

The association of outbreaks of AKAV infection with summer and autumn indicates that this virus is transmitted by arthropod vectors.[1,7] The biting midge of the genus *Culicoides* has been used as a model of AKAV transmission.[5,43,53,54] The more than 14,000 species in this genus are distributed globally, excluding Antarctica and New Zealand. The transmission of AKAV by *Culicoides* spp. differs according to species and locations. In Australia, AKAV is transmitted by *C. brevitarsis* Kieffer. Epidemiological studies showed that *C. brevitarsis* was active during the AKAV outbreak, and AKAV was isolated from this midge.[46,55] Due to suitable environmental conditions, *C. brevitarsis* is a common biting pest of domestic ruminants, which are widespread throughout the year in Northern Australia, mid-northern and northern coastal plains to the south coastal plains of New South Wales. Therefore, AKAV is endemic in these areas.[56–58] Occasionally, in epidemic areas, climate changes and windy conditions result in migration of infected midges to epidemic areas; however, the outbreak gradually subsides since infected midges cannot survive during the winter.[59–61] AKAV-neutralizing antibodies have been found in other host species, such as horse or camel, which are not common hosts of *C. brevitarsis* in endemic areas, suggesting that these animals are accidental hosts of *C. brevitarsis* or that this virus has other vectors.[58] In the southern parts of Japan, *C. oxystoma* is the principal vector of AKAV.[5,62] However, *C. brevitarsis* has been detected in the southern islands of Japan, leading to concerns of arbovirus outbreaks, such as those involving AKAV and other orthobunyaviruses.[63] In Oman, *Culicoides imicola* was reported to be an AKAV vector.[64] In Israel, the definitive vector remains unknown. It is possible that *Culicoides puncticollis* and *C. imicola* are vectors of AKAV; however, inoculation of AKAV to these midges via the oral route showed they were unable to replicate.[54,65] Moreover, AKAV can replicate in *Culicoides variipennis*, which is widespread in America, via oral and intrathoracic inoculation, while *Culicoides nubeculosus* is widespread in Europe but becomes infected via only intrathoracic inoculation in vitro.[53] However, there have been no reports of virus isolation from these midges in field surveys of these regions.

AKAV have been isolated from mosquitoes, such as *Aedes vexans* and *Culex tritaeniorhynchus* in Japan[3] or *Anopheles funestus* in Kenya.[42] However, experimental infection revealed that AKAV does not replicate in *Culex* and *Aedes* spp., which suggested that virus isolation from these vectors was derived from blood meal and not from viral replication in the mosquito itself.[55] Recently, AKAV was isolated from non-blooded *C. tritaeniorhynchus*, *Culex vishnui*, *Anopheles vagus*, and *Ochlerotatus* species in northern Vietnam.[9] Furthermore, AKAV propagated and maintained in *Aedes albopictus* cell culture as a persistent infection.[66] Thus, it remains possible that the virus can be transmitted by mosquito.

51.4 CLINICAL FEATURES AND PATHOGENESIS

51.4.1 CLINICAL SIGNS

The clinical signs of Akabane disease depend on the stage of pathological change. Severe signs are observed during the hydranencephaly and arthrogryposis stages (Figure 51.3).

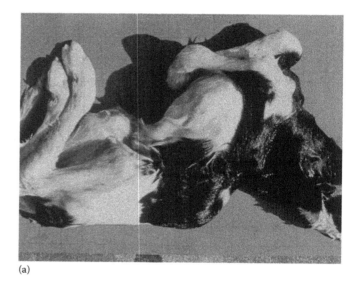

(a)

(b)

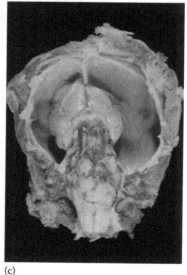

(c)

FIGURE 51.3 Clinical features of Akabane disease. (a) Newborn calf with severe arthrogryposis of fore- and hindlimbs and torticollis. (b) Cerebral defect of a naturally infected newborn calf. (c) Hydranencephaly and cerebral defect of an experimentally infected calf.

Infected pregnant cattle, sheep, and goats show dystocia, abortion, stillbirth, and premature birth while delayed labor can be observed with only pregnant goats.[67] Infected fetuses are generally dead during birth. Most live newborns show neurological problems, including the inability to stand, blindness, lack of a sucking reflex, nystagmus, and death within 3 days due to an inability to feed.[68,69] During the nonpurulent encephalomyelitis and myositis stages, the fetuses are usually born with mild to moderate neurological and muscular problems such as ataxia, muscle atrophy, paralysis, and blindness.[69]

With regard to AKAV infection in newborn and adults, results are generally transient viremia without any clinical signs.[70,71] However, infection with the highly pathogenic AKAV strain, Iriki, which was isolated from the cerebellum specimen of a calf, resulted in nonpurulent encephalitis.[72] Iriki-like strains can cause encephalomyelitis in young and adult animals.[11,41] Various clinical signs have been observed, including lassitude, anorexia, loss of milk yield, tremor, stretching of the legs, and ataxia. These are suggestive of a poor prognosis since there is typically no response to therapy. Therefore, all infected, symptomatic animals should be euthanized.

51.4.2 PATHOLOGY

Pathological examination of hydranencephaly and arthrogryposis shows disorders in two major organs, the central nervous and muscular systems. Gross lesions of the central nervous system present as a normal skull with hypoplasia in both sides of the cerebrum. The cranial cavity is replaced by a clear cerebrospinal fluid. The midbrain, cerebellum, medulla oblongata, and spinal cord are reduced in size.[70,71] The cystic cavity and small subcortical cysts in brain tissue can also be observed.[73] Skeletal deformities are present, including scoliosis, kyphosis, torticollis, and brachygnathism.[71,74] Skeletal muscle is atrophied and edematous, predominately in the limbs. Joints are fixed in flexion.[74] No macroscopic lesions were observed in other organs.

Histopathological lesions in the central nervous system show nerve cell degeneration. Most of the perivascular layer of nerve vessels was accumulated with plasma cells, histiocytes, and lymphocytes, so-called perivascular cuffs.[74] The ventral horn of the spinal cord also loses its motor neurons, gliosis, and perivascular cuffs,[75] whereas the muscle system shows muscle fiber degeneration and inflammatory cells infiltrate among degenerative fibers and in interstitial tissue.[76]

With regard to encephalitis in newborns and adults, no macroscopic lesions are identified in the central nervous or muscular systems. However, histopathological signs of nonpurulent encephalomyelitis are present,[77] comprising perivascular cuffs and nerve cell degeneration in the midbrain, cerebellopontine, and medulla oblongata, but rarely in the cerebrum and cerebellum.

51.4.3 PATHOGENESIS

The infection is initiated by intravenous inoculation of AKAV in the dermis using infected arthropod vectors. The virus has a short incubation period, and viremia is typically present 1–9 days after inoculation. However, the main cells responsible for viral multiplication remain unclear. AKAV-neutralizing antibody is detected 1–3 days after viremia and can be persistently detected for more than a year. During this period, the virus spreads to the brain, spinal cord, and skeletal muscle by a hematogenous route. However, it has been reported that there is also a tissue tropism of AKAV in the placenta, ovary, cornea, uvea, and lens fiber.[78,79]

The host immune system plays a crucial role in defense against AKAV infection. It has been suggested that both humoral and cellular immunity are responsible for the decrease or total elimination of the virus from blood vessels and infected tissues. Thus, animals with a mature immune system have the potential to resist AKAV infection,[71,80] whereas animals with an immature immune system, predominately the fetus, are susceptible to virus infection and exhibit virulent pathogenic lesions.[81]

The virulence pattern of Akabane disease predominately affected fetuses and caused congenital abnormalities including hydranencephaly and arthrogryposis, microencephaly, and porencephaly.[71] Transplacental vertical transmission is the major route of AKAV infection.[82] The pathological changes are closely related to the age at which the fetus is infected. The hydranencephaly and arthrogryposis lesions are produced in fetuses after exposure during an early stage of gestation, that is, 62–96 days in cattle[70] and 26–69 days in goats and sheep.[71,73,83]

AKAV pathogenesis is divided into the following three phases: the nonpurulent encephalomyelitis phase, polymyositis phase, and hydranencephaly with arthrogryposis phase (Figure 51.3). The nonpurulent encephalomyelitis phase is the primary pathological change during AKAV infection. Histopathological and immunohistochemistry studies have shown that nervous system damage due to AKAV is mediated by direct and indirect effects. The direct effects are caused by AKAV infection of neurons, resulting in cytolysis,[84] while the indirect effects are caused by cytotoxic cytokine responses resulting in inflammatory reactions, such as vascular damage.[74] AKAV causes inflammation in all parts of the central nervous system, such as the cerebrum, cerebellum, brainstem, and spinal cord. These inflammatory lesions typically develop to severe encephalitis (necrotizing encephalitis).

The polymyositis phase is the second pathological change during AKAV infection. There is evidence that this lesion is produced only during the myotubule and myofiber phases of muscular development in the fetus (first to middle stage of gestation) and is mediated by direct and indirect effects, resulting in abnormal myotubules and inflammation. Therefore, infection of AKAV after the late stages of gestation will not result in polymyositis.[76]

The hydranencephaly and arthrogryposis phase is the final stage of Akabane disease lesion development. Hydranencephaly and arthrogryposis are sequential changes of nonpurulent encephalomyelitis and polymyositis, respectively.[74] Hydranencephaly is produced by the circulating cerebrospinal fluid in necrosis areas during the encephalitis phase, while arthrogryposis and muscular atrophy are produced by the decrease in the number of ventral horn neurons that control muscle function and muscle fiber, which is abnormal due to polymyositis.

Hydranencephaly and arthrogryposis lesions are produced in the fetus following exposure to AKAV during the early to middle stages of gestation. If the fetus is exposed to the virus after the middle stage of gestation, the only pathological change produced is nonpurulent encephalomyetis.[41,71,73,85] Moreover, multifocal but not necrotizing encephalitis was produced during this stage. However, the differences in pathological lesions between the two groups remain unclear.

51.5 IDENTIFICATION AND DIAGNOSIS

Since Akabane disease is arthropod-borne, it spreads rapidly. Thus, diagnosis must be rapid and reliable in order to prevent transmission to noninfected animals.

A field diagnosis is based on clinical manifestation and detection of arthropod vector activity. Examination of postmortem changes by autopsy or necropsy, as well as histological examination, has been used. However, other orthobunyaviruses, including Aino and Peaton viruses, have been associated with bovine congenital abnormalities in Japan, and the disease itself is difficult to diagnose through clinical features alone.

Laboratory techniques for detection of AKAV can be divided into the following three categories: serological, viral isolation and identification, and viral antigen detection and identification. Serological analysis is a common and convenient method for AKAV detection due to its rapidity and high sensitivity. This method is generally used to diagnose viral infection, conduct epidemiological surveys, and determine immunity levels in vaccinated animals. Serum samples from a dam and its fetus are usually taken and subjected to serological testing. Alternatively, antibodies against AKAV can be detected in cerebrospinal and pericardial fluids.[86] Sera can be preserved long-term by freezing.

The serum-neutralization test (NT) is the most commonly used for serological diagnosis of AKAV infection.[71,87–89] The hemagglutination inhibition test,[90] complement fixation test (CFT),[91] hemolysis inhibition test,[40] and enzyme-linked immunosorbent assay (ELISA)[92,93] have also been used for serological diagnosis of AKAV infection. However, group-specific antibody-detection methods, such as CFT and ELISA, can detect cross-reactive antibodies against viruses of the same serogroup. These tests require correct interpretation because they detect immunoglobulin G (IgG), which can prolong positivity after virus exposure. To evaluate whether observed animals were recently infected, paired serum samples taken 1 week apart must show an increasing titer,[7,68] or IgM (a recent infection indicator) must be assayed simultaneously with IgG.[93]

Detection of neutralizing antibody in precolostral or fetal serum is reliable since cattle, sheep, and goats cannot transfer maternal antibodies to the fetus through the placenta; instead, newborn animals acquire antibodies from the dam via colostrum. Therefore, the presence of antibodies against AKAV in precolostral or fetal serum suggests that the fetus had been infected with the virus *in uterine*.

Virus isolation is a general technique that can be applied to identification and vaccine preparation. However, this technique is not appropriate in routine diagnosis work because it is slow and cannot be used for abnormally delivered calves due to the presence of antibody. In some cases, virus isolation was not successful, and other tests were positive for AKAV.[18,75,94] Defibrinated blood from sentinel cattle and central nervous tissue and skeletal muscle from aborted fetuses are typically used for virus isolation.

AKAV is commonly isolated using cell culture techniques. The cell types that are standard for AKAV isolation contain HmLu-1 cells[45] derived from hamster lung, BHK-21 cells from hamster kidney,[19] and Vero cells from African green monkey kidney,[78,80] due to their sensitivity to this virus based on the cytopathic effect (CPE) and plaque formation.[95] Moreover, experimental animals, such as suckling mice, hamsters,[3] or chicken embryos,[96] are commonly used for AKAV isolation.

Anti-AKAV antibody production is the key factor for immunological testing.[16] The techniques used for AKAV antigen detection include immunohistochemistry[18,41,77,85] and immunofluorescence[84] assays. Brain and skeletal muscle are typically subjected to histopathological examination. However, it is important to verify AKAV specificity using positive and negative controls due to the cross-reactivity among Simbu serogroup viruses.

Due to advancements in molecular biology technologies, amplification of AKAV genetic materials is one of the most valuable tools for confirmation of AKAV infection. Reverse-transcription polymerase chain-reaction (RT-PCR) as well as nested RT-PCR[97] has been proposed as the new standard for AKAV detection due to their rapidity and sensitivity. Moreover, multiplex RT-PCR[98,99] and quantitative real-time RT-PCR[99] facilitate rapid identification of viruses to the serotype level with improved accuracy.

51.6 TREATMENT AND PREVENTION

There is no practical treatment for AKAV infection because this virus causes congenital abnormalities in neonatal animals. Protection against Akabane disease requires vaccination and vector control.

Vaccination is the most practical prevention strategy for animals in endemic and epidemic areas. Development and use of vaccines against Akabane disease has been reported in Japan, Korea, and Australia. Inactivated vaccines have been used in Japan, Korea, and Australia,[15,100,101] while live-attenuated vaccines have been used only in Japan and South Korea.[15,102]

Inactivated and formalin-inactivated vaccines have been developed in Japan. Effectiveness was demonstrated after two intramuscular injections at 4-week intervals in the first immunization. NT titers were detected within 1 month of immunization, while NT levels became zero 12 months postimmunization. Therefore, an annual booster injection of inactivated vaccine was recommended. Moreover, the safety and effectiveness of the inactivated vaccine in pregnant cows and goats was found to be adequate, and the vaccine was deemed ready for general use in Japan.[100] An inactivated vaccine is also available in Australia. The vaccine is administrated by intramuscular injection on two occasions, 4 weeks apart, in heifers and cows prior to mating.[101] Recently, formalin-inactivated trivalent vaccines for Aino, Akabane, and Chuzan viruses have been developed in Japan and South Korea. This vaccine exhibits improved safety and immunogenicity profiles in mice, guinea pigs, and pregnant cows.[15]

Production of attenuated virus for live vaccine has been demonstrated in Japan. The attenuated strain was established using the cold temperature-adapted OBE-1 strain, which was isolated from a naturally infected bovine fetus during the 1974 outbreak, named the TS-C2 strain. This strain was attenuated by subjecting the OBE-1 strain to 20 passages in HmLu-1 cells at 30°C. The safety and immunogenicity of the TS-C2 strain in pregnant cows and ewes has been evaluated. The pregnant cows exhibited high immunogenicity within 4 weeks of inoculation. Moreover, no pathogenicity, pyrexia, leucopenia, or viremia was detected. Pregnant ewes also exhibited immunogenicity after inoculation; however, viremia was detected. Therefore, this vaccine candidate strain may be used in pregnant cows, but not in pregnant ewes for safety reasons.[102–104]

Although the vaccine has decreased the incidence of Akabane disease, especially in Australia, sporadic cases still occur in Japan and South Korea.[14] Recently, several variants of AKAV, the antigenic and pathogenic features of which differ from the classical strains of AKAV, such as OBE-1 and Iriki, have been isolated.[19] These extensive genetic variations may be responsible for vaccine failure. Therefore, an effective vaccine is still required. Furthermore, AKAV vaccination programs for calves may be considered in AKAV Iriki-like strain outbreak areas.

Molecular biology techniques play an essential role in modern vaccine development. Using reverse genetic manipulation, we obtained recombinant AKAV from cloned cDNA and exploited molecular analysis and vaccine development.[105,106] There is evidence that NSs protein acts as a virulence factor by suppressing the production of interferon-α/β,[32] and inoculation of a recombinant NSs-deleted AKAV strain resulted in a decreased mortality rate in mice. Therefore, this strain is considered a candidate live-attenuated vaccine.[105] However, the animals immunized with such a live vaccine will not produce anti-NSs antibody. Therefore, it may function as a molecular marker for differentiation of infected from vaccinated animals, which leads to novel Akabane eradication strategies.

Vector control is one of the most reliable methods of arbovirus disease prevention. This method includes the elimination of larvae and adult arthropods (biting midges and mosquitoes) using insecticides or the avoidance of exposure to arthropods by screening of animal sheds.[107] However, these vector control measures may be impractical or difficult to apply in the field since insecticides have a temporary effect and their excessive use has a negative impact on the environment.

51.7 CONCLUSIONS AND FUTURE PERSPECTIVES

In 2011, an outbreak of abortion, stillbirth, and birth at term of lambs, kids, and calves with neurologic signs and/or head, spine, or limb malformations occurred in several European countries. A novel orthobunyavirus sequence was detected by metagenomic analysis of the blood samples of affected animals at a farm near the city of Schmallenberg, Germany; the virus was subsequently isolated from the blood of a diseased cow using cell culture.[108] Therefore, the virus was named "Schmallenberg virus (SBV)." Based on phylogenetic analysis, the SBV genome was shown to be similar to that of Shamonda, Aino, and Akabane viruses and belonged to the Simbu serogroup. However, further genetic analyses of the entire genome sequences revealed that SBV may be a reassortant of Sathuperi and Shamonda viruses,[109] or an ancestor of Shamonda virus.[110]

Since the bunyavirus genome is segmented, new viruses have been produced in nature by genetic reassortment events among Simbu serogroup viruses.[109,111] Moreover, antigenic diversity among the orthobunyaviruses, as well as AKAV, has been demonstrated.[16] Emergence of a virus with novel antigenicity or pathogenicity will result in infected animals and their owners suffering severe damage, similar to that caused by SBV.

To control orthobunyavirus infections, further studies of the viruses, including AKAV, that belong to this genus are required to elucidate more precisely their pathogenicity. Moreover, vaccine development is important for disease prevention, but cannot be applied to emerging pathogens since vaccine development is a long-term process. However, molecular biological approaches using reverse genetics technology may be applicable at present.

REFERENCES

1. Blood, D.C. Arthrogryposis and hydranencephaly in newborn calves. *Aust Vet J* 32, 125–131 (1956).
2. Matsuyama, T. et al. Isolation of arbor viruses from mosquitoes collected at live-stock pens in Gumma Prefecture in 1959. *Jpn J Med Sci Biol* 13, 191–198 (1960).
3. Oya, A. et al. Akabane, a new arbor virus isolated in Japan. *Jpn J Med Sci Biol* 14, 101–108 (1961).
4. Doherty, R.L. Arboviruses of Australia. *Aust Vet J* 48, 172–180 (1972).
5. Kurogi, H. et al. Isolation of Akabane virus from the biting midge *Culicoides oxystoma* in Japan. *Vet Microbiol* 15, 243–248 (1987).
6. Hartley, W.J. et al. Serological evidence for the association of Akabane virus with epizootic bovine congenital arthrogryposis and hydranencephaly syndromes in New South Wales. *Aust Vet J* 51, 103–104 (1975).

7. Inaba, Y. et al. Akabane disease: Epizootic abortion, premature birth, stillbirth and congenital arthrgryposis–hydranencephaly in cattle, sheep and goats caused by Akabane virus.
Aust Vet J 51, 584–585 (1975).

8. Shimshony, S. An epizootic of Akabane disease in bovines,
ovines and caprines in Israel, 1969–70: Epidemiological
assessment. *Acta Morphol Acad Sci Hung* 28, 197–199 (1980).

9. Bryant, J.E. et al. Isolation of arboviruses from mosquitoes
collected in northern Vietnam. *Am J Trop Med Hyg* 73, 470–
473 (2005).

10. Chang, C.W. et al. Nucleotide sequencing of S-RNA segment
and sequence analysis of the nucleocapsid protein gene of the
newly isolated Akabane virus PT-17 strain. *Biochem Mol Biol
Int* 45, 979–987 (1998).

11. Liao, Y.K. et al. The isolation of Akabane virus (Iriki strain)
from calves in Taiwan. *J Basic Microbiol* 36, 33–39 (1996).

12. Al-Busaidy, S. et al. Neutralising antibodies to Akabane virus
in free-living wild animals in Africa. *Trop Anim Health Prod*
19, 197–202 (1987).

13. Taylor, W.P. and Mellor, P.S. The distribution of Akabane virus
in the Middle East. *Epidemiol Infect* 113, 175–185 (1994).

14. Inaba, Y. and Matsumoto, M. Akabane virus. In *Virus
Infections of Ruminants*, Vol. 3 (eds. Dinter, Z. and Morein, B.),
pp. 467–480 (Elsevier Science Publishers, Amsterdam, the
Netherlands, 1990).

15. Kim, Y.H. et al. Development of inactivated trivalent vaccine for the teratogenic Aino, Akabane and Chuzan viruses.
Biologicals 39, 152–157 (2011).

16. Akashi, H. and Inaba, Y. Antigenic diversity of Akabane virus
detected by monoclonal antibodies. *Virus Res* 47, 187–196
(1997).

17. Yamakawa, M. et al. Chronological and geographical variations in the small RNA segment of the teratogenic Akabane
virus. *Virus Res* 121, 84–92 (2006).

18. Kamata, H. et al. Encephalomyelitis of cattle caused by
Akabane virus in southern Japan in 2006. *J Comp Pathol* 140,
187–193 (2009).

19. Ogawa, Y. et al. Comparison of Akabane virus isolated from
sentinel cattle in Japan. *Vet Microbiol* 124, 16–24 (2007).

20. Calisher, C.E. History, classification, and taxonomy of viruses
in the family Bunyaviridae. In *The Bunyaviridae* (ed. Elliott,
R.M.), pp. 1–17 (Plenum Press, New York, 1996).

21. Cybinski, D.H. and St. George, T.D. A survey of antibody to
Aino virus in cattle and other species in Australia. *Aust Vet J*
54, 371–373 (1978).

22. Miura, Y. et al. Serological comparison of Aino and Samford
viruses in Simbu group of bunyaviruses. *Microbiol Immunol*
22, 651–654 (1978).

23. Kupferschmidt, K. Infectious disease. Scientists rush to find
clues on new animal virus. *Science* 335, 1028–1029 (2012).

24. Garigliany, M.M. et al. Schmallenberg virus in calf born at
term with porencephaly, Belgium. *Emerg Infect Dis* 18, 1005–
1006 (2012).

25. Bilk, S. et al. Organ distribution of Schmallenberg virus RNA
in malformed newborns. *Vet Microbiol* 159, 236–238 (2012).

26. Tesh, R.B. The emerging epidemiology of Venezuelan hemorrhagic fever and Oropouche fever in tropical South America.
Ann N Y Acad Sci 740, 129–137 (1994).

27. Ito, Y. et al. Electron microscopy of Akabane virus. *Acta Virol*
23, 198–202 (1979).

28. Schmaljohn, C.S. and Hooper, J.W. Bunyaviridae: The viruses
and their replication. In *Fields Virology*, vol. 2 (eds. Knipe,
D.M. and Howley, J.W.), pp. 1581–1602 (Lippincott Williams
& Wilkins, Philadelphia, PA, 2001).

29. Gonzalez-Scarano, F. et al. An avirulent G1 glycoprotein variant of La Crosse bunyavirus with defective fusion function.
J Virol 54, 757–763 (1985).

30. Acrani, G.O. et al. Apoptosis induced by Oropouche virus
infection in HeLa cells is dependent on virus protein expression. *Virus Res* 149, 56–63 (2010).

31. Bouloy, M. et al. Genetic evidence for an interferon-antagonistic function of rift valley fever virus nonstructural protein
NSs. *J Virol* 75, 1371–1377 (2001).

32. Weber, F. et al. Bunyamwera bunyavirus nonstructural protein
NSs counteracts the induction of alpha/beta interferon. *J Virol*
76, 7949–7955 (2002).

33. Lowen, A.C. and Elliott, R.M. Mutational analyses of the nonconserved sequences in the bunyamwera orthobunyavirus S
segment untranslated regions. *J Virol* 79, 12861–12870 (2005).

34. Hollidge, B.S. et al. Orthobunyavirus entry into neurons and
other mammalian cells occurs via clathrin-mediated endocytosis and requires trafficking into early endosomes. *J Virol* 86,
7988–8001 (2012).

35. Elliott, R.M. Emerging viruses: The Bunyaviridae. *Mol Med*
3, 572–577 (1997).

36. Fontana, J. et al. The unique architecture of bunyamwera virus
factories around the Golgi complex. *Cell Microbiol* 10, 2012–
2028 (2008).

37. Takahashi, E. et al. Physicochemical properties of Akabane
virus: A member of the Simbu arbovirus group of the family
bunyaviridae. *Vet Microbiol* 3, 45–54 (1978).

38. House, C. et al. Inactivation of viral agents in bovine serum by
gamma irradiation. *Can J Microbiol* 36, 737–740 (1990).

39. Goto, Y. et al. Improved hemagglutination of Simbu group
arboviruses with higher sodium chloride molarity diluent. *Vet
Microbiol* 1, 449–458 (1976).

40. Goto, Y. et al. Hemolytic activity of Akabane virus. *Vet
Microbiol* 4, 261–278 (1979).

41. Lee, J.K. et al. Encephalomyelitis associated with Akabane
virus infection in adult cows. *Vet Pathol* 39, 269–273 (2002).

42. Metselaar, D. and Robin, Y. Akabane virus isolated in Kenya.
Vet Rec 99, 86 (1976).

43. Doherty, R.L. et al. Virus strains isolated from arthropods during an epizootic of bovine ephemeral fever in Queensland.
Aust Vet J 48, 81–86 (1972).

44. Akashi, H. et al. Sequence determination and phylogenetic
analysis of the Akabane bunyavirus S RNA genome segment.
J Gen Virol 78, 2847–2851 (1997).

45. Kurogi, H. et al. Epizootic congenital arthrogryposis–hydranencephaly syndrome in cattle: Isolation of Akabane virus
from affected fetuses. *Arch Virol* 51, 67–74 (1976).

46. Della-Porta, A.J. et al. Congenital bovine epizootic arthrogryposis and hydranencephaly in Australia. Distribution of antibodies to Akabane virus in Australian Cattle after the 1974
epizootic. *Aust Vet J* 52, 496–501 (1976).

47. Parsonson, I.M. et al. Congenital abnormalities in foetal
lambs after inoculation of pregnant ewes with Akabane virus.
Aust Vet J 51, 585–586 (1975).

48. Huang, C.C. et al. Natural infections of pigs with Akabane
virus. *Vet Microbiol* 94, 1–11 (2003).

49. Yang, D.K. et al. Serosurveillance for Japanese encephalitis,
Akabane, and Aino viruses for thoroughbred horses in Korea.
J Vet Sci 9, 381–385 (2008).

50. Kurogi, H. et al. Pathogenicity of different strains of Akabane
virus for mice. *Natl Inst Anim Health Q (Tokyo)* 18, 1–7 (1978).

51. McPhee, D.A. et al. Teratogenicity of Australian Simbu serogroup and some other Bunyaviridae viruses: The embryonated
chicken egg as a model. *Infect Immun* 43, 413–420 (1984).

52. Andersen, A.A. and Campbell, C.H. Experimental placental transfer of Akabane virus in the hamster. *Am J Vet Res* 39, 301–304 (1978).

53. Jennings, M. and Mellor, P.S. *Culicoides*: Biological vectors of Akabane virus. *Vet Microbiol* 21, 125–131 (1989).

54. Mellor, P.S. et al. Infection of Israeli *Culicoides* with African horse sickness, blue tongue and Akabane viruses. *Acta Virol* 25, 401–407 (1981).

55. St George, T.D. et al. Isolations of Akabane virus from sentinel cattle and *Culicoides brevitarsis*. *Aust Vet J* 54, 558–561 (1978).

56. Murray, M.D. The seasonal abundance of female biting-midges, *Culicoides brevitarsis* Kieffer (Diptera, Ceratopogonidae), in Coastal South-Eastern Australia. *Aust J Zool* 39, 333–342 (1991).

57. Jagoe, S. et al. An outbreak of Akabane virus-induced abnormalities in calves after agistment in an endemic region. *Aust Vet J* 70, 56–58 (1993).

58. Cybinski, D.H. et al. Antibodies to Akabane virus in Australia. *Aust Vet J* 54, 1–3 (1978).

59. Murray, M.D. Akabane epizootics in New South Wales: Evidence for long-distance dispersal of the biting midge *Culicoides brevitarsis*. *Aust Vet J* 64, 305–308 (1987).

60. Bishop, A.L. et al. Models for the dispersal in Australia of the arbovirus vector, *Culicoides brevitarsis* Kieffer (Diptera: Ceratopogonidae). *Prev Vet Med* 47, 243–254 (2000).

61. Bishop, A.L. et al. Factors affecting the spread of *Culicoides brevitarsis* at the southern limit of distribution in eastern Australia. *Vet Ital* 40, 316–319 (2004).

62. Yanase, T. et al. Isolation of bovine arboviruses from *Culicoides* biting midges (Diptera: Ceratopogonidae) in southern Japan: 1985–2002. *J Med Entomol* 42, 63–67 (2005).

63. Yanase, T. et al. Detection of *Culicoides brevitarsis* activity in Kyushu. *J Vet Med Sci* 73, 1649–1652 (2011).

64. al-Busaidy, S.M. and Mellor, P.S. Isolation and identification of arboviruses from the Sultanate of Oman. *Epidemiol Infect* 106, 403–413 (1991).

65. Brenner, J. et al. Serological evidence of Akabane virus infection in northern Israel in 2001. *J Vet Med Sci* 66, 441–443 (2004).

66. Han, H.D. Propagation and persistent infection of Akabane virus in cultured mosquito cells. *Nihon Juigaku Zasshi* 43, 689–697 (1981).

67. Hashiguchi, Y. et al. Congenital abnormalities in newborn lambs following Akabane virus infection in pregnant ewes. *Natl Inst Anim Health Q (Tokyo)* 19, 1–11 (1979).

68. Haughey, K.G. et al. Akabane disease in sheep. *Aust Vet J* 65, 136–140 (1988).

69. Coverdale, O.R. et al. Congenital abnormalities in calves associated with Akabane virus and Aino virus. *Aust Vet J* 54, 151–152 (1978).

70. Kurogi, H. et al. Congenital abnormalities in newborn calves after inoculation of pregnant cows with Akabane virus. *Infect Immun* 17, 338–343 (1977).

71. Parsonson, I.M. et al. Congenital abnormalities in newborn lambs after infection of pregnant sheep with Akabane virus. *Infect Immun* 15, 254–262 (1977).

72. Miyazato, S. et al. Encephalitis of cattle caused by Iriki isolate, a new strain belonging to Akabane virus. *Nihon Juigaku Zasshi* 51, 128–136 (1989).

73. Narita, M. et al. The pathogenesis of congenital encephalopathies in sheep experimentally induced by Akabane virus. *J Comp Pathol* 89, 229–240 (1979).

74. Konno, S. et al. Akabane disease in cattle: Congenital abnormalities caused by viral infection. Spontaneous disease. *Vet Pathol* 19, 246–266 (1982).

75. Hartley, W.J. et al. Pathology of congenital bovine epizootic arthrogryposis and hydranencephaly and its relationship to Akabane virus. *Aust Vet J* 53, 319–325 (1977).

76. Konno, S. and Nakagawa, M. Akabane disease in cattle: Congenital abnormalities caused by viral infection. Experimental disease. *Vet Pathol* 19, 267–279 (1982).

77. Kono, R. et al. Bovine epizootic encephalomyelitis caused by Akabane virus in southern Japan. *BMC Vet Res* 4, 20 (2008).

78. Parsonson, I.M. et al. The consequences of infection of cattle with Akabane virus at the time of insemination. *J Comp Pathol* 91, 611–619 (1981).

79. Ushigusa, T. et al. A pathologic study on ocular disorders in calves in southern Kyushu, Japan. *J Vet Med Sci* 62, 147–152 (2000).

80. Della-Porta, A.J. et al. Akabane disease: Isolation of the virus from naturally infected ovine foetuses. *Aust Vet J* 53, 51–52 (1977).

81. McClure, S. et al. Maturation of immunological reactivity in the fetal lamb infected with Akabane virus. *J Comp Pathol* 99, 133–143 (1988).

82. Kirkland, P.D. et al. The development of Akabane virus-induced congenital abnormalities in cattle. *Vet Rec* 122, 582–586 (1988).

83. Kurogi, H. et al. Experimental infection of pregnant goats with Akabane virus. *Natl Inst Anim Health Q (Tokyo)* 17, 1–9 (1977).

84. Kitani, H. et al. Preferential infection of neuronal and astroglia cells by Akabane virus in primary cultures of fetal bovine brain. *Vet Microbiol* 73, 269–279 (2000).

85. Uchida, K. et al. Detection of Akabane viral antigens in spontaneous lymphohistiocytic encephalomyelitis in cattle. *J Vet Diagn Invest* 12, 518–524 (2000).

86. Kirkland, P.D. Akabane and bovine ephemeral fever virus infections. *Vet Clin North Am Food Anim Pract* 18, 501–514, viii–ix (2002).

87. Miura, Y. et al. Neutralizing antibody against Akabane virus in precolostral sera from calves with congenital arthrogryposis–hydranencephaly syndrome. *Arch Gesamte Virusforsch* 46, 377–380 (1974).

88. Sellers, R.F. and Herniman, K.A. Neutralising antibodies to Akabane virus in ruminants in Cyprus. *Trop Anim Health Prod* 13, 57–60 (1981).

89. Mohamed, M.E. et al. Akabane virus: Serological survey of antibodies in livestock in the Sudan. *Rev Elev Med Vet Pays Trop* 49, 285–288 (1996).

90. Goto, Y. et al. Hemagglutination-inhibition test applied to the study of Akabane virus infection in domestic animals. *Vet Microbiol* 3, 89–99 (1978).

91. Sato, K. et al. Appearance of slow-reacting and complement-requiring neutralizing antibody in cattle infected with Akabane virus. *Vet Microbiol* 14, 183–189 (1987).

92. Blacksell, S.D. et al. Rapid identification of Australian bunyavirus isolates belonging to the Simbu serogroup using indirect ELISA formats. *J Virol Methods* 66, 123–133 (1997).

93. Ungar-Waron, H. et al. ELISA test for the serodiagnosis of Akabane virus infection in cattle. *Trop Anim Health Prod* 21, 205–210 (1989).

94. Oem, J.K. et al. Bovine epizootic encephalomyelitis caused by Akabane virus infection in Korea. *J Comp Pathol* 147, 101–105 (2012).

95. Kurogi, H. et al. Development of Akabane virus and its immunogen in HmLu-1 cell culture. *Natl Inst Anim Health Q (Tokyo)* 17, 27–28 (1977).

96. Miah, A.H. and Spradbrow, P.B. The growth of Akabane virus in chicken embryos. *Res Vet Sci* 25, 253–254 (1978).

97. Akashi, H. et al. Detection and differentiation of Aino and Akabane Simbu serogroup bunyaviruses by nested polymerase chain reaction. *Arch Virol* 144, 2101–2109 (1999).

98. Ohashi, S. et al. Simultaneous detection of bovine arboviruses using single-tube multiplex reverse transcription–polymerase chain reaction. *J Virol Methods* 120, 79–85 (2004).

99. Stram, Y. et al. Detection and quantitation of Akabane and Aino viruses by multiplex real-time reverse-transcriptase PCR. *J Virol Methods* 116, 147–154 (2004).

100. Kurogi, H. et al. Development of inactivated vaccine for Akabane disease. *Natl Inst Anim Health Q (Tokyo)* 18, 97–108 (1978).

101. Charles, J.A. Akabane virus. *Vet Clin North Am Food Anim Pract* 10, 525–546 (1994).

102. Kurogi, H. et al. An attenuated strain of Akabane virus: A candidate for live virus vaccine. *Natl Inst Anim Health Q (Tokyo)* 19, 12–22 (1979).

103. Kurogi, H. et al. Immune response of various animals to Akabane disease live virus vaccine. *Natl Inst Anim Health Q (Tokyo)* 19, 23–31 (1979).

104. Hashiguchi, Y. et al. Responses of pregnant ewes inoculated with Akabane disease live virus vaccine. *Natl Inst Anim Health Q (Tokyo)* 21, 113–120 (1981).

105. Ogawa, Y. et al. Rescue of Akabane virus (family Bunyaviridae) entirely from cloned cDNAs by using RNA polymerase I. *J Gen Virol* 88, 3385–3390 (2007).

106. Ogawa, Y. et al. Sequence determination and functional analysis of the Akabane virus (family Bunyaviridae) L RNA segment. *Arch Virol* 152, 971–979 (2007).

107. Carpenter, S. et al. Control techniques for *Culicoides* biting midges and their application in the U.K. and northwestern Palaearctic. *Med Vet Entomol* 22, 175–187 (2008).

108. Hoffmann, B. et al. Novel orthobunyavirus in Cattle, Europe, 2011. *Emerg Infect Dis* 18, 469–472 (2012).

109. Yanase, T. et al. Genetic reassortment between Sathuperi and Shamonda viruses of the genus *Orthobunyavirus* in nature: Implications for their genetic relationship to Schmallenberg virus. *Arch Virol* 157, 1611–1616 (2012).

110. Goller, K.V. et al. Schmallenberg virus as possible ancestor of shamonda virus. *Emerg Infect Dis* 18, 1644–1646 (2012).

111. Akashi, H. et al. Antigenic and genetic comparisons of Japanese and Australian Simbu serogroup viruses: Evidence for the recovery of natural virus reassortants. *Virus Res* 50, 205–213 (1997).

52 Bluetongue Virus

Polly Roy and Bishnupriya Bhattacharya

CONTENTS

52.1 INTRODUCTION

Bluetongue virus (BTV), the causative agent of bluetongue (BT) disease, is an important pathogen of ruminants and can result in high morbidity and mortality in livestock resulting in significant economic loss.[1] BTV is transmitted between the vertebrate hosts by hematophagous arthropods belonging to the *Culicoides* genera,[2,3] and it replicates both in vertebrates and arthropods. Although previously the disease was restricted to the tropical and subtropical parts of the world, due to the change in global warming and the migration pattern of the vectors, BT is becoming a common disease of animals throughout the globe.

52.2 CLASSIFICATION AND MORPHOLOGY

BTV is a member of the *Orbivirus* genus in the Reoviridae family. Members of the Reoviridae family have nonenveloped, capsid structure where the structural proteins are arranged in concentric icosahedral shells comprising of one, two, or three protein layers. The *Orbivirus* genus is one of the nine genera within the Reoviridae family and is the only group of arboviruses (transmitted by arthropods) within the family. Although BTV shares some similarity with other members of the family (e.g., reoviruses and rotaviruses), there is a substantial difference in their detailed structure, physicochemical properties, replication cycle, pathogenesis, and epidemiology. Along with several unclassified isolates, 19 groups of orbiviruses have been established to date on the basis of their serological cross-reactivities. Due to its economic importance, BTV has been studied extensively at molecular, genetic, and structural level and as such serves as the model virus of the genus. BTV infects a number of wild and domestic ruminants (sheep, cattle, goats, deer, etc.), often causing high morbidity and mortality, particularly in sheep, although sometimes infection occurs with no apparent clinical symptoms. To date, there are at least 26 different serotypes of BTV (BTV-1, BTV-2, etc.) endemic in various parts in the world. The serotyping of BTV is not based on the sequence variation of the viral genes but is based on serological neutralization of virus particles. Other members of the *Orbivirus* genus that are closely related to BTV and transmitted by the same vectors are the African horse sickness virus (AHSV) and epizootic hemorrhagic disease virus (EHDV). Similar to the other members of the Reoviridae family, BTV has a segmented, double-stranded RNA (dsRNA) genome. The 10 dsRNA segments in BTV encode for 7 capsid proteins (4 structural and 3 enzymatic) and 3 or 4 nonstructural proteins. As summarized in Table 52.1, the complete sequence of each segment and the function of the respective proteins have been well established.[4]

The structural proteins of BTV particles (85 nm in diameter) are arranged in two concentric capsids enclosing the genome of 10 dsRNA segments. The outer capsid consists of two major proteins (VP2 and VP5), while the inner capsid (core) contains two major (VP3 and VP7) and three minor enzymatic proteins (VP1, VP4, and VP6).[5,6] While virion particles exhibit fuzzy appearance by negative staining methods,[6] unstained and unfixed particles show a well-defined outer surface. 3D cryo-electron microscopy (EM) reconstruction analysis has revealed that the intact virions have a well-ordered morphology with an icosahedral symmetry.[7] The outer capsid protein VP2 (110 kDa) forms 60 trimers, and each trimer sits on the surface of the inner core. VP2 trimers appear as "sail-shaped" spikes projecting 4 nm beyond the virion surface, and unlike other dsRNA viruses, the three tip domains of each trimer branch out from a central hub domain (Figure 52.1).[8] The top of the tip domain projecting upward from each monomer is rich in β-sheets suitable for cell surface attachment, while the lower hub domain has a distinct β-barrel fold with putative sialic acid–binding pocket.[8] Surrounding the VP2 hub domains are the 120 VP5 (60 kDa) trimers, which, in contrast to VP2 trimers, form globular structures and consist of predominant helices (Figure 52.1).[8] Strikingly, VP5 is

599

TABLE 52.1
BTV Coding Assignments

Genome Segment	Length (bp)	Proteins	Predicted Size (Da)	Estimated No. of Molecules per Virion[a]	Location in Virion Particle[b] (VP1–VP7)	Function
S1	3954	VP1	149,588	12	Inner core	RNA polymerase
S2	2926	VP2	111,112	180	Outer shell (exposed)	Receptor binding, virus entry, hemagglutinin-specific neutralization
S3	2772	VP3	103,344	120	Subcore layer (scaffold)	Forms scaffold for VP7 trimers, interacts with genomic RNA
S4	2011	VP4	76,433	~24	Inner core	Capping enzymes—guanylyltransferase, methyltransferases 1 and 2, RNA 5'-triphosphatase, inorganic pyrophosphatase NTPase
S5	1639	VP5	59,163	360	Outer shell (under surface)	Virus penetration, fusogenic
S6	1770	NS1	64,445	High level	Nonstructural	Tubules, upregulates the synthesis of viral proteins
S7	1156	VP7	38,548	780	Core surface layer	Group-specific, core entry to insect cells
S8	1123	NS2	40,999	NA	Nonstructural	Phosphorylated, forms cytoplasmic IBs, binds ssRNA, recruits RNA
S9	1046	VP6	35,750	60–72	Inner core	Binds ssRNA, dsRNA helicase, ATPase
		NS4		77–79	Nonstructural	Hypothesized to be involved in host–virus interaction
S10	822	NS3	25,572	NA	Nonstructural	Glycoproteins, membrane protein, aids virus trafficking and release
		NS3A	24,020	Low level		

Source: Adapted from Roy, P., in Knipe, D.M. et al. (eds.) *Fields Virology*, 4th ed., pp. 1735–1766, Lippincott-Raven, Philadelphia, PA, 2001.

structurally very similar to the fusion proteins of enveloped viruses (e.g., HIV, herpesviruses, vesicular stomatitis virus, and influenza virus), and this implies its involvement in membrane penetration.[8]

The outer capsid encloses the scaffolding core (75 nm in diameter), which also has an icosahedral symmetry and is composed of two concentric protein layers surrounding a central density of RNA and protein. The outer layer of the core consists of clusters of VP7 trimers (260 in total) that protrude out from the surface of the particle, giving the cores a bristly appearance.[9,10] The VP7 trimers are also arranged around distinctive 132 aqueous channels, some of which extend inward, penetrating through to the inner layer. These channels or pores in the core particles allow the viral mRNAs to extrude out of the particle and also allow the entrance of metabolites into the core particles (discussed later). While the VP7 (38 kDa) molecules on the core surface are covered by VP2 trimers,[7,8] the globular VP5 makes contacts with the sides of VP7 trimers that border the aqueous channels.[7,11] However, interactions between the trimers of VP5 with VP2 and VP7 are relatively weak,[8] which may be necessary to facilitate virus entry into the host cell (see later). The inner core layer that acts as the scaffold for VP7 deposition is solely made up of 60 dimers of VP3 (103 kDa).[10] Although VP3 is relatively featureless and exhibits a smooth, almost spherical shape, it has an angular appearance due to variations in thickness of the protein shell.[10,12]

The density in the central portion of the core structure contains the genomic RNA and the three enzymatic proteins: the polymerase protein VP1, the capping enzyme VP4, and a helicase protein VP6. Together, they form the viral transcription complex (TC).[9,10] The two larger proteins, VP1 (149.5 kDa) and VP4, are closely associated at the undersurface of VP3 layer, directly at the fivefold vertices of the core.[7] While the crystal structure of BTV VP1 is not known, a secondary structure-based 3D model has revealed that it has a cage-like structure with three distinct domains, similar to reovirus and rotavirus polymerases.[13] In contrast to the polymerase, the crystal structure of recombinant VP4 revealed that the capping enzyme is a unique elongated molecule with four discrete catalytic domains arranged in linear fashion.[14]

52.3 BIOLOGY AND REPLICATION

BTV initiates infection by binding to the surface of host cells. In mammalian cells, BTV attachment is mediated by the tip and hub domains of VP2.[8] Among the cellular receptors, BTV is known to bind to sialic acids present on the surface of erythrocyte and vertebrate cells.[15,16] In comparison, in insect cells, BTV core particles are highly infectious,[17] and the virus binds to the receptors through an RGD motif present on VP7, the outermost core protein.[18] Subsequent to receptor attachment in mammalian cells, the virus particles are rapidly internalized into endosomes through clathrin-dependent

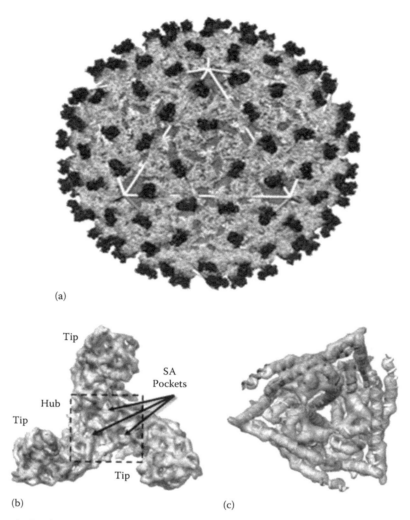

(a)

Tip

SA
Pockets

Hub

Tip

Tip

(b)

(c)

FIGURE 52.1 Cryo-EM analysis of BTV: (a) Cryo-EM structure of complete BTV-1 at 7 Å resolution. (b) Top view of VP2 triskelion formed by three VP2 monomers, each consisting of a tip domain and contributing a part of a hub domain. A putative SA-binding pocket is located at the hub domain of the triskelion. (c) Sixfold average density map of the VP5 trimer viewed from top with an embedded ribbon model. The density map reveals secondary structures demonstrating that VP5 contains many helices but only one β-sheet. (From Zhang, X. et al., *Proc. Natl. Acad. Sci. USA*, 107, 6292–6297, 2010.)

endocytotic pathway.[19] In the endosomes, the low pH induces the removal of VP2 and exposes VP5 to interact with the endosomal membranes to form pores, which further results in membrane destabilization.[8,20,21] Consequently, the outer capsid is removed, and the entire core is released into the cytoplasm where they remain intact to form transcriptionally active core particles by activating the TC in the cores.

The negative strands of dsRNA molecule present within the cores that serve as template for the synthesis of viral mRNAs are unwound by VP6 helicase.[22] Subsequently, the synthesis of positive-strand RNAs from the 3′ end of each of the negative-strand RNAs in the dsRNA molecule is initiated by VP1 polymerase.[23,24] The 5′ end of the newly synthesized positive-strand RNA molecules is then capped by VP4[25–27] and the progeny mRNAs that are extruded from the cores through the pores into the cytoplasm, where they serve as templates for translation in viral proteins.

In the cytoplasm of BTV-infected cells, a fibrillar network referred to as inclusion bodies (IBs) that are rich in

BTV transcripts and the phosphorylated nonstructural BTV protein NS2 develop around the infecting core particles.[28–30] Similar to other orbiviruses, BTV also induces the production of tubules composed of multimers of BTV nonstructural protein NS1[31,32] that upregulate the production of viral proteins.[33] The NS2 in the IBs recruits viral positive ssRNA[34–36] and VP3[37,38] to form the subcore structure that in turn associates with components of the viral transcriptase complex.[39,40] The VP3 subcore in the IBs serves as a scaffold for the addition of VP7 trimers, giving rise to more rigid and stable cores.[41,42] The successful development of reverse genetics (RG) system for BTV that resulted in the recovery of infectious virions from in vitro–generated BTV transcripts in transected cells[43] revealed that there are two stages in BTV replication: first, the assembly of replicase complex and core (primary replication), which is then followed by a secondary replication stage that results in the generation of mature virus particles.[44] In comparison to the in vivo RG system, a recently developed in vitro–based cell-free assembly system revealed that in the absence of NS2, the 3 TC proteins can interact with each other and the 10 single-stranded RNA (ssRNA) molecules prior to be surrounded by the VP3 and VP7 layers form mature core particles that were infectious for insect cells.[45] Since NS2 is needed for BTV maturation in infected cells but not during the assembly of core particles in vitro, it seems that the IBs formed in infected cells act as the concentrator of the core components to facilitate their assembly in the cytoplasm of infected cells.

Compared to the core proteins, the two outer capsid proteins VP2 and VP5 are not localized in the IBs.[38,46] Although it is still not clear where in the cytoplasm are the two outer capsid proteins attached to the core particles to form the mature viral particles, the association of both VP2 and VP5 to the intermediate filaments and lipid rafts, respectively, is believed to play an active role in virus assembly.[47,48] Both VP2 and VP5 also interact with BTV nonstructural protein NS3,[48,49] a protein that plays an important role in virus release and is the only glycosylated protein encoded by the virus.[50–52] In infected cells, NS3 is present in two forms, NS3 and NS3A, where the shorter form NS3A lacks the 13 residues from the N-terminal end of NS3. In infected cells, NS3 has been localized to intracellular organelles (Golgi complex and endoplasmic reticulum) and cellular membranes.[50–53] Similar to enveloped virus proteins, NS3 also contains late domains that interact with Tsg101,[52] a component of multivesicular bodies (MVBs). Further, NS3 also interacts with cellular protein p11 that forms a complex with annexin 2,[49] a member of the cellular exocytotic pathway. Subsequently, the RG-based virus recovery together with site-directed mutagenesis[43] not only confirmed NS3 as a key player in the interaction of viral particles with cellular proteins and in virus egress, but it has also highlighted that nonenveloped BTV utilizes the same cellular exocytotic pathway as some enveloped viruses.[54,55] Release of virions from the infected mammalian cells occurs predominantly as a result of cell lysis accompanied by cell death. However, virions may also release either via budding where the virus temporarily acquires an envelope[54,56] or via cell membrane destabilization mediated by the NS3 viroporin activity[53] or through extrusion in which virus particles move through a locally disrupted plasma membrane.[57] Released BTV particles and those remaining on the cell surface also often appear to retain an association with an underlying cortical layer of the cell cytoskeleton.[57]

52.4 EVOLUTION AND EPIDEMIOLOGY

Due to the presence of a segmented genome and on the basis of genetic compatibility, BTV serotypes reassort their genes during coinfection of their hosts. This process of virus evolution as opposed to accumulation of mutations, deletions, and insertions is well documented for a number of viruses with segmented genomes. While hybridization studies showed that the genes encoding the two outermost capsid proteins were the least conserved among all the BTV genes, phylogenetic analysis of BTV strains demonstrated that the geographic origin of the viruses (Asia/Australia, Americas/Caribbean, and Africa/Europe) determines the segregation of isolates into groups. In addition, although BTV serotypes evolved from a common ancestor, the gene pools of viruses on different continents have not only diverged significantly, but they have also evolved in close affiliation with competent *Culicoides* species, unique to that particular region (Table 52.2). Moreover, although the gene (segment 2) encoding for VP2, the outermost neutralizing antigen, is the least conserved, there is perfect correlation in the sequence of the gene belonging to 26 BTV serotypes.[59] As the sequence of segment 6 (encoding for VP5) has also been proven to be important for BTV serotype determination,[60,61] the phylogenetic characteristics of segments 2 and 6 are also important for serotype identification (Figure 52.2).

Geographically, while the northern- and southernmost boundaries of BTV have historically been 40° north and 35° south latitudes, respectively,[62] BTV recently has been documented to extend to 50° north.[1] Usually, the geographic distribution of BTV is categorized into three ecologic zones: *endemic*, *epidemic*, and *incursion*. *Endemic* zones are defined as typically tropical regions where BTV transmission occurs throughout the year and subclinical infection is common. In endemic zones, clinical disease generally occurs

TABLE 52.2
Global Distribution of BTV Serotypes and Primary Vector

Region	BTV Serotypes	Vector Species
Africa	1–19, 22, 24	C. imicola, C. bolitinos
Asia	1–4, 9, 12, 14–21, 23, 26	C. imicola, C. schultze grp, C. fulvus, C. actoni, C. wadai, C. brevitarsis, C. orientalis
Australia	1, 3, 9, 15, 16, 20, 21, 23	C. fulvus, C. actoni, C. wadai, C. brevitarsis
Europe	1, 2, 4, 8, 9, 16, 25	C. obsoletus, C. pulicaris
North America	2, 10, 11, 13, 17	C. sonorensis, C. insignis
South and Central America, Caribbean	1, 3, 4, 6, 8, 12, 17, 14	C. insignis

Source: Center for Emerging Issues, Brief overview of BT ecology, epidemiology, clinical, 2005.

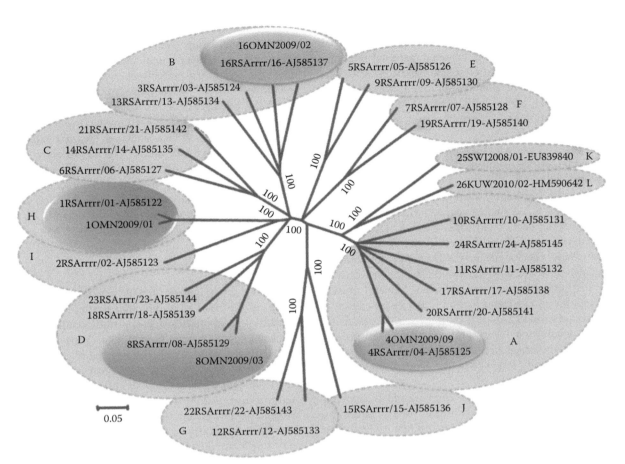

FIGURE 52.2 Neighbor-joining tree, showing relationships in Seg-2 between reference and field strains of BTV. (From Maan, N.S. et al., *PLoS One* 7, e32601, 2012.)

only in introduced, immunologically naïve susceptible species. *Epidemic* zones include temperate areas where outbreaks occur seasonally, generally in the late summer when vector populations peak. *Incursion* zones are areas that experience outbreaks infrequently when climatic conditions favor disease transmission by vectors. Currently, along with the Americas, Africa, Southern Asia, and Northern Australia, there has been a dramatic change in the distribution of BTV in Europe.[1] Along with Southern Europe (Italy, Portugal, and Spain), the virus has been reported in the Netherlands, Belgium, Germany, France, Czech Republic, Denmark, Luxembourg, Sweden, Norway, Switzerland, and the United Kingdom. More than global warming, an important contribution to the northward spread of BT disease has been the ability of novel European vectors (*Culicoides obsoletus* and *Culicoides pulicaris*) to acquire and transmit the disease. This is in contrast to the original *Culicoides imicola* vector, which is limited to North Africa and the Mediterranean.

52.5 CLINICAL FEATURES AND PATHOGENESIS

Although BTV infects most species of domestic and wild ruminants, the severity of resulting BT disease is mostly restricted to certain breeds of sheep and some species of deer.[63] The name "bluetongue" signifies the appearance of a dark blue and swollen tongue in some animals.[64] In 1902, the disease was mentioned as "a malarial catarrhal fever of sheep" but was renamed as "bluetongue" in 1905.[64] Currently, BT has been listed as a "notifiable disease" by the World Organization for Animal Health (OIE). In BT disease, the severity of clinical symptoms depends on the breed of sheep and the strain of virus. Hence, some of the European fine wool and mutton sheep breeds in unexposed areas are particularly susceptible to the disease.[65] Compared to subclinical infections observed in the indigenous breeds from endemic regions, fatality in infected sheep can vary between 30% and 50% in Northern Europe.[66] Although in cattle BT infections are usually subclinical,[67] the initial stages of new virus incursions into a serologically naive cattle population can cause the development of clinical symptoms.[68] Indeed during an outbreak of BTV-8 recently in Northern Europe, although less than 10% of cattle in affected areas developed clinical signs, there was a case fatality rate of up to 10% in these animals.[66] Cattle can also act as reservoir with prolonged viremia.

BTV infection of pregnant cattle and sheep can result in fetal infection, abortion, or congenital anomalies such as runting, blindness, deafness, hydranencephaly, arthrogryposis, campylognathia, and prognathia.[67,69–71] Congenital BTV infection occurring in late gestation usually does not result in congenital anomalies, although the newborn animal may circulate low levels of virus in their blood and have delayed immune responses. The interval between maternal infection and last detectable viremia in offspring has been as long as 20 weeks, suggesting a mechanism for long-term survival of BTV. However, such congenitally infected animals eventually clear the infection and are not immunologically tolerant or persistently infected.

Generally, the first symptom of BTV infection is the appearance of high temperature that usually lasts for 5–7 days in animals but can extend to 14 days in certain cases.[72] Subsequently, the respiratory rate of the infected animals also increases, and this is further accompanied by hyperemia and swelling of the buccal and nasal mucosa. The mucosal membrane of the gingival and oral cavity also hemorrhages, and at this stage, the tongue might swell up, turn blue in color, and also protrude from the mouth. Infected sheep also develops stiffness, lethargy, and anorexia. Skeletal muscle damage resulting in weakness and muscular degeneration is also observed in severe cases of BTV infection. Additionally, lesions also develop on the smooth muscles of esophagus and pharynx. This causes vomiting and results in pneumonia, which is frequently followed by the death of the infected animal.[72]

The widely accepted model of virus spread in infected vertebrate host has depicted that after being bitten by infected *Culicoides*, the virus particles migrate to the lymph nodes for replication. Subsequently, they are spread to spleen, thymus, and other lymph nodes where the particles become associated with circulating blood cells and might survive for weeks or months. Since BTV also binds to glycophorins[15,73] present on the surface of the blood cells, they may remain as invaginations of ovine and bovine erythrocytic cell membrane.[74] In this way, the virus particles protect themselves from the host defense mechanism (antibody, T cells) and also provide multiple opportunities for blood-sucking midges over an extended period of time. Prolonged BT viremias are particularly notable in cattle, which thus serve as virus reservoirs. After experimental inoculation of calves, infectious virus can be detected for up to 8 weeks, and viral nucleic acid can be detected in erythrocytes for up to 20 weeks. While BTV and the other orbiviruses such as EHDV and AHSV are known to infect leukocytes,[74–79] uncertainty exists regarding the role of leukocytes in virus-associated viremias[76,80] and the role of leukocytes in BTV infection of ruminant host.[81] In addition, although skin as an organ system has not been traditionally considered to be important for virus replication, recently, BTV has been observed to replicate in the skin of infected as early as 3 days postinfection.[82] The capability of BTV to infect dendritic cells and the potential use of these cells to disseminate the virus particles[83] indicated similarities of BTV with the spread of enveloped hemorrhagic disease-causing ssRNA viruses (Lassa, Rift Valley fever, Ebola, dengue, and yellow fever viruses) in human hosts.

The capability of BTV to replicate and damage endothelial cells (ECs), smooth muscles, and pericytes leads to a cascade of pathophysiological events that is characterized by capillary leakage, hemorrhage, and disseminated intravascular coagulation. While the severity of hemorrhages is potentially more severe and fatal in sheep, infected cattle develops subclinical infections with minimal endothelial infections.[67,69,72,84–86] Although traditionally, vascular injury in BT has been attributed to direct virus-mediated injury to ECs, the process of EC injury and dysfunction may also be due to the activities of virus-induced, host-derived inflammatory and vasoactive mediators.[84,86] Since recently, an in vitro study demonstrated that the adherens junctions of BTV-infected bovine EC monolayers remain intact throughout the course of virus infection and the cells become permeable mainly due to cell death, other factors linked to virus infection can also be responsible for this action.[87]

52.6 DIAGNOSIS

A preliminary diagnosis of BTV infection is usually undertaken on the basis of symptoms, postmortem reports, and epidemiological studies.[88] While traditionally, embryonated chicken eggs have been used to isolate the virus in laboratory setting,[88] BTV can also be isolated in cell lines of either mammalian (BHK-21, CPAE, or Vero) or insect origin (KC line from *Culicoides sonorensis* cells or the C6/36 line from *Aedes albopictus*).[89,90] Although sheep does provide a sensitive and reliable system for BTV isolation, it is only occasionally used, for example, in cases when a sample contains a very low virus titer.[88] Currently, molecular techniques such as reverse transcription polymerase chain reaction (RT-PCR) are increasingly being used in the laboratories as these techniques are not only far more rapid than the conventional detection systems but also more sensitive and robust means of detection of BTV RNA in blood or tissue samples as well as for serotyping.[59,91–97] In addition, commercial ELISA kits are widely available for detection of antibodies against BTV.[98,99]

52.7 TREATMENT AND PREVENTION

Vaccination is one of the important means adopted to control for BTV infection. Although BTV vaccines are usually serotype-specific and antibodies to one serotype do not cross-neutralize virus from another serotype, partial cross-protection can take place between vaccines that target closely related serotypes.[100] In addition, immune response is generated by both antibodies[101] and T-cell-mediated responses.[76,102,103] Since BTV genome is segmented and chances of reassortment of genome segments in hosts coinfected with more than one strain of BTV are very high, the ideal vaccine for BT should not only protect against as many virus serotypes as possible but should also not revert back to virulence nor recombine with circulating strains of the virus. In addition, the vaccine strain should be compatible with tests to distinguish between infected and vaccinated animals.

Two types of vaccines, inactivated and live attenuated, are currently available. Based on the successful passage of BTV in egg, the first live-attenuated vaccine was developed and used for protection against 15 BTV serotypes in

South Africa and other countries.[104–108] Until recently, the live-attenuated vaccines were the only commercially available BTV vaccines.[109] Although they are inexpensive and can provide good protection, they may not be effective in providing protection against infection with a heterologous BTV serotype. In addition, they also have other detrimental side effects such as clinical signs of BT, abortion, and reduced milk production[110] and also cause fetal malformation in vaccinated pregnant ewes.[69,111,112] As attenuated vaccine virus can also produce viremia lasting over 2 weeks in vaccinated sheep,[107] the possibility of vaccine virus infecting vectors is also present. The discovery of the attenuated vaccine strain of BTV in both unvaccinated sentinel cattle and insect vector in a field study in Italy[113] further highlighted the possible repercussions involving the use of attenuated BTV strains as vaccine for BT prevention and control. Such occurrences can either result in subsequent reversion to virulence or the production of recombinant progeny virus with novel properties after the reassortment of genes from wild and vaccine virus in the vaccinated animal or the vector.[107,114] Thus, the possible deleterious effects make the attenuated BTV strains not suitable as vaccine candidate for the eradication of BT disease.

Compared to attenuated vaccines, the more recent inactivated vaccines produced by treating BTV with beta-propiolactone,[115,116] gamma radiation,[117] or binary ethylenimine[111,118,119] are more effective in controlling BT. Although the production of inactivated BTV vaccines involves the maintenance of tight quality control so that each virus particle in a vaccine batch has been completely inactivated, if properly produced, they can induce reliable and protective immunity. Another drawback is the requirement of revaccination,[108] which further increases the cost of vaccination.

Other alternatives have also been developed for the protection of animals from BTV. Since VP2 by itself or in combination with other proteins can provide protection, one option was to vaccinate animals with purified recombinant viral proteins[120,121] expressed by baculovirus expression system, or the viral genes are vectored to the animals by either canarypox[122] or capripox[123] vector system. Baculovirus-generated virus-like particles (VLPs)[100,124,125] are an improved version of the recombinant protein approach to vaccinate animals. VLPs are complex structures that are made up of the four structural proteins (VP2, VP5, VP3, and VP7). Since VP2 in VLPs are present in exactly the same confirmation as in native viral particles,[7,8,11,126] they are highly immunogenic[127] and are known to elicit both B- and T-cell responses.[128] Studies have also shown that they are more immunogenic than VP2-based vaccination.[120] Further, the absence of BTV genome in VLPs negates the chances of reversion to virulence, reassortment, or incomplete inactivation. In addition, since VLPs do not contain any genes for the nonstructural proteins, the naturally infected animals showing antibody responses to the nonstructural proteins NS1 and/or NS2 can be easily differentiated from the vaccinated ones. Since the baculovirus used lacks a gene that is essential for the infection of its natural insect host, the recombinant virus also cannot replicate in the environment. Although the core-like

particles (CLPs) containing the highly conserved VP3 and VP7 can also be used for vaccine development, a study comparing VLP and CLP as vaccine candidates has shown that in comparison to VLP vaccination, CLP-vaccinated animals do not induce any neutralizing antibody response.[129] In addition, the lack of complete protection to sheep vaccinated by CLPs and the development of clinical manifestation of the disease and virus replication confirmed that the outer capsid proteins are needed to generate neutralizing antibody response needed for complete protection.[129] The main challenges of VLPs as a viable option for BTV vaccine development were that simple cloning strategies should be available that would quickly generate VLPs for multiple serotypes and that there should be cross protection between serotypes. To this effect, the preintegration of the conserved genes encoding for inner capsid proteins, followed by recombination with the outer capsid proteins of either BTV-2, BTV-4, or BTV-9, resulted in the development of VLPs that could protect vaccinated sheep from challenge with virulent virus belonging to the same serotype.[124] A separate study analyzing cross protection between serotypes by vaccination of Merino sheep with a cocktail of BTV VLPs for BTV-1 and BTV-4 or BTV-1, BTV-2, and BTV-8 demonstrated that the bivalent vaccine could protect against both virulent virus challenges.[125,130] An earlier study has also shown that VLP vaccines containing a cocktail of VLPs belonging to different serotypes can provide complete or partial protection against multiple serotypes of BTV.[131] Although the VLP as immunogen for vaccine development does need further analysis, since VLPs can be produced due to the high expression capabilities of baculovirus vectors (produced in serum-free medium) and can be purified using a one-step generic protocol based on the physical properties of the particle, they do offer advantages as potential vaccines over more conventional systems.

The RG system developed for BTV[43] has also been shown as an important tool for designing safe and efficacious attenuated vaccines. This system involves mutation of BTV genes, rescue of the mutated virus in permissive host cells, and pathogenesis study of the BTV mutants in the ruminant host. On the basis of pathogenic determinants of BTV, vaccine strains containing multiple attenuating mutations can then be designed and tested for protection in animals. The introduction of more than one attenuating mutation in the genome of the engineered strain and the presence of the attenuating mutations on several genome segments can reduce the risk of reversion to virulence and the regeneration of virulent BTV strain by reassortment with wild-type strains, respectively. Since it has been established that VP2 and VP5 from phylogenetically diverse serotypes can assemble on the conserved core proteins to create viable BTV strains,[43,127] it is possible to introduce the outer capsid proteins on a defined attenuated genetic background. Such an approach might be important for regions where protection from several cocirculating strains is needed. Indeed, highly attenuated disabled infectious single cycle (DISC) candidate vaccine strains have recently been generated. One such recently designed DISC vaccine strain of BTV-1 contains deletions in segment S9, the gene that encodes for VP6 protein.[132]

In addition, a defective BTV-8 strain was also made by reassorting the two RNA segments that encode the two outer capsid proteins (VP2 and VP5) of a highly pathogenic BTV-8 with the remaining eight RNA segments of one of the BTV-1 DISC viruses.[132] Both, BTV-1 and BTV-8 DISC, viruses were not only highly protective in sheep, but they are also compliant in differentiating infected from vaccinated animals (DIVA) vaccines.[132] Since the DISC strain might be missing one or more viral genes, it is easy to distinguish between vaccinated animals and those that have been exposed to infectious virus. Such information would permit an assessment on whether a country or region is BT-free, a decision that has important consequences for the trade of livestock.

Another means of controlling BTV involves reduction of midge populations to ineffective levels by agents of low toxicity to mammals[133–135] or to prevent vector attacks by stabling susceptible animals overnight since midges have nocturnal feeding habits.

52.8 CONCLUSION

In last two decades, impressive progress in BTV research has been achieved due to the development of technologies leading to a better understanding of BTV replication and life cycle in infected host cells. The generation of recombinant proteins and VLPs/CLPs has laid the essential foundation to studies leading to the characterization of the viral proteins and to understand the interactions between the BTV structural proteins during virus assembly. The recently developed RG system (both in vivo and in vitro) has not only made it possible to undertake an in-depth study in different aspects of BTV morphogenesis and life cycle in tissue culture systems, but it has also given impetus to the development of safe and effective vaccines against the virus. However, information on interactions between BTV and its hosts is still fragmentary. In particular, a longstanding problem in BTV research like in many other viruses is the relationship between infection and disease. What leads to extreme pathogenesis in one case but an almost silent infection in another? The broad host range of BTV represents a challenge in terms of understanding the interactions that lead to pathogenesis. Despite the relative simplicity of the virus, there are significant differences in the clinical severity of BTV infection of different species and between different breeds of the same species. The underlying basis of this substantial variation, the link between virus infection and BTV disease, will no doubt become a major focus in BTV research in future.

REFERENCES

1. Wilson, A.J. and Mellor, P. Bluetongue in Europe: Past, present and future. *Philos. Trans. R. Soc. Lond. B Biol. Sci.* 364, 2669–2681 (2009).
2. Du Toit, R.M. The transmission of bluetongue and horse sickness by *Culicoides*. *Onderstepoort J. Vet. Sci. Anim. Ind.* 19, 7–16 (1944).
3. Mellor, P.S. The replication of bluetongue virus in *Culicoides* vectors. *Curr. Top. Microbiol. Immunol.* 162, 143–161 (1990).
4. Roy, P. Orbiviruses, in Knipe, D.M. et al. (eds.) *Fields Virology*, 4th ed., pp. 1735–1766 (Lippincott–Raven, Philadelphia, PA, 2001).
5. Roy, P. Bluetongue virus genetics and genome structure. *Virus Res.* 13, 179–206 (1989).
6. Verwoerd, D.W. et al. Structure of the bluetongue virus capsid. *J. Virol.* 10, 783–794 (1972).
7. Nason, E. et al. Interactions between the inner and outer capsids of bluetongue virus. *J. Virol.* 78, 8059–8067 (2004).
8. Zhang, X. et al. Bluetongue virus coat protein VP2 contains a sialic acid-binding domain and VP5 has similarities to enveloped virus fusion proteins. *Proc. Natl. Acad. Sci. USA* 107, 6292–6297 (2010).
9. Prasad, B.V.V., Yamaguchi, S., and Roy, P. Three-dimensional structure of single-shelled bluetongue virus. *J. Virol.* 66, 2135–2142 (1992).
10. Grimes, J.M. et al. The atomic structure of the bluetongue virus core. *Nature* 395, 470–478 (1998).
11. Hewat, E.A., Booth, T.F., and Roy, P. Structure of bluetongue virus particles by cryoelectron microscopy. *J. Struct. Biol.* 109, 61–69 (1992).
12. Hewat, E.A. et al. Three-dimensional reconstruction of baculovirus expressed bluetongue virus core-like particles by cryo-electron microscopy. *Virology* 189, 10–20 (1992).
13. Wehrfritz, J.M. et al. Reconstitution of bluetongue virus polymerase activity from isolated domains based on a three-dimensional structural model. *Biopolymers* 86, 83–94 (2007).
14. Sutton, G. et al. Bluetongue virus VP4 is an RNA-capping assembly line. *Nat. Struct. Mol. Biol.* 14, 449–451 (2007).
15. Hassan, S.H. and Roy, P. Expression and functional characterization of bluetongue virus VP2 protein: Role in cell entry. *J. Virol.* 73, 9832–9842 (1999).
16. Bhattacharya, B. and Roy, P. Role of lipids on entry and exit of bluetongue virus, a complex non-enveloped virus. *Viruses* 2, 1218–1235 (2010).
17. Mertens, P.P. et al. Enhanced infectivity of modified bluetongue virus particles for two insect cell lines and for two *Culicoides* vector species. *Virology* 217, 582–593 (1996).
18. Tan, B. et al. RGD tripeptide of bluetongue virus VP7 protein is responsible for core attachment to *Culicoides* cells. *J. Virol.* 75, 3937–3947 (2001).
19. Forzan, M., Marsh, M., and Roy, P. Bluetongue virus entry into cells. *J. Virol.* 81, 4819–4827 (2007).
20. Hassan, S.H. et al. Expression and functional characterization of bluetongue virus VP5 protein: Role in cellular permeabilization. *J. Virol.* 75, 8356–8367 (2001).
21. Forzan, M., Wirblich, C., and Roy, P. A capsid protein of nonenveloped bluetongue virus exhibits membrane fusion activity. *Proc. Natl. Acad. Sci. USA* 101, 2100–2105 (2004).
22. Stauber, N. et al. Bluetongue virus VP6 protein binds ATP and exhibits an RNA-dependent ATPase function and a helicase activity that catalyze the unwinding of double-stranded RNA substrates. *J. Virol.* 71, 7220–7226 (1997).
23. Ramadevi, N., Kochan, G., and Roy, P. Polymerase complex of bluetongue virus in *Proceedings of the International Congress of Virology*, p. 35 (Sydney, New South Wales, Australia, 1999).
24. Boyce, M. et al. Purified recombinant bluetongue virus VP1 exhibits RNA replicase activity. *J. Virol.* 78, 3994–4002 (2004).
25. Mertens, P.P. et al. Analysis of guanyltransferase and transmethylase activities associated with bluetongue virus cores and recombinant baculovirus-expressed core-like particles, in Walton, T.E. and Osburn, B.M. (eds.) *Bluetongue, African Horse Sickness and Related Orbiviruses* (CRC Press, Boca Raton, FL, 1992).

26. Costas, J., Sutton, G., and Roy, P. Guanylyltransferase and RNA 5′-triphosphatase activities of the purified expressed VP4 protein of bluetongue virus. *J. Mol. Biol.* 280, 859–866 (1998).

27. Ramadevi, N. et al. Capping and methylation of mRNA by purified recombinant VP4 protein of bluetongue virus. *Proc. Natl. Acad. Sci. USA* 95, 13537–13542 (1998).

28. Huismans, H., van Dijk, A.A., and Bauskin, A.R. in vitro phosphorylation and purification of a nonstructural protein of bluetongue virus with affinity for single-stranded RNA. *J. Virol.* 61, 3589–3595 (1987).

29. Modrof, J., Lymperopoulos, K., and Roy, P. Phosphorylation of bluetongue virus nonstructural protein 2 is essential for formation of viral inclusion bodies. *J. Virol.* 79, 10023–10031 (2005).

30. Huismans, H. Protein synthesis in bluetongue virus-infected cells. *Virology* 92, 385–396 (1979).

31. Urakawa, T. and Roy, P. Bluetongue virus tubules made in insect cells by recombinant baculoviruses: Expression of the NS1 gene of bluetongue virus serotype 10. *J. Virol.* 62, 3919–3927 (1988).

32. Hewat, E.A. et al. 3-D Reconstruction of bluetongue virus tubules using cryoelectron microscopy. *J. Struct. Biol.* 108, 35–48 (1992).

33. Boyce, M., Celma, C.C., and Roy, P. Bluetongue virus nonstructural protein 1 is a positive regulator of viral protein synthesis. *Virol. J.* 9, 178 (2012).

34. Thomas, C.P., Booth, T.F., and Roy, P. Synthesis of bluetongue virus-encoded phosphoprotein and formation of inclusion bodies by recombinant baculovirus in insect cells: It binds the single-stranded RNA species. *J. Gen. Virol.* 71, 2073–2083 (1990).

35. Lymperopoulos, K. et al. Sequence specificity in the interaction of bluetongue virus non-structural protein 2 (NS2) with viral RNA. *J. Biol. Chem.* 278, 31722–31730 (2003).

36. Lymperopoulos, K. et al. Specific binding of bluetongue virus NS2 to different viral plus-strand RNAs. *Virology* 353, 17–26 (2006).

37. Kar, A.K., Iwatani, N., and Roy, P. Assembly and intracellular localization of the bluetongue virus core protein VP3. *J. Virol.* 79, 11487–11495 (2005).

38. Kar, A.K., Bhattacharya, B., and Roy, P. Bluetongue virus RNA binding protein NS2 is a modulator of viral replication and assembly. *BMC Mol. Biol.* 8, 4 (2007).

39. Loudon, P.T. and Roy, P. Assembly of five bluetongue virus proteins expressed by recombinant baculoviruses: Inclusion of the largest protein VP1 in the core and virus-like proteins. *Virology* 180, 798–802 (1991).

40. Kar, A.K., Ghosh, M., and Roy, P. Mapping the assembly of bluetongue virus scaffolding protein VP3. *Virology* 324, 387–399 (2004).

41. Limn, C.-H. et al. Functional dissection of the major structural protein of bluetongue virus: Identification of key residues within VP7 essential for capsid assembly. *J. Virol.* 74, 8658–8669 (2000).

42. Grimes, J. et al. The crystal structure of bluetongue virus VP7. *Nature* 373, 167–170 (1995).

43. Boyce, M., Celma, C.C., and Roy, P. Development of reverse genetics systems for bluetongue virus: Recovery of infectious virus from synthetic RNA transcripts. *J. Virol.* (2008).

44. Matsuo, E. and Roy, P. Bluetongue virus VP6 acts early in the replication cycle and can form the basis of chimeric virus formation. *J. Virol.* 83, 8842–8848 (2009).

45. Lourenco, S. and Roy, P. In vitro reconstitution of bluetongue virus infectious cores. *Proc. Natl. Acad. Sci. USA* 108, 13746–13751 (2011).

46. Hyatt, A.D. and Eaton, B.T. Ultrastructural distribution of the major capsid proteins within bluetongue virus and infected cells. *J. Gen. Virol.* 69, 805–815 (1988).

47. Bhattacharya, B., Noad, R.J., and Roy, P. Interaction between bluetongue virus outer capsid protein VP2 and vimentin is necessary for virus egress. *Virol. J.* 4, 7 (2007).

48. Bhattacharya, B. and Roy, P. Bluetongue virus outer capsid protein VP5 interacts with membrane lipid rafts via a SNARE domain. *J. Virol.* 27, 27 (2008).

49. Beaton, A.R. et al. The membrane trafficking protein calpactin forms a complex with bluetongue virus protein NS3 and mediates virus release. *Proc. Natl. Acad. Sci. USA* 99, 13154–13159 (2002).

50. Wu, X. et al. Multiple glycoproteins synthesized by the smallest RNA segment (S10) of bluetongue virus. *J. Virol.* 66, 7104–7112 (1992).

51. Hyatt, A.D., Zhao, Y., and Roy, P. Release of bluetongue virus-like particles from insect cells is mediated by BTV nonstructural protein NS3/NS3A. *Virology* 193, 592–603 (1993).

52. Wirblich, C., Bhattacharya, B., and Roy, P. Nonstructural protein 3 of bluetongue virus assists virus release by recruiting ESCRT-I protein Tsg101. *J. Virol.* 80, 460–473 (2006).

53. Han, Z. and Harty, R.N. The NS3 protein of bluetongue virus exhibits viroporin-like properties. *J. Biol. Chem.* 279, 43092–43097 (2004).

54. Celma, C.C. and Roy, P. A viral nonstructural protein regulates bluetongue virus trafficking and release. *J. Virol.* 83, 6806–6816 (2009).

55. Celma, C.C. and Roy, P. Interaction of calpactin light chain (S100A10/p11) and a viral NS protein is essential for intracellular trafficking of nonenveloped bluetongue virus. *J. Virol.* 85, 4783–4791 (2011).

56. Eaton, B.T., Hyatt, A.D., and Brookes, S.M. The replication of bluetongue virus. *Curr. Top. Microbiol. Immunol.* 162, 89–118 (1990).

57. Hyatt, A.D., Eaton, B.T., and Brookes, S.M. The release of bluetongue virus from infected cells and their superinfection by progeny virus. *Virology* 173, 21–34 (1989).

58. Center for emerging issues. Overview of the mediterranean basin bluetongue disease outbreak, 1998–2004 in brief overview of BT ecology, epidemiology, clinical (2005). Available at http://www.aphis.usda.gov/animal_health/emergingissues/downloads/bluetongue_med_basin_jan_05_update.pdf

59. Maan, N.S. et al. Identification and differentiation of the twenty six bluetongue virus serotypes by RT-PCR amplification of the serotype-specific genome segment 2. *PLoS One* 7, e32601 (2012).

60. Mertens, P.P. et al. A comparison of six different bluetongue virus isolates by cross-hybridization of the dsRNA genome segments. *Virology* 161, 438–447 (1987).

61. Mertens, P.P. et al. Analysis of the roles of bluetongue virus outer capsid proteins VP2 and VP5 in determination of virus serotype. *Virology* 170, 561–565 (1989).

62. Mellor, P.S. and Boorman, J. The transmission and geographical spread of African horse sickness and bluetongue viruses. *Ann. Trop. Med. Parasitol.* 89, 1–15 (1995).

63. Taylor, W.P. The epidemiology of bluetongue. *Rev. Sci. Tech. Off. Int. Epizoot.* 5, 351–356 (1986).

64. Spreull, J. Malarial catarrhal fever (bluetongue) of sheep in South Africa. *J. Comp. Pathol. Ther.* 18, 321–337 (1905).

65. Jeggo, M.J. et al. Virulence of bluetongue virus for British sheep. *Res. Vet. Sci.* 42, 24–28 (1987).

66. Darpel, K.E. et al. Clinical signs and pathology shown by British sheep and cattle infected with bluetongue virus serotype 8 derived from the 2006 outbreak in northern Europe. *Vet. Rec.* 161, 253–261 (2007).

67. MacLachlan, N.J. The pathogenesis and immunology of bluetongue virus infection of ruminants. *Comp. Immunol. Microbiol. Infect. Dis.* 17, 197–206 (1994).

68. Parsonson, I.M. *Pathology and Pathogenesis of BT Infections* (Springer-Verlag, Berlin, Germany, 1990).

69. Osburn, B.I. et al. Experimental viral-induced congenital encephalopathies. II. The pathogensis of bluetongue virus infection of fetal lambs. *Lab. Invest.* 25, 206–213 (1971).

70. Johnson, S.J. et al. Clinico-pathology of Australian bluetongue virus serotypes for sheep, in Walton, T.E. and Osburn, B.I. (eds.) *Bluetongue, African Horse Sickness and Related Orbiviruses*, pp. 737–743 (CRC Press, Boca Raton, FL, 1992).

71. Oberst, R.D. Viruses as teratogens. *Vet. Clin. North. Am. Food Anim. Pract.* 9, 23–31 (1993).

72. Erasmus, B.J. Bluetongue virus, in Dinter, Z. and Morein, B. (eds.) *Virus Infections of Ruminants*, Vol. 21, pp. 227–237 (Elsevier Biomedical, Amsterdam, the Netherlands, 1990).

73. Eaton, B.T. and Crameri, G.S. The site of bluetongue virus attachment to glycophorins from a number of animal erythrocytes. *J. Gen. Virol.* 70, 3347–3353 (1989).

74. Brewer, A.W. and MacLachlan, N.J. The pathogenesis of bluetongue virus infection of bovine blood cells in vitro: Ultrastructural characterization. *Arch. Virol.* 136, 287–298 (1994).

75. Whetter, L.E. et al. Bluetongue virus infection of bovine monocytes. *J. Gen. Virol.* 70, 1663–1676 (1989).

76. Stott, J.L. et al. Interaction of bluetongue virus with bovine lymphocytes. *J. Gen. Virol.* 71, 363–368 (1990).

77. Barratt-Boyes, S.M. and MacLachlan, N.J. Dynamics of viral spread in bluetongue virus infected calves. *Vet. Microbiol.* 40, 361–371 (1994).

78. Brodie, S.J. et al. The effects of pharmacological and lentivirus-induced immune suppression on orbivirus pathogenesis: Assessment of virus burden in blood monocytes and tissues by reverse transcription-in situ PCR. *J. Virol.* 72, 5599–5609 (1998).

79. Takamatsu, H. et al. A possible overwintering mechanism for bluetongue virus in the absence of the insect vector. *J. Gen. Virol.* 84, 227–235 (2003).

80. Barratt-Boyes, S.M. et al. Flow cytometric analysis of in vitro bluetongue virus infection of bovine blood mononuclear cells. *J. Gen. Virol.* 73, 1953–1960 (1992).

81. Lunt, R.A. et al. Cultured skin fibroblast cells derived from bluetongue virus-inoculated sheep and field-infected cattle are not a source of late and protracted recoverable virus. *J. Gen. Virol.* 87, 3661–3666 (2006).

82. Darpel, K.E. et al. Involvement of the skin during bluetongue virus infection and replication in the ruminant host. *Vet. Res.* 43, 40 (2012).

83. Hemati, B. et al. Bluetongue virus targets conventional dendritic cells in skin lymph. *J. Virol.* 83, 8789–8799 (2009).

84. DeMaula, C.D. et al. Infection kinetics, prostacyclin release and cytokine-mediated modulation of the mechanism of cell death during bluetongue virus infection of cultured ovine and bovine pulmonary artery and lung microvascular endothelial cells. *J. Gen. Virol.* 82, 787–794 (2001).

85. DeMaula, C.D. et al. Bluetongue virus-induced activation of primary bovine lung microvascular endothelial cells. *Vet. Immunol. Immunopathol.* 86, 147–157 (2002).

86. DeMaula, C.D. et al. The role of endothelial cell-derived inflammatory and vasoactive mediators in the pathogenesis of bluetongue. *Virology* 296, 330–337 (2002).

87. Drew, C.P. et al. Bluetongue virus infection alters the impedance of monolayers of bovine endothelial cells as a result of cell death. *Vet. Immunol. Immunopathol.* 136, 108–115 (2010).

88. Afshar, A. Bluetongue: Laboratory diagnosis. *Comp. Immunol. Microbiol. Infect. Dis.* 17, 221–242 (1994).

89. Wechsler, S.J. and McHolland, L.E. Susceptibilities of 14 cell lines to bluetongue virus infection. *J. Clin. Microbiol.* 26, 2324–2327 (1988).

90. Mecham, J. Detection and titration of bluetongue virus in *Culicoides* insect cell culture by an antigen-capture enzyme-linked immunosorbent assay. *J. Virol. Methods* 135, 269–271. (2006).

91. Katz, J. et al. Diagnostic analysis of the prolonged bluetongue virus RNA presence found in the blood of naturally infected cattle and experimentally infected sheep. *J. Vet. Diagn. Invest.* 6, 139–142 (1994).

92. MacLachlan, N.J. et al. Detection of bluetongue virus in the blood of inoculated calves: Comparison of virus isolation, PCR assay, and in vitro feeding of *Culicoides variipennis*. *Arch. Virol.* 136, 1–8 (1994).

93. Shaw, A.E. et al. Development and initial evaluation of a real-time RT-PCR assay to detect bluetongue virus genome segment 1. *J. Virol. Methods* 145, 115–126 (2007).

94. Toussaint, J.F. et al. Bluetongue virus detection by two real-time RT-qPCRs targeting two different genomic segments. *J. Virol. Methods* 140, 115–123 (2007).

95. van Rijn, P.A. et al. Sustained high-throughput polymerase chain reaction diagnostics during the European epidemic of bluetongue virus serotype 8. *J. Vet. Diagn. Invest.* 24, 469–478 (2012).

96. Vangeel, I. et al. Bluetongue sentinel surveillance program and cross-sectional serological survey in cattle in Belgium in 2010–2011. *Prev. Vet. Med.* 106, 235–243 (2012).

97. Falconi, C. et al. Evidence for BTV-4 circulation in free-ranging red deer (*Cervus elaphus*) in Cabañeros National Park, Spain. *Vet. Microbiol.* 159, 40–46 (2012).

98. Kramps, J.A. et al. Validation of a commercial ELISA for the detection of bluetongue virus (BTV)-specific antibodies in individual milk samples of Dutch dairy cows. *Vet. Microbiol.* 130, 80–87 (2008).

99. Mars, M.H. et al. Evaluation of an indirect ELISA for detection of antibodies in bulk milk against bluetongue virus infections in the Netherlands. *Vet. Microbiol.* 146, 209–214 (2010).

100. Roy P, B.D., LeBlois, H., and Erasmus, B.J. Long-lasting protection of sheep against bluetongue challenge after vaccination with virus-like particles: Evidence for homologous and partial heterologous protection. *Vaccine* 12, 805–811 (1994).

101. Jeggo, M.H., Wardley, R.C., and Taylor, W.P. Clinical and serological outcome following the simultaneous inoculation of three bluetongue virus types into sheep. *Res. Vet. Sci.* 37, 368–370 (1984).

102. Jeggo, M.H., Wardley, R.C., and Brownlie, J. A study of the role of cell-mediated immunity in bluetongue virus infection in sheep, using cellular adoptive transfer techniques. *Immunology* 52, 403–410 (1984).

103. Jeggo, M.H., Wardley, R.C., and Brownlie, J. Importance of ovine cytotoxic T cells in protection against bluetongue virus infection. *Prog. Clin. Biol. Res.* 178, 477–487 (1985).

104. Alexander, R.A. and Haig, D.A. The use of egg attenuated bluetongue virus in the production of a polyvalent vaccine for sheep: A propagation of the virus in sheep. *Onderstepport J. Vet. Sci. Anim. Ind.* 25, 3–15 (1951).

105. Hunter, P. and Modumo, J. A monovalent attenuated serotype 2 bluetongue virus vaccine confers homologous protection in sheep. *Onderstepoort J. Vet. Res.* 68, 331–333 (2001).

106. Lacetera, N. and Ronchi, B. Evaluation of antibody response and nonspecific lymphocyte blastogenesis following inoculation of a live attenuated bluetongue virus vaccine in goats. *Am. J. Vet. Res.* 65, 1331–1334 (2004).

107. Veronesi, E., Hamblin, C., and Mellor, P.S. Live attenuated bluetongue vaccine viruses in Dorset Poll sheep, before and after passage in vector midges (Diptera: Ceratopogonidae). *Vaccine* 23, 5509–5516 (2005).

108. Savini, G. et al. Assessment of efficacy of a bivalent BTV-2 and BTV-4 inactivated vaccine by vaccination and challenge in cattle. *Vet. Microbiol.* 133, 1–8 (2009).

109. Caporale, V. and Giovannini, A. Bluetongue control strategy, including recourse to vaccine: A critical review. *Rev. Sci. Tech.* 29, 573–591 (2010).

110. Savini, G., MacLachlan, N.J., Sanchez-Vizcaino, J.M., and Zientara, S. Vaccines against bluetongue in Europe. *Comp. Immunol. Microbiol. Infect. Dis.* 31, 101–120 (2008).

111. Schultz, G. and Delay, P.D. Losses in newborn lambs associated with bluetongue vaccination of pregnancy ewes. *J. Am. Vet. Med. Assoc.* 127, 224–226 (1955).

112. Young, S. and Cordy, D.R. An ovine fetal encephalopathy caused by bluetongue vaccine virus. *J. Neuropathol. Exp. Neurol.* 23, 635–642 (1964).

113. Ferrari, G. et al. Active circulation of bluetongue vaccine virus serotype-2 among unvaccinated cattle in central Italy. *Prev. Vet. Med.* 68, 103–113 (2005).

114. Batten, C.A. et al. A European field strain of bluetongue virus derived from two parental vaccine strains by genome segment reassortment. *Virus Res.* 137, 56–63 (2008).

115. Parker, J. et al. An experimental inactivated vaccine against bluetongue. *Vet. Rec.* 96, 284–287 (1975).

116. Savini, G. et al. An inactivated vaccine for the control of bluetongue virus serotype 16 infection in sheep in Italy. *Vet. Microbiol.* 124, 140–146 (2007).

117. Campbell, C.H. Immunogenicity of bluetongue virus inactivated by gamma irradiation. *Vaccine* 3, 401–406 (1985).

118. Berry, L.J. et al. Inactivated bluetongue virus vaccine in lambs: Differential serological responses related to breed. *Vet. Res. Commun.* 5, 289–293 (1982).

119. Ramakrishnan, M.A. et al. Immune responses and protective efficacy of binary ethylenimine (BEI)-inactivated bluetongue virus vaccines in sheep. *Vet. Res. Commun.* 30, 873–880 (2006).

120. Roy, P. et al. Recombinant virus vaccine for bluetongue disease in sheep. *J. Virol.* 64, 1998–2003 (1990).

121. Lobato, Z.I. et al. Antibody responses and protective immunity to recombinant vaccinia virus-expressed bluetongue virus antigens. *Vet. Immunol. Immunopathol.* 59, 293–309 (1997).

122. Boone, J.D. et al. Recombinant canarypox virus vaccine co-expressing genes encoding the VP2 and VP5 outer capsid proteins of bluetongue virus induces high level protection in sheep. *Vaccine* 25, 672–678 (2007).

123. Perrin, A. et al. Recombinant capripoxviruses expressing proteins of bluetongue virus: Evaluation of immune responses and protection in small ruminants. *Vaccine* 25, 6774–6783 (2007).

124. Stewart, M. et al. Validation of a novel approach for the rapid production of immunogenic virus-like particles for bluetongue virus. *Vaccine* 28, 3047–3054 (2010).

125. Pérez de Diego, A.C. et al. Characterization of protection afforded by a bivalent virus-like particle vaccine against bluetongue virus serotypes 1 and 4 in sheep. *PLoS One* 6, e26666 (2011).

126. Hewat, E.A., Booth, T.F., and Roy, P. Structure of correctly self-assembled bluetongue virus-like particles. *J. Struct. Biol.* 112, 183–191 (1994).

127. Loudon, P.T. et al. Expression of the outer capsid protein VP5 of two bluetongue viruses, and synthesis of chimeric double-shelled virus-like particles using combinations of recombinant baculoviruses. *Virology* 182, 793–801 (1991).

128. Noad, R. and Roy, P. Virus-like particles as immunogens. *Trends Microbiol.* 11, 438–444 (2003).

129. Stewart, M. et al. Protective efficacy of bluetongue virus-like and subvirus-like particles in sheep: Presence of the serotype-specific VP2, independent of its geographic lineage, is essential for protection. *Vaccine* 30, 2131–2139 (2012).

130. Stewart, M. et al. Bluetongue virus serotype 8 virus-like particles protect sheep against virulent virus infection as a single or multi-serotype cocktail immunogen. *Vaccine* 31, 553–558 (2013).

131. Roy, P., Callis, J., and Erasmus, B.J. Protection of sheep against Bluetongue disease after vaccination with core-like and virus-like particles: Evidence for homologous and partial heterologous proteins, in *Proceedings of the 97th Annual Meeting, United States Animal Health Association*, pp. 88–97 (Las Vegas, NV, 1994).

132. Matsuo, E. et al. Generation of replication-defective virus-based vaccines that confer full protection in sheep against virulent bluetongue virus challenge. *J. Virol.* 85, 10213–10221 (2011).

133. Schmahl, G. et al. Efficacy of Oxyfly on *Culicoides* species—The vectors of bluetongue virus and other insects. *Parasitol. Res.* 103, 1101–1103 (2008).

134. Schmahl, G. et al. Effects of permethrin (Flypor) and fenvalerate (Acadrex60, Arkofly) on *Culicoides* species—The vector of bluetongue virus. *Parasitol. Res.* 104, 815–820 (2009).

135. Schmahl, G. et al. Pilot study on deltamethrin treatment (Butox 7.5, Versatrine) of cattle and sheep against midges (*Culicoides* species, Ceratopogonidae). *Parasitol. Res.* 104, 809–813 (2009).

53 Bovine Spongiform Encephalopathy Agent

Akikazu Sakudo and Takashi Onodera

CONTENTS

53.1 INTRODUCTION

Prion diseases or transmissible spongiform encephalopathies (TSEs) are fatal neurological diseases that include Creutzfeldt–Jakob disease (CJD) in humans, scrapie in sheep and goats, bovine spongiform encephalopathy (BSE) in cattle, and chronic wasting disease (CWD) in cervids. A key event in prion diseases is the conversion of the cellular, host-encoded prion protein (PrP^C) to its abnormal isoform (PrP^{Sc}) predominantly in the central nervous system (CNS) of the infected host [1].

The diseases are transmissible under some circumstances, but unlike other transmissible disorders, prion diseases can also be caused by mutations in the host gene. The mechanism of prion spread among sheep and goats that develop natural scrapie is unknown. CWD, transmissible mink encephalopathy (TME), BSE, feline spongiform encephalopathy (FSE), and exotic ungulate encephalopathy (EUE) are all thought to occur after the consumption of prion-infected material. In this chapter, the fundamental aspects of BSE will be reviewed to help improve its risk assessment and control in the public health field.

53.2 CLASSIFICATION AND MORPHOLOGY

A hypothesis based on the structure of an agent can best be validated by structural and compositional analysis of the biologically active particles. Knowledge of both the chemical composition of the particle (percentage of protein, nucleic acid, carbohydrate, etc.) and the identity of the molecular components is crucial for evaluating hypotheses. While this concept appears to be obvious, applying it to the real world has proved to be problematic.

53.2.1 PROTEINS

By 1980, published reports indicated that scrapie agent was inactivated by proteases [2,3] or by treatment that modified or denatured proteins [4,5]. These studies demonstrated that the scrapie agent required a protein but did little to discriminate it structurally from other infectious particles [6]. The knowledge that a protein was involved undoubtedly motivated scientists to improve both the method for purifying the scrapie agent and techniques for identifying agent-specific proteins. Discovery of PrP^{Sc} and PrP^C initiated a period of remarkable progress in this field by providing the necessary technological breakthrough, that is, a physical marker for TSE agent/prions [7]. Numerous compelling observations made over the last three decades support the proposal that PrP^Sc is an essential component of TSE agents [8–11].

53.2.2 Nucleic Acids

The search for TSE-agent-specific nucleic acid seems to have declined over recent years. This could be due to the convincing nature of the negative results [12–15], which might have convinced some scientists that the protein hypothesis is valid. Alternatively, it may be that current techniques have been exhausted without solving the problem. If specific nucleic acids exist, its discovery would be of paramount importance for both practical and theoretical reasons. Several studies have ruled out large nucleic acids being associated with prion infection. However, no study has fully precluded the possibility of very small nucleic acids being involved in the infection process. The most informative results from direct research limit the putative nucleic acid from less than 80 nucleotides to 240 nucleotides in length [12,13]. This size range essentially eliminates viruses as a feasible model but does not exclude either the virino or prion hypothesis.

53.2.3 Ultrastructure

The inability to identify a TSE-specific agent was an important anomalous observation that contributed to the eventual crisis within this field of research. The discovery of scrapie-associated fibrils (SAFs) [16–18] provided the first ultrastructural observation that achieved some degree of consensus. SAFs were verified by several laboratories and found to be similar, though not identical, to prion rods that were identified shortly afterward [19,20]. A consensus on the issue of SAF/prion rod identity was not achieved, however, because the theoretical perspective of the respective research groups significantly shaped their interpretation of the observations. Some of the virus groups viewed SAF as a filamentous scrapie virus or later as a part of a tubulofilamentous nemavirus structure [21,22], while others viewed SAF as a part of the pathology. The prion group saw SAF and prion rods as either a related form of the same phenomenon (i.e., aggregate of PrP^Sc [23]) or as unrelated structures [24,25]. However, in both cases, these structures were considered unnecessary for prion biological activity. This is because neither rods nor SAFs are present after sonication or in liposome preparations even though high protein titers and PrP^Sc remained [26,27].

Other more conventional TSE agents have been proposed. Reports identifying specific virus particles or nucleoprotein complexes [28–30] have not been substantiated by others, and their relevance in terms of a plausible TSE agent remains in doubt.

53.3 BIOLOGY AND EPIDEMIOLOGY

53.3.1 Prion Strain

Three biological criteria are used to define agent strains: incubation times, distribution of vacuoles throughout the brain, and distribution of amyloid or PrP^Sc. There should be little doubt that different TSE isolates exhibit distinct phenotypes. Thus, the existence of more than one TSE agent phenotype in a single genetically identical host can be accepted as fact.

Prion proponents argue that different conformations of PrP^Sc can explain known strains. Indeed, specific examples to support this hypothesis have been reported in the literature [31,32]. Critics of the protein hypothesis cite the existence of at least 20 distinct strains and argue that different conformations of PrP^Sc cannot account for all of these variations. The prion proponents counter by using the following two arguments. First, most of the strains cited have not been characterized in a single inbred host, so the actual number of unique strains remains in doubt. PrP^Sc is actually a population of molecules having at least 400 different glycosylation variants. Therefore, combinations of protein conformation and different patterns of glycosylation could (at least numerically) account for all the strains known at this time.

53.3.2 Historical Aspect

Scrapie was thought of as a disease of sheep that did not infect humans, although its tissues were known to contain an infectious agent. Because there were no other known natural TSEs, when BSE first appeared, it was immediately thought to be derived from scrapie. Indeed, meat originating from sheep had been fed to cattle in the meal that they are to increase milk yield. Under the direction of the British Ministry of Agriculture, Fishery, and Foods (MAFF), a change had taken place in the way that this meal was prepared in the United Kingdom in the early 1980s. This change in policy coincided with the original infection in cattle.

A small farm in Surrey (United Kingdom) reported more than one cow developing a strange neurological disease. The cattle were immediately killed and their brains removed for examination. When it was found that the cattle had developed a previously unrecognized disease, the farmer wanted to publish the data but was advised not to by MAFF. It is now estimated that approximately 100 cattle may have developed BSE symptoms prior to 1987.

In 1987, Wells et al. published the data showing that a cow had developed a spongiform encephalopathy [33], although few extra details were given. MAFF soon realized that this was no straightforward disease and set up a committee to advise them on what action should be taken to avoid any risk to humans and cattle.

In April 1988, the UK government established the Southwood Committee to investigate the outbreak of BSE. That same year, the committee published a report stating that there would be minimal risk to humans as all infected cattle would be slaughtered. The response to the BSE crisis was to prevent all bovine material from entering the meal fed to cattle. This was brought into action in July 1988. The feed manufacturers were warned that this was going to happen several months in advance. The reporting of cases of BSE to MAFF was made obligatory, and farmers were compensated with half the value of a nonsick animal.

Since its first appearance in 1986, BSE has been registered at a clinical level in more than 180,000 bovine cattle in the United Kingdom, and it is believed that as many as one million undiagnosed infected cattle would have entered the human food chain [34]. This transmissible prion disease was transmitted from bovine tissue through consumption of contaminated meat-and-bone meal (MBM) [35]. MBM is used in animal feed in various proportions throughout most of the developed world. However, only a few European countries described BSE cases originating from within their borders following the initial outbreak in the United Kingdom. Epidemiological analysis revealed many years later that almost all European countries were affected by BSE. Between November 1986 and March 2012, 184,618 cases of BSE were confirmed in the United Kingdom. Since 1989, relatively small numbers of BSE cases (in total 6005) have also been reported in native cattle in Austria, Belgium, Czech Republic, Denmark, Finland, France, Germany, Greece, Ireland, Israel, Italy, Japan, Liechtenstein, Luxembourg, Netherlands, Poland, Portugal, Slovakia, Spain, and Switzerland. Since the introduction of monitoring programs to detect BSE in dead and slaughtered cattle, 12 countries have found their first native case (Austria, Czech Republic, Finland, Germany, Greece, Israel, Italy, Japan, Poland, Slovakia, Slovenia, and Spain). Small numbers of cases have also been reported in Canada, the Falkland Islands (Islas Malvinas), and Oman but solely in animals imported from the United Kingdom. Indeed, in the absence of large-scale studies, doubt remains as to whether many non-European countries are actually BSE-free [36].

In 1996, the British government announced several cases of a new form of CJD in young patients [37], called variant Creutzfeldt–Jakob disease (vCJD), caused by a BSE agent entering the body through the oral route [38]. Up to now, more than 180 cases of vCJD have been confirmed (with 173 in the United Kingdom) (Table 53.1). Many parameters remain unknown, including infective dose, susceptibility, and duration of the incubation period. Mathematical models have been developed to analyze the likely spread of the disease, but these are still imprecise. The initial models estimated between 100 and more than 100,000 cases of vCJD. However, more recent models suggest a total of a few hundred to a few thousand vCJD cases in the United Kingdom due to direct contamination of humans through consumption of beef products [39]. Moreover, the risk of secondary transmission from human to human has to be taken into account. Indeed, PrP^{res}, the disease-specific biochemical marker that is thought to be the infectious agent [40], has been found in peripheral lymphoid organs (e.g., tonsils, appendix) of vCJD patients [41]. The theoretical risk of iatrogenic transmission, either by blood transfusion, organ transplantation, or via contaminated surgical tools, could be higher than for sporadic CJD [42].

To prevent new cases of vCJD, the first measure has been to stop the consumption of potent infectious organs (i.e., CNS and some immune organs) [43], which correspond to 99% of total infectivity. Unfortunately, clinical signs of BSE appear late in the course of the disease, and clinical

TABLE 53.1

vCJD Current Data (April 2012)

Country	Total Number of Primary Cases (Number Alive)	Total Number of Secondary Cases: Blood Transfusion (Number Alive)	Cumulative Residence in the United Kingdom >6 Months during the Period 1980–1996
United Kingdom	173 (0)	3 (0)	176
France	25 (0)	—	1
Republic of Ireland	4 (0)	—	2
Italy	2 (0)	—	0
United States	3[a] (0)	—	2
Canada	2 (0)	—	1
Saudi Arabia	1 (0)	—	0
Japan	1[b] (0)	—	0
The Netherlands	3 (0)	—	0
Portugal	2 (0)	—	0
Spain	5 (0)	—	0
Taiwan	1 (0)	—	1

[a] The third US patient with vCJD was born and raised in Saudi Arabia and has lived permanently in the United States since late 2005. According to the US case report, the patient was most likely infected as a child when living in Saudi Arabia.

[b] The case from Japan had resided in the United Kingdom for 24 days in the period 1980–1996.

diagnosis is complicated by the absence of pathognomonic signatures [44]. Consequently, an infected animal could still enter the human food chain and present a risk to human health. For example, spinal cord could be incompletely removed, and dorsal root ganglia may be retained in vertebrae. Moreover, improvements in slaughtering practice did not prevent contamination of healthy materials by infected tissue.

In this context, the use of sensitive diagnostic tools was proposed to detect and remove infected animals. Initial epidemiological studies showed that a large number of previously undiagnosed infected animals could be detected by employing a sensitive test, even at the slaughterhouse. Systemic use of rapid and sensitive tests at abattoirs was then adopted within the European community and Japan from 2001 onwards in order to enforce consumer protection against the risk of contamination through the food chain.

53.3.3 Atypical BSE

The large-scale testing of livestock nervous tissues for the presence of PrP^{Sc} has lead to the recognition of two molecular signatures that are distinct from BSE. These were termed H-BSE and L-BSE (or BASE) (Table 53.2). Their PrP^{Sc} molecular signature differed from classical BSE in terms of the protease-resistant fragment size and glycopattern [45–47]. The experimental transmission of these cases to different lines of bovine PrP transgenic mice unambiguously demonstrated their infectious nature and confirmed their unique but distinctive strain type as compared to BSE [46,48–50]. These uncommon cases have been detected in aged asymptomatic cattle during systematic testing at the slaughterhouse. In France, a retrospective study of all TSE-positive cattle was

TABLE 53.2
Number of Atypical BSE Cases Worldwide

Country	H-Type	L-Type
Austria	0	2
Canada	1	1
Denmark	0	1
France	14	13
Germany	1	1
Ireland	1	0
Italy	0	4
Japan	0	1
Poland	2	8
Sweden	1	0
Switzerland	1	0
The Netherlands	1	2
United Kingdom	3	0
United States	2	0
Σ	27	33

60 cases worldwide

identified through the compulsory EU surveillance program (2001–2007) [51]. This study showed the following results:

1. All H-BSE and L-BSE cases detected by rapid tests were observed in animals over 8 years of age in either the fallen stock surveillance stream or in the abattoir.
2. No H-BSE and L-BSE were observed in the passive epidemio-surveillance network although, during retrospective interviews, the farmers and veterinarians for six of these animals reported clinical signs consistent with TSE in three fallen stock.
3. Frequency of H-BSE and L-BSE is, respectively, 0.35 and 0.41 cases per million adult cattle tested but increases to 1.9 and 1.7 cases per million in animals over 8 years of age.

The origin of these atypical BSE cases in cattle is currently unknown, as is the performance of the current active surveillance system for detecting H-BSE- and L-BSE-affected animals. Consequently, there is uncertainty about the true prevalence of these conditions within the stock of cattle. All atypical BSE cases identified in the European Union occur in animals born before the extended or real feed ban that was passed into law in January 2001 [44]. Hence, as with classical BSE, exposure of these animals to contaminated feed with low titers of TSE agents cannot be excluded. Equally, additional origins for these TSE forms must not be ruled out at this stage. In particular, the unusually old age of all H-BSE and L-BSE identified cases and their apparent low prevalence in the population could suggest that these atypical BSE forms arise spontaneously.

There are some data related to the peripheral distribution of L-BSE agent in cattle experimentally challenged by the intracerebral route. In a Japanese study, infectivity and PrP^{res} were detected in many peripheral nerves tested from mid-incubation onwards. PrP^{res} was not detected in lymphoid tissue [52]. In an Italian study, PrP^{res} was not detected in peripheral nerves [53], but the presence of infectivity in skeletal muscle, presumably linked to nervous structures, of natural cases was described [54].

Both L-BSE and H-BSE agents are able to propagate in experimentally challenged foreign species such as mouse, sheep, vole, primates, and hamster and in transgenic mice expressing heterologous (i.e., nonbovine PrP) sequences. Noticeably, the transmission barrier observed for the L-BSE agent was lower than that for classical BSE. In wild-type and transgenic mice expressing the VRQ allele of ovine PrP, the L-BSE agent acquired a phenotype indistinguishable from that observed by infection with the BSE agent [48–50]. More recently, transmission of H-BSE isolate originating from France and Poland to bovine PrP transgenic mice has been reported [55].

53.3.4 Atypical BSE Risk for Humans

Together these data indicate that there may be an etiological relationship between atypical and classical BSE. Intracerebral inoculation of brain from L-BSE-infected cattle to cynomolgus macaques induced a spongiform encephalopathy distinct in all aspects (clinical, lesional, and biochemical) from macaque BSE [56]. In terms of a primary passage through inoculation of the same amount of infected brain, incubation periods were shorter (23–25 months) than for BSE (38–40 months), suggesting that L-BSE may be more virulent than classical BSE for infecting primates.

In contrast, no clinical sign has been observed 72 months after intracerebral inoculation of brain from H-BSE, and recipient cynomolgus macaques remained healthy, suggesting a lower, if any, virulence of this agent for primates [56].

The intracerebral inoculation of L-BSE field isolates produced TSE disease in two lines of mice overexpressing human PrP (Met129), exhibiting a molecular phenotype distinct from classical BSE [57,58]. In one of them, the L-BSE agent appeared to propagate with no obvious transmission barrier: a 100% attack rate was observed on first passage, the incubation time was not reduced on subsequent passaging [56], and the L-type PrPSc biochemical signature was essentially conserved [57,58]. The latter appeared indistinguishable from that seen after experimental inoculation of MM2 sCJD in these mice [48]. These transmission features markedly differed from the low transmission efficiency of cattle BSE isolates to this [57,59] and other human PrP transgenic mouse lines [60].

H-type isolates failed to infect one line of "humanized" mice [57]. These mice overexpress human PrP and were inoculated intracranially with a low dilution inoculum, supporting the view that the transmission barrier of H-type BSE from cattle to humans might be quite high. The permissiveness of "humanized" transgenic mice expressing the valine allele at codon 129 (or hemizygous) to atypical BSE is still unknown.

Recently, another type of atypical BSE (Swiss 2011) was reported from Switzerland. Cows of 8 and 15 years of age were tested for BSE, which was confirmed by Western blotting experiments. The Prionics Western blot detected a similar three-band PrPres glycoprofile with molecular masses of roughly 16, 20, and 25 kDa for each animal, which were lower than the PrP protein bands detected in animals with classical BSE [61]. Combining Western blot analysis with an epitope mapping strategy, they ascertained that these animals displayed an N-terminally truncated PrPres that was different from classical BSE, L-BSE, and H-BSE. These findings raise the possibility that the cattle were affected by a prion disease not previously encountered before. In order to confirm this possibility, in vivo transmission studies using transgenic mice and cattle have been undertaken. The results from these ongoing studies will be used to assess the potential effect on disease control and public health.

53.4 CLINICAL FEATURES AND PATHOGENESIS

53.4.1 CONVENTIONAL MOUSE PASSAGES

BSE was recognized as a cattle TSE during the 1980s in the United Kingdom [62]. Early studies indicated that the BSE agent can only be found in the brain, spinal cord, and retinal

tissue of BSE-diseased cattle. Infectivity assessment in several tissues from orally inoculated cattle, using a bioassay based on RIII mice [63,64], revealed the presence of BSE infectivity in the CNS, optic nerve, retina, cervical, thoracic and trigeminal ganglia, and the facial and sciatic nerves, as well as in the distal ileum. The skeletal muscle, spleen, and other lymphatic tissues were shown to be free from detectable infectivity. More recently, infectivity was discovered on the tonsils from cattle killed 10 months after oral BSE challenge by intracerebral inoculation in cattle [65]. These results consistently show that BSE infectivity, after oral uptake, propagates only poorly in some intestinal lymphatic tissues (mainly Peyer's patches) and from there spread to the CNS, probably by an intraneural route via the peripheral nervous system [66,67].

These findings contrast to the way in which the scrapie agent spread in infected sheep, mice, and hamsters. In this case, the scrapie agent spreads via tissues such as spleen, other lymphatic tissues, and muscles, even during the preclinical stage [68–71]. However, BSE agent can be found in the lymphoreticular system and other peripheral tissues when transmitted to sheep or primates [72–75].

Experimental transmission to a common "reporter" species has highlighted the remarkable ability of the BSE agent to retain its biological properties after intermediate passage in a range of different hosts with distinct PrPC sequences. Initial studies involving transmission of various species of infected cattle to a panel of inbred mice expressing a or b mouse PrP allele suggested that cattle may have been infected by a single strain because incubation periods and distribution of spongiosis in the brain were uniform in each genotype, unlike that seen with scrapie or CJD isolates [76–78]. The two agent propagates were termed 301C and 301V, respectively. Strikingly, 301C and 301V were invariably obtained irrespective of the species infected by the BSE agent, either accidentally (cats, exotic ruminants, humans) or experimentally (sheep, goats, pigs, macaques) [76,77,79]. It remains unclear whether a more thorough adaptation of these species would lead to a conservation of the strain phenotype, given that continued passage in a new host can alter strain characteristics, for example, 301V and 301C [76]. A divergent evolution of the BSE agents has been reported following transmission to various lines of inbred mice, all carrying the PRNP-a allele [60]. Careful phenotypic comparison confirmed the presence of two distinct mouse strains resembling either the Chandler strain or mouse BSE [60,80]. These findings suggest that loci other than PrP might influence not only the susceptibility [80,81] but also the strain evolution. Puzzlingly, however, these results have not been reproduced in another study using the same panel of mice and different BSE isolates [50].

53.4.2 TRANSGENIC MOUSE PASSAGES

Experimental transmission of cattle BSE isolates to transgenic mice expressing a methionine (Met) allele of human PrP at codon 129 revealed very low transmission efficiency with a low attack rate and long incubation period, suggesting a strong transmission barrier for cattle BSE. This low BSE transmission efficiency to human PrP transgenic mice is occasionally accompanied by a strain shift allowing the appearance of an alternative, sporadic CJD-like phenotype in a proportion of mice [82]. The exact characteristics and further evolution of the vCJD epidemic still involve uncertainties owing to the prolonged incubation times. Nonetheless, this modeled high transmission barrier of human to cattle BSE might be an explanation for the currently low incidence of vCJD, given the high level of exposure to BSE during the "mad cow" crisis in the United Kingdom.

Sheep and goats are experimentally susceptible to BSE [83,84]. During the BSE epidemic, sheep and goats have been exposed to BSE-contaminated MBM, so BSE transmission to these species may have occurred. A recent study [85] evaluated human susceptibility to small ruminant-passaged BSE agents by inoculating two different transgenic mouse lines expressing the Met allele of human PrP at codon 129 (tg650 in INRA, France, and tg340 in INIA, Spain) with several sheep and goat BSE isolates and compared their transmission characteristics with those of cattle BSE. In this study, the transmission efficiency of cattle BSE isolates in both human PrP transgenic mouse models was low. When the sheep and goat BSE isolates were inoculated into human PrP transgenic mouse models, attack rates approaching 100% were observed from the primary passage onwards and mean incubation times [85].

53.4.3 Classical BSE Risk to Humans

These results clearly indicate that Met129 homozygous individuals are more susceptible to a sheep or goat BSE agent than to cattle BSE and that these agents might transmit with molecular and neuropathological properties indistinguishable from those of vCJD. The most recent EFSA opinion on BSE/TSE infectivity in small ruminant tissues [86] concluded with 95% confidence that the yearly number of BSE-infected animals entering the human food chain in the European Union ranges between 0 and 240 for sheep and between 0 and 381 for goats. Although these numbers are relatively low, the susceptibility of humans to a sheep BSE agent is thought to be higher than to bovine BSE. This has important ramifications for public health. Indeed, sheep might be a more dangerous source of BSE infectivity for man as compared to cattle because BSE-infected sheep PrP^{Sc} has been detected in peripheral organs such as tonsil, retropharyngeal lymph node, ileocecal and mesenteric lymph nodes [75,87,88]. Moreover, a mathematical model analysis was used to assess the human health risk from possible BSE infection derived from the British sheep flock. This study concluded that the risk to public health was higher from sheep BSE rather than cattle BSE [89]. Even more worryingly, transmission studies suggest that Met129 human PrP^{C} displays a preference for a BSE agent with an ovine, rather than a bovine, sequence. Thus, although few natural cases have been described and we cannot draw any definitive conclusions concerning the origin of vCJD, the risk of a potential goat and/or sheep BSE agent should not be underestimated.

Several transmission experiments in primate models were performed to assess the risk of BSE to human health. Lemur, marmosets, macaques, and squirrel monkeys developed spongiform encephalopathies after intracerebral inoculation of brains from BSE-infected cattle [73,90–92]. Secondary transmission to the same host (i.e., mice) of both macaque BSE and human vCJD induced similar lesional profiles. This observation provides additional evidence for the similarity between BSE and vCJD agents [79].

Subsequently, lemur and macaque models were used to demonstrate the transmissibility of BSE through the oral route [89,93]. In macaque, 5 g was sufficient to transmit the disease to one of two animals [93]. Risk of secondary transmission through transfusion was assessed in the same primate models: infectivity of blood components was demonstrated through intracerebral inoculation in lemurs [94], the intravenous route was demonstrated as an efficient way of transmission in macaques [74], and transmission was achieved through transfusion in this latter model [95].

Human case studies confirm previous findings regarding age-dependent susceptibility/exposure to infection [96,97], which those aged 10–20 years are at highest risk and those over ca. 40 years at much lower risk. Large-scale studies are underway to test tonsil and appendix tissue for the presence of PrP^{Sc} [98]. Although the results of these studies will provide useful information on the prevalence of late-stage infection, their interpretation will be limited by the relatively poor understanding of the sensitivity and specificity of the tests throughout disease incubation. The main priority therefore remains the development of a diagnostic test that is able to detect infection early in the incubation period and can be applied to large population samples, both in the United Kingdom and BSE-affected countries.

53.5 IDENTIFICATION AND DIAGNOSIS

Several biochemical methods have been used to diagnose BSE. This includes enzyme-linked immunosorbent assay (ELISA), Western blotting, and immunohistochemistry (IHC) (Figure 53.1). As an index of prion infection, the representative changes in the brain were detected in these assays. In ELISA and Western blotting, protease-resistant PrPs, which are accumulated in the brain and form deposits, were biochemically detected after treatment with proteinase K (PK). In the case of ELISA for BSE, a homogenate is usually prepared from the brain obex and treated with PK. The treated sample is applied to a microtiter plate for absorption then detected with anti-PrP antibody. This method is commercially exploited in the Bio-Rad Platelia BSE purification kit and Bio-Rad BSE detection kit (Bio-Rad, France), Enfer TSE kit (Abbott Laboratories, United States), and FRELISA BSE kit (Fujirebio, Inc., Japan). Although the extensively employed ELISA is a sensitive and high-throughput method, the large frequency of false-positives remains a problem. Therefore, if a result is positive, the ELISA must be repeated in order to validate the finding. If a positive result is obtained again, Western blotting and IHC are performed. The

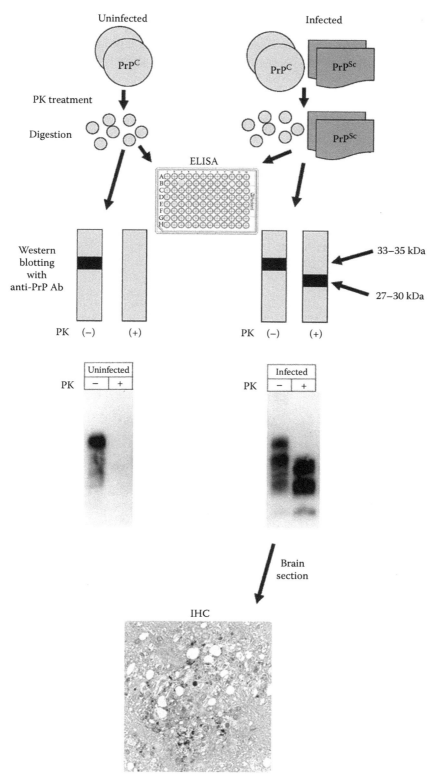

FIGURE 53.1 Diagnostic methods for prion infections. Brain tissue is used for the diagnosis of prion diseases. Therefore, the diagnosis is postmortem. Most methods of diagnosing prion diseases are based on the characteristics of PrP^{Sc}, which is resistant to PK. PK completely degrades PrP^C but only partially digests PrP^{Sc} because PrP^{Sc} forms protease-resistant aggregates. After the treatment, Western blotting and ELISA with anti-PrP antibody detect PK-resistant PrP, when PrP^{Sc} is included in the sample. As the N-terminal region of PrP^{Sc} is digested with PK (+), PK-resistant PrP (27–30 kDa) shows a shift to a lower molecular weight compared to untreated PrP (–) (33–35 kDa) in Western blotting. IHC is usually used for definitive diagnosis. In IHC, vacuolation and PrP deposits in brain are usually used as an index of prion diseases. (Modified from Sakudo, A. et al., *Int. J. Mol. Med.*, 27, 483, 2011, Figs. 3 and 4. With permission from Spandidos Publications Ltd.)

Western blotting uses a membrane to absorb PK-treated proteins separated by sodium dodecyl sulfate polyacrylamide gel electrophoresis (SDS-PAGE). After the absorption, PK-resistant PrP (PrP^{res}) in the membrane-bound proteins is detected with anti-PrP antibody. Importantly, Western blotting provides information on not only prion infections but also the mobility of peptides, which is influenced by the host genotype and prion strains [99,100]. In the case of IHC, the index for BSE is neuronal cell loss, astrocytosis, and vacuolation in addition to PrP accumulation (amyloid plaques). In the IHC analysis of brain sections, these changes are examined by light microscopy. Although vacuolation is also used as an index of prion infection, various combinations of prion strains with host species result in the accumulation of PrP without vacuolation in brain sections [101,102]. Recently, the protein misfolding cyclic amplification (PMCA) method

has been developed [103]. In this method, the index is the property of PrP^{Sc} converting PrP^{C} into PrP^{Sc}. This enables in vitro amplification of PK-resistant PrP (PrP^{res}) from a small quantity of PrP^{Sc} as seed by sequential cycles of incubation and sonication. Interestingly, levels of PrP^{res} amplified by this method correlated with the prion infectivity titer [104]. Furthermore, PMCA could detect prions in blood [105]. Thus, PMCA may be used to diagnose not only terminally diseased hamsters but also prion-infected presymptomatic hamsters [106]. This method has the highest sensitivity of any method for detecting PrP^{res} reported so far. Recently, PMCA has been modified, leading to recombinant PrP-PMCA (rPrP-PMCA) [107] and its combination with the quaking-induced conversion (QUIC) reaction [108,109]. Recently, other diagnostic methods for prion diseases have been developed (Table 53.3).

TABLE 53.3
Diagnostic Methods of Prion Infection

Method Name	Index of the Method	Procedures	References
Western blot	PK-resistant PrP	Detect PK-resistant PrP on the membrane.	[148]
ELISA	PK-resistant PrP	Detect PK-resistant PrP adsorbed on microtiter plates by anti-PrP antibody.	[149]
IHC	PK-resistant PrP	Immunostain tissue sections.	[150]
Bioassay	PK-resistant PrP, incubation time or infectivity titer	Transmission to mice.	[151]
Cell culture assay	PK-resistant PrP or infectivity titer	Transmission to cells.	[152]
Histoblot	PK-resistant PrP	Cryosection is blotted onto membrane before PK treatment and immunolabeling with anti-PrP.	[153]
Cell blot	PK-resistant PrP	Grow the cells on cover slip, directly transferred to membrane, and detect the PK-resistant PrP using anti-PrP antibody.	[154]
Slot blot	PK-resistant PrP	Filter the cell lysate through nitrocellulose membrane and detect the PK-resistant PrP using anti-PrP antibody.	[155]
PMCA	PK-resistant PrP	Amplification of misfolding protein by cycles of incubation and sonication.	[156]
CDI	PrP conformation	Specific antibody binding to denatured and native forms of PrP.	[157]
DELFIA	Insoluble PrP	Measure a percentage of the insoluble PrP extracted by two concentrations of guanidine.	[158]
Capillary gel electrophoresis	PK-resistant PrP	Competition between fluorescein-labeled synthetic PrP peptide and PrP present in samples is assayed by separation of free and antibody-peptide peaks using capillary electrophoresis.	[110,111]
FCS	Aggregation of prion protein	PrP aggregates were labeled by anti-PrP antibody tagged with fluorescent dyes, resulting in intensity fluorescent target, which were measured by dual-color fluorescence intensity distribution analysis.	[112]
Aptamer	PrP conformation	Use RNA aptamers that specifically recognize PrPC and/or PrPSc conformation.	[120]
FTIR spectroscopy	Alterations of spectral feature	Analyze FTIR spectra with chemometrics analysis.	[113]
Flow microbead immunoassay	PK-resistant PrP	Detect PK-resistant PrP using flow cytometry with anti-PrP antibody coupled with microbeads.	[159]
Paraffin-embedded tissue (PET) blot	PK-resistant PrP	PET section is collected on membrane for PK treatment and immunolabeling with anti-PrP antibody.	[160]
Surrogate marker	Change of expression level of 14-3-3 protein, erythroid-specific marker, or plasminogen	2D gel electrophoresis, differential display reverse-transcriptase PCR, or Western blotting of surrogate marker proteins for prion diseases.	[138–141]

Source: Modified from Sakudo, A. et al., *J. Vet. Med. Sci.*, 69, 329, 2007, with permission from the Japanese Society of Veterinary Science.
Note: PET, paraffin-embedded tissue; PMCA, protein misfolding cyclic amplification; CDI, conformation-dependent immunoassay; DELFIA, dissociation-enhanced lanthanide fluorescent immunoassay; FCS, fluorescence correlation spectroscopy; FTIR, Fourier transform infrared; ELISA, enzyme-linked immunosorbent assay; PK, proteinase K; PrP, prion protein; PrP^{C}, cellular isoform of PrP; PrP^{Sc}, abnormal isoform of PrP.

Capillary gel electrophoresis is combined with a competitive antibody-binding assay that uses a fluorescein-labeled synthetic PrP peptide and PrP [110,111]. The free and antibody-peptide peaks are separated by capillary electrophoresis and used for diagnosis.

In fluorescence correlation spectroscopy (FCS), single fluorescently labeled molecules in solution are detected [112]. By using this method, PrP^{Sc} can be labeled by the anti-PrP antibody or by conjugation with labeled recombinant PrP for diagnosis.

In Fourier transform infrared spectroscopy (FTIR), the difference of infrared spectra of samples between prion-infected and -noninfected animals is used for diagnosis [113]. Such a spectroscopy-based approach is promising for postmortem diagnosis when visible and near-infrared radiation (Vis-NIR) is used. Vis-NIR can be transmittable through the animal body and may facilitate the materialization of noninvasive diagnosis [114].

A detection tool for PrP^{Sc} has recently been developed. Some antibodies and aptamers are able to distinguish PrP^{Sc} from PrP^C. Antibodies having a conformational epitope of PrP^{Sc} were obtained by immunization of Tyr–Tyr–Arg peptide [115]. Antibodies 15B3, V5B2, mAb132, 3B7, 2C4, 1B12, and 3H6 can also recognize PrP^{Sc}-specific epitopes [116–119]. PrP^C and/or PrP^{Sc}-specific binding aptamers, which are RNA or DNA molecules specifically binding to a target protein, were reported [120–122]. PrP^{Sc}-specific antibodies and aptamers might not only serve as a diagnostic or analytical tool for investigating prion diseases but may also facilitate the development of useful therapeutic agents, that is, if they are found to inhibit the accumulation of PrP^{Sc}.

There are many genes and proteins with differential abundances in prion diseases [123]. Examples of deregulated genes associated with prion diseases include ATP-binding cassette, amyloid beta (A4) precursor-like protein1, apolipoprotein D and E, beta-2 microglobulin, CD9 molecule, clusterin, cystatin C, cathepsin S, glial fibrillary acidic protein, osteonectin, bone sialoprotein I, and early T-lymphocyte activation 1 [124–130]. Differentially expressed protein biomarkers for prion diseases include vitronectin, alpha-1-acid glycoprotein, alpha-1-antichymotrypsin, apolipoprotein B100, apolipoprotein E, cathelicidin antimicrobial peptide, clusterin, complement components C3 and C4, complement factors B and H, haptoglobin α-2 chain, histidine-rich glycoprotein, Ig gamma-2 chain C region, transthyretin, uroguanylin, vitronectin precursor, and α-2-macroglobulin [124–134]. Changes in response to prion infections have recently been employed as an index for prion diagnosis. For example, the level of urinary alpha-1-antichymotrypsin was dramatically increased in the urine of patients with sporadic CJD and other animal models of prion diseases [135]. Moreover, an increase of S-100β, cystatin C, and heart fatty acid binding protein (H-FABP) in the blood of CJD patients has been reported [136,137]. In particular, 14-3-3 [138,139], erythroid-specific marker [140], plasminogen [141], tau, phosphor-tau, S-100β, and neuron-specific enolase (NSE) are promising surrogate markers for prion diseases [123].

Taken together, prion diseases can usually be diagnosed by methods such as the ELISA, Western blotting, and immunohistochemical analysis [142]. Recently, there have been dramatic improvements in prion-sensitive diagnostic methods. Nonetheless, further developments of analytical methods to specifically and reliably detect prions are urgently needed.

53.6 TREATMENT AND PREVENTION

Prion agents are known to be highly resistant pathogens. Therefore, inactivation of prions requires appropriate treatment of surgical instruments used, for example, in craniotomy, spinal surgery, and ophthalmologic procedures. There have been four reported cases of iatrogenic CJD related to neurosurgical procedures, which occurred before the 1980s, but their causal relationships remain unclear. In addition, CDC suggested the risk of infection of CJD is relatively low but preventative measures should be put in place for specific procedures such as neurosurgery (CDC Guideline). The most important difference between prion and other pathogens is that prion has no associated nucleic acid [143]. Therefore, it cannot be inactivated by conventional sterilization procedures such as autoclaving (121°C, 20 min), exposure to UV or γ-ray irradiation [143]. Although treatment with alcohol, such as 70% ethanol, is effective for bacteria and enveloped viruses, there is no effect on prions. Inactivation of prions involves use of an autoclave under severe conditions (134°C, 18 min), NaOH (1 N, 20°C, 1 h), SDS (30%, 100°C, 10 min), and NaOCl (20,000 ppm, 20°C, 1 min) (Table 53.4). In contrast to other pathogens like viruses and bacteria, fixation using aldehyde slightly decreases the prion titer but this is insufficient. Therefore, for treatment of tissue sections of prion-infected animals, formic acid is recommended. Recently, the effectiveness of other treatments has been reported. Nonetheless, animal bioassays must be performed in order to fully demonstrate the effectiveness of these treatments on prion inactivation, which takes a considerable length of time (about 1 year or more). This makes it difficult to examine other forms of treatment given the cost and time-consuming nature of the tests. Currently, the following procedures for prion inactivation are recommended by the Japanese government and Society for Healthcare Epidemiology of America [144–146]: (1) Washing with appropriate detergents + SDS treatment (3%, 3–5 min), (2) treatment with alkaline detergents (80°C–93°C, 3–10 min) + autoclaving (134°C, 8–10 min), (3) washing with appropriate detergents + autoclaving (134°C, 18 min), and (4) washing with alkaline detergents (at a concentration and temperature according to instructions) + vaporized hydrogen peroxide gas plasma sterilization. Most importantly, a dried prion-infected apparatus is difficult to sterilize; therefore, prompt washing is essential [144]. This procedure should be followed by autoclaving at 134°C and vaporized hydrogen peroxide gas plasma sterilization in order to attain an assurance level of prion inactivation of less than 10^{-6}.

In December 1999, WHO made the following recommendations to reduce exposure to the BSE agent: (1) All countries

TABLE 53.4
Representatives of Effective Treatment on Inactivation of Prion

Treatment	References
NaOCl (20,000 ppm, 20°C, 1 h)	[143]
NaOH (1 N, 20°C, 1 h)	[143]
Autoclave under soaked condition in water (134°C, 18 min)	[143]
Alkaline detergent (1.6%, 43°C, 15 min)	[143]
Phenolic disinfectant (5%, 20°C, 30 min)	[143]
3% SDS, 100°C, 10 min	[162]
7 M Guanidine hydrochloride (room temperature, 2 h)	[162]
3 M Guanidine thiocyanate (room temperature, 2 h)	[162]
3 M Trichloroacetic acid (room temperature, 2 h)	[162]
60% Formic acid (room temperature, 2 h)	[162]
Radio-frequency gas plasma	[145]
Hydrogen peroxide gas plasma (STERRAD NX)	[145]
Chlorine (1000 ppm)	[145]
Copper (0.5 mmol/L) + hydrogen peroxide (100 mmol/L)	[145]
Sodium metaperiodate (0.01 M)	[145]
Quaternary ammonium compound	[145]
Peracetic acid (0.2%)	[145]
50% Phenol (room temperature, 2 h)	[162]
Enzymatic detergent (0.8%, 43°C, 5 min) + hydrogen peroxide gas plasma sterilization (1.5 mg/L, 25°C, 3 h)	[143]
Vaporized hydrogen peroxide (2 mg/L, 30°C, 3 cycles)	[143]

Note: SDS, sodium dodecyl sulfate.

Autoclaving with no soaking in water is insufficient for prion inactivation (dry conditions cause difficulty in inactivation) [143,163].

Enzymatic detergent (0.8%, 43°C, 5 min) + autoclave (121°C, 20 min), only enzymatic detergent (0.8%, 43°C, 5 min), only peroxyacetic acid (0.25%, 55°C, 12 min), only vaporized hydrogen peroxide gas (1.5 mg/L, 25°C, 3 h), or enzymatic detergent (0.8%, 43°C, 5 min) + vaporized hydrogen peroxide gas (1.5 mg/L, 25°C, 3 h) is insufficient for prion inactivation.

must prohibit the use of ruminant tissues in ruminant feed and must exclude tissues that are likely to contain the BSE agent from any animal or human food chain. BSE eradication was recommended during a WHO consultation held in December 1999. (2) All countries are encouraged to conduct risk assessment to determine if they are at risk for BSE in sheep and goats. It is advised that any tissue that may come from deer or elk with CWD (a transmissible spongiform disease of North American mule deer and elk) is not used in animal or human feed; however, at this time, there is no evidence to suggest that CWD in deer and elk can be transmitted to humans. (3) No infectivity has yet been detected in skeletal muscle tissue. Reassurance can be provided by removal of visible nervous and lymphatic tissue from meat (skeletal muscle). (4) Milk and milk products are considered safe. Tallow and gelatin are considered safe if prepared by a manufacturing process that has been shown experimentally to inactivate the transmissible agent and if prepared from specifically identified tissues or from cattle without risk of exposure to BSE. (5) Human veterinary vaccines prepared from bovine materials may carry the risk of transmission of animal TSE agents. The pharmaceutical industry should ideally avoid the use of bovine materials and materials from other animal species in which TSEs naturally occur. If absolutely necessary, bovine materials should be obtained from countries that have a surveillance system for BSE in place and that report either zero or sporadic cases of BSE. These precautions apply to the manufacture of cosmetics as well.

53.7 CONCLUSION

BSE or "mad cow disease" is a disease that affects the brain of cattle and humans. The disease was first diagnosed in the United Kingdom in 1986. Since then, the disease has occurred in many European countries as well as Japan, Canada, and the United States. Most of the reported cases of BSE (95%) have occurred in the United Kingdom.

Most cattle are infected when they orally ingest prion-contaminated "ruminant" MBM. This dietary supplement has been banned from feed since 1999 by WHO. A few cows may be able to pass BSE to their offspring, but animals do not infect each other by direct contact. BSE is typically a slow developing disease. Infected cattle appear normal for 2–8 years. As the disease develops, the brain is affected. Signs such as trembling, stumbling, and swaying and behavioral changes (e.g., nervousness, aggression, or frenzy) are seen.

Weight loss and a drop in milk production may be noted. Rarely, cattle with BSE will become suddenly ill with days. All cattle with BSE will die from the disease.

Humans who eat BSE-contaminated beef products can develop a disease called vCJD. Initial signs of vCJD include behavioral changes and abnormal sensations. As the disease progresses, incoordination and dementia develop, followed by coma and death. There is no cure for vCJD. Most people die within a year after signs occur. Most cases have been in people who lived in the United Kingdom during the BSE outbreak in the late 1980s.

Recent experimental evidence demonstrates that low level of TSE infectivity can be present in blood or blood fractions derived from experimental rodents and sheep models. Besides, three cases of vCJD were observed after the transfusion of contaminated blood in the United Kingdom. Of primary interest is the potential for plasma and biotechnology process to remove the human form of TSE. Owing to the biochemical similarities between animal and human PrP/infectivity, both species of PrP^{Sc} could be expected to partition in the same way during fractionation or purification. Nonetheless, it was important to demonstrate that the partitioning of animal PrP^{Sc} and human PrP^{CJD} was similar. At this moment, partitions of hamster PrP^{Sc}, sheep PrP^{Sc}, and human PrP^{CJD} were demonstrated in similar manner regardless of the partitioning step. PrP^{Sc} partitioning correlates with TSE infectivity partitioning. Rodent-adapted TSE infectivity partitioning is predictive of human CJD infectivity partitioning. The applying in vitro assays for the detection of PrP$^{Sc/CJD}$ is critical to identify the plasma and biotechnology manufacturing process steps that are capable of removing animal and human PrP$^{Sc/CJD}$. The PrP$^{Sc/CJD}$ assay such as PMCA could provide a practical means for performing rapid and thorough analysis of potential TSE removal [161]. Importantly, many separate partitioning experiments can be conducted in a relatively short time frame to establish the reproducibility of partitioning and removal steps.

ACKNOWLEDGMENTS

This work was supported in part by grants-in-aid for scientific research and grants-in-aid for Scientific Research on Innovative Areas (research in a proposed area) from the Ministry of Education, Culture, Sports, Science and Technology (MEXT). This work was also supported by grants-in-aid from the Research Committee of Prion Disease and Slow Virus Infection, the Ministry of Health, Labour and Welfare of Japan.

REFERENCES

1. Aguzzi, A. et al. Mammalian prion biology: One century of evolving concepts. *Cell*, 116, 313, 2004.
2. Cho, H.J. Requirement of a protein component for scrapie infectivity. *Intervirology*, 14, 213, 1980.
3. Prusiner, S.B. et al. Scrapie agent contains a hydrophobic protein. *Proc. Natl. Acad. Sci. USA*, 78, 6675, 1981.
4. Prusiner, S.B. Novel proteinaceous infectious particles cause scrapie. *Science*, 216, 136, 1982.
5. McKinley, M.P. et al. Reversible chemical modification of the scrapie agent. *Science*, 214, 1259, 1981.
6. Chandler, R.L. Encephalopathy in mice produced by inoculation with scrapie brain material. *Lancet*, 1, 1378, 1961.
7. Zlotnik, I. et al. The pathology of the brain of mice inoculated with tissues from scrapie sheep. *J. Comp. Pathol.*, 72, 360, 1962.
8. Bendheim, P.E. et al. The transmissible agent causing scrapie must contain more than protein: Against the proposition. *Rev. Med. Virol.*, 1, 139, 1991.
9. Somerville, R.A. The transmissible agent causing scrapie must contain more than protein: Against the proposition. *Rev. Med. Virol.*, 1, 131, 1991.
10. Somerville, R.A. Do prions exist? PBS/NOVA, http://pbs.org/wgbh/nova/madcow/prions.html, (accessed November 1, 2012) 2001.
11. Bolton, D.C. Do prions exist? PBS/NOVA, http://pbs.org/wgbh/nova/madcow/prions.html, (accessed November 1, 2012) 2001.
12. Kellings, K. et al. Analysis of nucleic acids in purified scrapie preparations. *Arch. Virol.*, 7, S215, 1993.
13. Meyer, N. et al. Search for a putative scrapie genome in purified prion fractions reveals a paucity of nucleic acids. *J. Gen. Virol.*, 72, 37, 1991.
14. Kellings, K. et al. Further analysis of nucleic acids in purified scrapie prion preparations by improved return refocusing gel electrophoresis. *J. Gen. Virol.*, 73, 1025, 1992.
15. Rinsner, D. et al. Prions and nucleic acids: Search for residual nucleic acids and screening for mutations in the *PrP*-gene. *Dev. Biol. Stand.*, 80, 173, 1993.
16. Merz, P.A. et al. Abnormal fibrils from scrapie-infected brain. *Acta Neuropathol. (Berl.)*, 54, 63, 1981.
17. Merz, P.A. et al. Scrapie-associated fibrils in Cteutzfeldt–Jakob disease. *Nature*, 306, 474, 1983.
18. Merz, P.A. et al. Infection-specific particles from the unconventional slow virus disease. *Science*, 225, 437, 1984.
19. Prusiner, S.B. et al. Scrapie prions aggregate to form amyloid-like birefringent rods. *Cell*, 35, 349, 1983.
20. Barry, R.A. et al. Antibodies to the scrapie protein decorate prion rods. *J. Immunol.*, 135, 603, 1985.
21. Narang, H.K. Evidence that single stranded-DNA wrapped around the tubulofilamentous particles termed "nomaviruses" is the genome of the scrapie agent. *Res. Virol.*, 149, 375, 1998.
22. Narang, H.K. The nature of the scrapie agent: The virus theory. *Proc. Soc. Exp. Biol. Med.*, 212, 208, 1996.
23. Carp, R.I. et al. The nature of unconventional slow infection agents remains a puzzle. *Alzheimer Dis. Assoc. Discord.*, 3, 79, 1989.
24. Gabizon, R. et al. Prion liposomes. *Biochem. J.*, 266, 1, 1990.
25. Prusiner, S.B. Molecular biology of prion disease. *Science*, 252, 1515, 1991.
26. McKinley, M.P. et al. Molecular characteristics of prion rods purified from scrapie-infected hamster brains. *J. Infect. Dis.*, 154, 110, 1986.
27. Gabizon, R. et al. Purified prion proteins and scrapie infectivity copartition into liposomes. *Proc. Natl. Acad. Sci. USA*, 84, 4017, 1987.
28. Field, E.J. et al. An electron-microscopic study of the scrapie mouse and rat: Further observations on "inclusion bodies" and virus-like particles. *J. Neurol. Sci.*, 17, 347, 1972.
29. Narang, H.K. et al. Tubulofilaments in negatively stained scrapie-infected brains: Relationship to scrapie-associated fibrils. *Proc. Natl. Acad. Sci. USA*, 84, 7730, 1987.

30. Narang, H.K. Scrapie associated tubulofilamentous particles in scrapie hamsters. *Intervirology*, 34, 105, 1992.

31. Bessen, R.A. et al. Distinct PrP properties suggest the molecular basis of strain variation in transmissible mink encephalopathy. *J. Virol.*, 68, 7859, 1994.

32. Telling, G.C. et al. Evidence for the conformation of the pathologic isoform of prion protein enciphering and propagating prion diversity. *Science*, 274, 2079, 1996.

33. Wells, G.A.H. et al. A novel progressive spongiform encephalopathy in cattle. *Vet. Rec.*, 121, 419, 1987.

34. Kimberlin, R.H. An overview of bovine spongiform encephalopathy. *Dev. Biol. Stand.*, 75, 75, 1991.

35. Nathanson, N. et al. Bovine spongiform encephalopathy (BSE): Causes and consequence of a common source epidemic. *Am. J. Epidemiol.*, 145, 959, 1997.

36. Matthews, D. BSE: A global update. *J. Appl. Microbiol.*, 94(Suppl), 120S, 2003.

37. Will, R.G. et al. A new variant of Creutzfeldt–Jakob diseases in UK. *Lancet*, 347, 921, 1996.

38. Hill, A.F. et al. The same prion strain causes vCJD and BSE. *Nature*, 389, 448, 1997.

39. Wroe, S.J. et al. Clinical presentation and pre-mortem diagnosis of variant–Creutzfeldt–Jakob disease associated with blood transfusion: A case report. *Lancet*, 368, 2061, 2006.

40. de Marco, M.F. et al. Large-scale immunohistochemical examination for lymphoreticular prion protein in tonsil specimens collected in Britain. *J. Pathol.*, 222, 380, 2010.

41. Ironside, J.W. et al. Retrospective study of prion–protein accumulation in tonsil and appendix tissues. *Lancet*, 355, 1693, 2000.

42. Shlomchik, M. et al. Neuroinvasion by a Creutzfeldt–Jakob disease agent in the absence of B cells and follicular dendritic cells. *Proc. Natl. Acad. Sci. USA*, 98, 9289, 2001.

43. Adkin, A. et al. Estimating the impact on the food chain of changing bovine spongiform encephalopathy (BSE) control measures: The BSE control model. *Prev. Vet. Med.*, 93, 170, 2010.

44. Ducrot, C. et al. Review on the epidemiology and dynamics of BSE epidemics. *Vet. Res.*, 39, 4, 2008.

45. Biacabe, A.G. et al. Distinct molecular phenotype in bovine prion diseases. *EMBO Rep.*, 5, 110, 2004.

46. Bushmann, A. et al. Atypical BSE in Germany—Proof on transmissibility and biochemical characterization. *Vet. Microbiol.*, 117, 103, 2006.

47. Casalone, C. et al. Identification of a second bovine amyloidotic spongiform encephalopathy: Molecular similarities with sporadic Creutzfeldt–Jakob disease. *Proc. Natl. Acad. Sci. USA*, 101, 3065, 2004.

48. Beringue, V. et al. A bovine prion acquires an epidemic bovine spongiform encephalopathy strain-like phenotype on interspecies transmission. *J. Neurosci.*, 27, 6965, 2007.

49. Beringue, V. et al. Isolation from cattle of a prion strain distinct from that causing bovine spongiform encephalopathy. *PLoS Pathogens*, 2, 956, 2006.

50. Capobianco, R. et al. Conversion of BASE prion strain into the BSE strain: The origin of BSE? *PLoS Pathogens*, 3, 2007.

51. Biacabe, A.G. et al. Atypical bovine spongiform encephalopathies, France, 2001–2007. *Emerg. Infect. Dis.*, 14, 298, 2008.

52. Iwamaru, Y. et al. Accumulation of L-type bovine prions in peripheral nerve tissues. *Emerg. Infect. Dis.*, 16, 1151, 2010.

53. Lombardi, G. et al. Intraspecies transmission of BASE induces clinical dullness and amyotrophic changes. *PLoS Pathogens*, 4, 5, 2008.

54. Suardi, S. et al. Infectivity in skeletal muscle of BASE-infected cattle. *Proceeding Prion 2009 Conference*, 47, 2009.

55. Espinosa, J.C. et al. Atypical H-type BSE infection in bovine-PrP transgenic mice let to the emergence of classical BSE strain features. *Prion*, 4, 137, 2010.

56. Comoy, E.E. et al. Atypical BSE (BASE) transmitted from asymptomatic aging cattle to primate. *PLoS One*, 3, e3017, 2008.

57. Beringue, V. et al. Transmission of atypical bovine prions to mice transgenic for human prion protein. *Emerg. Infect. Dis.*, 14, 1898, 2008.

58. Kong, Q. et al. Evaluation of the human transmission risk of an atypical bovine spongiform encephalopathy prion strain. *J. Virol.*, 82, 3697, 2008.

59. Beringue, V. et al. Prominent and persistent extraneural infection in human PrP transgenic mice infected with variant CJD. *PLoS One*, 3, 2008.

60. Asante, E.A. et al. BSE prions propagate as either variant CJD-like or sporadic CJD-like prion strains in transgenic mice expressing human prion protein. *EMBO J.*, 21, 6358, 2002.

61. Seuberlich, T. et al. Novel prion protein in BSE-affected cattle, Switzerland. *Emerg. Inf. Dis.*, 18, 158, 2012.

62. Wilesmith, J.W. et al. Bovine spongiform encephalopathy—Epidemiological studies. *Vet. Rec.*, 123, 638, 1988.

63. Wells, G.A.H. et al. Infectivity in the ileum of cattle challenged orally with bovine spongiform encephalopathy. *Vet. Rec.*, 135, 40, 1994.

64. Wells, G.A.H. et al. Preliminary observations on the pathogenesis of experimental bovine spongiform encephalopathy (BSE): An update. *Vet. Rec.*, 142, 103, 1998.

65. Welles, G.A.H. et al. Pathogenesis of experimental bovine spongiform encephalopathy: Preclinical infectivity in tonsil and observation on the distribution of lingual tonsil in slaughtered cattle. *Vet. Rec.*, 156, 401, 2005.

66. Buschmann, A. and Groschup, M.H. Highly bovine spongiform encephalopathy-sensitive transgenic mice confirm the essential restriction of infectivity to the nervous system in clinically diseased cattle. *J. Infect. Dis.*, 192, 934, 2005.

67. Epinosa, J.C. et al. Progression of prion infectivity in asymptomatic cattle after oral bovine spongiform encephalopathy challenge. *J. Gen. Virol.*, 88, 1379, 2007.

68. Bosque, P.J. et al. Prions in skeletal muscle. *Proc. Natl. Acad. Sci. USA*, 99, 3812, 2002.

69. Heggebo, R. et al. Detection of PrP^{Sc} in lymphoid tissues of lambs experimentally exposed to the scrapie agent. *J. Comp. Pathol.*, 128, 172, 2003.

70. Thomzig, A. et al. Widespread PrP^{Sc} accumulation in muscles of hamsters orally infected with scrapie. *EMBO Rep.*, 4, 530, 2003.

71. Thomzig, A. et al. Preclinical disposition of pathological prion protein PrP^{Sc} in muscles of hamsters orally exposed to scrapie. *J. Clin. Invest.*, 113, 1465, 2004.

72. Andreoletti, O. et al. Bovine spongiform encephalopathy agent in spleen from an ARR/ARR orally exposed sheep. *J. Gen. Virol.*, 87, 1043, 2006.

73. Bons, N. et al. Natural and experimental oral infection of nonhuman primates by bovine spongiform encephalopathy agents. *Proc. Natl. Acad. Sci. USA*, 96, 4046, 1999.

74. Herzog, C. et al. Tissue distribution of bovine spongiform encephalopathy agent in primates after intravenous or oral infection. *Lancet*, 363, 422, 2004.

75. Jeffrey, M. et al. Oral inoculation of sheep with the agent of bovine spongiform encephalopathy (BSE). 1. Onset and distribution of disease-specific PrP accumulation in brain and viscera. *J. Comp. Pathol.*, 124, 280, 2001.

76. Bruce, M.E. et al. Transmission of bovine spongiform encephalopathy and scrapie to mice—Strain variation and the species barrier. *Philos. Trans. R. Soc. Lond. Ser. B Biol. Sci.*, 343, 405, 1994.

77. Bruce, M.E. et al. Transmissions to mice indicate that "new variant" CJD is caused by BSE agent. *Nature*, 389, 498, 1997.

78. Green, R. et al. Primary isolation of encephalopathy agent based on a review the bovine spongiform in mice: Agent definition of 150 transmissions. *J. Comp. Pathol.*, 132, 117, 2005.

79. Lasmezas, C.I. et al. Adaptation of bovine spongiform encephalopathy agent to primates and comparison with Creutzfeldt–Jakob disease: Implication for human health. *Proc. Natl. Acad. Sci. USA*, 98, 4142, 2001.

80. Lloyd, S.E. et al. Characterization of two distinct prion strains derived from bovine spongiform encephalopathy transmission to inbred mice. *J. Gen. Virol.*, 85, 2471, 2004.

81. Lloyd, S.E. et al. Identification of multiple quantitative trait loci linked to prion disease incubation period in mice. *Proc. Natl. Acad. Sci. USA*, 98, 6279, 2001.

82. Bishop, M.T. et al. Predicting susceptibility and incubation time of human-to-human transmission of vCJD. *Lancet Neurol.*, 5, 393, 2006.

83. Foster, J.D. et al. Transmission of bovine spongiform encephalopathy to sheep and goats. *Vet. Rec.*, 133, 339, 1993.

84. Houston, E.F. et al. Clinical signs in sheep experimentally infected with scrapie and BSE. *Vet. Rec.*, 152, 333, 2003.

85. Padilla, D. et al. Sheep and goat BSE propagate more efficiently than cattle BSE in human PrP transgenic mice. *PLoS Pathogens*, 7, e1001319, 2011.

86. EFSA Panel on Biological Hazards (BIOHAZ). Scientific opinion on BSE/TSE infectivity in small ruminant tissues. *EFSA J.*, 8, 92, 2010.

87. Bellworthy, S.J. et al. Tissue distribution on bovine spongiform encephalopathy infectivity in Romney sheep up to the onset of clinical disease after oral challenge. *Vet. Rec.*, 156, 197, 2005.

88. Foster, J.D. et al. Clinical signs, histopathology and genetics of experimental transmission of BSE and natural scrapie to sheep and goats. *Vet. Rec.*, 148, 165, 2001.

89. Ferguson, N.M. et al. Estimating the human health risk from possible BSE infection of the British sheep flock. *Nature*, 415, 420, 2002.

90. Barker, H.F. et al. Experimental transmission of BSE and scrapie to the common marmoset. *Vet. Rec.*, 132, 403, 1993.

91. Lasmezas, C.I. et al. BSE transmission to macaque. *Nature*, 381, 743, 1996.

92. Williams, L. et al. Clinical, neuropathological and immunohistochemical features of sporadic and variant forms of Creutzfeldt–Jakob disease in the squirrel monkey (*Saimiri sciures*). *J. Gen. Virol.*, 88, 688, 2007.

93. Lasmezas, C.I. et al. Risk of oral infection with bovine spongiform encephalopathy agent in primates. *Lancet*, 365, 781, 2005.

94. Bons, N. et al. Brain and buffy coat transmission of bovine spongiform encephalopathy to the primate *Microcebus murinus*. *Transfusion*, 42, 513, 2002.

95. Comoy, E. et al. vCJD in primate: New insights for risk assessment. *Proceedings Prion 2008 Conference*, 21, 2008.

96. Ghani, A.C. et al. Predicted vCJD mortality in Great Britain. *Nature*, 406, 583, 2000.

97. Ghani, A.C. et al. Factors determining the pattern of the variant Creutzfeldt–Jakob disease (vCJD) epidemic in the UK. *Proc. R. Soc. Lond. B*, 270, 689, 2003.

98. Hilton, D.A. et al. Accumulation of prion protein in tonsil and appendix: Review of tissue samples. *Br. Med. J.*, 325, 633, 2002.

99. Pan, T. et al. Novel differences between two human prion strains revealed by two-dimensional gel electrophoresis. *J. Biol. Chem.*, 276, 37284, 2001.

100. Thuring, C.M. et al. Discrimination between scrapie and bovine spongiform encephalopathy in sheep by molecular size, immunoreactivity, and glycoprofile of prion protein. *J. Clin. Microbiol.*, 42, 972, 2004.

101. Iwata, N. et al. Distribution of PrP(Sc) in cattle with bovine spongiform encephalopathy slaughtered at abattoirs in Japan. *Jpn. J. Inf. Dis.*, 59, 100, 2006.

102. Orge, L. et al. Identification of putative atypical scrapie in sheep in Portugal. *J. Gen. Virol.*, 85, 3487, 2004.

103. Saborio, G.P., Permanne, B., and Soto, C. Sensitive detection of pathological prion protein by cyclic amplification of protein misfolding. *Nature*, 411, 810, 2001.

104. Castilla, J. et al. In vitro generation of infectious scrapie prions. *Cell*, 121, 195, 2005.

105. Thorne, L. and Terry, L.A. In vitro amplification of PrPSc derived from the brain and blood of sheep infected with scrapie. *J. Gen. Virol.*, 89, 3177, 2008.

106. Saa, P., Castilla, J., and Soto, C. Presymptomatic detection of prions in blood. *Science*, 313, 92, 2006.

107. Atarashi, R. et al. Simplified ultrasensitive prion detection by recombinant PrP conversion with shaking. *Nat. Methods*, 5, 211, 2008.

108. Atarashi, R. Recent advances in cell-free PrPSc amplification technique. *Protein Pept. Lett.*, 16, 256, 2009.

109. Wiliam, J.M., et al. Rapid end-point quantitation of prion seeding activity with sensitivity comparable to bioassays. *PLoS Pathog.*, 6, e1001217, 2010.

110. Jackman, R. and Schmerr, M.J. Analysis of the performance of antibody capture methods using fluorescent peptides with capillary zone electrophoresis with laser-induced fluorescence. *Electrophoresis*, 24, 892, 2003.

111. Schmerr, M.J. et al. Use of capillary electrophoresis and fluorescent labeled peptides to detect the abnormal prion protein in the blood of animals that are infected with a transmissible spongiform encephalopathy. *J. Chromatogr. A*, 853, 207, 1999.

112. Bieschke, J. et al. Ultrasensitive detection of pathological prion protein aggregates by dual-color scanning for intensely fluorescent targets. *Proc. Natl. Acad. Sci. USA*, 97, 5468, 2000.

113. Schmitt, J. et al. Identification of scrapie infection from blood serum by Fourier transform infrared spectroscopy. *Anal. Chem.*, 74, 3865, 2002.

114. Sakudo, A. et al., Near-infrared spectroscopy: Promising diagnostic tool for viral infections. *Biochem. Biophys. Res. Commun.*, 341, 279, 2006.

115. Paramithiotis, E. et al. A prion protein epitope selective for the pathologically misfolded conformation. *Nat. Med.*, 9, 893, 2003.

116. Korth, C. et al. Prion (PrPSc)-specific epitope defined by a monoclonal antibody. *Nature*, 390, 74, 1997.

117. Yamasaki, T. et al. Characterization of intracellular localization of PrP(Sc) in prion-infected cells using a mAb that recognizes the region consisting of aa 119–127 of mouse PrP. *J. Gen. Virol.*, 93, 668, 2012.

118. Ushiki-Kaku, Y. et al. Tracing conformational transition of abnormal prion proteins during interspecies transmission by using novel antibodies. *J. Biol. Chem.*, 285, 11931, 2010.

119. Curin Serbec, V. et al. Monoclonal antibody against a peptide of human prion protein discriminates between Creutzfeldt–Jakob's disease-affected and normal brain tissue. *J. Biol. Chem.*, 279, 3694, 2004.

120. Weiss, S. et al. RNA aptamers specifically interact with the prion protein PrP. *J. Virol.*, 71, 8790, 1997.

121. Rhie, A. et al. Characterization of 2′-fluoro-RNA aptamers that bind preferentially to disease-associated conformations of prion protein and inhibit conversion. *J. Biol. Chem.*, 278, 39697, 2003.

122. Sayer, N.M. et al. Structural determinants of conformationally selective, prion-binding aptamers. *J. Biol. Chem.*, 279, 13102, 2004.

123. Huzarewich, R.L., Siemens, C.G., and Booth, S.A. Application of "omics" to prion biomarker discovery. *J. Biomed. Biotechnol.*, 2010, 613504, 2010.

124. Sorensen, G. et al. Comprehensive transcriptional profiling of prion infection in mouse models reveals networks of responsive genes. *BMC Genomics*, 9, 114, 2008.

125. Xiang, W. et al. Identification of differentially expressed genes in scrapie-infected mouse brains by using global gene expression technology. *J. Virol.*, 78, 11051, 2004.

126. Riemer, C. et al. Gene expression profiling of scrapie-infected brain tissue. *Biochem. Biophys. Res. Commun.*, 323, 556, 2004.

127. Skinner, P.J. et al. Gene expression alterations in brains of mice infected with three strains of scrapie. *BMC Genomics*, 7, 114, 2006.

128. Dandoy-Dron, F. et al. Gene expression of scrapie: Cloning of a new scrapie-responsive gene and the identification of increased levels of seven other mRNA transcripts. *J. Biol. Chem.*, 273, 7691, 1998.

129. Rite, I. et al. Proteomic identification of biomarkers in the cerebrospinal fluid in a rat model of nigrostriatal dopaminergic degeneration. *J. Neurosci. Res.*, 85, 3607, 2007.

130. Brown, A.R. et al. Identification of up-regulated genes by array analysis in scrapie-infected mouse brains. *Neuropathol. Appl. Neurobiol.*, 30, 555, 2004.

131. Miele, G. et al. Urinary α1-antichymotrypsin: A biomarker of prion infection. *PLoS One*, 3, e3870, 2008.

132. Simon, S.L.R. et al. The identification of disease-induced biomarkers in the urine of BSE infected cattle. *Proteome Sci.*, 6, 23, 2008.

133. Sasaki, K. et al. Increased clusterin (apolipoprotein J) expression in human and mouse brains infected with transmissible spongiform encephalopathies. *Acta Neuropathol.*, 103, 199, 2002.

134. Guillaume, E. et al. A potential cerebrospinal fluid and plasmatic marker for the diagnosis of Creutzfeldt–Jakob disease. *Proteomics*, 3, 1495, 2003.

135. Miele, G. et al. Urinary alpha1-antichymotrypsin: A biomarker of prion infection. *PLoS One*, 3, e3870, 2008.

136. Otto, M. et al. Diagnosis of Creutzfeldt–Jakob disease by measurement of S100 protein in serum: Prospective case–control study. *BMJ*, 316, 577, 1998.

137. Guillaume, E. et al. A potential cerebrospinal fluid and plasmatic marker for the diagnosis of Creutzfeldt–Jakob disease. *Proteomics*, 3, 1495, 2003.

138. Zerr, I. et al. Diagnosis of Creutzfeldt–Jakob disease by two-dimensional gel electrophoresis of cerebrospinal fluid. *Lancet*, 348, 846, 1996.

139. Kenney, K. et al. An enzyme-linked immunosorbent assay to quantify 14-3-3 proteins in the cerebrospinal fluid of suspected Creutzfeldt–Jakob disease patients. *Ann. Neurol.*, 48, 395, 2000.

140. Miele, G., Manson, J., and Clinton, M. A novel erythroid-specific marker of transmissible spongiform encephalopathies. *Nat. Med.*, 7, 361, 2001.

141. Fischer, M.B. et al. Binding of disease-associated prion protein to plasminogen. *Nature*, 408, 479, 2000.

142. Gavier-Widén, D. et al. Diagnosis of transmissible spongiform encephalopathies in animals: A review. *J. Vet. Diagn. Invest.*, 17, 509, 2005.

143. Fichet, G. et al. Novel methods for disinfection of prion-contaminated medical devices. *Lancet*, 364, 521, 2004.

144. Mizusawa, H. and Kuroiwa, Y. Guideline for infection control of prion diseases, 2008. The Research Committee on Prion disease and Slow Virus Infection, Research on Measures for Intractable Diseases Health and Labour Sciences Research Grants, The Ministry of Health, Labour and Welfare, Japan. http://prion.umin.jp/guideline/index.html (accessed March 27, 2012).

145. Rutala, W.A. and Weber, D.J. Society for Healthcare Epidemiology of America. Guideline for disinfection and sterilization of prion-contaminated medical instruments. *Infect. Control Hosp. Epidemiol.*, 31, 107, 2010.

146. Siegel, J.D. et al. Guideline for isolation precautions: Preventing transmission of infectious agents in healthcare settings, 2007. http://www.cdc.gov/ncidod/dhqp/pdf/guidelines/isolation2007.pdf (accessed March 27, 2012).

147. Sakudo, A. et al. Fundamentals of prions and their inactivation (review). *Int. J. Mol. Med.*, 27, 483, 2011.

148. Inoue, Y. et al. Infection route-independent accumulation of splenic abnormal prion protein. *Jpn. J. Inf. Dis.*, 58, 78, 2005.

149. Grathwohl, K.U. et al. Sensitive enzyme-linked immunosorbent assay for detection of PrP(Sc) in crude tissue extracts from scrapie-affected mice. *J. Virol. Methods*, 64, 205, 1997.

150. McBride, P.A., Bruce, M.E., and Fraser, H. Immunostaining of scrapie cerebral amyloid plaques with antisera raised to scrapie-associated fibrils (SAF). *Neuropathol. Appl. Neurobiol.*, 14, 325, 1988.

151. Prusiner, S.B. et al. Measurement of the scrapie agent using an incubation time interval assay. *Ann. Neurol.*, 11, 353, 1982.

152. Klöhn, P.C. et al. A quantitative, highly sensitive cell-based infectivity assay for mouse scrapie prions. *Proc. Natl. Acad. Sci. USA*, 100, 11666, 2003.

153. Taraboulos, A. et al. Regional mapping of prion proteins in brain. *Proc. Natl. Acad. Sci. USA*, 89, 7620, 1992.

154. Bosque, P.J. and Prusiner, S.B. Cultured cell sublines highly susceptible to prion infection. *J. Virol.*, 74, 4377, 2000.

155. Winklhofer, K.F., Hartl, F.U., and Tatzelt, J. A sensitive filter retention assay for the detection of PrP(Sc) and the screening of anti-prion compounds. *FEBS Lett.*, 503, 41, 2001.

156. Castilla, J. et al. In vitro generation of infectious scrapie prions. *Cell*, 121, 195, 2005.

157. Safar, J. et al. Eight prion strains have PrP(Sc) molecules with different conformations. *Nat. Med.*, 4, 1157, 1998.

158. Barnard, G. et al. The measurement of prion protein in bovine brain tissue using differential extraction and DELFIA as a diagnostic test for BSE. *Luminescence*, 15, 357, 2000.

159. Murayama, Y. et al. Specific detection of prion antigenic determinants retained in bovine meat and bone meal by flow microbead immunoassay. *J. Appl. Microbiol.*, 101, 369, 2006.

160. Ritchie, D.L., Head, M.W., and Ironside, J.W. Advances in the detection of prion protein in peripheral tissues of variant Creutzfeldt–Jakob disease patients using paraffin-embedded tissue blotting. *Neuropathol. Appl. Neurobiol.*, 30, 360, 2004.

161. Sakudo, A. et al. Recent developments in prion disease research: Diagnostic tools and in vitro cell culture models. *J. Vet. Med. Sci.*, 69, 329, 2007.

162. Tateishi, J., Tashima, T., and Kitamoto, T. Practical methods for chemical inactivation of Creutzfeldt–Jakob disease pathogen. *Microbiol. Immunol.*, 35, 163, 1991.

163. Vadrot, C. and Darbord, J.C. Quantitative evaluation of prion inactivation comparing steam sterilization and chemical sterilants: Proposed method for test standardization. *J. Hosp. Infect.*, 64, 143, 2006.

54 Camelpox Virus

Vinayagamurthy Balamurugan, Gnanavel Venkatesan,
Veerakyathappa Bhanuprakash, and Raj Kumar Singh

CONTENTS

54.1 INTRODUCTION

Camelpox virus (CMLV) is a causative agent of an economically important contagious, often sporadic, and notifiable to Office International des Epizooties (World Organisation for Animal Health; OIE-WOAH) skin disease of camelids.[1] CMLV (family: Poxviridae, subfamily: Chordopoxvirinae, genus: *Orthopoxvirus* [OPV]) is closely related to variola virus (VARV) (the causative agent of smallpox) and was earlier thought to be a zoonotic agent, but so far, little evidence has been documented. Although camelpox has presumably existed for millennia, its causative agent was not isolated until the early 1970s, during the opening phase of the global smallpox eradication campaign.[2,3] The disease, restricted to camels, is enzootic in almost every region where camel rearing/breeding is practiced with the exception of Australia. According to the UN Food and Agriculture Organization (FAO), the total world camel population is ≈25 million (http://faostat.fao.org). The disease camelpox is confined to camel-rearing belts particularly in developing countries and causes economic impact due to considerable loss in terms of morbidity, mortality, loss of weight, and reduction in milk yield. Effective control of any disease warrants a prophylactic as well as a rapid, specific, and sensitive assay(s) for diagnosis and molecular epidemiological studies. The virus has gained attention from researchers due to its recent emergence with close genetic relatedness to VARV and carrying genes responsible for host immune-evasion mechanisms. The nature of occurrence, mode of transmission by contact, and difficulty in detection and identification make this agent a potential

biological or bioterror weapon like smallpox agent, creating panic and fear among common people. Considering the nature of the virus, an improved diagnostics and control methods would be of immense value to curtail the infection in the field. This chapter provides a comprehensive note on CMLV with particular reference to its classification, epidemiology, pathogenesis, biology of the disease, diagnostic approaches, and control measures with perspectives or future challenges.

54.2 CLASSIFICATION, MORPHOLOGY, PHYSICOCHEMICAL, AND GENOMIC PROPERTIES

CMLV, the causative agent of camelpox, belongs to the genus OPV and is one of eight genera of the subfamily Chordopoxvirinae of the family Poxviridae.[4] The other members of the genus include several pathogens of veterinary and zoonotic importance, namely, VARV, monkeypox virus (MPXV), vaccinia virus (VACV), buffalopox virus (BPXV, a variant of VACV), cowpox virus (CPXV), ectromelia virus (ECTV), rabbitpox virus (RPXV), taterapox virus, the North American OPVs (volepox virus, raccoonpox virus, and skunkpox virus), and an unclassified OPV species, Uasin Gishu disease virus.[1,5] Parapox- and papillomaviruses also cause a similar kind of infections in camelids. There are numerous CMLV strains that have been isolated from different outbreaks in different parts of the camel-rearing countries.[5]

The identification of CMLV agent caused an alarm when it was described as smallpox-like disease during smallpox eradication campaign,[6] which led to the discovery of the CMLV. CMLV is one of the least studied members of OPVs until recently. It is quiet difficult to distinguish from the prototypic VACV with respect to size, shape, structure, physicochemical properties and replication mechanism.[7,8] OPVs are large (250–350 nm), and the average size of CMLV virion estimated by electron microscopy (EM) is 224 × 389 nm.[4,9] A camelpox virion consists of an envelope, outer membrane, two lateral bodies, and a core. Like other OPVs, CMLV is brick-shaped and the outer membrane is covered with irregularly arranged tubular proteins. The growth kinetics in human embryonic lung (HEL) fibroblast cells indicates that CMLV is different from VACV and CPXV.[10]

CMLV is ether resistant but chloroform sensitive, whereas it is sensitive to both acidic (pH 3–5) and alkaline (pH 8.5–10) conditions.[11] Like other poxviruses, CMLV is susceptible to various disinfectants. The virus can be destroyed by either autoclaving or boiling for 10 min and ultraviolet rays in a few minutes.[12] The physicochemical properties of different CMLV strains with highlights of the differences existing between them are reported. For example, the strain Etha-78 is resistant to heat and chemical treatments while other strains exhibit phenotypes of sensitivity.[5] The CMLV hemagglutinates cockerel erythrocytes. In general, it is well recognized that poxvirus virions show high environmental stability and can remain contagious over several months. This feature is enhanced by the materials, namely, crusts, serum, blood, and other excretions in which the virus is released, and they can also show strong tolerance to high temperatures, pH, and chemicals.[13]

CMLV genome consists of a single linear double-stranded DNA molecule terminated by a hairpin loop that replicates in the cytoplasm.[4] The genome is Adenine and Thymine (AT) rich (66.9%) having cross-links that join the two DNA strands at both ends. The end of each DNA strand has long inverted tandem repeats that form single-stranded loops. The central region of the genome contains genes that are highly conserved among all sequenced OPVs.[14] Like other poxviruses, the genes are tightly packed with little noncoding sequences. Most genes at each terminus within 25 kb of genome are transcribed outwards towards the terminus and are variable in nature, coding for host range, virulence, and immunomodulation, whereas the genes within the center of the genome transcribed from either DNA strands are highly conserved and code for proteins involved in viral replication like RNA transcription, DNA replication, and virion assembly. The full-genome sequencing of two CMLV strains, one from Iran (CMLV-CMS) and another from Kazakhstan (CMLV-M96), has been completed.[7,15] It revealed that CMLV is variola's closest relative, sharing genes involved in basic replication and host-related functions, and probably, they may share a common ancestor.[7] A high degree of similarity in gene order, gene content, and amino acid composition in the region located between CMLV017 and CMLV184 (average 96% amino acid identity to VACV) confirms a close structural

and functional relationship between CMLV and other known OPVs. CMLV contains a unique ≈3 kb region, which encodes three open reading frames (ORFs) (CMLV185, CMLV186, CMLV187), which are absent in other OPVs. The molecular details about the genome structure and phylogenetic analysis of some selected genes indicate that CMLV is clearly distinct from VARV and VACV.

Genomic differences between CMLV and other OPVs are located in terminal regions where ORF colinearity and average amino acid identity decrease (82% to VACV) due to small and large nucleotide insertions, deletions, and translocations. Other notable differences are the deletion of a 14.5 kb region from the left end of the CMLV genome, which is present in CPXV, and the insertion of a 2.9 kb region (position 172,582–175,508), which is absent in VACV and CPXV.[16] CMLV contains 27 ORFs, which are absent in VARV, and conversely, CMLV lacks homologues of VARV, namely, interleukin (IL)-18 binding protein (D7L), ankyrin repeat protein (D8L), A-type inclusion (ATI)-like proteins (A26L and A27L), and ORFs of unknown function (A39L, A42L, B2L, B3L, B4L, B9R, and B10R). CMLV is similar to other OPVs in overall genome structure and composition, but CMLV genome lacks homologues of VARV C1L, E7L, A26L, A27L, A39L, A42R, B2L, B3L, and B4L and homologues of VACV K6L, A25L, A40R, A52R, and A53R genes.[16]

54.3 BIOLOGY AND EPIDEMIOLOGY

54.3.1 BIOLOGY

CMLV appears to share biological features with other OPVs mainly VARV. Both CMLV and VARV are restricted to a single host and induce a similar disease course.[9,17] CMLV was first regarded as an own OPV species in the mid-1970s, some time before phylogenetic analyses were conducted.[6] Earlier, CMLV was shown to share strong similarities with VARV as they both had a narrow host range and were indistinguishable in terms of pock formation on chorioallantoic membrane (CAM) in embryonated eggs, growth in cells, and low or absence of pathogenicity in various animal models.[6,18,19] Serological studies demonstrated the cross antigenicity among VACV, VARV, CPXV, and CMLV, but not with parapoxviruses (PPVs) and avipoxvirus infections.[6,11]

Camels have been used as animal models of camelpox infection, mainly for evaluating the efficacy of camelpox vaccines.[20–22] The two models of camelpox (moderate and fatal) have been successfully used to evaluate vaccine pathogenicity and efficacy. Experimental infections of guanacos (Lama) with CMLV have been successful, while natural infection of New World camelids has never been reported.[23] There have been several attempts to infect animals other than camels with CMLV in order to define its host range and develop animal models of camelpox since 1972, starting from large animals (horse and monkeys) to small laboratory animals (rabbits, rat, mice) including chicks and embryo. CMLV could exhibit different growth properties on CAM in embryonated eggs, in cell cultures, and also in laboratory animals. Many authors have compared the growth behavior of numerous CMLV strains in various cell cultures, animals, and embryo, which are reviewed by Duraffour and colleagues.[5] Most of the CMLV strains produce white pocks and are flat in shape, with sizes varying between 0.5 and 1.5 mm diameters at 37°C. Most of the CMLV strains did not induce pock proliferation, necrosis, or hemorrhage, in contrast to what was observed with VACV or CPXV.[11,24,25] In general, cells derived from camel, lamb, calf, pig, monkey, chicken, hamster, and mouse enable the propagation of CMLV strains. Both transformed and primary human cells are permissive to CMLV replication. However, cell monolayers derived from horse, rabbit, and dog lead to a poor replication of CMLV for most of the strains.[25] Other than camels, the species that have been infected successfully are monkeys and infant mice.[6] Immunocompetent adult mice (4–6 weeks old) have been reported to be resistant to CMLV challenge when injected by either intradermal or intracerebral or intranasal route.[26,27] However, suckling mice and athymic nude mice (2–3 weeks old) were susceptible or sensitive to a challenge with CMLV strain by intracerebral inoculation[11,24,28] and intranasal or subcutaneous inoculation,[29] respectively.

54.3.2 Epidemiology

Geographic distribution: According to the FAO, the total world camel population is approximately 25 million. In North Africa and western Asia, the majority are dromedary (single-hump) camels, while two-hump Bactrian camels are found in China, Mongolia, and other areas of East Asia. Camelpox is one of the most common contagious OPV diseases of the Old World (both *Camelus dromedarius* and *Camelus bactrianus*) and the New World camelids.[1] CMLV is considered to solely infect the Old World camelids.[17] The disease occurs throughout the camel-breeding areas of Africa, north of the equator, the Middle East, and Asia, as the camels are used for nomadic pastoralism, transportation, racing, and production of milk, wool, and meat purposes.[17,30] Although the causative agent was not isolated until 1970, camelpox is known in almost all camel-raising countries. Infections are commonly encountered in the herds of the nomadic pastoralists in the semidesert zones. It occurs in almost every country in which camel husbandry is practiced apart from the introduced dromedary camel in Australia and tylopods (llama and related species) in South America.[1] Of note, camelpox has never been reported in Australia, even though camel farming is practiced,[17] and has not been seen in feral camels in Australia, in wild Bactrian camels in China and Mongolia, or in New World camelids.[5] The disease has been reported initially in Punjab and Rajputana (India)[17,31] and later from many other countries. It has been reported in the Middle East (Bahrain Iran, Iraq, Saudi Arabia, UAE, and Yemen), in Asia (India, Afghanistan, and Pakistan), in Africa (Algeria, Egypt, Kenya, Mauretania, Niger, Somalia and Morocco, Ethiopia, Oman, Sudan), and in the southern parts of former USSR, where the disease is endemic.[5,20,27,32,33] Recently, Al-Ziabi and colleagues[34] have reported the first outbreak of camelpox in two provinces named Hama and Duma in Syria.

Risk factors: Clinically, camelpox occurs in the severe, generalized and the milder, localized forms and is accompanied with morbidity, mortality, and case fatality rate (CFR), which are variable depending on circulating virus in the herd.[34] The disease is socioeconomically significant as it incurs considerable loss in terms of morbidity, mortality, loss of weight, and reduction in milk yield.[27] The disease mostly affects young calves aged 2–3 years in a herd and due to waning of acquired immunity after 5–8 months,[35] with fatal severe form (generalized form) causing high mortality occasionally. In general, young camels (calves) under the age of 4 years and pregnant females appear more susceptible to camelpox.[34,36]

The disease is considered as endemic with a pattern of sporadic outbreaks, and various studies have demonstrated that the incidence of camelpox outbreaks increased during rainy seasons[17] with the appearance of more severe forms of the disease, while during the dry season, milder forms.[37,38] However, the involvement of arthropod vectors, which are abundant during rainy seasons, may exert a greater virus pressure on camel populations, which is supported by the isolation of CMLV from *Hyalomma dromedarii* ticks.[23,37,39] The incidence and CFR are mostly higher in male camels than females. The mortality in adult animals ranged from 10 to 28%, and in case of young animals, it is 25–100%. Further, the mortality is influenced by the presence of intercurrent diseases (like trypanosomosis), stress, the age and the nutritional status of the animal, and virus virulence. Abortion rates can reach even high (87%), as observed in Syria, albeit this high percentage might be explained by the absence of immunity as CMLV circulation had never been reported in this country earlier.[34] Outbreaks are often temporal due to the movement of camels for grazing and watering, and it results in mixing of the herds and the introduction of new camels into a herd.[40] A recent investigation of a CMLV outbreak in Eastern Saudi Arabia, by Yousif and Al-Naeem,[41] showed an atypical minute pock-like skin lesion (AMPL) that persisted on 42.9% of convalescent camels (8.8% of herd) for more than a year after the onset of clinical signs, and live CMLV was recovered from AMPL homogenates. They concluded that the small, often missed AMPL on infected animals or CMLV survival in the persistent skin lesions may play a key persistence mechanism in previously infected camel herds during interepizootic periods.

A high prevalence of antibodies to CMLV has also been reported.[17] Recovered animals become lifelong immune to re-infection and there is no chronic carrier state. The circulation of CMLV infections in herds has also been confirmed by seroepidemiologic studies that showed the presence of neutralizing antibodies in 9.8% of the animals in Libya[40] and in 9.14% in Saudi Arabia.[42] In contrast, a prevalence of neutralizing antibodies of 72.5% was measured in Sudan in an unvaccinated population following the outbreak during 1992–1994.[43] In general, the course and the outcome of camelpox may vary depending on the age, sex, and circulating CMLV strains differing in virulence.[34,36,44,45] Thus, the risk factors associated with higher incidence of camelpox include the average age of the animals (<4 years), the rainy season of the year, the introduction of new camels in a herd, and the common watering.[38]

Transmission: The CMLV is transmitted by either direct or indirect contact via a contaminated environment. The direct transmission occurs between infected and susceptible animals either by inhalation or through skin abrasion, although mechanical transmission may play a role.[17] The affected camels may shed virus through scab materials and secretions like milk, saliva, and ocular and nasal discharges[46] in the environment such as water, which becomes the source of infection for susceptible animals.[38] The dried scabs shed from the pox lesions may contain live virus particles at least for 4 months and contaminate the environment.[1] The role of an arthropod vector in the transmission of the disease has been suspected,[34] and the tick population during the rainy season is probably involved in the spread of the disease.[23,39] During an outbreak of camelpox in UAE, ticks collected from infected camels found to contain CMLV.[23] However, it is not clear whether ticks transmit CMLV mechanically or might be a true reservoir of the virus.[5] Among tick species, *H. dromedarii* have been found to be the predominant (90%) species infesting camels. Ticks seasonality with the highest infestations has been seen in Egypt, Sudan, and Saudi Arabia.[47,48] However, these reports did not reveal any correlations between tick's infestation in relation to temperature, relative humidity, or rainfall.[47,48] Further studies are needed to ensure the involvement of arthropods in the transmission of CMLV, but if confirmed, CMLV would be the first OPV to be transmitted via arthropods.[5] Moreover, other potential vectors such as biting flies and mosquitoes may also be involved. However, the role of these insects in mechanical spreading of the infection needs to be investigated as most of other poxviruses are transmitted mechanically. In general, like smallpox, camelpox is usually transmitted via airborne saliva droplets, but it can also spread through direct contact with skin lesions, and the virus can be transferred mechanically by ticks and other biting arthropods.

Zoonosis: CMLV could be a biowarfare agent. The threat CMLV poses to people whose well-being depends on the health of their camels makes the disease of considerable economic and public health importance.[5] Various observations suggested that human infections were rare, which may not be due to immunity induced via smallpox vaccination[19,49] as the cross-immunity exists between VACV, VARV, and CMLV. It is also proved based on a study conducted during the smallpox eradication campaign (1978–1979) in Somalia where the disease was endemic.[44] As a consequence, the immunological status of the individuals (smallpox vaccinated or exposed) might have been a bias for estimating the possible human cases. This disease is of less significance in developed countries as it incurs less threat to man or domestic animals, but it is a common cause of camel morbidity under traditional nomadic management system. CMLV is mostly host specific and was earlier thought as a zoonotic agent,[11] but until recently, little evidence has been documented from Somalia in smallpox-unvaccinated individuals[36,44] and from India in camel handlers or attendants.[50] Mild skin lesions in humans associated with camelpox have been reported[12] indicating that

camelpox may be of limited public health impact. Among the human cases, people drinking milk from camelpox-affected animals have been reported to develop ulcers on the lips and in the mouth, but these observations could not be visually or laboratory confirmed.[11] From the lesions (rashes) of a few of the herdsman, the involvement of CMLV could not be demonstrated.[36,44] However, under certain conditions, the virus could be pathogenic to man like cowpox and monkeypox[51] and is higher in immune-compromised individuals. No systematic epidemiological studies have been made on human cases due to the lack of immunological surveys for specific camelpox antibodies among unvaccinated herds.[40] During smallpox eradication campaign, it was thought that camelpox might represent a nonhuman reservoir of VARV, because the two causative agents were indistinguishable under certain laboratory conditions.[5] The self-limiting nature of human infection with CMLV suggests that it could be used as a live smallpox vaccine,[5] and from the historical records, it is interesting to note that centuries before Jenner developed vaccination, camelpox was circulating in Baluchistan (Southwest Asia), and it was described as effective as cowpox (inoculation of material from camelpox crusts was employed for that purpose in Iran) in protecting humans against smallpox.[52] This suggests that CMLV is able to elicit immune responses in humans that were strong enough to cross protect against VARV. Recently, the first conclusive evidence of zoonotic CMLV infection in humans (unvaccinated smallpox individuals) associated with outbreaks in dromedarian camels in the northwest region of India has been reported.[50] This outbreak involved camel handlers and attendants with clinical manifestations such as papules, vesicles, ulceration, and finally scabs over fingers and hands (Figure 54.1). The lesions remained localized, and there was no person-to-person transmission. On the basis of clinical and epidemiological features coupled with serological tests and molecular characterization of the causative agent, CMLV zoonosis was confirmed in three human cases.[50] Further epidemiological studies in regions endemic for camelpox are necessary to assess the circulation of CMLV, but currently, human infections do not seem to be of public health importance as described by Duraffour and colleagues.[5]

54.4 CLINICAL FEATURES AND PATHOGENESIS

54.4.1 CLINICAL FEATURES

The clinical manifestation of camelpox ranges from mild, local, inapparent infection confined over the skin to less common but severe systemic disease, depending on the CMLV strains involved in the infection.[17] The disease is characterized by an incubation period of 9–13 days with an initial rise in temperature, followed by enlarged lymph nodes, skin lesions, and prostration. The typical skin lesion/rash will pass through all the stages of pock-lesion progression, that is, the development of papules on labia, macules, papules, pustules, vesicles, and scabs.[5,23] Skin lesions appear 1–3 days after the onset of fever with erythematous macules

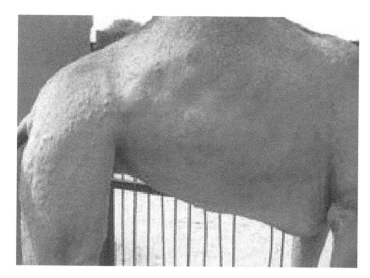

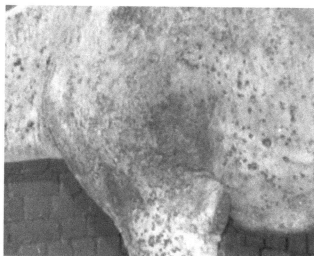

FIGURE 54.1   Clinical picture of camelpox. Infected camels exhibiting typical pox-like lesions on neck, shoulder, and other parts of the body.

to papules and vesicles and pustules and then crusts from ruptured pustules. In general, it takes 4–6 weeks for the lesion to heal. The infection is usually localized skin lesions, but occasionally, it leads to a generalized form, which is frequently seen in young animals aged 2–3 years in a herd associated with weaning and poor nutrition. Eruptions are mainly localized on the head, nostrils, the margins of the ears, and eyelids, as well as on the mucous membranes of the lips and the nose and also in the oral cavity. Later, lesions may extend to the neck, limbs, genitalia, mammary glands, and perineum or scrotum.[5] In contrast, lesions seen in the generalized form may spread over the body, particularly on the head and the limbs (Figure 54.2) with sometimes swellings on the neck and abdomen, and even multiple pox-like

lesions can be found on the mucous membranes of the mouth and on the respiratory and digestive tracts,[17] and the outcome of the disease is more likely fatal.[53] The animals may show salivation, anorexia, lacrimation, mucopurulent nasal discharge, and diarrhea. Pregnant animals may abort, and mortality in affected animals is due to septicemia caused by secondary bacterial infections like *Staphylococcus aureus*.[17,35] The size of the lesions in the lungs may vary between 0.5 and 1.3 cm in diameter and occasionally may reach up to 4–5 cm, and smaller lesions may have foci of hemorrhagic center. In contrast to smallpox, in which pustules occur only on the skin and the squamous epithelium of the oropharynx, severely ill camels also develop proliferative poxviral lesions in the bronchi and lungs.[54] Because oral

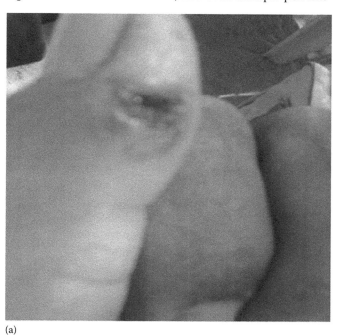

(a)

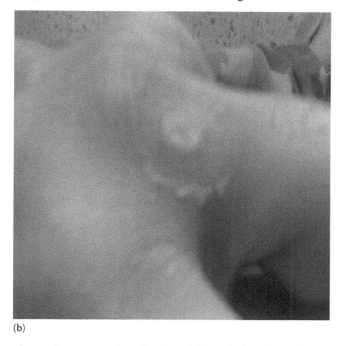

(b)

FIGURE 54.2   (a) Clinical lesions in infected camel handlers showing ulcerated open wound on thumb and (b) typical pock eruption on middle finger.

lesions severely impair the ability of young calves to feed, the CFR may reach moderate to high.

The histopathology of skin lesions reveals characteristic cytoplasmic swelling, vacuolation, and ballooning of the keratinocytes of the outer stratum spinosum of the epidermis.[10,55] The rupture of these cells produces vesicles and localized edema associated with perivascular cuffing of mononuclear cells, neutrophils, and eosinophils. Marked epithelial hyperplasia may occur at the borders of the skin lesions.[56] The lung lesions are characterized by hydropic degeneration, proliferation of bronchial epithelial cells associated with proliferative alveolitis and bronchiolitis infiltrated by macrophages, necrosis, and fibrosis leading to obliteration of normal architecture.[53] Immunohistochemical analyses of these lesions reveal numerous poxvirus antigen-positive cells in the bronchial epithelia. In camelpox infection, a rapid loss of body condition and marked reduction in the blood hemoglobin content, mean corpuscular hemoglobin concentration, packed cell volume, and red blood cell (RBC) and white blood cell (WBC) (particularly lymphocyte and neutrophil) counts have also been reported.

54.4.2 Pathogenesis

The CMLV enters commonly through the skin. However, the oronasal infection is also reported. After local replication and development of primary skin lesions, the virus spreads to local lymph nodes, which leads to a leukocyte-associated viremia, which may be associated with pyrexia. During this period, the virus can be isolated from various tissues, including the skin, turbinates, lungs, and also lymphoid organs. Widespread secondary skin lesions appear a few days after the onset of viremia, and new lesions continue to appear for 2–3 days; at that time, the viremia subsides. CMLV and VARV cause illness in a single host species and both viruses are distinguishable. CMLV has rarely caused disease in man. Similarly, VARV is unable to cause disease in camels, although camels immunized with VARV are resistant to subsequent infection with CMLV.[19] The injection of a large dose of VARV failed to cause disease in camels, but a tiny amount of material from a sick camel produced a severe febrile illness with a vesiculopustular rash that spreads to cohoused animals.[19] The virus has been found to be nonpathogenic to sheep, goats, rabbits, guinea pigs, rats, hamsters, and mice when inoculated by intradermal route.[27] The CMLV is host specific and does not infect other animal species, including cattle, sheep, and goats. Even the virus is nonpathogenic to sheep and cows in naturally, when they were in direct contact with infected camels.[34]

Camelpox infection can produce severe diseases, suggesting that CMLV may interfere with the host response to infection. Like other OPVs, CMLV encodes multiple genes that antagonize or affect the antiviral host immune response by interfering with the interferon (IFN) response, key proinflammatory cytokines (IL-1b, IL-18, and tumor necrosis factors [TNFs]), chemokines, and the complements.[5] A number of strategies used by the virus have been extensively reviewed.[57,58] The lack of small animal models for CMLV infection has greatly hampered the characterization of its immune-escape pathways, but the sequencing of two CMLV strains has brought additional knowledge on potential immunomodulatory proteins encoded by CMLV.[5] CMLV contains genes that can modulate or evade host immune responses, host cell apoptosis, and cell or tissue tropism. They are chemokine-binding protein, TNF receptor-II crmB, complement-binding protein, protein kinase inhibitors, STAT1 inhibitor,[59] serine proteinase inhibitors, CD47-like protein, IL-1/Toll-like receptor inhibitor,[60] IFN,[58] IFN-γ receptor, and IFN-α/IFN-β binding protein.[61] Similarly, CMLV encodes homologues of pox viral proteins of VACV, myxoma virus, and rabbit fibroma virus, which are known to affect virulence or host range. Several CMLV ORFs with potential host range and virulence functions are truncated or fragmented when compared to viral and cellular homologues. These smaller ORFs that have been annotated may or may not encode functional proteins. The ORFs 181R, 196R, 1L, 206R, and 2L 205R are predicted to encode or express CMLV soluble proteins that bind IFN-γ, IFN-α/IFN-β, CC chemokines, and TNF, respectively.[62] Proteins encoded by ORFs 31L, 188R, and 200R have similarity to serpins that have antifusion or antiapoptotic activity[63] involved in inflammation. To inhibit potent proinflammatory cytokine (IL-1b), CMLV could produce a viral soluble receptor encoded by three genes, CMLV193, CMLV194, and CMLV196, which are seen as separate parts of the VACV B16R IL-1 binding protein.[15] Proteins encoded by ORFs 32L and 55L are similar to VV proteins K3L and E3L that mediate resistance to IFN.[64] Protein 6L is closely related to an uncharacterized human protein of family UPF0005,[65] and possibly it regulates apoptosis in CMLV-infected cells.[5,16] Protein 201R contains a signal peptide and an Arg-Gly-Asp (RGD) motif, which mediates the binding of proteins to cell surface integrins and shows amino acid similarity to the C-terminal domain of OPV TNF receptors CrmB and CrmD.[66] VARV, CPXV, and CMLV encode soluble IFN-gRs that counteract the activity of the cytokine and possess broad species specificity. This novel property of the IFN-gR probably helped these OPVs to replicate in several species. The species specificity of the virus IFN-gR is also informative and indicates possible natural hosts of these viruses.[67] Recently, it has been demonstrated that CMLV expressed a novel protein inhibiting apoptosis (v-GAAP) and a novel virulence factor, the schlafen-like protein 176R-(v-slfn-57 kDa),[68,69] which is expressed both early and late in infection and play a role in the modulation of the innate and adaptive immune responses against pathogens.[70,71] In the case of IFN-α/IFN-β inhibition, CMLV secrete a protein with IFN-a binding activity (CMLV-CMS-252 with similarity to VACV B19R), but its inhibitory potency might be low.[72–74] To summarize the host cell interaction, CMLV may utilize several ways to alter or shut down the host immune response. Though these mechanisms have been described *in vitro*, they may reflect the *in vitro* situation and explain the pathogenicity of CMLV in the camel.[5]

54.5 DIAGNOSIS

The diagnosis of camelpox infection can be done based on clinical signs in affected animals. Following the appearance of clinical signs of the disease, tissue samples (skin or organ biopsies) are most useful to identify the infectious agent.[5] However, the confounding signs caused by contagious ecthyma (orf virus–PPV), papillomatosis, and insect bites demand camelpox to be differentiated from these infections using laboratory-based diagnostic methods. It is necessary to apply more than one method for confirmatory diagnosis as several diagnostic approaches are available.[16] Few complementary techniques might be advised for camelpox diagnosis, namely, transmission electron microscopy (TEM), virus isolation using cell culture, standard polymerase chain reaction (PCR) assays, immunohistochemistry, and demonstration of neutralizing antibodies.[5] The identity of the causative agent as CMLV must be confirmed by TEM, PCR, and/or sequencing.[75]

TEM and restriction enzyme analysis (REA) can be used to differentiate camelpox from other infections caused by OPV and PPVs.[14,34] TEM is a reliable and rapid method to demonstrate the presence of OPVs in scabs or tissue samples, although a relative high concentration of the virus in the sample is required.[5] This technique enables the differentiation between OPVs, which are brick-shaped, and PPVs, which are ovoid-shaped.[9] The camelpox antigen in infected scabs and pox lesions can be identified by using immunohistochemistry technique.[17] However, the isolation of the virus using embryonated eggs and various cell lines, namely, HeLa, GMK-AH1, WISH, Vero,[6] MA-104, and BHK-21 cells, as well as primary cell cultures like lamb testis and kidney, camel kidney, calf kidney, and chicken embryo fibroblast[11] could be the gold standard. Blood, serum, and homogenized scab or tissue

samples can be used to infect cell cultures. The infected cells should be monitored for cytopathic effects (CPE) for up to 10–12 days, which depend on the virus concentration. CPE include the formation of multinucleated syncytia and rounding, ballooning, and syncytia with degenerative changes (Figure 54.3b). Many different cell lines can be used; however, Vero, MA-104, or Dubca cells, in which the virus replicates easily, are preferred.[76] CAMs can also be used for the growth of CMLV,[8] but it is important to consider that the pocks produced by VARV and CMLV are indistinguishable.[6]

Similar to antigen detection methods, a battery of serological tests are available to detect CMLV antibodies. The conventional serological tests like hemagglutination, hemagglutination inhibition, neutralization,[77] complement fixation, fluorescent antibody, and indirect enzyme-linked immunosorbent assays (ELISA) have been described to detect CMLV antibodies.[11,24,51,78] Most of the conventional serological tests are time-consuming, labor-intensive, less sensitive, and slow and, therefore, not suitable for primary diagnosis but useful in secondary confirmatory testing and retrospective epidemiological studies.

To overcome the drawbacks associated with the aforementioned tests, the recent molecular biological tools and techniques like PCR, real-time PCR, and loop-mediated isothermal amplification (LAMP) assays have been used for the rapid and sensitive detection of CMLV DNA from clinical samples. DNA can be extracted from cell culture, clinical scab samples, or tissue material using numerous commercial kits. Recently, a reliable low-cost two-step extraction method has been developed for isolating CMLV DNA from skin samples.[79] The PCR techniques have been developed for targeting A-type inclusion body protein (ATIP) gene[80] and hemagglutinin (HA) gene[81,82] and used for specific detection of CMLV from skin biopsy and infected Vero cell cultures.[34]

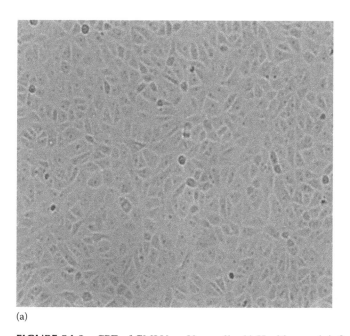

(a)

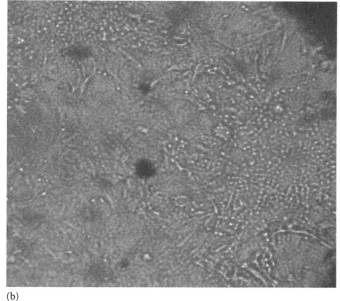

(b)

FIGURE 54.3 CPE of CMLV on Vero cells. (a) Healthy mock-infected cells. (b) CMLV-I (vaccine strain)-infected cells showing characteristic CPE, namely, rounding, ballooning, and syncytia with degenerative changes after 48 h postinfection.

The genus-specific PCR mentioned earlier yields product size specific for CMLV and thus can be differentiated from other OPVs. An extra step consisting of a *Bgl*II or *Xba*I restriction digestion allows the unequivocal identification of the virus species.[80,83] The HA-PCR amplicon *Taq*I restriction fragment length polymorphism (RFLP) permits to differentiate between OPV species, but species-specific primers within the HA ORF of OPVs have also been described.[81] Similarly, Huemer and colleagues[84] used A36R gene-based PCR followed by restriction enzyme digestion for differentiating OPVs including CMLVs.

Further, PCR strategies targeting HA gene[81] and B2L gene[85] have been developed for detection of CMLV and its differentiation from OPV and PPV infections in camels. In a similar direction, recently, a PCR assay based on the C18L gene (encoding ankyrin repeat protein) has been developed, which yields a specific amplicon of 243 bp in CMLV suspected cases.[86] This was employed successfully for the direct detection and differentiation of CMLV from other OPVs, PPVs, and capripoxviruses (CaPVs) in both cell culture samples and clinical specimens. Further, a duplex PCR based on the C18L and *DNA* polymerase (DNA pol) genes for specific and rapid detection and differentiation of CMLV from BPXV has also been developed.[86,87] These assays have the advantage of avoiding an extra step of restriction analysis. This will be an improved assay over the OPV-specific ATI or HA gene-based assays for the simultaneous detection and differentiation of CMLV. Similarly, Venkatesan and colleagues[88] described single-tube reaction for the detection and differentiation of CMLV from OPVs, CaPVs, and PPV in clinical samples. As an improvement over conventional PCR approaches, the real-time PCR techniques targeting A36R gene using fluorescence resonance energy transfer (FRET) method[89] and A13L, rpo18, and viral early transcription factor (VETF) genes[90] using melting curve analysis have been in use for rapid, highly sensitive, and specific

detection and quantitation of CMLV and other related OPVs. Recently, C18L gene-based real-time PCR based on SYBR green chemistry[86] (Figure 54.4a) and *Taq*Man hydrolysis probe[91] (Figure 54.4b) has been optimized for the specific detection of CMLV in clinical samples. Some of these methods are delineated in the OIE's Manual of Diagnostic Tests and Vaccines for Terrestrial Animals.[1] As a field applicable diagnostic tool, a simple, rapid, specific, and highly sensitive novel approach called as LAMP (Figure 54.5) has also been developed targeting C18L gene[92] and evaluated using field clinical samples. This assay appears to be a potential rapid and sensitive diagnostic tool for its application in less-equipped rural diagnostic laboratory settings. Different molecular diagnostics targeting different genes of CMLV for the detection and differentiation of CMLV with their advantages are summarized in Table 54.1.

54.6 TREATMENT, PREVENTION, IMMUNITY, AND ERADICATION

54.6.1 TREATMENT

Postexposure therapeutic approaches for camelpox infections are not mentioned in any of the literatures. However, application of antibiotics and administration of supplements may be useful to reduce the severity of the disease.[17] The use of antiviral drugs/agents may be of choice particularly in young camels as an alternative treatment. There are several classes of antiviral agents found useful for camelpox as applicable to other pox viral infections. There are potent antiviral molecules active *in vitro* and *in vivo* against poxviruses, including OPVs, and could be envisaged for the treatment of camelpox.[93,94] They include the molecules belonging to the acyclic nucleoside phosphonate (ANP) family, that is, cidofovir (Gilead, California, United States) and its lipid derivative CMX001 (Chimerix Inc., Durham, North Carolina, United

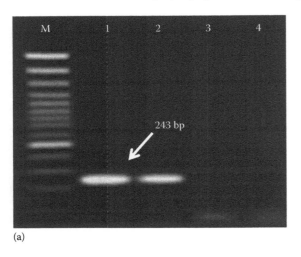
(a)

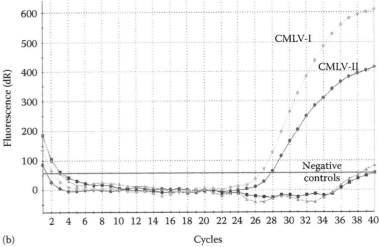
(b)

FIGURE 54.4 Diagnostic PCR of CMLV targeting C18L gene sequences. (a) Conventional PCR showing specific amplification in CMLV (Lane 1 and 2) and no such signal in BPXV and no template control (Lane 3 and 4). (b) *Taq*Man probe-based real-time PCR for rapid and sensitive detection of CMLV.

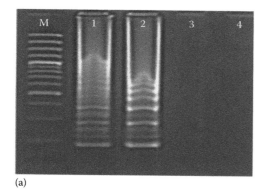

(a)

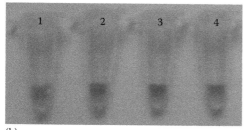

(b)

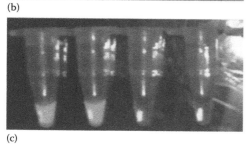

(c)

FIGURE 54.5 Diagnosis of CMLV by LAMP technique. (a) Specific detection of CMLV (Lane 1 and 2) and negative controls (Lane 3 and 4) by agarose gel analysis. (b) Naked eye colorimetric detection of LAMP reaction by hydroxyl naphthol blue (HNB) dye showing violet (negative—tube 3 and 4) to sky blue (positive—tube 1 and 2) color change. (c) SYBR green I dye showing change in color from original orange (negative—tube 3 and 4) to apple green (positive—tube 1 and 2).

States)[95,96] and the compound ST-246 (SIGA Inc., Corvallis, Oregon, United States).[97] Cidofovir and CMX001 are active against a broad range of DNA viruses including poxviruses. Both compounds target the viral DNA polymerase of OPVs and inhibit the functions.[98] These three drugs have gained investigational new drug (IND) status, allowing their emergency use for the treatment of life-threatening VACV infections.[5] The antiviral agents are cidofovir, ANPs[99], and other ANPs, which include HPMP-5-azaC, cHPMP-5-azaC, and HDE-cHPMP-5-azaC that represent promising candidates for treating poxvirus infections.[10,55] However, the development of resistance against CMLV and the OPVs was reported by *in vitro* cell culture method for cidofovir and its analog HPMPDAP, and therefore, a higher concentration of such drug is recommended for the treatment of animals.[99,100] Certain novel antiviral drugs are effective orally against pox viruses including CMLV targeting cellular enzymes (IMP dehydrogenase inhibitors, such as ribavirin, as well as the tyrosine kinase inhibitor STI-571, also called imatinib

mesylate or Gleevec) and viral enzymes including inhibitors of viral morphogenesis (TTP-6171), viral release (ST-246), and viral DNA synthesis (ANP analogs including HPMPC).[93] ST-246 is a potent inhibitor only of OPVs. It targets the protein F13L of VACV, which is required for the wrapping of intracellular mature viruses and the production of extracellular enveloped viruses.[97,101] Numerous studies have shown that ST-246 administered for 10–14 days at a dose of 100 mg/kg/day protects OPV-infected animals from disease development. In case of CMLV, the activity of the molecule (cidofovir, CMX001, and ST-246) has only been evaluated *in vitro* and are potent inhibitors of CMLV replication.[5,102] In mouse models of camelpox infection, cidofovir either formulated as cream or for systemic use protected animals from disease development and/or death. Nevertheless, CMX001 and ST-246 offer the advantage of being orally available, which may render them more attractive for veterinary use.[5] Among ANP drugs, cHPMPA, HPMP-5-azacytidine, HPMPDAP (2,6-diaminopurine derivative of cidofovir), HPMPO-DAPy, and HDE-HPMP-5-azaC are reported to have varying selectivity for CMLV isolates, namely, Iran (CML1) and Dubai (CML14) strains. Of these derivatives, HPMPO-DAPy is shown to have better selectivity index (SI) and potency with OPVs including CMLV when tested by *in vitro* cell culture.

54.6.2 Prevention and Control

The CMLV, which is reported to be closely related to variola virus at molecular level, warrants biosecurity and biosafety measures especially at borders to contain this transboundary and emerging disease. Because CMLV resembles variola in its dependence on a single host, the disease could potentially be eliminated through a combination of surveillance, vaccination, and quarantine.[5] In the early 1990s, Higgins and colleagues demonstrated that an outbreak could be halted by immunizing animals with human smallpox vaccine.[103] Research has been oriented towards the development of prophylactic methods to contain the spread of camelpox in enzootic countries. However, the development of camelpox vaccines has been initiated after the worldwide eradication of smallpox. At that time, the use of VACV as a prophylactic agent for other orthopoxviral diseases of animals was not recommended, most probably due to the potential danger to unvaccinated human contacts.[20] Because of the concern that VACV could accidentally be spread from recently inoculated camels to unvaccinated humans or to domestic or wild animals, researchers began to focus on developing attenuated CMLV vaccines that could only infect camels.[5] Nevertheless, it has been reported that a camelpox vaccine, developed in the former Soviet Union, was prepared from VACV.[20]

There is only little information on the production of vaccines against camelpox since its first inception about the concept of camelpox vaccine from former Soviet Union.[104] However, the details regarding strain, safety, and effectiveness are scanty. The knowledge of camelpox vaccine efficacy originates from field investigations using the commercialized CMLV-based vaccines. Of late, lactotherapy (scarification of

TABLE 54.1

Different Molecular Diagnostic Approaches for the Detection and Differentiation of CMLV

Serial No.	Diagnostic Tool/ Technique	Target Gene (Amplicon Size)	Features/Advantage(s)
1	Genus-specific single PCR	ATIP (881 bp)	Most widely used for diagnosis of OPVs[83]
		HA (1100 bp)	Commonly used PCR method earlier[82]
		C18L (477 bp)	Host range gene-targeted PCR[86]
		DNA Pol (96 bp)	Highly conserved and more sensitive for OPVs[86]
2	Species-specific single PCR	TNFR-II (270 bp)	Rapid and sensitive[33]
		C18L (243 bp)	Rapid and highly sensitive in direct detection of CMLV[86]
3	Restriction fragment PCR	HA and A36R	Rapid PCR followed by restriction enzyme digestion for differentiating OPVs including CMLV[81,84]
4	Duplex PCR	DNA Pol (96 bp) and C18L (243 bp)	Specific detection and simultaneous differentiation of CMLV from BPXV[86]
5	One-tube multiplex PCR	DNA Pol (CaPV, PPV, and OPV) and C18L (CMLV)	Single-tube reaction for detection and differentiation of CMLV from OPVs, CaPVs, and PPVs in clinical samples[88]
6	SYBR green real-time PCR	C18L (243 bp) with a specific Tm of 77.6°C for CMLV	Highly sensitive and rapid for detection and semiquantitation of CMLV[86]
		VETF, rpo18, and A13L genes	Detection of OPVs including CMLV based on melting curve analysis[90]
7	FRET real-time PCR	HA	Detection, differentiation, and quantification of VARV, MPXV, CPXV, CMLV, and VACV[89]
8	TaqMan hydrolysis probe real-time PCR	C18L (243 bp) with hydrolysis probe (Cy5-BHQ 2)	Highly specific and sensitive quantification of CMLV in clinical samples[91]
9	LAMP	C18L (198 bp)	Simple, rapid, specific, and sensitive detection of CMLV without the need of PCR and Q-PCR tools at rural field settings[92]

ATI, A-type inclusion protein; TNFR-II, tumor necrosis factor-binding protein receptor-II; C18L, ankyrin repeat protein; FRET, fluorescence resonance energy transfer; LAMP, loop-mediated isothermal amplification; VETF, virus early transcription factor; HA, hemagglutinin; OPV, orthopoxvirus; CaPV, capripoxvirus; PPV, parapoxvirus; BPXV, buffalopox virus; MPXV, monkeypox virus; VARV, variola virus; CPXV, cowpox virus; VACV, vaccinia virus.

a mixture of milk and camelpox-infected crusts) had been in use and practiced in Punjab (India), former USSR, and Bedouin to control camelpox. Some report on vaccines against camelpox using VACV strains in former USSR and the smallpox Elstree vaccine strain in former USSR and Bahrain have been found to cross protect against virulent CMLV.[5,16] A Saudi isolate of CMLV attenuated (Jouf-78 strain) in camel kidney cell cultures (CKCC) at passage level 78 was found to be safe (at $10^{6.8}$ TCID$_{50}$) and potent in camels (at 10^3–10^4 TCID$_{50}$) intradermally or subcutaneously[20] and commercially manufactured as vaccine (Orthovac[R]) by Jordanian Vaccine Company (JOVAC) and is being currently used in many countries. A similar kind of vaccine for camelpox has been in use in Egypt, Morocco, and Russia (apps. cfsph.iastate.edu). Further, in the UAE, Vero cell attenuated camelpox vaccine using isolate (strain CaPV298-2) of UAE[40,105] named as Ducapox[R] (Dubai camelpox attenuated vaccine) developed by Onderstepoort Biological Products (OBP) and produced by Highveld Biologicals, Republic of South Africa (RSA).[22] It confers immunity up to 6 years of vaccination in two animals. However, a booster vaccination is recommended in young animals vaccinated below 6 months. Further, the attenuated CMLV strain VD47/25, passaged 80 times in cell culture, has also been evaluated as camelpox vaccine in Mauritania.[21] Similarly, the formalin-inactivated aluminum hydroxide adjuvant camelpox vaccine (CMLV strain T8-1984) is available in Morocco and reported to give

protection only for 1 year.[1] It is manufactured and distributed by Biopharma and safe for young and adult camels and has been shown to induce CMLV neutralizing antibodies,[106] but it requires booster as well as annual vaccination for efficient protection. Both "Ducapox[R]" and inactivated camelpox vaccines were found safe in pregnant camels.[85] However, it is imperative to have a thermostabilizer of these poxviral vaccines, which would facilitate their use in hot, dry regions where the disease frequently occurs.

54.6.3 IMMUNITY

Immunity against camelpox is both humoral and cell mediated.[1] However, it is believed that circulating antibodies do not reflect the immune status of the animal.[17] Live attenuated vaccines provide protection for at least 6 years, probably longer,[22] whereas inactivated vaccine was reported to provide 1 year of protection.[1] The immune response in terms of antibody for both live and inactivated vaccines was observed on the second week post vaccination; and a booster dose increased antibody titers significantly. However, the inactivated vaccine induced only low level of antibody. Both types of vaccines are reported to produce a sound cell-mediated immune response when tested for delayed-type hypersensitivity. However, the live vaccine was relatively greater compared with the inactivated vaccine. Field trials showed that both vaccines induce a weak immune response in camels less than 6 months of age and older than 4 years due to the presence of maternal antibodies or the less-developed immune system in camel calves or due to the occurrence of preformed antibodies by vaccination in adult camels. Good immune responses were observed in 1–4-year-old camels as measured by serological tests.

The prevention and control of sporadic cases of camelpox infection in camel husbandry is of prime importance in developing countries like India. Similarly, attempts have been made in the Division of Virology, Indian Veterinary Research Institute, Mukteswar, India, to develop a live attenuated camelpox vaccine using CMLV-1 isolate in Vero cell line[16,27] in a similar line developed in Saudi Arabia. Further, considering the increased incidence of camelpox not only in camel but also in human,[50] studies on molecular epidemiology, specific diagnosis, and control measures are of paramount importance in reducing the circulation of CMLV in camels and also in humans as a public health aspect.

54.6.4 ERADICATION

The diagnostic tests and vaccines are needed for eradication of any infectious disease. Unfortunately, as with many public health problems, the challenge lies in bringing those tools to the affected animals. To eradicate camelpox, it would not be necessary to vaccinate all the camels. Instead, "ring vaccination" strategy could be employed, in which intensive surveillance to detect cases of disease (and differentiation from a clinically similar disease like contagious ecthyma), followed by vaccination of all surrounding contacts and continued monitoring to ensure that no more cases occurred.[5] Although, it has not been a recognized target for eradication efforts, and the toll of animal and human suffering from camelpox cannot be compared to the mass die-offs and famine caused by rinderpest, a deadly disease of cattle. Massive effort would not be required for camelpox, when compared to smallpox, rinderpest and poliomyelitis control and eradication programs, as it is confined to a specific region only. The fact that countries affected by camelpox include both wealthy nations, in which vaccines are in use, and some of the world's poorest countries like Somalia, where most of the infected animals are located, suggests that a program of mutual cooperation would be the best path for eradication. For a successful eradication campaign, they could recoup at least part of its cost by discontinuing their own programs of surveillance, diagnostic testing, and vaccination.[107] Researchers are now focusing on the goal of "one health" approach to combat the diseases of zoonotic and public health important.

54.7 PERSPECTIVES

The disease was considered inconsequential till recently, but it is considered emerging as an important public health problem during this decade and due to increased reported cases and outbreaks in camels. In the context, particular attention should be given to camelpox outbreaks, and identification of any human infections. Effective prevention and control

measures can be achieved through the use of proper diagnostic and prophylactic aids to curtail further spread as described for most of the zoonotic diseases. In general, several factors related to human activities, environmental changes, and virus could be the determinants of incidence and prevalence of the disease. To safeguard the public health from pathogens of zoonotic infections, application of skill, knowledge, and resources of veterinary public health is essential. Further, the control measures for emerging and reemerging pathogens are demanding, as there is population explosion. Novel, highly sensitive and specific techniques comprising genomics and proteomics along with conventional methods would be useful in the identification of emerging and reemerging pathogen or virus; thereby therapeutic/prophylactic/preventive measures would be applied on time. Intensified surveillance should be implemented to detect and restrain the spread of the disease. Research should focus on molecular biology of this virus so as to develop diagnostics and prophylactics in a modern way to combat this infection in short time.

Our present limited knowledge of CMLV prompts a series of challenging, interesting research questions that may drive further investigation as described by Duraffour and colleagues[5] to focus on the priorities for research on CMLV as follows. The different strains of CMLV may exist with varying virulence and pathogenicity, which could explain the moderate and generalized lethal forms of the disease. In this context, further sequencing of CMLV strains isolated from these two forms of cases could emphasize discrepancies in their gene content and allow the identification of genes associated with virulence or immune-evasion properties. Further analysis of virus pathogenesis and immune modulation should be carried out in relevance to other aspects of poxvirus biology. In-depth studies of CMLV strains may reveal the determinants of the narrow host tropism of CMLV, the duration of viral shedding, the spread from the initial site of infection, and the tissue tropism of the virus. It could be envisaged to evaluate their virulence in suckling mice or models in athymic nude mice. Further, in contrast to VACV and CPXV, the pathways used by CMLV to counteract host immune responses have to be studied in detail in order to know the CMLV ORFs with putative active immunoregulatory functions. Also, it would be interesting to identify which cytokines, chemokines, and INFs are deregulated *in vitro* and *in vivo* following CMLV exposure. Studies are required to confirm the speculation of the virus coevolved with its reservoir host, camels, and property to survive in hot temperatures. Epidemiological studies should be carried out to explore the circulation of CMLV among camels and possibly among sheep and goats, as previously reported.[42] If arthropod vectors prove to be involved in CMLV transmission, it would bring camelpox into a novel light, as the first vector-borne OPVs disease. In order to explore any role of arthropod vectors in the transmission of camelpox, arthropods that infest camels, such as ticks (*Hyalomma* and *Rhipicephalus*), suckling lice, flies, and fleas,[17] would be of great value to identify a potential vector for CMLV and might highlight animals other than camels that might be at risk of infection.[5]

Antiviral treatments may be beneficial for the management of camelpox outbreaks and/or in the event of vaccination failure. Further research should be oriented towards the determination of the long-term immunity conferred by CMLV-based vaccines including the development of thermostable vaccine. With few reported cases of camelpox infections in human and limited host range of virus in other animals and the fact that camels are the single reservoir host for camelpox, it may be envisaged, with a political will, to eradicate camelpox in the line of smallpox eradication.

REFERENCES

1. OIE Manual, *Manual of Diagnostic Tests and Vaccines for Terrestrial Animals*, 6th ed., Vol. 2, Chapter 2.7.14, OIE, Paris, France, 2008.
2. Sadykov, R.G., Cultivation of camelpox virus in chick embryos, In: *Virus Diseases of Agricultural Animals*, pp. 55, 1970. (In Russian)
3. Roslyakov, A.A., Comparative ultrastructure of viruses of camel pox, a pox-like disease of camels ("auzdik") and contagious ecthyma of sheep, *Vop. Virusol.*, 17, 26, 1972.
4. Moss, B., Poxviridae: The viruses and their replication, In: Knipe, D.M. and Howley, P.M. (eds.), *Fields Virology*, 5th ed., pp. 2905–2945 (Lippincott Williams & Wilkins, Philadelphia, PA), 2007.
5. Duraffour, S. et al., Camelpox, *Antivir. Res.*, 92(2), 167, 2011.
6. Baxby, D., Smallpox viruses from camels in Iran, *Lancet*, 7786, 1063, 1972.
7. Gubser, C. and Smith, G.L., The sequence of camelpox virus shows it is most closely related to variola virus, the cause of smallpox, *J. Gen. Virol.*, 83, 855, 2002.
8. Sheikh Ali, H.M., Khalafalla, A.I., and Nimir, A.H., Detection of camel pox and vaccinia viruses by polymerase chain reaction, *Trop. Anim. Health Prod.*, 41, 1637, 2009.
9. Damon, I., Poxviruses, In: Knipe, D.M. and Howley, P.M. (eds.), *Fields Virology*, 5th ed., pp. 2947–2975 (Lippincott Williams & Wilkins, Philadelphia, PA), 2007.
10. Duraffour, S. et al., Activities of several classes of acyclic nucleoside phosphonates against camelpox virus replication in different cell culture models, *Antimicrob. Agents Chemother.*, 51, 4410, 2007.
11. Davies, F.G., Mungai, J.N., and Shaw, T., Characteristics of a Kenyan camelpox virus, *J. Hyg. (Lond.)*, 7, 381, 1975.
12. Coetzer, J.A.W., Poxviridae, In: Coetzer, J.A.W. and Tustin, R.C. (eds.), *Infectious Diseases of Livestock*, 2nd ed., Vol. 2, pp. 1265–1267 (Oxford University Press, Cape Town, Southern Africa), 2004.
13. Rheinbaden, F.V., Gebel, J., Exner, M. et al., Environmental resistance, disinfection, and sterilization of poxviruses, In: Schmidt, A.A., Weber, A., and Mercer, O. (eds.), *Poxviruses*, pp. 397–405 (Birkhäuser Verlag, Basel, Switzerland), 2007.
14. Fenner, F., Wittek, R., and Dumbell, K.R., *The Orthopoxviruses* (Academic Press Inc., Harcourt Brace Jovanovich Publisher, New York), 1989.
15. Afonso, C.L. et al., The genome of camelpox virus, *Virology*, 295, 1, 2002.
16. Bhanuprakash, V. et al., Camelpox: Epidemiology, diagnosis and control measures, *Expert Rev. Anti-Infect. Ther.*, 8, 1187, 2010.
17. Wernery, U. and Kaaden, O.R., Camel pox. In: Wernery, U., and Kaaden, O.R. (eds.), *Infectious Diseases in Camelids*, 2nd edn., pp. 176–185 (Blackwell Science, Berlin, Vienna), 2002.
18. Baxby, D., Differentiation of smallpox and camelpox viruses in cultures of human and monkey cells, *J. Hyg. (Lond.)*, 72, 251, 1974.
19. Baxby, D. et al., A comparison of the response of camels to intradermal inoculation with camelpox and smallpox viruses, *Infect. Immun.*, 11, 617, 1975.
20. Hafez, S.M. et al., Development of a live cell culture camelpox vaccine, *Vaccine*, 8, 533, 1992.
21. Nguyen, B.V., Guerre, L., and Saint-Martin, G., Preliminary study of the safety and immunogenicity of the attenuated VD47/25 strain of camelpoxvirus, *Rev. Elev. Med. Vet. Pays Trop.*, 49, 189, 1996.
22. Wernery, U. and Zachariah, R., Experimental camel pox infection in vaccinated and unvaccinated dromedaries, *J. Vet. Med. Series B*, 46, 131, 1999.
23. Wernery, U., Meyer, H., and Pfeffer, M., Camel pox in the United Arab Emirates and its prevention, *J. Camel Pract. Res.*, 4, 135, 1997.
24. Tantawi, H.H., El Dahaby, H., and Fahmy, L.S., Comparative studies on poxvirus strains isolated from camels, *Acta Virol.*, 22, 451, 1978.
25. Renner-Muller, I.C., Meyer, H., and Munz, E., Characterization of camelpoxvirus isolates from Africa and Asia, *Vet. Microbiol.*, 45, 371, 1995.
26. Tscharke, D.C., Reading, P.C., and Smith, G.L., Dermal infection with vaccinia virus reveals roles for virus proteins not seen using other inoculation routes, *J. Gen. Virol.*, 83, 1977, 2002.
27. Bhanuprakash, V. et al., Isolation and characterization of Indian isolates of camel pox viruses, *Trop. Anim. Health. Prod.*, 42, 1271, 2010.
28. Otterbein, C.K. et al., *In vivo* and *in vitro* characterization of two camelpoxvirus isolates with decreased virulence, *Rev. Elev. Med. Vet. Pays Trop.*, 49, 114, 1996.
29. Duraffour, S. et al., Study of camelpox virus pathogenesis in athymic nude mice, *PLoS One*, 6, e21561, 2011.
30. Bett, B. et al., Using participatory epidemiological techniques to estimate the relative incidence and impact on livelihoods of livestock diseases amongst nomadic pastoralists in Turkana South District Kenya, *Prev. Vet. Med.*, 90, 194, 2009.
31. Leese, S., Two diseases of young camels, *J. Trop. Vet. Sci.*, 4, 1, 1909.
32. Chauhan, R.S. and Kaushik, R.K., Isolation of camelpox virus in India, *Brit. Vet. J.*, 143, 581, 1987.
33. Marodam, V. et al., Isolation and identification of camelpox virus, *Indian J. Anim. Sci.*, 76, 326, 2006.
34. Al-Ziabi, O., Nishikawa, H., and Meyer, H. The first outbreak of camelpox in Syria, *J. Vet. Med. Sci.*, 69, 541, 2007.
35. Nothelfer, H.B., Wernery, U., and Czerny, C.P., Camel pox: Antigen detection within skin lesions-immunohistochemistry as a simple method of etiological diagnosis, *J. Camel Pract. Res.*, 2, 119, 1995.
36. Kriz, B., A study of camelpox in Somalia, *J. Comp. Pathol.*, 92, 1, 1982.
37. Wernery, U., Kaaden, O.R., and Ali, M., Orthopox virus infections in dromedary camels in United Arab Emirates (U.A.E.) during winter season, *J. Camel Pract. Res.*, 4, 51, 1997.
38. Khalafalla, A.I. and Ali, Y.H., Observations on risk factors associated with viral diseases of camels in Sudan. In *Proceedings of the 12th International Conference of the Association of Institutions of Tropical Veterinary Medicine*, Montpellier, France, 20–22 August, pp. 101–105, 2007.

39. Wernery, U., Kinne, J., and Zachariah, R., Experimental camelpox infection in vaccinated and unvaccinated guanacos, *J. Camel Pract. Res.*, 7, 153, 2000.

40. Azwai, S.M. et al., Serology of *Orthopoxvirus cameli* infection in dromedary camels: Analysis by ELISA and western blotting, *Comp. Immunol. Microbiol.*, 19, 65, 1996.

41. Yousif, A.A. and Al-Naeem, A.A., Recovery and molecular characterization of live Camelpox virus from skin 12 months after onset of clinical signs reveals possible mechanism of virus persistence in herds, *Vet. Microbiol.*, 159, 320, 2012.

42. Housawi, F.M.T., Screening of domestic ruminants sera for the presence of anti-camel pox virus neutralizing antibodies, *Assiut Vet. Med. J.*, 53, 101, 2007.

43. Khalafalla, A.I., Mohamed, M.E.M., and Agab, H., Serological survey in camels of the Sudan for prevalence of antibodies to camelpox virus using ELISA technique, *J. Camel Pract. Res.*, 5, 197, 1998.

44. Jezek, Z., Kriz, B., and Rothbauer, V., Camelpox and its risk to the human population, *J. Hyg. Epid. Microb. Immunol.*, 27, 29, 1983.

45. Gitao, C.G., An investigation of camelpox outbreaks in two principal camel (*Camelus dromedarius*) rearing areas of Kenya, *Rev. Sci. Tech.*, 16, 841, 1997.

46. Ramyar, H. and Hessami, M., Isolation, cultivation and characterisation of camelpox virus, *J. Vet. Med. Series A.*, 19, 182, 1972.

47. Al-Khalifa, M.S., Khalil, G.M., and Diab, F.M., A two-year study of ticks infesting camels in Al-Kharj in Saudi Arabia, *Saudi J. Biol. Sci.*, 14, 211, 2007.

48. Elghali, A. and Hassan, S.M., Ticks (Acari: Ixodidae) infesting camels (*Camelus dromedarius*) in Northern Sudan, *Onderstepoort J. Vet. Res.*, 76, 177, 2009.

49. Falluji, M.M., Tantawi, H.H., and Shony, M.O., Isolation, identification and characterization of camelpox virus in Iraq, *J. Hyg. (Lond.)*, 83, 267, 1979.

50. Bera, B.C. et al., Zoonotic cases of camelpox infection in India, *Vet. Microbiol.*, 152, 29, 2011.

51. Marennikova, S.S., The results of examinations of wildlife monkeys for the presence of antibodies and viruses of the pox group, *Vopr. Virusol.*, 3, 321, 1975.

52. Tadjbakhsh, H., Traditional methods used for controlling animal diseases in Iran, *Rev. Sci. Tech.*, 13, 599, 1994.

53. Pfeffer, M. et al., Fatal form of camelpox virus infection, *Vet. J.*, 155, 107, 1998.

54. Kinne, J., Cooper, J.E., and Wernery, U., Pathological studies on camelpox lesions of the respiratory system in the United Arab Emirates (UAE), *J. Comp. Pathol.*, 118, 257, 1998.

55. Duraffour, S. et al., Activity of the antiorthopoxvirus compound ST-246 against vaccinia, cowpox and camelpox viruses in cell monolayers and organotypic raft cultures, *Antivir. Ther.*, 12, 1205, 2007.

56. Yager, J.A., Scott, D.W., and Wilcock, B.P., Viral diseases of the skin, In: Jubb, K.V.F., Kennedy, P.C., and Palmer, N. (eds.) *Pathology of Domestic Animals*, 4th ed., pp. 629–644 (Academic Press, San Diego, CA), 1991.

57. Nazarian, S.H. and McFadden, G., Immunomodulation by poxviruses, In: Schmidt, A.A., Weber, A., and Mercer, O. (eds.), *Poxviruses*, pp. 273–296 (Birkhäuser, Verlag, Basel, Switzerland), 2007.

58. Perdiguero, B. and Esteban, M., The interferon system and vaccinia virus evasion mechanisms, *J. Interferon Cytokine Res.*, 29, 581, 2009.

59. Najarro, P., Traktman, P., and Lewis, J.A., Vaccinia virus blocks gamma interferon signal transduction: Viral VH1 phosphatase reverses Stat1 activation, *J. Virol.*, 75, 3185, 2001.

60. Bowie, A. et al., A46R and A52R from vaccinia virus are antagonists of host IL-1 and toll-like receptor signalling, *Proc. Natl. Acad. Sci. USA*, 97, 10162, 2000.

61. Moss, B. and Shisler, J.L., Immunology 101 at poxvirus U: Immune evasion genes, *Semin. Immunol.*, 13, 59, 2001.

62. Alcami, A. et al., Vaccinia virus strains Lister, USSR and Evans express soluble and cell surface tumour necrosis factor's receptor, *J. Gen. Virol.*, 80, 949, 1999.

63. Turner, P.C., Musy, P.Y., and Moyer, R.W., Poxvirus serpins, In: G. McFadden (ed.), *Viroceptors, Virokines and Related Immune Modulators Encoded by DNA Viruses*, pp. 67–88 (R.G. Landes, Austin, TX), 1995.

64. Smith, G.L., Symons, J.A., and Alcami, A., Poxviruses: Interfering with interferon, *Semin. Virol.*, 8, 409, 1998.

65. Walter, L. et al., Identification of a novel conserved human gene, TEGT, *Genomics*, 28, 301, 1995.

66. Alcami, A. et al., Poxviruses: Capturing cytokines and chemokines, *Semin. Virol.*, 8, 419, 1998.

67. Alcami, A. and Smith, G.L., Vaccinia, cowpox, and camelpox viruses encode soluble gamma interferon receptors with novel broad species specificity, *J. Virol.*, 69, 4633, 1995.

68. Gubser, C. et al., A new inhibitor of apoptosis from vaccinia virus and eukaryotes, *PLoS Pathog.*, 3, e17, 2007.

69. Gubser, C. et al., Camelpox virus encodes a schlafen-like protein that affects orthopoxvirus virulence, *J. Gen. Virol.*, 88, 1667, 2007.

70. Eskra, L., Mathison, A., and Splitter, G., Microarray analysis of mRNA levels from RAW264.7 macrophages infected with *Brucella abortus*, *Infect. Immun.*, 71, 1125, 2003.

71. Geserick, P. et al., Modulation of T cell development and activation by novel members of the Schlafen (slfn) gene family harbouring an RNA helicase-like motif, *Int. Immunol.*, 16, 1535, 2004.

72. Symons, J.A., Alcami, A., and Smith, G.L., Vaccinia virus encodes a soluble type-I interferon receptor of novel structure and broad species-specificity, *Cell*, 81: 551–560. 1995.

73. Alcami, A., Symons, J.A., and Smith, G.L., The vaccinia virus soluble alpha/beta interferon (IFN) receptor binds to the cell surface and protects cells from the antiviral effects of IFN, *J. Virol.*, 74, 11230, 2000.

74. Montanuy, I., Alejo, A., and Alcami, A., Glycosaminoglycans mediate retention of the poxvirus type I interferon binding protein at the cell surface to locally block interferon antiviral responses, *FASEB J.*, 25, 1960, 2011.

75. Elliot, H. and Tuppurainen, E., Camelpox, *Manual of Diagnostic Tests and Vaccines for Terrestrial Animals*, p. 1177, OIE, Paris, France, 2010.

76. Pfeffer, M. et al., Diagnostic procedures for poxvirus infections in camelids, *J. Camel Pract. Res.*, 5, 189, 1998.

77. Boulter, E.A., The nature of the immune state produced by inactivated vaccinia virus in rabbits, *Am. J. Epidemiol.*, 94, 612, 1971.

78. Al Hendi, A.B. et al., A slow-spreading mild form of camelpox infection, *J. Vet. Med.*, 41, 71, 1994.

79. Yousif, A.A., Al-Naeem, A.A., and Al-Ali, M.A., A three-minute nonenzymatic extraction method for isolating PCR-quality Camelpox virus DNA from skin, *J. Virol. Methods*, 169, 138, 2010.

80. Meyer, H., Pfeffer, M., and Rziha, H.J., Sequence alterations within and downstream of the A-type inclusion protein genes allow differentiation of *Orthopoxvirus* species by polymerase chain reaction, *J. Virol.*, 75, 1975, 1994.

81. Ropp, S.L. et al., Polymerase chain reaction strategy for identification and differentiation of smallpox and other orthopoxviruses, *J. Clin. Microbiol.*, 33, 2069, 1995.

82. Damaso, C.R. et al., An emergent poxvirus from humans and cattle in Rio de Janeiro state: Cantagalo virus may derive from Brazilian small pox vaccine, *Virology*, 277, 439, 2000.

83. Meyer, H., Ropp, S.L., and Esposito, J.J., Gene for A-type inclusion body protein is useful for a polymerase chain reaction assay to differentiate orthopoxviruses, *J. Virol. Methods*, 64, 217, 1997.

84. Huemer, H.P, Hönlinger, B., and Höpfl, R., A simple restriction fragment PCR approach for discrimination of human pathogenic Old World animal Orthopoxvirus species, *Can. J. Microbiol.*, 54, 159, 2008.

85. Khalafalla, A.I. and El Dirdiri, G.A., Laboratory and field investigations of a live attenuated and an inactivated camelpox vaccine, *J. Camel Pract. Res.*, 10, 191, 2003.

86. Balamurugan, V. et al., A polymerase chain reaction strategy for the diagnosis of camel pox, *J. Vet. Diagn. Invest.*, 21, 231, 2009.

87. Singh, R.K. et al., Sequence analysis of C18L gene of Buffalopox virus: PCR strategy for specific detection and its differentiation from orthopoxviruses, *J. Virol. Methods*, 154, 146, 2008.

88. Venkatesan, G. et al., Multiplex PCR for detection and differentiation of Pox viruses, News letter, Indian Veterinary Research Institute, Izatnagar, UP, India, January–June, Vol. 31(1), p. 2, 2009.

89. Panning, M. et al., Rapid detection and differentiation of human pathogenic orthopox viruses by a fluorescence resonance energy transfer real-time PCR assay, *Clin. Chem.*, 50, 702, 2004.

90. Nitsche, A., Ellerbrok, H., and Pauli, G., Detection of orthopoxvirus DNA by real-time PCR and variola virus DNA by melting analysis, *J. Clin. Microbiol.*, 42, 1207, 2004.

91. Venkatesan, G. et al., TaqMan hydrolysis probe based real time PCR for detection and quantitation of camelpox virus in skin scabs, *J. Virol. Methods*, 181, 192, 2012.

92. Venkatesan, G. et al., Development of loop-mediated isothermal amplification assay for specific and rapid detection of camelpox virus in clinical samples, *J. Virol. Methods*, 183, 34, 2012.

93. Snoeck, R., Andrei, G., and De Clercq, E., Therapy of poxvirus infections, In: Mercer, A.A., Schmidt, A., and Weber, O. (eds.), *Poxviruses*, pp. 375–395 (Birkhauser, Basel, Switzerland), 2007.

94. Smee, D.F. et al., Progress in the discovery of compounds inhibiting orthopoxviruses in animal models, *Antivir. Chem. Chemother.*, 19, 115, 2008.

95. De Clercq, E. et al., Antiviral activity of phosphonylmethoxyalkyl derivatives of purine and pyrimidine, *Antivir. Res.*, 8, 261, 1987.

96. Kern, E.R. et al., Enhanced inhibition of orthopoxvirus replication *in vitro* by alkoxyalkyl esters of cidofovir and cyclic cidofovir, *Antimicrob. Agents Chemother.*, 46, 991, 2002.

97. Yang, G. et al., An orally bioavailable antipoxvirus compound (ST-246) inhibits extracellular virus formation and protects mice from lethal orthopoxvirus Challenge, *J. Virol.*, 79, 13139, 2005.

98. Andrei, G. et al., Cidofovir resistance in vaccinia virus is linked to diminished virulence in mice, *J. Virol.*, 80, 9391, 2006.

99. Smee, D.F. et al., Characterization of wild-type and cidofovir-resistant strains of camelpox, cowpox, monkeypox, and vaccinia viruses, *Antivir. Chem. Chemother.*, 46, 1329, 2002.

100. Duraffour, S., Selection and characterization of (S)-1-[3-hydroxy-2-(phosphonomethoxypropyl)-2,6-diaminopurine [HPMPDAP] resistant camelpox viruses, *Antivir. Res.*, 82, A67, 2009.

101. Duraffour, S. et al., Antiviral potency of ST-246 on the production of enveloped orthopoxviruses and characterization of ST-246 resistant vaccinia, cowpox and camelpox viruses, *Antivir. Res.*, 78, A29–A30, 2008.

102. Duraffour, S., Andrei, G., and Snoeck, R., Tecovirimat, a p37 envelope protein inhibitor for the treatment of smallpox infection, *Drugs*, 13, 181, 2010.

103. Higgins, A.J. et al., The epidemiology and control of an outbreak of camelpox in Bahrain, In: Allen, W.R. (ed.), *Proceedings of the First International Camel Conference*, pp. 101–104 (R&W Publications, Newmarket, Suffolk, U.K.), 1992.

104. Borisovich, Yu.F., Camel pox, In: F.M. Orlov (ed.), *Little-Known Contagious Diseases of Animals*, 2nd ed., pp. 32–42 (Izdatel'stvo Kolos, Moscow, Russia), 1973. Cited in *Vet. Bull.*, 44, 139, 1974.

105. Pfeffer, M. et al., Comparison of camelpox viruses isolated in Dubai, *Vet. Microbiol.*, 49, 135, 1996.

106. El Harrak, M. and Loutfi, C., La variole du dromadaire chez le jeune au Maroc. Isolement et identification du virus. Mise au point du vaccin et application à la prophylaxie, *Revue Elev. Med. Vet. Pays Trop.*, 53, 165, 2000.

107. Bary, M. and Babiuk, S., Camelpox: Target for eradication? *Antivir. Res.*, 92, 164, 2011.

55 Capripoxviruses

Veerakyathappa Bhanuprakash and M. Hosamani

CONTENTS

55.1 INTRODUCTION

Capripoxviruses are a group of poxviruses that include sheeppox virus (SPPV), goatpox virus (GTPV), and lumpy skin disease virus (LSDV). They cause sheeppox, goatpox, and lumpy skin disease. The target species for these viruses are sheep, goat, and cattle. Capripox diseases are transboundary, notifiable to Office Internationale des Epizooties (OIE), the world organization for animal health, and listed in select agent group. They are economically important in a country in which they are enzootic and restrict the international trade of animals and their byproducts. Goatpox and sheeppox are prevalent in Africa above the equator, Asia, and the Middle East, and occasional outbreaks occur in regions of Europe surrounding the Middle East. In contrary, LSD is enzootic in Africa and outbreaks have occurred in the Middle East surrounding Egypt.[1] Currently, there is a reduction in the distribution of these diseases, compared to earlier years. Yet, these viruses are expanding their territory with recent outbreaks of sheeppox and goatpox recorded in Vietnam, Mongolia, and Greece and outbreaks of LSD in Ethiopia, Egypt, and Israel. The legal and illegal trade of animals potentially contributes to their spread.[1]

55.2 CLASSIFICATION AND MORPHOLOGY

The genus *Capripoxvirus* (CaPV) is one of the eight genera of Chordopoxvirinae subfamily within the Poxviridae family. The genus comprises the economically important three viruses such as SPPV, GTPV, and LSDV. Sheeppox virus is the type species. Capripoxviruses are generally host specific and have specific geographic distribution. The viruses are physically, chemically, and morphologically similar to vaccinia virus (VACV). These viruses were originally classified on the basis of animal species from which they were isolated. Genomic studies indicate a close relationship between these viruses but they are phylogenetically different. Capripoxviruses are serologically indistinguishable and are able to cross-protect at varying degrees. Cross-infection between the species has been reported both naturally and experimentally, and it is possibly because of occurrence of genetic recombination reported among the members of capripoxvirus.[2]

The elementary bodies of SPPV were among the first to be seen with the light microscope.[3] Morphologically, the capripoxviruses resemble other poxviruses in several aspects. The virions are brick-shaped, enveloped with rounded ends. The virions have complex symmetry with or without an envelope, a surface membrane, a core, and lateral bodies. Each virion measures ~300 nm (length) × 270 nm (width) × 200 nm (height), and the surface is characterized by tubules. The appearance of the core is biconcave or cylindrical with two lateral bodies nested between the core or surface membranes. The core contains linear, double-stranded DNA and proteins organized in a nucleoprotein complex. Recent models indicate that the nucleoprotein complex might be cylindrical, folded at least twice along the long virion axis to form a Z-structure, which presents as three circles, arranged linearly when viewed as a section across the short axis.[4] During their life cycle, virions produce extracellular and intracellular particles, but the infection is initiated by extracellular virions. These viruses may be sequestered within inclusion bodies that are not occluded and typically contain one nucleocapsid. Virions mature naturally by budding through the membrane of the host cell. The viral genome encodes structural and nonstructural proteins. Lipids are located in the envelope and constitute to the tune of 4% by weight and are composed of glycolipids similar to host cell membranes. The lipids are host derived and synthesized *de novo*.[4]

55.3 BIOLOGY AND EPIDEMIOLOGY

Capripoxviruses contain a single, linear, double-stranded (ds) DNA with covalently closed ends, 154 kb in size.[4] The complete genome sequences of all the members of the genus *Capripoxvirus*, namely, SPPV, GTPV, and LSDV, are known.[5,6] The complete genome analysis indicate that the central region of the genome consists of coding sequences bound by two identical inverted terminal repeats (ITR). Complete genome sequencing of SPPV and GTPV revealed a total of 147 putative genes, including conserved poxvirus replicative, structural, and genes likely involved in virulence and host range; likewise, the genome of LSDV consists of 156 putative genes, which encode proteins involved in transcription and mRNA biogenesis, nucleotide metabolism, DNA replication, protein processing, virion structure and assembly, and viral virulence and host range.[5,6] SPPV and GTPV genomes are strikingly similar to each other, exhibiting 96% nucleotide identity over their entire length and they are closely related to LSDV with a nucleotide identity of 97%. In both SPPV and GTPV genomes, nine LSDV genes involved likely in virulence and host range are disrupted including a gene specific to LSDV (LSDV132) and genes similar to those coding for interleukin-1 receptor, myxoma virus M003.2 and M004.1 genes (two copies each), and vaccinia virus F11L, N2L, and K7L genes. The absence of these nine genes in SPPV and GTPV suggests a significant role for them in the bovine host range of LSDV. SPPV, GTPV, and LSDV genomes contain specific nucleotide differences, suggesting they are phylogenetically distinct.[6] The nucleotide composition of CaPV is A+T rich and it ranges from 73% to 75% and is uniformly distributed in all genomes. The notable changes include mutation or disruption of genes with predicted functions of virulence and host range. The ankyrin repeat proteins and kelch-like proteins are involved in virulence and host range. However, it is required to study the genomic sequences of more isolates of capripoxviruses to thoroughly understand the determinants of virulence, host specificity, and geographical distribution. The degree of species specificity can also be elucidated by comparative genomic analysis of SPPV, GTPV, and LSDV, and this would enable tracing of the origin of CaPV and their evolution[1].

The occurrences of sheeppox and goatpox diseases are associated with tangible and intangible losses. The tangible losses include productivity losses in terms of reduced milk yield, loss of weight, abortions, wool and hide damage, and increased susceptibility to pneumonia and fly strike.[7] Both morbidity and mortality can be very high, reaching up to 100% in naïve animals with significant case fatality rates.[8] Factors like host (age, sex, breed, nutritional, and immunological status), virus (strain, virulence, and pathogenicity), environment (micro and macro), poor management, feed scarcity, and inadequate veterinary services have direct influence on the epizootiology of sheeppox and goatpox.[9] In contrary, LSD is occasionally fatal in cattle with morbidity and mortality rates, respectively, of 10% and 1% in an affected herd. However, mortality rates more than 75% have also been reported.[10] Wide range of mortality because of LSD is possibly due to several factors including cattle breed, virus isolate, immune status of the animal, secondary infections, type of insect vector, and confusion over differential diagnosis of LSD and pseudo-LSD caused by Allerton herpes virus.[11] Sheeppox, goatpox, and LSD are the most important pox diseases of animals.[12] New outbreak of capripox in a disease-free region is immediately notifiable to OIE. The capripoxviruses are listed in the United States Department of Agriculture (USDA) national select agent registry.[1]

The distribution of sheeppox, goatpox, and LSD are clearly different. In the last half century, sheeppox and goatpox have been restricted to Asia and Africa, covering Africa north of the Equator, and into the Middle East, Turkey, and Asia including regions of the former Soviet Union, India, and China.[8,13] However, recently, they have extended into Vietnam and Mongolia in the east, with repeated incursions reported in Greece in southern Europe.[8]

The origin of LSD remains a mystery ever since it was identified from sheepox- and goatpox-free geographical regions. It was first identified in 1929 from sub-Saharan Africa and since then it spread to north and south of Africa. Currently, the disease is enzootic in Africa, though recent outbreaks have been reported from Egypt and Israel.[8] This clearly indicates that LSD is extending its territory beyond Africa to reach Middle East, Asia, and Europe.[14] Capripoxviruses are not present in north, central, or south America, East Asia (excluding Vietnam), and Australia. The spread of capripoxvirus diseases into new areas are mainly because of illegal animal trade[15] and inadequate veterinary services.[16] However, capripox-free countries have legislation in place to prevent transboundary spread of diseases. Biting flies have been implicated in the spread of capripoxviruses like the LSD outbreak that occurred in Israel during 1989.[17]

55.4 CLINICAL FEATURES AND PATHOGENESIS

Both sheeppox and goatpox are clinically alike.[18] The incubation period for sheeppox and goatpox ranges from 1 to 2 weeks with an initial rise in body temperature, increased respiratory and pulse rates, edema of the eyelids, inappetence, arched back, lacrimation, coughing, salivation, nasal discharge leading to crust formation, pneumonia, hypersensitivity, constipation, and scanty urine. Later, skin eruptions appear over the less hairy parts of the body and they undergo typical pox virus infection such as macular, papular, vesicular, pustular, and scab formation (Figure 55.1a through c). Scabs remain for 3–4 weeks and after healing cicatrix may persist. The lesions are distributed all over the hairless parts of the skin and mammary glands and mucous membranes of the respiratory and digestive tracts. The scabs shed by the recovered animals may remain infective for extended periods (many months). A fatal septicemia and pyemia may develop and the virus itself may result in the death of the animal during febrile phase of the disease in cases of hemorrhagic pocks. Sheeppox and goatpox affect both male and female animals of several breeds. The diseases are often generalized

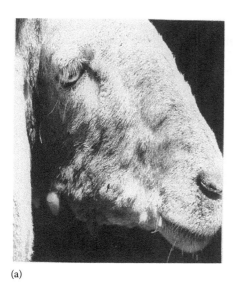
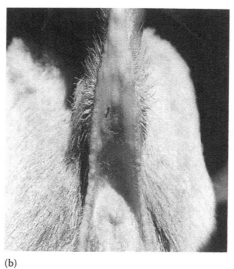

(a)   (b)   (c)

FIGURE 55.1 Photograph showing (a) typical sheeppox lesions on the face, (b) surface of the tail of a sheep and (c) goatpox lesions around the mouth of a goat.

in lambs and kids aged 4–5 months; whilst, the older animals are mildly affected. The suckling animals suffer severely, with lesions on oral and nasal mucosae. Mouth lesions constitute an important source of virus spread. Septicemia/pyemia due to secondary bacterial complications may lead to death. The latent brucella infection, tendovaginitis, orchitis, abortion, and peripheral paresis may be aggravated following sheeppox infection as reviewed.[1,8,19]

SPPV and GTPV are transmitted by the respiratory route during close contact and also through other mucous membranes or abraded skin. The transmission of SPPV and GTPV through semen or embryos has not been established. SPPV and GTPV can also be spread on fomites or mechanically by stable flies (insects) such as *Stomoxys calcitrans*, although the latter route may be uncommon. In case of sheeppox and goatpox, the postmortem lesions include pock lesions on lungs, pharynx, trachea, abomasums, kidney, testicle, tongue, hard and soft palate, esophagus, and intestines, an inflamed spleen, and all the lymph nodes in the body are enlarged. The epidermis, hypodermis, adnexa, collagen, elastin, and reticulin of the skin are affected. Microscopically, the epidermal changes include acanthuses, parakeratosis, hyperkeratosis, ballooning, and degeneration of proliferating epithelial cells. In the dermis, there will be cellular infiltration such as macrophages, neutrophils, lymphocytes, plasma cells, and occasionally eosinophils and vesiculitis leading to thrombosis, edema, and necrosis. In the lung, congestion, red hepatization, exudation, coagulative necrosis with inflammatory reaction and thickening of the interlobular septae, and the depletion of lymphocytes in spleen and liver are the characteristic changes as reviewed.[1,8,19]

In contrast to SPPV and GTPV, the LSDV mainly spreads mechanically by biting insects. The transmission by direct contact is insignificant. The LSDV affects cattle and the severity of the disease ranges from acute to subclinical. The incubation period ranges from 1 to 4 weeks following natural infection.[20] The disease is more severe in thin-skinned cattle such as Jersey, Guernsey, and African Sanga cattle like Fogera in Ethiopia. The predominant clinical sign of LSD is the formation of skin lesions all over the body, and after healing, scars remain and thereby damage the hide.[21] The characteristic signs of the disease include fever (some may remain nonfebrile), lymphadenopathy (especially subscapular and precrural), and nodules on the skin (1–5 cm in diameter) that progress into sitfasts and persist for many months. In severe cases, mucosae of eye, oral, and nasal cavities cause excessive lacrimation, salivation, and nasal discharges, respectively. All these secretions and even semen may contain LSDV.[1] Pox lesions may be noticed in the pharynx, larynx, trachea, lungs, and throughout the alimentary tract. Temporary or permanent infertility may occur in cows and bulls.[22] Several factors influence the disease progression as the rate of morbidity has a wide range of 3%–85%.[23] There is a clear cut difference in the exhibition of clinical signs following experimental infection of sheeppox, goatpox, and LSD. Following LSDV experimental infection, all cattle will not develop clinical signs[24]; in contrast, following sheeppox and goatpox infection of their respective natural hosts, there will be uniform range of responses. The severity of infection depends on virus isolates and all the three viruses are epitheliotropic.[21]

55.5 IDENTIFICATION AND DIAGNOSIS

Skin and visceral lesions are indicative of capripoxvirus infection but confirmatory diagnosis is achieved by specific laboratory techniques. A battery of serological, immunological, or nucleic acid–based tests is available for diagnosis of capripoxviruses with varying degrees of sensitivities. They are broadly classified in to (1) isolation of the virus and identification, (2) detection of CaPV antigen or CaPV-specific antibody, and (3) detection of virus-specific nucleic acid.

55.5.1 ISOLATION OF THE VIRUS AND IDENTIFICATION

Skin biopsies of early lesions can be used for virus isolation and histopathological and electron microscopic studies. The samples aspirated from lymph nodes are suitable for virus isolation. For histopathology, necropsy samples such as lungs, trachea, spleen, and rumen containing gross lesions are suitable. The samples intended for use in virus isolation should be shipped under wet ice if the shipment takes 48 h; if the shipment takes beyond 48 h, they can be shipped under dry ice. For histopathological work, samples should be preserved in 10% buffered formalin and not to be frozen. Serum samples should be taken from both acute and chronic cases. Electron microscopy (EM) can be used to identify capripoxviruses from skin lesions, but it will not differentiate between SPPV, GTPV, and LSDV. Neither it differentiates CaPVs from orthopoxviruses (OPVs) except with the application of specific immunological staining. *In vivo*, SPPV, GTPV, and LSDV can be propagated in sheep, goat, and cattle (natural hosts), respectively. Propagation of capripoxviruses in chorioallantoic membrane of embryonated chicken eggs varies significantly. Capripoxviruses do not grow in any of the laboratory animals.

For *in vitro* capripoxvirus isolation, skin lesions and oral and nasal swabs are the most useful specimens.[25] Capripoxviruses grow in a variety of sheep, goat, and cattle cells[26] including skin tissue, testes, thyroid, kidney, embryonic lung, and muscle cells of sheep origin; testes and kidney tissues of goat kid origin; and kidney, testes, fetal lung, and muscle cells of calf origin.[1,8] Some are successful in propagating SPPV in a few primary cells originating from pig also.[8] Presently, primary lamb testes or primary lamb kidney cells are the most commonly used cells for isolation. Lamb testicular tissue provides large number of fibroblasts for the growth of CaPV. The cytopathic effects in primary cells include rounding, tract formation, plaque formation, ballooning, nuclear vacuolation, chromatic formation, loss of continuity of the monolayer, and appearance of basophilic or acidophilic intracytoplasmic inclusion bodies. Nevertheless, primary cells are besotted with several disadvantages such as being difficult to establish, cell lot variation, and contamination with extraneous agents. However, recently a secondary cell line such as OA3.Ts has been evaluated as a replacement for primary cells.[27] Owing to several difficulties in establishing primary cell cultures, researchers have shifted to continuous cell lines. Some of the capripoxviruses have been successfully propagated in KEM-La, BHK_{21}, Vero, and MDBK cell lines.[8] Capripoxvirus isolation can be confirmed by immunostaining using anti-CaPV serum,[28] but it is not possible to differentiate between SPPV, GTPV, and LSDV. However, it is easy to visualize the capripoxvirus plaques after immunostaining.[27]

55.5.2 DETECTION OF CaPV ANTIGEN OR CaPV-SPECIFIC ANTIBODY

The literature describes several tests for the detection of CaPV antigen–antibody systems including agar gel precipitation test (AGPT), counterimmunoelectrophoresis (CIE), latex agglutination test, reverse phase passive hemagglutination test, single radial immunodiffusion test (SRID), single radial hemolysis test, and fluorescent antibody technique (FAT). These tests have their merits and demerits in terms of sensitivity, specificity, rapidity, and ease.[8,19] Current gold standard method is the neutralization test for determining anti-CaPV antibodies. But the test is laborious, is slow, and requires handling of live virus. Handling of live virus is not permitted in disease-free countries and also during posteradication phases. Lamb testes and lamb kidney primary cultures yield inconsistent results (virus breaks through due to variable susceptibility of cell cultures), and therefore, it is preferred to use cell lines like Vero. Recently, in order to increase the rapidity of the neutralization test, a recombinant capripoxvirus that expresses green fluorescent protein (GFP) has been evaluated in a virus neutralization assay by which the test can be completed within 48 h instead of 144 h.[29]

Various formats of enzyme-linked immunosorbent assays (ELISAs) have been reported for the detection of CaPV antigen or antibody such as immune-capture ELISA,[30] whole virus–based indirect ELISA for the detection of antibodies against GTPV,[31] ELISA based on inactivated CaPV antigen,[32] and recombinant P32 protein–based indirect ELISAs.[33,34] The results of latter two tests have been relatively comparable in terms of sensitivity and specificity with that of the neutralization test. ELISA based on inactivated CaPV antigen is impractical in routine diagnostic setup as the production of antigen is laborious and needs biocontainment facility.[1] However, as of today, there are no thoroughly validated ELISA for the detection of antibodies against SPPV, GTPV, and LSDV. Hence, it would be valuable to identify stable immunodominant antigen(s) of CaPV, so as to develop a recombinant protein–based ELISA with sensitivity and specificity comparable to the neutralization test. Western blotting is a specific and sensitive assay, but is laborious and difficult to interpret.[35]

Both humoral and cell-mediated immunity play important roles in the protection against CaPV infection. However, the longevity of humoral immune response to CaPV is unclear, but it is likely to be lifelong as that of vaccinia virus.[36] The other researchable area is the development of tests to differentiate infection from vaccination (DIVA), which is an absolute requirement during the posteradication phase of CaPV infection.

55.5.3 DETECTION OF VIRUS-SPECIFIC NUCLEIC ACID

To overcome one or the other drawbacks of aforementioned tests, highly sensitive molecular biology techniques are employed to detect nucleic acids of several viruses including CaPV. Among the techniques, conventional polymerase chain reaction (PCR),[33,37,38] PCR/PCR-restriction fragment length polymorphism (RFLP) for differentiation of CaPVs,[33,39,40] real-time PCRs based on various CaPV genes,[25,38,41] and loop-mediated isothermal amplification (LAMP) assays[42] are available for the detection of CaPV genetic material. The advantages of real-time PCRs are the rapidity, real-time quantification of PCR amplicons, and detection of reaction inhibitors. Similarly, the LAMP assay is simple, economical, and sensitive and can be used in rural settings and less-equipped laboratories.[42] Although PCRs have clear advantages, the PCR results should be confirmed by at least one additional test. Also PCRs in the form of multiplex formats have been developed using species-specific primers for the detection of CaPVs.[43] Similarly, a duplex PCR has been available for simultaneous detection of genetic material of CaPV and Orf virus (parapoxvirus).[44] The ultimate confirmation can be made by sequencing of one or several genes of individual CaPVs.

55.6 TREATMENT AND PREVENTION

The affected animals are to be isolated and maintained in an isolation facility. The palliative treatment is indicated along with the balanced diet feeding. The debilitated animals can be administered with 10% glucose saline intravenously. The lesions should be washed with 1:100–1:10,000 potassium permanganate lotion, and the application of 1:10 boric acid/mild antiseptic or antibiotic ointment topically with parenteral antibiotic administration to prevent secondary bacterial complications is warranted. A couple of nucleoside analogues have been identified as being effective against various pox viruses and may be useful as therapeutic agents in the treatment of CaPV infections in the near future.[8] Certain plant extracts have been found to be inhibitory to CaPV replication in vitro.[45]

Preventive measures largely include (1) the restriction of movement of animals and their products, (2) the decontamination of animal products, premises, and equipment, (3) quarantine, (4) test and slaughter, and (5) vaccination. The test and slaughter policy may not be feasible in all the countries due to issues such as social, economic, ethical, political, and regulatory concerns. Therefore, vaccination is the economically viable strategy to control CaPV infections. Accordingly, there have been several types of vaccination strategies including ovination, seroclavelization, sensitized virus, and inactivated vaccines. But these methods of immunization have one or other drawbacks like generalization and spread of infection, short duration of immunity, short shelf life, clogging of needles and syringes in cold conditions, large dosage and difficulty in transportation, necrosis and abscess formation, and mortality. Therefore, the research has been shifted to the use of live attenuated capripoxvirus vaccines. The animals that recover from capripoxvirus infection mount both humoral and cell-mediated immunity. Currently, live attenuated vaccines are in use in capripox enzootic countries. The attenuation of field isolates is done by multiple passages in cell cultures or embryonated chicken eggs. In some countries, homologous vaccines are used. However, Kenyan sheep and goatpox vaccine (KSG-1) is a naturally attenuated strain and displays characteristics that are intermediate between SPPV, GTPV, and LSDV.[46] Some vaccines may be safe in one host and too virulent to be used in another host. Although, most of the live attenuated vaccines work well, there have been reports of vaccine break down, reversion, short duration of protection, postvaccinal reactions, and low level of antibody production. Therefore, there is a great demand for the production of improved vaccines.[11] The postvaccinal reactions are attributed to virulent and host range genes, as raveled by complete genome sequencing of CaPVs. The proteins responsible for virulence include kelch-like and ankyrin repeat proteins.[6] Accordingly, CaPV strains like *Djelfa SPPV* isolate[47] and other isolates[48] with reduced virulence (safe) in target species have been developed.

The other forms of vaccines available for capripox infections are combined vaccines. The concept of combined vaccines is not new and is very well practiced in human medicine but is limited in veterinary medicine. The advantages of such vaccines are (1) overcoming the constraints of multiple injections, (2) improving timely vaccination coverage, (3) being economical, (4) reducing the cost of extra healthcare visits, and (5) facilitating the addition of new vaccines into immunization programs.[49] The combined vaccines for capripox include inactivated sheeppox with anthrax and polyvalent clostridial infections (braxy, infectious enterotoxaemia, lamb dysentery, and malignant edema)[50]; live attenuated SPPV-RF and peste des petites ruminants [PPR (Sungri/96)] combined vaccine[51]; formaldehyde inactivated recombinant epsilon toxin adjuvanted with aluminum hydroxide in combination with live attenuated sheeppox vaccine[49]; goatpox and PPR combined[52,53]; and simultaneous administration of live attenuated goatpox, PPR, and Orf vaccines.[54] The other strategy possibly enhances the immune response as well as solid protection against GTPV, PPR, and Orf challenge since Orf virus and more specifically some of the proteins (B2L) coded by the Orf virus have adjuvant effect as identified recently.[55]

There have been some efforts in developing subunit vaccines for capripox infections. P32 protein is a major immunodominant structural protein in CaPV and it induces neutralizing antibodies. There have been efforts to express the P32 in prokaryotic systems and utilize it as an immunizing agent for protection against CaPV.[56] Capripoxviruses have been the attractive candidate among vectored vaccines due to their restricted host range and thermostability. Accordingly, CaPVs have been in application as vectors for a variety of viral pathogens such as rabies,[57] peste des petits ruminants,[58–61] rinderpest,[62–67] rift valley fever virus,[68] and bluetongue.[69] The target gene in CaPV for generation of recombinant vaccines is *thymidine kinase (TK)*.[70] Capripoxvirus infections can be controlled, eradicated, and eliminated as they do not have carrier state, are nonzoonotic, possess conserved genome, have single serotype, they are nonpersistent, and have limited host range and are known to provide solid immunity.

55.7 CONCLUSIONS AND FUTURE PERSPECTIVES

It is very important to validate the currently available tests including various formats of PCRs, real-time PCR, LAMP, recombinant protein, and monoclonal antibody–based

ELISAs for rapid and accurate diagnosis and serosurveillance. It is also desired to identify the immunodominant domains of CaPVs so as to develop recombinant antigen–based ELISAs. The tool that will be of immense value is the development of tests to differentiate between the vaccinated from infected (DIVA), which is a requirement during posteradication phase of the disease.[71] These tools will be useful in responding to an outbreak, monitoring CaPV infections in enzootic areas, and to study the epidemiology of the disease. As the complete genome sequence of CaPVs is available, the factors like kelch-like and ankyrin repeat proteins are identified to be involved in the virulence of the virus. It is possible to develop less virulent and much safer CaPV by knocking out genes responsible for virulence. The spread of capripox infections into a new geographical area is a great concern which potentially occurs due to trade of infected animals and their products and movement of vectors. Naive population suffers severe morbidity and mortality. Test and slaughter policy is the best choice in a disease-free country. But, in an enzootic country, vaccination is the economically viable strategy to combat CaPV infection. Global eradication of disease would enable reduction in animal sufferings, improve the economic status of an enzootic country, and prevent the risk of spread of infection to disease-free countries.

REFERENCES

1. Babiuk, S. et al., Capripoxviruses: An emerging worldwide threat to sheep, goats and cattle, *Transbound. Emerg. Dis.*, 55, 263, 2008.
2. Kitching, P., Capripoxviruses. In: Brown, C. and Torres, A. (eds.), *Foreign Animal Diseases*, 7th edn. Boca Raton, FL: United States Animal Health Association, p. 189, 2008.
3. Borrel, A., Epithelioses infectieuses et epitheliomas. *Ann. de l' Instutut. Pasteur. de Lille, Paris*, 17, 81, 1903.
4. International Committee on Taxonomy of Viruses. Andrew MQ King, Elliot Lefkowitz, Michael J. Adams, Eric B. Carstens (eds.), *Virus Taxonomy: Ninth Report of the International Committee on Taxonomy of Viruses, Family: Poxviridae*. San Diego, CA: Academic Press, p. 291, 2011.
5. Tulman, E.R. et al., Genome of lumpy skin disease virus, *J. Virol.*, 75, 7122, 2001.
6. Tulman, E.R. et al., The genomes of sheeppox and goatpox viruses, *J. Virol.*, 76, 6054, 2002.
7. Yeruham, I. et al., Economic and epidemiological aspects of an outbreak of sheeppox in a dairy sheep flock, *Vet. Rec.*, 160, 236, 2007.
8. Bhanuprakash, V. et al., The current status of sheeppox disease, *Comp. Immunol. Microbiol. Infect. Dis.*, 29, 27, 2006.
9. Woldemeskel, M. and Ashenafi, H., Study on skin diseases in sheep from northern Ethiopia, *Dtsch. Tierarztl. Wochenschr.*, 110, 20, 2003.
10. Diesel, A.M. et al., The epizootiology of lumpy skin disease in South Africa. *Proceedings of 14th International Veterinary Congress*, London, U.K., vol. 2, p. 492, 1949.
11. Hunter, P. and Wallace, D., Lumpy skin disease in southern Africa: A review of the disease and aspects of control, *J. S. Afr. Vet. Assoc.*, 72, 68, 2001.
12. Carn, V.M., Control of capripoxvirus infections, *Vaccine*, 11, 1275, 1993.
13. Bhanuprakash, V. et al., An epidemiological study of sheep pox infection in Karnataka, India, *Sci. Tech. Rev.*, 24, 909, 2005.
14. Kitching, R.P. and Carn, V.M., Sheep pox and goat pox, *Office International des Epizooties Manual of Diagnostic Tests and Vaccines for Terrestrial Animals (Mammals, Birds and Bees)*, Paris, France: OIE, 2004.
15. Domenech, J. et al., Regional and international approaches on prevention and control of animal transboundary and emerging diseases, *Ann. N Y Acad. Sci.*, 1081, 90, 2006.
16. Rweyemamu, M. et al., Emerging diseases of Africa and the Middle East, *Ann. N Y Acad. Sci.*, 916, 61, 2000.
17. Yeruham, I. et al., Spread of lumpy skin disease in Israeli dairy herds, *Vet. Rec.*, 137, 91, 1995.
18. Kitching, R.P. and Taylor, W.P., Clinical and antigenic relationship between isolates of sheep and goat pox viruses, *Trop. Anim. Health Prod.*, 17, 64, 1985.
19. Rao, T.V.S. and Bandyopadhyay, S.K., A comprehensive review of goat pox and sheep pox and their diagnosis, *Anim. Health Res. Rev.*, 1, 127, 2000.
20. Coetzer, J.A.W., Lumpy skin disease. In: Coetzer, J.A.W. and R.C. Tustin (eds.), *Infectious Diseases of Livestock*, 2nd ed., pp. 1268–1276, Oxford, U.K.: University Press Southern Africa, 2004.
21. Davies, F.G., Lumpy skin disease, an African capripox virus disease of cattle. *Br. Vet. J.*, 147, 489, 1991.
22. Tuppurainen, E.S.M. and Oura, C.A.L., Lumpy skin disease: An emerging threat to Europe, the middle east and Asia, *Transbound. Emerg. Dis.*, 59, 40, 2012.
23. Woods, J.A., Lumpy skin disease—A review, *Trop. Anim. Health Prod.*, 20, 11, 1988.
24. Carn, V.M. and Kitching, R.P., The clinical response of cattle experimentally infected with lumpy skin disease (Neethling) virus, *Arch. Virol.*, 140, 503, 1995.
25. Bowden, T.R. et al., Capripoxvirus tissue tropism and shedding: A quantitative study in experimentally infected sheep and goats, *Virology*, 371, 380, 2008.
26. Binepal, Y.S. et al., Alternative cell lines for the propagation of lumpy skin disease virus, *Onderstepoort J. Vet. Res.*, 68, 151, 2001.
27. Babiuk, S. et al., Evaluation of an ovine testes cell line (OA3. Ts) for use in the propagation and detection of capripox virus and development of immunostaining technique for viral plaque visualization, *J. Vet. Diag. Invest.*, 5, 486, 2007.
28. Gulbahar, M.Y. et al., Immunohistochemical evaluation of inflammatory infiltrate in the skin and lung of lambs naturally infected with sheeppox virus, *Vet. Pathol.*, 43, 67, 2006.
29. Wallace, D.B. et al., Improved method for the generation and selection of homogeneous lumpy skin disease virus (SA-Neethling) recombinants, *J. Virol. Methods*, 146, 52, 2007.
30. Rao, T.V. et al., Evaluation of immunocapture ELISA for diagnosis of goat pox, *Acta Virol.*, 41, 345, 1997.
31. Bhanuprakash, V. et al., Detection of goat pox antibodies: Comparative efficacy of indirect ELISA and counterimmunoelectrophoresis, *J. Appl. Anim. Res.*, 30, 177, 2006.
32. Babiuk, S. et al., Detection of antibodies against capripoxviruses using an inactivated sheeppox virus ELISA, *Transbound. Emerg. Dis.*, 56, 132, 2009.
33. Heine, H.G. et al., A capripoxvirus detection PCR and antibody ELISA based on the major antigen P32, the homolog of the vaccinia virus H3L gene, *J. Immunol. Methods*, 227, 187, 1999.
34. Bhanot, V. et al., Expression of P32 protein of goatpox virus in Pichia pastoris and its potential use as a diagnostic antigen in ELISA, *J. Virol. Methods*, 162, 251, 2009.

35. Chand, P., Kitching, R.P., and Black, D.N., Western blot analysis of virus-specific antibody responses to capripoxvirus and contagious pustular dermatitis infections in sheep, *Epidemiol. Infect.*, 113, 377, 1994.

36. Hammarlund, E. et al., Duration of antiviral immunity after smallpox vaccination, *Nature Med.*, 9, 1131, 2003.

37. Mangana-Vougiouka, O. et al., Sheep poxvirus identification from clinical specimens by PCR, cell culture, immunofluorescence and agar gel immunoprecipitation assay, *Mol. Cell. Probes*, 14, 305, 2000.

38. Balamurugan, V. et al., Comparative efficacy of conventional and TaqMan polymerase chain reaction assays in the detection of capripoxviruses from clinical samples, *J. Vet. Diag. Invest.* 21, 225, 2009.

39. Hosamani, M. et al., Differentiation of sheep pox and goat poxviruses by sequence analysis and PCR-RFLP of P32 gene, *Virus Genes*, 29, 73, 2004.

40. Lamien, C.E. et al., Use of the Capripoxvirus homologue of Vaccinia virus 30 kDa RNA polymerase subunit (RPO30) gene as a novel diagnostic and genotyping target: Development of a classical PCR method to differentiate Goat poxvirus from Sheep poxvirus, *Vet. Microbiol.*, 149, 30, 2011.

41. Balinsky, C.A. et al., Rapid preclinical detection of sheeppox virus by a real-time PCR assay, *J. Clin. Microbiol.*, 46, 438, 2008.

42. Das, A., Babiuk, S., and McIntosh, M.T., Development of a loop-mediated isothermal amplification assay for rapid detection of capripoxviruses, *J. Clin. Microbiol.*, 50, 1613, 2012.

43. Orlova, E.S. et al., Differentiation of capripoxvirus species and strains by polymerase chain reaction, *Mol. Biol.*, (*Mosk*), 40, 158, 2006.

44. Zheng, M. et al., A duplex PCR assay for simultaneous detection and differentiation of Capripoxvirus and Orf virus, *Mol. Cell. Probes*, 21, 276, 2007.

45. Bhanuprakash, V. et al., In vitro antiviral activity of plant extracts on goatpox virus replication, *Indian J. Exp. Biol.*, 46, 120, 2008.

46. Kitching, R.P., Hammond, J.M., and Taylor, W.P., A single vaccine for the control of capripox infection in sheep and goats, *Res. Vet. Sci.*, 42, 53, 1987.

47. Achour, H.A. et al., Comparative study of the immunizing ability of some attenuated strains of sheep pox virus and of a sensitizing vaccine, *Sci. Tech. Rev.*, 19, 773, 2000.

48. Balinsky, C.A. et al., Sheeppox virus kelch-like gene SPPV-019 affects virus virulence, *J. Virol.*, 81, 11392, 2007.

49. Chandran, D. et al., Development of a recombinant epsilon toxoid vaccine against enterotoxemia and its use as a combination vaccine with live attenuated sheep pox virus against enterotoxemia and sheeppox, *Clin. Vaccine Immunol.*, 17, 1013, 2010.

50. Kadymov, R.A., Combined immunization of sheep against anthrax, sheep pox and clostridial infections, *Vet. Moscow*, 2, 50, 1975.

51. Chaudhary, S.S. et al., A Vero cell derived combined vaccine against sheep pox and Peste des Petits ruminants for sheep, *Vaccine*, 27, 2548, 2009.

52. Matrencher, A., Zoyem, N., and Diallo, A., Experimental study of a mixed vaccine against Peste des petits ruminants and capripox infection in goats in northern Cameroon, *Small Ruminant Res.*, 26, 39, 1997.

53. Hosamani, M. et al., A bivalent vaccine against goatpox and Peste des Petits ruminants induces protective immune response in goats, *Vaccine*, 24, 6058, 2006.

54. Bhanuprakash, V. et al., Protective immune response against PPR, goat pox and ORF viruses challenge in goats vaccinated with triple vaccine containing the thermo stable PPR, highly attenuated goatpox and attenuated ORF viruses. *XIX National Conference on Recent Trends in Viral Diseases Problems and Management*, March 11–13, 2010, Department of Virology, Venkateswara University, Tirupati, India. Indian Virological Society, p. 22, 2010.

55. Johnston, S.A. and McGuire, M.J., Use of parapox B2L protein to modify immune responses to administered antigens. Patent # US2004/0213807 A1, Filed Date: May 28, 2004. Publication Date: October 28, 2004.

56. Carn, V.M. et al., Protection of goats against capripox using a subunit vaccine, *Vet. Rec.*, 135, 434, 1994.

57. Aspden, K. et al., Immunogenicity of a recombinant lumpy skin disease virus (Neethling vaccine strain) expressing the rabies virus glycoprotein in cattle, *Vaccine*, 20, 2693, 2002.

58. Diallo, A. et al., Goat immune response to capripox vaccine expressing the hemagglutinin protein of *peste des petits ruminants*, *Ann. N Y Acad. Sci., USA*, 969, 88, 2002.

59. Berhe, G. et al., Development of a dual recombinant vaccine to protect small ruminants against peste-des-petits-ruminants virus and capripoxvirus infections, *J. Virol.*, 77, 1571, 2003.

60. Chen, W. et al., Recombinant goat poxvirus expressing PPRV H protein, *Chin. J. Biotechnol.*, 25, 496, 2009.

61. Chen, W. et al., A goat poxvirus-vectored peste-des-petits-ruminants vaccine induces long-lasting neutralization antibody to high levels in goats and sheep, *Vaccine*, 28, 4742, 2010.

62. Romero, C.H. et al., Single capripoxvirus recombinant vaccine for the protection of cattle against rinderpest and lumpy skin disease, *Vaccine*, 1, 737, 1993.

63. Romero, C.H. et al., Protection of cattle against rinderpest and lumpy skin disease with a recombinant Capri poxvirus expressing the fusion protein gene of rinderpest virus, *Vet. Rec.*, 135, 152, 1994.

64. Romero, C.H. et al., Recombinant capripoxvirus expressing the hemagglutinin protein gene of rinderpest virus: Protection of cattle against rinderpest and lumpy skin disease viruses, *Virology*, 204, 425, 1994.

65. Romero, C.H. et al., Protection of goats against peste des petits ruminants with recombinant capripoxviruses expressing the fusion and haemagglutinin protein genes of rinderpest virus, *Vaccine*, 13, 36, 1995.

66. Ngichabe, C.K. et al., Trial of a capripoxvirus-rinderpest recombinant vaccine in African cattle, *Epidemiol. Infect.*, 118, 63, 1997.

67. Ngichabe, C.K. et al., Long term immunity in African cattle vaccinated with a recombinant capripox-rinderpest virus vaccine, *Epidemiol. Infect.*, 128, 343, 2002.

68. Wallace, D.B. et al., Protective immune responses induced by different recombinant vaccine regimes to Rift Valley fever, *Vaccine*, 24, 7181, 2006.

69. Wade-Evans, A.M. et al., Expression of the major core structural protein (VP7) of bluetongue virus, by a recombinant capripoxvirus, provides partial protection of sheep against a virulent heterotypic bluetongue virus challenge, *Virology*, 220, 227, 1996.

70. Wallace, D.B. and Viljoen, G.J., Importance of thymidine kinase activity for normal growth of lumpy skin disease virus (SA-Neethling), *Arch. Virol.*, 147, 659, 2002.

71. Bhanuprakash, V., Hosamani, M., and Singh, R.K., Prospects of control and eradication of capripox from the Indian subcontinent: A perspective, *Antiviral Res.*, 91, 225, 2011.

56 Classical Swine Fever Virus

Benjamin Lamp and Till Rümenapf

CONTENTS

56.1 INTRODUCTION

With an estimated yield of 1.17×10^{12} kg in 2012 (*Source*: FAO Food Outlook), pork provides, after fish, the second largest source of animal protein worldwide. Main producing nations are China (46%), the United States (9.5%), and Germany (5.2%). As pork production increasingly occurs in large pig holdings, infection with devastating pathogens is becoming an imminent risk. The "big" OIE listed, notifiable viral epidemics in swine are food and mouth disease (FMD), Aujeszky's disease (or pseudorabies), and classical swine fever (CSF). Of increasing importance are porcine reproductive and respiratory disease (PRRS) and African swine fever (ASF).

CSF, formerly called hog cholera, is a highly contagious and economically important disease of domestic and wild pigs. The causative agent of CSF is classical swine fever virus (CSFV). Infections with highly virulent CSFV strains cause high morbidity and mortality in the affected population, while infections with less virulent CSFV strains may result in a more chronic course of disease. The severity of CSF varies not only with the strain but also with the age, immune status, and breed of affected pigs. A poor reproductive or fattening performance together with an accumulation of secondary bacterial infections might be the sole symptom of CSF in older pigs. Hence, similarity of symptoms to other diseases can make it challenging to diagnose CSF with certainty.

Once spread worldwide, the CSFV has been eradicated from domesticated swine in most developed countries. At the time, CSFV is distributed in Asia, South and Central America, some Caribbean islands, and Madagascar, but not the African continent. In order to ensure the CSFV-free status, CSFV-free countries prohibit vaccination and establish trade restrictions for living animals and pork from affected countries. CSFV-free countries apply a stamping out policy together with emergency vaccinations in case of virus outbreaks. Reintroduction of CSFV in these countries may have

a disastrous impact not only on meat production but also on international trade. During the 1997/1998 CSFV outbreak, about 400 herds were affected by CSFV in the Netherlands and approximately 12 million pigs were killed, causing an economically loss of around $2.3 billion.[1]

CSFV is a subject of official control, because the virus has a high risk of spread from the laboratory. Facilities working with CSFV should meet the requirements for containment group 3 pathogens determined by the risk assessment outlined in Chapter 1.1.3[2] of the OIEs Terrestrial Manual (Biosafety and Biosecurity in the Veterinary Microbiology Laboratory and Animal Facilities).

56.2 CLASSIFICATION AND MORPHOLOGY

The CSFV belongs to a large family of positive-strand RNA viruses, the Flaviviridae. This family currently consists of the three genera *Flavivirus*, *Pestivirus*, and *Hepacivirus* and will include the tentative genus *Pegivirus*, which has not been formally endorsed by the *International Committee on Taxonomy of Viruses*.[3]

CSFV is a member of the genus *Pestivirus* (from Latin pestis, "plague"). Natural host of CSFV are all members of the family Suidae, including various wild boar species, and all CSFV isolates belong to a single serotype displaying minor antigenic strain-specific variability. CSFV is closely related to the ruminant pestiviruses bovine virus diarrhea virus (BVDV-1 and BVDV-2) and border disease virus (BDV), which also might affect swine. A novel porcine pestivirus, which emerged in an Australian pig farm, was termed Bungowannah virus.[4] The classification of Bungowannah virus is still ongoing.

CSFV particles are small, spherical enveloped virions with an approximate diameter of 40–60 nm. An electron-dense inner core structure of about 30 nm has been described.[5] Due to the lipid bilayer membrane, they are readily inactivated by organic solvents, detergents, or heat.[6] Like other pestiviruses, CSFV is remarkably resistant against low pH,[7] and an incubation at pH 3.0 for 15 min does not result in a significant loss of infectivity.[8] Virus stability is extended by low temperatures, protein-rich environment, and humidity. CSF virions sediment at 100 Svedberg units and were concentrated at 1.10–1.12 g/mL in sucrose gradients.[9] In addition to the RNA genome and the lipid envelope, four structural proteins have been identified: the capsid protein (core) and three envelope glycoproteins (E^{rns}, E1, and E2).[10] Although the exact structure of CSFV virions is not known up until now, it is accepted that core protein and the viral RNA build up a ribonucleoprotein. Due to free cysteines, the structural proteins E^{rns}, E1, and E2 have been detected in the virion as disulfide-linked heterodimers (E^{rns}–E2 and E1–E2) as well as in homodimeric (E2–E2 and E^{rns}–E^{rns}) forms.[10–14] Mature E1 and E2 have a C-terminal transmembrane-spanning domain; E^{rns} is attached to the virions' surface by hydrophobic interactions, but it is also secreted from infected cells as a soluble protein.[15] E1, E2, and E^{rns} are essential for virion assembly.

56.3 BIOLOGY AND EPIDEMIOLOGY

The CSFV genome is a single-stranded RNA molecule with positive polarity, which spans approximately 12.5 kb lacking a 5′ cap and a 3′ poly(A) tract.[16] It encodes a single open reading frame (ORF) flanked by 5′ and 3′ untranslated regions (UTRs), the first containing conserved regions forming an intraribosomal entry site (IRES).[17] Cap-independent translation of the CSFV ORF resulted in a single large polyprotein of approximately 3900 amino acids, which is co- and posttranslationally processed into mature proteins by virus and host-encoded proteases.[18–20] The order of the cleavage products in the pestiviral polyprotein is NH$_2$-N^{pro}-C-E^{rns}-E1-E2-p7-NS2-NS3-NS4A-NS4B-NS5A-NS5B-COOH.[21] The four structural proteins, which made up the virion (C, E^{rns}, E1, and E2), are encoded at the 5′ end of the genome following an N-terminal autoproteinase (N^{pro}).[22] Within the pestiviral polyprotein, solely E^{rns}, E1, and E2 are glycosylated. Host signal peptidase is believed to cleave at the core/E^{rns}, E1/E2, E2/p7, and p7/NS2 sites.[15] The cellular signal peptide peptidase further processes a transmembrane region of the pestiviral core protein generating the C-terminus of its mature form.[19] The rear two-third of the genome encode eight additional nonstructural proteins (NSs) including a small transmembrane-spanning protein, termed p7.[20] An autoprotease activity within NS2 together with a cellular protease cofactor termed JIV (J-domain protein interacting with viral protein) is responsible for the cleavage between NS2 and NS3.[23] An incomplete cleavage between NS2 and NS3 leads to an accumulation of uncleaved NS2–3 molecules. NS3 and its cofactor NS4A cleave the viral polyprotein between Leu/Ser and Leu/Ala at four sites generating mature NS3, NS4A, NS4B, NS5A, and NS5B.[24] The cis-cleavage between NS3 and NS4A occurs cotranslational, while transcleavage processing of NS4–5 is incomplete. The NS4A/B–5A/B precursor and products of partial processing (NS5A/B, NS4A/B, and NS4B/5A) dominate against the mature nonstructural

proteins in CSFV-infected cells.[25] Pestiviral RNAs encoding solely the proteins NS3, NS4A, NS4B, and NS5B flanked by the cis-acting 5′ and 3′ NTR are still capable of genome replication, thus constituting a functional replicase complex.[26] The nonstructural protein precursor NS2–3 is essential for the generation of infectious virions,[27] while NS3, 4A, 4B, 5A, and 5B constitute the viral replicase. The enzymatic functions in the replicase have been elucidated for NS3 (serine protease, NTPase, helicase[28]), NS4A (protease cofactor[24]), and NS5B (RNA-dependent RNA polymerase[29]). Little is known about the function of NS4B and NS5A. NS5A is a phosphorylated RNA-binding protein[30] and NS4B might represent an NTPase.[31]

A unique feature of all pestiviruses is the appearance of two different biotypes, which have been characterized in detail for BVDV. A noncytopathogenic (ncp) biotype induces no apparent morphological changes in infected tissue culture cells and a cytopathogenic (cp) one. Field viruses usually belong to the ncp biotype, which is able to establish persistent infections in tissue culture cells and in the natural hosts. Cp variants of CSFVs have rarely been isolated always representing defective interfering particles with helper virus-dependent subgenomic viral RNAs. A genetically engineered infectious cp CSFV strain, which contains cp BVDV-derived sequences (BVDV-1, strain Cp8), was highly attenuated in pigs compared to the parental ncp CSFV.[32]

While bovine CD46 has been functionally characterized as a receptor for BVDV-1 and BVDV-2,[33] the cellular receptor of CSFV is so far not recognized. Adaptive mutations within E^{rns} that occur in cell culture passages have been linked to a heparin-binding phenotype.[34] Recombinant E2 of CSFV can block infection of CSFV as well as BVDV, indicating a common cellular receptor for all pestiviruses.[35] Furthermore, E2 of ruminant pestiviruses has been shown to determine the species preferences within cell cultures cells.[36] CSFV-pseudotyped HIV particles containing E1 and E2 of CSFV were sufficient to enter porcine cells.[37] The entry of BVDV has been characterized as clathrin-dependent endocytosis and endosomal fusion,[8] which is also believed for CSFV. Little information is available on the assembly and release of pestiviruses. However, there is good evidence for an assembly at membranes of the endoplasmic reticulum, maturation of virions within intracellular vesicles, and cellular exocytosis.[14,38]

Mutational analyses, applying reverse genetic methods, revealed that several regions within the CSFV genome influence viral virulence. The attenuation of CSFVs was achieved by sequence modifications of the structural proteins core,[9] E^{rns},[39] E1,[40] and E2[41] by deletion or abrogation of molecular determinants, such as glycosylation. Also, the exact molecular function is still under investigation; E^{rns} harbors an RNase activity preventing induction of cellular interferon induction after CSFV infection. Accordingly, an abrogation of the RNase activity within E^{rns} by point mutations (codon 297 and 346) affected CSFV's virulence.[42] E^{rns} has been crystallized and the structure of this unusual

RNase is solved using x-ray diffraction.[43] In addition, determinants of viral virulence have been identified in the nonstructural proteins, especially in N[pro].[44] The interaction of N[pro] with mediators of the innate immune system has been elucidated. N[pro] limits type I interferon induction by a direct interaction with interferon regulatory factor 3 (IRF3) and IRF7 in plasmacytoid dendritic cells.[45,46] However, virulence of CSFV strains was not reduced by specific elimination of IRF3 degradation signals.[47]

CSFV can be divided into three phylogenetic groups with three or four subgroups (1.1, 1.2, 1.3, 1.4; 2.1, 2.2, 2.3; 3.1, 3.2, 3.3, 3.4). These groups and subgroups have distinct geographical distribution patterns, but all of the groups have been isolated in Asia.[48] Group 1 is mainly affecting South and Central America as well as the Caribbean Sea, group 2 is found in the European Union, and group 3 is endemic in Asia.[49] CSFV has been eradicated from the United States, Canada, New Zealand, Australia, and most of western and central Europe. The situation is complicated by the fact that CSFV is endemic in domestic swine populations of Asia, Africa, South, and Central America and in wild boar populations in many parts of the world including Asia and several countries within Europe.

56.4 CLINICAL FEATURES AND PATHOGENESIS

CSF is highly contagious and can naturally infect all members of the family Suidae.[5] Peccaries, both *Tayassu tajacu* and *Tayassu pecari*, are also susceptible for CSFV.[50] Transmission occurs usually by oral and nasal routes but may take place by contact of skin lesions with virus-contaminated fomites. The primary sites of virus replication are the tonsils, the regional lymph nodes, and, via the peripheral blood, bone marrow and spleen. Monocytes and macrophages are the primary target cells of the CSFV. The virus prevents apoptosis of its host cells and blocks the antiviral effect of INF alpha. CSFV induces severe lymphopenia and most of the B cells are depleted, making CSFV an immunosuppressive virus. The apoptosis of uninfected endothelia cells leads to severe hemorrhages, unless the underlying mechanisms remain unknown.

All tissues, secretions, and excretions of infected pigs (such as oronasal secretions, lacrimal secretions, urine, feces, semen, and blood) contain the infectious agent. Infected pigs are the only known reservoir of CSFV. Virus shedding after infection starts before onset of clinical signs, and usually occurs throughout the course of disease. Chronically or persistently infected pigs may shed CSFV over several months. Transmission of CSFV occurs mainly by the oronasal route following direct exposure to infected animals. Spreads from primary infected herds often follows feeding of uncooked contaminated garbage. CSFV infections can also emerge after genital transmission or artificial insemination. CSFV is also spread mechanically by different insects, birds, and other wild or domesticated animals. Airborne transmission has been controversially discussed. Airborne CSFV infection seems possible over very short distances (max. 1 km);

also, exact scientific proof is still missing.[51] Infected sows may give birth to persistently infected piglets, which show no seroconversion.

CSFV incubation period ranges over 2–14 days and depends mainly on the CSFV strains virulence. The route of infection, the dose of the infectious particles, and the age of inoculated pigs may also influence incubation time. Overall, the clinical signs of CSF substantially vary with the respective virus strain. Highly virulent strains of CSFV cause an acute disease, while an infection with low-virulence strains results in a high percentage of chronic, mild, or even asymptomatic infections. In the past, highly virulent strains of CSFV dominated, but nowadays, CSFV strains of moderate virulence are more prevalent, probably due to combat programs after CSFV diagnoses. Adult pigs are less likely to develop severe symptoms than weaner pigs or piglets. Although breed-specific differences have been obvious, no CSFV-resistant breed is known.

Acute CSF is a severe hemorrhagic fever-like disease. Clinical symptoms include high fever (in piglets above 40°C), conjunctivitis, weakness, drowsiness, anorexia, constipation followed by a severe diarrhea, neurological disorders, and hemorrhages. Pigs may show incoordination, a weaving or staggering gait, and develop a typical posterior paresis followed by convulsions as symptoms of a nonsuppurative meningoencephalomyelitis. Respiratory signs and production of yellow bile-containing vomit is frequently seen in CSF outbreaks. Abdomen, ears, and tail typically show a purple cyanotic discoloration. Hemorrhages may occur in the skin and all inner organs. A severe leukopenia is usually the first symptom after CSF onset.[5] In acute CSF, histopathological lesions of organs might be inconspicuous or absent. Gross-pathological findings vary between different CSFV strains but consistently include lymphoid atrophy and growth retardation. A macroscopically visible atrophy of tonsils, thymus, and ileal Peyer's patches might be indicative for CSF in postmortem examination.[52] Typical CSF lesions include swollen marbled lymph nodes, serosal hemorrhages, and splenic infarctions. The necrotic ulcers in the mucosa of the gastrointestinal tract (buttons) are frequently observed in subacute cases of CSF. Neither these histopathological findings are pathognomonic for CSF nor should CSFV be excluded as a differential diagnosis due to a lack of the typical pathological lesions. Virtually, all pigs with acute CSF die within 1–3 weeks after onset of clinical signs.[5]

Moderately virulent strains of CSFV may cause a subacute form of CSF, mostly seen in older pigs. The symptoms resemble the clinical symptoms of the acute form of CSF but are less prominent. Fever persists for 2–3 weeks and a higher percentage of pigs survive the infection and recover. However, pigs suffering from subacute CSF die within a month or recover. Chronic CSF occurs after infection with low-virulence strains of CSFV or in partially immune herds. Chronic disease starts as a general illness with symptoms like anorexia, depression, elevated temperatures, leukopenia, and periods of constipation or diarrhea. In contrast to acute or subacute CSFV, the affected pigs improve after several weeks.

Following a period of a relatively undisturbed general condition, recurrent symptoms occur. Pigs show intermittent fever, anorexia, constipation or diarrhea, wasting or stunted growth, alopecia, and skin lesions. Chronic CSF leads to general immunosuppression and concurrent infections are frequently seen. The symptoms of chronic CSF can recurrent over weeks or months and often affect only a few pigs from the infected herd. At the end, a chronic CSF is always fatal.[5]

Clinical signs of CSF may not become evident in the herd within the first 2–3 weeks after introduction of a low-virulence CSFV. In breeding herds, a poor reproductive performance can be the first (and solely) clinical sign of CSF. Affected sows abort or give birth to stillborn, mummified, malformed, or dead piglets. An infection of pregnant sows between day 50 and 70 of gestation might induce a persistent infection of the fetuses. Such piglets may be born with a congenital tremor as a sign of disorders in the central nervous system. There are rare cases of asymptomatic piglets, which are persistently infected with CSFV and also persistently viremic. These animals exhibit conspicuous clinical symptoms after several months and develop a late onset form of disease. Typical symptoms include depression, stunted growth, dermatitis, diarrhea, conjunctivitis, ataxia, and posterior paresis. Persistent infected pigs usually survive for no more than 6 months.

Identification of CSFV epidemics in wild boar population is very challenging, because carcasses of dead animals as a main alert sign of epidemics were rarely found in the wild. Deceased wild animals were quickly eliminated by the predators and carcasses eaten by scavengers. Principal symptoms of CSFV in postmortem examination of wild boar have been described as round skin lesions, resembling scabies, and intestine ulcers.

Bovine pestiviruses might induce clinical symptoms similar to CSFV.[53] Differential diagnoses to CSF should also include other viral diseases, such as African swine fever (ASF), porcine dermatitis and nephropathy syndrome (PDNS), postweaning multisystemic wasting syndrome (PMWS), and multiple septicemic conditions. Septicemia is frequently seen in pigs following bacterial infections including salmonellosis (e.g., *Salmonella choleraesuis*), erysipelas, pasteurellosis, actinobacillosis (e.g., *Actinobacillus suis*), and *Haemophilus parasuis*. But all these viruses and bacteria might also cause concurrent infections of an underlying CSF infection.

56.5   IDENTIFICATION AND DIAGNOSIS

CSFV diagnostic is regulated by local authorities in many parts of the world.[54] CSFV is listed by the OIE; suspected cases were examined by specialized reference laboratories and confirmed outbreaks from all over the world were reported.

CSF is diagnosed by the detection of CSFV, its antigens, or nucleic acids in blood or tissue samples. CSFV antigens are usually detected by immunofluorescence or immunohistochemistry testing of tissue samples[55] or antigen-capture enzyme-linked immunosorbent assays (Ag ELISAs) from serum samples.[56] Direct fluorescent antibody tests (FATs)

on cryostat sections of organs from CSFV-affected pigs are commonly used for the detection of CSFV antigen. A panel of monoclonal antibodies (MAbs) can be used to determine between CSFV or non-CSFV pestivirus antigens.[57] Ag ELISAs targeting E^{rns} are a useful tool for herd screening, but do not allow a safely testing on single animal level. It is possible to isolate CSFV in different porcine cell lines, including SK-6 and PK-15 cells. Virus isolation from appropriate samples displays a sensitivity of up to 95%, but a negative result does not prove the absence of CSFV. Noncultivable virus particles might occur in the serum due to the presence of neutralizing antibodies. Infected cell cultures were identified by indirect immunofluorescence or immunoperoxidase staining using established monoclonal antibodies or specific antisera.[58] Today, polymerase chain reaction following reverse transcription (RT-PCRs) and real-time RT-PCR assays (rt RT-PCR) targeting the conserved 5′-NTR are used as in-house tests in most laboratories.[59] Total RNA or viral RNA is extracted from tissue or blood samples to detect the CSFV genomes. As a severe complication in diagnostics, ruminant pestiviruses frequently infect pigs and can lead to false-positive results in CSFV RT-PCR.[60] However, other pestiviruses are differentiated from CSFV by several discriminative RT-PCR systems or by nucleotide sequencing of resulting PCR fragments.[61] RT-PCR can be fully automated and performed within several hours. RT-PCR is believed to be the most sensitive method for CSFV diagnosis and viral genomes can be detected even if the animals are fully recovered.

Neutralization tests or immunoperoxidase procedures that use monoclonal antibodies exhibit a high specificity in differentiating CSFV from other pestiviruses and are therefore the gold standard for CSFV diagnostic.[57,62] The infectious agent, its RNA, and antigens are detectable in animals that die from the infection until the time of death, whether this is during an acute phase, during a chronic infection, or throughout persistent infection. Pigs might recover from infection and only have a transient viremia. In the field, whole blood, serum, and leucocytes samples provide a valuable tool for the confirmation of clinical CSF suspicion on herd level. However, samples from tonsils, spleen, and other lymphoid tissues are acknowledged as superior material for the early diagnostic of CSF.[52]

CSFV serology is used worldwide for diagnosis and surveillance. CSFV-specific antibodies appear usually from 1 to 3 weeks after infection targeting E2, E^{rns}, and NS3.[63,64] Pigs that recover from CSFV are protected by neutralizing antibodies for several years or for their whole lifetime. Fluorescent antibody virus neutralization (FAVN) assays using CSFV and related ruminant pestiviruses (BVDV, BDV) are the gold standard for the detection and discrimination of CSFV-specific antibodies. Commercially available ELISA systems, for example, competitive or blocking ELISAs, that use monoclonal antibodies targeting E2[65] or E^{rns}[66] can differentiate CSFV-specific antibodies from antibodies against ruminant pestiviruses. ELISA systems that depend on NS3 react with antibodies against all pestiviruses and are not able to differentiate a CSFV-specific immune response. ELISA systems are inexpensive compared to serum neutralization and represent a rapid tool for mass screening during outbreaks or for CSFV surveillance in negative regions.[67] Congenitally infected pigs are immunotolerant against CSFV proteins, persistently infected, and negative in serologically testing. Tracking down these persistently infected pigs is solely possible using techniques for direct CSFV detection, such as antigen ELISA, RT-PCR, or virus isolation.

56.6 TREATMENT AND PREVENTION

There is no specific treatment for CSF, but antibiotic therapy might reduce accompanying bacterial coinfections. Disease control is managed by immunization with live attenuated vaccines (e.g., C-strain, GPR strain) in countries with endemic CSFV in domestic swine populations, like many parts of Asia and South America. Countries can only cope with CSFV in wild boar populations, if they use strict biosecurity on pig farms to reduce the risk of infection. CSFV-free countries protect their status applying quarantines, movement bans, and good surveillance for imports from regions at risk. During CSFV outbreaks, all confirmed cases should be culled and all contact animals have to be slaughtered or subjected to suppressing immunization. Stables should be thoroughly disinfected and cleaned. The pig farms within a radius of half a kilometer surrounding the infected farm have the highest risk of CSFV infection. However, CSFV can spread over long distances by animal transportation and vector-based dissemination, so a culling of all pigs in the surrounding territories has been frequently practiced. In the European Union, a protection zone (3 km radius) and a surveillance zone (10 km radius) are established around each outbreak to restrict CSFV spread and pig movements. Furthermore, epidemiological investigations to trace the source of CSFV and its possible spread are carried out.[49] In CSFV-free areas, an emergence vaccination should exclusively be applied as a tool to assist in controlling of outbreaks.

Biosecurity measures und steady CSF serology are a prerequisite for the eradication of CSFV. Access to pig farms should be restricted. Separate clothing, entrance disinfection, isolation facilities, and hygiene measures can help to avoid CSFV introduction in pig farms. Recurrent serologic sampling in domestic swine population and wild boar is necessary to monitor for the potential reintroduction of CSFV.[68] The application of subunit vaccines against CSFV in domestic swine populations allows a differentiation of vaccinated and infected animals (DIVAs) using compatible diagnostic ELISA systems. Established DIVA systems for CSFV apply an E2 subunit vaccine and E^{rns} antigen or E^{rns}-specific antibody-detecting ELISAs.[69] E^{rns} antigen ELISA assays are based on the blocking activity of serum antibodies on the binding of E^{rns}-specific monoclonal antibodies and E^{rns}.[70] These tests can be fully automatized for high sample throughput and use serum or plasma.

Controlling CSFV in wild boar populations is more complicated than in domestic swine populations. The disease might persist over a long period of time or eventually become endemic in regions with CSFV-infected wild boar populations. Strategies for the control of CSF in wild boar usually include oral vaccination with attenuated live vaccines, like the C-strain.[71] Wild animal disease control via oral vaccination is labor intensive and requires safe and effective vaccines favorable with DIVA characteristics.

56.7 CONCLUSIONS

Available modified live vaccines induce a complete protection against the clinical course of CSFV within one week lasting for more than 1 year. The established vaccine strains are generally considered as safe. Neither adverse effects of multiple vaccinations nor reversions to pathogenic phenotypes have been reported. A partial protection and block of horizontal transmission was even observed if natural infection takes place simultaneously with C-strain vaccination. The use for oral applications has been proven, enabling a control of CSFV in wild boar populations.

Until now, a DIVA is not possible using classical modified live vaccines. A system using the E2 subunit vaccine and E^{rns} or NS3 antibody ELISAs is an elegant solution offering robust protection and DIVA characteristics. It is even safer than the attenuated CSFV strains but lacks the very quick onset of protection and can only be used parenterally. Furthermore, vertical and horizontal transmissions of CSFV have been observed after challenge experiments with E2 subunit–vaccinated animals due to the antigenic variation of E2. Chimeric pestiviruses, which consist of a BVDV backbone with a swap to CSFV E2 (CP7_E2alf)[72] or E1 and E2 (CP7_E1E2alf),[73] are promising live marker vaccine candidates for DIVA application in combination with E^{rns}-based ELISAs.[74] CSFV E^{rns}-specific antibodies are detectable within 10 days after CSFV infection with high sensitivity and can be differentiated from BVDV E^{rns}-specific antibodies.

The use of antiviral drugs has to be taken into consideration for the combat of emergency scenarios and numerous compounds have been shown to exert antiviral activity. As one example, VP32947 inhibits pestiviral replication due to specific interference with the enzymatic activity of the RNA-dependent polymerase (RdRp, NS5B).[75] Most of them, however, were developed to target hepatitis C virus and pestiviruses were taken as models. Nevertheless, the application of antivirals to combat a CSF outbreak is very much restricted by costs. Treated animals will almost certainly be banned from human consumption and thus represent no gain over culling.

REFERENCES

1. Stegeman, A. et al. The 1997–1998 epidemic of classical swine fever in the Netherlands. *Vet Microbiol* 73, 183–196 (2000).
2. http://www.oie.int/international-standard-setting/terrestrial-manual/access-online/
3. Simmonds, P. et al. Family Flaviviridae. In King, A. (ed.) *Virus Taxonomy. Ninth Report of the International Committee on Taxonomy of Viruses* (Academic Press, San Diego, CA, 2011).
4. Kirkland, P.D. et al. Identification of a novel virus in pigs—Bungowannah virus: A possible new species of pestivirus. *Virus Res* 129, 26–34 (2007).

5. Moennig, V., Floegel-Niesmann, G., and Greiser-Wilke, I. Clinical signs and epidemiology of classical swine fever: A review of new knowledge. *Vet J* 165, 11–20 (2003).

6. Rümenapf, T. et al. Structural proteins of hog cholera virus expressed by vaccinia virus: Further characterization and induction of protective immunity. *J Virol* 65, 589–597 (1991).

7. Liess, B. Hog cholera. In Paul, E. and Gibbs, J. (eds.) *Virus diseases of food Animals: A World Geography of Epidemiology and Control*, pp. 627–650 (Academic Press, London, U.K., 1981).

8. Krey, T., Thiel, H.J., and Rumenapf, T. Acid-resistant bovine pestivirus requires activation for pH-triggered fusion during entry. *J Virol* 79, 4191–4200 (2005).

9. Riedel, C. et al. The core protein of classical swine fever virus is dispensable for virus propagation in vitro. *PLoS Pathog* 8, e1002598 (2012).

10. Thiel, H.J. et al. Hog cholera virus: Molecular composition of virions from a pestivirus. *J Virol* 65, 4705–4712 (1991).

11. Konig, M. et al. Classical swine fever virus: Independent induction of protective immunity by two structural glycoproteins. *J Virol* 69, 6479–6486 (1995).

12. Weiland, E. et al. A second envelope glycoprotein mediates neutralization of a pestivirus, hog cholera virus. *J Virol* 66, 3677–3682 (1992).

13. Weiland, E. et al. Pestivirus glycoprotein which induces neutralizing antibodies forms part of a disulfide-linked heterodimer. *J Virol* 64, 3563–3569 (1990).

14. Weiland, F. et al. Localization of pestiviral envelope proteins E(rns) and E2 at the cell surface and on isolated particles. *J Gen Virol* 80, 1157–1165 (1999).

15. Rümenapf, T. et al. Processing of the envelope glycoproteins of pestiviruses. *J Virol* 67, 3288–3294 (1993).

16. Brock, K.V., Deng, R., and Riblet, S.M. Nucleotide sequencing of 5′ and 3′ termini of bovine viral diarrhea virus by RNA ligation and PCR. *J Virol Methods* 38, 39–46 (1992).

17. Pestova, T.V. et al. A prokaryotic-like mode of cytoplasmic eukaryotic ribosome binding to the initiation codon during internal translation initiation of hepatitis C and classical swine fever virus RNAs. *Genes Dev* 12, 67–83 (1998).

18. Rümenapf, T. et al. Molecular characterization of hog cholera virus. *Arch Virol Suppl* 3, 7–18 (1991).

19. Heimann, M. et al. Core protein of pestiviruses is processed at the C terminus by signal peptide peptidase. *J Virol* 80, 1915–1921 (2006).

20. Elbers, K. et al. Processing in the pestivirus E2-NS2 region: Identification of proteins p7 and E2p7. *J Virol* 70, 4131–4135 (1996).

21. Collett, M.S. et al. Proteins encoded by bovine viral diarrhea virus: The genomic organization of a pestivirus. *Virology* 165, 200–208 (1988).

22. Stark, R. et al. Processing of pestivirus polyprotein: Cleavage site between autoprotease and nucleocapsid protein of classical swine fever virus. *J Virol* 67, 7088–7095 (1993).

23. Lackner, T., Thiel, H.J., and Tautz, N. Dissection of a viral autoprotease elucidates a function of a cellular chaperone in proteolysis. *Proc Natl Acad Sci USA* 103, 1510–1515 (2006).

24. Tautz, N. et al. Serine protease of pestiviruses: Determination of cleavage sites. *J Virol* 71, 5415–5422 (1997).

25. Lamp, B. et al. Biosynthesis of classical swine fever virus nonstructural proteins. *J Virol* 85, 3607–3620 (2011).

26. Tautz, N. et al. Pathogenesis of mucosal disease: A cytopathogenic pestivirus generated by an internal deletion. *J Virol* 68, 3289–3297 (1994).

27. Agapov, E.V. et al. Uncleaved NS2-3 is required for production of infectious bovine viral diarrhea virus. *J Virol* 78, 2414–2425 (2004).

28. Xu, J. et al. Bovine viral diarrhea virus NS3 serine proteinase: Polyprotein cleavage sites, cofactor requirements, and molecular model of an enzyme essential for pestivirus replication. *J Virol* 71, 5312–5322 (1997).

29. Choi, K.H. et al. The structure of the RNA-dependent RNA polymerase from bovine viral diarrhea virus establishes the role of GTP in de novo initiation. *Proc Natl Acad Sci USA* 101, 4425–4430 (2004).

30. Reed, K.E., Gorbalenya, A.E., and Rice, C.M. The NS5A/NS5 proteins of viruses from three genera of the family Flaviviridae are phosphorylated by associated serine/threonine kinases. *J Virol* 72, 6199–6206 (1998).

31. Gladue, D.P. et al. Identification of an NTPase motif in classical swine fever virus NS4B protein. *Virology* 411, 41–49 (2011).

32. Gallei, A. et al. Cytopathogenicity of classical swine fever virus correlates with attenuation in the natural host. *J Virol* 82, 9717–9729 (2008).

33. Maurer, K. et al. CD46 is a cellular receptor for bovine viral diarrhea virus. *J Virol* 78, 1792–1799 (2004).

34. Hulst, M.M., van Gennip, H.G., and Moormann, R.J. Passage of classical swine fever virus in cultured swine kidney cells selects virus variants that bind to heparan sulfate due to a single amino acid change in envelope protein E(rns). *J Virol* 74, 9553–9561 (2000).

35. Hulst, M.M. and Moormann, R.J. Inhibition of pestivirus infection in cell culture by envelope proteins E(rns) and E2 of classical swine fever virus: E(rns) and E2 interact with different receptors. *J Gen Virol* 78, 2779–2787 (1997).

36. Liang, D. et al. The envelope glycoprotein E2 is a determinant of cell culture tropism in ruminant pestiviruses. *J Gen Virol* 84, 1269–1274 (2003).

37. Wang, Z. et al. Characterization of classical swine fever virus entry by using pseudotyped viruses: E1 and E2 are sufficient to mediate viral entry. *Virology* 330, 332–341 (2004).

38. Durantel, D. et al. Study of the mechanism of antiviral action of iminosugar derivatives against bovine viral diarrhea virus. *J Virol* 75, 8987–8998 (2001).

39. Sainz, I.F. et al. Removal of a N-linked glycosylation site of classical swine fever virus strain Brescia Erns glycoprotein affects virulence in swine. *Virology* 370, 122–129 (2008).

40. Fernandez-Sainz, I. et al. Alteration of the N-linked glycosylation condition in E1 glycoprotein of classical swine fever virus strain Brescia alters virulence in swine. *Virology* 386, 210–216 (2009).

41. Risatti, G.R. et al. Identification of a novel virulence determinant within the E2 structural glycoprotein of classical swine fever virus. *Virology* 355, 94–101 (2006).

42. Meyers, G., Saalmuller, A., and Buttner, M. Mutations abrogating the RNase activity in glycoprotein E(rns) of the pestivirus classical swine fever virus lead to virus attenuation. *J Virol* 73, 10224–10235 (1999).

43. Krey, T. et al. Crystal structure of the pestivirus envelope glycoprotein E(rns) and mechanistic analysis of its ribonuclease activity. *Structure* 20, 862–873 (2012).

44. Mayer, D., Hofmann, M.A., and Tratschin, J.D. Attenuation of classical swine fever virus by deletion of the viral N(pro) gene. *Vaccine* 22, 317–328 (2004).

45. Bauhofer, O. et al. Classical swine fever virus Npro interacts with interferon regulatory factor 3 and induces its proteasomal degradation. *J Virol* 81, 3087–3096 (2007).

46. Fiebach, A.R. et al. Classical swine fever virus N(pro) limits type I interferon induction in plasmacytoid dendritic cells by interacting with interferon regulatory factor 7. *J Virol* 85, 8002–8011 (2011).

47. Ruggli, N. et al. Classical swine fever virus can remain virulent after specific elimination of the interferon regulatory factor 3-degrading function of Npro. *J Virol* 83, 817–829 (2009).

48. Postel, A. et al. Classical swine fever virus isolates from Cuba form a new subgenotype 1.4. *Vet Microbiol* 161, 334–338 (2013).

49. Paton, D.J. et al. Genetic typing of classical swine fever virus. *Vet Microbiol* 73, 137–157 (2000).

50. Vargas Teran, M., Calcagno Ferrat, N., and Lubroth, J. Situation of classical swine fever and the epidemiologic and ecologic aspects affecting its distribution in the American continent. *Ann N Y Acad Sci* 1026, 54–64 (2004).

51. Weesendorp, E., Stegeman, A., and Loeffen, W.L. Quantification of classical swine fever virus in aerosols originating from pigs infected with strains of high, moderate or low virulence. *Vet Microbiol* 135, 222–230 (2009).

52. Lohse, L., Nielsen, J., and Uttenthal, A. Early pathogenesis of classical swine fever virus (CSFV) strains in Danish pigs. *Vet Microbiol* 159, 327–336 (2012).

53. Terpstra, C. Epizootiology of hog cholera. In Liess, B. (ed.) *Classical Swine Fever and Related Viral Infections*, pp. 201–232 (Martinus Nijhoff, Boston, MA, 1988).

54. Anonymous. Commission decision of February 1, 2002 approving a Diagnostic Manual establishing diagnostic procedures, sampling methods and criteria for evaluation of the laboratory tests for the confirmation of classical swine fever (2002/106/EC). *Official Journal of the European Union* L39/71 (2002).

55. Terpstra, C., Bloemraad, M., and Gielkens, A.L. The neutralizing peroxidase-linked assay for detection of antibody against swine fever virus. *Vet Microbiol* 9, 113–120 (1984).

56. Dewulf, J. et al. Analytical performance of several classical swine fever laboratory diagnostic techniques on live animals for detection of infection. *J Virol Methods* 119, 137–143 (2004).

57. Edwards, S., Moennig, V., and Wensvoort, G. The development of an international reference panel of monoclonal antibodies for the differentiation of hog cholera virus from other pestiviruses. *Vet Microbiol* 29, 101–108 (1991).

58. DeSmit, A.J., Terpstra, C., and Wensvoort, G. Comparison of viral isolation methods from whole blood or blood components for early diagnosis of CSF. In *Report on the Annual Meeting of National Swine Fever Laboratories*, Vol. DGVI/5848/95, pp. 21–22 (Commission of the European Communities, Brussels, Belgium, 1994).

59. Hoffmann, B. et al. Validation of a real-time RT-PCR assay for sensitive and specific detection of classical swine fever. *J Virol Methods* 130, 36–44 (2005).

60. Hurtado, A. et al. Genetic diversity of ruminant pestiviruses from Spain. *Virus Res* 92, 67–73 (2003).

61. Risatti, G. et al. Diagnostic evaluation of a real-time reverse transcriptase PCR assay for detection of classical swine fever virus. *J Clin Microbiol* 43, 468–471 (2005).

62. Wensvoort, G. Topographical and functional mapping of epitopes on hog cholera virus with monoclonal antibodies. *J Gen Virol* 70, 2865–2876 (1989).

63. Colijn, E.O., Bloemraad, M., and Wensvoort, G. An improved ELISA for the detection of serum antibodies directed against classical swine fever virus. *Vet Microbiol* 59, 15–25 (1997).

64. Lin, M., Trottier, E., and Mallory, M. Enzyme-linked immunosorbent assay based on a chimeric antigen bearing antigenic regions of structural proteins Erns and E2 for serodiagnosis of classical swine fever virus infection. *Clin Diagn Lab Immunol* 12, 877–881 (2005).

65. van Rijn, P.A. et al. Classical swine fever virus (CSFV) envelope glycoprotein E2 containing one structural antigenic unit protects pigs from lethal CSFV challenge. *J Gen Virol* 77, 2737–2745 (1996).

66. Langedijk, J.P. et al. Enzyme-linked immunosorbent assay using a virus type-specific peptide based on a subdomain of envelope protein E(rns) for serologic diagnosis of pestivirus infections in swine. *J Clin Microbiol* 39, 906–912 (2001).

67. de Smit, A.J. et al. Laboratory experience during the classical swine fever virus epizootic in the Netherlands in 1997–1998. *Vet Microbiol* 73, 197–208 (2000).

68. Blome, S. et al. Assessment of classical swine fever diagnostics and vaccine performance. *Rev Sci Tech* 25, 1025–1038 (2006).

69. Depner, K.R. et al. Classical swine fever (CSF) marker vaccine. Trial II. Challenge study in pregnant sows. *Vet Microbiol* 83, 107–120 (2001).

70. Floegel-Niesmann, G. Classical swine fever (CSF) marker vaccine. Trial III. Evaluation of discriminatory ELISAs. *Vet Microbiol* 83, 121–136 (2001).

71. Kaden, V. et al. An update on safety studies on the attenuated "RIEMSER Schweinepestoralvakzine" for vaccination of wild boar against classical swine fever. *Vet Microbiol* 143, 133–138 (2010).

72. Gabriel, C. et al. Towards licensing of CP7_E2alf as marker vaccine against classical swine fever-duration of immunity. *Vaccine* 30, 2928–2936 (2012).

73. Reimann, I. et al. Characterization of a new chimeric marker vaccine candidate with a mutated antigenic E2-epitope. *Vet Microbiol* 142, 45–50 (2010).

74. Aebischer, A., Muller, M., and Hofmann, M.A. Two newly developed E(rns)-based ELISAs allow the differentiation of classical swine fever virus-infected from marker-vaccinated animals and the discrimination of pestivirus antibodies. *Vet Microbiol* 161, 274–285 (2013).

75. Baginski, S.G. et al. Mechanism of action of a pestivirus antiviral compound. *Proc Natl Acad Sci USA* 97, 7981–7986 (2000).

57 Foot-and-Mouth Disease Virus

M. Hutber, E. Pilipcinec, and J. Bires

CONTENTS

57.1 INTRODUCTION

Foot-and-mouth disease (FMD), caused by a single-stranded RNA virus of the Picornaviridae family, is the most economically significant disease for animal farming. While adult animals infected with FMD usually recover from the disease and mortality rates remain low, morbidity within a susceptible population is invariably high and farm production losses can become significant. There is little economic benefit in maintaining national herds that become nonproductive, although the level at which production is considered economically useful or viable varies for any given geographical region. Regions that export animal products have historically included developed countries which remain free of FMD, and the international bans that are imposed upon any countries which lose their disease-free status can significantly affect trade. Consequently, disease-free regions usually become proactive in controlling FMD epidemics through a variety of control measures: these include vaccination, the culling of infected herds and flocks, and various biosecurity measures.

Conversely regions where FMD is endemic are generally unable to achieve disease-free status due to persistently high levels of viral challenge against their national herds and flocks. For the countries within these regions, primary objectives focus upon local protection of farming systems that yield high levels of animal products, and protection is typically achieved through regular prophylactic vaccination programmes, alongside biosecurity measures. Many underdeveloped countries lie within geographical regions of endemic FMD.

Semiendemic regions have historically adopted the traditional control measures employed by disease-free countries, particularly where the opportunities exist for regaining disease-free status; during epidemics within semiendemic regions, multiple vaccination administrations tend to be favored above slaughter control. Many developing countries lie within regions of semiendemic FMD.

57.2 CLASSIFICATION AND MORPHOLOGY

FMD is a highly contagious disease of cloven-hoofed animals and one of the most difficult to control.[1] Susceptibles include cattle, sheep, pigs, goats, and domestic buffalo, and the disease is additionally found in wild ruminants such as deer. Infection is characterized by fever and vesicles on the mouth and feet, as well as the udders of lactating animals. FMD is caused by a variety of strains of aphthovirus, in the family Picornaviridae, and there are seven immunologically distinct serotypes: O, A, C, SAT1, SAT2, SAT3, and ASIA1. Animals recovering from (or vaccinated against) a strain of one serotype still remain susceptible to challenges from strains of another serotype. This was demonstrated in Czechoslovakia (1954–1975) where 21 years of cumulative annual vaccinal protection against serotypes O and A did not protect susceptibles against a challenge from serotype C (Figure 57.1).

Within each serotype, there are a large number of strains that show differing degrees of antigenic diversity. This is particularly evident within the A serotype, in which vaccines prepared from one strain of serotype A virus may provide almost negligible immunity against another strain of serotype A. The genome of the FMD virus contains a single strand of RNA (approximately 8.2 kb) and similar to other RNA viruses has a high mutation rate: in conjunction with a plasticity of the major neutralizing sites on its surface, this explains the high antigenic variability.[2]

57.3 CLINICAL FEATURES AND PATHOGENESIS

The clinical severity of FMD is dependent upon the strain of virus, as well as the infecting dose, the species, and innate susceptibility of the host. Clinical signs are most apparent in high-yielding dairy cattle and intensively reared pigs, producing lesions that can be severe and debilitating. In adult sheep and goats, the clinical signs are typically mild and

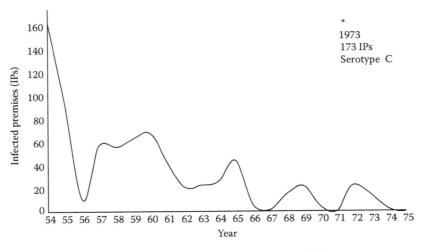

FIGURE 57.1 FMD prevalence among vaccinated livestock in Czechoslovakia (1954–1975).

transitory, such that they can be overlooked or mistaken by stockmen and veterinarians alike, even to the point of misdiagnosis.[3,4] Similarly, bovine FMD can also be misdiagnosed due to mild clinical signs.[5] The virus produces a high titer in epithelial cells, and lesions can usually be found on the hocks or elbows of pigs housed on hard flooring, where insults to legs are common. FMD virus will also destroy the replicating myocardial cells of young susceptible animals, producing high mortality through cardiomyopathy.[6,7]

The most common transmission route is via close contact spread between an infected and a susceptible animal, and large quantities of virus are excreted in exhaled breath. Cattle and sheep are particularly susceptible to infection by the aerosol route, requiring as little as 10 tissue culture infective doses 50 ($TCID_{50}$): with a larger respiratory volume, cattle are more susceptible to aerosol infection than sheep, while pigs are considerably less susceptible, requiring as much as 6000 $TCID_{50}$.[8] Nevertheless, all excretions and secretions from an infected animal will contain virus, so that infection may occur across damaged epithelium or via the aerosol route. This means that under favorable meteorological conditions, aerosol virus can potentially spread across considerable geographical distances, particularly where the source herd is porcine, and up to 3000 times more daily aerosol virus can be excreted during the acute stage of infection. In 1981, aerosol virus spread from pigs in Brittany (France) to cattle on the Island of Jersey and then onto the south coast of England at a total distance of 250 km.[9] Subsequently, preemptive culling was employed among pigs to prevent transmission across the south of England.

When an infected animal is slaughtered, the meat and organs will contain FMD virus, and if the carcass is frozen prior to rigor mortis, the virus will survive: should any of the infected products be fed to a susceptible species, for example in pigswill, an outbreak of FMD could occur.[10] For this reason, many countries retain strict regulations concerning the heat treatment of pigswill. FMD virus is particularly susceptible to inactivation outside a host through drying, high temperatures (>40°C in the field) or where the pH is <6 or >10. A carcass that is permitted to mature after slaughter (at 2°C for 24 h) produces a lactic acid buildup that will kill any virus in the meat through a reduction in pH to <6. The pH is not reduced in glandular tissue or bone marrow, although with additional safeguards it is still safe to import meat off the bone.[11] Milk from infectious animals also contains large quantities of live virus, and with inadequate heat treatment, calves and pigs can become infected. Semen from infected bulls and ova from infected cows may similarly be contaminated with live virus. Virus remains viable for days or weeks within a moist environment at neutral pH. Personnel handling infected animals may carry the live virus on hands, clothes, or in nasal passages, subsequently delivering infection to susceptible animals by close contact. Additional fomites include veterinary surgical instruments and artificial insemination equipment, all of which require suitable and timely cleansing and sterilization; outbreaks in Denmark (1982) and Italy (1993) were caused by fomite transmission. Milk tankers were also associated with the spread of virus during the 1967/1968 and 2001 UK epidemics, creating aerosols of virus-infected milk droplets during venting.[12] Recovered and vaccinated animals that have been exposed to a FMD challenge remain infected for a variable period of time. Cattle can carry the virus for over 3 years, while sheep may become carriers for 9 months and goats for 4 months: infection is carried in the epithelial cells of the pharynx, particularly the dorsal soft palate.[13,14] Carrier animals retain both a high level of neutralizing antibody within their sera and concurrently live virus in the pharynx, but the way in which the virus is protected from the host immune response is not understood.[15] Fifty percent of cattle and sheep exposed to a FMD challenge become disease carriers, where a carrier is defined by the recovery of live virus from the pharynx at 28+ days post infection. Oropharyngeal fluid from carrier animals can infect cattle and pigs experimentally, but to date, experimental transmission from carriers to susceptible animals has not been achieved.[16] Field evidence from Zimbabwe SAT2 outbreaks has indicated that carriers may initiate a new outbreak although difficulty in diagnosing carriers prohibits the international trade of recovered and vaccinated animals. This includes recovered porcine species that do not become carriers yet do carry FMD antibodies.[14,17]

57.4 IDENTIFICATION AND DIAGNOSIS

Diagnosis of FMD relies upon the identification and isolation of live FMD virus, although during an epidemic diagnosis may be based upon clinical signs alone or tracings to an infected premise (IP). There are a number of diagnostic tests that detect live virus, virus antigen, virus genome, and serological evidence for infection.[18] The perceived importance of the carrier animal has led to a development of tests that can distinguish between vaccination antibodies and infection antibodies.[14] FMD vaccine is produced from a suspension of whole virus, inactivated with aziridine and mixed with an oil or aluminum hydroxide/saponin adjuvant. Vaccinated animals produce no live virus and there is no expression of the eight viral nonstructural proteins (NSPs). Conversely, infected animals do replicate live virus and express the NSPs, producing antibodies to the structural proteins of the viral capsid as well as the NSPs. The presence of NSP antibodies to L, 2C, 3A, 3B, 3C, and 3D identifies a recovered animal. Although, the 3D NSP may contaminate the inactivated vaccine and yield antibodies, antibodies to 2C or the polyprotein 3ABC are reliable indicators of previous infection. These tests are not 100% sensitive in identifying vaccinated challenged carriers because some of these animals do not produce detectable antibodies to NSP, and yet they can carry FMD in the same manner as unvaccinated recovered animals. Until a test exists for carriers with full diagnostic sensitivity, then carriers will remain a constraint to trade.[14]

57.5 TREATMENT AND PREVENTION

The control of FMD within disease-free countries (United Kingdom, Australia, United States, etc.) is initially implemented by slaughter of all infected and susceptible in-contact animals, plus disinfection in and around IPs, intense surveillance for evidence of further spread, and a ban on the movement of susceptible animals.

Historically, some countries have employed disease control zones to regulate the national movement of livestock during periods of disease-free status. From the topography of the 1972–1973 epidemic in Czechoslovakia (Figure 57.2), it can be seen that IPs were restricted to two zones in the east of Slovakia and eight zones around Bratislava, while multiple clean zones were evidenced between the two foci. The demarcation of the two foci, and the restriction of infection to 10 out of 37 zones, is indicative of transmission abatement achieved through the use of disease control zones.

When an epidemic occurs within a disease-free region, a team of experienced veterinarians examines infected animals at the primary IP. Using estimations as to the age of disease lesions, and by tracing the recorded movements of animals on and off a farm, the team can advise as to the possible origin and potential spread of an epidemic.

A topical question during the UK 2001 FMD epidemic was whether or not to use vaccines as a supplementary or even primary means of disease control. A similar debate concerned the preemptive slaughter of in-contact herds geographically surrounding an IP and in combination with the slaughter of infected herds. The issues remain complex, with both advantages and disadvantages for each control method.

Disadvantages of Vaccination

1. Vaccinated animals are potential carriers of FMD.[19] While natural immunity produces subclinical infection among challenged sheep, vaccination suppresses clinical signs in cattle. Field data have confirmed that challenged vaccinated cattle with subclinical signs transmit FMD to other vaccinated animals and are therefore a greater threat to unvaccinated herds.[20] Consequently, regions vaccinating against FMD had been penalized in terms of international exports for a period of up to 12 months: more recently, respective ban lengths have been reduced to 3 months where vaccinated animals

B—Bratislava

FIGURE 57.2 Infected livestock within Slovakian movement control zones (1972–1973).

are culled and 6 months when vaccinated animals are permitted to live—vaccination inherently suppresses clinical signs, so subsequent slaughter control poses the smaller threat for exports.[21] Brownlie further debated the significance of a diagnostic test to distinguish between vaccinated animals and infected vaccinated animals and whether this would preserve the UK export status following vaccination.[21] Doel has stated that the likelihood of vaccine residual infectivity is low, such that unchallenged vaccinated animals are unlikely to become carriers.[1]

2. Vaccinated animals have been shown to succumb to rising challenge during the course of an outbreak. Therefore, vaccinated herds can still be vulnerable to high levels of challenge.[22]

3. Vaccines have been shown to be 81%–98% efficient, maintaining protective antibody titer levels for 2.5 months.[22] Hence, not all animals within a vaccinated herd are protected, and while current potent vaccines can curtail epidemics, they do not guarantee protection for any given herd. This in turn means that reinfection of vaccinated herds is possible, and waves of counterspread can occur between vaccinated herds.[23]

4. Implementation times for vaccine administration plus vaccine boost times to average herd protective levels (3–4 days) may be greater than the implementation times for preemptive slaughter. Hence, preemptive slaughter may require a shallower ring area for transmission abatement than ring vaccination.

Advantages of Vaccination

1. The same export ban (3 months) applies to a slaughter policy, as for a policy of vaccination with subsequent slaughter.[21] Animals can be vaccinated more rapidly than slaughtered (although, see 4 in *Disadvantages*), and the export ban could be preserved by culling the vaccinated animals at a later date.

2. A vaccination policy provides animal welfare advantages for the farming community (compared to preemptive slaughter control) and may consequently become a preferred option among the general public, albeit before an epidemic has reached its peak.

3. Initial slaughter of infected herds and flocks alone (rather than preemptive slaughter) during 2001 seemingly failed to reduce the effective contact rate (ECR) between herds to <1 and therefore did not appear to curtail the epidemic. However, this failure could have been inflated by subclinical transmission among sheep, where 75% of the animals culled on infected farms were sheep. A large number of infected animals that had failed to show clinical signs could have remained undiagnosed and perpetuated the epidemic. The factor that would thereby need to be addressed could be one of correctly diagnosing infected herds, rather than assuming

that the slaughter of infected herds alone was ineffective. It could thus be argued that the culling of infected herds did reduce the ECR <1, so long as a correct timely diagnosis was made, and the inter-herd transmission rate was not sufficiently high to require the implementation of any other control measure. Preemptive culling of in-contact herds or ring vaccination would, under these circumstances, be misdirecting control measures at clean herds. Moreover, Kitching has pointed out that among sheep, the infected undiagnosed flocks that were not culled or vaccinated could not sustain an adequate ECR to perpetuate the epidemic.[24] Therefore, it could be mooted that spatially increasing surveillance zones around IPs holding sheep would control disease spread effectively and without the need for vaccination or preemptive slaughter.

4. Implementation times for vaccine administration plus boost times to average herd protective levels (3–4 days) may be less than the implementation times for preemptive slaughter.[25] Under these conditions, preemptive slaughter would require a deeper ring area for transmission abatement than ring vaccination plus subsequent slaughter.

Advantages of Preemptive Slaughter

1. The curtailment of an epidemic requires the abatement of the ECR to <1, or put alternatively, the implementation speed of the control measures should exceed the transmission speed of the disease spread. Where speed = distance/time, then to control the epidemic, it becomes necessary to do the following:

   a. Either correctly estimate the required distance for disease control around each outbreak focus: overestimating the ring area for culling will result in clean herds and disease-free animals being slaughtered, while underestimating the appropriate ring area will enable the focus to spread radially outwards beyond the control ring. Or

   b. Decrease the implementation time for culling herds to a minimal level following initial diagnosis—the shorter the time for implementation, the smaller the required spatial ring for effective control around an IP or focus. This measure holds an advantage over 1a, in that decreasing the implementation time for the culling serves to reduce the required extent of the culling.

2. Preemptive slaughter of animals on premises that are contiguous with infected herds is equivalent to ring slaughter around a large epidemic focus. The difference between these two approaches principally lies in the scale of the control. The former divides an affected region into multiple foci and concentrates the control

towards places of maximum effect. A further advantage of the former approach is that the control remains within an affected region rather than at its boundaries. This therefore means that a higher level of protection is afforded to unaffected farms within a focus. Conversely, however, any significant level of misdiagnosis in the field will create a significantly higher level of culling directed at in-contact clean herds.

3. Preemptive culling reduces the number of herds within a region, diminishes the regional susceptible pool, and increases spatial barriers between farms. The overall effect of culling thus amounts to more than removing the viral challenge emerging from an infected farm.

4. While human error may occur both in the field and through inappropriate management decisions, the drastic nature of preemptive slaughter tends to ultimately move towards the eradication of an epidemic.

Disadvantages of Preemptive Slaughter

1. Timing of preemptive slaughter appears to be critical for its appropriate use. A tardy or lengthy implementation time allows established foci to expand, permits new foci to become established, and requires more extensive control measures to be adopted in the future.

2. FMD transmission via occult subclinical or preclinical undiagnosed animals can occur extensively before IP diagnosis, and culling preemptively around clinically affected herds and flocks would not necessarily remove subclinical disease. Hence, the diagnosis of infected animals and IPs should not always be based upon clinical signs alone.

3. The disease may have spread beyond the control area despite preemptive slaughter, particularly via wind-borne aerosol spread.[23] Under these circumstances, the culling of in-contact herds would become nonproductive.

4. The transmission rates for FMD vary according to both the infectious species excreting virus and the species of the in-contact susceptible animals—this adds additional variables when estimating the required zone for preemptive culling. A further variable develops as an epidemic progresses in that the level of subclinical disease can increase, particularly where vaccination programmes have been introduced as a control measure.

Slaughter control: A second question raised during the 2001 UK epidemic concerned the existence of a large FMD vaccine bank, held by the Institute for Animal Health, Pirbright. The vaccine bank was available to bank-member countries (United Kingdom, Eire, Finland, Malta, Norway, Sweden, Australia, New Zealand) at the onset of an epidemic. However, slaughter had been the preferred control option for most countries free of FMD, due to the complications of undiagnosed vaccinated carrier animals and subclinical disease produced by vaccination programmes.[15,19]

Cost-benefit analysis models have produced varied results with respect to the economic viability of slaughter compared to vaccinal control: some models have indicated that the cost of replacing culled herds is less than the cost of vaccination plus lost export revenues, while other models have economically favored vaccinal control, particularly in endemic areas.[26–31]

Regular vaccination: Vaccinal control using regular administrations is frequently used in areas where FMD is endemic and the prospects of control by slaughter are not good. Vaccination against FMD has been shown to be 81%–98% efficient with the most efficient vaccines providing the greatest rate of vaccinal boost.[22,32] While an efficacy of 81%–98% for vaccinated herds is not sufficient to protect a farm against every challenge, it is sufficiently high to limit outbreaks between farms and to ensure that an epidemic does not spread extensively across a region.

Rapid vaccination following the first diagnosed case on a farm (or postoutbreak vaccination [POV]) is as important for disease control as preoutbreak vaccination.[33] POV is relevant for herds that present a few animals with acute clinical signs at the onset of an outbreak. Waning vaccinal protection in these herds (due to a long period between vaccinations) fails to suppress any clinical signs in challenged animals, and hence, these signs are rapidly exhibited when the outbreak begins. Conversely, herds with strong vaccinal protection (due to a short period between vaccinations) suppress clinical signs in initially affected individuals, such that when clinical signs do eventually appear, a high number of animals exhibit them together. First-day incidence (FDI) in these herds is high and the initial clinical signs are mild.[20] Both POV and FDI are linearly correlated to prevalence or the final percentage of herd animals that become infected.

An entire herd on a given farm is usually vaccinated in order to provide suitable protection, due to the small spatial distances involved for possible transmission. A less effective way to use vaccines would be to partially vaccinate herds, for example, by selectively protecting young stock through vaccination, since young age groups exhibit a higher degree of mortality compared to older animals. However, this would not protect a herd against a highly contagious disease. The critical intervaccination period (CIP) for vaccinated herds is approximately 2.5 months, during which time the average antibody titer is maintained above the required level for protection but beyond which the titer levels for progressively more animals dip below protection.[22,34,35] The length of CIP, together with the timing of a vaccination programme, permits an assessment of the current immune status for any given herd: immune status is inherently an important parameter in monitoring the likely success of a viral challenge against a vaccinated herd. It is also noteworthy that revaccination sooner than 2.5 months appears to provide no additional advantage, in terms of reducing the final level of herd infection.

Strategic vaccination: The primary advantage of vaccination lies in the prophylactic protection offered to large numbers of susceptible animals. This is in contrast to the remedial action of removing infected or incubating animals by slaughter. A strategic barrier comprising vaccinated farms could be created around outbreak foci or could be used as an impediment to disease transmission into or out of a geographical region. For example, a band of vaccinated farms across the entrance to a peninsula could curtail spread into or from that area. A characteristic of vaccination programmes is that they remain in place during the course of an epidemic, acting as continual barriers to disease spread. Hence, areas of vaccinated herds when strategically placed could divide a region into protected subregions, thereby curtailing the spread of disease. Strategic herd vaccinations within a geographical region could significantly reduce the number of herds required to be vaccinated, and consequently, the percentage of the national herd that would require vaccination could be low.

An example of strategic vaccination was evident in Czechoslovakia from 1952 to 1974 (Table 57.1) where livestock were annually vaccinated within 20 km of the Czechoslovakian borders to the south and east of the country and within 10 km around cities, industrial centers, rendering plants, meat processing plants, slaughter houses, and holiday or recreational venues. In total, 600,000 cattle were vaccinated annually. During epidemics, pigs on suspected or confirmed IPs were culled, and livestock of susceptible species

TABLE 57.1
FMD Prevalence among Vaccinated Livestock in Czechoslovakia (1953–1975)

Year	No. of Infected Premises
1953	2264
1954	162
1955	92
1956	10
1957	58
1958	56
1959	63
1960	68
1961	42
1962	21
1963	23
1964	26
1965	44
1966	6
1967	1
1968	15
1969	22
1970	4
1971	1
1972	23
1973	173
1974	2
1975	0

TABLE 57.2
Prevaccination FMD Prevalence in Czechoslovakia (1938–1952)

Year	No. of Infected Premises
1938	212,173
1945	13,632
1952	40,114

were culled within protection zones around clusters of IPs. The cumulative effect of regular strategic vaccinations was to bring the number of annual IPs in Czechoslovakia down from 40,000 to 2 (Table 57.2).

From 1975 to 1991, blanket FMD vaccinations were administered annually across Czechoslovakia to all cattle, sheep, and goats, using bovine, ovine, and caprine Merial vaccines for serotypes A, O, and C. Merial porcine vaccines for serotypes A, O, and C were only administered to livestock within pig breeding farms. Czechoslovakia remained disease-free from FMD between 1975 and 1991. In 1991, annual blanket vaccinations were discontinued, and both the Czech and Slovak Republics have remained disease-free from 1991 to 2012.

Animal housing: The existence of natural and spatial barriers to FMD transmission demonstrates the possibility of strategic control and could be used when preemptive slaughter becomes inadequate. Hence, where geographical bottlenecks are exhibited within an affected region or topographic bottlenecks appear within a farm, these could be exploited to create divisions and subunits for abating transmission.[23] The creation of spatial barriers and hence divisions or isolating zones across a region (or within a farm) may be achieved by the use of animal housing. Such measures applied within herds not incubating the disease can protect farms and slow transmission between herds. Tinline concluded that animal housing would curtail FMD transmission for outbreaks in temperate climates.[36] Tinline's model simulations of the 1967–1968 UK epidemic predicted that ring-housing techniques would be more effective for the United Kingdom than ring vaccination programmes.

The implementation of animal housing at strategic points within regional geography or at appropriate bottlenecks within farm topographies, would not only decrease the probability of successful challenges against farm herds, but more importantly would decrease the level of virus leaving farms excreted by undiagnosed subclinical or incubating animals. Undiagnosed disease and infectious animals prior to slaughter become significant reasons for the continuation of an epidemic.

Biosecurity: Disease-preventive sanitary measures have been proposed for many years, but their effectiveness is largely dependent upon the vigilance applied to implementations at the farm level. Biosecurity measures at the farm level represent a source of self-assistance within areas of slaughter control and disseminate some of the administrative load during the course of an epidemic. In terms of sanitation, personnel movements are restricted both onto and within a farm, and recurrent disinfection of vehicles occurs alongside the use of disinfectants for personnel clothing, exterior and interior surfaces of housing, and feed coverings.[37] Biosecurity measures not only minimize the risk of infection reaching a farm but also reduce the likelihood of transmission across a farm to on-site susceptible animals: the control measures similarly reduce the risk of virus leaving a farm. Further measures restricting the level of virus excretion leaving farms include the division of animals into as many small housed groups as possible, with no intermixing of species—larger groups would permit the free intermixing of animals and FMD among pigs will heighten transmission. Farm buildings that are not used for the housing of susceptible animals have proven to be effective in curtailing FMD challenge.[23] Their positioning as physical barriers in the direction of a likely challenge will decrease the probability of herd infection. Using coverings or buildings for the storage of fodder and other feedstuffs will further reduce the likelihood of an outbreak. A feature of herd management that may minimize the transmission of FMD from wild animals to farmed species is the use of doubled boundary fencing around holdings. Such a measure provides a spatial barrier, where the effectiveness of the barrier is proportional to size.

57.6 BIOLOGY AND EPIDEMIOLOGY

During the 1967–1968 UK epidemic, slaughter control succeeded in controlling interherd transmission, but perhaps more importantly, an outbreak was not encountered within the United Kingdom for another 14 years.[38] Prior to 1967, FMD outbreaks had occurred more regularly within the United Kingdom. While the 1967–1968 UK epidemic was concentrated more towards the northwest of England, the 2001 UK epidemic rapidly exhibited a geographically diverse spread of outbreaks across the breadth and depth of the United Kingdom. For both epidemics, the spread of disease across the United Kingdom mirrored the spread of FMD within a single herd. The World Reference Laboratory for FMD collates outbreak studies worldwide and holds a large volume of data detailing spread across farms of up to 20,000 animals.[39] Similarities exist between the two UK epidemics and outbreaks for individual herds. Hence, a small number of source units (animals or herds) were initially infected, and this was followed by a period of very few (sometimes zero) units infected. Daily incidence from that point rose steadily to a peak and then declined, as predicted by an SIR (Susceptible-Infected-Recovered) model of spread.[40] It appears that UK livestock could be modeled as a single herd, with farms acting as units within the national herd. Both aerosol and contact spread (via livestock transport, fomites, or personnel) provided the means of disease transmission in a random fashion. Waves of spread were created by spatial barriers between farms and by natural geographical barriers or other curtailments, such that transmission was periodically but temporarily abated.[33,41] Counterwaves in different directions occurred when reinfection of susceptible units spread back across the same area.[23] These similarities

between intraherd and interherd spread allow administrators to gain insights into the epidemiology and control of disease within the national herd and flock, by examining disease spread at the intraherd level.

A useful method of assessing the relative merits and failings of a particular disease control policy (and whether or not to implement it) lies in the development of epidemiological models. The reliability of a model is dependent upon the quality of the data used for its construction and the use of appropriate construction methods: assumptions built into a model due to a lack of data will only diminish a model's usefulness. A model that has not been tested remains an untested construction and no one model should be relied upon in isolation from other models or proposals. If a model raises more questions than it answers, it is probably not fulfilling the important function of problem solving.

Various interherd models of FMD transmission have been based upon aerosol transmission.[9,42,43] These models are particularly useful in predicting the likely passage of challenging viral plumes in humid (RH >60%), cool, windy environmental conditions, with a high degree of cloud cover and reduced ultraviolet radiation. Under such circumstances, virus particles can survive for long periods and be carried for many miles, infecting farms in the direction of the plume(s) from source. However, in warmer, dry, sunnier, still conditions with outbreaks among sheep and cattle (rather than among high virus-excreting pigs), the aerosol models are less applicable. Under the latter conditions, a model of contact transmission becomes more relevant, where virus particles are carried between herds by the transport of livestock, fomites, or personnel and transport vehicles.[44,45]

One possible way of predicting the future course of an outbreak or epidemic across a farm or region (irrespective of meteorological conditions) lies in the monitoring of FDI and also first fortnight incidence (FFI). FDI is the initial number of animals with clinical signs or lesions or positive serological results on day 1 of an outbreak within a farm, while FFI is the number of infected farms or infected herds within a given region during the first fortnight of an epidemic.[20] FDI and FFI provide a mode of assessing the success of a viral challenge against a given farm or region and indicate the likely number of infectious animals (or herds) that have been excreting virus prior to culling. This in turn monitors the likely level of subsequent challenge against neighboring farms within a region, whether by aerosol or contact transmission. Theoretically, a high FDI (or high FFI) suggests a need for more rapid, extensive culling, while a low FDI (or low FFI) requires less extensive, preemptive slaughter. FDI and FFI are available at the start of an outbreak or epidemic and are indicative of how much virus has reached a given area as the disease spreads across a farm or region. FDI and FFI are correlated to herd and regional prevalence, respectively, whenever subclinical disease is indicated by the presentation of mild clinical signs. Alternative modeling parameters are available where the clinical signs are more acute.[20] FDI and FFI are accurate predictive modeling parameters for regions implementing slaughter and/or vaccinal control.

57.7 CONCLUSIONS AND FUTURE PERSPECTIVES

The control measures that a country uses to abate an epidemic of FMD (including vaccination) are to an extent predetermined by its epidemiological macroclimate. Changes to macroclimates require the coordination of long-term global disease control measures, and macroclimates can inherently determine where prophylactic vaccines are used, according to the following generality: endemic > semiendemic > disease-free. Accurate predictive models and cost-benefit analyses can indicate when, during the course of an epidemic, vaccines should or should not be used as a control measure against FMD. The decisions are usually economically based. Models that are not biologically accurate should not be used.

For endemic regions, CIP will determine when revaccination is used on a regular basis, while in disease-free regions, FFI will indicate when an epidemic is likely to be temporally lengthy and geographically widespread or the prevalence high, and therefore, vaccination should be considered. Semiendemic regions have been successfully protected against FMD challenges through annual blanket prophylaxis as well as annual strategic ring vaccinations.[46]

Modeling parameters such as FDI and FFI, alongside aerosol models and microclimatic conditions, can be used to determine how blanket, ring, and targeted vaccinations should or should not be used, either in prophylaxis or under emergency conditions. Models that are not biologically accurate should not be used.

Further conclusions can be drawn from the literature for various microclimatic factors relating to FMD.[46] In disease-free regions, emergency ring and targeted vaccination programmes that have successfully abated FMD outbreaks or epidemics have required concurrent culling of IPs. Vaccination to live offers welfare benefits while vaccination to cull does not. Ring and targeted vaccination probably offers no benefit above slaughter control in temporal or economic terms. Emergency blanket vaccinations have been shown to abate FMD epidemics, but any temporal or economic benefits over slaughter control remain controversial. Emergency blanket vaccinations may require multiple administrations in approaching maximal vaccinal immunity to abate epidemics. FMD vaccines have efficacy <100%. The likelihood of a successful vaccination programme diminishes with higher challenges, poorly chosen vaccines, extensively farmed livestock, and prolonged epidemics (with heightened levels of circulating virus). Regions demonstrating a change in disease status from semiendemic to disease-free have not been shown to require regular vaccination to maintain disease-free status.

In semiendemic regions, annual prophylactic ring and targeted vaccination programmes require multiple administrations and concurrent IP and preemptive culling to abate FMD epidemics. Annual prophylactic blanket and ring vaccinations of cattle, sheep, and goats can consistently prevent FMD epidemics in semiendemic regions. Annual prophylactic blanket vaccinations of cattle, sheep, and goats have not been shown to produce subclinical occult FMD.

In endemic regions, the CIP of 2.5 months is required to prevent FMD epidemics using prophylactic blanket vaccination. Regular (CIP < 4 months) prophylactic blanket revaccination can produce subclinical occult bovine FMD in 47% of herds.

Supplementary controls such as animal housing and topographical countrywide zoning can be used to abate disease transmission. They may or may not be practicable to implement for any given region.

ACKNOWLEDGMENT

The authors would wish to thank Dr. Hlinka (Stake Veterinary Administration, Slovakia) for his assistance in obtaining empirical data relating to FMD in Czechoslovakia.

REFERENCES

1. Doel TR (1996). Natural and vaccine-induced immunity to foot-and-mouth disease: The prospects for improved vaccines. *Revue Scientifique et Technique, Office International des Épizooties* 15(3), 883–911.
2. Domingo E, Martinez-Salas E, and Sobrino F (1985). The quasispecies (extremely heterogeneous) nature of viral RNA genome populations: Biological relevance—A review. *Gene* 40, 1–8.
3. De la Rue R, Watkins GH, and Watson PJ (2001). Idiopathic mouth ulcers in sheep (letter). *Veterinary Record* 149, 30–31.
4. Watson P (2002). The differential diagnosis of FMD in sheep in the UK in 2001. *State Veterinary Journal* 12, 20–24.
5. Kitching RP (2002). Clinical variation in foot and mouth disease: Cattle. *Revue Scientifique et Technique, Office International des Épizooties* 21, 499–504.
6. Kitching RP and Hughes GJ (2002). Clinical variation in foot and mouth disease: Sheep and goats. *Revue Scientifique et Technique, Office International des Épizooties* 21, 505–512.
7. Kitching RP and Alexandersen S (2002). Clinical variation in foot and mouth disease: Pigs. *Revue Scientifique et Technique, Office International des Épizooties* 21, 513–518.
8. Alexandersen S, Brotherhood L, and Donaldson AI (2002). Natural aerosol transmission of foot-and-mouth disease virus to pigs: Minimal infectious dose for strain O1 Lausanne. *Epidemiology and Infection* 128, 301–312.
9. Donaldson AL, Gloster J, Harvey LD, and Deans DH (1982). Use of prediction models to forecast and analyse airborne spread during the foot-and-mouth disease outbreaks in Brittany, Jersey and the Isle of Wight in 1981. *Veterinary Record* 110, 51–67.
10. Kitching RP (1998). A recent history of foot-and-mouth disease. *Journal of Comparative Pathology* 118, 89–108.
11. OIE (2001). *International Animal Health Code: Foot and Mouth Disease*, 10th ed. Office International des Épizooties, Paris, France, pp. 63–75 (Chapter 2.1.1).
12. Gibbens JC, Sharpe CE, Wilesmith JW et al. (2001). Descriptive epidemiology of the 2001 foot-and-mouth disease epidemic in Great Britain: The first five months. *Veterinary Record* 149, 729–743.
13. Zhang ZD and Kitching RP (2001). The localisation of persistent foot-and-mouth disease virus in the epithelial cells of the soft palate and pharynx. *Journal of Comparative Pathology* 124, 89–94.

14. Kitching RP (2002). Identification of foot and mouth disease virus carrier and subclinically infected animals and differentiation from vaccinated animals. *Revue Scientifique et Technique, Office International des Épizooties* 21, 531–538.

15. Salt JS (1993). The carrier state in foot-and-mouth disease, an immunological review. *British Veterinary Journal* 149, 207–223.

16. Van Bekkum JG (1973). The carrier state and foot and mouth disease. In: Pollard M (ed.), *Proceedings of the Second International Conference on Foot-and-Mouth Disease*. Gustav Stern Foundation Inc., New York, pp. 45–50.

17. Thomson GR (1996). The role of carriers in the transmission of foot and mouth disease. In: *Conference Proceedings of the 64th General Session of Office International des Épizooties*, Paris, France, pp. 87–103.

18. OIE (2001). *Manual of Standards for Diagnostic Tests and Vaccines: Foot and Mouth Disease*, 4th ed. Office International des Épizooties, Paris, France, pp. 77–92 (Chapter 2.1.1).

19. Kitching RP (1992). The excretion of foot-and-mouth disease virus by vaccinated animals. *State Veterinary Journal* 2, 7–11.

20. Hutber M, Kitching RP, and Conway DA (1999). Predicting the level of herd infection for outbreaks of foot-and-mouth disease in vaccinated herds. *Epidemiology and Infection* 122, 539–544.

21. Browlie J (2001). Strategic decisions to evaluate before implementing a vaccine programme in the face of a foot-and-mouth disease (FMD) outbreak. *Veterinary Record* 148, 358–360.

22. Hutber M, Kitching RP, and Conway DA (1998). Control of foot-and-mouth disease through vaccination and the removal of infected animals. *Tropical Animal Health and Production* 30(4), 217–227.

23. Hutber M and Kitching RP (2000). The role of management segregations in controlling foot-and-mouth disease. *Tropical Animal Health and Production* 32(5), 285–294.

24. Kitching RP (2001). SVS meeting memorandum April 20th 2001. *Sunday Times*, April 29, 2001.

25. Woolhouse M and Donaldson AI (2001). Managing foot-and-mouth. *Nature* 410, 515–516.

26. Carpenter TE and Thieme A (1979). A simulation modelling approach to measuring the economic effects of foot and mouth disease in beef and dairy cattle. In: *Proceedings of the Second International Symposium on Veterinary Epidemiology and Economics*, Canberra, Australia, pp. 511–516.

27. Chema S (1975). Vaccination as a method of foot-and-mouth disease control. An appraisal of the success achieved in Kenya, 1968–1973. *Bulletin de l'Office International des Épizooties* 83, 195–209.

28. Dijkhuizen AA (1996). Epidemiological and economic evaluation of foot-and-mouth disease control strategies in the Netherlands. *Netherlands Journal of Agricultural Science* 37, 1–12.

29. Garner MG and Lack MB (1995). An evaluation of alternate control strategies for foot-and-mouth disease in Australia: A regional approach. *Preventive Veterinary Medicine* 23, 9–32.

30. Power AP and Harris SA (1973). A cost benefit evaluation of alternative control policies for foot-and-mouth disease in Great Britain. *Journal of Agricultural Economics* 14(3), 573–579.

31. Thieme A (1982). Modelling the cost and benefits of foot and mouth disease control programs. In: *Third International Symposium on Veterinary Epidemiology and Economics*, Arlington, Virginia, USA, September 6–10, 1982. Veterinary Medicine Publishing, Edwardsville, IL, pp. 384–391.

32. Hingley PJ (1985). Problems in modelling responses of animals to foot-and-mouth disease vaccine. PhD thesis, Department of Applied Statistics, University of Reading, Reading, U.K.

33. Hutber M and Kitching RP (1996). The use of vector transition in the modelling of intra-herd foot-and-mouth disease. *Environmental and Ecological Statistics* 3, 245–255.

34. Hafez SM (1990). Studies on the control of foot-and-mouth disease in Saudi dairy farms. Report I, Ministry of Agriculture and Water, and King Abdul-Aziz City for Science and Technology, Riyadh, Saudi Arabia.

35. Hafez SM (1991). Studies on the control of foot-and-mouth disease in Saudi dairy farms. Report II, Ministry of Agriculture and Water, and King Abdul-Aziz City for Science and Technology, Riyadh, Saudi Arabia.

36. Tinline RR (1972). A simulation study of the 1967–68 FMD epizootic in Great Britain. PhD thesis, Bristol University, Bristol, U.K.

37. Ondrasovic M, Ondrasovicova O, Vargova M, and Sokol J (1994). *Animal Hygiene*. University of Veterinary Medicine, Kosice, Slovakia.

38. Haydon DT, Woolhouse MEJ, and Kitching RP (1997). An analysis of foot-and-mouth disease epidemics in the UK. *IMA Journal of Mathematics Applied in Medicine and Biology* 14, 1–9.

39. Hutber M (1997). Modelling the spread and maintenance of foot-and-mouth disease in a dairy herd. PhD thesis, The Business School, University of Hertfordshire, Hertfordshire, U.K.

40. Andersen RM and May RM (1991). *Infectious Diseases of Humans: Dynamics and Control*. Oxford University Press, Oxford, U.K.

41. Sanson RL (1992). The development of a decision support system for an animal disease emergency. PhD thesis, Massey University, Massey, New Zealand.

42. Gloster J, Blackall RM, Sellers RF, and Donaldson AI (1981). Forecasting the airborne spread of foot-and-mouth disease. *Veterinary Record* 108, 370–374.

43. Pech RP and McIlroy JC (1990). A model of the velocity of advance of foot and mouth disease in feral pigs. *Journal of Applied Ecology* 27, 635–650.

44. Hutber M, Kitching RP, and Pilipcinec E (2006). Predictions for the timing and use of culling or vaccination during a foot-and-mouth disease epidemic. *Research in Veterinary Science* 81(1), 31–36.

45. Hugh-Jones ME (1976). A simulation spatial model of the spread of foot and mouth disease through the primary movement of milk. *Journal of Hygiene* 77, 1–9.

46. Hutber M, Kitching RP, Fishwick J, and Bires J (2011). Foot-and-mouth disease: The question of implementing vaccinal control during an epidemic. *The Veterinary Journal* 188(1), 18–23.

58 Lumpy Skin Disease Virus

Estelle H. Venter

CONTENTS

58.1 INTRODUCTION

Lumpy skin disease (LSD) is an acute, subacute, or inapparent viral disease of cattle of all breeds.[1] It is characterized by fever, enlarged lymph nodes, and the formation of multiple firm, circumscribed nodules in the skin of affected animals and necrotic plaques in the mucous membranes mainly of the upper respiratory tract and oral cavity as well as generalized lymphadenopathy.[2] The disease is more severe in young animals, cows in the peak of lactation, and those individuals suffering from various stress factors, such as infectious and parasitic diseases or extreme weather conditions.[1]

The disease was listed in the Office International des Epizooties (OIE) "List A," which identifies diseases with the potential for rapid spread and ability to cause severe economic losses until the inception of a single disease list in 2006. The disease is included in the list of Notifiable Diseases of the World Organization of Animal Health (OIE). Criteria emphasized for inclusion include abortions, infertility problems in bulls and cows, loss of body mass, and a sharp drop in milk yield, which may cause considerable economic losses to farmers during outbreaks. Losses due to decreased milk production alone were estimated to be up to 40%–65% in an intensive milk-producing unit of 3000 cattle during an LSD outbreak in 2010 in the Sultanate of Oman.[3] The ulceration of nodular pox lesions leaves permanent scars that decrease the value of the hides.[4] Temporary loss of draft power may also occur, which remains important in rural regions.[5] In addition to direct losses due to the clinical disease, indirect costs caused by restrictions or a total ban on the international trade of live animals or animal products following an outbreak can be substantial. It acts as a hindrance to the introduction of new, high-productive breeds of cattle to endemic countries, and costs of control and eradication programs of LSD add to the financial burden.

The disease is caused by the lumpy skin disease virus (LSDV), a poxvirus with morbidity averaging 10% and mortality 1% in affected herds, although mortality rates over 75% have been recorded.[6] The reasons for the wide ranges in mortality following infection with LSDV are currently unknown but could be attributed to numerous factors that include the cattle breed, virus isolate, secondary bacterial infections, state of health of the animal, as well as the type of insect vector involved in transmission.[7]

The disease was mainly confined to Africa with only a few countries affected outside Africa, like in Israel as reported by Abraham and Zissman[8] and in the Middle East in Bahrain, Kuwait, Oman, and the United Arab Emirates.[9]

58.2 CLASSIFICATION AND MORPHOLOGY

The prototype strain of the LSDV is the Neethling strain, isolated in South Africa in 1944.[10] It is a large (300 nm) pleomorphic, double-stranded, unsegmented DNA virus that is classified in the *Capripoxvirus* genus within the subfamily Chordopoxvirinae (vertebrate poxviruses) of the family Poxviridae. It has only one serotype and is closely related to goatpox virus (GTPV) and sheeppox virus (SPPV), the only other members of the genus *Capripoxvirus*.[11] The virion contains numerous antigens most of which are shared by all the members of the same genus even though every species have their own specific polypeptides.[12,13] DNA analysis using restriction endonucleases on both field samples and vaccine strains showed 80% homology between strains of capripoxviruses.[14] The antigenic variation of field isolates of LSDV has not been reported.[15]

Mature LSDV particles have an oval profile and a larger lateral body than, for example, orthopox virions. Their average size is 320×260 nm.[12,13] Poxvirions are brick shaped or oval. Within the virion, there are over 100 polypeptides, which are arranged in a core, two lateral bodies, a membrane, and an envelope. The membrane and envelope are important structures for the interaction with the host cell. Mature virions that are released from the cell without cell disruption are enveloped. The envelope contains two layers of cellular lipids and several virus-specific polypeptides. Most of the virions released by the rupture of the host cell are therefore not enveloped. Both enveloped and nonenveloped virions are infectious.[12,13] The outer membrane is a lipoprotein bilayer that protects the core and lateral bodies. It has irregular arrangements of tubular protein called "filaments." The core is dumbbell shaped, and there are two lateral bodies of unknown nature. The core of the viruses contains proteins that include a transcriptase and several other enzymes.[12,13]

The 151 kbp genome of LSDV consists of a central coding region bounded by identical 2.4 kbp inverted terminal repeats and contains 156 putative genes. Genes of the central genomic region share a high degree of colinearity and amino acid identity (average of 65%) with genes of other known mammalian poxviruses, particularly suipoxvirus, yatapoxvirus, and leporipoxvirus. In terminal regions, colinearity is disrupted and poxvirus homologues are either absent or share a lower percentage of amino acid identity (average of 43%). Although LSDV is closely related to other members of the Chordopoxvirinae, it contains a unique complement of genes responsible for viral host range and virulence.[16]

The LSDV exhibits a remarkable degree of resistance to a range of environment conditions. Weiss[1] reported that the virus can survive for at least 33 days in skin lesions. More recently, Tuppurainen et al.,[17] using nucleic acid detection and virus isolation (VI) techniques in blood and skin of experimentally infected cattle, succeeded in isolating LSDV from skin lesions as long as 39 days postinfection and detected viral DNA in skin biopsies for up to 92 days postinfection. The LSDV survives for 2 h in a water bath at 55°C. It is phenol sensitive (2% for 15 min) and is susceptible to highly alkaline or acid pH solutions, for example, ether (20%), chloroform, and formalin (1%), and can be inactivated by sodium dodecyl sulfate (10%).[1]

58.3 BIOLOGY AND EPIDEMIOLOGY

A new skin disease of cattle was identified in northern Rhodesia (now Zambia) in 1929. Because the etiology of the disease was not known, it was thought to be caused by either an allergic reaction to insect bites[18] (Morris, 1931 cited by Weiss[1]) or plant poisoning (Le Roux, 1945 cited by Weiss[1]). At first, the disease was referred to as "pseudourticaria" or "lumpy disease." In October 1943, another outbreak of the disease occurred in Ngamiland, Bechuanaland Protectorate (Botswana) where it had not previously been described and was provisionally called "Ngamiland cattle disease."[19,20] In 1944, the disease spread to southern Rhodesia (Zimbabwe) (Houston, 1945, cited by Coetzer[2]) and to South Africa where it first appeared in the Transvaal (now divided into Gauteng, North West, and Limpopo provinces) (Thomas et al. 1945,

cited by Coetzer[22]). This economically devastating epidemic lasted until 1949, affecting some eight million cattle.[19,21]

The disease was now named "knopvelsiekte" or LSD.[21] The virus has spread throughout sub-Saharan Africa and its epidemiology is characterized by periodic outbreaks.[22,23] It was diagnosed in East Africa in Kenya in 1957,[24] Sudan in 1971,[25] Chad and Niger in 1973, Nigeria in 1974,[26] and Somalia in 1983. It has been reported in most countries in Africa and Madagascar.[12,23] In May 1988, the first outbreak occurred in Egypt in Ismailia, and despite all the control and eradication measures taken, the disease has become endemic.[27] The first outbreak in Egypt were reported by Ali et al.[27] and others and were confirmed by viral isolation, and this was reported by House et al.[28]. After 17 years of absence, a recent outbreak of LSD was reported in Egypt in 2006 by Babiuk[29] and El-Kholy.[30]

The first outbreak of LSD in Israel occurred in 1989. It was suggested that the disease spread from the Egyptian outbreak via insect vectors carried by wind or inside the vehicles of cattle merchants.[31] Ring vaccination program in a radius of 50 km of the outbreak, the slaughtering of all infected and in-contact cattle, sheep, and goats, together with the restriction of cattle movements eradicated the disease from Israel.[32] In June 2006, during the outbreak in Egypt, cases of LSD were again reported in Israel. At this stage, the Israeli authorities speculated that LSDV may have already been circulating in other Middle Eastern countries.[32]

Outbreaks or isolated cases of LSD have occurred in Bahrain in 1993, 1994, and 2003, the United Arab Emirates in 2000, Lebanon in 1993, Yemen in 1995, Kuwait in 1991, and Oman in 1984 and 2010.[3,33] Recent outbreaks (since 2006) in the Middle East occurred in Bahrain, Kuwait, Oman, and the United Arab Emirates.[3,9] The presence of LSDV in Saudi Arabia was never confirmed. There is a documented case of capripox infection of a captive-bred Arabian oryx (*Oryx leucoryx*) in Saudi Arabia.[34] This report, and other similar ones in the Middle East, has been classified as unconfirmed due to lack of VI.[2,29,35] This inability to isolate the virus and the possibility to cross-reactions between capripoxviruses raised concerns as to whether the Arabian oryx infection was due to GTPV or SPPV. In the past, capripoxvirus outbreaks have also been reported to be endemic in sheep and goats in Oman and Yemen.[36]

LSD occurs at regular intervals in most parts of South Africa. The epidemics of 1989/1990 and 2000/2001 in South Africa and most other countries in southern Africa were particularly severe and affected large numbers of cattle. Unfortunately, there are no accurate statistics on these epidemics.[22] More recently, LSD outbreaks were reported in 2010 in Eastern Cape, Mpumalanga, Limpopo, Free State, Gauteng, Western Cape, and North West Provinces of South Africa.[37]

LSD is most prevalent in the wet season or summer and along water courses,[25,38,39] although outbreaks may occur in the dry season.[20,39] In endemic areas, LSD has a tendency to disappear during the cold months, so-called overwintering of the disease.[22,23,40]

LSDV appears in areas free of SPPV and GTPV viruses, and it is not clear why SPPV and GTPV have not spread, for example, to the southern part of Africa. One speculation is that the ability of LSDV to provide heterologous cross-protection against SPPV and GTPV limits the spread of these diseases to the southern part of Africa. However, this is not consistent with the coexistence of LSDV, SPPV, and GTPV in countries such as Kenya.[29]

The spread of capripoxvirus into new areas is predominantly associated with the increase of animal movement through trade[41] as well as inadequate or breakdown of veterinary services.[42]

58.3.1 Host Specificity

With a few exceptions, capripoxviruses are highly host specific. LSD causes clinical disease only in cattle of all age groups. In Africa, imported *Bos taurus* breeds appear to be more susceptible than indigenous *Bos indicus* cattle.[22,23] However, even among groups of cattle of the same breed kept together under the same conditions, there is a large variation in the clinical reaction, which ranges from subclinical infection to death.[43] The failure of the virus to infect the whole herd probably depends on the virulence of the virus, immunological status of the host, and vector prevalence. While cattle are the definitive hosts, LSDV has been associated with an outbreak of capripox infection in Kenyan sheep[44] that was later confirmed by Kitching et al.[45]

The evidence of infection has been reported in some wildlife species but no clinical disease has been observed in nature. Antibodies to LSDV have been demonstrated in Arabian oryx (*Oryx leucoryx*)[34,46] in Saudi Arabia, Asian water buffalo (*Bubalis bubalis*) in Egypt,[16,27] black wildebeest (*Connochaetes gnou*), blue wildebeest (*Connochaetes taurinus*), eland (*Taurotragus oryx*), giraffe (*Giraffa camelopardalis*), impala (*Aepyceros melampus*), greater kudu (*Tragelaphus strepsiceros*), reedbuck (*Redunca arundinum*), springbok (*Antidorcas marsupialis*), and waterbuck (*Kobus ellipsiprymnus*) in Africa.[47–49] Giraffe and impala were proven to also be highly susceptible by experimental infection.[47] However, these authors did not observe clinical signs or viremia in two experimentally infected African buffalo (*Syncerus caffer*) calves. These negative results, in a very small population of African buffaloes, correspond to results later obtained by Hedger and Hamblin[48] and unpublished data of Howell and Coetzer cited by Coetzer.[2] However, seropositive African buffaloes have been detected in an LSDV-endemic area in Kenya[40] and the Kruger National Park, South Africa.[50] The experimental infection of LSDV in sheep and goats only causes a local erythematous swelling at the site of inoculation.[1]

These studies do not confirm that wildlife species play a significant role in the spread or maintenance of LSDV, and no strong evidence of a wildlife reservoir for capripoxviruses exists. Studies by Hosamani and coworkers[51] indicate that molecular markers for host specificity of capripoxviruses can be determined if sufficient virus strains and genes are analyzed.

58.3.2 Transmission

Direct contact between infected and susceptible animals is an inefficient route of infection of LSD.[22] It was demonstrated that under experimental conditions the disease was neither transmitted nor was immunity generated in naïve animals housed with infected animals in the absence of suitable insect vectors.[52] Thomas and coworkers[21] were the first to transmit the infectious agent by inoculating cattle with a suspension of skin nodules. In experimentally infected animals, only 50% are likely to develop clinical signs.[1,17,53]

It is currently believed that the most important mode of transmission of LSDV is likely to be through mechanical transmission of the virus by blood-feeding vectors.[1,54,55] Mechanical transmission, mainly by biting flies (*Stomoxys*, *Aedes*, and *Culex*), has been suspected to be responsible for disease transmission especially as the disease outbreaks are mostly seen during the rainy seasons and along water courses where there is increased insect activity.[1,20,31,52] Weather changes such as cold spells may adversely affect insect vector populations and thus reduce LSDV transmission,[22,40] although outbreaks during the dry season and winter months were also reported.[20,39]

The results of transmission studies by insects are not clear. Female *Aedes aegypti* and *Stomoxys calcitrans* have been demonstrated to have the ability to successfully transmit LSD,[54–57] and experimentally infected stable flies (*Stomoxys* spp.) can mechanically transmit capripoxvirus between sheep.[58] In other studies, however, LSDV transmission from infected to susceptible cattle using mosquitoes (*Anopheles stephensi*), stable flies (*Stomoxys calcitrans*), and biting midges (*Culicoides nubeculosus*) was not achieved.[57] Live LSDV has also been isolated from stable flies *Stomoxys calcitrans* and *Biomyia fasciata* after feeding on infected cattle as reported by Du Toit and Weiss in 1960.[1]

The variations in results may be due to low levels of viremia in the blood of infected animals that contribute to the inefficient transmission of LSDV by biting flies feeding on blood alone.[43]

These authors suggested that biting flies have to feed on skin lesions to obtain enough virus for transmission to take place.

The role of ticks in the transmission of the virus has been suspected[59] with some outbreaks being associated with the presence of ticks, especially *Amblyomma* spp., on affected animals.[25] Recently, Tuppurainen and coworkers[60] reported the potential role of ixodid ticks in the transmission of LSDV. The transovarial transmission of LSDV by *Rhipicephalus* (*Boophilus*) *decoloratus*, the mechanical or intrastadial transmission by *Rhipicephalus appendiculatus*, and mechanical/intrastadial and transstadial transmission by *Amblyomma hebraeum* have now been reported.[61,62]

Weiss[1] found that the virus persisted in skin lesions for more than 38 days, and more recently, viral DNA was demonstrated (by polymerase chain reaction [PCR]) in skin lesions for more than 90 days.[17] Infected saliva may contribute toward the spread of the disease,[20] but the disease is rarely transmitted to suckling calves through infected milk.[1] Shedding in semen may be prolonged; viral DNA (PCR) has been found in the semen of some bulls for at least 5 months after infection.[63] However, the role of semen in the transmission of the virus is not clear.[17,53,64,65] In a recent study, Annandale and coworkers (2012) showed that seminal transmission of LSDV in cattle is possible. Although the transmission can occur under experimental conditions, it has yet to be demonstrated whether this occurs naturally or when using semen containing amounts of virus typical of that shed by bulls. Although it was demonstrated that immunization prevents shedding of LSDV in semen,[66] it is not clear if the standard stepwise washing of embryos will successfully remove the virus.[67]

Animals can be infected experimentally by inoculation with material from skin nodules or blood or by ingestion of feed and water contaminated with saliva. The virus is very resistant to inactivation and can survive in desiccated crusts for up to 35 days.[1]

A carrier state in animals for LSD has not been demonstrated, and it has not been shown where the virus resides during the time of minimal or no insect activity (overwintering). The virus is maintained in forests or at forest edge situations, for example, in Kenya, in a fairly high rainfall zone at 1000–2500 mm.[40] The vertebrate hosts that might support the virus are wild buffaloes and domestic cattle populations. Subclinical infections of cattle might assist in the maintenance and the sporadic clinical cases that occur in most years and are thought to be accidental involvement in the maintenance cycle, in which it might not be necessary for any arthropod vector to be involved. Davies[40] was, however, of the opinion that circumstantial evidence has led to conclusions that an arthropod vector had to be involved in the rapid spread of the disease.[6,24]

The impact of global climate change and the evidence that insect vectors play a role in the transmission of LSDV (and speculated on as possible for SPPV and GTPV because of the very high virus concentrations in the skin) confirm the real risks of the establishment of LSD in the Middle East and Asia, as well as the further spread of the disease into other geographic regions, that is, Europe.

58.4 CLINICAL FEATURES AND PATHOGENESIS

LSDV causes an acute, subacute, or inapparent disease in cattle of all ages.[1,22,43] Reports of the clinical signs of both natural and experimentally infected animals have been reported.[1,2,10,20–22,43,63,65,68–71]

The duration of the natural incubation period is not well known, and 1 to 4 weeks was reported by Haig.[20] A biphasic febrile response occurs in experimentally infected animals from 2 to 4 weeks after exposure to the virus and animals remain febrile for 4 to 14 days postinfection.[20,43] Tuppurainen and coworkers[17] reported the incubation period in experimentally infected animals to be from 4 to 5 days and the viremic period could be determined to be from 1 to 12 days postinfection using VI and from 4 to 11 days postinfection

using PCR. The duration of the viremic period also did not correlate with the severity of the clinical disease. Carn and Kitching[43] experimentally infected cattle and observed generalized lesions 9 to 14 days postinfection and concluded that the development of generalized lesions did not seem to be dose related.

The morbidity rate varies widely, depending on the presence of insect vectors and host susceptibility, and ranges from 3% to 85% with an average of 10%. During the outbreak in Israel in 2006, up to 41.3% morbidity rates were reported.[32] The mortality rate is low in most cases (1%–3%) but has been as high as 40% in some outbreaks.[2] Unusually high mortality rates of 75%–85% have been reported,[6] but this remains unexplained.

Clinical disease is characterized by a biphasic febrile reaction that can reach 41°C. This may persist for 4–14 days. Clinical signs during this stage include disinclination to move, inappetence, salivation, lachrymation, and nasal discharge, which may be mucoid or mucopurulent. Lachrymation may be followed by conjunctivitis and, in some cases, by corneal opacity and blindness. The superficial lymph nodes, especially prescapular, precrural, and subparotid are usually markedly enlarged.[1,2,20,22,43,70,72]

Skin nodules are classical manifestations of LSD and have been well described. These nodules are usually widespread and may include the genitalia, udder, perineum, vulva, ears, legs, limbs, and skin around the head and neck.[1,2,17,22,70,73] Nodules on the skin and mucous membranes are usually developed within 2 days after onset of the febrile reaction. These nodules vary from 1 to 7 cm in diameter, and necrotic skin lesions may extend from the dermis and hypodermis into the surrounding tissues (Figure 58.1). Ulcerative lesions may appear on the conjunctiva, muzzle, nostrils, buccal mucosa, larynx, trachea, and abomasums (Figure 58.2). The lesions in the respiratory tract may result in primary or secondary pneumonia. Small nodules may resolve spontaneously, without any consequences, or may become ulcerated and sequestered. Although the nodules may exude serum initially, they develop a characteristic inverted conical zone of necrosis, which penetrates the epidermis and dermis, subcutaneous tissue, and sometimes the underlying muscle. These cores of necrotic material become separated from the adjacent skin and are called "sitfasts." Secondary bacterial infections are common within the necrotic cores. The scars may remain indefinitely thus rendering the hide worthless.[1,2,4,20,21,68,70,71,73,74]

The superficial lymph nodes become enlarged and edematous (Figure 58.1). In addition, feed intake decreases in affected cattle, milk yield can drop markedly, and animals may become emaciated. Inflammation and necrosis of the tendons or severe edema of the brisket and legs can result in lameness. Secondary bacterial infections can cause permanent damage to the tendons, joints, teats, and mammary glands. Bulls usually become temporarily infertile, but sometimes because of severe orchitis they may become permanently sterile. Pregnant cows may abort and

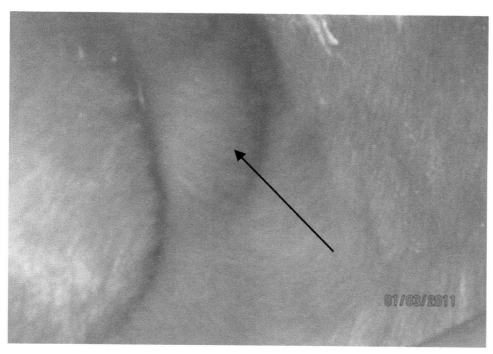

FIGURE 58.1   Enlarged lymph node.

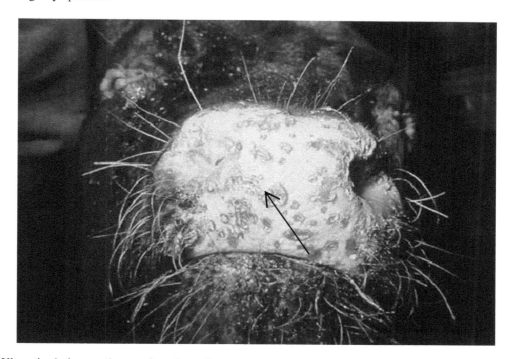

FIGURE 58.2   Ulcerative lesions on the muzzle and nostrils.

be in anestrus for several months.[1,2,70,71] A few animals die, but the majority slowly recover. Recovery can take several months, and some skin lesions may take a year or two to resolve. Deep holes or scars are often left in the skin (Figure 58.3a,b).[4]

Generally, poxviruses are epitheliotrophic and cause localized cutaneous (orf and pseudocowpox viruses) or systemic disease (LSDV and SPPV) involving the skin and many organs and tissues. Viral particles are enveloped when mature virus particles move to the Golgi complex; most particles are however nonenveloped and are released by cell disruption. Both enveloped and nonenveloped particles are infectious.[12,13] The pathogenesis is associated with both viral and host factors.[75] The control of virus replication by the host likely determines the clinical outcome. Viremia, likely cell associated,[56] also starts at the time of lesion occurrence and lasts until the time of seroconversion when host antibodies can neutralize the virus.

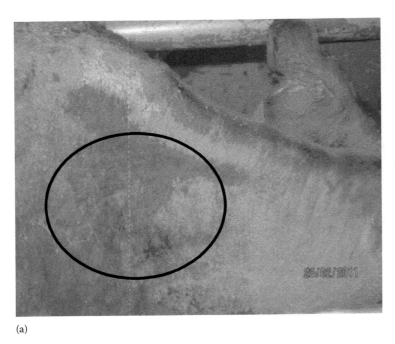

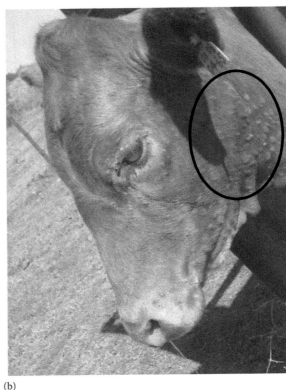

(a)   (b)

FIGURE 58.3   (a) and (b) Skin lesions on the neck.

Poxviruses usually enter the host through the skin or respiratory tract.[12,13] Initial multiplication of the LSDV occurs at the entry site. The cellular receptors for poxvirus entry are currently unknown,[76,77] but the functions of some of the virus-coded proteins have been determined, including some involved in the evasion of the host immune response.[78] The virion enters the host cell by fusion with the plasma membrane or by endocytosis. The transcriptase released from the core of the virion starts the formation of the mRNA within minutes after infection. The first polypeptides complete the uncoating of the core before the actual viral DNA synthesis begins 1.5–6 h after infection.[12,13] The replication of the LSDV is accompanied by the formation of round or irregularly shaped intracytoplasmic inclusion bodies.[10,70,79] There may be one or several inclusion bodies within one cell.[1]

In systemic infections, further viral replication takes place in the lymph nodes, followed by viremia and further viral multiplication in many different organs including the liver, spleen, and lungs.[13] The latter multiplication leads to the establishment of secondary viremia and subsequent infection and development of disseminated focal lesions in the skin.

The earliest description by Thomas and Maré[21] highlighted the histological changes. Their findings were confirmed by Prozesky and Barnard[70] who found that LSDV exerts its pathogenic effects by infiltrating a variety of cell types, including epithelial and endothelial cells, pericytes, and fibroblasts, resulting in lymphangitis and vasculitis. The latter report highlighted the difference in histological changes associated with acute or chronic lesions. During the acute stage, vasculitis and lymphangitis with concomitant thrombosis and infarction resulted in edema and necrosis.[70] The lesions were initially infiltrated by neutrophils and macrophages, and later on these cells were gradually replaced by lymphocytes, plasma cells, and macrophages, as well as fibroblasts.[70] The exact pathogenesis of the development of the lesions associated with LSD is not as well understood as the pathogenesis of SPPV.[12,13] It was established that sheeppox infection spreads in macrophages around the body and becomes localized in a variety of organs, including kidneys, testes, and lungs.[80] Coagulation necrosis is the result of thrombi in blood vessels. It is not known if a single cell type is responsible for the spread of LSDV around the body and its localization in various organs.

Nagi and coworkers[81] investigated cutaneous and testicular lesions in an outbreak of LSD in Egypt in 1989. He noted that diffuse degenerative and inflammatory changes could be observed in the seminiferous tubules and blood vessels. The seminiferous tubules were devoid of primary and secondary spermatocytes and of spermatids, although the spermatogonia and Sertoli cells appeared resistant. The author speculated that the possible infertility due to LSDV infection may be transient as the regeneration of the germinal epithelium depends mainly upon the persistence of spermatogonia and Sertoli cells, although extensive fibrosis may preclude the return to fertility. There are no other published reports on the pathology of LSDV in testicular tissue.[65]

Excretion in semen. Weiss[1] found that the virus was excreted in semen for 22 days after the fever reaction following experimental infection and that viral shedding in semen in some cases occurred in the absence of clinical evidence of infection. Although no further data exists as to the risk posed by semen in the transmission of the disease, the aforementioned information was sufficient to classify semen from countries where LSD occurs as a potential hazard.

Recent studies added additional data regarding the excretion of LSDV in semen. Irons and coworkers[63] and Tuppurainen et al.[17] demonstrated the excretion of the virus in semen for much longer than was previously shown. These authors demonstrated that the shedding of virus persisted despite termination of the viremic phase, seroconversion, and clinical recovery from the disease and that the virus could be detected in semen 159 days postinfection by using PCR. The effect of the virus on the quality of semen was reported by Irons et al.[63]. Modern detection methods of the virus in semen were reported by Bagla et al.,[53] and the use of PCR to detect the virus was described by Ireland and Binepal[82].

Annandale and coworkers[65] investigated the site of persistence in bulls shedding virus in semen for a period longer than 28 days. The development of lesions in the genital tract was studied and results compared to the presence of virus in different fractions of semen in an attempt to localize the virus to a specific site in the genital tract. They demonstrated the presence of viral DNA in all fractions of semen and in sheath wash samples from bulls that were shedding the virus in semen for protracted periods of time and concluded that the virus was not limited to a specific fraction of the ejaculate and that the testes and epididymides are most profoundly affected. Blood contamination is not responsible for the presence of viral particles in semen and the virus is not sperm associated. It is speculated that the ejaculate is contaminated with virus particles as they are shed from necrotic lesions in the genital tract.

The transmission of the virus to heifers and embryos through artificial insemination was recently published.[83] LSDV was detected by PCR, VI, or electron microscopy in blood, embryos, and organs of experimentally infected animals, and in heifers seroconverted by Day 27 postinsemination. If a highly infectious dose is used, lesions on mucous membranes, especially of the conjunctiva, nose, mouth, and reproductive tract, seemed more pronounced than what is normally seen in the field, where skin lesions are less outspoken. The high titer of virus used to spike semen in this trial precludes conclusions on the quantification of biosecurity risks associated with the presence of LSDV in bovine semen under field conditions.[83]

58.5 IDENTIFICATION AND DIAGNOSIS

A presumptive diagnosis of the disease can be made on clinical signs of fever, enlarged superficial lymph nodes, and the characteristic skin nodules. Differential diagnoses include pseudo-LSD/bovine herpes mammillitis, dermatophilosis, ringworm, insect or tick bites, besnoitiosis, ringworm,

Hypoderma bovis infestation, photosensitization, bovine papular stomatitis, urticaria, and cutaneous tuberculosis. However, most of these diseases can be distinguished from LSD by clinical signs, including the duration of the disease, as well as histopathology and other laboratory tests.[72]

The confirmation of LSD especially in a new area of infection requires VI and identification. Techniques that can be used include transmission electron microscopy, immunohistochemistry, VI in cell cultures, direct and indirect fluorescent antibody test (DFAT/IFAT), agar gel immunodiffusion and enzyme-linked immunosorbent assay (ELISA), Western blot, and serum neutralization test (SNT). Immunohistochemistry, using immunoperoxidase staining, can be used to visualize LSDV antigens in infected tissues.[29,43,64,65,73,84,85] This method is laborious, time consuming, and not a high-throughput assay and therefore not easily used to screen large animal populations.

Transmission electron microscopy is the most rapid diagnostic technique and permits reliable detection of LSDV particles in fresh or formalin-preserved samples.[86] It has been used in outbreaks[26,87] as well as experimental infections.[17,88] It has the advantage of not requiring specific reagents, which is not the case with serological and molecular tests.[89] However, access to a transmission electron microscope as well as a competent microscopist may not be available in most LSD-endemic countries.[86,90] Unlike serological and molecular tests, it is not suitable for primary screening of large number of samples. Furthermore, it cannot differentiate between SPPV, GTPV, and LSDV[91] or from the parapoxviruses that cause bovine papular stomatitis and pseudocowpox. Where orthopoxviruses are endemic in cattle,[92,93] transmission electron microscopy can only differentiate between these viruses and LSDV when specific immunological staining techniques are used.[29,94]

Generally, VI tends to be more sensitive than rapid antigen assays and less expensive than molecular tests.[95] The use of VI to detect LSDV and the cell lines used has been summarized in the literature.[10,29,79,96] Different cell types can be used and the virus grows well on primary lamb and calf adrenals, calf kidney and thyroid cells, fetal lamb and calf muscle cells, chicken embryo fibroblasts, adult vervet monkey kidney cells, and lamb testis cells.[97] The virus is commonly isolated using primary lamb kidney (LK) or primary lamb testis cells. Fetal lung, skin, muscle, and endothelial cells can also be used.[22,22,96] Primary cell culture of bovine dermis cells (BDCs) prepared from a fetal calf's ear was used to isolate LSDV.[17,53] Growth is indicated by the development of cytopathic effect (CPE), which may become evident after 4–10 days in most cell cultures.[22] An ovine testis cell line (OA3.Ts) for LSDV isolation was recently evaluated and the observed CPEs were similar to those obtained with the commonly used primary LK cells.[98] Distinct viral plaques indicative of LSDV could be detected in this cell line by immunostaining with capripoxvirus-specific antiserum. LSDV can be isolated from nodular skin lesions and ocular, nasal, and saliva swabs and buffy coat,[43] with skin lesions as well as nasal and oral swabs being the most useful,[99] although the use of VI techniques to isolate LSDV from semen is not very sensitive,[63]

Molecular diagnostic methods being used include conventional PCR,[17,35,82,90,100–103] real-time PCR,[85,99,104] and dot blot hybridization.[105] Studies have shown that real-time PCR detects capripoxvirus viremia (viral DNA) earlier than VI.[99] It is suggested that real-time PCR is more sensitive than conventional PCR in detecting capripoxviruses[106] and specifically for the detection of LSDV.[85] One disadvantage of PCR protocols is the detection of false positives due to reagent contamination with traces of capripox DNA from various sources.[82] A species-specific, real-time PCR method using dual hybridization probes has been developed for LSDV.[107] This PCR technique has been used to detect LSDV in semen and testicular tissue,[17,63,65] and it has been shown that PCR is much more sensitive than VI in detecting LSDV in semen as bovine semen is toxic to cell cultures.[17,53] Literature reported/described that PCR can be used for high-throughput work; although not cheaper than ELISA, it is more sensitive and can detect LSDV nucleic acid in skin samples postviremia.[85]

Serological tests include IFAT, SNT, ELISA, and immunoblotting (Western blotting). Cross-reactions with other poxviruses are seen in some of these assays.[15,108–111] Examinations using direct or indirect FAT may indicate the presence of LSD virus antigens especially in the early stages of the disease, but later nonspecific fluorescence may cause problems.[112] However, it is not possible to differentiate different members of the capripoxvirus group with direct or indirect FAT,[108] it is more time consuming than ELISAs,[111] and it is also less specific than SNT.[113]

The Western blot technique may be used to detect LSDV with reliable specificity and sensitivity; however, this assay is expensive and needs specialized equipment and training to be performed.[113,114] Agar gel immunodiffusion is also available, but cross-reactions occur in this test with bovine papular stomatitis and pseudocowpox virus.[9]

The SNT is the most specific serological test and gold standard for detecting antibodies against LSDV.[29,108,113] Although it can be used to perform retrospective serosurveillance, it is very time consuming. Since the immunity against LSDV is thought to be predominantly cell mediated, serological testing in general, and SNT in particular, are not sensitive enough to detect the presence of low levels of neutralizing antibodies. This should always be considered when interpreting results.[29,115] Therefore, a negative SNT result does not necessarily indicate that the animal has not been exposed to the virus.[113] The test also requires a live capripoxvirus, access to which is often not permitted in disease-free countries. An SNT utilizing a recombinant green fluorescent protein has been evaluated; preliminary indications are that it has decreased the length of time required for the detection of virus neutralization activity from 6 down to 2.[29,116]

Various ELISA protocols have been developed for use in detecting LSDV infection. Problems with these developments are the difficulty in producing sufficient volumes of inactivated whole virus and the instability of recombinant antigens.[117] An indirect ELISA (I-ELISA) was developed using an expressed recombinant capripoxvirus protein as antigen.[118] Carn later designed an antigen-trapping ELISA used to detect LSDV in the supernatant of cell cultures and skin biopsy samples.[84] The detector system was a guinea pig polyclonal antiserum raised against the recombinant capripoxvirus structural protein, P32. The advantages of this ELISA, especially in developing countries in Africa where LSD is endemic, include reduced costs, stability of reagents, as well as easy handling. It can also be used to differentiate between buffalopox virus and LSDV in water buffalo (*Bubalus bubalis*). Results of this ELISA correlated well with VI, though it was less sensitive. Another similar ELISA protocol developed in Australia using recombinant P32 protein as coating antigen[100] permits discrimination between capripox-, parapox- (orf), and orthopoxvirus (vaccinia virus) infections. These protocols use noninfectious antigens and can therefore be used in nonendemic countries.

Recently, Babiuk et al.[119] validated an ELISA that detects LSDV antibodies using an inactivated SPPV. Compared to SNT and Western blotting, it is easier to perform and is less time consuming. Unlike SNT, it does not require live LSDV and biosafety level (BSL)-3 facilities in LSD-free countries. Results can also be obtained within a day as opposed to the 6 days it takes to read the results from SNT. It compares well with SNT by detecting LSDV antibodies in experimentally tested cattle as early as 21 days postinfection. However, SNT proved to be slightly more specific. To avoid the cost and complex quality issues associated with producing the inactivated antigen used in this new ELISA, it is envisaged that these antigens will soon be replaced by recombinant immunodominant capripoxvirus proteins.[119] The development of recombinant protein ELISA is ongoing, and recently an ELISA based on a synthetic peptide targeting the major antigen P32 has been described for the detection of sheeppox and goatpox antibodies.[120] This ELISA has not been evaluated for the use of LSDV-positive cattle sera.

Samples to collect for laboratory diagnosis: Samples for VI and antigen-detection ELISA should be taken during the first week of symptoms, before neutralizing antibodies develop. Samples for PCR, blood ethylenediaminetetraacetic acid [EDTA] and/or skin lesions, can be collected after this time. In live animals, biopsy samples of skin nodules or lymph nodes can be used for VI and antigen detection. Scabs, nodular fluid, and skin scrapings may also be collected, specifically also for electron microscopy. Tissue and blood samples for VI and antigen detection should be kept chilled and shipped to the laboratory on ice. If the samples must be sent long distances without refrigeration, large pieces of tissue should be collected and the medium should contain 10% glycerol; the central part of the sample can be used for VI. The virus can be isolated from blood samples (collected into heparin or EDTA) during the early, viremic stage of the disease; VI from blood is unlikely to be successful after generalized lesions have been present for more than 4 days. Samples of lesions, including tissues from surrounding areas, should be submitted for histopathology. Acute and convalescent sera are collected for serology. At necropsy, samples for fixation in formalin should be taken from skin lesions and lesions in the respiratory and gastrointestinal tracts, and fresh samples should be taken and treated as described earlier.[17,63,70,74,113]

58.6 PREVENTION AND CONTROL

Import restrictions can help prevent the introduction of LSD. This disease is mainly spread to new areas by infected animals, but it could also be transmitted through contaminated hides and other products. Infected insects are suspected to have spread LSDV to new areas during some outbreaks. Outbreaks can be eradicated by quarantines, depopulation of infected and exposed animals, proper disposal of carcasses, cleaning and disinfection of the premises, and insect control. Vaccines were helpful in eradication during an outbreak in Israel. The virus is susceptible to ether (20%), chloroform, formalin (1%), and some detergents, as well as phenol (2% for 15 min). This virus can survive for long periods in the environment: up to 35 days in desiccated scabs and for at least 18 days in air-dried hides. Insecticides and repellents may also be helpful. Antibiotics are used to control secondary infections.[1,4,17]

The most widely used and viable means of controlling the disease in endemic countries is vaccination using highly effective attenuated virus vaccines. Four live attenuated strains of capripoxvirus have been used as vaccines specifically for the control of LSD.[121–123] These strains have been summarized in the OIE Terrestrial Manual[37,113]: a strain of Kenyan SPPV and GTPV passaged 18 times in lymphotoxin (LT) or fetal calf muscle cells, Yugoslavian RM 65 SPPV strain, Romanian SPPV strain, and LSDV strain from South Africa, passaged 60 times in lamb kidney cells and 20 times on the chorioallantoic membrane of embryonated chicken eggs. All strains of capripoxvirus, whether of bovine, ovine, or caprine origin, share a major neutralizing antigen, so that animals recovered from infection with one strain are resistant to infection with any other strain. Consequently, it is possible to protect cattle against LSD using strains of capripoxvirus derived from sheep or goats.[121] In 1989 and 1990, the Romanian strain of SPPV vaccine was used to help control the LSD outbreak in Egypt.[124] Control trials should however first be carried out prior to introducing a vaccine strain not usually used in cattle. Animals that recover from clinical capripoxvirus infection generate lifelong immunity, which protects against all capripoxvirus isolates.[115] Protection following vaccination with the South African LSDV strain is believed to be lifelong; however, experiences during outbreaks in 1990/1991 have challenged this evidence and more frequent vaccination is now recommended.[125] As immunity wanes, local capripoxvirus replication will occur at the site of inoculation, but the virus will not become generalized.[113]

The Neethling strain of the virus has been successfully used in a vaccine for the control of the disease in southern Africa.[1,2,7] This vaccine is developed by attenuation of a field isolate in tissue culture and on the chorioallantoic membranes of embryonated hens' eggs.[1] According to the manufacturers of the currently used vaccine, immunity to LSD starts developing 10 days after vaccination and reaches its peak after 21 days, though the vaccine does not necessarily confer absolute immunity to all vaccinated animals.[7] The possible reasons for vaccine failures have been investigated by Hunter and Wallace.[7]

Immunity to LSD is mainly cell mediated and forms a significant component in recovery from infection and in the long-term protection of an animal, but the production of antibodies is a useful indicator of the response to vaccination.[1,7,126] However, literature has shown the absence of detectable levels of antibodies after vaccination (using existing serological techniques) in some animals that were nevertheless immune to LSD when challenged.[1,66]

Although live attenuated vaccines work well, reports of vaccine breakdown, short duration of protection, and low levels of antibody as discussed earlier necessitate the need for improved vaccines.[7] New age vaccines using capripoxviruses have been designed[127] and experimentally tested as a vector for recombinant subunit vaccines, such as rabies,[128] rinderpest,[129,130] peste des petits ruminants,[131] and Rift Valley fever.[116]

58.7 CONCLUSIONS AND FUTURE PERSPECTIVES

Capripoxviruses cause pox of sheep and goats and LSD of cattle. These diseases are of economic significance to farmers in endemic regions and are a major constraint to international trade in livestock and their products. LSD is an acute, subacute, or inapparent disease of cattle of all breeds. This disease causes abortions, infertility problems in bulls and cows, loss of body mass, and a sharp drop in milk yield during outbreaks. The ulceration of nodular pox lesions leaves permanent scars that decrease the value of the hides.

LSD has the potential to become an emerging disease threat because of global climate change and changes in patterns of trade in animals and animal products. The distribution of the disease is considerably different from what it was 50 years ago, and outbreaks occurred outside the African continent in Madagascar, Israel (1989), and the Middle East (e.g., Kuwait 1991) and recently in United Arab Emirates (2000), Bahrain (2003), and Oman (2010). Increased legal and illegal trade in live animals provides the potential for further spread with, for instance, the possibility of LSD becoming firmly established in Asia.

It is evident that LSD has the ability to spread to new geographical areas and that naïve animal populations are at risk to suffer severe morbidity and mortality. It is therefore important to have the ability to monitor the disease by detecting the virus and antibodies using sensitive and reliable diagnostic techniques, including tests for the detection of cellular immunity. The improvement of the sensitivity and specificity of LSDV-specific antibody ELISA, especially for detection using wildlife sera, is essential for the successful control of the disease. Awareness programs and control measures should be put in place. This should include the development and implementation of new generation vaccines that can also be used in nonendemic countries. The study of the epidemiology of the disease and whether wildlife play a role as amplifying hosts in the maintenance of the disease should assist in the scientific knowledge of the transmission of the virus.

ACKNOWLEDGMENTS

I would like to thank my colleagues, collaborative researchers, and students that contributed to generate the research data used in this chapter:

Colleagues: Prof. JAW Coetzer, Dr. David Wallace, and Dr. Hein Stoltsz.

Students under my supervision or co-supervision: Prof. Peter Irons (PhD), Dr. Eeva Tuppurainen (MSc [VS]), Dr. Henry Annandale (MMedVet, Theriogenology), Dr. Victor Bagla (MSc [VS]), Dr. Uchi Osuagwuh (MSc [VS]), Dr. Shamsudeen Fagbo (MSc [VTD]), and Dr. Jimmy Lubinga (PhD).

Laboratory technologists: Ms. Karen Ebersohn, Ms. Rebone Mahlare, and Ms. Milana Troskie.

REFERENCES

1. Weiss, K.E. Lumpy skin disease virus. In: *Virology Monographs*, vol. 3, New York: Springer Verlag; pp. 111–131, 1968.

2. Coetzer, J.A.W. Lumpy skin disease. In: Coetzer, J.A.W. and Tustin, R.C., eds. *Infectious Diseases of Livestock with Special Reference to Southern Africa*, 2nd ed. Oxford, U.K.: Oxford University Press; pp. 1268–1276, 2004.

3. Kumar, S.M. An outbreak of lumpy skin disease in a Holstein dairy herd in Oman: A clinical report. *Asian J. Anim. Vet. Adv.* 6(8):851–859, 2011.

4. Green, H.F. Lumpy skin disease-its effect on hides and leather and a comparison in this respect with some other skin diseases. *Bull. Epizoot. Dis. Afr.* 7:63–94, 1959.

5. Gari, G., Bonnet, P., Roger, F., and Waret-Szkuta, A. Epidemiological aspects and financial impact of lumpy skin disease in Ethiopia. *Prev. Vet. Med.* 102(4):274–283, 2011.

6. Diesel, A.M. The epizootiology of lumpy skin disease in South Africa. Report. In: *Proceedings of the 14th International Veterinary Congress*, London, U.K., vol. 2, pp. 492–500, 1949.

7. Hunter, P. and Wallace, D. Lumpy skin disease in southern Africa: A review of the disease and aspects of control. *J. S. Afr. Vet. Assoc.* 72(2):68–71, 2001.

8. Abraham, A. and Zissman, A. Isolation of lumpy skin disease virus from cattle in Israel. *Isr. J. Vet. Med.* 46(1):20–23, 1991.

9. Anonymous. HANDISTATUS II. World Organization for Animal Health, 2008, http://www.oie.int (accessed on July, 2012).

10. Alexander, R.A., Plowright, W., and Haig, D.A. Cytopathogenic agents associated with lumpy-skin disease of cattle. *Bull. Epizoot. Dis. Afr.* 5:489–492, 1957.

11. Buller, R.M. et al. (eds.) *Virus Taxonomy: Eighth Report of the International Committee on the Taxonomy of Viruses*, 1st ed. Oxford, U.K.: Elsevier Academic Press; pp. 117–133, 2005.

12. Buller, R.M. and Fenner, F. Lumpy skin disease. In: Richman, D.D., Whitley, R.J., and Hayden, F.G., eds. *Poxviruses*, 3rd edn., Oxford: Elsevier Academic Press; pp. 387–408, 2009.

13. Anonymous. Chapter 7—Poxviridae. In: Maclachlan, N.J. and Dubovi, E.J., eds. *Fenner's Veterinary Virology*, 4th ed. San Diego, CA: Academic Press; pp. 151–165, 2011.

14. Black, D.N., Hammond, J.M., and Kitching, R.P. Genomic relationship between capripoxviruses. *Virus Res.* 5(2/3):277–292, 1986.

15. Davies, F.G. and Otema, C. Relationships of capripox viruses found in Kenya with two Middle Eastern strains and some orthopox viruses. *Res. Vet. Sci.* 31(2):253–255, 1981.

16. Tulman, E.R. et al. Genome of lumpy skin disease virus. *J. Virol.* 75(15):7122–7130, 2001.

17. Tuppurainen, E.S.M., Venter, E.H., and Coetzer, J.A.W. The detection of lumpy skin disease virus in samples of experimentally infected cattle using different diagnostic techniques. *Onderstepoort J. Vet. Res.* 72(2):153–164, 2005.

18. Macdonald, R.A.S. Annual report of the veterinary research officer for 1930. *Ann. Rep. Dept. Anim. Health* 1931:17–24, 1930.

19. Von Backstrom, U. Ngamiland cattle disease. Preliminary report on a new disease, the etiological agent being probably of an infectious nature. *J. S. Afr. Vet. Med. Assoc.* 16:29–35, 1945.

20. Haig, D.A. Lumpy skin disease. *Bull. Epizoot. Dis. Afr.* 5:421–430, 1957.

21. Thomas, A.D., Mare, C.v.E. Knop-velsiekte. *J. S. Afr. Vet. Med. Assoc.* 16:36–43, 1945.

22. Davies, F.G. Lumpy skin disease, an African capripox virus disease of cattle. *Br. Vet. J.* 147(6):489–503, 1991.

23. Davies, F.G. Lumpy skin disease of cattle: A growing problem in Africa and the near east. *World Anim. Rev.* 68(3):37–42, 1991.

24. MacOwan, K.D.S. Observations on the epizootiology of lumpy skin disease during the first year of its occurrence in Kenya. *Bull. Epizoot. Dis. Afr.* 7:7–20, 1959.

25. Ali, B.H. and Obeid, H.M. Investigation of the first outbreaks of lumpy skin disease in the Sudan. *Br. Vet. J.* 133(2):184–189, 1977.

26. Nawathe, D.R. et al. Lumpy skin disease in Nigeria. *Trop. Anim. Health Prod.* 10(1):49–54, 1978.

27. Ali, A.A. et al. Clinical and pathological studies on lumpy skin disease in Egypt. *Vet. Rec.* 127(22):549–550, 1990.

28. House, J.A. et al. The isolation of lumpy skin disease virus and bovine herpesvirus-4 from cattle in Egypt. *J. Vet. Diagn. Invest.* 2(2):111–115, 1990.

29. Babiuk, S. et al. Capripoxviruses: An emerging worldwide threat to sheep, goats and cattle. *Transbound. Emerg. Dis.* 55(7):263–272, 2008.

30. El-Kholy, A.A., Soliman, H.M.T., and Abdelrahman, K.A. Polymerase chain reaction for rapid diagnosis of a recent lumpy skin disease virus incursion to Egypt. *Arab. J. Biotechnol.* 11(2):293–302, 2008.

31. Yeruham, I. et al. Spread of lumpy skin disease in Israeli dairy herds. *Vet. Rec.* 137(4):91–93, 1995.

32. Brenner, J. et al. Appearance of skin lesions in cattle populations vaccinated against lumpy skin disease: Statutory challenge. *Vaccine* 27(10):1500–1503, 2009.

33. Shimshony, A. and Economides, P. Disease prevention and preparedness for animal health emergencies in the Middle East. *Rev. Sci. Tech.* 25(1):253–269, 2006.

34. Greth, A. et al. Capripoxvirus disease in an Arabian oryx (*Oryx leucoryx*) from Saudi Arabia. *J. Wildl. Dis.* 28(2):295–300, 1992.

35. Orlova, E.S. et al. Differentiation of capripoxvirus species and strains by polymerase chain reaction. *Mol. Biol. (New York)* 40(1):139–145, 2006.

36. Kitching, R.P., Mcgrane, J.J., and Taylor, W.P. Capripox in the Yemen-Arab-Republic and the Sultanate of Oman. *Trop. Anim. Health Prod.* 18(2):115–22 1986.

37. Anonymous. Lumpy skin disease—South Africa. World Organization for Animal Health, 2010, http://www.oie.int (accessed on August, 2012).

38. Fayed, A.A., Al-Gaabary, M.H., and Osman, S.A. Reappearance of lumpy skin disease (LSD) in Egypt. *Assiut Vet. Med. J.* 52(108):231–246, 2006.

39. Nawathe, D.R. et al. Some observations on the occurrence of lumpy skin disease in Nigeria. *Zentralbl. Vet. B* 29(1):31–36, 1982.

40. Davies, F.G. Observations on the epidemiology of lumpy skin disease in Kenya. *J. Hyg.* 88(1):95–102, 1982.

41. Domenech, J. et al. Regional and international approaches on prevention and control of animal transboundary and emerging diseases. *Ann. N.Y. Acad. Sci.* 1081:90–107, 2006.

42. Rweyemamu, M. et al. Emerging diseases of Africa and the Middle East. *Ann. N.Y. Acad. Sci.* 916:61–70, 2000.

43. Carn, V.M. and Kitching, R.P. The clinical response of cattle experimentally infected with lumpy skin disease (Neethling) virus. *Arch.Virol.* 140(3):503–513, 1995.

44. Burdin, M.L. and Prydie, J. Lumpy skin disease of cattle in Kenya. *Nature* 183:949–950, 1959.

45. Kitching, R.P., Bhat, P.P., and Black, D.N. The characterization of African strains of capripoxvirus. *Epidemiol. Infect.* 102(2):335–343, 1989.

46. Greth, A. et al. Serological survey for bovine bacterial and viral pathogens in captive Arabian oryx (*Oryx leucoryx* Pallas, 1776). *Rev. Sci. Tech.—Off. Int. Epizoot.* 11(4):1163–1168, 1992.

47. Young, E., Basson, P.A., and Weiss, K.E. Experimental infection of game animals with lumpy skin disease virus (prototype strain Neethling). *Onderstepoort J. Vet. Res.* 37:79–88, 1970.

48. Hedger, R.S. and Hamblin, C. Neutralising antibodies to lumpy skin disease virus in African wildlife. *Comp. Immunol. Microbiol. Infect. Dis.* 6(3):209–213, 1983.

49. Barnard, B.J.H. Antibodies against some viruses of domestic animals in southern African wild animals. *Onderstepoort J. Vet. Res.* 64(2):95–110, 1997.

50. Fagbo, S. Seroprevalence of Rift Valley fever and lumpy skin disease in African buffalo (*Syncerus caffer*) in the Kruger National and Hluhluwe-iMfolozi Parks, South Africa. MSc dissertation, University of Pretoria, http://upetd.up.ac.za/thesis/available/etd-10092012-161242/unrestricted/dissertation.pdf; 2012.

51. Hosamani, M. et al. Differentiation of sheep pox and goat poxviruses by sequence analysis and PCR-RFLP of P32 gene. *Virus Genes* 29(1):73–80, 2004.

52. Carn, V.M. and Kitching, R.P. An investigation of possible routes of transmission of lumpy skin disease virus (Neethling). *Epidemiol. Infect.* 114(1):219–226, 1995.

53. Bagla, V.P. et al. Elimination of toxicity and enhanced detection of lumpy skin disease virus on cell culture from experimentally infected bovine semen samples. *Onderstepoort J. Vet. Res.* 73(4):263–268, 2006.

54. Kitching, R.P. and Mellor, P.S. Insect transmission of capripoxvirus. *Res. Vet. Sci.* 40(2):255–258, 1986.

55. Chihota, C.M. et al. Mechanical transmission of lumpy skin disease virus by *Aedes aegypti* (Diptera: Culicidae). *Epidemiol. Infect.* 126(2):317–321, 2001.

56. Kitching, R.P. and Taylor, W.P. Transmission of capripoxvirus. *Res. Vet. Sci.* 39(2):196–199, 1985.

57. Chihota, C.M. et al. Attempted mechanical transmission of lumpy skin disease virus by biting insects. *Med. Vet. Entomol.* 17(3):294–300, 2003.

58. Mellor, P.S., Kitching, R.P., Wilkinson, P.J. Mechanical transmission of capripox virus and African swine fever virus by Stomoxys calcitrans. *Res. Vet. Sci.* 43(1):109–112, 1987.

59. Aiel, K. Lumpy skin disease. *Bovine Ovine* 22:2467, 2009.

60. Tuppurainen, E.S.M. et al. A potential role for ixodid (hard) tick vectors in the transmission of lumpy skin disease virus in cattle. *Transbound. Emerg. Dis.* 58(2):93–104, 2011.

61. Tuppurainen, E.S. et al. Mechanical transmission of lumpy skin disease virus by Rhipicephalus appendiculatus male ticks. *Epidemiol. Infect.* 1–6, 2012.

62. Lubinga, J. et al. Transstadial and mechanical transmission of lumpy skin disease virus by *Amblyomma hebraeum* ticks. *Epidemiol. Infect.* 2012; Accepted.

63. Irons, P.C., Tuppurainen, E.S.M., and Venter, E.H. Excretion of lumpy skin disease virus in bull semen. *Theriogenology* 63(5):1290–1297, 2005.

64. Bagla, V.P. The demonstration of lumpy skin disease virus in semen of experimentally infected bulls using different diagnostic techniques. MSc dissertation, University of Pretoria, Pretoria, South Africa, 2006. http://upetd.up.ac.za/thesis/available/etd-05272008-120446/unrestricted/dissertation.pdf.

65. Annandale, C.H. et al. Sites of persistence of lumpy skin disease virus in the genital tract of experimentally infected bulls. *Reprod. Domest. Anim.* 45(2):250–255, 2010.

66. Osuagwuh, U.I. et al. Absence of lumpy skin disease virus in semen of vaccinated bulls following vaccination and subsequent experimental infection. *Vaccine* 25(12):2238–2243, 2007.

67. Bielanski, A. Disinfection procedures for controlling microorganisms in the semen and embryos of humans and farm animals. *Theriogenology* 68(1):1–22, 2007.

68. Capstick, P.B. Lumpy skin disease-experimental infection. *Bull. Epizoot. Dis. Afr.* 7:51–62, 1959.

69. Capstick, P.B. et al. Protection of cattle against the "Neethling" type virus of lumpy skin disease. *Vet. Rec.* 71:422–423, 1959.

70. Prozesky, L. and Barnard, B.J.H. A study of the pathology of lumpy skin disease in cattle. *Onderstepoort J. Vet. Res.* 49(3):167–175, 1982.

71. Barnard, B.J.H. et al. Lumpy skin disease. In: Coetzer, J.A.W, Thomson, G.R., and Tustin, R.C., eds. *Infectious Diseases of Livestock with Special Reference to Southern Africa.* Cape Town, South Africa: Oxford University Press; pp. 604–612, 1994.

72. Anonymous. Lumpy skin disease technical disease card. World Organization for Animal Health, 2009, http://www.oie.int.

73. Tuppurainen, E.S.M. The detection of lumpy skin disease virus in samples of experimentally infected cattle using different diagnostic techniques. MSc dissertation, University of Pretoria, Pretoria, South Africa, 2004.

74. Annandale, C.H. Mechanisms by which lumpy skin disease virus is shed in semen of artificially infected bulls. MSc dissertation, University of Pretoria, Pretoria, South Africa, 2006.

75. Stanford, M.M. et al. Immunopathogenesis of poxvirus infections: Forecasting the impending storm. *Immunol. Cell Biol.* 85(2):93–102, 2007.

76. Moss, B. Poxvirus entry and membrane fusion. *Virology* 344(1):48–54, 2006.

77. Moss, B. Poxvirus cell entry: How many proteins does it take? *Viruses* 4(5):688–707, 2012.

78. Moss, B. and Shisler, J.L. Immunology 101 at poxvirus U: Immune evasion genes. *Semin. Immunol.* 13(1):59–66, 2001.

79. Prydie, J. and Coackley, W. Lumpy skin disease-tissue culture studies. *Bull. Epizoot. Dis. Afr.* 7:37–50, 1959.

80. Kitching, R.P. Sheeppox and goatpox. In: Coetzer, J.A.W. and Tustin, R.C., eds., *Infectious Diseases of Livestock with Special Reference to Southern Africa.* Oxford, U.K.: Oxford University Press; pp. 1277–1281, 2004.

81. Nagi, A.A. et al. Lumpy skin disease. I. Cutaneous and testicular lesions. *Assiut Vet. Med. J.* 23(45):90–99, 1990.

82. Ireland, D.C. and Binepal, Y.S. Improved detection of capripoxvirus in biopsy samples by PCR. *J. Virol. Methods* 74(1):1–7, 1998.

83. Annandale, C.H. et al. Seminal transmission of lumpy skin disease virus in heifers. *Transbound. Emerg. Dis.* 2012; In Press.

84. Carn, V.M. An antigen trapping ELISA for the detection of capripoxvirus in tissue culture supernatant and biopsy samples. *J. Virol. Methods* 51(1):95–102, 1995.

85. Babiuk, S. et al. Quantification of lumpy skin disease virus following experimental infection in cattle. *Transbound. Emerg. Dis.* 55(7):299–307, 2008.

86. Woods, J.A. Lumpy skin disease—A review. *Trop. Anim. Health Prod.* 20(1):11–17, 1988.

87. Khalafalla, A.I., Elamin, M.A.G., and Abbas, Z. Lumpy skin disease: Observations on the recent outbreaks of the disease in the Sudan. *Rev. Elev. Med. Vet. Pays. Trop.* 46(4):548–550, 1993.

88. Aspden, K. et al. Evaluation of lumpy skin disease virus, a capripoxvirus, as a replication-deficient vaccine vector. *J. Gen. Virol.* 84(8):1985–1996, 2003.

89. Goldsmith, C.S. and Miller, S.E. Modern uses of electron microscopy for detection of viruses. *Clin. Microbiol. Rev.* 22(4):552–563, 2009.

90. Min, Z. et al. Establishment of duplex PCR assay for quick differentiating capripoxvirus and orf viruses. *Vet. Sci. China* 37(11):931–934, 2007.

91. Kitching, R.P. and Smale, C. Comparison of the external dimensions of capripoxvirus isolates. *Res. Vet. Sci.* 41(3):425–427, 1986.

92. Yeruham, I. et al. Occurrence of cowpox-like lesions in cattle in Israel. *Rev. Elev. Med. Vet. Pays. Trop.* 49(4):299–302, 1996.

93. Singh, R.K. et al. Sequence analysis of C18L gene of buffalopox virus: PCR strategy for specific detection and differentiation of buffalopox from orthopoxviruses. *J. Virol. Methods* 154(1/2):146–153, 2008.

94. Gulbahar, M.Y. et al. Immunohistochemical evaluation of inflammatory infiltrate in the skin and lung of lambs naturally infected with sheeppox virus. *Vet. Pathol.* 43(1):67–75, 2006.

95. Leland, D.S. and Ginocchio, C.C. Role of cell culture for virus detection in the age of technology. *Clin. Microbiol. Rev.* 20(1):49–78, 2007.

96. Binepal, Y.S., Ongadi, F.A., and Chepkwony, J.C. Alternative cell lines for the propagation of lumpy skin disease virus. *Onderstepoort J. Vet. Res.* 68(2):151–153, 2001.

97. Davies, F.G. Characteristics of a virus causing a pox disease in sheep and goats in Kenya, with observations on the epidemiology and control. *J. Hyg.* 76(2):163–171, 1976.

98. Babiuk, S. et al. Evaluation of an ovine testis cell line (OA3. ts) for propagation of capripoxvirus isolates and development of an immunostaining technique for viral plaque visualization. *J. Vet. Diagn. Invest.* 19(5):486–491, 2007.

99. Bowden, T.R. et al. Capripoxvirus tissue tropism and shedding: A quantitative study in experimentally infected sheep and goats. *Virology* 371(2):380–393, 2008.

100. Heine, H.G. et al. A capripoxvirus detection PCR and antibody ELISA based on the major antigen P32, the homolog of the vaccinia virus H3L gene. *J. Immunol. Methods* 227(1–2):187–196, 1999.

101. Mangana-Vougiouka, O. et al. Sheep poxvirus identification by PCR in cell cultures. *J. Virol. Methods* 77(1):75–79, 1999.

102. Mangana-Vougiouka, O. et al. Sheep poxvirus identification from clinical specimens by PCR, cell culture, immunofluorescence and agar gel immunoprecipitation assay. *Mol. Cell. Probes* 14(5):305–310, 2000.

103. Stram, Y. et al. The use of lumpy skin disease virus genome termini for detection and phylogenetic analysis. *J. Virol. Methods* 151(2):225–229, 2008.

104. Balinsky, C.A. et al. Rapid preclinical detection of sheeppox virus by a real-time PCR assay. *J. Clin. Microbiol.* 46(2):438–442, 2008.

105. Awad, W.S. et al. Evaluation of different diagnostic methods for diagnosis of Lumpy skin disease in cows. *Trop. Anim. Health Prod.* 42(4):777–783, 2010.

106. Balamurugan, V. et al. Comparative efficacy of conventional and TaqMan polymerase chain reaction assays in the detection of capripoxviruses from clinical samples. *J. Vet. Diagn. Invest.* 21(2):225–231, 2009.

107. Lamien, C.E. et al. Real time PCR method for simultaneous detection, quantitation and differentiation of capripoxviruses. *J. Virol. Methods* 171(1):134–140, 2011.

108. Davies, F.G. and Otema, C. The antibody response in sheep infected with a Kenyan sheep and goat pox virus. *J. Comp. Pathol.* 88(2):205–210, 1978.

109. Davies, F.G. Possible role of wildlife as maintenance hosts for some African insect-borne virus diseases [bluetongue, ephemeral fever, African horse sickness, Rift Valley fever, lumpy skin disease, Nairobi sheep disease]. Wildlife disease research and economic development. In: *Proceedings of a Workshop Held in Kabete*, Kenya, September 8–9, 1980, pp. 24–27, 1981.

110. Davies, F.G. Lumpy skin disease. *Virus Diseases of Food Animals*, vol. II, pp. 751–764, 1981.

111. Gari, G. Evaluation of indirect fluorescent antibody test (IFAT) for the diagnosis and screening of lumpy skin disease using Bayesian method. *Vet. Microbiol.* 129(3–4):269–280, 2008.

112. Davies, F.G. et al. The laboratory diagnosis of lumpy skin disease. *Res. Vet. Sci.* 12:123–127, 1971.

113. Anonymous. Lumpy skin disease terrestrial manual. World Organization for Animal Health, 2010, http://www.oie.int.

114. Chand, P., Kitching, R.P., and Black, D.N. Western blood analysis of virus-specific antibody responses for capripox and contagious pustular dermatitis viral infections in sheep. *Epidemiol. Infect.* 113(2):377–385, 1994.

115. Kitching, R.P., Hammond, J.M., and Taylor, W.P. A single vaccine for the control of capripox infection in sheep and goats. *Res. Vet. Sci.* 42(1):53–60, 1987.

116. Wallace, D.B. et al. Improved method for the generation and selection of homogeneous lumpy skin disease virus (SA-Neethling) recombinants. *J. Virol. Methods* 146(1/2):52–60, 2007.

117. Bowden, T.R. et al. Detection of antibodies specific for sheeppox and goatpox viruses using recombinant capripoxvirus antigens in an indirect enzyme-linked immunosorbent assay. *J. Virol. Methods* 161(1):19–29, 2009.

118. Carn, V.M. et al. Use of a recombinant antigen in an indirect ELISA for detecting bovine antibody to capripoxvirus. *J. Virol. Methods* 49(3):285–294, 1994.

119. Babiuk, S. et al. Detection of antibodies against capripoxviruses using an inactivated sheeppox virus ELISA. *Transbound. Emerg. Dis.* 56(4):132–141, 2009.

120. Hong, T. et al. Serodiagnosis of sheeppox and goatpox using an indirect ELISA based on synthetic peptide targeting for the major antigen p32. *Virol. J.* 7: 245, 2010.

121. Coackley, W. and Capstick, P.B. Protection of cattle against lumpy skin disease. II. Factors affecting small scale production of a tissue culture propagated virus vaccine. *Res. Vet. Sci.* 2:369–374, 1961.

122. Carn, V.M. Control of capripoxvirus infections. *Vaccine* 11(13):1275–1279, 1993.

123. Brenner, J. et al. Lumpy skin disease (LSD) in a large dairy herd in Israel, June 2006. *Isr. J. Vet. Med.* 61(3/4):73–77, 2006.

124. Michael, A. et al. Control of lumpy skin disease outbreak in Egypt with Romanian sheep pox vaccine. *Assiut Vet. Med. J.* 36(71):173–180, 1997.

125. Kitching, R.P. Lumpy skin disease. In: OIE, ed., *Manual of Standards for Diagnostic Tests and Vaccines*. Paris, France: International Office of Epizootics, pp. 93–101, 1996.

126. Kitching, R.P. and Hammond, J. Poxvirus, infection and immunity. In: Roitt, I.M. and Delves, P.J., eds. *Encyclopaedia of Immunology*, vol. 3. London, U.K.: Academic Press; pp. 1261–1264, 1992.

127. Wallace, D.B. and Viljoen, G.J. Importance of thymidine kinase activity for normal growth of lumpy skin disease virus (SA-Neethling). *Arch. Virol.* 147(3):659–663, 2002.

128. Aspden, K. et al. Immunogenicity of a recombinant lumpy skin disease virus (Neethling vaccine strain) expressing the rabies virus glycoprotein in cattle. *Vaccine* 20(21/22):2693–2701, 2002.

129. Ngichabe, C.K. et al. Trial of a capripoxvirus-rinderpest recombinant vaccine in African cattle. *Epidemiol. Infect.* 118(1):63–70, 1997.

130. Ngichabe, C.K. et al. Long term immunity in African cattle vaccinated with a recombinant capripox-rinderpest virus vaccine. *Epidemiol. Infect.* 128(2):343–349, 2002.

131. Diallo, A. et al. Goat immune response to capripox vaccine expressing the hemagglutinin protein of peste des petits ruminants, 2002. In: *Proceedings of the Society for Tropical Veterinary Medicine and the Wildlife Diseases Association. Wildlife and Livestock, Disease and Sustainability: What Makes Sense?* Pilanesberg National Park, South Africa, July 22–27, 2001.

59 Malignant Catarrhal Fever Virus (Alcelaphine Herpesvirus Type 1)

George C. Russell

CONTENTS

59.1 INTRODUCTION

Malignant catarrhal fever (MCF) is a fatal generalized disease of ungulates that is characterized by systemic infiltration and proliferation of lymphocytes in many tissues, with extensive cytotoxic activity that appears to have some autoimmune characteristics.[1] MCF is caused by γ-herpesviruses of the genus *Macavirus* that share features of sequence and antigenicity. These viruses infect their natural hosts efficiently and without apparent disease, while contact with susceptible (usually ungulate) host species can lead to fatal MCF. Currently, four viruses have been associated with MCF in susceptible species, while other MCF viruses have been identified by a shared serological determinant and PCR-based sequence analysis.[2] Thus, alcelaphine herpesvirus 1 (AlHV-1) naturally infects wildebeest and is a cause of MCF in cattle in East and South Africa; ovine herpesvirus 2 (OvHV-2) infects domestic sheep and causes MCF in in-contact cattle and deer worldwide; and caprine herpesvirus 2 (CpHV-2) naturally infects goats and is associated with MCF cases in deer. The fourth virus shown to be associated with the disease has been termed "MCF virus of white-tailed deer" (MCFV-WTD) to reflect the susceptible host, while its natural host has yet to be identified.[3] In recent years, MCF has also been reported in pigs, initially in Scandinavia but more recently in a range of European countries. The clinical signs found in pigs were very similar to those seen in acutely affected cattle.

There is a clear range of susceptibility to MCF, with cattle species (*Bos taurus* and *Bos indicus*) being relatively resistant and experiencing sporadic disease. Most species of deer, bison (*Bison bison*) and water buffalo (*Bubalus bubalis*) are more susceptible, while Bali cattle (*Bos javanicus*) and Père David's deer (*Elaphurus davidianus*) are considered extremely susceptible.[4] The more resistant species, such as cattle, tend to have a longer course of disease with more obvious clinical signs, while the clinical signs are less dramatic and the course of disease is shorter in the more susceptible species. However, the usual end point in all MCF-affected animals is death.

A characteristic feature of MCF viruses is that the ability to propagate virus and the manifestation of MCF are entirely separated. Thus, cell-free virus is only propagated in reservoir hosts and is shed in mucous secretions, while MCF-susceptible hosts carry infected cells but do not produce infectious virus.[4] The consequence of this is that MCF does not spread among susceptible populations but can only be introduced by virus shed by reservoir hosts. Thus, MCF remains a sporadic condition that usually affects small numbers of individuals.[1]

Of the MCF viruses, AlHV-1 is the best characterized and remains the only MCF virus that can be propagated in cell culture. This chapter will therefore focus on the current state of knowledge for this virus but will draw in relevant insights from studies of other MCF viruses.

59.2 CLASSIFICATION AND MORPHOLOGY

MCF viruses are γ-herpesviruses of the genus *Macavirus* that share at least one serological determinant and sequence similarity within the DNA polymerase gene.[2] The genomes of AlHV-1 and OvHV-2 have been sequenced[5–7] and found to be collinear with other γ-herpesvirus genomes including human herpesvirus 8 (HHV-8; also known as Kaposi's sarcoma herpesvirus [KSHV][8]) and the prototype γ-herpesvirus herpesvirus saimiri (HVS[9]), allowing the assignment of function to many genes by similarity to known herpesvirus genes.[9] The sequenced MCF virus genomes comprise approximately 130 kilobase pairs (kbp) of moderate (~50%) GC content DNA, encoding about 72 genes, flanked by multiple tandem copies of high (>70%) GC content terminal repeat DNA (Figure 59.1). The terminal repeat units are 4.2 kbp in OvHV-2 and 1.1 kbp in AlHV-1 and lack obvious

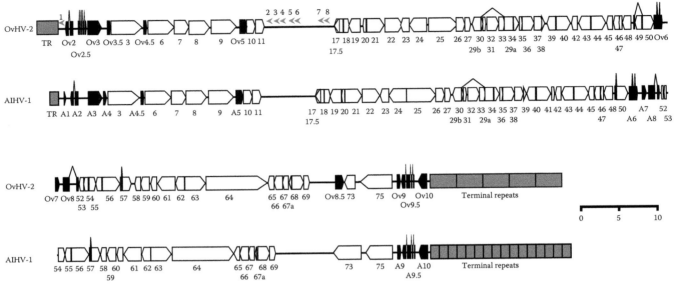

FIGURE 59.1 Schematic representation of the genomes of AlHV-1 and OvHV-2, illustrating the relative positions of the known genes. Each genome is depicted as a single black line with genes shown as block arrows to indicate the direction of transcription, filled white (conserved γ-herpesvirus genes) or black (MCF virus unique genes). The exons of spliced genes are joined by lines above the genes. Terminal repeat units are shown as gray boxes. Gene names, as published in References 5, 6, are given below the map of each genome, while the relative positions of microRNAs of OvHV-2[21] are indicated above that genome as numbered gray arrows. The black scale bar beside the aligned genomes is in kbp.

protein-coding potential. The infectious virus particle comprises a full-length viral genome contained within an icosahedral capsid, surrounded by a largely disordered tegument layer and enclosed within an envelope derived from cellular membranes studded with viral glycoproteins. The envelope proteins of AlHV-1 have been characterized by monoclonal antibody analysis of labeled virus particles,[10,11] while the composition of purified virions has been analyzed by proteomics.[12]

The moderate GC segment of each viral genome encodes all of the components of the mature virion in addition to a range of proteins required for viral DNA replication, virion assembly, and cellular regulation. About 60 genes in AlHV-1 and OvHV-2 have homologues within the γ-herpesvirus subfamily and are numbered by open reading frame (ORF) in accord with the scheme set out for HVS[9] (Figure 59.1). The remaining 12 genes are considered unique to MCF viruses and may have a role in the specific biology and pathology of these viruses. These genes are mainly found between the major groups of conserved genes, suggesting that there may be preferred sites for virus genome modification. The unique genes are numbered according to their position in the genome, with a prefix derived from the virus name (A1–A10 for AlHV-1; Ov2–Ov10 for OvHV-2) (Figure 59.1). The sequence of the OvHV-2 genome revealed that it carried homologues of all of the AlHV-1 unique genes, except A1 and A4, and also had additional unique genes that were named Ov2.5, Ov4.5, and Ov8.5 to reflect their positions between the other unique genes[6] (Figure 59.1). Bioinformatic analysis of the AlHV-1 sequence revealed the presence of a homologue of Ov4.5, termed A4.5, and a recent analysis of cDNA sequences has provided evidence for a previously

unannotated AlHV-1 unique gene—A9.5.[88] This analysis also suggested that the OvHV-2 genome carries a homologous gene, termed Ov9.5.

The potential contribution of the unique genes to the induction of pathology in MCF has made them a target for analysis. Biological functions have been assigned to several unique MCF virus genes by bioinformatic analysis,[5,6,13] and these assignments have been confirmed in a few genes that have been investigated.[14,15] However, some of the unique gene sequences do not have obvious homologues, making assignment of function difficult. Analysis of the predicted protein sequences using current bioinformatic tools may shed more light on the function of these gene products. This analysis is summarized in Table 59.1. In addition to analyses using standard bioinformatic tools (BLAST, PROSITE, NetNGlyc) with the latest database versions (September 2012), analysis was done using alignments of AlHV-1 and OvHV-2 unique gene homologues as query sequences for HHpred analysis (toolkit.tuebingen.mpg.de/hhpred). This approach uses profile hidden Markov models to search structural databases,[16] allowing similarities of both sequence and predicted secondary structure to be used to identify distant homologues of the unique genes as follows:

A1: This putative gene has no similarity to any known protein and contains no conserved sequence features. There remains no evidence for its expression as a protein.

A2/Ov2: These homologues contain a bZIP motif and show HHpred similarity to cellular transcription factors including C-FOS, C-JUN, and CCAAT/ enhancer-binding protein beta.

TABLE 59.1
Unique Genes in MCF Viruses and Their Predicted Function

AlHV-1 Gene[a]	OvHV-2 Gene[a]	Similarity (%)	BLASTp[b]	HHpred[c]	PROSITE[d]	NetNGlyc[e]	Function
A1	—	—	No hit	nd	No hit	0	Unknown
A2	Ov2	33	No hit	Proto-oncogene C-FOS	bZIP	0; 0	Transcription factor
—	Ov2.5	—	IL-10	IL-10	No hit	0	vIL-10
A3	Ov3	48	Semaphorin-7A	Semaphorin-7A	SEMA domain	10; 3	Semaphorin
—	Ov3.5	—	No hit	nd	No hit	1	Unknown
A4	—	—	No hit	nd	No hit	0	Unknown
A4.5	Ov4.5	51	BCL-2 homologues (γHV)	BCL-2-related protein A1	No hit	1; 1	BCL-2 family
A5	Ov5	46	GPCR (γHV)	Sphingosine 1-phosphate receptor 1	No hit	3; 5	G-protein-coupled receptor
A6	Ov6	27	No hit	Transcription factor ATF-4	No hit	0; 0	Transcription factor
A7	Ov7	45	BZLF2 (γHV)	Glycoprotein GP42 (HHV-4)	No hit	3; 1	Glycoprotein
A8	Ov8	50	No hit	Glycoprotein GP350 (HHV-4)	No hit	18; 12	Glycoprotein
—	Ov8.5	—	No hit	nd	No hit	0	Unknown, proline-rich
A9	Ov9	46	No hit	Apoptosis regulator BAX	No hit	0; 1	BCL-2 family
A9.5	Ov9.5	35	No hit	IL-4	No hit	8; 6	Secreted glycoprotein
A10	Ov10	15	No hit	No hit	No hit	0; 2	Unknown

[a] AlHV-1 and OvHV-2 unique genes are listed in order of their position in the genome, with clear homologues sharing the same line.

[b] BLASTp (www.ncbi.nlm.nih.gov/blast/Blast.cgi); *no hit* indicates that no homologue with e-value <0.01 was identified. Homologies with gene families found in several γ-herpesviruses (γHV) are indicated.

[c] HHpred (toolkit.tuebingen.mpg.de/hhpred); *no hit* indicates that no homologue with e-value <0.01 was identified; *nd* indicates that single sequences were not analyzed.

[d] PROSITE (prosite.expasy.org); *no hit* indicates that no significant similarity to any PROSITE domain was detected.

[e] NetNGlyc (www.cbs.dtu.dk/services/NetNGlyc); the number of predicted N-linked glycosylation sites with scores above the threshold (potential >0.5; jury agreement >7/9) in each protein sequence are given in order: AlHV-1 then OvHV-2.

A3/Ov3: The encoded proteins carry SEMA domains and appear to be most similar to semaphorin-7A, which is expressed on activated lymphocytes.[17]

Ov2.5: This protein has clear similarity to interleukin (IL-) 10 and shares the splicing pattern of the mammalian IL-10 gene. Functional analysis has shown that it has IL-10 activity.[14]

A4/Ov3.5: These genes have similar locations on the virus genomes, but are not considered homologues.[6] Although the protein sequences have only 20% overall amino acid similarity, the sequences have a core motif that is shared (58% over 19 residues), suggesting that they may be distant homologues. Both proteins are predicted to be glycosylated and have a signal sequence, suggesting they may be exported. Analyzed individually or together, no homologues could be detected.

A4.5/Ov4.5: These genes appear to encode a BCL-2 family member, potentially with prosurvival function, similar to BCL2A1. The sequences are also similar to the viral BCL-like proteins encoded by other γ-herpesviruses. A4.5-/Ov4.5-based protection from apoptosis might enhance survival of infected cells and it is interesting to note that BCL2A1 functions in inflammation to promote survival of leukocytes.[18]

A5/Ov5: These genes are similar in both sequence and predicted structure to G-protein-coupled receptors (GPCRs), particularly those encoded by other γ-herpesviruses. Interestingly, functional analysis of the A5 gene showed that it encoded a constitutively active GPCR that was not essential for growth in culture or induction of MCF.[15]

A6/Ov6: Although they are not predicted to have a bZIP motif, A6 and Ov6 appear to have similarity to the same family of transcription factors as A2/Ov2, sharing at least one homologue in common.

A7/Ov7: These proteins have similarity to the Epstein–Barr virus (HHV-4) glycoprotein BZLF2 (gp42), which mediates binding to MHC class II molecules, facilitating virus entry to B cells.[19] Like gp42, A7/Ov7 also shares structural similarity with a natural-killer receptor and is therefore likely to have a C-type lectin protein fold.

A8/Ov8: HHpred analysis identifies distant similarity between these proteins and the gp350 protein of Epstein–Barr virus. This viral glycoprotein binds CD21 and is involved in virus entry into B cells.

Ov8.5: The predicted sequence of Ov8.5 contains a high proportion of proline residues (24%) and comprises multiple blocks of a variable repeat sequence. The Ov8.5 sequence encompasses a direct repeat element that is essentially identical in both OvHV-2 isolates sequenced,[6,7] suggesting the region is conserved. Transcription has been detected across this region[20] but no other evidence for Ov8.5 gene function or expression is available.

A9/Ov9: Like A4.5 and Ov4.5, these genes have similarity to the BCL-2 family, although in this case, HHpred analysis suggests that the most similar protein is the proapoptotic protein BAX.

A9.5/Ov9.5: The putative A9.5 gene product was found to be encoded by spliced cDNAs observed in AlHV-1-infected tissues.[88] The splice sites and 170-residue ORF detected in A9.5 were found to be conserved in both sequenced isolates of OvHV-2,[6,7] suggesting the gene was conserved in this virus too. Interestingly, the Ov9.5 predicted polypeptide sequences were only 50% identical to each other and 30% identical to A9.5 but shared features including the positions of N-linked glycosylation sites and cysteine residues that suggest their function was conserved. HHpred analysis of the A9.5 and Ov9.5 sequences suggests structural similarity to the IL-4 family of cytokines.[88]

A10/Ov10: The conservation of an ORF of about 470 residues at the right end of the MCF viruses is the only evidence for the existence of the A10/Ov10 gene. The predicted protein sequences have little sequence similarity to each other (15%) (Table 59.1) and the current analysis found no similarity to known proteins (by BLASTp or HHpred) and no detected functional motifs, although previous analyses detected nuclear localization signals.[1] Ov10 is predicted to be glycosylated, while A10 is not (Table 59.1).

In addition to the currently annotated genes of MCF viruses, recent studies of OvHV-2 infected cells have identified eight putative microRNA species expressed by the virus,[21] all of which appeared to be abundant within the cells. The microRNAs were found in two regions of the genome (Figure 59.1), neither of which has coding capacity. Work to identify virus or host genes that may be regulated by these RNA species is ongoing.

59.3 BIOLOGY AND EPIDEMIOLOGY

MCF viruses are shed in the mucous secretions of infected individuals of the respective reservoir host species. For AlHV-1, wildebeest are infected at an early age in an intense epizootic that makes the calving season the most risky for potential MCF cases among cattle nearby.[22,23] Indeed, Massai pastoralists are known to move their cattle to poorer upland grazing to avoid the risk of MCF, despite the potential risks from other pathogens.[24,25] Similarly, OvHV-2 is shed in sheep nasal secretions[26] and efficiently infects lambs between 2 and 6 months. The lambing season has long been regarded as a high-risk period for in-contact cattle and deer, but research suggests that peak virus shedding by infected lambs occurs 6–9 months after birth.[27] While infection of reservoir hosts by MCF viruses appears efficient, there are no reports of clinical signs being caused by natural infection although minor clinical signs may go unnoticed. There have been reports that describe the disease following experimental infection of naïve sheep with OvHV-2, either by intravenous infusion of infected cells into lambs in utero[28] or by intranasal infection of adult sheep with high doses of virus shed in sheep nasal secretions.[29] In these cases, the animals appeared to develop MCF with characteristic clinical signs and histopathology, and although several of the lambs died,[28] two adult sheep recovered after a few weeks. Infection of sheep with lower doses of OvHV-2 led to infection in the absence of clinical signs.[30] Following initial infection, sheep and wildebeest remain infected for life and the virus is thought to become latent in a lymphocyte subset.

The cells that harbor the latent MCF virus have been identified in OvHV-2-infected sheep as a T cell subset.[31] This is in line with the identification of infected T cells from cattle and rabbits infected with OvHV-2 and AlHV-1,[32–34] suggesting that MCF may not be due to a change in the cell tropism of the viruses in MCF-susceptible species.[31]

Although virus-infected cell lines can be derived from MCF-affected lymphoid tissues,[34,35] only AlHV-1 can be propagated in cell culture.[36] Monolayers of a range of cell types are susceptible to infection by cells from AlHV-1-infected animals, producing a cytopathic effect. AlHV-1 remains cell-associated in culture over multiple serial passages but cell-free pathogenic virus can be derived by freeze-thaw lysis of infected cells after about passage 3.[37] Further passage in tissue culture leads to gradual attenuation of the virus for induction of MCF and an increasingly cell-free growth phenotype, associated with specific rearrangements of the virus genome.[38] The cell-free attenuated strain of AlHV-1 C500 has been maintained for over 1000 passages in cell culture without loss of infectivity or return to pathogenicity. This strain has been used as the basis for an attenuated virus vaccine.[37]

AlHV-1-infected wildebeest develop virus-neutralizing antibodies, both in the circulation and in nasal secretions.[39] OvHV-2-infected sheep also develop virus-specific antibody responses, but virus neutralization cannot be demonstrated because OvHV-2 cannot be propagated in culture. In sheep, it has been shown that virus shedding occurs in very short episodes.[40] These observations suggest that virus propagation in natural hosts may be limited by the host immune system and that virus-neutralizing antibodies may be an important component of control of MCF virus infection. An alternative suggestion is that latently infected T cells in the circulation are activated to replicate virus that can only infect nasal epithelia. This initiates a single round of synchronous virus replication in the nasal epithelium, with the progeny virus having a tropism for lung cells and being unable to infect local epithelial cell types. Thus, the shedding of virus is limited naturally by the cell tropism of virus propagated.[40]

MCF occurs when individuals of a susceptible species become infected with virus shed by a reservoir host. Experimental intranasal infection studies have demonstrated that, for OvHV-2 at least, there is a range of susceptibility for infection and a correlation between virus dose and the appearance of clinical signs.[29,30,41,42] Thus, sheep were infected without clinical signs with pooled sheep nasal secretions containing 10^3 OvHV-2 genome copies, while high doses of virus (>10^9 genome copies) induced fatal or transient MCF-like symptoms.[29] In bison, 5×10^4 OvHV-2 genome copies consistently induced disease,[41] while lower doses induced subclinical infection and occasional disease. Cattle were less susceptible than bison but both subclinical and clinical infections were demonstrated at 10^8 OvHV-2 genome copies.[43] Similarly, the experimental host, rabbit, could also be infected by nebulization of nasal secretions containing 10^7 OvHV-2 genome copies.[42]

The clinical signs of MCF can be varied, and several distinct manifestations of disease have been reported, including peracute, head and eye, alimentary, and neurological.[1] Most MCF cases in cattle present with fever, depression, and lymphadenopathy, while the common head and eye form is further characterized by nasal and ocular secretions, corneal opacity, skin lesions on the muzzle, and erosion of the gums, tongue, and palate. In the peracute form, either no clinical signs are detected or depression followed by diarrhea is seen, with death occurring within a few days of the first clinical signs. In the alimentary form, hemorrhagic diarrhea may also be found,[44] while nervous signs, ataxia, and blindness have been reported in the neurological form.[45] Although MCF is generally recognized as a disease with a high case fatality rate, reports of recovery from clinical MCF and chronic infection have also been published.[46–48] In addition, experimental and natural infection may also lead to subclinical infection.[41,43,49]

The development of a serological assay for MCF virus infection, based on a monoclonal antibody specific for an AlHV-1 antigen, has facilitated a number of epidemiological studies in both domesticated and wildlife species. These showed that most individuals within domestic and wild sheep species and muskox were infected and seropositive,[50,51] while the frequency of seropositivity in bison, cattle, deer, caribou, and moose was generally below 10%.[52,53] This seems to suggest that among MCF virus reservoir species, infection is efficient and subclinical, such that most animals become infected and seroconvert. In contrast, infection of MCF-susceptible species is less efficient and the resulting disease is often fatal, leading to a much lower frequency of seropositive animals. However, the indication that most susceptible species appear to have a subpopulation of MCF-seropositive animals[50,52,53] suggests that subclinical infection is part of the natural clinical spectrum of MCF. This view is supported by a longitudinal study of dairy cattle in which about half of the herd studied were seropositive for MCF at multiple timepoints.[54] Parallel PCR assays for OvHV-2 DNA were intermittently positive but none of the herd showed any signs of MCF.

The epidemiology of MCF is driven by the presence of reservoir species in close proximity to susceptible animals. The timing and magnitude of virus shedding and the proximity and numbers of reservoir species are important factors. Thus, Massai pastoralists avoid local grazing areas during the annual wildebeest migration and calving season, driving their cattle tens of kilometers to poorer upland grazing.[24,25] In contrast, cograzing of sheep and cattle is commonly seen in the United Kingdom, suggesting that there may be differences in the amount of virus shed, virus infectivity, or the susceptibility of cattle to AlHV-1 compared to OvHV-2.

59.4 CLINICAL FEATURES AND PATHOGENESIS

The clinical course of MCF varies with the susceptibility of the infected species. Highly susceptible species such as deer may show few symptoms other than fever before death within a few days, while hosts such as cattle have a more protracted course with a wider range of clinical signs over 1–2 weeks. MCF in cattle is characterized by fever, lymphadenopathy, and ocular and nasal discharge, with lesions in and around the mouth and muzzle and throughout the gastrointestinal tract. Occasional animals exhibit diarrhea. Histopathologic examination shows MCF is characterized by the infiltration and proliferation of lymphocytes in a range of tissues, leading to epithelial erosion, hyperplasia and necrosis of lymphoid organs, and interstitial accumulation of lymphoid cells in nonlymphoid tissues. Vasculitis is a common observation, found in many tissues but particularly in the brain, liver, and in the kidney, where lesions are visible on the surface of the cortex as pale foci of up to 5 mm in diameter.[4]

Studies of AlHV-1 and OvHV-2 infection of rabbits and cattle suggest that the disease is similar across species and viruses[32,55,56] but there are subtle differences in the specific pathology caused by AlHV-1 and OvHV-2.[56] More recent studies have shown that both viruses induce the proliferation of a cytotoxic subset of CD8-positive lymphocytes that appears to infiltrate multiple tissues from the circulation and that many of these T cells appear to be infected.[32,33,57,58] The involvement of unrestricted cytotoxicity in the pathogenesis of

MCF is supported by the observations that IL-15 is expressed in MCF lesions[59] and that IFN-γ is upregulated in tissues of rabbits and cattle infected with AlHV-1.[33,60–62]

Following infection with MCF virus by a natural (oronasal) route,[32,37] the virus is not detectable in the blood until just prior to the onset of clinical signs. It has been observed in sheep and in bison that OvHV-2 appears to replicate first in alveolar epithelial cells before infecting lymphocytes.[63,64] The observation that OvHV-2 derived from sheep nasal secretions appears to be unable to induce MCF following intravenous or intraperitoneal inoculation[42,49] but can induce MCF following intranasal administration[32,41,42,49,55] suggests that a phase of replication in the lung may be mandatory for natural infection.

This has not been observed for tissue-culture-propagated cell-free AlHV-1, which can cause MCF in cattle and rabbits following administration by either route.[37,88] This supports the view that the cell type in which the virus propagates may influence the tropism of the virus. The phenomenon of tropism switching, in which virus propagated in one cell type is infectious for a different cell type, has been cited as a potential explanation of the limited replication observed in bison lung and the limited nature of virus shedding episodes in sheep.[49,65] However, it should be noted that infectious cell-free virus has never been isolated from MCF-susceptible species, although cell-associated virus may be propagated from AlHV-1-infected tissues.

A recent study of MCF in bison described MCF-specific antibody responses, the appearance of MCF virus DNA in the blood and tissues, and clinical or histopathological signs in animals sacrificed at various times after infection with OvHV-2.[49] This confirmed studies in sheep that suggested an initial phase of replication in the lung[64] although local changes in immune response gene expression were limited to inflammatory-related genes,[49] suggesting differences in the control of OvHV-2 infection between sheep and bison. Following replication in the lung, the virus disseminated to the periphery, in line with results from other studies, and was found in T cell subsets that appear to be the main infected cell type.[32,57,58] Up to 10% of circulating T cells appear to be infected[57] in MCF-affected animals but there is conflicting evidence regarding virus gene expression patterns in peripheral tissues and lymphocytes. Studies of AlHV-1 infection of rabbits and OvHV-2 infection of cattle suggest that circulating lymphocytes and lymphoid tissues have a latent pattern of virus gene expression,[57,61] while analysis of tissues from OvHV-2-infected bison and rabbits suggests a less restricted pattern of virus gene expression, including markers of the lytic cycle.[49,66]

As clinical signs become evident, the viral load in peripheral blood mononuclear cells (PBMC) increases, as detected by real-time quantitative PCR,[67,68] and following necropsy, viral DNA can be detected by PCR in many tissues, including blood, liver, kidney, spleen, lymph node, brain, and cornea. This probably reflects the distribution of infected T cells via the bloodstream and the infiltration of these cells into multiple tissues where local cytotoxic activity is thought to induce tissue damage and ultimately death.[32,33,57,69]

Despite the detailed work that has been done, the role of infected and uninfected cells at lesion sites still remains to be elucidated. Previous work using in situ PCR and immunohistochemistry to identify the infected cells in MCF brain lesions showed that most of the infiltrating cells were CD8 positive and that many of these were infected.[70] This is in agreement with more recent studies in rabbits infected with a bioluminescent recombinant AlHV-1, which graphically illustrated the widespread distribution of infected cells and showed that they occurred in close proximity to MCF lesions.[69] These data confirm the involvement of infected cells in lesion development but still do not clearly show whether infected cells are the effectors of tissue damage or act as regulatory cells, inducing an effector phenotype in other cells recruited to lesions. Studies of MCF lesions that are able to simultaneously identify infected cells and effector cells or local tissue damage are needed to shed further light in this area. An improved understanding of the local extracellular milieu might also be of great interest, since studies from several groups have identified markers of clinical MCF including the depletion of IL-2 and the overexpression of IFN-γ and cytotoxic effectors such as perforin.[32,33,57,58,60,61,69] Detecting the expression of such molecules within tissue lesions might shed further light on their specific roles.

59.5 IDENTIFICATION AND DIAGNOSIS

Infection of reservoir host species with MCF viruses leads to the development of a virus-specific antibody response that is virus neutralizing, at least for AlHV-1 where it can be assayed.[4] In MCF-susceptible hosts, no virus-neutralizing antibody response is induced, but infected animals can produce antibodies that detect virus antigens by ELISA, indirect immunofluorescence, or Western blot. In addition, the AlHV-1 virus can be isolated from wildebeest and infected cattle, although virus from cattle is strictly cell-associated.[4] No other MCF viruses have been propagated in culture to date and this has limited both the analysis of their biology and the development of virus-specific diagnostic tests. Thus, in all diagnostic tests for MCF-specific antibodies, the antigens used are derived from AlHV-1 propagated in tissue culture.

The development of monoclonal antibodies against AlHV-1 antigens[11] has allowed the development of a competitive-inhibition (CI-) ELISA, which recognizes an epitope that appears to be common to MCF viruses. This assay[71] has been extensively used in the diagnosis of MCF in the United States and elsewhere. A direct ELISA, based on a detergent extract of AlHV-1-infected cells, was also found to be effective in detecting MCF virus infections,[72] while a quantitative ELISA based on an extract of AlHV-1 virions was used to analyze antibody responses to vaccination and challenge with AlHV-1.[73] It is of interest that intranasal challenge with OvHV-2 induced virus-specific antibodies in a dose-dependent manner in cattle and bison.[41,43] These studies demonstrated that animals challenged with lower doses of OvHV-2 could become subclinically infected without seroconverting. This contrasts with intranasal challenge experiments with a fatal dose of AlHV-1 in cattle[37,73] in which no unvaccinated, challenged animal developed antibody responses, detected either by ELISA or virus neutralization assay. This seems to be an effect of virus and species combination since parenteral infection of rabbits with either OvHV-2 or AlHV-1 induced the production of virus-specific antibodies. However, MCF-specific antibody responses are not consistently found in animals affected by sheep-associated MCF.[4,74,75] This may indicate that induction of disease can occur in the absence of a detectable antibody response. In some cases, this may be due to the speed of disease progression, particularly in highly susceptible species such as deer or bison, where peracute MCF may lead to death before an antibody response is detectable.[1]

The presence of infected cells in the blood and tissues of animals affected by MCF and the availability of sequence data from a number of MCF viruses have allowed the development of both standard and real-time PCR assays for MCF viruses that have proven useful in diagnostic testing.[67,68,76,77] These assays appear to be effective in MCF-affected animals, with sensitivity that compared well with CI-ELISA and histopathology,[53,78] but the nested PCR approach appears to be more sensitive than either a nonnested standard PCR or a real-time PCR assay. The power of PCR for the detection of clinical MCF cases may be a reflection of the pathology of MCF, with infected cells being frequent in the blood and infiltrating both lymphoid and nonlymphoid organs, allowing virus DNA to be detected in almost any tissue. In contrast, detection of MCF virus DNA in reservoir hosts can also be done by PCR, but low viral load in samples from latently infected animals may reduce the reliability of this approach for recognizing infected animals.[78] A pan-herpesvirus-nested PCR[79] has been used to amplify a region within the well-conserved DNA polymerase gene from a range of MCF-affected and reservoir host animals.[2] Analysis of the resulting DNA sequences and CI-ELISA testing of parallel serum samples demonstrated that the MCF viruses shared both the specific epitope recognized by the CI-ELISA and features of the DNA polymerase gene sequence. This work identified 10 putative MCF viruses, of which four had been associated with disease (AlHV-1, OvHV-2, CpHV-2, and MCFV-WTD[2]). The sensitivity and broad specificity of this consensus PCR assay gives it the potential to be the gold standard for PCR-based analysis of MCF virus infection.[79] The ease and sensitivity of PCR-based tests and the range of samples suitable for analysis has made this approach increasingly popular as a diagnostic tool. However, histopathology remains the definitive diagnostic test.[4] In the absence of histopathological evidence, a positive PCR or serological test would provide support to clinical observations for a diagnosis of MCF.

59.6 TREATMENT AND PREVENTION

Currently, there is no effective treatment or vaccine for MCF and the best approach to prevention remains the separation of reservoir hosts from susceptible species where possible. Recent studies have the potential to improve this situation. A candidate vaccine for AlHV-1 has been described, based on the use of the attenuated strain of AlHV-1 C500.[37] This live vaccine has been reported to provide significant protection against an intranasal challenge with a fatal dose of pathogenic AlHV-1 C500 and protection appears to remain for at least 6 months.[73] However, this approach has yet to be tested under field conditions.

Detailed analysis of MCF pathogenesis may also yield routes to treatment or rational therapy. For example, microarray analysis of cattle naturally infected with OvHV-2[61] has suggested that IL-2 depletion may be central to MCF, while a similar study of cattle experimentally infected with AlHV-1 identified immunological pathways and gene targets that were significantly upregulated.[60] These studies may be exploited by identifying agonists or inhibitors of selected host gene products that are potential therapeutic targets. Thus, aryl-substituted isobenzofuranones, inhibitors of the cytotoxic effector perforin that have been described in the context of immunosuppression and autoimmune disease,[80] may prevent the tissue damage seen in MCF. Similarly, interference with the cytokines and receptors that mediate the recruitment of activated T cells may be of benefit. The recent study of host gene expression in AlHV-1-induced MCF[60] suggested that ligands of CXCR3 might be involved in trafficking of activated T cells to lesion sites. Thus, the CXCR3 antagonist compound SCH 546738, which has been shown to inhibit chemotaxis and attenuate disease in a range of autoimmune models,[81] may have therapeutic value in MCF.

Similar examples may be found in studies of MCF pathogenesis, where inhibitors of implicated molecules such as IL-15[59] may be tested. Additionally, the recent development of a recombinant AlHV-1[82] allows the direct manipulation of specific virus genes[15] to identify candidates that might be essential for virus pathogenesis and therefore may be targets for specific inhibitors or antibody therapy.

However, in the absence of licensed vaccines or therapeutics, the best approach to the prevention of MCF is the separation of reservoir and susceptible hosts. The degree of separation required is difficult to calculate as it depends on multiple factors including the amount of virus shed by the reservoir species, the susceptibility of the target species, the number of each species in contact, the separation between the species, and the season. In both WA- and SA-MCF, contact during the calving/lambing period has been regarded as the greatest risk of infection. This may be especially true for WA-MCF where the influx of large numbers of wildebeest calving synchronously presents such a risk of MCF that pastoralist cattle are driven to poorer grazing where the risks of other diseases such as trypanosomiasis or east coast fever are increased, with consequent poor productivity.[24,25] In contrast, studies in lambs have shown that they may not become infected for several months,[27] suggesting that lambs shedding OvHV-2 may be a risk to susceptible species from 2 to 10 months of age. Virus shedding by adult sheep may also contribute significantly to the risk of MCF for in-contact animals,[83] while cases involving infection of large numbers of bison emphasize the risks of MCF for this particularly susceptible species.[84,85] Shared housing may compound the risk by concentrating the infectious material but even indirect contact between bison and sheep can lead to MCF[86] and cases of infection over several kilometers have also been reported.[87]

Thus, it is difficult to make precise recommendations for biosecurity purposes, but the strict separation of reservoir and susceptible species should form the basis for prevention of MCF, with no sharing of grazing, housing, feed, or water. There is no need to isolate MCF-affected animals, however, because they do not present an infection risk to other livestock. For valuable, highly susceptible species such as bison, deer, or exotic hoofstock in zoological collections, where airborne infection over large distances is possible, efforts should be made to ensure the greatest possible separation between reservoir and susceptible species. The prevailing wind direction and weather conditions should also be considered. For these animals, the development of an effective vaccine or therapy would be a major step forward.

59.7 CONCLUSIONS AND FUTURE PERSPECTIVES

MCF is a complex disease with a generally fatal outcome for susceptible species, yet infection of the reservoir hosts occurs without obvious clinical signs. In susceptible hosts, infected lymphocytes proliferate and spread systemically, while in reservoir hosts, infected lymphocytes spread systemically, but with little apparent proliferation. Thus, the ability of the host to control the proliferation of infected lymphocytes may define whether an animal develops MCF.

Much of the early work aimed at understanding the pathogenesis of MCF has been descriptive but recent studies are allowing specific hypotheses to be addressed. Thus, the function of virus genes can be analyzed in vivo using recombinant viruses; the consequences of MCF virus infection can be analyzed at the molecular level; tools have been developed to allow the identification and tracking of infected cells; and systems to test potential vaccines and therapeutics have been developed.

The development of intranasal challenge systems for both AlHV-1 and OvHV-2 provide a more natural approach to the analysis of pathogenesis and the evaluation of vaccines or therapies. For MCF induced by AlHV-1, progress has been made with the development of a vaccine that protects against experimental infection but this remains to be tested in the field and transferring this success to the more prevalent sheep-associated MCF will be a challenge. However, this experimental system allows the protective mechanisms of the vaccine to be investigated and may allow protective antigens to be characterized.

However, it is clear that the nature of the host is fundamental to disease and the analysis of the host response to MCF is likely to be a central component of future research. Published analyses of host immunological and transcriptional changes are likely to form the basis of more focused studies to elucidate the cell types and the activities that cause tissue damage during MCF. Analysis of the local cytokine environment is likely to be an important element of such studies, since

this will improve our understanding of the cell activation and trafficking that are clearly involved in lesion development.

The severity of MCF outbreaks can vary greatly, with differences in the clinical signs of disease and wide variation in the number of animals affected. Local circumstances are likely to influence the severity of specific outbreaks but there may also be virus-encoded determinants that affect the amount of virus shed or the pathogenesis of disease. There is a need for epidemiological studies to determine whether different strains of virus cause MCF outbreaks of distinct severity, in terms of clinical signs or number of animals affected. This should be underpinned by greater efforts to generate sequence data from multiple loci (or complete genomes) from clinical samples in order to clearly identify virus strains. This work may be facilitated by the increased availability of high-throughput sequencing, but the identification of loci that differ between a few fully sequenced genomes may provide clues to the best targets for molecular epidemiology projects.

In conclusion, MCF can have devastating effects in susceptible ungulates but is generally sporadic, limited by the inability of the causative viruses to propagate in the diseased host. Thus, animals suffering from MCF are truly *dead-end* hosts and the main approach to protection from MCF remains the separation of susceptible animals from those that shed virus.

REFERENCES

1. Russell, G.C., Stewart, J.P., and Haig, D.M., Malignant catarrhal fever: A review. *Vet. J.* 2009;179:324–335.
2. Li, H. et al., A novel subgroup of rhadinoviruses in ruminants. *J. Gen. Virol.* 2005;86:3021–3026.
3. Li, H. et al., Newly recognized herpesvirus causing malignant catarrhal fever in white-tailed deer (*Odocoileus virginianus*). *J. Clin. Microbiol.* 2000;38:1313–1318.
4. World Organisation for Animal Health (OIE), Malignant catarrhal fever, In *Manual of Diagnostic Tests and Vaccines for Terrestrial Animals*, 6th ed. Paris, France: OIE; 2008, pp. 779–788.
5. Ensser, A., Pflanz, R., and Fleckenstein, B., Primary structure of the alcelaphine herpesvirus 1 genome. *J. Virol.* 1997;71:6517–6525.
6. Hart, J. et al., Complete sequence and analysis of the ovine herpesvirus 2 genome. *J. Gen. Virol.* 2007;88:28–39.
7. Taus, N.S. et al., Comparison of ovine herpesvirus 2 genomes isolated from domestic sheep (*Ovis aries*) and a clinically affected cow (*Bos bovis*). *J. Gen. Virol.* 2007;88:40–45.
8. Russo, J.J. et al., Nucleotide sequence of the Kaposi sarcoma-associated herpesvirus (HHV8). *Proc. Natl. Acad. Sci. USA* 1996;93:14862–14867.
9. Albrecht, J.C. et al., Primary structure of the herpesvirus saimiri genome. *J. Virol.* 1992;66:5047–5058.
10. Adams, S.W. and Hutt-Fletcher, L.M., Characterization of envelope proteins of alcelaphine herpesvirus 1. *J. Virol.* 1990;64:3382–3390.
11. Li, H. et al., Identification and characterization of the major proteins of malignant catarrhal fever virus. *J. Gen. Virol.* 1995;76:123–129.
12. Dry, I. et al., Proteomic analysis of pathogenic and attenuated alcelaphine herpesvirus 1. *J. Virol.* 2008;82:5390–5397.

13. Coulter, L.J., Wright, H., and Reid, H.W., Molecular genomic characterization of the viruses of malignant catarrhal fever. *J. Comp. Pathol.* 2001;124:2–19.

14. Jayawardane, G. et al., A captured viral interleukin 10 gene with cellular exon structure. *J. Gen. Virol.* 2008;89:2447–2455.

15. Boudry, C. et al., The A5 gene of alcelaphine herpesvirus 1 encodes a constitutively active G-protein-coupled receptor that is non-essential for the induction of malignant catarrhal fever in rabbits. *J. Gen. Virol.* 2007;88:3224–3233.

16. Soding, J., Biegert, A., and Lupas, A.N., The HHpred interactive server for protein homology detection and structure prediction. *Nucleic Acids Res.* 2005;33:W244–W248.

17. Holmes, S. et al., Sema7A is a potent monocyte stimulator. *Scand. J. Immunol.* 2002;56:270–275.

18. Vogler, M., BCL2A1: The underdog in the BCL2 family. *Cell Death Differ.* 2012;19:67–74.

19. Shaw, P. et al., Characteristics of Epstein Barr virus envelope protein gp42. *Virus Genes* 2010;40:307–319.

20. Thonur, L. et al., Differential transcription of ovine herpesvirus 2 genes in lymphocytes from reservoir and susceptible species. *Virus Genes* 2006;32:27–35.

21. Levy, C.S. et al., Novel virus-encoded microRNA molecules expressed by ovine herpesvirus 2-immortalized bovine T-cells. *J. Gen. Virol.* 2012;93:150–154.

22. Mushi, E.Z., Rurangirwa, F.R., and Karstad, L., Shedding of malignant catarrhal fever virus by wildebeest calves. *Vet. Microbiol.* 1981;6:281–286.

23. Plowright, W., Malignant catarrhal fever. *Rev. Sci. Tech. Off. Int. Epiz.* 1986;5:897–918.

24. Cleaveland, S. et al., Assessing the impact of malignant catarrhal fever in Ngorongoro District, Tanzania. 2001. www.eldis.org/fulltext/cape_new/MCF_Maasai_Tanzania.pdf, accessed on September 25, 2012.

25. Bedelian, C., Nkedianye, D., and Herrero, M., Maasai perception of the impact and incidence of malignant catarrhal fever (MCF) in southern Kenya. *Prev. Vet. Med.* 2007;78:296–316.

26. Kim, O., Li, H., and Crawford, T.B., Demonstration of sheep-associated malignant catarrhal fever virions in sheep nasal secretions. *Virus Res.* 2003;98:117–122.

27. Li, H. et al., Transmission of ovine herpesvirus 2 in lambs. *J. Clin. Microbiol.* 1998;36:223–226.

28. Buxton, D. et al., Transmission of a malignant catarrhal fever-like syndrome to sheep—Preliminary experiments. *Res. Vet. Sci.* 1985;38:22–29.

29. Li, H. et al., Malignant catarrhal fever-like disease in sheep after intranasal inoculation with ovine herpesvirus-2. *J. Vet. Diagn. Invest.* 2005;17:171–175.

30. Taus, N.S. et al., Experimental infection of sheep with ovine herpesvirus 2 via aerosolization of nasal secretions. *J. Gen. Virol.* 2005;86:575–579.

31. Meier-Trummer, C.S., Ryf, B., and Ackermann, M., Identification of peripheral blood mononuclear cells targeted by Ovine herpesvirus-2 in sheep. *Vet. Microbiol.* 2010;141:199–207.

32. Li, H. et al., Characterization of ovine herpesvirus 2-induced malignant catarrhal fever in rabbits. *Vet. Microbiol.* 2011;150:270–277.

33. Dewals, B.G. and Vanderplasschen, A., Malignant catarrhal fever induced by Alcelaphine herpesvirus 1 is characterized by an expansion of activated CD3(+)CD8(+)CD4(−) T cells expressing a cytotoxic phenotype in both lymphoid and non-lymphoid tissues. *Vet. Res.* 2011;42:95.

34. Reid, H.W. et al., A cyto-toxic lymphocyte-T line propagated from a rabbit infected with sheep associated malignant catarrhal fever. *Res. Vet. Sci.* 1983;34:109–113.

35. Reid, H.W. et al., Isolation and characterization of lymphoblastoid-cells from cattle and deer affected with sheep-associated malignant catarrhal fever. *Res. Vet. Sci.* 1989;47:90–96.

36. Plowright, W., Ferris, R.D., and Scott, P.R., Blue wildebeest and the aetiological agent of malignant catarrhal fever. *Nature* 1960;188:1167–1169.

37. Haig, D.M. et al., An immunisation strategy for the protection of cattle against alcelaphine herpesvirus-1-induced malignant catarrhal fever. *Vaccine* 2008;26:4461–4468.

38. Wright, H. et al., Genome re-arrangements associated with loss of pathogenicity of the γ-herpesvirus alcelaphine herpesvirus-1. *Res. Vet. Sci.* 2003;75:163–168.

39. Plowright, W., Malignant catarrhal fever in East Africa III—Neutralizing antibody in free-living wildebeest. *Res. Vet. Sci.* 1967;8:129–136.

40. Li, H. et al., Shedding of ovine herpesvirus 2 in sheep nasal secretions: The predominant mode for transmission. *J. Clin. Microbiol.* 2004;42:5558–5564.

41. Gailbreath, K.L. et al., Experimental nebulization of American bison (*Bison bison*) with low doses of ovine herpesvirus 2 from sheep nasal secretions. *Vet. Microbiol.* 2010;143:389–393.

42. Gailbreath, K.L. et al., Experimental infection of rabbits with ovine herpesvirus 2 from sheep nasal secretions. *Vet. Microbiol.* 2008;132:65–73.

43. Taus, N.S. et al., Experimental aerosol infection of cattle (*Bos taurus*) with ovine herpesvirus 2 using nasal secretions from infected sheep. *Vet. Microbiol.* 2006;116:29–36.

44. Holliman, A. et al., Malignant catarrhal fever in cattle in the UK. *Vet. Rec.* 2007;161:494–495.

45. Mitchell, E. and Scholes, S., Unusual presentation of malignant catarrhal fever involving neurological disease in young calves. *Vet. Rec.* 2009;164:240–242.

46. OToole, D. et al., Chronic and recovered cases of sheep-associated malignant catarrhal fever in cattle. *Vet. Rec.* 1997;140:519–524.

47. Milne, E.M. and Reid, H.W., Recovery of a cow from malignant catarrhal fever. *Vet. Rec.* 1990;126:640–641.

48. Penny, C., Recovery of cattle from malignant catarrhal fever. *Vet. Rec.* 1998;142:227.

49. Cunha, C.W. et al., Ovine herpesvirus 2 infection in American bison: Virus and host dynamics in the development of sheep-associated malignant catarrhal fever. *Vet. Microbiol.* 2012;159:307–319.

50. Li, H. et al., Investigation of sheep-associated malignant catarrhal fever virus-infection in ruminants by PCR and competitive-inhibition enzyme-linked-immunosorbent-assay. *J. Clin. Microbiol.* 1995;33:2048–2053.

51. Zarnke, R.L., Li, H., and Crawford, T.B., Serum antibody prevalence of malignant catarrhal fever viruses in seven wildlife species from Alaska. *J. Wildl. Dis.* 2002;38:500–504.

52. Li, H. et al., Prevalence of antibody to malignant catarrhal fever virus in wild and domestic ruminants by competitive-inhibition ELISA. *J. Wildl. Dis.* 1996;32:437–443.

53. Frolich, K., Li, H., and Muller-Doblies, U., Serosurvey for antibodies to malignant catarrhal fever-associated viruses in free-living and captive cervids in Germany. *J. Wildl. Dis.* 1998;34:777–782.

54. Powers, J.G. et al., Evaluation of ovine herpesvirus type 2 infections, as detected by competitive inhibition ELISA and polymerase reaction assay in dairy cattle without clinical signs of malignant catarrhal fever. *J. Am. Vet. Med. Assoc.* 2005;227:606–611.

55. O'Toole, D. et al., Intra-nasal inoculation of American bison (*Bison bison*) with ovine herpesvirus-2 (OvHV-2) reliably reproduces malignant catarrhal fever. *Vet. Pathol.* 2007;44:655–662.

56. Anderson, I.E. et al., Immunohistochemical study of experimental malignant catarrhal fever in rabbits. *J. Comp. Pathol.* 2007;136:156–166.

57. Dewals, B. et al., Malignant catarrhal fever induced by alcelaphine herpesvirus 1 is associated with proliferation of CD8+ T cells supporting a latent infection. *PLoS ONE* 2008;3:e1627.

58. Nelson, D.D. et al., CD8(+)/perforin(+)/WC1(–) gamma delta T cells, not CD8(+) alpha beta T cells, infiltrate vasculitis lesions of American bison (*Bison bison*) with experimental sheep-associated malignant catarrhal fever. *Vet. Immunol. Immunopathol.* 2010;136:284–291.

59. Anderson, I.E. et al., Production and utilization of interleukin-15 in malignant catarrhal fever. *J. Comp. Pathol.* 2008;138:131–144.

60. Russell, G.C. et al., Host gene expression changes in cattle infected with Alcelaphine herpesvirus 1. *Virus Res.* 2012; 169:246–254.

61. Meier-Trummer, C.S. et al., Malignant catarrhal fever of cattle is associated with low abundance of IL-2 transcript and a predominantly latent profile of ovine herpesvirus 2 gene expression. *PLoS One* 2009;4:e6265.

62. Schock, A., Collins, R.A., and Reid, H.W., Phenotype, growth regulation and cytokine transcription in Ovine Herpesvirus-2 (OHV-2)-infected bovine T-cell lines. *Vet. Immunol. Immunopathol.* 1998;66:67–81.

63. Cunha, C.W. et al., Detection of ovine herpesvirus 2 major capsid gene transcripts as an indicator of virus replication in shedding sheep and clinically affected animals. *Virus Res.* 2008;132:69–75.

64. Taus, N.S. et al., Sheep (*Ovis aries*) airway epithelial cells support ovine herpesvirus 2 lytic replication in vivo. *Vet. Microbiol.* 2010;145:47–53.

65. Li, H. et al., Ovine herpesvirus 2 replicates initially in the lung of experimentally infected sheep. *J. Gen. Virol.* 2008;89:1699–1708.

66. Meier-Trummer, C.S. et al., Ovine herpesvirus 2 structural proteins in epithelial cells and M-cells of the appendix in rabbits with malignant catarrhal fever. *Vet. Microbiol.* 2009;137:235–242.

67. Traul, D.L. et al., A real-time PCR assay for measuring alcelaphine herpesvirus-1 DNA. *J. Virol. Methods* 2005;129:186–190.

68. Hussy, D. et al., Quantitative fluorogenic PCR assay for measuring ovine herpesvirus 2 replication in sheep. *Clin. Diagn. Lab. Immunol.* 2001;8:123–128.

69. Dewals, B. et al., Ex vivo bioluminescence detection of alcelaphine herpesvirus 1 infection during malignant catarrhal fever. *J. Virol.* 2011;85:6941–6954.

70. Simon, S. et al., The vascular lesions of a cow and bison with sheep-associated malignant catarrhal fever contain ovine herpesvirus 2-infected CD8(+) T lymphocytes. *J. Gen. Virol.* 2003;84:2009–2013.

71. Li, H. et al., A simpler, more sensitive competitive inhibition enzyme-linked immunosorbent assay for detection of antibody to malignant catarrhal fever viruses. *J. Vet. Diagn. Invest.* 2001;13:361–364.

72. Fraser, S.J. et al., Development of an enzyme-linked immunosorbent assay for the detection of antibodies against malignant catarrhal fever viruses in cattle serum. *Vet. Microbiol.* 2006;116(1–3):21–28.

73. Russell, G. et al., Duration of protective immunity and antibody responses in cattle immunised against alcelaphine herpesvirus-1-induced malignant catarrhal fever. *Vet. Res.* 2012;43:51.

74. Muller-Doblies, U.U. et al., Field validation of laboratory tests for clinical diagnosis of sheep-associated malignant catarrhal fever. *J. Clin. Microbiol.* 1998;36:2970–2972.

75. Rossiter, P.B., Antibodies to malignant catarrhal fever virus in cattle with non-wildebeest-associated malignant catarrhal fever. *J. Comp. Pathol.* 1983;93:93–97.

76. Flach, E.J. et al., Gamma herpesvirus carrier status of captive artiodactyls. *Res. Vet. Sci.* 2002;73:93–99.

77. Baxter, S.I.F. et al., PCR detection of the sheep-associated agent of malignant catarrhal fever. *Arch. Virol.* 1993;132:145–159.

78. Traul, D.L. et al., Validation of nonnested and real-time PCR for diagnosis of sheep-associated malignant catarrhal fever in clinical samples. *J. Vet. Diagn. Invest.* 2007;19:405–408.

79. VanDevanter, D.R. et al., Detection and analysis of diverse herpes-viral species by consensus primer PCR. *J. Clin. Microbiol.* 1996;34:1666–1671.

80. Spicer, J.A. et al., Inhibition of the pore-forming protein perforin by a series of aryl-substituted isobenzofuran-1(3H)-ones. *Bioorg. Med. Chem.* 2012;20:1319–1336.

81. Jenh, C.H. et al., A selective and potent CXCR3 antagonist SCH 546738 attenuates the development of autoimmune diseases and delays graft rejection. *BMC Immunol.* 2012;13:2.

82. Dewals, B. et al., Cloning of the genome of Alcelaphine herpesvirus 1 as an infectious and pathogenic bacterial artificial chromosome. *J. Gen. Virol.* 2006;87:509–517.

83. Li, H. et al., Levels of ovine herpesvirus 2 DNA in nasal secretions and blood of sheep: Implications for transmission. *Vet. Microbiol.* 2001;79:301–310.

84. O'Toole, D. et al., Malignant catarrhal fever in a bison (*Bison bison*) feedlot, 1993–2000. *J. Vet. Diagn. Invest.* 2002;14:183–193.

85. Li, H. et al., A devastating outbreak of malignant catarrhal fever in a bison feedlot. *J. Vet. Diagn. Invest.* 2006;18:119–123.

86. Berezowski, J.A. et al., An outbreak of sheep-associated malignant catarrhal fever in bison (*Bison bison*) after exposure to sheep at a public auction sale. *J. Vet. Diagn. Invest.* 2005;17:55–58.

87. Li, H. et al., Long distance spread of malignant catarrhal fever virus from feedlot lambs to ranch bison. *Can. Vet. J.* 2008;49:183–185.

88. Russell, G.C. et al., A novel spliced gene in Alcelaphine herpesvirus 1 encodes a glycoprotein which is secreted in vitro. *J. Gen. Virol.* 2013;94:2515.

60 Newcastle Disease Virus

Claudio L. Afonso and Patti J. Miller

CONTENTS

60.1 INTRODUCTION

Newcastle disease virus (NDV) is also known as avian paramyxovirus serotype-1 (APMV-1). While all NDV are referred to as APMV-1 and are of one serotype, only infections with virulent NDV (vNDV) cause Newcastle disease (ND). NDV strains are defined as virulent if they (1) have three or more basic amino acids in their fusion (F) cleavage sites (position 113–116 of the uncleaved F protein [F0]) with a phenylalanine at position 117 or (2) obtain an intracerebral pathogenicity index (ICPI) value of ≥0.7 in day-old chickens (*Gallus gallus*).[1] Failure to demonstrate multiple basic amino acids requires isolates to be tested for an ICPI value. There are three terms that are commonly used when discussing NDV isolates and the clinical disease they cause in chickens. Lentogens are nonvirulent NDV isolates that may cause no disease or minor respiratory infections. Mesogens and velogens are NDV isolates that lead to medium and severe clinical disease, respectively, that are both classified as vNDV. These three terms are most often used to describe the viruses based on clinical signs and lesions resulting from poultry infected with APMV-1 and were used more often before sequencing of viral genomes was not as readily available. NDV can infect over 200 species of birds and it is likely that all bird species are susceptible with mortalities of up to 100% possible in naïve birds of certain species. NDV is endemic in poultry in many countries of Africa and Asia, in smaller areas of Mexico, in Central and South America, and in cormorants in the United States and Canada. The virus continues to be endemic in pigeons worldwide since the third panzootic started in the 1970s in the Middle East. Infections with vNDV are reportable to the World Organisation for Animal Health (OIE) and,

if detected in poultry species, can lead to trade restrictions of poultry or poultry products. In 2011, 80 countries reported outbreaks of ND in poultry or wild birds (http://www.oie.int/ wahis_2/public/wahid.php/Diseaseinformation/statuslist). Member countries of the OIE are required to report within 24 h following the diagnosis of an ND outbreak. Strains of vNDV are agro-bioterrorism threat agents (http://www.fas. org/sgp/crs/terror/RL32521.pdf).

60.2 CLASSIFICATION AND MORPHOLOGY

NDV belong to the order Mononegavirales, family Paramyxoviridae, subfamily Paramyxovirinae, and genus Avulavirus.[2] They are single-stranded, nonsegmented, negative-sense RNA viruses that vary in shape from appearing filamentous with a diameter of 100 nm to round with a diameter of 100–500 nm, surrounded by a lipid bilayer envelope derived from the plasma membrane of the host cell in which the virus is grown.[3] Twenty to twenty-five percent of the w/w of the virus particle is derived from the host cell with an additional 6% w/w consisting of carbohydrate.[4] The genome exists in at least three different lengths (15, 186; 15, 192; or 15, 198),[5] follows the "rule of six," and is comprised of six genes (and six structural proteins) in 3′–5′ order: nucleocapsid (N), phosphoprotein (P), matrix (M), F, hemagglutinin–neuraminidase (HN), and the RNA-dependent RNA (large) polymerase (L). Editing of P produces at least one other protein, the V protein, which has anti-interferon properties. The tetrameric HN and trimeric F glycoproteins are 17 nm in length and are arranged on the outside of the virion. The HN protein allows the virus to attach to the host cell via a sialic acid receptor and the F protein facilitates the pH-independent fusion of the virus to the host membrane. The F protein is made in a precursor F0 form of which the N terminus has to be cleaved by a host cell protease into F1 and F2 pieces connected by a disulfide bond. The HN is produced in a precursor form, HN0, for some low-virulence NDV (loNDV) that also requires posttranslational cleavage for the carboxy end of the protein.[6] The M protein is found under the envelope. Inside of the virion, the RNA genome is bound tightly to the N protein and together they form the viral N complex. The P and L proteins are also bound to the N and together all three are important for replication.

60.3 BIOLOGY AND EPIDEMIOLOGY

60.3.1 ATTACHMENT, TRANSCRIPTION, AND REPLICATION

After the virus attaches to the host cell through the HN protein and fuses with the host cell membrane using the cleaved F protein, the viral N complex enters into the cytoplasm of the host cell where replication will occur. Besides direct fusion with the host cell membrane, endocytosis may also play a role in viral entry.[7] The polymerase (L) begins to transcribe the negative-sense genome at the 3′ end into positive-sense messenger RNA (mRNA) with 5′ caps and polyadenylated tails at the 3′ end for each gene.[3] The mRNAs of the genes nearer to the 3′ end are produced in higher amounts than those at the 5′ end. Soon after mRNAs are created, transcription switches to replication and the N mRNAs are used to produce the full-length positive strand antisense genome that will serve as a template for negative-sense genomic RNA.[8] Like all paramyxoviruses, productive virus infection will only occur when the total number of nucleotides in the RNA is a multiple of six.[8] Newly made viral proteins are organized by the M protein and will assemble at the plasma membrane and bud from the host cell.[9] The neuraminidase component of the HN protein is believed to prevent new virus particles from clumping together and/or from reinfecting the cell they just budded from, which would hinder their release and ability to infect new cells.[3] Furthermore, lipid rafts and cholesterol also appear to be critical for the assembly and release of ND virions from host cells.[10,11]

60.3.2 BIOLOGICAL PROPERTIES

While some variation can be seen in thermostability depending on the specific NDV isolate, many isolates will lose their infectivity within 10 min at 56°C, but some NDV isolates may remain infective up to 60–90 min[12] and the rare isolate may remain infective up to almost 5 h.[13] In addition, even though hemagglutination (HA) thermostability is often used to compare viruses,[14] the loss of HA does not always correlate with the loss of infectivity.[13] Virulent isolates have been found in infected tissues for up to 4 weeks and are able to remain viable in tap water and river water for multiple weeks.[15] Besides temperature, the ability of the virus to retain moisture and the amount of protein available to stabilize the virus, both of which depend on the material the virus is contained in,

also influence virus viability and explain why the virus could feasibly survive in the environment from one season to the next.[16] Notably, lyophilized NDV vaccine vials buried for greater than 20 years recovered from pharmaceutical dump sites have remained viable.[17] While NDV isolates can tolerate a wide pH range (2–11) for short time periods, they have maximum survivability when they are within the pH range of 5–9.[18] The ability of NDV to HA red blood cells (RBCs) and the rate that the virus elutes from RBCs were once commonly used to characterize NDV isolates before RNA isolation, PCR, and genome sequencing were available.[19] Some NDV variants adapted to and isolated from pigeons, pigeon paramyxovirus serotype-1 (PPMV-1), have lower HA titers and lower viral titers than other NDV isolates better adapted to replicate in chickens.[20,21]

60.3.3 RECOMBINATION

The factors affecting viral evolution are unknown. As with other RNA viruses, evidence suggests that most of the genetic changes in NDV occur as a result of evolutionary genomic drift due to the inability of the polymerase to proofread and correct errors in replication.[22,23] In addition, immune pressure may play a role as neutralizing antibodies remove isolates that bind to the antibodies, leaving the viruses with structural differences to replicate and potentially be passed to other birds or the environment.[24] Events compatible with recombination have been documented[22,25] and reports of recombinant NDV strains isolated from the field continue to be reported.[26] However, it is difficult to confirm if recombinant viruses are indeed the product of a natural event and not the result of laboratory contamination.[27,28]

When a dataset of 103 complete genomes was subjected to a recombinant analysis in our laboratory 14 of the 103 (13.5%) were found to be recombinant sequences. However, when the dataset containing 602 complete F gene sequences available in GenBank was evaluated only 15 of the 602 (2.5%) were discovered.[25] Because most of the recombination events found in GenBank sequences occur at different break points, it is possible that some of these events originated during the passage in eggs for samples containing more than one NDV or from other artifacts caused by the prevailing use of reverse transcription followed by PCR. Furthermore, there is only one documented case with documented evidence in which parent viruses may have produced a viable progeny through recombination.[22] Mixed infections with vaccine NDV and vNDV are not a rare occurrence and two NDV strains can infect the same cell, which is necessary for recombination to occur.[29] The role of recombination in NDV evolution needs to be further studied.

60.3.4 ANTIGENICITY

Differences in antigenicity have mostly been demonstrated with polyclonal antibodies that differentially affect HA-inhibition (HI) assays.[21] Monoclonal antibodies to the F protein[30] and to the HN protein also demonstrate differences in epitopes among isolates of different origins.[31] Even with recognized antigenic differences, NDV strains are considered to be of one serotype, because polyclonal antibodies induced by any NDV strain will neutralize any NDV.[1]

60.3.5 TRANSMISSION

NDV is spread horizontally from infected birds to susceptible birds through the ingestion or inhalation of viral particles from infected saliva, feces, or tissues.[32,33] While the amount of NDV needed to infect a bird will depend on the host species, the ND immune status of that bird, and the specific NDV strain, in general, a mean embryo infective dose of 10^3–10^4 is required to infect a naïve chicken.[34,35] Historically, ducks are more resistant to an infection with NDV compared to turkeys, and turkeys are more resistant than chickens.[36] It is important to note that different species may give significantly different results upon infection with the same virus, as was seen with a range of mortality after infection with a vNDV strain of 6%–66% mortality in five different duck species.[37] Because chicken embryos suffer high mortality rates after infection with a vNDV, it has been suggested that it is unlikely that vertical transmission would result in viable progeny, although there are reports of vNDV isolated from embryos and hatchlings.[38,39] However, high maternal antibody levels passed to progeny may decrease mortality and facilitate this type of transmission. In 2012, researchers from Israel confirmed vNDV isolated from allantoic fluid and intestinal content from embryonating eggs and from swab samples from hatched chicks that shed vNDV for at least 1 week (personal communication, Ruth Haddas, Shimon Perk, and Rosie Meir). The analysis continues to differentiate if vNDV was deposited through cracked eggs or penetrated the eggshell or if it was deposited vertically from the reproductive tract of the hen during egg production.

60.3.6 EPIDEMIOLOGY

NDV is distributed worldwide and its presence represents a huge reservoir and constant threat to all poultry industries and other endeavors that involve the raising or keeping of birds.[40] The evidence that NDV is continuously evolving at different rates in different areas of the world, leading to more diversity, is prevalent.[22] As a result, there is a need for continual surveillance and a better understanding of the epidemiology of the virus. The worldwide distribution and continuous evolution of NDV has allowed for the presence of large genetic variability among circulating NDV strains and has led scientist to group the viruses into two different classes, further divided into genotypes.[23] NDV strains of class I are predominantly found in wild birds and tend to be of low virulence, while viruses of class II are predominantly found in poultry and tend to be virulent. Two discordant systems have been used to classify NDV isolates generating confusion in the nomenclature.[41,42] However, in 2012, the genetic diversity of NDV was reevaluated and a unified classification system based on objective criteria to separate NDV was proposed.[25]

The new classification system is based on the utilization of the complete sequence of the F protein gene. Utilizing more than 602 sequence available in GenBank, the mean inter-populational evolutionary distance between all existing NDV genetic groups has been estimated and differences of 10% (at the nucleotide level) were proposed to be used as the cutoff value to assign new genotypes. This system grouped NDV strains of class I into a single genotype and viral strains of class II into fifteen genotypes (Figure 60.1 a through c). Class I comprised mainly viruses that have been isolated from waterfowl and shorebirds and occasionally from samples collected in live bird markets worldwide and captured wild birds.[23,43,44] Typically the genomes of class I viruses are the longest of the NDV isolates with 15, 198 nucleotides.[5] However, in 2012, class II viruses isolated from West Africa also contain a longer genome.[45] The overall phylogenetic distances between all viruses of class I are, in general, smaller than the distances between viruses of class II. However, it is not known if these reduced distances are a reflection of the worldwide diversity or simply due to lack of sampling among viruses of class I. The annual rate of change of virulent viruses could be as high as ten times the rate of change of loNDV, suggesting that other selective pressures such as vaccination may accelerate the rate of evolution of virulent viruses.[22] Despite a reduced genetic variability in viruses of class I, there are at least three distinct subgenotypes within this one genotype. Other strains of class I viruses not yet assigned into a subgenotype of genotype I have been isolated from chickens in Japan and ruddy turnstones in the United States (e.g., 142 RT from 2002) suggesting that the identification of additional genotypes or subgenotypes will continue.[25]

Analysis of class II viruses (Figure 60.1b) reveals a larger diversity in genotypes with fifteen genotypes clearly identified at first.[25] A 16th genotype was added with the discovery of vNDV that have been present in the Dominican Republic since the mid-1980s.[46] Class II viruses are present in both wild bird and poultry species; however, most vNDVs are isolated from poultry and are responsible for significant economic losses to the poultry industry worldwide.[47] While viruses of low virulence (loNDV), also called lentogenic, of class II still circulate in poultry (vaccine and wild strains), they are more often detected in wild birds. Genotypes I and II are the only genotypes that have representative loNDV circulating in wild birds, suggesting that these types of viruses may be the ancestral genotypes that gave rise to all virulent viruses. The loNDV strains from class II and genotypes I and II have been used as live vaccines against NDV for more than 60 years.[23] The loNDV strains of genotype I, subgenotype Ic, are currently widely distributed in Asia, and viruses of this subgenotype represent most loNDV strains isolated from chickens in China and Australia and one case in Malaysia. Viruses from subgenotype Ia (Figure 60.1c) also represent the only well-documented case of viruses that have mutated, becoming virulent, causing the 1998–2001 outbreak in Australia.[48] Newer isolates from subgenotype Ib have been isolated from migratory ducks and birds in China and Russia. However, an older Ib isolate, Ulster, was detected in chickens in Ireland in 1967 and was used later

FIGURE 60.1 (continued)

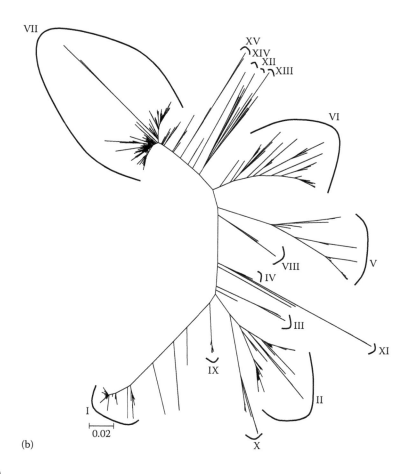

(b)

FIGURE 60.1 (continued)

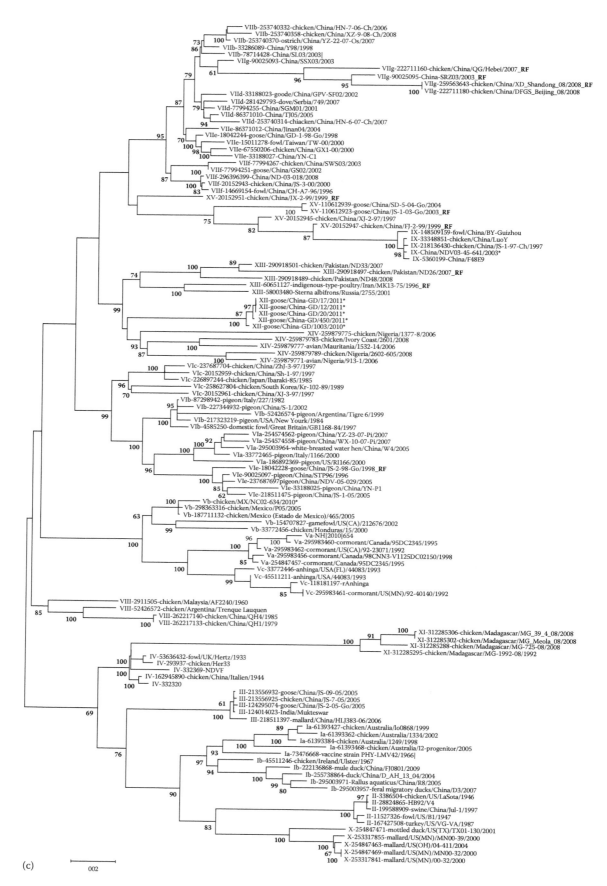

(c)

FIGURE 60.1 (continued)

as a vaccine. New loNDV strains of genotype X that are most similar to NDV strains of genotype II started to be identified in 2008 from a mallard, mottled duck, and northern pintail in the United States and from ducks and swans in Argentina.[25] The earliest vNDVs that have been genetically characterized from the United States, such as Texas GB/1948, are highly similar to the 1946 LaSota vaccine viruses (96% identity) suggesting that the original class II virulent viruses may have originated directly from loNDV through a change in the cleavage site of the F protein. Since 2008, there have been reports of ND outbreaks with genotype II vNDV strains in China and Egypt, which have not been detected circulating in poultry since vNDV of genotype V became prevalent in the 1970s in the United States, and genotype VII became prevalent in the 1980s.[49–52] Because most of these recent chicken vNDV isolates of genotype II are highly similar to the 1946 LaSota vaccine, it is hard to envision continuous evolution of NDV since 1946 with minor genetic changes. A more likely scenario is that loNDV vaccine-like viruses are escaping neutralization from infected vaccinated poultry and/or that mesogenic viruses of genotype II, which are still used as vaccines in some countries, are present in the environment.[53]

Some of the most widely distributed NDV strains from the virulent genotypes V, VI, and VII are also the most diverse. Therefore, these genotypes have been further divided in subgenotypes.[25] Viruses of genotype V are mainly distributed in the Americas with most frequent isolations in Mexico, Canada, and the United States. In Mexico, vNDVs are often detected in poultry despite intensive vaccination campaigns (subgenotype Vb), while in the United States, genotype V vNDVs only sporadically infect poultry (e.g., California 2002).[54,55] Subgenotype Va is comprised of vNDV strains isolated in North American cormorants where the virus seems to be maintained year to year with periodic outbreaks in newly hatched birds[56] but has spilled over into other bird populations.[57] Viruses of genotype VI were previously grouped into eight subgenotypes,[58] but a new classification system based on uniform genetic distances has established that genotype VI only comprises four subgenotypes.[25] Viruses of genotype VI are mainly isolated from pigeons,[59] but they occasionally spill over into chickens or other species.[60,61] These viruses are predominantly mesogenic and often do not cause mortality in chickens.[62] The viruses of genotype VI may contain a unique set of amino acid at the F protein cleavage site ([112]RRKKRF[117] instead of the most common [112]RRQKRF[117]). The role of these cleavage site motif differences on virulence is not known. Genotype VII comprises the largest number of NDV isolates and is the most diverse among all 16 NDV genotypes, which led to its classification into five subgenotypes, which were named as subgenotypes VIIb, VIId, VIIe, VIIf, and VIIg.[25] Newly identified genotypes XII through XIV are closely related to genotype VII.[25] Viruses of genotype XII caused the 2008 outbreak in Peru, and four related isolates have been identified from geese in live bird markets in China in 2011.[25,63] The relationship between the Peruvian and Chinese isolates is an epidemiological enigma that illustrates the tremendous risk to the poultry industry as a result of the ease of mobility of vNDV around the globe. The very close

---

FIGURE 60.1 (continued) Phylogenetic analysis of NDV. (a) Phylogenetic analysis based on 602 complete nucleotide sequence of the F gene of viruses representing NDV class I. The evolutionary history was inferred by using the maximum likelihood method based on the data-specific model. The tree with the highest log likelihood (−8604.4794) is shown. Initial tree(s) for the heuristic search was obtained automatically as follows. When the number of common sites was <100 or less than one-fourth of the total number of sites, the maximum parsimony method was used; otherwise, BIONJ method with maximum composite likelihood (MCL) distance matrix was used. A discrete gamma distribution was used to model evolutionary rate differences among sites (four categories [+G, parameter = 0.6182]). The rate variation model allowed for some sites to be evolutionarily invariable ([+I], 27.9736% sites). The tree is drawn to scale, with branch lengths measured in the number of substitutions per site. The analysis involved 111 nucleotide sequences. Codon positions included were the 1st, 2nd, 3rd, and noncoding. All positions containing gaps and missing data were eliminated. There were a total of 1662 positions in the final dataset. Evolutionary analyses were conducted in MEGA5. (b) Phylogenetic analysis based on the complete nucleotide sequence of the F gene of viruses representing NDV class II. The evolutionary history was inferred by using the maximum likelihood method based on the data-specific model. The tree with the highest log likelihood (−46323.5822) is shown. Initial tree(s) for the heuristic search was obtained automatically as follows. When the number of common sites was <100 or less than one-fourth of the total number of sites, the maximum parsimony method was used; otherwise, BIONJ method with MCL distance matrix was used. A discrete gamma distribution was used to model evolutionary rate differences among sites (four categories [+G, parameter = 0.8546]). The rate variation model allowed for some sites to be evolutionarily invariable ([+I], 8.7500% sites). The tree is drawn to scale, with branch lengths measured in the number of substitutions per site. The analysis involved 602 nucleotide sequences. Codon positions included were 1st + 2nd + 3rd + noncoding. All positions containing gaps and missing data were eliminated. There were a total of 1639 positions in the final dataset. Evolutionary analyses were conducted in MEGA5. (c) Phylogenetic analysis based on the complete nucleotide sequence of the F gene of viruses representing NDV class II. At least four sequences representing each NDV genotype were used in the analysis. The evolutionary history was inferred by using the maximum likelihood method based on the data-specific model. The tree with the highest log likelihood (−23299.0541) is shown. The percentage of trees in which the associated taxa are clustered together is shown next to the branches. Initial tree(s) for the heuristic search was obtained automatically as follows. When the number of common sites was <100 or less than one-fourth of the total number of sites, the maximum parsimony method was used; otherwise, BIONJ method with MCL distance matrix was used. A discrete gamma distribution was used to model evolutionary rate differences among sites (four categories [+G, parameter = 1.2175]). The rate variation model allowed for some sites to be evolutionarily invariable ([+I], 30.6963% sites). The tree is drawn to scale, with branch lengths measured in the number of substitutions per site. The analysis involved 123 nucleotide sequences. Codon positions included were 1st, 2nd, 3rd, and noncoding. All positions containing gaps and missing data were eliminated. There were a total of 1647 positions in the final dataset. Evolutionary analyses were conducted in MEGA5. (Data from Diego, D. et al., *Infect. Genet. Evol.*, 12, 1770, 2012.)

phylogenetic relationship between the viruses from Peru and China (>91% identity) suggests that the movement from Peru to China (or vice versa) may have occurred recently. Due to the large geographic separation between the countries and the lack of known migratory pathways between those two regions of the globe, it is possible that these viruses may have moved through human activity. Viruses of genotype VIII are not frequently reported; however, their distribution is wide with representatives in China, Argentina, Malaysia, and Africa.[5,25] While viruses from genotypes III, IV, and IX are still reported occasionally in China, they have not had a significant worldwide distribution compared to other vNDV. Genotype XI is confined to Madagascar where the virus has probably evolved from an older genotype.[25,64] Other recent genotypes include viruses that have only been reported in Africa (genotype XIV from 2006 to 2008); Pakistan, Russia, and Iran (genotype XIII, 1996–2008); or Dominican Republic (genotype XVI, 1986–2008) suggesting simultaneous evolution of the virus at different locations followed by migration and spread.[25,46]

Not surprisingly, genomic diversity among NDV isolates often corresponds with historical and geographical distribution. Genotypes I, II, III, IV, and IX emerged between the 1930s and 1960s and are considered "early" genotypes.[5] Genotype IV has rarely been isolated since the early 1980s except for India.[23,65] Genotype IX includes the first vNDV strain obtained in China in 1948, and viruses of this genotype are still circulating in Asia with sporadic outbreaks being reported in chickens and domestic ducks.[58,66,67] Genotypes V, VI, VII, VIII, and XI emerged after the 1960s and are considered "late" genotypes.[5] Genotypes V, VI, and VII contain only vNDV strains and are the predominant genotypes circulating worldwide.[22] Among these, genotype V viruses that cause a viscerotropic presentation with prominent lesion of organs emerged in the 1970s replacing virulent genotype II isolates that caused a neurotropic presentation with few gross lesions in chickens.[68] Genotype VI viruses have been isolated from multiple avian species since the 1960s and are still circulating at multiple locations worldwide.[59,61,69–71] Viruses of this genotype are particularly important because of their frequent association with doves and pigeons and the subsequent risk for introduction into poultry flocks due to their synanthropic tendencies.[69,72,73] Currently, genotype VII is the genotype most frequently associated with outbreaks of ND in the Middle East[74] and Asia,[75] and outbreaks caused by these viruses are of particular concern given that some strains have shown increased virulence in poultry, while others have expanded their host range and are now able to cause disease in geese.[76–78] Additionally, continued outbreaks of ND in South America (Venezuela and Colombia) have been attributed to a genotype VII virus, indicating that genotype VII viruses are spreading to other locations around the world.[79]

Genotypes X, XII, XIII, XIV, XV, and XVI comprise viruses with different biological properties and geographic origins. Newly defined genotype X, which contains viruses that have been previously classified into genotype IIa,[80] comprises viruses of low virulence that have been isolated from waterfowl and shorebirds in North America between

the late 1980s and the early 2000s. Genotype XII has been described earlier. Genotype XIV contains virulent viruses obtained in West and Central Africa between 2006 and 2008 (Figure 60.1c). These viruses have been previously classified by using the lineage system into lineage 7.[81] Genotype XV comprises viruses obtained from chickens and geese in China, which have been previously classified into subgenotype VIId (isolates XJ-2/97 and FJ-2/99) or VIIe (isolate JX-2/99).[82] The last genotype to be defined, genotype XVI, has also been described earlier.

The existence of reservoirs for vNDV is a question of interest that needs further investigation. Despite the existence of 16 genotypes containing vNDV strains, only two natural reservoirs where virus is maintained and shed from apparently healthy birds have been described. Cormorants harbor viruses of subgenotype Va and pigeons are reservoirs of viruses of genotype VI. Therefore, it is assumed that vNDV isolates from the other genotypes are maintained in vaccinated poultry that are able to shed vNDV into the environment to infect susceptible birds.[83] However, the recent detection of virulent viruses in geese[84] and white storks[85] demonstrates that there is a number of bird species that are susceptible to infection but do not show clinical signs,[86] suggesting that additional reservoirs are yet to be found in wild birds.[85,87]

Immune pressure from vaccination is likely to have an effect on viral evolution; however, the effect of vaccination on evolution has not been studied. The usage of billions of doses of vaccine worldwide (most commercial birds are vaccinated against NDV) combined with the continuous presence of vNDV in the environment, in some countries, suggests that the viruses that escape vaccination may be positively selected. The evidence showing the existence of viruses that escape vaccination has been presented[88–90]; however, vaccination studies have shown that live and inactivated vaccines that are currently used (e.g., LaSota, 1947) are still capable of preventing clinical disease and significantly reduce viral replication, but not as much as vaccines homologous to the challenge virus.[21,24,91] The increased virulence of present-day virulent viscerotropic viruses in comparison to the virulent but more attenuated neurotropic viruses that circulated in the 1940s suggests that vaccination may be selective for viruses of increased virulence. This is also another area of research that has not been fully developed.

The phylogenetic relatedness of current viruses from virulent genotypes circulating worldwide (genotypes V, VI, and VII) to virulent viruses circulating in previous years (genotypes III, IV, IX) strongly indicates that most currently vNDV have evolved from other vNDV that were circulating previously in poultry rather than from loNDV from wild birds or from vaccine strains. While a change in the cleavage from low to high virulence is possible, there is only one well-documented case showing that a change in the F cleavage site of a lentogenic virus mutated to produce a vNDV, causing an ND outbreak. The loNDV isolates of genotype I, class I from Australia (strain 99-1997 PR-32) in which the cleavage site changed from low virulence with a cleavage site of RRQGRF, to that of a vNDV with a cleavage site of RRQRRF (strain 99-1435) and an ICPI value of 1.8 and caused an outbreak in 1998.[48] In 2011, viruses of genotype II that resemble vaccine viruses were isolated in China (data not shown) and Egypt[52] and raise new concerns about the possibility of viruses of genotype II that have been thought to be very stable being able to mutate into virulent forms.

60.4 CLINICAL FEATURES AND PATHOGENESIS

The incubation period for NDV is usually 5 to 6 days but can vary from 3 to 4 days for experimental infections using susceptible birds and large challenge doses to 15 days when lower challenge doses or low levels of immunity are present. It will also depend on the species of the host and the virulence of the virus.[4] Clinical signs of ND in chickens may include a marked drop in egg production followed by depression, respiratory distress, hemorrhage and necrosis in multiple organs, neurological signs, and acute death.[4] None of these clinical signs are pathognomonic for ND, and other "rule outs" such as highly pathogenic avian influenza, infectious bronchitis virus, and infectious laryngotracheitis should be considered. While unvaccinated chickens may die acutely after infection without showing any signs of ND, vaccinated birds likely will show no signs, except a decrease in egg production, after being infected with a vNDV.[92] Juvenile cormorants and pigeons may succumb to disease, but older birds usually will not show signs of disease. Neurotropic vNDV will present differently than viscerotropic vNDV. Birds infected with neurotropic strains will present at first with a hypermetric gait while being bright and alert before developing a unilateral wing drop or leg paralysis. These birds will have little to no gross lesions upon necropsy. Birds infected with viscerotropic vNDV will become depressed, ruffled, and hot to the touch before becoming anorexic, listless, and hypothermic. These birds may also have neurotropic signs such as torticollis or tremors and will have multiple gross hemorrhagic lesions focusing on the lymphoid tissues throughout the body. Typically, birds infected will shed virus after infection no matter if they are vaccinated, with vNDV being shed in oral secretions first, peaking 4 days postinfection.[24] Shedding in feces occurs later and lasts longer and can be sporadic, especially in wild birds. Vaccination should decrease shedding two to three logs compared to nonvaccinated animals.[21]

60.5 IDENTIFICATION AND DIAGNOSIS

Molecular diagnostic tests are emerging as viable alternatives to classical diagnostics because they are easy to use, are inexpensive, and give rapid results. Most current rapid diagnostic methods are based on the use of reverse transcription followed by real-time PCR and those have been reviewed in 2009[93] and will not be discussed in detail here. Of the multiple molecular diagnostic methods developed, those based on real-time PCR seem to be the most accepted because of the speed, sensitivity, and specificity. Despite the availability of numerous molecular diagnostics for NDV, only one method[94] has undergone complete validation in the United States by the US Department of Agriculture (USDA) and is widely used worldwide, with other methods used regionally or for specific occasions. The method widely used in the United States uses two assays: the M gene assay, which is used to identify all NDV (APMV-1), and the F gene assay, which is used to identify only vNDV.[94] All assays that rely on primers and probes matching unknown viral genomes are limited in their capacity to detect new isolates depending on the rate of evolution or change the viruses incur over time. The M assay, in general, does not recognize viruses of class I with only 36% of these viruses identified as positive[44] and the F assay often fails to recognize viruses from pigeons (genotype VI)[59] and cormorants (genotype Va).[56,80] In 2010, NDV isolates from ND outbreaks affecting cormorants and gulls in the states of Minnesota, Massachusetts, Maine, New Hampshire, and Maryland were found to be closely related to the viruses that caused the ND outbreaks in Minnesota in 2008.[57] Similar to the results obtained with the 2008 isolates, the standard USDA F gene real-time reverse-transcription PCR (RRT-PCR) assay failed to detect the 2010 cormorant viruses, whereas all viruses were detected by a new cormorant-specific F gene RRT-PCR assay developed by the USDA as an alternate test.[56] NDV isolates from Pakistan also failed the USDA PCR M gene assay, designed to detect all NDV, and a new M gene test that detected all isolates was developed.[95] The USDA has not yet validated either of these new assays. Even if highly conserved regions are used, it is still possible to fail in detecting a strain because of the inherent genetic variation of NDV; therefore, constant surveillance and reassessing of current assays are always required. Panels containing multiple monoclonal antibodies to the HN protein were also commonly used to sort NDV isolates prior to genome sequencing.[96] However, with the ease of nucleotide sequencing, this technique is not used as often as it was.

Unfortunately, non-PCR-based assays (HA and HI assays) are not very sensitive and thus require amplification of the virus in specific pathogen-free embryonating chicken eggs (ECE). After amplification in eggs, the virus-infected allantoic fluid is mixed with chicken RBCs to see if the RBCs agglutinate. If they do agglutinate, the next step involves incubating the virus with serum containing NDV antibodies before mixing with the RBCs. If agglutination is inhibited, the diagnosis of NDV can be made. It is important to note that some NDV isolates do not HA well or at all. Pigeon isolates of genotype VI often do not HA well and cormorant isolates from the United States older than 2002 often do not HA at all. These pigeon and cormorant viruses will kill SPF ECE but should be passed at least twice in SPF ECE because vNDV adapted to species other than chickens may take longer to replicate. Thus, virus isolation assays are still considered the gold standard; unfortunately, they may be avoided due the time needed (up to 2 weeks for two passages to be completed) and the cost involved with obtaining SPF ECE. Therefore, there is still a need for a standardized approach to detect NDV.

60.6　TREATMENT AND PREVENTION

There is no treatment for ND. In the United States, infected poultry are culled to help contain the outbreak. For infected pigeons, cormorants, or exotic species, supportive care can be provided and the animal may survive. Neurological signs may or may not resolve. Biosecurity and vaccination continue to be the primary techniques used to prevent ND. Live NDV-based vaccines are the preferred method of vaccination because of the low cost and ease of application. Besides an increased capacity for inducing good mucosal immunity, live vaccines also have the advantage of having a low application cost due to mass application and an increased cellular immunity response, which reduces viral shedding. Oil emulsion vaccines formulated with inactivated virus-infected allantoic fluids are administered individually to each bird to provide long-lasting high antibody titers. However, withdrawal times between vaccination and slaughter reduce the ability to use these types of vaccines throughout the production period. Inactivated vaccines also lack the induction of mucosal immunity provided by live vaccines. Because all NDV are in one serotype, theoretically, any NDV can be used to make a vaccine. However, it has been shown that vaccines more similar to the challenge virus will decrease not only the amount of challenge virus shed from vaccinated birds but also the number of birds that shed the virus.[21,24] In addition, the same experiments show that some NDV strains are more antigenic than others, which can be observed by the HI antibody titer levels after equal amounts of vaccines are administered.

There are commercially available ND vaccines for chickens that use fowl pox or herpes virus of turkey as vectors for NDV F genes, and their use in the United States is becoming more popular. They have the benefit of not causing respiratory disease after vaccination, as live vaccines are able to do, but do take longer to reach protective antibody levels. For this reason, they may not be useful in areas were vNDV is endemic unless they are used with other types of ND vaccines. More information is needed on how these types of vaccines interact in situation with very high maternal antibodies or in species other than chickens. Racing pigeons are often vaccinated with killed ND vaccines formulated with inactivated a PPMV-1 strain.

60.7　CONCLUSIONS AND FUTURE PERSPECTIVES

In summary, ND, caused by virulent forms of APMV-1, remains a significant threat to poultry and susceptible wild birds species despite more than 60 years of intensive vaccination efforts worldwide. Control of the disease can be achieved by vaccination, but effective eradication in general requires culling of birds because there is no vaccine available capable of preventing viral replication. Furthermore, as virulent viruses normally replicate in immune animals, it is possible that vaccination may contribute to the evolution of virulent viruses. As there is an increased rate of evolution and there is an increased genetic diversity among virulent viruses isolated in poultry worldwide, the threat of ND is today as high as ever. The increased genetic diversity and worldwide mobility of circulating viruses creates challenges in diagnostics and vaccination.

Newly evolved and imported strains often fail detection using real-time PCR-based methods, and it is clear that vaccination cannot prevent virulent viruses from replicating and being shed in the environment. Indirect evidence demonstrating that current circulating velogenic viruses have the capacity to cause increased diseases in poultry and other species (goose) in comparison to mesogenic viruses circulating in wild birds and to older neurotropic viruses from the 1940s suggests that vaccination may actually contribute to increase the pathogenicity of circulating viruses. While updating the strains used in formulating vaccines will help with the amount of vNDV put back into the environment, ensuring that proper doses are applied to all of the birds in a flock is also crucial. Programs with intense vaccination protocols lead to hatchling with high maternal antibodies that require even higher live vaccine doses to account for the amount of vaccine that will be neutralized, and, therefore, these programs sometimes are not able to induce a high-enough antibody titer to decrease the amount of virus shed into the environment.

REFERENCES

1. OIE. Manual of diagnostic tests and vaccines for terrestrial animals: Mammals, birds and bees. In *Biological Standards Commission*, Vol. 1, Part 2, Chapter 2.03.14, pp. 1–19. World Organization for Animal Health, Paris, France, 2012.
2. Mayo, M.A. Virus taxonomy—Houston 2002. *Arch. Virol.* 147, 1071–1076 (2002).
3. Lamb, R. and Parks, G.D. Paramyxoviridae: The viruses and their replication. In *Fields Virology*, Vol. 1, Fields, B., Knipe, D.M., and Howley, P.M. (eds.), pp. 1449–1496. Lippincott Williams and Wilkins, Philadelphia, PA, 2007.
4. Alexander, D.J. and Senne, D.A. Newcastle disease, other avian paramyxoviruses, and pneumovirus infections. In *Diseases of Poultry*, Saif, Y.M. et al. (eds.), pp. 75–116. Iowa State University Press, Ames, IA, 2008.
5. Czegledi, A. et al. Third genome size category of avian paramyxovirus serotype 1 (Newcastle disease virus) and evolutionary implications. *Virus Res.* 120, 36–48 (2006).
6. Nagai, Y. and Klenk, H.D. Activation of precursors to both glycoproteins of Newcastle disease virus by proteolytic cleavage. *Virology* 77, 125–134 (1977).
7. Cantin, C., Holguera, J., Ferreira, L., Villar, E., and Munoz-Barroso, I. Newcastle disease virus may enter cells by caveolae-mediated endocytosis. *J. Gen. Virol.* 88, 559–569 (2007).
8. Peeters, B.P.H., Gruijthuijsen, Y.K., De Leeuw, O.S., and Gielkens, A.L.J. Genome replication of Newcastle disease virus: Involvement of the rule-of-six. *Arch. Virol.* 145, 1829–1845 (2000).
9. Harrison, M.S., Sakaguchi, T., and Schmitt, A.P. Paramyxovirus assembly and budding: Building particles that transmit infections. *Int. J. Biochem. Cell Biol.* 42, 1416–1429 (2010).

10. Laliberte, J.P., McGinnes, L.W., Peeples, M.E., and Morrison, T.G. Integrity of membrane lipid rafts is necessary for the ordered assembly and release of infectious Newcastle disease virus particles. *J. Virol.* 80, 10652–10662 (2006).

11. Martin, J.J., Holguera, J., Sanchez-Felipe, L., Villar, E., and Munoz-Barroso, I. Cholesterol dependence of Newcastle disease virus entry. *Biochim. Biophys. Acta* 1818, 753–761 (2012).

12. Lomniczi, B. Thermostability of Newcastle disease virus strains of different virulence. *Arch. Virol.* 47, 249–255 (1975).

13. Estola, T. Isolation of a Finnish Newcastle disease virus with an exceptionally high thermostability. *Avian Dis.* 18, 274–277 (1974).

14. King, D.J. and Seal, B.S. Biological and molecular characterization of Newcastle disease virus isolates from surveillance of live bird markets in the northeastern United States. *Avian Dis.* 41, 683–689 (1997).

15. Saber, M.S., Alfalluji, M., Siam, M.A., and Alobeidi, H. Survival of AG68V strain of Newcastle disease virus under certain local environmental conditions in Iraq. *J. Egypt. Vet. Med. Assoc.* 38, 73–82 (1978).

16. Olesiuk, O.M. Influence of environmental factors on viability of Newcastle disease virus. *Am. J. Vet. Res.* 43, 152–155 (1951).

17. Amendola, A. et al. Viable Newcastle disease vaccine strains in a pharmaceutical dump. *Emerg. Infect. Dis.* 13, 1901–1903 (2007).

18. Moses, H.E., Brandly, C.A., and Jones, E.E. The pH stability of viruses of Newcastle disease and fowl plague. *Science* 105, 477–479 (1947).

19. Spalatin, J., Hanson, R.P., and Beard, P.D. The hemagglutination-elution pattern as a marker in characterizing Newcastle disease virus. *Avian Dis.* 14, 542–549 (1970).

20. Stone, H.D. Efficacy of oil-emulsion vaccines prepared with pigeon paramyxovirus 1, Ulster, and LaSota Newcastle disease viruses. *Avian Dis.* 33, 157–162 (1989).

21. Miller, P.J., King, D.J., Afonso, C.L., and Suarez, D.L. Antigenic differences among Newcastle disease virus strains of different genotypes used in vaccine formulation affect viral shedding after a virulent challenge. *Vaccine* 25, 7238–7246 (2007).

22. Miller, P.J., Kim, L.M., Ip, H.S., and Afonso, C.L. Evolutionary dynamics of Newcastle disease virus. *Virology* 391, 64–72 (2009).

23. Miller, P.J., Decanini, E.L., and Afonso, C.L. Newcastle disease: Evolution of genotypes and the related diagnostic challenges. *Infect. Genet. Evol.* 10, 26–35 (2010).

24. Miller, P.J., Estevez, C., Yu, Q., Suarez, D.L., and King, D.J. Comparison of viral shedding following vaccination with inactivated and live Newcastle disease vaccines formulated with wild-type and recombinant viruses. *Avian Dis.* 53, 39–49 (2009).

25. Diel, D.G. et al. Genetic diversity of avian paramyxovirus type 1: Proposal for a unified nomenclature and classification system of Newcastle disease virus genotypes. *Infect. Genet. Evol.* 12, 1770–1779 (2012).

26. Yin, Y. et al. Molecular characterization of Newcastle disease viruses in Ostriches (Struthio camelus L.): Further evidences of recombination within avian paramyxovirus type 1. *Vet. Microbiol.* 149, 324–329 (2011).

27. Afonso, C.L. Not so fast on recombination analysis of Newcastle disease virus. *J. Virol.* 82, 9303 (2008).

28. Han, G.Z. and Worobey, M. Homologous recombination in negative sense RNA viruses. *Viruses* 3, 1358–1373 (2011).

29. Li, J. et al. Generation and characterization of a recombinant Newcastle disease virus expressing the red fluorescent protein for use in co-infection studies. *Virol. J.* 9, 227 (2012).

30. Choi, K.S., Lee, E.K., Jeon, W.J., and Kwon, J.H. Antigenic and immunogenic investigation of the virulence motif of the Newcastle disease virus fusion protein. *J. Vet. Sci.* 11, 205–211 (2010).

31. Russell, P.H. and Alexander, D.J. Antigenic variation of Newcastle disease virus strains detected by monoclonal antibodies. *Arch. Virol.* 75, 243–253 (1983).

32. Alexander, D.J. Newcastle disease: Methods of spread. In *Newcastle Disease*, Alexander, D.J. (ed.), pp. 256–272. Kluwer Academic Publishers, Dordrecht, the Netherlands, 1988.

33. Seal, B.S., King, D.J., Locke, D.P., Senne, D.A., and Jackwood, M.W. Phylogenetic relationships among highly virulent Newcastle disease virus isolates obtained from exotic birds and poultry from 1989 to 1996. *J. Clin. Microbiol.* 36, 1141–1145 (1998).

34. King, D.J. Influence of chicken breed on pathogenicity evaluation of velogenic neurotropic Newcastle disease virus isolates from cormorants and turkeys. *Avian Dis.* 40, 210–217 (1996).

35. Alexander, D.J. et al. Experimental assessment of the pathogenicity of the Newcastle disease viruses from outbreaks in Great Britain in 1997 for chickens and turkeys, and the protection afforded by vaccination. *Avian Pathol.* 28, 501–511 (1999).

36. Aldous, E.W. et al. Infection dynamics of highly pathogenic avian influenza and virulent avian paramyxovirus type 1 viruses in chickens, turkeys and ducks. *Avian Pathol.* 39, 265–273 (2010).

37. Shi, S.H. et al. Genomic sequence of an avian paramyxovirus type 1 strain isolated from Muscovy duck (*Cairina moschata*) in China. *Arch. Virol.* 156, 405–412 (2011).

38. Capua, I., Scacchia, M., Toscani, T., and Caporale, V. Unexpected isolation of virulent Newcastle disease virus from commercial embryonated fowls' eggs. *J. Vet. Med. B* 40, 609–612 (1993).

39. Roy, P. and Venugopalan, A.T. Unexpected Newcastle disease virus in day old commercial chicks and breeder hen. *Comp. Immunol. Microb.* 28, 277–285 (2005).

40. The World Bank. *World Livestock Disease Atlas: A Quantitative Analysis of Global Animal Health DataA (2006–2009)*, The International Bank for Reconstruction and Development/The World Bank and TAFS forum. Washington, DC, 2011. Available at http://www.oie.int/doc/en_document.php?numrec=4063003

41. Aldous, E.W., Mynn, J.K., Banks, J., and Alexander, D.J. A molecular epidemiological study of avian paramyxovirus type 1 (Newcastle disease virus) isolates by phylogenetic analysis of a partial nucleotide sequence of the fusion protein gene. *Avian Pathol.* 32, 239–257 (2003).

42. Herczeg, J. et al. Two novel genetic groups (VIIb and viii) responsible for recent Newcastle disease outbreaks in southern Africa, one (VIIb) of which reached southern Europe. *Arch. Virol.* 144, 2087–2099 (1999).

43. Kim, L.M., King, D.J., Suarez, D.L., Wong, C.W., and Afonso, C.L. Characterization of class I Newcastle disease virus isolates from Hong Kong live bird markets and detection using real-time reverse transcription-PCR. *J. Clin. Microbiol.* 45, 1310–1314 (2007).

44. Kim, L.M. et al. Phylogenetic diversity among low-virulence Newcastle disease viruses from waterfowl and shorebirds and comparison of genotype distributions to those of poultry-origin isolates. *J. Virol.* 81, 12641–12653 (2007).

45. Kim, S.H. et al. Complete genome sequence of a novel Newcastle disease virus strain isolated from a chicken in West Africa. *J. Virol.* 86, 11394–11395 (2012).

46. Courtney, S.C. et al. Highly divergent virulent isolates of Newcastle disease virus from the Dominican Republic are members of a new genotype that may have evolved unnoticed for over two decades. *J. Clin. Microbiol.* 51(2), 508–517 (2013).

47. Dundon, W.G. et al. Genetic data from avian influenza and avian paramyxoviruses generated by the European network of excellence (EPIZONE) between 2006 and 2011—Review and recommendations for surveillance. *Vet. Microbiol.* 154, 209–221 (2012).

48. Gould, A.R. et al. Virulent Newcastle disease in Australia: Molecular epidemiological analysis of viruses isolated prior to and during the outbreaks of 1998–2000. *Virus Res.* 77, 51–60 (2001).

49. Czegledi, A. et al. The occurrence of five major Newcastle disease virus genotypes (II, IV, V, VIand VIIb) in Bulgaria between 1959 and 1996. *Epidemiol. Infect.* 129, 679–688 (2002).

50. Tan, L.-T. et al. Molecular characterization of three new virulent Newcastle disease virus variants isolated from China. *J. Clin. Microbiol.* 46, 750–753 (2008).

51. Mohamed, M.H. et al. Complete genome sequence of a virulent Newcastle disease virus isolated from an outbreak in chickens in Egypt. *Virus Genes* 39, 234–237 (2009).

52. Mohamed, M.H., Kumar, S., Paldurai, A., and Samal, S.K. Sequence analysis of fusion protein gene of Newcastle disease virus isolated from outbreaks in Egypt during 2006. *Virol. J.* 8, 237 (2011).

53. Wu, S. et al. Genetic diversity of Newcastle disease viruses isolated from domestic poultry species in Eastern China during 2005–2008. *Arch. Virol.* 2, 253–261 (2010).

54. Kapczynski, D.R. and King, D.J. Protection of chickens against overt clinical disease and determination of viral shedding following vaccination with commercially available Newcastle disease virus vaccines upon challenge with highly virulent virus from the California 2002 exotic Newcastle disease outbreak. *Vaccine* 23, 3424–3433 (2005).

55. Absalon, A.E. et al. Complete genome sequence of a velogenic Newcastle disease virus isolated in Mexico. *Virus Genes* 45, 304–310 (2012).

56. Rue, C.A. et al. Evolutionary changes affecting rapid diagnostic of 2008 Newcastle disease viruses isolated from double-crested cormorants. *J. Clin. Microbiol.* 48, 2440–2448 (2010).

57. Diel, D.G. et al. Characterization of Newcastle disease viruses isolated from cormorant and gull species in the United States in 2010. *Avian Dis.* 56, 128–133 (2012).

58. Wang, Z. et al. Genotyping of Newcastle disease viruses isolated from 2002 to 2004 in China. *Ann. N. Y. Acad. Sci.* 1081, 228–239 (2006).

59. Kim, L.M. et al. Biological and phylogenetic characterization of pigeon paramyxovirus serotype-1 circulating in wild North American pigeons and doves. *J. Clin. Microbiol.* 46, 3303–3310 (2008).

60. Abolnik, C., Horner, R.F., Maharaj, R., and Viljoen, G.J. Characterization of a pigeon paramyxovirus (PPMV-1) isolated from chickens in South Africa. *Onderstepoort J. Vet. Res.* 71, 157–160 (2004).

61. Irvine, R.M. et al. Outbreak of Newcastle disease due to pigeon paramyxovirus type 1 in grey partridges (*Perdix perdix*) in Scotland in October 2006. *Vet. Rec.* 165, 531–535 (2009).

62. Dortmans, J.C., Koch, G., Rottier, P.J., and Peeters, B.P. Virulence of pigeon paramyxovirus type 1 does not always correlate with the cleavability of its fusion protein. *J. Gen. Virol.* 90(Pt 11), 2746–2750 (2009).

63. Diel, D.G. et al. Complete genome and clinicopathological characterization of a virulent Newcastle disease virus isolated from poultry in South America. *J. Clin. Microbiol.* 50, 378–387 (2011).

64. Maminiaina, O.F. et al. Newcastle disease virus in Madagascar: Identification of an original genotype possibly deriving from a died out ancestor of genotype IV. *PLoS One* 5, e13987 (2010).

65. Tirumurugaan, K.G. et al. Genotypic and pathotypic characterization of Newcastle disease viruses from India. *PLoS One* 6, e28414 (2011).

66. Qiu, X. et al. Entire genome sequence analysis of genotype IX Newcastle disease viruses reveals their early-genotype phylogenetic position and recent-genotype genome size. *Virol. J.* 8, 117 (2011).

67. Zhang, S. et al. Phylogenetic and pathotypical analysis of two virulent Newcastle disease viruses isolated from domestic ducks in China. *PLoS One* 6, e25000 (2011).

68. Utterback, W.W. and Schwartz, J.H. Epizootiology of velogenic viscerotropic Newcastle disease in southern California, 1971–1973. *J. Am. Vet. Med. Assoc.* 163, 1080–1088 (1973).

69. Alexander, D.J. Newcastle disease in the European Union 2000 to 2009. *Avian Pathol.* 40, 547–558 (2011).

70. Pikula, A., Smietanka, K., and Minta, Z. Antigenic and genetic characteristics of PPMV-1 isolated in Poland. In *Proceedings of the 17th World Veterinary Poultry Congress*, pp. 764–769. Cancun, Mexico, 2011.

71. Pchelkina, I.P. et al. Characteristics of pigeon paramyxovirus serotype-1 isolates (PPMV-1) from the Russian Federation from 2001 to 2009. *Avian Dis.* 57(1), 2–7 (2013).

72. Vidanovic, D. et al. Characterization of velogenic Newcastle disease viruses isolated from dead wild birds in Serbia during 2007. *J. Wildl. Dis.* 47, 433–441 (2011).

73. Abolnik, C. et al. Characterization of pigeon paramyxoviruses (Newcastle disease virus) isolated in South Africa from 2001 to 2006. *Onderstepoort J. Vet. Res.* 75, 147–152 (2008).

74. Munir, M., Zohari, S., Abbas, M., and Berg, M. Sequencing and analysis of the complete genome of Newcastle disease virus isolated from a commercial poultry farm in 2010. *Arch. Virol.* 157, 765–768 (2012).

75. Yi, J., Liu, C., Chen, B., and Wu, S. Molecular characterization of a virulent genotype VIId strain of Newcastle disease virus from farmed chickens in Shanghai. *Avian Dis.* 55, 279–284 (2011).

76. Liu, H. et al. Molecular epidemiological analysis of Newcastle disease virus isolated in china in 2005. *J. Virol. Methods* 140, 206–211 (2007).

77. Huang, Y., Wan, H.Q., Liu, H.Q., Wu, Y.T., and Liu, X.F. Genomic sequence of an isolate of Newcastle disease virus isolated from an outbreak in geese: A novel six nucleotide insertion in the non-coding region of the nucleoprotein gene. *Arch. Virol.* 149, 1445–1457 (2004).

78. Jinding, C., Ming, L., Tao, R., and Chaoan, X. A goose-sourced paramyxovirus isolated from Southern China. *Avian Dis.* 49, 170–173 (2005).

79. Perozo, F., Marcano, R., and Afonso, C.L. Biological and phylogenetic characterization of a genotype VII Newcastle disease virus from Venezuela: Efficacy of field vaccination. *J. Clin. Microbiol.* 50, 1204–1208 (2012).

80. Kim, L.M., Suarez, D.L., and Afonso, C.L. Detection of a broad range of class I and II Newcastle disease viruses using a multiplex real-time reverse transcription polymerase chain reaction assay. *J. Vet. Diagn. Invest.* 20, 414–425 (2008).

81. Cattoli, G. et al. Emergence of a new genetic lineage of Newcastle disease virus in West and Central Africa-Implications for diagnosis and control. *Vet. Microbiol.* 142, 168–178 (2010).

82. Liu, X.F., Wan, H.Q., Ni, X.X., Wu, Y.T., and Liu, W.B. Pathotypical and genotypical characterization of strains of Newcastle disease virus isolated from outbreaks in chicken and goose flocks in some regions of China during 1985–2001. *Arch. Virol.* 148, 1387–1403 (2003).

83. Ezeibe, M.C.O., Nwokike, E.C., Eze, J., and Eze, I.C. Detection and characterization of Newcastle disease virus from faeces of healthy free-roaming chickens in Nsukka, Nigeria. *Trop. Vet.* 24, 76–80 (2006).

84. Wan, H.Q., Chen, L.G., Wu, L.L., and Liu, X.F. Newcastle disease in geese: Natural occurrence and experimental infection. *Avian Pathol.* 33, 216–221 (2004).

85. Kaleta, E.F. and Kummerfeld, N. Isolation of herpesvirus and Newcastle disease virus from White Storks (*Ciconia ciconia*) maintained at four rehabilitation centres in northern Germany during 1983 to 2001 and failure to detect antibodies against avian influenza A viruses of subtypes H5 and H7 in these birds. *Avian Pathol.* 41, 383–389 (2012).

86. Roy, P., Venugopalan, A.T., Selvarangam, R., and Ramaswamy, V. Velogenic Newcastle disease virus in captive wild birds. *Trop. Anim. Health Prod.* 30, 299–303 (1998).

87. Zanetti, F., Berinstein, A., Pereda, A., Taboga, O., and Carrillo, E. Molecular characterization and phylogenetic analysis of Newcastle disease virus isolates from healthy wild birds. *Avian Dis.* 49, 546–550 (2005).

88. Cho, S.-H., Kim, S.-J., and Kwon, H.-J. Genomic sequence of an antigenic variant Newcastle disease virus isolated in Korea. *Virus Genes* 35, 293–302 (2007).

89. Cho, S.-H. et al. Variation of a Newcastle disease virus hemagglutinin-neuraminidase linear epitope. *J. Clin. Microbiol.* 46, 1541–1544 (2008).

90. Schat, K.A. and Baranowski, E. Animal vaccination and the evolution of viral pathogens. *Rev. Sci. Tech.* 26, 327–338 (2007).

91. Cornax, I., Miller, P.J., and Afonso, C.L. Characterization of live LaSota vaccine strain-induced protection in chickens upon early challenge with a virulent Newcastle disease virus of heterologous genotype. *Avian Dis.* 56, 464–470 (2012).

92. Bwala, D.G., Clift, S., Duncan, N.M., Bisschop, S.P., and Oludayo, F.F. Determination of the distribution of lentogenic vaccine and virulent Newcastle disease virus antigen in the oviduct of SPF and commercial hen using immunohistochemistry. *Res. Vet. Sci.* 93, 520–528 (2011).

93. Hoffmann, B. et al. A review of RT-PCR technologies used in veterinary virology and disease control: Sensitive and specific diagnosis of five livestock diseases notifiable to the World Organisation for Animal Health. *Vet. Microbiol.* 139, 1–23 (2009).

94. Wise, M.G. et al. Development of a real-time reverse-transcription PCR for detection of Newcastle disease virus RNA in clinical samples. *J. Clin. Microbiol.* 42, 329–338 (2004).

95. Khan, T.A. et al. Phylogenetic and biological characterization of Newcastle disease virus isolates from Pakistan. *J. Clin. Microbiol.* 48, 1892–1894 (2010).

96. Alexander, D.J. et al. Antigenic diversity and similarities detected in avian paramyxovirus type 1 (Newcastle disease virus) isolates using monoclonal antibodies. *Avian Pathol.* 26, 399–418 (1997).

61 Peste des Petits Ruminants Virus

Diallo Adama and Libeau Geneviève

CONTENTS

61.1 INTRODUCTION

Peste des petits ruminants (PPR) is a highly infectious transboundary animal disease that affects mainly sheep, goats, and small wild ruminants. This disease, caused by the peste des petits ruminants virus (PPRV), a virus of the *Morbillivirus* genus within the *Paramyxoviridae* family, is characterized clinically by ocular and nasal discharges, diarrhea, and erosive lesions of different mucous membranes. All these symptoms are similar to those of rinderpest (RP) that has recently been eradicated worldwide. An important distinguishing symptom that is found in acute PPR but which is absent in RP is bronchopneumonia. With morbidity and mortality rates that can be as high as 70%–80%, PPR is classified within the group of animal diseases that are notifiable to the Office International des Epizooties (OIE), the World Organization for Animal Health. It is considered as the main small ruminant disease in countries where it is endemic. Sheep and particularly goats (which are known as the cattle of the poor) contribute considerably to the cash income and nutrition of small farmers. So the control of a disease such as PPR, which is the main killer of those animals, is considered as an essential element in the fight for global food security and poverty alleviation.

61.2 CLASSIFICATION AND STRUCTURAL CHARACTERISTICS

61.2.1 CLASSIFICATION

Because of the close clinical resemblance between PPR and RP, and the strong cross-reaction between their causal agents, PPRV was considered for some time as a variant of the rinderpest virus (RPV) that was better adapted to small ruminants.[1] However, following the virus isolation in 1962[2] and serological and cross protection studies that were carried out in the 1970s, it was concluded that RPV and PPRV are two distinct but closely related viruses.[3,4] This distinction between these two ruminant viruses was further confirmed toward the end of the 1980s by a careful analysis of both epidemiological data, which indicated that both viruses were evolving independently in the field,[5] and data from biochemical studies.[6,7] Based on cross-reaction and cross protection studies, PPRV was classified in 1979[3] as part of the *Morbillivirus* genus within the *Paramyxoviridae* family along with RPV, the measles virus (MV), and the canine distemper virus (CDV). Later, toward the end of the 1980s, another group of emerging viruses that were seriously affecting aquatic mammals were added to this genus, namely, phocine distemper virus (PDV) in seals, dolphin morbillivirus (DMV) in dolphin, and porpoise morbillivirus (PMV) in porpoises.[8–10] Recently, based on genome sequence data, a morbillivirus of cat called feline morbillivirus (FmoPV) that is responsible for the tubulointerstitial nephritis in domestic cats has been identified.[11] Figure 61.1 shows the relationship between the different viruses of the group based on the amino acid sequence of their nucleocapsid protein. While it could be expected that the two ruminant morbilliviruses, PPRV and RPV, would be the most similar to each other within this group, the sequence analysis illustrated in Figure 61.1 shows that MV is, in fact, more closely related to RPV than PPRV as pointed out previously by Diallo et al.[7]

61.2.2 STRUCTURE OF THE PATHOGEN

PPRV, as all the members of the *Morbillivirus* genus, is composed of a nucleocapsid surrounded by an envelope.

It is a pleomorphic particle with a size varying between 150 and 700 nm, with a mean size of about 500 nm, bigger than that of the RPV particle that has a mean size of around 300 nm.[12,13] This size variation may be linked to the number of nucleocapsids that are incorporated into the virus particles. Rager et al.[14] reported that this number can be higher than 30 in the case of MV.

61.2.3 VIRUS NUCLEOCAPSID

The nucleocapsid is formed by the association of the viral genomic RNA with the nucleoprotein (N). Two other viral proteins are associated with this so-called ribonucleocapsid, namely, the phosphoprotein (P) and the RNA polymerase (L for large protein). The nucleocapsid appears as a tube that is about 1 μm long with a diameter at approximately 18 nm.[3,12] The genome is composed of a single-stranded RNA that

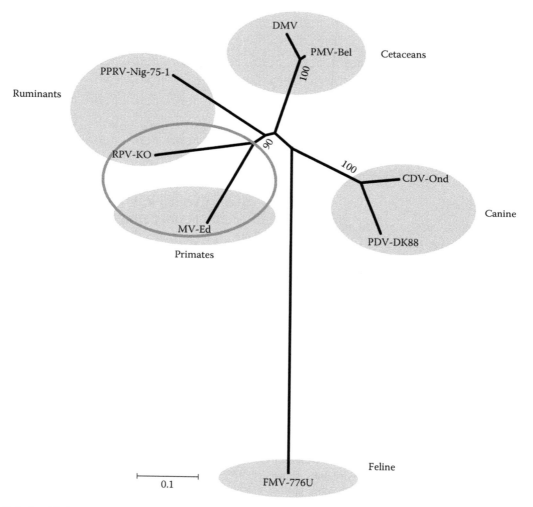

FIGURE 61.1 Relationship between viruses within the *Morbillivirus* genus based on the comparison of their full nucleoprotein sequences. The evolutionary history was inferred by using the maximum likelihood method based on the JTT matrix-based model. The bootstrap consensus tree inferred from 1000 replicates is taken to represent the evolutionary history of the taxa analyzed. Branches corresponding to partitions reproduced in less than 50% bootstrap replicates are collapsed. Initial tree(s) for the heuristic search was obtained automatically as follows. When the number of common sites was <100 or less than one fourth of the total number of sites, the maximum parsimony method was used; otherwise, BIONJ method with MCL distance matrix was used. The tree is drawn to scale, with branch lengths measured in the number of substitutions per site. Evolutionary analyses were conducted in MEGA5.

cannot serve as a matrix for direct translation into proteins and therefore needs to be transcribed into functional virus-specific positive-sense messenger RNAs. It is composed of 15,948 nucleotides,[15,16] longer than the genome of RPV that is made up of 15,882 nucleotides.[17] It contains six nonoverlapping transcriptional units corresponding to the six structural proteins: the N protein, the P protein, the matrix protein (M), the fusion protein (F), the hemagglutinin (H), and the RNA polymerase (L). The order of the transcription units on the genome starting from its 3′ end is N-P-M-F-H-L. The N gene is preceded by a small fragment of 52 nucleotides called the leader, while at the 5′ end, the L gene is followed by a trailer that is 37 nucleotides long. Genes are separated by a well-conserved nucleotide CTT triplet. Leader, trailer, and intergenic (IG) sequences CTT are never transcribed into RNA messengers. The leader and the trailer serve as attachment points of the RNA polymerase to the RNA. IG sequences serve as signals indicating the end of the gene under transcription to the RNA polymerase. This enzyme does not recognize them as stop signals during the replication step in which the full antigenomic or genomic RNA is synthesized. Both genomic and antigenomic RNAs are never naked. Indeed, as soon as their synthesis begins, they are wrapped by the nucleocapsid protein (N).[8]

The N protein is the most abundant of the viral proteins. It encapsidates the viral RNA and thereby protects it against nucleases. It is 525 amino acids long. The comparison of the PPRV amino acid sequence of the N protein with those of the other morbilliviruses has allowed the identification of a very well-conserved region in the middle of the protein and a C-terminal with very low homology.[7] Studies carried out on the N protein of MV have shown that the C-terminal of this protein is an intrinsically disordered domain, one of the characteristics of proteins that are able to interact with many partners through multiple sites.[18,19] This C-terminal fragment, located on the surface of the protein,[20] is involved in the interaction with the viral P protein in addition to some host cellular proteins involved in immune regulation.[19,21]

The P protein is the second protein associated with the nucleocapsid complex. For PPRV, this protein is 509 amino acids long, two residues more than the P protein of RPV.[22] It has a predicted molecular weight (MW) of 54.9 kDa based on its gene sequence, but it migrates in polyacrylamide gels with an apparent MW of 75 kDa [6] because of its high phosphate residue content that is added to the protein after its synthesis. P protein of negative strand viruses has at least two functions. The first is to act as a chaperone molecule for N in order to maintain the solubility of this protein in the cytoplasm thereby preventing it from auto-assembling or from associating nonspecifically with RNAs. The second role of the P protein is linked to its involvement in the RNA polymerase complex as a noncatalytic cofactor to the L protein.

The RNA polymerase (L) is the third viral protein of the nucleocapsid. With an identical length of 2183 amino acid residues in all of the morbilliviruses, L protein is the largest but least abundant of the viral proteins.[16] It constitutes the unique catalytic unit of the polymerase complex by assuming all enzymatic activities linked to the synthesis of the viral RNAs.

61.2.4 Virus Envelope

The first viral envelope protein in close contact with the nucleocapsid is the M protein. It is a basic protein composed of 335 amino acids [23] and is very well conserved within the morbilliviruses. It is generally believed that the M protein of paramyxoviruses forms an inner coat to the viral envelope serving as a bridge between the surface viral glycoproteins and the nucleocapsid core. However, following an electron cryotomographic study, Liljeroos et al.[24] showed that the M protein of MV forms helices coating the ribonucleocapsid rather than coating the inner leaflet of the membrane as previously thought. This complex that is formed within the cytoplasm is further transported to the host cell membrane in which the two other external envelope glycoproteins, the F and H proteins, are inserted. Whether its role is coating the nucleocapsid or forming an inner coat to the envelope, the M protein appears to play a central role in the formation of new virions that are liberated from the infected cell by budding.[25]

H and F proteins form spikes on the surface of the viral envelope. Both play crucial roles during the first steps of viral infection: the virus attaches to the membrane of the host cell to be infected via the H protein. Then via the F protein, the viral and the cell membranes are fused, a process that allows the delivery of the nucleocapsid into the cell cytoplasm. For viruses of the *Paramyxovirus* genus, the H protein possesses two biological functions: hemagglutination and neuraminidase activities. For long time, within the *Morbillivirus* genus, the hemagglutination activity had been identified only for MV H. However, studies carried out in the 1990s with PPRV showed that its protein H had both hemagglutinin and neuraminidase activities just like viruses of the *Paramyxovirus* genus.[26,27] The H of PPRV is composed of 609 amino acids. It is the protein to which virus-neutralizing antibodies produced in the infected host are directed against.

The second external protein of PPRV is the F protein that is composed of 546 amino acids similar to the F of RPV.[28] For all paramyxoviruses, F is synthesized as an inactive protein, F0, which is cleaved by cellular endopeptidases to give two fragments, F1 and F2, linked by disulfide bonds. This process liberates a hydrophobic peptide at the N-terminus of the F1 peptide called the fusion peptide because it is responsible for the fusion activity of the protein.[8] This cleavage is, therefore, essential for the infectivity of the virus. Certainly because H and F are proteins that are essential during the first step of viral infection and propagation in the host, they are the proteins to which the host protective immune response is directed.

61.2.5 Nonstructural Proteins

While the viral proteins described earlier are part of the virus particle, there are two other viral proteins, C and V, which are found only in infected cells and called nonstructural proteins. The synthesis of those two proteins is directed by the P gene. C is a small basic protein composed of 177 amino acid residues for both PPRV and RPV.[22] It is a result of the translation of the mRNA that directs the synthesis of the P protein but is the product of a different open reading frame (ORF). The second nonstructural protein of PPRV, the V protein, is a translation product of an mRNA that results from a nonfaithful transcription of the P gene: it differs from the P/C mRNA by an additional G that is inserted at nucleotide position 751 of the mRNA during the transcription process. This insertion results into a change in the ORF and thereby into the synthesis of a protein different from P. Both proteins share the fragment that corresponds to the translation of the mRNA before the insertion site. V is composed of 299 amino acid residues.[22] V and C are thought to be involved in the regulation of viral RNA synthesis.[8]

61.2.6   CHEMICAL AND PHYSICAL CHARACTERISTICS

PPRV is a fragile virus. A stability study that was carried out on cell culture viral suspensions revealed a half-life for the virus of 2.2 min at 56°C, 3.3 h at 37°C, 9.9 days at 4°C, and 24.2 days at −20°C.[29] The virus is stable in liquid suspension at pH values ranging from 5.8 to 9.5. As any enveloped virus, PPRV is inactivated by detergents and liquid solvents such as ether and chloroform.

61.3   CLINICAL FEATURES

PPR is primarily an acute respiratory disease. In general, the acute form of the disease begins within 5–6 days following infection with the onset of a dullness/depression in the affected animal, probably as one of the consequences of the fever. The animal becomes less interested in feed, and the ocular and oral membranes become congested. Next, ocular and nasal discharges appear that are initially serous but later become purulent. The discharges lead to the sticking together of parts of the eyelids or the partial blocking of the nostrils resulting in laborious breathing. The affected animal sometimes develops a moist and productive cough. In the oral cavity, discrete, tiny, and greyish necrotic foci develop over the reddish background. When the fever starts to drop about 4–5 days after the onset of the disease, necrotic spots appear that expand and coalesce resulting in extensive diphtheritic plaques. These lesions cover the membrane of the oral cavity and give the animal an unpleasant and fetid odor when it exhales. As the lesions disappear, irregular nonhemorrhagic erosive lesions remain. The lesions can also be found on the vulvar membrane and pregnant females may abort.[30–32] Along with the necrotic lesions, diarrhea also develops and it may be dysenteric. The severity of the diarrhea is correlated in many cases with the outcome of the disease. In 70%–80% of cases, animals will die within 10–12 days after the onset of the disease. Those animals that survive will recover fully within a week.

In the peracute form of the disease, 100% of affected animals will die. This form is mainly observed in young animals of more than 4 months old, animals which are no longer protected against PPR by maternal antibodies provided through the colostrum. After an incubation period of about 3 days following infection, the disease starts suddenly with high fever and a rectal temperature being between 40°C and 42°C. The animal is depressed and ceases eating. Different mucous membranes, and in particular those in the mouth and eyes, become significantly congested. One to two days after the onset of the disease, ocular and nasal discharges become apparent. Profuse diarrhea starts as the fever begins to decline. The animal will usually die within 5–6 days following the onset of disease symptoms.

If the dramatic impact of PPR on small ruminant productions is due to the high mortality rates of its superacute and acute forms, this disease can also occur in a subacute form in which all clinical signs are very mild and all animals recover within a week of the onset of symptoms. The diarrhea, which is slight, lasts for a maximum of 2–3 days and fever never exceeds 40°C. Ocular and nasal discharges are less abundant and only make crusts around the mouth and nostril orifices, symptoms that are similar to those of contagious ecthyma. In most cases, PPR in this form does not attract attention and may therefore remain unnoticed.

Following death after an acute case, the main and most striking findings during postmortem observation, in addition to the erosive–ulcerative and necrotic lesions in the mouth, are the following: mucopurulent or frothy exudates in the trachea and congestion of the respiratory and digestive tracks. The lung presents catarrhal or fibrinous bronchopneumonia, in particular, at the cardiac and apical lobes. The mesenteric lymph nodes may be swollen and congested. Linear hemorrhagic lesions are seen in the abomasum, the cecum, the colon, and along the folds of the rectum.[4,33–35] Due to their particular aspect, the lesions in the rectum are referred to a "zebra striping." The spleen is congested. The lymph nodes are also congested, edematous, and slightly enlarged. Necrosis may be present in the liver. Congestive and even necrotic lesions are sometimes seen in the kidneys. Histopathological examination reveals intracytoplasmic inclusion bodies. In addition, intranuclear viral eosinophilic inclusion bodies are also present in the cells of different tissues (lung, oral mucosa, kidney, and intestine epithelium), cells that may be fused to form syncytia. PPRV has been identified also in the brain even though no neurological disorders have yet been linked to this virus in small ruminants.[36,37]

61.4   PATHOGENESIS

Few studies investigating PPR pathogenesis have been undertaken to date. Most of our current understanding of this aspect of the disease derives mainly from results of studies on diseases caused by viruses closely related to PPRV (e.g., RP, measles and canine distemper) based on the assumption that their pathogenic processes are similar. The main route of infection by PPRV is by air through the respiratory tract like other morbilliviruses. In the case of the MV, it has long been thought that the first step of infection is the primary replication of the virus in the epithelial cells of the respiratory tract followed by a secondary amplification in the lymphoid tissues and dissemination throughout the whole organism. However, following the identification of the signaling lymphocyte activation molecule (CD150) (a protein expressed on lymphocyte and dendritic cell surface but not on epithelial cells) as the main receptor for morbillivirus wild types, including PPRV,[38–43] the model of host infection by MV was reexamined. The data that have been accumulated from these new studies strongly suggest that MV enters the host at the alveolar level by infecting macrophages and dendritic cells that then traffic the virus to bronchus-associated lymphoid tissue (BALT) and into the tracheobronchial lymph nodes. This infection results in local amplification of the virus and its subsequent systemic dissemination by viremia through infected CD150+ lymphocytes.[44] However, morbilliviruses are not only lymphotropic but are also epitheliotropic. As indicated earlier, PPRV is also found in different nonlymphoid tissues of the host such as lung, kidney, heart, and even the brain as is the case for MV and CDV. Thus, all these viruses use alternative receptor(s) present on epithelial cell surfaces to enable them to disseminate throughout the host. This other cell receptor for MV was identified in 2011: the protein Nectin-4.[45,46] It seems that both SLAM and Nectin-4 function equally well as morbillivirus receptors and that both are involved in the syncytia formation and cell-to-cell spread of the virus.[47] During morbillivirus infections, aerosols containing the viruses enter the upper respiratory tract and target dendritic cells and macrophages. These infected cells subsequently release the virus to the local lymph nodes where it can infect new populations of T and B cells expressing SLAM. Infected lymphocytes disseminate the virus throughout the host via the peripheral blood and lymphatic systems and then the epithelial cells of the respiratory, gastrointestinal, urinary, and endocrine systems via the Nectin-4 epithelial receptor.[47,48] So Nectin-4 functions as an "exit receptor," promoting amplification and subsequent release of the virus via different secretions. Both receptors are, therefore, required for the full pathogenesis of the virus.[49,50] This pathway of infection through the SLAM and Nectin-4 cell receptors has now been demonstrated for MV and CDV, and so it is most likely a valid pathway for too since it has now been demonstrated that Nectin-4 is also a receptor for PPRV.[125]

PPRV, like the other morbilliviruses, is a lymphotropic virus and it induces immunosuppression in the infected host. This immunosuppression, although transient, is profound and favors bacterial secondary infections that contribute to the severity of the disease.[51–53] In part, it is a result of the replication of the virus in the lymphoid cells and their subsequent destruction.[54,55] This may explain why the CDV clone that was engineered to make it blind for CD150 but at the same retaining its ability to binde to Nectin-4 spreads in the host but is not able anymore to induce immunosuppression.[50] In the case of the RPV, it has been demonstrated that the degree of the lymphopenia is in correlation with the virulence of the virus and the severity of the disease it causes.[56,57] However, there are data that indicate that the immunosuppression is not only the result of the destruction of lymphoid cells by the virus replication. Indeed, it has been demonstrated that the proliferation of freshly isolated, mitogen-stimulated bovine and caprine peripheral blood lymphocytes (PBLs) is inhibited by UV-inactivated RPV and PPRV.[58] It seems that the interaction of the two glycoproteins, F and H, with the lymphoid cell surface is involved in that inhibition.[58,59] Cell apoptosis was noted in goats infected by PPRV.[60] Morbillivirus N proteins, through their interactions with some cellular receptors, may be involved in that apoptosis and also in the inhibition of factors involved in host inflammation reactions.[61–64] In fact, many viral proteins contribute to the deregulation of the host immune response. In addition to the H, F, and N proteins indicated earlier, the two nonstructural proteins, C and V, inhibit the action of interferon.[65–67]

61.5 EPIDEMIOLOGY

61.5.1 HOST

Among domestic species, PPR primarily affects sheep and goats. In general, goats are more severely affected than sheep,[68] but some reports have highlighted cases of high mortality in sheep within small ruminants mixed flocks.[69,70] The seroprevalence to PPRV seems to differ notably among both species according to the field situation. Some authors observed that in mixed populations, the serological prevalence rate was higher in sheep than in goats, attributing this difference to the higher survival rate in sheep.[71,72] Others reported the opposite, arguing that goats were more susceptible than sheep and therefore had a higher probability of developing PPRV antibodies.[73–75] Indeed, it is difficult to fully interpret serological results obtained from field studies because, in addition to the host species and breeds, other factors such as breeding practices, animal densities, and trade may influence the susceptibility of the animal to infection.[76–79] Epidemiological surveillance studies carried out in different enzootic regions have revealed PPR seroprevalence in cattle, buffaloes, and camels.[71,72] This seroprevalence can be as high as 41% and 67% in the case of buffaloes and cattle, respectively, as reported in Pakistan.[80] Cattle are considered, and probably buffaloes also, as potential dead-end hosts for PPRV. It appears however that this virus, for reasons not yet elucidated, can occasionally overcome the innate resistance of these species and lead to the development of clinical signs. Mornet and collaborators[1] reported the development of clinical signs and a case fatality in calves experimentally infected with PPRV. In India, a case fatality rate of 96% was reported in domestic buffaloes (*Bubalus bubalis*) due to a virus that was isolated and identified as PPRV.[81] Clinical and epidemiological investigations coupled with laboratory results have led to the strong suspicion of a role of PPRV in the emergence of an epizootic disease in dromedary populations in the Horn of Africa. The prominent clinical signs observed in Sudan, similar to a previous outbreak reported in Ethiopia between 1995 and 1996, were respiratory syndrome, neurological signs, and abortion.[82–84] Viruses identified from sick dromedaries and small ruminants sharing the same grazing area were found phylogenetically identical.[85] Antelopes and other small wild ruminant species can also be severely affected resulting in high morbidity rates and deaths. Clinical signs were reported in different families of wild ungulates including Gazellinae, Tragelaphinae, and Caprinae subfamilies.[86–88] Some authors consider that in specific conditions, wildlife may play an important role in the epidemiology of PPR.[88]

61.5.2 MODE OF CONTAMINATION

PPR is highly contagious, with a potential for efficient and rapid spread. The sources of contamination are various excretions from sick animals: nasal and ocular discharges, saliva, feces, and urine.[3,89] Infected animals can excrete the virus at least 3 days before the onset of the disease.[90] From this infectious material, viral particles are released into the air and creating thereby aerosols that are inhaled by susceptible animals. Because PPRV is a very fragile virus, quickly inactivated at ambient temperature by UV from the sun and by desiccation, efficient virus transmission between animals is ensured only through close contacts between infected and susceptible animals. Such conditions are often met in and around shared watering holes and grazing sites. Live animal markets are ideal areas for virus transmission. For example, apparently healthy animals (but that are actually excreting the virus) that are brought to markets will be sources of contamination and dissemination of the disease. This is why in Muslim countries that are endemic for PPR, peaks of PPR outbreaks are recorded just after the religious festival of Eid during which the trade of sheep is dramatically increased. In sub-Saharan Africa, another period of high frequency in PPR outbreaks is the cold season (e.g., December–February), due to a better preservation of the virus under "cold" conditions and/or the tendency of animals to huddle closer to each other for warm.[76,91] In endemic countries, PPR outbreaks in the same area occur in cycles of 3 years. This is due to the fact that animals that recover from PPR are protected for life so PPRV can only be maintained in large populations if new susceptible hosts (e.g., newborns, transhumance, and newly

purchased animals) are available. Taking into consideration that the turnover of small ruminant flocks is around 3 years, a susceptible population to virus transmission is regenerated after 2–3 years.

61.5.3 CURRENT GEOGRAPHICAL DISTRIBUTION

Until the early 1970s, PPR was thought to be confined to West Africa. Since then, our understanding of the distribution of this disease is that it has steadily evolved in an apparent eastward manner from West Africa up to the Middle East and then Asia (Figure 61.2). The first PPR observation outside of West Africa was made in Sudan in 1972–1973.[92] In 1983, it appeared in the Arabian Peninsula.[87] As of the late 1980s, the disease's endemic regions have expanded into the Middle East and South Asia including China, Bhutan, and Vietnam.[93,94] The process of expansion into new uninfected territories has dramatically increased in Africa from 2005 to 2012 by spreading both northwards and southwards to cover all regions extending from North Africa to Tanzania, Democratic Republic of Congo (DRC), and Angola[74,85,93,95] (http://www.oie.int/wahis_2/public/wahid.php/Diseaseinformation/Diseasetimelines). PPRV strains that have been identified by different laboratories so far are divided into four phylogenetic lineages designated I–IV according to the sequence data derived from the N protein[96] or from the F protein genes.[97] In Figure 61.3, a phylogenetic tree is shown. It is derived from the partial N gene sequence of different isolates characterized to date. The lineage number I is the group of virus strains found in West Africa where the disease was first identified (Côte d'Ivoire) and also where the first virus isolation was made (Senegal). Lineage II is formed by a group of viruses that were initially found in Nigeria. Lineage III, which was first identified in East Africa, is shared between Africa and the Middle East on both sides of the Red Sea. Lineage IV, a unique lineage in Asia, covers a large area from Turkey to Southern Asia through the Arabian Peninsula. From the biological material submitted to

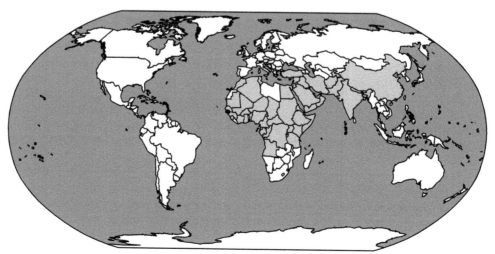

FIGURE 61.2 World map of PPR distribution. Countries that are indicated in light gray either have declared PPR to the OIE at least once or the information is from a publication. Although a country is entirely colored, the presence of PPR may be limited to only one or few regions.

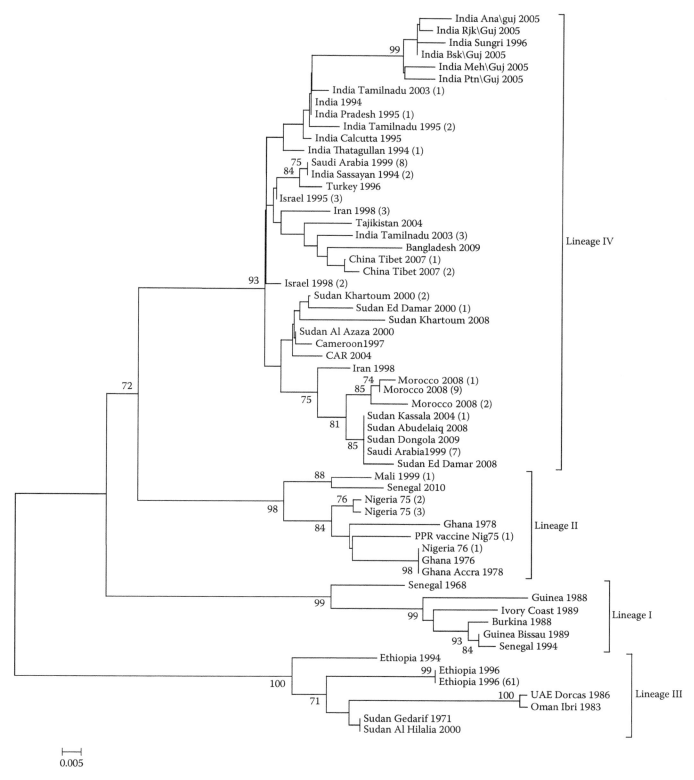

FIGURE 61.3 Phylogenetic tree of PPRV strains; tree drawn based on the comparison of partial nucleoprotein gene sequence of the different viruses. The tree was constructed using a weighted neighbor-joining method. It is drawn to scale, with branch lengths in the same units as those of the evolutionary distances used to infer the phylogenetic tree. Phylogenetic analyses were conducted using the Darwin package.

World Reference Laboratories (WRLs), it appears that recent emergences of PPR are accompanied by profound changes in the previous viral genotype distribution. This questions our knowledge of the previous distribution of lineages and thus requires a constant updating of virus data achieved through systematic epidemiological surveys. In an extraordinarily short period of time, lineage IV has become the predominant lineage in a core region of the African continent from Sudan to the Gulf of Guinea and the Mediterranean Sea.[85] Such a scenario is also seen in Senegal and Mauritania but involves lineage II viruses that were first detected in Nigeria, apparently slowly replacing or coexisting with lineage I viruses that were the only group detected in that region, from Senegal to Côte d'Ivoire.

61.6 DIAGNOSIS

For long time, PPR was overlooked in favor of many diseases (see Table 61.1) that have some similar clinical signs, in particular RP and pasteurellosis. In fact in many cases, the *Pasteurella* bacterial infection is associated with PPR. For the purpose of differential diagnosis, clinical diagnosis of PPR should be confirmed by laboratory testing.

61.6.1 ANTIBODY DETECTION

In the "OIE Terrestrial Manual," the prescribed test for serological diagnosis for international trade purposes is the virus neutralization test (VNT).[98,99] However, the most commonly used serological test is the ELISA technique. This test is perfectly suited to the analysis of large number of samples and is faster and less cumbersome to implement than the VNT. Competitive ELISA (c-ELISA) tests are recommended as an alternative to this test. The most widely used c-ELISAs are either the N or H antigen-based tests. For the current N-based c-ELISA,[100] the antigen is a recombinant PPRV N protein produced by a baculovirus in insect cells. For the H-based test developed by Anderson and McKay,[101] the antigen is a semipurified vaccine strain PPRV 75/1 virus. Both tests are available commercially. However, other ELISA tests, c-ELISA[102] or indirect ELISA,[103] have been developed. All these tests allow for the determination of the serological status of animals with respect to PPR but not the differentiation between vaccinated and infected animals.

61.6.2 VIRUS IDENTIFICATION

Antigen detection: A variety of tools have been developed, allowing for the direct detection of PPRV in tissue samples or swabs with good specificity and high sensitivity. These tests incorporate one or more monoclonal antibodies (Mabs) thereby improving specificity. The immunocapture ELISA developed by Libeau et al.[104] was a method of choice and an alternative to virus isolation. In addition, the rapid implementation of this test, combined with its simplicity, allowed its routine use in countries with the minimum of infrastructure. It is based on the detection of the N protein. Its detection limit value is comparable to that of current molecular diagnostic methods (i.e., 100.6 TCID50/well). A similar test was developed by Singh et al.[105] A chromatographic strip test device was also developed for PPRV to be used for viral antigen detection in eye swabs in field conditions.[106] A hemagglutination assay using chicken red blood cells has also been developed.[107]

Nucleic acid detection: A reverse transcription-polymerase chain reaction (RT-PCR) to amplify PPRV nucleic acid was developed in the mid-1990s. It remains an important diagnostic tool to positively identify the virus and perform subsequent phylogenetic characterization of isolates. Both the N and F genes are used as targets of the conventional RT-PCR.[108,109] The RT-PCR is also applicable directly on samples collected on filter paper and stored in absence of a cold chain.[110] Protocols for the quantitative RT-PCR (qRT-PCR) for the diagnosis of PPR have been published too.[111–113] The loop-mediated isothermal amplification (LAMP) technique has also been developed for the identification of PPRV nucleic acids,[114] a technique that may be suitable for field application.

61.6.3 VIRUS ISOLATION

Even if disease diagnosis in general has been dramatically improved with the development of ELISA and nucleic acid

TABLE 61.1
Clinical Differential Diagnosis of PPR with Some Other Small Ruminant Diseases

Disease	Nasal and Ocular Discharges	Mouth Lesions	Bronchopneumonia	Diarrhea
Oestrosis	++	−	−	−
ORF	−	+	−	−
Capripox disease	+	−	+	+
Contagious caprine pleuropneumonia (CCPP)	−	−	++	−
Blue tongue	+	+	−	−
Pasteurellosis	+	−	++	−
RP	++	++	−	++

detection technologies, pathogen isolation should remain the gold standard and be implemented whenever possible. For PPRV isolation, specimens should ideally be collected during the early phase of the disease (i.e., the prodromal or erosive phases) or from fresh carcasses. The most suitable pathological samples for PPRV isolation are ocular and nasal swabs and lung and lymph nodes, but other tissues such as intestine or white blood cells may also be used. Primary and secondary cultures of kidney cells of sheep or calf were used for many years for PPRV isolation.[68] Subsequently, cell lines that are easy to maintain in culture and address the various drawbacks of availability and quality of primary cells were then used. The African green monkey kidney (Vero) cell line is to date the most commonly used cell line for PPRV isolation.[4,68] In many cases, several blind passages are needed before a cytopathic effect (if any) is visible.[115] To improve on the efficiency of virus isolation, a monkey cell line expressing the PPRV receptor has been developed.[38] The development of this cell line was based on the demonstration that wild-type morbilliviruses preferentially use SLAM as a cell receptor.[39,42,43] The goat SLAM gene was cloned and the corresponding cDNA was integrated into the genome of the monkey cell line CV1. With this modified cell line, PPRV can now be isolated in less than 1 week.

61.7 TREATMENT AND PREVENTION

The treatment of animals affected by PPR involving the administration of anti-PPR serum or antibiotics in association with antidiarrheal medicines has been reported by some authors.[116,117] However, this approach may not be of practical use in the field given the low price of individual sheep and goats. So, presently, the control of PPR is ensured only through the implementation of effective prophylaxis. All sheep and goats of an affected stock should be under quarantine for at least 1 month after the last clinical case.[29] Animal movements need to be strictly controlled in the area of infection. Unfortunately, all these sanitary measures are difficult to maintain in countries where PPR is endemic. Therefore, the only effective way to control this disease is through vaccination. This takes advantage of the fact that recovery from morbillivirus infections is followed by a solid immune protective response. Very efficient PPR homologous vaccines have been developed and are commercially available.[118–120] They provide a lifelong immunity in the inoculated animals. Because PPR and capripox share almost the same geographical distributions, the F and H protein genes of PPRV have been inserted into the capripox virus genome to produce a dual thermostable recombinant vaccine to protect small ruminants against these two economical important diseases.[121,122]

61.8 CONCLUSIONS

As of 2000, the geographical distribution of PPR has rapidly expanded in Africa and Asia. It is now considered as the most important infectious disease of sheep and goats in endemic countries. In 2001, an international study for animal diseases prioritization in developing countries rated PPR in the list of the top 10 priority animal diseases to be considered for poverty alleviation in sub-Saharan Africa and South Asia.[123] This study clearly highlighted the economic importance of PPR and its impact on the livelihood of the poor in many developing countries. The increase in PPR outbreaks together with the rapid expansion of its endemic areas has now attracted many studies on this disease. Despite this increased interest in PPR, much has still to be done to understand the epidemiology of the disease, for example, the role of cattle and camels in the epidemiology of the disease, variation in the virulence of virus strains, and maintenance of the disease in an enzootic status. There is also a need to make available tools, vaccines, and tests that will enable differentiation between infected and vaccinated animals (DIVA vaccine) as well as tools that will improve the management of controls programs. Another area that deserves attention in the near future is the development of curative medicines in addition to vaccines as an improvement in PPR control. A promising approach is the possibility of blocking the expression of virus genes using RNA interference (RNAi). Preliminary studies carried out on this new technology to inhibit PPRV replication have provided promising results.[124]

Although much still needs to be learnt about the epidemiology of PPR and the development of suitable DIVA systems, the tools presently available for PPR control are, nevertheless, adequate to allow the implementation of an eradication program for this important disease similar to that which was successfully carried out for RP.

REFERENCES

1. Mornet, P. et al. La peste des petits ruminants en Afrique Occidentale Française. Ses rapports avec la peste bovine. *Rev. Elev. Méd. Vét. Pays Trop.*, 9, 313, 1956.
2. Gilbert, Y. and Monnier, J. Adaptation du virus de la peste des petits ruminants aux cultures cellulaires. *Rev. Elev. Med. Vet. Pays Trop.*, 15, 321, 1962.
3. Gibbs, E.P.J. et al. Classification of peste des petits ruminants virus as the fourth member of the genus morbillivirus. *Intervirology*, 11, 268, 1979.
4. Hamdy, F.M. et al. Etiology of stomatitis pneumoenteritis in Nigeria dwarf goats. *Can. J. Comp. Med.*, 40, 276, 1976.
5. Taylor, W.P. The distribution and epidemiology of peste des petits ruminants. *Prev. Vet. Med.*, 2, 157, 1984.
6. Diallo, A. et al. Comparison of proteins induced in cells infected with Rinderpest and peste des petits ruminants viruses. *J. Gen. Virol.*, 68, 2033, 1987.
7. Diallo, A. et al. Cloning of the nucleocapsid protein gene of the peste des petits ruminants virus: Relationship to other morbilliviruses. *J. Gen. Virol.*, 75, 233, 1994.
8. Barrett, T., Banyard, A.C., and Diallo, A. Molecular biology of the morbilliviruses. In: Thomas, B., Paul-Pierre, P., and William, P.T. (eds.). *Rinderpest and Peste des Petits Ruminants*. Oxford, U.K.: Academic Press, p. 31, 2006.
9. Kennedy, S. Morbillivirus infections in aquatic mammals. *J. Comp. Pathol.*, 119, 201, 1998.
10. Osterhaus, A.D.M.E. et al. Morbillivirus infections of aquatic mammals: Newly identified members of the genus. *Vet. Microbiol.*, 44, 219, 1995.

11. Woo, P.C. et al. Feline morbillivirus, a previously undescribed paramyxovirus associated with tubulointerstitial nephritis in domestic cats. *Proc. Natl. Acad. Sci. USA*, 109, 5435, 2012.

12. Bourdin, P. and Laurent-Vautier, A. Note sur la structure du virus de la peste des petits ruminants. *Rev. Elev. Méd. Vét. Pays Trop.*, 20, 383, 1967.

13. Durojaiye, O.A., Taylor, W.P., and Smale, C. The ultrastructure of peste des petits ruminants virus. *Zentralbl. Veterinärmed.*, 32, 460, 1985.

14. Rager, M. et al. Polyploid measles virus with hexameric genome length. *EMBO*, 21, 2364, 2002.

15. Chard, L.S. et al. Full genome sequences of two virulent strains of peste-des-petits ruminants virus, the Côte d'Ivoire 1989 and Nigeria 1976 strains. *Virus Res.*, 136, 192, 2008.

16. Minet, C. et al. Sequence analysis of the large (L) polymerase gene and trailer of the peste des petits ruminants virus vaccine strain Nigeria 75/1: Expression and use of the L protein in reverse genetics. *Virus Res.*, 145, 9, 2009.

17. Baron, M.D. et al. The genome sequence of the virulent Kabete "O" strain of rinderpest virus: Comparison with the derived vaccine. *J. Gen. Virol.*, 77, 3041, 1996.

18. Bourhis, J.M. et al. The C-terminal domain of measles virus nucleoprotein belongs to the class of intrinsically disordered proteins that fold upon binding to their physiological partner. *Virus Res.*, 99, 157, 2004.

19. Bourhis, J.M. et al. The intrinsically disordered C-terminal domain of the measles virus nucleoprotein interacts with the C-terminal domain of the phosphoprotein via two distinct sites and remains predominantly unfolded. *Protein Sci.*, 14, 1975, 2005.

20. Bodjo, S.C. et al. Mapping the peste des petits ruminants virus nucleoprotein: Identification of two domains involved in protein self-association. *Virus Res.*, 131, 23, 2008.

21. Couturier, M. et al. High affinity binding between Hsp70 and the C-terminal domain of the measles virus nucleoprotein requires an Hsp40 co-chaperone. *J. Mol. Recognit.*, 23, 301, 2010.

22. Mahapatra, M. et al. Sequence analysis of the phosphoprotein gene of peste des petits ruminants (PPR) virus: Editing of the gene transcript. *Virus Res.*, 96, 85, 2003.

23. Haffar, A. et al. The matrix protein gene sequence reveals close relationship between peste des petits ruminants virus (PPRV) and dolphin morbillivirus. *Virus Res.*, 64, 69–75, 1999.

24. Liljeroos, L. et al. Electron cryotomography of measles virus reveals how matrix protein coats the ribonucleocapsid within intact virions. *Proc. Natl. Acad. Sci. USA*, 108, 18085, 2011.

25. Peeples, M.E. Paramyxovirus M protein: Pulling it all together and taking it on the road. In: Kingsbury, D.W. (ed.). *The Paramyxoviruses*. New York: Plenum Press, p. 427, 1991.

26. Seth, S. and Shaila, M.S. The hemagglutinin–neuraminidase protein of peste des petits ruminants virus is biologically active when transiently expressed in mammalian cells. *Virus Res.*, 75, 169, 2001.

27. Wosu, L.O. Haemagglutination test for diagnosis of PPR diseases in goats with samples from live animals. *Small Rumin. Res.*, 5, 169, 1991.

28. Meyer, G. and Diallo, A. The nucleotide sequence of the fusion protein gene of the peste des petits ruminants virus: The long un-translated region in the 5'-end of the F-protein gene of morbilliviruses seems to be specific to each virus. *Virus Res.*, 37, 23, 1995.

29. Rossiter, P.B. Peste des petits ruminants. In: Coetzer, J.A.W. and Tustin, R.C. *Infectious Diseases of Livestock*, Vol. II. Cape Town, South Africa: Oxford University Press, pp. 660–672, 2004.

30. Abubakar, M., Ali, Q., and Khan, H.A. Prevalence and mortality rate of peste des petits ruminants (PPR): Possible association with abortion in goat. *Trop. Anim. Health Prod.*, 40, 317, 2008.

31. Kul, O. et al. Concurrent peste des petits ruminants virus and pestivirus infection in stillborn twin lambs. *Vet. Pathol.*, 45, 191, 2008.

32. Kulkarni, D.D. et al. Peste des petits ruminants in goats in India. *Vet Rec.*, 138, 187, 1996.

33. Aruni, A.W. et al. Histopathological study of a natural outbreak of peste des petits ruminants in goats of Tamilnadu. *Small Rumit. Res.*, 28, 233, 1998.

34. Bundza, A. et al. Experimental Peste des petits ruminant (goat plague) in goats and sheep. *Can. J. Vet. Res.*, 52, 46, 1988.

35. Obi, T.U. et al. Peste des petits ruminants (PPR) in goats in Nigeria: Clinical, microbiological and pathological features. *Zentralbl. Veterinärmed.*, 30, 751, 1983.

36. Toplu, N. Characteristic and non-characteristic pathological findings in peste des petits ruminants (PPR) of sheep in the Ege district of Turkey. *J. Comp. Pathol.*, 131, 135, 2004

37. Toplu, N., Oguzoglu, T.C., and Albayrak, H. Dual infection of foetal and neonatal small ruminants with border disease virus and peste des petits ruminants Virus (PPRV): Neuronal tropism of PPRV as a novel finding. *J. Comp. Pathol.*, 146, 289, 2012.

38. Adombi, C.M. et al. Monkey CV1 cell line expressing the sheep-goat SLAM protein: A highly sensitive cell line for the isolation of peste des petits ruminants virus from pathological specimens. *J. Virol. Methods*, 173, 306, 2011.

39. Baron, M.D. Wild-type Rinderpest virus uses SLAM (CD150) as its receptor. *J. Gen. Virol.*, 86, 1753, 2005.

40. Meng, X. et al. Tissue distribution and expression of signaling lymphocyte activation molecule receptor to peste des petits ruminant virus in goats detected by real-time PCR. *J. Mol. Histol.*, 42, 467, 2011.

41. Pawar, R.M., Dhinakar Raj, G., and Balachandran, C. Relationship between the level of signaling lymphocyte activation molecule mRNA and replication of peste-des-petits-ruminants virus in peripheral blood mononuclear cells of host animals. *Acta Virol.*, 52, 231, 2008.

42. Tatsuo, H. et al. SLAM (CDw150) is a cellular receptor for measles virus. *Nature*, 406, 893, 2000.

43. Yanagi, Y., Takeda, M., and Ohno, S. Measles virus: Cellular receptors, tropism and pathogenesis. *J. Gen. Virol.*, 87, 2767, 2006.

44. Lemon, K. et al. Early target cells of measles virus after aerosol infection of non-human primates. *PLoS Pathog.*, 7, e1001263, 2011.

45. Muhlebach, M.D. et al. Adherens junction protein nectin-4 is the epithelial receptor for measles virus. *Nature*, 480, 530, 2011.

46. Noyce, R.S. et al. Tumor cell marker PVRL4 (nectin 4) is an epithelial cell receptor for measles virus. *PLoS Pathog.*, 7, e1002240, 2011.

47. Noyce, R.S., Delpeut, S., and Richardson, C.D. Dog nectin-4 is an epithelial cell receptor for canine distemper virus that facilitates virus entry and syncytia formation. *Virology*, 436(1), 210, 2013.

48. Sato, H. et al. Morbillivirus receptors and tropism: Multiple pathways for infection. *Front. Microbiol.*, 3, 75, 2012.

49. Kato, S., Nagata, K., and Takeuchi, K. Cell tropism and pathogenesis of measles virus in monkeys. *Front. Microbiol.*, 3, 14, 2012.

50. Sawatsky, B. et al. Canine distemper virus epithelial cell infection is required for clinical disease but not for immunosuppression. *J. Virol.*, 86, 3658, 2012.

51. Jagtap, S.P. et al. Effect of immunosuppression on pathogenesis of peste des petits ruminants (PPR) virus infection in goats. *Microb. Pathog.*, 52, 217, 2012.

52. Rajak, K.K. et al. Experimental studies on immunosuppressive effects of peste des petits ruminants (PPR) virus in goats. *Comp. Immunol. Microbiol. Infect. Dis.*, 28, 287, 2005.

53. Schneider-Schaulies, S. et al. Measles virus induced immunosuppression: Targets and effector mechanisms. *Curr. Mol. Med.*, 1, 163, 2001.

54. Schobesberger, M. et al. Canine distemper virus-induced depletion of uninfected lymphocytes is associated with apoptosis. *Vet. Immunol. Immunopathol.*, 104, 33, 2005.

55. Von Messling, V., Milosevic, D., and Cattaneo, R. Tropism illuminated: Lymphocyte-based pathways blazed by lethal morbillivirus through the host immune system. *Proc. Natl. Acad. Sci. USA*, 101, 14216, 2004.

56. Rey Nores, J.E. and McCullough, K.C. Rinderpest virus isolates of different virulence vary in their capacity to infect bovine monocytes and macrophages. *J. Gen. Virol.*, 78, 1875, 1997.

57. Wohlsein, P. et al. Pathomorphological and immunohistological findings in cattle experimentally infected with rinderpest virus isolates of different pathogenicities. *Vet. Microbiol.*, 44, 141, 1995.

58. Heaney, J., Barrett, T., and Cosby, S.L. Inhibition of in vitro leukocyte proliferation by morbilliviruses. *J. Virol.*, 76, 3579, 2002.

59. Schlender, J. et al. Interaction of measles virus glycoproteins with the surface of uninfected peripheral blood lymphocytes induces immunosuppression in vitro. *Proc. Natl. Acad. Sci. USA*, 93, 13194, 1996.

60. Mondal, B. et al. Apoptosis induced by peste des petits ruminants virus in goat peripheral blood mononuclear cells. *Virus Res.*, 73, 113, 2001.

61. Kerdiles, Y.M. et al. Immunosuppression caused by measles virus: Role of viral proteins. *Rev. Med. Virol.*, 16, 49, 2005.

62. Kerdiles, Y.M. et al. Immunomodulatory properties of morbillivirus nucleoproteins. *Virol. Immunol.*, 19, 324, 2006.

63. Laine, D. et al. Measles virus (MV) nucleoprotein binds to a novel cell surface receptor distinct from FcRII via its C-terminal domain: Role in MV-induced immunosuppression. *J. Virol.*, 77, 11332, 2003.

64. Takayama, I. et al. The nucleocapsid protein of measles virus blocks host interferon response. *Virology*, 424, 45, 2012.

65. Nanda, S.K. and Baron, M.D. Rinderpest virus blocks type I and type II interferon action: Role of structural and nonstructural proteins. *J. Virol.*, 80, 7555, 2006.

66. Ohno, S. et al. Dissection of measles virus V protein in relation to its ability to block alpha/beta interferon signal transduction. *J. Gen. Virol.*, 85, 2991, 2004.

67. Shafer, J.A., Bellini, W.J., and Rota, P.A. The C protein of measles virus inhibits the type I interferon response. *Virology*, 315, 389, 2003.

68. Lefèvre, P.C. and Diallo, A. Peste des petits ruminants. *Rev. Sci. Tech.*, 9, 935, 1990.

69. Shaila, M.S. et al. Peste des petits ruminants of sheep in India. *Vet. Rec.*, 125, 602, 1989.

70. Yesilbağ, K. et al. Peste des petits ruminants outbreak in western Turkey. *Vet. Rec.*, 157, 260, 2005.

71. Abraham, G. et al. Antibody seroprevalences against peste des petits ruminants (PPR) virus in camels, cattle, goats and sheep in Ethiopia. *Prev. Vet. Med.*, 70, 51, 2005.

72. Özkul, A. et al. Distribution and host range of peste des petits ruminants virus in Turkey. *Emerg. Infect. Dis.*, 8, 709, 2002.

73. Al-Majali, A.M. et al. Seroprevalence of, and risk factors for, peste des petits ruminants in sheep and goats in Northern Jordan. *Prev. Vet. Med.*, 85, 1, 2008.

74. Ayari-Fakhfakh, E. et al. First serological investigation of peste-des-petits-ruminants and Rift Valley fever in Tunisia. *Vet. J.*, 187, 402, 2011.

75. Delil, F., Asfaw, Y., and Gebreegziabher, B. Prevalence of antibodies to peste des petits ruminants virus before and during outbreaks of the disease in Awash Fentale district, Afar, Ethiopia, *Trop. Anim. Health Prod.*, 44, 1329, 2012.

76. Abubakar, M. et al. Peste des petits ruminants virus (PPRV) infection: Its association with species, seasonal variations and geography. *Trop. Anim. Health Prod.*, 41, 1197, 2009.

77. Diop, M., Sarr, J., and Libeau, G. Evaluation of novel diagnostic tools for peste des petits ruminants virus in naturally infected goat herds. *Epidemiol. Infect.*, 133, 711, 2005.

78. Ezeokoli, D. et al. Clinical and epidemiological features of peste des petits ruminants in Sokoto red goats. *Rev. Elev. Méd. Vét. Pays Trop.*, 39, 269, 1986.

79. Hammouchi, M. et al. Experimental infection of alpine goats with a Moroccan strain of peste des petits ruminants virus (PPRV). *Vet. Microbiol.*, 160, 240, 2012.

80. Khan, H.A. et al. The detection of antibody against peste des petits ruminants virus in sheep, goats, cattle and buffaloes. *Trop. Anim. Health Prod.*, 40, 521, 2008.

81. Govindarajan, R. et al. Isolation of peste des petits ruminants virus from an outbreak in Indian buffalo (*Bubalus bubalis*). *Vet. Rec.*, 141, 573, 1997.

82. Megersa, B. et al. Epidemic characterization and modeling within herd transmission dynamics of an "emerging transboundary" camel disease epidemic in Ethiopia. *Trop. Anim. Health Prod.*, 44, 1643, 2012.

83. Khalafalla, A.I. et al. An outbreak of peste des petits ruminants (PPR) in camels in the Sudan. *Acta Trop.*, 116, 161, 2010.

84. Roger, F. et al. Investigation of a new pathological condition of camels in Ethiopia. *J. Camel Pract. Res.*, 2, 163, 2000.

85. Kwiatek, O. et al. Asian lineage of peste des petits ruminants virus, Africa. *Emerg. Infect. Dis.*, 17, 1223, 2011.

86. Bao, J. et al. Detection and genetic characterization of peste des petits ruminants virus in free-living bharals (*Pseudois nayaur*) in Tibet, China. *Res. Vet. Sci.*, 90, 238, 2011.

87. Furley, C.W., Taylor, W.P., and Obi, T.U. An outbreak of peste des petits ruminants in a zoological collection. *Vet. Rec.*, 121, 443, 1987.

88. Kinne, J. et al. Peste des petits ruminants in Arabian wildlife. *Epidemiol. Infect.*, 138, 1211, 2010.

89. Abegunde, A.A. and Adu, F. Excretion of the virus of peste des petits ruminants by goats. *Bull. Anim. Health. Prod. Afr.*, 25, 307, 1977.

90. Couacy-Hymann, E. et al. Early detection of viral excretion from experimentally infected goats with peste-des-petits ruminants virus. *Prev. Vet. Med.*, 78, 85, 2007.

91. Lancelot, R., Lesnoff, M., and McDermott, J.J. Use of Akaike information criteria for model selection and inference. An application to assess prevention of gastrointestinal parasitism and respiratory mortality of Guinean goats in Kolda, Senegal. *Prev. Vet. Med.*, 55, 217, 2002.

92. El Hag Ali, B. and Taylor, W.P. Isolation of peste des petits ruminants virus from Sudan. *Res. Vet. Sci.*, 36, 1, 1984.

93. Banyard, A.C. et al. Global distribution of peste des petits ruminants virus and prospects for improved diagnosis and control. *J. Gen. Virol.*, 91, 2885, 2010.

94. Maillard, J.C. et al. Examples of probable host-pathogen co-adaptation/co-evolution in isolated farmed animal populations in the mountainous regions of North Vietnam. *Ann. N. Y. Acad. Sci.*, 1149, 259, 2008.

95. De Nardi, M. et al. First evidence of peste des petits ruminants (PPR) virus circulation in Algeria (Sahrawi territories): Outbreak investigation and virus lineage identification. *Transbound. Emerg. Dis.*, 59, 214, 2012.

96. Kwiatek, O. et al. Peste des petits ruminants (PPR) outbreak in Tajikistan. *J. Comp. Pathol.*, 136, 111, 2007.

97. Dhar, P. et al. Recent epidemiology of peste des petits ruminants virus (PPRV). *Vet. Microbiol.*, 88, 153, 2002.

98. OIE. Manual of diagnostic tests and vaccines for terrestrial animals. http://www.oie.int/en/international-standard-setting/terrestrial-manual/access-online/, 2012.

99. Rossiter, P.B., Jessett, D.M., and Taylor, W.P. Microneutralisation systems for use with different strains of peste des petits ruminants and rinderpest virus. *Trop. Anim. Health Prod.*, 17, 75, 1985.

100. Libeau, G. et al. Development of a competitive ELISA for peste des petits ruminants virus antibody detection using a recombinant N protein. *Res. Vet. Sci.*, 58, 50, 1995.

101. Anderson, J. and McKay, J.A. The detection of antibodies against peste des petits ruminants virus in cattle, sheep and goats and the possible implications to rinderpest control programmes. *Epidemiol. Infect.*, 112, 225, 1994.

102. Singh, R.P. et al. Development of a monoclonal antibody based competitive-ELISA for detection and titration of antibodies to peste des petits ruminants (PPR) virus. *Vet. Microbiol.*, 98, 3, 2004.

103. Balamurugan, V. et al. Development of an indirect ELISA for the detection of antibodies against peste-des-petits-ruminants virus in small ruminants. *Vet. Res. Commun.*, 31, 355, 2007.

104. Libeau, G. et al. Rapid differential diagnosis of rinderpest and peste des petits ruminants using an immunocapture ELISA. *Vet. Rec.*, 134, 300, 1994.

105. Singh, R.P. et al. A sandwich-ELISA for the diagnosis of peste des petits ruminants (PPR) infection in small ruminants using anti-nucleocapsid protein monoclonal antibody. *Arch. Virol.*, 149, 2155, 2004.

106. Brüning-Richardson, A. et al. Improvement and development of rapid chromatographic strip-tests for the diagnosis of rinderpest and peste des petits ruminants viruses. *J. Virol. Methods*, 174, 42, 2011.

107. Ezeibe, M.C.O., Wosu, L.O., and Erumaka, I.G. Standardisation of the haemagglutination test for peste des petits ruminants (PPR). *Small Rumin. Res.*, 51, 269, 2004.

108. Couacy, E. et al. Rapid and sensitive detection of peste des petits ruminants virus by a polymerase chain reaction assay. *J. Virol. Methods*, 100, 17, 2002.

109. Forsyth, M.A. and Barrett, T. Evaluation of polymerase chain reaction for the detection and characterization of rinderpest and peste des petits ruminants viruses for epidemiological studies. *Virus Res.*, 39, 151, 1995.

110. Michaud, V. et al. Long-term storage at tropical temperature of dried-blood filter papers for detection and genotyping of RNA and DNA viruses by direct PCR. *J. Virol. Methods*, 146, 257, 2007.

111. Bao, J. et al. Development of one-step real-time RT-PCR assay for detection and quantitation of peste des petits ruminants virus. *J. Virol. Methods*, 148, 232, 2008.

112. Batten, C.A. et al. A real time RT-PCR assay for the specific detection of peste des petits ruminants virus. *J. Virol. Methods*, 171, 401, 2011.

113. Kwiatek, O. et al. Quantitative one-step real-time RT-PCR for the fast detection of the four genotypes of PPRV. *J. Virol. Methods*, 165, 168, 2010.

114. Li, L. et al. Rapid detection of peste des petits ruminants virus by a reverse transcription loop-mediated isothermal amplification assay. *J. Virol. Methods*, 170, 37, 2010.

115. Saliki, J.T. et al. Comparison of monoclonal antibody-based sandwich enzyme-linked immunosorbent assay and virus isolation for detection of peste des petits ruminants virus in goat tissues and secretions. *J. Clin. Microbiol.*, 32, 1349, 1994.

116. Akpan, M.O. et al. A comparison of three chemotherapeutic regimens in the management of naturally peste des petits ruminants (PPR) disease in West African dwarf goats. *Trop. Vet.*, 7, 87, 1999.

117. Anene, B.M., Ugochukwu, E.I., and Omamegbe, J.O. The appraisal of three different pharmaceutical regimes for the treatment of naturally occurring peste des petits ruminants (PPR) in goats. *Bull. Anim. Health Prod. Afr.*, 35, 1, 1987.

118. Diallo, A. et al. Atténuation d'une souche de virus de la peste des petits ruminants: Candidat pour un vaccin homologue vivant. *Rev. Elev. Méd. Vét. Pays Trop.*, 42, 311, 1989.

119. Diallo, A. et al. The threat of peste des petits ruminants: Progress in vaccine development for disease control. *Vaccine*, 25, 5591, 2007.

120. Sen, A. et al. Vaccines against peste des petits ruminants virus. *Expert Rev. Vaccines*, 9, 785, 2010.

121. Berhe, G. et al. Development of a dual recombinant vaccine to protect small ruminants against peste des petits ruminants and capripox infections. *J. Virol.*, 77, 1571, 2003.

122. Chen, W. et al. A goat poxvirus-vectored peste-des-petits-ruminants vaccine induces long-lasting neutralization antibody to high levels in goats and sheep. *Vaccine*, 28, 4742, 2010.

123. Perry, B.D. et al. *Investing in Animal Health Research to Alleviate Poverty.* Nairobi, Kenya: ILRI (International Livestock Research Institute), p. 148, 2002.

124. de Almeida, R. et al. Control of ruminant morbillivirus replication by small interfering RNA. *J. Gen. Virol.*, 88, 2307, 2007.

125. Birch, J. et al. Characterization of ovine Nectin-4, a novel peste des petits ruminants virus receptor. *J. Virol.*, 87, 4756, 2013.

62 Rinderpest Virus

Ashley C. Banyard, Anke Brüning-Richardson, and Satya Parida

CONTENTS

62.1 INTRODUCTION

Rinderpest, or "cattle plague," is a highly contagious viral disease that can affect all species of even-toed ungulates, order Artiodactyla. Until 2011, the causative agent of this disease, rinderpest virus (RPV), was the most feared viral infection of large ruminants causing severe mortalities across endemic regions. However, in 2011, this virus was declared as being eradicated following an extensive program of vaccination and serosurveillance.[1] Within the order Artiodactyla, rinderpest was able to infect animals including yak, giraffe, hippopotamus, antelope, cattle, and buffalo. The clinical manifestations of RPV infection could vary, depending both on the virus strain and the species infected, but RPV was generally characterized by certain classical symptoms. These included a sudden onset of fever leading to mucopurulent ocular and nasal discharges, necrosis, ulceration, and erosion of the mucosal lining of the oral cavity, nares, and digestive tract, the latter leading to severe diarrhea, dehydration, and death.[2] Infection by the same virus strain could vary from mild to acute and lethal depending on the species involved. Buffalo were extremely susceptible, and in some buffalo populations, infection with certain viral strains, which were mild in cattle, could exact morbidity and mortality rates approaching 90%.[3,4]

62.2 CLASSIFICATION AND MORPHOLOGY

RPV is a member of the order Mononegavirales, family Paramyxoviridae, subfamily Paramyxovirinae, genus *Morbillivirus*. All members of this order have single-stranded, nonsegmented RNA genomes with a negative polarity. Within the *Morbillivirus* genus are several well-characterized virus species including measles virus (MV), canine distemper virus (CDV), peste des petits ruminants virus (PPRV), phocine distemper virus (PDV), porpoise morbillivirus (PMV), dolphin morbillivirus (DMV), and morbilliviruses isolated from whale species termed cetacean morbilliviruses (CeMVs)[5] (Table 62.1 and Figure 62.1). Alongside this, several novel morbillivirus-like viruses have been discovered in domestic cats, feline morbillivirus (FeMV),[6] and numerous viruses in bats and rodents.[7]

Morbillivirus virions are normally spherical and approximately 150–300 nm in diameter, although pleomorphic variants are often seen in the electron microscope. They have a lipid-containing envelope that is derived from the host cell membrane on budding. These lipid envelopes contain spikelike glycoprotein protrusions of approximately 8–12 nm in length, spaced 7–10 nm apart, "studded" across their surfaces called peplomers.[8] Inside the lipid envelope, the virions contain a coiled helical nucleocapsid that is 13–18 nm in diameter and up to 1000 nm in length with a pitch of 5.5–7 nm depending on the genus.[9] The linear RNA genome of members of the *Morbillivirus* genus varies in length from 15 to 19 kb.[8]

The RPV genome consists of a single strand of RNA that is 15,882 nucleotides in length.[10] Each virion contains six structural proteins: the three nucleocapsid-associated proteins (N, an RNA-binding protein; P, a phosphoprotein; L, a large protein with characteristics of a polymerase protein); M, an unglycosylated matrix protein located under the virus envelope; and two glycosylated envelope proteins, namely, the fusion (F) and attachment (H) proteins. The two virus-encoded nonstructural proteins, C and V, are thought to play roles in the virus replication cycle. These two non-structural proteins may be essential for pathogenesis in vivo,[11] although they can be dispensed with in vitro.[12]

Alongside regions of the genome that code for proteins, the morbillivirus genome contains untranslated regions that flank each of the genes. These regions include the genome (GP) and antigenome (AGP) promoter regions that act as promoters of transcription and replication and untranslated regions positioned between each of the genes.[10,13]

TABLE 62.1

Characteristics of the Genus *Morbillivirus*

Virus	Principal Host	Other Susceptible Species	Distribution	Vaccines Available
RPV	Large ruminants	Subclinical infection of small ruminants/severe disease in wildlife species	Globally eradicated	Yes
PPRV	Small ruminants	Subclinical infection of large ruminants/severe disease in wildlife species	Primarily Africa and Asia, although incursions into Eastern European countries, e.g., Turkey	Yes
MV	Humans	Nonhuman primates	Global distribution—eradication campaign in progress	Yes
CDV	Dogs	Numerous hosts including, but not restricted to, members of Felidae, Hyaenidae, Mustelidae	Global distribution—not targeted for eradication	Yes
PDV	Seals	Mink and other terrestrial carnivores	Largely undefined	No
PMV	Porpoise	Other cetacean mammals	Global oceanic distribution	No
DMV	Dolphins	Other cetacean mammals	Global oceanic distribution	No
FeMV	Domestic cats	Currently unknown	Currently unknown	No

The latter contain both gene start and stop signals as well as an intergenic trinucleotide that is generally conserved within each morbillivirus species.

62.3 BIOLOGY AND EPIDEMIOLOGY

Rinderpest, like all other well-characterized morbilliviruses, is transmitted through contact with infected animals via droplet spread and contaminated fomites. Since the virus is relatively labile, it requires close contact for transmission and is easily controlled by zoo sanitary measures such as slaughter, disinfection, and quarantine. Historically, the topographical spread of the disease was generally a consequence of the development of livestock as a major trade product.[1]

The recognition of rinderpest as a distinct disease dates back to the European epizootic of A.D. 376–386 when the infection spread westwards from the Caspian basin or the Russian and Hungarian steppes.[14–16] Further invasions into Europe were often seen as a sequel to major military campaigns. In the eighteenth century in Western Europe, up to 200 million cattle were estimated to have succumbed to rinderpest, and as a consequence, widespread concern grew over the necessity to halt the spread of this cattle plague. This led to the establishment of the first veterinary schools in Europe.[15–17] During the nineteenth century, the disease continued to cause havoc and a major rinderpest outbreak occurred in Britain in 1865.[1] The twentieth century brought with it an increase in international trade in live cattle that disseminated the disease, causing new outbreaks in Europe as well as in the Americas.[16,18] Outbreaks such as these led to the establishment of the Paris-based Office International des Epizooties (OIE) to monitor the spread of rinderpest as well as other epizootic diseases.[19] The increasing trade in livestock on a global scale led to the occurrence of rinderpest in Australia by 1923,[20] but in both Australia and South America, the disease was quickly eradicated through the

swift application of quarantine and slaughter policies. Such policies also enabled the eradication of the disease from Western Europe. The last European outbreak occurred in 1949 with the importation of live antelope from East Africa into Rome zoo.[17] The continued prohibition of importation of live animals into Europe from enzootic areas has maintained Europe's freedom from this disease.

Historically, military campaigns and the unrestricted movement of wild game were contributing factors to its distribution and the spread of RPV from areas where rinderpest was enzootic proved disastrous. However, eradication campaigns managed to eliminate the threat of such epizootics in most countries. Despite its widespread occurrence in sub-Saharan Africa in the mid-1950s, rinderpest was restricted to a few well-defined regions of East Africa as a result of the Pan African Rinderpest Campaign (PARC) that worked to eradicate the disease between 1986 and 1999. Vaccination with a tissue culture attenuated vaccine, the Plowright rinderpest of bovine origin "Kabete" (RBOK) vaccine, proved a very effective method for controlling the spread of RPV. The efficacy of the vaccine was mainly due to the fact that there was only one serotype of the virus and the RBOK vaccine elicits lifelong immunity to the disease.[21] The PARC, mainly funded by the European Union (EU), was extended to include West and South Asia, where rinderpest was endemic up to the mid-1990s. A number of internationally coordinated and funded groups were then established to continue the efforts of previous campaigns to control and eventually eradicate rinderpest. Towards the end of the eradication campaign, rinderpest was thought to be restricted to one focus of infection in East Africa, the Kenya/Somalia ecosystem, while in Asia, Pakistan was the last stronghold but was considered free from the disease following extensive vaccination campaigns in 2003.[1] The combined efforts of the FAO in Rome, the Inter-African Bureau for Animal Resources (IBAR),

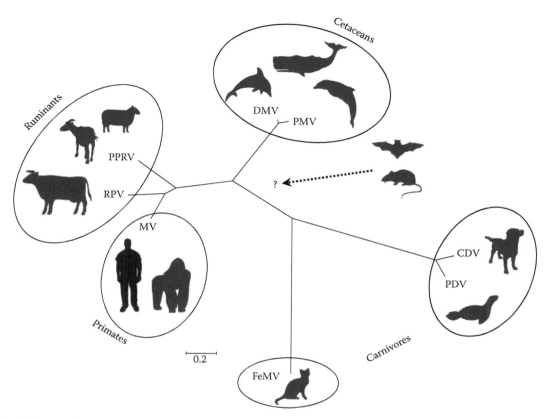

FIGURE 62.1 Phylogenetic analysis of the HA protein of morbilliviruses. All morbilliviruses for which full H gene data are available are included. The phylogeny was constructed using a minimum evolution amino/Poisson correction algorithm with 1000 bootstrap replicates using Mega4. The scale bar represents amino acid changes per position. Accession numbers for representative for each virus are RPV, X98291; PPRV, AY560591; CDV, AF305419; MV, AB012948; PDV, D10371; DMV, AJ608288; PMV, AY586537; and FeMV, JQ411016.

and the regional and national components of the Pan-African Programme for the Control of Epizootics (PACE), working under the coordination of the Global Rinderpest Eradication Programme (GREP), saw the eradication of rinderpest from the globe in 2010.[1] Comprehensive accounts of the historical aspects of rinderpest epidemiology, spread, and the gradual development of vaccines are reviewed in detail elsewhere.[22,23]

62.4 CLINICAL FEATURES AND PATHOGENESIS

Field strains of rinderpest varied greatly in their clinical manifestations from the severe classical form of the disease to mild or subclinical infection, but how such differences were genetically determined remains unknown.[24–29] An apparent adaptation to domestic livestock was seen towards the end of the eradication campaign. Isolates that circulated in a mild form in livestock appeared to cause severe disease in wildlife species, although the genetic drivers for this difference in pathogenicity remain unknown.[4] The course of a severe acute rinderpest infection was classically characterized by five distinct stages of disease. These included an incubation period, a prodromal fever, an erosive-mucosal phase, a diarrheic phase, and, in any surviving animals, a long period of convalescence.[30] The incubation period normally lasted between 3 and 5 days before a rise in temperature was seen. More virulent strains of rinderpest, such as the highly virulent Saudi/81

strain, caused a sharp increase in temperature 2–3 days post infection, although in general, the temperature rise was not seen until 3–5 days post infection. In contrast, the incubation period for mild strains could be as long as 2 weeks.[28,29] During this initial prodromal phase, the disease often went unnoticed, as clinical signs were not apparent, but as the temperature reaches its peak, more specific disease signs would generally begin to appear. Breathing would intensify and inappetence would occur as visible ocular and nasal surfaces became congested before "pinprick" lesions appeared in the oral cavity, which, depending on the strain of rinderpest present, could develop into highly ulcerative and erosive lesions.[21] These were generally observed on the lips, gums, corners of the mouth, tongue, and hard palate.[26,31] The severity of the lesions appeared to depend on the infecting virus strain. In the case of highly virulent strains, such as the Saudi/81 strain, the virus could cause the entire oral cavity to be covered in lesions where the skin was readily sloughed to expose a hemorrhagic layer of basal cells. This sloughing of necrotic cells produced a foul-smelling breath. However, other highly virulent strains, such as the Kabete "O" (KO) strain, did not cause such severe oral lesions. As the disease progressed, the animals often produced a frothy exudation from the mouth and both lacrimal and nasal secretions became mucopurulent in infections with highly virulent strains.[24–29,31,32] As the erosive-mucosal phase reached a peak, fever would subside, and diarrhea becomes severe. The feces contained mucous and epithelial cells as well as necrotic bloodied debris. The extreme oral trauma and diarrhea resulted in anorexia and severe dehydration. Such severe infections were invariably fatal, and before death, animals had sunken eyes, an arched back, and a lowered head. In fatal cases, animals would collapse and die 6–12 days after the onset of fever. In contrast, in nonfatal cases, the diarrhea would rapidly clear up after a week or so and the animals may slowly regain health.

62.5 IDENTIFICATION AND DIAGNOSIS

Field diagnosis of rinderpest was generally based on the observation of symptoms consistent with the three "Ds," discharge, diarrhea, and death. However, these features of infection were often only seen in late-stage disease and so early field observation of RPV was difficult. Indeed, numerous other viral diseases of livestock including peste des petits ruminants (PPR), foot and mouth disease (FMD), bovine viral diarrhea (BVD), bluetongue (BT), and malignant catarrhal fever (MCF) were all potential pathogens that could be confused with rinderpest at early time points post infection. Alongside this, the variation in the severity of disease from very mild to severe also hampered field diagnosis. For rinderpest to be diagnosed conclusively, laboratory testing was required.

However, the presence of rinderpest across the developing world made laboratory confirmation problematic as it often required the use of expensive laboratory equipment, highly skilled staff, and/or tissue culture facilities. Prior to the development of molecular techniques, antigen detection methods were used extensively across laboratories. Samples suitable for antigen detection included ocular and nasal secretions, gum debris, blood, and, following the death of an animal, postmortem tissues including tissue from lymph nodes, the tonsil, and spleen. However, not only did sampling need to occur during peak stages of disease, samples need rapid transportation to the laboratory for testing to ensure viability of the sample.

Several tools were developed to detect rinderpest antigen including the agar gel immunodiffusion (AGID) tests, the counterimmunoelectrophoresis (CIEP) test, direct and indirect immunoperoxidase tests, and different antigen detection ELISA tests.[30] For the AGID test, the development of a visible precipitate was used to detect antibody–antigen complex formation and give a diagnostic result. The AGID was easy to interpret, and commercially available diagnostic platforms made the test easy to acquire and perform and were first applied to RPV by White.[33] However, drawbacks to the test included its sensitivity alongside the inability to differentiate between RPV and PPRV antigen using this test. The CIEP test was a variation on the AGID test that utilized electric current to cause migration of antigen towards antibody within a gel, again to generate a precipitate upon the complex. Two CIEP tests were available in the late 1970s and early 1980s[34,35] with later adaptations being developed for field usage in 1984.[36] In relation to this form of test, both direct and indirect immunoperoxidase tests were developed

for use with biopsy material.[30] Further, immunofluorescent techniques were also developed where positive interactions following incubation with rinderpest-specific polyclonal and later monoclonal antibodies either directly or indirectly were visualized using light microscopy.[30] However, the provision of expensive equipment often precluded the use of microscope-based techniques where they were needed most.

Antigen-based ELISA techniques, immunocapture ELISAs, were later developed and offered a sensitive test that could detect virus antigen within a sample. Initially, this was developed using an RPV ELISA that was able to trap antigen within a suspect sample. The N protein-specific monoclonal antibody used was not, however, able to differentiate between RPV and PPRV, and so differential diagnosis could not be performed.[37,38] Several different sample types could be assessed using this platform including all secretions (ocular, nasal, and buccal) and tissues such as the lymph node, lung, and spleen. Later, the inclusion of two specific antibodies enabled the differential diagnosis of RPV and PPRV.[39,40]

While each of these antigen-based tests were of great utility in the diagnosis of RPV, during the eradication process, novel tools were developed that could be applied within the field for rapid detection of antigen within samples. This was particularly important following the cessation of mass vaccination where the circulation of natural virus was of concern and rapid "infield" tests were needed. Certainly, the rapidity of detection could prevent the dissemination of virus following emergency vaccination in areas where outbreaks occurred. In resource-limited settings, this was of particular importance. Rapid chromatographic strip tests were developed to overcome this issue and were found to be of great utility in the field through their robust and easy-to-use format, rapid result, and the requirement of only a small sample volume. The chromatographic strip test was based on the binding of RPV nucleocapsid protein in the sample to a rinderpest-specific antibody complexed with blue latex particles on the strip.[41] If a sample contained virus antigen, then the development of a blue line in the test result window would occur where any antigen–antibody complexes labeled with the blue latex particles were trapped by the rinderpest antibody. Since latex-bound antibody was applied in excess to the pad, unbound latex–antibody complexes travelled further to be trapped by a second antimouse antibody acting as an internal control. Interpretation of the test was simple, with two blue lines indicating RPV infection and one line indicating an RPV-negative result. The test was developed and commercialized[41,42] and was successfully used in the field in Pakistan[43] and Africa.[44] The test compared favorably to the antigen capture ELISA in terms of sensitivity. The simplicity of this test made it very useful for field diagnoses and the platform was further developed for the detection of the closely related PPRV.[42] Following the eradication of RPV, many believe that PPRV is a suitable target for eradication.[45,46] Alongside the successful application of this technology to the field, recent studies have shown that antigen trapped within these devices can be further probed back in the laboratory using molecular methods to detect viral nucleic acid.[47]

Where facilities were available in laboratories, virus isolation was considered a very important tool as not only did it enable diagnosis to be made, but it also generated live virus stocks that could be used as research tools. This method was considered a gold standard technique, especially in regions where the virus was not thought to be circulating. However, the major drawback to this approach was the necessity for expensive microbiological cabinets required for tissue culture work alongside the peripheral equipment needed to support cell culture facilities. Such equipment included incubators, centrifuges, filtration equipment, autoclaves, microscopes, and roller apparatus. Alongside this, highly skilled staff were needed to attempt growth of live virus from suspect samples. Early attempts to isolate live virus were performed in calf kidney cells that were derived from the perineum of a donor calf, and following dissection and processing, a primary cell culture could be established. Samples used to attempt virus isolation included blood samples and macerated tissue samples. Samples were left to incubate with the cell sheet for a two-week period during which, if live virus was present, characteristic cell fusions termed syncytia developed that were readily identified using microscopy (Figure 62.2). If syncytia did not develop, cells were blind passaged and monitored for a further period. A big drawback to this technique was the fact that the closely related PPRV was also able to generate syncytia in cell culture, precluding a differential diagnosis to be made between these two viruses.

During the 1980s, the advent of molecular techniques shifted the focus from detection of live virus and antigen to the detection of viral genomic material. Initially, cDNA probes specific for RPV were developed, although the sensitivity of such early molecular tests was poor.[48,49] The development of reverse transcription polymerase chain reaction (RT-PCR) revolutionized the detection of microbes across the globe. Forsyth and Barrett[50] developed the first RPV-specific RT-PCR assay based on primers that targeted either the P gene or the F gene. Not only did this assay enable specific differentiation between closely related morbilliviruses, but it also enabled the genetic typing of isolates by analysis of the sequence of resultant RT-PCR products. This enabled both current and historic samples to be compared phylogenetically, giving an indication of the origin of spread of the virus and determination of trade route by which the virus was introduced to a naive herd. This approach categorized RPV isolates into three distinct lineages that were largely based on geographical location of isolates (Figure 62.3). The utility of the RT-PCR was neatly demonstrated through the differentiation between vaccine virus and field isolates in infected animals.[51] Several more RT-PCR assays were developed for RPV during the final years of the eradication campaign with increased sensitivity and the application of newly developed commercial extraction techniques that enabled a rapid evaluation of suspect material.[52] The relevance of PCR techniques to the eradication program was high as these molecular tools gave mechanisms to rapidly type circulating virus. The detection of virus genetic material using "real-time technologies" further optimized the detection of nucleic acid during recent years and enabled specific detection of RPV.[53] However, drawbacks to these molecular techniques included both the need to sample animals during the acute stage of infection as often viral nucleic acid was only shed at high levels during this phase and

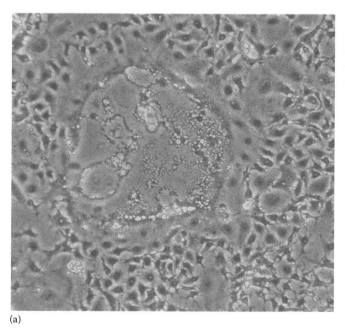

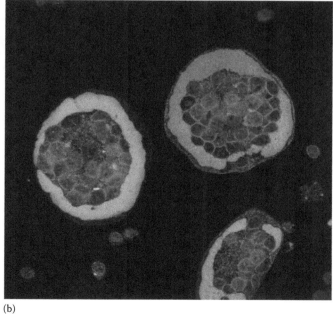

(a)   (b)

FIGURE 62.2 RPV in vitro: Syncytia formation in cell culture. (a) Light microscope image of a syncytia caused by rinderpest infection in vero cells. (b) Immunofluorescence on B95a cells of a recombinant RPV expressing GFP from within the polymerase gene. Infected syncytia can be clearly seen as accumulations of nuclei to produce giant cells. Uninfected cells can be seen as just individual nuclei. Within the syncytia the polymerase is present in "lakes" around the clusters of nuclei present in the centre of the fusion. The limits of the cells are defined by tubulin staining. (Images courtesy of Dr. Ashley C. Banyard and Dr. Paul Monaghan.)

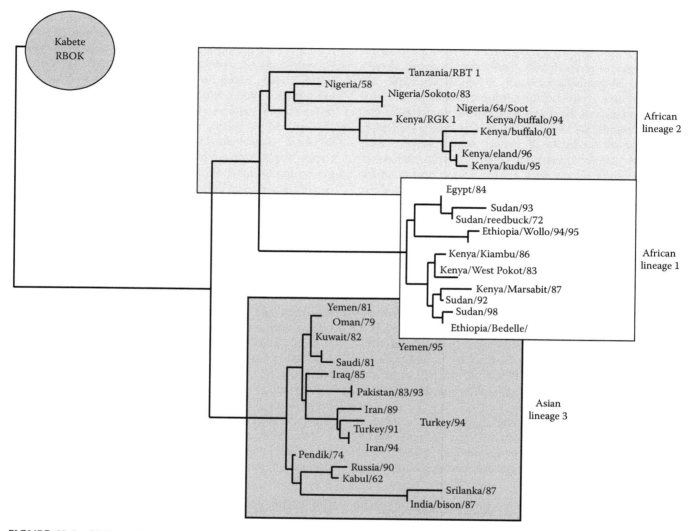

FIGURE 62.3 Phylogenetic analysis of rinderpest isolates. The phylogeny was constructed using the **PHYLIP DNADIST** and FITCH programs. (Adapted from Barrett, T. et al., *Rinderpest and Peste des Petits Ruminants Virus: Virus Plagues of Large and Small Ruminants*, Academic Press, London, U.K., p. 341, 2006.) Branch lengths are proportional to the genetic distances between virus isolates.

the requirement for expensive equipment and skilled staff to perform the assays. During the final stages of the eradication campaign, the former point was of great significance as it was believed that only mild strains of RPV were circulating, and so the absence of severe disease made the choice of animals to sample highly problematic. Furthermore, the availability of equipment and highly skilled staff precluded the local use of these techniques. However, the GREP established a network of laboratories taking receipt of samples that could then pass suspect materials to either national or world reference laboratories for analysis and/or confirmation of results using molecular techniques.

Alongside the detection of live virus, virus antigen, and viral genetic material, key to the eradication campaign was the detection of RPV-specific antibodies. Initially, virus neutralization tests[54] were developed for this purpose but faced the same drawbacks as virus isolation in tissue culture requiring both tissue culture facilities and highly skilled staff. The VNT was the OIE "gold standard" antibody detection test

for several years before the generation of ELISA techniques that proved to be both highly sensitive and specific and could be performed in resource-limited settings with comparative ease. The VNT test was first replaced by an indirect ELISA[55] and later by a competitive ELISA,[56] which could give a definitive result within a matter of hours. These assays were a significant improvement on the VNT that could take up to 2 weeks to give a definitive result.

With the development and successful application of safe, effective vaccines came the attempts to eradicate this devastating virus. In line with this, novel serological assays were developed to enable determination of herd immunity following vaccination. The initial ELISA format that depended on the indirect detection of rinderpest-specific antibodies was of great utility where postvaccinal herd immunity was desired. The highly effective tissue culture rinderpest vaccine (TCRV) (detailed in the following text), which induced lifelong immunity, was able to stimulate a high-titer antibody response among vaccinated animals. This fact was exploited for the development of an indirect ELISA for the detection of circulating antibodies both within uninfected animals that had survived natural infection and in vaccinated animals. This ELISA format depended on the reaction between antibodies present in a test sample and immobilized virus antigen within a 96-well plate format. The binding of antibodies in test sera bound to the immobilized antigen was detected following incubation with an HRP-conjugated secondary antibody and subsequent colorimetric readings on an ELISA plate reader. The presence of tightly regulated positive and negative controls enabled the assessment of the presence of RPV-specific antibodies within the test serum.[55,57] One principal drawback in this technique was the requirement for species-specific secondary antibodies that was particularly problematic where wildlife sera were being tested. This ELISA test was vastly improved by using monoclonal antibodies specific for the virus hemagglutinin (HA) protein that were used to compete with antibodies within suspect sera for virus antigen. The competitive ELISA test gave a readout following the addition of HRP-conjugated antibodies specific for the monoclonal antibody used to compete with antibodies in the test sera to give colorimetric readouts where the appearance of color signified a negative result, while if antibodies to RPV were present in the sample, less color or no color was visualized.[30] This new ELISA test found great utility during the eradication campaign and was essentially the tool that gave confidence both in the vaccination status of animals and in the lack of either vaccine-derived antibodies or antibodies developed following natural infection that enabled countries to declare freedom from RPV. This assay was made commercial and was supplied to the vast majority of countries across Africa and Asia during the eradication campaign. Numerous other ELISAs have been developed as alternative to the cH-ELISA, although in general, the eradication of RPV has precluded their use.[58-61]

62.6 TREATMENT AND PREVENTION

RPV was declared officially eradicated from the globe in 2011.[1] From a historical perspective, the mechanisms developed to treat and prevent rinderpest came from individuals now regarded as some of the most pioneering scientists to have lived. Initial attempts control the spread of RPV focused on efforts to cure sick animals but this approach met with little success. However, the discovery of the protective powers of serum taken from recovered animals at the end of the twentieth century drew attention to prevention, rather than cure, of rinderpest. Early techniques involved the inoculation of both virus and immune serum from convalescent animals and the preparation of hyperimmune serum in goats.[1] This was taken further with the application of immune serum, given either alone or in combination with infected blood. Later, Robert Koch demonstrated that serum from convalescent animals was only able to produce short-term passive immunity, whereas the combination of immune serum with virulent blood induced an active immunity. The "serum–virus simultaneous" method became widely used in Africa and India as

the most effective way of protection from rinderpest. While highly effective, this method had some important drawbacks including the induction of severe disease in young and weak animals as well as cattle in advanced pregnancy, the need of segregation of inoculated from uninoculated animals due to the potential infectivity of the former, the short shelf life of infectious blood, and the threat of spreading piroplasms onto the recipient animals.[51]

Such early attempts to transfer immunity were overcome by the development of inactivated or live attenuated RPV vaccines. The former were produced from infected bovine tissues including spleen, tonsil, or lymph node material that were chemically inactivated with various agents. These approaches enabled the eradication of the virus from several countries, including Iran, the Philippines, Sri Lanka, Thailand, and Russia.[62] However, the immunity produced by the inactivated vaccines was short-lived, and efforts were next focused on the generation of live attenuated virus strains, which were generated following multiple passages in various hosts. The first live attenuated vaccines were produced in goats and were considered not only cheap and efficacious, but more importantly, they did not transmit piroplasms, and one of them, Kenya/Kabete Attenuated Goat (KAG) vaccine, was able to induce lifelong immunity to rinderpest.[63] The shortage of goats in Japan and Korea was a main reason for using rabbits in RPV attenuation, but eventually, it resulted in the development of vaccines better suited to Asiatic cattle breeds.[64] This lapinized vaccine produced lifelong immunity although generation of inocula was difficult, with one rabbit only able to generate 600 doses of vaccine. This limiting factor led to the passage of this vaccine in goats to generate a lapinized/caprinized virus, which was then passaged in sheep to produce a vaccine. This final product was used in the eradication of rinderpest in China.[62] In Japan and Korea, the severe reactions of highly susceptible cattle following vaccination with the lapinized vaccine were overcome by the development of a strain adapted to grow in eggs. Interestingly, when using eggs, the level of attenuation of the avianized rinderpest strain was found to be correlated to passage level.[64]

The constant development of virus strains attenuated in various species helped to reduce the labor required for mass production of vaccines. However, the application of cell culture techniques vastly increased the scale up of vaccine manufacture. Following initial failures to grow goat-, egg-, or rabbit-adapted strains of rinderpest in chicken embryo fibroblasts, bovine kidney (BK) cells, or bovine embryonic kidney cells,[65,66] Plowright and Ferris reported the successful attenuation of virulent KO virus in primary calf kidney cell cultures[67] with virulence initially increasing upon passage before seeing a reduction in attenuation until the virus was completely attenuated and unable to cause disease even in the most susceptible breeds of cattle.[68] Although the TCRV was able to elicit neutralizing antibodies between 7 and 17 days, cattle were resistant to challenge as early as 4 days post vaccination. The experiments on the duration of protective immunity conferred by TCRV showed that vaccinated cattle withstood the challenge with virulent rinderpest several years following vaccination.[69,70] The vaccine recipients did not transmit the virus to in-contact animals, which correlated to the inability of the vaccine strain to replicate in the nasal or alimentary mucosae.[71] At the time of TCRV development, cattle less than one-year old were not included in vaccination campaigns with caprinized rinderpest strain due to severe reaction it caused in young animals.[72] Contrary to goat-attenuated vaccine, TCRV did not cause any adverse reaction in young or weak animals.

Many years after the generation of the TCRV, researchers determined the complete genome sequence of the vaccine strain and a virulent virus that, although passaged numerous times within the laboratory, was thought to be as close to the original KO virus as was available. The comparative sequence analysis of the vaccine and parent strains identified a total of 87 differing bases across the 15,882 nucleotide genome.[73] The mutations forced by tissue adaptation were stable and the virus did not revert to virulence even after several backpassages in cattle.[68,72,74] Extensive attempts to determine residues critical for virus attenuation were largely unsuccessful but highlighted the significance of mutations in several coding regions as well as the untranslated promoter regions.[13,75]

The TCRV vaccine proved to be a highly effective tool in the eradication of rinderpest, and as more and more areas were cleared of disease, the effort was diverted to serosurveillance. Since there is only one serotype of RPV, it was not possible to differentiate between vaccinated and infected animals using the OIE "gold standard" cH-ELISA test. Attention then shifted to the generation of a test that was able to *D*ifferentiate between *I*nfected and *V*accinated *A*nimals—the so-called DIVA concept. Despite the fact that RPV was successfully eradicated without the use of DIVA tools, several candidate vaccines and companion tests were developed including recombinant pox viruses expressing immunodominant glycoproteins or the N protein of RPV.[76–84] Alongside this approach, several recombinant versions of RPV were developed as potential DIVA vaccines using reverse genetics techniques. Candidate vaccines assessed included the generation of chimeric RPV with the glycoproteins of PPRV[85]; a chimeric RPV virus with the M, H, and F proteins derived from PPRV[86]; and a recombinant RPV where the N protein had been swapped with that of PPRV.[87]

The feasibility of generating a marker vaccine against RPV, through the addition of a foreign immunogen to its genome, was also examined. The first successful attempt was the development of recombinant RPV vaccines, which expressed intracellular or secreted forms of green fluorescent protein (GFP).[88] Although these vaccines elicited a strong protective antibody response against RPV, the response to the foreign antigen was poor.[88] A similar approach was also used with a virulent form of RPV to generate a research tool.[89] Alternative immunogens were assessed including a receptor site mutant form of the influenza virus HA gene and a membrane-anchored form of GFP gene.[90] These vaccines not only induced protective immunity against RPV but also elicited very good antibody response against respective marker proteins. Unfortunately, they are both positively marked and their use would only allow determination of vaccine cover in the field rather than enabling differentiation between vaccinated and infected animals.

62.7 CONCLUSION

The eradication of arguably the most devastating cattle disease ever known is one of the greatest successes in science and, alongside the eradication of smallpox, is the second viral pathogen to be globally eradicated. However, with eradication comes an increased awareness of the potential for use of the virus as a bioterror weapon. Reverse genetics techniques have enabled the generation of RPV and countless other viruses, from DNA copies of their genomes, meaning that even following sequestration of virus from laboratories around the globe, the potential for generation of the virus and use in such a way remains.

REFERENCES

1. Roeder, P.L. Rinderpest: The end of cattle plague. *Preventive Veterinary Medicine* 102, 98–106 (2011).
2. Barrett, T. and Rossiter, P.B. Rinderpest: The disease and its impact on humans and animals. In *Advances in Virus Research*, vol. 53, (eds. Karl, M., Frederick, A.M., and Aaron, J.S.) pp. 89–110 (Academic Press, San Diego, CA, 1999).
3. Barrett, T. et al. Rediscovery of the second African lineage of rinderpest virus: Its epidemiological significance. *The Veterinary Record* 142, 669–6711 (1998).
4. Kock, R.A. et al. Rinderpest epidemic in wild ruminants in Kenya 1993–1997. *Veterinary Record* 145, 275–283 (1999).
5. Banyard, A.C., Rima, B., and Barrett, T. The morbilliviruses. In *Rinderpest and Peste des Petits Ruminants—Virus Plagues of Large and Small Ruminants*, (eds. Barrett, T., Pastoret, P.-P., and Taylor, W.P.) pp. 13–26 (Elsevier, Oxford, U.K., 2006).
6. Woo, P.C. et al. Feline morbillivirus, a previously undescribed paramyxovirus associated with tubulointerstitial nephritis in domestic cats. *Proceedings of the National Academy of Sciences of the United States of America* 109, 5435–5440 (2012).
7. Drexler, J.F. et al. Bats host major mammalian paramyxoviruses. *Nature Communications* 3, 796 (2012).
8. Barrett, T., Subbarao, S.M., Belsham, G.J., and Mahy, B.W. The molecular biology of the morbilliviruses. In *The Paramyxoviruses*, (ed. Kingsbury, D.W.) pp. 83–102 (Plenum, New York, 1991).
9. Bhella, D., Ralph, A., and Yeo, R.P. Conformational flexibility in recombinant measles virus nucleocapsids visualised by cryo-negative stain electron microscopy and real-space helical reconstruction. *Journal of Molecular Biology* 340, 319–331 (2004).
10. Baron, M.D. and Barrett, T. The sequence of the N and L genes of rinderpest virus, and the 5′ and 3′ extra-genic sequences: The completion of the genome sequence of the virus. *Veterinary Microbiology* 44, 175–185 (1995).
11. Escoffier, C. et al. Nonstructural C protein is required for efficient measles virus replication in human peripheral blood cells. *Journal of Virology* 73, 1695–1698 (1999).

12. Baron, M.D. and Barrett, T. Rinderpest virus lacking the C and V proteins show specific defects in growth and transcription of viral RNAs. *Journal of Virology* 74, 2603–2611 (2000).

13. Banyard, A.C., Baron, M.D., and Barrett, T. A role for virus promoters in determining the pathogenesis of Rinderpest virus in cattle. *Journal of General Virology* 86, 1083–1092 (2005).

14. Curasson, G. *La Peste Bovine,* (Vigot Freres, Paris, France, 1932).

15. Henning, M.W. Rinderpest. In *Animal Diseases in South Africa,* (Central News Agency Ltd., Pretoria, South Africa, 1956).

16. Plowright, W. Rinderpest virus. *Monographs in Virology* 3, 25–110 (1968).

17. Scott, R.G. *Rinderpest and Peste des Petits Ruminants* (Academic Press, London, U.K., 1981).

18. Roberts, G.A. Rinderpest in Brazil. *Journal of the American Veterinary Medical Association* 60, 177–185 (1921).

19. Vittoz, R. Report of the director on the scientific and technical activities of the office international des epizooties from May 1962 to May 1963. *Bulletin De L'Office International des Epizooties* 60, 1443–1574 (1963).

20. Weston, E.A. Rinderpest in Australia. *Journal of the American Veterinary Medical Association* 66, 337–350 (1924).

21. Barrett, T. Rinderpest and distemper viruses. In *Encyclopaedia of Virology,* (Academic Press, London, U.K., 1994).

22. Spinage, C.A. *Cattle Plague: A History*, p. 770 (Kluwer Academic, London, U.K., 2003).

23. Barrett, T., Pastoret, P.-P., and Taylor, W.P. *Rinderpest and Peste des Petits Ruminants Virus: Virus Plagues of Large and Small Ruminants*, p. 341 (Academic Press, London, U.K., 2006).

24. Nawathe, D.R. and Lamorde, A.G. Towards global eradication of rinderpest. *Revue Scientifique et Technique Office International des Epizooites* 2, 77–91 (1983).

25. Wafula, J.S. and Kariuki, D.P. A recent outbreak of rinderpest in East Africa. *Tropical Animal Health and Production* 19, 173–176 (1987).

26. Brown, C.C. and Torres, A. Distribution of antigen in cattle infected with rinderpest virus. *Veterinary Pathology* 31, 194–200 (1994).

27. Wohlsein, P. et al. Pathomorphological and immunohistological findings in cattle experimentally infected with rinderpest virus isolates of different pathogenicity. *Veterinary Microbiology* 44, 141–149 (1995).

28. Taylor, W.P. Epidemiology and control of Rinderpest. *Revue Scientifique et Technique Office International des Epizooites* 5, 407–410 (1986).

29. Wamwayi, H.M., Fleming, M., and Barrett, T. Characterisation of African isolates of rinderpest virus. *Veterinary Microbiology* 44, 151–163 (1995).

30. Anderson, J., Barrett, T., and Scott, G.R. *FAO Animal Health Manual—Manual on the Diagnosis of Rinderpest*, 2nd ed., pp. 81–91 (Food and Agriculture Organization of the United Nations, Rome, Italy, 1996).

31. Liess, B. and Plowright, W. Studies on the pathogenesis of rinderpest in experimental cattle. I. Correlation of clinical signs, viraemia and virus excretion by various routes. *Journal of Hygiene* 62, 81–100 (1964).

32. Plowright, W. Studies on the pathogenesis of rinderpest in experimental cattle. II. Proliferation of the virus in different tissues following intranasal infection. *Journal of Hygiene* 62, 257–281 (1964).

33. White, G. A specific diffusable antigen of rinderpest virus demonstrated by the agar double-diffusion precipitation reaction. *Nature* 181, 1409 (1958).

34. Ali, B., El, H., and Lees, G.E. The application of immunoelectroprecipitation in the diagnosis of rinderpest. *The Bulletin of Animal Health and Production in Africa* 27, 1–6 (1979).

35. Rossiter, P.B. and Mushi, E.Z. Rapid detection of rinderpest virus antigens by counter-immunoelectrophoresis. *Tropical Animal Health and Production* 12, 209–216 (1980).

36. Rossiter, P.B. Notes on immunoprecipitin reactions with rinderpest virus. *Tropical Animal Health and Production* 17, 55–56 (1985).

37. Libeau, G., Diallo, A., Calvez, D., and Lefevre, P.C. A competitive ELISA using anti-N monoclonal antibodies for specific detection of rinderpest antibodies in cattle and small ruminants. *Veterinary Microbiology* 31, 147–160 (1992).

38. Libeau, G. and Lefevre, P.C. Comparison of rinderpest and peste des petits ruminants viruses using anti-nucleoprotein monoclonal antibodies. *Veterinary Microbiology* 25, 1–16 (1990).

39. Libeau, G., Diallo, A., Colas, F., and Guerre, L. Rapid differential diagnosis of rinderpest and peste des petits ruminants using an immunocapture ELISA. *Veterinary Record* 134, 300–304 (1994).

40. Libeau, G., Saliki, J.T., and Diallo, A. Caracterisation d'anticorps monoclonaux diriges contre le virus de la peste bovine et de la peste des petits ruminants: Identification d'epitopes conserves ou de specificite stricte sur la nucleoprotein. *Revue d'Elevage et de Medecine Veterinaire des Pays Tropicaux* 51, 181–190 (1997).

41. Bruning, A., Bellamy, K., Talbot, D., and Anderson, J. A rapid chromatographic strip test for the pen-side diagnosis of rinderpest virus. *Journal of Virological Methods* 81, 143–154 (1999).

42. Bruning-Richardson, A., Akerblom, L., Klingeborn, B., and Anderson, J. Improvement and development of rapid chromatographic strip-tests for the diagnosis of rinderpest and peste des petits ruminants viruses. *Journal of Virological Methods* 174, 42–46 (2011).

43. Hussain, M., Iqbal, M., Taylor, W.P., and Roeder, P.L. Pen-side test for the diagnosis of rinderpest in Pakistan. *Veterinary Record* 149, 300–302 (2001).

44. Wambura, P.N. et al. Diagnosis of rinderpest in Tanzania by a rapid chromatographic strip-test. *Tropical Animal Health and Production* 32, 141–145 (2000).

45. Anderson, J. et al. Rinderpest eradicated; what next? *The Veterinary Record* 169, 10–11 (2011).

46. Baron, M.D., Parida, S., and Oura, C.A. Peste des petits ruminants: A suitable candidate for eradication? *Veterinary Record* 169, 16–21 (2011).

47. Bruning-Richardson, A., Barrett, T., Garratt, J.C., and Anderson, J. The detection of rinderpest virus RNA extracted from a rapid chromatographic strip-test by RT-PCR. *Journal of Virological Methods* 173, 394–398 (2011).

48. Diallo, A., Barrett, T., Barbron, M., Subbarao, S.M., and Taylor, W.P. Differentiation of rinderpest and peste des petits ruminants viruses using specific cDNA clones. *Journal of Virological Methods* 23, 127–136 (1989).

49. Diallo, A., Libeau, G., Couacyhymann, E., and Barbron, M. Recent developments in the diagnosis of rinderpest and peste des petits ruminants. *Veterinary Microbiology* 44, 307–317 (1995).

50. Forsyth, M.A. and Barrett, T. Evaluation of polymerase chain reaction for the detection and characterisation of rinderpest and peste des petits ruminants viruses for epidemiological studies. *Virus Research* 39, 151–163 (1995).

51. Barrett, T. et al. The molecular-biology of rinderpest and peste-des-petits ruminants. *Annales De Medecine Veterinaire* 137, 77–85 (1993).

52. Couacy-Hymann, E., Bodjo, S.C., and Danho, T. Interference in the vaccination of cattle against rinderpest virus by antibodies against peste des petits ruminants (PPR) virus. *Vaccine* 24, 5679–5683 (2006).

53. Carrillo, C. et al. Specific detection of Rinderpest virus by real-time reverse transcription-PCR in preclinical and clinical samples from experimentally infected cattle. *Journal of Clinical Microbiology* 48, 4094–4101 (2010).

54. Plowright, W. and Ferris, R.D. Studies with rinderpest virus in tissue culture. III. The stability of cultured virus and its use in virus neutralization tests. *Archiv fur Die Gesamte Virusforschung* 11, 516–533 (1962).

55. Anderson, J., Rowe, L.W., Taylor, W.P., and Crowther, J.R. An enzyme-linked immunosorbent assay for the detection of IgG, IgA and IgM antibodies to rinderpest virus in experimentally infected cattle. *Research in Veterinary Science* 32, 242–247 (1982).

56. Anderson, J., McKay, J.A., and Butcher, R.N. The use of monoclonal antibodies in competitive ELISA for the detection of antibodies to rinderpest and peste des petits ruminants viruses: The seromonitoring of rinderpest throughout Africa, Phase One. in *Proceedings of the Final Research Coordination Meeting of the IAEA Rinderpest Control Projects*, (IAEA, Cote d'Ivoire, Vienna, Australia, 1991).

57. Anderson, J., Rowe, L., and Taylor, W. Use of an enzyme-linked immunosorbent assay for the detection of IgG antibodies to rinderpest virus in epidemiological surveys. *Research in Veterinary Science* 34, 77–81 (1983).

58. Singh, R.P., Sreenivasa, B.P., Dhar, P., Roy, R.N., and Bandyopadhyay, S.K. Development and evaluation of a monoclonal antibody based competitive enzyme-linked immunosorbent assay for the detection of rinderpest virus antibodies. *Revue Scientifique et Technique* 19, 754–763 (2000).

59. Khamehchian, S., Madani, R., Rasaee, M.J., Golchinfar, F., and Kargar, R. Development of 2 types of competitive enzyme-linked immunosorbent assay for detecting antibodies to the rinderpest virus using a monoclonal antibody for a specific region of the hemagglutinin protein. *Canadian Journal of Microbiology* 53, 720–726 (2007).

60. Renukaradhya, G.J., Suresh, K.B., Rajasekhar, M., and Shaila, M.S. Competitive enzyme-linked immunosorbent assay based on monoclonal antibody and recombinant hemagglutinin for serosurveillance of rinderpest virus. *Journal of Clinical Microbiology* 41, 943–947 (2003).

61. Choi, K.S. et al. Monoclonal antibody-based competitive ELISA for simultaneous detection of rinderpest virus and peste des petits ruminants virus antibodies. *Veterinary Microbiology* 96, 1–16 (2003).

62. Taylor, W.P., Roeder, P.L., and Rweyemamu, M.M. History of vaccines and vaccination In *Rinderpest and Peste des Petits Ruminants. Virus Plagues of Large and Small Ruminants*, (eds. Barrett, T., Pastoret, P.-P., and Taylor, W.P.) pp. 222–247 (Academic Press, London, U.K., 2006).

63. Brown, R.D. and Raschid, A. The duration of immunity following vaccination with caprinised virus in the field. Annual Report, East African Veterinary Research Organisation, 1956–1957, p. 21 (1958).

64. Nakamura, J. and Miyamoto, T. Avianization of lapinized rinderpest virus. *American Journal of Veterinary Research*, 14, 307–317 (1953).

65. Johnson, R.H. Rinderpest in tissue culture. I: Methods for virus production. *British Veterinary Journal* 118, 107–116 (1962).

66. Plowright, W. and Ferris, R.D. Studies with rinderpest virus in tissue culture. I. Growth and cytopathogenicity. *Journal of Comparative Pathology* 69, 152–172 (1959).

67. Plowright, W. and Ferris, R.D. Studies with rinderpest virus in tissue culture. II. Pathogenicity for cattle of culture-passaged virus. *Journal of Comparative Pathology* 69, 173–184 (1959).

68. Plowright, W. The application of monolayer tissue culture techniques in rinderpest research. II. The use of attenuated culture virus as a vaccine for cattle. *Bulletin Office International des Epizooties* 57, 253-276 (1962).

69. Plowright, W. The duration of immunity in cattle following inoculation of rinderpest cell culture vaccine. *The Journal of Hygiene (London)* 92, 285–296 (1984).

70. Rweyemamu, M.M., Reid, H.W., and Okuna, N. Observations on the behaviour of rinderpest virus in immune animals challenged intranasally. *Bulletin of Epizootic Diseases of Africa* 22, 1–9 (1974).

71. Taylor, W.P. and Plowright, W. Studies on the pathogenesis of rinderpest in experimental cattle. III. Proliferation of an attenuated strain in various tissues following subcutaneous inoculation. *The Journal of Hygiene (Cambridge)* 63, 263–275 (1965).

72. Johnson, R.H. Rinderpest in tissue culture. III: Use of the attenuated strain as a vaccine for cattle. *British Veterinary Journal* 118, 141–150 (1962).

73. Baron, M.D., Kamata, Y., Barras, V., Goatley, L., and Barrett, T. The genome sequence of the virulent Kabete "O" strain of rinderpest virus: Comparison with the derived vaccine. *Journal of General Virology* 77(Pt 12), 3041–3046 (1996).

74. Bansal, R.P., Chawla, S.K., Joshi, R.C., and Shukla, D.C. Studies on attenuated rinderpest vaccine of tissue culture origin. *Indian Journal of Animal Sciences* 44, 520–524 (1974).

75. Baron, M.D., Banyard, A.C., Parida, S., and Barrett, T. The Plowright vaccine strain of Rinderpest virus has attenuating mutations in most genes. *Journal of General Virology* 86, 1093–1101 (2005).

76. Yilma, T. et al. Protection of cattle against rinderpest with vaccinia virus recombinants expressing the HA or F gene. *Science* 242, 1058–1061 (1988).

77. Asano, K. et al. Immunological and virological characterization of improved construction of recombinant vaccinia virus expressing rinderpest virus hemagglutinin. *Archives of Virology* 116, 81–90 (1991).

78. Barrett, T., Belsham, G.J., Subbarao, S.M., and Evans, S.A. Immunization with a vaccinia recombinant expressing the F protein protects rabbits from challenge with a lethal dose of rinderpest virus. *Virology* 170, 11–18 (1989).

79. Belsham, G.J., Anderson, E.C., Murray, P.K., Anderson, J., and Barrett, T. Immune response and protection of cattle and pigs generated by a vaccinia virus recombinant expressing the F protein of rinderpest virus. *Veterinary Record* 124, 655–658 (1989).

80. Giavedoni, L., Jones, L., Mebus, C., and Yilma, T. A vaccinia virus double recombinant expressing the F and H genes of rinderpest virus protects cattle against rinderpest and causes no pock lesions. *Proceedings of the National Academy of Sciences the United States of America* 88, 8011–8015 (1991).

81. Romero, C.H. et al. Single capripoxvirus recombinant vaccine for the protection of cattle against rinderpest and lumpy skin disease. *Vaccine* 11, 737–742 (1993).

82. Tsukiyama, K. et al. Development of heat-stable recombinant rinderpest vaccine. *Archives of Virology* 107, 225–235 (1989).

83. Yamanouchi, K. and Barrett, T. Progress in the development of a heat-stable recombinant rinderpest vaccine using an attenuated vaccinia virus vector. *Revue Scientifique et Technique* 13, 721–735 (1994).

84. Yamanouchi, K. et al. Immunisation of cattle with a recombinant vaccinia vector expressing the haemagglutinin gene of rinderpest virus. *Veterinary Record* 132, 152–156 (1993).

85. Das, S.C., Baron, M.D., and Barrett, T. Recovery and characterization of a chimeric rinderpest virus with the glycoproteins of peste-des-petits-ruminants virus: Homologous F and H proteins are required for virus viability. *Journal of Virology* 74, 9039–9047 (2000).

86. Mahapatra, M., Parida, S., Baron, M.D., and Barrett, T. Matrix protein and glycoproteins F and H of Peste-des-petits-ruminants virus function better as a homologous complex. *Journal of General Virology* 87, 2021–2029 (2006).

87. Parida, S., Mahapatra, M., Kumar, S., Das, S.C., Baron, M.D., Anderson, J., and Barrett, T. Rescue of a chimeric Rinderpest virus with the nucleocapsid protein derived from peste-des-petits-ruminants virus: Use as a marker vaccine. *Journal of General Virology* 88(Pt 7), 2019–2027 (2007).

88. Walsh, E.P., Baron, M.D., Anderson, J., and Barrett, T. Development of a genetically marked recombinant rinderpest vaccine expressing green fluorescent protein. *Journal of General Virology* 81, 709–718 (2000).

89. Banyard, A.C., Simpson, J., Monaghan, P., and Barrett, T. Rinderpest virus expressing enhanced green fluorescent protein as a separate transcription unit retains pathogenicity for cattle. *Journal of General Virology* 91, 2918–2927 (2010).

90. Walsh, E.P. et al. Recombinant rinderpest vaccines expressing membrane-anchored proteins as genetic markers: Evidence of exclusion of marker protein from the virus envelope. *Journal of Virology* 74, 10165–10175 (2000).

63 Swine Vesicular Disease Virus

*Cristina Cano-Gómez, Paloma Fernández-Pacheco,
and Miguel Ángel Jiménez-Clavero*

CONTENTS

63.1 INTRODUCTION

Overview: Swine vesicular disease virus (SVDV) is an *Enterovirus* of the family Picornaviridae that infects pigs. In these hosts, it causes an infectious, contagious disease, called swine vesicular disease (SVD), characterized by the appearance of vesicles on the coronary bands of feet, heels, skin, snout, tongue, lips, and teats, along with fever. These symptoms are indistinguishable from those caused by foot-and-mouth disease virus (FMDV), vesicular stomatitis virus (VSV), and vesicular exanthema of swine virus (VESV), and hence its importance for animal health authorities. SVDV is closely related to coxsackievirus B5 (CV-B5),[1] a human virus, from which it is thought to derive by adaptation to pigs, possibly through changes in receptor usage.[2] Although originally considered as zoonotic,[3] public health risk is negligible as the infection of humans hasn't been reported recently and the late isolates of the virus have lost the ability to infect human cells[2] therefore decreasing its original zoonotic potential. Thus far, the disease, which was first described in Italy in 1966[4] has only been reported in farmed pig populations in Europe and far-East Asia. In Europe, since 1994 and after a period of relatively high activity and broader geographic distribution, the disease was effectively cleared from most countries, except from Italy, where, despite the application of intense eradication programs, SVD has been persistently reported causing occasional foci, and for this reason intense surveillance and eradication plans are in place.[5] The disease is also likely to be present in far-East Asia. The last reported case from this region occurred in Taiwan in 2000.

Biosecurity issues: In pigs, SVD is highly contagious and spreads rapidly by direct contact with infected animals and by environmental contamination. This ability to spread, together with its clinical resemblance to FMD, made SVD to be one of the diseases listed by the OIE (World Organization for Animal Health) in the *OIE Terrestrial Animal Health Code*.[6] OIE listed diseases are transmissible diseases that have the potential for very severe and rapid spread, irrespective of national borders (often referred to as "transboundary diseases"), and are of obligatory declaration to the OIE due to their serious socioeconomic or public health consequences and impact on the international trade of animals and animal products. As the importance of SVD relies on its similarity to FMD, all biosecurity measures regarding SVDV handling, including diagnosis involving biologically active virus, must follow those in force for FMDV handling, that is, OIE containment group-4 (and equivalent biocontainment levels like BSL-3, P3, etc.). This group includes "Organisms that cause severe human or animal disease, may represent a high risk of spread in the community or animal population and for which there is usually no effective prophylaxis or treatment."[6] However, SVD is not a severe disease by itself, and, as current diagnostic methods evolve, the capability of differentiating SVD from FMD is expected to be improved, even at the farm level, so the requirement for high biocontainment facilities and measures for SVDV work, as well as the inclusion of this disease in the OIE list, could be re-evaluated in the future.

63.2 CLASSIFICATION, MORPHOLOGY, AND BIOLOGY

63.2.1 VIRUS AND ITS GENOME

Taxonomically, SVDV is considered a subspecies within the species CV-B5. Nucleotide sequence identity between both viruses is 75%–85%.[7] From their antigenic cross-reactivity it was proposed as early as in 1973 that SVDV arose in pigs by interspecies transmission of CV-B5 from humans.[8,9] Later, the analysis of nucleotide sequences of 45 isolates from outbreaks of Europe and Asia between 1966 and 1993 allowed estimating the date of the divergence of SVDV from CV-B5 between 1945 and 1965.[10]

The nonenveloped virus capsid is composed of 60 protomers, each made up of one copy of each of the four structural proteins, VP1, VP2, VP3, and VP4, forming a T = 1 icosahedral shell of 25–30 nm diameter, enclosing the viral genome, a single-strand RNA molecule of positive sense composed of approximately 7400 nucleotides,[11–13] whose organization is essentially identical to that of poliovirus 1,

the virus type of the Picornaviridae family (Figure 63.1). Flanking a unique open reading frame encoding a precursor polyprotein of 2185 amino acid residues, there are two nontranslated regions, one at the 5′ end encompassing a highly structured region, the internal ribosome entry site (IRES), where the ribosome assembles to initiate the translation of the polyprotein, and one at the 3′ end also with a highly structured segment probably involved in RNA replication, followed by a chain of poly adenines (poly A) of variable length. The precursor polyprotein of SVDV is processed in the cytoplasm of the infected cell to yield 11 functional viral polypeptides. The first excision produces polypeptides P1, P2, and P3. After further processing, P1 gives rise to the structural proteins VP1–VP4, whereas P2 and P3 yield the nonstructural proteins (Figure 63.1). Among the nonstructural polypeptides there are two viral proteases (2A and 3C), one RNA polymerase (3D), the VPg, which is covalently bound to the 5′ end of the RNA molecule and participates in its replication, and other proteins whose functions are not well determined (2B, 2C, and 3A).

63.2.2 CAPSID AND ANTIGENIC STRUCTURE

In the mature virion, proteins VP1, VP2, and VP3 (33, 32, and 29 kDa, respectively)[14] are exposed on the surface, forming a compact proteic shell. These three capsid proteins share a tertiary structure, highly conserved among picornaviruses, that consists of a central hydrophobic core constituted by an eight stranded β-barrel. Conversely, VP4 (about 7.5 kDa) faces the inner side of the capsid, interacting with the viral RNA molecule.

Antigenically, SVDV shows two types of antigenic sites: (1) those defined by monoclonal antibody-resistant (MAR)

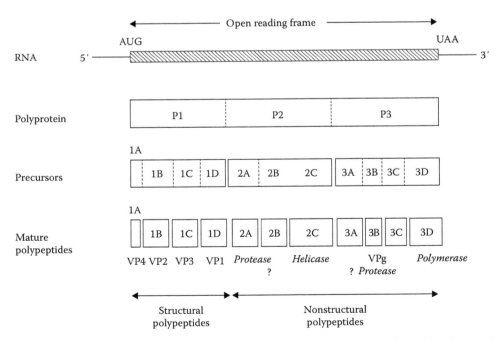

FIGURE 63.1 Genomic organization of SVDV and the post-translational processing of polyprotein products. Discontinuous lines indicate the sites of proteolytic cleavage.

mutant analyses, grouping in up to seven different neutralizing epitopes in the surface of the capsid,[15–17] and (2) those defined by pepscan analysis, essentially describing internal epitopes not involved in neutralization,[18] with one exception: one antigenic site at the N-terminal region of the VP1 protein, which is internal, but emerges to the surface upon virus interaction with the cell receptor, and is involved in neutralization.[19] The three-dimensional structure of SVDV capsid has been solved,[20,21] which enabled the fine mapping of the antigenic sites. A singular feature emerging from this study is the existence of a sphingosine molecule on each VP1 subunit, embedded in an internal cavity, known as the "hydrophobic pocket." This hydrophobic cavity has been identified in all enteroviruses studied, containing cellularly derived lipids, which are known as "pocket factors," which would stabilize the virus particles at the extracellular level.[22] Upon interaction with the receptor at the cell surface (see Section 63.5), these pocket factors are expelled, leading to unstable particles that, once internalized into the cell, undergo deep conformational changes, initiating virus uncoating and RNA release into the cytoplasm. Pocket factor analogues have been developed for use as antivirals[22] but, to date, no use in animal health is foreseen for these compounds, at least in the near future.

63.2.3 Virus Cell Interactions

Apparently all viruses belonging to the coxsackievirus B group, including SVDV, use a common surface molecule known as coxsackievirus-adenovirus receptor (CAR) as host cell receptor.[2,23] CAR is a 46 kDa transmembrane glycoprotein with two extracellular immunoglobulin-like domains, which also functions as an attachment molecule for adenovirus fiber proteins.[24] On the other hand, another surface molecule, decay accelerating factor (DAF, CD55), a 70 kDa protein involved in the control of homologous lysis by complement, has been postulated to act as a co-receptor used by some coxsackie B virus strains.[25–27] Evidence has been reported supporting also this role of DAF for SVDV but, interestingly, only earlier strains (It'66 and UK'72, isolated in Italy in 1966 and in the United Kingdom in 1972, respectively) are able to interact with DAF of human origin. By contrast, more recent strains (R1072, R1120, and SPA'93, isolated in Italy in the early 1990s and in Spain in 1993, respectively) have lost their ability to interact with human DAF, and have not replaced this interaction with the, otherwise different, porcine homologous version of the human DAF molecule.[2] This finding suggests that in its adaptation to swine from humans, SVDV has lost certain capacity to interact with human cells. In fact recent SVDV isolates, though able to interact with CAR, have lost the capacity of infecting human HeLa cells, still present in earlier isolates.[2] This phenotypic change has been suggested to involve a lower tropism for human cells and, consequently, a reduced zoonotic potential for SVDV.[28]

In addition to binding to cell surface glycoproteins, the interaction of many viruses with the cell surface is mediated in many instances by sulfated glycosaminoglycans (GAGs), particularly heparan sulfate proteoglycans.[29] Among the picornaviruses, FMDV–heparin/heparan sulfate interaction has been attributed to an adaptation to cell cultures.[30] In SVDV, interaction with heparan sulfate GAG has been shown to mediate the attachment to the host cell, indicating that this molecule could act as receptor or co-receptor for this virus.[31] The importance of this interaction in the process of cell attachment and entry remains to be determined.

The mechanism of cell entry of SVDV is not fully characterized. Similarly to other picornaviruses, the capsid of SVDV undergoes a series of conformational changes upon interaction with its receptor, leading to a structurally and antigenically altered particle named "A particle." In this process, the N-terminus of VP1 becomes externalized and VP4 is lost. As noted earlier, this process uncovers a cryptic epitope which is involved in neutralization.[19] Those "A particles" are the most abundant form of the intracellular virus in the early stages of infection, and seem to be a necessary intermediate preceding the uncoating and release of RNA into the cytoplasm in picornaviruses.

63.2.4 Genetic and Antigenic Variation

SVDV can be seen as a relatively recently emerged virus variant of CV-B5, which adapted to pigs,[1,7,10] providing a good example of the obvious but sometimes overlooked concept that adaptations and crossing the species barrier can occur either to or from humans, and that the human being is just one more species in this regard.[32] Serologically, all isolates constitute a single serotype. However, in its short history, the virus has evolved into at least four recognized antigenic variants that can be distinguished by different monoclonal antibodies, correlating with four genetically distinct groups: group 1 comprises the first isolate (It'66); group 2 includes European and Asian SVD viruses isolated in the 1970s and early 1980s; group 3 is constituted by Italian isolates of the 1980s and 1990s, and group 4, SVDV isolates circulating in the late 1980s and early 1990s in a broad territory of Europe.[33] As mentioned earlier, in recent years SVDV foci have been almost exclusively reported in Italy, where an intensive surveillance program is in place. As a result of this surveillance, a number of SVDV isolates have been studied, all of which belong to antigenic and genetic group 4,[5] while no representative of antigenic/genetic groups 1–3 has been isolated in the last 20 years, a fact that might indicate a possible extinction of these three groups.

63.3 EPIDEMIOLOGY

63.3.1 Geographic Range of Disease Occurrence

The first occurrence of SVD was reported in Lombardy, Italy, in 1966, where it was clinically recognized as FMD but physical and chemical analysis showed that the causative agent was an enterovirus originally termed "porcine enterovirus,"[4] though eventually it was renamed as "swine vesicular disease virus." The next outbreak was reported in

Hong Kong in 1971, in an FMD vaccine trial.[34] Since then, SVD has been reported in different European and Asian countries (Table 63.1). More recently, besides Italy, where SVD reoccurs every year, outbreaks of SVD have occurred in Portugal in 2003–2004 (Leiria district) and in 2007 (Beja district).[35] Currently, the European Union countries have been declared disease-free, except Southern Italy, where SVD is endemic despite eradication campaigns in place in this country since 1995.[5,36,37] Africa, North and South America, and Oceania have remained free of the disease, but in various Asian countries the disease is likely to be present,[36,38] the last case of SVD from the far-East Asia being reported from Taiwan in 2000.[36]

63.3.2 HOST RANGE

SVDV is considered to be highly contagious in pigs of all ages with variable morbidity, which can be as high as 80%–100% in individual pens, although it depends on a wide range of factors, including virulence of the viral strain involved, type

of farm, and time elapsed between infection and detection, among others. Mortality is negligible. The susceptibility to natural infections has only been reported in swine, either Euro-Asian pigs or American one-toed pigs.[39] Infections in nonswine species have also been reported: relatively high virus titers have been detected in the pharynx of sheep (but not in cattle) kept in close contact with SVD-infected pigs. Such sheep developed significant levels of neutralizing antibodies, indicating that they were infected, so the virus managed to replicate in their tissues, but without showing clinical signs. Despite this finding, sheep do not seem to play any role in the transmission of the disease.[40] Experimental infection has been induced in 1-day old mice, being lethal to newborn mice.[4,41] Although SVDV is closely related to human CV-B5, and few cases of human illness have been reported caused by infections with earlier strains of SVDV, affecting people working in laboratories handling (and exposed to) high virus titers and/or infected pigs,[3] public health risk is not considered since (as mentioned in Sections 63.1 and 63.5) the infection of human hasn't been reported recently and apparently and the virus has lost the ability to infect human cells, decreasing its original zoonotic potential.[2]

63.3.3 DISEASE TRANSMISSION AND VIRUS INACTIVATION

The disease is transmitted mainly by direct contact with infected animals and their feces (fecal–oral route), fomites, and contaminated food. The virus resists fermentation and smoking processes used for conservation of certain food products, but is inactivated either after heat-treatment (above 69°C) of pork products[42] or after treatment of cured ham for 1 year.[43] SVDV-contaminated environments have been shown to be as infectious as direct inoculation or contact with infected pigs. By contrast, airborne spread by aerosols from the lesions and skin is not an efficient way of SVDV transmission.[44,45] Nevertheless, inefficient disinfection of vehicles used in transport of pigs constitutes another important source of infection and disease dissemination from one farm to another.[37] A collective open drainage system or regular movements of infected pigs between pens increase the likelihood of spread within the farm, leading to the consideration of SVD as a "pen disease" rather than a "farm disease."[46]

The notable persistence and particular high stability of SVD virus particles play an essential role in the epidemiology of SVD.[47] The virus is resistant to most common alkaline or acidic disinfectants as well as to detergents and organic solvents such as ether and chloroform. Unaffected by desiccation and freezing, it can survive between 4 and 11 months over a pH range of 2.5–12 and temperature range of 12°C to −20°C. The inactivation of SVD virus is achieved by heating at 56°C for 1 h, or by using disinfectants such as 1% sodium hydroxide (pH 12.4), 2% formaldehyde, and even 70% alcohol (allowing appropriate contact times).[48]

Once the infection is acquired, pigs secrete SVDV both orally and nasally, the secretion ceasing normally within 2 weeks. The virus can be excreted in large amounts in feces, starting up to 48 h before the onset of clinical signs, and lasting

TABLE 63.1
Occurrence of SVD around the World

Country	Years of Report	Year of Most Recent Report
Europe		
Austria	1972–1976, 1978	1979
Belgium	1973, 1979, 1992	1993
Bulgaria	1971	1971
France	1973–1975, 1982	1983
Germany	1973–1977, 1979–1982	1985
Greece	1979	1979
Holland	1975, 1992	1994
Italy	1966, 1972–1984, 1988–1989, 1991–2010	2011
Malta	1975	1978
Poland	1972	1973
Portugal	1995, 2003–2004	2007
Romania	1972–1973	1987
Russia	1972	1976
Spain	1993	1993
Switzerland	1973	1975
The Netherlands	1975	1994
Ukraine	1972	1977
United Kingdom	1972–1977, 1979–1981	1982
Asia		
China	1971–1977, 1979–1981, 1984–1985, 1887–1989	1991
Japan	1973	1975
Macau	1989	1989
Taiwan	1997–1998	2000

usually around 10 days.[49] Nevertheless, occasionally, excretion of the virus is observed for more time, even up to 70 days.[50] Furthermore, in certain circumstances (probably influenced by stress conditions) fecal excretion of the virus is resumed after cessation, and in this type of cases the presence of the virus in feces has been observed up to 126 days after the initial infection.[51] Although the virus found in tissues of infected pigs might bring on the contamination of pork products and introduce the disease agent into the food chain, (particularly when the disease is not immediately recognized), in countries where this practice is not allowed, swill feeding is generally regarded as unimportant in the epidemiology of SVD.[37]

Generally, the highest viral shedding occurs from vesicles within the first week after the infection[45] and the virus does not persist in the tissues longer than 28 days after the infection. However, persistent infection of pigs has been observed experimentally using viral isolates of European origin, in which the virus remains in tissues for longer than 3 months after the infection,[51,52] although it has proven difficult to reproduce these results due to the low frequency and lack of consistency observed for the phenomenon of SVDV persistency.[53] The epidemiological significance of persistent SVDV infection in pigs is uncertain. Experimental evidence indicates that in stress conditions like mixing of pigs from different pens, reactivation of latent virus infections for a short period of time is possible. However, to what extent this latency gives rise to epidemiologically relevant sources of infection remains unknown.

63.4 PATHOGENESIS AND IMMUNITY

This virus can enter the body through either broken skin lesions or mucous membranes of the digestive tract by ingestion of excreta or secretions from infected pigs.[54–56] The incubation period is usually 2–7 days, but it can be longer if the dose of the virus is small. Viral replication starts in the primary site of infection, usually the gastrointestinal tract, spreading rapidly via the lymphatic system into the bloodstream at 1–2 days postinfection (dpi). This occurs evenly in pigs experimentally infected by direct inoculation of the virus as well as in pigs infected through contact with a contaminated environment.[45] The highest virus titers are found in vesicles and tissues as lesions develop, decreasing as the antibody response rises, after the first week of clinical disease. The presence of circulating neutralizing antibodies, observed as early as 4 dpi, provides long lasting protection against future reinfection. The tropism for epithelial tissues is well known but virus titers in the myocardium, brain, kidney, and spleen are higher than the titers in plasma or serum, being probably also sites of replication.[45,54,55] Similarly, the virus can be isolated from lymph nodes which may contain high viral titres[57] possibly either due to virus drainage from tissues or to local replication, or both.

Immunohistochemistry and in situ hybridization have been used to detect SVDV in epithelial and dermal cells, and in infected tissues,[58] but more research must be implemented in order to know which cells are involved in SVDV replication, which would help to identify the mechanism behind the host tropism of the virus.[59]

As noted earlier, the development of neutralizing antibodies in the early stages of the infection is of paramount importance to impair virus dissemination throughout the body and limit the disease. IgM class antibodies dominate the earliest steps in the humoral immune response to SVDV. At 8–12 dpi a switch to IgG occurs,[46,60] which provides a long-living immune protection to further SVDV infections. As noted earlier, only one serotype of SVDV is recognized, which means that prior infection or vaccination with a given SVDV strain will protect against further infection with any other SVDV variant.

63.5 CLINICAL SIGNS

The importance of SVD relies on the fact that it is clinically indistinguishable from FMD, VS, and VES.[61] The differential diagnosis based on the observation of clinical signs is not feasible, and thus it requires conducting specific laboratory investigations (see Section 63.6).

SVD courses most often as an acute, self-limited infection with most perceptible clinical signs being the appearance of vesicles on limbs, tongue, and snout. However, the infection may remain unnoticed since it is often asymptomatic. Disease outcome may vary in appearance, and this variation poses some difficulties in the initial diagnosis. Affected pigs may show inappetence or even anorexia for 1 or 2 days, although this is not a consistent finding. Within 1–5 days after exposure, body temperature may rise to 41°C for 2–3 days; however, in experimental infection with European SVDV, pyrexia has not been observed.[47] Initially, blanching areas in the epithelium precede the formation of vesicles around the coronary bands and interdigital spaces, and on the skin of the lower parts of the limbs, particularly at pressure points such as the knees. Vesicles may also appear on the snout, specifically on the dorsal surface, on the lips, tongue, and mammary glands, being less common in the oral cavity. Rupture of vesicles releases a straw-colored fluid (vesicular fluid) rich in infectious virus, and reveals a shallow ulcer with a red hemorrhagic base. In a few instances punctiform ulcerative lesions have been found on the lower lips, but such lesions are not characteristic of the disease. Extension of the lesions on the coronary band may arise, and the hoof wall may separate from the underlying tissues, but the complete detachment of the hoof, as occurs with FMD, is uncommon.

Neurological signs have been observed, though rarely, in experimentally infected animals. They were reported as nonsuppurative meningitis and panencephalomyelitis, principally affecting the midst and forebrain. These neurological signs may include trembling, unsteady gait, back arching, and chorea (rhythmic jerking).[54,62] Isolation from an aborted fetus suggests that abortion may occur, but there is no evidence that the pregnant sow is relevant for the perpetuation of the disease and dissemination of the infection. Experimentally, abortion is not typically seen and the virus is unlikely to infect or cross the placenta.[63] Other disease signs include lameness, salivation, and moderate general weakening. Normally, infected pigs recover after the acute phase of the disease within 2–3 weeks, with all the skin lesions healed completely. Frequently, the only evidence of infection in the pig after careful examination is a dark horizontal line on the hooves where the growth was temporarily interrupted.

SVD may be subclinical, mild, or severe, depending on the virulence of the strain and husbandry conditions. Regarding virulence, differences in pathogenicity between different strains have been assigned to specific regions and even single point mutations in the virus genome. Differences in pathogenicity between the virulent J1'73 and the avirulent H/3'76 strains were mapped in a region comprising nucleotides 2233–3368 of the viral RNA, corresponding to the C-terminus of VP3, the whole VP1, and the N-terminus of 2A viral protease.[38] Moreover, two different amino acid changes within this region, that is, glycine at the VP1-132 and proline at the 2A-20 positions, were identified as determinants of the high virulent phenotype. With regard to husbandry conditions, more severe lesions are seen when pigs are housed on concrete, particularly damp concrete, than on straw bedding or in grass. In addition, young pigs tend to show more severe clinical signs than older pigs. The only postmortem lesions of SVD are the vesicles that can be seen in live pigs.

63.6 DIAGNOSIS

As SVD is clinically indistinguishable from FMD and other vesicular diseases of pigs (hence its status as OIE-listed disease), the correct differential diagnosis is laboratory-based and must be carried out using specific laboratory tests available. The standards for SVD diagnosis are described in the OIE *Manual of Diagnostic Test and Vaccines for Terrestrial Animals*.[61] Any vesicular condition in pigs may be caused by FMD; therefore, any sign of vesicular disease in pigs should be considered infected with FMDV until proven otherwise. Once suspicion of FMD is discarded, the diagnosis of SVD requires the facilities of a specialized laboratory, with appropriate biosecurity and biocontention measures (see Section 63.1). Countries lacking such facility should send samples for investigation to an OIE Reference Laboratory for SVD.[61]

The laboratory diagnosis of SVD requires either the isolation of the virus and/or detection of its genome, or the demonstration of specific antibodies, or a combination of both.[47] A brief account of the current techniques and methods for the diagnosis of SVD is given in the following, finalizing with an overview of recent advances in the area.

63.6.1 SAMPLES

Samples of vesicular fluids and portions of vesicular lesion tissues are collected separately and aseptically in sterile containers. Fecal samples and serum from animals with or without lesions are collected for virus isolation, reverse transcription polymerase chain reaction (RT-PCR) analysis, and serological tests. Serum samples can be used for routine disease surveillance or export certification. Lymph nodes, thyroid and adrenal glands, kidney, spleen, and heart tissues can be collected from slaughtered animals. Independent sets of sterile instruments should be used for sampling of each animal.

63.6.2 IDENTIFICATION OF THE AGENT

Viral antigen can be detected by indirect sandwich enzyme-linked immunosorbent assay (ELISA) if sufficient antigen is present on the sample.[64] The antigen ELISA is the fastest and simplest test for SVDV detection in epithelium homogenates, where virus concentration is high enough to give a positive result in this test. However, the main drawback of this technique is its low sensitivity, making it unsuitable for the analysis of other types of samples. In particular, more sensitive tests are required for detection of SVDV in fecal samples, often the only available sample for virological diagnosis, since very frequently the infection courses without the development of vesicles/skin lesions. Several highly specific and sensitive RT-PCR techniques have been developed to detect SVDV RNA either in conventional or in real-time format, some of them enabling the simultaneous differentiation from FMDV and VSV.[49,59,65–69]

The virus can be isolated from epithelium of skin lesions, fecal homogenates, and vesicular fluids. For this, the clarified homogenates are inoculated on monolayers of IB-RS-2 cells[70] or other susceptible porcine cells (SK6, PK-15), and these are examined daily for the development of cytopathic effect (CPE), for 2–3 days. The supernatant of cell culture is harvested and virus identification is performed by ELISA or RT-PCR. The virus, if present in feces, is often found in small amounts, so its isolation from this type of sample often requires 2–3 successive blind passages and may be affected by other interfering enteric viruses which are highly prevalent in pig fecal samples.[71,72] If CPE is not observed after three blind passages, the sample is considered negative in virus isolation.

Virus isolation is the reference *gold standard* for virus detection and identification, and is used as a confirmatory method if a positive antigen ELISA or RT-PCR result is not associated with the detection of clinical signs of disease, the detection of seropositive pigs, or a direct epidemiological connection with a confirmed outbreak. If any inoculated culture subsequently develops a CPE, the demonstration of SVD antigen by ELISA or by RT-PCR will suffice to make a positive diagnosis.[61]

63.6.3 SEROLOGICAL TESTS

Virus neutralization test (VNT) and ELISA are the techniques more commonly used for the detection of antibodies raised in response to SVDV infection. Various ELISA tests have been developed[60,73–75] including formats able to specifically differentiate IgM and IgG class immunoglobulins involved in SVDV binding.[60] Some of them are available commercially. While ELISA is the most appropriate technique for screening and has been shown to be efficient in large-scale serosurveillance programes,[57] VNT is currently the *gold standard* assay for confirmatory detection of antibodies to

SVD virus.[76] It is used to confirm the results after ELISA screening. Neutralizing antibodies can be detected as early as 3–4 days after experimental infection. The VNT is a laborious technique which requires 2–3 days to obtain the results. It is a quantitative test which is performed using IB-RS-2 cells or other susceptible porcine cells. The test measures the ability of serial dilutions of a given serum sample to neutralize (block) the infection of the cells by a known amount of infectious virus, previously titrated. It also needs including a reference antiserum of known titer, control cells, and infectious virus, which is used as antigen in the test.

63.6.4 Singleton Reactors

Singleton reactors (SRs) are sera from individual pigs serologically positive for SVDV, but which have shown no clinical signs and for which there is neither a history of the disease on the holding nor contact with a known outbreak.[47] These SRs are very infrequent. In serological surveillance programs for SVD about 1 in 220 tested samples yields positive by standard monoclonal antibody-competition (MAC)-ELISA, and the gold standard VNT detects 1–3 SRs every 1000 sera tested. This low but consistent frequency of false positive reactions is a common problem, provided that surveillance programs for SVD often analyze thousands of samples in some instances. The presence of an SR in a herd imposes restriction of movements to the farm in which it is found, leading to disruption of control measures and eradication programs, and imposes quarantine restrictions to farms and bans for international trade.[47] The SR can be differentiated from true seropositive samples of infected pigs by resampling of the positive animals and their cohorts.[61] IgM class immunoglobulin is responsible for most, if not all, SR reactions. The problem of SR can be reduced by applying a combination of VNT, MAC-ELISA, and isotype specific ELISA.[77] The factors giving rise to SR sera are unknown; however, it seems likely that this condition is not due to a specific immune response to SVDV or other related viruses, but to other responses resulting in a nonspecific IgM-mediated reaction.

63.6.5 Recent Advances in SVD Diagnosis

In the recent years a number of new techniques have been incorporated to the range of SVDV diagnostic tools. These new developments are mainly addressed to improve virus detection and identification, and particularly differentiation from FMDV, the "key issue" in SVDV differential diagnosis. Two main strategies have been explored. On the one hand, multiplexing methods, enabling the simultaneous detection of a wide range of pathogens, including SVDV and FMDV, but also others that are considered in the differential diagnosis of these diseases; on the other hand, "on-site" or portable devices are sought, aiming at enabling rapid virus detection and identification at the farm level, thus saving time from the first suspicion of the disease to its confirmation and the application of control measures.

A recent development of the first kind is the nucleic acid-based multiplexed assay for detection of FMDV, and ruling-out for six other animal diseases that cause vesicular or ulcerative lesions in cattle, sheep, and swine (including SVD), based on a combination of magnetic bead-flow cytometry technology (LUMINEX) and RT-PCR.[78] Another approach in this regard uses solid-phase (chip) microarray technology combined with RT-PCR to address multiplexing of the whole range of vesicular disease viruses.[79] This method enables the efficient detection and differentiation, including correct molecular serotyping of all known serotypes of FMDV, of the two known serotypes of VSV and SVDV simultaneously.

As for the second strategy, a lateral flow device ("penside test") has been developed, which showed a good performance in the detection of SVDV in vesicular fluids and cell culture passage-derived supernatants.[80] Portable devices allowing "on-site" RT-PCR analysis have been developed and tested for their performance on detecting FMDV genome,[81] which could be suitable also for detection of other vesicular diseases, including SVDV, in the same setting, provided that there are RT-PCR techniques (specifically, fluorescence-based real-time formats are of particular utility in this regard) of enough sensitivity and specificity available. As explained in Section 63.6.2, there are a number of real-time RT-PCR techniques available for SVDV diagnostic, and their adaptation to portable devices should not be a problem. On the other hand, loop-mediated isothermal amplification (LAMP) assays provide some advantages in this regard with respect to classical or real-time RT-PCR methods, since they are faster and do not rely on expensive equipment such as portable thermal cyclers, providing robust performance in modestly equipped laboratories, as can be the case for field stations or mobile diagnostic units. A LAMP assay for the detection of SVDV has been recently developed.[82]

63.7 CONTROL AND PROPHYLAXIS

SVD does not cause severe production losses, and improved biosecurity plays a major role in controlling the disease. The main sanitary measures in place are stamping out, restriction of pig movements (protection and surveillance zones), cleaning and disinfection, restrictions on swill feeding and on importation of pig products from SVD-affected regions.[61] The often asymptomatic course of the infection poses a risk at importing livestock from countries where SVD status is not regularly assessed.

Vaccines against SVD are not available commercially, although experimental studies have been conducted to develop vaccines using different approaches.[83–87] In agreement with the status of SVD as an OIE-listed disease, where control is mainly achieved by stamping out and other sanitary measures, vaccination is disregarded, due to the general concern about the presence of specific SVDV antibodies in pigs, which would be raised after vaccination. Any attempt to develop a vaccine useful in the control of SVD has to consider the need for efficient methods to discriminate vaccinated from infected animals (DIVA tests) accompanying the vaccine.

There is currently no treatment for SVD, although antiviral activity of pocket factor analogues in infections caused by enteroviruses (see Section 63.2.2) might open a valuable control strategy that merits exploration in the future.

63.8 CONCLUSIONS

SVDV is the causative agent of a disease of pigs whose importance relies on its resemblance to FMD. This fact imposes severe restrictions on the movements of pigs for which convincing evidence of being free of the virus (or of antibodies raised to it) is not provided. This poses a significant burden for the trade of these animals, and for this reason the disease may cause severe economic losses for the producers, even though it is in fact a mild disease from the clinical point of view. Differentiation from FMD, though not possible clinically, is feasible if appropriate diagnostic tests are applied. Improvement of the diagnostic techniques is making differential diagnosis of vesicular diseases more and more affordable, feasible, and easy, and nowadays portable devices allow a rapid and accurate differentiation of SVDV from FMDV infections on-site. As these tests become affordable and availability of competent laboratory services becomes more and more accessible, the constraint originally imposed on SVD due to its similarity to FMD will lose its sense. This, together with the fact that in recent years circulating SVDV strains are predominantly asymptomatic, makes it necessary to rethink the measures currently in place for the control of SVD.

REFERENCES

1. Zhang, G., Wilsden, G., Knowles, N.J., and McCauley, J.W. Complete nucleotide sequence of a coxsackie B5 virus and its relationship to swine vesicular disease virus. *J Gen Virol* 74, 845–853 (1993).
2. Jimenez-Clavero, M.A., Escribano-Romero, E., Ley, V., and Spiller, O.B. More recent swine vesicular disease virus isolates retain binding to coxsackie-adenovirus receptor, but have lost the ability to bind human decay-accelerating factor (CD55). *J Gen Virol* 86, 1369–1377 (2005).
3. Knowles, N.J. and Sellers, R.F. Swine vesicular disease. In Beran, G.W. (ed.), *Handbook of Zoonoses, Section B: Viral*, pp. 437–444 (CRC Press Inc., Boca Raton, FL, 1994).
4. Nardelli, L. et al. A foot and mouth disease syndrome in pigs caused by an enterovirus. *Nature* 219, 1275–1276 (1968).
5. Bellini, S., Santucci, U., Zanardi, G., Brocchi, E., and Marabelli, R. Swine vesicular disease surveillance and eradication activities in Italy. *Rev Sci Tech* 26, 585–593 (2007).
6. OIE. *Terrestrial Animal Health Code* (OIE World Organisation for Animal Health, Paris, France, 2011).
7. Knowles, N.J. and McCauley, J.W. Coxsackievirus B5 and the relationship to swine vesicular disease virus. *Curr Top Microbiol Immunol* 223, 153–167 (1997).
8. Graves, J.H. Serological relationship of swine vesicular disease virus and Coxsackie B5 virus. *Nature* 245, 314–315 (1973).
9. Brown, F., Talbot, P., and Burrows, R. Antigenic differences between isolates of swine vesicular disease virus and their relationship to Coxsackie B5 virus. *Nature* 245, 315–316 (1973).
10. Zhang, G., Haydon, D.T., Knowles, N.J., and McCauley, J.W. Molecular evolution of swine vesicular disease virus. *J Gen Virol* 80, 639–651 (1999).
11. Inoue, T., Suzuki, T., and Sekiguchi, K. The complete nucleotide sequence of swine vesicular disease virus. *J Gen Virol* 70, 919–934 (1989).
12. Inoue, T., Yamaguchi, S., Kanno, T., Sugita, S., and Saeki, T. The complete nucleotide sequence of a pathogenic swine vesicular disease virus isolated in Japan (J1'73) and phylogenetic analysis. *Nucleic Acids Res* 21, 3896 (1993).
13. Seechurn, P., Knowles, N.J., and McCauley, J.W. The complete nucleotide sequence of a pathogenic swine vesicular disease virus. *Virus Res* 16, 255–274 (1990).
14. Tsuda, T., Tokui, T., and Onodera, T. Induction of neutralizing antibodies by structural proteins VP1 and VP2 of swine vesicular disease virus. *Nihon Juigaku Zasshi* 49, 129–132 (1987).
15. Kanno, T., Inoue, T., Wang, Y., Sarai, A., and Yamaguchi, S. Identification of the location of antigenic sites of swine vesicular disease virus with neutralization-resistant mutants. *J Gen Virol* 76, 3099–3106 (1995).
16. Nijhar, S.K. et al. Identification of neutralizing epitopes on a European strain of swine vesicular disease virus. *J Gen Virol* 80, 277–282 (1999).
17. Borrego, B., Carra, E., Garcia-Ranea, J.A., and Brocchi, E. Characterization of neutralization sites on the circulating variant of swine vesicular disease virus (SVDV): A new site is shared by SVDV and the related coxsackie B5 virus. *J Gen Virol* 83, 35–44 (2002).
18. Jimenez-Clavero, M.A., Douglas, A., Lavery, T., Garcia-Ranea, J.A., and Ley, V. Immune recognition of swine vesicular disease virus structural proteins: Novel antigenic regions that are not exposed in the capsid. *Virology* 270, 76–83 (2000).
19. Jimenez-Clavero, M.A., Escribano-Romero, E., Douglas, A.J., and Ley, V. The N-terminal region of the VP1 protein of swine vesicular disease virus contains a neutralization site that arises upon cell attachment and is involved in viral entry. *J Virol* 75, 1044–1047 (2001).
20. Verdaguer, N., Jimenez-Clavero, M.A., Fita, I., and Ley, V. Structure of swine vesicular disease virus: Mapping of changes occurring during adaptation of human coxsackie B5 virus to infect swine. *J Virol* 77, 9780–9789 (2003).
21. Fry, E.E. et al. Crystal structure of Swine vesicular disease virus and implications for host adaptation. *J Virol* 77, 5475–5486 (2003).
22. Tuthill, T.J., Groppelli, E., Hogle, J.M., and Rowlands, D.J. Picornaviruses. *Curr Top Microbiol Immunol* 343, 43–89 (2010).
23. Martino, T.A. et al. The coxsackie-adenovirus receptor (CAR) is used by reference strains and clinical isolates representing all six serotypes of coxsackievirus group B and by swine vesicular disease virus. *Virology* 271, 99–108 (2000).
24. Coyne, C.B. and Bergelson, J.M. CAR: A virus receptor within the tight junction. *Adv Drug Deliv Rev* 57, 869–882 (2005).
25. Bergelson, J.M. et al. Coxsackievirus B3 adapted to growth in RD cells binds to decay-accelerating factor (CD55). *J Virol* 69, 1903–1906 (1995).
26. Martino, T.A. et al. Cardiovirulent coxsackieviruses and the decay-accelerating factor (CD55) receptor. *Virology* 244, 302–314 (1998).
27. Shafren, D.R. et al. Coxsackieviruses B1, B3, and B5 use decay accelerating factor as a receptor for cell attachment. *J Virol* 69, 3873–3877 (1995).

28. Alexandersen, S. et al. Picornaviruses. In Zimmerman, J., Karriker, L., Ramirez, A., Schwartz, K., and Stevenson, G. (eds.), *Diseases of Swine*, pp. 587–620 (John Wiley & Sons, Chichester, U.K., 2012).

29. Fears, C.Y. and Woods, A. The role of syndecans in disease and wound healing. *Matrix Biol* 25, 443–456 (2006).

30. Fry, E.E. et al. Structure of foot-and-mouth disease virus serotype A10 61 alone and complexed with oligosaccharide receptor: Receptor conservation in the face of antigenic variation. *J Gen Virol* 86, 1909–1920 (2005).

31. Escribano-Romero, E., Jimenez-Clavero, M.A., Gomes, P., Garcia-Ranea, J.A., and Ley, V. Heparan sulphate mediates swine vesicular disease virus attachment to the host cell. *J Gen Virol* 85, 653–663 (2004).

32. Jimenez-Clavero, M.A. Animal viral diseases and global change: Bluetongue and West Nile fever as paradigms. *Front Genet* 3, 105 (2012).

33. Brocchi, E. et al. Molecular epidemiology of recent outbreaks of swine vesicular disease: Two genetically and antigenically distinct variants in Europe, 1987–1994. *Epidemiol Infect* 118, 51–61 (1997).

34. Mowat, G.N., Darbyshire, J.H., and Huntley, J.F. Differentiation of a vesicular disease of pigs in Hong Kong from foot-and-mouth disease. *Vet Rec* 90, 618–621 (1972).

35. Knowles, N.J. et al. Reappearance of swine vesicular disease virus in Portugal. *Vet Rec* 161, 71 (2007).

36. Bellini, S., Alborali, L., Zanardi, G., Bonazza, V., and Brocchi, E. Swine vesicular disease in northern Italy: Diffusion through densely populated pig areas. *Rev Sci Tech* 29, 639–648 (2010).

37. EFSA. Scientific opinion on swine vesicular disease and vesicular stomatitis. *EFSA J* 10, 2631 (2012).

38. Kanno, T. et al. Mapping the genetic determinants of pathogenicity and plaque phenotype in swine vesicular disease virus. *J Virol* 73, 2710–2716 (1999).

39. Wilder, F.W. et al. Susceptibility of one-toed pig to certain diseases exotic to the United States. In *Proceedings of the 78th Annual Meeting of the United States Animal Health Association*, pp. 195–199 (1974).

40. Burrows, R., Mann, J.A., Goodridge, D., and Chapman, W.G. Swine vesicular disease: Attempts to transmit infection to cattle and sheep. *J Hyg (Lond)* 73, 101–107 (1974).

41. Kadoi, K. The propagation of a strain of swine vesicular disease virus in one-day-old mice. *Nihon Juigaku Zasshi* 45, 821–823 (1983).

42. McKercher, P.D., Morgan, D.O., McVicar, J.W., and Shuot, N.J. Thermal processing to inactivate viruses in meat products. *Proc Annu Meet U S Anim Health Assoc* 84, 320–328 (1980).

43. Mebus, C. et al. Survival of several porcine viruses in different Spanish dry-cured meat products. *Food Chem* 59, 555–559 (1997).

44. Herniman, K.A., Medhurst, P.M., Wilson, J.N., and Sellers, R.F. The action of heat, chemicals and disinfectants on swine vesicular disease virus. *Vet Rec* 93, 620–624 (1973).

45. Dekker, A., Moonen, P., de Boer-Luijtze, E.A., and Terpstra, C. Pathogenesis of swine vesicular disease after exposure of pigs to an infected environment. *Vet Microbiol* 45, 243–250 (1995).

46. Dekker, A., van Hemert-Kluitenberg, F., Baars, C., and Terpstra, C. Isotype specific ELISAs to detect antibodies against swine vesicular disease virus and their use in epidemiology. *Epidemiol Infect* 128, 277–284 (2002).

47. Lin, F. and Kitching, R.P. Swine vesicular disease: An overview. *Vet J* 160, 192–201 (2000).

48. Terpstra, C. Swine vesicular disease in the Netherlands. *Tijdschr Diergeneeskd* 117, 623–626 (1992).

49. Reid, S.M., Ferris, N.P., Hutchings, G.H., King, D.P., and Alexandersen, S. Evaluation of real-time reverse transcription polymerase chain reaction assays for the detection of swine vesicular disease virus. *J Virol Methods* 116, 169–176 (2004).

50. Niedbalski, W. Application of different diagnostic methods for the detection of SVDV infection in pigs. *Bull Vet Inst Pulawy* 43, 11–18 (1999).

51. Lin, F., Mackay, D.K., and Knowles, N.J. The persistence of swine vesicular disease virus infection in pigs. *Epidemiol Infect* 121, 459–472 (1998).

52. Lahellec, M. and Gourreau, J.M. Maladie vésiculeuse du porc: étude anatomo-pathologique. *Ann Rech Vet* 6, 179–186 (1975).

53. Lin, F., Mackay, D.K., Knowles, N.J., and Kitching, R.P. Persistent infection is a rare sequel following infection of pigs with swine vesicular disease virus. *Epidemiol Infect* 127, 135–145 (2001).

54. Chu, R.M., Moore, D.M., and Conroy, J.D. Experimental swine vesicular disease, pathology and immunofluorescence studies. *Can J Comp Med* 43, 29–38 (1979).

55. Lai, S.S., McKercher, P.D., Moore, D.M., and Gillespie, J.H. Pathogenesis of swine vesicular disease in pigs. *Am J Vet Res* 40, 463–468 (1979).

56. Mann, J.A. and Hutchings, G.H. Swine vesicular disease: Pathways of infection. *J Hyg (Lond)* 84, 355–363 (1980).

57. Dekker, A. Swine vesicular disease, studies on pathogenesis, diagnosis, and epizootiology: A review. *Vet Q* 22, 189–192 (2000).

58. Mulder, W.A. et al. Detection of early infection of swine vesicular disease virus in porcine cells and skin sections. A comparison of immunohistochemistry and in-situ hybridization. *J Virol Methods* 68, 169–175 (1997).

59. Lin, F., Mackay, D.K., and Knowles, N.J. Detection of swine vesicular disease virus RNA by reverse transcription-polymerase chain reaction. *J Virol Methods* 65, 111–121 (1997).

60. Brocchi, E., Berlinzani, A., Gamba, D., and De Simone, F. Development of two novel monoclonal antibody-based ELISAs for the detection of antibodies and the identification of swine isotypes against swine vesicular disease virus. *J Virol Methods* 52, 155–167 (1995).

61. OIE. Swine vesicular disease. In *Manual of Diagnostic Tests & Vaccines for Terrestrial Animals*, pp. 1139–1145 (World Organisation for Animal Health, Paris, France, 2008).

62. Lenghaus, C., Mann, J.A., Done, J.T., and Bradley, R. Neuropathology of experimental swine vesicular disease in pigs. *Res Vet Sci* 21, 19–27 (1976).

63. Watson, W.A. Swine vesicular disease in Great Britain. *Can Vet J* 22, 195–200 (1981).

64. Ferris, N.P. and Dawson, M. Routine application of enzyme-linked immunosorbent assay in comparison with complement fixation for the diagnosis of foot-and-mouth and swine vesicular diseases. *Vet Microbiol* 16, 201–209 (1988).

65. Nunez, J.I. et al. A RT-PCR assay for the differential diagnosis of vesicular viral diseases of swine. *J Virol Methods* 72, 227–235 (1998).

66. Fernandez, J. et al. Rapid and differential diagnosis of foot-and-mouth disease, swine vesicular disease, and vesicular stomatitis by a new multiplex RT-PCR assay. *J Virol Methods* 147, 301–311 (2008).

67. Rasmussen, T.B., Uttenthal, A., and Aguero, M. Detection of three porcine vesicular viruses using multiplex real-time primer-probe energy transfer. *J Virol Methods* 134, 176–182 (2006).

68. Hakhverdyan, M., Rasmussen, T.B., Thoren, P., Uttenthal, A., and Belak, S. Development of a real-time PCR assay based on primer-probe energy transfer for the detection of swine vesicular disease virus. *Arch Virol* 151, 2365–2376 (2006).

69. McMenamy, M.J. et al. Development of a minor groove binder assay for real-time one-step RT-PCR detection of swine vesicular disease virus. *J Virol Methods* 171, 219–224 (2011).

70. De Castro, M.P. Behaviour of the foot and mouth disease virus in cell cultures: Susceptibility of the IB-RS-2 cell line. *Arq Inst Biol (Sao Paulo)* 31, 155–166 (1964).

71. Buitrago, D. et al. A survey of porcine picornaviruses and adenoviruses in fecal samples in Spain. *J Vet Diagn Invest* 22, 763–766 (2010).

72. Cano-Gomez, C. et al. Analyzing the genetic diversity of teschoviruses in Spanish pig populations using complete VP1 sequences. *Infect Genet Evol* 11, 2144–2150 (2011).

73. Hamblin, C. and Crowther, J.R. A rapid enzyme-linked immunosorbent assay for the serological confirmation of swine vesicular disease. *Br Vet J* 138, 247–252 (1982).

74. Armstrong, R.M. and Barnett, I.T. An enzyme-linked immunosorbent assay (ELISA) for the detection and quantification of antibodies against swine vesicular disease virus (SVDV). *J Virol Methods* 25, 71–79 (1989).

75. Dekker, A., Moonen, P.L., and Terpstra, C. Validation of a screening liquid phase blocking ELISA for swine vesicular disease. *J Virol Methods* 51, 343–348 (1995).

76. Golding, S.M., Hedger, R.S., and Talbot, P. Radial immuno-diffusion and serum-neutralisation techniques for the assay of antibodies to swine vesicular disease. *Res Vet Sci* 20, 142–147 (1976).

77. De Clercq, K. Reduction of singleton reactors against swine vesicular disease virus by a combination of virus neutralisation test, monoclonal antibody-based competitive ELISA and isotype specific ELISA. *J Virol Methods* 70, 7–18 (1998).

78. Lenhoff, R.J. et al. Multiplexed molecular assay for rapid exclusion of foot-and-mouth disease. *J Virol Methods* 153, 61–69 (2008).

79. Lung, O. et al. Multiplex RT-PCR detection and microarray typing of vesicular disease viruses. *J Virol Methods* 175, 236–245 (2011).

80. Ferris, N.P. et al. Development and laboratory evaluation of a lateral flow device for the detection of swine vesicular disease virus in clinical samples. *J Virol Methods* 163, 477–480 (2010).

81. Madi, M. et al. Rapid detection of foot-and-mouth disease virus using a field-portable nucleic acid extraction and real-time PCR amplification platform. *Vet J* 193, 67–72 (2012).

82. Blomstrom, A.L. et al. A one-step reverse transcriptase loop-mediated isothermal amplification assay for simple and rapid detection of swine vesicular disease virus. *J Virol Methods* 147, 188–193 (2008).

83. Mowat, G.N., Prince, M.J., Spier, R.E., and Staple, R.F. Preliminary studies on the development of a swine vesicular disease vaccine. *Arch Gesamte Virusforsch* 44, 350–360 (1974).

84. Delagneau, J.F., Guerche, J., Adamowicz, P., and Prunet, P. Maladie vésiculeuse du porc: propriétés physicochimiques et immunogènes de la souche France 1/73. *Ann Microbiol (Paris)* 125, 559–574 (1974).

85. Gourreau, J.M., Dhennin, L., and Labie, J. Mise au point d'un vaccin a virus inactivé contre la maladie vesiculeuse du porc. *Rec Med Vet* 151, 85–89 (1975).

86. McKercher, P.D. and Graves, J.H. A mixed vaccine for swine: An aid for control of foot-and-mouth and swine vesicular diseases. *Bol Centro Panameric Fiebre Aftosa* 23/24, 37–49 (1976).

87. Jimenez-Clavero, M.A., Escribano-Romero, E., Sanchez-Vizcaino, J.M., and Ley, V. Molecular cloning, expression and immunological analysis of the capsid precursor polypeptide (P1) from swine vesicular disease virus. *Virus Res* 57, 16370 (1998).

64 Vesicular Stomatitis Virus

M.D. Salman and B.J. McCluskey

CONTENTS

64.1 INTRODUCTION

Vesicular stomatitis (VS) is a disease of the western hemisphere with geographical areas throughout South America, Central America, and Mexico considered endemic. Sporadic outbreaks of VS occur in the United States primarily in the southwestern states and must be considered one of the most enigmatic diseases of the western hemisphere. Throughout the second half of the last century, many bright minds worked diligently to answer difficult questions surrounding this disease. Several virologic, pathologic, and epidemiologic questions remain unanswered about this disease whose effect on individual animals is generally innocuous but whose political effects are substantial.

64.2 CLASSIFICATION AND MORPHOLOGY

Vesicular stomatitis viruses (VSV) are members of the family Rhabdoviridae, which includes viruses that infect vertebrates, invertebrates, and several plant species. The viruses of this family known to infect mammals are classified into two genera, *Lyssavirus* and *Vesiculovirus*. Rabies is the most well-characterized and most devastating virus of the *Lyssavirus* genus, while VSV are the prototype viruses of the *Vesiculovirus* genus.[69] VSV are bullet-shaped and generally 180 nm long and 75 nm wide.[72] The nucleocapsid or ribonucleoprotein core (RNP) and lipoprotein envelope surrounding the RNP are the two major structural components of VSV. Extending from the outer surface of the envelope are spike-like projections. The genomic structure of VSV is a single strand of negative-sense RNA and is composed of five genes, N, P, M, G, and L, representing the nucleocapsid protein, phosphoprotein (a component of the viral RNA polymerase), matrix protein, glycoprotein, and the large protein (a component of the viral RNA polymerase), respectively. The RNA genome is 11,161 nucleotides long and is transcribed by the RNA-dependent RNA polymerase composed of the L and P proteins. The polymerase generates five monocistronic, capped, and polyadenylated mRNAs. Cellular translational mechanisms produce the five structural proteins of an intact virion. Studies have shown that the synthesis of gene transcripts follows the order of the genes in the genome and there is a gradient in the amount of the transcripts, which follows this same order (i.e., N > P > M > G > L).[5] The shift from transcription to replication is hypothesized to occur when large quantities of the N protein bind to the nascent leader RNA and prevents termination at the leader-N protein junction. Thus, continuous passage of the RNA polymerase down the genome occurs, resulting in the production of a positive strand, which in turn is replicated by the RNA-dependent RNA polymerase to generate the negative-sense RNA to be packaged in the virion.

As is typical of RNA polymerases, their infidelity results in many nucleotide substitutions, and researchers have suggested that with the established error rates of the polymerase, one could expect every clone in a VSV population to differ from one another at a number of nucleotide positions.[64] It was further suggested that many of these base substitutions result in lethal mutations and subsequent production of noninfectious virions. The evolutionary pattern of VSV suggests adaptations to geographical or ecological pressures, and thus in some endemic areas, virus genomes appear to remain relatively stable.[56] Although there are many members of the *Vesiculovirus* genus, two are of particular interest to animal health, vesicular stomatitis virus-New Jersey serotype (VSV-NJ) and vesicular stomatitis virus-Indiana serotype (VSV-IN). These two viruses are similar in size and morphology but generate distinct neutralizing antibodies in infected animals. Thus, although considered distinct viruses, they are often distinguished only by terming one serotype New Jersey and the other serotype Indiana.[48] Other members of the *Vesiculovirus* genus include Cocal, Jurona, Carajas, Maraba, Piry, Calchaqui, Yug Bogdanovac, Isfahan, Chandipura, Perinet, and Porton-S.[69] Cocal and Alagoas are subtypes of VSV-IN and have been associated with vesicular disease in animals in South America. Piry, Chandipura, and Isfahan produced only mild lesions in experimentally infected animals.[73] The remaining *Vesiculoviruses* have been isolated from arthropods, mammals, or both, but are not associated with the disease.[69]

64.3 BIOLOGY AND EPIDEMIOLOGY

64.3.1 NATURAL HISTORY

Accounts of VS were first reported in literature in 1916 when horses in Denver, Colorado, stockyards showed the clinical signs typical of VSV infection. Later that summer, an outbreak of VS in cattle and horses in the San Luis Valley of Colorado occurred. In this same year, cases occurred in horses in Illinois, Iowa, Kansas, Missouri Montana, Nebraska, South Dakota, Utah, and Wyoming.[30] In 1925, a trainload of healthy cattle arrived in Indiana from Kansas City. These cattle were dispersed to individual farms and soon after developed lesions of the tongue and oral mucosa, then the disease spread to horses in the area. The infectious agent was isolated and termed the vesicular stomatitis Indiana strain.[16] In 1926, an extensive outbreak of VS occurred in New Jersey where approximately 750 cattle on 33 farms were affected. The disease appeared only in very few horses. The agent again was isolated and found to be distinct from the Indiana strain isolated in the previous year. This new strain of VSV was termed vesicular stomatitis New Jersey strain. Over the past 100 years since its identification, VS has occurred sporadically throughout the United States. Only states in New England appear to have been spared incursions of VS.

In 1995, 1162 investigations were conducted in 42 states during a VS outbreak. VS was confirmed in six states including Arizona (1 premises), Colorado (165 premises), New Mexico (186 premises), Texas (1 premises), Utah (6 premises), and Wyoming (8 premises). Overall, 78% of the positive premises-housed horses were positive for VS,

22% of premises-housed cattle were positive for VS, and there was one VS-positive llama. All cases where virus isolation was successful were due to the VSV-NJ serotype.[7,42] During the 1997 outbreak, 689 total investigations for suspected VS occurred in 40 states. There were a total of 380 (55%) premises identified as housing animals positive for VS in four states: Arizona, Colorado, New Mexico, and Utah. Nationwide, horses comprised 704 of 802 (88%) of the examinations conducted for suspect VS, and 362 of 374 (97%) positive premises had horses diagnosed as positive for VS. Cattle comprised 78 of 802 (10%) of the examinations conducted, and 12 of 374 (3%) positive premises had cattle diagnosed with VS. The remaining 2% of examinations were conducted in sheep, goats, swine, llamas, elk, and one dog. None of these species were positive for VS. Both VSV-NJ and VSV-IN were isolated from clinical cases.

In 1998, 130 of 232 (6%) investigations nationwide were positive for VS. A total of four states were affected including Arizona (15 positive premises), Colorado (102 positive premises), New Mexico (12 positive premises), and Texas (1 positive premises). Premises where an equid was the species positive for VS represented 99% of all positive premises. Only one premise was identified as housing cattle with VS, and this occurred in only one cow. All of the VSV isolated during this outbreak were VSV-IN.

Over 750 premises, between 2004 and 2006, housed affected animals in nine states (Arizona, Colorado, Idaho, Montana, Nebraska, New Mexico, Texas, Utah, and Wyoming).[19] A total of five premises in 2009 were affected, two premises in Texas, and three in New Mexico. All five premises had only horses affected.[35] In 2010, four premises in Arizona with only horses affected were reported.[36]

Until recently, VS was considered endemic on Ossabaw Island, Georgia, where cattle, raccoons, white-tailed deer, horses, and feral swine were seropositive to VSV-NJ.[24,61,62] Recent serologic test results of white-tailed deer and feral swine and the failure to isolate VSV-NJ from sand flies on the island suggest that the virus is no longer present.[38]

As mentioned previously, VS is a disease of the western hemisphere. Although limited, there are reports of the endemic nature of VSV and other vesiculoviruses from Argentina,[9] Brazil,[22] Columbia,[65] and other South American countries.[3] Extensive research conducted in Costa Rica has shown the endemic nature of the virus and the disease.[56,57,67]

64.3.2 EPIDEMIOLOGY

64.3.2.1 Transmission

Outbreaks of VS in the southwestern United States typically begin in the late spring or early summer. Subsequently, premises with livestock positive for VS then are diagnosed throughout the summer with the last cases usually identified in the late fall. Also typical is the northward progression of recognized VS cases over time. Index cases for outbreaks in the United States usually are identified in southern New Mexico or Arizona. It has been suggested that both the temporal and spatial characteristics of VS outbreaks are associated with arthropod abundance with identification of new cases occurring as warmer weather induces insect hatching and ceases when cold weather predominates, inhibiting vector hatches.

During the VS outbreak in 1982–1983, disease entered California through the transport of infected cattle purchased in Idaho.[29] Additional evidence from investigative reports suggests that VSV can be moved to new locations through infected animals. For example, the only case of VS infection identified in Texas during the 1995 outbreak was due to movement of a horse from an area in New Mexico that was experiencing increased VSV activity into Texas. It was apparent that this horse had become infected in New Mexico and then subsequently exhibited clinical signs of VS infection after being moved back to Texas.

Direct contact transmission was observed when pigs were experimentally inoculated with VSV-NJ. Serologically naive pigs were housed in direct contact with pigs that were experimentally inoculated with VSV-NJ by routes that would simulate contact or mechanical transmission.[63] In a second experiment, pigs infected with VSV-NJ by contact were housed with additional naive pigs. Pigs were monitored and sampled daily for clinical disease and virus isolation, and serologic testing was performed before and after infection or contact. Contact transmission developed only when vesicular lesions were evident and contact pigs shed virus as early as 1 day after contact. Transmission was lesion-dependent, that is transmission only occurred when the infected pigs had visible lesions. Contact transmission was efficient, with infections ranging from subclinical to clinical with development of vesicular lesions.

64.3.2.2 Vectors (Arthropods)

The evidence for arthropod transmission of VSV is most compelling for sand flies (*Lutzomyia shannoni*)[8,11–14] and black flies (Diptera: Simuliidae). Other species of insects may also be competent biologic or mechanical vectors of VSV. Table 64.1 contains species of arthropods from which VSV has been isolated.

Black flies are competent experimental vectors for VSV-NJ, and this virus has been isolated from Simuliidae trapped in the wild during outbreaks of VS.[17,25] *Simulium vittatum* (black fly) females intrathoracically infected with virus transmit infectious virus in their saliva after 10 days.[17] Efficient transmission of VSV-NJ occurs between infected and noninfected black flies cofeeding on nonviremic deer mice, suggesting that black flies could act as a transfer vector between nonviremic vertebrate hosts and domestic livestock.[44–46] Recent experiments with domestic cattle showed transmission from infected to noninfected black flies when feeding simultaneously and in close proximity on the same animal. Uninfected flies physically separated from infected flies by up to 11 cm were able to acquire the virus.[60] These experiments are critical in explaining how VSV may be maintained and transmitted without a viremic reservoir host.

The flight range of potential VSV insect vectors vary, but none would be adequate to explain the often large distances observed between both individual and clusters of infected premises. The analysis of backward wind trajectories during the VS outbreaks in 1982 and 1985 suggests the feasibility of infected insects being transported for long distances on wind currents and subsequently landing on noninfected premises many miles away.[59]

64.3.2.3 Reservoirs

Arboviruses generally use vertebrates as reservoirs for transmission via arthropods.[21] A vertebrate reservoir normally would experience a viremia during which time they are infective for hematophagous insects. Viremia has not been observed in livestock species that exhibit clinical signs. Bats (*Myotis lucifugus*) were inoculated subcutaneously with Coca VSV to determine their potential as maintenance hosts of VSV. Inoculated bats were shown to remain viremic for 10 days when housed at 22°C, and those kept in hibernation conditions were viremic for 16 days.[18] It was not determined if these periods of viremia were due to virus replication or merely persistence. Persistent infections of VSV-IN were established in immunocompetent Syrian hamsters with virus recoverable up to 8 months after infection.[6] Virus was isolated from brain, spleen, and liver homogenates in these experiments.[26] Other experimental work with hamsters revealed viral RNA in the brain, cerebellum, spleen, liver, kidney, and lung 2 months after infection and in central nervous system tissues at 10 and 12 months postinfection.[6] Infectious virus, however, was not recovered from any experimental animals in this study. The pathogenesis of VSV-NJ was investigated in deer mice (*Peromyscus maniculatus*), a potential reservoir species in the southwestern United States. Virus was inoculated experimentally into mice and was identified by immunohistochemistry in central nervous system tissues and the heart for up to 5 days postinoculation.[15] Blood clearance experiments in wild mammals, chickens, pigeons, and other birds from Panama indicated that the virus was cleared in all animal species by 120 min postinoculation.[66]

TABLE 64.1
Insect Genera from Which Vesicular Stomatitis Viruses Have Been Isolated

Genus	Common Name	Transmission
Tabanus	Horsefly	Yes
Chrysops	Deerfly	Yes
Aedes	Mosquito	Yes
Culex	Mosquito	Yes
Culicoides	Biting midge	Yes
Musca	Housefly	No
Hippelates	Eye gnats	No
Simulium	Blackfly	Yes
Lutzomyia	Sand fly	Yes
Stomoxys	Stable fly	Yes

TABLE 64.2

Wild Vertebrates Identified to Have Antibodies to VSV

Species	Common Name
Alouatta villosa	Howler monkey
Antilocapra americana	Pronghorn antelope
Antilope cervicapra	Black buck
Aotus trivirgatus	Night monkey
Artibus spp.	Fruit bat
Baiomys taylori	Pygmy mouse
Bassaricyon gabbii	Olingo
Bradypus infuscatus	Sloth
Canis latrans	Coyote
Cervus elaphus	Elk
Coendu rothschildi	Porcupine
Dasypus novemcinctus	Armadillo
Didelphis virginiana	Opossum
Felis rufus	Bobcat
Lepus californicus	Jackrabbit
Lynx rufus	Lynx
Meleagris gallopavo	Wild turkey
Mus musculus	House mouse
Mephitis mephitis	Skunk
Myocastor coypus	Nutria
Neotoma mexicana	Wood rat
Odocoileus virginianus	White-tailed deer
Odocoileus hemionus	Mule deer
Ovis canadensis	Bighorn sheep
Peromyscus maniculatus	Deer mouse
Procyon lotor	Raccoon
Saguinus geoffroyi	Marmoset

Serologic evidence of exposure to VSV has been shown in many vertebrate species. Table 64.2 lists select species from which antibodies to one or both VSV serotypes have been detected.

64.3.2.4 Risk Factors

A cross-sectional study of 348 farms conducted in Costa Rica found that cattle residing in areas between 500 and 1500 m in premontane or lower montane moist forest had a higher risk of seropositivity to VSV-NJ as compared with cattle at lower elevations. Cattle residing at 0–500 m and less than 2 m of annual rainfall (tropical dry forest) were also at higher risk of seropositivity to VSV-NJ.[4] No factors were associated with increased risk of seropositivity to VSV-IN. A prospective case-control study conducted on 22 Costa Rican dairy farms evaluated cow, farm, and ecological risk factors of clinical VS. Affected cattle were generally older, with 7-year-old cows having the highest age-specific incidence rate. Clinical disease also was associated with cows in lactation and with higher acute antibody titers to VSV-IN. Farm factors associated with clinical disease included the presence of poultry and a longer calving interval on the farm. Two ecological factors were forced into the multivariate models, the reported presence of sand flies and the farm's location

in forest land. The two forced ecological variables were the only ones found to be associated significantly with clinical disease.[67]

A case-control study was designed to identify management factors affecting the risk of animals developing VS in the southwestern United States.[37] Horses, cattle, and sheep with suspected VS on 395 premises in Arizona, Colorado, New Mexico, and Utah were included in the study. Data were collected during the VS outbreak in 1997 with cases defined as those premises completing a questionnaire and having at least one animal confirmed positive for VS. Control premises were those investigating and completing a questionnaire but on which animals were tested for VS and then determined to be negative for VS. Results indicated that animals with access to a shelter or barn had a reduced risk of developing VS. This effect was more pronounced for equine premises. Risk of developing disease was increased where animals had access to pasture. On all premises where owners reported increased insect populations and where animals were housed less than 0.25 miles from a source of running water odds of developing VS were increased (insects odds ratio of 2.5, 95% confidence interval [CI] 1.47–4.47; running water odds ratio of 2.6, 95% CI 1.32–5.0).

A more recent case-control study supported these findings suggesting that insect control and spending time in shelters decreased the odds for VS infection and that premises with grassland or pasture or that had a body of water were at higher risk.[19] A retrospective case-control study was designed to determine potential risk factors for VS in Colorado in 1995.[47] Data were collected on 52 premises that had VS-positive animals and 33 that did not have VS-positive animals during the 1995 VS outbreak and 8 premises that were in the vicinity of premises with VS-positive animals during the 1995 outbreak. Premises level and animal level data were collected including management practices and ecological variables. Premises that had at least one seropositive animal in 1996 were significantly more likely to be case premises than control premises. For case premises, there was an association between serologic status of the animals in 1996 and their clinical disease status in 1995. There were no significant risk factors on the premises level or animal level identified in this study.

As previously mentioned, some studies have shown that animals located in specific ecologic zones were more likely to be seropositive for VSV than animals in other zones. It was suggested from these studies that habitat requirements of either reservoirs or arthropod vectors were the factors associated with zone preference. An extensive study of human inhabitants of rural Central America found that antibodies to VSV-NJ were associated with persons living at elevations between 350 and 649 m, with relatively dry climate, low-density vegetation, and seasonal alterations in ground moisture. Similar risks were found for VSV-IN with the addition of increased risk in moist, high-density tree cover habitats.

64.3.2.5 Economic Impact

Financial impacts of VS on livestock producers have been reported during a number of outbreaks in the United States.

An outbreak on an Alabama dairy in 1962 resulted in net losses of $13,889 on sale of cows, $15,000–$20,000 in milk production, and $10,000–$14,000 in milk quota.[20] A cost of $253.31 per clinical case was calculated for 13 Colorado dairies affected by VS in 1982.[1] The greatest loss was due to cows culled (46.6% of the total loss). Two California dairies affected by VS in 1982 sustained losses of $202/cow for one dairy and $97/cow for the other dairy for total losses of $225,000 for both dairies over a two-month period.[28] Again, culling accounted for the highest percentage (56%) of the loss. Another dairy affected by VS in 1982, located in Idaho, sustained total losses of approximately $50,000.[40] Losses were attributed to involuntary culling, secondary bacterial infections and death, lost milk production, and losses due to early dry-off of cows. Information concerning financial effects of VS on 16 Colorado beef ranches was investigated following the 1995 outbreak.[32] Median financial loss was $7,818/ranch, and mean financial loss was $15,565/ranch for total losses for the 16 ranches approaching $250,000. Financial losses in these beef herds primarily were attributed to increased culling rates, death of pregnant cows, loss of income from calves, and costs of additional labor during the outbreaks.

64.3.2.6 Zoonotic Potential

Three people were infected with VSV through exposure to experimentally infected animals while working in laboratories.[31] Fever, general malaise, and muscle pain were signs similar in all three individuals. Mild stomatitis was observed in two of the three individuals. In all three, recoveries were completed and rapid. Although no virus was isolated from any of these individuals, high neutralizing antibody titers to VSV-NJ were present in all three. Two of the individuals handled experimentally infected cattle, and the third was splashed with virus-containing material while harvesting infected allantoic fluids.

In the 1950s, laboratory workers at the Agricultural Research Laboratory in Beltsville, Maryland, were tested routinely for VSV complement-fixing antibodies.[51] A summary of this work indicated that VS in humans appears as an acute, self-limiting infection with signs similar to influenza. Overall, 96% of laboratory workers and animal handlers tested had positive titers to VSV, while only 57% of the seropositive individuals could recall having clinical signs.

An investigation of owners and handlers of infected cattle was conducted during an outbreak of VS in 1965.[23] A total of 41 persons were interviewed and had specimens collected for virus isolation and serologic testing. Eight persons had serologic evidence of exposure to VSV, and all eight had lived or worked on ranches where cattle were confirmed to have been infected with VSV-IN. Fever, general malaise, myalgia, nausea, and pharyngitis were observed. Vesicular lesions of the gums occurred in two people. Similarly, a study was conducted of veterinarians, research workers, and regulatory personnel who were exposed to VSV during an outbreak in Colorado in 1982.[53] Neutralizing antibody prevalence was higher in exposed persons with clinical signs than in those without a history of clinical illness. Higher risk of seropositivity was found for individuals who examined the oral cavity of infected animals and had open wounds on hands or arms and for those examining horses rather than cattle. Overall, infectivity of humans was low.

A survey of approximately 20,000 Central Americans, living in areas endemic for VSV, revealed an overall prevalence of neutralizing antibodies to VSV-NJ and VSV-IN of 48% and 18% respectively.

64.4 CLINICAL FEATURES AND PATHOGENESIS

Infection of epithelium with VSV induces intercellular edema in the malpighian layer and the epithelial cells become separated by vacuolar cavities.[54] Cellular necrosis is concomitant with edema with cells shrinking in size but not undergoing lysis. The infiltration of inflammatory cells, including granulocytes and monocytes, eventually results in cellular lysis. Vesicles develop when the necrotic, edematous mucosa breaks free from underlying tissue forming a cavity filled with cellular exudates. The separation occurs at the basal layer. Vesicle formation through intercellular edema, cellular necrosis, and inflammatory cell infiltration generally occurs within 48 h following experimental inoculation, and vesicles may disappear through seepage of edematous fluid soon after this time.

The clinical signs of VS infection occur in cattle, horses, and swine but rarely in llamas. Signs follow a typical viral incubation period of 3–7 days with an initial febrile period followed by ptyalism in cattle and horses.[39] Cattle often can be heard smacking their lips or seen immersing their mouths in water troughs without drinking. Lesions of the oral mucosa include raised, blanched, and rarely fluid-filled vesicles. The dorsal lingual surface often is affected, but the gingival surfaces, palate, and mucocutaneous junctions may also exhibit lesions.[52] Vesicles are very short-lived and rupture leaving ulcerations and erosions. Lesions often coalesce to form large denuded areas of oral mucosa with the presence of epithelial tags.

Vesicular and/or ulcerative lesions outside of the oral mucosa occur on the snout of pigs, teats of cattle, and coronary bands of pigs, cattle, and horses. Teat lesions are not as common as oral lesions but in cattle may be associated with severe mastitis. Milk production may drop dramatically due to both the inability of the animals to eat and mastitis. Lesions of the feet typically manifest as a coronitis with edema and inflammation extending from the coronary band proximally up the lower leg. Hoof lesions in swine often result in sloughing of the hoof wall. A very common lesion observed by diagnosticians during VS outbreaks in the United States yet rarely described in the literature is crusting or scabbing lesions of the muzzle, ventral abdominal wall, prepuce, and udder of affected horses. These lesions typically start as discrete, small (approximately 1 cm) erosions that quickly coalesce so that large crusted or scabbed areas are observed. These lesions are often associated with feeding sites of hematophagous arthropods. VSVs have been isolated from samples collected from these crusting or scabbing lesions.

Subclinical infections are common in livestock during outbreaks of VS. One study reported disease prevalence of 44.7% in horses on 17 premises but a seroprevalence of 61.0%. Only 4.5% of cattle on the 17 premises investigated showed clinical signs, while 67.6% were seropositive to VSV-NJ. In another study, also conducted during the 1982 outbreak in the southwestern United States, disease prevalence ranged from 0% to 30%, but the percent seropositive ranged from 14% to 100%.[70]

64.5 IDENTIFICATION AND DIAGNOSIS

64.5.1 CURRENT LABORATORY TESTS

Detection of antibodies to VSV can be accomplished through application of the serum neutralization test (SNT), complement fixation test (CFT), and enzyme-linked immunosorbent assays (ELISA). The SNT has been considered the standard serologic test for VS antibodies for many years. The World Organisation for Animal Health (OIE) recognizes the neutralization test as a prescribed test for international trade.[58] Samples with detectable antibody greater than 1:40 are considered positive for international trade purposes.[2] The CFT also is recognized as a prescribed test for international trade purposes, and samples with titers greater than 1:5 are considered positive. Most recently, the competitive enzyme-linked immunosorbent assay (cELISA) has become the serologic test of choice for screening purposes during outbreaks of VS in the United States. The cELISA is considered a prescribed test for international trade by the OIE with a sample considered positive if the absorbance is greater than or equal to 50% of the absorbance of the diluent control. An ELISA (mcELISA) capable of detecting the IgM class of antibody to VSV was developed following the 1982 outbreak of VS in the southwestern United States.[68] The assay was shown to be capable of identifying recent exposure to VSV. This assay is not a prescribed test for international trade as determined by the OIE.

A comparison of these four tests (SNT, CFT, cELISA, mcELISA) by examination of experimentally inoculated animals indicated that the cELISA performed comparably to the SNT. The relative sensitivity and specificity of the cELISA to the SNT was 88% and 99% respectively. Positive mcELISA and cELISA responses consistently appeared 1–2 days prior to CFT seroconversion, but all animals reverted to mcELISA negative status by 49 days post-exposure. The current serologic diagnostic testing scheme employed during outbreaks in the United States is to screen samples with the cELISA for antibodies to either serotype of virus with those considered positive on these assays further tested by the SNT and CFT.

For virus isolation, vesicular fluid, epithelial tags, or swabs from fresh lesions are the ideal diagnostic sample. VSVs are propagated easily in cell culture as evident by their use in a variety of basic virologic studies. Samples can be inoculated onto an assortment of cell types including Vero, BHK-1, and IB-RS-2 cell cultures. VSVs will cause cytopathic effects in all three cell types.[58] Fluorescent antibody staining using conjugates specific for VSV-NJ and VSV-IN can be employed for serotype differentiation.

The detection of genomic sequences of VSV may be used to identify the presence of virus in tissue or swab samples. Many polymerase chain reaction (PCR) assays to detect various genomic sequences of VSV have been developed but are used primarily at this time for research purposes.[33,34,41,49,55]

64.5.2 DIFFERENTIAL DIAGNOSIS

In cattle, the differential diagnoses include foot and mouth disease (FMD), foot rot, and chemical or thermal burns. The oral lesions can also be similar to those of infectious bovine rhinopneumonitis, bovine viral diarrhea, malignant catarrhal fever, rinderpest, and epizootic hemorrhagic disease. In pigs, FMD, swine vesicular disease (SVD), vesicular exanthema of swine (VES), foot rot, and chemical and thermal burns should be considered. Bluetongue, contagious ecthyma, lip and leg ulceration, and foot rot are among the differentials in sheep. Toxic and mechanical causes of ulcers and erosions should be considered in horses.[43]

64.6 PREVENTION

The concept of using an unmodified viable VSV as a vaccine was proposed first in the 1920s.[50] Cattle inoculated intramuscularly with VSV did not develop lesions and when challenged with virus locally, were resistant. A field trial of attenuated virus vaccine was conducted in a large dairy herd in Panama. Cattle were vaccinated over a three-year period, 1962–1964, and then serum antibody titers as well as prevalence of disease were followed until 1966.[40] Protection for up to 4 years after vaccination was reported.

A special license was obtained from the United States Department of Agriculture in 1967 to produce and sell an attenuated VSV-NJ lyophilized vaccine, although sale of this vaccine in the United States was discontinued in 1972.[40] This same vaccine, however, was used in Guatemala for many years with reported success.

A commercially available inactivated VSV-NJ vaccine was used in Colorado during the 1985 outbreak, but serologic data regarding its immunogenicity and efficacy were not available. A field trial to examine the humoral responses to this vaccine was conducted in a 350 cow dairy.[27] Two doses of this formalin-inactivated cell culture-derived vaccine were administered intramuscularly 30 days apart to lactating and nonlactating adult cattle in this herd. Geometric mean titers peaked at 1:530 by 21 days after the second vaccination and declined to a geometric mean titer of 1:65 by 175 days postvaccination. The lack of detectable antibody in the control group of cattle indicated that exposure to wild-type virus did not occur and therefore the efficacy of the vaccine could not be ascertained. A similar study was conducted by the author during the 1995 VSV outbreak. Three commercial dairies approved to use an autogenous inactivated virus vaccine produced from a 1995 isolate of virus were enlisted in a field trial. Serum samples were collected from all cattle in the study prior to vaccination, and all were determined to be free of antibodies to both VSV-NJ and VSV-IN. Two doses of vaccine were administered 14 days apart, and all cattle in the study subsequently were bled three times. On all operations, all vaccinated cattle generated serum neutralizing antibodies to VSV-NJ, but antibody titers waned quickly to low levels by 250 days postvaccination. There was no indication that wild-type virus infected livestock on these operations or on operations near them so that determining whether cattle were protected with the vaccine was not accomplished.

A DNA vaccine that expressed the glycoprotein gene of VSV-NJ was evaluated for neutralizing antibody responses in mice, calves, and horses.[10] The vaccine elicited antibody titers in individuals from all species, but the level of antibody required to afford protection to challenge was not determined.

During the 1990s outbreaks in the United States, gene expression of VSV-NJ was altered through the rearrangement of the viral genome. Translocation of the N gene attenuated the virus to increasing extents and reduced lethality in mice without reductions in the ability to generate protective immunity.[71] This may provide a novel approach to the development of an attenuated VSV vaccine that will not elicit clinical disease but may afford adequate protection.

64.7   CONCLUSION

VS is a viral disease with a wide range of susceptible host species. The disease, however, is limited in its geographical distribution to the Americas. The clinical manifestations and pathology do not differ broadly between susceptible species with limited morbidity and mortality. The etiological VSVs are members of the family Rhabdoviridae, which includes viruses that infect vertebrates, invertebrates, and several plant species. Two major strains of VSV are common to cause the infection. Insects are suggested to play a role in the transmission of VSV, but the mechanism is not clear. Experimental studies have shown that the virus can replicate in insects. Humans can become infected mainly through direct contacts with infected animals or the virus, but disease is considered mild and self-limiting.

Serological tests are usually performed for the detection of the infection; however, viral isolation or molecular techniques are available for confirmation of the viral presence and their strains. The differential diagnoses of the infection in cattle include FMD, foot rot, and chemical or thermal burns. The oral lesions can also be similar to those infections that can cause mucosal infections such as FMD. Toxic and mechanical causes of ulcers and erosions should be considered as part of the differential diagnosis in horses.

The disease is considered one of economic importance due to trade barriers as its clinical manifestations are similar to FMD.

REFERENCES

1. Alderink, F.J. 1984. Vesicular stomatitis epidemic in Colorado: Clinical observations and financial losses reported by Dairymen. *Prev. Vet. Med.* 3, 29–44.
2. Alvarado, J.F., Dolz, G., Herrero, M.V., McCluskey, B.J., Salman, M.D. 2002. Comparison of the serum neutralization test and a competitive enzyme-linked immunosorbent assay for the detection of antibodies to vesicular stomatitis virus New Jersey and vesicular stomatitis virus Indiana. *J. Vet. Diag. Invest.* 14, 204–242.
3. Astudillo, V.M., Estupinan, A.J., Rosenberg F.J., da Silva A.J.M., Dora, J.F.P., Urbina, M., Tamayo, H., Lora, J.Q., Morrero J.C. 1984. *Epidemiological Study of Vesicular Stomatitis in South America. Proceedings of the International. Conference of Vesicular Stomatitis*, Mexico City, Mexico, September 1984, pp. 23–83.
4. Atwill, E.R., Rodriguez, L.L., Hird, D.W. 1996. Environmental and host factors associated with seropositivity to New Jersey and Indiana vesicular stomatitis. *Prev. Vet. Med.* 15, 303–314.
5. Banerjee, A.K. 1987. The transcription complex of vesicular stomatitis virus. *Cell* 48, 363–364.
6. Barrera, J.C., Letchworth, G.J. 1996. Persistence of vesicular stomatitis virus New Jersey RNA in convalescent hamsters. *Virology* 219, 453–464.
7. Bridges, V.E., McCluskey, B.J., Salman, M.D., Hurd, H.S., Dick, J. 1997. Review of the 1995 vesicular stomatitis outbreak in the western United States. *J. Am. Vet. Med. Assoc.* 211, 556–560.
8. Brinson, F.J., Hagan, D.V., Comer, J.A. et al. 1992. Seasonal abundance of Lutzomyia shannoni (Diptera: Psychodidae) on Ossabaw Island, Georgia. *J. Med. Entomol.* 29, 178–182.
9. Calisher, C.H., Monath, T.P., Sabattini, M.S., Mitchell, C.J., Lazuick, J.S., Tesh, R.B., Cropp, C.B. 1987. A newly recognized vesiculovirus, Calchaqui virus, and subtypes of Melao and Maguari viruses from Argentina, with serologic evidence for infections of humans and horses. *Am. J. Trop. Med. Hyg.* 36, 114–119.
10. Cantlon, J.D., Gordy, P.W., Bowen, R.A. 2000. Immune responses in mice, cattle and horses to a DNA vaccine for vesicular stomatitis. *Vaccine* 18, 2368–2374.
11. Comer, J.A., Irby, W.S., Kavanaugh, D.M. 1994. Hosts of Lutzomyia shannoni (Diptera: Psychodidae) in relation to vesicular stomatitis virus on Ossabaw Island, Georgia, United States. *Med. Vet. Entomol.* 8, 325–330.
12. Comer, J.A., Kavanaugh, D.M., Stallknecht, D.E. et al. 1993. Effect of forest type on the distribution of Lutzomyia shannoni (Diptera: Psychodidae) and vesicular stomatitis virus on Ossabaw Island, Georgia. *J. Med. Entomol.* 30, 555–560.
13. Comer, J.A., Stallknecht, D.E., Nettles, V.F. 1995. Incompetence of domestic pigs as amplifying hosts of vesicular stomatitis virus for Lutzomyia shannoni (Diptera: Psychodidae). *J. Med. Entomol.* 32, 741–744.
14. Comer, J.A., Stallknecht, D.E., Nettles, V.F. 1995. Incompetence of white-tailed deer as amplifying hosts of vesicular stomatitis virus for Lutzomyia shannoni (Diptera: Psychodidae). *J. Med. Entomol.* 32, 738–740.
15. Cornish, T.E., Stallknecht, D.E., Brown, C.C., Seal, B.S., Howerth, E.W. 2001. Pathogenesis of experimental vesicular stomatitis virus (New Jersey serotype) infection in the deer mouse (*Peromyscus maniculatus*). *Vet. Pathol.* 38, 396–406.
16. Cotton, W.E. 1927. Vesicular stomatitis. *Vet. Med.* 22, 169–175.
17. Cupp, E.W., Mare, C.J., Cupp, M.S. et al. 1992. Biological transmission of vesicular stomatitis virus (New Jersey) by *Simulium vittatum* (Diptera: Simuliidae). *J. Med. Entomol.* 29, 137–140.

18. Donaldson, A.I. 1970. Bats as possible maintenance hosts for vesicular stomatitis virus. *Am. J. Epidemiol.* 92, 132–136.

19. Duarte, P.C., Morley, P.S., Traub-Dargatz, J.L. et al. 2008. Factors associated with vesicular stomatitis in animals in the western United States. *J. Am. Vet. Med. Assoc.* 232, 249–256.

20. Ellis, E.M., Kendall, H.E. 1964. The public health and economic effects of vesicular stomatitis in a herd of dairy cattle. *J. Am. Vet. Med. Assoc.* 144, 377–380.

21. Fenner, F. Bachman, P.A., Gibbs, E.P.J. et al. 1987. *Veterinary Virology.* Academic Press, Orlando, FL, p. 299.

22. Fernandez, A.A., Sondahl, M.S. 1985. Antigenic and immunogenic characterization of various strains of the Indiana serotype of vesicular stomatitis isolated in Brazil. Bol. Centro Panamericano Fiebre Aftosa. 51, 27-30.

23. Fields, B.N., Hawkins, K. 1967. Human infection with the virus of vesicular stomatitis during an epizootic. *N. Engl. J. Med.* 277, 989–994.

24. Fletcher, W.O., Stallknecht, D.E., Jenney, E.W. 1985. Serologic surveillance for vesicular stomatitis virus on Ossabaw Island, Georgia. *J. Wildl. Dis.* 21, 100–104.

25. Francy, D.B., Moore, C.G., Smith, G.C. et al. 1988. Epizootic vesicular stomatitis in Colorado, 1982: Isolation of virus from insects collected along the northern Colorado Rocky Mountain front range. *J. Med. Entomol.* 25, 343–347.

26. Fultz, P.N., Shadduck, J.A., Kang, C.Y., Streilein, J.W. 1982. Vesicular stomatitis virus can establish persistent infections in Syrian hamsters. *J. Gen. Virol.* 63, 493–497.

27. Gearhart, M.A., Webb, P.A., Knight, A.P., Salman, M.D., Smith, J.A., Erickson, G.A. 1987. Serum neutralizing antibody titers in dairy cattle administered an inactivated vesicular stomatitis virus vaccine. *J. Am. Vet. Med. Assoc.* 191, 819–822.

28. Goodger, W.J., Thurmond, M., Nehay, J., Mitchell, J., Smith, P. 1985. Economic impact of an epizootic of bovine vesicular stomatitis in California. *J. Am. Vet. Med. Assoc.* 186, 370–373.

29. Hansen, D.E., Thurmond, M.C., Thorburn, M. 1985. Factors associated with the spread of clinical vesicular stomatitis in California dairy cattle. *Am. J. Vet. Res.* 46, 789–795.

30. Hanson, R.P. 1952. The natural history of vesicular stomatitis. *Bacteriol. Rev.* 16, 179–204.

31. Hanson, R.P., Rasmussen, A.F., Brandly, C.A., Brown, J.W. 1950. Human infection with the virus of vesicular stomatitis. *J. Lab. Clin. Med.* 36, 754–758.

32. Hayek, A.M., McCluskey, B.J., Chavez, G.T., Salman, M.D. 1998. Financial impact of the 1995 outbreak of vesicular stomatitis on 16 beef ranches in Colorado. *J. Am. Vet. Med. Assoc.* 212, 820–823.

33. Hofner, M.C., Carpenter, W.C., Ferris, N.P. et al. 1994. A hemi-nested PCR assay for the detection and identification of vesicular stomatitis virus nucleic acid. *J. Virol. Methods* 50, 11–20.

34. Hole, K., Velazques-Salinas, L., Clavijo, A. 2010. Improvement and optimization of a multiplex real-time reverse transcription polymerase chain reaction assay for the detection and typing of vesicular stomatitis virus. *J. Vet. Diagn. Invest.* 22, 428–433.

35. http://www.aphis.usda.gov/vs/nahss/equine/vsv/vsv2009_final.htm, accessed on 12/22/2011.

36. http://www.aphis.usda.gov/vs/nahss/equine/vsv/vsv2010_final.htm, accessed on 12/22/2011.

37. Hurd, H.S., McCluskey, B.J., Mumford, E.L. 1999. Management factors affecting the risk for vesicular stomatitis in livestock operations in the western United States. *J. Am. Vet. Med. Assoc.* 215, 1263–1268.

38. Killmaster, L.F., Stallknecht, D.E., Howerth, E.W. et al. 2011. Apparent disappearance of vesicular stomatitis New Jersey virus from Ossabaw Island, Georgia. *Vector Borne Zoonotic Dis.* 11, 559–565.

39. Knight, A.P., Messer, N.T. 1983. Vesicular stomatitis. *Comp. Cont. Ed.* 5, 2–6.

40. Lauerman, L.H., Hanson, R.P. 1968. Live vesicular stomatitis virus vaccines. *Proc. US Livestock Sanitary Assoc.* 72, 591–597.

41. Magnuson, R., Triantis, J., Rodriguez, L. et al. 2003. A multiplex, single tube RT-PCR for detection and differentiation of vesicular stomatitis virus serotypes in culture supernatants and mosquitoes. *J. Vet. Diagn. Invest.* 15, 561–567.

42. McCluskey, B.J., Hurd, H.S., Mumford, E.L. 1999. Review of the 1997 outbreak of vesicular stomatitis in the western United States. *J. Am. Vet. Med. Assoc.* 215, 1259–1262.

43. Mead, D.G., Murphy, M.D., Howerth, E.W. et al. 2002. Biological transmission of vesicular stomatitis virus, New Jersey serotype, to domestic pigs by black flies. In *Proceedings of the American Association of Veterinary Laboratory Diagnosticians*, St. Louis, MO, p. 90.

44. Mead, D.G., Murphy, M.D., Rodriguez, L.L. et al. 2003. Biological transmission of vesicular stomatitis virus, New Jersey serotype, to horses by black flies. In *Proceedings of the American Association of Veterinary Laboratory Diagnosticians*, San Diego, CA.

45. Mead, D.G., Ramberg, F.B., Besselsen, D.G. et al. 2000. Transmission of vesicular stomatitis virus from infected to noninfected black flies co-feeding on nonviremic deer mice. *Science* 287, 485–487.

46. Mumford, E.L., McCluskey, B.J., Traub-Dargatz, J.L., Schmitt, B.J., Salman, M.D. 1988. Public veterinary medicine: Public health. Serologic evaluation of vesicular stomatitis virus exposure in horses and cattle in 1996. *J. Am. Vet. Med. Assoc.* 213, 1265–1269.

47. Nichol, S.T., Rowe, J.E., Fitch, W.M. 1993. Punctuated equilibrium and positive Darwinian evolution in vesicular stomatitis virus. *Evolution* 90, 10424–10428.

48. Nunez, J.I., Blanco, E., Hernandez, T. et al. 1998. A RT-PCR assay for the differential diagnosis of vesicular viral diseases of swine. *J. Virol. Methods* 72, 227–235.

49. Olitsky, P.K., Traum, J., Schoening, H.W. 1928. Report of foot and mouth disease. Commission of the U.S. Department of Agriculture, Washington, DC. USDA Technical Bulletins No. 76.

50. Patterson, W.C., Mott, L.O., Jenney, E.W. 1958. A study of vesicular stomatitis in man. *J. Am. Vet. Med. Assoc.* 19, 57–62.

51. Reif, J.S. 1994. Vesicular stomatitis. In: Beran, G.S. (ed.), *Handbook of Zoonoses*. CRC Press, Boca Raton, FL, pp. 171–179.

52. Reif, J.S., Webb, P.A., Monath, T.P., Emerson, J.K., Poland, J.D., Kemp, G.E., Cholas, G. 1987. Epizootic vesicular stomatitis in Colorado, 1982: Infection in occupational risk groups. *Am. J. Trop. Med. Hyg.* 36, 177–182.

53. Ribelin, W.E. 1958. The cytopathogenesis of vesicular stomatitis virus infection in cattle. *Am. J. Vet. Res.* 19, 66–73.

54. Rodriguez, L.L., Letchworth, G.J., Spiropoulou, C.F. et al. 1993. Rapid detection of vesicular stomatitis virus New Jersey serotype in clinical samples by using polymerase chain reaction. *J. Clin. Microbiol.* 31, 2016–2020.

55. Rodriguez, L.L., Fitch, W.M., Nichol, S.T. 1996. Ecological factors rather than temporal factors dominate the evolution of vesicular stomatitis virus. *Proc. Natl. Acad. Sci. USA* 93, 13030–13035.

56. Rodriguez, L.L., Vernon, S., Morales, A.I., Letchworth, G.J. 1990. Serological monitoring of vesicular stomatitis New Jersey virus in enzootic regions of Costa Rica. *Am. J. Trop. Med. Hyg.* 42, 272–281.

57. Schmitt, B. 2000. Vesicular stomatitis. In *OIE Manual of Standards for Diagnostic Test and Vaccines*, 4th ed. pp. 93–99. OIE, Paris, France.

58. Sellers, R.F., Maarouf, A.R. 1990. Trajectory analysis of winds and vesicular stomatitis in North America, 1982–1985. *Epidemiol. Infect.* 104, 313–328.

59. Smith, P.F., Howerth, E.W. Carter, D. et al. 2011. Domestic cattle as a non-conventional amplifying host of vesicular stomatitis New Jersey Virus. *Med. Vet. Entomol.* 25, 184–191.

60. Stallknecht, D.E., Nettles, V.F., Fletcher, W.O. et al. 1985. Enzootic vesicular stomatitis New Jersey type in an insular feral swine population. *Am. J. Epidemiol.* 122, 876–883.

61. Stallknecht, D.E. 2000. VSV-NJ on Ossabaw Island, Georgia: The truth is out there. *Ann. N.Y. Acad. Sci.* 916, 431–436.

62. Stallknecht, D.E., Perzak, D.E., Bauer, L.D., Murphy, M.D. Howerth, E.W. 2001. Contact transmission of vesicular stomatitis virus New Jersey in pigs. *Am. J. Vet. Res.* 62, 516–520.

63. Steinhauer, D.A., de la Torre, J.C., Holland, J.J. 1989. High nucleotide substitution error frequencies in clonal pools of vesicular stomatitis virus. *J. Virol.* 63, 2063–2071.

64. Tesh, R.B., Boshell, J., Modi, G.B., Morales, A., Young, D.G., Corredor, A., Ferro de Carrasquilla, C. et al. 1987. Natural infection of humans, animals, and phlebotomine sand flies with the Alagoas serotype of vesicular stomatitis virus in Colombia. *Am. J. Trop. Med. Hyg.* 36, 653–661.

65. Tesh, R.B., Peralta, P.H., Johnson, K.M. 1969. Ecologic studies of vesicular stomatitis virus. I. Prevalence of infection among animals and humans living in an area of endemic VSV activity. *Am. J. Epidemiol.* 90, 255–261.

66. Vanleeuwen, J.A., Rodriguez, L.L., Waltner-Toews, D. 1995. Cow, farm, and ecologic risk factors of clinical vesicular stomatitis on Costa Rican dairy farms. *Am. J. Trop. Med. Hyg.* 53, 342–350.

67. Vernon, S.D., Webb, P.A. 1985. Recent vesicular stomatitis virus infection detected by immunoglobulin M antibody capture enzyme-linked immunosorbent assay. *J. Clin. Microbiol.* 22, 582–586.

68. Wagner, R.R., Rose, J.K. 1996. Rhabdoviridae: The viruses and their replication. *Fields Virol.* 3, 1121–1135.

69. Walton, T.E., Webb, P.A., Kramer, W.L., Smith, G.C., Davis, T., Holbrook, F.R., Moore, C.G. et al. 1987. Epizootic vesicular stomatitis in Colorado, 1982: epidemiologic and entomologic studies. *Am. J. Trop. Med. Hyg.* 36, 166–176.

70. Wertz, G.W., Perepelitsa, V.P., Ball, L.A. 1998. Gene rearrangement attenuates expression and lethality of a nonsegmented negative strand RNA virus. *Proc. Natl. Acad. Sci. USA* 95, 3501–3506.

71. White, D.O., Fenner, F.J. 1994. Rhabdoviridae. In: White, D.O., Fenner, F.J. (eds.), *Medical Virology*, 4th ed. Academic Press, San Diego, CA, pp. 476–477.

72. Wilks, C.R., House, J.A. 1986. Susceptibility of various animals to the vesiculoviruses Isfahan and Chandipura. *J. Hyg. (Lond.)* 97, 359–368.

73. Wilson, W.C., Letchworth, G.J., Herrero, M.V. et al. 2009. Field evaluation of a multiplex real-time reverse-transcription polymerase chain reaction assay for detection of vesicular stomatitis virus. *J. Vet. Diagn. Inv.* 21, 179–186.

Section VI

Microbes Affecting Animals: Bacteria

65 *Ehrlichia ruminantium*

Nathalie Vachiéry, Isabel Marcelino, Dominique Martinez, and Thierry Lefrançois

CONTENTS

65.1 INTRODUCTION

Heartwater, caused by *Ehrlichia ruminantium,* is one of the most important tick-borne diseases of livestock in Africa. It is a major obstacle for the introduction of high-producing animals into Africa,[1] causing mortality rates up to 90%. For instance, in cattle, a mortality rate of 60% due to heartwater is common, and in merino sheep, it may rise up to 80%.[2] For the Southern African Development Community (Angola, Botswana, Malawi, Mozambique, South Africa, Swaziland, Tanzania, and Zimbabwe), the losses are estimated around 47.6 millions of dollars per year. These economical losses are mainly due to mortality, diminution of productivity in farming systems, and cost of treatment (use of antibiotics and acaricides).

Heartwater occurs in nearly all the sub-Saharan countries of Africa where *Amblyomma* spp. ticks are present and in the surrounding islands Madagascar, Reunion, Mauritius, Zanzibar, the Comoros Islands, and São Tomé. The disease is also reported in the Caribbean (Guadeloupe, Marie-Galante, and Antigua), from where it threatens the American mainland.[3–5] All domestic and wild ruminants can be infected, but the former appears to be the most susceptible. Indigenous domestic ruminants are usually more resistant to the disease. Ruminants remain the main target of the pathogen; however, human and canine cases potentially due to *E. ruminantium* have been reported in South Africa.[6,7]

The average natural incubation period is 2–3 weeks, but it can vary from 10 days to 1 month. In most cases, heartwater is an acute febrile disease, with a sudden rise in body temperature. The animal is restless, walks in circles, makes sucking movements, stands rigidly with tremors of the superficial muscles, and finally, the animal falls to the ground, pedaling before dying. The most common macroscopic lesions are hydropericardium, hydrothorax, and pulmonary oedema. A clinical diagnosis of heartwater is based on the presence of *Amblyomma* vectors, nervous signs, and presence of transudates in the pericardium and thorax on *postmortem* examination. Nowadays, efficient molecular diagnostics are available for a reliable heartwater diagnosis.

Four immunization strategies are currently available: the "infection and treatment" method and 3 experimental vaccines based on in vitro–attenuated[8] or chemically inactivated *E. ruminantium*[9] and recombinant DNA and/or proteins.[10,11] The "infection and treatment" method is based on the injection of blood infected with live Ball 3 organisms, followed by antibiotic treatment at the beginning of hyperthermia. Although it is the only commercially available vaccine, its use is limited to South Africa.[12] After recovery, the vaccinated animals have a long-lasting immunity against homologous strains. Nevertheless, this method has many drawbacks such as the requirement of a cold chain, the potential transmission of other pathogens, and cannot be used worldwide.[13] The attenuated vaccine is very effective against homologous and heterologous challenges but, as for the previous vaccine, has major concerns related to vaccine storage, and there is also a risk of reversion to virulence after injection in the animals. Moreover, few attenuated strains are currently available. The inactivated vaccine has major advantages including the availability of a complete industrial process for the production, purification, and formulation of large amounts of *E. ruminantium* at low cost.[14,15] Although this vaccine protects the vaccinated animals against death, it does not protect against the disease, leading to high levels of morbidity. This inactivated vaccine is also effective against homologous and some heterologous challenges and can include several strains of *E. ruminantium* to improve vaccine efficacy. Recombinant DNA-/protein-based vaccines are also being developed, but major improvements are still required. Globally, the major bottleneck in developing an effective vaccine against heartwater is related to the high antigenic differences between strains, even in restricted areas, therefore, hampering the finding of protective antigens.[16,17] Heartwater will therefore remain a disease of major economical importance in susceptible countries until a safer, more cost-effective, and better-defined live or subunit vaccine becomes available.

For this reason, additional research on heartwater disease is essential. Nowadays, global and integrative high-throughput approaches such as transcriptomics and proteomics are being used to increase the knowledge on *E. ruminantium* pathogenesis. Molecular epidemiological studies are also being performed worldwide to study *E. ruminantium* diversity.

65.2 CLASSIFICATION AND MORPHOLOGY

Traditional rickettsial taxonomy assigned *Cowdria ruminantium* as the sole member of the genus *Cowdria* in the tribe *Ehrlichieae*. This was one of the three tribes within the family *Rickettsiaceae* in the order *Rickettsiales*, which initially encompassed all intracellular bacteria but from which

the *Chlamydiae* were later removed.[18] The obligate intracellular nature of *E. ruminantium*, coupled with morphological features suggestive of a Chlamydia-like life cycle, led to confusion as to its position in the Ehrlichial hierarchy.[1] In 2001, Dumler et al.[19] defined after *16S ribosomal DNA* and *groESL* gene comparisons that all members of the tribes *Ehrlichieae* and *Wolbachieae* had to be transferred to the family *Anaplasmataceae* and that the family *Rickettsiaceae* had to be eliminated. In *Anaplasmataceae*, four genera are present: *Anaplasma*, *Ehrlichia*, *Wolbachia*, and *Neorickettsia*. The genus *Ehrlichia* includes now *E. ruminantium* (formerly *Cowdria ruminantium*), *E. chaffeensis*, and *E. canis* and excludes *Anaplasma bovis*, *A. marginale*, *A. centrale*, and *A. phagocytophilum* (all belonging now to genus *Anaplasma*) and the genus *Neorickettsia* (including *Ehrlichia sennetsu* and *E. risticii*).

E. ruminantium is a small pleomorphic gram-negative obligate intracellular bacterium (0.2–2.5 μm). *E. ruminantium* colonies called morula are found in the cytoplasm of vascular endothelial cells of ruminants[20,21] and to a lesser extent in neutrophils.[22,23] Each growing colony is contained inside a vacuole-like structure without fusing with other vacuoles or with lysosomes. Inside host cells such as bovine endothelial cells, *E. ruminantium* colonies are sometimes arranged in grapefruit and close to the nucleus inside the host endothelial cell (Figure 65.1a).

65.3 BIOLOGY AND EPIDEMIOLOGY

65.3.1 *EHRLICHIA RUMINANTIUM* LIFE CYCLE

Characterization of *E. ruminantium* in host cells was possible in 1985, when the first in vitro cultivation of the organism in a calf endothelial cell line was described.[24] Many endothelial cell lines from cattle, sheep, goats, wild African mammals,[25] human,[26] and murine origins as well as myeloid and monocytic lineages have been examined to improve and standardize the in vitro cultivation.[27] Several studies have shown that *E. ruminantium* is able to infect tick cell lines[28,29] as well as cells from nonendothelial origin such as Chinese hamster ovary cells (CHO) and baby hamster kidney cells (BHK).[30,31]

In host cells, *E. ruminantium* has developmental stages similar to *Chlamydia* species.[32] The deduced biphasic developmental cycle is characterized by two morphologically distinct forms, the elementary bodies (EB) and the reticulate bodies (RBs) (Figure 65.1a and b).[32] EBs are small (0.2–0.5 μm in diameter) extracellular infectious forms of the bacterium, and after cell colonization, they reside within intracytoplasmic inclusions (morula) where they convert into the larger (0.75–2.5 μm) intracellular noninfectious, metabolically active RBs.[33,34] The RBs multiply by binary fission,[33] rapidly filling the inclusion, which expands in size. RBs recondense back into EBs toward the end of the cycle and are then released from the host cell (Figure 65.2). Transmission electron microscope studies of in vitro-cultivated organisms demonstrated the presence of intracellular RBs 2–4 days after infection and intermediate bodies 4–5 days after infection.

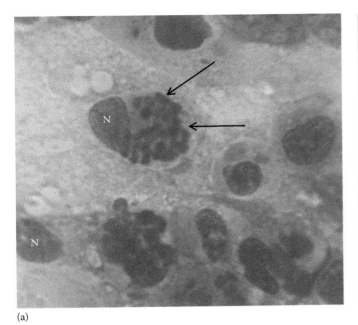

(a)

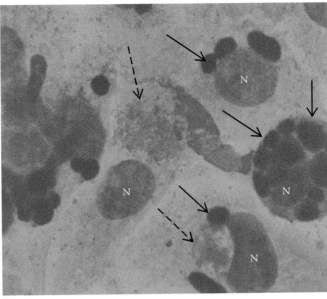

(b)

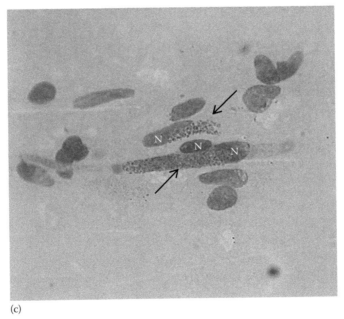

(c)

FIGURE 65.1 *Ehrlichia ruminantium* organisms in endothelial cells. (a) Banan 112 strain morula (full arrow) in bovine aortic endothelial cells (magnification: 1000×). (b) Gardel strain morula (full arrow) and elementary bodies (EBs) (dashed arrow) in bovine aortic endothelial cells. (c) Gardel strain morula in brain capillary smear (magnification: 400×). "N" stands for host cell nucleus.

Large numbers of EBs are observed after rupture of endothelial cells 5–6 days after infection.[32] The relation between the stage of development and time postinfection depends on the strain and its adaptation to *in vitro* conditions. For instance, the virulent *E. ruminantium* Gardel strain (isolated in Guadeloupe, FWI) lyses 5 days postinfection, whereas the attenuated phenotype lyses after 4 days.

E. ruminantium development cycle and its infectivity within the tick are poorly understood. It is thought that after an infected blood meal, initial replication of the organism takes place in the intestinal epithelium of the tick and that the salivary glands eventually become parasitized

(Figure 65.2).[35] Transmission of the parasite to the vertebrate host probably takes place either by regurgitation of their gut contents or through the saliva of the tick while feeding. The minimum period required for transmission of the parasite after tick attachment is between 27 and 38 h in nymphs and 21 and 75 h in adults.[36]

65.3.2 GEOGRAPHIC DISTRIBUTION

Heartwater is widespread in sub-Saharan Africa. Countries like Lesotho, Somalia, southern Angola, Botswana, Namibia, and western and south-central South Africa have

Amblyomma variegatum

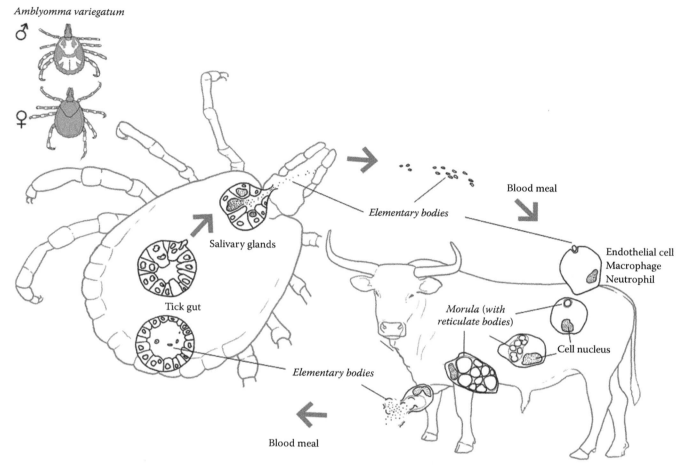

FIGURE 65.2 *Ehrlichia ruminantium* life cycle. *E. ruminantium* organisms initially develop in the gut epithelial cells of ticks and subsequently invade and develop in the salivary gland cells of the vector. The vertebrate host is infected via salivary glands of the tick during blood meal. In the vertebrate host, *E. ruminantium* proliferate in vascular endothelial cells, neutrophils, and macrophages, presenting a biphasic developmental cycle with two morphologically distinct forms, the EB and the RB. Organisms enter cells as elementary bodies through a process resembling phagocytosis and divide by binary fission within intracytoplasmic vacuoles resulting in large colonies of RB (morula). *Amblyomma* ticks are three-host ticks and become infected during the larval and nymphal stages when they feed on infected hosts. Nymphal and adult ticks transmit *E. ruminantium* to susceptible hosts without losing their infective condition. (From Marcelino, I. et al. *J Proteomics.* 75(14):4232-50. 2012.)

not been threatened by heartwater since their climate is unsuitable for *Amblyomma* ticks.[21] Elsewhere, heartwater is present in the Comoros Islands, Mayotte, Madagascar, the Mascarenes, Reunion, and Mauritius, where the major vector, *A. variegatum*, is established. *A. variegatum* is also present in the Caribbean islands, and heartwater is endemic in Guadeloupe, Antigua, and Marie-Galante. The improvement of molecular diagnostic allows confirming the presence of *E. ruminantium* in different countries but also permit to study *E. ruminantium* prevalence in ticks in Burkina Faso,[37] the Gambia,[38] Nigeria (personal communications, Dr. Maxwell Opara), Uganda, Guadeloupe, and Antigua.[3]

65.3.3 Species Affected

Small ruminants, goat and sheep, are more susceptible to heartwater than cattle. Moreover, the susceptibility to heartwater varies according to the different breeds of domestic ruminants. *Bos indicus* (zebu) breeds are generally more

resistant than European breeds.[1] A wide variety of wild ruminant species may become infected with *E. ruminantium*, but so far only the bleskbok, black wildebeest, helmeted guinea fowl, leopard tortoise, and scrub hare have been reported to develop clinical disease.[39–41] Knowledge of the susceptibility of wild ruminants to heartwater is particularly important where farmers wish to reintroduce ruminant game species into heartwater endemic areas. Wild ruminants also play a role as sources of infection for ticks, particularly in those areas where stringent tick control in domestic animals is practiced.[42] *E. ruminantium* can also infect ferrets and mice[39]; however, the pathogenicity of the different strains of *E. ruminantium* to mice varies significantly.

Cases of possible human and canine infection by *E. ruminantium* have been reported in South Africa[7,43] but only genetic typing was done with no confirmatory isolation and further genetic characterization. No other human cases have been described in the literature until now.

65.3.4 Transmission

E. ruminantium can be transmitted by at least 10 species of *Amblyomma* ticks: *A. variegatum* (the major vector in Africa and in the Caribbean), *A. hebraeum*, *A. pomposum*, *A. gemma*, *A. lepidum*, *A. tholloni*, *A. sparsum*, *A. astrion*, *A. cohaerens*, and *A. marmoreum*. Two North American species, *A. maculatum* and *A. cajennense*, can also transmit *E. ruminantium* in the laboratory experimental conditions, but neither has been implicated in natural infections. These ticks are three-host ticks and become infected during the larval and nymphal stages when they feed on infected domestic and wild ruminants. Nymphae and adult ticks transmit *E. ruminantium* to susceptible hosts without losing the infection. Intrastadial transmission has been demonstrated,[44] but transovarial transmission has only been reported once under laboratory conditions and probably does not occur in nature.[45]

65.4 CLINICAL FEATURES AND PATHOGENESIS

65.4.1 Clinical Signs

The incubation period in natural infections is usually 2 weeks, but can be up to 1 month. The course of the disease may range from peracute to mild depending on age, immune status, breed, and virulence of *E. ruminantium* strain.[46] The clinical signs may include a sudden fever (up to 42°C), severe respiratory distress, hyperesthesia, lacrimation, terminal convulsions, and sudden death. Occasionally, animals also have diarrhea. Animals with the acute form of heartwater usually die within a week after the onset of the disease. The peracute form of heartwater is relatively rare. In typical cases, animals show nervous symptoms such as rapid blinking of the eyes, hypersensitivity to touch, and once recumbent, they exhibit pedaling movements.[2,46] In these conditions, the recovery is rare.

65.4.2 Pathology and Histopathology

At necropsy, the characteristic lesion of heartwater is hydropericardium (origin of the name "heartwater"), with straw-colored to reddish pericardial fluid, and it is more pronounced in sheep and goats than in cattle. Other common lesions include splenomegaly, pulmonary and mediastinal edema, hydrothorax, and ascites. Histopathological examination reveals variable numbers of *E. ruminantium* colonies discernible in the cytoplasm of endothelial cells (Figure 65.1), particularly those of the brain capillaries and lungs.[46–48] Transmission electron microscopical studies of the lung lesions in sheep and goats reveal the presence of minor cytopathic changes in endothelial cells. Apart from mild swelling of mitochondria and endoplasmic reticulum, no other changes occur in most parasitized alveolar endothelial cells.

65.4.3 Pathogenesis

The pathogenesis of the disease is still poorly understood, but the following hypotheses have been proposed. After infection of the host with *E. ruminantium*, initial replication of the organisms appears to take place in reticulo-endothelial cells and macrophages in the regional lymph nodes. From here, the organisms are disseminated via the blood stream to invade endothelial cells of blood vessels of various organs where further multiplication occurs.[49] The rickettsemia coincides with the onset of fever. There is an increased vascular permeability allowing the seepage of plasma proteins, which results in transudation through the serous membranes with resultant tissue oedema[50] and effusion into body cavities. Brain oedema is responsible for the nervous signs. Hydropericardium contributes to cardiac dysfunction during the terminal stages of the disease, and progressive pulmonary edema and hydrothorax result in asphyxiation.[51,52] The pathogenesis of vascular permeability remains speculative as the intracytoplasmic development of the organisms seems to have little detectable cytopathic effect upon the endothelial cells,[53] and there is also no apparent correlation between the number of parasitized cells in the pulmonary blood vessels and the severity of the pulmonary edema.[54] It has been proposed that an endotoxin[55] and increased cerebrospinal fluid pressure[50] play a role in the development of lung edema.

65.5 IDENTIFICATION AND DIAGNOSIS

The traditional methods of diagnosis of heartwater include recognition of clinical signs (described earlier) and identification of the pathogen through microscopic examination of brain smears or bacterium isolation from infected blood or tick homogenates. Serological assays such as indirect fluorescent antibody (IFA) test, enzyme-linked immunosorbent assays (ELISA), and immunoblotting (Western blotting) were developed but had poor sensitivity and specificity. The *E. ruminantium* diagnostic is now mainly based on molecular diagnostics using nested and real-time polymerase chain reaction (PCR) methods.

65.5.1 Identification of the Agent

Classical method: Brain smear staining: The confirmation of a diagnosis based on clinical signs and postmortem lesions requires the demonstration of the organisms in the cytoplasm of endothelial cells of blood vessels. Typical colonies of *E. ruminantium* can be observed in brain smears made after death. Brain smears are air dried, fixed with methanol, and stained with eosin and blue methylene (RAL555). *E. ruminantium* occurs as clumps of reddish-purple to blue, coccoid to pleomorphic organisms in the cytoplasm of capillary endothelial cells close to the nucleus (Figure 65.1c). Colonies can be difficult to find in samples from animals treated with antibiotics or from animals with peracute disease or depending on the *E. ruminantium* strain. Colonies are still visible 2 days after death in a brain sample stored at room temperature (20°C–25°C) and up to 34 days in a brain sample stored in a refrigerator at 4°C. Technical expertise is required to differentiate *E. ruminantium* colonies from other hemoparasites (*Babesia bovis*), certain blood cells (thrombocytes, granulocytes), normal subcellular structures (mitochondria, mast cell granules), or stain artifacts (stain precipitates).

Isolation of Ehrlichia ruminantium: The isolation of *E. ruminantium* strains is possible by intravenously inoculation of infected blood from reacting animals or tick homogenate into a susceptible animal or to engorged field ticks on susceptible animals. During hyperthermia, blood is collected and incubated in vitro on endothelial cells from umbilical cord, aorta, or the pulmonary artery of different ruminant species (cattle, goat, sheep). *E. ruminantium* isolation should be encouraged as it would allow the phenotypic/genetic characterization of strains present in a region and thereafter improve vaccination programs.

65.5.2 SEROLOGICAL TESTS

Two serological diagnostic tests based on the detection of antibodies against a major antigenic *E. ruminantium* protein (MAP1) are available: a competitive ELISA MAP1 and an indirect ELISA using a fraction of MAP1 protein, MAP1-B.[56,57] These assays display cross-reaction with other *Ehrlichia* species specifically with *E. chaffeensis* and *E. muris*. The indirect MAP1-B ELISA is used routinely at the regional OIE reference laboratory. ELISAs against heartwater are suitable for prevalence studies at herd level but cannot be used either for diagnostic purposes on clinical cases or to evaluate the epidemiological status of imported animals. Indeed, there is a 15 day delay in seroconversion after animal infection, and the seropositivity period lasts several weeks for bovines and less than 6 months for small ruminants.

65.5.3 MOLECULAR DIAGNOSIS

Molecular diagnostics are the gold methods for the diagnosis of heartwater. For the past 10 years, there was an important improvement in the development of molecular tools for the diagnostic of heartwater and the genetic typing of the different strains of *E. ruminantium*.

65.5.3.1 PCR and Nested PCR

The molecular method consisting in PCR amplification of a *pCS20* fragment gene specific of *E. ruminantium* followed by membrane hybridization was first developed.[58] Low levels of infection in animals and in ticks that fed on carrier animals are detected by PCR, while a hybridization reaction with the *pCS20* probe alone (without PCR first) usually remains negative.[59] Experimentally, the detection limit of the conventional PCR assay was found to be between 10 and 100 organisms per sample, whereas it was between 1 and 10 organisms after PCR/hybridization. Thirty seven strains from all endemic areas were detected by PCR/hybridization with a high specificity (98%). However, the sensitivity of the PCR assay is variable depending on the nature of the sample and *E. ruminantium* load.[60]

Two nested PCR assays were then developed to enhance detection limits for samples with low levels of rickettsemia and herewith avoid the use of a hybridization step.[61,62] Both assays also use the *pCS20* region as the target sequence. The assay developed by Mahan and co-workers uses two external primers U24 and L24 for the primary reaction and then the primers AB 128 and AB 129 for the nested reaction.[62] The sensitivity of detection of this assay is one organism per sample. The nested PCR test developed by Martinez and co-workers[61] uses a pair of external primers comprising the AB128 sense primer together with an antisense primer called AB130. These amplify a 413 bp fragment used as a template in a second-round PCR using AB128 and AB129 as internal primers. The nested PCR shows an average detection limit of six organisms per sample. The *pCS20* nested PCR allowed regular detection of *E. ruminantium* organisms from ticks, blood, brain, and lungs from infected animals, whether the samples are processed fresh, after freezing, or preservation in 70% ethanol.

The diagnosis of heartwater using molecular diagnostic also showed to be much more sensitive than the one based on examination of brain smears from dead ruminants. Indeed, when brain smear samples were analyzed using these two methodologies, a substantial improvement in the detection of positive animals for heartwater was obtained, increasing the detection level from 75% (brain smear observation) to 97% using *pCS20* nested PCR.[63] The range of strain detection was increased by the use of primers including universal nucleotides AB128′ AB130′, and AB129′, and this method is used routinely for *E. ruminantium* detection in field samples, especially in ticks.[4,63] The detection of *E. ruminantium* by nested PCR is possible in the blood of animals 1 or 2 days before hyperthermia and during the hyperthermia period but not on asymptomatic animals. PCR-based methods appear to be more reliable in detecting infection in ticks, and this could have epidemiological value in determining the *E. ruminantium* prevalence in ticks and the geographical distribution of *E. ruminantium*.

A nested PCR targeting the entire *map1* polymorphic gene was developed in parallel in order to type the strains by restriction fragment length polymorphism (RFLP) or sequencing of the amplification fragment directly from the *pCS20* positive samples.[61] Its detection limit was evaluated at around 60 organisms per sample. This method cannot be used for diagnosis due to polymorphic characteristic of the *map1* targeted gene.

In general, the major drawback of the nested PCRs is the possibility to get cross contaminations due to the additional round of PCR. To limit this issue, technical cares must be taken.

65.5.3.2 Quantitative Real-Time PCR

Several quantitative real-time PCRs have been developed for the detection of *E. ruminantium* targeting *map1*, *map1-1*, and *pCS20*.[64–66] These methods have been described for the detection and quantitative determination of *E. ruminantium* organisms either for its growth kinetics in the blood of experimentally infected sheep (during the hyperthermia reaction period[65]) or in vitro.[14,15,34,64,67] Real-time PCR targeting *map1* and *map1-1* polymorphic genes was tested only on six strains and therefore cannot be used for diagnostic. Another real-time PCR assay targeting *pCS20* region has a sensitivity level similar to the nested PCR, but it was tested so far only on 15 different strains; screening on additional strains should be performed before using it as a diagnostic tool. The use of real-time PCR for heartwater diagnosis will be also technically advantageous, since only one round of PCR per sample is required, limiting the cross-contamination issues observed for the nested PCR.

65.5.3.3 Typing Methods

The genetic characterization and structure of *E. ruminantium* population at regional scale is essential in order to select potential vaccinal strains. The genetic typing of strains was previously done using RFLP on the polymorphic gene *map1* after PCR amplification.[63,68] Based on the genome analysis of two different strains Gardel and Welgevonden, truncated and unique coding sequences specific of strains have been identified. This analysis allows the development of a differential strain-specific diagnosis using nested PCRs targeting six unique and four truncated coding DNA sequences (CDS).[69] New multilocus methods such as multilocus sequence typing (MLST)[37,63] and multilocus variable number of tandem repeated sequence analysis (MLVA)[70] were recently adapted and validated for *E. ruminantium*. Two studies on restricted areas in Burkina Faso demonstrated the presence of several different *E. ruminantium* clusters and identified one strain population in stasis and another in clonal expansion. New studies are currently done at larger scale using both MLST and MLVA methods.

65.6 TREATMENT AND PREVENTION

65.6.1 CHEMOTHERAPY

Infected animals can be treated with antibiotic (short- or long-acting tetracyclines according to recommended dosage) during the early febrile stages, and the treatment confers long-lasting immunity.[75] The main problem is the timing of treatment, as in general, the sick animals display visual symptoms when it is too late to treat.

Prevention in goats and cattle can be performed by tetracycline treatment at different times after ruminant introduction in pastures. The animals are treated simultaneously and indiscriminately with antibiotics whether a febrile reaction occurs or not. Antibiotic treatments are sufficient to protect them from contracting heartwater, while at the same time allowing them to develop a natural immunity.[71,72] The success of this method is dependent on all the animals becoming naturally infected with heartwater during the time that they are protected by the drug.

65.6.2 VECTOR CONTROL

Heartwater is usually introduced into free areas by infected animals, including subclinical carriers, or by ticks. Therefore, sustained, intensive tick control measures may, under certain conditions, succeed in preventing outbreaks of heartwater, even in endemic areas. The disease can be controlled successfully if all the animals on the farm are dipped in acaricide baths regularly throughout the year and if there are limited number of game and birds hosting ticks. Intensive dipping programs (high-frequency dipping) have a major drawback as it can induce tick resistance to the dipping compound. An integrated tick control strategy taking into account the recent data on the heterogeneous drop-off rhythm of *A. variegatum* nymphs has been proposed to reduce pasture infestation by adult ticks.[73] Modeling of *Amblyomma* population dynamics based on biotic and abiotic parameters is being develop.[74] These models will allow obtaining habitat suitability maps and therefore test different control strategies.

65.6.3 VACCINATION

Four immunization strategies against heartwater are currently available: the "infection and treatment" method based on blood infected with live *E. ruminantium* and three candidate vaccines based on in vitro attenuated or chemically inactivated in vitro grown bacteria and recombinant DNA/protein. The main problem for the development of an effective vaccine for worldwide application resides on the presence of numerous strains in the field with high genetic diversity, being necessary to choose the vaccinal strain genotype(s) according to the region.

65.6.3.1 "Infection and Treatment" Method

Field observations and experiments under laboratory conditions have shown that cattle, sheep, and goats are capable of developing a protective immunity against heartwater after surviving a virulent infection. In South Africa, this led to the development of an "infection and treatment" type of immunization where animals are injected with blood infected with fully virulent *E. ruminantium* organisms of the Ball 3 strain and are subsequently treated with tetracyclines to prevent disease.[75,76] Despite the low cross protection of the Ball 3 blood vaccine strain against other *E. ruminantium* strains and the fact that this is an expensive and dangerous methodology, it has been the only commercially available vaccine for over 50 years.[77] Whenever large numbers of commercial livestock are introduced to heartwater endemic regions, the prevention method described earlier is also used.

65.6.3.2 Live Attenuated Vaccine

In the early 1990s, an attenuated strain of *E. ruminantium* (Senegal strain) was prepared as a live vaccine by serial passage in vitro in endothelial cells.[78] This attenuated strain, while providing immunity to homologous challenge, was nevertheless not fully efficient to provide cross-protection against other virulent strains.[79] Another *E. ruminantium* strain from Guadeloupe (Gardel strain) has also been attenuated in vitro, after 200 passages. This strain also proved to provide a good protection against heterologous challenge with other strains.[80] Zweygarth and co-workers successfully attenuated the virulent Welgevonden strain of *E. ruminantium* by 50 continuous passages in a canine macrophage–monocyte cell line.[30] The use of live attenuated vaccines is nevertheless limited since cross-protection against different isolates is not complete. The main disadvantage of attenuated vaccines is the possible reversion to virulence and the need for storage in liquid nitrogen until used.

65.6.3.3 Inactivated Vaccine

Martinez and co-workers developed an inactivated vaccine against heartwater, based on bovine endothelial cell culture-derived *E. ruminantium* organisms that are chemically inactivated or lysed.[9,81,82] Nowadays, the development of a fully scalable bioprocess for the production[14] and purification[67] of *E. ruminantium* and the optimization of antigen storage conditions (−20°C or refrigerated)[15] allowed to produce large number of vaccine doses for widespread application. Recently, Marcelino and co-workers developed a ready-to-use vaccine that could be easily used in the field and even withstand up to 3–4 days at 37°C before injection.[83] As soon as regional isolates are available in culture after isolation, it could be possible to produce an inactivated vaccine including a cocktail of regional strains. The main difficulty is to choose the strains, which could protect against other circulating strains. The choice will depend on genetic characteristics and markers, which are not yet defined. The main inconvenient of the inactivated vaccines is the observation of animal morbidity during challenge.

65.6.3.4 Recombinant Vaccine

Besides the increased safety and reduced price, the use of recombinant vaccines could permit the correct presentation of the antigen after endogenous processing leading to a long-lasting immunity. To develop such vaccine, it is nonetheless necessary to identify *E. ruminantium* antigens that would induce a protective immune response. The *map1* gene was cloned and tested as a naked-DNA vaccine in a mouse model system.[84] Other genes such as groE operon (*groES* and *groEL*)[85] and *cpg 1*[86] have also been cloned and tested as a recombinant DNA vaccine to protect animals against death due to heartwater infection. Subunit vaccines using denatured *E. ruminantium* have also been tested, although no protection was acquired.[87] The use of four open reading frames (ORFs) DNA/recombinant proteins prime/boost resulted in a 100% survival rate after homologous needle challenge but lower survival rates (20% of protection) during field tick challenge.[88] Recently, a prime/boost vaccination trial using the polymorphic *cpg1* gene and the recombinant protein also resulted in complete protection of vaccinated animals after homologous challenge; no trials with heterologous strains have yet been performed.[89] Due to the polymorphic property of *cpg1*, a cocktail of representative *CpG1* from different strains should be included in the vaccine before any field trial. Although recombinant vaccines look promising under experimental conditions, results during field trials have been less successful. Moreover, simple intramuscular immunization is not enough to induce protection, and the use of a gene gun is necessary for prime DNA injection, which is not suitable for large vaccination campaign.

65.7 "OMICS" APPROACHES FOR IMPROVED UNDERSTANDING OF *E. RUMINANTIUM* PATHOGENESIS

Global "omics" approaches (genomics, proteomics, transcriptomics, and metabolomics) in a systems biology context are becoming key tools to increase knowledge on the biology of infectious diseases, specially to improve knowledge of the complex host–vector–pathogen interactions.[90] The sequencing of three *E. ruminantium* phenotypes[91–93] paves the way of using "omics" approaches for this pathogen.

In 2010, Emboule and co-workers optimized the selective capture of transcribed sequences (SCOTS) methodology to successfully capture *E. ruminantium* mRNAs, avoiding the contaminants of host cell origin and eliminating rRNA, which accounts for 80% of total RNA encountered.[94] This method is essential to perform transcriptomic studies on the intracellular form of the bacterium (RB) avoiding host cell contaminants. Pruneau and colleagues[95] have recently determined the genome-wide transcriptional profile of *E. ruminantium* replicating inside bovine aortic endothelial cells using cDNA microarrays. Interestingly, over 50 genes were found to have differential expression levels between RBs and EBs. A high number of genes involved in metabolism, nutrient exchange, and defense mechanisms, including those involved in resistance to oxidative stress, were significantly induced in RBs, indicating an active metabolism of *E. ruminantium* inside host cells (for bacterial growth inside vacuoles) and the need to protect themselves against host cell defense mechanisms. Finally, the authors demonstrate that the transcription factor *dksA,* known to induce virulence in other microorganisms, is overexpressed in the infectious form of *E. ruminantium.*

Marcelino and co-workers used a two dimensional electrophoresis (2-DE) analysis coupled to mass spectrometry (MALDI-TOF–TOF) to establish the first 2DE proteome map of *E. ruminantium* cultivated in endothelial cells.[96] Interestingly, among the 64 spots identified, only four proteins that belong to the MAP1 family were identified; the other proteins detected were mainly related to energy, amino acid, and general metabolism (26%); to protein turnover, chaperones, and survival (21%); and to information processes (14%) or classified as hypothetical proteins (23%). Interestingly, 25% of the detected proteins were found to be isoforms suggesting that posttranslational modifications might be important in EB for the regulation of cellular processes such as host cell recognition, signaling, and metabolism and in determining antigenicity as previously observed in other *Rickettsiales.*

65.8 CONCLUSIONS AND FUTURE PERSPECTIVES

The vaccination strategies developed so far have proven not to be fully effective due to genetic and antigenic diversity of *E. ruminantium*. At the moment, the experimental inactivated vaccine is the most suitable for large-scale application, with an optimized industrial process available and its ability to include several strains within the vaccine to design an appropriate regional vaccine. To improve the vaccine efficacy, it will be necessary to isolate in vitro several strains from each susceptible geographic region to study their ability for protection; genotyping of protective strains will be also crucial to identify genetic markers linked to clusters of protection. More globally, it is essential to perform molecular epidemiology studies to evaluate the structuration of strains in order to design regional vaccines.

On the other hand, further studies are required to better understand *E. ruminantium* pathogenesis in order to identify protective antigens and elaborate next-generation vaccines. New breakthroughs in vaccine research are increasingly reliant on novel "omics" approaches such as genomics, proteomics, transcriptomics, and other less known "omics" such as metabolomics, immunomics, and vaccinomics.[97] These "omics" approaches will deepen our understanding on the following: (1) *E. ruminantium* pathogenesis and attenuation mechanisms, (2) *E. ruminantium* host subversion mechanisms, and (3) the key biological processes leading to protective immunity. These high-throughput technologies will also significantly contribute to overcome knowledge gaps on the role of key parasite molecules involved in cell invasion, adhesion, and tick transmission and surely revolutionize the capacity for discovering potential candidate vaccines, such as proteins involved in protective immune response, tick feeding, or parasite development. These researches will contribute to find new treatments or next-generation vaccines.

REFERENCES

1. Uilenberg, G. Heartwater (*Cowdria ruminantium* infection): Current status. *Adv Vet Sci Comp Med* 27, 427–480 (1983).
2. OIE. Heartwater. Institute for International Cooperation in Animal Biologies (OIE) collaborating center, College of Veterinary Medicine (Iowa State University, Ames, IA, 2005).
3. Vachiery, N. et al. *Amblyomma variegatum* ticks and heartwater on three Caribbean Islands. *Ann N Y Acad Sci* 1149, 191–195 (2008).
4. Molia, S. et al. *Amblyomma variegatum* in cattle in Marie Galante, French Antilles: Prevalence, control measures, and infection by *Ehrlichia ruminantium*. *Vet Parasitol* 153, 338–346 (2008).
5. Kasari, T.R. et al. Recognition of the threat of *Ehrlichia ruminantium* infection in domestic and wild ruminants in the continental United States. *J Am Vet Med Assoc* 237, 520–530 (2010).
6. Allsopp, M.T. and Allsopp, B.A. Novel *Ehrlichia* genotype detected in dogs in South Africa. *J Clin Microbiol* 39, 4204–4207 (2001).
7. Allsopp, M.T., Louw, M., and Meyer, E.C. *Ehrlichia ruminantium*: An emerging human pathogen? *Ann N Y Acad Sci* 1063, 358–360 (2005).
8. Zweygarth, E. et al. An attenuated *Ehrlichia ruminantium* (Welgevonden stock) vaccine protects small ruminants against virulent heartwater challenge. *Vaccine* 23, 1695–1702 (2005).
9. Martinez, D. et al. Protection of goats against heartwater acquired by immunisation with inactivated elementary bodies of *Cowdria ruminantium*. *Vet Immunol Immunopathol* 41, 153–163 (1994).
10. Simbi, B.H. et al. Evaluation of *E. ruminantium* genes in DBA/2 mice as potential DNA vaccine candidates for control of heartwater. *Ann N Y Acad Sci* 1078, 424–437 (2006).
11. Sebatjane, S.I. et al. *In vitro* and in vivo evaluation of five low molecular weight proteins of *Ehrlichia ruminantium* as potential vaccine components. *Vet Immunol Immunopathol* 137, 217–225 (2010).
12. Allsopp, B.A. Trends in the control of heartwater. *Onderstepoort J Vet Res* 76, 81–88 (2009).
13. Shkap, V. et al. Attenuated vaccines for tropical theileriosis, babesiosis and heartwater: The continuing necessity. *Trends Parasitol* 23, 420–426 (2007).
14. Marcelino, I. et al. Process development for the mass production of *Ehrlichia ruminantium*. *Vaccine* 24, 1716–1725 (2006).
15. Marcelino, I. et al. Effect of the purification process and the storage conditions on the efficacy of an inactivated vaccine against heartwater. *Vaccine* 25, 4903–4913 (2007).
16. Allsopp, M.T. and Allsopp, B.A. Extensive genetic recombination occurs in the field between different genotypes of *Ehrlichia ruminantium*. *Vet Microbiol* 124, 58–65 (2007).
17. Barbet, A.F., Byrom, B., and Mahan, S.M. Diversity of *Ehrlichia ruminantium* major antigenic protein 1–2 in field isolates and infected sheep. *Infect Immun* 77, 2304–2310 (2009).
18. Weiss, E. and Moulder, J. The rickettsias and chlamydias. In: Holt, J. (ed.) *Bergey's Manual of Systematic Bacteriology*, pp. 687–739 (Williams & Wilkins, Baltimore, MD, 1984).
19. Dumler, J.S. et al. Reorganization of genera in the families Rickettsiaceae and Anaplasmataceae in the order Rickettsiales: Unification of some species of *Ehrlichia* with *Anaplasma*, *Cowdria* with *Ehrlichia* and *Ehrlichia* with *Neorickettsia*, descriptions of six new species combinations and designation of *Ehrlichia equi* and "HGE agent" as subjective synonyms of *Ehrlichia phagocytophila*. *Int J Syst Evol Microbiol* 51, 2145–2165 (2001).
20. Prozesky, L., Bezuidenhout, J.D., and Paterson, C.L. Heartwater: An *in vitro* study of the ultrastructure of *Cowdria ruminantium*. *Onderstepoort J Vet Res* 53, 153–159 (1986).
21. Yunker, C.E. Heartwater in sheep and goats: A review. *Onderstepoort J Vet Res* 63, 159–170 (1996).
22. Logan, L.L. et al. The development of *Cowdria ruminantium* in neutrophils. *Onderstepoort J Vet Res* 54, 197–204 (1987).
23. Jongejan, F. and Thielemans, M.J. Identification of an immunodominant antigenically conserved 32-kilodalton protein from *Cowdria ruminantium*. *Infect Immun* 57, 3243–3246 (1989).
24. Bezuidenhout, J.D., Paterson, C.L., and Barnard, B.J. *In vitro* cultivation of *Cowdria ruminantium*. *Onderstepoort J Vet Res* 52, 113–120 (1985).
25. Smith, G.E. et al. Growth of *Cowdria ruminantium* in tissue culture endothelial cell lines from wild African mammals. *J Wildl Dis* 34, 297–304 (1998).
26. Totte, P. et al. Bovine and human endothelial cell growth on collagen microspheres and their infection with the rickettsia *Cowdria ruminantium*: Prospects for cells and vaccine production. *Rev Elev Med Vet Pays Trop* 46, 153–156 (1993).
27. Yunker, C.E. Current status of *in vitro* cultivation of Cowdria ruminantium. *Vet Parasitol* 57, 205–211 (1995).
28. Bell-Sakyi, L. et al. Growth of *Cowdria ruminantium*, the causative agent of heartwater, in a tick cell line. *J Clin Microbiol* 38, 1238–1240 (2000).
29. Bell-Sakyi, L. *Ehrlichia ruminantium* grows in cell lines from four ixodid tick genera. *J Comp Pathol* 130, 285–293 (2004).

30. Zweygarth, E. and Josemans, A.I. Continuous *in vitro* propagation of *Cowdria ruminantium* (Welgevonden stock) in a canine macrophage-monocyte cell line. *Onderstepoort J Vet Res* 68, 155–157 (2001).

31. Zweygarth, E. and Josemans, A.I. *In vitro* infection by *Ehrlichia ruminantium* of baby hamster kidney (BHK), Chinese hamster ovary (CHO-K1) and Madin Darby bovine kidney (MDBK) cells. *Onderstepoort J Vet Res* 70, 165–168 (2003).

32. Jongejan, F. et al. The tick-borne rickettsia *Cowdria ruminantium* has a *Chlamydia*-like developmental cycle. *Onderstepoort J Vet Res* 58, 227–237 (1991).

33. Prozesky, L. Heartwater. The morphology of *Cowdria ruminantium* and its staining characteristics in the vertebrate host and *in vitro*. *Onderstepoort J Vet Res* 54, 173–176 (1987).

34. Marcelino, I. et al. Characterization of *Ehrlichia ruminantium* replication and release kinetics in endothelial cell cultures. *Vet Microbiol* 110, 87–96 (2005).

35. Kocan, K.M. and Bezuidenhout, J.D. Morphology and development of *Cowdria ruminantium* in Amblyomma ticks. *Onderstepoort J Vet Res* 54, 177–182 (1987).

36. Bezuidenhout, J. *DVSc*. Sekere aspekte van hartwateroordraging, voorkoms van die organisme in bosluise en in vitro kweking. University of Pretoria, Pretoria, South Africa (1988).

37. Adakal, H. et al. MLST scheme of *Ehrlichia ruminantium*: Genomic stasis and recombination in strains from Burkina-Faso. *Infect Genet Evol* 9, 1320–1328 (2009).

38. Faburay, B. et al. Point seroprevalence survey of *Ehrlichia ruminantium* infection in small ruminants in The Gambia. *Clin Diagn Lab Immunol* 12, 508–512 (2005).

39. Oberem, P.T. and Bezuidenhout, J.D. Heartwater in hosts other than domestic ruminants. *Onderstepoort J Vet Res* 54, 271–275 (1987).

40. Kock, N.D. et al. Detection of *Cowdria ruminantium* in blood and bone marrow samples from clinically normal, free-ranging Zimbabwean wild ungulates. *J Clin Microbiol* 33, 2501–2504 (1995).

41. Peter, T.F. et al. Susceptibility and carrier status of impala, sable, and tsessebe for *Cowdria ruminantium* infection (heartwater). *J Parasitol* 85, 468–472 (1999).

42. Peter, T.F. et al. Cowdria ruminantium infection in ticks in the Kruger National Park. *Vet Rec* 145, 304–307 (1999).

43. Allsopp, M.T., Louw, M., and Meyer, E.C. *Ehrlichia ruminantium*—An emerging human pathogen. *S Afr Med J* 95, 541 (2005).

44. Andrew, H.R. and Norval, R.A. The role of males of the bont tick (*Amblyomma hebraeum*) in the transmission of *Cowdria ruminantium* (heartwater). *Vet Parasitol* 34, 15–23 (1989).

45. Bezuidenhout, J.D. and Jacobsz, C.J. Proof of transovarial transmission of *Cowdria ruminantium* by Amblyomma hebraeum. *Onderstepoort J Vet Res* 53, 31–34 (1986).

46. Van de Pypekamp, H.E. and Prozesky, L. Heartwater. An overview of the clinical signs, susceptibility and differential diagnoses of the disease in domestic ruminants. *Onderstepoort J Vet Res* 54, 263–266 (1987).

47. Prozesky, L. The pathology of heartwater. III. A review. *Onderstepoort J Vet Res* 54, 281–286 (1987).

48. Van Amstel, S.R. et al. The clinical pathology and pathophysiology of heartwater: A review. *Onderstepoort J Vet Res* 54, 287–290 (1987).

49. Du Plessis, J.L. Pathogenesis of heartwater. I. *Cowdria ruminantium* in the lymph nodes of domestic ruminants. *Onderstepoort J Vet Res* 37, 89–95 (1970).

50. Brown, C.C. and Skowronek, A.J. Histologic and immuno-chemical study of the pathogenesis of heartwater (*Cowdria ruminantium* infection) in goats and mice. *Am J Vet Res* 51, 1476–1480 (1990).

51. Uilenberg, G. Studies on cowdriosis in Madagascar. I. *Rev Elev Med Vet Pays Trop* 24, 239–249 (1971).

52. Owen, N.C. et al. Physiopathological features of heartwater in sheep. *J S Afr Vet Assoc* 44, 397–403 (1973).

53. Pienaar, J.G. Electron microscopy of *Cowdria (Rickettsia) ruminantium* (Cowdry, 1926) in the endothelial cells of the vertebrate host. *Onderstepoort J Vet Res* 37, 67–78 (1970).

54. Prozesky, L. and Du Plessis, J.L. The pathology of heartwater. II. A study of the lung lesions in sheep and goats infected with the Ball3 strain of *Cowdria ruminantium*. *Onderstepoort J Vet Res* 52, 81–85 (1985).

55. Amstel, S.V. et al. The presence of endotoxin activity in cases of experimentally-induced heartwater in sheep. *Onderstepoort J Vet Res* 55, 217–220 (1988).

56. van Vliet, A.H. et al. Use of a specific immunogenic region on the *Cowdria ruminantium* MAP1 protein in a serological assay. *J Clin Microbiol* 33, 2405–2410 (1995).

57. Katz, J.B. et al. Development and evaluation of a recombinant antigen, monoclonal antibody-based competitive ELISA for heartwater serodiagnosis. *J Vet Diagn Invest* 9, 130–135 (1997).

58. Mahan, S.M. et al. A cloned DNA probe for *Cowdria ruminantium* hybridizes with eight heartwater strains and detects infected sheep. *J Clin Microbiol* 30, 981–986 (1992).

59. Peter, T.F. et al. Development and evaluation of PCR assay for detection of low levels of *Cowdria ruminantium* infection in *Amblyomma* ticks not detected by DNA probe. *J Clin Microbiol* 33, 166–172 (1995).

60. Peter, T.F. et al. Detection of the agent of heartwater, *Cowdria ruminantium*, in *Amblyomma* ticks by PCR: Validation and application of the assay to field ticks. *J Clin Microbiol* 38, 1539–1544 (2000).

61. Martinez, D. et al. Nested PCR for detection and genotyping of *Ehrlichia ruminantium*: Use in genetic diversity analysis. *Ann N Y Acad Sci* 1026, 106–113 (2004).

62. Mahan, S.M., Simbi, B.H., and Burridge, M.J. The pCS20 PCR assay for *Ehrlichia ruminantium* does not cross-react with the novel deer ehrlichial agent found in white-tailed deer in the United States of America. *Onderstepoort J Vet Res* 71, 99–105 (2004).

63. Adakal, H. et al. Efficiency of inactivated vaccines against heartwater in Burkina Faso: Impact of *Ehrlichia ruminantium* genetic diversity. *Vaccine* 28, 4573–4580 (2010).

64. Peixoto, C.C. et al. Quantification of *Ehrlichia ruminantium* by real time PCR. *Vet Microbiol* 107, 273–278 (2005).

65. Postigo, M. et al. Kinetics of experimental infection of sheep with *Ehrlichia ruminantium* cultivated in tick and mammalian cell lines. *Exp Appl Acarol* 28, 187–193 (2002).

66. Steyn, H.C. et al. A quantitative real-time PCR assay for *Ehrlichia ruminantium* using pCS20. *Vet Microbiol* 131, 258–265 (2008).

67. Peixoto, C. et al. Purification by membrane technology of an intracellular *Ehrlichia ruminantium* candidate vaccine against heartwater. *Process Biochem* 42, 1084–1089 (2007).

68. Faburay, B. et al. Immunisation of sheep against heartwater in The Gambia using inactivated and attenuated *Ehrlichia ruminantium* vaccines. *Vaccine* 25, 7939–7947 (2007).

69. Vachiery, N. et al. Differential strain-specific diagnosis of the heartwater agent: *Ehrlichia ruminantium*. *Infect Genet Evol* 8, 459–466 (2008).

70. Pilet, H. et al. A new typing technique for the Rickettsiales *Ehrlichia ruminantium*: Multiple-locus variable number tandem repeat analysis. *J Microbiol Methods* 88, 205–211 (2012).

71. Gruss, B. A practical approach to the control of heartwater in the Angora goat and certain sheep breeds in the Eastern Cape coastal region. In: *Tick Biology and Control*, pp. 135–136 (Tick Research Unit, Rhodes University, Grahamstown, South Africa, 1981).

72. Purnell, R.E. Development of a prophylactic regime using Terramycin/LA to assist in the introduction of susceptible cattle into heartwater endemic areas of Africa. *Onderstepoort J Vet Res* 54, 509–512 (1987).

73. Stachurski, F. and Adakal, H. Exploiting the heterogeneous drop-off rhythm of *Amblyomma variegatum* nymphs to reduce pasture infestation by adult ticks. *Parasitology* 137, 1129–1137 (2010).

74. Porphyre, T. et al. How do temperature and humidity affect ticks population dynamics? A global sensitivity analysis of the tick *Amblyomma variegatum* climatic niche. *Ecol Modelling* (submitted).

75. du Plessis, J.L. and Bezuidenhout, J.D. Investigations on the natural and acquired resistance of cattle to artificial infection with *Cowdria ruminantium*. *J S Afr Vet Assoc* 50, 334–348 (1979).

76. Du Plessis, J.L. and Malan, L. The block method of vaccination against heartwater. *Onderstepoort J Vet Res* 54, 493–495 (1987).

77. Du Plessis, J.L. et al. The heterogeneity of *Cowdria ruminantium* stocks: Cross-immunity and serology in sheep and pathogenicity to mice. *Onderstepoort J Vet Res* 56, 195–201 (1989).

78. Jongejan, F. Protective immunity to heartwater (*Cowdria ruminantium* infection) is acquired after vaccination with *in vitro*-attenuated rickettsiae. *Infect Immun* 59, 729–731 (1991).

79. Jongejan, F. et al. Vaccination against heartwater using *in vitro* attenuated *Cowdria ruminantium* organisms. *Rev Elev Med Vet Pays Trop* 46, 223–227 (1993).

80. Martinez, D. Analysis of the immune response of ruminants to Cowdria ruminantium infection - development of an inactivated vaccine. PhD thesis, Utrecht University, Utrecht, the Netherlands (1997).

81. Martinez, D. et al. Comparative efficacy of Freund's and Montanide ISA50 adjuvants for the immunisation of goats against heartwater with inactivated *Cowdria ruminantium*. *Vet Parasitol* 67, 175–184 (1996).

82. Mahan, S.M. et al. The inactivated *Cowdria ruminantium* vaccine for heartwater protects against heterologous strains and against laboratory and field tick challenge. *Vaccine* 16, 1203–1211 (1998).

83. Marcelino, I. et al. Stability and efficacy of a ready-to-use inactivated vaccine against heartwater using Montanide ISA water-in-oil emulsions prepared with different emulsifying techniques (submitted).

84. Nyika, A. et al. A DNA vaccine protects mice against the rickettsial agent *Cowdria ruminantium*. *Parasite Immunol* 20, 111–119 (1998).

85. Lally, N.C. et al. The *Cowdria ruminantium* groE operon. *Microbiology* 141, 2091–2100 (1995).

86. Louw, E. et al. Sequencing of a 15-kb *Ehrlichia ruminantium* clone and evaluation of the cpg1 open reading frame for protection against heartwater. *Ann N Y Acad Sci* 969, 147–150 (2002).

87. van Vliet, A.H. et al. Cloning and partial characterization of the Cr32 gene of *Cowdria ruminantium*. *Rev Elev Med Vet Pays Trop* 46, 167–170 (1993).

88. Pretorius, A. et al. A heterologous prime/boost immunisation strategy protects against virulent *E. ruminantium* Welgevonden needle challenge but not against tick challenge. *Vaccine* 26, 4363–4371 (2008).

89. Pretorius, A. et al. Studies of a polymorphic *Ehrlichia ruminantium* gene for use as a component of a recombinant vaccine against heartwater. *Vaccine* 28, 3531–3539 (2010).

90. Marcelino, I. et al. Tick-borne diseases in cattle: Applications of proteomics to develop new generation vaccines. *J Proteomics* 75, 4232–4250 (2012).

91. Frutos, R. et al. Comparative genomic analysis of three strains of *Ehrlichia ruminantium* reveals an active process of genome size plasticity. *J Bacteriol* 188, 2533–2542 (2006).

92. Frutos, R. et al. Comparative genomics of three strains of *Ehrlichia ruminantium*: A review. *Ann N Y Acad Sci* 1081, 417–433 (2006).

93. Collins, N.E. et al. The genome of the heartwater agent *Ehrlichia ruminantium* contains multiple tandem repeats of actively variable copy number. *Proc Natl Acad Sci USA* 102, 838–843 (2005).

94. Emboule, L. et al. Innovative approach for transcriptomic analysis of obligate intracellular pathogen: Selective capture of transcribed sequences of *Ehrlichia ruminantium*. *BMC Mol Biol* 10, 111 (2009).

95. Pruneau, L. et al. Global gene expression profiling of *Ehrlichia ruminantium* at different stages of development. *FEMS Immunol Med Microbiol* 64, 66–73 (2012).

96. Marcelino, I. et al. Proteomic analyses of *Ehrlichia ruminantium* highlight differential expression of MAP1-family proteins. *Vet Microbiol* 156, 305–314 (2012).

97. Bagnoli, F. et al. Designing the next generation of vaccines for global public health. *Omics* 15, 545–566 (2011).

66 *Mycoplasma mycoides* subsp. *mycoides* and *Mycoplasma capricolum* subsp. *capripneumoniae*

F. Thiaucourt and L. Manso-Silván

CONTENTS

66.1 INTRODUCTION

Two mycoplasmas belonging to the so-called "mycoides cluster," *Mycoplasma mycoides* subsp. *mycoides* "small colony" (MmmSC) and *Mycoplasma capricolum* subsp. *capripneumoniae* (Mccp), are responsible for causing contagious bovine pleuropneumonia (CBPP) and contagious caprine pleuropneumonia (CCPP), respectively. These organisms induce acute or subacute disease in ruminants characterized by unilateral serofibrinous pleuropneumonia with severe pleural effusion.

66.1.1 CBPP HISTORY

Old descriptions of cattle diseases made by Roman scholars or during the sixteenth century by Gallo and Estienne are too imprecise to conclude that they were referring to CBPP. Even the more recent publication by Scheuchzer in 1732[1] certainly did not refer to CBPP, and others made at that time could be

relating to rinderpest rather than to CBPP. In fact, the earliest unambiguous description of CBPP may be that of B. de Haller in 1772.[2]

Interestingly, the molecular dating of the most recent common ancestor to a representative set of MmmSC strains indicated that the appearance of this agent must have happened around 1700.[3] Genomic comparisons with the other members of the mycoides cluster showed that the closest relative to MmmSC is another subspecies, *M. mycoides* subsp. *capri*, which is usually found in goats. It has been hypothesized that MmmSC resulted from a recent adaptation of this goat pathogen to a new host, cattle.[4]

The geographic distribution of CBPP expanded gradually in Europe during the first half of the nineteenth century. It was well established around Paris in France in 1798 and appeared in Belgium in 1827, the Northern German States in 1830, Great Britain in 1841, Catalonia in 1846, and Sweden in 1847.[5] This expansion was clearly linked to live animal trade.

During the second half of the nineteenth century, CBPP gained a worldwide expansion and was considered then, in the preantibiotic era, one of most important diseases for cattle, together with rinderpest. North America was infected in 1843 while South America remained free of the disease. The colony of the Cape of Good Hope was contaminated in 1853 and Australia shortly after, in 1858. Apparently, Asia was contaminated some time later, through imports of milking cattle from Australia in 1910.[6] CBPP spread in the southern part of Africa is well known from historical accounts, starting from its introduction through Friesian bulls imported from the Netherlands at Mossel Bay in 1853. The disease progressed to the north by trek oxen and reached the Huila province in Angola in the early 1880s. It caused heavy losses to cattle raising and even starvation in some tribes such as the Xhosa.[7] There are no historical accounts on CBPP introduction in the rest of the African continent and some authors have hypothesized that it had been present "from time immemorial."[8] However, recent genetic analysis of MmmSC strains of African origin indicated that CBPP must have been imported at the beginning of the nineteenth century, at a time when CBPP was well known to occur in Europe.[3]

Owing to its importance, strict control measures were put in place in many countries to eradicate CBPP at the end of the nineteenth century. These measures were based on clinical identification of affected herds, slaughter, and animal movement control. Some countries also used the "inoculation" technique described by Willems to protect cattle herds.[9] However, eradication was achieved only when strict slaughter policies were implemented. This allowed, for example, the United States to be declared CBPP-free in 1892, Great Britain in 1896, France in 1902, Austria in 1919, and South Africa in 1924. However, CBPP persisted as an enzootic disease in many sub-Saharan African countries, where the disease was controlled by vaccination. It benefitted from international efforts to eradicate rinderpest, as combined vaccination campaigns were put in place. Rinderpest was finally eradicated but CBPP persisted, notably because of the short duration of protection of CBPP vaccines, while rinderpest vaccines elicit lifelong protection. The halt in the international aid to fight rinderpest resulted in an expansion of CBPP in the African continent coupled with an increase in the number of outbreaks in the enzootic zones.

The CBPP situation in Europe during the twentieth century was characterized by sporadic outbreaks occurring in the southern part of the continent at 10–20-year intervals,[10] the last outbreaks being recorded in Portugal in 1999. It was clearly shown by molecular data that these outbreaks were resurgences of the disease rather than the result of importation from infected zones.[11] This emphasized the need to better understand the reasons of eradication campaign failures in this continent.

66.1.2 CCPP History

CCPP history can be subdivided into three periods. The first period (1873–1890) corresponds to the first clinical descriptions of the disease. P. Thomas was the first to describe CCPP very precisely in 1873 in Algeria, where local herdsmen knew it under the name "Bou Frida" by reference to the fact that only one lung was affected.[12] Thomas recorded the cyclic appearance of CCPP at 5–8-year intervals and also its appearance during the winter season. This led people to believe that this disease was in relation to climate rather than to an infectious agent, although goat owners knew that a recovered goat was never affected again. Very shortly afterward, CCPP was introduced into the "colony of the Cape of Good Hope" in 1881 with the import of angora goats from Turkey.[13] D. Hutcheon rapidly concluded that the disease was contagious and promoted a control strategy based on the slaughter of affected goats and the inoculation of the others. At that time, inoculation meant injecting infectious material by different routes to induce protection. After 8 years of efforts, the disease was eradicated from the southern part of Africa.[14]

The second period (1890–1976) is a period of confusion for CCPP. The reasons for this confusion are the very fastidious nature of the causative organism, Mccp, and the fact that goats can be affected by a variety of bacterial pathogens, notably mycoplasmas of the mycoides cluster, which can induce similar symptoms and lesions. Most of the CCPP descriptions that were made during that period in Europe were certainly referring to other diseases and especially to what is now known as the "contagious agalaxia syndrome."[15] One exception, though, was the description of CCPP made by Greek authors in 1928[16] after the importation of goats from mainland Turkey during the war at "Smyrne" (now Izmir) in 1920–1922. In Africa some publications indicate that the disease was certainly present in the eastern part of the continent in 1912,[17] in Kenya in 1929,[18] and in Eritrea in 1932.[19] The disease may also have been present in India, as noted by Walker in 1914[20] and Longley in 1940.[21] Efforts to isolate the causative organism were met with failure, although various bacteria were isolated. The isolation of a mycoplasma from a disease that resembled CCPP in Turkey created more confusion, as CCPP was certainly confused with the agalaxia syndrome at that time. The real CCPP agent was finally isolated and characterized in 1976 by McOwan and Minette in Kenya under the name "type F38."[22]

The third period extends from 1976 until today, with a very slow and progressive recognition of the real distribution and impact of the disease. CCPP was evidenced mostly in East, Central, and North Africa as well as the Middle East, although its real distribution was suspected to be much wider. The presence of CCPP was finally confirmed in Turkey, in the Thrace region in 2005[23] and in the East of the country in 2009.[24] This last finding was a clear indication that the disease was certainly present in Asia and this was confirmed for Tadjikistan in 2010[25] and then for China in 2011.[26] The fact that CCPP can affect wildlife is also a recent finding[27] that may have important consequences in terms of wildlife conservation or introduction risks.

66.2　CLASSIFICATION AND MORPHOLOGY

66.2.1　Phylogeny and Taxonomy

MmmSC and Mccp are wall-less bacteria belonging to the class *Mollicutes*, order *Mycoplasmatales*, family *Mycoplasmataceae*, and genus *Mycoplasma*. The *Mollicutes* (*mollis*, soft and *cutis*, skin, in Latin), trivially referred to as mycoplasmas, constitute the smallest organisms capable of autonomous replication. They are characterized by their small size and lack of a cell wall, which renders them pleomorphic and intrinsically resistant to beta-lactam antibiotics. Although the absence of a cell wall makes them technically Gram-negative, they actually constitute a distinct lineage within the Gram-positive bacteria with low G+C content.

More precisely, MmmSC and Mccp belong to the mycoides cluster[28,29] which includes three other ruminant pathogens: *M. mycoides* subsp. *capri*, *M. capricolum* subsp. *capricolum*, and *Mycoplasma leachii*. In addition, the pathogen *Mycoplasma putrefaciens* and the related saprophytic species *Mycoplasma cottewii* and *Mycoplasma yeatsii* are phylogenetically related to this cluster (Table 66.1). *M. mycoides* subsp. *capri*, *M. capricolum* subsp. *capricolum*, and *M. putrefaciens* are involved in a complex syndrome known as "contagious agalactia," or MAKePS (for mastitis, arthritis, keratoconjunctivitis, pneumonia, and septicemia) in small ruminants.[15] Finally, *M. leachii* causes mastitis and polyarthritis in cattle.[30]

The CCPP agent was known for many years as the "F38-type mycoplasma," in reference to the type strain F38[T]. This taxon, first isolated in 1976,[22] remained unassigned until 1993, when sufficient evidence was gathered to support a subspecies relationship of this agent with *M. capricolum* subsp. *capricolum*.[31] As for the CBPP agent, it was the first mycoplasma to be isolated[32] and it represents the type for the class *Mollicutes*. For decades, MmmSC was considered just the "small colony" biotype of *M. mycoides* subsp. *mycoides*, together with the "large colony" biotype. The two taxa had

been associated, separate from *M. mycoides* subsp. *capri*, due to serological cross-reactions. However, numerous investigations suggested that the "large colony" strains were more closely related to *M. mycoides* subsp. *capri* than to MmmSC. Recently, a thorough phylogenetic analysis based on five housekeeping gene sequences[33] prompted an amendment of the taxonomy of the mycoides cluster and the "large colony" strains were finally united to *M. mycoides* subsp. *capri*, leaving the "small colony" strains as sole representatives of *M. mycoides* subsp. *mycoides*.[29] The phylogeny of the mycoides cluster[28] (Figure 66.1) reveals the presence of two subclusters; one comprising MmmSC and *M. mycoides* subsp. *capri* and another one including Mccp, *M. capricolum* subsp. *capricolum*, and *M. leachii*. *M. putrefaciens*, *M. cottewii*, and *M. yeatsii* are separated but closely related to this cluster, as well as other reported mycoplasma strains isolated from wildlife that remain unassigned (*M.* sp. in Figure 66.1).

66.2.2　Morphology

MmmSC and Mccp display similar morphology. They are pleomorphic, nonspiral, and nonmotile and they both show a tendency to produce filamentous growth. MmmSC and Mccp cells are surrounded by an exopolysaccharide layer or pseudocapsule.

66.3　BIOLOGY AND EPIDEMIOLOGY

66.3.1　Biology

Mycoplasmas are obligate parasites that live in relatively unchanging niches requiring little adaptive capability. These wall-less bacteria with a very low G+C% genomes evolved from more conventional progenitors in the *Firmicutes* taxon by a process of massive genome reduction. The absence of a cell wall makes them quite fragile outside their ecological niche. They usually survive in mucosal tissues, where they can find appropriate nutrients to compensate for the loss of various biosynthetic pathways and where they find the appropriate temperature and humidity to allow them to multiply.

66.3.2　Epidemiology

CBPP. MmmSC affects the species of the *Bos* genus, cattle, zebu, and yak, as well as the water buffalo (*Bubalus bubalis*). In cattle, there seems to be a variation of susceptibility according to breed. Younger animals seem to be less affected and to present arthritis lesions more frequently,[34] while older animals present usually the classical lung lesions. CBPP is not a zoonosis.

MmmSC has been isolated from lungs or other organs of small ruminants, although these animals are not susceptible to the disease.[35] The role of small ruminants is certainly negligible but their possible implication in the long-term undetected persistence of MmmSC strains remains to be ascertained.

Individual susceptibility may vary according to climatic conditions or intercurrent infections that may modify the

---

TABLE 66.1

"*Mycoides* Cluster" and Related *Mycoplasma* Species

Designation	Host Species	Disease
Mycoides cluster		
M. mycoides subsp. *mycoides* (MmmSC)	Bovine	CBPP
M. mycoides subsp. *capri*	Caprine, ovine	MAKePS
M. capricolum subsp. *capricolum*	Caprine, ovine	MAKePS
M. capricolum subsp. *capripneumoniae* (Mccp)	Caprine, wildlife	CCPP
M. leachii	Bovine	Mastitis, arthritis
Related species		
M. putrefaciens	Caprine, ovine	MAKePS
M. cottewii	Caprine, ovine	(Saprophytic)
M. yeatsii	Caprine, ovine	(Saprophytic)

Note: The former *M. mycoides* subsp. *mycoides* LC biotype has been included in the *M. mycoides* subsp. *capri* subspecies.

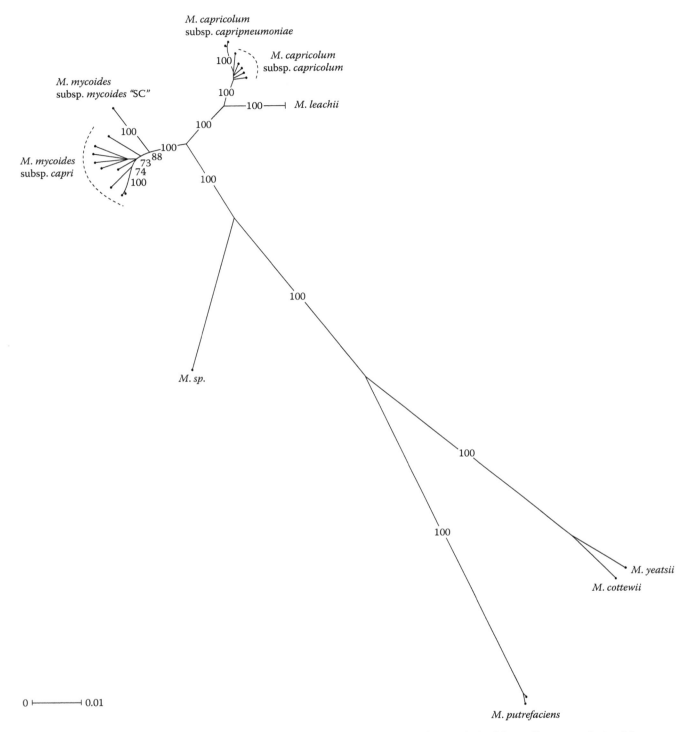

FIGURE 66.1 Phylogeny of the "mycoides cluster" and related species. Phylogenetic tree derived from distance analysis of five concatenated protein-coding partial gene sequences (*fusA*, *glpQ*, *gyrB*, *lepA*, and *rpoB*). The tree was constructed using the neighbor-joining algorithm. Bootstrap percentage values were calculated from 500 resamplings and all values over 70% are displayed. The scale bar shows the equivalent distance to 1 substitution per 100 nucleotide positions.

immune response of the affected animals. Immunosuppressed animals are less susceptible to CBPP.

CBPP is now present mostly in sub-Saharan Africa[36] and it seems to be expanding gradually to the south from Tanzania toward Zambia and from Cameroon toward Congo and Gabon (Figure 66.2). The presence of the disease in the Middle East and Asia is not well established as official declarations rely mostly on clinical signs. No outbreak has been detected in Europe since 1999.

The real importance of the disease in enzootic regions is difficult to establish. There are no incentives for declaration and owners usually treat their animals with antibiotics as

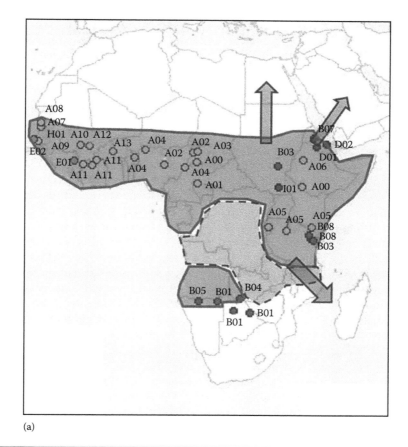

(a)

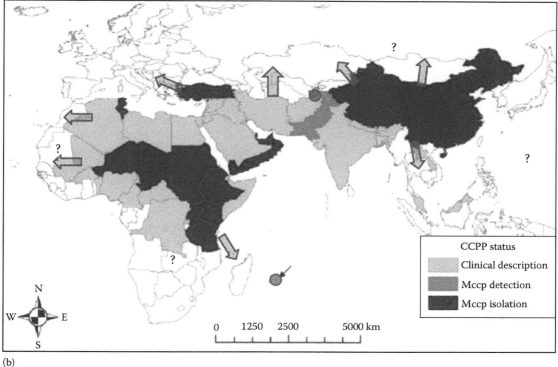

(b)

FIGURE 66.2 Distribution and extension risks for (a) CBPP and (b) CCPP in 2012: (a) CBPP is present (continuous line) or suspected (dotted line) in all the intertropical regions in Africa. Arrows indicate probable risks of expansion. MmmSC subtypes are indicated (From Yaya, A. et al., *Vet. Res.*, 39, 14, 2008). (b) CCPP is present in Africa and Asia where it has been confirmed by isolation of Mccp (dark color) or by direct detection by PCR (medium gray color). It is highly suspected in neighboring countries based on clinical descriptions (light gray color). Arrows indicate probable risks of expansion.

soon as they observe respiratory symptoms without knowing if they are caused by CBPP or not. Recent studies in a pastoralist area in Africa, using a participatory approach, have shown that CBPP ranked among the most important cattle diseases according to their owners.[37]

CBPP is always transmitted by live animals. There is no indirect transmission by fomites or by meat products. There are some reports showing the presence of MmmSC in the semen of affected bulls.[38] The possible role of semen in the transmission of CBPP has never been established. Incubation lasts about 30 days (min 8–max 180) and MmmSC excretion can occur during this period (up to 40 days), while animals do not display any symptoms or any sero-conversion. Excretion is at its maximum during the acute stage of the disease (15 days), when animals cough, excreting infected droplets. It appears that animals in the chronic stage of the disease (1 month up to 2 years) can still excrete mycoplasmas, although it is difficult to evaluate its intensity and periodicity. However, this prolonged excretion may be responsible for the long-term persistence of the disease.[39] This infectivity may last up to 2 years post infection.

CBPP being transmitted by live animals and by direct contact only, disease transmission is clearly dependent on cattle density and animal movements. This has a clear impact on the kind of preventive measures that can be effective. Transhumance may favor the contact between infected and naive herds, especially at watering spots while gathering all the animals of a herd in a kraal at night may favor a rapid spread of the disease within the herd.

There are very few data concerning the influence of antibiotic treatments on mycoplasma shedding. These treatments do not seem to increase the number of chronic carriers[40] but they certainly do not allow a complete cure, as MmmSC can be isolated from recovered treated animals.[41]

Molecular epidemiology of CBPP. Various typing tools can be used to subtype MmmSC strains. These tools take advantage of molecular events that generate polymorphisms in the MmmSC genome. For example, the insertion and duplication of "insertion sequences" can be evidenced by Southern blot techniques,[42,43] while the duplication or deletion of repeated elements (variable number of tandem repeats) may be evidenced by polymerase chain reaction (PCR) amplification and evaluation of the fragment length.[44] However, the most robust technique may be the multilocus sequence analysis (MLSA) that detects various kinds of genetic variations including single nucleotide polymorphisms (SNPs).[11] In other bacteria, this technique is usually performed on housekeeping genes and is called multilocus sequence typing (MLST). However, MmmSC genomes are too conserved to allow the detection of SNPs in a small number of housekeeping genes (typically seven or eight). Hence, more variable genes or noncoding DNA were targeted for MmmSC, allowing a clear correlation between MLSA types, geographic locations of the strains, and trading routes. MLSA also showed that MmmSC strains that had been isolated in Europe up to 1999 in Portugal were indigenous to Europe and may have resulted from the undetected persistence of MmmSC strains in between two CBPP outbreaks.

The MLSA technique was refined recently by using an extended set of genes (63 instead of 8)[3] thanks to the use of "next-generation sequencing" techniques. Results showed notably that all the recent European MmmSC strains, isolated in Italy, France, Spain, and Portugal during the two CBPP events in the 1980s and 1990s, originated from a single ancestor. The reasons for this unnoticed persistence of MmmSC strains, which are responsible for periodic CBPP reoccurrence, are still unknown. This could be due to the persistence of strains with lower pathogenicity that could suddenly regain their original virulence. This could also be due to the persistence of MmmSC strains in an unusual, unsusceptible, host and the transmission to cattle when conditions may favor this event.

CCPP. Mccp affects mostly goats and there is no breed or age variation of susceptibility. However, wild ruminants can also be affected such as Nubian ibex, Laristan mouflon, wild goat, beira antelope, or gerenuk.[27] This susceptibility has been shown in animals kept in captivity as it is obviously more difficult to confirm the disease in wildlife. At one stage, CCPP was also suspected in the markhors in Tadjikistan. Lung lesions were very similar to CCPP lesions but another mycoplasma species (*M. capricolum* subsp. *capricolum*) was isolated.[45] CCPP is not a zoonosis.

Due to the very fastidious nature of Mccp, it is difficult to establish if it can be found in other organs or species. There are some reports of Mccp being isolated from sheep from nasal swabs or lungs. Unlike Mccp, many other mycoplasma species such as *Mycoplasma agalactiae* or other mycoplasmas of the mycoides cluster can be found as saprophytes in the ear canal of goats.[46,47]

As for CBPP, climate and other infections seem to have a direct influence on the course of the disease. In North Africa, CCPP was observed during the winter, while in Oman, the disease seems to occur at the end of the dry season. In China, CCPP is often associated with pox infections.[48]

CCPP distribution is now better known and it certainly exists within the whole region (polygon) delimited by Tunisia–Niger–Tanzania–Oman–Pakistan–China–Turkey, where CCPP has been isolated and characterized (Figure 66.2). The full extension of the disease, outside these boundaries, is difficult to ascertain. It has been recently imported in Maurice Island with goats originating from East Africa.

There are very few field data allowing a precise epidemiological study of CCPP. In Tunisia, the disease was evidenced in 1981 and it affected mostly imported goats of European origin. The disease seemed to extend then to Libya. Since then, there have been no reports of CCPP in that region although no control measures were put in place. CCPP was shortly afterward evidenced in Chad in 1987 from a lesion found at the slaughterhouse,[49] while there were no field reports of any disease affecting goats. However, CCPP caused heavy losses in the same country in 1995. Similarly, CCPP was suspected in Ethiopia in 1984 at the Sudanese border but no confirmation was obtained at that time. CCPP struck Ethiopia again in 1991 as a consequence of animal movements from neighboring Somalia.[50] The disease rapidly spread in the goat herds of nomads living in the Rift Valley, causing very heavy losses. Since then, the disease has occurred in an enzootic form, similar to what is observed in East African countries such as Kenya, Tanzania, and Uganda.

CCPP is transmitted by live animals only. However, there are no precise data regarding the possible excretion before symptoms appear in the infected animals. Unlike for CBPP, CCPP-affected animals in the chronic stage of the disease do not display sequestered lesions. However, this does not prevent a possible long-term carrier status.

The recent demonstration that CCPP could affect wildlife raises new concerns and questions. One possibility is that wildlife is only recently being contaminated as goats are now encroaching on land that was previously occupied only by wildlife. Alternatively, wild ungulates may constitute a reservoir for pathogenic mycoplasmas.

Molecular epidemiology of CCPP: Typing tools for Mccp strains have mostly focused on the detection of SNPs. The first techniques were based on the polymorphisms observed among the two 16S rRNA genes found in the genomes of the mycoplasmas of the mycoides cluster.[51] More recently, an MLSA technique was developed for Mccp strains.[25] Based on the sequencing of eight loci, it allowed a much more precise typing of Mccp strains, which were distributed in five groups that correlated well with the geographic origin of the strains. MLSA showed that the strains isolated in China and Tajikistan constitute a separate group, indicating that CCPP may have existed in the Asian continent for quite some time. It also confirmed the presence of strains of two different lineages in East Africa.

The main limitation to these studies is the paucity of Mccp strains for analysis. This is due to the fastidious nature of Mccp, which makes isolation very difficult, and also to the lack of incentive to study goat diseases in many countries.

66.4   CLINICAL FEATURES AND PATHOGENESIS

66.4.1   CLINICAL AND PATHOLOGICAL FEATURES

CBPP and CCPP present similar clinical and pathological features. Both diseases show strict tissue tropism, with lesions located exclusively in the thoracic cavity and consisting in severe, exudative pleuropneumonia, associated with fibrinous pleurisy. In both cases, almost invariably, only one lung is affected. However, the thoracic localization of CBPP appears to be merely related to the natural entry route, as similar lesions can be observed when MmmSC cultures are inoculated artificially by other routes. A good example is the invasive edematous lesion, known as Willem's reaction, that is observed when MmmSC is inoculated subcutaneously. No such reaction is observed with the CCPP agent.

Clinical signs and lesions are similar and their severity depends on whether the diseases present in an acute or subacute–chronic form.

The classical clinical picture is observed in the acute forms that develop when the etiologic agents infect naive populations. The first signs are reluctance to move, fever, and anorexia. Gradually, respiratory symptoms become more prominent. They are characterized by a fast and painful respiration and an intermittent, violent cough. In the final stages, the animals stand with their forelegs wide apart, head and neck extended, and mouth open. Death may quickly follow as a result of respiratory distress or heart failure. Otherwise, the animals recover gradually, though they may still carry and excrete the etiologic agent for months. Subacute or chronic forms of these diseases may be observed in enzootic regions. In these forms, the clinical signs are similar, but milder, and consist in transient, slight fever and occasional coughing, which may only be noticeable following exercise.

In the acute forms of both diseases, large amounts of pleural exudate are found in the thoracic cavity, with caseous fibrinous deposits in the pleura. In CBPP, a typical interstitial pneumonia is observed, characterized by enlarged interlobular septa, which are distended by the presence of extensive edema. This serous fluid separates lung lobules at different stages of hepatization that vary from light to deep red, yellow, or gray, giving a marbled appearance to the lungs. These characteristic lesions are not found in CCPP, where affected lungs may be totally hepatized and present a fine, granular texture and a color that varies from purple to gray. In the chronic forms of both diseases, the pleural fluid disappears and pleural adhesions are prominent but the characteristic encapsulated necrotic lesions, known as sequestra, that are observed in CBPP are absent in CCPP.

66.4.2   PATHOGENESIS

The pathogenesis is not well understood, but both diseases are characterized by a massive inflammatory reaction that may be induced directly by toxic mycoplasma components or indirectly, through the release of proinflammatory cytokines. No classical virulence factors such as adhesins or toxins have been identified, and virulence has been attributed to surface or secreted components and intrinsic metabolic functions.

Several MmmSC components have been proposed as virulence attributes and a few hypotheses have been suggested to explain the pathogenesis of CBPP. Lipoproteins are considered important triggers of the inflammatory process by mediating adhesion and stimulating the release of proinflammatory cytokines. The immunodominant lipoprotein LppQ of MmmSC is assumed to have superantigen-like properties and appears to play an adverse reaction in vaccination. But the best described mechanism for induction of cellular damage by MmmSC is based on the metabolism of glycerol. The enzyme L-α-glycerophosphate oxidase (glpO) plays a central role in this pathway.[52] It catalyzes deoxidation of glycerol-3-phosphate into dihydroxyacetone phosphate (DHAP), resulting in the release of hydrogen peroxide (H_2O_2), which induces host cell damage and inflammation. An intimate contact between the mycoplasma and the host cell is required for the translocation of H_2O_2 and reactive oxygen species.[53] However, the mechanisms involved in adhesion are still to be elucidated. This and other factors may explain the fact that vaccinal strains produce large amounts of H_2O_2.

The role of exopolysaccharides in disease initiation and persistence has long been speculated. A direct cytopathic effect of the capsular polysaccharide (galactan) of MmmSC has been reported,[54] and its role in autoimmunity has been proposed as an explanation of the pathogenesis of CBPP: immunecomplexes formed due to antigenic relatedness between mycoplasma and lung galactan precipitate producing tissue damage and therefore allowing mycoplasma invasion. The galactan has also been associated with persistence and dissemination of MmmSC in the host, presumably through protection from the bactericidal activity of complement and other host defense mechanisms.[55] Furthermore, MmmSC has been shown to produce biofilms, which confer resistance to many stresses such as heat, detergent, and peroxide, as well as increased persistence.[56]

The role of variable surface proteins in evasion of the immune response has been widely described in mycoplasmas. The protein Vmm of MmmSC has been shown to undergo reversible phase variation at high frequency.[57] However, it is not an immunodominant antigen and its function and role in the pathogenesis of CBPP are still to be elucidated.

For what concerns the pathogenesis of CCPP, very little information is available and hypotheses are mainly drawn from comparison with CBPP. The only proposed virulence factor of Mccp is the exopolysaccharide,[58] which may play a similar role to that of MmmSC.

66.5 IDENTIFICATION AND DIAGNOSIS

Specific diagnosis of CBPP and CCPP is crucial, as these important transboundary diseases are listed by the World Organisation for Animal Health—Organisation International des Epizooties (OIE) as notifiable diseases. Clinical and pathological features can be highly evocative in the acute form of these diseases, though the classical presentation is rarely observed in practice due to insidious disease circulation or concomitant infections. Differential diagnosis with pulmonary diseases such as acute pasteurellosis or other mycoplasmosis with lung tropism may be required. In addition, CBPP sequestra may be confused with actinobacillosis lesions or old hydatid cysts. In any case, laboratory diagnosis is required.

66.5.1 DIRECT DETECTION

Direct diagnosis by culture and isolation of the etiologic agents is still mandatory for absolute confirmation by the OIE. Furthermore, isolation allows the detailed characterization of strains (e.g., genome sequencing) and is essential for the generation of reference culture collections. The preferred samples for isolation are pleural fluid, hepatized lung lesions (preferably in the interface between healthy and diseased tissue), and tracheobronchial and mediastinal lymph nodes from acute disease.

The CBPP and CCPP agents can be grown in adequate media such as the modified Hayflick's medium incubated at 37°C with 5% CO_2 and maximum humidity. However,

while MmmSC may grow relatively well, Mccp is extremely fastidious and isolation trials are often unsuccessful. In any case, isolation may be compromised by poor conservation of samples, bacterial overgrowth (including other mycoplasmas) or antibiotic treatments. Once isolated, the members of the mycoides cluster may be difficult to identify by classical techniques based on cultural properties and biochemical and serological reactions. MmmSC and Mccp present several characteristic cultural features. They display filamentous growth in broth culture and, when shaken, the sediment rises forming characteristic "silky swirls." On solid medium, they produce small colonies (particularly in the case of Mccp) with a fried egg appearance. A few biochemical tests classically proposed for mycoplasmas (glucose fermentation, arginin hydrolysis, tetrazolium chloride reduction, and phosphatase activity) may also help orient the diagnosis but they do not allow the discrimination between members of the mycoides cluster, and serological confirmation by growth inhibition or immunofluorescence is required. However, serodiagnosis is often hampered by cross-reactions between cluster members. Cross-reactions are frequent between MmmSC and "large colony" strains of *M. mycoides* subsp. *capri*[59] and between Mccp and *M. leachii*.[60] In addition, these techniques are time- and labor-consuming, and their sensitivity is low.

Because of these difficulties, direct detection of the etiological agents in clinical material is a very useful alternative for the confirmation of CBPP and CCPP outbreaks. Since the 1990s, the use of specific PCR tests has made detection and identification of these two organisms much more sensitive and reliable. These assays may be even performed from samples dried on filter paper, which can be easily transported without cold-chain requirements. PCR-based techniques have been described for the identification of MmmSC and Mccp based on the amplification of "CAP21 sequence" or 16S rDNA gene fragments, followed by restriction enzyme digestion of the amplicons.[61,62] Specific PCR assays have also been designed for MmmSC[63] and Mccp,[64] which allow direct detection of the etiological agents. PCR can also be used to identify the vaccine strain T1 specifically.[65] Still, conventional PCR requires post-PCR processing (i.e., agarose gel electrophoresis) for visualization of the results, increasing the time of analysis and, most importantly, the risk of contamination. Real-time, quantitative PCR assays based on SYBR Green detection have been developed for both MmmSC and Mccp,[66,67] whereas Taqman probes have been described only for MmmSC.[68,69] These assays do not require post-PCR processing, reducing greatly the risk of contamination. Furthermore, they allow quantification of the target DNA in the samples and afford increased sensitivity. They constitute the assays of choice for rapid disease confirmation in CBPP-free countries.

66.5.2 DIAGNOSIS BY SEROLOGICAL METHODS

During the past decades, complement fixation test (CFT) has been the method of choice for the detection of antibodies to

CBPP[70] and CCPP. It is still a prescribed method for the OIE for both diseases. The main advantage of this test is that it can be performed quite easily with very little equipment,[71] for example, in mobile laboratories. It was widely used for the CBPP eradication campaigns in Australia and more recently in Italy and Portugal. However, it also has a number of drawbacks. In the case of CBPP, it lacks specificity and the percentage of false positive reactions may reach 2%. As it detects mostly IgM, its sensitivity may rapidly decline over time and it is therefore more suitable for the early detection of CBPP cases in an epizootic context rather than for prevalence studies in enzootic situations. Besides, this technique requires a good laboratory experience and trained personnel. In the case of CCPP, the CFT is even less specific. Goats are very often infected by mycoplasmas of the mycoides cluster, which may induce very high cross-reactions, as the antigen used for the CFT is a crude concentrate of mycoplasmas.

Since the 1990s, some enzyme-linked immunosorbent assays (ELISAs) have been developed. The tests that are still available are competitive ELISAs (cELISAs) which have been designed to obtain an improved specificity thanks to the competition with a monoclonal antibody (MAb). The cELISAs for CBPP, developed with MAb 117/5,[72] and CCPP, with MAb 4.52,[73] proved highly specific. The precocity of detection proved slightly retarded as compared to the CFT (1 week) but cELISAs detected more animals in the chronic stage of the disease.[74] Thanks to the very high specificity of this test, it can be used for herd positivity detection and the sensitivity is increased when selecting animals showing suspicious symptoms in the previous months. The vaccines used in Africa to control CBPP are live vaccines that do not induce an important and persistent seroconversion, lasting less than 3 months. Hence, cELISA and CFT can still be used to detect CBPP outbreaks. On the contrary, CCPP vaccines are made of inactivated antigens and adjuvant (saponin) and induce long-lasting serological responses. The cELISA can then be used for controlling the efficacy of vaccination campaigns but no more for the detection of outbreaks. The cELISA for CBPP is a prescribed test for the OIE.

Immunoblotting tests have been used extensively in Europe for the confirmation of CBPP cases when the prevalence of CBPP was very low, as CFT often yielded false positive results. The test relies on the detection of five specific bands. It is quite cumbersome to perform and requires a careful preparation of antigen and skilled personnel.[75]

The slide agglutination test, based on whole colored antigen, was developed long ago to provide a rapid confirmation of CBPP outbreaks in the field.[76] The test performs well when a significant number of collected sera induce a strong agglutination result. However, it lacks specificity. More recently, agglutination tests based on latex beads sensitized with mycoplasma-excreted polysaccharides have been developed to detect MmmSC and Mccp.[77] Their exact sensitivity and specificity are not well known but they can clearly detect affected herds.

66.6 TREATMENT AND PREVENTION

66.6.1 ANTIBIOTHERAPY

Mycoplasmas are intrinsically resistant to penicillins and other antibiotics acting on cell wall synthesis and they become rapidly resistant to aminosides, such as streptomycin. However, they are sensitive to many other antibiotics that may be used to treat CBPP and CCPP.[78] The most effective are tetracyclines; macrolides, such as spiramycin, erythromycin, and tulathromycin; and fluoroquinolones. Antibiotherapy results in reduction of clinical symptoms and pathology but it does not assure complete clearance of the infection, so animals may act as chronic, asymptomatic carriers. Furthermore, as antibioresistance is considered one of the most important threats to human health, there is a global trend to reduce the use of antibiotics. For this reason, antibiotherapy may only be recommended as part of well-designed eradication strategies, coupled to prophylactic measures.

66.6.2 PROPHYLAXIS

As the epidemiology of these diseases is quite simple, based exclusively on direct transmission by contact between infected and susceptible animals, standard methods for disease control may be applied. Sanitary measures constitute the most effective strategy in disease-free countries, and the first golden rule is the prohibition of importation of live animals from infected or suspected zones. Furthermore, CBPP and CCPP have been eradicated from zones, countries, and entire continents by a combination of strategies including drastic stamping-out policies and movement restrictions. Slaughtering of infected and in-contact animals is still recommended to neutralize new foci in disease-free countries. However, restriction of animal movements in regions that practice extensive nomadic livestock herding is not feasible and the massive slaughter of infected animals or herds is out of the question in countries where the diseases are enzootic. A medical prophylaxis based on the use of vaccines is the preferred alternative to reduce the prevalence and limit the expansion of CBPP and CCPP in these regions.

Inoculation procedures for the prevention of CBPP and CCPP are very old, particularly for CBPP. Indeed, the first inoculations for CBPP control were performed before the discovery of the etiologic agent.[79] They consisted in the insertion of lung pieces or pleural exudates from CBPP-infected animals into healthy cattle. In the 1850s in Europe, Willems proposed the dense connective tissue at the tip of the tail as inoculation site to avoid the extensive edematous reactions, known as "Willems' reactions," that were otherwise generated by subcutaneous inoculations into the trunk or limbs. Fulani herdsmen from Senegal developed a similar method but they chose the bridge of the nose as inoculation site. This resulted in a massive inflammatory reaction, often causing the formation of a "third horn," which was mistaken by French zoologists as a trait of a new bovine race that they named *Bos triceros*.[8] Similar procedures have been implemented in goats for the prevention of CCPP. In the 1880s, following the introduction of the disease in South Africa from Turkey, Hutcheon "vaccinated" goats by inoculating infected lung extracts subcutaneously.[14]

Since then, many different preparations have been proposed for the prevention of CBPP and CCPP, based either on live, attenuated strains or on inactivated antigens.

For CBPP control, several inactivated vaccine preparations have been tested experimentally, though in the field, live, attenuated vaccines have been used for decades.[79] These vaccines were developed from MmmSC strains displaying reduced virulence upon isolation, which were attenuated by serial passage in heterologous tissue and/or culture medium so as to allow subcutaneous inoculations in cattle without the development of Willems' reactions. The best known examples are strain V5 from Australia[80] and strains KH3J[81] and T1[82] from Africa.

Strain V5 originated from an Australian isolate that was passaged in broth and directly used as a liquid vaccine to be administered at the tip of the tail. It was successfully used in Australia as part of the eradication strategy in this country, allowing a dramatic reduction of prevalence before more drastic stamping-out measures were put in place.[83] Vaccine strain KH3J, from southern Sudan, was passaged up to 85 times in vitro.[84] Although no pathogenic effects were recorded upon subcutaneous inoculation, KH3J was abandoned due to low and short duration of immunity and it was replaced by vaccines derived from strain T1, which are still extensively used. Strain T1, from Tanzania, was attenuated by 44 passages in embryonated eggs followed by several passages in broth, resulting in vaccine strain T1/44.[82] T1SR, a streptomycin-resistant variant, was obtained after three additional passages in broth with increasing streptomycin concentrations to be used in combination with a rinderpest vaccine that contained this antibiotic. This variant was successfully used in West Africa during the massive rinderpest vaccination campaigns but it has been abandoned since the eradication of rinderpest in these countries.

T1 vaccines have been shown to be effective in experimental trials[85] and after extensive use in the field. The main advantage of these vaccines is their relatively low production costs. The fact that they only induce a transient seroconversion (generally no longer than 3 months) can be turned into an advantage, as it allows the detection of outbreaks, but it hampers seromonitoring of vaccination campaigns. Other drawbacks are their thermolability and the fact that only limited protection is afforded by a single administration, while the duration of this immunity is rather short (6 months for T1SR and up to 1 year for T1/44). However, T1SR is completely safe, whereas T1/44 presents some residual virulence and Willems' reactions are occasionally observed.[86]

On the other hand, CCPP vaccines used in the field consist in inactivated, adjuvanted, whole mycoplasmas. Preparations adjuvanted with saponin[87] have been experimentally tested. A modified version of the latter has been used for years in Kenya, and seems to afford good and relatively long-lasting protection. The main advantages of these vaccines are their thermostability and their compatibility with antibiotic

treatments. As they induce a seroconversion, the efficacy of vaccination campaigns may be monitored, though this may hamper outbreak detection. The main drawback is their high production costs.

66.7 CONCLUSIONS AND FUTURE PERSPECTIVES

66.7.1 CBPP

At the end of the 1980s, CBPP had its smallest distribution, thanks to coordinated eradication efforts that had taken place during the 1960s and 1970s. It persisted only in sub-Saharan Africa, where its economical importance was also minimized by internationally funded vaccination campaigns. The final and global eradication of this disease seemed reasonably achievable.

However, 30 years later, CBPP is progressively expanding. It is now present, or highly suspected, in Africa in the whole region comprised between the two tropics. This expansion results from various factors. Funds allocated to fight animal diseases by African developing countries are very scarce. In parallel, international funds are now targeting other threatening diseases such as foot and mouth disease or avian influenza. Finally, a number of animal health economists are now considering CBPP as a private good, which can be managed at the farm level, rather than a transboundary disease that has to be tackled by the states. As a consequence, antibiotic treatment of CBPP cases is no longer considered a taboo, as it may reduce the farmers' economic losses.

The renewed interest in antibiotic treatments for CBPP is in fact raising new concerns. Although these treatments do not seem to increase the percentage of chronic carriers, they do not allow a complete cure and live mycoplasmas may persist in the treated animals. It is also of common knowledge that antibiotics are not properly administered in many countries where CBPP persists. The quality of the drugs may not be adequately controlled, vials may be stored incorrectly, and requested dosage and regimen not properly followed. Such improper management of antibiotic use may have very detrimental effects, promoting the emergence of antibioresistance in pathogenic bacteria. This may favor the persistence of MmmSC-infected animals that may be difficult to identify.

With CBPP being present in such a vast majority of African states, the risk for neighboring regions has never been so high. North Africa, the Indian Ocean, and the Middle East present the highest risk due to imports or smuggling of live animals.

In Europe, the periodic reoccurrence of CBPP at 10–20 years' interval is also a source of concern. Till now there is no scientific or technical explanation for the unnoticed persistence of infection in between two outbreaks. The rarity of such events is an indication that the biological events that may be at the source of such outbreaks have a very low frequency, which makes them even more difficult to identify. Two hypotheses have been raised: the persistence of a pathogenic MmmSC strain in a nonbovine reservoir or the persistence of a low pathogenic MmmSC strain in cattle with a reversion to virulence at some point of time. Surveillance should therefore focus on the regular and systematic isolation or detection of mycoplasma strains that circulate in small and large ruminants. This also calls for the systematic bacterial analysis of CBPP-like lesions which are observed at the slaughterhouses.

Finally, until complete eradication is achieved, CBPP should continue to be considered as a real threat. Although eradication was achieved in many countries or continents in the past, the situation prevailing in many African countries today is preventing the same successful strategies to be applied there. The most limiting factor for eradication is the short immunity afforded by the present CBPP vaccines, added to the inability to control animal movements and to implement stamping-out policies. Alternative and more innovative strategies must therefore be put in place to fight CBPP. The first priority should focus on the generation of new vaccines inducing longer immunity. The second one should be to design multivalent vaccines allowing a drastic reduction of the vaccination costs and permitting international funds targeting other diseases to benefit CBPP control. Finally, a controlled use of antibiotics may be an alternative to slaughter. A strategy combining vaccination and antibiotic treatments may therefore pave the way for CBPP eradication. What is certain is that, if nothing is done, CBPP will certainly continue its expansion.

66.7.2 CCPP

What was suspected for a long time has now been confirmed: CCPP is a disease having a worldwide distribution extending from Central Africa to China. Its distribution may even be greater but this awaits confirmation. The American and Australian continents are free of the disease and protected by their isolation, the low number of goats, and the ban on importation of live animals. Europe is also free of CCPP but should be considered at risk. Goat populations are quite numerous in countries north of the Mediterranean basin such as Spain, France, Italy, and Greece, where goat cheese is highly appreciated and part of a cultural heritage together with olive oil. CCPP has already been evidenced in the European part of Turkey which is bordering the European Union (Bulgaria and Greece), calling for renewed surveillance efforts.

In fact, the recent development of new CCPP-specific diagnostic tools such as real-time PCR and competitive ELISA may ease these surveillance measures. However, the main limitation to this exercise is the lack of awareness concerning CCPP among veterinary services, which are more concerned with rapidly emerging diseases, and identified threats such as vector-borne viral infections (e.g., Bluetongue or Schmallenberg) or Peste des Petits Ruminants. CCPP may be difficult to spot and may remain unnoticed for quite a long time as goats are very often infected by other mycoplasmas of the mycoides cluster which may induce CCPP-like lesions. Due to the fastidious nature of Mccp isolation, attempts will certainly be met with failure and CCPP will be spotted only if specific tools are used. Accordingly, efforts should be made to improve veterinary services and diagnostic laboratory awareness.

A striking feature of CCPP is that wildlife may be affected by this disease. In fact, it is difficult to establish which species are really susceptible. Field observations are scarce for obvious reasons. However, this finding has two main consequences:

1. CCPP should be considered a threat to many endangered wild ungulate species. Great care should therefore be given to avoid contact between small ruminants and wildlife, as difficult as this may be.
2. Wildlife may be considered as a potential reservoir for pathogenic mycoplasmas, including Mccp. The exchange of wild animals for conservation purposes could in fact favor the emergence of mycoplasma diseases. This threat is quite insidious as ruminants are known to harbor pathogenic mycoplasmas in their ear canal without showing any clinical signs.

Owing to the distribution of the disease and its insidious nature, its eradication is an unlikely prospect. CCPP vaccines are quite effective but expensive to produce. In fact, there is a paucity of vaccine producers and the demand exceeds the offer. Until the development of the competitive ELISA, there was also a lack of efficient tools to evaluate the quality of the available products. The recognition of the wide distribution and economical impact of CCPP may be an incentive for more vaccine producers to develop CCPP vaccines. However, there is an urgent need for improved and cheaper products for these vaccines to be used on a large scale and this will be a difficult task due to Mccp fastidiousness.

It is therefore quite likely that antibiotics will be used to treat CCPP cases. As for CBPP, the uncontrolled use of antibiotics is a real cause of concern.

REFERENCES

1. Scheuchzer, J.J. *Fliegender Zungen-Krebs, eine Vieh-Seuche, welche anno 1732. die Eydgenössische Lande ergriffen*, p. 60 (Heidegger, Zürich, Switzerland, 1732).
2. de Haller, B. Mémoire sur la contagion parmi le bétail, mis au jour pour l'instruction du public, p. 32 (Berne, Switzerland, 1773).
3. Dupuy, V. et al. Evolutionary history of contagious bovine pleuropneumonia using next generation sequencing of *Mycoplasma mycoides* subsp. *Mycoides* "Small Colony". *PLoS One* 7, e46821 (2012).
4. Thiaucourt, F. et al. *Mycoplasma mycoides*, from "mycoides Small Colony" to "capri". A microevolutionary perspective. *BMC Genomics* 12, 114 (2011).
5. Curasson, G. Péripneumonie bovine. In: G. Curasson (ed.), *Traité de pathologie exotique vétérinaire et comparée*, vol. II, pp. 276–353 (Vigot Frères, Paris, France, 1942).
6. Hutyra, F., Marek, J., and Manniger, R. Contagious pleuropneumonia of cattle. In Greig, R. (ed.), *Special Pathology and Therapeutics of the Diseases of Domestic Animals*, vol. 1, pp. 455–471 (Baillère, Tindall, Cox, London, U.K. 1949).
7. Peires, J.B. *The Dead Will Arise*, p. 348 (Ravan, Johannesburg, South Africa, 1989).

8. de Rochebrune, A.T. Sur le Bos triceros Rochbr. et sur l' inoculation préventive de la péripneumonie épizootique par les Maures et les Pouls de la Sénégambie. *C. R. Acad. Sci. Hebd. Seances Acad. Sci. D* 100, 658–660 (1885).

9. Willems, L. Mémoires sur la pleuropneumonie épizootique du gros bétail. *Recl. Méd. Vét. Pratique (Maisons Alfort)* 3, 401–434 (1852).

10. Regalla, J. et al. Manifestation and epidemiology of contagious bovine pleuropneumonia in Europe. *Rev. Sci. Tech.* 15, 1309–1329 (1996).

11. Yaya, A. et al. Genotyping of *Mycoplasma mycoides* subsp. *mycoides* SC by multilocus sequence analysis allows molecular epidemiology of contagious bovine pleuropneumonia. *Vet. Res.* 39, 14 (2008).

12. Thomas, P. *Rapport médical sur le Bou Frida* (Jourdan, A., Alger, Algeria, 1873).

13. Hutcheon, D. Contagious pleuro-pneumonia in angora goats. *Vet. J.* 13, 171–180 (1881).

14. Hutcheon, D. Contagious pleuropneumonia in goats at Cape Colony, South Africa. *Vet. J.* 29, 399–404 (1889).

15. Thiaucourt, F. and Bolske, G. Contagious caprine pleuropneumonia and other pulmonary mycoplasmoses of sheep and goats. *Rev. Sci. Tech.* 15, 1397–1414 (1996).

16. Melanidi, C. and Stylianopoulos, M. La pleuropneumonie contagieuse des chèvres en Grèce. *Rev. Gén. Méd. Vét.* 37, 490–493 (1928).

17. Schellhase, W. Ein Beitrag zur Kenntnis der ansteckenden Lungenbrustfellentzundung der Ziegen in Deutsch-Ostafrica. *Zeit. Infect. Haustiere* 12, 70–83 (1912).

18. Mettam, R.W.M. Contagious pleuro-pneumonia of goats in East Africa. In *Pan African Veterinary Conference*, pp. 173–178, Pretoria, South Africa, 1929).

19. Pirani, A. Sulla pleuropomonite infettivo-contagiosa delle capre in Eritrea. *Profilassi* 5, 170–175 (1932).

20. Walker, G.K. Pleuro-pneumonia of goats in the Kangra district, Punjab, India. *J. Comp. Path. Ther.* 27, 68–71 (1914).

21. Longley, E.O. Contagious pleuro-pneumonia of goats. *Ind. J. Vet. Sci. Anim. Husb.* 10, 127–197 (1940).

22. MacOwan, K.J. and Minette, J.E. A mycoplasma from acute contagious caprine pleuropneumonia in Kenya. *Trop. Anim. Health Prod.* 8, 91–95 (1976).

23. Ozdemir, U. et al. Contagious caprine pleuropneumonia in the Thrace region of Turkey. *Vet. Rec.* 156, 286–287 (2005).

24. Çetinkaya, B. et al. Detection of contagious caprine pleuropneumonia in East Turkey. *Rev. Sci. Tech. OIE* 28, 1037–1044 (2009).

25. Manso-Silvan, L. et al. Multi-locus sequence analysis of *Mycoplasma capricolum* subsp. *capripneumoniae* for the molecular epidemiology of contagious caprine pleuropneumonia. *Vet. Res.* 42, 86 (2011).

26. Chu, Y. et al. Genome sequence of *Mycoplasma capricolum* subsp. *capripneumoniae* strain M1601. *J. Bacteriol.* 193, 6098–6099 (2011).

27. Arif, A. et al. Contagious caprine pleuropneumonia outbreak in captive wild ungulates at Al Wabra Wildlife Preservation, State of Qatar. *J. Zoo Wildl. Med.* 38, 93–96 (2007).

28. Cottew, G.S. et al. Taxonomy of the *Mycoplasma mycoides* cluster. *Isr. J. Med. Sci.* 23, 632–635 (1987).

29. Manso-Silvan, L. et al. *Mycoplasma leachii* sp. nov. as a new species designation for *Mycoplasma* sp. bovine group 7 of Leach, and reclassification of *Mycoplasma mycoides* subsp. *mycoides* LC as a serovar of *Mycoplasma mycoides* subsp. *capri*. *Int. J. Syst. Evol. Microbiol.* 59, 1353–1358 (2009).

30. Leach, R.H. Comparative studies of mycoplasma of bovine origin. *Ann. N.Y. Acad. Sci.* 143, 305–316 (1967).

31. Bonnet, F. et al. DNA relatedness between field isolates of *Mycoplasma* F38 group, the agent of contagious caprine pleuropneumonia, and strains of *Mycoplasma capricolum*. *Int. J. Syst. Bacteriol.* 43, 597–602 (1993).

32. Nocard, E. et al. Le microbe de la péripneumonie. *Ann. Inst. Pasteur (Paris)* 12, 240–262 (1898).

33. Manso-Silván, L., Perrier, X., and Thiaucourt, F. Phylogeny of the *Mycoplasma mycoides* cluster based on analysis of five conserved protein-coding sequences: Consequences in taxonomy. *Int. J. Syst. Evol. Microbiol.* 57, 2247–2258 (2007).

34. Masiga, W.N. and Windsor, R.S. Some evidence of an age susceptibility to contagious bovine pleuropneumonia. *Res. Vet. Sci.* 24, 328–333 (1978).

35. Brandao, E. Isolation and identification of *Mycoplasma mycoides* subspecies *mycoides* SC strains in sheep and goats. *Vet. Rec.* 136, 98–99 (1995).

36. Masiga, W.N., Domenech, J., and Windsor, R.S. Manifestation and epidemiology of contagious bovine pleuropneumonia in Africa. *Rev. Sci. Tech.* 15, 1283–1308 (1996).

37. Shiferaw, T.J., Moses, K., and Manyahilishal, K.E. Participatory appraisal of foot and mouth disease in the Afar pastoral area, northeast Ethiopia: Implications for understanding disease ecology and control strategy. *Trop. Anim. Health Prod.* 42, 193–201 (2010).

38. Stradaioli, G. et al. *Mycoplasma mycoides* subsp. *mycoides* SC identification by PCR in sperm of seminal vesiculitis-affected bulls. *Vet. Res.* 30, 457–466 (1999).

39. Windsor, R.S. and Masiga, W.N. Investigations into the role of carrier animals in the spread of contagious bovine pleuropneumonia. *Res. Vet. Sci.* 23, 224–229 (1977).

40. Niang, M. et al. Experimental studies on long-acting tetracycline treatment in the development of sequestra in contagious bovine pleuropneumonia-infected cattle. *J. Vet. Med. Anim. Health* 2, 35–45 (2010).

41. Huebschle, O.J. et al. Danofloxacin (Advocin) reduces the spread of contagious bovine pleuropneumonia to healthy in-contact cattle. *Res. Vet. Sci.* 81, 304–309 (2006).

42. Cheng, X. et al. Insertion element IS1296 in *Mycoplasma mycoides* subsp. *mycoides* small colony identifies a European clonal line distinct from African and Australian strains. *Microbiology* 141, 3221–3228 (1995).

43. Vilei, E.M. et al. Genomic and antigenic differences between the European and African/Australian clusters of *Mycoplasma mycoides* subsp. *mycoides* SC. *Microbiology* 146, 477–486 (2000).

44. McAuliffe, L., Ayling, R.D., and Nicholas, R.A. Identification and characterization of variable-number tandem-repeat markers for the molecular epidemiological analysis of *Mycoplasma mycoides* subspecies *mycoides* SC. *FEMS Microbiol. Lett.* 276, 181–188 (2007).

45. Ostrowski, S. et al. Fatal outbreak of *Mycoplasma capricolum* pneumonia in endangered markhors. *Emerg. Infect. Dis.* 17, 2338–2341 (2011).

46. Cottew, G.S. and Yeats, F.R. Mycoplasmas and mites in the ears of clinically normal goats. *Aust. Vet. J.* 59, 77–81 (1982).

47. DaMassa, A.J. and Brooks, D.L. The external ear canal of goats and other animals as a mycoplasma habitat. *Small Ruminant Res.* 4, 85–93 (1991).

48. Chu, Y. et al. Molecular detection of a mixed infection of Goatpox virus, Orf virus, and *Mycoplasma capricolum* subsp. *capripneumoniae* in goats. *J. Vet. Diagn. Invest.* 23, 786–789 (2011).

49. Lefevre, P.C. et al. *Mycoplasma* species F 38 isolated in Chad. *Vet. Rec.* 121, 575–576 (1987).

50. Thiaucourt, F. et al. Contagious caprine pleuropneumonia in Ethiopia. *Vet. Rec.* 131, 585 (1992).

51. Pettersson, B. et al. Molecular evolution of *Mycoplasma capricolum* subsp. *capripneumoniae* strains, based on polymorphisms in the 16S rRNA genes. *J. Bacteriol.* 180, 2350–2358 (1998).

52. Pilo, P., Frey, J., and Vilei, E.M. Molecular mechanisms of pathogenicity of *Mycoplasma mycoides* subsp. *mycoides* SC. *Vet. J.* 174, 513–521 (2007).

53. Bischof, D.F. et al. Cytotoxicity of *Mycoplasma mycoides* subsp. *mycoides* small colony type to bovine epithelial cells. *Infect. Immun.* 76, 263–269 (2008).

54. Buttery, S.H., Lloyd, L.C., and Titchen, D.A. Acute respiratory, circulatory and pathological changes in the calf after intravenous injections of the galactan from *Mycoplasma mycoides* subsp. *mycoides*. *J. Med. Microbiol.* 9, 379–391 (1976).

55. Lloyd, L.C., Buttery, S.H., and Hudson, J.R. The effect of the galactan and other antigens of *Mycoplasma mycoides* var. *mycoides* on experimental infection with that organism in cattle. *J. Med. Microbiol.* 4, 425–439 (1971).

56. McAuliffe, L. et al. Biofilm-grown *Mycoplasma mycoides* subsp. *mycoides* SC exhibit both phenotypic and genotypic variation compared with planktonic cells. *Vet. Microbiol.* 129, 315–324 (2008).

57. Persson, A. et al. Variable surface protein Vmm of *Mycoplasma mycoides* subsp. *mycoides* small colony type. *J. Bacteriol.* 184, 3712–3722 (2002).

58. Rurangirwa, F.R. et al. Composition of a polysaccharide from mycoplasma (F-38) recognised by antibodies from goats with contagious pleuropneumonia. *Res. Vet. Sci.* 42, 175–178 (1987).

59. Cottew, G.S. and Yeats, F.R. Subdivision of *Mycoplasma mycoides* subsp. *mycoides* from cattle and goats into two types. *Aust. Vet. J.* 54, 293–296 (1978).

60. ter Laak, E.A. Identification of mycoplasma F38 biotype. *Vet. Rec.* 129, 295 (1991).

61. Bashiruddin, J.B. et al. Use of the polymerase chain reaction to detect mycoplasma DNA in cattle with contagious bovine pleuropneumonia. *Vet. Rec.* 134, 240–241 (1994).

62. Bascunana, C.R. et al. Characterization of the 16S rRNA genes from *Mycoplasma* sp. strain F38 and development of an identification system based on PCR. *J. Bacteriol.* 176, 2577–2586 (1994).

63. Dedieu, L., Mady, V., and Lefevre, P.C. Development of a selective polymerase chain reaction assay for the detection of *Mycoplasma mycoides* subsp. *mycoides* S.C. (contagious bovine pleuropneumonia agent). *Vet. Microbiol.* 42, 327–339 (1994).

64. Woubit, S. et al. A specific PCR for the identification of *Mycoplasma capricolum* subsp. *capripneumoniae*, the causative agent of contagious caprine pleuropneumonia (CCPP). *Vet. Microbiol.* 104, 125–132 (2004).

65. Lorenzon, S. et al. Specific PCR identification of the T1 vaccine strains for contagious bovine pleuropneumonia. *Mol. Cell. Probes* 14, 205–210 (2000).

66. Fitzmaurice, J. et al. Real-time polymerase chain reaction assays for the detection of members of the mycoides cluster. *N.Z. Vet. J.* 56, 40–47 (2008).

67. Lorenzon, S., Manso-Silvan, L., and Thiaucourt, F. Specific real-time PCR assays for the detection and quantification of *Mycoplasma mycoides* subsp. *mycoides* SC and *Mycoplasma capricolum* subsp. *capripneumoniae*. *Mol. Cell. Probes* 22, 324–328 (2008).

68. Gorton, T.S. et al. Development of real-time diagnostic assays specific for *Mycoplasma mycoides* subspecies *mycoides* Small Colony. *Vet. Microbiol.* 111, 51–58 (2005).

69. Schnee, C. et al. Assessment of a novel multiplex real-time PCR assay for the detection of the CBPP agent *Mycoplasma mycoides* subsp. *mycoides* SC through experimental infection in cattle. *BMC Vet. Res.* 7, 47 (2011).

70. Campbell, A.D. and Turner, A.W. Studies on contagious bovine pleuropneumonia of cattle. II. A complement fixation test for the diagnosis of contagious bovine pleuropneumonia. Its use in experimental investigations and in the control of disease. *Bull. Counc. Sci. Ind. Res.* 97, 11–52 (1936).

71. Turner, A.W. and Campbell, A.D. A note on the application of the complement fixation test to the control of bovine pleuropneumonia. *Aust. Vet. J.* 13, 183–186 (1937).

72. Le Goff, C. and Thiaucourt, F. A competitive ELISA for the specific diagnosis of contagious bovine pleuropneumonia (CBPP). *Vet. Microbiol.* 60, 179–191 (1998).

73. Thiaucourt, F. et al. The use of monoclonal antibodies in the diagnosis of contagious caprine pleuropneumonia (CCPP). *Vet. Microbiol.* 41, 191–203 (1994).

74. Muuka, G. et al. Comparison of complement fixation test, competitive ELISA and LppQ ELISA with post-mortem findings in the diagnosis of contagious bovine pleuropneumonia (CBPP). *Trop. Anim. Health Prod.* 43, 1057–1062 (2011).

75. Schubert, E. et al. Serological testing of cattle experimentally infected with *Mycoplasma mycoides* subsp. *mycoides* Small Colony using four different tests reveals a variety of seroconversion patterns. *BMC Vet. Res.* 7, 72, 11 (2011).

76. Lindley, E.P. The rapid slide agglutination test in the control of contagious bovine pleuropneumonia in Nigeria. *Bull. Epiz. Dis. Afr.* 6, 373–375 (1959).

77. March, J.B., Gammack, C., and Nicholas, R. Rapid detection of contagious caprine pleuropneumonia using a *Mycoplasma capricolum* subsp. *capripneumoniae* capsular polysaccharide-specific antigen detection latex agglutination test. *J. Clin. Microbiol.* 38, 4152–4159 (2000).

78. Ayling, R.D. et al. Assessing the in vitro effectiveness of antimicrobials against *Mycoplasma mycoides* subsp. *mycoides* small-colony type to reduce contagious bovine pleuropneumonia infection. *Antimicrob. Agents Chemother.* 49, 5162–5165 (2005).

79. Provost, A. et al. Contagious bovine pleuropneumonia. *Rev. Sci. Tech. Off. Int. Epiz.* 6, 625–679 (1987).

80. Hudson, J.R. Contagious bovine pleuropneumonia. The keeping properties of the V5 vaccine used in Australia. *Aust. Vet. J.* 44, 123–129 (1968).

81. Hudson, R. Contagious bovine pleuropneumonia: The immunizing value of the attenuated strain KH3J. *Aust. Vet. J.* 41, 43–49 (1965).

82. Sheriff, D. and Piercy, S.E. Experiments with an avianised strain of the organism of contagious bovine pleuropneumonia. *Vet. Rec.* 64, 615–621 (1952).

83. Newton, L.G. Contagious bovine pleuropneumonia in Australia: Some historic highlights from entry to eradication. *Aust. Vet. J.* 69, 306–317 (1992).

84. Doutre, M. Valeur de l'immunité conférée par deux vaccins lyophilisés préparés à l'aide des souches KH3J et T1. *Bull. Off. Int. Epizoot.* 72, 103–129 (1969).

85. Gilbert, F.R. et al. The efficacy of T1 strain broth vaccine against contagious bovine pleuropneumonia: In-contact trials carried out six and twelve months after primary vaccination. *Vet. Rec.* 86, 29–33 (1970).

86. Lindley, E.P. Experiences with a lyophilised contagious bovine pleuropneumonia vaccine in the Ivory coast. *Trop. Anim. Health Prod.* 3, 32–42 (1971).

87. Rurangirwa, F.R. et al. Vaccination against contagious caprine pleuropneumonia caused by F38. *Isr. J. Med. Sci.* 23, 641–643 (1987).

67 *Rathayibacter toxicus*

Ian T. Riley, Jeremy G. Allen, and Martin J. Barbetti

CONTENTS

67.1 INTRODUCTION

Rathayibacter toxicus is a Gram-positive, coryneform bacterium that infests the foliage and floral structures of grasses and is economically important because of its toxin-producing capability. The bacterium is transferred from infested soils into plants by plant-parasitic nematodes of the genus *Anguina*, which produce foliar or seedgalls in several species of grasses. *R. toxicus* can proliferate within the nematode galls, killing the nematodes and completely filling the lumen of the gall (forming a bacterial gall). Often, bacterial growth in an inflorescence is sufficient to cause oozing from the floral structures as a yellow bacterial slime (gummosis). Ingestion of the toxins (glycolipids known as corynetoxins) by grazing animals results in poisoning that is often fatal.

Given that *R. toxicus* is an organism that both colonizes plants and affects animals through the toxins it produces, the following discussion considers these aspects as distinct but interlinked. For example, managing bacterial colonization of the host plant is important to minimize exposure of animals to the toxins but is quite different from the management of animals exposed to the toxins to minimize clinical effects and economic loss. The impact of *R. toxicus* and its vector on the growth and reproduction of the plant host is minimal compared to most economically important plant pathogens. For this reason, after the description of *R. toxicus* as a nematode-vectored, plant-colonizing organism, this chapter will focus on its toxicity. This toxicosis, corynetoxin poisoning, is known as annual ryegrass toxicity or flood plains staggers depending on the host and vector involved.

The tripartite associations *R. toxicus* forms with host grass and vector nematode differ between regions. Details of these associations including the host and vector, name given to poisoning of livestock, geographic region, occurrence of outbreaks, and current status are summarized in Table 67.1. This chapter cannot cover all details of these complex relationships, the contributing factors in each circumstance, and the economic impacts, so the reader is directed to relevant reviews [1–4] for additional information.

67.2 CLASSIFICATION AND MORPHOLOGY

Kingdom: Bacteria
Phylum: Actinobacteria
Class: Actinobacteria
Order: Actinomycetales
Family: Microbacteriaceae Park et al. 1995
Genus: *Rathayibacter* Zgurskaya et al. 1993
Species: *Rathayibacter toxicus* (Riley and Ophel 1992) Sasaki et al. 1998

R. toxicus was initially described as *Clavibacter toxicus* [5] with the transfer to the genus *Rathayibacter* proposed by Sasaki et al. [6]. *Rathayibacter*, when first proposed by Zgurskaya et al. [7], consisted of three species of plant-associated coryneform bacteria, *Rathayibacter iranicus*, *Rathayibacter tritici*, and *Rathayibacter rathayi*, all of which had been placed in the genus *Clavibacter* [8]. The genus represents a consistent cluster within the Microbacteriaceae phylogenetic branch [6,9,10] with the following chemotaxonomic properties: peptidoglycan of the B type based on 2,4-diaminobutyric acid (L-isomer); major menaquinone of MK-10; phosphatidylglycerol and diphosphatidylglycerol as principal phospholipids; and predominantly saturated anteiso- and isomethyl-branched fatty acids [6,7].

Before the application of chemotaxonomic and molecular methods, species of *Rathayibacter* were classified on the basis of physiological tests, host range, and the nature of their host infection [11]. Subsequently, *Rathayibacter* spp. were shown to differ in their cellular protein patterns [12], cell-wall sugar composition [7,8], allozyme pattern [13], enzyme profiles [14], polyamine content [15], and most recently, 16S rDNA analysis [16]. Dorofeeva et al. [16] proposed a further two species, *Rathayibacter caricis* and *Rathayibacter festucae*, giving the genus six named species. However, *R. toxicus* is the only member of its genus to produce toxins and cause economic impact on animal industries.

Cells of *R. toxicus* are non-spore-forming, nonmotile, coryneform rods of 0.6–0.75 by 1.5 μm, including a capsule that is 0.08–0.14 μm thick [17]. Surface colonies on agar are typically convex, smooth, entire, mucoid, and glistening, with light- to mid-yellow pigmentation [5]. In old growth (>3 weeks), pale yellow variants are common and form streaks or convex protrusions, and they maintain their pigmentation in subsequent cultures [5].

67.3 BIOLOGY AND EPIDEMIOLOGY

R. toxicus is a nematode-associated bacterium. It adheres to the cuticle of the second-stage juveniles (J2) of certain *Anguina* spp. allowing it to be carried into the host plant where it colonizes intraplant spaces and galls initiated by the vector nematode (Figure 67.1). In order to understand the biology and epidemiology of the bacterium, it is important to first describe the life history of its vector nematodes. A schematic of the *R. toxicus* disease cycle is provided in Figure 67.2, showing the development of the bacterium and toxin relative to that of the nematode and host.

In nature, there are two identified vectors of *R. toxicus*, *Anguina funesta* in *Lolium rigidum* (annual ryegrass) and *Anguina paludicola* in *Lachnagrostis filiformis* (annual blowngrass) and *Polypogon monspeliensis* (annual beard grass). Another anguinid vector of a toxigenic bacteria found in *Festuca nigrescens* (Table 67.1) has not been identified, and the bacterium is considered to be related but not identical to *R. toxicus*. The biology of this anguinid has not been studied in detail, but is likely to be similar to *A. funesta*. In addition to the vectors found in nature, inoculation experiments (both direct to nematode juveniles in vitro and to nematode-inoculated host plants grown in pots outdoors) have shown that other *Anguina* spp. can act as vectors for *R. toxicus* (Table 67.1).

Here, the life history of the two identified vectors is described briefly; for a more detailed description of these, see the recent review by Riley and Barbetti [4]. In addition, the potential for *R. toxicus* to form novel host/vector associations is further discussed (Section 67.3.3), as this is an additional reason why *R. toxicus* could be regarded as a security sensitive microbe.

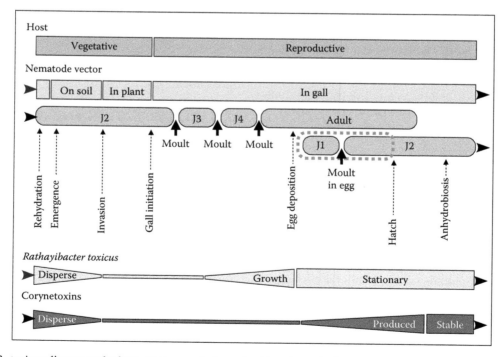

FIGURE 67.1 *R. toxicus* disease cycle; host grass, nematode vector (*Anguina* sp.), *R. toxicus*, and corynetoxin development over a full growing season. In the case of *R. toxicus* in *L. rigidum* with *A. funesta* as the vector in a Mediterranean climate zone, this would be from the autumn rains through early summer. *Note*: not to scale and relative timings indicative only.

FIGURE 67.2 *R. toxicus* in *L. rigidum*. From left to right: (1) nematode gall fully colonized by *R. toxicus*, (2) *A. funesta* gall, (3) *L. rigidum* seed, and (4) *L. rigidum* diaspore. The diaspore normally contains a single seed (3), which can be replaced by either gall type, (1) or (2), and still appear superficially the same.

TABLE 67.1

Tripartite Associations of *Rathayibacter toxicus* with Host Grass and Vector Nematode, and Where Applicable, the Name Currently Given to the Consequent Corynetoxin Poisoning of Livestock, Geographic Region, and Period of Occurrence of Outbreaks of Toxicity, and Current Status of Outbreaks

Host Grass	Vector Nematode	Corynetoxin Poisoning	Geographic Region	Occurrence	Current Status	References
Festuca nigrescens[a]	Undetermined	—	Oregon, United States	1939–1960s	No recent outbreaks	[70–72]
Lolium rigidum	*Anguina funesta*	Annual ryegrass toxicity	South Australia, Australia	1956–present	Continuing infrequent outbreaks	[18,73,74]
			Western Australia, Australia	1968–present	Continuing annual outbreaks	[66,75,76]
			South Africa	1979/1980, 2009	Rare (or unreported) outbreaks	[77,78]
Polypogon monspeliensis	*Anguina paludicola*	Floodplain staggers	South Australia, Australia	c. 1970	No recent outbreaks	[33,79,80]
Lachnagrostis filiformis	*A. paludicola*	Floodplain staggers	New South Wales, Australia	1990/1991	No recent outbreaks	[33,37,80,81]
Triticum aestivum	*Anguina tritici*	—	—	—	Experimental plant inoculation only	[19]
Ehrharta longiflora	*Anguina australis*	—	—	—	Experimental plant inoculation only	[41]

[a] The toxins in this association were similar but not identical to corynetoxins, so it is possible that the bacterium was a regional variant of *R. toxicus* or a closely related species [69].

67.3.1 *RATHAYIBACTER TOXICUS* IN ASSOCIATION WITH *ANGUINA FUNESTA*

R. toxicus was first known from its association with *A. funesta*, a seedgall nematode that induces galls in the ovaries of *L. rigidum* and some closely related grasses [18,19]. The main host, *L. rigidum*, is an important pasture grass as well as a major crop weed [20] across southern Australia. In this tripartite association of bacterium, nematode, and pasture grass, grazing animals are exposed to sufficient of the bacterial toxins for economic losses to have occurred and for the cause to have been investigated. As the bacterium adheres to J2 of *A. funesta*, the survival, dispersal, and invasive stage of the nematode, this is an appropriate stage from which to describe this association [21–25].

The J2 that carry *R. toxicus* are initially contained in the nematode galls that replace the seed within the diaspores of the host. The diaspores, with either seed or galls, will remain where produced, or be dispersed over short distances by wind and surface water flow, or over long distances in hay and harvested pasture seed or as a contaminant in grain. Given the size and structure of the diaspores is not altered, the same dispersal mechanisms operate on nematode galls and host seed, allowing for perpetuation of the life cycle. With the arrival of the autumn rains in the Mediterranean climate zone of southern Australia, the anhydrobiotic J2 rehydrate and emerge from the galls. The J2 can survive through several hydration/dehydration cycles [26]. On a damp soil surface, the J2 will actively seek a host plant to invade [27], probably assisted by rain splash. The J2 congregate at the apex of infested tillers and initiate galls by invading and modifying developing host ovaries. Within the gall, the J2 feed and molt through two further juvenile stages to adults (females and males, usually a few but up to 20 per gall [28]). Following mating, the females deposit up to several hundred eggs per gall [28]. Within the eggs, the nematodes develop and molt to J2. Hatching occurs within the gall with the hatched J2 maturing physiologically to survive desiccation. The life cycle is complete by late spring to early summer, coinciding with the maturity but before senescence of the host.

When *R. toxicus* is carried into the host inflorescence and nematode galls by adhering to the nematode cuticle, the resulting colonization can range from individual nematode galls through to the whole inflorescence with yellow gummosis [17]. The development of the nematode within colonized galls is inhibited, and in fully colonized galls, no eggs are produced. In the colonized galls, *R. toxicus* produces potent glycolipid toxins known as corynetoxins [29] that inhibit *N*-glycosylation of proteins and are often fatal to livestock [30]. The bacterium within the galls or colonized plant tissue will survive over summer to be dispersed by the autumn rains and again adhere to *A. funesta* J2. The potential for the bacterium to persist in the environment as a soil or phytoplane organism in the absence of a nematode vector has not been established.

Corynetoxins are produced by the bacterium late in the disease cycle, apparently as a stationary phase product.

A bacteriophage (useful in bacterial typing, see Section 67.5.1) also increases within the bacterial cell mass late in the disease cycle. It is possible that these two processes are interlinked, as in culture the phage appears to be involved in toxin production [31]. However, the assay of toxin and phage concentrations in field samples did not confirm this relationship [32].

R. toxicus in *L. rigidum* is widespread in South Australia (SA) and Western Australia, but it does not occur as an economic problem throughout the full range of the host or the nematode. Neither has the nematode been found across the full range of its known alternative hosts such as *Lolium perenne* (perennial ryegrass) and *Festuca* spp. [18], which are widely grown in the higher rainfall areas of southern Australia. It is possible that further spread is still to occur, but it is also likely that various environmental factors constrain this spread. Riley and Barbetti [4] suggested that the timing and intensity of autumn rains, time of host germination, host density, interplant competition, and grazing frequency and intensity could be important factors. For the populations of the vector nematode to develop and persist, nematode invasion is best synchronized with initiation of host ovaries [24], with the growing season of sufficient length and land management practices such that infested tillers survive through to maturity. In the case of *L. rigidum*, its occurrence as a major weed in crops has facilitated the latter. Additionally, *Dilophospora alopecuri*, a fungal antagonist of many *Anguina* spp., is now widely established in southern Australia, which might also be limiting further spread of *R. toxicus* in *L. rigidum* as an economic problem.

Beyond southern Australia, *R. toxicus* in *L. rigidum* has only been recorded in South Africa (Table 67.1). It is considered that that population arose from infested seed imported from Australia. Riley and Barbetti [4] speculated that in Australia, both *R. toxicus* and its vector are naturalized exotics and that they have not come to the attention of science in their native range as the special conditions leading to high population densities and economic problems do not occur.

67.3.2 *RATHAYIBACTER TOXICUS* IN ASSOCIATION WITH *ANGUINA PALUDICOLA*

In Australia, *R. toxicus* is also found in association with *A. paludicola* in two host grasses, *L. filiformis* (annual blowngrass) and *P. monspeliensis* (annual beard grass) [33]. Although there are similarities to the association of the bacterium with *A. funesta*, the adaptation of *A. paludicola* to survive in seasonally inundated ecosystems makes for some important differences. The summary in the succeeding text highlights these differences; for more details, see the reports of Bourke et al. [34], Bryden et al. [35], Bertozzi and McKay [36], Davis et al. [37], and Bertozzi [38].

The biology of *A. paludicola* and its association with *R. toxicus* in *P. monspeliensis* has been studied in detail [38]. In some poorly drained pastures of southeast SA,

P. monspeliensis becomes dominant with flooding in wet years, whereas *Hordeum leporinum* (barley grass) is dominant in dry years. *A. paludicola* is adapted to colonizing its host under flooded conditions. It induces galls in apical meristems and panicle branches as well as ovary initials, produces two generations of galls per year, emerges rapidly from rehydrated galls to invade the host, and can survive some degree of inundation.

With flooding that can follow autumn rains in the southeast of SA, the ability of *A. paludicola* to emerge, invade, and induce galls quickly is important. Host invasion can happen within 24 h compared to up to 1 month for *A. funesta* [18]. Galls are initiated in a vegetative apex rather than waiting for floral initiation, as in *A. funesta*, allowing *A. paludicola* to complete a life cycle before midwinter. These galls provide a second cohort of juveniles to invade new tillers that develop after flooding recedes in the spring. In the spring life cycle, seedgalls and a minor proportion of panicle branch galls are formed. After gall initiation, the development from J2 to adults takes up to 14 days, which is similar to *A. funesta* [18]. In contrast to *A. funesta*, *A. paludicola* induces enlargement of infested florets. The final stages of the life cycle of *A. paludicola* parallel that of *A. funesta*, with the mature galls containing anhydrobiotic J2 that survive the summer to emerge with the autumn rains of the next year.

R. toxicus has been observed to colonize the three gall types (apical, panicle, and seedgalls) induced in *P. monspeliensis*, as well as cause gummosis of the entire inflorescence. Corynetoxins at concentrations lethal to livestock are produced [39] with the resultant poisoning called floodplain staggers.

Much less is known of *R. toxicus* and *A. paludicola* in *L. filiformis*. The nematode forms apical and seedgalls. High populations of the host, nematode, and bacterium and lethal concentrations of corynetoxins are only known to have occurred in the upper catchment of the Darling River, New South Wales (NSW) in 1991, after several seasons with winter rainfall or floods (conditions that have not occurred since).

In the southeast of SA, *A. paludicola* was detected in over a third of sites with *P. monspeliensis* during a 1991–1992 survey [38]. In NSW, during surveys in 1993 and 1996, the nematode was found more widely in *L. filiformis* than during the 1991 outbreaks, but still within the same general area. Only a single infested population of *P. monspeliensis* has been found in NSW. Although the vector nematode exhibits host specificity (which could be due to physiological specialization of the local nematode ecotype or an interaction of host and its environment), *R. toxicus* appears not to be host specific, but rather appears to be dependent on the presence of a suitable vector nematode.

67.3.3 POTENTIAL FOR NOVEL ASSOCIATIONS

Given that in nature *R. toxicus* colonizes three main grass hosts and seems to be limited by its vector nematode more so than the host, it is feasible that novel associations could develop. This was tested experimentally (Table 67.1) by adding *A. funesta* galls colonized with *R. toxicus* to (1) *Triticum*

aestivum (wheat) grown in pots also inoculated with *Anguina tritici* [40] and (2) *Ehrharta longiflora* (annual veldtgrass) with *Anguina australis* [41]. In both cases, *R. toxicus* was able to colonize the hosts and produce toxin. Also, in vitro inoculation of nematodes found that *R. toxicus* can adhere to *Anguina* spp. with known ranges allopatric to that of the bacterium [42]. A third line of evidence is the occasional record of *R. toxicus* colonizing grasses other than *L. rigidum* in conjunction with infected *L. rigidum* populations. In dense populations, it appears that the nematode will inadvertently carry the bacterium into nonhost (i.e., for the nematode) grasses in which it can grow (e.g., *Austrodanthonia caespitosa*, *Avena sativa*, *Hordeum* sp., [43,44]). Although *R. toxicus* has only become a widespread, serious, and persistent economic problem under particular circumstances that appear to be unique to Australia, the evidence indicates that new associations could develop following its introduction to other areas. However, economic impacts are only likely to occur if the new circumstances allow the bacterium, nematode, and host to develop through to maturity with moderate to high population densities over a series of years and where livestock are exposed to the toxins.

The poisoning of livestock in Oregon, United States, from 1939 for about two decades (Table 67.1) by toxins similar to corynetoxins produced by a bacterium closely related to *R. toxicus* (or a regional variant thereof) highlights the potential for not only different vectors but also new bacterial genotypes.

67.4 CLINICAL FEATURES AND PATHOGENESIS

Clinical signs of corynetoxin poisoning in livestock (summarized by Allen [45] from Purcell [46], Bourke [47], Cheeke [48]) appear abruptly, usually following some external stimulation. They include tremors, ataxia, adoption of a wide-based stance, stumbling and falling over, nystagmus, convulsions while recumbent, and death. Animals may appear to recover between episodes of ataxia and convulsions. Sheep often exhibit a high stepping gait with their forelimbs, while the head is held high. When forced to run, some sheep have a stiff-legged gait. Some cattle appear disorientated and wander aimlessly between episodes of convulsions. Signs may appear as soon as 4 days after introduction to toxic pasture or feed. Animals may continue to exhibit neurological signs and die for up to 10 days after removal from toxic feed. Ewes may abort.

Corynetoxins have been found to affect sheep, cattle, horses, donkeys, pigs, guinea pigs, rats, mice, and chickens [46] either when incidentally or deliberately exposed. The effects of corynetoxins are cumulative, so the total lethal dose is the same whether given as a single dose or as repeated smaller doses up to 2 months apart [49].

Most cases of corynetoxin poisoning occur in late spring to early summer as the host grass inflorescences mature and toxins reach their maximum concentrations. Although the toxins are stable and concentrations in colonized nematode galls do not decline in summer, the incidence of field outbreaks declines. The cause for this decline in incidence has not been determined, but could involve changes in grazing behavior of the livestock and/or reduced availability of the toxic material as the plant residues lodge and break up and toxic galls fall to the ground. Toxic material in harvested hay or feed grain remains toxic and can cause poisoning whenever consumed by livestock [50].

Postmortem examination usually reveals a pale-tan- to yellowish-colored liver, and the carcass may be mildly icteric. Small hemorrhages are commonly seen in the gallbladder and heart, and less commonly in the rumen, small intestine, kidney, and lymph nodes throughout the body. Elevated liver enzymes in the blood indicate hepatic damage caused by the toxins. Despite significant neurological signs, microscopic changes in the brain are mild, indistinct, and often absent [3].

67.5 IDENTIFICATION AND DIAGNOSIS

67.5.1 BACTERIAL CHARACTERIZATION

Description of *Rathayibacter toxicus* (after Riley and Ophel [5], and Sasaki et al. [6]): Cells are obligately aerobic, Gram-positive, non-spore-forming, nonmotile, coryneform rods of 0.6–0.75 by 1.5 μm, including a capsule of 0.08–0.14 μm in thickness. Surface colonies on agar are convex, smooth, entire, mucoid, and glistening, with light- to mid-yellow pigmentation. In old growth (>3 weeks), pale yellow variants are common and form streaks or convex protrusions, and they maintain their pigmentation in subsequent cultures. From cultured inoculum, single colonies take 5–6 days to appear, and from rehydrated anhydrobiotic material obtained from colonized ryegrass, colonies take 8–14 days to appear. Colonies are often variable in size because the adhesive properties of the mucoid capsule make dispersal to single cells difficult, especially with cells obtained from solid media and colonized plants. Optimal temperature for growth is 26°C, with no growth at 37°C. In the cell wall, peptidoglycans have amino acids, D-glutamic acid, glycine, D-alanine, and L-DAB, at a molar ratio of 1:1:1:2. The predominant isoprenoid quinone is MK-10. Acid is produced from galactose, mannose, and xylose. Acetate, glutamate, malonate, succinate, and tartrate are not utilized. It is positive for H_2S production from cysteine but negative for hydrolysis of starch and gelatin. The type strain (ICMP 9525 = NCPPB 3552) was isolated from ryegrass in SA.

Rathayibacter toxicus in culture: Isolation of *R. toxicus* from infected plant material is subject to contamination issues because this bacterium is relatively slow growing. The presumptive identification by an experienced operator can be made on colony growth rate, morphology, and color. However, there is no selective medium, and the confirmation of identification by biochemical or physiological tests is not possible. Diagnosis based on menaquinone composition of the 2,4-diaminobutyric acid in the cell walls as proposed by Sasaki et al. [6] is not a practical option for most laboratories. For routine confirmation of the identity of putative

R. toxicus strains, bacteriophage typing [51] or specific immunoassays [43,52,53] are available.

Rathayibacter toxicus in infected plants: The ability to detect and quantify *R. toxicus* in infected plant material is useful for predicting the risk of corynetoxin poisoning in the field [54], for monitoring the effectiveness of management strategies [55], and for confirmation of corynetoxin poisoning of livestock. Initially, this was done by looking for bacterial gummosis and bacterially colonized nematode galls in the field, but for more reliable and quantitative results, grass inflorescences were collected, dried, threshed, cleaned by sieving and air aspiration, and examined with transmitted light. Trained operators were able to distinguish galls from normal seed within the diaspores and the number per unit weight used as an estimate of the population density. This method is still used for detecting nematode galls for phytosanitary certification of seed for export. However, for *R. toxicus* detection and quantification, a monoclonal-based enzyme-linked immunoassay is routinely used [52,53]. This assay is applied to oaten hay for export from Australia [56], as ryegrass is a common contaminant. Since the establishment of testing hay for export in 1996, there has been no incident of corynetoxin poisoning arising from exported hay.

67.5.2 CORYNETOXIN POISONING

Description of corynetoxins (after Finnie [3]): *R. toxicus* produces a unique mixture of 16 closely related glycolipid toxins, designated corynetoxins, which belong to the tunicamycin group of nucleoside antibiotics collectively referred to as tunicaminyluracil antibiotics. Tunicamycin, originally isolated from *Streptomyces lysosuperificus*, is structurally similar to corynetoxins and produces the same biological effects. Streptovirudin, mycospocidin, and antibiotics 24010 and MM19290 are other closely related antibiotics. Tunicaminyluracil antibiotics have a common structure of uracil, tunicamine (a C_{11} amino sugar), and N-acetylglucosamine (Figure 67.3). Attached to the tunicamine by an amide link are fatty acids of variable chain length and terminal branching. The mixture of variants is diagnostic of their microbial source.

Corynetoxins are produced by the bacterium late in the disease cycle, apparently as a stationary phase product. The bacteriophage, mentioned in Section 67.5.1, also increases within the bacterial mass at this time. It is possible that these two processes are interlinked.

Identification of corynetoxins: Reverse-phase high-performance liquid chromatography (HPLC) analytical methods using UV and/or mass spectrometry (MS) [57,58] can be used to detect corynetoxins from extracted plant material infected or contaminated with *R. toxicus*. However, this method requires a concentration step to achieve adequate sensitivity and is not suitable for high-throughput routine diagnosis [59].

For more rapid detection, enzyme-linked immunosorbent assays (ELISAs) based on antibodies against corynetoxins produced in sheep injected with an immunogenic conjugate prepared by chemically bonding a toxin-related hapten to

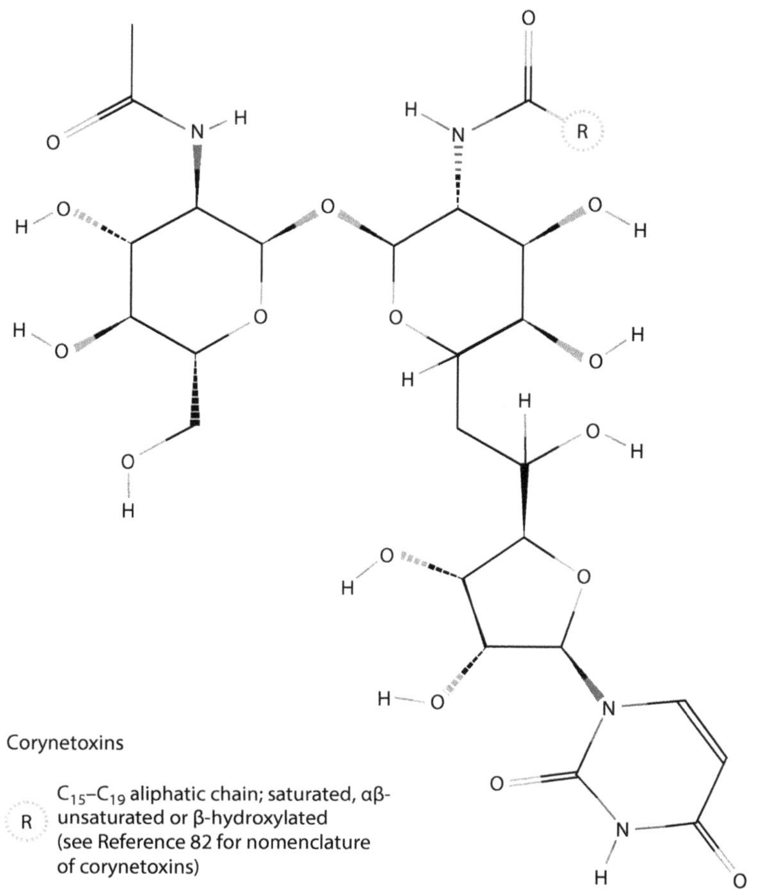

FIGURE 67.3 Structure of corynetoxins produced by *R. toxicus*. (Modified image from National Center for Biotechnology Information, US National Library of Medicine, Bethesda, MD.)

fetal calf serum proteins using 1-ethyl-3-(3-dimethylamino-propyl) carbodiimide as the linker [60,61] have been developed. Although technologically a significant advancement, a corynetoxin ELISA service has not been provided by any diagnostic laboratory because it is more complex and slower to perform, and thereby more costly than the ELISA for *R. toxicus*. In spite of the fact that corynetoxins are produced by *R. toxicus* only as the host grass matures, the bacterial assay provides a workable surrogate for a direct toxin assay.

Corynetoxin poisoning: Although corynetoxin poisoning results in a characteristically severe, episodic, and frequently fatal neurological disorder in affected animals, neither the clinical signs nor the associated neuropathological changes (which can be mild, indistinct, or absent) are diagnostic on their own [3]. The effects of corynetoxins can be more evident in liver damage, but again, this is largely nonspecific. Electron microscope examination of liver tissue can reveal unusual concretions in distended and/or dilated rough endoplasmic reticulum, a change that is possibly more indicative of corynetoxin poisoning [3]. The toxins inhibit *N*-acetylglucosamine-1-phosphate transferase, an enzyme involved in the initial step of biosynthesis of *N*-glycoproteins [30]. Given glycoproteins are important in cell membranes and cell interactions, the lack of any highly

specific pathology caused by corynetoxins is consistent with their impact on a fundamental process in animal physiology.

Without a definitive diagnostic feature, the confirmation of corynetoxin poisoning routinely relies on the detection of *R. toxicus* in the feed source, either as colonized plant material or by immunoassay. The development of an immunocapture MS assay for carbohydrate-deficient transferrin in serum is reported to have prospects for application as a high-throughput clinical assessment of corynetoxins exposure in humans or livestock [62].

67.6 TREATMENT AND PREVENTION

67.6.1 CARE AND TREATMENT OF AFFECTED LIVESTOCK

There is no satisfactory treatment for corynetoxin poisoning that can be applied in the field. Both an antidote [63,64] and a vaccine [61,65] have been developed, and although representing great scientific achievement, they have not provided practical treatment options. The application of the antidote is constrained by not being effective in animals exposed to doses of corynetoxins well in excess of the lethal dose and the fact that most outbreaks of corynetoxin poisoning are not detected sufficiently early to prevent high-dose exposure.

In addition, the stresses caused by mustering the animals and administering the antidote can accelerate and exacerbate development of signs and death. The vaccine requires a series of inoculations to give useful immunity and it is not known what level of protection would be provided against high-dose exposure. Although both might have application for high-value animals under intensive management, a more cost-effective way is to protect such animals by reducing the risk of exposure to toxic feed.

The only recommended option for livestock affected by corynetoxin poisoning is to immediately remove them from the toxic pasture or to remove the source of toxic feed. When moving affected animals, stress is to be minimized. To achieve best recovery, the affected animals are provided with good-quality feed, water, and shade and any disturbance kept to a minimum. Even when recovery is apparent, the animals should not be reexposed to even sublethal concentrations of toxin within that season.

67.6.2 CONTROL OF *RATHAYIBACTER TOXICUS*, ITS HOSTS, AND VECTORS IN THE FIELD

To protect livestock from corynetoxin poisoning, it would be ideal to ensure feed is produced in fields free of *R. toxicus* or at least with population densities unlikely to cause clinical or subclinical effects. In this tripartite relationship, it could be achieved by control of the grass host, vector nematode, or the bacterium. With weeds having more options for agronomic and chemical control than nematodes or bacteria in agricultural systems, grass host control has been the primary, but not the only, approach to reducing exposure of livestock to corynetoxins. However, in Australia, corynetoxin poisoning occurs when the hosts become dominant, and this presents a challenge for control.

L. rigidum, in the context in which it is a host for *R. toxicus*, is a major weed of dryland field crops in a ley farming system and the dominant grass in the pasture phase. Consequently, most outbreaks of corynetoxin poisoning occur in the pasture phase [66]. Ryegrass control in these systems is problematic, especially with adoption of reduced or zero tillage systems and the development of herbicide resistance. Also, control of *L. rigidum* to an economic threshold of crop production might not be adequate for *A. funesta* and *R. toxicus* control, and a reduced population of a heavily infected host might result in higher toxin concentrations in total feed on offer and an increased risk to livestock grazing crop residues. A full review of *L. rigidum* control is not possible here, but it does represent an important component of integrated control of *R. toxicus*.

In the case of floodplain staggers in the southeast of SA, the host (*P. monspeliensis*) becomes dominant under flood conditions. In recent years, much of this low-lying land has been drained, and there have been a series of years with below average rainfall, which together have provided control. However, while the causal organisms are unlikely to have been eliminated, given the scale of the drainage system, it is now unlikely that floodplain staggers will ever reoccur to any significant extent in the region. In NSW, the host (*L. filiformis*) occurs in rangelands subject to spasmodic seasonal flooding and conditions that led to it becoming dominant are rare and not economically controlled.

Controlling the vector nematode, and the bacterium more directly, is done by removing infected *L. rigidum* tillers in the pasture phase. The first cohort of tillers produced by the host are the most likely to be colonized and can be removed before toxin is produced or nematodes reach their survival stage by mowing, crash grazing, or spray topping with contact herbicides. To be effective, these operations must be applied before anthesis, which is earlier than would normally be applied for seed set control. Consequently, this might not prevent seed set from later tillering.

Another approach has been the development of ryegrass cultivars resistant to *A. funesta* [2]. Again, this approach is constrained by the difficulty of controlling the well-adapted local annual ryegrass ecotypes [4].

In addition to agronomic methods, natural decline in the outbreaks of corynetoxin poisoning in WA prompted a search for a biocontrol agent [67]. The nematode-vectored, plant pathogenic fungus, *D. alopecuri*, was considered to be the likely cause of the decline because of its known biology, its high incidence in areas where outbreaks had declined, and its apparent absence from areas with no decline [67]. Barbetti and Riley [68] demonstrated the potential for field application of *D. alopecuri* to reduce *R. toxicus* colonization of *L. rigidum*. The fungus was commercially cultured and distributed to outbreak areas where it did not occur. The fungus colonizes the host inflorescence more aggressively, thus reducing bacterial colonization and nematode reproduction.

67.6.3 MINIMIZING EXPOSURE OF LIVESTOCK

Since corynetoxin poisoning has no practical treatment, and in outbreaks many animals will have consumed a lethal dose (including asymptomatic animals on the same feed source) by the time signs are observed, it is important to adopt strategies that minimize the exposure of livestock to the toxins. Testing of standing and harvested fodder for *R. toxicus* by either visual examination for galls or by immunoassay provides an indication of risk on which grazing/feeding decisions can be made. For landowners, testing over a series of years will provide information on spatial and temporal variation of the causal organisms and the effectiveness of efforts to achieve their control. Fields with the greatest risk can be grazed before the toxin develops or later in the summer when the risk tends to decline.

Also, during the late autumn to early summer when the risk is greatest, livestock should be monitored daily, forcing the animals to run, which induces the expression of signs. On any indication of poisoning, the exposed animals need to be moved to a pasture or other feed source known to be free of the toxins. Identification of a safe field and/or feed source is best established by prior testing, rather than at the time the decision is needed. During the period when the area of infested *L. rigidum* was increasing most rapidly across southern Australia, a significant proportion of outbreaks of corynetoxin poisoning occurred on farms not previously known to be infested because stockowners were not inclined to preemptively adopt the aforementioned strategies.

67.7 CONCLUSIONS AND FUTURE PERSPECTIVES

R. toxicus, through its nematode-assisted colonization of pasture grasses and production of corynetoxins, created a significant economic problem in Australia, and to a minor extent in South Africa. The bacterium is slow growing, does not readily produce toxin in culture, and is reliant on a suitable nematode vector infesting a host grass population within a context that allows both to reach maturity. As an economic problem in Australia, it has occurred in particular (and, in some cases, transient or now historic) circumstances that have facilitated large and protected populations of its vectors and hosts. The importance of this bacterium has declined in recent decades with better understanding and management by stockowners, assisted by the development of natural antagonism.

R. toxicus, and the corynetoxins it produces, represents both security and biosecurity (phytosanitary) risks because the bacterium can colonize grasses used for animal and human consumption, and corynetoxins are cumulative, highly toxic chemicals (affecting a fundamental physiological process common to all animals) for which there is no effective treatment. However, these properties of corynetoxins are not unique and apply to tunicaminyluracil antibiotics from any microbial source. Commercially available tunicaminyluracil antibiotics are expensive, which may be a reflection of limited demand in a niche market, but more likely because production is technically difficult. *R. toxicus* does not readily produce toxins in culture, so it has not become a microbe of choice for production of these antibiotics.

As a biosecurity (phytosanitary) concern, *R. toxicus* does not yet have a global distribution and its potential introduction with its vector into suitable host populations represents a risk. Further, given that the *R. toxicus* is not specific to a particular host grass or vector nematode (*Anguina* sp.), its introduction to new areas could lead to the establishment of novel associations of economic importance. However, this risk should not be considered high because, if its history in Australia is indicative, economic problems only develop in highly specific circumstances. Also, several decades of research on the toxicology, plant pathology, and field management of the disease complex means that progress to resolution of a problem in any new context would most likely be much less protracted. However, considering that the signs and pathology of corynetoxin poisoning are not unique, initial investigations are likely to be misdirected with delays in correct diagnosis impeding the introduction of control procedures to minimize the impact of the disease.

Therefore, it is imperative that veterinary practitioners, as they are likely to be the first to encounter cases of corynetoxin poisoning in any new context, as well as plant toxicologists and pasture scientists, be familiar with corynetoxin poisoning. It is recommended that national authorities responsible for biosecurity and incursion management develop and disseminate within their jurisdiction information on early detection and rapid response to suspected instances of corynetoxin poisoning.

REFERENCES

1. Frahn, J.L. et al., Structure of the corynetoxins, metabolites of *Corynebacterium rathayi* responsible for toxicity of annual ryegrass (*Lolium rigidum*) pastures. *Aust. J. Chem.* 37, 165, 1984.
2. McKay, A.C. and Ophel, K.M. Toxigenic *Clavibacter/Anguina* associations infecting grass seedheads. *Ann. Rev. Phytopathol.* 31, 153, 1993.
3. Finnie, J.W., Review of corynetoxins poisoning of livestock, a neurological disorder produced by a nematode-bacterium complex. *Aust. Vet. J.* 84, 271, 2006.
4. Riley, I.T. and Barbetti, M.J. Australian anguinids: Their agricultural impact and control. *Australas. Plant Pathol.* 37, 289, 2008.
5. Riley, I.T. and Ophel K.M., *Clavibacter toxicus* sp. nov., the toxigenic bacterium responsible for annual ryegrass toxicity in Australia. *Int. J. Syst. Bacteriol.* 42, 92, 1992.
6. Sasaki, J. et al., Taxonomic significance of 2,4- diaminobutyric acid isomers in the cell wall peptidoglycan of actinomycetes and reclassification of *Clavibacter toxicus* as *Rathayibacter toxicus* comb. nov. *Int. J. Syst. Bacteriol.* 48, 403, 1998.
7. Zgurskaya, H.I. et al., *Rathayibacter* gen. nov., including the species *Rathayibacter rathayi* comb. nov., *Rathayibacter tritici* comb. nov., *Rathayibacter iranicus* comb. nov., and six strains from annual grasses. *Int. J. Syst. Bacteriol.* 43, 143, 1993.
8. Davis, M.J. et al., *Clavibacter*: A new genus containing some phytopathogenic coryneform bacteria, including *Clavibacter xyli* subsp. *xyli* sp. nov., subsp. nov. and *Clavibacter xyli* subsp. *cynodontis* subsp. nov., pathogens that cause ratoon stunting disease of sugarcane and Bermudagrass stunting disease. *Int. J. Syst. Bacteriol.* 34, 107, 1984.
9. Rainey, F. et al., Further evidence for the phylogenetic coherence of actinomycetes with group B-peptidoglycan and evidence for the phylogenetic intermixing of the genera *Microbacterium* and *Aureobacterium* as determined by 16s rDNA analysis. *FEMS Microbiol. Lett.* 118, 135, 1994.
10. Takeuchi, M. and Yokota, A., Phylogenetic analysis of the genus *Microbacterium* based on 16S rRNA gene sequences. *FEMS Microbiol. Lett.* 124, 11, 1994.
11. Dye, D.W. and Kemp, W.J., A taxonomic study of plant pathogenic *Corynebacterium* species. *N. Z. J. Agric. Res.* 20, 563, 1977.
12. Carlson, R.R. and Vidaver, A.K., Taxonomy of *Corynebacterium* plant pathogens, including a new pathogen of wheat, based on polyacrylamide gel electrophoresis of cellular proteins. *Int. J. Syst. Bacteriol.* 32, 315, 1982.
13. Riley, I.T. et al., Genetic analysis of plant pathogenic bacteria in the genus *Clavibacter* using allozyme electrophoresis. *J. Gen. Microbiol.* 134, 3025, 1988.
14. De Bruyne, E. et al., Enzymatic relatedness amongst phytopathogenic coryneform bacteria and its potential use for their identification. *Syst. Appl. Microbiol.* 15, 393, 1992.

15. Altenburger, P. et al., Polyamine distribution in actinomycetes with group B peptidoglycan and species of the genera *Brevibacterium, Corynebacterium,* and *Tsukamurella. Int. J. Syst. Bacteriol.* 47, 270, 1997.

16. Dorofeeva, L.V. et al., *Rathayibacter caricis* sp. nov. and *Rathayibacter festucae* sp. nov., isolated from the phyllosphere of *Carex* sp. and the leaf gall induced by the nematode *Anguina graminis* on *Festuca rubra* L., respectively. *Inter. J. Syst. Evol. Microbiol.* 52, 1917, 2002.

17. Bird, A.F. and Stynes, B.A., The morphology of a *Corynebacterium* sp. parasitic on annual rye grass. *Phytopathology* 67, 828, 1977.

18. Price, P.C. et al., Annual ryegrass toxicity: Parasitism of *Lolium rigidum* by a seed gall forming nematode (*Anguina* sp.). *Ann. Appl. Biol.* 91, 359, 1979.

19. Riley, I.T., *Vulpia myuros* and the annual ryegrass toxicity organisms, *Anguina funesta,* and *Clavibacter toxicus. Fund. Appl. Nematol.* 18, 595, 1995.

20. Gill, G.S., Why annual ryegrass is a problem in Australian agriculture. *Plant Prot. Q.* 11(Suppl. 1), 193, 1996.

21. Bird, A.F. and Stynes, B.A., The life cycle of *Anguina agrostis*: Embryogenesis. *Int. J. Parasitol.* 11, 23, 1981.

22. Bird, A.F. and Stynes, B.A., The life cycle of *Anguina agrostis*: Post- embryonic growth of the second stage larva. *Int. J. Parasitol.* 11, 243, 1981.

23. Bird, A.F. and Stynes, B.A., The life cycle of *Anguina agrostis*: Development in host plant. *Int. J. Parasitol.* 11, 431, 1981.

24. McKay, A.C. et al., Ecological field studies on *Anguina funesta,* the vector in annual ryegrass toxicity. *Aust. J. Agric. Res.* 32, 917, 1981.

25. McKay, A.C., Investigations to develop methods to control the nematode associated with annual ryegrass toxicity. PhD thesis, The University of Adelaide, Adelaide, South Australia, Australia, 1985.

26. Preston, C.M. and Bird, A.F. Physiological and morphological changes associated with recovery from anabiosis in the dauer larva of the nematode *Anguina agrostis. Parasitology* 95, 125, 1987.

27. Riley, I.T. and McKay, A.C., Invasion of some grasses by *Anguina funesta* (Nematoda: Anguinidae) juveniles. *Nematologica* 37, 447, 1991.

28. Riley, I.T. and Bertozzi, T., Variation in sex ratios in four *Anguina* (Nematoda: Anguinidae) species. *Trans. R. Soc. S. Aust.* 128, 43, 2004.

29. Edgar, J.A. et al., Corynetoxins, causative agents of annual ryegrass toxicity; their identification as tunicamycin group antibiotics. *J. Chem. Soc. Chem. Commun.* 4, 222, 1982.

30. Jago, M.V. et al., Inhibition of glycosylation by corynetoxin, the causative agent of annual ryegrass toxicity: A comparison with tunicamycin. *Chem. Biol. Interact.* 45, 223, 1983.

31. Ophel, K.M. et al., Association of bacteriophage particles with toxin production by *Clavibacter toxicus,* the causal agent of annual ryegrass toxicity. *Mol. Plant Pathol.* 83, 676, 1993.

32. Kowalski, M.C. et al., Development and application of polymerase chain reaction-based assays for *Rathayibacter toxicus* and a bacteriophage associated with annual ryegrass (*Lolium rigidum*) toxicity. *Aust. J. Exp. Agric.* 47, 177, 2007.

33. Bertozzi, T. and Davies, K.A., *Anguina paludicola* sp. n. (Tylenchida: Anguinidae): The nematode associated with *Rathayibacter toxicus* infection in *Polypogon monspeliensis* and *Lachnagrostis filiformis* in Australia. *Zootaxa* 2060, 33, 2009.

34. Bourke, C.A. et al., Flood plain staggers, a tunicaminyluracil toxicosis of cattle in northern New South Wales. *Aust. Vet. J.* 69, 228, 1992.

35. Bryden, W.L. et al., Corynetoxicosis of livestock: A nematode-bacterium disease complex associated with different grasses. In Colegate, S.M. and Dorling, P.R. (eds.), *Plant-Associated Toxins: Agricultural, Phytochemical and Ecological Aspects.* CAB International, Wallingford, U.K., p. 410, 1994.

36. Bertozzi, T. and McKay, A.C., Incidence on *Polypogon monspeliensis* of *Clavibacter toxicus* and *Anguina* sp., the organisms associated with 'flood plain staggers' in South Australia. *Aust. J. Exp. Agric.* 35, 567, 1995.

37. Davis, E.O. et al., Clinical, pathological and epidemiological aspects of flood plain staggers, a corynetoxicosis of livestock grazing *Agrostis avenacea. Aust. Vet. J.* 72, 187, 1995.

38. Bertozzi, T., Biology and control of the anguinid nematode associated with flood plain staggers. PhD thesis, The University of Adelaide, Adelaide, South Australia, Australia, 2003.

39. Edgar, J.A. et al., Identification of corynetoxins as the cause of poisoning associated with annual beardgrass (*Polypogon monspeliensis* (L.) Desf.) and blown grass (*Agrostis avenacea* C. Gemelin). In Colegate, S.M. and Dorling, P.R. (eds.), *Plant-Associated Toxins: Agricultural, Phytochemical and Ecological Aspects.* CAB International, Wallingford, U.K., p. 393, 1994.

40. Riley, I.T., *Anguina tritici* is a potential vector of *Clavibacter toxicus. Australas. Plant Pathol.* 21, 147, 1992.

41. Riley, I.T. et al., *Anguina australis,* a vector for *Rathayibacter toxicus* in *Ehrharta longiflora. Australas. Plant Pathol.* 30, 171, 2001.

42. Riley, I.T and McKay, A.C., Specificity of the adhesion of some plant pathogenic microorganisms to the cuticle of nematodes in the genus *Anguina* (Nematoda: Anguinidae). *Nematologica* 36, 90, 1990.

43. Riley, I.T., Serological relationships between strains of coryneform bacteria responsible for annual ryegrass toxicity and other plant pathogenic corynebacteria. *Int. J. Syst. Bacteriol.* 37, 153, 1987.

44. Chatel, D.L., Research on the nematode-bacterial association of annual ryegrass responsible for annual ryegrass toxicity, 1978 to 1982. Western Australian Department of Agriculture, Perth, Western Australia, Division of Plant Industries Technical Report, No. 54, p. 38, 1992.

45. Allen, J., Annual ryegrass toxicity—An animal disease caused by toxins produced by a bacterial plant pathogen. *Microbiol. Aust.* 33, 18, 2012.

46. Purcell, D.A. Annual ryegrass toxicity: A review. In Seawright, A.A. et al. (eds.), *Plant Toxicology.* Queensland Poisonous Plants Committee, Brisbane, Queensland, Australia, p. 553, 1985.

47. Bourke, C.A., Tunicaminyluracil toxicity, an emerging problem in livestock fed grass or cereal products. In Colegate, S.M. and Dorling, P.R. (eds.), *Plant-Associated Toxins: Agricultural, Phytochemical and Ecological Aspects.* CAB International, Wallingford, U.K., p. 399, 1994.

48. Cheeke, P.R., *Natural Toxicants in Feeds, Forages, and Poisonous Plants,* 2nd ed. Interstate Publishers Inc., Danville, IL, p. 260, 1998.

49. Jago, M.V. and Culvenor, C.C.J. Tunicamycin and corynetoxin poisoning in sheep. *Aust. Vet. J.* 64, 232, 1987.

50. Allen, J.G., Annual ryegrass toxicity. *Proceedings of ARGT Workshop.* Department of Agriculture Western Australia, Wooroloo, Western Australia, Australia, p. 1, 2002.

51. Riley, I.T. and Gooden J.M., Bacteriophage specific for the *Clavibacter* sp. associated with annual ryegrass toxicity. *Lett. Appl. Microbiol.* 12, 158, 1991.

52. Masters, A.M. et al., An enzyme-linked immunosorbent assay for the detection of *Rathayibacter toxicus*, the bacterium involved in annual ryegrass toxicity, in hay. *Aust. J. Agric. Res.* 57, 731, 2006.

53. Masters, A.M. et al., Improvements to the immunoassay for detection of *Rathayibacter toxicus* in hay. *Crop Past. Sci.* 62, 523, 2011.

54. McKay, A.C. and Riley, I.T. Sampling ryegrass to assess the risk of annual ryegrass toxicity. *Aust. Vet. J.* 70, 241, 1993.

55. Riley, I.T., Paddock sampling for management of annual ryegrass toxicity. *J. Agric. West. Aust.* 33, 51, 1992.

56. Anon., Standard for minimising the risk of corynetoxin contamination of hay and straw for export. Australian Quarantine and Inspection Service, Department of Agriculture Fisheries and Forestry Australian Government. (http://www.daff. gov.au/aqis/export/plants-plant-products/corynetoxin, last reviewed April 23, 2007), 2007.

57. Cockrum, P.A. and Edgar, J.A., Rapid estimation of corynetoxins in bacterial galls from annual ryegrass (*Lolium rigidum* Gaudin) by high-performance liquid chromatography. *Aust. J. Agric. Res.* 36, 35, 1985.

58. Anderton, N. et al., The identification of corynetoxin-like tunicaminyluracil-glycolipids in nematode galls in *Festuca nigrescens* from North America and New Zealand. In Acamovic, T. et al. (eds.), *Poisonous Plants and Related Toxins*. CABI Publishing, Wallingford, U.K., p. 204, 2003.

59. Than, K.A. et al., Plant-associated toxins in animal feed: Screening and confirmation assay development. *Anim. Feed Sci. Technol.* 121, 5, 2005.

60. Than., K.A. et al., Development of a vaccine against annual ryegrass toxicity. In Garland, T. and Barr, A.C. (eds.), *Toxic Plants and Other Natural Toxicants*. CAB International, Wallingford, U.K., p. 165, 1998.

61. Than, K.A. et al., Analysis of corynetoxins: A comparative study of an indirect competitive ELISA and HPLC. In Acamovic, T. et al. (eds.), *Poisonous Plants and Related Toxins*. CAB International, Wallingford, U.K., p. 402, 2004.

62. Penno, M.A.S. et al., Detection and measurement of carbohydrate deficient transferrin in serum using immuno-capture mass spectrometry: Diagnostic applications for annual ryegrass toxicity and corynetoxin exposure. *Res. Vet. Sci.* 93, 611, 2012.

63. Stewart, P.L. et al., Protective effects of cyclodextrins on tunicaminyluracil toxicity. In Garland, T. and Barr, A.C. (eds.), *Plant Toxins and Other Natural Toxicants*. CAB International, Wallingford, U.K., p. 179, 1998.

64. Allen, J.G. et al. Annual ryegrass toxicity in sheep is not prevented by administration of cyclodextrin via controlled release devices. In Riet-Correa, F. et al. (eds.), *Poisoning by Plants, Mycotoxins and Related Toxins*. CAB International, Wallingford, U.K., p. 331, 2011.

65. Than, K.A. et al., Development of an immunoassay for corynetoxins. In Garland, T. and Barr, A.C. (eds.), *Toxic Plants and Other Natural Toxicants*. CAB International, Wallingford, U.K., p. 49, 1998.

66. Stynes, B.A. and Wise, J.L., The distribution and importance of annual ryegrass toxicity in Western Australia and its occurrence in relation to cropping rotations and cultural practices. *Aust. J. Agric. Res.* 31, 557, 1980.

67. Riley, I.T., *Dilophospora alopecuri* and decline in annual ryegrass toxicity in Western Australia. *Aust. J. Agric. Res.* 45, 841, 1994.

68. Barbetti, M.J. and Riley, I.T., Field application of *Dilophospora alopecuri* to manage annual ryegrass toxicity caused by *Rathayibacter toxicus*. *Plant Dis.* 90, 229, 2006.

69. Riley, I.T. et al., Poisoning of livestock in Oregon in the 1940s to 1960s attributed to corynetoxins produced by *Rathayibacter* in nematode galls in chewing fescue (*Festuca nigrescens*). *Vet. Hum. Toxicol.* 45, 160, 2003.

70. Haag, J.R., Toxicity of nematode infested Chewing's fescue seed. *Science* 102, 406, 1945.

71. Shaw, J.N. and Muth, O.H., Some types of forage poisoning in Oregon cattle and sheep. *J. Am. Vet. Med. Assoc.* 114, 315, 1949.

72. Galloway, J.H., Grass seed nematode poisoning in livestock. *J. Am. Vet. Med. Assoc.* 139, 1212, 1961.

73. McIntosh, G.H. et al., Toxicity of parasitised Wimmera ryegrass, *Lolium rigidum*, for sheep and cattle. *Aust. Vet. J.* 43, 349, 1967.

74. Fisher, J.M., Annual ryegrass toxicity. Biennial Report of the Waite Agricultural Research Institute Adelaide, South Australia, 1976–1977. 1977.

75. Gwyn, R. and Hadlow, A.J., Toxicity syndrome in sheep grazing Wimmera ryegrass in Western Australia. *Aust. Vet. J.* 47, 408, 1971.

76. Pink, B.N., Ryegrass toxicity in Western Australia February 1989, Australian Bureau of Statistics, Canberra, Australia, Catalogue No. 7421.5, 1989.

77. Schneider, D.J., First report of annual ryegrass toxicity in the Republic of South Africa. *Onderstepoort J. Vet. Res.* 48, 251, 1981.

78. Grewar, J.D. et al., Annual ryegrass toxicity in Thoroughbred horses in Ceres in the Western Cape Province, South Africa. *J. S. Afr. Vet. Med. Assoc.* 80, 220, 2009.

79. Finnie, J.W. Corynetoxin poisoning in sheep in the southeast of South Australia associated with annual beard grass (*Polypogon monspeliensis*). *Aust. Vet. J.* 68, 370, 1991.

80. McKay, A.C. et al., Livestock death associated with *Clavibacter toxicus/Anguina* sp. Infection in seedheads of *Agrostis avenacea* and *Polypogon monspeliensis*. *Plant Dis.* 77, 635, 1993.

81. Bryden, W.L. et al., Flood-plain staggers—An intoxication in cattle due to the ingestion of blown grass (*Agrostis avenacea*). *Proc. Nutr. Soc. Aust.* 16, 240, 1991.

82. Cockrum, P.A. and Edgar, J.A., High-performance liquid chromatographic comparison of the tunicaminyluracil-based antibiotics corynetoxin, tunicamycin, streptovirudin and MM 19290. *J. Chrom.* 268, 245–254, 1983.

Section VII

Microbes Affecting Plants

68 *Candidatus* Liberibacter

Kirsten S. Pelz-Stelinski and Lukasz L. Stelinski

CONTENTS

68.1 INTRODUCTION

Bacteria in the genus *Candidatus (Ca.)* Liberibacter consist of several distinct species of psyllid-borne plant pathogens that also serve as endosymbionts of their insect vectors. *Ca.* Liberibacter spp. are phloem-limited, gram-negative alpha-Proteobacteria in the family Rhizobiaceae. As indicated by the *Candidatus* designation, the bacteria in this genus have not been cultivated in a sustained manner. Three species of *Ca.* Liberibacter, *Ca.* Liberibacter asiaticus (Las), *Ca.* Liberibacter americanus (Lam), and *Ca.* Liberibacter africanus (Laf), are found in a variety of mostly rutaceous host plants and are the putative causal agents of huanglongbing (HLB), otherwise known as greening disease of citrus. Las is associated with HLB throughout Asia and the Americas, while Lam and Laf are only found in association with the disease in Brazil and Africa, respectively. A fourth *Ca.* Liberibacter species, *Ca.* Liberibacter solanacearum (Lso) (also known as *Ca.* Liberibacter psyllaurous), is found in association with solanaceous plants and is the probable causal agent of Zebra chip disease in potatoes.[1]

68.2 CLASSIFICATION, MORPHOLOGY, AND GENOME FEATURES

Classification: *Ca.* Liberibacter species are phloem-restricted bacteria that to date have not been cultivated in sustained pure culture[2]; however, limited success has been reported for Las.[3] The bacteria are members of the alpha subdivision of the class *Proteobacteria*.[4,5] Phylogenetic analysis of 16S rDNA places *Ca.* Liberibacter spp. in a distinct subgroup of the α-*Proteobacteria*, which shares only 86% homology with its closest relatives in the alpha-2 subgroup. Gene sequence analyses indicate some geographic variation among Las isolates, which may correspond to phenotypic adaptations to local environments.[6–8]

Morphology: *Ca.* Liberibacter spp. occur as filamentous (rod-shaped) or spherical structures. The filamentous form ranges from 0.6 to 1.2 μm in length and 0.2 to 0.3 μm in width, and the round form from 0.1 to 0.5 μm in diameter.[3,9–13] The gram-negative bacteria possess an outer cell membrane with a thin peptidoglycan layer but lack pili and flagella.[4]

Genome features: The genome sequence of Las has recently been obtained through metagenomic analysis of DNA extracted from a single Las-infected psyllid (Psy62).[14] The 1.23 Mb circular genome of Las is characterized by a high percentage of genes contributing to its virulence, including cell motility and active transport genes. This intracellular plant pathogen and insect symbiont also lacks genes common to free-living bacteria, including those encoding type III and type IV secretion systems, or plant cell wall and extracellular degradation enzymes found in type II secretion systems. Similarly, the 1.26 Mb metagenomic sequence of Lso, isolated from potato psyllid DNA, exhibits a reduced genome consistent with its role as an intracellular plant pathogen and/or insect symbiont.[15] The inability to maintain *Ca.* Liberibacter spp. in a sustained culture has limited efforts to understand the contribution of putative virulence genes, including those encoding ABC transporters and the type I secretion system, to the pathogenicity of these organisms.[16] Based on shared sequence similarity with *pSymA*-encoded proteins of the plant symbiotic bacterium, *Sinorhizobium meliloti*, several Las genes have been identified that are predicted to encode proteins important for host plant colonization.[17] Two bacteriophage genomes found in association with the Las genome may also contribute to the pathogenicity of the bacterium to host plants.[18] In addition, genes associated with zinc uptake homologous to the *znuABC* systems found in other bacterial genera are present in Las, which may underlie the resemblance of HLB symptoms to the appearance of plants deficient in this micronutrient.[19]

68.3 DISTRIBUTION, VECTORS, AND TRANSMISSION

Distribution: The distribution of *Ca.* Liberibacter spp. has increased in recent years, spreading throughout the primary citrus-growing areas of the Americas. Las and Lam occur concurrently in Brazil. Las has also been detected in Mexico, Belize, and throughout the citrus-growing regions of the United States.[20] Las and Laf are endemic to Asia and Africa, respectively.[21] Lso has been confirmed in New Zealand, Mexico, the United States, and Central America.[22,23]

Vectors: *Ca.* Liberibacter spp. are vector-borne pathogens transmitted by several species of psyllids (Hemiptera: Psyllidae). The Asian citrus psyllid (ACP), *Diaphorina citri* Kuwayama, is the vector of Las and Lam throughout Asia and the Americas, while the African citrus psyllid, *Trioza erytreae* (Del Guercio), is the only known vector of Laf in Africa. Lso is transmitted to susceptible solanaceous hosts by the potato/tomato psyllid, *Bactericera cockerelli*.[24]

Transmission: The ACP, *D. citri*, is the primary vector of Las, the presumed causal agent of HLB, which is the leading economic disease of citrus worldwide. The psyllids can acquire the pathogen from the phloem tissue of infected plants as adults and nymphs[25,26] and subsequently infect healthy plants by feeding. There is evidence that the pathogen can also be transferred to the eggs of infected females transovarially.[26] In addition, adult females can acquire the bacteria from infected males during mating.[13] Currently in Florida, the average infection rate of HLB in citrus groves is estimated to be 1.6%, reaching up to 100% in the southern and eastern parts of the state[27] and will potentially increase given the mobility of the adult vector[28,29] and the large acreage (ca. 57,650 ha) of abandoned citrus groves in the state (USDA-NASS 2011).

There is no evidence supporting the transmission of Las or Laf through citrus seed.[30,31] Seedlings propagated from the seeds of infected plants remain free of Las infection when subjected to multiple tests by quantitative polymerase chain reaction (PCR) assays over a 3-year period.[30] However, the pathogen has been detected in the seed coats of several citrus varieties "Sanguinelli" sweet orange and "Conners" grapefruit.[32] In contrast, Lso is readily transmitted by seed tubers used to replant potato crops.[33]

68.4 SYMPTOMS OF INFECTION

The symptoms of HLB caused by the bacteria Las, Lam, and Laf during the early phase of infection include leaf yellowing and asymmetrical chlorosis, or "blotchy mottle." Yellowing may occur on a single shoot or an entire branch and may easily be mistaken for nutrient deficiencies. As the disease progresses, yellowing may spread throughout the tree canopy followed by twig dieback. The resulting decline in tree productivity is compounded by the occurrence of small, lopsided, discolored fruit of bitter flavor that often drop prematurely and are unfit for juice and fresh fruit markets.[11] Stunted root systems and twig dieback in infected trees occur due to starch accumulation within leaves, resulting in the constriction of the phloem elements that ultimately results in tree decline and potential death. The bacteria appear to affect all citrus, although the severity of symptom expression may differ among species. Historically, replanting has been the primary strategy for maintaining the viability of infected groves as the affected trees become weakened 5–6 years postinfection.

Potato crops afflicted with Zebra chip disease associated with Lso are distinguished by chlorotic, scorched, or curled foliage with browning of the vascular tissue. Swollen nodes may also be present. Potatoes are rendered unfit for potato chip production due to the conversion of starch in the harvested potatoes to water-soluble sugar during the cooking process, resulting in the cosmetically unappealing, characteristic brown stripes reflected in the disease namesake.[34] Infected tomato fruit may be asymptomatic or may be small and misshapen.

68.5 IDENTIFICATION AND DIAGNOSIS

Although the symptomology described earlier can be used to diagnose HLB disease in citrus, these symptoms often resemble those associated with nutritional deficiencies or other disorders. In addition, infected plants may be asymptomatic; thus, identification and diagnosis of *Ca.* Liberibacter infection in citrus by PCR analysis of plant samples is the most reliable method for detecting the presence of pathogen DNA. Conventional PCR assays may be used for the detection of *Ca.* Liberibacter spp.[5,22]; however, a more accurate, sensitive method for detection and identification is quantitative PCR using a TaqMan technology.[35–37] Due to the amount of time and expense associated with PCR, an iodine-based starch test may be used as a faster, reduced cost alternative for diagnosis.[38,39] The reaction of iodine with starch, which accumulates in the leaves of HLB-afflicted plants,[40] results in a characteristic black color change. Although less accurate than PCR analysis, this method is useful for rapid diagnosis of HLB in field settings.

68.6 MANAGEMENT

Pathogen management: Since the discovery of HLB, the management of the disease in commercial citrus has primarily focused on three approaches: (1) use of clean planting material for new groves and replacement trees within established groves obtained from certified nurseries (FDACS 2008), (2) reduction of *D. citri* populations in groves using insecticides, and (3) removal (or "rouging") of infected trees to remove inoculum sources. Control of *D. citri* with insecticides is believed to decrease the rate of HLB spread within and between groves,[41] but this costs growers an additional $1000 or more per acre annually.[27] Suppression of ACP populations typically consists of foliar (6–9/year) and soil (2–3/year) applications of insecticides.[42,43] This aggressive management of *D. citri* through multiple applications of insecticides within a single growing season not only increases the chance of insecticide resistance development but also greatly increases the cost of citrus production. In some parts of Florida, United States, insecticide resistance levels have reached 30–35-fold as compared with known susceptible populations for some of the most commonly used insecticides, including neonicotinoids and organophosphates.[44] Resistance levels are not yet at the point where product failure would occur, but considering the short time span within which they developed and continued intense use of a limited number of registered products, it is important to explore compounds with different modes of action for *D. citri* management to prevent future product failures due to resistance.

A recent program to increase effectiveness of *D. citri* management and possibly help reduce costs of HLB management is the development of Citrus Health Management Areas (CHMAs). Participating growers coordinate *D. citri* control sprays and chemical selection to reduce psyllid movement between groves and target their applications to critical times of the year, such as the dormant winter season.[41] In addition to insecticide costs, tree removal and replacement costs may be substantial. Recent guidelines by Spann et al.[45] suggest that tree removal may not be an economically sound strategy, depending on the percentage of HLB-infected trees, size and age of the grove, and the aggressiveness of psyllid management tactics. When infection levels become high, it is no longer economically feasible to replace HLB-infected trees. Furthermore, due to the long lag period between inoculation with Las and the appearance of symptoms on the tree, many more trees may be infected in a grove than are apparent by visual symptoms.[46]

Nutritional management: Alternatives to tree removal are being sought that would maintain existing trees in production even when infected with HLB. One such method that has been attempted is the application of supplemental nutrition to prevent and/or reverse disease symptoms. The symptoms of HLB infection are similar to nutrient deficiencies. It is known that trees infected with Las are deficient in Ca, Mg, Mn, and B.[47] Although researchers in China reported that fertilizer and micronutrient applications had no discernable effect on HLB management, results from studies in Florida suggest that there may be potential for these applications to prolong the productivity of groves infected with HLB for 3–5 years or longer.[45,48]

The use of nutritional supplements as a tactic for extending productivity of Las-infected trees originated with a Florida grower who had a high rate of infection in his grove soon after HLB was first discovered in the state. Rather than replacing much of the grove, this grower implemented a tree health program based on nutritional sprays and systemic acquired resistance (SAR) inducing materials applied 3 times/year to coincide with flushing cycles of citrus. Currently, trees in experimental plots in this grove have a 92% or greater HLB infection rate. However, productivity and fruit quality remains within economic and industry standards.[45,48] Many growers with HLB in their groves have adopted some kind of supplemental nutritional program and there are a number of products and application regimes utilized.[49] However, psyllid management is still believed to be essential in groves managed with supplemental nutrient sprays since many more sources of Las inoculum are present as opposed to groves in which infected tree removal is practiced. It is believed that repeated inoculation of trees by infected psyllids will decrease tree life span in proportion to the number of such inoculations. Also, the nutrition-based management technique has not yet been proven sustainable in the long term through scientific experiments. Therefore, an intensive insecticide spray plan is necessary to mitigate effects of HLB.

68.7 CONCLUSIONS

Ca. Liberibacter spp. are insect-transmitted bacterial pathogens that cause plant decline and reduced production of plant fruit and quality and may cause eventual plant death. These species of bacteria have significant negative effects on production of commercial tree fruit and vegetables worldwide. Current management strategies focus on elimination of vector populations, removal of pathogen inoculum, or maintenance of infected plants through nutritional supplements to mitigate the negative effects of disease symptoms on plant growth. Future management strategies are focusing on the development of resistant or tolerant plant varieties and/or engineering of vectors that are unable to transmit the pathogens for mass release in the field.[20]

ACKNOWLEDGMENT

Kirsten S. Pelz-Stelinski and Lukasz L. Stelinski thank the Citrus Research and Development Foundation for funding.

REFERENCES

1. Liefting, L.W. et al. A new '*Candidatus Liberibacter*' species associated with diseases of Solanaceous crops. *Plant Dis.* 93: 208–214, 2009.
2. Davis, M. et al. Co-cultivation of '*Candidatus Liberibacter asiaticus*' with actinobacteria from citrus with huanglongbing. *Plant Dis.* 11: 1547–1550, 2008.
3. Sechler, A. et al. Cultivation of '*Candidatus Liberibacter asiaticus*', '*L. africanus*', and '*L. americanus*' associated with huanglongbing. *Phytopathology* 99: 480–486, 2009.
4. Garnier, M., Danel, N., and Bové, J.M. The greening organism is a gram-negative bacterium, pp. 115–124. In: *Proceedings of the 9th International Conference of Citrus Virologists*, Garnsey, S.M., Timmers, L.W., and Dodds, J.A. (eds.). University of California, Riverside, CA, 1984.
5. Jagoueix, S., Bové, J.M., and Garnier, M. The phloem-limited bacterium of greening disease of citrus is a member of alpha subdivision of Proteobacteria. *Int. J. Syst. Bacteriol.* 44: 379–386, 1994.
6. Hu, W.Z. et al. Diversity of the omp gene in *Candidatus Liberibacter asiaticus* in China. *J. Plant Pathol.* 93: 211–214, 2011.

7. Katoh, H. et al. Differentiation of Indian, East Timorese, Papuan and Floridian *Candidatus Liberibacter asiaticus'* isolates on the basis of simple sequence repeat and single nucleotide polymorphism profiles at 25 loci. *Ann. Appl. Biol.* 160: 291–297, 2012.

8. Wang, X. et al. Molecular characterization of a mosaic locus in the genome of '*Candidatus Liberibacter asiaticus'*. *BMC Microbiol.* 12: 18, 2012.

9. Laflèche, D. and Bové, J.M. Structures de type mycoplasma dans les feuilles d'orangers atteints de la maladie du greening. *C. R. Acad. Sci., Paris* 270: 1915–1917, 1970.

10. Laflèche, D. and Bové, J.M. Mycoplasmes dans les agrumes atteints de "greening", de "stubborn" ou de maladies similaires. *Fruits* 25: 455–465, 1970.

11. Bové, J.M. Huanglongbing: A destructive, newly-emerging, century-old disease of citrus. *J. Plant Pathol.* 88: 7–37, 2006.

12. Hajivand, S. et al. Ultrastructures of *Candidatus Liberibacter asiaticus* and its damage in huanglongbing (HLB) infected citrus. *Afr. J. Biotechnol.* 9: 5897–5901, 2010.

13. Mann, R.S. et al. Sexual transmission of a plant pathogenic bacterium, *Candidatus Liberibacter asiaticus*, between conspecific insect vectors during mating. *PLoS One* 6: e29197, 2011.

14. Duan, Z.Y. et al. Complete genome sequence of citrus huanglongbing bacterium, '*Candidatus Liberibacter asiaticus'* obtained through metagenomics. *Mol. Plant Microbe Interact.* 22: 1011–1120, 2009.

15. Lin, H. et al. The complete genome sequence of '*Candidatus Liberibacter solanacearum'*, the bacterium associated with potato Zebra Chip Disease. *PloS One* 6: e19135, 2011.

16. Kuykendall, L.D., Shao, J.Y., and Hartung, J.S. '*Ca. Liberibacter asiaticus'* proteins orthologous with *pSymA*-encoded proteins of *Sinorhizobium meliloti*: Hypothetical roles in plant host interaction. *PLoS One* 7: e38725, 2012.

17. Zhang, S. et al. '*Ca.* Liberibacter asiaticus' carries an excision plasmid prophage and a chromosomally integrated prophage that becomes lytic in plant infections. *Mol. Plant-Microbe Interact.* 24: 458–468, 2011.

18. Vahling-Armstrong, C.M. et al. Two plant bacteria, *S. meliloti* and *Ca.* Liberibacter asiaticus share functional *znuABC* homologues that encode for a high affinity zinc uptake system. *PLoS One* 7: e37340, 2012.

19. Li, W. et al. The ABC transporters in *Candidatus Liberibacter asiaticus*. *Proteins: Struct. Funct. Bioinf.* 80: 2614–2628, 2012.

20. Grafton-Cardwell, E. E., Stelinski, L.L., and Stansly, P.A. Biology and management of Asian citrus psyllid, vector of the huanglongbing pathogens. *Annu. Rev. Entomol.* 58: 413–432, 2013.

21. Texeira, D.C. et al. First report of a huanglongbing-like disease of citrus in Sao Paolo State, Brazil, and association of a new *Liberibacter* species, "*Candidatus Liberibacter americanus*", with the disease. *Plant Dis.* 89: 107, 2005.

22. Liefting, L.W. et al. A new '*Candidatus Liberibacter'* species in *Solanum tuberosum* in New Zealand. *Plant Dis.* 92: 1474, 2008.

23. Wen, A. et al. Detection, distribution, and genetic variability of '*Candidatus Liberibacter'* species associated with zebra complex disease of potato in North America. *Plant Dis.* 93: 1102–1115, 2009.

24. Hansen, A.K. et al. A new huanglongbing species, '*Candidatus Liberibacter psyllaurous*,' found to infect tomato and potato, is vectored by the psyllid *Bactericera cockerelli* (Sulc). *Appl. Environ. Microbiol.* 74: 5862–5865, 2008.

25. Xu, C.F. et al. Further study of the transmission of citrus huanglongbing by a psyllid, *Diaphorina citri* Kuwayama. In: *Proceedings of the 10th Conference of the International Organization of Citrus Virolologists*, Timmer, L.W., Garnsey, S.M., and Navarro, L. (eds.). Riverside, CA: IOCV, 1988, pp. 243–248.

26. Pelz-Stelinski, K.S., Brlansky, R.H., and Rogers, M.E. Transmission of *Candidatus Liberibacter asiaticus* by the Asian citrus psyllid, *Diaphorina citri*. *J. Econ. Entomol.* 103: 1531–1541, 2010.

27. Morris, A. and Muraro, R. Economic evaluation of citrus greening management and control strategies. Gainesville, FL: University of Florida, IFAS/CREC. EDIS Publication #FE712, 2009.

28. Boina D.R. et al. Quantifying dispersal of *Diaphorina citri* (Hemiptera: Psyllidae) by immunomarking and potential impact of unmanaged groves on commercial citrus management. *Environ. Entomol.* 38: 1250–1258, 2009.

29. Tiwari, A. et al. Incidence of *Candidatus Liberibacter asiaticus* infection in abandoned citrus occurring in proximity to commercially managed groves. *J. Econ. Entomol.* 103: 1972–1978, 2010.

30. Hartung, J.S. et al. Lack of evidence for transmission of '*Candidatus' Liberibacter asiaticus* through true citrus seed. *Plant Dis.* 94: 1200–1205, 2010.

31. van Vuuren, S.P., Cook, G., and Pietersen, G. Lack of evidence for seed transmission of '*Candidatus Liberibacter africanus'* associated with greening (huanglongbing) in citrus in South Africa. *Plant Dis.* 95: 1026–1026, 2011.

32. Hilf, M.E. Colonization of citrus seed coats by *Candidatus Liberibacter asiaticus*: Implications for seed transmission of the bacterium. *Phytopathology* 10: 1242–1250, 2011.

33. Pitman, A. et al. Tuber transmission of '*Candidatus Liberibacter solanacearum'* and its association with zebra chip on potato in New Zealand. *Eur. J. Plant Pathol.* 129: 389–398, 2011.

34. Duroy A.N. et al. LC-MS Analysis of phenolic compounds in tubers showing zebra chip symptoms. *Am. J. Potato Res.* 86: 88–95, 2009.

35. Li W., Hartung, J.S., and Levy, L. Quantitative real-time PCR for detection and identification of *Candidatus Liberibacter* species associated with huanglongbing. *J. Microbiol. Methods* 66: 104–115, 2006.

36. Li W., Hartung, J.S., and Levy, L. Evaluation of DNA amplification methods for improved detection of '*Candidatus Liberibacter* species' associated with citrus huanglongbing. *Plant Dis.* 91: 51–58, 2007.

37. Li, W. et al. Multiplex real-time PCR for detection, identification and quantification of '*Candidatus Liberibacter solanacearum'* in potato plants with zebra chip. *J. Microbiol. Methods* 78: 59–65, 2009.

38. Hong, L.T.T. and Truc, N.T.N. Iodine reaction quick detection of huanglongbing disease. *Proceedings of the 2003 Annual Workshop of JIRCAS Mekong Delta Project*, pp. 1–11, 2003.

39. Etxeberria, E. et al. An iodine-based starch test to assist in selecting leaves for HLB testing. UF/IFAS EDIS HS375, 2007.

40. Etxeberria, E. et al. Anatomical distribution of abnormally high levels of starch in HLB-affected Valencia orange trees. *Physiol. Mol. Plant Pathol.* 74: 76–83, 2009.

41. Rogers, M.E., Stansly, P.A., and Stelinski, L.L. Asian citrus psyllid and citrus leafminer. In: *Florida Citrus Pest Management Guide, University of Florida*, Rogers, M.E., Dewdney, M.M., and Spann, T.M. (eds.). IFAS extension Publication No. SP-43, 2011, pp. 43–51.

42. Tiwari, S., Pelz-Stelinski, K.S., and Stelinski, L.L. Effect of *Candidatus Liberibacter asiaticus* infection on susceptibility of Asian citrus psyllid, *Diaphorina citri*, to selected insecticides. *Pest Manag. Sci.* 67: 94–99, 2011.

43. Tiwari, S. et al. Characterization of five CYP4 genes from Asian citrus psyllid and their expression levels in *Candidatus Liberibacter asiaticus* infected and uninfected psyllids. *Insect Mol. Biol.* 20: 733–744, 2011.

44. Tiwari, S. et al. Insecticide resistance in field populations of Asian citrus psyllid in Florida. *Pest Manag. Sci.* 67: 1258–1268, 2011.

45. Spann, T.M. et al. Foliar nutrition for HLB. *Citrus Ind.* 92: 6–10, 2011.

46. Gottwald, T.R. Current epidemiological understanding of citrus huanglongbing. *Annu. Rev. Phytopathol.* 48: 119–139, 2010.

47. Spann, T.M. and Schumann, A.W., Citrus greening-associated nutrient deficiency. *Citrus Ind.* 90: 14–15, 2009.

48. Rouse, B. et al. Monitoring trees infected with huanglongbing in a commercial grove receiving nutritional/SAR foliar sprays in southwest Florida. *Proc. Florida State Hort. Soc.* 123: 118–120, 2010.

49. Roka, F.M. Defining the economic "tipping point" in the management of citrus greening: Follow the standard protocol or shift to an enhanced foliar nutritional program. *Proceedings of the 18th International Farm Management Congress.* Canterbury, New Zealand, 2011.

69 *Peronosclerospora philippinensis* and Related Species

László Kredics, László Galgóczy, and Csaba Vágvölgyi

CONTENTS

69.1 INTRODUCTION

Maize is among the most important crops of the world. As exotic downy mildew species are capable of causing severe losses to maize production, they represent significant biosecurity threats. To date, eight downy mildew species have been reported to be pathogenic on maize, three of which have been reported in the United States: *Peronosclerospora sorghi* (downy mildew of sorghum and maize), *Sclerophthora macrospora* (crazy top), and *Sclerophthora graminicola* (green ear downy mildew).[1] *Peronosclerospora philippinensis* (W. Weston), C. G. Shaw 1978, is an oomycetous fungus that causes the Philippine downy mildew (PDM) disease of maize. Although PDM has not been reported in the United States and the western hemisphere, this disease is known to cause significant crop losses in Asia and the Philippines. On farms across the Philippines, the annual yield losses due to PDM on maize have been 40%–60%, but in the case of sweet corn, losses of 100% have also been reported.[1–3] The national yield loss in the 1974–1975 crop year was estimated to be about 8%, which was corresponding to US $23 million.[2] The disease is generally less severe in India, but losses up to 60% have been reported.[4,5] Data available on yield losses in the case of *P. philippinensis* infection of sugarcane are limited; however, losses of 35% of the harvested and extracted sugar have been noted.[2] The most severe disease occurs in tropical climates and in areas where the annual amount of rainfall is between 154 and 307 mm.[2]

Along with the closely related species *P. sacchari* (which was supposed to be conspecific with *P. philippinensis* by some authors),[6–8] *P. philippinensis* is included in the select agent list of the Animal and Plant Health Inspection Service (APHIS), US Department of Agriculture (USDA), as a bioterror agent because it has the potential to pose a severe threat to the production of maize, which is one of the most important crops in the United States where it is grown as a monoculture in the North American corn belt.[9] A third member of the genus *Peronosclerospora*, *P. maydis*, is also listed by APHIS as a regulated plant pest.[10,11] The aforementioned species, along with *P. spontanea* and *P. sorghi*, are also listed as high risk plant pathogens in Australia that should be carefully monitored and quarantined.[12] Three *Peronosclerospora* species, *P. philippinensis*, *P. sacchari*, and *P. sorghi* are included in the Target List for Plant Pathogens of North Australia Quarantine Strategy (NAQS) as having the potential for significant adverse impact on agriculture.[13]

69.2 CLASSIFICATION AND MORPHOLOGY

The causative agent of the PDM of maize was originally described as *Sclerospora philippinensis* by Weston in 1920.[14] The *Peronosclerospora* subgenus of *Sclerospora*[15] was raised to the generic rank in 1978 by Shaw[16] based on the production of true conidia which always germinate by a germ tube. *S. philippinensis* along with seven other former species of *Sclerospora* (*S. dichanthiicola*, *S. maydis*, *S. miscanthi*, *S. sacchari*, *S. spontanea*, *S. sorghi*, and *S. westonii*) were transferred to the genus *Peronosclerospora*, and *P. sacchari* was defined as the type species of the genus.[15] Besides *P. philippinensis* (W. Weston) C.G. Shaw, 12 further valid species have been accepted up to date within the genus *Peronosclerospora*: *P. australiensis* R.G. Shivas, Ryley, Telle, Liberato & Thines, *P. dichanthiicola* (Thirum. & Naras.) C.G. Shaw (attacking *Sorghum* spp. and maize), *P. eriochloae* Ryley & Langdon (found on *Eriochloa pseudoacrotricha*), *P. heteropogonis* Siradhana, Dange, Rathore & S.D. Singh (causal agent of the Rajasthan downy mildew of maize), *P. maydis* (Racib.) C.G. Shaw (causal agent of the Java downy mildew or sleepy disease of maize), *P. miscanthi* (T. Miyake) C.G. Shaw, *P. noblei* (W. Weston) C.G. Shaw (attacking wild sorghum), *P. sacchari* (T. Miyake) Shirai & Hara (causal agent of the sugarcane downy mildew), *P. sargae* R.G. Shivas, M.J. Ryley, Telle & Thines (attacking *Sorghum* spp.), *P. sorghi* (W. Weston & Uppal) C.G. Shaw (causal agent of the sorghum downy mildew), *P. spontanea* (W. Weston) C. G. Shaw (causal agent of the spontaneum downy mildew), and *P. westonii* (Sriniv. Naras. & Thirum.) C.G. Shaw.[17]

The graminicolous downy mildew genera *Sclerospora* and *Peronosclerospora* were known as members of the Peronosporaceae family within the Peronosporales order until 1984, when they were moved to the new order Sclerosporales by Dick et al.[18] on the basis of their parasitism on graminaceous hosts, a thickened, sclerified oogonial wall, and more or less plerotic oospores. Two families were suggested within this new order, the Sclerosporaceae including the genera *Peronosclerospora* and *Sclerospora*, and the Verrucalvaceae including the genera *Sclerophthora* and *Verrucalvus*. Later the Sclerosporales order was removed from the Peronosporomycetidae and placed in the Saprolegniomycetidae.[19] However, results of molecular phylogenetic investigations have not supported the aforementioned classifications and clearly placed *Sclerospora* and *Peronosclerospora* within Peronosporaceae.[20,21] In the year 2013 the recently accepted taxonomic position of the genus *Peronosclerospora* is the following: family Peronosporaceae, order Peronosporales, phylum Oomycota, kingdom Chromista.[22]

Mycelia of the fungus are 8 μm wide, irregularly constricted and inflated, growing intercellularly in all parts of the host plant except from the root.[14] Haustoria are simple, 8 × 2 μm in size, vesiculiform to subdigitate. Conidiophores (150–400 μm × 15–26 μm) are dichotomously branched 2 to 4 times; they develop through the stomata during the night dew or moist conditions. Conidia borne on the sterigmata are 17–21 × 27–39 μm, hyaline, elongate ovoid to round cylindrical, slightly rounded at the apex. The germination follows always by a germ tube. The average diameter of oogonia is 22.9 μm; they have smooth wall with frequently adhering fragments of antheridial cell or oogonial stalk. Oospores are rarely occurring, regularly spherical, central to eccentric, 15.5–22.5 μm in diameter with an average of 19.2. Their wall is 2–4 μm thick, and they have a homogenous, finely granular content.[14]

69.3 BIOLOGY AND EPIDEMIOLOGY

Although *P. philippinensis* is known to produce an overwintering spore form (oospore), its role in the lifecycle has not been established. Production of *P. philippinensis* conidia requires darkness and night temperatures ranging from 21°C to 26°C, accompanied by free moisture: the presence of at least a thin film of water on the infected leaf surface is necessary.[3] The most intensive conidium production was found to occur at 90%–94% relative humidity and 23°C–24°C for 6–8 h.[9] Development of conidiophores and conidia, dissemination as well as inoculation and penetration of the host plants take place during the night time.[9] The dew deposit begins around 7–8 p.m. Conidiophore initials appear in stomata around

10–11 p.m. when the surface is already wet.[9] Conidiophores develop until midnight, which is followed by the development of conidia until 1 a.m. The discharge and dispersal of conidia begins around 1 a.m., reaches its maximum at 2–3 a.m., and lessens between 3 and 4 a.m. Conidia are forcibly discharged, but only 1–2 mm, and usually all conidia on the conidiophore are ejected at once.[9] The conidium and the sterigmata keep bulging out until the basal septum of the conidium and the apical septum of the sterigmata bulge enough, enabling the conidium to be shot away. Germination of conidia begins between 3 and 4 a.m. and continues until 5 a.m., which is followed by the drying of conidiophores at 5–6 a.m.[9]

Conidial production can continue for over 2 months on maize plants, but even up to 8 months on wild sugarcane.[9] Conidia produced on nearby infected maize plants or other hosts such as susceptible grass species or sugarcane represent the source of primary infection in maize.[3] The PDM disease can spread locally via raindrops and wind.[1] Wind dispersal results in localized spread among fields in a given geographical area;[3] however, serious storms, for example, typhoons, may spread the disease over several miles.[9] Splashing water from dew and rain as well as insects probably play only a minor role in dissemination of PDM.[9] Seed transmission of *P. philippinensis* may also occur, as the fungus becomes established within the seed as a mycelium in the pericarp, the embryo, and the endosperm.[23] However, there are no external symptoms on the seed and its quality is not affected. In the case of another member of the genus, *P. sorghi*, it has been clearly demonstrated that seed transmission will not occur if the moisture content of the grain or seed is below 14%.[24] PDM can be introduced to new locations also by the movement of infected plant tissues.[1] Infection rates by airborne conidia are high at temperatures above 16°C.[23] Before germination, conidia keep swelling and their size and shape keeps changing.[9] Germination takes place on maize plants in less than 1 h;[9] during this process the conidia produce one or more germ tubes.[9] Optimal temperatures for germination are between 20°C and 24°C; however, germination can take place also at temperatures as low as 6.5°C.[9] Although appressoria are usually formed, penetration without appressoria may also occur. Germ tubes invade through the stomata and develop into a mycelium in the mesophyll.[9] Mycelia are most abundant in the leaf regions showing chlorotic symptoms, but they are also found in leaf areas without any discoloration.[9] Hyphae grow intercellularly; the cell walls are penetrated by haustoria which enter the lumen, which, however, does not result in collapsing of the host cells.[9] The virulence of different isolates may vary substantially.[23]

PDM is considered as a monocyclic disease, which means that *P. philippinensis* completes only one or a few growth cycles within one growing season; however, in the tropics, this pathogen has many characteristics of polycyclic organisms which complete a large number of cycles in a season and cause devastating epidemics.[9] This can be explained with the situation that in certain areas of the Philippines (e.g., Mindanao), maize is planted all around the year, which results in the coexistence of maize plantations that are widely differing in development stage.[9] An additional problem is that other susceptible hosts (e.g., wild sugarcane) capable of producing large numbers of conidia under appropriate environmental conditions are abundant in or near the maize fields.[9] This special epidemiological situation may result in infections and serious crop damages all around the year.

The confirmed geographical distribution of *P. philippinensis* includes China, India, Indonesia, Nepal, Pakistan, the Philippines, and Thailand.[3] Besides maize (*Zea mays*) and teosinte (*Euchlaena luxurians*),[14] the natural host range of *P. philippinensis* includes common oat (*Avena sativa* L.),[25] sugarcane (*Saccharum officinarum* L.),[25] sorghum species (*Sorghum bicolor* Moench,[14] *Sorghum propinquum* (Kunth) Hitchc),[25] wild sorghum (*Sorghum arundinaceum* (Willd.) Stapf),[26] Johnsongrass (*Sorghum halepense* (L.) Pers),[25] silver grass (*Miscanthus japonicus* Andress),[27] and wild sugarcane (*Saccharum spontaneum* L.).[27] Bonde and Peterson[7] found that the host range of *P. philippinensis* is nearly identical to that of *P. sacchari*. In order to evaluate the threat of *P. philippinensis* to U.S. agriculture, they performed artificial inoculation experiments on 72 plant species from 22 genera to examine whether this downy mildew pathogen has any potential alternative hosts which enable overwintering and may serve as a reservoir to infect maize. From the tribe Andropogoneae, 19 species from 6 genera (*Andropogon*, *Bothriochloa*, *Eulalia*, *Saccharum*, *Schizachyrium*, and *Sorghum*), while from the tribe Maydeae, 4 species from 2 genera (*Tripsacum* and *Zea*) proved to be susceptible to *P. philippinensis*. Among the potential hosts, big bluestem (*Andropogon gerardii*) and little bluestem (*Schizachyrium scoparium*) are common perennial grasses in the prairie states of the United States, thereby representing potential reservoirs of *P. philippinensis*. The authors also compared the host ranges of *P. philippinensis* and *P. sacchari*, which were found remarkably similar, suggesting a very close phylogenetic relationship between the two species.[7]

69.4 INFECTION FEATURES AND PATHOGENESIS

P. philippinensis can infect two-leaf stage and older maize plants up to 3 weeks of age.[1] Four to six-week-old plants are almost immune.[9] Based on controlled inoculation experiments, whorl is the most susceptible part of maize seedlings.[9] The first symptoms of PDM typically appear as chlorotic stripes or overall yellowing at the first true and successive leaves as early as 9 days after planting.[23,28] All leaves of the plant may show these characteristic symptoms of long, yellow chlorotic stripes. Conidiation is more frequent on the lower leaf surface, but it can occur on both sides.[23] The clearest indication of PDM is a downy covering on the lower surface of the leaves, which is the site of conidium production and the source for spreading of the infection further to maize plants.[28] *P. philippinensis* also invades the shoot apex and the stem, but without any visible external symptoms.[23] After becoming established, the fungus can survive in the apex until the death of the plant.[9] In the case of young plants

the infection may result in stunting and death. As the plant ages, older leaves may narrow, become abnormally erect, and appear somewhat dried out.[28] Systemic infections can cause tassel malformation, interrupted ear formation, and reduction of pollen production.[28]

The severity of the disease is highly influenced by environmental factors like temperature, humidity, wind, light, and dew.[29–31]

69.5 IDENTIFICATION AND DIAGNOSIS

There is an urgent need for rapid and accurate methods for the identification of *P. philippinensis* in order to be able to prevent the introduction of this serious maize pathogen to new regions like the United States.[8] Although the differentiation of PDM from the maize and sorghum pathogen *P. sorghi* is microscopically possible based on clear differences in the conidiophore morphology,[3] oospores are not known, missing, or rare in the case of certain *Peronosclerospora* species (e.g., *P. maydis*, *P. philippinensis*, and *P. spontanea*). Additionally, the conidium morphology was found to be highly dependent on the environmental conditions, like temperature as well as the host species and cultivar,[25,32] which makes the morphology-based identification of *P. philippinensis* at the species level and its differentiation from other closely related species of the genus very difficult.[15] *P. philippinensis* and *P. sacchari* were shown to be very similar morphologically.[6] According to the original species description, the only differences between them are that *P. philippinensis* cannot form oospores and it does not attack sugarcane;[14] however, in a later publication it was demonstrated that *P. philippinensis* can also colonize sugarcane.[7] Duck et al.[34] reported that these two species are indistinguishable from each other on the basis of sporulation, both of them producing 4,000–10,000 spores per cm^2 of infected leaf area. Furthermore, similarly to *P. sacchari*, *P. philippinensis* was shown to have less systemic infection with dew periods at the lower temperatures of 10°C–16°C.[30]

PDM cannot be diagnosed solely based on the early field symptoms, as physiological conditions and a series of plant pathogens including other downy mildews result in similar—in the case of *P. sacchari* almost identical—symptoms.[3]

In 1984, Bonde et al.[35] suggested that isoenzyme analysis of a series of enzymes (aspartate aminotransferase, fumarase, glucose phosphate isomerase, glutamate pyruvate transaminase, glutathione reductase, isocitrate dehydrogenase, leucine aminopeptidase, malate dehydrogenase, peptidase with glycilleucine, peptidase with leucyl-leucyl-leucine, phosphogluconate dehydrogenase, and superoxide dismutase) by horizontal starch gel electrophoresis may be a useful tool for distinguishing species of the genus *Peronosclerospora* and elucidating their phylogenetic relationships. Besides six cultures of *P. sorghi*, the examined strains included two cultures of *P. philippinensis* and two cultures of *P. sacchari*. Based on the results of the study the authors concluded that *P. sacchari* in Taiwan may be essentially the same pathogen as *P. philippinensis* in the Philippines. In a subsequent study, Micales et al.[8] included further 17 enzymes (acid phosphatase, aconitase, adenylate kinase, alpha-glycerophosphate dehydrogenase, diaphorase, esterase, fructose diphosphatase, glucose-6-phosphate dehydrogenase, glucokinase, β-glucosidase, glutamate dehydrogenase, glyceraldehyde-3-phosphate dehydrogenase, mannitol dehydrogenase, mannose phosphate isomerase, peptidase with phenylalanyl-proline, triose phosphate isomerase, and xanthine dehydrogenase), an aminopeptidase assay, and a larger set of *Peronosclerospora* strains (16 *P. sorghii*, 8 *P. sacchari*, 8 *P. philippinensis*, and 2 *P. maydis*) to evaluate the discriminatory power of isoenzyme analysis within the genus. The aminopeptidase assay applied was not able to differentiate between *P. sorghi*, *P. sacchari*, and *P. philippinensis* due to intraspecific variations. Isoenzyme analysis was shown to have a higher discriminatory power within the genus; however, *P. sacchari* and *P. philippinensis* were found to exhibit identical banding patterns for 22 out of the 26 enzymes tested, and therefore the authors suggested that they may represent the same species.[8]

DNA-based molecular methods have already shown the potential to be applicable for differentiation of *Peronosclerospora* species. Yao et al.[36] identified an A–T rich DNA region from a *P. sorghi* library which hybridizes only to DNA of *P. sorghi* or to DNA from *P. sorghi*-infected leaves and not to DNA samples from *P. philippinensis*, *P. sacchari*, and *P. maydis*. After an outbreak of sorghum downy mildew in Texas, Perumal et al.[37] identified simple sequence repeat (SSR) markers useful for sorghum downy mildew (*P. sorghi*) and related species after high throughput sequencing of a microsatellite-enriched library prepared from a *P. sorghi* pathotype by using a simplified biotinylated oligonucleotide capture protocol. For the 55 cloned microsatellite sequences, 54 SSR primer pairs were designed, out of which 15 amplified SSR amplicons are unique to *P. philippinensis*, thereby being applicable for the development of molecular diagnostic tests. This study also revealed a narrow genetic distance between *P. philippinensis* and *P. sacchari*; however, the unique identity of *P. philippinensis* from other species was clearly depicted.[37]

69.6 TREATMENT AND PREVENTION

Breeding maize varieties resistant to *P. philippinensis* proved to be a useful option to prevent PDM. In order to perform a successful breeding program for PDM resistance, information is needed about the nature and inheritance of the resistance traits.[38] Crosses were performed between seven susceptible and five resistant maize lines by Gomez et al.[39] Based on the results it was concluded that the resistance behavior could be due to partial dominance and that the reaction of the disease is under the control of a few gene pairs only.[39] Carañgal et al.[40] performed crosses between introduced susceptible and native resistant varieties, selected a number of tolerant lines, and suggested that resistance to PDM has a polygenic nature. The host resistance to PDM of maize was studied in a diallel cross of three susceptible, two resistant, and four intermediate inbred lines by Mochizuki et al.,[41] who concluded that the resistance to PDM is governed by a few factor pairs

only and that it is conditioned by dominant genes. Yamada and Aday[42] examined the usefulness of local Philippine varieties as genetic resources for developing maize varieties resistant to PDM by performing artificial inoculations, and concluded that the susceptibility was governed by recessive genes. Fertilizer conditions affecting the susceptibility to PDM were also studied in resistant and susceptible varieties of maize, and high nitrogen levels were found to increase the susceptibility of susceptible cultivars, although they had less effect on the resistance of resistant cultivars.[43] A screening test for maize varieties resistant to PDM was also developed.[44] Dahlan and Aday.[38] studied the genetic variances of maize resistance to PDM and found that the additive genetic variance was a major component of the total genetic variance, especially under moderate disease incidence. Kaneko and Aday[45] studied a PDM-resistant and a susceptible maize variety, and their F1 generation hybrid, and found that the changes in percentage of infection depended on the age of the seedlings at the time of inoculation. The mode of inheritance of resistance was found to change from complete to partial dominance. The authors suggested that the resistance is governed by a polygenic system with a threshold nature.[45] Ebron and Raymundo[46] found that resistant inbred lines show an extended duration of local infection, a delayed onset of systemic infection, and a slow rate of downy mildew development, and correlated these epidemiological components of resistance with one another. De Leon et al.[47] studied the inheritance of maize resistance to PDM using resistant and susceptible homozygous derivatives of three maize populations by artificial inoculation, and found that the resistance is polygenically inherited and is mainly controlled by additive gene effects. George et al.[48] analyzed quantitative trait loci (QTLs) involved in resistance to the important downy mildew pathogens of maize including *P. philippinensis*, and identified six genomic regions on chromosomes 1, 2, 6, 7, and 10 involved in downy mildew resistance. The results of the study suggested that the expression of the QTLs may be dependent on the environment. Furthermore, the authors identified SSR markers tightly linked to a strong, stable QTL on chromosome 6, which have the potential to be used in marker-assisted selection.[48] Although maize varieties resistant to *P. philippinensis* have been developed in Asia, the maize hybrids recently cultivated in the United States are highly susceptible to PDM, and American breeding lines which are highly resistant to *P. sorghi* are also generally susceptible to *P. philippinensis*.[37]

Cultural practices suggested to have a value in controlling PDM include the elimination of weed hosts, manipulating the planting dates, regulating plant density (wider spacing or interplanting maize with other crops which enables better air drainage and faster drying of the plants), removing infected plants as soon as possible in order to reduce inoculum level, and planting maize only once a year, not continuosly.[9]

Regarding the possibilities of chemical control, significant efforts have been made in the Philippines to develop appropriate control strategies: 63 fungicides and 8 antibiotics were tested until 1965.[49] Fentin hydroxide and maneb were reported to provide effective control of PDM.[49,50] Metalaxyl—a compound showing specific toxicity to fungi in the Peronosporales—applied as a foliar spray or in the form of the metalaxyl-based seed-dressing fungicide Ridomit (Ciba-Geigy2) is also effective for the control of PDM.[31,51] Molina and Exconde[51] reported that metalaxyl-treated seeds were not infected regardless of rainfall frequency, while 97.6% of the untreated seedlings were infected. Other possible chemical control options mentioned by Magill et al.[3] are Duter and Dithane M-45, Dexon, Cela, azoxystrobin, chlorothalonil, and mefanoxam. Control of PDM could potentially be accomplished with a seed treatment with mefanoxam or metalaxyl.[3]

An environment-friendly alternative of chemical control could be the application of biocontrol agents. Although there are no reports to date about studies aimed at the biological control of PDM, efforts were made to develop biocontrol strategies against *P. sorghi*, the causative agent of maize downy mildew in India.[52] The disease could be significantly controlled both under greenhouse and field conditions by foliar and seed treatment of a talc-based formulation based on a *Pseudomonas fluorescens* strain. Furthermore, leaf extracts of the neem tree (*Azadirachta indica*) and the Chilean mesquite (*Prosophis chilensis*) proved similarly effective as the pseudomonads in controlling maize downy mildew; however, the plant extracts were less inhibitory to conidial germination than *P. fluorescens* or metalaxyl. The tested biological treatments were able to suppress *P. sorghi* sporulation on diseased leaves. The control mechanism was suggested to be associated with enhanced induction of defense enzymes (e.g., β-1,3-glucanase and chitinase). Under field conditions, the application of the studied biocontrol strategies resulted in the reduction of disease incidence and increased yield. The promising results of this study suggest that bacteria and plant extracts may also be potential candidates for the biological control of *P. philippinensis*.[52]

69.7   CONCLUSION

Although the distribution of *P. philippinensis* is currently limited to Asia, there is a considerable risk that this highly destructive disease may eventually spread to other tropical and temperate maize-growing regions.[9] Regulatory protocols are in place to prevent the introduction of this serious maize pathogen to the United States.[1] A recovery plan was produced for PDM (as well as for the brown stripe downy mildew of maize caused by *Sclerophthora rayssiae* var. *zeae*)[3] as part of the National Plant Disease Recovery System (NPDRS) called for in Homeland Security Presidential Directive Number 9 (HSPD-9), which suggests the development of a coordinated framework for survey, detection, and prediction of maize downy mildews. The recovery plan defines the need for a surveillance and monitoring program which can provide timely information about the incidence and severity of maize downy mildew in the United States, a web-based system for information management of monitoring observations, forecasts, and decision criteria to stakeholders, and prediction modeling for the spread and establishment of maize downy mildew. The development of species-specific molecular detection tools, compilation of digital images of disease symptoms and pathogen morphology on relevant hosts and their distribution, the compilation of bibliography of downy mildew related literature, the gathering of current information on industry activities related to the testing of recently developed fungicides (e.g., strobilurins, azoles) for maize downy mildew control in Asia, the development of geophytopathological models for predicting the establishment and spread potential of *P. philippinensis* in the United States, and the organization of training courses on management of downy mildew diseases and fungicide resistance are among the priorities of this recovery plan.[3]

ACKNOWLEDGMENT

Authorial activity of László Galgóczy and Csaba Vágvölgyi was supported in the frame of TÁMOP 4.2.4. A/2-11-1-2012-0001.

REFERENCES

1. Byrne, J.M. and Hammerschmidt, R. Biology and diagnosis of Philippine downy mildew. In: *NPDN National Meeting 2007*, poster presentation, www.plantmanagementnetwork. org/proceedings/npdn/2007/posters/05DownyMildew. pdf, 2007 (accessed on November 4, 2013)..
2. Exconde, O.R. and Raymundo, A.D. Yield loss caused by Philippine corn downy mildew. *Philipp Agric*. 58, 115, 1974.
3. Magill, C. et al. Recovery plan for Philippine downy mildew and brown stripe downy mildew of corn caused by *Peronosclerospora philippinensis* and *Sclerophthora rayssiae* var. *zeae*, respectively, www.ars.usda.gov/SP2UserFiles/Place/00000000/opmp/Corn%20Downy%20Mildew%20 09-18-06.pdf, 2006.
4. Bonde, M.R. Epidemiology of downy mildew diseases of maize, sorghum and pearl millet. *Trop Pest Manag*. 28, 49, 1982.
5. Payak, M.M. Downy mildews of maize in India. *Trop Agric Res*. 8, 13, 1975.
6. Bonde, M.R., Duck, N.B., and Peterson, G.L. Morphological comparison of *Peronosclerospora sacchari* from Taiwan with *P. philippinensis* from the Philippines. *Phytopathology*. 74, 755, 1984.
7. Bonde, M.R. and Peterson, G.L. Comparison of host ranges of *Peronosclerospora philippinensis* and *P. sacchari*. *Phytopathology*. 73, 875, 1983.
8. Micales, J.A., Bonde, M.R., and Peterson, G.L. Isozyme analysis and aminopeptidase activities within the genus *Peronosclerospora*. *Phytopathology*. 78, 1396, 1988.
9. Thurston, H.D. *Tropical Plant Diseases*. The American Phytopathological Society, St. Paul, MN, 1984.
10. Cline, E.T. and Farr, D.F. Synopsis of fungi listed as regulated plant pests by the USDA Animal and Plant Health Inspection Service: Notes on nomenclature, disease, plant hosts, and geographic distribution. *Plant Health Progr*. doi:10.1094/ PHP-2006-0505-01-DG, 2006.
11. Rossman, A.Y. et al. Evaluating the threat posed by fungi on the APHIS list of regulated plant pests. *Plant Health Progr*. doi:10.1094/PHP-2006-0505-01-PS, 2006.
12. Telle, S. et al. Molecular phylogenetic analysis of *Peronosclerospora* (Oomycetes) reveals cryptic species and genetically distinct species parasitic to maize. *Eur J Plant Pathol*. 130, 521, 2011.

13. Floyd, R. Review of Australia's Quarantine Function: Question on notice. http://www.aphref.aph.gov.au/house/committee/jcpaa/aqis/submissions/sub49.pdf 2002 (accessed on November 4, 2013).

14. Weston, W.H.J. Philippine downy mildew of maize. *J Agric Res.* 19, 97, 1920.

15. Ito, S. Kleine Notizen über parasitische Pilze Japans. *Bot Mag Tokyo.* 27, 217, 1913.

16. Shaw, C.G. *Peronosclerospora* species and other downy mildews of the Gramineae. *Mycologia.* 70, 594, 1978.

17. Shivas, R.G. et al. *Peronosclerospora australiensis* sp. nov. and *Peronosclerospora sargae* sp. nov., two newly recognised downy mildews in northern Australia, and their biosecurity implications. *Aust Plant Pathol.* 41, 125, 2012.

18 Dick, M.W., Wong, P.T.W., and Clark, G. The identity of the oomycete causing "Kikuyu Yellows", with a reclassification of the downy mildews. *Bot J Linn Soc.* 89, 171, 1984.

19. Dick, M.W. et al. A new genus of the *Verrucalvaceae (Oomycetes). Bot J Linn Soc.* 99, 97, 1989.

20. Göker, M. et al. How do obligate parasites evolve? A multigene phylogenetic analysis of downy mildews. *Fungal Genet Biol.* 44, 105, 2007.

21. Hudspeth, D.S.S., Stenger, D., and Hudspeth, M.E.S. A *cox2* phylogenetic hypothesis for the downy mildews and white rusts. *Fungal Divers.* 13, 47, 2003.

22. CAB International. *Index Fungorum.* Online. CAB International, Wallingford, U.K., http://www.indexfungorum.org/Names/NamesRecord.asp?RecordID=118331, 2013.

23. Jepson, S.B. Philippine downy mildew of corn. Oregon State University Extension Service, http://www.science.oregon-state.edu/bpp/Plant_Clinic/Disease_sheets/Philippine%20downy%20mildew%20of%20corn.pdf, 2008 (accessed on November 4, 2013).

24. Adenle, V.O. and Cardwell, K.F. Seed transmission of *Peronosclerospora sorghi*, causal agent of maize downy mildew in Nigeria. *Plant Pathol.* 49, 628, 2000.

25. Exconde, O.R., Elec, J.V., and Advincula, B.A. Host range of *Sclerospora philippinensis* Weston in the Philippines. *Philipp Agric.* 52, 175, 1968.

26. Storey, H.H. and McClean, A.P.D. A note upon the conidial *Sclerospora* of maize in South Africa. *Phytopathology.* 20, 107, 1930.

27. Weston, W.H.J. Another conidial *Sclerospora* of Philippine maize. *J Agric Res.* 20, 669, 1921.

28. White, D.G. (ed.) 1999. *Compendium of Corn Diseases.* APS Press, St. Paul, MN.

29. Semangoen, H. Studies on downy mildew of maize in Indonesia, with special reference to the perrenation of the fungus. *Indian Phytopathol.* 23, 307, 1970.

30. Barredo, F.C. and Exconde, O.R. Incidence of Philippine corn downy mildew as affected by inoculum suscept and environment. *Philipp Agric.* 57, 232, 1973.

31. Exconde, O.R. and Molina, A.B. Ridomit (Ciba-Geigy2), a seed-dressing fungicide for the control of Philippine corn downy mildew. *Philipp J Crop Sci.* 3, 60, 1978.

32. Leu, L.S. Effects of temperature on conidial size and sporulation of *Sclerospora sacchari. Plant Prot Bull Taiwan.* 15, 106, 1973.

33. Bonde, M.R. et al. Effect of temperature on conidial germination and systemic infection of maize by *Peronosclerospora* species. *Phytopathology.* 82, 104, 1992.

34. Duck, N.B. et al. Sporulation of *Peronosclerospora sorghi, P. sacchari* and *P. philippinensis* on maize. *Phytopathology.* 77, 438, 1987.

35. Bonde, M.R. et al. Isozyme analysis to differentiate species of *Peronosclerospora* causing downy mildews of maize. *Phytopathology.* 74, 1278, 1984.

36. Yao, C.L., Magill, C.W., and Frederiksen, R.A. Use of an A–T-rich DNA clone for identification and detection of *Peronosclerospora sorghi. Appl Environ Microbiol.* 57, 2027, 1991.

37. Perumal, R. et al. Simple sequence repeat markers useful for sorghum downy mildew (*Peronosclerospora sorghi*) and related species. *BMC Genet.* 9, 77, 2008.

38. Dahlan, M.M. and Aday, B.A. Estimation of genetic variances of resistance to downy mildew (*Sclerospora philippinensis* Weston) in a composite corn variety. *Philipp J Crop Sci.* 3, 5, 1978.

39. Gomez, A.A. et al. Preliminary studies on the inheritance of the reaction of corn downy mildew disease. *Philipp Agric.* 47, 113, 1963.

40. Carañgal, V., Claudio, M., and Sumayao, M. Breeding for resistance to maize downy mildew caused by *Sclerospora philippinensis* in the Philippines. *Indian Phytopathol.* 23, 285, 1970.

41. Mochizuki, N., Carañgal, V.R., and Aday, B.A. Diallel analysis of host resistance to Philippine downy mildew of maize caused by *Sclerospora philippinensis. JARQ.* 8, 184, 1974.

42. Yamada, M. and Aday, B. Usefulness of local varieties for developing resistant varieties to Philippine downy mildew disease. *Maize Genet Coop Newslett.* 51, 68, 1977.

43. Yamada, M. and Aday, B.A. Fertilizer conditions affecting susceptibility to downy mildew diseases, *Sclerospora philippinensis* Weston, in resistant and susceptible materials of maize. *Ann Phytopathol Soc Jpn.* 43, 291, 1977.

44. Yamada, M. and Aday, B.A. Development of screening test for resistant materials to Philippine downy mildew disease of maize caused by *Sclerospora philippinensis* Weston. *Jpn J Breed.* 27, 131, 1977.

45. Kaneko, K. and Aday, B.A. Inheritance of resistance to Philippine downy mildew of maize, *Peronosclerospora philippinensis. Crop Sci.* 20, 590, 1980.

46. Ebron, L.A. and Raymundo, A.D. Quantitative resistance to Philippine corn downy mildew caused by *Peronosclerospora philippinensis* (Weston) Shaw. *Philipp Agric.* 70, 217, 1987.

47. De Leon, C. et al. Genetics of resistance to Philippine downy mildew in three maize populations. *Indian J Genet.* 4, 406, 1993.

48. George, M.L. et al. Identification of QTLs conferring resistance to downy mildews of maize in Asia. *Theor Appl Genet.* 107, 544, 2003.

49. Exconde, O.R. Chemical control of maize downy mildew. pp. 157. In: *Symposium on Downy Mildew of Maize.* Tropical Agriculture Research Series No. 8, Tropical Agriculture Research Center, Ministry of Agriculture and Forestry, Tokyo, 1975.

50. Exconde, O.R. Philippine corn downy mildew: Assessment of present: Knowledge and future research needs. *Kasetsart J.* 10, 94, 1976.

51. Molina, A.B. and Exconde, O.R. Efficacy of Apron 35 SD (MetaLaxyl) against Philippine corn downy mildew 1. Effects of seed dressing methods and rainfall frequency. *Philipp Agric.* 64, 99, 1981.

52. Kamalakannan, A. and Shanmugam, V. Management approaches of maize downy mildew using biocontrol agents and plant extracts. *Acta Phytopathol Entomol Hung.* 44, 255, 2009.

70 *Phoma glycinicola*

Glen L. Hartman

CONTENTS

70.1 INTRODUCTION

Red leaf blotch of soybean [*Glycine max* (L.) Merr.] is the common name for the disease caused by the fungal pathogen *Phoma glycinicola* Gruyter and Boerema. Other names for red leaf blotch include Dactuliophora leaf spot, Pyrenochaeta leaf blotch, and Pyrenochaeta leaf spot. The disease was first reported in Africa in 1957[1] and is currently a threat to production in Central and Southern African countries where it is endemic.[2]

P. glycinicola is a culturable, soilborne fungus.[3] It is known to infect soybean and *Neonotonia wightii* (Arnott) Lackey, a perennial legume that inhabits the woodlands and grasslands of Southern Africa.[1,4] The fungus is likely to have other hosts and was shown experimentally to infect many other legumes.[3] The fungus is known to occur in Cameroon, Ethiopia, Malawi, Nigeria, Republic of Congo, Rwanda, Uganda, Zambia, and Zimbabwe. There was one report of the fungus occurring in samples collected in Bolivia in 1982.[5] Other reports have not corroborated this find. The pathogen has not been reported in the United States.

Most of the published literature about red leaf blotch are from studies completed in Zambia and Zimbabwe, where yield losses of up to 50% were reported.[2,6] Statistics on yield losses from other affected areas generally are not available. If the fungus/disease were found in the United States, it could become economically important, because of its potential to cause significant yield losses and the importance of the soybean crop to the agro-industry in the United States. The soybean production area in the United States has increased from 6.1 million hectares planted in 1950 to 30.6 million hectares planted in 2012 with an estimated worth of $35.8 billion in 2011.[7] If the fungus became widespread in the United States and severe epidemics followed, the states that produce the most soybeans would suffer the highest economic losses (Table 70.1). Significant losses in soybean yield beyond current levels, caused, by the introduction *P. glycinicola* into leading soybean-producing countries like Brazil and the United States would have implications for food security because of our dependence on the soybean crop, directly and indirectly, for food products.[8]

70.2 CLASSIFICATION AND MORPHOLOGY

The fungal pathogen causing red leaf blotch has undergone numerous name changes. Originally, it was named *Pyrenochaeta glycines* Stewart, based on the pycnidial stage,[1] and *Dactuliophora glycines* Leakey, based on the sclerotial stage.[4] In 1986, both stages, the pycnidial and sclerotial, were observed in herbarium specimens linking the two epithets to the same fungus.[9] In 1988, a new genus and species, *Dactuliochaeata glycines* (Stewart) Hartman and Sinclair, was established to accommodate *P. glycines* and its synanamorph, *D. glycines*.[10] Recently, the fungus was classified as a *Phoma* species and renamed *P. glycinicola*.[11] The fungus is unique among the species in *Phoma*, because only *P. glycinicola* produces well-defined, melonized sclerotia that on their own can be infectious or can produce pycnidia on their surface, which then produce infectious conidia.[10] The genome of the fungus has not been sequenced nor have specific gene sequences been generated to compare this fungus with other *Phoma* species. It is quite probable that *P. glycinicola* will be reclassified again because it has unique features unlike any other *Phoma* species once its genome is sequenced and compared to other *Phoma* and *Phoma*-like species.

The fungus produces conidiogenous cells that are monophialidic and ampulliform formed from the inner cells of the pycnidial wall.[10] Pycnidiospores are ellipsoidal, one-celled, and 4–8 μm long by 1–3 μm wide. Sclerotia range in size from 96 to 357 μm in diameter and are mostly spherical, dark brown to black, and covered with setae that are 5–36 μm in length.

TABLE 70.1

Soybean Production for the Top Five States in the United States in 2012

State	Harvested Hectares (Millions)
Iowa	3.7
Illinois	3.6
Minnesota	2.8
Missouri	2.1
Indiana	2.1

Source: USDA-NASS, National Agricultural Statistics Service, 2012, http://www.nass.usda.gov, verified November 2, 2012.

70.3 BIOLOGY AND EPIDEMIOLOGY

Sclerotia of *P. glycinicola* reside in the upper soil matrix associated with decaying leaf litter from either infected soybeans or other legume hosts (Figure 70.1).[2] Infection occurs when rain splashes soilborne sclerotia or conidia from pycnidia onto leaf surfaces, where they germinate and infect plants.[2] After infection, the fungus causes leaf lesions and blotches that vary in size but can increase to 2 cm in diameter. Heavily diseased leaves senesce prematurely, and when foliage from diseased plants drop, newly formed sclerotia and pycnidia are released back into the soil, where they overseason and serve as the initial inoculum for the next cycle. Sclerotia germinate to form either pycnidia, secondary sclerotia, or infectious hyphae (Figure 70.2). Little is known about the overseasoning of pycnidia in leaf litter or in soil.

The combination of *P. glycinicola* survival and its potential host range is an important factor that may affect disease epidemics should the fungus enter and become established in the United States. Part of the establishment would depend on the survival of sclerotia. There have been only a few biological studies that have examined this fungus in detail. One study showed that there was up to 19 sclerotia per gram of soil.[3] In addition, it showed that sclerotia kept at 5°C for 18 months or heat-treated at 100°C still germinated at rates of 90% and 22%, respectively, indicating that sclerotia are effective survival units.[3] Another biological factor to consider along with the density and survival of sclerotia is conidial germination. Temperature between 20°C and 25°C were optimal for germination, but conidia did not germinate when incubated at 5°C or 35°C for over 12 h, indicating that conidia are not long-lived in more extreme temperatures.[3] In culture, *P. glycinicola* produce pycnidia and sclerotia abundantly on media containing asparagine or casein hydrolysate, respectively. Cultures grown on culture media with greater than 1% sodium chloride do not produce fully developed pycnidia.

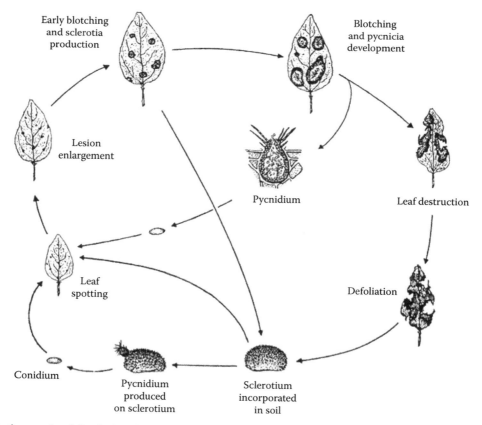

FIGURE 70.1 Infections cycle of *P. glycinicola*, the causal fungus of red leaf blotch of soybean. (Reprinted with permission from the American Phytopathological Society.)

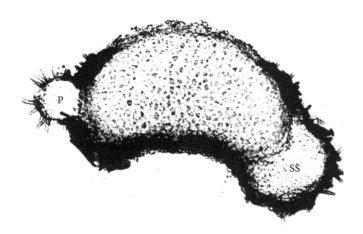

FIGURE 70.2 Cross section of a germinating sclerotium of *Phoma glycinicola* showing the development of a pycnidium (P) and of a secondary sclerotium (SS). (Reprinted with permission from the Mycological Society of America.)

Under field conditions or in nature, *P. glycinicola* is known to infect soybean and *N. wightii*, but there is a good possibility that more legumes are susceptible under field conditions or in natural ecosystems. It was shown that leaf disks of ten *Glycine* spp. and six other legumes (cowpea, kudzu, lentil, lima bean, common pea, pigeon pea, and winter vetch) when inoculated with *P. glycinicola* became infected.[3] It is possible that *P. glycinicola* would attack other important commercial crops in the United States, such as alfalfa, beans, and peanuts, along with other forage and timber legumes. The potential that *P. glycinicola* has a broader host range than is currently known will have to be considered if and when the fungus is discovered in the United States.

Local spread of the *P. glycinicola* occurs in rain showers and when animal or human activities transport the fungal propagules among plants and fields. Sclerotia, residing in soil, are the primary source of inoculum. There are no studies on the transport of sclerotia among fields, but it is reasonable to assume that they are transported similarly into other soilborne pathogens. Secondary spread via conidia is not well understood.[2] The conidia are watersplashed onto leaves just as other *Phoma* species that cause plant diseases.[11] Long-distance transport of these conidia has not been studied, and no evidence exists that the pathogen propagules are seedborne or airborne. Long-distance spread could occur through transport of untreated plant material, via debris accompanying seed from infected fields, or through the movement of contaminated soil.

The economic impact of red leaf blotch to soybean production in the United States has not been determined in part because threshold levels are not known. A risk assessment, compiled by the United States Department of Agriculture (USDA) Animal and Plant Health Inspection Service Plan Protection Quarantine, indicated that the highest risk of establishment in the United States is in the Mississippi River Valley, parts of the eastern Midwest, and the Mid-Atlantic coast.[12] In terms of potential yield losses, studies have shown that 50% yield losses occurred in Zambia,[13] which may also serve as an estimation for yield losses in fields in the United States if the fungus becomes established without any mitigation. However, other factors like predisposition of plants to the fungus and the spread of the various propagules (conidia, pycnidia, and sclerotia) by water, wind, and other means are not well understood. Additional biological data are needed to accurately predict short- and long-distance dispersal of *P. glycinicola* before accurate disease forecasting models become reality.

70.4 CLINICAL FEATURES AND PATHOGENESIS

The US Agricultural Bioterrorism Protection Act of 2002 specifies requirements for possession, use, and transfer of organisms listed as select agents or toxins, such as *P. glycinicola*. Once an unregistered diagnostic laboratory identifies a presumptive select agent, the Agriculture Select Agent Program must be notified and the select agent must either be destroyed or transferred with prior approval of the Agriculture Select Agent Program. If a diagnostic laboratory held part of a screened sample (or culture) for voucher purposes, and the sample forwarded to the USDA Beltsville Laboratory was identified as positive for a select agent, then the USDA Beltsville Laboratory would notify both the Agriculture Select Agent Program and the sending diagnostic laboratory that a select agent has been identified. Clarification of these requirements and other information related to adherence to the select agent regulations is available at the Agriculture Select Agent Program website (www.aphis. usda.gov/programs/ag_selectagent/). *P. glycinicola* went on the list in 2009, and in that same year, a recovery plan that included a weather-based pest risk map was developed by the USDA Animal and Plant Health Inspection Service.[12]

All isolates of the *P. glycinicola* appear to be culturable and pathogenic.[3] There is no information available to indicate that any isolates or strains of the fungus have any specialization that would make them more virulent or aggressive than other isolates, so response procedures based on the first detection of *P. glycinicola* in the United States would not be strain dependent.

70.5 IDENTIFICATION AND DIAGNOSIS

P. glycinicola causes similar symptoms on both of its primary hosts, soybean and *N. wightii*. Characteristic lesions develop on foliage, petioles, pods, and stems of soybeans. Initial symptoms include lesions that appear first on unifoliolate leaves associated with primary veins (Figure 70.3). At this early stage of infection, the disease may appear to look like *Alternaria* leaf spot, brown spot, or target spot, and caution is needed before confirming that a plant is infected with the red leaf blotch pathogen until the fungal structures form and are identified by expert mycologists that are familiar with this fungus. As the disease progresses, more characteristic lesions develop on trifoliolate leaves, appearing as dark red spots on the upper leaf surfaces and similar reddish brown

FIGURE 70.3 (a) Red leaf blotch lesions on a young soybean leaf, predominantly along the veins of an upper leaf surface. (b) Lesions spreading on a lower leaf surface and (c) coalescing into larger lesions. (d) An upper leaf surface showing advanced blotching and (e) lesions on petioles, pods, and stem. (Reprinted with permission from the American Phytopathological Society.)

spots with dark borders on the lower surfaces (Figure 70.3). The fungus also causes symptoms on petioles, stems, and pods (Figure 70.3).

The National Plant Disease Diagnostic Network will be crucial in evaluating the movement and spread of this disease in the event this pathogen is discovered in the United States. To fulfill this role, educational materials developed by the National Plant Disease Diagnostic Network are available so that the disease can be recognized from samples that are received in the diagnostic network. High-quality closeup photos of diseased soybean leaves and microscopic photos of the pathogen have been supplied to the National Plant Diagnostic Network. A standard operating procedure to aid preliminary identification has been developed.

Currently, the diagnosis of red leaf blotch and the identification of the pathogen are limited to a small group of scientists that have firsthand experience working with this pathogen and the disease it causes. The USDA Cooperative Agricultural Pest Survey does not have a mandate to look for red leaf blotch during the course of their disease and pest surveys. No methods have been developed to confirm a positive molecular identification of *P. glycinicola*. An international cooperative partnership to identify samples and extract DNA would provide the necessary DNA samples to complete the molecular diagnosis of the fungus.

70.6 TREATMENT AND PREVENTION

There are many possible treatments and preventative techniques to manage red leaf blotch; however, most of these have not been studied in detail. What follows are some general statements that may be useful in the overall management of red leaf blotch.

Biological and cultural control: There are no biological control agents available at the present time nor information on the role of cultural practices in suppression of *P. glycinicola*. Whether biological control agents and/or cultural practices that are effective against other soybean pathogens will be effective against red leaf blotch is not known. Since *P. glycinicola* relies on sclerotia to survive, new or existing biological control agents that infect sclerotia may be used to reduce active populations of the fungus. Likewise, certain cultural practice like rotations with nonhost crops and/or specific plowing techniques may be effective at reducing the sclerotial density in the soil.

Chemical control: Soil fumigant may be effective in reducing the survival of sclerotia in the soil, but limitations on fumigants and economical feasibility could limit this as a practical management tool. Foliar fungicides would most likely be effective and it was shown that fentin acetate was effective in controlling of red leaf blotch in Africa.[6,13] Fentin acetate is no longer in production, but there are many other foliar fungicides with different chemistries that are registered for use on soybeans in the United States. The currently registered fungicides in the United States need to be tested for their efficacy against *P. glycinicola* preferably before the fungus is found in the United States.

Eradication: Early detection and destruction of infected/infested plant material may make eradication possible in the United States if there is a single entry point into a known area with limited spread. This should be followed by planting nonhost crops, like corn, rice, or wheat, and extensive monitoring over time to check for recurrence of the disease. Eradication may not work if multiple entries have occurred or if the spread is extensive.

Germplasm: In 1982 and 1984, a report showed that US-grown commercial cultivars were susceptible to this disease in field tests in Zambia and Zimbabwe.[5] Host resistance may be a viable option in managing red leaf blotch if this disease enters and persists in the United States. There are over 16,000 soybean germplasm accessions listed the USDA soybean germplasm bank,[14] and the vast majority of these accessions have not been evaluated for resistance to *P. glycinicola*.

Recommended disease management strategy: The recommended disease management strategy for red leaf blotch once it is found in the United States will depend upon the extent of spread and severity associated with the find. If the first detection indicates that it is a small local outbreak, then the destruction of infected material may make eradication possible. Subsequent planting of nonhost crops may help to keep the disease from reappearing. Conversely, if the distribution extends beyond a point source (multiple fields or states), then disease management will rely on emergency-labeled fungicides. While additional fungicides registrations may fill some gaps and allow a crop to be produced, the costs associated with producing that crop will increase. Furthermore, if a susceptible crop is grown the following season, further disease monitoring followed by implementation of a fungicide plan may be warranted after positive confirmation of the disease. Over time, long-term studies evaluating other aspects of disease suppression and management (tillage and rotation studies) will be essential as will be the discovery of sources of host resistance and incorporation of the resistance genes into commercial soybean cultivars.

Current and future surveys: There are no general surveys monitoring for red leaf blotch in the United States and there are no plans for any specific surveys for red leaf blotch. It has been recommended that the USDA Pest Information Platform for Extension and Education (www.sbr.ipmpipe.org/cgi-bin/sbr/public.cgi), an early warning system to monitor for and forecast the occurrence of soybean rust and other diseases, be adapted to include the detection of red leaf blotch.[12] Soybean leaf samples collected by a network of scouts are submitted to diagnostic laboratories and analyzed for the presence of soybean rust, soybean aphid, and other diseases. Results are entered in the Pest Information Platform for Extension and Education database. The online system could be expanded to provide updates and recommendations for red leaf blotch should it be found in the United States.

Generally, once the detection of a select pathogen is confirmed by a USDA Animal Plant Health Inspection Service Plant Protection and Quarantine–recognized laboratory, they, in cooperation with the State Department of Agriculture, are responsible for the deployment of teams of experts and survey personnel to the site of the initial detection. The teams initiate delimiting surveys and recommend actions that may include regulatory measures to quarantine infested or potentially infested production areas to prevent infected material from moving. Control measures may include host removal and destruction and/or ensuring adherence to required sanitary practices. The results of the delimiting survey determine if the disease is generally distributed throughout commercial and noncommercial plant hosts in an area, or if the disease is isolated. If isolated, then there is a possibility of eradicating the pathogen by destroying all infected plants prior to production and deposition of sclerotia. If the detection is late, fields must be removed from production indefinitely. Research on sclerotial/pycnidial longevity is needed to address how long a field or location is to remain in quarantine or under surveillance. Animal Plant Health Inspection Service has the authority to impose quarantines and regulatory requirements to control and prevent both importation and interstate movement of quarantine diseases or regulated articles, and works in conjunction with state regulatory actions that restrict intrastate movement. Any disease mitigation strategy would need to be coordinated among federal, state, and local regulatory officials. Soybean growers may be compensated by the USDA Risk Management Agency for losses caused by red leaf blotch if the producer can verify that available control measures were applied. If there are no effective control measures available or there are insufficient amounts of chemicals available for effective control, resulting losses in production would be compensated.

Prevention/exclusion: The chance of *P. glycinicola* being discovered in the United States is relatively low, because there is little, if any, import of soybean seed and associated debris from infested areas in Africa. Thus, phytosanitary regulations are not needed. If the pathogen were discovered in the United States, spread could be slowed and possibly controlled by fungicides, but probably not eradicated. Since *P. glycinicola* is listed in the Agriculture Select Agent Program, continued exclusion of this disease through normal quarantine activities is an initial step in prevention of the pathogen from entering the United States.

70.7 CONCLUSIONS AND FUTURE PERSPECTIVES

The countries that currently are most affected by red leaf blotch are in tropical and subtropical regions of Africa. In the United States, almost all soybean-growing states are in the temperate zone with the majority of the soybean production being in the Midwest. Information about *P. glycinicola* survival in temperate climates does not exist, but laboratory studies have shown that sclerotia of *P. glycinicola* survive at low temperatures, indicating that this disease could be a threat to soybean production anywhere in the United States.

If red leaf blotch were discovered in the United States, diagnosticians, cooperative extension experts, growers, and others would need to be trained on its identification and know how to distinguish red leaf blotch from other diseases that resemble it.[15] Early detection relies on accurate information, and early detection is absolutely necessary to provide the time needed for successful eradication of the disease from a localized area. Should there be a positive identification of red leaf blotch, cooperative extension and research experts will need to further educate growers about the threat posed by *P. glycinicola* and the cultural and chemical practices to manage its impact.

Educational and extension priorities will need to focus on the development of training materials on detection, monitoring, and management of red leaf blotch that can be distributed to diagnosticians, soybean growers, crop advisors, and industry. Clinicians will need to have hands-on training since they are likely to get numerous samples through the National Plant Disease Diagnostic network that will need to be diagnosed. Information gleaned from research on disease management, like fungicide guidelines, will need to be disseminated at meetings, workshops, and other venues. Current programs like the USDA Pest Information Platform for Extension and Education may have to incorporate red leaf blotch into their field monitoring system.

Research in for the future will need to focus on factors that will help answer the following questions. Can effective measures be implemented to control, confine, or destroy the pathogen in the field once it is discovered in the United States? Is *P. glycinicola* likely to survive a period of winter in the United States? Will current fungicides be efficacious and will current application methods of fungicides be effective to manage red leaf blotch? Will epidemiology studies be able to determine predisposition factors involved in disease development and to evaluate disease gradients to measure short- and long-term spread of the pathogen? Can predictive models be developed to predict potential epidemics of *P. glycinicola* based on research about pathogen survival, spread, and alternative hosts in the United States? Will sources of resistance in soybean be discovered that are effective against invasion of *P. glycinicola*? Will effective resistance genes be easy to cross into elite breeding stock for the development of resistant commercial soybean cultivars?

In conclusion, the introduction of *P. glycinicola* could potentially cause soybean production losses not only to the United States but also in other soybean-producing countries, like Brazil, where the disease is not known to occur. In the United States, because of the current precautions and potential rapid response by the Animal Plant Health Inspection Service, the impact of an accidental point source introduction would likely be relatively minor with no or little economic impact. If the area of infestation were greater than a local infestation, then recovery time and economic impact may be much greater. In both cases, one of the short-term management options would be the application of fungicides. If the pathogen persists in importance and/or the research community becomes proactive, then the long-term management scheme will rely more on host plant resistance. In any event, there has been relatively little threat of the pathogen entering the United States since the discovery of the fungus causing red leaf blotch in 1957. Unless there are some drastic changes in trade or unlawful activities, it remains unlikely that the fungus will be discovered in the United States anytime soon.

REFERENCES

1. Stewart, R.B., An undescribed species of *Pyrenochaeta* on soybeans, *Mycologia*, 49, 115, 1957.
2. Hartman, G.L. et al., Red leaf blotch of soybeans, *Plant Dis.*, 71, 113, 1987.
3. Hartman, G.L. and Sinclair, J.B., Cultural studies on *Dactuliochaeta glycines*, the causal agent of red leaf blotch of soybeans, *Plant Dis.*, 76, 847, 1992.
4. Leakey, C.L.A., Dactuliophora, a new genus of *Mycelia Sterilia* from tropical Africa, *Trans. Br. Mycol. Soc.*, 47, 341, 1964.
5. Sinclair, J.B., Threats to production in the tropics: Red leaf blotch and leaf rust, *Plant Dis.*, 73, 604, 1989.
6. Datnoff, L.E., Sinclair, J.B., and Naik, D.M., Effect of red leaf blotch on soybean yields in Zambia, *Plant Dis.*, 71, 132, 1987.
7. USDA-NASS, National Agricultural Statistics Service, 2012. http://www.nass.usda.gov (verified November 2, 2012).
8. Hartman, G.L., West, E., and Herman, T., Crops that feed the world 2. Soybean-worldwide production, use, and constraints caused by pathogens and pests, *Food Secur.*, 3, 5, 2011.
9. Datnoff, L.E. et al., *Dactuliophora glycines*, a sclerotial state of *Pyrenochaeta glycines*, *Trans. Br. Mycol. Soc.*, 87, 297, 1986.
10. Hartman, G.L. and Sinclair, J.B., *Dactuliochaeta*, a new genus for the fungus causing red leaf blotch of soybeans, *Mycologia*, 80, 696, 1988.
11. Boerema, G.H. et al., *Phoma Identification Manual: Differentiation of Specific and Infra-Specific Taxa in Culture*, CABI Publishing, Wallingford, U.K., 2004, 470pp.
12. Hartman, G.L. et al., Recovery plan for red leaf blotch of soybean caused by *Phoma glycinicola*, Government Publication/Report. http://www.ars.usda.gov/SP2UserFiles/Place/00000000/opmp/Soybean RLB FINAL July 2009.pdf, 4, 2009.
13. Hartman, G.L. and Sinclair, J.B., Red leaf blotch (*Dactuliochaeta glycines*) of soybeans (*Glycine max*) and its relationship to yield, *Plant Pathol.*, 45, 332, 1996.
14. USDA-ARS, National Genetic Resources Program. Germplasm Resources Information Network—(GRIN). [Online Database] National Germplasm Resources Laboratory, Beltsville, MD. Available at: http://www.ars-grin.gov/cgi-bin/npgs/acc/display.pl? 1444837 (November 5, 2012).
15. Hartman, G.L., Sinclair, J.B., and Rupe, J.C. (eds.), *Compendium of Soybean Diseases*, American Phytopathological Society, St. Paul, MN, 1999, 99pp.

71 *Ralstonia solanacearum* Race 3, Biovar 2

Mark Fegan and Alan Christopher Hayward

CONTENTS

71.1 INTRODUCTION

Ralstonia solanacearum is a soilborne plant pathogen, widely distributed in the tropics and subtropics, which colonizes the xylem, causing bacterial wilt in a wide range of economically important host plants. Within this diverse species, *R. solanacearum* race 3, biovar 2 (R3bv2), the principal cause of brown rot of potato, is uniquely adapted to pathogenicity at high elevation in the tropics and cool-temperate environments in the northern and southern hemispheres. R3bv2 has the propensity for distribution in latently infected potato tubers as well as on susceptible geranium cuttings exposed to contaminated irrigation water during propagation. This agent is considered to be a high risk to seed potato production worldwide. In the European Union, R3bv2 has an A2 phytosanitary categorization[1] and is listed as a quarantine pathogen in Canada and as a select agent under the Agricultural Bioterrorism Protection Act of 2002 in the United States.[2]

R. solanacearum includes many strains differing in host range and host preference and, on this basis, was divided into three "races."[3,4] This work was the first to recognize the unique epidemiological attributes of race 3. *R. solanacearum* was first classified into four biovars based on the ability to oxidize three disaccharides, lactose, maltose, and cellobiose, and to utilize three hexose alcohols, mannitol, sorbitol, and dulcitol.[5] Later work in China has led to the designation of a fifth[6] and a sixth[7] biovar. Race 3 strains possess the biovar 2 phenotype in which the disaccharides are oxidized and none of the sugar alcohols are utilized. For historical reasons, the designations "race 3" and "biovar 2" continue to

be used although the race and biovar classification systems have now been superseded by the results of later molecular work.[8] Based upon the recognition that *R. solanacearum* is composed of four groups of strains termed phylotypes, it has been proposed to subdivide the species *R. solanacearum sensu lato* into two new species, *Ralstonia sequeirae* for phylotypes I and III combined and *Ralstonia haywardii* for phylotype IV.[9] The type strain of the species is K60, which possesses the biovar 1 phenotype; accordingly, the original species name remains with phylotype 2. R3bv2 is a phenotypically and genetically homogeneous subgroup of phylotype II (cf. 71.3).

71.2 CULTURAL, MORPHOLOGICAL, AND PHYSIOLOGICAL PROPERTIES

R3bv2 shares many features in common with members of the genus *Pseudomonas sensu lato*. R3bv2 is an aerobic, non-spore-forming, Gram-negative rod-shaped bacterium with one to several polar flagella enabling swimming motility in a liquid medium. One isolate 017a was shown to produce a dense polar tuft of pili (fimbriae)[10] enabling spreading growth and twitching motility on solid surfaces.

The following phenotypic properties are consistently present or give positive reactions in both R3bv2 and *R. sequeirae*: accumulation of poly-beta-hydroxybutyrate inclusions, production of a loose extracellular polysaccharide slime on glucose-containing medium, oxidative metabolism of glucose and other sugars, solubility in 3% KOH, Kovacs' oxidase and catalase, citrate utilization, and growth in nutrient medium containing 1% NaCl. Most isolates of R3bv2 reduce nitrate to nitrite.

Several isolates from Chile and Colombia lack this capacity; none are capable of producing gas from nitrate.[11] Properties universally absent include production of a fluorescent pigment in a low iron-containing medium, arginine dihydrolase, gelatine liquefaction, starch hydrolysis, aesculin hydrolysis, levan production, and growth in nutrient medium containing 2% NaCl.[12] R3bv2 does not grow at 4°C or 40°C and has an optimum temperature in culture of 27°C, in contrast to *R. sequeirae*, which has an optimum temperature in the range of 30°C–35°C.

Under certain conditions in culture, colonies of R3bv2 spontaneously undergo a change from fluidal to nonfluidal morphology, accompanied by reduced pathogenicity and other phenotypic properties. This phenomenon is known as phenotype conversion.[13,14]

R3bv2 is less versatile than *R. sequeirae* in the range of carbon-containing substrates that can be used as sole source of carbon and of energy in a defined medium.[15] The relative lack of metabolic versatility in R3bv2 is consistent with pathogenic specialization to potato and a few other host plants relative to the wide host range of *R. sequeirae*.

The average percent G + C content of the genome of strain UW551, an isolate of R3bv2 from Geranium, was determined to be 64.5%; by comparison, the genome of GMI1000 representing *R. sequeirae* is 67% G + C.[16]

71.3 PHYLOGENY, PHYLOTYPES, AND SEQUEVARS

R. solanacearum belongs to the *Burkholderia* group within the class β-proteobacteria and is phenotypically and genetically heterogeneous.[12] The closest relatives of *R. solanacearum* are *Ralstonia syzygii* and the blood disease bacterium (BDB).[17] A name change has recently been proposed for *R. syzygii* and the BDB as *R. haywardii* subsp. *syzygii* and *R. haywardii* subsp. *celebensis*, respectively.[9]

Since the late 1980s, a variety of molecular methods of greatly improved discriminatory power, in comparison to the phenotypically based race and biovar classification systems,

have been used. These have led to the understanding that *R. solanacearum* is a species complex, a heterogeneous group of related but genetically distinct strains.[8,16,18] The subdivision of the species into four phylotypes on the basis of sequencing of conserved genes, and into sequevars according to differences in sequence of the endoglucanase gene,[8] is a system now widely used and accepted. The complexity of the species complex first became clear in the work of Cook and colleagues[19,20] who used restriction fragment length polymorphism (RFLP)-based genotyping to divide *R. solanacearum* strains into 46 multilocus genotypes (MLGs), which clustered into two major groups of strains. The presence of these two genetic groups was confirmed by several other investigations employing molecular methods[17,21,22] and a further two groups were identified.[18,23,24] Each of the four groups (phylotypes) corresponds roughly with the geographic origin of the strains contained within each phylotype. Phylotype I contains strains primarily from Asia, phylotype II strains primarily from the Americas, phylotype III strains from Africa and surrounding islands, and phylotype IV includes strains isolated primarily from Indonesia (including *R. syzygii* and the BDB).[8] Subsequently, phylotype II has been subdivided into two subclusters termed IIA and IIB.[25] The classification of *R. solanacearum* into the four phylotypes has been confirmed by the use of multilocus sequence typing (MLST),[26] comparative genomic hybridization data,[27–29] and whole-genome sequence comparisons.[9,29] The phylotyping scheme has been further refined by the MLST analysis of Wicker et al.[26] who identified a number of clades within phylotypes II and IV. Within phylotype II, four clades were recognized, two in phylotype IIA (clades 2 and 3) and two in phylotype IIB (clades 4 and 5). Within the phylotyping classification scheme, as extended by the work of Wicker et al.,[26] *R. solanacearum* R3bv2 strains causing brown rot of potato belong to phylotype IIB, clade 5, sequevars 1 and 2 (Figure 71.1). *R. solanacearum* phylotype IIB sequevar 1 strains have been spread worldwide, whereas sequevar 2 strains appear to be restricted in their distribution. Cellier

	Ralstonia solanacearum Species Complex					
Phylotype[a]	Phylotype I	Phylotype III	Phylotype IIA	Phylotype IIB	Phylotype IV	
Origin[a]	Asia	Africa	Americas	Americas	Indonesia	
Clades[b]	1	6	2	3	4	5	7	8
Sequevars[b]						1	2		
Proposed taxonomic reclassification[c]	*Ralstonia sequeirae*		*Ralstonia solanacearum*			*Ralstonia haywardii*	

FIGURE 71.1 Position of *R. solanacearum* R3bv2 (marked in gray highlight) within the *R. solanacearum* species complex. [a] As described by Fegan and Prior[8]; [b] as described by Wicker et al.[26]; [c] as described by Remenant et al.[9]

TABLE 71.1

Examples of the Global Dispersal of *R. solanacearum* R3bv2 on Plant Material

Introduction	Source of Plant Material	References
To southern Sweden on potatoes for factory processing (1972–1977)	Mediterranean region	Olsson [33]
To Mauritius since about 2005 on imported ware potatoes	Market potatoes planted as seed	Khoodoo et al. [43]
To Europe on cuttings of *Pelargonium zonale*	Cuttings from nurseries in Kenya for propagation in Belgium, Germany, and the Netherlands	Janse et al. [41,93]
To United States on cuttings of *Pelargonium zonale*	Cuttings from nurseries in Guatemala for propagation in Wisconsin (1999), Pennsylvania, Delaware, and Connecticut	Williamson et al. [94], Kim et al. [95]

et al.[30] discovered that only phylotype IIB sequevar 1 strains maintain a high level of virulence for potato under low temperature, whereas phylotype IIB sequevar 2 strains dramatically decreased in virulence under cold temperatures. It may be speculated that this adaptation maybe the reason that the *R. solanacearum* R3bv2 strains that have been spread worldwide belong to phylotype IIB sequevar 1 and not sequevar 2.

R. solanacearum R3bv2 phylotype IIB sequevar 1 strains are recognized as being clonal.[24,30] Indeed, repetitive element PCR (rep-PCR) has found that all R3bv2 isolates have the same or very similar rep-PCR profiles.[7,31] However, not all phylotype IIB sequevar 1 strains belong to biovar 2. Fegan et al.[31] identified strains of *R. solanacearum* biovar 1 that have the same rep-PCR pattern as R3bv2 strains. These *R. solanacearum* biovar 1 strains also have a similar pathogenic phenotype to *R. solanacearum* R3bv2 strains[32] and belong to phylotype IIA sequevar 1 (Fegan, unpublished data). Therefore, although these strains are biochemically not biovar 2, in terms of pathogenicity and genotype, they are equivalent to R3bv2 strains. This calls into question the designation of *R. solanacearum* strains as R3bv2 in the select agents scheme; although these strains appear to correspond to race 3, they are not biovar 2. These strains are also identified as R3bv2 strains using molecular detection methods (see Section 71.7.3).

Although *R. solanacearum* R3bv2 phylotype IIB sequevar 1 strains are considered to be clonal, recent comparative genomic hybridization studies have identified a degree of variation within these strains.[27] Cellier et al.[27] have been able to distinguish between *R. solanacearum* R3bv2 phylotype IIB sequevar 1 strains, associated with European outbreaks of brown rot, of Andean and African origin.

71.4 EPIDEMIOLOGY, HOST RANGE, AND GEOGRAPHICAL DISTRIBUTION

71.4.1 EPIDEMIOLOGY AND HOST RANGE

There is a consensus that R3bv2 evolved in South America near the center of origin of the tuber-bearing *Solanum* species and, from there, was distributed worldwide on latently

infected seed or ware potatoes or on potatoes for processing (Table 71.1). After introduction into an area, the fate of the pathogen depends on capacity to survive in soil, in the rhizosphere or root system of secondary host plants, and on survival in water, in reservoirs, and arterial water courses.[32a] The importance of secondary weed hosts and their role in the contamination of river water used for irrigation was clearly established by work carried out in Sweden.[33] Imported potatoes infested with R3bv2 were processed at a factory located adjacent to the Pinnan river in southern Sweden. Untreated effluent from the factory flowed into the river, and at points along the river, the adventitious roots of the riparian semi-aquatic weed known as bittersweet (*Solanum dulcamara*) became infected by the contaminated factory effluent. Although the population of R3bv2 in river water declined, bacterial exudate from the adventitious roots maintained the population of R3bv2. River water used to irrigate potato crops growing adjacent to the river led to dissemination of the disease. There have been several other instances of the involvement of contaminated river water from factory effluent and of the involvement of *S. dulcamara* in other parts of western Europe including the Netherlands and the United Kingdom.[32a,34] The experience in Sweden has led to a search for other riparian weeds that can serve a comparable role to that of *S. dulcamara* in north western Europe. There has also been an effort to more clearly define the distinction between host and nonhost plants so that the selection of plants used in crop rotation after brown rot infestation can have a better scientific basis.[35]

Survival of R3bv2 in soils is dependent on many biotic and abiotic factors and suppressive soils have been identified.[36] Fallow periods of 6 months to 3 years have been found to be adequate for eradication of the pathogen in infested soils. In tests carried out in the Netherlands and in Egypt, the decline of populations of R3bv2 was faster in sandy soils than in clay soils.[37] Tomlinson et al.[38] confirmed the faster rate of decline in Egyptian sandy soils than in clay soils and also showed that temperature was the most important factor affecting survival. Survival was generally longer at 15°C than at 4°C, 28°C, or 35°C. There is limited survival of the pathogen in compost. Survival was for less than 2 weeks in infected plant material or in infested soil samples incorporated into compost heaps.[38] R3bv2 was eradicated from infested potato waste after 6 h at 52°C of anaerobic digestion at low concentration and after 12 h at high concentration of the pathogen.[39]

The most important crop plant hosts of R3bv2 are potato and tomato. Severe losses on tomato have occurred when tomato is planted in the same soil in which brown rot has occurred. The pathogen has been reported occasionally on eggplant (*Solanum melongena*) and on pepper (*Capsicum annuum*) and on solanaceous weeds such as *Solanum nigrum*, and in north, western, and southern Europe on *S. dulcamara*. There are also numerous nonsolanaceous weeds that have been reported as latent (symptomless) carriers of infection,[40–42] as well as other plants in which R3bv2 is maintained in the rhizosphere. In Mauritius, Khoodoo et al.[43] have identified weeds that promote the persistence of R3bv2 in soil, notably *Oxalis latifolia* (broadleaf wood sorrel), a common weed in potato fields in Mauritius, which was latently infected. Solanaceous weeds including *Solanum americanum* and the wild tomato (*Lycopersicon pimpinellifolium*) were also identified as alternative hosts. In the Netherlands, Janse et al.[41] used two artificial inoculation methods, drenching of soil with inoculum without wounding of the root system and stem inoculation, to test the pathogenicity of a potato isolate and a *Pelargonium* isolate on a range of plants. Both isolates were pathogenic to potato, eggplant, and *S. nigrum*. *Pelargonium zonale*, tomato, and *Portulaca oleracea* all showed typical wilt symptoms by either method of inoculation.

Attempts have been made to more clearly define the distinction between host and nonhost plants[35,44] in vitro and in pot plants and in the glasshouse. Alvarez et al.[35] provided data based on histological studies on 20 plant species chosen because of their potential use in crop rotation. One month postinoculation, sections of root and stems were analyzed to locate the pathogen on surfaces, in cortex, and/or xylem. Plants in which the xylem was colonized were classified as hosts, those without xylem colonization as nonhosts, and hosts generally infected in a few xylem vessels or occasionally in all xylem bundles were classified as tolerant. The latter included some cabbage, kidney bean, and rutabaga cultivars and the weed *S. dulcamara*. Nonhosts were the cultivars tested of alfalfa, barley, black radish, carrot, celery, colocynth, fennel, fiber flax, field bean, field pea, horseradish, maize, and zucchini. The authors stress the preliminary nature of their findings because of the variability of the results within a plant family, notably in the Brassicaceae including susceptible, tolerant, and nonhosts and the Leguminosae including either tolerant or nonhosts. Cultivars of a single species may differ in response, and the aggressiveness of the strain of R3bv2 may also affect the results.

There is ample evidence of the dispersal of R3bv2 by river and flood water contaminated by the runoff from infested fields.[32a,34] Members of the *R. solanacearum* species complex, as well as many other microbial species, survive for many years as suspensions in sterile water held at temperatures of 15°C–20°C.[45,46] In natural waterways, survival is for much shorter periods because of competition from other microflora.

Álvarez et al.[47] showed that lytic bacteriophages, indigenous protozoa, and predatory bacteria were all involved in decline in population density of R3bv2 in Spanish rivers. A similar decline in population in Egyptian irrigation canal water was attributed to an antagonistic microflora.[48] *R. solanacearum* R3bv2 has also been shown to enter a viable but nonculturable (VBNC) state when subjected to unfavorable environmental conditions in both soil and water habitats.[49–51] The VBNC state allows the pathogen to survive in the environment in a dormant form and may also explain the population decline in water.

71.4.2 GEOGRAPHICAL DISTRIBUTION

There are many historical records of outbreaks of brown rot of potato known to be caused by R3bv2, particularly in Europe and the Mediterranean region. In several instances, there have been no later reports of the disease and the pathogen is assumed to have declined to extinction. There are also reports of brown rot of potato in which no or limited diagnostic studies were carried out, and many others in which no distinction was made between brown rot caused members of the *R. solanacearum* species complex differing in biovar and sequevar from R3bv2. The current status of brown rot is not known in many countries with historical records of disease occurrence. In the following account, emphasis is given to recent investigations using advanced diagnostic methods.

R3bv2 does not occur in Canada or the United States on potato but is present in Mexico and many parts of Central and South America. In Guatemala, the presence of brown rot attributed to R3bv2 was confirmed at altitudes higher than 1700 m above sea level with a mean temperature of 4°C–20°C. All isolates were of sequevar 1 and indistinguishable from other potato isolates from Costa Rica and Mexico as well as from *Pelargonium* in Kenya.[52] Brown rot of potato caused by R3bv2 occurs in Argentina, Colombia, Chile, Peru, and Venezuela and is endemic in many areas. Most recently, the disease has been found to be widespread in potato-growing regions of Uruguay.[53] In Brazil, R3bv2 is present in all potato-producing regions as the most prevalent strain.[54]

In Africa, brown rot of potato caused by R3bv2 is present in Burundi, Egypt, Ethiopia,[55] Kenya, Reunion, Rwanda, South Africa, Tanzania, Uganda, and Zambia. In West Africa, R3bv2 is present in Cameroon but not in Côte d'Ivoire.[56] In Mauritius, where other member of the *R. solanacearum* species complex are endemic and cause disease of solanaceous crops including potato, brown rot of potato caused by R3bv2 appears to be a recent introduction possibly on imported ware potatoes. There have been major losses in three growing seasons since 2005.[43]

In Oceania, R3bv2 occurs in Papua New Guinea.[57] In Australia, there have been disease outbreaks prior to and since 1990 in Queensland, New South Wales, Victoria, and South Australia but not in Tasmania. The disease does not occur in Western Australia. An outbreak in several locations southwest of Perth in 1967–1968, attributed to the use of a common contaminated seed source, was eradicated.[58]

In Asia, brown rot caused by R3bv2 is present in China, India (West Bengal), Indonesia, Iran, the Philippines,[59] Nepal, and Thailand. A recent extensive survey of infection in potato fields in the warm and temperate regions of southwestern Japan has shown that brown rot caused by other members of the *R. solanacearum* species complex is common, whereas R3bv2 is absent.[60] The authors attribute this to the fact that in Japan, only domestically produced seed potatoes have been distributed for cultivation. The import of seed potatoes from outside Japan has been strictly prohibited by a plant protection law established in 1951. In contrast to the situation in Japan, R3bv2 is found in all of the four main potato-growing regions of China, to an overwhelming extent in high latitudes and highlands of lower latitude regions, but not limited to cool-temperate zones and tropical highlands. Several R3bv2 isolates have been obtained from the lowland plains in China suggesting that the dissemination of R3bv2 may be associated with movement of symptomless latently infected seed potato tubers from upland regions. As recently as the late 1970s, bacterial wilt of potato was observed only in mountain areas of Hunan and Sichuan province but has now spread to more than 10 provinces. Other members of the *R. solanacearum* species complex were found on potato and other hosts exclusively on the plains south of the Yellow River.[61]

Sporadic outbreaks of potato brown rot caused by R3bv2 have been reported since 1989 in restricted areas of 9 of the 15 member states of the European Community (Belgium, France, Germany, Greece, Italy, the Netherlands, Portugal, Spain, and the United Kingdom), and more recently, there have been isolated potato brown rot outbreaks in Hungary and Turkey.[34] The outbreak of brown rot in Sweden (1972–1977) has been well documented.[33,62] Eradication has been achieved by removal of infected *S. dulcamara* over a 5-year period from within 30 km of a contaminated water course. Eradication has probably been successfully achieved in other countries but awaits confirmation.[62a]

All of the outbreaks of potato brown rot in Europe and the Mediterranean region have been attributed to R3bv2, with a single exception. In May 2007, potato plants showing symptoms of brown rot were collected from some potato fields in the Baixo Mondego region of central Portugal, and the pathogen was identified as *R. solanacearum* biovar 1 on the basis of both biochemical tests and DNA-based methods.[40,63]

71.5 DISEASE SYMPTOMS

On potato, the aboveground symptoms of rapid wilting of leaves and stems are most evident at the warmest time of the day, with recovery as conditions become cooler; eventually, plants fail to recover, become yellow and brown, necrotic, and die. There may be a brown discoloration of the stem above the soil and epinasty of the petioles may occur. When stems are cut, a white slimy mass of bacterial ooze is expressed from the vascular tissue. Threads of bacterial ooze are observed within 30–60 s after suspension of portions of stem in water. On potato tubers under optimum conditions for disease expression, bacterial ooze emerges from both the eyes and from the stolon end of the detached tuber. In heavy clay soils, soil adheres to the eyes. Cut tubers exhibit a brown discoloration and necrosis of the vascular ring; a creamy bacterial exudate is expressed on standing or when the tuber is squeezed (Figure 71.2).

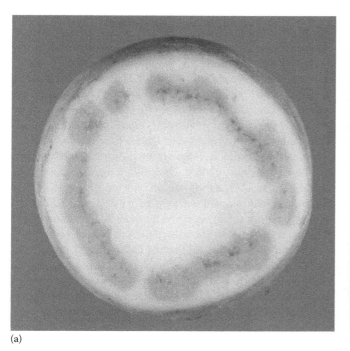

(a)

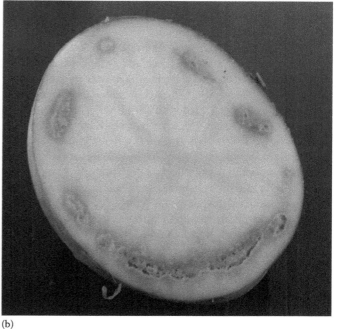

(b)

FIGURE 71.2 Potato tuber infected with *R. solanacearum* R3bv2 exhibits a brown discoloration and necrosis of the vascular ring. (a) Water soaking extending from the vascular tissue. (b) Cavitation of the vascular ring. (Courtesy of Dr. Julian Smith, The Food and Environment Research Agency, York, England.)

Under cool growing conditions, symptoms may be absent aboveground and harvested tubers show no symptoms. The tubers harvested from infected plants may or may not show symptoms and plants without wilt symptoms may sometimes produce diseased tubers. Latent infection is a common feature of brown rot disease when the growing conditions are suboptimal for disease expression. Several weed hosts that become latently infected, or less commonly show symptoms, are known (cf. 71.4) and are an important factor in epidemiology. Symptomless latent infections have also been reported in geranium.[64]

71.6 INFECTION AND PATHOGENESIS

Entry into a host plant is usually through a wound. Histological studies have shown that R3bv2 enters the potato plant from the soil through root extremities and at the point where lateral roots emerge as well as through cracks in the developing endodermis. Wounds are also made by nematodes and through agricultural practices. Chemotaxis and swimming motility probably play a role in the concentration of the pathogen at these sites.[13] In tubers, the eyes and stolon heel ends are the main entrance sites.[45,49] After invasion of a susceptible host, the pathogen multiplies and moves systemically within the plant before symptoms of bacterial wilt occur. Studies on the histology of infection of R3bv2 on potato[13] have been complemented by parallel work with tomato as host and other strains of R. solanacearum.[65] These studies show that the multiplication of the pathogen occurs in the root cortex for 2–6 days before the endodermis is breached. The intercellular spaces within the vascular cylinder are colonized, the xylem vessels are invaded, and there is rapid migration through the vascular tissue.[65] Contrasting results have been obtained under different test conditions.[66]

There is a very extensive literature on the biochemical and genetic basis of virulence and pathogenicity in the R. solanacearum species complex, most derived from work with strains K60 and AW, two closely related representatives of phylotype II, biovar 1, or with GMI1000, a representative of phylotype I, biovar 3.[66] Although similar work has not been done with R3bv2, it is unlikely that the mechanisms involved are greatly different. Virulence factors identified include extracellular polysaccharides; the protein appendages, flagella and pili (fimbriae), involved in swimming and twitching motility, respectively; plant cell wall–degrading enzymes; and plant hormones and secondary metabolites such as siderophores. Production of these virulence factors is regulated by a complex sensory network.[67] Spontaneous loss of virulence and ability to produce the acidic extracellular heteropolysaccharide EPS1 and other properties of R. solanacearum in culture, a phenomenon known as phenotype conversion (PC), appears to be general in all representatives of the species complex and is consistently found in R3bv2.

71.7 DIAGNOSTIC METHODS

R. solanacearum can be detected directly in plant material using a variety of screening tests including the production of bacterial ooze from cut surfaces of symptomatic plants, culture using a semiselective media, and immunodiagnostic and molecular methods. Similarly, R. solanacearum can be identified in environmental samples such as water or soil by means of microbiological, immunodiagnostic, and molecular methods. Even when immunodiagnostic or molecular methods are used to directly identify the presence of R. solanacearum in plant and environmental samples, culturing is recommended to confirm pathogen viability and provide a pure strain to confirm identity and perform pathogenicity assays.[66]

71.7.1 CULTURE

Isolation of R. solanacearum from freshly collected symptomatic plants on TZC medium[68] is relatively easy due to the high numbers of the pathogen present in the tissues. However, isolation from latently infected plant material, where pathogen numbers are low, and from environmental samples such as soil, where other faster growing microorganisms dominate the microbial flora, can be difficult.[66] Several selective media have been developed to aid in the isolation of R. solanacearum from these difficult sample types; the most commonly used of these are SMSA[69] and modified SMSA.[70] To accurately identify cultured R. solanacearum, a combination of traditional biochemical tests (as indicated earlier), race determination,[4] or molecular testing (see Section 71.7.3) should be completed.

71.7.2 IMMUNODIAGNOSTIC METHODS

Immunodiagnostic methods are fast, reliable, and relatively low-cost methods for the initial screening of material for the presence of R. solanacearum.[71,72] However, these tests are species specific and are not able to discriminate R. solanacearum R3bv2 from other R. solanacearum strains. The simplest and most rapid immunodiagnostic kits available are dipsticks and lateral flow devices that are commercially available from various sources such as Agdia (Elkhart, Indiana, United States), Neogen Europe Ltd. (Auchincruive, Ayr, Scotland), and Pocket Diagnostic (Sand Hutton Applied Innovation Campus, York, United Kingdom). A number of these are accepted methods for the detection of R. solanacearum R3bv2 by the USDA-APHIS.

Enzyme-linked immunosorbent assays (ELISAs) are relatively inexpensive, rapid, and robust and kits are commercially available from Agdia (Elkhart, Indiana, United States) and Neogen Europe Ltd. (Auchincruive, Ayr, Scotland). However, ELISA protocols using polyclonal antibodies are, at best, able to detect fewer than 10^4 bacteria/mL of sample and tests employing monoclonal antibodies are even less sensitive.[71] Direct ELISA, indirect ELISA, and double antibody sandwich (DAS) assays have been developed and employed to detect *R. solanacearum* R3bv2 in plant material, soil, and water. Priou et al.[73] developed a variation to the ELISA protocol where samples can be bound to a nitrocellulose membrane and be stored for long periods prior to conducting the ELISA. Prior enrichment of a sample can improve the sensitivity of the test by the use of selective or nonselective media. Priou et al.[73] also developed a postenrichment DAS-ELISA protocol employing selective enrichment in modified SMSA prior to running the DAS-ELISA on the enriched sample. This procedure was able to detect approximately 100 cfu/g in naturally infested soil, and the protocol has also proven useful in the detection of *R. solanacearum* R3bv2 in symptomless latently infected potato stems and tubers.[72] Other immunodiagnostic methods such as immunofluorescent antibody staining and immunofluorescence-colony staining have also been successfully employed to detect *R. solanacearum* R3bv2 in infected potato tubers[70] and soil.[74]

71.7.3 NUCLEIC ACID-BASED METHODS

Many nucleic acid-based assays have been developed for the detection of *R. solanacearum* R3bv2 in plant and environmental samples. Overall, depending on the target sequence chosen, nucleic acid-based methodologies are more sensitive and potentially more specific than serological-based techniques but, like serological methods, cannot distinguish viable from nonviable cells. Only a few molecular detection assays employed to detect *R. solanacearum* R3bv2 are specific for R3bv2 with most detecting other strains of *R. solanacearum*. For example, tests targeting the ribosomal RNA genes are only able to identify the species or phylotype.[75,76] The discrimination of R3bv2 from other subgroups of *R. solanacearum* is imperative as many different *R. solanacearum* strains may infect potato, including race 1 strains belonging to biovars 2T, 3, and 4. It is especially important to be able to specifically identify R3bv2 strains in the United States due to the organism being on the select agent list. Fegan et al.[31] described the first R3bv2-specific PCR test, which targeted a region of the R3bv2 genome identified by competitive hybridization. This PCR has been commonly used in conjunction with rapid serological test strips for the detection of *R. solanacearum* R3bv2.[77] The sequence of the DNA fragment, on which the test developed by Fegan et al.[31] is based, has also been used to develop two real-time PCR assays.[78,79] The specificity of the PCR assays based upon this DNA fragment has been questioned due to the proximity of the fragment to phage or prophage sequences indicating that the fragment may be laterally transferred.[80] Indeed, the PCR test developed by Fegan et al.[31] has been shown to detect *R. solanacearum* strains belonging to biovar 1. However, these biovar 1 strains have been shown to be genetically, biochemically, and pathogenically similar to *R. solanacearum* R3bv2 strains but lack the ability to oxidize the three disaccharides that differentiate biovars 1 and 2.[31,32]

Using an in silico bioinformatic approach, Kubota et al.[80] identified four putative R3bv2-specific genomic loci (RRSL_2401, RRSL_2403, RRSL_1249, and RRSL_2208). Conventional PCR tests developed to loci RRSL_2401 and RRSL_2403 amplified template DNA from all R3bv2 strains tested, and PCR tests developed from the remaining two loci (RRSL_1249 and RRSL_2208) amplified all but one of the R3bv2 strains tested. The two loci RRSL_2401 and RRSL_2403 are located adjacent to the fragment identified by Fegan et al.[31] Although these fragments were also associated with putative phage-associated genes, one of these regions has proven to be a stable part of all R3bv2 genomes tested and not present in other *R. solanacearum* strains.[80] One of these R3bv2-specific genes has recently been used to develop a real-time PCR test.[81]

PCR tests are susceptible to inhibition by compounds that are introduced into the test along with the sample. For example, extracts from potato tubers have been found to interfere with PCR[70,78,79] leading to the production of false negative test results. Endogenous and exogenous internal positive controls have been employed[79,82] as a means to confirm the presence of inhibitors of the reaction and therefore validating negative results, as well as confirming positive results. Various methods have been used to reduce the effect of inhibitory substances. Simply diluting the template 10- to 100-fold has been shown to increase the success of PCR[79] but this also has the effect of also decreasing the sensitivity of the PCR. Addition of substances to neutralize the inhibitors of PCR[83] has also been successfully employed as has preenrichment of the sample prior to PCR.[78,84] Immunomagnetic separation (IMS) and magnetic capture hybridization (MCH) have also been used to separate the bacterial cells (IMS) or DNA (MHC) away from the inhibitory substances followed by R3bv2-specific real-time PCR.[81]

Isothermal amplification technologies have also been employed to detect *R. solanacearum* R3bv2. Loop-mediated isothermal amplification (LAMP) protocols, which are able to rapidly synthesize large amounts of target DNA at a constant temperature,[85] have been developed for the specific detection of *R. solanacearum* R3bv2 strains using three of the R3bv2 loci described earlier (RRSL_2403, RRSL_1249, and RRSL_2208).[77] Due to the isothermal nature of the reaction, LAMP is more amenable to the development of in-field tests than PCR.[80] Another isothermal amplification technique, AmpliDet RNA, has also been used to detect *R. solanacearum* in potato tuber extracts.[86] However, this technique, based upon nucleic acid sequence-based amplification (NASBA), is not specific for R3bv2.

71.8 DISEASE PREVENTION AND MANAGEMENT

The variety of habitats into which the disease has been introduced in the tropics, subtropics, regions with a Mediterranean climate, and to higher latitudes is very great. In general, the prospects for exclusion and eradication are greater in the cooler regions at higher latitudes, exemplified by the experience in Sweden,[62] and more difficult at higher elevations in the tropics and subtropics where brown rot of potato caused by R3bv2 has become established. In these regions, different strategies have been developed to limit or possibly to eradicate the disease. Integrated management is an approach applicable in many circumstances. French[36] has listed seventeen factors that influence the efficacy of control, each given a rating of 1–7; the higher the number, the more the factor is judged to contribute to good control of the disease. The ratings will differ according to site-specific factors.

In Egypt, brown rot of potato caused by R3bv2 is endemic in the Nile Delta, and ware potato exports from this area to the European Union have often been contaminated with the pathogen. This problem has been overcome by relocating potato production from the Nile Delta to pest-free areas in or on the fringes of the desert, areas with no history of potato production that can be shown to be free of the pathogen in weeds, soil, and surface, and arterial waterways.[38] In Australia, there are many examples of the introduction of brown rot caused by R3bv2 on uncertified seed potatoes into pristine land with no previous history of potato production. After the disease outbreak, potato production is either abandoned at that site and moved to an alternative location or the pathogen is eradicated by a suitable pasture rotation of a minimum of 2½ years accompanied by removal of volunteer potato plants and alternative weed hosts such as S. nigrum.[36] In most instances, there has not been work to confirm the absence of the pathogen from these previously infested soils, but there is a strong presumption that R3bv2 has declined to extinction.

Early detection of disease development on a field scale plays a role in disease management. At ground level, disease becomes visually detectable when extensive damage to the crop has already occurred. Chávez et al.[87] have used a remote sensing diagnostic method to detect physiological and morphological changes in plants caused by the infection. A multispectral signal, enhanced by multifractal analysis, detected both symptomatic and latently infected plants, matching the results of ELISA tests in the laboratory to a high degree. The capacity of the method to detect asymptomatic latent infections indicates its potential as a monitoring tool for the control of bacterial wilt in potato crops.

In the Netherlands, biological soil disinfestation (BSD) was effective in reducing soil populations of R3bv2 under three test conditions including in a naturally infested commercial field. BSD also significantly reduced survival of R3bv2 in buried potato tubers and promoted decomposition of superficially buried potatoes remaining after harvest, effectively destroying an important inoculum reservoir of R3bv2.[88] BSD involves the induction of anaerobic soil conditions by increasing microbial respiration through addition of fresh organic amendments and by limiting oxygen availability by covering with airtight plastic sheeting. BSD is a promising management strategy for potato brown rot in countries where the disease is endemic.

Soil amendments with manure or compost may be an indirect way of stimulating biological control. Comparative studies in Dutch and Egyptian soils showed that addition of cow manure, an ammonia-producing amendment, can reduce populations of R3bv2, whereas addition of compost and organic amendments did not necessarily result in enhanced decline of the pathogen.[37,89] In addition, there has been some research to find specific biological agents effective in the management of bacterial wilt.[90] The efficacy of *Stenotrophomonas maltophilia* as a biological control agent for brown rot of potato has been tested in parallel in the Netherlands and Egypt with contrasting results. *S. maltophilia* applied directly to soil or by bacterization of potato eyepieces significantly suppressed potato brown rot in Egyptian clay soil but not in Dutch clay soil. The authors concluded that this agent may be useful for the control of brown rot in the Nile Delta of Egypt, where the bacterium was originally isolated, but recognize that the approval of the use of *S. maltophilia*, a bacterium associated with various illnesses in humans, will require regulatory approval.[90]

There have been many attempts to control bacterial wilts of potato and tomato caused by the *R. solanacearum* species by the application of toxic chemicals including fumigants to the soil with very limited success.[45] Foliar sprays and applications of antibiotics have been ineffective. Various bactericides have been tested for their efficacy in protecting geranium plants (*Pelargonium hortorum*) from infection by R3bv2. Potassium salts of phosphorous acid were found to be effective in protecting plants when applied as drench, an effect attributed to the active portion, phosphorous acid, acting as a bacteriostatic agent in the soil. The plants were not protected from aboveground infection on wounded surfaces.[91]

71.9 CONCLUSION AND FUTURE PERSPECTIVES

R. solanacearum R3bv2 presents a major constraint to potato production worldwide. In the developing world, brown rot of potato poses a major food security problem, whereas in the developed world, the pathogen is the subject of quarantine and select agent legislation[1,2] and disease incursion on previously disease-free land has led to legal proceedings.[34]

Understanding of the diversity of the *R. solanacearum* species complex and the position of R3bv2 strains within the species complex has helped in the understanding of the evolution of the pathogen and the development of targeted molecular tests. Recent genomic sequencing of representative strains of the *R. solanacearum* species complex, guided in part by our understanding of the diversity of the pathogen, has improved our knowledge of the genetic mechanisms of pathogenicity of *R. solanacearum*,[92] and it is hoped that these methods will also help improve our understanding of the ecological adaptability of the pathogen.

The ecology of *R. solanacearum* R3bv2 is complex as the pathogen is found in three different environments: soil, water, and the plant host. While we have an understanding of the host pathogen interaction, we are still only beginning to understand how *R. solanacearum* R3bv2 strains interact with the soil and water environments. An understanding of these interactions has implications for the development of control methods. While in the developed world the need for adequate quarantine procedures and purchase of disease-free planting material are well understood, and relatively easily implemented, this is not the case in the developing world. Farmers in the tropics and subtropics are often too poor to purchase high-quality seed, if such seed is available; often there is replanting in land already infested with R3bv2. It is a major challenge for the future to improve disease management in these circumstances.

REFERENCES

1. Anon. Diagnostic protocols for regulated pests. European and Mediterranean plant protection organisation. *Bulletin OEPP/ EPPO Bulletin* 34, 173–178 (2004).
2. Lambert, C.D. Agricultural bioterrorism protection act of 2002: Possession, use and transfer of biological agents and toxins; interim and final rule. (7 CFR Part 331). *Federal Register* 67, 76908–76938 (2002).
3. Buddenhagen, I.W. and Kelman, A. Biological and physiological aspects of bacterial wilt caused by *Pseudomonas solanacearum*. *Annual Review of Phytopathology* 2, 203–230 (1964).
4. Buddenhagen, I.W., Sequeira, L., and Kelman, A. Designation of races in *Pseudomonas solanacearum*. *Phytopathology* 52, 726 (1962).
5. Hayward, A.C. Characteristics of *Pseudomonas solanacearum*. *Journal of Applied Bacteriology* 27, 265–277 (1964).
6. Granada, G.A. and Sequeira, L. A new selective medium for *Pseudomonas solanacearum*. *Plant Disease* 67, 1084–1088 (1983).
7. Xue, Q.Y. et al. Genetic diversity of *Ralstonia solanacearum* strains from China assessed by PCR-based fingerprints to unravel host plant- and site-dependent distribution patterns. *FEMS Microbiology Ecology* 75, 507–519 (2011).
8. Fegan, M. and Prior, P. How complex is the *Ralstonia solanacearum* species complex? In *Bacterial Wilt Disease and the Ralstonia solanacearum Species Complex*, (eds.) Allen, C., Prior, P., and Hayward, A.C., pp. 449–461. APS Press, St. Paul, MN (2005).
9. Remenant, B. et al. *Ralstonia syzygii*, the blood disease bacterium and some Asian *R. solanacearum* strains form a single genomic species despite divergent lifestyles. *PLoS One* 6, e24356. http://www.plosone.org/article/info%3Adoi%2F10.1371%2Fjournal.pone.0024356 (2011).
10. Henrichsen, J. and Blom, J. Examination of fimbriation of some gram-negative rods with and without twitching and gliding motility. *Acta Pathologica et Microbiologica Scandinavica, Section B* 83, 161–170 (1975).
11. Hayward, A.C., El-Nashaar, H.M., Nydegger, U., and De Lindo, L. Variation in nitrate metabolism in biovars of *Pseudomonas solanacearum*. *Journal of Applied Bacteriology* 69, 269–280 (1990).
12. Hayward, A.C. Systematics and phylogeny of *Pseudomonas solanacearum* and related bacteria. In *Bacterial Wilt: The Disease and Its Causative Agent Pseudomonas solanacearum*, (eds.) Hayward, A.C. and Hartman, G.L., pp. 123–136. CAB International, Wallingford, U.K. (1994).

13. Álvarez, B., Biosca, E.G., and López, M.M. On the life of *Ralstonia solanacearum*, a destructive bacterial plant pathogen. In *Current Research, Technology and Education in Applied Microbiology and Microbial Biotechnology*, Vol. 1, (ed.) Méndez-Vilas, A., pp. 267–275. Formatex, Badajoz, Spain (2010).
14. Denny, T.P. et al. Phenotype conversion of *Pseudomonas solanacearum*: Its molecular basis and potential function. In *Bacterial Wilt: The Disease and Its Causative Agent Pseudomonas solanacearum*, (eds.) Hayward, A.C. and Hartman, G.L., pp. 137–144. CAB International, Wallingford, U.K. (1994).
15. Palleroni, N.J. and Doudoroff, M. Phenotypic characterization and deoxyribonucleic acid homologies of *Pseudomonas solanacearum*. *Journal of Bacteriology* 107, 690–696 (1971).
16. Gabriel, D.W. et al. Identification of open reading frames unique to a select agent: *Ralstonia solanacearum* race 3 biovar 2. *Molecular Plant-Microbe Interactions* 19, 69–79 (2006).
17. Taghavi, M. et al. Analysis of the phylogenetic relationships of strains of *Burkholderia solanacearum*, *Pseudomonas syzygii*, and the blood disease bacterium of banana based on 16S rRNA gene sequences. *International Journal of Systematic Bacteriology* 46, 10–15 (1996).
18. Fegan, M. et al. Phylogeny, diversity and molecular diagnostics of *Ralstonia solanacearum*. In *Bacterial Wilt Disease: Molecular and Ecological Aspects*, (eds.) Prior, P., Allen, C., and Elphinstone, J., pp. 19–23. INRA Editions, Paris, France (1998).
19. Cook, D., Barlow, E., and Sequeira, L. Genetic diversity of *Pseudomonas solanacearum*: Detection of restriction fragment length polymorphisms with DNA probes that specify virulence and the hypersensitive response. *Molecular Plant-Microbe Interactions* 2, 113–121 (1989).
20. Cook, D. and Sequeira, L. Genetic diversity of *Pseudomonas solanacearum*: Detection of restriction fragment length polymorphisms with DNA probes that specify virulence and the hypersensitive response. In *Bacterial Wilt: The Disease and Its Causative Agent, Pseudomonas solanacearum*, (eds.) Hayward, A.C. and Hartman, G.L., pp. 77–93. CAB International, Wallingford, U.K. (1994).
21. Gillings, M., Fahy, P., and Davies, C. Restriction analysis of an amplified polygalacturonase gene fragment differentiates strains of the phytopathogenic bacterium *Pseudomonas solanacearum*. *Letters in Applied Microbiology* 17, 44–48 (1993).
22. Seal, S.E., Jackson, L.A., and Daniels, M.J. Use of tRNA consensus primers to indicate subgroups of *Pseudomonas solanacearum* by polymerase chain reaction amplification. *Applied and Environmental Microbiology* 58, 3759–3761 (1992).
23. Poussier, S. et al. Partial sequencing of the *hrpB* and endoglucanase genes confirms and expands the known diversity within the *Ralstonia solanacearum* species complex. *Systematic and Applied Microbiology* 23, 479–486 (2000).
24. Poussier, S. et al. Genetic diversity of *Ralstonia solanacearum* as assessed by PCR-RFLP of the *hrp* gene region, AFLP and 16S rRNA sequence analysis, and identification of an African subdivision. *Microbiology* 146, 1679–1692 (2000).
25. Fegan, M. and Prior, P. Diverse members of the *Ralstonia solanacearum* species complex cause bacterial wilts of banana. *Australasian Plant Pathology* 35, 93–101 (2006).
26. Wicker, E. et al. Contrasting recombination patterns and demographic histories of the plant pathogen *Ralstonia solanacearum* inferred from MLSA. *ISME Journal* 6, 961–974 (2011).

27. Cellier, G. et al. Phylogeny and population structure of brown rot- and Moko disease-causing strains of *Ralstonia solanacearum* phylotype II. *Applied and Environmental Microbiology* 78, 2367–2375 (2012).

28. Guidot, A. et al. Genomic structure and phylogeny of the plant pathogen *Ralstonia solanacearum* inferred from gene distribution analysis. *Journal of Bacteriology* 189, 377–387 (2007).

29. Remenant, B. et al. Genomes of three tomato pathogens within the *Ralstonia solanacearum* species complex reveal significant evolutionary divergence. *BMC Genomics* 11, 379. http://www.biomedcentral.com/1471-2164/11/379 (2010).

30. Cellier, G. and Prior, P. Deciphering phenotypic diversity of *Ralstonia solanacearum* strains pathogenic to potato. *Phytopathology* 100, 1250–1261 (2010).

31. Fegan, M. et al. Development of a diagnostic test based on the polymerase chain reaction (PCR) to identify strains of *Ralstonia solanacearum* exhibiting the biovar 2 genotype. In *Bacterial Wilt Disease: Molecular and Ecological Aspects*, (eds.) Prior, P., Allen, C., and Elphinstone, J., pp. 34–43. Springer, Berlin, Germany (1998).

32. Marin, J.E. and El-Nashaar, H.M. Pathogenicity and new phenotypes of *Pseudomonas solanacearum* from Peru. In *Bacterial Wilt: ACIAR Proceedings No. 45*, pp. 78–84. Australian Centre for International Agricultural Research, Canberra, Australian Capital Territory, Australia (1993).

32a. Parkinson, N. et al. Application of variable-number tandem-repeat typing to discriminate *Ralstonia solanacearum* strains associated with English watercourses and disease outbreaks. *Applied and Environmental Microbiology* 79, 6016–6022 (2013).

33. Olsson, K. Experience of brown rot caused by *Pseudomonas solanacearum* (Smith) Smith in Sweden. *EPPO Bulletin/Bulletin OEPP* 6, 199–207 (1976).

34. Elphinstone, J.G. The current bacterial wilt situation: A global perspective. In *Bacterial Wilt: The Disease and the Ralstonia solanacearum Species Complex*, (eds.) Allen, C., Prior, P., and Hayward, A.C., pp. 9–28. American Phytopathological Society Press, St. Paul, MN (2005).

35. Álvarez, B. et al. Comparative behaviour of *Ralstonia solanacearum* biovar 2 in diverse plant species. *Phytopathology* 98, 59–68 (2008).

36. French, E.R. Strategies for integrated control of bacterial wilt of potatoes. In *Bacterial Wilt: The Disease and Its Causative Agent Pseudomonas solanacearum*, (eds.) Hayward, A.C. and Hartman, G.L., pp. 199–209. CAB International, Wallingford, U.K. (1994).

37. Messiha, N.A.S. et al. Effects of soil type, management type and soil amendments on the survival of the potato brown rot bacterium *Ralstonia solanacearum*. *Applied Soil Ecology* 43, 206–215 (2009).

38. Tomlinson, D.L. et al. Limited survival of *Ralstonia solanacearum* race 3 in bulk soils and composts from Egypt. *European Journal of Plant Pathology* 131, 197–209 (2011).

39. Ryckeboer, J., Cops, S., and Coosemans, J. The fate of plant pathogens and seeds during anaerobic digestion and aerobic composting of source separated household wastes. *Compost Science and Utilization* 10, 204–216 (2002).

40. Cruz, L. et al. Molecular epidemiology of *Ralstonia solanacearum* strains from plants and environmental sources in Portugal. *European Journal of Plant Pathology* 133, 687–706 (2012).

41. Janse, J.D. et al. Introduction to Europe of *Ralstonia solanacearum* biovar 2, race 3 in *Pelargonium zonale* cuttings from Kenya. In *Bacterial Wilt: The Disease and the Ralstonia solanacearum Species Complex*, (eds.) Allen, C., Prior, P., and Hayward, A.C., pp. 81–94. American Phytopathological Society Press, St. Paul, MN (2005).

42. Pradhanang, P.M., Elphinstone, J.G., and Fox, R.T.V. Identification of crop and weed hosts of *Ralstonia solanacearum* biovar 2 in the hills of Nepal. *Plant Pathology* 49, 403–413 (2000).

43. Khoodoo, M.H.R., Ganoo, E.S., and Saumtally, A.S. Molecular characterization and epidemiology of *Ralstonia solanacearum* race 3 biovar 2 causing brown rot of potato in Mauritius. *Journal of Phytopathology* 158, 503–512 (2010).

44. Orozco-Miranda, E.F., Takatsu, A., and Uesugi, C.H. Colonization of the roots of weeds cultivated in vitro and in pots by *Ralstonia solanacearum*, biovars 1, 2 and 3. *Fitopatologia Brasileira* 29, 121–127 (2004).

45. Kelman, A. *The Bacterial Wilt Caused by Pseudomonas solanacearum. A Literature Review and Bibliography*, p. 194. North Carolina Agricultural Experiment Station Technical Bulletin 99, Raleigh, NC (1953).

46. Liao, C.H. and Shollenberger, L.M. Survivability and long-term preservation of bacteria in water and in phosphate-buffered saline. *Letters in Applied Microbiology* 37, 45–50 (2003).

47. Álvarez, B., López, M.M., and Biosca, E.G. Influence of native microbiota on survival of *Ralstonia solanacearum* phylotype II in river water microcosms. *Applied and Environmental Microbiology* 73, 7210–7217 (2007).

48. Tomlinson, D.L. et al. Recovery of *Ralstonia solanacearum* from canal water in traditional potato-growing areas of Egypt but not from designated Pest-Free Areas (PFAs). *European Journal of Plant Pathology* 125, 589–601 (2009).

49. Álvarez, B., López, M.M., and Biosca, E.G. Survival strategies and pathogenicity of *Ralstonia solanacearum* phylotype II subjected to prolonged starvation in environmental water microcosms. *Microbiology* 154, 3590–3598 (2008).

50. Van Elsas, J.D. et al. Effects of ecological factors on the survival and physiology of *Ralstonia solanacearum* bv. 2 in irrigation water. *Canadian Journal of Microbiology* 47, 842–854 (2001).

51. Van Overbeek, L.S. et al. The low-temperature-induced viable-but-nonculturable state affects the virulence of *Ralstonia solanacearum* biovar 2. *Phytopathology* 94, 463–469 (2004).

52. Sanchez Perez, A. et al. Diversity and distribution of *Ralstonia solanacearum* strains in Guatemala and rare occurrence of tomato fruit infection. *Plant Pathology* 57, 320–331 (2008).

53. Siri, M.I., Sanabria, A., and Pianzzola, M.J. Genetic diversity and aggressiveness of *Ralstonia solanacearum* strains causing bacterial wilt of potato in Uruguay. *Plant Disease* 95, 1292–1301 (2011).

54. Santana, B.G. et al. Diversity of Brazilian biovar 2 strains of *Ralstonia solanacearum*. *Journal of General Plant Pathology* 78, 190–200 (2012).

55. Lemessa, F. and Zeller, W. Isolation and characterisation of *Ralstonia solanacearum* strains from Solanaceae crops in Ethiopia. *Journal of Basic Microbiology* 47, 40–49 (2007).

56. N'Guessan, C.A. et al. So near and yet so far: The specific case of *Ralstonia solanacearum* populations from Côte d'Ivoire in Africa. *Phytopathology* 102, 733–740 (2012).

57. Tomlinson, D.L. and Gunther, M.T. Bacterial wilt in Papua New Guinea. In *Bacterial Wilt Disease in Asia and the South Pacific*, (ed.) Persley, G.J., pp. 35–39. Australian Centre for International Agricultural Research, Canberra, Australian Capital Territory, Australia (1986).

58. Hayward, A.C. Biotypes of *Pseudomonas solanacearum* in Australia. *Australian Plant Pathology Society Newsletter* 4, 9–11 (1975).

59. Villa, J.E. et al. Phylogenetic relationships of *Ralstonia solanacearum* species complex strains from Asia and other continents based on 16S rDNA, endoglucanase, and hrpB gene sequences. *Journal of General Plant Pathology* 71, 39–46 (2005).

60. Horita, M. et al. Analysis of genetic and biological characters of Japanese potato strains of *Ralstonia solanacearum*. *Journal of General Plant Pathology* 76, 196–207 (2010).

61. Xu, J. et al. Genetic diversity of *Ralstonia solanacearum* strains from China. *European Journal of Plant Pathology* 125, 641–653 (2009).

62. Persson, P. Successful eradication of *Ralstonia solanacearum* from Sweden. *Bulletin OEPP/EPPO Bulletin* 28, 113–119 (1998).

62a. Janse, J.D. Review on brown rot (*Ralstonia solanacearum* race 3, biovar 2, phylotype IIB) epidemiology and control in the Netherlands since 1995: A success story of integrated pest management. *Journal of Plant Pathology* 94, 257–272 (2012).

63. Cruz, L. et al. H. *Ralstonia solanacearum* biovar 1 associated with a new outbreak of potato brown rot in Portugal. *Phytopathologia Mediterranea* 47, 87–91 (2008).

64. Swanson, J.K. et al. Detection of latent infections of *Ralstonia solanacearum* race 3 biovar 2 in geranium. *Plant Disease* 91, 828–834 (2007).

65. Vasse, J., Frey, P., and Trigalet, A. Microscopic studies of intercellular infection and protoxylem invasion of tomato roots by *Pseudomonas solanacearum*. *Molecular Plant-Microbe Interactions* 8, 241–251 (1995).

66. Denny, T.P. Plant pathogenic *Ralstonia* species. In *Plant-Associated Bacteria*, (ed.) Gnanamanickam, S.S., pp. 573–644. Springer, Dordrecht, the Netherlands (2006).

67. Schell, M.A. Control of virulence and pathogenicity genes of *Ralstonia solanacearum* by an elaborate sensory network. *Annual Review of Phytopathology* 38, 263–292 (2000).

68. Kelman, A. The relationship of pathogenicity of *Pseudomonas solanacearum* to colony appearance in a tetrazolium medium. *Phytopathology* 44, 693–695 (1954).

69. Engelbrecht, M.C. Modification of a semi-selective medium for the isolation and quantification of *Pseudomonas solanacearum*. *ACIAR Bacterial Wilt Newsletter* 10, 3–5 (1994).

70. Elphinstone, J.G. et al. Sensitivity of different methods for the detection of *Ralstonia solanacearum* in potato tuber extracts. *EPPO Bulletin* 26, 663–678 (1996).

71. Pradhanang, P.M., Elphinstone, J.G., and Fox, R.T.V. Sensitive detection of *Ralstonia solanacearum* in soil: A comparison of different detection techniques. *Plant Pathology* 49, 414–422 (2000).

72. Priou, S. et al. Detection of *Ralstonia solanacearum* (biovar 2A) in stems of symptomless plants before harvest of the potato crop using post-enrichment DAS-ELISA. *Plant Pathology* 59, 59–67 (2010).

73. Priou, S., Gutarra, L., and Aley, P. An improved enrichment broth for the sensitive detection of *Ralstonia solanacearum* (biovars 1 and 2A) in soil using DAS-ELISA. *Plant Pathology* 55, 36–45 (2006).

74. Van Der Wolf, J.M. et al. Immunofluorescence colony-staining (IFC) for detection and quantification of *Ralstonia* (*Pseudomonas*) *solanacearum* biovar 2 (race 3) in soil and verification of positive results by PCR and dilution plating. *European Journal of Plant Pathology* 106, 123–133 (2000).

75. Pastrik, K.H. and Maiss, E. Detection of *Ralstonia solanacearum* in potato tubers by polymerase chain reaction. *Journal of Phytopathology* 148, 619–626 (2000).

76. Seal, S.E. et al. Determination of *Ralstonia* (*Pseudomonas*) *solanacearum* rDNA subgroups by PCR tests. *Plant Pathology* 48, 115–120 (1999).

77. Kubota, R. et al. In silico genomic subtraction guides development of highly accurate, DNA-based diagnostics for *Ralstonia solanacearum* race 3 biovar 2 and blood disease bacterium. *Journal of General Plant Pathology* 77, 182–193 (2011).

78. Ozakman, M. and Schaad, N.W. A real-time BIO-PCR assay for detection of *Ralstonia solanacearum* race 3, biovar 2, in asymptomatic potato tubers. *Canadian Journal of Plant Pathology* 25, 232–239 (2003).

79. Weller, S.A. et al. Detection of *Ralstonia solanacearum* strains with a quantitative, multiplex, real-time, fluorogenic PCR (TaqMan) assay. *Applied and Environmental Microbiology* 66, 2853–2858 (2000).

80. Kubota, R. et al. Non-instrumented nucleic acid amplification (NINA) for rapid detection of *Ralstonia solanacearum* race 3 biovar 2. *Biological Engineering Transactions* 4, 69–80 (2011).

81. Ha, Y. et al. A rapid, sensitive assay for *Ralstonia solanacearum* race 3 biovar 2 in plant and soil samples using magnetic beads and real-time PCR. *Plant Disease* 96, 258–264 (2012).

82. Smith, D.S. and De Boer, S.H. Implementation of an artificial reaction control in a TaqMan method for PCR detection of *Ralstonia solanacearum* race 3 biovar 2. *European Journal of Plant Pathology* 124, 405–412 (2009).

83. Poussier, S. et al. Evaluation of procedures for reliable PCR detection of *Ralstonia solanacearum* in common natural substrates. *Journal of Microbiological Methods* 51, 349–359 (2002).

84. Weller, S.A. et al. Detection of *Ralstonia solanacearum* from potato tissue by post-enrichment TaqMan PCR. *EPPO Bulletin* 30, 381–383 (2000).

85. Kubota, R. et al. Detection of *Ralstonia solanacearum* by loop-mediated isothermal amplification. *Phytopathology* 98, 1045–1051 (2008).

86. Van Der Wolf, J.M. et al. Specific detection of *Ralstonia solanacearum* 16S rRNA sequences by AmpliDet RNA. *European Journal of Plant Pathology* 110, 25–33 (2004).

87. Chávez, P. et al. Detection of bacterial wilt infection caused by *Ralstonia solanacearum* in potato (*Solanum tuberosum* L.) through multifractal analysis applied to remotely sensed data. *Precision Agriculture* 13, 236–255 (2012).

88. Messiha, N.A.S. et al. Biological Soil Disinfestation (BSD), a new control method for potato brown rot, caused by *Ralstonia solanacearum*; race 3 biovar 2. *European Journal of Plant Pathology* 117, 403–415 (2007).

89. Messiha, N.A.S. et al. Potato brown rot incidence and severity under different management and amendment regimes in different soil types. *European Journal of Plant Pathology* 119, 367–381 (2007).

90. Messiha, N.A.S. et al. *Stenotrophomonas maltophilia*: A new potential biocontrol agent of *Ralstonia solanacearum*, causal agent of potato brown rot. *European Journal of Plant Pathology* 118, 211–225 (2007).

91. Norman, D.J. et al. Control of bacterial wilt of geranium with phosphorous acid. *Plant Disease* 90, 798–802 (2006).

92. Genin, S. and Denny, T.P. Pathogenomics of the *Ralstonia solanacearum* Species Complex. *Annual Review of Phytopathology* 50, 67–89 (2012).

93. Janse, J.D. et al. Introduction to Europe of *Ralstonia solanacearum* biovar 2, race 3 in Pelargonium zonale cuttings. *Journal of Plant Pathology* 86, 147–155 (2004).

94. Williamson, L. et al. *Ralstonia solanacearum* race 3, biovar 2 strains isolated from geranium are pathogenic on potato. *Plant Disease* 86, 987–991 (2002).

95. Kim, S.H. et al. *Ralstonia solanacearum* race 3, biovar 2, the causal agent of brown rot of potato, identified in geraniums in Pennsylvania, Delaware and Connecticut. *Plant Disease* 87, 450 (2003).

72 *Sclerophthora rayssiae* var. zeae

László Galgóczy, László Kredics, Máté Virágh, and Csaba Vágvölgyi

CONTENTS

72.1 INTRODUCTION

Agriculture is a tempting target for bioterrorists because of the economic and social impact it exerts. Countries relying on agriculture to feed their people and having large sectors of economy based on agriculture are especially vulnerable to bioterrorists' attacks that target their agriculturally important plants.[1] Corn is an important grain crop worldwide, especially in the United States where approximately 30 million hectares of land is planted with this cereal every year.[2] An oomycetous fungus, *Sclerophthora rayssiae* var. *zeae*, causes the brown stripe downy mildew (BSDM) disease of corn and results in enormous yield losses in tropical Asia (e.g., 20%–90% in India). Although this plant pathogenic fungus is not present outside Asia, *S. rayssiae* var. *zeae* is included in the select agent list of the US Department of Agriculture's Animal and Plant Health Inspection Service (USDA APHIS) as a bioterror agent because it has the potential to pose a severe threat to global plant health and economy.[2,3]

72.2 CLASSIFICATION AND MORPHOLOGY

The genus *Sclerophthora* is classified in the family Peronosporaceae, order Peronosporales, class Oomycota, kingdom Chromista.[4] With >500 recognized species, the class Oomycota (Peronosporomycetes) includes the so-called water molds and downy mildews. Similar to fungi, as the members of the class Oomycota (meaning egg fungi, which refers to the large round oogonia or structures containing the female gametes) produce the filamentous threads, have the ability to obtain nutrients via absorption, and reproduce via spores; they have been considered traditionally as part of the kingdom Fungi (true fungi, including Ascomycota or ascomycetes, Basidiomycota or basidiomycetes, Chytridiomycota or chytrids, Zygomycota or zygomycetes, and Glomeromycota).

However, the Oomycota demonstrate notable ultrastructural and biochemical differences from the true fungi. For example, the cell wall of oomycetes is composed of beta glucans and cellulose instead of chitin as in the true fungi; the Oomycota produce motile zoospores with two flagella that are oriented posteriorly (whiplash flagellum) and anteriorly (tinsel flagellum), while some true fungi (e.g., the Chytridiomycota) produce motile zoospores with flagella of the posterior whiplash type only; the vegetative cells of the Oomycota generally consist of coenocytic hyphae (hyphae without septa or cross-walls) with diploid nuclei in comparison with the true fungi whose mycelium tends to be divided into cells by cross-walls, with each cell containing one, two, or more haploid nuclei; the Oomycota possess mitochondria with tubular cristae and protoplasmic and nuclear-associated microtubules, rather than flattened mitochondrial cristae as seen in the true fungi. Recent molecular evidence indicates that the Oomycota share a common ancestor with the other members of the heterokont algae or Chromista and thus belong to the kingdom Chromista.[3]

Sclerophthora rayssiae (synonym: *Sclerophthora rayssiae* var. *rayssiae*) was first described by Kenneth et al.[5] Subsequently, *Sclerophthora rayssiae* var. *zeae* was identified by Payak and Renfro[6] as a potential fungal pathogen of some plants belonging to the Poaceae.[6] Besides *Sclerophthora rayssiae*, other species in the genus *Sclerophthora* include *Sclerophthora butleri*, *Sclerophthora cryophila*, *Sclerophthora farlowii*, *Sclerophthora lolii*, *Sclerophthora macrospora*, and *Sclerophthora northii*.

S. rayssiae var. *zeae* is an obligate plant parasite and its mycelium occurs only in the mesophyll of the leaf blades and sheaths. This fungus forms irregular hyphae, which are lobulate rather than tubular. Hyaline oogonia (33.0–44.5 μm) with one or two paragynous antheridia are thin-walled and produce pleurotic, spherical or subspherical, brownish, smooth-walled hyaline oospores (29–37 μm). Sporangiophores are short, hyaline, ovate, obclavate, elliptic or cylindrical; smooth-walled, papillate sporangia (29–66.5 × 18.5–26 μm) are produced sympodially in groups of two to six in a basipetal succession on them. Sporangia have a raised truncate, rounded or tapering poroid apex and they are caducous, with a persistent, straight or cuneate peduncle. Sporangia contain four to eight hyaline and spherical zoospores (7.5–11.0 μm).[2,6,7]

72.3 BIOLOGY AND EPIDEMIOLOGY

S. rayssiae var. *zeae* is an obligate plant pathogenic fungus with sexual and asexual reproduction. This fungus can survive and overwinter via oospores (primary inocula, sexual reproduction) in infected seeds, plant debris, or in the soil.[3] After indirect germination in leaf mesophyll or substomatal cavities the oospores produce sporangiophores with sporangia (secondary inocula, asexual reproduction) that contain four to eight zoospores. Less frequently the sporangium germinates directly and produces a germ tube which can penetrate into the leaves and repeat the cycle. More frequently sporangia disperse via wind, water splash, or physical contact and initiate new infections through escape of their zoospores.[2]

The viability of oospores can remain for 3–5 years in air-dried leaf tissue,[8,9] but the infected seeds dried to ≤14% moisture and stored for four or more weeks cannot transmit the disease.[7] The optimal moisture and the temperature are essential for sporangia production and germination; furthermore film of water is also required. However, sporangia are produced over a wide temperature range (18°C–30°C), but the optimum for abundant production is 22°C–25°C and the germination optimum via zoospores is at 20°C–22°C, and periods of high moisture are also required.[2] In optimal environmental conditions sporangia appear in infected leaves within 3 h and their second generation arises after 9 h.[9] The produced zoospores can germinate between 15°C and 30°C.[2]

The primary host of *S. rayssiae* var. *zeae* is maize (*Zea mays*). Additionally, *Digitaria sanguinalis* (L.) Scop. and *Digitaria bicornus* (Lam.) Roem. & Schult. also can be infected by the fungus, but the *S. rayssiae* var. *zeae* strains isolated from *D. bicornus* (Lam.) Roem. & Schult. did not cause infection on maize.[2,10] Artificial inoculation experiments revealed that several unspecified species of millets, cereals, and grasses are not susceptible to the fungus.[9] *Zea mays* may be infected with *S. rayssiae* var. *zeae* when the seeds carry the plant debris with viable oospores or the oospores and mycelia are within the embryo.[11,12] It was observed that infected leaf debris is more important in the initiation of new infections compared to that when the disease can become established on seedlings grown from infected seed or from seed transmission.[8,11,12] The optimal temperature of the soil (28°C–32.5°C) is important for the development of the disease. The young plants are more susceptible to the infection than the older ones.[2] Under optimal conditions the generation time of sporangia from oospores is rapid; the sporangia appear after 10 days. The sporangial

releases occur in huge amounts in the afternoon of sunny days when high moisture is present.[9] The produced sporangia are dispersed short distance (1–1.65 m from an infected field) in wind, rain drops, or via physical contact, but the long transport via wind is not common.[2,9] At the optimum temperature range (21°C–26°C) of sporangial infection, no more than 10 days are required for rapid spreading of the disease throughout the crop, but the pathogen does not systemically infect the plant.[2]

BSDM caused by *S. rayssiae* var. *zeae* was first reported in India in 1962, with the causative organism identified in 1967. It later spread throughout the country and also appeared in Myanmar, Nepal, Pakistan, and Thailand.[2,6] This disease is most severe in areas of India with high rainfall (100–200 cm/year), but also occurs and causes moderate or light crop losses in the areas with 60–100 and 50–70 cm rainfall per year.[2] BSDM is common in the high rainfall regions of India (20%–100% incidence)[8] and causes 20%–90% grain yield reduction, but it is dependent on the annual rainfall. The losses also depend on the region where the BSDM appeared, the type of the maize cultivar, the life stage of the plant, and when and how severely the tissue is affected. The greatest crop losses were observed in fields with abundant summer moisture and with warm soil. Based on a previous susceptibility survey among different genotypes of Indian maize, 29% of them were susceptible and the remaining proved to be moderately to highly resistant to the disease.[12] The young plants are more susceptible and the loss may be total when three-quarters or more of the foliage is affected prior to flowering and the ear formation is either totally suppressed or markedly attenuated.[2,6,9]

72.4 INFECTION FEATURES AND PATHOGENESIS

BSDM is not a systemic disease, only the leaves of the plant are affected. The young plants are most susceptible to the disease than the older ones, and the susceptibility decreases with the age of the plant. The symptom of BSDM is observable in the early growing period and is maintained through the whole season.[13]

S. rayssiae var. *zeae* causes leaf lesions with well-defined margins, which are parallel and delimited by the veins. The stripes are 3–7 mm narrow, chlorotic, and yellowish at the first stage of infection, but later they become reddish to purple in some maize genotypes. The lateral development of the lesions causes severe striping and blotching.[2,6,13] *S. rayssiae* var. *zeae* can sporulate on both sides of the lesions and show greyish-white woolly growth, and the occurrence of sporangia when the lesions become necrotic. Appearance of the oospores is limited to only the necrotic tissues in the leaf mesophyll or beneath the stomata, and is not in the vascular tissue.[2,13]

The leaf lesion symptom in the greatest degree appears on the lower leaves first: these leaves become pale-brown and burnt and the severely affected ones may be shed prematurely. The BSDM may be associated with suppressed seed development when the disease occurs prior the flowering, and smaller seed size or early plant death results. The vegetative and floral tissues are not affected by BSDM, malformation is not observed in them, and the affected leaves do not shed.[11,13]

72.5 IDENTIFICATION AND DIAGNOSIS

As *S. rayssiae* var. *zeae* is an obligate plant pathogenic fungus, it does not grow on artificial media in spite of the fact that the successful cultivation of *Sclerophthora* was reported.[2,14] The diagnosis and identification is only possible via the macroscopic investigation of the infected plant materials and microscopic examination of the diseased tissues; serological and/or molecular methods for specific pathogen detection have not been published yet.[2]

The following morphological features are important for the identification of *S. rayssiae* var. *zeae*: sporangiophores are short, determinate, and are produced from hyphae in the substomatal cavities; sporangia arise in basipetal succession, forming sympodially in groups of 2–6; typically appear hyaline, ovate, obclavate, elliptic or cylindrical, smooth-walled, papillate with a projecting truncate, rounded or tapering poroid apex; are caducous, with a persistent, straight or cuneate peduncle; range from $18.5–26.0 \times 29.0–66.5$ μm in size; possess lens-shaped pores through which zoospores or cytoplasm may escape; and contain 4–8 zoospores; zoospores are hyaline, spherical, varying from 7.5 to 11.0 μm in diameter; oogonia are hyaline to light, 33.0–44.5 μm in diameter, thin-walled, with one or two paragynous antheridia; and oospores are pleurotic, spherical or subspherical, ranging from 29.5 to 37.0 μm in size; hyaline, with one prominent oil globule; whose cell walls are smooth, glistening, about 4 μm in thickness, confluent with the oogonial wall.[2] In particular, its conidiophore structure and dimension as well as spore (conidia) shape and size are of diagnostic value. Morphological guides are available for the exact identification of this pathogen.[6,7]

To assist the diagnosis of *S. rayssiae* var. *zeae*, chlorotic symptomatic tissue may be placed into a moist chamber and incubated at 22°C–25°C for 3–9 h for enhanced sporangia production. Oogonia and oospores in necrotic tissue can be visualized by clearing the leaf tissues in 2% sodium or potassium hydroxide solution at 45°C–50°C, washing in several changes of distilled water, then staining with 0.1% cotton blue (methyl blue) in 50% glycerol for up to 20 min at 45°C–50°C. If confirmed with BSDM, all tissues in possession as well as supplies and materials contaminated during examination of the suspect plant material must be destroyed or sterilized by autoclaving at a minimum of 103.4 kPa, 121°C, for 20 min.[2]

BSDM may be mistaken for diseases caused by other *Sclerophthora* spp. that occur on corn and by other downy mildew genera (*Peronosclerospora* and *Sclerospora*).[13] The crazy top and sorghum downy mildew caused by *Sclerophthora macrospora* and *Peronosclerospora sorghi*, respectively, show the most similar symptoms to BSDM. However, in contrast to *S. rayssiae* var. *zeae*, *S. macrospora*

causes leaf or floral malformation or distortion; its sporangia and oospores are larger and the oospores develop mainly in vascular bundles. Systemic infection is observable in the case of *P. sorghi*, which causes irregular whitish yellow stripes on leaves. Chlorotic blade base is observable on the affected leaves. The stripes do not change their color to purple and red in the later stage of the infection.[2]

72.6 TREATMENT AND PREVENTION

Two main methods are available for treatment and prevention of BSDM caused by *S. rayssiae* var. *zeae*: the chemical control applies fungicides on the seeds and leaves; the nonchemical control uses different varieties of maize.

It was observed that the severity of BSDM was reduced when mancozeb (0.3%) or triphenyltin chloride or triphenyltin acetate spray was applied on the leaves.[15–17] Excellent disease control was achieved when seed treatment with metalaxyl (4 g/kg up to 30th day after planting) was combined with one foliar application of this pesticide (225 ppm on the 30th day after planting).[18] Based on this study, the application of metalaxyl-based fungicide on the seeds and leaves has been recommended for treatment of BSDM up to date.

The use of resistant varieties of maize prevents the development of BSDM and the spreading of *S. rayssiae* var. *zeae* on and throughout the corn fields. Mass selection of maize variants under epiphytotic conditions could result in a significant improvement for resistance to BSDM.[19] Changes in the (Ca + Mg)/K ratio could improve the resistance of various crops to different downy mildews.[20]

Beyond the aforementioned protection possibilities, there are other management principles recommended by the International Maize and Wheat Improvement Center (http://maizedoctor.cimmyt.org/index.php) for the prevention of BSDM: planting early in the season, increasing row and plant spacing, avoiding rotation or simultaneous cultivation of maize with alternate hosts of downy mildew, rouging out the infected plants at early stage, destructing the plant debris by deep plowing and other methods, planting when soil temperature is below 20°C and when seed has low moisture content (<9%) before planting, controlling weeds to increase aeration within the crop, and reducing moisture levels in the soil.

72.7 CONCLUSION

S. rayssiae var. *zeae* is an oomycetous fungus that causes the BSDM.[21–23] Although BSDM due to *S. rayssiae* var. *zeae* is limited only to the tropical countries of Asia and this pathogen infects corn only, it could be a potential weapon in the hands of bioterrorists due to the wide ecological features of the fungus allowing its dispersion into other countries of the world. Furthermore, after a possible infection of the endangered country, the financial losses would be substantial due to the enormous yield losses of corn (e.g., it may reach billions of USD in the United States).[2] These are the reasons why this fungus is worth considering, and why the development of a rapid, reliable molecular technique for its identification is needed. With recent advances in the study of oomycete genetics and genomics and ongoing efforts,[24–27] it can be envisaged that sensitive and precise methods for *S. rayssiae* var. *zeae* detection and effective measures for its control and prevention will become a reality in the foreseeable future.

ACKNOWLEDGMENT

Authorial activity of László Galgóczy and Csaba Vágvölgyi was supported in the frame of TÁMOP 4.2.4. A/2-11-1-2012-0001.

REFERENCES

1. Budowle, B., Murch, R., and Chakraborty, R., Microbial forensics: The next forensic challenge, *Int. J. Legal Med.*, 119, 317, 2005.
2. Putnam, M.L., Brown stripe downy mildew (*Sclerophthora rayssiae* var. *zeae*) of maize, *Plant Health Progr.*, doi:10.1094/PHP-2007-1108-01-DG, http://www.plantmanagementnetwork.org/pub/php/diagnostic-guide/2007/stripe/, 2007.
3. Fry, W.E. and Grünwald, N.J., Introduction to oomycetes. *Plant Health Instr.*, doi:10.1094/PHI-I-2010-1207-01, http://www.apsnet.org/edcenter/intropp/PathogenGroups/Pages/IntroOomycetes.aspx, 2010.
4. CAB International, *Index Fungorum*, Online, CAB International, Wallingford, U.K., http://www.indexfungorum.org/names/NamesRecord.asp?RecordID = 353844, 2012.
5. Kenneth, R.G., Koltin, Y., and Wahl, I., Barley diseases newly found in Israel, *Bull. Torrey Bot. Club*, 91, 185, 1964.
6. Payak, M.M. and Renfro, B.L., A new downy mildew disease of maize, *Phytopathology*, 57, 394, 1967.
7. Smith, D.R. and Renfro, B.L., Downy mildews. In: White, D.G. (ed.) *Compendium of Corn Diseases*, 3rd ed., St. Paul, MN: APS Press, p. 8, 1999.
8. Singh, J.P., Infectivity and survival of oospores of *Sclerophthora rayssiae* var. *zeae*, *Indian J. Exp. Biol.*, 9, 530, 1971.
9. Singh, J.P., Renfro, B.L., and Payak, M.M., Studies on the epidemiology and control of brown stripe downy mildew of maize (*Sclerophthora rayssiae* var. *zeae*), *Indian Phytopathol.*, 23, 194, 1970.
10. Bains, S.S. et al., Role of *Digitaria sanguinalis* in outbreaks of brown stripe downy mildew of maize, *Plant Dis. Rep.*, 62, 143, 1978.
11. Lal, S. and Prasad, T., Detection and management of seed-borne nature of downy mildew diseases of maize, *Seeds Farms*, 15, 35, 1989.
12. Singh, R.S., Joshi, M.M., and Chaube, H.S., Further evidence of the seedborne nature of maize downy mildews and their possible control with chemicals, *Plant Dis. Rep.*, 52, 446, 1967.
13. Sullivan, M. and Jones, E., Corn commodity-based survey guideline, USDA APHIS, Plant Protection and Quarantine, Center for Plant Health Science and Technology, 2010, http://caps.ceris.purdue.edu/survey/manual/corn_guidelines.
14. Tokura, R., Axenic or artificial culture of the downy mildew fungi of gramineous plants, *Trop. Agric. Res. Ser.*, 8, 57, 1975.
15. Lal, S., Brown stripe and sugarcane downy mildews of maize: Germplasm evolution, resistance breeding and chemical control, *Trop. Agric. Res. Ser.*, 8, 235, 1975.
16. Lal, S. et al., Field evaluation of some systemic and nonsystemic fungicides for the control of brown stripe downy mildew of maize, *Pesticides*, 10, 28, 1976.
17. Nene, Y.L. and Saxena, S.C., Studies on the fungicidal control of downy mildew of maize caused by *Sclerophthora rayssiae* var. *zeae*, *Indian Phytopathol.*, 23, 216, 1970.
18. Lal, S., Saxena, S.C., and Upadhyay, R.N., Control of brown stripe downy mildew of maize by metalaxyl, *Plant Dis.*, 64, 874, 1980.
19. Khehra, A.S. et al., Selection for brown stripe downy mildew resistance in maize, *Euphytica*, 30, 393, 1981.
20. Bains, S.S., Jhooty, J.S., and Sharma, N.K., The relation between cation-ratio and host-resistance to certain downy mildew and root-knot diseases, *Plant Soil*, 81, 69, 1984.
21. Green, A., Brown stripe downy mildew. *Sclerophthora rayssiae* var. *zeae*, *PNKTO* 55, 1, 1984.
22. Thakur, R.P. and Mathur, K., Downy mildews of India, *Crop Prot.*, 21, 333, 2002.
23. Spencer-Phillips, P.T.N., Gisi, U., and Lebeda, A. (eds.) *Advances in Downy Mildew Research*, vol. 1, Dordrecht, the Netherlands: Kluwer Academic Publishers/Springer, 2002.
24. Riethmüller, A. et al., Phylogenetic relationships of the downy mildews (Peronosporales) and related groups based on nuclear large subunit ribosomal DNA sequences, *Mycologia*, 94, 834, 2002.
25. Lebeda, A., Spencer-Phillips, P.T.N., and Cooke, B.M., (eds.) *The Downy Mildews—Genetics, Molecular Biology and Control*, Issue 1, Dordrecht, the Netherlands: Springer, 2008.
26. Thines, M. et al., Phylogenetic relationships of graminicolous downy mildews based on *cox2* sequence data, *Mycol. Res.*, 112, 345, 2008.
27. Lamour, K. et al., Taxonomy and phylogeny of the downy mildews (Peronosporaceae). In Lamour, K. and Kamoun, S. (eds.) *Oomycete Genetics and Genomics: Diversity, Interactions, and Research Tools*, Hoboken, NJ: John Wiley & Sons, Inc., 2009.

73 *Synchytrium endobioticum*

Jarosław Przetakiewicz

CONTENTS

73.1 INTRODUCTION

Synchytrium endobioticum (Shilb.) Perc. is a fungus which causes potato wart disease known also as black wart, warty disease, potato tumor, black cancer, or black scab. The fungus is a member of genus *Synchytrium*, which contains about 200 species. While all *Synchytrium* species are parasites, *S. endobioticum* is the most important economically and phytosanitarily. The main host of *S. endobioticum* is potato, but the fungus is able to infect other species of the genus *Solanum* (covering annual and perennial plants).[1] Although *S. endobioticum* originated from Andean zone in South America, thanks to the popularity of the potatoes, it is distributed worldwide and occurs almost around the world. First found in 1876 in the United Kingdom,[2,3] *S. endobioticum* was only described by Shilberszky in 1896.[9] According to Hampson,[3] thousands of outbreaks had been reported in England by 1919. Similar cases had been also documented in Germany in the 1920s–1940s.[4] Soon, most countries in Europe had reported outbreaks of potato wart disease.[1] Currently, *S. endobioticum* is described in 15 European countries.[5] The geographical distribution of this pathogen includes all European and Mediterranean Plant Protection Organization (EPPO) regions (except a few countries), Asia, North and South America, as well as Oceania (New Zealand).[6]

S. endobioticum is the obligate biotrophic pathogen of a plant which does not produce hyphae. Its cell wall is produced only by winter sporangia, which allow survival in soil for dozens of years. For the production and release of zoospores as well as biflagellated zygotes, free water is essential. Upon infection, potatoes develop galls. Successful infections, however, occur only at young sprouts of susceptible potato cultivars.[7] After infection, *S. endobioticum* produces two different kinds of sporangia in the galls. Summer sporangia known as sori have a thin cell wall and form sporangia from where haploid zoospores emerge and steadily reinfect the host tissue like sprouts, tubers, eye tubers, stolons, and roots only in tomato.[8] Under appropriate conditions, haploid zoospores develop into diploid zygotes in a process known as isogamy, and the latter are able to infect host cells and form winter sporangia (resting sporangia) with thicker cell walls. Resting sporangia are embedded deeper into the host tissue than the sori (which is always on the surface), and are designed for longer survival in the soil.

73.2 CLASSIFICATION AND MORPHOLOGY

In the older classification, this species has been included in the Kingdom Protista but now *S. endobioticum* (Schilberszky) Perecival (synonyms: *Chrysophlyctis endobiotica* (Schilberszky), *Synchytrium solani* [Massee]) belongs to the Kingdom Fungi,[10] phylum Chytridiomycota, order Chytridiales, family Synchytriaceae, genus *Synchytrium*, and species *S. endobioticum*. Traditionally, *S. endobioticum* has been placed in the subgenus *Mesochytrium*[11] but on the basis of the model of germination it was transferred to the subgenus *Microsynchytrium*.[12] The latter observations were confirmed subsequently by Lange and Olson.[13a]

The most characteristic stage of *S. endobioticum* is golden brown winter (resting) sporangia (Figure 73.1). They are usually spherical to ovoid in shape and 24–75 µm in diameter with a thick-walled (triple wall) structure, which is ornamented with irregularly shaped wing-like protrusions.[13b] The sporangium wall is composed among other things of C18.0, C18.1, C18.2, C20.0, and C20.4 fatty acids. A large portion of the sporangium wall lipids contain wax esters with branched chains. This is the reason that the sporangium wall can serve as a protective device for continued viability of *S. endobioticum* in the soil.[14] Transmission electron microscopy revealed that the cell wall was composed of exospores—derived from the host potato, a mesospore and endospore composed of the vesicle wall and the sporangial wall.[15] The other electron microscope observation of the developing wall revealed microfibrils with discrete orientation and incorporation of a precursor of chitin was restricted to the periphery of the developing wall of the winter sporangium. The mature wall

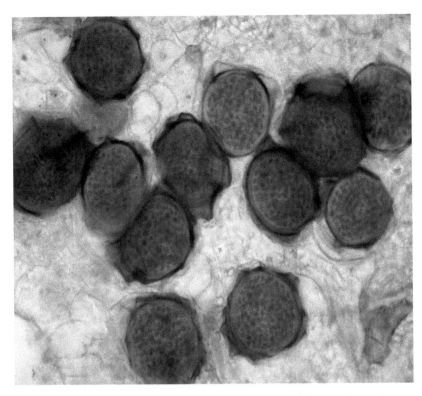

FIGURE 73.1   Ripped winter (resting) sporangia of *S. endobioticum*, pathotype 2(Ch1) in wart tissue of variety Desirée. There is a visible thick-walled structure, which is ornamented with irregularly shaped wing-like protrusions and homogeneous, granular, grayish contents (lipid bodies). Magnification 400×.

of winter sporangia with compact layers of orient chitin microfibrils serve as a protective cover, which may be related to its survival characteristics.[16] The cytoplasmic organization of the resting spores was described by Lange and Olson.[13b] The cytoplasm of the resting sporangium contains a large number of closely packed lipid bodies and irregular electron-dense bodies, which are interspaced with fine channels of cytoplasm.[13b]

The next characteristic is disease symptoms of *S. endobioticum* on the host (potato). The typical symptoms of potato wart disease on tubers are the proliferating warts, which may vary markedly in form but are primarily spherical to irregular.[17] The infection originates from eye tissue, but may expand to stolons, entire young tubers, and, in favorable conditions, stems, foliage, and even flowers.[1] The aboveground warts are green which in time become dark to black. Underground warts are white, yellow, or purple to brown. The color is dependent on the potato variety. If the variety produces purple sprouts, then the warts will be purple too. The warts are similar to cauliflower and they usually have the same color but are sometimes compared to walnut kernel. The warts proliferate to galls. Galls vary markedly in form[3] but are primarily spheroid. Although the typical disease symptoms of *S. endobioticum* are described as tumor-like structures/outgrowth,[18,19] galls are primarily parenchymatous and phytoteratological and not phytooncological, and should therefore not be referred to as tumors.[3] Warts differ in size from pea-sized nodules to the size of a fist. Warts maintained

under laboratory conditions (for inoculum production) can reach 220 g from one eye/sprout. The size of warts depends not only on environmental conditions (cool summer, wet soil, etc.) but on potato variety as well. In an extremely susceptible variety, specified pathotypes of *S. endobioticum* can differ markedly up to very big sizes producing winter sporangia in the last stage of growth. In the slightly susceptible variety, specified pathotypes of *S. endobioticum* are able to produce very small warts or only weakly influence proliferation of host tissue, producing winter sporangia in the first stage of infection. The winter sporangia were visible 13 days after inoculation of slightly susceptible varieties. In uncomfortable conditions (high temperature or dry soil), warts stop growing and begin to produce winter spores.

73.3   BIOLOGY AND EPIDEMIOLOGY

S. endobioticum as an obligate biotrophic parasite passages from one vegetative generation to the next by winter (resting) sporangia as a result of generative reproduction. Thick-walled winter sporangia can survive till the next season (vegetative season of host). The wall structure of the winter sporangia influences their longevity. Malec[20] cited some researches in which winter sporangia was found to survive up to over a dozen of years. According to Hartman and Akeley,[21] it was possible to find potato wart disease after 20 years since the first detection. In general, winter sporangia can remain viable in the range of 10–40 years or more and are found at depths of up to 50 cm.[2,3,17,22,23,75] According to McDonnell and Kavanagh,[24] the winter spores of *S. endobioticum* can survive for 31 years in field conditions. This longevity might be affected by soil types and climate.[25] The soil types associated with the subarid region are less suitable for the pathogen than those associated with the damp, cool region. In the grassland, winter sporangia can remain viable longer than in arable soil, where the oxygenation is higher.[20]

Under favorable conditions (temperature at 8°C–24°C, high moisture of soil), part of the resting spores germinate and release uninucleate haploid zoospores. They use flagellum to move in the soli water to find the host cell—epidemic cells of young organs of potato like eye, sprout young tuber, or stolons but never roots. In the case of tomato, it is possible to infect both roots and stems. The warts on the roots are the size of a pea or smaller. Microscopic examinations revealed that root warts contain numerous winter sporangia.[8] These become greatly enlarged and the enclosed fungus forms a short-lived, quickly reproducing stage, the summer sporangium, from which numerous zoospores are rapidly discharged and reinfect surrounding cells (secondary infection), which again produce summer sporangia. The presence of parasites inside the epidemic cell causes its hypertrophy and their influence on surrounding cells causes their hypertrophy too. Inside, the infected host cell develops prosorus and then sorus. One sorus contains 1–9 sporangia (average 3–7) and one sporangium contains 200–300 motile zoospores.[1] The vegetation cycle can be repeated as long as infection conditions are favorable, which leads to quick growth of gall. The cells around those penetrated also swell and the tissue proliferates, producing a characteristic cauliflower appearance.

Usually, most handbooks inform us that springtime is the best term for germination of winter sporangia. In practice, a part of winter sporangia are able to germinate directly from maturated warts without any dormancy period. Through the preparation of microscopic slides from matured warts, it is clear that the germination process of winter sporangia begins by showing visible vacuoles in the central part of the spores, before the warts start decaying and rotting. It was observed that on matured warts, new warts with summer sporangia appear after 4–6 weeks. The fresh warts are a result of secondary infection due to germination of winter sporangia. Additionally, it is possible to get fresh warts after 4 weeks of incubation with winter sporangia from matured macerated warts (to release winter sporangia). These observations suggest the possibility of closing the life cycle of *S. endobioticum* two or more times in one vegetation season of the host.

Under certain conditions of stress, such as water shortage, the zoospores may fuse into pairs to form a zygote; the host cell in which it forms does not swell but divides. The host cell wall remains closely attached, forming an outer layer to the resistant, thick-walled winter sporangium. This matures and is released into the soil from rotting warts. Winter sporangia can remain viable for at least 30 years and are found at depths of up to 50 cm. The disease can spread in infected seed tubers which may have incipient warts that pass undetected, or in infested soil attached to tubers. The sporangia resist digestion by animals, and can thus spread through feces.

Many pathotypes of the fungus exist, defined by their virulence on differential potato cultivars. Pathotype 1 (European race 1) is the most common in the EPPO region and, in addition, the only pathotype occurring in most countries. Other pathotypes, numbered up to 18,[26] occur mainly in the rainy mountainous areas of central and eastern Europe (Alps, Carpathians), for example, in the Czech Republic, Germany,[2,72] Poland,[71] and the former USSR.[73] They persist mainly in small garden potato plots and not in commercial potato crops.[27]

73.4 CLINICAL FEATURES AND PATHOGENESIS

The fungus *S. endobioticum* is not lethal for humans and animals. Casual eating by humans or feeding by animal warts does not kill the winter sporangia. Moreover, passage of the warts through the digestive system stimulates winter sporangia to germinate quicker. In experiments, when winter sporangia of *S. endobioticum* were fed to water snails, it caused eroding of exospores (one of the layers of the cell wall) and an increase in the germination rate (production of vesicles).[15] In other experiments, earthworms facilitate the small-scale dissemination of wart disease of potato by soil ingestion and ejection of ingested materials in other places.[28] On the large scale, the pathogen was distributed by humans. This species is a classic example of the distribution of a disease by man first within Europe, thence to North America, and then to New Zealand and other distant lands.[29]

However, according to some hypotheses, the emergence of new, more virulent pathotypes is caused by growing resistant varieties, which are characterized by a low level of resistance. This hypothesis was proven by Malec,[30,31] who showed that repeated passages selected a more virulent pathotype. Initially it was believed that *S. endobioticum* is morphologically and biologically homogeneous, and does not differ in terms of parasitic properties. The first report of the existence of diversity of races appeared in 1941,[32] when the evidence of physiological specialization of *S. endobioticum* and more virulent pathotype detected in Thuringia in Gießübel—pathotype 2 (G1). In Europe, more than 40 pathotypes of *S. endobioticum* have already been identified.[33]

73.5 IDENTIFICATION AND DIAGNOSIS

Contrary to most of the potato pathogens, the symptoms of potato wart disease are not visible on plants, although there may be a reduction in plant vigor.[6] Due to the biotrophic character of *S. endobioticum* the host is not killed by the fungus and the underground symptoms may not be visible until harvest (Figure 73.2). Only in very suitable conditions the symptoms might be visible above the ground: stem, foliage, and in extreme conditions on inflorescences. They are usually green in color with malformations of infected parts of plants, which are similar to broccoli. In time, they become dark to black-colored and turn into a decaying mass. In most

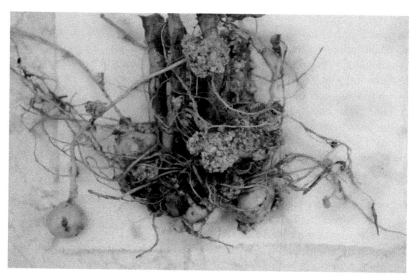

FIGURE 73.2   Potato wart disease on stems, stolons, tubers, and tuber eyes directly after harvesting.

cases, the symptoms are visible on underground parts of potato, that is, on stolons, stems, bulbs, and eyes of matured tubers. In soil, the high content of winter sporangia (above 500 per gram) may lead to infection of all eyes of seed potato and develop only warts without any emergences of potato. In very suitable conditions, one might find only warts and no tubers except seed potato. They can vary from pea size up to fist size. Usually, it is possible to find tubers, tubers with warts, and stolons with warts (Figure 73.2). The tubers that look healthy usually have infected eyes and during storage or in the next vegetation season, potato wart disease can reoccur. Underground warts are cauliflower-like, and white to purple in color. In the case of rains during the cultivation season, which rinse off the soil from the warts, they turn green after exposure to light. Even after damage/breakup of infected plants by animals or men, whole or part of warts do not die and can develop further with time, producing summer and winter sporangia. This is possible because wart tissue includes supply of starch, which is used for the development of pathogen. Usually at the end of the vegetation season, when *S. endobioticum* produces winter sporangia, the warts become brittle. In this stage, they can disintegrate, rot, and decay to release winter sporangia. In the meantime, warts turn dark and during inspection it is possible to mistake them with clod of soil or even with stones.

Detection of winter sporangia requires the use of laboratory equipment. Several methods have been described for inspection of soil. In 1925, Mary D. Glynne described in detail the method for extracting winter sporangia from soil.[34] Now, in EPPO Region Standards, PM 3/59 (soil tests and descheduling of previously infested plots)[35] and PM 7/28 (diagnostic protocols for regulated pests)[17] are respected for detection of winter sporangia from soil. These standards were based on methods of Mygind,[36] Pratt,[22] Potocek,[38] and Laidlaw.[39] All of them concern direct examination of winter sporangia in soil. Recently, for improving reliability and reducing the cost of testing, modifications of direct examination using different chemicals,[37,40] Hendrickx centrifuge,[41] or polymerase chain reaction (PCR) techniques have been reported.[42–44,70] In general, these methods aim to detect winter sporangia of *S. endobioticum* species and distinguish between viable and dead organisms. The details of morphology of winter sporangia were described in Section 73.3. This information is required in the case of direct detection using light microscopy. For bioassay methods or field tests, it is essential to know the pathotype of *S. endobioticum* involved in the original infestation. In the case of a new focus, extremely susceptible varieties to all known pathotypes of *S. endobioticum* are required.

73.6   TREATMENT AND PREVENTION

In general, prevention of potato wart disease by using chemicals is not possible. However, more than 400 compounds have been tested and the chemical structure of 149 compounds investigated against *S. endobioticum*. Most of these chemicals were found to be as biotoxic and phytotoxic, and thus not helpful for the control of *S. endobioticum*.[3,45] Several biological treatments, such as the use of chitin, urea, organic fertilizers, and crab shell meal have been proposed but without expected success.[9] Contrary to the long quarantine of other pathogens such as *Globodera rostochiensis* (yellow potato cyst nematode), the winter sporangia express high resistance to physical factors like high temperature. Viable winter sporangia of *S. endobioticum* could be extracted from samples for 70 days at 30°C–40°C, composting for 21 days at 50°C–55°C, and after composting for 12 days at 60°C–65°C. Moreover, the viable resting spores could be extracted after pasteurization for 90 min at 70°C and heating in a water bath at 80°C and in a dry oven at 90°C for 8 h.[46]

The only strategies to confine the disease are strict quarantine and phytosanitary measures and the cultivation of resistant varieties of potato.[19] Now the pathogen is classified as an IAII quarantine pest in the EU Directive 2000/29/EC.[47]

The first acts were issued in 1908 in Ireland, and a few years later in Scotland, England, and Germany.[48] The first legislation was primarily to prohibit the importation from other countries of potatoes infected by *S. endobioticum*. At that time, the acts did not take into account the possibility of the spread of the disease within the country. This contributed to an increasing number of outbreaks in such a short time in many countries.[1] As a consequence of the wide spread of potato wart disease was the development of research on the biological control of the fungus. The biggest discovery was finding resistant varieties of potato to *S. endobioticum* among cultivated ones. Systematic studies on resistance of potato cultivars to *S. endobioticum* started in England in 1909.[1] The field tests allowed for selection of resistant genotypes to *S. endobioticum*. In 1941,[32,49] new pathotypes of *S. endobioticum* were discovered for the first time: pathotype 2(G1) in Giessübel (Turyngia, Germany) and pathotype SB in the region of Carpathian (former Czechoslovakia). Although the second one is now extinct, the first one is still noticed in the fields in Germany[50] and has also been dragged into the Netherlands, the Czech Republic,[33] and even into Canada.[3] At least 40 different pathotypes of *S. endobioticum* have been recorded worldwide since 1941.[33] New pathotypes are still being discovered.[18] Polish studies have shown that potato varieties which are weakly resistant or slightly susceptible to appropriate pathotypes of *S. endobioticum* might play an important role in the development of new, more virulent pathotypes.[51] A heterogeneous population of a definite pathotype of *S. endobioticum* is a subject of selection for virulence by infecting the potato varieties that are weakly resistant to the infection by that pathotype of *S. endobioticum*. The crosses between selected individuals and genetic recombination may contribute to the creation of new pathotypes of the fungus that may be able to break the resistance of a definite potato variety and strongly infect it.[31] Methods used by the European Union Member States to assess the resistance of potato cultivars to wart disease are not harmonized. Members of the European Union should notify their list of potato cultivars resistant to *S. endobioticum* to the Commission.[52] It should be noted that some cultivars stated as resistant in one country are not recognized as resistant in other and vice versa. This is a result of applying different methods of resistance assessment to *S. endobioticum*.[62a] The directive of the European Union 69/464[52] only indicates that the resistant varieties are those that respond to pathogen infection in such a way that there is no risk of secondary infection. The directive also does not specify which method should be used to determine the lack of secondary infection. Many more clarifications have been included in the EPPO Diagnostic Protocol PM 7/28,[17] where the resistance assessment methods and pathotype identification of *S. endobioticum* are described: Spieckermann[53] and Glynne–Lemmerzahl[34,54,55] and field tests. The EPPO Diagnostic Protocol PM 3/59[35] also recommends Potocek's tube tests.[38] As a result, different countries apply one of these methods or others not included in the EPPO Diagnostic Protocols.

For example, in the Scottish Agricultural Science Agency, sprouts of potato are infected by zoospores of *S. endobioticum* obtained from germinated winter sporangia of the fungus.[56,69] In the Netherlands, in the Spieckermann test the winter sporangia are used too, but as inoculum. The winter sporangia in the meantime germinate during incubation spores with sprouts of the host[53]; in Germany and Poland, the Glynne–Lemmerzahl method is used for scoring of potato genotypes.[57–61,62a,63,74] The Glynne–Lemmerzahl method seems to be one of the proper methods to characterize potato resistance to *S. endobioticum* because of its highly reliability in comparison to field, pot, or Spieckermann tests. This method was used and modified in Scotland[64] and compared with the Spieckermann test.[56,65] The conclusions indicated that resistance assessment to *S. endobioticum* using the Spieckermann test does not allow for distinguishing resistant cultivars from those that respond with a low level of resistance (slightly susceptible). The application of the Glynne–Lemmerzahl method allows for this distinguishability and this method should therefore be applied in mass tests.[62b]

73.7 CONCLUSIONS AND FUTURE PERSPECTIVES

Currently, *S. endobioticum* is classified as a quarantine organism in most countries of the world that have very strict regulations concerning the movement of plant materials, especially from countries where this pathogen is recorded. Although potato wart disease seems to be in remission in most countries of Europe, in some countries with warmer continental climate new foci appear with this pathogen.[19,66a,66b,67] This may suggest adaptation of *S. endobioticum* to warmer and/ or dryer climate as well as the lack of adequate control in countries where the disease has not been present before. The lack of control is manifested by the widespread cultivation of susceptible varieties to pathotype 1 (D1) of *S. endobioticum* and importation of infected seed tubers of potato. The other problem is bioterrorism; the *S. endobioticum* could be taken into consideration as a biological weapon as a threat to potato. *S. endobioticum* is included in the list of particularly dangerous plant pathogens listed in the ABPA Act and in the HSDP-9 Directive in the United States.[68] On the other hand, there are some varieties of potato which seem to be resistant to all known pathotypes or at least to most of the known pathotypes of *S. endobioticum*. These varieties as well as germplasm of wild species of *Solanum* will be the source of resistance to pathotypes of *S. endobioticum* in breeding programs.

The other problem concerns unification of pathotype identification and resistance assessment. There is an urgent need to harmonize these issues. Comparative studies of two different test protocols were conducted in 2009 and 2010 by a total of five German, Polish, and Dutch laboratories.[76] Currently, an international project Euphresco II (acronym SENDO) has been realized with the following motive: "Development of a new differential set of potato cultivars for the identification of pathotypes of *Synchytrium endobioticum*." This project includes participants from different countries, not only from the European Union.

REFERENCES

1. Malec, K., Potato wart (*Synchytrium endobioticum*) Schilb Perc., *Z. Prac. Inst. Ziem. Bonin.*, 1, 21, 1983.
2. Langerfeld, E., *Synchytrium endobioticum*, A comprehensive account of the potato wart pathogen from literature reports, *Mitt. Biol. Bundesanst. Land- Forstw.*, 219, 1, 1984.
3. Hampson, M.C., History, biology and control of potato wart disease in Canada, *Can. J. Plant Pathol.*, 15, 223, 1993.
4. Stachewicz, H., 100 years of potato wart disease its distribution and current importance, *Nachrichtenbl Pflanzenschutz DDR*, 43, 109, 1989.
5. EPPO, *Pest Quarantine Database*, Version 4.4, EPPO, Paris, France, 2005.
6. EPPO/CAB, *Synchytrium endobioticum*, In: *Quarantine Pests for Europe*, 2nd ed., CAB International, Wallingford, U.K., 1996.
7. Karling, J.S., *Synchytrium*, Academic Press, New York, 1964.
8. Hampson, M., Infection of additional hosts of *Synchytrium endobioticum*, the causal agent of potato wart disease: 2. Tomato, tobacco and species of *Capsicastrum, Datura, Physalis* and *Schizanthus, Can. Plant Dis. Surv.*, 59, 3, 6, 1979.
9. Schilberszky, K., Ein neuer Schorf-parasit der Kartoffelknollen, *Ber Deut. Bot. Ger.*, 14, 36, 1896.
10. Alexopoulos, C.J., Mims, C.W., and Blackwel, M.I., In: Alexopoulos, C.J. (ed.), *Introductory Mycology*, John Wiley & Sons, Inc., New York, 1996.
11. Curtis, K.M., The life history and cytology of *Synchytrium endobioticum*, the cause of wart disease in potato, *Philos Trans. R. Soc. Lond. Ser.*, B210, 409, 1921.
12. Kole, A.P., Resting-spore germination in *Synchytrium endobioticum*, *Neth. J. Plant Pathol.*, 71, 72, 1965.
13a. Lange, L. and Olson, L.W., Germination and parasitation of the resting sporangia of *Synchytrium endobioticum*, *Protoplasma*, 106, 69, 1981.
13b. Lange, L. and Olson, L.W., Development of the zoosporangia of *Synchytrium endobioticum*, the causal agent of potato wart disease, *Protoplasma*, 106, 97, 1981.
14. Bal, A.K., Dey, A.C., and Hampson, M.C., Resting sporangium of *Synchytrium endobioticum*: Its structure and composition of the lipids and fatty acids, *Arch. Microbiol.*, 140, 178, 1984.
15. Hampson, M., Yang, A.F., and Bal, A.K., Ultrastructure of *Synchytrium endobioticum* resting pores and enhancement of germination using snails, *Mycologia*, 86, 733, 1994.
16. Murphy, A.M., Bal, A.K., and Hampson, M.C., Incorporation of uridine diphospho-N-acetyl-D-glucosamine in the resting sporangium wall of *Synchytrium endobioticum*, *Experientia*, 38, 244, 1981.
17. EPPO, Diagnostic protocols for regulated pests, PM 7/28(1), *Synchytrium endobioticum*, *Bull. OEPP/EPPO Bull.*, 34, 213, 2004.
18. Çakir, E. et al., Identification of pathotypes of *Synchytrium endobioticum* found in infested fields in Turkey, *Bull. OEPP/EPPO Bull.*, 39, 175, 2009.
19. Ballvora, A. et al., Multiple alleles for resistance and susceptibility modulate the defense response in the interaction of tetraploid potato (*Solanum tuberosum*) with *Synchytrium endobioticum* pathotypes 1, 2, 6 and 18, *Theor. Appl. Genet.*, 123, 1281, 2011.
20. Malec, K., Viability of resting sporangia of the fungus *Synchytrium endobioticum* (Schilb.) Perc, in soil under natural conditions, *Biul. Inst. Ziem.*, 23, 87, 1979.
21. Hartman, R.E. and Akeley, R.W., Potato wart in America, *Am. Potato J.*, 32, 317, 1944.
22. Pratt, M.A., A wet-sieving and flotation technique for the detection of resting sporangia of Synchytrium endobioticum in soil, *Ann. Appl. Biol.*, 82, 21, 1976.
23. Langerfeld, E. and Stachewicz, H., Pathotypes of potato wart in West and East Germany, *Gesunde Pflanzen*, 45, 9, 1993.
24. McDonnell, M.B. and Kavanagh, J.A., Studies on *Synchytrium endobioticum* (Schilb.) Perc, In Ireland, *J. Life Sci. R. Dublin Soc.*, 1, 177, 1980.
25. Bojnansky, V., The effect of soil type on the development and severity of potato wart disease, *E. Potato J.*, 11, 100, 1986.
26. EPPO, Report of the 2nd meeting of the EPPO Panel on potato wart disease, EPPO Document No. 5205, 1982.
27. Bojnansky, V., Potato wart pathotypes in Europe from an ecological point of view, *Bull. OEPP/EPPO Bull.*, 14, 141, 1984.
28. Hampson, M.C. and Coombes, J.W., Pathogenesis of *Synchytrium endobioticum* VII. Earthworms as vectors of wart disease of potato, *Plant Soil*, 116, 147, 1989.
29. Stakman, E.C. and Harrar, J.G., *Principles of Plant Pathology*, Ronald Press, New York, 1957.
30. Malec, K., Changes in the virulence of fungus *Synchytrium endobioticum* Schilb. Perc. depending on the degree of susceptibility of potato cultivars and to the date of infection, *Hodowla Roślin, Aklimatyzacja i Nasiennictwo*, 7, 25, 1963.
31. Malec, K., Investigations on the occurrence of new, highly virulent biotypes of *Synchytrium endobioticum* (Schilb.) Perc., *Biul. Inst. Ziem.*, 14, 131, 1974.
32. Braun, H., Biologische Spezialisierung bei *Synchytrium endobioticum* (Vorläufige Mitteilung), *Z. Pflkrankh., U. Pflsch.*, 5, H., 11, 481, 1942.
33. Baayen, R.P. et al., History of potato wart disease in Europe—A proposal for harmonisation in defining pathotypes, *Eur. J. Plant Pathol.*, 116, 21, 2006.
34. Glynne, M.D., Infection experiments with wart disease of potato *Synchytrium endobioticum* (Schilb.), *Perc, Ann. Appl. Biol.*, 12, 34, 1925.
35. EPPO, EPPO standards PM 3/59 *Synchytrium endobioticum*: Soil tests and descheduling of previously infested plots, *Bull. OEPP/EPPO Bull.*, 29, 225, 1999.
36. Mygind, H., Examination of soil samples for potato wart sporangia, *Acta Agric. Scand.*, 11, 114, 1961.
37. Zelya, A.G. and Melnik, P.E., Detection methods for *Synchytrium endobioticum*, *Bull. OEPP/EPPO Bull.*, 28, 543, 1998.
38. Potocek, J., Quantitative determination of resting zoosporangia of the potato wart pathogen in soil samples, *Ochrana Rostlin*, 13, 251, 1977.
39. Laidlaw, W.M.R., A method for the detection of the resting sporangia of potato wart disease (*Synchytrium endobioticum*) in the soil of old outbreak sites, *Potato Res.*, 28, 223, 1985.
40. van Leeuwen, G.C.M. et al., Direct examination of soil for sporangia of *Synchytrium endobioticum* using chloroform, calcium chloride and zinc sulphate as extraction reagents, *Bull. OEPP/EPPO Bull.*, 35, 25, 2005.
41. Wander, J.G.N. et al., A novel technique using the Hendrickx centrifuge for extracting winter sporangia of *Synchytrium endobioticum* from soil, *Eur. J. Plant Pathol.*, 119, 165, 2007.
42. Niepold, F. and Stachewicz, H., PCR-detection of *Synchytrium endobioticum*, *Z. Pflanzenkr. Pflanzenschutz*, 111, 313, 2004.
43. van den Boogert, P.H.J.F. et al., Development of PCR-based detection methods for the quarantine phytopathogen *Synchytrium endobioticum*, causal agent of potato wart disease, *Eur. J. Plant Pathol.*, 113, 47, 2005.

44. van Gent-Pelzer, M.P.E., Krijger, M., and Bonants, P.J.M., Improved real-time PCR assay for detection of the quarantine potato pathogen *Synchytrium endobioticum* in zonal centrifuge extracts from soil and in plants, *Eur. J. Plant Pathol.*, 126, 129, 2010.
45. Hampson, M.C., Screening systemic fungicides for potato wart disease, *Can. Plant Dis. Surv.*, 57, 75, 1977.
46. Steinmöller, S. et al., Effects of sanitation processes on survival of *Synchytrium endobioticum* and *Globodera rostochiensis*, *Eur. J. Plant Pathol.*, 133, 753, 2012.
47. EU, Council Directive 2000/29 of 8 May 2000 on protective measures against the introduction into the community of organisms harmful to plants or plant products and against their spread within the community, *Off. J. Eur. Comm.*, L169, 112, 2000.
48. Grabowski, L., Potato wart/*Synchytrium endobioticum* Perc. in Poland. *Choroby i szkodniki roślin*, 2, 1, 1925.
49. Blattny, C., Preliminary note on the races of *Synchytrium endobioticum*, *Pres. Ann. Akad. Techechoslov, Agr.*, 17, 40, 1942.
50. Flath, K., Potato wart disease, *S. endobioticum* in Germany: 1. Official variety testing—Experience and results 2. Identification and occurrence of pathotypes, *EPPO Workshop Associated to the Panel on Phytosanitary Measures for Potato, Exchange of Experiences on Testing for Synchytrium endobioticum*, February 2, 2012, Moscow, Russia, 2012.
51. Malec, K., Research on virulence of potato wart disease caused by *Synchytrium endobioticum* Schilb., Percival, *Biul. IHAR*, 6, 75, 1959.
52. EU, Council Directive of December 8, 1969 on control of potato wart disease, *Off. J. Eur. Comm.*, L323, 561, 1969.
53. Spieckermann, A. and Kothoff, P., Testing potatoes for wart resistance, *DLP*, 51, 114, 1924.
54. Lemmerzahl, J., A new simplified infection procedure for testing potato cultivars for wart resistance, *Züchter*, 2, 288, 1930.
55. Noble, M. and Glynne, M.D., Wart disease of potatoes, *FAO Plant Protect. Bull.*, 18, 125, 1970.
56. Browning, I.A. and Darling, M., Development of potato wart susceptibility testing in Scotland, *Potato Res.*, 38, 363, 1995.
57. Hille, A., Assessment of potato cultivars for their behaviour towards *Synchytrium endobioticum*, the cause of potato wart disease, *Nachrichtenbl. Deut. Pflanzenschutzd*, 17, 137, 1965.
58. Stachewicz, H., Identification of pathotypes of the potato wart pathogen *Synchytrium endobioticum* by means of test cultivars. *Archiv. f Phyt. Pflan*, 16, 1, 1980.
59. Langerfeld, E. and Stachewicz, H., Assessment of varietal reactions to potato wart (*Synchytrium endobioticum*) in Germany, *Bull. OEPP/EPPO Bull.*, 24, 793, 1994.
60. Malec, K., Investigation method of potato breeding materials towards cancer resistance. *Biuletyn Instytutu Ziemniaka*, 5–6, 143, 1965.
61. Malec, K., Modifications of the method of testing the potato breeding stocks for the resistance to potato wart *Synchytrium endobioticum* (Schilb) Perc., *Biul. Inst. Ziem.*, 10, 5, 1972.
62a. Przetakiewicz, J., Assessment of the resistance of potato cultivars to *Synchytrium endobioticum* (Schilb). Perc. in Poland, *Bull. OEPP/EPPO Bull.*, 38, 211, 2008.
62b. Przetakiewicz, J., Comparison of two methods for assessing the rate of infection of potato sprouts by pathotype 1(D1) of *Synchytrium endobioticum* (Schilb.) Perc., *Biuletyn IHAR*, 248, 67, 2008.
63. Przetakiewicz, J., Suggestions for changes in the Polish scale used to evaluate the resistance of potato cultivars to potato wart disease according to the EPPO Diagnostic Protocol PM 7/28, *Biuletyn IHAR*, 254, 169, 2009.
64. Sharma, R. and Cammack, R.H., A modification of the Glynne-Lemmerzahl method for tastin resistance of potato varieties to wart disease, *Synchytrium endobioticum* (Shilb.) Perc., *Potato Res.*, 19, 165, 1976.
65. Przetakiewicz, J. and Kopera, K., The comparison of usefulness the Glynne-Lemmerzahl and the Spieckermann methods to assess resistance of potato (*Solanum tuberosum* L.) to *Synchytrium endobioticum* (Schilb.) Perc. pathotype 1(D1) in mass tests, *Biuletyn IHAR*, 243, 235, 2007.
66a. Çakir, E., First report of potato wart disease in Turkey, *Plant Pathol.*, 54, 584, 2005.
66b. Çakir, E. et al., Potato wart disease survey in Turkey, *Bull. OEPP/EPPO Bull.*, 35, 489, 2005.
67. Dimitrova, L. et al., Occurrence of potato wart disease (*Synchytrium endobioticum*) in Bulgaria: Identification of pathotype(s) present, *Bull. OEPP/EPPO Bull.*, 41, 195, 2012.
68. USDA-APHIS, Agricultural bioterrorism protection act 2002 listing of biological agents and requirements and procedures for notification of possession, *Fed. Reg.*, 67, 52283, 2002.
69. Browning, I.A., A comparison of laboratory and field reaction of a range of potato cultivars to infection with *Synchytrium endobioticum* (Schilb.) Perc., *Potato Res.*, 38, 281, 1995.
70. Brugmans, B. et al., Exploitation of a marker dense linkage map of potato for positional cloning of a wart disease resistance gene, *Theor. Appl. Genet.*, 112, 269, 2006.
71. Grabowski, L. and Leszczenko, P., Prevalence of potato wart disease and the progress of study on resistance of potato varieties against *Synchytrium endobioticum* (Schilb.) Perc., *Pr. Wydz. Chorób Rośl. PINGW, Bydgoszcz*, 10, 3, 1931.
72. Langerfeld, E., Stachewicz, H., and Rintelen, J., Pathotypes of *Synchytrium endobioticum* in Germany, *Bull. OEPP/EPPO Bull.*, 24, 799, 1994.
73. Hampson, M.C., Research on potato wart disease in the U.S.S.R.—A literature review. *Can. Plant Dis. Surv.*, 58, 7, 1979.
74. Malec, K., Methodology of testing potato breeding materials for wart resistance used in the laboratory of testing for resistance to quarantine diseases and pests, Institute for Potato Research, *Biul. Inst. Ziem.*, 25, 125, 1980.
75. Hartman, R.E., Potato wart eradication program in Pennsylvania, *Am. Potato J.*, 32, 317, 1956.
76. Flath, K. et al., Interlaboratory tests of the Glynne-Lemmerzahl method, *EPPO Workshop Associated to the Panel on Phytosanitary Measures for Potato, Exchange of Experiences on Testing for Synchytrium endobioticum*, February 2, 2012, Moscow, Russia 2012.

Paul E. Roffey and Michelle E. Gahan

CONTENTS

74.1 INTRODUCTION

Xanthomonas oryzae is the most important bacterial pathogen of rice. Two closely related pathovars, namely, *X. oryzae* pv. *oryzae* (Xoo) and *X. oryzae* pv. *oryzicola* (Xoc), cause the rice diseases bacterial leaf blight (BLB) and bacterial leaf streak (BLS), respectively. A number of plant species are vulnerable to infection by *X. oryzae*; however, the host species in which it has the greatest economic impact is rice.[1]

Rice is a grass that belongs to the genus *Oryza*. There are over 20 species of rice but only two commercially cultivated species, *Oryza sativa* (Asian rice) and *Oryza glaberrima* (African rice).[2] Of these, *O. glaberrima* accounts for less than 1% of the world's rice production.[3] Asian rice (*O. sativa*) is one of the world's oldest and most important food crops being the primary food source for more than half of the world's population.[4] *O. sativa* was domesticated from its wild ancestor *Oryza rufipogon* approximately 10,000 years ago and has been subjected to artificial selection to improve crop characteristics ever since.[5] However, it is in the last 50 years, during the so-called green revolution, that the most significant advances in crop improvement have occurred through intensive research to develop cultivars with enhanced yield, resistance to disease and pests, increased tolerance to harsh environments and low nutrient conditions, reduced water requirements, increased nutritional content, and shorter growing times.[6] Despite world production increasing over 300% from 1961 to 2010 (from 215 to 696 million tonnes),[7] it will still need to increase by more than 60% on top of current production to meet the needs of the world population in

2025.[6] Hence, the management of rice disease is essential if world demand for rice is to be maintained.

BLB is one of the most serious diseases of rice and the worst of the two diseases caused by *X. oryzae*, causing losses of yield generally ranging from 20% to 50% depending on weather, location, and rice variety but has been reported to be as high as 74%.[8] The impact of BLS typically ranges from 0% to 20% loss of yield depending on weather, location, and rice variety but has been as high as 32%.[9]

BLB was first reported by farmers in southern Japan in 1884.[1] It was not until 1911 that the disease was attributed to a bacterium, then called *Bacillus oryzae*, which was subsequently classified as *X. oryzae* in 1922.[10,11] By this time, BLB was common throughout southern Japan. Its incidence in northern Japan and other Asian countries remained sporadic until the 1960s and 1970s when BLB devastated crops in nearly every Asian country. This outbreak coincided with the adoption of new, high-yielding rice varieties that had inadequate disease resistance.[10,12] In the years following, cooperative international efforts developed resistant strains of rice to replace the susceptible varieties, and by the mid-1980s, incidence had declined to current levels. In spite of the use of resistant rice varieties, losses of 15%–20% are still common throughout the Asian countries.[12] BLB first emerged in Mali in Africa in 1979 and quickly spread throughout the rice-growing regions of northwest Africa. Its spread also correlated with intensification of rice agriculture and an expansion of Asian rice varieties.[13] Interestingly, the African Xoo strains are genetically distinct from the Asian strains suggesting these arose from a different evolutionary history and most probably from native *X. oryzae* strains.[14] Currently, BLB is prevalent in tropical and temperate areas of Asia and Africa and has been reported in Australia, Latin America, and the Caribbean.[9]

BLS was first discovered in the Philippines in 1918. At that time, it was incorrectly characterized as bacterial blight. It was not until 1957 that the causal agent was distinguished as being different from that of bacterial blight and given the name *Xanthomonas oryzicola*.[9] BLS is largely restricted to tropical and subtropical regions of Asia, northwest Africa, and northern Australia and is in epidemic proportions in China.[1,9]

The most effective and economical strategy for controlling BLB and BLS is through the ongoing development of new resistant varieties of rice, with the aim of keeping ahead of bacterial adaptation to resistance.[9] Interestingly, climate change may well present the next great challenge to rice production as high temperature and high humidity are factors that are favorable to BLB disease development while being unfavorable to the expression of rice genes that confer resistance to disease. Most of the current rice resistance genes lose efficacy at higher temperatures thereby increasing the vulnerability of resistant crops to disease,[13] a factor that will need to be considered in the development of future resistant strains of rice.

74.2   CLASSIFICATION AND MORPHOLOGY

Since *X. oryzae* pv. *oryzae* and *X. oryzae* pv. *oryzicola* were first described, both have passed through several changes of classification and names. First isolated in 1911 as the causative agent of BLB, Xoo was originally classified as *B. oryzae* and then subsequently renamed *Pseudomonas oryzae* and then *X. oryzae* in 1922.[10,11] In 1978, it was reclassified as *Xanthomonas campestris* pv. *oryzae*.[15] Other synonyms include *Xanthomonas itoana*, *Xanthomonas kresek*, and *Xanthomonas translucens* f.sp. *oryzae*.[1,9] Xoc, first documented in 1918 as the causative agent of BLS, was originally misidentified as Xoo. In 1957, it was distinguished from Xoo and given the name *X. oryzicola*.[16] In 1964, it was reclassified as *X. translucens* f.sp. *oryzae*.[17] and then reclassified again to *X. campestris* pv. *oryzicola* in 1978.[15] Another synonym that has been used is *X. translucens* f.sp. *oryzicola*.[1] In 1990, both pathovars were reclassified to their current taxonomic status as *X. oryzae* pv. *oryzae* and *X. oryzae* pv. *oryzicola*.[11]

X. oryzae is a Gram-negative straight rod-shaped bacterium, approximately 0.8–2.0 μm long and 0.4–0.6 μm wide. All *Xanthomonas* species are motile by a single polar flagellum and are obligate aerobes, catalase positive, and oxidase weak or negative. *X. oryzae* bacteria produce H_2S, are lysed in 3% KOH, are lipase positive, and are negative for urease, indole, cytochrome oxidase, 2-ketogluconate formation, egg yolk hydrolysis, and nitrate reduction.[11] Bacterial growth occurs on D-fructose, D-galactose, D-xylose, D-glucose, sucrose, cellobiose, trehalose, and sodium-containing media such as sodium fumarate, sodium L-lactate, sodium oxaloacetate, and sodium succinate. There is no growth on D-ribose,

L-rhamnose, L-arabinose, lactose, raffinose, inulin, ethanol, *n*-propanol, methanol, ethanediol, *meso*-inositol, sorbitol, dulcitol, adenine, cytosine, guanine, thymine, sodium–potassium tartrate, sodium glyoxylate, 2-ketogluconic acid, tannic acid, *p*-hydroxybenzoic acid, hydroquinone, resorcinol, amygdalin, phloroglucinol, arbutin, or esculin.[11] Acid is formed from D-glucose, D-fructose, D-galactose, trehalose, and cellobiose, but not from D-ribose, L-rhamnose, salicin, sorbitol, maltose, lactose, *meso*-inositol, adonitol, dulcitol, or inulin. There is no acidification of litmus milk. Bacteria do not grow in 3% NaCl or at 4°C or 35°C but there is weak growth at 32°C and 10°C. Esculin, TWEEN 40, and TWEEN 80 are hydrolyzed. Growth is inhibited in media containing 0.001% (wt/vol) tetracycline hydrochloride, 0.1% triphenyl–tetrazolium chloride, 0.005% chloramphenicol, 0.001% novobiocin, or 0.001% doxycycline. However, bacteria are able to grow in the presence of 0.001% trimethoprim or 0.005% metronidazole.[11]

The key biochemical tests that distinguish the pathovars are β-glucosidase, tyrosinase, phenylalanine deaminase, sensitivity to 0.001% cupric nitrate, acetoin production, growth on L-alanine as the sole carbon source, and growth on 0.2% vitamin-free casamino acids (Table 74.1).[1,18,19]

Although colony morphology will vary depending on the choice of media, in general, colonies are yellow in color due to the presence of xanthomonadin, a membrane-bound pigment. On nutrient agar, colonies of *Xoo* are slow growing, opaque, circular, smooth, convex, and, depending on the age of the colonies, pale yellow to straw in color. *Xoc* colonies, although faster growing than *Xoo*, still are slow growing, circular, smooth, entire, convex, and mucoid. They appear whitish at first then become straw to pale yellow in color.[1] It should be noted that as both pathovars are slow growing, they can be overgrown by faster-growing contaminants confounding identification by colony morphology.

Pathovars *oryzae* and *oryzicola* are clearly distinguished by host symptoms. Numerous phenotypic and genotypic studies support this distinction; these include nutritional, metabolic and biochemical studies,[19,20] serotyping,[21,22] fatty acid analysis,[23,24] protein profiling,[19,25] restriction digestion patterns,[26] phylogenetic gene sequence comparisons,[27,28] plasmid profiling,[29] repetitive sequence-based polymerase chain reaction (PCR),[30,31] other PCR-based techniques,[32–34] and whole genome comparisons.[35,36] The vast majority of the Asian isolates reside within the two recognized pathovars, Xoo or Xoc, whereas the African and United States isolates appear to reside within separate groups that may be representative of distinct pathovars.[14,27]

74.3   BIOLOGY AND EPIDEMIOLOGY

Xoo typically infects the rice leaf through hydathodes, as well as through broken trichomes (i.e., surface hairs) or wounds in the roots or leaves. Hydathodes are natural openings composed of a series of water pores situated at the leaf tips and margins. Water is excreted through the hydathodes early

TABLE 74.1

Selected Biochemical Tests for Identification and Differentiation of *X. oryzae* Pathovars

Test	*X. oryzae* pv. *oryzae*	*X. oryzae* pv. *oryzicola*	*X. campestris* (Type Species)
Esculin hydrolysis	+	+	+
Catalase	+	+	+
Lipase	+	+	+
Oxidase	−	−	−
Lecithinase	−	−	+
Urease	−	−	−
Growth on sodium propionate as a carbon source	−	−	+
Starch hydrolysis	−	+	
β-Glucosidase	−	+	
Phenylalanine deaminase	−	+	
Tyrosinase	−	+	
Sensitivity to 0.001% cupric acid	+	−	
Acetoin production	−	+	
Growth on 0.2% vitamin-free casamino acids	−	+	
Growth on L-alanine as sole carbon source	−	+	

Note: +, positive; −, negative.

in the morning when water is abundant and transpiration is minimal. As the stomata open and transpiration accelerates, these water droplets may be drawn back into the leaf thereby providing a vehicle through which bacteria present on the surface of the leaf can be carried into the leaf.[9,37,38] Wounds are the other major mode of entry of bacterial cells into the leaf. Indeed, in endemic areas, severe outbreaks of BLB commonly follow wind storms, heavy rain, and/or hail.[9] Once inside the leaf, cells multiply in the intercellular spaces of the parenchymal tissue directly underlying the hydathode, then invade the vascular bundle thereby spreading throughout the plant via the xylem. In the xylem vessels, the cells have a preference for the lumen and do not multiply within the xylem parenchyma, phloem, mestome sheath, and vascular bundle sheath.[39] Within a few days, the bacterial cells and extracellular polysaccharide (EPS) fills the xylem, which then oozes out from the hydathodes forming beads of exudate on the surface of the leaf.[38,39] The exudate is a characteristic sign of the disease, as well as a common source of secondary infection.[9] Lesions typically extend from the point of origin down the leaf veins, eventually merging with others to form larger lesions that extend down from the leaf tip (Figure 74.1).

Xoc typically enters leaves through stomata or through wounds. Stomata are pores on the surface of the leaf that regulate the exchange of CO_2, O_2, and water vapor into and out of the leaf. Once inside the leaf, cells multiply in the substomatal space and then colonize the intercellular spaces of the parenchymal tissue directly underlying the stomata. Xylem veins act as a barrier as infected areas expand causing neighboring lesions to coalesce resulting in larger lesions

in the form of streaks that run lengthwise along the leaf (Figure 74.2). The infection does not spread systemically and so remains a disease of the leaf. At the late stages of infection, the bacteria can enter the xylem vessels where there is reduced multiplication compared to *Xoo*. Lesions discharge a thin yellow bacterial ooze via the stomata, which forms a source for the spread of disease to neighboring plants.[9,10] Like Xoo, wounds are the other major mode of entry of bacterial cells into the leaf and increases in incidence are seen following wind storms, heavy rain, and/or hail.[9]

FIGURE 74.1 Bacterial leaf blight at mature crop stage. (Courtesy of Dr. Seong-Sook Han, National Institute of Crop Science, South Korea.)

FIGURE 74.2 Bacterial leaf streak at mature crop stage. (Courtesy of Dr. Casiana Vera Cruz, International Rice Research Institute, Philippines.)

With the exception of movement by windstorms, both pathovars only move short distances in infected crops.[9] The main vehicle for spread within and between fields is contaminated water where bacterium-laden ooze from infected plants is the primary source of inoculum. Long-distance dispersal requires human input, such as distribution of contaminated seeds and infected seedlings and contaminated soil, stubble, and rice straw.[1] Overlapping crops of different ages and monocrop successions are the major factors in the spread and propagation of disease as these provide a continuum of disease.[40] *X. oryzae* is not well suited to long-term survival outside of a living tissue. Xoo is known to survive on rice hulls for several months after harvest, although seed transmission to subsequent crops is yet to be clearly demonstrated.[10] Indeed, seedling infection occurs rarely in nurseries suggesting incidence through infected seed is unlikely as a major source of transmission.[41] Survival for up to 5 months on dried stubble and straw has been demonstrated, but when left moist on the surface of soil or in the high-humidity conditions typical of the tropics, the organism dies within 2 months.[40] Similarly, survival in paddy water and soil rarely extends beyond 2 months[42] and is not considered an important source of inoculum.[9] Studies have shown that susceptible grass weeds and other *Sativa* species can serve as hosts for the survival of *X. oryzae* between crop cycles thereby providing inoculum for the carry over to successive crops.[40] In the absence of weeds, the fallowing of fields remains the most reliable mechanism for breaking the cycle of infection in endemic areas.

74.4 CLINICAL FEATURES AND PATHOGENESIS

74.4.1 CLINICAL FEATURES

Due to differences in the mode of infection, the clinical features of Xoo and Xoc are distinguishable. Xoo infection causes BLB and Kresek disease. Symptoms in common include exudates, pale yellow leaves and stunted growth. BLB is a vascular disease characterized by wavy elongated lesions. These lesions initially present as small water-soaked lesions near the tip and margins of fully developed leaves and then rapidly enlarge in length and width along the veins, forming a yellow lesion with a wavy margin along the lead edges (see Figure 74.1). The whole leaf may become affected, and lesions can occur on one or both sides of the leaf. The diseased area then becomes whitish or grayish, following which the leaf dies. Symptoms are usually observed at the tillering stage with disease incidence increasing with plant growth and peaking at the flowering stage.[1,10] Kresek disease, a more severe form of the disease, arises due to systemic spread of the bacteria and is common in young plants and during the tillering stage brought about by damage to roots or leaves during transplantation at the seedling stage. Kresek is characterized by leaf wilting, following which leaves roll up, turn grey/green, and shrivel. Entire plants die; however, if they do survive, plants look yellowish and are stunted, and in these plants, the bacteria are found in the internodes and crowns of infected stems, but not in the leaves themselves.[1,10]

Xoc is the causative agent of BLS, which can occur at any growth stage of the host. BLS is characterized by narrow, interveinal water-soaked translucent lesions, which appear as streaks of various lengths (see Figure 74.2). These lesions enlarge, darken (yellowish orange to brown depending on the cultivar), may coalesce to form patches, and later become necrotic. Tiny amber/yellow droplets of bacterial exudate are often present on the lesions.[1,10] Both BLB and BLS can occur in the same leaf.

74.4.2 PATHOGENESIS

X. oryzae produces a wide range of virulence factors, which enable them to adhere to the plant surface, invade the intracellular spaces, multiply, acquire nutrients, and evade or counteract the plant defenses. Groups of genes associated with the pathogenesis of *X. oryzae* pathovars include effector or avirulence genes (avr), hypersensitive response and pathogenicity (hrp) genes, and genes associated with production of EPSs and enzymes or cell wall degradation. Genes are also present for fimbriae and outer membrane proteins associated with adhesion and for flagellar biosynthesis and chemotaxis that enable motility.[43] *X. oryzae* pathovars have virulence genes involved in lipopolysaccharide (LPS) production, a component of the bacterial cell wall that blocks antimicrobials by restricting membrane permeability and protects bacteria against environmental stresses, and the *rpf* genes which is a two-component regulatory system that regulates virulence factors including biofilm formation.[44–46] Of the two *X. oryzae* pathovars, the majority of research into elucidating the pathogenic mechanisms has focused on Xoo as this is the more pathogenic of the two.

For pathogenicity of both pathovars, a functioning bacterial type III secretion system, encoded by hrp genes located in a pathogenicity island, is essential. Type III secretion systems are highly conserved among Gram-negative plant and animal pathogens where they function like a syringe to deliver proteins, known as effectors, into the host cells. Many of the type III effectors function as virulence factors, having either enzymatic or transcription-like activities that can regulate the expression of host genes, modulate host defense signaling pathways, interfere with host cell functions, and/or modify or degrade host proteins. This enhances the pathogenicity of the microorganism and induces disease in the host. The importance of the type III effectors in the pathogenicity of *X. oryzae* is indicated by studies showing that bacteria with mutations in the type III secretion system have impaired growth *in planta* and fail to cause disease in susceptible rice cultivars.[47,48] Type III effectors that contribute to virulence include members of the AvrBs3 gene family including PthA, PthXo1, PthXo7, Avrb6, and AvrXa7. Members of the AvrBs3 family are found in many *Xanthomonas* species where they play a role in manipulating plant transcription. AvrBs3 effectors have eukaryotic nuclear localization signal motifs that mediate nuclear import and a highly conserved repeated 34-amino acid residue consensus sequence that mediates protein dimerization and DNA binding. These effectors also contain an acidic activation domain in the C terminus, which mediates the activation of plant gene expression (for a review of type III effectors, see [49,50]). PthXo7 has been found to contribute to lesion length and bacterial growth,[51] AvrXa7 has been found to contribute to bacterial aggressiveness and lesion development,[47] and PthXo1 induces expression of the rice susceptibility gene *Os8N3*.[52] For both Xoo and Xoc, manipulation of host transcription and interference with host defenses through type III effectors are an important mechanism in pathogenesis.

X. oryzae pathovars also contain a general secretory pathway known as a type II secretion system that secretes extracellular enzymes, including cell-wall-degrading enzymes and toxins associated with virulence. These include cellulases, proteases, polygalacturonases, lipases, pectin-degrading enzymes, and xylanases. The latter is particularly important as *Xoo* multiplies and spreads in the xylem vessels where there is an abundance of xylan. Type II secretion mutants exhibit a loss in virulence and reduced pathogenicity due to a blockage in the secretion of these enzymes from the bacteria.[53] Another group of enzymes associated with virulence are the antioxidant scavenging enzymes (e.g., catalase and ascorbate peroxidase), which have been shown to counteract the production of active oxygen species such as O_2 and H_2O_2 produced by plants as a defense mechanism.[54] Other enzymes that play a role in virulence are phytase, encoded by the *phyA*

gene and involved in the degradation of phytic acid, a plant phosphate storage molecule,[55] and shikimate dehydrogenase, encoded by *aroE*, an enzyme involved in the aromatic amino acid biosynthesis pathway.[56] Interestingly, using DNA microarray analysis, mutations in the *aroK* gene, another member of the aromatic amino acid biosynthesis pathway, has been shown to affect pigment production and virulence.[57]

Both *X. oryzae* pathovars produce copious amounts of EPSs, otherwise known as xanthan gum, which is responsible for the mucoid phenotype of the colonies. EPS is a polymer of repeating pentamer units composed of two mannose subunits, two glucose subunits, and one glucuronic subunit. EPS plays a critical role in facilitating adhesion of the bacteria to host cell surfaces, protecting bacteria from environmental stresses such as dehydration and toxic compounds, and contributes to plant wilting by blocking water flow in xylem vessels.[58] Production of EPS is controlled by the *gum* operon, which is highly conserved among *Xanthomonas* species, with mutations in the *gum* operon having been shown to cause a loss of EPS production and virulence.[59,60]

Quorum sensing allows bacteria to sense their population density and synchronize individual activity, enabling them to act cooperatively. This provides a competitive advantage over their hosts. Xoo employs two different types of quorum-sensing factors, diffusible signal molecules (DSF) and activator of XA21-mediated immunity protein (Ax21). Ax21-mediated quorum sensing has been shown to mediate biofilm formation, motility, and virulence,[61] and DSF regulates virulence factors including production of EPS and xylanase.[62]

Infection of rice by Xoo has been shown to be associated with activation of the *Xa13* host resistance gene and copper transporter proteins (COPT1 and COPT5), which modulate copper redistribution in rice, specifically the removal of copper from the xylem.[63] Due to its antimicrobial properties, copper is an important element in a number of agricultural pesticides, and so the ability to remove copper from the xylem vessels can promote bacterial multiplication and disease spread. Another player in the pathogenicity of Xoo is the PhoP–PhoQ two-component regulatory system, which is found in other Gram-negative bacteria such as *Salmonella* and regulates a wide variety of biological processes in response to environmental fluctuations. This plays a role in promoting bacterial survival and multiplication by enabling Xoo to tolerate acidic conditions, provides resistance to antimicrobials, and controls regulation of the type III secretion system.[64]

74.5 IDENTIFICATION AND DIAGNOSIS

Identification can be undertaken based on the clinical features with the yellow exudates, and thinner translucent streaks, distinguishing BLS from BLB in which the streaks are opaque. However, in the later stages of infection, where both are characterized by withering and leaf death, if a symptomatic infection is suspected, or if screening of young plants/seeds is required, the bacteria must first be isolated from plant or seed material. Following isolation, the two pathovars can be identified and distinguished using a range of methods including morphological examination and biochemical tests (see Section 74.2) as well as serological, molecular, protein, and pathogenicity tests.

74.5.1 ISOLATION

The direct plating method and liquid assay method are commonly employed to isolate bacteria from infected plant material. In the direct plating method, sections of leaf tissues or seeds, preferably symptomatic, are surface-sterilized with 2% sodium hypochlorite, washed with distilled water, dried by blotting, and then plated directly onto media and incubated at 25°C–28°C. Suitable media include 1% dextrose nutrient agar, Wakimoto agar, peptone sucrose agar, growth factor agar, nutrient broth yeast extract agar medium, and nutrient agar.[1,18] In the liquid assay method, the plant material and/or seed is surfaced-sterilized and washed, as for the direct planting method, and then macerated in distilled water, and the resulting suspension is plated out.

74.5.2 SEROLOGICAL IDENTIFICATION

Serological identification provides a useful method of identification, particularly in mixed cultures. Monoclonal antibodies have been produced for Xoo and Xoc.[21] These pathovar-specific antibodies can be utilized to identify and distinguish the pathovars from each other and also from nonpathogenic xanthomonads in enzyme-linked immunosorbent assays (ELISA), immunofluorescence assays, or immunoelectron microscopy. The use of monoclonal antibodies, in combination with growth on semiselective media, has been shown to improve identification.[65] A commercial ELISA kit (Agdia), which can be used on plant tissue and bacterial cultures, is available to detect *Xanthomonas* species including Xoo.

74.5.3 MOLECULAR IDENTIFICATION

DNA–DNA hybridization studies show an 80% homology between the two pathovars[11] and a guanine plus cytosine (G + C) base % composition of 64.6% for Xoo and 65% for Xoc.[19] Restriction fragment length polymorphism (RFLP) analysis of the genetic relationship between the two pathovars, using a 2.4 kb *EcoR I-Hind III* fragment isolated from Xoo containing a highly repetitive sequence as a probe, has been used for pathovar identification.[66] This study found that the proportion of mismatch fragments between Xoo and Xoc was 0.84–0.93, whereas between Xoo strains, this was 0.08–0.36, and between Xoc strains, it was 0.08–0.69.

Sequencing of bacterial genomes followed by bioinformatic analysis has enabled the identification of unique genomic regions, which can be targeted for identification. PCR has been widely used as this provides a quick, sensitive method for identification, particularly when real-time PCR

is employed, which negates the need for prior culturing and can detect bacteria in nonsymptomatic plant material. When designing primers for identification, it is essential that these are capable of distinguishing *X. oryzae* pathovars of diverse geographical origins from other plant bacterial pathogens as well as differentiating Xoo from Xoc. Assays that detect *X. oryzae* pathovars have been developed targeting repeated elements,[14,30] insertion sequences,[67] and the 16S–23S ribosomal intergenic spacer region.[68,69] Pathovar-specific PCRs targeting a putative membrane protein[70] and membrane fusion protein[71] for Xoc and an *rhs* family gene[32] and the siderophore receptor gene[72] for Xoo have also been developed, which have been proposed to be more specific and differentiate pathovars. A multiplex PCR assay, which uses four sets of primers to produce amplicons of difference sizes, has also been developed, which is specific for *X. oryzae* and distinguishes Xoo from Xoc.[33]

The genomes of a number of the *X. oryzae* strains have already been sequenced. With the introduction of new sequencing platforms and improvements in methodology, not only is the cost of sequencing falling, turnaround time is continually improving thereby increasing the feasibility of identification by sequencing. However, knowledge in bioinformatics will be required to interpret the sequencing data, which may limit the widespread adaptation of this identification approach.

74.5.4 OTHER METHODS OF IDENTIFICATION

Sodium dodecyl sulfate polyacrylamide gel electrophoresis (SDS-PAGE) protein patterns have been used to differentiate Xoo and Xoc with the pathovars displaying distinct protein patterns.[19,73] Fatty acid profiling has also been used for identification, and although very similar, the fatty acid profiles of Xoo and Xoc do differ in the relative amounts of 12:0 iso3OH, 12:0 3OH, and 15:0 anteiso fatty acids.[74]

74.5.5 PATHOGENICITY TESTS

In addition to identification of the isolates, pathogenicity testing is undertaken on susceptible rice cultivars. For Xoo, 30–45-day-old International Rice (IR) strains IR24 or IR8 are recommended and for Xoc IR24 or IR50. Leaf clipping and spray inoculation methods are available and for both methods, following inoculation, plants are maintained under greenhouse conditions commonly at 28°C–32°C with high humidity with a 12 h light/dark cycle.[1] The leaf clipping method, originally developed by Kauffman et al.,[75] involves cutting 2–3 cm of the tips of 30–40 leaves of rice plants while immersed in the bacterial suspension, or alternatively the cut leaves can be sprayed with the bacterial culture. Plants are observed for symptoms, and lesion lengths measured, starting 48–72 h post inoculation and up to 15 days.[1] The spray inoculation method, developed by Cottyn et al.,[76] involves evenly spraying whole plants with the bacterial suspension and examining for lesions 10 days after inoculation. Development of BLB or BLS lesions confirms pathogenicity.

74.6 TREATMENT AND PREVENTION

Due to the disease severity, greater emphasis is placed on the treatment and prevention of Xoo infection than Xoc. Although there is little discussion of treatments and/or prevention in the literature for Xoc, in general, measures taken against Xoo should also be effective against Xoc. Control methods include seed testing, crop management, prohibiting import of rice seeds from countries with endemic infections, chemical and biological control, disease forecasting, and the use of resistant cultivars of rice. Crop management practices useful for control include burning fields prior to planting to remove the previous seasons' material, minimizing the cultivation of overlapping crops of different ages in adjacent fields, fallowing of fields for a minimum of 2 months, removing weeds to reduce habitats for the bacteria, proper plant spacing, and using fertilizers that avoid excess nitrogen as this can favor disease development by promoting vegetative growth of the plant.

Chemical control has been employed for a number of decades with strategies focused on the use of chemicals for eliminating bacteria on seeds, for treating seedlings prior to transplantation, and for field control. Initial control involved the use of Bordeaux, a mixture of hydrated lime and copper sulfate, as well as mercury and copper compounds and antibiotics including streptomycin and its derivates. Streptomycin has been used extensively in agriculture for controlling plant disease since its introduction in 1955; however, antibiotic-resistant strains emerged shortly after. More recently, Park et al.[77] have purified two compounds (bottromycin A2 and dunaimycin D3S) from *Streptomyces bottropensis*, which were effective against Xoo *in vitro*. Further greenhouse studies are required to determine the effectiveness of these compounds *in vivo*. Fumigation of seedbeds prior to sowing with formalin and chlorinating irrigation water has also been shown to have an effect. Additional chemicals used include L-chloramphenicol, dithianon, nickel-dimethyldithiocarbamate, fentiazon, probenazole, tecloftalam, and phenazine oxide. Some chemicals, such as probenazole, can be applied to the rice paddy water before and after seedling transplantation, and others, such as tecloftalam and phenazine oxide, are sprayed directly onto the plants (for a review of chemical methods, see [9,78]). The success of controlling the disease with chemical measures is debated, with many arguing that there is no truly effective chemical treatment. To be successful, this strategy will require the use of a chemical that is effective at a low concentration, is not harmful to the environment or toxic, is capable of translocating through the plant, and is stable and long lasting once administered so it will not require multiple treatments. Additionally, a reliable forecasting system to determine the ideal time of application for effective control is required, and issues surrounding the development of drug- and antibiotic-resistant strains will need to be addressed.

Biological control has been explored as an environmentally friendly alternative to chemical control. Bacterial antagonists, such as *Pseudomonas*,[79,80] *Bacillus*,[81] and *Streptomyces*[82] strains, have been shown to inhibit growth of Xoo *in vitro*. Research has also focused on the use of natural antimicrobial agents to control plant bacterial diseases with the argument being that these are nontoxic and nonpolluting and will not accumulate in the food chain. Essential oil and extracts of *Cleistocalyx operculatus*[83] and *Metasequoia glyptostroboides*[84] have both shown efficacy at controlling *Xanthomonas* species including Xoo.

Sequencing of the bacterial genome has opened the door to the development of rationally designed antibacterial drugs. Targets that are being evaluated include leucine aminopeptidase (LAP), a highly conserved essential enzyme encoded by the *pepA* gene[85]; β-ketoacyl-ACP synthase III (KASIII), an important enzyme involved in fatty acid biosynthesis encoded by the *fabH* gene[86]; and the diffusible signal factor synthase RpfF protein, which is involved in quorum sensing.[87] These studies, designed to select appropriate targets based on the genome sequence, are the first steps in the drug development process, and now further studies are required to develop and test the efficacy of these and other antibacterial drugs. Studies identifying essential bacterial virulence factors may also reveal novel targets for drug development.

The most effective and economically feasible way to control disease is the use of host resistance cultivars of rice. More than 30 rice resistance (R) genes have been identified. R genes, which recognize corresponding effector proteins from the invading bacteria, can be introduced into a variety of commercial rice varieties either individually or in combination to control disease. It is proposed that R proteins act as receptors, which bind specifically to a ligand produced by the pathogen directly or indirectly by an avr gene in a relationship referred to as the gene-for-gene hypothesis.[88] R genes are given the prefix Xa and can be grouped into two major classes: receptor kinase (RLK) and nucleotide-binding site leucine-rich repeat (NBS-LRR) (for a review, see [50]). There are also minor classes of resistance genes that have been identified. The RLK class can respond to a variety of effectors due to its multi-domain structure consisting of an extracellular LRR receptor domain, a transmembrane domain, and an intracellular kinase domain. Two examples of RLKs are *Xa21* and *Xa26*, which confer broad resistance to many strains of Xoo.[50] *Xa21*, the first R gene to be cloned, is reported to have introgressed into rice from the related species *Oryza longistaminata*.[89] The NBS-LRR class of genes is represented by *Xa1*, whose expression is induced by pathogen infection and wounding with expression products localizing in the cytoplasm.[90] *Xa1* only confers narrow resistance, for example, while it has been shown to be effective against isolates of Xoo from Japan, it is not effective against most strains from the Philippines.[50] Another member of the NBS-LRR class is *Rxo1*, from maize, which, when transferred to rice, has been shown to confer broad resistance to Xoc isolates.[91] However, long-term and large-scale cultivation of resistant rice cultivars has resulted in the evolution of new bacterial races that are resistant to the R genes. There is variation in the time taken for the bacteria to develop resistance,[92–94] and this has been shown to correlate with the evolutionary fitness cost to the bacteria to develop resistance.[94] When using resistant strains of rice, it is important to determine the durability of resistance, geographical appropriateness, and population structure of the bacteria. Therefore, specific regional breeding and deployment programs will be required. Studies have shown that the development of bacterial resistance can be delayed by pyramiding multiple resistance genes into rice cultivars.[95,96] Environmental conditions of the region will also need to be considered as it has been shown that most of the rice BLB resistance genes, except *Xa7,* lose efficacy at high temperatures,[93] which has particular implications for countries such as Africa.

74.7 CONCLUSIONS AND FUTURE PERSPECTIVES

As a pathogen of rice, *X. oryzae* presents a very significant threat to global food security. Rice is the world's most important food crop, being the primary source of food for over a third of the world's population; hence, the potential impact of this bacterium on food supply cannot be understated. *X. oryzae* is considered a quarantine pest in most rice-growing countries, and most have restrictions in place to limit spread via the movement of rice plants, rice seed, and rice products. In the United States, *X. oryzae* is listed as a select agent necessitating strict biosecurity measures to prevent the introduction of new virulent strains or the spread of endemic strains to areas that are not endemic.[97] In Europe, *X. oryzae* is listed by the European and Mediterranean Plant Protection Organization (EPPO) as a quarantine pest necessitating regulation in its member countries,[98] and globally, the Australia Group emphasizes its importance as a plant pathogen for export control.[99]

It is unlikely the importance of this bacterium in rice production will diminish in the future as the demand for rice continues to grow. Indeed, it is estimated that world production will need to increase by more than 60% on top of current levels to meet the needs of the world population in 2025.[6] Hence, the management of rice diseases is essential if world demand for rice is to be maintained, and the management of *X. oryzae* is paramount to achieving this objective.

REFERENCES

1. OEPP/EPPO. *Xanthomonas oryzae. EPPO Bull.* 37, 543–553 (2007).
2. Sweeney, M. and McCouch, S. The complex history of the domestication of rice. *Ann. Bot.* 100, 951–957 (2007).
3. Agnoun, Y. et al. The African rice *Oryza glaberrima* Steud: Knowledge distribution and prospects. *Int. J. Biol.* 4, 158–180 (2012).
4. International Rice Genome Sequencing Project. The map-based sequence of the rice genome. *Nature* 436, 793–800 (2005).
5. Asano, K. et al. Artificial selection for a green revolution gene during *japonica* rice domestication. *Proc. Natl. Acad. Sci. USA* 108, 11034–11039 (2011).
6. Yang, J. and Zhang, J. Crop management techniques to enhance harvest index in rice. *J. Exp. Bot.* 61, 3177–3189 (2010).
7. Food and Agriculture Organization of the United Nations. FAOSTAT. Retrieved December 3, 2012, from http://faostat.fao.org
8. Reddy, A.P.K. et al. Relationship of bacterial leaf blight severity to grain yield of rice. *Phytopathology* 69, 967–969 (1979).
9. Nino-Liu, D.O., Ronald, P.C., and Bogdanove, A.J. *Xanthomonas oryzae* pathovars: Model pathogens of a model crop. *Mol. Plant Pathol.* 7, 303–324 (2006).
10. Mew, T.W. et al. Focus on bacterial-blight of rice. *Plant Dis.* 77, 5–12 (1993).
11. Swings, J. et al. Reclassification of the causal agents of bacterial-blight (*Xanthomonas campestris* pv. *oryzae*) and bacterial leaf streak (*Xanthomonas campestris* pv. *oryzicola*) of rice as pathovars of *Xanthomonas oryzae* (Ex Ishiyama 1922) Sp-Nov, Nom-Rev. *Int. J. Syst. Bacteriol.* 40, 309–311 (1990).
12. Iyer-Pascuzzi, A.S. and McCouch, S.R. Recessive resistance genes and the *Oryza sativa-Xanthomonas oryzae* pv. *oryzae* pathosystem. *Mol. Plant. Microbe Interact.* 20, 731–739 (2007).
13. Verdier, V., Vera Cruz, C., and Leach, J.E. Controlling rice bacterial blight in Africa: Needs and prospects. *J. Biotechnol.* 159, 320–328 (2012).
14. Gonzalez, C. et al. Molecular and pathotypic characterization of new *Xanthomonas oryzae* strains from West Africa. *Mol. Plant. Microbe Interact.* 20, 534–546 (2007).
15. Young, J.M. et al. Proposed nomenclature and classification for plant pathogenic bacteria. *N. Z. J. Agric. Res.* 21, 153–177 (1978).
16. Fang, C.T. et al. A comparison of the rice bacterial leaf blight organism with the bacterial leaf streak organisms of rice and Leersia Hexandra Swartz. *Acta Phytopathol. Sin.* 3, 99–124 (1957).
17. Bradbury, J.F. Nomenclature of the bacterial leaf streak pathogen of rice. *Int. J. Syst. Bacteriol.* 21, 72 (1971).
18. Reddy, O.R. and Ou, S.H. Differentiation of *Xanthomonas translucens* f.sp. *oryzicola* (Fang et al.) Bradbury, the leaf-streak pathogen, from *Xanthomonas oryzae* (Uyeda and Ishiyama) Dowson, the blight pathogen of rice, by enzymatic tests. *Int. J. Syst. Bacteriol.* 24, 450–452 (1974).
19. Vera Cruz, C. et al. Differentiation between *Xanthomonas curnpestri* pv. *oryzae, Xanthomonas campestris* pv. *oryzicola* and the bacterial "brown blotch" pathogen on rice by numerical analysis of phenotypic features and protein gel electrophoregrams. *J. Gen. Microbiol.* 130, 2983–2999 (1984).
20. Van den Mooter, M. and Swings, J. Numerical analysis of 295 phenotypic features of 266 *Xanthomonas* strains and related strains and an improved taxonomy of the genus. *Int. J. Syst. Bacteriol.* 40, 348–369 (1990).
21. Benedict, A.A. et al. Pathovar-specific monoclonal-antibodies for *Xanthomonas campestris* pv. *oryzae* and for *Xanthomonas campestris* pv. *oryzicola. Phytopathology* 79, 322–328 (1989).
22. Mahanta, I.C. and Addy, S.K. Serological specificity of *Xanthomonas oryzae,* incitant of bacterial-blight of rice. *Int. J. Syst. Bacteriol.* 27, 383–385 (1977).
23. Vauterin, L., Yang, P., and Swings, J. Utilization of fatty acid methyl esters for the differentiation of new *Xanthomonas* species. *Int. J. Syst. Bacteriol.* 46, 298–304 (1996).
24. Yang, P. et al. Application of fatty-acid methyl-esters for the taxonomic analysis of the genus *Xanthomonas. Syst. Appl. Microbiol.* 16, 47–71 (1993).
25. Vauterin, L., Swings, J., and Kersters, K. Grouping of *Xanthomonas campestris* pathovars by SDS-PAGE of proteins. *J. Gen. Microbiol.* 137, 1677–1687 (1991).
26. Berthier, Y. et al. Characterization of *Xanthomonas campestris* pathovars by rRNA gene restriction patterns. *Appl. Environ. Microbiol.* 59, 851–859 (1993).

27. Triplett, L.R. et al. Genomic analysis of *Xanthomonas oryzae* isolates from rice grown in the United States reveals substantial divergence from known *X. oryzae* pathovars. *Appl. Environ. Microbiol.* 77, 3930–3937 (2011).

28. Parkinson, N. et al. Phylogenetic structure of *Xanthomonas* determined by comparison of *gyrB* sequences. *Int. J. Syst. Evol. Microbiol.* 59, 264–274 (2009).

29. Xu, G.W. and Gonzalez, C.F. Plasmid, genomic, and bacteriocin diversity in United States strains of *Xanthomonas campestris* pv. *oryzae*. *Phytopathology* 81, 628–631 (1991).

30. Louws, F.J. et al. Specific genomic fingerprints of phytopathogenic *Xanthomonas* and *Pseudomonas* pathovars and strains generated with repetitive sequences and PCR. *Appl. Environ. Microbiol.* 60, 2286–2295 (1994).

31. Rademaker, J.L. et al. A comprehensive species to strain taxonomic framework for *xanthomonas*. *Phytopathology* 95, 1098–1111 (2005).

32. Cho, M.S. et al. Sensitive and specific detection of *Xanthomonas oryzae* pv. *oryzae* by real-time bio-PCR using pathovar-specific primers based on an rhs family gene. *Plant Dis.* 95, 589–594 (2011).

33. Lang, J.M. et al. Genomics-based diagnostic marker development for *Xanthomonas oryzae* pv. *oryzae* and *X. oryzae* pv. *oryzicola*. *Plant Dis.* 94, 311–319 (2010).

34. Rademaker, J.L. et al. Comparison of AFLP and rep-PCR genomic fingerprinting with DNA-DNA homology studies: *Xanthomonas* as a model system. *Int. J. Syst. Evol. Microbiol.* 50(2), 665–677 (2000).

35. Bogdanove, A.J. et al. Two new complete genome sequences offer insight into host and tissue specificity of plant pathogenic *Xanthomonas* spp. *J. Bacteriol.* 193, 5450–5464 (2011).

36. Salzberg, S.L. et al. Genome sequence and rapid evolution of the rice pathogen *Xanthomonas oryzae* pv. *oryzae* PXO99A. *BMC Genomics* 9, 204 (2008).

37. Huang, J.S. and De Cleene, M. How rice plants are infected by *Xanthomonas campestris* pv. *oryzae*. In *Proceedings of the International Workshop on Bacterial Blight of Rice*, pp. 31–42. International Rice Research Institute, Manila, Philippines (1988).

38. Mew, T.W., Mew, I.P.C., and Huang, J.S. Scanning electron-microscopy of virulent and avirulent strains of *Xanthomonas campestris* pv. *oryzae* on rice leaves. *Phytopathology* 74, 635–641 (1984).

39. Noda, T. and Kaku, H. Growth of *Xanthomonas oryzae* pv. *oryzae* in planta and in guttation fluid of rice. *Ann. Phytopathol. Soc. Jpn.* 65, 9–14 (1999).

40. Reddy, R. and Shang-zhi, Y. Survival of *Xanthomonas campestris* pv. *oryzae*, the causal organism of bacterial blight of rice. In *Proceedings of the International Workshop on Bacterial Blight of Rice*, pp. 65–78. International Rice Research Institute, Manila, Philippines (1988).

41. Mew, T.W., Unnamalai, N., and Baraoidan, M.R. Does rice seed transmit the bacterial blight pathogen? In *Proceedings of the International Workshop on Bacterial Blight of Rice*, pp. 55–63. International Rice Research Institute, Manila, Philippines (1988).

42. Hsieh, S.P.Y. and Buddenhagen, I.W. Survival of tropical *Xanthomonas-oryzae* in relation to substrate, temperature, and humidity. *Phytopathology* 65, 513–519 (1975).

43. Lee, B.M. et al. The genome sequence of *Xanthomonas oryzae* pathovar *oryzae* KACC10331, the bacterial blight pathogen of rice. *Nucleic Acids Res.* 33, 577–586 (2005).

44. Wang, L. et al. Novel candidate virulence factors in rice pathogen *Xanthomonas oryzae* pv. *oryzicola* as revealed by mutational analysis. *Appl. Environ. Microbiol.* 73, 8023–8027 (2007).

45. Wang, J.C. et al. Genome-wide identification of pathogenicity genes in *Xanthomonas oryzae* pv. *oryzae* by transposon mutagenesis. *Plant Pathol.* 57, 1136–1145 (2008).

46. Jeong, K.S. et al. Virulence reduction and differing regulation of virulence genes in rpf mutants of *Xanthomonas oryzae* pv. *oryzae*. *Plant Pathol. J.* 24, 143–151 (2008).

47. Bai, J.F. et al. *Xanthomonas oryzae* pv. *oryzae* avirulence genes contribute differently and specifically to pathogen aggressiveness. *Mol. Plant. Microbe Interact.* 13, 1322–1329 (2000).

48. Yang, B. and White, F.F. Diverse members of the AvrBs3/PthA family of type III effectors are major virulence determinants in bacterial blight disease of rice. *Mol. Plant. Microbe Interact.* 17, 1192–1200 (2004).

49. Kay, S. and Bonas, U. How *Xanthomonas* type III effectors manipulate the host plant. *Curr. Opin. Microbiol.* 12, 37–43 (2009).

50. White, F.F. and Yang, B. Host and pathogen factors controlling the rice-*Xanthomonas oryzae* interaction. *Plant Physiol.* 150, 1677–1686 (2009).

51. Sugio, A. et al. Two type III effector genes of *Xanthomonas oryzae* pv. *oryzae* control the induction of the host genes OsTFIIAgamma1 and OsTFX1 during bacterial blight of rice. *Proc. Natl. Acad. Sci. USA* 104, 10720–10725 (2007).

52. Yang, B., Sugio, A., and White, F.F. Os8N3 is a host disease-susceptibility gene for bacterial blight of rice. *Proc. Natl. Acad. Sci. USA* 103, 10503–10508 (2006).

53. Ray, S.K., Rajeshwari, R., and Sonti, R.V. Mutants of *Xanthomonas oryzae* pv. *oryzae* deficient in general secretory pathway are virulence deficient and unable to secrete xylanase. *Mol. Plant. Microbe Interact.* 13, 394–401 (2000).

54. Chithrashree, C. and Srinivas, C. Role of antioxidant scavenging enzymes and extracellular polysaccharide in pathogenicity of rice bacterial blight pathogen *Xanthomonas oryzae* pv. *oryzae*. *Afr. J. Biotechnol.* 11, 13186–13193 (2012).

55. Chatterjee, S., Sankaranarayanan, R., and Sonti, R.V. PhyA, a secreted protein of *Xanthomonas oryzae* pv. *oryzae*, is required for optimum virulence and growth on phytic acid as a sole phosphate source. *Mol. Plant. Microbe Interact.* 16, 973–982 (2003).

56. Goel, A.K., Rajagopal, L., and Sonti, R.V. Pigment and virulence deficiencies associated with mutations in the *aroE* gene of *Xanthomonas oryzae* pv. *oryzae*. *Appl. Environ. Microbiol.* 67, 245–250 (2001).

57. Park, Y.J. et al. Virulence analysis and gene expression profiling of the pigment-deficient mutant of *Xanthomonas oryzae* pathovar *oryzae*. *FEMS Microbiol. Lett.* 301, 149–155 (2009).

58. Denny, T.P. Involvement of bacterial polysaccharides in plant pathogenesis. *Annu. Rev. Phytopathol.* 33, 173–197 (1995).

59. Dharmapuri, S. and Sonti, R.V. A transposon insertion in the gumG homologue of *Xanthomonas oryzae* pv. *oryzae* causes loss of extracellular polysaccharide production and virulence. *FEMS Microbiol. Lett.* 179, 53–59 (1999).

60. Kim, S.Y. et al. Mutational analysis of the gum gene cluster required for xanthan biosynthesis in *Xanthomonas oryzae* pv. *oryzae*. *Biotechnol. Lett.* 31, 265–270 (2009).

61. Han, S.W. et al. Small protein-mediated quorum sensing in a Gram-negative bacterium. *PLoS One* 6, e29192 (2011).

62. He, Y.W. et al. Rice bacterial blight pathogen *Xanthomonas oryzae* pv. *oryzae* produces multiple DSF-family signals in regulation of virulence factor production. *BMC Microbiol.* 10, 187 (2010).

63. Yuan, M. et al. The bacterial pathogen *Xanthomonas oryzae* overcomes rice defenses by regulating host copper redistribution. *Plant Cell* 22, 3164–3176 (2010).

64. Lee, S.W. et al. The *Xanthomonas oryzae* pv. *oryzae* PhoPQ two-component system is required for AvrXA21 activity, hrpG expression, and virulence. *J. Bacteriol.* 190, 2183–2197 (2008).

65. Gnanamanickam, S.S. et al. Problems in detection of *Xanthomonas oryzae* pv. *oryzae* in rice seed and potential for improvement using monoclonal antibodies. *Plant Dis.* 78, 173–178 (1994).

66. Leach, J.E. et al. A repetitive DNA-sequence differentiates *Xanthomonas campestris* pv. *oryzae* from other pathovars of *Xanthomonas campestris*. *Mol. Plant. Microbe Interact.* 3, 238–246 (1990).

67. Sakthivel, N., Mortensen, C.N., and Mathur, S.B. Detection of *Xanthomonas oryzae* pv. *oryzae* in artificially inoculated and naturally infected rice seeds and plants by molecular techniques. *Appl. Microbiol. Biotechnol.* 56, 435–441 (2001).

68. Adachi, N. and Oku, T. PCR-mediated detection of *Xanthomonas oryzae* pv. *oryzae* by amplification of the 16–23S rDNA spacer region sequence. *J. Gen. Plant Pathol.* 66, 303–309 (2000).

69. Kim, H.M. and Song, W.Y. Characterization of ribosomal RNA intergenic spacer region of several seedborne bacterial pathogens of rice. *Seed Sci. Technol.* 24, 571–580 (1996).

70. Kang, M.J. et al. Quantitative in planta PCR assay for specific detection of *Xanthomonas oryzae* pv. *oryzicola* using putative membrane protein based primer set. *Crop Protect.* 40, 22–27 (2012).

71. Kang, M.J. et al. Specific detection of *Xanthomonas oryzae* pv. *oryzicola* in infected rice plant by use of PCR assay targeting a membrane fusion protein gene. *J. Microbiol. Biotechnol.* 18, 1492–1495 (2008).

72. Zhao, W.J. et al. Detection of *Xanthomonas oryzae* pv. *oryzae* in seeds using a specific TaqMan probe. *Mol. Biotechnol.* 35, 119–127 (2007).

73. Kersters, K. et al. Protein electrophoresis and DNA:DNA hybridizations of xanthomonads from grasses and cereals. *EPPO Bull.* 19, 51–55 (1989).

74. Stead, D.E. Grouping of *Xanthomonas campestris* pathovars of cereals and grasses by fatty acid profiling. *EPPO Bull.* 19, 57–68 (1989).

75. Kauffman, H.E. et al. An improved technique for evaluating resistance of rice varieties of *Xanthomonas oryzae*. *Plant Dis. Rep.* 57, 537–541 (1973).

76. Cottyn, B., Cerez, M.T., and Mew, T.W. Bacteria. In *A Manual of Rice Seed Health Testing*, Mew, T.W. and Mistra, J.K. (eds.), pp. 322–328. IRRI, Manila, Philippines (1994).

77. Park, S.B. et al. Screening and identification of antimicrobial compounds from *Streptomyces bottropensis* suppressing rice bacterial blight. *J. Microbiol. Biotechnol.* 21, 1236–1242 (2011).

78. Devadath, S. Chemical control of bacterial blight of rice. In *Proceedings of the International Workshop on Bacterial Blight of Rice*, pp. 89–98. International Rice Research Institute, Manila, Philippines (1988).

79. Gnanamanickam, S.S. et al. An overview of bacterial blight disease of rice and strategies for its management. *Curr. Sci.* 77, 1435–1444 (1999).

80. Sivamani, E., Anuratha, C.S., and Gnanamanickam, S.S. Toxicity of *Pseudomonas fluorescens* towards bacterial plant-pathogens of banana (*Pseudomonas solanacearum*) and rice (*Xanthomonas campestris* pv. *oryzae*). *Curr. Sci.* 56, 547–548 (1987).

81. Lin, D. et al. A 3.1-kb genomic fragment of *Bacillus subtilis* encodes the protein inhibiting growth of *Xanthomonas oryzae* pv. *oryzae*. *J. Appl. Microbiol.* 91, 1044–1050 (2001).

82. Ndonde, M.J.M. and Semu, E. Preliminary characterization of some *Streptomyces* species from four Tanzanian soils and their antimicrobial potential against selected plant and animal pathogenic bacteria. *World J. Microbiol. Biotechnol.* 16, 595–599 (2000).

83. Bajpai, V.K. et al. Antibacterial activity of essential oil and extracts of *Cleistocalyx operculatus* buds against the bacteria of *Xanthomonas* spp. *J. Am. Oil Chem. Soc.* 87, 1341–1349 (2010).

84. Bajpai, V.K., Cho, M.J., and Kang, S.C. Control of plant pathogenic bacteria of *Xanthomonas* spp. by the essential oil and extracts of *Metasequoia glyptostroboides* Miki ex Hu in vitro and in vivo. *J. Phytopathol.* 158, 479–486 (2010).

85. Huynh, K.H. et al. Cloning, expression, crystallization and preliminary X-ray crystallographic analysis of leucine aminopeptidase (LAP) from the pepA gene of *Xanthomonas oryzae* pv. *oryzae*. *Acta Crystallogr. F.* 65, 952–955 (2009).

86. Huynh, K.H. et al. Cloning, expression, crystallization and preliminary X-ray crystallographic analysis of beta-ketoacyl-ACP synthase III (FabH) from *Xanthomonas oryzae* pv. *oryzae*. *Acta Crystallogr. F* 65, 460–462 (2009).

87. Reddy, V.S. et al. *In silico* model of DSF synthase RpfF protein from *Xanthomonas oryzae* pv. *Oryzae*: A novel target for bacterial blight of rice disease. *Bioinformation* 8, 504–507 (2012).

88. Flor, H.H. Current status of the gene-for-gene concept. *Annu. Rev. Phytopathol.* 9, 275–296 (1971).

89. Song, W.Y. et al. A receptor kinase-like protein encoded by the rice disease resistance gene, Xa21. *Science* 270, 1804–1806 (1995).

90. Yoshimura, S. et al. Expression of Xa1, a bacterial blight-resistance gene in rice, is induced by bacterial inoculation. *Proc. Natl. Acad. Sci. USA* 95, 1663–1668 (1998).

91. Zhao, B. et al. A maize resistance gene functions against bacterial streak disease in rice. *Proc. Natl. Acad. Sci. USA* 102, 15383–15388 (2005).

92. Mew, T.W., Cruz, C.M.V., and Medalla, E.S. Changes in race frequency of *Xanthomonas oryzae* pv. *oryzae* in response to rice cultivars planted in the Philippines. *Plant Dis.* 76, 1029–1032 (1992).

93. Webb, K.M. et al. A benefit of high temperature: Increased effectiveness of a rice bacterial blight disease resistance gene. *New Phytol.* 185, 568–576 (2010).

94. Vera Cruz, C.M. et al. Predicting durability of a disease resistance gene based on an assessment of the fitness loss and epidemiological consequences of avirulence gene mutation. *Proc. Natl. Acad. Sci. USA* 97, 13500–13505 (2000).

95. Kottapalli, K.R., Lakshmi Narasu, M., and Jena, K.K. Effective strategy for pyramiding three bacterial blight resistance genes into fine grain rice cultivar, Samba Mahsuri, using sequence tagged site markers. *Biotechnol. Lett.* 32, 989–996 (2010).

96. Huang, N. et al. Pyramiding of bacterial blight resistance genes in rice: Marker-assisted selection using RFLP and PCR. *Theor. Appl. Genet.* 95, 313–320 (1997).

97. National Select Agent Registry. Select agents and toxins list. Retrieved December 17, 2012, from http://www.select-agents.gov/select%20agents%20and%20Toxins%20list.html

98. European and Mediterranean Plant Protection Organization. EPPO A1 list of pests recommended for regulation as quarantine pests. Retrieved December 17, 2012, from http://www.eppo.int/QUARANTINE/listA1.htm

99. The Australia Group. List of plant pathogens for export control. Retrieved December 17, 2012, from http://www.australiagroup.net/en/plants.html

75 *Xylella fastidiosa*

Rodrigo P.P. Almeida, Helvécio D. Coletta-Filho, and João R.S. Lopes

CONTENTS

75.1 INTRODUCTION

The United States Department of Agriculture currently lists fewer than a dozen plant-associated microbial species as select agents. This list is a by-product of the Agricultural Bioterrorism Protection Act, which is part of the larger Public Health Security and Bioterrorism Preparedness Response Act of 2002. Although there is discussion within the academic community about the relative risk of taxa in that list [1], work within the United States on any of these taxa is highly regulated, even if the organism is already established in specific regions where research is being conducted and immediately needed to reduce pathogen spread. Among the taxa in this list is the citrus variegated chlorosis (CVC) strain of the plant pathogenic bacterium *Xylella fastidiosa*. Although we will focus this chapter on the phylogenetic group of *X. fastidiosa* causing CVC (subsp. *pauca*), general aspects of *X. fastidiosa* classification, detection, and biology require background that includes other subspecies, which are included here as necessary.

CVC History: In 1987, sweet orange plants (*Citrus sinensis*) in commercial orchards located in the northwest region of São Paulo state, Brazil, were found to be diseased with previously unknown symptoms. Initial hypotheses on the causes of this new disease included nutritional deficiencies, the emergence of a novel virus, and the introduction of the etiological agent of the disease known as Huanglongbing [2]. The first years after its discovery were followed by increased disease spread, which could not be controlled effectively as its etiology was unknown. Initial spatiotemporal analyses of the epidemics indicated that a contagious and likely vector-borne pathogen was associated with the disease [3]. Early work demonstrated that grafting of tissue from symptomatic plants resulted in transmission of the etiological agent, and microscopic examination showed bacteria colonizing the xylem vessels of infected plants [4].

The fulfillment of Koch's postulates by independent groups around 1993 [5,6] identified the bacterium *X. fastidiosa* as the etiological agent of the disease, by then named CVC. Research expanded into various directions, from the identification of insect vectors to epidemiological studies and breeding programs. The majority of those studies were conducted by local researchers and published in Portuguese in Brazilian journals. Vector transmission of the bacterium in citrus was first reported in 1996 [7]. Epidemiological studies showed that vectors played a major role spreading the bacterium within and between citrus orchards, along with planting of infected nursery trees that was responsible for long-distance dispersal [8,9]. Nursery trees can easily become infected with *X. fastidiosa* by grafting of infected plant tissue or naturally by vectors if produced in open fields where CVC is endemic. As consequence of these findings, a disease management program to prevent pathogen introduction and secondary spread was established based on the production and planting of healthy nursery plants, as well as eradication or pruning of diseased trees and vector control with insecticides in citrus orchards. In addition, all nursery plant production in São Paulo state changed from the open-field system to a certified program with screen houses, which became mandatory in 2003 [10].

Today, CVC is endemic throughout the citrus regions of São Paulo state, as well as all other Brazilian states that have sweet orange planted over large areas. According to recent surveys of disease incidence (www.fundecitrus.com.br), approximately 40% of the 200 million sweet orange plants in São Paulo show CVC symptoms. CVC is apparently restricted to Brazil and the neighboring countries of Argentina and Paraguay [11], although we expect the disease is present elsewhere in South America. If introduced in other important citrus-growing regions of the world, CVC could have large economic and social impacts. Economic losses due to CVC in Brazil, including only yield loss due to yearly tree removal, were estimated in 120 million dollars per year [2]. A citrus disease with CVC-like symptoms and associated with *X. fastidiosa* was found in citrus trees used as shade plants in coffee plantations in Costa Rica [12]. However, *X. fastidiosa* strains associated with the disease in Costa Rica are genetically distinct from those causing CVC in Brazil, suggesting that new strains pathogenic to citrus or other crops may evolve independently [13].

75.2 CLASSIFICATION AND MORPHOLOGY

75.2.1 CLASSIFICATION

The species *X. fastidiosa* is a gammaproteobacterium within the family *Xanthomonadaceae*, order Xanthomonadales. It is the sole species in the genus *Xylella*, although the only genetic group known to occur outside of the Americas may represent another species based on currently available data [14]. The species forms a monophyletic group within *Xanthomonadaceae*, with the plant-associated *Xanthomonas* spp. as its phylogenetically closest taxa [15]. All taxa in the species share similar biological and genetic characteristics, the main distinction among phylogenetic groups is the host plant in which they cause disease. The general congruence between host range and phylogenetic placement is the basis for *X. fastidiosa* infraspecies classification.

There are four subspecies of *X. fastidiosa*: subsp. *fastidiosa, multiplex, sandyi,* and *pauca*. This classification is based on DNA relatedness [16] and multilocus sequence typing (MLST) results [17,18]. In addition to biological differences among the subspecies, work has also indicated that they represent allopatric populations with evidence of recent long-distance dispersal. Subsp. *pauca* is limited to South America, while subsp. *fastidiosa* occurs in North America but is much more diverse within Central America, suggesting the latter is its center of origin [13,19]. The subsp. *multiplex* and *sandyi* have only been found in North America, except for a population of subsp. *multiplex* colonizing plums in Brazil, which is thought to have been introduced in the 1900s via contaminated plant material [20]. Because vegetative propagation through grafting is widely used for most long-lived perennial *X. fastidiosa* hosts, transportation of live plant tissue is a common practice in the various agricultural industries affected by this pathogen, eventually increasing its geographic distribution.

Because DNA–DNA reassociation assays are not available to most research groups, MLST has become the benchmark for *X. fastidiosa* classification. The current scheme uses sequence data from seven housekeeping genes under neutral selection [21]. The use of multiple genes is especially important because the species is naturally competent [22] and high rates of intra- and intersubspecies homologous recombination have been observed in field populations [13,19]. Therefore, once an isolate has been identified as belonging to the species *X. fastidiosa*, MLST is highly recommended for the necessary phylogenetic resolution that is also biologically informative. Although there are well-recognized and robust intrasubspecies phylogenetic groups that are also biologically distinct, those are currently largely discriminated based on the host plants in which they cause disease [13,19,21].

Isolates causing CVC belong to the subsp. *pauca*, to which also belong isolates causing coffee leaf scorch (also known as coffee stem atrophy; [19]). Although there is evidence of recombination between these groups, MLST effectively resolves them into two distinct clusters. Typing schemes using various other markers have yielded inconclusive results, which would be expected for recombining populations, and are discouraged for identification purposes. It should be noted that CVC isolates have been shown to cause disease in grape (*Vitis vinifera*; [23]), while isolates from coffee plants in Brazil were not able to cause disease in citrus or vice versa [19,24]. However, given the high recombination rates observed in *X. fastidiosa* populations and what appears to be the capacity of different phylogenetic groups to converge and cause disease in the same host plant, the introduction of additional genetic diversity into new populations (e.g., subsp. *pauca* into North America) carries risks that go beyond one specific plant species of economic importance.

75.2.2 MORPHOLOGY

X. fastidiosa are single, aflagellate, rod-shaped cells, with a rippled cell wall and estimated dimensions of 0.25–0.5 µm in diameter and 0.9–4.0 µm in length [25–27]. These estimates were obtained from different subspecies and media conditions, in addition to cells colonizing plant tissue. Estimates of cell size from *X. fastidiosa* colonizing insects are not available, but scanning and transmission electron microscopy indicates that cell size is similar in that environment [28–30]. Although cells are devoid of flagella, they possess both short (type I pilus) and long fimbriae (type IV pilus); in fact, it has been shown that type IV pili are responsible for twitching motility in *X. fastidiosa* [31,32]. These traits are also supported by genomic sequences [33,34].

75.3 BIOLOGY AND EPIDEMIOLOGY

The epidemiology of *X. fastidiosa* diseases is dependent on a variety of ecological, biotic, and abiotic factors and differs significantly from disease to disease; sometimes, the same disease may have different epidemiology if vectors, for example, are different [35]. Despite these differences, basic aspects of the biology of *X. fastidiosa* are reasonably similar for representatives of its subspecies.

75.3.1 DUAL-HOST LIFESTYLE

X. fastidiosa is primarily considered a plant pathogen, despite the fact that it successfully colonizes two very distinct hosts: plants and insect vectors. In fact, colonization of both hosts is required for dissemination of the bacterium in the landscape; it is unfortunate that most research has so far focused on plants as hosts, as insects are equally important. The economic importance of *X. fastidiosa* diseases also obscures the fact that it most likely evolved to be a harmless endophyte; the bacterium is capable of multiplying and moving within a wide host range but causes disease in very few hosts in a specific manner [36]. Furthermore, work on its pathogenicity mechanisms led to the conclusion that *X. fastidiosa* regulates its gene expression in a cell density-dependent manner, essentially turning off its plant colonization machinery when in high density [37]. This counterintuitive scenario is explained by the fact that traits necessary for insect colonization, and consequently plant-to-plant transmission, are only expressed at high cell densities. Because insect vectors discriminate against symptomatic plants [38], symptom expression is expected to decrease transmission rates, effectively reducing pathogen fitness and disease spread.

75.3.2 GENOME

The genome of *X. fastidiosa* is reduced compared to its sister clade *Xanthomonas* spp. For the species *X. fastidiosa*, there are a limited number of genomes available, but they are reasonably conserved and have high degrees of overall sequence similarity [39]. The genome is approximately 2.6 Mb in size, has no evidence of codon bias, and, unlike *Xanthomonas* spp., does not harbor a type III secretion system [33]. This is intriguing as host specificity is a strong characteristic of *X. fastidiosa* phylogenetic groups. Yet the fact that *X. fastidiosa* does not appear to interact with living cells of insects

or plants, but rather colonizes surfaces devoid of cells, may explain why this secretion system was disposed of during lineage splitting from *Xanthomonas*.

The genome of CVC isolate 9a5c shares 98% of genes with Pierce's disease (PD) of grapevine isolate, suggesting similar metabolites and pathogenesis pathways [34], but with many genomic rearrangements as consequence of phage integrations. The CVC genome has the polygalacturonase (required for the degradation of pit membranes and intervessel migration) gene truncated, which may explain the low aggressiveness of CVC-causing *X. fastidiosa* compared to PD.

75.3.3 PLANT COLONIZATION

CVC exclusively affects sweet orange tress under natural conditions in Brazil. Work with more than 200 accessions of *C. sinensis* failed to detect any resistant or tolerant variety to *X. fastidiosa*, but different degrees of susceptibility were observed [40]. Some mandarins (*C. sinensis* cv. Carvalhais, Emperor, Wilking, and Tankan), sour orange (*C. aurantium* L.), tangelos (*C. sinensis* cv. Page, Swanee, and Williams), and tangors (*C. sinensis* x *C. reticulata* cv. Dweet, Hansen, Ortanique, Temple, and Umatilla) are also susceptible to CVC. On the other hand, some varieties of mandarins (*C. reticulata*), acid lime (*C. aurantifolia*), lemon (*C. limon*), grapefruit (*C. paradisi*), pummelo (*C. grandis*), tangor, kumquats, and *Poncirus trifoliata* present high tolerance and resistance to the disease [40,41]. Therefore, within the genus *Citrus*, there is a broad spectrum of resistance to CVC, which has been used in breeding programs [41,42]. Information regarding other noncitrus natural hosts for *X. fastidiosa* subsp. *pauca* is limited. Artificial (mechanical) inoculation of *Catharanthus roseus* (Madagascar periwinkle) and *Nicotiana tabacum* (tobacco) showed susceptibility to *X. fastidiosa* and its potential to be used as model plant species [43,44]. Under natural conditions, *X. fastidiosa* was unevenly found in 10 out of 23 species of weeds sampled in two groves affected by CVC [45]. However, those weedy hosts do not appear to be important to the epidemiology of the disease as pathogen reservoirs.

Experimental work on the infection and colonization of citrus plants by *X. fastidiosa* has shown variable results, which appears to be a consequence of environmental conditions, host tissue, and bacterial inocula. The method of pricking plant tissues with entomological needles through a droplet of concentrated bacterial suspension (10^8–10^9 UFC/mL) placed on plant stems or petioles [46] is commonly used for mechanical inoculation of *X. fastidiosa* in controlled assays. Six-month-old nursery plants were less prompt to infection and *X. fastidiosa* colonization following pinpricking inoculation than plants of the same age obtained from seeds (juvenile tissue), which showed more evident CVC symptoms (FAA Mourão, personal communication). Like other *X. fastidiosa*, the virulence of CVC-causing isolates was affected by successive passage on culture medium resulting in lower infection rate, poor host colonization, and migration of bacterial cells in the plant [47].

75.3.4 DISEASE SPREAD

Similarly to other *X. fastidiosa*, the CVC pathogen can be transmitted by grafting if infected plant material is used and by insect vectors under field conditions. In the beginning of the CVC outbreak in São Paulo state and before 2003, transmission by infected plant material, that is, nursery plants or vegetative material used for grafting, was probably the main mode of CVC spread to areas far from the initial foci in northern areas of the state, including other Brazilian states. We believe that the following factors were important for this nonintentional spreading. The first one is the long incubation period required for symptom expression, which varies from about 6 months to years, depending on environmental conditions. The second is that the bacterium can be transmitted from plant material taken from infected but yet asymptomatic plants used for grafting. In 2003, when production of healthy nursery trees under vector-proof screen houses became mandatory, the tree-to-tree transmission of *X. fastidiosa* by vectors is the major, if not only, form of bacterial spread in São Paulo state.

75.3.5 INSECT VECTORS AND TRANSMISSION

Vectors are required for natural *X. fastidiosa* dissemination. Therefore, a robust understanding of vector ecology is necessary for the development of management practices. Xylem-sap-feeding sharpshooter leafhoppers (Hemiptera, Cicadellidae) and spittlebugs (Hemiptera, Cercopidae) are vectors of *X. fastidiosa*; sharpshooters are considered of greater economic importance and epidemiological relevance [48,49]. General characteristics of *X. fastidiosa* transmission appear to be universal in the sense that the biology of the process seems to be shared by all vector species and pathogen subspecies [50]. In fact, Frazier [51] proposed that all species in the family Cicadellidae are potential vectors because of their habit to feed in plant xylem vessels, which are colonized by *X. fastidiosa*. So far, Frazier's assessment of vector specificity has been correct.

Although some spittlebugs have been shown to transmit *X. fastidiosa* under experimental conditions, the great majority of known vectors of this bacterium are sharpshooter leafhoppers, which are specially fit to transmit *X. fastidiosa* because of their diversity in natural and agricultural ecosystems, polyphagy, mobility, and specialization on xylem-sap feeding [35,50]. Although most of sharpshooter leafhoppers are neotropical, particularly in the tribe Proconiini, many species are present in North America, Africa, Asia, and Australia. Species composition can be very rich in some agricultural systems, particularly those with a higher diversity of trees, shrubs, and herbaceous plants in surrounding areas or between crop rows, which may serve as sharpshooter hosts [52]. Sharpshooter species are generally polyphagous; oviposition and nymphal development occurs on selected hosts, but adults are quite mobile and usually feed on a wide range of plants of various botanical families and growth habits. Because *X. fastidiosa* can be persistently transmitted for life after acquisition by sharpshooter adults [50], vector polyphagy allows the bacterium the opportunity to exploit several host plant species. In addition, vector specialization on xylem-sap feeding optimizes the chances of acquisition and inoculation of this xylem-limited bacterium. Feeding behavior studies show that sharpshooters spend most of the time on plants with their stylets in the xylem vessels [53].

Citrus orchards in tropical and subtropical regions are usually rich in sharpshooter species, not only because citrus is an adequate feeding and developmental host for some of them, but also because of the diversity of host plants of various growth habits in the ground vegetation or in adjacent areas. In São Paulo, more than 20 species of sharpshooters have been described inhabiting citrus orchards [52,54]. Some of them are abundant on weeds in the ground cover and accidentally found on citrus trees, whereas others show opposite distribution or are more commonly found on trees or shrubs in adjacent natural vegetation (e.g., woods and swamps). Thirteen out of 17 sharpshooters tested have been confirmed as vectors of the CVC strain of *X. fastidiosa* [7,55–57]. Because of the characteristics of CVC epidemiology and the relevance of tree-to-tree transmission (secondary spread) in citrus orchards, sharpshooters that more often visit citrus trees are considered the most important vectors. This is the case for leafhopper species *Acrogonia citrina*, *Bucephalogonia xanthophis*, *Dilobopterus costalimai*, and *Oncometopia facialis*. It should be noted, however, that sharpshooter species composition and abundance on citrus orchards vary among regions because of differences in climate, vegetation types, and host plants [57]. Thus, surveys of sharpshooter species and studies aimed at identification of key vectors are necessary in other regions or countries where CVC emerges as new disease.

75.3.6 MECHANISMS OF TRANSMISSION BY VECTORS

X. fastidiosa transmission by vectors has only been studied in detail with the PD system in California. However, observations appear to be applicable for other diseases, including CVC. Transmission occurs in a noncirculative manner, with bacteria colonizing the cuticular surface of the mouthparts of sharpshooter vectors [58]. This cuticle is part of the exoskeleton of insects and is shed at each molt; therefore, although nymphs are capable of transmitting *X. fastidiosa*, they lose inoculum at each molt. Adults, which do not molt, retain *X. fastidiosa* for life. The regions of the foregut colonized by *X. fastidiosa* are named precibarium and cibarium, which are posterior to the maxillary stylets and found "inside" the head of vectors. The maxillary stylets, which are not colonized by *X. fastidiosa*, form a straw-like canal that penetrates plant tissue, through which xylem sap is sucked into the insect's gut. Cells colonizing the foregut form a biofilm that is subject to rapid fluid flow (estimated at 8 cm/s; [28]) and frequent turbulence (once a second; [59]) due to a pumping system responsible for intake of up to 1000 times the insect's body weight daily [60]. The mechanism of pathogen inoculation into plants is yet to be understood.

X. fastidiosa acquisition and inoculation efficiencies increase with vector plant access time, up to 2–4 days [58,61].

Transmission is context dependent, with plant–pathogen–vector interactions strongly affecting overall efficiency [62]; in fact, even vector within-plant tissue preference has been shown to affect efficiency [63]. The major factors shown to affect the efficiency with which sharpshooters transmit *X. fastidiosa* from plant to plant are bacterial populations within the host functioning as a source of the pathogen [64]. Because there is no evidence of specificity between vector species and *X. fastidiosa* genotype, experiments need to be performed to estimate transmission rates when new vector–pathogen combinations are of importance. For example, transmission of *X. fastidiosa* from grapevines in California can reach efficiencies approaching 100% over a four-day period [58], while estimates range from ~1% to 30% with vectors spreading CVC [57,65]. In other words, data on transmission efficiency from one system are not transferable to another.

75.3.7 PATHOGEN POPULATION STRUCTURE

Coletta-Filho and Machado [66] showed that populations of *X. fastidiosa* were geographically structured by using samples collected from plants grown in five different geographic regions in São Paulo state. Bacterial populations were found to be genetically different from each other, indicating spatial structure and limited gene flow among populations. Contrary to the effect of geographic origin on genetic structure of *X. fastidiosa*, no relationship was observed between pathogen genetic diversity and *C. sinensis* varieties from which isolates/populations were obtained, suggesting that host responses to infection were not selecting for specific pathogen genotypes [67]. More recently, Coletta-Filho and Almeida (unpublished data) showed that different *X. fastidiosa* genotypes were found colonizing 4-cm-long branch fragments from *C. sinensis* trees; consequently, different genotypes were found within individual trees. These studies indicate that, at the population level, *X. fastidiosa* causing CVC may be very diverse but that populations are structured in space. In addition, no relationship between pathogen and host plant genotype was found, suggesting that any isolate causing CVC in Brazil should be considered of high risk if introduced into other sweet orange-producing regions.

75.3.8 EPIDEMIOLOGY

Disease progress is faster during spring and summer than in autumn and winter seasons [9], alternating periods of rapid and slow rates of increase in the proportion of diseased plants, which are best explained by sigmoid-shaped models such as Gompertz and logistic [3]. Spring and summer also appear to be periods of higher rates of *X. fastidiosa* transmission, mainly because of the higher vector populations observed on citrus orchards during these two seasons. Sharpshooters show strong preference for citrus flushes [68], which are usually more numerous and vigorous during the rainy season (spring and summer).

Diseased plants show a patchy distribution in citrus orchards of São Paulo state [3,9], as expected for a vector-borne contagious pathogen. Analyses of disease foci structure and dynamics in affected orchards showed coalescence of foci at higher incidences (>30%) of diseased plants, indicating that tree-to-tree transmission (secondary spread) by vectors takes place [8]. No significant CVC aggregation is observed within citrus rows, suggesting that movement of machines during mechanical or cultural practices have no effect on disease spread [69]. A clear edge effect is often observed, with initial foci appearing near the borders with older orchards, showing that previously infected orchards represent major sources of inoculum for primary spread.

The epidemiology of *Xylella* diseases may change dramatically if vector species with different host plant preferences, feeding habits, and dispersal abilities are introduced. An example is PD in the Central Valley of California, for which only primary spread was observed until the 1980s, presumably promoted by grass-feeding sharpshooters that accidentally landed on grapes [70]. The situation changed after introduction in California of the highly polyphagous, abundant, and mobile glassy-winged sharpshooter, *Homalodisca vitripennis*, which colonizes grapes and was able to promote rapid vine-to-vine spread of the pathogen and exponential increase in PD progress [35,50]. Therefore, the epidemiological characteristics of CVC if introduced in regions with other vector species, host plants, and environmental conditions may be different from those reported in Brazil.

75.4 PATHOGENESIS

The mechanisms of *X. fastidiosa* virulence are not entirely understood, but the development of a xylem-limited biofilm leading to vessel occlusion and subsequent reduced water conductance (water stress) is the leading hypothesis to explain disease symptoms [71]. In the specific case of CVC, other hypotheses have been proposed, including the possibility that toxins secreted by *X. fastidiosa* may affect host physiology, such as alterations to the photosynthetic machinery [72]. *C. sinensis* responses to *X. fastidiosa* infection include those typical of water deficit symptoms such as decreased of photosynthesis, transpirations, stomatal conductance, and water potential [73]. Nitrogen metabolism is highly affected in CVC symptomatic plant as observed by the imbalance of enzymes like glutamine synthetase and proteases [74]. However, it is unclear if the negative effect of nitrogen metabolism is a direct consequence of pathogen presence or a physiological response of plant to the water stress. No disturbance on hormones (auxin and abscisic acid) was observed in CVC symptomatic leaves [75].

75.5 IDENTIFICATION AND DIAGNOSIS OF CVC

75.5.1 DISEASE SYMPTOMS

Unlike the majority of *X. fastidiosa* diseases, CVC symptoms do not include scorched leaves. Typically, irregular chlorosis evolves in mature leaves recognized by interveinal

FIGURE 75.1 (a) Plant with CVC symptoms on the left and healthy plant on the right. (b) CVC foliar symptoms, including necrotic spots surrounded by yellowing leaf tissue. (c) Difference in fruit size between healthy (left) and infected (right) plants. (d) Photograph of *D. costalimai*, a species of sharpshooter leafhoppers in Brazil that are vectors of *X. fastidiosa* causing CVC.

yellowing on the upper side of leaf and corresponding brownish gumlike material over the side (Figure 75.1). Later on, brown spots coalesce and necrosis becomes evident, eventually leading to leaves dropping from branches. Zinc- and ironlike deficiency can be frequently observed in the affected leaves. Stunted trees show twig dieback and fruits reduce in size and harden, becoming unsuitable for the juice industry as well as for the fresh fruit market (Figure 75.1). Severely infected plants do not die but become economically nonproductive.

75.5.2 BACTERIAL CULTURE

The PW or PWG media [64,76] are well suited for bacterial isolation from CVC symptomatic tissues (petioles or branches). On these media, small (~0.30 mm of diameter), white, and convex colonies are observed under a dissecting microscope after approximately 10 days of growth at 27°C–30°C. Other media like BCYE, CS20, and PD2 [77–79] also support cell growth, but it may take over 20 days for colonies to be observable.

75.5.3 SEROLOGICAL ASSAYS

To our knowledge, no monoclonal antibody was developed to specifically recognize *X. fastidiosa* subsp. *pauca*. All the serology-based methods used for diagnosis of *X. fastidiosa* use polyclonal antibodies that recognize other subspecies pathogenic to hosts like grapevine, mulberry, almond, elm, plum, ragweed, and periwinkle [5]. Protocols based on serological approaches like DAS-ELISA and dot immunoblotting assay (DIBA) are detailed as described in the EPPO standard protocols for regulated pests at http://www.eppo.int.

75.5.4 MOLECULAR TOOLS

As consequence of popularization of DNA-based techniques, PCR detection of *X. fastidiosa* is now routine. The most useful primer set to recognize any *X. fastidiosa* is the RST31/RST33 [80]. The primer set CVC-1 and 272-2int is specific to CVC isolates and works well [81], although it should be noted that this set is also known to amplify DNA from isolates infecting coffee in Brazil, which are not pathogenic to

C. sinensis (Coletta-Filho, unpublished data). Both primer sets are recommended by the EPPO standard protocols. Based on our experience, the RST primer set is more sensitive compared to other sets that have been developed and are available in the literature. A TaqMan® real-time quantitative PCR protocol is also available for detection and has the benefit of not amplifying isolates originated from coffee plants [82]. As discussed earlier, although individual loci may be used for *X. fastidiosa* diagnostics, the use of multiple loci (MLST) for typing purposes provides a more robust placement of isolates within this species [21] and for biological inferences such as host range to be made.

75.6 TREATMENT AND PREVENTION

CVC management uses an integrated strategy that involves the principles of exclusion, eradication, and protection. Growers in Brazil have implemented several preventive control measures for CVC management, including (1) planting of certified nursery trees; (2) pruning of affected branches in mildly symptomatic trees and removal of very symptomatic plants, both practiced with the objective of removing inoculum from orchards; and (3) spraying of insecticides to control vector populations [83].

75.6.1 PLANTING OF CERTIFIED NURSERY TREES

The use of healthy nursery plants is one of the most important strategies for management of CVC and other citrus diseases. As consequence of the significant increase of CVC in São Paulo at the end of the 1990s, the production of certified citrus nursery trees within vector-proof screen houses has become mandatory in that state since January 2003. According to the law, all the steps involved in nursery plant production must be carried out under protected conditions, including growing of rootstock seedlings for grafting, bud stick sources, and storage of grafted plants [10]. This certified program has contributed to a significant reduction of CVC incidence in citrus tress younger than 2 years old, as well as an increase of fruit production by 21% (average of 8 years) compared to artificially inoculated plants [84].

75.6.2 DISEASED TREE REMOVAL OR PRUNING

In areas where the disease is already established, frequent inspections should be done in citrus orchards for detection of diseased trees, especially during summer and fall when CVC symptoms become more evident. The sooner the disease is identified in the orchard and the diseased branches or trees are totally removed, the lower the probability that CVC will become endemic in the orchard. The pruning strategy used for CVC management has been done with the objective of removing a tree section that is colonized by bacteria, consequently eliminating a source of inoculum and disease spread within plants. Pruning of symptomatic plant material is successful only if done in trees older than 3 years with CVC symptoms present in only a few leaves [85]. In this case,

the branch must be cut at least 70 cm below symptomatic leaves. For trees with symptomatic leaves throughout the tree's canopy or with symptomatic fruits, pruning is not feasible because bacteria are already systemic, including in the basal trunk of the tree. In this case, and for trees younger than 3 years of age with any degree of disease, plants should be immediately removed from orchards.

75.6.3 VECTOR CONTROL

Vector control in affected orchards is another important measure for CVC management because sharpshooters can acquire the pathogen from either symptomatic or asymptomatic infected citrus plants and spread it to other trees within and between orchards. Vector control with insecticides (e.g., pyrethroids, organophosphates, and neonicotinoids) is widely used by citrus growers in São Paulo state for this purpose. Insecticides can be applied via soil or on the basal portion of the tree trunk for systemic action or sprayed on the tree canopy for contact action. Systemic effects of insecticide applications are only obtained in nursery plants and trees up to 3 years old and during the rainy season. For older trees, only contact action by insecticide sprays is effective against the vectors. During the drier months of the year (April–September), insecticide sprays are required for all plant ages. Insecticide treatments should be done throughout the year to prevent pathogen transmission in young plants up to 3 years old, which can become systemically infected by *X. fastidiosa* soon after vector inoculation. Older orchards in affected areas should be sprayed when sharpshooter vectors are trapped by yellow sticky cards placed at the height of 1.8 m on the tree canopy (at least 1 card/ha), during periodic samplings. A larger number of traps should be used in the orchard borders to detect vector immigration, especially in borders facing other citrus orchards or natural habitats of the sharpshooters such as woody vegetation and swamps.

75.7 CONCLUSIONS AND FUTURE PERSPECTIVES

CVC is an important disease in South America, and its emergence 25 years ago has resulted in significant changes to sweet orange production in that region, primarily in São Paulo state, Brazil. The introduction of *X. fastidiosa* subsp. *pauca* isolates causing CVC into citrus-growing regions such as the United States and Europe could have devastating consequences. That is especially true for the United States, where large populations of insect vectors are already established in the states of Florida and California, which are responsible for the bulk of citrus production in the country. Aggressive large-scale management practices proven to be successful in Brazil are labor intensive and costly and may not be economically feasible in the United States. In addition, awareness of the CVC-like disease in Costa Rica [12] must be increased due to its potential threat, although little is known about the biology or geographic distribution of that specific pathogen.

The risks due to the introduction of CVC-causing *X. fastidiosa* subsp. *pauca* into the United States, if accidental or deliberate, are substantial. Because vectors are established on citrus in Florida and California and because there is no *X. fastidiosa*–vector specificity required for transmission, it is likely that the disease would spread very quickly. Furthermore, because disease symptoms may take more than one season to develop, while vector acquisition of the pathogen is possible from asymptomatic trees, the proportion of infected trees in an orchard may be much larger than the number of symptomatic ones. The best strategy available for countries without this pathogen is to have aggressive legislation and quarantine efforts to avoid its introduction. Once introduced, we do not believe it can be eradicated and that resources would be better used trying to reduce the speed with which it moves in space. That is especially true for the United States, as this pathogen has a very wide host range and would be present on alternative hosts, and vectors would assure its spread throughout orchards and the landscape.

REFERENCES

1. Young, J.M. et al., Plant-pathogenic bacteria as biological weapons—Real threats? *Phytopathology*, 98, 1060, 2008.
2. Bové, J.M. and Ayres, A.J., Etiology of three recent diseases of citrus in São Paulo State, sudden death, variegated chlorosis and Huanglongbing, *IUBMB Life*, 59, 346, 2007.
3. Gottwald, T.R. et al., Preliminary spatial and temporal analysis of citrus variegated chlorosis (CVC) in São Paulo, Brazil, *Proceedings of the 12th Conference of the International Organization of Citrus Virology*. P. Moreno, J. V. Da Graca and L. W. Timmer (eds.). Riverside, CA, p. 327, 1993.
4. Rossetti, V. et al., Présence de bactéries dans le xylème d'orangers atteints de chlorose variégé, une nouvelle maladie des agrumes au Bresil, Comptes rendus de l'Académie des sciences. Série 3, *Sciences de la vie*, 30, 345, 1990.
5. Chang, C.J. et al., Culture and serological detection of *Xylella fastidiosa*, the xylem-limited bacterium associated with citrus variegated chlorosis disease, *Curr. Microbiol.*, 27, 137, 1993.
6. Hartung, J.S. et al., Citrus variegated chlorosis bacterium: Axenic culture, pathogenicity, and serological relationships with other strains of *Xylella fastidiosa*, *Phytopathology*, 84, 591, 1994.
7. Roberto, S.R. et al., Transmissão de *Xylella fastidiosa* pelas cigarrinhas *Dilobopterus costalimai*, *Acrogonia terminalis* e *Oncometopia facialis* em citros, *Fitopatol. Bras.*, 21, 517, 1996.
8. Laranjeira, F.F., Bergamin, F.A., and Amorim, L., Dynamics and structure of citrus variegated chlorosis (CVC) foci, *Fitopatol. Bras.*, 23, 36, 1998.
9. Roberto, S.R., Farias, P.R.S., and Bergamin Filho, A., Geostatistical analysis of spatial dynamic of citrus variegated chlorosis, *Fitopatol. Bras.*, 27, 599, 2002.
10. Carvalho, S.A., Regulamentação atual da Agência de Defesa Agropecuária para a produção, estocagem, comércio, transporte e plantio de mudas cítricas no Estado de São Paulo, *Laranja*, 24, 199, 2003.
11. De Souza, A.A. et al., Citrus response to *Xylella fastidiosa* infection, the causal agent of Citrus Variegated Chlorosis, *Tree Forest Sci. Biotechnol.*, S2, 73, 2009.
12. Aguilar, E. et al., First report of *Xylella fastidiosa* infecting citrus in Costa Rica, *Plant Dis.*, 89, 687, 2005.
13. Nunney, L. et al., Population genomic analysis of a bacterial plant pathogen: Novel insight into the origin of Pierce's disease of grapevine in the US, *PLoS ONE* 5, e15488. 2010.
14. Su, C.C. et al., Specific characters of 16S rRNA gene and 16S–23S rRNA internal transcribed spacer sequences of *Xylella fastidiosa* pear leaf scorch strains, *Eur. J. Plant Pathol.*, 132, 203, 2012.
15. Pieretti, I. et al., The complete genome sequence of *Xanthomonas albilineans* provides new insights into the reductive genome evolution of the xylem-limited *Xanthomonadaceae*, *BMC Genomics*, 10, 616, 2009.
16. Schaad, N.W. et al., *Xylella fastidiosa* subspecies: *X. fastidiosa* subsp. *piercei*, subsp. nov., *X. fastidiosa* subsp. *multiplex* subsp. nov., and *X. fastidiosa* subsp. *pauca* subsp. nov., *Syst. Appl. Microbiol.*, 27, 290, 2004.
17. Schuenzel, E.L. et al., A multigene phylogenetic study of clonal diversity and divergence in North American strains of the plant pathogen *Xylella fastidiosa*, *Appl. Environ. Microbiol.*, 71, 3832, 2005.
18. Scally, M. et al., Multilocus sequence type system for the plant pathogen *Xylella fastidiosa* and relative contributions of recombination and point mutation to clonal diversity, *Appl. Environ. Microbiol.*, 71, 8491, 2005.
19. Almeida, R.P.P. et al., Genetic structure and biology of *Xylella fastidiosa* causing disease in citrus and coffee in Brazil, *Appl. Environ. Microbiol.*, 74, 3690, 2008.
20. Nunes, L.R. et al., Microarray analyses of *Xylella fastidiosa* provide evidence of coordinated transcription control of laterally transferred elements, *Genome Res.*, 13, 570, 2003.
21. Yuan, X. et al., Multilocus sequence typing of *Xylella fastidiosa* causing Pierce's disease and oleander leaf scorch in the United States, *Phytopathology*, 100, 601, 2010.
22. Kung, S.H. and Almeida, R.P.P., Natural competence and recombination in the plant pathogen *Xylella fastidiosa*, *Appl. Environ. Microbiol.*, 77, 5278, 2011.
23. Li, W.B. et al., Citrus and coffee strains of *Xylella fastidiosa* induce Pierce's disease in grapevine, *Plant Dis.*, 86, 1206, 2002.
24. Prado, S.S. et al., Host colonization differences between citrus and coffee isolates of *Xylella fastidiosa* in reciprocal inoculation, *Sci. Agric.*, 65, 251, 2008.
25. Davis, M.J., Purcell, A.H., and Thomson, S.V., Pierce's disease of grapevines: Isolation of the causal bacterium, *Science*, 199, 75, 1978.
26. Wells, J.M. et al., *Xylella fastidiosa*, new-genus, new-species gram negative xylem-limited fastidious plant bacteria related to *Xanthomonas* spp., *Int. J. Syst. Bacteriol.*, 37, 136, 1987.
27. Chagas, C.M., Rossetti, V., and Beretta, M.J.G., Electron microscopy studies of a xylem-limited bacterium in sweet orange affected with citrus variegated chlorosis disease in Brazil, *J. Phytopathol.*, 134, 306, 1992.
28. Purcell, A.H., Finlay, A.H., and McLean, D.L., Pierce's disease bacterium: Mechanism of transmission by leafhopper vectors, *Science*, 206, 839, 1979.
29. Brlansky, R.H. et al., Colonization of the sharpshooter vectors, *Oncometopia nigricans* and *Homalodisca coagulata*, by xylem-limited bacteria, *Phytopathology*, 73, 530, 1983.
30. Almeida, R.P.P. and Purcell, A.H., Patterns of *Xylella fastidiosa* colonization on the precibarium of leafhopper vectors relative to transmission to plants, *Ann. Entomol. Soc. Am.*, 99, 884, 2006.

31. Meng, Y. et al., Upstream migration of *Xylella fastidiosa* via pilus-driven twitching motility, *J. Bacteriol.*, 187, 5560, 2005.

32. De La Fuente, L., Burr, T.J., and Hoch, H.C., Mutations in type I and type IV pilus biosynthetic genes affect twitching motility rates in *Xylella fastidiosa*, *J. Bacteriol.*, 189, 7507, 2007.

33. Simpson, A.J. et al., The genome sequence of the plant pathogen *Xylella fastidiosa*, *Nature*, 406, 151, 2000.

34. Van Sluys, M.A. et al., Comparative analyses of the complete genome sequences of Pierce's disease and citrus variegated chlorosis strains of *Xylella fastidiosa*, *J. Bacteriol.*, 185, 1018, 2003.

35. Redak, R.A. et al., The biology of xylem fluid-feeding insect vectors of *Xylella fastidiosa* and their relation to disease epidemiology, *Annu. Rev. Entomol.*, 49, 243, 2004.

36. Chatterjee, S., Almeida, R.P.P., and Lindow, S.E., Living in two worlds: The plant and insect lifestyles of *Xylella fastidiosa*, *Annu. Rev. Phytopathol.*, 46, 243, 2008.

37. Newman, K.L. et al., Cell-cell signaling controls *Xylella fastidiosa* interactions with both insects and plants, *Proc. Natl. Acad. Sci. USA*, 101, 1737, 2004.

38. Daugherty, M.P. et al., Vector preference for hosts differing in infection status: Sharpshooter movement and *Xylella fastidiosa* transmission, *Ecol. Entomol.*, 36, 654, 2011.

39. Bhattacharyya, A. et al., Whole-genome comparative analysis of three phytopathogenic *Xylella fastidiosa* strains, *Proc. Natl. Acad. Sci. USA*, 99, 12403, 2002.

40. Laranjeira, F.F. et al., Cultivares e espécies cítricas hospedeiras de *Xylella fastidiosa* em condições de campo, *Fitopatol. Bras.*, 23, 147, 1998.

41. Coletta-Filho, H.D. et al., Analysis of resistance to *Xylella fastidiosa* within a hybrid population of Pera sweet orange × Murcott tangor, *Plant Pathol.*, 56, 661, 2007.

42. Garcia, A.L. et al., Citrus responses to *Xylella fastidiosa* infection, *Plant Dis.*, 96, 1245, 2012.

43. Lopes, S.A. et al., *Nicotiana tabacum* as an experimental host for the study of plant-*Xylella fastidiosa* interactions, *Plant Dis.*, 84, 827, 2000.

44. Monteiro, P.B. et al., *Catharanthus roseus*, an experimental host plant for the citrus strain of *Xylella fastidiosa*, *Plant Dis.*, 85, 246, 2001.

45. Lopes, S.A. et al., Weeds as alternative hosts of the citrus, coffee, and plum strains of *Xylella fastidiosa* in Brazil, *Plant Dis.*, 87, 544, 2003.

46. Almeida, R.P.P. et al., Multiplication and movement of a citrus strain of *Xylella fastidiosa* within sweet orange, *Plant Dis.*, 85, 382, 2001.

47. Souza, A.A. et al., Analysis of gene expression in two growth states of *Xylella fastidiosa* and its relationship with pathogenicity, *Mol. Plant Microbe. Interact.*, 16, 867, 2003.

48. Severin, H.H.P., Transmission of the virus of Pierce's disease by leafhoppers, *Hilgardia*, 19, 190, 1949.

49. Severin, H.H.P., Spittle-insect vectors of Pierce's disease virus. II. Life history and virus transmission, *Hilgardia*, 19, 357, 1950.

50. Almeida, R.P.P. et al., Vector transmission of *Xylella fastidiosa*: Applying fundamental knowledge to generate disease management strategies, *Ann. Entomol. Soc. Am.*, 98, 775, 2005.

51. Frazier, N.W., Xylem viruses and their insect vectors, *Proceedings International Conference on Virus and Vector on Perennial Hosts, with Special Reference to Vitis*, W.B. Hewitt (eds.), University of California, Division of Agricultural Sciences, Davis, CA, p. 91, 1965.

52. Giustolin, T.A. et al., Diversidade de Hemiptera Auchenorrhyncha em citros, café e fragmento de floresta nativa do Estado de São Paulo, *Neotrop. Entomol.*, 38, 834, 2009.

53. Miranda, M.P. et al., Characterization of electrical penetration graphs of *Bucephalogonia xanthophis*, a vector of *Xylella fastidiosa* in citrus, *Entomol. Exp. Appl.*, 130, 35, 2009.

54. Yamamoto, P.T. and Gravena, S., Espécies e abundância de cigarrinhas e psilídeos (Homoptera) em pomares cítricos, *Ann. Soc. Entomol. Bras.*, 29, 169, 2000.

55. Krugner, R. et al., Transmission efficiency of *Xylella fastidiosa* to citrus by sharpshooters and identification of two new vector species, *Proceedings of the 14th Conference of the International Organization of Citrus Virologists*. J.V. da Graça, R.F. Lee, R.K. Yokomi (eds.). IOCV c/o Dept. Plant Pathology, University of California-Riverside, Riverside, CA, p. 423, 2000.

56. Yamamoto, P.T. et al., Transmissão de *Xylella fastidiosa* por cigarrinhas *Acrogonia virescens* e *Homalodisca ignorata* (Hemiptera: Cicadellidae) em plantas cítricas, *Summa Phytopathol.*, 28, 178, 2002.

57. Lopes, J.R.S. and Krugner, R., Transmission ecology and epidemiology of the citrus variegated chlorosis strain of *Xylella fastidiosa*, *Vector-Mediated Transmission of Plant Pathogens*, J.K. Brown (ed.), APS Press, Saint Paul, MN, 2014 (accepted).

58. Purcell, A.H. and Finlay, A.H., Evidence for noncirculative transmission of Pierce's disease bacterium by sharpshooter leafhoppers, *Phytopathology*, 69, 393, 1979.

59. Dugravot, S. et al., Correlations of cibarial muscle activities of *Homalodisca* spp. sharpshooters (Hemiptera: Cicadellidae) with EPG ingestion waveform and excretion, *J. Insect Physiol.*, 54, 1467, 2008.

60. Mittler, T.E., Water tensions in plants—An entomological perspective, *Ann. Entomol. Soc. Am.*, 60, 1074, 1967.

61. Almeida, R.P.P. and Purcell, A.H., Transmission of *Xylella fastidiosa* to grapevines by *Homalodisca coagulata* (Hemiptera, Cicadellidae), *J. Econ. Entomol.*, 96, 265, 2003.

62. Lopes, J.R.S., Context-dependent transmission of a generalist plant pathogen: Host species and pathogen strain mediate insect vector competence, *Entomol. Exp. Appl.*, 131, 216, 2009.

63. Daugherty, M.P., Lopes, J.R.S., and Almeida, R.P.P., Vector within-host feeding preference mediates transmission of a heterogeneously distributed pathogen, *Ecol. Entomol.*, 35, 360, 2010.

64. Hill, B.L. and Purcell, A.H., Acquisition and retention of *Xylella fastidiosa* by an efficient vector, *Graphocephala atropunctata*, *Phytopathology*, 85, 209, 1995.

65. Marucci, R.C., Lopes, J.R.S., and Cavichioli, R.R., Transmission efficiency of *Xylella fastidiosa* by sharpshooters (Hemiptera: Cicadellidae) in coffee and citrus, *J. Econ. Entomol.*, 101, 1114, 2008.

66. Coletta-Filho, H.D. and Machado, M.A., Geographical genetic structure of *Xylella fastidiosa* from citrus in São Paulo State, Brazil, *Phytopathology*, 93, 28, 2003.

67. Coletta-Filho, H.D. and Machado, M.A., Evaluation of the genetic structure of *Xylella fastidiosa* populations from different *Citrus sinensis* varieties, *Appl. Environ. Microbiol.*, 68, 3731, 2002.

68. Marucci, R.C. et al., Feeding site preference of *Dilobopterus costalimai* Young and *Oncometopia facialis* (Signoret) (Hemiptera: Cicadellidae) on citrus plants, *Neotrop. Entomol.*, 33, 759, 2004.

69. Laranjeira, F.F. et al., Dinâmica espacial da clorose variegada dos citros em três regiões do Estado de São Paulo, *Fitopatol. Bras.*, 29, 56, 2004.

70. Purcell, A.H. and Frazier, N.W., Habitats and dispersal of the principal leafhopper vectors of Pierce's disease in the San-Joaquin Valley California USA, *Hilgardia*, 53, 1, 1985.

71. Hopkins, D.L. and Purcell, A.H., *Xylella fastidiosa*: Cause of Pierce's disease of grapevine and other emergent diseases, *Plant Dis.*, 86, 1056, 2002.

72. Ribeiro, R.V., Machado, E.C., and Oliveira, R.F., Early photosynthesis responses of sweet orange plants inoculated with *Xylella fastidiosa*, *Physiol. Mol. Plant Pathol.*, 62, 167, 2003.

73. Ribeiro, R.V., Machado, E.C., and Oliveira, R.F., Growth- and leaf-temperature effects on photosynthesis of sweet Orange seedlings infected with *Xylella fastidiosa*, *Plant Pathol.* 53, 334, 2004.

74. Purcino, R.P. et al., *Xylella fastidiosa* disturbs nitrogen metabolism and causes a stress response in sweet orange *Citrus sinensis* cv. Pera, *J. Exp. Bot.*, 58, 2733, 2007.

75. Gomes, M.M.A. et al., Abscisic acid and indole-3-acetic acid contents in orange trees infected by *Xylella fastidiosa* and submitted to cycles of water stress, *Plant Growth Regul.*, 39, 263, 2003.

76. Davis, M.J. et al., Axenic culture of the bacteria associated with phony disease of peach and plum leaf scald, *Curr. Microbiol.*, 6, 309, 1981.

77. Chang, C.J. and Walker, J.T., Bacterial leaf scorch of northern red oak: Isolation, cultivation, and pathogenicity of a xylem-limited bacterium, *Plant Dis.*, 72, 730, 1988.

78. Davis, M.J., Purcell, A.H., and Thomson, S.V., Isolation media for Pierce's disease bacterium, *Phytopathology*, 70, 425, 1980.

79. Wells, J.M. et al., Medium for isolation and growth of bacteria associated with plum leaf scald and phony peach diseases, *Appl. Environ. Microbiol.*, 42, 357, 1981.

80. Minsavage, G.V. et al., Development of a polymerase chain-reaction protocol for detection of *Xylella fastidiosa* in plant tissue, *Phytopathology*, 84, 456, 1994.

81. Pooler, M. R. and J. S. Hartung, Specific identification of strains of *Xylella fastidiosa* causing citrus variegated chlorosis disease using sequence characterized amplified regions, *Phytopathology*, 85, 1157, 1995.

82. Oliveira, A.C. et al., Quantification of *Xylella fastidiosa* from citrus trees by real-time polymerase chain reaction assays, *Phytopathology*, 92, 1048, 2002.

83. Laranjeira, F.F. et al., Prevalence, incidence and distribution of citrus variegated chlorosis in Bahia, Brazil, *Trop. Plant Pathol.*, 33, 339, 2008.

84. Goncalves, F.P. et al., Role of healthy nursery plants in orange yield during eight years of Citrus Variegated Chlorosis epidemics, *Sci. Horticult.*, 129, 343, 2011.

85. Coletta-Filho, H.D. et al., Distribution of *Xylella fastidiosa* within sweet orange trees, influence of age and level of symptom expression of citrus variegated chlorosis, *Proceedings of 14th International Organization of Citrus Virologists*, Riverside, CA, p. 243, 2000.

Index

Milton Keynes UK
Ingram Content Group UK Ltd.
UKHW050131071024
449327UK00029B/2535

9 780367 378745